www.brookscole.com

www.brookscole.com is the World Wide Web site for Thomson Brooks/Cole and is your direct source to dozens of online resources.

At *www.brookscole.com* you can find out about supplements, demonstration software, and student resources. You can also send email to many of our authors and preview new publications and exciting new technologies.

www.brookscole.com
Changing the way the world learns®

In memory of Nancy B. Kinsell and Robert C. Kinsell, whose dedication to their family and their country has been an inspiration to me.

AST

To Don Rubo, whom I admire and respect

RDG

Books in the Tussy and Gustafson Series

In paperback:

Basic Mathematics for College Students Third Edition
Student edition: ISBN 0-534-42223-3
Instructor's edition: ISBN 0-534-42224-1

Prealgebra Third Edition
Student edition: ISBN 0-534-40280-1
Instructor's edition: ISBN 0-534-40281-X

Developmental Mathematics for College Students Second Edition
Student edition: ISBN 0-534-99776-7
Instructor's edition: ISBN 0-534-99775-5

Introductory Algebra Third Edition
Student edition: ISBN 0-534-40735-8
Instructor's edition: ISBN 0-534-40736-6

Intermediate Algebra Third Edition
Student edition: ISBN 0-534-49394-7
Instructor's edition: ISBN 0-534-49395-5

Basic Geometry for College Students
ISBN 0-534-39180-X

In hardcover:

Elementary Algebra Third Edition
Student edition: ISBN 0-534-41914-3
Instructor's edition: ISBN 0-534-41915-1

Intermediate Algebra Third Edition
Student edition: ISBN 0-534-41923-2
Instructor's edition: ISBN 0-534-41924-0

Elementary and Intermediate Algebra Third Edition
Student edition: ISBN 0-534-41932-1
Instructor's edition: ISBN 0-534-41933-X

SECOND EDITION

Developmental Mathematics for College Students

Alan S. Tussy

Citrus College

R. David Gustafson

Rock Valley College

Australia • Brazil • Canada • Mexico • Singapore • Spain • United Kingdom • United States

Developmental Mathematics for College Students, Second Edition
Alan S. Tussy and R. David Gustafson

Executive Editor: Jennifer Laugier
Development Editor: Kirsten Markson
Assistant Editor: Rebecca Subity
Editorial Assistant: Christina Ho
Technology Project Manager: Sarah Woicicki
Marketing Manager: Greta Kleinert
Marketing Assistant: Brian R. Smith
Marketing Communications Manager: Bryan Vann
Project Manager, Editorial Production: Hal Humphrey
Art Director: Vernon Boes
Print Buyer: Barbara Britton
Permissions Editor: Kiely Sisk
Production Service: Helen Walden
Text Designer: Diane Beasley
Art Editor: Helen Walden
Photo Researcher: Helen Walden
Copy Editor: Carol Reitz
Illustrator: Lori Heckelman/LHI Technical Illustration
Cover Designer: Cheryl Carrington
Cover Image: Ian Cartwright/Getty Images
Cover Printer: Quebecor World/Dubuque
Compositor: G&S Book Services
Printer: Quebecor World/Dubuque

© 2006 Thomson Brooks/Cole, a part of The Thomson Corporation. Thomson, the Star logo, and Brooks/Cole are trademarks used herein under license.

ALL RIGHTS RESERVED. No part of this work covered by the copyright hereon may be reproduced or used in any form or by any means — graphic, electronic, or mechanical, including photocopying, recording, taping, web distribution, information storage and retrieval systems, or in any other manner — without the written permission of the publisher.

Printed in the United States of America
1 2 3 4 5 6 7 08 07 06 05

For more information about our products, contact us at:
Thomson Learning Academic Resource Center
1-800-423-0563

For permission to use material from this text or product, submit a request online at **http://www.thomsonrights.com.**

Any additional questions about permissions can be submitted by email to **thomsonrights@thomson.com.**

Library of Congress Control Number: 2005925785
Student Edition: ISBN 0-534-99776-7

Thomson Higher Education
10 Davis Drive
Belmont, CA 94002-3098
USA

CONTENTS

3 Fractions and Mixed Numbers 155

4 Decimals 237

5 Percent 305

6 Descriptive statistics 361

7 Introduction to Geometry 393

8 Algebraic Expressions, Equations, and Inequalities 475

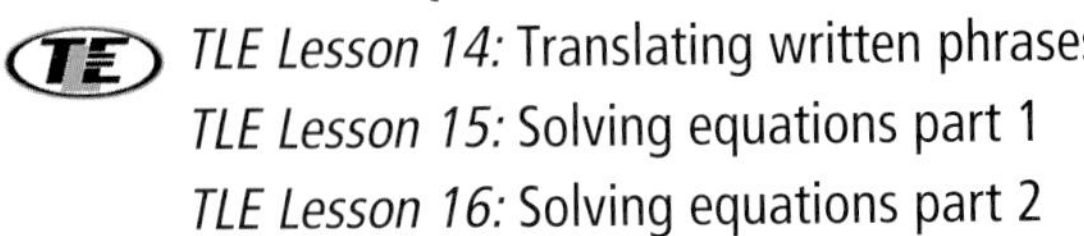

9 Graphs, Linear Equations, and Functions 559

10 Exponents and polynomials 653

11 Factoring and Quadratic Equations 729

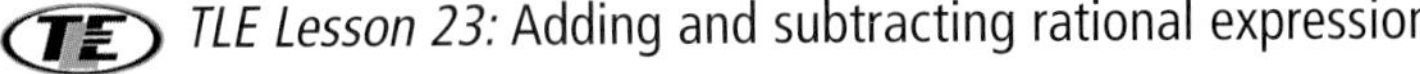

12 Rational Expressions and Equations 789

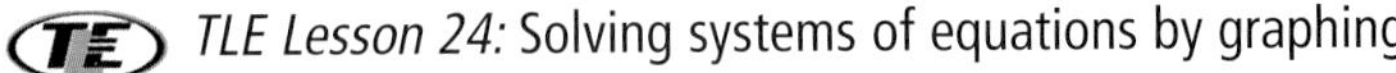

13 Solving Systems of Equations and Inequalities 866

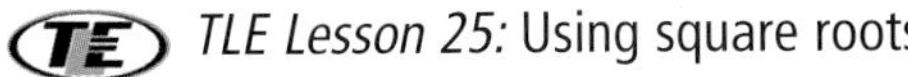

14 Roots and Radicals 937

15 Quadratic Equations 1013

Appendices A-1

PREFACE

Features of the text

A Blend of the Traditional and Reform Approaches

Developmental Mathematics for College Students, Second Edition, employs a variety of instructional methods that reflect the recommendations for NCTM and AMATYC. You will find extensive opportunities for skill practice and the well-defined pedagogy that are hallmarks of a traditional approach. You will also find an emphasis on reasoning, modeling, and communications skills that are a part of today's mathematics reform movement.

Thorough Coverage of Arithmetic

This book provides thorough coverage of the arithmetic of whole numbers, fractions, and decimals. Other topics traditionally taught in an arithmetic course are also included, such as percent, ratio and proportion, and measurement.

Arithmetic and Algebra Are Integrated Throughout

To prepare students for introductory algebra, this book provides a thorough review of arithmetic while introducing basic algebra topics. For example, Chapter 1 covers whole-number arithmetic, but it also introduces the concept of variable, develops the geometric formulas for perimeter and area, and shows how to evaluate numerical expressions. In Chapter 1, we also lay the groundwork for rectangular coordinate graphing and solve some simple equations. Additionally, we establish a five-step problem solving strategy and use it to solve real-life problems.

Competency with Signed Numbers

The rules for adding, subtracting, multiplying, and dividing integers are first introduced in Chapter 2. Students apply these rules again in Chapter 3 with signed fractions, and again in Chapter 4 with signed decimals.

We feel that this spiral approach is superior to that of introducing signed numbers in a single chapter at, or near, the end of the arithmetic review. Revisiting these rules in several different contexts builds a thorough understanding of signed numbers. This pays great dividends when the student makes the transition to the introductory algebra part of the text.

Philosophy of the Text

The second edition retains the basic philosophy of the first edition. However, we have made several improvements as a direct result of the comments and suggestions we received from instructors and students. Our goal has been to make the book more enjoyable to read, easier to understand, and more relevant to students.

Proven Features of the Text

- The authors' five-step problem-solving strategy teaches students to analyze the problem, form an equation, solve the equation, state the conclusion, and check the result. In a step-by-step manner, this approach clarifies the thought process and mathematical skills necessary to solve a wide variety of problems. As a result, students' confidence is increased and their problem-solving abilities are strengthened.
- STUDY SETS are found at the end of every section and feature a unique organization, tailored to improve students' ability to read, write, and communicate mathematical ideas, thereby approaching topics from a wide variety of perspectives. Each comprehensive STUDY SET is divided into seven types of problems: Vocabulary, Concepts, Notation, Practice, Applications, Writing, and Review.
  - VOCABULARY, NOTATION, and WRITING problems help students improve their ability to read, write, and communicate mathematical ideas.
  - CONCEPT problems reinforce major ideas through exploration and foster independent thinking and the ability to interpret graphs and data.
  - PRACTICE problems provide the necessary drill for mastery while the APPLICATIONS problems provide opportunities for students to deal with real-life situations. Each STUDY SET concludes with a Review section that consists of problems randomly selected from previous sections.
- SELF CHECK problems, adjacent to most worked examples, reinforce concepts and build confidence. The answer to each Self Check follows the problem to give instant feedback.
- The KEY CONCEPT section is a one-page review, found at the end of each chapter, that revisits the importance of the role the concept plays in the overall picture.
- REAL-LIFE APPLICATIONS are presented from many disciplines, including science, business, economics, manufacturing, entertainment, history, art, music, and mathematics.
- CALCULATOR SNAPSHOT sections introduce keystrokes and show how scientific calculators can be used to solve application problems. For those who do not wish to use calculators, this feature can be omitted.
- CUMULATIVE REVIEW EXERCISES at the end of every chapter help students retain what they have learned in previous sections.

New to This Edition

The second edition has several new features that will make it even more relevant and useful to students.

New Chapter Openers with TLE Labs: TLE (The Learning Equation) is interactive courseware that uses a guided inquiry approach to teaching developmental mathematics concepts. Each chapter opener integrates TLE Labs with the topics of the chapter, which offers an opportunity to seamlessly integrate TLE courseware into your course. For more information on TLE, instructors can see the Preview at the beginning of their book.

Check Your Knowledge: New pretests at the beginning of each chapter are helpful in gauging a student's knowledge base for the upcoming chapter. Instructors can assign the pretest to see how well students are prepared for the material in the chapter, and customize subsequent lessons based on the needs of their students. Students can also take the pretest themselves as a warm-up for the chapter and to structure review. Answers to the pretests appear at the back of the book.

Study Skills Workshops: This complete mini-course in mathematics study skills helps students and instructors tackle the problem of inadequate study habits in an organized series of lessons. Each chapter opens with a one-page Study Skills Workshop sequenced to address relevant study skill issues as the students move through the course. For example, students learn how to use a calendar to schedule study times in the first lesson, learn how to use study groups around the time of the midterm exam, and learn how to effectively study for the final examination in one of the last lessons. This helpful reference can be used in the classroom or assigned as homework.

Think It Through: Each chapter contains one or two THINK IT THROUGH features that make the connection between mathematics and student life. These problems are student-relevant and apply mathematics skills from the chapter to a real-life situation. Topics include tuition costs, statistics about college life, and many more topics directly connected to the student experience.

Colorful New Design: A colorful new design makes the book more visually appealing and better organizes information on the page.

Changes in Content

Many sections have been rewritten for better clarity. Many applications have been updated to use more recent data. Several topics requested by users have also been included, such as solving percent problems by using proportions and properties of parallelograms.

Ancillary Resources

For detailed information on the ancillary resources available for this text, instructors can refer to the Preview section at the beginning of their book.

Calculators

For those instructors who wish to use calculators as part of the instruction, the text includes a CALCULATOR SNAPSHOT feature. In the arithmetic part of the text, this feature introduces keystrokes and shows how scientific calculators can be used to solve applications problems. In the algebra part of the text, this feature gives keystrokes for both scientific and graphing calculators. Some Study Sets include problems that are to be solved using a calculator. These problems are indicated by a calculator logo.

Instructors who do not wish to introduce calculators can skip that material without interrupting the flow of ideas.

Acknowledgments

We are grateful to the following people who reviewed this manuscript and the other manuscripts in the paperback series at various stages of development. They all had valuable suggestions that have been incorporated into the text.

The following people reviewed the earlier editions in the series:

Linda Beattie
Western New Mexico University

Julia Brown
Atlantic Community College

Linda Clay
Albuquerque TVI

John Coburn
Saint Louis Community College–Florissant Valley

Sally Copeland
Johnson County Community College

Ben Cornelius
Oregon Institute of Technology

James Edmondson
Santa Barbara Community College

David L. Fama
Germanna Community College

Barbara Gentry
Parkland College

Laurie Hoecherl
Kishwaukee College

Judith Jones
Valencia Community College

Therese Jones
Amarillo College

Joanne Juedes
University of Wisconsin–Marathon County

Dennis Kimzey
Rogue Community College

Sally Lesik
Holyoke Community College

Elizabeth Morrison
Valencia Community College

Jan Alicia Nettler
Holyoke Community College

Scott Perkins
Lake–Sumter Community College

Angela Peterson
Portland Community College

J. Doug Richey
Northeast Texas Community College

Angelo Segalla
Orange Coast College

June Strohm
Pennsylvania State Community College–DuBois

Rita Sturgeon
San Bernardino Valley College

Jo Anne Temple
Texas Technical University

Sharon Testone
Onondaga Community College

Marilyn Treder
Rochester Community College

Thomas Vanden Eynden
Thomas More College

The following people reviewed the books in this series in preparation for the current revisions:

Cedric E. Atkins
Mott Community College

William D. Barcus
SUNY, Stony Brook

Kathy Bernunzio
Portland Community College

Girish Budhwar
United Tribes Technical College

Sharon Camner
Pierce College–Fort Steilacoom

Robin Carter
Citrus College

Ann Corbeil
Massasoit Community College

Carolyn Detmer
Seminole Community College

Maggie Flint
Northeast State Technical Community College

Charles Ford
Shasta College

Michael Heeren
Hamilton College

Monica C. Kurth
Scott Community College

Sandra Lofstock
St. Petersberg College–Tarpon Springs Center

Marge Palaniuk
United Tribes Technical College

Jane Pinnow
University of Wisconsin–Parkside

Eric Sims
Art Institute of Dallas

Annette Squires
Palomar College

Lee Ann Spahr
Durham Technical Community College

John Strasser
Scottsdale Community College

Stuart Swain
University of Maine at Machias

Celeste M. Teluk
D'Youville College

Sven Trenholm
Herkeimer County Community College

Stephen Whittle
Augusta State University

Mary Lou Wogan
Klamath Community College

Without the talents and dedication of the editorial, marketing, and production staff of Brooks/Cole, this revision of *Developmental Mathematics* could not have been so well accomplished. We express our sincere appreciation for the hard work of Bob Pirtle, Jennifer Laugier, Helen Walden, Lori Heckelman, Vernon Boes, Diane Beasley, Sarah Woicicki, Greta Kleinert, Jessica Bothwell, Bryan Vann, Kirsten Markson, Rebecca Subity, Hal Humphrey, Christine Davis, and G & S Typesetters for their help in creating the book. Special thanks to David Casey of Citrus College for his hard work on the pretests and to Sheila Pisa for writing the excellent Study Skills Workshops.

Alan S. Tussy
R. David Gustafson

For the Student

Success in Mathematics

To be successful in mathematics, you need to know how to study it. The following checklist will help you develop your own personal strategy to study and learn the material. The suggestions below require some time and self-discipline on your part, but it will be worth the effort. This will help you get the most out of the course.

As you read each of the following statements, place a check mark in the box if you can truthfully answer Yes. If you can't answer Yes, think of what you might do to make the suggestion part of your personal study plan. You should go over this checklist several times during the semester to be sure you are following it.

Preparing for the Class

- ❑ I have made a commitment to myself to give this course my best effort.
- ❑ I have the proper materials: a pencil with an eraser, paper, a notebook, a ruler, a calculator, and a calendar or day planner.
- ❑ I am willing to spend a minimum of two hours doing homework for every hour of class.
- ❑ I will try to work on this subject every day.
- ❑ I have a copy of the class syllabus. I understand the requirements of the course and how I will be graded.
- ❑ I have scheduled a free hour after the class to give me time to review my notes and begin the homework assignment.

Class Participation

- ❑ I know my instructor's name.
- ❑ I will regularly attend the class sessions and be on time.
- ❑ When I am absent, I will find out what the class studied, get a copy of any notes or handouts, and make up the work that was assigned when I was gone.

- ❑ I will sit where I can hear the instructor and see the board.
- ❑ I will pay attention in class and take careful notes.
- ❑ I will ask the instructor questions when I don't understand the material.
- ❑ When tests, quizzes, or homework papers are passed back and discussed in class, I will write down the correct solutions for the problems I missed so that I can learn from my mistakes.

Study Sessions

- ❑ I will find a comfortable and quiet place to study.
- ❑ I realize that reading a math book is different from reading a newspaper or a novel. Quite often, it will take more than one reading to understand the material.
- ❑ After studying an example in the textbook, I will work the accompanying Self Check.
- ❑ I will begin the homework assignment only after reading the assigned section.
- ❑ I will try to use the mathematical vocabulary mentioned in the book and used by my instructor when I am writing or talking about the topics studied in this course.
- ❑ I will look for opportunities to explain the material to others.
- ❑ I will check all my answers to the problems with those provided in the back of the book (or with the *Student Solutions Manual*) and resolve any differences.
- ❑ My homework will be organized and neat. My solutions will show all the necessary steps.
- ❑ I will work some review problems every day.
- ❑ After completing the homework assignment, I will read the next section to prepare for the coming class session.
- ❑ I will keep a notebook containing my class notes, homework papers, quizzes, tests, and any handouts — all in order by date.

Special Help

- ❑ I know my instructor's office hours and am willing to go in to ask for help.
- ❑ I have formed a study group with classmates that meets regularly to discuss the material and work on problems.
- ❑ When I need additional explanation of a topic, I use the tutorial videos and the interactive CD, as well as the Web site.
- ❑ I make use of extra tutorial assistance that my school offers for mathematics courses.
- ❑ I have purchased the *Student Solutions Manual* that accompanies this text, and I use it.

To follow each of these suggestions will take time. It takes a lot of practice to learn mathematics, just as with any other skill.

No doubt, you will sometimes become frustrated along the way. This is natural. When it occurs, take a break and come back to the material after you have had time to clear your thoughts. Keep in mind that the skills and discipline you learn in this course will help make for a brighter future. Good luck!

iLrn Tutorial Quick Start Guide

iLrn Can Help You Succeed in Math

iLrn™ is an online program that facilitates math learning by providing resources and practice to help you succeed in your math course. Your instructor chose to use iLrn because it provides online opportunities for learning (Explanations found by clicking

Read Book), practice (Exercises), and evaluating (Quizzes). It also gives you a way to keep track of your own progress and manage your assignments.

The mathematical notation in iLrn is the same as that you see in your textbooks, in class, and when using other math tools like a graphing calculator. iLrn can also help you run calculations, plot graphs, enter expressions, and grasp difficult concepts. You will encounter various problem types as you work through iLrn, all of which are designed to strengthen your skills and engage you in learning in different ways.

Logging in to 1Pass

Registering with the PIN Code on the 1Pass Card *Situation:* Your instructor has not given you a PIN code for an online course, but you have a textbook with a 1Pass PIN code. With 1Pass, you have one simple PIN code access to all media resources associated with your textbook. Please refer to your 1Pass card for a complete list of those resources.

Initial Log-in

To access your web gateway through 1Pass:

1. Check the outside of your textbook to see if there is an additional 1Pass card.
2. Take this card (and the additional 1Pass card if appropriate) and go to http://1pass.thomson.com.
3. Type in your 1Pass access code (or codes).
4. Follow the directions on the screen to set up your personal username and password.
5. Click through to launch your personal portal.
6. Access the media resources associated with your text . . . all the resources are just one click away.
7. Record your username and password for future visits and be sure to use the same username for all Thomson Learning resources.

For tech support, contact us at 1 (800) 423-0563.

> You will be asked to enter a valid e-mail address and password. Save your password in a safe place. You will need them to log in the next time you use 1Pass. Only your e-mail address and password will allow you to reenter 1Pass.

Subsequent Log-in

1. Go to **http://1pass.thomson.com.**
2. Type your e-mail address and password (see boxed information above) in the "Existing Users" box; then click on **Login.**

Navigating through Your iLrn Tutorial

To navigate between chapters and sections, use the drop-down menu below the top navigation bar. This will give you access to the study activities available for each section.

The view of a tutorial in iLrn looks like this.

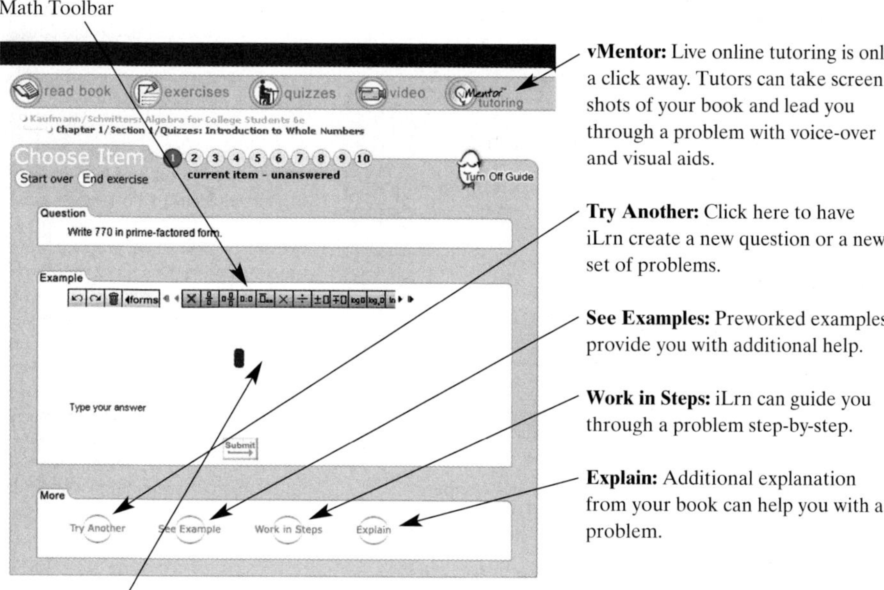
Math Toolbar
read book
exercises
quizzes
video
vMentor tutoring
Chapter 1/Section 1/Quizzes: Introduction to Whole Numbers
Choose Item
Start over
End exercise
current item - unanswered
Turn Off Guide
Question
Write 770 in prime-factored form.
Example
Type your answer
Submit
More
Try Another
See Example
Work in Steps
Explain
vMentor: Live online tutoring is only a click away. Tutors can take screen shots of your book and lead you through a problem with voice-over and visual aids.
Try Another: Click here to have iLrn create a new question or a new set of problems.
See Examples: Preworked examples provide you with additional help.
Work in Steps: iLrn can guide you through a problem step-by-step.
Explain: Additional explanation from your book can help you with a problem.

Type your answer here.

Online Tutoring with vMentor

Access to iLrn also means access to online tutors and support through vMentor™, which provides live homework help and tutorials. To access vMentor while you are working in the Exercises or Tutorial areas in iLrn, click on the **vMentor Tutoring** button at the top right of the navigation bar above the problem or exercise.

Next, click on the **vMentor** button; you will be taken to a Web page that lists the steps for entering a vMentor classroom. If you are a first-time user of vMentor, you might need to download Java software before entering the class for the first class. You can either take an Orientation Session or log in to a vClass from the links at the bottom of the opening screen.

All vMentor Tutoring is done through a vClass, an Internet-based virtual classroom that features two-way audio, a shared whiteboard, chat, messaging, and experienced tutors.

You can access vMentor Sunday through Thursday, as follows:

5 p.m. to 9 p.m. Pacific Time

6 p.m. to 10 p.m. Mountain Time

7 p.m. to 11 p.m. Central Time

8 p.m. to midnight Eastern Time

If you need additional help using vMentor, you can access the Participant Quick Reference Guide at this Web site: **http://www.elluminate.com/support/docs/Elive_Participant_Quick_Reference_Guide_6.0.pdf.**

Interact with TLE Online Labs

Use TLE Online Labs to explore and reinforce key concepts introduced in this text. These electronic labs give you access to additional instruction and practice problems, so you can explore each concept interactively, at your own pace. Not only will you be better prepared, but also you will perform better in the class overall.

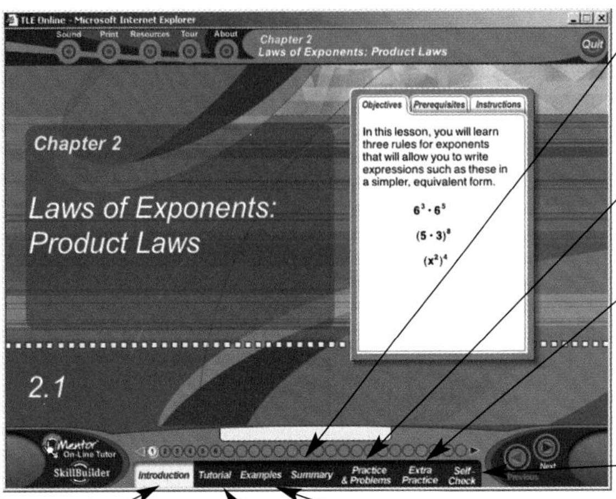

Summary: The Summary revisits the problem presented in the Introduction and encourages you to apply the mathematics you learned in the Tutorial and Examples.

Practice & Problems: Practice & Problems presents up to 25 questions organized in four or five categories.

Extra Practice: Extra practice presents questions like those in the Examples. After each question you have the option to try again, see the answer, see a sample solution, or try another question of the same type.

Self-Check: Self-Check presents up to 10 dynamically generated questions. To complete a lesson you must obtain the minimum standard (about 70%).

Introduction: Each lesson opens with objectives and prerequisites and provides brief instructions on using TLE.

Tutorial: The Tutorial provides the main instruction for the lesson. Hint and Success Tips teach strategies that can be used to solve the problem.

Examples: The examples expand on what you learned in the Tutorial. A hidden picture is progressively revealed as you complete each example.

APPLICATIONS INDEX

Examples that are applications are shown with boldface page numbers.
Exercises that are applications are shown with lightface page numbers.

CHAPTER 1

Whole Numbers

Getty Images

TLE Office managers play an important role in many businesses. They oversee the day-to-day activities of a company, making sure that the business runs smoothly and efficiently. To be an effective office manager, one needs excellent organizational, planning, and communication skills. Strong mathematical skills are also necessary to perform such job responsibilities as scheduling meetings, managing payroll and budgets, and designing office workspace layouts.

To learn more about the use of mathematics in the business world, visit The Learning Equation on the Internet at http://tle.brookscole.com. (The log-in instructions are in the Preface.) For Chapter 1, the online lessons are:

- *TLE* Lesson 1: Whole Numbers
- *TLE* Lesson 2: Order of Operations
- *TLE* Lesson 3: Solving Equations

Check Your Knowledge

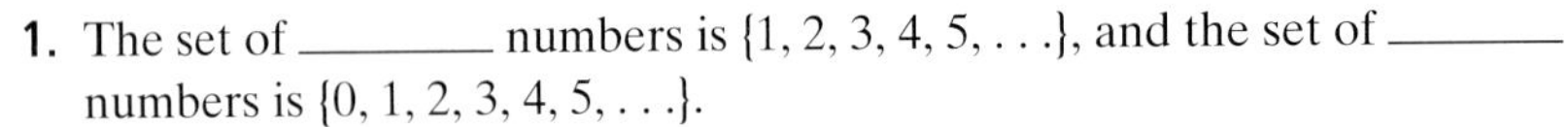

1. The set of ________ numbers is {1, 2, 3, 4, 5, . . .}, and the set of ________ numbers is {0, 1, 2, 3, 4, 5, . . .}.

2. Numbers that are to be multiplied are called ________. The result of a multiplication problem is called a ________. The answer to a division problem is called the ________.

3. The property that guarantees that we can add two numbers in either order is called the ____________ property of addition. The property that allows us to group numbers in an addition in any way we wish is called the ____________ property of addition.

4. A ________ number is a whole number, greater than 1, that has only 1 and itself as factors.

5. Write 7,343 in expanded notation.

6. Round 27,450 to the nearest hundred.

Number of persons (1–10) vs. Blood types: A, B, O, AB

PROBLEM 7

Refer to the data in the table.

Blood type	A	B	O	AB
Number of persons	5	7	9	4

7. Use the data to make a bar graph.

8. Use the data to make a line graph.

9. Place one of the symbols $<$ or $>$ in the blank to make a true statement: 91 ___ 19

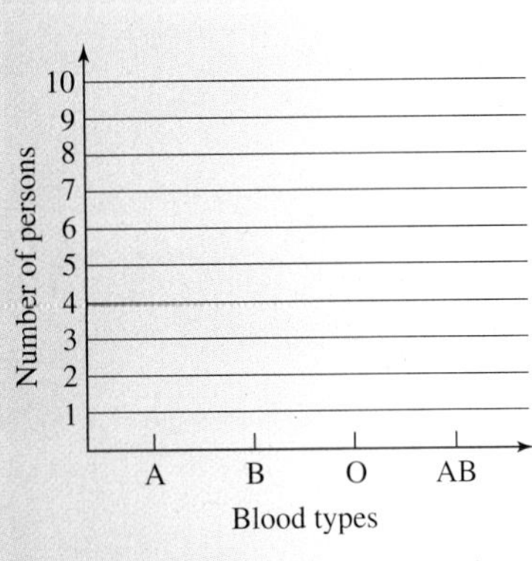

PROBLEM 8

Perform each operation.

10. Add: $\begin{array}{r} 3{,}742 \\ +\ 1{,}379 \\ \hline \end{array}$

11. Subtract 289 from 347.

12. On Wednesday, the high temperature was 72°. The high temperature rose 9° on Thursday and fell 12° on Friday. What was the high temperature on Friday?

13. Multiply: $\begin{array}{r} 432 \\ \times\ 57 \\ \hline \end{array}$

14. Divide: $79\overline{)4{,}537}$

15. Find the perimeter and the area of a rectangle that is 13 ft wide and 19 ft long.

16. Find the prime factorization of 950.

Evaluate the expressions.

17 $3 + 4 - 2$

18. $3 + 4 \cdot 25$

19. $\dfrac{(4^3 - 2) + 7}{5(2 + 4) - 7}$

20. $3 \cdot 7 - 2\,[10 - 3\,(5 - 2)]$

Solve each equation and check the result.

21. $x - 2 = 1$

22. $y + 3 = 5$

23. $4n = 64$

24. $\dfrac{z}{4} = 12$

25. Joan scored 95, 85, 73, 62, and 0 on five math quizzes. Find her mean (average) quiz score.

26. Let a variable represent the unknown quantity. Then write and solve an equation to answer the following question:

After moving his shop, a barber lost 24 regular customers. If he has 65 regular customers left, how many did he have before the move?

Getty Images

Study Skills Workshop

GET ORGANIZED!

Sometimes students who have had difficulty learning math in the past think that their problem is not being born with the ability to "do math." This isn't true! Learning math is a skill and, much like learning to play a musical instrument, takes daily, organized practice. Below are some strategies that will get you off to a good start.

Attend Class. Attending class every meeting is one of the most important things you can do to succeed. Your instructor explains material, gives examples to support the text, and information about topics that are not in your book, or may make announcements regarding homework assignments and test dates. Getting to know at least a few of your classmates is also important to your success. Find a classmate or two on whom you can depend for information, who can help with homework, or with whom you can form a study group.

Make a Calendar. Because daily practice is so important in learning math, it is a good idea to set up a calendar that lists all of your time commitments and time for studying and doing your homework. A general rule for how much study time to budget is to allot two hours outside of class for every lecture hour. If your class meets for three hours per week, plan on six hours per week for homework and study.

Gather Needed Materials. All math classes require textbooks, notebooks, pencils (with big erasers!), and usually as much scrap paper as you can gather. If you are not sure that you have everything you need, check with your instructor. Ideally, you should have your materials by your second class meeting and bring them to every class meeting after that. Additional materials that may be of use outside of class are the online tutorial program iLrn (www.iLrn.com) and the Video Skillbuilder CD-ROM that is packaged with your textbook.

What Does Your Instructor Expect From You? Your instructor's syllabus is documentation of his or her expectations. Many times your instructor will detail in the syllabus how your grade is determined, when office hours are held, and where you can get help outside of class. Read the syllabus thoroughly and make sure you understand all that is required.

ASSIGNMENT

1. Download a calendar online at series.brookscole.com/tussypaperback or make up your own calendar that includes class and study times for each course you are taking as well as times for work and other essential activities (e.g., church activities, social obligations, time with children). You may want to schedule additional time to study a week before a test. Also include time for physical exercise and rest—important for reducing the effects of stress that school brings.
2. Download and print out the Course Information Sheet online at series.brookscole.com/tussypaperback or make a list of the following items:
   a. Instructor name, office location and hours, phone number, e-mail address
   b. Test dates, if scheduled
   c. The work that determines your course grade and how grades are calculated
3. Write down names, phone numbers, and e-mail addresses of at least two classmates.
4. Does your school have tutorial services or a math lab/learning center? If so, where are they located? What are their hours of operation?

In this chapter, we begin our mathematics study by examining the procedures used to solve problems that involve whole numbers.

1.1 An Introduction to the Whole Numbers

- Sets of numbers • Place value • Expanded notation • Graphing on the number line
- Ordering of the whole numbers • Rounding whole numbers • Tables and graphs

In this section, we will discuss the natural numbers and the whole numbers. These numbers are used to answer questions such as How many?, How fast?, How heavy?, and How far?

- The movie *Saving Private Ryan* won 5 Academy Awards.
- The speed limit on interstate highways in Wisconsin is 65 mph.
- The Statue of Liberty weighs 225 tons.
- The driving distance between Chicago and Houston is 1,067 miles.

Sets of numbers

A **set** is a collection of objects. Two sets in mathematics are the natural numbers (the numbers that we count with) and the whole numbers. When writing a set, we use **braces** { } to enclose its **members** (or **elements**).

The set of natural numbers

$$\{1, 2, 3, 4, 5, 6, 7, 8, 9, 10, 11, 12, \dots\}$$

The set of whole numbers

$$\{0, 1, 2, 3, 4, 5, 6, 7, 8, 9, 10, 11, 12, \dots\}$$

The three dots at the ends of the lists above indicate that the sets continue on forever. There is no largest natural number or whole number.

Since every natural number is also a whole number, we say that the set of natural numbers is a **subset** of the set of whole numbers. However, not all whole numbers are natural numbers, because 0 is a whole number but not a natural number.

Place value

When we express a whole number with a *numeral* containing the *digits* 0, 1, 2, 3, 4, 5, 6, 7, 8, 9, we say that we have written the number in **standard notation.** The position of a digit in a numeral determines its value. In the numeral 325, the 5 is in the *ones column,* the 2 is in the *tens column,* and the 3 is in the *hundreds column.*

3 2 5

Hundreds column (3) — Tens column (2) — Ones column (5)

To make a numeral easy to read, we use commas to separate its digits into groups of three, called **periods.** Each period has a name, such as *ones, thousands, millions,* and so on. The following table shows the place value of each digit in the numeral 345,576,402,897,415, which is read as

three hundred forty-five trillion, five hundred seventy-six billion, four hundred two million, eight hundred ninety-seven thousand, four hundred fifteen

345 trillion			576 billion			402 million			897 thousand			4 hundred fifteen		
3	4	5	5	7	6	4	0	2	8	9	7	4	1	5
Trillions			Billions			Millions			Thousands			Ones		
Hundreds	Tens	Ones	Hundreds	Tens	Ones	Hundreds	Tens	Ones	Hundreds	Tens	Ones	Hundreds	Tens	Ones

As we move to the left in this table, the place value of each column is 10 times greater than the column to its right. This is why we call our number system a *base-10 number system.*

EXAMPLE 1 **TV news.** In 2003, there were 73,365,880 basic cable subscribers in the United States. Which digit in 73,365,880 tells the number of hundreds?

Solution In 73,365,880, the hundreds column is the third column from the right. The digit 8 tells the number of hundreds.

Self Check 1
In 2003, there were 158,722,000 cellular telephone subscribers in the United States. Which digit in 158,722,000 tells the number of ten thousands?

Answer 2

Expanded notation

In the numeral 6,352, the digit 6 is in the thousands column, 3 is in the hundreds column, 5 is in the tens column, and 2 is in the ones (or units) column. The meaning of 6,352 becomes clear when we write it in **expanded notation.**

6 thousands + 3 hundreds + 5 tens + 2 ones

We read the numeral 6,352 as "six thousand, three hundred fifty-two."

EXAMPLE 2 Write each number in expanded notation: **a.** 63,427 and **b.** 1,251,609.

Solution

a. 6 ten thousands + 3 thousands + 4 hundreds + 2 tens + 7 ones

We read this number as "sixty-three thousand, four hundred twenty-seven."

b. 1 million + 2 hundred thousands + 5 ten thousands + 1 thousand + 6 hundreds + 0 tens + 9 ones

Since 0 tens is zero, the expanded notation can also be written as

1 million + 2 hundred thousands + 5 ten thousands + 1 thousand + 6 hundreds + 9 ones

We read this number as "one million, two hundred fifty-one thousand, six hundred nine."

Self Check 2
Write 808,413 in expanded notation.

Answer 8 hundred thousands + 8 thousands + 4 hundreds + 1 ten + 3 ones. Read as "eight hundred eight thousand, four hundred thirteen."

Self Check 3
Write seventy-six thousand three in standard notation.

Answer 76,003

EXAMPLE 3 Write twenty-three thousand forty in standard notation.

Solution In expanded notation, the number is written as

2 ten thousands + 3 thousands + 4 tens   There are 0 hundreds and 0 ones.

In standard notation, this is written as 23,040.

Graphing on the number line

Whole numbers can be represented by points on the **number line.** The number line is a horizontal or vertical line that is used to represent numbers graphically. Like a ruler, the number line is straight and has uniform markings. (See Figure 1-1.) To construct the number line, we begin on the left with a point on the line representing the number 0. This point is called the **origin.** We then proceed to the right, drawing equally spaced marks and labeling them with whole numbers that increase progressively in size. The arrowhead at the right indicates that the number line continues forever.

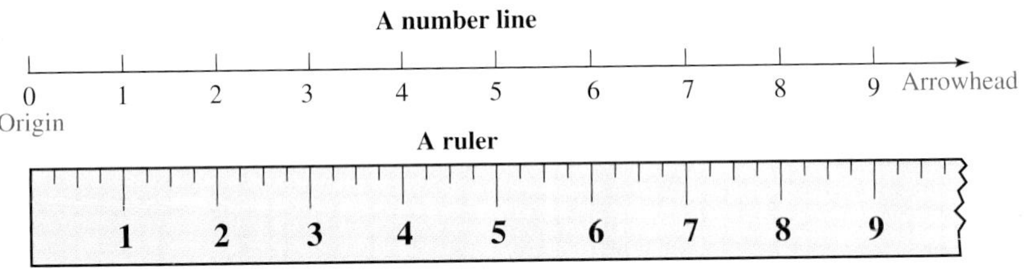

FIGURE 1-1

Using a process known as **graphing,** we can represent a single number or a set of numbers on the number line. *The graph of a number* is the point on the number line that corresponds to that number. *To graph a number* means to locate its position on the number line and then to highlight it using a dot. Figure 1-2 shows the graphs of the whole numbers 5 and 8.

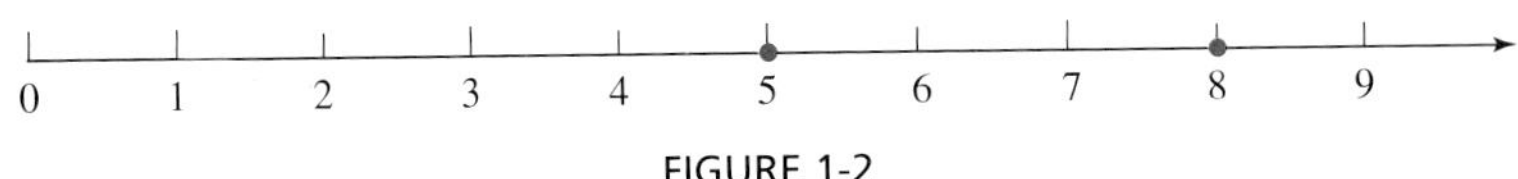

FIGURE 1-2

Ordering of the whole numbers

As we move to the right on the number line, the numbers get larger. Because 8 lies to the right of 5, we say that 8 is greater than 5. The **inequality symbol** $>$ ("is greater than") can be used to write this fact.

$8 > 5$ Read as "8 is greater than 5."

Since $8 > 5$, it is also true that $5 < 8$. (Read as "5 is less than 8.")

! COMMENT To distinguish between these two inequality symbols, remember that the inequality symbol always points to the smaller of the two numbers involved.

$8 > 5$ $5 < 8$

Points to the smaller number

EXAMPLE 4 Place an < or > symbol in the box to make a true statement: **a.** 3 □ 7 and **b.** 18 □ 16.

Solution

a. Since 3 is to the left of 7 on the number line, $3 < 7$.

b. Since 18 is to the right of 16 on the number line, $18 > 16$.

Self Check 4
Place an < or > symbol in the box to make a true statement.
a. 12 □ 4 **b.** 7 □ 10

Answers **a.** >, **b.** <

Rounding whole numbers

When we don't need exact results, we often round numbers. For example, when a teacher with 36 students in his class orders 40 textbooks, he has rounded the actual number to the *nearest ten,* because 36 is closer to 40 than it is to 30.

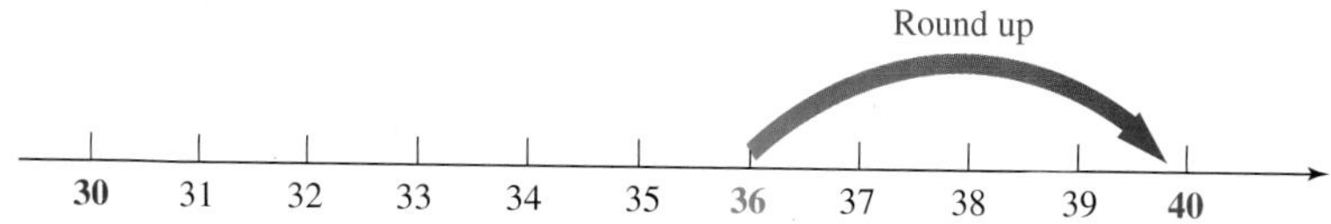

When a geologist says that the height of Alaska's Mount McKinley is "about 20,300 feet," she has rounded to the *nearest hundred,* because its actual height of 20,320 feet is closer to 20,300 than it is to 20,400.

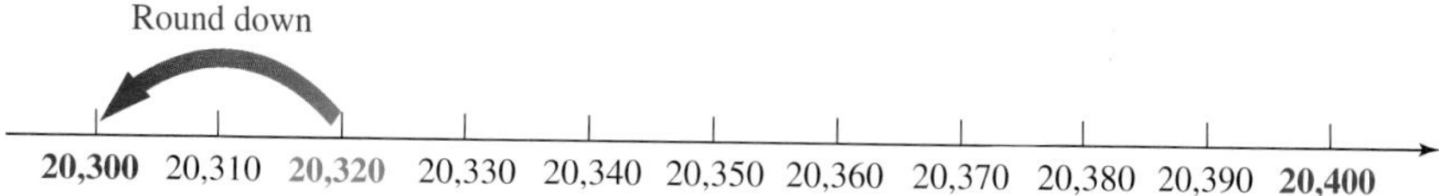

To round a whole number, we follow an established set of rules. To round a number to the nearest ten, for example, we begin by locating the **rounding digit** in the tens column. If the **test digit** to the right of that column (the digit in the ones column) is 5 or greater, we *round up* by increasing the tens digit by 1 and placing a 0 in the ones column. If the test digit is less than 5, we *round down* by leaving the tens digit unchanged and placing a 0 in the ones column.

EXAMPLE 5 Round each number to the nearest ten: **a.** 3,764 and **b.** 12,087.

Solution

a. We find the rounding digit in the tens column, which is 6.

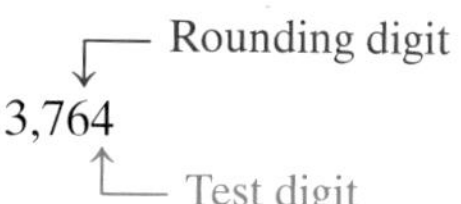

We then look at the test digit to the right of 6, the 4 in the ones column. Since $4 < 5$, we round down by leaving the 6 unchanged and replacing the test digit with 0. The rounded answer is 3,760.

b. We find the rounding digit in the tens column, which is 8.

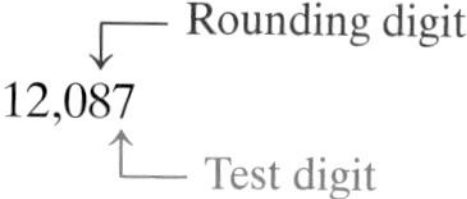

Self Check 5
Round each number to the nearest ten:
a. 35,642
b. 3,756

We then look at the test digit to the right of 8, the 7 in the ones column. Because $7 > 5$, we round up by adding 1 to 8 and replacing the test digit with 0. The rounded answer is 12,090.

Answers **a.** 35,640, **b.** 3,760

A similar procedure is used to round numbers to the nearest hundred, the nearest thousand, the nearest ten thousand, and so on.

Rounding a whole number

1. To round a number to a certain place, locate the rounding digit in that place.
2. Look at the test digit to the right of the rounding digit.
3. If the test digit is 5 or greater, round up by adding 1 to the rounding digit and changing all of the digits to the right of the rounding digit to 0.

   If the test digit is less than 5, round down by keeping the rounding digit and changing all of the digits to the right of the rounding digit to 0.

EXAMPLE 6 Round 7,960 to the nearest hundred.

Solution First, we find the rounding digit in the hundreds column. It is 9.

Rounding digit
7,960
Test digit

We then look at the 6 to the right of 9. Because $6 > 5$, we round up and increase 9 in the hundreds column by 1. Since the 9 in the hundreds column represents 900, increasing 9 by 1 represents increasing 900 to 1,000. Thus, we replace the 9 with a 0 and add 1 to the 7 in the thousands column. Finally, we replace the two rightmost digits with 0's. The rounded answer is 8,000.

Self Check 6
Round 365,283 to the nearest hundred.

Answer 365,300

EXAMPLE 7 **U.S. cities.** In 2003, Denver was the nation's 26th largest city. Round the 2003 population of Denver given in Figure 1-3 **a.** to the nearest thousand and **b.** to the nearest ten thousand.

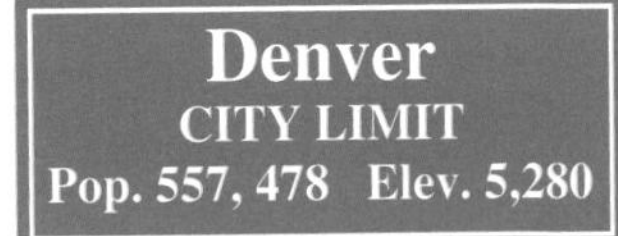

FIGURE 1-3

Solution

a. The rounding digit in the thousands column is 7.
The test digit, 4, is less than 5, so we round down. To the nearest thousand, Denver's population in 2003 was 557,000.

b. The rounding digit in the ten thousands column is 5. The test digit, 7, is greater than 5, so we round up. To the nearest ten thousand, Denver's population in 2003 was 560,000.

Self Check 7
Round the elevation of Denver **a.** to the nearest hundred feet and **b.** to the nearest thousand feet.

Answers **a.** 5,300 ft, **b.** 5,000 ft

Tables and graphs

The table in Figure 1-4(a) on the next page is an example of the use of whole numbers. It shows the number of women elected to the United States House of Representatives in the congressional elections held every two years from 1996 to 2004.

Table

Year	1996	1998	2000	2002	2004
Number of women elected	54	56	59	60	65

(a)

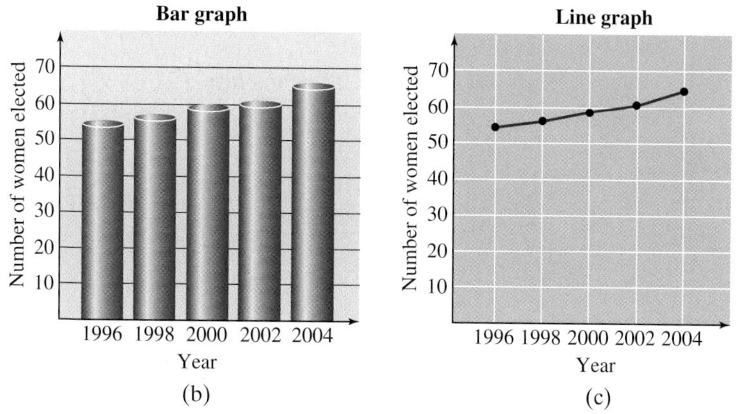

Bar graph

(b)

Line graph

(c)

FIGURE 1-4

In Figure 1-4(b), the election results are presented in a **bar graph.** The horizontal axis is labeled "Year" and scaled in units of 2 years. The vertical axis is labeled "Number of women elected" and scaled in units of 10. The bar directly over each year extends to a height indicating the number of women elected to Congress that year.

Another way to present the information in the table is with a **line graph.** Instead of using a bar to denote the number of women elected, we use a dot drawn at the correct height. After drawing data points for 1996, 1998, 2000, 2002, and 2004, we connect the points with line segments to create the line graph in Figure 1-4(c).

Re-entry Students

THINK IT THROUGH

"A re-entry student is considered one who is the age of 25 or older, or those students that have had a break in their academic work for 5 years or more. Nationally, this group of students is growing at an astounding rate."

Student Life and Leadership Department, University Union, Cal Poly University, San Luis Obispo

Some common concerns expressed by adult students considering returning to school are listed below in Column I. Match each concern to an encouraging reply in Column II.

Column I

1. I'm too old to learn.
2. I don't have the time.
3. I didn't do well in school the first time around. I don't think a college would accept me.
4. I'm afraid I won't fit in.
5. I don't have the money to pay for college.

Column II

a. Many students qualify for some type of financial aid.

b. Taking even a single class puts you one step closer to your educational goal.

c. There's no evidence that older students can't learn as well as younger ones.

d. More than 41% of the students in college are older than 25.

e. Typically, community colleges and career schools have an open admissions policy.

Adapted from *Common Concerns for Adult Students,* Minnesota Higher Education Services Office

Section 1.1 STUDY SET

VOCABULARY *Fill in the blanks.*

1. A _____ is a collection of objects.
2. The set of ________ numbers is $\{1, 2, 3, 4, 5, \ldots\}$, and the set of ________ numbers is $\{0, 1, 2, 3, 4, 5, \ldots\}$.
3. When 297 is written as 2 hundreds + 9 tens + 7 ones, it is written in __________ notation.
4. If we ________ 627 to the nearest ten, we get 630.
5. Using a process known as graphing, we can represent whole numbers as points on the ________ line.
6. The symbols $>$ and $<$ are __________ symbols.

CONCEPTS *Consider the numeral 57,634.*

7. What digit is in the tens column?
8. What digit is in the thousands column?
9. What digit is in the hundreds column?
10. What digit is in the ten thousands column?
11. What set of numbers is obtained when 0 is combined with the natural numbers?
12. Place the numbers 25, 17, 37, 15, 45 in order, from smallest to largest.
13. Graph: 1, 3, 5, and 7.

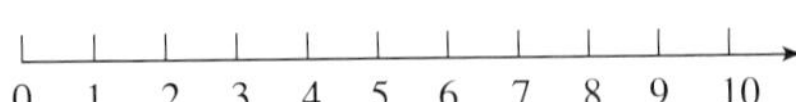

14. Graph: 0, 2, 4, 6, and 8.

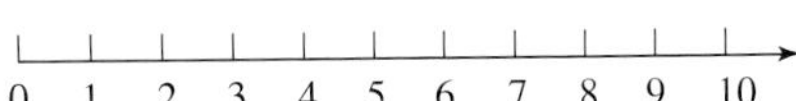

15. Graph: the whole numbers less than 6.

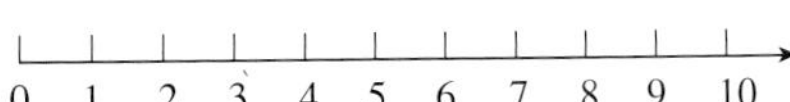

16. Graph: the whole numbers between 2 and 8.

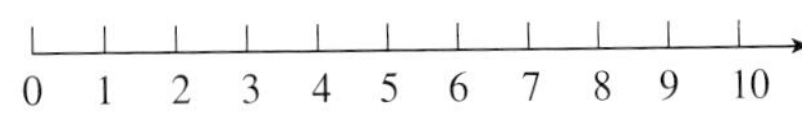

Place an $>$ or $<$ symbol in the box to make a true statement.

17. 47 □ 41
18. 53 □ 67
19. 309 □ 300
20. 841 □ 814
21. 2,052 □ 2,502
22. 999 □ 998
23. Since $4 < 7$, it is also true that 7 □ 4.
24. Since $9 > 0$, it is also true that 0 □ 9.

NOTATION *Fill in the blanks.*

25. The symbols { }, called ________, are used when writing a set.
26. The symbol $>$ means ____ ________ ______, and the symbol $<$ means ____ _____ ______.

PRACTICE *Write each number in expanded notation and then write it in words.*

27. 245
28. 508
29. 3,609
30. 3,960
31. 32,500
32. 73,009
33. 104,401
34. 570,003

Write each number in standard notation.

35. 4 hundreds + 2 tens + 5 ones
36. 7 hundreds + 7 tens + 7 ones
37. 2 thousands + 7 hundreds + 3 tens + 6 ones
38. 7 billions + 3 hundreds + 5 tens
39. Four hundred fifty-six
40. Three thousand seven hundred thirty-seven
41. Twenty-seven thousand five hundred ninety-eight
42. Seven million, four hundred fifty-two thousand, eight hundred sixty
43. Nine thousand one hundred thirteen
44. Nine hundred thirty
45. Ten million, seven hundred thousand, five hundred six
46. Eighty-six thousand four hundred twelve

Round 79,593 to the nearest . . .

47. ten

48. hundred

49. thousand

50. ten thousand

Round 5,925,830 to the nearest . . .

51. thousand

52. ten thousand

53. hundred thousand

54. million

Round $419,161 to the nearest . . .

55. $10

56. $100

57. $1,000

58. $10,000

APPLICATIONS

59. EATING HABITS The following list shows the ten countries with the largest per-person annual consumption of meat. Construct a two-column table that presents the data in order, beginning with the largest per-person consumption. (The abbreviation "lb" means "pounds.")

Australia: 239 lb
Austria: 229 lb
Canada: 211 lb
Cyprus: 236 lb
Denmark: 219 lb
New Zealand: 259 lb
Saint Lucia: 222 lb
Spain: 211 lb
Uruguay: 230 lb
United States: 261 lb

60. PRESIDENTS The following list shows the ten youngest U.S. presidents and their ages (in years/days) when they took office. Construct a two-column table that presents the data in order, beginning with the youngest president.

C. Arthur 50 yr/350 days
G. Cleveland 47 yr/351 days
W. Clinton 46 yr/154 days
M. Filmore 50 yr/184 days
J. Garfield 49 yr/105 days
U. Grant 46 yr/236 days
J. Kennedy 43 yr/236 days
F. Pierce 48 yr/101 days
J. Polk 49 yr/122 days
T. Roosevelt 42 yr/322 days

61. MISSIONS TO MARS The United States, Russia, Europe, and Japan have launched Mars space probes. The graph in the next column shows the success rate of the missions, by decade.

a. What decade had the greatest number of successful or partially successful missions? How many?

b. What decade had the greatest number of unsuccessful missions? How many?

c. Which decade had the greatest number of missions? How many?

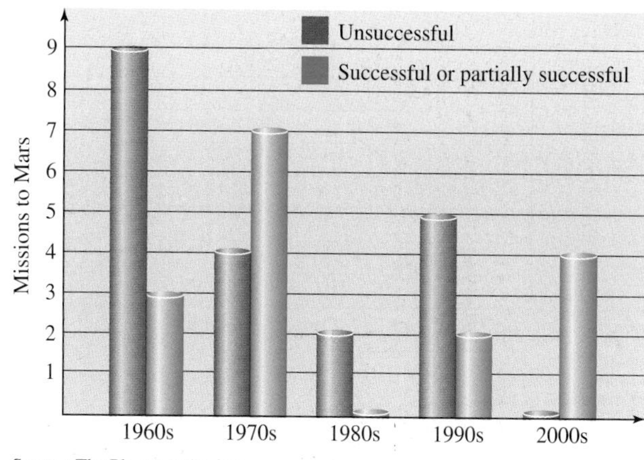

Source: The Planetary Society

62. BANKING The illustration shows the number of banks that were closed or taken over by federal agencies during the years 1935–1995.

a. During what two time spans was there an upsurge in bank failures?

b. In what year were there the most bank failures? Estimate the number of banks that failed that year.

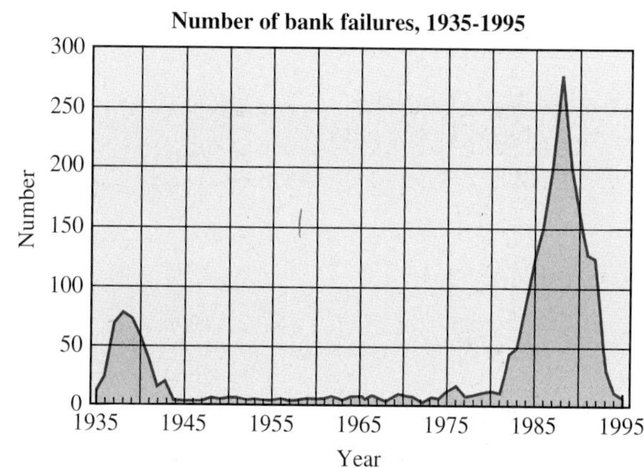

Source: FDIC Division of Research and Statistics

63. ENERGY RESERVES Construct a bar graph (see the next page) using the data shown in the table.

NATURAL GAS RESERVES, 2003 (IN TRILLION CUBIC FEET)	
United States	187
Venezuela	148
Canada	60
Argentina	27
Mexico	9

Source: *Oil and Gas Journal*

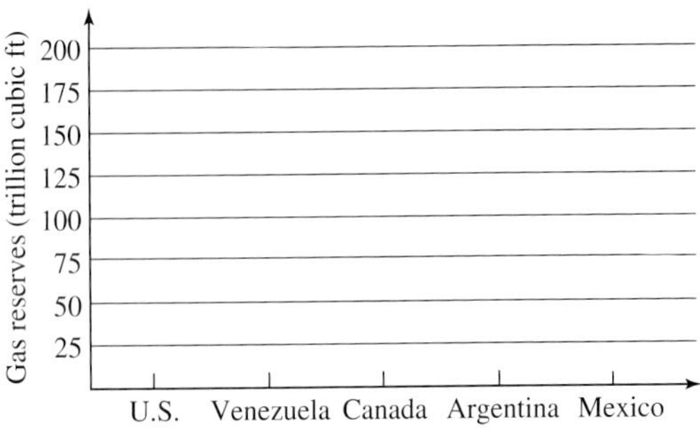

64. ENERGY RESERVES Construct a line graph using the data in the table on the previous page.

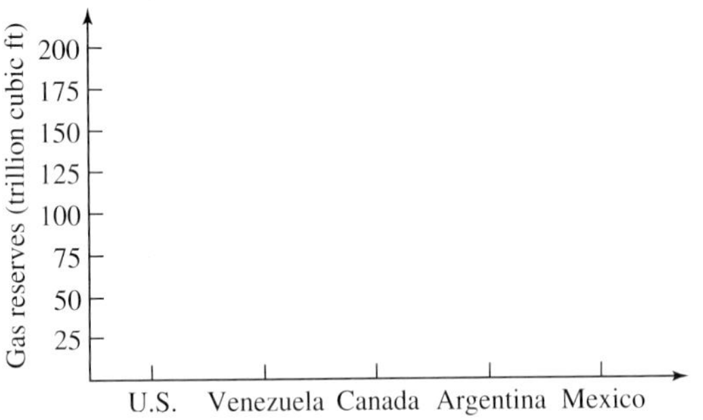

65. COFFEE Construct a line graph using the data in the table.

STARBUCKS LOCATIONS

Year	Number
1997	1,412
1998	1,886
1999	2,135
2000	3,501
2001	4,709
2002	5,886
2003	7,225
2004	8,337

Source: Starbucks Company

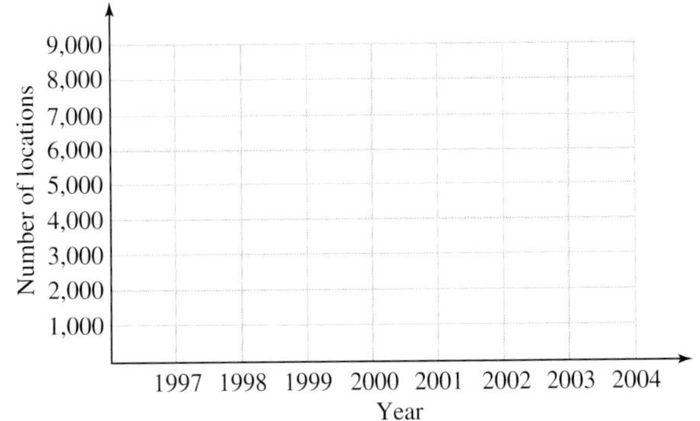

66. COFFEE Construct a bar graph using the data in the table in Exercise 65.

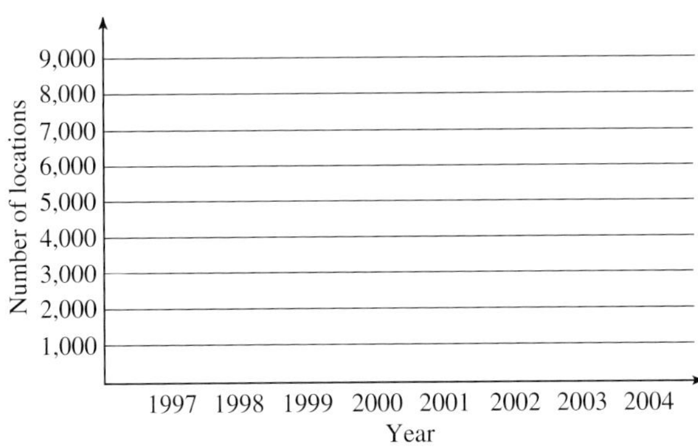

67. Complete each check by writing the amount in words on the proper line.

a.

No. 201 March 9, 20 05

Payable to Davis Chevrolet $ 15,601.00

_______________ DOLLARS

Don Smith

45-365-02

b.

No. 7890 Aug. 12, 20 05

Payable to Dr. Anderson $ 3,433.00

_______________ DOLLARS

Juan Decito

45-828-02

68. ANNOUNCEMENTS One style used when printing formal invitations and announcements is to write all numbers in words. Use this style to write each of the following phrases.

a. This diploma awarded this 27th day of June 2004.

b. The suggested contribution for the fundraiser is \$850 a plate, or an entire table may be purchased for \$5,250.

69. EDITING Edit this excerpt from a history text by circling all numbers written in words and rewriting them using digits.

Abraham Lincoln was elected with a total of one million, eight hundred sixty-five thousand, five hundred ninety-three votes — four hundred eighty-two thousand, eight hundred eighty more than the runner-up, Stephen Douglas. He was assassinated after having served a total of one thousand five hundred three days in office. Lincoln's Gettysburg Address, a mere two hundred sixty-nine words long, was delivered at the battle site where forty-three thousand four hundred forty-nine casualties occurred.

70. READING METERS The amount of electricity used in a household is measured by a meter in kilowatt-hours (kwh). Determine the reading on the meter shown in the illustration. (When the pointer is between two numbers, read the lower number.)

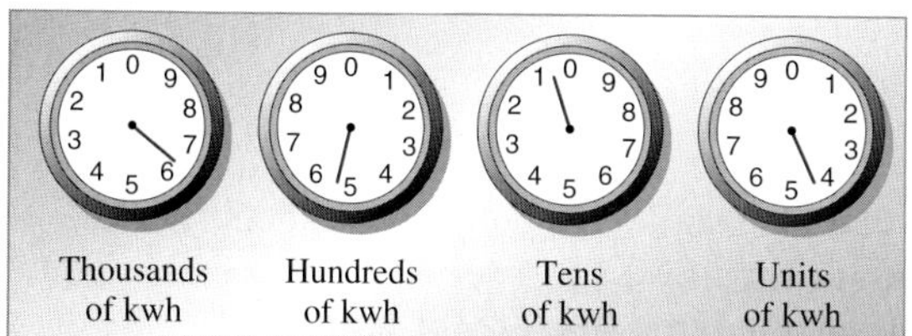

71. SPEED OF LIGHT The speed of light in a vacuum is 299,792,458 meters per second. Round this number

a. to the nearest hundred thousand meters per second.

b. to the nearest million meters per second.

72. CLOUDS Draw a vertical number line scaled from 0 to 40,000 feet, in units of 5,000 feet. Graph each cloud type given in the table at the proper altitude.

Cloud type	Altitude (ft)
Altocumulus	21,000
Cirrocumulus	37,000
Cirrus	38,000
Cumulonimbus	15,000
Cumulus	8,000
Stratocumulus	9,000
Stratus	4,000

WRITING

73. Explain why the natural numbers are called the counting numbers.

74. Explain how you would round 687 to the nearest ten.

75. The houses in a new subdivision are priced "in the low 130s." What does this mean?

76. A million is a thousand thousands. Explain why this is so.

1.2 Adding and Subtracting Whole Numbers

- Properties of addition • Adding whole numbers
- Perimeter of a rectangle and a square • Subtracting whole numbers
- Combinations of operations

Mastering addition and subtraction of whole numbers enables us to solve many problems from geometry, business, and science. For example, to find the distance around a rectangle, we need to add the lengths of its four sides. To prepare an annual budget, we need to add separate line items. To find the difference between two temperatures, we need to subtract one from the other.

Properties of addition

Addition is the process of finding the total of two (or more) numbers. It can be illustrated using the number line. (See Figure 1-5 on the next page.) For example, to compute $4 + 5$, we begin at 0 and draw an arrow 4 units long, extending to the right. This represents 4. From the tip of that arrow, we draw another arrow 5 units long, also extending to the right. The second arrow points to 9. This result corresponds to the addition fact $4 + 5 = 9$, where 4 and 5 are called **addends** or **terms,** and 9 is called the **sum.**

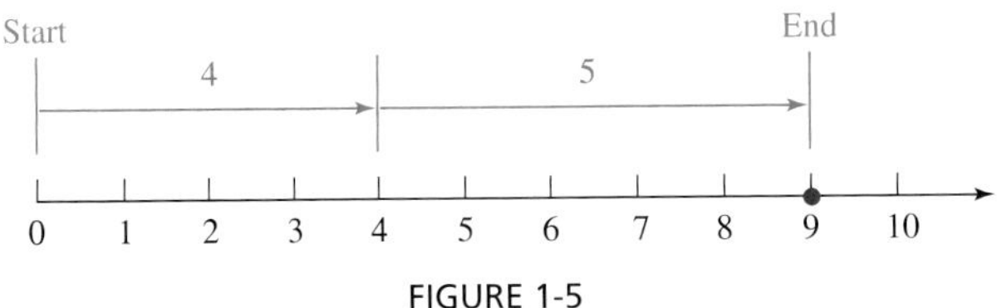

FIGURE 1-5

We have used a number line to find that $4 + 5 = 9$. If we add 4 and 5 in the opposite order, Figure 1-6 shows that we get the same result: $5 + 4 = 9$.

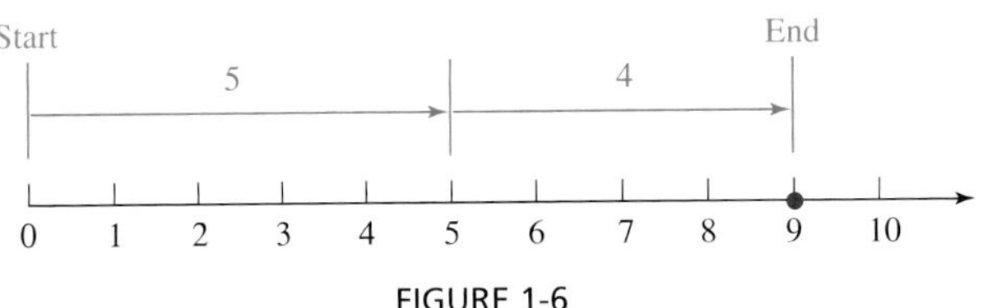

FIGURE 1-6

These examples illustrate that two whole numbers can be added in either order to get the same sum. The order in which we add two numbers does not affect the result. This property is called the **commutative property of addition.** To state the commutative property of addition concisely, we can use variables.

Variables

A **variable** is a letter that is used to stand for a number.

We now use the variables a and b to state the commutative property of addition.

Commutative property of addition

If a and b represent numbers, then

$$a + b = b + a$$

To find the sum of three whole numbers, we add two of them and then add the third to that result. In the following examples, we add $3 + 4 + 2$ in two ways. We will use the **grouping symbols** ( ), called **parentheses,** to show this. We must perform the operation within parentheses first.

Method 1: Group 3 + 4

$(\mathbf{3 + 4}) + 2 = \mathbf{7} + 2$	Because of the parentheses, add 3 and 4 first to get 7.
$= 9$	Then add 7 and 2 to get 9.

Method 2: Group 4 + 2

$3 + (\mathbf{4 + 2}) = 3 + \mathbf{6}$	Because of the parentheses, add 4 and 2 to get 6.
$= 9$	Then add 3 and 6 to get 9.

Either way, the sum is 9. It does not matter how we group (or associate) numbers in addition. This property is called the **associative property of addition.**

Associative property of addition

If a, b, and c represent numbers, then

$$(a + b) + c = a + (b + c)$$

Whenever we add 0 to a number, the number remains the same. For example,

$$3 + 0 = 3, \qquad 5 + 0 = 5, \qquad \text{and} \qquad 9 + 0 = 9$$

These examples suggest the **addition property of 0.**

Addition property of 0

If a represents any number, then

$$a + 0 = a \qquad \text{and} \qquad 0 + a = a$$

EXAMPLE 1 Find each sum: **a.** $8 + 9$ and $9 + 8$, **b.** $5 + (1 + 8)$ and $(5 + 1) + 8$, and **c.** $(3 + 0) + 4$.

Solution

a. $8 + 9 = 17$ and $9 + 8 = 17$ The results are the same.

b. In each case, we perform the addition within parentheses first.

$$5 + (\mathbf{1 + 8}) = 5 + \mathbf{9} \qquad (\mathbf{5 + 1}) + 8 = \mathbf{6} + 8$$
$$= 14 \qquad\qquad = 14 \qquad \text{The results are the same.}$$

c. $(\mathbf{3 + 0}) + 4 = \mathbf{3} + 4$ Perform the addition within parentheses first: $3 + 0 = 3$.
$= 7$

Self Check 1

Find each sum:

a. $6 + 7$ and $7 + 6$

b. $2 + (6 + 3)$ and $(2 + 6) + 3$

c. $3 + (0 + 4)$

Answers **a.** 13, 13, **b.** 11, 11, **c.** 7

Adding whole numbers

We can add whole numbers greater than 10 by using a vertical format that adds digits with the same place value. Because the additions in each column often exceed 9, it is sometimes necessary to *carry* the excess to the next column to the left. For example, to add 27 and 15, we write the numerals with the digits of the same place value aligned vertically.

$$\begin{array}{r} 27 \\ +15 \\ \hline \end{array}$$ This is called vertical format.

We begin by adding the digits in the ones column: $7 + 5 = 12$. Because $12 =$ 1 ten and 2 ones, we place a 2 in the ones column of the answer and carry 1 to the tens column.

$$\begin{array}{r} {}^{1} \\ 2\mathbf{7} \\ +1\mathbf{5} \\ \hline \mathbf{2} \end{array}$$ Add the digits in the ones column: $7 + 5 = 12$. Carry 1 (shown in blue) to the tens column.

Then we add the digits in the tens column.

$$\begin{array}{r} {}^{1} \\ 27 \\ +15 \\ \hline 42 \end{array}$$ Add 1, 2, and 1. Place the result, 4, in the tens column of the answer.

Thus, $27 + 15 = 42$.

Self Check 2
Add: 675 + 1,497.

EXAMPLE 2 Add: 9,834 + 692.

Solution We write the numerals with their corresponding digits aligned vertically. Then we add the numbers, one column at a time, working from right to left.

$$\begin{array}{r} 9{,}83\mathbf{4} \\ +\ \ 69\mathbf{2} \\ \hline \mathbf{6} \end{array}$$
Add the digits in the ones column and place the result in the ones column of the answer.

$$\begin{array}{r} ^{1}\ \ \\ 9{,}8\mathbf{3}4 \\ +\ \ 6\mathbf{9}2 \\ \hline \mathbf{2}6 \end{array}$$
Add the digits in the tens column. The result, 12, exceeds 9. Place the 2 in the tens column of the answer and carry 1 (shown in blue) to the hundreds column.

$$\begin{array}{r} ^{1\,1}\ \ \ \ \\ 9{,}834 \\ +\ \ 692 \\ \hline 526 \end{array}$$
Add the digits in the hundreds column. Since the result, 15, exceeds 9, place the 5 in the hundreds column of the answer and carry 1 (shown in green) to the thousands column.

$$\begin{array}{r} ^{1\,1}\ \ \ \ \\ 9{,}834 \\ +\ \ \ \ 692 \\ \hline \mathbf{10}{,}526 \end{array}$$
Since the sum of the digits in the thousands column is 10, write 0 in the thousands column and 1 in the ten thousands column of the answer.

Thus, 9,834 + 692 = 10,526.

Answer 2,172

To see if the result in Example 2 is reasonable, we can **estimate** the answer. We know that 9,834 is a little less than 10,000, and 692 is a little less than 700. We estimate that the answer will be a little less than 10,000 + 700, or 10,700. An answer of 10,526 is reasonable. Estimation is discussed in more detail later in this chapter.

Words such as *increase, gain, credit, up, forward, rise, in the future, and to the right* are used to indicate addition.

EXAMPLE 3 **Calculating temperatures.**

At noon, the temperature in Helena, Montana, was 31°. By 1:00 P.M., the temperature had increased 5°, and by 2:00 P.M., it had risen another 7°. Find the temperature at 2:00 P.M.

Solution To the temperature at noon, we add the two increases.

$$31 + 5 + 7$$

The two additions are done working from left to right.

$$\begin{aligned} \mathbf{31 + 5} + 7 &= \mathbf{36} + 7 \\ &= 43 \end{aligned}$$

The temperature at 2:00 P.M. was 43°.

Self Check 4
By 1700, the populations of the four colonies were New Hampshire 5,000, New York 19,100, Massachusetts 55,900, and Virginia 58,600. Find the total population.

EXAMPLE 4 **History.**

The populations of four American colonies in 1630 are shown in Figure 1-7 on the next page. Find the total population.

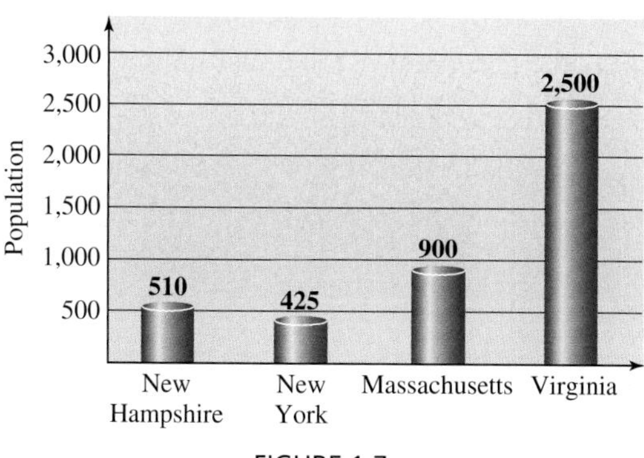

FIGURE 1-7

Solution The word *total* indicates that we must add the populations of the colonies.

$$\begin{array}{r} {}^{2}510 \\ 425 \\ 900 \\ +2,500 \\ \hline 4,335 \end{array}$$

Align the numerals vertically. Add the digits, one column at a time, working from right to left.

The total population was 4,335.

Answer 138,600

Perimeter of a rectangle and a square

Figure 1-8(a) is an example of a four-sided figure called a **rectangle.** Either of the longer sides of a rectangle is called its **length** and either of the shorter sides is called its **width.** Together, the length and width are called the **dimensions** of the rectangle. For any rectangle, opposite sides have the same measure.

When all four of the sides of a rectangle have the same measure, we call the rectangle a **square.** An example of a square is shown in Figure 1-8(b).

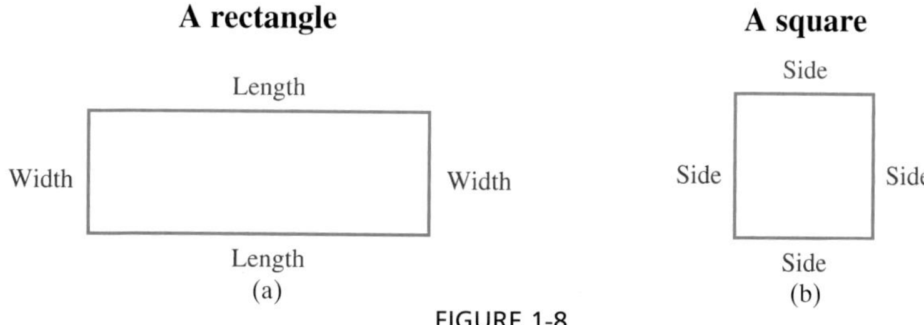

FIGURE 1-8

The distance around a rectangle or a square is called its **perimeter.** To find the perimeter of a rectangle, we add the lengths of its four sides.

The perimeter of a rectangle	=	length	+	length	+	width	+	width

To find the perimeter of a square, we add the lengths of its four sides.

The perimeter of a square	=	side	+	side	+	side	+	side

Self Check 5

A Monopoly game board is square, with sides 19 inches long. Find the perimeter of the board.

Answer 76 in.

EXAMPLE 5 Find the perimeter of the dollar bill shown in Figure 1-9.

Solution To find the perimeter of the rectangular-shaped bill, we add the lengths of its four sides.

Width = 65 mm

Length = 156 mm

mm stands for millimeters

FIGURE 1-9

$$\begin{array}{r} {\scriptstyle 2\,2} \\ 156 \\ 156 \\ 65 \\ +\ \underline{65} \\ 442 \end{array}$$

The perimeter is 442 mm.

To see whether this result is reasonable, we estimate the answer. Because the rectangle is about 150 mm by 70 mm, its perimeter is approximately 150 + 150 + 70 + 70, or 440 mm. An answer of 442 mm is reasonable.

Subtracting whole numbers

Subtraction is the process of finding the difference between two numbers. It can be illustrated using the number line. (See Figure 1-10.) For example, to compute $9 - 4$, we begin at 0 and draw an arrow 9 units long, extending to the right. From the tip of that arrow, we draw another arrow 4 units long, but extending to the left. (This represents taking away 4.) The second arrow points to 5, indicating that $9 - 4 = 5$. In this subtraction fact, 9 is called the **minuend,** 4 is called the **subtrahend,** and 5 is called the **difference.**

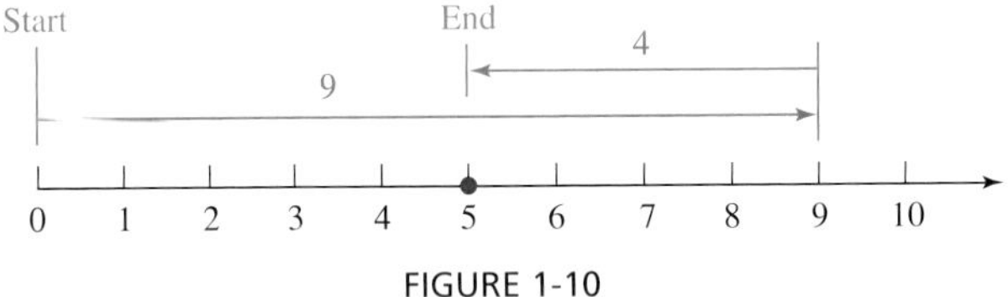

FIGURE 1-10

With whole numbers, we cannot subtract in the opposite order and find the difference $4 - 9$, because we cannot take away 9 objects from 4 objects. Since subtraction of whole numbers cannot be done in either order, **subtraction is not commutative.**

Subtraction is not associative either, because if we group in different ways, we get different answers. For example,

$$(\mathbf{9 - 5}) - 1 = \mathbf{4} - 1 \qquad \text{but} \qquad 9 - (\mathbf{5 - 1}) = 9 - \mathbf{4}$$
$$= 3 \qquad\qquad\qquad\qquad = 5$$

Self Check 6

Perform the subtractions: $8 - (5 - 2)$ and $(8 - 5) - 2$.

Answers 5, 1

EXAMPLE 6 Perform the subtractions: $9 - (6 - 3)$ and $(9 - 6) - 3$.

Solution In each case, we perform the subtraction within parentheses first.

$$9 - (\mathbf{6 - 3}) = 9 - \mathbf{3} \qquad (\mathbf{9 - 6}) - 3 = \mathbf{3} - 3$$
$$= 6 \qquad\qquad\qquad = 0$$

Note that the results are different.

Whole numbers can be subtracted using a vertical format. Because subtractions often require subtracting a larger digit from a smaller digit, we may need to *borrow.* For example, to subtract 15 from 32, we write the minuend, 32, and the subtrahend, 15, in a vertical format, aligning the digits with the same place value.

$$\begin{array}{r} 32 \\ -15 \\ \hline \end{array}$$

Write the numerals in a column, with corresponding digits aligned vertically.

Since 5 can't be subtracted from 2, we borrow from the tens column of 32.

$$\begin{array}{r} {}^{2}\,{}^{12} \\ \not{3}\;\not{2} \\ -1\;\mathbf{5} \\ \hline \mathbf{7} \end{array}$$

To subtract in the ones column, borrow 1 ten from the tens column. We show this by drawing a slash through the 3 and writing a 2 above it. Add 10 to the 2 in the ones column, which gives 12. Then subtract: 12 − 5 = 7.

$$\begin{array}{r} {}^{2}\,{}^{12} \\ \not{3}\;\not{2} \\ -\mathbf{1}\;5 \\ \hline \mathbf{1}\;7 \end{array}$$

Subtract in the tens column: 2 − 1 = 1.

Thus, 32 − 15 = 17. To check the result, we add the difference, 17, and the subtrahend, 15. We should obtain the minuend, 32.

Check:

$$\begin{array}{r} {}^{1} \\ 17 \\ +15 \\ \hline 32 \end{array}$$

EXAMPLE 7 Subtract 576 from 2,021.

Solution The number to be subtracted is 576. This sentence translates to 2,021 − 576. In vertical format, we have

$$\begin{array}{r} 2{,}021 \\ -\;\;576 \\ \hline \end{array}$$

Write the numerals in a column, with the digits of the same place value aligned vertically.

$$\begin{array}{r} {}^{1}\;{}^{11} \\ 2{,}0\;\not{2}\;\not{1} \\ -\;5\;7\;\mathbf{6} \\ \hline \mathbf{5} \end{array}$$

To subtract in the ones column, borrow 1 ten from the tens column and add it to the ones column. Then subtract: 11 − 6 = 5.

Since we can't subtract 7 from 1 in the tens column, we borrow. Because there is a 0 in the hundreds column of 2,021, we must borrow from the thousands column. We can take 1 thousand from the thousands column (leaving 1 thousand behind) and write it as 10 hundreds, placing a 10 in the hundreds column. From these 10 hundreds, we take 1 hundred (leaving 9 hundreds behind) and think of it as 10 tens. We add these 10 tens to the 1 ten that is already in the tens column to get 11 tens. From these 11 tens, we subtract 7 tens: 11 − 7 = 4.

$$\begin{array}{r} {}^{9} \\ {}^{1}\;\not{10}\;\mathbf{11}\;11 \\ \not{2},\not{0}\,\not{2}\,\not{1} \\ -\;5\;\mathbf{7}\;6 \\ \hline \mathbf{4}\;5 \end{array}$$

To subtract in the tens column, borrow 10 hundreds from the thousands digit and add it to the hundreds digit. Borrow 10 tens from the hundreds digit and add it to the tens digit. Then subtract: 11 − 7 = 4.

$$\begin{array}{r} {}^{9} \\ {}^{1}\;\not{10}\;11\;11 \\ \not{2},\not{0}\,\not{2}\,\not{1} \\ -\;\mathbf{5}\;7\;6 \\ \hline \mathbf{4}\;4\;5 \end{array}$$

Subtract in the hundreds column: 9 − 5 = 4.

Self Check 7

Subtract 1,445 from 2,021. Then check the result using addition.

$$\begin{array}{r} \overset{1}{\cancel{2}},\overset{\overset{9}{\cancel{10}}}{\cancel{0}}\ \overset{11}{\cancel{2}}\ \overset{11}{\cancel{1}} \\ -\quad 5\ 7\ 6 \\ \hline \mathbf{1},\ 4\ 4\ 5 \end{array}$$ Subtract in the thousands column: $1 - 0 = 1$.

Thus, $2{,}021 - 576 = 1{,}445$. Check the result using addition.

Answer 576; $576 + 1{,}445 = 2{,}021$

Words such as *minus, decrease, loss, debit, down, backward, fall, reduce, in the past,* and *to the left* indicate subtraction.

EXAMPLE 8 Vehicle crashes. In 2000, the number of motor vehicle traffic crashes in the United States was 6,394,000. That number declined in 2001, dropping by 71,000. In 2002, it fell by an additional 7,000. How many motor vehicle traffic crashes were there in 2002?

Solution The words *dropping* and *fell* indicate subtraction. We can show the calculations necessary to solve this example in a single expression, as shown below. The two subtractions are done working from left to right.

$$\begin{aligned} \mathbf{6{,}394{,}000} - \mathbf{71{,}000} - 7{,}000 &= \mathbf{6{,}323{,}000} - 7{,}000 \\ &= 6{,}316{,}000 \end{aligned}$$

In 2002, there were 6,316,000 motor vehicle traffic crashes in the United States.

To answer questions about *how much more* or *how many more,* subtraction can be used.

Combinations of operations

Additions and subtractions often appear in the same problem. It is important to read the problem carefully, locate the useful information, and organize it correctly.

Self Check 9

One share of ABC Corporation stock cost $75. The price fell $7 per share. However, it recovered and rose $13 per share. What is its current price?

Answer $81

EXAMPLE 9 Bus passengers. Twenty-seven people were riding a bus on Route 47. At the Seventh Street stop, 16 riders got off the bus and 5 got on. How many riders were left on the bus?

Solution The route and street number are not important. The phrase *got off the bus* indicates subtraction, and the phrase *got on* indicates addition. The number of riders on the bus can be found by calculating $27 - 16 + 5$. Working from left to right, we have

$$\begin{aligned} \mathbf{27} - \mathbf{16} + 5 &= \mathbf{11} + 5 \\ &= 16 \end{aligned}$$

There were 16 riders left on the bus.

! COMMENT When making the calculation in Example 9, we must perform the subtraction first. If the addition is done first, we obtain an incorrect answer of 6. For expressions containing addition and subtraction, perform them as they occur from *left to right.*

$$\begin{aligned} 27 - 16 + 5 &= 27 - 21 \\ &= 6 \end{aligned}$$

Calculators

CALCULATOR SNAPSHOT

A calculator can be helpful when you are checking an answer or performing a tedious computation. Before making regular use of one, make sure that you have mastered the fundamentals of arithmetic.

Several brands of calculators are available. For specific details about the operation of your calculator, please consult the owner's manual.

To check the addition done in Example 4 using a scientific calculator, we enter these numbers and press these keys.

510 [+] 425 [+] 900 [+] 2500 [=] 4335

The display shows that in 1630, the total population of the four colonies was 4,335.

We can use a scientific calculator to check the subtraction performed in Example 8 by entering these numbers and pressing these keys.

17126 [−] 937 [−] 253 [=] 15936

The display shows that there were 15,936 alcohol-related traffic deaths in 1998.

Section 1.2 STUDY SET

VOCABULARY *Fill in the blanks.*

1. When two numbers are added, the result is called a ______. The numbers that are to be added are called ________.
2. A ________ is a letter that stands for a number.
3. When two numbers are added, the result is called a ______.
4. The figure on the left is an example of a ________. The figure on the right is an example of a ________.

5. When two numbers are subtracted, the result is called a ________. In a subtraction problem, the ________ is subtracted from the ________.
6. The property that guarantees that we can add two numbers in either order and get the same sum is called the ________ property of addition.
7. The property that allows us to group numbers in an addition in any way we want is called the ________ property of addition.
8. The distance around a rectangle (or a square) is called its ________.

CONCEPTS *What property of addition is shown?*

9. $3 + 4 = 4 + 3$
10. $(3 + 4) + 5 = 3 + (4 + 5)$
11. $7 + (8 + 2) = (7 + 8) + 2$
12. $(8 + 5) + 1 = 1 + (8 + 5)$
13. a. Use the variables x and y to write the commutative property of addition.

    b. Use the variables x, y, and z to write the associative property of addition.
14. Show how to check the result:
$$\begin{array}{r} 74 \\ -29 \\ \hline 45 \end{array}$$
15. Fill in the blank: Any number added to ___ stays the same.
16. a. In calculating $12 + (8 + 5)$, which numbers should be added first?

    b. In calculating $60 - 15 + 4$, which operation should be performed first?
17. What addition fact is illustrated below?

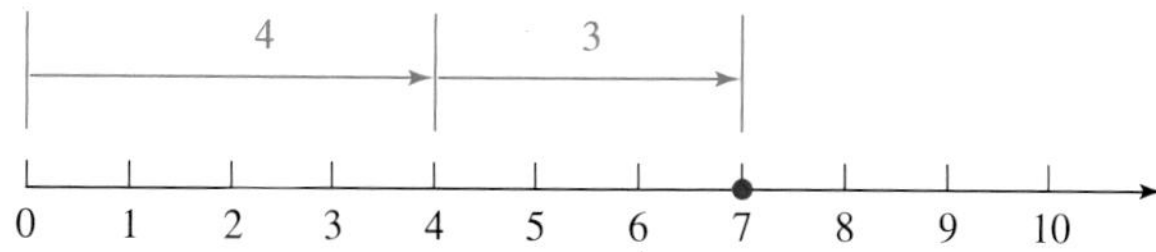

18. What subtraction fact is illustrated below?

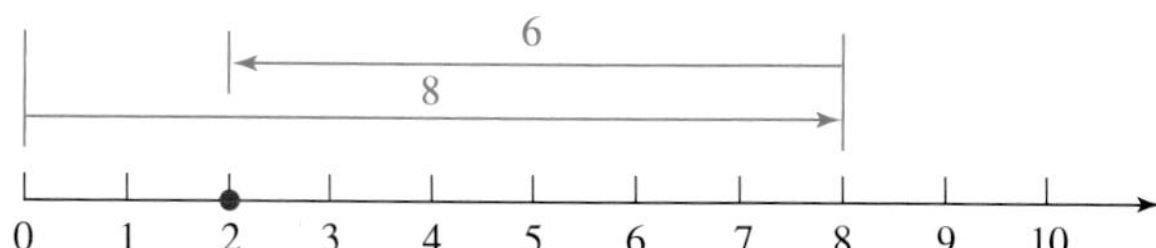

NOTATION *Fill in the blanks.*

19. The grouping symbols ( ) are called ________.
20. The minus sign − means ________.

Complete each solution.

21. (36 + 11) + 5 = ___ + 5
 = ___

22. 12 + (15 + 2) = 12 + ___
 = ___

PRACTICE *Perform each addition.*

23. 25 + 13
24. 47 + 12
25. 156 + 305
26. 647 + 38
27. 19 + 39 + 53
28. 27 + 16 + 48
29. (95 + 16) + 39
30. 832 + (97 + 27)
31. 25 + (321 + 17)
32. (4,231 + 213) + 5,234

33. 632 + 347

34. 423 + 570

35. 1,372 + 613

36. 2,477 + 693

37. 6,427 + 3,573

38. 3,567 + 8,778

39. 8,539 + 7,368

40. 5,799 + 6,879

41. 1,246 + 578 + 37

42. 4,689 + 3,422 + 26

43. 3,156 + 1,578 + 578

44. 2,379 + 4,779 + 2,339

Find the perimeter of each rectangle or square.

45. 

46. 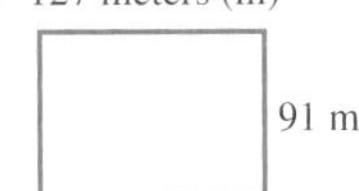

47. 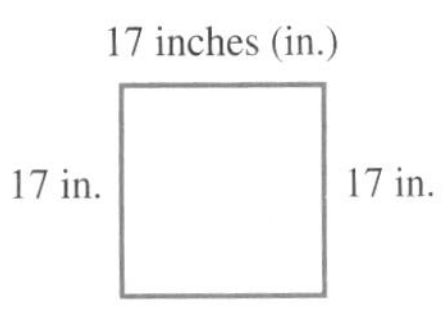

48. 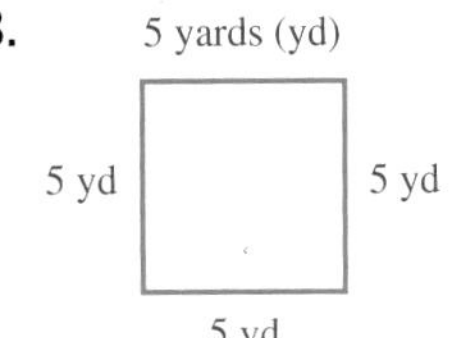

Perform each subtraction.

49. 17 − 14
50. 42 − 31
51. 39 − 14
52. 45 − 32
53. 174 − 71
54. 257 − 155
55. 633 − (598 − 30)
56. 600 − (497 − 60)
57. 160 − 15 − 4
58. 498 − 17 − 162
59. 29 − 17 − 12
60. 53 − 26 − 27
57. 160 − 15 − 4
58. 498 − 17 − 162
59. 29 − 17 − 12
60. 53 − 26 − 27
61. Subtract 343 from 367.
62. Subtract 122 from 224.
63. Subtract 305 from 423.
64. Subtract 270 from 330.

65. 1,537 − 579

66. 2,470 − 863

67. 4,267 − 2,578

68. 7,356 − 3,578

69. 17,246 − 6,789

70. 34,510 − 27,593

71. 15,700 − 15,397

72. 35,021 − 23,999

Perform the computations.

73. 43 − 12 + 9
74. 59 − 16 + 2
75. 120 + 30 − 40
76. 600 + 99 − 54

APPLICATIONS

77. TAXIS For a 17-mile trip, Wanda paid the taxi driver \$23. If \$5 was a tip, how much was the fare?

78. SPACE FLIGHTS Astronaut Walter Schirra's first space flight orbited the Earth 6 times and lasted 9 hours. His second flight orbited the Earth 16 times and lasted 26 hours. How long was Schirra in space?

79. DOW JONES AVERAGE How much did the Dow rise on the day described by the graph?

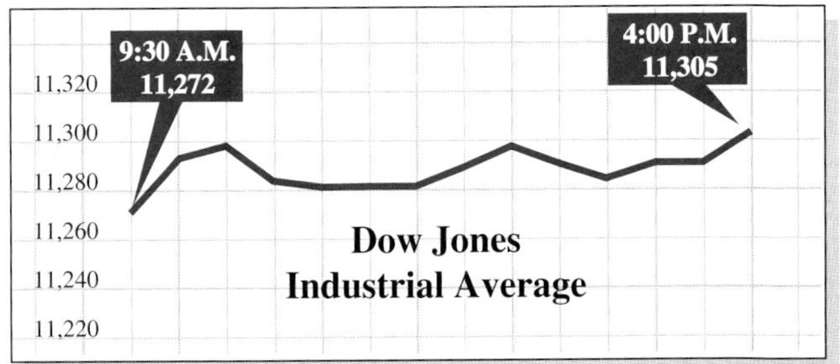

80. JEWELRY Gold melts at about 1,947° F. The melting point of silver is 183° F lower. What is the melting point of silver?

81. BANKING A savings account contained $370. After a deposit of $40 and a withdrawal of $197, how much is in the account?

82. TRAVEL A student wants to make the 2,221-mile trip from Detroit to Seattle in three days. If she drives 751 miles on the first day and 875 miles on the second day, how far must she travel on the third day?

83. TAX DEDUCTIONS For tax purposes, a woman kept the mileage records shown on the right. Find the total number of miles that she drove in the first 6 months of the year.

Month	Miles driven
January	2,345
February	1,712
March	1,778
April	445
May	1,003
June	2,774

84. COMPANY BUDGETS A department head prepared an annual budget with the line items shown on the right. Find the projected number of dollars to be spent.

Line item	Amount
Equipment	$17,242
Utilities	$5,443
Travel	$2,775
Supplies	$10,553
Development	$3,225
Maintenance	$1,075

Refer to the following table. To use this salary schedule, note that the annual salary of a third-year teacher with 15 units of course work beyond a Bachelor's degree is $30,887 (Step 3/Column 2).

Teachers' Salary Schedule			
Years teaching	Column 1: B.D.	Column 2: B.D. + 15	Column 3: B.D. + 30
Step 1	$26,785	$28,243	$29,701
Step 2	$28,107	$29,565	$31,023
Step 3	$29,429	$30,887	$32,345
Step 4	$30,751	$32,209	$33,667
Step 5	$32,073	$33,531	$34,989

85. INCOME How much money will a new teacher make in his first five years of teaching if he begins at

a. Step 1/Column 1?

b. Step 1/Column 3?

86. PAY INCREASES If a teacher is now on Step 2/Column 2, how much more money will she make next year when she

a. gains one year of teaching experience?

b. completes 15 units of course work?

87. BLUEPRINTS Find the length of the house shown in the blueprint.

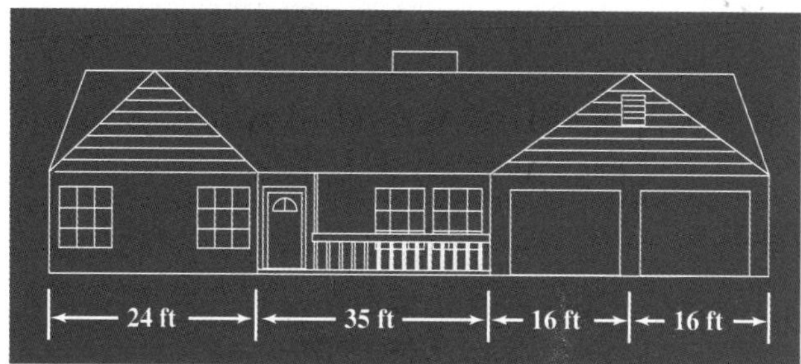

88. MACHINERY Find the length of the motor on the machine shown below (cm means centimeters).

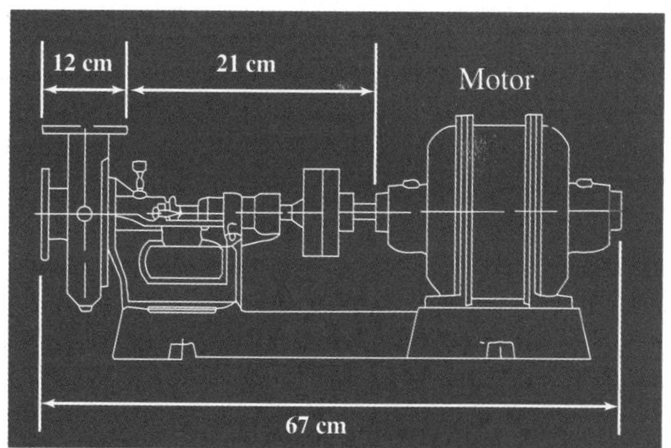

89. CANDY The graph below shows U.S. candy sales in 2003 during four holiday periods. Find the sum of these seasonal candy sales.

Source: National Confectioners Association

90. DALMATIANS See the graph below. Between which two years was the drop in registrations the greatest? What was that drop?

Number of new Dalmatians registered with the American Kennel Club

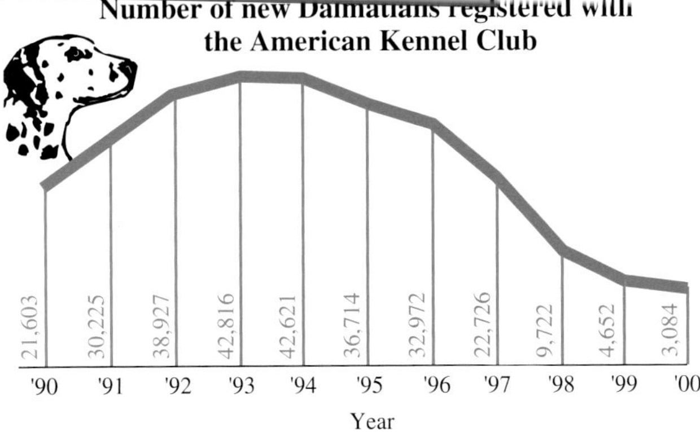

91. CITY FLAGS To decorate a city flag, yellow fringe is to be sewn around its outside edges, as shown. The fringe comes on long spools and is sold by the inch. How many inches of fringe must be purchased to complete the project?

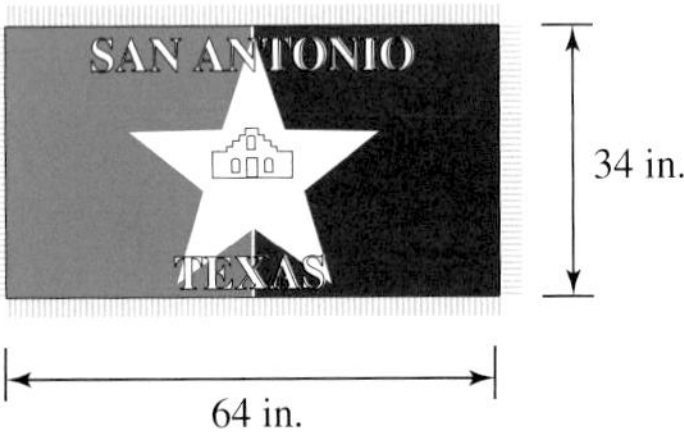

92. BOXING How much padded rope is needed to create the square boxing ring below if each side is 24 feet long?

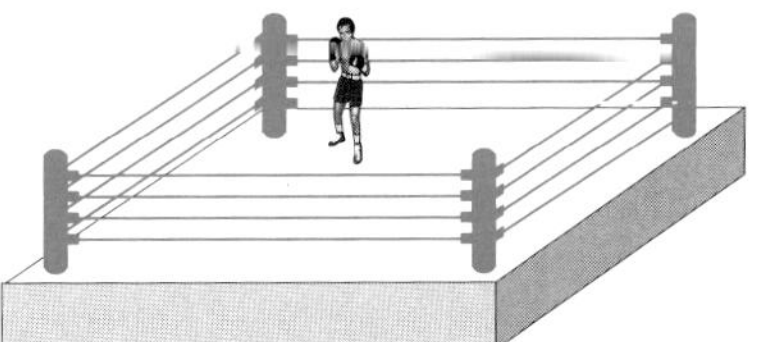

WRITING

93. Explain why the operation of addition is commutative.

94. Explain how addition can be used to check subtraction.

REVIEW *Write each numeral in expanded notation.*

95. 3,125

96. 60,037

Round 6,354,784 to the specified place.

97. Nearest ten

98. Nearest hundred

99. Nearest ten thousand

100. Nearest hundred thousand

1.3 Multiplying and Dividing Whole Numbers

- Properties of multiplication • Multiplying whole numbers
- Finding the area of a rectangle • Division • Properties of division
- Dividing whole numbers

Mastering multiplication and division of whole numbers enables us to find areas of geometric figures and to solve many business and transportation problems. For example, to find the area of a rectangle, we need to multiply its length by its width. To figure a paycheck, we need to multiply the number of hours worked by the hourly rate of pay. To calculate the fuel economy of a bus, we need to divide the miles it travels by the number of gallons of gas that are used.

Properties of multiplication

Three symbols are used to indicate multiplication.

Symbols that are used for multiplication

Symbol		Example
$\times$	times symbol	4×5 or $\begin{array}{r} 4 \\ \underline{\times 5} \end{array}$
$\cdot$	raised dot	$4 \cdot 5$
( )	parentheses	(4)(5) or 4(5) or (4)5

Recall that a variable is a letter that stands for a number. We often multiply a variable by another number or multiply a variable by another variable. When we do this, we don't need to use a symbol for multiplication.

$5a$ means $5 \cdot a$, ab means $a \cdot b$, and xyz means $x \cdot y \cdot z$

! COMMENT In this book, we seldom use the $\times$ symbol, because it can be confused with the letter x.

Multiplication is repeated addition. For example, $4 \cdot 5$ means the sum of four 5's.

$$4 \cdot 5 = \overbrace{5 + 5 + 5 + 5}^{\text{The sum of four 5's}}$$
$$= 20$$

In the above multiplication, the result of 20 is called a **product.** The numbers that were multiplied (4 and 5) are called **factors.**

$$\begin{array}{ccccc} \text{Factor} & & \text{Factor} & & \text{Product} \\ \downarrow & & \downarrow & & \downarrow \\ 4 & \cdot & 5 & = & 20 \end{array}$$

The multiplication $5 \cdot 4$ means the sum of five 4's.

$$5 \cdot 4 = \overbrace{4 + 4 + 4 + 4 + 4}^{\text{The sum of five 4's}}$$
$$= 20$$

We see that $4 \cdot 5 = 20$ and $5 \cdot 4 = 20$. The results are the same. These examples illustrate that the order in which we multiply two numbers does not affect the result. This property is called the **commutative property of multiplication.**

Commutative property of multiplication

If a and b represent numbers, then

$a \cdot b = b \cdot a$ or, more simply, $ab = ba$

Table 1-1 on the next page summarizes the basic multiplication facts.

- To find the product of 6 and 8 using the table, we find the intersection of the **6**th row and the **8**th column. The product is 48.
- To find the product of 8 and 6, we find the intersection of the **8**th row and the **6**th column. Once again, the product is 48.

In the table, we see that the set of answers above and the set of answers below the diagonal line in bold print are identical. This further illustrates that multiplication is commutative.

·	0	1	2	3	4	5	6	7	8	9
0	**0**	0	0	0	0	0	0	0	0	0
1	0	**1**	2	3	4	5	6	7	8	9
2	0	2	**4**	6	8	10	12	14	16	18
3	0	3	6	**9**	12	15	18	21	24	27
4	0	4	8	12	**16**	20	24	28	32	36
5	0	5	10	15	20	**25**	30	35	40	45
6	0	6	12	18	24	30	**36**	42	48	54
7	0	7	14	21	28	35	42	**49**	56	63
8	0	8	16	24	32	40	48	56	**64**	72
9	0	9	18	27	36	45	54	63	72	**81**

TABLE 1-1

From the table, we see that whenever we multiply a number by 0, the product is 0. For example,

$0 \cdot 5 = 0, \quad 0 \cdot 8 = 0, \quad \text{and} \quad 9 \cdot 0 = 0$

We also see that whenever we multiply a number by 1, the number remains the same. For example,

$3 \cdot 1 = 3, \quad 7 \cdot 1 = 7, \quad \text{and} \quad 1 \cdot 9 = 9$

These examples suggest the multiplication properties of 0 and 1.

Multiplication properties of 0 and 1

If a represents any number, then

$a \cdot 0 = 0 \quad \text{and} \quad 0 \cdot a = 0$

$a \cdot 1 = a \quad \text{and} \quad 1 \cdot a = a$

Application problems that involve repeated addition are often more easily solved using multiplication.

Self Check 1

At a rate of $8 per hour, how much will a school bus driver earn if she works from 8:00 A.M. until noon?

Answer $32

EXAMPLE 1 Computing daily wages. Raul worked an 8-hour day at an hourly rate of $9. How much money did he earn?

Solution For each of the 8 hours, Raul earned $9. His total pay for the day is the sum of eight 9's: 9 + 9 + 9 + 9 + 9 + 9 + 9 + 9. This repeated addition can be calculated by multiplication.

Total wages $= 8 \cdot 9$

$= 72$ See the multiplication table.

Raul earned $72.

To multiply three numbers, we first multiply two of them and then multiply that result by the third number. In the following examples, we multiply $3 \cdot 2 \cdot 4$ in two ways. The parentheses show us which multiplication to perform first.

Method 1: Group 3 · 2

$(\mathbf{3 \cdot 2}) \cdot 4 = \mathbf{6} \cdot 4$ Multiply 3 and 2 to get 6.

$= 24$ Then multiply 6 and 4 to get 24.

Method 2: Group 2 · 4

$3 \cdot (\mathbf{2} \cdot \mathbf{4}) = 3 \cdot \mathbf{8}$ Multiply 2 and 4 to get 8.

$= 24$ Then multiply 3 and 8 to get 24.

The answers are the same. This illustrates that changing the grouping when multiplying numbers does not affect the result. This property is called the **associative property of multiplication.**

Associative property of multiplication

If a, b, and c represent numbers, then

$(a \cdot b) \cdot c = a \cdot (b \cdot c)$ or, more simply, $(ab)c = a(bc)$

Multiplying whole numbers

To find the product 8 · 47, it is inconvenient to add up eight 47's. Instead, we find the product by a multiplication process.

$$\begin{array}{r} 47 \\ \times \quad 8 \\ \hline \end{array}$$
Write the factors in a column, with the corresponding digits aligned vertically.

$$\begin{array}{r} {}^{5} \\ 4\mathbf{7} \\ \times \quad \mathbf{8} \\ \hline \mathbf{6} \end{array}$$
Multiply 7 by 8. The product is 56. Place 6 in the ones column of the answer and carry 5 (in blue) to the tens column.

$$\begin{array}{r} {}^{5} \\ \mathbf{4}7 \\ \times \quad \mathbf{8} \\ \hline \mathbf{37}6 \end{array}$$
Multiply 4 by 8. The product is 32. To the 32, add the carried 5 to get 37. Place the 7 in the tens column and the 3 in the hundreds column of the answer.

The product is 376.

To find the product 23 · 435, we use the multiplication process. Because $23 = 20 + 3$, we multiply 435 by 20 and by 3 and then add the products. To do this, we write the factors in a column, with the corresponding digits aligned vertically. We then begin the process by multiplying 435 by 3.

$$\begin{array}{r} {}^{1} \\ 43\mathbf{5} \\ \times \ 2\mathbf{3} \\ \hline \mathbf{5} \end{array}$$
Multiply 5 by 3. The product is 15. Place 5 in the ones column and carry 1 (in blue) to the tens column.

$$\begin{array}{r} {}^{1\,1} \\ 4\mathbf{3}5 \\ \times \ 2\mathbf{3} \\ \hline \mathbf{0}5 \end{array}$$
Multiply 3 by 3. The product is 9. To the 9, add the carried 1 to get 10. Place the 0 in the tens column and carry the 1 (in green) to the hundreds column.

$$\begin{array}{r} {}^{1\,1} \\ \mathbf{4}35 \\ \times \ 2\mathbf{3} \\ \hline \mathbf{13}05 \end{array}$$
Multiply 4 by 3. The product is 12. Add the 12 to the carried 1 to get 13. Write 13.

We continue by multiplying 435 by 2 tens, or 20.

$$\begin{array}{r} {}^{1} \\ 43\mathbf{5} \\ \times \ \mathbf{2}3 \\ \hline 1305 \\ \mathbf{0} \end{array}$$
Multiply 5 by 2. The product is 10. Write 0 in the tens column and carry 1 (in purple).

$$\begin{array}{r} {}^{1} \\ 4\mathbf{3}5 \\ \times \ \mathbf{2}3 \\ \hline 1305 \\ \mathbf{7}0 \end{array}$$
Multiply 3 by 2. The product is 6. Add 6 to the carried 1 to get 7. Write the 7. There is no carry.

$$\begin{array}{r} \overset{1}{\mathbf{4}}35 \\ \times\ \mathbf{2}3 \\ \hline 1305 \\ \mathbf{8}70 \end{array}$$

Multiply 4 by 2. The product is 8. There is no carry to add. Write the 8.

$$\begin{array}{r} 435 \\ \times\ 23 \\ \hline 1305 \\ 870 \\ \hline 10005 \end{array}$$

Draw another line beneath the two completed rows. Add the two rows. This sum gives the product of 435 and 23.

Thus, $23 \cdot 435 = 10{,}005$.

Self Check 2

For highway driving, how far can the Explorer travel on a tank of gas?

EXAMPLE 2 **Mileage.** Specifications for a Ford Explorer 4 × 4 are shown in the table below. For city driving, how far can it travel on a tank of gas? (The abbreviation "mpg" means "miles per gallon.")

Engine	4.0 L V6
Fuel capacity	21 gal
Fuel economy (mpg)	15 city/19 hwy

Solution For city driving, each of the 21 gallons of gas that the tank holds enables the Explorer to go 15 miles. The total distance it can travel is the sum of twenty-one 15's. This can be calculated by multiplication: $21 \cdot 15$.

$$\begin{array}{r} \overset{1}{1}5 \\ \times 21 \\ \hline 15 \\ 30 \\ \hline 315 \end{array}$$

For city driving, the Explorer can go 315 miles on a tank of gas.

Answer 399 mi

EXAMPLE 3 **Calculating production.** The labor force of an electronics firm works two 8-hour shifts each day and manufactures 53 television sets each hour. Find how many sets will be manufactured in 5 days.

Solution The number of TV sets manufactured in 5 days is given by the product.

2 shifts per day		8 hr per shift		53 sets per hr		5 days	
↓		↓		↓		↓	
2	·	8	·	53	·	5	This could also be written 2(8)(53)(5).

We perform the multiplications working from left to right.

$$\begin{aligned} \mathbf{2}\cdot\mathbf{8}\cdot 53\cdot 5 &= \mathbf{16}\cdot 53\cdot 5 && \text{Multiply 2 and 8.} \\ &= 848\cdot 5 && \text{Multiply 16 and 53.} \\ &= 4{,}240 \end{aligned}$$

So 4,240 television sets will be manufactured in 5 days.

Checking an answer

CALCULATOR SNAPSHOT

We can use a scientific calculator to check the multiplication performed in Example 3. To find the product $2 \cdot 8 \cdot 53 \cdot 5$, we enter these numbers and press these keys.

2 [×] 8 [×] 53 [×] 5 [=]   4240

The display verifies that the multiplication was done correctly in Example 3.

We can use multiplication to count objects arranged in rectangular patterns. For example, the following display on the left below shows a rectangular array consisting of 5 rows of 7 stars. The product $5 \cdot 7$, or 35, indicates the total number of stars.

Because multiplication is commutative, the array on the right below, consisting of 7 rows of 5 stars, contains the same number of stars.

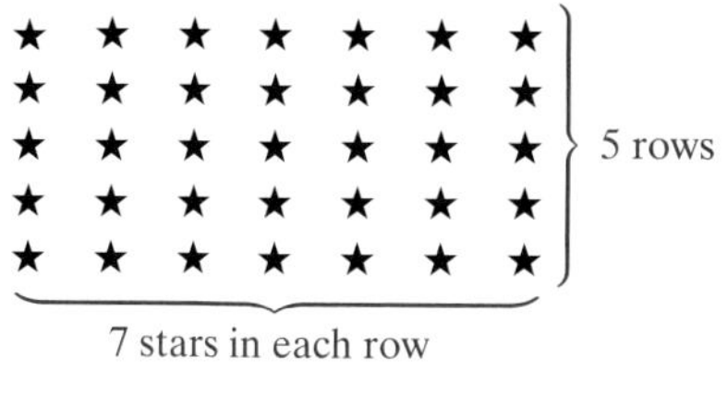

5 rows of 7 stars is 35 stars: $5 \cdot 7 = 35$.

★ ★ ★ ★ ★
★ ★ ★ ★ ★
★ ★ ★ ★ ★
★ ★ ★ ★ ★   7 rows
★ ★ ★ ★ ★
★ ★ ★ ★ ★
★ ★ ★ ★ ★

5 stars in each row

7 rows of 5 stars is 35 stars: $7 \cdot 5 = 35$.

EXAMPLE 4 **Computer science.** To draw graphics on a computer screen, a computer controls each *pixel* (one dot on the screen). See Figure 1-11. If a computer graphics image is 800 pixels wide and 600 pixels high, how many pixels does the computer control?

FIGURE 1-11

Solution The graphics image is a rectangular array of pixels. Each of its 600 rows consists of 800 pixels. The total number of pixels is the product of 600 and 800.

$600 \cdot 800 = 480{,}000$ This could be written as 600(800).

The computer controls 480,000 pixels.

Self Check 4
On a color monitor, each of the pixels can be red, green, or blue. How many colored pixels does the computer control?

Answer 1,440,000

Finding the area of a rectangle

One important application of multiplication is finding the area of a rectangle. The **area of a rectangle** is the measure of the amount of surface it encloses. Area is measured in square units, such as square inches (denoted as in.^{2}) or square centimeters (denoted as cm^2). (See Figure 1-12 on the next page.)

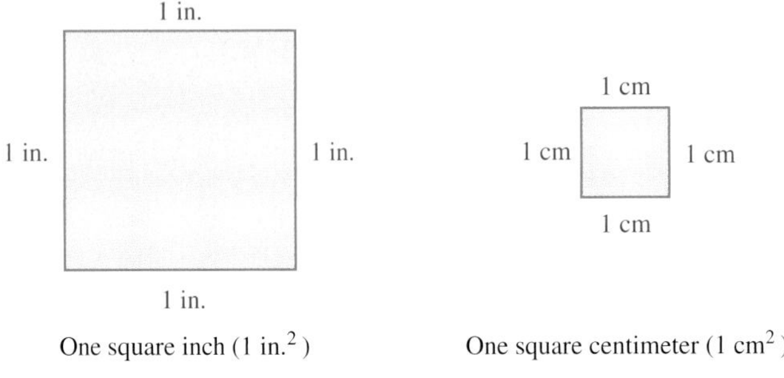

FIGURE 1-12

The rectangle in Figure 1-13 has a length of 5 centimeters and a width of 3 centimeters. Each small square covers an area of 1 square centimeter (1 cm^2). The small squares form a rectangular pattern, with 3 rows of 5 squares.

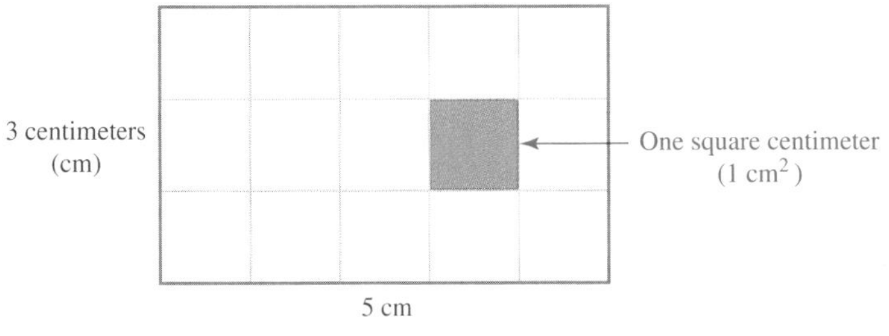

FIGURE 1-13

Because there are $5 \cdot 3$, or 15, small squares, the area of the rectangle is 15 cm^2. This suggests that the area of any rectangle is the product of its length and its width.

$$\text{Area of a rectangle} = \text{length} \cdot \text{width}$$

Using the variables l and w to represent the length and width, we can write this formula in simpler form.

Area of a rectangle

The area A of a rectangle is the product of the rectangle's length l and its width w.

$$A = l \cdot w \quad \text{or, more simply,} \quad A = lw$$

Self Check 5

Find the area of a 9-inch-by-12-inch sheet of paper.

Answer 108 in.^{2}

EXAMPLE 5 Gift wrap. When completely unrolled, a long sheet of gift wrapping paper has the dimensions shown in Figure 1-14. How many square feet of gift wrap are on the roll?

FIGURE 1-14

Solution To find the number of square feet of paper, we need to find the area of the rectangle shown in the figure.

$A = lw$ This is the formula for the area of a rectangle.

$= \mathbf{12} \cdot \mathbf{3}$ Replace l with 12 and w with 3.

$= 36$ Perform the multiplication.

There are 36 square feet (ft^2) of wrapping paper on the roll.

! COMMENT Remember that the perimeter of a rectangle is the distance around it. The area of a rectangle is a measure of the surface it encloses.

Division

If $12 is distributed equally among 4 people, we must divide to see that each person receives $3.

$$4\overline{)12}\ \text{with quotient } 3$$

Three symbols can be used to indicate division.

Symbols that are used for division

Symbol		Example
÷	division symbol	$12 \div 4$
$\overline{)}$	long division	$4\overline{)12}$
—	fraction bar	$\frac{12}{4}$

In a division, the number that is being divided is called the **dividend,** the number that we are dividing by is called the **divisor,** and the answer is called the **quotient.**

$$\text{Dividend} \div \text{divisor} = \text{quotient} \qquad \begin{array}{r}\text{quotient}\\ \text{Divisor}\overline{)\text{dividend}}\end{array} \qquad \frac{\text{Dividend}}{\text{Divisor}} = \text{quotient}$$

Division can be thought of as repeated subtraction. To divide 12 by 4 is to ask, How many 4's can be subtracted from 12? Exactly three 4's can be subtracted from 12 to get 0.

$$12 \overbrace{- 4 - 4 - 4}^{\text{Three 4's}} = 0$$

Thus, $12 \div 4 = 3$.

Division is also related to multiplication.

$$\frac{12}{4} = 3 \quad \text{because} \quad 4 \cdot 3 = 12 \qquad \text{and} \qquad \frac{20}{5} = 4 \quad \text{because} \quad 5 \cdot 4 = 20$$

Properties of division

We will now consider three types of division that involve zero. In the first case, we will examine a division of zero; in the second, a division by zero; in the third, a division of zero by zero.

Division statement	**Related multiplication statement**	**Result**
$\frac{0}{2} = ?$	$2(?) = 0$ ↑ This must be 0 if the product is to be 0.	$\frac{0}{2} = \mathbf{0}$

$\frac{2}{0} = ?$ $\qquad 0(?) = 2$ $\qquad$ There is no quotient.

$\uparrow$ There is no number that gives 2 when multiplied by 0.

$\frac{0}{0} = ?$ $\qquad 0(?) = 0$ $\qquad$ Any number can be the quotient.

$\uparrow$ Any number times 0 is 0.

We see that $\frac{0}{2} = 0$. This suggests that the quotient of 0 divided by any nonzero number is 0. Since $\frac{2}{0}$ does not have a quotient, we say that division of 2 by 0 is *undefined.* In general, division of any nonzero number by 0 is undefined. Since $\frac{0}{0}$ can be any number, we say that $\frac{0}{0}$ is undetermined.

Division with 0

1. If a represents any nonzero number, $\frac{0}{a} = 0$.
2. If a represents any nonzero number, $\frac{a}{0}$ is undefined.
3. $\frac{0}{0}$ is undetermined.

The example $\frac{12}{1} = 12$ illustrates that *any number divided by 1 is the number itself.* The example $\frac{12}{12} = 1$ illustrates that *any number (except 0) divided by itself is 1.*

Division properties

If a represents any number,

$$\frac{a}{1} = a \quad \text{and} \quad \frac{a}{a} = 1 \text{ (provided that } a \neq 0)$$

Read $\neq$ as "is not equal to."

Dividing whole numbers

We can use a process called **long division** to divide whole numbers. To divide 832 by 23, for example, we proceed as follows.

Quotient →

Divisor → $23\overline{)832}$ (Dividend)

Place the divisor and the dividend as indicated. The quotient will appear above the long division symbol.

We will find the quotient using the following division process.

$$\begin{array}{r} 4 \\ 23\overline{)832} \end{array}$$

Ask: How many times will 23 divide 83? Because an estimate is 4, place 4 in the tens column of the quotient.

$$\begin{array}{r} 4 \\ 23\overline{)832} \\ 92 \end{array}$$

Multiply $23 \cdot 4$ and place the answer, 92, under the 83. Because 92 is larger than 83, our estimate of 4 for the tens column of the quotient was too large.

$$\begin{array}{r} 3 \\ 23\overline{)832} \\ 69\downarrow \\ \hline 142 \end{array}$$

Revise the estimate of the quotient to be 3. Multiply $23 \cdot 3$ to get 69, place 69 under the 83, draw a line, and subtract.

Bring down the 2 in the ones column.

```
     3 7
2 3)8 3 2
    6 9
    1 4 2
    1 6 1
```

Ask: How many times will 23 divide 142? The answer is approximately 7. Place 7 in the ones column of the quotient. Multiply $23 \cdot 7$ to get 161. Place 161 under 142. Because 161 is larger than 142, the estimate of 7 is too large.

```
     3 6
2 3)8 3 2
    6 9
    1 4 2
    1 3 8
        4
```

Revise the estimate of the quotient to be 6. Multiply: $23 \cdot 6 = 138$.

Place 138 under 142 and subtract.

The quotient is 36, and the leftover 4 is the **remainder.** We can write this result as 36 R 4.

To check the result of a division, we multiply the divisor by the quotient and then add the remainder. The result should be the dividend.

Check: Quotient $\cdot$ divisor + remainder = dividend

$$36 \cdot 23 + 4 = 832$$

$$828 + 4 = 832$$

$$832 = 832$$

Applications that involve forming groups can often be solved using division.

EXAMPLE 6 Soup kitchens. A soup kitchen plans to feed 1,990 people. Because of space limitations, only 165 people can be served at one time. How many seatings will be necessary to feed everyone? How many will be served at the last seating?

Solution The 1,990 people can be fed 165 at a time. To find the number of seatings, we divide.

```
       12
165)1,990
    1 65↓
     340
     330
      10
```

The quotient is 12, and the remainder is 10. Thirteen seatings will be needed: 12 full-capacity seatings and one partial seating to serve the remaining 10 people.

Self Check 6

Each gram of fat in a meal provides 9 calories. A fast-food meal contains 243 calories from fat. How many grams of fat does the meal contain?

Answer 27

Retailing

CALCULATOR SNAPSHOT

A salesperson sold a number of calculators for $17 each, and her total sales were $1,819. To find the number of calculators she sold, we must divide the total sales by the cost of each calculator. We can use a calculator to evaluate $1{,}819 \div 17$ by entering these numbers and pressing these keys.

1819 [÷] 17 [=]   107

The salesperson sold 107 calculators.

Section 1.3 STUDY SET

VOCABULARY *Fill in the blanks.*

1. ______________ is repeated addition.
2. Numbers that are to be multiplied are called ________. The result of a multiplication is called a ________.
3. The statement $ab = ba$ expresses the ____________ property of multiplication. The statement $(ab)c = a(bc)$ expresses the ___________ property of multiplication.
4. If a square measures 1 inch on each side, its area is 1 ________ ________.
5. In a division, the dividend is divided by the ________. The result of a division is called a __________.
6. The ______ of a rectangle is the amount of surface it encloses.

CONCEPTS

7. Write $8 + 8 + 8 + 8$ as a multiplication.
8. **a.** Use the variables x and y to write the commutative property of multiplication.

   b. Use the variables x, y, and z to write the associative property of multiplication.
9. How do we find the amount of surface enclosed by a rectangle?
10. Determine whether *perimeter* or *area* is the concept that should be applied to find each of the following.

    a. The amount of floor space to be carpeted

    b. The amount of clear glass to be tinted

    c. The amount of lace needed to trim the sides of a handkerchief
11. Perform each multiplication.

    a. $1 \cdot 25$ **b.** $62(1)$

    c. $10 \cdot 0$ **d.** $0(4)$
12. Perform each division.

    a. $25 \div 1$ **b.** $\frac{7}{1}$

    c. $\frac{0}{1}$ **d.** $\frac{5}{0}$

    e. $\frac{0}{0}$ **f.** $\frac{0}{2,757}$

13. Write a multiplication statement that finds the number of red squares.

14. Consider

$$\begin{array}{r} 12 \\ 15\overline{)182} \\ \underline{15} \\ 32 \\ \underline{30} \\ 2 \end{array}$$

Fill in the blanks: $12 \cdot \_\_ + \_\_ = \_\_$.

NOTATION

15. **a.** Write three symbols that are used for multiplication.

    b. Write three symbols that are used for division.
16. Write each multiplication in simpler form.

    a. $8 \cdot x$ **b.** $l \cdot w$
17. What does ft^2 mean?
18. Draw a figure having an area of 1 square inch.

PRACTICE *Perform each multiplication.*

19. $12 \cdot 7$
20. $15 \cdot 8$
21. $27(12)$
22. $35(17)$
23. $9 \cdot (4 \cdot 5)$
24. $(3 \cdot 5) \cdot 12$
25. $5 \cdot 7 \cdot 3$
26. $7 \cdot 6 \cdot 8$
27. $\begin{array}{r} 99 \\ \times 77 \\ \hline \end{array}$
28. $\begin{array}{r} 73 \\ \times 59 \\ \hline \end{array}$
29. $\begin{array}{r} 20 \\ \times 53 \\ \hline \end{array}$
30. $\begin{array}{r} 78 \\ \times 20 \\ \hline \end{array}$
31. $\begin{array}{r} 112 \\ \times\ 23 \\ \hline \end{array}$
32. $\begin{array}{r} 232 \\ \times\ 53 \\ \hline \end{array}$
33. $\begin{array}{r} 207 \\ \times\ 97 \\ \hline \end{array}$
34. $\begin{array}{r} 768 \\ \times\ 70 \\ \hline \end{array}$
35. $13,456 \cdot 217$
36. $17,456 \cdot 257$
37. $3,302 \cdot 358$
38. $123,112 \cdot 46$

Find the area of each rectangle or square.

39. 

40. 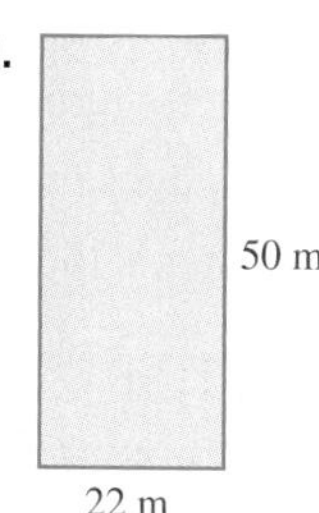

41. 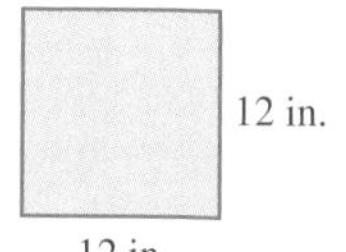

42. 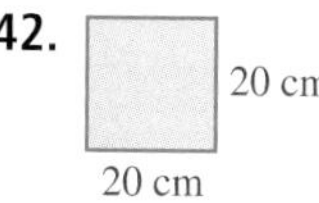

Perform each division.

43. $40 \div 5$

44. $40 \div 8$

45. $42 \div 14$

46. $65 \div 13$

47. $132 \div 11$

48. $132 \div 12$

49. $\frac{221}{17}$

50. $\frac{221}{13}$

51. $13\overline{)949}$

52. $73\overline{)949}$

53. $33\overline{)1,353}$

54. $41\overline{)1,353}$

55. $39\overline{)7,995}$

56. $71\overline{)7,313}$

57. $29\overline{)6,090}$

58. $13\overline{)7,410}$

Perform each division and give the quotient and the remainder.

59. $31\overline{)273}$

60. $25\overline{)290}$

61. $37\overline{)743}$

62. $79\overline{)931}$

63. $42\overline{)1,273}$

64. $83\overline{)3,280}$

65. $57\overline{)1,795}$

66. $99\overline{)9,876}$

APPLICATIONS

67. WAGES A cook worked 12 hours at $11 per hour. How much did she earn?

68. CHESSBOARDS A chessboard consists of 8 rows, with 8 squares in each row. How many squares are on a chessboard?

69. FINDING DISTANCE A car with a tank that holds 14 gallons of gasoline goes 29 miles on 1 gallon. How far can the car go on a full tank?

70. RENTING APARTMENTS Mia owns an apartment building with 18 units. Each unit generates a monthly income of $450. Find her total monthly income.

71. CONCERTS A jazz quartet gave two concerts in each of 37 cities. Approximately 1,700 fans attended each concert. How many people heard the group?

72. CEREAL A cereal maker advertises "Two cups of raisins in every box." Find the number of cups of raisins in a case of 36 boxes of cereal.

73. ORANGE JUICE It takes 13 oranges to make one can of orange juice. Find the number of oranges used to make a case of 24 cans.

74. ROOM CAPACITY A college lecture hall has 17 rows of 33 seats. A sign on the wall reads, "Occupancy by more than 570 persons is prohibited." If the seats are filled and there is one instructor, is the college breaking the rule?

75. ELEVATORS There are 14 people in an elevator with a capacity of 2,000 pounds. If the average weight of a person on the elevator is 150 pounds, is the elevator overloaded?

76. CHANGING UNITS There are 12 inches in 1 foot. How many inches are in 80 feet?

77. WORD PROCESSING A student used the option shown in the illustration when typing a report. How many entries will the table hold?

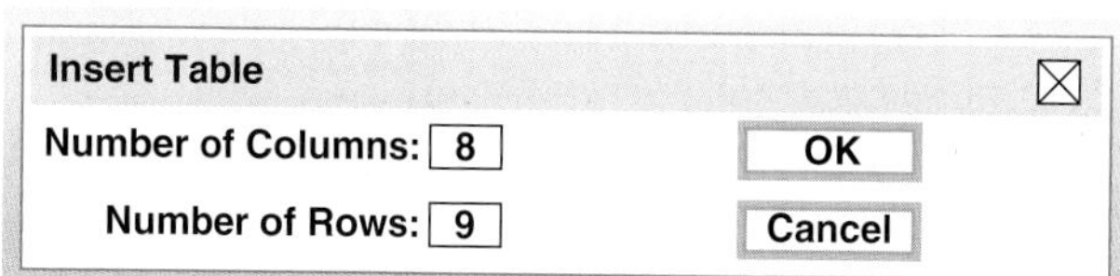

78. PRESCRIPTIONS How many tablets should a pharmacist put in the container shown on the right?

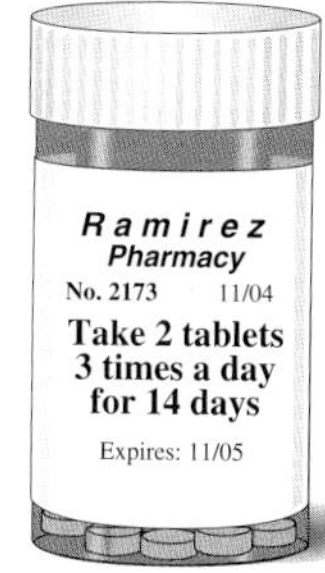

79. DISTRIBUTING MILK A first-grade class received 73 half-pint cartons of milk to distribute evenly to the 23 students. How many cartons were left over?

80. LIFT SYSTEMS If the bus shown below weighs 58,000 pounds, how much weight is on each jack?

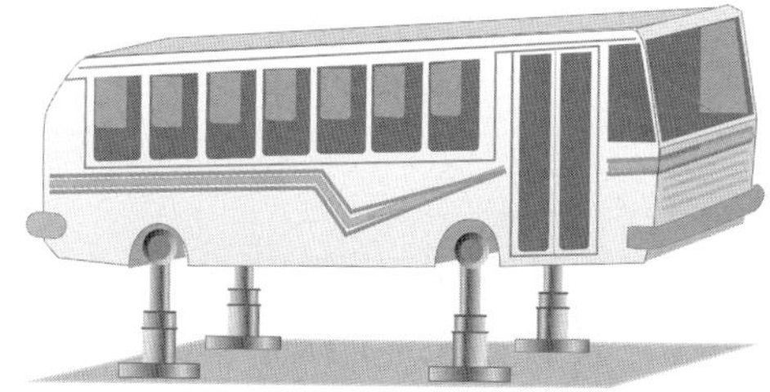

81. MILEAGE A touring rock group travels in a bus that has a range of 700 miles on one tank (140 gallons) of gasoline. How far can the bus travel on 1 gallon of gas?

82. RUNNING Brian runs 7 miles each day. In how many days will Brian run 371 miles?

83. How many feet more than 2 miles is 11,000 feet? (*Hint:* 5,280 feet = 1 mile.)

84. DOUGHNUTS How many dozen doughnuts must be ordered for a meeting if 156 people are expected to attend and each person will be served one doughnut?

85. PRICE OF TEXTBOOKS An author knows that her publisher received $954,193 on the sale of 23,273 textbooks. What is the price of each book?

86. WATER DISCHARGES The Susquehanna River discharges 38,200 cubic feet of water per second into the Chesapeake Bay. How long will it take for the river to discharge 1,719,000 cubic feet?

87. VOLLEYBALL LEAGUES A total of 216 girls tried out for a city volleyball program. How many girls should be put on each team roster if the following requirements must be met?

- All the teams are to have the same number of players.
- A reasonable number of players on a team is 7 to 10.
- For scheduling purposes, there must be an even number of teams.

88. AREA OF WYOMING The state of Wyoming is approximately rectangular-shaped with dimensions 360 miles long and 270 miles wide. Find its perimeter and its area.

89. COMPARING ROOMS Which has the greater area: a rectangular room that is 14 feet by 17 feet or a square room that is 16 feet on each side? Which has the greater perimeter?

90. MATTRESSES A queen-size mattress measures 60 inches by 80 inches, and a full-size mattress measures 54 inches by 75 inches. How much more sleeping surface is there on a queen-size mattress?

91. GARDENING A rectangular garden is 27 feet long and 19 feet wide. A path in the garden uses 125 square feet of space. How many square feet are left for planting?

92. TENNIS See the illustration.

a. Find the number of square feet of court area a singles tennis player must defend.

b. Do the same for a doubles player.

c. What is the difference between the two results?

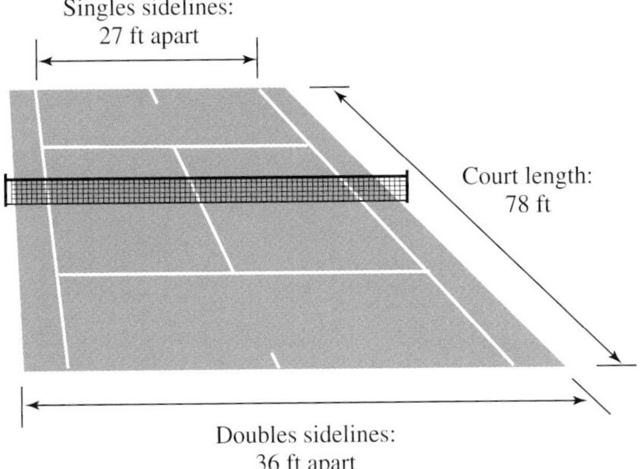

WRITING

93. Explain why the division of two numbers is not commutative.

94. Explain the difference between what perimeter measures and what area measures.

95. Explain the difference between 1 foot and 1 square foot.

96. When two numbers are multiplied, the result is 0. What conclusion can be drawn about the numbers?

REVIEW

97. Consider 372,856. What digit is in the hundreds column?

98. Round 45,995 to the nearest thousand.

99. Add: 357, 39, and 476.

100. DISCOUNTS A car, originally priced at $17,550, is being sold for $13,970. By how many dollars has the price been decreased?

ESTIMATION

In the previous two sections, we have used **estimation** as a means of checking the reasonableness of an answer. We now take a more in-depth look at the process of estimating.

Estimation is used to find an *approximate* answer to a problem. Estimates can be helpful in two ways. First, they serve as an accuracy check that can detect major computational errors. If an answer does not seem reasonable when compared to the estimate, the original problem should be reworked. Second, some situations call for only an approximate answer rather than the exact answer.

There are several ways to estimate, but there is one overriding theme of all the methods: The numbers in the problem are simplified so that the computation can be made easily and quickly. The first method we will study uses what is called **front-end rounding.** Each number is rounded to its largest place value, so that all but the first digit of each number is 0.

EXAMPLE 1 Estimating sums, differences, and products.

a. Estimate the sum: 3,714 + 2,489 + 781 + 5,500 + 303.

Solution Use front-end rounding.

$$\begin{array}{rcr}
3{,}714 & \longrightarrow & 4{,}000 \\
2{,}489 & \longrightarrow & 2{,}000 \\
781 & \longrightarrow & 800 \\
5{,}500 & \longrightarrow & 6{,}000 \\
+\ 303 & \longrightarrow & +\ 300 \\
\hline
 & & 13{,}100
\end{array}$$

Each number is rounded to its largest place value. All but the first digit is 0.

The estimate is 13,100.

If we compute 3,714 + 2,489 + 781 + 5,500 + 303, the sum is 12,787. We can see that our estimate is close; it's just 313 more than 12,787. This example illustrates the tradeoff when using estimation: The calculations are easier to perform and they take less time, but the answers are not exact.

b. Estimate the difference: 46,721 − 13,208.

Solution Use front-end rounding.

$$\begin{array}{rcr}
46{,}721 & \longrightarrow & 50{,}000 \\
-13{,}208 & \longrightarrow & -10{,}000 \\
\hline
 & & 40{,}000
\end{array}$$

Only the first digit is nonzero.

The estimate is 40,000.

c. Estimate the product: $334 \cdot 59$.

Solution Use front-end rounding.

$$\begin{array}{rcr}
334 & \longrightarrow & 300 \\
\times\ 59 & \longrightarrow & \times\ 60 \\
\hline
 & & 18{,}000
\end{array}$$

334 rounds to 300, 59 rounds to 60.

The estimate is 18,000.

Self Check 1

a. Estimate the sum:

$$\begin{array}{r}
6{,}780 \\
3{,}278 \\
566 \\
4{,}230 \\
+1{,}923 \\
\hline
\end{array}$$

b. Estimate the difference:

$$\begin{array}{r}
89{,}070 \\
-15{,}331 \\
\hline
\end{array}$$

c. Estimate the product:

$$\begin{array}{r}
707 \\
\times 251 \\
\hline
\end{array}$$

Answers **a.** 16,600, **b.** 70,000, **c.** 210,000

To estimate quotients, we will use a method that approximates both the dividend and the divisor so that they will divide easily. With this method, some insight and intuition are needed. There is one rule of thumb for this method: If possible, round both numbers up or both numbers down.

Self Check 2
Estimate: $33{,}642 \div 42$.

EXAMPLE 2 Estimating quotients. Estimate the quotient: $170{,}715 \div 57$.

Solution Both numbers are rounded up. The division can then be done in your head.

The dividend is approximately

$$170{,}715 \div 57 \qquad 180{,}000 \div 60 = 3{,}000$$

The divisor is approximately

The estimate is 3,000.

Answer 800

STUDY SET *Use front-end rounding to find an estimate to check the reasonableness of each answer. Write yes if it appears reasonable and no if it does not.*

1.
```
 25,405
 11,222
  8,909
  1,076
 14,595
+33,999
 73,206
```

2.
```
 568,334
− 31,225
 497,109
```

3.
```
   451
 ×  73
39,923
```

4.
```
   616
 ×  98
60,368
```

Use estimation to check the reasonableness of each answer.

5. $57{,}238 \div 28 = 200$

6. $322\overline{)13{,}202}$ with quotient 41

Use an estimation procedure to answer each problem.

7. CAMPAIGNING The number of miles flown each day by a politician on a campaign swing are shown here. Estimate the number of miles she flew during this time.

Day 1	3,546 miles
Day 2	567
Day 3	1,203
Day 4	342
Day 5	2,699

8. SHOPPING MALLS The total sales income for a downtown mall in its first three years in operation are shown here.

2000	$5,234,301
2001	$2,898,655
2002	$6,343,433

Estimate the difference in income for 2001 and 2002 as compared to the first year, 2000.

9. GOLF COURSES Estimate the number of bags of grass seed needed to plant a fairway whose area is 86,625 square feet if the seed in each bag covers 2,850 square feet.

10. CENSUS Estimate the total population of the ten largest counties in the United States as of 2003.

LARGEST COUNTIES, BY POPULATION	
1. Los Angeles, CA	9,871,506
2. Cook, IL	5,351,552
3. Harris, TX	3,596,086
4. Maricopa, AZ	3,389,260
5. Orange, CA	2,957,766
6. San Diego, CA	2,930,886
7. Kings, NY	2,472,523
8. Miami-Dade, FL	2,341,167
9. Dallas, TX	2,284,096
10. Queens, NY	2,225,486

Source: Bureau of the Census

11. CURRENCY Estimate the number of $5 bills in circulation as of June 30, 2004, if the total value of the currency was $9,373,288,075.

12. CORPORATIONS In 2003, General Motors Corporation had sales of $185,524,000,000. Approximately how many times larger was this than the 2003 sales of IBM, which were $89,000,000,000?

1.4 Prime Factors and Exponents

- Factoring whole numbers • Even and odd whole numbers • Prime numbers
- Composite numbers • Finding prime factorizations with the tree method
- Exponents • Finding prime factorizations with the division method

In this section, we will describe how to represent whole numbers in alternative forms. The procedures used to find these forms involve multiplication and division. We will then discuss exponents, a shortcut way to represent repeated multiplication.

Factoring whole numbers

The statement $3 \cdot 2 = 6$ has two parts: the numbers that are being multiplied, and the answer. The numbers that are being multiplied are *factors,* and the answer is the *product.* We say that 3 and 2 are factors of 6.

Factors

Numbers that are multiplied together are called **factors.**

EXAMPLE 1 Find the factors of 12.

Solution We need to find the possible ways that we can multiply two whole numbers to get a product of 12.

$$1 \cdot 12 = 12, \qquad 2 \cdot 6 = 12, \qquad \text{and} \qquad 3 \cdot 4 = 12$$

In order from least to greatest, the factors of 12 are 1, 2, 3, 4, 6, and 12.

Self Check 1
Find the factors of 20.

Answer 1, 2, 4, 5, 10, and 20

Example 1 shows that 1, 2, 3, 4, 6, and 12 are the factors of 12. This observation was established by using multiplication facts. Each of these factors is related to 12 by division as well. Each of them divides 12, leaving a remainder of 0. Because of this fact, we say that 12 is *divisible* by each of its factors. When a division ends with a remainder of 0, we say that the division comes out even or that one of the numbers divides the other *exactly*.

Divisibility

One number is divisible by another if, when we divide them, the remainder is 0.

When we say that 3 is a factor of 6, we are using the word *factor* as a noun. The word *factor* is also used as a verb.

Factoring a whole number

To **factor** a whole number means to express it as the product of other whole numbers.

Self Check 2
Factor 18 using **a.** two factors and **b.** three factors.

Answers **a.** $1 \cdot 18, 2 \cdot 9, 3 \cdot 6$, **b.** $2 \cdot 3 \cdot 3$

EXAMPLE 2 Factor 40 using **a.** two factors and **b.** three factors.

Solution

a. There are several possibilities.

$$40 = 1 \cdot 40, \quad 40 = 2 \cdot 20, \quad 40 = 4 \cdot 10, \quad \text{or} \quad 40 = 5 \cdot 8$$

b. Again, there are several possibilities. Two of them are

$$40 = 5 \cdot 4 \cdot 2 \quad \text{and} \quad 40 = 2 \cdot 2 \cdot 10$$

Even and odd whole numbers

Even and odd whole numbers

If a whole number is divisible by 2, it is called an **even** number.

If a whole number is not divisible by 2, it is called an **odd** number.

The even whole numbers are the numbers

0, 2, 4, 6, 8, 10, 12, 14, 16, 18, . . .

The odd whole numbers are the numbers

1, 3, 5, 7, 9, 11, 13, 15, 17, 19, . . .

There are infinitely many even and infinitely many odd whole numbers.

Prime numbers

Self Check 3
Find the factors of 23.

Answer 1 and 23

EXAMPLE 3 Find the factors of 17.

Solution

$$1 \cdot 17 = 17$$

The only factors of 17 are 1 and 17.

In Example 3 and its Self Check, we saw that the only factors of 17 are 1 and 17, and the only factors of 23 are 1 and 23. Numbers that have only two factors, 1 and the number itself, are called **prime numbers.**

Prime numbers

A **prime number** is a whole number, greater than 1, that has only 1 and itself as factors.

The prime numbers are the numbers

2, 3, 5, 7, 11, 13, 17, 19, 23, 29, 31, . . .

The dots at the end of the list indicate that there are infinitely many prime numbers.

Note that the only even prime number is 2. Any other even whole number is divisible by 2 and thus has 2 as a factor, in addition to 1 and itself. Also note that not all odd whole numbers are prime numbers. For example, since 15 has factors of 1, 3, 5, and 15, it is not a prime number.

Composite numbers

The set of whole numbers contains many prime numbers. It also contains many numbers that are not prime.

Composite numbers

The **composite numbers** are whole numbers, greater than 1, that are not prime.

The composite numbers are the numbers

4, 6, 8, 9, 10, 12, 14, 15, 16, 18, . . .

The three dots at the end of the list indicate that there are infinitely many composite numbers.

EXAMPLE 4 **a.** Is 37 a prime number? **b.** Is 45 a prime number?

Solution

a. Since 37 is a whole number greater than 1 and its only factors are 1 and 37, it is prime.

b. The factors of 45 are 1, 3, 5, 9, 15, and 45. Since there are factors other than 1 and 45, 45 is not prime. It is a composite number.

Self Check 4

a. Is 57 a prime number?

b. Is 39 a prime number?

Answers **a.** no, **b.** no

! COMMENT The numbers 0 and 1 are neither prime nor composite, because neither is a whole number greater than 1.

Finding prime factorizations with the tree method

Every composite number can be formed by multiplying a specific combination of prime numbers. The process of finding that combination is called **prime factorization.**

Prime factorization

To find the **prime factorization** of a whole number means to write it as the product of only prime numbers.

One method for finding the prime factorization of a number is called the **tree method.** We use the tree method in the diagrams below to find the prime factorization of 90 in two ways.

1. Factor 90 as $9 \cdot 10$.
2. Factor 9 and 10.
3. The process is complete when only prime numbers appear.

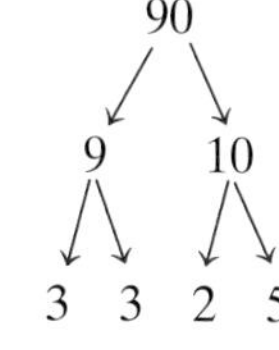

1. Factor 90 as $6 \cdot 15$.
2. Factor 6 and 15.
3. The process is complete when only prime numbers appear.

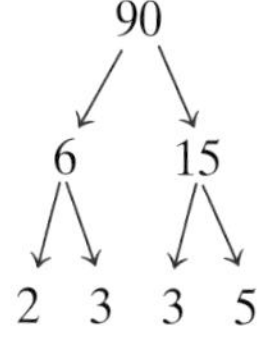

In either case, the prime factors are $2 \cdot 3 \cdot 3 \cdot 5$. Thus, the prime-factored form of 90 is $2 \cdot 3 \cdot 3 \cdot 5$. As we have seen, it does not matter how we factor 90. We will always get the same set of prime factors. No other combination of prime factors will multiply together and produce 90. This example illustrates an important fact about composite numbers.

Fundamental theorem of arithmetic

Any composite number has exactly one set of prime factors.

Self Check 5
Use a factor tree to find the prime factorization of 120.

Answer $2 \cdot 2 \cdot 2 \cdot 3 \cdot 5$

EXAMPLE 5 Use a factor tree to find the prime factorization of 210.

Solution

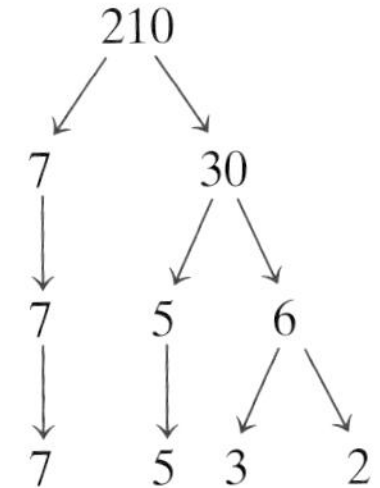

Factor 210 as $7 \cdot 30$.

Bring down the 7. Factor 30 as $5 \cdot 6$.

Bring down the 7 and the 5. Factor 6 as $3 \cdot 2$.

The prime factorization of 210 is $7 \cdot 5 \cdot 3 \cdot 2$. Writing the prime factors in order, from least to greatest, we have $210 = 2 \cdot 3 \cdot 5 \cdot 7$.

Exponents

In the Self Check of Example 5, we saw that the prime factorization of 120 is $2 \cdot 2 \cdot 2 \cdot 3 \cdot 5$. Because this factorization has three factors of 2, we call 2 a *repeated factor*. To express a repeated factor, we can use an **exponent.**

Exponent and base

An **exponent** is used to indicate repeated multiplication. It tells how many times the **base** is used as a factor.

The exponent is 3.

$$\underbrace{2 \cdot 2 \cdot 2}_{\text{Repeated factors}} = 2^3$$

The base is 2.

Read 2^3 as "2 to the third power" or "2 cubed."

The prime factorization of 120 can be written in a more compact form using exponents: $2 \cdot 2 \cdot 2 \cdot 3 \cdot 5 = 2^3 \cdot 3 \cdot 5$.

In the **exponential expression** a^n, a is the base and n is the exponent. The expression is called a **power of *a*.**

Self Check 6
Use exponents to write each prime factorization:
a. $3 \cdot 3 \cdot 7$ **b.** $5(5)(7)(7)$
c. $2 \cdot 2 \cdot 2 \cdot 3 \cdot 3 \cdot 5$

Answers **a.** $3^2 \cdot 7$, **b.** $5^2(7^2)$, **c.** $2^3 \cdot 3^2 \cdot 5$

EXAMPLE 6 Use exponents to write each prime factorization: **a.** $5 \cdot 5 \cdot 5$, **b.** $7 \cdot 7 \cdot 11$, and **c.** $2(2)(2)(2)(3)(3)(3)$.

Solution

a. $5 \cdot 5 \cdot 5 = 5^3$ — 5 is used as a factor 3 times.

b. $7 \cdot 7 \cdot 11 = 7^2 \cdot 11$ — 7 is used as a factor 2 times.

c. $2(2)(2)(2)(3)(3)(3) = 2^4(3^3)$ — 2 is used as a factor 4 times, and 3 is used as a factor 3 times.

EXAMPLE 7 Find each power: **a.** 7^2, **b.** 2^5, **c.** 10^4, and **d.** 6^1.

Solution

a. $7^2 = 7 \cdot 7 = 49$ — Read 7^2 as "7 to the second power" or "7 squared." Write the base 7 as a factor two times.

b. $2^5 = 2 \cdot 2 \cdot 2 \cdot 2 \cdot 2 = 32$ — Read 2^5 as "2 to the fifth power." Write the base 2 as a factor five times.

c. $10^4 = 10 \cdot 10 \cdot 10 \cdot 10 = 10{,}000$ — Read 10^4 as "10 to the fourth power." Write the base 10 as a factor four times.

d. $6^1 = 6$ — Read 6^1 as "6 to the first power." Write the base 6 once.

Self Check 7

Which of the numbers 3^5, 4^4, and 5^3 is the largest?

Answer $4^4 = 256$

! COMMENT Note that 2^5 means $2 \cdot 2 \cdot 2 \cdot 2 \cdot 2$. It does not mean $2 \cdot 5$. That is, $2^5 = 32$ and $2 \cdot 5 = 10$.

EXAMPLE 8 The prime factorization of a number is $2^3 \cdot 3^4 \cdot 5$. What is the number?

Solution To find the number, we find the value of each power and then perform the multiplication.

$$\mathbf{2^3 \cdot 3^4} \cdot 5 = \mathbf{8 \cdot 81} \cdot 5 \quad 2^3 = 8 \text{ and } 3^4 = 81.$$
$$= 648 \cdot 5 \quad \text{Perform the multiplications, working from left to right.}$$
$$= 3{,}240$$

The number is 3,240.

Self Check 8

The prime factorization of a number is $3 \cdot 5^2 \cdot 7$. What is the number?

Answer 525

Bacterial growth

CALCULATOR SNAPSHOT

At the end of one hour, a culture contains two bacteria. Suppose the number of bacteria doubles every hour thereafter. Use exponents to determine how many bacteria the culture will contain after 24 hours.

We can use Table 1-2 to help model the situation. From the table, we see a pattern developing: The number of bacteria in the culture after 24 hours will be 2^{24}. We can evaluate this exponential expression using the exponential key [y^x] on a scientific calculator ([x^y] on some models).

To find the value of 2^{24}, we enter these numbers and press these keys.

Time	Number of bacteria
1 hr	$2 = 2^1$
2 hr	$4 = 2^2$
3 hr	$8 = 2^3$
4 hr	$16 = 2^4$
24 hr	$? = 2^{24}$

TABLE 1-2

2 [y^x] 24 [=] — 16777216

Since $2^{24} = 16{,}777{,}216$, there will be 16,777,216 bacteria after 24 hours.

Finding prime factorizations with the division method

We can also find the prime factorization of a whole number by division. For example, to find the prime factorization of 363, we begin the division method by choosing the *smallest* prime number that will divide the given number exactly. We continue this "inverted division" process until the result of the division is a prime number.

Step 1: The prime number 2 doesn't divide 363 exactly, but 3 does. The result is 121, which is not prime. We continue the division process.

$$\begin{array}{r} 3\underline{|363} \\ 121 \end{array}$$

Step 2: Next, we choose the smallest prime number that will divide 121. The primes 2, 3, 5, and 7 don't divide 121 exactly, but 11 does. The result is 11, which is prime. We are done.

$$\begin{array}{r} \mathbf{3}\underline{|363} \\ \mathbf{11}\underline{|121} \\ \mathbf{11} \end{array}$$

$363 = \mathbf{3 \cdot 11 \cdot 11}$

Using exponents, we can write the prime factorization of 363 as $3 \cdot 11^2$.

Self Check 9

Use the division method to find the prime factorization of 108. Use exponents to express the result.

Answer $2^2 \cdot 3^3$

EXAMPLE 9 Use the division method to find the prime factorization of 100. Use exponents to express the result.

Solution

2 divides 100 exactly. The result is 50, which is not prime. ⟶ $2\underline{|100}$
2 divides 50 exactly. The result is 25, which is not prime. ⟶ $2\underline{|50}$
5 divides 25 exactly. The result is 5, which is prime. We are done. ⟶ $5\underline{|25}$
5

The prime factorization of 100 is $2 \cdot 2 \cdot 5 \cdot 5$ or $2^2 \cdot 5^2$.

! COMMENT In Example 9, it would be incorrect to begin the division process with

$10\underline{|100}$

because 10 is not a prime number.

Section 1.4 STUDY SET

VOCABULARY *Fill in the blanks.*

1. Numbers that are multiplied together are called ________.
2. One number is ________ by another if the remainder is 0 when they are divided. When a division ends with a remainder of 0, we say that one of the numbers divides the other ________.
3. To ________ a whole number means to express it as the product of other whole numbers.
4. A ________ number is a whole number, greater than 1, that has only 1 and itself as factors.
5. Whole numbers, greater than 1, that are not prime numbers are called ________ numbers.
6. An ________ whole number is exactly divisible by 2. An ________ whole number is not exactly divisible by 2.
7. To prime factor a number means to write it as a product of only ________ numbers.
8. An ________ is used to represent repeated multiplication.
9. In the exponential expression 6^4, 6 is called the ________ and 4 is called the ________.
10. Another way to say "5 to the second power" is 5 ________. Another way to say "7 to the third power" is 7 ________.

CONCEPTS

11. Write 27 as the product of two factors.
12. Write 30 as the product of three factors.

13. The complete list of the factors of a whole number is given. What is the number?

a. 2, 4, 22, 44, 11, 1

b. 20, 1, 25, 100, 2, 4, 5, 50, 10

14. a. Find the factors of 24.

b. Find the prime factorization of 24.

15. Find the factors of each number.

a. 11

b. 23

c. 37

d. From the results obtained in parts a–c, what can be said about 11, 23, and 37?

16. Suppose a number is divisible by 10. Is 10 a factor of the number?

17. If 4 is a factor of a whole number, will 4 divide the number exactly?

18. Give examples of whole numbers that have 11 as a factor.

The prime factorization of a whole number is given. Find the number.

19. $2 \cdot 3 \cdot 3 \cdot 5$

20. $3^3 \cdot 2$

21. $11^2 \cdot 5$

22. $2 \cdot 2 \cdot 2 \cdot 7$

23. Can we change the order of the base and the exponent in an exponential expression and obtain the same result? In other words, does $3^2 = 2^3$?

24. Find the prime factors of 20 and 35. What prime factor do they have in common?

25. Find the prime factors of 20 and 50. What prime factors do they have in common?

26. Find the prime factors of 30 and 165. What prime factors do they have in common?

27. Find the prime factors of 30 and 242. What prime factor do they have in common?

28. Find 1^2, 1^3, and 1^4. From the results, what can be said about any power of 1?

29. Finish the process of prime factoring 150. Compare the results.

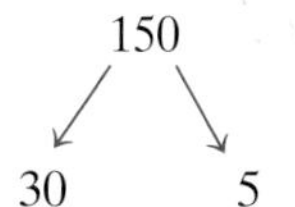

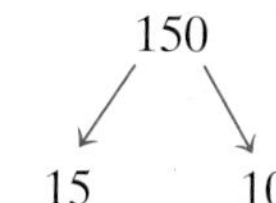

30. Find three whole numbers, less than 10, that would fit at the top of this tree diagram.

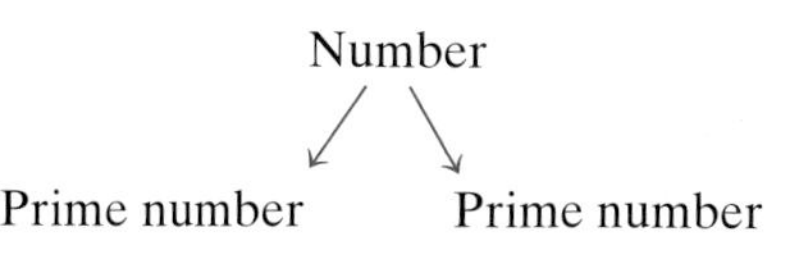

31. Complete the table.

Product of the factors of 12	Sum of the factors of 12
$1 \cdot 12$	
$2 \cdot 6$	
$3 \cdot 4$	

32. Consider 1, 4, 9, 16, 25, 36, 49, 64, 81, 100. Of the numbers listed, which is the *largest* factor of

a. 18 **b.** 24 **c.** 50

33. When using the division method to find the prime factorization of an even number, what is an obvious choice with which to start the division process?

34. When using the division method to find the prime factorization of a number ending in 5, what is an obvious choice with which to start the division process?

NOTATION *Write the repeated multiplication represented by each expression.*

35. 7^3

36. 8^4

37. 3^5

38. 4^6

39. $5^2(11)$

40. $2^3 \cdot 3^2$

Simplify each expression.

41. 10^1

42. 2^1

Use exponents to write each expression in simpler form.

43. $2 \cdot 2 \cdot 2 \cdot 2 \cdot 2$

44. $3 \cdot 3 \cdot 3 \cdot 3 \cdot 3 \cdot 3$

45. $5 \cdot 5 \cdot 5 \cdot 5$

46. $9 \cdot 9 \cdot 9$

47. $4(4)(5)(5)$

48. $12 \cdot 12 \cdot 12 \cdot 16$

PRACTICE *Find the factors of each whole number.*

49. 10

50. 6

51. 40

52. 75

53. 18

54. 32

55. 44

56. 65

57. 77

58. 81

59. 100

60. 441

Write each number in prime-factored form.

61. 39

62. 20

63. 99

64. 105

65. 162

66. 400

67. 220

68. 126

69. 64

70. 243

71. 147

72. 98

Evaluate each exponential expression.

73. 3^4

74. 5^3

75. 2^5

76. 10^5

77. 12^2

78. 7^3

79. 8^4

80. 9^5

81. $3^2(2^3)$

82. $3^3(4^2)$

83. $2^3 \cdot 3^3 \cdot 4^2$

84. $3^2 \cdot 4^3 \cdot 5^2$

85. 234^3

86. 51^4

87. $23^2 \cdot 13^3$

88. $12^3 \cdot 15^2$

APPLICATIONS

89. PERFECT NUMBERS A whole number is called a **perfect number** when the sum of its factors that are less than the number equals the number. For example, 6 is a perfect number, because $1 + 2 + 3 = 6$. Find the factors of 28. Then use addition to show that 28 is also a perfect number.

90. CRYPTOGRAPHY Information is often transmitted in code. Many codes involve writing products of large primes, because they are difficult to factor. To see how difficult, try finding two prime factors of 7,663. (*Hint:* Both primes are greater than 70.)

91. LIGHT The illustration shows that the light energy that passes through the first unit of area, 1 yard away from the bulb, spreads out as it travels away from the source. How much area does that energy cover 2 yards, 3 yards, and 4 yards from the bulb? Express each answer using exponents.

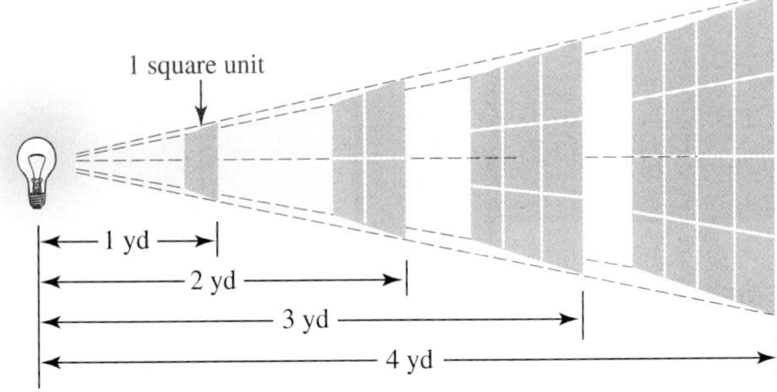

92. CELL DIVISION After one hour, a cell has divided to form another cell. In another hour, these two cells have divided so that four cells exist. In another hour, these four cells divide so that eight exist.

a. How many cells exist at the end of the fourth hour?

b. The number of cells that exist after each division can be found using an exponential expression. What is the base?

c. Use a calculator to find the number of cells after 12 hours.

WRITING

93. Explain how to test a number to see whether it is prime.

94. Explain how to test a number to see whether it is even.

95. Explain the difference between the *factors* of a number and the *prime factorization* of the number.

96. Explain why it would be incorrect to say that the area of the square shown on the right is 25^2 ft. How should we express its area?

REVIEW

97. Round 230,999 to the nearest thousand.

98. Write the set of whole numbers.

99. What is $0 \div 15$?

100. Multiply: $15 \cdot (6 \cdot 9)$.

101. What is the formula for the area of a rectangle?

102. MARCHING BANDS When a university band lines up in eight rows of 15 musicians, there are five musicians left over. How many band members are there?

1.5 Order of Operations

- Order of operations
- Evaluating expressions with no grouping symbols
- Evaluating expressions with grouping symbols
- The arithmetic mean (average)

Punctuation marks, such as commas, quotations, and periods, serve an important purpose when writing compositions. They determine the way in which sentences are to be read and interpreted. To read and interpret mathematical expressions correctly, we must use an agreed-upon set of priority rules for the *order of operations*.

Order of operations

Suppose you are asked to contact a friend if you see a certain type of watch for sale while you are traveling in Europe. While in Switzerland, you spot the watch and send the following e-mail message.

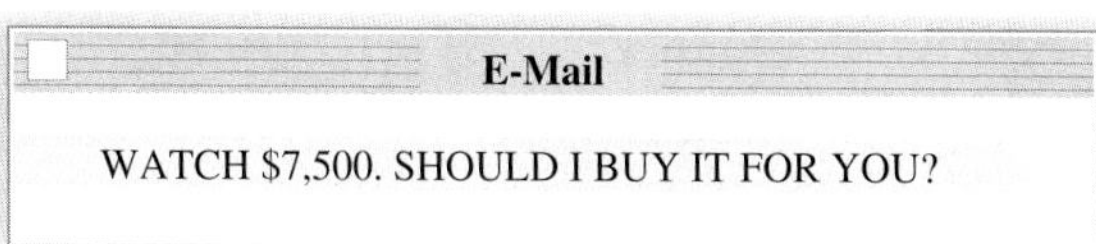

E-Mail
WATCH $7,500. SHOULD I BUY IT FOR YOU?

The next day, you get this response from your friend.

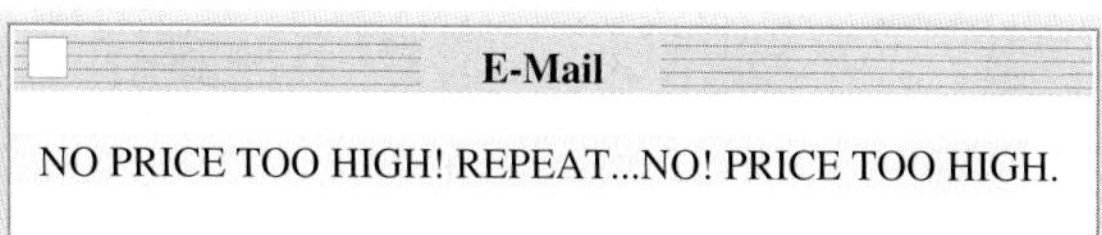

E-Mail
NO PRICE TOO HIGH! REPEAT...NO! PRICE TOO HIGH.

Something is wrong. One statement says to buy the watch at any price. The other says not to buy it, because it's too expensive. The placement of the exclamation point makes us read these statements differently, which results in different interpretations.

When we read a mathematical statement, the same kind of confusion is possible. For example, we consider

$3 + 2 \cdot 5$

The above expression contains two operations: addition and multiplication. We can *evaluate* it (find its value) in two ways. We can perform the addition first and then the multiplication. Or we can perform the multiplication first and then the addition. However, we get different results.

$\mathbf{3 + 2} \cdot 5 = \mathbf{5} \cdot 5$	Add first: $3 + 2 = 5$.	$3 + \mathbf{2 \cdot 5} = 3 + \mathbf{10}$	Multiply first: $2 \cdot 5 = 10$.
$= 25$	Multiply 5 and 5.	$= 13$	Add 3 and 10.

↑ Different results ↑

If we don't establish an order of operations, the expression $3 + 2 \cdot 5$ will have two different answers. To avoid this possibility, we evaluate expressions in the following order.

Order of operations

1. Perform all calculations within parentheses and other grouping symbols following the order listed in steps 2–4, working from the innermost pair to the outermost pair.
2. Evaluate all exponential expressions.
3. Perform all multiplications and divisions as they occur from left to right.
4. Perform all additions and subtractions as they occur from left to right.

When all grouping symbols have been removed, repeat steps 2–4 to complete the calculation.

If a fraction bar is present, evaluate the expression above the bar (called the **numerator**) and the expression below the bar (called the **denominator**) separately. Then perform the division indicated by the fraction bar, if possible.

To evaluate $3 + 2 \cdot 5$ correctly, we must apply the rules for the order of operations. Since there are no parentheses and there are no exponents, we perform the multiplication first and then the addition.

Ignore the addition for now → and perform the multiplication first: $2 \cdot 5 = 10$.

$$3 + \mathbf{2 \cdot 5} = 3 + \mathbf{10}$$

$$= 13 \quad \text{Now perform the addition.}$$

Using the rules for the order of operations, we see that the correct answer is 13.

Evaluating expressions with no grouping symbols

Self Check 1
Evaluate: $4 \cdot 3^3 - 6$.

Answer 102

EXAMPLE 1 Evaluate: $2 \cdot 4^2 - 8$.

Solution Since the expression does not contain any grouping symbols, we begin with step 2 of the rules for the order of operations.

$2 \cdot \mathbf{4^2} - 8 = 2 \cdot \mathbf{16} - 8$ Evaluate the exponential expression: $4^2 = 16$.

$= 32 - 8$ Perform the multiplication: $2 \cdot 16 = 32$.

$= 24$ Perform the subtraction.

Self Check 2
Evaluate: $10 - 2 \cdot 3 + 24$.

Answer 28

EXAMPLE 2 Evaluate: $8 - 3 \cdot 2 + 16$.

Solution Since the expression does not contain grouping symbols and since there are no powers to find, we look for multiplications or divisions to perform.

$8 - \mathbf{3 \cdot 2} + 16 = 8 - \mathbf{6} + 16$ Perform the multiplication: $3 \cdot 2 = 6$.

$= 2 + 16$ Working from left to right, perform the subtraction: $8 - 6 = 2$.

$= 18$ Perform the addition.

! COMMENT Some students incorrectly think that additions are always performed before subtractions. As Example 2 shows, this is not true. Working from left to right, we perform the additions or subtractions *in the order in which they occur.* The same is true for multiplications and divisions.

Self Check 3
Evaluate: $36 \div 9 + 4(2)3$.

Answer 28

EXAMPLE 3 Evaluate: $192 \div 6 - 5(3)2$.

Solution Although this expression contains parentheses, there are no calculations to perform within them. Since there are no powers, we perform multiplications and divisions as they are encountered from left to right.

$\mathbf{192 \div 6} - 5(3)2 = \mathbf{32} - 5(3)2$ Working from left to right, perform the division: $192 \div 6 = 32$.

$= 32 - 15(2)$ Working from left to right, perform the multiplication: $5(3) = 15$.

$= 32 - 30$ Perform the multiplication: $15(2) = 30$.

$= 2$ Perform the subtraction.

EXAMPLE 4 **Phone bills.** Figure 1–15 shows the rates for international telephone calls charged by a 10-10 long-distance company. A businesswoman calls Germany for 20 minutes, South Korea for 5 minutes, and Mexico City for 35 minutes. What is the total cost of the calls?

All rates are per minute.	
Canada	10¢
Germany	23¢
Jamaica	68¢
Mexico City	42¢
South Korea	29¢

FIGURE 1-15

Solution We can find the cost of a call (in cents) by multiplying the rate charged per minute by the length of the call (in minutes). To find the total cost, we add the costs of the three calls.

The cost of the call to Germany		The cost of the call to South Korea		The cost of the call to Mexico City
↓		↓		↓
23(20)	+	29(5)	+	42(35)

To evaluate this expression, we apply the rules for the order of operations.

$$23(20) + 29(5) + 42(35) = 460 + 145 + 1{,}470 \quad \text{Perform the multiplications.}$$
$$= 2{,}075 \quad \text{Perform the additions.}$$

The total cost of the calls is 2,075 cents, or \$20.75.

Evaluating expressions with grouping symbols

Grouping symbols serve as mathematical punctuation marks. They help determine the order in which an expression is to be evaluated. Examples of grouping symbols are parentheses ( ), brackets [ ], and the fraction bar —.

In the next example, we have two similar-looking expressions. However, because of the parentheses, we evaluate them in a different order.

EXAMPLE 5 Evaluate each expression: **a.** $12 - 3 + 5$ and **b.** $12 - (3 + 5)$.

Solution

a. We perform the additions and subtractions as they occur, from left to right.

$$\mathbf{12 - 3} + 5 = \mathbf{9} + 5 \quad \text{Perform the subtraction: } 12 - 3 = 9.$$
$$= 14 \quad \text{Perform the addition.}$$

b. This expression contains parentheses. We must perform the calculation within the parentheses first.

$$12 - (\mathbf{3 + 5}) = 12 - \mathbf{8} \quad \text{Perform the addition: } 3 + 5 = 8.$$
$$= 4 \quad \text{Perform the subtraction.}$$

Self Check 5
Evaluate each expression:
a. $20 - 7 + 6$
b. $20 - (7 + 6)$

Answer **a.** 19, **b.** 7

EXAMPLE 6 Evaluate: $(2 + 6)^3$.

Solution We begin by performing the calculation within the parentheses.

$$(\mathbf{2 + 6})^3 = \mathbf{8}^3 \quad \text{Perform the addition.}$$
$$= 512 \quad \text{Evaluate the exponential expression: } 8^3 = 8 \cdot 8 \cdot 8 = 512.$$

Self Check 6
Evaluate: $(1 + 3)^4$.

Answer 256

Self Check 7
Evaluate: $50 - 4(12 - 5 \cdot 2)$.

Answer 42

EXAMPLE 7 Evaluate: $5 + 2(13 - 5 \cdot 2)$.

Solution This expression contains grouping symbols. We will apply the rules for the order of operations within the parentheses first, to evaluate $13 - 5 \cdot 2$.

$5 + 2(13 - \mathbf{5 \cdot 2}) = 5 + 2(13 - \mathbf{10})$	Perform the multiplication within the parentheses.
$= 5 + 2(3)$	Perform the subtraction within the parentheses.
$= 5 + 6$	Perform the multiplication: $2(3) = 6$.
$= 11$	Perform the addition.

Sometimes an expression contains two or more sets of grouping symbols. Since it can be confusing to read an expression such as $16 + 2(14 - 3(5 - 2))$, we often use brackets in place of the second pair of parentheses.

$$16 + 2[14 - 3(5 - 2)]$$

If an expression contains more than one pair of grouping symbols, we always begin by working within the innermost pair and then work to the outermost pair.

Innermost parentheses
↓ ↓
$16 + 2[14 - 3(5 - 2)]$
↑ ↑
Outermost brackets

Self Check 8
Evaluate: $140 - 7[4 + 3(6 - 2)]$.

Answer 28

EXAMPLE 8 Evaluate: $16 + 6[14 - 3(5 - 2)]$.

Solution

$16 + 6[14 - 3(\mathbf{5 - 2})] = 16 + 6[14 - 3(\mathbf{3})]$	Perform the subtraction within the parentheses.
$= 16 + 6(14 - 9)$	Perform the multiplication within the brackets. Since only one set of grouping symbols is needed, write $14 - 9$ within parentheses.
$= 16 + 6(5)$	Perform the subtraction within the parentheses.
$= 16 + 30$	Perform the multiplication: $6(5) = 30$.
$= 46$	Perform the addition.

Self Check 9
Evaluate: $\frac{3(14) - 6}{2(3^2)}$.

EXAMPLE 9 Evaluate: $\frac{2(13) - 2}{3(2^3)}$.

Solution A fraction bar is a grouping symbol. We evaluate the numerator and denominator separately and then perform the indicated division.

$$\frac{2(13) - 2}{3(2^3)} = \frac{26 - 2}{3(8)}$$ In the numerator, perform the multiplication. In the denominator, perform the calculation within the parentheses.

$$= \frac{24}{24}$$ In the numerator, perform the subtraction. In the denominator, perform the multiplication.

$$= 1$$ Perform the division.

Answer 2

The arithmetic mean (average)

The **arithmetic mean,** or **average,** of several numbers is a value around which the numbers are grouped. It gives you an indication of the center of the set of numbers. When finding the mean of a set of numbers, we usually need to apply the rules for the order of operations.

Finding an arithmetic mean

To find the mean of a set of scores, divide the sum of the scores by the number of scores.

EXAMPLE 10 **Basketball.** In 1998, the Lady Vols of the University of Tennessee won the women's basketball championship, capping a perfect 39-0 season. Find their average margin of victory in their last four tournament games shown below.

Regional	Regional final	Semifinal	Championship
Beat Rutgers by 32 points	Beat North Carolina by 6 points	Beat Arkansas by 28 points	Beat Louisiana Tech by 18 points

Solution To find the average margin of victory, add the margins of victory and divide by 4.

$$\text{Average} = \frac{32 + 6 + 28 + 18}{4}$$

$$= \frac{84}{4}$$

$$= 21$$

Their average margin of victory was 21 points.

Self Check 10

Syracuse University won the 2003 NCAA men's basketball championship. Find their average margin of victory in their six tournament games, which they won by 11, 12, 1, 16, 11, and 3 points.

Answer 9 points

Order of operations and parentheses

CALCULATOR SNAPSHOT

Scientific calculators have the rules for order of operations built in. Even so, some evaluations require the use of a left parenthesis key [( ] and a right parenthesis key [ )]. For example, to evaluate $\frac{240}{20 - 15}$, we enter these numbers and press these keys.

240 [÷] [( ] 20 [−] 15 [ )] [=]

48

THINK IT THROUGH Preparing for Class

"Only about 13% of full-time students spend more than 25 hours a week preparing for class, the approximate number that faculty members say is needed to do well in college." The National Survey of Student Engagement Annual Report 2003

The National Survey of Student Engagement 2003 Annual Report questioned thousands of full-time college students about their weekly activities. Use the given clues to determine the results of the survey shown below.

Full-time Student Time Usage per Week

Activity	**Time per week**
Preparing for class....................	14 hours
Working on-campus or off-campus......	Four hours less than the time spent preparing for class
Participating in co-curricular activities...	Half as many hours as the time spent working on-campus or off-campus
Relaxing and socializing..............	Two hours more than twice the time spent participating in co-curricular activities
Providing care for dependents..........	Three hours less than one-half of the time spent relaxing and socializing
Commuting to class..................	One hour more than the time spent providing care for dependents

Section 1.5 STUDY SET

VOCABULARY *Fill in the blanks.*

1. The grouping symbols ( ) are called ____________, and the symbols [ ] are called __________.
2. The expression above a fraction bar is called the ____________. The expression below a fraction bar is called the ______________.
3. To _________ the expression $2 + 5 \cdot 4$ means to find its value.
4. To find the _______________ of several values, we add the values and divide by the number of values.

CONCEPTS

5. Consider $5(2)^2 - 1$. How many operations need to be performed to evaluate the expression? List them in the order in which they should be performed.
6. Consider $15 - 3 + (5 \cdot 2)^3$. How many operations need to be performed to evaluate this expression? List them in the order in which they should be performed.
7. Consider $\frac{5 + 5(7)}{2 + (8 - 4)}$. In the numerator, what operation should be done first? In the denominator, what operation should be done first?
8. In the expression $\frac{3 - 5(2)}{5(2) + 4}$, the bar is a grouping symbol. What does it separate?
9. Explain the difference between $2 \cdot 3^2$ and $(2 \cdot 3)^2$.
10. Use brackets to write $2(12 - (5 + 4))$ in better form.

NOTATION *Complete each solution to evaluate the expression.*

11. $28 - 5(2)^2 = 28 - 5(\quad)$
$= 28 - \quad$
$= \quad$

12. $2 + (5 + 6 \cdot 2) = 2 + (5 + \quad)$
$= 2 + \quad$
$= \quad$

13. $[4(2 + 7)] - 6 = [4(9)] - 6$
$= \square - 6$
$= \square$

14. $\dfrac{5(3) + 12}{9 - 6} = \dfrac{\square + 12}{\square}$
$= \dfrac{\square}{\square}$
$= \square$

PRACTICE *Evaluate each expression.*

15. $7 + 4 \cdot 5$
16. $10 - 2 \cdot 2$
17. $2 + 3(0)$
18. $5(0) + 8$
19. $20 - 10 + 5$
20. $80 - 5 + 4$
21. $25 \div 5 \cdot 5$
22. $6 \div 2 \cdot 3$
23. $7(5) - 5(6)$
24. $4 \cdot 2 + 2 \cdot 4$
25. $4^2 + 3^2$
26. $12^2 - 5^2$
27. $2 \cdot 3^2$
28. $3^3 \cdot 5$
29. $3 + 2 \cdot 3^4 \cdot 5$
30. $3 \cdot 2^3 \cdot 4 - 12$
31. $5 \cdot 10^3 + 2 \cdot 10^2 + 3 \cdot 10^1 + 9$
32. $8 \cdot 10^3 + 0 \cdot 10^2 + 7 \cdot 10^1 + 4$
33. $3(2)^2 - 4(2) + 12$
34. $5(1)^3 + (1)^2 + 2(1) - 6$
35. $(8 - 6)^2 + (4 - 3)^2$
36. $(2 + 1)^2 + (3 + 2)^2$
37. $60 - \left(6 + \dfrac{40}{8}\right)$
38. $7 + \left(5^3 - \dfrac{200}{2}\right)$
39. $6 + 2(5 + 4)$
40. $3(5 + 1) + 7$
41. $3 + 5(6 - 4)$
42. $7(9 - 2) - 1$
43. $(7 - 4)^2 + 1$
44. $(9 - 5)^3 + 8$
45. $6^3 - (10 + 8)$
46. $5^2 - (9 + 3)$
47. $50 - 2(4)^2$
48. $30 + 2(3)^3$
49. $16^2 - 4(2)(5)$
50. $8^2 - 4(3)(1)$
51. $39 - 5(6) + 9 - 1$
52. $15 - 3(2) - 4 + 3$
53. $(18 - 12)^3 - 5^2$
54. $(9 - 2)^2 - 3^3$
55. $2(10 - 3^2) + 1$
56. $1 + 3(18 - 4^2)$
57. $6 + \dfrac{25}{5} + 6(3)$
58. $15 - \dfrac{24}{6} + 8 \cdot 2$
59. $3\left(\dfrac{18}{3}\right) - 2(2)$
60. $2\left(\dfrac{12}{3}\right) + 3(5)$
61. $(2 \cdot 6 - 4)^2$
62. $2(6 - 4)^2$
63. $4[50 - (3^3 - 5^2)]$
64. $6[15 + (5 \cdot 2^2)]$
65. $80 - 2[12 - (5 + 4)]$
66. $15 + 5[12 - (2^2 + 4)]$
67. $2[100 - (5 + 4)] - 45$
68. $8[6(6) - 6^2] + 4(5)$
69. $\dfrac{10 + 5}{6 - 1}$
70. $\dfrac{18 + 12}{2(3)}$
71. $\dfrac{5^2 + 17}{6 - 2^2}$
72. $\dfrac{3^2 - 2^2}{(3 - 2)^2}$
73. $\dfrac{(3 + 5)^2 + 2}{2(8 - 5)}$
74. $\dfrac{25 - (2 \cdot 3 - 1)}{2 \cdot 9 - 8}$
75. $\dfrac{(5 - 3)^2 + 2}{4^2 - (8 + 2)}$
76. $\dfrac{(4^3 - 2) + 7}{5(2 + 4) - 7}$
77. $12{,}985 - (1{,}800 + 689)$
78. $\dfrac{897 - 655}{88 - 77}$
79. $3{,}245 - 25(16 - 12)^2$
80. $\dfrac{24^2 - 4^2}{22 + 58}$

APPLICATIONS *In Problems 81–86, write an expression to solve each problem. Then evaluate the expression.*

81. BUYING GROCERIES At the supermarket, Carlos has 2 cases of soda, 4 bags of potato chips, and 2 cans of dip in his cart. Each case of soda costs \$6, each bag of chips costs \$2, and each can of dip costs \$1. Find the total cost of the groceries.

82. JUDGING The scores received by a junior diver are as follows.

5	2	4	6	3	4

The formula for computing the overall score for the dive is as follows:

1. Throw out the lowest score.
2. Throw out the highest score.
3. Divide the sum of the remaining scores by 4.

Find the diver's score.

83. BANKING When a customer deposits cash, a teller must complete a "currency count" on the back of the deposit slip. In the illustration, what is the total amount of cash being deposited?

Currency count, for financial use only			
24	x 1's		
—	x 2's		
6	x 5's		
10	x 10's		
12	x 20's		
2	x 50's		
1	x 100's		
	TOTAL \$		

84. WRAPPING GIFTS How much ribbon is needed to wrap the package shown if 15 inches of ribbon are needed to make the bow?

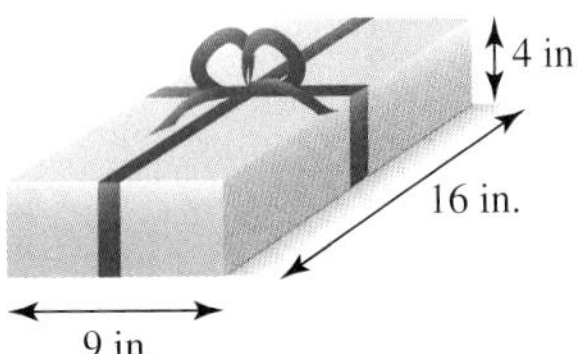

85. SCRABBLE Illustration (a) shows part of the game board before and Illustration (b) shows it after the words *brick* and *aphid* were played. Determine the scoring for each word. (The number on each tile gives the point value of the letter.)

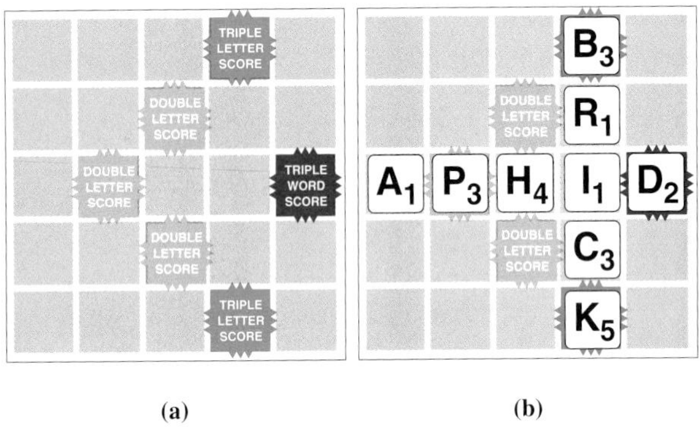

86. THE GETTYSBURG ADDRESS Here is an excerpt from Abraham Lincoln's Gettysburg Address.

Fourscore and seven years ago, our fathers brought forth on this continent a new nation, conceived in liberty, and dedicated to the proposition that all men are created equal.

Lincoln's comments refer to the year 1776, when the United States declared its independence. If a score is 20 years, in what year did Lincoln deliver the Gettysburg Address?

87. CLIMATE One December week, the temperatures in Honolulu, Hawaii, were 75°, 80°, 83°, 80°, 77°, 72°, and 86°. Find the week's average (mean) temperature.

88. GRADES In a psychology class, a student had test scores of 94, 85, 81, 77, and 89. He also overslept, missed the final exam, and received a 0 on it. What was his test average in the class?

89. NATURAL NUMBERS What is the average (mean) of the first nine natural numbers?

90. ENERGY USAGE Find the average number of therms of natural gas used per month. Then draw a dashed line across the graph in the next column showing the average.

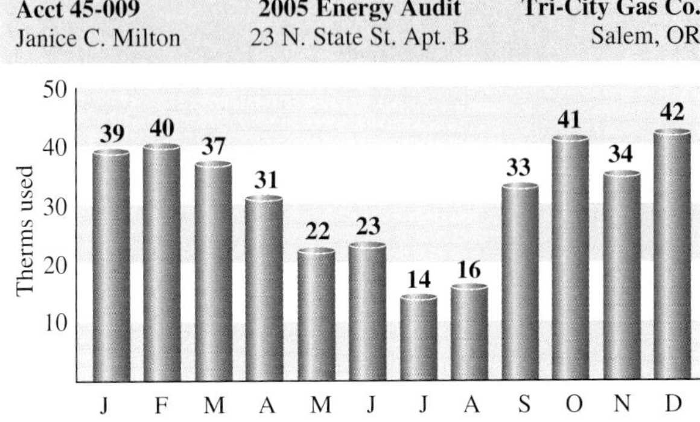

91. FAST FOOD The table shows the sandwiches Subway advertises as its low-fat menu. What is the average (mean) number of calories for the group of sandwiches?

6-inch subs	Calories	Fat (g)
Veggie Delite	237	3
Turkey Breast	289	4
Turkey Breast & Ham	295	5
Ham	302	5
Roast Beef	303	5
Subway Club	312	5
Roasted Chicken Breast	348	6

92. TV RATINGS The list below shows the number of people watching *Who Wants to Be a Millionaire*? on five weeknights in November 1999. How large was the average audience?

Monday	26,800,000
Tuesday	24,900,000
Wednesday	22,900,000
Thursday	25,900,000
Friday	21,900,000

WRITING

93. Explain why rules for the order of operations are necessary.

94. Explain the difference between the steps used to evaluate $5 \cdot 2^3$ and $(5 \cdot 2)^3$.

95. Explain the process of finding the mean of a large group of numbers. What does an average tell you?

96. What does it mean when we say to perform all additions and subtractions *as they occur from left to right*?

REVIEW *Perform the operations.*

97. $\begin{array}{r} 4{,}029 \\ +3{,}271 \\ \hline \end{array}$

98. $\begin{array}{r} 4{,}263 \\ -3{,}764 \\ \hline \end{array}$

99. $\begin{array}{r} 417 \\ \times\ 23 \\ \hline \end{array}$

100. $82\overline{)50{,}430}$

1.6 Solving Equations by Addition and Subtraction

- Equations • Checking solutions • Solving equations
- Problem solving with equations

The language of mathematics is *algebra.* The word *algebra* comes from the title of a book written by the Arabian mathematician Al-Khowarazmi around A.D. 800. Its title, *Ihm aljabr wa'l muqabalah,* means restoration and reduction, a process then used to solve equations. In this section, we will begin discussing equations, one of the most powerful ideas in algebra.

Equations

An **equation** is a statement indicating that two expressions are equal. Some examples of equations are

$x + 5 = 21,\quad 16 + 5 = 21,\quad \text{and}\quad 10 + 5 = 21$

Equations

Equations are mathematical sentences that contain an = symbol.

In the equation $x + 5 = 21$, the expression $x + 5$ is called the **left-hand side,** and 21 is called the **right-hand side.** The letter x is the **variable** (or the **unknown**).

An equation can be true or false. For example, $16 + 5 = 21$ is a true equation, whereas $10 + 5 = 21$ is a false equation. An equation containing a variable can be true or false, depending upon the value of the variable. If x is 16, the equation $x + 5 = 21$ is true, because

$\mathbf{16} + 5 = 21$ Substitute 16 for x.

However, this equation is false for all other values of x.

Any number that makes an equation true when substituted for its variable is said to *satisfy* the equation. Such numbers are called **solutions.** Because 16 is the only number that satisfies $x + 5 = 21$, it is the only solution of the equation.

Checking solutions

EXAMPLE 1 Verify that 18 is a solution of the equation $x - 3 = 15$.

Solution We substitute 18 for x in the equation and verify that both sides of the equation are equal.

Self Check 1
Is 8 a solution of $x + 17 = 25$?

$x - 3 = 15$	This is the given equation.
$\mathbf{18} - 3 \stackrel{?}{=} 15$	Substitute 18 for x. Read $\stackrel{?}{=}$ as "is possibly equal to."
$15 = 15$	Perform the subtraction.

Since $15 = 15$ is a true equation, 18 is a solution of $x - 3 = 15$.

Answer yes

Self Check 2
Is 5 a solution of $20 = y - 17$?

EXAMPLE 2 Is 23 a solution of $32 = y + 10$?

Solution We substitute 23 for y and simplify.

$32 = y + 10$	This is the given equation.
$32 \stackrel{?}{=} \mathbf{23} + 10$	Substitute 23 for y.
$32 \neq 33$	Perform the addition.

Since the left-hand and right-hand sides are not equal, 23 is not a solution of $32 = y + 10$.

Answer no

Solving equations

Since the solution of an equation is usually not given, we must develop a process to find it. This process is called *solving the equation.* To develop an understanding of the properties and procedures used to solve an equation, we will examine $x + 2 = 5$ and make some observations as we solve it in a practical way.

We can think of the scales shown in Figure 1-16(a) as representing the equation $x + 2 = 5$. The weight (in grams) on the left-hand side of the scales is $x + 2$, and the weight (in grams) on the right-hand side is 5. Because these weights are equal, the scales are in balance. To find x, we need to isolate it. That can be accomplished by removing 2 grams from the left-hand side of the scales. Common sense tells us that we must also remove 2 grams from the right-hand side if the scales are to remain in balance. In Figure 1-16(b), we can see that x grams are balanced by 3 grams. We say that we have *solved* the equation and that the *solution* is 3.

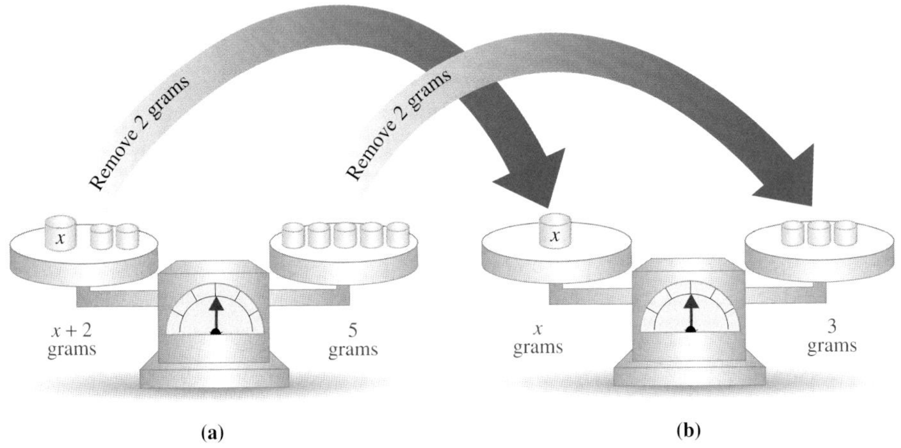

FIGURE 1-16

From this example, we can make observations about solving an equation.

- To find the value of x, we needed to isolate it on the left-hand side of the scales.
- To isolate x, we had to undo the addition of 2 grams. This was accomplished by subtracting 2 grams from the left-hand side.

- We wanted the scales to remain in balance. When we subtracted 2 grams from the left-hand side, we subtracted the same amount from the right-hand side.

The observations suggest a property of equality: *If the same quantity is subtracted from equal quantities, the results will be equal quantities.* We can express this property in symbols.

Subtraction property of equality

Let a, b, and c represent numbers.

If $a = b$, then $a - c = b - c$.

When we use this property, the resulting equation will be equivalent to the original equation.

Equivalent equations

Two equations are **equivalent equations** when they have the same solutions.

In the previous example, we found that $x + 2 = 5$ is equivalent to $x = 3$. This is true because these equations have the same solution, 3.

We now show how to solve $x + 2 = 5$ using an algebraic approach.

EXAMPLE 3 Solve: $x + 2 = 5$.

Solution To isolate x on the left-hand side of the equation, we undo the addition of 2 by subtracting 2 from both sides of the equation.

$x + 2 = 5$	This is the equation to solve.
$x + 2 \mathbf{- 2} = 5 \mathbf{- 2}$	Subtract 2 from both sides.
$x = 3$	On the left-hand side, subtracting 2 undoes the addition of 2 and leaves x. On the right-hand side, $5 - 2 = 3$.

We check by substituting 3 for x in the original equation and simplifying. If 3 is the solution, we will obtain a true statement.

Check: $\mathbf{x} + 2 = 5$	This is the original equation.
$\mathbf{3} + 2 \stackrel{?}{=} 5$	Substitute 3 for x.
$5 = 5$	Perform the addition.

Since the resulting equation is true, 3 is the solution.

Self Check 3

Solve $x + 7 = 14$ and check the result.

Answer 7

A second property that we will use to solve equations involves addition. It is based on the following idea: *If the same quantity is added to equal quantities, the results will be equal quantities.* In symbols, we have the following property.

Addition property of equality

Let a, b, and c represent numbers.

If $a = b$, then $a + c = b + c$.

We can think of the scales shown in Figure 1-17(a) on the next page as representing the equation $x - 2 = 3$. To find x, we need to use the addition property of equality and add 2 grams of weight to each side. The scales will remain in balance. From the scales in Figure 1-17(b), we can see that x grams are balanced by 5 grams. The solution of $x - 2 = 3$ is therefore 5.

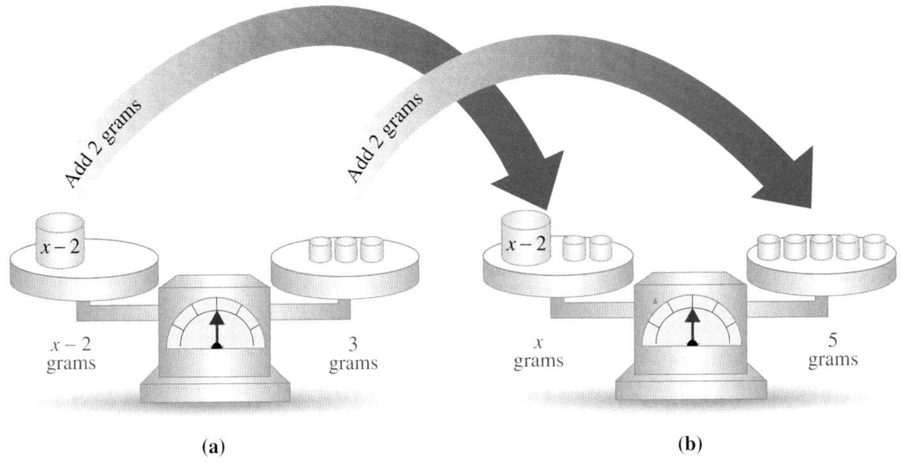

FIGURE 1-17

To solve $x - 2 = 3$ algebraically, we apply the addition property of equality. We can isolate x on the left-hand side of the equation by adding 2 to both sides.

$x - 2 = 3$	This is the equation to solve.
$x - 2 + \mathbf{2} = 3 + \mathbf{2}$	To undo the subtraction of 2, add 2 to both sides.
$x = 5$	On the left-hand side, adding 2 undoes the subtraction of 2 and leaves x. On the right-hand side, $3 + 2 = 5$.

To check this result, we substitute 5 for x in the original equation and simplify.

Check: $\mathbf{x} - 2 = 3$	This is the original equation.
$\mathbf{5} - 2 \stackrel{?}{=} 3$	Substitute 5 for x.
$3 = 3$	Perform the subtraction.

Since this is a true statement, 5 is the solution.

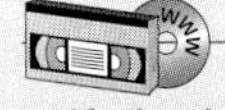

Self Check 4

Solve $75 = b - 38$ and check the result.

EXAMPLE 4 Solve: $19 = y - 7$.

Solution To isolate the variable y on the right-hand side, we use the addition property of equality. We can undo the subtraction of 7 by adding 7 to both sides.

$19 = y - 7$	This is the equation to solve.
$19 + \mathbf{7} = y - 7 + \mathbf{7}$	Add 7 to both sides.
$26 = y$	On the left-hand side, $19 + 7 = 26$. On the right-hand side, adding 7 undoes the subtraction of 7 and leaves y.
$y = 26$	When we state a solution, it is common practice to write the variable first. If $26 = y$, then $y = 26$.

We check by substituting 26 for y in the original equation and simplifying.

Check: $19 = \mathbf{y} - 7$	This is the original equation.
$19 \stackrel{?}{=} \mathbf{26} - 7$	Substitute 26 for y.
$19 = 19$	Perform the subtraction.

Since this is a true statement, 26 is the solution.

Answer 113

Problem solving with equations

The key to problem solving is to understand the problem and then to devise a plan for solving it. The following list of steps provides a good strategy to follow.

Strategy for problem solving

1. **Analyze the problem** by reading it carefully to understand the given facts. What information is given? What vocabulary is given? What are you asked to find? Often a diagram will help you visualize the facts of a problem.
2. **Form an equation** by picking a variable to represent the quantity to be found. Then express all other unknown quantities as expressions involving that variable. Key words or phrases can be helpful. Finally, translate the words of the problem into an equation.
3. **Solve the equation.**
4. **State the conclusion.**
5. **Check the result** in the words of the problem.

We will now use this five-step strategy to solve problems. The purpose of the following examples is to help you learn the strategy, even though you can probably solve these examples without it.

EXAMPLE 5 **Financial data.** Figure 1-18 shows the 1999 quarterly net income for Nike, the athletic shoe company. What was the company's total net income for 1999?

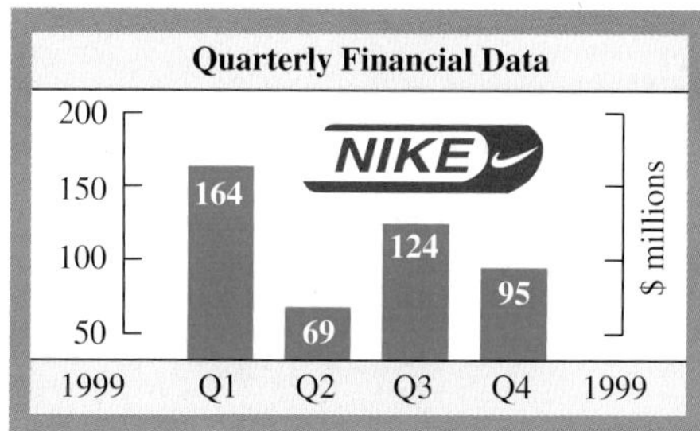

FIGURE 1-18

Analyze the problem

- We are given the net income for each quarter.
- We are asked to find the total net income.

Form an equation

Let n = the total net income for 1999. To form an equation involving n, we look for a key word or phrase in the problem.

Key word: *total* **Translation:** *addition*

Now we translate the words of the problem into an equation.

Total net income	is	1st qtr. net income	plus	2nd qtr. net income	plus	3rd qtr. net income	plus	4th qtr. net income.
n	$=$	164	$+$	69	$+$	124	$+$	95

Solve the equation

$$\begin{aligned} n &= 164 + 69 + 124 + 95 && \text{We are working in millions of dollars.} \\ &= 452 && \text{Perform the additions.} \end{aligned}$$

State the conclusion

Nike's total net income was \$452 million.

Check the result

We can check the result by estimation. To estimate, we round the net income from each quarter and add.

$$160 + 70 + 120 + 100 = 450$$

The answer, 452, is reasonable.

EXAMPLE 6 **Small business.** Last year a hairdresser lost 17 customers who moved away. If he now has 73 customers, how many did he have originally?

Analyze the problem

- We know that he started with some unknown number of customers, and after 17 moved away, 73 were left.
- We are asked to find the number of customers he had before any moved away.

Form an equation

We can let c = the original number of customers. To form an equation involving c, we look for a key word or phrase in the problem.

Key phrase: *moved away* **Translation:** *subtraction*

Now we translate the words of the problem into an equation.

The original number of customers	minus	17	is	the remaining number of customers.
c	$-$	17	$=$	73

Solve the equation

$$\begin{aligned} c - 17 &= 73 \\ c - 17 + \mathbf{17} &= 73 + \mathbf{17} && \text{To undo the subtraction of 17, add 17 to both sides.} \\ c &= 90 && \text{Simplify each side of the equation.} \end{aligned}$$

State the conclusion

He originally had 90 customers.

Check the result

The hairdresser had 90 customers. After losing 17, he has $90 - 17$, or 73 left. The answer, 90, checks.

EXAMPLE 7 **Buying a house.** Sue wants to buy a house that costs \$87,000. Since she has only \$15,000 for a down payment, she will have to borrow some additional money by taking a mortgage. How much will she have to borrow?

Analyze the problem

- The house costs \$87,000.
- Sue has \$15,000 for a down payment.
- We must find how much money she needs to borrow.

Form an equation

We can let x = the money that she needs to borrow. To form an equation involving x, we look for a key word or phrase in the problem.

Key phrase: *borrow some additional money* **Translation:** *addition*

Now we translate the words of the problem into an equation.

The amount Sue has	plus	the amount she borrows	is	the total cost of the house.
15,000	$+$	x	$=$	87,000

Solve the equation

$$15{,}000 + x = 87{,}000$$

$$15{,}000 + x - \mathbf{15{,}000} = 87{,}000 - \mathbf{15{,}000}$$ To undo the addition of 15,000, subtract 15,000 from both sides.

$$x = 72{,}000$$ Perform the subtractions.

State the conclusion

She must borrow \$72,000.

Check the result

With a \$72,000 mortgage, she will have \$15,000 + \$72,000, which is the \$87,000 that is necessary to buy the house. The answer, 72,000, checks.

Section 1.6 STUDY SET

VOCABULARY *Fill in the blanks.*

1. An equation is a statement that two expressions are ______. An equation contains an ____ symbol.
2. A ______ of an equation is a number that satisfies the equation.
3. To ______ the solution of an equation, we substitute the value for the variable in the original equation and see whether the result is a true statement.
4. A letter that is used to represent a number is called a ______.
5. ______ equations have exactly the same solutions.
6. To solve an equation, we ______ the variable on one side of the equals symbol.

CONCEPTS *Complete the rules.*

7. If $x = y$ and c is any number, then $x + c =$ ____ $+$ ____.
8. If $x = y$ and c is any number, then $x - c =$ ____ $-$ ____.

9. In $x + 6 = 10$, what operation is performed on the variable? How do we undo that operation to isolate the variable?

10. In $9 = y - 5$, what operation is performed on the variable? How do we undo that operation to isolate the variable?

NOTATION *Complete each solution to solve the given equation.*

11.
$$
\begin{aligned}
x + 8 &= 24 \\
x + 8 - \underline{\quad} &= 24 - \underline{\quad} \\
x &= 16
\end{aligned}
$$

Check:
$$
\begin{aligned}
x + 8 &= 24 \\
\underline{\quad} + 8 &\;\underline{\quad}\; 24 \\
\underline{\quad} &= 24
\end{aligned}
$$

So ___ is the solution.

12.
$$
\begin{aligned}
x - 8 &= 24 \\
x - 8 + \underline{\quad} &= 24 + \underline{\quad} \\
x &= 32
\end{aligned}
$$

Check:
$$
\begin{aligned}
x - 8 &= 24 \\
\underline{\quad} - 8 &\stackrel{?}{=} 24 \\
\underline{\quad} &= 24
\end{aligned}
$$

So ___ is the solution.

PRACTICE *Determine whether each statement is an equation.*

13. $x = 2$

14. $y - 3$

15. $7x < 8$

16. $7 + x = 2$

17. $x + y = 0$

18. $3 - 3y > 2$

19. $1 + 1 = 3$

20. $5 = a + 2$

For each equation, is the given number a solution?

21. $x + 2 = 3;\ 1$

22. $x - 2 = 4;\ 6$

23. $a - 7 = 0;\ 7$

24. $x + 4 = 4;\ 0$

25. $8 - y = y;\ 5$

26. $10 - c = c;\ 5$

27. $x + 32 = 0;\ 16$

28. $x - 1 = 0;\ 4$

29. $z + 7 = z;\ 7$

30. $n - 9 = n;\ 9$

31. $x = x;\ 0$

32. $x = 2;\ 0$

Use the addition or subtraction property of equality to solve each equation. **Check each answer.**

33. $x - 7 = 3$

34. $y - 11 = 7$

35. $a - 2 = 5$

36. $z - 3 = 9$

37. $1 = b - 2$

38. $0 = t - 1$

39. $x - 4 = 0$

40. $c - 3 = 0$

41. $y - 7 = 6$

42. $a - 2 = 4$

43. $70 = x - 5$

44. $66 = b - 6$

45. $312 = x - 428$

46. $x - 307 = 113$

47. $x - 117 = 222$

48. $y - 27 = 317$

49. $x + 9 = 12$

50. $x + 3 = 9$

51. $y + 7 = 12$

52. $c + 11 = 22$

53. $t + 19 = 28$

54. $s + 45 = 84$

55. $23 + x = 33$

56. $34 + y = 34$

57. $5 = 4 + c$

58. $41 = 23 + x$

59. $99 = r + 43$

60. $92 = r + 37$

61. $512 = x + 428$

62. $x + 307 = 513$

63. $x + 117 = 222$

64. $y + 38 = 321$

65. $3 + x = 7$

66. $b - 4 = 8$

67. $y - 5 = 7$

68. $z + 9 = 23$

69. $4 + a = 12$

70. $5 + x = 13$

71. $x - 13 = 34$

72. $x - 23 = 19$

APPLICATIONS *Complete each solution.*

73. ARCHAEOLOGY A 1,700-year-old manuscript is 425 years older than the clay jar in which it was found. How old is the jar?

Analyze the problem

- The manuscript is ________ old.
- The manuscript is ________ older than the jar.
- We are asked to find ________________.

Form an equation Since we want to find the age of the jar, we can let x = ________________. Now we look for a key word or phrase in the problem.

Key phrase: ____________

Translation: ___________

Now we translate the words of the problem into an equation.

___	is	425	plus	the age of the jar.
___	=	425	+	___

Solve the equation

$$
\begin{aligned}
\underline{\quad} &= 425 + x \\
1{,}700 - \underline{\quad} &= 425 + x - \underline{\quad} \\
\underline{\quad} &= x
\end{aligned}
$$

State the conclusion ________________.

Check the result If the jar is ___ years old, then the manuscript is $1{,}275 + 425 =$ ___ years old. The answer checks.

74. BANKING After a student wrote a \$1,500 check to pay for a car, he had a balance of \$750 in his account. How much did he have in the account before he wrote the check?

Analyze the problem

- A ______ check was written.
- The balance became ______.
- We are asked to find ______________________________.

Form an equation Since we want to find his balance before he wrote the check, we let x = ______________________________

Now we look for a key word or phrase in the problem.

Key phrase: ____________

Translation: ____________

Now we translate the words of the problem into an equation.

The original balance in the account	minus	\$1,500	is	\$750.
x	$-$		$=$	750

Solve the equation

$$\quad - 1{,}500 = 750$$
$$x - 1{,}500 + \quad = 750 + \quad$$
$$x = \quad$$

State the conclusion ______________________________

Check the result The original balance was ______. After writing a check, he had a balance of \$2,250 − \$1,500, or ______. The answer checks.

Let a variable represent the unknown quantity. Then write and solve an equation to answer the question.

75. ELECTIONS The illustration shows the votes received by the three major candidates running for President of the United States in 1996. Find the total number of votes cast for them.

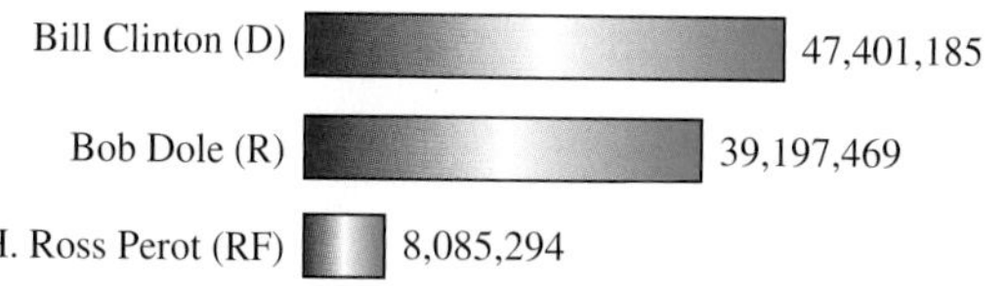

76. HIT RECORDS The oldest artist to have a number 1 single was Louis Armstrong at age 67, with *Hello Dolly.* The youngest artist to have the number 1 single was 12-year-old Jimmy Boyd, with *I Saw Mommy Kissing Santa Claus.* What is the difference in their ages?

77. PARTY INVITATIONS Three of Mia's party invitations were lost in the mail, but 59 were delivered. How many invitations did she send?

78. HEARING PROTECTION The sound intensity of a jet engine is 110 decibels. What noise level will an airplane mechanic experience if the ear plugs she is wearing reduce the sound intensity by 29 decibels?

79. FAST FOODS The franchise fee and startup costs for a Taco Bell restaurant are \$287,000. If an entrepreneur has \$68,500 to invest, how much money will she need to borrow to open her own Taco Bell restaurant?

80. BUYING GOLF CLUBS A man needs \$345 for a new set of golf clubs. How much more money does he need if he now has \$317?

81. CELEBRITY EARNINGS *Forbes* magazine estimates that in 2003, Celine Dion earned \$28 million. If this was \$152 million less than Oprah Winfrey's earnings, how much did Oprah earn in 2003?

82. HELP WANTED From the following ad from the classified section of a newspaper, determine the value of the benefit package. (\$45 K means \$45,000.)

★ACCOUNTS PAYABLE★
2-3 yrs exp as supervisor. Degree a +. High vol company. Good pay, \$45K & xlnt benefits; total compensation worth \$52K. Fax resume.

83. POWER OUTAGES The electrical system in a building automatically shuts down when the meter shown reads 85. By how much must the current reading increase to cause the system to shut down?

84. VIDEO GAMES After a week of playing Sega's *Sonic Adventure,* a boy scored 11,053 points in one game—an improvement of 9,485 points over the very first time he played. What was his score for his first game?

85. AUTO REPAIRS A woman paid $29 less to have her car fixed at a muffler shop than she would have paid at a gas station. At the gas station, she would have paid $219. How much did she pay to have her car fixed?

86. RIDING BUSES A man had to wait 20 minutes for a bus today. Three days ago, he had to wait 15 minutes longer than he did today, because four buses passed by without stopping. How long did he wait three days ago?

WRITING

87. Explain what it means for a number to satisfy an equation.

88. Explain how to tell whether a number is a solution of an equation.

89. Explain what Figure 1-16 (page 56) is trying to show.

90. Explain what Figure 1-17 (page 58) is trying to show.

91. When solving equations, we *isolate* the variable. Write a sentence in which the word *isolate* is used in a different context.

92. Think of a number. Add 8 to it. Now subtract 8 from that result. Explain why we will always obtain the original number.

REVIEW

93. Round 325,784 to the nearest ten

94. Find the power: 1^5.

95. Evaluate: $2 \cdot 3^2 \cdot 5$.

96. Represent $4 + 4 + 4$ as a multiplication.

97. Evaluate: $8 - 2(3) + 1^3$.

98. Write 1,055 in words.

1.7 Solving Equations by Division and Multiplication

- The division property of equality
- The multiplication property of equality
- Problem solving with equations

In the previous section, we solved equations of the forms

$$x - 4 = 10 \qquad \text{and} \qquad x + 5 = 16$$

by using the addition and subtraction properties of equality. In this section, we will learn how to solve equations of the forms

$$2x = 8 \qquad \text{and} \qquad \frac{x}{3} = 25$$

by using the division and multiplication properties of equality.

The division property of equality

To solve many equations, we must divide both sides of the equation by the same nonzero number. The resulting equation will be equivalent to the original one. This idea is summed up in the division property of equality: *If equal quantities are divided by the same nonzero quantity, the results will be equal quantities.*

Division property of equality

Let a, b, and c represent numbers and c is not 0.

If $a = b$, then $\frac{a}{c} = \frac{b}{c}$.

We will now consider how to solve the equation $2x = 8$. Recall that $2x$ means $2 \cdot x$. Therefore, the given equation can be rewritten as $2 \cdot x = 8$. We can think of the scales in Figure 1-19(a) as representing the equation $2 \cdot x = 8$. The weight (in grams) on the left-hand side of the scales is $2 \cdot x$, and the weight (in grams) on the right-hand side is 8. Because these weights are equal, the scales are in balance. To find x, we need to isolate it. That can be accomplished by using the division property of equality to remove half of the weight from each side. The scales will remain in balance. From the scales shown in Figure 1-19(b), we see that x grams are balanced by 4 grams.

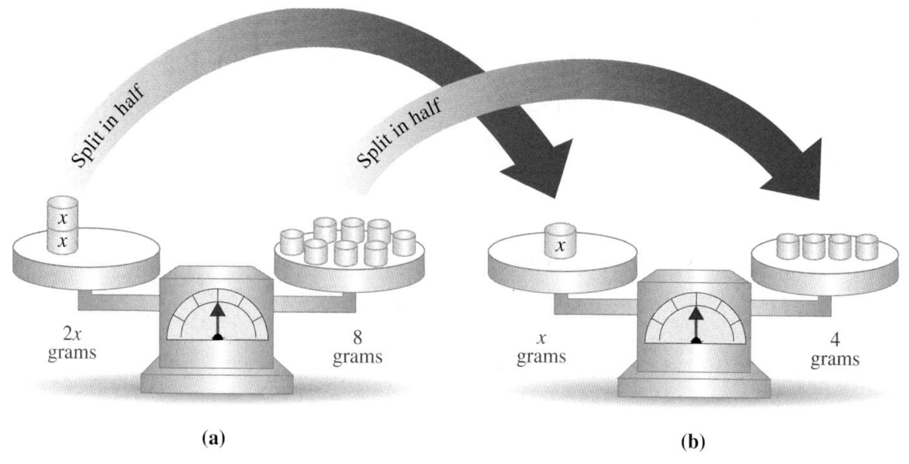

FIGURE 1-19

We now show how to solve $2x = 8$ using an algebraic approach.

EXAMPLE 1 Solve: $2x = 8$.

Solution Recall that $2x = 8$ means $2 \cdot x = 8$. To isolate x on the left-hand side of the equation, we undo the multiplication by 2 by dividing both sides of the equation by 2.

$2x = 8$ This is the equation to solve.

$\frac{2x}{\mathbf{2}} = \frac{8}{\mathbf{2}}$ To undo the multiplication by 2, divide both sides by 2.

$x = 4$ When x is multiplied by 2 and that product is then divided by 2, the result is x. Perform the division: $8 \div 2 = 4$.

To check this result, we substitute 4 for x in $2x = 8$.

Check: $2x = 8$ This is the original equation.

$2 \cdot \mathbf{4} \stackrel{?}{=} 8$ Substitute 4 for x.

$8 = 8$ Perform the multiplication.

Since $8 = 8$ is a true statement, 4 is the solution.

Self Check 1
Solve $17x = 153$ and check the result.

Answer 9

The multiplication property of equality

We can also multiply both sides of an equation by the same nonzero number to get an equivalent equation. This idea is summed up in the multiplication property of equality: *If equal quantities are multiplied by the same nonzero quantity, the results will be equal quantities.*

Multiplication property of equality

Let a, b, and c represent numbers and c is not 0.

If $a = b$, then $c \cdot a = c \cdot b$ or, more simply, $ca = cb$.

We can think of the scales shown in Figure 1-20(a) as representing the equation $\frac{x}{3} = 25$. The weight on the left-hand side of the scales is $\frac{x}{3}$ grams, and the weight on the right-hand side is 25 grams. Because these weights are equal, the scales are in balance. To find x, we can use the multiplication property of equality to triple (or multiply by 3) the weight on each side. The scales will remain in balance. From the scales shown in Figure 1-20(b), we can see that x grams are balanced by 75 grams.

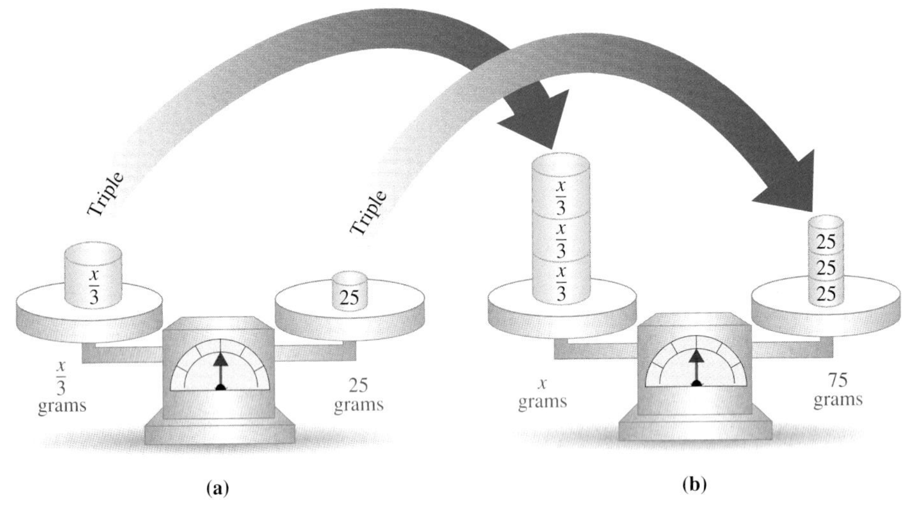

FIGURE 1-20

We now show how to solve $\frac{x}{3} = 25$ using an algebraic approach.

Self Check 2

Solve $\dfrac{x}{12} = 24$ and check the result.

Answer 288

EXAMPLE 2 Solve: $\dfrac{x}{3} = 25$.

Solution To isolate x on the left-hand side of the equation, we undo the division of the variable by 3 by multiplying both sides by 3.

$\dfrac{x}{3} = 25$ This is the equation to solve.

$\mathbf{3} \cdot \dfrac{x}{3} = \mathbf{3} \cdot 25$ To undo the division by 3, multiply both sides by 3.

$x = 75$ When x is divided by 3 and that quotient is then multiplied by 3, the result is x. Perform the multiplication: $3 \cdot 25 = 75$.

Check: $\dfrac{x}{3} = 25$ This is the original equation.

$\dfrac{\mathbf{75}}{3} \stackrel{?}{=} 25$ Substitute 75 for x.

$25 = 25$ Perform the division: $75 \div 3 = 25$. The answer checks.

Problem solving with equations

As before, we can use equations to solve problems. Remember that the purpose of these early examples is to help you learn the strategy, even though you can probably solve the examples without it.

EXAMPLE 3 Buying electronics. The owner of an apartment complex bought six television sets that were on sale for \$499 each. What was the total cost?

Analyze the problem

- 6 television sets were bought.
- They cost \$499 each.
- We are asked to find the total cost.

Form an equation

We let c = the total cost of the TVs. To form an equation, we look for a key word or phrase in the problem. We can add 499 six times or multiply 499 by 6. Since it is easier, we will multiply.

Key phrase: *six TVs, each costing \$499* **Translation:** *multiplication*

Now we translate the words of the problem into an equation.

The total number of TVs	multiplied by	the cost of each TV	is	the total cost.
6	$\cdot$	499	$=$	c

Solve the equation

$6 \cdot 499 = c$

$2{,}994 = c$ Perform the multiplication.

State the conclusion

The total cost will be \$2,994.

Check the result

We can check by estimation. Since each TV costs a little less than \$500, we would expect the total cost to be a little less than $6 \cdot \$500$, or \$3,000. An answer of \$2,994 is reasonable.

EXAMPLE 4 Splitting an inheritance. If seven brothers inherit \$343,000 and split the money evenly, how much will each brother get?

Analyze the problem

- There are 7 brothers.
- They split \$343,000 evenly.
- We are asked to find how much each brother will get.

Form an equation

We can let g = the number of dollars each brother will get. To form an equation, we look for a key word or phrase in the problem.

Key phrase: *split the money evenly* **Translation:** *division*

Now we translate the words of the problem into an equation.

The total amount of the inheritance	divided by	the number of brothers	is	the share each brother will get.
343,000	$\div$	7	$=$	g

Solve the equation

$\frac{343{,}000}{7} = g$ $\quad$ $343{,}000 \div 7$ can be written as $\frac{343{,}000}{7}$.

$49{,}000 = g$ $\quad$ Perform the division.

State the conclusion

Each brother will get $49,000.

Check the result

If we multiply $49,000 by 7, we get $343,000.

EXAMPLE 5 **Traffic violations.** For a speeding ticket, a motorist had to pay a fine of $592. The violation occurred on a stretch of highway posted with special signs like the one shown in Figure 1-21. What would the fine have been if such signs were not posted?

TRAFFIC FINES DOUBLED IN CONSTRUCTION ZONE

FIGURE 1-21

Analyze the problem

- The motorist was fined $592.
- The fine was double what it would normally have been.
- We are asked to find what the fine would have been, had the area not been posted.

Form an equation

We can let f = the amount that the fine would normally have been. To form an equation, we look for a key word or phrase in the problem or analysis.

Key word: *double* **Translation:** *multiply by 2*

Now we translate the words of the problem into an equation.

Two	times	the normal speeding fine	is	the new fine.
2	$\cdot$	f	$=$	592

Solve the equation

$2f = 592$ $\quad$ Write $2 \cdot f$ as $2f$.

$\frac{2f}{\mathbf{2}} = \frac{592}{\mathbf{2}}$ $\quad$ To undo the multiplication by 2, divide both sides by 2.

$f = 296$ $\quad$ Perform the division: $592 \div 2 = 296$.

State the conclusion

The fine would normally have been \$296.

Check the result

If we double \$296, we get 2(\$296) = \$592. The answer checks.

EXAMPLE 6 **Entertainment costs.** A five-piece band worked on New Year's Eve. If each player earned \$120, what fee did the band charge?

Analyze the problem

- There were 5 players in the band.
- Each player made \$120.
- We are asked to find the band's fee. We know that the fee divided by the number of players will give each person's share.

Form an equation

We can let f = the band's fee. To form an equation, we look for a key word or phrase. In this case, we find it in the analysis of the problem.

Key phrase: *divided by* **Translation:** *division*

Now we translate the words of the problem into an equation.

The band's fee	divided by	the number in the band	is	each person's share.
f	$\div$	5	=	120

Solve the equation

$$\frac{f}{5} = 120 \quad \text{Write } f \div 5 \text{ as } \frac{f}{5}.$$

$$\mathbf{5} \cdot \frac{f}{5} = \mathbf{5} \cdot 120 \quad \text{To undo the division by 5, multiply both sides by 5.}$$

$$f = 600 \quad \text{Perform the multiplication: } 5 \cdot 120 = 600.$$

State the conclusion

The band's fee was \$600.

Check the result

If we divide \$600 by 5, we get each person's share: \$120.

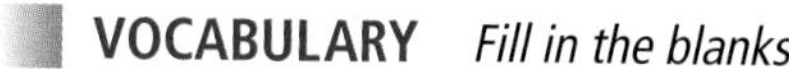

Section 1.7 STUDY SET

VOCABULARY *Fill in the blanks.*

1. According to the ________ property of equality: If equal quantities are divided by the same nonzero quantity, the results will be equal quantities.

2. According to the ____________ property of equality: If equal quantities are multiplied by the same nonzero quantity, the results will be equal quantities.

CONCEPTS *Fill in the blanks.*

3. If we multiply x by 6 and then divide that product by 6, the result is ___.

4. If we divide x by 8 and then multiply that quotient by 8, the result is ___.

5. If $x = y$, then $\frac{x}{z} = \frac{\quad}{\quad}$ where $z \neq 0$.

6. If $x = y$, then $zx = $ ___ where $z \neq 0$.

7. In the equation $4t = 40$, what operation is being performed on the variable? How do we undo it?

8. In the equation $\frac{t}{15} = 1$, what operation is being performed on the variable? How do we undo it?

9. Name the first step in solving each of the following equations.

a. $x + 5 = 10$ **b.** $x - 5 = 10$

c. $5x = 10$ **d.** $\frac{x}{5} = 10$

10. For each of the following equations, check the given possible solution.

a. $16 = t - 8$; 33

b. $16 = t + 8$; 8

c. $16 = 8t$; 128

d. $16 = \frac{t}{8}$; 2

NOTATION *Complete each solution to solve the given equation.*

11. $3x = 12$

$$\frac{3x}{\quad} = \frac{12}{\quad}$$

$x = 4$

Check: $3x = 12$

$3 \cdot \quad \stackrel{?}{=} 12$

$\quad = 12$

So ___ is the solution.

12. $\frac{x}{5} = 9$

$$\quad \cdot \frac{x}{5} = \quad \cdot 9$$

$x = 45$

Check: $\frac{x}{5} = 9$

$\frac{\quad}{5} \stackrel{?}{=} 9$

$\quad = 9$

So ___ is the solution.

PRACTICE *Use the division or the multiplication property of equality to solve each equation.* **Check each answer.**

13. $3x = 3$ **14.** $5x = 5$

15. $2x = 192$ **16.** $4x = 120$

17. $17y = 51$ **18.** $19y = 76$

19. $34y = 204$ **20.** $18y = 90$

21. $100 = 100x$ **22.** $35 = 35y$

23. $16 = 8r$ **24.** $44 = 11m$

25. $\frac{x}{7} = 2$ **26.** $\frac{x}{12} = 4$

27. $\frac{y}{14} = 3$ **28.** $\frac{y}{13} = 5$

29. $\frac{a}{15} = 5$ **30.** $\frac{b}{25} = 5$

31. $\frac{c}{13} = 3$ **32.** $\frac{d}{100} = 11$

33. $1 = \frac{x}{50}$ **34.** $1 = \frac{x}{25}$

35. $7 = \frac{t}{7}$ **36.** $4 = \frac{m}{4}$

37. $9z = 90$ **38.** $3z = 6$

39. $7x = 21$ **40.** $13x = 52$

41. $86 = 43t$ **42.** $288 = 96t$

43. $21s = 21$ **44.** $31x = 155$

45. $\frac{d}{20} = 2$ **46.** $\frac{x}{16} = 4$

47. $400 = \frac{t}{3}$ **48.** $250 = \frac{y}{2}$

APPLICATIONS *Complete each solution.*

49. THE NOBEL PRIZE In 1998, three Americans, Louis Ignarro, Robert Furchgott, and Dr. Fred Murad, were awarded the Nobel Prize for Medicine. They shared the prize money. If each person received \$318,500, what was the Nobel Prize cash award?

Analyze the problem

- ___ people shared the cash award.
- Each person received ______.
- We are asked to find the ____________________.

Form an equation

Since we want to find what the Nobel Prize cash award was, we let c = ____________________. To form an equation, we look for a key word or phrase in the problem.

Key phrase: ____________________

Translation: ________

Now we translate the words of the problem into an equation.

The Nobel Prize cash award	divided by	the number of recipients	was	$318,500.
___	÷	3	=	318,500

Solve the equation

$$\frac{x}{3} = 318{,}500$$

$$\_\_ \cdot \frac{x}{3} = \_\_ \cdot 318{,}500$$

$$x = \_\_\_$$

State the conclusion

____

Check the result

If we divide the Nobel Prize cash award by 3, we have $\frac{\_\_\_}{3}$ = ______. This was the amount each person received. The answer checks.

50. INVESTING An investor has watched the value of his portfolio double in the last 12 months. If the current value of his portfolio is $274,552, what was its value one year ago?

Analyze the problem

- The value of the portfolio ______ in 12 months.
- The current value is ______.
- We must find ____________________.

Form an equation

We can let x = ____________________.

We now look for a key word or phrase in the problem.

Key phrase: ________

Translation: ________

Now we translate the words of the problem into an equation.

2	times	the value of the portfolio one year ago	is	the current value of the portfolio.
2	·	___	=	274,552

Solve the equation

$$2x = \_\_\_$$

$$\frac{2x}{\_\_} = \frac{274{,}552}{\_\_}$$

$$x = \_\_\_$$

State the conclusion

____

Check the result

If the value of the portfolio one year ago was ______ and it doubled, its current value would be ______. The answer checks.

Let a variable represent the unknown quantity. Then write and solve an equation to answer the question.

51. SPEED READING An advertisement for a speed reading program claimed that successful completion of the course could triple a person's reading rate. If Alicia can currently read 130 words a minute, at what rate can she expect to read after taking the classes?

52. COST OVERRUNS Lengthy delays and skyrocketing costs caused a rapid-transit construction project to go over budget by a factor of 10. The final audit showed the project costing $540 million. What was the initial cost estimate?

53. STAMPS Large sheets of commemorative stamps honoring Marilyn Monroe are to be printed. On each sheet, there are 112 stamps, with 8 stamps per row. How many rows of stamps are on a sheet?

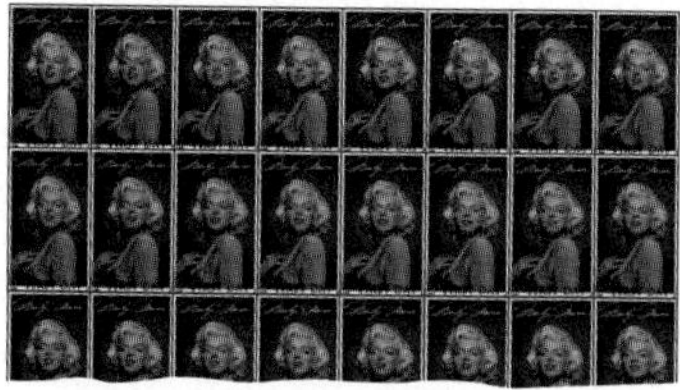

54. SPREADSHEETS The grid shown below is a Microsoft Excel spreadsheet. The rows are labeled with numbers, and the columns are labeled with letters. Each empty box of the grid is called a *cell.* Suppose a certain project calls for a spreadsheet with 294 cells, using columns A through F. How many rows will need to be used?

Microsoft Excel-Book 1

File Edit View Insert Format Tools

	A	B	C	D	E	F
1						
2						
3						
4						
5						
6						
7						
8						

Sheet 1 Sheet 2 Sheet 3 Sheet 4 Sheet 5

55. PHYSICAL EDUCATION A high school PE teacher had the students in her class form three-person teams for a basketball tournament. Thirty-two teams participated in the tournament. How many students were in the PE class?

56. LOTTO WINNERS The grocery store employees listed below pooled their money to buy $120 worth of lottery tickets each week, with the understanding they would split the prize equally if they happened to win. One week they did have the winning ticket and won $480,000. What was each employee's share of the winnings?

Sam M. Adler	Ronda Pellman	Manny Fernando
Lorrie Jenkins	Tom Sato	Sam Lin
Kiem Nguyen	H. R. Kinsella	Tejal Neeraj
Virginia Ortiz	Libby Sellez	Alicia Wen

57. ANIMAL SHELTERS The number of phone calls to an animal shelter quadrupled after the evening news aired a segment explaining the services the shelter offered. Before the publicity, the shelter received 8 calls a day. How many calls did the shelter receive each day after being featured on the news?

58. OPEN HOUSES The attendance at an elementary school open house was only half of what the principal had expected. If 120 people visited the school that evening, how many had she expected to attend?

59. GRAVITY The weight of an object on Earth is 6 times greater than what it is on the moon. The following weighing situation took place on the Earth. If it took place on the moon, what weight would the scale register?

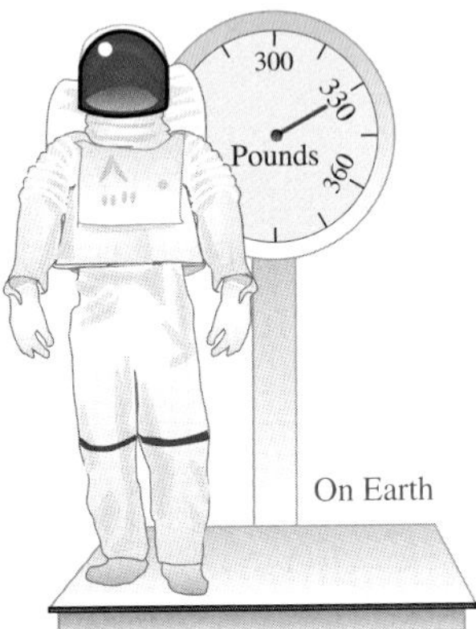

60. INFOMERCIALS The number of orders received each week by a company selling skin care products increased fivefold after a Hollywood celebrity was added to the company's infomercial. After adding the celebrity, the company received abou 175 orders each week. How many orders were received each week before the celebrity took part?

WRITING

61. Explain what Figure 1-19 (page 65) is trying to show.

62. Explain what Figure 1-20 (page 66) is trying to show.

63. What does it mean to solve an equation?

64. Think of a number. Double it. Now divide it by 2. Explain why you always obtain the original number.

REVIEW

65. Find the perimeter of a rectangle with sides measuring 8 cm and 16 cm.

66. Find the area of a rectangle with sides measuring 23 inches and 37 inches.

67. Find the prime factorization of 120.

68. Find the prime factorization of 150.

69. Evaluate: $3^2 \cdot 2^3$.

70. Evaluate: $5 + 6 \cdot 3$.

71. FUEL ECONOMY Five basic models of automobiles made by Saturn have city mileage ratings of 24, 22, 28, 29, and 27 miles per gallon. What is the average (mean) city mileage for the five models?

72. Solve: $x - 4 = 20$.

KEY CONCEPT

Variables

One of the objectives of this course is for you to become comfortable working with **variables.** You will recall that a variable is a letter that stands for a number.

The application problems of Sections 1.6 and 1.7 were solved with the help of a variable. In these problems, we let the variable represent an unknown quantity such as the number of customers a hairdresser used to have, the age of a jar, and the cash award given a Nobel Prize winner. We then wrote an equation to describe the situation mathematically and solved the equation to find the value represented by the variable.

Suppose that you are going to solve the following problems. What quantity should be represented by a variable? State your response in the form "Let x = ..."

1. The monthly cost to lease a van is \$120 less than the monthly cost to buy it. To buy it, the monthly payments are \$290. How much does it cost to lease the van each month?
2. One piece of pipe is 10 feet longer than another. Together, their lengths total 24 feet. How long is the shorter piece of pipe?
3. The length of a rectangular field is 50 feet. What is its width if it has a perimeter of 200 feet?
4. If one hose can fill a vat in 2 hours and another can fill it in 3 hours, how long will it take to fill the vat if both hoses are used?
5. Find the distance traveled by a motorist in 3 hours if her average speed was 55 mph.
6. In what year was a couple married if their 50th anniversary was in 2004?

Variables can also be used to state properties of mathematics in a concise, "shorthand" notation. State each property using mathematical symbols and the given variable(s).

7. Use the variables a and b to state that two numbers can be added in either order to get the same sum.
8. Use the variable x to state that when 0 is subtracted from a number, the result is the same number.
9. Use the variable b to state that the result when dividing a number by 1 is the same number.
10. Use the variable x to show that the sum of a number and 1 is greater than the number.
11. Using the variable n, state the fact that when 1 is subtracted from any number, the difference is less than the number.
12. State the fact that the product of any number and 0 is 0, using the variable a.
13. Use the variables r, s, and t to state that the way we group three numbers when adding them does not affect the answer.
14. Using the variable n, state the fact that when a number is multiplied by 1, the result is the number.

ACCENT ON TEAMWORK

SECTION 1.1

PLACE VALUE Have each student in your group bring a calculator to class so that you can examine several different models. For each model, determine the largest number (if there is one) that can be entered on the display of the calculator. Then press the appropriate calculator keys to add 1 to that number. What does the display show?

LARGE NUMBERS Bill Gates, founder of Microsoft Corporation, is said to be a billionaire. How many millions make 1 billion?

SECTION 1.2

READING THE PROBLEM In reading Example 9 of Section 1.2, you will notice that it contains several facts that are not used in the solution of the problem. Have each person in your group write a similar problem that requires careful reading to extract the useful information. Then have each person share his or her problem with the other students in the group.

SECTION 1.3

DIVISIBILITY TESTS Certain tests can help us decide whether one whole number is divisible by another.

- A number is divisible by 2 if the last digit of the number is 0, 2, 4, 6, or 8.
- A number is divisible by 3 if the sum of the digits is divisible by 3.
- A number is divisible by 4 if the number formed by the last two digits is divisible by 4.
- A number is divisible by 5 if the last digit of the number is 0 or 5.
- A number is divisible by 6 if the last digit of the number is 0, 2, 4, 6, or 8 and the sum of the digits is divisible by 3.
- A number is divisible by 8 if the number formed by the last three digits is divisible by 8.
- A number is divisible by 9 if the sum of the digits is divisible by 9.
- A number is divisible by 10 if the last digit of the number is 0.
- Determine whether each number is divisible by 2, 3, 4, 5, 6, 8, 9, and/or 10.

  a. 660 **b.** 2,526
  c. 11,523 **d.** 79,503
  e. 135,405 **f.** 4,444,440

SECTION 1.4

COMMON FACTORS The prime factorizations of 36 and 126 are shown below. The prime factors that are common to 36 and 126 (highlighted in color) are 2, 3, and 3.

$$36 = \mathbf{2} \cdot 2 \cdot \mathbf{3} \cdot \mathbf{3}$$
$$126 = \mathbf{2} \cdot \mathbf{3} \cdot \mathbf{3} \cdot 7$$

Find the common prime factors for each of the following pairs of numbers.

a. 25, 45 **b.** 24, 60
c. 18, 45 **d.** 40, 112
e. 180, 210 **f.** 242, 198

SECTION 1.5

ORDER OF OPERATIONS Consider the expression

$$5 + 8 \cdot 2^3 - 3 \cdot 2$$

Insert a set of parentheses somewhere in the expression so that, when it is evaluated, you obtain

a. 63 **b.** 132
c. 21 **d.** 127

SECTION 1.6

SOLVING EQUATIONS Borrow a scale and some weights from the chemistry department. Use them as part of a class presentation to explain how the subtraction property of equality is used to solve the equation $x + 2 = 5$. See the discussion and Figure 1-16 on page 56 for some suggestions on how to do this.

SECTION 1.7

FORMING EQUATIONS Reread Example 4 in Section 1.7. This problem could have been solved by forming an equation involving the operation of multiplication instead of the operation of division.

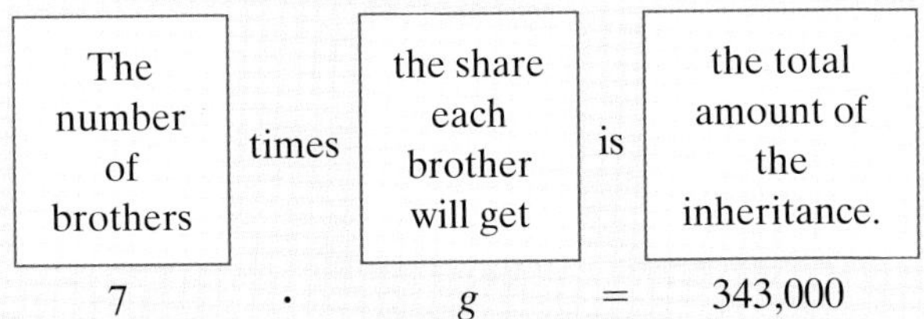

For Examples 5 and 6 in Section 1.7, write another equation that could be used to solve the problem. Then solve the equation and state the result.

CHAPTER REVIEW

SECTION 1.1 An Introduction to the Whole Numbers

CONCEPTS

A *set* is a collection of objects. The set of *natural numbers* is

$\{1, 2, 3, 4, 5, \ldots\}$

The set of *whole numbers* is

$\{0, 1, 2, 3, 4, 5, \ldots\}$

REVIEW EXERCISES

Graph each set.

1. The natural numbers less than 5

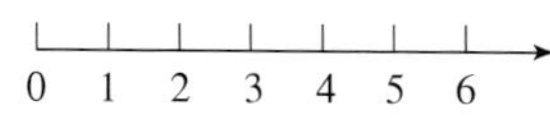

2. The whole numbers between 0 and 3

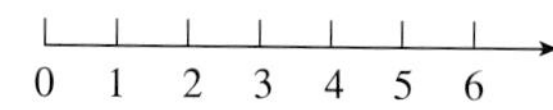

Whole numbers are often used in tables, bar graphs, and line graphs.

FARMING The table below shows the size of the average U.S. farm (in acres) for the period 1940–2000, in 20-year increments.

Year	1940	1960	1980	2000
Average size (acres)	174	297	426	432

3. Construct a bar graph of the data.

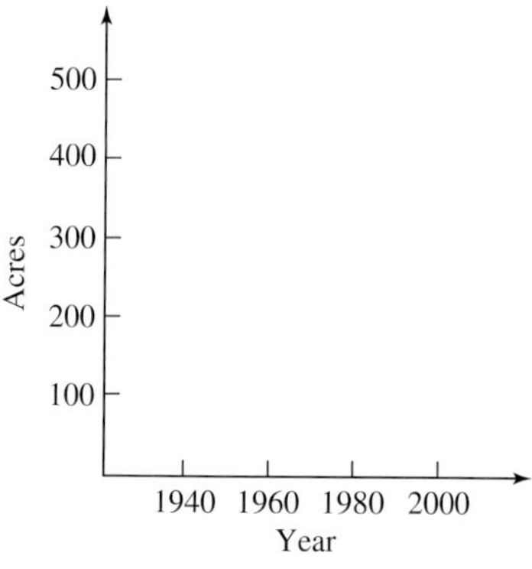

4. Construct a line graph of the data.

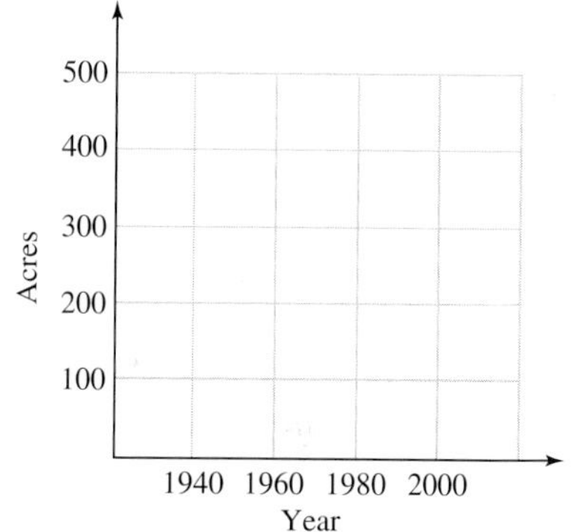

The digits in a whole number have *place value.*

Consider the number 2,365,720. Which digit is in the given column?

5. Ten thousands
6. Hundreds

A whole number is written in *expanded notation* when its digits are written with their place values.

Write each number in expanded notation.

7. 570,302
8. 37,309,054

We use the digits 0, 1, 2, 3, 4, 5, 6, 7, 8, and 9 to write a number in *standard notation.*

Write each number in standard notation.

9. 3 thousands + 2 hundreds + 7 ones
10. Twenty-three million, two hundred fifty-three thousand, four hundred twelve
11. Sixteen billion

The symbol < means "is less than." The symbol > means "is greater than."

Place an < or > symbol in the box to make a true statement.

12. 9 ☐ 7

13. 3 ☐ 5

To give approximate answers, we often use *rounded numbers.*

Round 2,507,348 to the specified place.

14. Nearest hundred

15. Nearest ten thousand

16. Nearest ten

17. Nearest hundred thousand

SECTION 1.2 Adding and Subtracting Whole Numbers

Addition is the process of finding the total of two (or more) numbers. Do additions within parentheses first.

Find each sum.

18. 56 + 22

19. 137 + 0

20. 15 + (27 + 13)

21. 82 + 17 + 50

22. (111 + 222) + 444

23. 0 + 2,332

The *commutative* and *associative properties of addition:*

$a + b = b + a$

$(a + b) + c = a + (b + c)$

Perform each addition.

24. $\begin{array}{r} 236 \\ +\underline{282} \end{array}$

25. $\begin{array}{r} 5{,}345 \\ +\underline{655} \end{array}$

26. 135 + 213 + 615 + 47

27. 4,447 + 7,478 + 13,061

What property of addition is shown?

28. 12 + 8 = 8 + 12

29. 12 + (8 + 2) = (12 + 8) + 2

The *perimeter* of a rectangle or a square is the distance around it.

30. Find the perimeter of square with sides 24 inches long.

Subtraction is the process of finding the difference between two numbers.

Perform each subtraction.

31. 18 − 5

32. 9 − (7 − 2)

33. 22 − 5 − 6

34. Subtract 5,177 from 5,231.

35. $\begin{array}{r} 343 \\ -\underline{269} \end{array}$

36. $\begin{array}{r} 17{,}800 \\ -\underline{15{,}725} \end{array}$

Give the addition or subtraction fact that is illustrated by each figure.

37.

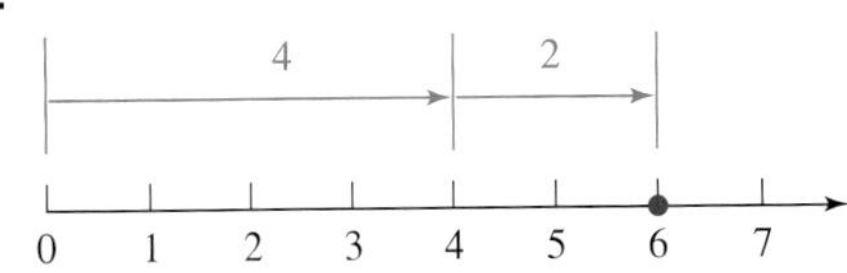

38.

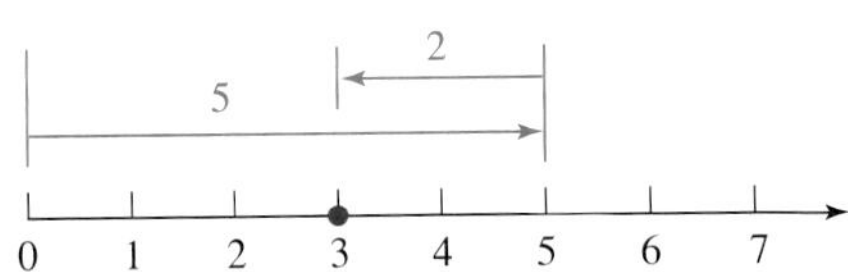

39. TRAVEL A direct flight from Omaha to San Francisco costs $237. Another flight with one stop in Reno costs $192. How much can be saved by taking the less expensive flight?

40. SAVINGS ACCOUNTS A savings account contains $931. If the owner deposits $271 and makes withdrawals of $37 and $380, find the final balance.

41. REBATES The price of a new Honda Civic was advertised in a newspaper as $21,991*. A note at the bottom of the ad read, "*Reflects $1,550 factory rebate." What was the car's original sticker price?

SECTION 1.3 *Multiplying and Dividing Whole Numbers*

Multiplication is repeated addition. For example,

$$4 \cdot 6 = \overbrace{6 + 6 + 6 + 6}^{\text{The sum of four 6's}} = 24$$

The result, 24, is called the *product,* and the 4 and 6 are called *factors.*

The *commutative* and *associative properties of multiplication:*

$a \cdot b = b \cdot a$

$(a \cdot b) \cdot c = a \cdot (b \cdot c)$

The *area A of a rectangle* is the product of its length l and its width w.

$A = l \cdot w$

Perform each multiplication.

42. $8 \cdot 7$

43. $7(8)$

44. $8 \cdot 0$

45. $7 \cdot 1$

46. $10 \cdot 8 \cdot 7$

47. $5 \cdot (7 \cdot 6)$

48. $157 \cdot 21$

49. $3{,}723 \cdot 48$

50. $\begin{array}{r} 356 \\ \times \underline{\ \ 89} \end{array}$

51. $\begin{array}{r} 5{,}624 \\ \times \underline{\ \ \ 81} \end{array}$

What property of multiplication is shown?

52. $12 \cdot (8 \cdot 2) = (12 \cdot 8) \cdot 2$

53. $12 \cdot 8 = 8 \cdot 12$

54. WAGES If a math tutor worked for 38 hours and was paid $9 per hour, how much did she earn?

55. HORSESHOES Find the perimeter and the area of the rectangular horseshoe court shown below.

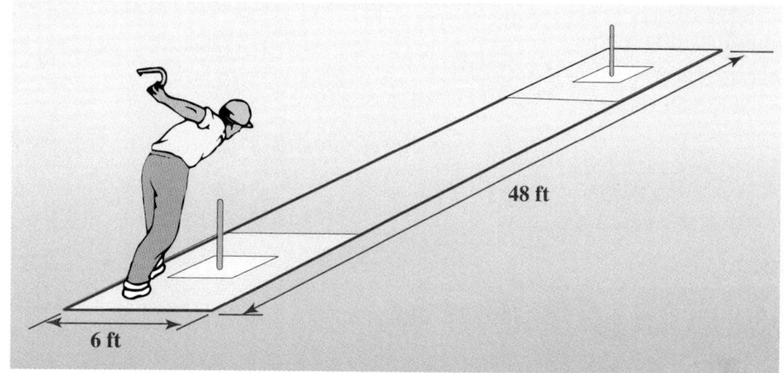

56. PACKAGING There are 12 eggs in one dozen, and 12 dozen in one gross. How many eggs are in a shipment of 5 gross?

Division is an operation that determines how many times a number (the *divisor*) is contained in another number (the *dividend*). *Remember that you can never divide by 0.*

Perform each division, if possible.

57. $\frac{6}{3}$

58. $\frac{15}{1}$

59. $73 \div 0$

60. $\frac{0}{8}$

61. $357 \div 17$

62. $1{,}443 \div 39$

63. $21\overline{)405}$

64. $54\overline{)1{,}269}$

65. TREATS If 745 candies are divided equally among 45 children, how many will each child receive? How many candies will be left over?

66. COPIES An elementary school teacher had copies of a 3-page social studies test made at Quick Copy Center. She was charged for 84 sheets of paper. How many copies of the test were made?

SECTION 1.4 *Prime Factors and Exponents*

Numbers that are multiplied together are called *factors.*

Find all of the factors of each number.

67. 18

68. 25

A *prime number* is a whole number, greater than 1, that has only 1 and itself as factors. Whole numbers greater than 1 that are not prime are called *composite numbers.*

Identify each number as a prime, composite, or neither.

69. 31

70. 100

71. 1

72. 0

73. 125

74. 47

Whole numbers divisible by 2 are *even* numbers. Whole numbers not divisible by 2 are *odd* numbers.

Identify each number as an even or odd number.

75. 171

76. 214

77. 0

78. 1

The *prime factorization* of a whole number is the product of its prime factors.

Find the prime factorization of each number.

79. 42

80. 375

An *exponent* is used to indicate repeated multiplication. In the *exponential expression* a^n, a is the base and n is the exponent.

Write each expression using exponents.

81. $6 \cdot 6 \cdot 6 \cdot 6$

82. $5 \cdot 5 \cdot 5 \cdot 13 \cdot 13$

Evaluate each expression.

83. 5^3

84. 11^2

85. $2^3 \cdot 5^2$

86. $2^2 \cdot 3^3 \cdot 5^2$

SECTION 1.5 *Order of Operations*

Perform mathematical operations in the following order:

1. Perform all calculations within parantheses and other grouping symbols.
2. Evaluate all exponential expressions.
3. Perform all multiplications and divisions in order from left to right.
4. Perform all additions and subtractions in order from left to right.

To evaluate an expression containing grouping symbols, perform all calculations within each pair of grouping symbols, working from the innermost pair to the outermost pair.

Evaluate each expression.

87. $13 + 12 \cdot 3$

88. $35 - 15 \div 5$

89. $(13 + 12)3$

90. $(8 - 2)^2$

91. $8 \cdot 5 - 4 \div 2$

92. $8 \cdot (5 - 4 \div 2)$

93. $2 + 3(10 - 4 \cdot 2)$

94. $4(20 - 5 \cdot 3 + 2) - 4$

95. $3^3\left(\frac{12}{6}\right) - 1^4$

96. $\frac{12 + 3 \cdot 7}{5^2 - 14}$

97. $7 + 3[10 - 3(4 - 2)]$

98. $5 + 2[(15 - 3 \cdot 4) - 2]$

99. DICE GAMES Write an expression that finds the total on all the dice shown below. Then evaluate the expression.

The *arithmetic mean* (average) is a value around which numbers are grouped.

100. YAHTZEE Find the player's average (mean) score for the 6 games.

yahtzee
SCORE CARD

Game #1	Game #2	Game #3	Game #4	Game #5	Game #6
159	244	184	240	166	213

SECTION 1.6 Solving Equations by Addition and Subtraction

An *equation* is a statement that two expressions are equal.

A *variable* is a letter that stands for a number.

Two equations with exactly the same solutions are called *equivalent equations.*

To solve an equation, isolate the variable on one side of the equation by undoing the operation performed on it.

If the same number is added to (or subtracted from) both sides of an equation, an equivalent equation results.

If $a = b$, then $a + c = b + c$.

If $a = b$, then $a - c = b - c$.

Problem-solving strategy:

1. Analyze the problem.
2. Form an equation.
3. Solve the equation.
4. State the conclusion.
5. Check the result.

Determine whether the given number is a solution of the equation. Explain.

101. $x + 2 = 13$; 5

102. $x - 3 = 1$; 4

Identify the variable in each equation.

103. $y - 12 = 50$

104. $114 = 4 - t$

Solve the equation and check the result.

105. $x - 7 = 2$

106. $x - 11 = 20$

107. $225 = y - 115$

108. $101 = p - 32$

109. $x + 9 = 18$

110. $b + 12 = 26$

111. $175 = p + 55$

112. $212 = m + 207$

113. $x - 7 = 0$

114. $x + 15 = 1{,}000$

Let a variable represent the unknown quantity. Then write and solve an equation to answer the question.

115. FINANCING A newly married couple made a \$25,500 down payment on a \$122,750 house. How much did they need to borrow?

116. DOCTOR'S PATIENTS After moving his office, a doctor lost 13 patients. If he had 172 patients left, how many did he have originally?

SECTION 1.7 *Solving Equations by Division and Multiplication*

If both sides of an equation are divided by (or multiplied by) the same nonzero number, an equivalent equation results.

If $a = b$, then $\frac{a}{c} = \frac{b}{c}$

If $a = b$, then $a \cdot c = b \cdot c$

Solve the equation and check the result.

117. $3x = 12$

118. $15y = 45$

119. $105 = 5r$

120. $224 = 16q$

121. $\frac{x}{7} = 3$

122. $\frac{a}{3} = 12$

123. $15 = \frac{s}{21}$

124. $25 = \frac{d}{17}$

125. $12x = 12$

126. $\frac{x}{12} = 12$

Let a variable represent the unknown quantity. Then write and solve an equation to answer the question.

127. CARPENTRY If you cut a 72-inch board into three equal pieces, how long will each piece be? Disregard any loss due to cutting.

128. JEWELRY Four sisters split the cost of a gold chain evenly. How much did the chain cost if each sister's share was $32?

CHAPTER 1 TEST

1. Graph the whole numbers less than 5.

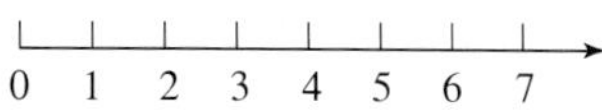

2. Write "five thousand two hundred sixty-six" in expanded notation.

3. Write "7 thousands + 5 hundreds + 7 ones" in standard notation.

4. Round 34,752,341 to the nearest million.

In Problems 5–6, refer to the data in the table.

Lot number	1	2	3	4
Defective bolts	7	10	5	15

5. Use the data to make a bar graph.

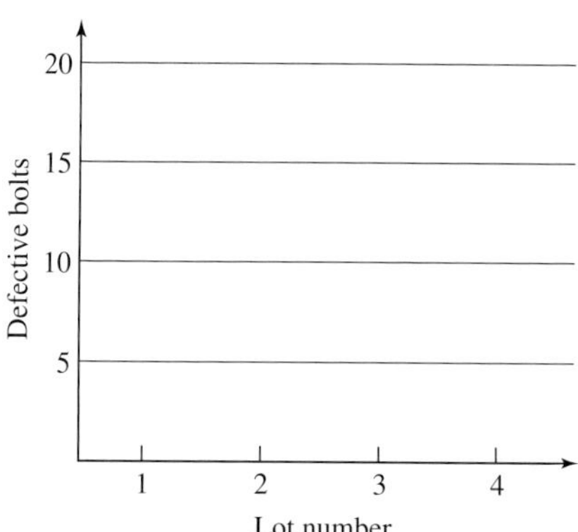

6. Use the data to make a line graph.

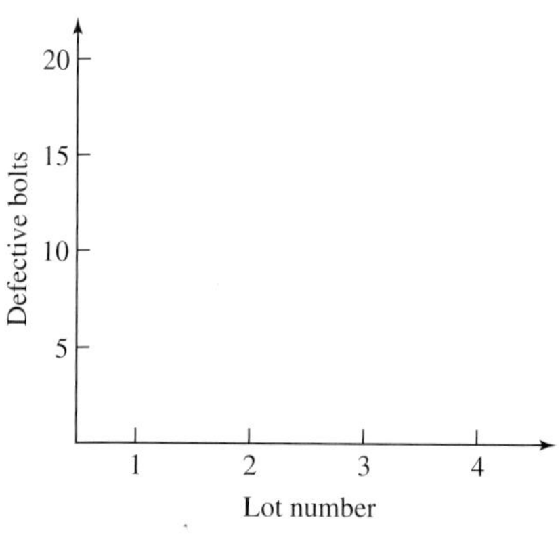

Place an < or > symbol in the box to make a true statement.

7. 15 ☐ 10

8. 1,247 ☐ 1,427

9. Add: 327 + 435 + 123 + 606.

10. Subtract 287 from 535.

11. Add: $\begin{array}{r} 44{,}526 \\ +13{,}579 \\ \hline \end{array}$

12. Subtract: $\begin{array}{r} 4{,}521 \\ -3{,}579 \\ \hline \end{array}$

13. STOCKS On Tuesday, a share of KBJ Company was selling at \$73. The price rose \$12 on Wednesday and fell \$9 on Thursday. Find its price on Thursday.

14. List the factors of 20 in order, from least to greatest.

Perform each operation.

15. Multiply: $\begin{array}{r} 53 \\ \times\ 8 \\ \hline \end{array}$

16. Multiply: 367(73).

17. Divide: $63\overline{)4{,}536}$.

18. Divide $73\overline{)8{,}379}$.

19. Perform each operation, if possible.

    a. $15 \cdot 0$

    b. $\dfrac{0}{15}$

20. What property is illustrated by each statement?

    a. $18 \cdot (9 \cdot 40) = (18 \cdot 9) \cdot 40$

    b. $23{,}999 + 1 = 1 + 23{,}999$

21. FURNITURE SALES See the following ad. Find the perimeter of the rectangular space under the tent. Then fill in the blank in the advertisement.

The Greatest Parking Lot Tent
Sale in Our History!
? square feet
in our "outdoor showroom"
2 DAYS ONLY!
Thomastown Home
Furnishings
105 ft
75 ft

22. PRINTOUTS A computer printout can list 74 student names on one page of paper. If 3,451 student names are to be printed, how many names will appear on the last page of the printout?

23. COLLECTIBLES There are 12 baseball cards in every pack. There are 24 packs in every box. There are 12 boxes in every case. How many cards are in a case?

24. Find the prime factorization of 252.

25. Evaluate: $9 + 4 \cdot 5$.

26. Evaluate: $\dfrac{3 \cdot 4^2 - 2^2}{(2 - 1)^3}$.

27. Evaluate: $10 + 2[12 - 2(6 - 4)]$.

28. GRADES A student scored 73, 52, and 70 on three exams and received 0 on two missed exams. Find his average (mean) score.

29. Is 3 a solution of the equation $x + 13 = 16$? Explain why or why not.

Solve each equation. **Check the result.**

30. $100 = x + 1$

31. $y - 12 = 18$

32. $5t = 55$

33. $\dfrac{q}{3} = 27$

Let a variable represent the unknown quantity. Then write and solve an equation to answer the question.

34. PARKING After many student complaints, a college decided to commit funds to double the number of parking spaces on campus. This increase would bring the total number of spaces up to 6,200. How many parking spaces does the college have at this time?

35. LIBRARIES A library building is 6 years shy of its 200th birthday. How old is the building at this time?

36. Explain what it means to *solve* an equation.

CHAPTER 2

The Integers

Getty Images

There are few things more breathtaking than a star-filled sky on a clear night. Stars come in different colors, sizes, shapes, and ages. You have probably noticed that some stars are bright and others are faint. Astronomers have developed a scale to classify the relative brightness of stars. The most brilliant stars (including the sun) are assigned negative number magnitudes while positive number magnitudes are assigned to the dimmer stars (and planets).

To learn more about positive and negative numbers, visit The Learning Equation on the Internet at http://tle.brookscole.com. (The log-in instructions are in the Preface.) For Chapter 2, the online lessons are:

- *TLE* Lesson 4: An Introduction to Integers
- *TLE* Lesson 5: Adding Integers
- *TLE* Lesson 6: Subtracting Integers

Check Your Knowledge

1. The ________ value of a number is the distance between the number and zero on a number line.
2. When 0 is added to a number, the number remains the same. We call 0 the additive ________.
3. Two numbers that are the same distance from 0 on the number line, but on opposite sides of it, are called ________________.
4. The product of two integers with ________ signs is negative.
5. Insert one of the symbols > or < in the blank to make the statement true: $-15 \quad -16$.
6. Find the mean (average) of the temperatures shown in the table on the left.

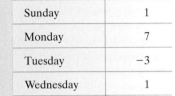

Daily high temperatures	
Sunday	1
Monday	7
Tuesday	−3
Wednesday	1
Thursday	0
Friday	−1
Saturday	2

7. Add:
   a. $-27 + 13$
   b. $12 + (-12)$
   c. $(-2) + (-2) + (-2) + (-2)$
   d. $(-4 + 7) + [1 + (-6) + 4]$
8. Subtract:
   a. $7 - 13$ b. $3 - (-3)$ c. $0 - 5 - 7$
9. Find each product.
   a. $3(-3)$ b. $-5(-20)$ c. $(-3)(-2)(-4)$
10. Write the related multiplication statement for $\frac{-18}{3} = -6$.
11. Find each quotient, if possible.
   a. $\frac{36}{-9}$ b. $\frac{-900}{-30}$ c. $\frac{-3}{2 + 1 - 3}$
12. Evaluate each expression.
   a. $-(-7)$ b. $-|-7|$ c. $3|-6 + 2|$
13. Evaluate each power.
   a. -3^2 b. $(-3)^2$ c. $(3 - 4)^2$
14. Evaluate each expression.
   a. $24 \div 3 \cdot 2$ b. $6 + \frac{12}{-4} - 4^2$
   c. $-5 - 2[7 - (-3)(-2)^3]$ d. $\frac{-3^3 + (-4)(-6)}{-3(3 - 5)^2 + 9}$
15. The price of a certain computer dropped from $620 to $500 in six months. How much did the price drop per month?
16. On one lie detector test, a burglar scored −19, which indicates deception. However, on a second test, he scored −4, which is inconclusive. Find the difference in the scores.
17. Let a variable represent the unknown quantity. Then write and solve an equation to answer the following question.
    After Michelle deposited $175 in her checking account, it was still $55 overdrawn. What was the account balance before the deposit?
18. Explain what is meant when we say that subtraction is the same as adding the opposite.

Study Skills Workshop

ATTITUDES, PAST EXPERIENCE, AND LEARNING STYLES

Getty

Your personality will affect the way you learn math; some traits will work toward your benefit and some may not. The key is to maximize your strengths, minimize weaknesses, and find learning methods that will best suit your personal style.

Attitudes. How do you feel about math? How do you feel about taking control of your learning environment? What do you think of your chance for success in this class? If you respond to these questions with answers like: "Ugh! I'll never learn math!" or "My math teachers just never taught me well," you may be setting yourself up for failure. Try to alter your attitude slightly. Constructive thinking and positive self-talk may bring good results. Instead of telling yourself "I'm just stupid when it comes to math," try saying "Math is not my strongest subject, but I will work hard at it and learn new things this time." Know that in truly learning any new concept there are bound to be frustrations, but they can be overcome with support, strategy, and hard work.

Past Experience. Good or bad experiences in previous math classes may affect the way you view math now. Many students who have had negative math experiences in the past may feel anxiety and stress in the present. If you have high levels of stress or feel helpless when dealing with numbers, you may have math anxiety. These counterproductive feelings can be overcome with extra preparation, support services, relaxation techniques, and even hypnotherapy.

Learning Styles. What type of learner are you? The answer to this question will help you determine how you study, how you do your homework, and maybe even where you choose to sit in class. For example, visual-verbal learners learn best by reading and writing, so a good strategy for them in studying is to rewrite notes and examples. However, audio learners learn best by listening, so making audiotapes of important concepts may be their best study strategy. Kinesthetic learners like to move and do things with their hands, so incorporating the use of games or puzzles (often called manipulatives) in studying may be helpful.

ASSIGNMENT

1. Take a learning skills inventory test such as that found online at http://www.metamath.com/multiple/multiple_choice_questions.cgi to determine what type of learner you are. Once you have determined your learning style, write a plan on how will you use this information to help you succeed in your class.
2. Describe your past experiences in math courses. Have they generally been good or bad? Why? If you feel you have math anxiety, visit your school's counseling office and schedule an appointment to find ways of dealing with this problem.

In this chapter, we introduce the concept of negative number and explore an extension of the set of whole numbers, called the integers.

2.1 An Introduction to the Integers

- The integers • Extending the number line • More on inequality
- Absolute value • The opposite of a number • The – symbol

We have seen that whole numbers can be used to describe many situations that arise in everyday life. For example, we can use whole numbers to express temperatures above zero, the balance in a checking account, or an altitude above sea level. However, we cannot use whole numbers to express temperatures below zero, the balance in a checking account that is overdrawn, or how far an object is below sea level. (See Figure 2-1.)

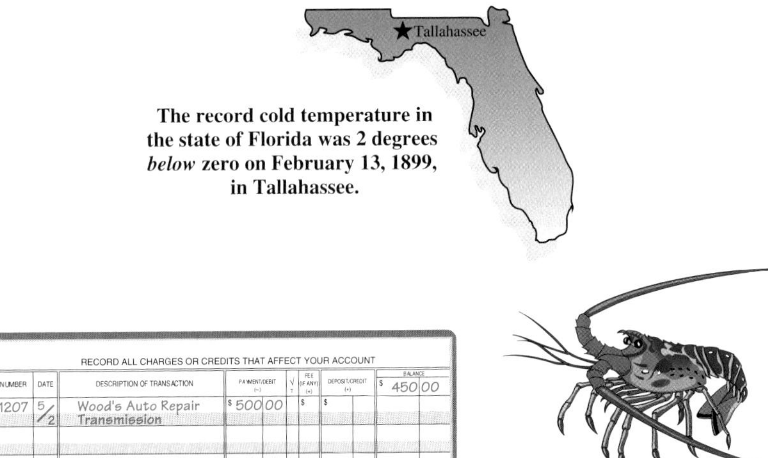

The record cold temperature in the state of Florida was 2 degrees *below* zero on February 13, 1899, in Tallahassee.

A check for $500 was written when there was only $450 in the account. The checking account is *overdrawn*.

The American lobster is found off the East Coast of North America at depths as much as 600 feet *below* sea level.

FIGURE 2-1

In this section, we see how negative numbers can be used to describe these three situations as well as many others.

The integers

To describe a temperature of 2 degrees (2°) above zero, a balance of $50, or 600 feet above sea level, we can use numbers called **positive numbers.** All positive numbers are greater than 0, and we can write them using a **positive sign** + or no sign at all.

In words	**In symbols**	**Read as**
2 degrees above zero	+2 or 2	positive two
A balance of $50	+50 or 50	positive fifty
600 feet above sea level	+600 or 600	positive six hundred

To describe a temperature of 2 degrees below zero, $50 overdrawn, or 600 feet below sea level, we need to use negative numbers. **Negative numbers** are numbers less than 0, and they are written using a **negative sign** −.

In words	In symbols	Read as
2 degrees below zero	-2	negative two
\$50 overdrawn	-50	negative fifty
600 feet below sea level	-600	negative six hundred

Positive and negative numbers

Positive numbers are greater than 0. **Negative numbers** are less than 0.

! COMMENT Zero is neither positive nor negative.

We often call positive and negative numbers **signed numbers.** The first three of the following signed numbers are positive, and the last three are negative.

$$+12, \quad +26, \quad 515, \quad -12, \quad -26, \quad \text{and} \quad -515$$

The collection of positive whole numbers, the negatives of the whole numbers, and 0 is called the set of **integers** (read as "in-ti-jers").

The set of integers

$$\{\ldots, -5, -4, -3, -2, -1, 0, 1, 2, 3, 4, 5, \ldots\}$$

Since every natural number is an integer, we say that the set of natural numbers is a subset of the integers. See Figure 2-2. Since every whole number is an integer, we say that the set of whole numbers is a **subset** of the integers.

$$\text{The set of integers} \rightarrow \{\ldots, -7, -6, -5, -4, -3, -2, -1, \underbrace{0, \overbrace{1, 2, 3, 4, 5, 6, 7, \ldots}^{\text{The set of natural numbers}}}_{\text{The set of whole numbers}}\}$$

FIGURE 2-2

! COMMENT Note that the negative integers and 0 are not natural numbers. Also note that negative integers are not whole numbers.

Extending the number line

In Section 1.1, we introduced the number line. We can use an extension of the number line to learn about negative numbers.

Negative numbers can be represented on a number line by extending the line to the left. Beginning at the origin (the 0 point), we move to the left, marking equally spaced points as shown in Figure 2-3. As we move to the right on the number line, the values of the numbers increase. As we move to the left, the values of the numbers decrease.

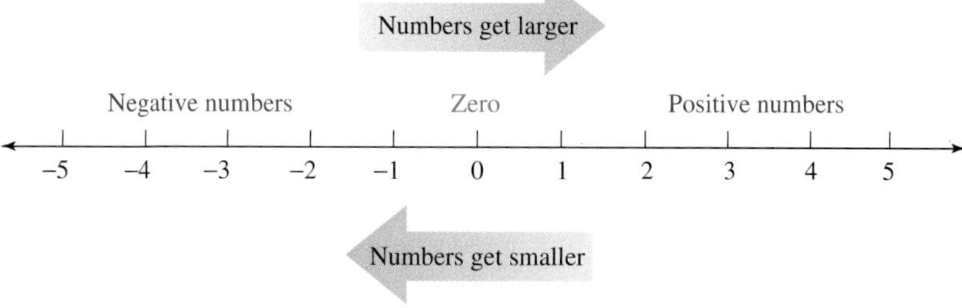

FIGURE 2-3

The thermometer shown in Figure 2-4(a) is an example of a vertical number line. It is scaled in degrees and shows a temperature of $-10°$. The time line shown in Figure 2-4(b) is an example of a horizontal number line. It is scaled in increments of 500 years.

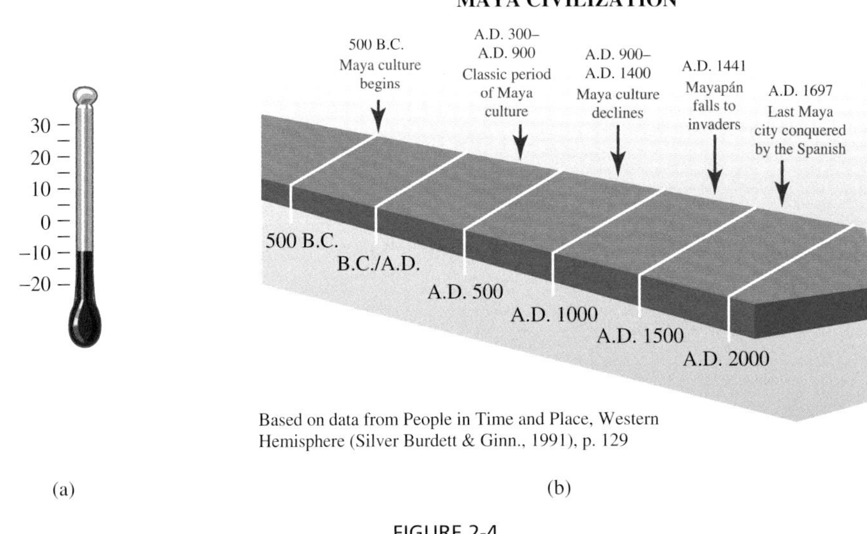

(a) (b)

FIGURE 2-4

Self Check 1

On the number line, graph -4, -2, 1, and 3.

Answer

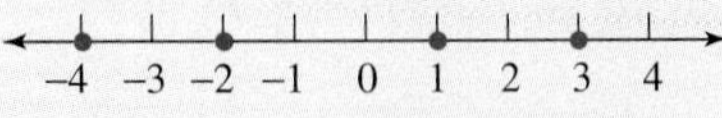

EXAMPLE 1 On the number line, graph -3, -1, 2, and 4.

Solution To graph each integer, we locate its position on the number line and draw a dot.

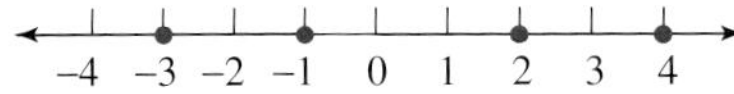

By extending the number line to include negative numbers, we can represent more situations using bar graphs and line graphs. For example, the bar graph shown in Figure 2-5 illustrates the annual profits *and losses* of Toys R Us over a nine-year period. Note that the profit in 2004 was \$88 million and that the loss in 1999 was \$132 million.

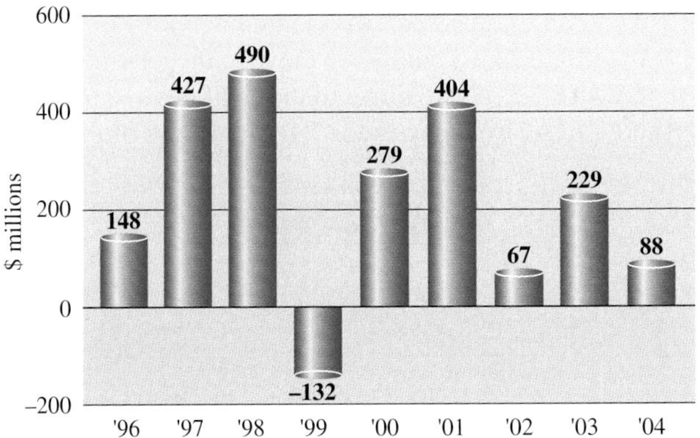

FIGURE 2-5

Credit Card Debt

THINK IT THROUGH

"The most dangerous pitfall for many college students is the overuse of credit cards. Many banks do their best to entice new card holders with low or zero-interest cards." Gary Schatsky, certified financial planner

Which numbers on the credit card statement below are actually debts and, therefore, could be represented using negative numbers?

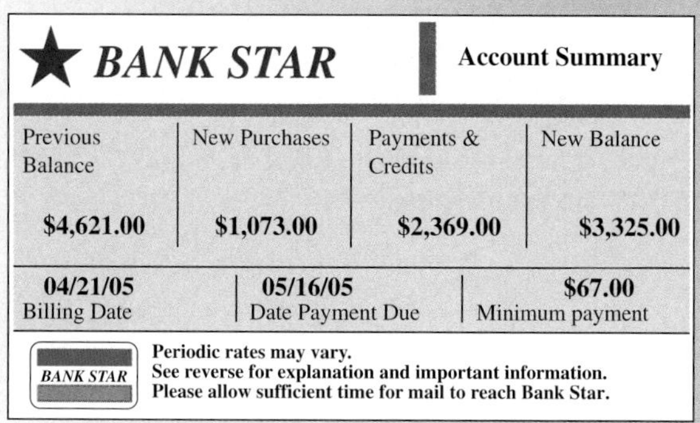

BANK STAR — **Account Summary**

Previous Balance	New Purchases	Payments & Credits	New Balance
\$4,621.00	**\$1,073.00**	**\$2,369.00**	**\$3,325.00**

04/21/05 Billing Date	**05/16/05** Date Payment Due	**\$67.00** Minimum payment

BANK STAR

Periodic rates may vary.
See reverse for explanation and important information.
Please allow sufficient time for mail to reach Bank Star.

More on inequality

Recall that the symbol $<$ means "is less than" and that $>$ means "is greater than." Figure 2-6 shows the graph of the integers -2 and 1. Since -2 is to the left of 1 on the number line, $-2 < 1$. Since $-2 < 1$, it is also true that $1 > -2$.

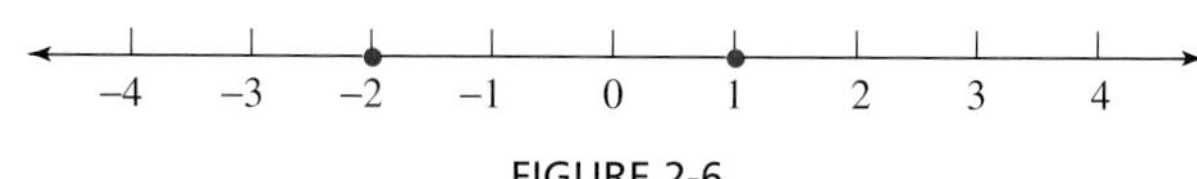

FIGURE 2-6

EXAMPLE 2 Use one of the symbols $>$ or $<$ to make each statement true:
a. $4 \ \square \ -5$ and **b.** $-4 \ \square \ -2$.

Solution

a. Since 4 is to the right of -5 on the number line, $4 > -5$.

b. Since -4 is to the left of -2 on the number line, $-4 < -2$.

Self Check 2
Use one of the symbols $>$ or $<$ to make each statement true:
a. $6 \ \square \ -6$
b. $-6 \ \square \ -5$

Answers **a.** $>$, **b.** $<$

Three other commonly used inequality symbols are the "is not equal to" symbol $\neq$, the "is less than or equal to" symbol $\leq$, and the "is greater than or equal to" symbol $\geq$.

$5 \neq 2$	Read as "5 is not equal to 2."
$6 \leq 10$	Read as "6 is less than or equal to 10." This statement is true, because $6 < 10$.
$12 \leq 12$	Read as "12 is less than or equal to 12." This statement is true, because $12 = 12$.
$17 \geq 15$	Read as "17 is greater than or equal to 15." This statement is true, because $17 > 15$.
$20 \geq 20$	Read as "20 is greater than or equal to 20." This statement is true, because $20 = 20$.

Self Check 3
Use an inequality symbol to write "30 is less than or equal to 35."

Answer $30 \leq 35$

EXAMPLE 3 Use an inequality symbol to write **a.** "8 is not equal to 5" and **b.** "50 is greater than or equal to 40."

Solution

a. $8 \neq 5$

b. $50 \geq 40$

Absolute value

Using the number line, we can see that the numbers 3 and -3 are both a distance of 3 units away from 0, as shown in Figure 2-7.

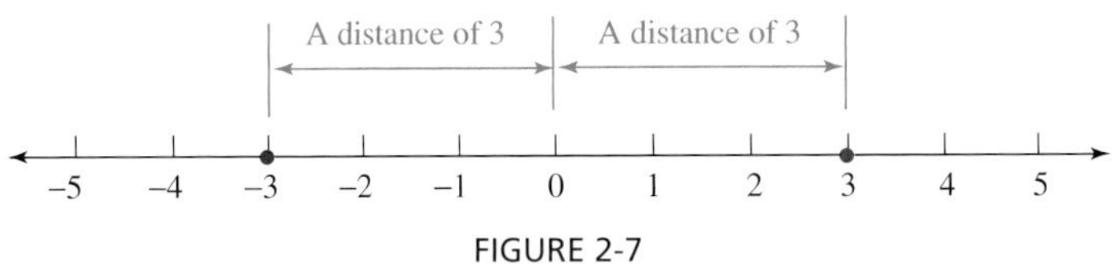

FIGURE 2-7

The **absolute value** of a number gives the distance between the number and 0 on the number line. To indicate absolute value, the number is inserted between two vertical bars, called the **absolute value symbol.** For example, we can write $|-3| = 3$. This is read as "The absolute value of negative 3 is 3," and it tells us that the distance between -3 and 0 on the number line is 3 units. From Figure 2-7, we also see that $|3| = 3$.

Absolute value

The **absolute value** of a number is the distance on the number line between the number and 0.

! COMMENT Absolute value expresses distance. The absolute value of a number is always positive or 0. It is never negative.

Self Check 4
Evaluate each expression:
a. $|-9|$
b. $|4|$

Answers **a.** 9, **b.** 4

EXAMPLE 4 Evaluate each expression: **a.** $|8|$, **b.** $|-5|$, and **c.** $|0|$.

Solution

a. On the number line, the distance between 8 and 0 is 8. Therefore,

$$|8| = 8$$

b. On the number line, the distance between -5 and 0 is 5. Therefore,

$$|-5| = 5$$

c. On the number line, the distance between 0 and 0 is 0. Therefore,

$$|0| = 0$$

The opposite of a number

Opposites or negatives

Two numbers that are the same distance from 0 on the number line, but on opposite sides of it, are called **opposites** or **negatives.**

Figure 2-8 shows that for each natural number on the number line, there is a corresponding natural number, called its *opposite,* to the left of 0. For example, we see that 3 and -3 are opposites, as are -5 and 5. Note that 0 is its own opposite.

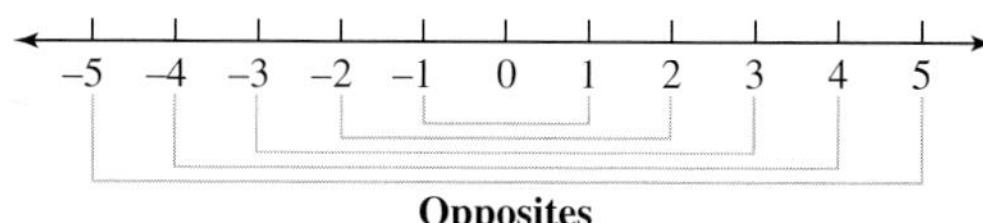

FIGURE 2-8

To write the opposite of a number, a $-$ symbol is used. For example, the opposite of 5 is -5 (read as "negative 5"). Parentheses are needed to express the opposite of a negative number. The opposite of -5 is written as $-(-5)$. Since 5 and -5 are the same distance from 0, the opposite of -5 is 5. Therefore, $-(-5) = 5$. This leads to the following conclusion.

The double negative rule

If a is any number, then

$$-(-a) = a$$

In words, this rule says that *the opposite of the negative of a number is that number.*

Number	Opposite	
57	-57	Read as "negative fifty-seven."
-8	$-(-8) = 8$	Read as "the opposite of negative eight." Apply the double negative rule.
0	$-0 = 0$	The opposite of 0 is 0.

The concept of opposite can also be applied to an absolute value. For example, the opposite of the absolute value of -8 can be written as $-|-8|$. Think of this as a two-step process. Find the absolute value first, and then attach a $-$ to that result.

First, find the absolute value.

$$-|-8| = -8$$

Then attach a $-$ sign.

EXAMPLE 5 Simplify each expression: **a.** $-(-44)$ and **b.** $-|-225|$.

Solution

a. $-(-44)$ means the opposite of -44. Since the opposite of -44 is 44, we write

$$-(-44) = 44$$

b. $-|-225|$ means the opposite of $|-225|$. Since $|-225| = 225$, and the opposite of 225 is -225, we write

$$-|-225| = -225$$

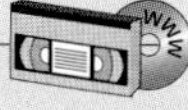

Self Check 5

Simplify each expression:
a. $-(-1)$ and **b.** $-|-99|$.

Answers **a.** 1, **b.** -99

The – symbol

The – symbol is used to indicate a negative number, the opposite of a number, and the operation of subtraction. The key to interpreting the – symbol correctly is to examine the context in which it is used.

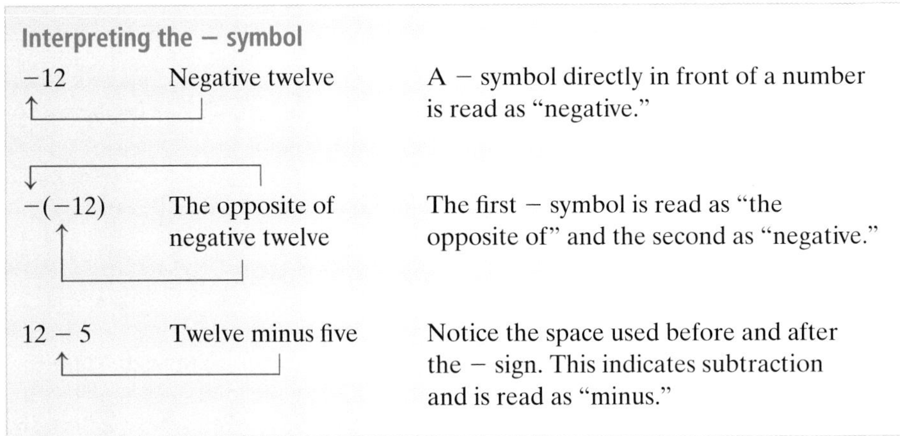

Interpreting the – symbol

-12	Negative twelve	A – symbol directly in front of a number is read as "negative."
$-(-12)$	The opposite of negative twelve	The first – symbol is read as "the opposite of" and the second as "negative."
$12 - 5$	Twelve minus five	Notice the space used before and after the – sign. This indicates subtraction and is read as "minus."

Section 2.1 STUDY SET

VOCABULARY *Fill in the blanks.*

1. Numbers can be represented by points equally spaced on the number ______.
2. The point on the number line representing 0 is called the ______.
3. To ______ a number means to locate it on the number line and highlight it with a dot.
4. The graph of a number is the point on the number ______ that represents that number.
5. The symbols $>$ and $<$ are called ______ symbols.
6. ______ numbers are less than 0.
7. The ______ ______ of a number is the distance between the number and 0 on the number line.
8. Two numbers that are the same distance from 0 on the number line, but on opposite sides of it, are called ______.
9. $\{\ldots, -3, -2, -1, 0, 1, 2, 3, \ldots\}$ is called the set of ______.
10. The double negative rule states that the negative of the ______ of a number is that number. If a is any number, then $-(-a) = \square$.

CONCEPTS

11. Refer to each graph and use an inequality symbol $>$ or $<$ to make a true statement.

a. −2 0 2

$-2 \;\square\; 2$

b. −1 0

$0 \;\square\; -1$

c. −1 0 1

$-1 \;\square\; 0$ and $1 \;\square\; 0$

12. Determine what is wrong with each number line.

a.
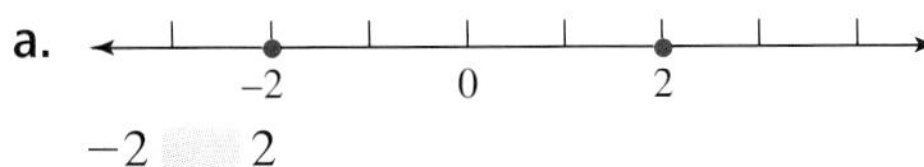

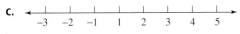

b.
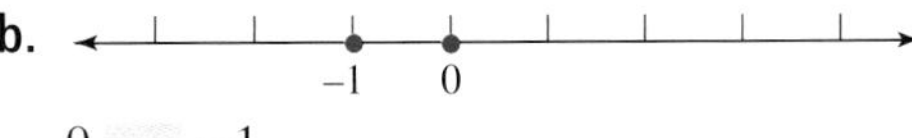

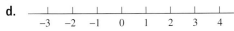

c.
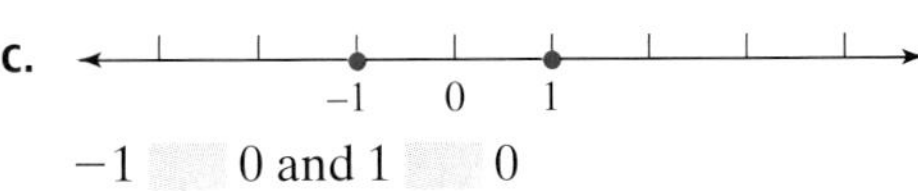

d.
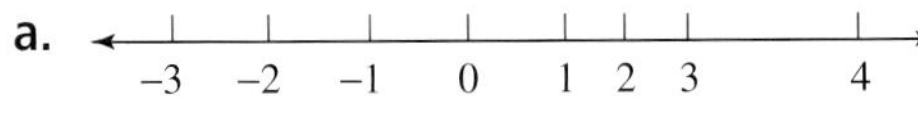

13. Does every number on the number line have an opposite?

14. Is the absolute value of a number always positive?

15. Which of the following contains a minus sign: $15 - 8$, $-(-15)$, or -15?

16. Is there a number that is both greater than 10 and less than 10 at the same time?

17. Express the fact $12 < 15$ using the $>$ symbol.

18. Express the fact $5 > 4$ using the $<$ symbol.

19. Represent each of these situations using a signed number.

 a. \$225 overdrawn

 b. 10 seconds before liftoff

 c. 3 degrees below normal

 d. A deficit of \$12,000

 e. A racehorse finished 2 lengths behind the leader.

20. Represent each of these situations using a signed number, and then describe its opposite in words.

 a. A trade surplus of \$3 million

 b. A bacteria count 70 more than the standard

 c. A profit of \$67

 d. A business \$1 million in the "black"

 e. 20 units over their quota

21. If a number is less than 0, what type of number must it be?

22. If a number is greater than 0, what type of number must it be?

23. On the number line, what number is 3 units to the right of -7?

24. On the number line, what number is 4 units to the left of 2?

25. Name two numbers on the number line that are a distance of 5 away from -3.

26. Name two numbers on the number line that are a distance of 4 away from 3.

27. Which number is closer to -3 on the number line: 2 or -7?

28. Which number is farther from 1 on the number line: -5 or 8?

29. Give examples of the $-$ symbol used in three different ways.

30. What is the opposite of 0?

NOTATION

31. Translate each phrase to mathematical symbols.

 a. The opposite of negative eight

 b. The absolute value of negative eight

 c. Eight minus eight

 d. The opposite of the absolute value of negative eight

32. Write the set of integers.

PRACTICE *Simplify each expression.*

33. $|9|$
34. $|12|$
35. $|-8|$
36. $|-1|$
37. $|-14|$
38. $|-85|$
39. $-|20|$
40. $-|110|$
41. $-|-6|$
42. $|0|$
43. $|203|$
44. $-|-11|$
45. -0
46. $-|0|$
47. $-(-11)$
48. $-(-1)$
49. $-(-4)$
50. $-(-9)$
51. $-(-12)$
52. $-(-25)$

Graph each set of numbers on the number line.

53. $\{-3, 0, 3, 4, -1\}$

54. $\{-4, -1, 2, 5, 1\}$

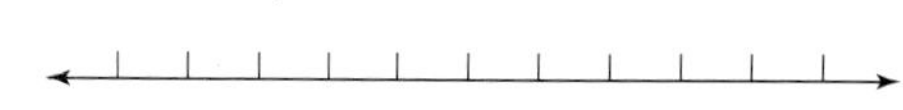

55. The opposite of -3, the opposite of 5, and the absolute value of -2

56. The absolute value of 3, the opposite of 3, and the number that is 1 less than -3

Insert one of the symbols $>$, $<$, or $=$ in the blank to make a true statement.

57. -5 ☐ 5
58. 0 ☐ -1
59. -12 ☐ -6
60. -6 ☐ -7
61. -10 ☐ -11
62. -11 ☐ -20
63. $|-2|$ ☐ 0
64. $|-30|$ ☐ -40

Insert one of the symbols ≥ or ≤ in the blank to make a true statement.

65. -14 ___ -15

66. -77 ___ -76

67. 210 ___ 210

68. 37 ___ 37

69. $-1{,}255$ ___ $-(-1{,}254)$

70. 0 ___ -3

71. $-|-3|$ ___ 4

72. $-|-163|$ ___ -150

APPLICATIONS

73. FLIGHT OF A BALL A boy throws a ball from the top of a building, as shown. At the instant he does this, his friend starts a stopwatch and keeps track of the time as the ball rises to a peak and then falls to the ground. Use the vertical number line to complete the table by finding the position of the ball at each specified time.

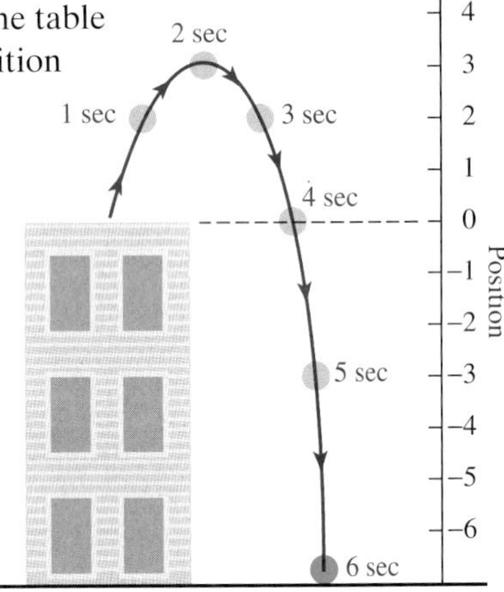

Time	Position of ball
1 sec	
2 sec	
3 sec	
4 sec	
5 sec	
6 sec	

74. SHOOTING GALLERIES At an amusement park, a shooting gallery contains moving ducks. The path of one duck is shown, along with the time it takes the duck to reach certain positions on the gallery wall. Complete the table using the horizontal number line in the illustration.

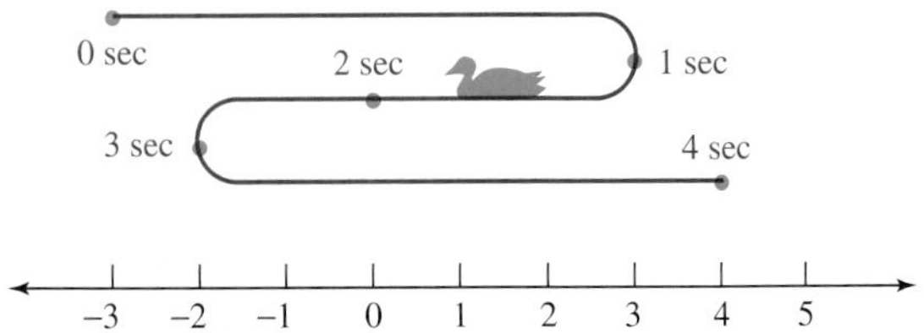

Time	Position of duck
0 sec	
1 sec	
2 sec	
3 sec	
4 sec	

75. TECHNOLOGY The readout from a testing device is shown. It is important to know the height of each of the three peaks and the depth of each of the three valleys. Use the vertical number line to find these numbers.

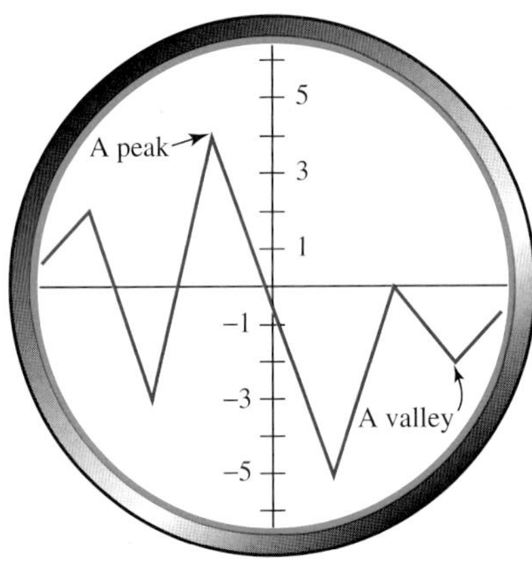

76. FLOODING A week of daily reports listing the height of a river in comparison to flood stage is given in the table. Complete the bar graph in the illustration.

Flood stage report	
Sun.	2 ft below
Mon.	3 ft over
Tue.	4 ft over
Wed.	2 ft over
Thu.	1 ft below
Fri.	3 ft below
Sat.	4 ft below

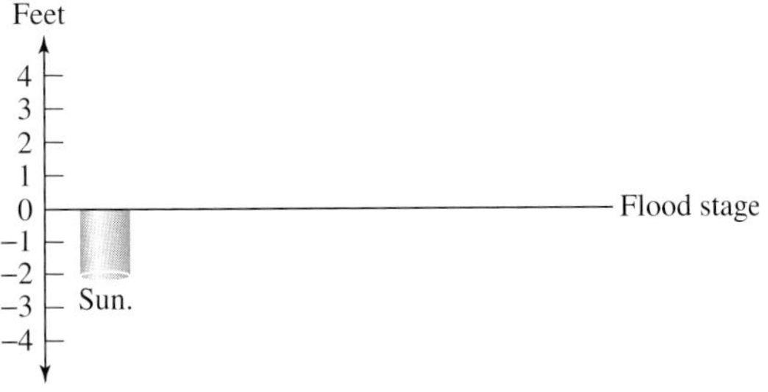

77. GOLF In golf, *par* is the standard number of strokes considered necessary on a given hole. A score of -2 indicates that a golfer used 2 strokes less than par. A score of $+2$ means 2 more strokes than par were used. In the illustration, each golf ball represents the score of a professional golfer on the 16th hole of a certain course.

a. What score was shot most often on this hole?

b. What was the best score on this hole?

c. Explain why this hole appears to be too easy for a professional golfer.

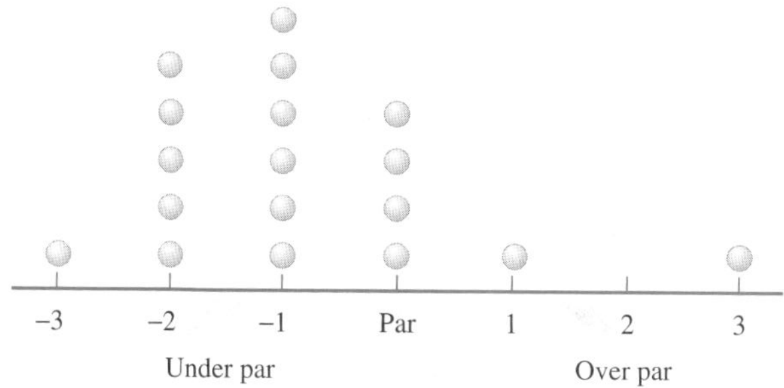

78. PAYCHECKS Examine the items listed on the following paycheck stub. Then write two columns on your paper — one headed "positive" and the other "negative." List each item under the appropriate heading.

Tom Dryden Dec. 04		Christmas bonus	$100
Gross pay	$2,000	**Reductions**	
Overtime	$300	Retirement	$200
Deductions		**Taxes**	
Union dues	$30	Federal withholding	$160
U.S. Bonds	$100	State withholding	$35

79. WEATHER MAPS The illustration shows the predicted Fahrenheit temperatures for a day in mid-January.

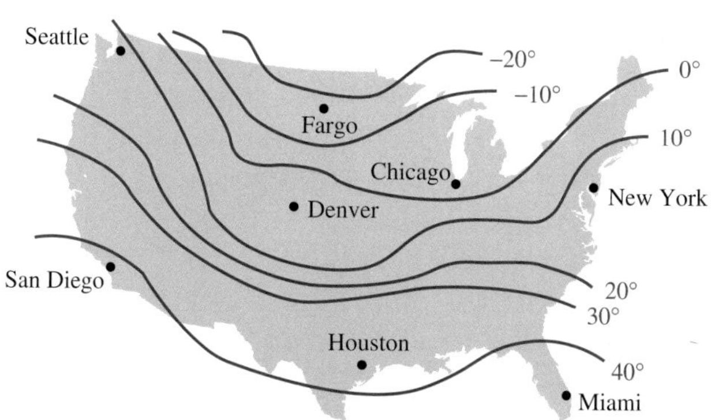

a. What is the temperature range for the region including Fargo, North Dakota?

b. According to the prediction, what is the warmest it should get in Houston?

c. According to this prediction, what is the coldest it should get in Seattle?

80. PROFITS/LOSSES The graph in the illustration shows the net income of Apple Computer Inc. for the years 2000–2004.

a. In what year did the company suffer a loss? Estimate each loss.

b. In what year did Apple have the greatest profit? Estimate it.

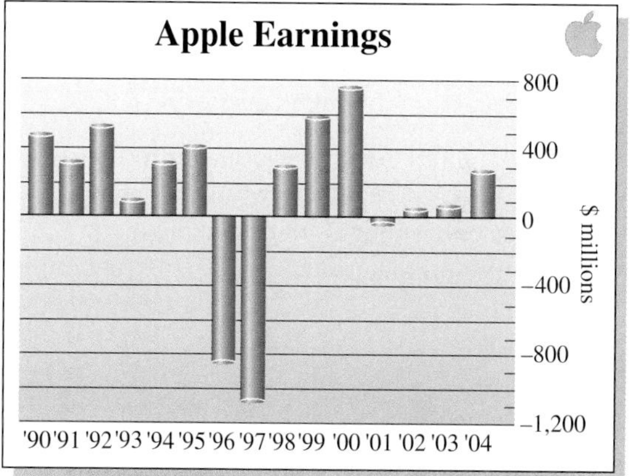

Source: Morningstar.com

81. HISTORY Number lines can be used to display historical data. Some important world events are shown on the time line in the illustration.

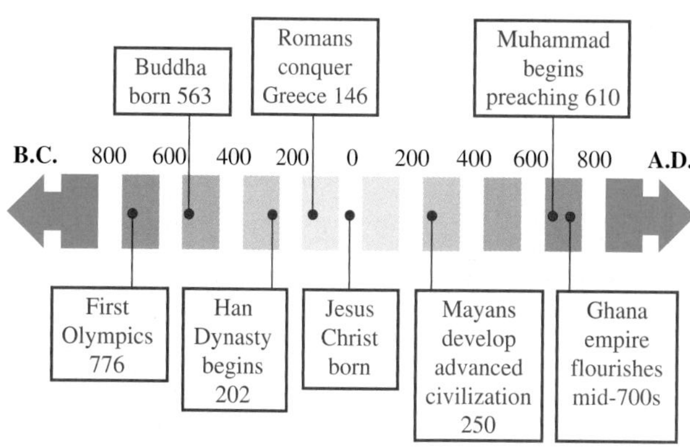

a. What basic unit is used to scale this time line?

b. What can be thought of as positive numbers?

c. What can be thought of as negative numbers?

d. What important event distinguishes the positive from the negative numbers?

82. ASTRONOMY Astronomers use a type of number line called the *apparent magnitude scale* to denote the brightness of objects in the sky. The brighter an object appears to an observer on Earth, the more negative is its apparent magnitude. Graph each of the following on the scale in the illustration.

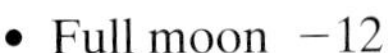

- Full moon −12
- Pluto +15
- Sirius (a bright star) −2
- Sun −26
- Venus −4
- Visual limit of binoculars +10
- Visual limit of large telescope +20
- Visual limit of naked eye +6

83. LINE GRAPHS Each thermometer in the illustration gives the daily high temperature in degrees Fahrenheit. Plot each daily high temperature on the grid and then construct a line graph.

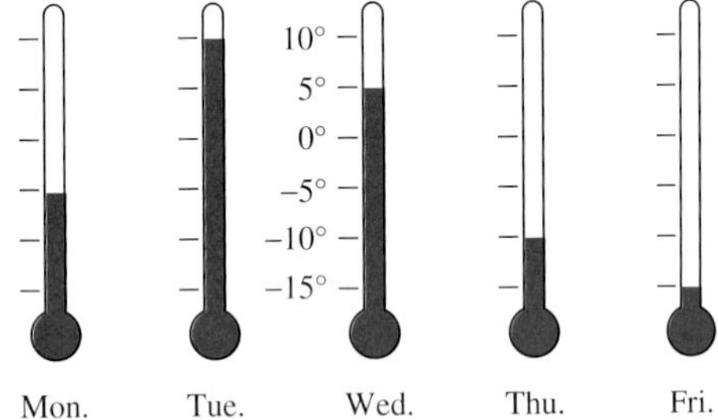

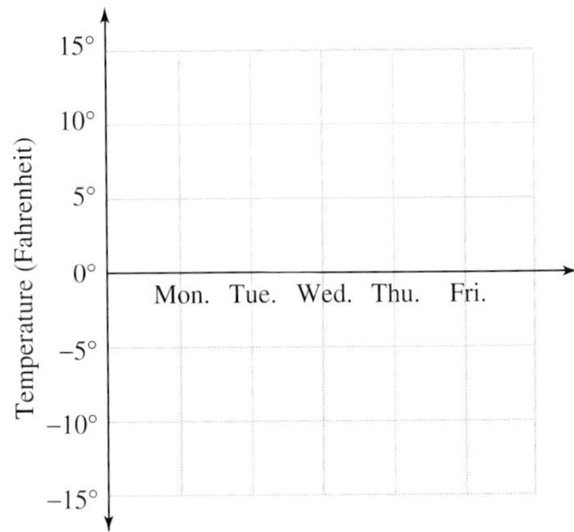

84. GARDENING The illustration in the next column shows the depths at which the bottoms of various types of flower bulbs should be planted. (The symbol ″ represents inches.)

a. At what depth should a tulip bulb be planted?

b. How much deeper are hyacinth bulbs planted than gladiolus bulbs?

c. Which bulb must be planted the deepest? How deep?

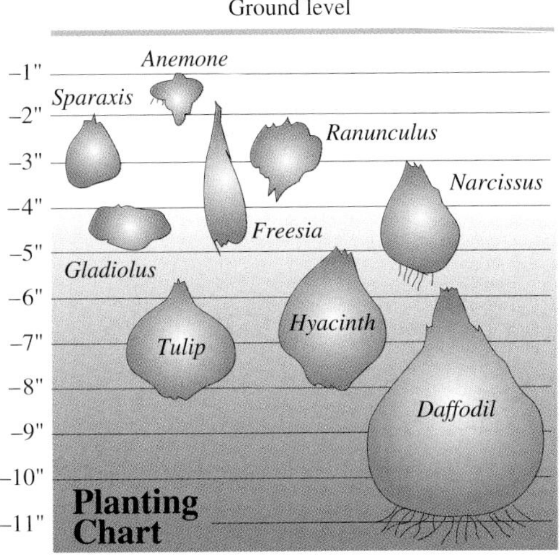

WRITING

85. Explain the concept of the opposite of a number.

86. What real-life situation do you think gave rise to the concept of a negative number?

87. Explain why the absolute value of a number is never negative.

88. Give an example of the use of the number line that you have seen in another course.

89. DIVING Divers use the terms *positive buoyancy, neutral buoyancy,* and *negative buoyancy* as shown. What do you think each of these terms means?

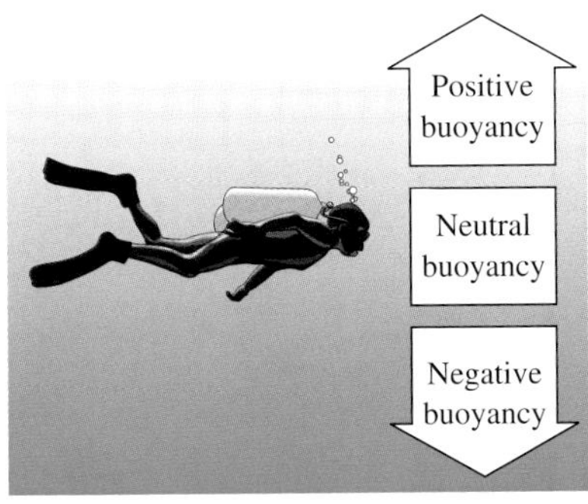

90. NEW ORLEANS The city of New Orleans, Louisiana, lies largely below sea level. Find out why the city is not under water.

REVIEW

91. Round 23,456 to the nearest hundred.

92. Evaluate: $19 - 2 \cdot 3$.

93. Subtract 2,081 from 2,842.

94. Divide 345 by 15.

95. Give the name of the property illustrated here:

$$(13 \cdot 2) \cdot 5 = 13 \cdot (2 \cdot 5)$$

96. Write four times five using three different notations.

2.2 Adding Integers

- Adding two integers with the same sign
- Adding two integers with different signs
- The addition property of zero
- The additive inverse of a number

A dramatic change in temperature occurred in 1943 in Spearfish, South Dakota. On January 22, at 7:30 A.M., the temperature was −4°F. In just two minutes, the temperature rose 49 degrees! To calculate the temperature at 7:32 A.M., we need to add 49 to −4.

$$-4 + 49$$

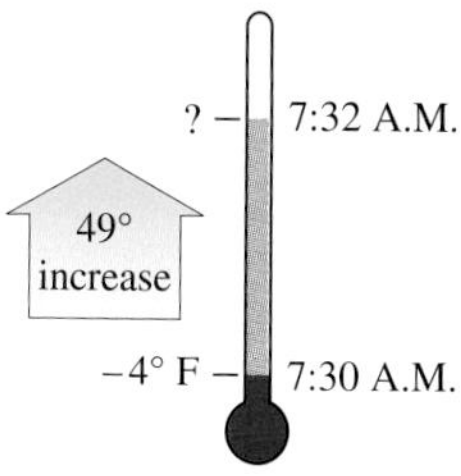

To perform this addition, we must know how to add positive and negative integers. In this section, we develop rules to help us make such calculations.

Adding two integers with the same sign

4 + 3
both positive

To explain addition of signed numbers, we can use the number line. (See Figure 2-9.) To compute $4 + 3$, we begin at the **origin** (the point labeled 0) and draw an arrow 4 units long, pointing to the right. This represents positive 4. From that point, we draw an arrow 3 units long, pointing to the right, to represent positive 3. The second arrow points to the answer. Therefore, $4 + 3 = 7$.

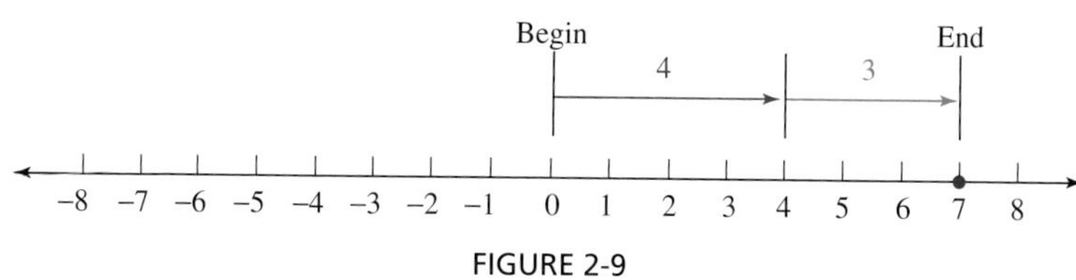

FIGURE 2-9

To check our work, let's think of the problem in terms of money. If you had $4 and earned $3 more, you would have a total of $7.

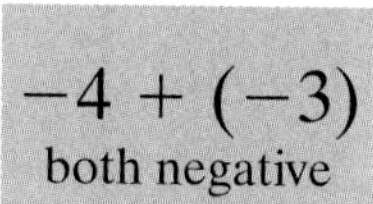

To compute $-4 + (-3)$, we begin at the origin and draw an arrow 4 units long, pointing to the left. (See Figure 2-10.) This represents −4. From there, we draw an arrow 3 units long, pointing to the left, to represent −3. The second arrow points to the answer: $-4 + (-3) = -7$.

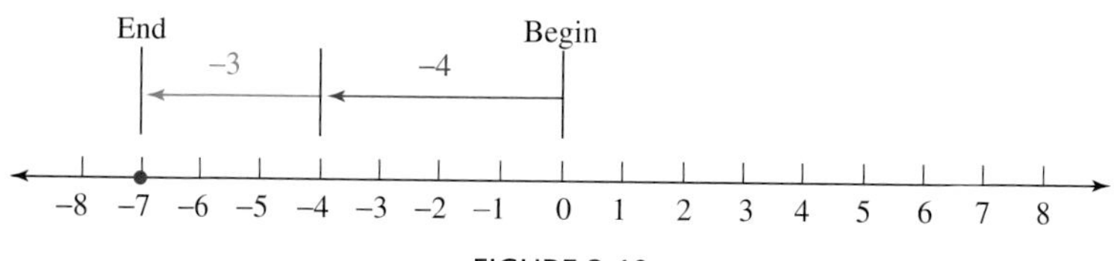

FIGURE 2-10

Let's think of this problem in terms of money. If you had a debt of \$4 (negative 4) and then took on \$3 more debt (negative 3), you would be in debt \$7 (negative 7).

Here are some observations about the process of adding two numbers that have the same sign, using the number line.

- Both arrows point in the same direction and "build" on each other.
- The answer has the same sign as the two numbers being added.

4	+	3	=	7		−4	+	(−3)	=	−7
Positive	+	positive	=	positive answer		Negative	+	negative	=	negative answer

These observations suggest the following rule.

Adding two integers with the same sign

To add two integers with the same sign, add their absolute values and attach their common sign to the sum. If both integers are positive, their sum is positive. If both integers are negative, their sum is negative.

! COMMENT When writing additions that involve signed numbers, write negative numbers within parentheses to separate the negative sign − from the plus sign +.

$9 + (-4)$ $\quad$ ~~$9 + -4$~~ $\quad$ and $\quad$ $-9 + (-4)$ $\quad$ ~~$-9 + -4$~~

EXAMPLE 1 Find the sum: $-9 + (-4)$.

Solution

Step 1: To add two integers with the same sign, we first add the absolute values of each of the integers. Since $|-9| = 9$ and $|-4| = 4$, we begin by adding 9 and 4.

$$9 + 4 = 13$$

Step 2: We then attach the common sign (which is negative) to this result. Therefore,

$$-9 + (-4) = -13$$

↑ Make the answer negative.

After some practice, you will be able to do this kind of problem in your head. It will not be necessary to show all the steps as we have done here.

Self Check 2

Find the sum:

a. $7 + 5$

b. $-300 + (-100)$

Answers **a.** 12, **b.** −400

EXAMPLE 2 Find the sum: **a.** $6 + 4$ and **b.** $-80 + (-60)$.

Solution

a. Since both integers are positive, the answer is positive.

$$6 + 4 = 10$$

b. Since both integers are negative, the answer is negative.

$$-80 + (-60) = -140$$

Adding two integers with different signs

$4 + (-3)$
one positive,
one negative

To compute $4 + (-3)$, we start at the origin and draw an arrow 4 units long, pointing to the right. (See Figure 2-11.) This represents positive 4. From there, we draw an arrow 3 units long, pointing to the left, to represent -3. The second arrow points to the answer: $4 + (-3) = 1$.

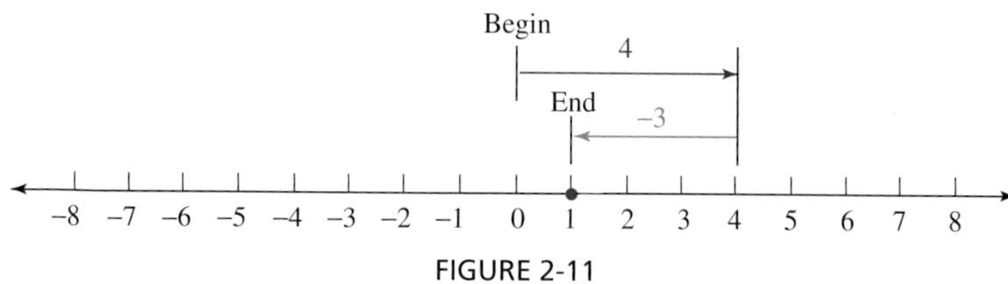

FIGURE 2-11

In terms of money, if you had $4 (positive 4) and then took on a debt of $3 (negative 3), you would have $1 (positive 1).

$-4 + 3$
one positive,
one negative

The problem $-4 + 3$ can be illustrated by drawing an arrow 4 units long from the origin, pointing to the left. (See Figure 2-12.) This represents -4. From there, we draw an arrow 3 units long, pointing to the right, to represent positive 3. The second arrow points to the answer: $-4 + 3 = -1$.

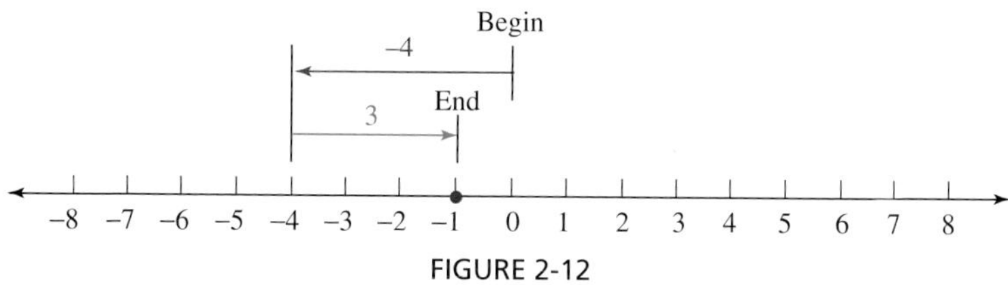

FIGURE 2-12

This problem can be thought of as owing $4 (negative 4) and then paying back $3 (positive 3). You will still owe $1 (negative 1).

The last two examples lead us to some observations about adding two integers with different signs, using the number line.

- The arrows representing the integers point in opposite directions.
- The longer of the two arrows determines the sign of the answer.

These observations suggest the following rule.

Adding two integers with different signs

To add two integers with different signs, subtract their absolute values, the smaller from the larger. Then attach to that result the sign of the integer with the larger absolute value.

EXAMPLE 3 Find the sum: $5 + (-7)$.

Solution

Step 1: To add two integers with different signs, we first subtract the smaller absolute value from the larger absolute value. Since $|5|$, which is 5, is smaller than $|-7|$, which is 7, we begin by subtracting 5 from 7.

$$7 - 5 = 2$$

Step 2: -7 has the larger absolute value, so we attach a negative sign to the result from step 1. Therefore,

$$5 + (-7) = -2$$

Make the answer negative.

Self Check 4

Find the sum:

a. $-2 + 7$

b. $6 + (-9)$

Answers **a.** 5, **b.** -3

EXAMPLE 4 Find the sum: **a.** $-8 + 5$ and **b.** $11 + (-5)$.

Solution

a. Since -8 has the larger absolute value, the answer is negative.

$-8 + 5 = -3$ Because the signs of the numbers are different, subtract their absolute values, 5 from 8, to get 3. Attach a negative sign to that result.

b. Since 11 has the larger absolute value, the answer is positive.

$11 + (-5) = 6$ Subtract the absolute values, 5 from 11, to get 6. The answer is positive.

THINK IT THROUGH Cash Flow

"College can be trial by fire — a test of how to cope with pressure, freedom, distractions, and a flood of credit card offers. It's easy to get into a cycle of overspending and unnecessary debt as a student." *Planning for College,* Wells Fargo Bank

If your income is less than your expenses, you have a *negative* cash flow. A negative cash flow can be a red flag that you should increase your income and/or reduce your expenses. Which of the following activities can increase income and which can decrease expenses?

- Buy generic or store-brand items.
- Get training and/or more education.
- Use your student ID to get discounts at stores, events, etc.
- Work more hours.
- Turn a hobby or skill into a money-making business.
- Tutor young students.
- Stop expensive habits, like smoking, buying snacks every day, etc.
- Attend free activities and free or discounted days at local attractions.
- Sell rarely used items, like an old CD player.
- Compare the prices of at least three products or at three stores before buying.

Based on the *Building Financial Skills* by National Endowment for Financial Education.

EXAMPLE 5 **Temperature change.** At the beginning of this section, we learned that at 7:30 A.M. on January 22, 1943, in Spearfish, South Dakota, the temperature was $-4°$. The temperature then rose 49 degrees in just two minutes. What was the temperature at 7:32 A.M.?

Solution The phrase *temperature rose 49 degrees* indicates addition. We need to add 49 to -4.

$-4 + 49 = 45$ Subtract the smaller absolute value, 4, from the larger absolute value, 49. The sum is positive.

At 7:32 A.M., the temperature was 45°F.

EXAMPLE 6 Add: $-3 + 5 + (-12) + 2$.

Solution This expression contains four integers. We add them, working from left to right.

$$-3 + 5 + (-12) + 2 = 2 + (-12) + 2 \quad \text{Add: } -3 + 5 = 2.$$
$$= -10 + 2 \quad \text{Add: } 2 + (-12) = -10.$$
$$= -8$$

Self Check 6
Add: $-12 + 8 + (-6) + 1$.

Answer -9

An alternative approach to problems like Example 6 is to add all the positive numbers, add all the negative numbers, and then add those results.

EXAMPLE 7 Find the sum: $-3 + 5 + (-12) + 2$.

Solution We can use the commutative property of addition to reorder the numbers and use the associative property of addition to group the positives together and the negatives together.

$$-3 + 5 + (-12) + 2 = \mathbf{5} + \mathbf{2} + (\mathbf{-3}) + (\mathbf{-12}) \quad \text{Reorder the numbers.}$$
$$= (\mathbf{5} + \mathbf{2}) + [(\mathbf{-3}) + (\mathbf{-12})] \quad \text{Group the positives. Group the negatives.}$$

We perform the operations inside the grouping symbols first.

$$(\mathbf{5} + \mathbf{2}) + [(\mathbf{-3}) + (\mathbf{-12})] = \mathbf{7} + (\mathbf{-15}) \quad \text{Add the positives. Add the negatives.}$$
$$= -8 \quad \text{Add the numbers with different signs.}$$

Self Check 7
Find the sum:

$$-12 + 8 + (-6) + 1$$

Answer -9

EXAMPLE 8 Evaluate: $[-1 + (-5)] + (-7 + 5)$.

Solution By the rules for the order of operations, we must perform the operations within the grouping symbols first.

$$[\mathbf{-1} + (\mathbf{-5})] + (\mathbf{-7} + \mathbf{5}) = \mathbf{-6} + (\mathbf{-2}) \quad \text{Perform the addition within the brackets. Perform the addition within parentheses.}$$
$$= -8 \quad \text{Add the numbers with the same sign.}$$

Self Check 8
Evaluate:

$(-6 + -8) + [10 + (-17)]$.

Answer -21

CALCULATOR SNAPSHOT **Entering negative numbers**

Nigeria is the United States' second largest trading partner in Africa. To calculate the 2002 U.S. trade balance with Nigeria, we add the \$1,057,700,000 worth of exports to Nigeria (considered positive) to the \$5,945,300,000 worth of imports from Nigeria (considered negative). We can use a scientific calculator to perform the addition: $1,057,700,000 + (-5,945,300,000)$.

- We do not have to do anything special to enter a positive number. When we key in 1,057,700,000, a positive number is entered.
- To enter $-5,945,300,000$, we press the change-of-sign key [+/−] *after entering* 5,945,300,000. Note that the change-of-sign key is different from the subtraction key [−].

1057700000 [+] 5945300000 [+/−] [=] -4887600000

In 2002, the United States had a trade balance of $-\$4,887,600,000$ with Nigeria. Because the result is negative, it is called a *trade deficit.*

The addition property of zero

When 0 is added to a number, the number remains the same. For example, $5 + 0 = 5$, and $0 + (-4) = -4$. Because of this, we call 0 the **additive identity.**

Addition property of 0

For any number a,

$$a + 0 = a \quad \text{and} \quad 0 + a = a$$

In words, this property states that *the sum of any number and 0 is that number.*

The additive inverse of a number

A second fact concerning 0 and the operation of addition can be demonstrated by considering the sum of a number and its opposite. To illustrate this, we use the number line in Figure 2-13 to add 6 and its opposite, -6. We see that $6 + (-6) = 0$.

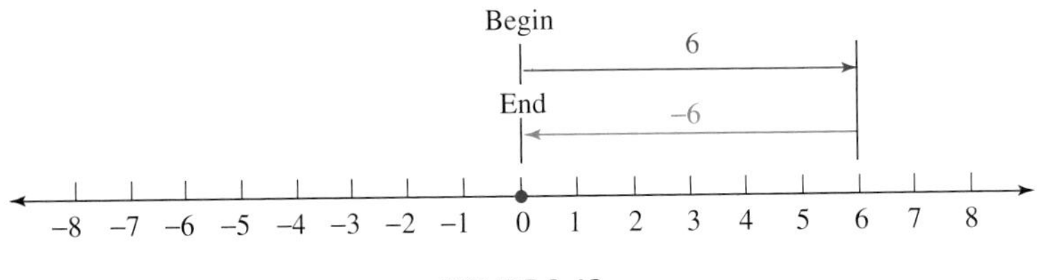

FIGURE 2-13

If the sum of two numbers is 0, the numbers are said to be **additive inverses** of each other. Since $6 + (-6) = 0$, we say that 6 and -6 are additive inverses.

We can now classify a pair of numbers such as 6 and -6 in three ways: as opposites, negatives, or additive inverses.

The additive inverse of a number

For any numbers a and b, if $a + b = 0$, then a and b are called **additive inverses.** That is, two numbers are said to be additive inverses if their sum is 0.

EXAMPLE 9 What is the additive inverse of -3? Justify your result.

Solution The additive inverse of -3 is its opposite, 3. To justify the result, we add and show that the sum is 0.

$$-3 + 3 = 0$$

Self Check 9
What is the additive inverse of 12? Justify your result.

Answer -12; $12 + (-12) = 0$

Section 2.2 STUDY SET

VOCABULARY *Fill in the blanks.*

1. When 0 is added to a number, the number remains the same. We call 0 the additive ________.
2. Since $-5 + 5 = 0$, we say that 5 is the additive ________ of -5. We can also say that 5 and -5 are ________.

CONCEPTS *Find each answer using the number line.*

3. $-3 + 6$
4. $-3 + (-2)$
5. $-5 + 3$
6. $-1 + (-3)$
7. **a.** Is the sum of two positive integers always positive?
   b. Is the sum of two negative integers always negative?
8. **a.** What is the sum of a number and its additive inverse?
   b. What is the sum of a number and its opposite?
9. Find each absolute value.
   a. $|-7|$ **b.** $|10|$
10. If the sum of two numbers is 0, what can be said about the numbers?

Fill in the blanks.

11. To add two integers with unlike signs, ________ their absolute values, the smaller from the larger. Then attach to that result the sign of the number with the ________ absolute value.
12. To add two integers with like signs, add their ________ values and attach their common ______ to the sum.

NOTATION *Complete each solution to evaluate the expression.*

13. $-16 + (-2) + (-1) = \square + (-1)$
    $= \square$
14. $-8 + (-2) + 6 = \square + 6$
    $= \square$
15. $(-3 + 8) + (-3) = \square + (-3)$
    $= \square$
16. $-5 + [2 + (-9)] = -5 + (\square)$
    $= \square$
17. Explain why the expression $-6 + -5$ is not written correctly. How should it be written?
18. What mathematical symbol is suggested when the word *sum* is used?

PRACTICE *Find the additive inverse of each number.*

19. -11
20. 9
21. -23
22. -43
23. 0
24. 1
25. 99
26. 250

Find each sum.

27. $-6 + (-3)$
28. $-2 + (-3)$
29. $-5 + (-5)$
30. $-8 + (-8)$
31. $-6 + 7$
32. $-2 + 4$
33. $-15 + 8$
34. $-18 + 10$
35. $20 + (-40)$
36. $25 + (-10)$
37. $30 + (-15)$
38. $8 + (-20)$
39. $-1 + 9$
40. $-2 + 7$
41. $-7 + 9$
42. $-3 + 6$
43. $5 + (-15)$
44. $16 + (-26)$
45. $24 + (-15)$
46. $-4 + 14$
47. $35 + (-27)$
48. $46 + (-73)$
49. $24 + (-45)$
50. $-65 + 31$

Evaluate each expression.

51. $-2 + 6 + (-1)$
52. $4 + (-3) + (-2)$
53. $-9 + 1 + (-2)$
54. $5 + 4 + (-6)$
55. $6 + (-4) + (-13) + 7$
56. $8 + (-5) + (-10) + 6$
57. $9 + (-3) + 5 + (-4)$
58. $-3 + 7 + 1 + (-4)$
59. Find the sum of -6, -7, and -8.
60. Find the sum of -11, -12, and -13.

Find each sum.

61. $-7 + 0$
62. $6 + 0$
63. $9 + 0$
64. $0 + (-15)$
65. $-4 + 4$
66. $18 + (-18)$
67. $2 + (-2)$
68. $-10 + 10$
69. What number must be added to -5 to obtain 0?
70. What number must be added to 8 to obtain 0?

Evaluate each expression.

71. $2 + (-10 + 8)$
72. $(-9 + 12) + (-4)$
73. $(-4 + 8) + (-11 + 4)$
74. $(-12 + 6) + (-6 + 8)$
75. $[-3 + (-4)] + (-5 + 2)$
76. $[9 + (-10)] + (-7 + 9)$
77. $[6 + (-4)] + [8 + (-11)]$
78. $[5 + (-8)] + [9 + (-15)]$
79. $-2 + [-8 + (-7)]$
80. $-8 + [-5 + (-2)]$
81. $789 + (-9{,}135)$
82. $2{,}701 + (-4{,}089)$
83. $-675 + (-456) + 99$
84. $-9{,}750 + (-780) + 2{,}345$

APPLICATIONS *Use signed numbers to help answer each question.*

85. G FORCES As a fighter pilot dives and loops, different forces are exerted on the body, just like the forces you experience when riding on a roller coaster. Some of the forces, called G's, are positive and some are negative. The force of gravity, 1G, is constant. Complete the diagram in the next column.

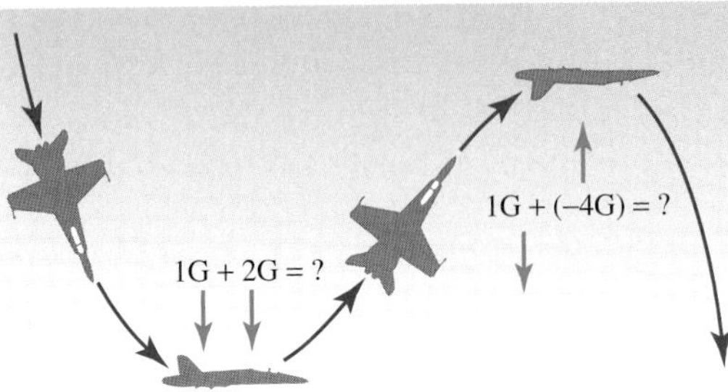

86. CHEMISTRY The first several steps of a chemistry lab experiment are listed here. The experiment begins with a compound that is stored at −40° F.

    Step 1: Raise the temperature of the compound 200°.

    Step 2: Add sulfur and then raise the temperature 10°.

    Step 3: Add 10 milliliters of water, stir, and raise the temperature 25°.

    What is the resulting temperature of the mixture after step 3?

87. CASH FLOW The maintenance costs, utilities, and taxes on a duplex are \$900 per month. The owner of the apartments receives monthly rental payments of \$450 and \$380. Does this investment produce a positive cash flow each month?

88. JOGGING A businessman's lunchtime workout includes jogging up ten stories of stairs in his high-rise office building. If he starts on the fourth level below ground in the underground parking garage, on what story of the building will he finish his workout?

89. HEALTH Find the point total for the six risk factors (in blue) on the medical questionnaire below. Then use the table at the bottom of the form to determine the risk of contracting heart disease for the man whose responses are shown.

Age		**Total Cholesterol**	
Age	Points	Reading	Points
35	−4	280	3
Cholesterol		**Blood Pressure**	
HDL	Points	Systolic/Diastolic	Points
62	−3	124/100	3
Diabetic		**Smoker**	
	Points		Points
Yes	4	Yes	2

10-Year Heart Disease Risk

Total Points	**Risk**	Total Points	**Risk**
−2 or less	1%	5	4%
−1 to 1	2%	6	6%
2 to 3	3%	7	6%
4	4%	8	7%

Source: National Heart, Lung, and Blood Institute

90. SPREADSHEETS Monthly rain totals for four counties are listed in the spreadsheet shown below. The **−1** entered in cell B1 means that the rain total for Suffolk County for a certain month was 1 inch below average. We can analyze this data by asking the computer to perform various operations.

a. To ask the computer to add the numbers in cells C1, C2, C3, and C4, we type SUM(C1:C4). Find this sum.

b. Find SUM(B4:F4).

	A	B	C	D	E	F
1	Suffolk	**−1**	−1	0	+1	+1
2	Marin	0	−2	+1	+1	−1
3	Logan	−1	+1	+2	+1	+1
4	Tipton	−2	−2	+1	−1	−3

91. ATOMS An atom is composed of protons, neutrons, and electrons. A proton has a positive charge (represented by +1), a neutron has no charge, and an electron has a negative charge (−1). Two simple models of atoms are shown. What is the net charge of each atom?

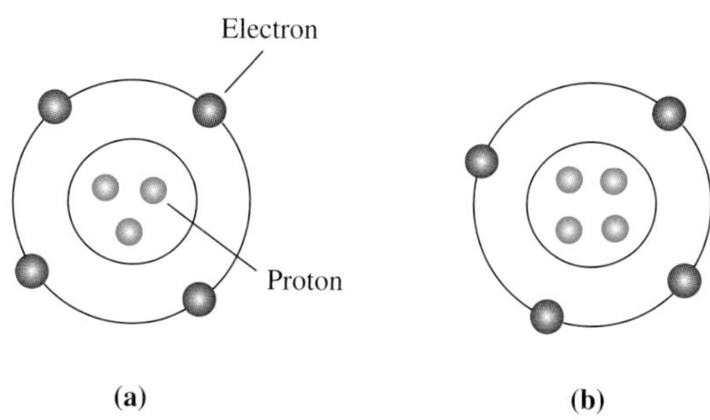

92. POLITICAL POLLS Six months before a general election, the incumbent senator found himself trailing the challenger by 18 points. To overtake his opponent, the campaign staff decided to use a four-part strategy. Each part of this plan is shown below, with the anticipated point gain.

1. Intense TV ad blitz +10
2. Ask for union endorsement +2
3. Voter mailing +3
4. Get-out-the-vote campaign +1

With these gains, will the incumbent overtake the challenger on election day?

93. FLOODING After a heavy rainstorm, a river that had been 4 feet under flood stage rose 11 feet in a 48-hour period. Find the height of the river after the storm in comparison to flood stage.

94. MILITARY SCIENCE During a battle, an army retreated 1,500 meters, regrouped, and advanced 3,500 meters. The next day, it had to retreat 1,250 meters. Find the army's net gain.

95. AIRLINES The graph below shows the annual net income for Delta Air Lines during the years 2000–2003. Estimate the company's total net income over this span of four years.

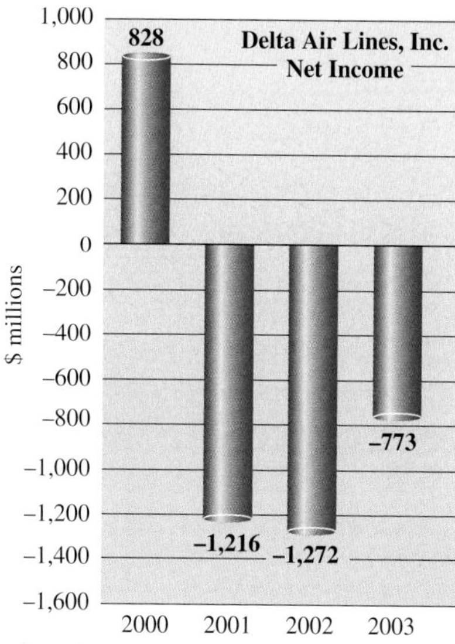

Source: Hoover's Online

96. ACCOUNTING On a financial balance sheet, debts (considered negative numbers) are denoted within parentheses. Assets (considered positive numbers) are written without parentheses. What is the 2004 fund balance for the preschool whose financial records are shown?

Community Care Preschool Balance Sheet, June 2004	
Fund balances	
Classroom supplies	$ 5,889
Emergency needs	927
Holiday program	(2,928)
Insurance	1,645
Janitorial	(894)
Licensing	715
Maintenance	(6,321)
BALANCE	?

WRITING

97. Is the sum of a positive and a negative number always positive? Explain why or why not.

98. How do you explain the fact that when asked to *add* −4 and 8, we must actually *subtract* to obtain the result?

99. Why is the sum of two negative numbers a negative number?

100. Write an application problem that will require adding -50 and -60.

REVIEW

101. Find the area of the rectangle shown below.

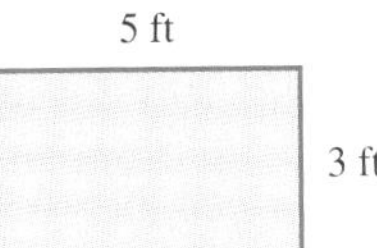

102. Find the perimeter of the rectangle in Exercise 101.

103. A car with a tank that holds 15 gallons of gasoline goes 25 miles on 1 gallon. How far can it go on a full tank?

104. What property is illustrated by the statement $5 \cdot 15 = 15 \cdot 5$?

105. Prime factor 125. Use exponents to express the result.

106. Perform the division: $\frac{144}{12}$.

2.3 Subtracting Integers

- Adding the opposite • Order of operations • Applications of subtraction

In this section, we study another way to think about subtraction. This new procedure is helpful when subtraction problems involve negative numbers.

Adding the opposite

The subtraction problem $6 - 4$ can be thought of as taking away 4 from 6. We can use the number line to illustrate this. (See Figure 2-14.) Beginning at the origin, we draw an arrow of length 6 units in the positive direction. From that point, we move back 4 units to the left. The answer, called the **difference,** is 2.

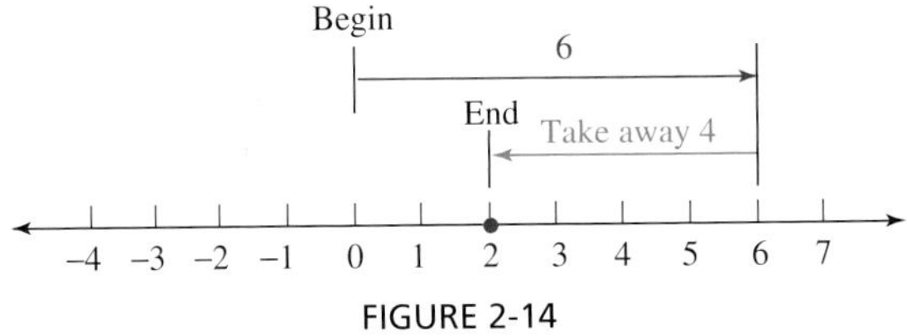

FIGURE 2-14

The work shown in Figure 2-14 looks like the illustration for the *addition* problem $6 + (-4) = 2$, shown in Figure 2-15.

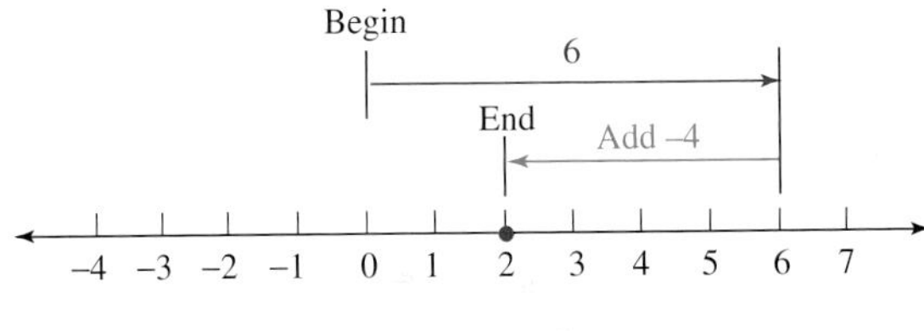

FIGURE 2-15

In the first problem, $6 - 4$, we subtracted 4 from 6. In the second, $6 + (-4)$, we added -4 (which is the opposite of 4) to 6. In each case, the result was 2.

Subtracting → $6 - 4 = 2$

Adding the opposite of 4 → $6 + (-4) = 2$

The same result

This observation helps to justify the following rule for subtraction.

Rule for subtraction

If a and b are any numbers, then

$$a - b = a + (-b)$$

In words, this rule states that *to subtract two integers, add the opposite of the second integer to the first integer.*

The rule for subtraction is also stated as follows: *subtraction is the same as adding the opposite of the number to be subtracted.*

You won't need to use this rule for every subtraction problem. For example, $6 - 4$ is obviously 2; it does not need to be rewritten as adding the opposite. But for more complicated problems such as $-6 - 4$ or $3 - (-5)$, where the result is not obvious, the subtraction rule will be quite helpful.

EXAMPLE 1 Subtract: $-6 - 4$.

Solution The number to be subtracted is 4. Applying the subtraction rule, we write

$-6 - 4 = -6 + (-4)$ Write the subtraction as an addition of the opposite of 4, which is -4. Write -4 within parentheses.

$= -10$ To add -6 and -4, apply the rule for adding two negative numbers.

To check the result, we add the difference, -10, and the subtrahend, 4. We should obtain the minuend, -6.

$$-10 + 4 = -6$$

The answer, -10, checks.

Self Check 1

Find $-2 - 3$ and check the result.

Answer -5

EXAMPLE 2 Subtract: $3 - (-5)$.

Solution The number being subtracted is -5.

$3 - (-5) = 3 + 5$ Write the subtraction as an addition of the opposite of -5, which is 5.

$= 8$ Perform the addition.

Self Check 2

Subtract: $3 - (-2)$.

Answer 5

Self Check 3
Subtract -8 from -3.

Answer 5

EXAMPLE 3 Subtract -3 from -8.

Solution The number being subtracted is -3, so we write it after -8.

$-8 - (-3) = -8 + 3$	Add the opposite of -3, which is 3.
$= -5$	Perform the addition.

Remember that any subtraction problem can be rewritten as an equivalent addition. We just add the opposite of the number that is to be subtracted.

Subtraction can be written as addition . . .

$$4 - 8 = 4 + (-8) = -4$$
$$4 - (-8) = 4 + 8 = 12$$
$$-4 - 8 = -4 + (-8) = -12$$
$$-4 - (-8) = -4 + 8 = 4$$

of the opposite of the number to be subtracted.

Order of operations

Expressions can contain repeated subtraction or subtraction in combination with grouping symbols. To work these problems, we apply the rules for the order of operations, listed on page 47.

Self Check 4
Evaluate: $-3 - 5 - (-1)$.

Answer -7

EXAMPLE 4 Evaluate: $-1 - (-2) - 10$.

Solution This problem involves two subtractions. We work from left to right, rewriting each subtraction as an addition of the opposite.

$-1 - (-2) - 10 = -1 + 2 + (-10)$	Add the opposite of -2, which is 2. Add the opposite of 10, which is -10. Write -10 in parentheses.
$= 1 + (-10)$	Work from left to right. Add $-1 + 2$.
$= -9$	Perform the addition.

Self Check 5
Evaluate: $-2 - (-6 - 5)$.

Answer 9

EXAMPLE 5 Evaluate: $-8 - (-2 - 2)$.

Solution We must perform the subtraction within the parentheses first.

$-8 - (-2 - 2) = -8 - [-2 + (-2)]$	Add the opposite of 2, which is -2. Since -2 must be written within parentheses, we write $-2 + (-2)$ within brackets.
$= -8 - (-4)$	Add -2 and -2. Since only one set of grouping symbols is now needed, we write -4 within parentheses.
$= -8 + 4$	Add the opposite of -4, which is 4.
$= -4$	Perform the addition.

Applications of subtraction

Things are constantly changing in our daily lives. The temperature, the amount of money we have in the bank, and our ages are examples. In mathematics, the operation of subtraction is used to measure change.

! COMMENT In general, to find the change in a quantity, we subtract the earlier value from the later value.

EXAMPLE 6 Change of water level. On Monday, the water level in a city storage tank was 6 feet above normal. By Friday, the level had fallen to a mark 4 feet below normal. Find the change in the water level from Monday to Friday. (See Figure 2-16.)

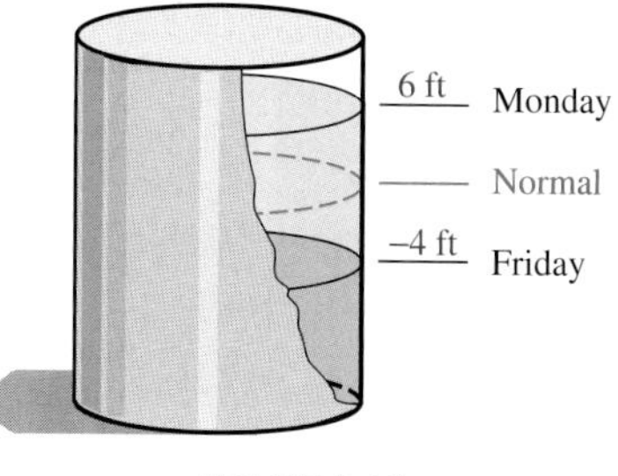

FIGURE 2-16

Solution We use subtraction to find the amount of change. The water levels of 4 feet below normal (the later value) and 6 feet above normal (the earlier value) can be represented by -4 and 6, respectively.

The water level Friday	minus	the water level Monday	is	the change in the water level.

$-4 - 6 = -4 + (-6)$ Add the opposite of 6, which is -6.

$= -10$ Perform the addition. The negative result indicates that the water level fell.

The water level fell 10 feet from Monday to Friday.

In the next example, the number line serves as a mathematical model of a real-life situation. You will see how the operation of subtraction can be used to find the distance between two points on the number line.

EXAMPLE 7 Artillery. In a practice session, an artillery group fired two rounds at a target. The first landed 65 yards short of the target, and the second landed 50 yards past it. (See Figure 2-17.) How far apart were the two impact points?

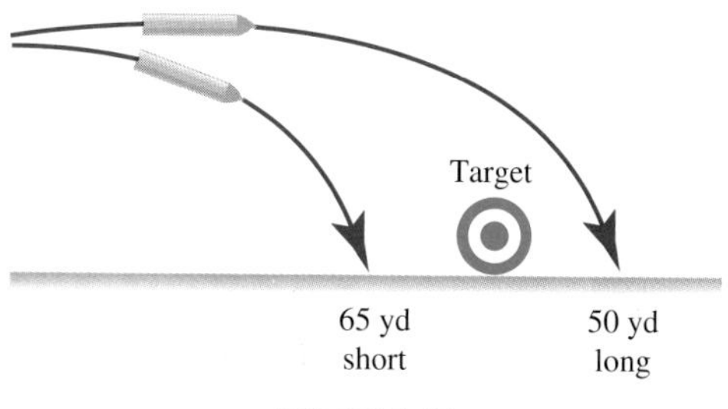

FIGURE 2-17

Solution We can use the number line to model this situation. The target is the origin. The words *short of the target* indicate a negative number, and the words *past it* indicate a positive number. Therefore, we graph the impact points at −65 and 50 in Figure 2-18.

FIGURE 2-18

The phrase *how far apart* tells us to subtract.

The position of the long shot	minus	the position of the short shot	is	the distance between the impact points.

$$
\begin{aligned}
50 - (-65) &= 50 + 65 &&\text{Add the opposite of } -65.\\
&= 115 &&\text{Perform the addition.}
\end{aligned}
$$

The impact points are 115 yards apart.

CALCULATOR SNAPSHOT Subtraction with negative numbers

The world's highest peak is Mount Everest in the Himalayas. The greatest ocean depth yet measured lies in the Mariana Trench near the island of Guam in the western Pacific. (See Figure 2-19) To find the range between the highest peak and the greatest depth, we must subtract.

29,035 − (−36,025)

To perform this subtraction using a scientific calculator, we enter these numbers and press these keys.

29035 [−] 36025 [+/−] [=] 65060

The range is 65,060 feet between the highest peak and the lowest depth. We could have applied the subtraction rule to write 29,035 − (−36,025) as 29,035 + 36,025 before using the calculator.

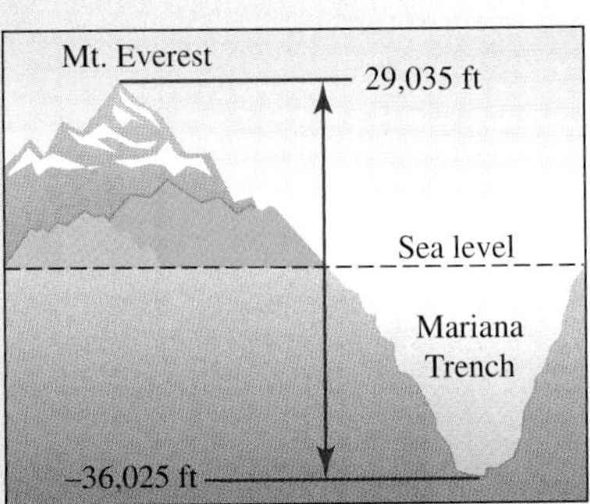

FIGURE 2-19

Section 2.3 STUDY SET

VOCABULARY *Fill in the blanks.*

1. The answer to a subtraction problem is called the ________.
2. Two numbers that are the same distance from 0 on the number line, but on opposite sides of it, are called ________.

CONCEPTS *Fill in the blanks.*

3. Subtraction is the same as ________ the ________ of the number to be subtracted.
4. Subtracting 3 is the same as adding ____.
5. Subtracting -6 is the same as adding ____.
6. The opposite of -8 is ____.
7. For any numbers a and b, $a - b = a +$ ____.
8. **a.** $2 - 7 = 2 +$ ____
   b. $2 - (-7) = 2 +$ ____
   c. $-2 - 7 = -2 +$ ____
   d. $-2 - (-7) = -2 +$ ____
9. After using parentheses as grouping symbols, if another set of grouping symbols is needed, we use ________.
10. We can find the ________ in a quantity by subtracting the earlier value from the later value.
11. Write this problem using mathematical symbols: negative eight minus negative four.
12. Write this problem using mathematical symbols: negative eight subtracted from negative four.
13. Find the distance between -4 and 3 on the number line.
14. Find the distance between -10 and 1 on the number line.
15. Is subtracting 3 from 8 the same as subtracting 8 from 3? Explain.
16. Evaluate each expression.
    a. $-2 - 0$ **b.** $0 - (-2)$

NOTATION *Complete each solution to evaluate each expression.*

17. $1 - 3 - (-2) = 1 + (\ \ \ ) + 2$
    $= -2 + \ \_\_\_$
    $= \ \_\_\_$

18. $-6 + 5 - (-5) = -6 + 5 + \ \_\_\_$
    $= \ \_\_\_ + 5$
    $= \ \_\_\_$

19. $(-8 - 2) - (-6) = [-8 + (\ \ \ )] - (-6)$
    $= \ \_\_\_ - (-6)$
    $= -10 + \ \_\_\_$
    $= \ \_\_\_$

20. $-5 - (-1 - 4) = -5 - [-1 + (\ \ \ )]$
    $= -5 - (\ \ \ )$
    $= -5 + \ \_\_\_$
    $= \ \_\_\_$

PRACTICE *Find each difference.*

21. $8 - (-1)$
22. $3 - (-8)$
23. $-4 - 9$
24. $-7 - 6$
25. $-5 - 5$
26. $-7 - 7$
27. $-5 - (-4)$
28. $-9 - (-1)$
29. $-1 - (-1)$
30. $-4 - (-3)$
31. $-2 - (-10)$
32. $-6 - (-12)$
33. $0 - (-5)$
34. $0 - 8$
35. $0 - 4$
36. $0 - (-6)$
37. $-2 - 2$
38. $-3 - 3$
39. $-10 - 10$
40. $4 - 4$
41. $9 - 9$
42. $4 - (-4)$
43. $-3 - (-3)$
44. $-5 - (-5)$

Evaluate each expression.

45. $-4 - (-4) - 15$
46. $-3 - (-3) - 10$
47. $-3 - 3 - 3$
48. $-1 - 1 - 1$
49. $5 - 9 - (-7)$
50. $6 - 8 - (-4)$
51. $10 - 9 - (-8)$
52. $16 - 14 - (-9)$
53. $-1 - (-3) - 4$
54. $-2 - 4 - (-1)$
55. $-5 - 8 - (-3)$
56. $-6 - 5 - (-1)$
57. $(-6 - 5) - 3$
58. $(-2 - 1) - 5$
59. $(6 - 4) - (1 - 2)$
60. $(5 - 3) - (4 - 6)$
61. $-9 - (6 - 7)$
62. $-3 - (6 - 12)$
63. $-8 - [4 - (-6)]$
64. $-1 - [5 - (-2)]$
65. $[-4 + (-8)] - (-6)$
66. $[-5 + (-4)] - (-2)$

67. Subtract -3 from 7.

68. Subtract 8 from -2.

69. Subtract -6 from -10.

70. Subtract -4 from -9.

Use a calculator to perform each subtraction.

71. $-1{,}557 - 890$

72. $-345 - (-789)$

73. $20{,}007 - (-496)$

74. $-979 - (-44{,}879)$

75. $-162 - (-789) - 2{,}303$

76. $-787 - 1{,}654 - (-232)$

APPLICATIONS *Use signed numbers to help answer each question.*

77. SCUBA DIVING A diver jumps from his boat into the water and descends 50 feet. He pauses to check his equipment and then descends an additional 70 feet. Use a signed number to represent the diver's final depth.

78. TEMPERATURE CHANGE Rashawn flew from his New York home to Hawaii for a week of vacation. He left blizzard conditions and a temperature of $-6°$, and stepped off the airplane into 85° weather. What temperature change did he experience?

79. READING PROGRAMS In a state reading test administered at the start of a school year, an elementary school's performance was 23 points below the county average. The principal immediately began a special tutorial program. At the end of the school year, retesting showed the students to be only 7 points below the average. How many points did the school's reading score improve over the year?

80. SUBMARINES A submarine was traveling 2,000 feet below the ocean's surface when the radar system warned of an impending collision with another sub. The captain ordered the navigator to dive an additional 200 feet and then level off. Find the depth of the submarine after the dive.

81. AMPERAGE During normal operation, the ammeter on a car reads $+5$. If the headlights, which draw a current of 7 amps, and the radio, which draws a current of 6 amps, are both turned on, what number will the ammeter register?

82. GIN RUMMY After a losing round, a card player must subtract the value of each of the cards left in his hand from his previous point total of 21. If face cards are counted as 10 points, what is his new score?

83. GEOGRAPHY Death Valley, California, is the lowest land point in the United States, at 282 feet below sea level. The lowest land point on the Earth is the Dead Sea, which is 1,348 feet below sea level. How much lower is the Dead Sea than Death Valley?

84. LIE DETECTOR TESTS On one lie detector test, a burglar scored -18, which indicates deception. However, on a second test, he scored -1, which is inconclusive. Find the difference in the scores.

85. FOOTBALL A college football team records the outcome of each of its plays during a game on a stat sheet. Find the net gain (or loss) after the 3rd play.

Down	Play	Result
1st	Run	Lost 1 yd
2nd	Pass – sack!	Lost 6 yd
Penalty	Delay of game	Lost 5 yd
3rd	Pass	Gained 8 yd
4th	Punt	—

86. ACCOUNTING Complete the balance sheet below. Then determine the overall financial condition of the company by subtracting the total liabilities from the total assets.

Walker Corporation
Balance Sheet 2001

Assets				
Cash	$11	1	0	9
Supplies	7	8	6	2
Land	67	5	4	3
Total assets	$			
Liabilities				
Accounts payable	$79	0	3	7
Income taxes	20	1	8	1
Total liabilities	$			

87. DIVING A diver jumps from a platform. After she hits the water, her momentum takes her to the bottom of the pool.

a. Use the number line and signed numbers to model this situation. Show the top of the platform, the water level as 0, and the bottom of the pool.

b. Find the total length of the dive from the top of the platform to the bottom of the pool.

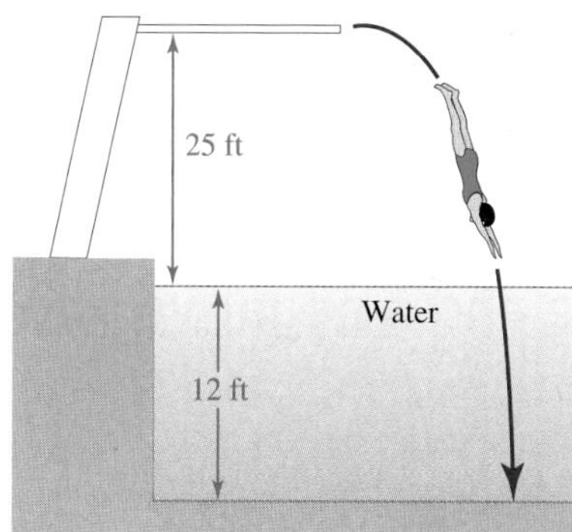

88. TEMPERATURE EXTREMES The highest and lowest temperatures ever recorded in several cities are shown below. List the cities in order, from the largest to smallest range in temperature extremes.

	Extreme temperatures	
City	**Highest**	**Lowest**
Atlantic City, NJ	106	−11
Barrow, AK	79	−56
Kansas City, MO	109	−23
Norfolk, VA	104	−3
Portland, ME	103	−39

89. CHECKING ACCOUNTS Michael has \$1,303 in his checking account. Can he pay his car insurance premium of \$676, his utility bills of \$121, and his rent of \$750 without having to make another deposit? Explain.

90. HISTORY Two of the greatest Greek mathematicians were Archimedes (287–212 B.C.) and Pythagoras (569–500 B.C.). How many years apart were they born?

WRITING

91. Explain what is meant when we say that subtraction is the same as addition of the opposite.

92. Give an example showing that it is possible to subtract something from nothing.

93. Explain how to check the result: $-7 - 4 = -11$.

94. Explain why students don't need to change every subtraction they encounter to an addition of the opposite. Give some examples.

REVIEW

95. Round 5,989 to the nearest ten.

96. Round 5,999 to the nearest hundred.

97. List the factors of 20.

98. It takes 13 oranges to make one can of orange juice. Find the number of oranges used to make 12 cans.

99. Evaluate: $12^2 - (5 - 4)^2$.

100. What property does the following illustrate?

$$15 + 12 = 12 + 15$$

101. Solve: $5x = 15$.

102. Solve: $x + 5 = 15$.

2.4 Multiplying Integers

- Multiplying two positive integers • Multiplying a positive and a negative integer
- Multiplying a negative and a positive integer • Multiplying by zero
- Multiplying two negative integers • Powers of integers

We now turn our attention to multiplication of integers. When we multiply two nonzero integers, the first factor can be positive or negative. The same is true for the second factor. This means that there are four possible combinations to consider.

Positive · positive	Positive · negative
Negative · positive	Negative · negative

In this section, we discuss these four combinations and use our observations to establish rules for multiplying two integers.

Multiplying two positive integers

4(3)
like signs
both positive

We begin by considering the product of two positive integers, 4(3). Since both factors are positive, we say that they have *like* signs. In Chapter 1, we learned that multiplication is repeated addition. Therefore, 4(3) represents the sum of four 3's.

$4(3) = 3 + 3 + 3 + 3$ Multiplication is repeated addition. Write 3 as an addend four times.

$4(3) = 12$ The result is 12, which is a positive number.

This result suggests that *the product of two positive integers is positive.*

Multiplying a positive and a negative integer

4(−3)
unlike signs
one positive, one negative

Next, we consider $4(-3)$. This is the product of a positive and a negative integer. The signs of these factors are *unlike.* According to the definition of multiplication, $4(-3)$ means that we are to add -3 four times.

$4(-3) = (-3) + (-3) + (-3) + (-3)$ Use the definition of multiplication. Write -3 as an addend four times.

$4(-3) = (-6) + (-3) + (-3)$ Work from left to right. Apply the rule for adding two negative numbers.

$4(-3) = (-9) + (-3)$ Work from left to right. Apply the rule for adding two negative numbers.

$4(-3) = -12$ Perform the addition.

This result is -12, which suggests that *the product of a positive integer and a negative integer is negative.*

Multiplying a negative and a positive integer

−3(4)
unlike signs
one negative, one positive

To develop a rule for multiplying a negative and a positive integer, we consider $-3(4)$. Notice that the factors have *unlike* signs. Because of the commutative property of multiplication, the answer to $-3(4)$ will be the same as the answer to $4(-3)$. We know that $4(-3) = -12$ from the previous discussion, so $-3(4) = -12$. This suggests that *the product of a negative integer and a positive integer is negative.*

Putting the results of the last two cases together leads us to the rule for multiplying two integers with unlike signs.

Multiplying two integers with unlike signs

To multiply a positive integer and a negative integer, or a negative integer and a positive integer, multiply their absolute values. Then make the answer negative.

EXAMPLE 1 Find each product: **a.** $7(-5)$, **b.** $20(-8)$, and **c.** $-8 \cdot 5$.

Solution To multiply integers with unlike signs, we multiply their absolute values and make the product negative.

a. $7(-5) = -35$ Multiply the absolute values, 7 and 5, to get 35. Then make the answer negative.

b. $20(-8) = -160$ Multiply the absolute values, 20 and 8, to get 160. Then make the answer negative.

c. $-8 \cdot 5 = -40$ Multiply the absolute values, 8 and 5, to get 40. Then make the answer negative.

Self Check 1
Find each product:
a. $2(-6)$
b. $30(-2)$
c. $-15 \cdot 2$

Answers **a.** -12, **b.** -60, **c.** -30

! COMMENT When writing multiplication involving signed numbers, do not write a negative sign – next to a raised dot · (the multiplication symbol). Instead, use parentheses to show the multiplication.

$6(-2)$ ~~$6 \cdot -2$~~ and $-6(-2)$ ~~$-6 \cdot -2$~~

Multiplying by zero

Before we can develop a rule for multiplying two negative integers, we need to examine multiplication by 0. If 4(3) means that we are to find the sum of four 3's, then $0(-3)$ means that we are to find the sum of zero -3's. Obviously, the sum is 0. Thus, $0(-3) = 0$.

The commutative property of multiplication guarantees that we can change the order of the factors in the multiplication problem without affecting the result.

$$(-3)(0) = 0(-3) = 0$$

↑ Change the order of the factors. ↑ The result is still 0.

We see that the order in which we write the factors 0 and -3 doesn't matter—their product is 0. This example suggests that *the product of any number and 0 is 0.*

Multiplying by 0

If a is any number, then

$$a \cdot 0 = 0 \qquad 0 \cdot a = 0$$

EXAMPLE 2 Find $-12 \cdot 0$.

Solution Since the product of any number and 0 is 0, we have

$$-12 \cdot 0 = 0$$

Self Check 2
Find $0(-56)$.

Answer 0

Multiplying two negative integers

$-3(-4)$
like signs
both negative

To develop a rule for multiplying two negative integers, we consider the pattern displayed on the next page. There, we multiply -4 by a series of factors that decrease by 1. After determining each product, we graph each product on the number line (Figure 2-20). See if you can determine the answers to the last three multiplication problems by examining the pattern of answers leading up to them.

This factor decreases by 1 as you read down the column. ↓ Look for a pattern here. ↓

$$4(-4) = -16$$
$$3(-4) = -12$$
$$2(-4) = -8$$
$$1(-4) = -4$$
$$0(-4) = 0$$
$$-1(-4) = ?$$
$$-2(-4) = ?$$
$$-3(-4) = ?$$

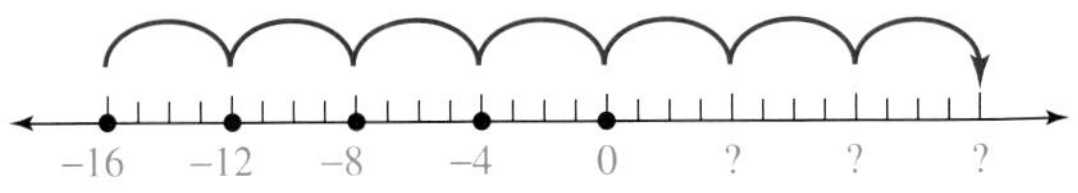

−16 −12 −8 −4 0 ? ? ?

FIGURE 2-20

From the pattern, we see that

$$-1(-4) = 4$$
$$-2(-4) = 8$$
$$-3(-4) = 12$$

↑ ↑ ↑
For two negative factors, the product is a positive.

These results suggest that *the product of two negative integers is positive.* Earlier in this section, we saw that the product of two positive integers is also positive. This leads to the following conclusion.

Multiplying two integers with like signs

To multiply two positive integers, or two negative integers, multiply their absolute values. The answer is positive.

Self Check 3
Find each product:
a. $-9(-7)$
b. $-12(-2)$

Answers a. 63, **b.** 24

EXAMPLE 3 Find each product: a. $-5(-9)$ and b. $-8(-10)$.

Solution To multiply two negative integers, we multiply their absolute values and make the result positive.

a. $-5(-9) = 45$ Multiply the absolute values, 5 and 9, to get 45. The answer is positive.

b. $-8(-10) = 80$ Multiply the absolute values, 8 and 10, to get 80. The answer is positive.

We now summarize the rules for multiplying two integers.

Multiplying two integers

To multiply two integers, multiply their absolute values.

1. The product of two integers with *like* signs is positive.
2. The product of two integers with *unlike* signs is negative.

We can use a calculator to multiply signed numbers.

Multiplication with negative numbers

CALCULATOR SNAPSHOT

At Thanksgiving time, a large supermarket chain offered customers a free turkey with every grocery purchase of \$100 or more. Each turkey cost the store \$8, and 10,976 people took advantage of the offer. Since each of the 10,976 turkeys given away represented a loss of \$8 (which can be expressed as -8 dollars), the company lost $10{,}976(-8)$ dollars. To find this product, we enter these numbers and press these keys on a scientific calculator.

10976 [×] 8 [+/−] [=]   [-87808]

The negative result indicates that with this promotion, the supermarket chain gave away \$87,808 in turkeys.

EXAMPLE 4 Multiply: **a.** $-3(-2)(-6)(-5)$ and **b.** $-2(-4)(-5)$.

Solution

a. $-3(-2)(-6)(-5) = 6(-6)(-5)$ Work from left to right: $-3(-2) = 6$.

$= -36(-5)$ Work from left to right: $6(-6) = -36$.

$= 180$

b. $-2(-4)(-5) = 8(-5)$ Work from left to right: $-2(-4) = 8$.

$= -40$

Self Check 4
Multiply: **a.** $-1(-2)(-5)$ and **b.** $-2(-7)(-1)(-2)$.

Answers **a.** -10, **b.** 28

Example 4, part a, illustrates that

A product is positive when there are an even number of negative factors.

Part b illustrates that

A product is negative when there are an odd number of negative factors.

Powers of integers

Recall that exponential expressions are used to represent repeated multiplication. For example, 2 to the third power, or 2^3, is a shorthand way of writing $2 \cdot 2 \cdot 2$. In this expression, 3 is the exponent and the base is positive 2. In the next example, we evaluate exponential expressions with bases that are negative numbers.

EXAMPLE 5 Find each power: **a.** $(-2)^4$ and **b.** $(-5)^3$.

Solution

a. $(-2)^4 = (-2)(-2)(-2)(-2)$ Write -2 as a factor 4 times.

$= 4(-2)(-2)$ Work from left to right. Multiply -2 and -2 to get 4.

$= -8(-2)$ Work from left to right. Multiply 4 and -2 to get -8.

$= 16$ Perform the multiplication.

b. $(-5)^3 = (-5)(-5)(-5)$ Write -5 as a factor 3 times.

$= 25(-5)$ Work from left to right. Multiply -5 and -5 to get 25.

$= -125$ Perform the multiplication.

Self Check 5
Find each power:
a. $(-3)^4$
b. $(-4)^3$

Answers **a.** 81, **b.** -64

In Example 5, part a, -2 was raised to an even power, and the answer was positive. In part b, another negative number, -5, was raised to an odd power, and the answer was negative. These results suggest a general rule.

Even and odd powers of a negative integer

When a negative integer is raised to an even power, the result is positive.

When a negative integer is raised to an odd power, the result is negative.

Self Check 6

Find the power: $(-1)^8$.

Answer 1

EXAMPLE 6 Find the power: $(-1)^5$.

Solution We have a negative integer raised to an odd power. The result will be negative.

$$\begin{aligned}(-1)^5 &= (-1)(-1)(-1)(-1)(-1)\\ &= -1\end{aligned}$$

! COMMENT Although the expressions -3^2 and $(-3)^2$ look similar, they are not the same. In -3^2, the base is 3 and the exponent 2. The $-$ sign in front of 3^2 means the opposite of 3^2. In $(-3)^2$, the base is -3 and the exponent is 2. When we evaluate them, it becomes clear that they are not equivalent.

-3^2 represents *the opposite of* 3^2.

$-3^2 = -(3 \cdot 3)$ Write 3 as a factor 2 times.

$= -9$ Multiply within the parentheses first.

$(-3)^2$ represents $(-3)(-3)$.

$(-3)^2 = (-3)(-3)$ Write -3 as a factor 2 times.

$= 9$ The product of two negative numbers is positive.

Notice that the results are different.

Self Check 7

Evaluate: **a.** -4^2 and **b.** $(-4)^2$.

Answers **a.** -16, **b.** 16

EXAMPLE 7 Evaluate: a. -2^2 and b. $(-2)^2$.

Solution

a. $-2^2 = -(2 \cdot 2)$ Since 2 is the base, write 2 as a factor two times.

$= -4$ Perform the multiplication within the parentheses.

b. $(-2)^2 = (-2)(-2)$ The base is -2. Write it as a factor twice.

$= 4$ The signs are like, so the product is positive.

CALCULATOR SNAPSHOT Raising a negative number to a power

Negative numbers can be raised to a power using a scientific calculator. We use the change-of-sign key [+/−] and the power key [y^x] (on the some calculators, [x^y]). For example, to evaluate $(-5)^6$, we enter these numbers and press these keys.

5 [+/−] [y^x] 6 [=]   15625

The result is 15,625.

Some scientific calculators require parentheses when entering a negative base raised to a power.

Section 2.4 STUDY SET

VOCABULARY *Fill in the blanks.*

1. In the multiplication $-5(-4)$, the integers -5 and -4, which are being multiplied, are called ________. The answer, 20, is called the ________.
2. The definition of multiplication tells us that $3(-4)$ represents repeated ________ $-4 + (-4) + (-4)$.
3. In the expression -3^5, ▢ is the base and 5 is the ________.
4. In the expression $(-3)^5$, ▢ is the base and ▢ is the exponent.

CONCEPTS *Fill in the blanks.*

5. The product of two integers with ________ signs is negative.
6. The product of two integers with like signs is ________.
7. The ____________ property of multiplication implies that $-2(-3) = -3(-2)$.
8. The product of 0 and any number is ▢.
9. Find $-1(9)$. In general, what is the result when we multiply a positive number by -1?
10. Find $-1(-9)$. In general, what is the result when we multiply a negative number by -1?
11. When we multiply two integers, there are four possible combinations of signs. List each of them.
12. When we multiply two integers, there are four possible combinations of signs. How can they be grouped into two categories?
13. If each of the following powers were evaluated, what would be the *sign* of the result?
    a. $(-5)^{13}$ b. $(-3)^{20}$
14. A student claimed, "A positive and a negative is negative." What is wrong with this statement?
15. Find each absolute value.
    a. $|-3|$ b. $|12|$
    c. $|-5|$ d. $|9|$
    e. $|10|$ f. $|-25|$
16. Find each product and then graph it on a number line. What is the distance between every two products?

$$2(-2),\quad 1(-2),\quad 0(-2),\quad -1(-2),\quad -2(-2)$$

17. a. Complete the table.

Problem	Number of negative factors	Answer
$-2(-2)$		
$-2(-2)(-2)(-2)$		
$-2(-2)(-2)(-2)(-2)(-2)$		

    b. The answers entered in the table help to justify the following rule: The product of an ______ number of negative integers is positive.

18. a. Complete the table.

Problem	Number of negative factors	Answer
$-2(-2)(-2)$		
$-2(-2)(-2)(-2)(-2)$		
$-2(-2)(-2)(-2)(-2)(-2)(-2)$		

    b. The answers entered in the table help to justify the following rule: The product of an ______ number of negative integers is negative.

NOTATION *Complete each solution to evaluate the expression.*

19. $-3(-2)(-4) = \square(-4)$
    $= \square$
20. $(-3)^4 = (-3)(-3)(-3)\square$
    $= \square(-3)(-3)$
    $= \square(-3)$
    $= \square$
21. Explain why the expression below is not written correctly. How should it be written?

$$-6 \cdot -5$$

22. Translate to mathematical symbols.
    a. the product of negative three and negative two
    b. negative five, squared
    c. the opposite of five squared

PRACTICE *Find each product.*

23. $-9(-6)$
24. $-5(-5)$
25. $-3 \cdot 5$
26. $-6 \cdot 4$

27. $12(-3)$
28. $11(-4)$
29. $(-8)(-7)$
30. $(-9)(-3)$
31. $(-2)10$
32. $(-3)8$
33. $-40 \cdot 3$
34. $-50 \cdot 2$
35. $-8(0)$
36. $0(-27)$
37. $-1(-6)$
38. $-1(-8)$
39. $-7(-1)$
40. $-5(-1)$
41. $1(-23)$
42. $-35(1)$

Evaluate each expression.

43. $-6(-4)(-2)$
44. $-3(-2)(-3)$
45. $5(-2)(-4)$
46. $3(-3)(3)$
47. $2(3)(-5)$
48. $6(2)(-2)$
49. $6(-5)(2)$
50. $4(-2)(2)$
51. $(-1)(-1)(-1)$
52. $(-1)(-1)(-1)(-1)$
53. $-2(-3)(3)(-1)$
54. $5(-2)(3)(-1)$
55. $3(-4)(0)$
56. $-7(-9)(0)$
57. $-2(0)(-10)$
58. $-6(0)(-12)$
59. Find the product of -6 and the opposite of 10.
60. Find the product of the opposite of 9 and the opposite of 8.

Find each power.

61. $(-4)^2$
62. $(-6)^2$
63. $(-5)^3$
64. $(-6)^3$
65. $(-2)^3$
66. $(-4)^3$
67. $(-9)^2$
68. $(-10)^2$
69. $(-1)^5$
70. $(-1)^6$
71. $(-1)^8$
72. $(-1)^9$

Evaluate each expression.

73. $(-7)^2$ and -7^2
74. $(-5)^2$ and -5^2
75. -12^2 and $(-12)^2$
76. -11^2 and $(-11)^2$

Use a calculator to evaluate each expression.

77. $-76(787)$
78. $407(-32)$
79. $(-81)^4$
80. $(-6)^5$
81. $(-32)(-12)(-67)$
82. $(-56)(-9)(-23)$
83. $(-25)^4$
84. $(-41)^5$

APPLICATIONS *Use signed numbers to help solve each problem.*

85. DIETING After giving a patient a physical exam, a physician felt that the patient should begin a diet. Two options were discussed.

	Plan #1	Plan #2
Length	10 weeks	14 weeks
Daily exercise	1 hr	30 min
Weight loss per week	3 lb	2 lb

a. Find the expected weight loss from each diet plan. Express each answer as a signed number.

b. With which plan should the patient expect to lose more weight? Explain why the patient might not choose it.

86. INVENTORIES A spreadsheet is used to record inventory losses at a warehouse. The items, their cost, and the number missing are listed in the table.

	A	B	C	D
1	Item	Cost	Number of units	$ losses
2	CD	$5	−11	
3	TV	$200	−2	
4	Radio	$20	−4	

a. What instruction should be given to find the total losses for each type of item? Find each of those losses and fill in column D.

b. What instruction should be given to find the *total* inventory losses for the warehouse? Find this number.

87. MAGNIFICATION Using an electronic testing device, a mechanic can check the emissions of a car. The results of the test are displayed on a screen.

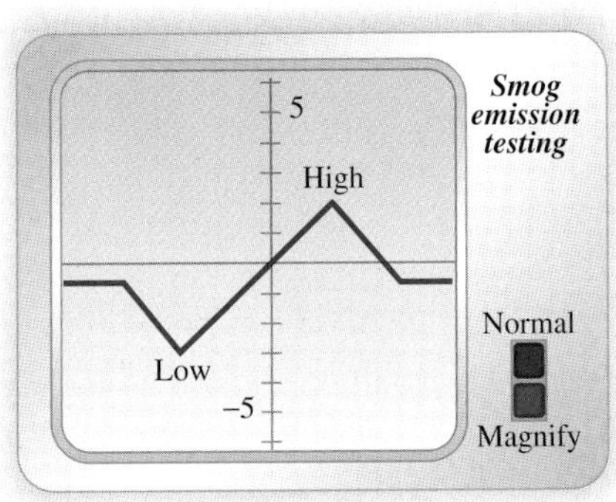

a. Find the high and low values for this test as shown on the screen.

b. By switching a setting on the monitor, the picture on the screen can be magnified. What would be the new high and new low if every value were doubled?

88. LIGHT Sunlight is a mixture of all colors. When sunlight passes through water, the water absorbs different colors at different rates, as shown.

a. Use a signed number to represent the depth to which red light penetrates water.

b. Green light penetrates 4 times deeper than red light. How deep is this?

c. Blue light penetrates 3 times deeper than orange light. How deep is this?

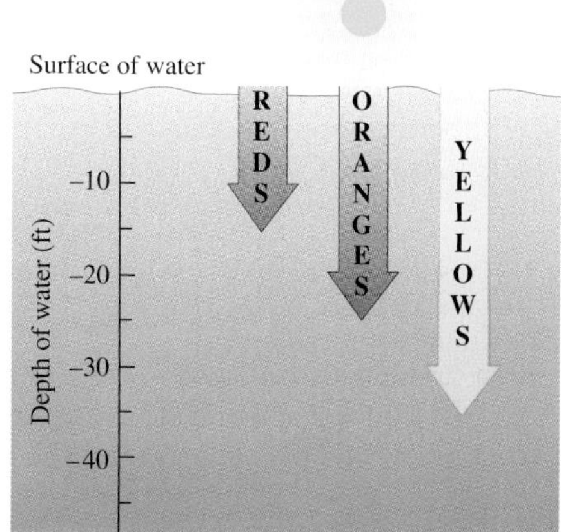

89. TEMPERATURE CHANGE A farmer, worried about his fruit trees suffering frost damage, calls the weather service for temperature information. He is told that temperatures will be decreasing approximately 4° every hour for the next five hours. What signed number represents the total change in temperature expected over the next five hours?

90. DEPRECIATION For each of the last four years, a businesswoman has filed a $200 depreciation allowance on her income tax return, for an office computer system. What signed number represents the total amount of depreciation written off over the four-year period?

91. EROSION A levee protects a town in a low-lying area from flooding. According to geologists, the banks of the levee are eroding at a rate of 2 feet per year. If something isn't done to correct the problem, what signed number indicates how much of the levee will erode during the next decade?

92. DECK SUPPORTS After a winter storm, a homeowner has an engineering firm inspect his damaged deck. Their report concludes that the original pilings were not anchored deep enough, by a factor of 3. What signed number represents the depth to which the pilings should have been sunk?

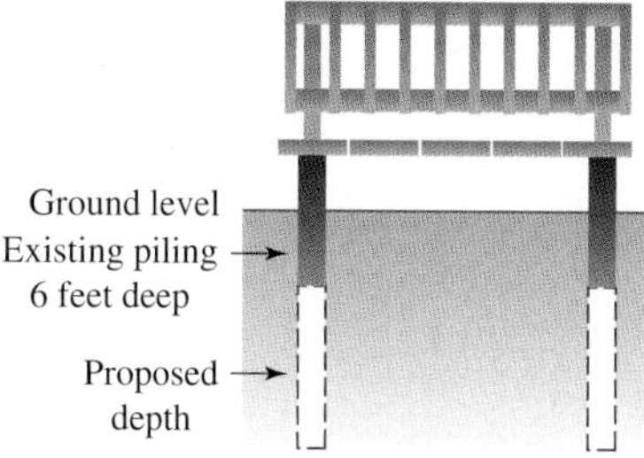

93. WOMEN'S NATIONAL BASKETBALL ASSOCIATION The average attendance for the WNBA Houston Comets is 8,110 a game. Suppose the team gives a sports bag, costing $3, to everyone attending a game. What signed number expresses the financial loss from this promotional giveaway?

94. HEALTH CARE A health care provider for a company estimates that 75 hours per week are lost by employees suffering from stress-related or preventable illness. In a 52-week year, how many hours are lost? Use a signed number to answer.

WRITING

95. If a product contains an even number of negative factors, how do we know that the result will be positive?

96. Explain why the product of a positive number and a negative number is negative, using $5(-3)$ as an example.

97. Explain why the result is the opposite of the original number when a number is multiplied by -1.

98. Can you think of any number that yields a nonzero result when it is multiplied by 0? Explain your response.

REVIEW

99. The prime factorization of a number is $3^2 \cdot 5$. What is the number?

100. Solve: $\frac{y}{8} = 10$.

101. The enrollment at a college went from 10,200 to 12,300 in one year. What was the increase in enrollment?

102. Find the perimeter of a square with sides 6 yards long.

103. What does the symbol $<$ mean?

104. List the first ten prime numbers.

2.5 Dividing Integers

- The relationship between multiplication and division
- Rules for dividing integers • Division and zero

In this section, we develop rules for division of integers, just as we did for multiplication of integers. We also consider two types of division involving 0.

The relationship between multiplication and division

Every division fact containing three numbers has a related multiplication fact involving the same three numbers. For example,

$\frac{6}{3} = 2$ because $3(2) = 6$ Remember that in the division statement, 6 is the *dividend,* 3 is the *divisor,* and 2 is the *quotient.*

$\frac{20}{5} = 4$ because $5(4) = 20$

Rules for dividing integers

We now use the relationship between multiplication and division to help develop rules for dividing integers. There are four cases to consider.

Case 1: In the first case, a positive integer is divided by a positive integer. From years of experience, we already know that the result is positive. Therefore, *the quotient of two positive integers is positive.*

Case 2: Next, we consider the quotient of two negative integers. As an example, consider the division $\frac{-12}{-2} = ?$. We can do this division by examining its related multiplication statement, $-2(?) = -12$. Our objective is to find the number that should replace the question mark. To do this, we use the rules for multiplying integers, introduced in the previous section.

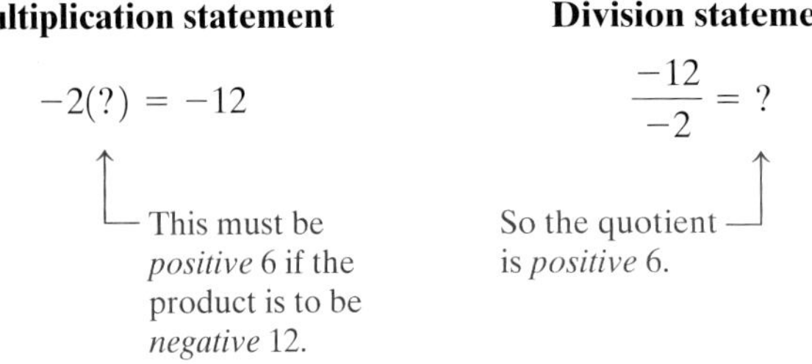

Therefore, $\frac{-12}{-2} = 6$. From this example, we can see that *the quotient of two negative integers is positive.*

Case 3: The third case we examine is the quotient of a positive integer and a negative integer. Let's consider $\frac{12}{-2} = ?$. Its equivalent multiplication statement is $-2(?) = 12$.

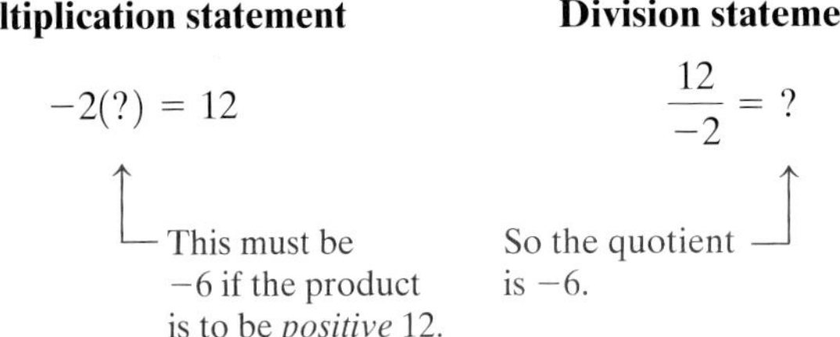

Multiplication statement
$-2(?) = 12$
This must be -6 if the product is to be *positive* 12.
Division statement
$\frac{12}{-2} = ?$
So the quotient is -6.

Therefore, $\frac{12}{-2} = -6$. This result shows that *the quotient of a positive integer and a negative integer is negative.*

Case 4: Finally, to find the quotient of a negative integer and a positive integer, let's consider $\frac{-12}{2} = ?$. Its equivalent multiplication statement is $2(?) = -12$.

Multiplication statement	**Division statement**
$2(?) = -12$	$\frac{-12}{2} = ?$
This must be -6 if the product is to be -12.	So the quotient is -6.

Therefore, $\frac{-12}{2} = -6$. From this example, we can see that *the quotient of a negative integer and a positive integer is negative.*

We now summarize the results from the previous discussion.

Dividing two integers

To divide two integers, divide their absolute values.

1. The quotient of two integers with *like* signs is positive.
2. The quotient of two integers with *unlike* signs is negative.

The rules for dividing integers are similar to those for multiplying integers.

EXAMPLE 1 Find each quotient: **a.** $\frac{-35}{7}$ and **b.** $\frac{20}{-5}$.

Solution To divide integers with unlike signs, we find the quotient of their absolute values and make the quotient negative.

a. $\frac{-35}{7} = -5$ Divide the absolute values, 35 by 7, to get 5. The quotient is negative.

To check the result, we multiply the divisor, 7, and the quotient, -5. We should obtain the dividend, -35.

$7(-5) = -35$

The answer, -5, checks.

b. $\frac{20}{-5} = -4$ Divide the absolute values, 20 by 5, to get 4. The quotient is negative.

Self Check 1

Find each quotient and check the result:

a. $\frac{-45}{5}$

b. $\frac{60}{-20}$

Answers **a.** -9, **b.** -3

EXAMPLE 2 Divide: $\frac{-12}{-3}$.

Solution The integers have like signs. The quotient will be positive.

$\frac{-12}{-3} = 4$ Divide the absolute values, 12 by 3, to get 4. The quotient is positive.

Self Check 2

Divide: $\frac{-21}{-3}$.

Answer 7

EXAMPLE 3 **Price reductions.** Over the course of a year, a retailer reduced the price of a television set by an equal amount each month, because it was not selling. By the end of the year, the cost was \$132 less than at the beginning of the year. How much did the price fall each month?

Solution We label the drop in price of \$132 for the year as -132. It occurred in 12 equal reductions. This indicates division.

$\frac{-132}{12} = -11$ The quotient of a negative number and a positive number is negative.

The drop in price each month was \$11.

Division and zero

To review the concept of division of 0, we look at $\frac{0}{-2} = ?$. The related multiplication statement is $-2(?) = 0$.

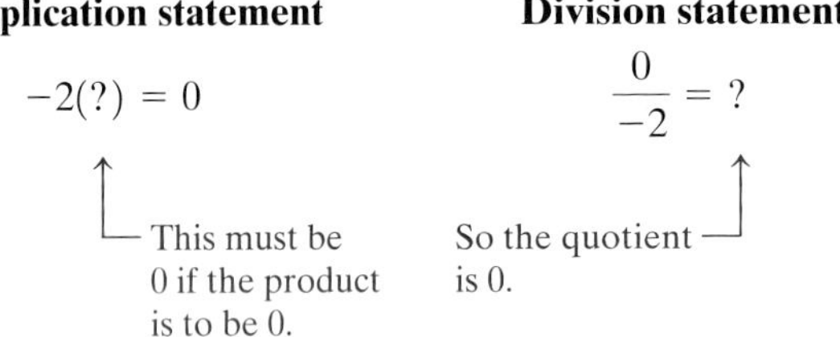

Therefore, $\frac{0}{-2} = 0$. This example suggests that *the quotient of 0 divided by any nonzero number is 0.*

To review division by 0, let's look at $\frac{-2}{0} = ?$. The related multiplication statement is $0(?) = -2$.

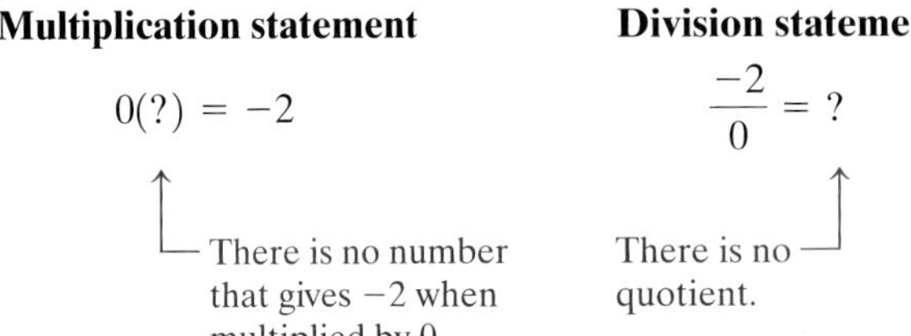

Multiplication statement
Division statement
$0(?) = -2$
$\frac{-2}{0} = ?$
There is no number that gives -2 when multiplied by 0.
There is no quotient.

Therefore, $\frac{-2}{0}$ does not have an answer. We say that division by 0 is undefined. This example suggests that *the quotient of any number divided by 0 is undefined.*

Division with 0

1. If a represents any nonzero number, $\frac{0}{a} = 0$.
2. If a represents any nonzero number, $\frac{a}{0}$ is undefined.
3. $\frac{0}{0}$ is undetermined.

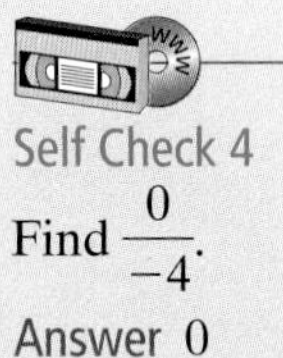

Self Check 4
Find $\frac{0}{-4}$.
Answer 0

EXAMPLE 4 Find $\frac{-4}{0}$, if possible.

Solution Since $\frac{-4}{0}$ is division by 0, the division is undefined.

Division with negative numbers

CALCULATOR SNAPSHOT

The Bureau of Labor Statistics estimated that the United States lost 146,400 jobs in the manufacturing sector of the economy in 2003. Because the jobs were lost, we write this as −146,400. To find the average number of manufacturing jobs lost each month, we divide: $\frac{-146,400}{12}$. To perform this division using a scientific calculator, we enter these numbers and press these keys.

146400 [+/−] [÷] 12 [=]   -12200

The average number of manufacturing jobs lost each month in 2003 was 12,200.

Section 2.5 STUDY SET

VOCABULARY *Fill in the blanks.*

1. In $\frac{-27}{3} = -9$, the number −9 is called the ________, and the number 3 is the ________.
2. Division by 0 is ________. Division ____ 0 by a nonzero number is 0.
3. The ________ ________ of a number is the distance between it and 0 on the number line.
4. {. . . , −4, −3, −2, −1, 0, 1, 2, 3, 4, . . .} is the set of ________.
5. The quotient of two negative integers is ________.
6. The quotient of a negative integer and a positive integer is ________.

CONCEPTS

7. Write the related multiplication statement for $\frac{-25}{5} = -5$.
8. Write the related multiplication statement for $\frac{0}{-15} = 0$.
9. Show that there is no answer for $\frac{-6}{0}$ by writing the related multiplication statement.
10. Find the value of $\frac{0}{5}$.
11. Write a related division statement for $5(-4) = -20$.
12. How do the rules for multiplying integers compare with the rules for dividing integers?
13. Determine whether each statement is always true, sometimes true, or never true.
    a. The product of a positive integer and a negative integer is negative.
    b. The sum of a positive integer and a negative integer is negative.
    c. The quotient of a positive integer and a negative integer is negative.
14. Determine whether each statement is always true, sometimes true, or never true.
    a. The product of two negative integers is positive.
    b. The sum of two negative integers is negative.
    c. The quotient of two negative integers is negative.

PRACTICE *Find each quotient, if possible.*

15. $\frac{-14}{2}$
16. $\frac{-10}{5}$
17. $\frac{-8}{-4}$
18. $\frac{-12}{-3}$
19. $\frac{-25}{-5}$
20. $\frac{-36}{-12}$
21. $\frac{-45}{-15}$
22. $\frac{-81}{-9}$
23. $\frac{40}{-2}$
24. $\frac{35}{-7}$
25. $\frac{50}{-25}$
26. $\frac{80}{-40}$

27. $\frac{0}{-16}$

28. $\frac{0}{-6}$

29. $\frac{-6}{0}$

30. $\frac{-8}{0}$

31. $\frac{-5}{1}$

32. $\frac{-9}{1}$

33. $-5 \div (-5)$

34. $-11 \div (-11)$

35. $\frac{-9}{9}$

36. $\frac{-15}{15}$

37. $\frac{-10}{-1}$

38. $\frac{-12}{-1}$

39. $\frac{-100}{25}$

40. $\frac{-100}{50}$

41. $\frac{75}{-25}$

42. $\frac{300}{-100}$

43. $\frac{-500}{-100}$

44. $\frac{-60}{-30}$

45. $\frac{-200}{50}$

46. $\frac{-500}{100}$

47. Find the quotient of -45 and 9.

48. Find the quotient of -36 and -4.

49. Divide 8 by -2.

50. Divide -16 by -8.

Use a calculator to perform each division.

51. $\frac{-13{,}550}{25}$

52. $\frac{-3{,}876}{-19}$

53. $\frac{272}{-17}$

54. $\frac{-6{,}776}{-77}$

APPLICATIONS *Use signed numbers to help solve each problem.*

55. TEMPERATURE DROP During a five-hour period, the temperature steadily dropped as shown. What was the average change in the temperature per hour over this five-hour time span?

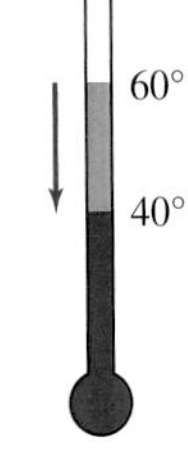

56. PRICE DROPS Over a three-month period, the price of a DVR steadily fell as shown. What was the average monthly change in the price of the DVR over this period?

57. SUBMARINE DIVES In a series of three equal dives, a submarine is programmed to reach a depth of 3,030 feet below the ocean surface. What signed number describes how deep each of the three dives will be?

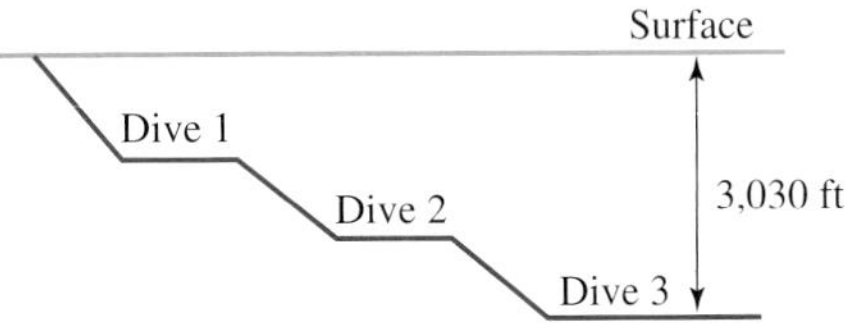

58. GRAND CANYON A mule train is to travel from a stable on the rim of the Grand Canyon to a camp on the canyon floor, approximately 5,000 feet below the rim. If the guide wants the mules to be rested after every 1,000 feet of descent, how many stops will be made on the trip?

59. BASEBALL TRADES At the midway point of the season, a baseball team finds itself 12 games behind the league leader. Team management decides to trade for a talented hitter, in hopes of making up at least half of the deficit in the standings by the end of the year. Where in the league standings does management expect to finish at season's end?

60. BUDGET DEFICITS A politician proposed a two-year plan for cutting a county's $20-million budget deficit, as shown. If this plan is put into effect, what will be the change in the county's financial status in two years?

	Plan	Prediction
1st year	Raise taxes, drop subsidy programs	Will cut deficit in half
2nd year	Search out waste and fraud	Will cut remaining deficit in half

61. MARKDOWNS The owner of a clothing store decides to reduce the price on a line of jeans that are not selling. She feels she can afford to lose $300 of projected income on these pants. By how much can she mark down each of the 20 pairs of jeans?

62. WATER RESERVOIRS Over a week's time, engineers at a city water reservoir released enough water to lower the water level 35 feet. On average, how much did the water level change each day during this period?

63. PAY CUTS In a cost-cutting effort, a business decides to lower expenditures on salaries by $9,135,000. To do this, all of the 5,250 employees will have their salaries reduced by an equal dollar amount. How big a pay cut will each employee experience?

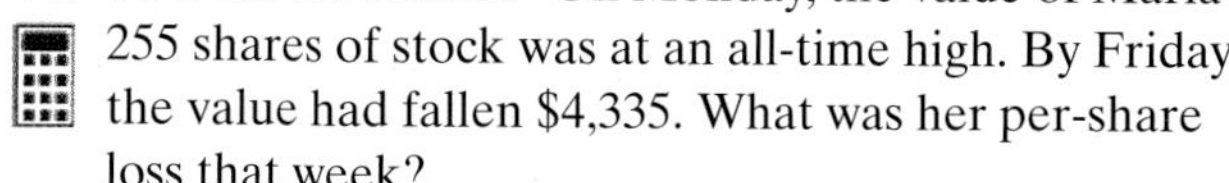

64. STOCK MARKET On Monday, the value of Maria's 255 shares of stock was at an all-time high. By Friday, the value had fallen \$4,335. What was her per-share loss that week?

WRITING

65. Explain why the quotient of two negative numbers is positive.

66. Think of a real-life situation that could be represented by $\frac{0}{4}$. Explain why the answer would be 0.

67. Using a specific example, explain how multiplication can be used as a check for division.

68. Explain what it means when we say that division by 0 is undefined.

REVIEW

69. Evaluate: $3\left(\frac{18}{3}\right)^2 - 2(2)$.

70. List the set of whole numbers.

71. Find the prime factorization of 210.

72. The statement $(4 + 8) + 10 = 4 + (8 + 10)$ illustrates what property?

73. Is the following true? $17 \geq 17$

74. Does $8 - 2 = 2 - 8$?

75. Solve: $99 = r - 43$.

76. Sharif has scores of 55, 70, 80, and 75 on four mathematics tests. What is his mean (average) score?

2.6 Order of Operations and Estimation

• Order of operations • Absolute value • Estimation

In this section, we evaluate expressions involving more than one operation. To do this, we apply the rules for the order of operations and the rules for working with integers. We also continue the discussion of estimating an answer. Estimation can be used when you need a quick indication of the size of the actual answer to a calculation.

Order of operations

In Section 1.5, we introduced the following rules for the order of operations: an agreed-upon sequence of steps for completing the operations of arithmetic.

Order of operations

1. Perform all calculations within parentheses and other grouping symbols in the following order listed in steps 2–4, working from innermost pair to outermost pair.
2. Evaluate all exponential expressions.
3. Perform all multiplications and divisions as they occur from left to right.
4. Perform all additions and subtractions as they occur from left to right.

When all grouping symbols have been removed, repeat steps 2–4 to complete the calculation.

If a fraction bar is present, evaluate the expression above the bar (the *numerator*) and the expression below the bar (the *denominator*) separately. Then perform the division indicated by the fraction bar, if possible.

EXAMPLE 1 Evaluate: $-4(-3)^2 - (-2)$.

Solution This expression contains the operations of multiplication, raising to a power, and subtraction. Since there are no calculations to be made *within* parentheses, we begin by evaluating the exponential expression.

Self Check 1
Evaluate: $-5(-2)^2 - (-6)$.

$$\begin{aligned} -4(\mathbf{-3})^2 - (-2) &= -4(\mathbf{9}) - (-2) && \text{Evaluate the exponential expression: } (-3)^2 = 9.\\ &= -36 - (-2) && \text{Perform the multiplication: } -4(9) = -36.\\ &= -36 + 2 && \text{To subtract, add the opposite of } -2.\\ &= -34 && \text{Perform the addition.} \end{aligned}$$

Answer -14

Self Check 2
Evaluate: $4(2) + (-4)(-3)(-2)$.

Answer -16

EXAMPLE 2 Evaluate: $2(3) + (-5)(-3)(-2)$.

Solution This expression contains the operations of multiplication and addition. By the rules for the order of operations, we perform the multiplications first.

$$\begin{aligned} 2(3) + (-5)(-3)(-2) &= 6 + (-30) && \text{Perform the multiplications from left to right.}\\ &= -24 && \text{Perform the addition.} \end{aligned}$$

Self Check 3
Evaluate: $45 \div (-5)3$.

Answer -27

EXAMPLE 3 Evaluate: $40 \div (-4)5$.

Solution This expression contains the operations of division and multiplication. We perform the divisions and multiplications as they occur from left to right.

$$\begin{aligned} \mathbf{40 \div (-4)}5 &= \mathbf{-10} \cdot 5 && \text{Perform the division first: } 40 \div (-4) = -10.\\ &= -50 && \text{Perform the multiplication.} \end{aligned}$$

Self Check 4
Evaluate: $-3^2 - (-3)^2$.

Answer -18

EXAMPLE 4 Evaluate: $-2^2 - (-2)^2$.

Solution This expression contains the operations of raising to a power and subtraction. We are to find the powers first. (Recall that -2^2 means the *opposite* of 2^2.)

$$\begin{aligned} -2^2 - (-2)^2 &= -4 - 4 && \text{Find the powers: } -2^2 = -4 \text{ and } (-2)^2 = 4.\\ &= -8 && \text{Perform the subtraction.} \end{aligned}$$

Self Check 5
Evaluate: $-18 + 4(-7 + 9)$.

Answer -10

EXAMPLE 5 Evaluate: $-15 + 3(-4 + 7)$.

Solution

$$\begin{aligned} -15 + 3(\mathbf{-4 + 7}) &= -15 + 3(\mathbf{3}) && \text{Perform the addition within the parentheses.}\\ &= -15 + 9 && \text{Perform the multiplication: } 3(3) = 9.\\ &= -6 && \text{Perform the addition.} \end{aligned}$$

Self Check 6
Evaluate: $-2 + (9 - 3)^2$.

Answer 34

EXAMPLE 6 Evaluate: $-10 + (8 - 4)^2$.

Solution By the rules for the order of operations, we must perform the operation within the parentheses first.

$$\begin{aligned} -10 + (\mathbf{8 - 4})^2 &= -10 + (\mathbf{4})^2 && \text{Perform the subtraction within the parentheses: } 8 - 4 = 4.\\ &= -10 + 16 && \text{Evaluate the exponential expression: } (4)^2 = 16.\\ &= 6 && \text{Perform the addition.} \end{aligned}$$

EXAMPLE 7 Evaluate: $\frac{-20 + 3(-5)}{(-4)^2 - 21}$.

Solution We first evaluate the expressions in the numerator and the denominator, separately.

$\frac{-20 + 3(-5)}{(-4)^2 - 21} = \frac{-20 + (-15)}{16 - 21}$ In the numerator, perform the multiplication: $3(-5) = -15$. In the denominator, evaluate the power: $(-4)^2 = 16$.

$= \frac{-35}{-5}$ In the numerator, add: $-20 + (-15) = -35$. In the denominator, subtract: $16 - 21 = -5$.

$= 7$ Perform the division.

Self Check 7

Evaluate: $\frac{-9 + 6(-4)}{(-5)^2 - 28}$.

Answer 11

EXAMPLE 8 Evaluate: $-5[-1 + (2 - 8)^2]$.

Solution We begin by working within the innermost pair of grouping symbols and work to the outermost pair.

$-5[-1 + (2 - 8)^2] = -5[-1 + (-6)^2]$ Perform the subtraction within the parentheses.

$= -5(-1 + 36)$ Evaluate the power within the brackets.

$= -5(35)$ Perform the addition within the parentheses.

$= -175$ Perform the multiplication.

Self Check 8

Evaluate: $-4[-2 + (5 - 9)^2]$.

Answer -56

Absolute value

You will recall that the absolute value of a number is the distance between the number and 0 on the number line. Earlier in this chapter, we evaluated simple absolute value expressions such as $|-3|$ and $|10|$. Absolute value symbols are also used in combination with more complicated expressions, such as $|-4(3)|$ and $|-6 + 1|$. When we apply the rules for the order of operations to evaluate these expressions, *the absolute value symbols are considered to be grouping symbols,* and any operations within them are to be completed first.

EXAMPLE 9 Find each absolute value: **a.** $|-4(3)|$ and **b.** $|-6 + 1|$.

Solution We perform the operations within the absolute value symbols first.

a. $|-4(3)| = |-12|$ Perform the multiplication within the absolute value symbol: $-4(3) = -12$.

$= 12$ Find the absolute value of -12.

b. $|-6 + 1| = |-5|$ Perform the addition within the absolute value symbol: $-6 + 1 = -5$.

$= 5$ Find the absolute value of -5.

Self Check 9

Find each absolute value:

a. $|(-6)(5)|$

b. $|-3 + (-26)|$

Answers **a.** 30, **b.** 29

! COMMENT Just as $-5(8)$ means $-5 \cdot 8$, the expression $-5|8|$ (read as "negative 5 times the absolute value of 8") means $-5 \cdot |8|$. To evaluate such an expression, we find the absolute value first and then multiply.

$$
\begin{aligned}
-5|8| &= -5 \cdot 8 && \text{Find the absolute value: } |8| = 8. \\
&= -40 && \text{Perform the multiplication.}
\end{aligned}
$$

Self Check 10
Evaluate: $7 - 5|-1 - 6|$.

Answer -28

EXAMPLE 10 Evaluate: $8 - 4|-6 - 2|$.

Solution We perform the operation within the absolute value symbol first.

$$
\begin{aligned}
8 - 4|\mathbf{-6 - 2}| &= 8 - 4|\mathbf{-8}| && \text{Perform the subtraction within the absolute value symbol: } -6 - 2 = -8. \\
&= 8 - 4(8) && \text{Find the absolute value: } |-8| = 8. \\
&= 8 - 32 && \text{Perform the multiplication: } 4(8) = 32. \\
&= -24 && \text{Perform the subtraction.}
\end{aligned}
$$

Estimation

Recall that the idea behind estimation is to simplify calculations by using rounded numbers that are close to the actual values in the problem. When an exact answer is not necessary and a quick approximation will do, we can use estimation.

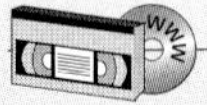

EXAMPLE 11 **The stock market.** The Dow Jones Industrial Average is announced at the end of each trading day to give investors an indication of how the New York Stock Exchange performed. A positive number indicates good performance, while a negative number indicates poor performance. Estimate the net gain or loss of points in the Dow for the week shown in Figure 2-21.

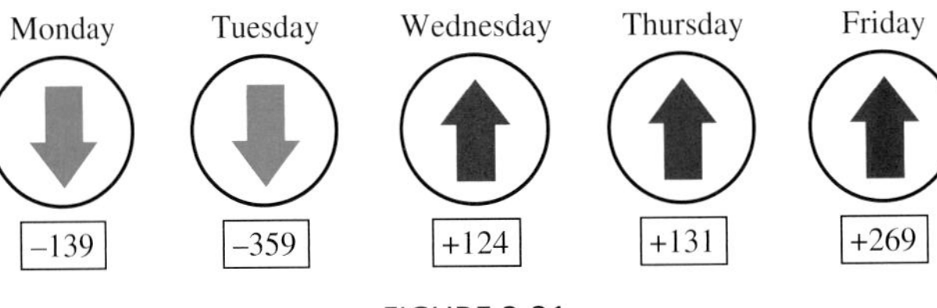

FIGURE 2-21

Solution We will approximate each of these numbers. For example, -139 is close to -140, and $+124$ is close to $+120$. To estimate the net gain or loss, we add the approximations.

$$
\begin{aligned}
-140 + (-360) + 120 + 130 + 270 &= -500 + 520 && \text{Add positive and negative numbers separately to get subtotals.} \\
&= 20 && \text{Perform the addition.}
\end{aligned}
$$

This estimate tells us that there was a gain of approximately 20 points in the Dow.

Section 2.6 STUDY SET

VOCABULARY *Fill in the blanks.*

1. When asked to evaluate expressions that contain more than one operation, we should apply the rules for the _______ of operations.
2. In situations where an exact answer is not needed, an approximation or _________ is a quick way of obtaining a rough idea of the size of the actual answer.
3. Absolute value symbols, parentheses, and brackets are types of _________ symbols.
4. If an expression involves two sets of grouping symbols, always begin working within the _________ symbols and then work to the _________.

CONCEPTS

5. Consider $5(-2)^2 - 1$. How many operations need to be performed to evaluate this expression? List them in the order in which they should be performed.
6. Consider $15 - 3 + (-5 \cdot 2)^3$. How many operations need to be performed to evaluate this expression? List them in the order in which they should be performed.
7. Consider $\frac{5 + 5(7)}{2 + (4 - 8)}$. In the numerator, what operation should be performed first? In the denominator, what operation should be performed first?
8. In the expression $4 + 2(-7 - 1)$, how many operations need to be performed? List them in the order in which they should be performed.
9. Explain the difference between -3^2 and $(-3)^2$.
10. In the expression $-2 \cdot 3^2$, what operation should be performed first?

NOTATION *Complete each solution to evaluate the expression.*

11. $-8 - 5(-2)^2 = -8 - 5(\square)$
$= -8 - \square$
$= -8 + (\square)$
$= \square$

12. $2 + (5 - 6 \cdot 2) = 2 + (5 - \square)$
$= 2 + [5 + (\square)]$
$= 2 + (\square)$
$= \square$

13. $[-4(2 + 7)] - 6 = [-4(\square)] - 6$
$= \square - 6$
$= \square$

14. $\frac{|-9 + (-3)|}{9 - 6} = \frac{|\square|}{3}$
$= \frac{\square}{3}$
$= \square$

PRACTICE *Evaluate each expression.*

15. $(-3)^2 - 4^2$
16. $-7 + 4 \cdot 5$
17. $3^2 - 4(-2)(-1)$
18. $2^3 - 3^3$
19. $(2 - 5)(5 + 2)$
20. $-3(2)^2 4$
21. $-10 - 2^2$
22. $-50 - 3^3$
23. $\frac{-6 - 8}{2}$
24. $\frac{-6 - 6}{-2 - 2}$
25. $\frac{-5 - 5}{2}$
26. $\frac{-7 - (-3)}{2 - 4}$
27. $-12 \div (-2)2$
28. $-60(-2) \div 3$
29. $-16 - 4 \div (-2)$
30. $-24 + 4 \div (-2)$
31. $|-5(-6)|$
32. $|-7 - 9|$
33. $|-4 - (-6)|$
34. $|-2 + 6 - 5|$
35. $5|3|$
36. $5|4|$
37. $-6|-7|$
38. $-6|-4|$
39. $(7 - 5)^2 - (1 - 4)^2$
40. $5^2 - (-9 - 3)$
41. $-1(2^2 - 2 + 1^2)$
42. $(-7 - 4)^2 - (-1)$
43. $-50 - 2(-3)^3$
44. $(-2)^3 - (-3)(-2)$
45. $-6^2 + 6^2$
46. $-9^2 + 9^2$
47. $3\left(\frac{-18}{3}\right) - 2(-2)$
48. $2\left(\frac{-12}{3}\right) + 3(-5)$
49. $6 + \frac{25}{-5} + 6 \cdot 3$
50. $-5 - \frac{24}{6} + 8(-2)$

51. $\dfrac{1 - 3^2}{-2}$

52. $\dfrac{-3 - (-7)}{2^2 - 3}$

53. $\dfrac{-4(-5) - 2}{-6}$

54. $\dfrac{(-6)^2 - 1}{-4 - 3}$

55. $-3\left(\dfrac{32}{-4}\right) - (-1)^5$

56. $-5\left(\dfrac{16}{-4}\right) - (-1)^4$

57. $6(2^3)(-1)$

58. $2(3^3)(-2)$

59. $2 + 3[5 - (1 - 10)]$

60. $12 - 2[1 - (-8 + 2)]$

61. $-7(2 - 3 \cdot 5)$

62. $-4(1 + 3 \cdot 5)$

63. $-[6 - (1 - 4)^2]$

64. $-[9 - (9 - 12)^2]$

65. $15 + (-3 \cdot 4 - 8)$

66. $11 + (-2 \cdot 2 + 3)$

67. $|-3 \cdot 4 + (-5)|$

68. $|-8 \cdot 5 - 2 \cdot 5|$

69. $|(-5)^2 - 2 \cdot 7|$

70. $|8 \div (-2) - 5|$

71. $-2 + |6 - 4^2|$

72. $-3 - 4|6 - 7|$

73. $2|1 - 8| \cdot |-8|$

74. $2(5) - 6(|-3|)^2$

75. $-2(-34)^2 - (-605)$

75. $11 - (-15)(24)^2$

77. $-60 - \dfrac{1{,}620}{-36}$

78. $\dfrac{2^5 - 4^6}{-42 + 58}$

79. $-30 + (7 - 2)^2$

80. $-11 + (4 - 1)^2$

81. $(3 - 4)^2 - (3 - 9)^2$

82. $(5 - 9)^2 - (2 - 10)^2$

Make an estimate.

83. $-379 + (-103) + 287$

84. $\dfrac{-67 - 9}{-18}$

85. $-39 \cdot 8$

86. $-568 - (-227)$

87. $-3{,}887 + (-5{,}106)$

88. $-333(-4)$

89. $\dfrac{6{,}267}{-5}$

90. $-36 + (-78) + 59 + (-4)$

APPLICATIONS

91. TESTING In an effort to discourage her students from guessing on multiple-choice tests, a professor uses the grading scale shown in the table in the next column. If unsure of an answer, a student does best to skip the question, because incorrect responses are penalized very heavily. Find the test score of a student who gets 12 correct and 3 wrong and leaves 5 questions blank.

Response	Value
Correct	+3
Incorrect	−4
Left blank	−1

92. THE FEDERAL BUDGET See below. Suppose you were hired to write a speech for a politician who wanted to highlight the improvement in the federal government's finances during 1990s. Would it be better for the politician to refer to the average budget deficit/surplus for the last half, or for the last four years of that decade? Explain your reasoning.

U.S. Budget Deficit/Surplus
($ billions)

Deficit	Year	Surplus
−164	1995	
−107	1996	
−22	1997	
	1998	+70
	1999	+123

93. SCOUTING REPORTS The illustration shows a football coach how successful his opponent was running a "28 pitch" the last time the two teams met. What was the opponent's average gain with this play?

Play: 28 pitch

Gain 16 yd	Gain 10 yd	Loss 2 yd	No gain
Gain 4 yd	Loss 4 yd	TD Gain 66 yd	Loss 2 yd

94. SPREADSHEETS The table shows the data from a chemistry experiment in spreadsheet form. To obtain a result, the chemist needs to add the values in row 1, double that sum, and then divide that number by the smallest value in column C. What is the final result of these calculations?

	A	B	C	D
1	12	−5	6	−2
2	15	4	5	−4
3	6	4	−2	8

Estimate the answer to each question.

95. OIL PRICES The price per barrel of crude oil fluctuates with supply and demand. It can rise and fall quickly. The line graph shows how many cents the price per barrel rose or fell each day for a week. For example, on Monday the price rose 68 cents, and on Tuesday it rose an additional 91 cents. Estimate the net gain or loss in the value of a barrel of crude oil for the week.

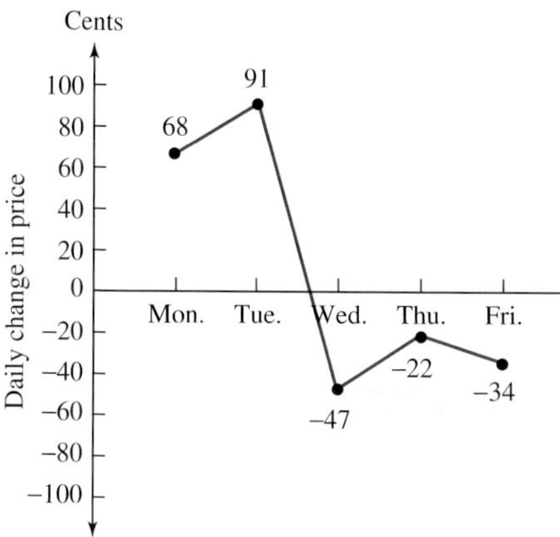

96. ESTIMATION Quickly determine a reasonable estimate of the exact answer in each of the following situations.

a. A diver, swimming at a depth of 34 feet below sea level, spots a sunken ship beneath him. He dives down another 57 feet to reach it. What is the depth of the sunken ship?

b. A dental hygiene company offers a money-back guarantee on its tooth whitener kit. When the kit is returned by a dissatisfied customer, the company loses the $11 it cost to produce it, because it cannot be resold. How much money has the company lost because of this return policy if 56 kits have been mailed back by customers?

c. A tram line makes a 7,561-foot descent from a mountaintop in 18 equal stages. How much does it descend in each stage?

WRITING

97. When evaluating expressions, why are rules for the order of operations necessary?

98. In the rules for the order of operations, what does the phrase *as they occur from left to right* mean?

99. Name a situation in daily life where you use estimation.

100. List some advantages and some disadvantages of the process of estimation.

REVIEW

101. Round 5,456 to the nearest thousand.

102. Evaluate: $6^2 - (10 - 8)$.

103. How do we find the perimeter of a rectangle?

104. Is this statement true or false? "When measuring perimeter, we use square units."

105. An elevator has a weight capacity of 1,000 pounds. Seven people, with an average weight of 140 pounds, are in it. Is it overloaded?

106. List the factors of 36.

2.7 Solving Equations Involving Integers

- The properties of equality • Solving equations involving $-x$
- Combinations of operations • Problem solving with equations

In this section, we revisit the topic of solving equations. The equations that we will be solving involve negative numbers, and some of the solutions are negative numbers as well. The section concludes with an application problem.

The properties of equality

Recall that the addition property of equality states: *If the same number is added to equal quantities, the results will be equal quantities.* When working with negative numbers, we can use this property in a new way.

To solve $x + (-8) = -10$, we need to isolate x on the left-hand side of the equation. We can do this by adding 8 to both sides.

$$x + (-8) = -10$$

$$x + (-8) \mathbf{+ 8} = -10 \mathbf{+ 8}$$ Use the addition property of equality. To undo the addition of -8, add 8 to both sides.

$$x = -2$$ Perform the additions: $(-8) + 8 = 0$ and $-10 + 8 = -2$.

From this example, we see that to undo addition, we can *add the opposite* of the number that is added to the variable.

Self Check 1
Solve $-6 + y = -8$ and check the result.

Answer -2

EXAMPLE 1 Solve: $-4 + x = -12$.

Solution We want to isolate x on one side of the equation. We can do this by adding the opposite of -4 to both sides.

$$-4 + x = -12$$

$$-4 + x \mathbf{+ 4} = -12 \mathbf{+ 4}$$ To undo the addition of -4, add 4 to both sides.

$$x = -8$$ Perform the additions: $-4 + 4 = 0$ and $-12 + 4 = -8$.

To check, we substitute -8 for x in the original equation and then simplify.

$$-4 + \mathbf{x} = -12$$ This is the original equation.

$$-4 + (\mathbf{-8}) \stackrel{?}{=} -12$$ Substitute -8 for x.

$$-12 = -12$$ Perform the addition: $-4 + (-8) = -12$.

Since the statement $-12 = -12$ is true, the solution is -8.

Recall that the subtraction property of equality states: *If the same quantity is subtracted from equal quantities, the results will be equal quantities.*

Self Check 2
Solve: $c + 4 = -3$.

Answer -7

EXAMPLE 2 Solve: $x + 16 = -8$.

Solution

$$x + 16 = -8$$

$$x + 16 \mathbf{- 16} = -8 \mathbf{- 16}$$ To undo the addition of 16, subtract 16 from both sides.

$$x = -8 + (-16)$$ Simplify the left side of the equation. On the right side, write the subtraction as addition of the opposite.

$$x = -24$$ Perform the addition.

Check the result.

Self Check 3
Solve $-2 + 8 = y + 3(-4)$ and check the result.

EXAMPLE 3 Solve: $-3 + 7 = h + 11(-2)$.

Solution The expressions on either side of the equation should be simplified before we use the properties of equality. There is an addition to perform on the left-hand side and a multiplication to perform on the right-hand side of the equation.

$$-3 + 7 = h + 11(-2)$$

$$4 = h + (-22)$$ Perform the addition: $-3 + 7 = 4$. Perform the multiplication: $11(-2) = -22$.

$$\mathbf{4 + 22} = h + (-22) \mathbf{+ 22}$$ To undo the addition of -22, add 22 to both sides.

$$26 = h$$ Simplify: $4 + 22 = 26$ and $(-22) + 22 = 0$.

$$h = 26$$ Since $26 = h$, $h = 26$.

Check:

$-3 + 7 = \mathbf{h} + 11(-2)$	This is the original equation.
$-3 + 7 \stackrel{?}{=} \mathbf{26} + 11(-2)$	Substitute 26 for *h*.
$4 \stackrel{?}{=} 26 + (-22)$	Simplify both sides.
$4 = 4$	Perform the addition.

Answer 18

Recall that the division property of equality states: *If equal quantities are divided by the same nonzero quantity, the results will be equal quantities.*

EXAMPLE 4 Solve: **a.** $-3x = 15$ and **b.** $-16 = -4y$.

Self Check 4

Solve each equation and check the result:

a. $-7k = 28$

b. $-40 = -8f$

Solution

a. Recall that $-3x$ indicates multiplication: $-3 \cdot x$. We must undo the multiplication of x by -3. To do this, we divide both sides of the equation by -3.

$-3x = 15$	This is the original equation.
$\frac{-3x}{\mathbf{-3}} = \frac{15}{\mathbf{-3}}$	Divide both sides by -3.
$x = -5$	-3 times x, divided by -3, is x. On the right-hand side, perform the division: $15 \div (-3) = -5$.

Check:

$-3x = 15$	This is the original equation.
$-3(\mathbf{-5}) \stackrel{?}{=} 15$	Substitute -5 for x.
$15 = 15$	Perform the multiplication: $-3(-5) = 15$.

b. $-16 = -4y$

$\frac{-16}{\mathbf{-4}} = \frac{-4y}{\mathbf{-4}}$	To undo the multiplication by -4, divide both sides by -4.
$4 = y$	Perform the divisions.
$y = 4$	If $4 = y$, then $y = 4$.

Check the result.

Answers **a.** -4, **b.** 5

Recall that the multiplication property of equality states: *If equal quantities are multiplied by the same nonzero quantity, the results will be equal quantities.*

EXAMPLE 5 Solve: $\frac{x}{-5} = -10$.

Self Check 5

Solve $\frac{t}{-3} = 4$ and check the result.

Solution In this equation, x is being divided by -5. To undo this division, we multiply both sides of the equation by -5.

$\frac{x}{-5} = -10$	
$\mathbf{-5}\left(\frac{x}{-5}\right) = \mathbf{-5}(-10)$	Use the multiplication property of equality. Multiply both sides by -5.
$x = 50$	When x is divided by -5 and then multiplied by -5, the result is x. Perform the multiplication: $-5(-10) = 50$.

Check:

$$\frac{x}{-5} = -10 \quad \text{This is the original equation.}$$

$$\frac{\mathbf{50}}{-5} \stackrel{?}{=} -10 \quad \text{Substitute 50 for } x.$$

$$-10 = -10 \quad \text{Perform the division.}$$

The solution is 50.

Answer -12

Solving equations involving $-x$

Consider the equation $-x = 3$. The variable x is not isolated, because there is a $-$ sign in front of it. The notation $-x$ means -1 times x. Therefore, the equation $-x = 3$ can be rewritten as $-1x = 3$. To isolate the variable, we can either multiply both sides by -1 or divide both sides by -1.

$-x = 3$		$-x = 3$	
$-1x = 3$	$-x = -1x$.	$-1x = 3$	$-x = -1x$.
$(\mathbf{-1})(-1x) = (\mathbf{-1})3$	Multiply both sides by -1.	$\frac{-1x}{\mathbf{-1}} = \frac{3}{\mathbf{-1}}$	Divide both sides by -1.
$1x = -3$	$-1(-1) = 1$.	$x = -3$	$-1x \div (-1) = x$; $3 \div (-1) = -3$.
$x = -3$	$1x = x$.		

Self Check 6

Solve $-h = -10$ and check the result.

EXAMPLE 6 Solve: $-x = -9$.

Solution

$$-x = -9$$

$$-1x = -9 \quad -x = -1x.$$

$$\mathbf{-1}(-1x) = \mathbf{-1}(-9) \quad \text{Multiply both sides by } -1.$$

$$x = 9 \quad \text{Perform the multiplications: } -1(-1x) = x \text{ and } -1(-9) = 9.$$

Check:

$$-x = -9 \quad \text{This is the original equation.}$$

$$-(\mathbf{9}) \stackrel{?}{=} -9 \quad \text{Substitute 9 for } x.$$

$$-9 = -9$$

The solution is 9.

Answer 10

Combinations of operations

In the previous examples, each equation was solved by using a single property of equality. If the equation is more complicated, it is usually necessary to use several properties of equality to solve it. For example, consider the equation $2x + 5 = 9$ and the operations performed on x.

$$2x + 5 = 9$$

The variable is multiplied by 2. (arrow to $2x$)

Then 5 is added. (arrow to 5)

To solve this equation, we use the rules for the order of operations in reverse.

- First, use the subtraction property of equality to undo the addition of 5.
- Second, apply the division property of equality to undo the multiplication by 2.

$$
\begin{aligned}
2x + 5 &= 9 \\
2x + 5 - \mathbf{5} &= 9 - \mathbf{5} && \text{To undo the addition of 5, subtract 5 from both sides.} \\
2x &= 4 && \text{Perform the subtraction on each side: } 5 - 5 = 0 \text{ and } 9 - 5 = 4. \\
\frac{2x}{\mathbf{2}} &= \frac{4}{\mathbf{2}} && \text{To undo the multiplication by 2, divide both sides by 2.} \\
x &= 2 && \text{Perform the divisions.}
\end{aligned}
$$

EXAMPLE 7 Solve: $-4x - 5 = 15$.

Self Check 7
Solve $-6b - 1 = 11$ and check the result.

Solution The operations performed on x are multiplication by -4 and subtraction of 5. We undo these operations in the reverse order.

$$
\begin{aligned}
-4x - 5 &= 15 \\
-4x - 5 + \mathbf{5} &= 15 + \mathbf{5} && \text{Add 5 to both sides.} \\
-4x &= 20 && \text{Perform the addition on each side: } -5 + 5 = 0 \text{ and } 15 + 5 = 20. \\
\frac{-4x}{\mathbf{-4}} &= \frac{20}{\mathbf{-4}} && \text{Divide both sides by } -4. \\
x &= -5 && \text{Perform the divisions.}
\end{aligned}
$$

Check:

$$
\begin{aligned}
-4x - 5 &= 15 && \text{This is the original equation.} \\
-4(\mathbf{-5}) - 5 &\stackrel{?}{=} 15 && \text{Substitute } -5 \text{ for } x. \\
20 - 5 &\stackrel{?}{=} 15 && \text{Perform the multiplication: } -4(-5) = 20. \\
15 &= 15 && \text{Perform the subtraction.}
\end{aligned}
$$

The solution is -5.

Answer -2

EXAMPLE 8 Solve: $2 - 3p = -1$.

Self Check 8
Solve: $6 - 8k = -34$.

Solution We begin by writing the subtraction on the left-hand side as addition of the opposite.

$$
\begin{aligned}
2 - 3p &= -1 \\
-3p + 2 &= -1 && \text{Rewrite the left-hand side of the equation.} \\
-3p + 2 - \mathbf{2} &= -1 - \mathbf{2} && \text{To undo the addition of 2 on the left-hand side of the equation, subtract 2 from both sides.} \\
-3p &= -3 && \text{Simplify.} \\
\frac{-3p}{\mathbf{-3}} &= \frac{-3}{\mathbf{-3}} && \text{To undo the multiplication by } -3, \text{ divide both sides by } -3. \\
p &= 1 && \text{Perform the divisions.}
\end{aligned}
$$

Check this result in the *original* equation.

Answer 5

EXAMPLE 9 Solve: $\frac{y}{-2} - 6 = -18$.

Self Check 9
Solve: $\frac{m}{-8} - 10 = -14$.

Solution The operations performed on y are division by -2 and subtraction of 6. We undo these operations in the opposite order.

$$\frac{y}{-2} - 6 = -18$$

$$\frac{y}{-2} - 6 + \mathbf{6} = -18 + \mathbf{6}$$ To undo the subtraction of 6, add 6 to both sides.

$$\frac{y}{-2} = -12$$ Simplify both sides of the equation.

$$\mathbf{-2}\left(\frac{y}{-2}\right) = \mathbf{-2}(-12)$$ To undo the division by −2, multiply both sides by −2.

$$y = 24$$ Perform the multiplications.

Check this result in the original equation.

Answer 32

Problem solving with equations

EXAMPLE 10 **Water management.** One day, enough water was released from a reservoir to lower the water level 17 feet to a reading of 33 feet below capacity. What was the water level reading before the release?

Analyze the problem

Figure 2-22 illustrates the given information and what we are asked to find.

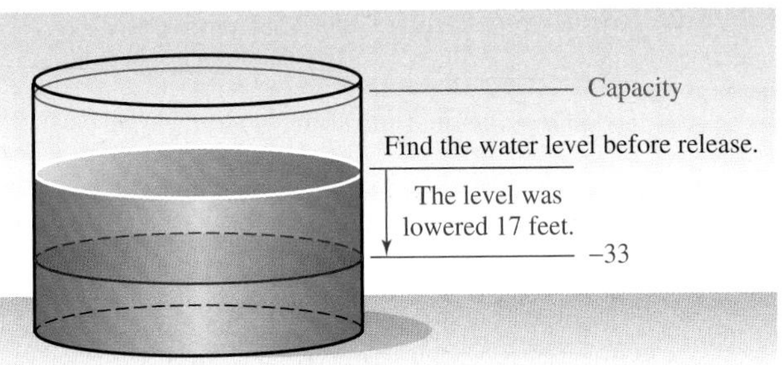

FIGURE 2-22

Form an equation

Let x = the water level reading before the release.

The water level reading before the release	minus	the number of feet the water level was lowered	is	the new water level reading.
x	$-$	17	$=$	-33

Solve the equation

$$x - 17 = -33$$

$$x - 17 + \mathbf{17} = -33 + \mathbf{17}$$ To undo the subtraction of 17, add 17 to both sides.

$$x = -16$$ Perform the additions: $-17 + 17 = 0$ and $-33 + 17 = -16$.

State the conclusion

The water level reading before the release was −16 feet.

Check the result

If the water level reading was initially −16 feet and was then lowered 17 feet, the new reading would be $-16 - 17 = -16 + (-17) = -33$ feet. The answer checks.

Section 2.7 STUDY SET

VOCABULARY *Fill in the blanks.*

1. To ______ an equation, we isolate the variable on one side of the = symbol.
2. When solving an ________, the objective is to find all values of the variable that will make the equation true.

CONCEPTS

3. If we multiply x by -3 and then divide that product by -3, what is the result?
4. If we divide x by -4 and then multiply that quotient by -4, what is the result?
5. In the equation $x + 3 = 10$, we can isolate x in two ways. Find the missing numbers.
   a. $x + 3 - 3 = 10 - \square$
   b. $x + 3 + (-3) = 10 + \square$
6. In the equation $12 + c = 10$, we can isolate c in two ways. Find the missing numbers.
   a. $12 + c - \square = 10 - \square$
   b. $12 + c + (\square) = 10 + (\square)$
7. Determine what operations are performed on the variable x and the order in which they occur.
   a. $-2x = -100$
   b. $-6 + x = -9$
   c. $-4x - 8 = 12$
   d. $-1 = -6 + (-5x)$
8. Determine what operations are performed on the variable x and the order in which they occur.
   a. $\dfrac{x}{-4} - 8 = 50$
   b. $-16 = -5 + \dfrac{x}{-3}$

Fill in the blanks.

9. When solving the equation $t - 4 = -8 - 2$, it is best to ________ the right-hand side of the equation first before undoing any operations performed on the variable.
10. To solve the equation $-2x - 4 = 6$, we first undo the ________ of 4. Then we undo the ________ by -2.
11. When solving an equation, we isolate the variable by undoing the operations performed on it in the ________ order.
12. To solve $-x = 6$, we can multiply or divide both sides of the equation by $\square$.
13. When solving each of these equations, which operation should you undo first?
    a. $-2x - 3 = -19$
    b. $-6 + \dfrac{h}{-3} = -14$
14. When solving each of these equations, which operation should you undo first?
    a. $5 + (-9x) = -1$
    b. $-16 = -9 + \dfrac{t}{7}$

NOTATION *Complete each solution to solve the equation.*

15.
$$\begin{aligned} y + (-7) &= -16 + 3 \\ y + (-7) &= \square \\ y + (-7) + 7 &= -13 + \square \\ y &= -6 \end{aligned}$$

16.
$$\begin{aligned} x - (-4) &= -1 - 5 \\ x + 4 &= -1 + \square \\ x + 4 &= -6 \\ x + 4 - 4 &= -6 - \square \\ x &= -6 + (-4) \\ x &= -10 \end{aligned}$$

17.
$$\begin{aligned} -13 &= -4y - 1 \\ -13 + \square &= -4y - 1 + \square \\ \square &= -4y \\ \frac{-12}{\square} &= \frac{-4y}{\square} \\ \square &= y \\ y &= 3 \end{aligned}$$

18.
$$\begin{aligned} 1 &= \frac{m}{-5} + 6 \\ 1 - \square &= \frac{m}{-5} + 6 - \square \\ -5 &= \frac{m}{-5} \\ \square(-5) &= \square\left(\frac{m}{-5}\right) \\ \square &= m \\ m &= 25 \end{aligned}$$

19. What does $-10x$ mean?
20. What does $\dfrac{x}{-8}$ mean?

PRACTICE *Determine whether the given number is a solution of the equation.*

21. $-3x - 4 = 2;\ -2$
22. $\frac{x}{-2} + 5 = -10;\ 20$
23. $-x + 8 = -4;\ 4$
24. $-3 + 2x = -3;\ 0$

Solve each equation. **Check each result.**

25. $x + 6 = -12$
26. $y + 1 = -4$
27. $-6 + m = -20$
28. $-12 + r = -19$
29. $-5 + 3 = -7 + f$
30. $-10 + 4 = -9 + t$
31. $h - 8 = -9$
32. $x - 1 = -7$
33. $0 = y + 9$
34. $0 = t + 5$
35. $r - (-7) = -1 - 6$
36. $x - (-1) = -4 - 3$
37. $t - 4 = -8 - (-2)$
38. $r - 1 = -3 - (-4)$
39. $x - 5 = -5$
40. $r - 4 = -4$
41. $-2s = 16$
42. $-3t = 9$
43. $-5t = -25$
44. $-6m = -60$
45. $-2 + (-4) = -3n$
46. $-10 + (-2) = -4x$
47. $-9h = -3(-3)$
48. $-6k = -2(-3)$
49. $\frac{t}{-3} = -2$
50. $\frac{w}{-4} = -5$
51. $0 = \frac{y}{8}$
52. $0 = \frac{h}{7}$
53. $\frac{x}{-2} = -6 + 3$
54. $\frac{a}{-5} = -7 + 6$
55. $\frac{x}{4} = -5 - 8$
56. $\frac{r}{2} = -5 - 1$
57. $2y + 8 = -6$
58. $5y + 1 = -9$
59. $-21 = 4h - 5$
60. $-22 = 7l - 8$
61. $-3v + 1 = 16$
62. $-4e + 4 = 24$
63. $8 = -3x + 2$
64. $15 = -2x + (-11)$
65. $-35 = 5 - 4x$
66. $12 = -9 - 3x$
67. $4 - 5x = 34$
68. $15 - 6x = 21$
69. $-5 - 6 - 5x = 4$
70. $-7 - 5 - 7x = 16$
71. $4 - 6x = -5 - 9$
72. $8 - 2d = -5 - 5$
73. $\frac{h}{-6} + 4 = 5$
74. $\frac{p}{-3} + 3 = 8$
75. $-2(4) = \frac{t}{-6} + 1$
76. $-2(5) = \frac{y}{-3} + 3$
77. $0 = 6 + \frac{c}{-5}$
78. $0 = -6 + \frac{s}{-3}$
79. $-1 = -8 + \frac{h}{-2}$
80. $-5 = 4 + \frac{g}{-4}$
81. $2x + 3(0) = -6$
82. $3x - 4(0) = -12$
83. $2(0) - 2y = 4$
84. $5(0) - 2y = 10$
85. $-x = 8$
86. $-y = 12$
87. $-15 = -k$
88. $-4 = -p$

APPLICATIONS *Complete each solution.*

89. SHARKS During a research project, a diver inside a shark cage made observations at a depth of 120 feet. For a second set of observations, the cage was raised to a depth of 75 feet. How many feet was the cage raised between observations?

Analyze the problem

- The first observations were at −120 ft.
- The next observations were at ____ ft.
- We must find ______________________.

Form an equation

Let x = ______________________

Key word: *raised* **Translation:** ________

Translate the words of the problem into an equation.

The first position of the cage	plus	the amount the cage was raised	is	the second position of the cage.
−120	+		=	

Solve the equation

$\quad + x = \quad$

$-120 + x + \quad = -75 + \quad$

$x = 45$

State the conclusion

____

Check the result

If we add the number of feet the cage was raised to the first position, we get $-120 + \quad = -75$. The answer checks.

90. TRACK TIMES In one race, an athlete's time for the mile was 7 seconds under the school record. In a second race, her time continued to drop, to 16 seconds under the old school record. How much time did she drop between the first and second races?

Analyze the problem

- The 1st race was 7 sec under the record.
- The 2nd race was ____ sec under the record.
- We must find

____

Form an equation

Let x = ______________________

Key phrase: *dropped* **Translation:** __________

Translate the words of the problem into an equation.

First race performance	minus	amount of time dropped	is	second race performance.
-7	$-$		$=$	

Solve the equation

$$\quad - x = \quad$$
$$-7 - x + \quad = -16 + \quad$$
$$-x = \quad$$
$$x = 9$$

State the conclusion

____

Check the result

If we subtract the time she dropped in the second race from the time dropped in the first race, we get $-7 - \quad = -16$. The answer checks.

Let a variable represent the unknown quantity. Then write and solve an equation to answer the question.

91. MARKET SHARE After its first year of business, a manufacturer of smoke detectors found its market share 43 points behind the industry leader. Five years later, it trailed the leader by only 9 points. How many points of market share did the company pick up over this five-year span?

92. WEATHER FORECASTS The weather forecast for Fairbanks, Alaska warned listeners that the daytime high temperature of 2° below zero would drop to a nighttime low of 28° below. What was the overnight change in temperature?

93. CHECKING ACCOUNTS After he made a deposit of $220, a student's account was still $215 overdrawn. What was his checking account balance before the deposit?

94. FOOTBALL During the first half of a football game, a team ran for a total of 43 yards. After a dismal second half, they ended the game with a total of -8 yards rushing. What was their rushing total in the second half?

95. POLLS Six months before an election, a political candidate was 31 points behind in the polls. Two days before the election, polls showed that his support had skyrocketed; he was now only 2 points behind. How much support had he gained over the six-month period?

96. HORSE RACING At the midway point of a 6-furlong horse race, the long shot was 3 lengths ahead of the pre-race favorite. In the last half of the race, the long shot lost ground and eventually finished 6 lengths behind the favorite. By how many lengths did the long shot lose to the favorite during the last half of the race?

97. PRICE REDUCTIONS Over the past year, the price of a video game player has dropped each month. If the price fell $60 this year, how much did the price drop each month on average?

98. REBATES A store decided that it could afford to lose some money to promote a new line of sunglasses. A $9 rebate was offered to each customer purchasing these sunglasses. If the rebate program resulted in a loss of $225 for the store, how many customers took advantage of the offer?

99. DREDGING A HARBOR In order to handle larger vessels, port officials are having a harbor deepened by dredging. The harbor bottom is already 47 feet below sea level. After the dredging, the bottom will be 65 feet below sea level. How many feet must be dredged out?

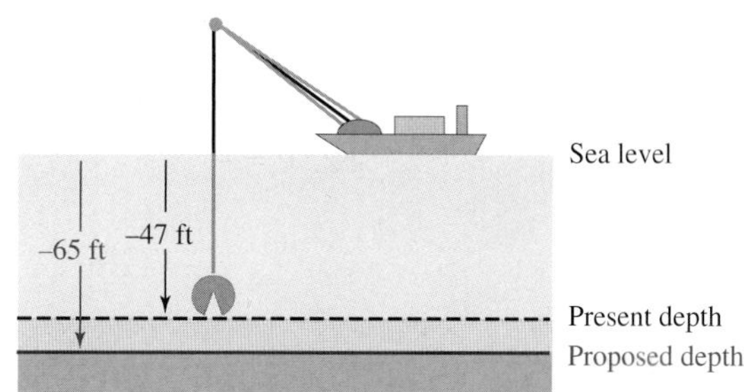

100. ROLLER COASTERS The end of a roller coaster ride consists of a steep plunge from a peak 145 feet above ground level. The car then comes to a screeching halt in a cave that is 25 feet below ground level. How many feet does the roller coaster drop at the end of the ride shown in the illustration below?

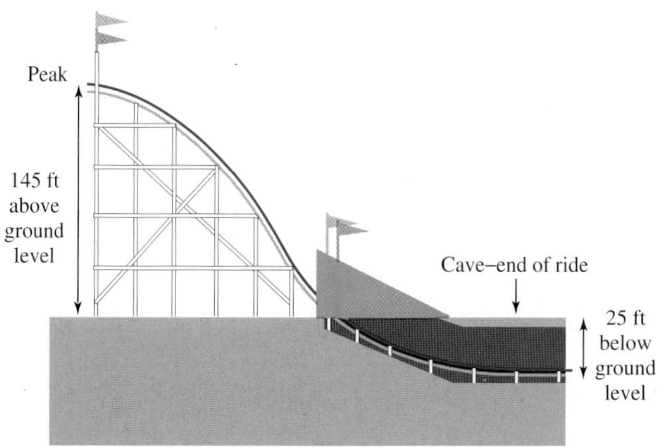

101. INTERNATIONAL TIME ZONES The world is divided into 24 times zones. Each zone is one hour ahead of or behind its neighboring zones. In the portion of the world time zone map shown, we see that Tokyo is in zone +9. What time zone is Seattle in if it is 17 hours behind Tokyo?

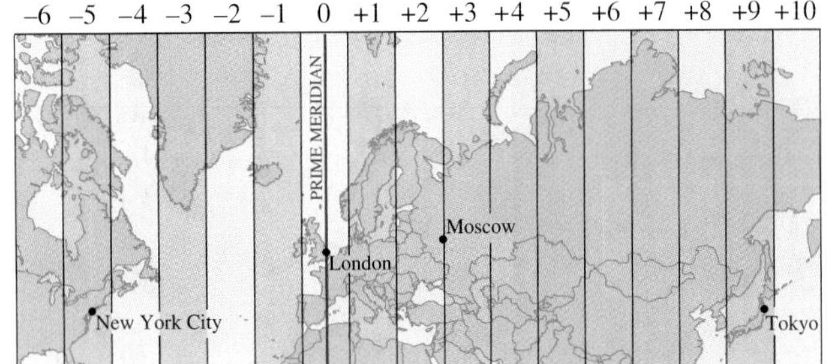

102. PROFITS AND LOSSES In its first year of business, a nursery suffered a loss due to frost damage, ending the year \$11,560 in the red. In the second year, it made a sizable profit. If the total profit for the first two years in business was \$32,090, how much profit was made the second year?

WRITING

103. Explain why the variable is not isolated in the equation $-x = 10$.

104. Explain how to check the result after solving an equation.

REVIEW

105. Write 5^6 without using exponents.

106. Give the definition of an even whole number.

107. Solve: $7 + 3y = 43$.

108. How can the addition $2 + 2 + 2 + 2 + 2$ be represented using multiplication?

109. Write $16 \div 8$ using a fraction bar.

110. To evaluate the expression $5(6) - 3 + 2$, list the operations in the order in which they must be performed.

KEY CONCEPT

Signed Numbers

In algebra, we work with both positive and negative numbers. We study negative numbers because they are necessary to describe many situations in daily life.

Represent each of these situations using a signed number.

1. Stocks fell 5 points.
2. The river was 12 feet over flood stage.
3. 30 seconds before going on the air
4. A business \$6 million in the red
5. 10 degrees above normal
6. The year 2000 B.C.
7. \$205 overdrawn
8. 14 units under their quota

The number line can be used to illustrate positive and negative numbers.

9. On this number line, label the location of the positive integers and the negative integers.

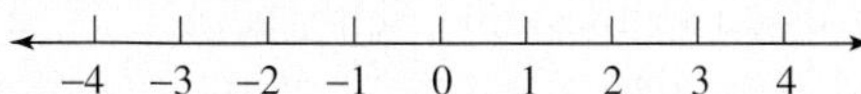

10. On this number line, graph 2 and its opposite.

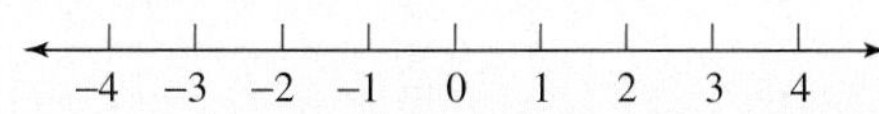

11. Two numbers, x and y, are graphed on this number line. What can you say about their relative sizes?

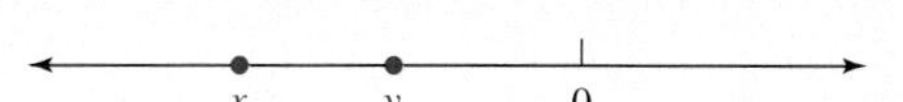

12. The absolute value of -3, written $|-3|$, is the distance between -3 and 0 on a number line. Show this distance on the number line.

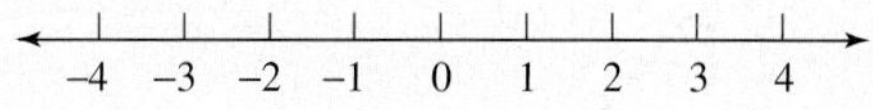

In the space provided, summarize how addition, multiplication, and division are performed with two integers having like signs and with two integers having unlike signs. Then explain the method that is used to subtract integers.

13. **Addition**

    Like signs:

    Unlike signs:

14. **Multiplication**

    Like signs:

    Unlike signs:

15. **Division**

    Like signs:

    Unlike signs:

16. **Subtraction with integers**

ACCENT ON TEAMWORK

SECTION 2.1

OPPOSITES Have everyone in the class get in pairs. The object of the game is for one member of a team, using one-word clues, to get the other member to say each of the words in Column I. This is done by giving your partner a word clue having the opposite meaning. For example, if you want your partner to say the word *up*, give the clue *down*. Go through all of the words in Column I. Keep track of how many words are guessed correctly.

Then switch assignments. The person who first gave the clues now receives the clues. Go through all of the words in Column II.

Column I		**Column II**	
below	surplus	win	positive
over	overdrawn	gain	increase
ahead	profit	forward	debt
before	deduct	retreat	withdrawal
less	liabilities	rise	accelerate

SECTION 2.2

ADDING INTEGERS To illustrate how to add $-5 + 3$, think of a hole 5 feet deep (-5) which then has 3 feet of dirt (3) added to it. The illustration shows that the resulting hole would be 2 feet deep (-2). So $-5 + 3 = -2$.

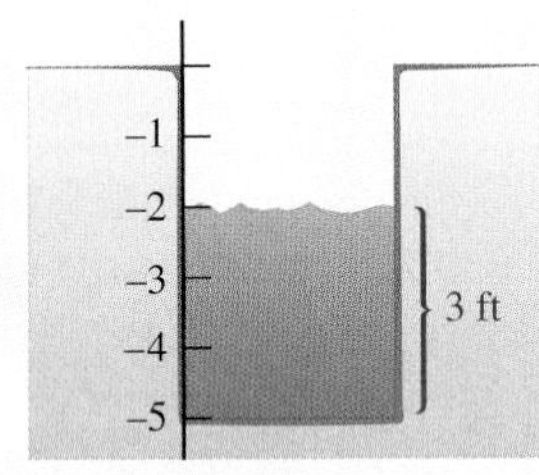

Draw a similar picture to help find each sum.

a. $-4 + 1$ **b.** $-5 + 4$
c. $-6 + 6$ **d.** $-3 + (-1)$
e. $-3 + (-3)$ **f.** $-3 + 5$

SECTION 2.3

SUBTRACTING INTEGERS Write a subtraction problem in which the difference of two negative numbers is

a. a positive number. **b.** a negative number.

SECTION 2.4

MULTIPLYING INTEGERS

a. Complete the multiplication table that follows.
b. Construct another table with a top row of -1 through -10 and a first column of 1 through 10.
c. Construct a table with a top row of -1 through -10 and a first column of -1 through -10.

·	1	2	3	4	5	6	7	8	9	10
−1										
−2										
−3										
−4										
−5										
−6										
−7										
−8										
−9										
−10										

SECTION 2.5

OPERATIONS WITH TWO INTEGERS For each operation listed in the table, tell whether the answer is always positive, always negative, or may be positive or negative.

Signs of the two integers	Add	Subtract	Multiply	Divide
Both positive				
Both negative				
One positive, one negative				

SECTION 2.6

ESTIMATION Estimate the answer to each problem.

a. $405 - 567$ **b.** $-2{,}564 - 2{,}456$
c. $989 - 898$ **d.** $-23{,}250 + 22{,}750$
e. $56(-87)$ **f.** $-40 - 30 - 45$
g. $608 \div (-2)$ **h.** $-94 + 90 - 45$

SECTION 2.7

SOLVING EQUATIONS What is wrong with each solution?

a. Solve: $2x + 4 = 10$.

$$2x + 4 = 10$$
$$\frac{2x}{2} + 4 = \frac{10}{2}$$
$$x + 4 = 5$$
$$x + 4 - 4 = 5 - 4$$
$$x = 1$$

b. Solve: $2x + 4 = 10$.

$$2x + 4 = 10$$
$$2x + 4 - 4 = 10$$
$$2x = 10$$
$$\frac{2x}{2} = \frac{10}{2}$$
$$x = 5$$

CHAPTER REVIEW

SECTION 2.1 An Introduction to the Integers

CONCEPTS

The *number line* is a horizontal or vertical line used to represent numbers graphically.

A *negative number* is less than 0. A *positive number* is greater than 0.

Integers:
$\{\ldots, -2, -1, 0, 1, 2, \ldots\}$

Inequality symbols:

$>$ is greater than

$<$ is less than

$\geq$ is greater than or equal to

$\leq$ is less than or equal to

The *absolute value* of a number is the distance between it and 0 on the number line.

On the number line, two numbers the same distance away from 0, but on different sides of it, are called *opposites.*

REVIEW EXERCISES

Graph each set of numbers.

1. $\{-3, -1, 0, 4\}$

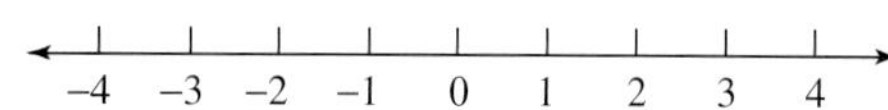

2. The integers greater than -3 but less than 4

Insert one of the symbols $>$ or $<$ in the blank to make a true statement.

3. $-7 \; \square \; 0$

4. $-20 \; \square \; -19$

Insert one of the symbols $\geq$ or $\leq$ in the blank to make a true statement.

5. $|-16| \; \square \; -16$

6. $56 \; \square \; 56$

7. WATER PRESSURE Salt water exerts a pressure of 14.7 pounds per square inch at a depth of 33 feet. Express the depth using a signed number.

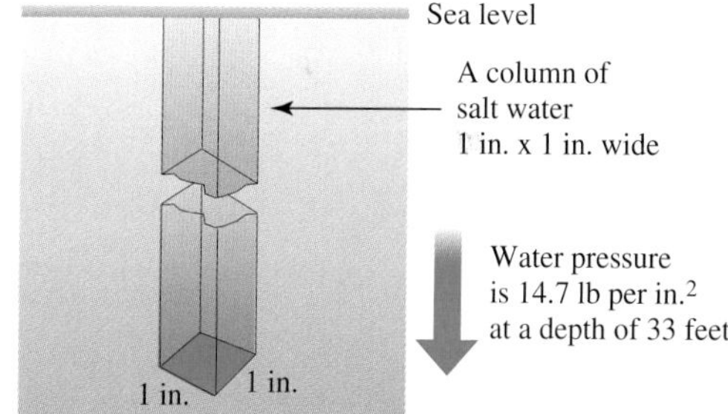

Represent each of these situations using a signed number.

8. A deficit of \$1,200

9. 10 seconds before going on the air

Evaluate each expression.

10. $|-4|$

11. $|0|$

12. $|-43|$

13. $-|12|$

Explain the meaning of each red − sign.

14. -5

15. $-(-5)$

16. $-(-5)$

17. $5 - (-5)$

The opposite of the opposite of a number is that number.

Find each of the following.

18. $-(-12)$

19. The opposite of 8

20. The opposite of -8

21. -0

SECTION 2.2 *Adding Integers*

To add two integers with *like signs,* add their absolute values and attach their common sign to that sum.

To add two integers with *unlike signs,* subtract their absolute values, the smaller from the larger. Attach the sign of the number with the larger absolute value to that result.

Use a number line to find each sum.

22. $4 + (-2)$

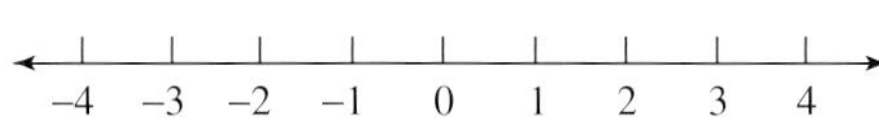

23. $-1 + (-3)$

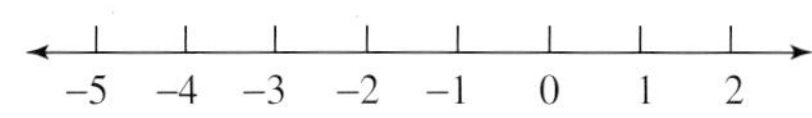

Add.

24. $-6 + (-4)$

25. $-23 + (-60)$

26. $-1 + (-4) + (-3)$

27. $-4 + 3$

28. $-28 + 140$

29. $9 + (-20)$

30. $3 + (-2) + (-4)$

31. $(-2 + 1) + [(-5) + 4]$

Addition property of 0: If a is any number, then

$a + 0 = a$ and $0 + a = a$

If $a + b = 0$, then a and b are called *additive inverses.*

Add.

32. $-4 + 0$

33. $0 + (-20)$

34. $-8 + 8$

35. $73 + (-73)$

Give the additive inverse of each number.

36. -11

37. 4

38. DROUGHT During a drought, the water level in a reservoir fell to a point 100 feet below normal. After two rainy months, it rose a total of 35 feet. How far below normal was the water level after the rain?

SECTION 2.3 *Subtracting Integers*

Rule for subtraction: If a and b are any numbers, then

$a - b = a + (-b)$

Subtract.

39. $5 - 8$

40. $-9 - 12$

41. $-4 - (-8)$

42. $-6 - 106$

43. $-8 - (-2)$

44. $7 - 1$

45. $0 - 37$

46. $0 - (-30)$

47. Fill in the blanks: Subtracting a number is the same as ________ the ________ of that number.

Evaluate each expression.

48. $-9 - 7 + 12$

49. $7 - [(-6) - 2]$

50. $1 - (2 - 7)$

51. $-12 - (6 - 10)$

52. Subtract 27 from -50.

53. Evaluate: $2 - [-(-3)]$.

54. GOLD MINING Some miners discovered a small vein of gold at a depth of 150 feet. This prompted them to continue their exploration. After descending another 75 feet, they came upon a much larger find. Use a signed number to represent the depth of the second discovery.

To find the *change* in a quantity, subtract the earlier value from the later value.

55. RECORD TEMPERATURES The lowest and highest recorded temperatures for Alaska and Virginia are shown here. For each state, find the difference in temperature between the record high and low.

Alaska:	Low $-80°$ Jan. 23, 1971	Virginia:	Low $-30°$ Jan. 22, 1985
	High 100° June 27, 1915		High 110° July 15, 1954

SECTION 2.4 *Multiplying Integers*

The product of two integers with *like signs* is positive. The product of two integers with *unlike signs* is negative.

Multiply.

56. $-9 \cdot 5$

57. $-3(-6)$

58. $7(-2)$

59. $(-8)(-47)$

60. $-20 \cdot 5$

61. $-1(-1)$

62. $-1(25)$

63. $(5)(-30)$

64. $(-6)(-2)(-3)$

65. $4(-3)3$

66. $0(-7)$

67. $(-1)(-1)(-1)(-1)$

68. TAX DEFICITS A state agency's prediction of a tax shortfall proved to be two times worse than the actual deficit of \$3 million. The federal prediction of the same shortfall was even more inaccurate — three times the amount of the actual deficit. Complete the illustration, which summarizes these incorrect forecasts.

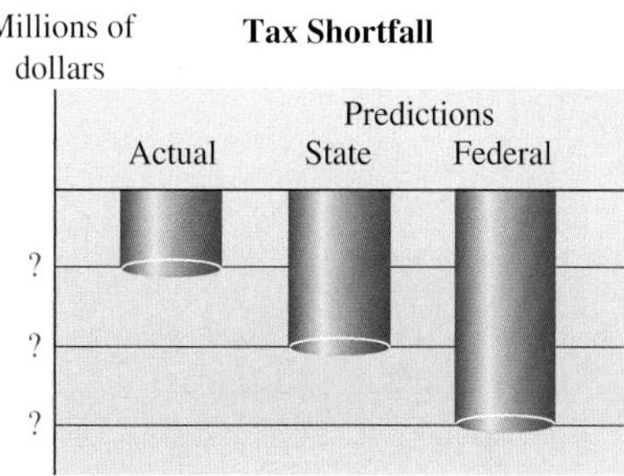

An *exponent* is used to represent repeated multiplication.

Find each power.

69. $(-5)^2$

70. $(-2)^5$

71. $(-8)^2$

72. $(-4)^3$

When a negative integer is raised to an *even* power, the result is positive. When it is raised to an *odd* power, the result is negative.

73. When $(-5)^9$ is evaluated, will the result be positive or negative?

74. Explain the difference between -2^2 and $(-2)^2$ and then evaluate each.

SECTION 2.5 Dividing Integers

75. Fill in the blanks: We know that $\frac{-15}{5} = -3$ because $\quad(\quad) = \quad$.

The quotient of two numbers with *like* signs is positive.

The quotient of two integers with *unlike* signs is negative.

Divide.

76. $\frac{-14}{7}$

77. $\frac{25}{-5}$

78. $-64 \div 8$

79. $\frac{-202}{-2}$

If a is any nonzero number,

$$\frac{0}{a} = 0$$

Division by 0 is undefined.

Find each quotient, if possible.

80. $\frac{0}{-5}$

81. $\frac{-4}{0}$

82. $\frac{-673}{-673}$

83. $\frac{-10}{-1}$

84. PRODUCTION TIME Because of improved production procedures, the time needed to produce an electronic component dropped by 12 minutes over the past six months. If the drop in production time was uniform, how much did it change each month over this period?

SECTION 2.6 Order of Operations and Estimation

The rules for the order of operations:

1. Perform all calculations within parentheses and other grouping symbols.
2. Evaluate all exponential expressions.
3. Perform all the multiplications and divisions, working from left to right.
4. Perform all the additions and subtractions, working from left to right.

Always work from the *innermost* set of grouping symbols to the *outermost* set.

A fraction bar is a grouping symbol.

Evaluate each expression.

85. $2 + 4(-6)$

86. $7 - (-2)^2 + 1$

87. $2 - 5(4) + (-25)$

88. $-3(-2)^3 - 16$

89. $-2(5)(-4) + \frac{|-9|}{3^2}$

90. $-4^2 + (-4)^2$

91. $-12 - (8 - 9)^2$

92. $7|-8| - 2(3)(4)$

93. $-4\left(\frac{15}{-3}\right) - 2^3$

94. $-20 + 2(12 - 5 \cdot 2)$

95. $-20 + 2[12 - (-7 + 5)^2]$

96. $8 - |-3 \cdot 4 + 5|$

97. $\frac{10 + (-6)}{-3 - 1}$

98. $\frac{3(-6) - 11 + 1}{4^2 - 3^2}$

An *estimation* is an approximation that gives a quick idea of what the actual answer would be.

Estimate each answer.

99. $-89 + 57 + (-42)$

100. $\frac{-507}{-26}$

101. $(-681)(9)$

102. $317 - (-775)$

SECTION 2.7 *Solving Equations Involving Integers*

To *solve an equation* means to find all values of the variable that, when substituted into the original equation, make the equation a true statement.

Is −4 a solution of the equation? Explain why or why not.

103. $2x + 6 = -2$

104. $6 + \frac{x}{2} = -4$

To solve an equation, we undo the operations in the reverse order from that in which they were performed on the variable. The objective is to isolate the variable.

Solve each equation. **Check the result.**

105. $t + (-8) = -18$

106. $\frac{x}{-3} = -4$

107. $y + 8 = 0$

108. $-7m = -28$

109. $-x = -15$

110. $4 = -y$

111. $-5t + 1 = -14$

112. $3(2) = 2 - 2x$

113. $\frac{x}{-4} - 5 = -1 - 1$

114. $c - (-5) = 5$

The five-step *problem-solving strategy* can be used when solving application problems.

1. Analyze the problem.
2. Form an equation.
3. Solve the equation.
4. State the conclusion.
5. Check the result.

Let a variable represent the unknown quantity. Then write and solve an equation to answer the question.

115. CHECKING ACCOUNTS After Lee made a deposit of \$85 into his checking account, it was still \$47 overdrawn. What was his account balance before making the deposit?

116. WIND-CHILL FACTOR If the wind is blowing at 25 miles per hour, an air temperature of 5° below zero will feel like 51° below zero. Find the perceived change in temperature that is caused by the wind.

117. CREDIT CARD PROMOTIONS During the holidays, a store offered an \$8 gift certificate to any customer applying for its credit card. If this promotion cost the company \$968, how many customers applied for credit?

118. BANK FAILURES When a group of 7 investors decided to acquire a failing bank, each had to assume an equal share of the bank's total indebtedness, which was \$57,400. How much debt did each investor assume?

CHAPTER 2 TEST

1. Insert one of the symbols $>$ or $<$ in the blank to make the statement true.
   a. $-8 \quad -9$
   b. $-8 \quad |-8|$
   c. The opposite of 5 $\quad$ 0

2. List the integers.

3. SCHOOL ENROLLMENTS According to the projections in the table, which high school will face the greatest shortage in the year 2010?

High schools with shortage of classroom seats by 2010	
Sylmar	−669
San Fernando	−1,630
Monroe	−2,488
Cleveland	−350
Canoga Park	−586
Polytechnic	−2,379
Van Nuys	−1,690
Reseda	−462
North Hollywood	−1,004
Hollywood	−774

4. Use a number line to find the sum: $-3 + (-2)$.

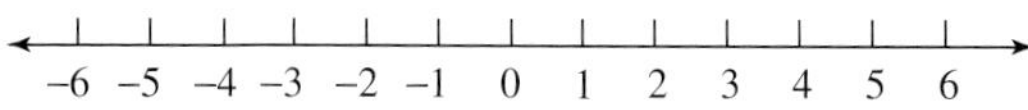

5. Add.
   a. $-65 + 31$
   b. $-17 + (-17)$
   c. $[6 + (-4)] + [-6 + (-4)]$

6. Subtract.
   a. $-7 - 6$
   b. $-7 - (-6)$
   c. $0 - 15$
   d. $-60 - 50 - 40$

7. Find each product.
   a. $-10 \cdot 7$
   b. $-4(-2)(-6)$
   c. $(-2)(-2)(-2)(-2)$
   d. $-55(0)$

8. Write the related multiplication statement for $\frac{-20}{-4} = 5$.

9. Find each quotient, if possible.
   a. $\frac{-32}{4}$
   b. $\frac{8}{6 - 6}$
   c. $\frac{-5}{1}$
   d. $\frac{0}{-6}$

10. BUSINESS TAKEOVERS Six businessmen are contemplating taking over a company that has potential, but they must pay off the debt taken on by the company over the past three quarters. If they plan equal ownership, how much will each have to contribute to pay off the debt?

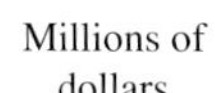

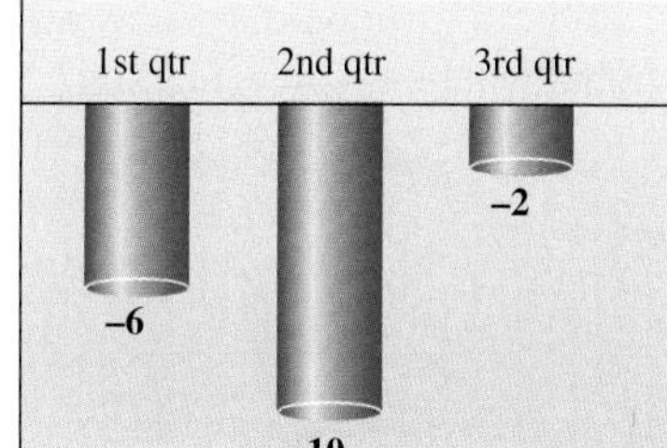

11. GEOGRAPHY The lowest point on the African continent is the Qattarah Depression in the Sahara Desert, 436 feet below sea level. The lowest point on the North American continent is Death Valley, California, 282 feet below sea level. Find the difference in these elevations.

12. Evaluate each expression.
    a. $-(-6)$
    b. $|-7|$
    c. $-|-9 + 3|$
    d. $2|-66|$

13. Find each power.

a. $(-4)^2$ b. -4^2

c. $(-4 + 3)^5$

Evaluate each expression.

14. $-18 \div 2 \cdot 3$

15. $4 - (-3)^2 + 6$

16. $-3 + \left(\frac{-16}{4}\right) - 3^3$

17. $-10 + 2[6 - (-2)^2(-5)]$

18. $\dfrac{4(-6) - 2^2}{-3 - 4}$

19. TEMPERATURE CHANGE In a lab, the temperature of a fluid was reduced 6° per hour for 12 hours. What signed number represents the change in temperature?

20. GAUGES Many automobiles have an ammeter like that shown. If the headlights, which draw a current of 8 amps, and the radio, which draws a current of 4 amps, are both turned on, which way will the arrow move? What will be the new reading?

21. CARD GAMES After the first round of a card game, Tommy had a score of 8. When he lost the second round, he had to deduct the value of the cards left in his hand from his first-round score. (See the illustration below.) What was his score after two rounds of the game? For scoring, face cards are counted as 10 points and aces as 1 point.

22. Is -10 a solution of $\frac{x}{5} - 16 = -14$?

Solve each equation.

23. $c - (-7) = -8$

24. $6 - x = -10$

25. $\frac{x}{-4} = 10$

26. $-6x = 0$

27. $3x + (-7) = -11 + (-11)$

28. $-5 = -6a + 7$

29. $\frac{x}{-2} + 3 = (-2)(-6)$

Let a variable represent the unknown quantity. Then write and solve an equation to answer each question.

30. CHECKING ACCOUNTS After making a deposit of $225, a student's account was still $19 overdrawn. What was her balance before the deposit?

31. HOSPITAL CAPACITY One morning, the number of beds occupied by patients in a hospital was 3 under capacity. By afternoon, the number of unoccupied beds was 21. If no new patients were admitted, how many patients were released to go home?

32. Multiplication means repeated addition. Use this fact to show that the product of a positive and a negative number, such as $5(-4)$, is negative.

33. Explain why the absolute value of a number can never be negative.

34. Is the inequality $12 \geq 12$ true or false? Explain why.

CHAPTERS 1–2 CUMULATIVE REVIEW EXERCISES

Consider the numbers in the set $\{-2, -1, 0, 1, 2\frac{3}{2}, 5, 9\}$.

1. List each natural number.

2. List each whole number.

3. List each negative number.

4. List each integer.

Consider the number 7,326,549.

5. Which digit is in the thousands column?

6. Which digit is in the hundred thousands column?

7. Round to the nearest hundred.

8. Round to the nearest ten thousand.

9. BIDS A school district received the bids shown in the table. Which company should be awarded the contract?

Citrus Unified School District Bid 02-9899 CABLING AND CONDUIT INSTALLATION	
Datatel	$2,189,413
Walton Electric	$2,201,999
Advanced Telecorp	$2,175,081
CRF Cable	$2,174,999
Clark & Sons	$2,175,801

10. NUCLEAR POWER The table gives the number of nuclear power plants in the United States for the years 1978–2003, in five-year increments. Construct a bar graph using the data.

Year	1978	1983	1988	1993	1998	2003
Plants	70	80	108	109	104	104

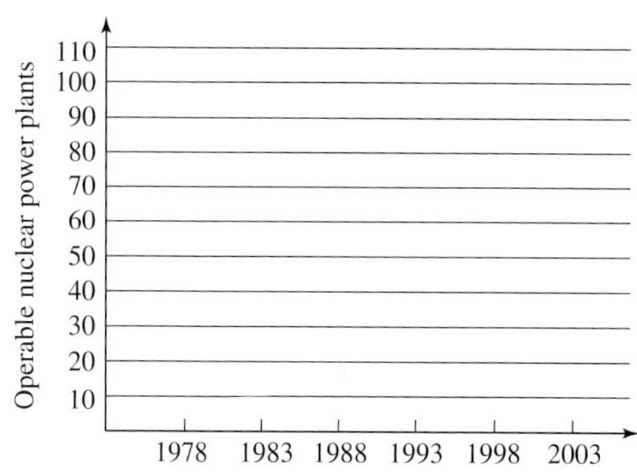

Source: *The World Almanac, 2005*

Perform each operation.

11. $237 + 549$

12. $6{,}375 - 2{,}569$

13. $\begin{array}{r} 5{,}369 \\ -\ \underline{685} \end{array}$

14. $\begin{array}{r} 7{,}899 \\ \underline{+5{,}237} \end{array}$

15. Find the perimeter and the area of the rectangular garden.

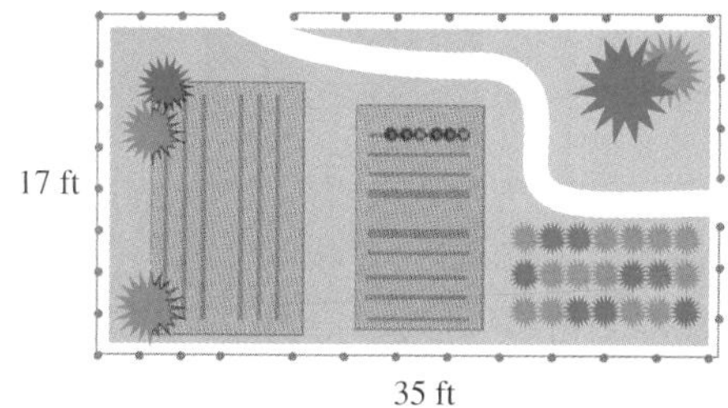

16. In a shipment of 147 pieces of furniture, 27 pieces were sofas, 55 were leather chairs, and the rest were wooden chairs. Find the number of wooden chairs.

Perform each operation.

17. $435 \cdot 27$

18. $1{,}261 \div 97$

19. $\begin{array}{r} 4{,}587 \\ \times\ \underline{67} \end{array}$

20. $38\overline{)17{,}746}$

21. SHIPPING There are 12 tennis balls in one dozen, and 12 dozen in one gross. How many tennis balls are there in a shipment of 12 gross?

22. Find all of the factors of 18.

Identify each number as a prime, a composite, an even, or an odd.

23. 17

24. 18

25. 0

26. 1

27. Find the prime factorization of 504.

28. Write the expression $11 \cdot 11 \cdot 11 \cdot 11$ using an exponent.

Evaluate each expression.

29. $5^2 \cdot 7$

30. $16 + 2[14 - 3(5 - 4)^2]$

31. $25 + 5 \cdot 5$

32. $\dfrac{16 - 2 \cdot 3}{2 + (9 - 6)}$

33. SPEED CHECKS A traffic officer monitored several cars on a city street. She found that the speeds of the cars were as follows:

38, 42, 36, 38, 48, 44

On average, were the drivers obeying the 40-mph speed limit?

34. Determine whether 6 is a solution of the equation $3x - 2 = 16$. Explain why or why not.

Solve each equation and check the result.

35. $50 = x + 37$

36. $a - 12 = 41$

37. $5p = 135$

38. $\dfrac{y}{8} = 3$

Graph each set on the number line.

39. $\{-2, -1, 0, 2\}$

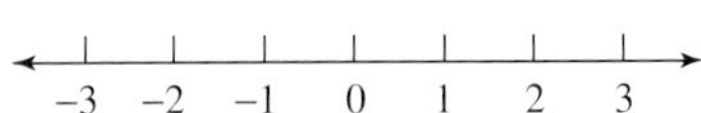

40. The integers greater than -4 but less than 2

–4 –3 –2 –1 0 1 2

41. True or false: $-17 < -16$.

42. Evaluate: 3^2 and -3^2.

Evaluate each expression.

43. $-2 + (-3)$

44. $-15 + 10 + (-9)$

45. $-3 - 5$

46. $-15^2 - 2\,|-3|$

47. $(-8)(-3)$

48. $5(-7)^3$

49. $\dfrac{-14}{-7}$

50. $\dfrac{450}{-9}$

51. $5 + (-3)(-7)$

52. $-20 + 2[12 - 5(-2)(-1)]$

53. $\dfrac{10 - (-5)}{1 - 2 \cdot 3}$

54. $\dfrac{3(-6) - 10}{3^2 - 4^2}$

Solve each equation. Check the result.

55. $-5t + 1 = -14$

56. $\dfrac{x}{-3} - 2 = -2(-2)$

57. BUYING A BUSINESS When 12 investors decided to buy a bankrupt company, they agreed to assume equal shares of the company's debt of $1,512,444. How much was each person's share?

58. THE MOON The difference in the maximum and the minimum temperatures on the moon's surface is 540° F. The maximum temperature, which occurs at lunar noon, is 261° F. Find the minimum temperature, which occurs just before lunar dawn.

CHAPTER 3

Fractions and Mixed Numbers

Getty Images

TLE It has been said that "A penny saved is a penny earned." Smart shoppers certainly know this is true. A lot of money can be saved by waiting for big-ticket items, such as appliances, furniture, and jewelry, to go on sale before purchasing them. Many stores advertise from 1/3 to 1/2 off on these items at various times during the year. In such cases, an informed shopper needs a working knowledge of fractions to calculate discounts and sale prices.

To learn more about fractions and discount shopping, visit The Learning Equation on the Internet at http://tle.brookscole.com. (The log-in instructions are in the Preface.) For Chapter 3, the online lessons are:

- *TLE* Lesson 7: Adding and Subtracting Fractions
- *TLE* Lesson 8: Multiplying Fractions and Mixed Numbers

Check Your Knowledge

1. For the fraction $\frac{3}{4}$, 3 is the __________ and 4 is the __________.
2. A fraction is said to be in ________ terms if the only factor common to the numerator and denominator is 1.
3. Fractions that represent the same amount, such as $\frac{2}{3}$ and $\frac{4}{6}$, are called __________ fractions.
4. Two numbers are called __________ if their product is 1.
5. A ________ number such as $5\frac{3}{4}$ contains a whole-number part and a fractional part.
6. What fractional part of a day is 10 hours? What is the remaining fractional part of the day?
7. Simplify: $\dfrac{24}{30}$.
8. Multiply: $\dfrac{1}{2}\left(\dfrac{3}{4}\right)$.
9. Divide: $-3\dfrac{3}{8} \div 2\dfrac{1}{4}$.
10. Add: $\dfrac{2}{5} + \dfrac{7}{8}$.
11. Subtract: $\dfrac{2}{3} - \dfrac{1}{4}$.
12. Express $\dfrac{5}{6}$ as an equivalent fraction with denominator 36.
13. Graph: $-\dfrac{9}{8}, \dfrac{1}{9}$, and $1\dfrac{1}{12}$.

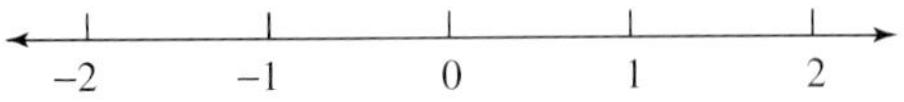

14. Find the area of a triangle with base of length $\dfrac{7}{4}$ cm and height $\dfrac{5}{7}$ cm.
15. Add: $123\dfrac{4}{5} + 189\dfrac{2}{3}$.
16. Subtract: $13\dfrac{1}{3} - 8\dfrac{7}{8}$.
17. Evaluate: $\left(\dfrac{5}{4}\right)^2 + \left(\dfrac{2}{3} - 2\dfrac{1}{6}\right)$.
18. Simplify each complex fraction.
    a. $\dfrac{\frac{3}{5}}{\frac{1}{3}}$
    b. $\dfrac{\frac{1}{5} - \frac{1}{6}}{\frac{1}{4} - \frac{1}{3}}$
19. Solve each equation.
    a. $-\dfrac{x}{6} = -12$
    b. $\dfrac{x}{2} - \dfrac{2}{3} = 1$
20. Four-fifths of the people questioned at a beach were using sun screen. Twelve of those surveyed were not using sun screen. How many people took part in the survey?
21. Josie and five friends purchased ten lottery tickets together. If one of their tickets is chosen, the prize is $\$10\frac{1}{2}$ million. What would Josie's share of the prize be if they won?
22. A corporate executive negotiated a seven-year employment contract providing for equal annual payments. The executive received 3 million for the first two years of the contract term. How much is she due for the rest of the contract term?

Study Skills Workshop

LISTENING IN CLASS AND TAKING NOTES

Attending all of your class meetings is crucial to your success in a math course. Your instructor will be giving explanations and examples that may not be found in your textbook, as well as other information about your course (test dates, homework assignments, etc.). Because this information is not found anywhere else, and because it is normal to forget material shortly after it is presented, it is important that you keep a written record of what occurred in your class. (*Note:* Auditory learners may also want to keep a recorded version of class notes. Ask your instructor for permission to record lectures.)

Listening in Class. Listening in class is different from listening to your favorite CDs or MP3 files because it requires that you be an *active* listener. It is usually impossible to write down everything that your instructor says in class, but you want to get what's important. In order to catch the significant material, you should be waiting for cues from your instructor: pauses in lectures and statements such as "This is really important" or "This is a question that shows up frequently on tests" are indications that you should be paying special attention. You should be listening with a pencil in hand, ready to record these examples, definitions, or concepts.

Taking Notes. Usually when giving a lecture, your instructor will tell you the important chapter(s) and section(s) in your textbook. Bring your textbook to every class and open it to the section that your instructor is covering. If possible, find out what section you will be covering before your next lecture and read this section *before* class. As you read, try to identify which terms and definitions your instructor will think are important. When your instructor refers to definitions that are found in the text, it is not necessary for you to write them down in entirety, but just jot the term being defined and the page number on which it appears. If topics are being discussed that are not in the text, record them using abbreviations and phrases rather than complete sentences.

Don't worry about making your class notes really neat; you should be "reworking" your notes after class (this will be discussed in more detail in the next Study Skills Workshop). However, you should organize your notes as much as possible as you write them. Write down the examples your instructor gives in step-by-step detail. Label examples and definitions that your instructor covers in detail as "key concepts." If you miss a step, or if you don't understand a step in the example, ask your instructor to explain. The examples should look a good deal like the problems that you will see in your homework assignment, so it is really important that you write them down as completely as possible. As soon as possible after class, sit down with a classmate and compare notes, to see whether either one of you missed anything.

ASSIGNMENT

1. Find out from your instructor which section(s) will be covered in your next class. Pre-read the section(s) and make a list of terms and definitions that you predict your instructor will think are most important.
2. In your next class, bring your textbook and keep it open to the sections being covered. If your instructor mentions a term or definition that is found in your text, write the term and the page number on which it appears in your notes. Write every example that your instructor gives, making sure that you have included all the steps.
3. Find at least one classmate with whom you can review notes. Make an appointment to compare your class notes as soon as possible after the class. Did you find differences in your notes?

Whole numbers are used to count objects. When we need to represent parts of a whole, fractions can be used.

3.1 The Fundamental Property of Fractions

- Basic facts about fractions • Equivalent fractions • Simplifying a fraction
- Expressing a fraction in higher terms

There is no better place to start a study of fractions than with *the fundamental property of fractions.* This property is the foundation for two fundamental procedures that are used when working with fractions. But first, we review some basic facts about fractions.

Basic facts about fractions

1. A Fraction Can Be Used to Indicate Equal Parts of a Whole. In our everyday lives, we often deal with parts of a whole. For example, we talk about parts of an hour, parts of an inch, and parts of a pound.

half of an hour three-eighths of an inch of rain a quarter-pound hamburger

2. A Fraction Is Composed of a Numerator, a Denominator, and a Fraction Bar.

$$\text{Fraction bar} \longrightarrow \frac{3 \longleftarrow \text{Numerator}}{4 \longleftarrow \text{Denominator}}$$

The denominator (in this case, 4) tells us that a whole was divided into four equal parts. The numerator tells us that we are considering three of those equal parts.

3. Fractions Can Be Proper or Improper. If the numerator of a fraction is less than its denominator, the fraction is called a **proper fraction.** A proper fraction is less than 1. Fractions whose numerators are greater than or equal to their denominators are called **improper fractions.**

Proper fractions	**Improper fractions**
$\frac{1}{4}, \frac{2}{3},$ and $\frac{98}{99}$	$\frac{7}{2}, \frac{98}{97}, \frac{16}{16},$ and $\frac{5}{1}$

4. The Denominator of a Fraction Cannot Be 0. $\frac{7}{0}, \frac{23}{0}$, and $\frac{0}{0}$ are meaningless expressions. (Recall that $\frac{7}{0}, \frac{23}{0}$, and $\frac{0}{0}$ represent *division* by 0, and a number cannot be divided by 0.) However, $\frac{0}{7} = 0$ and $\frac{0}{23} = 0$.

5. Fractions Can Be Negative. There are times when a negative fraction is needed to describe a quantity. For example, if an earthquake causes a road to sink one-half inch, the amount of movement can be represented by $-\frac{1}{2}$ inch.

Negative fractions can be written in three ways. The negative sign can appear in the numerator, in the denominator, or in front of the fraction.

$$\frac{-1}{2} = \frac{1}{-2} = -\frac{1}{2} \qquad \frac{-15}{8} = \frac{15}{-8} = -\frac{15}{8}$$

Negative fractions

If a and b represent positive numbers,

$$\frac{-a}{b} = \frac{a}{-b} = -\frac{a}{b}$$

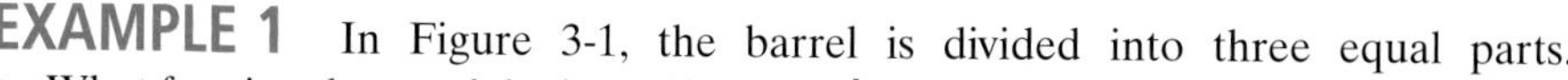

EXAMPLE 1 In Figure 3-1, the barrel is divided into three equal parts. **a.** What fractional part of the barrel is full? **b.** What fractional part is empty?

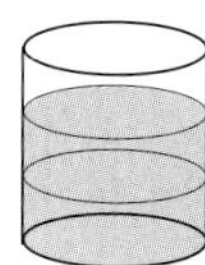

FIGURE 3-1

Solution

a. Two of the three parts are full. Therefore, the barrel is $\frac{2}{3}$ full.

b. One of the three equal parts is not filled. The barrel is $\frac{1}{3}$ empty.

The fractions $\frac{2}{3}$ and $\frac{1}{3}$ are both proper fractions.

Self Check 1

a. According to the calendar below, what fractional part of the month has passed? **b.** What fractional part remains?

DECEMBER

X	X	X	X	X	X	X
X	X	X	X	12	13	14
15	16	17	18	19	20	21
22	23	24	25	26	27	28
29	30	31				

Answers **a.** $\frac{11}{31}$, **b.** $\frac{20}{31}$

Fractions are often referred to as **rational numbers.** All integers are rational numbers, because every integer can be written as a fraction with a denominator of 1. For example,

$$2 = \frac{2}{1}, \qquad -5 = \frac{-5}{1}, \qquad \text{and} \qquad 0 = \frac{0}{1}$$

Since every integer is also a rational number, the integers are a subset of the rational numbers.

! COMMENT Not all rational numbers are integers. For example, the rational number $\frac{7}{8}$ is not an integer.

Equivalent fractions

Fractions can look different but still represent the same number. For example, let's divide the rectangle in Figure 3-2(a) in two ways. In Figure 3-2(b), we divide it into halves (2 equal-sized parts). In Figure 3-2(c), we divide it into fourths (4 equal-sized parts). Notice that one-half of the figure is the same size as two-fourths of the figure.

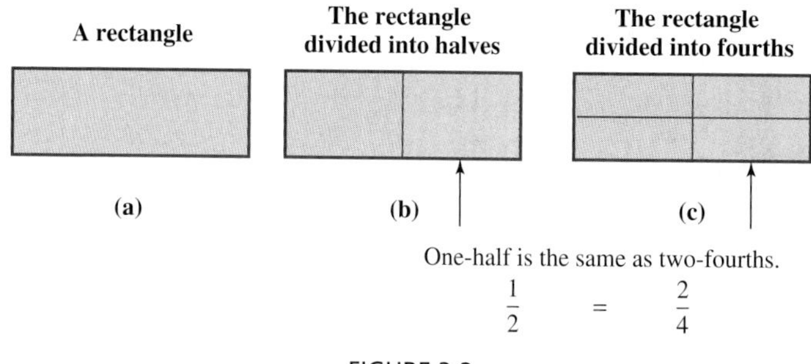

FIGURE 3-2

The fractions $\frac{1}{2}$ and $\frac{2}{4}$ look different, but Figure 3-2 shows that they represent the same amount. We say that they are **equivalent fractions.**

Equivalent fractions

Two fractions are **equivalent** if they represent the same number.

Simplifying a fraction

If we replace a fraction with an equivalent fraction that contains smaller numbers, we are **simplifying** or **reducing the fraction.** To simplify a fraction, we use the following fact.

The fundamental property of fractions

Multiplying or dividing the numerator and the denominator of a fraction by the same nonzero number does not change the value of the fraction. In symbols, for all numbers a, b, and x (provided b and x are not 0),

$$\frac{a}{b} = \frac{a \cdot x}{b \cdot x} \quad \text{and} \quad \frac{a}{b} = \frac{a \div x}{b \div x}$$

For example, we consider $\frac{24}{28}$. It is apparent that 24 and 28 have a common factor 4. By the fundamental property of fractions, we can divide the numerator and denominator of this fraction by 4.

$$\frac{24}{28} = \frac{24 \div \mathbf{4}}{28 \div \mathbf{4}}$$ Divide the numerator by 4. Divide the denominator by 4.

$$= \frac{6}{7}$$ Perform each division: $24 \div 4 = 6$ and $28 \div 4 = 7$.

Thus, $\frac{24}{28} = \frac{6}{7}$. We say that $\frac{24}{28}$ and $\frac{6}{7}$ are equivalent fractions, because they represent the same number.

In practice, we show the previous simplification in a slightly different way.

$$\frac{24}{28} = \frac{\mathbf{4} \cdot 6}{\mathbf{4} \cdot 7}$$ Once you see a common factor of the numerator and the denominator, factor each of them so that it shows. In this case, 24 and 28 share a common factor 4.

$$= \frac{\overset{1}{\cancel{4}} \cdot 6}{\underset{1}{\cancel{4}} \cdot 7}$$ Divide the numerator and the denominator by 4 by drawing slashes through the common factors. Use small 1's to represent the result of each division of 4 by 4. Note that $\frac{4}{4} = 1$.

$$= \frac{6}{7}$$ Multiply in the numerator and in the denominator: $1 \cdot 6 = 6$ and $1 \cdot 7 = 7$.

In the second step of the previous simplification, we say that we *divided out the common factor 4.*

Simplifying a fraction

We can **simplify** a fraction by factoring its numerator and denominator and then dividing out all common factors in the numerator and denominator.

When a fraction can be simplified no further, we say that it is written in **lowest terms.**

Lowest terms

A fraction is in **lowest terms** if the only factor common to the numerator and denominator is 1.

EXAMPLE 2 Simplify to lowest terms: $\frac{25}{75}$.

Solution The numerator and the denominator have a common factor 25.

$$\frac{25}{75} = \frac{25 \cdot 1}{3 \cdot 25}$$ Factor 25 as 25 · 1 and 75 as 3 · 25.

$$= \frac{\overset{1}{\cancel{25}} \cdot 1}{3 \cdot \underset{1}{\cancel{25}}}$$ Divide out the common factor 25. Note that $\frac{25}{25} = 1$.

$$= \frac{1}{3}$$ Multiply in the numerator and in the denominator: $1 \cdot 1 = 1$ and $1 \cdot 3 = 3$.

Self Check 2

Simplify to lowest terms: $\frac{60}{80}$.

Answer $\frac{3}{4}$

EXAMPLE 3 Simplify: $\frac{90}{126}$.

Solution To find the common factors that will divide out, we prime factor 90 and 126.

$$\frac{90}{126} = \frac{2 \cdot 3 \cdot 3 \cdot 5}{2 \cdot 3 \cdot 3 \cdot 7}$$ Use the tree method or the division method to prime factor 90 and 126: $90 = 2 \cdot 3 \cdot 3 \cdot 5$ and $126 = 2 \cdot 3 \cdot 3 \cdot 7$.

$$= \frac{\overset{1}{\cancel{2}} \cdot \overset{1}{\cancel{3}} \cdot \overset{1}{\cancel{3}} \cdot 5}{\underset{1}{\cancel{2}} \cdot \underset{1}{\cancel{3}} \cdot \underset{1}{\cancel{3}} \cdot 7}$$ Divide out the common factors 3, 3, and 2.

$$= \frac{5}{7}$$ Multiply in the numerator and in the denominator.

We can also simplify this fraction by noting that the numerator and denominator have a common factor 18.

$$\frac{90}{126} = \frac{5 \cdot \overset{1}{\cancel{18}}}{7 \cdot \underset{1}{\cancel{18}}}$$ Factor 90 as 5 · 18 and 126 as 7 · 18. Divide out the common factor 18.

$$= \frac{5}{7}$$

Self Check 3

Simplify: $\frac{42}{150}$.

Answer $\frac{7}{25}$

! COMMENT Negative fractions are simplified in the same way as positive fractions. Just remember to write a negative sign − in each step of the solution.

$$-\frac{45}{72} = -\frac{5 \cdot \overset{1}{\cancel{9}}}{8 \cdot \underset{1}{\cancel{9}}} = -\frac{5}{8}$$

Expressing a fraction in higher terms

It is sometimes necessary to replace a fraction with an equivalent fraction that involves larger numbers or more complex terms. This is called **expressing the fraction in higher terms** or **building** the fraction.

For example, to write $\frac{3}{8}$ as an equivalent fraction with a denominator 40, we can use the fundamental property of fractions and multiply the numerator and denominator by 5.

$$\frac{3}{8} = \frac{3 \cdot \mathbf{5}}{8 \cdot \mathbf{5}}$$ Multiply the numerator by 5. Multiply the denominator by 5.

$$= \frac{15}{40}$$ Perform the multiplications in the numerator and in the denominator.

Therefore, $\frac{3}{8} = \frac{15}{40}$.

EXAMPLE 4 Write $\frac{5}{7}$ as an equivalent fraction with a denominator 28.

Solution We need to multiply the denominator by 4 to obtain 28. By the fundamental property of fractions, we must multiply the numerator by 4 as well.

$$\frac{5}{7} = \frac{5 \cdot \mathbf{4}}{7 \cdot \mathbf{4}}$$ Multiply the numerator and denominator by 4.

$$= \frac{20}{28}$$ Perform the multiplication in the numerator and in the denominator.

Self Check 4
Write $\frac{2}{3}$ as an equivalent fraction with a denominator 24.

Answer $\frac{16}{24}$

EXAMPLE 5 Write 4 as a fraction with a denominator 6.

Solution First, express 4 as a fraction: $4 = \frac{4}{1}$. To obtain a denominator 6, we need to multiply the numerator and denominator by 6.

$$\frac{4}{1} = \frac{4 \cdot \mathbf{6}}{1 \cdot \mathbf{6}}$$

$$= \frac{24}{6}$$ Perform each multiplication: $4 \cdot 6 = 24$ and $1 \cdot 6 = 6$.

Self Check 5
Write 5 as a fraction with a denominator 3.

Answer $\frac{15}{3}$

Section 3.1 STUDY SET

VOCABULARY *Fill in the blanks.*

1. For the fraction $\frac{7}{8}$, 7 is the ________ and 8 is the ________.
2. When we express 15 as $3 \cdot 5$, we say that we have ________ 15.
3. A ________ fraction is less than 1.
4. A fraction is said to be in ________ terms if the only factor common to the numerator and denominator is 1.
5. Two fractions are ________ if they have the same value.
6. A ________ can be used to indicate the number of equal parts of a whole.
7. Multiplying the numerator and denominator of a fraction by a number to obtain an equivalent fraction that involves larger numbers is called expressing the fraction in ________ terms or ________ the fraction.
8. We can ________ a fraction that is not in lowest terms by applying the fundamental property of fractions. We ________ out common factors of the numerator and denominator.

CONCEPTS

9. What common factor (other than 1) do the numerator and the denominator have?

a. $\frac{2}{16}$ **b.** $\frac{6}{9}$

c. $\frac{10}{15}$ **d.** $\frac{14}{35}$

10. Given: $\frac{15}{35} = \frac{3 \cdot \overset{1}{\cancel{5}}}{\underset{1}{\cancel{5}} \cdot 7}$. In this work, what do the slashes and small 1's mean?

11. What concept studied in this section is shown?

12. Why can't we say that $\frac{2}{5}$ of the figure is shaded?

13. a. Explain the difference in the two approaches used to simplify $\frac{20}{28}$.

$$\frac{\overset{1}{\cancel{4}} \cdot 5}{\underset{1}{\cancel{4}} \cdot 7} \quad \text{and} \quad \frac{\overset{1}{\cancel{2}} \cdot \overset{1}{\cancel{2}} \cdot 5}{\underset{1}{\cancel{2}} \cdot \underset{1}{\cancel{2}} \cdot 7}$$

b. Are the results the same?

14. What concept studied in this section does this statement illustrate?

$$\frac{5}{10} = \frac{4}{8} = \frac{3}{6} = \frac{2}{4} = \frac{1}{2}$$

15. Why isn't this a valid application of the fundamental property of fractions?

$$\frac{10}{11} = \frac{2+8}{2+9} = \frac{\overset{1}{\cancel{2}} + 8}{\underset{1}{\cancel{2}} + 9} = \frac{9}{10}$$

16. Write the fraction $\frac{7}{-8}$ in two other ways.

17. Write as a fraction.

a. 8 **b.** -25

18. Fill in the blanks in the following solution to write $\frac{5}{9}$ as an equivalent fraction with denominator 27.

$$\frac{5 \cdot \square}{9 \cdot \square} = \frac{15}{27}$$

NOTATION *Complete each solution to simplify each fraction.*

19. $\frac{18}{24} = \frac{2 \cdot \square \cdot 3}{2 \cdot 2 \cdot \square \cdot 3}$

$= \frac{\overset{1}{\cancel{\square}} \cdot 3 \cdot \overset{1}{\cancel{\square}}}{\underset{1}{\cancel{\square}} \cdot 2 \cdot 2 \cdot \underset{1}{\cancel{\square}}}$

$= \frac{3}{4}$

20. $\frac{60}{90} = \frac{2 \cdot \square}{3 \cdot \square}$

$= \frac{\square \cdot \overset{1}{\cancel{30}}}{\square \cdot \underset{1}{\cancel{30}}}$

$= \frac{2}{3}$

PRACTICE *Simplify each fraction to lowest terms, if possible.*

21. $\frac{3}{9}$ **22.** $\frac{5}{20}$

23. $\frac{7}{21}$ **24.** $\frac{6}{30}$

25. $\frac{20}{30}$ **26.** $\frac{12}{30}$

27. $\frac{15}{6}$ **28.** $\frac{24}{16}$

29. $-\frac{28}{56}$ **30.** $-\frac{45}{54}$

31. $-\frac{90}{105}$ **32.** $-\frac{26}{78}$

33. $\frac{60}{108}$ **34.** $\frac{75}{125}$

35. $\frac{180}{210}$ **36.** $\frac{76}{28}$

37. $\frac{55}{67}$ **38.** $\frac{41}{51}$

39. $\frac{36}{96}$ **40.** $\frac{48}{120}$

41. $\frac{25}{35}$ **42.** $\frac{16}{20}$

43. $\frac{12}{15}$ **44.** $\frac{10}{15}$

45. $\frac{6}{7}$ **46.** $\frac{4}{5}$

47. $\frac{7}{8}$ **48.** $\frac{10}{21}$

49. $-\frac{10}{30}$

50. $-\frac{14}{28}$

51. $\frac{15}{25}$

52. $\frac{16}{24}$

53. $\frac{35}{28}$

54. $\frac{35}{25}$

55. $\frac{56}{28}$

56. $\frac{32}{8}$

Write each fraction as an equivalent fraction with the indicated denominator.

57. $\frac{7}{8}$, denominator 40

58. $\frac{3}{4}$, denominator 24

59. $\frac{4}{5}$, denominator 35

60. $\frac{5}{7}$, denominator 49

61. $\frac{5}{6}$, denominator 54

62. $\frac{11}{16}$, denominator 32

63. $\frac{1}{2}$, denominator 30

64. $\frac{1}{3}$, denominator 60

65. $\frac{2}{7}$, denominator 14

66. $\frac{3}{10}$, denominator 50

67. $\frac{9}{10}$, denominator 60

68. $\frac{2}{3}$, denominator 27

69. $\frac{5}{4}$, denominator 20

70. $\frac{9}{4}$, denominator 44

71. $\frac{2}{15}$, denominator 45

72. $\frac{5}{12}$, denominator 36

Write each number as a fraction with the indicated denominator.

73. 3 as fifths

74. 4 as thirds

75. 6 as eighths

76. 3 as sixths

77. 4 as ninths

78. 7 as fourths

79. −2 as halves

80. −10 as ninths

APPLICATIONS *Use fractions to answer each question.*

81. COMMUTING How much of the commute from home to work has the motorist in the illustration made?

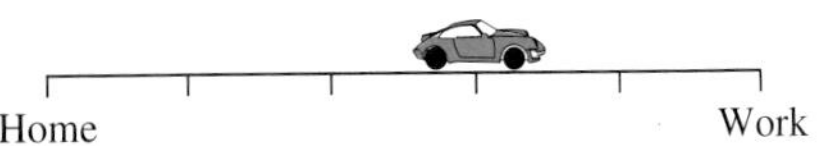

82. TIME CLOCKS For each clock, how much of the hour has passed?

a. 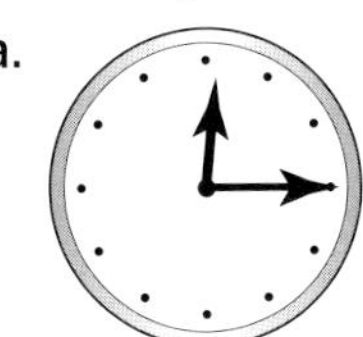

b. 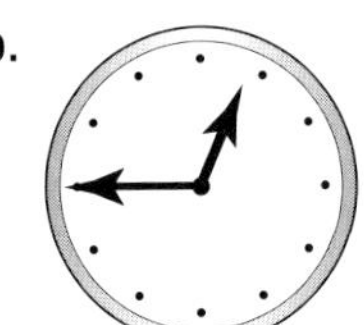

c. 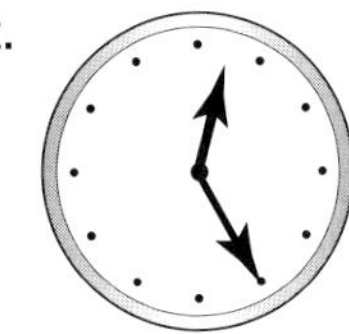

d. 

83. SINKHOLES The illustration shows a side view of a depression in the sidewalk near a sinkhole. Describe the movement of the sidewalk using a signed number. (On the tape measure, 1 inch is divided into 16 equal parts.)

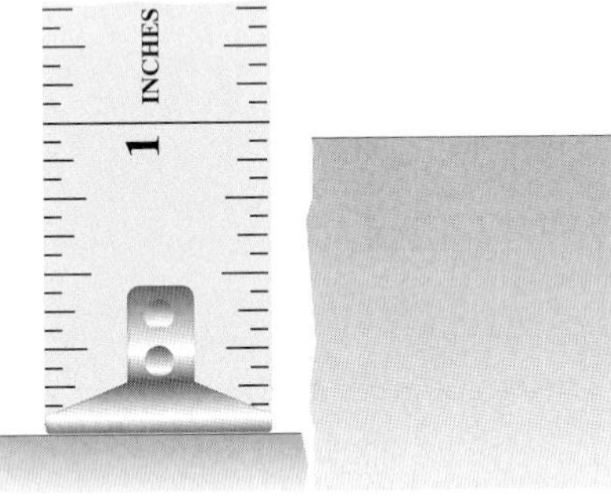

84. POLITICAL PARTIES The illustration shows the political party affiliation of the governors of the 50 states, as of January 1, 2000.

a. What fraction were Democrats?

b. What fraction were Republicans?

c. What fraction were neither Democrat nor Republican?

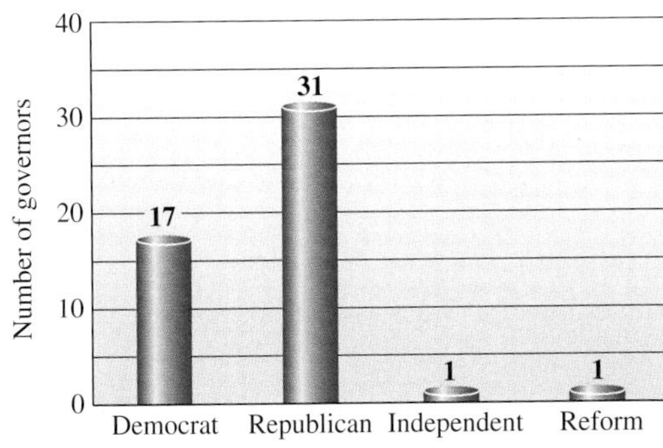

85. PERSONNEL RECORDS Complete the table by finding the amount of the job that will be completed by each person working alone for the given number of hours.

Name	Total time to complete the job alone	Time worked alone	Amount of job completed
Bob	10 hours	7 hours	
Ali	8 hours	1 hour	

86. GAS TANKS How full does the gauge indicate the gas tank is? How much of the tank has been used?

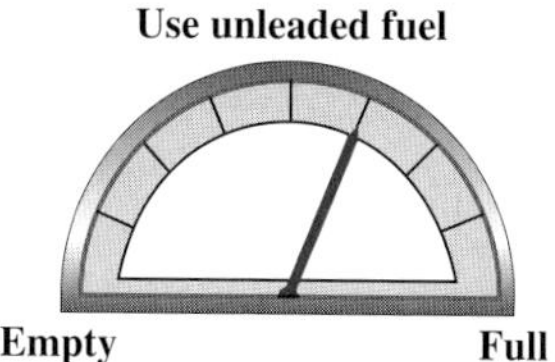

87. MUSIC The illustration shows the finger position needed to produce a length of string (from the bridge to the fingertip) that gives low C on a violin. To play other notes, fractions of that length are used. Locate these finger positions.

a. $\frac{1}{2}$ of the length gives middle C.

b. $\frac{3}{4}$ of the length gives F above low C.

c. $\frac{2}{3}$ of the length gives G.

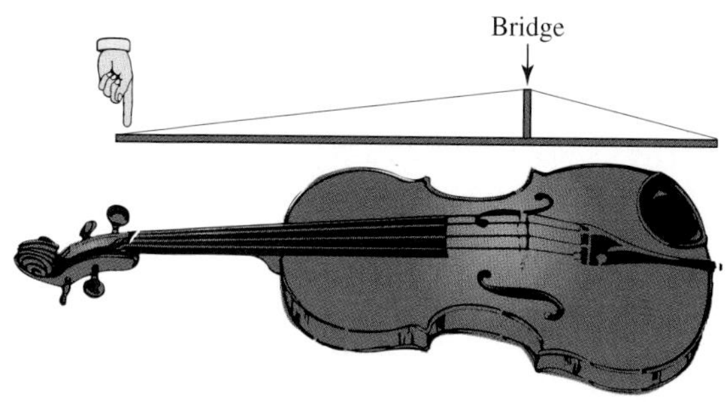

88. RULERS On the ruler, determine how many spaces are between the numbers 0 and 1. Then determine to what number the arrow is pointing.

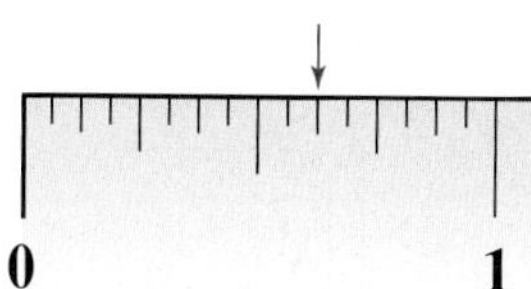

89. MACHINERY The operator of a machine is to turn the dial shown below from setting A to setting B. Express this instruction in two different ways, using fractions of one complete revolution.

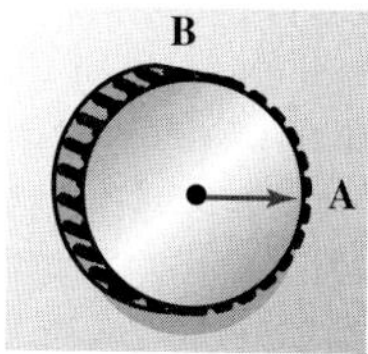

90. EARTH'S ROTATION The Earth rotates about its vertical axis once every 24 hours.

a. What is the significance of $\frac{1}{24}$ of a rotation to us on Earth?

b. What significance does $\frac{24}{24}$ of a revolution have?

91. SUPERMARKET DISPLAYS The amount of space to be given each type of snack food in a supermarket display case is expressed as a fraction. Complete the model of the display, showing where the adjustable shelves should be located, and label where each snack food should be stocked.

$\frac{3}{8}$: potato chips

$\frac{2}{8}$: peanuts

$\frac{1}{8}$: pretzels

$\frac{2}{8}$: tortilla chips

SNACKS

92. MEDICAL CENTERS Hospital designers have located a nurse's station at the center of a circular building. Use the circle graph to show how to divide the surrounding office space so that each medical department has the proper fractional amount allocated to it. Label each department.

$\frac{2}{12}$: Radiology

$\frac{5}{12}$: Pediatrics

$\frac{1}{12}$: Laboratory

$\frac{3}{12}$: Orthopedics

$\frac{1}{12}$: Pharmacy

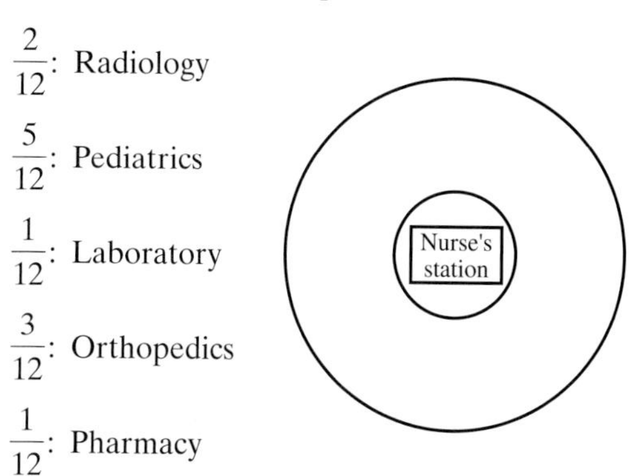

93. CAMERAS When the shutter of a camera stays open longer than $\frac{1}{125}$ second, any movement of the camera will probably blur the picture. With this in mind, if a photographer is taking a picture of a fast-moving object, should she select a shutter speed of $\frac{1}{60}$ or $\frac{1}{250}$?

94. GROSS DOMESTIC PRODUCT The GDP is the official measure of the size of the U.S. economy. It represents the market value of all goods and services that have been bought during a given period of time. The GDP for the second quarter of 2004 is listed below. What is meant by the phrase *second quarter of 2004?*

Second quarter of 2004	\$11,657,000,000

WRITING

95. Explain the concept of equivalent fractions.
96. What does it mean for a fraction to be in lowest terms?
97. Explain the difference between three-fourths and three-fifths of a pizza.
98. Explain both parts of the fundamental property of fractions.

REVIEW

99. Solve: $-5x + 1 = 16$.
100. Solve: $\frac{y}{2} + 2 = 4$.
101. Round 564,112 to the nearest thousand.
102. Give the definition of a prime number.

3.2 Multiplying Fractions

- Multiplying fractions
- Simplifying when multiplying fractions
- Powers of a fraction
- Applications

In the next three sections, we will discuss how to add, subtract, multiply, and divide fractions. We begin with the operation of multiplication.

Multiplying fractions

Suppose that a television network is going to take out a full-page ad to publicize its fall lineup of shows. The prime-time shows are to get $\frac{3}{5}$ of the ad space and daytime programming the remainder. Of the space devoted to prime time, $\frac{1}{2}$ is to be used to promote weekend programs. How much of the newspaper page will be used to advertise weekend prime-time programs?

The ad for the weekend prime-time shows will occupy $\frac{1}{2}$ of $\frac{3}{5}$ of the page. This can be expressed as $\frac{1}{2} \cdot \frac{3}{5}$. We can calculate $\frac{1}{2} \cdot \frac{3}{5}$ using a three-step process, illustrated below.

Step 1: We divide the page into fifths and shade three of them. This represents the fraction $\frac{3}{5}$, the amount of the page used to advertise prime-time shows.

Prime-time ad space

Step 2: Next, we find $\frac{1}{2}$ of the shaded part of the page by dividing the page into halves, using a vertical line.

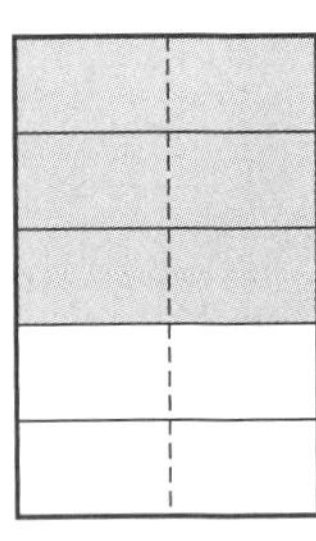

Step 3: Finally, we highlight (in purple) $\frac{1}{2}$ of the shaded parts determined in step 2. The highlighted parts are 3 out of 10 or $\frac{3}{10}$ of the page. They represent the amount of the page used to advertise the weekend prime-time shows. This leads us to the conclusion that $\frac{1}{2} \cdot \frac{3}{5} = \frac{3}{10}$.

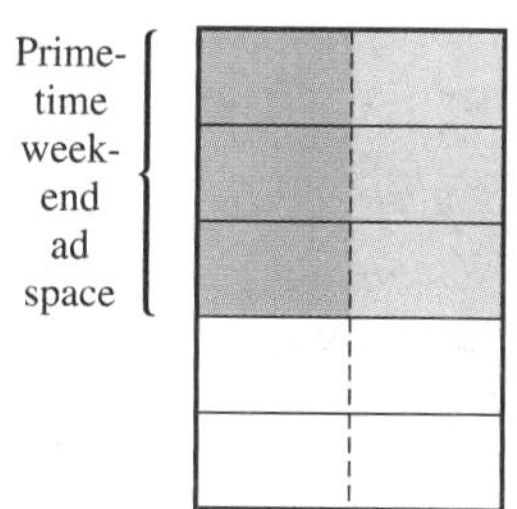

Two observations can be made from this result.

- The numerator of the answer is the product of the numerators of the original fractions.

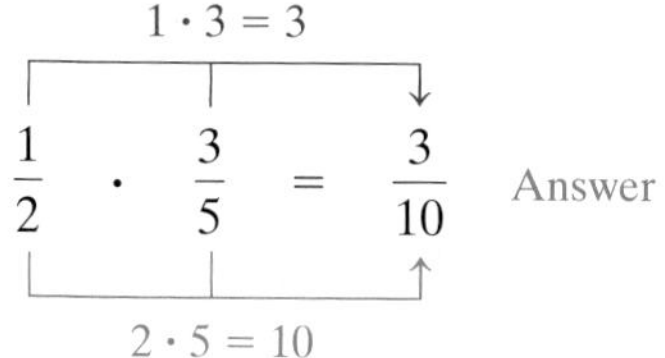

- The denominator of the answer is the product of the denominators of the original fractions.

These observations suggest the following rule for multiplying two fractions.

Multiplying fractions

To multiply two fractions, multiply the numerators and multiply the denominators. In symbols, if a, b, c, and d represent numbers and b and d are not 0,

$$\frac{a}{b} \cdot \frac{c}{d} = \frac{a \cdot c}{b \cdot d}$$

EXAMPLE 1 Multiply: $\frac{7}{8} \cdot \frac{3}{5}$.

Solution

$$\frac{7}{8} \cdot \frac{3}{5} = \frac{7 \cdot 3}{8 \cdot 5}$$ Multiply the numerators and multiply the denominators.

$$= \frac{21}{40}$$ Perform the multiplications: $7 \cdot 3 = 21$ and $8 \cdot 5 = 40$.

Self Check 1

Multiply: $\frac{5}{9} \cdot \frac{2}{3}$.

Answer $\frac{10}{27}$

The rules for multiplying integers also hold for multiplying fractions. When we multiply two fractions with *like* signs, the product is positive. When we multiply two fractions with *unlike* signs, the product is negative.

EXAMPLE 2 Multiply: $-\frac{3}{4}\left(\frac{1}{8}\right)$.

Solution

$$-\frac{3}{4}\left(\frac{1}{8}\right) = -\frac{3 \cdot 1}{4 \cdot 8}$$ Multiply the numerators and multiply the denominators. Since the fractions have unlike signs, the product is negative.

$$= -\frac{3}{32}$$ Perform the multiplications: $3 \cdot 1 = 3$ and $4 \cdot 8 = 32$.

Self Check 2

Multiply: $\frac{5}{6}\left(-\frac{1}{3}\right)$.

Answer $-\frac{5}{18}$

Simplifying when multiplying fractions

After multiplying two fractions, we should simplify the result, if possible.

Self Check 3

Multiply: $\frac{6}{25} \cdot \frac{5}{6}$.

EXAMPLE 3 Multiply: $\frac{5}{8} \cdot \frac{4}{5}$.

Solution

$\frac{5}{8} \cdot \frac{4}{5} = \frac{5 \cdot 4}{8 \cdot 5}$ Multiply the numerators and multiply the denominators.

$= \frac{5 \cdot 4}{4 \cdot 2 \cdot 5}$ In the denominator, factor 8 as $4 \cdot 2$ so that we can simplify the fraction.

$= \frac{\overset{1}{\cancel{5}} \cdot \overset{1}{\cancel{4}}}{\underset{1}{\cancel{4}} \cdot 2 \cdot \underset{1}{\cancel{5}}}$ Divide out the common factors 4 and 5.

$= \frac{1}{2}$ Perform the multiplications: $1 \cdot 1 = 1$ and $1 \cdot 2 \cdot 1 = 2$.

Answer $\frac{1}{5}$

EXAMPLE 4 Multiply: $45\left(-\frac{1}{14}\right)\left(-\frac{7}{10}\right)$.

Solution This expression involves three factors, and one of them is an integer. When multiplying fractions and integers, we express each integer as a fraction with a denominator 1.

Recall that a product is positive when there are an even number of negative factors. Since $45\left(-\frac{1}{14}\right)\left(-\frac{7}{10}\right)$ has two negative factors, the product is positive.

$45\left(-\frac{1}{14}\right)\left(-\frac{7}{10}\right) = \frac{45}{1}\left(\frac{1}{14}\right)\left(\frac{7}{10}\right)$ Write 45 as a fraction: $45 = \frac{45}{1}$. Since the product has an even number of negative factors, the answer is positive. We can drop both − signs and continue.

$= \frac{\mathbf{45} \cdot 1 \cdot 7}{1 \cdot \mathbf{14} \cdot \mathbf{10}}$ Multiply the numerators and multiply the denominators.

$= \frac{\mathbf{3 \cdot 3 \cdot 5} \cdot 1 \cdot 7}{1 \cdot \mathbf{2 \cdot 7} \cdot \mathbf{2 \cdot 5}}$ Factor 45 as $3 \cdot 3 \cdot 5$, factor 14 as $2 \cdot 7$, and factor 10 as $2 \cdot 5$.

$= \frac{3 \cdot 3 \cdot \overset{1}{\cancel{5}} \cdot 1 \cdot \overset{1}{\cancel{7}}}{1 \cdot 2 \cdot \underset{1}{\cancel{7}} \cdot 2 \cdot \underset{1}{\cancel{5}}}$ Divide out the common factors 5 and 7.

$= \frac{9}{4}$ Multiply in the numerator and in the denominator.

! COMMENT The answer in Example 4 was $\frac{9}{4}$, an improper fraction. In arithmetic, we often write such fractions as mixed numbers. In algebra, it is often more useful to leave $\frac{9}{4}$ in this form. We will discuss improper fractions and mixed numbers in more detail in Section 3.5.

In the next example, we multiply an expression containing a variable by a fraction.

EXAMPLE 5 Multiply: $\frac{1}{4}(4y)$.

Solution

$$\frac{1}{4}(4y) = \frac{1}{4} \cdot \frac{4y}{1}$$ Write $4y$ as a fraction: $4y = \frac{4y}{1}$.

$$= \frac{1 \cdot 4 \cdot y}{4 \cdot 1}$$ Multiply the numerators and multiply the denominators.

$$= \frac{1 \cdot \overset{1}{\cancel{4}} \cdot y}{\underset{1}{\cancel{4}} \cdot 1}$$ Divide out the common factor 4.

$$= \frac{y}{1}$$ Multiply in the numerator and in the denominator.

$$= y$$ Simplify: $\frac{y}{1} = y$.

Self Check 5

Multiply: $\frac{1}{5} \cdot 5m$.

Answer m

To multiply $\frac{1}{2}$ and x, we can express the product as $\frac{1}{2}x$, or we can use the concept of multiplying fractions to write it in a different form.

$$\frac{1}{2} \cdot x = \frac{1}{2} \cdot \frac{x}{1}$$ Write x as a fraction: $x = \frac{x}{1}$.

$$= \frac{1 \cdot x}{2 \cdot 1}$$ Multiply the numerators and multiply the denominators.

$$= \frac{x}{2}$$ Multiply in the numerator and in the denominator.

The product of $\frac{1}{2}$ and x can be expressed as $\frac{1}{2}x$ or $\frac{x}{2}$. Similarly, $\frac{3}{4}t = \frac{3t}{4}$ and $-\frac{5}{16}y = -\frac{5y}{16}$.

Powers of a fraction

If the base of an exponential expression is a fraction, the exponent tells us how many times to write that fraction as a factor. For example,

$$\left(\frac{2}{3}\right)^2 = \frac{2}{3} \cdot \frac{2}{3} = \frac{2 \cdot 2}{3 \cdot 3} = \frac{4}{9}$$ $\frac{2}{3}$ is used as a factor 2 times.

EXAMPLE 6 Find the power: $\left(-\frac{4}{5}\right)^2$.

Solution Exponents are used to indicate repeated multiplication.

$$\left(-\frac{4}{5}\right)^2 = \left(-\frac{4}{5}\right)\left(-\frac{4}{5}\right)$$ Write $-\frac{4}{5}$ as a factor 2 times.

$$= \frac{4 \cdot 4}{5 \cdot 5}$$ The product of two fractions with like signs is positive. Multiply the numerators and multiply the denominators.

$$= \frac{16}{25}$$ Multiply in the numerator and in the denominator.

Self Check 6

Find the power: $\left(-\frac{3}{4}\right)^3$.

Answer $-\frac{27}{64}$

Applications

EXAMPLE 7 **House of Representatives.** In the United States House of Representatives, a bill was introduced that would require a $\frac{3}{5}$ vote of the 435 members to authorize any tax increase. Under this requirement, how many representatives would have to vote for a tax increase before it could become law?

Solution

$$\frac{3}{5} \text{ of } 435 = \frac{3}{5} \cdot \frac{435}{1}$$ Here, the word *of* means to multiply. Write 435 as a fraction: $435 = \frac{435}{1}$.

$$= \frac{3 \cdot 435}{5 \cdot 1}$$ Multiply the numerators and multiply the denominators.

$$= \frac{3 \cdot 3 \cdot 5 \cdot 29}{5 \cdot 1}$$ Prime factor 435 as $3 \cdot 5 \cdot 29$.

$$= \frac{3 \cdot 3 \cdot \overset{1}{\cancel{5}} \cdot 29}{\underset{1}{\cancel{5}} \cdot 1}$$ Divide out the common factor 5.

$$= \frac{261}{1}$$ Multiply in the numerators: $3 \cdot 3 \cdot 1 \cdot 29 = 261$. Multiply in the denominator: $1 \cdot 1 = 1$.

$$= 261$$ Simplify: $\frac{261}{1} = 261$.

It would take 261 representatives voting in favor to pass a tax increase.

As Figure 3-3 shows, a triangle has three sides. The length of the base of the triangle can be represented by the letter b and the height by the letter h. The height of a triangle is always perpendicular (makes a square corner) to the base. This is denoted by the symbol ⌉.

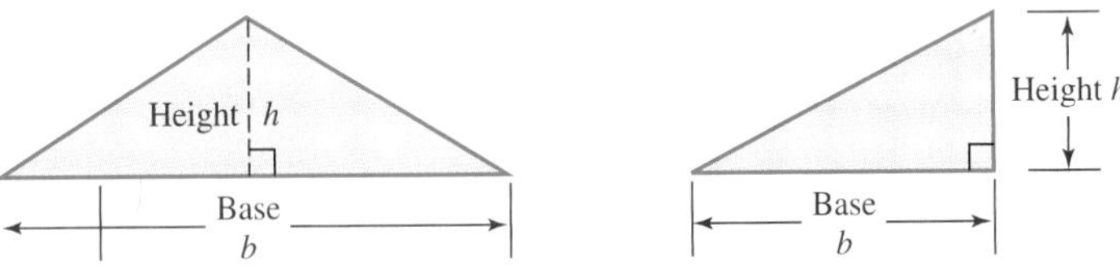

FIGURE 3-3

Recall that the area of a figure is the amount of surface that it encloses. The area of a triangle can be found by using the following formula.

Area of a triangle

The area A of a triangle is one-half the product of its base b and its height h.

$$\text{Area} = \frac{1}{2}(\text{base})(\text{height}) \quad \text{or} \quad A = \frac{1}{2}bh$$

EXAMPLE 8 **Geography.** Approximate the area of Virginia using the triangle in Figure 3-4.

Solution We will approximate the area of the state by finding the area of the triangle.

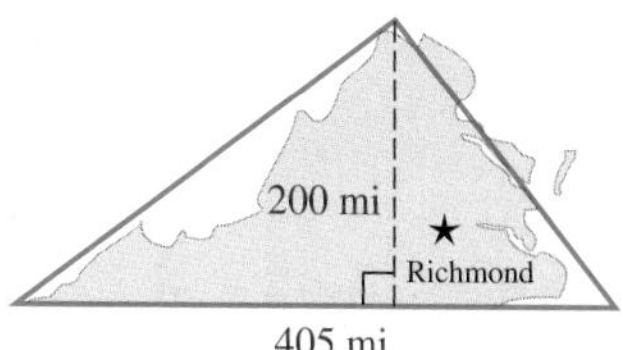

FIGURE 3-4

$A = \frac{1}{2}bh$	This is the formula for the area of a triangle.
$= \frac{1}{2}(405)(200)$	Substitute 405 for b and 200 for h.
$= \frac{1}{2}\left(\frac{405}{1}\right)\left(\frac{200}{1}\right)$	Write 405 and 200 as fractions.
$= \frac{1 \cdot 405 \cdot 200}{2 \cdot 1 \cdot 1}$	Multiply the numerators. Multiply the denominators.
$= \frac{1 \cdot 405 \cdot \overset{1}{\cancel{2}} \cdot 100}{\underset{1}{\cancel{2}} \cdot 1 \cdot 1}$	Factor 200 as $2 \cdot 100$. Then divide out the common factor 2.
$= 40{,}500$	Multiply in the numerator: $405 \cdot 100 = 40{,}500$.

The area of Virginia is approximately 40,500 square miles.

Section 3.2 STUDY SET

VOCABULARY *Fill in the blanks.*

1. The word *of* in mathematics usually means ________.

2. The ______ of a triangle is the amount of surface that it encloses.

3. The result of a multiplication problem is called the ________.

4. To ________ a fraction means to divide out common factors of the numerator and denominator.

5. In the formula for the area of a triangle, $A = \frac{1}{2}bh$, b stands for the length of the ______ and h stands for the _______.

6. A ________ is an equation that mathematically describes a known relationship between two or more variables.

CONCEPTS

7. Find the result when multiplying $\frac{a}{b} \cdot \frac{c}{d}$.

8. Write each of the following as fractions.

a. 4

b. -3

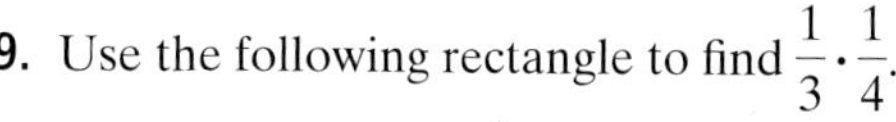

9. Use the following rectangle to find $\frac{1}{3} \cdot \frac{1}{4}$.

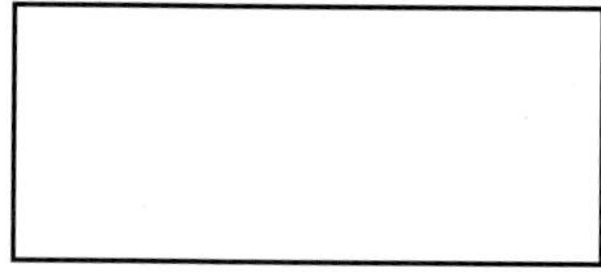

a. Using vertical lines, divide the given rectangle into four equal parts and lightly shade one of them. What fractional part of the rectangle did you shade?

b. To find $\frac{1}{3}$ of the shaded portion, use two horizontal lines to divide the given rectangle into three equal parts and lightly shade one of them. Into how many equal parts is the rectangle now divided? How many parts have been shaded twice? What is $\frac{1}{3} \cdot \frac{1}{4}$?

10. In the following solution, what mistake did the student make that caused him to work with such large numbers?

$$\frac{44}{63} \cdot \frac{27}{55} = \frac{44 \cdot 27}{63 \cdot 55}$$
$$= \frac{1{,}188}{3{,}465}$$

11. a. Is the product of two numbers with unlike signs positive or negative?
    b. Is the product of two numbers with like signs positive or negative?
12. a. Multiply $\frac{9}{10}$ and 20.
    b. When we multiply two numbers, is the product always larger than both those numbers?
13. Determine whether each statement is true or false.
    a. $\frac{1}{2}x = \frac{x}{2}$
    b. $\frac{2t}{3} = \frac{2}{3}t$
    c. $-\frac{3}{8}a = -\frac{3}{8a}$
    d. $\frac{-4e}{7} = -\frac{4e}{7}$
14. What is the numerator of the result for the multiplication problem shown here?

$$\frac{4}{15} \cdot \frac{3}{4} = \frac{\overset{1}{\cancel{4}} \cdot \overset{1}{\cancel{3}}}{5 \cdot \underset{1}{\cancel{3}} \cdot \underset{1}{\cancel{4}}}$$

NOTATION *Complete each solution.*

15. $$\frac{5}{8} \cdot \frac{7}{15} = \frac{5 \cdot \square}{8 \cdot \square}$$
$$= \frac{5 \cdot 7}{8 \cdot \square \cdot 5}$$
$$= \frac{\overset{1}{\cancel{5}} \cdot 7}{8 \cdot \square \cdot \underset{1}{\cancel{5}}}$$
$$= \frac{7}{24}$$

16. $$\frac{7}{12} \cdot \frac{4}{21} = \frac{7 \cdot 4}{\square \cdot \square}$$
$$= \frac{7 \cdot 4}{\square \cdot 4 \cdot 3 \cdot \square}$$
$$= \frac{\overset{1}{\cancel{7}} \cdot \overset{1}{\cancel{4}}}{3 \cdot \underset{\square}{\cancel{4}} \cdot 3 \cdot \underset{\square}{\cancel{7}}}$$
$$= \frac{1}{9}$$

PRACTICE *Multiply. Write all answers in lowest terms.*

17. $\frac{1}{4} \cdot \frac{1}{2}$
18. $\frac{1}{3} \cdot \frac{1}{5}$
19. $\frac{3}{8} \cdot \frac{7}{16}$
20. $\frac{5}{9} \cdot \frac{2}{7}$
21. $\frac{2}{3} \cdot \frac{6}{7}$
22. $\frac{5}{12} \cdot \frac{3}{4}$
23. $\frac{14}{15} \cdot \frac{11}{8}$
24. $\frac{5}{16} \cdot \frac{8}{3}$
25. $-\frac{15}{24} \cdot \frac{8}{25}$
26. $-\frac{20}{21} \cdot \frac{7}{16}$
27. $\left(-\frac{11}{21}\right)\left(-\frac{14}{33}\right)$
28. $\left(-\frac{16}{35}\right)\left(-\frac{25}{48}\right)$
29. $\frac{7}{10}\left(\frac{20}{21}\right)$
30. $\left(\frac{7}{6}\right)\frac{9}{49}$
31. $\frac{3}{4} \cdot \frac{4}{3}$
32. $\frac{4}{5} \cdot \frac{5}{4}$
33. $\frac{1}{3} \cdot \frac{15}{16} \cdot \frac{4}{25}$
34. $\frac{3}{15} \cdot \frac{15}{7} \cdot \frac{14}{27}$
35. $\left(\frac{2}{3}\right)\left(-\frac{1}{16}\right)\left(-\frac{4}{5}\right)$
36. $\left(\frac{3}{8}\right)\left(-\frac{2}{3}\right)\left(-\frac{12}{27}\right)$
37. $\frac{5}{6} \cdot 18$
38. $6\left(-\frac{2}{3}\right)$
39. $15\left(-\frac{4}{5}\right)$
40. $-2\left(-\frac{7}{8}\right)$
41. $\frac{5x}{12} \cdot \frac{1}{6}$
42. $\frac{2t}{3} \cdot \frac{7}{8}$
43. $\frac{b}{12} \cdot \frac{3}{10}$
44. $\frac{5c}{8} \cdot \frac{1}{15}$
45. $\frac{1}{3} \cdot 3d$
46. $\frac{1}{16} \cdot 16x$
47. $\frac{2}{3} \cdot \frac{3s}{2}$
48. $\frac{3}{5} \cdot \frac{5h}{3}$
49. $-\frac{5}{6} \cdot \frac{6}{5}c$
50. $-\frac{5}{7} \cdot \frac{7}{5}w$
51. $\frac{x}{2} \cdot \frac{4}{9}$
52. $\frac{c}{5} \cdot \frac{25}{36}$

Multiply and express the product in two ways.

53. $\frac{5}{6} \cdot x$
54. $\frac{2}{3} \cdot y$
55. $-\frac{8}{9} \cdot v$
56. $-\frac{7}{6} \cdot m$

Find each power.

57. $\left(\frac{2}{3}\right)^2$
58. $\left(\frac{3}{5}\right)^2$
59. $\left(-\frac{5}{9}\right)^2$
60. $\left(-\frac{5}{6}\right)^2$
61. $\left(\frac{4}{3}\right)^2$
62. $\left(\frac{3}{2}\right)^2$
63. $\left(-\frac{3}{4}\right)^3$
64. $\left(-\frac{2}{5}\right)^3$

65. Complete the multiplication table of fractions.

$\cdot$	$\frac{1}{2}$	$\frac{1}{3}$	$\frac{1}{4}$	$\frac{1}{5}$	$\frac{1}{6}$
$\frac{1}{2}$					
$\frac{1}{3}$					
$\frac{1}{4}$					
$\frac{1}{5}$					
$\frac{1}{6}$					

66. Complete the table by finding the original fraction.

Original fraction squared	Original fraction
$\frac{1}{9}$	
$\frac{1}{100}$	
$\frac{4}{25}$	
$\frac{16}{49}$	
$\frac{81}{36}$	
$\frac{9}{121}$	

Find the area of each triangle.

67.

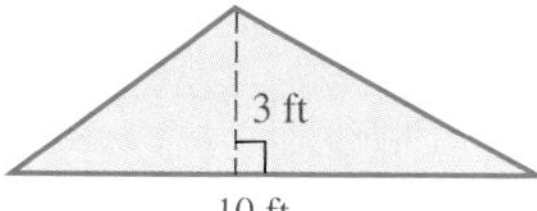

68.

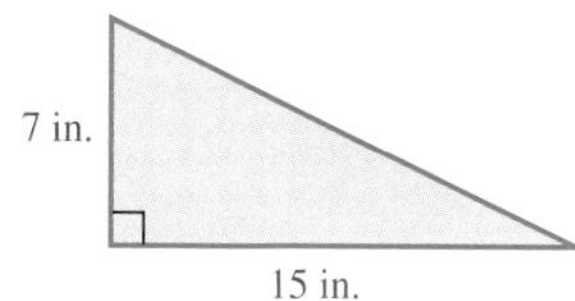

69.

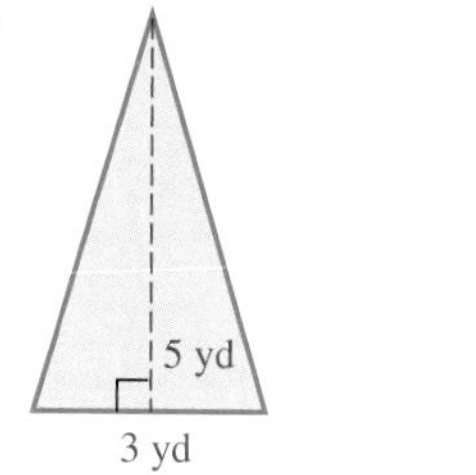

70.

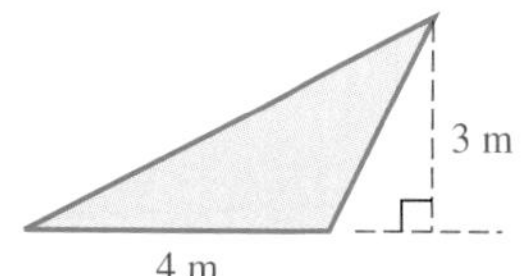

APPLICATIONS

71. THE CONSTITUTION Article V of the United States Constitution requires a two-thirds vote of the House of Representatives to propose a constitutional amendment. The House has 435 members. Find the number of votes needed to meet this requirement.

72. GENETICS Gregor Mendel (1822–1884), an Augustinian monk, is credited with developing a heredity model that became the foundation of modern genetics. In his experiments, he crossed purple-flowered plants with white-flowered plants and found that $\frac{3}{4}$ of the third-generation offspring plants had purple flowers and $\frac{1}{4}$ had white flowers. According to this concept, when the group of third-generation offspring plants shown below flower, how many will have purple flowers?

73. TENNIS BALLS A tennis ball is dropped from a height of 54 inches. Each time it hits the ground, it rebounds one-third of the previous height it fell. Find the three missing rebound heights in the illustration.

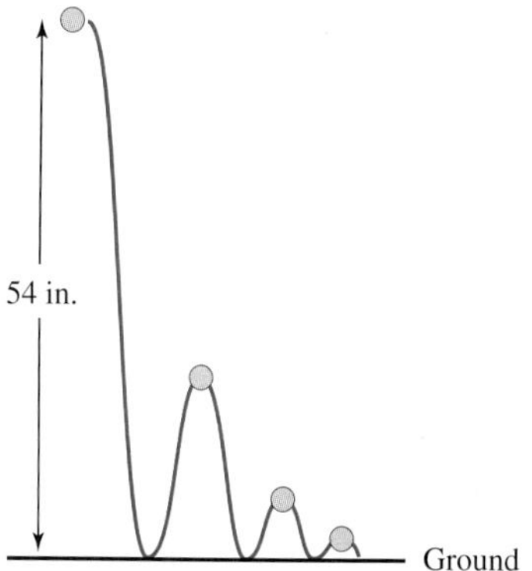

74. ELECTIONS The following illustration shows the final election returns for a city bond measure.

a. How many votes were cast?

b. Find two-thirds of the number of votes cast.

c. Did the bond measure pass?

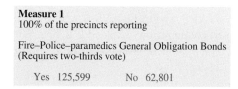

Measure 1
100% of the precincts reporting

Fire–Police–paramedics General Obligation Bonds
(Requires two-thirds vote)

Yes 125,599 No 62,801

75. COOKING Use the recipe below, along with the concept of multiplication with fractions, to find how much sugar and molasses are needed to make one dozen cookies.

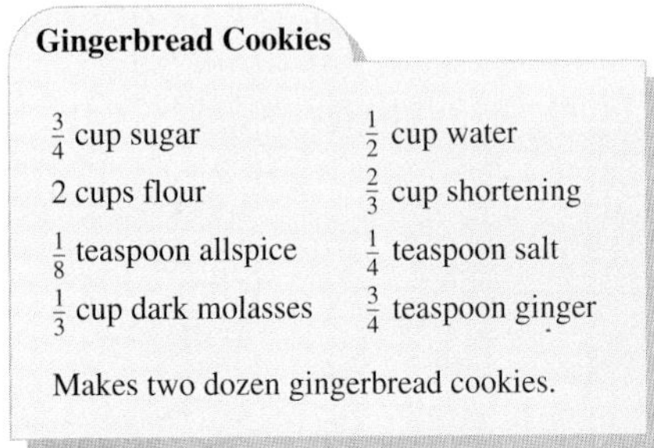

Gingerbread Cookies
3/4 cup sugar
2 cups flour
1/8 teaspoon allspice
1/3 cup dark molasses
1/2 cup water
2/3 cup shortening
1/4 teaspoon salt
3/4 teaspoon ginger
Makes two dozen gingerbread cookies.

76. THE EARTH'S SURFACE The surface of the Earth covers an area of approximately 196,800,000 square miles. About $\frac{3}{4}$ of that area is covered by water. Find the number of square miles of the surface covered by water.

77. BOTANY In an experiment, monthly growth rates of three types of plants doubled when nitrogen was added to the soil. Complete the illustration by charting the improved growth rate next to each normal growth rate.

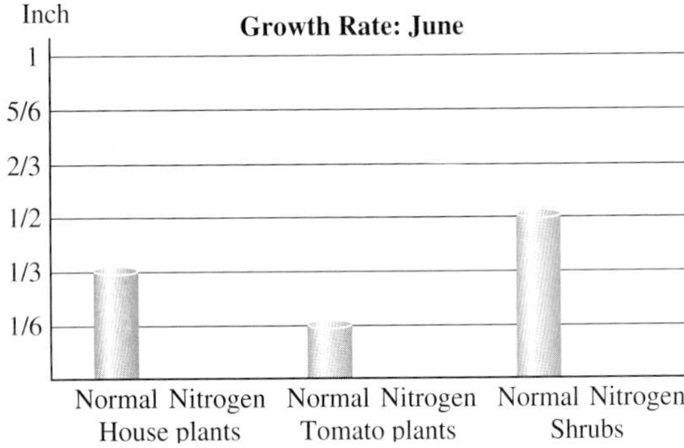

78. STAMPS The best designs in a contest to create a wildlife stamp are shown. To save on paper costs, the postal service has decided to choose the stamp that has the smaller area. Which one is that?

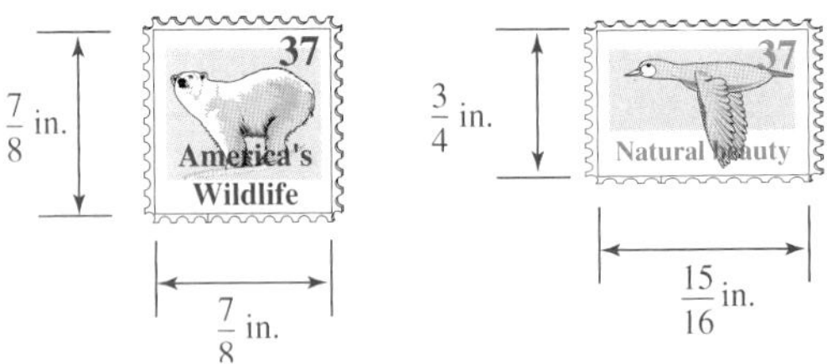

79. THE STARS AND STRIPES The illustration in the next column shows a folded U.S. flag. When it is placed on a table as part of an exhibit, how much area will it occupy?

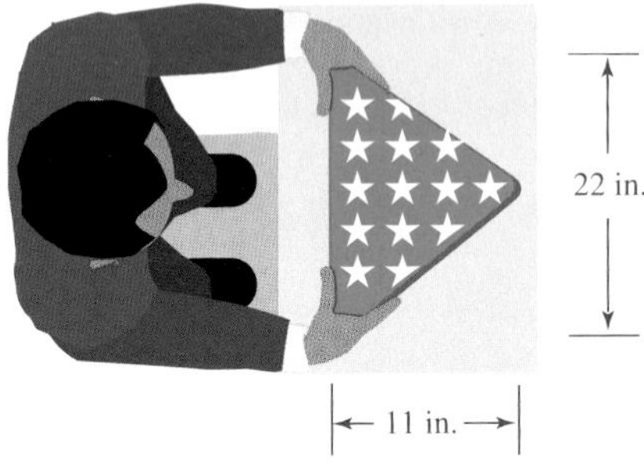

80. WINDSURFING Estimate the area of the sail on the windsurfing board shown below.

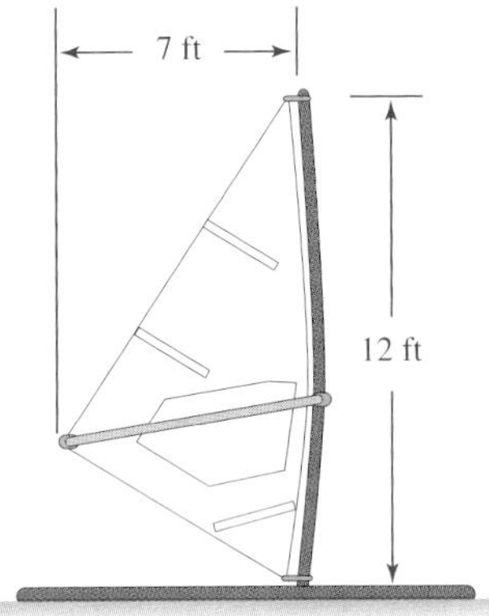

81. TILE DESIGNS A design for bathroom tile is shown. Find the amount of area on a tile that is blue.

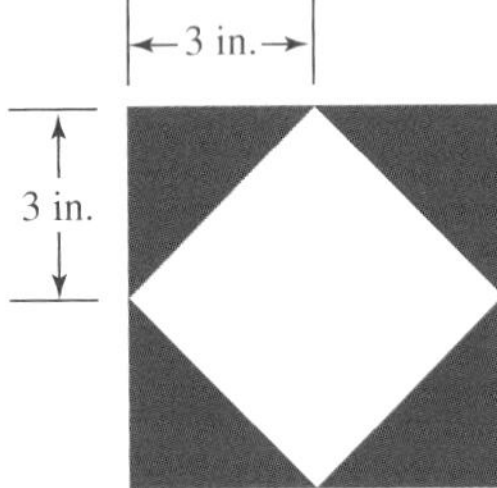

82. GEOGRAPHY Estimate the area of New Hampshire, using the triangle shown below.

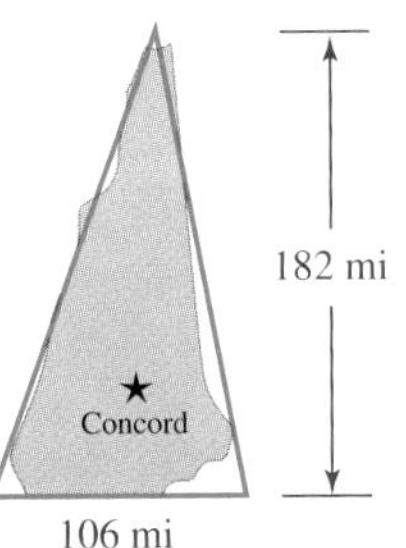

WRITING

83. In mathematics, the word *of* usually means multiply. Give three real-life examples of this usage.

84. Explain how you could multiply the number 5 and another number and obtain an answer that is less than 5.

85. MAJORITIES The definition of the word *majority* is as follows: "a number greater than one-half of the total." Explain what it means when a teacher says, "A majority of the class voted to postpone the test until Monday." Give an example.

86. What does area measure?

REVIEW

87. Round 987,459 to the nearest thousand.

88. Solve: $3x + 1 = 7$.

89. Is -6 a solution of $2x + 6 = 6$?

90. Find the prime factorization of 125.

3.3 Dividing Fractions

• Division with fractions • Reciprocals • A rule for dividing fractions

In this section, we will discuss how to divide fractions. We will examine problems involving positive and negative fractions as well as fractions containing variables. The skills you learned in Section 3.2 will be useful in this section.

Division with fractions

Suppose that the manager of a candy store buys large bars of chocolate and divides each one into four equal parts to sell. How many fourths can be obtained from 5 bars?

We are asking: How many $\frac{1}{4}$'s are there in 5? To answer the question, we need to use the operation of division. We can represent this division as $5 \div \frac{1}{4}$. (See Figure 3-5.)

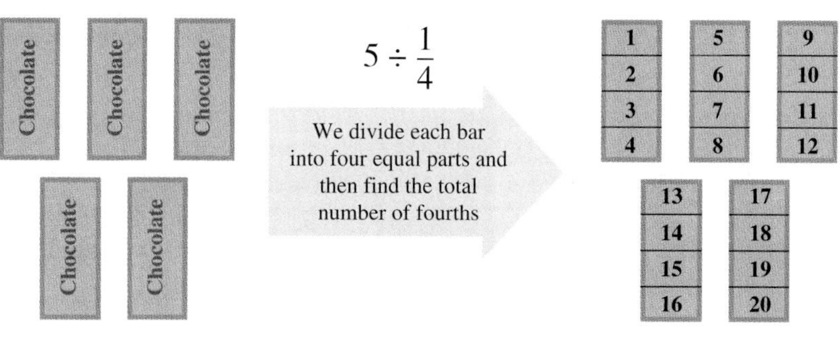

FIGURE 3-5

There are 20 fourths in the 5 bars of chocolate. This result leads to the following observations.

- This division problem involves a fraction: $5 \div \frac{1}{4}$.
- Although we were asked to find $5 \div \frac{1}{4}$, we solved the problem using *multiplication* instead of *division:* $5 \cdot 4 = 20$.

Later in this section, we will see that these observations suggest a rule for dividing fractions. But before we can discuss that rule, we need to introduce a new term.

Reciprocals

Division with fractions involves working with **reciprocals.** To present the concept of reciprocal, we consider the problem $\frac{7}{8} \cdot \frac{8}{7}$.

$$\frac{7}{8} \cdot \frac{8}{7} = \frac{7 \cdot 8}{8 \cdot 7}$$ Multiply the numerators and multiply the denominators.

$$= \frac{\overset{1}{\cancel{7}} \cdot \overset{1}{\cancel{8}}}{\underset{1}{\cancel{8}} \cdot \underset{1}{\cancel{7}}}$$ Divide out the common factors of 7 and 8.

$$= \frac{1}{1}$$ Multiply in the numerator and in the denominator.

$$= 1$$ $\frac{1}{1} = 1$.

The product of $\frac{7}{8}$ and $\frac{8}{7}$ is 1.

Whenever the product of two numbers is 1, we say that those numbers are *reciprocals.* Therefore, $\frac{7}{8}$ and $\frac{8}{7}$ are reciprocals. To find the reciprocal of a fraction, *we invert the numerator and the denominator.*

Reciprocals

Two numbers are called **reciprocals** if their product is 1.

! COMMENT Zero does not have a reciprocal, because the product of 0 and a number can never be 1.

Self Check 1

For each number, find its reciprocal and show that their product is 1:

a. $\frac{3}{5}$

b. $-\frac{5}{6}$

c. 8

Answers **a.** $\frac{5}{3}$, **b.** $-\frac{6}{5}$ **c.** $\frac{1}{8}$

EXAMPLE 1 For each number, find its reciprocal and show that their product is 1: **a.** $\frac{2}{3}$, **b.** $-\frac{3}{4}$, and **c.** 5.

Solution

a. The reciprocal of $\frac{2}{3}$ is $\frac{3}{2}$.

$$\frac{2}{3} \cdot \frac{3}{2} = \frac{\overset{1}{\cancel{2}} \cdot \overset{1}{\cancel{3}}}{\underset{1}{\cancel{3}} \cdot \underset{1}{\cancel{2}}} = 1$$

b. The reciprocal of $-\frac{3}{4}$ is $-\frac{4}{3}$.

$$-\frac{3}{4}\left(-\frac{4}{3}\right) = \frac{\overset{1}{\cancel{3}} \cdot \overset{1}{\cancel{4}}}{\underset{1}{\cancel{4}} \cdot \underset{1}{\cancel{3}}} = 1$$ The product of two fractions with like signs is positive.

c. $5 = \frac{5}{1}$, so the reciprocal of 5 is $\frac{1}{5}$.

$$5 \cdot \frac{1}{5} = \frac{5}{1} \cdot \frac{1}{5} = \frac{\overset{1}{\cancel{5}} \cdot 1}{1 \cdot \underset{1}{\cancel{5}}} = 1$$

A rule for dividing fractions

In the candy store example, we saw that we can find $5 \div \frac{1}{4}$ by computing $5 \cdot 4$. That is, division by $\frac{1}{4}$ (a fraction) is the same as multiplication by 4 (the reciprocal of $\frac{1}{4}$).

$$5 \div \frac{1}{4} = 5 \cdot 4$$

This observation suggests a general rule for dividing fractions.

Dividing fractions

To divide fractions, multiply the first fraction by the reciprocal of the second fraction. In symbols, if a, b, c, and d represent numbers and b, c, and d are not 0, then

$$\frac{a}{b} \div \frac{c}{d} = \frac{a}{b} \cdot \frac{d}{c}$$

For example, to find $\frac{5}{7} \div \frac{3}{4}$, we multiply the first fraction by the reciprocal of the second.

Change the division to multiplication.

$$\frac{5}{7} \div \frac{3}{4} = \frac{5}{7} \cdot \frac{4}{3}$$

The reciprocal of $\frac{3}{4}$ is $\frac{4}{3}$.

$$= \frac{5 \cdot 4}{7 \cdot 3}$$ Multiply the numerators. Multiply the denominators.

$$= \frac{20}{21}$$ Multiply in the numerator and in the denominator: $5 \cdot 4 = 20$ and $7 \cdot 3 = 21$.

Therefore, $\frac{5}{7} \div \frac{3}{4} = \frac{20}{21}$.

EXAMPLE 2 Divide: $\frac{1}{3} \div \frac{4}{5}$.

Solution

$$\frac{1}{3} \div \frac{4}{5} = \frac{1}{3} \cdot \frac{5}{4}$$ Multiply $\frac{1}{3}$ by the reciprocal of $\frac{4}{5}$, which is $\frac{5}{4}$.

$$= \frac{1 \cdot 5}{3 \cdot 4}$$ Multiply the numerators and multiply the denominators.

$$= \frac{5}{12}$$ Multiply in the numerator and in the denominator.

Self Check 2

Divide: $\frac{2}{3} \div \frac{7}{8}$.

Answer $\frac{16}{21}$

Self Check 3

Divide: $\frac{4}{5} \div \frac{8}{25}$.

Answer $\frac{5}{2}$

EXAMPLE 3 Divide: $\frac{9}{16} \div \frac{3}{20}$.

Solution

$$\frac{9}{16} \div \frac{3}{20} = \frac{9}{16} \cdot \frac{20}{3}$$ Multiply $\frac{9}{16}$ by the reciprocal of $\frac{3}{20}$, which is $\frac{20}{3}$.

$$= \frac{9 \cdot 20}{16 \cdot 3}$$ Multiply the numerators and multiply the denominators.

$$= \frac{\overset{1}{\cancel{3}} \cdot 3 \cdot 5 \cdot \overset{1}{\cancel{4}}}{\underset{1}{\cancel{4}} \cdot 4 \cdot \underset{1}{\cancel{3}}}$$ Factor 9 as $3 \cdot 3$, factor 20 as $4 \cdot 5$, and factor 16 as $4 \cdot 4$. Then divide out the common factors 3 and 4.

$$= \frac{15}{4}$$ Multiply in the numerator and in the denominator.

EXAMPLE 4 **Surfboard designs.** Most surfboards are made of foam plastic covered with several layers of fiberglass to keep them water-tight. How many layers are needed to build up a finish three-eighths of an inch thick if each layer of fiberglass has a thickness of one-sixteenth of an inch?

Solution We need to know how many one-sixteenths there are in three-eighths. To answer this question, we will use division and find $\frac{3}{8} \div \frac{1}{16}$.

$$\frac{3}{8} \div \frac{1}{16} = \frac{3}{8} \cdot \frac{16}{1}$$ Multiply $\frac{3}{8}$ by the reciprocal of $\frac{1}{16}$, which is $\frac{16}{1}$.

$$= \frac{3 \cdot 16}{8 \cdot 1}$$ Multiply the numerators and multiply the denominators.

$$= \frac{3 \cdot 2 \cdot \overset{1}{\cancel{8}}}{\underset{1}{\cancel{8}} \cdot 1}$$ Factor 16 as $2 \cdot 8$. Then divide out the common factor 8.

$$= \frac{6}{1}$$ Multiply in the numerator and in the denominator.

$$= 6$$ Simplify: $\frac{6}{1} = 6$.

The number of layers of fiberglass to be applied is 6.

Self Check 5

Divide: $\frac{2}{3} \div \left(-\frac{7}{6}\right)$

Answer $-\frac{4}{7}$

EXAMPLE 5 Divide: $\frac{1}{6} \div \left(-\frac{1}{18}\right)$.

Solution When working with divisions involving negative fractions, we use the same rules as for multiplying numbers with like or unlike signs.

$$\frac{1}{6} \div \left(-\frac{1}{18}\right) = \frac{1}{6}\left(-\frac{18}{1}\right)$$ Multiply $\frac{1}{6}$ by the reciprocal of $-\frac{1}{18}$, which is $-\frac{18}{1}$.

$$= -\frac{1 \cdot 18}{6 \cdot 1}$$ The product of two fractions with unlike signs is negative. Multiply the numerators and multiply the denominators.

$$= -\frac{1 \cdot 3 \cdot \overset{1}{\cancel{6}}}{\underset{1}{\cancel{6}} \cdot 1}$$ Factor 18 as $3 \cdot 6$. Then divide out the common factor 6.

$$= -\frac{3}{1}$$ Multiply in the numerator and in the denominator.

$$= -3$$ Simplify: $\frac{3}{1} = 3$.

EXAMPLE 6 Divide: $-\frac{21}{36} \div (-3)$.

Solution

$$-\frac{21}{36} \div (-3) = -\frac{21}{36}\left(-\frac{1}{3}\right)$$ Multiply $-\frac{21}{36}$ by the reciprocal of -3, which is $-\frac{1}{3}$.

$$= \frac{21 \cdot 1}{36 \cdot 3}$$ The product of two fractions with like signs is positive. Multiply the numerators and multiply the denominators.

$$= \frac{3 \cdot 7 \cdot 1}{36 \cdot 3}$$ Factor 21 as $3 \cdot 7$.

$$= \frac{\overset{1}{\not{3}} \cdot 7 \cdot 1}{36 \cdot \underset{1}{\not{3}}}$$ Divide out the common factor 3.

$$= \frac{7}{36}$$ Multiply in the numerator and in the denominator.

Self Check 6

Divide: $-\frac{35}{16} \div (-7)$.

Answer $\frac{5}{16}$

Section 3.3 STUDY SET

VOCABULARY *Fill in the blanks.*

1. Two numbers are called __________ if their product is 1.
2. The result of a division problem is called the ________.

CONCEPTS

3. Complete this statement:

$$\frac{1}{2} \div \frac{2}{3} = \square \cdot \square$$

4. Find the reciprocal of each number.

   a. $\frac{2}{5}$ b. -3

5. Using horizontal lines, divide each rectangle into thirds. What division problem does this illustrate? What is the quotient of that problem?

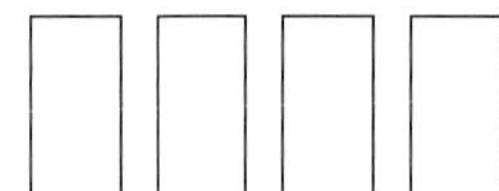

6. Using horizontal lines, divide each rectangle into fifths. What division problem does this illustrate? What is the quotient of that problem?

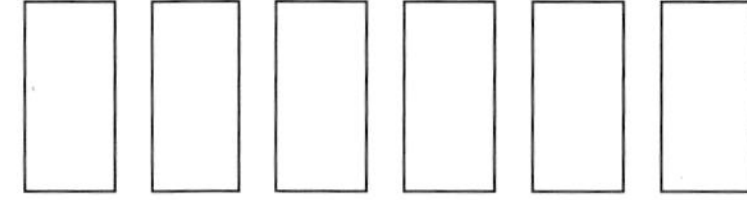

7. Multiply $\frac{4}{5}$ and its reciprocal. What is the result?
8. Multiply $\frac{7}{6}$ and its reciprocal. What is the result?
9. a. Divide: $15 \div 3$.

   b. Rewrite $15 \div 3$ as multiplication by the reciprocal and find the result.

   c. Complete this statement: Division by 3 is the same as multiplication by $\square$.
10. a. Divide: $10 \div \frac{1}{5}$.

    b. Complete this statement: Division by $\frac{1}{5}$ is the same as multiplication by $\square$.

NOTATION *Complete each solution.*

11. $$\frac{25}{36} \div \frac{10}{9} = \frac{25}{36} \cdot \frac{\square}{\square}$$
$$= \frac{25 \cdot \square}{36 \cdot \square}$$
$$= \frac{5 \cdot \square \cdot 9}{4 \cdot 9 \cdot 2 \cdot 5}$$
$$= \frac{\overset{1}{\not{\square}} \cdot 5 \cdot \overset{1}{\not{\square}}}{4 \cdot \underset{1}{\not{\square}} \cdot 2 \cdot \underset{1}{\not{\square}}}$$
$$= \frac{5}{8}$$

12. $\frac{4}{9} \div \frac{8}{27} = \frac{4}{9} \cdot \frac{\square}{\square}$

$= \frac{4 \cdot \square}{9 \cdot \square}$

$= \frac{4 \cdot 3 \cdot 9}{9 \cdot 2 \cdot 4}$

$= \frac{\overset{1}{\cancel{\square}} \cdot 3 \cdot \overset{1}{\cancel{\square}}}{\underset{1}{\cancel{\square}} \cdot 2 \cdot \underset{1}{\cancel{\square}}}$

$= \frac{3}{2}$

PRACTICE *Find each quotient.*

13. $\frac{1}{2} \div \frac{3}{5}$

14. $\frac{5}{7} \div \frac{5}{6}$

15. $\frac{3}{16} \div \frac{1}{9}$

16. $\frac{5}{8} \div \frac{2}{9}$

17. $\frac{4}{5} \div \frac{4}{5}$

18. $\frac{2}{3} \div \frac{2}{3}$

19. $\left(-\frac{7}{4}\right) \div \left(-\frac{21}{8}\right)$

20. $\left(-\frac{15}{16}\right) \div \left(-\frac{5}{8}\right)$

21. $3 \div \frac{1}{12}$

22. $9 \div \frac{3}{4}$

23. $120 \div \frac{12}{5}$

24. $360 \div \frac{36}{5}$

25. $-\frac{4}{5} \div (-6)$

26. $-\frac{7}{8} \div (-14)$

27. $\frac{15}{16} \div 180$

28. $\frac{7}{8} \div 210$

29. $-\frac{9}{10} \div \frac{4}{15}$

30. $-\frac{3}{4} \div \frac{3}{2}$

31. $\frac{9}{10} \div \left(-\frac{3}{25}\right)$

32. $\frac{11}{16} \div \left(-\frac{9}{16}\right)$

33. $-\frac{1}{8} \div 8$

34. $-\frac{1}{15} \div 15$

35. $\frac{15}{32} \div \frac{15}{32}$

36. $-\frac{1}{64} \div \left(-\frac{1}{64}\right)$

37. $\frac{4}{5} \div \frac{3}{2}$

38. $\frac{2}{3} \div \frac{3}{2}$

39. $\frac{1}{8} \div \frac{3}{4}$

40. $\frac{1}{9} \div \frac{3}{5}$

41. $\frac{13}{16} \div \frac{1}{2}$

42. $\frac{7}{8} \div \frac{6}{7}$

43. $-\frac{15}{32} \div \frac{3}{4}$

44. $-\frac{7}{10} \div \frac{4}{5}$

APPLICATIONS

45. MARATHONS Each lap around a stadium track is $\frac{1}{4}$ mile. How many laps would a runner have to complete to get a 26-mile workout?

46. COOKING A recipe calls for $\frac{3}{4}$ cup of flour, and the only measuring container you have holds $\frac{1}{8}$ cup. How many $\frac{1}{8}$ cups of flour would you need to add to follow the recipe?

47. LASER TECHNOLOGIES Using a laser, a technician slices thin pieces of aluminum off the end of a rod that is $\frac{7}{8}$ inch long. How many $\frac{1}{64}$-inch-wide slices can be cut from this rod?

48. FURNITURE A production process applies several layers of a clear acrylic coat to outdoor furniture to help protect it from the weather. If each protective coat is $\frac{3}{32}$ inch thick, how many applications will be needed to build up $\frac{3}{8}$ inch of clear finish?

49. UNDERGROUND CABLES In the illustration, which construction proposal will require the fewest days to install underground TV cable from the broadcasting station to the subdivision?

Proposal	*Amount of cable installed per day*	*Comments*
Route 1	$\frac{3}{5}$ of a mile	Longer than Route 2
Route 2	$\frac{2}{5}$ of a mile	Terrain very rocky

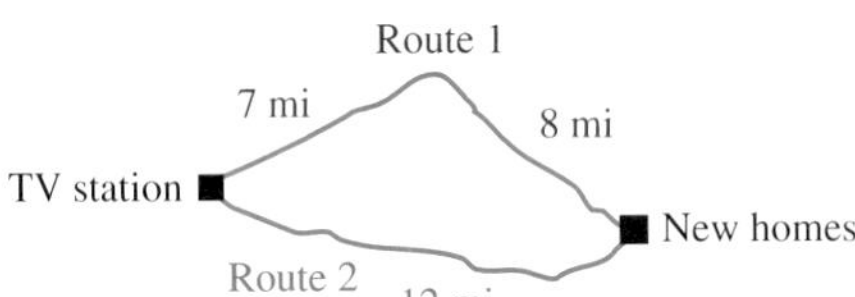

50. PRODUCTION PLANNING The materials used to make a pillow are shown. Examine the inventory list to decide how many pillows can be manufactured in one production run with the materials in stock.

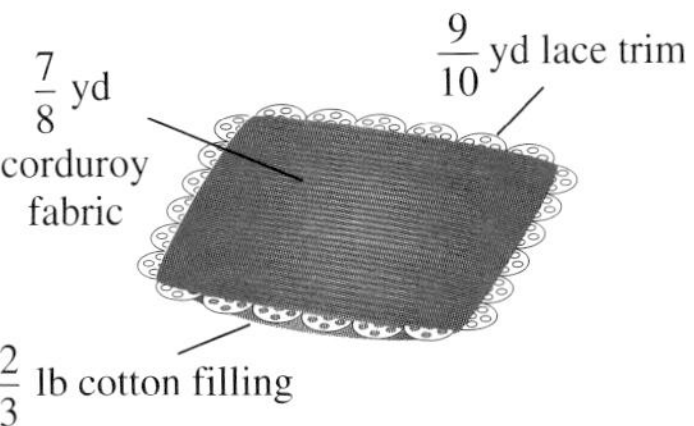

Factory Inventory List

Materials	*Amount in stock*
Lace trim	135 yd
Corduroy fabric	154 yd
Cotton filling	98 lb

51. 3 × 5 CARDS Ninety 3 × 5 cards are shown stacked next to a ruler.
    a. Into how many parts is 1 inch divided on the ruler?
    b. How thick is the stack of cards?
    c. How thick is one 3 × 5 card?

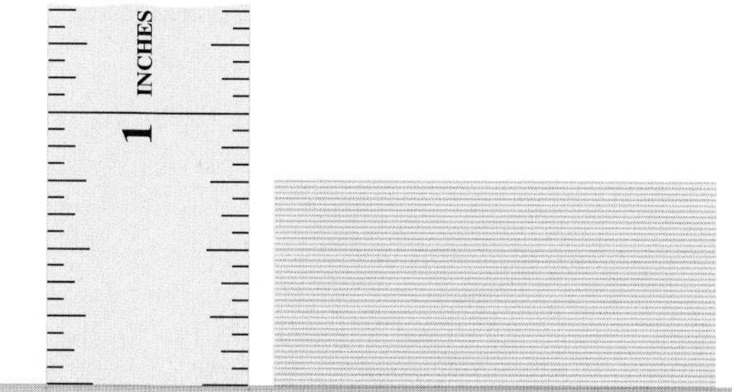

52. COMPUTER PRINTERS The illustration shows how the letter E is formed by a dot matrix printer. What is the height of a dot?

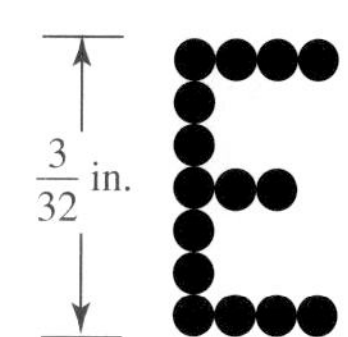

53. FORESTRY A set of forestry maps divides the 6,284 acres of an old-growth forest into $\frac{4}{5}$-acre sections. How many sections do the maps contain?

54. HARDWARE A hardware chain purchases large amounts of nails and packages them in $\frac{9}{16}$-pound bags for sale. How many of these bags of nails can be obtained from 2,871 pounds of nails?

WRITING

55. Explain how to divide two fractions.
56. Explain why 0 does not have a reciprocal.
57. Write an application problem that could be solved by finding $10 \div \frac{1}{5}$.
58. Explain why dividing a fraction by 2 is the same as finding $\frac{1}{2}$ of it.

REVIEW

59. Solve: $4x - 2 = -18$.
60. Solve: $\frac{x}{3} - 1 = 4$.
61. Divide: $14\overline{)378}$.
62. Add: $12{,}346 + 78 + 599 + 6{,}787$.
63. True or false: If equal amounts are subtracted from the numerator and the denominator of a fraction, the result will be an equivalent fraction.
64. Graph each of these numbers on a number line: $-2, 0, |-4|$, and the opposite of 1.

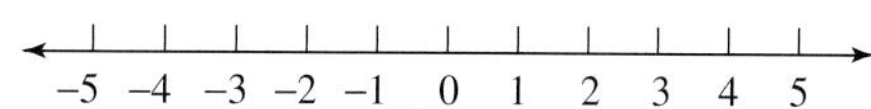

65. Round 637,512 to the nearest hundred.
66. Define the word *variable.*

3.4 Adding and Subtracting Fractions

- Fractions with the same denominator
- Fractions with different denominators
- Finding the LCD
- Comparing fractions

In arithmetic and algebra, *we can only add or subtract objects that are similar.* For example, we can add dollars to dollars, but we cannot add dollars to oranges. This concept is important when adding or subtracting fractions.

Fractions with the same denominator

Consider the problem $\frac{3}{5} + \frac{1}{5}$. When we write it in words, it is apparent that we are adding similar objects.

three-**fifths** + one-**fifth**

Similar objects

Because the denominators of $\frac{3}{5}$ and $\frac{1}{5}$ are the same, we say that they have a **common denominator.** Since the fractions have a common denominator, we can add them. Figure 3-6 illustrates the addition process.

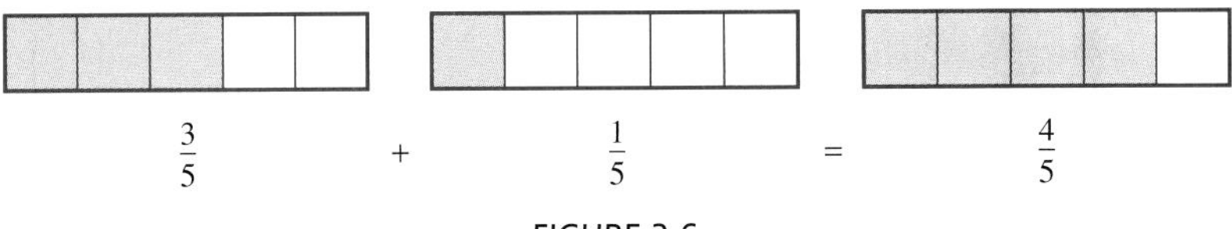

$$\frac{3}{5} + \frac{1}{5} = \frac{4}{5}$$

FIGURE 3-6

We can make some observations about the addition shown in the figure.

The *sum* of the numerators is the numerator of the answer.

$$\frac{3}{5} + \frac{1}{5} = \frac{4}{5}$$

The answer is a fraction that has the *same* denominator as the two fractions that were added.

These observations suggest the following rule.

Adding or subtracting fractions with the same denominators

To add (or subtract) fractions with the same denominators, add (or subtract) their numerators and write that result over the common denominator. Simplify the result, if possible.

In symbols, if a, b, and c represent numbers and c is not 0, then

$$\frac{a}{c} + \frac{b}{c} = \frac{a+b}{c} \quad \text{and} \quad \frac{a}{c} - \frac{b}{c} = \frac{a-b}{c}$$

Self Check 1

Add: $\frac{5}{12} + \frac{1}{12}$.

Answer $\frac{1}{2}$

EXAMPLE 1 Add: $\frac{1}{8} + \frac{5}{8}$.

Solution

$\frac{1}{8} + \frac{5}{8} = \frac{1+5}{8}$ Add the numerators. Write the sum over the common denominator 8.

$= \frac{6}{8}$ Perform the addition: $1 + 5 = 6$. The fraction can be simplified.

$= \frac{\overset{1}{\cancel{2}} \cdot 3}{\underset{1}{\cancel{2}} \cdot 4}$ Factor 6 as $2 \cdot 3$ and 8 as $2 \cdot 4$. Divide out the common factor 2.

$= \frac{3}{4}$ Multiply in the numerator and in the denominator.

EXAMPLE 2 Subtract: $-\frac{7}{3} - \left(-\frac{2}{3}\right)$.

Solution

$-\frac{7}{3} - \left(-\frac{2}{3}\right) = -\frac{7}{3} + \frac{2}{3}$ Add the opposite of $-\frac{2}{3}$, which is $\frac{2}{3}$.

$= \frac{-7}{3} + \frac{2}{3}$ Write $-\frac{7}{3}$ as $\frac{-7}{3}$.

$= \frac{-7+2}{3}$ Add the numerators. Write the sum over the common denominator 3.

$= \frac{-5}{3}$ Perform the addition: $-7 + 2 = -5$.

$= -\frac{5}{3}$ Rewrite the fraction: $\frac{-5}{3} = -\frac{5}{3}$.

Self Check 2

Subtract: $-\frac{9}{11} - \left(-\frac{3}{11}\right)$.

Answer $-\frac{6}{11}$

Fractions with different denominators

Now we consider the problem $\frac{3}{5} + \frac{1}{3}$. Since the denominators are different, we cannot add these fractions in their present form.

three-**fifths** + one-**third**

↑ Not similar objects ↑

To add these fractions, we need to find a common denominator. The smallest common denominator (called the **least** or **lowest common denominator**) usually is the easiest common denominator to work with.

Least common denominator

The **least common denominator (LCD)** for a set of fractions is the smallest number each denominator will divide exactly.

In the problem $\frac{3}{5} + \frac{1}{3}$, the denominators are 5 and 3. The numbers 5 and 3 divide many numbers exactly; 30, 45, and 60, to name a few. But the smallest number that 5 and 3 divide exactly is 15. This is the LCD. We will now build each fraction into a fraction with a denominator of 15 by using the fundamental property of fractions.

$\frac{3}{5} + \frac{1}{3} = \frac{3 \cdot \mathbf{3}}{5 \cdot \mathbf{3}} + \frac{1 \cdot 5}{3 \cdot 5}$

We need to multiply this denominator by 3 to obtain 15. We need to multiply this denominator by 5 to obtain 15.

$= \frac{9}{15} + \frac{5}{15}$ Perform the multiplications in the numerators and denominators.

$= \frac{9+5}{15}$ Add the numerators and write the sum over the common denominator 15.

$= \frac{14}{15}$ Perform the addition: $9 + 5 = 14$.

Figure 3-7 shows $\frac{3}{5}$ and $\frac{1}{3}$ expressed as equivalent fractions with a denominator of 15. Once the denominators are the same, the fractions can be added easily.

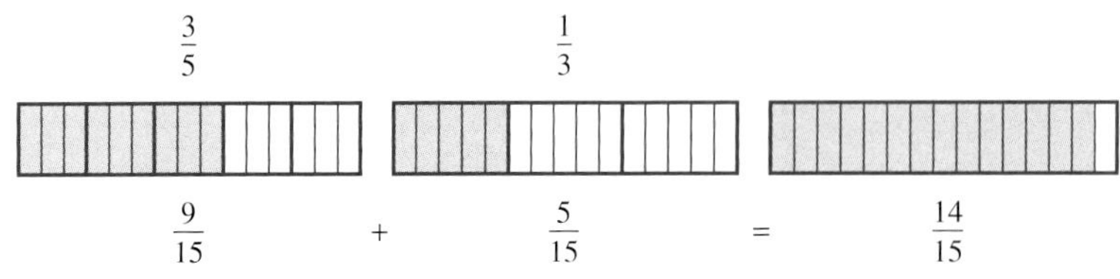

FIGURE 3-7

To add or subtract fractions with different denominators, we follow these steps.

Adding or subtracting fractions with different denominators

1. Find the LCD.
2. Express each fraction as an equivalent fraction with a denominator that is the LCD.
3. Add or subtract the resulting fractions. Simplify the result if possible.

Self Check 3

Add: $\frac{1}{2} + \frac{2}{5}$.

Answer $\frac{9}{10}$

EXAMPLE 3 Add: $\frac{1}{8} + \frac{2}{3}$.

Solution Since the smallest number the denominators 8 and 3 divide exactly is 24, the LCD is 24.

$$\frac{1}{8} + \frac{2}{3} = \frac{1 \cdot \mathbf{3}}{8 \cdot \mathbf{3}} + \frac{2 \cdot \mathbf{8}}{3 \cdot \mathbf{8}}$$ Express each fraction in terms of 24ths.

$$= \frac{3}{24} + \frac{16}{24}$$ Perform the multiplications in the numerators and denominators.

$$= \frac{3 + 16}{24}$$ Add the numerators and write the sum over the common denominator 24.

$$= \frac{19}{24}$$ Perform the addition in the numerator: $3 + 16 = 19$.

Self Check 4

Add: $-6 + \frac{2}{9}$.

Answer $-\frac{52}{9}$

EXAMPLE 4 Add: $-5 + \frac{1}{4}$.

Solution

$$-5 + \frac{1}{4} = \frac{-5}{1} + \frac{1}{4}$$ Write -5 as $\frac{-5}{1}$. The smallest number that 1 and 4 divide exactly is 4, so the LCD is 4.

$$= \frac{-5 \cdot \mathbf{4}}{1 \cdot \mathbf{4}} + \frac{1}{4}$$ Express $\frac{-5}{1}$ in terms of 4ths.

$$= \frac{-20}{4} + \frac{1}{4}$$ Multiply in the numerator and in the denominator.

$$= \frac{-20 + 1}{4}$$ Write the sum of the numerators over the common denominator 4.

$$= \frac{-19}{4}$$ Perform the addition: $-20 + 1 = -19$.

$$= -\frac{19}{4}$$ Rewrite: $\frac{-19}{4} = -\frac{19}{4}$.

Budgets

THINK IT THROUGH

"Putting together a budget is crucial if you don't want to spend your way into serious problems. You're also developing a habit that can serve you well throughout your life." Liz Pulliam Weston, MSN Money

The circle graph below shows a suggested budget for new college graduates as recommended by Springboard, a nonprofit consumer credit counseling service. What fraction of net take-home pay should be spent on housing?

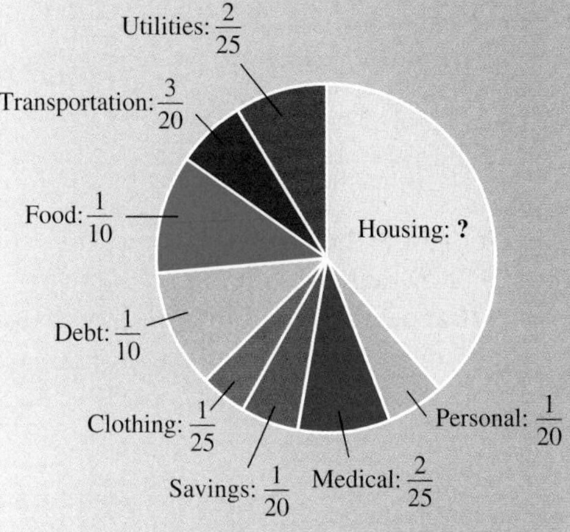

Finding the LCD

When we add or subtract fractions with different denominators, the LCD is not always obvious. We will now develop two methods for finding the LCD of a set of fractions. As an example, let's find the LCD of $\frac{3}{8}$ and $\frac{1}{10}$.

Method 1: A **multiple** of a number is the product of that number and a natural number. The multiples of 8 and the multiples of 10 are shown below.

Multiples of 8	**Multiples of 10**
$8 \cdot 1 = 8$	$10 \cdot 1 = 10$
$8 \cdot 2 = 16$	$10 \cdot 2 = 20$
$8 \cdot 3 = 24$	$10 \cdot 3 = 30$
$8 \cdot 4 = 32$	$10 \cdot 4 = \mathbf{40}$
$8 \cdot 5 = \mathbf{40}$	$10 \cdot 5 = 50$
$8 \cdot 6 = 48$	$10 \cdot 6 = 60$
$8 \cdot 7 = 56$	$10 \cdot 7 = 70$
$8 \cdot 8 = 64$	$10 \cdot 8 = \mathbf{80}$
$8 \cdot 9 = 72$	$10 \cdot 9 = 90$
$8 \cdot 10 = \mathbf{80}$	$10 \cdot 10 = 100$

The multiples that the lists have in common are highlighted in red.

The smallest multiple common to both lists is 40. It is the smallest number that 8 and 10 divide exactly. Therefore, 40 is the LCD of $\frac{3}{8}$ and $\frac{1}{10}$. These observations suggest a method for finding the LCD of a set of fractions.

Finding the LCD by finding multiples

1. List the multiples of each denominator.
2. The smallest multiple common to the lists found in step 1 is the LCD of the fractions.

Method 2: If the LCD for $\frac{3}{8}$ and $\frac{1}{10}$ is a number that 8 and 10 divide exactly, the prime factorization of the LCD must include the prime factorization of 8 (which is $2 \cdot 2 \cdot 2$) and the prime factorization of 10 (which is $5 \cdot 2$). The smallest number that meets both of these requirements is $2 \cdot 2 \cdot 2 \cdot 5$. Therefore, the LCD is $2 \cdot 2 \cdot 2 \cdot 5 = 40$.

$$\left.\begin{array}{r} 8 = 2 \cdot 2 \cdot 2 \\ 10 = 5 \cdot 2 \end{array}\right\} \text{LCD} = \overbrace{2 \cdot 2 \cdot 2}^{\text{The prime factorization of 8}} \cdot 5 = 40$$

with $2 \cdot 5$ marked as "The prime factorization of 10".

In the prime factorization of 8, the factor 2 appears three times. It appears three times in the product $(2 \cdot 2 \cdot 2 \cdot 5)$ that gives the LCD. In the prime factorization of 10, the factor 5 appears once. It appears once in the product that gives the LCD. These observations suggest another method for finding the LCD of a set of fractions.

Finding the LCD using prime factorization

1. Prime factor each denominator.
2. The LCD is a product of prime factors, where each factor is used the greatest number of times it appears in any one factorization found in step 1.

Self Check 5

Subtract: $\frac{33}{35} - \frac{11}{14}$.

Answer $\frac{11}{70}$

EXAMPLE 5 Subtract: $\frac{19}{21} - \frac{5}{18}$.

Solution We use prime factorization to find the LCD.

Step 1: Find the prime factorization of each denominator.

$21 = 7 \cdot 3$

$18 = 3 \cdot 3 \cdot 2$

Step 2: The factors 7, 3, and 2 appear in the prime factorizations.

The greatest number of times 7 appears in any one factorization is once. The greatest number of times 3 appears in any one factorization is twice. The greatest number of times 2 appears in any one factorization is once.

$$\text{LCD} = 7 \cdot 3 \cdot 3 \cdot 2 = 126$$

$\frac{19}{21} - \frac{5}{18} = \frac{19 \cdot \mathbf{6}}{21 \cdot \mathbf{6}} - \frac{5 \cdot 7}{18 \cdot 7}$ Express each fraction in terms of 126ths.

$= \frac{114}{126} - \frac{35}{126}$ Perform the multiplications in the numerators and in the denominators.

$= \frac{114 - 35}{126}$ Write the difference of the numerators over the common denominator 126.

$= \frac{79}{126}$ Perform the subtraction: $114 - 35 = 79$.

EXAMPLE 6 Television viewing habits. Students on a college campus were asked to estimate to the nearest hour how much television they watched each day. The results are given in the pie chart in Figure 3-8. For example, the chart tells us that $\frac{1}{4}$ of those responding watched 1 hour per day. Find the fraction of the student body watching from 0 to 2 hours daily.

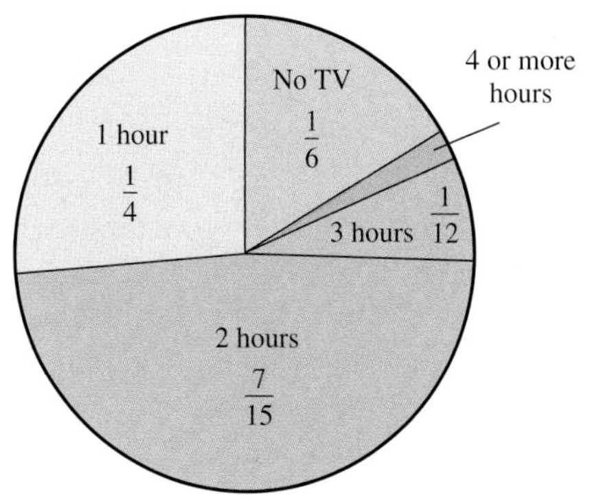

FIGURE 3-8

Solution To answer this question, we need to add $\frac{1}{6}$, $\frac{1}{4}$, and $\frac{7}{15}$. To find the LCD, we prime factor each of the denominators.

$$\left.\begin{aligned} 6 &= 3 \cdot 2 \\ 4 &= 2 \cdot 2 \\ 15 &= 5 \cdot 3 \end{aligned}\right\} \text{LCD} = 5 \cdot 3 \cdot 2 \cdot 2 = 60$$

In any one factorization, the greatest number of times 5 appears is once, the greatest number of times 3 appears is once, and the greatest number of times 2 appears is twice.

$$\frac{1}{6} + \frac{1}{4} + \frac{7}{15} = \frac{1 \cdot \mathbf{10}}{6 \cdot \mathbf{10}} + \frac{1 \cdot \mathbf{15}}{4 \cdot \mathbf{15}} + \frac{7 \cdot \mathbf{4}}{15 \cdot \mathbf{4}}$$

Express each fraction in terms of 60ths.

$$= \frac{10}{60} + \frac{15}{60} + \frac{28}{60}$$

Perform the multiplication in the numerators and denominators.

$$= \frac{10 + 15 + 28}{60}$$

Add the numerators. Write the sum over the common denominator 60.

$$= \frac{53}{60}$$

Perform the addition: $10 + 15 + 28 = 53$.

The fraction of the student body watching 0 to 2 hours of television daily is $\frac{53}{60}$.

Comparing fractions

If fractions have the same denominator, the fraction with the larger numerator is the larger fraction. If their denominators are different, we need to write the fractions with a common denominator before we can make a comparison.

> **Comparing fractions**
>
> To compare unlike fractions, write the fractions as equivalent fractions with the same denominator — preferably the LCD. Then compare their numerators. The fraction with the larger numerator is the larger fraction.

EXAMPLE 7 Which fraction is larger: $\frac{5}{6}$ or $\frac{7}{8}$?

Solution To compare these fractions, we express each with the LCD of 24.

$$\frac{5}{6} = \frac{5 \cdot \mathbf{4}}{6 \cdot \mathbf{4}} \qquad \frac{7}{8} = \frac{7 \cdot \mathbf{3}}{8 \cdot \mathbf{3}}$$

Express each fraction in terms of 24ths.

$$= \frac{20}{24} \qquad = \frac{21}{24}$$

Perform the multiplications in the numerators and denominators.

Next, we compare the numerators. Since $21 > 20$, we conclude that $\frac{21}{24}$ is greater than $\frac{20}{24}$. Thus, $\frac{7}{8} > \frac{5}{6}$.

Self Check 7
Which fraction is larger: $\frac{7}{12}$ or $\frac{3}{5}$?

Answer $\frac{3}{5}$

Section 3.4 STUDY SET

VOCABULARY *Fill in the blanks.*

1. The ______ common denominator for a set of fractions is the smallest number each denominator will divide exactly.
2. __________ fractions, such as $\frac{1}{2}$ and $\frac{2}{4}$, are fractions that represent the same amount.
3. To express a fraction in _______ terms, we multiply the numerator and denominator by the same number.
4. _________ a fraction is the process of multiplying the numerator and the denominator of the fraction by the same number.

CONCEPTS

5. The rule for adding fractions is

$$\frac{a}{c} + \frac{b}{c} = \frac{a+b}{c} \quad \text{where } c \neq 0$$

Fill in the blanks: This rule tells us how to add fractions having like ____________. To find the sum, we add the ___________ and then write that result over the _________ denominator.

6. **a.** Add the indicated fractions.

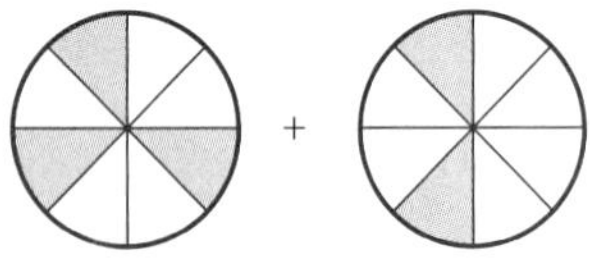

b. Subtract the indicated fractions.

7. Why must we do some preliminary work before doing the following addition?

$$\frac{2}{9} + \frac{2}{5}$$

8. Why must we do some preliminary work before doing the following subtraction?

$$\frac{5}{6} - \frac{5}{18}$$

9. By what are the numerator and the denominator of the following fraction being multiplied?

$$\frac{5 \cdot 4}{6 \cdot 4}$$

10. Consider $\frac{3}{4}$. By what should we multiply the numerator and denominator of this fraction to express it as an equivalent fraction with the given denominator?

a. 12 **b.** 36

11. Consider the following prime factorizations.

$$24 = 2 \cdot 2 \cdot 2 \cdot 3$$
$$90 = 2 \cdot 3 \cdot 3 \cdot 5$$

For any one factorization, what is the greatest number of times

a. a 5 appears?

b. a 3 appears?

c. a 2 appears?

12. **a.** List the first ten multiples of 9 and the first ten multiples of 12.

b. What is the LCM of 9 and 12?

13. The denominators of two fractions involved in a subtraction problem have the prime-factored forms $2 \cdot 2 \cdot 5$ and $2 \cdot 3 \cdot 5$. What is the LCD for the fractions?

14. The denominators of three fractions involved in a subtraction problem have the prime-factored forms $2 \cdot 2 \cdot 5$, $2 \cdot 3 \cdot 5$, and $2 \cdot 3 \cdot 3 \cdot 5$. What is the LCD for the fractions?

15. **a.** Divide the figure on the left into fourths and shade one part. Divide the figure on the right into thirds and shade one part. Which shaded part is larger?

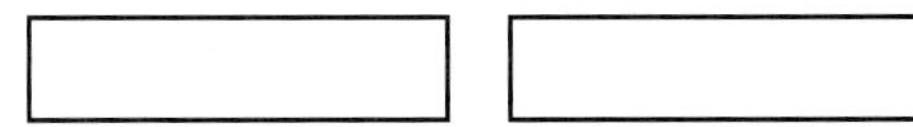

b. Express the shaded part of each figure in part a as a fraction. Show that one of those fractions is larger than the other by expressing both in terms of a common denominator and then comparing them.

16. Place an $>$ or $<$ symbol in the blank to make a true statement.

a. $\frac{32}{35}$ ☐ $\frac{31}{35}$ **b.** $\frac{7}{8}$ ☐ $\frac{31}{32}$

NOTATION *Complete each solution.*

17. $\frac{2}{5} + \frac{1}{3} = \frac{2 \cdot \square}{5 \cdot \square} + \frac{1 \cdot 5}{3 \cdot 5}$

$= \frac{\square}{15} + \frac{\square}{15}$

$= \frac{\square + \square}{15}$

$= \frac{11}{15}$

18. $\frac{7}{8} - \frac{2}{3} = \frac{7 \cdot 3}{\square \cdot 3} - \frac{2 \cdot 8}{\square \cdot 8}$

$= \frac{21}{\square} - \frac{16}{\square}$

$= \frac{21 - 16}{\square}$

$= \frac{5}{24}$

PRACTICE *The denominators of two fractions are given. Find the lowest common denominator.*

19. 18, 6
20. 15, 3
21. 8, 6
22. 10, 4
23. 8, 20
24. 14, 21
25. 15, 12
26. 25, 30

Perform each operation. Simplify when necessary.

27. $\frac{3}{7} + \frac{1}{7}$
28. $\frac{16}{25} - \frac{9}{25}$
29. $\frac{37}{103} - \frac{17}{103}$
30. $\frac{54}{53} - \frac{52}{53}$
31. $\frac{11}{25} - \frac{1}{25}$
32. $\frac{7}{8} - \frac{1}{8}$
33. $\frac{5}{7} + \frac{3}{7}$
34. $\frac{17}{11} - \frac{12}{11}$
35. $\frac{1}{4} + \frac{3}{8}$
36. $\frac{2}{3} + \frac{1}{6}$
37. $\frac{13}{20} - \frac{1}{5}$
38. $\frac{71}{100} - \frac{1}{10}$
39. $\frac{4}{5} + \frac{2}{3}$
40. $\frac{1}{4} + \frac{2}{3}$
41. $\frac{1}{8} + \frac{2}{7}$
42. $\frac{1}{6} + \frac{5}{9}$
43. $\frac{3}{4} - \frac{2}{3}$
44. $\frac{4}{5} - \frac{1}{6}$
45. $\frac{5}{6} - \frac{3}{4}$
46. $\frac{7}{8} - \frac{5}{6}$
47. $\frac{16}{25} - \left(-\frac{3}{10}\right)$
48. $\frac{3}{8} - \left(-\frac{1}{6}\right)$
49. $-\frac{7}{16} + \frac{1}{4}$
50. $-\frac{17}{20} + \frac{4}{5}$
51. $\frac{1}{12} - \frac{3}{4}$
52. $\frac{11}{60} - \frac{13}{20}$
53. $-\frac{5}{8} - \frac{1}{3}$
54. $-\frac{7}{20} - \frac{1}{5}$
55. $-3 + \frac{2}{5}$
56. $-6 + \frac{5}{8}$
57. $-\frac{3}{4} - 5$
58. $-2 - \frac{7}{8}$
59. $\frac{1}{3} + \frac{1}{4} + \frac{1}{5}$
60. $\frac{1}{10} + \frac{1}{8} + \frac{1}{5}$
61. $-\frac{2}{3} + \frac{5}{4} + \frac{1}{6}$
62. $-\frac{3}{4} + \frac{3}{8} + \frac{7}{6}$
63. $\frac{5}{24} + \frac{3}{16}$
64. $\frac{17}{20} - \frac{4}{15}$
65. $-\frac{11}{15} - \frac{2}{9}$
66. $-\frac{19}{18} - \frac{5}{12}$
67. $\frac{7}{25} + \frac{1}{15}$
68. $\frac{11}{20} - \frac{1}{8}$
69. $\frac{4}{27} + \frac{1}{6}$
70. $\frac{8}{9} - \frac{7}{12}$

71. Find the difference of $\frac{11}{60}$ and $\frac{2}{45}$.

72. Find the sum of $\frac{9}{48}$ and $\frac{7}{40}$.

73. Subtract $\frac{5}{12}$ from $\frac{2}{15}$.

74. Find the sum of $\frac{11}{24}$ and $\frac{7}{36}$ increased by $\frac{5}{48}$.

APPLICATIONS

75. BOTANY To assess the effects of smog on tree development, botanists cut down a pine tree and measured the width of the growth rings for the last two years (see the illustration on the next page).

a. What was the growth over this two-year period?

b. What is the difference in the widths of the two rings?

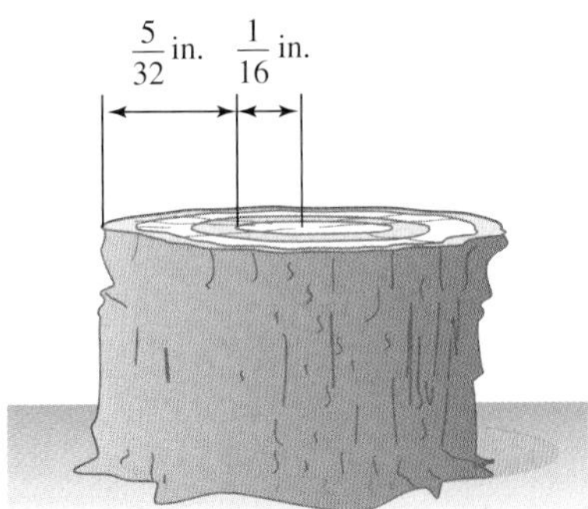

76. MAGAZINE LAYOUTS The page design for a magazine cover includes a blank strip at the top, called a header, and a blank strip at the bottom of the page, called a footer. In the illustration, how much page length is lost because of the header and footer?

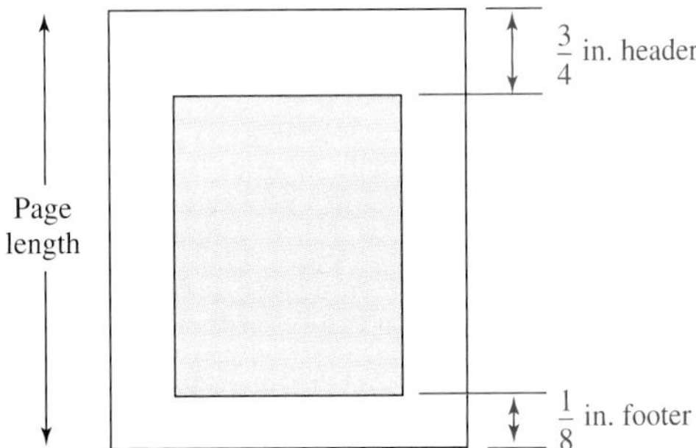

77. FAMILY DINNER A family bought two large pizzas for dinner. Several pieces of each pizza were not eaten, as shown. How much pizza was left? Could the family have been fed with just one pizza?

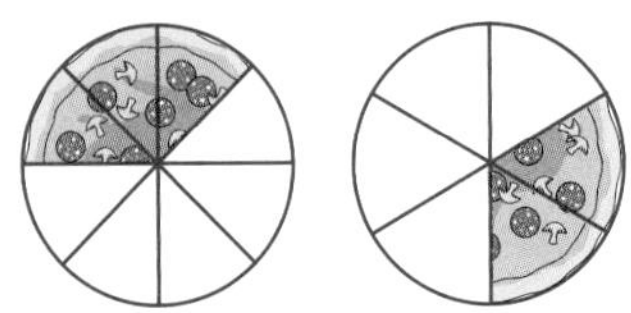

78. GASOLINE BARRELS The contents of two identical-sized barrels are shown. If they are poured into an empty third barrel that is the same size, how much of the third barrel will they fill?

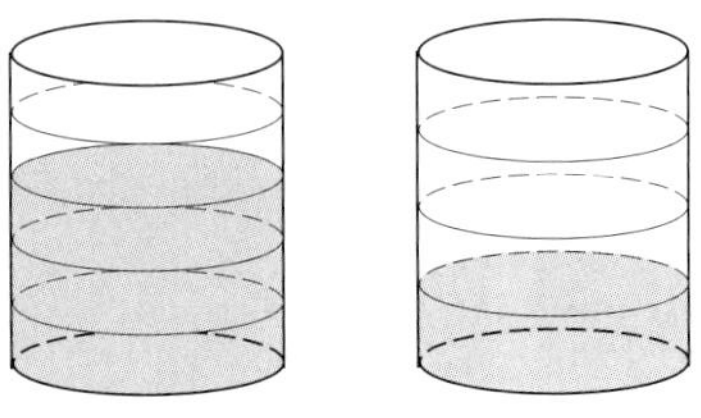

79. WEIGHTS AND MEASURES A consumer protection agency verifies the accuracy of butcher shop scales by placing a known three-quarter-pound weight on the scale and then comparing that to the scale's readout. According to the illustration, by how much is this scale off? Does it result in undercharging or overcharging customers on their meat purchases?

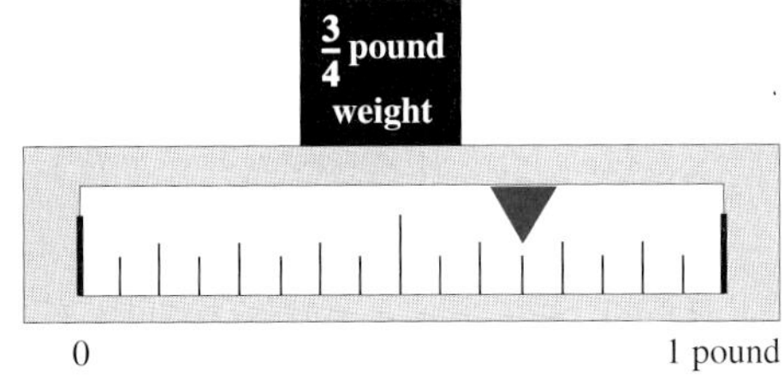

80. WRENCHES A mechanic hangs his wrenches above a tool bench in order of narrowest to widest. What is the proper order of the wrenches in the illustration?

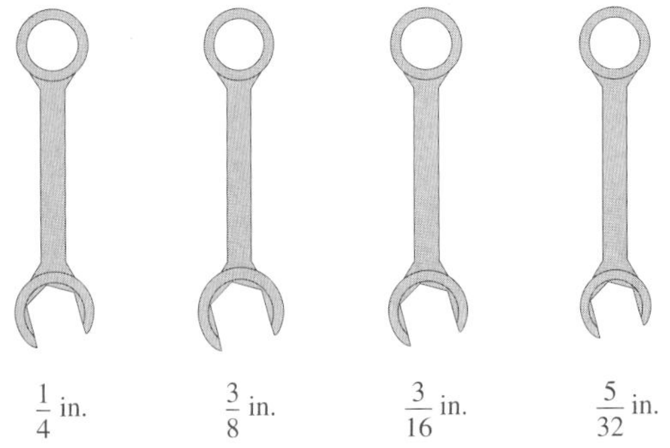

81. HIKING The illustration shows the length of each part of a three-part hike. Rank the lengths from longest to shortest.

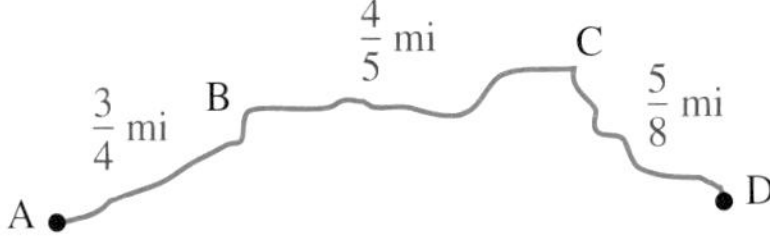

82. FIGURE DRAWING As an aid in drawing the human body, artists divide the body into three parts. Each part is then expressed as a fraction of the total body height. For example, the torso is $\frac{4}{15}$ of the body height. What fraction of body height is the head?

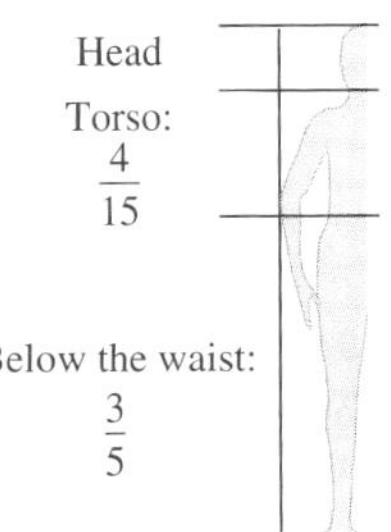

83. STUDY HABITS College students taking a full load were asked to give the average number of hours they studied each day. The results are shown in the pie chart. What fraction of the students study 2 hours or more daily?

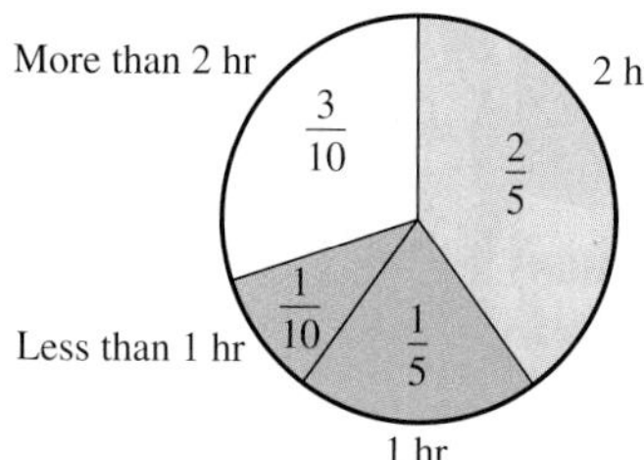

84. MUSICAL NOTES The notes used in music have fractional values. Their names and the symbols used to represent them are shown in Illustration (a). In common time, the values of the notes in each measure must add up to 1. Is the measure in Illustration (b) complete?

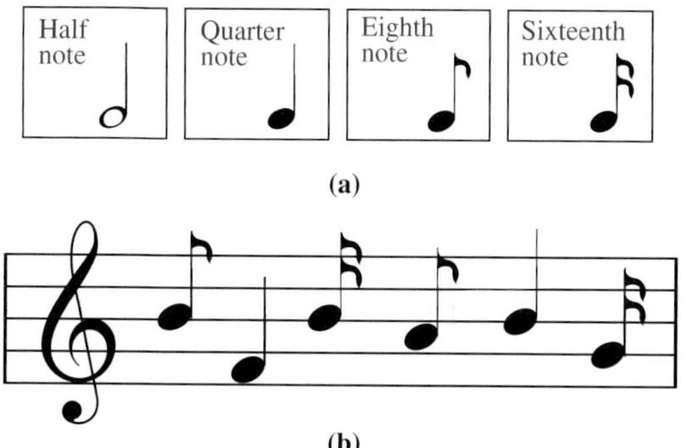

(a)

(b)

85. GARAGE DOOR OPENERS What is the difference in strength between a $\frac{1}{3}$-hp and a $\frac{1}{2}$-hp garage door opener?

86. DELIVERY TRUCKS A truck can safely carry a one-ton load. Should it be used to deliver one-half ton of sand, one-third ton of gravel, and one-fifth ton of cement in one trip to a job site?

WRITING

87. How are the procedures for expressing a fraction in higher terms and simplifying a fraction to lowest terms similar, and how are they different?

88. Given two fractions, how do we find their lowest common denominator?

89. How do we compare the sizes of two fractions with different denominators?

90. What is the difference between a common denominator and the lowest common denominator?

REVIEW

91. Find the prime factorization of 20.

92. Solve: $3x + 1 = -5$.

93. What is the formula for finding the area of a rectangle?

94. What is the formula for finding the perimeter of a rectangle?

The LCM and the GCF

As we have seen, the **multiples** of a number can be found by multiplying it successively by 1, 2, 3, 4, 5, and so on. The multiples of 4 and the multiples of 6 are shown below.

$1 \cdot 4 = 4$	$1 \cdot 6 = 6$	
$2 \cdot 4 = 8$	$2 \cdot 6 = \mathbf{12}$	
$3 \cdot 4 = \mathbf{12}$	$3 \cdot 6 = 18$	
$4 \cdot 4 = 16$	$4 \cdot 6 = \mathbf{24}$	Common multiples of 4 and 6 are highlighted in red.
$5 \cdot 4 = 20$	$5 \cdot 6 = 30$	
$6 \cdot 4 = \mathbf{24}$	$6 \cdot 6 = \mathbf{36}$	
$7 \cdot 4 = 28$	$7 \cdot 6 = 42$	
$8 \cdot 4 = 32$	$8 \cdot 6 = 48$	
$9 \cdot 4 = \mathbf{36}$	$9 \cdot 6 = 54$	

Because 12 is the smallest number that is a multiple of both 4 and 6, it is called the **least common multiple (LCM)** of 4 and 6.

Making lists like those shown above can be tedious. A more efficient method to find the least common multiple of several numbers is as follows.

Finding the least common multiple

1. Write each of the numbers in prime-factored form.
2. The least common multiple is a product of prime factors, where each prime factor is used the greatest number of times it appears in any one factorization found in step 1.

Self Check 1
Find the LCM of 18 and 84.

EXAMPLE 1 Least common multiple. Find the LCM of 24 and 36.

Solution

Step 1: First, we find the prime factorizations of 24 and 36.

$$24 = 2 \cdot 2 \cdot 2 \cdot 3$$
$$36 = 2 \cdot 2 \cdot 3 \cdot 3$$

Step 2: The prime factorizations of 24 and 36 contain the prime factors 3 and 2. We use each of these factors the greatest number of times it appears in any one factorization.

The greatest number of times 3 appears in any one factorization is two times.

The greatest number of times 2 appears in any one factorization is three times.

$$\text{LCD} = 3 \cdot 3 \cdot 2 \cdot 2 \cdot 2 = 72$$

The least common multiple of 24 and 36 is 72.

Answer 252

Because 2 divides 36 exactly and because 2 divides 120 exactly, 2 is called a **common factor** of 36 and 120.

$$\frac{36}{2} = 18 \qquad \frac{120}{2} = 60$$

The numbers 36 and 120 have other common factors, such as 3 and 6. The **greatest common factor (GCF)** of 36 and 120 is the largest number that is a factor of both. We follow these steps to find the greatest common factor of several numbers.

Finding the greatest common factor

1. Write each of the numbers in prime-factored form.
2. The greatest common factor is the product of the prime factors that are common to the factorizations found in step 1. If the numbers have no factors in common, the GCF is 1.

Self Check 2
Find the GCF of 60 and 150.

EXAMPLE 2 Greatest common factor. Find the GCF of 36 and 120.

Solution

Step 1: We find the prime factorizations of 36 and 120.

$$36 = \mathbf{2 \cdot 2 \cdot 3} \cdot 3$$
$$120 = \mathbf{2 \cdot 2} \cdot 2 \cdot \mathbf{3} \cdot 5$$

Step 2: One factor of 3 (highlighted in red) and two factors of 2 (highlighted in blue) are common to the factorizations of 36 and 120. To find the GCF, we form their product.

$$\text{GCF} = \mathbf{2} \cdot \mathbf{2} \cdot \mathbf{3} = 12$$

The greatest common factor of 36 and 120 is 12.

Answer 30

STUDY SET *The LCM and the GCF*

Find the least common multiple of the given numbers.

1. 3, 5
2. 7, 11
3. 8, 14
4. 8, 12
5. 14, 21
6. 16, 20
7. 6, 18
8. 3, 9
9. 44, 60
10. 36, 60
11. 100, 120
12. 120, 180
13. 6, 24, 36
14. 6, 10, 18
15. 18, 54, 63
16. 16, 30, 84

Find the greatest common factor of the given numbers.

17. 6, 9
18. 8, 12
19. 22, 33
20. 15, 20
21. 16, 20
22. 18, 24
23. 25, 100
24. 16, 80
25. 100, 120
26. 120, 180
27. 48, 108
28. 60, 96
29. 18, 24, 36
30. 30, 50, 90
31. 18, 54, 63
32. 28, 42, 84

33. NURSING A nurse, working in an intensive care unit, has to check a patient's vital signs every 45 minutes. Another nurse has to give the same patient his medication every 2 hours. If both nurses are in the patient's room together now, how long will it be until they are once again in the room together?

34. BARBECUES A certain brand of hot dogs comes in packages of 10. A certain brand of hot dog buns comes in packages of 12. For a family reunion barbecue, how many packages of hot dogs and how many packages of hot dog buns should be purchased so that no hot dogs and no buns are wasted?

3.5 Multiplying and Dividing Mixed Numbers

- Mixed numbers • Writing mixed numbers as improper fractions
- Writing improper fractions as mixed numbers • Graphing fractions and mixed numbers
- Multiplying and dividing mixed numbers

In the next two sections, we will show how to add, subtract, multiply, and divide *mixed numbers.* These numbers are widely used in daily life. Here are a few examples.

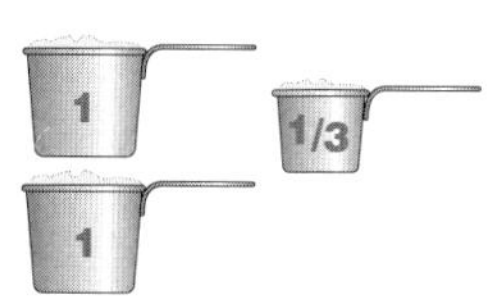

The recipe calls for $2\frac{1}{3}$ cups of flour.

It took $3\frac{3}{4}$ hours to paint the living room.

The entrance to the park is $1\frac{1}{2}$ miles away.

Mixed numbers

A **mixed number** is the *sum* of a whole number and a proper fraction. For example, $3\frac{3}{4}$ is a mixed number.

$$3\frac{3}{4} \quad = \quad 3 \quad + \quad \frac{3}{4}$$

↑	↑	↑
Mixed number	Whole number	Proper fraction

! COMMENT Note that $3\frac{3}{4}$ means $3 + \frac{3}{4}$, even though the + symbol is not written. Don't confuse $3\frac{3}{4}$ with $3 \cdot \frac{3}{4}$ or $3(\frac{3}{4})$, which indicate the multiplication of 3 and $\frac{3}{4}$.

In this section, we will work with negative as well as positive mixed numbers. For example, the negative mixed number $-4\frac{3}{4}$ could be used to represent $4\frac{3}{4}$ feet below sea level. We think of $-4\frac{3}{4}$ as $-4 - \frac{3}{4}$.

Writing mixed numbers as improper fractions

To see that mixed numbers are related to improper fractions, consider $3\frac{3}{4}$. To write $3\frac{3}{4}$ as an improper fraction, we need to find out how many *fourths* it represents. One way is to use the fundamental property of fractions.

$3\frac{3}{4} = 3 + \frac{3}{4}$	Write the mixed number $3\frac{3}{4}$ as a sum.
$= \frac{3}{1} + \frac{3}{4}$	Write 2 as a fraction: $3 = \frac{3}{1}$.
$= \frac{3 \cdot \mathbf{4}}{1 \cdot \mathbf{4}} + \frac{3}{4}$	Express $\frac{3}{1}$ as a fraction with denominator 4.
$= \frac{12}{4} + \frac{3}{4}$	Perform the multiplications in the numerator and denominator.
$= \frac{15}{4}$	Add the numerators: $12 + 3 = 15$. Write the sum over the common denominator, 4.

Thus, $3\frac{3}{4} = \frac{15}{4}$.

We can obtain the same result with far less work. To change $3\frac{3}{4}$ to an improper fraction, we simply multiply 3 by 4 and add 3 to get the numerator, and keep the denominator of 4.

$$3\frac{3}{4} = \frac{3(4) + 3}{4} = \frac{12 + 3}{4} = \frac{15}{4}$$

This example illustrates the following general rule.

Writing a mixed number as an improper fraction

To write a mixed number as an improper fraction, multiply the whole-number part by the denominator of the fraction and add the result to the numerator. Write this sum over the denominator.

EXAMPLE 1 Write the mixed number $5\frac{1}{6}$ as an improper fraction.

Solution

$5\frac{1}{6} = \frac{5(6) + 1}{6}$ Multiply 5 by the denominator 6. Add the numerator 1. Write this sum over the denominator 6.

$= \frac{30 + 1}{6}$ Perform the multiplication: $5(6) = 30$.

$= \frac{31}{6}$ Perform the addition: $30 + 1 = 31$.

Self Check 1
Write the mixed number $3\frac{3}{8}$ as an improper fraction.

Answer $\frac{27}{8}$

To write a negative mixed number in fractional form, we temporarily ignore the − sign and use the method shown in Example 1 on the positive mixed number. Once that procedure is completed, we write a − sign in front of the result. For example, $-3\frac{1}{4} = -\frac{13}{4}$.

Writing improper fractions as mixed numbers

To write an improper fraction as a mixed number, we must find two things: the *whole-number part* and the *fractional part* of the mixed number. To develop a procedure to do this, let's consider the improper fraction $\frac{7}{3}$. To find the number of groups of 3 in 7, we can divide 7 by 3. This will find the whole-number part of the mixed number. The remainder is the numerator of the fractional part of the mixed number.

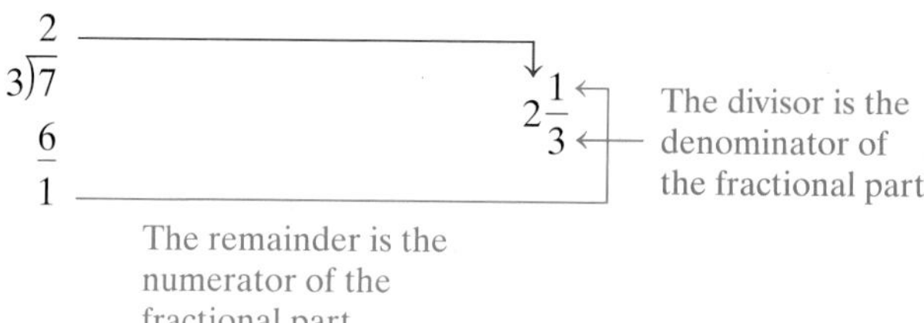

This example suggests the following general rule.

> **Writing an improper fraction as a mixed number**
> To write an improper fraction as a mixed number, divide the numerator by the denominator to obtain the whole-number part. The remainder over the divisor is the fractional part.

EXAMPLE 2 Write $\frac{29}{6}$ as a mixed number.

Solution

```
  4    Divide the numerator by the denominator.
6)29
  24
   5   The remainder is 5.
```

Thus, $\frac{29}{6} = 4\frac{5}{6}$.

Self Check 2
Write $\frac{43}{5}$ as a mixed number.

Answer $8\frac{3}{5}$

THINK IT THROUGH Sleep

"Lack of sleep among college students is an old problem, but one that appears to be getting worse, according to some national surveys." CBS news.com

A 2001 study of over 1,500 college students found that they averaged six hours and forty minutes of sleep per night (Holy Cross Health Survey). James Maas, a professor at Cornell University and author of *Power Sleep,* recommends eight hours of sleep per night for college students. Find the difference between Professor Maas's recommendation and the survey average for college students. Express the result as a mixed number.

Graphing fractions and mixed numbers

Earlier, we graphed whole numbers and integers on a number line. Fractions and mixed numbers can also be graphed on a number line.

EXAMPLE 3 Graph $-2\frac{3}{4}, -1\frac{1}{2}, -\frac{1}{8},$ and $\frac{13}{5}$ on a number line.

Solution To help locate the graph of each number, we make some observations.

- Since $-2\frac{3}{4} < -2$, the graph of $-2\frac{3}{4}$ is to the left of -2.
- The number $-1\frac{1}{2}$ is between -1 and -2.
- The number $-\frac{1}{8}$ is less than 0.
- Expressed as a mixed number, $\frac{13}{5} = 2\frac{3}{5}$.

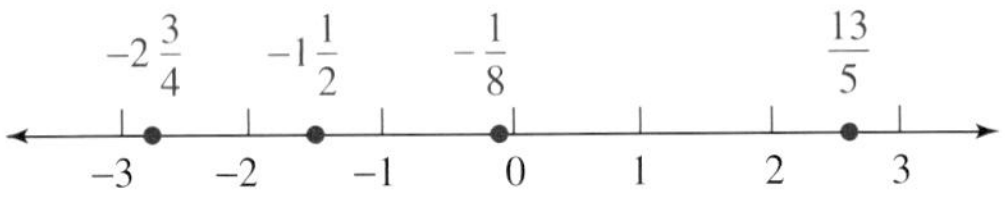

Self Check 3

Graph: $-1\frac{7}{8}, -\frac{2}{3},$ and $\frac{9}{4}$.

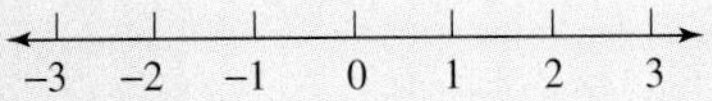

Answer

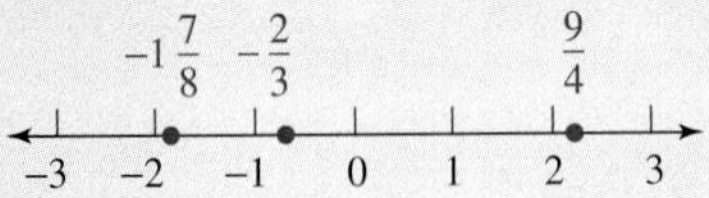

Multiplying and dividing mixed numbers

Multiplying and dividing mixed numbers

To multiply or divide mixed numbers, first change the mixed numbers to improper fractions. Then do the multiplication or division of the fractions.

EXAMPLE 4 Multiply: $5\frac{1}{5} \cdot 1\frac{2}{13}$.

Solution

$$5\frac{1}{5} \cdot 1\frac{2}{13} = \frac{26}{5} \cdot \frac{15}{13} \quad \text{Write each mixed number as an improper fraction.}$$

$$= \frac{26 \cdot 15}{5 \cdot 13} \quad \text{Multiply the numerators and multiply the denominators.}$$

Self Check 4

Multiply: $9\frac{3}{5} \cdot 3\frac{3}{4}$.

$= \frac{2 \cdot 13 \cdot 3 \cdot 5}{5 \cdot 13}$ Factor 26 as $2 \cdot 13$ and 15 as $3 \cdot 5$.

$= \frac{2 \cdot \overset{1}{\cancel{13}} \cdot 3 \cdot \overset{1}{\cancel{5}}}{\underset{1}{\cancel{5}} \cdot \underset{1}{\cancel{13}}}$ Divide out the common factors 13 and 5.

$= \frac{6}{1}$ Multiply in the numerator and denominator.

$= 6$ Simplify: $\frac{6}{1} = 6$.

Answer 36

EXAMPLE 5 Divide: $-3\frac{3}{8} \div 2\frac{1}{4}$.

Self Check 5

Divide: $3\frac{4}{15} \div \left(-2\frac{1}{10}\right)$.

Solution

$-3\frac{3}{8} \div 2\frac{1}{4} = -\frac{27}{8} \div \frac{9}{4}$ Write each mixed number as an improper fraction.

$= -\frac{27}{8} \cdot \frac{4}{9}$ Multiply by the reciprocal of $\frac{9}{4}$.

$= -\frac{27 \cdot 4}{8 \cdot 9}$ The product of two fractions with unlike signs is negative. Multiply the numerators and multiply the denominators.

$= -\frac{3 \cdot \overset{1}{\cancel{9}} \cdot \overset{1}{\cancel{4}}}{2 \cdot \underset{1}{\cancel{4}} \cdot \underset{1}{\cancel{9}}}$ Factor 27 as $3 \cdot 9$ and 8 as $2 \cdot 4$. Divide out the common factors 9 and 4.

$= -\frac{3}{2}$ Multiply in the numerator and denominator.

$= -1\frac{1}{2}$ Write $-\frac{3}{2}$ as a mixed number.

Answer $-1\frac{5}{9}$

EXAMPLE 6 **Government grants.** If $\$12\frac{1}{2}$ million is to be divided equally among five cities to fund recreation programs, how much will each city receive?

Solution To find the amount received by each city, we divide the grant money by 5.

$12\frac{1}{2} \div 5 = \frac{25}{2} \div \frac{5}{1}$ Write $12\frac{1}{2}$ as an improper fraction, and write 5 as a fraction.

$= \frac{25}{2} \cdot \frac{1}{5}$ Multiply by the reciprocal of $\frac{5}{1}$.

$= \frac{25 \cdot 1}{2 \cdot 5}$ Multiply the numerators and multiply the denominators.

$= \frac{\overset{1}{\cancel{5}} \cdot 5 \cdot 1}{2 \cdot \underset{1}{\cancel{5}}}$ Factor 25 as $5 \cdot 5$. Divide out the common factor 5.

$= \frac{5}{2}$ Multiply in the numerator and denominator.

$= 2\frac{1}{2}$ Write $\frac{5}{2}$ as a mixed number.

Each city will receive $\$2\frac{1}{2}$ million.

Section 3.5 STUDY SET

VOCABULARY *Fill in the blanks.*

1. A ________ number is the sum of a whole number and a proper fraction.
2. An __________ fraction is a fraction with a numerator that is greater than or equal to its denominator.
3. To _______ a number means to locate its position on the number line and highlight it with a dot.
4. Multiplying or dividing the __________ and ___________ of a fraction by the same nonzero number does not change the value of the fraction.

CONCEPTS

5. What signed number could be used to describe each situation?
   a. A temperature of five and one-half degrees below zero
   b. A sprinkler pipe that is $6\frac{7}{8}$ inches under the sidewalk
6. What signed number could be used to describe each situation?
   a. A rain total two and three-tenths of an inch lower than the average.
   b. Three and one-half minutes before liftoff
7. a. In the illustration, the divisions on the face of the meter represent fractions. What value is the arrow registering?
   b. If the arrow moves four marks to the left, what value will it register?

8. a. In the illustration, the divisions on the face of the meter represent fractions. What value is the arrow registering?
   b. If the arrow moves up one mark, what value will it register?

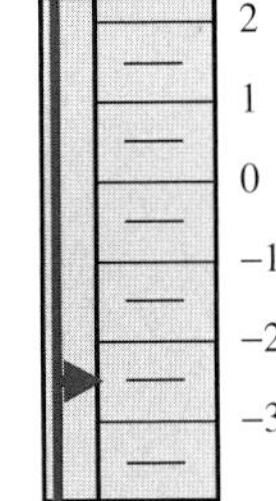

9. What fractions have been graphed on the number line?

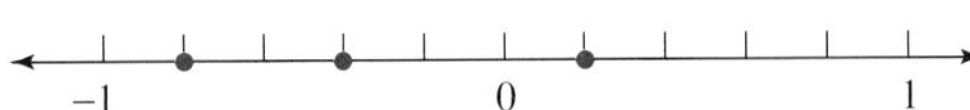

10. What mixed numbers have been graphed on the number line?

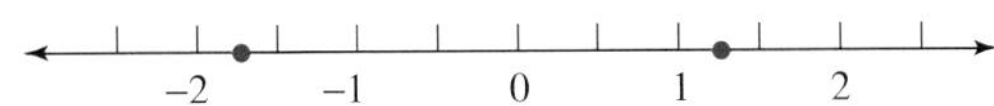

11. DIVING Complete the description of the following dive by filling in the blank with a mixed number.

Forward   somersaults

12. PRODUCT LABELING The label below uses mixed numbers. Write each one as an improper fraction.

Laundry Basket
1 3/4 Bushel
•Easy-grip rim is reinforced to handle the biggest loads
23 1/4" L X 18 7/8" W X 10 1/2" H

13. Draw a picture that represents $\frac{17}{8}$ pizzas.
14. a. What mixed number is represented below?
    b. What improper fraction is shown below?

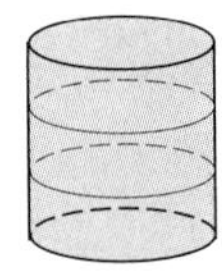 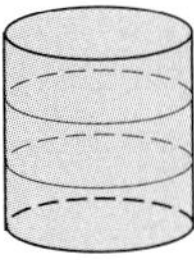 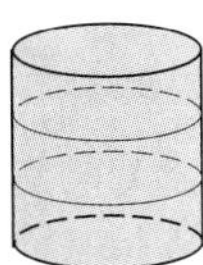 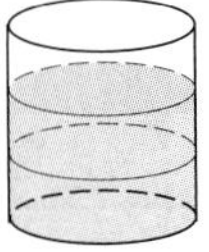

NOTATION *Complete each solution.*

15. $-5\frac{1}{4}\cdot 1\frac{1}{7} = -\frac{21}{4}\cdot\frac{\square}{7}$

$= -\frac{21\cdot\square}{4\cdot 7}$

$= -\frac{3\cdot\overset{1}{\cancel{7}}\cdot 2\cdot\overset{1}{\cancel{\square}}}{\underset{1}{\cancel{\square}}\cdot\underset{1}{\cancel{7}}}$

$= -\frac{\square}{1}$

$= -6$

16. $-5\frac{5}{6}\div 2\frac{1}{12} = -\frac{\square}{6}\div\frac{25}{12}$

$= -\frac{35}{6}\cdot\frac{12}{\square}$

$= -\frac{35\cdot 12}{6\cdot\square}$

$= -\frac{\overset{1}{\cancel{5}}\cdot\square\cdot 2\cdot\overset{1}{\cancel{6}}}{\underset{1}{\cancel{6}}\cdot\underset{1}{\cancel{5}}\cdot\square}$

$= -\frac{\square}{5}$

$= -2\frac{4}{5}$

PRACTICE *Write each improper fraction as a mixed number. Simplify the result, if possible.*

17. $\frac{15}{4}$

18. $\frac{41}{6}$

19. $\frac{29}{5}$

20. $\frac{29}{3}$

21. $-\frac{20}{6}$

22. $-\frac{28}{8}$

23. $\frac{127}{12}$

24. $\frac{197}{16}$

Write each mixed number as an improper fraction.

25. $6\frac{1}{2}$

26. $8\frac{2}{3}$

27. $20\frac{4}{5}$

28. $15\frac{3}{8}$

29. $-6\frac{2}{9}$

30. $-7\frac{1}{12}$

31. $200\frac{2}{3}$

32. $90\frac{5}{6}$

Graph each set of numbers on the number line.

33. $\left\{-2\frac{8}{9}, 1\frac{2}{3}, \frac{16}{5}\right\}$

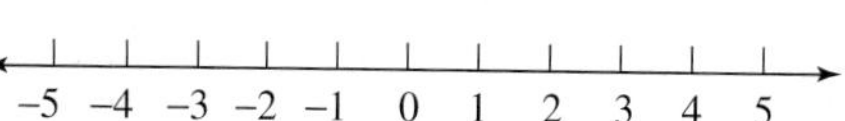

34. $\left\{-\frac{3}{4}, -3\frac{1}{4}, \frac{5}{2}\right\}$

−5 −4 −3 −2 −1 0 1 2 3 4 5

35. $\left\{3\frac{1}{7}, -\frac{98}{99}, -\frac{10}{3}\right\}$

−5 −4 −3 −2 −1 0 1 2 3 4 5

36. $\left\{-2\frac{1}{5}, \frac{4}{5}, -\frac{11}{3}\right\}$

−5 −4 −3 −2 −1 0 1 2 3 4 5

Multiply.

37. $1\frac{2}{3}\cdot 2\frac{1}{7}$

38. $2\frac{3}{5}\cdot 1\frac{2}{3}$

39. $-7\frac{1}{2}\left(-1\frac{2}{5}\right)$

40. $-4\frac{1}{8}\left(-1\frac{7}{9}\right)$

41. $3\frac{1}{16}\cdot 4\frac{4}{7}$

42. $5\frac{3}{5}\cdot 1\frac{11}{14}$

43. $-6\cdot 2\frac{7}{24}$

44. $-7\cdot 1\frac{3}{28}$

45. $2\frac{1}{2}\left(-3\frac{1}{3}\right)$

46. $\left(-3\frac{1}{4}\right)\left(1\frac{1}{5}\right)$

47. $2\frac{5}{8}\cdot\frac{5}{27}$

48. $3\frac{1}{9}\cdot\frac{3}{32}$

49. Find the product of $1\frac{2}{3}$, 6, and $-\frac{1}{8}$.

50. Find the product of $-\frac{5}{6}$, -8, and $-2\frac{1}{10}$.

Evaluate each power.

51. $\left(1\frac{2}{3}\right)^2$

52. $\left(3\frac{1}{2}\right)^2$

53. $\left(-1\frac{1}{3}\right)^3$

54. $\left(-1\frac{1}{5}\right)^3$

Divide.

55. $3\frac{1}{3} \div 1\frac{5}{6}$

56. $3\frac{3}{4} \div 5\frac{1}{3}$

57. $-6\frac{3}{5} \div 7\frac{1}{3}$

58. $-4\frac{1}{4} \div 4\frac{1}{2}$

59. $-20\frac{1}{4} \div \left(-1\frac{11}{16}\right)$

60. $-2\frac{7}{10} \div \left(-1\frac{1}{14}\right)$

61. $6\frac{1}{4} \div 20$

62. $4\frac{2}{5} \div 11$

63. $1\frac{2}{3} \div \left(-2\frac{1}{2}\right)$

64. $2\frac{1}{2} \div \left(-1\frac{5}{8}\right)$

65. $8 \div 3\frac{1}{5}$

66. $15 \div 3\frac{1}{3}$

67. Find the quotient of $-4\frac{1}{2}$ and $2\frac{1}{4}$.

68. Find the quotient of 25 and $-10\frac{5}{7}$.

APPLICATIONS

69. CALORIES A company advertises that its mints contain only $3\frac{1}{5}$ calories apiece. What is the calorie intake if you eat an entire package of 20 mints?

70. CEMENT MIXERS A cement mixer can carry $9\frac{1}{2}$ cubic yards of concrete. If it makes 8 trips to a job site, how much concrete will be delivered to the site?

71. SHOPPING In the illustration, what is the cost of buying the fruit in the scale?

72. FRAMES How much molding is needed to make the square picture frame below?.

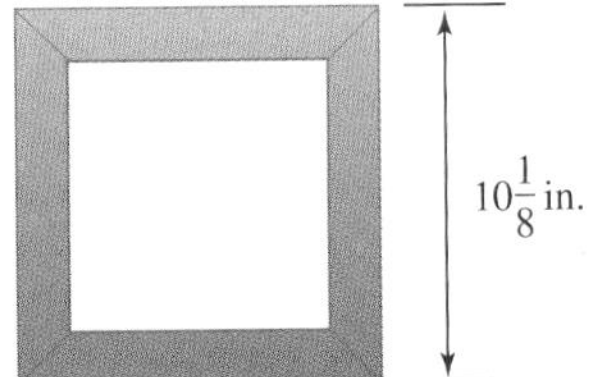

73. SUBDIVISIONS A developer donated to the county 100 of the 1,000 acres of land she owned. She divided the remaining acreage into $1\frac{1}{3}$-acre lots. How many lots were created?

74. CATERING How many people can be served $\frac{1}{3}$-pound hamburgers if a caterer purchases 200 pounds of ground beef?

75. GRAPH PAPER Mathematicians use specially marked paper, called *graph paper,* when drawing figures. It is made up of $\frac{1}{4}$-inch squares. Find the length and width of the following piece of graph paper.

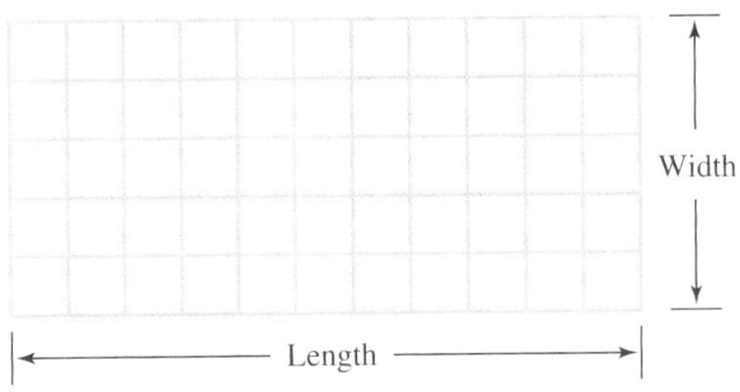

76. LUMBER As shown in the following illustration, 2-by-4's from the lumber yard do not really have dimensions of 2 inches by 4 inches. How wide and how high is the stack of 2-by-4's?

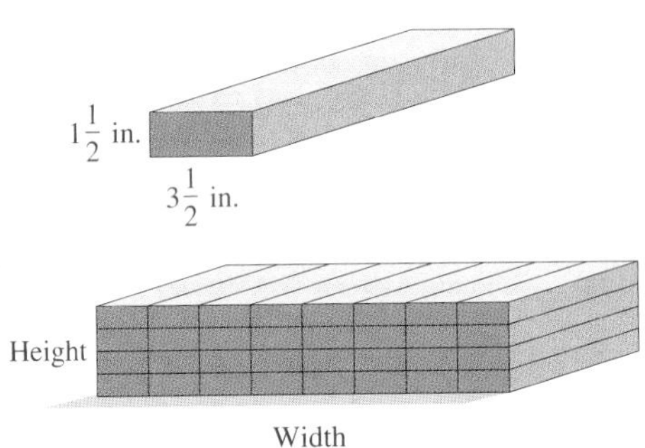

77. EMERGENCY EXIT The following sign marks the emergency exit on a school bus. Find the area of the sign.

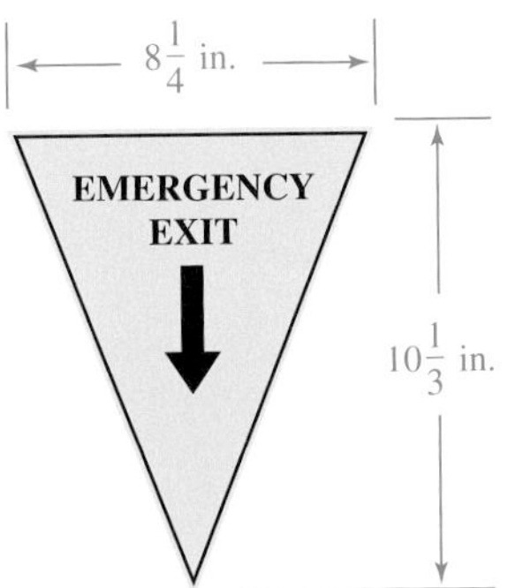

78. HORSE RACING The race tracks on which thoroughbred horses run are marked off in $\frac{1}{8}$-mile-long segments called furlongs. How many furlongs are there in a $1\frac{1}{16}$-mile race?

79. FIRE ESCAPES The fire escape stairway in an office building is shown. Each riser is $7\frac{1}{2}$ inches high. If each floor is 105 inches high and the building is 43 stories tall, how many steps are in the stairway?

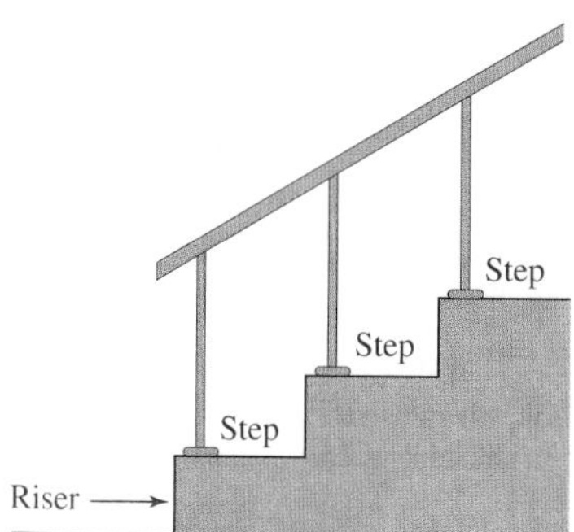

80. LICENSE PLATES Find the area of the license plate shown below.

81. SHOPPING ON THE INTERNET A mother is ordering a pair of jeans for her daughter from the screen shown. If the daughter's height is $60\frac{3}{4}$ in. and her waist is $24\frac{1}{2}$ in., on what size and what cut should the mother point and click?

Girl's jeans- regular cut

Size	7	8	10	12	14	16
Height	50-52	52-54	54-56	$56\frac{1}{4}$-$58\frac{1}{2}$	59-61	61-62
Waist	$22\frac{1}{4}$-$22\frac{3}{4}$	$22\frac{3}{4}$-$23\frac{1}{4}$	$23\frac{3}{4}$-$24\frac{1}{4}$	$24\frac{3}{4}$-$25\frac{1}{4}$	$25\frac{3}{4}$-$26\frac{1}{4}$	$26\frac{1}{4}$-28

Girl's jeans- slim cut

Size	7	8	10	12	14	16
Height	50-52	52-54	54-56	$56\frac{1}{2}$-$58\frac{1}{2}$	59-61	61-62
Waist	$20\frac{3}{4}$-$21\frac{1}{4}$	$21\frac{1}{4}$-$21\frac{3}{4}$	$22\frac{1}{4}$-$22\frac{3}{4}$	$23\frac{1}{4}$-$23\frac{3}{4}$	$24\frac{1}{4}$-$24\frac{3}{4}$	25-$26\frac{1}{2}$

To order:
Point arrow to proper size/cut and click

82. SEWING Use the following table to determine the number of yards of fabric needed:

a. to make a size 16 top if the fabric to be used is 60 inches wide.

b. to make size 18 pants if the fabric to be used is 45 inches wide.

8767 Pattern
stitch'n save
by McCall's

Front

SIZES	8	10	12	14	16	18	20	
Top								
45"	$2\frac{1}{4}$	$2\frac{3}{8}$	$2\frac{3}{8}$	$2\frac{3}{8}$	$2\frac{1}{2}$	$2\frac{5}{8}$	$2\frac{3}{4}$	**Yds**
60"	2	2	$2\frac{1}{8}$	$2\frac{1}{8}$	$2\frac{1}{8}$	$2\frac{1}{8}$	$2\frac{1}{8}$	
Pants								
45"	$2\frac{5}{8}$	$2\frac{5}{8}$	$2\frac{5}{8}$	$2\frac{5}{8}$	$2\frac{5}{8}$	$2\frac{5}{8}$	$2\frac{5}{8}$	**Yds**
60"	$1\frac{3}{4}$	2	$2\frac{1}{4}$	$2\frac{1}{4}$	$2\frac{1}{4}$	$2\frac{1}{4}$	$2\frac{1}{2}$	

WRITING

83. Explain the difference between $2\frac{3}{4}$ and $2(\frac{3}{4})$.

84. Give three examples of how you use mixed numbers in daily life.

85. Explain the procedure used to write an improper fraction as a mixed number.

86. Explain the procedure used to multiply two mixed numbers.

REVIEW

87. Evaluate: $3^2 \cdot 2^3$.

88. If a represents a number, then $a \cdot 0 = \square$.

89. Write $8 + 8 + 8 + 8$ as a multiplication.

90. If a square measures 1 inch on each side, find its area?

91. Solve: $\frac{x}{2} = -12$.

92. In the formula $A = lw$, what do l and w represent?

3.6 Adding and Subtracting Mixed Numbers

- Adding mixed numbers
- Adding mixed numbers in vertical form
- Subtracting mixed numbers

In this section, we will discuss methods for adding and subtracting mixed numbers. The first method works well when the whole-number parts of the mixed numbers are small. The second method works well when the whole-number parts of the mixed numbers are large. The third method uses columns as a way to organize the work.

Adding mixed numbers

We can add mixed numbers by writing them as improper fractions. To do so, we follow these steps.

Adding mixed numbers: method 1

1. Write each mixed number as an improper fraction.
2. Write each improper fraction as an equivalent fraction with a denominator that is the LCD.
3. Add the fractions.
4. Change the result to a mixed number if desired.

Self Check 1

Add: $3\frac{2}{3} + 1\frac{1}{5}$.

Answer $4\frac{13}{15}$

EXAMPLE 1 Add: $4\frac{1}{6} + 2\frac{3}{4}$.

Solution

$4\frac{1}{6} + 2\frac{3}{4} = \frac{25}{6} + \frac{11}{4}$ Write each mixed number as an improper fraction: $4\frac{1}{6} = \frac{25}{6}$ and $2\frac{3}{4} = \frac{11}{4}$.

By inspection, we see that the common denominator is 12.

$= \frac{25 \cdot \mathbf{2}}{6 \cdot \mathbf{2}} + \frac{11 \cdot 3}{4 \cdot 3}$ Write each fraction as a fraction with a denominator of 12.

$= \frac{50}{12} + \frac{33}{12}$ Perform the multiplications in the numerators and denominators.

$= \frac{83}{12}$ Add the numerators: $50 + 33 = 83$. Write the sum over the common denominator 12.

$= 6\frac{11}{12}$ Write the improper fraction as a mixed number: $\frac{83}{12} = 6\frac{11}{12}$.

We can also add mixed numbers by adding their whole-number parts and their fractional parts. To do so, we follow these steps.

Adding mixed numbers: method 2

1. Write each mixed number as the sum of a whole number and a fraction.
2. Use the commutative property of addition to write the whole numbers together and the fractions together.
3. Add the whole numbers and the fractions separately.
4. Write the result as a mixed number if necessary.

EXAMPLE 2 Find the sum: $168\frac{3}{4} + 85\frac{1}{5}$.

Solution

$$168\frac{3}{4} + 85\frac{1}{5} = 168 + \frac{3}{4} + 85 + \frac{1}{5}$$ Write each mixed number as the sum of a whole number and a fraction.

$$= 168 + 85 + \frac{3}{4} + \frac{1}{5}$$ Use the commutative property of addition to change the order of the addition.

$$= 253 + \frac{3}{4} + \frac{1}{5}$$ Add the whole numbers: $168 + 85 = 253$.

$$= 253 + \frac{3 \cdot \mathbf{5}}{4 \cdot \mathbf{5}} + \frac{1 \cdot \mathbf{4}}{5 \cdot \mathbf{4}}$$ Write each fraction as a fraction with denominator 20.

$$= 253 + \frac{15}{20} + \frac{4}{20}$$ Multiply in the numerators and denominators.

$$= 253 + \frac{19}{20}$$ Add the numerators and write the sum over the common denominator 20.

$$= 253\frac{19}{20}$$ Write the sum as a mixed number.

Self Check 2

Find the sum: $275\frac{1}{6} + 81\frac{3}{5}$.

Answer $356\frac{23}{30}$

! COMMENT If we use method 1 to add the mixed numbers in Example 2, the numbers we encounter are cumbersome. As expected, the result is the same: $253\frac{19}{20}$.

$$168\frac{3}{4} + 85\frac{1}{5} = \frac{675}{4} + \frac{426}{5}$$ Write $168\frac{3}{4}$ and $85\frac{1}{5}$ as improper fractions.

$$= \frac{675 \cdot \mathbf{5}}{4 \cdot \mathbf{5}} + \frac{426 \cdot \mathbf{4}}{5 \cdot \mathbf{4}}$$ The LCD is 20.

$$= \frac{3{,}375}{20} + \frac{1{,}704}{20}$$

$$= \frac{5{,}079}{20}$$

$$= 253\frac{19}{20}$$

Generally speaking, the larger the whole-number parts of the mixed numbers get, the more difficult it becomes to add those mixed numbers using method 1.

Adding mixed numbers in vertical form

By working in columns, we can use a third method to add mixed numbers. The strategy is the same as in Example 2: Add whole numbers to whole numbers and fractions to fractions.

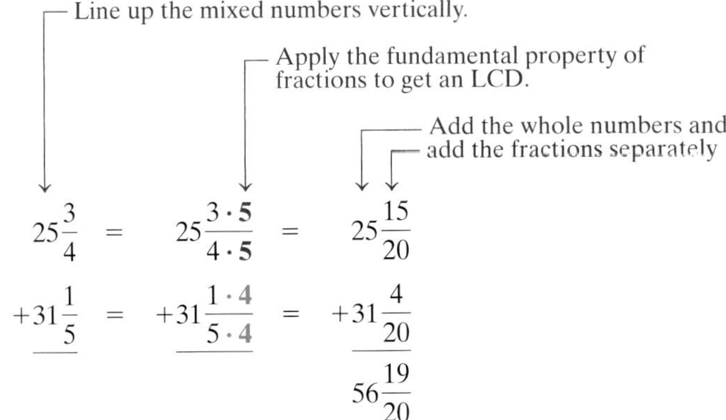

Line up the mixed numbers vertically.
Apply the fundamental property of fractions to get an LCD.
Add the whole numbers and add the fractions separately
$25\frac{3}{4} = 25\frac{3 \cdot 5}{4 \cdot 5} = 25\frac{15}{20}$
$+31\frac{1}{5} = +31\frac{1 \cdot 4}{5 \cdot 4} = +31\frac{4}{20}$
$56\frac{19}{20}$

EXAMPLE 3 Suspension bridges. Find the total length of cable that must be ordered if cables a, d, and e of the suspension bridge in Figure 3-9 are to be replaced. (See the table on the right.)

Bridge Specifications

Cable	a	b	c
Length (feet)	$75\frac{1}{12}$	$54\frac{1}{6}$	$43\frac{1}{4}$

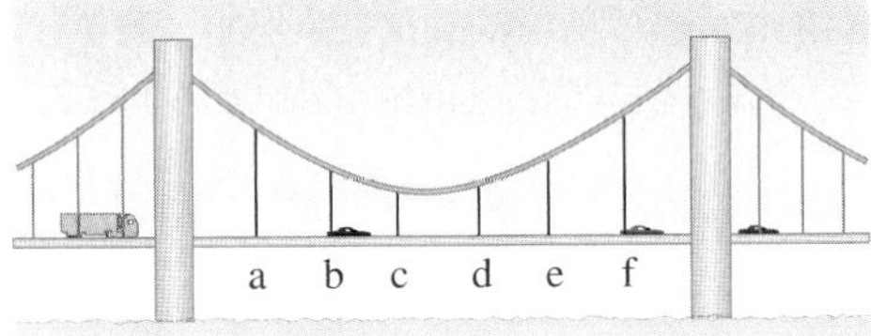

FIGURE 3-9

Solution To find the total length of cable to be ordered, we add the lengths of cables a, d, and e. Because of the symmetric design, cables e and b and cables d and c are the same length.

Length of cable a	plus	length of cable d (or cable c)	plus	length of cable e (or cable b)	equals	the total length needed.
$75\frac{1}{12}$	$+$	$43\frac{1}{4}$	$+$	$54\frac{1}{6}$	$=$	total length

We add the mixed numbers using a vertical format.

$$75\frac{1}{12} = 75\frac{1}{12} = 75\frac{1}{12}$$

$$43\frac{1}{4} = 43\frac{1 \cdot 3}{4 \cdot 3} = 43\frac{3}{12}$$

$$+54\frac{1}{6} = +54\frac{1 \cdot 2}{6 \cdot 2} = +\ 54\frac{2}{12}$$

$$172\frac{6}{12} = 172\frac{1}{2} \quad \text{Simplify: } \frac{6}{12} = \frac{1}{2}.$$

The total length of cable needed for the replacement is $172\frac{1}{2}$ feet.

When we add mixed numbers, the sum of the fractions sometimes yields an improper fraction, as in the next example.

EXAMPLE 4 Add: $45\frac{2}{3} + 96\frac{4}{5}$.

Solution

$$\begin{array}{rcrcr} 45\frac{2}{3} & = & 45\frac{2 \cdot \mathbf{5}}{3 \cdot \mathbf{5}} & = & 45\frac{10}{15} \\ +96\frac{4}{5} & = & +96\frac{4 \cdot \mathbf{3}}{5 \cdot \mathbf{3}} & = & +\ 96\frac{12}{15} \\ \hline & & & & 141\frac{22}{15} \end{array}$$

The whole-number part of the answer ↑ ↑ The fractional part of the answer is an improper fraction.

Now we write the improper fraction as a mixed number.

$$141\frac{22}{15} = 141 + \frac{\mathbf{22}}{\mathbf{15}} = 141 + \mathbf{1}\frac{\mathbf{7}}{\mathbf{15}} = 142\frac{7}{15}$$

Self Check 4

Add: $76\frac{11}{12} + 49\frac{5}{8}$.

Answer $126\frac{13}{24}$

Subtracting mixed numbers

Subtracting mixed numbers is similar to adding mixed numbers.

EXAMPLE 5 **Cooking.** How much butter is left in a 10-pound tub if $2\frac{2}{3}$ pounds are used for a wedding cake?

Solution The phrase "How much is left?" suggests subtraction.

$10 - 2\frac{2}{3} = \frac{10}{1} - \frac{8}{3}$ Write 10 as a fraction: $10 = \frac{10}{1}$. Write $2\frac{2}{3}$ as $\frac{8}{3}$.

By inspection, we see that the LCD is 3.

$\frac{10}{1} - 2\frac{2}{3} = \frac{10 \cdot \mathbf{3}}{1 \cdot \mathbf{3}} - \frac{8}{3}$ Write the first fraction with a denominator 3.

$= \frac{30}{3} - \frac{8}{3}$ Perform the multiplications in the first fraction.

$= \frac{30 - 8}{3}$ Subtract the numerators and write the difference over the common denominator.

$= \frac{22}{3}$ Perform the subtraction: $30 - 8 = 22$.

$= 7\frac{1}{3}$ Write $\frac{22}{3}$ as a mixed number.

There are $7\frac{1}{3}$ pounds of butter left in the tub.

In the next example, $\frac{1}{5}$ is less than $\frac{2}{3}$. Because of this, we have to borrow.

Self Check 6

Subtract: $101\frac{3}{4} - 79\frac{15}{16}$.

Answer $21\frac{13}{16}$

EXAMPLE 6 Subtract: $34\frac{1}{5} - 11\frac{2}{3}$.

Solution We will use the vertical form to subtract. The LCD is 15, so we write each fraction as a fraction with a denominator 15.

$$\begin{array}{rcrcr} 34\frac{1}{5} & = & 34\frac{1\cdot\mathbf{3}}{5\cdot\mathbf{3}} & = & 34\frac{3}{15} \\ -11\frac{2}{3} & = & -11\frac{2\cdot 5}{3\cdot 5} & = & -11\frac{10}{15} \end{array}$$

Since $\frac{10}{15}$ is larger than $\frac{3}{15}$, borrow 1 (in the form of $\frac{15}{15}$) from 34 and add it to $\frac{3}{15}$ to obtain $33\frac{3}{15} + \frac{15}{15} = 33\frac{18}{15}$. Then we subtract the fractions and the whole numbers separately.

$$\begin{array}{rcr} 33\frac{3}{15} + \frac{15}{15} & = & 33\frac{18}{15} \\ -11\frac{10}{15} & = & -11\frac{10}{15} \\ & & 22\frac{8}{15} \end{array}$$

Self Check 7

Subtract: $2{,}300 - 129\frac{31}{32}$.

Answer $2{,}170\frac{1}{32}$

EXAMPLE 7 Subtract: $419 - 53\frac{11}{16}$.

Solution We align the numbers vertically and borrow 1 (in the form of $\frac{16}{16}$) from 419. Then we subtract the fractions and subtract the whole numbers separately.

$$\begin{array}{rcr} 419 & = & 418\frac{16}{16} \\ -\ 53\frac{11}{16} & = & -\ 53\frac{11}{16} \\ & & 365\frac{5}{16} \end{array}$$

Section 3.6 STUDY SET

VOCABULARY *Fill in the blanks.*

1. By the ____________ property of addition, we can add numbers in any order.
2. A ________ number such as $1\frac{7}{8}$ contains a whole-number part and a fractional part.
3. Consider

$$\begin{array}{rcl} 80\frac{1}{3} & = & 79\frac{1}{3} + \frac{3}{3} \\ -24\frac{2}{3} & = & -24\frac{2}{3} \end{array}$$

To do the subtraction, we ________ 1 in the form of $\frac{3}{3}$.

4. The letters LCD stand for ______ _________ ____________.

CONCEPTS

5. **a.** For $76\frac{3}{4}$, list the whole-number part and the fractional part.
   b. Write $76\frac{3}{4}$ as a sum.
6. Use the commutative property of addition to get the whole numbers together.

$$14 + \frac{5}{6} + 53 + \frac{1}{6}$$

7. What property is being highlighted here?

$$\begin{array}{r} 25\frac{3\cdot\mathbf{5}}{4\cdot\mathbf{5}} \\ +31\frac{1\cdot 4}{5\cdot 4} \end{array}$$

8. a. The denominators of two fractions, expressed in prime-factored form, are $5 \cdot 2$ and $5 \cdot 3$. Find the LCD for the fractions.

b. The denominators of three fractions, in prime-factored form, are $3 \cdot 5$, $2 \cdot 3$, and $3 \cdot 3$. Find the LCD for the fractions.

9. Simplify.

a. $9\frac{17}{16}$ **b.** $1{,}288\frac{7}{3}$

c. $16\frac{12}{8}$ **d.** $45\frac{24}{20}$

10. Consider

$$\begin{array}{r} 108\frac{1}{3} \\ -\ 99\frac{2}{3} \\ \hline \end{array}$$

a. Explain why we will have to borrow if we subtract the mixed numbers in this way.

b. In what form will we borrow a 1 from 108?

NOTATION *Complete each solution.*

11.
$$\begin{aligned} 70\frac{3}{5} + 39\frac{2}{7} &= \square + \frac{3}{5} + \square + \frac{2}{7} \\ &= \square + \square + \frac{3}{5} + \frac{2}{7} \\ &= 109 + \frac{3}{5} + \frac{2}{7} \\ &= 109 + \frac{3 \cdot \square}{5 \cdot \square} + \frac{2 \cdot \square}{7 \cdot \square} \\ &= 109 + \frac{21}{\square} + \frac{10}{\square} \\ &= 109 + \frac{\square}{35} \\ &= 109\frac{31}{35} \end{aligned}$$

12.
$$\begin{array}{rcl} 67\frac{3}{8} &=& 67\frac{3 \cdot \square}{8 \cdot \square} = \\ -23\frac{2}{3} &=& -23\frac{2 \cdot \square}{3 \cdot \square} = \\ \hline \end{array}$$

$$\begin{array}{rcccl} 67\frac{9}{24} &=& \square\frac{9}{24} + \frac{\square}{\square} &=& 66\frac{\square}{24} \\ -23\frac{16}{24} &=& -23\frac{16}{24} &=& -23\frac{16}{24} \\ \hline & & & & 43\frac{\square}{24} \end{array}$$

PRACTICE *Find each sum or difference.*

13. $2\frac{1}{5} + 2\frac{1}{5}$

14. $3\frac{1}{3} + 2\frac{1}{3}$

15. $8\frac{2}{7} - 3\frac{1}{7}$

16. $9\frac{5}{11} - 6\frac{2}{11}$

17. $3\frac{1}{4} + 4\frac{1}{4}$

18. $2\frac{1}{8} + 3\frac{3}{8}$

19. $4\frac{1}{6} + 1\frac{1}{5}$

20. $2\frac{2}{5} + 3\frac{1}{4}$

21. $2\frac{1}{2} - 1\frac{1}{4}$

22. $13\frac{5}{6} - 4\frac{2}{3}$

23. $2\frac{5}{6} - 1\frac{3}{8}$

24. $4\frac{5}{9} - 2\frac{1}{6}$

25. $5\frac{1}{2} + 3\frac{4}{5}$

26. $6\frac{1}{2} + 2\frac{2}{3}$

27. $7\frac{1}{2} - 4\frac{1}{7}$

28. $5\frac{3}{4} - 1\frac{3}{7}$

29. $56\frac{2}{5} + 73\frac{1}{3}$

30. $44\frac{3}{8} + 66\frac{1}{5}$

31. $380\frac{1}{6} + 17\frac{1}{4}$

32. $103\frac{1}{2} + 210\frac{2}{5}$

33. $228\frac{5}{9} + 44\frac{2}{3}$

34. $161\frac{7}{8} + 19\frac{1}{3}$

35. $778\frac{5}{7} - 155\frac{1}{3}$

36. $339\frac{1}{2} - 218\frac{3}{16}$

37. $140\frac{5}{6} - 129\frac{4}{5}$

38. $291\frac{1}{4} - 289\frac{1}{12}$

39. $422\frac{13}{16} - 321\frac{3}{8}$

40. $378\frac{3}{4} - 277\frac{5}{8}$

Find each difference.

41. $16\frac{1}{4} - 13\frac{3}{4}$

42. $40\frac{1}{7} - 19\frac{6}{7}$

43. $76\frac{1}{6} - 49\frac{7}{8}$

44. $101\frac{1}{4} - 70\frac{1}{2}$

45. $140\frac{3}{16} - 129\frac{3}{4}$

46. $211\frac{1}{3} - 8\frac{3}{4}$

47. $334\frac{1}{9} - 13\frac{5}{6}$

48. $442\frac{1}{8} - 429\frac{2}{3}$

Find the sum or difference.

49. $7 - \frac{2}{3}$

50. $6 - \frac{1}{8}$

51. $9 - 8\frac{3}{4}$

52. $11 - 10\frac{4}{5}$

53. $4\frac{1}{7} - \frac{4}{5}$

54. $5\frac{1}{10} - \frac{4}{5}$

55. $6\frac{5}{8} - 3$

56. $10\frac{1}{2} - 6$

57. $\frac{7}{3} + 2$

58. $\frac{9}{7} + 3$

59. $2 + 1\frac{7}{8}$

60. $3\frac{3}{4} + 5$

Find each sum.

61. $12\frac{1}{2} + 5\frac{3}{4} + 35\frac{1}{6}$

62. $31\frac{1}{3} + 20\frac{2}{5} + 10\frac{1}{15}$

63. $58\frac{7}{8} + 340 + 61\frac{1}{4}$

64. $191 + 233\frac{1}{16} + 16\frac{5}{8}$

Find each sum or difference.

65. $-3\frac{3}{4} + \left(-1\frac{1}{2}\right)$

66. $-3\frac{2}{3} + \left(-1\frac{4}{5}\right)$

67. $-4\frac{5}{8} - 1\frac{1}{4}$

68. $-2\frac{1}{16} - 3\frac{7}{8}$

APPLICATIONS

69. FREEWAY TRAVEL A freeway exit sign is shown. How far apart are the Citrus Ave. and Grand Ave. exits?

Citrus Ave. $\frac{3}{4}$ mi

Grand Ave. $3\frac{1}{2}$ mi

70. BASKETBALL See the illustration. What is the difference in height between the tallest and the shortest starting players?

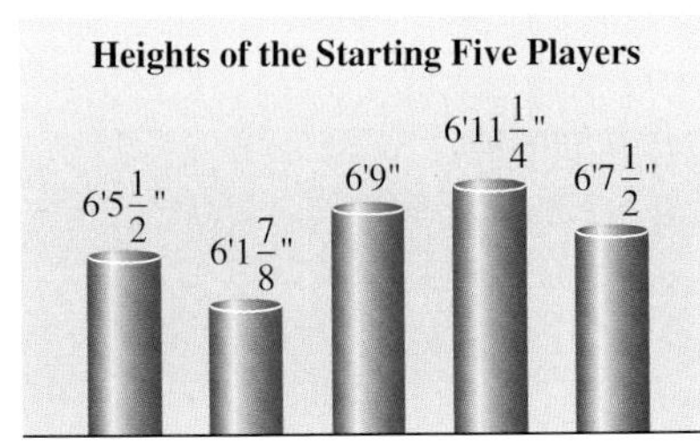

71. TRAIL MIX A camper doubles up on the amount of sunflower seeds called for in the following recipe. How much trail mix will the adjusted recipe yield?

Trail Mix

A healthy snack–great for camping trips

$2\frac{3}{4}$ cups peanuts
$\frac{1}{2}$ cup sunflower seeds
$\frac{2}{3}$ cup raisins
$\frac{1}{3}$ cup coconut
$2\frac{2}{3}$ cups oat flakes
$\frac{1}{4}$ cup pretzels

72. AIR TRAVEL A businesswoman's flight leaves Los Angeles at 8 A.M. and arrives in Seattle at 9:45 A.M.

a. Express the duration of the flight as a mixed number.

b. Upon arrival, she boards a commuter plane at 11:15 A.M., arriving at her final destination at 11:45 A.M. Express the length of this flight as a fraction.

c. Find the total time of these two flights.

73. HOSE REPAIRS To repair a bad connector, Ming Lin removes $1\frac{1}{2}$ feet from the end of a 50-foot garden hose. How long is the hose after the repair?

74. SEWING To make some draperies, Liz needs $12\frac{1}{4}$ yards of material for the den and $8\frac{1}{2}$ yards for the living room. If the material comes only in 21-yard bolts, how much will be left over after both sets of draperies are completed?

75. SHIPPING A passenger ship and a cargo ship leave San Diego harbor at midnight. During the first hour, the passenger ship travels south at $16\frac{1}{2}$ miles per hour, while the cargo ship is traveling north at a rate of $5\frac{1}{5}$ miles per hour.

a. Complete the following table.

b. How far apart are the ships at 1:00 A.M.?

	Rate (mph) ·	Time traveling (hr) =	Distance traveled (mi)
Passenger ship		1	
Cargo ship		1	

76. HARDWARE Refer to the illustration below. How long should the threaded part of the bolt be?

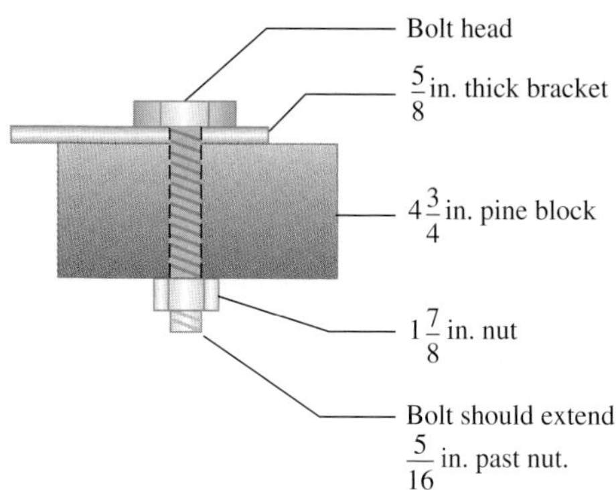

77. SERVICE STATIONS Use the service station sign to answer the following questions.

a. What is the difference in price between the least and most expensive types of gasoline at the self-service pump?

b. How much more is the cost per gallon for full service?

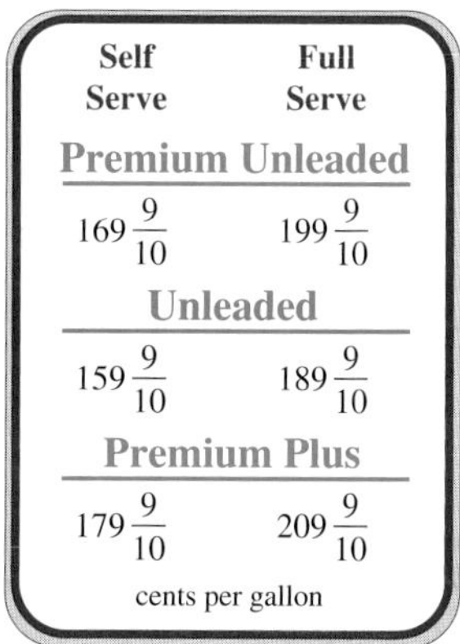

Self Serve
Full Serve
Premium Unleaded
$169\frac{9}{10}$
$199\frac{9}{10}$
Unleaded
$159\frac{9}{10}$
$189\frac{9}{10}$
Premium Plus
$179\frac{9}{10}$
$209\frac{9}{10}$
cents per gallon

78. SEPTUPLETS On November 19, 1997, at Iowa Methodist Medical Center, Bobbie McCaughey gave birth to seven babies. From the information, shown below, find the combined birthweights of the babies.

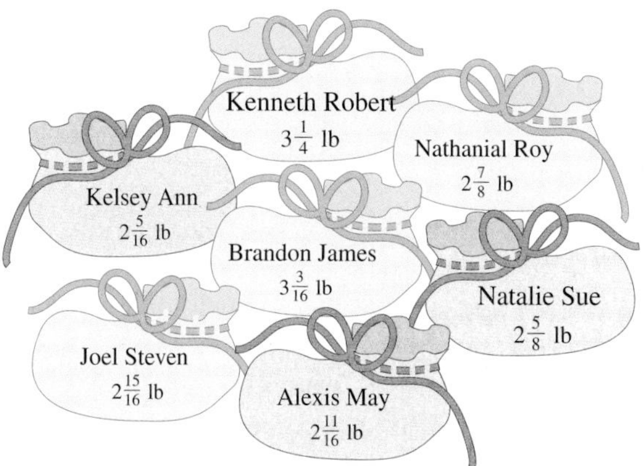

79. WATER SLIDES An amusement park added a new section to a water slide to create a slide $311\frac{5}{12}$ feet long. How long was the slide before the addition?

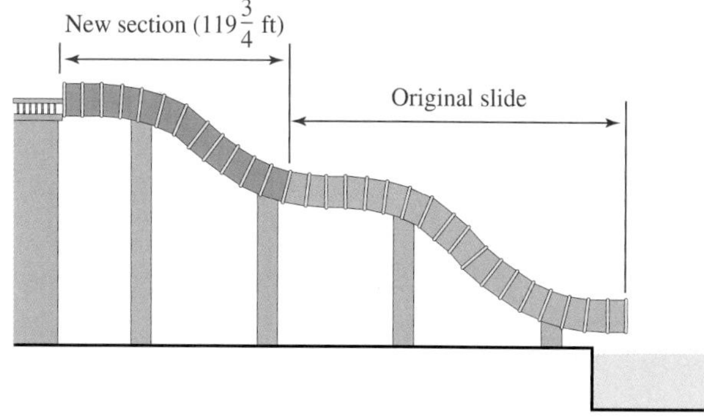

80. JEWELRY A jeweler is to cut the following 7-inch-long gold braid into three pieces. He aligns a 6-inch-long ruler directly below the braid and makes the proper cuts. Find the length of piece 2 of the braid.

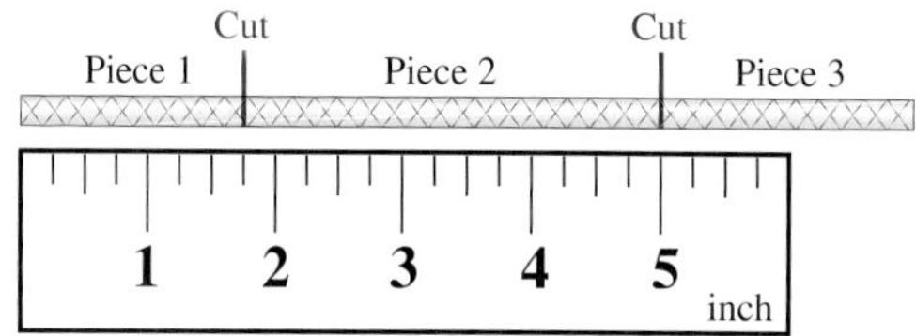

WRITING

81. Of the methods studied to add mixed numbers, which do you like better, and why?

82. When subtracting mixed numbers, when is borrowing necessary? How is it done?

83. Explain how to add $1\frac{3}{8}$ and $2\frac{1}{4}$ if we write each one as an improper fraction.

84. Explain the process of simplifying $12\frac{16}{5}$.

REVIEW

85. Solve: $2x - 1 = 13$.

86. Multiply: $-3(-4)(-5)$.

87. Subtract: $-2 - (-8)$.

88. Evaluate: $(-2)^3(3)^2$.

89. Subtract: $-2 - 8$.

90. Evaluate: $|-12|$.

3.7 Order of Operations and Complex Fractions

- Order of operations • Complex fractions • Simplifying complex fractions

In this section, we will evaluate expressions involving fractions and mixed numbers. We will also discuss complex fractions and the methods that are used to simplify them.

Order of operations

The rules for the order of operations are used to evaluate numerical expressions that involve more than one operation.

EXAMPLE 1 Evaluate: $\frac{3}{4} + \frac{5}{3}\left(-\frac{1}{2}\right)^3$.

Self Check 1

Evaluate: $\frac{7}{8} + \frac{3}{2}\left(-\frac{1}{4}\right)^2$.

Answer $\frac{31}{32}$

Solution The expression involves the operations of raising to a power, multiplication, and addition. By the rules for the order of operations, we must evaluate the power first, the multiplication second, and the addition last.

$$\frac{3}{4} + \frac{5}{3}\left(-\frac{\mathbf{1}}{\mathbf{2}}\right)^3 = \frac{3}{4} + \frac{5}{3}\left(-\frac{\mathbf{1}}{\mathbf{8}}\right)$$ Evaluate the power: $\left(-\frac{1}{2}\right)^3 = -\frac{1}{8}$.

$$= \frac{3}{4} + \left(-\frac{5}{24}\right)$$ Perform the multiplication: $\frac{5}{3}\left(-\frac{1}{8}\right) = -\frac{5}{24}$.

$$= \frac{3 \cdot 6}{4 \cdot 6} + \left(-\frac{5}{24}\right)$$ The LCD is 24. Write the first fraction as a fraction with denominator 24.

$$= \frac{18}{24} + \left(-\frac{5}{24}\right)$$ Multiply in the numerator: $3 \cdot 6 = 18$. Multiply in the denominator: $4 \cdot 6 = 24$.

$$= \frac{13}{24}$$ Add the numerators: $18 + (-5) = 13$. Write the sum over the common denominator.

If an expression contains grouping symbols, we do the operations within the grouping symbols first.

EXAMPLE 2 Evaluate: $\left(\frac{7}{8} - \frac{1}{4}\right) \div \left(-2\frac{3}{16}\right)$.

Solution

$$\left(\frac{7}{8} - \frac{1}{4}\right) \div \left(-2\frac{3}{16}\right) = \left(\frac{7}{8} - \frac{\mathbf{1 \cdot 2}}{\mathbf{4 \cdot 2}}\right) \div \left(-2\frac{3}{16}\right)$$ Within the first set of parentheses, write $\frac{1}{4}$ as a fraction with denominator 8.

$$= \left(\frac{7}{8} - \frac{2}{8}\right) \div \left(-2\frac{3}{16}\right)$$ Multiply in the numerator: of $1 \cdot 2 = 2$. Multiply in the denominator: $4 \cdot 2 = 8$.

$$= \frac{5}{8} \div \left(-2\frac{3}{16}\right)$$ Subtract the numerators and write the difference over the common denominator: $7 - 2 = 5$.

$= \frac{5}{8} \div \left(-\frac{35}{16}\right)$ Write the mixed number as an improper fraction.

$= \frac{5}{8}\left(-\frac{16}{35}\right)$ Multiply by the reciprocal of $-\frac{35}{16}$.

$= -\frac{5 \cdot 16}{8 \cdot 35}$ The product of two fractions with unlike signs is negative. Multiply the numerators and multiply the denominators.

$= -\frac{\overset{1}{\cancel{5}} \cdot 2 \cdot \overset{1}{\cancel{8}}}{\underset{1}{\cancel{8}} \cdot \underset{1}{\cancel{5}} \cdot 7}$ Factor 16 as $2 \cdot 8$ and factor 35 as $5 \cdot 7$. Divide out the common factors 8 and 5.

$= -\frac{2}{7}$ Multiply in the numerator: $1 \cdot 2 \cdot 1 = 2$. Multiply in the denominator: $1 \cdot 1 \cdot 7 = 7$.

EXAMPLE 3 **Masonry.** To build a wall, a mason will use blocks that are $5\frac{3}{4}$ inches high, held together with $\frac{3}{8}$-inch-thick layers of mortar. (See Figure 3-10.) If the plans call for 8 layers of blocks, what will be the height of the wall when completed?

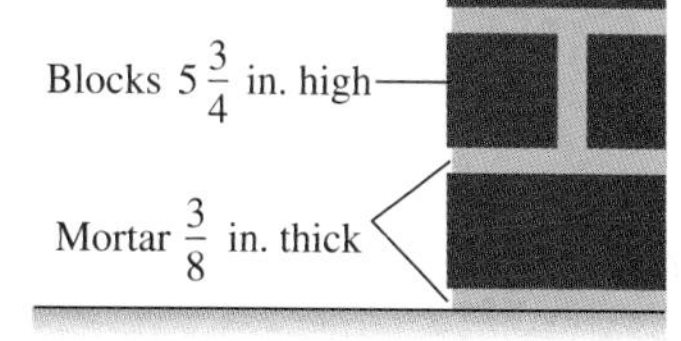

FIGURE 3-10

Solution To find the height, we must consider 8 layers of blocks and 8 layers of mortar. We will compute the height contributed by one block and one layer of mortar and then multiply that result by 8.

8	times	(height of 1 block	plus	height of 1 layer of mortar)	equals	height of block wall.
8		($5\frac{3}{4}$	+	$\frac{3}{8}$)	=	height of wall

$8\left(5\frac{3}{4} + \frac{3}{8}\right) = 8\left(\frac{23}{4} + \frac{3}{8}\right)$ Write $5\frac{3}{4}$ as the improper fraction $\frac{23}{4}$.

$= 8\left(\frac{23 \cdot \mathbf{2}}{4 \cdot \mathbf{2}} + \frac{3}{8}\right)$ Express $\frac{23}{4}$ in terms of 8ths.

$= 8\left(\frac{46}{8} + \frac{3}{8}\right)$ Perform the multiplication in the numerator and denominator.

$= \frac{8}{1}\left(\frac{49}{8}\right)$ Write 8 as $\frac{8}{1}$. Write the sum of the numerators over the common denominator 8.

$= \frac{\overset{1}{\cancel{8}} \cdot 49}{\underset{1}{\cancel{8}}}$ Multiply the numerators and the denominators. Divide out the common factor 8.

$= 49$ Simplify: $\frac{49}{1} = 49$.

The wall will be 49 inches high.

Complex fractions

Fractions whose numerators, denominators, or both contain fractions are called *complex fractions*. Here is an example.

A fraction in the numerator ⟶ $\dfrac{\frac{3}{4}}{\frac{7}{8}}$ ⟵ The main fraction bar

A fraction in the denominator ⟶

Complex fraction

A **complex fraction** is a fraction whose numerator or denominator, or both, contain one or more fractions or mixed numbers.

Here are more examples of complex fractions.

$$\dfrac{-\frac{1}{4}-\frac{4}{5}}{2\frac{4}{5}} \quad \longleftarrow \text{Numerator, Main fraction bar, Denominator} \longrightarrow \quad \dfrac{\frac{1}{3}+\frac{1}{4}}{\frac{1}{3}-\frac{1}{4}}$$

Simplifying complex fractions

To *simplify* complex fractions means to express them as fractions in simplified form.

Simplifying a complex fraction: method 1

Write the numerator and the denominator of the complex fraction as single fractions. Then perform the indicated division of the two fractions and simplify.

Method 1 is based on the fact that the main fraction bar of the complex fraction indicates division.

$\dfrac{\frac{1}{4}}{\frac{2}{5}}$ ⟵ The main fraction bar means "divide the fraction in the numerator by the fraction in the denominator." ⟶ $\frac{1}{4} \div \frac{2}{5}$

Self Check 4

Simplify: $\dfrac{\frac{1}{6}}{\frac{3}{8}}$.

EXAMPLE 4 Simplify: $\dfrac{\frac{1}{4}}{\frac{2}{5}}$.

Solution Since the numerator and the denominator of this complex fraction are single fractions, we can do the indicated division.

$$\frac{\frac{1}{4}}{\frac{2}{5}} = \frac{1}{4} \div \frac{2}{5}$$ Express the complex fraction as an equivalent division problem.

$$= \frac{1}{4} \cdot \frac{5}{2}$$ Multiply by the reciprocal of $\frac{2}{5}$.

$$= \frac{1 \cdot 5}{4 \cdot 2}$$ Multiply the numerators and multiply the denominators.

$$= \frac{5}{8}$$

Answer $\frac{4}{9}$

A second method is based on the fundamental property of fractions.

Simplifying a complex fraction: method 2

Multiply the numerator and the denominator of the complex fraction by the LCD of all the fractions that appear in its numerator and denominator. Then simplify.

EXAMPLE 5 Simplify: $\dfrac{-\frac{1}{4} + \frac{2}{5}}{\frac{1}{2} - \frac{4}{5}}$.

Solution Examine the numerator and the denominator of the complex fraction. The fractions involved have denominators of 4, 5, and 2. The LCD of these fractions is 20.

$$\frac{-\frac{1}{4} + \frac{2}{5}}{\frac{1}{2} - \frac{4}{5}} = \frac{20\left(-\frac{1}{4} + \frac{2}{5}\right)}{20\left(\frac{1}{2} - \frac{4}{5}\right)}$$ Use the fundamental property of fractions. Multiply the numerator and the denominator of the complex fraction by 20. Note how parentheses are used to show this.

$$= \frac{20\left(-\frac{1}{4}\right) + 20\left(\frac{2}{5}\right)}{20\left(\frac{1}{2}\right) - 20\left(\frac{4}{5}\right)}$$ Use the distributive property in the numerator and in the denominator.

$$= \frac{-5 + 8}{10 - 16}$$ Perform the multiplications by 20.

$$= \frac{3}{-6}$$ Perform the addition in the numerator and the subtraction in the denominator.

$$= -\frac{1}{2}$$ Simplify.

EXAMPLE 6 Simplify: $\dfrac{7 - \frac{2}{3}}{4\frac{5}{6}}$.

Solution Examine the numerator and the denominator of the complex fraction. The fractions have denominators of 3 and 6. The LCD of these fractions is 6.

Self Check 6

Simplify: $\dfrac{5 - \frac{3}{4}}{1\frac{7}{8}}$.

$$\frac{7-\frac{2}{3}}{4\frac{5}{6}} = \frac{7-\frac{2}{3}}{\frac{29}{6}}$$ Express $4\frac{5}{6}$ as an improper fraction.

$$= \frac{6\left(7-\frac{2}{3}\right)}{6\left(\frac{29}{6}\right)}$$ Use the fundamental property of fractions. Multiply the numerator and the denominator of the complex fraction by the LCD, 6.

$$= \frac{6(7)-6\left(\frac{2}{3}\right)}{6\left(\frac{29}{6}\right)}$$ Use the distributive property in the numerator. Distribute the multiplication by 6.

$$= \frac{42-4}{29}$$ Perform the multiplications by 6.

$$= \frac{38}{29}$$ Perform the subtraction in the numerator.

$$= 1\frac{9}{29}$$ Write $\frac{38}{29}$ as a mixed number.

Answer $2\frac{4}{15}$

Section 3.7 STUDY SET

VOCABULARY *Fill in the blanks.*

1. $\frac{\frac{1}{2}}{\frac{3}{4}}$ is a ________ fraction.

2. To evaluate an algebraic expression, we ________ specific numbers for the variables in the expression and simplify.

CONCEPTS

3. What division is represented by the complex fraction?

$$\frac{\frac{2}{3}}{\frac{1}{5}}$$

4. Write the division as a complex fraction.

$$-\frac{7}{8} \div \frac{3}{4}$$

5. What is the common denominator of all the fractions in the complex fraction?

$$\frac{\frac{2}{3}-\frac{1}{5}}{\frac{1}{3}+\frac{4}{5}}$$

6. What is the common denominator of all the fractions in the complex fraction?

$$\frac{\frac{1}{8}-\frac{3}{16}}{-5\frac{3}{4}}.$$

7. When the complex fraction is simplified, will the result be positive or negative?

$$\frac{-\frac{2}{3}}{\frac{3}{4}}$$

8. What property is being applied?

$$\frac{1+\frac{1}{11}}{\frac{1}{2}} = \frac{22\left(1+\frac{1}{11}\right)}{22\left(\frac{1}{2}\right)}$$

9. What is the LCD of fractions with the denominators 6, 4, and 5?

10. What operations are involved in the numerical expression?

$$5\left(6\frac{1}{3}\right) + \left(-\frac{1}{4}\right)^2$$

NOTATION *Complete each solution to simplify the complex fraction.*

11.
$$\frac{\frac{1}{8}}{\frac{3}{4}} = \frac{1}{8} \div \square$$
$$= \frac{1}{8} \cdot \square$$
$$= \frac{1 \cdot \square}{8 \cdot 3}$$
$$= \frac{1 \cdot \overset{1}{\cancel{4}}}{2 \cdot \underset{1}{\cancel{}} \cdot 3}$$
$$= \frac{1}{6}$$

12.
$$\frac{\frac{1}{6}+\frac{1}{5}}{-\frac{1}{15}} = \frac{30\left(\frac{1}{6}+\frac{1}{5}\right)}{\square\left(-\frac{1}{15}\right)}$$
$$= \frac{\square\left(\frac{1}{6}\right) + \square\left(\frac{1}{5}\right)}{30\left(-\frac{1}{15}\right)}$$
$$= \frac{5+6}{\square}$$
$$= \frac{\square}{-2}$$
$$= -5\frac{1}{2}$$

PRACTICE *Evaluate each expression.*

13. $\frac{2}{3}\left(-\frac{1}{4}\right) + \frac{1}{2}$

14. $-\frac{7}{8} - \left(\frac{1}{8}\right)\left(\frac{2}{3}\right)$

15. $\frac{4}{5} - \left(-\frac{1}{3}\right)^2$

16. $-\frac{3}{16} - \left(-\frac{1}{2}\right)^3$

17. $-4\left(-\frac{1}{5}\right) - \left(\frac{1}{4}\right)\left(-\frac{1}{2}\right)$

18. $(-3)\left(-\frac{2}{3}\right) - (-4)\left(-\frac{3}{4}\right)$

19. $1\frac{3}{5}\left(\frac{1}{2}\right)^2\left(\frac{3}{4}\right)$

20. $2\frac{3}{5}\left(-\frac{1}{3}\right)^2\left(\frac{1}{2}\right)$

21. $\frac{7}{8} - \left(\frac{4}{5} + 1\frac{3}{4}\right)$

22. $\left(\frac{5}{4}\right)^2 + \left(\frac{2}{3} - 2\frac{1}{6}\right)$

23. $\left(\frac{9}{20} \div 2\frac{2}{5}\right) + \left(\frac{3}{4}\right)^2$

24. $\left(1\frac{2}{3} \cdot 15\right) + \left(\frac{7}{9} \div \frac{7}{81}\right)$

25. $\left(-\frac{3}{4} \cdot \frac{9}{16}\right) + \left(\frac{1}{2} - \frac{1}{8}\right)$

26. $\left(\frac{8}{5} - 1\frac{1}{3}\right) - \left(-\frac{4}{5} \cdot 10\right)$

27. $\left|\frac{2}{3} - \frac{9}{10}\right| \div \left(-\frac{1}{5}\right)$

28. $\left|-\frac{3}{16} \div 2\frac{1}{4}\right| + \left(-2\frac{1}{8}\right)$

29. $\left(2 - \frac{1}{2}\right)^2 + \left(2 + \frac{1}{2}\right)^2$

30. $\left(1 - \frac{3}{4}\right)\left(1 + \frac{3}{4}\right)$

Find one-half of the given number and square that result. Express the answer as an improper fraction.

31. -7

32. -5

33. $\frac{11}{2}$

34. $\frac{7}{3}$

Find the perimeter of each figure.

35.

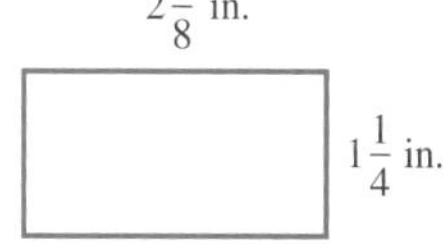

36.

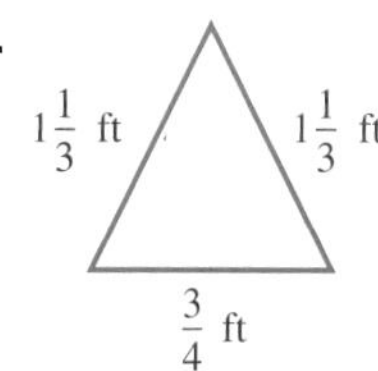

Simplify each complex fraction.

37. $\dfrac{\frac{2}{3}}{\frac{4}{5}}$

38. $\dfrac{\frac{3}{5}}{\frac{9}{25}}$

39. $\dfrac{-\frac{14}{15}}{\frac{7}{10}}$

40. $\dfrac{\frac{5}{27}}{-\frac{5}{9}}$

41. $\dfrac{5}{\frac{10}{21}}$

42. $\dfrac{6}{\frac{3}{8}}$

43. $\dfrac{-\frac{5}{6}}{-1\frac{7}{8}}$

44. $\dfrac{-\frac{4}{3}}{-2\frac{5}{6}}$

45. $\dfrac{\frac{1}{2}+\frac{1}{4}}{\frac{1}{2}-\frac{1}{4}}$

46. $\dfrac{\frac{1}{3}+\frac{1}{4}}{\frac{1}{3}-\frac{1}{4}}$

47. $\dfrac{\frac{3}{8}+\frac{1}{4}}{\frac{3}{8}-\frac{1}{4}}$

48. $\dfrac{\frac{2}{5}+\frac{1}{4}}{\frac{2}{5}-\frac{1}{4}}$

49. $\dfrac{\frac{1}{5}+3}{-\frac{4}{25}}$

50. $\dfrac{-5-\frac{1}{3}}{\frac{1}{6}+\frac{2}{3}}$

51. $\dfrac{5\frac{1}{2}}{-\frac{1}{4}+\frac{3}{4}}$

52. $\dfrac{4\frac{1}{4}}{\frac{2}{3}+\left(-\frac{1}{6}\right)}$

53. $\dfrac{\frac{1}{5}-\left(-\frac{1}{4}\right)}{\frac{1}{4}+\frac{4}{5}}$

54. $\dfrac{\frac{1}{8}-\left(-\frac{1}{2}\right)}{\frac{1}{4}+\frac{3}{8}}$

55. $\dfrac{\frac{1}{3}+\left(-\frac{5}{6}\right)}{1\frac{1}{3}}$

56. $\dfrac{\frac{3}{7}+\left(-\frac{1}{2}\right)}{1\frac{3}{4}}$

APPLICATIONS

57. SANDWICH SHOPS A sandwich shop sells a $\frac{1}{2}$-pound club sandwich, made up of turkey meat and ham. The owner buys the turkey in $1\frac{3}{4}$-pound packages and the ham in $2\frac{1}{2}$-pound packages. If he mixes a package of each of the meats together, how many sandwiches can he make from the mixture?

58. SKIN CREAMS Using a formula of $\frac{1}{2}$ ounce of sun block, $\frac{2}{3}$ ounce of moisturizing cream, and $\frac{3}{4}$ ounce of lanolin, a beautician mixes her own brand of skin cream. She packages it in $\frac{1}{4}$-ounce tubes. How many tubes can be produced using this formula?

59. PHYSICAL FITNESS Two people begin their workouts from the same point on a bike path and travel in opposite directions, as shown. How far apart are they in $1\frac{1}{2}$ hours? Use the table to help organize your work.

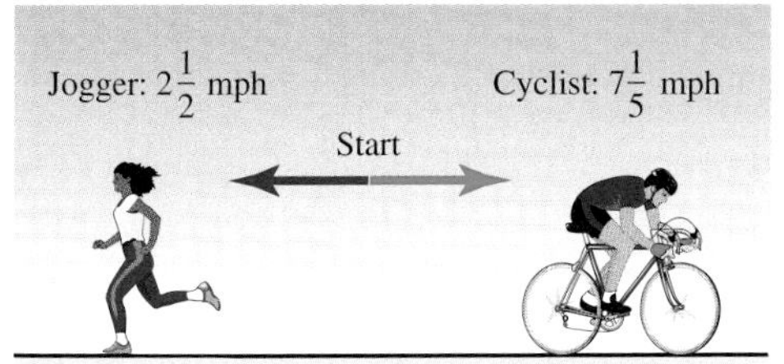

	Rate (mph)	Time (hr)	Distance (mi)
Jogger			
Cyclist			

60. SLEEP The illustration compares the amount of sleep a 1-month-old baby got to the $15\frac{1}{2}$-hour daily requirement recommended by Children's Hospital of Orange County, California. For the week, how far below the baseline was the baby's daily average?

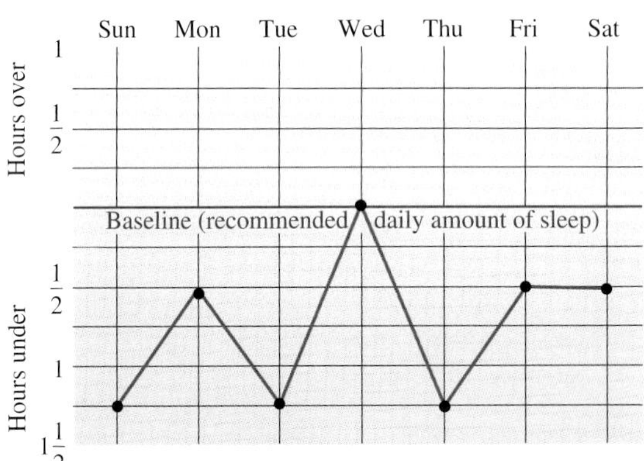

61. POSTAGE RATES Can the following ad package be mailed for the 1-ounce rate?

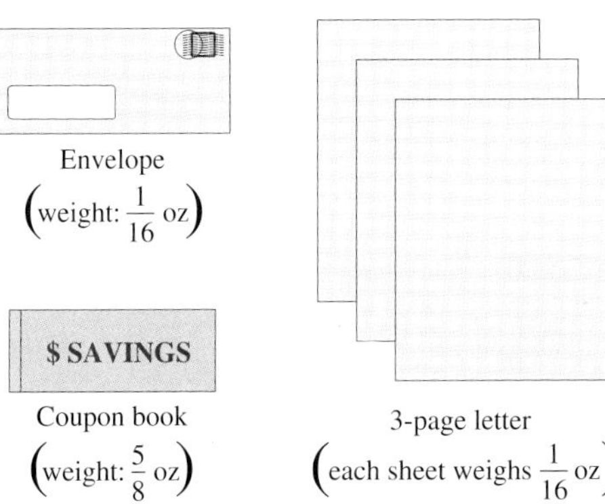

62. PLYWOOD To manufacture a sheet of plywood, several layers of thin laminate are glued together, as shown. Then an exterior finish is affixed to the top and bottom. How thick is the finished product?

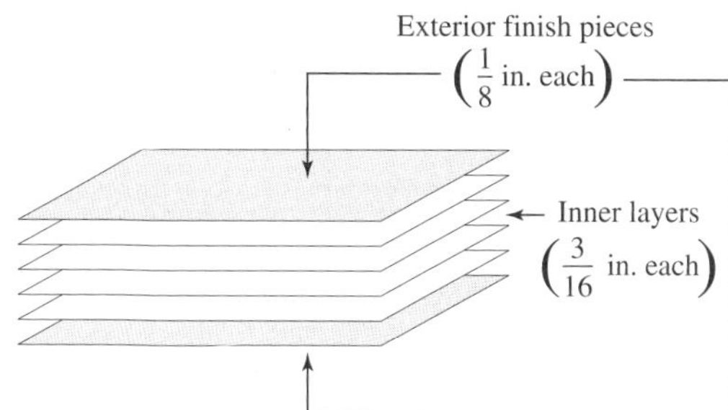

63. PHYSICAL THERAPY After back surgery, a patient undertook a walking program to rehabilitate her back muscles, as specified in the table. What was the total distance she walked over this three-week period?

Week	Distance per day
#1	$\frac{1}{4}$ mile
#2	$\frac{1}{2}$ mile
#3	$\frac{3}{4}$ mile

64. READING PROGRAMS To improve reading skills, elementary-school children read silently at the end of the school day for $\frac{1}{4}$ hour on Mondays and for $\frac{1}{2}$ hour on Fridays. For the month of January, how many total hours did the children read silently in class?

JANUARY						
S	M	T	W	T	F	S
	1	2	3	4	5	6
7	8	9	10	11	12	13
14	15	16	17	18	19	20
21	22	23	24	25	26	27
28	29	30	31			

65. AMUSEMENT PARKS At the end of a ride at an amusement park, a boat splashes into a pool of water. The time (in seconds) that it takes two pipes to refill the pool is given by

$$\frac{1}{\frac{1}{10} + \frac{1}{15}}$$

Find this time.

66. HIKING A scout troop plans to hike from the campground to Glenn Peak. Since the terrain is steep, they plan to stop and rest after every $\frac{2}{3}$ mile. With this plan, how many parts will there be to this hike?

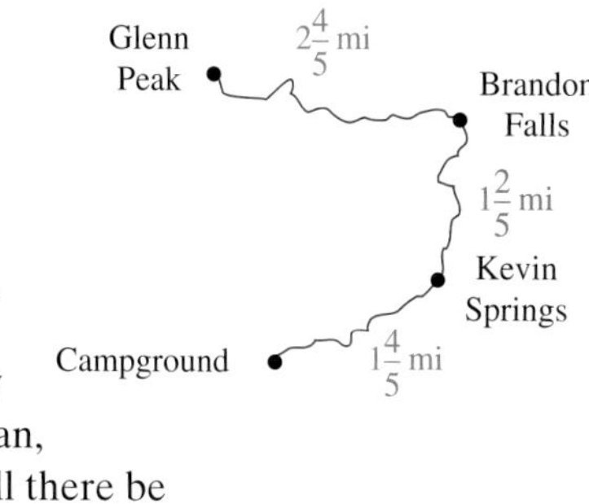

WRITING

67. What is a complex fraction?

68. Explain method 1 for simplifying complex fractions.

69. Write an application problem using a complex fraction, and then solve it.

70. Explain method 2 for simplifying complex fractions.

REVIEW

71. Evaluate: $\frac{2(15) + 6}{2 \cdot 3^2}$.

72. Find the area of a rectangle that is 6 feet by 12 feet.

73. Solve: $-5x + 3 = 28$.

74. List the factors of 24.

75. Evaluate: $2 + 3[-3 - (-4 - 1)]$.

76. What is the sign of the quotient of two numbers with unlike signs?

3.8 Solving Equations Containing Fractions

- Using reciprocals to solve equations
- An alternate method
- The addition and subtraction properties of equality
- Problem solving with equations

In this section, we will discuss how to solve both equations containing fractions and equations whose solutions are fractions. We will make use of the properties of equality and several concepts from this chapter.

Using reciprocals to solve equations

In the equation $\frac{3}{4}x = 5$, the variable is multiplied by $\frac{3}{4}$. To undo this multiplication and isolate the variable, we can use the multiplication property of equality and multiply both sides of the equation by the reciprocal of $\frac{3}{4}$.

$\mathbf{\frac{4}{3}}\left(\frac{3}{4}x\right) = \mathbf{\frac{4}{3}}(5)$ Multiply both sides by the reciprocal of $\frac{3}{4}$, which is $\frac{4}{3}$.

$\left(\frac{4}{3} \cdot \frac{3}{4}\right)x = \frac{4}{3} \cdot \frac{5}{1}$ Use the associative property of multiplication to regroup the factors. Write 5 as $\frac{5}{1}$.

$\frac{\overset{1}{\cancel{4}} \cdot \overset{1}{\cancel{3}}}{\underset{1}{\cancel{3}} \cdot \underset{1}{\cancel{4}}}x = \frac{4 \cdot 5}{3 \cdot 1}$ Multiply the numerators and multiply the denominators. On the left, divide out the common factors 4 and 3.

$1x = \frac{20}{3}$ Multiply in the numerators and in the denominators.

$x = \frac{20}{3}$ Simplify: $1x = x$.

In algebra, we usually leave a solution to an equation as an improper fraction rather than converting it to a mixed number.

COMMENT We can write expressions such as $\frac{4a}{5}$ and $\frac{-9h}{16}$ in an equivalent form so that the fractional coefficients are more evident.

$$\frac{4a}{5} = \mathbf{\frac{4}{5}}a \qquad \frac{-9h}{16} = \mathbf{-\frac{9}{16}}h$$

Self Check 1

Solve $-\frac{3}{2}t = 15$ and check the result.

Answer -10

EXAMPLE 1 Solve: $-\frac{7}{8}k = 21$.

Solution The coefficient of the variable is $-\frac{7}{8}$. To isolate k, we multiply both sides of the equation by the reciprocal of $-\frac{7}{8}$.

$-\frac{7}{8}k = 21$

$\mathbf{-\frac{8}{7}}\left(-\frac{7}{8}k\right) = \mathbf{-\frac{8}{7}}(21)$ Multiply both sides by the reciprocal of $-\frac{7}{8}$, which is $-\frac{8}{7}$.

$1k = -\frac{8}{7}(21)$ The product of a number and its reciprocal is 1: $-\frac{8}{7}(-\frac{7}{8}) = 1$.

$k = -\frac{8}{7} \cdot \frac{21}{1}$ Simplify: $1k = k$. Write 21 as $\frac{21}{1}$.

$k = -\frac{8 \cdot 3 \cdot \overset{1}{\cancel{7}}}{\underset{1}{\cancel{7}} \cdot 1}$ The product of two numbers with unlike signs is negative. Multiply the numerators and the denominators. Prime factor 21 and then divide out the common factor 7.

$k = -24$ Multiply in the numerator and the denominator.

Check the result by substituting -24 for k in the original equation.

An alternate method

Another method of solving equations such as $\frac{3}{4}x = 5$ uses two steps to isolate the variable. In this method, we consider the variable to be multiplied by 3 and divided by 4. Then, in reverse order, we undo these operations.

$$\mathbf{4}\left(\frac{3}{4}x\right) = \mathbf{4}(5)$$ To undo the division by 4, multiply both sides by 4.

$$\left(4 \cdot \frac{3}{4}\right)x = 4(5)$$ Use the associative property to regroup the factors.

$$\left(\frac{\overset{1}{\cancel{4}} \cdot 3}{1 \cdot \underset{1}{\cancel{4}}}\right)x = 4(5)$$ Write 4 as $\frac{4}{1}$, multiply the numerators and the denominators, and divide out the common factor 4.

$$3x = 20$$ Multiply in the numerator and the denominator.

$$\frac{3x}{\mathbf{3}} = \frac{20}{\mathbf{3}}$$ To undo the multiplication by 3, divide both sides by 3.

$$x = \frac{20}{3}$$

EXAMPLE 2 Solve: $\frac{3}{5}h = -9$.

Solution

$$\frac{3}{5}h = -9$$

$$\mathbf{5}\left(\frac{3}{5}h\right) = \mathbf{5}(-9)$$ To undo the division by 5, multiply both sides by 5.

$$3h = -45$$ Perform the multiplications.

$$\frac{3h}{\mathbf{3}} = -\frac{45}{\mathbf{3}}$$ To undo the multiplication by 3, divide both sides by 3.

$$h = -15$$ Perform the divisions.

Self Check 2

Solve: $\frac{5}{9}t = -10$.

Answer -18

The addition and subtraction properties of equality

The addition property of equality enables us to add the same number to both sides of an equation and obtain an equivalent equation. In the next example, we will use this property to help solve an equation that contains fractions.

EXAMPLE 3 Solve: $y - \frac{15}{32} = \frac{1}{32}$.

Solution To isolate y on the left-hand side, we need to undo the subtraction of $\frac{15}{32}$.

$$y - \frac{15}{32} = \frac{1}{32}$$

$$y - \frac{15}{32} + \mathbf{\frac{15}{32}} = \frac{1}{32} + \mathbf{\frac{15}{32}}$$ Add $\frac{15}{32}$ to both sides.

$$y = \frac{16}{32}$$ Simplify: $-\frac{15}{32} + \frac{15}{32} = 0$ and $\frac{1}{32} + \frac{15}{32} = \frac{16}{32}$.

$$y = \frac{1}{2}$$ Simplify the fraction: $\frac{16}{32} = \frac{\overset{1}{\cancel{16}} \cdot 1}{2 \cdot \underset{1}{\cancel{16}}} = \frac{1}{2}$.

Self Check 3

Solve: $\frac{11}{16} = a - \frac{1}{16}$.

Answer $\frac{3}{4}$

Self Check 4

Solve: $y + \frac{1}{5} = \frac{2}{3}$.

EXAMPLE 4 Solve: $x + \frac{1}{6} = \frac{3}{4}$.

Solution In this equation, $\frac{1}{6}$ is added to x. We undo this operation by subtracting $\frac{1}{6}$ from both sides.

$$x + \frac{1}{6} = \frac{3}{4}$$

$$x + \frac{1}{6} - \mathbf{\frac{1}{6}} = \frac{3}{4} - \mathbf{\frac{1}{6}} \quad \text{Subtract } \frac{1}{6} \text{ from both sides.}$$

$$x = \frac{3}{4} - \frac{1}{6} \quad \text{Perform the subtraction: } \frac{1}{6} - \frac{1}{6} = 0.$$

$$x = \frac{3 \cdot \mathbf{3}}{4 \cdot \mathbf{3}} - \frac{1 \cdot \mathbf{2}}{6 \cdot \mathbf{2}} \quad \text{Use the fundamental property of fractions to write each fraction in terms of the LCD, which is 12.}$$

$$x = \frac{9}{12} - \frac{2}{12} \quad \text{Perform the multiplications in the numerators and denominators.}$$

$$x = \frac{7}{12} \quad \text{Subtract the fractions.}$$

Answers $\frac{7}{15}$

Problem solving with equations

EXAMPLE 5 **Native Americans.** The United States Constitution requires a population count, called a *census,* to be taken every ten years. In the 2000 census, the population of the Navajo tribe was 298,000. This was about two-fifths of the population of the largest Native American tribe, the Cherokee. What was the population of the Cherokee tribe in 2000?

Analyze the problem

- In 2000, the population of the Navajo tribe was 298,000.
- The population of the Navajo tribe was $\frac{2}{5}$ the population of the Cherokee tribe.
- Find the population of the Cherokee tribe in 2000.

Form an equation

Let x = the population of the Cherokee tribe. Next, we look for a key word or phrase in the problem.

Key phrase: two-fifths of **Translation:** multiply by $\frac{2}{5}$

The population of the Navajo tribe	was	$\frac{2}{5}$	of	the population of the Cherokee tribe.
298,000	=	$\frac{2}{5}$	$\cdot$	x

Solve the equation

$$298{,}000 = \frac{2}{5}x$$

$$\mathbf{\frac{5}{2}}(298{,}000) = \mathbf{\frac{5}{2}}\left(\frac{2}{5}x\right) \quad \text{To isolate } x \text{ on the right-hand side, multiply both sides by the reciprocal of } \frac{2}{5}.$$

$$745{,}000 = x$$

On the left-hand side, $\frac{5}{2}(298{,}000) = \frac{1{,}490{,}000}{2} = 745{,}000$. On the right-hand side, $\frac{5}{2}(\frac{2}{5}x) = 1x = x$.

Solve the conclusion

In 2000, the population of the Cherokee tribe was about 745,000.

Check the result

Using a fraction to compare the two populations, we have

$$\frac{298{,}000}{745{,}000} = \frac{298}{745} = \frac{2 \cdot 149}{5 \cdot 149} = \frac{2}{5}$$

The answer checks.

EXAMPLE 6 **Reading a book.** A student has read $\frac{2}{5}$ of a book and wants to have read $\frac{3}{4}$ of it by tomorrow morning. How much more of the book must she read?

Analyze the problem

- A student has read $\frac{2}{5}$ of a book.
- She wants to have read $\frac{3}{4}$ of it by tomorrow morning.
- The amount she has read plus the amount she needs to read equals $\frac{3}{4}$ of the book.
- Find how much more of the book she needs to read.

Form an equation

We let x = the part of the book that needs to be read. Then we look for a key word or phrase.

Key phrase: plus **Translation:** add

The part that has been read	plus	The part that needs to be read	is	$\frac{3}{4}$ of the book.
$\frac{2}{5}$	$+$	x	$=$	$\frac{3}{4}$

Solve the equation

$$\frac{2}{5} + x = \frac{3}{4}$$

$$\frac{2}{5} - \mathbf{\frac{2}{5}} + x = \frac{3}{4} - \mathbf{\frac{2}{5}}$$ To undo the addition by $\frac{2}{5}$, subtract $\frac{2}{5}$ from both sides.

$$x = \frac{3}{4} - \frac{2}{5}$$ $\frac{2}{5} - \frac{2}{5} = 0$.

$$x = \frac{3 \cdot \mathbf{5}}{4 \cdot \mathbf{5}} - \frac{2 \cdot 4}{5 \cdot 4}$$ Apply the fundamental property of fractions to get an LCD 20.

$$x = \frac{15}{20} - \frac{8}{20}$$ Perform the multiplications in the numerators and the denominators.

$$x = \frac{7}{20}$$ Subtract the fractions.

State the conclusion

The student must read $\frac{7}{20}$ more of the book.

Check the result

The student has read $\frac{2}{5}$ of the book. If she reads $\frac{7}{20}$ more, she will have read

$$\frac{2}{5} + \frac{7}{20} = \frac{2 \cdot 4}{5 \cdot 4} + \frac{7}{20} = \frac{8}{20} + \frac{7}{20} = \frac{15}{20} = \frac{3}{4}$$

of the book. The answer checks.

Section 3.8 STUDY SET

VOCABULARY *Fill in the blanks.*

1. To find the __________ of a fraction, invert the numerator and the denominator.
2. In the expression $\frac{5}{12}x$, x is called a __________.
3. The ______ __________ ____________ of a set of fractions is the smallest number each denominator will divide exactly.
4. A __________ of an equation, when substituted into that equation, makes a true statement.

CONCEPTS

5. Is 40 a solution of $\frac{5}{8}x = 25$? Explain.
6. Give the reciprocal of each number.
   a. $\frac{7}{9}$ b. $-\frac{1}{2}$
7. What is the result when a number is multiplied by its reciprocal?
8. Perform each multiplication.
   a. $\frac{3}{2}\left(\frac{2}{3}x\right)$ b. $-\frac{16}{15}\left(-\frac{15}{16}t\right)$
   c. $25\left(\frac{2}{5}\right)$ d. $16\left(\frac{3}{8}\right)$
9. Translate to mathematical symbols.
   a. Four-fifths of the population p
   b. One-quarter of the time t
10. Explain two ways in which the variable x can be isolated: $\frac{2}{3}x = -4$.

NOTATION *Complete each solution to solve the equation.*

11. $\frac{7}{8}x = 21$

$\square\left(\frac{7}{8}x\right) = \square(21)$

$x = 24$

12. $h + \frac{1}{2} = \frac{2}{3}$

$h + \frac{1}{2} - \square = \frac{2}{3} - \square$

$h = \square$

13. Determine whether each statement is true or false.
   a. $\frac{1}{2}x = \frac{x}{2}$ b. $\frac{1}{8}y = 8y$
   c. $-\frac{1}{2}x = \frac{-x}{2} = \frac{x}{-2}$ d. $\frac{7p}{8} = \frac{7}{8}p$
14. Write the product of $\frac{4}{7}$ and x in two ways.

PRACTICE *Solve each equation.*

15. $\frac{4}{7}x = 16$
16. $\frac{2}{3}y = 30$
17. $\frac{7}{8}t = -28$
18. $\frac{5}{6}c = -25$
19. $-\frac{3}{5}h = 4$
20. $-\frac{5}{6}f = -2$

21. $\frac{2}{3}x = \frac{4}{5}$

22. $\frac{5}{8}y = \frac{10}{11}$

23. $\frac{2}{5}y = 0$

24. $\frac{4}{9}x = 0$

25. $-\frac{5c}{6} = -25$

26. $-\frac{7t}{4} = -35$

27. $\frac{5f}{7} = -2$

28. $\frac{3h}{5} = -35$

29. $\frac{5}{8}y = \frac{1}{10}$

30. $\frac{1}{16}x = \frac{5}{24}$

31. $2x + 1 = 0$

32. $3y - 1 = 0$

33. $5x - 1 = 1$

34. $4c + 1 = -2$

35. $x - \frac{1}{9} = \frac{7}{9}$

36. $x + \frac{1}{3} = \frac{2}{3}$

37. $x + \frac{1}{9} = \frac{4}{9}$

38. $x - \frac{1}{6} = \frac{1}{6}$

39. $x - \frac{1}{6} = \frac{2}{9}$

40. $y - \frac{1}{3} = \frac{4}{5}$

41. $y + \frac{7}{8} = \frac{1}{4}$

42. $t + \frac{5}{6} = \frac{1}{8}$

43. $\frac{5}{4} + t = \frac{1}{4}$

44. $\frac{2}{3} + y = \frac{4}{3}$

45. $x + \frac{3}{4} = -\frac{1}{2}$

46. $y - \frac{5}{6} = \frac{1}{3}$

47. $\frac{-x}{4} + 1 = 10$

48. $\frac{-y}{6} - 1 = 5$

49. $2x - \frac{1}{2} = \frac{1}{3}$

50. $3y - \frac{2}{5} = \frac{1}{8}$

51. $\frac{1}{2}x - \frac{1}{9} = \frac{1}{3}$

52. $\frac{1}{4}y - \frac{2}{3} = \frac{1}{2}$

53. $5 + \frac{x}{3} = \frac{1}{2}$

54. $4 + \frac{y}{2} = \frac{3}{5}$

55. $\frac{2}{5}x + 1 = \frac{1}{3}$

56. $\frac{2}{3}y + 2 = \frac{1}{5}$

57. $\frac{x}{3} + \frac{1}{4} = -2$

58. $\frac{5}{6} + \frac{y}{4} = -1$

59. $4 + \frac{s}{3} = 8$

60. $6 + \frac{y}{5} = 1$

61. $\frac{5h}{6} - 8 = 12$

62. $\frac{6a}{7} - 1 = 11$

63. $-4 + 9 + \frac{5t}{12} = 0$

64. $-4 + 10 + \frac{3y}{8} = 0$

65. $-3 - 2 + \frac{4x}{15} = 0$

66. $-1 - 9 + \frac{2y}{15} = 0$

APPLICATIONS *Complete each solution.*

67. TRANSMISSION REPAIRS A repair shop found that $\frac{1}{3}$ of its customers with transmission problems needed a new transmission. If the shop installed 32 new transmissions last year, how many customers did the shop have last year?

Analyze the problem

- Only ____ of the customers needed new transmissions.
- The shop installed ____ new transmissions last year.
- Find the number of ________ the shop had last year.

Form an equation

Let x = ________________.

Key phrase: *one-third of*
Translation: ________

$\frac{1}{3}$ of the number of customers last year	was	32.
____	=	32

Solve the equation

$$\frac{1}{3}x = 32$$

$$\square\left(\frac{1}{3}x\right) = \square(32)$$

$$x = \square$$

State the conclusion ________________

Check the result If we find $\frac{1}{3}$ of 96, we get ____. The answer checks.

68. CATTLE RANCHING A rancher is preparing to fence in a rectangular grazing area next to a $\frac{3}{4}$-mile-long lake. He has determined that $1\frac{1}{2}$ square miles of land are needed to ensure that overgrazing does not occur. How wide should this grazing area be?

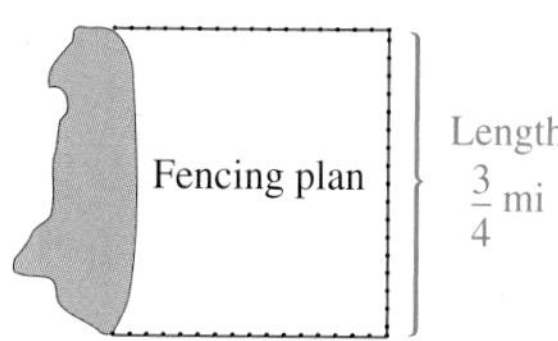

Analyze the problem

- The grazing area is ____ $= \frac{3}{2}$ square miles.
- The length of the rectangle is $\frac{3}{4}$ mile.
- Find the ________ of the grazing area.

Form an equation

Let w = ____________________.

Key word: *area* **Translation:** A =

The area of the rectangle	is	the length times the width.
	=	

Solve the equation

$$\frac{3}{2} = \frac{3}{4}w$$

$$\left(\frac{3}{2}\right) = \left(\frac{3}{4}w\right)$$

$$= w$$

State the conclusion

____

Check the result If we multiply the length and the width of the rectangular area, we get $\frac{3}{4} \cdot 2 = \frac{3}{2} = 1\frac{1}{2}$ square miles. The answer checks.

Choose a variable to represent the unknown. Then write and solve an equation to answer each question.

69. TOOTH DEVELOPMENT During a checkup, a pediatrician found that only four-fifths of a child's baby teeth had emerged. The mother counted 16 teeth in the child's mouth. How many baby teeth will the child eventually have?

70. GENETICS Bean plants with inflated pods were cross-bred with bean plants with constricted pods. Of the offspring plants, three-fourths had inflated pods and one-fourth had constricted pods. If 244 offspring plants had constricted pods, how many offspring plants resulted from the cross-breeding experiment?

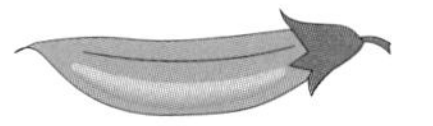

Inflated pod

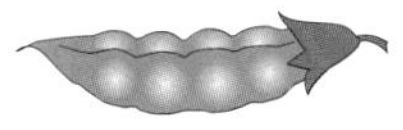

Constricted pod

71. TELEPHONE BOOKS A telephone book consists of the white pages and the yellow pages. Two-thirds of the book consists of the white pages; the white pages number 300. Find the total number of pages in the telephone book.

72. BROADWAY MUSICALS A theater usher at a Broadway musical finds that seven-eighths of the patrons attending a performance, which is 350 people, are in their seats by show time. If the show is always a complete sellout, how many seats does the theater have?

73. HOME SALES In less than a month, three-quarters of the homes in a new subdivision were purchased. This left only 9 homes to be sold. How many homes are there in the subdivision? (*Hint:* First determine what fractional part of the homes in the subdivision were not yet sold.)

74. WEDDING GUESTS Of those invited to a wedding, three-tenths were friends of the bride. The friends of the groom numbered 84. How many people were invited to the wedding? (*Hint:* First determine what fractional part of the people invited to the wedding were friends of the groom.)

75. SAFETY REQUIREMENTS In developing taillights for an automobile, designers must be aware of a safety standard that requires an area of 30 square inches to be visible from behind the vehicle. If the designers want the taillights to be $3\frac{3}{4}$ inches high, how wide must they be to meet safety standards?

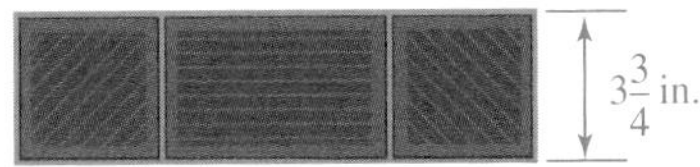

76. GRAPHIC ARTS A design for a yearbook is shown. The page is divided into 12 parts. The parts that are shaded will contain pictures, and the remainder of the squares will contain copy. If the pictures are to cover an area of 100 square inches, how many square inches are there on the page?

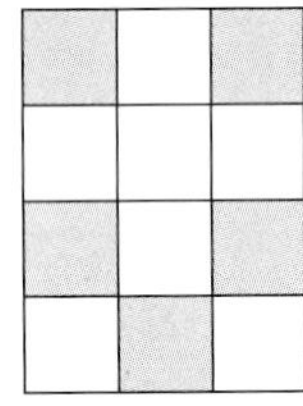

77. TEXTBOOKS An editor has reviewed $\frac{1}{3}$ of an algebra textbook. She wants to have reviewed $\frac{3}{4}$ of the book by the end of the week. How much more of the textbook must she review?

78. PAINTING A crew has painted $\frac{1}{4}$ of an apartment building. They want to have painted $\frac{7}{8}$ of the building by the end of the day. How much more of the apartment building must they paint?

WRITING

79. What does it mean to isolate the variable when solving an equation?

80. Explain how to determine whether -30 is a solution of $\frac{5}{6}x = -25$.

81. Which method, the reciprocal method or the two-step method, would you use to solve the equation $\frac{5}{16}t = 15$? Why?

82. Use an example to show why dividing by a number is the same as multiplying by its reciprocal.

REVIEW

83. Evaluate: $(-2)^5$.

84. Evaluate: $\frac{-4 - 8}{3}$.

85. Solve: $3x - 2 = 7$.

86. Evaluate: $(-1)^{15}$.

87. Round 12,590,767 to the nearest million.

88. In the expression $(-4)^6$, what do we call -4 and what do we call 6?

KEY CONCEPT

The Fundamental Property of Fractions

The **fundamental property of fractions** states that multiplying or dividing the numerator and the denominator of a fraction by the same nonzero number does not change the value of the fraction. This property is used to simplify fractions and to express fractions in higher terms. The following problems review both procedures. Complete each solution.

1. Simplify: $\frac{15}{25}$.

Step 1: The numerator and the denominator share a common factor of ___.

Step 2: Apply the fundamental property of fractions. Divide the numerator and the denominator by the common factor ___.

$$\frac{15}{25} = \frac{15 \div \_\_}{25 \div \_\_}$$

Step 3: Perform the divisions to simplify the fraction.

$$= \frac{3}{\_\_}$$

2. In practice, we often show the simplifying process described in Problem 1 in a different form.

Step 1: Factor 15 as 3 · ___ and 25 as ___ · 5.

$$\frac{15}{25} = \frac{3 \cdot \_\_}{\_\_ \cdot 5}$$

Step 2: The slashes and small 1's indicate that the numerator and the denominator have been divided by ___.

$$= \frac{3 \cdot \overset{1}{\cancel{5}}}{\underset{1}{\cancel{5}} \cdot 5}$$

Step 3: Multiply in the numerator and the denominator.

$$= \frac{\_\_}{5}$$

3. When adding or subtracting fractions and mixed numbers, we often need to express a fraction in higher terms. This is called building the fraction. Express $\frac{1}{5}$ as a fraction with denominator 35.

Step 1: We must multiply 5 by ___ to obtain 35.

Step 2: Use the fundamental property of fractions. Multiply the numerator and the denominator by ___.

$$\frac{1}{5} = \frac{1 \cdot \_\_}{5 \cdot \_\_}$$

Step 3: Multiply in the numerator and the denominator.

$$= \frac{\_\_}{35}$$

ACCENT ON TEAMWORK

SECTION 3.1

EQUIVALENT FRACTIONS Complete the labeling of each number line using fractions with the same denominator.

a. 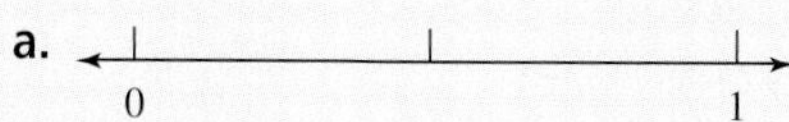

b. 0 1

c. 0 1

d. 0 1

FRACTIONS Give everyone in your group a strip of paper that is the same length. Determine ways to fold the strip of paper into

a. fourths
b. eighths
c. thirds
d. sixths

SECTION 3.2

MULTIPLICATION When we multiply 2 and 4, the answer is greater than 2 and greater than 4. Is this always the case? Is the product of two numbers always greater than either of the two numbers? Explain your answer.

POWERS When we square the number 4, the answer is greater than 4. Is the square of a number always greater than the number? Explain your answer.

SECTION 3.3

DIVIDING SNACKS Devise a way to divide seven brownies equally among six people.

SECTION 3.4

ADDING FRACTIONS Without actually doing the addition, explain why $\frac{3}{7} + \frac{1}{4}$ must be less than 1 and why $\frac{4}{7} + \frac{3}{4}$ must be greater than 1.

COMPARING FRACTIONS

a. When 1 is added to the numerator of a fraction, is the result greater than or less than the original fraction? Explain your reasoning.
b. When 1 is added to the denominator of a fraction, is the result greater than or less than the original fraction? Explain your reasoning.

COMPARING FRACTIONS Think of a fraction. Add 1 to its numerator and add 1 to its denominator. Is the resulting fraction greater than, less than, or equal to the original fraction? Explain your reasoning.

SECTION 3.5

DIVISION WITH MIXED NUMBERS Division can be thought of as repeated subtraction. Use this concept to solve the following problem.

$5\frac{1}{4}$ yards of ribbon needs to be cut into pieces that are $\frac{3}{4}$ of a yard long to form bows. How many bows can be made?

SECTION 3.6

MIXED NUMBERS Two mixed numbers, A and B, are graphed below. Estimate where on the number line the graph of $A + B$ would lie.

A B
0 1 2 3 4 5 6 7

SECTION 3.7

COMPLEX FRACTIONS Write a problem that could be solved by simplifying the complex fraction.

$$\frac{\frac{7}{8}}{\frac{3}{4}}$$

SECTION 3.8

SOLVING EQUATIONS

a. Solve the equation $\frac{3}{4}x = 15$. Undo the multiplication by $\frac{3}{4}$ by dividing both sides by $\frac{3}{4}$.
b. Do the same for $-\frac{7}{8}x = 21$.

CHAPTER REVIEW

SECTION 3.1 The Fundamental Property of Fractions

CONCEPTS

Fractions can be used to indicate equal parts of a whole.

A fraction is composed of a *numerator, a denominator,* and a *fraction bar.*

If a and b are positive numbers,

$$\frac{-a}{b} = \frac{a}{-b} = -\frac{a}{b} \quad (b \neq 0)$$

Equivalent fractions represent the same number.

The *fundamental property of fractions:* Dividing the numerator and denominator of a fraction by the same nonzero number does not change the value of the fraction.

To *simplify* a fraction that is not in lowest terms, divide the numerator and denominator by the same number.

A fraction is in *lowest terms* if the only factor common to the numerator and denominator is 1.

The fundamental property of fractions: Multiplying the numerator and denominator of a fraction by a nonzero number does not change its value.

$$\frac{a}{b} = \frac{a \cdot x}{b \cdot x} \quad (b \neq 0, x \neq 0)$$

Expressing a fraction in higher terms results in an equivalent fraction that involves larger numbers or more complex terms.

REVIEW EXERCISES

1. If a woman gets seven hours of sleep each night, what part of a whole day does she spend sleeping?

2. In the illustration, why can't we say that $\frac{3}{4}$ of the figure is shaded?

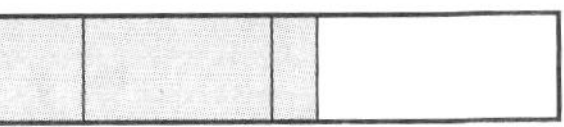

3. Write the fraction $\frac{2}{-3}$ in two other ways.

4. What concept about fractions does the illustration demonstrate?

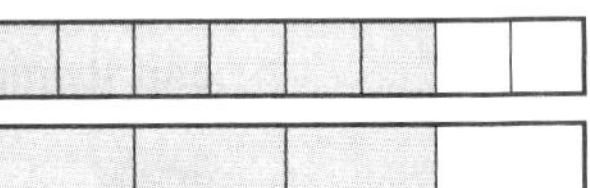

5. Explain the procedure shown here.

$$\frac{4}{6} = \frac{4 \div 2}{6 \div 2} = \frac{2}{3}$$

6. Explain what the slashes and the 1's mean.

$$\frac{4}{6} = \frac{\overset{1}{\cancel{2}} \cdot 2}{\underset{1}{\cancel{2}} \cdot 3} = \frac{2}{3}$$

Simplify each fraction to lowest terms.

7. $\frac{15}{45}$ **8.** $\frac{20}{48}$ **9.** $-\frac{63}{84}$ **10.** $\frac{66}{108}$

11. Explain what is being done and why it is valid.

$$\frac{5}{8} = \frac{5 \cdot 2}{8 \cdot 2} = \frac{10}{16}$$

Write each fraction or whole number with the indicated denominator (shown in red).

12. $\frac{2}{3}$, 18 **13.** $-\frac{3}{8}$, 16 **14.** $\frac{7}{15}$, 45 **15.** 4, 9

SECTION 3.2 *Multiplying Fractions*

To *multiply two fractions,* multiply their numerators and multiply their denominators.

$$\frac{a}{b} \cdot \frac{c}{d} = \frac{a \cdot c}{b \cdot d}$$

Multiply.

16. $\frac{1}{2} \cdot \frac{1}{3}$ **17.** $\frac{2}{5}\left(-\frac{7}{9}\right)$ **18.** $\frac{9}{16} \cdot \frac{20}{27}$ **19.** $\frac{5}{6} \cdot \frac{1}{3} \cdot \frac{18}{25}$

20. $\frac{3}{5} \cdot 7$ **21.** $-4\left(-\frac{9}{16}\right)$ **22.** $3\left(\frac{1}{3}\right)$ **23.** $-\frac{6}{7}\left(-\frac{7}{6}\right)$

Determine whether each statement is true or false.

24. $\frac{3}{4}x = \frac{3x}{4}$ **25.** $-\frac{5}{9}e = -\frac{5}{9e}$

Multiply.

26. $\frac{3}{5} \cdot \frac{10}{27}$ **27.** $-\frac{2}{3}\left(\frac{4}{7}s\right)$ **28.** $\frac{4}{9} \cdot \frac{3}{28}$ **29.** $9m\left(-\frac{5}{81}\right)$

An *exponent* indicates repeated multiplication.

Evaluate each power.

30. $\left(\frac{3}{4}\right)^2$ **31.** $\left(-\frac{5}{2}\right)^3$ **32.** $\left(\frac{2}{3}\right)^2$ **33.** $\left(-\frac{2}{5}\right)^3$

In mathematics, the word *of* usually means multiply.

34. GRAVITY ON THE MOON Objects on the moon weigh only one-sixth as much as on Earth. How much will an astronaut weigh on the moon if he weighs 180 pounds on Earth?

The *area of a triangle:*

$$A = \frac{1}{2}bh$$

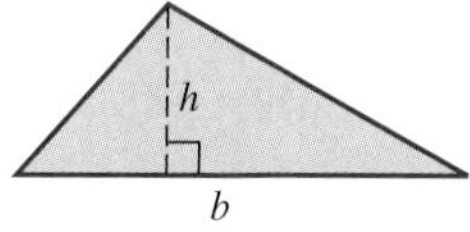

35. Find the area of the triangular sign.

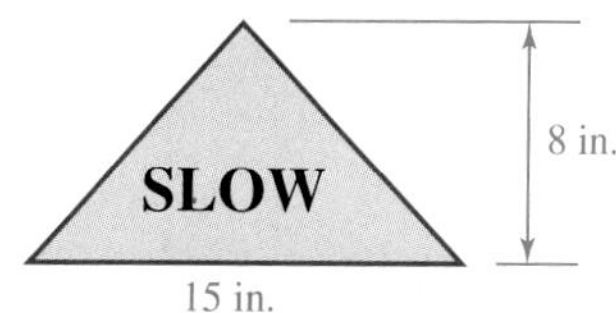

SECTION 3.3 *Dividing Fractions*

Two numbers are called *reciprocals* if their product is 1.

Find the reciprocal of each number.

36. $\frac{1}{8}$ **37.** $-\frac{11}{12}$ **38.** -6 **39.** 200

To *divide two fractions,* multiply the first by the reciprocal of the second.

$$\frac{a}{b} \div \frac{c}{d} = \frac{a}{b} \cdot \frac{d}{c}$$

Divide.

40. $\frac{1}{6} \div \frac{11}{25}$ **41.** $-\frac{7}{8} \div \frac{1}{4}$ **42.** $-\frac{15}{16} \div (-10)$

43. $8 \div \frac{16}{5}$ **44.** $\frac{1}{8} \div \frac{1}{4}$ **45.** $\frac{4}{5} \div \frac{1}{2}$

46. GOLD COINS How many $\frac{1}{16}$-ounce coins can be cast from a $\frac{3}{4}$-ounce bar of gold?

SECTION 3.4 *Adding and Subtracting Fractions*

To add (or subtract) fractions with like denominators, add (or subtract) their numerators and write the result over the common denominator.

$$\frac{a}{c} + \frac{b}{c} = \frac{a + b}{c}$$

$$\frac{a}{c} - \frac{b}{c} = \frac{a - b}{c}$$

Add or subtract.

47. $\frac{2}{7} + \frac{3}{7}$ **48.** $-\frac{3}{5} - \frac{3}{5}$ **49.** $\frac{3}{8} - \frac{1}{8}$ **50.** $\frac{7}{8} + \frac{7}{8}$

51. Explain why we cannot immediately add $\frac{1}{2} + \frac{2}{3}$ without doing some preliminary work.

The *LCD* must include the set of prime factors of each of the denominators.

52. Use prime factorization to find the least common denominator for fractions with denominators 45 and 30.

To add or subtract fractions with unlike denominators, we must first express them as equivalent fractions with the same denominator, preferably the LCD.

Add or subtract.

53. $\frac{1}{6} + \frac{2}{3}$ **54.** $\frac{2}{5} + \left(-\frac{3}{8}\right)$

55. $-\frac{3}{8} - \frac{5}{6}$ **56.** $3 - \frac{1}{7}$

57. $\frac{13}{6} - 6$ **58.** $\frac{1}{3} + \frac{1}{4} + \frac{1}{5}$

59. MACHINE SHOPS How much must be milled off the $\frac{3}{4}$-inch-thick steel rod so that the collar will slip over the end of it?

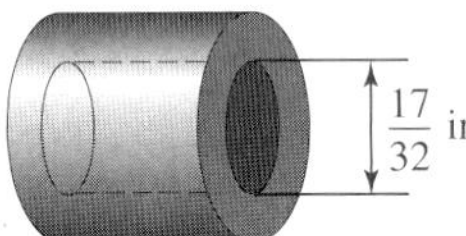

Steel rod

To *compare fractions,* write them as equivalent fractions with the same denominator. Then the fraction with the larger numerator will be the larger fraction.

60. TELEMARKETING In the first hour of work, a telemarketer made 2 sales out of 9 telephone calls. In the second hour, she made 3 sales out of 11 calls. During which hour was the rate of sales to calls better?

SECTION 3.5 *Multiplying and Dividing Mixed Numbers*

A *mixed number* is the sum of its whole-number part and its fractional part.

61. What mixed number is represented in the illustration?

62. What improper fraction is represented in the illustration?

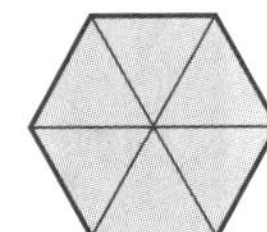 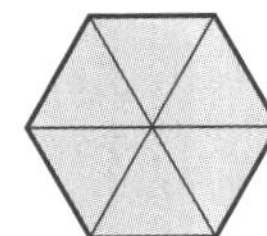 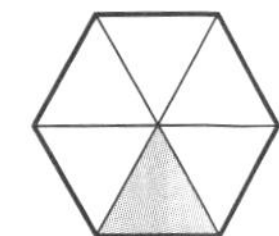

To change an *improper fraction* to a mixed number, divide the numerator by the denominator to obtain the whole-number part. Write the remainder over the denominator for the fractional part.

Express each improper fraction as a mixed number or a whole number.

63. $\frac{16}{5}$ **64.** $-\frac{47}{12}$ **65.** $\frac{6}{6}$ **66.** $\frac{14}{6}$

To change a mixed number to an improper fraction, multiply the whole number by the denominator and add the result to the numerator. Write this sum over the denominator.

Write each mixed number as an improper fraction.

67. $9\frac{3}{8}$ **68.** $-2\frac{1}{5}$ **69.** $100\frac{1}{2}$ **70.** $1\frac{99}{100}$

71. Graph: $-2\frac{2}{3}$, $\frac{8}{9}$, and $\frac{59}{24}$.

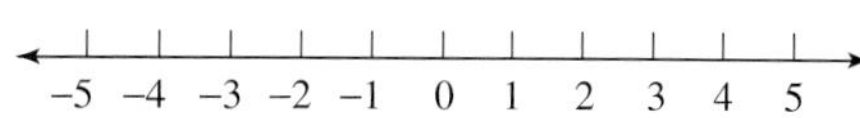

To *multiply* or *divide mixed numbers,* change the mixed numbers to improper fractions and then perform the operations as usual.

Multiply or divide. Write answers as mixed numbers when appropriate.

72. $-5\frac{1}{4} \cdot \frac{2}{35}$ **73.** $\left(-3\frac{1}{2}\right) \div \left(-3\frac{2}{3}\right)$

74. $\left(-6\frac{2}{3}\right)(-6)$ **75.** $-8 \div 3\frac{1}{5}$

76. CAMERA TRIPODS The three legs of a tripod can be extended to become $5\frac{1}{2}$ times their original length. If each leg is $8\frac{3}{4}$ inches long when collapsed, how long will a leg become when it is completely extended?

SECTION 3.6 *Adding and Subtracting Mixed Numbers*

To add (or subtract) mixed numbers, we can change each to an improper fraction and use the method of Section 3.4.

Add or subtract.

77. $1\frac{3}{8} + 2\frac{1}{5}$ **78.** $3\frac{1}{2} + 2\frac{2}{3}$

79. $2\frac{5}{6} - 1\frac{3}{4}$ **80.** $3\frac{7}{16} - 2\frac{1}{8}$

To add mixed numbers, we can add the whole numbers and the fractions separately.

81. PAINTING SUPPLIES In a project to restore a house, painters used $10\frac{3}{4}$ gallons of primer, $21\frac{1}{2}$ gallons of latex paint, and $7\frac{2}{3}$ gallons of enamel. Find the total number of gallons of paint used.

Vertical form can be used to add or subtract mixed numbers.

Add or subtract.

82. $\begin{array}{r} 133\frac{1}{9} \\ +\ 49\frac{1}{6} \\ \hline \end{array}$ **83.** $\begin{array}{r} 98\frac{11}{20} \\ +\ 14\frac{3}{5} \\ \hline \end{array}$

84. $\begin{array}{r} 50\frac{5}{8} \\ -19\frac{1}{6} \\ \hline \end{array}$ **85.** $\begin{array}{r} 375\frac{3}{4} \\ -\ 59 \\ \hline \end{array}$

If the fraction being subtracted is larger than the first fraction, we need to *borrow* from the whole number.

Subtract.

86. $23\frac{1}{3} - 2\frac{5}{6}$ **87.** $39 - 4\frac{5}{8}$

SECTION 3.7 *Order of Operations and Complex Fractions*

A *complex fraction* is a fraction whose numerator or denominator, or both, contain one or more fractions or mixed numbers.

To simplify a complex fraction, *Method 1:* The main fraction bar of a complex fraction indicates division.

Method 2: Multiply the numerator and denominator of the complex fraction by the LCD of all the fractions that appear in it.

Evaluate each numerical expression.

88. $\frac{3}{4} + \left(-\frac{1}{3}\right)^2\left(\frac{5}{4}\right)$

89. $\left(\frac{2}{3} \div \frac{16}{9}\right) - \left(1\frac{2}{3} \cdot \frac{1}{15}\right)$

Simplify each complex fraction.

90. $\dfrac{\frac{3}{5}}{-\frac{17}{20}}$

91. $\dfrac{\frac{2}{3} - \frac{1}{6}}{-\frac{3}{4} - \frac{1}{2}}$

SECTION 3.8 *Solving Equations Containing Fractions*

Solve each equation. Check the result.

92. $\frac{2}{3}x = 16$

93. $-\frac{7s}{4} = -49$

94. $\frac{y}{5} = -\frac{1}{15}$

95. $2x - 3 = 8$

96. $\frac{c}{3} - \frac{3}{8} = 2$

97. $\frac{5h}{9} - 1 = -3$

98. $4 - \frac{d}{4} = 0$

99. $\frac{t}{10} - \frac{2}{3} = \frac{1}{5}$

100. TEXTBOOKS In writing a history text, the author decided to devote two-thirds of the book to events prior to World War II. The remainder of the book deals with history after the war. If pre-World War II history is covered in 220 pages, how many pages does the textbook have?

CHAPTER 3 TEST

1. See the illustration.
   a. What fractional part of the plant is above ground?
   b. What fractional part of the plant is below ground?

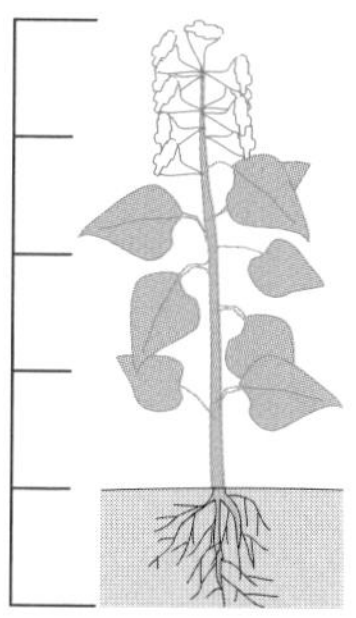

2. Simplify each fraction.

   a. $\frac{27}{36}$
   b. $\frac{72}{180}$

3. Multiply: $-\frac{3}{4}\left(\frac{1}{5}\right)$.

4. COFFEE DRINKERS Of 100 adults surveyed, $\frac{2}{5}$ said they started off their morning with a cup of coffee. Of the 100, how many would this be?

5. Divide: $\frac{4}{3} \div \frac{1}{9}$.

6. Subtract: $\frac{1}{6} - \frac{4}{5}$.

7. Express $\frac{7}{8}$ as an equivalent fraction with denominator 24.

8. Graph: $2\frac{4}{5}$, $-1\frac{1}{7}$, and $\frac{7}{6}$.

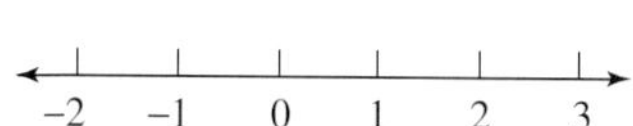

9. SPORTS CONTRACTS A basketball player signed a nine-year contract for $\$13\frac{1}{2}$ million. How much is this per year?

10. Add: $157\frac{5}{9} + 103\frac{3}{4}$.

11. Subtract: $67\frac{1}{4} - 29\frac{5}{6}$.

12. BOXING When Oscar De La Hoya fought Pernell Whitaker, the "Tale of the Tape" shown below appeared in the sports section of many newspapers. What was the difference in the fighters'
    a. weights?
    b. chests (expanded)?
    c. waists?

Tale of the Tape

De La Hoya		Whitaker
24 yr	Age	33 yr
146 1/2 lb	Weight	146 1/2 lb
5-11	Height	5-6
72 in.	Reach	69 in.
39 in.	Chest (Normal)	37 in.
42 1/4 in.	Chest (Expanded)	39 1/2 in.
31 3/4 in.	Waist	28 in.

13. Add: $-\frac{3}{7} + 2$.

14. SEWING When cutting material for a $10\frac{1}{2}$-inch-wide placemat, a seamstress allows $\frac{5}{8}$ inch at each end for a hem. How wide should the material be cut?

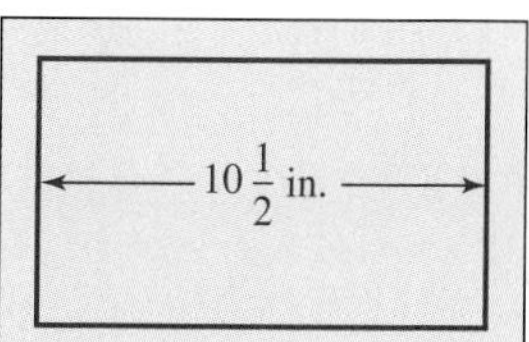

15. Find the perimeter and the area of the triangle shown below.

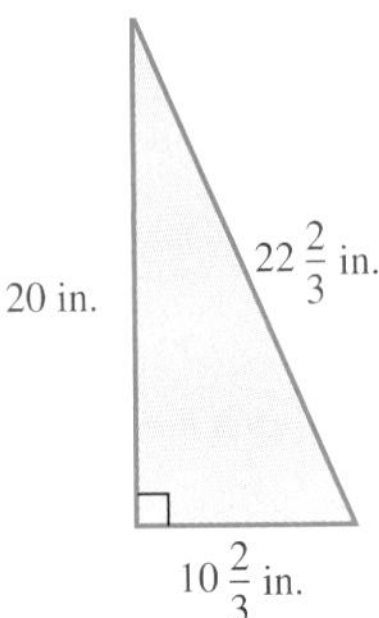

16. NUTRITION A box of Tic Tacs contains 40 of the $1\frac{1}{2}$-calorie flavored breath mints. How many calories are there in a box of Tic Tacs?

17. COOKING How many servings are there in an 8-pound roast, if the suggested serving size is $\frac{2}{3}$ pound?

18. Evaluate:

$$\left(\frac{2}{3}\cdot\frac{5}{16}\right) - \left(-1\frac{3}{5} \div 4\frac{4}{5}\right)$$

19. Simplify the complex fraction.

$$\frac{-\frac{5}{6}}{\frac{7}{8}}$$

20. Simplify the complex fraction.

$$\frac{\frac{1}{2}+\frac{1}{3}}{-\frac{1}{6}-\frac{1}{3}}$$

21. Is 6 a solution of $\frac{5}{2}x = 15$? Explain.

Solve each equation.

22. $\frac{x}{3} = 14$

23. $-\frac{5}{2}t = 18$

24. $6x - 4 = -3$

25. $y + \frac{9}{16} = \frac{11}{16}$

26. $\frac{x}{6} - \frac{2}{3} = \frac{1}{12}$

27. JOB APPLICANTS Three-fourths of the applicants for a position had previous experience. If 144 people applied, how many had previous experience? How many did not have previous experience?

28. What are the parts of a fraction? What does a fraction represent?

29. Explain what is meant when we say, "The product of any number and its reciprocal is 1."

30. Explain what mathematical concept is being shown.

a. $\frac{6}{8} = \frac{\overset{1}{\cancel{2}}\cdot 3}{\underset{1}{\cancel{2}}\cdot 4} = \frac{3}{4}$

b. 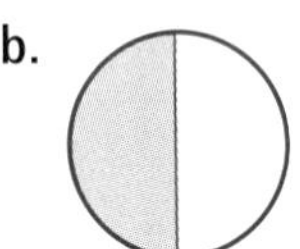 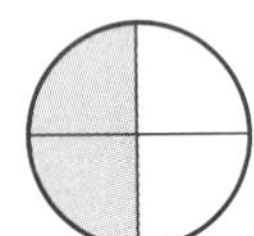

c. $\frac{3}{5} = \frac{3\cdot 4}{5\cdot 4} = \frac{12}{20}$

CHAPTERS 1–3 CUMULATIVE REVIEW EXERCISES

Consider the number 5,434,679.

1. Round to the nearest hundred.
2. Round to the nearest ten thousand.
3. THE STOCK MARKET The graph below shows the performance of the Dow Jones Industrial Average on the last trading day of 1999. Estimate the highest mark that the market reached. At what time during the day did that occur?

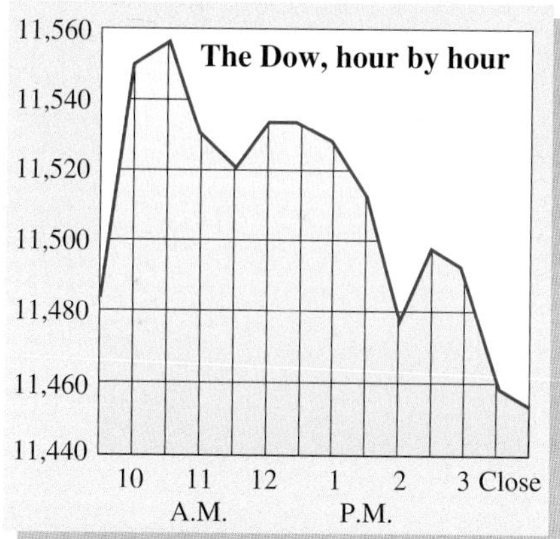

4. BANKS One of the world's largest banks, with total assets of \$691,920,300,000, is the Bank of Tokyo–Mitsubishi Ltd., Japan. In what place value column is the digit 6 located?

Perform each operation.

5. $\begin{array}{r} 4{,}679 \\ +3{,}457 \\ \hline \end{array}$

6. $\begin{array}{r} 7{,}897 \\ -4{,}378 \\ \hline \end{array}$

7. $\begin{array}{r} 5{,}345 \\ \times\quad 56 \\ \hline \end{array}$

8. $35\overline{)34{,}685}$

Refer to the rectangular swimming pool shown below.

9. Find the perimeter of the pool.
10. Find the area of the pool's surface.

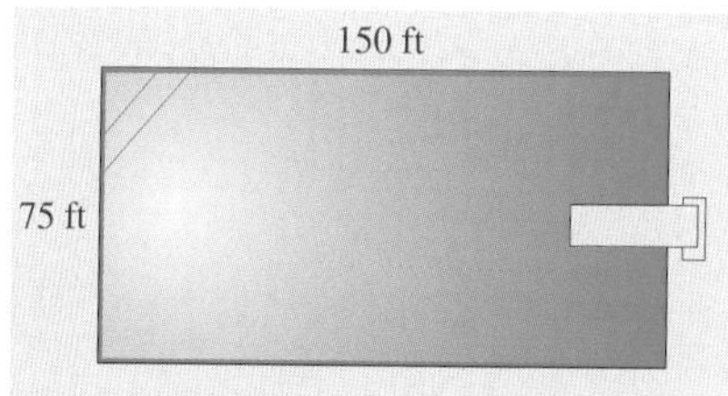

Find the prime factorization of each number.

11. 84
12. 450
13. 360
14. 3,600

Evaluate each expression.

15. $6 + (-2)(-5)$
16. $(-2)^3 - 3^3$
17. $\dfrac{2(-7) + 3(2)}{2(-2)}$
18. $\dfrac{2(3^2 - 4^2)}{-2(3) - 1}$

Solve each equation and check the result.

19. $3x + 2 = -13$
20. $-5z - 7 = 18$
21. $\dfrac{y}{4} - 1 = -5$
22. $\dfrac{n}{5} + 1 = 0$
23. Is -15 a solution of $2x + 45 = 30$?
24. OBSERVATION HOURS To get a Master's degree in educational psychology, a student must have 100 hours of observation time at a clinic. If the student has already observed for 37 hours, how many 3-hour shifts must he observe to complete the requirement?

Simplify each fraction.

25. $\dfrac{21}{28}$
26. $\dfrac{40}{16}$

Perform each operation.

27. $\dfrac{6}{5}\left(-\dfrac{2}{3}\right)$
28. $\dfrac{14}{8} \div \dfrac{7}{2}$
29. $\dfrac{2}{3} + \dfrac{3}{4}$
30. $\dfrac{4}{3} - \dfrac{3}{5}$

Write each mixed number as an improper fraction.

31. $3\frac{5}{6}$

32. $-6\frac{5}{8}$

Perform each operation.

33. $4\frac{2}{3} + 5\frac{1}{4}$

34. $14\frac{2}{5} - 8\frac{2}{3}$

35. FIRE HAZARDS Two terminals in an electrical switch were so close that electricity could jump the gap and start a fire. The following illustration shows a newly designed switch that will keep this from happening. By how much was the distance between the ground terminal and the hot terminal increased?

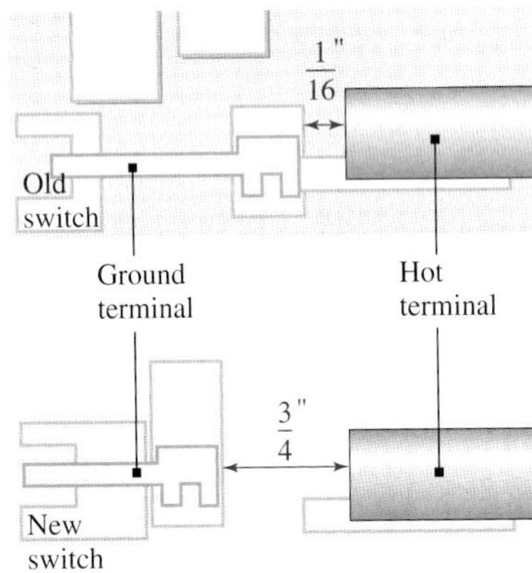

36. SHAVING Advertisements claim that a shaving lotion for men cuts shaving time by a third. If it normally takes a man 90 seconds to shave, how much time will he *save* if he uses the special lotion? If he uses the special lotion, how long will it take him to shave?

Simplify each expression.

37. $\left(\frac{1}{4} - \frac{7}{8}\right) \div \left(-2\frac{3}{16}\right)$

38. $\dfrac{\frac{2}{3} - 7}{4\frac{5}{6}}$

Solve each equation and check the result.

39. $x + \frac{1}{5} = -\frac{14}{15}$

40. $3 = \frac{5}{8}x + \frac{1}{2}$

41. $\frac{2}{3}x = -10$

42. $3y - 8 = 0$

43. Explain the difference between an *expression* and an *equation.*

44. What is a variable?

CHAPTER 4

Decimals

CORBIS

TLE Sports fans love to talk facts, figures, and trivia. Many sports records are listed as decimal numbers. For example, did you know that over his fifteen-year career in the NBA, Michael Jordan averaged 30.1 points per game. Jerry Rice, considered by many to be the best receiver ever to play in the NFL, has averaged 14.8 yards per catch. Swimmer Janet Evans of the United States still holds the world record in the 800-meter freestyle. Her time of 8 minutes and 16.22 seconds was set in 1989 and it has stood for over 15 years!

To learn more about decimals and how they are used in sports, visit The Learning Equation on the Internet at http://tle.brookscole.com. (The log-in instructions are in the Preface.) For Chapter 4, the online lessons are:

- *TLE* Lesson 9: Decimal Conversions
- *TLE* Lesson 10: Ordering Numbers

Check Your Knowledge

1. To multiply decimals, multiply them as if they were whole numbers. The number of decimal places in the product is the same as the ______ of the decimal places in the factors.
2. When we find what number is squared to obtain a given number, we are finding the square ______ of the given number.
3. Write 0.084 as a fraction.
4. Brittany purchased a CD player for \$39.95, headphones for \$17.95, and two CDs for \$13.95 each. What was the total cost of her purchases?
5. Perform each operation in your head.

   a. $354{,}278.2 \div 1{,}000$ b. $2.000478 \cdot 10{,}000$
6. A rectangular table is 3.5 ft long and 2.25 ft wide.

   a. What is its area?   b. What is its perimeter?
7. Evaluate: $2.8 + (1.2)(-0.5)^2$.
8. Write each fraction as a decimal. Use an overbar, if necessary.

   a. $\frac{3}{20}$ b. $\frac{5}{8}$ c. $\frac{1}{9}$
9. Divide, and round your answer to the nearest tenth: $\frac{25.736}{16.3}$.
10. Graph $-\frac{3}{8}$ and $\frac{1}{4}$ on the number line.
    Label each point using its decimal equivalent.

    −2 −1 0 1 2
11. Find the exact answer: $\frac{1}{6} + 0.25$.
12. Individual grades on a quiz were 5, 8, 3, 5, 7, 9, 4, 9, 10, 7, 9. Find the average quiz score. Round to one decimal place.
13. Graph $-\sqrt{3}$ and $\sqrt{2}$ on the number line. Label each point with its decimal approximation rounded to two decimal places.

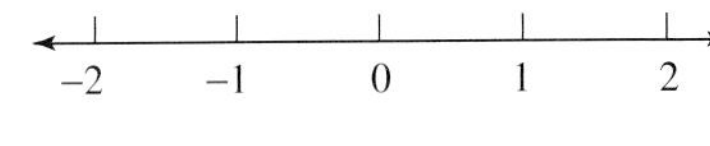

14. Evaluate each expression.

    a. $\sqrt{16} + 5\sqrt{9}$ b. $\sqrt{\frac{36}{25}} - \sqrt{\frac{1}{16}}$

    c. $\sqrt{0.16} - \sqrt{0.49}$
15. Insert the proper symbol $<$ or $>$ to make each statement true.

    a. -2.7 ☐ -2.75 b. $\sqrt{2}$ ☐ 2 c. $0.\overline{3}$ ☐ 0.3
16. Mindy drove 342 miles in 6.5 hours. What was her average speed in miles per hour? Round to the nearest tenth.

Solve each equation.

17. $\frac{c}{2.3} = -4.1$ 18. $0.5x + 1.5 = -3$ 19. $3x - 3.3 - 2.3 = -3.2$
20. It costs a business \$50 a month plus 4 cents per copy made to rent a color copier. If they have budgeted \$75 a month for copier expenses, how many color copies can they make each month?

Study Skills Workshop

HOMEWORK

Doing a thorough job with your homework is one of the most important things you can do to be successful in your class. Sitting in class and listening to the lecture will help you place concepts in short-term memory, but in order to do well on tests and in subsequent classes, you want to put these concepts in long-term memory. When done correctly, homework assignments will help with this.

Are You Giving Yourself Enough Time. Recall that in the first Study Skills Workshop assignment you made a study calendar that scheduled two hours of study and homework for every hour that you spend in class. If you are not adhering to this schedule, make changes to ensure that you can spend enough time outside of class to learn new material. Make sure you spread this time out over a period of at least five days per week, rather than in one or two long sessions.

Before You Start Your Homework Problems. In the Study Skills Workshop for Chapter 6, your assignment is to rework your notes. Always rework the notes that relate to your homework before starting your assignment. After reworking your notes, read the sections in the textbook that pertain to the homework problems, looking especially at the examples. With a pencil in hand, rework the examples, trying to understand each step. Keep a list of anything that you don't understand, both in your notes and in the textbook examples.

Doing Your Homework Problems. Once you have reviewed your notes and the textbook examples, you should be able to successfully manage the bulk of your homework assignment easily. When you work on your homework, keep your textbook and notes handy so that you can refer to them if necessary. If you have trouble with a homework question, look through your textbook and notes to see if you can identify an example that is similar to the homework question. Apply the same strategy to your homework problem. If there are places where you get stuck, add these to your questions list.

Before Your Homework Is Due. At least one day before your assignment is due, seek help with your question list. You can get help by contacting a classmate for assistance, making an appointment with a tutor, or visiting your instructor during office hours. Make sure to bring your list and try to pinpoint exactly where in the process you got stuck.

ASSIGNMENT

1. Review your study calendar. Are you following it? If not, what changes can you make to adhere to the rule: two hours of homework and study for every hour of class?
2. Find five homework problems in your assignment that are similar to the examples in your textbook. List each problem along with its matching example. Were there any homework problems in your assignment that didn't have an example that was similar?
3. Make a list of homework problems that you had questions about or didn't know how to do. See your tutor or your instructor during office hours with your list of problems and ask him or her to work through these problems with you.

Decimals provide another way to represent fractions and mixed numbers. They are often used in measurement, because it is easy to put decimals in order and to compare them.

4.1 An Introduction to Decimals

- Decimals • The place value system for decimal numbers
- Reading and writing decimals • Comparing decimals • Rounding

This section introduces the **decimal numeration system**— an extension of the place value system that we used with whole numbers.

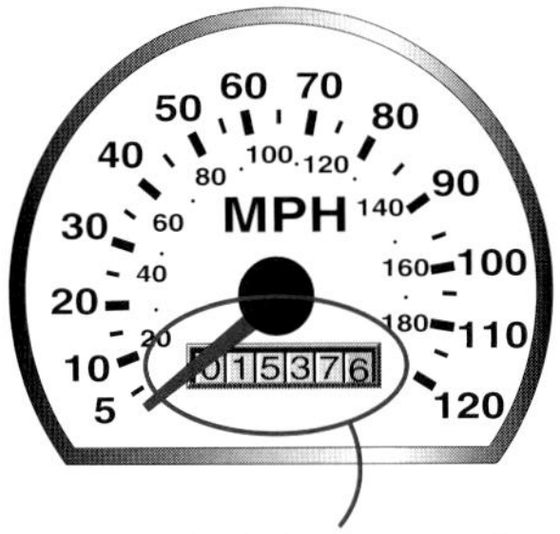

We can use the decimal numeration system to express the car's mileage. The odometer reads 1,537.6 miles.

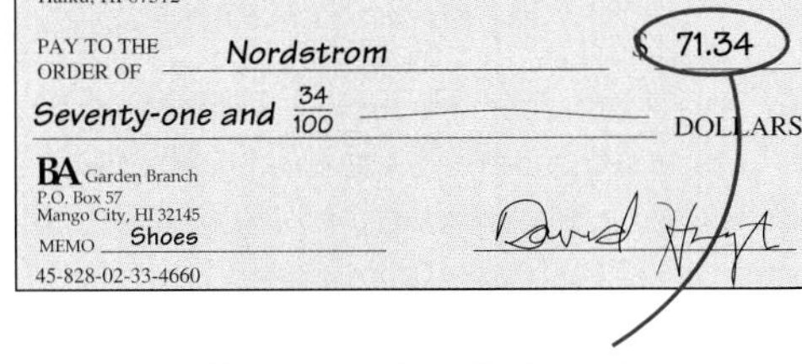
David Hoyt
612 Lelani
Haiku, HI 67512

Feb. 21, 20 05

PAY TO THE ORDER OF Nordstrom $ 71.34

Seventy-one and 34/100 DOLLARS

BA Garden Branch
P.O. Box 57
Mango City, HI 32145

MEMO Shoes

45-828-02-33-4660

The amount of the check is written using the decimal numeration system.

Decimals

Like fraction notation, decimal notation is used to denote a part of a whole. However, when writing a number in decimal notation, we don't use a fraction bar, nor is a denominator shown. A number written in decimal notation is often called a **decimal.**

In Figure 4-1, a rectangle is divided into 10 equal parts. One-tenth of the figure is shaded. We can use either the fraction $\frac{1}{10}$ or the decimal 0.1 to describe the amount of the figure that is shaded. Both are read as "one-tenth."

$$\frac{1}{10} = 0.1$$

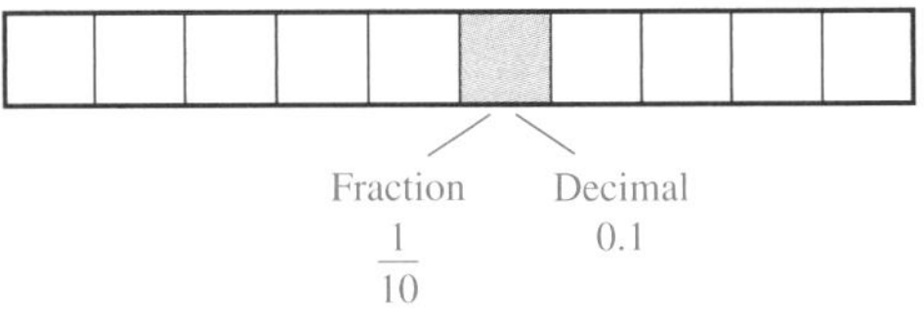

FIGURE 4-1

In Figure 4-2, a square is divided into 100 equal parts. One of the 100 parts is shaded. The amount of the figure that is shaded can be represented by the fraction $\frac{1}{100}$ or by the decimal 0.01. Both are read as "one one-hundredth."

$$\frac{1}{100} = 0.01$$

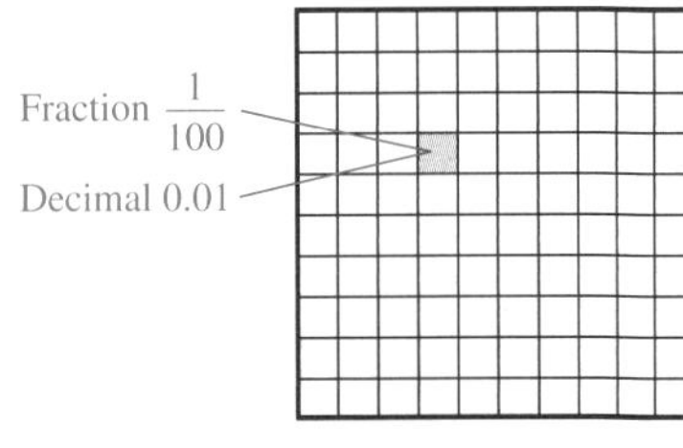

FIGURE 4-2

The place value system for decimal numbers

Decimal numbers are written by placing digits (0, 1, 2, 3, 4, 5, 6, 7, 8, 9) into place value columns that are separated by a **decimal point.** (See Figure 4-3.) The place value names of all the columns to the right of the decimal point end in "th." The "th" tells us that the value of the column is a fraction whose denominator is a power of 10. Columns to the left of the decimal point have a value greater than or equal to 1; columns to the right of the decimal point have a value less than 1. We can show the value represented by each digit of a decimal by using **expanded notation.**

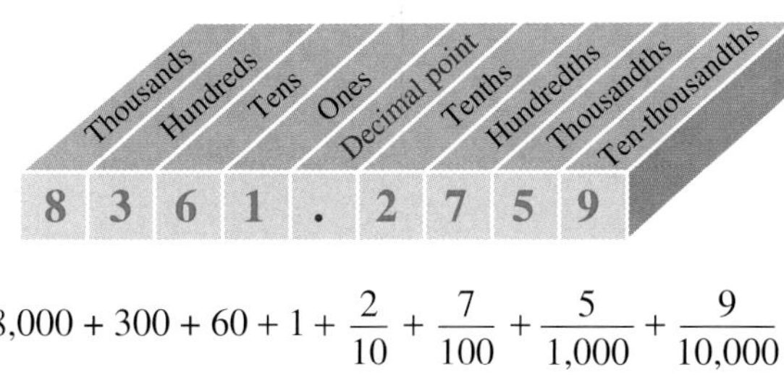

$$8{,}000 + 300 + 60 + 1 + \frac{2}{10} + \frac{7}{100} + \frac{5}{1{,}000} + \frac{9}{10{,}000}$$

Expanded notation

FIGURE 4-3

Decimal points are used to separate the whole-number part of a decimal from its fractional part.

12 . 37

Whole-number part — Decimal point — Fractional part

When there is no whole-number part of a decimal, we can show that by entering a zero to the left of the decimal point.

.85 = 0.85

↑ No whole-number part ↑ Enter a zero here, if desired.

We can write a whole number in decimal notation by placing a decimal point to its right and then adding a zero, or zeros, to the right of the decimal point.

99 = 99.0 = 99.00

↑ A whole number ↑ ↑ Place a decimal point here and enter a zero, or zeros, to the right of it.

Writing additional zeros to the right of the decimal point *following the last digit* does not change the value of the decimal. Deleting additional zeros to the right of the decimal point following the last digit does not change the value of the decimal.

12.37 = 12.370 = 12.3700

↑ ↑ These additional zeros do not change the value of the decimal.

Reading and writing decimals

The decimal 12.37 can be read as "twelve point three seven." Another way of reading a decimal states the whole-number part first and then the fractional part.

Reading a decimal

1. Look to the left of the decimal point and say the name of the whole number.
2. The decimal point is then read as "and."
3. Say the fractional part of the decimal as a whole number followed by the name of the place value column of the digit that is farthest to the right.

When we use this procedure, here is the other way to read 12.37.

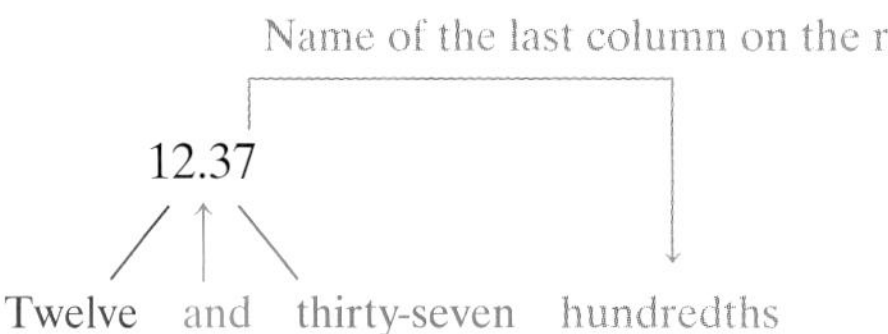

When we read a decimal in this way, it is easy to write it in words and as a mixed number.

Decimal	Words	Mixed number
12.37	Twelve and thirty-seven hundredths	$12\frac{37}{100}$

Self Check 1

Write each decimal in words and then as a mixed number.

a. Sputnik 1, the first artificial satellite, weighed 184.3 pounds.

b. The planet Mercury makes one revolution every 87.9687 days.

Answers a. One hundred eighty-four and three tenths, or $184\frac{3}{10}$ b. Eighty-seven and nine thousand six hundred eighty-seven ten-thousandths, or $87\frac{9,687}{10,000}$

EXAMPLE 1 World records. Write each decimal in words and then as a fraction or mixed number. **Do not simplify the fraction.**

a. According to the *Guinness Book of World Records*, the fastest qualifying speed at the Indianapolis 500 was 236.986 mph by Aire Luyendyk in 1996.

b. The smallest freshwater fish is the dwarf pygmy goby, found in the Philippines. Adult males weigh 0.00014 ounce.

Solution

a. The whole number part of **236.**986 to the left of the decimal point is 236. The fractional part, stated as a whole number, is 986. The digit the farthest to the right is 6 and it is in the thousandths column. Thus,

236.986 is two hundred thirty-six and nine hundred eighty-six thousandths, or $236\frac{986}{1,000}$.

b. The whole number part of **0.**00014 to the left of the decimal point is 0. The fractional part, stated as a whole number, is 14. The digit the farthest to the right is 4 and it is in the hundred-thousandths column. Thus,

0.00014 is fourteen hundred-thousandths, or $\frac{14}{100,000}$.

Decimals can be negative. For example, a record low temperature of $-128.6°$ F was recorded in Vostok, Antarctica, on July 21, 1983. This is read as "negative one hundred twenty-eight and six tenths." Written as a mixed number, it is $-128\frac{6}{10}$.

Comparing decimals

The relative sizes of a set of decimals can be determined by scanning their place value columns from left to right, column by column, looking for a difference in the digits. For example,

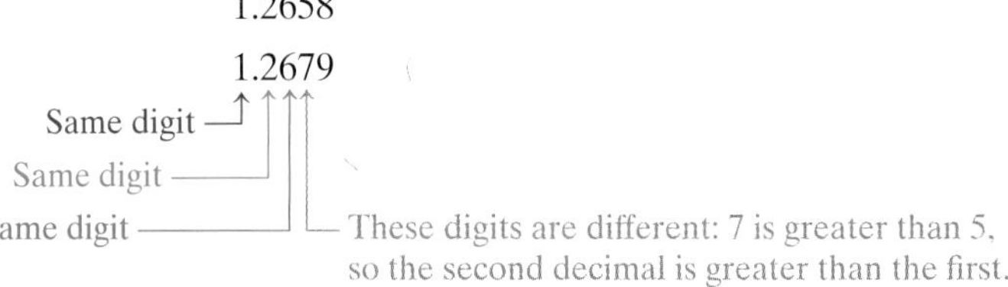

Thus, 1.2679 is greater than 1.2658. We write $1.2679 > 1.2658$.

Comparing positive decimals

1. Make sure both numbers have the same number of decimal places to the right of the decimal point. Write any additional zeros necessary to achieve this.
2. Compare the digits of each decimal, column by column, working from left to right.
3. When two digits differ, the decimal with the greater digit is the greater number.

EXAMPLE 2 Which is greater: 54.9 or 54.929?

Solution

54.900 Write two zeros after 9 so that both decimals have the same number of digits to the right of the decimal point.

54.929
  ↑

Working from left to right, we find this is the first column in which the digits differ. Since 2 is greater than 0, we can conclude that 54.929 > 54.9.

Self Check 2
Which is greater: 113.7 or 113.657?

Answer 113.7

Comparing negative decimals

1. Make sure both numbers have the same number of decimal places to the right of the decimal point. Write any additional zeros necessary to achieve this.
2. Compare the digits of each decimal, column by column, working from left to right.
3. When two digits differ, the decimal with the smaller digit is the greater number.

EXAMPLE 3 Which is greater: −10.45 or −10.419?

Solution

−10.450 Write a zero after 5 to help in the comparison.

−10.419
  ↑

As we work from left to right, this is the first column in which the digits differ. Since 1 is less than 5, we conclude that −10.419 > −10.45.

Self Check 3
Which is greater: −703.8 or −703.78?

Answer −703.78

Self Check 4

Graph: -1.1, -0.6, 0.8, and 1.9.

Answer

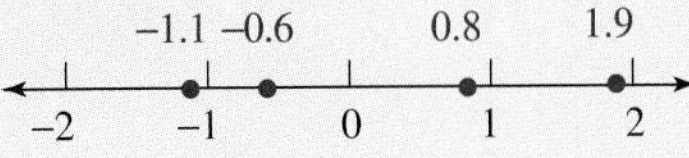

EXAMPLE 4 Graph: -1.8, -1.23, -0.3, and 1.89.

Solution To graph each decimal, we locate its position on the number line and draw a dot. Since -1.8 is to the left of -1.23, we can write $-1.8 < -1.23$.

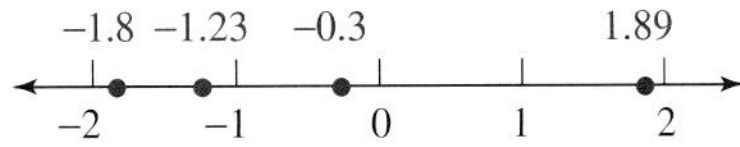

Rounding

When working with decimals, we often round answers to a specific number of decimal places.

Rounding a decimal

1. To round a decimal to a specified decimal place, locate the digit in that place. Call it the *rounding digit.*
2. Look at the *test digit* to the right of the rounding digit.
3. If the test digit is 5 or greater, round up by adding 1 to the rounding digit and dropping all the digits to its right. If the test digit is less than 5, round down by keeping the rounding digit and dropping all the digits to its right.

EXAMPLE 5 **Chemistry.** In a chemistry class, a student uses a balance to weigh a compound. The digital readout on the scale shows 1.2387 grams. Round this decimal to the nearest thousandth of a gram.

Solution We are asked to round to the nearest thousandth.

Add 1 to the 8.

1.2387

The rounding digit (8) — The test digit (7) is 5 or greater. Therefore, add 1 to the rounding digit and drop all other digits to its right.

The compound weighs approximately 1.239 grams.

Self Check 6

Round each decimal to the indicated place value:

a. -708.522 to the nearest tenth

b. 9.1198 to the nearest thousandth

EXAMPLE 6 Round each decimal to the indicated place value: **a.** -645.13 to the nearest tenth and **b.** 33.097 to the nearest hundredth.

Solution

a. -645.13

Rounding digit (1) — Since the test digit (3) is less than 5, drop it and all the digits to its right.

The result is -645.1.

b. 33.097

Rounding digit — Since the test digit is greater than 5, we add 1 to 9 and drop all the digits to the right.

33.09 (10) Adding a 1 to the 9 requires that we carry a 1 to the tenths column.

When we are asked to round to the nearest hundredth, we must have a digit in the hundredths column, even if it is a zero. Therefore, the result is 33.10.

Answers **a.** −708.5, **b.** 9.120

Section 4.1 STUDY SET

VOCABULARY *Fill in the blanks.*

1. Give the name of each place value column.

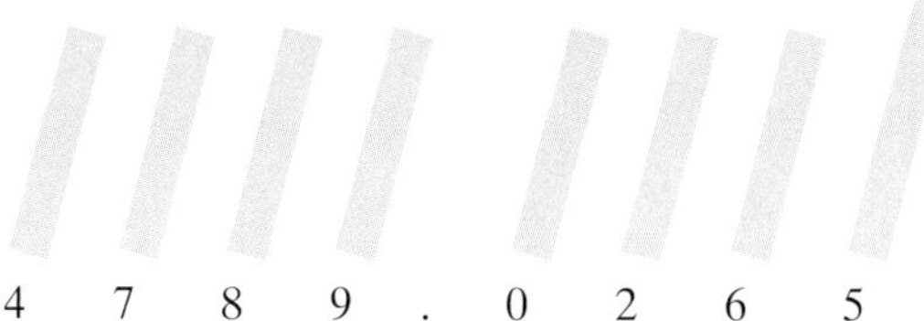

4 7 8 9 . 0 2 6 5

2. We can show the value represented by each digit of the decimal 98.6213 by using ________ notation.

$$98.6213 = 90 + 8 + \frac{6}{10} + \frac{2}{100} + \frac{1}{1,000} + \frac{3}{10,000}$$

3. We can approximate a decimal number using the process called ________.

4. When we read 2.37, the decimal point can be read as "______" or "______."

CONCEPTS

5. Consider the decimal 32.415.
   - **a.** Write the decimal in words.
   - **b.** What is its whole-number part?
   - **c.** What is its fractional part?
   - **d.** Write the decimal in expanded notation.

6. Write $400 + 20 + 8 + \frac{9}{10} + \frac{6}{100}$ as a decimal.

7. Graph: $\frac{7}{10}$, −0.7, $-3\frac{1}{100}$, and 3.01.

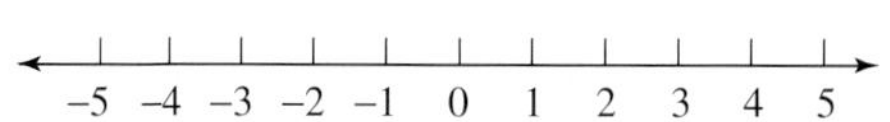

8. Graph: −1.21, −3.29, and −4.25.

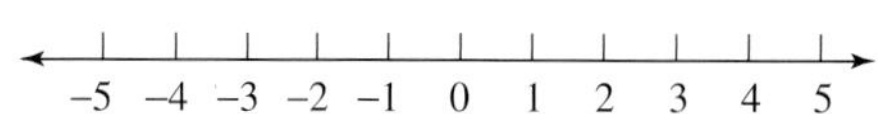

9. Determine whether the statement is true or false.
   - **a.** 0.9 = 0.90
   - **b.** 1.260 = 1.206
   - **c.** −1.2800 = −1.280
   - **d.** 0.001 = .0010

10. Write each fraction as a decimal.
   - **a.** $\frac{9}{10}$
   - **b.** $\frac{63}{100}$
   - **c.** $\frac{111}{1,000}$
   - **d.** $\frac{27}{10,000}$

11. Represent the shaded part of the square as a fraction and a decimal.

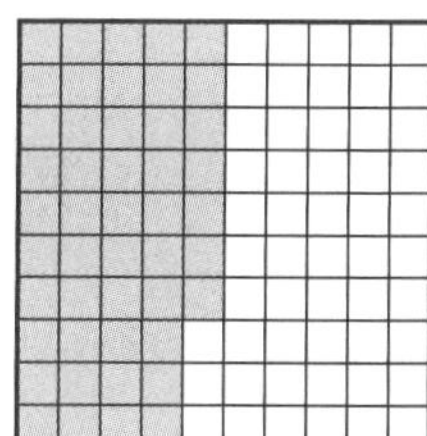

12. Represent the shaded part of the rectangle using a fraction and a decimal.

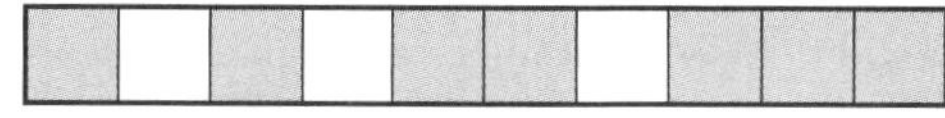

13. The line segment shown below is 1 inch long. Show a length of 0.3 inch on it.

14. Read the meter on the right. What decimal is indicated by the arrow?

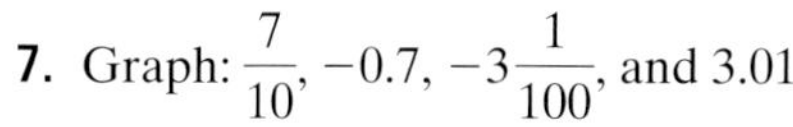

NOTATION

15. Construct a decimal number by writing
0 in the tenths column,
4 in the thousandths column,
1 in the tens column,
9 in the thousands column,
8 in the hundreds column,
2 in the hundredths column,
5 in the ten-thousandths column, and
6 in the ones column.

16. Represent each situation using a signed number.
- **a.** A deficit of \$15,600.55
- **b.** A river 6.25 feet under flood stage
- **c.** A state budget \$6.4 million in the red
- **d.** 3.9 degrees below zero
- **e.** 17.5 seconds prior to liftoff
- **f.** A checking account overdrawn by \$33.45

PRACTICE *Write each decimal in words and as a fraction or mixed number.*

17. 50.1 **18.** 0.73

19. −0.0137 **20.** −76.09

21. 304.0003 **22.** 68.91

23. −72.493 **24.** −31.5013

Write each decimal using numbers.

25. Negative thirty-nine hundredths

26. Negative twenty-seven and forty-four hundredths

27. Six and one hundred eighty-seven thousandths

28. Ten and fifty-six ten-thousandths

Round each decimal to the nearest tenth.

29. 506.098 **30.** 0.441

31. 2.718218 **32.** 3,987.8911

Round each decimal to the nearest hundredth.

33. −0.137 **34.** −808.0897

35. 33.0032 **36.** 64.0059

Round each decimal to the nearest thousandth.

37. 3.14159 **38.** 16.0995

39. 1.414213 **40.** 2,300.9998

Round each decimal to the nearest whole number.

41. 38.901 **42.** 405.64

43. 2,988.399 **44.** 10,453.27

Round each amount to the value indicated.

45. \$3,090.28
- **a.** Nearest dollar
- **b.** Nearest ten cents

46. \$289.73
- **a.** Nearest dollar
- **b.** Nearest ten cents

Fill in the blanks with the proper symbol (<, >, or =).

47. −23.45 ☐ −23.1 **48.** −301.98 ☐ −302.45

49. −.065 ☐ −.066 **50.** −3.99 ☐ −3.9888

Arrange the decimals in order, from least to greatest.

51. 132.64, 132.6499, 132.6401

52. 0.007, 0.00697, 0.00689

APPLICATIONS

53. WRITING CHECKS Complete the check below by writing in the amount, using a decimal.

Ellen Russell
455 Santa Clara Ave.
Parker, CO 25413
April 14, 20 05
PAY TO THE ORDER OF Citicorp $
One thousand twenty-five and 78/100 DOLLARS
BA Downtown Branch
P.O. Box 2456
Colorado Springs, CO 23712
MEMO Mortgage
Ellen Russell
45-828-02-33-4660

54. MONEY We use a decimal point when working with dollars, but the decimal point is not necessary when working with cents. For each dollar amount in the table, give the equivalent amount expressed as cents.

Dollars	Cents
\$0.50	
\$0.05	
\$0.55	
\$5.00	
\$0.01	

55. INJECTIONS A syringe is shown below. Use an arrow to show to what point the syringe should be filled if a 0.38-cc dose of medication is to be administered. ("cc" stands for "cubic centimeters.")

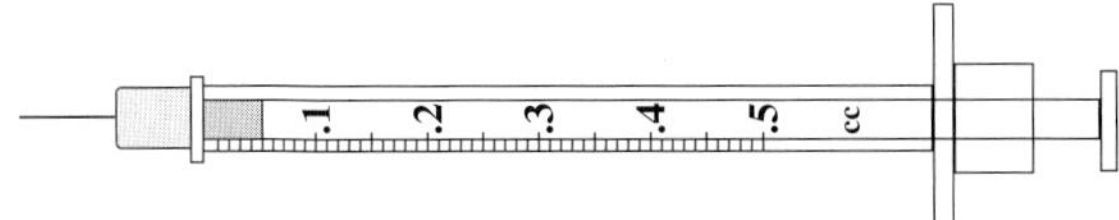

56. LASERS The laser used in laser vision correction is so precise that each pulse can remove 39 millionths of an inch of tissue in 12 billionths of a second. Write each of these numbers as decimals.

57. THE METRIC SYSTEM The metric system is widely used in science to measure length (meters), weight (grams), and capacity (liters). Round each decimal to the nearest hundredth.

a. 1 ft is 0.3048 meter.

b. 1 mi is 1,609.344 meters.

c. 1 lb is 453.59237 grams.

d. 1 gal is 3.785306 liters.

58. WORLD RECORDS As of October 2004, four American women held individual world records in swimming. Their times are given below in the form *minutes: seconds*. Round each to the nearest tenth of a second.

100-meter backstroke	Natalie Coughlin	0:59.58
200-meter breaststroke	Amanda Beard	2:22.44
400-meter freestyle	Janet Evans	4:03.85
800-meter freestyle	Janet Evans	8:16.22
1,500-meter freestyle	Janet Evans	15:52.10

59. GEOLOGY Geologists classify types of soil according to the grain size of the particles that make up the soil. The four major classifications are shown below. Complete the table by classifying each sample.

Clay	0.00008 in. and under
Silt	0.00008 in. to 0.002 in.
Sand	0.002 in. to 0.08 in.
Granule	0.08 in. to 0.15 in.

Sample	Location	Size (in.)	Classification
A	riverbank	0.009	
B	pond	0.0007	
C	NE corner	0.095	
D	dry lake	0.00003	

60. MICROSCOPES A microscope used in a lab is capable of viewing structures that range in size from 0.1 to 0.0001 centimeter. Which of the structures listed below would be visible through this microscope?

Structure	Size (in cm)
bacterium	0.00011
plant cell	0.015
virus	0.000017
animal cell	0.00093
asbestos fiber	0.0002

61. AIR QUALITY The following table shows the cities with the highest one-hour concentrations of ozone (in parts per million) during the summer of 1999. Rank the cities in order, beginning with the city with the highest reading.

Crestline, California	0.170
Galveston, Texas	0.176
Houston, Texas	0.202
Texas City, Texas	0.206
Westport, Connecticut	0.188
White Plains, New York	0.171

Source: *Los Angeles Times* (August 18, 1999)

62. DEWEY DECIMAL SYSTEM A system for classifying books in a library is the Dewey Decimal System. Books on the same subject are grouped together by number. For example, books about the arts are assigned numbers between 700 and 799. When stacked on the shelves, the books are to be in numerical order, from left to right. How should the titles in the illustration be rearranged to be in the proper order?

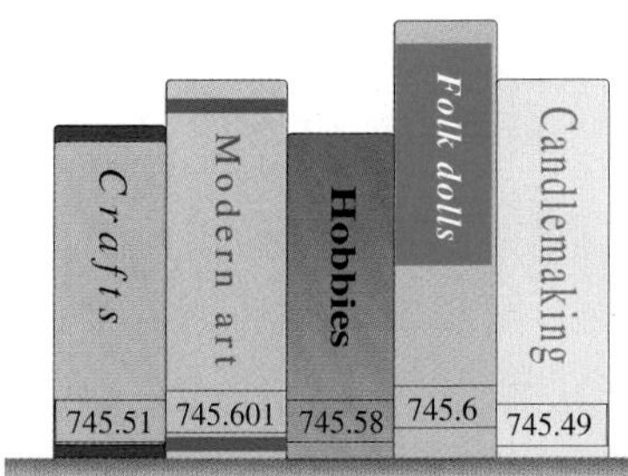

63. THE OLYMPICS The results of the women's all-around gymnastic competition in the 2004 Athens Olympic Games are shown in the following table. Which gymnasts won the gold, silver, and bronze medals?

Name	Country	Score
Nan Zhang	China	38.049
Ana Pavlova	Russia	38.024
Nicoleta Sofronie	Romania	37.948
Carly Patterson	U.S.A.	38.387
Svetlana Khorkina	Russia	38.211
Irina Yarotska	Ukraine	37.687

64. TUNEUPS The six spark plugs from the engine of a Nissan Quest were removed, and the spark plug gap was checked. If vehicle specifications call for the gap to be from 0.031 to 0.035 inch, which of the plugs should be replaced?

Cylinder 1: 0.035 in.
Cylinder 2: 0.029 in.
Cylinder 3: 0.033 in.
Cylinder 4: 0.039 in.
Cylinder 5: 0.031 in.
Cylinder 6: 0.032 in.

Spark plug gap

65. E-COMMERCE The gain (or loss) in value of one share of Amazon.com stock is shown in the graph below for eleven quarters. (For accounting purposes, a year is divided into four quarters.)

a. In what quarter, of what year, was there the greatest gain? Estimate the gain.

b. In what quarter, of what year, was there the greatest loss? Estimate the loss.

Loss Per Share

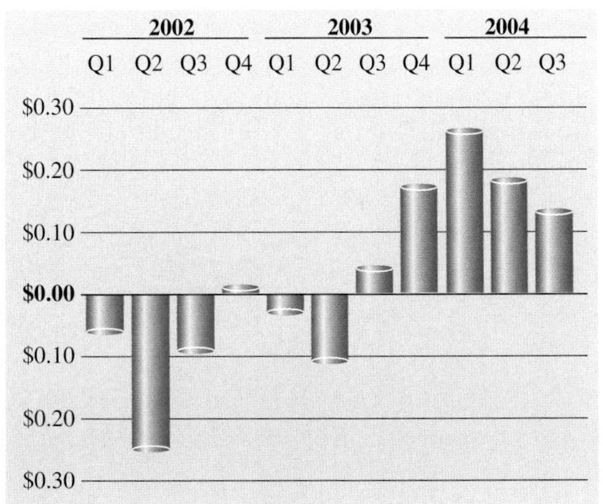

Source: Amazon.com

66. GASOLINE PRICES Refer to the data in the table. Then construct a line graph showing the annual national average retail price per gallon for unleaded regular gasoline for the years 1997 through 2003 (according to *The World Almanac 2005*).

Year	1997	1998	1999	2000	2001	2002	2003
Price (¢)	123.4	105.9	116.5	151.0	146.1	135.8	159.1

WRITING

67. Explain the difference between ten and a tenth.

68. "The more digits a number contains, the larger it is." Is this statement true? Explain your response.

69. How are fractions and decimals related?

70. Explain the benefits of a monetary system that is based on decimals instead of fractions.

71. The illustration shows an unusual notation some service stations use to express the price of a gallon of gasoline. Explain what this notation means. Then express each price using decimal notation.

REGULAR $225\frac{9}{10}$ UNLEADED $235\frac{9}{10}$ UNLEADED + $245\frac{9}{10}$

72. Write a definition for each of these words.

decade *decathlon* *decimal*

REVIEW

73. Add: $75\frac{3}{4} + 88\frac{4}{5}$.

74. Multiply: $\frac{2}{15}\left(-\frac{5}{4}\right)$.

75. Solve: $8y + 8 = 9$.

76. Find the area of a triangle with base 16 in. and height 9 in.

77. Express the fraction $\frac{2}{3}$ as an equivalent fraction with a denominator of 36.

78. Add: $-2 + (-3) + 4$.

79. Subtract: $-15 - (-6)$.

80. Subtract: $\frac{3}{4} - \frac{2}{3}$.

4.2 Adding and Subtracting Decimals

• Adding decimals • Subtracting decimals • Adding and subtracting signed decimals

To add or subtract objects, they must be similar. The federal income tax form shown in Figure 4-4 has a vertical line to ensure that dollars are added to dollars and cents added to cents. In this section, we will show how decimals are added and subtracted using this type of vertical column format.

Form 1040EZ	Department of the Treasury—Internal Revenue Service **Income Tax Return for Single and Joint Filers With No Dependents 2003**			
Income **Attach Form(s) W-2 here.** Enclose but do not attach any payment.	1 Total wages, salaries, and tips. This should be shown in box 1 of your form(s) W-2.	1	21,056	89
	2 Taxable interest. If the total is over $1,500, you cannot use Form 1040EZ.	2	42	06
	3 Unemployment compensation and Alaska Permanent Fund dividends (see page 14).	3	200	00
	4 Add lines 1, 2, and 3. This is your **adjusted gross income.**	4	21,298	95

FIGURE 4-4

Adding decimals

When adding decimals, we line up the columns so that ones are added to ones, tenths are added to tenths, hundredths are added to hundredths, and so on. As an example, consider the following problem.

Line up the columns and the decimal points vertically. Then add the numbers.

$$\begin{array}{r} 12.140 \\ 3.026 \\ 4.000 \\ +\ 0.700 \\ \hline 19.866 \end{array}$$

Write the decimal point in the result directly under the decimal points in the problem.

Adding decimals

1. Line up the decimal points, using the vertical column format.
2. Add the numbers as you would add whole numbers.
3. Write the decimal point in the result directly below the decimal points in the problem.

EXAMPLE 1 Add: 1.903 + 0.6 + 8 + 0.78.

Solution

$$\begin{array}{r} \overset{2}{1}.903 \\ 0.600 \\ 8.000 \\ +\ 0.780 \\ \hline 11.283 \end{array}$$

To make the addition by columns easier, write two zeros after 6, a decimal point and three zeros after 8, and one zero after 0.78.

Carry a 2 (shown in blue) to the ones column.

The result is 11.283.

Self Check 1
Add: 0.07 + 35 + 0.888 + 4.1.

Answer 40.058

CALCULATOR SNAPSHOT Preventing heart attacks

The bar graph in Figure 4-5 shows the number of grams of fiber in a standard serving of each of several foods. It is believed that men can significantly cut their risk of heart attack by eating at least 25 grams of fiber a day. Does this diet meet or exceed the 25-gram requirement?

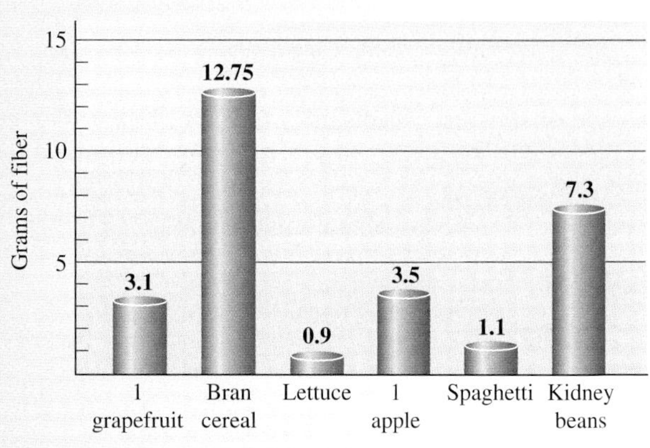

FIGURE 4-5

To find the total fiber intake, we will add the fiber content of each of the foods. We can use a scientific calculator to add the decimals.

3.1 [+] 12.75 [+] .9 [+] 3.5 [+] 1.1 [+] 7.3 [=]   28.65

Since 28.65 > 25, this diet exceeds the daily fiber requirement of 25 grams.

Subtracting decimals

To subtract decimals, we line up the decimal points and corresponding columns so that we subtract like objects—tenths from tenths, hundredths from hundredths, and so on.

Subtracting decimals

1. Line up the decimal points using the vertical column format.
2. Subtract the numbers as you would subtract whole numbers.
3. Write the decimal point in the result directly below the decimal points in the problem.

EXAMPLE 2 Subtract: **a.** 279.6 − 138.7 and **b.** 15.4 − 13.059.

Solution

a.

$$\begin{array}{r} \scriptstyle 8\ 16 \\ 27\not{9}.\not{6} \\ -138.7 \\ \hline 140.9 \end{array}$$

To subtract in the tenths column, borrow 1 one in the form of 10 tenths from the ones column. Add 10 to the 6 in the tenths column, which gives 16 (shown in blue).

b.

$$\begin{array}{r} \scriptstyle 9 \quad \\ \scriptstyle 3\ \not{10}\ 10 \\ 15.\not{4}\ \not{0}\ \not{0} \\ -13.0\ 5\ 9 \\ \hline 2.3\ 4\ 1 \end{array}$$

Add two zeros to the right of 15.4 to make borrowing easier. First, borrow from the tenths column; then borrow from the hundredths column.

Self Check 2
Subtract:
a. 382.5 − 227.1
b. 30.1 − 27.122

Answers **a.** 155.4, **b.** 2.978

EXAMPLE 3 **Conditioning programs.** A 350-pound football player lost 15.7 pounds during the first week of practice. During the second week, he gained 4.9 pounds. Find his weight after the first two weeks of practice.

Solution The word *lost* indicates subtraction. The word *gained* indicates addition.

Beginning weight	minus	first week weight loss	plus	second week weight gain	equals	weight after two weeks of practice.

$$\mathbf{350 - 15.7} + 4.9 = \mathbf{334.3} + 4.9$$ Working from left to right, perform the subtraction first: $350 - 15.7 = 334.3$.

$$= 339.2$$ Perform the addition.

The player's weight is 339.2 pounds after two weeks of practice.

Weather balloons

CALCULATOR SNAPSHOT

A giant weather balloon is made of neoprene, a flexible rubberized substance, that has an uninflated thickness of 0.011 inch. When the balloon is inflated with helium, the thickness becomes 0.0018 inch.

To find the change in thickness, we need to subtract. We can use a scientific calculator to subtract the decimals.

.011 [−] .0018 [=] [0.0092]

After the balloon is inflated, the neoprene loses 0.0092 of an inch in thickness.

Adding and subtracting signed decimals

To add signed decimals, we use the same rules that we used for adding integers.

Adding two decimals

With like signs: Add their absolute values and attach their common sign to the sum.

With unlike signs: Subtract their absolute values (the smaller from the larger) and attach the sign of the number with the larger absolute value to the sum.

EXAMPLE 4 Add: $-6.1 + (-4.7)$.

Solution Since the decimals are both negative, we add their absolute values and attach a negative sign to the result.

$$-6.1 + (-4.7) = -10.8$$ Add the absolute values, 6.1 and 4.7, to get 10.8. Use their common sign.

Self Check 4
Add: $-5.04 + (-2.32)$.

Answer -7.36

Self Check 5
Add: $-21.4 + 16.75$.

Answer -4.65

EXAMPLE 5 Add: $5.35 + (-12.9)$.

Solution In this example, the signs are unlike. Since -12.9 has the larger absolute value, we subtract 5.35 from 12.9 to get 7.55, and attach a negative sign to the result.

$$5.35 + (-12.9) = -7.55$$

Self Check 6
Subtract: $-1.18 - 2.88$

Answer -4.06

EXAMPLE 6 Subtract: $-4.3 - 5.2$.

Solution To subtract signed decimals, we can add the opposite of the decimal that is being subtracted.

$-4.3 - 5.2 = -4.3 + (-5.2)$ Add the opposite of 5.2, which is -5.2.

$= -9.5$ Add the absolute values, 4.3 and 5.2, to get 9.5. Attach a negative sign to the result.

Self Check 7
Subtract: $-2.56 - (-4.4)$.

Answer 1.84

EXAMPLE 7 Subtract: $-8.37 - (-16.2)$.

Solution

$-8.37 - (-16.2) = -8.37 + 16.2$ Add the opposite of -16.2, which is 16.2.

$= 7.83$ Subtract the smaller absolute value from the larger, 8.37 from 16.2, to get 7.83. Since 16.2 has the larger absolute value, the result is positive.

Self Check 8
Evaluate: $-4.9 - (-1.2 + 5.6)$.

Answer -9.3

EXAMPLE 8 Evaluate: $-12.2 - (-14.5 + 3.8)$.

Solution We perform the addition within the grouping symbols first.

$-12.2 - (\mathbf{-14.5 + 3.8}) = -12.2 - (\mathbf{-10.7})$ Perform the addition: $-14.5 + 3.8 = -10.7$.

$= -12.2 + 10.7$ Add the opposite of -10.7.

$= -1.5$ Perform the addition.

Section 4.2 STUDY SET

VOCABULARY *Fill in the blanks.*

1. The answer to an addition problem is called the ______.
2. The answer to a subtraction problem is called the ______.

CONCEPTS

3. To subtract signed decimals, add the ______ of the decimal that is being subtracted.

4. a. Add: 0.3 + 0.17.

b. Write 0.3 and 0.17 as fractions. Find a common denominator for the fractions and add them.

c. Express your answer to part b as a decimal.

d. Compare your answers from part a and part c.

NOTATION

5. Every whole number has an unwritten decimal _______ to its right.

6. In the subtraction problem below, we must borrow. How much is borrowed from the 3, and in what form is it borrowed?

$$\begin{array}{r} \overset{2\ 11}{29.\cancel{3}\,\cancel{1}} \\ -25.1\,6 \\ \hline \end{array}$$

PRACTICE *Perform each operation.*

7. $\begin{array}{r} 32.5 \\ +\ 7.4 \\ \hline \end{array}$

8. $\begin{array}{r} 6.3 \\ +13.5 \\ \hline \end{array}$

9. $\begin{array}{r} 21.6 \\ +33.12 \\ \hline \end{array}$

10. $\begin{array}{r} 19.4 \\ +31.95 \\ \hline \end{array}$

11. 12 + 3.9

12. 0.01 + 3.6

13. 0.03034 + 0.2003

14. 19.9 + 19.9

15. 247.9 + 40 + 0.56

16. 0.0053 + 1.78 + 6

17. 45 + 9.9 + 0.12 + 3.02

18. 505.01 + 23 + 0.989 + 12.07

19. $\begin{array}{r} 12.98 \\ -\ 3.45 \\ \hline \end{array}$

20. $\begin{array}{r} 1.6 \\ -0.16 \\ \hline \end{array}$

21. $\begin{array}{r} 78.1 \\ -\ 7.81 \\ \hline \end{array}$

22. $\begin{array}{r} 202.234 \\ -\ 19.34 \\ \hline \end{array}$

23. 5 − 0.023

24. 30 − 11.98

25. 24 − 23.81

26. 7.001 − 5.9

27. −45.6 + 34.7

28. −19.04 + 2.4

29. 46.09 + (−7.8)

30. 34.7 + (−30.1)

31. −7.8 + (−6.5)

32. −5.78 + (−33.1)

33. −0.0045 + (−0.031)

34. −90.09 + (−0.087)

35. −9.5 − 7.1

36. −7.08 − 14.3

37. 30.03 − (−17.88)

38. 143.3 − (−64.01)

39. −2.002 − (−4.6)

40. −0.005 −(−8)

41. −7 − (−18.01)

42. −63.04 − (−8.911)

Evaluate each expression. Remember to perform the operations within grouping symbols first.

43. (3.4 − 6.6) + 7.3

44. 3.4 − (6.6 + 7.3)

45. (−9.1 − 6.05) − (−51)

46. −9.1 − (−6.05) + 51

47. 16 − (67.2 + 6.27)

48. −43 − (0.032 − 0.045)

49. (−7.2 + 6.3) − (−3.1 − 4)

50. 2.3 + [2.4 − (2.5 −2.6)]

51. |−14.1 + 6.9| + 8

52. 15 − |−2.3 + (−2.4)|

53. Find the sum of *two and forty-three hundredths* and *five and six tenths.*

54. Find the difference of *nineteen hundredths* and *six thousandths.*

APPLICATIONS

55. SPORTS PAGES In the sports pages of any newspaper, decimal numbers are used quite often.

a. "German bobsledders set a world record today with a final run of 53.03, finishing ahead of the Italian team by only fourteen thousandths of a second." What was the time for the Italian bobsled team?

b. "The women's figure skating title was decided by only thirty-three hundredths of a point." If the winner's point total was 102.71, what was the second-place finisher's total?

56. NURSING The following table shows a patient's health chart. A nurse failed to fill in certain portions. (98.6° Fahrenheit is considered normal.) Complete the table.

Day of week	Patient's temperature	How much above normal
Monday	99.7°	
Tuesday		2.5°
Wednesday	98.6°	
Thursday	100.0°	
Friday		0.9

57. VEHICLE SPECIFICATIONS Certain dimensions of a compact car are shown. Find the wheelbase of the car.

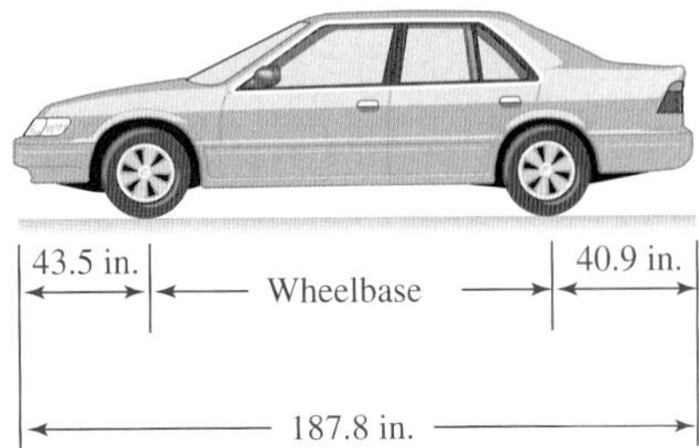

58. pH SCALE The pH scale shown below is used to measure the strength of acids and bases in chemistry. Find the difference in pH readings between

a. bleach and stomach acid.

b. ammonia and coffee.

c. blood and coffee.

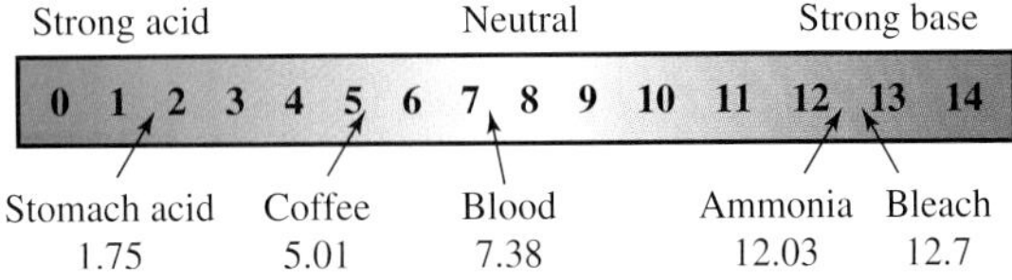

59. BAROMETRIC PRESSURE Barometric pressure readings are recorded on the following weather map. In a low-pressure area (L on the map), the weather is often stormy. The weather is usually fair in a high-pressure area (H). What is the difference in readings between the areas of highest and lowest pressure? In what part of the country would you expect the weather to be fair?

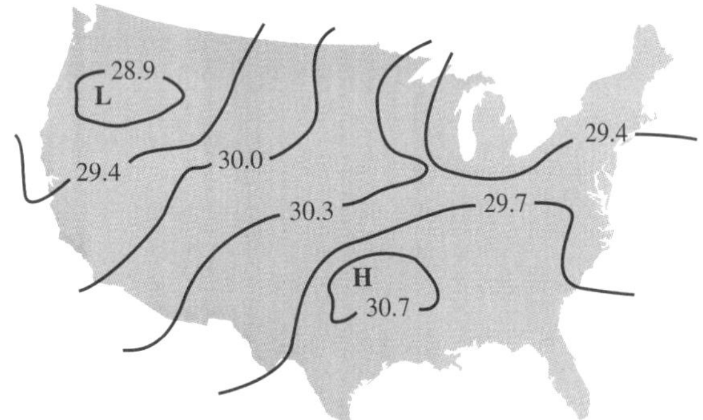

60. QUALITY CONTROL An electronics company has strict specifications for silicon chips used in a computer. The company will install only chips that are within 0.05 centimeter of the specified thickness. The table gives that specification for two types of chip. Fill in the blanks to complete the chart.

Chip type	Thickness specification	Acceptable range	
		Low	High
A	0.78 cm		
B	0.643 cm		

61. OFFSHORE DRILLING A company needs to construct a pipeline from an offshore oil well to a refinery located on the coast. Company engineers have come up with two plans for consideration, as shown. Use the information in the illustration to complete the table.

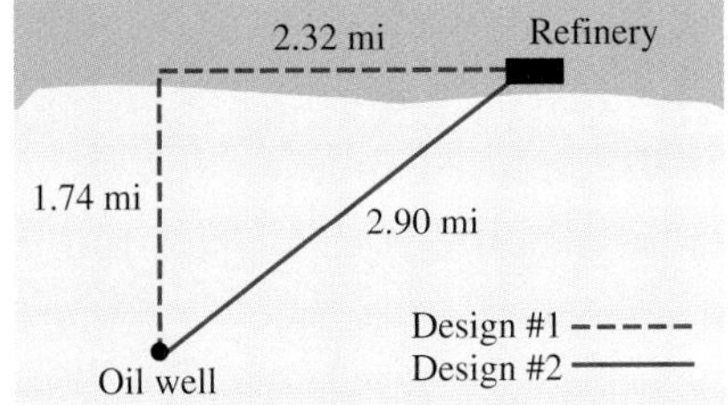

	Pipe underwater (mi)	Pipe underground (mi)	Total pipe (mi)
Design 1			
Design 2			

62. TELEVISION The following illustration shows the six most-watched television shows of all time (excluding Super Bowl games).

a. What was the combined total audience of all six shows?

b. How many more people watched the last episode of "MASH" than watched the last episode of "Seinfeld"?

c. How many more people would have had to watch the last "Seinfeld" to move it into a tie for fifth place?

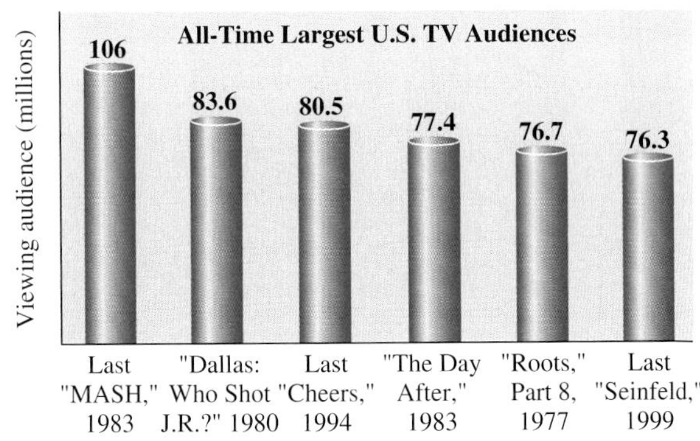

Source: Nielson Media Research

63. RECORDHOLDERS The late Florence Griffith-Joyner of the United States still holds the world record in the 100-meter sprint: 10.49 seconds. Jodie Henry of Australia holds the world record in the 100-meter freestyle swim. Jodie's time is 53.52. How much faster did Griffith-Joyner run the 100 meters than Henry swam it?

64. FLIGHT PATH See the illustration. Find the added distance a plane must travel to avoid flying through the storm.

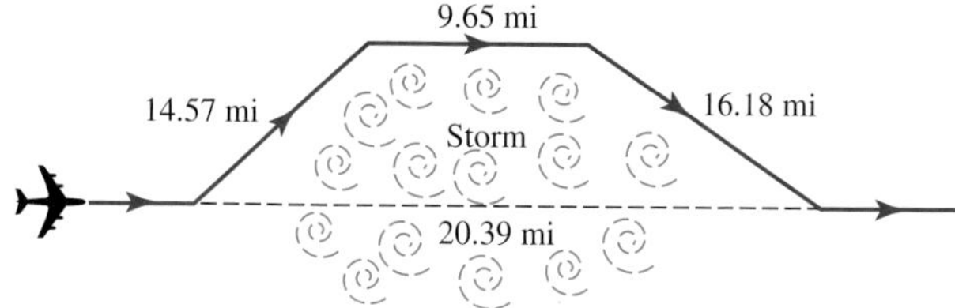

65. DEPOSIT SLIPS A deposit slip for a savings account is shown below. Find the subtotal and the total deposit.

Deposit

Cash	242	50
Checks (properly endorsed)	116	10
	47	93
Total from reverse side	359	16
Subtotal		
Less cash	25	00
Total deposit		

66. MOTION Forces such as water current or wind can increase or decrease the speed of an object in motion. Find the speed of each object.

a. An airplane's speed in still air is 450 mph, and it has a tail wind of 35.5 mph helping it along.

b. A man can paddle a canoe at 5 mph in still water, but he is going upstream. The speed of the current against him is 1.5 mph.

67. THE HOME SHOPPING NETWORK The illustration shows a description of a cookware set that was sold on television.

a. Find the difference between the manufacturer's suggested retail price (MSRP) and the sale price.

b. Including shipping and handling (S & H), how much will the cookware set cost?

Item 229-442	
Continental 9-piece Cookware Set	
Stainless steel	
MSRP	$149.79
HSN Price	$59.85
On Sale	**$47.85**
S & H	$7.95

68. RETAILING Complete the table by filling in the retail price of each appliance, given its cost to the dealer and the store markup.

Item	Cost	Markup	Retail price
Refrigerator	$510.80	$105.00	
Washing machine	$189.50	$55.50	
Dryer	$163.99	$60.75	

Evaluate each expression.

69. 2,367.909 + 5,789.0253

70. 0.00786 + 0.3423

71. 9,000.09 − 7,067.445

72. 1 − 0.004999

73. 3,434.768 − (908 − 2.3 + .0098)

74. 12 − (0.723 + 3.05611)

WRITING

75. Explain why we line up the decimal points and corresponding columns when adding decimals.

76. Explain why we can write additional zeros to the right of a decimal such as 7.89 without affecting its value.

77. Explain what is wrong with the work shown below.

$$\begin{array}{r} 203.56 \\ 37 \\ +\ \ 0.43 \\ \hline 204.36 \end{array}$$

78. Consider the addition

$$\begin{array}{r} \overset{2}{2}3.7 \\ 41.9 \\ +12.8 \\ \hline 78.4 \end{array}$$

Explain the meaning of the small 2 written above the ones column.

REVIEW

79. Add: $44\frac{3}{8} + 66\frac{1}{5}$.

80. Simplify: $\dfrac{-\frac{3}{4}}{\frac{5}{16}}$.

81. Multiply: $\dfrac{-15}{26} \cdot 1\dfrac{4}{9}$.

82. Evaluate: $2 + 5[-2 - (6 + 1)]$.

4.3 Multiplying Decimals

- Multiplying decimals • Multiplying decimals by powers of 10
- Multiplying signed decimals • Order of operations

We now focus on the operation of multiplication. First, we develop a method used to multiply decimals. Then we use that method to evaluate expressions and to solve problems involving decimals.

Multiplying decimals

To show how to multiply decimals, we examine the multiplication $0.3 \cdot 0.17$, finding the product in a roundabout way. First, we will write 0.3 and 0.17 as fractions and multiply them. Then we will express the resulting fraction as a decimal.

$$0.3(0.17) = \frac{3}{10} \cdot \frac{17}{100} \quad \text{Express 0.3 and 0.17 as fractions.}$$

$$= \frac{3 \cdot 17}{10 \cdot 100} \quad \text{Multiply the numerators and multiply the denominators.}$$

$$= \frac{51}{1{,}000} \quad \text{Multiply in the numerator and denominator.}$$

$$= 0.051 \quad \text{Write } \tfrac{51}{1{,}000} \text{ as a decimal.}$$

From this example, we can make observations about multiplying decimals.

- The digits in the answer are found by multiplying 3 and 17.

  $0.3 \cdot 0.17 = 0.051$

  $3 \cdot 17 = 51$

- The answer has 3 decimal places. The *sum* of the number of decimal places in the factors 0.3 and 0.17 is also 3.

  $0.3 \cdot 0.17 = 0.051$

  1 decimal place 2 decimal places 3 decimal places

  These observations suggest the following rule for multiplying decimals.

Multiplying decimals

To multiply two decimals:

1. Multiply the decimals as if they were whole numbers.
2. Find the total number of decimal places in both factors.
3. Place the decimal point in the result so that the answer has the same number of decimal places as the total found in step 2.

Self Check 1

Mutiply: $2.74 \cdot 4.3$.

EXAMPLE 1 Multiply: $5.9 \cdot 3.4$.

Solution We temporarily ignore the decimal points and multiply the decimals as if they were whole numbers. Initially, we think of this problem as 59 times 34.

$$\begin{array}{r} 59 \\ \times\ 34 \\ \hline 236 \\ 177 \\ \hline 2006 \end{array}$$

To place the decimal point in the product, we find the total number of digits to the right of the decimal points in the factors.

$$\begin{array}{r} 5.9 \\ \times\ 3.4 \\ \hline 236 \\ 177 \\ \hline 20.06 \end{array} \quad \begin{array}{l} \leftarrow 1 \text{ decimal place} \\ \leftarrow 1 \text{ decimal place} \end{array} \Big\} \text{ The answer will have } 1 + 1 = 2 \text{ decimal places.}$$

Locate the decimal point so that the answer has 2 decimal places.

Answer 11.782

When we multiply decimals, it is not necessary to line up the decimal points, as the next example illustrates.

EXAMPLE 2 Multiply: 1.3(0.005).

Self Check 2
Multiply: (0.0002)7.2.

Solution We multiply 13 by 5.

$$\begin{array}{r} 1.3 \\ \times 0.005 \\ \hline 65 \end{array} \quad \begin{array}{l} \leftarrow 1 \text{ decimal place} \\ \leftarrow 3 \text{ decimal places} \end{array} \Big\} \text{ The answer will have } 1 + 3 = 4 \text{ decimal places.}$$

We then place the decimal point in the result.

$$\begin{array}{r} 1.3 \\ \times\ 0.005 \\ \hline 0.\mathbf{00}65 \end{array}$$

Add 2 placeholder zeros and position the decimal point so that the product has 4 decimal places.

Answer 0.00144

Heating costs

CALCULATOR SNAPSHOT

When billing a household, a gas company converts the amount of natural gas used into units of heat energy called *therms*. The number of therms used by a household in one month and the cost per therm are shown below.

Customer charge . . . . 39 therms @ $0.72264

To find the total charges for the month, we multiply the number of therms by the cost per therm: 39 · 0.72264.

39 [×] .72264 [=] 28.18296

Rounding to the nearest cent, we see that the total charge is $28.18.

Self Check 3
Multiply: 178(2.7).

Answer 480.6

EXAMPLE 3 Multiply: 234(3.1).

Solution

```
   234   ← No decimal places  }
×  3.1   ← 1 decimal place    }  The answer will have 0 + 1 = 1 decimal place.
  23 4
 702
 725.4   Locate the decimal point so that the answer has 1 decimal place.
```

Multiplying decimals by powers of 10

The numbers 10, 100, and 1,000 are called *powers of 10,* because they are the results when we evaluate 10^1, 10^2, and 10^3, respectively. To develop a rule to determine the product when multiplying a decimal and a power of 10, we will multiply 8.675 by three different powers of 10.

Multiply: 8.675 · **10**	Multiply: 8.675 · **100**	Multiply: 8.675 · **1,000**
8.675 × 10 0000 8675 86.750	8.675 × 100 0000 0000 8675 867.500	8.675 × 1000 0000 0000 0000 8675 8675.000
The answer is 86.75.	The answer is 867.5	The answer is 8,675.

We can make observations about the results.

- In each case, the answer contains the same digits as the factor 8.675.
- When we inspect the answers, the decimal point in the first factor 8.675 appears to be moved to the right by the multiplication process. The number of decimal places it moves depends on the power of 10 by which 8.675 is multiplied.

One zero in 10	Two zeros in 100	Three zeros in 1,000
8.675 · 10 = 86.75	8.675 · 100 = 867.5	8.675 · 1,000 = 8675.
It moves one place to the right.	It moves two places to the right.	It moves three places to the right.

These observations suggest the following rule.

Multiplying a decimal by a power of 10

To multiply a decimal by a power of 10, move the decimal point to the right the same number of places as there are zeros in the power of 10.

Self Check 4
Find each product:
a. 0.721 · 100
b. 6.08(1,000)

EXAMPLE 4 Find each product: **a.** 2.81 · 10 and **b.** 0.076 · 10,000.

Solution

a. 2.81 · 10 = 28.1 Since 10 has 1 zero, move the decimal point 1 place to the right.

b. $0.076 \cdot 10{,}000 = 0760.$ Since 10,000 has 4 zeros, move the decimal point 4 places to the right. Write a placeholder zero (shown in blue).

$= 760$

Answers **a.** 72.1, **b.** 6,080

EXAMPLE 5 **Tachometers.** A tachometer indicates the engine speed of a vehicle, in revolutions per minute (rpm). What engine speed is indicated by the tachometer in Figure 4-6?

FIGURE 4-6

Solution The needle is pointing to 4.5. The notation "RPM × 1000" on the tachometer instructs us to multiply 4.5 by 1,000 to find the engine speed.

$4.5 \cdot 1{,}000 = 4500.$ Since 1,000 has 3 zeros, move the decimal point 3 places to the right. Write two placeholder zeros.

$= 4{,}500$

The engine speed is 4,500 rpm.

Overtime

THINK IT THROUGH

"Employees covered by the Fair Labor Standards Act must receive overtime pay for hours worked in excess of 40 in a workweek of at least 1.5 times their regular rates of pay." U.S. Department of Labor

The map of the United States shown below is divided into nine regions. The average hourly wage for private industry workers in each region is also listed in the legend below the map. Determine the average hourly wage for the region where you live. Then calculate the corresponding average hourly overtime wage for that region.

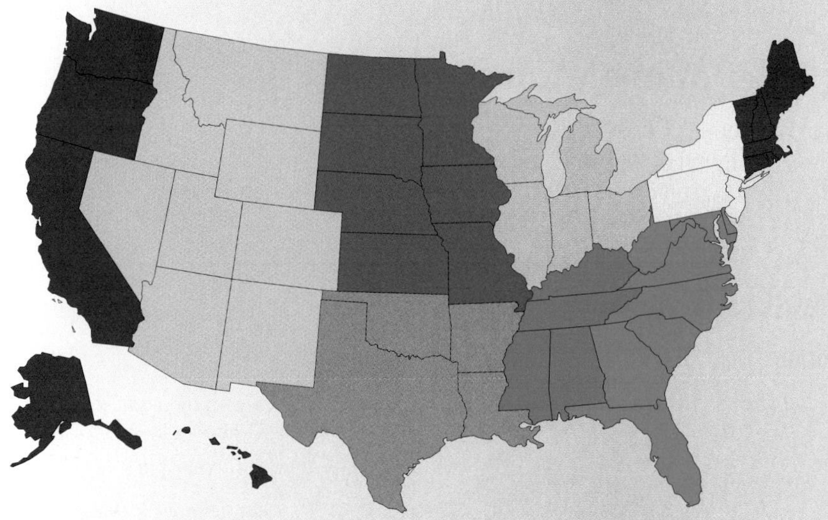

West North Central: \$16.30	East South Central: \$13.97	New England: \$20.10
Mountain: \$15.65	East North Central: \$17.16	Middle Atlantic: \$19.08
Pacific: \$19.11	West South Central: \$15.22	South Atlantic: \$15.88

Multiplying signed decimals

Recall that the product of two numbers with like signs is positive, and the product of two numbers with unlike signs is negative.

Self Check 6
Multiply:
a. $6.6(-5.5)$
b. $(-44.968)(-100)$

Answers **a.** -36.3, **b.** $4{,}496.8$

EXAMPLE 6 Multiply: **a.** $-1.8(4.5)$ and **b.** $(-1{,}000)(-59.08)$.

Solution

a. Since the decimals have unlike signs, their product is negative.

$-1.8(4.5) = -8.1$ Multiply the absolute values, 1.8 and 4.5, to get 8.1. Make the result negative.

b. Since the decimals have like signs, their product is positive.

$(-1{,}000)(-59.08) = 59{,}080$ Multiply the absolute values, 1,000 and 59.08. Since 1,000 has 3 zeros, move the decimal point 3 places to the right. Write a placeholder zero.

Self Check 7
Evaluate:
a. $(-1.3)^2$
b. $(0.09)^2$

Answers **a.** 1.69, **b.** 0.0081

EXAMPLE 7 Evaluate: **a.** $(2.4)^2$ and **b.** $(-0.05)^2$.

Solution

a. $(2.4)^2 = (2.4)(2.4)$ Write 2.4 as a factor 2 times.

$= 5.76$ Perform the multiplication.

b. $(-0.05)^2 = (-0.05)(-0.05)$ Write -0.05 as a factor 2 times.

$= 0.0025$ Perform the multiplication. The product of two decimals with like signs is positive.

Order of operations

In the remaining examples, we use the rules for the order of operations to evaluate expressions involving decimals.

Self Check 8
Evaluate:
$-2|-4.4 + 5.6| + (-0.8)^2$.

Answer -1.76

EXAMPLE 8 Evaluate: $-(0.6)^2 + 5|-3.6 + 1.9|$.

Solution

$-(0.6)^2 + 5|-3.6 + 1.9| = -(0.6)^2 + 5|-1.7|$ Perform the addition within the absolute value symbols.

$= -(0.6)^2 + 5(1.7)$ Simplify: $|-1.7| = 1.7$.

$= -0.36 + 5(1.7)$ Find the power: $(0.6)^2 = 0.36$.

$= -0.36 + 8.5$ Perform the multiplication: $5(1.7) = 8.5$.

$= 8.14$ Perform the addition.

EXAMPLE 9 **Weekly earnings.** A cashier's workweek is 40 hours. After his daily shift is over, he can work overtime at a rate 1.5 times his regular rate of $7.50 per hour. How much money will he earn in a week if he works 6 hours of overtime?

Solution First, we need to find his overtime rate, which is 1.5 times his regular rate of $7.50 per hour.

$1.5(7.50) = 11.25$

His overtime rate is $11.25 per hour.

To find his total weekly earnings, we use the following fact.

The regular rate	times	40 hours	plus	the overtime rate	times	overtime hours worked	equals	his total earnings.

$7.50(40) + 11.25(6) = 300 + 67.50$ Perform the multiplications.

$= 367.50$ Perform the addition.

The cashier's earnings for the week are $367.50.

Section 4.3 STUDY SET

VOCABULARY *Fill in the blanks.*

1. In the multiplication problem $2.89 \cdot 15.7$, the numbers 2.89 and 15.7 are called ________. The answer, 45.373, is called the ________.
2. Numbers such as 10, 100, and 1,000 are called ________ of 10.

CONCEPTS *Fill in the blanks.*

3. To multiply decimals, multiply them as if they were ________ numbers. The number of decimal places in the product is the same as the ______ of the decimal places in the factors.
4. To multiply a decimal by a power of 10, move the decimal point to the ______ the same number of decimal places as the number of ______ in the power of 10.
5. **a.** Multiply $\frac{3}{10}$ and $\frac{7}{100}$.

   b. Now write both fractions from part a as decimals. Multiply them in that form. Compare your results from parts a and b.
6. **a.** Multiply 0.11 and 0.3.

   b. Now write both decimals in part a as fractions. Multiply them in that form. Compare your results from parts a and b.

NOTATION

7. Suppose that the result of multiplying two decimals is 2.300. Write this result in simpler form.
8. When we move the decimal point to the right, does the decimal number get larger or smaller?

PRACTICE *Perform each multiplication.*

9. $(0.4)(0.2)$
10. $(0.2)(0.3)$
11. $(-0.5)(0.3)$
12. $(0.6)(-0.7)$
13. $(1.4)(0.7)$
14. $(2.1)(0.4)$
15. $(0.08)(0.9)$
16. $(0.003)(0.9)$
17. $(-5.6)(-2.2)$
18. $(-7.1)(-4.1)$
19. $(-4.9)(0.001)$
20. $(0.001)(-7.09)$
21. $(-0.35)(0.24)$
22. $(-0.85)(0.42)$
23. $(-2.13)(4.05)$
24. $(3.06)(-1.82)$
25. $16 \cdot 0.6$
26. $24 \cdot 0.8$
27. $-7(8.1)$
28. $-5(4.7)$
29. $0.04(306)$
30. $0.02(417)$
31. $60.61(-0.3)$
32. $-70.07 \cdot 0.6$
33. $-0.2(0.3)(-0.4)$
34. $-0.1(-2.2)(0.5)$
35. $5.5(10)(-0.3)$
36. $6.2(100)(-0.8)$
37. $4.2 \cdot 10$
38. $10 \cdot 7.1$

39. $67.164 \cdot 100$
40. $708.199 \cdot 100$
41. $-0.056(10)$
42. $-100(0.0897)$
43. $1,000(8.05)$
44. $23.7(1,000)$
45. $0.098(10,000)$
46. $3.63(10,000)$
47. $-0.2 \cdot 1,000$
48. $-1,000 \cdot 1.9$

Complete each table.

49.

Decimal	Its square
0.1	
0.2	
0.3	
0.4	
0.5	
0.6	
0.7	
0.8	
0.9	

50.

Decimal	Its cube
0.1	
0.2	
0.3	
0.4	
0.5	
0.6	
0.7	
0.8	
0.9	

Find each power.

51. $(1.2)^2$
52. $(2.3)^2$
53. $(-1.3)^2$
54. $(-2.5)^2$

Evaluate each expression.

55. $-4.6(23.4 - 19.6)$
56. $6.9(9.8 - 8.9)$
57. $(-0.2)^2 + 2(7.1)$
58. $(-6.3)(3) - (1.2)^2$
59. $(-0.7 - 0.5)(2.4 - 3.1)$
60. $(-8.1 - 7.8)(0.3 + 0.7)$
61. $(0.5 + 0.6)^2(-3.2)$
62. $(-5.1)(4.9 - 3.4)^2$
63. $|-2.6| \cdot |-7.2|$
64. $4|-3.1| + 5|-5.5|$
65. $|-2.6 - 6.7|^2$
66. $-3|-8.16 + 9.9|$

APPLICATIONS

67. CONCERT SEATING Two types of tickets were sold for a concert. Floor seating cost \$12.50 a ticket, and balcony seats were \$15.75.

a. Complete the table in the next column and find the receipts from each type of ticket.

b. Find the total receipts from the sale of both types of tickets.

Ticket type	Price	Number sold	Receipts
Floor		1,000	
Balcony		100	

68. CITY PLANNING In the city map shown below, the streets form a grid. The lines are 0.35 mile apart. Find the distance of each trip.

a. The airport to the Convention Center

b. City Hall to the Convention Center

c. The airport to City Hall

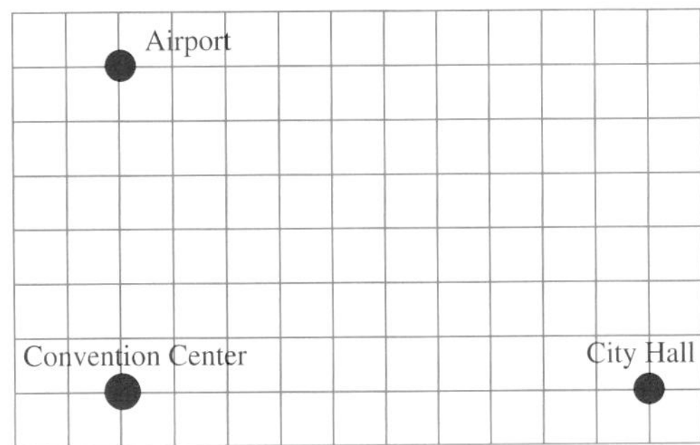

69. STORM DAMAGE After a rainstorm, the saturated ground under a hilltop house began to give way. A survey team noted that the house dropped 0.57 inch initially. In the next two weeks, the house fell 0.09 inch per week. How far did the house fall during this three-week period?

70. WATER USAGE In May, the water level of a reservoir reached its high mark for the year. During the summer months, as water usage increased, the level dropped. In the months of May and June, it fell 4.3 feet each month. In August, because of high temperatures, it fell another 8.7 feet. By September, how far below the year's high mark had the water level fallen?

71. WEIGHTLIFTING The barbell is evenly loaded with iron plates. How much plate weight is loaded on the barbell?

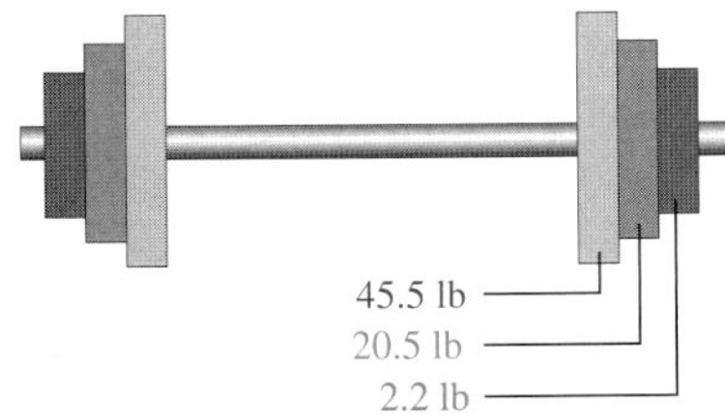

72. PLUMBING BILLS In the following illustration, an invoice for plumbing work is torn. What is the charge for the 4 hours of work? What is the total charge?

Carter Plumbing 100 W. Dalton Ave.		Invoice #210
Standard service charge 4 hr @ $40.55/hr		$25.75
Total		

73. BAKERY SUPPLIES A bakery buys various types of nuts as ingredients for cookies. Complete the table by filling in the cost of each purchase.

Type of nut	Price per pound	Pounds	Cost
Almonds	$5.95	16	
Walnuts	$4.95	25	
Peanuts	$3.85	20	

74. RETROFITS The illustration shows the width of the three columns of an existing freeway overpass. A computer analysis indicates that each column needs to be increased in width by a factor of 1.4 to ensure stability during an earthquake. According to the analysis, how wide should each of the columns be?

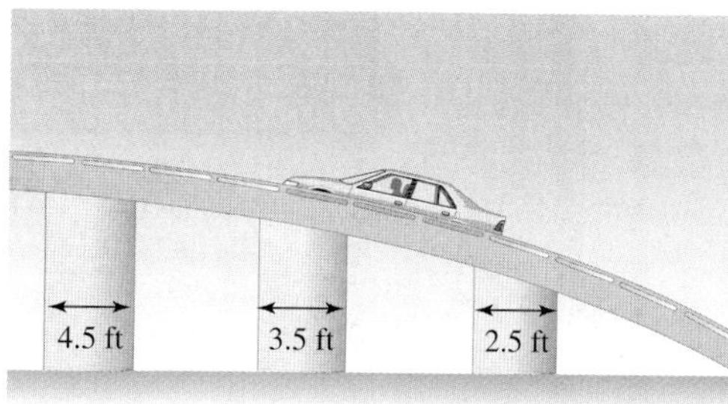

75. POOL CONSTRUCTION Long bricks, called *coping,* can be used to outline the edge of a swimming pool. How many meters of coping will be needed in the construction of the swimming pool shown?

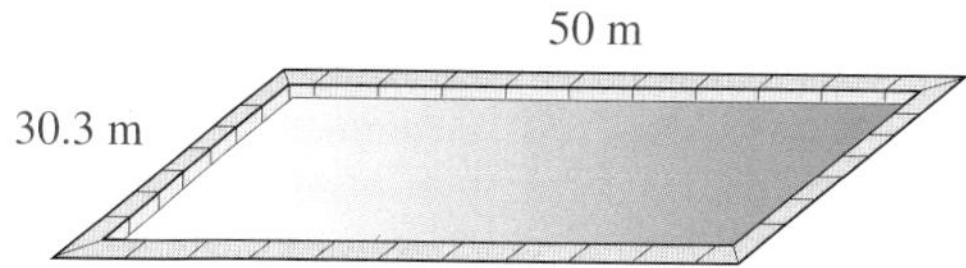

76. SOCCER A soccer goal measures 24 feet wide by 8 feet high. Major league soccer officials are proposing to increase its width by 1.5 feet and increase its height by 0.75 foot.

a. What is the area of the goal opening now?

b. What would it be if their proposal is adopted?

c. How much area would be added?

77. BIOLOGY DNA is found in cells. It is referred to as the genetic "blueprint." In humans, it determines such traits as eye color, hair color, and height. A model of DNA appears on the right. If Å = 0.000000004 inch, determine the three dimensions shown in the illustration.

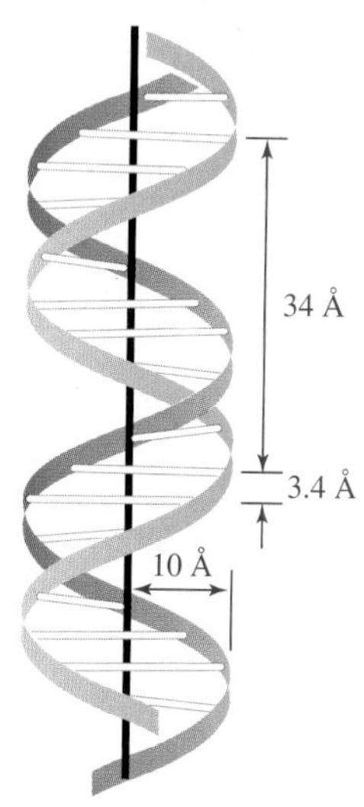

78. TACHOMETERS See the illustration.

a. To what number is the tachometer needle pointing? Give your estimate in decimal form.

b. What engine speed (in rpm) does the tachometer indicate?

Use a calculator to answer each problem.

79. $(-9.0089 + 10.0087)(15.3)$

80. $(-4.32)^3 - 78.969$

81. $(18.18 + 6.61)^2 + (5 - 9.09)^2$

82. $304 - 3.780876(100)$

83. ELECTRIC BILLS When billing a household, a utility company charges for the number of kilowatt-hours used. A kilowatt-hour (kwh) is a standard measure of electricity. If the cost of 1 kwh is $0.14277, what is the electric bill for a household using 719 kwh in a month? Round the answer to the nearest cent.

84. UTILITY TAXES Some gas companies are required to tax the number of therms used each month by the customer. What are the taxes collected on a monthly usage of 31 therms if the tax rate is $0.00566 per therm? Round the answer to the nearest cent.

WRITING

85. Explain how to determine where to place the decimal point in the answer when multiplying two decimals.

86. List the similarities and differences between whole-number multiplication and decimal multiplication.

87. What is a decimal place?

88. What is the purpose of the rules for the order of operations?

REVIEW

89. Solve: $7x + 5 = 54$.

90. Multiply: $3\frac{1}{3}\left(-1\frac{4}{5}\right)$.

91. Write this notation in words: $|-3|$.

92. What is the LCD of fractions with denominators of 4, 5, and 6?

93. Simplify: $-\frac{8}{8}$.

94. Find one-half of 7 and square the result.

4.4 Dividing Decimals

- Dividing a decimal by a whole number
- Divisors that are decimals
- Rounding when dividing
- Dividing decimals by powers of 10
- Order of operations

In Chapter 1, we used a process called long division to divide whole numbers.

Long division form

$$\begin{array}{rrl} & 2 & \leftarrow \text{Quotient} \\ \text{Divisor} \rightarrow & 5\overline{)10} & \leftarrow \text{Dividend} \\ & \underline{10} & \\ & 0 & \leftarrow \text{Remainder} \end{array}$$

In this section, we will consider division problems in which the divisor, the dividend, or both are decimals.

Dividing a decimal by a whole number

To use long division to divide 47 by 10, we proceed as follows.

$$\begin{array}{r} 4\frac{7}{10} \\ 10\overline{)47} \\ \underline{40} \\ 7 \end{array}$$

Here the result is written in quotient + $\frac{\text{remainder}}{\text{divisor}}$ form.

To perform this same division using decimals, we write 47 as 47.0 and divide as we would divide whole numbers.

$$\begin{array}{r} 4.7 \\ 10\overline{)47.0} \\ \underline{40\downarrow} \\ 7\,0 \\ \underline{7\,0} \\ 0 \end{array}$$

Note that the decimal point in the result is placed directly above the decimal point in the dividend.

After subtracting 40 from 47, bring down the 0.

Since $4\frac{7}{10} = 4.7$, either method gives the same result. The second part of this discussion suggests the following method for dividing a decimal by a whole number.

Dividing a decimal by a whole number

1. Write the problem in long division form.
2. Divide as if working with whole numbers.
3. Write the decimal point in the result directly above the decimal point in the dividend. If necessary, additional zeros can be written to the right of the dividend to allow the division to proceed.

EXAMPLE 1 Divide: 71.68 ÷ 28.

Solution

```
      .
28)71.68
```

Since we are dividing by a whole number, write the decimal point in the answer directly above the decimal point in the dividend.

```
    2.56
28)71.68
   56 ↓|
   15 6
   14 0↓
    1 68
    1 68
       0
```

Divide as if working with whole numbers.

The remainder is 0.

The answer is 2.56.

We can check this result by multiplying the divisor by the quotient; their product should equal the dividend. Since $2.56 \cdot 28 = 71.68$, the result is correct.

Self Check 1

Divide: 101.44 ÷ 32.

Answer 3.17

EXAMPLE 2 Divide: 19.2 ÷ 5.

Solution

```
   3.8
5)19.2
  15↓
   4 2
   4 0
     2
```

All the digits in the dividend have been used, but the remainder is not 0.

We can write a zero to the right of 2 in the dividend and continue the division process. Recall that writing additional zeros to the right of the decimal point does not change the value of the decimal.

```
   3.84
5)19.20
  15  |
   4 2|
   4 0↓
     20
     20
      0
```

Write a zero to the right of the 2 and bring it down.

Continue to divide.

The remainder is 0.

The answer is 3.84.

Check: $3.84 \cdot 5 = 19.2$.

Self Check 2

Divide: 3.4 ÷ 4.

Answer 0.85

Divisors that are decimals

When the divisor is a decimal, we change it to a whole number and proceed as in division of whole numbers. To illustrate, we consider the problem $0.36\overline{)0.2592}$, where the divisor is a decimal. First, we express the division in another form.

$0.36\overline{)0.2592}$ can be represented by $\dfrac{0.2592}{0.36}$

To write the divisor, 0.36, as a whole number, its decimal point needs to be moved 2 places to the right. This can be accomplished by multiplying it by 100. However, if the denominator of the fraction is multiplied by 100, the numerator must also be multiplied by 100 so that the fraction maintains the same value.

$\dfrac{0.2592}{0.36} = \dfrac{0.2592 \cdot \mathbf{100}}{0.36 \cdot \mathbf{100}}$ Multiply numerator and denominator by 100.

$= \dfrac{25.92}{36}$ Multiplying by 100 moves both decimal points 2 places to the right.

This fraction represents the division problem $36\overline{)25.92}$. From this result, we can make the following observations.

- The division problem $0.36\overline{)0.2592}$ is equivalent to $36\overline{)25.92}$. That is, they have the same answer.
- The decimal points in *both* the divisor and the dividend of the first division problem have been moved 2 decimal places to the right to create the second division problem.

$0.36\overline{)0.2592}$ becomes $36\overline{)25.92}$

These observations suggest the following rule for division with decimals.

Dividing with a decimal divisor

1. Move the decimal point of the divisor so that it becomes a whole number.
2. Move the decimal point of the dividend the same number of places to the right.
3. Divide as if working with whole numbers. Write the decimal point in the answer directly above the decimal point in the dividend.

Self Check 3

Divide: $\dfrac{0.6045}{0.65}$.

EXAMPLE 3 Divide: $\dfrac{0.2592}{0.36}$.

Solution

$0\,36\overline{)0\,25.92}$ Move the decimal point 2 places to the right in the divisor and dividend.

$$\begin{array}{r} 0.72 \\ 36\overline{)25.92} \\ \underline{25\,2} \\ 72 \\ \underline{72} \\ 0 \end{array}$$

Now divide as with whole numbers. Write the decimal point in the answer directly above the decimal point in the dividend.

The result is 0.72.

Check: $0.72 \cdot 36 = 25.92$.

Answer 0.93

Rounding when dividing

In Example 3, the division process stopped after we obtained a zero from the second subtraction. Sometimes when we divide, the subtractions never give a zero remainder, and the division process continues forever. In such cases, we can round the result.

EXAMPLE 4 Divide: $\frac{2.35}{0.7}$. Round to the nearest hundredth.

Solution Using long division form, we have $0.7\overline{)2.35}$.

$0\,7\overline{)2\,3.5}$ To write the divisor as a whole number, move the decimal point 1 place to the right. Do the same for the dividend. Place the decimal point in the answer directly above the decimal point in the dividend.

$7\overline{)23.500}$ To round to the hundredths column, divide to the thousandths column. Write two zeros to the right of the dividend.

$$\begin{array}{r} 3.357 \\ 7\overline{)23.500} \\ \underline{21} \\ 2\,5 \\ \underline{2\,1} \\ 40 \\ \underline{35} \\ 50 \\ \underline{49} \\ 1 \end{array}$$

After dividing to the thousandths column, round to the hundredths column. The rounding digit is 5. The test digit is 7.

To the nearest hundredth, the answer is 3.36.

The nucleus of a cell — CALCULATOR SNAPSHOT

The nucleus of a cell contains vital information about the cell in the form of DNA. A typical animal cell has a nucleus that is only 0.00023622 inch across. How many nuclei would have to be laid end-to-end to extend to a length of 1 inch?

To find how many 0.00023622-inch lengths there are in 1 inch, we must use division: 1 ÷ 0.00023622.

1 [÷] .00023622 [=]   [4233.3418]

It takes approximately 4,233 nuclei laid end-to-end to extend to a length of 1 inch.

Dividing decimals by powers of 10

To develop a set of rules for division by a power of 10, we consider the problem 8.13 ÷ 10.

$$\begin{array}{r} 0.813 \\ 10\overline{)8.130} \\ \underline{0} \\ 8\,1 \\ \underline{8\,0} \\ 13 \\ \underline{10} \\ 30 \\ \underline{30} \\ 0 \end{array}$$

Write a zero to the right of the 3.

We note that the quotient, 0.813, and the dividend, 8.13, are the same except for the location of the decimal points. The quotient can be easily obtained by moving the decimal point of the dividend 1 place to the left. This observation suggests the following rule for dividing a decimal by a power of 10.

Dividing a decimal by a power of 10

To divide a decimal by a positive power of 10, move the decimal point to the left the same number of places as there are zeros in the power of 10.

Self Check 5
Find the quotient:
a. $721.3 \div 100$
b. $\frac{1.07}{1,000}$

Answers **a.** 7.213, **b.** 0.00107

EXAMPLE 5 Find the quotient: **a.** $16.74 \div 10$ and **b.** $8.6 \div 10,000$.

Solution

a. $16.74 \div 10 = 1.674$ Since 10 has 1 zero, move the decimal point 1 place to the left.

b. $8.6 \div 10,000 = .00086$ Since 10,000 has 4 zeros, move the decimal point 4 places to the left. Write 3 placeholder zeros.

$= 0.00086$

Order of operations

In the next example, we will use the rules for the order of operations to evaluate an expression that involves division by a decimal.

Self Check 6
Evaluate: $\frac{2.7756 + 3(-0.63)}{-0.8}$.

Answer -1.107

EXAMPLE 6 Evaluate: $\frac{2(0.351) + 0.5592}{-0.4}$.

Solution

$\frac{2(0.351) + 0.5592}{-0.4} = \frac{0.702 + 0.5592}{-0.4}$ Perform the multiplication first: $2(0.351) = 0.702$.

$= \frac{1.2612}{-0.4}$ Perform the addition: $0.702 + 0.5592 = 1.2612$.

$= -3.153$ Perform the division. The quotient of two numbers with unlike signs is negative.

Self Check 7
Use the data below to find the average yearly rainfall for St. Louis for the years 1995–2004. Rainfall totals are in inches.

1995: 41.68	1996: 43.67
1997: 31.23	1998: 43.62
1999: 34.06	2000: 37.37
2001: 35.29	2002: 40.95
2003: 46.06	2004: 42.27

Answer 39.62 in.

EXAMPLE 7 **Rainfall.** Use the data in the table to find the average seasonal rainfall for Los Angeles for the years 1999–2004.

Seasonal Rainfall, Los Angeles Civic Center

Season (July 1– June 30)	1999–2000	2000–2001	2001–2002	2002–2003	2003–2004
Total inches of rainfall	11.57	17.94	4.42	16.42	9.25

Source: *Los Angeles Almanac*

Solution To find the average seasonal rainfall, we add the rainfall totals and divide by the number of seasons, which is 5.

$$\text{Average} = \frac{11.57 + 17.94 + 4.42 + 16.42 + 9.25}{5}$$

$$= \frac{59.6}{5}$$

$$= 11.92 \quad \text{Perform the division.}$$

The average seasonal rainfall in Los Angeles for the years 1999–2004 was 11.92 in.

GPA

THINK IT THROUGH

"In considering all of the factors that are important to employers as they recruit students in colleges and universities nationwide, college major, grade point average, and work-related experience usually rise to the top of the list."
Mary D. Feduccia, Ph.D., Career Services Director, Louisiana State University

A grade point average (GPA) is a weighted average based on the grades received and the number of units (credit hours) taken. A GPA for one semester (or term) is defined as:

the quotient of the sum of the grade points earned for each class and the sum of the number of units taken. The number of grade points earned for a class is the product of the number of units assigned to the class and the value of the grade received in the class.

1. Use the table of grade values on the top right to compute the GPA for the student whose semester grade report is shown. Round to the nearest hundredth.
2. If you were enrolled in school last semester (or term), list the classes taken, units assigned, and grades received in the same format as shown in the sample grade report below on the bottom right. Then calculate your GPA.

Grade	Value
A	4
B	3
C	2
D	1
F	0

Class	Units	Grade
Geology	4	C
Algebra	5	A
Psychology	3	C
Spanish	2	B

Section 4.4 STUDY SET

VOCABULARY *Fill in the blanks.*

1. In the division $2.5\overline{)4.075} = 1.63$, the decimal 4.075 is called the ________, the decimal 2.5 is the ________, and 1.63 is the ________.
2. In $\frac{33.6}{0.3}$, the fraction ______ indicates division.

CONCEPTS *Fill in the blanks.*

3. To divide by a decimal, move the decimal point of the divisor so that it becomes a ________ number. The decimal point of the dividend is then moved the same number of places to the ________. The decimal point in the quotient is written directly ________ the decimal point in the dividend.
4. To divide a decimal by a power of 10, move the decimal point to the ______ the same number of decimal places as the number of zeros in the power of 10.
5. Is this statement true or false?

   $45 = 45.0 = 45.000$
6. When a decimal is divided by 10, is the answer smaller or larger than the original number?
7. To complete the division $7.8\overline{)14.562}$, the decimal points of the divisor and dividend are moved 1 place to the right. This is equivalent to multiplying the numerator and the denominator of $\frac{14.562}{7.8}$ by what number?
8. **a.** When dividing decimals with like signs, what is the sign of the quotient?

   b. When dividing decimals with unlike signs, what is the sign of the quotient?

9. How can we check the result of this division?

$$\frac{1.917}{0.9} = 2.13$$

10. When rounding a decimal to the hundredths column, what other column must we look at first?

11. A student performed the division

$$4.6\overline{)9.522}$$

and obtained the answer 2.07. Without doing the division, check this result. Is it correct?

12. In the division problem below, explain why we can write the additional zeros (shown in red) after 5.

$$\begin{array}{r} 0.3 \\ 16\overline{)5.5\mathbf{0000}} \end{array}$$

NOTATION

13. Explain what the arrows are illustrating.

$$4\,67\overline{)32\,08.7}$$

14. What is this arrow illustrating?

$$\begin{array}{r} 0.7 \\ 4\overline{)3.100} \\ 2\,8\downarrow \\ 30 \end{array}$$

PRACTICE *Perform each division.*

15. $8\overline{)36}$

16. $4\overline{)10}$

17. $-39 \div 4$

18. $-26 \div 8$

19. $49.6 \div 8$

20. $23.5 \div 5$

21. $9\overline{)288.9}$

22. $6\overline{)337.8}$

23. $(-14.76) \div (-6)$

24. $(-13.41) \div (-9)$

25. $\frac{-55.02}{7}$

26. $\frac{-24.24}{8}$

27. $45\overline{)119.7}$

28. $41\overline{)146.37}$

29. $250.95 \div 35$

30. $241.86 \div 29$

31. $41.6 \div 0.32$

32. $31.8 \div 0.15$

33. $(-199.5) \div (-0.19)$

34. $(-2{,}381.6) \div (-0.26)$

35. $\frac{0.0102}{0.017}$

36. $\frac{0.0092}{0.023}$

37. $\frac{0.0186}{0.031}$

38. $\frac{0.416}{0.52}$

Divide and round each result to the nearest tenth.

39. $3\overline{)16}$

40. $7\overline{)20}$

41. $-5.714 \div 2.4$

42. $-21.21 \div 3.8$

Divide and round each result to the nearest hundredth.

43. $12.243 \div 0.9$

44. $13.441 \div 0.6$

45. $0.04\overline{)0.03164}$

46. $0.08\overline{)0.02201}$

Perform each division mentally.

47. $7.895 \div 100$

48. $23.05 \div 10$

49. $0.064 \div (-100)$

50. $0.0043 \div (-10)$

51. $1000\overline{)34.8}$

52. $100\overline{)678.9}$

53. $\frac{45.04}{10}$

54. $\frac{22.32}{100}$

Evaluate each expression. Round each result to the nearest hundredth.

55. $\frac{-1.2 - 3.4}{3(1.6)}$

56. $\frac{(-1.3)^2 + 6.7}{-0.9}$

57. $\frac{40.7(-5.3)}{0.4 - 0.61}$

58. $\frac{(0.5)^2 - (0.3)^2}{0.005 + 0.1}$

Evaluate each expression. If an answer is not exact, round it to the nearest hundredth.

59. $\frac{5(48.38 - 32)}{9}$

60. $\frac{5(19.94 - 32)}{9}$

61. $\frac{6.7 - 0.3^2 + 1.6}{0.3^3}$

62. $\frac{3.6 - (-1.5)}{0.5(-1.5) - 0.4(3.6)}$

APPLICATIONS

63. BUTCHER SHOPS A meat slicer is designed to trim 0.05-inch-thick pieces from a sausage. If the sausage is 14 inches long, how many slices will result?

64. COMPUTERS A computer can do an arithmetic computation in 0.00003 second. How many of these computations could it do in 60 seconds?

65. HIKING Use the information below to find the time of arrival for the hiker.

66. ELECTRONICS A volume control is shown below. If the distance between the Low and High settings is 21 cm, how far apart are the equally spaced volume settings?

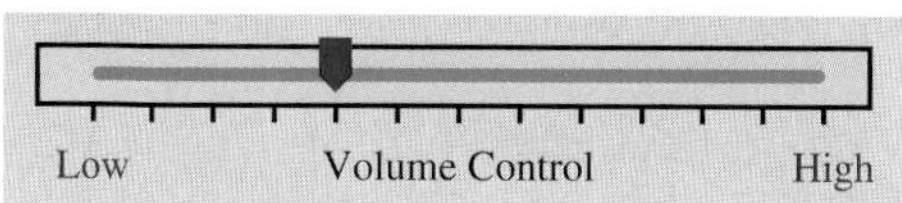

67. SPRAY BOTTLES Production planners have found that each squeeze of the trigger of a spray bottle emits 0.015 ounce of liquid. How many squeezes would there be in an 8.5-ounce bottle?

68. CAR LOANS See the following loan statement. How many more monthly payments must be made to pay off the loan?

American Finance Company	June
Monthly payment:	Paid to date: \$547.30
\$42.10	Loan balance: \$631.50

69. HOURLY PAY The illustration shows the average hours worked and the average weekly earnings of U.S. production workers in 1998 and 2003. What did the average production worker earn per hour in 1998 and in 2003? Round to the nearest cent.

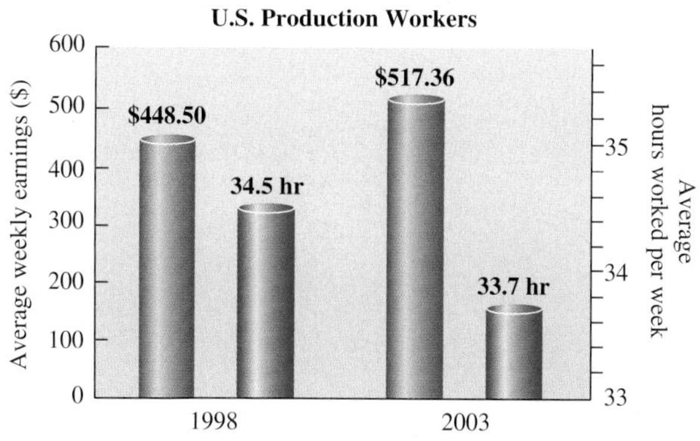

Source: *The World Almanac 2005*

70. TRAVEL The illustration shows the annual number of person-trips of 50 miles or more (one way) for the years 1999–2003, as estimated by the Travel Industry Association of America. Find the average number of trips per year for this period.

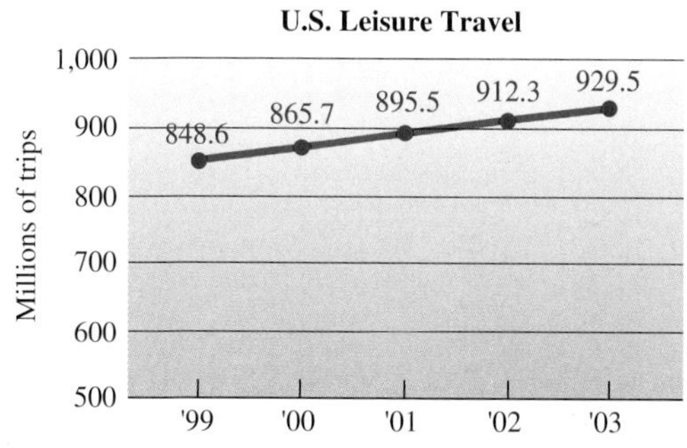

71. OIL WELLS Geologists have mapped out the substances through which engineers must drill to reach an oil deposit. (See the illustration.) What is the average depth that must be drilled each week if this is to be a four-week project?

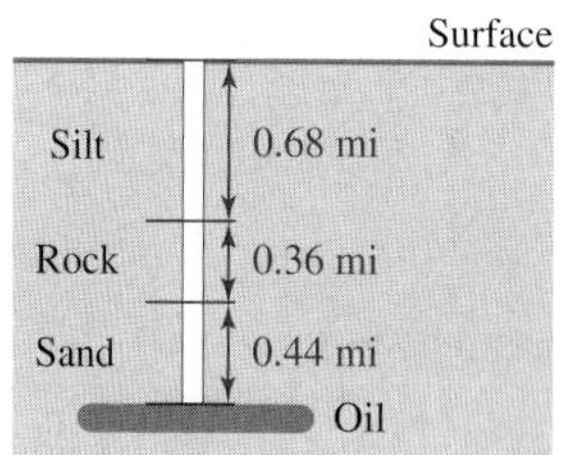

72. INDY 500 The illustration shows the first row of the starting grid for the 2004 Indianapolis 500 automobile race. The drivers' speeds on a qualifying run were used to rank them in this order. What was the average qualifying speed for the drivers in the first row?

B. Rice	D. Wheldon	D. Franchitti
222.024 mph	221.524 mph	221.471 mph

Evaluate each expression. Round to the nearest hundredth.

73. $\dfrac{8.6 + 7.99 + (4.05)^2}{4.56}$

74. $\dfrac{0.33 + (-0.67)(1.3)^3}{0.0019}$

75. $\left(\dfrac{45.9098}{-234.12}\right)^2 - 4$

76. $\left(\dfrac{6.0007}{3.002}\right) - \left(\dfrac{78.8}{12.45}\right)$

WRITING

77. Explain the process used to divide two numbers when both the divisor and the dividend are decimals.

78. Explain why we must sometimes use rounding when writing the answer to a division problem.

79. The division $0.5\overline{)2.005}$ is equivalent to $5\overline{)20.05}$. Explain what *equivalent* means in this case.

80. In $3\overline{)0.7}$, why can additional zeros be placed to the right of 0.7 without affecting the outcome?

REVIEW

81. Simplify the complex fraction: $\dfrac{\frac{7}{8}}{\frac{3}{4}}$.

82. Express the fraction $\frac{3}{4}$ as an equivalent fraction with a denominator of 36.

83. List the set of integers.

84. Round 57,205 to the nearest hundred.

85. Solve: $-\dfrac{3}{4}A = -9$.

86. Evaluate: $\left(\dfrac{1}{2}\right)^3 - \left(\dfrac{1}{2}\right)^2$.

Estimation

In this section, we will use estimation procedures to approximate the answers to addition, subtraction, multiplication, and division problems involving decimals. You will recall that we use rounding when estimating to help simplify the computations so that they can be performed quickly and easily.

Self Check 1

a. Estimate to the nearest ten the sum of 526.93 and 284.03.

b. Estimate using front-end rounding: 512.33 − 36.47.

Answers **a.** 810, **b.** 460

EXAMPLE 1 Estimating sums and differences.

a. Estimate to the nearest ten: 261.76 + 432.94.

b. Estimate using front-end rounding: 381.77 − 57.01.

Solution

a. We round each number to the nearest ten.

$$\begin{array}{ccc} 261.76 & + & 432.94 \\ \downarrow & & \downarrow \\ 260 & + & 430 \end{array} = 690$$

261.76 rounds to 260, and 432.94 rounds to 430.

The estimate is 690. If we compute 261.76 + 432.94, the sum is 694.7. We can see that our estimate is close; it's just 4.7 less than 694.7. Example 1, part a, illustrates the tradeoff when using estimation. The calculations are easier to perform and they take less time, but the answers are not exact.

b. We use front-end rounding.

$$\begin{array}{ccc} 381.77 & - & 57.01 \\ \downarrow & & \downarrow \\ 400 & - & 60 \end{array} = 340$$

Each number is rounded to its largest place value: 381.77 to the nearest hundred and 57.01 to the nearest ten.

The estimate is 340.

Self Check 2

a. Estimate the product:

$42 \cdot 17.65$

b. Estimate the product:

$182 \cdot 24.04$

c. Estimate the product:

$979.3 \cdot 2.3485$

Answers **a.** 800, **b.** 4,000, **c.** 2,348.5

EXAMPLE 2 Estimating products.

Estimate each product: **a.** $6.41 \cdot 27$, **b.** $5.2 \cdot 13.91$, and **c.** $0.124 \cdot 98.6$.

Solution

a. We use front-end rounding.

$$6.41 \cdot 27 \approx 6 \cdot 30$$

The symbol ≈ means "is approximately equal to."

The estimate is 180.

b. We use front-end rounding.

$$5.2 \cdot 13.91 \approx 5 \cdot 10$$

The estimate is 50.

c. Notice that $98.6 \approx 100$.

$$0.124 \cdot 98.6 \approx 0.124 \cdot 100$$

To multiply a decimal by 100, move the decimal point 2 places to the right.

The estimate is 12.4.

When estimating a quotient, we round the divisor and the dividend so that they will divide evenly. Try to round both numbers up or both numbers down.

EXAMPLE 3 **Estimating quotients.** Estimate: $246.03 \div 4.31$.

Solution 4.31 is close to 4. A multiple of 4 close to 246.03 is 240. (Note that both the divisor and dividend were rounded down.)

$246.03 \div 4.31 \approx 240 \div 4$ Perform the division in your head.

The estimate is 60.

Self Check 3
Estimate: $6{,}429.6 \div 7.19$.

Answer 900

STUDY SET *Use the following information about refrigerators to estimate the answers to each question. Remember that answers may vary, depending on the rounding method used.*

Deluxe model	**Standard model**	**Economy model**
Price: $978.88	Price: $739.99	Price: $599.95
Capacity: 25.2 cubic feet	Capacity: 20.6 cubic feet	Capacity: 18.8 cubic feet
Energy cost: $6.79 a month	Energy cost: $5.61 a month	Energy cost: $4.39 a month

1. How much more expensive is the deluxe model than the standard model?
2. A couple wants to buy two standard models, one for themselves and one for their newly married son and daughter-in-law. What is the total cost?
3. How much less storage capacity does the economy model have than the standard model?
4. The owner of a duplex apartment wants to purchase a standard model for one unit and an economy model for the other. What will be the total cost?
5. A stadium manager has a budget of $20,000 to furnish the luxury boxes at a football stadium with refrigerators. How many standard models can she purchase for this amount?
6. How many more cubic feet of storage do you get with the deluxe model compared to the economy model?
7. Three roommates are planning on purchasing the deluxe model and splitting the cost evenly. How much will each have to pay?
8. What is the energy cost per year to run the deluxe model?
9. If you make a $220 down payment on the standard model, how much of the cost is left to finance?
10. The economy model can be expected to last for 10 years. What would be the total energy cost over that period?

Estimate the answer to each problem. Does the result on the calculator display seem reasonable?

11. $25.9 + 345.1 + 0.09$ [347.78]
12. $8{,}345.889 - 345.6$ [8000.289]
13. $42{,}090.8 + 3{,}303.09$ [45393.89]
14. $10.007 - 0.626$ [3.747]
15. $9.8(8.8)$ [86.24]
16. $\frac{24.56}{2.2}$ [1.116363636]
17. $53 \cdot 5.61$ [241.23]
18. $89.11 \div 22.707$ [39.24340]

4.5 Fractions and Decimals

- Writing fractions as equivalent decimals • Repeating decimals
- Rounding repeating decimals • Graphing fractions and decimals
- Problems involving fractions and decimals

In this section, we will further investigate the relationship between fractions and decimals.

Writing fractions as equivalent decimals

To write $\frac{5}{8}$ as a decimal, we use the fact that $\frac{5}{8}$ indicates the division $5 \div 8$. We can convert $\frac{5}{8}$ to decimal form by performing the division.

$$\begin{array}{r} .625 \\ 8\overline{)5.000} \\ \underline{4\,8} \\ 20 \\ \underline{16} \\ 40 \\ \underline{40} \\ 0 \end{array}$$

Write a decimal point and additional zeros to the right of 5.

0 ← The remainder is 0.

Thus, $\frac{5}{8} = 0.625$.

Writing a fraction as a decimal

To write a fraction as a decimal, divide the numerator of the fraction by its denominator.

EXAMPLE 1 Write $\frac{3}{4}$ as a decimal.

Solution We divide the numerator by the denominator.

$$\begin{array}{r} .75 \\ 4\overline{)3.00} \\ \underline{2\,8} \\ 20 \\ \underline{20} \\ 0 \end{array}$$

Write a decimal point and two zeros to the right of 3.

0 ← The remainder is 0.

Thus, $\frac{3}{4} = 0.75$.

Self Check 1

Write $\frac{3}{16}$ as a decimal.

Answer 0.1875

In Example 1, the division process ended because a remainder of zero was obtained. In this case, we call the quotient, 0.75, a **terminating decimal.**

Repeating decimals

Sometimes, when we are finding a decimal equivalent of a fraction, the division process never gives a remainder of zero. In this case, the result is a **repeating decimal.** Examples of repeating decimals are 0.4444. . . and 1.373737. . . . The three dots tell us that a block of digits repeats in the pattern shown. Repeating decimals can be written using a bar over the repeating block of digits. For example, 0.4444. . . can be written as $0.\overline{4}$, and 1.373737. . . can be written as $1.\overline{37}$.

! COMMENT When using an overbar to write a repeating decimal, use the fewest digits necessary to show the repeating block of digits.

$0.333\ldots = 0.\overline{333}$ $\qquad 6.7454545\ldots = 6.7\overline{454}$

$0.333\ldots = 0.\overline{3}$ $\qquad 6.7454545\ldots = 6.7\overline{45}$

EXAMPLE 2 Write $\frac{5}{12}$ as a decimal.

Self Check 2
Write $\frac{4}{11}$ as a decimal.

Solution We use division to find the decimal equivalent.

$$\begin{array}{r} .4166 \\ 12\overline{)5.0000} \\ \underline{4\,8} \\ 20 \\ \underline{12} \\ 80 \\ \underline{72} \\ 80 \\ \underline{72} \\ 8 \end{array}$$

Write a decimal point and four zeros to the right of 5.

It is apparent that 8 will continue to reappear as the remainder. Therefore, 6 will continue to reappear in the quotient. Since the repeating pattern is now clear, we can stop the division.

Thus, $\frac{5}{12} = 0.41\overline{6}$.

Answer $0.\overline{36}$

Every fraction can be written as either a terminating decimal or a repeating decimal. For this reason, the set of fractions (**rational numbers**) forms a subset of the set of decimals called the set of **real numbers.** The set of real numbers corresponds to *all* points on a number line.

Not all decimals are terminating or repeating decimals. For example,

$0.2020020002\ldots$

does not terminate, and it has no repeating block of digits. This decimal cannot be written as a fraction with an integer numerator and a nonzero integer denominator. Thus, it is not a rational number. It is an example from the set of **irrational numbers.**

Rounding repeating decimals

When a fraction is written in decimal form, the result is either a terminating or a repeating decimal. Repeating decimals are often rounded to a specified place value.

EXAMPLE 3 Write $\frac{1}{3}$ as a decimal and round to the nearest hundredth.

Solution First, we divide the numerator by the denominator to find the decimal equivalent of $\frac{1}{3}$.

$$\begin{array}{r} 0.333 \\ 3\overline{)1.000} \\ \underline{9} \\ 10 \\ \underline{9} \\ 10 \\ \underline{9} \\ 1 \end{array}$$

Write a decimal point and additional zeros to the right of 1.

We see that the division process never gives a remainder of zero. When we write $\frac{1}{3}$ in decimal form, the result is the repeating decimal $0.333\ldots = 0.\overline{3}$.

To find a decimal approximation of $\frac{1}{3}$ to the nearest hundredth, we proceed as follows.

0.333. . . — Round 0.333 to the nearest hundredth by examining the test digit in the thousandths column.

Since 3 is less than 5, we round down, and we have

$\frac{1}{3} \approx 0.33$

Read $\approx$ as "is approximately equal to."

Self Check 4
Write $\frac{7}{24}$ as a decimal and round to the nearest thousandth.

Answer 0.292

EXAMPLE 4 Write $\frac{2}{7}$ as a decimal and round to the nearest thousandth.

Solution

$$\begin{array}{r} .2857 \\ 7\overline{)2.0000} \\ \underline{1\,4} \\ 60 \\ \underline{56} \\ 40 \\ \underline{35} \\ 50 \\ \underline{49} \\ 1 \end{array}$$

Write a decimal point and additional zeros to the right of 2.

To round to the thousandths column, we must divide to the ten thousandths column.

0.2857 — Round 0.2857 to the nearest thousandth by examining the test digit in the ten thousandths column.

Since 7 is greater than 5, we round up, and $\frac{2}{7} \approx 0.286$.

CALCULATOR SNAPSHOT The fixed-point key

After performing a calculation, a scientific calculator can round the result to a given decimal place. This is done using the *fixed-point key.* As we did in Example 4, let's find the decimal equivalent of $\frac{2}{7}$ and round to the nearest thousandth. This time, we will use a calculator.

First, we set the calculator to round to the third decimal place (thousandths) by pressing [FIX] 3. Then we press

2 [÷] 7 [=] 0.286

Thus, $\frac{2}{7} \approx 0.286$. To round to the nearest tenth, we would fix 1; to round to the nearest hundredth, we would fix 2, and so on.

If your calculator does not have a fixed-point key, see the owner's manual.

Self Check 5
Write $8\frac{19}{20}$ in decimal form.

EXAMPLE 5 Write $5\frac{3}{8}$ in decimal form.

Solution To write a mixed number in decimal form, recall that a mixed number is made up of a whole-number part and a fractional part. Since we can write $5\frac{3}{8}$ as $5 + \frac{3}{8}$, we need only consider how to write $\frac{3}{8}$ as a decimal.

$$\begin{array}{r} .375 \\ 8\overline{)3.000} \\ \underline{2\,4} \\ 60 \\ \underline{56} \\ 40 \\ \underline{40} \\ 0 \end{array}$$

Write a decimal point and three zeros to the right of 3.

Thus, $5\frac{3}{8} = 5 + \frac{3}{8} = 5 + 0.375 = 5.375$. We would obtain the same result if we changed $5\frac{3}{8}$ to the improper fraction $\frac{43}{8}$ and divided 43 by 8.

Answer 8.95

Graphing fractions and decimals

The number line can be used to show the relationship between fractions and their respective decimal equivalents. Figure 4-7 shows some commonly used fractions that have terminating decimal equivalents. For example, we see from the graph that $\frac{13}{16} = 0.8125$.

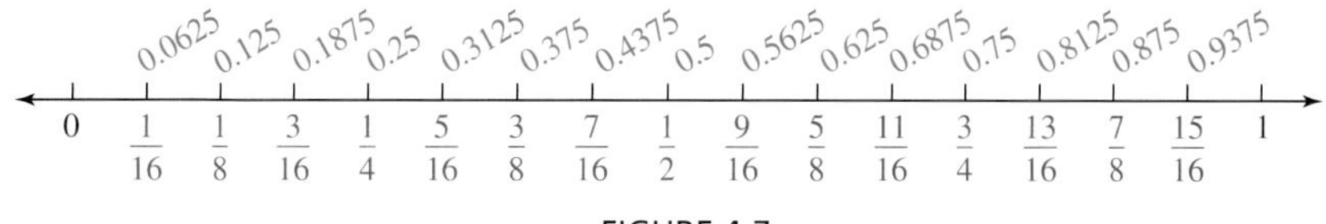

FIGURE 4-7

The number line in Figure 4-8 shows some commonly used fractions that have repeating decimal equivalents.

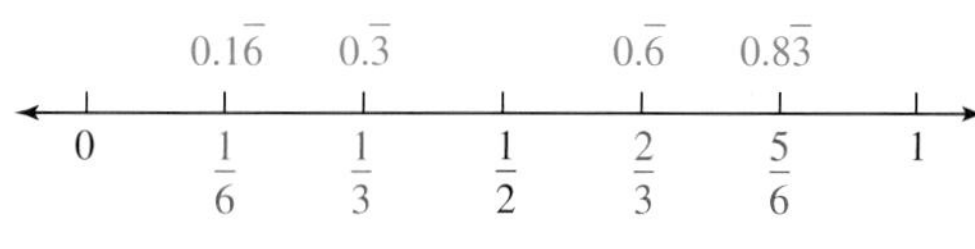

FIGURE 4-8

Problems involving fractions and decimals

Numerical expressions can contain both fractions and decimals. In the following examples, we show how different methods can be used to solve problems of this type.

EXAMPLE 6 Evaluate $\dfrac{1}{3} + 0.27$ by working in terms of fractions.

Solution We write 0.27 as a fraction and add it to $\dfrac{1}{3}$.

$\dfrac{1}{3} + 0.27 = \dfrac{1}{3} + \dfrac{27}{100}$ Replace 0.27 with $\dfrac{27}{100}$.

$= \dfrac{1 \cdot \mathbf{100}}{3 \cdot \mathbf{100}} + \dfrac{27 \cdot \mathbf{3}}{100 \cdot \mathbf{3}}$ The LCD for $\frac{1}{3}$ and $\frac{27}{100}$ is 300. Express each fraction in terms of 300ths.

$= \dfrac{100}{300} + \dfrac{81}{300}$ Multiply in the numerators and in the denominators.

$= \dfrac{181}{300}$ Add the numerators and write the sum over the common denominator, 300.

Self Check 6

Evaluate by working in terms of fractions: $0.53 - \frac{1}{6}$.

Answer $\dfrac{109}{300}$

Self Check 7
Evaluate by working in terms of decimals: $0.53 - \frac{1}{6}$.

Answer 0.36

EXAMPLE 7

Evaluate $\frac{1}{3} + 0.27$ by working in terms of decimals.

Solution We have seen that the decimal equivalent of $\frac{1}{3}$ is the repeating decimal 0.333. . . . To add $\frac{1}{3}$ to 0.27, we round 0.333. . . to the nearest hundredth: $\frac{1}{3} \approx 0.33$.

$$\frac{1}{3} + 0.27 \approx 0.33 + 0.27 \quad \text{Approximate } \frac{1}{3} \text{ with the decimal 0.33.}$$
$$\approx 0.60 \quad \text{Perform the addition.}$$

In the previous two examples, we evaluated $\frac{1}{3} + 0.27$ in different ways. In Example 6, we obtained the exact answer, $\frac{181}{300}$. In Example 7, we obtained an approximation, 0.60. It is apparent that the results are in agreement when we write $\frac{181}{300}$ in decimal form: $\frac{181}{300} = 0.60333\ldots$.

Self Check 8
Perform the operations: $(-0.6)^2 + (2.3)\left(\frac{1}{8}\right)$.

Answer 0.6475

EXAMPLE 8

Perform the operations: $\left(\frac{4}{5}\right)(1.35) + (0.5)^2$.

Solution It appears simplest to work in terms of decimals. We use division to find the decimal equivalent of $\frac{4}{5}$.

$$\begin{array}{r} .8 \\ 5\overline{)4.0} \\ \underline{4\,0} \\ 0 \end{array}$$
Write a decimal point and one zero to the right of the 4.

Now we use the rules for the order of operations to evaluate the given expression.

$$\left(\frac{4}{5}\right)(1.35) + (0.5)^2 = (\mathbf{0.8})(1.35) + (0.5)^2 \quad \text{Replace } \tfrac{4}{5} \text{ with its decimal equivalent, 0.8.}$$
$$= (0.8)(1.35) + 0.25 \quad \text{Find the power: } (0.5)^2 = 0.25.$$
$$= 1.08 + 0.25 \quad \text{Perform the multiplication: } (0.8)(1.35) = 1.08.$$
$$= 1.33 \quad \text{Perform the addition.}$$

EXAMPLE 9

Shopping. During a trip to the grocery store, a shopper purchased $\frac{3}{4}$ pound of fruit, priced at \$0.88 a pound, and $\frac{1}{3}$ pound of fresh-ground coffee, selling for \$6.60 a pound. Find the total cost of these items.

Solution To find the cost of each item, we multiply the amount purchased by its unit price. Then we add the two individual costs to obtain the total cost.

Cost of fruit	plus	cost of coffee	equals	total cost.
$\left(\frac{3}{4}\right)(0.88)$	$+$	$\left(\frac{1}{3}\right)(6.60)$	$=$	total cost

Because 0.88 is divisible by 4 and 6.60 is divisible by 3, we can work with the decimals and fractions in this form; no conversion is necessary.

$$\left(\frac{3}{4}\right)(0.88) + \left(\frac{1}{3}\right)(6.60) = \left(\frac{3}{4}\right)\left(\frac{0.88}{1}\right) + \left(\frac{1}{3}\right)\left(\frac{6.60}{1}\right)$$ Express 0.88 as $\frac{0.88}{1}$ and 6.60 as $\frac{6.60}{1}$. Multiply the numerators and the denominators.

$$= \frac{2.64}{4} + \frac{6.60}{3}$$

$$= 0.66 + 2.20$$ Perform each division.

$$= 2.86$$ Perform the addition.

The total cost of the items is \$2.86.

Section 4.5 STUDY SET

VOCABULARY *Fill in the blanks.*

1. The decimal form of the fraction $\frac{1}{3}$ is a ________ decimal, which is written $0.\overline{3}$ or 0.3333. . . .
2. The decimal form of the fraction $\frac{2}{5}$ is a ________ decimal, which is written 0.4.
3. The ________ equivalent of $\frac{1}{16}$ is 0.0625.
4. We read $\approx$ as "is ________ equal to."

CONCEPTS

5. **a.** What division is indicated by $\frac{7}{8}$?
   b. Fill in the blank: To write a fraction as a decimal, divide the ________ of the fraction by its denominator.
6. Insert the proper symbol < or > in the blank.
   a. $0.\overline{6}$ ☐ 0.7 **b.** $0.\overline{6}$ ☐ 0.6
7. When we round 0.272727. . . to the nearest hundredth, is the result larger or smaller than the original number?
8. Write each decimal in fraction form.
   a. 0.7 **b.** 0.77
9. Graph: $1\frac{3}{4}$, -0.75, $0.\overline{6}$, and $-3.8\overline{3}$.

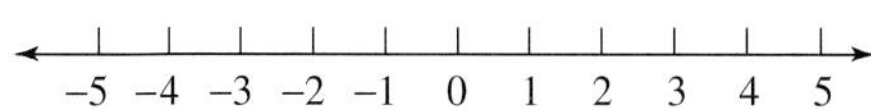

10. Graph: $2\frac{7}{8}$, -2.375, $0.\overline{3}$, and $4.1\overline{6}$.

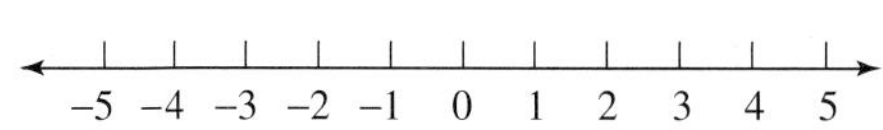

11. Determine whether each statement is true or false.
    a. $\frac{1}{3} = 0.3$ **b.** $\frac{3}{4} = 0.75$
    c. $20\frac{1}{2} = 20.5$ **d.** $\frac{1}{16} = 0.1\overline{6}$
12. When evaluating the expression $0.25 + \left(2.3 + \frac{2}{5}\right)^2$, would it be easier to work in terms of fractions or in terms of decimals?

NOTATION

13. Examine the color portion of the long division below.
    a. Will the remainder ever be zero?
    b. What can be deduced about the decimal equivalent of $\frac{5}{6}$?

$$\begin{array}{r} .833 \\ 6\overline{)5.000} \\ \underline{4\,8} \\ \mathbf{20} \\ \underline{\mathbf{18}} \\ \mathbf{20} \end{array}$$

14. Write each repeating decimal using an overbar.
    a. 0.888. . . **b.** 0.323232. . .
    c. 0.56333. . . **d.** 0.8898989. . .

PRACTICE *Write each fraction in decimal form.*

15. $\frac{1}{2}$
16. $\frac{1}{4}$
17. $-\frac{5}{8}$
18. $-\frac{3}{5}$

19. $\frac{9}{16}$ **20.** $\frac{3}{32}$

21. $-\frac{17}{32}$ **22.** $-\frac{15}{16}$

23. $\frac{11}{20}$ **24.** $\frac{19}{25}$

25. $\frac{31}{40}$ **26.** $\frac{17}{20}$

27. $-\frac{3}{200}$ **28.** $-\frac{21}{50}$

29. $\frac{1}{500}$ **30.** $\frac{1}{250}$

Write each fraction in decimal form. Use an overbar.

31. $\frac{2}{3}$ **32.** $\frac{7}{9}$

33. $\frac{5}{11}$ **34.** $\frac{4}{15}$

35. $-\frac{7}{12}$ **36.** $-\frac{17}{22}$

37. $\frac{1}{30}$ **38.** $\frac{1}{60}$

Write each fraction in decimal form. Round to the nearest hundredth.

39. $\frac{7}{30}$ **40.** $\frac{14}{15}$

41. $\frac{17}{45}$ **42.** $\frac{8}{9}$

Write each fraction in decimal form. Round to the nearest thousandth.

43. $\frac{5}{33}$ **44.** $\frac{5}{12}$

45. $\frac{10}{27}$ **46.** $\frac{17}{21}$

Write each fraction in decimal form. Round to the nearest hundredth.

47. $\frac{4}{3}$ **48.** $\frac{10}{9}$

49. $-\frac{34}{11}$ **50.** $-\frac{25}{12}$

Write each mixed number in decimal form. Round to the nearest hundredth when the result is a repeating decimal.

51. $3\frac{3}{4}$ **52.** $5\frac{4}{5}$

53. $-8\frac{2}{3}$ **54.** $-1\frac{7}{9}$

55. $12\frac{11}{16}$ **56.** $32\frac{1}{8}$

57. $203\frac{11}{15}$ **58.** $568\frac{23}{30}$

Fill in the correct symbol (< or >) to make a true statement. (Hint: Express each number as a decimal.)

59. $\frac{7}{8}$ ___ 0.895 **60.** 4.56 ___ $4\frac{2}{5}$

61. $-\frac{11}{20}$ ___ $-0.\overline{4}$ **62.** $-9.0\overline{9}$ ___ $-9\frac{1}{11}$

Evaluate each expression. Work in terms of fractions.

63. $\frac{1}{9} + 0.3$ **64.** $\frac{2}{3} + 0.1$

65. $0.9 - \frac{7}{12}$ **66.** $0.99 - \frac{5}{6}$

67. $\frac{5}{11}(0.3)$ **68.** $(0.9)\left(\frac{1}{27}\right)$

69. $\frac{1}{3}\left(-\frac{1}{15}\right)(0.5)$ **70.** $(-0.4)\left(\frac{5}{18}\right)\left(-\frac{1}{3}\right)$

71. $\frac{1}{4}(0.25) + \frac{15}{16}$ **72.** $\frac{2}{5}(0.02) - (0.04)$

Evaluate each expression to the nearest hundredth.

73. $0.24 + \frac{1}{3}$ **74.** $0.02 + \frac{5}{6}$

75. $5.69 - \frac{5}{12}$ **76.** $3.19 - \frac{2}{3}$

77. $\frac{3}{4}(0.43) - \frac{1}{12}$ **78.** $-\frac{2}{5}(0.33) + 0.45$

Evaluate each expression. Work in terms of decimals.

79. $(3.5 + 6.7)\left(-\frac{1}{4}\right)$ **80.** $\left(-\frac{5}{8}\right)(5.3 - 3.9)$

81. $\left(\frac{1}{5}\right)^2(1.7)$ **82.** $(2.35)\left(\frac{2}{5}\right)^2$

83. $7.5 - (0.78)\left(\frac{1}{2}\right)$ **84.** $8.1 - \left(\frac{3}{4}\right)(0.12)$

85. $\frac{3}{8}(-3.2) + (4.5)\left(-\frac{1}{9}\right)$

86. $(-0.8)\left(\frac{1}{4}\right) + \left(\frac{1}{3}\right)(0.39)$

Evaluate each expression. Round to the nearest hundredth.

87. $\dfrac{3\frac{1}{5} + 2\frac{1}{2}}{5.69} + 3\frac{1}{4}$

88. $4\frac{2}{3} - \dfrac{2.7 - \frac{7}{8}}{0.12}$

Write each fraction in decimal form.

89. $\frac{23}{101}$

90. $\frac{1}{99}$

91. $\frac{2,046}{55}$

92. $-\frac{11}{128}$

APPLICATIONS

93. DRAFTING The architect's scale below has several measuring edges. The edge marked 16 divides each inch into 16 equal parts. Find the decimal form for each fractional part of 1 inch that is highlighted on the scale.

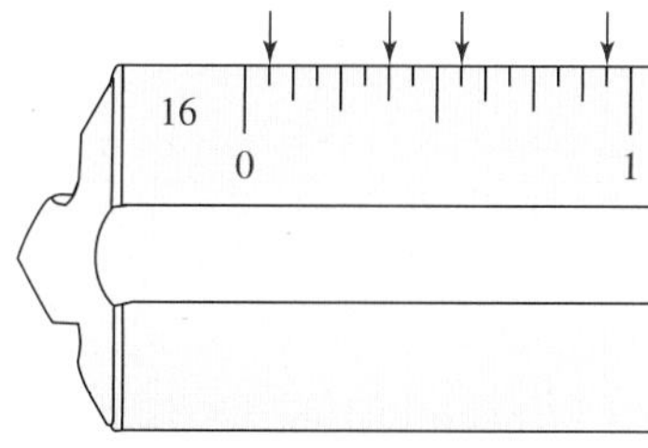

94. FREEWAY SIGNS The freeway sign below gives the number of miles to the next three exits. Convert the mileages to decimal notation.

BARRANCA AVE.	$\frac{3}{4}$ mi
210 FREEWAY	$2\frac{1}{4}$ mi
ADA ST.	$3\frac{1}{2}$ mi

95. GARDENING Two brands of replacement line for a lawn trimmer are labeled in different ways as shown in the next column. On one package, the line's thickness is expressed as a decimal; on the other, as a fraction. Which line is thicker?

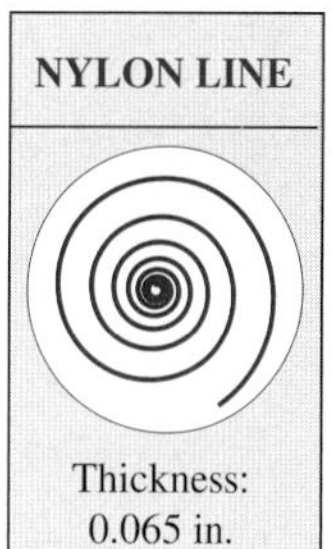

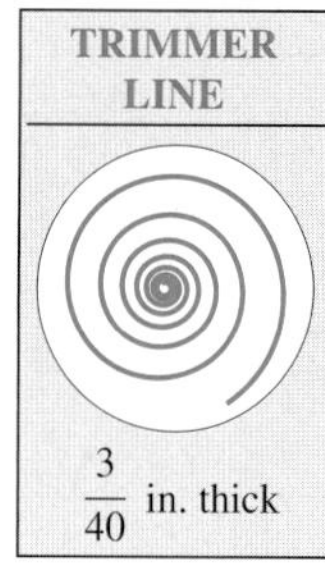

96. AUTO MECHANICS While doing a tuneup, a mechanic checks the gap on one of the spark plugs of a car to be sure it is firing correctly. The owner's manual states that the gap should be $\frac{2}{125}$ inch. The gauge the mechanic uses to check the gap is in decimal notation; it registers 0.025 inch. Is the spark plug gap too large or too small?

97. HORSE RACING In thoroughbred racing, the time a horse takes to run a given distance is measured using fifths of a second. For example, 55^2 (read "fifty-five and two") means $55\frac{2}{5}$ seconds. The illustration lists four split times for a horse. Express the times in decimal form.

Speedy Flight	Turfway Park, Ky	3-year-old			
17 May 97	$1\frac{1}{16}$ mile	$:23^2$	$:23^4$	$:24^1$	$:32^3$

98. GEOLOGY A geologist weighed a rock sample at the site where it was discovered and found it to weigh $17\frac{7}{8}$ lb. Later, a more accurate digital scale in the laboratory gave the weight as 17.671 lb. What is the difference in the two measurements?

99. WINDOW REPLACEMENTS The amount of sunlight that comes into a room depends on the area of the windows in the room. What is the area of the window shown below?

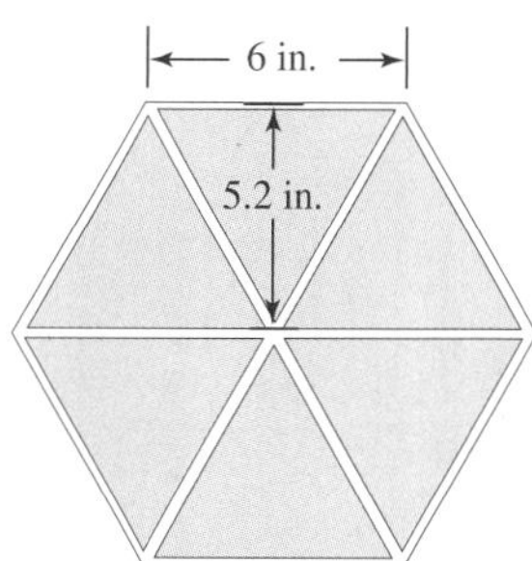

100. FIRE CONTAINMENT A command post asked each of three fire crews to estimate the length of the fire line they were fighting. Their reports came back in different forms, as indicated in the illustration on the next page. Find the perimeter of the fire.

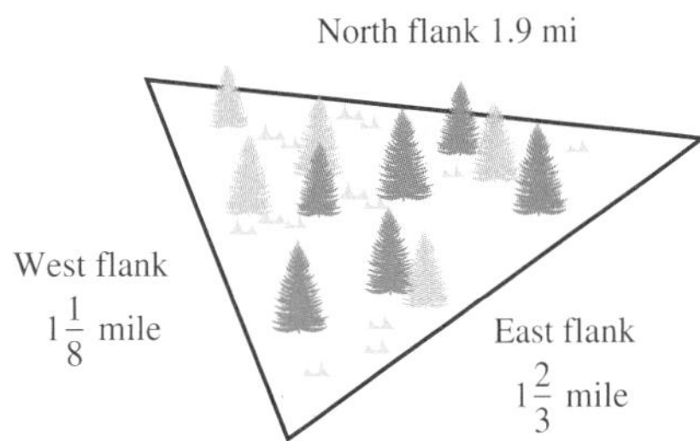

WRITING

101. Explain the procedure used to write a fraction in decimal form.
102. Compare and contrast the two numbers 0.5 and $0.\overline{5}$.
103. A student represented the repeating decimal 0.1333. . . as $0.1\overline{333}$. Is this correct? Explain why or why not.
104. Is 0.10100100010000 . . . a repeating decimal? Explain why or why not.

REVIEW

105. Add: $-2 + (-3) + 10 + (-6)$.
106. Evaluate: $-3 + 2[-3 + (2 - 7)]$.
107. Add: $\frac{5}{2} + \frac{2}{3}$.
108. Subtract: $\frac{3}{7} - \frac{1}{5}$.
109. Multiply: $\frac{3}{4} \cdot \frac{8}{9}$.
110. Divide: $\frac{4}{5} \div \frac{9}{10}$.

4.6 Solving Equations Containing Decimals

- Solving equations
- Problem solving with equations

Solving equations

Recall that the addition and subtraction properties of equality allow us to add the same number to or subtract the same number from both sides of an equation.

Self Check 1
Solve each equation:
a. $4.6 + t = 15.7$
b. $-1.24 = r - 0.04$

Answers **a.** 11.1, **b.** −1.2

EXAMPLE 1 Solve each equation: **a.** $x + 3.5 = 7.8$ and **b.** $y - 1.23 = -4.52$.

Solution

a. To isolate x, we undo the addition of 3.5 by subtracting 3.5 from both sides of the equation.

$$x + 3.5 = 7.8$$
$$x + 3.5 - \mathbf{3.5} = 7.8 - \mathbf{3.5} \quad \text{Subtract 3.5 from both sides.}$$
$$x = 4.3 \quad \text{Simplify.}$$

b. To isolate y, we undo the subtraction of 1.23 by adding 1.23 to both sides of the equation.

$$y - 1.23 = -4.52$$
$$y - 1.23 + \mathbf{1.23} = -4.52 + \mathbf{1.23} \quad \text{Add 1.23 to both sides.}$$
$$y = -3.29 \quad \text{Simplify.}$$

Check each result.

The multiplication property of equality states that we can multiply both sides of an equation by the same nonzero number.

EXAMPLE 2 Solve: $\frac{m}{2} = -24.8$.

Solution To isolate m, we undo the division by 2 by multiplying both sides of the equation by 2.

$$\frac{m}{2} = -24.8$$

$$\mathbf{2}\left(\frac{m}{2}\right) = \mathbf{2}(-24.8) \quad \text{Multiply both sides by 2.}$$

$$m = -49.6 \quad \text{Perform the multiplications.}$$

Check the result.

Self Check 2

Solve: $\frac{y}{3} = -13.11$.

Answer -39.33

The division property of equality states that we can divide both sides of an equation by the same nonzero number.

EXAMPLE 3 Solve: $-4.6x = -9.66$.

Solution To isolate x, we undo the multiplication by -4.6 by dividing by -4.6.

$$-4.6x = -9.66$$

$$\frac{-4.6x}{\mathbf{-4.6}} = \frac{-9.66}{\mathbf{-4.6}} \quad \text{Divide both sides by } -4.6.$$

$$x = 2.1 \quad \text{Perform the divisions.}$$

Check the result.

Self Check 3

Solve: $-22.32 = -3.1m$.

Answer 7.2

Sometimes, more than one property must be used to solve an equation. In the next example, we use the addition property of equality and the division property of equality.

EXAMPLE 4 Solve: $8.1y - 6.04 = -13.33$.

Solution The left-hand side involves a multiplication and a subtraction. To solve the equation, we must undo these operations, but in the opposite order. We begin by undoing the subtraction.

$$8.1y - 6.04 = -13.33$$

$$8.1y - 6.04 + \mathbf{6.04} = -13.33 + \mathbf{6.04} \quad \text{To undo the subtraction of 6.04, add 6.04 to both sides.}$$

$$8.1y = -7.29 \quad \text{Simplify: } -13.33 + 6.04 = -7.29.$$

$$\frac{8.1y}{\mathbf{8.1}} = \frac{-7.29}{\mathbf{8.1}} \quad \text{To undo the multiplication of 8.1, divide both sides by 8.1.}$$

$$y = -0.9 \quad \text{Perform the divisions.}$$

Check:

$$8.1y - 6.04 = -13.33$$

$$8.1(\mathbf{-0.9}) - 6.04 \stackrel{?}{=} -13.33 \quad \text{Substitute } -0.9 \text{ for } y.$$

$$-7.29 - 6.04 \stackrel{?}{=} -13.33 \quad \text{Perform the multiplication: } 8.1(-0.9) = -7.29.$$

$$-13.33 = -13.33 \quad \text{Perform the subtraction by adding the opposite: } -7.29 - 6.04 = -7.29 + (-6.04) = -13.33.$$

The solution is -0.9.

Self Check 4

Solve: $-4.2h + 3.14 = 1.88$.

Answer 0.3

Problem solving with equations

EXAMPLE 5 **Business expenses.** A business owner decides to rent a copy machine (Figure 4-9) instead of buying one. Under the rental agreement, the company is charged \$65 per month plus 2¢ for every copy made. If the business has budgeted \$125 for copier expenses each month, how many copies can be made before exceeding the budget?

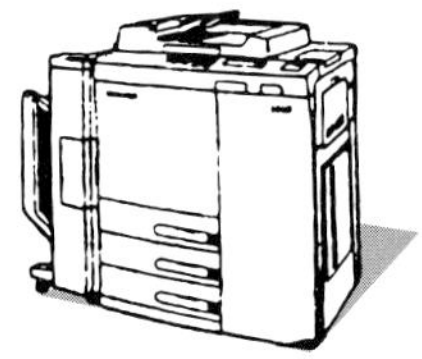

FIGURE 4-9

Analyze the problem

- The basic rental charge is \$65 a month.
- There is a 2¢ charge for each copy made.
- \$125 is budgeted for copier expenses each month.
- We must find the maximum number of copies that can be made each month.

Form an equation

Let x = the maximum number of copies that can be made. We can write the amount budgeted for copier expenses in two ways.

The basic fee	plus	the cost of the copies	is	the amount budgeted each month.

We can find the total cost of the copies by multiplying the cost per copy by the maximum number of copies that can be made. Notice that the costs are expressed in terms of dollars and cents. We need to work in terms of one unit, so we write 2¢ as \$0.02 and work in terms of dollars.

65	plus	0.02	·	the maximum number of copies made	is	125.
65	+	0.02	·	x	=	125

Solve the equation

$$65 + 0.02x = 125$$

$$65 + 0.02x - \mathbf{65} = 125 - \mathbf{65}$$ To undo the addition of 65, subtract 65 from both sides.

$$0.02x = 60$$ Simplify.

$$\frac{0.02x}{\mathbf{0.02}} = \frac{60}{\mathbf{0.02}}$$ To undo the multiplication by 0.02, divide both sides by 0.02.

$$x = 3{,}000$$ Perform the divisions.

State the conclusion

The business can make up to 3,000 copies each month without exceeding its budget.

Check the result

If we multiply the cost per copy and the maximum number of copies, we get $\$0.02 \cdot 3{,}000 = \$60$. Then we add the \$65 monthly fee: $\$60 + \$65 = \$125$. The answer checks.

Section 4.6 STUDY SET

VOCABULARY *Fill in the blanks.*

1. To ______ an equation, we isolate the variable on one side of the equals symbol.
2. To ______ the solution of an equation, we substitute the value for the variable in the original equation and see whether the result is a true statement.
3. When we add the same number to both sides of an equation, we are using the ________ property of equality.
4. A ________ is a letter that is used to stand for a number.

CONCEPTS

5. Show that 1.7 is a solution of $2.1x - 6.3 = -2.73$ by checking it.
6. Show that 0.04 is a solution of $\frac{y}{2} + 0.7 = 0.72$ by checking it.

NOTATION *Complete each solution to solve the equation.*

7. $$0.6s - 2.3 = -1.82$$
$$0.6s - 2.3 + \square = -1.82 + \square$$
$$\square = 0.48$$
$$\frac{0.6s}{\square} = \frac{0.48}{\square}$$
$$s = 0.8$$

8. $$\frac{x}{2} + 1 = -5.2$$
$$\frac{x}{2} + 1 - \square = -5.2 - \square$$
$$\square = -6.2$$
$$2\left(\frac{x}{2}\right) = 2(\square)$$
$$x = -12.4$$

PRACTICE *Solve each equation. Check the result.*

9. $x + 8.1 = 9.8$
10. $6.75 + y = 8.99$
11. $7.08 = t - 0.03$
12. $14.1 = k - 13.1$
13. $-5.6 + h = -17.1$
14. $-0.05 + x = -1.25$
15. $7.75 = t - (-7.85)$
16. $3.33 = y - (-5.55)$
17. $2x = -8.72$
18. $3y = -12.63$
19. $-3.51 = -2.7x$
20. $-1.65 = -0.5f$
21. $\frac{x}{2.04} = -4$
22. $\frac{y}{2.22} = -6$
23. $\frac{-x}{5.1} = -4.4$
24. $\frac{-t}{8.1} = -3$
25. $\frac{1}{3}x = -7.06$
26. $\frac{1}{5}x = -3.02$
27. $\frac{x}{100} = 0.004$
28. $\frac{y}{1{,}000} = 0.0606$
29. $2x + 7.8 = 3.4$
30. $3x - 1.2 = -4.8$
31. $-0.8 = 5y + 9.2$
32. $-9.9 = 6t + 14.1$
33. $0.3x - 2.1 = 7.2$
34. $0.4a + 3.3 = -5.1$
35. $-1.5b + 2.7 = 1.2$
36. $-2.1x - 3.1 = 5.3$
37. $0.9a - 6 = -5.73$
38. $-0.6t + 4 = 3.46$

APPLICATIONS *Complete each solution.*

39. PETITION DRIVES On weekends, a college student works for a political organization, collecting signatures for a petition drive. Her pay is \$15 a day plus 30 cents for each signature she obtains. How many signatures does she have to collect to make \$60 a day?

Analyze the problem

- Her base pay is ☐ dollars a day.
- She makes ☐ cents for each signature.
- She wants to make ☐ dollars a day.
- Find the number of ☐ she needs to get.

Form an equation Let x = ______________________

We need to work in terms of the same units, so we write 30 cents as ☐.

If we multiply the pay per signature by the number of signatures, we get the money she makes just from collecting signatures. Therefore, ☐ = total amount (in dollars) made from collecting signatures.

We can express the money she earns in a day in two ways.

Base pay	+	0.30	·	the number of signatures	is	60.
15	+	☐			=	60

Solve the equation

$15 + \_\_\_\_ = \_\_\_\_$

$\_\_\_\_ = 45$

$x = 150$

State the conclusion

____

Check the result

If she collects ____ signatures, she will make 0.30 · ____ = ____ dollars from signatures. If we add this to \$15, we get \$60. The answer checks.

40. HIGHWAY CONSTRUCTION A 12.8-mile highway is in its third and final year of construction. In the first year, 2.3 miles of the highway were completed. In the second year, 4.9 miles were finished. How many more miles of the highway need to be completed?

Analyze the problem

- The planned highway is ____ miles long.
- The 1st year, ____ miles were completed.
- The 2nd year, ____ miles were completed.
- Find the number of ____ yet to be completed.

A diagram will help us understand the problem.

12.8-mi highway

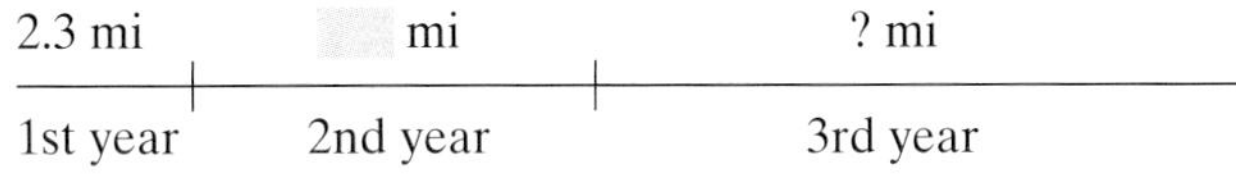

Form an equation

Let x = ____________________________________

We can express the length of the highway in two ways.

Miles 1st year	+	miles 2nd year	+	the number of miles yet to be completed	is	12.8.
____	+	4.9	+	____	=	12.8

Solve the equation

$2.3 + 4.9 + \_\_\_\_ = 12.8$

$\_\_\_\_ + \_\_\_\_ = 12.8$

$x = \_\_\_\_$

State the conclusion

____

Check the result

Add: ____ + ____ + ____ = 12.8. The answer checks.

Choose a variable to represent the unknown. Then write and solve an equation to answer the question.

41. DISASTER RELIEF After hurricane damage estimated at \$27.9 million, a county looked to three sources for relief. Local agencies contributed \$6.8 million toward the cleanup. A state emergency fund offered another \$12.5 million. When applying for federal government help, how much should the county ask for?

42. TELETHONS Midway through a telethon, the donations had reached \$16.7 million. How much more was donated in the second half of the program if the final total pledged was \$30 million?

43. GRADE POINT AVERAGES After receiving her grades for the fall semester, a college student noticed that her overall GPA had dropped by 0.18. If her new GPA was 3.09, what was her GPA at the beginning of the fall semester?

44. MONTHLY PAYMENTS A food dehydrator offered on a home shopping channel can be purchased by making 3 equal monthly payments. If the price is \$113.25, how much is each monthly payment?

45. POINTS PER GAME As a senior, a college basketball player's scoring average was double that of her junior season. If she averaged 21.4 points a game as a senior, how many did she average as a junior?

46. NUTRITION One 3-ounce serving of broiled ground beef has 7 grams of saturated fat. This is 14 times the amount of saturated fat in 1 cup of cooked crab meat. How many grams of saturated fat are in 1 cup of cooked crab meat?

47. FUEL EFFICIENCY Each year, the Federal Highway Administration determines the number of vehicle-miles traveled in the country and divides it by the amount of fuel consumed to get an average miles per gallon (mpg). The illustration shows how the figure has changed over the years to reach a high of 16.7 mpg in 1998. What was the average miles per gallon in 1960?

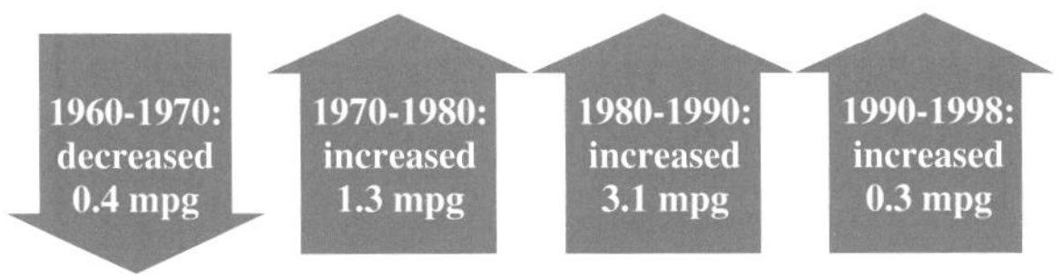

48. RATINGS REPORTS The following illustration shows the prime-time television ratings for the week of January 3, 2000. If the Fox network ratings had been $\frac{1}{2}$ point higher, there would have been a three-way tie for second place. What prime-time rating did Fox have that week?

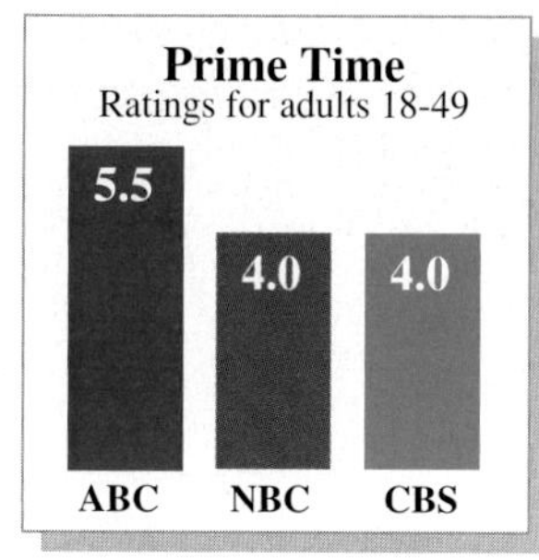

Source: Neilsen Media Research

49. CALLIGRAPHY A city honors its citizen of the year with a framed certificate. A calligrapher charges \$20 for the frame and then 15 cents a word for writing out the proclamation. If the city charter prohibits gifts in excess of \$50, what is the maximum number of words that can be printed on the award?

50. HELIUM BALLOONS The organizer of a jog-a-thon wants an archway of balloons constructed at the finish line of the race. A company charges a \$100 setup fee and then 8 cents for every balloon. How many balloons will be used if \$300 is spent for the decoration?

WRITING

51. Did you encounter any differences in solving equations containing decimals as compared to solving equations containing only integers? Explain your answer.

52. Explain how to verify that 29.2 is a solution of $0.2x - 0.16 = 5.68$.

REVIEW

53. Add: $-\frac{2}{3} + \frac{3}{4}$.

54. Add: $-\frac{2}{3} + \frac{1}{5}$.

55. Divide: $\frac{7}{8} \div \frac{13}{16}$.

56. Multiply: $2\frac{1}{3} \cdot 4\frac{1}{2}$.

57. Evaluate: $\frac{-3 - 3}{-3 + 4}$.

58. Write a complex fraction using $-\frac{4}{5}$ and $\frac{1}{5}$. Then simplify it.

59. Solve: $\frac{6}{5}x = 10$.

60. Solve: $\frac{2}{3}x + 1 = 7$.

4.7 Square Roots

- Square roots • Evaluating numerical expressions containing radicals
- Square roots of fractions and decimals • Using a calculator to find square roots
- Approximating square roots

There are six basic operations of arithmetic. We have seen the relationships between addition and subtraction and between multiplication and division. In this section, we will explore the relationship between raising a number to a power and finding a root. Decimals will play an important role in this discussion.

Square roots

When we raise a number to the second power, we are squaring it, or finding its **square.**

The square of 6 is 36, because $6^2 = 36$.

The square of -6 is 36, because $(-6)^2 = 36$.

The **square root** of a given number is a number whose square is the given number. For example, the square roots of 36 are 6 and -6, because either number, when squared, yields 36. We can express this concept using symbols.

Square root

A number b is the **square root** of a if $b^2 = a$.

Self Check 1

Find the square roots of 64.

Answers 8 and -8

EXAMPLE 1 Find the square roots of 49.

Solution Ask yourself, What number was squared to obtain 49? The two answers are

$$7^2 = 49 \quad \text{and} \quad (-7)^2 = 49$$

Thus, 7 and -7 are the square roots of 49.

In Example 1, we saw that 49 has two square roots—one positive and one negative. The symbol $\sqrt{\ }$ is called a **radical symbol** and is used to indicate a positive square root.

When a number, called the **radicand,** is written under a radical symbol, we have a **radical expression.** Some examples of radical expressions are

$$\sqrt{36} \qquad \sqrt{100} \qquad \sqrt{144} \qquad \sqrt{81}$$

In the radical expression $\sqrt{36}$, 36 is called the radicand.

To evaluate (or simplify) a radical expression, we need to find the positive square root of the radicand. For example, if we evaluate $\sqrt{36}$ (read as "the square root of 36"), the result is

$$\sqrt{36} = 6$$

because $6^2 = 36$. The negative square root of 36 is denoted $-\sqrt{36}$, and we have

$$-\sqrt{36} = -6$$

Read as "the negative square root of 36 is -6" or "the opposite of the square root of 36 is -6."

Self Check 2

Simplify each expression:

a. $\sqrt{144}$ and **b.** $-\sqrt{81}$.

Answers **a.** 12, **b.** -9

EXAMPLE 2 Simplify: **a.** $\sqrt{81}$ and **b.** $-\sqrt{100}$.

Solution

a. $\sqrt{81}$ means the positive square root of 81.

$$\sqrt{81} = 9, \text{ because } 9^2 = 81$$

b. $-\sqrt{100}$ means the opposite (or negative) of the square root of 100. Since $\sqrt{100} = 10$, we have

$$-\sqrt{100} = -10$$

! COMMENT Radical expressions such as

$$\sqrt{-36} \qquad \sqrt{-100} \qquad \sqrt{-144} \qquad \sqrt{-81}$$

do not represent real numbers. This is because there are no real numbers that, when squared, give a negative number.

Be careful to note the difference between expressions such as $-\sqrt{36}$ and $\sqrt{-36}$. We have seen that $-\sqrt{36}$ is a real number: $-\sqrt{36} = -6$. On the other hand, $\sqrt{-36}$ is not a real number.

Evaluating numerical expressions containing radicals

Numerical expressions can contain radical expressions. When applying the rules for the order of operations, we treat a radical expression as we would a power.

EXAMPLE 3 Evaluate: **a.** $\sqrt{64} + \sqrt{9}$ and **b.** $-\sqrt{25} - \sqrt{4}$.

Solution

a. $\sqrt{64} + \sqrt{9} = 8 + 3$ Evaluate each radical expression first.

$= 11$ Perform the addition.

b. $-\sqrt{25} - \sqrt{4} = -5 - 2$ Evaluate each radical expression first.

$= -7$ Perform the subtraction.

Self Check 3
Evaluate:
a. $\sqrt{121} + \sqrt{1}$ and
b. $-\sqrt{9} - \sqrt{16}$.

Answers **a.** 12, **b.** -7

EXAMPLE 4 Evaluate: **a.** $6\sqrt{100}$ and **b.** $-5\sqrt{16} + 3\sqrt{9}$.

Solution

a. We note that $6\sqrt{100}$ means $6 \cdot \sqrt{100}$.

$6\sqrt{100} = 6(10)$ Simplify the radical first.

$= 60$ Perform the multiplication.

b. $-5\sqrt{16} + 3\sqrt{9} = -5(4) + 3(3)$ Simplify each radical first.

$= -20 + 9$ Perform the multiplications.

$= -11$ Perform the addition.

Self Check 4
Evaluate each expression:
a. $8\sqrt{64}$ and
b. $-6\sqrt{25} + 2\sqrt{36}$.

Answers **a.** 64, **b.** -18

Square roots of fractions and decimals

So far, we have found square roots of whole numbers. We can also find square roots of fractions and decimals.

EXAMPLE 5 Find each square root: **a.** $\sqrt{\frac{25}{64}}$ and **b.** $\sqrt{0.81}$.

Solution

a. $\sqrt{\frac{25}{64}} = \frac{5}{8}$, because $\left(\frac{5}{8}\right)^2 = \frac{25}{64}$.

b. $\sqrt{0.81} = 0.9$, because $(0.9)^2 = 0.81$.

Self Check 5
Find each square root:
a. $\sqrt{\frac{16}{49}}$ and **b.** $\sqrt{0.04}$.

Answers **a.** $\frac{4}{7}$, **b.** 0.2

Using a calculator to find square roots

We can also use a calculator to find square roots.

CALCULATOR SNAPSHOT

Finding a square root

We use the $\boxed{\sqrt{\ }}$ key (square root key) on a scientific calculator to find square roots. For example, to find $\sqrt{729}$, we enter these numbers and press these keys.

729 $\boxed{\sqrt{\ }}$ 27

We have found that $\sqrt{729} = 27$. To check this result, we need to square 27. This can be done by entering the numbers 2 and 7 and pressing the $\boxed{x^2}$ key. We obtain 729. Thus, 27 is the square root of 729.

Approximating square roots

Numbers whose square roots are whole numbers are called **perfect squares.** The perfect squares that are less than or equal to 100 are

0, 1, 4, 9, 16, 25, 36, 49, 64, 81, 100

To find the square root of a number that is not a perfect square, we can use a calculator. For example, to find $\sqrt{17}$, we enter these numbers and press the square root key.

17 $\sqrt{\ }$

The display reads 4.123105626. This result is not exact, because $\sqrt{17}$ is a **nonterminating decimal** that never repeats. $\sqrt{17}$ is an *irrational number.* Together, the rational and the irrational numbers form the set of *real numbers.* Rounding to the nearest thousandth, we have

$$\sqrt{17} = 4.123$$

EXAMPLE 6 Use a scientific calculator to find each square root. Round to the nearest hundredth.

a. $\sqrt{373}$, **b.** $\sqrt{56.2}$, and **c.** $\sqrt{0.0045}$.

Solution

a. From the calculator, we get $\sqrt{373} \approx 19.31320792$. Rounded to the nearest hundredth, $\sqrt{373}$ is 19.31.

b. From the calculator, we get $\sqrt{56.2} \approx 7.496665926$. Rounded to the nearest hundredth, $\sqrt{56.2}$ is 7.50.

c. From the calculator, we get $\sqrt{0.0045} \approx 0.067082039$. Rounded to the nearest hundredth, $\sqrt{0.0045}$ is 0.07.

Self Check 6

Use a scientific calculator to find each square root. Round to the nearest hundredth.

a. $\sqrt{607.8}$

b. $\sqrt{0.076}$

Answers **a.** 24.65, **b.** 0.28

Section 4.7 STUDY SET

VOCABULARY *Fill in the blanks.*

1. When we find what number is squared to obtain a given number, we are finding the square ______ of the given number.
2. Whole numbers such as 25, 36, and 49 are called ________ squares because their square roots are whole numbers.
3. The symbol $\sqrt{\ }$ is called a ________ symbol. It indicates that we are to find a ________ square root.
4. The decimal number that represents $\sqrt{17}$ is a ______________ decimal — it never ends.
5. In $\sqrt{26}$, the number 26 is called the _________.
6. The symbol $\approx$ means ____ _____________ _______ ____.

CONCEPTS *Fill in the blanks.*

7. The square of 5 is ▢, because $(5)^2 =$ ▢.
8. The square of $\frac{1}{4}$ is ▢, because $\left(\frac{1}{4}\right)^2 =$ ▢.
9. $\sqrt{49} = 7$, because ▢ $= 49$.
10. $\sqrt{4} = 2$, because ▢ $= 4$.
11. $\sqrt{\frac{9}{16}} =$ ▢, because $\left(\frac{3}{4}\right)^2 = \frac{9}{16}$.
12. $\sqrt{0.16} =$ ▢, because $(0.4)^2 = 0.16$.

13. Without evaluating the following square roots, write them in order, from smallest to largest: $\sqrt{23}$, $\sqrt{11}$, $\sqrt{27}$, $\sqrt{6}$.
14. Without evaluating the following square roots, write them in order from smallest to largest: $-\sqrt{13}$, $-\sqrt{5}$, $-\sqrt{17}$, $-\sqrt{37}$.
15. Find each square root.
    a. $\sqrt{1}$ b. $\sqrt{0}$
16. Multiplication can be thought of as the opposite of division. What is the opposite of finding the square root of a number?

Use a calculator.

17. a. Use a calculator to approximate $\sqrt{6}$ to the nearest tenth.
    b. Square the result from part a.
    c. Find the difference between 6 and the answer to part b.
18. a. Use a calculator to approximate $\sqrt{6}$ to the nearest hundredth.
    b. Square the result from part a.
    c. Find the difference between the answer to part b and 6.
19. Graph: $\sqrt{9}$ and $-\sqrt{5}$.

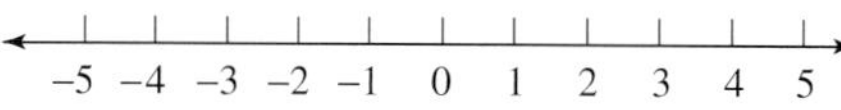

20. Graph: $-\sqrt{3}$ and $\sqrt{7}$.

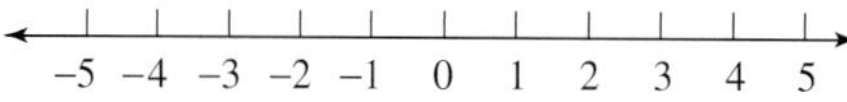

21. Between what two whole numbers would each square root be located when graphed on the number line?
    a. $\sqrt{19}$ b. $\sqrt{87}$
22. Between what two whole numbers would each square root be located when graphed on the number line?
    a. $\sqrt{33}$ b. $\sqrt{50}$

NOTATION *Complete each solution to evaluate the expression.*

23. $-\sqrt{49} + \sqrt{64} =$ ▢ $+$ ▢
    $= 1$
24. $2\sqrt{100} - 5\sqrt{25} = 2($ ▢ $) - 5($ ▢ $)$
    $=$ ▢ $- 25$
    $= -5$

PRACTICE *Evaluate each expression without using a calculator.*

25. $\sqrt{16}$
26. $\sqrt{64}$
27. $-\sqrt{121}$
28. $-\sqrt{144}$
29. $-\sqrt{0.49}$
30. $-\sqrt{0.64}$
31. $\sqrt{0.25}$
32. $\sqrt{0.36}$
33. $\sqrt{0.09}$
34. $\sqrt{0.01}$
35. $-\sqrt{\frac{1}{81}}$
36. $-\sqrt{\frac{1}{4}}$
37. $-\sqrt{\frac{16}{9}}$
38. $-\sqrt{\frac{64}{25}}$
39. $\sqrt{\frac{4}{25}}$
40. $\sqrt{\frac{36}{121}}$

41. $5\sqrt{36} + 1$

42. $2 + 6\sqrt{16}$

43. $-4\sqrt{36} + 2\sqrt{4}$

44. $-6\sqrt{81} + 5\sqrt{1}$

45. $\sqrt{\frac{1}{16}} - \sqrt{\frac{9}{25}}$

46. $\sqrt{\frac{25}{9}} - \sqrt{\frac{64}{81}}$

47. $5(\sqrt{49})(-2)$

48. $(-\sqrt{64})(-2)(3)$

49. $\sqrt{0.04} + 2.36$

50. $\sqrt{0.25} + 4.7$

51. $-3\sqrt{1.44}$

52. $-2\sqrt{1.21}$

Use a calculator to complete each square root table. Round to the nearest thousandth when an answer is not exact.

53.

Number	Square root
1	
2	
3	
4	
5	
6	
7	
8	
9	
10	

54.

Number	Square root
10	
20	
30	
40	
50	
60	
70	
80	
90	
100	

Use a calculator to evaluate each of the following.

55. $\sqrt{1{,}369}$

56. $\sqrt{841}$

57. $\sqrt{3{,}721}$

58. $\sqrt{5{,}625}$

Use a calculator to approximate each of the following to the nearest hundredth.

59. $\sqrt{15}$

60. $\sqrt{51}$

61. $\sqrt{66}$

62. $\sqrt{204}$

Use a calculator to approximate each of the following to the nearest thousandth.

63. $\sqrt{24.05}$

64. $\sqrt{70.69}$

65. $-\sqrt{11.1}$

66. $\sqrt{0.145}$

Use a calculator to evaluate each expression. If an answer is not exact, round to the nearest ten thousandth.

67. $\sqrt{24{,}000{,}201}$

68. $-\sqrt{4.012009}$

69. $-\sqrt{0.00111}$

70. $\sqrt{\frac{27}{44}}$

APPLICATIONS *Square roots have been used to express various lengths. Solve each problem by evaluating any square roots. You may need to use a calculator. Round to the nearest tenth, if necessary.*

71. CARPENTRY Find the length of the slanted side of each roof truss.

a.

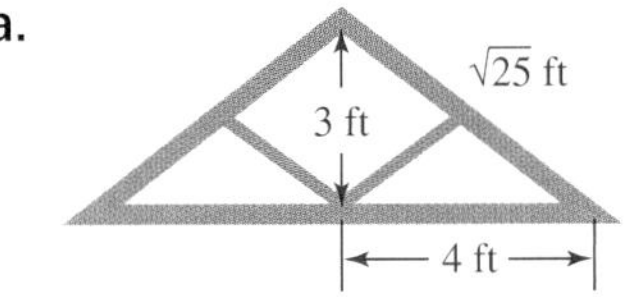

b.

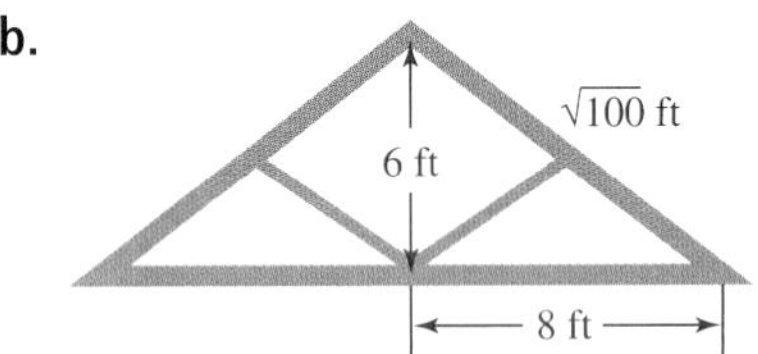

72. RADIO ANTENNAS How far from the base of the following antenna is each guy wire anchored to the ground? (The measurements are in feet.)

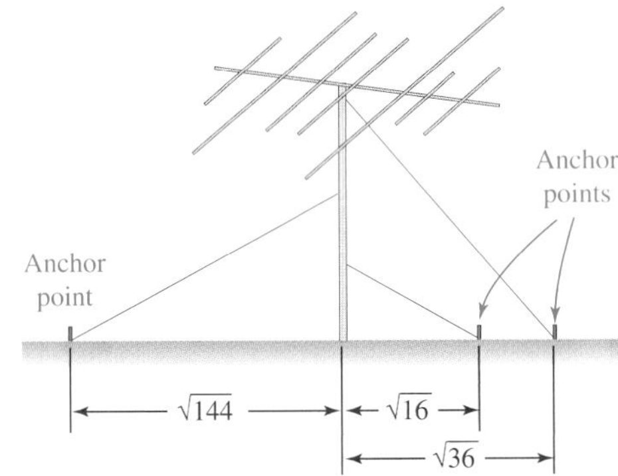

73. BASEBALL The illustration shows some dimensions of a major league baseball field. How far is it from home plate to second base?

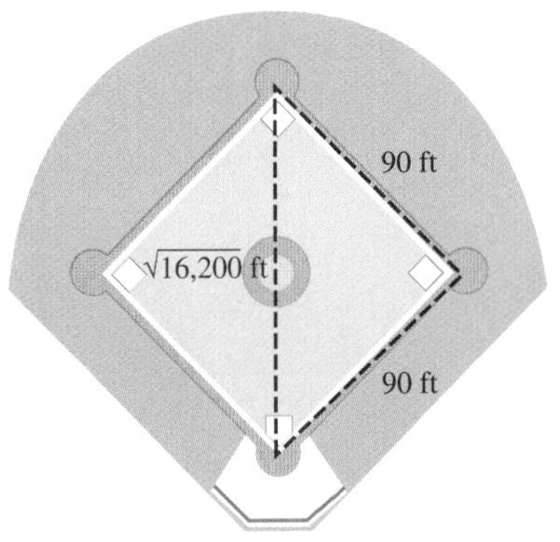

74. SURVEYING Use the imaginary triangles set up by a surveyor to find the length of each lake. (The measurements are in meters.)

a. Length: $\sqrt{318,096}$

b. Length: $\sqrt{93,025}$

75. BIG-SCREEN TELEVISIONS The picture screen on a television set is measured diagonally. What size screen is shown?

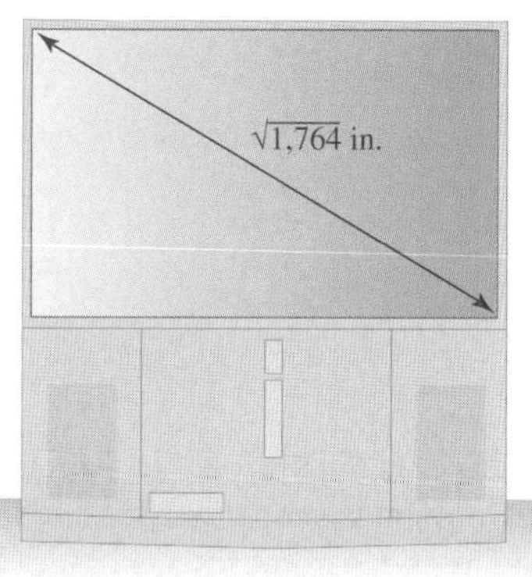

76. LADDERS A painter's ladder is shown. How long are the legs of the ladder?

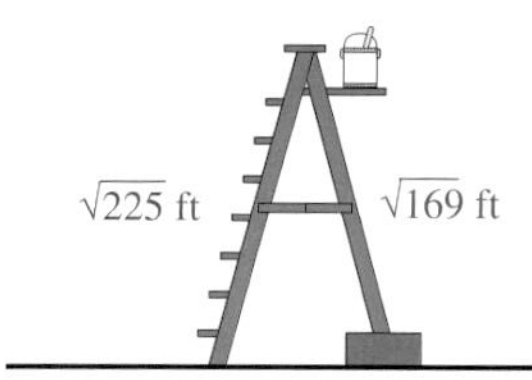

WRITING

77. When asked to find $\sqrt{16}$, a student answered 8. Explain his misunderstanding of square root.

78. Explain the difference between the square and the square root of a number.

79. What is a nonterminating decimal? Use an example in your explanation.

80. What do you think might be meant by the term *cube root?*

81. Explain why $\sqrt{-4}$ does not represent a real number.

82. Is there a difference between $-\sqrt{25}$ and $\sqrt{-25}$? Explain.

REVIEW

83. When solving the equation $2x - 5 = 11$, what operations must be undone in order to isolate the variable?

84. Simplify: $\dfrac{\frac{-2}{3}}{8}$.

85. Evaluate: $5(-2)^2 - \dfrac{16}{4}$.

86. Subtract: $\dfrac{5}{8} - \dfrac{3}{4}$.

87. Divide: $\dfrac{5}{8} \div \dfrac{3}{4}$.

88. List the set of whole numbers.

89. Solve: $8 + \dfrac{a}{5} = 14$.

90. Insert the proper symbol, $<$ or $>$, in the blank to make a true statement: -15 ☐ -14.

KEY CONCEPT

The Real Numbers

A **real number** is any number that can be expressed as a decimal. The set of real numbers corresponds to all points on a number line. All of the types of numbers that we have discussed in this book are real numbers. As we have seen, the set of real numbers is made up of several subsets of numbers.

If possible, list the numbers that belong to each set. If it is not possible to list them, define the set in words.

1. Natural numbers

2. Whole numbers

3. Integers

4. Rational numbers

5. Irrational numbers

This diagram shows how the set of real numbers is made up of two distinct sets: the rational and the irrational numbers. Since every natural number is a whole number, we show the set of natural numbers included in the whole numbers. Because every whole number is an integer, the whole numbers are shown contained in the integers. Since every integer is a rational number, we show the integers included in the rational numbers.

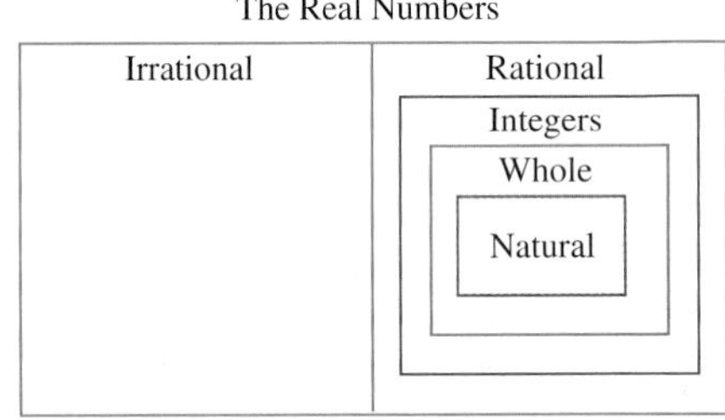

Determine whether each statement is true or false.

6. Every integer is a real number.

7. Every fraction can be written as a terminating decimal.

8. Every real number is a whole number.

9. Some irrational numbers are integers.

10. Some rational numbers are natural numbers.

11. No numbers are both rational and irrational numbers.

12. All real numbers can be graphed on a number line.

13. The set of whole numbers is a subset of the irrational numbers.

14. All decimals either terminate or repeat.

15. Every natural number is an integer.

16. List the numbers in the set $\{-2, -1.2, -\frac{7}{8}, 0, 1\frac{2}{3}, 2.75, \sqrt{23}, 10, 1.161661666\ldots\}$ that are

a. natural numbers

b. whole numbers

c. integers

d. rational numbers

e. irrational numbers

f. real numbers

ACCENT ON TEAMWORK

SECTION 4.1

ROUNDING

a. Find all the numbers that have three digits that round to 4.7.

b. Find all the numbers that have four digits that round to 8.09.

c. Find all numbers that have two digits that round to 0.0.

SECTION 4.2

VISUAL MODELS Shade a grid like the one in the illustration to compute each addition or subtraction.

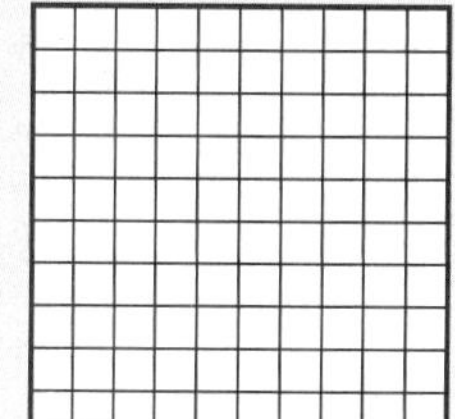

a. $0.62 + 0.24$ **b.** $0.45 - 0.41$

c. $0.21 + 0.29$ **d.** $0.98 - 0.18$

e. $0.2 + 0.17$ **f.** $0.57 - 0.3$

SECTION 4.3

SEQUENCES Multiplication by 2 is used to form the terms of the sequence 3, 6, 12, 24, 48, 96, ... That is, to form the second term (which is 6), we multiply the first term (which is 3) by 2. To get the third term (which is 12), we multiply the second term by 2. To get the fourth term, we multiply the third term by 2, and so on.

What multiplication is used to form the terms of each of the following sequences?

a. 0.2134, 2.134, 21.34, 213.4, 2,134, 21,340, . . .

b. 0.00005, 0.005, 0.5, 50, 5,000, . . .

c. 3, 0.9, 0.27, 0.081, 0.0243, 0.00729, . . .

d. 0.7, 0.07, 0.007, 0.0007, 0.00007, 0.000007, . . .

SECTION 4.4

Read the Think-It-Through feature on page 269 to learn how to compute a semester grade point average (GPA).

A student, taking the classes shown on the grade report below, had a semester GPA of 2.6. Determine what letter grade the student received in Spanish II.

Course no.	Course title	Units	Grade
101	Intro. Accounting	5.0	C
201	Intro. Psychology	3.0	A
102	Spanish II	4.0	?
142	Swimming	1.0	D
080	Keyboarding	2.0	C

SECTION 4.5

EQUIVALENT DECIMALS A student was asked to write several fractions as decimals. His answers, which are all incorrect, are shown below. What was he doing wrong?

$$\frac{2}{5} = 2.5 \qquad \frac{4}{15} = 3.75 \qquad \frac{3}{4} = 1.\overline{3}$$

SECTION 4.6

SOLVING EQUATIONS If an equation contains decimals, we can multiply both sides by a power of 10 to clear the equation of decimals. Multiply both sides of each equation by the appropriate power of 10; then solve it.

a. $0.6x = 3.6$

b. $0.8x + 0.4 = 5.2$

c. $0.62x + 1.24 = 5.58$

SECTION 4.7

A SPIRAL OF ROOTS To do this project, you will need a blank piece of paper, a ruler, a 3×5 index card, and a pencil. Begin by drawing a triangle with two sides 1 inch long, as shown. Use the corner of the 3×5 card to help draw the "sharp corner" (90-degree angle) of the triangle. Then draw the dashed blue line to complete the first triangle. It is $\sqrt{2}$ inches long.

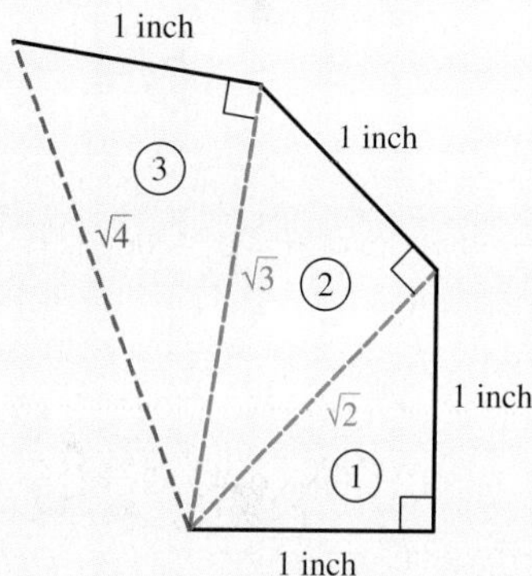

Next, create a second triangle using one side of the first triangle and drawing another side 1 inch long as shown. Complete the second triangle by drawing the dashed green line. It is $\sqrt{3}$ inches long. Draw a third triangle in a similar fashion. The dashed purple line is $\sqrt{4} = 2$ inches long. Draw a fourth triangle, a fifth triangle, and so on. If the pattern continues, what is the length of the dashed side of each new triangle?

EVALUATING SQUARE ROOTS A student was asked to evaluate the three square root expressions shown below. Examine his answers and then explain what he is doing wrong.

$$\sqrt{16} = 8 \qquad \sqrt{64} = 32 \qquad \sqrt{100} = 50$$

CHAPTER REVIEW

SECTION 4.1 An Introduction to Decimals

CONCEPTS

Decimal notation is used to denote part of a whole.

Expanded notation is used to show the value represented by each digit in the *decimal numeration system.*

$5.6791 =$

$5 + \frac{6}{10} + \frac{7}{100} + \frac{9}{1{,}000} + \frac{1}{10{,}000}$

To express a decimal in words, say:

1. the whole number to the left of the decimal point;
2. "and" for the decimal point;
3. the whole number to the right of the decimal point, followed by the name of the last place-value column on the right.

To compare the size of two decimals, compare the digits of each decimal, column by column, working from left to right.

A decimal point and additional zeros may be written to the right of a whole number.

REVIEW EXERCISES

1. Represent the amount of the square that is shaded, using a decimal and a fraction.

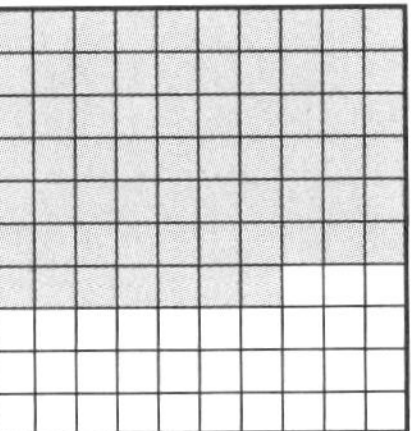

2. In the illustration, shade 0.8 of the rectangle.

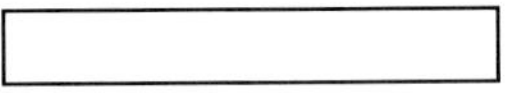

3. Write 16.4523 in expanded notation.

Write each decimal in words and then as a fraction or mixed number.

4. 2.3

5. -15.59

6. 0.0601

7. 0.00001

8. Graph: 1.55, -0.8, and -2.7.

–5 –4 –3 –2 –1 0 1 2 3 4 5

9. VALEDICTORIANS At the end of the school year, the five students listed were in the running to be class valedictorian. Rank the students in order by GPA, beginning with the valedictorian.

Name	GPA
Diaz, Cielo	3.9809
Chou, Wendy	3.9808
Washington, Shelly	3.9865
Gerbac, Lance	3.899
Singh, Amani	3.9713

10. True or false: $78 = 78.0$.

Place the proper symbol $<$, $>$, or $=$ in the blank to make a true statement.

11. 4.5 ☐ 4.6

12. -2.35 ☐ -2.53

13. 10.90 ☐ 10.9

14. 0.027894 ☐ 0.034

To round a decimal, locate the rounding digit and the test digit.

1. If the test digit is a number less than 5, drop it and all digits to the right of the rounding digit.
2. If the test digit is 5 or larger, add 1 to the rounding digit and drop all digits to its right.

Round each decimal to the specified place-value column.

15. 4.578: hundredths

16. 3,706.0895: thousandths

17. −0.0614: tenths

18. 88.12: tenths

SECTION 4.2 *Adding and Subtracting Decimals*

To add (or subtract) decimals:

1. Line up their decimal points.
2. Add (or subtract) as you would with whole numbers.
3. Write the decimal point in the result directly below the decimal points in the problem.

Perform each addition or subtraction.

19. $19.5 + 34.4 + 12.8$

20. $3.4 + 6.78 + 35 + 0.008$

21. $68.47 - 53.3$

22. $45.08 - 17.37$

Evaluate each expression.

23. $-16.1 + 8.4$

24. $-4.8 - (-7.9)$

25. $-3.55 + (-1.25)$

26. $-15.1 - 13.99$

27. $-8.8 + (-7.3 - 9.5)$

28. $(5 - 0.096) - (-0.035)$

29. SALE PRICES A calculator normally sells for $52.20. If it is being discounted $3.99, what is the sale price?

30. MICROWAVE OVENS A microwave oven is shown. How tall is the window?

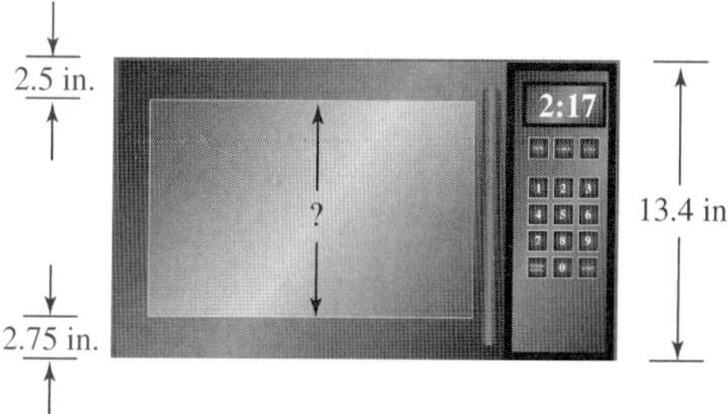

SECTION 4.3 *Multiplying Decimals*

To multiply decimals:

1. Multiply as if working with whole numbers.
2. Place the decimal point in the result so that the answer has the same number of decimal places as the total number of decimal places in the factors.

Perform each multiplication.

31. $(-0.6)(0.4)$

32. $2.3 \cdot 0.9$

33. $5.5(-3.1)$

34. $32.45(6.1)$

35. $(-0.003)(-0.02)$

36. $7 \cdot 0.6$

To multiply a decimal by a power of 10, move the decimal point to the right the same number of places as there are zeros in the power of 10.

Perform each multiplication in your head.

37. $1{,}000(90.1452)$

38. $(-10)(-2.897)(100)$

Exponents are used to represent repeated multiplication.

Find each power.

39. $(0.2)^2$ **40.** $(-0.15)^2$ **41.** $(3.3)^2$ **42.** $(0.1)^3$

Evaluate each expression.

43. $(0.6 + 0.7)^2 - 12.3$ **44.** $3(7.8) + 2(1.1)^2$

To evaluate an algebraic expression, substitute specific numbers for the variables in the expression and apply the rules for the order of operations.

45. $2(3.14)(4)^2 - 8.1$

46. WORD PROCESSORS The Page Setup screen for a word processor is shown. Find the area that can be filled with text on an 8.5-inch-by-11-inch piece of paper if the margins are set as shown.

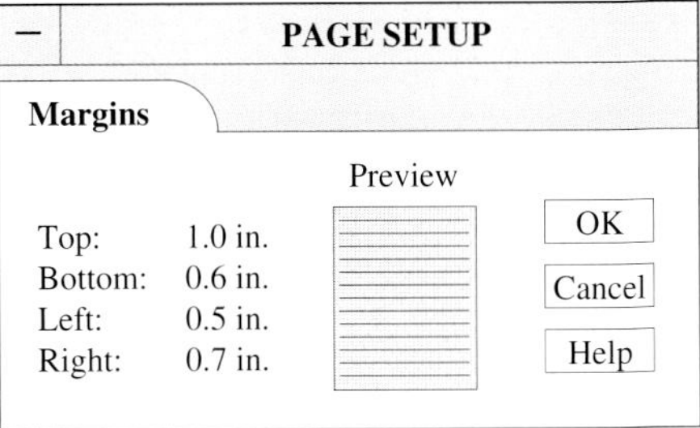

47. AUTO PAINTING A manufacturer uses a three-part process to finish the exterior of the cars it produces.

Step 1: A 0.03-inch-thick rust-prevention undercoat is applied.

Step 2: Three layers of color coat, each 0.015 inch thick, are sprayed on.

Step 3: The finish is then buffed down, losing 0.005 inch of its thickness.

What is the resulting thickness of the automobile's finish?

SECTION 4.4 *Dividing Decimals*

To divide a decimal by a whole number:

1. Divide as if working with whole numbers.
2. Write the decimal point in the result directly above the decimal point in the dividend.

Perform each division.

48. $12\overline{)15}$ **49.** $-41.8 \div 4$

50. $\dfrac{-29.67}{-23}$ **51.** $24.618 \div 6$

52. $12.47 \div (-4.3)$ **53.** $\dfrac{0.0742}{1.4}$

54. $\dfrac{15.75}{0.25}$ **55.** $\dfrac{-0.03726}{-0.046}$

To divide by a decimal:

1. Move the decimal point in the divisor so that it becomes a whole number.
2. Move the decimal point in the dividend the same number of places to the right.
3. Use the process for dividing a decimal by a whole number.

Divide and round each result to the nearest tenth.

56. $78.98 \div 6.1$ **57.** $\dfrac{-5.338}{0.008}$

58. Evaluate $\dfrac{5(68.4 - 32)}{9}$ and round to the nearest hundredth.

59. THANKSGIVING DINNER The cost of purchasing the ingredients for a Thanksgiving turkey dinner for a family of 5 was \$41.70. What was the cost of the dinner per person?

To divide a decimal by a power of 10, move the decimal point to the left the same number of places as there are zeros in the power of 10.

Perform each division in your head.

60. $89.76 \div 100$

61. $\dfrac{0.0112}{-10}$

62. Evaluate: $\dfrac{(1.4)^2 + 2(4.6)}{0.5 + 0.3}$

63. SERVING SIZE The illustration shows the package labeling on a box of children's cereal. Use the information given to find the number of servings.

Nutrition Facts	
Serving size	1.1 ounce
Servings per container	?
Package weight	15.5 ounces

64. TELESCOPES To change the position of a focusing mirror on a telescope, an adjustment knob is used. The mirror moves 0.025 inch with each revolution of the knob. The mirror needs to be moved 0.2375 inch to improve the sharpness of the image. How many revolutions of the adjustment knob does this require?

SECTION 4.5 *Fractions and Decimals*

To write a fraction as a decimal, divide the numerator by the denominator.

Write each fraction in decimal form.

65. $\dfrac{7}{8}$

66. $-\dfrac{2}{5}$

67. $-\dfrac{9}{16}$

68. $\dfrac{3}{50}$

We obtain either a *terminating* or a *repeating* decimal when using division to write a fraction as a decimal.

Write each fraction in decimal form. Use an overbar.

69. $\dfrac{6}{11}$

70. $-\dfrac{2}{3}$

An overbar can be used instead of the three dots . . . to represent the repeating pattern in a repeating decimal.

Write each fraction in decimal form. Round to the nearest hundredth.

71. $\dfrac{19}{33}$

72. $\dfrac{31}{30}$

Place the proper symbol $<$ or $>$ in the blank to make a true statement.

73. $\dfrac{13}{25}$ ☐ 0.499

74. $-0.\overline{26}$ ☐ $-\dfrac{4}{15}$

75. Graph $1\frac{1}{8}$, $-\frac{1}{3}$, $2\frac{3}{4}$, and $-\frac{9}{10}$ on the number line. Label each point using the decimal equivalent of the fraction or mixed number.

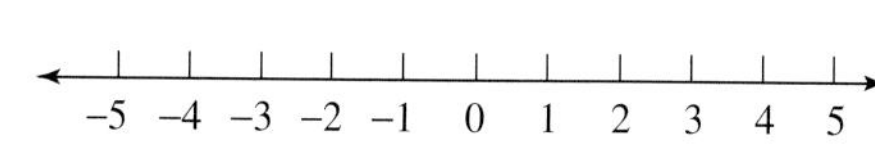

Evaluate each expression. Find the exact answer.

76. $\dfrac{1}{3} + 0.4$

77. $\dfrac{4}{5}(-7.8)$

78. $\dfrac{1}{2}(9.7 + 8.9)(10)$

79. $\dfrac{1}{3}(3.14)(3)^2(4.2)$

80. Evaluate: $\frac{4}{3}(3.14)(2)^3$. Round the result to the nearest hundredth.

81. ROADSIDE EMERGENCIES In case of trouble, truckers carry reflectors to be placed on the highway shoulder to warn approaching cars of a stalled vehicle. What is the area of the triangular reflector shown?

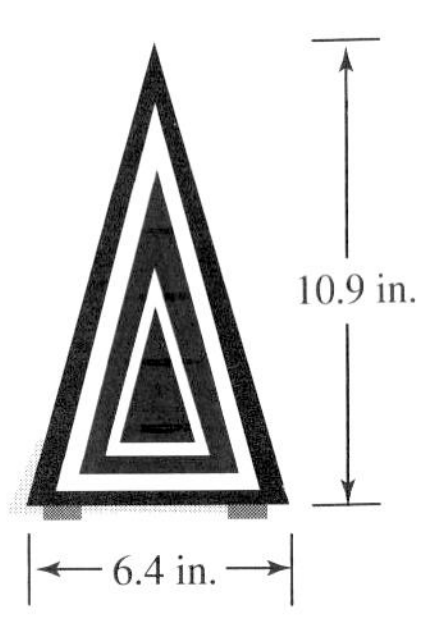

SECTION 4.6 Solving Equations Containing Decimals

The five-step problem-solving strategy:

1. Analyze the problem.
2. Form an equation.
3. Solve the equation.
4. State the conclusion.
5. Check the result.

Solve each equation.

82. $y + 12.4 = -6.01$

83. $0.23 + x = 5$

84. $\frac{x}{1.78} = -3$

85. $-1.61b = -27.37$

86. Is -1.1 a solution of $-1.3 = 1.2r + 0.02$?

87. BOWLING If it costs \$1.45 to rent shoes and 95 cents a game to use a lane, how many games can be bowled for \$10?

SECTION 4.7 Square Roots

The number b is a *square root* of a if $b^2 = a$.

88. Fill in the blanks.

a. The symbol $\sqrt{\ }$ is called a ________ symbol.

b. $\sqrt{64} = 8$, because ____ $= 64$.

A *radical symbol* $\sqrt{\ }$ is used to indicate a positive square root. The square root of a *perfect square* is a whole number.

Evaluate each expression without using a calculator.

89. $\sqrt{49}$

90. $-\sqrt{16}$

91. $\sqrt{100}$

92. $\sqrt{0.09}$

93. $\sqrt{\frac{64}{25}}$

94. $\sqrt{0.81}$

95. $-\sqrt{\frac{1}{36}}$

96. $\sqrt{0}$

97. Between what two whole numbers would $\sqrt{83}$ be located when graphed on a number line?

A square root can be approximated using a calculator.

98. Use a calculator to approximate $\sqrt{11}$. Round to the nearest tenth. Now square the approximation. How close is it to 11?

99. Graph each square root on the number line: $\sqrt{3}$, $-\sqrt{2}$, and $\sqrt{0}$.

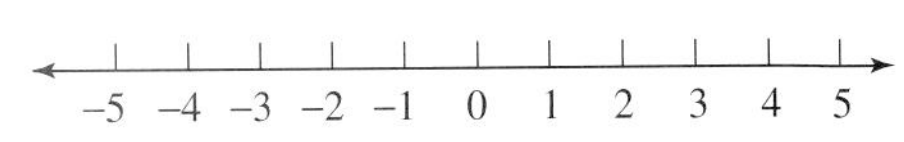

When evaluating an expression containing square roots, treat a radical as you would a power when applying the rules for the order of operations.

Evaluate each expression without using a calculator.

100. $-3\sqrt{100}$

101. $5\sqrt{0.25}$

102. $-3\sqrt{49} - \sqrt{36}$

103. $\sqrt{\frac{9}{100}} + \sqrt{1.44}$

104. Use a calculator to approximate $\sqrt{19}$ to the nearest hundredth.

CHAPTER 4 TEST

1. Express the amount of the square that is shaded, using a fraction and a decimal.

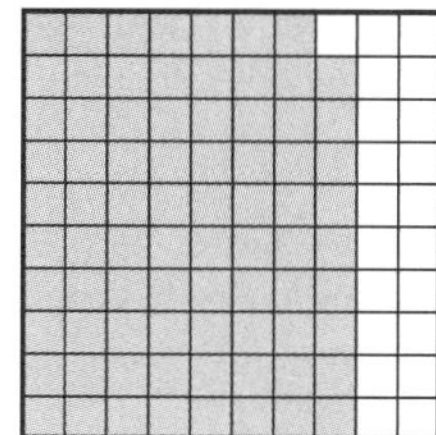

2. WATER PURITY A county health department sampled the pollution content of tap water in five cities, with the results shown. Rank the cities in order, from dirtiest tap water to cleanest.

City	Pollution, parts per million
Monroe	0.0909
Covington	0.0899
Paston	0.0901
Cadia	0.0890
Selway	0.1001

3. Write each decimal in words and then as a fraction or mixed number. **Do not simplify the fraction.**
   a. SKATEBOARDING Gary Hardwick of Carlsbad, California, set the skateboard speed record of 62.55 mph in 1998.

   b. MONEY A dime weighs 0.08013 ounce.

4. Round to the nearest thousandth: 33.0495.
5. SKATING RECEIPTS At an ice-skating complex, receipts on Friday were \$30.25 for indoor skating and \$62.25 for outdoor skating. On Saturday, the corresponding amounts were \$40.50 and \$75.75. Find the total receipts for the two days.
6. Perform each operation in your head.
   a. $567.909 \div 1{,}000$ b. $0.00458 \cdot 100$
7. EARTHQUAKE FAULT LINES After an earthquake, geologists found that the ground on the west side of the fault line had dropped 0.83 inch. The next week, a strong aftershock caused the same area to sink 0.19 inch deeper. How far did the ground on the west side of the fault drop because of the seismic activity?

Perform each operation.

8. $2 + 4.56 + 0.89 + 3.3$
9. $45.2 - 39.079$
10. $(0.32)^2$
11. $-6.7(-2.1)$
12. NEW YORK CITY Central Park, which lies in the middle of Manhattan, is the city's best-known park. If it is 2.5 miles long and 0.5 mile wide, what is its area?

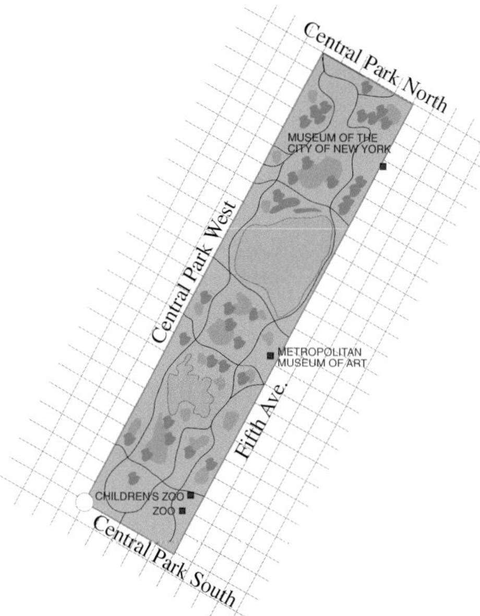

13. TELEPHONE BOOKS To print a telephone book, 565 sheets of paper were used. If the book is 2.3 inches thick, what is the thickness of each sheet of paper? (Round to the nearest thousandth of an inch.)
14. Evaluate: $4.1 - (3.2)(0.4)^2$.
15. Write each fraction as a decimal.
   a. $\frac{17}{50}$ b. $\frac{5}{12}$
16. Perform the division and round to the nearest hundredth: $\frac{12.146}{-5.3}$.
17. Divide: $11\overline{)13}$.

18. Graph $\frac{3}{8}$ and $-\frac{4}{5}$ on the number line. Label each point using the decimal equivalent of the given fractions.

19. Find the exact answer: $\frac{2}{3} + 0.7$.

Solve each equation.

20. $-2.4t = 16.8$

21. $-0.008 + x = 6$

22. $-0.53 = 0.0225 + 1.3x$.

23. CHEMISTRY In a lab experiment, a chemist mixed three compounds together to form a mixture weighing 4.37 g. Later, she discovered that she had forgotten to record the weight of compound C in her notes. Find the weight of compound C used in the experiment.

	Weight
Compound A	1.86 g
Compound B	2.09 g
Compound C	?
Mixture total	4.37 g

24. WEDDING COSTS A printer charges a setup fee of \$24 and then 95 cents for each wedding announcement printed (tax included). If a couple has budgeted \$100 for printing costs, how many announcements can they have made?

25. Fill in the blank: $\sqrt{144} = 12$, because $\quad = 144$.

26. Graph: $\sqrt{2}$ and $-\sqrt{5}$.

−5 −4 −3 −2 −1 0 1 2 3 4 5

Evaluate each expression.

27. $-2\sqrt{25} + 3\sqrt{49}$

28. $\sqrt{\frac{1}{36}} - \sqrt{\frac{1}{25}}$

Insert the proper symbol < or > to make a true statement.

29. $-6.78 \quad -6.79$

30. $\frac{3}{8} \quad 0.3$

31. $\sqrt{\frac{16}{81}} \quad \frac{16}{81}$

32. $0.45 \quad 0.\overline{45}$

Find each square root.

33. $-\sqrt{0.04}$

34. $\sqrt{1.69}$

35. Although the decimal 3.2999 contains more digits than 3.3, it is smaller than 3.3. Explain why this is so.

36. What is a repeating decimal? Give an example.

CHAPTERS 1–4 CUMULATIVE REVIEW EXERCISES

1. THE EXECUTIVE BRANCH The annual salaries for the President and the Vice President of the United States are \$400,000 and \$203,000, respectively. How much more money does the President make than the Vice President during a four-year term?

2. Use the variables x, y, and z to write the associative property of addition.

3. Divide: $43\overline{)1,161}$.

4. How many thousands are there in one million?

5. Find the prime factorization of 220.

6. List the factors of 20, from least to greatest.

7. List the set of whole numbers.

8. Add: $-8 + (-5)$.

9. Fill in the blank to make the statement true: Subtraction is the same as ______ the opposite.

10. Complete the solution to evaluate the expression.

$$\begin{aligned}(-6)^2 - 2(5 - 4 \cdot 2) &= (-6)^2 - 2(5 - \square) \\ &= (-6)^2 - 2(\square) \\ &= \square - 2(-3) \\ &= 36 - (\square) \\ &= 36 + \square \\ &= 42\end{aligned}$$

11. Consider the division statement $\frac{-15}{-5} = 3$. What is its related multiplication statement?

12. Find the power: $(-1)^5$.

13. Solve: $8 - 2d = -5 - 5$.

14. Solve: $0 = 6 + \frac{c}{-5}$.

15. Evaluate: $|-7(5)|$.

16. What is the opposite of -102?

17. Round 3.60745 to the nearest hundredth.

18. CHECKING ACCOUNTS After a deposit of \$995, a student's checking account was still \$105 overdrawn. What was the balance in the account before the deposit?

19. Solve: $7x - 38 = -3$.

20. What fraction of the stripes in the flag are white?

21. Although the fractions listed below look different, they all represent the same value. What concept does this illustrate?

$$\frac{1}{2} = \frac{2}{4} = \frac{3}{6} = \frac{4}{8} = \frac{5}{10} = \frac{6}{12}$$

22. Simplify: $\frac{90}{126}$.

Perform the operations.

23. $\frac{3}{8} \cdot \frac{7}{16}$

24. $-\frac{15}{8} \div \frac{10}{1}$

25. $\frac{4}{3} + \frac{2}{7}$

26. $-4\frac{1}{4}\left(-4\frac{1}{2}\right)$

27. $76\frac{1}{6} - 49\frac{7}{8}$

28. $\dfrac{\frac{5}{27}}{-\frac{5}{9}}$

29. Solve: $\frac{2}{3}y = -30$.

30. Solve: $\frac{d}{6} - \frac{2}{3} = \frac{2}{3}$.

31. KITES Find the area of the kite.

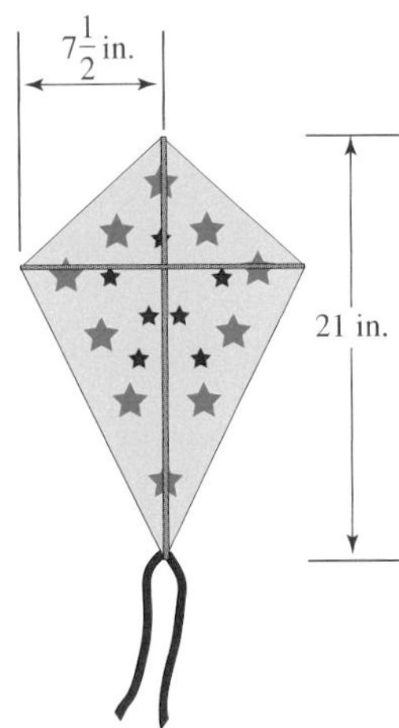

32. Graph: $\{-3\frac{1}{4}, 0.75, -1.5, -\frac{9}{8}, 3.8, \sqrt{4}\}$.

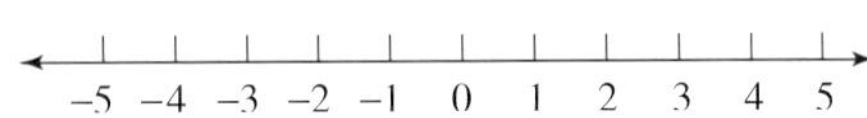

33. GLASS Some electronic and medical equipment uses glass that is only 0.00098 inch thick. Round this number to the nearest thousandth.

34. Place the proper symbol > or < in the box to make the statement true.

356.1978 ☐ 356.22

Perform the operations.

35. $-1.8(4.52)$

36. $\dfrac{-21.28}{-3.8}$

37. $56.012(100)$

38. $\dfrac{0.897}{10{,}000}$

39. Evaluate: $-9.1 - (-6.05 - 51)$.

40. WEEKLY SCHEDULES Refer to the illustration. Determine the number of hours during a week that a typical adult spends watching television.

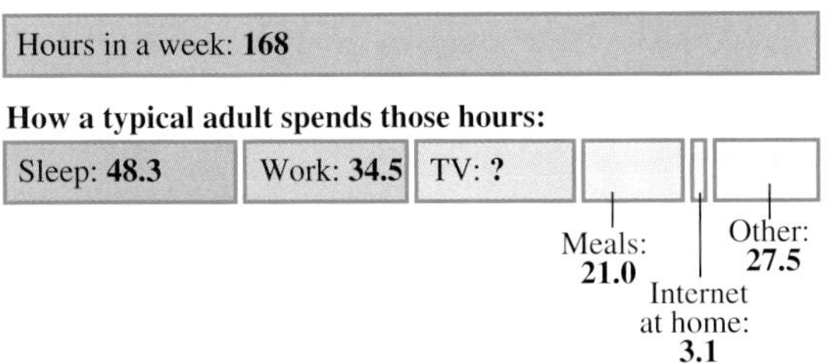

Source: National Sleep Foundation and the U.S. Bureau of Statistics

41. LITERATURE The novel *Fahrenheit 451,* by Ray Bradbury, is a story about censorship and book burning. Convert 451° F to degrees Celsius by evaluating $\dfrac{4(451 - 32)}{9}$. Round to the nearest tenth of a degree.

42. Write $\dfrac{5}{12}$ as a decimal. Use an overbar.

43. Solve: $-3 = 0.5t + 1.5$.

44. CONCESSIONAIRES At a ballpark, a vendor is paid \$22 a game plus 35¢ for each bag of peanuts she sells. How many bags of peanuts must she sell to make \$50 a game?

45. Evaluate: $-4\sqrt{36} + 2\sqrt{81}$.

46. Find the exact answer: $\dfrac{1}{3} + 0.4$.

CHAPTER 5

Percent

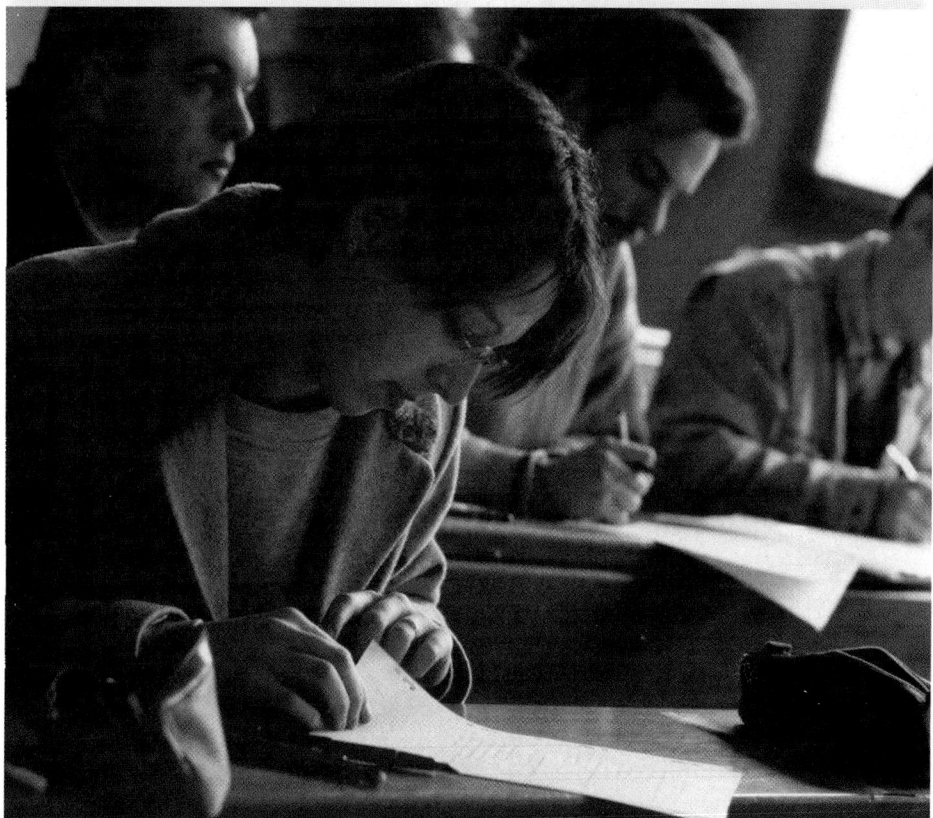

CORBIS

Percents are commonly used to present numerical information. The word percent comes from the Latin phrase *per centum*, which means parts per one hundred. Many instructors use a percent grading scale when evaluating their students' work. For example, if a student correctly answers 85 out of 100 true/false questions on a history exam, the student's grade on the exam can be expressed as 85%.

To learn more about percent and its many applications, visit The Learning Equation on the Internet at http://tle.brookscole.com. (The log-in instructions are in the Preface.) For Chapter 5, the online lesson is:

- *TLE* Lesson 11: Percent

Check Your Knowledge

1. Percent means parts per ________ ________.
2. In the statement "40 is 50% of 80," 40 is the ________, 50% is the ________, and 80 is the ______.
3. The difference between the original price and the sale price of an item is called the ________.
4. In banking, the original amount of money borrowed or deposited is known as the ________.
5. ________ interest is interest paid on the principal and the accumulated interest. Interest computed on only the original principal is called ________ interest.
6. Change each fraction to a decimal and to a percent.

   a. $\frac{3}{4}$ b. $\frac{5}{8}$ c. $\frac{29}{20}$

7. Change each decimal to a percent and to a fraction.

   a. 0.35 b. 3.98 c. 0.105

8. Change each percent to a decimal and to a fraction.

   a. 25% b. 200% c. 0.5%

9. If a glass is 65% full, what percent of the glass is empty?
10. Find the exact percent equivalent for each fraction.

    a. $\frac{3}{8}$ b. $\frac{5}{6}$

11. Change $\frac{2}{3}$ to a percent.

    a. Give the exact percent.

    b. Round your answer to part a to the nearest tenth of a percent.

12. What is 65% of 500?
13. What percent of 200 is 34?
14. 13 is 25% of what number?
15. What percent of 50 is 125?
16. A pen is normally priced at $14.95. The sale price is 20% off the normal price. What are the amount of the discount and the sale price?
17. A CD player is on sale for $24.95, which is $5.00 off the regular price. Find the regular price and the discount rate. Round to the nearest one percent.
18. If the sales tax rate is 8.25%, what is the price of an item when the sales tax amount is $2.47? Round to the nearest cent.
19. Find the simple interest on a $2,000.00 savings account for one year if the interest rate is 2.3%.
20. Find the account balance after two years on a $1,500.00 investment with earnings of 5.6% compounded quarterly.
21. Only 3 of 46 parking spaces in a parking lot were not taken. What percent of the parking lot spaces were filled? Round to the nearest one percent.
22. If Sheila borrows $1,000 with simple annual interest at 18%, what will be the payoff amount if she pays off the loan after 3 months?
23. If Leslie wants to tip 15% on a $35.40 meal, what will the total cost be, including the tip?
24. A quiz has 15 questions. Assuming that the questions are weighted equally, how many questions must George answer correctly to score at least 85%?

Getty Images

Study Skills Workshop

HELP OUTSIDE OF CLASS

Have you ever had the experience of understanding everything your instructor is saying in class, only to go home, try a homework problem, and be completely stumped? This is a common complaint among math students. The key to success is to take care of these problems before you go on to tackle new material. Below are some suggestions for finding help outside of your classroom.

Instructor Office Hours. Your instructor may hold office hours during the week to help students with questions. Usually these hours are listed in your syllabus, and you do not need to make appointments to see your instructor at these times. Remember to bring a list of questions that are giving you trouble and try to pinpoint exactly where in the process you are getting stuck. This will help your instructor answer your questions efficiently and effectively.

Tutorial Centers. Many colleges have tutorial centers where students can meet with a tutor, either one-on-one or in a group of students taking the same class. Tutorial centers usually offer their services for free and have regular hours of operation. When you visit your tutor, bring your list of questions that detail where in the process you're having difficulty.

Math Labs. Some colleges have math labs or learning centers where students can drop in at their convenience to have their math questions answered or where they can hang out and work on their homework. If this is available at your college, try to organize your study calendar to spend some time there doing your homework.

Study Groups. Study groups are groups of classmates who meet outside of class to discuss homework problems or study for tests. Study groups work best when they are relatively small (no more than four members), when they meet regularly, and when they follow these guidelines:

- Members should have attempted all homework problems before meeting.
- No one person in the group should be responsible for doing all of the work or all of the explaining.
- The group should meet in a place where members can spread out and talk, not in a quiet area of the library.
- Members should practice verbalizing and explaining processes and concepts to others in the study group. The best way to really learn a topic is by teaching it to someone else.

ASSIGNMENT

1. List your instructor's office hours and location. Next, pay a visit to your instructor during his or her office this week, even if you don't have any homework questions.
2. List the hours that your college's tutorial center is open, the location of the center, and how to make an appointment with a tutor.
3. Find out whether your college has a math lab or learning center. If so, list its hours of operation, location, and rules.
4. Find at least two other students who can meet for a study group. Plan to meet two days before your next homework assignment is due and follow the guidelines given above. After your group has met, evaluate how well it worked. Is there anything you might do to make it better the next time?

Percents are based on the number 100. They offer us a standardized way to measure and describe many situations in our daily lives.

5.1 Percents, Decimals, and Fractions

- The meaning of percent
- Changing a percent to a fraction
- Changing a percent to a decimal
- Changing a decimal to a percent
- Changing a fraction to a percent

Percents are a popular way to present numeric information. Stores use them to advertise discounts, manufacturers use them to describe the content of their products, and banks use them to list interest rates for loans and savings accounts. Newspapers are full of statistics presented in percent form. In this section, we introduce percent and show how fractions, decimals, and percents are interrelated.

The meaning of percent

A percent tells us the number of parts per 100. You can think of a percent as the *numerator* of a fraction that has a denominator of 100.

Percent

Percent means parts per one hundred.

In Figure 5-1, 93 out of 100 equal-sized squares are shaded. Thus, $\frac{93}{100}$ or 93 percent of the figure is shaded. The word *percent* can be written using the symbol %, so 93% of Figure 5-1 is shaded.

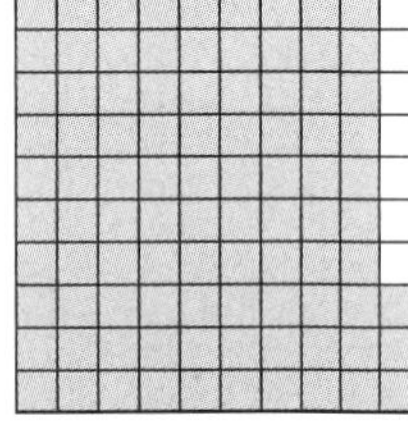

FIGURE 5-1

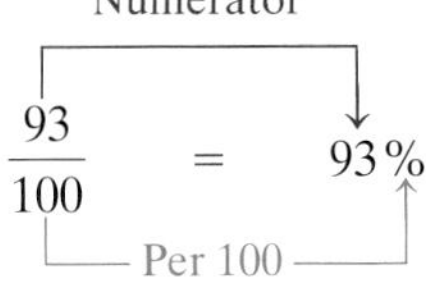

If the entire grid in Figure 5-1 had been shaded, we would say that 100 out of the 100 squares, or 100%, was shaded. Using this fact, we can determine what percent of the figure is *not* shaded by subtracting the percent of the figure that is shaded from 100%.

$$100\% - 93\% = 7\%$$

So 7% of Figure 5-1 is not shaded.

Changing a percent to a fraction

To change a percent to an equivalent fraction, we use the definition of percent.

Changing a percent to a fraction

To change a percent to a fraction, drop the % symbol and write the given number over 100. Then simplify the fraction, if possible.

EXAMPLE 1 **Earth's atmosphere.** The chemical makeup of the Earth's atmosphere is 78% nitrogen, 21% oxygen, and 1% other gases. Write each percent as a fraction.

Solution We begin with nitrogen.

$78\% = \frac{78}{100}$ Use the definition of percent: 78% means 78 parts per one hundred. This fraction can be simplified.

$= \frac{\overset{1}{\cancel{2}} \cdot 39}{\underset{1}{\cancel{2}} \cdot 50}$ Factor 78 as $2 \cdot 39$ and 100 as $2 \cdot 50$. Divide out the common factor of 2.

$= \frac{39}{50}$ Perform the multiplication in the numerator and in the denominator.

Nitrogen makes up $\frac{78}{100}$, or $\frac{39}{50}$, of the Earth's atmosphere.

Oxygen makes up 21% or $\frac{21}{100}$ of the Earth's atmosphere. Other gases make up 1% or $\frac{1}{100}$ of the atmosphere.

Self Check 1

An average watermelon is 92% water. Write this percent as a fraction.

Answer $\frac{23}{25}$

EXAMPLE 2 **Unions.** In 2003, 12.9% of the U.S. labor force belonged to a union. Write this percent as a fraction.

Solution

$12.9\% = \frac{12.9}{100}$ Drop the % symbol and write 12.9 over 100.

$= \frac{12.9 \cdot \mathbf{10}}{100 \cdot \mathbf{10}}$ To obtain a whole number in the numerator, multiply by 10. This will move the decimal point 1 place to the right. Multiply the denominator by 10 as well.

$= \frac{129}{1{,}000}$ Perform the multiplication in the numerator and in the denominator.

Thus, $12.9\% = \frac{129}{1{,}000}$. This means that 129 out of every 1,000 workers in the U.S. labor force belonged to a union in 2003.

Self Check 2

In 2002, 13.3% of the U.S. labor force belonged to a union. Write this percent as a fraction.

Answer $\frac{133}{1{,}000}$

Self Check 3

Write $83\frac{1}{3}\%$ as a fraction.

Answer $\frac{5}{6}$

EXAMPLE 3 Write $66\frac{2}{3}\%$ as a fraction.

Solution

$$66\frac{2}{3}\% = \frac{66\frac{2}{3}}{100}$$ Drop the % symbol and write $66\frac{2}{3}$ over 100.

$$= 66\frac{2}{3} \div 100$$ The fraction bar indicates division.

$$= \frac{200}{3} \cdot \frac{1}{100}$$ Change $66\frac{2}{3}$ to an improper fraction and then multiply by the reciprocal of 100.

$$= \frac{2 \cdot 100 \cdot 1}{3 \cdot 100}$$ Multiply the numerators and the denominators. Factor 200 as $2 \cdot 100$.

$$= \frac{2 \cdot \overset{1}{\cancel{100}} \cdot 1}{3 \cdot \underset{1}{\cancel{100}}}$$ Divide out the common factor of 100.

$$= \frac{2}{3}$$ Multiply in the numerator. Multiply in the denominator.

Changing a percent to a decimal

To write a percent as a decimal, recall that a percent can be written as a fraction with denominator 100, and that a denominator of 100 indicates division by 100.

Consider 14.25%, which means 14.25 parts per 100.

$$14.25\% = \frac{14.25}{100}$$ Use the definition of percent: write 14.25 over 100.

$$= 14.25 \div 100$$ The fraction bar indicates division.

$$= 0.14\,25$$ To divide a decimal by 100, move the decimal point 2 places to the left.

$$= 0.1425$$

This example suggests the following procedure.

Changing a percent to a decimal

To change a percent to a decimal, drop the % symbol and divide by 100 by moving the decimal point 2 places to the left.

Self Check 4

What percent of all music sold is produced on LPs (long-playing vinyl record albums)? Write the percent as a decimal.

EXAMPLE 4 **The music industry.** Figure 5-2 shows that the compact disc has become the format of choice among most consumers. What percent of all music sold is produced on CDs? Write the percent as a decimal.

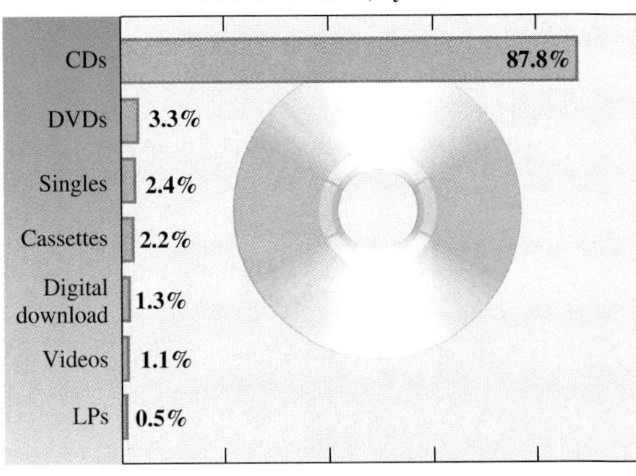

Source: *The Recording Industry Association of America*

FIGURE 5-2

Solution From the graph, we see that 87.8% of all music sold is produced on CDs. To write 87.8% as a decimal, we proceed as follows.

87.8% = .87 8	Drop the % symbol and divide by 100 by moving the decimal point 2 places to the left.
= 0.878	Write a 0 to the left of the decimal point.

Answer 0.5, 0.005

EXAMPLE 5 Write 310% as a decimal.

Solution The whole number 310 has an understood decimal point to the right of 0.

310% = 310.0%	Write a decimal point and a 0 to the right of 310.
= 3.10 0	Drop the % symbol and divide by 100 by moving the decimal point 2 places to the left.
= 3.100	
= 3.1	Drop the unnecessary zeros to the right of the 1.

Self Check 5

Write 600% as a decimal.

Answer 6

EXAMPLE 6 **Oregon.** The population of Oregon is approximately $1\frac{1}{4}\%$ of the population of the United States. Write this percent as a decimal.

Solution To change a percent to a decimal, we drop the % symbol and divide by 100 by moving the decimal point 2 places to the left. In this case, however, there is no decimal point in $1\frac{1}{4}\%$ to move. Since $1\frac{1}{4} = 1 + \frac{1}{4}$, and since the decimal equivalent of $\frac{1}{4}$ is 0.25, we can write $1\frac{1}{4}\%$ in an equivalent form as 1.25%.

$1\frac{1}{4}\% = 1.25\%$	Write $1\frac{1}{4}$ as 1.25.
= 0.01 25	Drop the % symbol and divide by 100 by moving the decimal point 2 places to the left.
= 0.0125	

Self Check 6

Write $15\frac{3}{4}\%$ as a decimal.

Answer 0.1575

Changing a decimal to a percent

To change a percent to a decimal, we drop the % symbol and move the decimal point 2 places to the left. To write a decimal as a percent, we move the decimal point 2 places to the right and insert a % symbol.

Changing a decimal to a percent

To change a decimal to a percent, multiply the decimal by 100 by moving the decimal point 2 places to the right, and then insert a % symbol.

Self Check 7
Write 0.5343 as a percent.

Answer 53.43%

EXAMPLE 7 Geography. Land areas make up 0.291 of the Earth's surface. Write this decimal as a percent.

Solution

$0.291 = 0\,29.1\%$ Multiply the decimal by 100 by moving the decimal point 2 places to the right, and then insert a % symbol.

$= 29.1\%$

Changing a fraction to a percent

We will use a two-step process to change a fraction to a percent. First, we write the fraction as a decimal. Then we change that decimal to a percent.

Fraction ⟶ Decimal ⟶ Percent

Changing a fraction to a percent

To change a fraction to a percent:

1. Write the fraction as a decimal by dividing its numerator by its denominator.
2. Multiply the decimal by 100 by moving the decimal point 2 places to the right.
3. Insert a % symbol.

Self Check 8
Write $\frac{7}{8}$ as a percent.

Answer 87.5%

EXAMPLE 8 Television. The highest-rated television show of all time was the episode of "M*A*S*H" that aired on February 28, 1983. Surveys found that three out of every five American households watched this show. Express the rating as a percent.

Solution 3 out of 5 we can express as $\frac{3}{5}$. We need to change this fraction to a decimal.

$$\begin{array}{r} 0.6 \\ 5\overline{)3.0} \\ \underline{3\,0} \\ 0 \end{array}$$

Write 3 as 3.0 and then divide the numerator by the denominator.

$\frac{3}{5} = 0.6$ The result is a terminating decimal.

$0.6 = 0\,60.\%$ Write a placeholder 0 to the right of the 6. Multiply the decimal by 100 by moving the decimal point 2 places to the right, and then insert a % symbol.

$= 60\%$

So 60% of American households watched the episode of "M*A*S*H."

In Example 8, the result of the division was a terminating decimal. Sometimes when we change a fraction to a decimal, the result of the division is a repeating decimal.

EXAMPLE 9 Write $\frac{5}{6}$ as a percent.

Solution The first step is to change $\frac{5}{6}$ to a decimal.

```
  0.8333
6)5.0000
  4 8
    20
    18
     20
     18
      20
```

Write 5 as 5.0000. Divide the numerator by the denominator.

$\frac{5}{6} = 0.8333\ldots$ The result is a repeating decimal.

$= 0\,83.33\ldots\%$ Change 0.8333. . . to a percent. Multiply the decimal by 100 by moving the decimal point 2 places to the right, and then insert a % symbol.

$= 83.33\ldots\%$ 83.333. . . is a repeating decimal.

We must now decide whether we want an approximation or an exact answer. For an approximation, we can round 83.333. . .% to a specific place value. For an exact answer, we can represent the repeating part of the decimal using an equivalent fraction.

Approximation

$\frac{5}{6} = 83.33\ldots\%$

$\approx 83.3\%$ Round to the nearest tenth.

$\frac{5}{6} \approx 83.3\%$

Exact answer

$\frac{5}{6} = 83.3333\ldots\%$

$= 83\frac{1}{3}\%$ Use the fraction $\frac{1}{3}$ to represent .333. . . .

$\frac{5}{6} = 83\frac{1}{3}\%$

Some percents occur so frequently that it is useful to memorize their fractional and decimal equivalents.

Percent	Decimal	Fraction
1%	0.01	$\frac{1}{100}$
10%	0.1	$\frac{1}{10}$
20%	0.2	$\frac{1}{5}$
25%	0.25	$\frac{1}{4}$

Percent	Decimal	Fraction
$33\frac{1}{3}\%$	0.3333. . .	$\frac{1}{3}$
50%	0.5	$\frac{1}{2}$
$66\frac{2}{3}\%$	0.6666. . .	$\frac{2}{3}$
75%	0.75	$\frac{3}{4}$

Section 5.1 STUDY SET

VOCABULARY *Fill in the blanks.*

1. ________ means parts per one hundred.
2. When we change a fraction to a decimal, the result is either a ____________ or a repeating decimal.

CONCEPTS *Fill in the blanks.*

3. To write a percent as a fraction, drop the % symbol and write the given number over ______. Then ________ the fraction, if possible.

4. To change a percent to a decimal, drop the % symbol and divide by 100 by moving the decimal point 2 places to the _____.
5. To change a decimal to a percent, multiply the decimal by 100 by moving the decimal point 2 places to the ______, and then insert a % symbol.
6. To write a fraction as a percent, first write the fraction as a ________. Then multiply the decimal by 100 by moving the decimal point 2 places to the ______, and insert a % symbol.

NOTATION

7. a. See the illustration. Express the amount of the figure that is shaded as a decimal, a percent, and a fraction.
   b. What percent of the figure is not shaded?

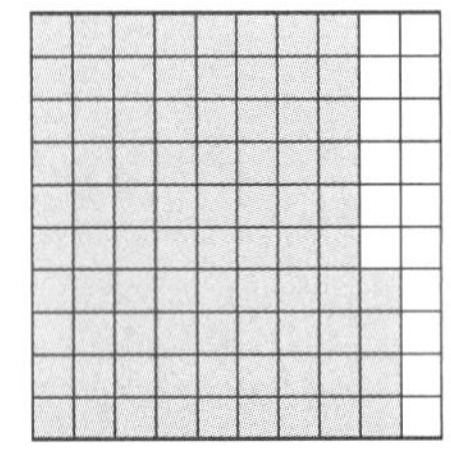

8. In the illustration below, each set of 100 squares represents 100%. What percent is shaded?

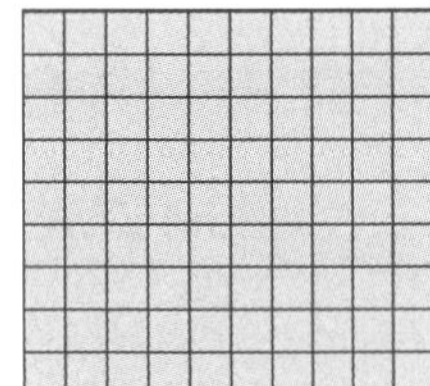

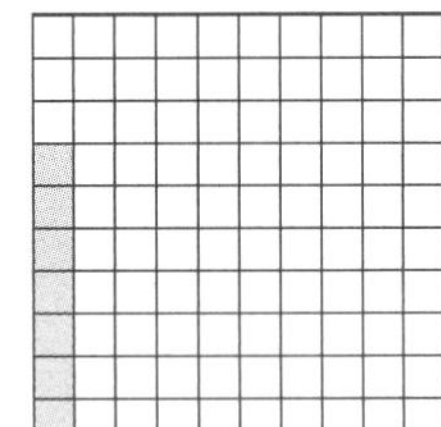

PRACTICE *Change each percent to a fraction. Simplify when necessary.*

9. 17%
10. 31%
11. 5%
12. 4%
13. 60%
14. 40%
15. 125%
16. 210%
17. $\frac{2}{3}\%$
18. $\frac{1}{5}\%$
19. $5\frac{1}{4}\%$
20. $6\frac{3}{4}\%$
21. 0.6%
22. 0.5%
23. 1.9%
24. 2.3%

Change each percent to a decimal.

25. 19%
26. 83%
27. 6%
28. 2%
29. 40.8%
30. 34.2%
31. 250%
32. 600%
33. 0.79%
34. 0.01%
35. $\frac{1}{4}\%$
36. $8\frac{1}{5}\%$

Change each decimal to a percent.

37. 0.93
38. 0.44
39. 0.612
40. 0.727
41. 0.0314
42. 0.0021
43. 8.43
44. 7.03
45. 50
46. 3
47. 9.1
48. 8.7

Change each fraction to a percent.

49. $\frac{17}{100}$
50. $\frac{29}{100}$
51. $\frac{4}{25}$
52. $\frac{47}{50}$
53. $\frac{2}{5}$
54. $\frac{21}{50}$
55. $\frac{21}{20}$
56. $\frac{33}{20}$
57. $\frac{5}{8}$
58. $\frac{3}{8}$
59. $\frac{3}{16}$
60. $\frac{1}{32}$

Find the exact equivalent percent for each fraction.

61. $\frac{2}{3}$
62. $\frac{1}{6}$
63. $\frac{1}{12}$
64. $\frac{4}{3}$

Express each of the given fractions as a percent. Round to the nearest hundredth of a percent.

65. $\frac{1}{9}$
66. $\frac{2}{3}$
67. $\frac{5}{9}$
68. $\frac{7}{3}$

APPLICATIONS

69. U.N. SECURITY COUNCIL The United Nations has 191 members. The United States, Russia, the United Kingdom, France, and China, along with 10 other nations, make up the Security Council.
   a. What fraction of the members of the United Nations belong to the Security Council?
   b. Write your answer to part a in percent form. (Round to the nearest one percent.)

70. ECONOMIC FORECASTS One economic indicator of the national economy is the number of orders placed by manufacturers. One month, the number of orders rose one-fourth of one percent.
   a. Write this using a % symbol.
   b. Express it as a fraction.
   c. Express it as a decimal.

71. PIANO KEYS Of the 88 keys on a piano, 36 are black.
   a. What fraction of the keys are black?
   b. What percent of the keys are black? (Round to the nearest one percent.)

72. INTEREST RATES Write as a decimal the interest rate associated with each of these accounts.
   a. Home loan: 7.75%
   b. Savings account: 5%
   c. Credit card: 14.25%

73. THE HUMAN SPINE The human spine consists of a group of bones (vertebrae).
   a. What fraction of the vertebrae are lumbar?
   b. What percent of the vertebrae are lumbar? (Round to the nearest one percent.)
   c. What percent of the vertebrae are cervical? (Round to the nearest one percent.)

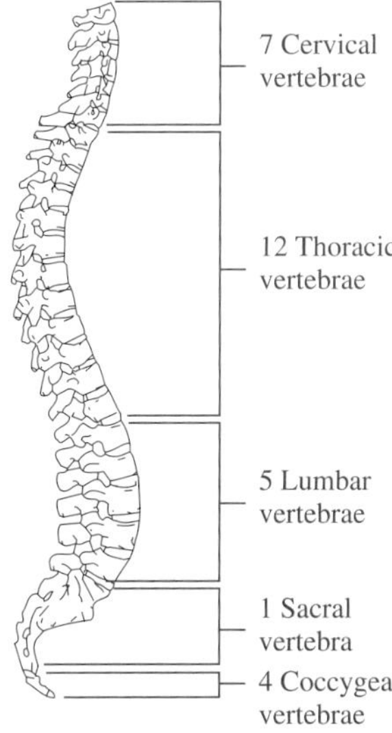

74. REGIONS OF THE COUNTRY The continental United States is divided into seven regions.
   a. What percent of the 50 states are in the Rocky Mountain region?
   b. What percent of the 50 states are in the Midwestern region?
   c. What percent of the 50 states are not located in any of the seven regions shown here?

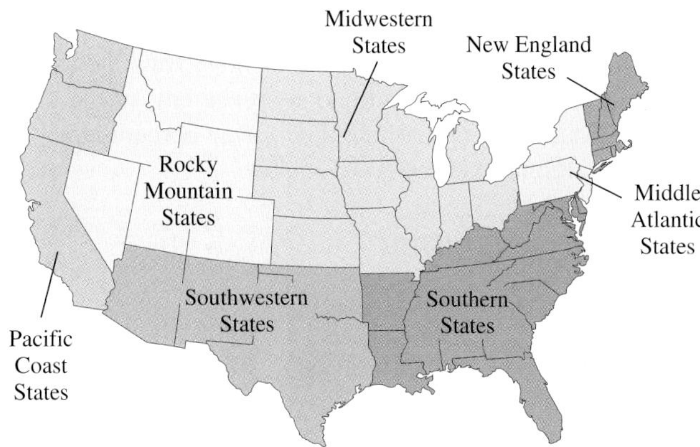

75. STEEP GRADES Sometimes, signs are used to warn truckers when they are approaching a steep grade on the highway. For a 5% grade, how many feet does the road rise over a 100-foot run?

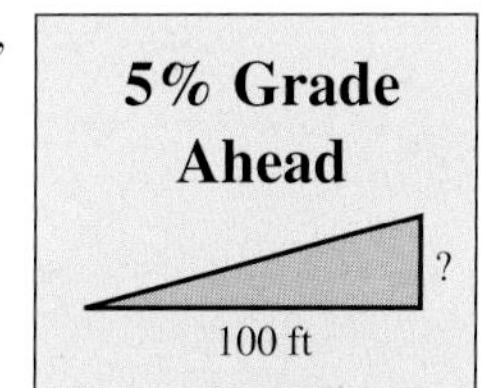

76. COMPANY LOGOS In the illustration, what part of the company's logo is red? Express your answer as a percent, a fraction, and a decimal. **Do not round.**

Recycling Industries Inc.

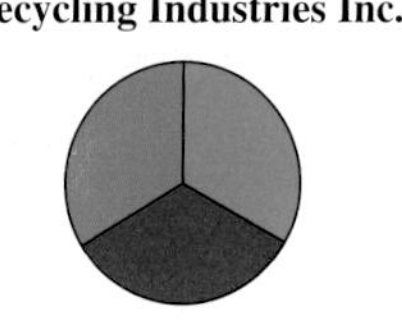

77. SOAP Ivory soap claims to be $99\frac{44}{100}$% pure. Write this percent as a decimal.

78. DRUNK DRIVING In most states, it is illegal to drive with a blood alcohol concentration of 0.08% or more. Change this percent to a fraction. **Do not simplify.** Explain what the numerator and the denominator of the fraction represent.

79. BASKETBALL In the following table, we see that Chicago has won 60 of 67, or $\frac{60}{67}$ of its games. In what form is the team's winning percentage presented in the newspaper? Express it as a percent.

Eastern conference			
Team	W	L	Pct.
Chicago	60	7	.896

80. WON-LOST RECORDS In sports, when a team wins as many as it loses, it is said to be playing "500 ball." Examine the following table and explain the significance of the number 500.

Eastern conference			
Team	W	L	Pct.
Orlando	33	33	.500

81. HUMAN SKIN The illustration shows roughly what percent each section of the body represents of the total skin area. Determine the missing percent, and then complete the bar graph below.

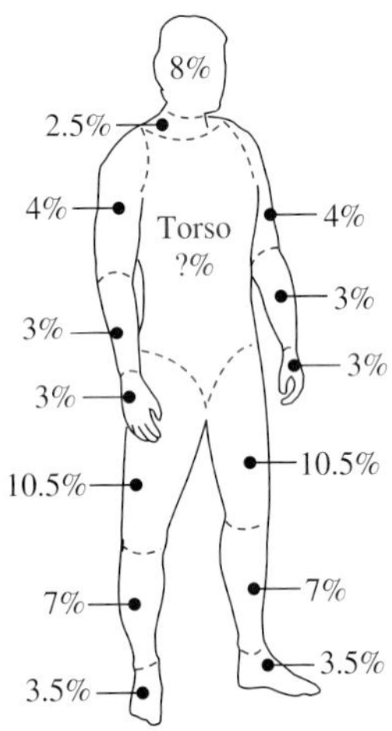

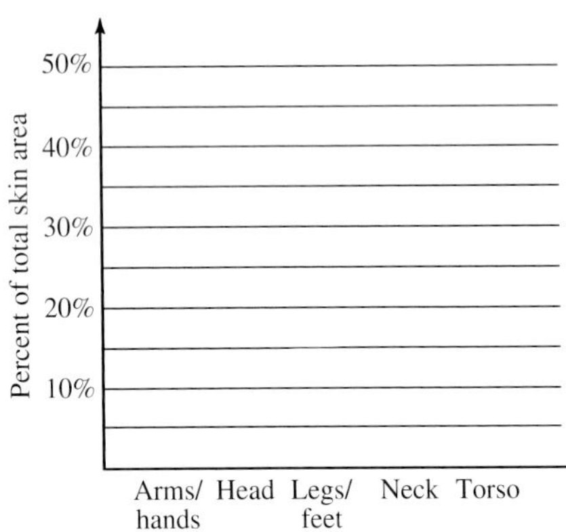

82. RAP MUSIC The following table shows what percent rap/hip-hop music sales were of total U.S. dollar sales of recorded music for the years 1997–2003. On the illustration in the next column, construct a line graph using the given data.

1997	1998	1999	2000	2001	2002	2003
10.1%	9.7%	10.8%	12.9%	11.4%	13.8%	13.3%

Source: *The Recording Industry Association of America*

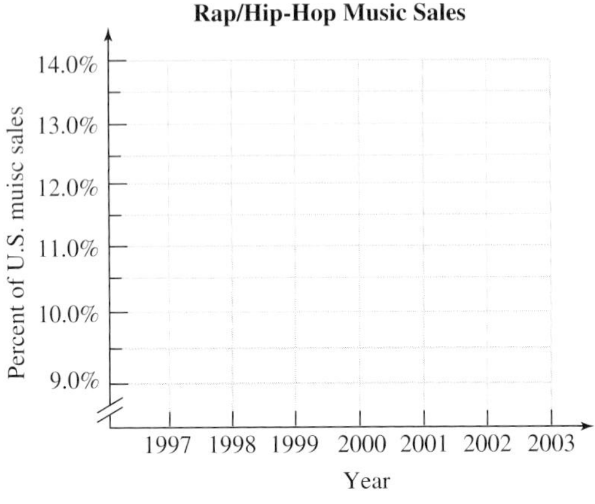

83. CHARITIES A 1998 fact sheet released by the American Red Cross stated, "For the past three fiscal years, an average of 92 cents of every dollar spent by the Red Cross went to programs and services to help those in need." What percent of the money spent by the Red Cross went to programs and services?

84. TAXES Santa Anita Thoroughbred Racetrack in Arcadia, California, has to pay a one-third of 1% tax on all the money wagered at the track. Write the percent as a fraction.

A calculator will be helpful to solve these problems.

85. BIRTHDAYS If the day of your birthday represents $\frac{1}{365}$ of a year, what percent of the year is it? Round to the nearest hundredth of a percent.

86. POPULATION As a fraction, each resident of the United States represents approximately $\frac{1}{295,000,000}$ of the population. Express this as a percent. Round to one nonzero digit.

WRITING

87. If you were writing advertising, which form do you think would attract more customers: "25% off" or "$\frac{1}{4}$ off"? Explain.

88. Many coaches ask their players to give a 110% effort during practices and games. What do you think this means? Is it possible?

89. Explain how to change a fraction to a percent.

90. Explain how an amusement park could have an attendance that is 103% of capacity.

91. CHAMPIONS Muhammad Ali won 92% of his professional boxing matches. Does that mean he had exactly 100 fights and won 92 of them? Explain.

92. CALCULATORS To change the fraction $\frac{15}{16}$ to a percent, a student used a calculator to divide 15 by 16. The display is shown below.

0.9375

Now what keys should the student press to change this decimal to a percent?

REVIEW

93. Solve: $-\frac{2}{3}x = -6$.

94. Add: $\frac{1}{3} + \frac{1}{4} + \frac{1}{2}$.

95. Subtract: $\frac{7}{11} - \frac{2}{9}$.

96. Find the area of a square with a side that is 4 feet long.

97. Add: $3.875 + 23.2$.

98. Subtract: $41 - 10.287$.

5.2 Solving Percent Problems

- Percent problems • Finding the amount • Finding the percent • Finding the base
- Restating the problem • An alternative approach: the percent formula • Circle graphs

Percent problems occur in three forms. In this section, we will study a single procedure that can be used to solve all three types. It involves the equation-solving skills that we studied earlier.

Percent problems

The articles on the front page of the newspaper in Figure 5-3 suggest three types of percent problems.

- In the labor article, if we want to know how many union members voted to accept the new offer, we would ask:

  What number is 84% of 500?

- In the article on drinking water, if we want to know what percent of the wells are safe, we would ask:

  38 is what percent of 40?

- In the article on new appointees, if we want to know how many examiners are on the State Board, we would ask:

  6 is 75% of what number?

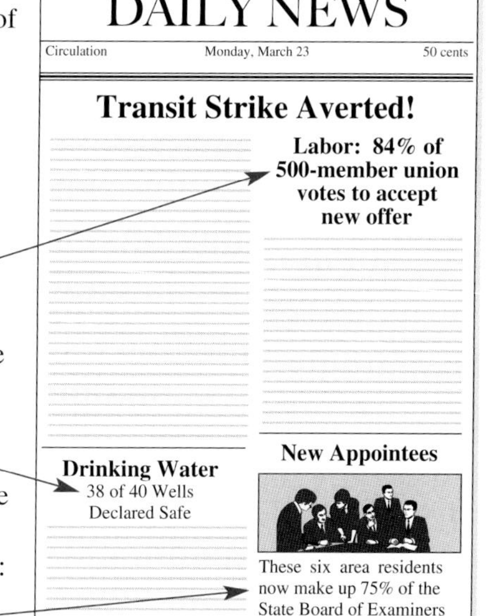
DAILY NEWS

Circulation Monday, March 23 50 cents

Transit Strike Averted!

Labor: 84% of 500-member union votes to accept new offer

Drinking Water
38 of 40 Wells Declared Safe

New Appointees

These six area residents now make up 75% of the State Board of Examiners

FIGURE 5-3

These percent problems have features in common.

- Each problem contains the word *is*. Here, *is* can be translated to an = symbol.
- Each of the problems contains a phrase such as *what number* or *what percent*. In other words, there is an unknown quantity that can be represented by a variable.
- Each problem contains the word *of*. In this context, *of* means multiply.

These observations suggest that each of the percent problems can be translated into an equation. The equation, called a **percent equation,** will contain a variable, and the operation of multiplication will be involved.

Finding the amount

To solve the labor union problem, we translate the words into an equation and then solve it.

What number	is	84%	of	500?	
↓	↓	↓	↓	↓	
x	$=$	84%	$\cdot$	500	Translate to mathematical symbols.

$x = 0.84 \cdot 500$ Change 84% to a decimal: 84% = 0.84.

$= 420$ Perform the multiplication.

We have found that 420 is 84% of 500. That is, 420 union members voted to accept the new offer.

! COMMENT When solving percent equations, always write the percent as a decimal or a fraction before performing any calculations. For example, in the previous problem, we wrote 84% as 0.84 before multiplying by 500.

Percent problems involve a comparison of numbers or quantities. In the statement "420 is 84% of 500," the number 420 is called the **amount,** 84% is the **percent,** and 500 is called the **base.** Think of the base as the standard of comparison—it represents the whole of some quantity. The amount is a part of the base, but it can exceed the base when the percent is more than 100%. The percent, of course, has the % symbol.

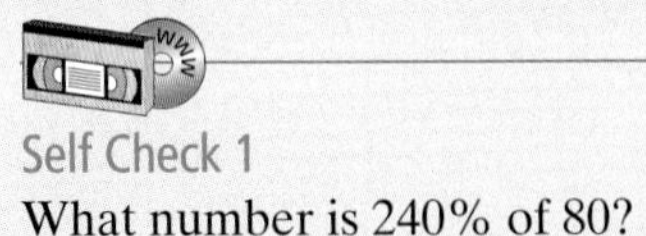

Self Check 1
What number is 240% of 80?

EXAMPLE 1 What number is 160% of 15.8?

Solution First, we translate the words into an equation.

What number	is	160%	of	15.8?	
↓		↓		↓	
x	$=$	160%	$\cdot$	15.8	x is the amount, 160% is the percent, and 15.8 is the base.

Then we solve the equation.

$x = 1.6 \cdot 15.8$ Change 160% to a decimal: 160% = 1.6.

$= 25.28$ Perform the multiplication.

Thus, 25.28 is 160% of 15.8.

Answer 192

Finding the percent

In the drinking water problem, we must find the percent. Once again, we translate the words of the problem into an equation and solve it.

38	is	what percent	of	40?
↓		↓		↓
38	=	x	$\cdot$	40

38 is the amount, x is the percent, and 40 is the base.

$38 = 40x$ Use the commutative property of multiplication to write $x \cdot 40$ as $40x$.

$\frac{38}{\mathbf{40}} = \frac{40x}{\mathbf{40}}$ To undo the multiplication by 40, divide both sides by 40.

$0.95 = x$ Perform the divisions.

$x = 0.95$ Since $0.95 = x$, $x = 0.95$.

$x = 95\%$ To change a decimal to a percent, multiply the decimal by 100 by moving the decimal point 2 places to the right, and then insert a % symbol.

Thus, 38 is 95% of 40. That is, 95% of the wells referred to in the article were declared safe.

EXAMPLE 2 14 is what percent of 32?

Solution First, we translate the words into an equation.

14	is	what percent	of	32?
↓		↓		↓
14	=	x	$\cdot$	32

14 is the amount, x is the percent, and 32 is the base.

Then we solve the equation.

$14 = 32x$ Rewrite the right-hand side: $x \cdot 32 = 32x$.

$\frac{14}{\mathbf{32}} = \frac{32x}{\mathbf{32}}$ To undo the multiplication by 32, divide both sides by 32.

$0.4375 = x$ Perform the divisions.

$43.75\% = x$ Change 0.4375 to a percent. Multiply the decimal by 100 by moving the decimal point 2 places to the right, and then insert a % symbol.

Thus, 14 is 43.75% of 32.

Self Check 2
9 is what percent of 16?

Answer 56.25%

Cost of an air bag

CALCULATOR SNAPSHOT

An air bag is estimated to add an additional $500 to the cost of a car. What percent of the $16,295 sticker price is the cost of the air bag?

First, we translate the words into an equation.

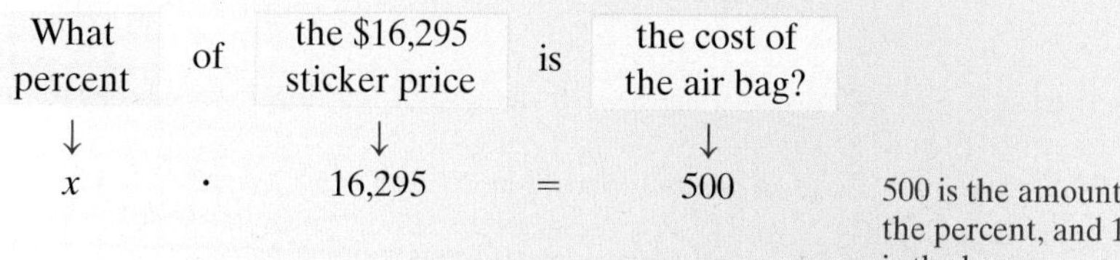

What percent	of	the $16,295 sticker price	is	the cost of the air bag?
↓		↓		↓
x	$\cdot$	16,295	=	500

500 is the amount, x is the percent, and 16,295 is the base.

Then we solve the equation.

$$16{,}295x = 500 \qquad x \cdot 16{,}295 = 16{,}295x.$$

$$\frac{16{,}295x}{\mathbf{16{,}295}} = \frac{500}{\mathbf{16{,}295}} \qquad \text{To undo the multiplication by 16,295, divide both sides by 16,295.}$$

$$x = \frac{500}{16{,}295}$$

To perform the division using a calculator, enter these numbers and press these keys.

500 [÷] 16295 [=] 0.030684259

This display gives the answer in decimal form. To change it to a percent, we multiply the result by 100 and insert a % symbol. This moves the decimal point 2 places to the right. If we round to the nearest tenth of a percent, the cost of the air bag is about 3.1% of the sticker price.

Finding the base

In the problem about the State Board of Examiners, we must find the base. As before, we translate the words of the problem into an equation and solve it.

6	is	75%	of	what number?	
↓		↓		↓	
6	=	75%	·	x	6 is the amount, 75% is the percent, and x is the base.

$$6 = 0.75x \qquad \text{Change 75\% to 0.75.}$$

$$\frac{6}{\mathbf{0.75}} = \frac{0.75x}{\mathbf{0.75}} \qquad \text{To undo the multiplication by 0.75, divide both sides by 0.75.}$$

$$8 = x \qquad \text{Perform the divisions.}$$

Thus, 6 is 75% of 8. That is, there are 8 examiners on the State Board.

Self Check 3
150 is $66\frac{2}{3}\%$ of what number?

EXAMPLE 3 31.5 is $33\frac{1}{3}\%$ of what number?

Solution

31.5	is	$33\frac{1}{3}\%$	of	what number?	
↓		↓		↓	
31.5	=	$33\frac{1}{3}\%$	·	x	31.5 is the amount, $33\frac{1}{3}\%$ is the percent, and x is the base.

In this case the computations can be made easier by changing the percent to a fraction instead of to a decimal. We write $33\frac{1}{3}\%$ as a fraction and proceed as follows.

$$31.5 = \frac{1}{3} \cdot x \qquad 33\frac{1}{3}\% = \frac{1}{3}.$$

$$\mathbf{3} \cdot 31.5 = \mathbf{3} \cdot \frac{1}{3}x \qquad \text{To isolate } x \text{ on the right-hand side, multiply both sides by 3.}$$

$$94.5 = x \qquad \text{Perform the multiplications: } 3 \cdot 31.5 = 94.5 \text{ and } 3 \cdot \tfrac{1}{3} = 1.$$

Thus, 31.5 is $33\frac{1}{3}\%$ of 94.5.

Answer 225

Restating the problem

Not all percent problems are presented in the form we have been studying. In Example 4, we must examine the given information carefully so that we can restate the problem in the familiar form.

EXAMPLE 4 **Housing.** In an apartment complex, 110 of the units are currently being rented. This represents an 88% occupancy rate. How many units are there in the complex?

Solution An occupancy rate of 88% means that 88% of the units are occupied. We restate the problem in the form we have been studying.

110	is	88%	of	what number?	
↓		↓		↓	
110	=	88%	·	x	110 is the amount, 88% is the percent, and x is the base.

Now we solve the equation.

$110 = 0.88x$ Change 88% to a decimal: 88% = 0.88.

$\frac{110}{\mathbf{0.88}} = \frac{0.88x}{\mathbf{0.88}}$ To undo the multiplication by 0.88, divide both sides by 0.88.

$125 = x$ Perform the divisions.

The complex has 125 units.

An alternative approach: the percent formula

In any percent problem, the relationship between the amount, the percent, and the base is as follows: *Amount is percent of base.* This relationship is shown in the **percent formula.**

The percent formula

Amount = percent · base

The percent formula can be used as an alternative way to solve percent problems. With this method, we need to identify the *amount* (the part that is compared to the whole), the *percent* (indicated by the % symbol or the word *percent*), and the *base* (the whole of some quantity, usually following the word *of*).

EXAMPLE 5 What number is 160% of 15.8?

Solution In this example, the percent is 160% and the base is 15.8, the number following the word *of.* We can let A stand for the amount and use the percent formula.

Amount	=	percent	·	base	
↓		↓		↓	
A	=	160%	·	15.8	Substitute 160% for the percent and 15.8 for the base.

Self Check 5

What number is 240% of 80?

The statement $A = 160\% \cdot 15.8$ is an equation, with the amount A being the unknown. We can find the unknown amount by multiplication.

$A = 1.6 \cdot 15.8$ Change 160% to a decimal: 160% = 1.6.

$= 25.28$ Perform the multiplication.

Answer 192

Thus, 25.28 is 160% of 15.8. Note that we got the same result in Example 1.

Self Check 6

9 is what percent of 16?

Answer 56.25%

EXAMPLE 6 14 is what percent of 32?

Solution In this example, 14 is the amount and 32 is the base. Once again, we use the percent formula and let p stand for the percent.

Amount	=	percent	·	base	
↓		↓		↓	
14	=	p	·	32	Substitute 14 for the amount and 32 for the base.

The statement $14 = p \cdot 32$ is an equation, with the percent p being the unknown. We can find the unknown percent by division.

$14 = p \cdot 32$ This is the equation to solve.

$14 = 32p$ Rewrite the right-hand side: $p \cdot 32 = 32p$.

$\frac{14}{\mathbf{32}} = \frac{32p}{\mathbf{32}}$ To undo the multiplication by 32, divide both sides by 32.

$0.4375 = p$ Perform the divisions: $\frac{14}{32} = 0.4375$.

$p = 43.75\%$ To change the decimal to a percent, multiply the decimal by 100 by moving the decimal point 2 places to the right, and then insert a % symbol.

Thus, 14 is 43.75% of 32. Note that we got the same result in Example 2.

Self Check 7

150 is $66\frac{2}{3}\%$ of what number?

EXAMPLE 7 31.5 is $33\frac{1}{3}\%$ of what number?

Solution In this example, 31.5 is the amount and $33\frac{1}{3}\%$ is the percent. To find the base (which we will call b), we form an equation using the percent formula.

Amount	=	percent	·	base	
↓		↓		↓	
31.5	=	$33\frac{1}{3}\%$	·	b	Substitute 31.5 for the amount and $33\frac{1}{3}\%$ for the percent.

The statement $31.5 = 33\frac{1}{3}\% \cdot b$ is an equation, with the base b being the unknown. We can find the unknown base by multiplication.

$31.5 = 33\frac{1}{3}\% \cdot b$ This is the equation to solve.

$31.5 = \frac{1}{3}b$ $33\frac{1}{3}\% = \frac{33\frac{1}{3}}{100} = \frac{1}{3}$.

$\mathbf{3} \cdot 31.5 = \mathbf{3} \cdot \frac{1}{3}b$ To isolate b on the right-hand side, multiply both sides by 3.

$94.5 = b$ Perform the multiplication: $31.5 \cdot 3 = 94.5$.

Thus, 31.5 is $33\frac{1}{3}\%$ of 94.5. Note that we got the same result in Example 3.

Answer 225

Solving Percent Problems Using Proportions

A **proportion** is a statement that two fractions are equal. Some examples of proportions are

$$\frac{2}{5} = \frac{4}{10}, \quad \frac{3}{7} = \frac{9}{21}, \quad \text{and} \quad \frac{a}{b} = \frac{c}{d} \quad \text{where } b \text{ and } d \text{ are not } 0$$

In the proportion $\frac{a}{b} = \frac{c}{d}$, a and d are called the **extremes** and b and c are called the **means.**

means $\frac{a}{b} = \frac{c}{d}$ **extremes**

The first two examples above illustrate the following property of proportions.

Property of Proportions

In a proportion, the product of the means is always equal to the product of the extremes.

extremes / **means** $\frac{2}{5} = \frac{4}{10}$ $2(10) = 5(4) = 20$ We call the products 2(10) and 5(4) **cross products.**

extremes / **means** $\frac{3}{7} = \frac{9}{21}$ $3(21) = 7(9) = 63$ We call the products 3(21) and 7(9) **cross products.**

From the previous section, we know that a percent is the numerator of a fraction with a denominator of 100. For example,

$$7\% = \frac{7}{100}, \quad 33\% = \frac{33}{100}, \quad \text{and} \quad 525\% = \frac{525}{100}$$

Because of this fact, we can solve percent problems by using the following proportion.

$$\begin{array}{r} \text{percent} \longrightarrow \\ 100 \longrightarrow \end{array} \frac{p}{100} = \frac{a}{b} \begin{array}{l} \longleftarrow \text{amount} \\ \longleftarrow \text{base} \end{array}$$

To illustrate the method, we will solve the same three examples shown in the section.

EXAMPLE 1 **Finding the amount.** What number is 160% of 15.8?

Solution

In this problem, 160 is the percent p and 15.8 is the base b. To find the amount, we substitute these values into the percent proportion and solve for a.

$$\frac{p}{100} = \frac{a}{b} \quad \text{This is the percent formula.}$$

$$\frac{160}{100} = \frac{a}{15.8} \quad \text{Substitute 160 for } p \text{ and 15.8 for } b.$$

$$160(15.8) = 100a \quad \text{In a proportion, the cross products are equal.}$$

$$2{,}528 = 100a \quad \text{Multiply.}$$

$$25.28 = a \quad \text{Divide both sides by 100.}$$

Thus, 160% of 15.8 is 25.28.

EXAMPLE 2 **Finding the percent.** 14 is what percent of 32?

Solution

In this problem, 14 is the amount a and 32 is the base b. To find the percent, we substitute these values into the percent proportion and solve for p.

$$\frac{p}{100} = \frac{a}{b} \quad \text{This is the percent formula.}$$

$$\frac{p}{100} = \frac{14}{32} \quad \text{Substitute 14 for } a \text{ and 32 for } b.$$

$$32(p) = 100(14) \quad \text{In a proportion, the cross products are equal.}$$

$$32p = 1{,}400 \quad \text{Multiply.}$$

$$p = 43.75 \quad \text{Divide both sides by 32.}$$

Thus, 14 is 43.75% of 32.

EXAMPLE 3 **Finding the base.** 31.5 is $33\frac{1}{3}$% of what number?

Solution

In this problem, $33\frac{1}{3}$% is the percent p and 31.5 is the amount a. To find the base, we substitute these values into the percent proportion and solve for b.

$$\frac{p}{100} = \frac{a}{b} \quad \text{This is the percent formula.}$$

$$\frac{33\frac{1}{3}}{100} = \frac{31.5}{b} \quad \text{Substitute } 33\tfrac{1}{3}\% \text{ for } p \text{ and 31.5 for } a.$$

$$33\tfrac{1}{3}(b) = 100(31.5) \quad \text{In a proportion, the cross products are equal.}$$

$$33\tfrac{1}{3}b = 3{,}150 \quad \text{Multiply.}$$

$$b = 94.5 \quad \text{Divide both sides by } 33\tfrac{1}{3}.$$

Thus, 31.5 is $33\frac{1}{3}$% of 94.5.

STUDY SET

1. Find 240% of 80.
2. 9 is what percent of 16?
3. 150 is $66\frac{2}{3}$% of what number?
4. 25 is what percent of 80?
5. 120 is 80% of what number?
6. Find 95% of 300.
7. 37 is what percent of 25?
8. 563.2 is 110% of what number?
9. 525 is 75% of what number?
10. 864 is what percent of 1,200?

Circle graphs

Percents are used with **circle graphs,** or **pie charts,** as a way of presenting data for comparison. In Figure 5-4, the entire circle represents the total amount of electricity generated in the United States in 2003. The pie-shaped pieces of the graph show the relative sizes of the energy sources used to generate the electricity. For example, we see that the greatest amount of electricity (51%) was generated from coal. Note that if we add the percents from all categories (51% + 3% + 7% + 16% + 20% + 3%), the sum is 100%.

The 100 tick marks equally spaced around the circle serve as a visual aid when constructing a circle graph. For example, to represent hydropower as 7%, a line was drawn from the center of the circle to a tick mark. Then we counted off 7 ticks and drew a second line from the center to that tick to complete the pie-shaped wedge.

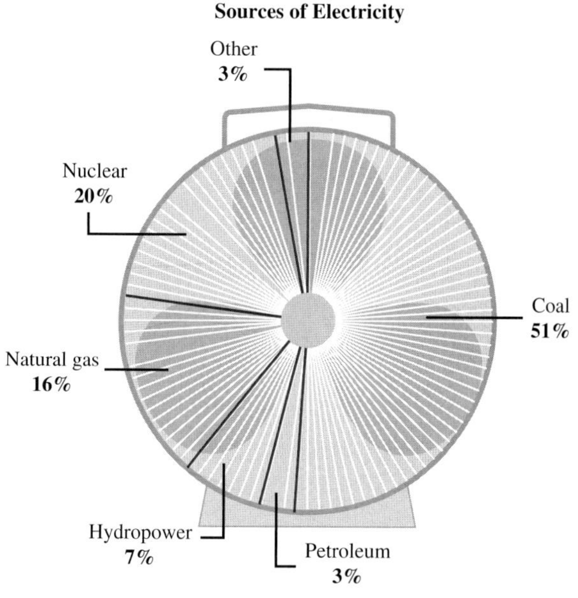

FIGURE 5-4

EXAMPLE 8 Presidential elections.

Results from the 2004 presidential election are shown in Figure 5-5 on the right. Use the information to find the number of states won by George W. Bush.

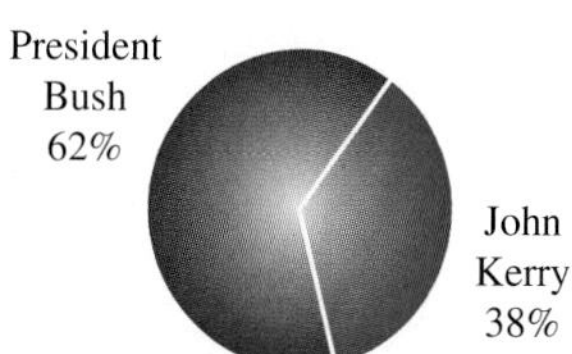

2004 Presidential Election
States won by each candidate

FIGURE 5-5

Solution The circle graph shows that George W. Bush was victorious in 62% of the 50 states. Here, the percent is 62% and the base is 50. We use the percent formula and solve for the amount.

Amount	=	percent	·	base	
A	=	62%	·	50	Substitute 62% for the percent and 50 for the base.

$A = 0.62 \cdot 50$ Change 62% to a decimal: 62% = 0.62. This is the equation to solve.

$A = 31$ Perform the multiplication.

George W. Bush won 31 states in the 2004 presidential election.

THINK IT THROUGH Community College Students

"Community Colleges are centers of educational opportunity. More than 100 years ago, this unique, American invention put publicly funded higher education at close-to-home facilities and initiated a practice of welcoming all who desire to learn, regardless of wealth, heritage or previous academic experience. Today, the community college of continues the process of making higher education available to a maximum number of people at 1,166 public and independent community colleges." The American Association of Community Colleges (AACC)

Over 33,500 students responded to the 2002 Community College Survey of Student Engagement. Several of the results are shown below. Study each circle graph and then complete its legend.

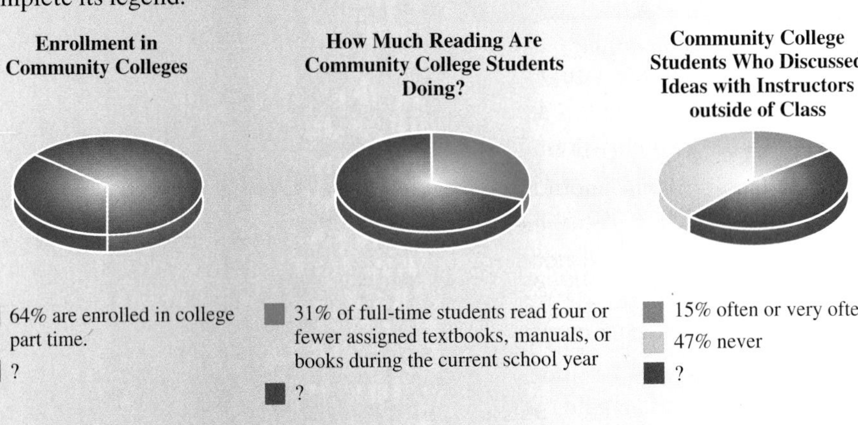
Enrollment in Community Colleges
64% are enrolled in college part time.
?
How Much Reading Are Community College Students Doing?
31% of full-time students read four or fewer assigned textbooks, manuals, or books during the current school year
?
Community College Students Who Discussed Ideas with Instructors outside of Class
15% often or very often
47% never
?

Section 5.2 STUDY SET

VOCABULARY *Fill in the blanks.*

1. In a circle ________, pie-shaped wedges are used to show the division of a whole quantity into its component parts.

2. In the statement "45 is 90% of 50," 45 is the ________, 90% is the ________, and 50 is the ________.

CONCEPTS *Translate each sentence into a percent equation.* ***Do not solve the equation.***

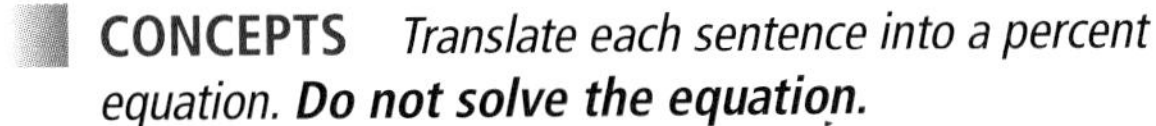

3. What number is 10% of 50?
4. 16 is 55% of what number?
5. 48 is what percent of 47?
6. 12 is what percent of 20?

7. When we compute with percents, the percent must be changed to a decimal or a fraction. Change each percent to a decimal.

   a. 12% b. 5.6%

   c. 125% d. $\frac{1}{4}$%

8. When we compute with percents, the percent must be changed to a decimal or a fraction. Change each percent to a fraction.

   a. $33\frac{1}{3}$% b. $66\frac{2}{3}$%

   c. $16\frac{2}{3}$% d. $83\frac{1}{3}$%

9. Without doing the calculation, determine whether 120% of 55 is more than 55 or less than 55.

10. Without doing the calculation, determine whether 12% of 55 is more than 55 or less than 55.

11. Solve each of the following problems in your head.

    a. What is 100% of 25?

    b. What percent of 132 is 132?

    c. What number is 87% of 100?

12. To solve the problem

    15 is what percent of 75?

    a student wrote a percent equation, solved it, and obtained $x = 0.2$. For her answer, the student wrote

    15 is 0.2% of 75.

    Explain her error.

13. E-MAIL The circle graph shows the types of e-mail messages a typical Internet user receives. *Spam* is the name given "junk" e-mail that is sent to a large number of people to promote products or services. What percent of e-mail messages is spam?

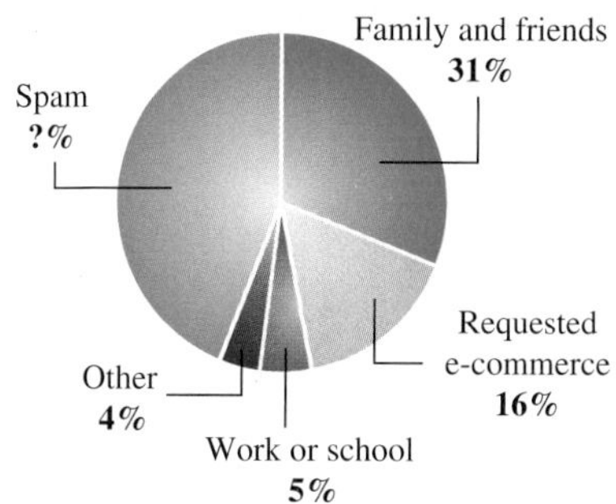

Source: *USA Today*, October 31, 2003

14. HOUSING In the last quarter of 2000, approximately 105.5 million housing units in the United States were occupied. Use the data in the circle graph in the next column to determine what percent were owner-occupied.

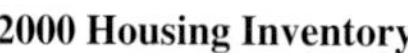
2000 Housing Inventory

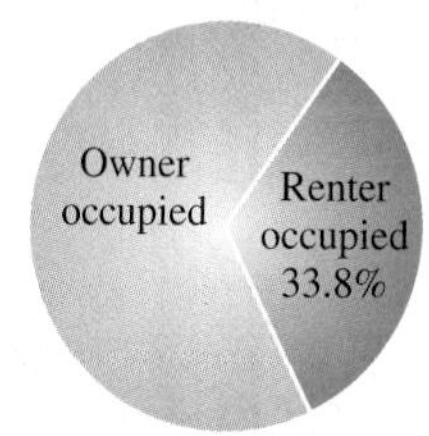

Source: *The U.S. Census Bureau*

NOTATION

15. How is each of the following words or phrases translated in this section?

    a. of

    b. is

    c. what number

16. a. Write the repeating decimal shown in the calculator display as a percent. Use an overbar.

    0.456666666

    b. Round your answer to part a to the nearest hundredth of a percent.

    c. Write your answer to part a using a fraction.

PRACTICE *Solve each problem by solving a percent equation.*

17. What number is 36% of 250?
18. What number is 82% of 300?
19. 16 is what percent of 20?
20. 13 is what percent of 25?
21. 7.8 is 12% of what number?
22. 39.6 is 44% of what number?
23. What number is 0.8% of 12?
24. What number is 5.6% of 4,040?
25. 0.5 is what percent of 40,000?
26. 0.3 is what percent of 15?
27. 3.3 is 7.5% of what number?
28. 8.4 is 20% of what number?
29. Find $7\frac{1}{4}$% of 600.
30. Find $1\frac{3}{4}$% of 800.
31. 102% of 105 is what number?
32. 210% of 66 is what number?
33. $33\frac{1}{3}$% of what number is 33?

34. $66\frac{2}{3}\%$ of what number is 28?

35. $9\frac{1}{2}\%$ of what number is 5.7?

36. $\frac{1}{2}\%$ of what number is 5,000?

37. What percent of 8,000 is 2,500?

38. What percent of 3,200 is 1,400?

Use a circle graph to illustrate the given data. A circle divided into 100 sections is provided to aid in the graphing process.

39. ENERGY Complete the graph to show what percent of the total U.S. energy produced was provided by each source in 2003.

Renewable	12%
Nuclear	11%
Coal	31%
Natural gas	29%
Petroleum	17%

Source: *Energy Information Administration*

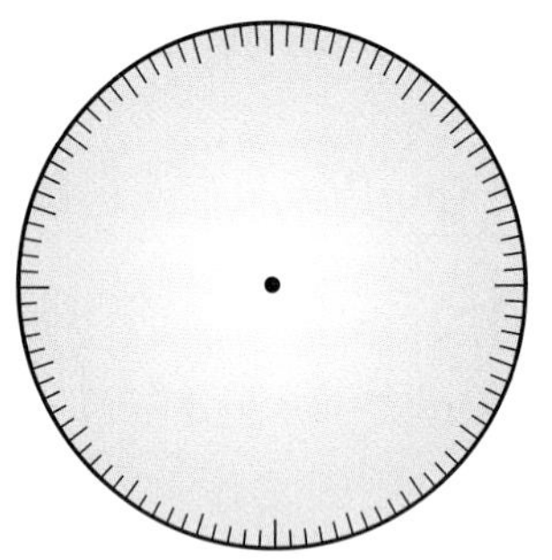

Source: Energy Information Administration

40. GREENHOUSE EFFECT Complete the graph to show what percent of the total U.S. emissions from human activities in 2002 came from each greenhouse gas.

Carbon dioxide	83%
Nitrous oxide	6%
Methane	9%
PFCs, HFCs	2%

Source: *The World Almanac 2005*

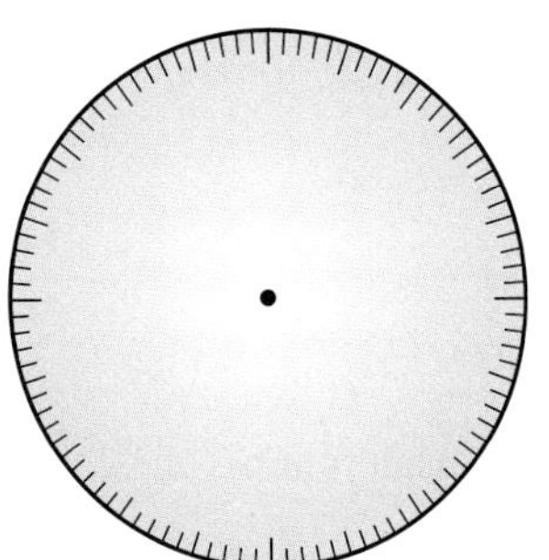

APPLICATIONS

41. CHILD CARE After the first day of registration, 84 children had been enrolled in a new day care center. That represented 70% of the available slots. What was the maximum number of children the center could enroll?

42. RACING PROGRAMS One month before a stock car race, the sale of ads for the official race program was slow. Only 12 pages, or just 60% of the available pages, had been sold. What was the total number of pages devoted to advertising in the program?

43. GOVERNMENT SPENDING The illustration shows the breakdown of federal spending for fiscal year 2003. If the total spending was approximately \$1,800 billion, how many dollars were spent on Social Security, Medicare, and other retirement programs?

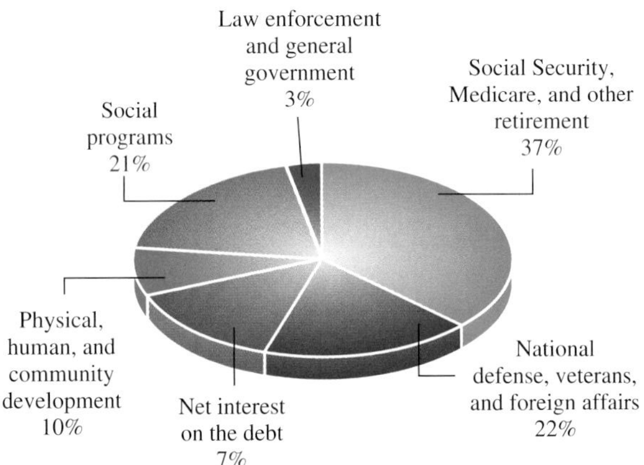

Source: 2004 Federal Income Tax Form 1040

44. GOVERNMENT INCOME Complete the table by finding what percent of total federal government income each source provided in 2003. Round to the nearest percent. Then complete the circle graph on the next page.

Total income, fiscal year 2003: \$2,200 billion		
Source of income	**Amount**	**Percent of total**
Social Security, Medicare, unemployment taxes	\$726 billion	
Personal income taxes	\$814 billion	
Corporate income taxes	\$132 billion	
Excise, estate, customs taxes	\$154 billion	
Borrowing to cover deficit	\$374 billion	

Source: 2003 Federal Income Tax Form

2003 Federal Income Sources

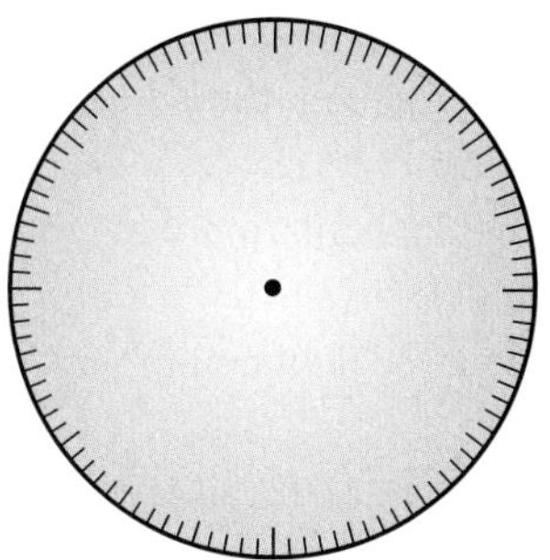

45. THE INTERNET The message at the bottom of the computer screen indicates that 24% of the 50K bytes of information that the user has decided to view have been downloaded to her computer. How many more bytes of information must be downloaded? (50K stands for 50,000.)

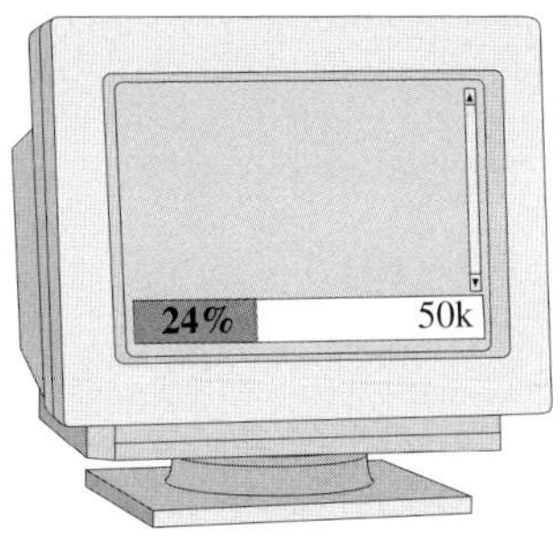

46. REBATES A long-distance telephone company offered its customers a rebate of 20% of the cost of all long-distance calls made in the month of July. One customer's calls are listed in the table. What amount will this customer receive in the form of a rebate?

Date	Time	Place called	Min.	Amount
Jul 4	3:48 P.M.	Denver	47	$3.80
Jul 9	12:00 P.M.	Detroit	68	$7.50
Jul 20	8:59 A.M.	San Diego	70	$9.45

47. PRODUCT PROMOTIONS To promote sales, a free 6-ounce bottle of shampoo is packaged with every large bottle. Use the information on the package to find how many ounces of shampoo the large bottle contains.

48. NUTRITION FACTS The nutrition label on a package of corn chips is shown (g stands for grams and mg for milligrams).

a. How many milligrams of sodium are in one serving of chips?

b. According to the label, what percent of the daily value of sodium is this?

c. What daily value of sodium intake is deemed healthy?

Nutrition Facts
Serving Size: 1 oz. (28g/About 29 chips)
Servings Per Container: About 11

Amount Per Serving

Calories 160	Calories from Fat 90
	% Daily Value
Total fat 10g	**15%**
Saturated fat 1.5 g	**7%**
Cholesterol 0mg	**0%**
Sodium 240mg	**12%**
Total carbohydrate 15g	**5%**
Dietary fiber 1g	**4%**
Sugars less than 1g	
Protein 2g	

49. DRIVER'S LICENSES On the written part of his driving test, a man answered 28 out of 40 questions correctly. If 70% correct is passing, did he pass the test?

50. THE ALPHABET What percent of the English alphabet do the vowels a, e, i, o, and u make up? (Round to the nearest one percent.)

51. MIXTURES Complete the table to find the number of gallons of sulfuric acid in each of two storage tanks.

	Gallons of solution in tank	% sulfuric acid	Gallons of sulfuric acid in tank
Tank 1	60	50%	
Tank 2	40	30%	

52. CUSTOMER GUARANTEES To assure its customers of low prices, the Home Club offers a "10% Plus" guarantee. If the customer finds the same item selling for less somewhere else, he or she receives the difference in price, plus 10% of the difference. A woman bought miniblinds at the Home Club for $120 but later saw the same blinds on sale for $98 at another store. How much can she expect to be reimbursed?

53. MAKING COPIES The zoom key on the control panel of a copier programs is to print a magnified or reduced copy of the original document. If the zoom is set at 180% and the original document contains type that is 1.5 inches tall, what will be the height of the type on the copy?

54. MAKING COPIES The zoom setting for a copier is entered as a decimal: 0.98. Express it as a percent and find the resulting type size on the copy if the original has type 2 inches in height.

55. INSURANCE The cost to repair a car after a collision was \$4,000. The automobile insurance policy paid the entire bill except for a \$200 deductible, which the driver paid. What percent of the cost did he pay?

56. FLOOR SPACE A house has 1,200 square feet on the first floor and 800 square feet on the second floor. What percent of the square footage of the house is on the first floor?

57. MAJORITIES In Los Angeles City Council races, if no candidate receives more than 50% of the vote, a runoff election is held between the first- and second-place finishers. From the election results in the table, determine whether there must be a runoff election for District 10.

City council	District 10
Nate Holden	8,501
Madison T. Shockley	3,614
Scott Suh	2,630
Marsha Brown	2,432

58. PORTS In 2002, the busiest port in the United States was the Port of South Louisiana, which handled 216,396,497 tons of goods. Of that amount, 124,908,067 tons were domestic goods and 91,488,430 tons were foreign. What percent of the total was domestic? Round to the nearest tenth of a percent.

WRITING

59. Explain the relationship in a percent problem between the amount, the percent, and the base.

60. Write a real-life situation that could be described by "9 is what percent of 20?"

61. Explain why 150% of a number is more than the number.

62. Explain why "Find 9% of 100" is an easy problem to solve.

REVIEW

63. Add: $2.78 + 6 + 9.09 + 0.3$.

64. Evaluate: $\sqrt{64} + 3\sqrt{9}$.

65. On a number line, which number is closer to 5: 4.9 or 5.001?

66. Multiply: $34.5464 \cdot 1{,}000$.

67. Find the power: $(0.2)^3$.

68. Solve: $0.4x + 1.2 = -7.8$.

5.3 Applications of Percent

• Taxes • Commissions • Percent of increase or decrease • Discounts

In this section, we discuss applications of percent. Three of them (taxes, commissions, and discounts) are directly related to purchasing. A solid understanding of these concepts will make you a better consumer. The fourth application uses percent to describe increases or decreases of such things as unemployment and grocery store sales.

Taxes

The sales receipt in Figure 5-6 on the next page gives a detailed account of what items were purchased, how many of each were purchased, and the price of each item.

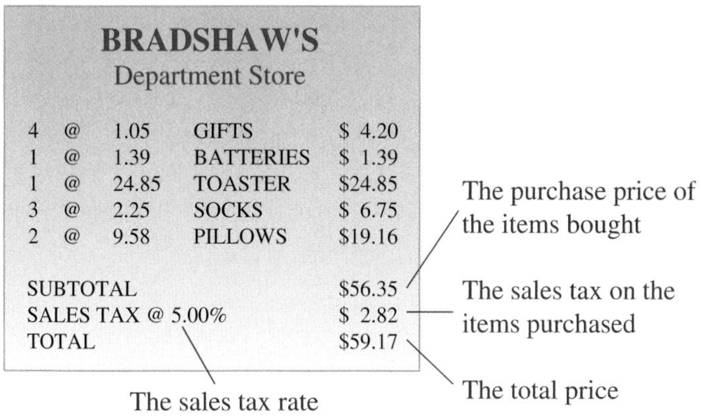

BRADSHAW'S
Department Store

4	@	1.05	GIFTS	\$ 4.20
1	@	1.39	BATTERIES	\$ 1.39
1	@	24.85	TOASTER	\$24.85
3	@	2.25	SOCKS	\$ 6.75
2	@	9.58	PILLOWS	\$19.16
SUBTOTAL				\$56.35
SALES TAX @ 5.00%				\$ 2.82
TOTAL				\$59.17

The purchase price of the items bought

The sales tax on the items purchased

The total price

The sales tax rate

FIGURE 5-6

The receipt shows that the \$56.35 purchase price (labeled *subtotal*) was taxed at a **rate** of 5%. Sales tax of \$2.82 was charged. The sales tax was then added to the subtotal to get the total price of \$59.17.

Finding the total price

Total price = purchase price + sales tax

In Example 1, we verify that the amount of sales tax shown on the receipt in Figure 5-6 is correct.

EXAMPLE 1 **Sales tax.** Find the sales tax on a purchase of \$56.35 if the sales tax rate is 5%.

Solution First we write the problem so that we can translate it into an equation. The rate is 5%. We are to find the amount of the tax.

What number	is	5%	of	56.35?
x	$=$	5%	$\cdot$	56.35

$x = 0.05 \cdot 56.35$ Change 5% to a decimal: 5% = 0.05.

$= 2.8175$ Perform the multiplication.

Rounding to the nearest cent (hundredths), we find that the sales tax would be \$2.82. The sales receipt in Figure 5-6 is correct.

Self Check 1
What would the sales tax be if the \$56.35 purchase were made in Texas, which has a 6.25% state sales tax?

Answer \$3.52

In addition to sales tax, we pay many other types of taxes in our daily lives. Income tax, gasoline tax, and Social Security tax are just a few.

EXAMPLE 2 **Withholding tax.** A waitress found that \$11.04 was deducted from her weekly gross earnings of \$240 for federal income tax. What withholding tax rate was used?

Solution First, we write the problem in a form that can be translated into an equation. We need to find the tax rate.

Self Check 2
A tax of \$5,250 had to be paid on an inheritance of \$15,000. What is the inheritance tax rate?

11.04	is	what percent	of	240?
11.04	=	x	$\cdot$	240

$11.04 = 240x$ Rewrite the right-hand side: $x \cdot 240 = 240x$.

$\frac{11.04}{\mathbf{240}} = \frac{240x}{\mathbf{240}}$ To undo the multiplication by 240, divide both sides by 240.

$0.046 = x$ Perform the divisions.

$4.6\% = x$ Change 0.046 to a percent.

Answer 35%

The withholding tax rate was 4.6%.

Commissions

Instead of working for a salary or getting paid at an hourly rate, many salespeople are paid on **commission.** They earn an amount based on the goods or services they sell.

Self Check 3

An insurance salesperson receives a 4.1% commission on each $120 premium paid by a client. What is the amount of the commission on this premium?

Answer $4.92

EXAMPLE 3 **Commissions.** The commission rate for a salesperson at an appliance store is 16.5%. Find his commission from the sale of a refrigerator costing $499.95.

Solution We write the problem so that it can be translated into an equation. We are to find the amount of the commission.

What number	is	16.5%	of	499.95?
x	=	16.5%	$\cdot$	499.95

$x = 0.165 \cdot 499.95$ Change 16.5% to a decimal: 16.5% = 0.165.

$= 82.49175$ Use a calculator to perform the multiplication.

Rounding to the nearest cent (hundredth), we find that the commission is $82.49.

Percent of increase or decrease

Percents can be used to describe how a quantity has changed. For example, consider Figure 5-7, which compares the number of hours of work it took the average U.S. worker to earn enough to buy a dishwasher in 1950 and 1998.

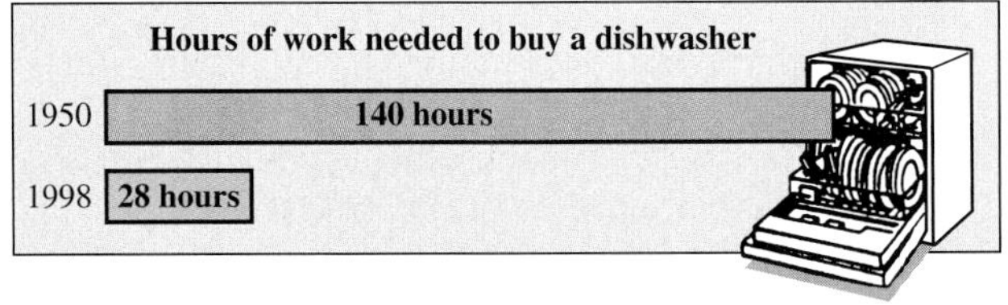

Source: Federal Reserve Bank of Dallas

FIGURE 5-7

From the figure, we see that the number of hours an average American had to work in order to buy a dishwasher has decreased over the years. To describe this decrease using a percent, we first subtract to find the amount of the decrease.

$140 - 28 = 112$ Subtract the hours of work needed in 1998 from the hours of work needed in 1950.

Next, we find what percent of the original number of hours of work needed in 1950 this difference represents.

112	is	what percent	of	140?
112	=	x	$\cdot$	140

$112 = 140x$ Rewrite the right-hand side: $x \cdot 140 = 140x$.

$\frac{112}{\mathbf{140}} = \frac{140x}{\mathbf{140}}$ To undo the multiplication by 140, divide both sides by 140.

$0.8 = x$ Perform the divisions.

$80\% = x$ Change 0.8 to a percent.

From 1950 to 1998, there was an 80% decrease in the number of hours it took the average U.S. worker to earn enough to buy a dishwasher.

Finding the percent of increase or decrease

To find the percent of increase or decrease:

1. Subtract the smaller number from the larger to find the amount of increase or decrease.
2. Find what percent the difference is of the original amount.

EXAMPLE 4 **Auctions.** A 1996 auction included an oak rocking chair used by President John F. Kennedy in the Oval Office. The chair, originally valued at $5,000, sold for $453,500. Find the percent of increase in the value of the rocking chair.

Solution First, we find the amount of increase.

$453{,}500 - 5{,}000 = 448{,}500$ Subtract the original value from the price paid at auction.

The rocking chair increased in value by $448,500. Next, we find what percent of the original value the increase represents.

448,500	is	what percent	of	5,000?
448,500	=	x	$\cdot$	5,000

$448{,}500 = 5{,}000x$ Rewrite the right-hand side: $x \cdot 5{,}000 = 5{,}000x$.

$\frac{448{,}500}{\mathbf{5{,}000}} = \frac{5{,}000x}{\mathbf{5{,}000}}$ To undo the multiplication by 5,000, divide both sides by 5,000.

$89.7 = x$ Perform the divisions.

$8{,}970\% = x$ Change 89.7 to a percent.

The Kennedy rocking chair increased in value by an amazing 8,970%.

Self Check 4
In one school district, the number of home-schooled children increased from 15 to 150 in 4 years. Find the percent of increase.

Answer 900%

EXAMPLE 5
Population decline. Norfolk, Virginia, experienced a decrease in population over the ten-year period from 1990 to 2000. Use the information in Figure 5-8 to determine the population of Norfolk in 2000.

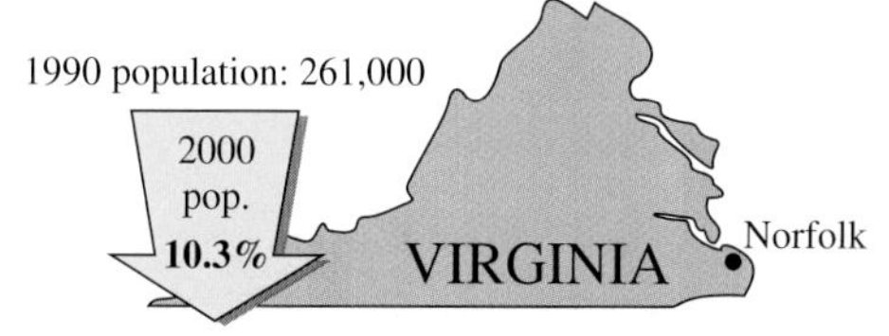

Source: U.S. Bureau of the Census (1999)

FIGURE 5-8

Solution In 1990, the population was 261,000. We are told that the number fell, and we need to find out by how much. To do so, we solve the following percent formula.

$$\text{Amount} = \text{percent} \cdot \text{base}$$

$$A = 10.3\% \cdot 261{,}000$$

$A = 0.103 \cdot 261{,}000$ Change 10.3% to a decimal: 10.3% = 0.103.

$A = 26{,}883$ Perform the multiplication.

From 1990 to 2000, the population decreased by 26,883. To find the city's population in 2000, we subtract the decrease from the population in 1990.

$$261{,}000 - 26{,}883 = 234{,}117$$

In 2000, the population of Norfolk, Virginia, was 234,117.

We can solve this problem in another way. If the population of Norfolk decreased by 10.3%, then the population in 2000 was 100% − 10.3% = 89.7% of the population in 1990. Using this approach, we can find the 2000 population directly by solving the following percent formula.

$$\text{Amount} = \text{percent} \cdot \text{base}$$

$$A = 89.7\% \cdot 261{,}000$$

$A = 0.897 \cdot 261{,}000$ Change 89.7% to a decimal: 89.7% = 0.897.

$A = 234{,}117$ Perform the multiplication.

As before, we see that the population of Norfolk in 2000 was 234,117.

THINK IT THROUGH Fastest Growing Occupations

"With the Baby Boomer generation aging, more medical professionals will be required to manage the health needs of the elderly."

Jonathan Stanewick, Compensation Analyst, Salary.com, 2004

The table below shows predictions by the U.S. Department of Labor, Bureau of Labor Statistics, of the fastest growing occupations for the years 2002–2012. Find the percent increase in the number of jobs for each occupation. Which occupation has the greatest percent increase?

Occupation	Number of jobs in 2002	Predicted number of jobs in 2012	Education or training required
Home health aide	580,000	859,000	On-the-job training or AA degree
Medical assistant	365,000	579,000	On-the-job training or AA degree
Physician assistant	63,000	94,000	Bachelor's or Master's degree
Social service assistant	305,000	454,000	On-the-job training or AA degree
Systems, data analyst	186,000	292,000	Bachelor's degree

Discounts

The difference between the original price and the sale price of an item is called the **discount.** If the discount is expressed as a percent of the selling price, it is called the **rate of discount.** We will use the information in the advertisement shown in Figure 5-9 to discuss how to find a discount and how to find a discount rate.

FIGURE 5-9

EXAMPLE 6 **Shoe sales.** Find the amount of the discount on the pair of men's basketball shoes shown in Figure 5-9. Then find the sale price.

Solution To find the discount, we find 25% of the regular price, \$59.80.

What number	is	25%	of	59.80?
x	$=$	25%	$\cdot$	59.80

$x = 0.25 \cdot 59.80$ Change 25% to a decimal: 25% = 0.25.

$= 14.95$ Perform the multiplication.

The discount is \$14.95. To find the sale price, we subtract the amount of the discount from the regular price.

$59.80 - 14.95 = 44.85$

The sale price of the men's basketball shoes is \$44.85.

Self Check 6
Sunglasses, regularly selling for \$15.40, are discounted 15%. Find the sale price.

Answer \$13.09

In Example 6, we used the following formula to find the sale price.

> **Finding the sale price**
>
> Sale price = original price − discount

EXAMPLE 7 **Discounts.** What is the rate of discount on the ladies' aerobic shoes advertised in Figure 5-9?

Solution We can think of this as a percent-of-decrease problem. We first compute the amount of the discount. This decrease in price is found using subtraction.

$39.99 - 21.99 = 18$

The shoes are discounted \$18. Now we find what percent of the original price the discount is.

18	is	what percent	of	39.99?
18	$=$	x	$\cdot$	39.99

Self Check 7
An early-bird special at a restaurant offers a \$10.99 prime rib dinner for only \$7.95 if it is ordered before 6 P.M. Find the rate of discount. Round to the nearest one percent.

$$18 = 39.99x \quad x \cdot 39.99 = 39.99x.$$

$$\frac{18}{\mathbf{39.99}} = \frac{39.99x}{\mathbf{39.99}} \quad \text{To undo the multiplication by 39.99, divide both sides by 39.99.}$$

$$0.450113 \approx x \quad \text{Perform the division.}$$

$$45.0113\% \approx x \quad \text{Change 0.450113 to a percent.}$$

Rounded to the nearest one percent, the discount rate is 45%.

Answer 28%

Section 5.3 STUDY SET

VOCABULARY *Fill in the blanks.*

1. Some salespeople are paid on __________. It is based on a percent of the total dollar amount of the goods or services they sell.
2. When we use percent to describe how a quantity has increased when compared to its original value, we are finding the percent of ________.
3. The difference between the original price and the sale price of an item is called the ________.
4. The ______ of a sales tax is expressed as a percent.

CONCEPTS

5. Fill in the blanks: To find the percent decrease, ________ the smaller number from the larger number to find the amount of decrease. Then find what percent that difference is of the ________ amount.
6. NEWSPAPERS The table below shows how the circulations of two daily newspapers changed from 1997 to 2003.

CIRCULATION		
	Miami Herald	USA Today
1997	356,803	1,629,665
2003	315,850	2,154,539

Source: *The World Almanac 2005*

a. What was the *amount of decrease* of the *Miami Herald*'s circulation?

b. What was the *amount of increase* of *USA Today*'s circulation?

APPLICATIONS *Solve each problem. If a percent answer is not exact, round to the nearest one percent.*

7. SALES TAXES The state sales tax rate in Utah is 4.75%. Find the sales tax on a dining room set that sells for $900.
8. SALES TAXES Find the sales tax on a pair of jeans costing $40 if they are purchased in Arkansas, which has a sales tax rate of 4.625%.
9. ROOM TAXES After checking out of a hotel, a man noticed that the hotel bill included an additional charge labeled *room tax.* If the price of the room was $129 plus a room tax of $10.32, find the room tax rate.
10. EXCISE TAXES While examining her monthly telephone bill, a woman noticed an additional charge of $1.24 labeled *federal excise tax.* If the basic service charges for that billing period were $42, what is the federal excise tax rate?
11. SALES RECEIPTS Complete the following sales receipt by finding the subtotal, the sales tax, and the total.

NURSERY CENTER
Your one-stop garden supply

3 @ 2.99	PLANTING MIX	$ 8.97
1 @ 9.87	GROUND COVER	$ 9.87
2 @ 14.25	SHRUBS	$28.50
	SUBTOTAL	$
	SALES TAX @ 6.00%	$
	TOTAL	$

12. SALES RECEIPTS Complete the following sales receipt by finding the prices, the subtotal, the sales tax, and the total.

McCOY'S FURNITURE

1 @ 450.00	SOFA	$
2 @ 90.00	END TABLES	$
1 @ 350.00	LOVE SEAT	$
	SUBTOTAL	$
	SALES TAX @ 4.20%	$
	TOTAL	$

13. SALES TAX In order to raise more revenue, some states raise the sales tax rate. How much additional money will be collected on the sale of a $15,000 car if the sales tax rate is raised 1%?

14. FOREIGN TRAVEL Value-added tax is a consumer tax imposed on goods and services. Currently, there are VAT systems in place all around the world. (The United States is one of the few industrialized nations not using a value-added tax system.) Complete the table by determining the VAT a traveler would pay in each country on a dinner costing $20.95.

Country	VAT rate	Tax on a $20.95 dinner
Canada	7%	
Germany	16%	
England	17.5%	
Sweden	25%	

15. PAYCHECKS Use the information on the paycheck stub to find the tax rate for the federal withholding, worker's compensation, Medicare, and Social Security taxes that were deducted from the gross pay.

6286244	
Issue date: 03-27-05	
GROSS PAY	$360.00
TAXES	
FED. TAX	$ 28.80
WORK. COMP.	$ 4.32
MEDICARE	$ 5.22
SOCIAL SECURITY	$ 22.32
NET PAY	$299.34

16. GASOLINE TAXES In one state, a gallon of unleaded gasoline sells for $1.89. This price includes federal and state taxes that total approximately $0.54. Therefore, the price of a gallon of gasoline, before taxes, is about $1.35. What is the tax rate on gasoline?

17. OVERTIME Factory management wants to reduce the number of overtime hours by 25%. If the total number of overtime hours is 480 this month, what is the target number of overtime hours for next month?

18. COST-OF-LIVING INCREASES If a woman making $32,000 a year receives a cost-of-living increase of 2.4%, how much is her raise? What is her new salary?

19. REDUCED CALORIES A company advertised its new, improved chips as having 36% fewer calories per serving than the original style. How many calories are in a serving of the new chips if a serving of the original style contained 150 calories?

20. POLICE FORCE A police department plans to increase its 80-person force by 5%. How many additional officers will be hired? What will be the new size of the department?

21. ENDANGERED SPECIES The illustration shows the total number of endangered and threatened plant and animal species for each of the years 1993–1999, as determined by the U.S. Fish and Wildlife Service. When was there a decline in the total? To the nearest percent, find the percent of decrease in the total for that period.

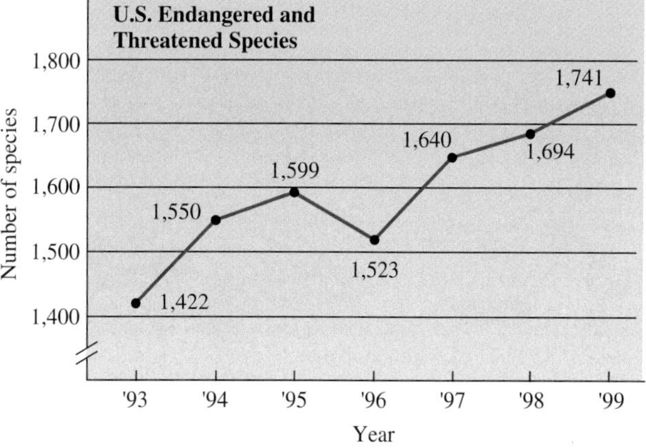

Source: *The New York Times 2000 Almanac*

22. CROP DAMAGE After flooding damaged much of the crop, the cost of a head of lettuce jumped from $0.99 to $2.20. What percent of increase is this?

23. CAR INSURANCE A student paid a car insurance premium of $400 every three months. Then the premium dropped to $360, because she qualified for a good-student discount. What was the percent of decrease in the premium?

24. BUS PASSES To increase the number of riders, a bus company reduced the price of a monthly pass from $112 to $98. What was the percent of decrease?

25. LAKE SHORELINES Because of a heavy spring runoff, the shoreline of a lake increased from 5.8 miles to 7.6 miles. What was the percent of increase in the shoreline?

26. BASEBALL The illustration shows the path of a baseball hit 110 mph, with a launch angle of 35 degrees, at sea level and at Coors Field, home of the Colorado Rockies. What is the percent of increase in the distance the ball travels at Coors Field?

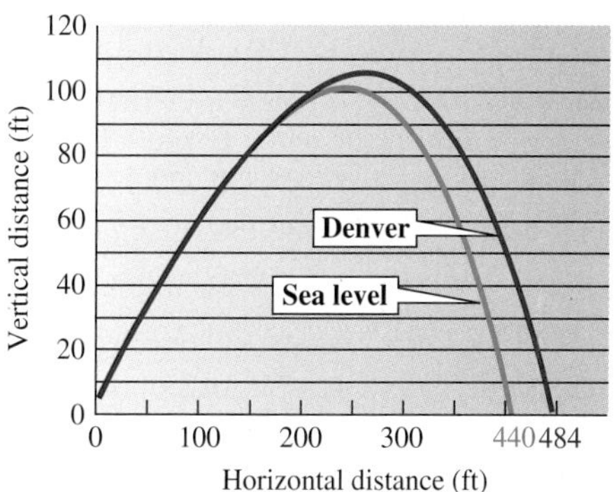

Source: *Los Angeles Times*

27. EARTH MOVING The illustration shows the typical soil volume change during earth moving. (One cubic yard of soil fits in a cube that is 1 yard long, 1 yard wide, and 1 yard high.)

   a. Find the percent of increase in the soil volume as it goes through step 1 of the process.

   b. Find the percent of decrease in the soil volume as it goes through step 2 of the process.

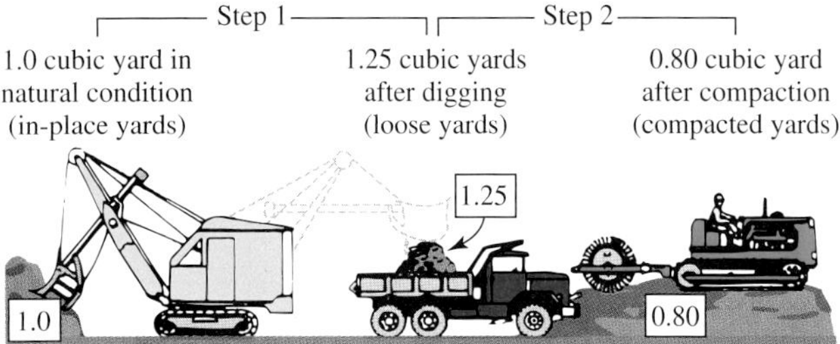

Source: U.S. Department of the Army

28. PARKING The management of a mall has decided to increase the parking area. The plans are shown. What will be the percent of increase in the parking area once the project is completed?

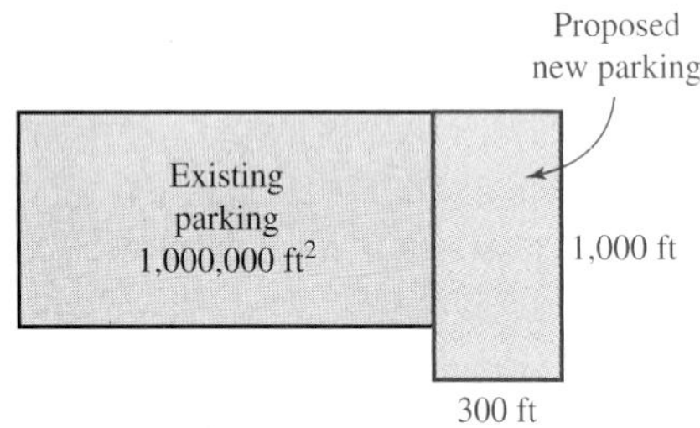

29. REAL ESTATE After selling a house for \$198,500, a real estate agent split the 6% commission with another agent. How much did each person receive?

30. MEDICAL SUPPLIES A salesperson for a medical supplies company is paid a commission of 9% for orders under \$8,000. For orders exceeding \$8,000, she receives an additional 2% in commission on the total amount. What is her commission on a sale of \$14,600?

31. SPORTS AGENTS A sports agent charges her clients a fee to represent them during contract negotiations. The fee is based on a percent of the contract amount. If the agent earned \$37,500 when her client signed a \$2,500,000 professional football contract, what rate did she charge for her services?

32. ART GALLERIES An art gallery displays paintings for artists and receives a commission from the artist when a painting is sold. What is the commission rate if a gallery received \$135.30 when a painting was sold for \$820?

33. CONCERT PARKING A concert promoter gets $33\frac{1}{3}\%$ of the revenue the arena receives from its parking concession the night of the performance. How much can the promoter make if 6,000 cars are anticipated and parking costs \$6 a car?

34. KITCHENWARE A homemaker invited her neighbors to a kitchenware party to show off cookware and utensils. As party hostess, she received 12% of the total sales. How much was purchased if she received \$41.76 for hosting the party?

35. WATCHES Find the regular price and the rate of discount for the watch that is on sale.

36. STEREOS Find the regular price and the rate of discount for the stereo system that is on sale.

37. RINGS What does a ring regularly sell for if it has been discounted 20% and is on sale for \$149.99? (*Hint:* The ring is selling for 80% of its regular price.)

38. BLINDS What do vinyl blinds regularly sell for if they have been discounted 55% and are on sale for \$49.50? (*Hint:* The blinds are selling for 45% of their regular price.)

39. DVRs What are the sale price and the discount rate for a digital video recorder with remote that regularly sells for \$399.97 and is being discounted \$50?

40. CAMCORDERS What are the sale price and the discount rate for a camcorder that regularly sells for \$559.97 and is being discounted \$80?

41. REBATES Find the discount, the discount rate, and the reduced price for a case of motor oil if a shopper receives the manufacturer's rebate mentioned in the following ad.

GXT
MOTOR OIL
Regular price $15.48/case
Mfr's rebate: $3.60

42. COUPONS Find the discount, the discount rate, and the reduced price for a box of cereal that normally sells for \$3.29 if a shopper presents the following coupon at a store that doubles the value of the coupon.

43. SHOPPING Determine the Home Shopping Network price of the ring described below if it sells it for 55% off of the retail price. Ignore shipping and handling costs.

Item 169-117
2.75 lb ctw
10K
Blue Topaz
Ring
6, 7, 8, 9, 10
Retail value \$170
HSN Price
\$??.??
S&H \$5.95

44. INFOMERCIALS The host of a TV infomercial says that the suggested retail price of a rotisserie grill is \$249.95 and that it is now offered "for just 4 easy payments of only \$39.95." What is the discount, and what is the discount rate?

WRITING

45. List the pros and cons of working on commission.
46. In Example 6, explain why you get the correct answer for the sale price by finding 75% of the regular price.
47. Explain the difference between a tax and a tax rate.
48. Explain how to find the sale price of an item if you know the regular price and the discount rate.

REVIEW

49. Multiply: $-5(-5)(-2)$.
50. Solve: $4x + 24 = 12$.
51. Evaluate: $-4 - (-7)$.
52. Evaluate: $|-5 - 8|$.
53. A store clerk earns \$12.50 an hour. How much will she earn in a 40-hour week?
54. Add: $\frac{3}{5} + \frac{4}{7}$.
55. Subtract: $\frac{4}{9} - \frac{3}{5}$.
56. Multiply: $\frac{7}{9} \cdot \frac{6}{11}$.
57. Divide: $\frac{6}{7} \div \frac{3}{5}$.
58. Simplify: $\frac{\frac{3}{5}}{\frac{2}{7}}$.

Estimation

We will now discuss some estimation methods that can be used when working with percent. To begin, we consider a way to find 10% of a number quickly. Recall that 10% of a number is found by multiplying the number by 10% or 0.1. When multiplying a number by 0.1, we simply move the decimal point 1 place to the left to find the result.

EXAMPLE 1 **10% of a number.** Find 10% of 234.

Solution To find 10% of 234, move the decimal point 1 place to the left.

$$234 = 23.4\,0$$

Thus, 10% of 234 is 23.4, or approximately 23.

To find 15% of a number, first find 10% of the number. Then find half of that to obtain the other 5%. Finally, add the two results.

EXAMPLE 2 Estimating 15% of a number. Estimate 15% of 78.

Solution

10% of 78 is 7.8, or about 8. $\longrightarrow$ 8
Add half of 8 to get the other 5%. $\longrightarrow$ + 4

$$\begin{array}{r} 8 \\ +\ \underline{4} \\ 12 \end{array}$$

Thus, 15% of 78 is approximately 12.

To find 20% of a number, first find 10% of it and then double that result. A similar procedure can be used when working with any multiple of 10%.

EXAMPLE 3 Estimating 20% of a number. Estimate 20% of 3,234.15.

Solution 10% of 3,234.15 is 323.415 or about 323. To find 20%, double that.

Thus, 20% of 3,234.15 is approximately 646.

EXAMPLE 4 1% of a number. Find 1% of 0.8.

Solution To find 1% of a number, multiply it by 0.01, because 1% = 0.01. When multiplying a number by 0.01, simply move the decimal point 2 places to the left to find the result.

$$0.8 = .00\,8$$
$$= 0.008$$

Thus, 1% of 0.8 is 0.008.

EXAMPLE 5 50% of a number. Find 50% of 2,800,000,000.

Solution To find 50% of a number means to find $\frac{1}{2}$ of that number. To find one-half of a number, simply divide it by 2. Thus, 50% of 2,800,000,000 is 2,800,000,000 ÷ 2 = 1,400,000,000.

To find 25% of a number, first find 50% of it and then divide that result by 2.

EXAMPLE 6 **Estimating 25% of a number.** Estimate 25% of 16,813.

Solution 16,813 is about 16,800. Half of that is 8,400. Thus, 50% of 16,813 is approximately 8,400.

To estimate 25% of 16,813, divide 8,400 by 2. Thus, 25% of 16,813 is approximately 4,200.

100% of a number is the number itself. To find 200% of a number, double the number.

EXAMPLE 7 **Estimating 200% of a number.** Estimate 200% of 65.198.

Solution 65.198 is about 65. To find 200% of 65, double it. Thus, 200% of 65.198 is approximately $65 \cdot 2$ or 130.

STUDY SET *Estimate each answer.*

1. COLLEGE COURSES 20% of the 815 students attending a small college were enrolled in a science course. How many students is this?
2. SPECIAL OFFERS In the grocery store, a 65-ounce bottle of window cleaner was marked "25% free." How many ounces are free?
3. DISCOUNTS By how much is the price of a VCR discounted if the regular price of $196.88 is reduced by 30%?
4. TIPPING A restaurant tip is normally 15% of the cost of the meal. Find the tip on a dinner costing $38.64.
5. FIRE DAMAGE An insurance company paid 50% of the $107,809 it cost to rebuild a home that was destroyed by fire. How much did the insurance company pay?
6. SAFETY INSPECTIONS Of the 2,580 vehicles inspected at a safety checkpoint, 10% had code violations. How many cars had code violations?
7. WEIGHTLIFTING A 158-pound weightlifter can bench press 200% of his body weight. How many pounds can he bench press?
8. TESTING On a 120-question true/false test, 5% of a student's answers were wrong. How many questions did she miss?
9. TRAFFIC STUDIES According to an electronic traffic monitor, 20% of the 650 motorists who passed it were speeding. How many of these motorists were speeding?
10. SELLING HOMES A homeowner has been told she will recoup 70% of her $5,000 investment if she paints her home before selling it. What is the potential payback if she paints her home?

Approximate the percent and then estimate each answer.

11. NO-SHOWS The attendance at a seminar was only 31% of what the organizers had anticipated. If 68 people were expected, how many actually attended the seminar?
12. "A" STUDENTS Of the 900 students in a school, 16% were on the principal's honor roll. How many students were on the honor roll?
13. INTERNET SURVEY The illustration shows an online survey question. How many people voted yes?

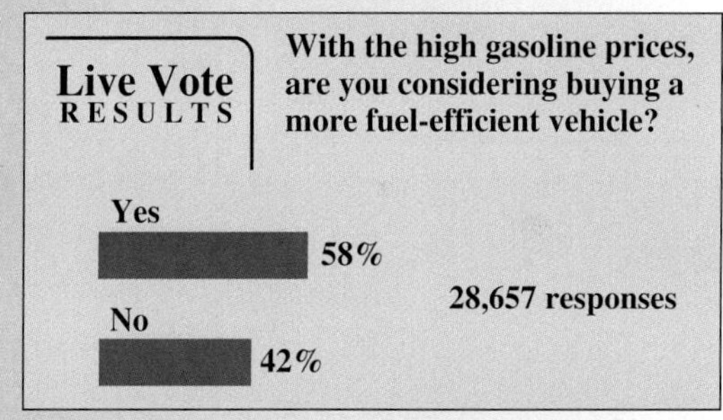

14. MEDICARE The Medicare payroll tax rate is 1.45%. How much Medicare tax will be deducted from a paycheck of $596?
15. VOTING On election day, 48% of the 6,200 workers at the polls were volunteers. How many volunteers helped with the election?
16. BUDGETS Each department at a college was asked to cut its budget by 21%. By how much money should the mathematics department budget be reduced if it is currently $4,515?

5.4 Interest

• Simple interest • Compound interest

When money is borrowed, the lender expects to be paid back the amount of the loan plus an additional charge for the use of the money. The additional charge is called **interest.** When money is deposited in a bank, the depositor is paid for the use of the money. The money the deposit earns is also called interest. In general, interest is money that is paid for the use of money.

Simple interest

Interest is calculated in one of two ways: either as **simple interest** or as **compound interest.** We will begin by discussing simple interest. First, we need to introduce some key terms associated with borrowing or lending money.

Principal: the amount of money that is invested, deposited, or borrowed.

Interest rate: a percent that is used to calculate the amount of interest to be paid. It is usually expressed as an annual (yearly) rate.

Time: the length of time (usually in years) that the money is invested, deposited, or borrowed.

The amount of interest to be paid depends on the principal, the rate, and the time. That is why all three are usually mentioned in advertisements for bank accounts, investments, and loans. (See Figure 5-10.)

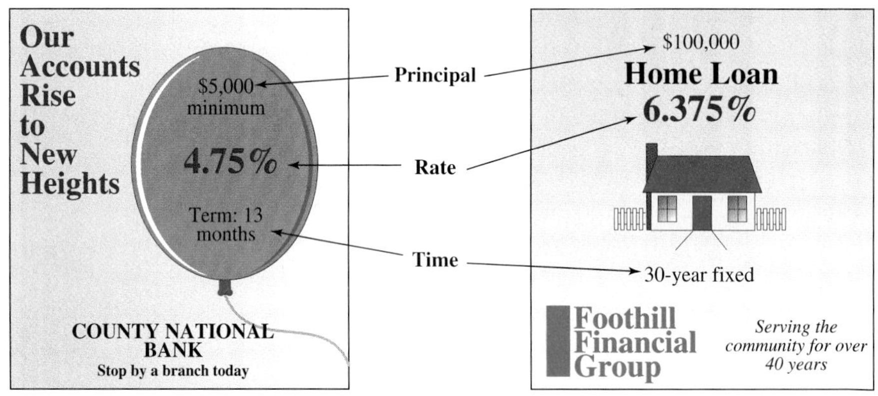

FIGURE 5-10

Simple interest is interest earned on the original principal. It is found by using a formula.

Simple interest formula

Interest = principal · rate · time

or

$I = Prt$

where the rate r is expressed as an annual rate and the time t is expressed in years.

EXAMPLE 1 \$3,000 is invested for 1 year at a rate of 5%. How much interest is earned?

Solution We will use the formula $I = Prt$ to calculate the interest earned. The principal is \$3,000, the interest rate is 5% (or 0.05), and the time is 1 year.

$P = 3{,}000 \qquad r = 5\% = 0.05 \qquad t = 1$ year

$I = \mathbf{Prt}$	Write the interest formula.
$= \mathbf{3{,}000 \cdot 0.05 \cdot 1}$	Substitute the values for P, r, and t.
$= 150$	Perform the multiplication.

The interest earned in 1 year is \$150.

The information given in this problem and the result can be presented in a table.

Principal	Rate	Time	Interest earned
\$3,000	5%	1 year	\$150

Self Check 1

If \$4,200 is invested for 2 years at a rate of 4% annual interest, how much interest is earned?

Answer \$336

When we use the formula $I = Prt$, the time must be expressed in years. If the time is given in days or months, we rewrite it as a fractional part of a year. For example, a 30-day investment lasts $\frac{30}{365}$ of a year, since there are 365 days in a year. For a 6-month loan, we express the time as $\frac{6}{12}$ or $\frac{1}{2}$ of a year, since there are 12 months in a year.

EXAMPLE 2 To start a carpet-cleaning business, a couple borrows \$5,500 to purchase equipment and supplies. If the loan has a 14% interest rate, how much must they repay at the end of the 90-day period?

Solution First, we find the amount of interest paid on the loan. We must rewrite the time (90-day period) as a fractional part of a 365-day year.

$P = 5{,}500 \qquad r = 14\% = 0.14 \qquad t = \frac{90}{365}$

$I = \mathbf{Prt}$	Write the interest formula.
$= \mathbf{5{,}500 \cdot 0.14 \cdot \frac{90}{365}}$	Substitute the values for P, r, and t.
$= \frac{5{,}500}{1} \cdot \frac{0.14}{1} \cdot \frac{90}{365}$	Write 5,500 and 0.14 as fractions.
$= \frac{69{,}300}{365}$	Use a calculator to multiply the numerators. Multiply the denominators.
≈ 189.86	Use a calculator to perform the division. Round to the nearest cent.

The interest on the loan is \$189.86. To find how much they must pay back, we add the principal and the interest.

$5{,}500 + 189.86 = 5{,}689.86$

The couple must pay back \$5,689.86 at the end of 90 days.

Self Check 2

How much must be repaid if \$3,200 is borrowed at a rate of 15% for 120 days?

Answer \$3,357.81

Compound interest

Most savings accounts pay **compound interest** rather than simple interest. Compound interest is interest paid on accumulated interest. To illustrate this concept, suppose that \$2,000 is deposited in a savings account at a rate of 5% for 1 year. We can use the formula $I = Prt$ to calculate the interest earned at the end of 1 year.

$$\begin{aligned} I &= Prt \\ &= 2{,}000 \cdot 0.05 \cdot 1 && \text{Substitute for } P, r, \text{ and } t. \\ &= 100 && \text{Perform the multiplication.} \end{aligned}$$

Interest of \$100 was earned. At the end of the first year, the account contains the interest (\$100) plus the original principal (\$2,000), for a balance of \$2,100.

Suppose that the money remains in the savings account for another year at the same interest rate. For the second year, interest will be paid on a principal of \$2,100. That is, during the second year, we earn *interest on the interest* as well as on the original \$2,000 principal. Using $I = Prt$, we can find the interest earned in the second year.

$$\begin{aligned} I &= Prt \\ &= 2{,}100 \cdot 0.05 \cdot 1 && \text{Substitute for } P, r, \text{ and } t. \\ &= 105 && \text{Perform the multiplication.} \end{aligned}$$

In the second year, \$105 of interest is earned. The account now contains that interest plus the \$2,100 principal, for a total of \$2,205.

As Figure 5-11 shows, we calculated the simple interest two times to find the compound interest.

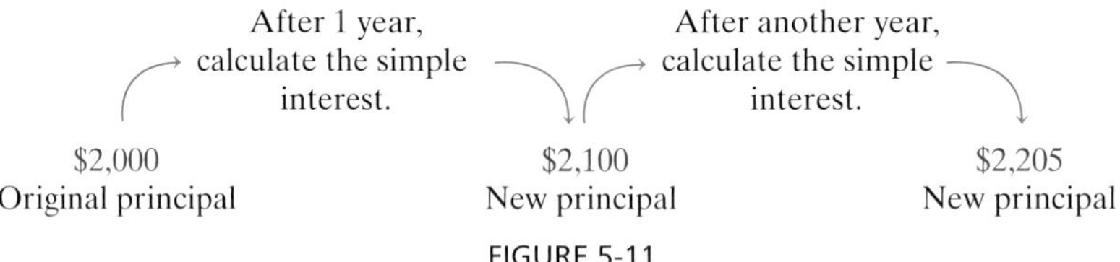

FIGURE 5-11

If we compute the *simple interest* on \$2,000, at 5% for 2 years, the interest earned is $I = 2{,}000 \cdot 0.05 \cdot 2 = 200$. Thus, the account balance would be \$2,200. Comparing the balances, the account earning compound interest will contain \$5 more than the account earning simple interest.

In the previous example, the interest was calculated at the end of each year, or **annually.** When compounding, we can compute the interest in other time increments, such as **semiannually** (twice a year), **quarterly** (four times a year), or even **daily.**

EXAMPLE 3 Compound interest. As a gift for her newborn granddaughter, a grandmother opens a \$1,000 savings account in the baby's name. The interest rate is 4.2%, compounded quarterly. Find the amount of money the child will have in the bank on her first birthday.

Solution If the interest is compounded quarterly, the interest will be computed four times in one year. To find the amount of interest \$1,000 will earn in the first quarter of the year, we use the simple interest formula, where t is $\frac{1}{4}$ of a year.

Interest earned in the first quarter

$$P = 1{,}000 \qquad r = 4.2\% = 0.042 \qquad t = \frac{1}{4}$$

$$\begin{aligned} I &= 1{,}000 \cdot 0.042 \cdot \frac{1}{4} \\ &= \$10.50 \end{aligned}$$

The interest earned in the first quarter is \$10.50. This now becomes part of the principal for the second quarter.

$$\$1{,}000 + \$10.50 = \$1{,}010.50$$

To find the amount of interest \$1,010.50 will earn in the second quarter of the year, we use the simple interest formula, where t is again $\frac{1}{4}$ of a year.

$$P = \mathbf{1{,}010.50} \qquad r = \mathbf{0.042} \qquad t = \frac{\mathbf{1}}{\mathbf{4}}$$

$$I = \mathbf{1{,}010.50} \cdot \mathbf{0.042} \cdot \frac{\mathbf{1}}{\mathbf{4}}$$

$$\approx \$10.61 \quad \text{(Rounded)}$$

The interest earned in the second quarter is \$10.61. This becomes part of the principal for the third quarter.

\$1,010.50 + \$10.61 = \$1,021.11

To find the interest \$1,021.11 will earn in the third quarter of the year, we proceed as follows.

$$P = \mathbf{1{,}021.11} \qquad r = \mathbf{0.042} \qquad t = \frac{\mathbf{1}}{\mathbf{4}}$$

$$I = \mathbf{1{,}021.11} \cdot \mathbf{0.042} \cdot \frac{\mathbf{1}}{\mathbf{4}}$$

$$\approx \$10.72 \quad \text{(Rounded)}$$

The interest earned in the third quarter is \$10.72. This now becomes part of the principal for the fourth quarter.

\$1,021.11 + \$10.72 = \$1,031.83

To find the interest \$1,031.83 will earn in the fourth quarter, we again use the simple interest formula.

$$P = \mathbf{1{,}031.83} \qquad r = \mathbf{0.042} \qquad t = \frac{\mathbf{1}}{\mathbf{4}}$$

$$I = \mathbf{1{,}031.83} \cdot \mathbf{0.042} \cdot \frac{\mathbf{1}}{\mathbf{4}}$$

$$\approx \$10.83 \quad \text{(Rounded)}$$

The interest earned in the fourth quarter is \$10.83. Adding this to the existing principal, we get

\$1,031.83 + \$10.83 = \$1,042.66

The amount that has accumulated in the account after four quarters, or 1 year, is \$1,042.66.

Computing compound interest by hand is tedious. The **compound interest formula** can be used to find the total amount of money that an account will contain at the end of the term.

Compound interest formula

The total amount A in an account can be found using the formula

$$A = P\left(1 + \frac{r}{n}\right)^{nt}$$

where P is the principal, r is the annual interest rate expressed as a decimal, t is the length of time in years, and n is the number of compoundings in one year.

A calculator is often helpful in solving compound interest problems.

CALCULATOR SNAPSHOT — Compound interest

A businessman invests $9,250 at 7.6% interest, to be compounded monthly. To find what the investment will be worth in 3 years, we use the compound interest formula with the following values.

$P = \$9{,}250, \quad r = 7.6\% = 0.076, \quad t = 3 \text{ years}, \quad n = 12 \text{ times a year (monthly)}$

We apply the compound interest formula.

$$A = P\left(1 + \frac{r}{n}\right)^{nt} \qquad \text{Write the compound interest formula.}$$

$$= \mathbf{9{,}250}\left(1 + \frac{\mathbf{0.076}}{\mathbf{12}}\right)^{\mathbf{12(3)}} \qquad \text{Substitute the values of } P, r, t, \text{ and } n.$$

$$= 9{,}250\left(1 + \frac{0.076}{12}\right)^{36} \qquad \text{Simplify the exponent: } 12(3) = 36.$$

To evaluate the expression on the right-hand side of the equation, we enter these numbers and press these keys.

9250 [×] [(] 1 [+] .076 [÷] 12 [)] [y^x] 36 [=]   11610.43875

Rounded to the nearest cent, the amount in the account after 3 years will be $11,610.44.

If your calculator does not have parenthesis keys, calculate the sum within the parentheses first. Then find the power. Finally, multiply by 9,250.

Self Check 4
Find the amount of interest $25,000 will earn in 10 years if it is deposited in an account at 5.99% interest, compounded daily.

Answer $20,505.20

EXAMPLE 4 A man deposited $50,000 in a long-term account at 6.8% interest, compounded daily. How much money will he be able to withdraw in 7 years if the principal is to remain in the bank?

Solution "Compounded daily" means that compounding will be done 365 times in a year.

$P = \$50{,}000 \qquad r = 6.8\% = 0.068 \qquad t = 7 \text{ years} \qquad n = 365 \text{ times a year}$

$$A = P\left(1 + \frac{r}{n}\right)^{nt} \qquad \text{Write the compound interest formula.}$$

$$= \mathbf{50{,}000}\left(1 + \frac{\mathbf{0.068}}{\mathbf{365}}\right)^{365(7)} \qquad \text{Substitute the values of } P, r, t, \text{ and } n.$$

$$= 50{,}000\left(1 + \frac{0.068}{365}\right)^{2{,}555} \qquad \text{Perform the multiplication: } 365(7) = 2{,}555.$$

$$\approx 80{,}477.58 \qquad \text{Use a calculator. Round to the nearest cent.}$$

The account will contain $80,477.58 at the end of 7 years. To find the amount the man can withdraw, we subtract.

$80{,}477.58 - 50{,}000 = 30{,}477.58$

The man can withdraw $30,477.58 without having to touch the $50,000 principal.

Section 5.4 STUDY SET

VOCABULARY *Fill in the blanks.*

1. In banking, the original amount of money borrowed or deposited is known as the ________.
2. Borrowers pay ________ to lenders for the use of their money.
3. The percent that is used to calculate the amount of interest to be paid is called the ________ rate.
4. ________ interest is interest paid on accumulated interest.
5. Interest computed on only the original principal is called ________ interest.
6. Percent means parts per ________.

CONCEPTS

7. When we do calculations with percents, they must be changed to decimals or fractions. Change each percent to a decimal.

   a. 7% b. 9.8% c. $6\frac{1}{4}\%$

8. Express each of the following as a fraction of a year. Simplify the fraction.

   a. 6 months b. 90 days
   c. 120 days d. 1 month

9. Complete the table by finding the simple interest earned.

Principal	Rate	Time	Interest earned
\$10,000	6%	3 years	

10. Determine how many times a year the interest on a savings account is calculated if the interest is compounded

    a. semiannually b. quarterly
    c. daily d. monthly

11. a. What concept studied in this section is illustrated by the following diagram?
    b. What was the original principal?
    c. How many times was the interest found?
    d. How much interest was earned on the first compounding?
    e. For how long was the money invested?

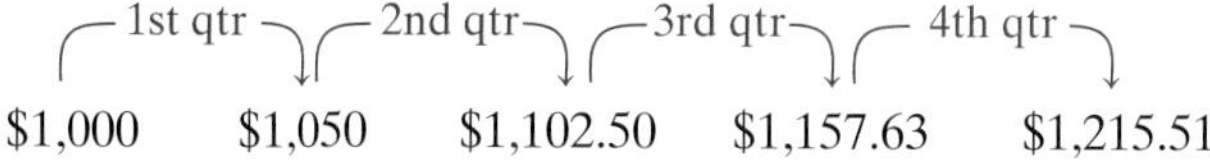

12. \$3,000 is deposited in a savings account that earns 10% interest compounded annually. Complete the series of calculations shown in the illustration to find how much money will be in the account at the end of 2 years.

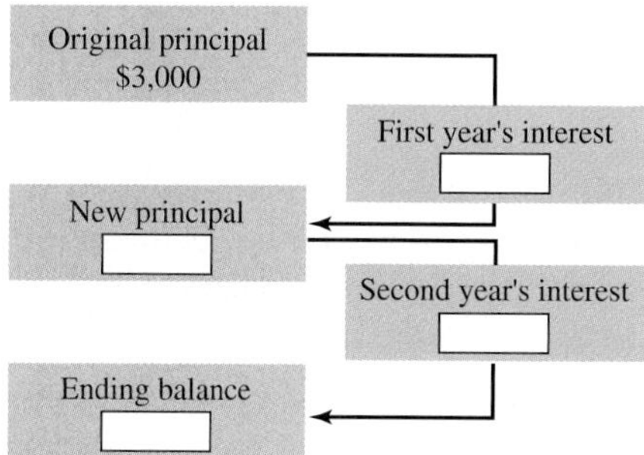

NOTATION

13. In the formula $I = Prt$, what operations are indicated by Prt?
14. In the formula $A = P\left(1 + \frac{r}{n}\right)^{nt}$, how many operations must be performed to find A?

APPLICATIONS *In the following problems, use simple interest.*

15. RETIREMENT INCOME A retiree invests \$5,000 in a savings plan that pays 6% per year. What will the account balance be at the end of the first year?
16. INVESTMENTS A developer promised a return of 8% annual interest on an investment of \$15,000 in her company. How much could an investor expect to make in the first year?
17. REMODELING A homeowner borrows \$8,000 to pay for a kitchen remodeling project. The terms of the loan are 9.2% annual interest and repayment in 2 years. How much interest will be paid on the loan?
18. CREDIT UNIONS A farmer borrowed \$7,000 from a credit union. The money was loaned at 8.8% annual interest for 18 months. How much money did the credit union charge him for the use of the money?
19. MEETING PAYROLLS In order to meet end-of-the-month payroll obligations, a small business had to borrow \$4,200 for 30 days. How much did the business have to repay if the interest rate was 18%?

20. CAR LOANS To purchase a car, a man takes out a loan for \$2,000. If the interest rate is 9% per year, how much interest will he have to pay at the end of the 120-day loan period?

21. SAVINGS ACCOUNTS Find the interest earned on \$10,000 at $7\frac{1}{4}\%$ for 2 years. Use the table to organize your work.

P	*r*	*t*	*I*

22. TUITION A student borrows \$300 from an educational fund to pay for books for spring semester. If the loan is for 45 days at $3\frac{1}{2}\%$ annual interest, what will the student owe at the end of the loan period?

23. LOAN APPLICATIONS Complete the loan application below.

Loan Application Worksheet

1. Amount of loan (principal) *\$1,200.00*
2. Length of loan (time) *2 YEARS*
3. Annual percentage rate *8%*
4. Interest charged ____________
5. Total amount to be repaid ____________
6. Check method of repayment:
   ☐ 1 lump sum ☑ monthly payments

Borrower agrees to pay *24* equal payments of ________ to repay loan.

24. LOAN APPLICATIONS Complete the loan application below.

Loan Application Worksheet

1. Amount of loan (principal) *\$810.00*
2. Length of loan (time) *9 mos.*
3. Annual percentage rate *12%*
4. Interest charged ____________
5. Total amount to be repaid ____________
6. Check method of repayment:
   ☐ 1 lump sum ☑ monthly payments

Borrower agrees to pay *9* equal payments of ________ to repay loan.

25. LOW-INTEREST LOANS An underdeveloped country receives a low-interest loan from a bank to finance the construction of a water treatment plant. What must the country pay back at the end of 2 years if the loan is for \$18 million at 2.3%?

26. CITY REDEVELOPMENT A city is awarded a low-interest loan to help renovate the downtown business district. The \$40-million loan, at 1.75%, must be repaid in $2\frac{1}{2}$ years. How much interest will the city have to pay?

A calculator will be helpful in solving these problems.

27. COMPOUNDING ANNUALLY If \$600 is invested in an account that earns 8%, compounded annually, what will the account balance be after 3 years?

28. COMPOUNDING SEMIANNUALLY If \$600 is invested in an account that earns annual interest of 8%, compounded semiannually, what will the account balance be at the end of 3 years?

29. COLLEGE FUNDS A ninth-grade student opens a savings account that locks her money in for 4 years at an annual rate of 6%, compounded daily. If the initial deposit is \$1,000, how much money will be in the account when she begins college in four years?

30. CERTIFICATES OF DEPOSIT A 3-year certificate of deposit pays an annual rate of 5%, compounded daily. The maximum allowable deposit is \$90,000. What is the most interest a depositor can earn from the CD?

31. TAX REFUNDS A couple deposits an income tax refund check of \$545 in an account paying an annual rate of 4.6%, compounded daily. What will the size of the account be at the end of 1 year?

32. INHERITANCES After receiving an inheritance of \$11,000, a man deposits the money in an account paying an annual rate of 7.2%, compounded daily. How much money will be in the account at the end of 1 year?

33. LOTTERIES Suppose you won \$500,000 in the lottery and deposited the money in a savings account that paid an annual rate of 6% interest, compounded daily. How much interest would you earn each year?

34. CASH GIFTS After receiving a \$250,000 cash gift, a university decides to deposit the money in an account paying an annual rate of 5.88%, compounded quarterly. How much money will the account contain in 5 years?

WRITING

35. What is the difference between simple and compound interest?
36. Explain: *Interest is the amount of money paid for the use of money.*
37. On some accounts, banks charge a penalty if the depositor withdraws the money before the end of the term. Why would a bank do this?
38. Explain why it is better for a depositor to open a savings account that pays 5% interest, compounded daily, than one that pays 5% interest, compounded monthly.

REVIEW

39. Evaluate: $\sqrt{\frac{1}{4}}$.
40. Find the power: $\left(\frac{1}{4}\right)^2$.
41. Round 23.045 to the nearest tenth.
42. Round 23.045 to the nearest hundredth.
43. Solve: $\frac{2}{3}x = -2$.
44. Divide: $-12\frac{1}{2} \div 5$.
45. Multiply: $8\frac{1}{3} \cdot 6$.
46. Evaluate: $(0.2)^2 - (0.3)^2$.

KEY CONCEPT

Percent

Since the word *percent* means *parts per one hundred,* we can think of a percent as the numerator of a fraction that has a denominator of 100. To write a percent as a fraction, we drop the percent symbol and write the given number over 100.

Finish each convension.

$67\% = \frac{\square}{100}$ $\quad 56\% = \frac{56}{\square} = \frac{14}{25}$ $\quad 0.05\% = \frac{\square}{100} = \frac{5}{10{,}000} = \frac{1}{\square}$

To change a percent to a decimal, we drop the percent symbol and move the decimal point two places to the left.

Finish each conversion.

$67\% = \square$ $\quad 56\% = \square$ $\quad 0.05\% = \square$

To write a fraction as a percent, we write the fraction as a decimal and then move the decimal point two places to the right and insert a percent symbol.

Finish each conversion.

$\frac{3}{4} = 0.75 = \square$ $\qquad \frac{4}{5} = \square = 80\%$

$\frac{5}{8} = \square = 62.5\%$ $\qquad \frac{25}{4} = 6.25 = \square$

To solve problems involving percent, we use the percent formula.

Amount = percent · base

Solve each problem.

1. Find 32% of 620.
2. 300 is what percent of 500?
3. 25 is 40% of what number?
4. Find 125% of 850.
5. 106.25 is what percent of 625?
6. 163.84 is 32% of what number?

Percents are used to compute interest. If I is the interest, P is the principal, r is the annual rate (or percent), and t is the length of time in years, the formula for simple interest is

$$I = Prt$$

7. Find the amount of interest that will be earned if $10,000 is invested for 5 years at 6% annual interest.

If A is the amount, P is the principal, r is the annual rate of interest, t is the length of time in years, and n is the number of compoundings in one year, the formula for compound interest is

$$A = P\left(1 + \frac{r}{n}\right)^{nt}$$

8. Find the amount of interest that will be earned if $10,000 is invested for 5 years at 6% annual interest, compounded quarterly.

ACCENT ON TEAMWORK

SECTION 5.1

M & M'S   Give each member of your group a bag of M & M's candies.

a. Determine what percent of the total number of M & M's in your bag are yellow. Do the same for each of the other colors. Enter the results in the table. (Round to the nearest one percent.)

M & M's color	Percent
Yellow	
Brown	
Green	
Red	
Blue	

b. Present the data in the table using the circle graph. Compare your graph to the graphs made by the other members of your group. Do the colors occur in the same percentages in each of the bags?

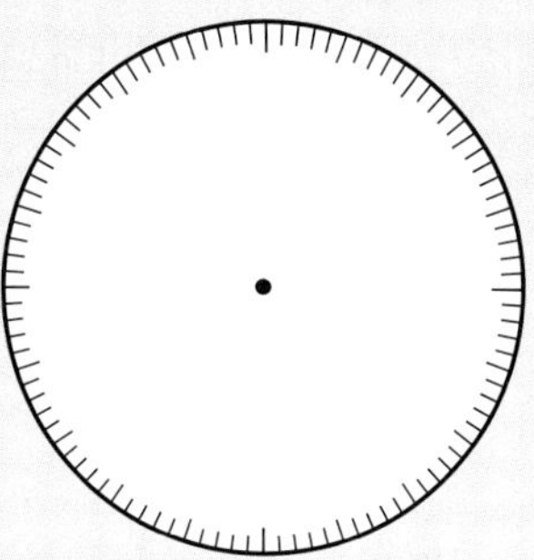

SECTION 5.2

NUTRITION   Have each person in your group bring in a nutrition label like the one shown and write the name of the food product on the back. Have the members of the group exchange labels. With the label that you receive, determine what percent of the total calories come from fat.

The USDA recommends that no more than 30% of a person's daily calories should come from fat. Which products exceed the recommendation?

Nutrition Facts
Serving Size 1 meal

Amount Per Serving

Calories 560	Calories from Fat 190
	% Daily Value
Total fat 21g	32%
Saturated fat 9 g	43%
Cholesterol 60mg	20%
Sodium 2110mg	88%
Total carbohydrate 67g	22%
Dietary fiber 7g	29%
Sugars less than 25g	
Protein 27g	

SECTION 5.3

ENROLLMENTS   From your school's admissions office, get the enrollment figures for the last ten years. Calculate the percent of increase (or decrease) in enrollment for each of the following periods.

- Ten years ago to the present
- Five years ago to the present
- One year ago to the present

NEWSPAPER ADS   Have each person in your group find a newspaper advertisement for some item that is on sale. The ad should include only two of the four details listed below.

- The regular price
- The sale price
- The discount
- The discount rate

For example, the following ad gives the regular price and the sale price, but it doesn't give the discount or the discount rate.

Have the members of your group exchange ads. Determine the two missing details on the ad that you receive. In your group, which item had the highest discount rate?

SECTION 5.4

INTEREST RATES   Recall that interest is money that the borrower pays to the lender for the use of the money. The amount of interest that the borrower must pay depends on the interest rate charged by the lender.

Have members of your group call banks, savings and loans, credit unions, and other financial services to get the lending rates for various types of loans. (See the yellow pages of the phone book.)

Find out what rate is charged by credit cards such as VISA, department stores, and gasoline companies. List the interest rates in order, from greatest to least, and present your findings to the class.

CHAPTER REVIEW

SECTION 5.1 *Percents, Decimals, and Fractions*

CONCEPTS

Percent means parts per one hundred.

REVIEW EXERCISES

Express the amount of each figure that is shaded as a percent, as a decimal, and as a fraction. Each set of squares represents 100%.

1. 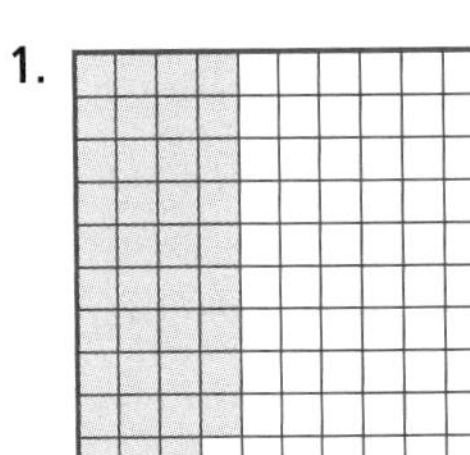

2. 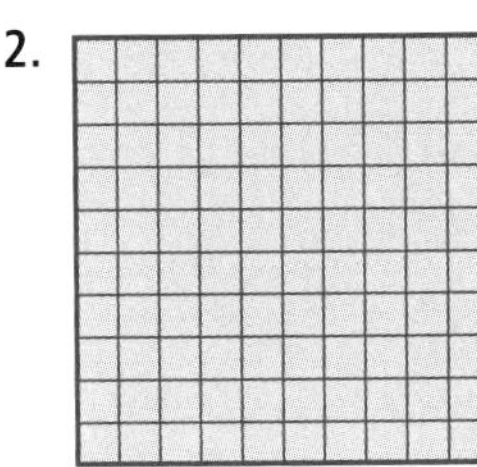 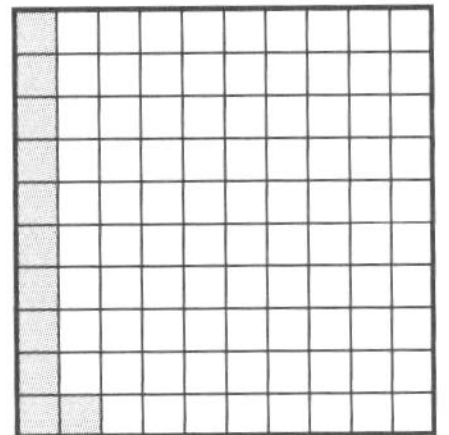

3. In Problem 1, what percent of the figure is not shaded?

To change a percent to a fraction, drop the % symbol and put the given number over 100.

Change each percent to a fraction.

4. 15%
5. 120%
6. $9\frac{1}{4}\%$
7. 0.1%

To change a percent to a decimal, drop the % symbol and divide by 100 by moving the decimal point 2 places to the left.

Change each percent to a decimal.

8. 27%
9. 8%
10. 155%
11. $1\frac{4}{5}\%$

To change a decimal to a percent, multiply the decimal by 100 by moving the decimal point 2 places to the right, and then insert a % symbol.

Change each decimal to a percent.

12. 0.83
13. 0.625
14. 0.051
15. 6

To change a fraction to a percent, write the fraction as a decimal by dividing its numerator by its denominator. Multiply the decimal by 100 by moving the decimal point 2 places to the right, and then insert a % symbol.

Change each fraction to a percent.

16. $\frac{1}{2}$
17. $\frac{4}{5}$
18. $\frac{7}{8}$
19. $\frac{1}{16}$

Find the exact percent equivalent for each fraction.

20. $\frac{1}{3}$
21. $\frac{5}{6}$

Change each fraction to a percent. Round to the nearest hundredth of a percent.

22. $\frac{5}{9}$
23. $\frac{8}{3}$

24. BILL OF RIGHTS There are 27 amendments to the Constitution of the United States. The first ten are known as the Bill of Rights. What percent of the amendments were adopted after the Bill of Rights? (Round to the nearest one percent.)
25. Express one-tenth of one percent as a fraction.

SECTION 5.2 *Solving Percent Problems*

The percent formula:

Amount = percent · base

We can translate a percent problem from words into an equation. A *variable* is used to stand for the unknown number; *is* can be translated to an = symbol; and *of* means multiply.

26. Identify the amount, the base, and the percent in the statement "15 is $33\frac{1}{3}$% of 45."

27. Translate the given sentence into a percent equation.

What number	is	32%	of	96?

Solve each percent problem.

28. What number is 40% of 500?

29. 16% of what number is 20?

30. 1.4 is what percent of 80?

31. $66\frac{2}{3}$% of 3,150 is what number?

32. Find 220% of 55.

33. What is 0.05% of 60,000?

34. RACING The nitro–methane fuel mixture used to power some experimental cars is 96% nitro and 4% methane. How many gallons of each fuel component are needed to fill a 15-gallon fuel tank?

35. HOME SALES After the first day on the market, 51 homes in a new subdivision had already sold. This was 75% of the total number of homes available. How many homes were originally for sale?

36. HURRICANES In a mobile home park, 96 of the 110 trailers were either damaged or destroyed by hurricane winds. What percent is this? (Round to the nearest one percent.)

37. TIPPING The cost of dinner for a family of five at a restaurant was $36.20. Find the amount of the tip if it should be 15% of the cost of dinner.

A *circle graph* is a way of presenting data for comparison. The sizes of the segments of the circle indicate the percents of the whole represented by each category.

38. AIR POLLUTION Complete the circle graph to show the given data.

Sources of carbon monoxide air pollution	
Transportation vehicles	63%
Fuel combustion in homes, offices, electrical plants	12%
Industrial processes	8%
Solid-waste disposal	3%
Miscellaneous	14%

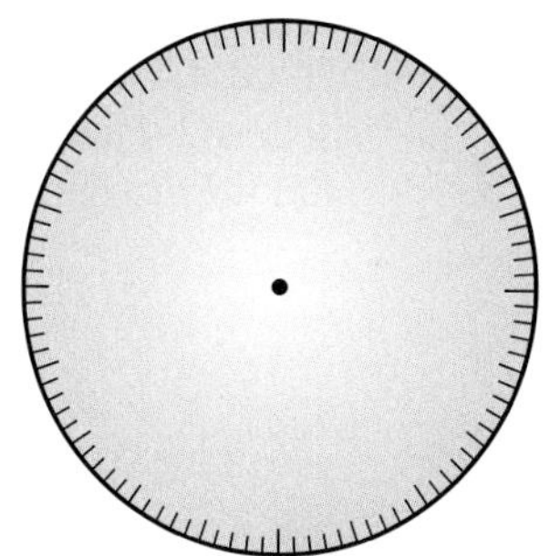

39. EARTH'S SURFACE The surface of the Earth is approximately 196,800,000 square miles. Use the graph to determine the number of square miles of the Earth's surface that are covered with water.

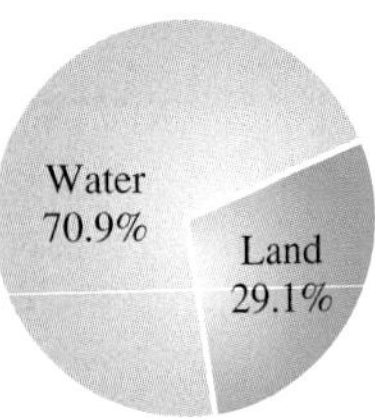

SECTION 5.3 *Applications of Percent*

To find the total price of an item:

Total price = purchase price + sales tax

40. SALES RECEIPTS Complete the sales receipt.

CAMERA CENTER	
35mm Canon Camera	\$59.99
SUBTOTAL	\$59.99
SALES TAX @ 5.5%	
TOTAL	

41. TAX RATES Find the sales tax rate if the sales tax is \$492 on the purchase of an automobile priced at \$12,300.

Commission is based on a percent of the total dollar amount of the goods or services sold.

42. COMMISSIONS If the commission rate is 6%, find the commission earned by an appliance salesperson who sells a washing machine for \$369.97 and a dryer for \$299.97.

To find *percent of increase or decrease:*

1. Subtract the smaller number from the larger to find the amount of increase or decrease.
2. Find what percent the difference is of the original amount.

43. Fill in the blank: Always find the percent of increase or decrease of a quantity with respect to the ________ amount.

44. TROOP SIZE The size of a peacekeeping force was increased from 10,000 to 12,500 troops. What percent of increase is this?

45. GAS MILEAGE Experimenting with a new brand of gasoline in her truck, a woman found that the gas mileage fell from 18.8 to 17.0 miles per gallon. What percent of decrease is this? (Round to the nearest tenth of a percent.)

The difference between the original price and the sale price of an item is called the *discount.*

To find the *sale price:*

Sale price = original price − discount

46. TOOL CHESTS Use the information in the ad to find the discount, the original price, and the discount rate on the tool chest.

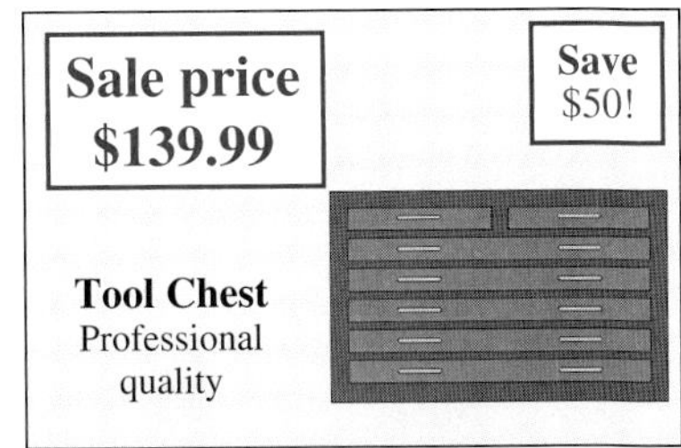

SECTION 5.4 Interest

Simple interest is interest earned on the original principal and is found using the formula

$$I = Prt$$

where P is the principal, r is the annual interest rate, and t is the length of time in years.

47. Find the interest earned on $6,000 invested at 8% per year for 2 years. Use the following chart to organize your work.

P	r	t	I

48. CODE VIOLATIONS A business was ordered to correct safety code violations in a production plant. To pay for the needed corrections, the company borrowed $10,000 at 12.5% for 90 days. Find the total amount that had to be paid after 90 days.

Compound interest is interest earned on interest.

49. MONTHLY PAYMENTS A couple borrows $1,500 for 1 year at $7\frac{3}{4}$% and decides to repay the loan by making 12 equal monthly payments. How much will each monthly payment be?

50. Find the amount of money that will be in a savings account at the end of 1 year if $2,000 is the initial deposit and the annual interest rate of 7% is compounded semiannually. (*Hint:* Find the simple interest twice.)

51. Find the amount that will be in a savings account at the end of 3 years if a deposit of $5,000 earns interest at an annual rate of $6\frac{1}{2}$%, compounded daily.

The compound interest formula:

$$A = P\left(1 + \frac{r}{n}\right)^{nt}$$

where A is the amount in the account, P is the principal, r is the annual interest rate, n is the number of compoundings in one year, and t is the length of time in years.

52. CASH GRANTS Each year a cash grant is given to a deserving college student. The grant consists of the interest earned that year on a $500,000 savings account. What is the cash award for the year if the money is invested at an annual rate of 8.3%, compounded daily?

CHAPTER 5 TEST

1. Express the amount of the figure that is shaded as a percent, as a fraction, and as a decimal.

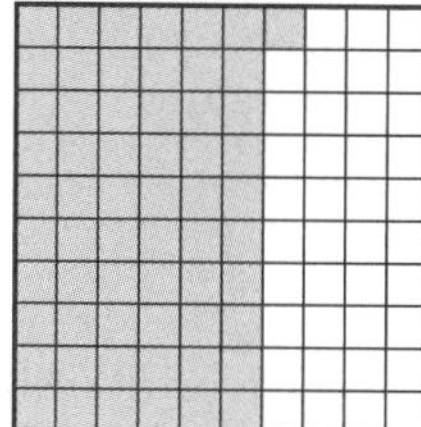

2. In the illustration, each set of 100 squares represents 100%. Express as a percent the amount of the figure that is shaded. Then express that percent as a fraction and as a decimal.

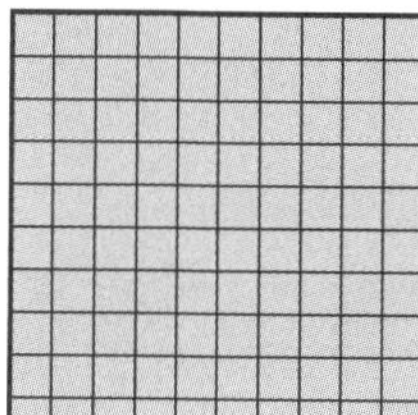

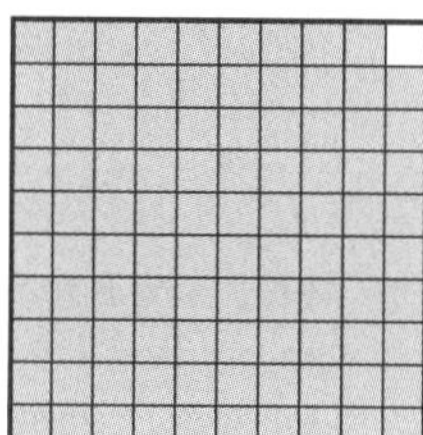

3. Change each percent to a decimal.

   a. 67% b. 12.3% c. $9\frac{3}{4}\%$

4. Change each fraction to a percent.

   a. $\frac{1}{4}$ b. $\frac{5}{8}$ c. $\frac{3}{25}$

5. Change each decimal to a percent.

   a. 0.19 b. 3.47 c. 0.005

6. Change each percent to a fraction.

   a. 55% b. 0.01% c. 125%

7. Change $\frac{7}{30}$ to a percent. Round to the nearest hundredth of a percent.

8. WEATHER REPORTS A weather reporter states that there is a 40% chance of rain. What are the chances that it will not rain?

9. Find the exact percent equivalent for the fraction $\frac{2}{3}$.

10. Find the exact percent equivalent for the fraction $\frac{1}{4}$.

11. SHRINKAGE Refer to the label on a pair of jeans.

    a. How much length will be lost due to shrinkage?

    b. What will be the resulting length?

WAIST	INSEAM
33	34

Expect shrinkage of approximately **3%** in length after the jeans are washed.

12. 65 is what percent of 1,000?

13. TIPPING Find the amount of a 15% tip on a meal costing $25.40.

14. FUGITIVES As of October 2004, 450 of the 479 fugitives who have appeared on the FBI's Ten Most Wanted list have been apprehended. What percent is this? Round to the nearest tenth of a percent.

15. SWIMMING WORKOUTS A swimmer was able to complete 18 laps before a shoulder injury forced him to stop. This was only 20% of a typical workout. How many laps does he normally complete during a workout?

16. COLLEGE EMPLOYEES The 700 employees at a community college fall into three major categories, as shown. How many employees are in administration?

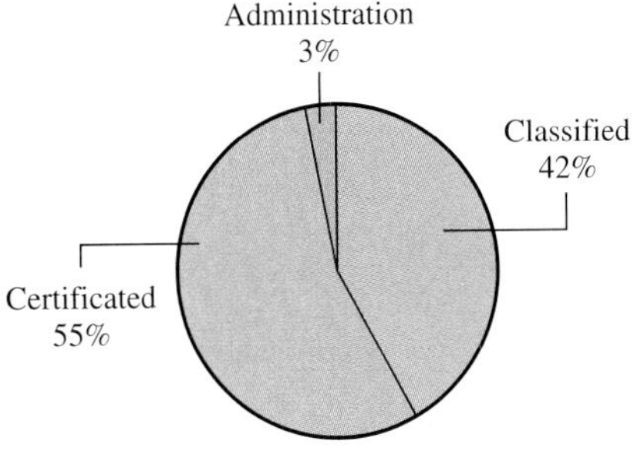

17. Find 24% of 600.

18. HAIRCUTS The illustration shows the number of minutes it took the average U.S. worker to earn enough to pay for a man's haircut in 1950 and 1998. Find the percent of decrease, to the nearest one percent.

19. INSURANCE An insurance salesperson receives a 4% commission on the annual premium of any policy she sells. Find her commission on a homeowner's policy if the premium is $898.

20. COST-OF-LIVING INCREASES A teacher earning $40,000 just received a cost-of-living increase of 3.6%. What is the teacher's new salary?

21. CAR WAX SALE A car waxing kit, regularly priced at $14.95, is on sale for $3 off. What are the sale price, the discount, and the rate?

22. POPULATION INCREASES After a new freeway was completed, the population of a city it passed through increased from 12,808 to 15,565 in two years. What percent of increase is this? (Round to the nearest one percent.)

23. Find the simple interest on a loan of $3,000 at 5% per year for 1 year.

24. Find the amount of interest earned on an investment of $24,000 paying an annual rate of 6.4% interest, compounded daily for 3 years.

25. POLITICAL ADS Explain what is unclear about the ad shown below.

Re-elect
Sal Berchetto
for District Attorney
Berchetto has a proven record of bringing crime down to 37%.
A man of integrity and experience

26. In Section 5.4, we discussed *interest.* What is interest?

CHAPTERS 1–5 CUMULATIVE REVIEW EXERCISES

1. SHAQUILLE O'NEAL Use the data in the table to complete the line graph that shows the growth of Shaquille O'Neal, the Miami Heat center.

Age (yr)	4	6	8	10	12	16	21	28
Weight (lb)	56	82	108	139	192	265	302	315

Based on data from *Los Angeles Times*

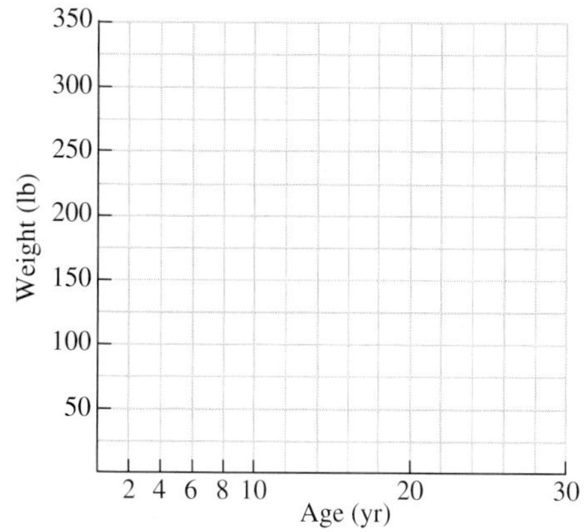

2. State the commutative property of multiplication.

3. a. Find the factors of 40.

   b. Find the prime factorization of 40.

4. AUTO INSURANCE See the premium comparison below. What is the average (mean) six-month insurance premium for the companies listed?

Allstate	$2,672	Mercury	$1,370
Auto Club	$1,680	State Farm	$2,737
Farmers	$2,485	20th Century	$1,692

Criteria: Six-month premium. Husband, 45, drives a 1995 Explorer, 12,000 annual miles. Wife, 43, drives a 1996 Dodge Caravan, 12,000 annual miles. Son, 17, is an occasional operator. All have clean driving records.

5. PAINTING A square tarp has sides 8 feet long. When it is laid out on a floor, how much area will it cover?

6. Evaluate: $-12 - (-5)$.

7. Evaluate: $12 - 2[-8 - 2^4(-1)]$.

8. Evaluate: $|-55|$.

9. Solve: $6 = 2 - 2x$.

10. Solve: $3x - 4 = 2$.

11. FRUIT STORAGE Evaluate $\frac{5(59 - 32)}{9}$ to determine the missing Celsius temperature on the label on the box of bananas shown.

12. SPELLING What fraction of the letters in the word *Mississippi* are vowels?

13. Simplify: $\frac{10}{15}$.

14. Simplify: $\frac{24}{60}$.

Perform the operations.

15. $-\frac{16}{35} \cdot \frac{25}{48}$

16. $4\frac{2}{5} \div 11$

17. $\frac{4}{3} + \frac{2}{7}$

18. $34\frac{1}{9} - 13\frac{5}{6}$

19. Solve: $\frac{5}{6}y = -25$.

20. Solve: $\frac{y}{6} - 2 = 1$.

Perform the operations.

21. $78.1 - 7.81$

22. $2.13(-4.05)$

23. $0.752(1,000)$

24. $\frac{241.86}{2.9}$

25. Evaluate $\frac{3.6 - (-1.5)}{0.5(1.5) - 0.4(3.6)}$. Round to the nearest hundredth.

26. Round 452.0298 to the nearest thousandth.

27. Write $\frac{11}{15}$ as a decimal. Use an overbar.

28. Solve: $\frac{y}{2.22} = -5$.

29. Evaluate: $3\sqrt{81} - 8\sqrt{49}$.

30. LABOR COSTS On the repair bill shown, one line cannot be read. How many hours of labor did it take to repair the car?

***Brian Wood* Auto Repair**

Parts.. $175.00
Total labor (at $35 an hour)..........................
Total.. $297.50

31. Complete the table.

Percent	Decimal	Fraction
	0.29	
47.3%		
		$\frac{7}{8}$

32. 16% of what number is 20?

33. 16% of 400 is what number?

34. 800 is what percent of 10,000?

35. TIPPING Complete the sales draft below if a 15% tip, rounded up to the next dollar, is to be left for the waiter.

STEAK STAMPEDE
Bloomington, MN
Server #12\ AT

VISA	67463777288
NAME	DALTON/ LIZ
AMOUNT	$75.18
GRATUITY $__________	
TOTAL $__________	

36. GENEALOGY Through an extensive computer search, a genealogist determined that worldwide, 180 out of every 10 million people had his last name. What percent is this?

37. SAVINGS ACCOUNT Find the simple interest earned on $10,000 at $7\frac{1}{4}$% for 2 years.

38. Explain why 100% of 50 is 50.

CHAPTER 6

Descriptive Statistics

CORBIS

Every ten years, the federal government conducts a survey, called a *census*, to gather information about the population of the United States. Across the country, members of households, as well as individuals, are asked to complete a questionnaire that asks their age, gender, and race. The Census Bureau collects the data and then uses a branch of mathematics called *statistics* to examine the information. A census report is published that presents the results in tables and graphs.

To learn more about statistics, visit The Learning Equation on the Internet at http://tle.brookscole.com. (The log-in instructions are in the Preface.) For Chapter 6, the online lesson is:

- *TLE* Lesson 12: Statistics

Check Your Knowledge

1. A histogram is a type of ______ graph.
2. A __________ is like a bar graph, but the bars are composed of pictures, where each picture represents a quantity.
3. A __________ polygon is a special line graph formed from a histogram.
4. The sum of several values divided by the number of values is called the ______ of the distribution.
5. The ________ of several values written in increasing order is the middle value.
6. The value that appears most often in a distribution is called the _______ of the distribution.

Refer to the graph, which depicts the blood types of a sample of individuals.

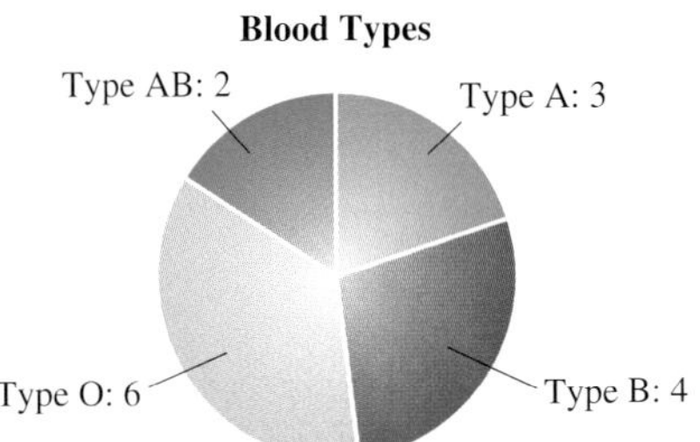

7. What type of graph is shown?
8. How many individuals have blood type B?
9. How many individuals are in the sample group?
10. What percent of the group have blood type A?
11. Which blood type group is the largest? The smallest?

Refer to the graph on the right, which depicts a class grade distribution.

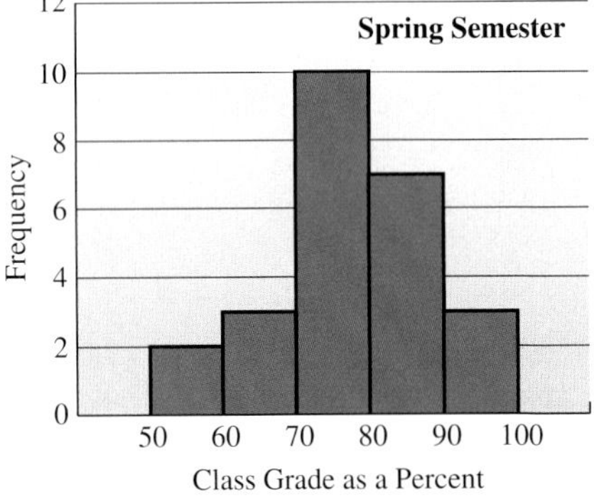

12. What type of graph is shown?
13. How many individuals have earned grades in the 60–70 range?
14. How many students are in the class?
15. What percent of the group earned grades in the 70–80 range?
16. How many students earned at least 80?
17. What percent of the class has grades below a 70?

A group of college men were surveyed to determine how many pairs of shoes they own. The results of the survey were:

5, 8, 3, 5, 7, 9, 4, 9, 10, 7, 9

18. Find the mean number of pairs owned. Round to the nearest one pair.
19. Find the median number of pairs owned.
20. Find the mode of the list.
21. Find the range of the list.
22. Which of the above three measures of central tendency would be affected by the removal of 10 from the list?
23. Which of the above three measures of central tendency would be affected by the removal of 3 from the list?

Study Skills Workshop

REWORKING NOTES AS A STUDY AID

Getty Images

In the last Study Skills Workshop, you learned the importance of taking notes. In this workshop, you will learn to rework your notes to use in studying. *Reworking notes* simply means that you look over class notes, fill in any missing information, and then put the information in a new, easy-to-study format.

Identify Concepts. As soon as possible after class, sit down with your notes and see whether you can identify the key concepts that were discussed in your lecture. Key concepts are usually identified by definitions, rules, or formulas.

Find Examples. Once you have identified a key concept, find all of the examples that were used to illustrate that concept, and make sure you understand them.

Choose a Format for Reworking. You can rework your notes using index cards, a study sheet divided into three columns, or audio form.

- *Index card format:* Make a tab for your cards that states a key concept. On the front of the first index card, write the concept. On the back of the card, write the definition or formula; you will want to memorize this. Using an index card for each example, write the example on the front of the card. On the back, work each step of the example, stating the reason for each step. Repeat this process for every concept learned. Store your index cards in a box, like those used for recipes, and group cards by concept.
- *Three-column study sheet:* Divide an $8\frac{1}{2}$-by-11-inch sheet of paper into three columns. Label the leftmost column "Concept," the middle column "Example," and the rightmost column "Steps and Reasons." Rewrite your notes, filling in the columns on your study sheet as labeled.
- *Audiotape or CD:* Using a recorder, state a key concept. Pause for a few seconds and then state the definition or formula. State an example and verbally explain each step. Continue to state examples and steps. Repeat for the remaining concepts.

Use the Reworked Notes for Studying.

- *Index cards:* Look at the concept on the first card in a group. Attempt to state the definition or the rule. If you are unable to do so, look at the back and read it. Keep doing this until you can recite from memory. Then see whether you can work the problem without any help. Continue to rework the problem until you can do it without looking at the solution.
- *Three-column study sheet:* Go through the concepts and examples, covering the work in the rightmost column while you attempt to work the problem in the middle. If you get stuck, uncover the work to look at the steps. Continue to rework the problem until you can do it without looking at the solution.
- *Audiotape:* In the pause between each phase of the problem, see whether you can recite the steps before they are revealed on the tape. Repeat until you can recite all of the steps on your own.

ASSIGNMENT

1. Determine which format is the best one for you to use to rework your notes. If you think of a different format, describe it.
2. Rework your notes from the last lecture in your chosen format.
3. Compare your reworked notes with a classmate who has chosen the same format. Then compare with someone who has chosen a different format. Are you satisfied with your reworked notes? If not, analyze what is not working and try to fix it.
4. Ask your instructor or a tutor for an evaluation of your reworked notes.

Newspapers and magazines often present information in the form of graphs and tables. In this chapter, we show how information can be obtained by reading graphs. We then discuss three measures of central tendency: the mean, the median, and the mode.

6.1 Reading Graphs and Tables

- Reading data from tables • Reading bar graphs • Reading pictographs
- Reading circle graphs • Reading line graphs • Reading histograms and frequency polygons

It is often said that a picture is worth a thousand words. In this section, we show how to read information from mathematical pictures called *graphs.*

Reading data from tables

The **table** in Figure 6-1(a), the **bar graph** in Figure 6-1(b), and the **circle graph** or **pie chart** in Figure 6-1(c) all show the results of a survey of viewers' opinions. In the bar graph, the length of each bar represents the percent of responses in each category. In the circle graph, the size of each region represents the percent of responses. The two graphs tell the story more quickly and more clearly than the table of numbers.

Ratings of Prime-Time News Coverage

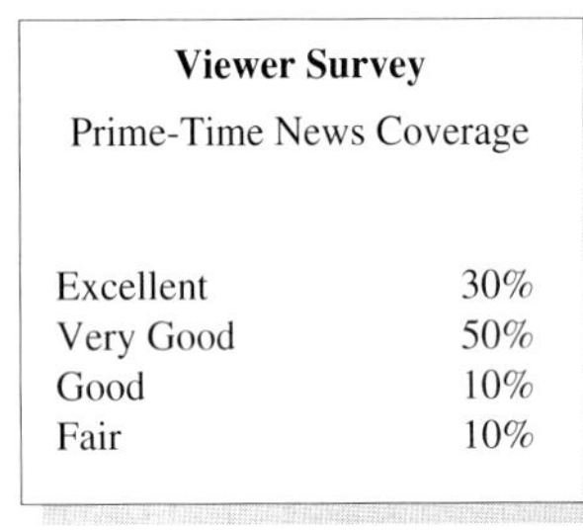

Viewer Survey Prime-Time News Coverage	
Excellent	30%
Very Good	50%
Good	10%
Fair	10%

(a)

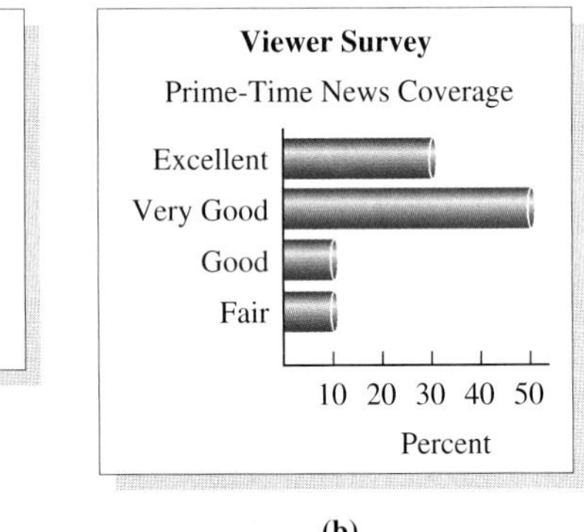

(b)

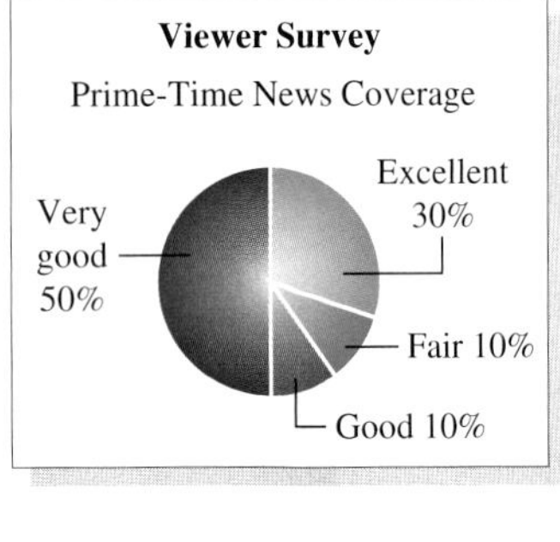

(c)

FIGURE 6-1

It is easy to see from either graph that the largest percent of those surveyed rated the programming *very good,* and that the responses *good* and *fair* were tied for last. The same information is available in the table of Figure 6-1(a), but it is not as easy to see at a glance.

Data are often presented in tables, with information organized in rows and columns. To read a table, we must find the intersection of the row and the column that contains the needed information.

Postal rates (in 2004) for priority mail appear in Table 6-1 on the next page. To find the cost of mailing an $8\frac{1}{2}$-pound package by priority mail to postal zone 4, we find the *row* of the postage table for a package that does not exceed 9 pounds. We then find the *column* for zone 4. At the intersection of this row and this column, we read the number 11.70. This means that it would cost $11.70 to mail the package.

Postage Rates for Priority Mail 2004						
Weight Not Over (pounds)	Zones Local, 1, 2, & 3	Zone 4	Zone 5	Zone 6	Zone 7	Zone 8
1	$3.85	$3.85	$3.85	$3.85	$3.85	$3.85
2	3.95	4.55	4.90	5.05	5.40	5.75
3	4.75	6.05	6.85	7.15	7.85	8.55
4	5.30	7.05	8.05	8.50	9.45	10.35
5	5.85	8.00	9.30	9.85	11.00	12.15
6	6.30	8.85	9.90	10.05	11.30	12.30
7	6.80	9.80	10.65	11.00	12.55	14.05
8	7.35	10.75	11.45	11.95	13.80	15.75
9	7.90	**11.70**	12.20	12.90	15.05	17.50
10	8.40	12.60	13.00	14.00	16.30	19.20
11	8.95	13.35	13.75	15.15	17.55	20.90
12	9.50	14.05	14.50	16.30	18.80	22.65

TABLE 6-1

Reading bar graphs

EXAMPLE 1 Income.
The bar graph in Figure 6-2 shows the total income generated by three sectors of the economy in each of three years. The height of each bar, representing income in billions of dollars, is measured on the scale on the vertical *axis.* The years appear on the horizontal axis. Read the graph to answer the following questions.

a. What income was generated by retail sales in 1990?

b. Which sector of the economy consistently generated the most income?

c. By what amount did income from the wholesale sector increase from 1980 through 2000?

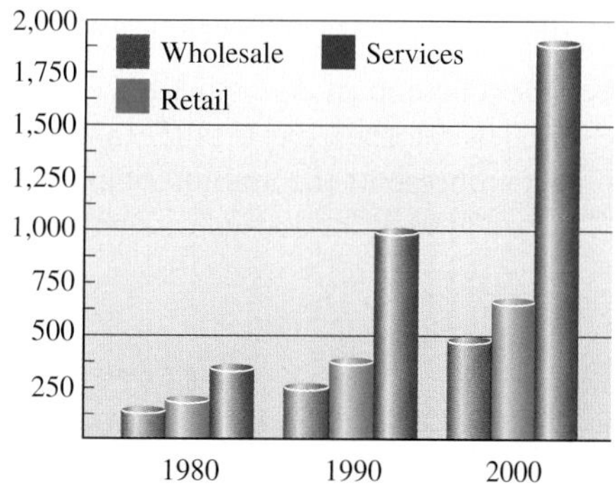

FIGURE 6-2

Self Check 1
What income was generated by the services sector in 1990?

Solution

a. The second group of bars indicates income in 1990, and the middle bar of that group shows sales in the retail sector. Since the vertical axis is scaled in units of \$125 billion, the height of that bar is approximately $250 + 125 = 375$, which represents \$375 billion.

b. In each group, the rightmost bar is the tallest. That bar, according to the key, represents income from the services sector of the economy. Therefore, services consistently generated the most income.

c. According to the color key, the leftmost bar in each group shows income from the wholesale sector. That sector generated about $125 billion in 1980 and $500 billion in 2000. The amount of increase in income is the difference of these two quantities.

$$\$500 \text{ billion} - \$125 \text{ billion} = \$375 \text{ billion}$$

Wholesale income increased by $375 billion between 1980 and 2000.

Answer $1,000 billion

Self Check 2
Which model has shown the greatest decrease in sales?

EXAMPLE 2 Automobile sales. The bar graph in Figure 6-3 shows the number of cars of various models purchased in Dale County for two consecutive years.

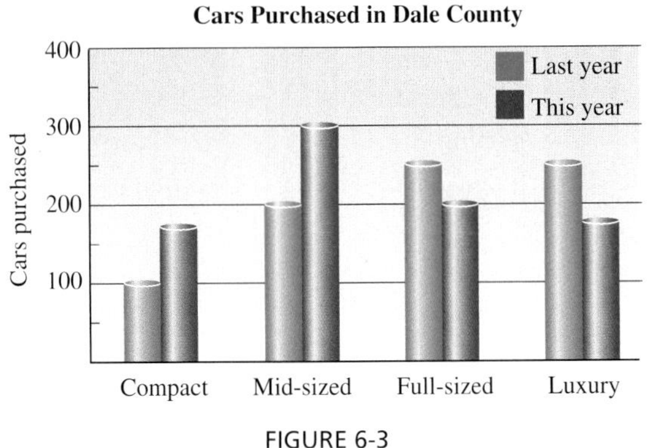

FIGURE 6-3

a. Which models have shown a decrease in sales?

b. Which model showed the greatest increase in sales?

Solution

a. In each pair, the bar on the left gives last year's sales. The bar on the right represents this year's sales. Only for full-sized and luxury cars is the left bar taller than the right bar. This means that full-sized and luxury cars have decreased in sales.

b. Sales of compact and mid-sized cars have increased over last year, because for these models the right bar is taller than the left bar. The difference in the heights of the bars represents the amount of increase. That increase is greater for mid-sized cars. Of all models, mid-sized cars have shown the greatest increase in sales.

Answer luxury cars

Reading pictographs

A **pictograph** is like a bar graph, but the bars are composed of pictures, where each picture represents a quantity. In Figure 6-4, each picture represents 50 pizzas ordered during exam week. The top bar contains three complete pizzas and one partial pizza. This indicates that the men in the men's residence hall ordered $3 \cdot 50$, or 150 pizzas, plus approximately $\frac{1}{4}$ of 50, or about 13 pizzas. This totals 163 pizzas. The women in the women's residence hall ordered $4\frac{1}{2} \cdot 50$, or 225 pizzas.

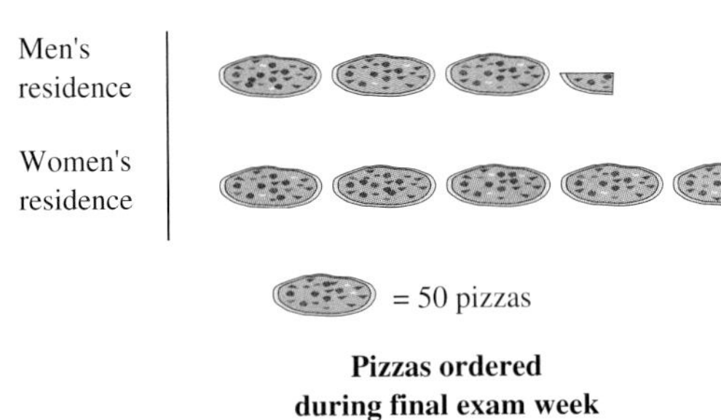

Pizzas ordered during final exam week

FIGURE 6-4

Reading circle graphs

EXAMPLE 3 **Gold production.** The circle graph in Figure 6-5 gives information about world gold production. The entire circle represents the world's total production, and the sizes of the segments of the circle represent the parts of that total contributed by various nations and regions. Use the graph to answer the following questions.

a. What percent of the total was the combined production of the United States and Canada?

b. What percent of the total production came from sources other than those listed?

c. If the world's total production of gold was 56.3 million ounces during the year of the survey, how many ounces did Australia produce?

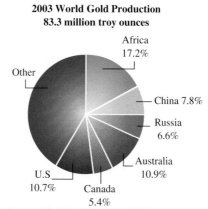

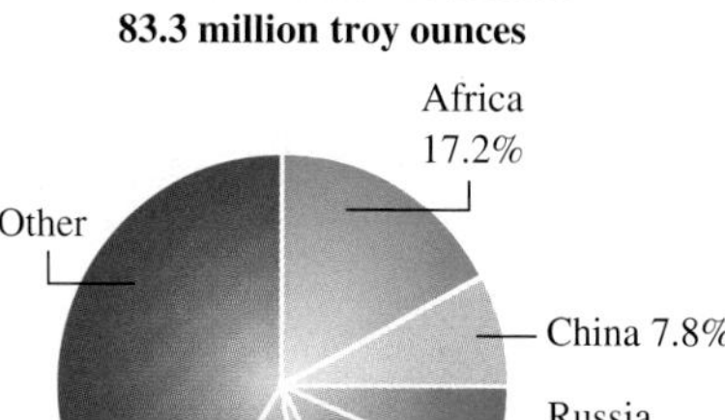

Source: *The World Almanac 2005*

FIGURE 6-5

Self Check 3
To the nearest tenth of a million, how many ounces of gold did Russia produce?

Solution

a. According to the graph, the United States produced 10.7% and Canada produced 5.4% of the total. Together, they produced (10.7 + 5.4)%, or 16.1% of the total.

b. To find the percent of gold produced by countries that are not listed, we add the contributions of all the listed sources and subtract that total from 100%.

$$\begin{aligned} 100\% - (17.2\% + 7.8\% + 6.6\% + 10.9\% + 5.4\% + 10.7\%) &= 100\% - 58.6\% \\ &= 41.4\% \end{aligned}$$

The countries that are not listed produced 41.4% of the world's total production of gold.

c. From the graph, we see that Australia produced 10.9% of the world's gold. Since the world total was 83.3 million ounces, Australia's share (in millions of ounces) was

$$\begin{aligned} 10.9\% \text{ of } 83.3 &= (0.109)(83.3) \\ &= 9.0797 \end{aligned}$$

Rounded to the nearest tenth of a million, Australia produced 9.1 million ounces of gold.

Answer 5.5 million ounces

Reading line graphs

Another graph, called a **line graph,** is used to show how quantities change with time. From such a graph, we can determine when a quantity is increasing and when it is decreasing.

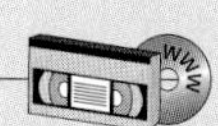

EXAMPLE 4 **Automobile production.** The line graph in Figure 6-6 on the next page shows how U.S. automobile production has changed since 1900. Look at the graph and answer the following questions.

a. How many automobiles were manufactured in 1940?

b. How many were manufactured in 1950?

c. Over which 20-year span did automobile production increase most rapidly?

Self Check 4
How many more cars were produced in 1960 than in 1940?

d. When did production decrease?

e. Why is a broken line used for a portion of the graph?

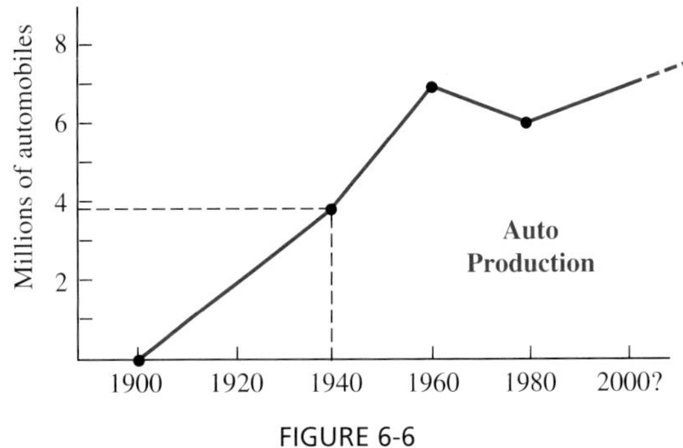

FIGURE 6-6

Solution

a. To find the number of autos produced in 1940, we follow the dashed line from the label 1940 straight up to the graph and then directly over to the scale. There, we read about 3.8. Since the scale indicates millions of automobiles, approximately 3.8 million autos were produced in 1940.

b. To find the number of autos produced in 1950, we find the point halfway between 1940 and 1960. From there, we move up to the graph and then sideways to the scale, where we read about 5. Approximately 5 million autos were manufactured in 1950.

c. Since the upward tilt of the graph is the greatest between 1940 and 1960, auto production increased most rapidly in those years.

d. Between 1960 and 1980, the graph drops, indicating that auto production decreased during that period.

e. Because beyond the year 2000 was still in the future when the graph was made, the production levels were *projections,* and the broken line indicates that the numbers are only estimates.

Answer about 3.2 million

Self Check 5
In Figure 6-7, what is train 1 doing at time D?

EXAMPLE 5 The graph in Figure 6-7 shows the movements of two trains. The horizontal axis represents time, and the vertical axis represents the distance that the trains have traveled.

a. How are the trains moving at time A?

b. At what time (A, B, C, D, or E) are both trains stopped?

c. At what times have both trains gone the same distance?

Train 1
Train 2
Distance
A B C D E
Time

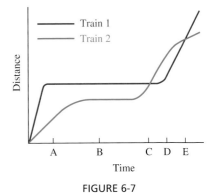

FIGURE 6-7

Solution The movement of train 1 is represented by the red line, and that of train 2 is represented by the blue line.

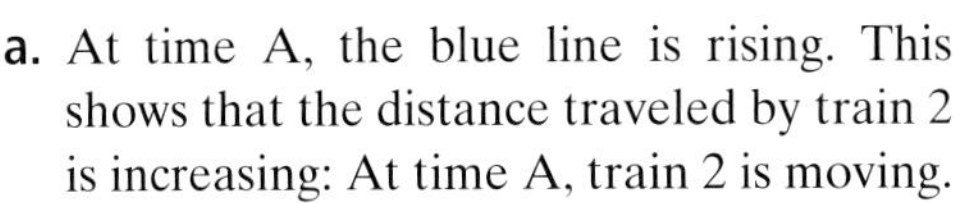

a. At time A, the blue line is rising. This shows that the distance traveled by train 2 is increasing: At time A, train 2 is moving. At time A, the red line is horizontal. This indicates that the distance traveled by train 1 is not changing: At time A, train 1 is stopped.

b. To find the time at which both trains are stopped, we find the time at which both the red and the blue lines are horizontal. At time B, both trains are stopped.

c. At any time, the height of a line gives the distance a train has traveled. Both trains have traveled the same distance whenever the two lines are the same height — that is, at any time when the lines intersect. This occurs at times C and E.

Answer Train 1, which had been stopped, is beginning to move.

Reading histograms and frequency polygons

A pharmaceutical company is sponsoring a series of reruns of old Westerns. The marketing department must choose from three advertisements.

1. Children talking about Chipmunk Vitamins
2. A college student catching a quick breakfast and a TurboPill Vitamin
3. A grandmother talking about Seniors Vitamins

A survey of the viewing audience records the age of each viewer, counting the number in the 6-to-15-year-old age group, the 16-to-25-year-old age group, and so on. The graph of the data is displayed in a special type of bar graph called a **histogram** as shown in Figure 6-8. The vertical axis, labeled *Frequency,* indicates the number of viewers in each age group. For example, the histogram shows that 105 viewers are in the 36-to-45-year-old age group.

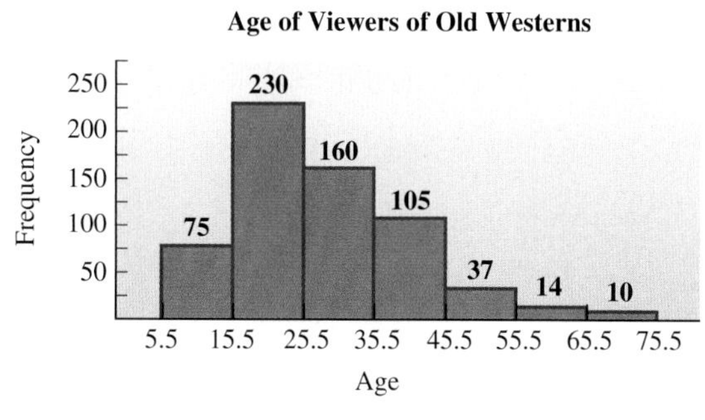

FIGURE 6-8

A histogram is a bar graph with three important features.

1. The bars of a histogram touch.
2. Data values never fall at the edge of a bar.
3. The widths of each bar are equal and represent a range of values.

The width of each bar in Figure 6-8 represents an age span of 10 years. Since most viewers are in the 16-to-25-year-old age group, the marketing department decides to advertise TurboPills in commercials that appeal to active young adults.

EXAMPLE 6 **Carry-on luggage.** An airline weighs the carry-on luggage of 2,260 passengers. See the histogram in Figure 6-9.

a. How many passengers carried luggage in the 8-to-11-pound range?

b. How many carried luggage in the 12-to-19-pound range?

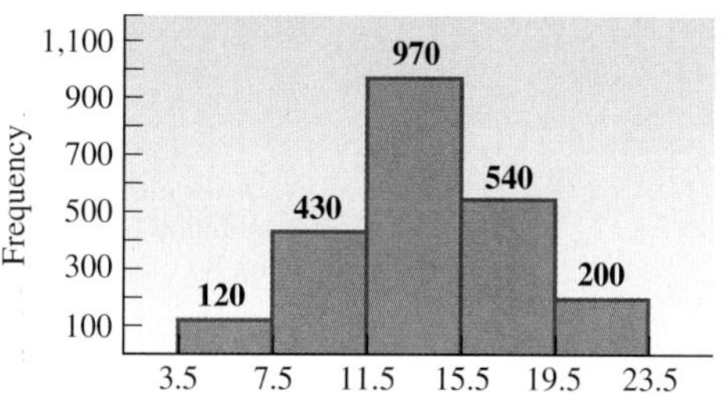

FIGURE 6-9

Solution

a. The second bar, with edges at 7.5 and 11.5 pounds, corresponds to the 8-to-11-pound range. Use the height of the bar (or the number written there) to determine that 430 passengers carried such luggage.

b. The 12-to-19-pound range is covered by two bars. The total number of passengers with luggage in this range is 970 + 540, or 1,510.

A special line graph, called a **frequency polygon,** can be constructed from the histogram in Figure 6-9 by joining the center points at the top of each bar. (See Figure 6-10.) On the horizontal axis, we write the coordinate of the middle value of each bar. After erasing the bars, we get the frequency polygon shown in Figure 6-11.

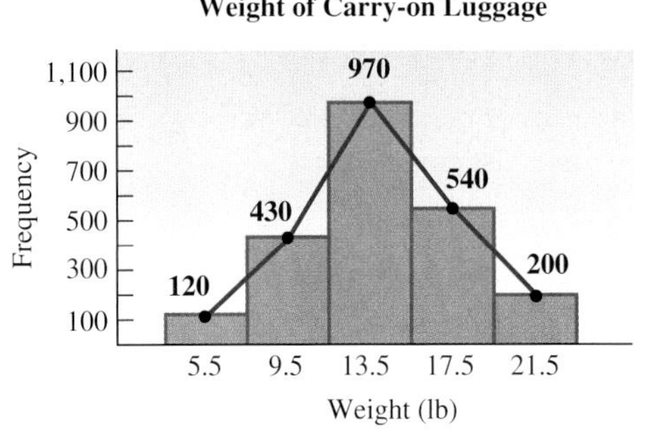

FIGURE 6-10

FIGURE 6-11

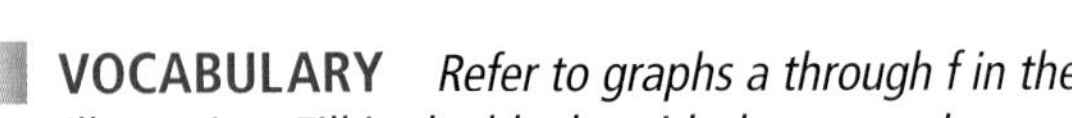

Section 6.1 STUDY SET

VOCABULARY *Refer to graphs a through f in the illustration. Fill in the blanks with the correct letter.*

1. Graph ____ is a bar graph.
2. Graph ____ is a circle graph.
3. Graph ____ is a pictograph.
4. Graph ____ is a line graph.
5. Graph ____ is a histogram.
6. Graph ____ is a frequency polygon.

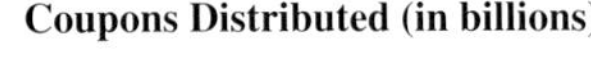

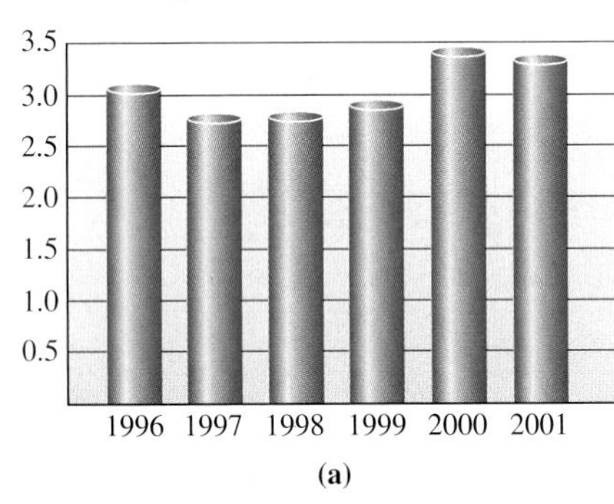

(a)

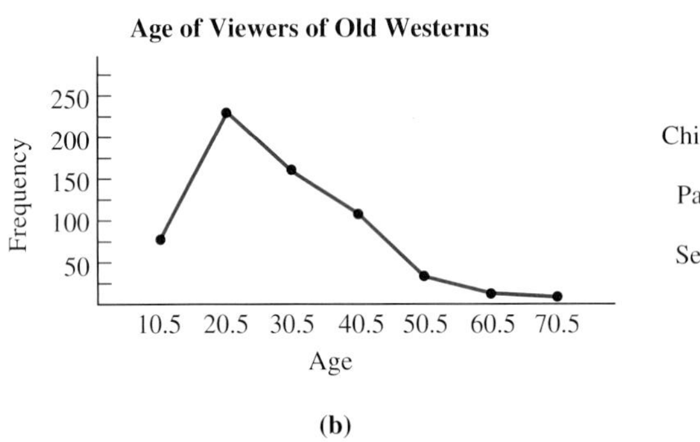

(b)

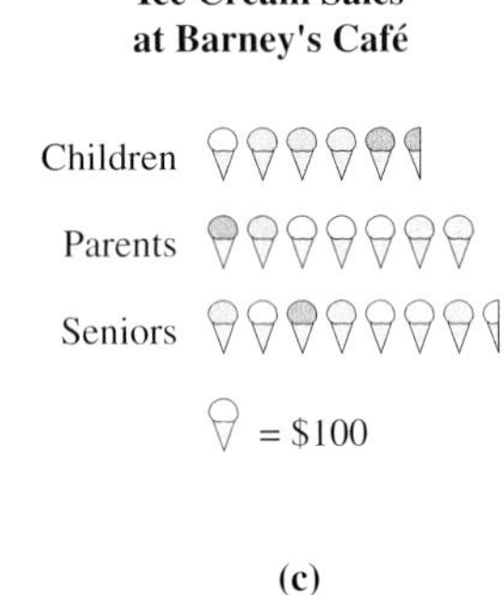

(c)

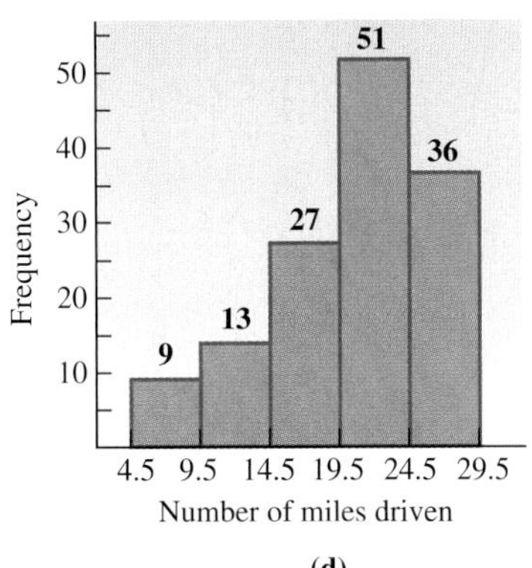

(d)

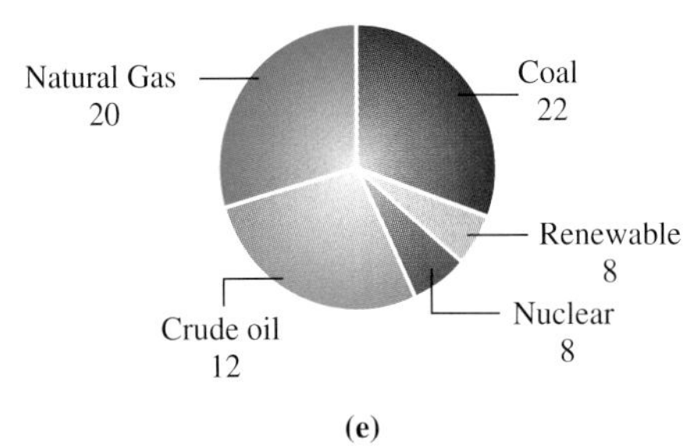

(e)

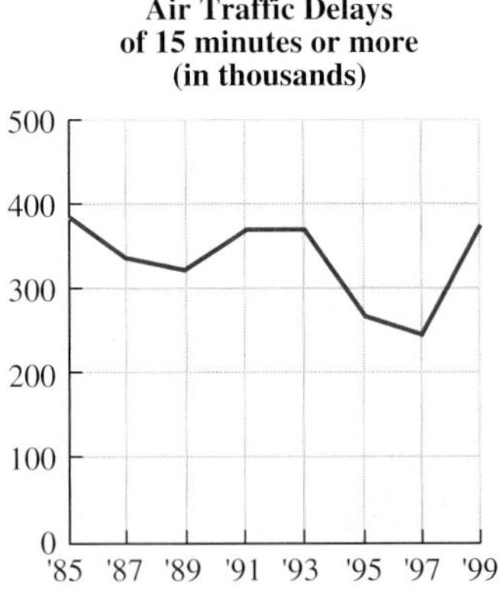

(f)

CONCEPTS *Fill in the blanks.*

7. The ______ of a histogram touch. The widths of the bars of a histogram are ______ and represent a range of values.
8. A ______ polygon can be constructed from a histogram by joining the center points at the top of each bar.

APPLICATIONS *Refer to the postal rate table, Table 6-1, on page 365.*

9. PRIORITY MAIL Find the cost of using priority mail to send a package weighing $7\frac{1}{4}$ pounds to zone 3.
10. PRIORITY MAIL Find the cost of sending (priority mail) a package weighing $2\frac{1}{4}$ pounds to zone 5.
11. COMPARING POSTAGE Juan wants to send a package weighing 6 pounds 1 ounce to a friend living in zone 2. Fourth-class postage would be $2.79. How much could he save by sending the package fourth class instead of priority mail?
12. SENDING TWO PACKAGES Jenny wants to send a birthday gift and an anniversary gift to her brother, who lives in zone 6. One package weighs 2 pounds 9 ounces, and the other weighs 3 pounds 8 ounces. If she uses priority mail, how much will she save by sending both gifts as one package instead of two? (*Hint:* 16 ounces = 1 pound.)

Refer to the federal income tax tables.

13. FILING A JOINT RETURN Raul has an adjusted income of $57,100, is married, and files jointly. Compute his tax.
14. FILING A SINGLE RETURN Herb is single and has an adjusted income of $79,250. Compute his tax.
15. TAX-SAVING STRATEGIES Angelina is single and has an adjusted income of $53,000. If she gets married, she will gain other deductions that will reduce her income by $2,000, and she can file a joint return. How much will she save in tax by getting married?
16. FILING STATUS A man with an adjusted income of $53,000 married a woman with an adjusted income of $75,000. They filed a joint return. Would they have saved on their taxes if they had both stayed single?

Revised 2003 Tax Rate Schedules

	If TAXABLE INCOME		The TAX is		
	THEN				
	Is Over	But Not Over	This Amount	Plus This %	Of the Excess Over
SCHEDULE X —					
Single	$0	$7,000	$0.00	10%	$0.00
	$7,000	$28,400	$700.00	15%	$7,000
	$28,400	$68,800	$3,910.00	25%	$28,400
	$68,800	$143,500	$14,010.00	28%	$68,800
	$143,500	$311,950	$34,926.00	33%	$143,500
	$311,950	—	$90,514.50	35%	$311,950
SCHEDULE Y-1 —					
Married Filing Jointly or Qualifying Widow(er)	$0	$14,000	$0.00	10%	$0.00
	$14,000	$56,800	$1,400.00	15%	$14,000
	$56,800	$114,650	$7,820.00	25%	$56,800
	$114,650	$174,700	$22,282.50	28%	$114,650
	$174,700	$311,950	$39,096.50	33%	$174,700
	$311,950	—	$84,389.00	35%	$311,950

Refer to the following graph.

17. Which source supplied the least amount of energy in 1975?

18. Which energy source remained essentially unchanged between 1975 and 2000?

19. What percent of electrical energy was produced by crude oil in 1975?

20. Which sources provided about 10% of the U.S. energy needs in 2000?

21. Which source supplied the greatest amount of energy in 2000?

22. On which sources of energy does the U.S. rely more heavily since 1975?

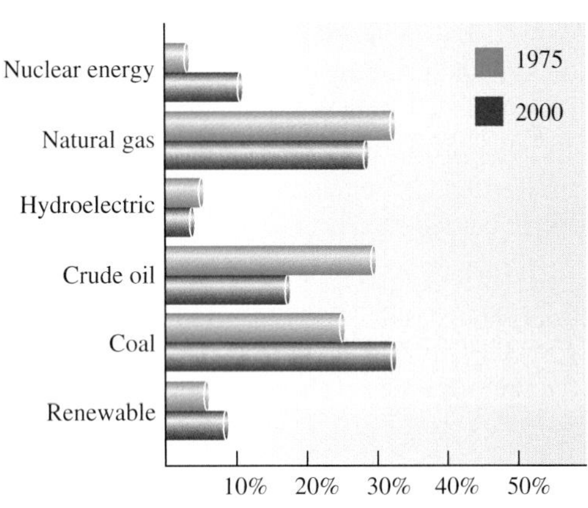

Refer to the following graph.

23. The world production of lead in 1970 was approximately equal to the production of zinc in another year. In what other year was that?

24. The world production of zinc in 1990 was approximately equal to the production of lead in another year. In what other year was that?

25. In what year was the production of zinc less than one-half that of lead?

26. In what year was the production of zinc more than twice that of lead?

27. By how many metric tons did the production of zinc increase between 1970 and 1980?

28. By how many metric tons did the production of lead decrease between 1980 and 1990?

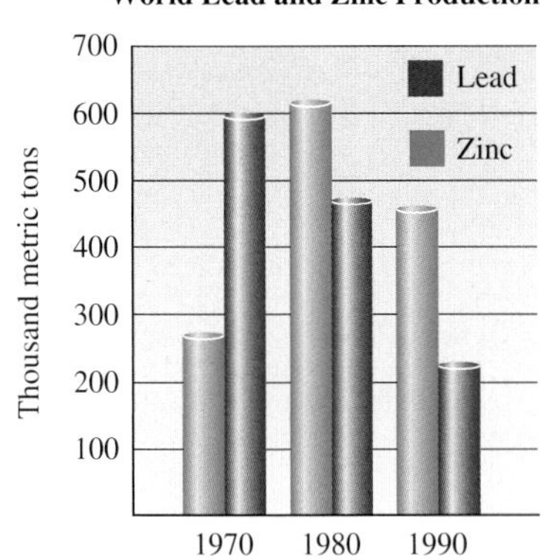

Refer to the following graph.

29. In which categories of moving violations have arrests decreased since last month?

30. Last month, which violation occurred most often?

31. This month, which violation occurred least often?

32. Which violation has shown the greatest decrease in number of arrests since last month?

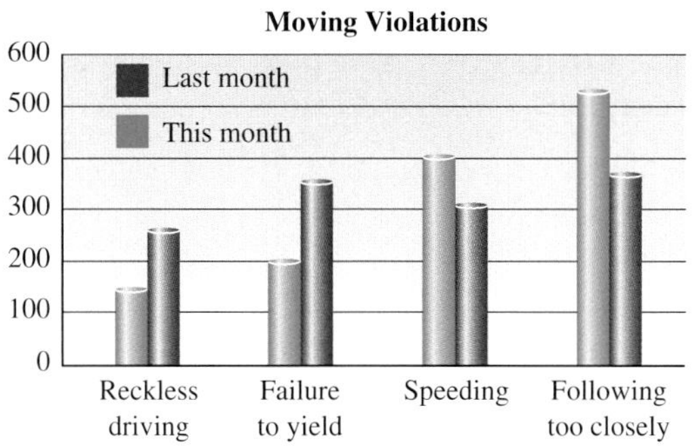

Refer to the pictograph.

33. Which group (children, parents, or seniors) spent the most money on ice cream at Barney's Café?

34. How much money did parents spend on ice cream?

35. How much more money did seniors spend than parents?

36. How much more money did seniors spend than children?

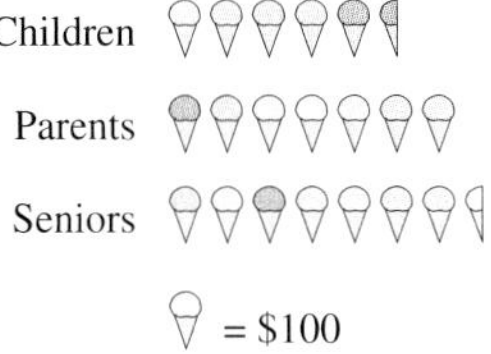

Refer to the circle graph on the next page.

37. Two of the seven languages considered are spoken by groups of about the same size. Which languages are they?

38. Of the languages in the graph, which is spoken by the greatest number of people?

39. Do more people speak Russian or English?

40. What percent of the world's population speak Russian or English?

41. What percent of the world's population speak a language other than these seven?

42. What percent of the world's population do not speak either French or German?

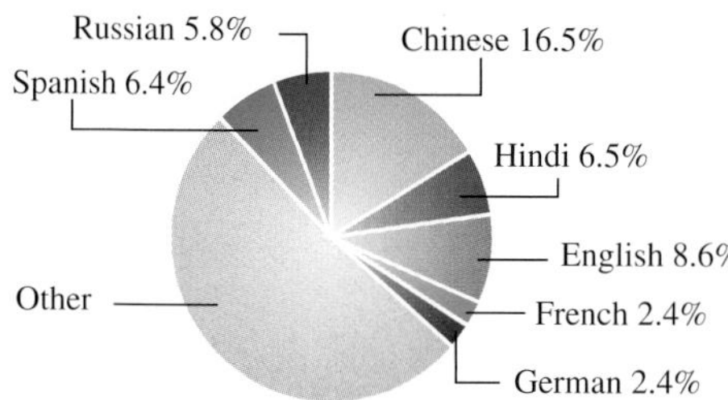

Refer to the following graph.

43. What percent of total energy production is from nuclear energy?

44. What percent of energy production comes from renewable sources?

45. What percent of total energy production comes from coal and crude oil combined?

46. By what *percent* does energy produced from coal exceed that produced from crude oil?

47. By what *percent* does energy produced from coal exceed that produced from nuclear sources?

48. If production of nuclear energy tripled in the next 10 years and other sources remained the same, what percent of total energy sources would nuclear energy be?

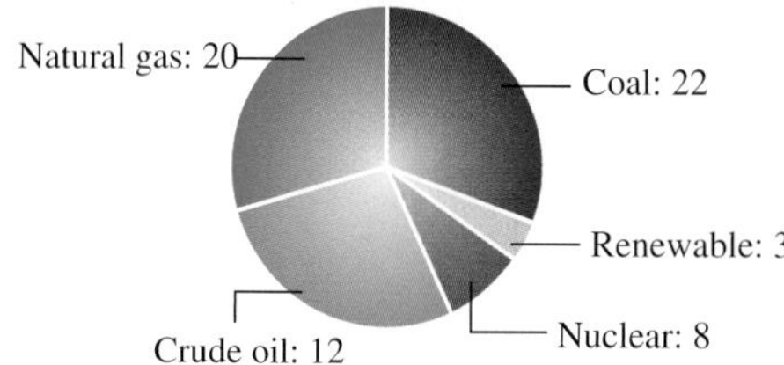

Refer to the line graph in the next column.

49. What were the average weekly earnings in mining for the year 1975?

50. What were the average weekly earnings in construction for the year 1980?

51. In the period between 1982 and 1984, which salary was increasing most rapidly?

52. In approximately what year did miners begin to earn more than construction workers?

53. In the period from 1970 to 1995, which workers received the greatest increase in wages?

54. In what five-year interval did wages in mining increase most rapidly?

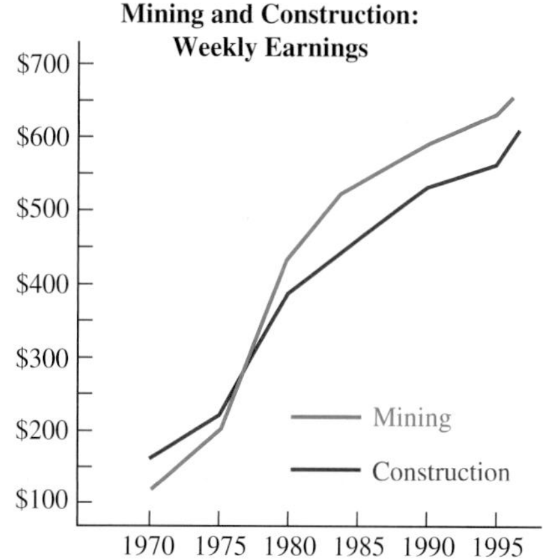

Refer to the following line graph.

55. Which runner ran faster at the start of the race?

56. Which runner stopped to rest first?

57. Which runner dropped the baton and had to go back to get it?

58. At what times (A, B, C, or D) was runner 1 stopped and runner 2 running?

59. Describe what was happening at time D.

60. Which runner won the race?

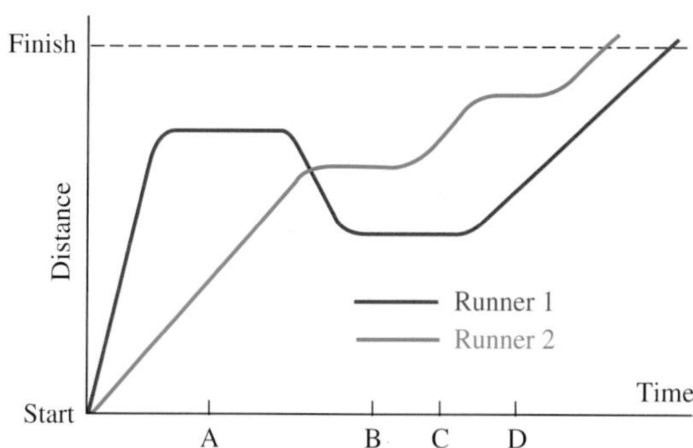

61. COMMUTING MILES An insurance company has collected data on the number of miles its employees drive to and from work. The data are presented in the histogram on the next page. How many employees commute between 14.5 and 19.5 miles per week?

62. COMMUTING The employees of a marketing firm were surveyed to determine the number of miles that they commute to work each week. How many employees commute 14 miles or less per week?

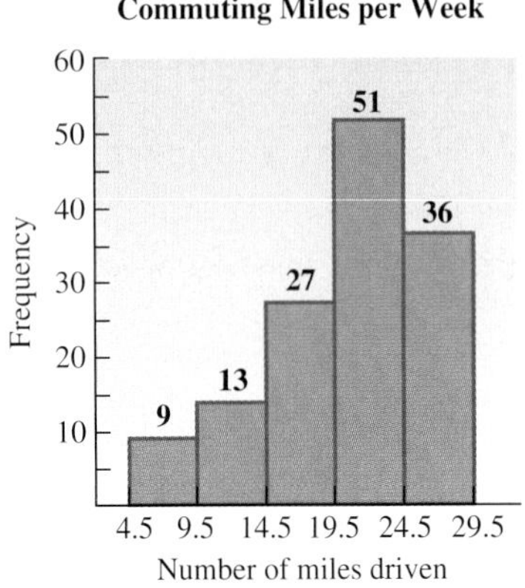

63. NIGHT SHIFT STAFFING A hospital administrator surveyed the medical staff to determine the number of room calls during the night. She constructed the frequency polygon below. On how many nights were there about 30 room calls?

64. NIGHT SHIFT STAFFING Refer to Problem 63 and the graph below. On how many nights were there about 60 room calls?

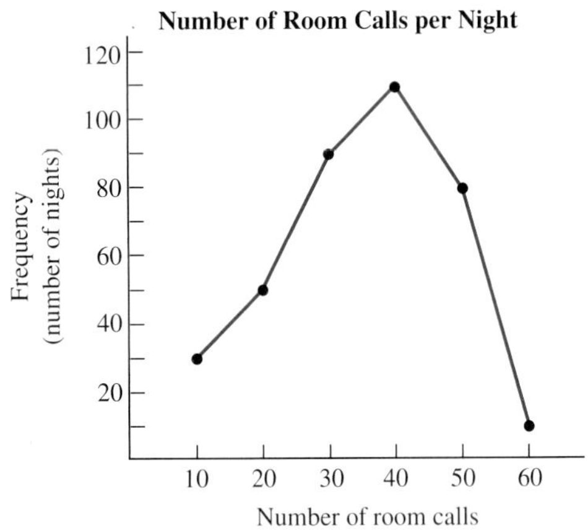

65. MAKING BAR GRAPHS Use the data in the table below to make a bar graph showing the number of U.S. farms in the years 1950 through 2000.

66. MAKING LINE GRAPHS Use the data in the table to make a line graph showing the average acreage of U.S. farms for the years 1950 through 2000.

Year	Number of U.S. farms (in millions)	Average size of U.S. farms (acres)
1950	5.6	213
1960	4.0	297
1970	2.9	374
1980	2.4	426
1990	2.1	460
2000	2.2	434

Source: U.S. Dept. of Agriculture

67. MAKING LINE GRAPHS The coupon in the table provides savings for shoppers. Make a line graph that relates the original price (in dollars, on the horizontal axis) to the sale price (on the vertical axis).

SAVE!	On purchases of
\$10	\$100–\$250
\$25	\$250–\$500
\$50	over \$500

68. MAKING HISTOGRAMS To study the effect of fluoride in preventing tooth decay, researchers counted the number of fillings in the teeth of 28 patients and recorded these results:

3, 7, 11, 21, 16, 22 18, 8, 12, 3, 7, 2, 8, 19, 12,
19, 12, 10, 13, 10, 14, 15, 14, 14, 9, 10, 12, 13

Tally the results by completing the table. Then make a histogram. The first bar extends from 0.5 to 5.5, the second bar from 5.5 to 10.5, and so on.

Number of fillings	Frequency
1–5	
6–10	
11–15	
16–20	
21–25	

WRITING *Write a paragraph using your own words.*

69. What kind of presentation (table, bar graph, line graph, pie chart, pictograph, or histogram) is most appropriate for displaying each type of information?

- The percent of students, classified by major
- The percent of biology majors each year since 1970
- The number of hours spent studying for finals
- Various ethnic populations of the ten largest cities
- Average annual salary of corporate executives for ten major industries

Explain your choices.

70. A histogram is a special type of bar graph. Explain.

REVIEW *Perform the operations.*

71. $5 - 3 \cdot 4$

72. $3(6 - 9) + 4$

73. $\left(\frac{1}{2} + \frac{1}{3}\right)^2$

74. $5^2 + |6 - 10|$

75. Write the prime numbers between 10 and 30.

76. Write the first ten composite numbers.

77. Write the even numbers less than 6 that are not prime.

78. Write the prime numbers between 0 and 10.

6.2 Mean, Median, and Mode

- The mean (the arithmetic average) • The median • The mode • The range

Graphs are not the only way of describing *distributions* (lists) of numbers compactly. We can often find *one* number that represents the center of all of the numbers in a collection of data. We have already seen one such typical number: the *mean,* or the *average.* There are two others: the *median* and the *mode.* In this section, we discuss these three measures of **central tendency.**

The mean (the arithmetic average)

A student has taken five tests this semester, scoring 87, 73, 89, 92, and 84. To find out how well she is doing, she calculates the **mean,** or the **arithmetic average,** of these grades, by finding the sum of the grades and then dividing by 5.

$$\text{Mean score} = \frac{87 + 73 + 89 + 92 + 84}{5}$$

$$= \frac{425}{5}$$

$$= 85$$

The mean score is 85. Some exams were better and some were worse, but 85 is a good indication of her performance in the class.

The mean (arithmetic average)

The **mean,** or the **arithmetic average,** of several values is given by the formula

$$\text{Mean (or average)} = \frac{\text{sum of the values}}{\text{number of values}}$$

EXAMPLE 1 **Store sales.** The week's sales in three departments of the Tog Shoppe are given in Table 6-2 on the next page. Find the mean of the daily sales in the women's department for this week.

Self Check 1

Find the mean daily sales in all three departments of the Tog Shoppe on Wednesday.

	Men's department	Women's department	Children's department
Monday	$2,315	$3,135	$1,110
Tuesday	2,020	2,310	890
Wednesday	1,100	3,206	1,020
Thursday	2,000	2,115	880
Friday	955	1,570	1,010
Saturday	850	2,100	1,000

TABLE 6-2

Solution Use a calculator to add the sales in the women's department for the week. Then divide the sum of those six values by 6.

$$\text{Mean sales in the women's department} = \frac{3{,}135 + 2{,}310 + 3{,}206 + 2{,}115 + 1{,}570 + 2{,}100}{6}$$
$$= \frac{14{,}436}{6}$$
$$= 2{,}406$$

Answer $1,775.33

The mean of the week's daily sales in the women's department is $2,406.

CALCULATOR SNAPSHOT Finding the mean

Most scientific calculators do statistical calculations and can easily find the mean of a set of numbers. To use a statistical calculator in statistical mode to find the mean in Example 1, try these keystrokes:

- Set the calculator to statistical mode.
- Reset the calculator to clear the *statistical registers.*
- Enter each number, followed by the [Σ+] key instead of the [+] key. That is, enter 3,135, press [Σ+], enter 2,310, press [Σ+], and so on.
- When all data are entered, find the mean by pressing the [$\bar{x}$] key. You may need to press [2nd] first. The mean is 2,406.

Because keystrokes vary among calculator brands, you might have to check the owner's manual if these instructions don't work.

Self Check 2
If Bob drove 3,360 miles in February 2005, how many miles did he drive per day, on average?

EXAMPLE 2 **Driving.** In January, Bob drove a total of 4,805 miles. On the average, how many miles did he drive per day?

Solution To find the average number of miles driven per day, we divide the total number of miles by the number of days. Because there are 31 days in January, we divide 4,805 by 31.

$$\begin{aligned}\text{Average number of miles per day} &= \frac{\text{total miles driven}}{\text{number of days}}\\ &= \frac{4{,}805}{31}\\ &= 155\end{aligned}$$

On average, Bob drove 155 miles per day.

Answer 120

The median

The mean is not always representative of the values in a list. For example, suppose that the weekly earnings of four workers in a small business are \$280, \$300, \$380, and \$240, and the owner of the company pays himself \$5,000. The mean salary is

$$\begin{aligned}\text{Mean salary} &= \frac{280 + 300 + 380 + 240 + 5{,}000}{5}\\ &= \frac{6{,}200}{5}\\ &= 1{,}240\end{aligned}$$

The owner could say, "Our employees earn an average of \$1,240 per week." Clearly, the mean does not fairly represent the typical worker's salary.

A better measure of the company's typical salary is the *median:* the salary in the middle when all the numbers are arranged by size.

240 280 300 380 5,000

↑ The middle salary (300)

The typical worker earns \$300 per week, far less than the mean salary.

If there is an even number of values in a list, there is no middle value. In that case, the median is the mean of the two numbers closest to the middle. For example, there is no middle number in the list 2, 5, 6, 8, 13, 17. The two numbers closest to the middle are 6 and 8. The median is the mean of 6 and 8, which is $\frac{6+8}{2}$, or 7.

The median

The **median** of several values is the middle value. To find the median:

1. Arrange the values in increasing order.
2. If there is an odd number of values, the median is the value in the middle.
3. If there is an even number of values, the median is the average of the two values that are closest to the middle.

EXAMPLE 3 **Grades.** On an exam, there were three scores of 59, four scores of 77, and scores of 43, 47, 53, 60, 68, 82, and 97. Find the median score.

Solution We arrange the 14 scores in increasing order.

43 47 53 59 59 59 **60 68** 77 77 77 77 82 97

Self Check 3

Find the median of these values: 7.5, 2.1, 9.8, 5.3, 6.2.

Since there is an even number of scores, the median is the mean of the two scores closest to the middle: the 60 and the 68.

The median is $\frac{60 + 68}{2}$, or 64.

Answer 6.2

The mode

A hardware store displays 20 outdoor thermometers. Twelve of them read 68°, and the other eight have different readings. To choose an accurate thermometer, should we choose one with a reading that is closest to the *mean* of all 20, or to their *median?* Neither. Instead, we should choose one of the 12 that all read the same, figuring that any of those that agree will likely be correct.

By choosing that temperature that appears most often, we have chosen the *mode* of the 20 numbers.

The mode

The **mode** of several values is the single value that occurs most often. The mode of several values is also called the **modal value.**

Self Check 4
Find the mode of these values: 2, 3, 4, 6, 2, 4, 3, 4, 3, 4, 2, 5

EXAMPLE 4 Find the mode of these values: 3, 6, 5, 7, 3, 7, 2, 4, 3, 5, 3, 7, 8, 7, 3, 7, 6, 3, 4.

Solution To find the mode of the numbers in the list, we make a chart of the distinct numbers that appear and make tally marks to record the number of times they occur.

2	3	4	5	6	7	8
/	~~////~~ /	//	//	//	~~////~~	/

Because 3 occurs more times than any other number, it is the mode.

Answer 4

The range

Another measure that is used to describe a collection of numbers is the **range.** It indicates the spread of the data.

The range

The **range** of a list of values is the difference between the largest and smallest values.

Self Check 5
Consider the following set of measurements:

1.7, 1.2, 1.8, 1.6, 1.2

Find **a.** the mean, **b.** the median, **c.** the mode, and **d.** the range.

EXAMPLE 5 **Machinist's tools.** The diameters (distances across) of eight stainless steel bearings were found using the vernier calipers shown in Figure 6-12. Find **a.** the mean, **b.** the median, **c.** the mode, and **d.** the range of the set of measurements listed below.

3.43 cm, 3.25 cm, 3.48 cm, 3.39 cm, 3.54 cm, 3.48 cm, 3.23 cm, 3.24 cm

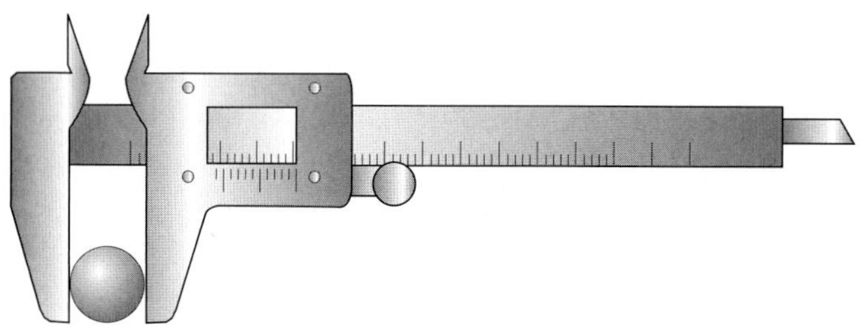

FIGURE 6-12

Solution

a. To find the mean, we add the measurements and divide by the number of values, which is 8.

$$\text{Mean} = \frac{3.43 + 3.25 + 3.48 + 3.39 + 3.54 + 3.48 + 3.23 + 3.24}{8} = 3.38 \text{ cm}$$

b. To find the median, we first arrange the measurements in increasing order.

3.23, 3.24, 3.25, 3.39, 3.43, 3.48, 3.48, 3.54

Because there is an even number of measurements, the median will be the sum of the middle two values (3.39 and 3.43) divided by 2.

$$\text{Median} = \frac{3.39 + 3.43}{2} = \frac{6.82}{2} = 3.41 \text{ cm}$$

c. Since the measurement 3.48 cm occurs most often, it is the mode.

d. In part b, we see that the smallest value is 3.23 and the largest value is 3.54. To find the range, we subtract the smallest value from the largest value.

$$\text{Range} = 3.54 - 3.23 = 0.31 \text{ cm}$$

Answers **a.** 1.5, **b.** 1.6, **c.** 1.2, **d.** 0.6

The Value of an Education

THINK IT THROUGH

"Additional education makes workers more productive and enables them to increase their earnings." Virginia Governor, Mark R. Warner, 2004

As college costs increase, some people wonder if it is worth it to spend years working toward a degree when that same time could be spent earning money. The following median income data makes it clear that, over time, additional education is well worth the investment. Use the given facts to complete the bar graph.

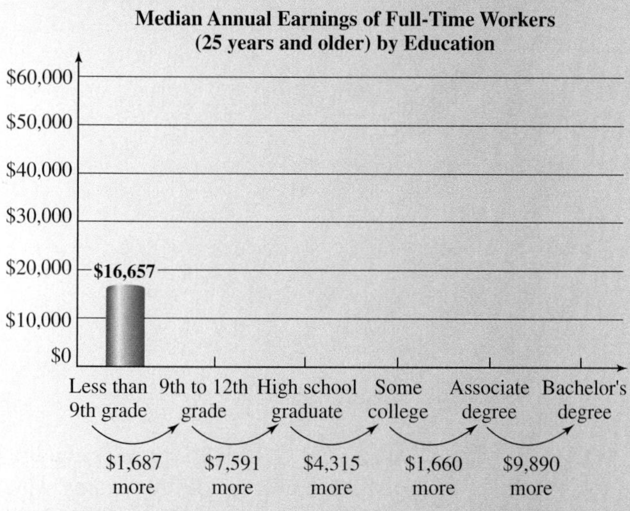

Section 6.2 STUDY SET

VOCABULARY *Fill in the blanks.*

1. The sum of the values in a distribution of numbers divided by the number of values in the distribution is called the _______ of the distribution.
2. The value that appears most often in a distribution is called the _______ of the distribution.
3. The ________ of several values arranged in increasing order is the middle value.
4. The _______ of a list of values is the difference between the largest and smallest values.

CONCEPTS *Fill in the blanks.*

5. The mean of several values is given by

$$\text{Mean} = \frac{\text{the sum of the values}}{}$$

6. Consider the following list of values:

4, 5, 5, 6, 8, 9, 9, 15

a. $\text{Median} = \frac{\square + \square}{2} = \square$

b. The range of the list is $\square$.

PRACTICE *Find the mean of each list of data.*

7. 3, 4, 7, 7, 8, 11, 16
8. 13, 15, 17, 17, 15, 13
9. 5, 9, 12, 35, 37, 45, 60, 77
10. 0, 0, 3, 4, 7, 9, 12
11. 15, 7, 12, 19, 27, 17, 19, 35, 20
12. 45, 67, 42, 35, 86, 52, 91, 102

Find the median of each list of data.

13. 2, 5, 9, 9, 9, 17, 29
14. 16, 18, 27, 29, 35, 47
15. 4, 7, 2, 11, 5, 4, 9, 17
16. 0, 0, 3, 4, 0, 0, 3, 4, 5
17. 18, 17, 2, 9, 21, 23, 21, 2
18. 5, 13, 5, 23, 43, 56, 32, 45

Find the mode (if any) of each list of data.

19. 3, 5, 7, 3, 5, 4, 6, 7, 2, 3, 1, 4
20. 12, 12, 17, 17, 12, 13, 17, 12
21. 5, 9, 12, 35, 37, 45, 60
22. 0, 3, 0, 2, 7, 0, 6, 0, 3, 4, 2, 0
23. 23.1, 22.7, 23.5, 22.7, 34.2, 22.7
24. $\frac{1}{2}, \frac{1}{3}, \frac{1}{3}, 2, \frac{1}{2}, 2, \frac{1}{5}, \frac{1}{2}, 5, \frac{1}{3}$

APPLICATIONS

25. SOFT DRINKS A survey of soft-drink machines indicates the following prices for a can (in cents): 50, 60, 50, 50, 70, 75, 50, 45, 50, 50, 65, 75, 60, 75, 100, 50, 80, 75. Find the mean price of a soft drink.
26. COMPUTER SUPPLIES Several computer stores reported differing prices for toner cartridges for a laser printer (in dollars): 51, 55, 73, 75, 72, 70, 53, 59, 75. Find the mean price of a toner cartridge.
27. SOFT DRINKS Find the median price for a soft drink. (See Exercise 25.)
28. COMPUTER SUPPLIES Find the median price for a toner catridge. (See Exercise 26.)
29. SOFT DRINKS Find the modal price for a soft drink. (See Exercise 25.)
30. COMPUTER SUPPLIES Find the mode of the prices for a toner cartridge. (See Exercise 26.)
31. CHANGING TEMPERATURES Temperatures are recorded at hourly intervals, as listed in the table. Find the average temperature of the period from midnight to 11:00 A.M.

Time	Temperature	Time	Temperature
12:00 A.M.	53	12:00 noon	71
1:00	53	1:00 P.M.	75
2:00	57	2:00	77
3:00	58	3:00	77
4:00	59	4:00	79
5:00	59	5:00	72
6:00	60	6:00	70
7:00	62	7:00	64
8:00	64	8:00	61
9:00	66	9:00	59
10:00	68	10:00	53
11:00	70	11:00	51

32. SEMESTER GRADES Frank's algebra grade is based on the average of four exams, which count equally. His grades are 75, 80, 90, and 85. Find his average.

33. AVERAGE TEMPERATURES Find the average temperature for the 24-hour period. (See Exercise 31.)

34. FINAL EXAMS If Frank's professor decided to count the fourth examination double, what would Frank's average be? (See Exercise 32.)

35. FLEET MILEAGE An insurance company's sales force uses 37 cars. Last June, those cars logged a total of 98,790 miles. On the average, how many miles did each car travel that month?

36. BUDGETING FOR GROCERIES The Hinrichs family spent $519 on groceries last April. On the average, how much did they spend each day?

37. DAILY MILEAGE Find the average number of miles driven daily for each car in Exercise 35.

38. GROCERY COSTS See Exercise 36. The Hinrichs family has five members. What is the average spent for groceries for one family member for one day?

39. EXAM AVERAGES Roberto received the same score on each of five exams, and his mean score is 85. Find his median score and his modal score.

40. EXAM SCORES The scores on the first exam of the students in a history class were 57, 59, 61, 63, 63, 63, 87, 89, 95, 99, and 100. Kia got a score of 70 and claims that "70 is better than average." Which of the three measures of central tendency is she better than: the mean, the median, or the mode?

41. COMPARING GRADES A student received scores of 37, 53, and 78 on three quizzes. His sister received scores of 53, 57, and 58. Who had the better average? Whose grades were more consistent?

42. What is the average of all of the integers from -100 to 100, inclusive?

43. OCTUPLETS In December 1998, Nkem Chukwu gave birth to eight babies in Texas Children's Hospital. Find the mean, the median, and the range of their birth weights.

Ebuka (girl) 24 oz	Odera (girl) 11.2 oz
Chidi (girl) 27 oz	Ikem (boy) 17.5 oz
Echerem (girl) 28 oz	Jioke (boy) 28.5 oz
Chima (girl) 26 oz	Gorom (girl) 18 oz

44. ICE SKATING Listed below are Tara Lipinski's artistic impression scores for the long program of the women's figure skating competition at the 1998 Winter Olympics. Find the mean, median, mode, and range. Round to the nearest tenth.

Australia	5.8	Germany	5.8	Ukraine	5.9
Hungary	5.8	U.S.	5.8	Poland	5.8
Austria	5.9	Russia	5.9	France	5.9

45. COMPARISON SHOPPING A survey of grocery stores found the price of a 15-ounce box of Cheerios cereal ranging from $3.89 to $4.39. (See below.) What are the mean, median, mode, and range of the prices listed?

$4.29	$3.89	$4.29	$4.09	$4.24	$3.99
$3.98	$4.19	$4.19	$4.39	$3.97	$4.29

46. EARTHQUAKES The magnitudes of 1999's major earthquakes are listed below. Find the mean, median, mode, and range. Round to the nearest tenth.

1/19/99	New Ireland, Papua New Guinea	7.0
2/6/99	Santa Cruz Islands, S. Pacific Sea	7.3
3/4/99	Celebes Sea, Indonesia	7.1
4/5/99	New Britain, Papua New Guinea	7.4
4/8/99	E. Russia/N.E. China border	7.1
5/10/99	New Britain, Papua New Guinea	7.1
5/16/99	New Britain, Papua New Guinea	7.1
8/17/99	Izmit region, western Turkey	7.4
9/21/99	Taiwan	7.6
9/30/99	Oaxaca, Mexico	7.4
11/12/99	Bolu Province, northwest Turkey	7.2

47. FUEL EFFICIENCY The ten most fuel-efficient cars in 2002, based on manufacturer's estimated city and highway average miles per gallon (mpg), are shown in the table. Find the mean, median, and mode of both sets of data.

Model	mpg city/hwy
Honda Insight	61/68
Toyota Prius	52/45
Honda Civic Hybrid	47/51
VW Jetta Wagon	42/50
VW Golf	42/49
VW Jetta Sedan	42/49
VW Beetle	42/49
Honda Civic Coupe	36/44
Toyota Echo	34/41
Chevy Prizm	32/41

Source: edmonds.com

48. SPORT FISHING The report shown on the next page lists the fishing conditions at Pyramid Lake for a Saturday in January. Find the median, the mode, and the range of the weights of the striped bass caught at the lake.

Pyramid Lake— Some striped bass are biting but are on the small side. Striking jigs and plastic worms. Water is cold: 38°. Weights of fish caught (lb): 6, 9, 4, 7, 4, 3, 3, 5, 6, 9, 4, 5, 8, 13, 4, 5, 4, 6, 9

49. Refer to the data in the table.
   a. Find the mean. Round to the nearest tenth of a percent.
   b. Find the median.
   c. Find the mode.
   d. Find the range.

U.S. Unemployment Rate (1990–2002) in percent

1990	1991	1992	1993	1994	1995	1996
5.6	6.8	7.5	6.9	6.1	5.6	5.4
1997	**1998**	**1999**	**2000**	**2001**	**2002**	
4.9	4.5	4.2	4.0	4.7	5.8	

Source: U.S. Department of Labor

50. SCHOLASTIC APTITUDE TEST The mean SAT verbal test scores of college-bound seniors for the years 1993–2003 are listed below. Find the mean, median, mode, and range. Round to the nearest one point.

1993	1994	1995	1996	1997	1998
500	499	504	505	505	505
1999	**2000**	**2001**	**2002**	**2003**	
505	505	506	504	507	

Source: *The World Almanac 2004*

WRITING

51. Explain how to find the mean, the median, and the mode of several numbers.

52. The mean, median, and mode are used to measure the central tendency of a list of numbers. What is meant by central tendency?

REVIEW

53. Find the prime factorization of 81.

54. Find the LCD for two fractions whose denominators are 36 and 81.

Perform each operation.

55. $\frac{3}{4} \cdot \frac{2}{9}$

56. $\frac{2}{15} \div \frac{4}{5}$

57. $\frac{18}{5} + \frac{12}{5}$

58. $\frac{7}{12} - \frac{5}{12}$

59. $\frac{8}{5} + \frac{3}{10}$

60. $\frac{5}{6} - \frac{1}{12}$

KEY CONCEPT

Mean, Median, and Mode

To indicate the center of a distribution of numbers, we can use the mean, the median, or the mode.

- The **mean** of a distribution is the sum of the values in the distribution divided by the number of values in the distribution.

$$\text{Mean} = \frac{\text{sum of the values in the distribution}}{\text{number of values in the distribution}}$$

- The **median** of several values written in increasing order is the middle value. Just as many scores are above the median as are below it. If there are an even number of values in the distribution, the median is the mean of the two values that are closest to the middle.
- The **mode** of a distribution is the value that occurs most often.
- The **range** of a list of values is the difference between the largest and smallest values.

Consider the following distribution: 3, 7, 4, 12, 15, 23, 17, 21, 15, 20.

1. Calculate the mean.

2. Find the median.

3. Find the mode.

4. Find the range.

5. Are the mean, the median, and the mode the same number?

Consider the distribution 4, 2, 6, 8, 6, 10.

6. Calculate the mean.

7. Find the median.

8. Find the mode.

9. Find the range.

10. Are the mean, the median, and the mode the same number?

Construct a distribution with the following characteristics.

11. The mean is greater than the mode.

12. The mean is less than the median.

13. The mode is less than the median.

14. The mode is greater than the median.

ACCENT ON TEAMWORK

SECTION 6.1

DAILY HIGH AND LOW TEMPERATURES Make a bar graph that shows the daily high and low temperatures for your city for a two-week period. You can find this information in a newspaper. From your graph, answer the following questions.

a. What was the highest high temperature?

b. What was the lowest high temperature?

c. What was the highest low temperature?

d. What was the lowest low temperature?

e. What was the difference between the highest high temperature and the lowest low temperature?

f. Were any trends apparent from the graph?

SECTION 6.2

MEAN, MEDIAN, AND MODE

1. Find the mean, median, and mode of the following set of values.

2.3 2.3 3.6 3.8 4.5

a. Is the mean of the set of values one of the values in the set?

b. Is the median of the set of values one of the values in the set?

c. Is the mode of the set of values one of the values in the set?

2. Construct a set of values (not all the same number) such that

$$\text{mean} = \text{median} = \text{mode}$$

3. Construct a set of values such that

$$\text{mean} < \text{median} < \text{mode}$$

4. Construct a set of values such that

$$\text{mean} > \text{median} > \text{mode}$$

CHAPTER REVIEW

SECTION 6.1 *Reading Graphs and Tables*

CONCEPTS

Numerical information can be presented in the form of tables, bar graphs, pictographs, circle graphs, and line graphs.

REVIEW EXERCISES

Refer to the table.

1. WIND-CHILL TEMPERATURES Find the wind-chill temperature on a 10° F day when a 15-mph wind is blowing.
2. WIND SPEEDS The wind-chill temperature is −25° F, and the outdoor temperature is 15° F. How fast is the wind blowing?

Determining the Wind-Chill Temperature

Wind speed	Actual temperature													
	35° F	30° F	25° F	20° F	15° F	10° F	5° F	0° F	−5° F	−10° F	−15° F	−20° F	−25° F	−30° F
5 mph	33°	27°	21°	16°	12°	7°	0°	−5°	−10°	−15°	−21°	−26°	−31°	−36°
10 mph	22	16	10	3	−3	−9	−15	−22	−27	−34	−40	−46	−52	−58
15 mph	16	9	−2	−5	−11	−18	−25	−31	−38	−45	−51	−58	−65	−72
20 mph	12	4	−3	−10	−17	−24	−31	−39	−46	−53	−60	−67	−74	−81
25 mph	8	1	−7	−15	−22	−29	−36	−44	−51	−59	−66	−74	−81	−88
30 mph	6	−2	−10	−18	−25	−33	−41	−49	−56	−64	−71	−79	−86	−93
35 mph	4	−4	−12	−20	−27	−35	−43	−52	−58	−67	−74	−82	−89	−97
40 mph	3	−5	−13	−21	−29	−37	−45	−53	−60	−69	−76	−84	−92	−100
45 mph	2	−6	−14	−22	−30	−38	−46	−54	−62	−70	−78	−85	−93	−102

Refer to the graph below.

3. How many coupons were distributed in 2000?
4. Between what years did the number of coupons distributed remain essentially unchanged?
5. Between what two years did the number of coupons distributed increase the most?
6. Between what two years was there the greatest decrease in the number of coupons distributed?

Coupons Distributed (in billions)

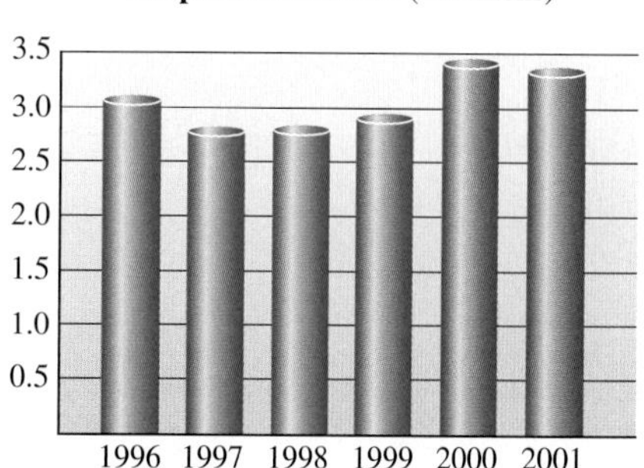

Refer to the graph below.

7. How many eggs were produced in Wisconsin in 1985?
8. How many eggs were produced in Nebraska in 1987?
9. In what year was the egg production of Wisconsin equal to that of Nebraska?
10. What was the total egg production of Wisconsin and Nebraska in 1988?

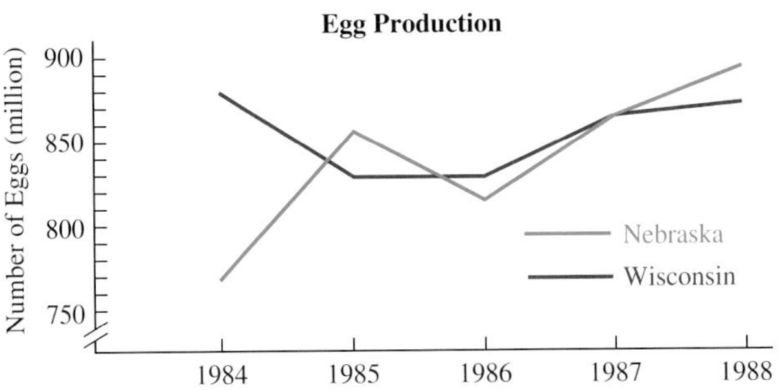

A *histogram* is a bar graph with these features:

1. The bars of the histogram touch.
2. Data values never fall at the edge of a bar.
3. The widths of the bars are equal and represent a range of values.

A *frequency polygon* is a special line graph formed from a histogram.

Refer to the graph below.

11. A survey of the television viewing habits of 320 households produced the following histogram. How many households watch between 6 and 15 hours of TV each week?
12. How many households watch 11 hours or more each week?

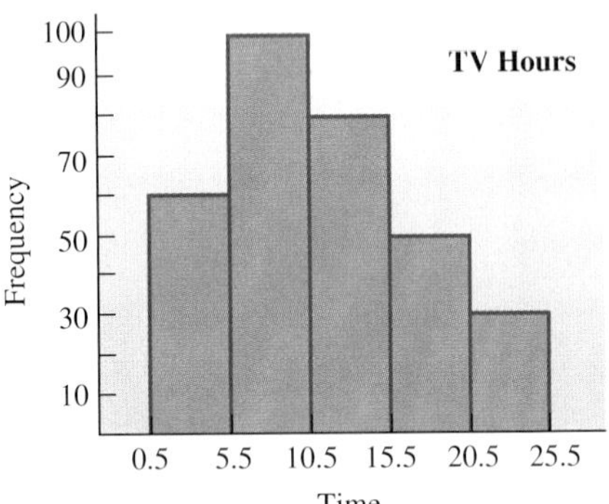

SECTION 6.2 Mean, Median, and Mode

The *mean* (or average) is given by the formula

$$\text{Mean} = \frac{\text{sum of the values}}{\text{number of values}}$$

13. EARNING AN A Jose worked hard this semester, earning grades of 87, 92, 97, 100, 100, 98, 90, and 98. If he needs a 95 average to earn an A in the class, did he make it?

To find the *median* of several values:

1. Arrange the values in increasing order.
2. If there is an odd number of values, the median is the value in the middle.
3. If there is an even number of values, the median is the average of the two values that are closest to the middle.

14. GRADE SUMMARIES The students in a mathematics class had final averages of 43, 83, 40, 100, 40, 36, 75, 39, and 100. When asked how well her students did, their teacher answered, "43 was typical." What measure was the teacher using?

15. PRETZEL PACKAGING Samples of SnacPak pretzels were weighed to find out whether the package claim "Net weight 1.2 ounces" was accurate. The tally appears in the table. Find the modal weight.

Weights of SnacPak Pretzels

Ounces	Number
0.9	1
1.0	6
1.1	18
1.2	23
1.3	2

The *mode* of several numbers is the single value that occurs most often.

The *range* of a list of values is the difference between the largest and smallest values.

16. Find the mean weight of the samples in Exercise 15.

17. BLOOD SAMPLES A medical laboratory technician examined a blood sample under a microscope and measured the sizes (in microns) of the white blood cells. The data are listed below. Find the mean, median, mode, and range.

7.8 6.9 7.9 6.7 6.8 8.0 7.2 6.9 7.5

18. TOBACCO SETTLEMENTS In November 1998, the country's four largest tobacco companies reached an agreement with 46 states to pay $206.4 billion to cover public health costs related to smoking. The payments to each of the New England states are shown below. Find the median payment.

Connecticut	$3.63 billion	New Hampshire	$1.3 billion
Maine	$1.5 billion	Rhode Island	$1.4 billion
Massachusetts	$8.0 billion	Vermont	$0.81 billion

CHAPTER 6 TEST

Refer to the graph below. Keeping one prisoner for one month costs $2,266.

1. How much money is spent monthly, per prisoner, to pay the prison staff?

2. How much money is spent monthly, per prisoner, on office costs?

3. What percent of the monthly allotment is spent on one prisoner's food?

4. What percent of the monthly allotment is spent on one prisoner's recreation and training?

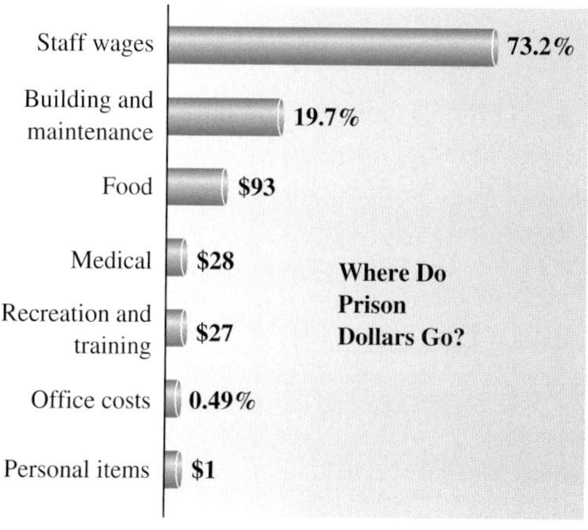

Refer to the graph below.

5. Approximately what percent of all employees are in the food and clothing industries?

6. Among workers in food and clothing, 2.4 million are in food and the rest are in clothing. What percent of all workers are in clothing?

Employees in Industry
(in millions)

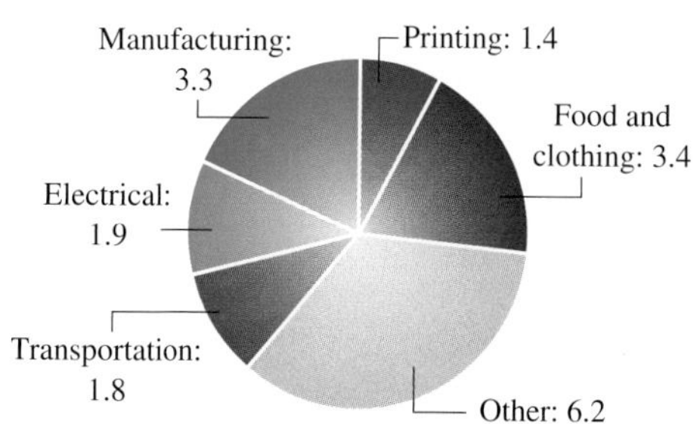

Refer to the line graph below.

7. How many air traffic delays occurred in 1995?

8. Which year was worst for air traffic delays?

9. In 1999, about 2% of the air traffic delays were due to controller equipment glitches. How many flights were delayed for that reason?

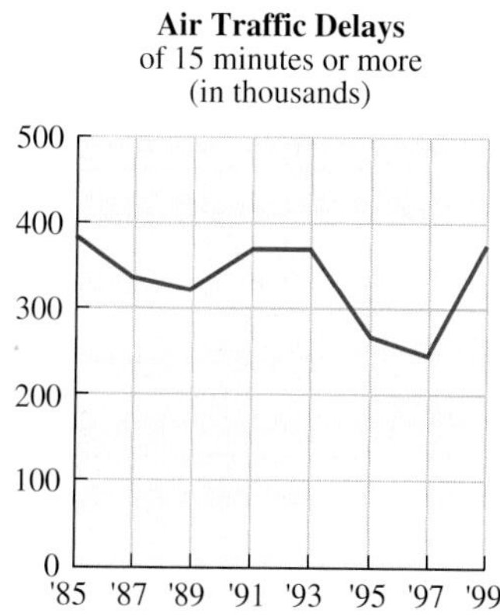

10. Refer to the circle graph below. What is the missing percent for the weather delays?

Causes of Air Traffic Delays

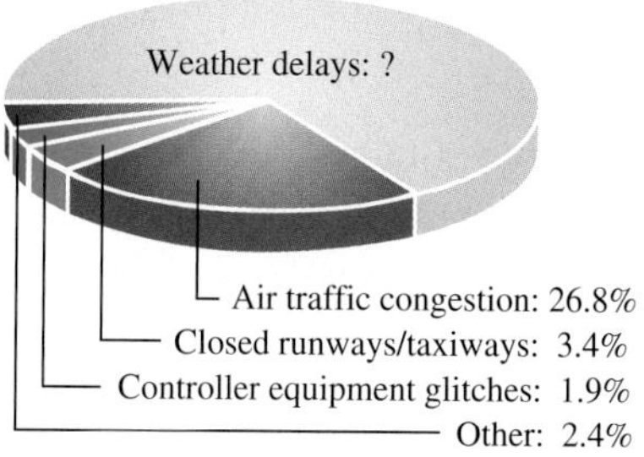

Refer to the illustration and choose the best answer from the following statements.

A. Both bicyclists are moving, and bicyclist 1 is faster than 2.

B. Both bicyclists are moving, and bicyclist 2 is faster than 1.

C. Bicyclist 1 is stopped, and bicyclist 2 is not.

D. Bicyclist 2 is stopped, and bicyclist 1 is not.

E. Both bicyclists are stopped.

11. Indicate what is happening at time A.

12. Indicate what is happening at time B.

13. Indicate what is happening at time C.

14. Which bicyclist won the race?

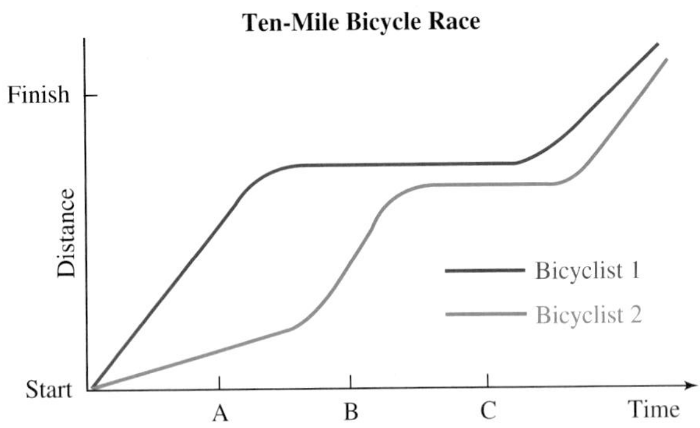

Refer to this information: The hours served last month by the individual volunteers at a homeless shelter were 4, 6, 8, 2, 8, 10, 11, 9, 5, 12, 5, 18, 7, 5, 1, and 9.

15. Find the median of the hours of volunteer service.

16. Find the mean of the hours of volunteer service.

17. Find the mode of the hours of volunteer service.

18. If the one value of 18 were removed from the list, which would be most affected: the mean or the median?

19. RATINGS The seven top-rated cable television programs for the week of February 8–14 are given below. What are the mean, median, mode, and range of the ratings? Round to the nearest tenth.

Show/day/time/network	Rating
1. "WCW Monday," Mon. 9 p.m., TNT	4.5
2. "WCW Monday," Mon. 10 p.m., TNT	4.4
3. "WCW Monday," Mon. 8 p.m., TNT	3.9
4. "WWF Special," Sat. 9 p.m., USA	3.6
5. "WWF Wrestling," Sun. 7 p.m., USA	3.1
6. "Dog Show," Tues. 8 p.m., USA	3.1
7. "WWF Special," Sat. 8 p.m., USA	2.9

20. STATISTICS The bar graph has an asterisk * that refers readers to a note at the bottom. In your own words, complete the explanation of the term *median.*

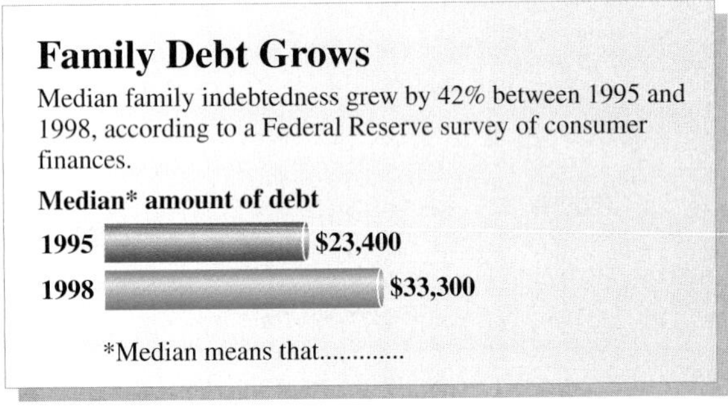

Family Debt Grows
Median family indebtedness grew by 42% between 1995 and 1998, according to a Federal Reserve survey of consumer finances.
Median* amount of debt
1995
$23,400
1998
$33,300
*Median means that............

Source: *Los Angeles Times* (February 1, 2000)

CHAPTERS 1–6 CUMULATIVE REVIEW EXERCISES

1. GASOLINE In 1999, gasoline consumption in the United States was three hundred fifty-eight million, six hundred thousand gallons a day. Write this number in standard notation.

2. Round 49,999 to the nearest thousand.

Perform each operation.

3. $\begin{array}{r} 38{,}908 \\ +15{,}696 \\ \hline \end{array}$

4. $\begin{array}{r} 9{,}700 \\ -5{,}491 \\ \hline \end{array}$

5. $\begin{array}{r} 345 \\ \times\ 67 \\ \hline \end{array}$

6. $23\overline{)2{,}001}$

7. Explain how to check the following result using addition.

$$\begin{array}{r} 1{,}142 \\ -\ 459 \\ \hline 683 \end{array}$$

8. THE VIETNAMESE CALENDAR An animal represents each Vietnamese lunar year. Recent Years of the Cat are listed below. If the cycle continues, what year will be the next Year of the Cat?

1915 1927 1939 1951 1963 1975 1987 1999

9. Consider the multiplication statement $4 \cdot 5 = 20$. Show that multiplication is repeated addition.

10. ROOM DIVIDERS Four pieces of plywood, each 22 inches wide and 62 inches high, are to be covered with fabric, front and back, to make the room divider shown. How many square inches of fabric will be used?

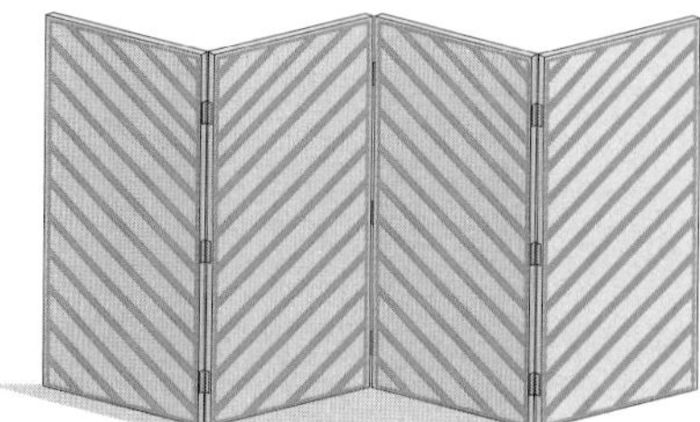

11. **a.** Find the factors of 18.
    b. Find the prime factorization of 18.

12. List the first ten prime numbers.

13. Why isn't 27 a prime number?

14. Evaluate: $(9 - 2)^2 - 3^3$.

15. Divide: $\dfrac{-315}{-1}$.

16. Simplify: $-(-6)$.

17. Graph the integers greater than -3 but less than 4.

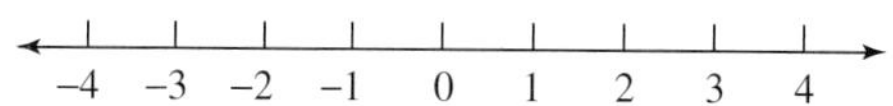

18. Find the absolute value: $|-5|$.

19. Is the statement $-12 > -10$ true or false?

20. ANNUAL NET INCOME Use the following data for the Polaroid Corporation to construct a line graph.

Year	1995	1996	1997	1998	1999
Total net income ($ millions)	−139	15	−127	−51	9

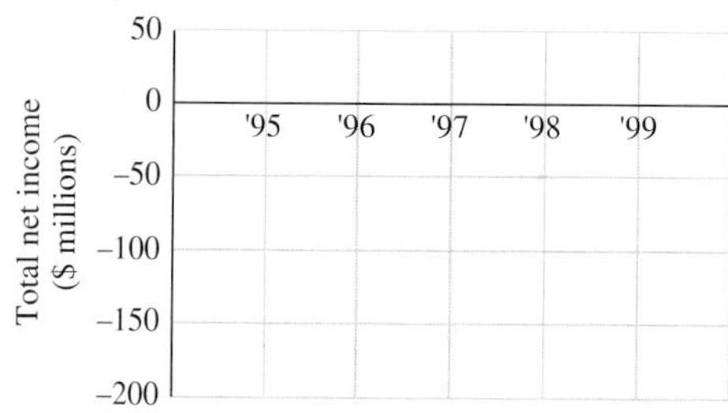

Perform each operation.

21. $-25 + 5$

22. $25 - (-5)$

23. $-25(5)(-1)$

24. $\dfrac{-25}{-5}$

25. Evaluate: $\dfrac{(-6)^2 - 1^5}{-4 - 3}$.

26. Evaluate: $-3 + 3(-4 - 4 \cdot 2)^2$.

27. Evaluate: -3^2 and $(-3)^2$.

28. PLANETS Mercury orbits closer to the sun than does any other planet. Temperatures on Mercury can get as high as 810° F and as low as −290° F. What is the temperature range?

29. 360 is 45% of what number?

30. 90 is what percent of 600?

31. Write an expression illustrating division by 0 and an expression illustrating division of 0. Which is undefined?

32. Add: $\dfrac{1}{2} + \dfrac{2}{3}$.

33. Subtract: $\dfrac{1}{2} - \dfrac{2}{3}$.

34. TENNIS Find the length of the handle on the tennis racquet shown.

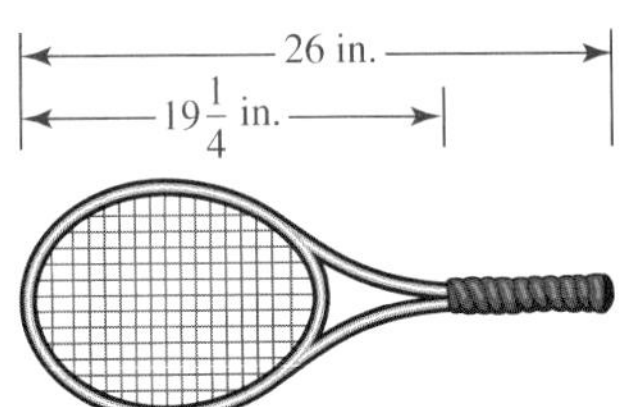

35. Multiply: $\dfrac{4}{5} \cdot \dfrac{2}{7}$.

36. Divide: $2\dfrac{4}{5} \div 2\dfrac{2}{3}$.

37. Complete the table.

	Rate (mph)	Time (hr)	Distance traveled (mi)
Truck	55	4	

38. Multiply: $3.45 \cdot 100$.

39. Multiply: $(0.31)(2.4)$.

40. Divide: $0.72\overline{)536.4}$.

41. Change $\dfrac{8}{11}$ to a decimal.

42. CLASS TIME In a chemistry course, students spend a total of 300 minutes in lab and lecture each week. If $\dfrac{7}{15}$ of the time is spent in lab each week, how many minutes are spent in lecture each week?

43. WEEKLY SCHEDULES Refer to the illustration. Determine the number of hours during a week that an adult spends, on average, on the Internet at home.

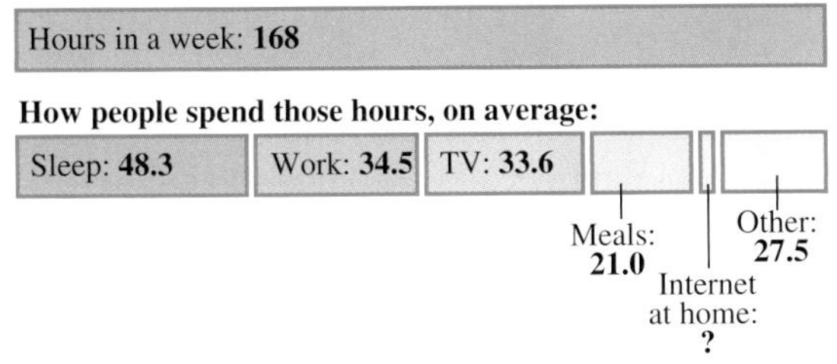

Based on data from the National Sleep Foundation and the United States Bureau of Statistics

44. TEAM GPA The grade point averages of the players on a badminton team are listed below. Find the mean, median, mode, and range of the team's GPAs.

3.04	4.00	2.75	3.23	3.87	2.20
3.02	2.25	2.99	2.56	3.58	2.75

CHAPTER 7

An Introduction to Geometry

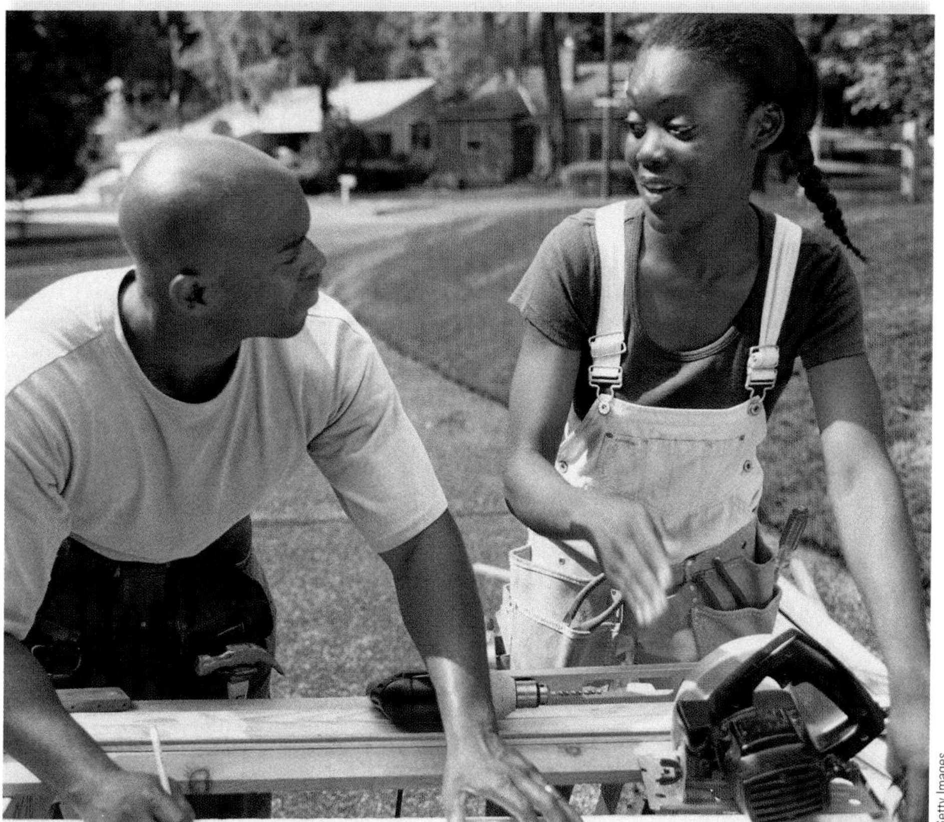

Getty Images

Many people enjoy tackling home-improvement projects themselves. These projects run more smoothly and are more cost efficient if careful planning is done in advance. This planning often requires some mathematics, and in particular, geometry. For example, to purchase the correct amount of materials to fence a yard, you need to know how to calculate *perimeter*. If you are painting a bedroom, you need to calculate the total *area* of the wall surfaces to determine the number of gallons of paint to buy.

To learn more about perimeter and area, visit The Learning Equation on the Internet at http://tle.brookscole.com. (The log-in instructions are in the Preface.) For Chapter 7, the online lesson is:

- *TLE* Lesson 13: Perimeter and Area

Check Your Knowledge

1. __________ lines do not intersect. If two lines intersect and form right angles, they are __________.
2. A polygon with four sides is called a __________. A __________ is a polygon with three sides.
3. A triangle that includes a 90° angle is called a ________ triangle. The longest side of such a triangle is the __________.
4. Triangles that are the same size and the same shape are called __________ triangles. If two triangles are the same shape but not necessarily the same size, they are called ________ triangles.
5. The distance around a polygon is called the __________. The measure of the surface enclosed by a polygon is called its ______.
6. A segment drawn from the center of a circle to a point on the circle is called a ________. The distance around a circle is called its __________. A __________ is a chord that passes through the center of a circle.
7. The space contained within a geometric solid is called its ________.
8. Match the descriptions with the angles.
   a. 37° ____ I. right angle
   b. 90° ____ II. straight angle
   c. 125° ____ III. acute angle
   d. 180° ____ IV. obtuse angle

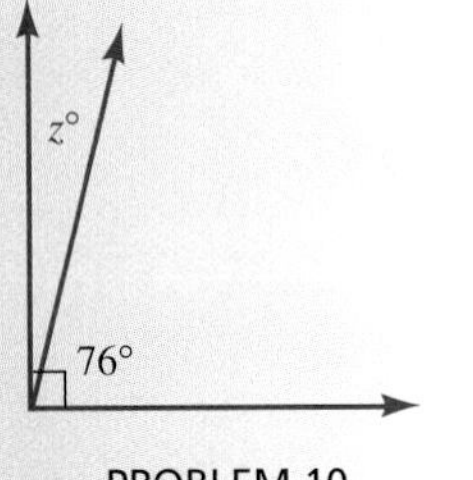

PROBLEM 10

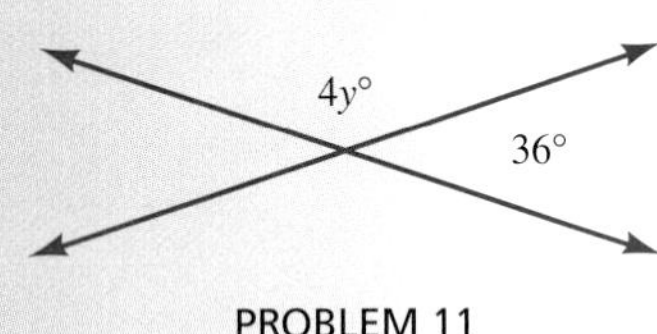

PROBLEM 11

9. In the notation $\angle ABC$, which point is the vertex of the angle?
10. Refer to the illustration on the left. Find z.
11. Refer to the illustration on the left. Find y.
12. Find the supplement of an angle measuring 119°.
13. Refer to the illustration, in which lines ℓ_1 and ℓ_2 are parallel.
    a. Find m($\angle 1$).
    b. Find m($\angle 2$).
    c. Find m($\angle 3$).
    d. Find m($\angle 4$).
    e. Find x.

50° $(x + 55)°$ ℓ_1 ℓ_2 1 2 3 4

14. If the measure of one angle of a triangle is 45° and the measure of another is 55°, what is the measure of the third angle?
15. A rectangle is 12 ft long and 5 ft wide.
    a. Find the length of the diagonal of the rectangle.
    b. Find the area of the rectangle.
16. Find the area of a triangle with base 4.5 in. and height 7.8 in.
17. Find the area of a circle if the diameter is 10 ft.
18. Find the volume of a sphere that is 4 ft in diameter.
19. Find the volume of a rectangular solid with length 5 ft, width 4 ft, and height 10 ft.
20. Find the volume of a cylinder 8 ft in diameter and 10 ft tall.

Study Skills Workshop

STUDYING FOR TESTS

Getty Images

Doing homework regularly is important, but tests require another type of strategy for preparation. Before you take a test, it is important that you have done your best to commit the important concepts to memory.

How Much Time Should I Devote to Preparing for a Test? The time needed to prepare for a test will vary according to how well you have understood and memorized the basic concepts. Plan to prepare at least four days in advance of your test and schedule this on your study calendar.

How Do I Prepare? **Four days before the test** Know exactly what material the test will cover. Then imagine that you could bring one $8\frac{1}{2}'' \times 11''$ sheet of paper to the test. What would you write on that sheet? Go through each section (and your reworked notes) to identify the key rules and definitions to include on your study sheet. Keep this paper with you all the time until the test, and review it whenever you can.

Three days before the test Go through your reworked notes and find the examples that your instructor gave you in class. Add any examples you might have missed to your paper and continue to look at it whenever you have time.

Two days before the test Use a chapter test in your textbook, or make up your own practice test by choosing a sampling of problems from your text or reworked notes. You can also take a quiz on chapter and section material online using the iLrn Web site rather than create your own test. This site is located at www.iLrn.com, and the quizzes are found in the "Tutorial" section. Try to include the same number of questions that your real test will have. Choose problems that have a solution that can be checked. Then, *with your book closed,* take a *timed* trial. Don't be upset if you can't do everything perfectly; it's only a trial. When you are done, check your answers. Be honest with yourself! Make a list of the topics that were difficult and add these to your study sheet.

One day before the test Get help, if necessary, with yesterday's problems from your instructor during office hours, from a tutor at the tutorial center, or from a classmate. Practice your trial test again, without books or notes. Go back over any problem you didn't get correct. Get plenty of rest the night before your test.

Test day Review your study sheet, if you have time. Focus on how well you have prepared and relax as much as possible. When taking your test, complete the problems that you are sure of first. Skip the problems that you don't understand right away, and return to them later. Be aware of time throughout the test so that you don't spend too long on any one problem.

ASSIGNMENT

1. Four days before your test, prepare your study sheet.
2. Three days before your test, go through all of the examples in your reworked notes.
3. Two days before your test, make a written practice test and time youreslf on it. Then check your answers.
4. The day before the test, fix the problems that you did incorrectly. Use your textbook, notes, classmate, tutor, or instructor for help. Review all of your practice problems until you can do them without any help. Gather the materials that you will need for your test and get enough rest.
5. On test day, arrive in class a few minutes early. Look over your study sheet if you have time. Relax as much as possible, knowing that you have prepared well.

Geometry comes from the Greek words geo *(meaning earth) and* metron *(meaning measure).*

7.1 Some Basic Definitions

- Points, lines, and planes
- Angles
- Adjacent and vertical angles
- Complementary and supplementary angles

In this chapter, we study two-dimensional geometric figures such as rectangles and circles. In daily life, it is often necessary to find the perimeter or area of one of these figures. For example, to find the amount of fencing that is needed to enclose a circular garden, we must find the perimeter of a circle (called its *circumference*). To find the amount of paint needed to paint a room, we must find the area of its four rectangular walls.

We also study three-dimensional figures such as cylinders and spheres. To find the amount of space enclosed within these figures, we must find their volumes.

Points, lines, and planes

Geometry is based on three words: **point, line,** and **plane.** Although we will make no attempt to define these words formally, we can think of a point as a geometric figure that has position but no length, width, or depth. Points are always labeled with capital letters. Point A is shown in Figure 7-1(a).

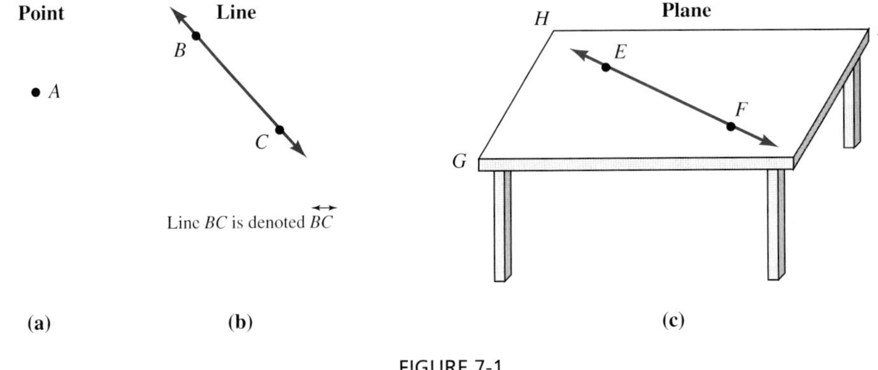

FIGURE 7-1

A line is infinitely long but has no width or depth. Figure 7-1(b) shows line BC, passing through points B and C. A plane is a flat surface, like a table top, that has length and width but no depth. In Figure 7-1(c), line EF lies in the plane GHI.

As Figure 7-1(b) illustrates, points B and C determine exactly one line, the line BC. In Figure 7-1(c), the points E and F determine exactly one line, the line EF. In general, any two points determine exactly one line.

Other geometric figures can be created by using parts or combinations of points, lines, and planes.

> **Line segment**
>
> The **line segment** AB, denoted as $\overline{AB}$, is the part of a line that consists of points A and B and all points in between. (See Figure 7-2 on the next page.) Points A and B are the **endpoints** of the segment.

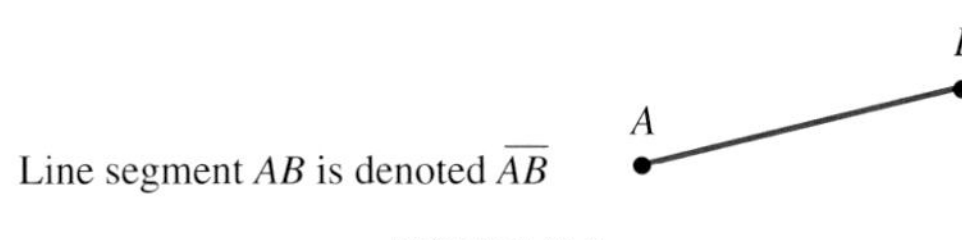

FIGURE 7-2

Since the **midpoint** of a line segment divides the segment into two parts of equal length, we say that the midpoint **bisects** the line segment. In Figure 7-3, M is the midpoint of segment AB, because the measure of $\overline{AM}$, denoted as $m(\overline{AM})$, is equal to the measure of $\overline{MB}$, denoted as $m(\overline{MB})$.

$$m(\overline{AM}) = 4 - 1$$
$$= 3$$

and

$$m(\overline{MB}) = 7 - 4$$
$$= 3$$

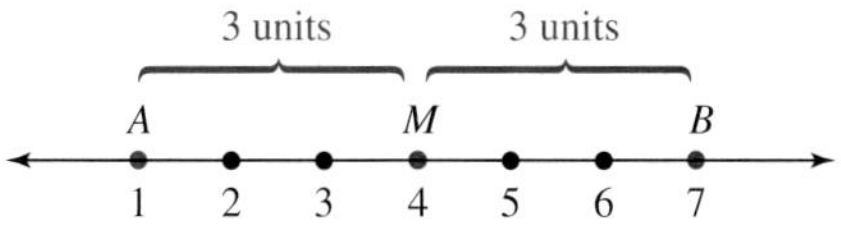

3 units 3 units
A M B
1 2 3 4 5 6 7

FIGURE 7-3

Since the measure of both segments is 3 units, M bisects segment AB and $m(\overline{AM}) = m(\overline{MB})$.

When two line segments have the same measure, we say that they are **congruent.** Since $m(\overline{AM}) = m(\overline{MB})$, we can write

$\overline{AM} \cong \overline{MB}$ Read $\cong$ as "is congruent to."

Another geometric figure is the *ray,* as shown in Figure 7-4.

Ray

A **ray** is the part of a line that begins at some point (say, A) and continues forever in one direction. Point A is the **endpoint** of the ray.

Ray AB is denoted as $\overrightarrow{AB}$. The endpoint of the ray is always listed first.

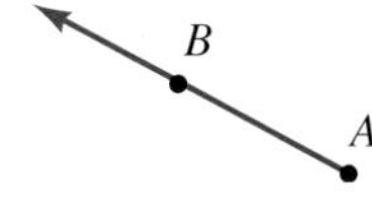

FIGURE 7-4

Angles

Angle

An **angle** is a figure formed by two rays with a common endpoint. The common endpoint is called the **vertex,** and the rays are called **sides.**

The angle in Figure 7-5 can be denoted as

$\angle BAC$, $\angle CAB$, $\angle A$, or $\angle 1$ The symbol $\angle$ means angle.

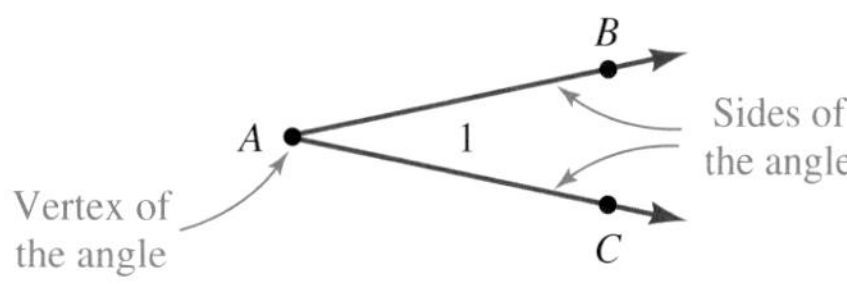

FIGURE 7-5

! COMMENT When using three letters to name an angle, be sure the letter name of the vertex is the middle letter.

One unit of measurement of an angle is the **degree.** It is $\frac{1}{360}$ of a full revolution. We can use a **protractor** to measure angles in degrees. (See Figure 7-6.)

Angle	Measure in degrees
$\angle ABC$	30°
$\angle ABD$	60°
$\angle ABE$	110°
$\angle ABF$	150°
$\angle ABG$	180°

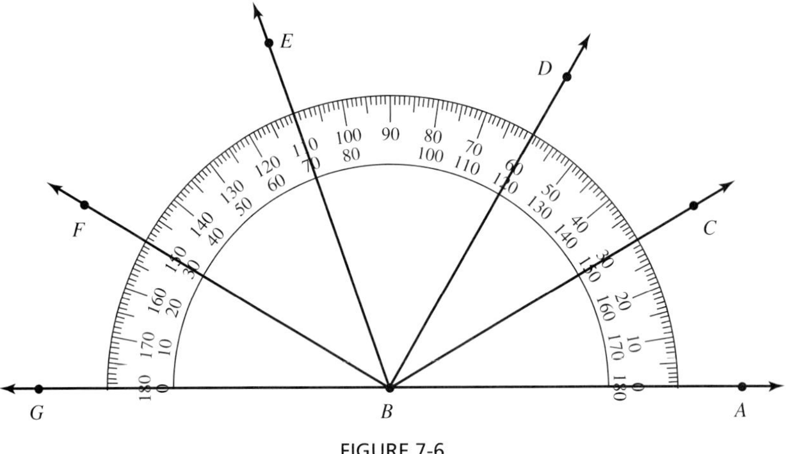

FIGURE 7-6

If we read the protractor from left to right, we can see that the measure of $\angle GBF$ (denoted as m($\angle GBF$)) is 30°.

When two angles have the same measure, we say that they are congruent. Since m($\angle ABC$) = 30° and m($\angle GBF$) = 30°, we can write

$$\angle ABC \cong \angle GBF$$

We classify angles according to their measure, as in Figure 7-7.

Classification of angles

Acute angles: Angles whose measures are greater than 0° but less than 90°

Right angles: Angles whose measures are 90°

Obtuse angles: Angles whose measures are greater than 90° but less than 180°

Straight angles: Angles whose measures are 180°

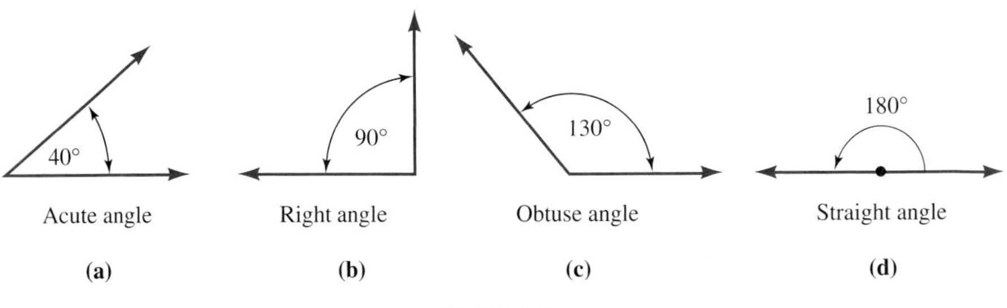

FIGURE 7-7

EXAMPLE 1 Classify each angle in Figure 7-8 as an acute angle, a right angle, an obtuse angle, or a straight angle.

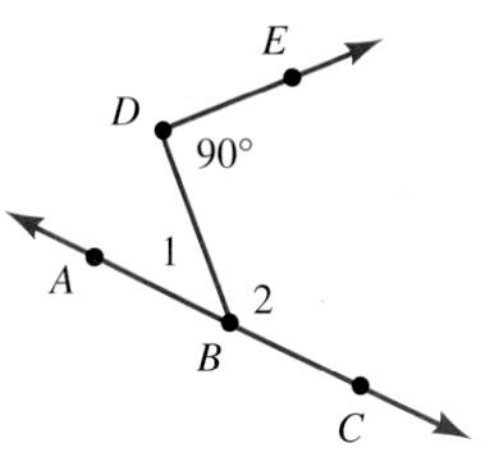

FIGURE 7-8

Solution

Since $m(\angle 1) < 90°$, it is an acute angle.

Since $m(\angle 2) > 90°$, but less than 180°, it is an obtuse angle.

Since $m(\angle BDE) = 90°$, it is a right angle.

Since $m(\angle ABC) = 180°$, it is a straight angle.

Adjacent and vertical angles

Two angles that have a common vertex and are side-by-side are called **adjacent angles.**

EXAMPLE 2 Two angles with measures of $x°$ and 35° are adjacent angles. Use the information in Figure 7-9 to find x.

Solution We can use algebra to solve this problem. Since the sum of the measures of the angles is 80°, we have

$$x + 35 = 80$$

$$x + 35 - \mathbf{35} = 80 - \mathbf{35}$$ To undo the addition of 35, subtract 35 from both sides.

$$x = 45$$ Perform the subtractions: $35 - 35 = 0$ and $80 - 35 = 45$.

FIGURE 7-9

Thus, x is 45.

Self Check 2

In the figure below, find x.

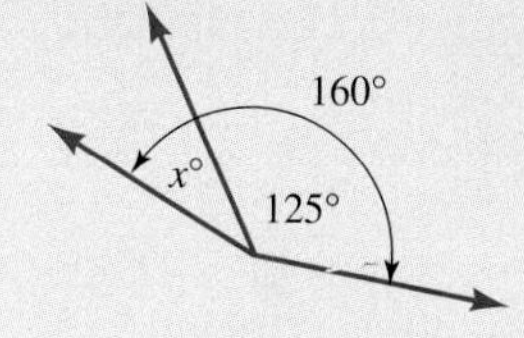

Answer 35

When two lines intersect, pairs of nonadjacent angles are called **vertical angles.** In Figure 7-10(a), lines l_1 (read as "line l sub 1") and l_2 (read as "line l sub 2") intersect. $\angle 1$ and $\angle 3$ are vertical angles, as are $\angle 2$ and $\angle 4$.

To illustrate that vertical angles always have the same measure, we refer to Figure 7-10(b) with angles having measures of $x°$, $y°$, and 30°. Since the measure of any straight angle is 180°, we have

$$\mathbf{30} + x = 180 \quad \text{and} \quad \mathbf{30} + y = 180$$

$$x = 150 \qquad\qquad y = 150$$ To undo the addition of 30, subtract 30 from both sides.

Since x and y are both 150, $x = y$.

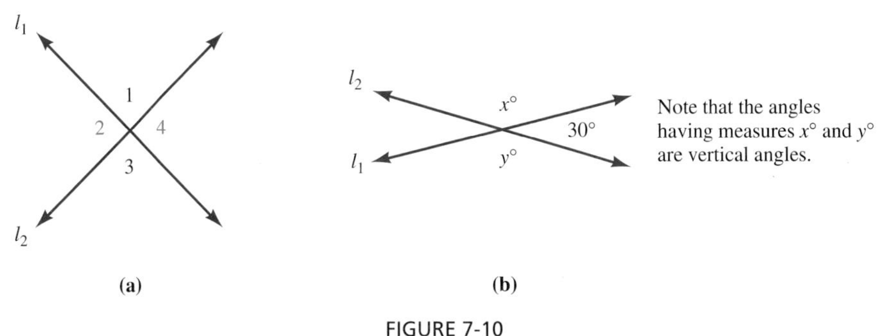

FIGURE 7-10

Property of vertical angles

Vertical angles are congruent (have the same measure).

Self Check 3

In Figure 7-11, find:

a. m(∠2)

b. m(∠4)

Answers **a.** 100°, **b.** 30°

EXAMPLE 3 In Figure 7-11, find: **a.** m(∠1) and **b.** m(∠3).

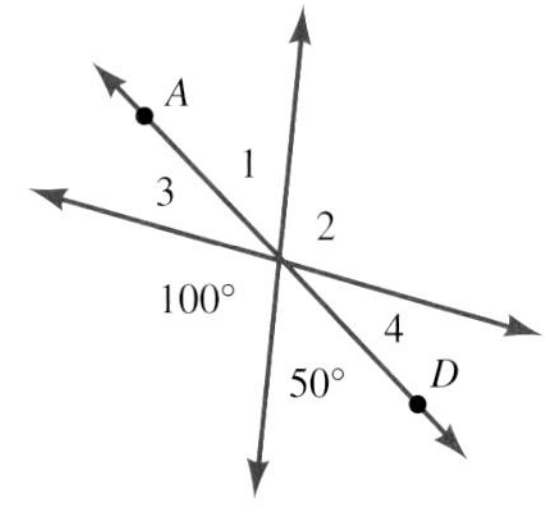

FIGURE 7-11

Solution

a. The 50° angle and ∠1 are vertical angles. Since vertical angles are congruent, m(∠1) = 50°.

b. Since AD is a line, the sum of the measures of ∠3, the 100° angle, and the 50° angle is 180°. If m(∠3) = x, we have

$$x + 100 + 50 = 180$$

$$x + 150 = 180 \quad \text{Perform the addition: } 100 + 50 = 150.$$

$$x = 30 \quad \text{Subtract 150 from both sides.}$$

Thus, m(∠3) = 30°.

Self Check 4

In the figure below, find y.

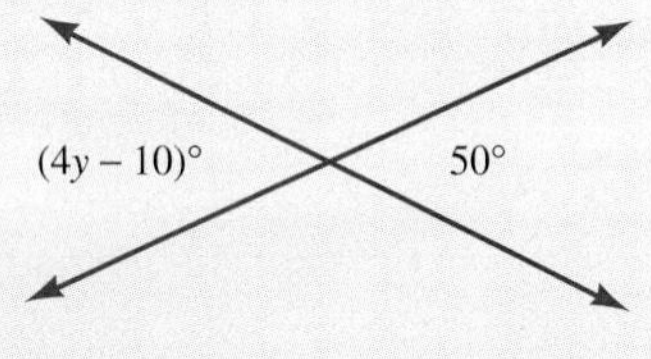

Answer 15

EXAMPLE 4 In Figure 7-12, find x.

FIGURE 7-12

Solution Since the angles are vertical angles, they have equal measures.

$$4x - 20 = 120$$

$$4x = 140 \quad \text{To undo the subtraction of 20, add 20 to both sides.}$$

$$x = 35 \quad \text{To undo the multiplication by 4, divide both sides by 4.}$$

Thus, x is 35.

Complementary and supplementary angles

Complementary and supplementary angles

Two angles are **complementary angles** when the sum of their measures is 90°.

Two angles are **supplementary angles** when the sum of their measures is 180°.

Angles with measures of 60° and 30° are examples of complementary angles, because the sum of their measures is 90°. Each angle is the complement of the other.

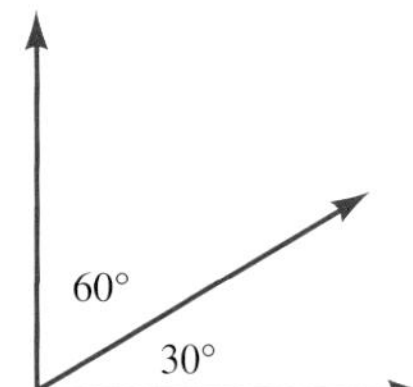

Angles of 130° and 50° are supplementary, because the sum of their measures is 180°. Each angle is the supplement of the other.

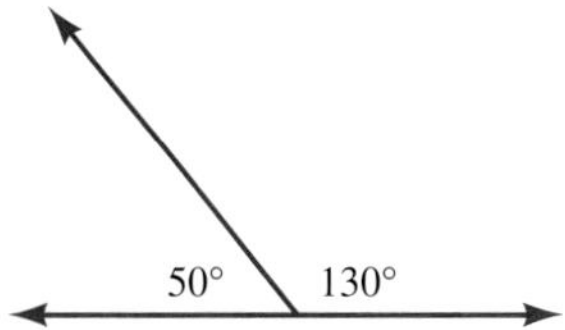

! COMMENT The definition of supplementary angles requires that the sum of two angles be 180°. Three angles of 40°, 60°, and 80° are not supplementary even though their sum is 180°.

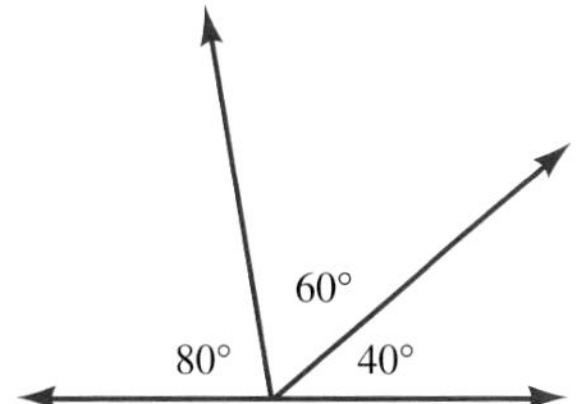

EXAMPLE 5

a. Find the complement of a 35° angle.

b. Find the supplement of a 105° angle.

Solution

a. See Figure 7-13. Let x represent the complement of the 35° angle. Since the angles are complementary, we have

FIGURE 7-13

$x + 35 = 90$ The sum of the angles' measures must be 90°.

$x = 55$ To undo the addition of 35, subtract 35 from both sides.

The complement of 35° is 55°.

b. See Figure 7-14. Let y represent the supplement of the 105° angle. Since the angles are supplementary, we have

$y + 105 = 180$ The sum of the angles' measures must be 180°.

$y = 75$ To undo the addition of 105, subtract 105 from both sides.

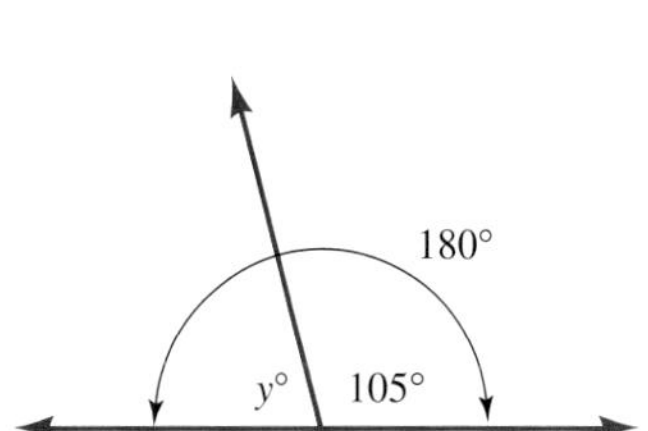

FIGURE 7-14

The supplement of 105° is 75°.

Self Check 5

a. Find the complement of a 50° angle.

b. Find the supplement of a 50° angle.

Answers **a.** 40°, **b.** 130°

Section 7.1 STUDY SET

VOCABULARY *Fill in the blanks.*

1. A line ________ has two endpoints.
2. Two points __________ at most one line.
3. A _________ divides a line segment into two parts of equal length.
4. An angle is measured in ________.
5. A __________ is used to measure angles.
6. An ______ angle is less than 90°.
7. A ______ angle measures 90°.
8. An _______ angle is greater than 90° but less than 180°.
9. The measure of a straight angle is ____.
10. Adjacent angles have the same vertex and are __________.
11. The sum of two ____________ angles is 180°.
12. The sum of two complementary angles is ____.

CONCEPTS *Refer to the illustration and determine whether each statement is true. If a statement is false, explain why.*

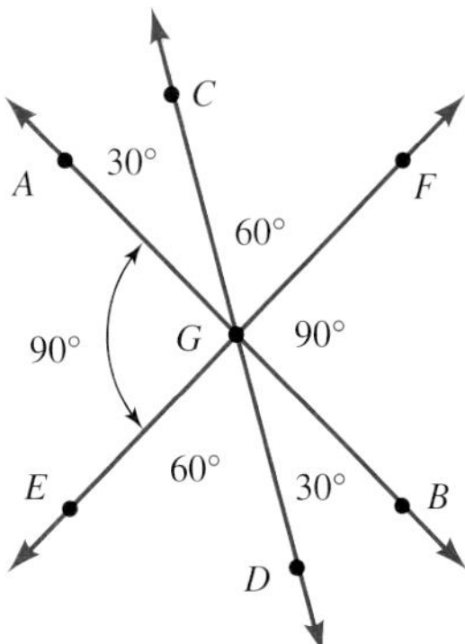

13. $\overrightarrow{GF}$ has point G as its endpoint.
14. $\overline{AG}$ has no endpoints.
15. Line CD has three endpoints.
16. Point D is the vertex of $\angle DGB$.
17. $m(\angle AGC) = m(\angle BGD)$
18. $\angle AGF \cong \angle BGE$
19. $\angle FGB \cong \angle EGA$
20. $\angle AGC$ and $\angle CGF$ are adjacent angles.

Refer to the illustration above and determine whether each angle is an acute angle, a right angle, an obtuse angle, or a straight angle.

21. $\angle AGC$
22. $\angle EGA$
23. $\angle FGD$
24. $\angle BGA$
25. $\angle BGE$
26. $\angle AGD$
27. $\angle DGC$
28. $\angle DGB$

Refer to the illustration below and determine whether each statement is true. If a statement is false, explain why.

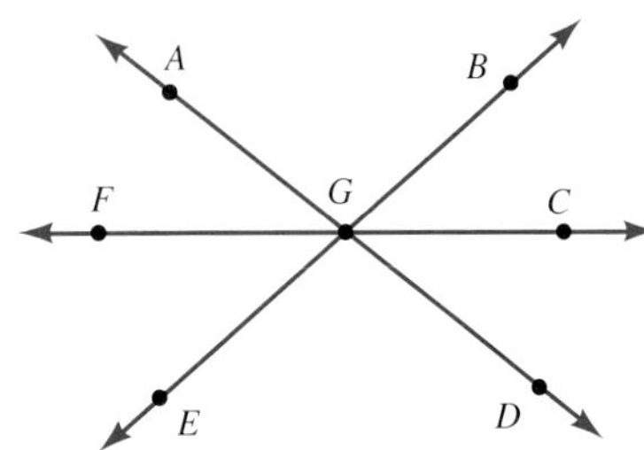

29. $\angle AGF$ and $\angle DGC$ are vertical angles.
30. $\angle FGE$ and $\angle BGA$ are vertical angles.
31. $m(\angle AGB) = m(\angle BGC)$
32. $\angle AGC \cong \angle DGF$

Refer to the illustration below and determine whether each pair of angles are congruent.

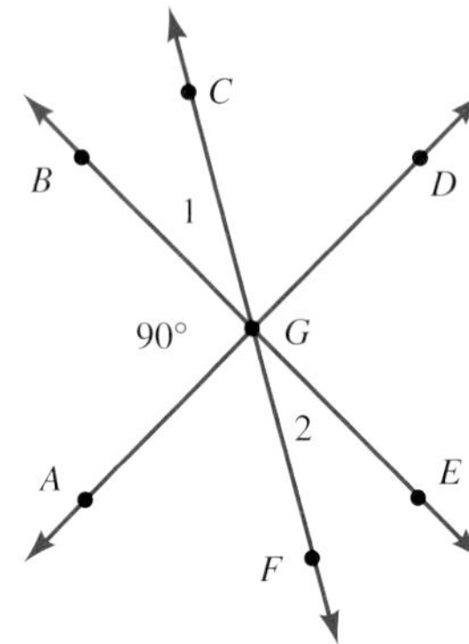

33. $\angle 1$ and $\angle 2$
34. $\angle FGB$ and $\angle CGE$
35. $\angle AGB$ and $\angle DGE$
36. $\angle CGD$ and $\angle CGB$
37. $\angle AGF$ and $\angle FGE$
38. $\angle AGB$ and $\angle BGD$

Refer to the illustration above and determine whether each statement is true.

39. $\angle 1$ and $\angle CGD$ are adjacent angles.
40. $\angle 2$ and $\angle 1$ are adjacent angles.
41. $\angle FGA$ and $\angle AGC$ are supplementary.
42. $\angle AGB$ and $\angle BGC$ are complementary.
43. $\angle AGF$ and $\angle 2$ are complementary.
44. $\angle AGB$ and $\angle EGD$ are supplementary.
45. $\angle EGD$ and $\angle DGB$ are supplementary.
46. $\angle DGC$ and $\angle AGF$ are complementary.

NOTATION *Fill in the blanks.*

47. The symbol $\angle$ means ______.
48. The symbol $\overline{AB}$ is read as "____________ AB."
49. The symbol $\overrightarrow{AB}$ is read as "______ AB."
50. The symbol   is read as "is congruent to."

PRACTICE *Refer to the illustration below and find the length of each segment.*

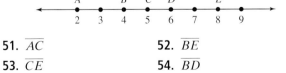

51. $\overline{AC}$
52. $\overline{BE}$
53. $\overline{CE}$
54. $\overline{BD}$
55. $\overline{CD}$
56. $\overline{DE}$

Refer to the illustration above and find each midpoint.

57. Find the midpoint of $\overline{AD}$.
58. Find the midpoint of $\overline{BE}$.

Use a protractor to measure each angle.

59. **60.**

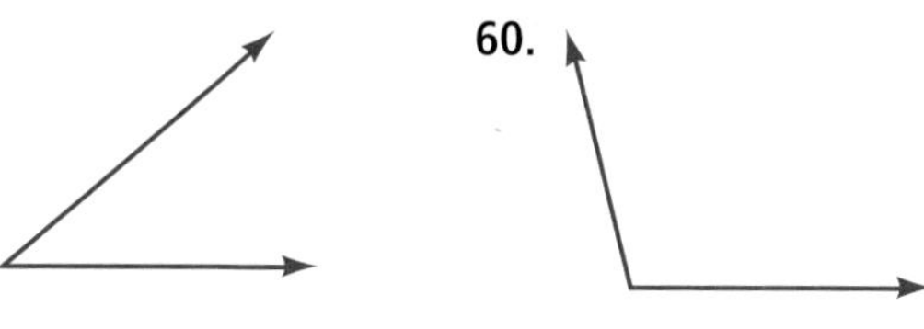

61. **62.**

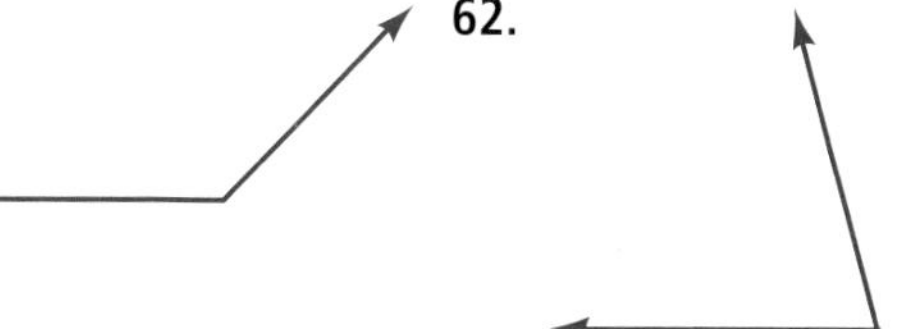

Find x.

63.

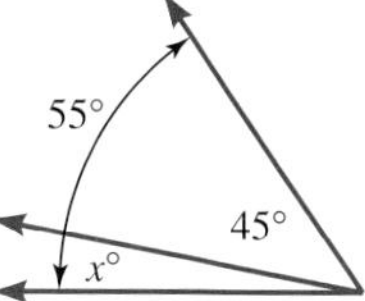

64.

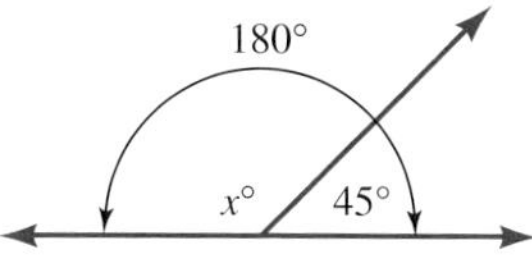

65.

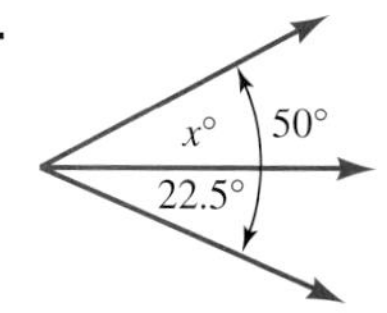

66.

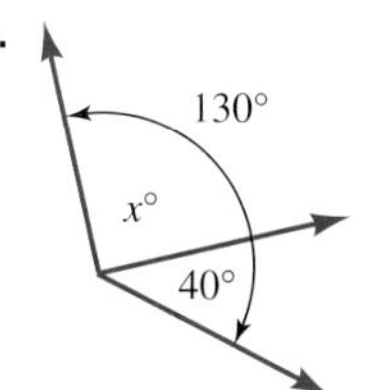

67.

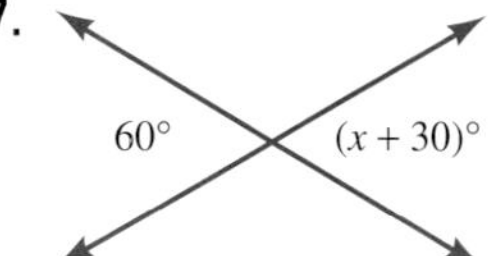

68.

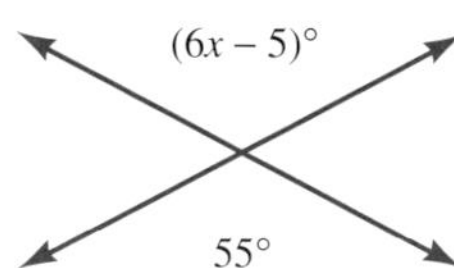

69.

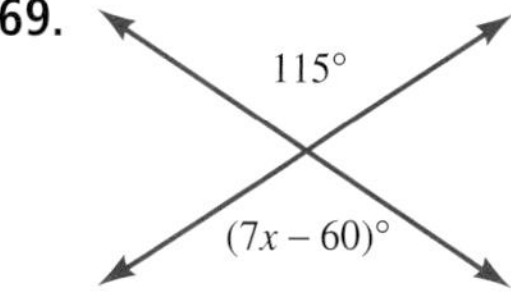

70.

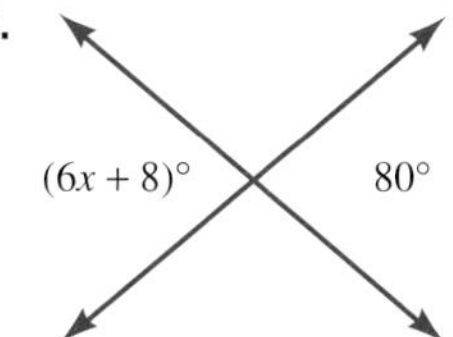

Let x represent the unknown angle measure. Draw a diagram, write an appropriate equation, and solve it for x.

71. Find the complement of a 30° angle.

72. Find the supplement of a 30° angle.

73. Find the supplement of a 105° angle.

74. Find the complement of a 75° angle.

Refer to the illustration below in which $m(\angle 1) = 50°$. *Find the measure of each angle or sum of angles.*

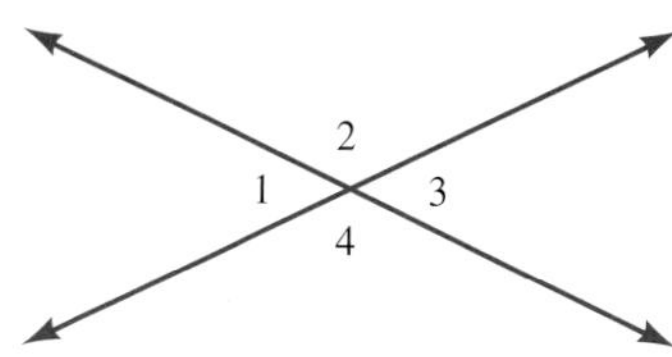

75. $\angle 4$

76. $\angle 3$

77. $m(\angle 1) + m(\angle 2) + m(\angle 3)$

78. $m(\angle 2) + m(\angle 4)$

Refer to the illustration below, in which $m(\angle 1) + m(\angle 3) + m(\angle 4) = 180°$, $\angle 3 \cong \angle 4$, *and* $\angle 4 \cong \angle 5$. *Find the measure of each angle.*

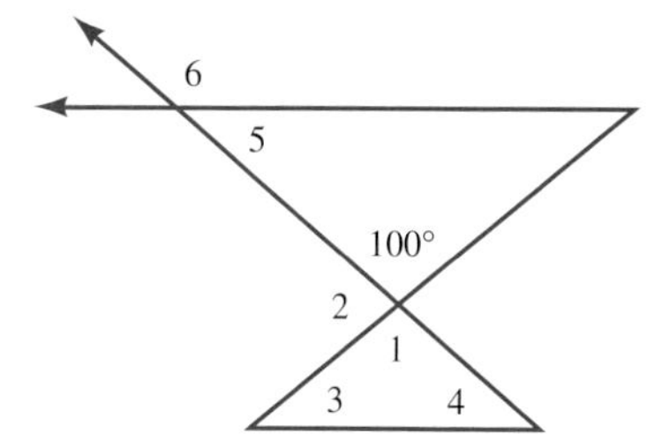

79. $\angle 1$

80. $\angle 2$

81. $\angle 3$

82. $\angle 6$

APPLICATIONS

83. BASEBALL Use the following definition to draw the strike zone for the player shown.

> *The strike zone is that area over home plate the upper limit of which is a horizontal line at the midpoint between the top of the shoulders and the top of the uniform pants and the lower level is a line at the hollow beneath the kneecap.*

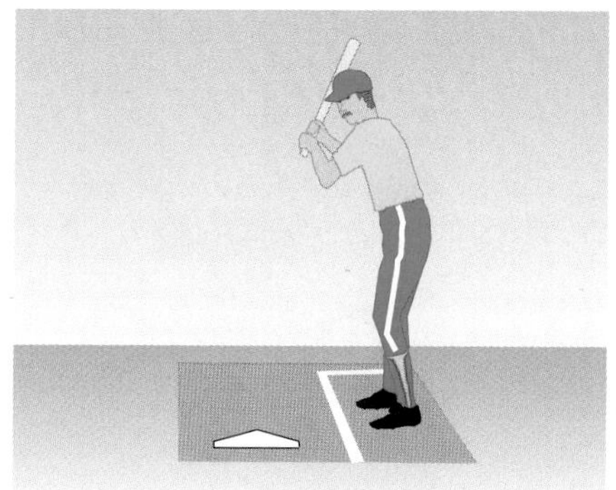

84. PHYSICS The illustration shows a 15-pound block that is suspended with two ropes, one of which is horizontal. Classify each numbered angle in the illustration as acute, obtuse, or right.

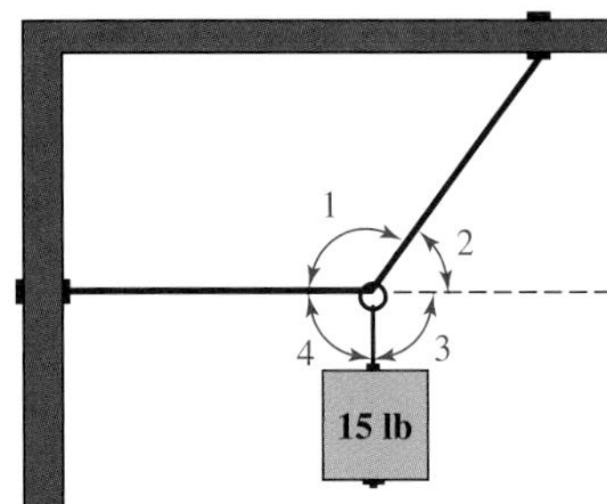

85. SYNTHESIZERS Refer to the illustration. Find x and y.

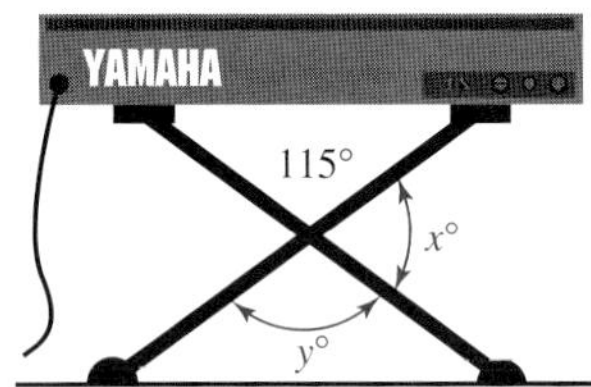

86. AVIATION How many degrees from the horizontal position are the wings of the airplane?

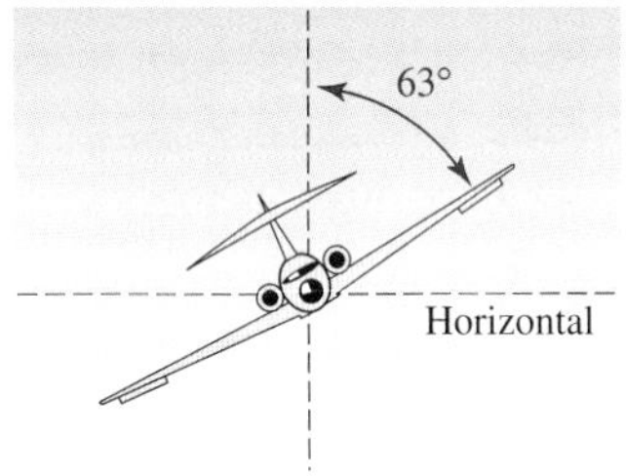

87. GARDENING What angle does the handle of the lawn mower make with the ground?

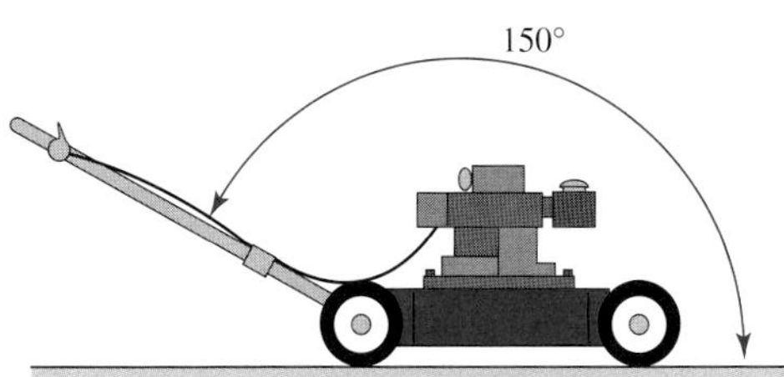

88. MUSICAL INSTRUMENTS Suppose that you are a beginning band teacher describing the correct posture needed to play various instruments. Use the following diagrams to approximate the angle measure at which each instrument should be held in relation to the student's body: **a.** flute **b.** clarinet **c.** trumpet

a.

b.

c.

WRITING

89. PHRASES Explain what you think each of these phrases means. How is geometry involved?

a. The president did a complete 180-degree flip on the subject of a tax cut.

b. The rollerblader did a "360" as she jumped off the ramp.

90. In the statements below, the ° symbol is used in two different ways. Explain the difference.

85° F and $m(\angle A) = 85°$

91. What is a protractor?

92. Explain the difference between a ray and a line segment.

93. Explain why an angle measuring 105° cannot have a complement.

94. Explain why an angle measuring 210° cannot have a supplement.

REVIEW

95. Find: 2^4.

96. Add: $\frac{1}{2} + \frac{2}{3} + \frac{3}{4}$.

97. Subtract: $\frac{3}{4} - \frac{1}{8} - \frac{1}{3}$.

98. Multiply: $\frac{5}{8} \cdot \frac{2}{15} \cdot \frac{6}{5}$.

99. Divide: $\frac{12}{17} \div \frac{4}{34}$.

100. What is 7% of 7?

7.2 Parallel and Perpendicular Lines

• Parallel and perpendicular lines • Transversals and angles • Properties of parallel lines

In this section, we consider *parallel* and *perpendicular* lines. Since parallel lines are always the same distance apart, the railroad tracks shown in Figure 7-15(a) illustrate one application of parallel lines. Figure 7-15(b) shows one of the events of men's gymnastics, the parallel bars. Since perpendicular lines meet and form right angles, the monument and the ground shown in Figure 7-15(c) illustrate one application of perpendicular lines.

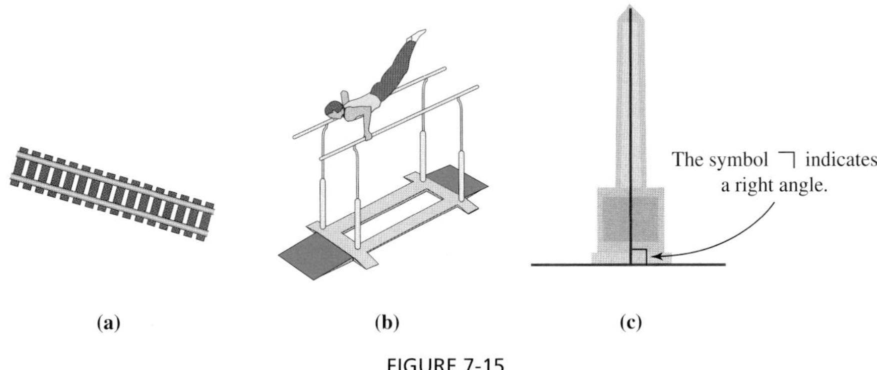

FIGURE 7-15

Parallel and perpendicular lines

If two lines lie in the same plane, they are called **coplanar.** Two coplanar lines that do not intersect are called **parallel lines.** See Figure 7-16(a).

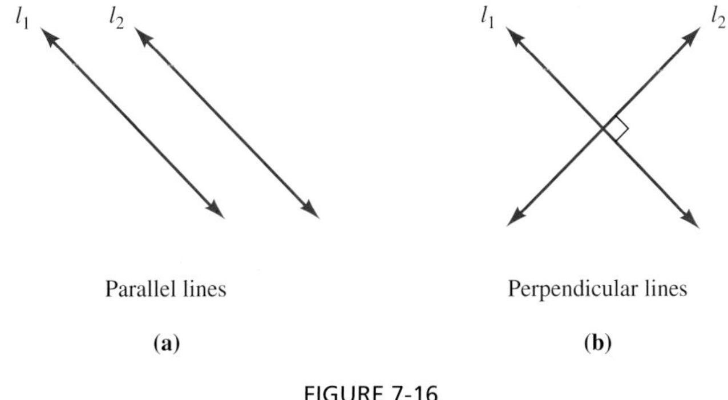

FIGURE 7-16

Parallel lines

Parallel lines are coplanar lines that do not intersect.

If lines l_1 (read as "l sub 1") and l_2 (read as "l sub 2") are parallel, we can write $l_1 \parallel l_2$, where the symbol $\parallel$ is read as "is parallel to."

Perpendicular lines

Perpendicular lines are lines that intersect and form right angles.

In Figure 7-16(b), $l_1 \perp l_2$, where the symbol $\perp$ is read as "is perpendicular to."

Transversals and angles

A line that intersects two or more coplanar lines is called a **transversal.** For example, line l_1 in Figure 7-17 is a transversal intersecting lines l_2, l_3, and l_4.

When two lines are cut by a transversal, the following types of angles are formed.

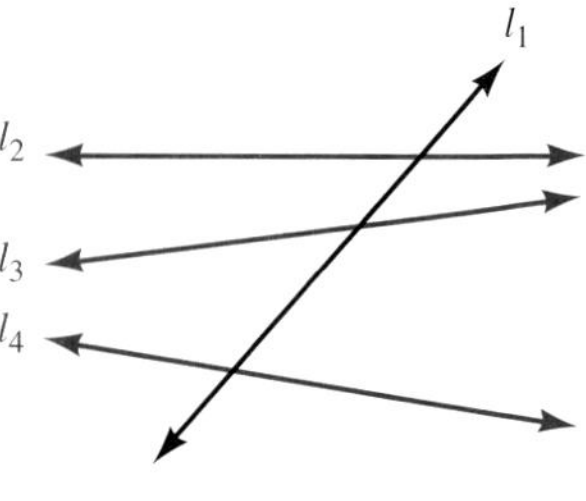

FIGURE 7-17

Alternate interior angles:

$\angle 4$ and $\angle 5$

$\angle 3$ and $\angle 6$

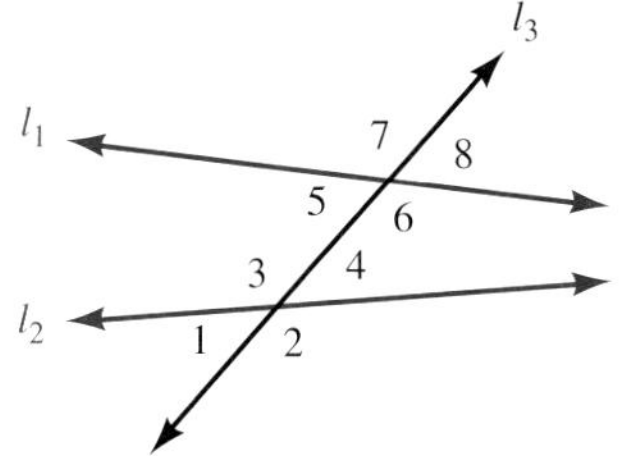

Corresponding angles:

$\angle 1$ and $\angle 5$

$\angle 3$ and $\angle 7$

$\angle 2$ and $\angle 6$

$\angle 4$ and $\angle 8$

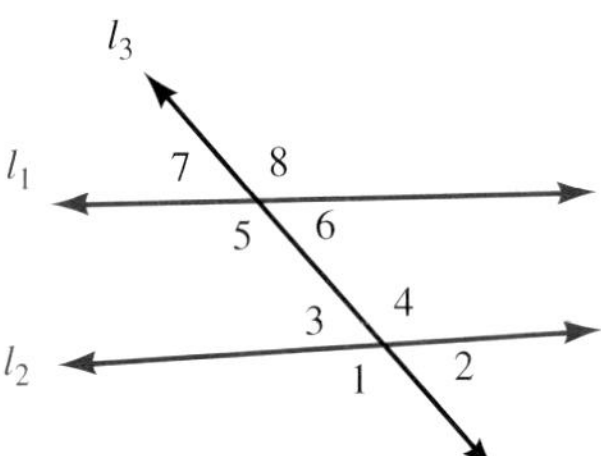

Interior angles:

$\angle 3$, $\angle 4$, $\angle 5$, and $\angle 6$

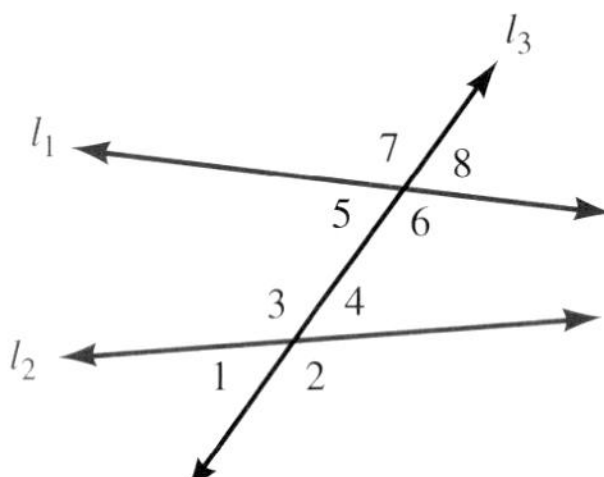

EXAMPLE 1 In Figure 7-18, identify **a.** all pairs of alternate interior angles, **b.** all pairs of corresponding angles, and **c.** all interior angles.

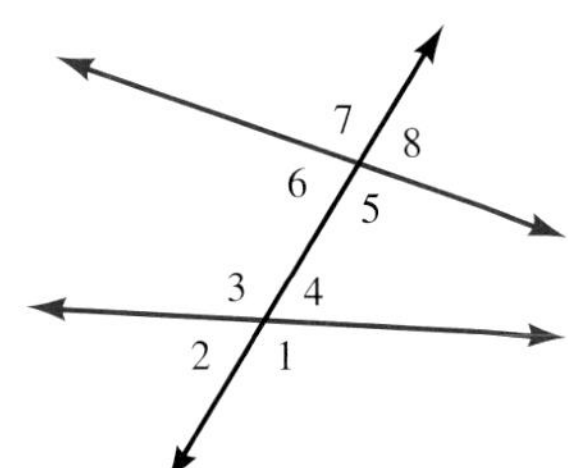

FIGURE 7-18

Solution

a. Pairs of alternate interior angles are

$\angle 3$ and $\angle 5$, $\angle 4$ and $\angle 6$

b. Pairs of corresponding angles are

$\angle 1$ and $\angle 5$, $\angle 4$ and $\angle 8$, $\angle 2$ and $\angle 6$, $\angle 3$ and $\angle 7$

c. Interior angles are

$\angle 3$, $\angle 4$, $\angle 5$, and $\angle 6$

Properties of parallel lines

1. If two parallel lines are cut by a transversal, alternate interior angles are congruent. (See Figure 7-19.) If $l_1 \parallel l_2$, then $\angle 2 \cong \angle 4$ and $\angle 1 \cong \angle 3$.
2. If two parallel lines are cut by a transversal, corresponding angles are congruent. (See Figure 7-20.) If $l_1 \parallel l_2$, then $\angle 1 \cong \angle 5$, $\angle 3 \cong \angle 7$, $\angle 2 \cong \angle 6$, and $\angle 4 \cong \angle 8$.

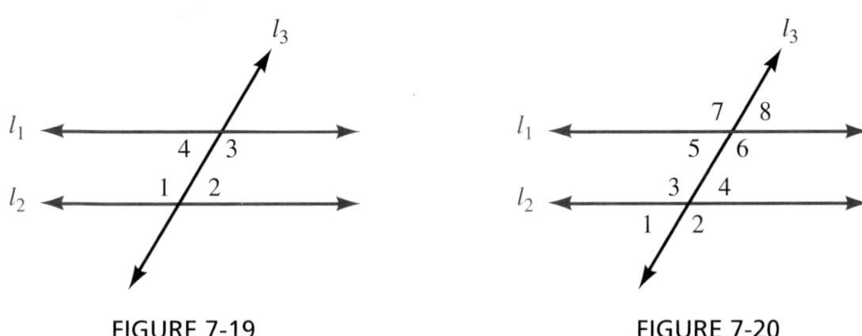

FIGURE 7-19 FIGURE 7-20

3. If two parallel lines are cut by a transversal, interior angles on the same side of the transversal are supplementary. (See Figure 7-21.) If $l_1 \parallel l_2$, then $\angle 1$ is supplementary to $\angle 2$ and $\angle 4$ is supplementary to $\angle 3$.
4. If a transversal is perpendicular to one of two parallel lines, it is also perpendicular to the other line. (See Figure 7-22.) If $l_1 \parallel l_2$ and $l_3 \perp l_1$, then $l_3 \perp l_2$.

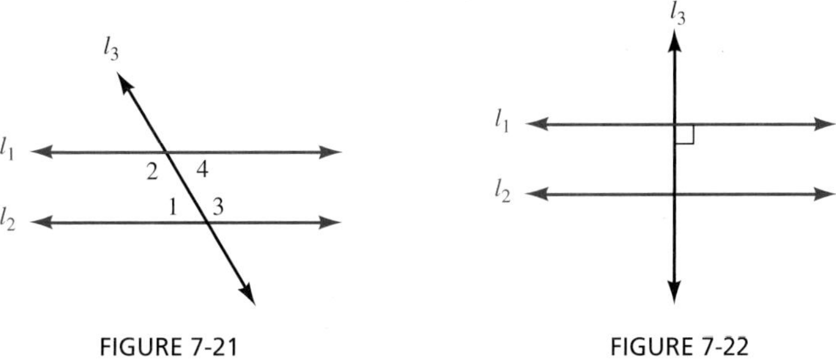

FIGURE 7-21 FIGURE 7-22

5. If two lines are parallel to a third line, they are parallel to each other. (See Figure 7-23.) If $l_1 \parallel l_2$ and $l_1 \parallel l_3$, then $l_2 \parallel l_3$.

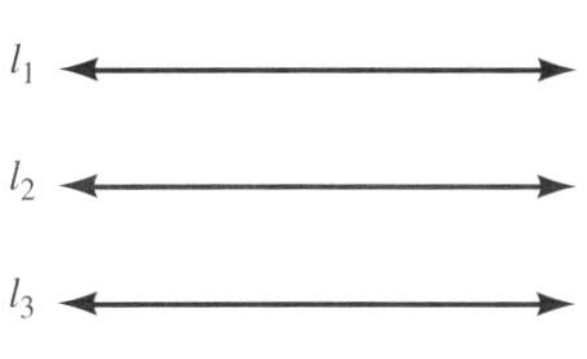

FIGURE 7-23

EXAMPLE 2 See Figure 7-24. If $l_1 \parallel l_2$ and $m(\angle 3) = 120°$, find the measures of the other angles.

Solution

$m(\angle 1) = 60°$ $\angle 3$ and $\angle 1$ are supplementary.

$m(\angle 2) = 120°$ Vertical angles are congruent: $m(\angle 2) = m(\angle 3)$.

$m(\angle 4) = 60°$ Vertical angles are congruent: $m(\angle 4) = m(\angle 1)$.

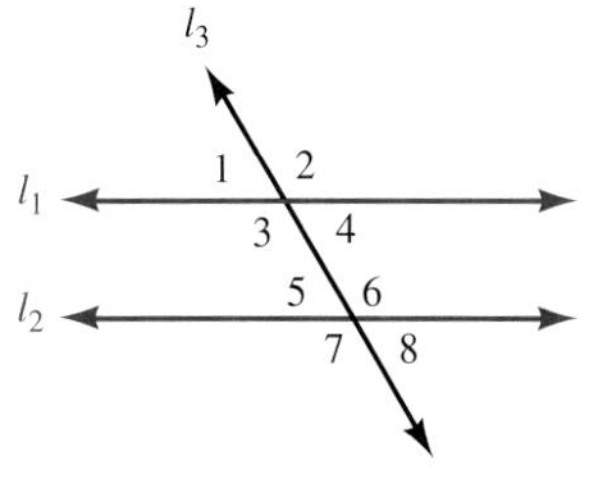

FIGURE 7-24

Self Check 2
If $l_1 \parallel l_2$ and $m(\angle 8) = 50°$, find the measures of the other angles. (See Figure 7-24.)

$m(\angle 5) = 60°$	If two parallel lines are cut by a transversal, alternate interior angles are congruent: $m(\angle 5) = m(\angle 4)$.
$m(\angle 6) = 120°$	If two parallel lines are cut by a transversal, alternate interior angles are congruent: $m(\angle 6) = m(\angle 3)$.
$m(\angle 7) = 120°$	Vertical angles are congruent: $m(\angle 7) = m(\angle 6)$.
$m(\angle 8) = 60°$	Vertical angles are congruent: $m(\angle 8) = m(\angle 5)$.

Answers $m(\angle 5) = 50°$, $m(\angle 7) = 130°$, $m(\angle 6) = 130°$, $m(\angle 3) = 130°$, $m(\angle 4) = 50°$, $m(\angle 1) = 50°$, $m(\angle 2) = 130°$

EXAMPLE 3 See Figure 7-25. If $\overline{AB} \parallel \overline{DE}$, which pairs of angles are congruent?

Solution Since $\overline{AB} \parallel \overline{DE}$, corresponding angles are congruent. So we have

$$\angle A \cong \angle 1 \quad \text{and} \quad \angle B \cong \angle 2$$

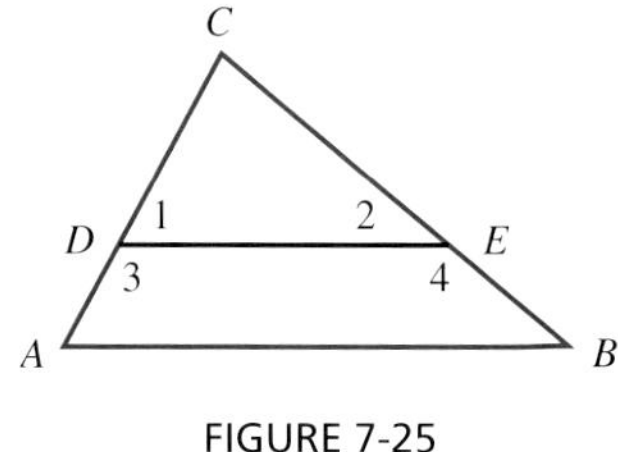

FIGURE 7-25

Self Check 4
In the figure below, $l_1 \parallel l_2$. Find y.

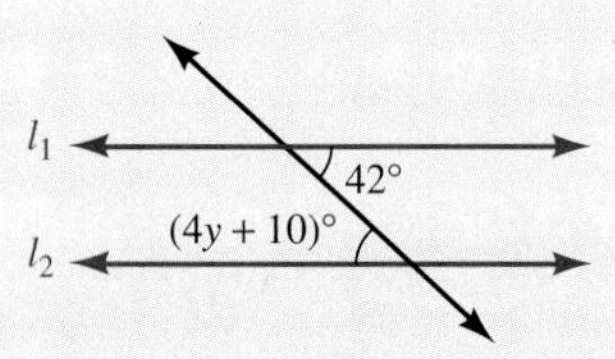

Answer 8

EXAMPLE 4 In Figure 7-26, $l_1 \parallel l_2$. Find x.

Solution The angles involving x are corresponding angles. Since $l_1 \parallel l_2$, all pairs of corresponding angles are congruent.

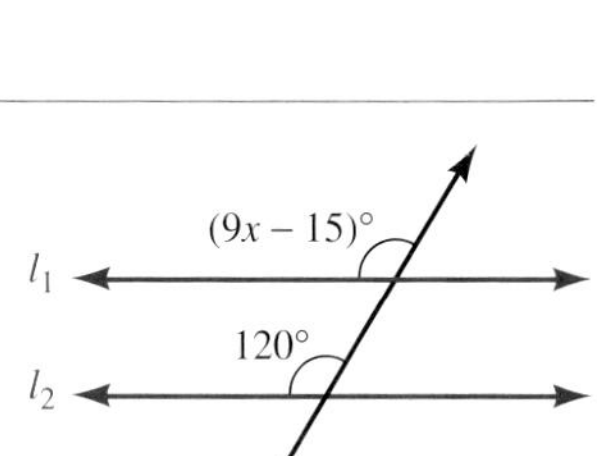

FIGURE 7-26

$9x - 15 = 120$	The angle measures are equal.
$9x = 135$	To undo the subtraction of 15, add 15 to both sides.
$x = 15$	To undo the multiplication by 9, divide both sides by 9.

Thus, x is 15.

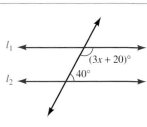

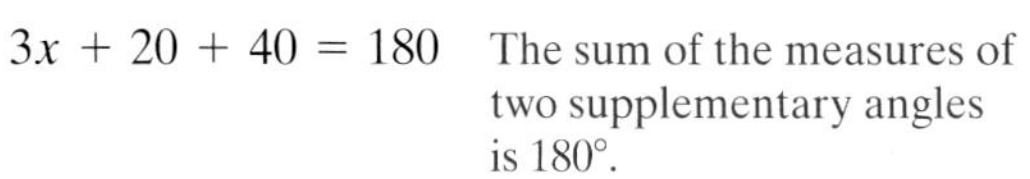

EXAMPLE 5 In Figure 7-27, $l_1 \parallel l_2$. Find x.

Solution Since the angles are interior angles on the same side of the transversal, they are supplementary.

$3x + 20 + 40 = 180$	The sum of the measures of two supplementary angles is 180°.
$3x + 60 = 180$	$20 + 40 = 60$.
$3x = 120$	To undo the addition of 60, subtract 60 from both sides.
$x = 40$	To undo the multiplication by 3, divide both sides by 3.

FIGURE 7-27

Thus, x is 40.

Section 7.2 STUDY SET

VOCABULARY *Fill in the blanks.*

1. Two lines in the same plane are ________.
2. ________ lines do not intersect.
3. If two lines intersect and form right angles, they are ________.
4. A ________ intersects two or more coplanar lines.
5. In the following illustration, $\angle 4$ and $\angle 6$ are ________ interior angles.
6. In the following illustration, $\angle 2$ and $\angle 6$ are ________ angles.

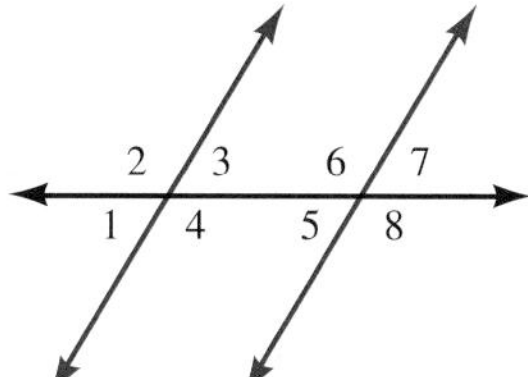

CONCEPTS

7. Which pairs of angles in the illustration above are alternate interior angles?
8. Which pairs of angles in the illustration above are corresponding angles?
9. Which angles shown in the illustration above are interior angles?
10. In the following illustration, $l_1 \parallel l_2$. What can you conclude about l_1 and l_3?

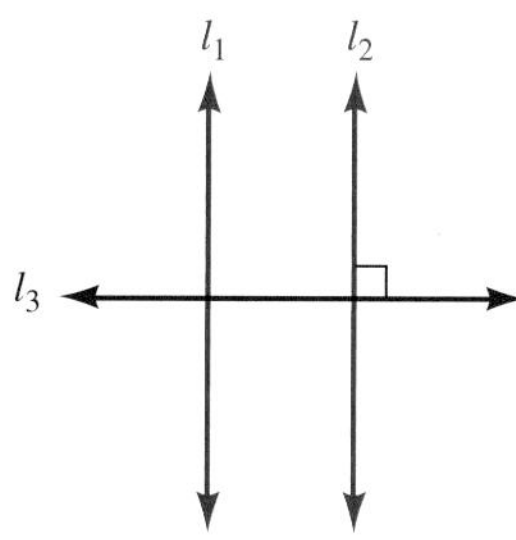

11. In the illustration, $l_1 \parallel l_2$ and $l_2 \parallel l_3$. What can you conclude about l_1 and l_3?

12. In the following illustration, $\overline{AB} \parallel \overline{DE}$. What pairs of angles are congruent?

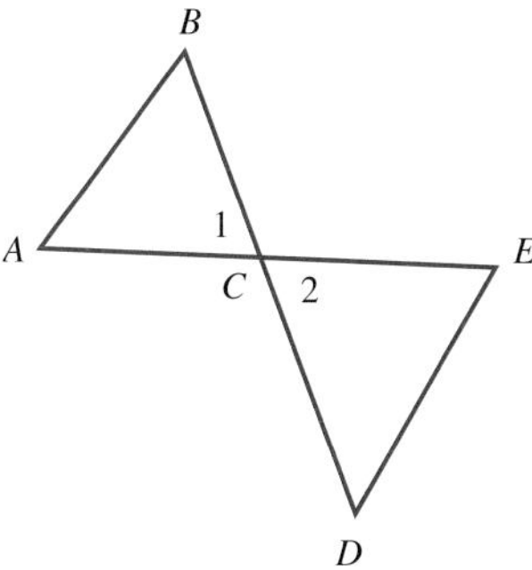

NOTATION *Fill in the blanks.*

13. The symbol ⅂ indicates ________.
14. The symbol $\parallel$ is read as "________."
15. The symbol $\perp$ is read as "________."
16. The symbol l_1 is read as "________."

PRACTICE

17. In the illustration, $l_1 \parallel l_2$ and $m(\angle 4) = 130°$. Find the measures of the other angles.

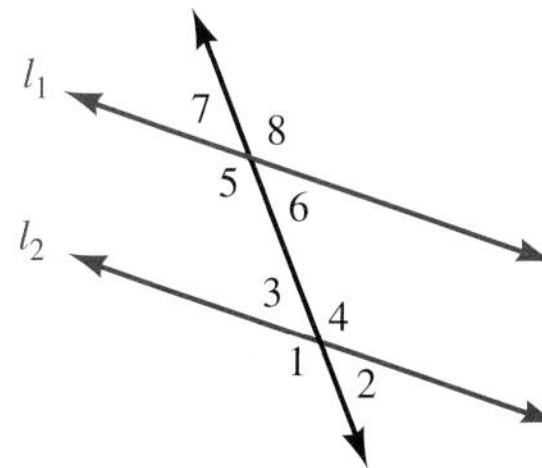

18. In the illustration, $l_1 \parallel l_2$ and $m(\angle 2) = 40°$. Find the measures of the other angles.

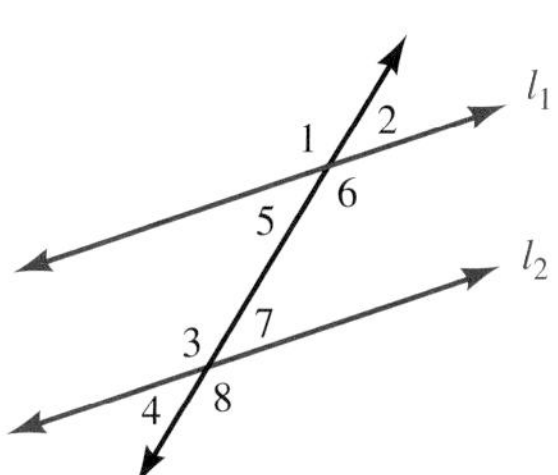

19. In the illustration, $l_1 \parallel \overrightarrow{AB}$. Find the measure of each angle.

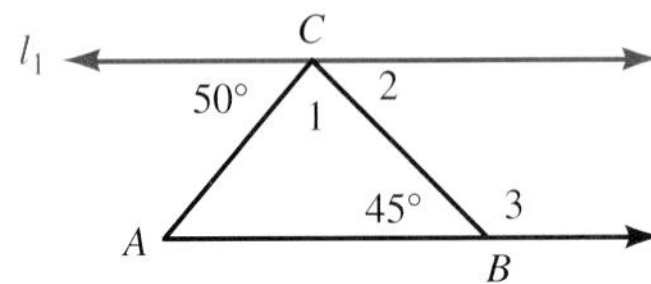

20. In the illustration, $\overline{AB} \parallel \overline{DE}$. Find m(∠$B$), m(∠$E$), and m(∠1).

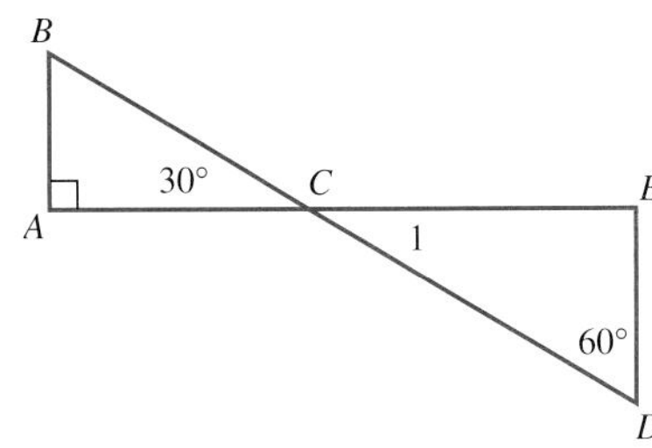

Find x given that $l_1 \parallel l_2$.

21.

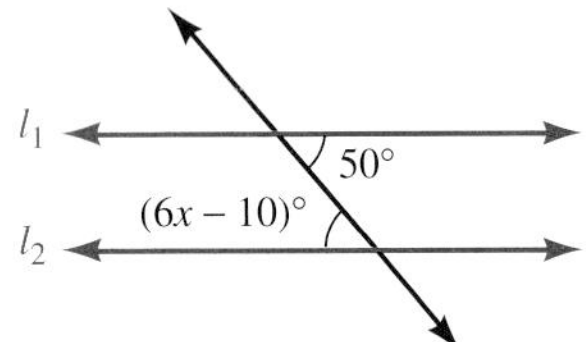

22.

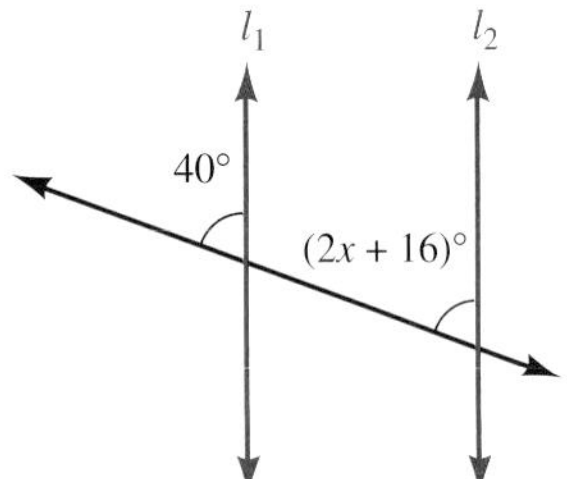

23.

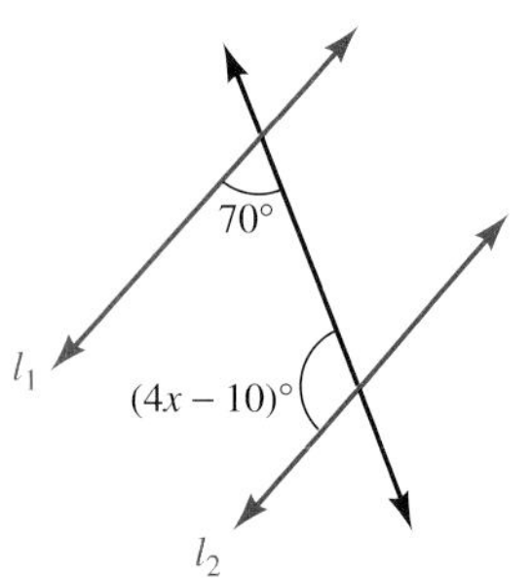

24.

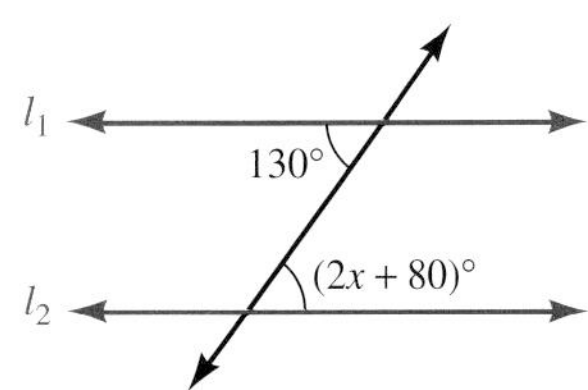

Find x.

25. $l_1 \parallel \overrightarrow{AC}$

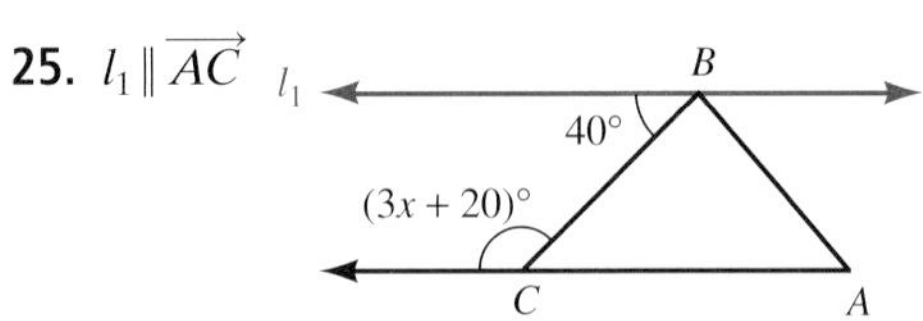

26. $\overline{AB} \parallel \overline{DE}$

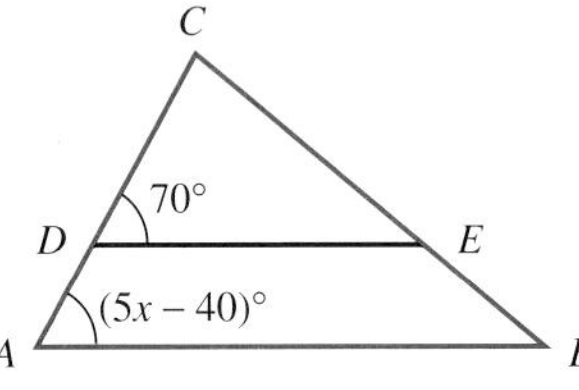

27. $\overline{AB} \parallel \overline{DE}$

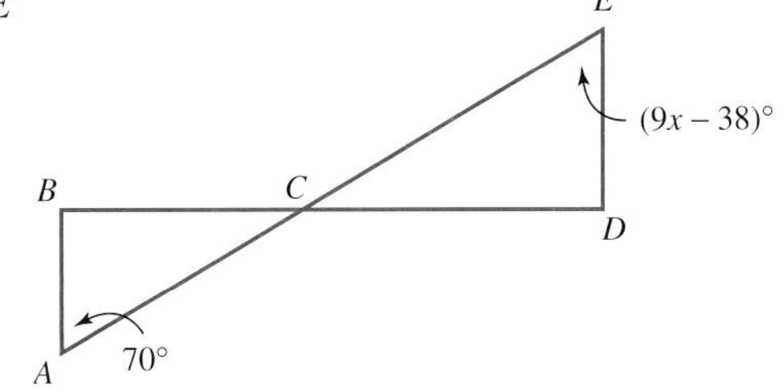

28. $\overline{AC} \parallel \overline{BD}$

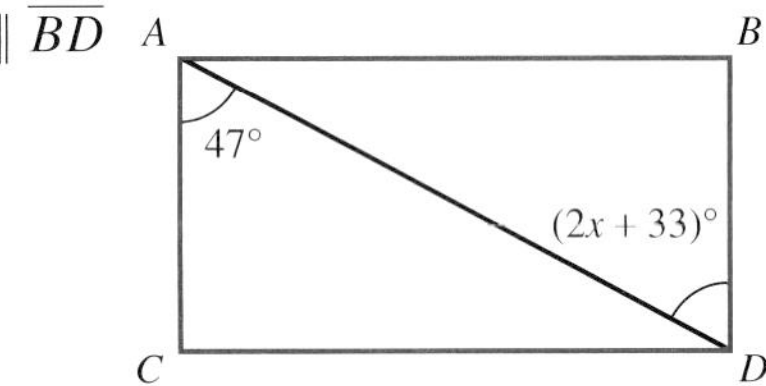

APPLICATIONS

29. CONSTRUCTING PYRAMIDS The Egyptians used a device called a **plummet** to determine whether stones were properly leveled. A plummet, shown in the illustration, is made up of an A-frame and a plumb bob suspended from the peak of the frame. How could a builder use a plummet to tell that the stone on the left is not level and that the stones on the right are level?

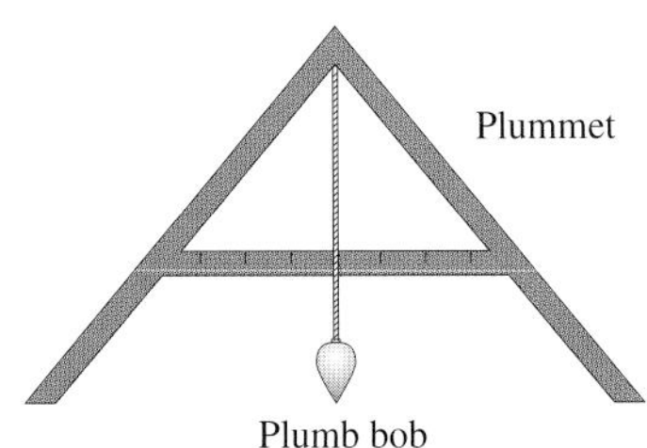

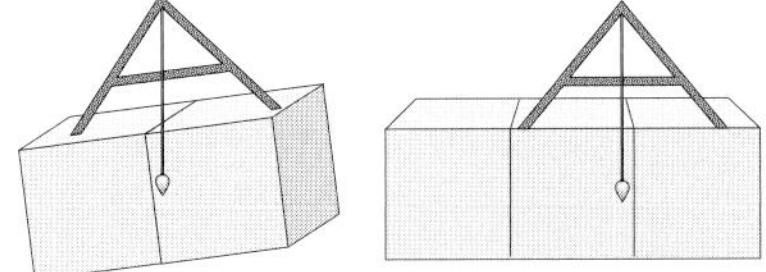

30. DIAGRAMMING SENTENCES English instructors have their students diagram sentences to help teach proper sentence structure. The illustration is a diagram of the sentence *The cave was rather dark and damp.* Point out pairs of parallel and perpendicular lines used in the diagram.

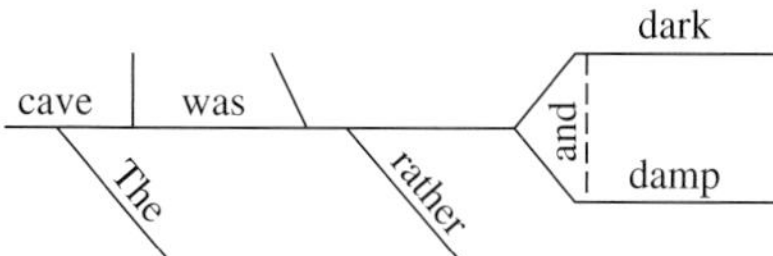

31. LOGOS Point out any perpendicular lines that can be found on the BMW company logo shown below.

32. PAINTING SIGNS For many sign painters, the most difficult letter to paint is a capital E, because of all of the right angles involved. How many right angles are there?

33. HANGING WALLPAPER Explain why the concepts of perpendicular and parallel are both important when hanging wallpaper.

34. TOOLS What geometric concepts are seen in the design of the rake?

WRITING

35. PARKING DESIGNS Using terms from this chapter, write a paragraph describing the parking layout shown in the next column.

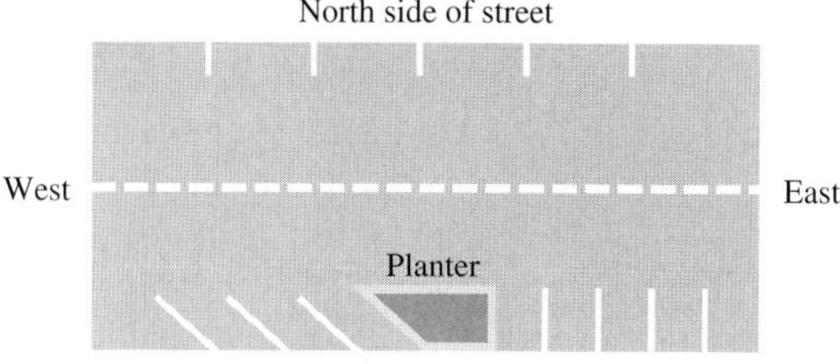

36. In your own words, explain what is meant by each of the following sentences.
    a. The hikers were told that the path *parallels* the river.
    b. John's quick rise to fame and fortune *paralleled* that of his older brother.
    c. The judge stated that the case that was before her court was without *parallel.*

37. Why do you think that $\angle 4$ and $\angle 6$ shown in the illustration for Problem 5 are called alternate interior angles?

38. Why do you think that $\angle 4$ and $\angle 8$ shown in the illustration for Problem 5 are called corresponding angles?

39. Are pairs of alternate interior angles always congruent? Explain.

40. Are pairs of interior angles always supplementary? Explain.

REVIEW

41. Find 60% of 120.

42. 80% of what number is 400?

43. What percent of 500 is 225?

44. Simplify: $3.45 + 7.37 \cdot 2.98$.

45. Is every whole number an integer?

46. Multiply: $2\frac{1}{5} \cdot 4\frac{3}{7}$.

7.3 Polygons

- Polygons • Triangles • Properties of isosceles triangles
- The sum of the measures of the angles of a triangle • Quadrilaterals
- Properties of rectangles • The sum of the measures of the angles of a polygon

In this section, we discuss figures called *polygons.* We see these shapes every day. For example, the walls in most buildings are rectangular. We also see rectangular shapes in doors, windows, and sheets of paper.

The gable ends of many houses are triangular, as are the sides of the Great Pyramid in Egypt. Triangular shapes are especially important because triangles are rigid and contribute strength and stability to walls and towers.

The designs in tile or linoleum floors often use the shapes of a pentagon or a hexagon. Stop signs are in the shape of an octagon.

Polygons

Polygon

A **polygon** is a closed geometric figure with at least three line segments for its sides.

The figures in Figure 7-28 are **polygons.** They are classified according to the number of sides they have. The points where the sides intersect are called **vertices.** If a polygon has sides that are all the same length and angles that have the same measure, we call it a **regular polygon.**

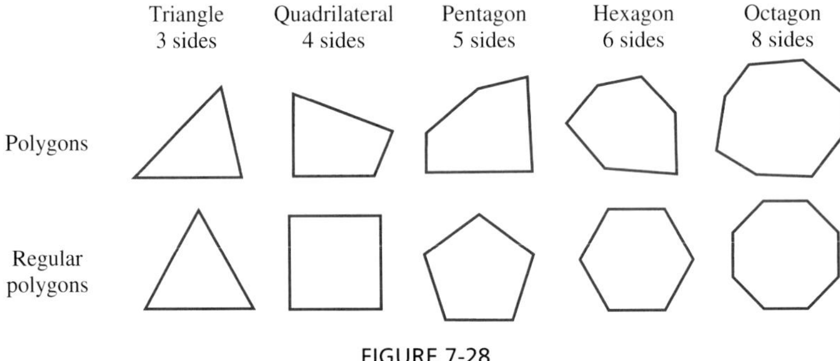

FIGURE 7-28

Self Check 1

Give the number of vertices of

a. a quadrilateral

b. a pentagon

Answers **a.** 4, **b.** 5

EXAMPLE 1 Give the number of vertices of **a.** a triangle and **b.** a hexagon.

Solution

a. From Figure 7-28, we see that a triangle has three angles and therefore three vertices.

b. From Figure 7-28, we see that a hexagon has six angles and therefore six vertices.

From the results of Example 1, we see that the number of vertices of a polygon is equal to the number of its sides.

Triangles

A **triangle** is a polygon with three sides. Figure 7-29 illustrates some common triangles. The slashes on the sides of a triangle indicate which sides are of equal length.

Equilateral triangle (all sides equal length)

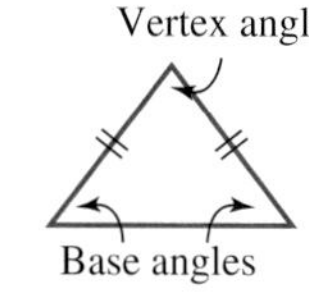

Isosceles triangle (at least two sides of equal length)

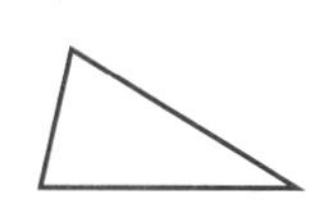

Scalene triangle (no sides equal length)

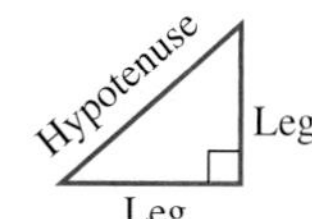

Right triangle (has a right angle)

FIGURE 7-29

! COMMENT Since equilateral triangles have at least two sides of equal length, they are also isosceles. However, isosceles triangles are not necessarily equilateral.

Since every angle of an equilateral triangle has the same measure, an equilateral triangle is also **equiangular.**

In an isosceles triangle, the angles opposite the sides of equal length are called **base angles,** the sides of equal length form the **vertex angle,** and the third side is called the **base.**

The longest side of a right triangle is called the **hypotenuse,** and the other two sides are called **legs.** The hypotenuse of a right triangle is always opposite the 90° angle.

Properties of isosceles triangles

1. Base angles of an isosceles triangle are congruent.
2. If two angles in a triangle are congruent, the sides opposite the angles have the same length, and the triangle is isosceles.

EXAMPLE 2 Is the triangle in Figure 7-30 an isosceles triangle?

Solution $\angle A$ and $\angle B$ are angles of the triangle. Since $m(\angle A) = m(\angle B)$, we know that $m(\overline{AC}) = m(\overline{BC})$ and that $\triangle ABC$ (read as "triangle ABC") is isosceles.

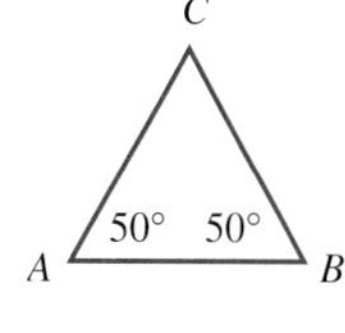

FIGURE 7-30

Self Check 2
In the figure below, $l_1 \parallel \overline{AB}$. Is the triangle an isosceles triangle?

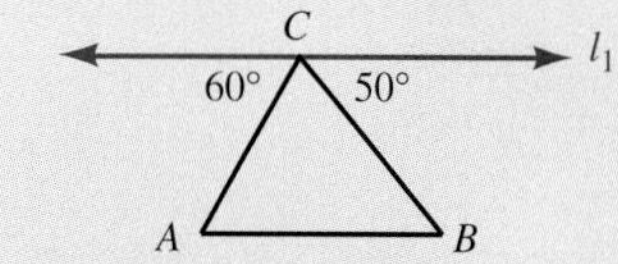

Answer no

The sum of the measures of the angles of a triangle

If you draw several triangles and carefully measure each angle with a protractor, you will find that the sum of the angle measures in each triangle is 180°.

Angles of a triangle
The sum of the angle measures of any triangle is 180°.

Self Check 3
In the figure below, find y.

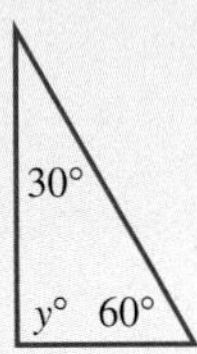

Answer 90

EXAMPLE 3 See Figure 7-31. Find x.

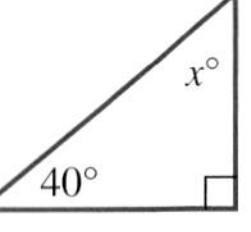

FIGURE 7-31

Solution Since the sum of the angle measures of any triangle is 180°, we have

$x + 40 + 90 = 180$

$x + 130 = 180$ Perform the addition: $40 + 90 = 130$.

$x = 50$ To undo the addition of 130, subtract 130 from both sides.

Thus, $x = 50$.

EXAMPLE 4 See Figure 7-32. If one base angle of an isosceles triangle measures 70°, how large is the vertex angle?

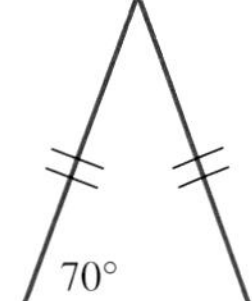

FIGURE 7-32

Solution Since one of the base angles measures 70°, so does the other. If we let x represent the measure of the vertex angle, we have

$x + 70 + 70 = 180$ The sum of the measures of the angles of a triangle is 180°.

$x + 140 = 180$ Perform the addition: $70 + 70 = 140$.

$x = 40$ To undo the addition of 140, subtract 140 from both sides.

The vertex angle measures 40°.

Quadrilaterals

A **quadrilateral** is a polygon with four sides. Some common quadrilaterals are shown in Figure 7-33.

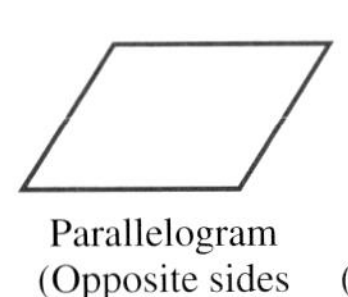

Parallelogram (Opposite sides parallel)

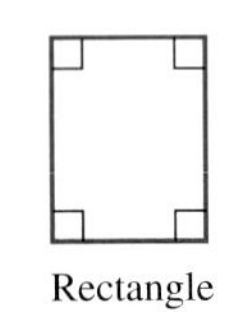

Rectangle (Parallelogram with four right angles)

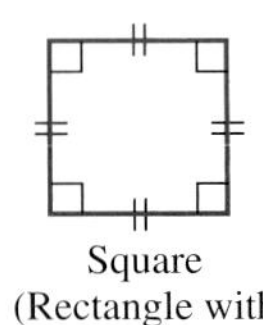

Square (Rectangle with sides of equal length)

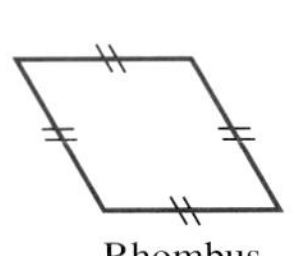

Rhombus (Parallelogram with sides of equal length)

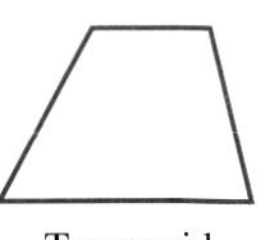

Trapezoid (Exactly two sides parallel)

FIGURE 7-33

A basic quadrilateral is the parallelogram.

Properties of parallelograms

1. The opposite sides of a parallelogram are parallel.
2. The opposite sides of a parallelogram are congruent.
3. The opposite angles of a parallelogram are congruent.
4. The consecutive angles of a parallelogram are supplementary.
5. The diagonals of a parallelogram bisect each other.
6. The diagonals of a parallelogram divide the parallelogram into two triangles with the same size and the same shape.

EXAMPLE 5 Refer to parallelogram $ABCD$. If $m(\angle A) = 60°$, $m(\overline{AB}) = 8$ cm, and $m(\overline{AD}) = 6$, find the measures of the other angles and sides.

$m(\angle B) = 120°$ by Property 4

$m(\angle C) = 60°$ by Property 3

$m(\angle D) = 120°$ by Property 4

$m(\overline{BC}) = 6$ cm by Property 2

$m(\overline{DC}) = 8$ cm by Property 2

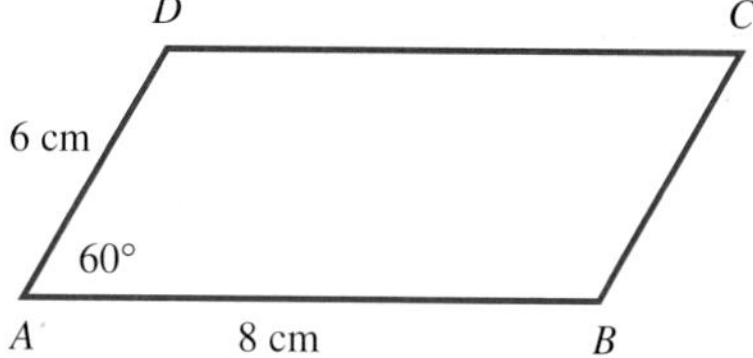

Self Check 5

Refer to parallelogram $ABCD$, where $m(\overline{DE}) = 5$ cm. Find $m(\overline{EB})$.

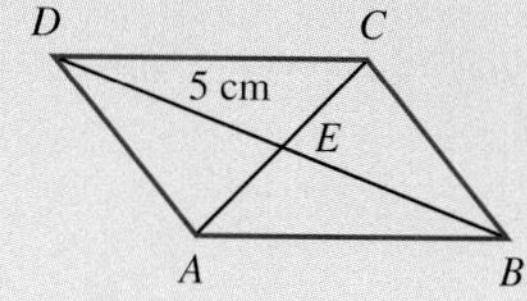

Answer 5 cm

If a parallelogram has one right angle, it is called a **rectangle.** A rectangle has all of the properties of a parallelogram plus a few more.

Properties of rectangles

1. The opposite sides of a rectangle are parallel.
2. The opposite sides of a rectangle are congruent.
3. All angles of a rectangle are right angles.
4. The diagonals of a rectangle bisect each other and are congruent.
5. If the diagonals of a parallelogram are congruent, the parallelogram is a rectangle.
6. The diagonals of a rectangle divide the rectangle into two triangles with the same size and the same shape.

EXAMPLE 6 **Squaring a foundation.** A carpenter intends to build a shed with an 8-by-12-foot base. How can he make sure that the rectangular foundation is square?

Solution See Figure 7-34. The carpenter can use a tape measure to find the lengths of diagonals AC and BD. If these diagonals are of equal length, the figure will be a rectangle and have four right angles. Then the foundation will be square.

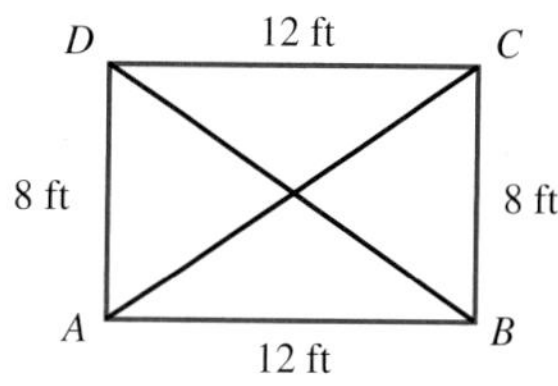

FIGURE 7-34

EXAMPLE 7 In rectangle $ABCD$ (Figure 7-35), the length of $\overline{AC}$ is 20 centimeters. Find each measure: **a.** $m(\overline{BD})$, **b.** $m(\angle 1)$, and **c.** $m(\angle 2)$.

FIGURE 7-35

Solution

a. Since the diagonals of a rectangle are of equal length, $m(\overline{BD})$ is also 20 centimeters.

b. We let $m(\angle 1) = x$. Then, since the angles of a rectangle are right angles, we have

$$x + 30 = 90$$

$$x = 60 \quad \text{To undo the addition of 30, subtract 30 from both sides.}$$

Thus, $m(\angle 1) = 60°$.

Self Check 7

In rectangle $ABCD$ (Figure 7-35), the length of $\overline{DC}$ is 16 centimeters. Find each measure:

a. $m(\overline{AB})$

b. $m(\angle 3)$

c. $m(\angle 4)$

c. We let $m(\angle 2) = y$. Then, since the sum of the angle measures of a triangle is 180°, we have

$$30 + 30 + y = 180$$
$$60 + y = 180 \quad \text{Simplify: } 30 + 30 = 60.$$
$$y = 120 \quad \text{To undo the addition of 60, subtract 60 from both sides.}$$

Thus, $m(\angle 2) = 120°$.

Answers **a.** 16 cm, **b.** 120°, **c.** 60°

The parallel sides of a trapezoid are called **bases,** the nonparallel sides are called **legs,** and the angles on either side of a base are called **base angles.** If the nonparallel sides are the same length, the trapezoid is an **isosceles trapezoid.** In an isosceles trapezoid, the base angles are congruent.

EXAMPLE 8 **Cross section of a drainage ditch.** A cross section of a drainage ditch (Figure 7-36) is an isosceles trapezoid with $\overline{AB} \parallel \overline{CD}$. Find x and y.

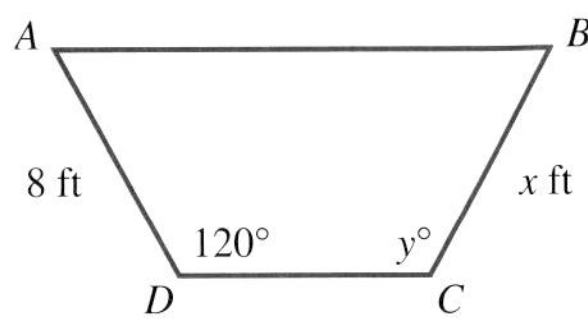

FIGURE 7-36

Solution Since the figure is an isosceles trapezoid, its nonparallel sides have the same length. So $m(\overline{AD})$ and $m(\overline{BC})$ are equal, and $x = 8$.

Since the base angles of an isosceles trapezoid are congruent, $m(\angle D) = m(\angle C)$. Thus, y is 120.

The sum of the measures of the angles of a polygon

We have seen that the sum of the angle measures of any triangle is 180°. Since a polygon with n sides can be divided into $n - 2$ triangles, the sum of the angle measures of the polygon in degrees is $(n - 2)180$.

Angles of a polygon

The sum S, in degrees, of the measures of the angles of a polygon with n sides is given by the formula

$$S = (n - 2)180 \quad \text{or} \quad S = 180n - 360$$

EXAMPLE 9 Find the sum of the angle measures of a pentagon.

Solution Since a pentagon has 5 sides, we substitute 5 for n in the formula and simplify.

$$S = (n - 2)180$$
$$= (5 - 2)180 \quad \text{Substitute 5 for } n.$$
$$= (3)180 \quad \text{Perform the subtraction within the parentheses.}$$
$$= 540$$

The sum of the angles of a pentagon is 540°.

Self Check 9
Find the sum of the angle measures of a quadrilateral.

Answer 360°

EXAMPLE 10 The sum of the measures of the angles of a polygon is 1,080°. Find the number of sides the polygon has.

Solution To find the number of sides the polygon has, we substitute 1,080 for S in the formula $S = 180n - 360$ and then solve for n.

$$S = 180n - 360$$

$$\mathbf{1{,}080} = 180n - 360 \quad \text{Substitute 1,080 for } S.$$

$$1{,}080 + \mathbf{360} = 180n - 360 + \mathbf{360} \quad \text{To undo the subtraction of 360, add 360 to both sides.}$$

$$1{,}440 = 180n \quad \text{Simplify.}$$

$$\frac{1{,}440}{\mathbf{180}} = \frac{180n}{\mathbf{180}} \quad \text{To undo the multiplication of 180, divide both sides by 180.}$$

$$8 = n \quad \text{Perform the division: } \frac{1{,}440}{180} = 8.$$

The polygon has 8 sides. It is an octagon.

Self Check 10
The sum of the measures of the angles of a polygon is 720°. Find the number of sides the polygon has.

Answer 6

Section 7.3 STUDY SET

VOCABULARY *Fill in the blanks.*

1. A ________ polygon has sides that are all the same length and angles that all have the same measure.
2. A polygon with four sides is called a ____________. A ________ is a polygon with three sides.
3. A ________ is a polygon with six sides.
4. A polygon with five sides is called a ________.
5. An eight-sided polygon is an ________.
6. The points where the sides of a polygon intersect are called ________.
7. A triangle with three sides of equal length is called an __________ triangle.
8. An ________ triangle has two sides of equal length.
9. The longest side of a right triangle is the __________.
10. The ______ angles of an isosceles triangle have the same measure.
11. A ____________ with a right angle is a rectangle.
12. A rectangle with all sides of equal length is a _______.
13. A ________ is a parallelogram with four sides of equal length.
14. A ________ has two sides that are parallel and two sides that are not parallel.
15. The legs of an ________ trapezoid have the same length.
16. The ________ of a polygon is the distance around it.

CONCEPTS *Give the number of sides each polygon has and classify it as a triangle, quadrilateral, pentagon, hexagon, or octagon. Then give the number of vertices it has.*

17.

18.

19.

20.

21.

22.

23.

24.

Classify each triangle as an equilateral triangle, an isosceles triangle, a scalene triangle, or a right triangle.

25. 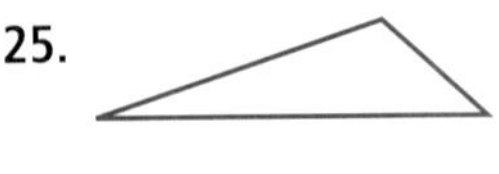

26. 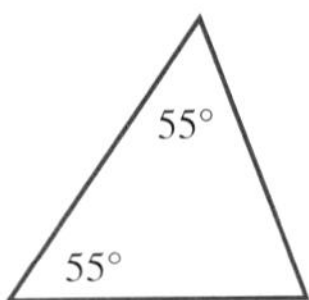

27. 

28. 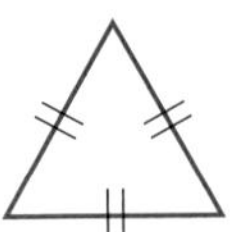

29. 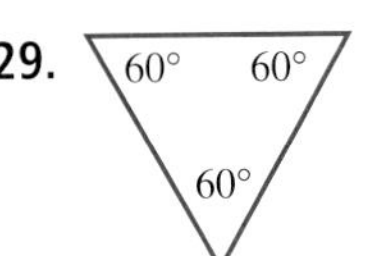

30. 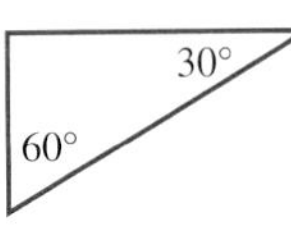

31. 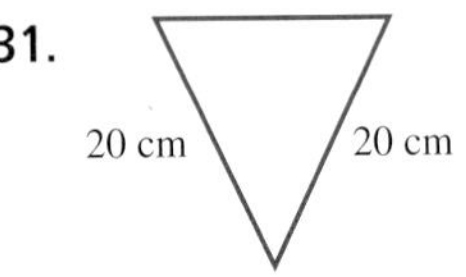

32. 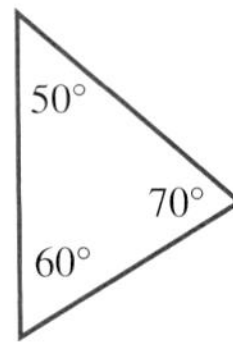

Classify each quadrilateral as a rectangle, a square, a rhombus, or a trapezoid. More than one name can be used for some figures.

33. 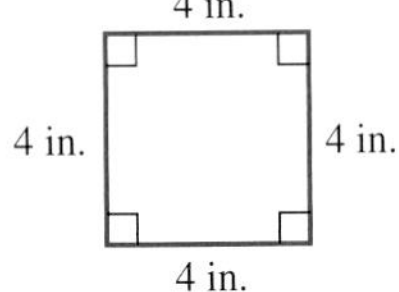

34. 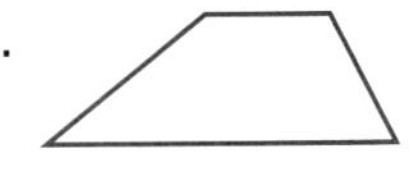

35. 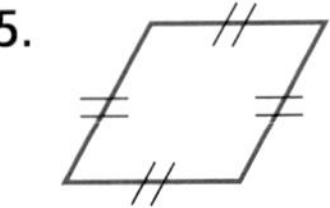

36. 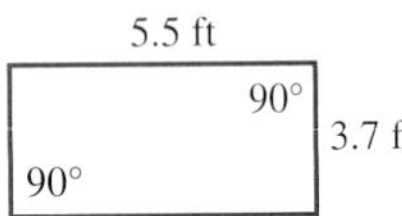

37. 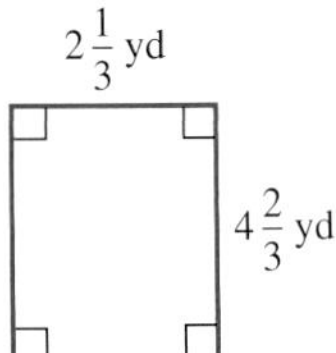

38. 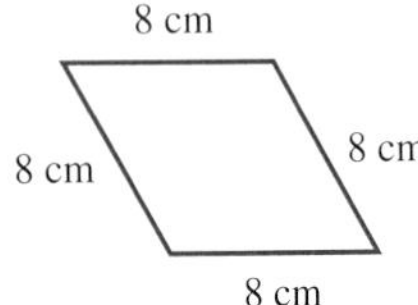

39. 

40. 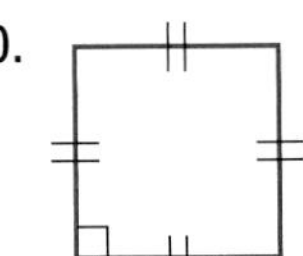

NOTATION *Fill in the blanks.*

41. The symbol △ means ________.

42. The symbol m(∠1) means the ________ of angle 1.

PRACTICE *See the following illustration. The measures of two angles of △ABC are given. Find the measure of the third angle.*

43. $m(\angle A) = 30°$ and $m(\angle B) = 60°$
$m(\angle C) =$ ______.

44. $m(\angle A) = 45°$ and $m(\angle C) = 105°$
$m(\angle B) =$ ______.

45. $m(\angle B) = 100°$ and $m(\angle A) = 35°$
$m(\angle C) =$ ______.

46. $m(\angle B) = 33°$ and $m(\angle C) = 77°$
$m(\angle A) =$ ______.

47. $m(\angle A) = 25.5°$ and $m(\angle B) = 63.8°$
$m(\angle C) =$ ______.

48. $m(\angle B) = 67.25°$ and
$m(\angle C) = 72.5°$
$m(\angle A) =$ ______.

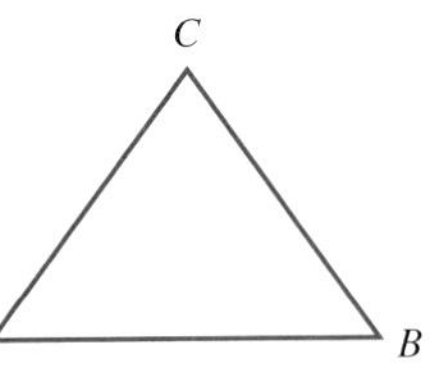

Refer to parallelogram ABCD shown below.

49. $m(\angle A) =$ ______

50. $m(\angle D) =$ ______

51. $m(\overline{AD}) =$ ______

52. $\overline{AB} \parallel$ ______

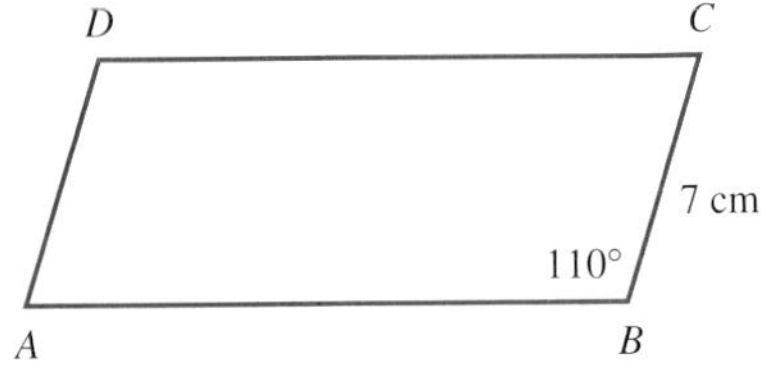

Refer to rectangle ABCD, shown below.

53. $m(\angle 1) =$ ______.

54. $m(\angle 3) =$ ______.

55. $m(\angle 2) =$ ______.

56. If $m(\overline{AC})$ is 8 cm, then $m(\overline{BD}) =$ ______.

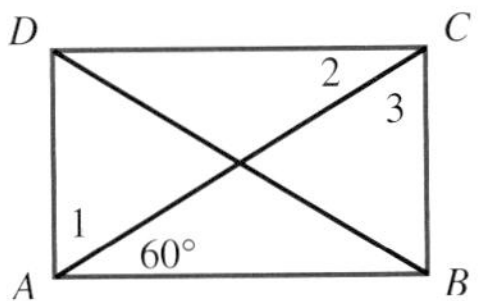

Find the sum of the angle measures of each polygon.

57. A hexagon

58. An octagon

59. A decagon (10 sides)

60. A dodecagon (12 sides)

Find the number of sides a polygon has if the sum of its angle measures is the given number.

61. 900° **62.** 1,260°
63. 2,160° **64.** 3,600°

APPLICATIONS

65. Give three uses of triangles in everyday life.

66. Give three uses of rectangles in everyday life.

67. Give three uses of squares in everyday life.

68. Give a use of a trapezoid in everyday life.

69. POLYGONS IN NATURE As we see in illustration (a), a starfish has the approximate shape of a pentagon. What approximate polygon shape do you see in each of the other objects? **b.** Lemon **c.** Chili pepper **d.** Apple

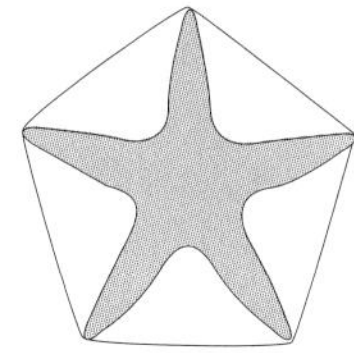
(a)

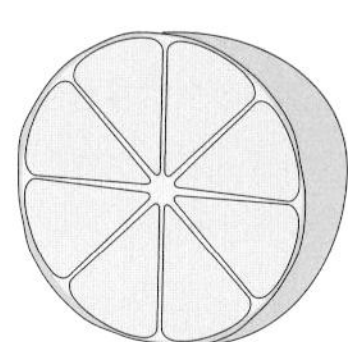
(b)

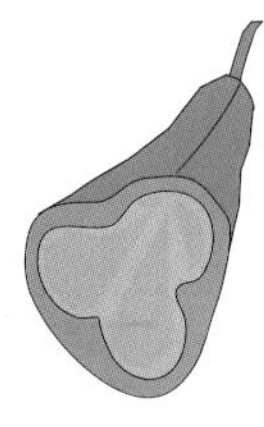
(c)

(d)

70. FLOWCHARTS A flowchart shows a sequence of steps to be performed by a computer to solve a given problem. When designing a flowchart, the programmer uses a set of standardized symbols to represent various operations to be performed by the computer. Locate a rectangle, a rhombus, and a parallelogram in the flowchart shown.

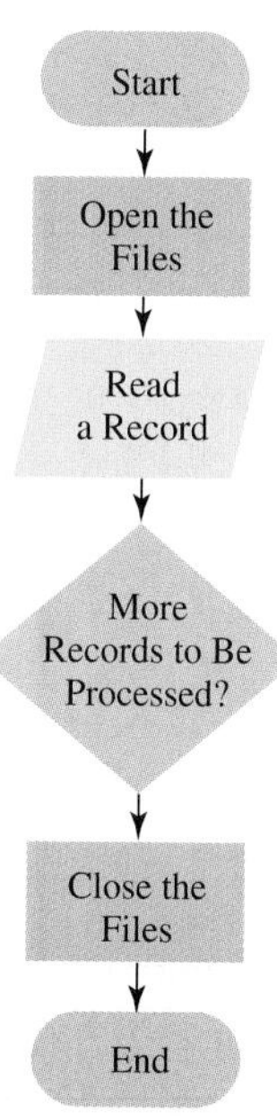

71. CHEMISTRY Polygons are used to represent the chemical structure of compounds graphically. In the illustration, what types of polygons are used to represent methylprednisolone, the active ingredient in an antiinflammatory medication?

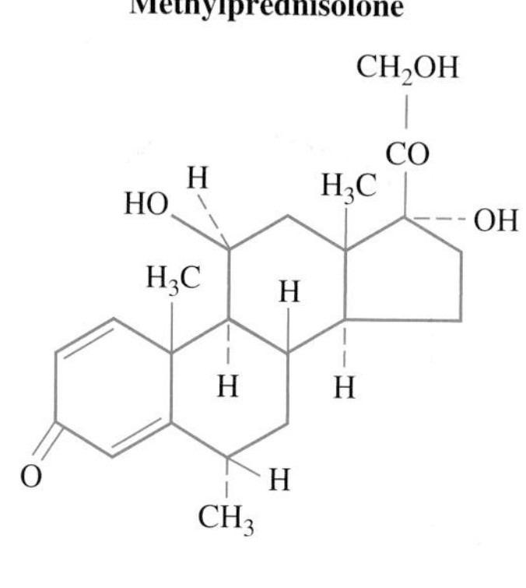

72. PODIUMS What polygon describes the shape of the upper portion of the podium?

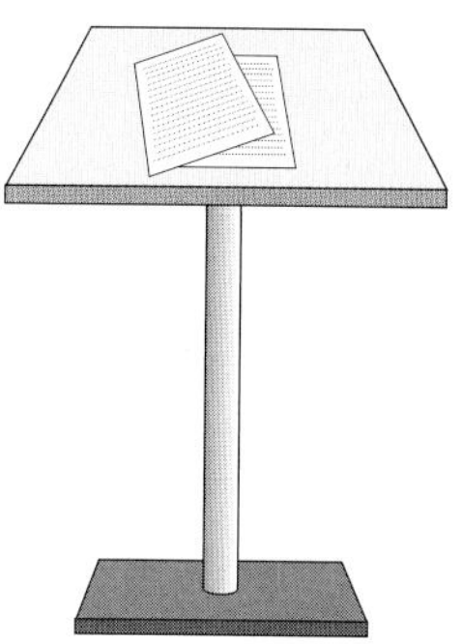

73. EASELS Show how two of the legs of the easel form the equal sides of an isosceles triangle.

74. AUTOMOBILE JACKS Refer to the illustration. Show that no matter how high the jack is raised, it always forms two isosceles triangles.

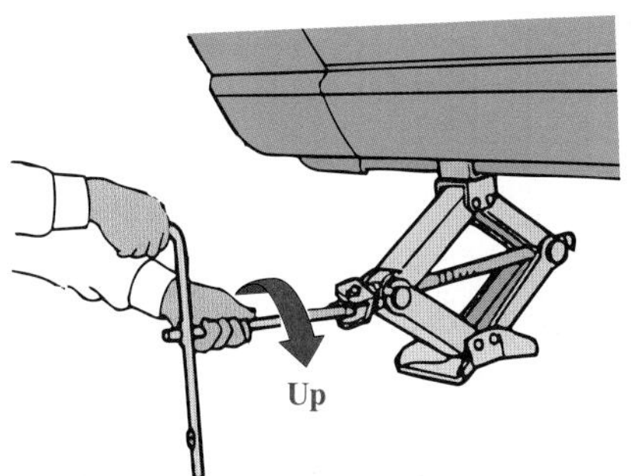

WRITING

75. Explain why a square is a rectangle.
76. Explain why a trapezoid is not a parallelogram.

REVIEW

77. Find 20% of 110.
78. Find 15% of 50.
79. What percent of 200 is 80?
80. 20% of what number is 500?
81. Simplify: $0.85 \div 2(0.25)$.
82. FIRST AID When checking an accident victim's pulse, a paramedic counted 13 beats during a 15-second span. How many beats would be expected in 60 seconds?

7.4 Properties of Triangles

- Congruent triangles
- The Pythagorean theorem

Proportions and triangles are often used to measure distances indirectly. For example, by using a proportion, Eratosthenes (275–195 B.C.) was able to estimate the circumference of the Earth with remarkable accuracy. On a sunny day, we can use properties of similar triangles to calculate the height of a tree while staying safely on the ground. By using a theorem proved by the Greek mathematician Pythagoras (about 500 B.C.), we can calculate the length of the third side of a right triangle whenever we know the lengths of two sides.

Congruent triangles

Triangles that have the same area and the same shape are called **congruent triangles.** In Figure 7-37, triangles ABC and DEF are congruent.

$\triangle ABC \cong \triangle DEF$ Read as "Triangle ABC is congruent to triangle DEF."

Corresponding angles and corresponding sides of congruent triangles are called **corresponding parts.** The notation $\triangle ABC \cong \triangle DEF$ shows which vertices are corresponding parts.

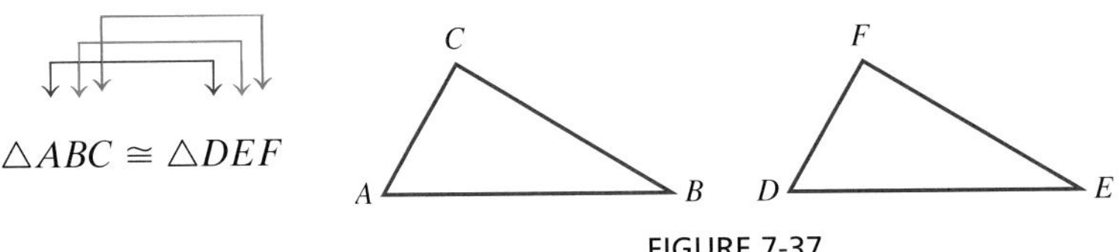

FIGURE 7-37

Corresponding parts of congruent triangles always have the same measure. For the congruent triangles shown in Figure 7-37,

$$m(\angle A) = m(\angle D), \quad m(\angle B) = m(\angle E), \quad m(\angle C) = m(\angle F),$$
$$m(\overline{BC}) = m(\overline{EF}), \quad m(\overline{AC}) = m(\overline{DF}), \quad m(\overline{AB}) = m(\overline{DE})$$

EXAMPLE 1 Name the corresponding parts of the congruent triangles in Figure 7-38.

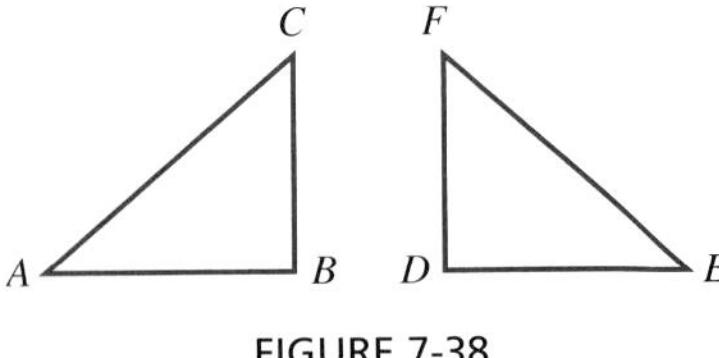

FIGURE 7-38

Solution The corresponding angles are

$$\angle A \text{ and } \angle E, \quad \angle B \text{ and } \angle D, \quad \angle C \text{ and } \angle F$$

Since corresponding sides are always opposite corresponding angles, the corresponding sides are

$\overline{BC}$ and $\overline{DF}$, $\overline{AC}$ and $\overline{EF}$, $\overline{AB}$ and $\overline{ED}$

We will discuss three ways of showing that two triangles are congruent.

SSS property

If three sides of one triangle are congruent to three sides of a second triangle, the triangles are congruent.

The triangles in Figure 7-39 are congruent because of the SSS property.

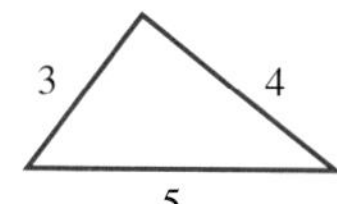

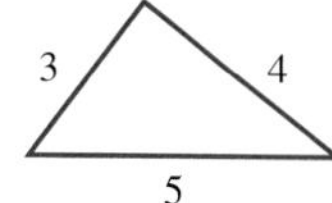

FIGURE 7-39

SAS property

If two sides and the angle between them in one triangle are congruent, respectively, to two sides and the angle between them in a second triangle, the triangles are congruent.

The triangles in Figure 7-40 are congruent because of the SAS property.

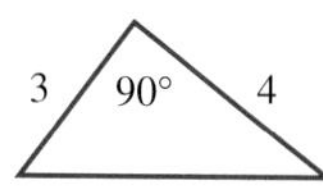

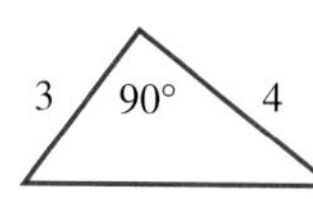

FIGURE 7-40

ASA property

If two angles and the side between them in one triangle are congruent, respectively, to two angles and the side between them in a second triangle, the triangles are congruent.

The triangles in Figure 7-41 are congruent because of the ASA property.

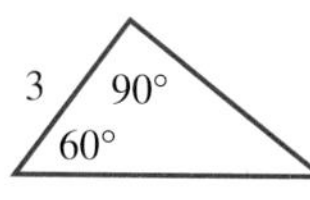

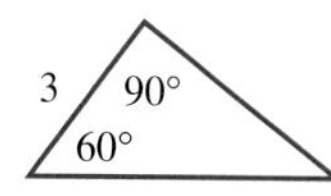

FIGURE 7-41

! COMMENT There is no SSA property. To illustrate this, consider the triangles in Figure 7-42. Two sides and an angle of $\triangle ABC$ are congruent to two sides and an angle of $\triangle DEF$. But the congruent angle is not between the congruent sides.

We refer to this situation as SSA. Obviously, the triangles are not congruent, because they have different areas.

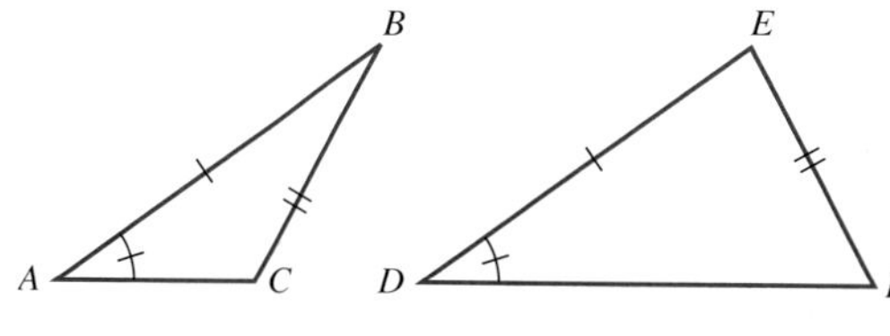

FIGURE 7-42

The slash marks indicate congruent parts. That is, the sides with one slash are the same length, the sides with two slashes are the same length, and the angles with one slash have the same measure.

EXAMPLE 2 Explain why the triangles in Figure 7-43 are congruent.

Solution Since vertical angles are congruent,

$$m(\angle 1) = m(\angle 2)$$

From the figure, we see that

$$m(\overline{AC}) = m(\overline{EC}) \quad \text{and} \quad m(\overline{BC}) = m(\overline{DC})$$

Since two sides and the angle between them in one triangle are congruent, respectively, to two sides and the angle between them in a second triangle, $\triangle ABC \cong \triangle EDC$ by the SAS property.

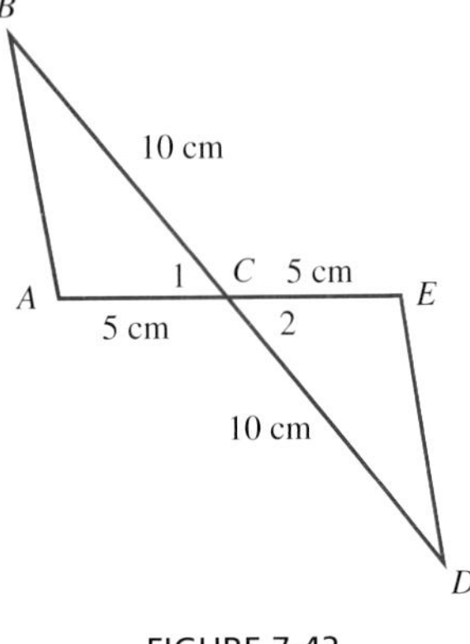

FIGURE 7-43

The Pythagorean theorem

In the movie *The Wizard of Oz,* the scarecrow was in search of a brain. To prove that he had found one, he tried to recite the Pythagorean theorem. In words, the Pythagoream theorem can be stated as:

In a right triangle, the square of the hypotenuse is equal to the sum of squares of the other two sides.

Pythagorean theorem

If the length of the hypotenuse of a right triangle is c and the lengths of its legs are a and b, then

$$a^2 + b^2 = c^2$$

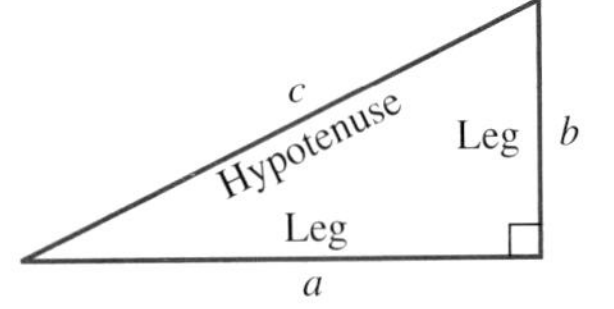

Self Check 3

A 26-foot ladder rests against the side of a building. If the base of the ladder is 10 feet from the wall, how far up the side of the building does the ladder reach?

EXAMPLE 3 Constructing a high-ropes adventure course. A builder of a high-ropes adventure course wants to secure the pole shown in Figure 7-44 by attaching a cable from the anchor stake 8 feet from its base to a point 6 feet up the pole. How long should the cable be?

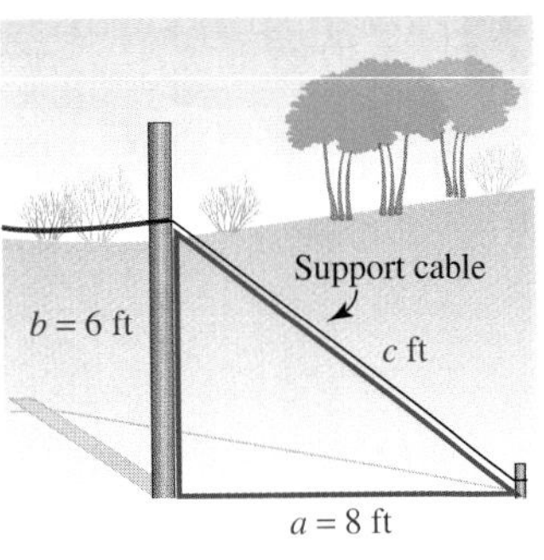

FIGURE 7-44

Solution The support cable, the pole, and the ground form a right triangle. If we let c represent the length of the cable (the hypotenuse), then we can use the Pythagorean theorem with $a = 8$ and $b = 6$ to find c.

$c^2 = a^2 + b^2$ This is the Pythagorean theorem.

$c^2 = 8^2 + 6^2$ Substitute 8 for a and 6 for b.

$c^2 = 64 + 36$ Evaluate the exponential expressions.

$c^2 = 100$ Simplify the right-hand side.

To find c, we must find a number that, when squared, is 100. There are two such numbers, one positive and one negative; they are the square roots of 100. Since c represents the length of a support cable, c cannot be negative. For this reason, we need only find the positive square root of 100 to get c.

$c^2 = 100$ This is the equation to solve.

$c = \sqrt{100}$ The symbol $\sqrt{}$ is used to indicate the positive square root of a number.

$c = 10$ $\sqrt{100} = 10$, because $10^2 = 100$.

The support cable should be 10 feet long.

Answer 24 ft

Finding the width of a television screen

CALCULATOR SNAPSHOT

The size of a television screen is the diagonal measure of its rectangular screen. (See Figure 7-45.) To find the width of a 27-inch screen that is 17 inches high, we use the Pythagorean theorem with $c = 27$ and $b = 17$.

$$c^2 = a^2 + b^2$$
$$\mathbf{27}^2 = a^2 + \mathbf{17}^2$$
$$27^2 - 17^2 = a^2$$

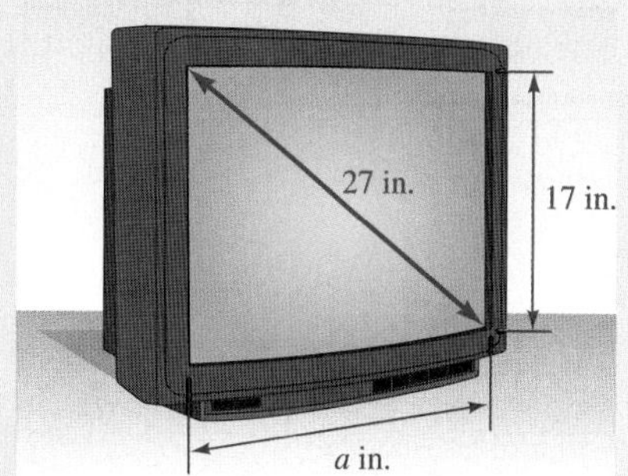

FIGURE 7-45

The variable a represents the width of a television screen, so it must be positive. To find a, we find the positive square root of the result when 17^2 is subtracted from 27^2. Using a radical symbol to indicate this, we have

$$\sqrt{27^2 - 17^2} = a$$

We can evaluate the expression on the left-hand side by entering these numbers and pressing these keys.

[( ] 27 [x^2] [−] 17 [x^2] [ )] [$\sqrt{}$] 20.97617696

To the nearest inch, the width of the television screen is 21 inches.

It is also true that

If the square of one side of a triangle is equal to the sum of the squares of the other two sides, the triangle is a right triangle.

EXAMPLE 4 Is a triangle with sides of 5, 12, and 13 meters a right triangle?

Solution We can use the Pythagorean theorem to answer this question. Since the longest side of the triangle is 13 meters, we must substitute 13 for c. It doesn't matter which of the two remaining side lengths we substitute for a and which we substitute for b.

$c^2 = a^2 + b^2$ This is the Pythagorean theorem.

$\mathbf{13}^2 \stackrel{?}{=} \mathbf{5}^2 + \mathbf{12}^2$ Substitute 13 for c, 5 for a, and 12 for b.

$169 \stackrel{?}{=} 25 + 144$ Evaluate the exponential expressions.

$169 = 169$ Simplify the right-hand side.

Since the square of the longest side is equal to the sum of the squares of the other two sides, the triangle is a right triangle.

Self Check 4

Is a triangle with sides of 9, 40, and 41 meters a right triangle?

Answer yes

Self Check 5
Is a triangle with sides of 4, 5, and 6 inches a right triangle?

Answer no

EXAMPLE 5 Is a triangle with sides of 2, 2, and 3 feet a right triangle?

Solution We check to see whether the square of the longest side is equal to the sum of the squares of the other two sides.

$c^2 = a^2 + b^2$	This is the Pythagorean theorem.
$3^2 \stackrel{?}{=} 2^2 + 2^2$	Substitute 3 for c, 1 for a, and 2 for b.
$9 \stackrel{?}{=} 4 + 4$	Evaluate the exponential expressions.
$9 \neq 8$	Simplify the right-hand side.

Since the square of the longest side is not equal to the sum of the squares of the other two sides, the triangle is not a right triangle.

Section 7.4 STUDY SET

VOCABULARY *Fill in the blanks.*

1.  triangles are the same size and the same shape.
2. All ____________ parts of congruent triangles have the same measure.
3. If a triangle has an angle that measures 90°, it is called a ______ triangle.
4. The ____________ is the longest side of a right triangle.

CONCEPTS *Determine whether each statement is true. If a statement is false, tell why.*

5. If three sides of one triangle are the same length as three sides of a second triangle, the triangles are congruent.
6. If two sides of one triangle are the same length as two sides of a second triangle, the triangles are congruent.
7. If two sides and an angle of one triangle are congruent, respectively, to two sides and an angle of a second triangle, the triangles are congruent.
8. If two angles and the side between them in one triangle are congruent, respectively, to two angles and the side between them in a second triangle, the triangles are congruent.
9. Are the triangles shown on the right congruent?

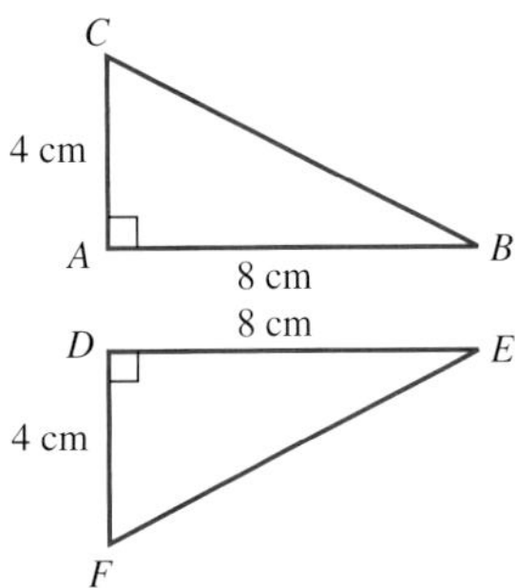

10. Are the triangles shown below congruent?

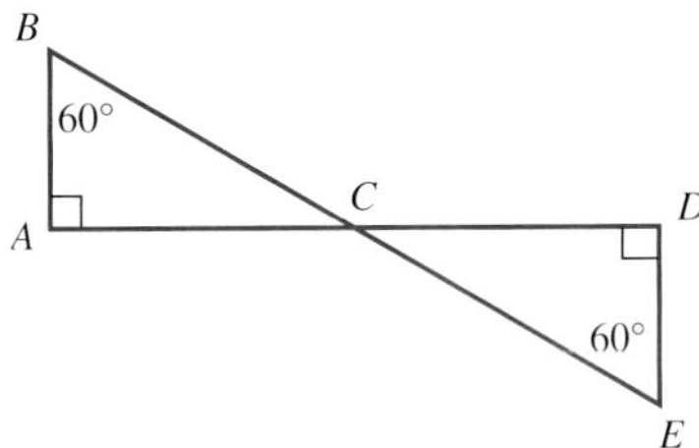

11. The Pythagorean theorem states that for a right triangle, $c^2 = a^2 + b^2$. What do the variables a, b, and c represent?
12. A triangle has sides of length 3, 4, and 5 centimeters. Substitute the lengths into $c^2 = a^2 + b^2$ and show that a true statement results. From the result, what can we conclude about the triangle?

NOTATION *Fill in the blanks.*

13. The symbol ≅ is read as "____ __________ ____."
14. The symbol $m(\angle A)$ is read as "_____ _________ ____ angle A."

PRACTICE *Name the corresponding parts of the congruent triangles.*

15. Refer to the illustration.
$\overline{AC}$ corresponds to ____.
$\overline{DE}$ corresponds to ____.
$\overline{BC}$ corresponds to ____.
$\angle A$ corresponds to ____.
$\angle E$ corresponds to ____.
$\angle F$ corresponds to ____.

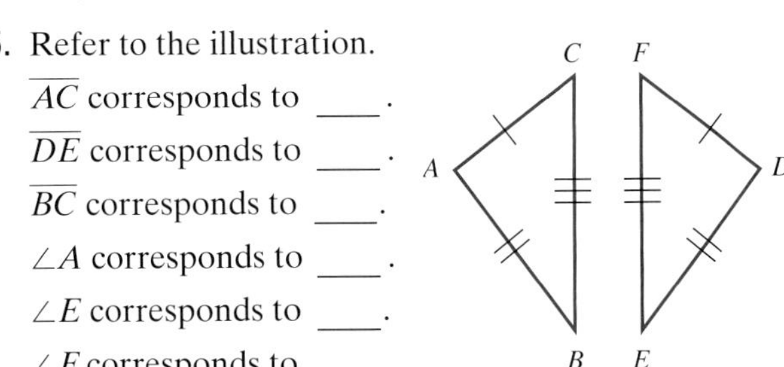

16. Refer to the illustration.

$\overline{AB}$ corresponds to ____.
$\overline{EC}$ corresponds to ____.
$\overline{AC}$ corresponds to ____.
$\angle D$ corresponds to ____.
$\angle B$ corresponds to ____.
$\angle 1$ corresponds to ____.

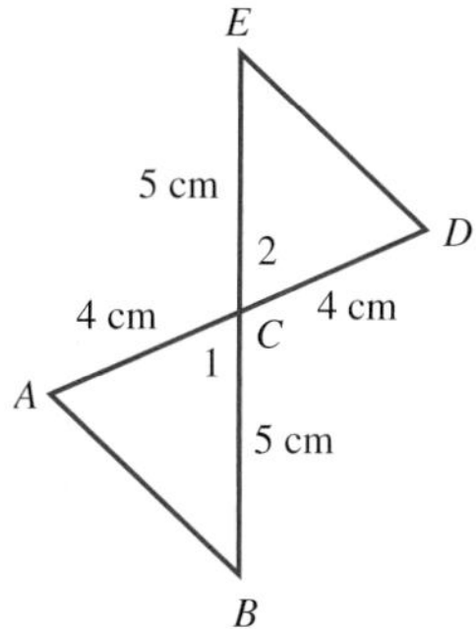

Determine whether each pair of triangles is congruent. If they are explain why.

17.

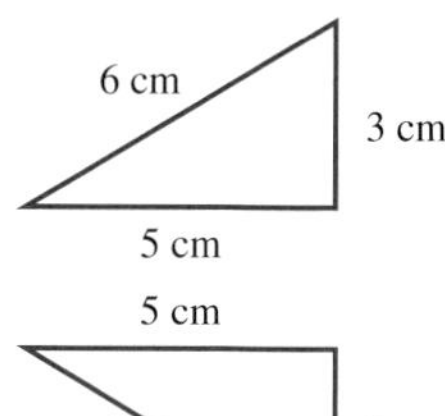

18.

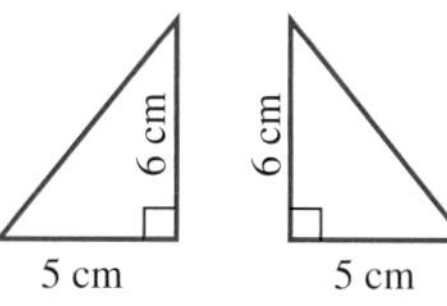

19.

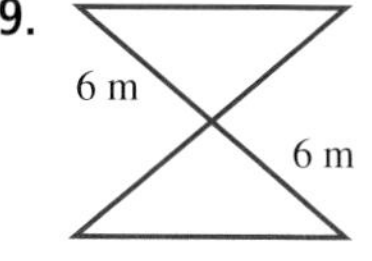

20.

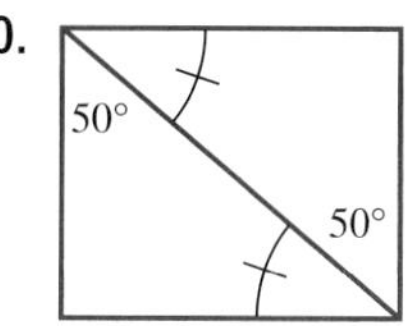

21.

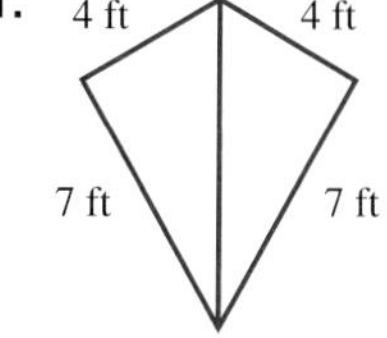

22.

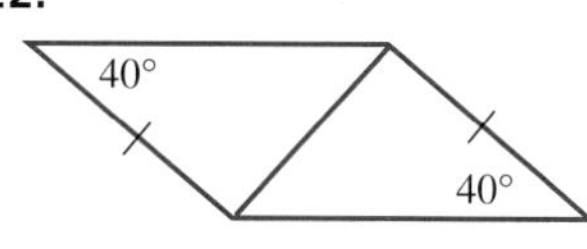

23.

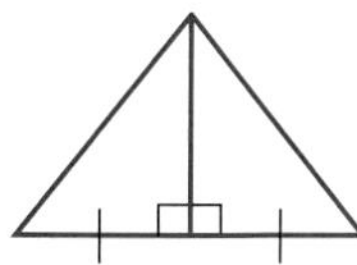

24.

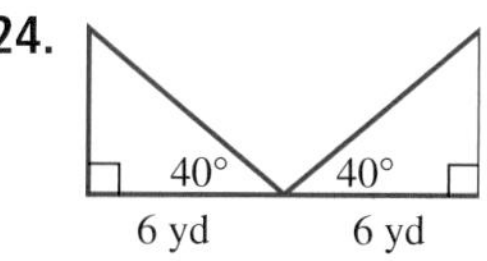

Find x.

25.

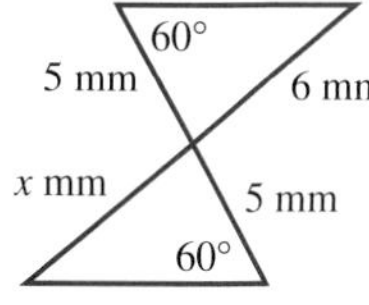

26.

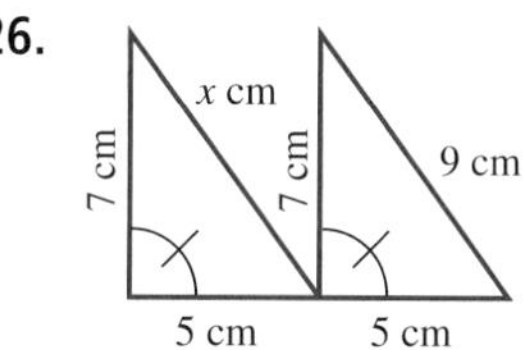

27.

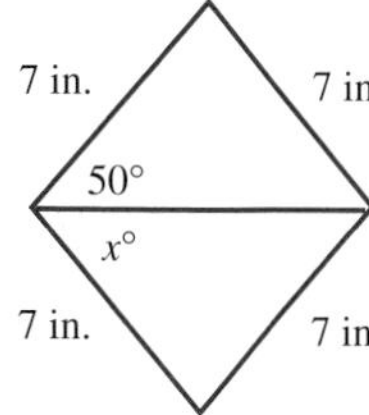

28.

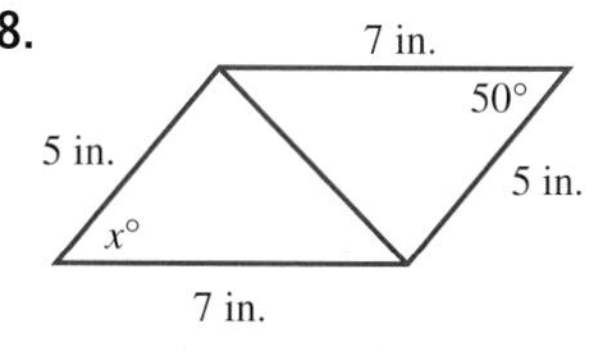

Refer to the illustration below and find the length of the unknown side.

29. $a = 3$ and $b = 4$. Find c.

30. $a = 12$ and $b = 5$. Find c.

31. $a = 15$ and $c = 17$. Find b.

32. $b = 45$ and $c = 53$. Find a.

33. $a = 5$ and $c = 9$. Find b.

34. $a = 1$ and $b = 7$. Find c.

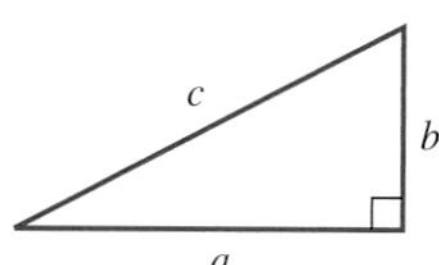

The lengths of the three sides of a triangle are given. Determine whether the triangle is a right triangle.

35. 8, 15, 17

36. 6, 8, 10

37. 7, 24, 26

38. 9, 39, 40

APPLICATIONS *Solve each problem. If an answer is not exact, give the answer to the nearest tenth.*

39. ADJUSTING LADDERS A 20-foot ladder reaches a window 16 feet above the ground. How far from the wall is the base of the ladder?

40. LENGTH OF GUY WIRES A 30-foot tower is to be fastened by three guy wires attached to the top of the tower and to the ground at positions 20 feet from its base. How much wire is needed?

41. PICTURE FRAMES After gluing and nailing two pieces of picture frame molding together, a frame maker checks her work by making a diagonal measurement. If the sides of the frame form a right angle, what measurement should the frame maker read on the yardstick?

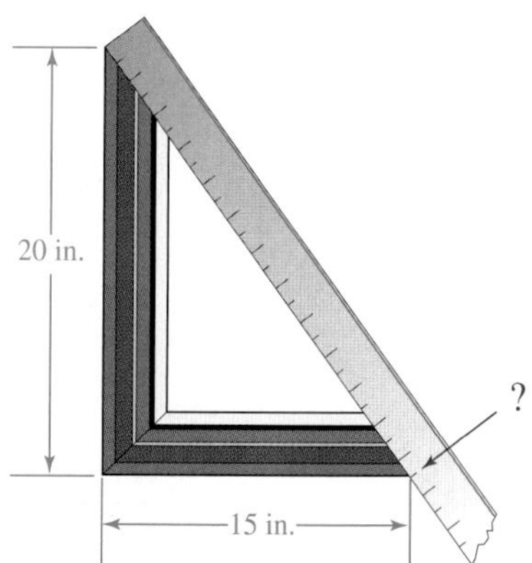

42. CARPENTRY The gable end of the roof shown is divided in half by a vertical brace, 8 feet in height. Find the length of the roof line.

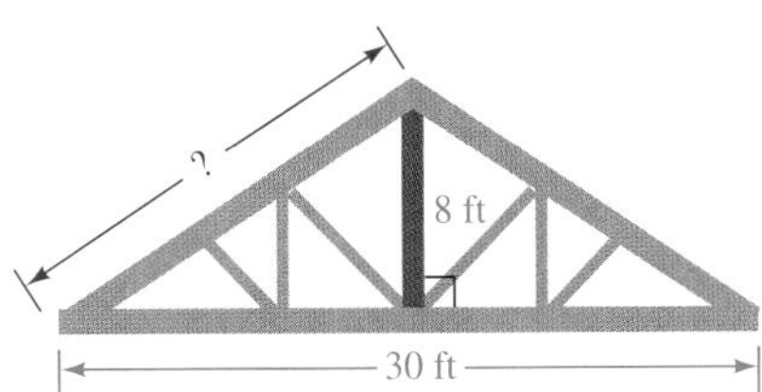

43. BASEBALL A baseball diamond is a square with each side 90 feet long. How far is it from home plate to second base?

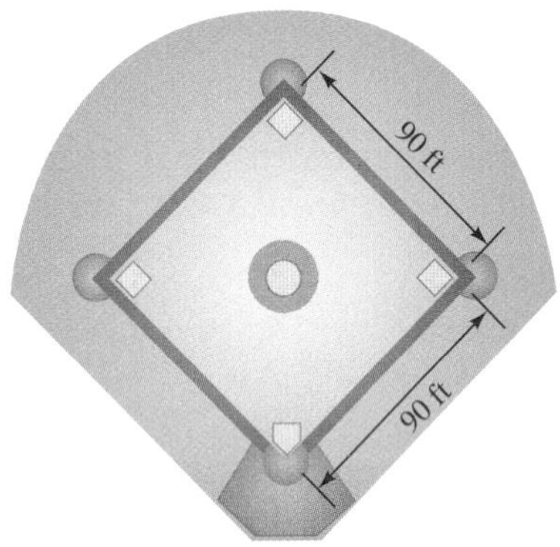

44. TELEVISION What size is the television screen shown?

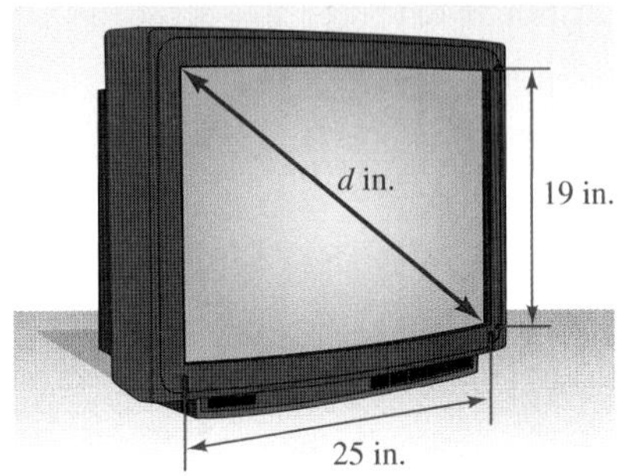

WRITING

45. Explain the Pythagorean theorem.

46. Explain the procedure used to solve the equation $c^2 = 64$. (Assume that c is positive.)

REVIEW *Estimate the answer to each problem.*

47. $\dfrac{0.95 \cdot 3.89}{2.997}$

48. 21% of 42

49. 32% of 60

50. $\dfrac{4.966 + 5.001}{2.994}$

51. 49.5% of 18.1

52. 98.7% of 0.03

7.5 Perimeters and Areas of Polygons

- Perimeters of polygons • Perimeters of figures that are combinations of polygons
- Areas of polygons • Areas of figures that are combinations of polygons

In this section, we discuss how to find perimeters and areas of polygons. Finding perimeters is important when estimating the cost of fencing or the cost of woodwork in a house. Finding areas is important when calculating the cost of carpeting, the cost of painting a house, or the cost of fertilizing a yard.

Perimeters of polygons

Recall that the **perimeter** of a polygon is the distance around it. Since a square has four sides of equal length s, its perimeter P is $s + s + s + s$, or $4s$.

Perimeter of a square

If a square has a side of length s, its perimeter P is given by the formula

$$P = 4s$$

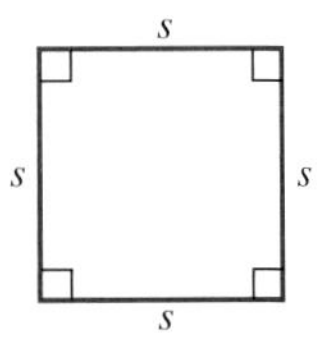

EXAMPLE 1 Find the perimeter of a square whose sides are 7.5 meters long.

Solution The perimeter of a square is given by the formula $P = 4s$. We substitute 7.5 for s and simplify.

$$P = 4s$$
$$= 4(\mathbf{7.5})$$
$$= 30$$

The perimeter is 30 meters.

Self Check 1
Find the perimeter of a square whose sides are 23.75 centimeters long.

Answer 95 cm

Since a rectangle has two lengths l and two widths w, its perimeter P is $l + l + w + w$, or $2l + 2w$.

Perimeter of a rectangle
If a rectangle has length l and width w, its perimeter P is given by the formula

$$P = 2l + 2w$$

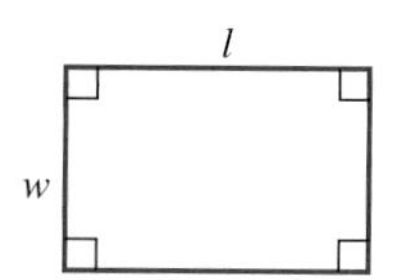

EXAMPLE 2 Find the perimeter of the rectangle in Figure 7-46.

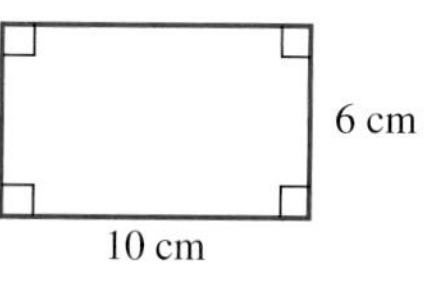

FIGURE 7-46

Solution The perimeter is given by the formula $P = 2l + 2w$. We substitute 10 for l and 6 for w and simplify.

$$P = 2l + 2w$$
$$= 2(\mathbf{10}) + 2(6)$$
$$= 20 + 12$$
$$= 32$$

The perimeter is 32 centimeters.

Self Check 2
Find the perimeter of the isosceles trapezoid below.

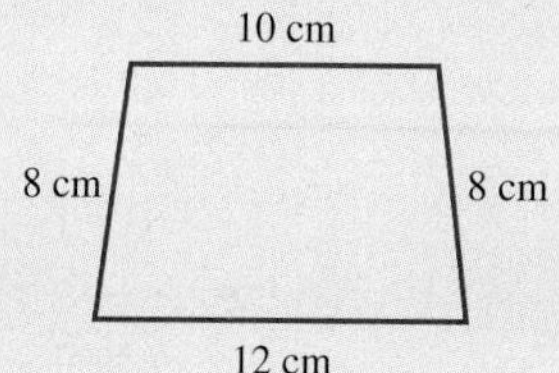

Answer 38 cm

EXAMPLE 3 Find the perimeter of the rectangle in Figure 7-47, in meters.

Solution Since 1 meter = 100 centimeters, we can convert 80 centimeters to meters by multiplying by 1, written in the form $\frac{1\text{ m}}{100\text{ cm}}$.

$$80\text{ cm} = 80\text{ cm} \cdot \frac{1\text{ m}}{100\text{ cm}} \quad \text{Multiply by 1: } \frac{1\text{ m}}{100\text{ cm}} = 1.$$

$$= \frac{80}{100}\text{ m} \quad \text{The units of centimeters divide out.}$$

$$= 0.8\text{ m} \quad \text{Divide by 100 by moving the decimal point 2 places to the left.}$$

3 m

80 cm

FIGURE 7-47

Self Check 3
Find the perimeter of the triangle below, in inches.

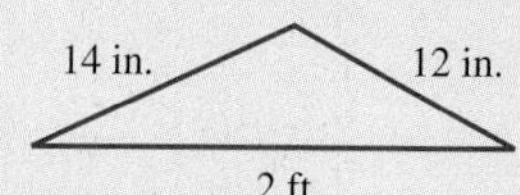

We can now substitute 3 for l and 0.8 for w to get

$$\begin{aligned} P &= 2l + 2w \\ &= 2(\mathbf{3}) + 2(\mathbf{0.8}) \\ &= 6 + 1.6 \\ &= 7.6 \end{aligned}$$

The perimeter is 7.6 meters.

Answer 50 in.

EXAMPLE 4 The perimeter of the isosceles triangle in Figure 7-48 is 50 meters. Find the length of its base.

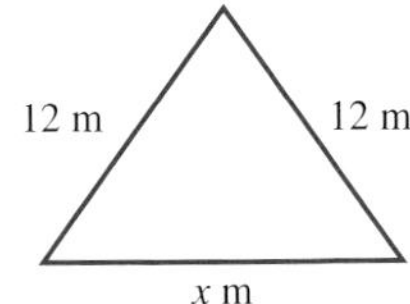

FIGURE 7-48

Solution Two sides are 12 meters long, and the perimeter is 50 meters. If x represents the length of the base, we have

$$\begin{aligned} 12 + 12 + x &= 50 \\ 24 + x &= 50 && \text{Simplify: } 12 + 12 = 24. \\ 24 + x - \mathbf{24} &= 50 - \mathbf{24} && \text{To undo the addition of 24, subtract 24 from both sides.} \\ x &= 26 \end{aligned}$$

The length of the base is 26 meters.

Self Check 4
The perimeter of an isosceles triangle is 60 meters. If one of its sides of equal length is 15 meters long, how long is its base?

Answer 30 m

Perimeters of figures that are combinations of polygons

CALCULATOR SNAPSHOT Perimeter of a figure

See Figure 7-49. To find the perimeter, we need to know the values of x and y. Since the figure is a combination of two rectangles, we can use a calculator to see that

$$x = 20.25 - 10.17 = 10.08 \quad \text{and} \quad y = 12.5 - 4.75 = 7.75$$

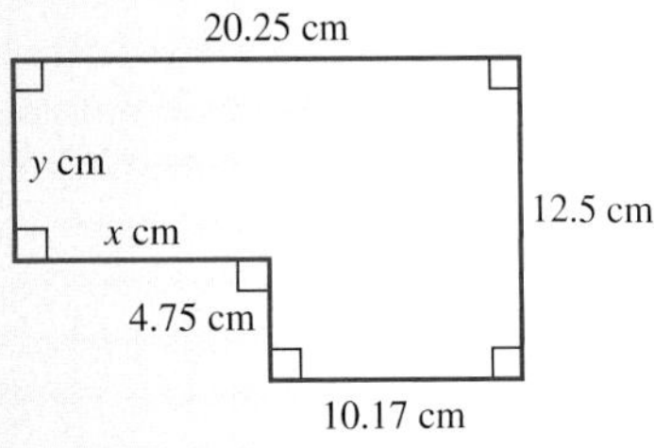

FIGURE 7-49

The perimeter P of the figure is

$$\begin{aligned} P &= 20.25 + 12.5 + 10.17 + 4.75 + x + y \\ &= 20.25 + 12.5 + 10.17 + 4.75 + \mathbf{10.08} + \mathbf{7.75} \end{aligned}$$

We can use a calculator to evaluate the expression on the right-hand side by entering these numbers and pressing these keys.

20.25 [+] 12.5 [+] 10.17 [+] 4.75 [+] 10.08 [+] 7.75 [=]

65.5

The perimeter is 65.5 centimeters.

Areas of polygons

Recall that the **area** of a polygon is the measure of the amount of surface it encloses. Area is measured in square units, such as square inches or square centimeters. See Figure 7-50.

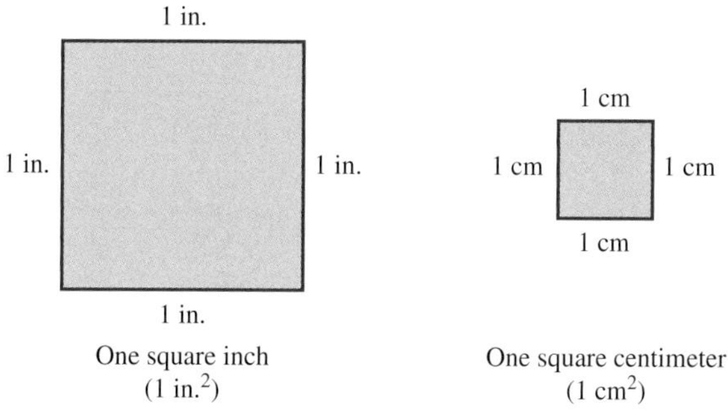

FIGURE 7-50

In everyday life, we often use areas. For example,

- To carpet a room, we buy square yards.
- A can of paint will cover a certain number of square feet.
- To measure vast amounts of land, we often use square miles.
- We buy house roofing by the "square." One square is 100 square feet.

The rectangle shown in Figure 7-51 has a length of 10 centimeters and a width of 3 centimeters. If we divide the rectangle into squares as shown in the figure, each square represents an area of 1 square centimeter—a surface enclosed by a square measuring 1 centimeter on each side. Because there are 3 rows with 10 squares in each row, there are 30 squares. Since the rectangle encloses a surface area of 30 squares, its area is 30 square centimeters, often written as 30 cm^2.

This example illustrates that to find the area of a rectangle, we multiply its length by its width.

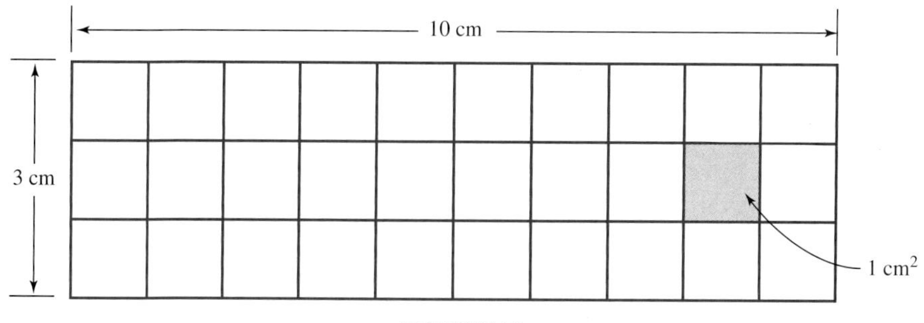

FIGURE 7-51

! COMMENT Do not confuse the concepts of perimeter and area. Perimeter is the distance around a polygon. It is measured in linear units, such as centimeters, feet, or miles. Area is a measure of the surface enclosed within a polygon. It is measured in square units, such as square centimeters, square feet, or square miles.

In practice, we do not find areas by counting squares in a figure. Instead, we use formulas for finding areas of geometric figures, as shown in Table 7-1 on the next page.

Figure	Name	Formula for area
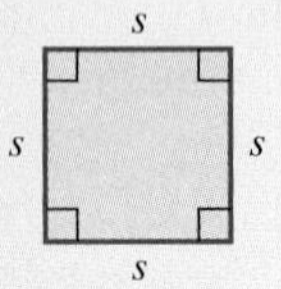	Square	$A = s^2$, where s is the length of one side.
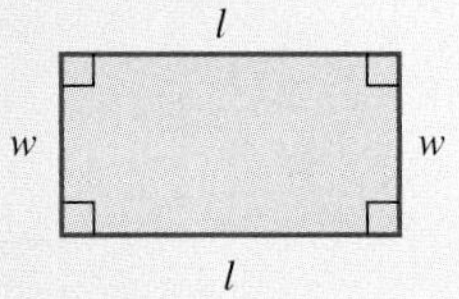	Rectangle	$A = lw$, where l is the length and w is the width.
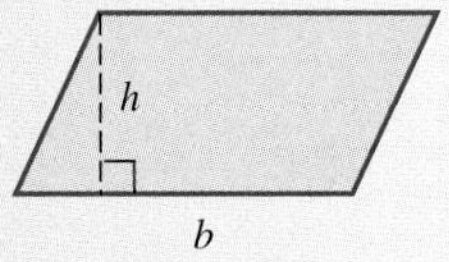	Parallelogram	$A = bh$, where b is the length of the base and h is the height. (A height is always perpendicular to the base.)
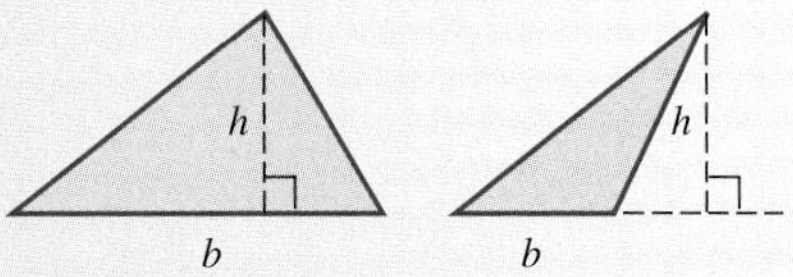	Triangle	$A = \frac{1}{2}bh$, where b is the length of the base and h is the height. The segment perpendicular to the base and representing the height is called an **altitude.**
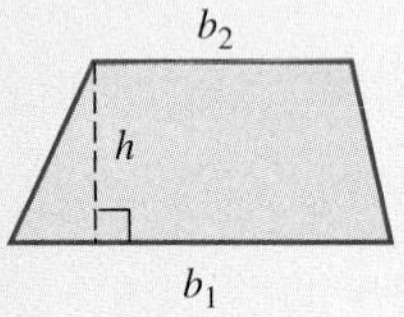	Trapezoid	$A = \frac{1}{2}h(b_1 + b_2)$, where h is the height of the trapezoid and b_1 and b_2 represent the lengths of the bases.

TABLE 7-1

EXAMPLE 5 Find the area of the square in Figure 7-52.

Solution We can see that the length of one side of the square is 15 centimeters. We can find its area by using the formula $A = s^2$ and substituting 15 for s.

$$A = s^2$$

$$= (\mathbf{15})^2 \quad \text{Substitute 15 for } s.$$

$$= 225 \quad \text{Evaluate the exponential expression: } 15 \cdot 15 = 225.$$

The area of the square is 225 cm^2.

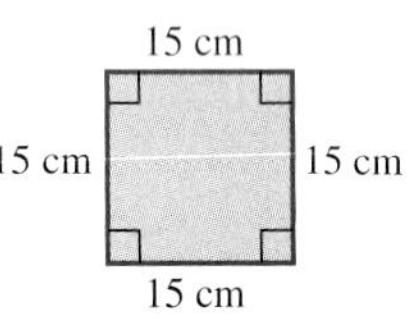

FIGURE 7-52

Self Check 5

Find the area of the square shown below.

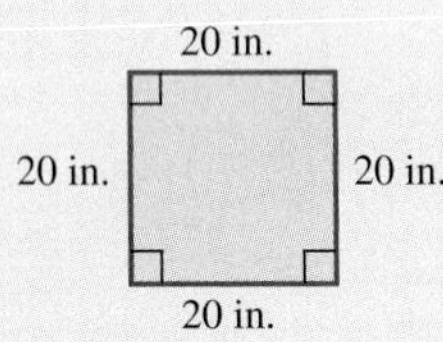

Answer 400 $in.^2$

EXAMPLE 6 Find the number of square feet in 1 square yard. (See Figure 7-53.)

Solution Since 3 feet = 1 yard, each side of 1 square yard is 3 feet long.

$$1\ yd^2 = (\mathbf{1}\ yd)^2$$

$$= (\mathbf{3}\ ft)^2 \quad \text{Substitute 3 feet for 1 yard.}$$

$$= 9\ ft^2 \quad (3\ ft)^2 = (3\ ft)(3\ ft) = 9\ ft^2.$$

There are 9 square feet in 1 square yard.

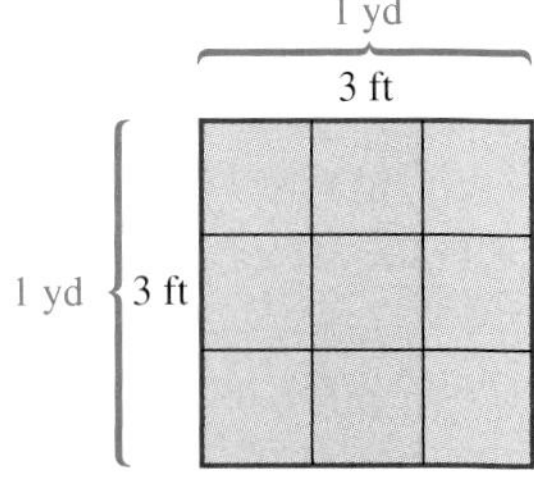

FIGURE 7-53

Self Check 6

Find the number of square centimeters in 1 square meter.

Answer 10,000 cm^2

EXAMPLE 7 **Women's sports.** Field hockey is a team sport in which players use sticks to try to hit a ball into their opponents' goal. Find the area of the rectangular field shown in Figure 7-54. Give the answer in square feet.

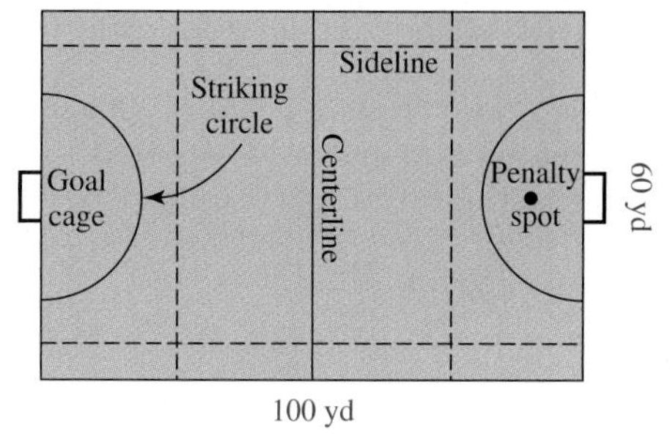

FIGURE 7-54

Solution To find the area in square yards, we substitute 100 for l and 60 for w in the formula for the area of a rectangle, and simplify.

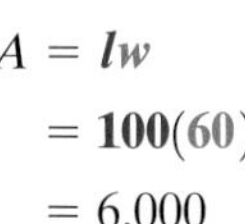

$$\begin{aligned} A &= lw \\ &= \mathbf{100}(\mathbf{60}) \\ &= 6{,}000 \end{aligned}$$

The area is 6,000 square yards. Since there are 9 square feet per square yard, we can convert this number to square feet by multiplying by 1, written in the form $\frac{9 \text{ ft}^2}{1 \text{ yd}^2}$.

$$\begin{aligned} 6{,}000 \text{ yd}^2 &= 6{,}000 \text{ yd}^2 \cdot \frac{9 \text{ ft}^2}{1 \text{ yd}^2} && \text{Multiply by the unit conversion factor: } \frac{9 \text{ ft}^2}{1 \text{ yd}^2}. \\ &= 6{,}000 \cdot 9 \text{ ft}^2 && \text{The units of square yards divide out.} \\ &= 54{,}000 \text{ ft}^2 && \text{Multiply: } 6{,}000 \cdot 9 = 54{,}000. \end{aligned}$$

The area of the field is 54,000 ft^2.

Self Check 7
Find the area in square inches of a rectangle with dimensions of 6 inches by 2 feet.

Answer 144 in.^{2}

Dorm Rooms

THINK IT THROUGH

"The United States has more than 4,000 colleges and universities, with 2 million students living in college dorms." Washingtonpost.com, 2004

The average dormitory room in a residence hall has about 180 square feet of floor space. The rooms are usually furnished with the following items having the given dimensions:

- 2 extra-long twin beds (39 in. wide × 80 in. long × 24 in. high)
- 2 dressers (18 in. wide × 36 in. long × 48 in. high)
- 2 bookcases (12 in. wide × 24 in. long × 40 in. high)
- 2 desks (24 in. wide × 48 in. long × 28 in. high)

How many square feet of floor space are left?

EXAMPLE 8 Find the area of the parallelogram in Figure 7-55.

FIGURE 7-55

Solution The length of the base of the parallelogram is

$$5 \text{ feet} + 25 \text{ feet} = 30 \text{ feet}$$

The height is 12 feet. To find the area, we substitute 30 for b and 12 for h in the formula for the area of a parallelogram and simplify.

$$\begin{aligned} A &= bh \\ &= \mathbf{30}(\mathbf{12}) \\ &= 360 \end{aligned}$$

The area of the parallelogram is 360 ft^2.

Self Check 8
Find the area of the parallelogram below.

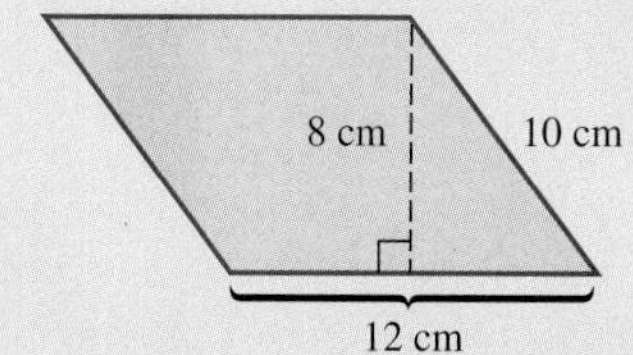

Answer 96 cm^2

Self Check 9

Find the area of the triangle below.

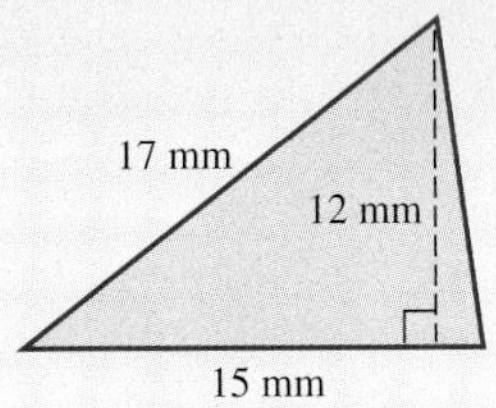

Answer 90 mm^2

EXAMPLE 9 Find the area of the triangle in Figure 7-56.

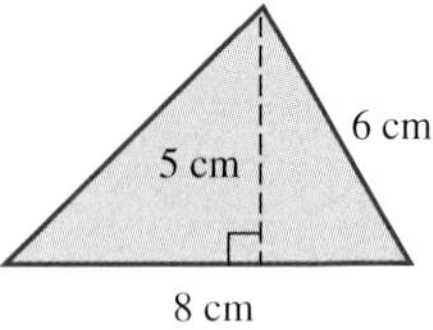

FIGURE 7-56

Solution We substitute 8 for b and 5 for h in the formula for the area of a triangle, and simplify. (The side having length 6 cm is additional information that is not used to find the area.)

$$A = \frac{1}{2}\boldsymbol{bh}$$

$$= \frac{1}{2}(\mathbf{8})(\mathbf{5}) \quad \text{The length of the base is 8 cm. The height is 5 cm.}$$

$$= 4(5) \quad \text{Perform the multiplication: } \frac{1}{2}(8) = 4.$$

$$= 20$$

The area of the triangle is 20 cm^2.

EXAMPLE 10 Find the area of the triangle in Figure 7-57.

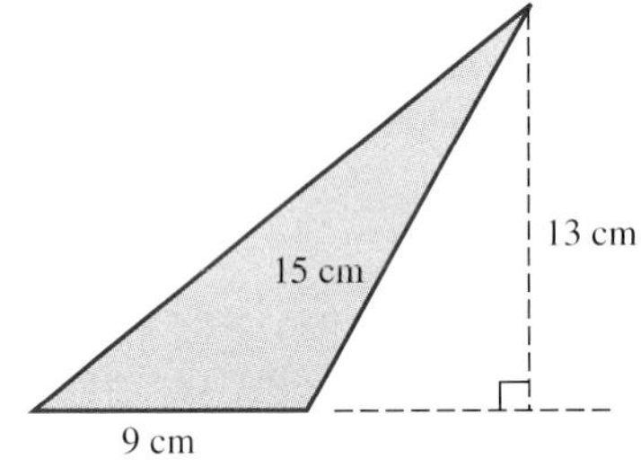

FIGURE 7-57

Solution In this case, the altitude falls outside the triangle.

$$A = \frac{1}{2}\boldsymbol{bh}$$

$$= \frac{1}{2}(\mathbf{9})(\mathbf{13}) \quad \text{Substitute 9 for } b \text{ and 13 for } h.$$

$$= \frac{1}{2}\left(\frac{9}{1}\right)\left(\frac{13}{1}\right) \quad \text{Write 9 as } \frac{9}{1} \text{ and 13 as } \frac{13}{1}.$$

$$= \frac{117}{2} \quad \text{Multiply the fractions.}$$

$$= 58.5 \quad \text{Perform the division.}$$

The area of the triangle is 58.5 cm^2.

EXAMPLE 11 Find the area of the trapezoid in Figure 7-58.

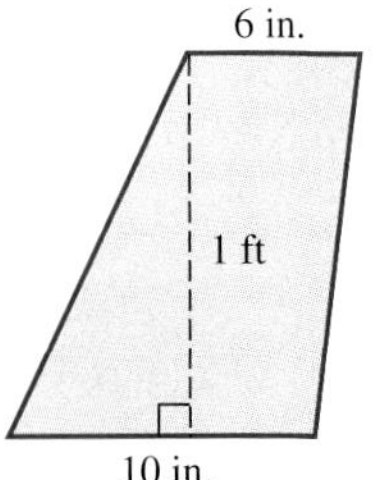

FIGURE 7-58

Solution In this example, $b_1 = 10$ and $b_2 = 6$. It is incorrect to say that $h = 1$, because the height of 1 foot must be expressed as 12 inches to be consistent with the units of the bases. Thus, we substitute 10 for b_1, 6 for b_2, and 12 for h in the formula for finding the area of a trapezoid and simplify.

$$A = \frac{1}{2}h(b_1 + b_2)$$

$= \frac{1}{2}(12)(10 + 6)$ The length of the lower base is 10 in. The length of the upper base is 6 in. The height is 12 in.

$= \frac{1}{2}(12)(16)$ Perform the addition within the parentheses.

$= 6(16)$ Perform the multiplication: $\frac{1}{2}(12) = 6$.

$= 96$

The area of the trapezoid is 96 in.^{2}.

Self Check 11

Find the area of the trapezoid below.

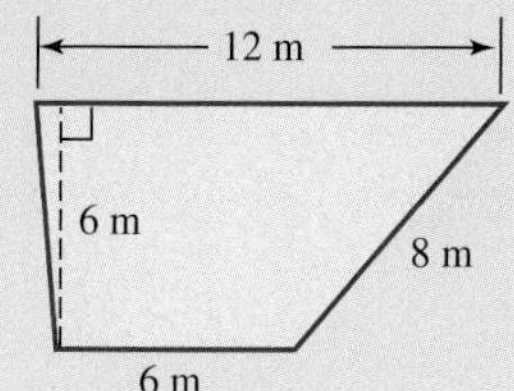

Answer 54 m^2

Areas of figures that are combinations of polygons

EXAMPLE 12 Carpeting a room. A living room/dining room area has the floor plan shown in Figure 7-59. If carpet costs \$29 per square yard, including pad and installation, how much will it cost to carpet the room? (Assume no waste.)

FIGURE 7-59

Solution First we must find the total area of the living room and the dining room:

$$A_{\text{total}} = A_{\text{living room}} + A_{\text{dining room}}$$

Since $\overline{CF}$ divides the space into two rectangles, the areas of the living room and the dining room are found by multiplying their respective lengths and widths.

$$\begin{aligned}\text{Area of living room} &= lw\\ &= 7(4)\\ &= 28\end{aligned}$$

The area of the living room is 28 yd^2.

To find the area of the dining room, we find its length by subtracting 4 yards from 9 yards to obtain 5 yards. We note that its width is 4 yards.

$$\begin{aligned}\text{Area of dining room} &= lw\\ &= 5(4)\\ &= 20\end{aligned}$$

The area of the dining room is 20 yd^2.

The total area to be carpeted is the sum of these two areas.

$$\begin{aligned}A_{\text{total}} &= A_{\text{living room}} + A_{\text{dining room}}\\ &= 28\text{ yd}^2 + 20\text{ yd}^2\\ &= 48\text{ yd}^2\end{aligned}$$

At \$29 per square yard, the cost to carpet the room will be $48 \cdot \$29$, or \$1,392.

EXAMPLE 13 **Area of one side of a tent.** Find the area of one side of the tent in Figure 7-60.

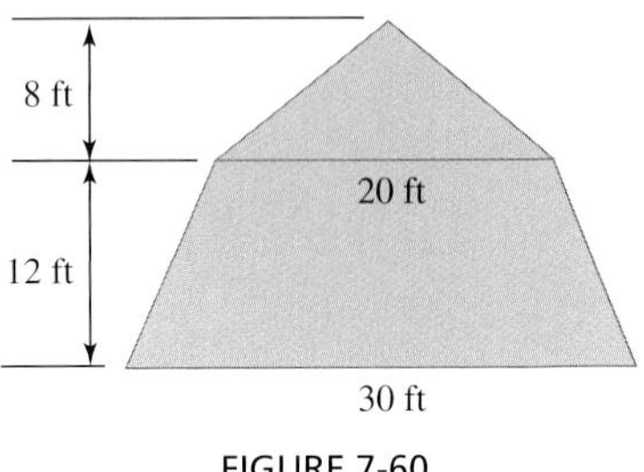

FIGURE 7-60

Solution Each side is a combination of a trapezoid and a triangle. Since the bases of each trapezoid are 30 feet and 20 feet and the height is 12 feet, we substitute 30 for b_1, 20 for b_2, and 12 for h into the formula for the area of a trapezoid.

$$\begin{aligned} A_{\text{trap.}} &= \frac{1}{2}h(b_1 + b_2) \\ &= \frac{1}{2}(12)(30 + 20) \\ &= 6(50) \\ &= 300 \end{aligned}$$

The area of the trapezoid is 300 ft^2.

Since the triangle has a base of 20 feet and a height of 8 feet, we substitute 20 for b and 8 for h in the formula for the area of a triangle.

$$\begin{aligned} A_{\text{triangle}} &= \frac{1}{2}bh \\ &= \frac{1}{2}(20)(8) \\ &= 80 \end{aligned}$$

The area of the triangle is 80 ft^2.

The total area of one side of the tent is

$$\begin{aligned} A_{\text{total}} &= A_{\text{trap.}} + A_{\text{triangle}} \\ &= 300 \text{ ft}^2 + 80 \text{ ft}^2 \\ &= 380 \text{ ft}^2 \end{aligned}$$

The total area is 380 ft^2.

Section 7.5 STUDY SET

VOCABULARY *Fill in the blanks.*

1. The distance around a polygon is called the ________.
2. The perimeter of a polygon is measured in ________ units.
3. The measure of the surface enclosed by a polygon is called its ________.
4. If each side of a square measures 1 foot, the area enclosed by the square is 1 ________ foot.
5. The area of a polygon is measured in ________ units.
6. The segment that represents the height of a triangle is called an ________.

CONCEPTS *Sketch and label each of the figures described.*

7. Two different rectangles, each having a perimeter of 40 in.
8. Two different rectangles, each having an area of 40 $in.^2$

9. A square with an area of 25 m^2
10. A square with a perimeter of 20 m
11. A parallelogram with an area of 15 yd^2
12. A triangle with an area of 20 ft^2
13. A figure consisting of a combination of two rectangles whose total area is 80 ft^2
14. A figure consisting of a combination of a rectangle and a square whose total area is 164 ft^2

NOTATION *Fill in the blanks*

15. The formula for the perimeter of a square is $P =$ ___.
16. The formula for the perimeter of a rectangle is $P =$ ___.
17. The symbol 1 $in.^2$ means one ________ ________.
18. One square meter is expressed as 1 m ___.
19. The formula for the area of a square is $A =$ ___.
20. The formula for the area of a rectangle is $A =$ ___.
21. The formula $A = \frac{1}{2}bh$ gives the area of a ________.
22. The formula $A = \frac{1}{2}h(b_1 + b_2)$ gives the area of a ________.

PRACTICE *Find the perimeter of each figure.*

23.

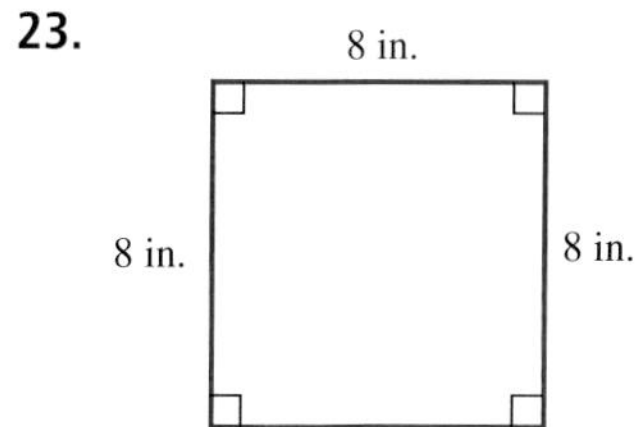

24.

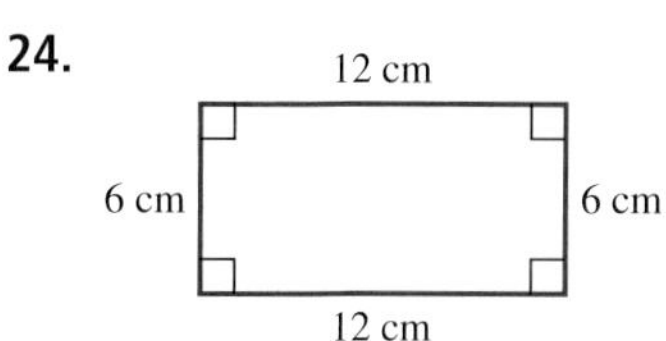

25.

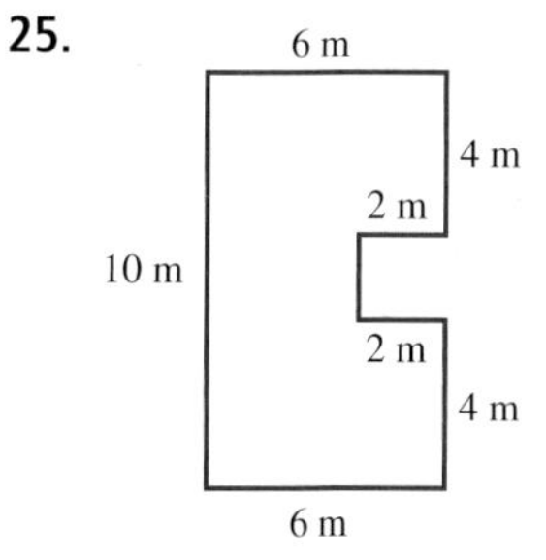

26.

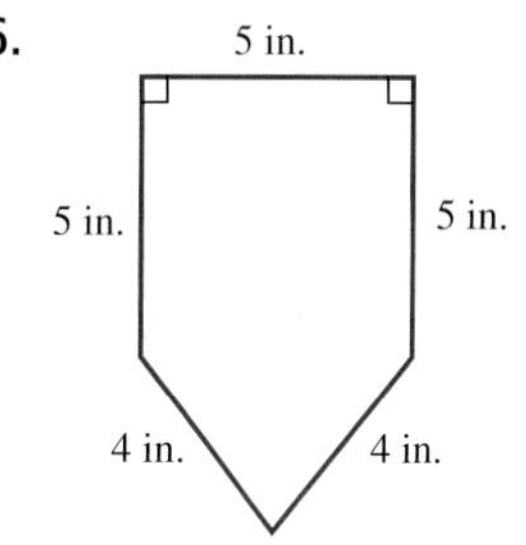

27.

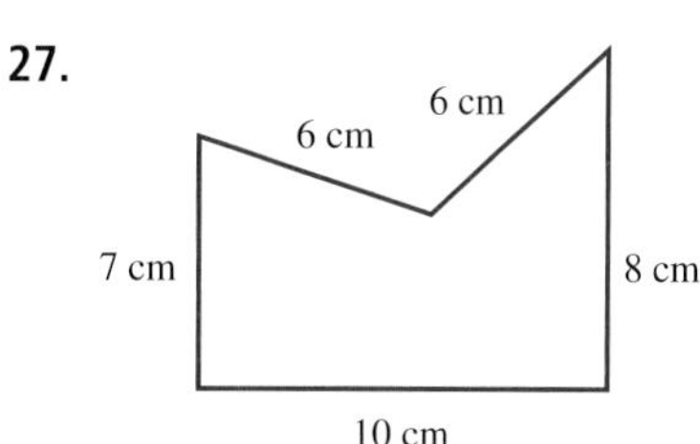

28.

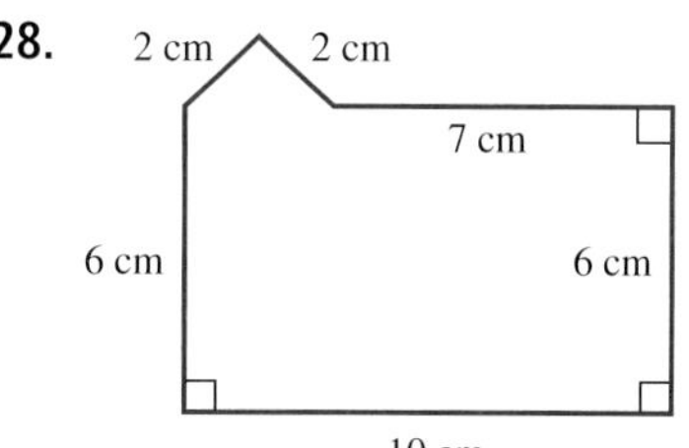

Solve each problem.

29. Find the perimeter of an isosceles triangle with a base of length 21 centimeters and sides of length 32 centimeters.
30. The perimeter of an isosceles triangle is 80 meters. If the length of one side is 22 meters, how long is the base?
31. The perimeter of an equilateral triangle is 85 feet. Find the length of each side.
32. An isosceles triangle with sides of 49.3 inches has a perimeter of 121.7 inches. Find the length of the base.

Find the area of the shaded part of each figure.

33.

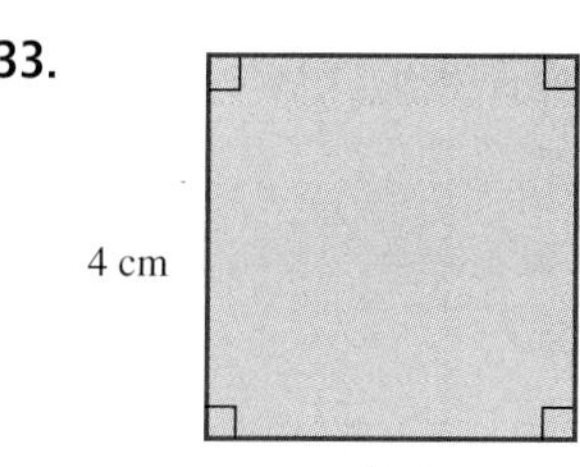

34.

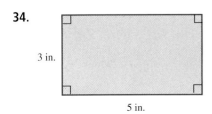

35.

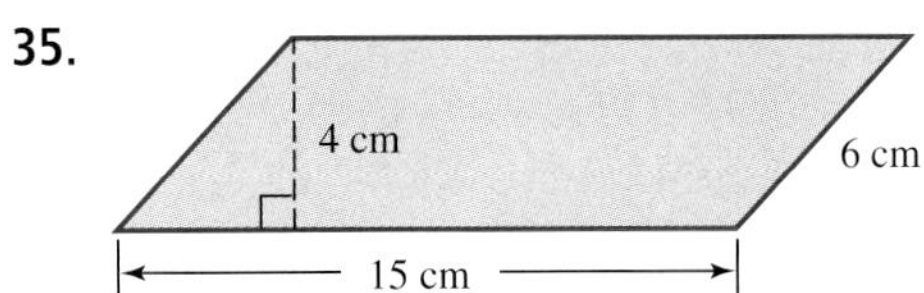

36.

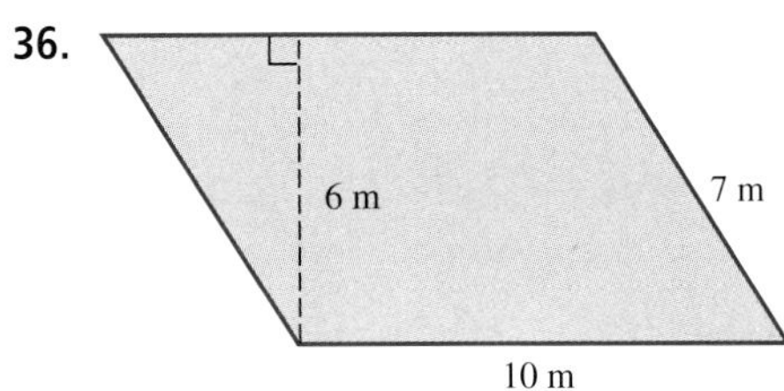

37.

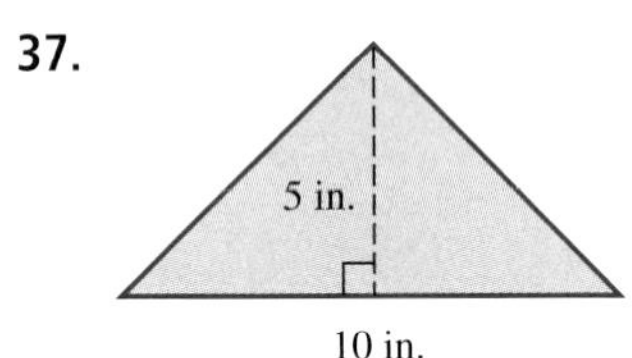

38.

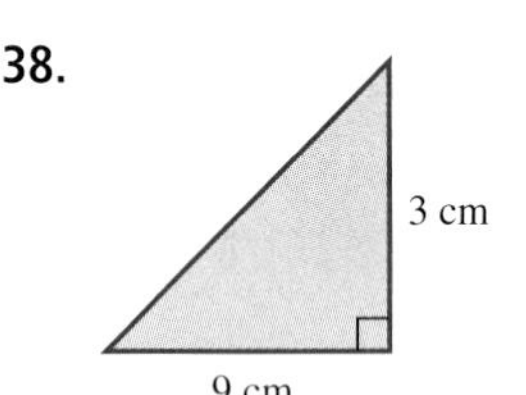

39.

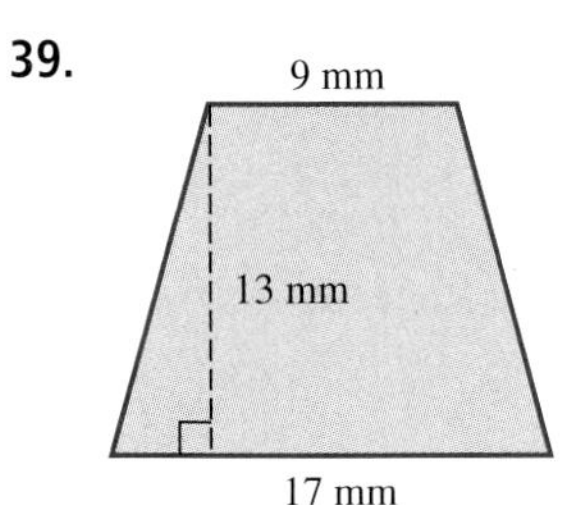

40.

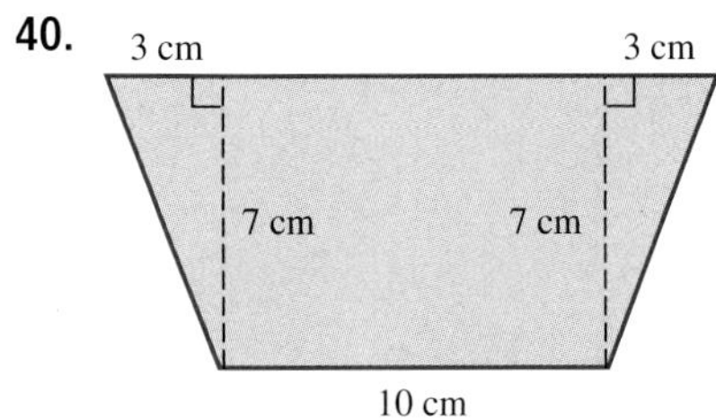

41.

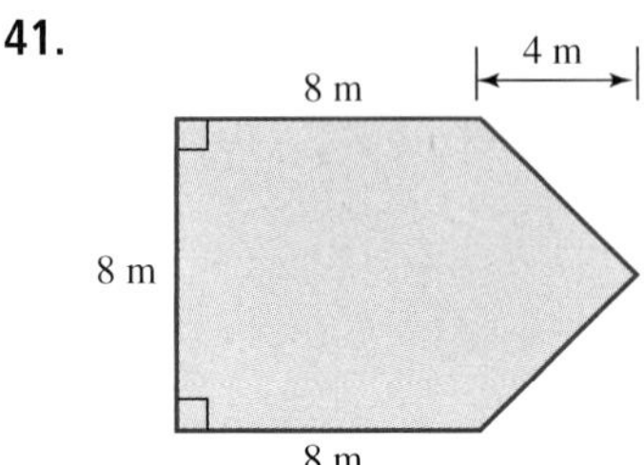

42.

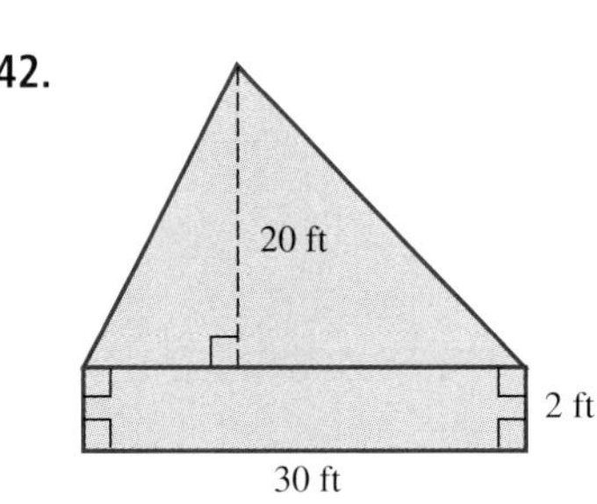

43.

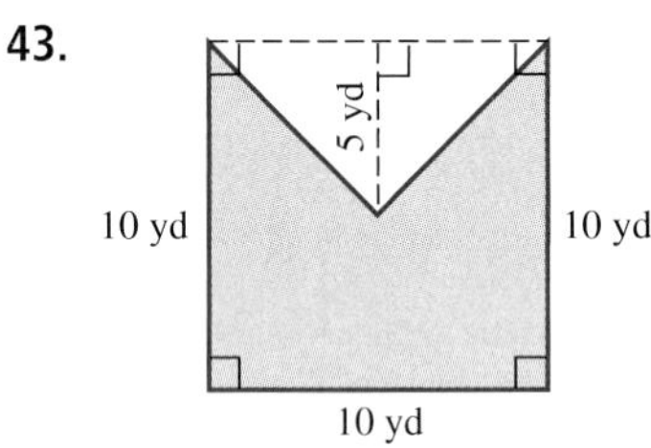

44.

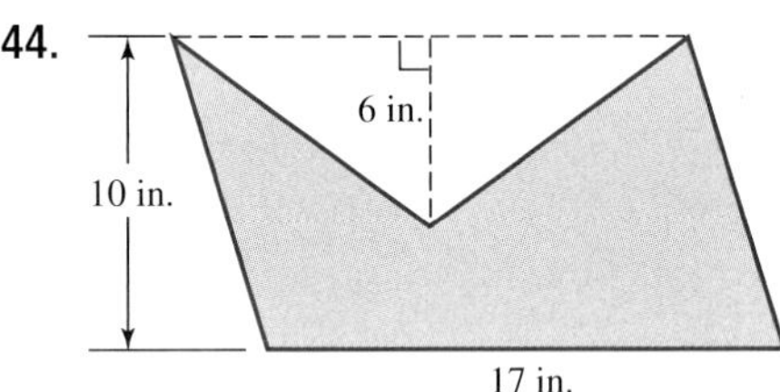

45.

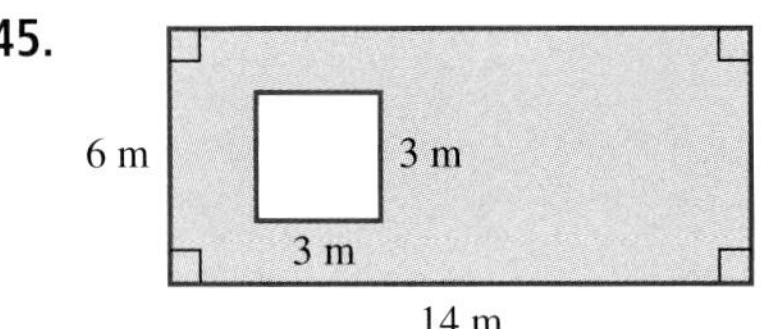

46.

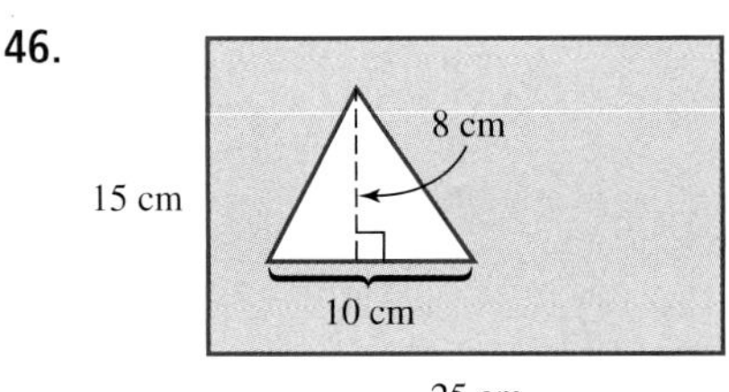

47. How many square inches are in 1 square foot?

48. How many square inches are in 1 square yard?

APPLICATIONS

49. FENCING YARDS A man wants to enclose a rectangular yard with fencing that costs $12.50 a foot, including installation. Find the cost of enclosing the yard if its dimensions are 110 ft by 85 ft.

50. FRAMING PICTURES Find the cost of framing a rectangular picture with dimensions of 24 inches by 30 inches if framing material costs $8.46 per foot, including matting.

51. PLANTING SCREENS A woman wants to plant a pine-tree screen around three sides of her backyard. If she plants the trees 3 feet apart, how many trees will she need?

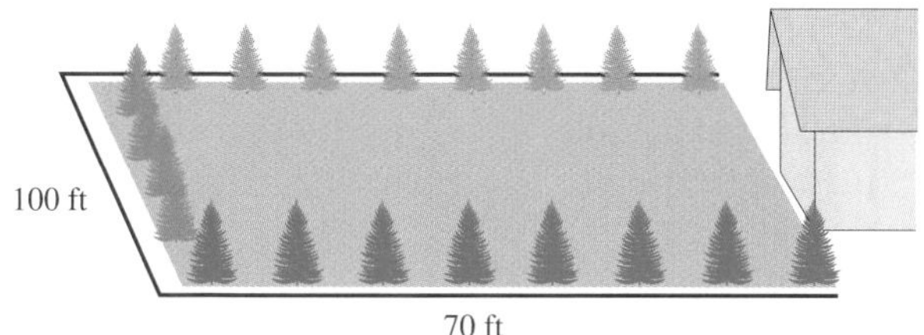

52. PLANTING MARIGOLDS A gardener wants to plant a border of marigolds around the garden shown to keep out rabbits. How many plants will she need if she allows 6 inches between plants?

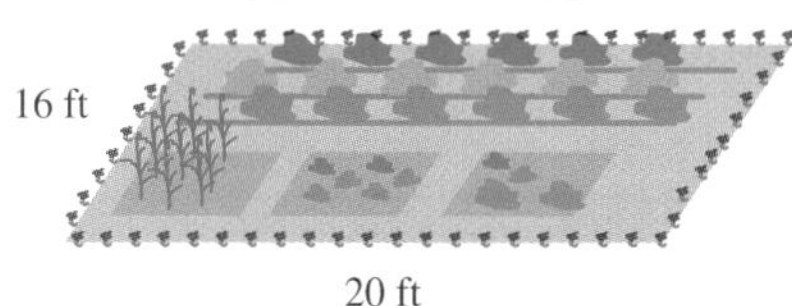

53. BUYING A FLOOR Which is more expensive: a ceramic-tile floor costing $3.75 per square foot or linoleum costing $34.95 per square yard?

54. BUYING A FLOOR Which is cheaper: a hardwood floor costing $5.95 per square foot or a carpeted floor costing $37.50 per square yard?

55. CARPETING A ROOM A rectangular room is 24 feet long and 15 feet wide. At $30 per square yard, how much will it cost to carpet the room? (Assume no waste.)

56. CARPETING A ROOM A rectangular living room measures 30 by 18 feet. At $32 per square yard, how much will it cost to carpet the room? (Assume no waste.)

57. TILING A FLOOR A rectangular basement room measures 14 by 20 feet. Vinyl floor tiles that are 1 ft^2 cost $1.29 each. How much will the tile cost to cover the floor? (Disregard any waste.)

58. PAINTING A BARN The north wall of a barn is a rectangle 23 feet high and 72 feet long. There are five windows in the wall, each 4 by 6 feet. If a gallon of paint will cover 300 ft^2, how many gallons of paint must the painter buy to paint the wall?

59. MAKING A SAIL If nylon is $12 per square yard, how much would the fabric cost to make a triangular sail with a base of 12 feet and a height of 24 feet?

60. PAINTING A GABLE The gable end of a warehouse is an isosceles triangle with a height of 4 yards and a base of 23 yards. It will require one coat of primer and one coat of finish to paint the triangle. Primer costs $17 per gallon, and the finish paint costs $23 per gallon. If one gallon covers 300 square feet, how much will it cost to paint the gable, excluding labor?

61. GEOGRAPHY Use the dimensions of the trapezoid that is superimposed over the state of Nevada to estimate the area of the "Silver State."

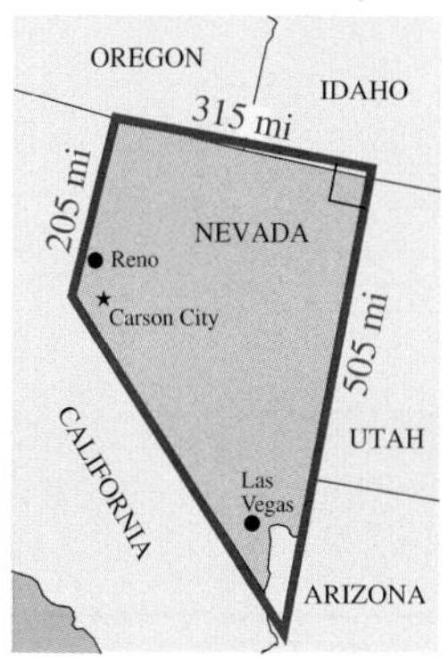

62. SWIMMING POOLS A swimming pool has the shape shown. How many square meters of plastic sheeting will be needed to cover the pool? How much will the sheeting cost if it is $2.95 per square meter? (Assume no waste.)

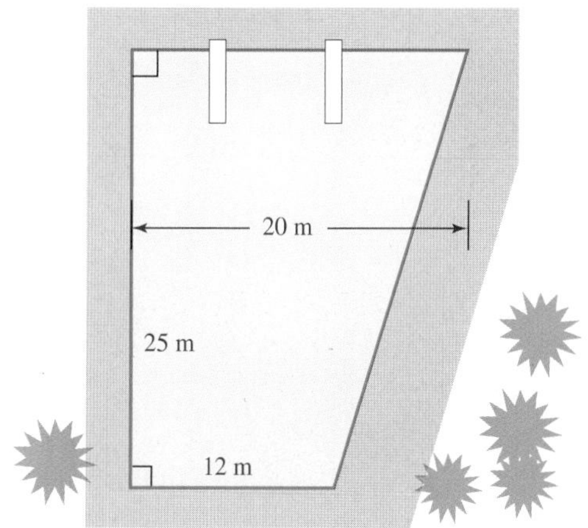

63. CARPENTRY How many sheets of 4-foot-by-8-foot sheetrock are needed to drywall the inside walls on the first floor of the barn shown? (Assume that the carpenters will cover each wall entirely and then cut out areas for the doors and windows.)

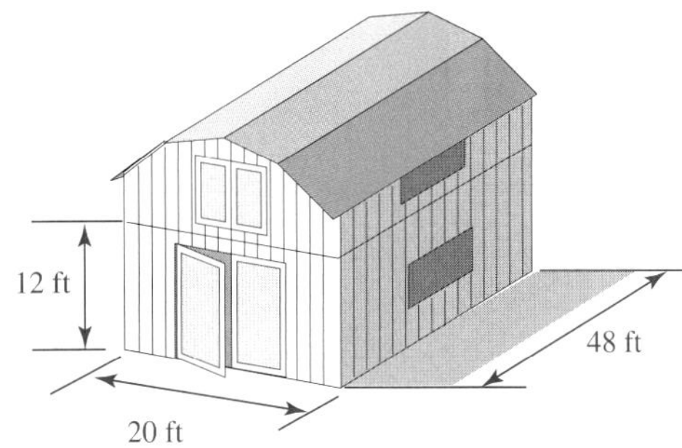

64. CARPENTRY If it costs \$90 per square foot to build a one-story home in northern Wisconsin, estimate the cost of building the house with the floor plan shown.

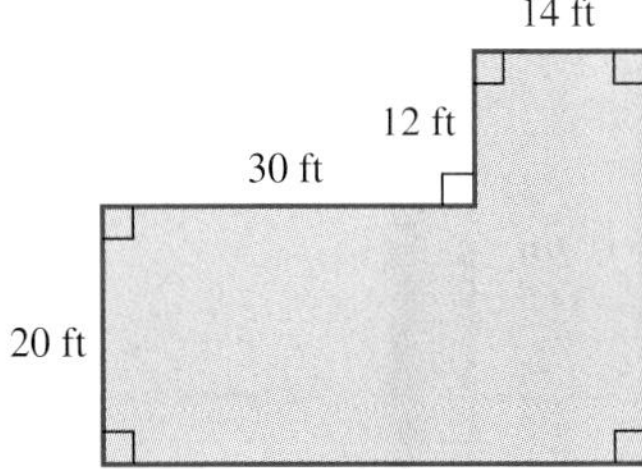

65. DRIVING SAFETY The illustration in the next column shows the areas on a highway that a truck driver cannot see in the truck's rear view mirrors. Use the scale to determine the approximate dimensions of each blind spot. Then estimate the area of each of them.

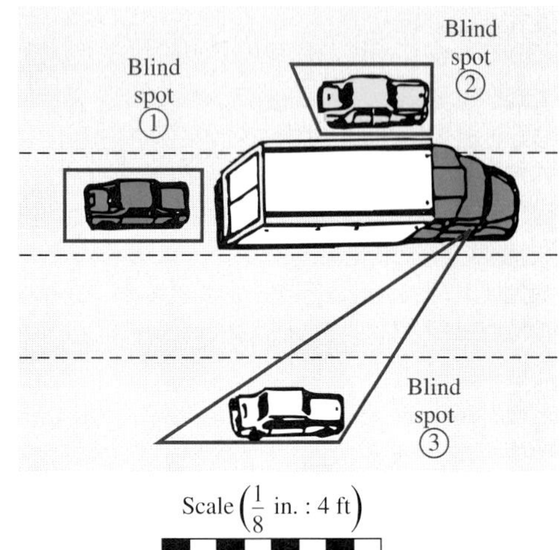

66. ESTIMATING AREA Estimate the area of the sole plate of the iron by thinking of it as a combination of a trapezoid and a triangle.

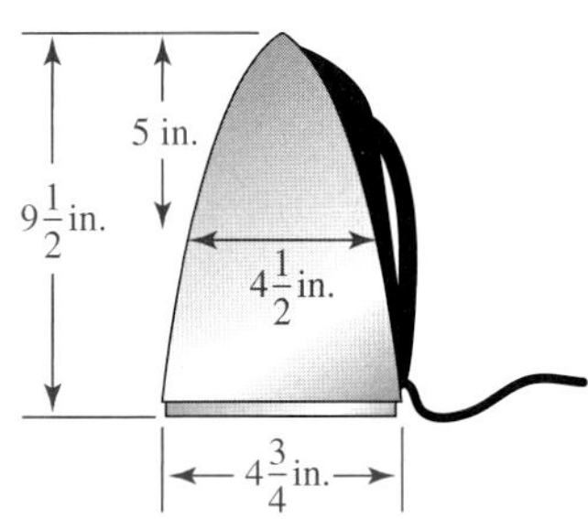

WRITING

67. Explain the difference between perimeter and area.

68. Why is it necessary that area be measured in square units?

REVIEW *Perform the calculations. Write all improper fractions as mixed numbers.*

69. $\frac{3}{4} + \frac{2}{3}$

70. $\frac{7}{8} - \frac{2}{3}$

71. $3\frac{3}{4} + 2\frac{1}{3}$

72. $7\frac{5}{8} - 2\frac{5}{6}$

73. $7\frac{1}{2} \div 5\frac{2}{5}$

74. $5\frac{3}{4} \cdot 2\frac{5}{6}$

7.6 Circles

• Circles • Circumference of a circle • Area of a circle

In this section, we discuss circles, one of the most useful geometric figures. In fact, the discovery of fire and the circular wheel were two of the most important events in the history of the human race.

Circles

Circle

A **circle** is the set of all points in a plane that lie a fixed distance from a point called its **center.**

A segment drawn from the center of a circle to a point on the circle is called a **radius.** (The plural of *radius* is *radii.*) From the definition, it follows that all radii of the same circle are the same length.

A **chord** of a circle is a line segment connecting two points on the circle. A **diameter** is a chord that passes through the center of the circle. Since a diameter D of a circle is twice as long as a radius r, we have

$$D = 2r$$

Each of the previous definitions is illustrated in Figure 7-61, in which O is the center of the circle.

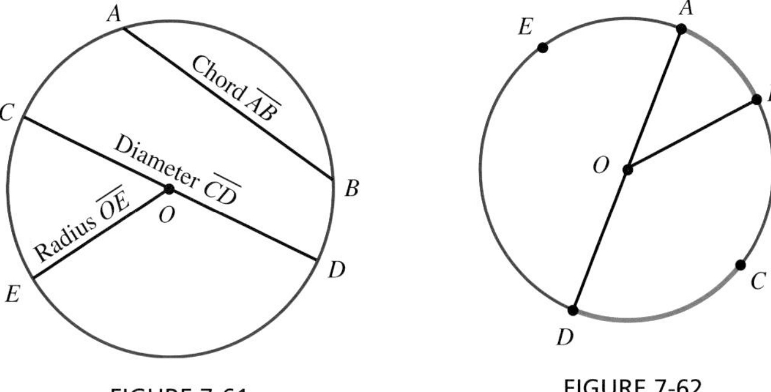

FIGURE 7-61

FIGURE 7-62

Any part of a circle is called an **arc.** In Figure 7-62, the part of the circle from point A to point B is $\overset{\frown}{AB}$, read as "arc AB." $\overset{\frown}{CD}$ is the part of the circle from point C to point D. An arc that is half of a circle is a **semicircle.**

Semicircle

A **semicircle** is an arc of a circle whose endpoints are the endpoints of a diameter.

If point O is the center of the circle in Figure 7-62, $\overline{AD}$ is a diameter and $\overset{\frown}{AED}$ is a semicircle. The middle letter E is used to distinguish semicircle $\overset{\frown}{AED}$ from semicircle $\overset{\frown}{ABCD}$.

An arc that is shorter than a semicircle is a **minor arc.** An arc that is longer than a semicircle is a **major arc.** In Figure 7-62,

$\overset{\frown}{AB}$ is a minor arc and $\overset{\frown}{ABCDE}$ is a major arc

Circumference of a circle

Since early history, mathematicians have known that the ratio of the distance around a circle (the circumference) divided by the length of its diameter is approximately 3. First Kings, Chapter 7, of the Bible describes a round bronze tank that was 10 cubits from rim to rim and 30 cubits in circumference, and $\frac{30}{10} = 3$. Today, we have a better value for this ratio, known as π (pi). If C is the circumference of a circle and D is the length of its diameter, then

$$\pi = \frac{C}{D}, \quad \text{where } \pi = 3.141592653589\ldots$$

$\frac{22}{7}$ and 3.14 are often used as estimates of π.

If we multiply both sides of $\pi = \frac{C}{D}$ by D, we have the following formula.

Circumference of a circle

The circumference of a circle is given by the formula

$C = \pi D$ where C is the circumference and D is the length of the diameter

Since a diameter of a circle is twice as long as a radius r, we can substitute $2r$ for D in the formula $C = \pi D$ to obtain another formula for the circumference C.

$$C = 2\pi r$$

Self Check 1

To the nearest tenth, find the circumference of a circle that has a radius of 12 meters.

Answer 75.4 m

EXAMPLE 1

Find the circumference of a circle that has a diameter of 10 centimeters. (See Figure 7-63.)

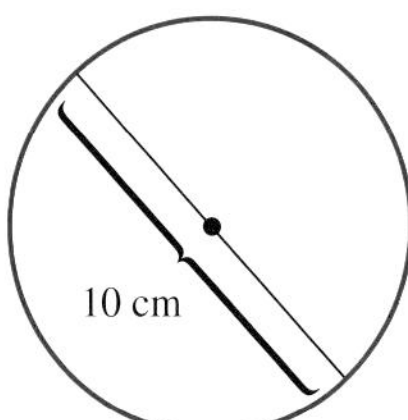

FIGURE 7-63

Solution We substitute 10 for D in the formula for the circumference of a circle.

$$\begin{aligned} C &= \pi \mathbf{D} \\ &= \pi(\mathbf{10}) \\ &\approx 3.14(10) \quad \text{Replace } \pi \text{ with an approximation: } \pi \approx 3.14. \\ &\approx 31.4 \end{aligned}$$

The circumference is approximately 31.4 centimeters.

CALCULATOR SNAPSHOT

Calculating revolutions of a tire

When the $\boxed{\pi}$ key on a scientific calculator is pressed (on some models, the $\boxed{\text{2nd}}$ key must be pressed first), an approximation of π is displayed. To illustrate how to use this key, consider the following problem. How many times does a 15-inch tire revolve when a car makes a 25-mile trip?

We first find the circumference of the tire.

$$\begin{aligned} C &= \pi \mathbf{D} \\ &= \pi(\mathbf{15}) \quad \text{Substitute 15 for } D\text{, the diameter of the tire.} \\ &= 15\pi \quad \text{Normally, we rewrite a product such as } \pi(15) \text{ so that } \pi \text{ is the second factor.} \end{aligned}$$

The circumference of the tire is 15π inches.

We then change 25 miles to inches using two unit conversion factors.

$$\frac{25}{1} \text{ miles} \cdot \frac{5{,}280 \text{ feet}}{1 \text{ mile}} \cdot \frac{12 \text{ inches}}{1 \text{ foot}} = 25(5{,}280)(12) \text{ in.}$$

The total distance of the trip is 25(5,280)(12) inches.

Finally, we divide the total distance of the trip by the circumference of the tire to get

$$\text{The number of revolutions of the tire} = \frac{25(5{,}280)(12)}{15\pi}$$

To approximate the value of $\frac{25(5{,}280)(12)}{15\pi}$ using a scientific calculator, we enter

$\boxed{(}$ 25 $\boxed{\times}$ 5280 $\boxed{\times}$ 12 $\boxed{)}$ $\boxed{\div}$ $\boxed{(}$ 15 $\boxed{\times}$ $\boxed{\pi}$ $\boxed{)}$ $\boxed{=}$ $\boxed{33613.52398}$

The tire makes about 33,614 revolutions.

EXAMPLE 2 **Architecture.** A Norman window is constructed by adding a semicircular window to the top of a rectangular window. Find the perimeter of the Norman window shown in Figure 7-64.

FIGURE 7-64

Solution The window is a combination of a rectangle and a semicircle. The perimeter of the rectangular part is

$$P_{\text{rectangular part}} = 8 + 6 + 8 = 22 \quad \text{Add only 3 sides.}$$

The perimeter of the semicircle is one-half the circumference of a circle that has a 6-foot diameter.

$$\begin{aligned} P_{\text{semicircle}} &= \frac{1}{2}\pi \boldsymbol{D} \\ &= \frac{1}{2}\pi(\mathbf{6}) && \text{Substitute 6 for } D. \\ &\approx 9.424777961 && \text{Use a calculator.} \end{aligned}$$

The total perimeter is the sum of the two parts.

$$\begin{aligned} P_{\text{total}} &\approx 22 + 9.424777961 \\ &\approx 31.424777961 \end{aligned}$$

To the nearest hundredth, the perimeter of the window is 31.42 feet.

Area of a circle

If we divide the circle shown in Figure 7-65(a) into an even number of pie-shaped pieces and then rearrange them as shown in Figure 7-65(b), we have a figure that looks like a parallelogram. The figure has a base that is one-half the circumference of the circle, and its height is about the same length as a radius of the circle.

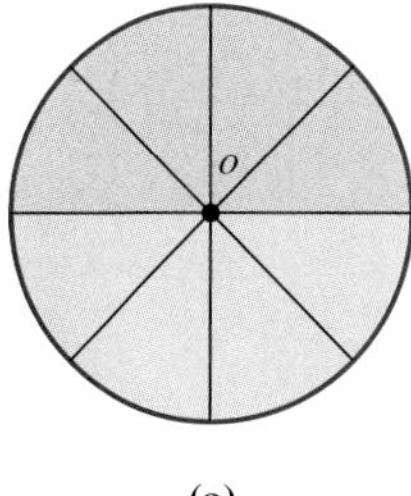

(a)

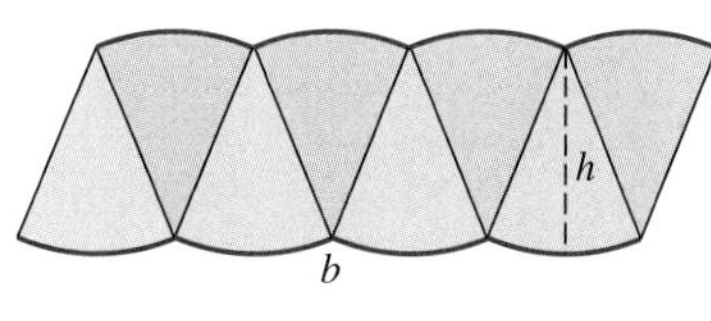

(b)

FIGURE 7-65

If we divide the circle into more and more pie-shaped pieces, the figure will look more and more like a parallelogram, and we can find its area by using the formula for the area of a parallelogram.

$$A = bh$$

$$= \frac{1}{2}Cr \quad \text{Substitute } \frac{1}{2} \text{ of the circumference for } b \text{, and } r \text{ for the height.}$$

$$= \frac{1}{2}(2\pi r)r \quad \text{Make a substitution: } C = 2\pi r.$$

$$= \pi r^2 \quad \text{Simplify: } \tfrac{1}{2} \cdot 2 = 1 \text{ and } r \cdot r = r^2.$$

Area of a circle

The **area of a circle** with radius r is given by the formula

$$A = \pi r^2$$

Self Check 3

To the nearest tenth, find the area of a circle with a diameter of 12 feet.

Answer 113.1 ft^2.

EXAMPLE 3 To the nearest tenth, find the area of the circle in Figure 7-66.

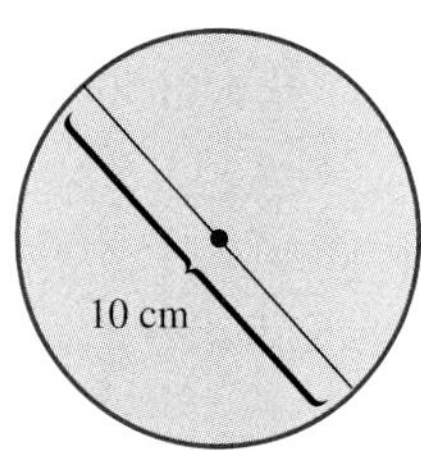

FIGURE 7-66

Solution Since the length of the diameter is 10 centimeters and the length of a diameter is twice the length of a radius, the length of the radius is 5 centimeters. To find the area of the circle, we substitute 5 for r in the formula for the area of a circle.

$$A = \pi r^2$$
$$= \pi(5)^2$$
$$= 25\pi$$
$$\approx 78.53981634 \quad \text{Use a calculator.}$$

To the nearest tenth, the area is 78.5 cm^2.

CALCULATOR SNAPSHOT Painting a helicopter pad

Orange paint is available in gallon containers at $19 each, and each gallon will cover 375 ft^2. To calculate how much the paint will cost to cover a circular helicopter pad 60 feet in diameter, we first calculate the area of the helicopter pad.

$$A = \pi r^2$$
$$= \pi(30)^2 \quad \text{Substitute one-half of 60 for } r.$$
$$= 30^2\pi$$

The area of the pad is $30^2\pi$ ft^2. Since each gallon of paint will cover 375 ft^2, we can find the number of gallons of paint needed by dividing $30^2\pi$ by 375.

$$\text{Number of gallons needed} = \frac{30^2\pi}{375}$$

To approximate the value of $\frac{30^2\pi}{375}$ using a scientific calculator, we enter

30 [x^2] [×] [π] [=] [÷] 375 [=] 7.539822369

The result is approximately 7.54. Because paint comes only in full gallons, the painter will need to purchase 8 gallons. The cost of the paint will be 8($19), or $152.

EXAMPLE 4 Find the shaded area in Figure 7-67.

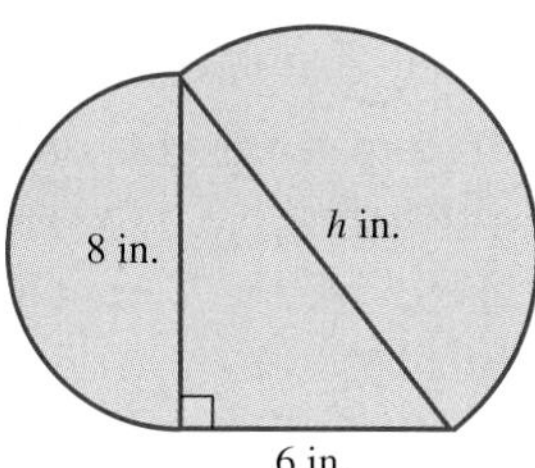

FIGURE 7-67

Solution The figure is a combination of a triangle and two semicircles. By the Pythagorean theorem, the hypotenuse h of the right triangle is

$$h = \sqrt{6^2 + 8^2} = \sqrt{36 + 64} = \sqrt{100} = 10$$

The area of the triangle is

$$A_{\text{right triangle}} = \frac{1}{2}\boldsymbol{bh} = \frac{1}{2}(\mathbf{6})(\mathbf{8}) = 3(8) = 24$$

The area enclosed by the smaller semicircle is

$$A_{\text{smaller semicircle}} = \frac{1}{2}\pi \boldsymbol{r}^2 = \frac{1}{2}\pi(\mathbf{4})^2 = \frac{1}{2}\pi(16) = 8\pi$$

The area enclosed by the larger semicircle is

$$A_{\text{larger semicircle}} = \frac{1}{2}\pi \boldsymbol{r}^2 = \frac{1}{2}\pi(\mathbf{5})^2 = \frac{1}{2}\pi(25) = 12.5\pi$$

The total area is

$$A_{\text{total}} = 24 + 8\pi + 12.5\pi \approx 88.4026494 \quad \text{Use a calculator.}$$

To the nearest hundredth, the area is 88.40 in.^{2}.

Section 7.6 STUDY SET

VOCABULARY *Fill in the blanks.*

1. A segment drawn from the center of a circle to a point on the circle is called a ______.
2. A segment joining two points on a circle is called a ______.
3. A ________ is a chord that passes through the center of a circle.
4. An arc that is one-half of a complete circle is a ________.
5. An arc that is shorter than a semicircle is called a ______ arc.
6. An arc that is longer than a semicircle is called a ______ arc.
7. The distance around a circle is called its ____________.
8. The surface enclosed by a circle is called its _____.

CONCEPTS *Refer to the illustration.*

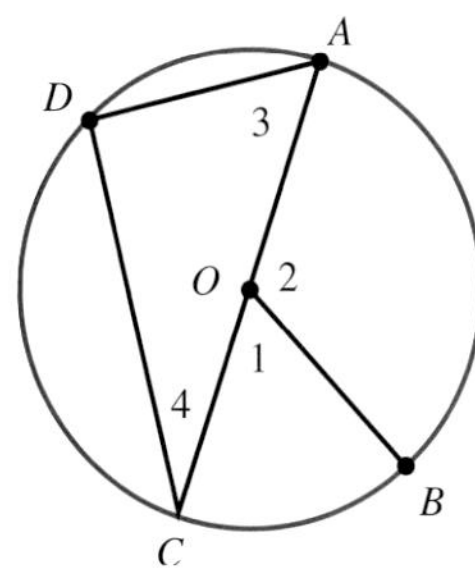

9. Name each radius.
10. Name a diameter.
11. Name each chord.
12. Name each minor arc.
13. Name each semicircle.
14. Name each major arc.

15. If you know the radius of a circle, how can you find its diameter?

16. If you know the diameter of a circle, how can you find its radius?

17. Suppose the two legs of the compass shown on the right are adjusted so that the distance between the pointed ends is 1 inch. Then a circle is drawn.

a. What will the radius of the circle be?

b. What will the diameter of the circle be?

c. What will the circumference of the circle be?

d. What will the area of the circle be?

18. Suppose we find the distance around a can and the distance across the can using a measuring tape, as shown. Then we make a comparison, in the form of a ratio:

$$\frac{\text{the distance around the can}}{\text{the distance across the top of the can}}$$

After we do the indicated division, the result will be close to what number?

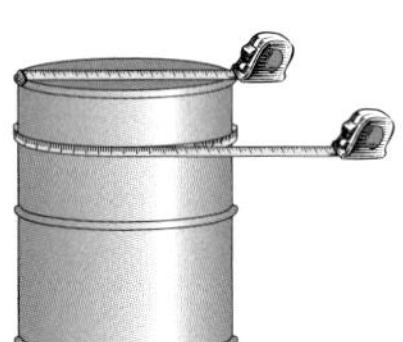

19. When evaluating $\pi(6)^2$, what operation should be performed first?

20. Round $\pi = 3.141592653589\ldots$ to the nearest hundredth.

NOTATION *Fill in the blanks.*

21. The symbol $\widehat{AB}$ is read as ____ ____.

22. To the nearest hundredth, the value of π is ____.

23. The formula for the circumference of a circle is $C =$ ____ or $C = 2\pi$ ____.

24. The formula $A = \pi r^2$ gives the area of a ____.

25. If C is the circumference of a circle and D is its diameter, then $\frac{C}{D} =$ ____.

26. If D is the diameter of a circle and r is its radius, then $D =$ ____ r.

27. Write $\pi(8)$ in a better form.

28. What does $2\pi r$ mean?

PRACTICE *Solve each problem. Answers may vary slightly depending on which approximation of π is used.*

29. To the nearest hundredth, find the circumference of a circle that has a diameter of 12 inches.

30. To the nearest hundredth, find the circumference of a circle that has a radius of 20 feet.

31. Find the diameter of a circle that has a circumference of 36π meters.

32. Find the radius of a circle that has a circumference of 50π meters.

Find the perimeter of each figure to the nearest hundredth.

33. **34.**

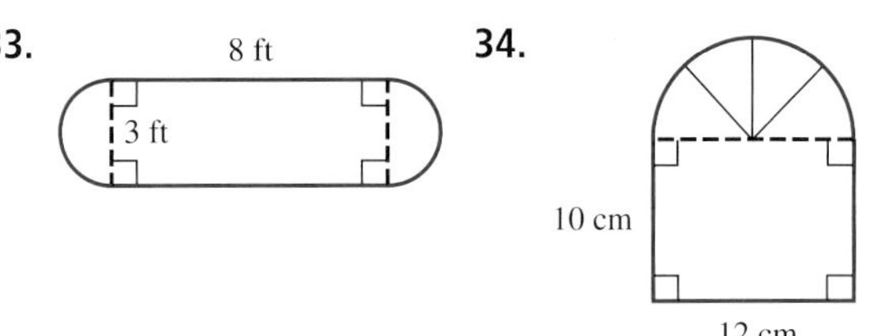

35. **36.**

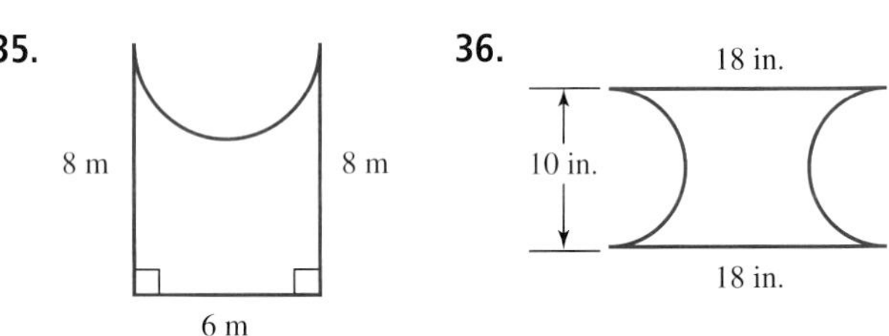

Find the area of each circle to the nearest tenth.

37. **38.**

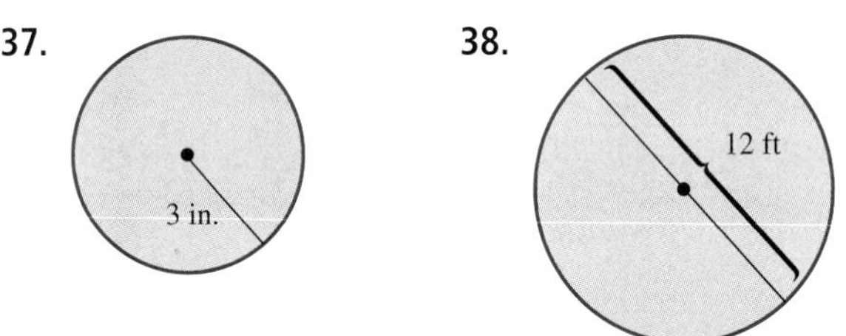

Find the total area of each figure to the nearest tenth.

39.

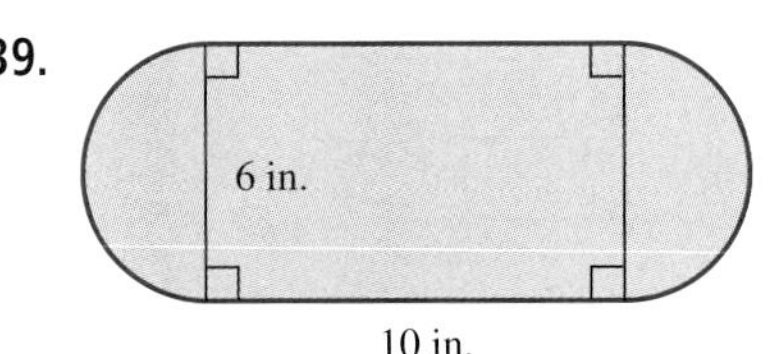

40.

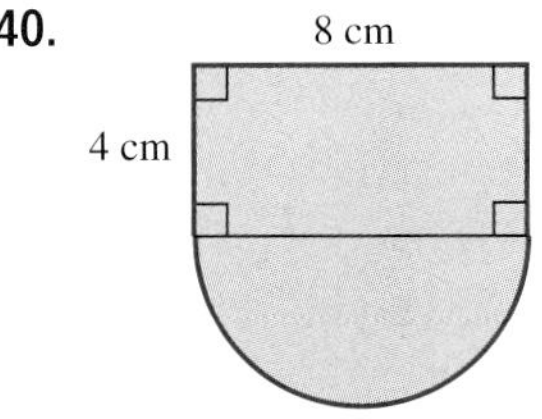

41. 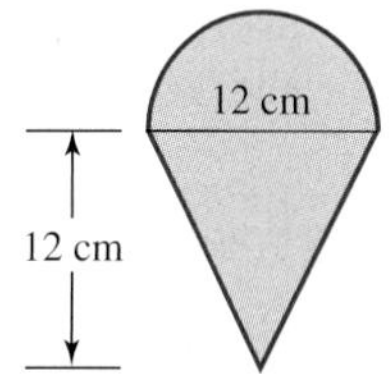

42. 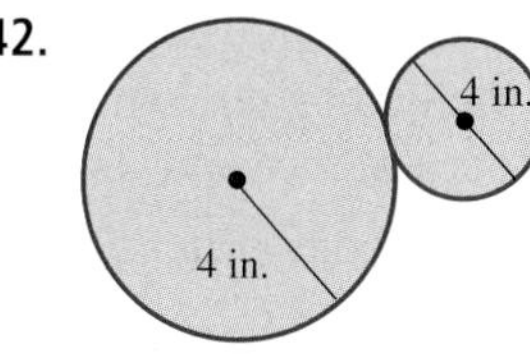

Find the area of each shaded region to the nearest tenth.

43. 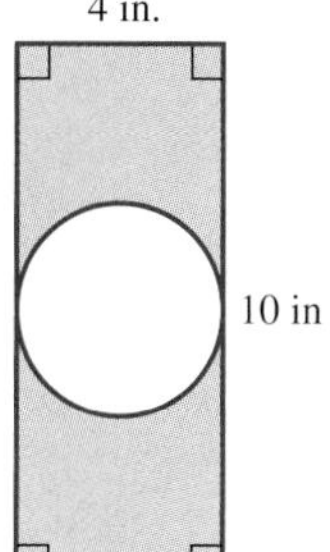

44. 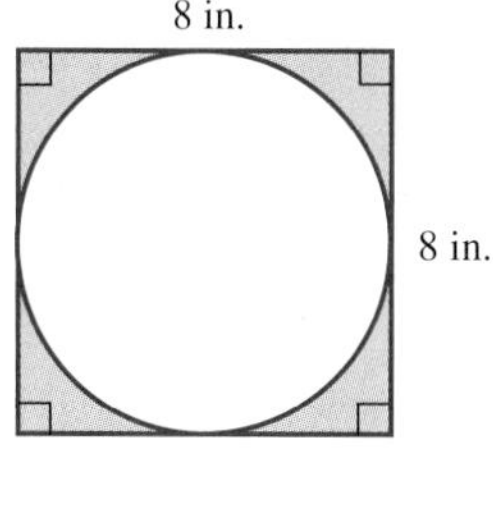

45. 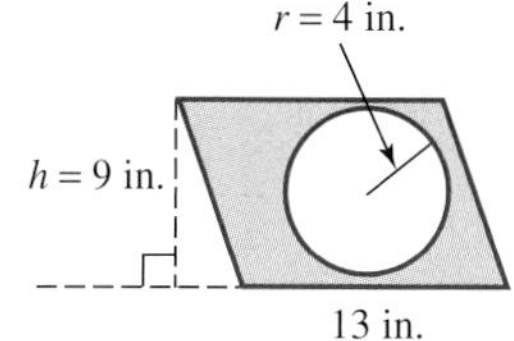

46. 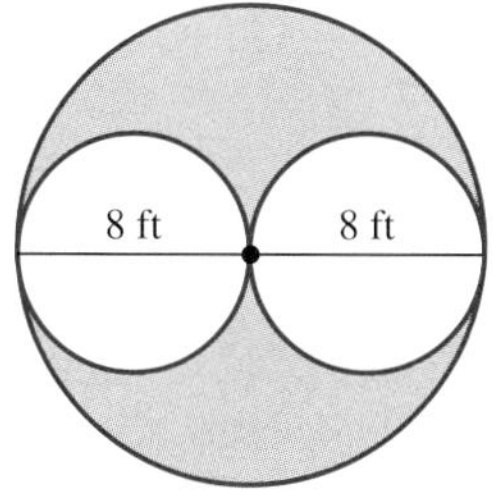

APPLICATIONS *Give each answer to the nearest hundredth. Answers may vary slightly depending on which approximation of π is used.*

47. AREA OF ROUND LAKE Round Lake has a circular shoreline that is 2 miles in diameter. Find the area of the lake.

48. HELICOPTERS How far does a point on the tip of a rotor blade travel when it makes one complete revolution?

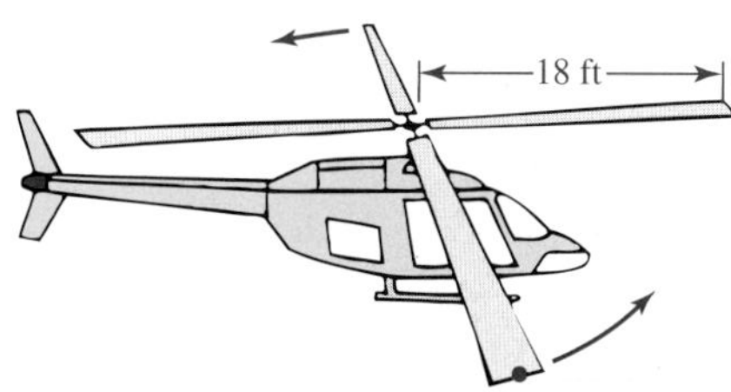

49. GIANT SEQUOIAS The largest sequoia tree is the General Sherman Tree in Sequoia National Park in California. In fact, it is considered to be the largest living thing in the world. According to the *Guinness Book of World Records,* it has a circumference of 102.6 feet, measured $4\frac{1}{2}$ feet above the ground. What is the diameter of the tree at that height?

50. TRAMPOLINES The distance from the center of the trampoline to the edge of its steel frame is 7 feet. The protective padding covering the springs is 15 inches wide. Find the area of the circular jumping surface of the trampoline, in square feet.

51. JOGGING Joan wants to jog 10 miles on a circular track $\frac{1}{4}$ mile in diameter. How many times must she circle the track?

52. FIXING THE ROTUNDA The rotunda at a state capitol is a circular area 100 feet in diameter. The legislature wishes to appropriate money to have the floor of the rotunda tiled. The lowest bid is $83 per square yard, including installation. How much must the legislature spend?

53. BANDING THE EARTH A steel band is drawn tightly about Earth's equator. The band is then loosened by increasing its length by 10 feet, and the resulting slack is distributed evenly along the band's entire length. How far above Earth's surface is the band? (*Hint:* You don't need to know Earth's circumference.)

54. CONCENTRIC CIRCLES Two circles are called **concentric circles** if they have the same center. Find the area of the band between two concentric circles if their diameters are 10 centimeters and 6 centimeters.

55. ARCHERY See the illustration. Find the area of the entire target and the bull's eye. What percent of the area of the target is the bull's eye?

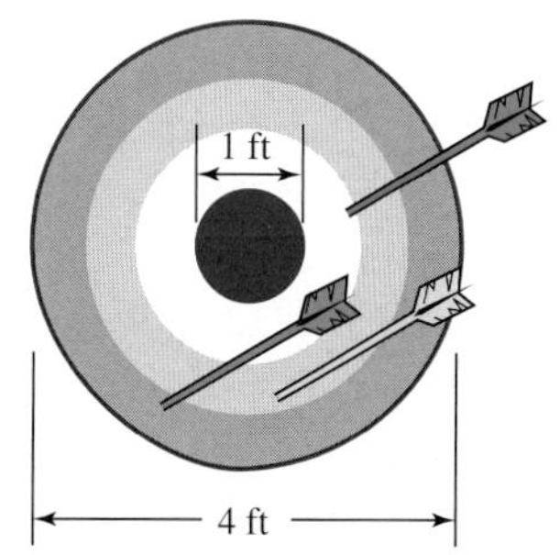

56. LANDSCAPE DESIGNS See the illustration. How much of the lawn does not get watered by the sprinklers at the center of each circle?

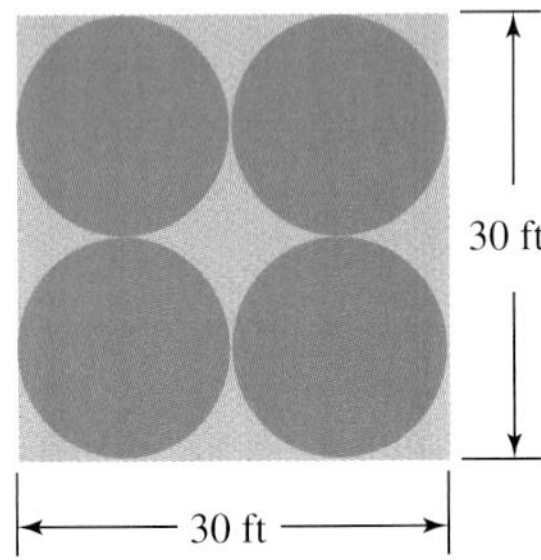

WRITING

57. Explain what is meant by the circumference of a circle.
58. Explain what is meant by the area of a circle.
59. Explain the meaning of π.
60. Distinguish between a major arc and a minor arc.
61. Explain what it means for a car to have a small turning radius.
62. The word *circumference* means the distance around a circle. In your own words, explain what is meant by each of the following sentences.
    a. A boat owner's dream was to *circumnavigate* the globe.
    b. The teenager's parents felt that he was always trying to *circumvent* the rules.
    c. The class was shown a picture of a circle *circumscribed* about an equilateral triangle.

REVIEW

63. Change $\frac{9}{10}$ to a percent.
64. Change $\frac{7}{8}$ to a percent.
65. How many sides does a pentagon have?
66. What is the sum of the measures of the angles of a triangle?

7.7 Surface Area and Volume

- Volumes of solids • Surface areas of rectangular solids
- Volumes and surface areas of spheres • Volumes of cylinders
- Volumes of cones • Volumes of pyramids

In this section, we discuss a measure of capacity called **volume.** Volumes are measured in cubic units, such as cubic inches, cubic yards, or cubic centimeters. For example,

- We buy gravel or topsoil by the cubic yard.
- We measure the capacity of a refrigerator in cubic feet.
- We often measure amounts of medicine in cubic centimeters.

We also discuss surface area. The ability to compute surface area is necessary to solve problems such as calculating the amount of material necessary to make a cardboard box or a plastic beach ball.

Volumes of solids

A **rectangular solid** and a **cube** are two common geometric solids. (See Figure 7-68 on the next page.)

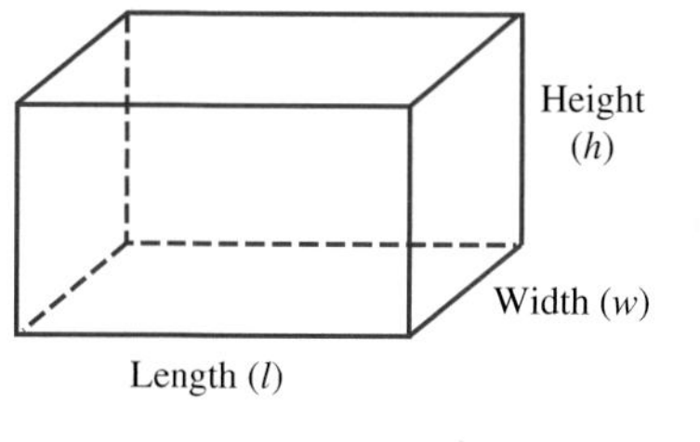

A rectangular solid

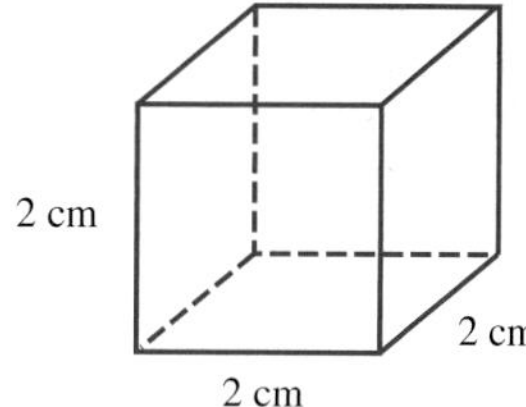

A cube

FIGURE 7-68

The **volume** of a rectangular solid is a measure of the space it encloses. Two common units of volume are cubic inches (in.^{3}) and cubic centimeters (cm^3). (See Figure 7-69.)

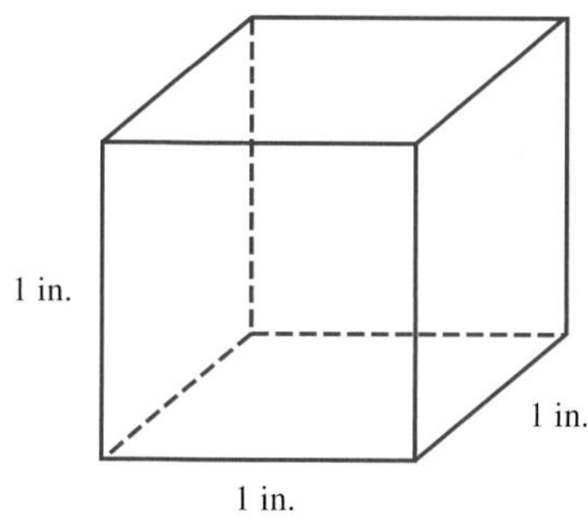

1 cubic inch (1 in.^{3})

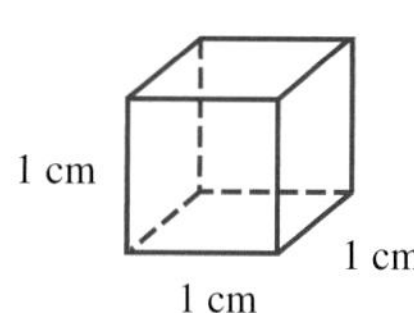

1 cubic centimeter (1 cm^3)

FIGURE 7-69

If we divide the rectangular solid shown in Figure 7-70 into cubes, each cube represents a volume of 1 cm^3. Because there are 2 levels with 12 cubes on each level, the volume of the rectangular solid is 24 cm^3.

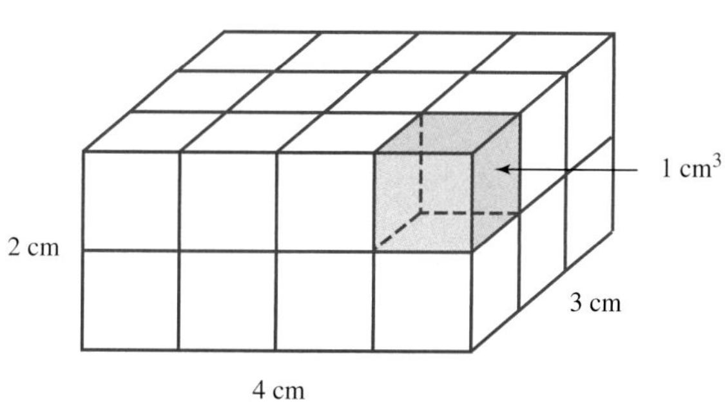

FIGURE 7-70

In practice, we do not find volumes by counting cubes. Instead, we use the formulas shown in Table 7-2.

Figure	Name	Volume	Figure	Name	Volume
	Cube	$V = s^3$		Cylinder	$V = \pi r^2 h$ or $V = Bh$*
	Rectangular solid	$V = lwh$		Cone	$V = \frac{1}{3}\pi r^2 h$ or $V = \frac{1}{3}Bh$*

*B represents the area of the base that is shaded in the figure.

(*continued*)

Figure	Name	Volume	Figure	Name	Volume
h	Prism	$V = Bh^*$	h	Pyramid	$V = \frac{1}{3}Bh^*$
r	Sphere	$V = \frac{4}{3}\pi r^3$			

$^*B$ represents the area of the base that is shaded in the figure.

TABLE 7-2

! COMMENT The height of a geometric solid is always measured along a line perpendicular to its base. In each of the solids in Figure 7-71, h is the height.

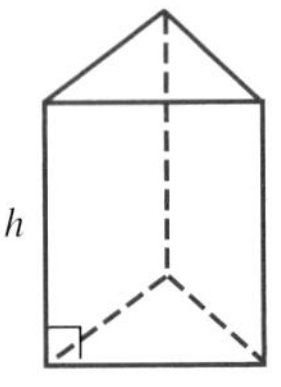

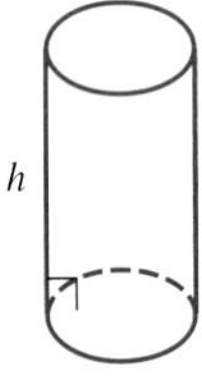

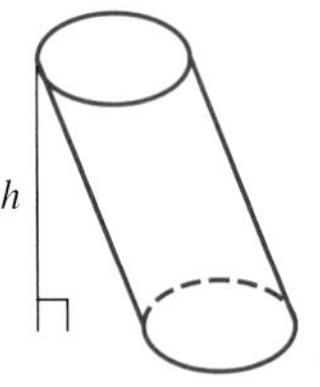

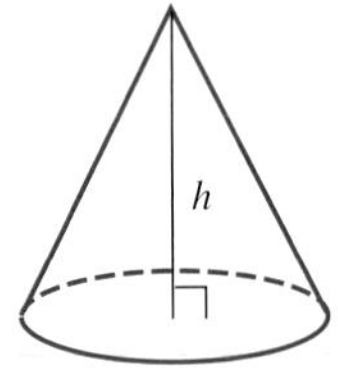

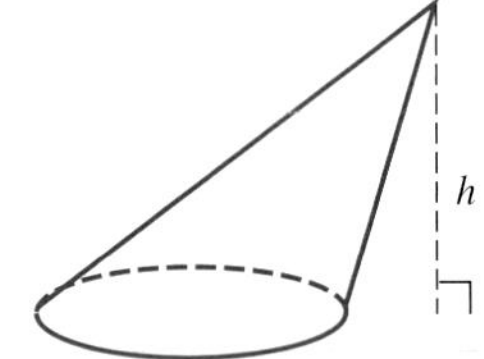

FIGURE 7-71

Self Check 1
How many cubic centimeters are in 1 cubic meter?

Answer 1,000,000 cm^3

EXAMPLE 1 How many cubic inches are there in 1 cubic foot? (See Figure 7-72.)

Solution Since a cubic foot is a cube with each side measuring 1 foot, each side also measures 12 inches. Thus, the volume in cubic inches is

$$V = s^3 \quad \text{This is the formula for the volume of a cube.}$$
$$= (\mathbf{12})^3 \quad \text{Substitute 12 for } s.$$
$$= 1{,}728$$

There are 1,728 cubic inches in 1 cubic foot.

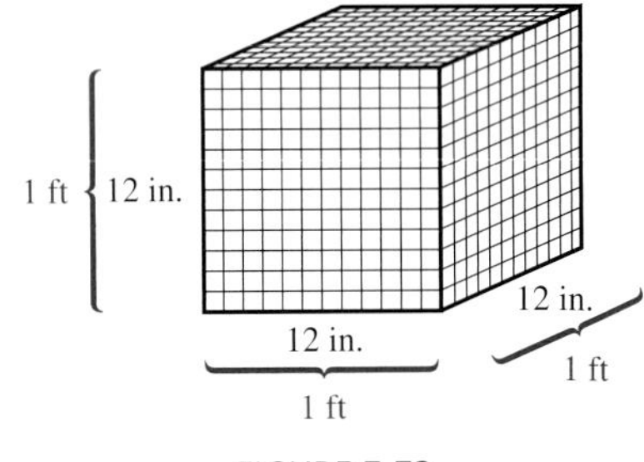

FIGURE 7-72

Self Check 2
Find the volume of a rectangular solid with dimensions of 8 by 12 by 20 meters.

EXAMPLE 2 **Volume of an oil storage tank.** An oil storage tank is in the form of a rectangular solid with dimensions of 17 by 10 by 8 feet. (See Figure 7-73.) Find its volume.

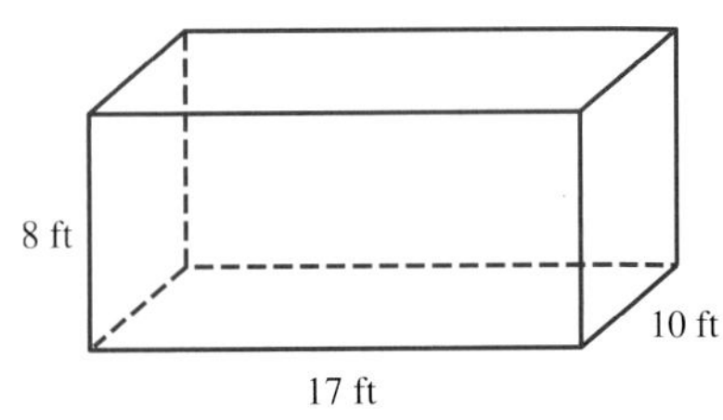

FIGURE 7-73

Solution To find the volume, we substitute 17 for l, 10 for w, and 8 for h in the formula $V = lwh$ and simplify.

$$\begin{aligned} V &= lwh \\ &= 17(10)(8) \\ &= 1{,}360 \end{aligned}$$

The volume is 1,360 ft³.

Answer 1,920 m³

EXAMPLE 3 Volume of a triangular prism.

Find the volume of the triangular prism in Figure 7-74.

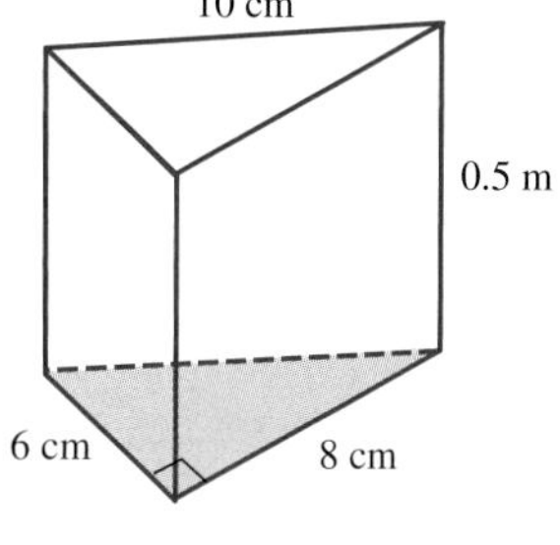

FIGURE 7-74

Solution The volume of the prism is the area of its base multiplied by its height. Since there are 100 centimeters in 1 meter, the height in centimeters is

$$\begin{aligned} 0.5 \text{ m} &= 0.5(\mathbf{1\ m}) \\ &= 0.5(\mathbf{100\ cm}) && \text{Substitute 100 centimeters for 1 meter.} \\ &= 50 \text{ cm} \end{aligned}$$

The area of the triangular base is $\frac{1}{2}(6)(8) = 24$ square centimeters. The height of the prism is 50 centimeters. Substituting into the formula for the volume of a prism, we have

$$\begin{aligned} V &= Bh \\ &= 24(50) \\ &= 1{,}200 \end{aligned}$$

The volume of the prism is 1,200 cm³.

Self Check 3

Find the volume of the triangular prism below.

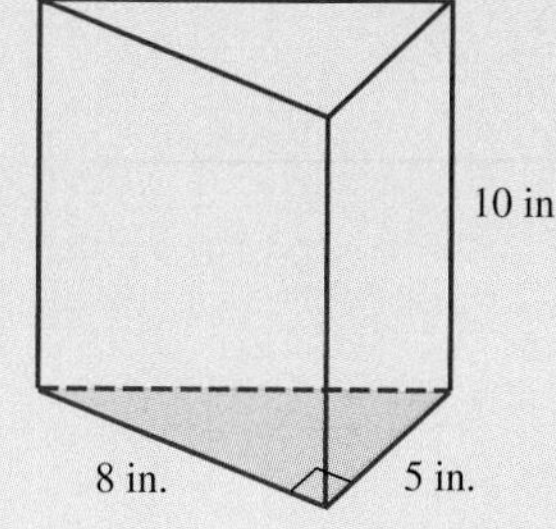

Answer 200 in.³

Surface areas of rectangular solids

The **surface area** of a rectangular solid is the sum of the areas of its six faces. Figure 7-75 shows how we can unfold the faces of a cardboard box to derive a formula for its surface area (SA).

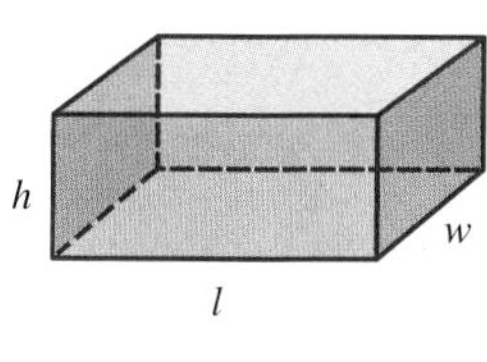

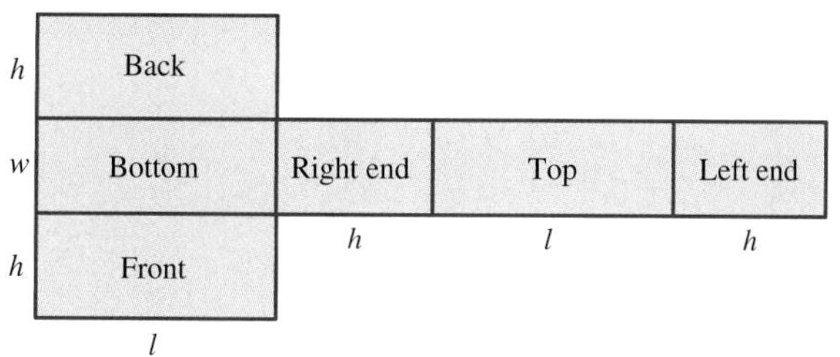

FIGURE 7-75

$$\begin{aligned} SA &= A_{\text{bottom}} + A_{\text{back}} + A_{\text{front}} + A_{\text{right end}} + A_{\text{top}} + A_{\text{left end}} \\ &= lw + lh + lh + hw + lw + hw \\ &= 2lw + 2lh + 2hw && \text{Combine like terms.} \end{aligned}$$

Surface area of a rectangular solid

The surface area of a rectangular solid is given by the formula

$$SA = 2lw + 2lh + 2hw$$

where l is the length, w is the width, and h is the height.

Self Check 4
Find the surface area of a rectangular solid with dimensions of 8 by 12 by 20 meters.

Answer 992 m^2

EXAMPLE 4 Surface area of an oil tank. An oil storage tank is in the form of a rectangular solid with dimensions of 17 by 10 by 8 feet. (See Figure 7-76.) Find the surface area of the tank.

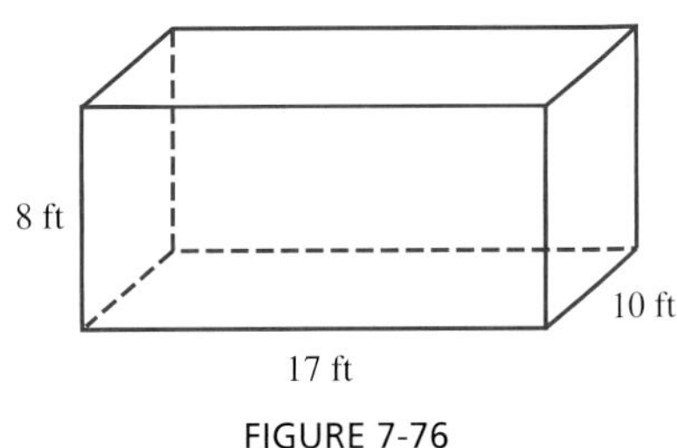

FIGURE 7-76

Solution To find the surface area, we substitute 17 for l, 10 for w, and 8 for h in the formula for surface area and simplify.

$$\begin{aligned} SA &= 2lw + 2lh + 2hw \\ &= 2(\mathbf{17})(\mathbf{10}) + 2(\mathbf{17})(\mathbf{8}) + 2(\mathbf{8})(\mathbf{10}) \\ &= 340 + 272 + 160 \\ &= 772 \end{aligned}$$

The surface area is 772 ft^2.

Volumes and surface areas of spheres

A **sphere** is a hollow, round ball. (See Figure 7-77.) The points on a sphere all lie at a fixed distance r from a point called its *center.* A segment drawn from the center of a sphere to a point on the sphere is called a *radius.*

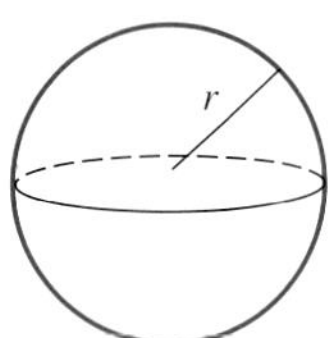

FIGURE 7-77

CALCULATOR SNAPSHOT

Filling a water tank

See Figure 7-78. To calculate how many cubic feet of water are needed to fill a spherical water tank with a radius of 15 feet, we substitute 15 for r in the formula for the volume of a sphere.

$$\begin{aligned} V &= \frac{4}{3}\pi r^3 \\ &= \frac{4}{3}\pi(\mathbf{15})^3 \end{aligned}$$

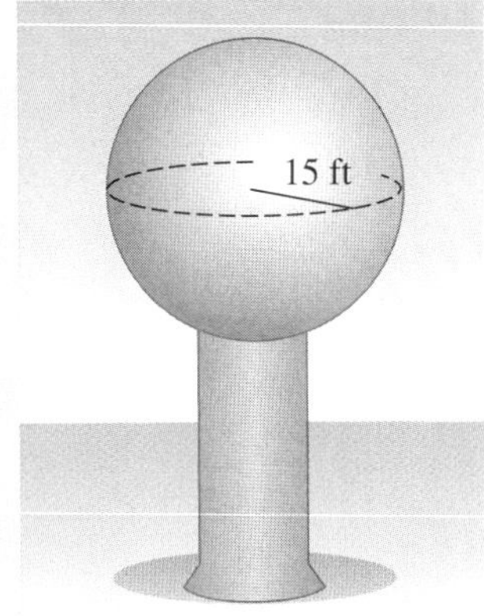

FIGURE 7-78

To approximate the value of $\frac{4}{3}\pi(15)^3$ using a scientific calculator, we enter

15 $\boxed{y^x}$ 3 $\boxed{=}$ $\boxed{\times}$ 4 $\boxed{\div}$ 3 $\boxed{=}$ $\boxed{\times}$ $\boxed{\pi}$ $\boxed{=}$ 14137.16694

To the nearest tenth, $14{,}137.2 \text{ ft}^3$ of water are needed to fill the tank.

There is a formula to find the surface area of a sphere.

Surface area of a sphere

The surface area of a sphere with radius r is given by the formula

$$SA = 4\pi r^2$$

EXAMPLE 5 **Manufacturing beach balls.** A beach ball is to have a diameter of 16 inches. (See Figure 7-79.) How many square inches of material will be needed to make the ball? (Disregard any waste.)

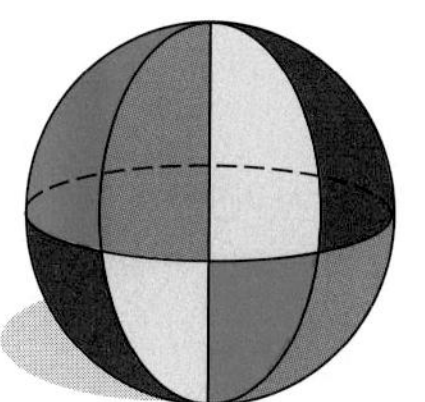

FIGURE 7-79

Solution Since a radius r of the ball is one-half the diameter, $r = 8$ inches. We can now substitute 8 for r in the formula for the surface area of a sphere.

$$\begin{aligned} SA &= 4\pi r^2 \\ &= 4\pi(\mathbf{8})^2 \\ &= 4\pi(64) \\ &= 256\pi && \text{Simplify: } 4 \cdot 64 = 256. \\ &\approx 804.2477193 && \text{Use a calculator.} \end{aligned}$$

A little more than 804 in.^2 of material is needed to make the ball.

Volumes of cylinders

A **cylinder** is a hollow figure like a piece of pipe. (See Figure 7-80.)

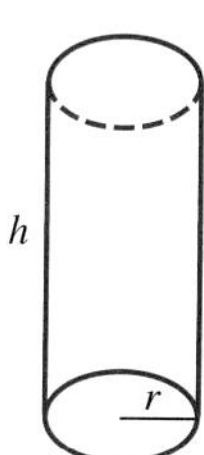

FIGURE 7-80

EXAMPLE 6 Find the volume of the cylinder in Figure 7-81.

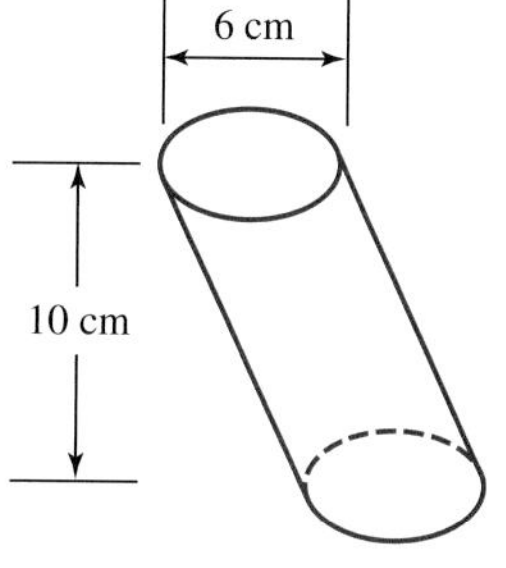

FIGURE 7-81

Solution Since a radius is one-half the diameter of the circular base, $r = 3$ cm. From the figure, we see that the height of the cylinder is 10 cm. So we can substitute 3 for r and 10 for h in the formula for the volume of a cylinder.

$$\begin{aligned} V &= \pi r^2 h \\ &= \pi(\mathbf{3})^2(\mathbf{10}) \\ &= 90\pi && \text{Simplify: } (3)^2(10) = 9(10) = 90. \\ &\approx 282.7433388 && \text{Use a calculator.} \end{aligned}$$

To the nearest hundredth, the volume of the cylinder is 282.74 cm^3.

CALCULATOR SNAPSHOT Volume of a silo

A silo is a structure used for storing grain. The silo in Figure 7-82 is a cylinder 50 feet tall topped with a **hemisphere** (a half-sphere). To find the volume of the silo, we add the volume of the cylinder to the volume of the dome.

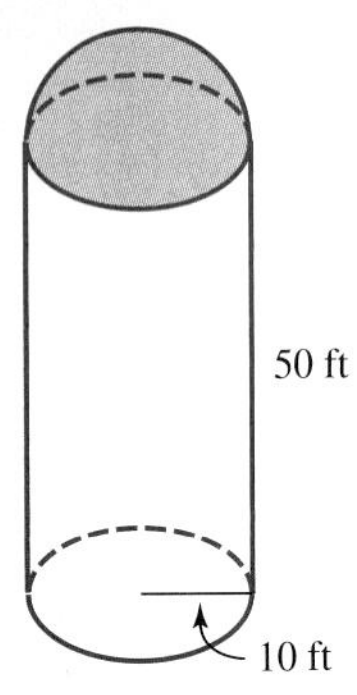

FIGURE 7-82

$$\text{Volume}_{\text{cylinder}} + \text{volume}_{\text{dome}} = (\text{area}_{\text{cylinder's base}})(\text{height}_{\text{cylinder}}) + \frac{1}{2}(\text{volume}_{\text{sphere}})$$

$$= \pi r^2 h + \frac{1}{2}\left(\frac{4}{3}\pi r^3\right)$$

$$= \pi r^2 h + \frac{2\pi r^3}{3} \qquad \frac{1}{2}\left(\frac{4}{3}\pi r^3\right) = \frac{1}{2}\cdot\frac{4}{3}\pi r^3 = \frac{4}{6}\pi r^3 = \frac{2\pi r^3}{3}$$

$$= \pi(10)^2(50) + \frac{2\pi(10)^3}{3} \qquad \text{Substitute 10 for } r \text{ and 50 for } h.$$

To approximate the value of $\pi(10)^2(50) + \frac{2\pi(10)^3}{3}$ using a scientific calculator, we enter

[π] [×] 10 [x^2] [×] 50 [=] [+] [(] 2 [×] [π] [×] 10 [y^x] 3
[÷] 3 [)] [=]

17802.35837

The volume of the silo is approximately 17,802 ft^3.

EXAMPLE 7 **Machining a block of metal.** See Figure 7-83. Find the volume that is left when the hole is drilled through the metal block.

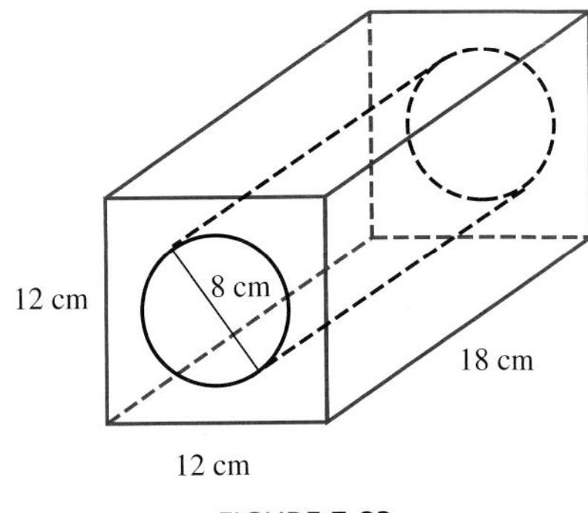

FIGURE 7-83

Solution We must find the volume of the rectangular solid and then subtract the volume of the cylinder. We will think of the rectangular solid and the cylinder as lying on their sides. Thus, the height is 18 cm when we find each volume.

$$V_{\text{rect. solid}} = lwh$$
$$= 12(12)(18)$$
$$= 2{,}592$$

$$V_{\text{cylinder}} = \pi r^2 h$$
$$= \pi(4)^2(18)$$
$$= 288\pi \qquad \text{Simplify: } (4)^2(18) = 16(18) = 288.$$
$$\approx 904.7786842 \qquad \text{Use a calculator.}$$

$$
\begin{aligned}
V_{\text{drilled block}} &= V_{\text{rect. solid}} - V_{\text{cylinder}} \\
&\approx 2{,}592 - 904.7786842 \\
&\approx 1{,}687.221316 \qquad \text{Use a calculator.}
\end{aligned}
$$

To the nearest hundredth, the volume is 1,687.22 cm^3.

Volumes of cones

Two **cones** are shown in Figure 7-84. Each cone has a height h and a radius r, which is the radius of the circular base.

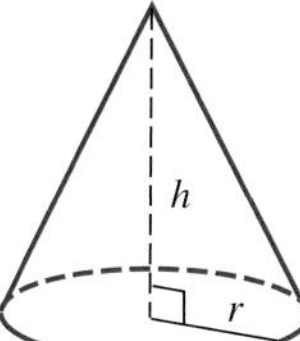

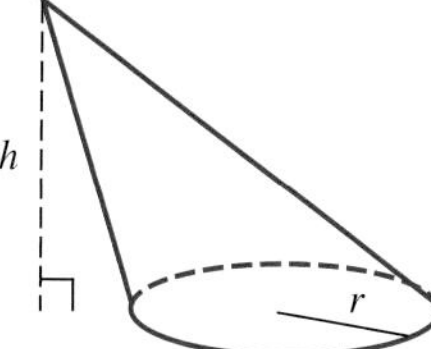

FIGURE 7-84

EXAMPLE 8 To the nearest tenth, find the volume of the cone in Figure 7-85.

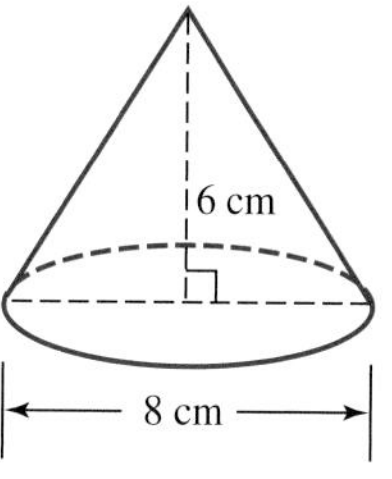

FIGURE 7-85

Solution Since the radius is one-half the diameter, $r = 4$ cm. We then substitute 4 for r and 6 for h in the formula for the volume of a cone.

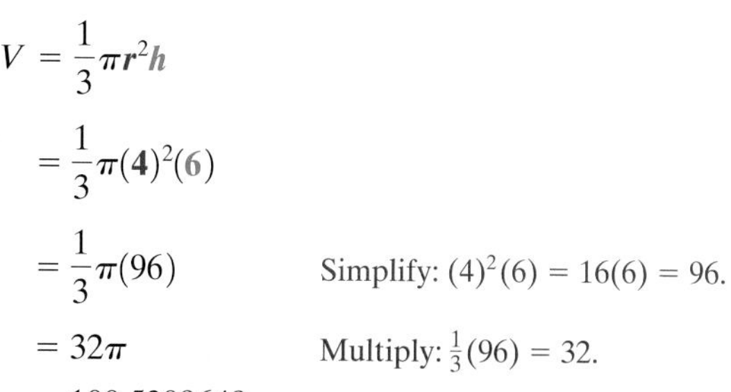

$$
\begin{aligned}
V &= \frac{1}{3}\pi r^2 h \\
&= \frac{1}{3}\pi(4)^2(6) \\
&= \frac{1}{3}\pi(96) \qquad \text{Simplify: } (4)^2(6) = 16(6) = 96. \\
&= 32\pi \qquad \text{Multiply: } \tfrac{1}{3}(96) = 32. \\
&\approx 100.5309649
\end{aligned}
$$

To the nearest tenth, the volume is 100.5 cubic centimeters.

Volumes of pyramids

Two **pyramids** with a height h are shown in Figure 7-86.

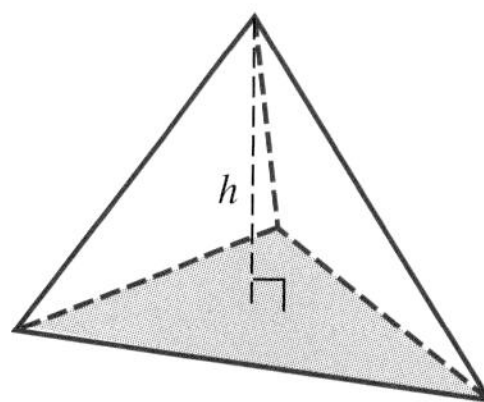

The base is a triangle.

(a)

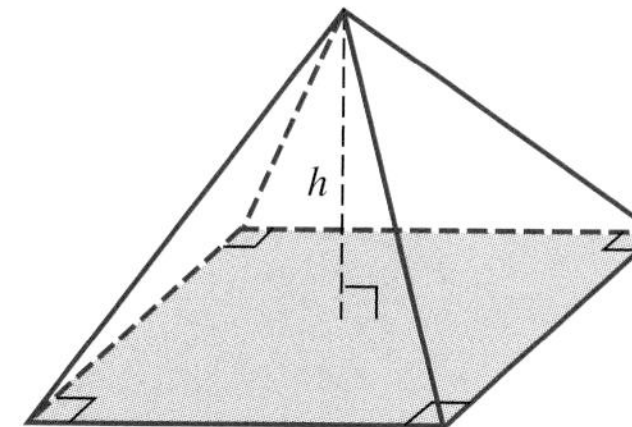

The base is a square.

(b)

FIGURE 7-86

Self Check 9
Find the volume of the pyramid shown below.

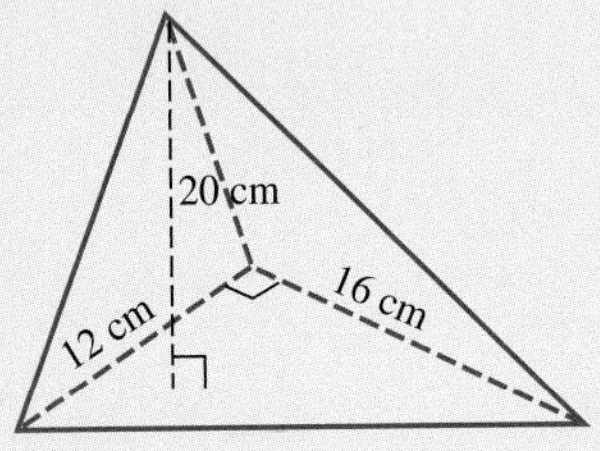

Answer 640 cm^3

EXAMPLE 9 Find the volume of a pyramid that has a square base with each side 6 meters long and a height of 9 meters.

Solution Since the base is a square with each side 6 meters long, the area of the base is 6^2 m^2, or 36 m^2. We can then substitute 36 for the area of the base and 9 for the height in the formula for the volume of a pyramid.

$$V = \frac{1}{3}\mathbf{B}\mathbf{h}$$

$$= \frac{1}{3}(\mathbf{36})(\mathbf{9})$$

$$= 12(9) \qquad \text{Multiply: } \tfrac{1}{3}(36) = 12.$$

$$= 108$$

The volume of the pyramid is 108 m^3.

Section 7.7 STUDY SET

VOCABULARY *Fill in the blanks.*

1. The space contained within a geometric solid is called its ________.
2. A ___________ solid is like a hollow shoe box.
3. A ______ is a rectangular solid with all sides of equal length.
4. The volume of a cube with each side 1 inch long is 1 ______ inch.
5. The ________ area of a rectangular solid is the sum of the areas of its faces.
6. The point that is equidistant from every point on a sphere is its ________.
7. A ________ is a hollow figure like a drinking straw.
8. A ___________ is one-half of a sphere.
9. A ______ looks like a witch's pointed hat.
10. A figure that has a polygon for its base and that rises to a point is called a ________.

CONCEPTS *Write the formula used for finding the volume of each solid.*

11. A rectangular solid
12. A prism
13. A sphere
14. A cylinder
15. A cone
16. A pyramid
17. Write the formula for finding the surface area of a rectangular solid.
18. Write the formula for finding the surface area of a sphere.
19. How many cubic feet are in 1 cubic yard?
20. How many cubic inches are in 1 cubic yard?
21. How many cubic decimeters are in 1 cubic meter?
22. How many cubic millimeters are in 1 cubic centimeter?

Which geometric concept (perimeter, circumference, area, volume, or surface area) should be applied to find each of the following?

23. **a.** The size of a room to be air conditioned
    b. The amount of land in a national park
    c. The amount of space in a refrigerator freezer
    d. The amount of cardboard in a shoe box
    e. The distance around a checkerboard
    f. The amount of material used to make a basketball
24. **a.** The amount of cloth in a car cover
    b. The size of a trunk of a car
    c. The amount of paper used for a postage stamp
    d. The amount of storage in a cedar chest
    e. The amount of beach available for sunbathing
    f. The distance the tip of a propeller travels

25. In the following illustration, the unit of measurement of length that was used to draw the figure was the inch.

a. What is the volume of the figure?

b. What is the area of the front of the figure?

c. What is the area of the base of the figure?

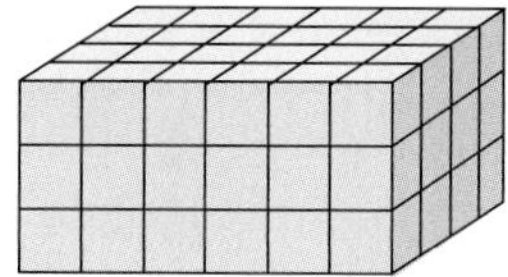

26. The cardboard box shown is a cube. Suppose the six faces were unfolded to lie flat on a table. Draw a picture of what this would look like.

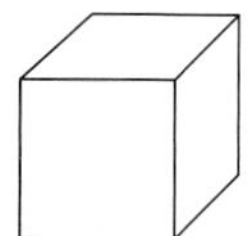

NOTATION *Fill in the blanks.*

27. The notation 1 in.^{3} is read as one ______ ______.

28. One cubic centimeter is represented as 1 cm__.

PRACTICE *Find the volume of each solid. If an answer is not exact, round to the nearest hundredth. (Answers may vary slightly, depending on which approximation of π is used.)*

29. A rectangular solid with dimensions of 3 by 4 by 5 centimeters

30. A rectangular solid with dimensions of 5 by 8 by 10 meters

31. A prism whose base is a right triangle with legs 3 and 4 meters long and whose height is 8 meters

32. A prism whose base is a right triangle with legs 5 and 12 feet long and whose height is 10 feet

33. A sphere with a radius of 9 inches

34. A sphere with a diameter of 10 feet

35. A cylinder with a height of 12 meters and a circular base with a radius of 6 meters

36. A cylinder with a height of 4 meters and a circular base with a diameter of 18 meters

37. A cone with a height of 12 centimeters and a circular base with a diameter of 10 centimeters

38. A cone with a height of 3 inches and a circular base with a radius of 4 inches

39. A pyramid with a square base 10 meters on each side and a height of 12 meters

40. A pyramid with a square base 6 inches on each side and a height of 4 inches

Find the surface area of each solid. If an answer is not exact, round to the nearest hundredth.

41. A rectangular solid with dimensions of 3 by 4 by 5 centimeters

42. A cube with a side 5 centimeters long

43. A sphere with a radius of 10 inches

44. A sphere with a diameter of 12 meters

Find the volume of each figure. If an answer is not exact, round to the nearest hundredth. (Answers may vary slightly, depending on which approximation of π is used.)

45.

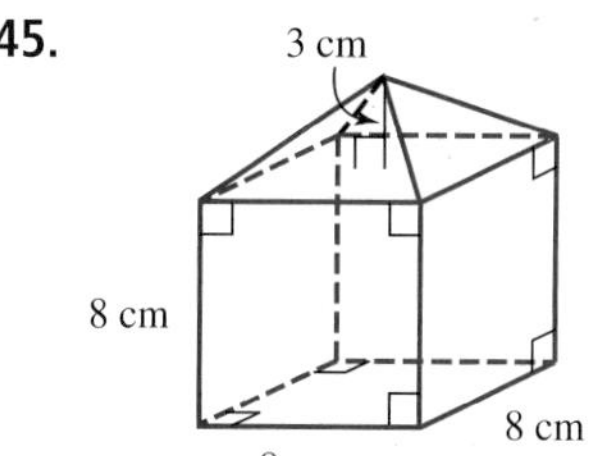

46.

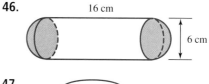

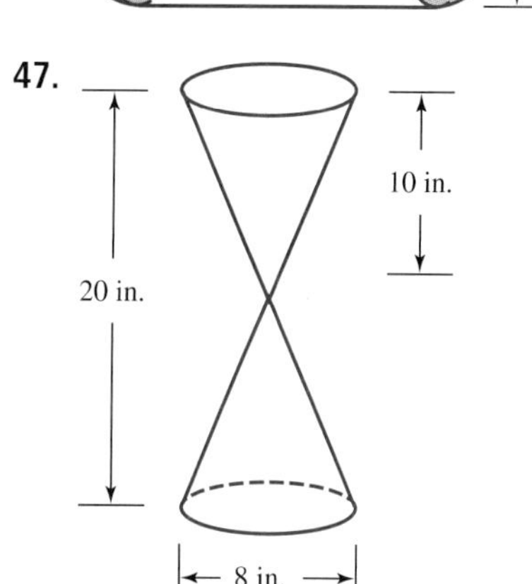

47.

20 in.
10 in.
8 in.

48.

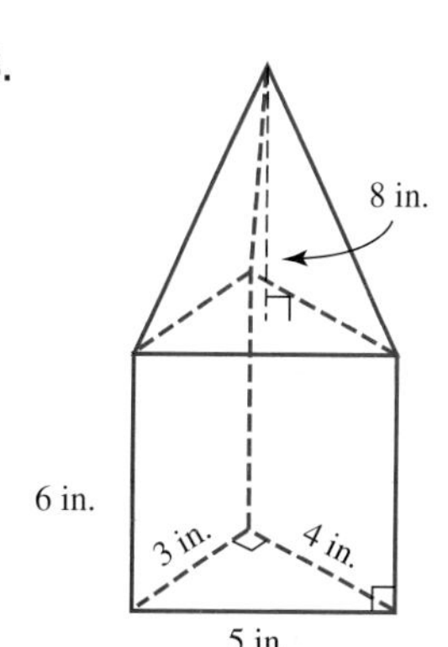

APPLICATIONS *Solve each problem. If an answer is not exact, round to the nearest hundredth.*

49. VOLUME OF A SUGAR CUBE A sugar cube is $\frac{1}{2}$ inch on each edge. How much volume does it occupy?

50. VOLUME OF A CLASSROOM A classroom is 40 feet long, 30 feet wide, and 9 feet high. Find the number of cubic feet of air in the room.

51. WATER HEATERS Complete the ad for the high-efficiency water heater shown.

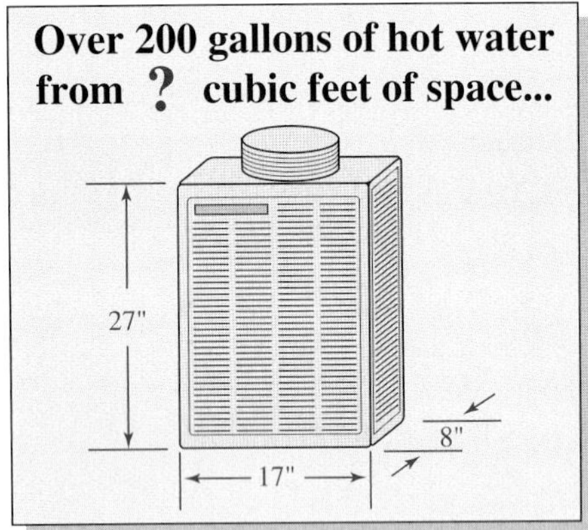

52. REFRIGERATOR CAPACITY The largest refrigerator advertised in a J. C. Penney catalog has a capacity of 25.2 cubic feet. How many cubic inches is this?

53. VOLUME OF AN OIL TANK A cylindrical oil tank has a diameter of 6 feet and a length of 7 feet. Find the volume of the tank.

54. VOLUME OF A DESSERT A restaurant serves pudding in a conical dish that has a diameter of 3 inches. If the dish is 4 inches deep, how many cubic inches of pudding are in each dish?

55. HOT-AIR BALLOONS The lifting power of a spherical balloon depends on its volume. How many cubic feet of gas will a balloon hold if it is 40 feet in diameter?

56. VOLUME OF A CEREAL BOX A box of cereal measures 3 by 8 by 10 inches. The manufacturer plans to market a smaller box that measures $2\frac{1}{2}$ by 7 by 8 inches. By how much will the volume be reduced?

57. ENGINES The *compression ratio* of an engine is the volume in one cylinder with the piston at bottom-dead-center (B.D.C.), divided by the volume with the piston at top-dead-center (T.D.C.). From the data given in the illustration in the next column, what is the compression ratio of the engine? Use a colon to express your answer as a ratio.

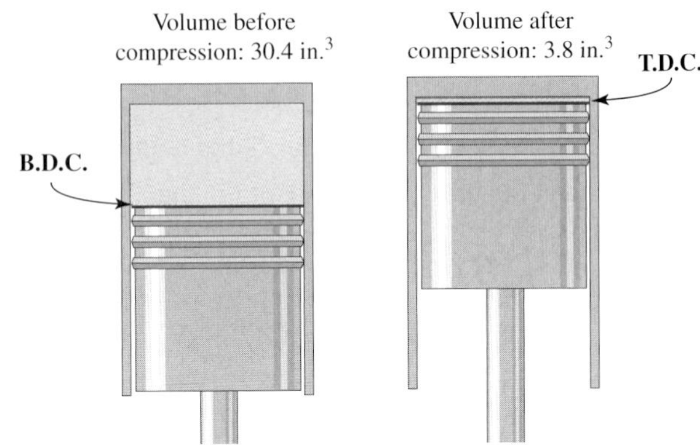

58. LINT REMOVERS The illustration shows a handy gadget; it uses a cylinder of sheets of sticky paper that can be rolled over clothing and furniture to pick up lint and pet hair. After the paper is full, that sheet is peeled away to expose another sheet of sticky paper. Find the area of the first sheet by using the formula $LSA = 2\pi rh$, where LSA represents the lateral surface area of the cylinder.

WRITING

59. What is meant by the *volume* of a cube?

60. What is meant by the *surface area* of a cube?

61. Are the units used to measure area different from the units used to measure volume? Explain.

62. The dimensions (length, width, and height) of one rectangular solid are entirely different numbers from the dimensions of another rectangular solid. Would it be possible for the rectangular solids to have the same volume? Explain.

REVIEW

63. Evaluate: $-5(5 - 2)^2 + 3$.

64. BUYING PENCILS Carlos bought 6 pencils at \$0.60 each and a notebook for \$1.25. He gave the clerk a \$5 bill. How much change did he receive?

65. Solve: $\frac{x}{-4} = \frac{1}{4}$.

66. 38 is what percent of 40?

KEY CONCEPT

Formulas

A **formula** is a mathematical expression that is used to express a relationship between quantities. We have studied formulas used in mathematics, business, geometry, and science.

Write a formula describing the mathematical relationship between the given quantities.

1. Distance traveled (d), rate traveled (r), time traveling at that rate (t)
2. Sale price (s), original price (p), discount (d)
3. Perimeter of a rectangle (P), length of the rectangle (l), width of the rectangle (w)
4. Amount of simple interest earned (I), principal (P), interest rate (r), time the money is invested (t)

Use a formula to solve each problem.

5. Find the area (A) of the triangular lot.

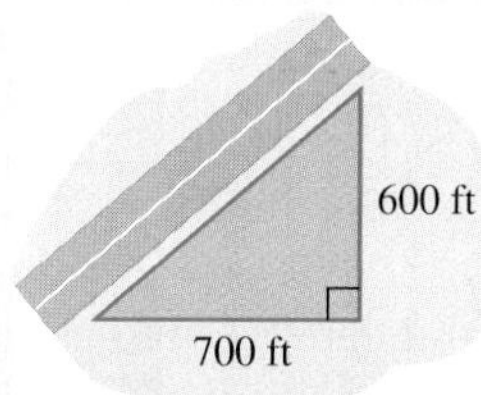

6. Find the volume (V) of the ice chest.

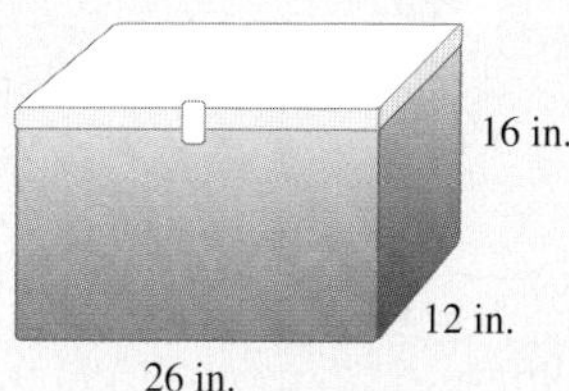

7. Find the retail price (p) of a cookware set that costs the store owner \$45.50 and is marked up \$35.
8. Find the profit (p) made by a school T-shirt sale if revenue was \$14,500 and costs were \$10,200.
9. Find the distance (d) that a rock falls in 3 seconds after being dropped from the edge of a cliff.
10. Find the temperature in degrees Celsius (C) if the temperature in degrees Fahrenheit is 59.

Sometimes we use the same formula to answer several related questions. The results can be displayed in a table.

11. Find the interest earned by each account.

Type of account	Principal	Annual rate earned	Time invested	Interest earned
Savings	\$5,000	5%	3 yr	
Passbook	\$2,250	2%	1 yr	
Trust fund	\$10,000	6.25%	10 yr	

ACCENT ON TEAMWORK

SECTION 7.1

WRITING DIGITS In the illustration, the digit 1 is drawn using one angle, and the digit 2 is drawn using two angles. Draw the digit 3 using three angles, the digit 4 using four angles, and so on for all of the digits up to and including 9.

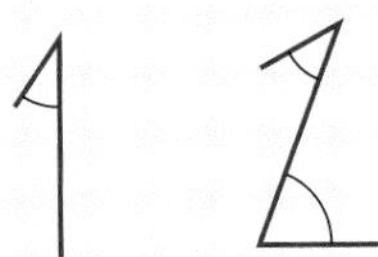

SECTION 7.2

CONSTRUCTIONS

Step 1: See Illustration (a). Using a straightedge, draw $\overline{AB}$. Then place the sharp point of a compass at A and draw an arc.

Step 2: With the same compass setting, place the sharp point at B. As shown in Illustration (b), draw another arc that intersects the arc from step 1 at two points. Label these points C and D.

Step 3: Using a straightedge, draw a line through points C and D. Label the point where line CD intersects $\overline{AB}$ as point E. Does $m(\overline{AE}) = m(\overline{EB})$?

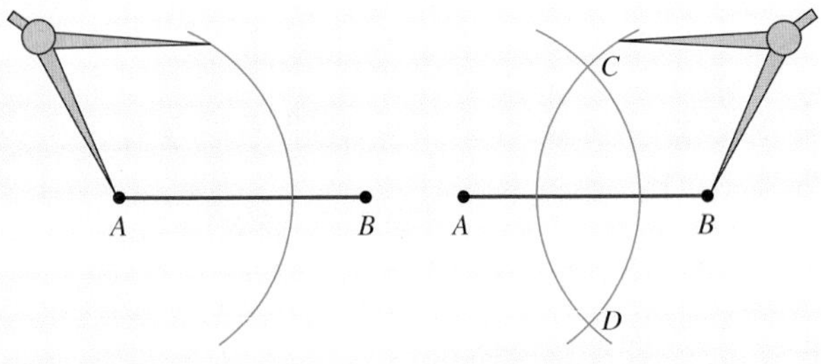

SECTION 7.3

TANGRAM A tangram is a puzzle in which geometric shapes are arranged to form other shapes. Cut out the pieces in the illustration. Assemble them so that they form a square. There should be no gaps, overlaps, or holes.

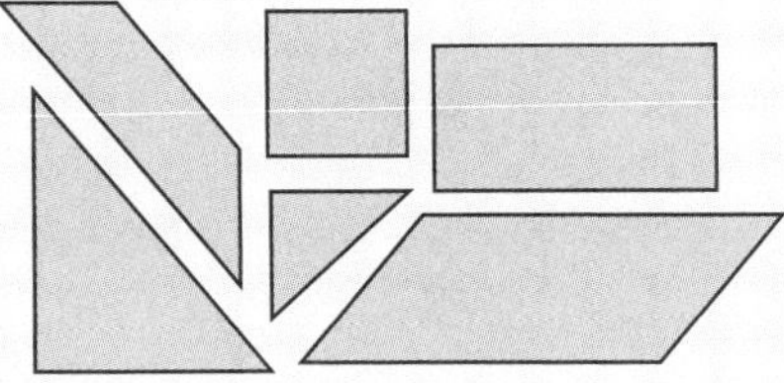

SECTION 7.4

CONGRUENT TRIANGLES Draw a triangle on a piece of paper. Then measure the lengths of its sides (with a ruler) and the angle measures (with a protractor). Choose a combination of any three measurements and tell them to your partner. Are the given facts sufficient for your partner to construct a triangle congruent to yours?

SECTION 7.5

AREA Find the area of the shaded figure on the square grid.

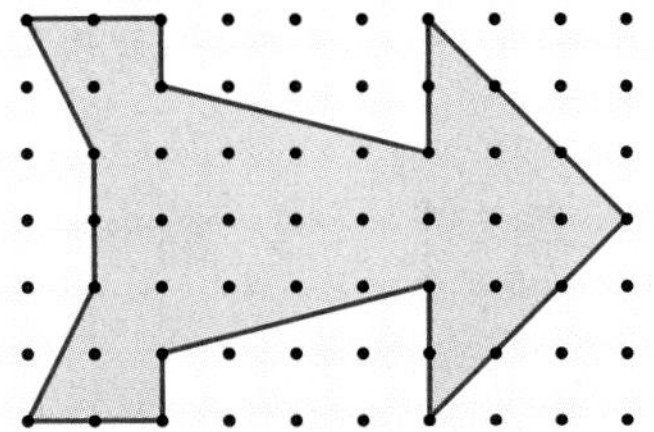

SECTION 7.6

PI Carefully measure the circumference and the diameter of different-size circles. Record the measurements in a table like the one below. Then use a calculator to find $\frac{C}{D}$. The result should be a number close to π.

Object	Circumference	Diameter	$\frac{C}{D}$
Jar	$3\frac{1}{2}$ in.	$1\frac{1}{8}$ in.	3.11

SECTION 7.7

PYRAMIDS Cut out, fold, and glue together the pattern shown. Estimate the volume and surface area of the pyramid.

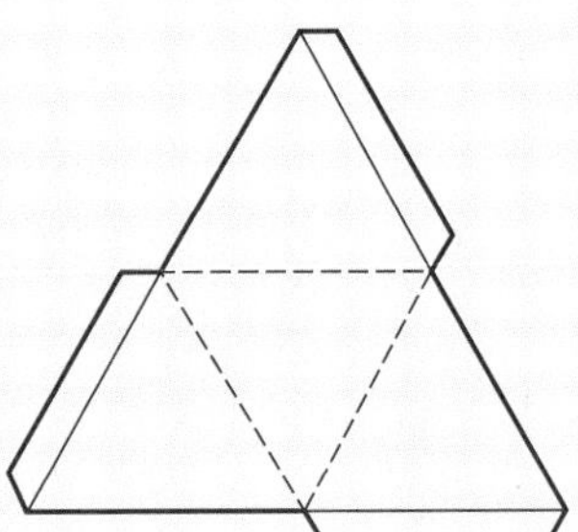

CHAPTER REVIEW

SECTION 7.1 Some Basic Definitions

CONCEPTS

In geometry, we study *points, lines,* and *planes.*

A *line segment* is a part of a line with two endpoints. A *ray* is a part of a line with one endpoint.

An *angle* is a figure formed by two rays with a common endpoint. The common endpoint is called the *vertex* of the angle.

A *protractor* is used to find the measure of an angle.

An *acute angle* is greater than 0° but less than 90°. A *right angle* measures 90°. An *obtuse angle* is greater than 90° but less than 180°. A *straight angle* measures 180°.

Two angles that have the same vertex and are side-by-side are called *adjacent angles.*

REVIEW EXERCISES

1. In the illustration, identify a point, a line, and a plane.

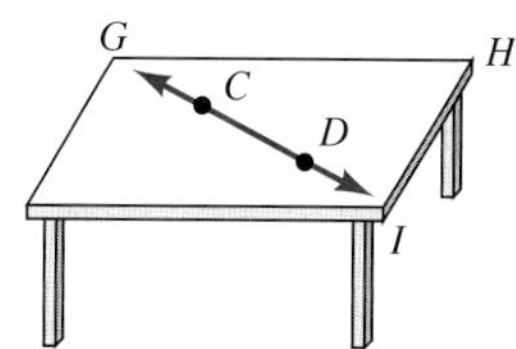

2. In the illustration, find $m(\overline{AB})$.

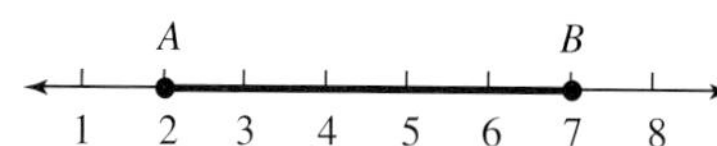

3. In the illustration below, give four ways to name the angle.

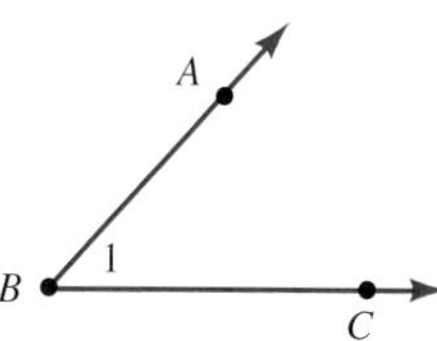

4. In the illustration above, use a protractor to find the measure of the angle.

5. In the illustration below, identify each acute angle, right angle, obtuse angle, and straight angle.

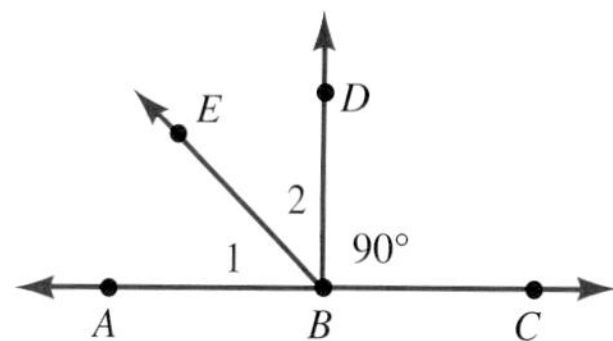

The measures of several angles are given. Identify each angle as an acute angle, a right angle, an obtuse angle, or a straight angle.

6. $m(\angle A) = 150°$

7. $m(\angle B) = 90°$

8. $m(\angle C) = 180°$

9. $m(\angle D) = 25°$

10. The two angles shown are adjacent angles. Find x.

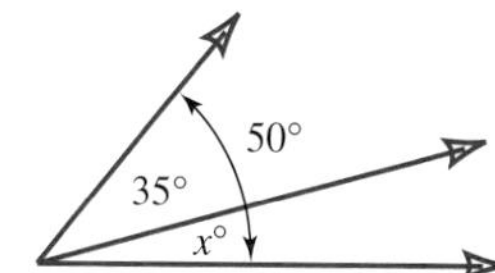

11. Line AB is shown. Find y.

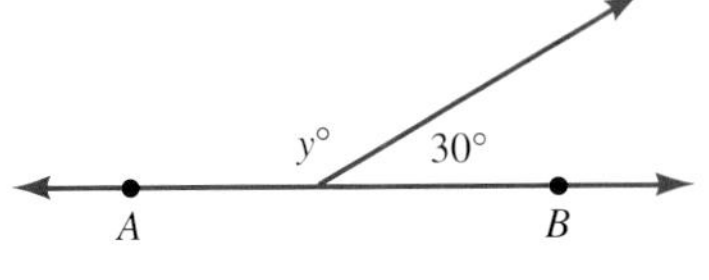

When two lines intersect, pairs of nonadjacent angles are called *vertical angles.*

Vertical angles have the same measure.

12. Find **a.** m(∠1) and **b.** m(∠2).

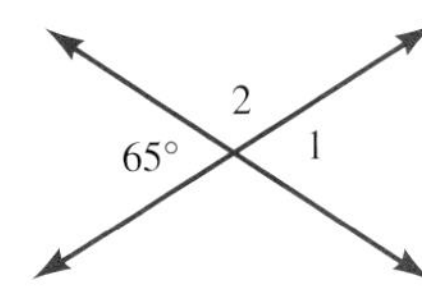

If the sum of two angles is 90°, the angles are *complementary.* If the sum of two angles is 180°, the angles are *supplementary.*

13. Find the complement of an angle that measures 50°.
14. Find the supplement of an angle that measures 140°.
15. Are angles measuring 30°, 60°, and 90° supplementary?

SECTION 7.2 Parallel and Perpendicular Lines

Parallel lines do not intersect. *Perpendicular* lines intersect and make right angles.

A line that intersects two or more *coplanar* lines is called a *transversal.*

16. Which part of the illustration represents parallel lines?

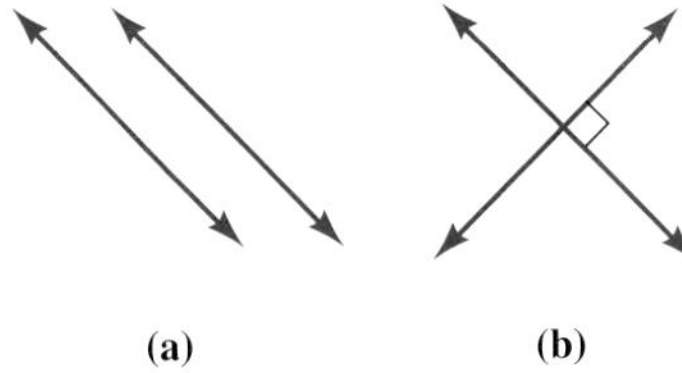

When a transversal intersects two coplanar lines, *alternate interior angles* and *corresponding angles* are formed.

17. Identify all pairs of alternate interior angles shown in the illustration below.
18. Identify all pairs of corresponding angles shown in the illustration below.
19. Identify all pairs of vertical angles shown in the illustration below.

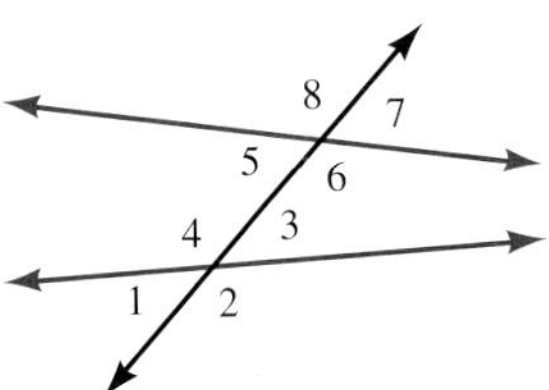

If two parallel lines are cut by a transversal.

1. alternate interior angles are congruent (have equal measures).
2. corresponding angles are congruent.
3. interior angles on the same side of the transversal are supplementary.

20. In the illustration below, $l_1 \parallel l_2$. Find the measure of each angle.

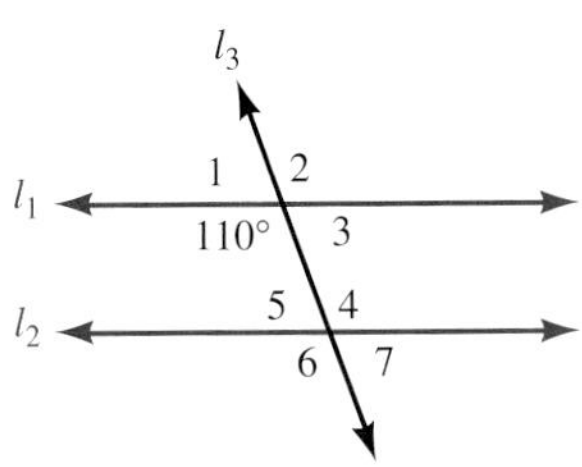

21. In the illustration below, $\overline{DC} \parallel \overline{AB}$. Find the measure of each angle.

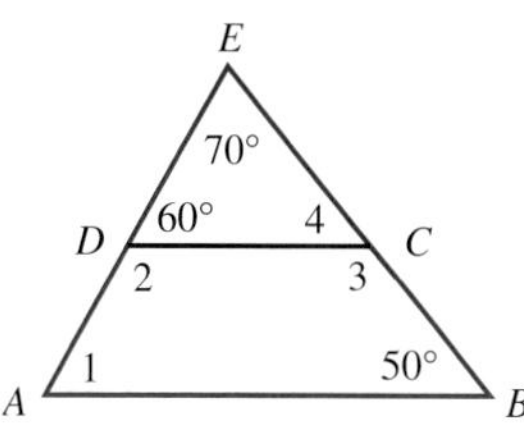

22. In Illustration (a), $l_1 \parallel l_2$. Find x.

23. In Illustration (b), $l_1 \parallel l_2$. Find x.

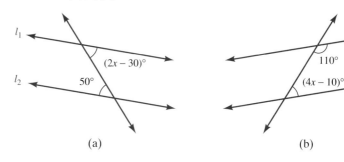

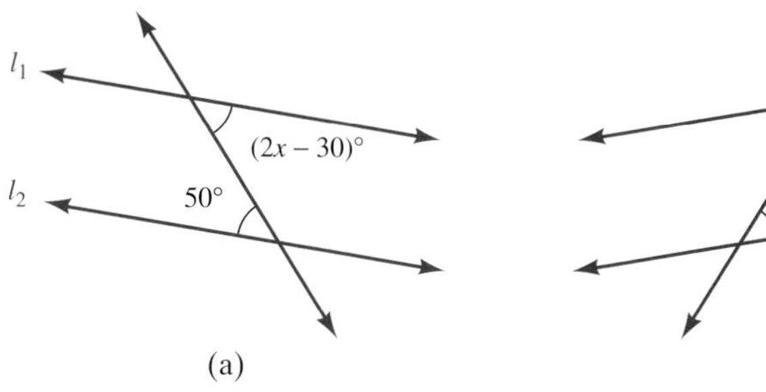

SECTION 7.3 Polygons

A *polygon* is a closed geometric figure. The points at which the sides intersect are called *vertices.* A *regular polygon* has sides that are all the same length and angles that are all the same measure.

Polygons are classified as follows:

Number of sides	Name
3	triangle
4	quadrilateral
5	pentagon
6	hexagon
8	octagon

An *equilateral triangle* has three sides of equal length.
An *isosceles triangle* has at least two sides of equal length.
A *scalene triangle* has no sides of equal length.
A *right triangle* has one right angle.

Identify each polygon as a triangle, a quadrilateral, a pentagon, a hexagon, or an octagon.

24. 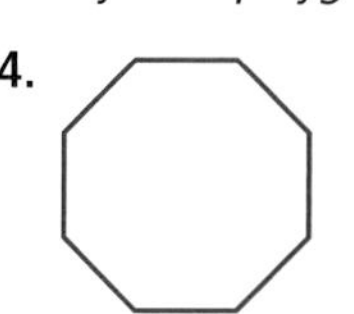

25. 

26. 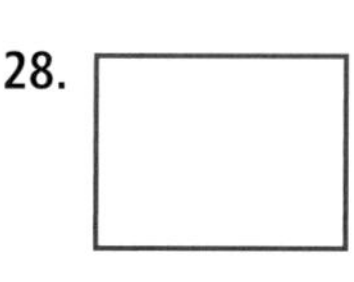

27.

28.

Give the number of vertices in each polygon.

29. Triangle

30. Quadrilateral

31. Octagon

32. Hexagon

Classify each of the triangles as an equilateral triangle, an isosceles triangle, a scalene triangle, or a right triangle.

33. 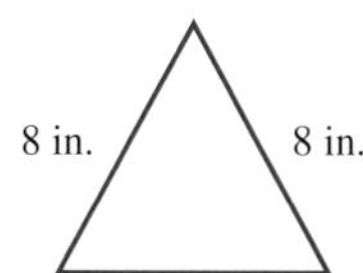

34. 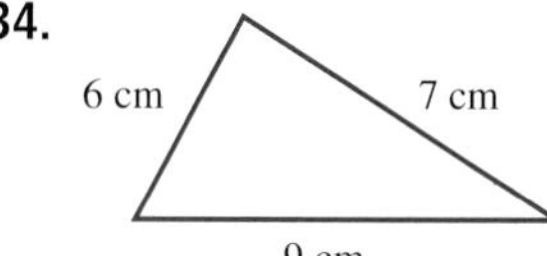

In an isosceles triangle, the angles opposite the sides of equal length are called *base angles.* The third angle is called the *vertex angle.* The third side is called the *base.*

Properties of isosceles triangles:

1. The base angles are congruent.
2. If two angles in a triangle are congruent, the sides opposite the angles are congruent, and the triangle is isosceles.

The sum of the measures of the angles of any triangle is 180°.

35.

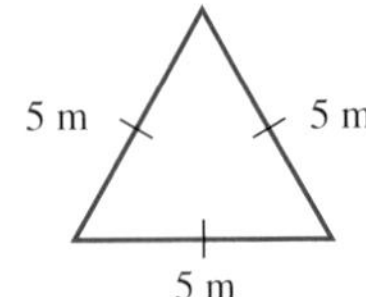

36.

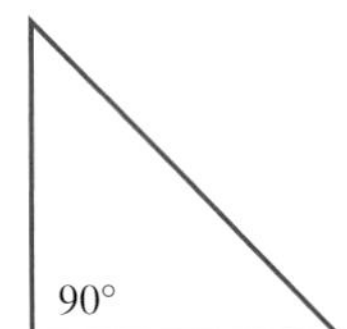

Determine whether each triangle is isosceles.

37.

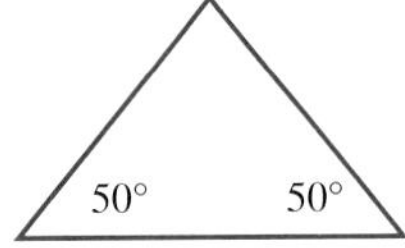

38.

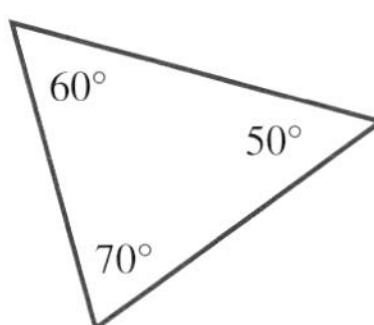

In each triangle, find x.

39.

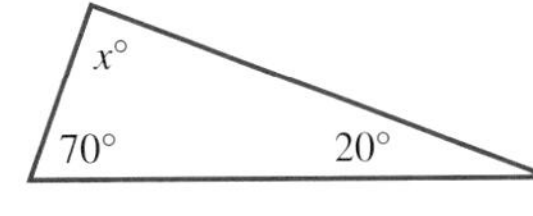

40.

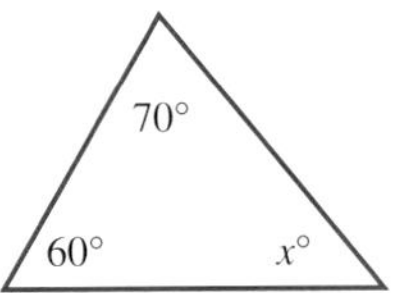

41. If one base angle of an isosceles triangle measures 65°, how large is the vertex angle?

42. If one base angle of an isosceles triangle measures 60°, what can you conclude about the triangle?

Quadrilaterals are classified as follows:

Property	Name
Opposite sides parallel	parallel-ogram
Parallelogram with four right angles	rectangle
Rectangle with all sides equal	square
Parallelogram with sides of equal length	rhombus
Exactly two sides parallel	trapezoid

Classify each quadrilateral as a parallelogram, a rectangle, a square, a rhombus, or a trapezoid.

43.

44.

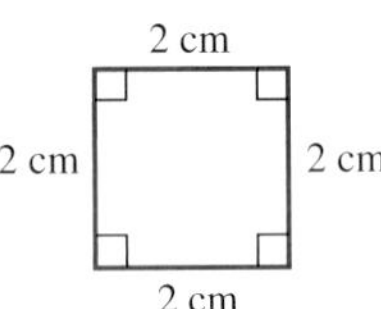

45.

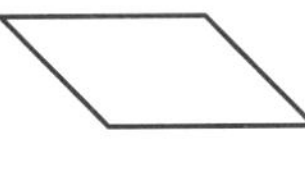

46.

47.

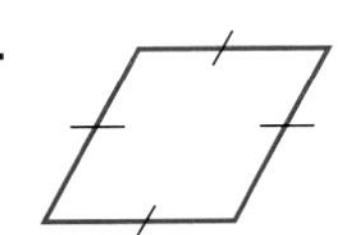

48.

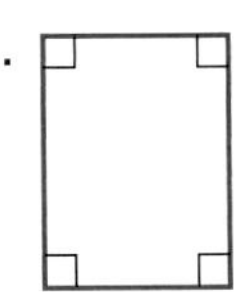

Properties of parallelograms

1. The opposite sides of a parallelogram are parallel.
2. The opposite sides of a parallelogram are congruent.
3. The opposite angles of a parallelogram are congruent.

In parallelogram ABCD, shown below, $m(\angle A) = 135°$ *and* $m(\overline{AD}) = 11$ cm. *Find each measure.*

49. $m(\angle B)$

50. $m(\overline{BC})$

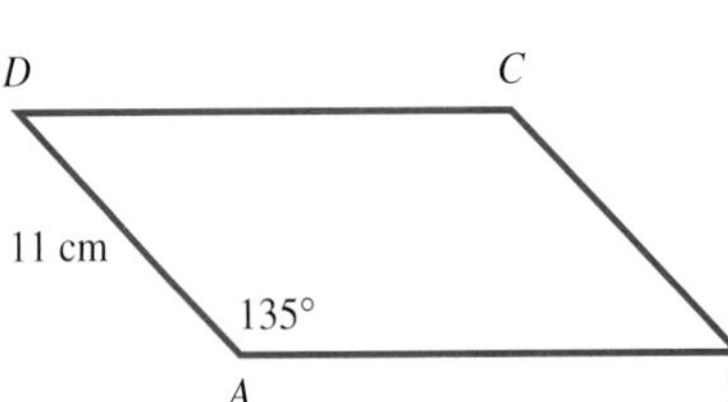

4. The consecutive angles of a parallelogram are supplementary.
5. The diagonals of a parallelogram bisect each other.
6. The diagonals of a parallelogram divide the parallelogram into two triangles with the same size and the same shape.

In the illustration below, the length of diagonal $\overline{AC}$ of rectangle ABCD is 15 centimeters. Find each measure.

51. $m(\overline{BD})$ **52.** $m(\angle 1)$ **53.** $m(\angle 2)$

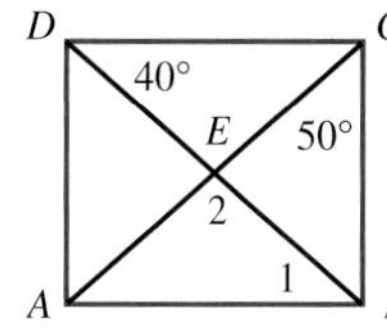

Properties of rectangles:

1. All angles are right angles.
2. Opposite sides are parallel.
3. Opposite sides are of equal length.
4. Diagonals are of equal length.
5. If the diagonals of a parallelogram are of equal length, the parallelogram is a rectangle.

In the illustration below, ABCD is a rectangle. Classify each statement as true or false.

54. $m(\overline{AB}) = m(\overline{DC})$

55. $m(\overline{AD}) = m(\overline{DC})$

56. Triangle ABE is isosceles.

57. $m(\overline{AC}) = m(\overline{BD})$

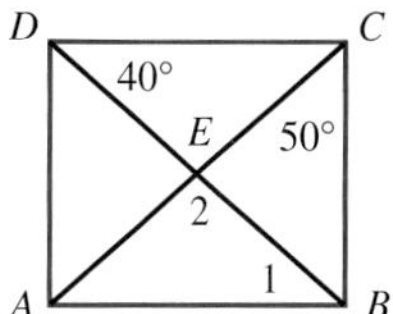

The parallel sides of a trapezoid are called *bases.* The nonparallel sides are called *legs.* If the legs of a trapezoid are of equal length, it is *isosceles.* In an isosceles trapezoid, the angles opposite the sides of equal length are *base angles,* and they are congruent.

In the illustration, ABCD is an isosceles trapezoid. Find each measure.

58. $m(\angle B)$ **59.** $m(\angle C)$

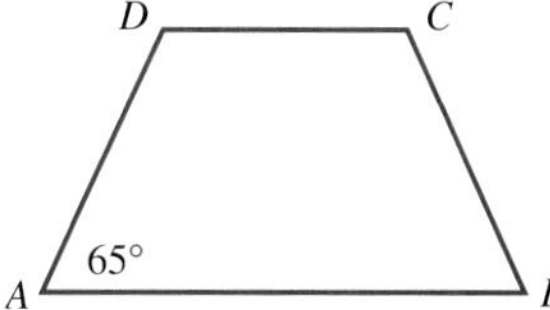

The sum of the measures of the angles of a polygon (in degrees) is given by the formula

$$S = (n - 2)180$$

or

$$S = 180n - 360$$

Find the sum of the angle measures of each polygon.

60. Quadrilateral **61.** Hexagon

SECTION 7.4 Properties of Triangles

If two triangles have the same size and the same shape, they are *congruent triangles.*

Corresponding parts of congruent triangles have the same measure.

See the illustration. Complete the list of corresponding parts.

62. $\angle A$ corresponds to ______.

63. $\angle B$ corresponds to ______.

64. $\angle C$ corresponds to ______.

65. $\overline{AC}$ corresponds to ______.

66. $\overline{AB}$ corresponds to ______.

67. $\overline{BC}$ corresponds to ______.

Three ways to show that two triangles are congruent are

1. the SSS property
2. the SAS property
3. the ASA property

Determine whether the triangles in each pair are congruent. If they are, explain why.

68.

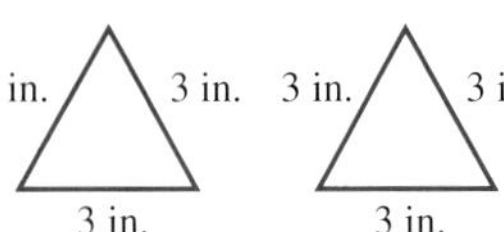

69.

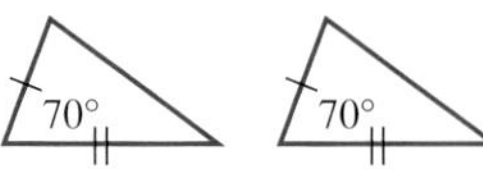

70.

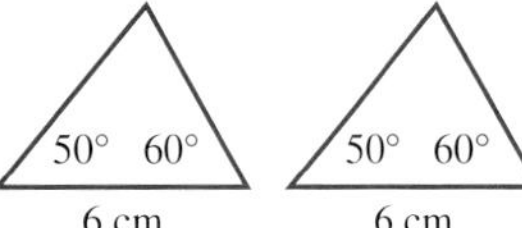

71.

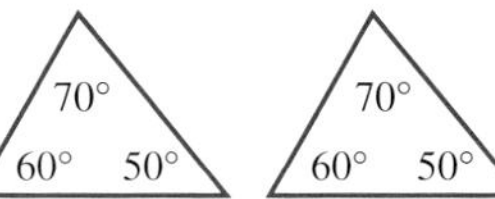

The Pythagorean theorem: If the length of the *hypotenuse* of a right triangle is c, and the lengths of its legs are a and b, then

$$a^2 + b^2 = c^2$$

Refer to the illustration and find the length of the unknown side.

72. If $a = 5$ and $b = 12$, find c.

73. If $a = 8$ and $c = 17$, find b.

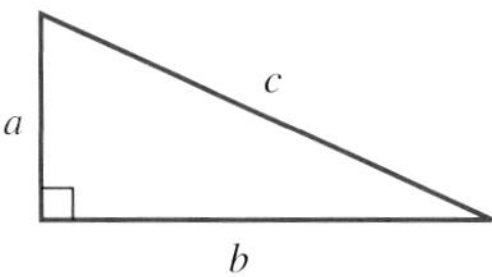

74. To the nearest tenth, find the height of the television screen shown.

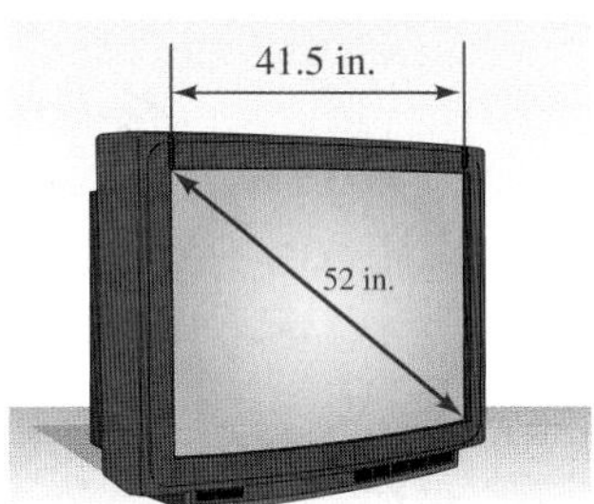

SECTION 7.5 Perimeters and Areas of Polygons

The *perimeter* of a polygon is the distance around it.

75. Find the perimeter of a square with sides 18 inches long.

76. Find the perimeter of a rectangle that is 3 meters long and 1.5 meters wide.

Find the perimeter of each polygon.

77.

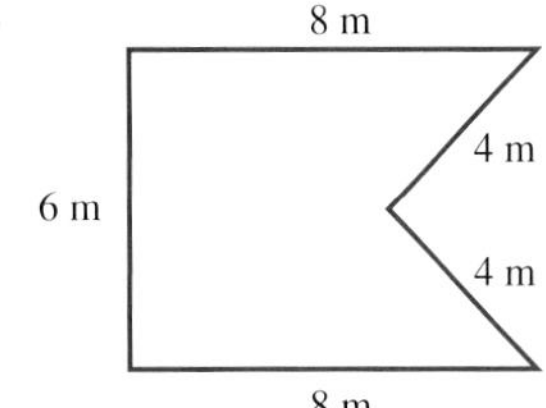

78.

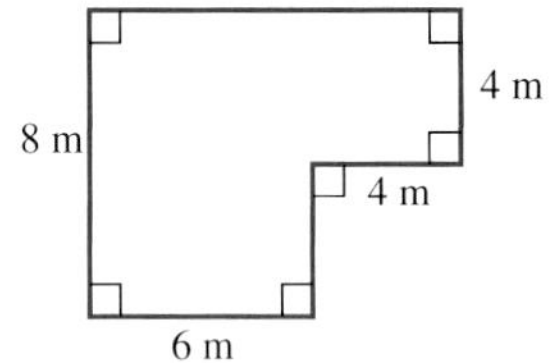

The *area* of a polygon is the measure of the surface it encloses.

Formulas for area:

Figure	Area
Square	$A = s^2$
Rectangle	$A = lw$
Parallel-ogram	$A = bh$
Triangle	$A = \frac{1}{2}bh$
Trapezoid	$A = \frac{1}{2}h(b_1 + b_2)$

Find the area of each polygon.

79.

80.

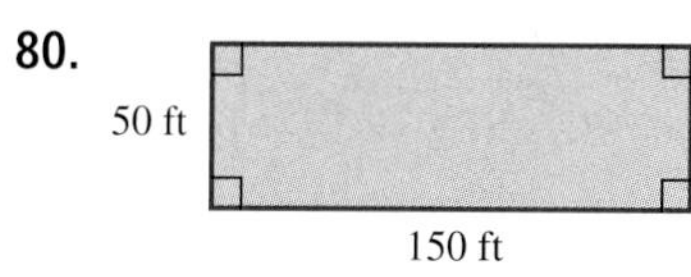

81.

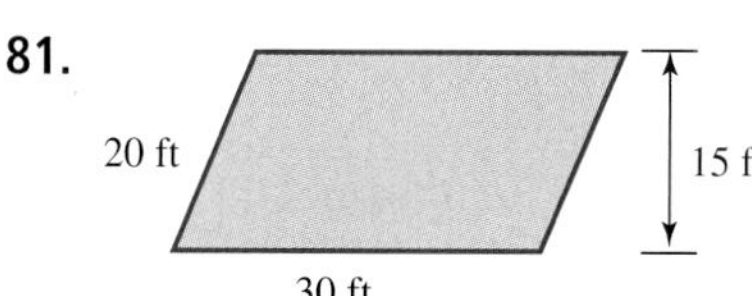

82.

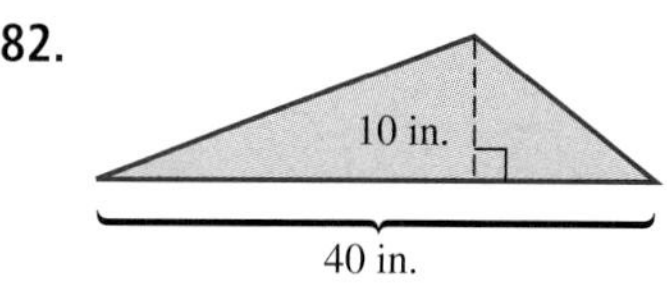

83.

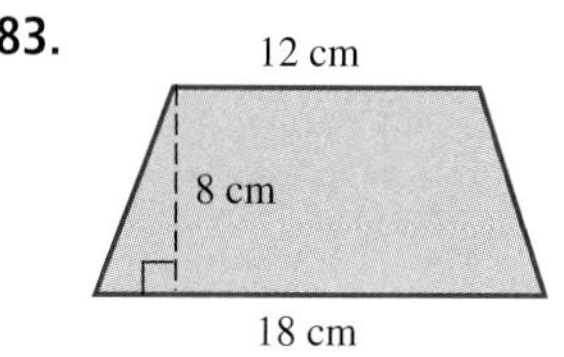

84.

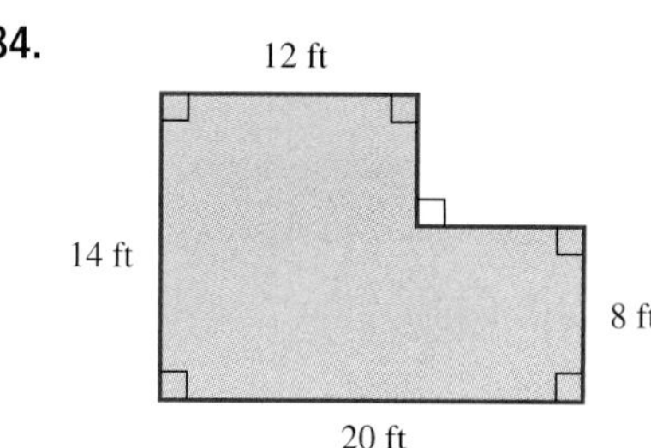

85.

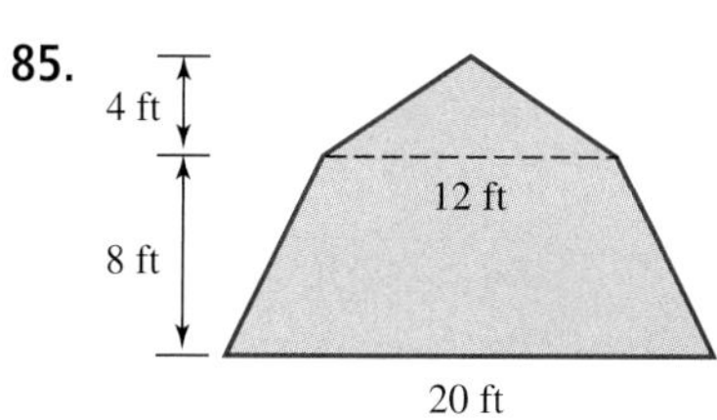

86.

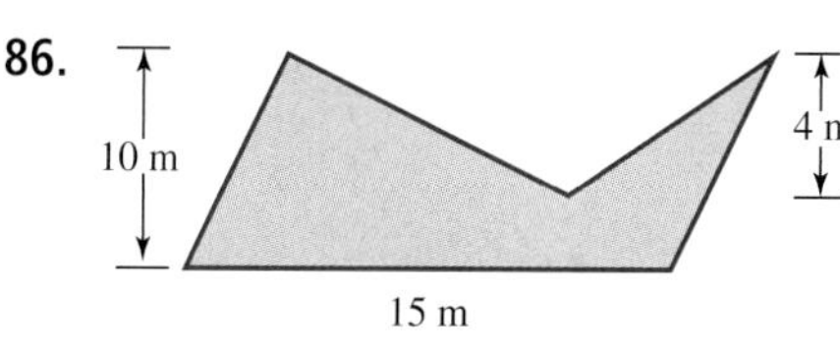

87. How many square feet are there in 1 square yard?

88. How many square inches are in 1 square foot?

SECTION 7.6 *Circles*

A *circle* is the set of all points in a plane that lie a fixed distance from a point called its *center.* The fixed distance is the circle's *radius.*

A *chord* of a circle is a line segment connecting two points on the circle.

Refer to the illustration.

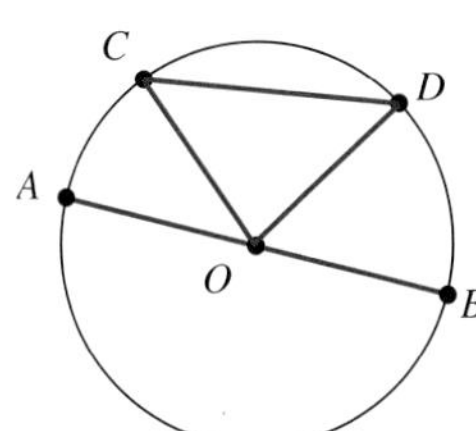

89. Name each chord.

90. Name each diameter.

91. Name each radius.

92. Name the center.

A *diameter* is a chord that passes through the circle's center.

The *circumference* (perimeter) of a circle is given by the formulas

$C = \pi D$ or $C = 2\pi r$

where $\pi = 3.14159\ldots$.

The *area* of a circle is given by the formula

$A = \pi r^2$

Find each answer to the nearest tenth.

93. Find the circumference of a circle with a diameter of 21 centimeters.

94. Find the perimeter of the figure shown.

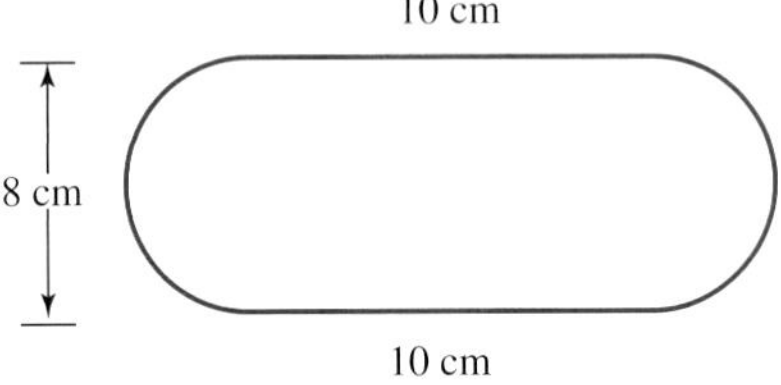

95. Find the area of a circle with a diameter of 18 inches.

96. Find the area of the figure shown in above.

SECTION 7.7 Surface Area and Volume

The *volume* of a solid is a measure of the space it occupies.

Figure	Volume
Cube	$V = s^3$
Rectangular solid	$V = lwh$
Prism	$V = Bh^*$
Sphere	$V = \frac{4}{3}\pi r^3$
Cylinder	$V = \pi r^2 h$
Cone	$V = \frac{1}{3}\pi r^2 h$
Pyramid	$V = \frac{1}{3}Bh^*$

*B represents the area of the base.

Find the volume of each solid to the nearest unit.

97.

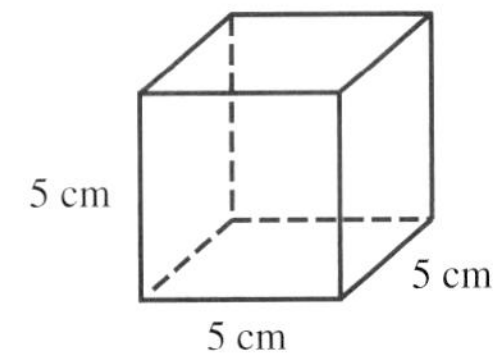

98.

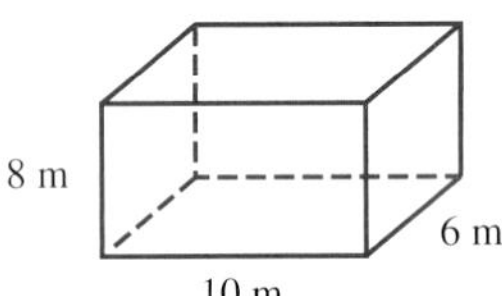

99.

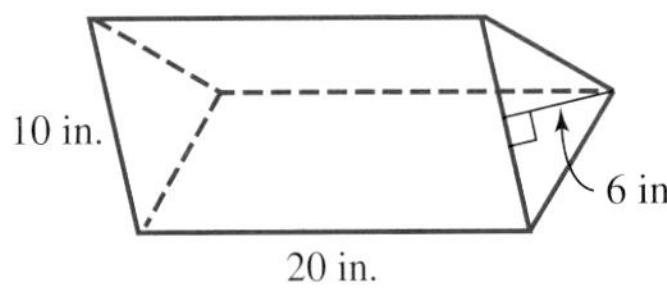

100.

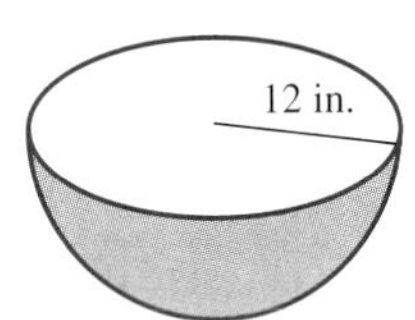

101.

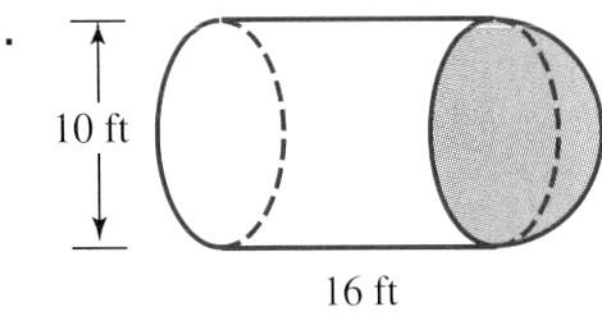

102.

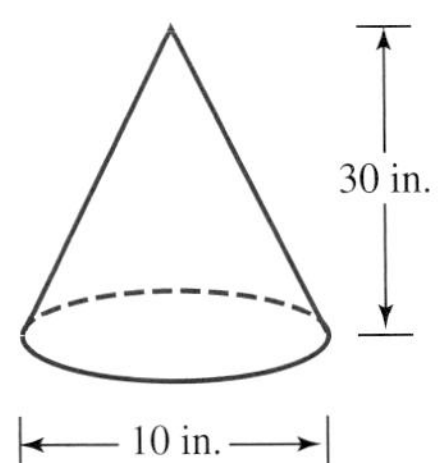

The *surface area* of a rectangular solid is the sum of the areas of its six faces.

The surface area of a sphere is given by the formula

$SA = 4\pi r^2$

103.

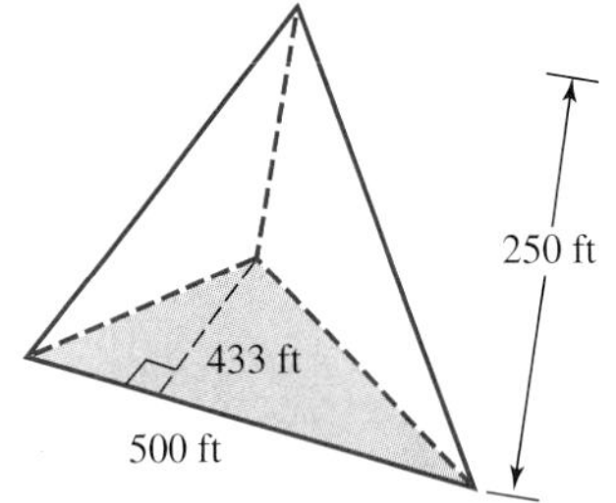

104.

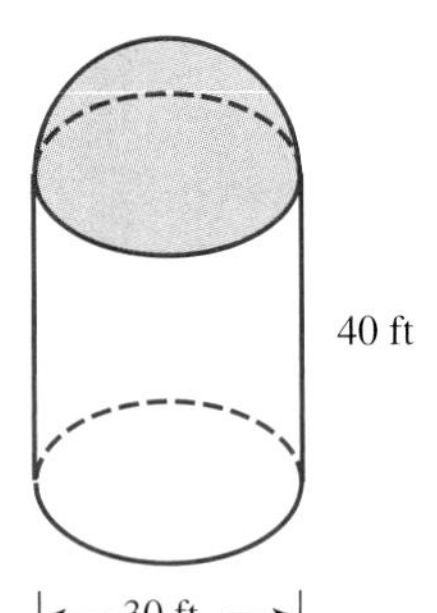

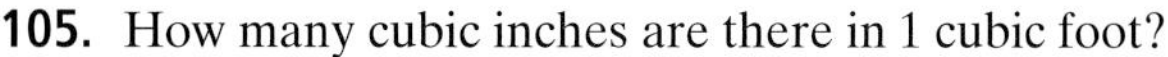

105. How many cubic inches are there in 1 cubic foot?

106. How many cubic feet are there in 2 cubic yards?

To the nearest tenth, find the surface area of each solid.

107.

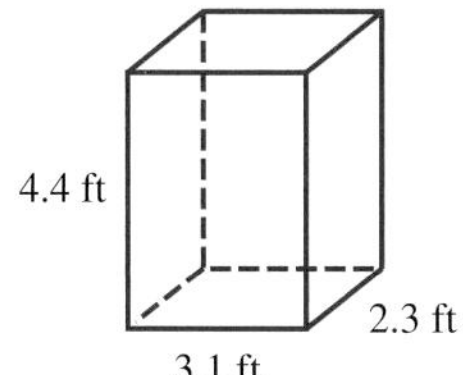

108.

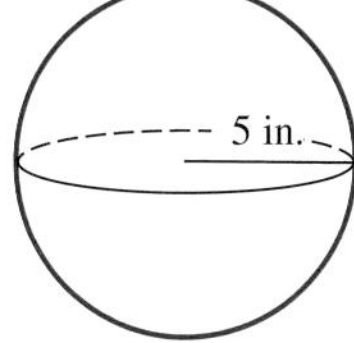

CHAPTER 7 TEST

1. Find $m(\overline{AB})$,

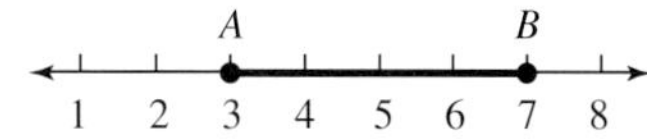

2. Which point is the vertex of $\angle ABC$?

Determine whether each statement is true or false.

3. An angle of 47° is an acute angle.

4. An angle of 90° is a straight angle.

5. An angle of 180° is a right angle.

6. An angle of 132° is an obtuse angle.

7. Find x.

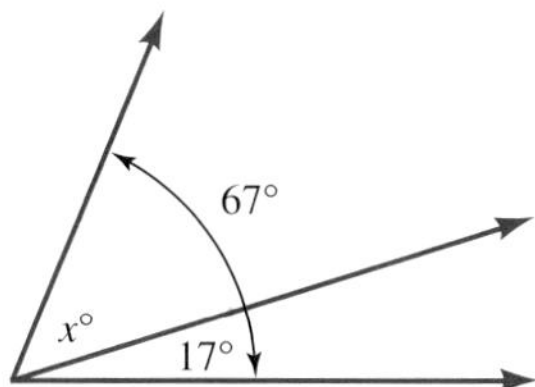

8. Find y.

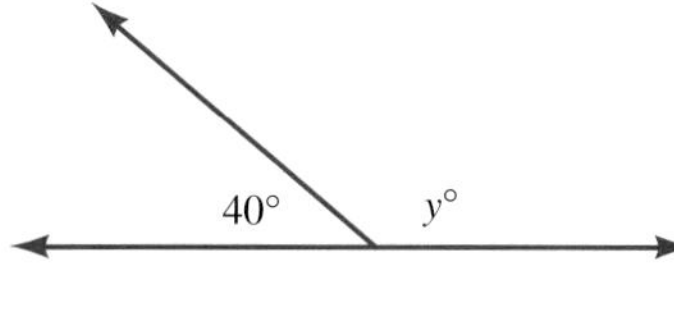

9. Find y.

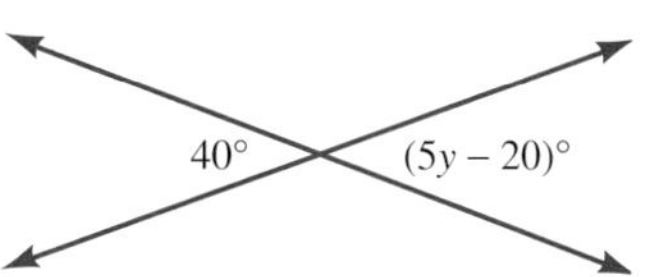

10. CALLIGRAPHY The illustration shows how the tip of the pen should be held at a 45° angle to the horizontal. What is x?

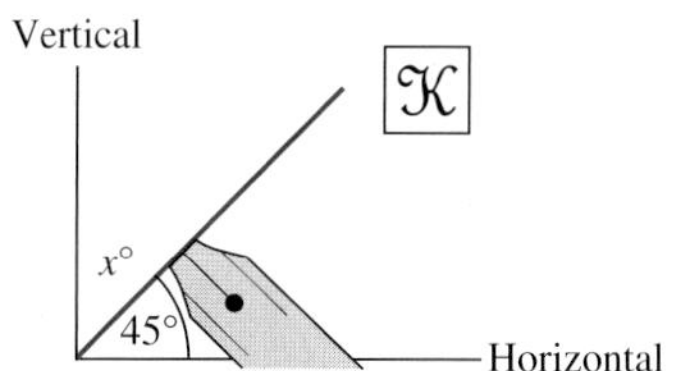

11. Find the complement of an angle measuring 67°.

12. Find the supplement of an angle measuring 117°.

Refer to the illustration below, in which $l_1 \parallel l_2$.

13. $m(\angle 1) =$ ______.

14. $m(\angle 2) =$ ______.

15. $m(\angle 3) =$ ______.

16. Find x.

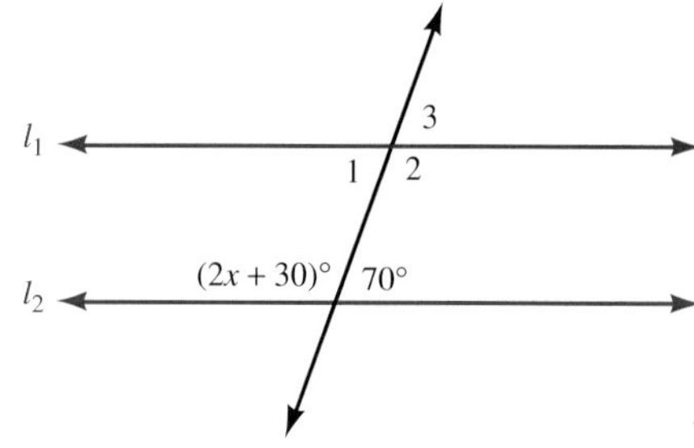

17. Complete the table.

Polygon	Number of sides
Triangle	
Quadrilateral	
Hexagon	
Pentagon	
Octagon	

18. Complete the table about triangles.

Property	Kind of triangle
All sides of equal length	
No sides of equal length	
Two sides of equal length	

Refer to the illustration below.

19. Find $m(\angle A)$.

20. Find $m(\angle C)$.

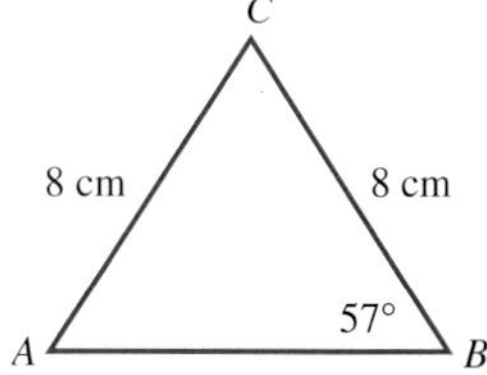

21. If the measures of two angles in a triangle are 65° and 85°, find the measure of the third angle.

22. Find the sum of the measures of the angles in a decagon (a ten-sided polygon).

23. In the illustration, $ABCD$ is a rectangle. Name three pairs of segments with equal lengths.

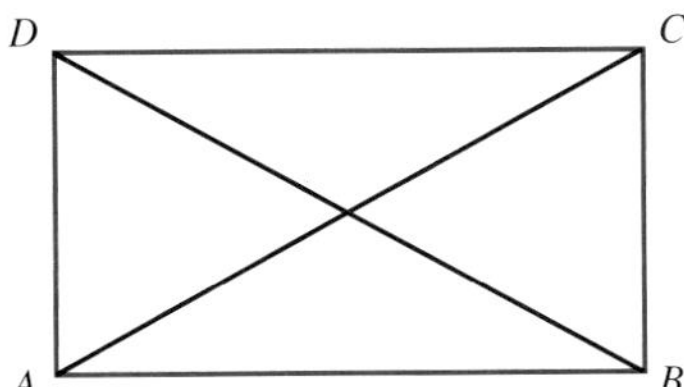

24. In the illustration, $ABCD$ is an isosceles trapezoid. Find x.

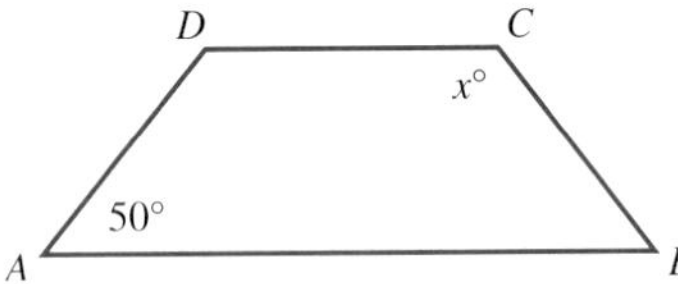

Refer to the illustration, in which $\triangle ABC \cong \triangle DEF$.

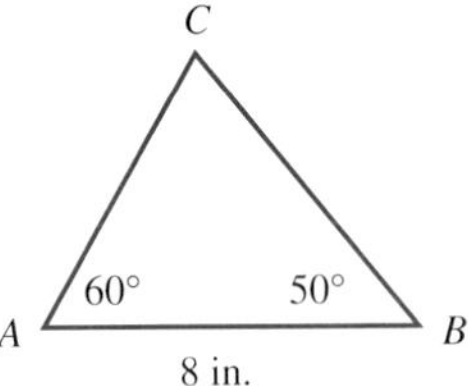

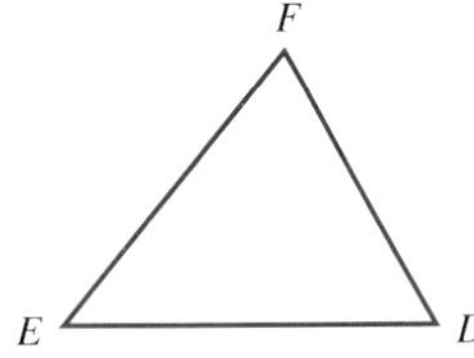

25. Find m($\overline{DE}$).

26. Find m($\angle E$).

Give each answer to the nearest tenth.

27. A baseball diamond is a square with each side 90 feet long. What is the straight-line distance from third base to first base?

28. Find the area of a triangle with a base 44.5 centimeters long and a height of 17.6 centimeters.

29. Find the area of a trapezoid with a height of 6 feet and bases that are 12.2 feet and 15.7 feet long.

30. THE OLYMPICS Steel rod is to be bent to form the interlocking rings of the Olympic Games symbol. How many feet of steel rod will be needed to make the symbol if the diameter of each ring is to be 6 feet?

31. Find the area of a circle with a diameter that is 6 feet long.

32. Find the volume of a rectangular solid with dimensions 4.3 by 5.7 by 6.5 meters.

33. Find the volume of a sphere that is 8 meters in diameter.

34. Find the volume of a 10-foot-tall pyramid that has a rectangular base 5 feet long and 4 feet wide.

35. Give a real-life example in which the concept of perimeter is used. Do the same for area and for volume. Be sure to discuss the type of units used in each case.

36. Draw a cube. Explain how to find its surface area.

CHAPTERS 1–7 CUMULATIVE REVIEW EXERCISES

1. AMUSEMENT PARKS Use the data in the table to construct a bar graph on the illustration.

Fatal accidents on amusement park rides										
Year	'93	'94	'95	'96	'97	'98	'99	'00	'01	'02
Number	4	2	4	3	4	7	6	1	3	2

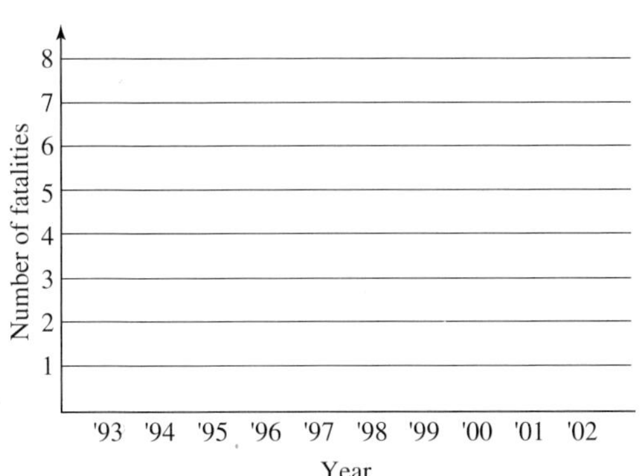

Source: U.S. Product Consumer Safety Commission

2. USED CARS The following ad appeared in *The Car Trader.* (O.B.O. means "or best offer.") If offers of \$8,750, \$8,875, \$8,900, \$8,850, \$8,800, \$7,995, \$8,995, and \$8,925 were received, what was the selling price of the car?

> 1969 Ford Mustang. New tires
> Must sell!!!! \$10,500 O.B.O.

3. Subtract: 35,021 − 23,999.

4. Divide: 1,353 ÷ 41.

5. Round 2,109,567 to the nearest thousand.

6. Prime factor 220.

7. Find all the factors of 24.

8. List the set of integers.

9. Evaluate: $-10(-2) - 2^3 + 1$.

10. Evaluate: $5 - 3[4^2 - (1 + 5 \cdot 2)]$.

11. Evaluate: $|-6 - (-3)|$.

12. Simplify: $\dfrac{2(2) + 3(-3)}{-4 - (-3)}$.

Solve each equation. **Check each result.**

13. $-x + 2 = 13$

14. $-2t - 7 = 13$

15. $4 + \dfrac{x}{5} - 6 = -1$

16. $4x - 40 = -20$

17. Simplify: $\dfrac{35}{28}$.

18. Add: $45\dfrac{2}{3} + 96\dfrac{4}{5}$.

19. Subtract: $\dfrac{3}{4} - \dfrac{3}{5}$.

20. BAKING A 5-pound bag of all-purpose flour contains $17\frac{1}{2}$ cups. A baker uses $3\frac{3}{4}$ cups. How much flour is left?

21. Multiply: $-\dfrac{6}{25}\left(2\dfrac{7}{24}\right)$.

22. Divide: $\dfrac{15}{8} \div \dfrac{45}{8}$.

23. What is the reciprocal of $\dfrac{9}{8}$?

24. Write $7\dfrac{1}{2}$ as an improper fraction.

25. PET MEDICATION A pet owner was told to use an eye dropper to administer medication to his sick kitten. The cup shown below contains 8 doses of the medication. Determine the size of a single dose.

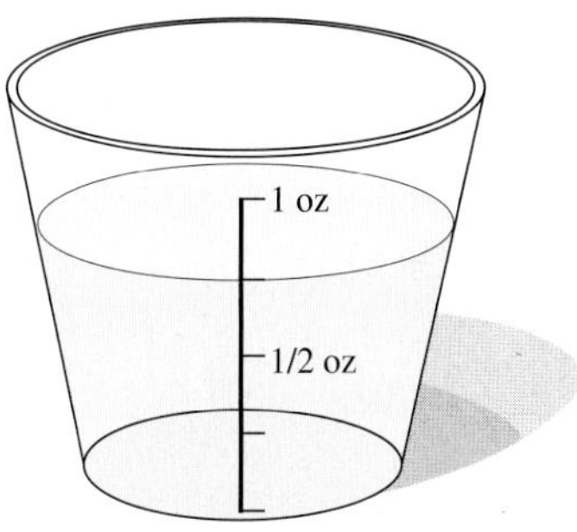

26. Evaluate: $\dfrac{3}{4} + \left(-\dfrac{1}{3}\right)^2\left(\dfrac{5}{4}\right)$.

27. Simplify: $\dfrac{7 - \dfrac{2}{3}}{4\dfrac{5}{6}}$.

28. GLOBAL WARMING The graph below shows the annual mean global temperature change as measured by satellites orbiting the Earth.

a. When was the greatest rise in temperature recorded? What was it?

b. When was the greatest decline in temperature recorded? What was it?

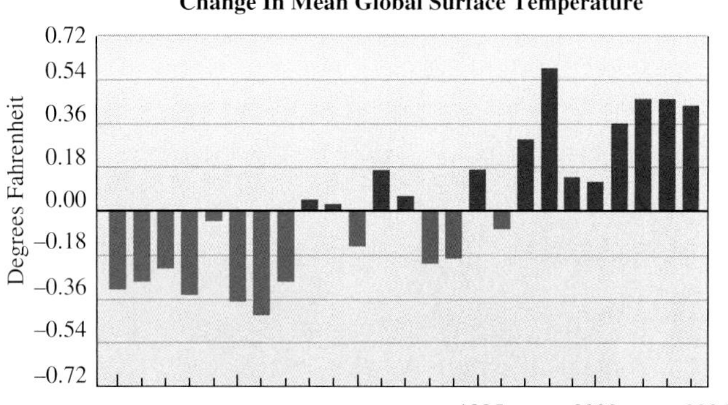

Source: National Oceanic and Atmospheric Administration

29. Graph each member of the set on the number line:
$\left\{-4\frac{5}{8}, \sqrt{17}, 2.89, \frac{2}{3}, -0.1, -\sqrt{9}, \frac{3}{2}\right\}$

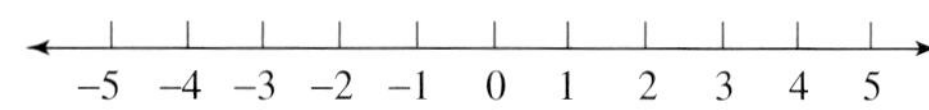

30. Round the number pi to the nearest ten thousandth: $\pi = 3.141592654. . . .$

31. Place the proper symbol (> or <) in the blank: 154.34 ☐ 154.33999.

32. Add: $3.4 + 106.78 + 35 + 0.008$.

33. Multiply: $-5.5(-3.1)$.

34. Multiply: $(89.9708)(1{,}000)$.

35. Divide: $\frac{0.0742}{1.4}$.

36. Evaluate: $-8.8 + (-7.3 - 9.5)$.

37. Evaluate: $\frac{7}{8}(9.7 + 15.8)$.

38. Change $\frac{2}{15}$ to a decimal.

39. Evaluate $\frac{(-1.3)^2 + 6.7}{-0.9}$ and round to the nearest hundredth.

40. DECORATIONS A mother has budgeted \$20 for decorations for her daughter's birthday party. She decides to buy a tank of helium for \$15.15 and some balloons. If the balloons sell for 5 cents apiece, how many balloons can she buy?

41. Find the square root of 100.

42. Evaluate: $2\sqrt{121} - 3\sqrt{64}$.

43. Evaluate: $\sqrt{\frac{49}{81}}$.

44. What percent of the figure is shaded? What percent is not shaded?

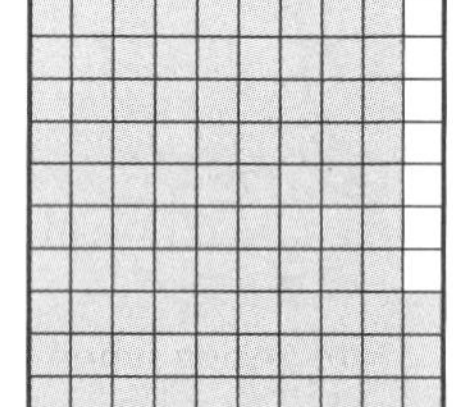

45. What number is 15% of 450?

46. 24.6 is 20.5% of what number?

47. Complete the table.

Percent	Decimal	Fraction
57%		
	0.001	
		$\frac{1}{3}$

48. STUDENT GOVERNMENT In an election for Student Body President, 560 votes were cast. Stan Cisneros received 308 votes, and Amy Huang-Sims received 252 votes. Use a circle graph to show the percent of the vote received by each candidate.

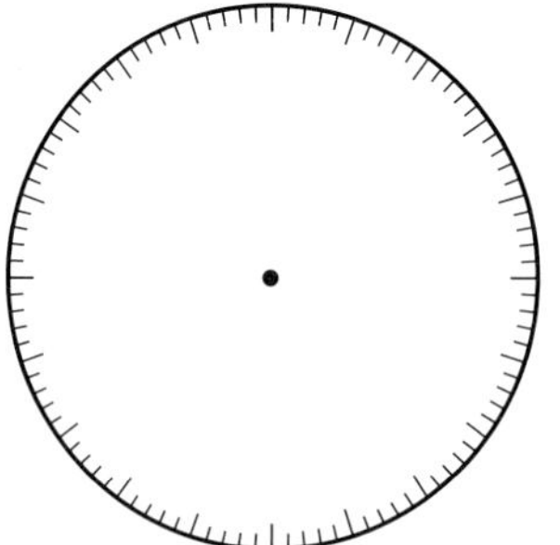

49. SHOPPING What is the regular price of the calculator shown below?

50. SALES TAX If the sales tax rate is $6\frac{1}{4}\%$, how much sales tax will be added to the price of a new car selling for $18,550?

51. COLLECTIBLES A German Hummel porcelain figurine, which was originally purchased for $125, was sold by a collector ten years later for $750. What was the percent increase in the value of the figurine?

52. PAYING OFF LOANS To pay for tuition, a college student borrows $1,500 for two months. If the annual interest rate is 9%, how much will the student have to repay when the loan comes due?

53. RETIREMENT When he got married, a man invested $5,000 in an account that guaranteed to pay 8% interest, compounded monthly, for 50 years. At the end of 50 years, how much will his account be worth?

54. How many degrees are in a right angle?

55. How many degrees are in an acute angle?

56. Find the supplement of an angle of 105°.

57. Find the complement of an angle of 75°.

Refer to the illustration, in which $l_1 \parallel l_2$. Find the measure of each angle.

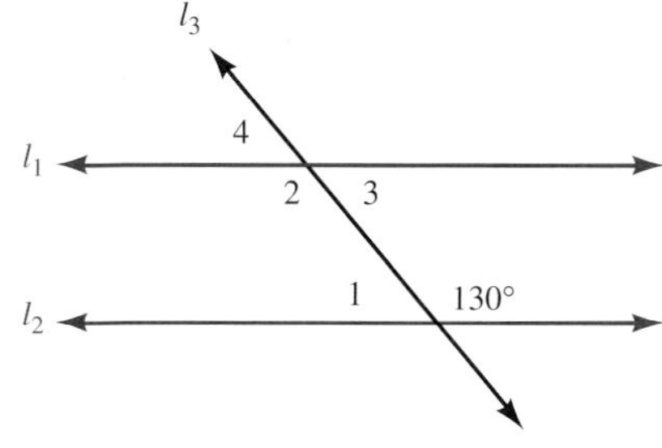

58. m(∠1) 59. m(∠3)

60. m(∠2) 61. m(∠4)

Refer to the illustration, in which $\overrightarrow{AB} \parallel \overline{DE}$ and $m(\overline{AC}) = m(\overline{BC})$. Find the measure of each angle.

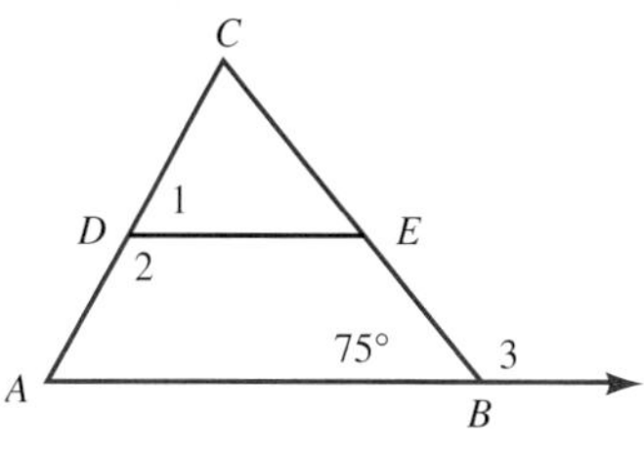

62. m(∠1) 63. m(∠C)

64. m(∠2) 65. m(∠3)

66. JAVELIN THROW Determine x and y.

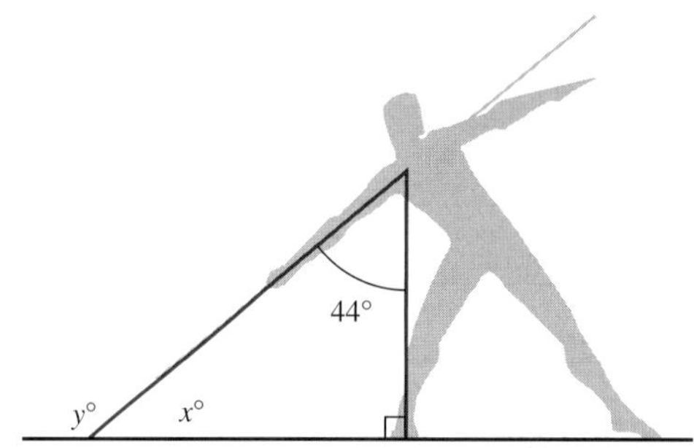

67. Find the sum of the angles of a pentagon.

68. If two sides of a right triangle measure 5 meters and 12 meters, how long is the hypotenuse?

If an answer is not exact, round to the nearest hundredth.

69. Find the perimeter and area of a rectangle with dimensions of 9 meters by 12 meters.

70. Find the area of a triangle with a base that is 14 feet long and an altitude of 18 feet.

71. Find the area of a trapezoid that has bases that are 12 inches and 14 inches long and a height of 7 inches.

72. Find the circumference and area of a circle with a diameter of 14 centimeters.

73. Find the area of the shaded region, which is created using two semicircles.

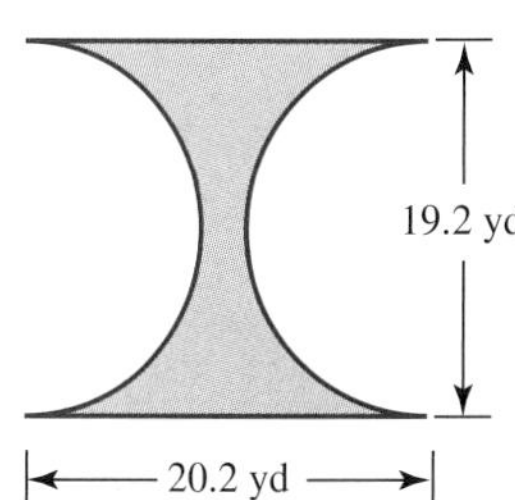

74. Find the volume of a rectangular solid with dimensions of 5 meters by 6 meters by 7 meters.

75. Find the volume of a sphere with a diameter of 10 inches.

76. Find the volume of a cone that has a circular base 8 meters in diameter and a height of 9 meters.

77. Find the volume of a cylindrical pipe that is 20 feet long and 6 inches in diameter.

78. Find the surface area of a block of ice that is in the shape of a rectangular solid with dimensions 15 in. $\times$ 24 in. $\times$ 18 in.

CHAPTER 8

Algebraic Expressions, Equations, and Inequalities

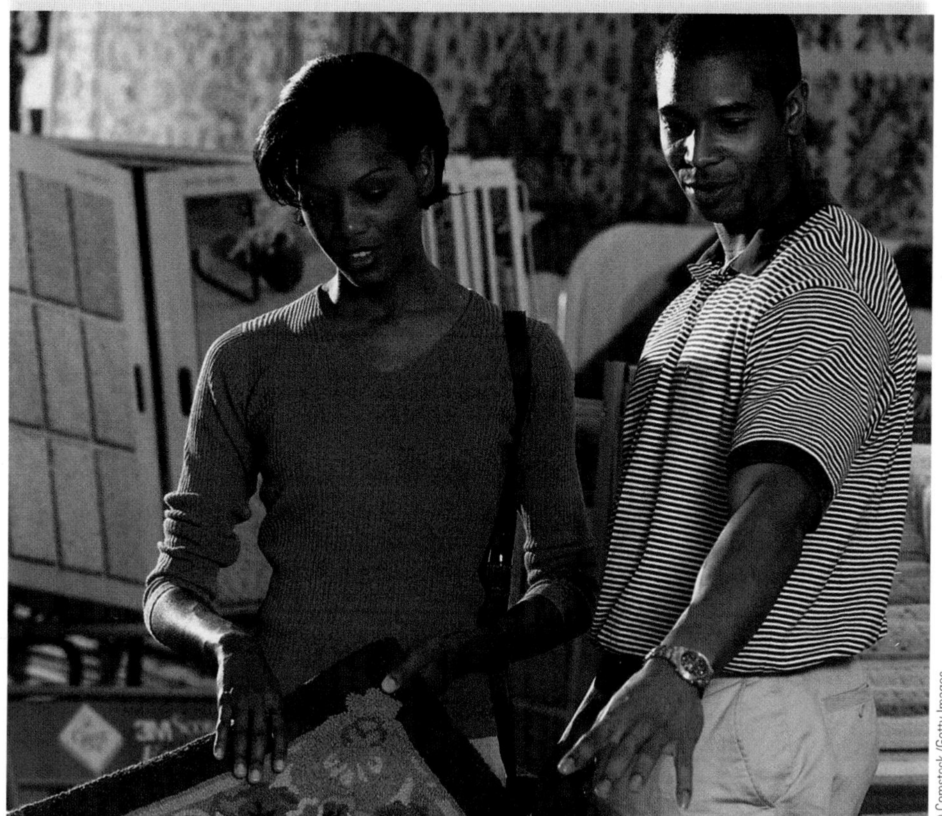

© Comstock/Getty Images

TLE When shopping for big-ticket items such as appliances, automobiles, or home furnishings, we often have to stay within a budget. Besides price, the quality and dependability of an item must also be considered. The problem-solving skills that we will discuss can help you make wise decisions when making large purchases such as these.

To learn more about the use of algebra in the marketplace, visit *The Learning Equation* on the Internet at http://tle.brookscole.com. (Log-in instructions are in the Preface.) For Chapter 8, the online lessons are:

- *TLE* Lesson 14: Translating Written Phrases
- *TLE* Lesson 15: Solving Equations Part 1
- *TLE* Lesson 16: Solving Equations Part 2

Check Your Knowledge

Evaluate each expression for $x = 2$ and $y = -3$.

1. $x + y^2$

2. $\frac{x + y}{x - y}$

Remove the parentheses.

3. $6(x + y - 2)$

4. $-2(x - y - 2)$

Simplify:

5. $\frac{3}{5}(15x)$

6. $3(a + 4) - 7(3a - 2)$

Solve each equation:

7. $-5x = 10$

8. $\frac{x}{6} = 12$

9. $-0.2x + 5 = -2$

10. $-12 - (7x + 4) = 6$

11. $\frac{x}{2} - \frac{x}{3} = 1$

12. $2x + 7 = 5x - 2$

13. Solve for y: $2x - y = 3$.

14. Find the perimeter of a rectangle with a length of 8.5 cm and a width of 12.5 cm.

15. A compact disc has radius 6 centimeters. Find the circumference of the disc. Round to one decimal place.

16. George decides to invest part of his $1,000 graduation gift at 10% annual interest and the remainder at 4% annual interest. How much did he invest at each rate if he receives $76 interest in 1 year?

17. How many pints of a 5% cleaning solution must be mixed with 5 pints of a 9% cleaning solution to give a 7% solution?

18. Find the area of a triangle with base 10 feet and height 5 feet.

19. Find the volume of a concrete block that is $17\frac{1}{2}$ inches long, $5\frac{1}{2}$ inches wide, and $7\frac{1}{2}$ inches tall.

20. The distance from Pasadena to Sacramento, California, is approximately 386 miles. What is Sarah's average speed to the nearest mile per hour if she makes the trip in 6 hours?

21. If Michael leaves San Francisco for Los Angeles at noon going 45 miles per hour, and Jo leaves one hour later traveling at 60 miles per hour, at what time will Jo overtake Michael?

Solve each inequality, graph the solution, and write the solution in interval notation.

22. $2x - 5 \geq 3$

23. $5 - 2x \leq 3$

24. $-2 < 3(x - 2) + 1 \leq 7$

Study Skills Workshop

AFTER THE TEST

CORBIS

When your test is returned to you, don't just file it and forget it. Remember that you might see this material again on your final exam! Use this as an opportunity to learn a little more about both the material on the test and the effectiveness of your test preparation.

Review Your Test. Go back over the problems that were either partially or wholly incorrect on your test and make sure that you understand how to do them correctly. If you need help, seek assistance from your instructor, a tutor, or a classmate. Math is a sequential process, so this is a really important step: many concepts that appeared on this test will be used in material that will be covered later. Also, check that your points were totaled correctly and that you received a correct score on your test. Even instructors occasionally make arithmetic mistakes! If you find such errors, your teacher will appreciate you pointing this out after class or during office hours.

Review Your Homework and Study Habits. This is a little more difficult, because it requires being as objective as possible about your personal practices. Some questions that you might ask yourself are: Am I allowing enough time for homework? Am I understanding my homework assignments? Am I getting help with homework problems that are causing me difficulty? Did I allow enough time to prepare for my test? Was there anything that I should have done differently to prepare? Was there something on the test that I didn't anticipate? If so, what can I do to minimize surprises the next time? Are there things that I would like to add to my test preparation list?

ASSIGNMENT

1. Correct any test problem for which you did not receive full credit. Turn this in to your instructor.
2. Write a short analysis of your homework and study habits. What things are you doing that are working well? What things need improvement? What might work better the next time you are preparing for a test?

In this chapter, we will learn how to write, evaluate, and simplify algebraic expressions. We will then use these skills to solve equations and inequalities.

8.1 Algebraic Expressions

- Translating from words to mathematical symbols
- Writing algebraic expressions to represent unknown quantities • Looking for hidden operations
- Number and value • Evaluating algebraic expressions • Making tables

Since problems in algebra are often presented in words, the ability to interpret what you read is important. In this section, we will introduce several strategies that will help you translate English words into mathematical symbols. We will begin by discussing the most fundamental skill — locating key words and phrases that represent four operations of arithmetic.

Translating from words to mathematical symbols

An **algebraic expression** is a collection of numbers and/or variables that are combined by using the operations of arithmetic. In the following tables, we list some *key words* and *key phrases* that are used to represent addition, subtraction, multiplication, and division, and we show how they can be translated to form algebraic expressions.

Addition

The phrase	Translates to the algebraic expression
the *sum* of p and 15	$p + 15$
10 *plus* c	$10 + c$
5 *added to* a	$a + 5$
4 *more than* r	$r + 4$
8 *greater than* A	$A + 8$
S *increased by* 100	$S + 100$
exceeds L *by* 20	$L + 20$

Subtraction

The phrase	Translates to the algebraic expression
the difference of 30 and k	$30 - k$
1,000 *minus* R	$1{,}000 - R$
15 *less than* w	$w - 15$
r *decreased by* 5	$r - 5$
T *reduced by* 80	$T - 80$
7 *subtracted from* s	$s - 7$
2,000 *less* c	$2{,}000 - c$

Multiplication

The phrase	Translates to the algebraic expression
the *product* of 60 and h	$60h$
10 *times* A	$10A$
twice w	$2w$
$\frac{1}{2}$ *of* t	$\frac{1}{2}t$

Division

The phrase	Translates to the algebraic expression
the *quotient* of B and 5	$\frac{B}{5}$
T *divided by* 50	$\frac{T}{50}$
the *ratio* of h to 3	$\frac{h}{3}$
n *split into* 8 equal parts	$\frac{n}{8}$

! COMMENT The phrase *greater than* is used to indicate addition. The phrase *is greater than* refers to the symbol >. A similar comment applies to the phrases *less than* and *is less than.*

EXAMPLE 1 Write each phrase as an algebraic expression: **a.** the sum of n and 12, **b.** the product of z and 60, and **c.** 6 less than z.

Solution

a. Key word: *sum* **Translation:** add

The phrase translates to

$$n + 12$$

b. Key word: *product* **Translation:** multiply

The variable z is to be multiplied by 60. So we have: $z \cdot 60$. This can be written as

$$60z$$

c. Key phrase: *less than* **Translation:** subtract

Since z is to be made less, we will subtract 6 from it. When we translate to mathematical symbols, we must reverse the order in which 6 and x appear in the given phrase.

6 less than z

$z - 6$

Self Check 1
Write each phrase as an algebraic expression:

a. $\frac{3}{4}$ of k

b. A split into 5 equal parts

c. Eighty subtracted from n

Answers **a.** $\frac{3}{4}k$, **b.** $\frac{A}{5}$, **c.** $n - 80$

! COMMENT Be careful when translating a subtraction. For example, we have seen that 6 *less than* z translates to $z - 6$. It would be incorrect to translate it as $6 - z$, because $6 - z$ and $z - 6$ are not the same. The phrase z *less than six* translates to $6 - z$.

Writing algebraic expressions to represent unknown quantities

When solving problems, we often begin by letting a variable stand for an unknown quantity. Frequently, a problem will contain a second unknown but related quantity, which can be described using an algebraic expression involving the original variable.

EXAMPLE 2 A butcher trims 4 ounces of fat from a roast that originally weighed x ounces. Write an algebraic expression that represents the weight of the roast after it is trimmed.

Solution We let x = the original weight of the roast (in ounces).

Key word: *trimmed* **Translation:** subtract

After 4 ounces of fat have been trimmed, the weight of the roast is $(x - 4)$ ounces.

Self Check 2
When a secretary rides the bus to work, it takes her m minutes. If she drives her own car, her travel time exceeds this by 15 minutes. How can we represent the time it takes her to get to work by car?

Answer $(m + 15)$ minutes

Self Check 3
A handyman estimates that it will take the same amount of time to sand as it will to paint some kitchen cabinets. If the entire job takes x hours, how can we express the time it will take him to do the painting?

Answer $\frac{x}{2}$ hours

EXAMPLE 3 The swimming pool in Figure 8-1 is x feet wide. If it is to be sectioned into 8 equally wide swimming lanes, write an algebraic expression that represents the width of each lane.

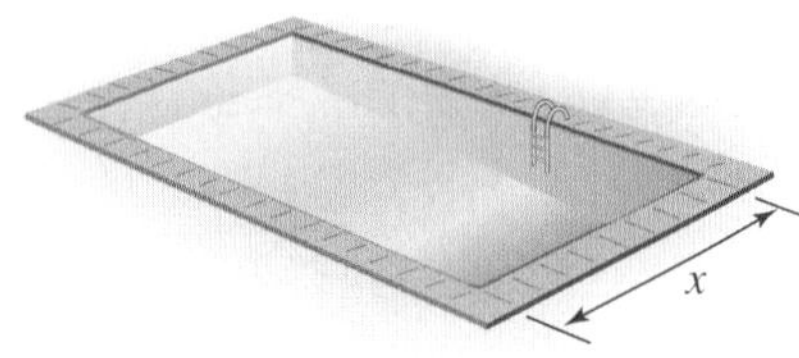

FIGURE 8-1

Solution We let x = the width of the swimming pool (in feet).

Key phrase: *sectioned into 8 equally wide lanes* **Translation:** divide

The width of each lane is $\frac{x}{8}$ feet.

When we are solving problems, the variable to be used is rarely specified. We must decide what the unknown quantities are and how they will be represented using variables. The following examples illustrate how to approach these situations.

Self Check 4
The McDonald's Chicken Deluxe sandwich has 5 fewer grams of fat than the Quarter Pounder hamburger. Express the number of grams of fat in each sandwich, using one variable.

Answer x = the number of grams of fat in the hamburger; $x - 5$ = the number of grams of fat in the chicken sandwich

EXAMPLE 4 The value of a collectible doll is three times that of an antique toy truck. Express the value of each, using one variable.

Solution There are two unknown quantities. Since the doll's value is related to the truck's value, we will let x = the value of the toy truck in dollars.

Key phrase: *3 times* **Translation:** multiply by 3

The value of the doll is $\$3x$.

! COMMENT A variable is used to represent an unknown number. Therefore, in Example 4, it would be incorrect to write, "Let x = toy truck," because the truck is not a number. We need to write, "Let x = the *value* of the toy truck."

Self Check 5
Part of a \$900 donation to a preschool was designated to go to the scholarship fund, the remainder to the building fund. Choose a variable to represent the amount donated to one of the funds. Write an expression for the amount donated to the other fund.

Answers s = amount donated to scholarship fund in dollars; $900 - s$ = amount donated to building fund

EXAMPLE 5 A 10-inch-long paintbrush has two parts: a handle and bristles. Choose a variable to represent the length of one of the parts. Then write an expression that represents the length of the other part.

Solution A drawing is helpful in explaining this problem. If the entire paintbrush is 10 inches long, and if we let h represent the length of the handle, then the bristles are $(10 - h)$ inches long.

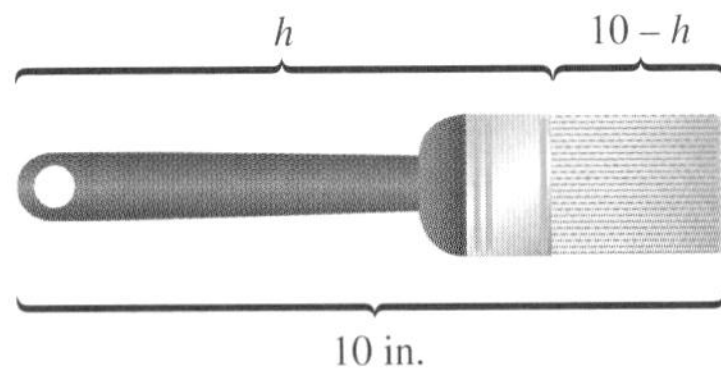

EXAMPLE 6 In the second semester, student enrollment in a retraining program at a college was 32 more than twice that of the first semester. Express the student enrollment in the program each semester, using one variable.

Solution Since the second-semester enrollment is expressed in terms of the first-semester enrollment, we let $x =$ the enrollment in the first semester.

Key phrase: *more than* **Translation:** add

Key word: *twice* **Translation:** multiply by 2

The enrollment for the second semester is $2x + 32$.

Self Check 6

The number of votes received by the incumbent in an election was 55 fewer than three times the number the challenger received. Express the number of votes received by each candidate, using one variable.

Answers $x =$ the number of votes received by the challenger; $3x - 55 =$ the number of votes received by the incumbent

Looking for hidden operations

When analyzing problems, we aren't always given key words or key phrases to help establish what mathematical operation to use. Sometimes a careful reading of the problem is needed to determine the hidden operations.

EXAMPLE 7 Disneyland, located in Anaheim, California, was in operation 16 years before the opening of Walt Disney World, in Orlando, Florida. Euro Disney, in Paris, France, was constructed 21 years after Disney World. Use algebraic expressions to express the ages (in years) of each of these Disney attractions.

Solution The ages of Disneyland and Euro Disney are both related to the age of Walt Disney World. Therefore, we will let $x =$ the age of Walt Disney World.

In carefully reading the problem, we find that Disneyland was built 16 years *before* Disney World, so its age is more than that of Disney World.

Key phrase: *more than* **Translation:** add

In years, the age of Disneyland is $x + 16$. Euro Disney was built 21 years *after* Disney World, so its age is less than that of Disney World.

Key phrase: *less than* **Translation:** subtract

In years, the age of Euro Disney is $x - 21$. The results are summarized in Table 8-1.

Attraction	Age
Disneyland	$x + 16$
Disney World	x
Euro Disney	$x - 21$

TABLE 8-1

EXAMPLE 8 How many months are in x years?

Solution Since there are no key words, we must carefully analyze the problem to write an expression that represents the number of months in x years. It is often helpful to consider some specific cases. For example, let's calculate the number of

Self Check 8
Complete the table. How many days is h hours?

Number of hours	Number of days
24	
48	
72	
h	

Answers $1, 2, 3; \frac{h}{24}$

months in 1 year, 2 years, and 3 years. When we write the results in a table, a pattern is apparent.

Number of years	Number of months
1	12
2	24
3	36
x	$12x$

We multiply the number of years by 12 to find the number of months.

Therefore, if $x =$ the number of years, the number of months is $12 \cdot x$ or $12x$.

Number and value

Some problems deal with quantities that have value. In these problems, we must distinguish between *the number of* and *the value of* the unknown quantity. For example, to find the value of 3 quarters, we multiply the number of quarters by the value (in cents) of one quarter. Therefore, the value of 3 quarters is $3 \cdot 25¢ = 75¢$.

The same distinction must be made if the number is unknown. For example, the value of n nickels is not $n¢$. The value of n nickels is $n \cdot 5¢ = (5n)¢$. For problems of this type, we will use the relationship

Number · value = total value

Self Check 9
Find the value of
a. six \$50 savings bonds
b. t \$100 savings bonds
c. $(x - 4)$ \$1,000 savings bonds

Answers **a.** \$300, **b.** \$100t, **c.** \$1,000$(x - 4)$

EXAMPLE 9 Suppose a roll of paper towels sells for 79¢. Find the cost of **a.** five rolls of paper towels, **b.** x rolls of paper towels, and **c.** $x + 1$ rolls of paper towels.

Solution In each case, we will multiply the number of rolls of paper towels by the value of one roll (79¢) to find the total cost.

a. The cost of 5 rolls of paper towels is $5 \cdot 79¢ = 395¢$, or \$3.95.

b. The cost of x rolls of paper towels is $x \cdot 79¢ = (79 \cdot x)¢ = (79x)¢$.

c. The cost of $x + 1$ rolls of paper towels is $(x + 1) \cdot 79¢ = 79 \cdot (x + 1)¢ = 79(x + 1)¢$.

Evaluating algebraic expressions

To **evaluate an algebraic expression,** we replace each variable with a given number value. (When we replace a variable with a number, we say we are **substituting** for the variable.) Then we do the necessary calculations following the rules for the order of operations. For example, to evaluate $x^2 - 2x + 1$ for $x = 3$, we begin by substituting 3 for x.

$$\mathbf{x}^2 - 2\mathbf{x} + 1 = \mathbf{3}^2 - 2(\mathbf{3}) + 1 \quad \text{Substitute 3 for } x.$$
$$= 9 - 2(3) + 1 \quad \text{Evaluate the exponential expression: } 3^2 = 9.$$
$$= 9 - 6 + 1 \quad \text{Perform the multiplication: } 2(3) = 6.$$
$$= 4 \quad \text{Working left to right, perform the subtraction and then the addition.}$$

We say that 4 is the **value** of this expression when $x = 3$.

! COMMENT When replacing a variable with its numerical value, use parentheses around the replacement number to avoid possible misinterpretation. For example, when substituting 5 for x in $2x + 1$, we show the multiplication using parentheses: $2(5) + 1$. If we don't show the multiplication, we could misread the expression as $25 + 1$.

EXAMPLE 10 Evaluate: **a.** $-y$ and **b.** $-3(y + x^2)$ for $x = 3$ and $y = -4$.

Solution

a. $-y = -(-4)$ Substitute -4 for y.

$= 4$ The opposite of -4 is 4.

b. $-3(y + x^2) = -3(-4 + 3^2)$ Substitute 3 for x and -4 for y.

$= -3(-4 + 9)$ Work within the parentheses first. Evaluate the exponential expression.

$= -3(5)$ Perform the addition within the parentheses.

$= -15$

Self Check 10
Evaluate: **a.** $-x$ and **b.** $5(x - y)$ for $x = -2$ and $y = 3$.

Answers **a.** 2, **b.** -25

EXAMPLE 11 **Surface area of a swim fin.** Divers use swim fins because they provide a much larger surface area to push against the water than do bare feet. Consequently, the diver can swim faster wearing them. In Figure 8-2, we see that the fin is in the shape of a trapezoid. The algebraic expression $\frac{1}{2}h(b + d)$ gives the area of a trapezoid, where h is the height and b and d are the lengths of the lower and upper bases, respectively. To find the area of the fin shown here, we evaluate the algebraic expression for $h = 14$, $b = 3.5$, and $d = 8.5$.

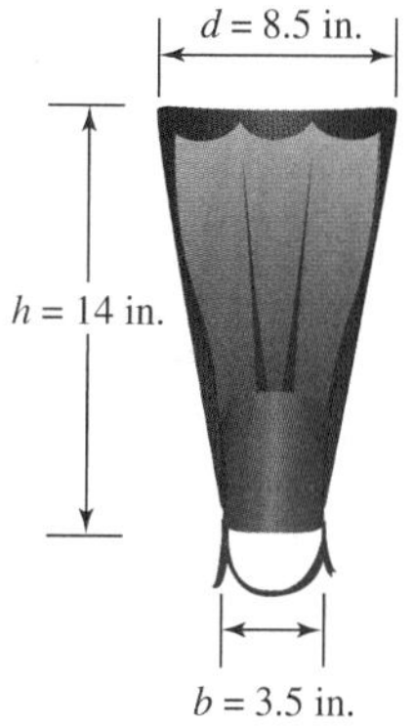

$\frac{1}{2}h(b + d) = \frac{1}{2}(14)(3.5 + 8.5)$ Substitute 14 for h, 3.5 for b, and 8.5 for d.

$= \frac{1}{2}(14)(12)$ Perform the addition within the parentheses.

$= 7(12)$ Work from left to right: $\frac{1}{2}(14) = 7$.

$= 84$

FIGURE 8-2

The fin has an area of 84 square inches.

EXAMPLE 12 **Temperature conversion.** The expression $\frac{9C + 160}{5}$ converts a temperature in degrees Celsius (represented by C) to a temperature in degrees Fahrenheit. Convert $-170°$ C, the coldest temperature on the moon, to degrees Fahrenheit.

Solution To convert $-170°$ C to degrees Fahrenheit, we evaluate the algebraic expression for $C = -170$.

Self Check 12
On January 22, 1943, the temperature in Spearfish, South Dakota changed from $-20°$ C to $7.2°$ C in two minutes. Convert $-20°$ C to degrees Fahrenheit.

$$\frac{9C + 160}{5} = \frac{9(\mathbf{-170}) + 160}{5}$$ Substitute -170 for C.

$$= \frac{-1{,}530 + 160}{5}$$ Perform the multiplication.

$$= \frac{-1{,}370}{5}$$ Perform the addition.

$$= -274$$ Perform the division.

In degrees Fahrenheit, the coldest temperature on the moon is $-274°$.

Answer $-4°$ F

CALCULATOR SNAPSHOT

Evaluating algebraic expressions

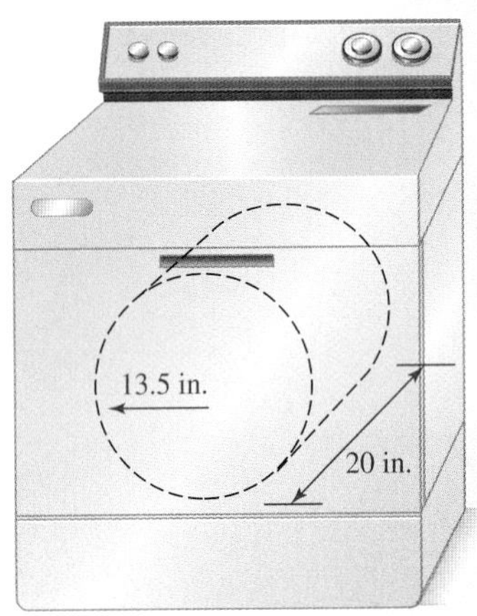

FIGURE 8-3

The rotating drum of a clothes dryer is a cylinder. (See Figure 8-3.) To find the capacity of the dryer, we can find its volume by evaluating the algebraic expression $\pi r^2 h$, where r represents the radius and h represents the height of the drum. (Here, the cylinder is lying on its side). If we substitute 13.5 for r and 20 for h, we obtain $\pi(13.5)^2(20)$. Using a scientific calculator, we can evaluate the expression by entering these numbers and pressing these keys.

[π] [×] 13.5 [x^2] [×] 20 [=] 11451.10522

Using a graphing calculator, we can evaluate the expression by entering these numbers and pressing these keys.

[2nd] [π] [×] 13.5 [x^2] [×] 20 [ENTER]

π*13.5²*20
11451.10522

To the nearest cubic inch, the capacity of the dryer is 11,451 in.^{3}.

Making tables

Self Check 13

In Example 13, suppose the height of the rocket is given by $112t - 16t^2$. Complete the table to find out how many seconds after launch it would hit the ground.

t	$112t - 16t^2$
1	
3	
5	
7	

EXAMPLE 13 **Ballistics.** If a toy rocket is shot into the air with an initial velocity of 80 feet per second, its height (in feet) after t seconds in flight is given by the algebraic expression

$$80t - 16t^2$$

How many seconds after the launch will it hit the ground?

Solution We can substitute positive values for t, the time in flight, until we find the one that gives a height of 0. At that time, the rocket will be on the ground. We will begin by finding the height after the rocket has been in flight for 1 second ($t = 1$) and record the result in a table.

$$80t - 16t^2 = 80(\mathbf{1}) - 16(\mathbf{1})^2$$ Substitute 1 for t.

$$= 64$$

After 1 second in flight, the height of the rocket is 64 feet. We continue to pick more values of t until we find out when the height is 0.

As we evaluate $80t - 16t^2$ for various values of t, we can show the results in a **table of values.** In the column headed "t," we list each value of the variable to be used in the evaluations. In the column headed "$80t - 16t^2$," we write the result of each evaluation.

t	$80t - 16t^2$
1	64
2	96
3	96
4	64
5	0

Evaluate for $t = 2$:
$80t - 16t^2 = 80(2) - 16(2)^2 = 96$
Evaluate for $t = 3$:
$80t - 16t^2 = 80(3) - 16(3)^2 = 96$
Evaluate for $t = 4$:
$80t - 16t^2 = 80(4) - 16(4)^2 = 64$
Evaluate for $t = 5$:
$80t - 16t^2 = 80(5) - 16(5)^2 = 0$

Since the height of the rocket is 0 when $t = 5$, the rocket will hit the ground in 5 seconds.

Answer 7 sec (the heights are 96, 192, 160, and 0)

The two columns of a table of values are sometimes headed with the terms **input** and **output,** as shown in the table below. The t-values are the inputs into the expression $80t - 16t^2$, and the resulting values are thought of as the outputs.

Input	Output
1	64
2	96
3	96
4	64
5	0

Section 8.1 STUDY SET

VOCABULARY *Fill in the blanks.*

1. To ________ an algebraic expression, we substitute the values for the variables and then apply the rules for the order of operations.
2. Variables and/or numbers can be combined with the operation symbols of addition, subtraction, multiplication, and division to create algebraic ________.
3. $2x + 5$ is an example of an algebraic ________, whereas $2x + 5 = 7$ is an example of an ________.
4. When we evaluate an algebraic expression, such as $5x - 8$, for several values of x, we can keep track of the results in an input/output ______.

CONCEPTS

5. Write two algebraic expressions that contain the variable x and the numbers 6 and 20.
6. **a.** Complete the table to determine how many days are in w weeks.

Number of weeks	Number of days
1	
2	
3	
w	

b. Complete the table to answer this question: s seconds is how many minutes?

Number of seconds	Number of minutes
60	
120	
180	
s	

7. When evaluating $3x - 6$ for $x = 4$, what misunderstanding can occur if we don't write parentheses around 4 when it is substituted for the variable?

8. If the knife in the illustration is 12 inches long, how long is the blade?

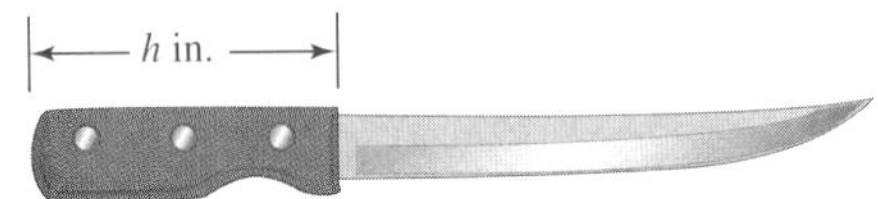

9. a. In the illustration, the weight of the van is 500 pounds less than twice the weight of the car. Express the weight of the van and the car using the variable x.

b. If the actual weight of the car is 2,000 pounds, what is the weight of the van?

10. See the illustration.

a. If we let b represent the length of the beam, write an algebraic expression for the length of the pipe.

b. If we let p represent the length of the pipe, write an algebraic expression for the length of the beam.

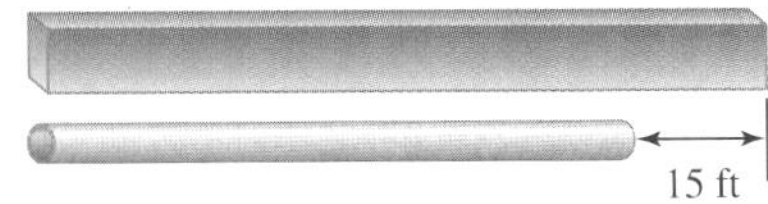

11. Complete the table.

Type of coin	Number ·	Value in cents =	Total value in cents
Nickel	6		
Dime	d		
Half dollar	$x + 5$		

12. If $x = -9$, find the value of

a. $-x$ **b.** $-(-x)$ **c.** $-x^2$ **d.** $(-x)^2$

NOTATION *Complete each solution.*

13. Evaluate the expression $9a - a^2$ for $a = 5$.

$$\begin{aligned} 9a - a^2 &= 9(\quad) - (\quad)^2 \\ &= 9(5) - \quad \\ &= \quad - 25 \\ &= 20 \end{aligned}$$

14. Evaluate $\frac{4x^2 - 3y}{9(x - y)}$ when $x = 4$ and $y = -3$.

$$\begin{aligned} \frac{4x^2 - 3y}{9(x - y)} &= \frac{4(\mathbf{4})^2 - 3(\mathbf{-3})}{9[\mathbf{4} - (\mathbf{-3})]} \\ &= \frac{4(\quad) - 3(\quad)}{9(\quad)} \\ &= \frac{(\quad) - (\quad)}{\quad} \\ &= \frac{73}{63} \end{aligned}$$

PRACTICE *Translate each phrase to an algebraic expression. If no variable is given, use x as the variable.*

15. The sum of the length l and 15

16. The difference of a number and 10

17. The product of a number and 50

18. Three-fourths of the population p

19. The ratio of the amount won w and lost l

20. The tax t added to c

21. P increased by p

22. 21 less than the total height h

23. The square of k minus 2,005

24. s subtracted from S

25. J reduced by 500

26. Twice the attendance a

27. 1,000 split n equal ways

28. Exceeds the cost c by 25,000

29. 90 more than the current price p

30. 64 divided by the cube of y

31. The total of 35, h, and 300

32. x decreased by 17

33. 680 fewer than the entire population p

34. Triple the number of expected participants

35. The product of d and 4, decreased by 15

36. Forty-five more than the quotient of y and 6

37. Twice the sum of 200 and t

38. The square of the quantity 14 less than x

39. The absolute value of the difference of a and 2

40. The absolute value of a, decreased by 2

If n represents a number, write a word description of each algebraic expression. (Answers may vary.)

41. $n - 7$

42. $n^2 + 7$

43. $7n + 4$

44. $3(n + 1)$

45. How many minutes are there in **a.** 5 hours and **b.** h hours?

46. A woman watches television x hours a day. Express the number of hours she watches TV **a.** in a week and **b.** in a year.

47. **a.** How many feet are in y yards?
b. How many yards are in f feet?

48. A sales clerk earns \$$x$ an hour. How much does he earn in **a.** an 8-hour day and **b.** a 40-hour week?

49. If a car rental agency charges 29¢ a mile, express the rental fee if a car is driven x miles.

50. A model's skirt is x inches long. The designer then lets the hem down 2 inches. How can we express the length (in inches) of the altered skirt?

51. A soft-drink manufacturer produced c cans of cola during the morning shift. Write an expression for how many six-packs of cola can be assembled from the morning shift's production.

52. The tag on a new pair of 36-inch-long jeans warns that after washing, they will shrink x inches in length. Express the length (in inches) of the jeans after they are washed.

53. A caravan of b cars, each carrying 5 people, traveled to the state capital for a political rally. Express how many people were in the car caravan.

54. A caterer always prepares food for 10 more people than the order specifies. If p people are to attend a reception, write an expression for the number of people she should prepare for.

55. Tickets to a circus cost \$5 each. Express how much tickets will cost for a family of x people if they also pay for two of their neighbors.

56. If each egg is worth e¢, express the value (in cents) of a dozen eggs.

Complete each table.

57.

x	$x^3 - 1$
0	
−1	
−3	

58.

g	$g^2 - 7g + 1$
0	
7	
−10	

59.

s	$\frac{5s + 36}{s}$
1	
6	
−12	

60.

a	$2{,}500a + a^3$
2	
4	
−5	

61.

Input x	Output $2x - \frac{x}{2}$
100	
−300	

62.

Input x	Output $\frac{x}{3} + \frac{x}{4}$
12	
−36	

63.

x	$(x + 1)(x + 5)$
−1	
−5	
−6	

64.

x	$\frac{1}{x + 8}$
−7	
−9	
−8	

Evaluate each expression, given that $x = 3$, $y = -2$, and $z = -4$.

65. $3y^2 - 6y - 4$

66. $-z^2 - z - 12$

67. $(3 + x)y$

68. $(4 + z)y$

69. $(x + y)^2 - |z + y|$

70. $[(z - 1)(z + 1)]^2$

71. $(4x)^2 + 3y^2$

72. $4x^2 + (3y)^2$

73. $-\dfrac{2x + y^3}{y + 2z}$

74. $-\dfrac{2z^2 - y}{2x - y^2}$

Evaluate each expression for the given values of the variables.

75. $b^2 - 4ac$ for $a = -1, b = 5$, and $c = -2$

76. $(x - a)^2 + (y - b)^2$ for $x = -2, y = 1, a = 5$, and $b = -3$

77. $a^2 + 2ab + b^2$ for $a = -5$ and $b = -1$

78. $\dfrac{x - a}{y - b}$ for $x = -2, y = 1, a = 5$, and $b = 2$

79. $\dfrac{n}{2}[2a + (n - 1)d]$ for $n = 10, a = -4$, and $d = 6$

80. $\dfrac{a(1 - r^n)}{1 - r}$ for $a = -5, r = 2$, and $n = 3$

81. $\dfrac{a^2 + b^2}{2}$ for $a = 1.8$ and $b = -7.6$

82. $(y^3 - 52y^2)^2$ for $y = 55$

APPLICATIONS

83. ROCKETRY The algebraic expression $64t - 16t^2$ gives the height of a toy rocket (in feet) t seconds after being launched. Find the height of the rocket for each of the times shown in the table.

t	h
0	
0.5	
1	
1.5	
2	
2.5	
3	
3.5	
4	

84. GROWING SOD To determine the number of square feet of sod *remaining* in a field after filling an order, the manager of a sod farm uses the expression $20{,}000 - 3s$ (where s is the number of 1-foot-by-3-foot strips the customer has ordered). To sod a soccer field, a city orders 7,000 strips of sod. Evaluate the expression for this value of s and explain the result.

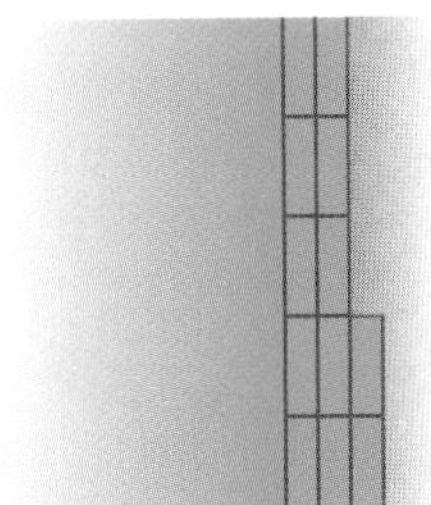

1-ft-by-3-ft strips of sod, cut and ready to be loaded on a truck for delivery

85. ANTIFREEZE The expression

$$\frac{5(F - 32)}{9}$$

converts a temperature in degrees Fahrenheit (given as F) to degrees Celsius. Convert the temperatures listed on the container of antifreeze shown in the illustration to degrees Celsius. Round to the nearest degree.

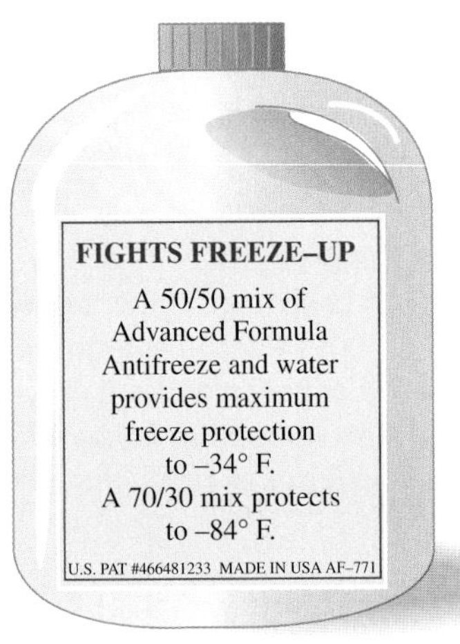

FIGHTS FREEZE–UP
A 50/50 mix of Advanced Formula Antifreeze and water provides maximum freeze protection to –34° F.
A 70/30 mix protects to –84° F.
U.S. PAT #466481233 MADE IN USA AF–771

86. TEMPERATURE ON MARS On Mars, maximum summer temperatures can reach 20° C. However, daily temperatures average −33° C. Convert each of these temperatures to degrees Fahrenheit. Round to the nearest degree.

87. The utility knife blade shown in the illustration is in the shape of a trapezoid. Find the area of the front face of the blade. (See Example 11 on page 483 for the expression that gives the area of a trapezoid.)

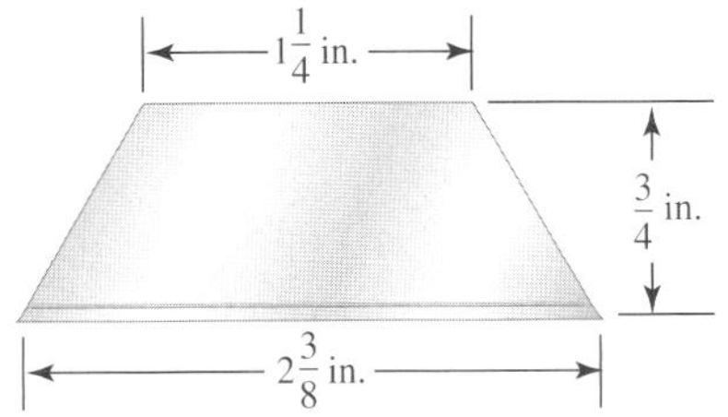

88. TRUMPET MUTES The expression

$$\pi[b^2 + d^2 + (b + d)s]$$

can be used to find the total surface area of the trumpet mute shown in the illustration. Evaluate the expression for the given dimensions to find the number of square inches of cardboard (to the nearest tenth) used to make the mute.

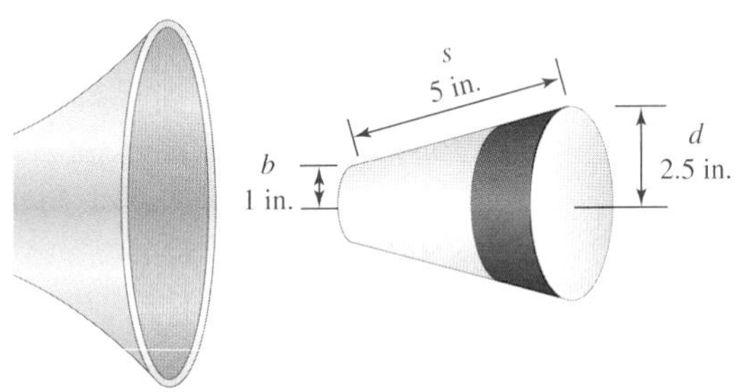

89. LANDSCAPING A grass strip is to be planted around a tree, as shown in the illustration. Find the number of square feet of sod to order by evaluating the expression $\pi(R^2 - r^2)$. Round to the nearest square foot.

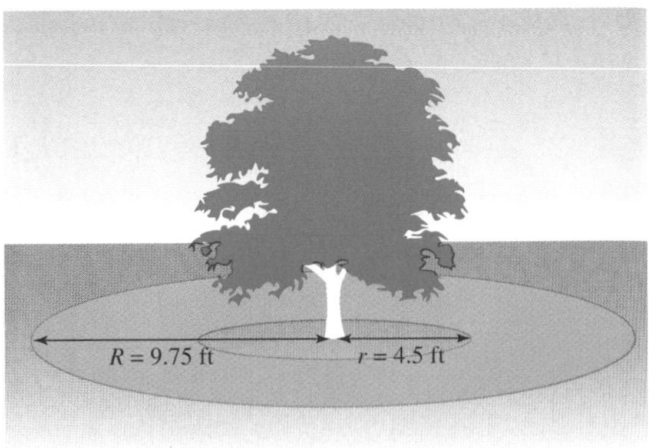

90. ENERGY CONSERVATION A fiberglass blanket wrapped around a water heater helps prevent heat loss. Find the number of square feet of heater surface the blanket covers by evaluating the algebraic expression $2\pi rh$, where r is the radius and h is the height. Round to the nearest square foot.

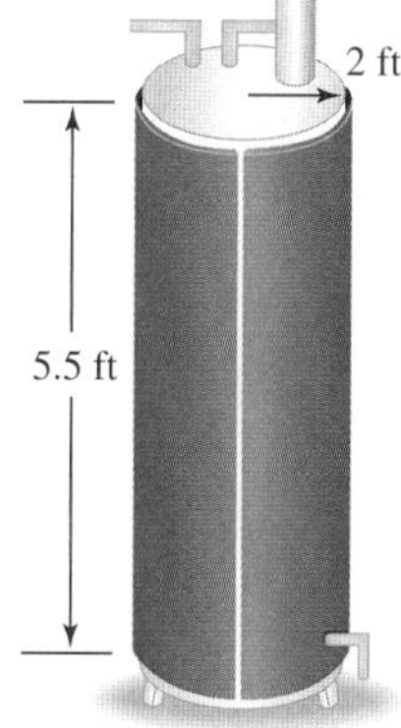

WRITING

91. What is an algebraic expression? Give some examples.

92. What is a variable? How are variables used in this section?

93. In this section, we substituted a number for a variable. List some other uses of the word *substitute* that you encounter in everyday life.

94. Explain why d dimes are not worth d cents.

REVIEW

95. Simplify: -0.

96. Is the statement $-5 > -4$ true or false?

97. Evaluate: $|-\frac{2}{3}|$.

98. Evaluate: $2^3 \cdot 3^2$.

99. Write $5 \cdot 5 \cdot 5 \cdot 5$ in exponential form.

100. Evaluate: $15 + 2[15 - (12 - 10)]$.

101. Find the mean (average) of the three test scores 84, 93, and 72.

102. Fill in the blanks: In the multiplication statement $5 \cdot x = 5x$, 5 and x are called ________, and $5x$ is called the ________.

8.2 Simplifying Algebraic Expressions

- Simplifying algebraic expressions involving multiplication • The distributive property
- The vocabulary of algebraic expressions • Like terms • Combining like terms

In arithmetic, we often replace one numerical expression with another expression that is equivalent and simpler in form. For example, when we express $\frac{40}{80}$ as $\frac{1}{2}$, we say we have *simplified* $\frac{40}{80}$. In algebra, we often simplify algebraic expressions. To **simplify an algebraic expression,** we use properties of algebra to write the expression in an equivalent, less complicated form.

Simplifying algebraic expressions involving multiplication

Two properties that are often used to simplify algebraic expressions are the associative and commutative properties of multiplication. Recall that the associative property of multiplication enables us to change the grouping of factors involved in a multiplication. The commutative property of multiplication enables us to change the order of the factors.

As an example, let's consider the expression $8(4x)$ and simplify it as follows:

$8(4x) = 8 \cdot (4 \cdot x)$ — $4x = 4 \cdot x$.

$= (8 \cdot 4) \cdot x$ — Use the associative property of multiplication to group 4 with 8 instead of with x.

$= 32x$ — Perform the multiplication within the parentheses: $8 \cdot 4 = 32$.

Since $8(4x) = 32x$, we say that $8(4x)$ simplifies to $32x$. To verify that $8(4x)$ and $32x$ are **equivalent expressions** (represent the same number), we can evaluate each expression for several choices of x. For each value of x, the results should be the same.

If $x = 10$		If $x = -3$	
$8(4x) = 8[4(\mathbf{10})]$	$32x = 32(\mathbf{10})$	$8(4x) = 8[4(\mathbf{-3})]$	$32x = 32(\mathbf{-3})$
$= 8(40)$	$= 320$	$= 8(-12)$	$= -96$
$= 320$		$= -96$	

Self Check 1

Simplify each expression:

a. $9 \cdot 6s$

b. $8(\frac{7}{8}h)$

c. $21p(-3q)$

d. $-4(6m)(-2m)$

Answer a. $54s$, b. $7h$, c. $-63pq$, d. $48m^2$

EXAMPLE 1 Simplify each expression: **a.** $15a(-7)$, **b.** $5(\frac{4}{5}x)$, **c.** $-5r(-6s)$, and **d.** $3(7p)(-5p)$.

Solution

a. $15a(-7) = 15(-7)a$ Use the commutative property of multiplication to change the order of the factors.

$= -105a$ Working left to right, perform the multiplications.

b. $5\left(\frac{4}{5}x\right) = \left(5 \cdot \frac{4}{5}\right)x$ Use the associative property of multiplication to group the numbers.

$= 4x$ Multiply: $5 \cdot \frac{4}{5} = \frac{5}{1} \cdot \frac{4}{5} = \frac{\overset{1}{\cancel{5}} \cdot 4}{1 \cdot \underset{1}{\cancel{5}}} = 4$.

c. We note that the expression contains two variables.

$-5r(-6s) = [-5(-6)][r \cdot s]$ Use the commutative and associative properties of multiplication to group the numbers and group the variables.

$= 30rs$ Perform the multiplications within the brackets: $-5(-6) = 30$ and $r \cdot s = rs$.

d. $3(7p)(-5p) = [3(7)(-5)](p \cdot p)$ Use the commutative and associative properties of multiplication to change the order and to regroup the factors.

$= -105p^2$ Perform the multiplication within the grouping symbols: $3(7)(-5) = -105$ and $p \cdot p = p^2$.

! COMMENT Note that $p \cdot p$ can be written with an exponent as p^2. Just as $4 \cdot 4 = 4^2$ and $7 \cdot 7 = 7^2$, the expression $p \cdot p = p^2$.

The distributive property

To introduce the **distributive property,** we will consider the expression $4(5 + 3)$, which can be evaluated in two ways.

Method 1. Rules for the order of operations: We compute the sum within the parentheses first.

$4(\mathbf{5 + 3}) = 4(\mathbf{8})$ Perform the addition within the parentheses first.

$= 32$ Perform the multiplication.

Method 2. The distributive property: We multiply both 5 and 3 by 4, and then we add the results.

$$4(5 + 3) = 4(5) + 4(3) \quad \text{Distribute the multiplication by 4.}$$
$$= 20 + 12 \quad \text{Perform the multiplications.}$$
$$= 32 \quad \text{Perform the addition.}$$

Notice that each method gives a result of 32.

We can interpret the distributive property geometrically. Figure 8-4 shows three rectangles that are divided into squares. Since the area of the rectangle on the left-hand side of the equals sign can be found by multiplying its width by its length, its area is $4(5 + 3)$ square units. We can evaluate this expression, or we can count squares; either way, we see that the area is 32 square units.

The area shown on the right-hand side is the sum of the areas of two rectangles: $4(5) + 4(3)$. Either by evaluating this expression or by counting squares, we see that this area is also 32 square units. Therefore,

$$4(5 + 3) = 4(5) + 4(3)$$

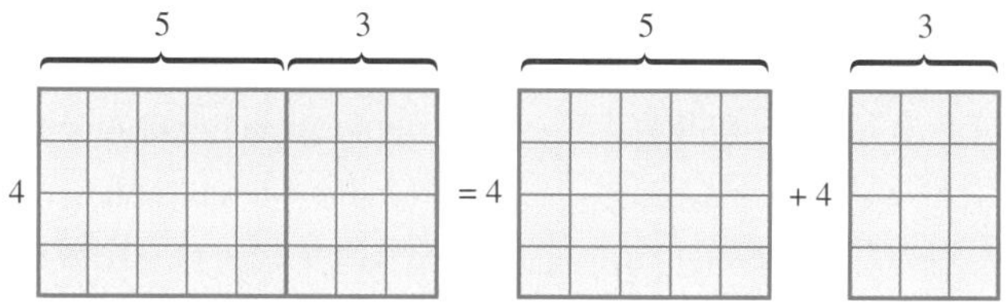

FIGURE 8-4

Figure 8-5 shows the general case where the width is a and the length is $b + c$.

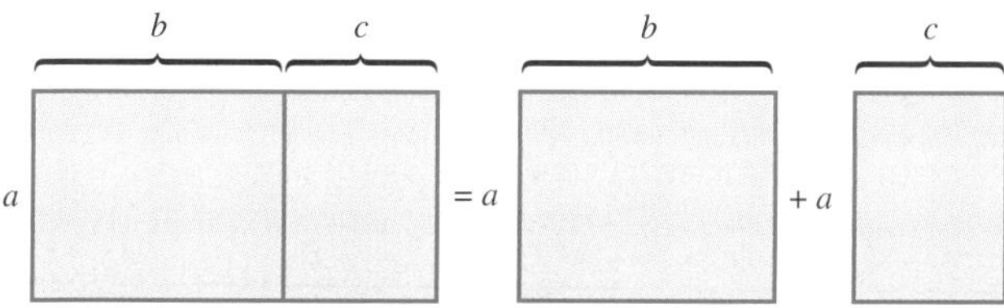

FIGURE 8-5

Using Figure 8-5 as a basis, we can now state the distributive property in symbols.

The distributive property

If a, b, and c represent real numbers, then

$$a(b + c) = ab + ac$$

Since subtraction is the same as adding the opposite, the distributive property also holds for subtraction.

The distributive property

If a, b, and c represent real numbers, then

$$a(b - c) = ab - ac$$

We can use the distributive property to remove the parentheses when an expression is multiplied by a quantity. For example, to remove the parentheses in $5(x + 2)$, we distribute the factor of 5 that is outside the parentheses over the addition of x and 2, and add the products.

$$\mathbf{5}(x + 2) = \mathbf{5}(x) + \mathbf{5}(2)$$ Distribute the multiplication by 5.

$$= 5x + 10$$ Perform the multiplications.

! COMMENT Since the expression $5(x + 2)$ contains parentheses, some students are tempted to perform the addition within the parentheses first. However, we cannot add x and 2, because we do not know the value of x. Instead, we should multiply $x + 2$ by 5, which requires the use of the distributive property.

Self Check 2
Use the distributive property to remove parentheses:
a. $5(p + 2)$
b. $4(t - 1)$
c. $-8(2x - 4)$
d. $p(p - 5)$

Answers **a.** $5p + 10$, **b.** $4t - 4$ **c.** $-16x + 32$, **d.** $p^2 - 5p$

EXAMPLE 2 Use the distributive property to remove parentheses: **a.** $3(x - 8)$, **b.** $-12(a + 1)$, **c.** $-6(-3y - 8)$, and **d.** $x(x + 2)$.

Solution

a. $\mathbf{3}(x - 8) = \mathbf{3}(x) - \mathbf{3}(8)$ Distribute the multiplication by 3.

$= 3x - 24$ Perform the multiplications.

b. $\mathbf{-12}(a + 1) = \mathbf{-12}(a) + (\mathbf{-12})(1)$ Distribute the multiplication by -12.

$= -12a + (-12)$ Perform the multiplications.

$= -12a - 12$ Write the addition of -12 as subtraction of 12.

c. $\mathbf{-6}(-3y - 8) = \mathbf{-6}(-3y) - (\mathbf{-6})(8)$ Distribute the multiplication by -6.

$= 18y - (-48)$ Perform the multiplications.

$= 18y + 48$ Add the opposite of -48, which is 48.

d. $\mathbf{x}(x + 2) = \mathbf{x}(x) + \mathbf{x}(2)$ Distribute the multiplication by x.

$= x^2 + 2x$ Perform the multiplications: $x(x) = x^2$ and $x(2) = 2x$.

! COMMENT The fact that an expression contains parentheses does not necessarily mean that the distributive property can be applied. For example, the distributive property does not apply to the expressions

$6(5x)$ or $6(-7 \cdot y)$ Here a product is multiplied by 6. Simplifying, we have $6(5x) = 30x$ and $6(-7 \cdot y) = -42y$.

However, the distributive property does apply to the expressions

$6(5 + x)$ or $6(-7 - y)$ Here a sum and a difference are multiplied by 6. Distributing the 6, we have $6(5 + x) = 30 + 6x$ and $6(-7 - y) = -42 - 6y$.

To use the distributive property to simplify $-(x + 10)$, we note that the negative sign in front of the parentheses represents -1.

The $-$ sign represents -1.

$-(x + 10) = \mathbf{-1}(x + 10)$

$= \mathbf{-1}(x) + (\mathbf{-1})(10)$ Distribute the multiplication by -1.

$= -x + (-10)$ Multiply: $-1(x) = -x$ and $(-1)(10) = -10$.

$= -x - 10$ Write the addition of -10 as a subtraction.

EXAMPLE 3 Simplify: $-(-12 - 3p)$.

Solution

$-(-12 - 3p)$

$= \mathbf{-1}(-12 - 3p)$ Change the − sign in front of the parentheses to −1.

$= \mathbf{-1}(-12) - (\mathbf{-1})(3p)$ Distribute the multiplication by −1.

$= 12 - (-3p)$ Multiply: $-1(-12) = 12$ and $(-1)(3p) = -3p$.

$= 12 + 3p$ To subtract $-3p$, add the opposite of $-3p$, which is $3p$.

Self Check 3
Simplify: $-(-5x + 18)$.

Answer $5x - 18$

Since multiplication is commutative, we can write the distributive property in the following forms.

$(b + c)a = ba + ca \qquad (b - c)a = ba - ca$

EXAMPLE 4 Multiply: $(6x + 4y)\frac{1}{2}$.

Solution

$(6x + 4y)\frac{\mathbf{1}}{\mathbf{2}} = (6x)\frac{\mathbf{1}}{\mathbf{2}} + (4y)\frac{\mathbf{1}}{\mathbf{2}}$ Distribute the multiplication by $\frac{1}{2}$.

$= 3x + 2y$ Multiply: $(6x)\frac{1}{2} = (6 \cdot \frac{1}{2})x = 3x$ and $(4y)\frac{1}{2} = (4 \cdot \frac{1}{2})y = 2y$.

Self Check 4
Multiply: $(-6x - 24y)\frac{1}{3}$.

Answer $-2x - 8y$

The distributive property can be extended to situations in which there are more than two terms within parentheses.

The extended distributive property
If a, b, c, and d represent real numbers, then

$a(b + c + d) = ab + ac + ad$ and $a(b - c - d) = ab - ac - ad$

EXAMPLE 5 Remove parentheses: $-0.3(3a - 4b + 7)$.

Solution

$\mathbf{-0.3}(3a - 4b + 7)$

$= \mathbf{-0.3}(3a) - (\mathbf{-0.3})(4b) + (\mathbf{-0.3})(7)$ Distribute the multiplication by −0.3.

$= -0.9a - (-1.2b) + (-2.1)$ Perform the three multiplications.

$= -0.9a + 1.2b + (-2.1)$ To substract $-1.2b$, add its opposite, which is $1.2b$.

$= -0.9a + 1.2b - 2.1$ Write the addition of −2.1 as a subtraction.

Self Check 5
Remove parentheses:
$-0.7(2r + 5s - 8)$

Answer $-1.4r - 3.5s + 5.6$

The vocabulary of algebraic expressions

Addition signs separate algebraic expressions into parts called **terms.** The expression $5x + 8$ contains two terms, $5x$ and 8.

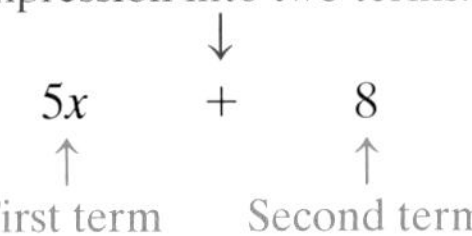
The + sign separates the
expression into two terms.
5x + 8
First term
Second term

A term can be

- a number. Examples are 8, 98.6, and -45.
- a variable, or a product of variables (which may be raised to powers). Examples are x, s^3, rt, and a^2bc^4.
- a product of a number and one or more variables (which may be raised to powers). Examples are $-35x$, $\frac{1}{2}bh$, and πr^2h.

Since subtraction can be expressed as addition of the opposite, the expression $6x - 5$ can be written in the equivalent form $6x + (-5)$. We can then see that $6x - 5$ contains two terms, $6x$ and -5.

Self Check 6
List the terms in each expression:
a. $\frac{1}{3}Bh$, **b.** $3p + 5q - 1.2$, and **c.** $b^2 - 4ac$

Answers
a. $\frac{1}{3}Bh$, **b.** $3p, 5q, -1.2$, **c.** $b^2, -4ac$

EXAMPLE 6 List the terms in each expression: **a.** $-4p + 7 + 5p$, **b.** $-12r^2st$, and **c.** $y^3 + 8y^2 - 3y - 24$.

Solution

a. $-4p + 7 + 5p$ has three terms: $-4p$, 7, and $5p$.

b. The expression $-12r^2st$ has one term: $-12r^2st$.

c. $y^3 + 8y^2 - 3y - 24$ can be written as $y^3 + 8y^2 + (-3y) + (-24)$. It contains four terms: y^3, $8y^2$, $-3y$, and -24.

In a term that is the product of a number and one or more variables, the number is called the **numerical coefficient,** or simply the **coefficient.** In the expression $5x$, the coefficient is 5, and x is the variable part. Other examples are shown in Table 8-2.

Term	Coefficient	Variable part
$8y^2$	8	y^2
$-0.9pq$	-0.9	pq
$\frac{3}{4}b$	$\frac{3}{4}$	b
$-\frac{x}{6}$	$-\frac{1}{6}$	x
x	1	x
$-t$	-1	t
15	15	none

TABLE 8-2

Notice that when there is no number in front of a variable, the term has an *implied coefficient* of 1. When there is a negative sign in front of the variable, the term has an *implied coefficient* of -1. For example, $-t = -1t$.

EXAMPLE 7 Identify the coefficient and the variable part of each term in $-7x^2 + 3x - 6$.

Solution

Term	Coefficient	Variable part
$-7x^2$	-7	x^2
$3x$	3	x
-6	-6	none

Self Check 7
Identify the coefficient of each term in $p^3 - 12p^2 + 3p - 4$.

Answers $1, -12, 3, -4$

! COMMENT It is important to be able to distinguish between a *term* of an expression and a *factor* of a term. Terms are separated by + symbols. Factors are numbers and/or variables that are multiplied together. For example, x is a term of the expression $18 + x$, because x and 18 are separated by a + symbol. In the expression $18x + 9$, x is a factor of the term $18x$, because x and 18 are multiplied together.

Like terms

The expression $5p + 7q - 3p + 12$, which can be written $5p + 7q + (-3p) + 12$, contains four terms, $5p$, $7q$, $-3p$, and 12. Since the variable of $5p$ and $-3p$ are the same, we say that these terms are **like** or **similar terms.**

Like terms (similar terms)

Like terms (or **similar terms**) are terms with exactly the same variables raised to exactly the same powers. Any numbers (called **constants**) in an expression are considered to be like terms.

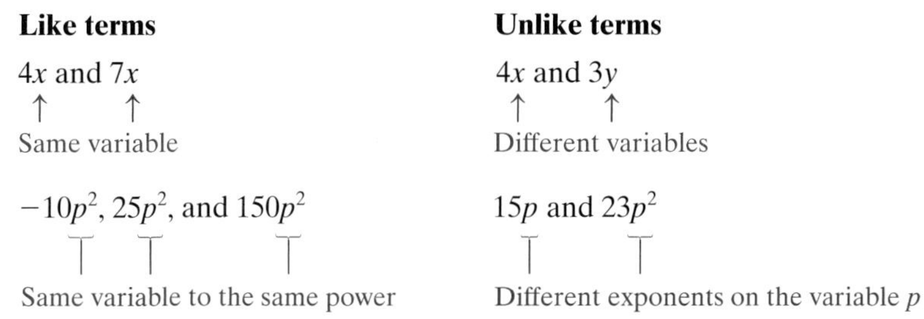

! COMMENT When looking for like terms, don't look at the coefficients of the terms. Consider only their variable parts.

EXAMPLE 8 List like terms: **a.** $7r + 5 + 3r$, **b.** $x^4 - 6x^2 - 5$, and **c.** $-7m + 7 - 2 + m$.

Solution

a. $7r + 5 + 3r$ contains the like terms $7r$ and $3r$.

b. $x^4 - 6x^2 - 5$ contains no like terms.

c. $-7m + 7 - 2 + m$ contains two pairs of like terms: $-7m$ and m are like terms, and the constants, 7 and -2, are like terms.

Self Check 8
List like terms:
a. $5x - 2y + 7y$
b. $-5pq + 17p - 12q - 2pq$

Answers **a.** $-2y$ and $7y$, **b.** $-5pq$ and $-2pq$

Combining like terms

If we are to add (or subtract) objects, they must have the same units. For example, we can add dollars to dollars and inches to inches, but we cannot add dollars to inches. The same is true when we work with terms of an algebraic expression. They can be added or subtracted only when they are like terms.

This expression can be simplified, because it contains like terms.	This expression cannot be simplified, because its terms are not like terms.
$3x + 4x$	$3x + 4y$
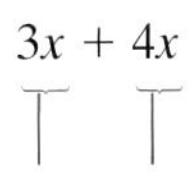	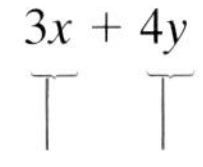
Like terms The variable parts are identical.	Unlike terms The variable parts are not identical.

To simplify an expression containing like terms, we use the distributive property. For example, we can simplify $3x + 4x$ as follows:

$3x + 4x = (\mathbf{3} + \mathbf{4})x$ Use the distributive property.

$= \mathbf{7}x$ Perform the addition within the parentheses: $3 + 4 = 7$.

We have simplified the expression $3x + 4x$ by **combining like terms.** The result is the equivalent expression $7x$. This example suggests the following general rule.

> **Combining like terms**
>
> To add or subtract like terms, combine their coefficients and keep the same variables with the same exponents.

Self Check 9
Simplify by combining like terms:

a. $5n + (-8n)$

b. $-1.2a^3 + (1.4a^3)$

Answers **a.** $-3n$, **b.** $0.2a^3$

EXAMPLE 9 Simplify by combining like terms: **a.** $-8p + (-12p)$ and **b.** $0.5s^2 - 0.3s^2$.

Solution

a. $-8p + (-12p) = -20p$ Add the coefficients of the like terms: $-8 + (-12) = -20$. Keep the variable p.

b. $0.5s^2 - 0.3s^2 = 0.2s^2$ Subtract: $0.5 - 0.3 = 0.2$. Keep the variable part s^2.

Self Check 10
Simplify: $8R + 7r - 14R - 21r$.

Answer $-6R - 14r$

EXAMPLE 10 Simplify: $7P - 8p - 12P + 25p$.

Solution The uppercase P and the lowercase p are different variables. We can use the commutative property of addition to write like terms next to each other.

$7P - 8p - 12P + 25p$

$= 7P + (-8p) + (-12P) + 25p$ Rewrite each subtraction as the addition of the opposite.

$= 7P + (-12P) + (-8p) + 25p$ Use the commutative property of addition to write the like terms together.

$= -5P + 17p$ Combine like terms: $7P + (-12P) = -5P$ and $-8p + 25p = 17p$.

The expression in Example 10 contained two sets of like terms, and we rearranged the terms so that like terms were next to each other. With practice, you will be able to combine like terms without having to write them next to each other.

EXAMPLE 11 Simplify: $4(x + 5) - 3(2x - 4)$.

Solution

$4(x + 5) - 3(2x - 4)$

$= 4x + 20 - 6x + 12$ Use the distributive property twice.

$= -2x + 32$ Combine like terms: $4x - 6x = -2x$ and $20 + 12 = 32$.

Self Check 11
Simplify: $-5(y - 4) + 2(4y + 6)$.

Answer $3y + 32$

Section 8.2 STUDY SET

VOCABULARY *Fill in the blanks.*

1. To ________ an algebraic expression, we use properties of algebra to write the expression in a less complicated form.
2. A ______ is a number or a product of a number and one or more variables.
3. In the term $3x^2$, the number factor 3 is called the __________ and x^2 is called the _________ part.
4. Two terms with exactly the same variables and exponents are called ______ terms.
5. We can use the distributive property to ________ the parentheses in the expression $2(x + 8)$.
6. The ___________ property of multiplication enables us to change the order of the factors involved in a multiplication.

CONCEPTS

7. What property does the statement $a(b + c) = ab + ac$ illustrate?
8. Complete this statement:

   $a(b + c + d) =$ ______

9. The illustration shows an application of the distributive property. Fill in the blanks.

   $2(\ \ + \ \ ) \quad = \quad 2(\ \ ) \quad + \quad 2(\ \ )$

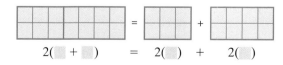

10. Complete the table.

Term	Coefficient	Variable part
$6m$		
$-75t$		
w		
$\frac{1}{2}bh$		

11. Fill in the blanks.
    a. $2(x + 4) = 2x \ \square \ 8$
    b. $2(x - 4) = 2x \ \square \ 8$
    c. $-2(x + 4) = -2x \ \square \ 8$
    d. $-2(x - 4) = -2x \ \square \ 8$
    e. $-2(-x + 4) = 2x \ \square \ 8$
    f. $-2(-x - 4) = 2x \ \square \ 8$
12. Complete this statement: To add or subtract like terms, combine their __________ and keep the same variables and __________.
13. A board was cut into two pieces, as shown. Add the lengths of the two pieces. How long was the original board?

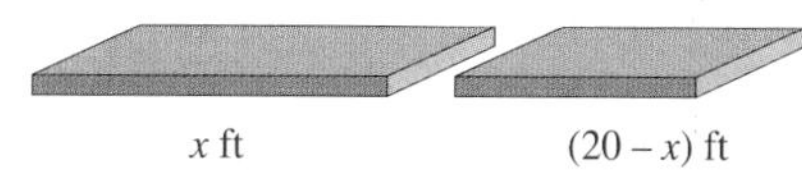

x ft $\qquad (20 - x)$ ft

14. Let x = the number of miles driven on the first day of a 2-day driving trip. Translate the verbal model to mathematical symbols, and simplify by combining like terms.

the miles driven day 1	plus	100 miles more than the miles driven day 1

15. a. Two angles are called **complementary angles** if the sum of their measures is 90°. Add the measures of the angles in illustration (a). Are they complementary angles?

b. Two angles are called **supplementary angles** if the sum of their measures is 180°. Add the measures of the angles in illustration (b). Are they supplementary angles?

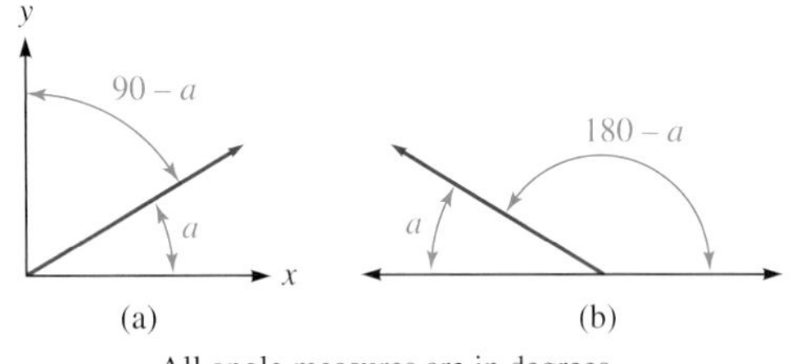

All angle measures are in degrees.

16. Simplify each expression, if possible.

a. $5(2x)$ and $5 + 2x$

b. $6(-7x)$ and $6 - 7x$

c. $2(3x)(3)$ and $2 + 3x + 3$

d. $x \cdot x$ and $x + x$

NOTATION *Complete each solution.*

17. $7(a + 2) = \square(a) + \square(2)$
$= 7a + \square$

18. $6(b - 5) + 12b + 7 = 6(\square) - 6(\square) + 12b + 7$
$= 6b - \square + 12b + 7$
$= 6b + \square b - \square + 7$
$= \square b - 23$

19. a. Are $2K$ and $3k$ like terms?

b. Are $-d$ and d like terms?

20. Fill in the blank.

$-(x + 10) = -\square(x + 10)$

21. Write each expression using fewer symbols.

a. $5x - (-1)$

b. $16t + (-6)$

22. In the following table, a student's answers to five homework problems are compared to the answers in the back of the book. Are the answers equivalent?

Student's answer	Book's answer	Equivalent?
$10x$	$10 + x$	
$3 + y$	$y + 3$	
$5 - 8a$	$8a - 5$	
$3(x) + 4$	$3(x + 4)$	
$2x$	x^2	

PRACTICE *Simplify each expression.*

23. $9(7m)$

24. $12n(8)$

25. $5(-7q)$

26. $-7(5t)$

27. $12\left(\frac{5}{12}x\right)$

28. $15\left(\frac{4}{15}w\right)$

29. $8\left(\frac{3}{4}y\right)$

30. $27\left(\frac{2}{3}x\right)$

31. $(-5p)(-4b)$

32. $(-7d)(-7c)$

33. $-5(4r)(-2r)$

34. $7t(-4t)(-2)$

Use the distributive property to remove parentheses.

35. $5(x + 3)$

36. $4(x + 2)$

37. $-2(b - 1)$

38. $-7(p - 5)$

39. $(3t - 2)8$

40. $(2q + 1)9$

41. $(2y - 1)6$

42. $(3w - 5)5$

43. $0.4(x - 4)$

44. $-2.2(2q + 1)$

45. $-\frac{2}{3}(3w - 6)$

46. $\frac{1}{2}(2y - 8)$

47. $r(r - 10)$

48. $h(h + 4)$

49. $-(x - 7)$

50. $-(y + 1)$

51. $17(2x - y + 2)$

52. $-12(3a + 2b - 1)$

53. $-(-14 + 3p - t)$

54. $-(-x - y + 5)$

55. Identify the coefficient of each term.

a. $-b$

b. $-9.9x^3$

c. $\frac{1}{4}x$

d. $-\frac{2x}{3}$

56. Determine whether the variable x is used as a factor or as a term.

a. $24 - x$ **b.** $24x$

c. $24 + 3x$ **d.** $x - 12$

Identify the coefficient of each term.

57. $-5r + 4s$ **58.** $2m + n - 3m + 2n$

59. $-15r^2s$ **60.** $4b^2 - 5b + 6$

61. $50a + 2$ **62.** $a^2 - ab + b^2$

63. $x^3 - 125$ **64.** $-2.55x + 1.8$

Simplify each expression by combining like terms.

65. $3x + 17x$ **66.** $12y - 15y$

67. $8x^2 - 5x^2$ **68.** $17x^2 + 3x^2$

69. $-4x + 4x$ **70.** $-16y + 16y$

71. $-7b + 7b$ **72.** $-2c + 2c$

73. $a + a + a$ **74.** $t - t - t - t$

75. $0 - 3x$ **76.** $0 - 4a$

77. $0 - (-t)$ **78.** $0 - (-2y)$

79. $3x + 5x - 7x$ **80.** $-y + 3y + 2y$

81. $-13x^2 + 2x^2 - 5x^2$ **82.** $-8x^3 - x^3 + 2x^3$

83. $1.8h - 0.7h$ **84.** $-5.7m + 4.3m$

85. $\frac{3}{5}t + \frac{1}{5}t$ **86.** $\frac{3}{16}x - \frac{5}{16}x$

87. $-0.2r - (-0.6r)$ **88.** $-1.1m - (-2.4m)$

89. $2z + 5(z - 3)$ **90.** $12(m + 11) - 11$

91. $-(c + 7) - 2(c - 3)$ **92.** $-(z + 2) + 5(3 - z)$

93. $2x + 4(X - x) + 3X$ **94.** $3p - 6(p + z) + p$

95. $(a + 2) - (a - b)$ **96.** $3z + 2(Z - z) + Z$

97. $x(x + 3) - 3x^2$ **98.** $2x + x(x - 3)$

APPLICATIONS

99. THE AMERICAN RED CROSS In 1891, Clara Barton founded the Red Cross. Its symbol is a white flag bearing a red cross. If each side of the cross in the illustration has length x, write an algebraic expression for the perimeter (the total distance around the outside) of the cross.

100. BILLIARDS Billiard tables vary in size, but all tables are twice as long as they are wide.

a. If the following billiard table is x feet wide, write an expression involving x that represents its length.

b. Write an expression for the perimeter of the table.

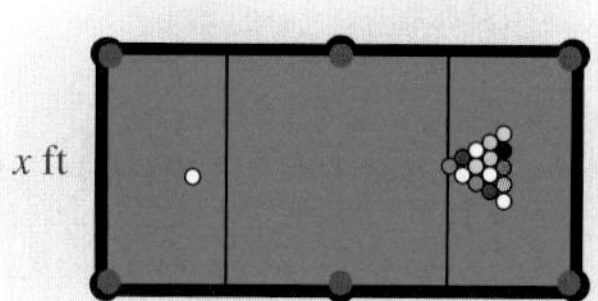

101. PING-PONG Write an expression for the perimeter of the table shown in the illustration.

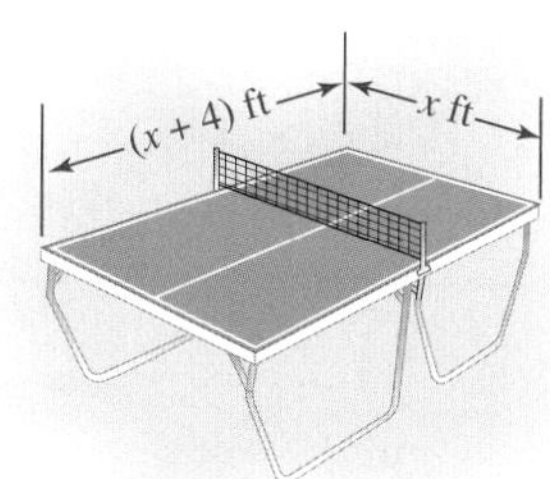

102. SEWING Write an expression for the length of the yellow trim needed to outline a pennant with the given side lengths.

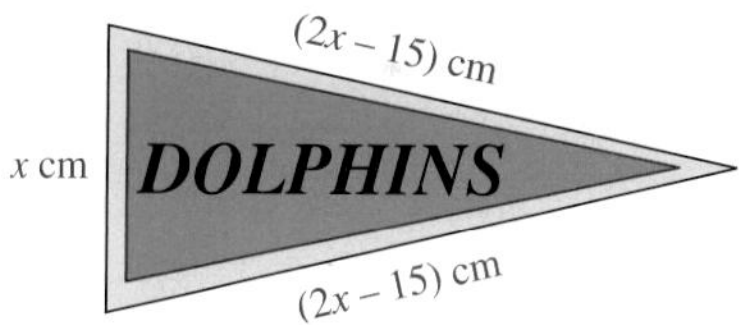

WRITING

103. Explain why the distributive property applies to $2(3 + x)$ but not to $2(3x)$.

104. Explain why $3x^2y$ and $5x^2y$ are like terms, and explain why $3x^2y$ and $5xy^2$ are not like terms.

105. Distinguish between a *factor* and a *term* of an algebraic expression. Give examples.

106. Describe how to combine like terms.

REVIEW *Evaluate each expression for $x = -3$, $y = -5$, and $z = 0$.*

107. $x^2z(y^3 - z)$ **108.** $|y^3 - z|$

109. $\dfrac{x - y^2}{2y - 1 + x}$ **110.** $\dfrac{2y + 1}{x} - x$

8.3 Solving Equations

- Solving equations by using more than one property of equality
- Simplifying expressions to solve equations • Identities and contradictions

We have solved simple equations using the properties of equality. In this section, we will solve more-complicated equations. Our objective is to develop a general strategy that can be used to solve any kind of linear equation.

Solving equations by using more than one property of equality

Recall the following properties of equality:

- If the same quantity is added to (or subtracted from) equal quantities, the results will be equal quantities.
- If equal quantities are multiplied (or divided) by the same nonzero quantity, the results will be equal quantities.

We have already solved many simple equations using one of the properties listed above. For example, to solve $x + 6 = 10$, we isolate x by subtracting 6 from both sides.

$$x + 6 = 10$$

$x + 6 - \mathbf{6} = 10 - \mathbf{6}$ To undo the addition of 6, subtract 6 from both sides.

$x = 4$ Perform the subtractions. The solution is 4.

To solve $2x = 10$, we isolate x by dividing both sides by 2.

$$2x = 10$$

$\frac{2x}{\mathbf{2}} = \frac{10}{\mathbf{2}}$ To undo the multiplication by 2, divide both sides by 2.

$x = 5$ Perform the divisions. The solution is 5.

Sometimes several properties of equality must be applied in succession to solve an equation. For example, on the left-hand side of, $2x + 6 = 10$, the variable x is first multiplied by 2, and then 6 is added to that product. To isolate x, we use the rules for the order of operations in reverse. First, we undo the addition of 6, and then we undo the multiplication by 2.

$$2x + 6 = 10$$

$2x + 6 - \mathbf{6} = 10 - \mathbf{6}$ To undo the addition of 6, subtract 6 from both sides.

$2x = 4$ Perform the subtractions.

$\frac{2x}{\mathbf{2}} = \frac{4}{\mathbf{2}}$ To undo the multiplication by 2, divide both sides by 2.

$x = 2$ Perform the divisions. The solution is 2.

Self Check 1
Solve and check:
$8x - 13 = 43$

EXAMPLE 1 Solve: $-12x + 5 = 17$.

Solution On the left-hand side of the equation, x is multiplied by -12, and then 5 is added to that product. To isolate x, we undo these operations in the opposite order.

- To undo the addition of 5, we subtract 5 from both sides.
- To undo the multiplication by -12, we divide both sides by -12.

$$-12x + 5 = 17$$

$-12x + 5 - \mathbf{5} = 17 - \mathbf{5}$ Subtract 5 from both sides.

$-12x = 12$ Subtract: $5 - 5 = 0$ and $17 - 5 = 12$.

$\frac{-12x}{\mathbf{-12}} = \frac{12}{\mathbf{-12}}$ Divide both sides by -12.

$x = -1$ Perform the divisions.

Check: $-12x + 5 = 17$ This is the original equation.

$-12(\mathbf{-1}) + 5 \stackrel{?}{=} 17$ Substitute -1 for x.

$12 + 5 \stackrel{?}{=} 17$ Multiply.

$17 = 17$ Perform the addition.

Since the statement $17 = 17$ is true, -1 is the solution.

Answer 7

EXAMPLE 2 Solve: $\frac{2x}{3} = -6$.

Self Check 2
Solve and check: $\frac{7h}{16} = -14$

Solution On the left-hand side, x is multiplied by 2, and then that product is divided by 3. To solve this equation, we must undo these operations in the opposite order.

- To undo the division of 3, we multiply both sides by 3.
- To undo the multiplication by 2, we divide both sides by 2.

$$\frac{2x}{3} = -6$$

$\mathbf{3}\left(\frac{2x}{3}\right) = \mathbf{3}(-6)$ Multiply both sides by 3.

$2x = -18$ On the left-hand side: $3\left(\frac{2x}{3}\right) = \frac{\overset{1}{\cancel{3}}}{1}\left(\frac{2x}{\underset{1}{\cancel{3}}}\right) = 2x$.

$\frac{2x}{\mathbf{2}} = \frac{-18}{\mathbf{2}}$ Divide both sides by 2.

$x = -9$ Perform the divisions.

Check: $\frac{2x}{3} = -6$ This is the original equation.

$\frac{2(\mathbf{-9})}{3} \stackrel{?}{=} -6$ Substitute -9 for x.

$\frac{-18}{3} \stackrel{?}{=} -6$ Multiply.

$-6 = -6$ Perform the division.

Since we obtain a true statement, -9 is the solution.

Answer -32

Another approach can be used to solve the equation from Example 2. To isolate the variable, we will use the fact that the product of a number and its **reciprocal,** or **multiplicative inverse,** is 1. Since $\frac{2x}{3} = \frac{2}{3}x$, the equation can be rewritten as

$$\frac{2}{3}x = -6$$

To isolate x, we multiply both sides by $\frac{3}{2}$ the reciprocal (multiplicative inverse) of $\frac{2}{3}$.

$\frac{\mathbf{3}}{\mathbf{2}}\left(\frac{2}{3}x\right) = \frac{\mathbf{3}}{\mathbf{2}}(-6)$ The coefficient of x is $\frac{2}{3}$. Multiply both sides by the reciprocal of $\frac{2}{3}$, which is $\frac{3}{2}$.

$\left(\frac{3}{2} \cdot \frac{2}{3}\right)x = \frac{3}{2}(-6)$ On the left-hand side, apply the associative property of multiplication to regroup factors.

$1x = -9$ Perform the multiplications: $\frac{3}{2} \cdot \frac{2}{3} = 1$ and $\frac{3}{2}(-6) = -9$.

$x = -9$ Simplify: $1x = x$.

Self Check 3

Solve: $\frac{7}{12}a - 6 = -27$.

EXAMPLE 3 Solve: $-\frac{5}{8}m - 2 = -12$.

Solution The coefficient of m is the fraction $-\frac{5}{8}$. We proceed as follows.

- To undo the subtraction of 2, we add 2 to both sides.
- To undo the multiplication by $-\frac{5}{8}$, we multiply both sides by its reciprocal, $-\frac{8}{5}$.

$-\frac{5}{8}m - 2 = -12$

$-\frac{5}{8}m - 2 + \mathbf{2} = -12 + \mathbf{2}$ Add 2 to both sides.

$-\frac{5}{8}m = -10$ Perform the additions: $-2 + 2 = 0$ and $-12 + 2 = -10$.

$-\frac{\mathbf{8}}{\mathbf{5}}\left(-\frac{5}{8}m\right) = -\frac{\mathbf{8}}{\mathbf{5}}(-10)$ Multiply both sides by $-\frac{8}{5}$.

$m = 16$ On the left-hand side: $-\frac{8}{5}\left(-\frac{5}{8}\right)m = 1m = m$.

On the right-hand side: $-\frac{8}{5}(-10) = \frac{8 \cdot 2 \cdot \overset{1}{\cancel{5}}}{\underset{1}{\cancel{5}}} = 16$.

Check that 16 is the solution.

Answer -36

Self Check 4

Solve: $-6.6 - m = -2.7$.

EXAMPLE 4 Solve: $-0.2 = -0.8 - y$.

Solution To solve the equation, we begin by eliminating -0.8 from the right-hand side by adding 0.8 to both sides.

$-0.2 = -0.8 - y$

$-0.2 + \mathbf{0.8} = -0.8 - y + \mathbf{0.8}$ Add 0.8 to both sides.

$0.6 = -y$ Perform the additions: $-0.2 + 0.8 = 0.6$ and $-0.8 + 0.8 = 0$.

Since the term $-y$ has an understood coefficient of -1, the equation can be rewritten as $0.6 = -1y$. To isolate y, either multiply both sides or divide both sides by -1.

$0.6 = -1y$ Write $-y$ as $-1y$.

$\frac{0.6}{\mathbf{-1}} = \frac{-1y}{\mathbf{-1}}$ To undo the multiplication by -1, divide both sides by -1.

$-0.6 = y$ Perform the divisions.

$y = -0.6$

Verify that -0.6 satisfies the equation.

Answer -3.9

Simplifying expressions to solve equations

EXAMPLE 5 Solve: $3(k + 1) - 5k = 0$.

Solution In this example, we must use the distributive property to remove parentheses.

$3(k + 1) - 5k = 0$	
$3k + 3 - 5k = 0$	Distribute the multiplication by 3.
$-2k + 3 = 0$	Combine like terms: $3k - 5k = -2k$.
$-2k + 3 - \mathbf{3} = 0 - \mathbf{3}$	To undo the addition of 3, subtract 3 from both sides.
$-2k = -3$	Perform the subtractions: $3 - 3 = 0$ and $0 - 3 = -3$.
$\frac{-2k}{\mathbf{-2}} = \frac{-3}{\mathbf{-2}}$	To undo the multiplication by -2, divide both sides by -2.
$k = \frac{3}{2}$	Simplify: $\frac{-3}{-2} = \frac{3}{2}$.

Check:

$3(k + 1) - 5k = 0$	This is the original equation.
$3\left(\frac{3}{2} + 1\right) - 5\left(\frac{3}{2}\right) \stackrel{?}{=} 0$	Substitute $\frac{3}{2}$ for k.
$3\left(\frac{3}{2} + \frac{2}{2}\right) - 5\left(\frac{3}{2}\right) \stackrel{?}{=} 0$	To add $\frac{3}{2}$ and 1, write 1 as $\frac{2}{2}$.
$3\left(\frac{5}{2}\right) - 5\left(\frac{3}{2}\right) \stackrel{?}{=} 0$	Perform the addition within the parentheses.
$\frac{15}{2} - \frac{15}{2} \stackrel{?}{=} 0$	Perform the multiplications.
$0 = 0$	Perform the subtraction.

Self Check 5
Solve:
$-5(x - 3) + 3x = 11$

Answer 2

EXAMPLE 6 Solve: $3x - 15 = 4x + 36$.

Solution To solve for x, all the terms containing x must be on the same side of the equation. We can eliminate $3x$ from the left-hand side by subtracting $3x$ from both sides.

$3x - 15 = 4x + 36$	
$3x - 15 - \mathbf{3x} = 4x + 36 - \mathbf{3x}$	Subtract $3x$ from both sides.
$-15 = x + 36$	Combine like terms: $3x - 3x = 0$ and $4x - 3x = x$.
$-15 - \mathbf{36} = x + 36 - \mathbf{36}$	To undo the addition of 36, subtract 36 from both sides.
$-51 = x$	Perform the subtractions.
$x = -51$	

Check:

$3x - 15 = 4x + 36$	This is the original equation.
$3(\mathbf{-51}) - 15 \stackrel{?}{=} 4(\mathbf{-51}) + 36$	Substitute -51 for x.
$-153 - 15 \stackrel{?}{=} -204 + 36$	Perform the multiplications.
$-168 = -168$	Simplify each side.

Self Check 6
Solve:
$30 + 6n = 4n - 2$

Answer -16

Self Check 7
Solve:
$\frac{x}{4} + \frac{1}{2} = -\frac{1}{8}$

EXAMPLE 7 Solve: $\frac{x}{6} - \frac{5}{2} = -\frac{1}{3}$.

Solution Since integers are easier to work with, we will clear the equation of the fractions by multiplying both sides by the least common denominator (LCD), which is 6.

$$\frac{x}{6} - \frac{5}{2} = -\frac{1}{3}$$

$$\mathbf{6}\left(\frac{x}{6} - \frac{5}{2}\right) = \mathbf{6}\left(-\frac{1}{3}\right)$$ 6 is the smallest number that each denominator will divide exactly. Multiply both sides by 6 to clear the equation of the fractions.

$$6\left(\frac{x}{6}\right) - 6\left(\frac{5}{2}\right) = 6\left(-\frac{1}{3}\right)$$ On the left-hand side, distribute the multiplication by 6.

$$x - 15 = -2$$ Perform each multiplication by 6. Note that the resulting equation does not contain any fractions.

$$x - 15 + \mathbf{15} = -2 + \mathbf{15}$$ To undo the subtraction of 15, add 15 to both sides.

$$x = 13$$ Perform the additions.

Verify that 13 satisfies the equation.

Answer $-\frac{5}{2}$

The preceding examples suggest the following strategy for solving basic equations.

Strategy for solving equations

1. Clear the equation of fractions.
2. Use the distributive property to remove parentheses, if necessary.
3. Combine like terms, if necessary.
4. Undo the operations of addition and subtraction to get the variables on one side and the constant terms on the other.
5. Undo the operations of multiplication and division to isolate the variable.
6. Check the result.

Self Check 8
Solve:
$\frac{3x + 23}{7} = x + 5$

EXAMPLE 8 Solve: $\frac{3x + 11}{5} = x + 3$.

Solution

$$\frac{3x + 11}{5} = x + 3$$

$$\mathbf{5}\left(\frac{3x + 11}{5}\right) = \mathbf{5}(x + 3)$$ Clear the equation of the fraction by multiplying both sides by 5.

$$3x + 11 = 5x + 15$$ On the left-hand side, simplify: $\frac{\overset{1}{\cancel{5}}}{1}\left(\frac{3x + 11}{\underset{1}{\cancel{5}}}\right)$.

On the right-hand side, distribute the multiplication by 5.

$$3x + 11 - \mathbf{11} = 5x + 15 - \mathbf{11}$$ Subtract 11 from both sides.

$$3x = 5x + 4$$ Perform the subtractions.

$$3x - \mathbf{5x} = 5x + 4 - \mathbf{5x}$$ To eliminate $5x$ from the right-hand side, subtract $5x$ from both sides.

$$-2x = 4$$ Combine like terms: $3x - 5x = -2x$ and $5x - 5x = 0$.

$$\frac{-2x}{\mathbf{-2}} = \frac{4}{\mathbf{-2}}$$ To undo the multiplication by -2, divide both sides by -2.

$$x = -2$$ Perform the divisions.

Verify that -2 satisfies the equation.

Answer -3

! COMMENT Remember that when you multiply one side of an equation by a nonzero number, you must multiply the other side of the equation by the same number.

Identities and contradictions

Equations in which some numbers satisfy the equation and others don't are called **conditional equations.** The equations in Examples 1–8 are conditional equations.

An equation that is true for all values of its variable is called an **identity.**

$x + x = 2x$ This is an identity, because it is true for all values of x.

An equation that is not true for any values of its variable is called a **contradiction.** Such equations are said to have no solution. An example is

$x = x + 1$ No number is 1 greater than itself.

EXAMPLE 9 Solve: $3(x + 8) + 5x = 2(12 + 4x)$.

Solution

$$3(x + 8) + 5x = 2(12 + 4x)$$

$$3x + 24 + 5x = 24 + 8x$$ On each side of the equation, use the distributive property.

$$8x + 24 = 24 + 8x$$ Combine like terms.

$$8x + 24 - \mathbf{8x} = 24 + 8x - \mathbf{8x}$$ Subtract $8x$ from both sides.

$$24 = 24$$ Combine like terms: $8x - 8x = 0$.

The terms involving x drop out and the result, $24 = 24$, is true. This means that any number substituted for x in the original equation will yield a true statement. Therefore, every real number is a solution and this equation is an identity.

Self Check 9
Solve:
$3(x + 5) - 4(x + 4) = -x - 1$

Answer all real numbers; this equation is an identity

EXAMPLE 10 Solve: $3(d + 7) - d = 2(d + 10)$.

Solution

$$3(d + 7) - d = 2(d + 10)$$

$$3d + 21 - d = 2d + 20$$ Use the distributive property.

$$2d + 21 = 2d + 20$$ Combine like terms.

$$2d + 21 - \mathbf{2d} = 2d + 20 - \mathbf{2d}$$ Subtract $2d$ from both sides.

$$21 = 20$$ Combine like terms.

In this case, the terms involving d drop out. Since the result $21 = 20$ is false, the original equation has no solution. It is a contradiction.

Self Check 10
Solve:
$-4(c - 3) + 2c = 2(10 - c)$

Answer No solution. This equation is a contradiction.

Section 8.3 STUDY SET

VOCABULARY *Fill in the blanks.*

1. An ________ is a statement that two quantities are equal.
2. To solve an equation, we must ________ the variable on one side of the equation.
3. If a number is a solution of an equation, the number is said to ________ the equation.
4. In $2(x - 7)$, "to remove parentheses" means to apply the ________ property.
5. The product of a number and its ________ is 1.
6. An equation that is true for all values of its variable is called an ________. An equation that is not true for any values of its variables is called a ________.

CONCEPTS *Fill in the blanks.*

7. To solve the equation $2x - 7 = 21$, we first undo the ________ of 7 by adding 7 to both sides. We then undo the ________ by 2 by dividing both sides by 2.
8. To solve the equation $\frac{x}{2} + 3 = 5$, we first undo the ________ of 3 by subtracting 3 from both sides. We then undo the ________ by 2 by multiplying both sides by 2.
9. To solve the equation $\frac{x}{2} + 3 = 5$, we first undo the ________ of 3 by subtracting 3 from both sides. We then undo the ________ by 2 by multiplying both sides by 2.
10. To solve $\frac{s}{3} + \frac{1}{4} = -\frac{1}{2}$, we can clear the equation of the fractions by ________ both sides of the equation by 12.
11. One method of solving $-\frac{4}{5}x = 8$ is to multiply both sides of the equation by the reciprocal of $-\frac{4}{5}$. What is the reciprocal of $-\frac{4}{5}$?
12. a. Combine like terms on the left-hand side of $6x - 8 - 8x = -24$.

    b. Combine like terms on the right-hand side of $5a + 1 = 9a + 16 + a$.

    c. Combine like terms on both sides of $12 - 3r + 5r = -8 - r - 2$.
13. What is the LCD for the fractions in the equation $\frac{x}{3} - \frac{4}{5} = \frac{1}{2}$?
14. Complete the three multiplications necessary to clear the given equation of fractions.

$$\frac{2}{3} - \frac{b}{2} = -\frac{4}{3}$$
$$6\left(\frac{2}{3} - \frac{b}{2}\right) = 6\left(-\frac{4}{3}\right)$$
$$6\left(\frac{2}{3}\right) - 6\left(\frac{b}{2}\right) = 6\left(-\frac{4}{3}\right)$$
$$\square - \square = \square$$

15. a. Simplify: $3x + 5 - x$.

    b. Solve: $3x + 5 - x = 9$.

    c. Evaluate $3x + 5 - x$ for $x = 9$.

    d. Check: is -1 a solution of $3x + 5 - x = 9$?
16. a. Simplify: $3(x - 4) - 4x$.

    b. Solve: $3(x - 4) - 4x = 0$.

    c. Evaluate $3(x - 4) - 4x$ for $x = 0$.

    d. Check: Is -1 a solution of $3(x - 4) - 4x = 0$?

NOTATION *Complete each solution.*

17.
$$2x - 7 = 21$$
$$2x - 7 + \square = 21 + \square$$
$$2x = \square$$
$$\frac{2x}{\square} = \frac{28}{\square}$$
$$x = 14$$

18.
$$\frac{x}{2} + 3 = 5$$
$$\frac{x}{2} + 3 - \square = 5 - \square$$
$$\frac{x}{2} = \square$$
$$\square\left(\frac{x}{2}\right) = \square(2)$$
$$x = 4$$

19. Fill in the blanks.

    a. $-x = \square x$.   b. $\frac{3x}{5} = \frac{\square}{\square}x$.

    c. If $-31 = x$, then $x = \square$
20. When checking a solution of an equation, the symbol $\stackrel{?}{=}$ is used. What does it mean?

PRACTICE *Solve each equation and check the result.*

21. $2x + 5 = 17$
22. $3x - 5 = 13$
23. $5q - 2 = 1$
24. $4p + 3 = 2$
25. $0.6 = 4.1 - x$
26. $1.2 - x = -1.7$
27. $-g = -4$
28. $-u = -20$
29. $-8 - 3c = 0$
30. $-5 - 2d = 0$
31. $\frac{5}{6}k = 10$
32. $\frac{2c}{5} = 2$
33. $-\frac{t}{3} + 2 = 6$
34. $-\frac{x}{5} - 5 = -12$
35. $\frac{2x}{3} - 2 = 4$
36. $\frac{2}{5}y + 3 = 9$
37. $\frac{x + 5}{3} = 11$
38. $\frac{x + 2}{13} = 3$
39. $\frac{y - 2}{7} = -3$
40. $\frac{x - 7}{3} = -1$
41. $2(-3) + 4y = 14$
42. $4(-1) + 3y = 8$
43. $-2x - 4(1) = -6$
44. $-5x - 3(5) = 0$
45. $3(x + 2) - x = 12$
46. $2(x - 4) + x = 7$
47. $-3(2y - 2) - y = 5$
48. $-(3a + 1) + a = 2$
49. $0 - 2y = 8$
50. $0 - 7x = -21$
51. $60r - 50 = 15r - 5$
52. $100f - 75 = 50f + 75$
53. $5x + 7.2 = 4x$
54. $3x + 2.5 = 2x$
55. $15x = x$
56. $-7y = -8y$
57. $4 + \frac{y}{2} = \frac{3}{5}$
58. $5 + \frac{x}{3} = \frac{1}{2}$
59. $\frac{1}{3} + \frac{c}{5} = -\frac{3}{2}$
60. $\frac{1}{2} + \frac{x}{5} = \frac{3}{4}$
61. $\frac{y}{6} + \frac{y}{4} = -1$
62. $\frac{x}{3} + \frac{x}{4} = -2$
63. $-\frac{2}{9} = \frac{5x}{6} - \frac{1}{3}$
64. $\frac{2}{3} = -\frac{2x}{3} + \frac{3}{4}$
65. $\frac{1}{2}x - \frac{1}{9} = \frac{1}{3}$
66. $\frac{1}{4}y - \frac{2}{3} = \frac{1}{2}$
67. $\frac{2}{5}x + 1 = \frac{1}{3} + x$
68. $\frac{2}{3}y + 2 = \frac{1}{5} + y$
69. $3(a + 2) = 2(a - 7)$
70. $9(t - 1) = 6(t + 2) - t$
71. $9(x + 11) + 5(13 - x) = 0$
72. $3(x + 15) + 4(11 - x) = 0$
73. $\frac{3t - 21}{2} = t - 6$
74. $\frac{2t - 18}{3} = t - 8$
75. $\frac{10 - 5s}{3} = s + 6$
76. $\frac{40 - 8s}{5} = -2s$
77. $2 - 3(x - 5) = 4(x - 1)$
78. $2 - (4x + 7) = 3 + 2(x + 2)$

Solve each equation. If it is an identity or a contradiction, so indicate.

79. $8x + 3(2 - x) = 5(x + 2) - 4$
80. $5(x + 2) = 5x - 2$
81. $-3(s + 2) = -2(s + 4) - s$
82. $21(b - 1) + 3 = 3(7b - 6)$
83. $2(3z + 4) = 2(3z - 2) + 13$
84. $x + 7 = \frac{2x + 6}{2} + 4$
85. $4(y - 3) - y = 3(y - 4)$
86. $5(x + 3) - 3x = 2(x + 8)$

Solve each equation.

87. $1.73x = -4.952 - 2.27x$
88. $\frac{h}{709} - 23{,}898 = -19{,}678$
89. $20(x - 3.7) = 32{,}832$
90. $9.35 - 1.4y = 7.32 + 1.5y$

WRITING

91. Explain the difference between *simplifying* an expression and *solving* an equation. Give some examples.
92. To solve $3x - 4 = 5x + 1$, one student began by subtracting $3x$ from both sides. Another student solved the same equation by first subtracting $5x$ from both sides. Will the students get the same solution? Explain why or why not.
93. What does it mean to clear an equation such as $\frac{1}{4} + \frac{x}{2} = \frac{3}{8}$ of the fractions?
94. Explain the error.

Solve: $2x + 4 = 30$.

$$\frac{2x}{2} + 4 = \frac{30}{2}$$
$$x + 4 = 15$$
$$x + 4 - 4 = 15 - 4$$
$$x = 11$$

REVIEW

95. Simplify: $-(-8)$.
96. Subtract: $-8 - (-8)$.
97. Multiply: $-8(-8)$.
98. Add: $\frac{1}{8} + \frac{1}{8}$.
99. Multiply: $\frac{1}{8} \cdot \frac{1}{8}$.
100. Divide: $\frac{0.8}{8}$.
101. Simplify: $8x + 8 + 8x - 8$.
102. Evaluate: -1^8.

8.4 Formulas

- Formulas from business
- Formulas from science
- Formulas from geometry
- Solving formulas

A **formula** is an equation that is used to state a relationship between two or more variables. Formulas are used in many fields: economics, physical education, anthropology, biology, automotive repair, and nursing, to name a few. In this section, we will consider formulas from business, science, and geometry.

Formulas from business

A formula to find the retail price: To make a profit, a merchant must sell a product for more than he or she paid for it. The price at which the merchant sells the product, called the **retail price,** is the sum of what the item cost the merchant plus the **markup.**

Retail price = cost + markup

Using r to represent the retail price, c the wholesale cost, and m the markup, we can write this formula as

$$\boxed{r = c + m}$$

As an example, suppose a jeweler purchases a gold ring at a wholesale jewelry mart for \$612.50. Then she sets the price of the ring at \$837.95 for sale in her store. We can find the markup on the ring as follows:

$$r = c + m$$

$$\mathbf{837.95} = \mathbf{612.50} + m \quad \text{Substitute 837.95 for } r \text{ and 612.50 for } c.$$

$$837.95 - \mathbf{612.50} = 612.50 + m - \mathbf{612.50} \quad \text{To undo the addition of 612.50, subtract 612.50 from both sides.}$$

$$225.45 = m \quad \text{Perform the subtractions.}$$

The markup on the ring is \$225.45.

A formula for profit: The **profit** a business makes is the difference between the **revenue** (the money it takes in) and the costs.

Profit = revenue − costs

Using p to represent the profit, r the revenue, and c the costs, we can write this formula as

$$\boxed{p = r - c}$$

Self Check 1
A PTA spaghetti dinner made a profit of \$275.50. If the cost to host the dinner was \$1,235, how much revenue did it generate?

EXAMPLE 1 Charitable giving. In 2001, the Salvation Army collected \$2.31 billion. Of that amount, \$1.92 billion went directly to the support of its programs. What were the 2001 administrative costs of the organization?

Solution The charity collected \$2.31 billion in revenue. We can think of the \$1.92 billion that was spent on programs as profit. We need to find the administrative costs, c.

$$p = r - c \quad \text{This is the formula for profit.}$$

$$\mathbf{1.92} = \mathbf{2.31} - c \quad \text{Substitute 1.92 for } p \text{ and 2.31 for } r.$$

$$1.92 - \mathbf{2.31} = 2.31 - c - \mathbf{2.31}$$ To eliminate 2.31, subtract 2.31 from both sides.

$$-0.39 = -c$$ Subtract: $1.92 - 2.31 = -0.39$ and $2.31 - 2.31 = 0$.

$$\frac{-0.39}{\mathbf{-1}} = \frac{-c}{\mathbf{-1}}$$ Since $-c = -1c$, divide (or multiply) both sides by -1.

$$0.39 = c$$ Perform the divisions.

In 2001, the Salvation Army had administrative costs of \$0.39 billion.

Answer \$1,510.50

A formula for simple interest: When money is borrowed, the lender expects to be paid back the amount of the loan plus an additional charge for the use of the money. The additional charge is called **interest.** When money is deposited in a bank, the depositor is paid for the use of the money. The money the deposit earns is also called interest. In general, interest is the money that is paid for the use of money.

Interest is calculated in two ways: as **simple interest** or as **compound interest.** To find simple interest, we use the formula

Interest = principal · rate · time

Using I to represent the simple interest, P the principal (the amount of money that is invested, deposited, or borrowed), r the annual interest rate, and t the length of time in years, we can write the formula as

$$I = Prt$$

EXAMPLE 2 **Retirement income.** One year after investing \$15,000 in a mini-mall development, a retired couple received a check for \$1,125 in interest. What interest rate did their money earn that year?

Solution The couple invested \$15,000 (the principal) for 1 year (the time) and made \$1,125 (the interest). We need to find the annual interest rate.

$$I = Prt$$ This is the formula for simple interest.

$$\mathbf{1{,}125} = \mathbf{15{,}000}r(\mathbf{1})$$ Substitute 1,125 for I, 15,000 for P, and 1 for t.

$$1{,}125 = 15{,}000r$$ Simplify the right-hand side.

$$\frac{1{,}125}{\mathbf{15{,}000}} = \frac{15{,}000r}{\mathbf{15{,}000}}$$ To solve for r, undo the multiplication by 15,000 by dividing both sides by 15,000.

$$0.075 = r$$ Perform the divisions.

$$7.5\% = r$$ To write 0.075 as a percent, multiply 0.075 by 100 by moving the decimal point two places to the right and insert a % symbol.

The couple received an annual rate of 7.5% that year.

Self Check 2

A father lent his daughter \$12,200 at a 2% annual simple interest rate for a down payment on a house. If the interest on the loan amounted to \$610, for how long was the loan?

Answer 2.5 years

Formulas from science

A formula for distance traveled: If we know the average rate (speed) at which we will be traveling and the time we will be traveling at that rate, we can find the distance traveled by using the formula

Distance = rate · time

Using d to represent the distance, r the average rate (speed), and t the time, we can write this formula as

$$d = rt$$

! COMMENT When using this formula, the units must be the same. For example, if the rate is given in miles per hour, the time must be expressed in hours.

Self Check 3
An elevator in a building travels at an average rate of 288 feet per minute. How long will it take the elevator to climb 30 stories, a distance of 360 feet?

Answer 1.25 minutes

EXAMPLE 3 Finding the rate. As they migrate from the Bering Sea to Baja California, gray whales swim for about 20 hours each day, covering a distance of approximately 70 miles. Estimate their average swimming rate in miles per hour (mph).

Solution Since the distance d is 70 miles and the time t is 20 hours, we substitute 70 for d and 20 for t in the formula $d = rt$, and then solve for r.

$$d = rt$$

$$\mathbf{70} = r(\mathbf{20}) \quad \text{Substitute 70 for } d \text{ and 20 for } t.$$

$$\frac{70}{\mathbf{20}} = \frac{20r}{\mathbf{20}} \quad \text{To undo the multiplication by 20, divide both sides by 20.}$$

$$3.5 = r \quad \text{Perform the divisions.}$$

The whales' average swimming rate is 3.5 mph.

A formula for converting degrees Fahrenheit to degrees Celsius: Many marquees, like the one shown in Figure 8-6, flash two temperature readings, one in degrees Fahrenheit and one in degrees Celsius. The Fahrenheit scale is used in the American system of measurement. The Celsius scale is used in the metric system. The formula that relates a Fahrenheit temperature F to a Celsius temperature C is:

$$C = \frac{5(F - 32)}{9}$$

CITY SAVINGS
TEMP 30°C

FIGURE 8-6

Self Check 4
Change $-175°$ C, the temperature on Saturn, to degrees Fahrenheit.

EXAMPLE 4 Change the temperature reading on the sign in Figure 8-6 to degrees Fahrenheit.

Solution Since the temperature C in degrees Celsius is 30°, we substitute 30 for C in the formula and solve for F.

$$C = \frac{5(F - 32)}{9}$$

$$\mathbf{30} = \frac{5(F - 32)}{9} \quad \text{Substitute 30 for } C.$$

$$\mathbf{9}(30) = \mathbf{9}\left[\frac{5(F - 32)}{9}\right] \quad \text{To clear the equation of the fraction, multiply both sides by 9.}$$

$$270 = 5(F - 32) \quad \text{Simplify: } 9(30) = 270 \text{ and } \frac{\overset{1}{\cancel{9}}}{1}\left[\frac{5(F - 32)}{\underset{1}{\cancel{9}}}\right] = 5(F - 32).$$

$$270 = 5F - 5(32)$$ Distribute the multiplication by 5.

$$270 = 5F - 160$$ Multiply: $5(32) = 160$.

$$270 + \mathbf{160} = 5F - 160 + \mathbf{160}$$ To undo the subtraction of 160, add 160 to both sides.

$$430 = 5F$$ Simplify: $270 + 160 = 430$ and $-160 + 160 = 0$.

$$\frac{430}{\mathbf{5}} = \frac{5F}{\mathbf{5}}$$ To undo the multiplication by 5, divide both sides by 5.

$$86 = F$$ Perform the divisions.

Thus, 30° C is equivalent to 86° F.

Answer $-283°$ F

Formulas from geometry

The **perimeter** of a geometric figure is the distance around it. Perimeter is measured in linear units, such as inches, feet, yards, and meters. The **area** of the figure is the amount of surface that it encloses. Area is measured in square units, such as square inches, square feet, square yards, and square meters (denoted as in.^{2}, ft^2, yd^2, and m^2, respectively). Table 8-3 shows the formulas for the perimeter P and area A of several geometric figures.

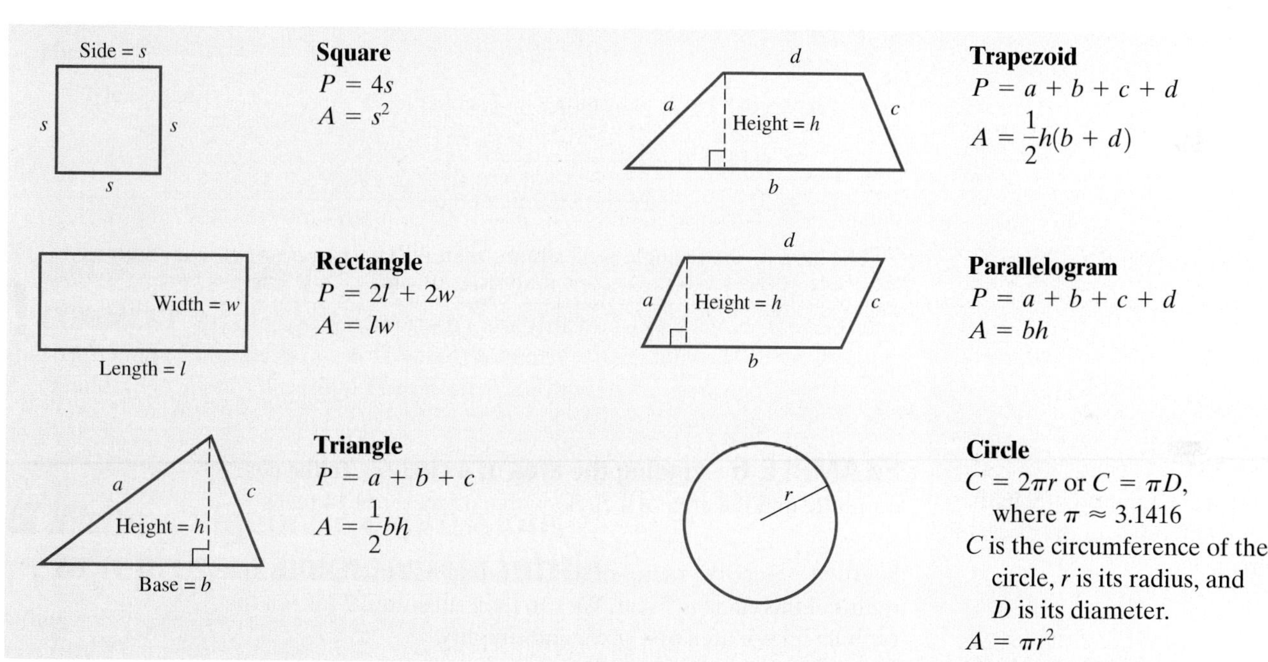

TABLE 8-3

A **circle** (see Figure 8-7) is the set of all points in a plane that are a fixed distance from a point called its **center.** A segment drawn from the center of a circle to a point on the circle is called a **radius.** Since a **diameter** of a circle is a segment passing through the center that joins two points on the circle, the diameter D of a circle is twice as long as its radius r.

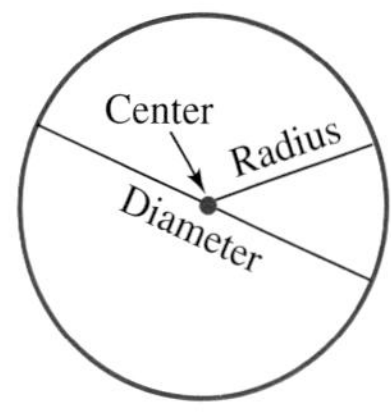

FIGURE 8-7

$$\mathbf{D = 2r}$$

The perimeter of a circle is called its **circumference.** The formulas for the circumference of a circle are

$$\mathbf{C = 2\pi r} \quad \text{or} \quad \mathbf{C = \pi D}$$

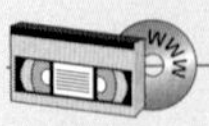

Self Check 5

a. The flag of Eritrea, a country in east Africa, is shown below. What is the perimeter of the flag?

b. Find the area of the red triangular region of the flag.

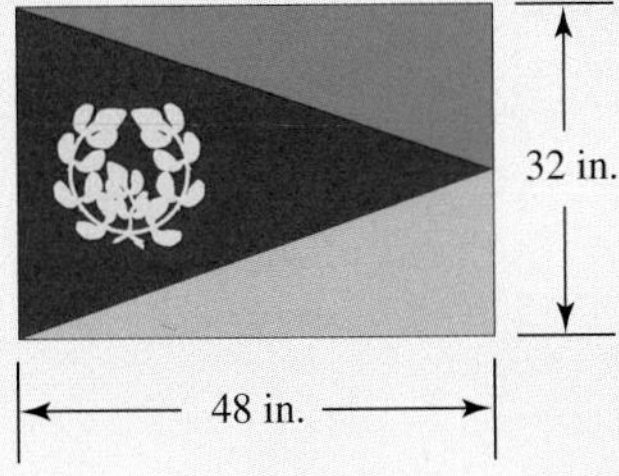

Answers **a.** 160 in., **b.** 768 in.^{2}

EXAMPLE 5 Finding perimeters and areas. Find **a.** the perimeter of a square with sides 6 inches long and **b.** the area of a triangle with base 8 meters and height 13 meters.

Solution

a. The perimeter of a square is given by the formula $P = 4s$, where P is the perimeter and s is the length of one side. Since the sides of the square are 6 inches long, we substitute 6 for s and simplify.

$$P = 4s$$
$$P = 4(\mathbf{6}) \quad \text{Substitute 6 for } s.$$
$$= 24 \quad \text{Perform the multiplication.}$$

The perimeter of the square is 24 inches.

b. The area of a triangle is given by the formula $A = \frac{1}{2}bh$. Since the base of the triangle is 8 meters and the height is 13 meters, we substitute 8 for b and 13 for h and simplify.

$$A = \frac{1}{2}bh$$
$$A = \frac{1}{2}(\mathbf{8})(\mathbf{13}) \quad \text{Substitute 8 for } b \text{ and 13 for } h.$$
$$= 4(13) \quad \tfrac{1}{2}(8) = \tfrac{8}{2} = 4.$$
$$= 52 \quad \text{Perform the multiplication.}$$

The area of the triangle is 52 square meters. This can be written as 52 m^2.

Self Check 6

To the nearest hundredth, find the circumference of the circle of Example 6.

Answer 43.98 ft

EXAMPLE 6 Finding the area of a circle. To the nearest tenth, find the area of a circle with a diameter of 14 feet.

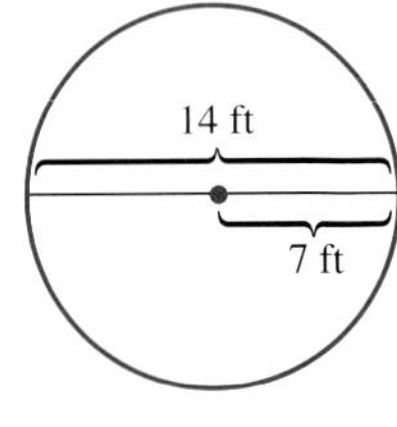

Solution Since the radius of a circle is one-half its diameter, the radius of this circle is 7 feet. We can then substitute 7 for r in the formula for the area of a circle and simplify.

$$A = \pi r^2$$
$$A = \pi(\mathbf{7})^2$$
$$= 49\pi \quad \text{First, evaluate the exponential expression: } 7^2 = 49.$$
$$\approx 153.93804 \quad \text{Use a calculator to multiply. Enter these numbers and press these keys on a scientific calculator: } 49 \times \pi =.$$

To the nearest tenth, the area is 153.9 ft^2.

The **volume** of a three-dimensional geometric solid is the amount of space it encloses. Table 8-4 on the next page shows the formula for the volume V of several solids. Volume is measured in cubic units, such as cubic inches, cubic feet, and cubic meters (denoted as in.^{3}, ft^3, and m^3, respectively).

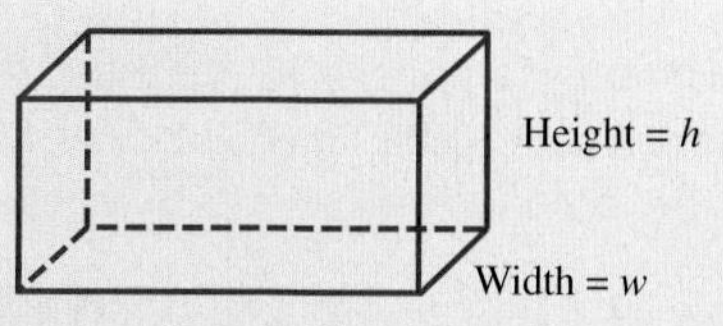

Rectangular solid
$V = lwh$

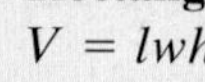

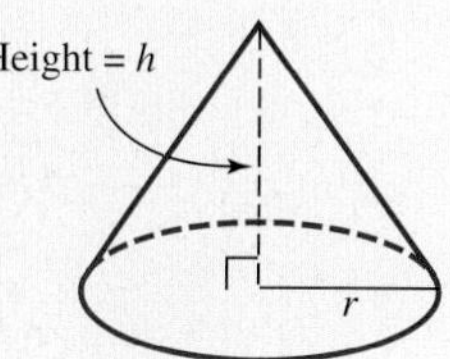

Cone
$V = \frac{1}{3}\pi r^2 h$

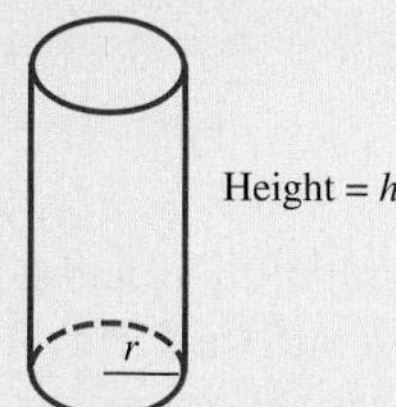

Cylinder
$V = \pi r^2 h$

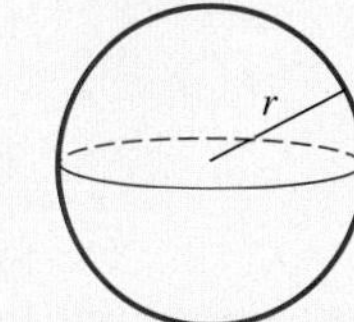

Sphere
$V = \frac{4}{3}\pi r^3$

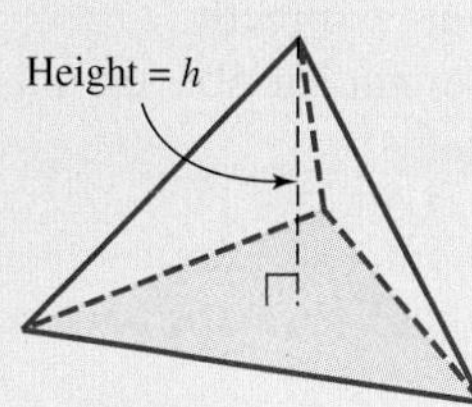

Pyramid
$V = \frac{1}{3}Bh^*$

$^*B$ represents the area of the base

TABLE 8-4

EXAMPLE 7 **Finding volumes.** To the nearest tenth, find the volume of each figure.

a.

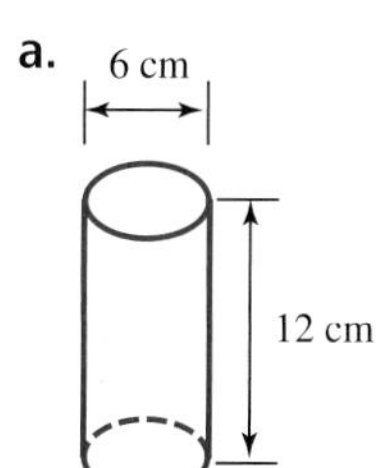

b.

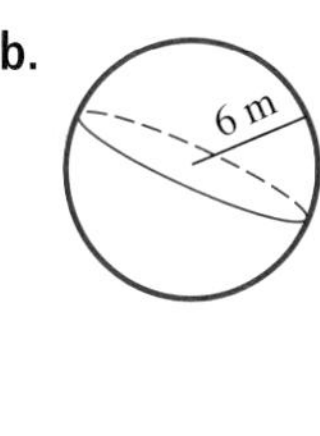

Self Check 7
Find the volume of each figure:
a. a rectangular solid with length 7 inches, width 12 inches, and height 15 inches and
b. a cone whose base has radius 12 meters and whose height is 9 meters. Give the answer to the nearest tenth.

Solution

a. To find the volume of a cylinder, we use the formula $V = \pi r^2 h$, where r is the radius and h is the height. Since the radius of a circle is one-half its diameter, the radius of the cylinder is $\frac{1}{2}(6\text{cm}) = 3\text{cm}$. The height of the cylinder is 12 cm. We substitute 3 for r and 12 for h in the formula for volume and proceed as follows.

$V = \pi r^2 h$	This is the formula for the volume of a cylinder.
$V = \pi(\mathbf{3})^2(\mathbf{12})$	Substitute 3 for r and 12 for h.
$= \pi(9)(12)$	Evaluate the exponential expression: $(3)^2 = 9$.
$= 108\pi$	Multiply: $9(12) = 108$.
≈ 339.2920066	Use a calculator.

To the nearest tenth, the volume is 339.3 cubic centimeters. This can be written as 339.3 cm^3.

b. To find the volume of the sphere, we substitute 6 for r in the formula for the volume of a sphere and proceed as follows.

$V = \frac{4}{3}\pi r^3$	This is the formula for the volume of a sphere.
$V = \frac{4}{3}\pi(\mathbf{6})^3$	Since the radius is 6 m, substitute 6 for r.

$$= \frac{4}{3}\pi(216) \quad (6)^3 = 6 \cdot 6 \cdot 6 = 216.$$

$$= 288\pi \quad \text{Multiply: } \tfrac{4}{3}(216) = \tfrac{4(216)}{3} = \tfrac{864}{3} = 288.$$

$$\approx 904.7786842 \quad \text{Use a calculator.}$$

To the nearest tenth, the volume is 904.8 m^3.

Answers **a.** 1,260 $in.^3$, **b.** 1,357.2 m^3

Solving formulas

Suppose we wish to find the bases of several triangles whose areas and heights are known. It could be tedious to substitute values for A and h into the formula and then repeatedly solve the formula for b. A better way is to solve the formula $A = \frac{1}{2}bh$ for b first, and then substitute values for A and h and compute b directly.

To **solve an equation for a variable** means to isolate that variable on one side of the equation, with all other quantities on the opposite side.

Self Check 8
Solve $V = lwh$ for w.

EXAMPLE 8 Solve $A = \frac{1}{2}bh$ for b.

Solution To solve for b, we must isolate b on one side of the equation.

$$A = \frac{1}{2}bh$$

$$\mathbf{2}A = \mathbf{2} \cdot \frac{1}{2}bh \quad \text{To clear the equation of the fraction, multiply both sides by 2.}$$

$$2A = bh \quad \text{Simplify: } 2 \cdot \tfrac{1}{2} = \tfrac{2}{2} = 1.$$

$$\frac{2A}{\mathbf{h}} = \frac{bh}{\mathbf{h}} \quad \text{To undo the multiplication by } h\text{, divide both sides by } h.$$

$$\frac{2A}{h} = b \quad \text{On the right-hand side, divide out the common factor of } h\text{: } \frac{b\not{h}}{\not{h}} = b.$$

$$b = \frac{2A}{h} \quad \text{Reverse the sides of the equation to write } b \text{ on the left.}$$

Answer $w = \frac{V}{lh}$

Self Check 9
Solve $P = 2l + 2w$ for w.

EXAMPLE 9 Solve $P = 2l + 2w$ for l.

Solution To solve for l, we must isolate l on one side of the equation.

$$P = 2l + 2w$$

$$P - \mathbf{2w} = 2l + 2w - \mathbf{2w} \quad \text{To undo the addition of } 2w\text{, subtract } 2w \text{ from both sides.}$$

$$P - 2w = 2l \quad \text{Combine like terms: } 2w - 2w = 0.$$

$$\frac{P - 2w}{\mathbf{2}} = \frac{2l}{\mathbf{2}} \quad \text{To undo the multiplication by 2, divide both sides by 2.}$$

$$\frac{P - 2w}{2} = l \quad \text{Simplify the right-hand side.}$$

We can write the result as $l = \frac{P - 2w}{2}$.

Answer $w = \frac{P - 2l}{2}$

EXAMPLE 10 Solve $2y - 4 = 3x$ for y.

Solution

$2y - 4 = 3x$	This is the given equation.
$2y - 4 + \mathbf{4} = 3x + \mathbf{4}$	To undo the subtraction of 4, add 4 to both sides.
$2y = 3x + 4$	On the left-hand side, simplify: $-4 + 4 = 0$.
$\frac{2y}{\mathbf{2}} = \frac{3x + 4}{\mathbf{2}}$	To undo the multiplication by 2, divide both sides by 2.
$y = \frac{3x}{2} + \frac{4}{2}$	On the right-hand side, rewrite $\frac{3x+4}{2}$ as the sum of two fractions with like denominators, $\frac{3x}{2}$ and $\frac{4}{2}$.
$y = \frac{3}{2}x + 2$	Write $\frac{3x}{2}$ as $\frac{3}{2}x$. Simplify: $\frac{4}{2} = 2$.

Self Check 10

Solve $3y + 12 = x$ for y.

Answer $y = \frac{1}{3}x - 4$

EXAMPLE 11 Solve $V = \pi r^2 h$ for r^2.

Solution We want to isolate r^2 on one side of the equation.

$V = \pi r^2 h$	
$\frac{V}{\boldsymbol{\pi h}} = \frac{\pi r^2 h}{\boldsymbol{\pi h}}$	To undo the multiplication by π and h on the right-hand side, divide both sides by πh.
$\frac{V}{\pi h} = r^2$	On the right-hand side, divide out π and h: $\frac{\overset{1}{\cancel{\pi}} r^2 \overset{1}{\cancel{h}}}{\underset{1}{\cancel{\pi}}\underset{1}{\cancel{h}}} = r^2$.
$r^2 = \frac{V}{\pi h}$	Reverse the sides of the equation so that r^2 is on the left.

Self Check 11

Solve $a^2 + b^2 = c^2$ for b^2.

Answer $b^2 = c^2 - a^2$

Section 8.4 STUDY SET

VOCABULARY *Fill in the blanks.*

1. A ________ is an equation that is used to state a relationship between two or more variables.
2. The ________ of a three-dimensional geometric solid is the amount of space it encloses.
3. The distance around a geometric figure is called its ________.
4. A ________ is the set of all points in a plane that are a fixed distance from a point called its center.
5. A segment drawn from the center of a circle to a point on the circle is called a ________.
6. The amount of surface that is enclosed by a geometric figure is called its ________.
7. The perimeter of a circle is called its ________.
8. A segment passing through the center of a circle and connecting two points on the circle is called a ________.

CONCEPTS

9. Use variables to write the formula relating the following:
   - **a.** Time, distance, rate
   - **b.** Markup, retail price, cost
   - **c.** Costs, revenue, profit
   - **d.** Interest rate, time, interest, principal
   - **e.** Circumference, radius

10. Complete the table.

Principal	· Rate	· Time	= Interest
\$2,500	5%	2 yr	
\$15,000	4.8%	1 yr	

11. Complete the table to find how far light and sound travel in 60 seconds. (*Hint:* mi/sec means miles per second.)

	Rate	· Time	= Distance
Light	186,282 mi/sec	60 sec	
Sound	1,088 ft/sec	60 sec	

12. Give the name of each figure.

a. 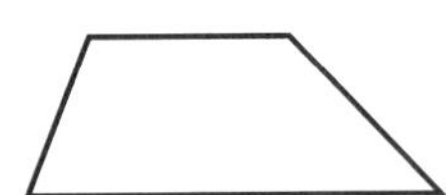

b. 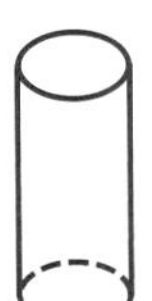

c. 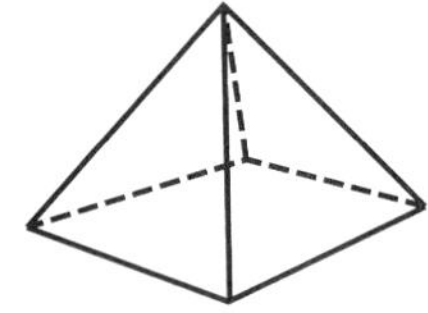

d. 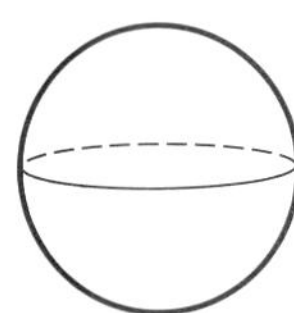

e. 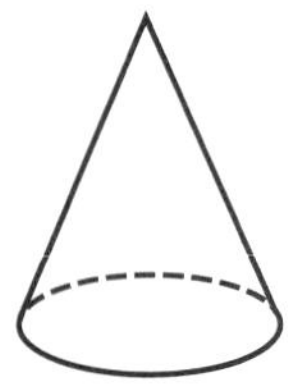

f. 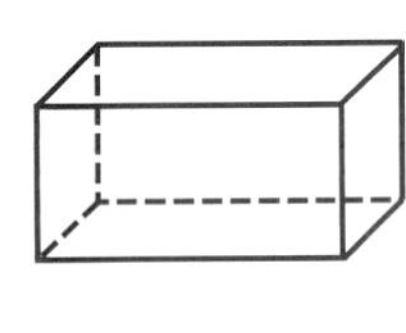

13. Determine which geometric concept — perimeter, circumference, area, or volume — should be used to find the following:

a. The amount of storage in a freezer

b. How far a bicycle tire rolls in one revolution

c. The amount of land making up the Sahara Desert

d. The distance around a Monopoly game board

14. Determine which unit of measurement — ft, ft^2, or ft^3 — would be appropriate when finding the following:

a. The amount of storage inside a safe

b. The ground covered by a sleeping bag lying on the floor

c. The distance the tip of an airplane propeller travels in one revolution

d. The size of the trunk of a car

15. Write an expression for the area of the rectangle.

16. Write an expression for the area of the triangle.

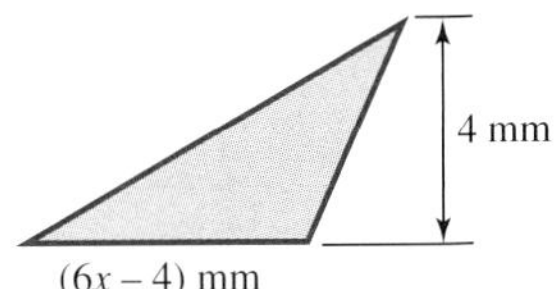

NOTATION *Complete each solution.*

17. Solve $V = \frac{1}{3}Bh$ for B.

$$V = \frac{1}{3}Bh$$
$$\square(V) = \square\left(\frac{1}{3}\right)Bh$$
$$3V = \square$$
$$\frac{3V}{\square} = \frac{Bh}{\square}$$
$$\frac{3V}{h} = \square$$
$$B = \frac{3V}{h}$$

18. Solve $Ax + By = C$ for y.

$$Ax + By = C$$
$$Ax + By - \square = C - \square$$
$$\square = C - Ax$$
$$\frac{By}{\square} = \frac{C - Ax}{\square}$$
$$y = \frac{C - Ax}{B}$$

19. a. Approximate π to the nearest hundredth.

b. What does 98π mean?

c. In the formula for the volume of a cylinder, $V = \pi r^2 h$, what does r represent? What does h represent?

20. a. What does ft^2 mean?

b. What does $in.^3$ mean?

PRACTICE *Use a formula to solve each problem.*

21. SWIMMING In 1930, a man swam down the Mississippi River from Minneapolis to New Orleans, a total of 1,826 miles. He was in the water for 742 hours. To the nearest tenth, what was his average swimming rate?

22. THE ROSE PARADE Rose Parade floats travel down the 5.5-mile-long parade route at a rate of 2.5 mph. How long will it take a float to complete the parade if there are no delays?

23. HOLLYWOOD Figures for the summer of 1998 showed that the movie *Saving Private Ryan* had U.S. box-office receipts of $190 million. What were the production costs to make the movie if, at that time, the studio had made a $125 million profit?

24. SERVICE CLUBS After expenses of $55.15 were paid, a Rotary Club donated $875.85 in proceeds from a pancake breakfast to a local health clinic. How much did the pancake breakfast gross?

25. ENTREPRENEURS To start a mobile dog-grooming service, a woman borrowed $2,500. If the loan was for 2 years and the amount of interest was $175, what simple interest rate was she charged?

26. BANKING Three years after opening an account that paid 6.45% annually, a depositor withdrew the $3,483 in interest earned. How much money was left in the account?

27. METALLURGY Change 2,212° C, the temperature at which silver boils, to degrees Fahrenheit. Round to the nearest degree.

28. LOW TEMPERATURES Cryobiologists freeze living matter to preserve it for future use. They can work with temperatures as low as −270° C. Change this to degrees Fahrenheit.

29. VALENTINE'S DAY Find the markup on a dozen roses if a florist buys them wholesale for $12.95 and sells them for $37.50.

30. STICKER PRICES The factory invoice for a minivan shows that the dealer paid $16,264.55 for the vehicle. If the sticker price of the van is $18,202, how much over factory invoice is the sticker price?

31. YO-YOS How far does a yo-yo travel during one revolution of the "around the world" trick if the length of the string is 21 inches?

32. HORSE TRAINING A horse trots in a perfect circle around its trainer at the end of a 28-foot-long rope. How far does the horse travel as it circles the trainer once?

Solve each formula for the given variable.

33. $E = IR$; for R
34. $d = rt$; for t
35. $V = lwh$; for w
36. $I = Prt$; for r
37. $C = 2\pi r$; for r
38. $V = \pi r^2 h$; for h
39. $a + b + c = 180$; for a
40. $P = a + b + c$; for b
41. $y = mx + b$; for x
42. $P = 2l + 2w$; for l
43. $A = P + Prt$; for t
44. $S = 2\pi rh + 2\pi r^2$; for h
45. $V = \frac{1}{3}\pi r^2 h$; for h
46. $K = \frac{1}{2}mv^2$; for m
47. $x = \frac{a + b}{2}$; for b
48. $A = \frac{a + b + c}{3}$; for c
49. $D = \frac{C - s}{n}$; for s
50. $2E = \frac{T - t}{9}$; for t
51. $E = mc^2$; for c^2
52. $s = 4\pi r^2$; for r^2
53. $c^2 = a^2 + b^2$; for a^2
54. $Kg = \frac{wv^2}{2}$; for v^2
55. $A = \frac{1}{2}h(b + d)$; for b
56. $h = vt + 16t^2$; for t^2
57. $3y - 9 = x$; for y
58. $5y - 25 = x$; for y
59. $4y + 16 = -3x$; for y
60. $6y + 12 = -5x$; for y

APPLICATIONS

61. PROPERTIES OF WATER The boiling point and the freezing point of water are to be given in both degrees Celsius and degrees Fahrenheit on the thermometer in the illustration. Find the missing degree measures.

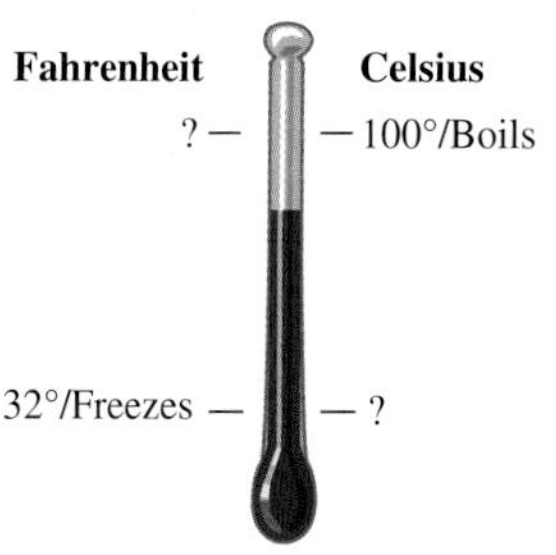

62. SPEED LIMITS Several state speed limits for trucks are shown. At each of these speeds, how far would a truck travel in $2\frac{1}{2}$ hours?

63. AVON PRODUCTS, INC. Complete the financial statement shown below.

Quarterly financials Income statement (dollar amounts in millions except per share amounts)	Quarter ending Dec 02	Quarter ending Sep 02
Revenue	1,854.1	1,463.4
Cost of goods sold	679.5	506.5
Gross profit		

Based on data from Hoover's Online

64. CREDIT CARDS The finance charge section of a person's credit card statement says, "annual percentage rate (APR) is 19.8%." Determine how much finance charges (interest) the card owner would have to pay if the account's average balance for the year was $2,500.

65. CARPENTRY Find the perimeter and area of the truss.

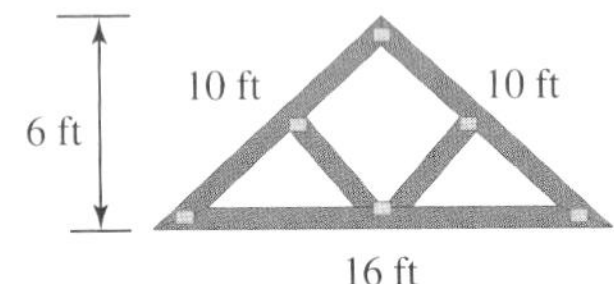

66. CAMPERS Find the area of the window of the camper shell.

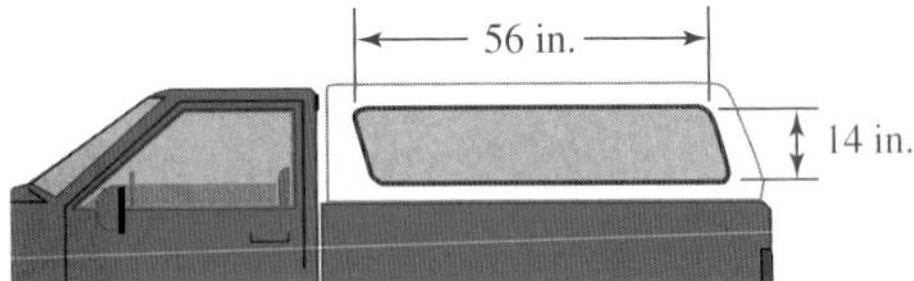

67. ARCHERY To the nearest tenth, find the circumference and area of the target.

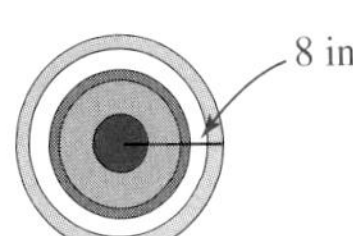

68. GEOGRAPHY The circumference of the Earth is about 25,000 miles. Find its diameter to the nearest mile.

69. LANDSCAPING Find the perimeter and the area of the redwood trellis.

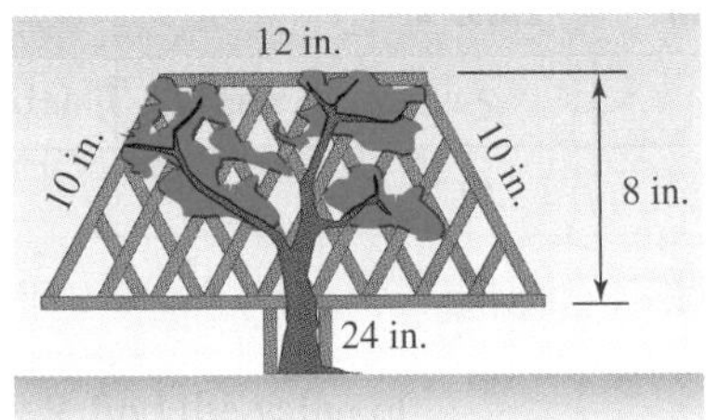

70. HAMSTER HABITATS Find the amount of space in the plastic tube.

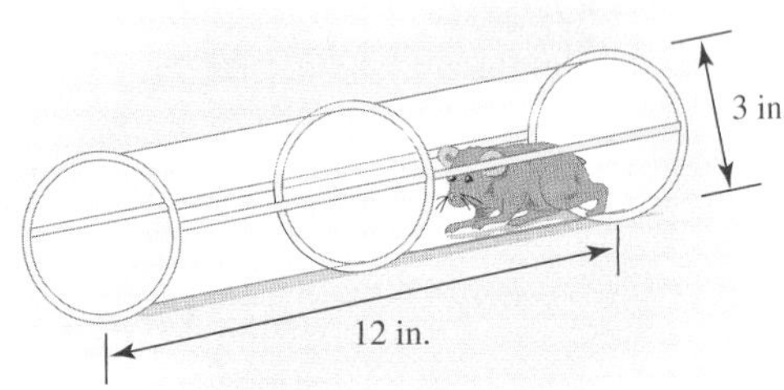

71. THE WALL The Vietnam Veterans Memorial is a black granite wall recognizing the more than 58,000 Americans who lost their lives or remain missing. A diagram of the wall is shown. Find the total area of the two triangular-shaped surfaces on which the names are inscribed.

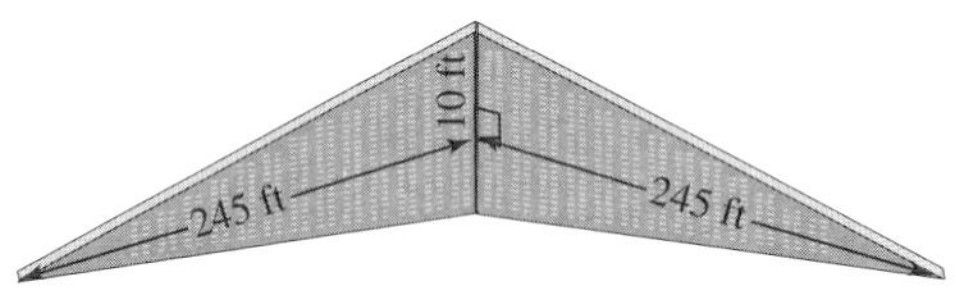

72. SIGNAGE Find the perimeter and area of the service station sign.

73. RUBBER MEETS THE ROAD A sport truck tire has the road surface footprint shown in the illustration. Estimate the perimeter and area of the tire's footprint.

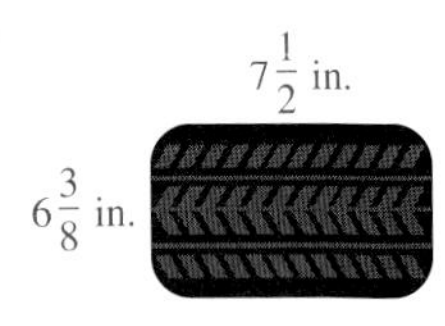

74. SOFTBALL The strike zone in fast-pitch softball is the region over home plate that is between the batter's armpit and the top of her knees. Find the area of the strike zone.

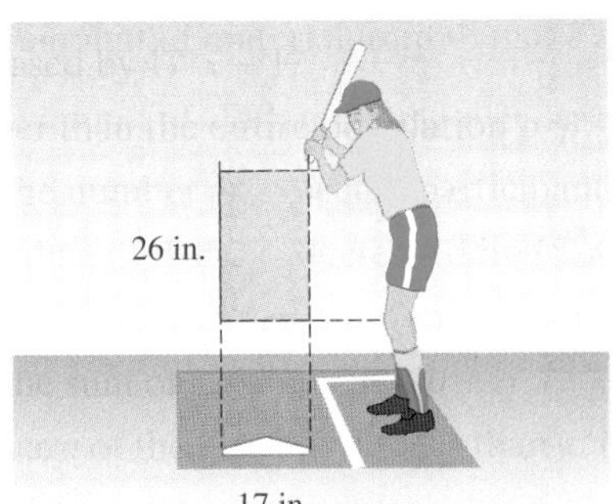

75. FIREWOOD The dimensions of a cord of firewood are shown in the illustration. Find the area on which the wood is stacked and the volume the cord of firewood occupies.

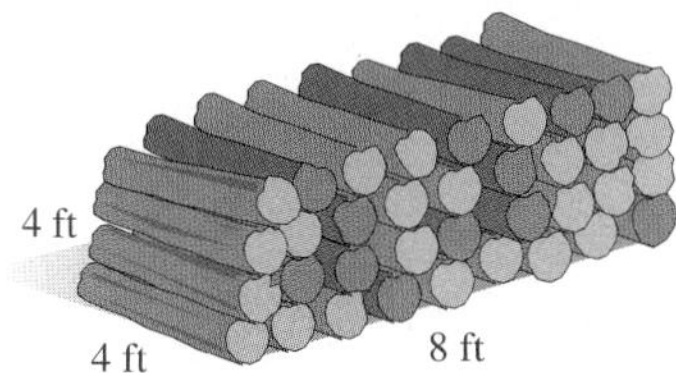

76. TEEPEES The teepees constructed by the Blackfoot Indians were cone-shaped tents made of long poles and animal hide, about 10 feet high and about 15 feet across at the ground. Estimate the volume of a teepee with these dimensions, to the nearest cubic foot.

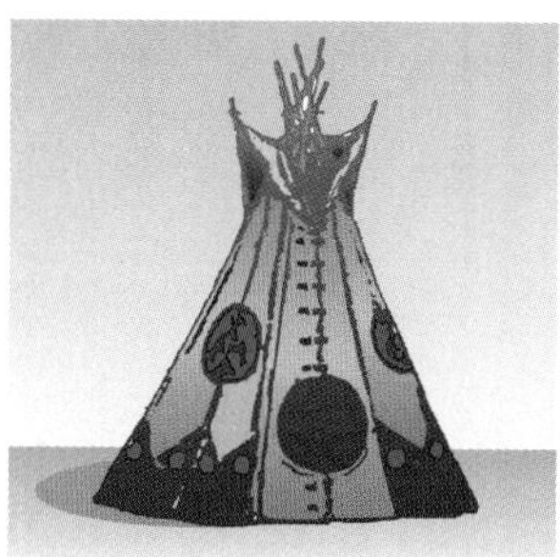

77. IGLOOS During long journeys, some Canadian Inuit (Eskimos) built winter houses of snow blocks piled in the dome shape shown. Estimate the volume of an igloo having an interior height of 5.5 feet to the nearest cubic foot.

78. PYRAMIDS The Great Pyramid at Giza in northern Egypt is one of the most famous works of architecture in the world. Use the information in the illustration to find the volume to the nearest cubic foot.

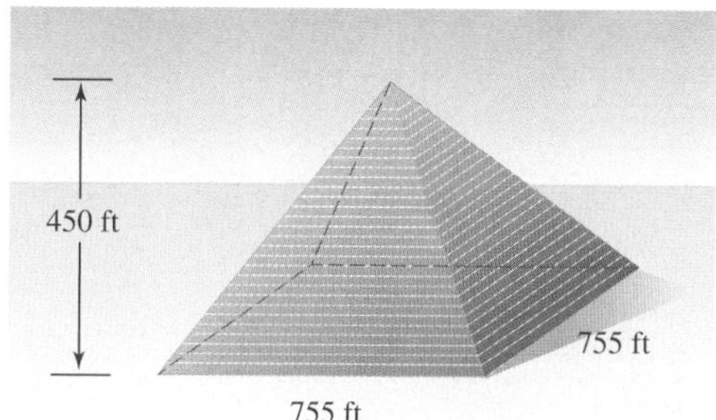

79. BARBECUING Use the fact that the fish is 18 inches long to find the area of the barbecue grill to the nearest square inch.

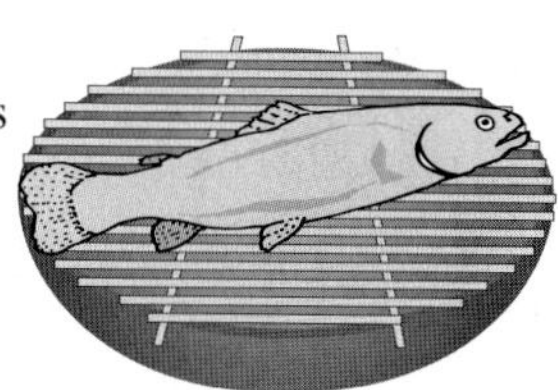

80. SKATEBOARDING A half-pipe ramp used for skateboarding is in the shape of a semicircle with a radius of 8 feet. To the nearest tenth of a foot, what is the length of the arc that the skateboarder travels on the ramp?

81. GEOMETRY The measure a of an interior angle of a regular polygon with n sides is given by the formula

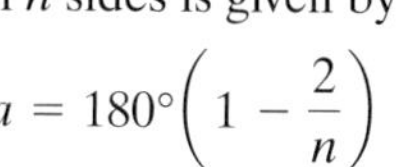

$$a = 180°\left(1 - \frac{2}{n}\right)$$

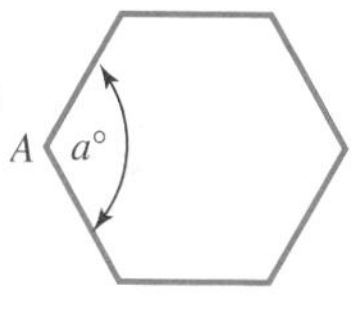

Solve the formula for n. How many sides does a regular polygon have if an interior angle is 108°? (*Hint:* Distribute first.)

82. THERMODYNAMICS The Gibbs free-energy function is given by $G = U - TS + pV$. Solve this formula for the pressure p.

WRITING

83. The formula $P = 2l + 2w$ is also an equation, but an equation such as $2x + 3 = 5$ is not a formula. What equations do you think should be called formulas?

84. Explain what it means to solve the equation $P = 2l + 2w$ for w.

85. After solving $A = B + C + D$ for B, a student compared her answer with that at the back of the textbook.

Student's answer: $B = A - C - D$

Book's answer: $B = A - D - C$

Could this problem have two different-looking answers? Explain.

86. Suppose the volume of a cylinder is 28 cubic feet. Explain why it is incorrect to express the volume as 28^3 ft.

REVIEW

87. Find 82% of 168.

88. 29.05 is what percent of 415?

89. What percent of 200 is 30?

90. SHOPPING A woman bought a coat for $98.95 and some gloves for $7.95. If the sales tax was 6%, how much did the purchase cost her?

8.5 Problem Solving

- Finding more than one unknown
- Solving number–value problems
- Solving uniform motion problems
- Solving geometric problems
- Solving investment problems
- Solving mixture problems

In this section, we will solve different types of problems using the five-step problem-solving strategy.

Finding more than one unknown

EXAMPLE 1 **California coastline.** The first part of California's 17-Mile Drive scenic tour, shown in Figure 8-8, begins at the Pacific Grove entrance and continues to Seal Rock. It is 1 mile longer than the second part of the drive, which extends from Seal Rock to the Lone Cypress. The final part of the tour winds through the hills of the Monterey Peninsula, eventually returning to the entrance. This part of the drive is 1 mile longer than four times the length of the second part. How long is each of the three parts of 17-Mile Drive?

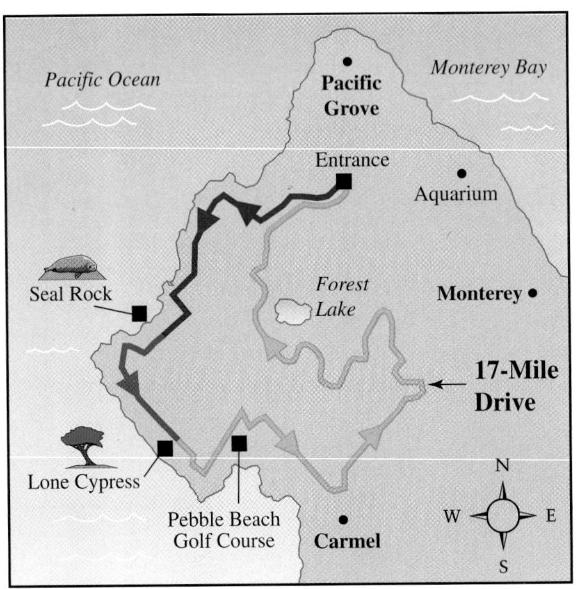

FIGURE 8-8

Solution

Analyze the problem

In Figure 8-9, we straighten out 17-Mile Drive so it can be modeled with a line segment. The drive is composed of three parts. We need to find the length of each part.

Form an equation

Since the lengths of the first part and of the third part of the scenic drive are related to the length of the second part, we will let x represent the length of that part. We then express the other lengths in terms of that variable.

x = the length of the second part of the drive.

$x + 1$ = the length of the first part of the drive.

$4x + 1$ = the length of the third part of the drive.

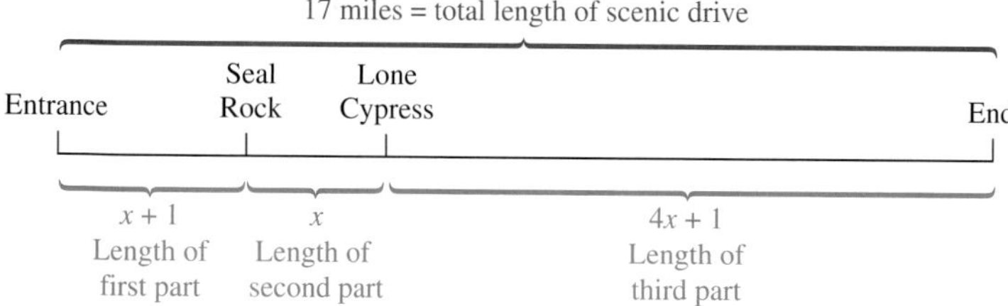

FIGURE 8-9

The sum of the lengths of the three parts of the drive must be 17 miles.

The length of part 1	plus	the length of part 2	plus	the length of part 3	equals	the total length.
$x + 1$	$+$	x	$+$	$4x + 1$	$=$	17

Solve the equation

$$x + 1 + x + 4x + 1 = 17$$

$$6x + 2 = 17 \quad \text{Combine like terms: } x + x + 4x = 6x \text{ and } 1 + 1 = 2.$$

$$6x = 15 \quad \text{To undo the addition of 2, subtract 2 from both sides.}$$

$$\frac{6x}{6} = \frac{15}{6} \quad \text{To undo the multiplication by 6, divide both sides by 6.}$$

$$x = 2.5 \quad \text{Perform the divisions.}$$

Recall that x represents the length of the second part of the drive. To find the lengths of the first and third parts, we evaluate the expressions $x + 1$ and $4x + 1$ for $x = 2.5$.

First part of drive

$$x + 1 = \mathbf{2.5} + 1$$
$$= 3.5$$

Third part of drive

$$4x + 1 = 4(\mathbf{2.5}) + 1 \quad \text{Substitute 2.5 for } x.$$
$$= 10 + 1$$
$$= 11$$

State the conclusion

The first part of the drive is 3.5 miles long, the second part is 2.5 miles long, and the third part is 11 miles long.

Check the result

Because the sum of 3.5 miles, 2.5 miles, and 11 miles is 17 miles, the answers check.

Solving geometric problems

EXAMPLE 2 **Gardens.** A gardener wants to use 62 feet of fencing to enclose a rectangular-shaped garden. Find the length and width of the garden if its length is to be 4 feet longer than twice its width.

Solution

Analyze the problem

We can make a sketch of the garden, as shown in Figure 8-10. We know that its length is to be 4 feet longer than twice its width. We also know that its perimeter is to be 62 feet.

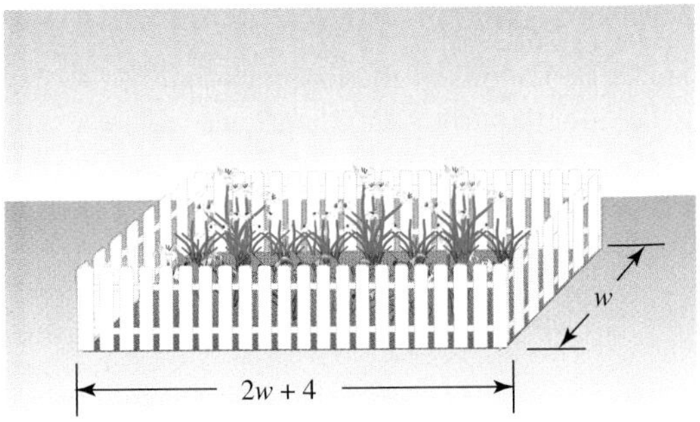

FIGURE 8-10

Form an equation

If we let w = the width of the garden, then $2w + 4$ = its length. Since the formula for the perimeter of a rectangle is $P = 2l + 2w$, the perimeter of the garden is $2(2w + 4) + 2w$, which is also 62. This fact enables us to form the equation.

2	times	the length	plus	2	times	the width	is	the perimeter.
2	$\cdot$	$(2w + 4)$	$+$	2	$\cdot$	w	$=$	62

Solve the equation

$$2(2w + 4) + 2w = 62$$

$$4w + 8 + 2w = 62 \quad \text{Use the distributive property to remove parentheses.}$$

$$6w + 8 = 62 \quad \text{Combine like terms: } 4w + 2w = 6w.$$

$$6w = 54 \quad \text{To undo the addition of 8, subtract 8 from both sides.}$$

$$w = 9 \quad \text{To undo the multiplication by 6, divide both sides by 6.}$$

State the conclusion

The width of the garden is 9 feet. Since $2w + 4 = 2(\mathbf{9}) + 4 = 22$, the length is 22 feet.

Check the result

If the garden has a width of 9 feet and a length of 22 feet, its length is 4 feet longer than twice the width $(2 \cdot 9 + 4 = 22)$. Since its perimeter is $(2 \cdot 22 + 2 \cdot 9)$ feet = 62 feet, the answers check.

EXAMPLE 3 Isosceles triangles.

If the vertex angle of an isosceles triangle is 56°, find the measure of each base angle.

Solution

Analyze the problem

An **isosceles triangle** has two sides of equal length, which meet to form the **vertex angle.** In this case, the measurement of the vertex angle is 56°. We can sketch the triangle as shown in Figure 8-11. The **base angles** opposite the equal sides are also equal. We need to find their measure.

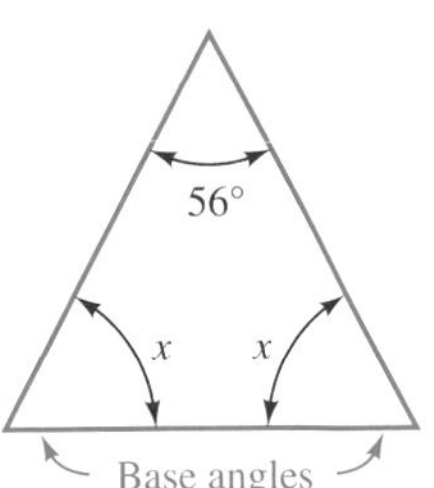

FIGURE 8-11

Form an equation

If we let x = the measure of one base angle, the measure of the other base angle is also x. Since the sum of the angles of any triangle is 180°, the sum of the base angles and the vertex angle is 180°. We can use this fact to form the equation.

One base angle	plus	the other base angle	plus	the vertex angle	is	180°.
x	$+$	x	$+$	56	$=$	180

Solve the equation

$x + x + 56 = 180$

$2x + 56 = 180$ Combine like terms: $x + x = 2x$.

$2x = 124$ To undo the addition of 56, subtract 56 from both sides.

$x = 62$ To undo the multiplication by 2, divide both sides by 2.

State the conclusion

The measure of each base angle is 62°.

Check the result

The measure of each base angle is 62°, and the vertex angle measures 56°. Since $62° + 62° + 56° = 180°$, the answer checks.

Solving number–value problems

Some problems deal with quantities that have a monetary value. In these problems, we must distinguish between the *number of* and the *value of* the unknown quantity. For problems of this type, we will use the relationship

Number · value = total value

EXAMPLE 4 **Restaurants.** A restaurant owner wants to purchase some new tables, chairs, and dinner plates for the dining area of her establishment. She plans to buy four chairs and four plates for each new table. She also needs 20 additional plates to keep in case of breakage. If a table costs \$100, a chair \$50, and a plate \$5, how many of each can she buy if she takes out a loan for \$6,500 to pay for the new items?

Solution

Analyze the problem

We know the *value* of each item: Tables cost \$100, chairs cost \$50, and plates cost \$5 each. We need to find the *number* of tables, chairs, and plates she can purchase for \$6,500.

Form an equation

The number of chairs and plates she needs depends on the number of tables she buys. So we let t = the number of tables to be purchased. Since every table requires 4 chairs and 4 plates, she needs to order $4t$ chairs. Because an additional 20 plates are needed, she should order $4t + 20$ plates. The total value of each purchase is the *product* of the number of items bought and the price, or value, of each item.

Item	Number ·	Value =	Total value
Tables	t	100	$100t$
Chairs	$4t$	50	$50(4t)$
Plates	$4t + 20$	5	$5(4t + 20)$

The total purchase can be expressed in two ways.

The value of the tables	+	the value of the chairs	+	the value of the plates	is	the total value of the purchase.
$100t$	+	$50(4t)$	+	$5(4t + 20)$	=	6,500

Solve the equation

$$100t + 50(4t) + 5(4t + 20) = 6{,}500$$

$$100t + 200t + 20t + 100 = 6{,}500 \quad \text{Perform the multiplications.}$$

$$320t + 100 = 6{,}500 \quad \text{Combine like terms.}$$

$$320t = 6{,}400 \quad \text{Subtract 100 from both sides.}$$

$$t = 20 \quad \text{Divide both sides by 320.}$$

State the conclusion

The purchases are summarized as follows:

Item	Number	·	Value	=	Total value
Table	$t = 20$		$100		$2,000
Chairs	$4t = 80$		$50		$4,000
Plates	$4t + 20 = 100$		$5		$500
Total					$6,500

Check the result

Because the total purchase is $6,500, the answer checks.

THINK IT THROUGH Federal Minimum Wage

According to Current Population Survey estimates for 2002, about 570,000 people were reported earning exactly $5.15, the prevailing Federal minimum wage.

U.S. Department of Labor, Bureau of Labor Statistics

A federal hourly minimum wage was first established in 1938. Over the years, Congress has periodically raised it. The current $5.15-an-hour federal minimum wage is 15 cents more than 20 times the initial minimum wage set in 1938. What was the first minimum wage?

Some states have passed laws establishing a minimum wage that is above the federal level. Washington state, by 1 cent, tops Alaska as the state with the highest minimum wage. If the sum of their minimum wages is $14.31, what is the minimum wage of Washington and of Alaska?

Solving investment problems

To find the amount of simple interest I an investment earns, we use the formula

$$I = Prt$$

where P is the principal, r is the annual rate, and t is the time in years.

EXAMPLE 5 Paying tuition. A college student invested the \$12,000 inheritance he received and decided to use the annual interest earned to pay his yearly tuition costs of \$945. The highest rate offered by a savings and loan at that time was 6% annual simple interest. At this rate, he could not earn the needed \$945, so he invested some of the money in a riskier, but more lucrative, investment offering a 9% return. How much did he invest at each rate?

Solution

Analyze the problem

We know that \$12,000 was invested for 1 year at two rates: 6% and 9%. We are asked to find the amount invested at each rate so that the total return would be \$945.

Form an equation

Let $x =$ the amount invested at 6%. Then $12{,}000 - x =$ the amount invested at 9%.

If \$$x$ (the principal P) is invested at 6% (the rate r), the interest earned in 1 year is Pr or \$$0.06x$. At 9%, the rest of the inheritance money, \$$(12{,}000 - x)$, would earn \$$0.09(12{,}000 - x)$ interest. These facts are summarized in the following table.

	P	$\cdot$ r	$\cdot$ t	$=$ I
Savings and loan	x	0.06	1	$0.06x$
Riskier investment	$12{,}000 - x$	0.09	1	$0.09(12{,}000 - x)$

We can use the information in the last column of the table to form an equation.

The interest earned at 6%	plus	the interest earned at 9%	is	the total interest.
$0.06x$	$+$	$0.09(12{,}000 - x)$	$=$	945

Solve the equation

$$0.06x + 0.09(12{,}000 - x) = 945$$

$$\mathbf{100}[0.06x + 0.09(12{,}000 - x)] = \mathbf{100}(945)$$ Multiply both sides by 100 to clear the equation of decimals.

$$\mathbf{100}(0.06x) + \mathbf{100}(0.09)(12{,}000 - x) = 100(945)$$ Distribute the multiplication by 100.

$$6x + 9(12{,}000 - x) = 94{,}500$$ Multiply by 100.

$$6x + 108{,}000 - 9x = 94{,}500$$ Use the distributive property.

$$-3x + 108{,}000 = 94{,}500$$ Combine like terms.

$$-3x = -13{,}500$$ Subtract 108,000 from both sides.

$$x = 4{,}500$$ Divide both sides by -3.

State the conclusion

The student invested \$4,500 at 6% and $\$12{,}000 - \$4{,}500 = \$7{,}500$ at 9%.

Check the result

The first investment earned 6% of \$4,500, or \$270. The second earned 9% of \$7,500, or \$675. The total return was $\$270 + \$675 = \$945$. The answers check.

Solving uniform motion problems

If we know the rate r at which we will be traveling and the time t we will be traveling at that rate, we can find the distance d traveled by using the formula

$$d = rt$$

EXAMPLE 6 **Coast Guard rescues.** A cargo ship, heading into port, radios the Coast Guard that it has engine trouble and that its speed has dropped to 3 knots (3 nautical miles per hour). Immediately, a Coast Guard cutter leaves the port and speeds at a rate of 25 knots directly toward the disabled ship, which is 21 nautical miles away. How long will it take the Coast Guard cutter to reach the cargo ship?

Solution

Analyze the problem

The diagram in Figure 8-12(a) shows the situation.

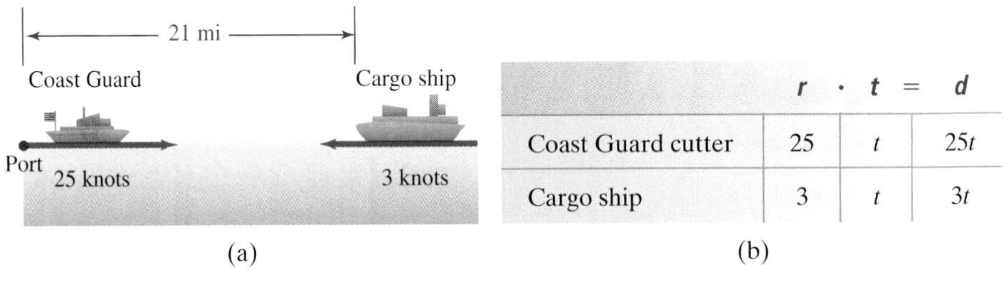

(a)

	r	$\cdot\ t$	$= d$
Coast Guard cutter	25	t	$25t$
Cargo ship	3	t	$3t$

(b)

FIGURE 8-12

We know the *rate* of each ship (25 knots and 3 knots), and we know that they must close a *distance* of 21 nautical miles between them. We don't know the *time* it will take them to do this.

Form an equation

Let t = the time it takes for the ships to meet. Using $d = rt$, we find that $25t$ represents the distance traveled by the Coast Guard cutter and $3t$ represents the distance traveled by the cargo ship. This information is recorded in the table in Figure 8-12(b). We can use the information in the last column of the table to form an equation.

The distance the Coast Guard cutter travels	plus	the distance the cargo ship travels	is	the initial distance between the two ships
$25t$	$+$	$3t$	$=$	21

Solve the equation

$$25t + 3t = 21$$

$$28t = 21 \quad \text{Combine like terms.}$$

$$t = \frac{21}{28} \quad \text{Divide both sides by 28.}$$

$$t = \frac{3}{4} \quad \text{Simplify the fraction: } \frac{21}{28} = \frac{3 \cdot \overset{1}{\cancel{7}}}{4 \cdot \underset{1}{\cancel{7}}} = \frac{3}{4}.$$

State the conclusion

The ships will meet in $\frac{3}{4}$ hour, or 45 minutes.

Check the result

In $\frac{3}{4}$ hour, the Coast Guard cutter travels $25 \cdot \frac{3}{4} = \frac{75}{4}$ nautical miles, and the cargo ship travels $3 \cdot \frac{3}{4} = \frac{9}{4}$ nautical miles. Together, they travel $\frac{75}{4} + \frac{9}{4} = \frac{84}{4} = 21$ nautical miles. Since this is the initial distance between the ships, the answer checks.

Solving mixture problems

We now discuss how to solve two types of mixture problems. In the first type, a *liquid mixture* of a desired strength is made from two solutions with different concentrations.

EXAMPLE 7 **Mixing solutions.** A chemistry experiment calls for a 30% sulfuric acid solution. If the lab supply room has only 50% and 20% sulfuric acid solutions on hand, how much of each should be mixed to obtain 12 liters of a 30% acid solution?

Solution
Analyze the problem

We must find how much of the 50% solution and how much of the 20% solution is needed to obtain 12 liters of a 30% acid solution.

Form an equation

If x = the numbers of liters (L) of the 50% solution used in the mixture, the remaining $(12 - x)$ liters must be the 20% solution. See Figure 8-13(a). Only 50% of the x liters, and only 20% of the $(12 - x)$ liters, is pure sulfuric acid. The total of these amounts is also the amount of acid in the final mixture, which is 30% of 12 liters. This information is shown in the chart in Figure 8-13(b).

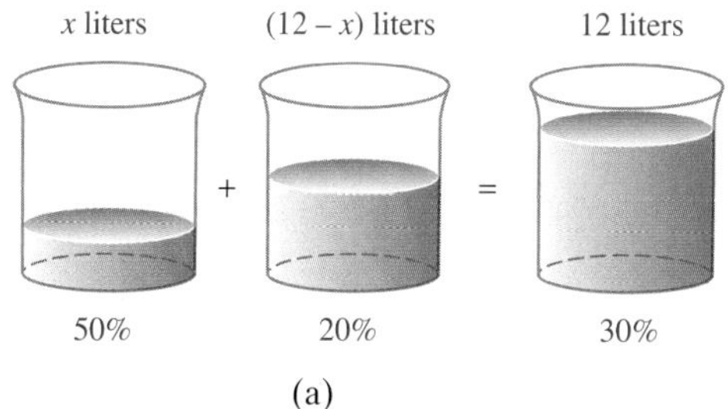

(a)

Solution	% acid ·	Liters =	Amount of acid
50% solution	x	0.50	$0.50x$
20% solution	$12 - x$	0.20	$0.20(12 - x)$
30% mixture	12	0.30	$0.30(12)$

(b)

FIGURE 8-13

We can use the information in the last column of the table to form an equation.

The acid in the 50% solution	plus	the acid in the 20% solution	equals	the acid in the final mixture.
50% of x	$+$	20% of $(12 - x)$	$=$	30% of 12

Solve the equation

$0.50x + 0.20(12 - x) = 0.30(12)$ — 50% = 0.50, 20% = 0.20, and 30% = 0.30.

$5x + 2(12 - x) = 3(12)$ — Multiply both sides by 10 to clear the equation of decimals.

$5x + 24 - 2x = 36$ — Distribute the multiplication by 2.

$3x + 24 = 36$ — Combine like terms.

$3x = 12$ — Subtract 24 from both sides.

$x = 4$ — Divide both sides by 3.

State the conclusion

The mixture will contain 4 liters of 50% solution and $12 - 4 = 8$ liters of 20% solution.

Check the result

Verify that this answer checks.

In the next example, a *dry mixture* of a specified value is created from two differently priced components.

EXAMPLE 8 **Snack foods.** Because fancy cashews priced at \$9 per pound were not selling, a produce clerk decided to combine them with less expensive peanuts and sell the mixture for \$7 per pound. How many pounds of peanuts, selling at \$6 per pound, should be mixed with 50 pounds of cashews to obtain such a mixture?

Solution

Analyze the problem

We know the value of the cashews (\$9 per pound) and the peanuts (\$6 per pound). We also know that 50 pounds of cashews are to be mixed with an unknown number of pounds of peanuts to obtain a mixture worth \$7 per pound.

Form an equation

To solve this problem, we use the formula $v = pn$, where v is value, p is the price per pound, and n is the number of pounds.

Suppose that x pounds of peanuts are used in the mixture. At \$6 per pound, they are worth \$6x. At \$9 per pound, the 50 pounds of cashews are worth $\$9 \cdot 50 = \$450$. Their combined value will be $\$(6x + 450)$. We also know that the mixture weighs $(50 + x)$ pounds. At \$7 per pound, that mixture will be worth $\$7(50 + x)$. This information is shown in the table in Figure 8-14.

	p ·	n =	v
Peanuts	6	x	$6x$
Cashews	9	50	450
Mixture	7	$50 + x$	$7(50 + x)$

FIGURE 8-14

We can use the information in the last column of the table to form an equation.

The value of the peanuts	plus	the value of the cashews	equals	the value of the mixture.
$6x$	$+$	450	$=$	$7(50 + x)$

Solve the equation

$$6x + 450 = 7(50 + x)$$

$$6x + 450 = 350 + 7x \quad \text{Distribute the multiplication by 7.}$$

$$100 = x \quad \text{Subtract } 6x \text{ and 350 from both sides.}$$

State the conclusion

Thus, 100 pounds of peanuts should be used in the mixture.

Check the result

The value of 100 pounds of peanuts at \$6 per pound is \$600
The value of 50 pounds of cashews at \$9 per pound is \$450
The value of the mixture is \$1,050

The value of 150 pounds of the mixture at \$7 per pound is also \$1,050. The answer checks.

Section 8.5 STUDY SET

VOCABULARY *Fill in the blanks.*

1. The ________ of a triangle or a rectangle is the distance around it.
2. An ________ triangle is a triangle with two sides of the same length.
3. The equal sides of an isosceles triangle meet to form the ________ angle.
4. The angles opposite the equal sides in an isosceles triangle are called ______ angles, and they have equal measures.

CONCEPTS

5. PLUMBING A plumber wants to cut a 17-foot pipe into three sections. The longest section is to be three times as long as the shortest, and the middle-sized section is to be 2 feet longer than the shortest.

   a. Complete the diagram below.

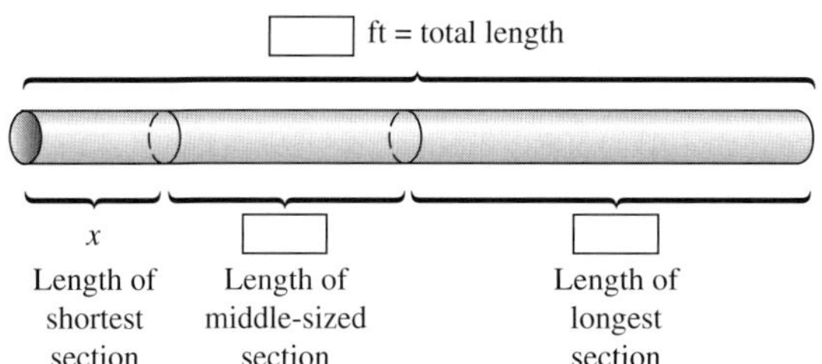

   b. To solve this problem, an equation is formed, it is solved, and it is found that $x = 3$. How long is each section of pipe?

6. What is the sum of the measures of the angles of any triangle?
7. Use a ruler to draw an isosceles triangle with sides 3 inches long and a base 2 inches long. Label the vertex and the base angles.
8. a. Complete the table, which shows the inventory of nylon brushes that a paint store carries.

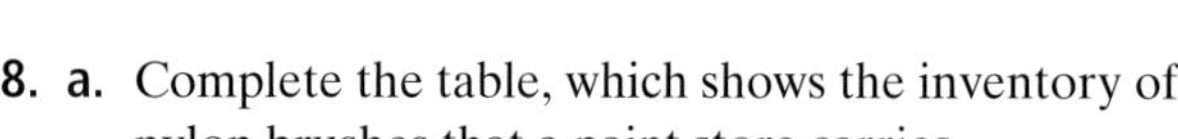

	Number	· Value	= Total value
1 inch brush	$\frac{x}{2}$	\$4	
2 inch brush	x	\$5	
3 inch brush	$x + 10$	\$7	

   b. Which type of brush does the store have the largest number of?

   c. What is the least expensive brush?

   d. What is the total value of the inventory of nylon brushes?

9. In the advertisement, what are the principal, the rate, and the time for the investment opportunity shown?

Invest in Mini Malls!
Builder seeks daring people who want to earn big \$\$\$\$\$\$. In just 1 year, you will earn a gigantic 14% on an investment of only \$30,000! Call now.

10. a. Complete the table, which gives the details about two investments that were made by a retired couple.

	P	· r	· t	= I
Certificate of deposit	x	0.04	1	
Brother-in-law's business	$2x$	0.06	1	

   b. How much more money was invested in the brother-in-law's business than in the certificate of deposit?

   c. What is the total amount of interest the couple will make from these investments?

11. COMMUTERS When a husband and wife leave for work, they drive in opposite directions. Their average speeds are different. However, their drives last the same amount of time. Complete the table, which gives the details of each person's morning commute.

	r	$\cdot$ t	$=$ d
Husband	35 mph	t hr	
Wife	45 mph		

12. Two oil and vinegar salad dressings are combined to make a new mixture. Complete the table.

	Amount	$\cdot$ Strength	$=$ Pure vinegar
Strong	x	0.06	
Weak		0.03	
Mixture	10	0.05	

13. See the illustration.

a. How many gallons of acid are there in the second barrel?

b. Suppose the contents of the two barrels are poured into an empty third barrel. How many gallons of liquid will the third barrel contain?

c. Estimate the concentration of the solution in the third barrel: 19%, 32%, or 43% acid.

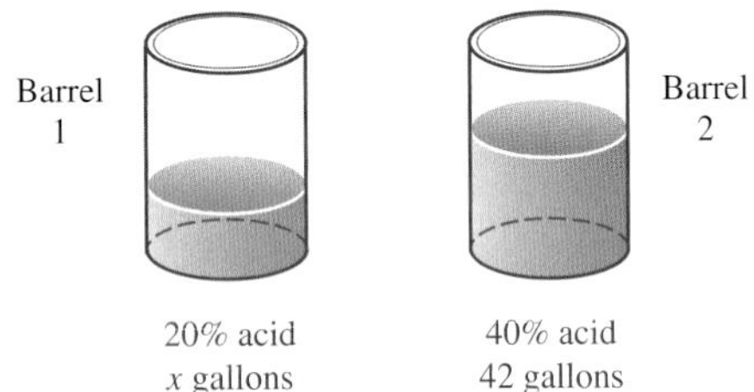

14. Complete the table, which gives the details about the ingredients in a box of breakfast cereal.

	Amount (oz)	Value ($/oz)	Total Value
Blueberries	x	$0.38	
Bran Flakes	14	$0.08	
Blueberries & Bran Flakes Cereal	$14 + x$	$0.21	

PRACTICE *Solve each equation by first clearing it of decimals.*

15. $0.08x + 0.07(15{,}000 - x) = 1{,}110$

16. $0.108x + 0.07(16{,}000 - x) = 1{,}500$

17. Two angles are called **complementary angles** when the sum of their measures is 90°. Find the measures of the complementary angles shown in the illustration.

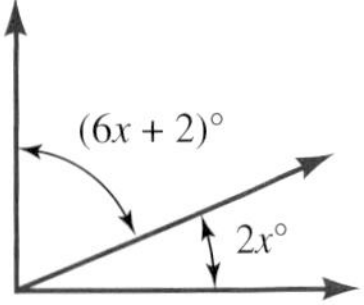

18. Two angles are called **supplementary angles** when the sum of their measures is 180°. Find the measures of the supplementary angles shown in the illustration.

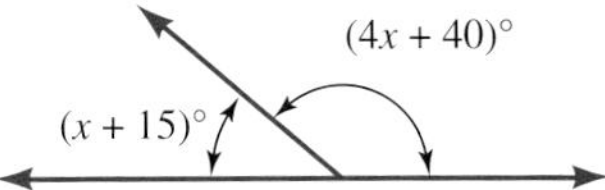

19. In the illustration, two lines intersect to form **vertical angles.** Use the fact that vertical angles have the same measure to find x.

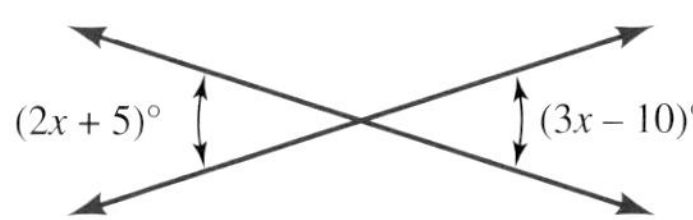

20. Find the measures of the vertical angles shown. (See Exercise 19.)

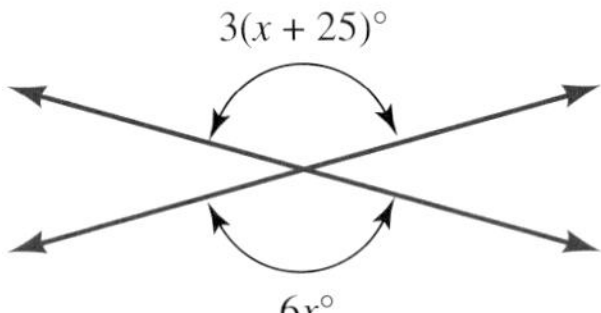

APPLICATIONS

21. CARPENTRY The 12-foot board in the illustration has been cut into two sections, one twice as long as the other. How long is each section?

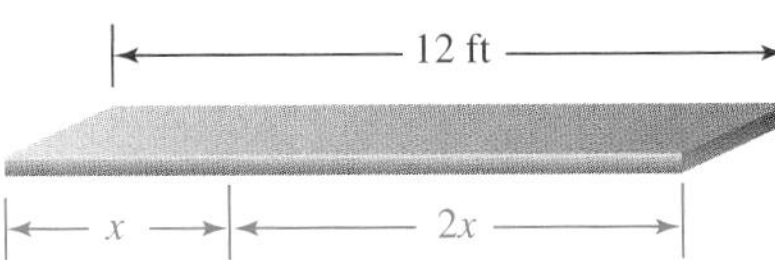

22. ROBOTICS The robotic arm shown will extend a total distance of 18 feet. Find the length of each section.

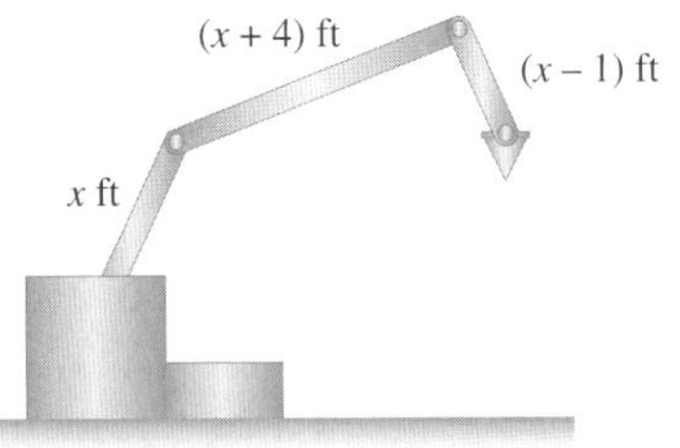

23. SOLAR HEATING One solar panel in the illustration is 3.4 feet wider than the other. Find the width of each panel.

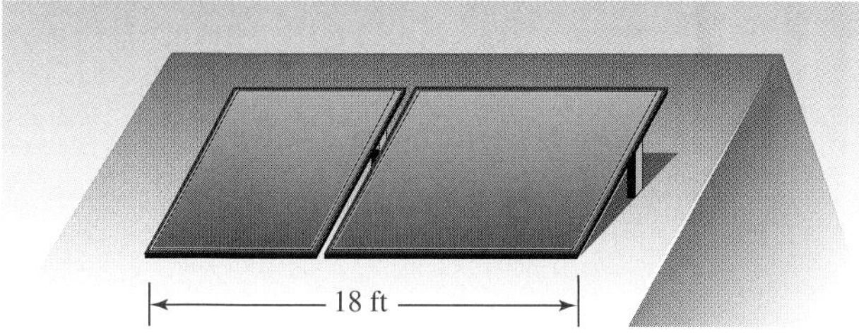

24. PLUMBING A 20-foot pipe has been cut into two sections, one 3 times as long as the other. How long is each section?

25. TOURING A rock group plans to travel for a total of 38 weeks, making three major concert tours. They will be in Japan for 4 more weeks than they will be in Australia. Their stay in Sweden will be 2 weeks less than that in Australia. How many weeks will they be in each country?

26. PUBLISHER'S INVENTORY A novel can be purchased in a hardcover edition for \$15.95 or in paperback for \$4.95. The publisher printed 11 times as many paperbacks as hardcover books. A total of 114,000 books were printed. How many of each type were printed?

27. COUNTING CALORIES A slice of pie with a scoop of ice cream has 850 calories. The calories in the pie alone are 100 more than twice the calories in the ice cream alone. How many calories are in each food?

28. WASTE DISPOSAL Two tanks hold a total of 45 gallons of a toxic solvent. One tank holds 6 gallons more than twice the amount in the other. How many gallons does each tank hold?

29. ACCOUNTING Determine the 2002 net income of Sears, Roebuck and Co. from the data in the graph. (Source: Hoover's Online Internet service.)

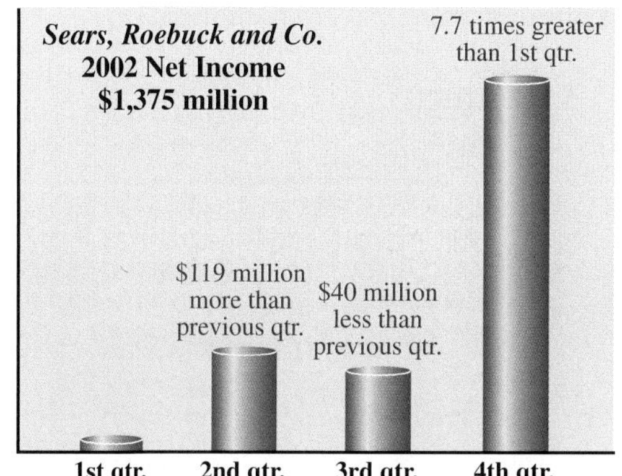

30. LOCKS The three numbers of the combination for the lock shown are **consecutive integers,** and their sum is 81. (Consecutive integers follow each other, like 7, 8, 9.) Complete the instructions below that will open the lock. (*Hint:* If x represents the smallest integer, $x + 1$ represents the next integer, and $x + 2$ represents the largest integer.)

Spin dial to the right one complete revolution to ___. Turn to the left to ___. Turn to the right to ___, and lift the handle.

31. TRUSSES The following truss is in the form of an isosceles triangle. Each of the two equal sides is 4 feet less than the third side. If the perimeter is 25 feet, find the lengths of the sides.

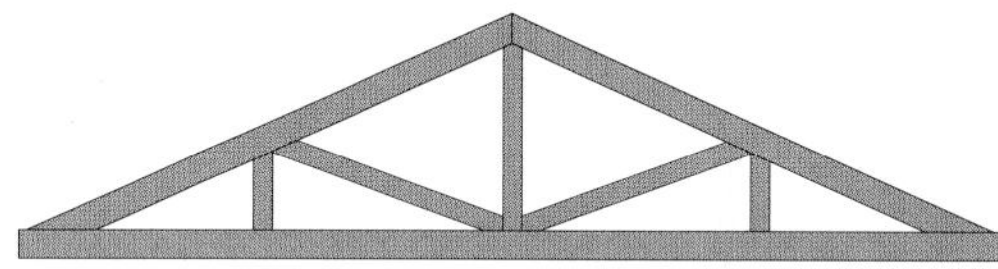

32. FIRST AID The sling shown in the illustration is in the shape of an isosceles triangle with a perimeter of 144 inches. The longest side of the sling is 18 inches longer than either of the other two sides. Find the lengths of each side.

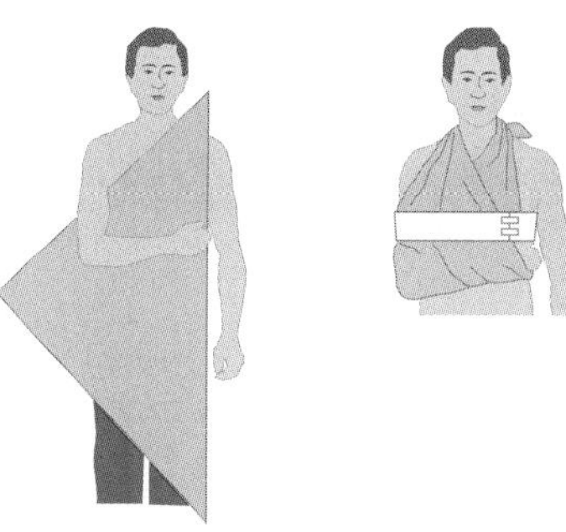

33. SWIMMING POOLS The seawater Orthlieb Pool in Casablanca, Morocco is the largest swimming pool in the world. With a perimeter of 1,110 meters, this rectangular-shaped pool has a length that is 30 meters more than 6 times its width. Find its dimensions.

34. ART The *Mona Lisa* was completed by Leonardo da Vinci in 1506. The length of the picture is 11.75 inches less than twice the width. If the perimeter of the picture is 102.5 inches, find its dimensions.

35. GUY WIRES The two guy wires shown form an isosceles triangle. Each of the base angles of the triangle is 4 times the third angle (the vertex angle a). Find the measure of the vertex angle.

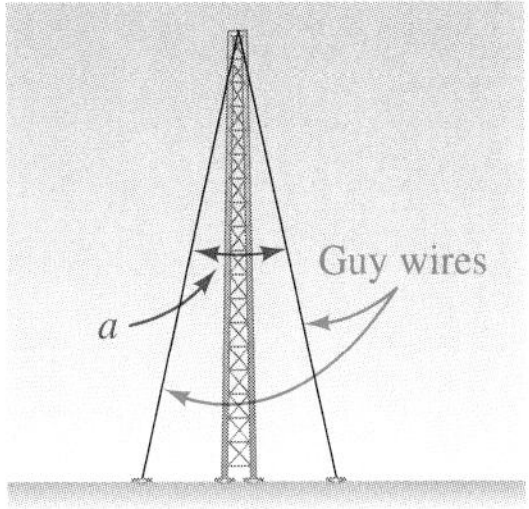

36. MOUNTAIN BIKES For the bicycle frame in the illustration, the angle that the horizontal crossbar makes with the seat support is 15° less than twice the angle at the steering column. The angle at the pedal gear is 25° more than the angle at the steering column. Find these angle measures.

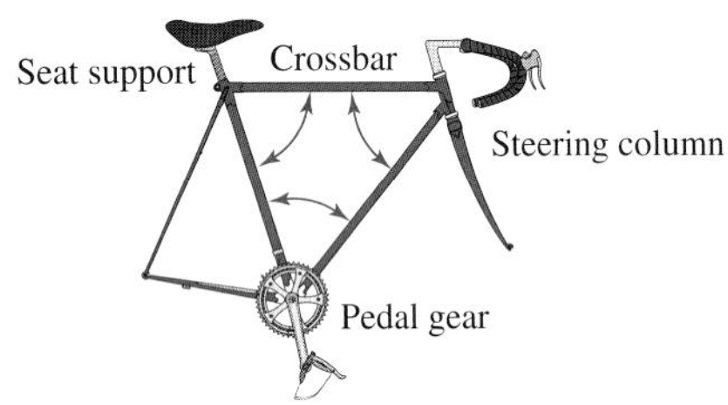

37. WAREHOUSING COSTS A store warehouses 40 more portables than big-screen TV sets, and 15 more consoles than big-screen sets. Storage costs for the different TV sets are shown in the table. If storage costs $276 per month, how many big-screen sets are in stock?

Type of TV	Monthly cost
Portable	$1.50
Console	$4.00
Big-screen	$7.50

38. APARTMENT RENTALS The owners of an apartment building rent 1-, 2-, and 3-bedroom units. They rent equal numbers of each, with the monthly rents given in the table. If the total monthly income is $36,550, how many of each type of unit are there?

Unit	Rent
One-bedroom	$550
Two-bedroom	$700
Three-bedroom	$900

39. SOFTWARE SALES Three software applications are priced as shown the table. Spreadsheet and database programs sold in equal numbers, but 15 more word processing applications were sold than the other two combined. If the three applications generated sales of $72,000, how many spreadsheets were sold?

Software	Price
Spreadsheet	$150
Database	$195
Word processing	$210

40. INVENTORIES With summer approaching, the number of air conditioners sold is expected to be double that of stoves and refrigerators combined. Stoves sell for $350, refrigerators for $450, and air conditioners for $500, and sales of $56,000 are expected. If stoves and refrigerators sell in equal numbers, how many of each appliance should be stocked?

41. TAXES On January 2, 2003, Terrell Washington opened two savings accounts. At the end of the year his bank mailed him the form shown below for income tax purposes. If a total of $15,000 was initially deposited and if no further deposits or withdrawals were made, how much money was originally deposited in account number 721?

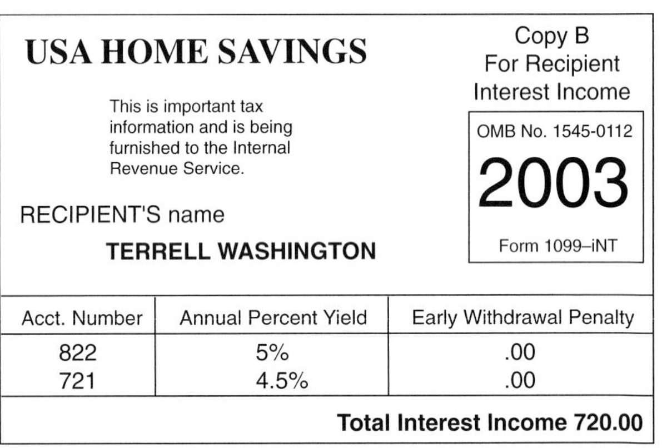
USA HOME SAVINGS

This is important tax information and is being furnished to the Internal Revenue Service.

Copy B
For Recipient
Interest Income

OMB No. 1545-0112
2003
Form 1099–iNT

RECIPIENT'S name
TERRELL WASHINGTON

Acct. Number	Annual Percent Yield	Early Withdrawal Penalty
822	5%	.00
721	4.5%	.00

Total Interest Income 720.00

42. PRESENTATIONS A financial planner recommends a plan for a client who has $65,000 to invest. At the end of the presentation, the client asks, "How much will be invested at each rate?" Answer this question using the given information.

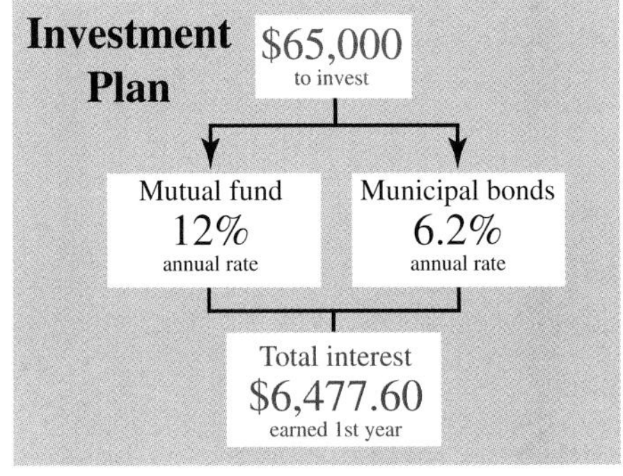

43. INVESTMENTS Equal amounts are invested in each of three accounts paying 7%, 8%, and 10.5% annually. If one year's combined interest income is $1,249.50, how much is invested in each account?

44. RETIREMENT A professor wants to supplement her retirement income with investment interest. If she invests $28,000 at 6% interest, how much more would she have to invest at 7% to achieve a goal of $3,500 per year in supplemental income?

45. FINANCIAL PLANNING A plumber has a choice of two investment plans:

- An insured fund that pays 11% interest
- A risky investment that pays a 13% return

If the same amount invested at the higher rate would generate an extra $150 per year, how much does the plumber have to invest?

46. INVESTMENTS The amount of annual interest earned by $8,000 invested at a certain rate is $200 less than $12,000 would earn at a rate 1% lower. At what rate is the $8,000 invested?

47. TORNADOES During a storm, two teams of scientists leave a university at the same time in specially designed vans to search for tornadoes. The first team travels east at 20 mph and the second travels west at 25 mph, as shown. If their radios have a range of up to 90 miles, how long will it be before they lose radio contact?

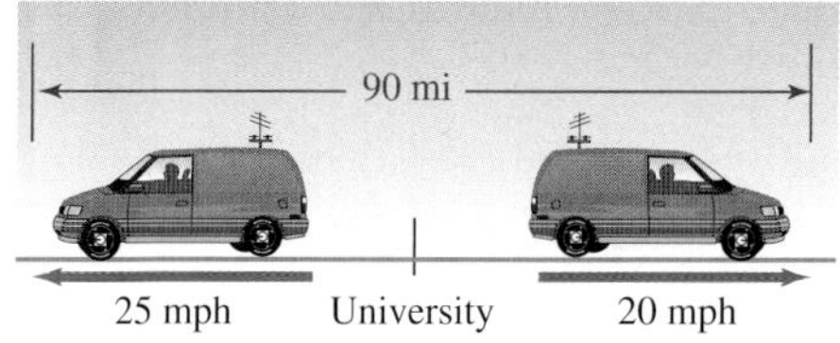

48. SEARCH AND RESCUE Two search-and-rescue teams leave base at the same time looking for a lost boy. The first team, on foot, heads north at 2 mph and the other, on horseback, south at 4 mph. How long will it take them to search a distance of 21 miles between them?

49. SPEED OF TRAINS Two trains are 330 miles apart, and their speeds differ by 20 mph. Find the speed of each train if they are traveling toward each other and will meet in 3 hours.

50. AVERAGE SPEED A car averaged 40 mph for part of a trip and 50 mph for the remainder. If the 5-hour trip covered 210 miles, for how long did the car average 40 mph?

51. AIR TRAFFIC CONTROL An airliner leaves Berlin, Germany, headed for Montreal, Canada, flying at an average speed of 450 mph. At the same time, an airliner leaves Montreal headed for Berlin, averaging 500 mph. If the airports are 3,800 miles apart, when will the air traffic controllers have to make the pilots aware that the planes are passing each other?

52. ROAD TRIPS A bus, carrying the members of a marching band, and a truck, carrying their instruments, leave a high school at the same time. The bus travels at 65 mph and the truck at 55 mph. In how many hours will they be 75 miles apart?

53. SALT SOLUTIONS How many gallons of a 3% salt solution must be mixed with 50 gallons of a 7% solution to obtain a 5% solution?

54. MAKING CHEESE To make low-fat cottage cheese, milk containing 4% butterfat is mixed with 10 gallons of milk containing 1% butterfat to obtain a mixture containing 2% butterfat. How many gallons of the richer milk must be used?

55. ANTISEPTIC SOLUTIONS A nurse wants to add water to 30 ounces of a 10% solution of benzalkonium chloride to dilute it to an 8% solution. How much water must she add?

56. PHOTOGRAPHIC CHEMICALS A photographer wishes to mix 2 liters of a 5% acetic acid solution with a 10% solution to get a 7% solution. How many liters of 10% solution must be added?

57. MIXING FUELS How many gallons of fuel costing $1.15 per gallon must be mixed with 20 gallons of a fuel costing $0.85 per gallon to obtain a mixture costing $1 per gallon?

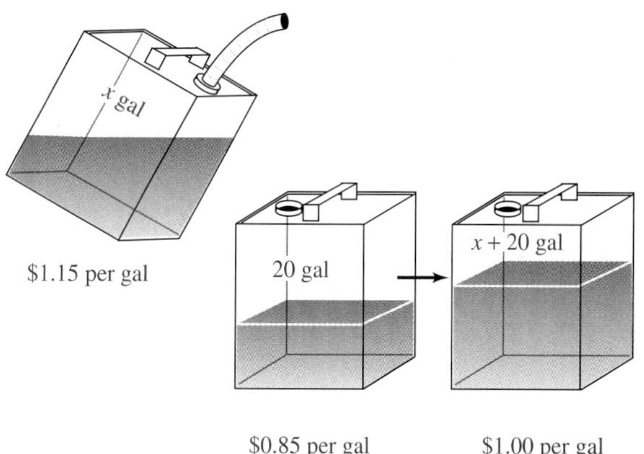

58. MIXING PAINT Paint costing $19 per gallon is to be mixed with 5 gallons of a $3-per-gallon thinner to make a paint that can be sold for $14 per gallon. How much paint will be produced?

59. MIXING CANDY Lemon drops worth $1.90 per pound are to be mixed with jelly beans that cost $1.20 per pound to make 100 pounds of a mixture worth $1.48 per pound. How many pounds of each candy should be used?

60. MIXING CANDY Twenty pounds of lemon drops are to be mixed with cherry chews to make a mixture that will sell for \$1.80 per pound. How much of the more expensive candy should be used? See the table.

Candy	Price per pound
Peppermint patties	\$1.35
Lemon drops	\$1.70
Licorice lumps	\$1.95
Cherry chews	\$2.00

61. BLENDING COFFEE A store sells regular coffee for \$4 a pound and gourmet coffee for \$7 a pound. To get rid of 40 pounds of the gourmet coffee, a shopkeeper makes a blend to put on sale for \$5 a pound. How many pounds of regular coffee should he use?

62. BLENDING LAWN SEED A store sells bluegrass seed for \$6 per pound and ryegrass seed for \$3 per pound. How much ryegrass must be mixed with 100 pounds of bluegrass to obtain a blend that will sell for \$5 per pound?

WRITING

63. Create a mixture problem of your own, and solve it.

64. Use an example to explain the difference between the quantity and the value of the materials being combined in a mixture problem.

65. A car travels at 60 mph for 15 minutes. Why can't we multiply the rate, 60, and the time, 15, to find the distance traveled by the car?

66. Create a geometry problem that could be answered by solving the equation $2w + 2(w + 5) = 26$.

REVIEW *Use the distributive property to remove parentheses.*

67. $-25(2x - 5)$

68. $-12(3a + 4b - 32)$

69. $-(-3x - 3)$

70. $\frac{1}{2}(4b - 8)$

Combine like terms.

71. $8p - 9q + 11p + 20q$

72. $-5(t - 120) - 7(t + 5)$

8.6 Inequalities

- Inequality symbols • Graphing inequalities • Interval notation • Solving inequalities
- Graphing compound inequalities • Solving compound inequalities • An application

Inequalities are expressions indicating that two quantities are not necessarily equal. They appear in many situations:

- An airplane is rated to fly at altitudes that are less than 36,000 feet.
- To melt ice, the temperature must be greater than 32° F.
- To earn a B, I need a final exam score of at least 80%.

Inequality symbols

We can use **inequality symbols** to show that two expressions are not equal.

Inequality symbols

$\neq$	means	"is not equal to"
$<$	means	"is less than"
$>$	means	"is greater than"
$\leq$	means	"is less than or equal to"
$\geq$	means	"is greater than or equal to"

EXAMPLE 1 Write each inequality in words: **a.** $6 \neq 9$, **b.** $8 > 4$, **c.** $12 \geq 0$, **d.** $5 \leq 5$.

Solution

a. $6 \neq 9$ is read as "6 is not equal to 9."

b. $8 > 4$ is read as "8 is greater than 4."

Self Check 1
Write each inequality in words: **a.** $15 < 20$, **b.** $y \geq 9$, **c.** $10 \geq 1$, and **d.** $30 \leq 30$.

c. $12 \geq 0$ is read as "12 is greater than or equal to 0." This is true, because $12 > 0$.

d. $5 \leq 5$ is read as "5 is less than or equal to 5." This is true, because $5 = 5$.

Answers **a.** 15 is less than 20. **b.** y is greater than or equal to 9. **c.** 10 is greater than or equal to 1. **d.** 30 is less than or equal to 30.

If two numbers are graphed on a number line, the one to the right is the greater. For example, from Figure 8-15, we see that $-1 > -4$, because -1 lies to the right of -4.

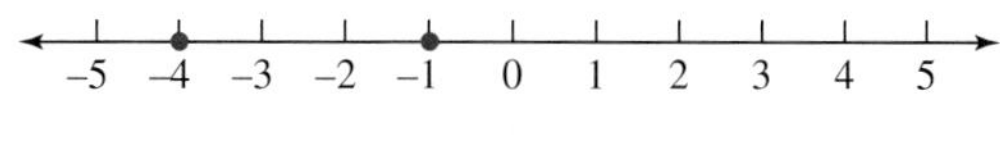

FIGURE 8-15

Inequalities can be written so that the inequality symbol points in the opposite direction. For example, the following statements both indicate that 27 is a smaller number than 32.

$27 < 32$ (27 is less than 32) and $32 > 27$ (32 is greater than 27)

The following statements both indicate that 9 is greater than or equal to 6.

$9 \geq 6$ (9 is greater than or equal to 6) and $6 \leq 9$ (6 is less than or equal to 9)

Variables can be used with inequality symbols to show mathematical relationships. For example, consider the statement, "You must be taller than 54 inches to ride the roller coaster." If we let h represent a person's height in inches, then to ride the roller coaster, $h > 54$ inches.

EXAMPLE 2 Express the following situation using an inequality symbol: "The occupancy of the dining room cannot exceed 200 people."

Solution If p represents the number of people that can occupy the room, then p cannot be greater than (exceed) 200. Another way to state this is that p must be *less than or equal to* 200.

$p \leq 200$

Self Check 2

Express the following statement using an inequality symbol: "The thermostat on the pool heater is set so that the water temperature t is at least 72°."

Answer $t \geq 72$

Graphing inequalities

Graphs of inequalities involving real numbers are **intervals** on the number line. For example, two versions of the graph of all real numbers x such that $x > -3$ are shown in Figure 8-16.

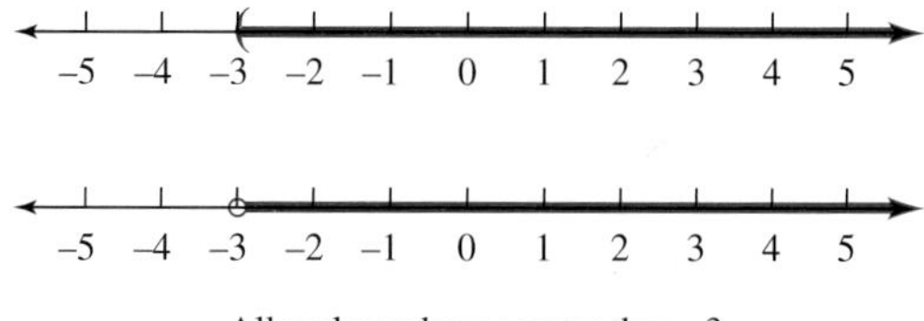

FIGURE 8-16

The red arrow pointing to the right shows that all numbers to the right of -3 are in the graph. The **parenthesis** (or open circle) at -3 indicates that -3 is not in the graph.

Interval notation

The interval shown in Figure 8-16 can be expressed in **interval notation** as $(-3, \infty)$. Again, the first parenthesis indicates that -3 is not included in the interval. The infinity symbol ∞ does not represent a number. It indicates that the interval continues on forever to the right.

Figure 8-17 shows two versions of the graph of $x \leq 2$. The thick red arrow pointing to the left shows that all numbers to the left of 2 are in the graph. The **bracket** (or closed circle) at 2 indicates that 2 is included in the graph. We can express this interval as $(-\infty, 2]$. Here the bracket indicates that 2 is included in the interval.

From now on, we will use parentheses or brackets when graphing intervals, because they are consistent with interval notation.

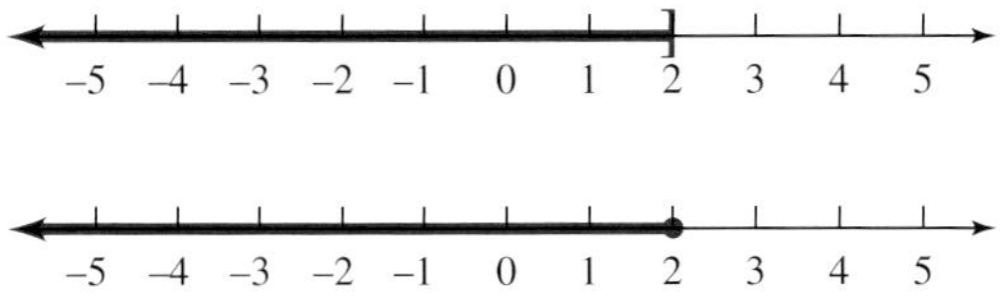

All real numbers less than or equal to 2

FIGURE 8-17

Self Check 3
What inequality is represented by each graph?

a. 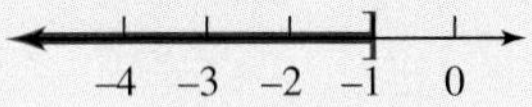

b. 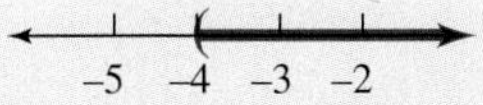

Answers **a.** $x \leq -1$, $(-\infty, -1]$
b. $x > -4$, $(-4, \infty)$

EXAMPLE 3 What inequality is represented by each graph?

a. 

b. 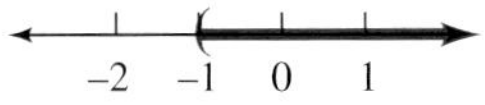

Solution

a. This is the interval $(-\infty, 3]$, which consists of all real numbers less than or equal to 3. The inequality is $x \leq 3$.

b. This is the interval $(-1, \infty)$, consisting of all real numbers greater than -1. The inequality is $x > -1$.

Solving inequalities

A **solution of an inequality** is any number that makes the inequality true. For example, 2 is a solution of $x \leq 3$, because $2 \leq 3$.

To solve more complicated inequalities, we will use the addition, subtraction, multiplication, and division properties of inequality. When we use one of these properties, the resulting inequality will always be equivalent to the original one.

Addition and subtraction properties of inequality

For real numbers a, b, and c,

If $a < b$, then $a + c < b + c$.
If $a < b$, then $a - c < b - c$.

Similar statements can be made for the symbols $>$, $\leq$, and $\geq$.

The **addition property of inequality** can be stated this way:

If a quantity is added to both sides of an inequality, the resulting inequality will have the same direction as the original one.

The **subtraction property of inequality** can be stated this way:

If a quantity is subtracted from both sides of an inequality, the resulting inequality will have the same direction as the original one.

EXAMPLE 4 Solve $x + 3 > 2$ and graph its solution.

Solution To isolate the x on the left-hand side of the $>$ sign, we proceed as we would when solving equations.

$x + 3 > 2$

$x + 3 - \mathbf{3} > 2 - \mathbf{3}$ To undo the addition of 3, subtract 3 from both sides.

$x > -1$ Subtract: $3 - 3 = 0$ and $2 - 3 = 2 + (-3) = -1$.

All real numbers greater than -1 are solutions of $x + 3 > 2$. This means the inequality has *infinitely many* solutions. The graph of the solutions (see Figure 8-18) includes all points to the right of -1 but does not include -1. Expressed as an interval, this is $(-1, \infty)$.

−3 −2 −1 0 1 2 3

FIGURE 8-18

Since the solution contains infinitely many numbers, we cannot check to see whether each of them satisfies the original inequality. As an informal check, we pick several numbers in the graph, such as 1 and 30, substitute each number for x in the inequality, and see whether it satisfies the inequality.

$x + 3 > 2$

$\mathbf{1} + 3 \overset{?}{>} 2$ Substitute 1 for x.

$4 > 2$ Perform the addition.

$x + 3 > 2$

$\mathbf{30} + 3 \overset{?}{>} 2$ Substitute 30 for x.

$33 > 2$ Perform the addition.

Since $4 > 2$, we know that 1 satisfies the inequality. Since $33 > 2$, we know that 30 satisfies the inequality. The result $(x > -1)$ appears to be correct.

Self Check 4
Solve $x - 3 \le -2$ and graph its solution. Then use interval notation to write the solution.

Answer $x \le 1, (-\infty, 1]$

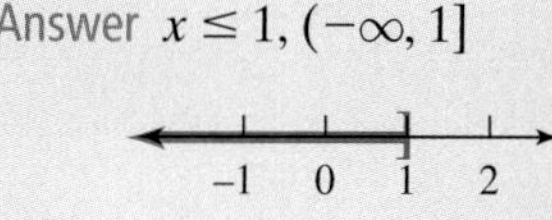

If both sides of the inequality $2 < 5$ are multiplied by a *positive* number, such as 3, another true inequality results.

$2 < 5$

$\mathbf{3} \cdot 2 < \mathbf{3} \cdot 5$ Multiply both sides by 3.

$6 < 15$ Perform the multiplications: $3 \cdot 2 = 6$ and $3 \cdot 5 = 15$.

However, if we multiply both sides of $2 < 5$ by a *negative* number, such as -3, the direction of the inequality symbol must be reversed to produce another true inequality.

$2 < 5$

$\mathbf{-3} \cdot 2 > \mathbf{-3} \cdot 5$ Multiply both sides by the negative number -3 and reverse the direction of the inequality.

$-6 > -15$ Perform the multiplications: $-3 \cdot 2 = -6$ and $-3 \cdot 5 = -15$.

The inequality $-6 > -15$ is true because -6 is to the right of -15 on the number line.

Multiplication and division properties of inequalities

For real numbers a, b, and c,

If $a < b$ and $c > 0$, then $ac < bc$.
If $a < b$ and $c < 0$, then $ac > bc$.
If $a < b$ and $c > 0$, then $\frac{a}{c} < \frac{b}{c}$.
If $a < b$ and $c < 0$, then $\frac{a}{c} > \frac{b}{c}$.

Similar statements can be made for the symbols $>$, $\leq$, and $\geq$.

The **multiplication property of inequality** can be stated this way:

If both sides of an inequality are multiplied by the same positive number, the resulting inequality will have the same direction as the original one.

If both sides of an inequality are multiplied by the same negative number, the resulting inequality will have the opposite direction from the original one.

The **division property of inequality** can be stated this way:

If both sides of an inequality are divided by the same positive number, the resulting inequality will have the same direction as the original one.

If both sides of an inequality are divided by the same negative number, the resulting inequality will have the opposite direction from the original one.

Self Check 5

Solve $2x - 7 > -13$ and graph the solution. Then use interval notation to write the solution.

EXAMPLE 5 Solve $-5 \geq 3x + 7$ and graph the solution.

Solution

$-5 \geq 3x + 7$

$-5 - \mathbf{7} \geq 3x + 7 - \mathbf{7}$ To undo the addition of 7, subtract 7 from both sides.

$-12 \geq 3x$ Subtract: $-5 - 7 = -5 + (-7) = -12$ and $7 - 7 = 0$.

$\frac{-12}{\mathbf{3}} \geq \frac{3x}{\mathbf{3}}$ To undo the multiplication by 3, divide both sides by 3.

$-4 \geq x$ Perform the divisions.

It is common practice to present a solution such as $-4 \geq x$ in an equivalent form with the variable on the left-hand side. If -4 is greater than or equal to x, then x must be less than or equal to -4, and we can write the solution as

$x \leq -4$

The graph (shown in Figure 8-19) consists of all real numbers less than or equal to -4. Using interval notation, we have $(-\infty, -4]$.

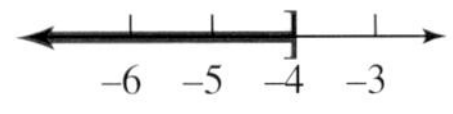

FIGURE 8-19

To check, we can pick several numbers in the graph, such as -6 and -20, and see whether each one satisfies the inequality.

For $x = -6$		For $x = -20$
$-5 \geq 3x + 7$		$-5 \geq 3x + 7$
$-5 \overset{?}{\geq} 3(\mathbf{-6}) + 7$	Substitute -6 for x.	$-5 \overset{?}{\geq} 3(\mathbf{-20}) + 7$
$-5 \overset{?}{\geq} -18 + 7$	Multiply.	$-5 \overset{?}{\geq} -60 + 7$
$-5 \geq -11$	Add.	$-5 \geq -53$

Since $-5 \geq -11$, we know that -6 satisfies the inequality. Since $-5 \geq -53$, we know that -20 satisfies the inequality. The result, $x \leq -4$, appears to be correct.

Answer $x > -3, (-3, \infty)$

EXAMPLE 6 Solve $5 - 3x < 14$ and graph the solution.

Solution

$$5 - 3x < 14$$

$$5 - 3x - 5 < 14 - 5$$ To isolate $-3x$ on the left-hand side, subtract 5 from both sides.

$$-3x < 9$$ Subtract: $5 - 5 = 0$ and $14 - 5 = 9$.

$$\frac{-3x}{-3} > \frac{9}{-3}$$ To undo the multiplication by -3, divide both sides by -3. Since we are dividing by a negative number, we reverse the direction of the $<$ symbol.

$$x > -3$$

The graph is shown in Figure 8-20. This is the interval $(-3, \infty)$, which consists of all real numbers greater than -3.

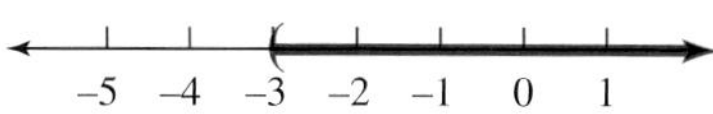

FIGURE 8-20

Check the result.

Self Check 6
Solve $-2x - 5 \geq -11$ and graph the solution. Then use interval notation to write the solution.

Answer $x \leq 3, (-\infty, 3]$

EXAMPLE 7 Solve $\frac{x}{-15} \geq -6$ and graph the solution.

Solution

$$\frac{x}{-15} \geq -6$$

$$-15\left(\frac{x}{-15}\right) \leq -15(-6)$$ To undo the division by -15, multiply both sides by -15. Since we are multiplying by a negative number, we reverse the direction of the $\geq$ symbol.

$$x \leq 90$$ Perform the multiplications.

The graph is shown in Figure 8-21. This is the interval $(-\infty, 90]$, which consists of all real numbers less than or equal to 90.

FIGURE 8-21

Self Check 7
Solve $\frac{h}{-20} < 10$ and graph the solution.

Answer $h > -200, (-200, \infty)$

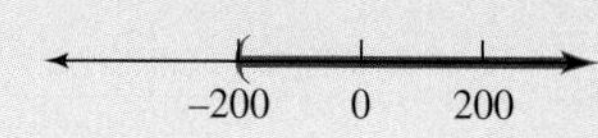

! COMMENT Remember that if both sides of an inequality are multiplied or divided by a negative number, the direction of the inequality symbol must be reversed.

EXAMPLE 8 Solve $5(x + 1) \leq 2(x - 3)$ and graph the solution.

Solution

$$5(x + 1) \leq 2(x - 3)$$

$$5x + 5 \leq 2x - 6$$ Use the distributive property on both sides of the inequality.

Self Check 8
Solve $3(x - 2) > -(x + 1)$ and graph the solution. Then use interval notation to write the solution.

$5x + 5 - \mathbf{2x} \le 2x - 6 - \mathbf{2x}$	To eliminate $2x$ from the right side, subtract $2x$ from both sides.
$3x + 5 \le -6$	Combine like terms on both sides.
$3x + 5 - \mathbf{5} \le -6 - \mathbf{5}$	To undo the addition of 5, subtract 5 from both sides.
$3x \le -11$	Perform the subtractions.
$\frac{3x}{\mathbf{3}} \le \frac{-11}{\mathbf{3}}$	To undo the multiplication by 3, divide both sides by 3.
$x \le -\frac{11}{3}$	

The graph is shown in Figure 8-22. This is the interval $(-\infty, -\frac{11}{3}]$, which consists of all real numbers less than or equal to $-\frac{11}{3}$. We note that $-\frac{11}{3} = -3\frac{2}{3}$.

FIGURE 8-22

Check the result.

Answer $x > \frac{5}{4}$, $\left(\frac{5}{4}, \infty\right)$

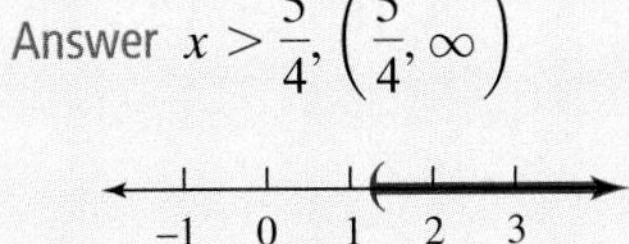

Graphing compound inequalities

Two inequalities can be combined into a **compound inequality** to indicate that numbers lie *between* two fixed values. For example, $-2 < x < 3$ is a combination of

$$-2 < x \quad \text{and} \quad x < 3$$

It indicates that x is greater than -2 and that x is also less than 3. The solution of $-2 < x < 3$ consists of all numbers that lie *between* -2 and 3. The graph of this interval appears in Figure 8-23. We can express this interval as $(-2, 3)$.

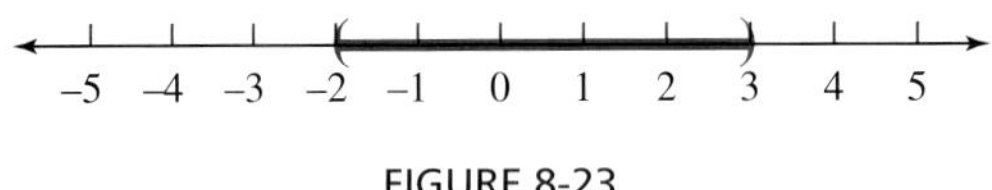

FIGURE 8-23

Self Check 9

What inequality is represented by the following graph?

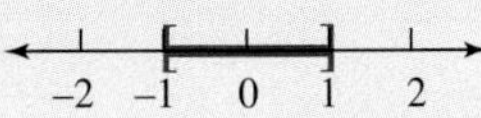

Answer $-1 \le x \le 1$. This is the interval $[-1, 1]$.

EXAMPLE 9 What inequality is represented by the following graph?

Solution

$1 < x \le 5$. This is the interval $(1, 5]$.

Self Check 10

Graph the interval $-2 \le x < 1$. Then use interval notation to write the solution.

Answer $[-2, 1)$

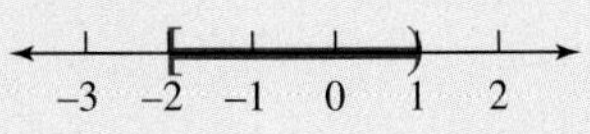

EXAMPLE 10 Graph the interval $-4 < x \le 0$.

Solution The interval $-4 < x \le 0$ consists of all real numbers between -4 and 0, including 0. The graph appears in Figure 8-24. This is the interval $(-4, 0]$.

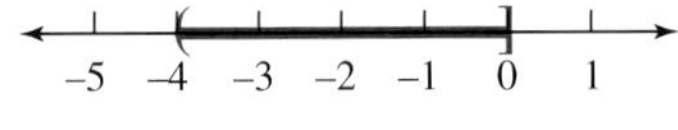

FIGURE 8-24

To check, we pick a number, such as -2, in the graph and see whether it satisfies the inequality. Since $-4 < -2 \le 0$, the answer appears to be correct.

Solving compound inequalities

To solve compound inequalities, we use the same methods we used for solving equations. However, instead of applying the properties of equality to both sides of an equation, we will apply the properties of inequality to all three parts of the inequality.

EXAMPLE 11 Solve $-4 < 2(x - 1) \leq 4$ and graph the solution.

Solution

$-4 < 2(x - 1) \leq 4$	
$-4 < 2x - 2 \leq 4$	Distribute the multiplication by 2.
$-4 + \mathbf{2} < 2x - 2 + \mathbf{2} \leq 4 + \mathbf{2}$	To undo the subtraction of 2, add 2 to all three parts.
$-2 < 2x \leq 6$	Perform the additions.
$\frac{-2}{\mathbf{2}} < \frac{2x}{\mathbf{2}} \leq \frac{6}{\mathbf{2}}$	To isolate x, we undo the multiplication by 2 by dividing all three parts by 2.
$-1 < x \leq 3$	

The graph of the solution appears in Figure 8-25. This is the interval $(-1, 3]$.

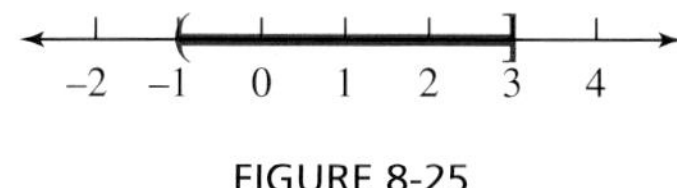

FIGURE 8-25

Check the solution.

Self Check 11
Solve $-6 \leq 3(x + 2) \leq 6$ and graph the solution. Then use interval notation to write the solution.

Answer $-4 \leq x \leq 0$, $[-4, 0]$

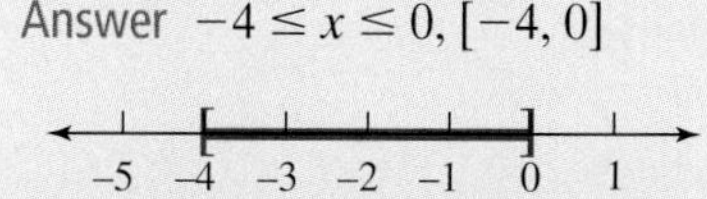

An application

When solving problems, phrases such as "not more than," "at least," or "should exceed" suggest that an *inequality* should be written instead of an *equation*.

EXAMPLE 12 **Grades.** A student has scores of 72%, 74%, and 78% on three exams. What percent score does he need on the last exam to earn no less than a grade of B (80%)?

Solution

Analyze the problem

We know three of the student's scores. We are to find what he must score on the last exam to earn at least a B grade.

Form an inequality

We can let x represent the score on the fourth (and last) exam. To find the average grade, we add the four scores and divide by 4. To earn no less than a grade of B, the student's average must be greater than or equal to 80%.

The average of the four grades	must be greater than or equal to	80.
$\frac{72 + 74 + 78 + x}{4}$	$\geq$	80

Solve the inequality

We can solve this inequality for x.

$\frac{224 + x}{4} \geq 80$ Simplify the numerator: $72 + 74 + 78 = 224$.

$224 + x \geq 320$ To clear the inequality of the fraction, multiply both sides by 4.

$x \geq 96$ To undo the addition of 224, subtract 224 from both sides.

State the conclusion

To earn a B, the student must score 96% or better on the last exam. Of course, the student cannot score higher than 100%. The graph appears in Figure 8-26. This is the interval [96, 100].

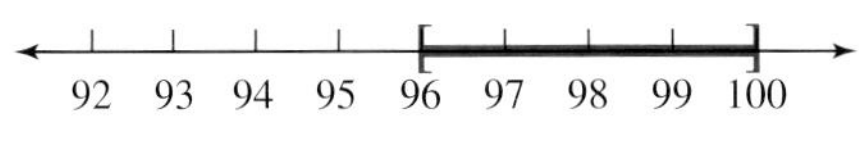

FIGURE 8-26

Check the result

Pick some numbers in the interval, and verify that the average of the four scores will be 80% or greater.

Section 8.6 STUDY SET

VOCABULARY *Fill in the blanks.*

1. An expression containing one of the symbols $>$, $<$, $\geq$, $\leq$, or $\neq$ is called an ________.
2. Graphs of inequalities involving real numbers are called ________ on the number line.
3. A ________ of an inequality is any real number that makes the inequality true.
4. The inequality $-4 < x \leq 12$ is an example of a ________ inequality.

CONCEPTS

5. Determine whether each statement is true or false.
   a. $35 \geq 34$ b. $5.61 \geq 5.61$
   c. $-16 \leq -17$ d. $0 \leq -2\frac{1}{8}$
   e. $\frac{3}{4} \leq 0.75$ f. $-0.6 \geq -0.5$
6. Determine whether each number is a solution of $3x + 7 < 4x - 2$.
   a. 12 b. -6
   c. 0 d. 9

Fill in the blanks.

7. If a quantity is added to or subtracted from both sides of an inequality, the resulting inequality will have the ________ direction as the original one.
8. If both sides of an inequality are multiplied or divided by a positive number, the resulting inequality will have the ________ direction as the original one.
9. If both sides of an inequality are multiplied or divided by a negative number, the resulting inequality will have the ________ direction from the original one.
10. To solve compound inequalities, the properties of inequalities are applied to all ________ parts of the inequality.
11. The solution of an inequality is graphed below.

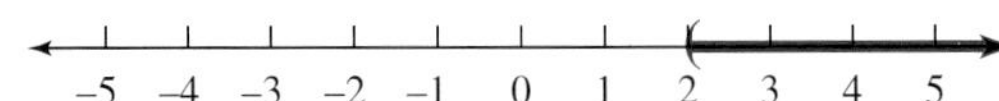

   a. If 3 is substituted for the variable in the inequality, will a true or a false statement result?
   b. If -3 is substituted for the variable in the inequality, will a true or a false statement result?

12. The solution of a compound inequality is graphed below.

a. If 3 is substituted for the variable in the inequality, will a true or a false statement result?

b. If -3 is substituted for the variable in the inequality, will a true or a false statement result?

13. Solve the inequality $2x - 4 > 12$, and give the solution:

a. in words

b. using a graph

c. using interval notation

14. Solve the compound inequality $-4 < 2x < 12$, and give the solution:

a. in words

b. using a graph

c. using interval notation

NOTATION *Fill in the blanks.*

15. The symbol $<$ means "____ ____ ____," and the symbol $>$ means "____ ____ ____."

16. The symbol $\geq$ means "____ ____ ____ or equal to," and the symbol $\leq$ means "is less than ____ ____ ____."

17. The symbol $\neq$ means "____ ____ ____ ____."

18. In the interval $[4, 8)$, the endpoint 4 is ______, but the endpoint 8 is not included.

19. Suppose you solve an inequality and obtain $-2 < x$. Rewrite this inequality so that x is on the left-hand side.

20. Explain what is wrong with the compound inequality $8 < x < -1$.

Write each inequality so the inequality symbol points in the opposite direction.

21. $17 \geq -2$

22. $-32 < -10$

Complete the solution to solve each inequality.

23.

$$4x - 5 \geq 7$$
$$4x - 5 + \square \geq 7 + \square$$
$$4x \geq \square$$
$$\frac{4x}{\square} \geq \frac{12}{\square}$$
$$x \geq 3$$

24.

$$\frac{-x}{2} + 4 < 5$$
$$\frac{-x}{2} + 4 - \square < 5 - \square$$
$$\frac{-x}{2} < \square$$
$$\square\left(\frac{-x}{2}\right) < \square(1)$$
$$\square < 2$$
$$\frac{-x}{\square} \ \square \ \frac{2}{-1}$$
$$x > -2$$

PRACTICE *Graph each inequality. Then write the solution using interval notation.*

25. $x < 5$

26. $x \geq -2$

27. $-3 < x \leq 1$

28. $-1 \leq x \leq 3$

Write the inequality that is represented by each graph. Then write the solution in interval notation.

29.  (graph: shaded to the left, open endpoint at -1)

30. 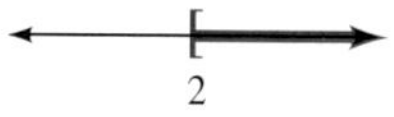 (graph: shaded to the right, closed endpoint at 2)

31.  (graph: open endpoint at -7, closed endpoint at 2)

32. (graph: closed endpoint at -3, closed endpoint at 1)

Solve each inequality, graph the solution, and write the solution in interval notation.

33. $x + 2 > 5$

34. $x + 5 \geq 2$

35. $-x - 3 \leq 7$

36. $-x - 9 > 3$

37. $3 + x < 2$

38. $5 + x \geq 3$

39. $2x - 0.3 \leq 0.5$

40. $-3x - 0.5 < 0.4$

41. $-3x - 7 > -1$

42. $-5x + 7 \leq 12$

43. $-4x + 6 > 17$

44. $7x - 1 > 5$

45. $\frac{y}{4} + 1 \leq -9$

46. $\frac{r}{8} - 7 \geq -8$

47. $-\frac{1}{2}n \geq -1$

48. $-\frac{1}{3}t \leq -3$

49. $\frac{x}{-42} - 1 > -1$

50. $\frac{a}{-25} + 3 < 3$

51. $\frac{2}{3}x \geq 2$

52. $\frac{3}{4}x < 3$

53. $-\frac{7}{8}x \leq 21$

54. $-\frac{3}{16}x \geq -9$

55. $2x + 9 \leq x + 8$

56. $3x + 7 \leq 4x - 2$

57. $9x + 13 \geq 8x$

58. $7x - 16 < 6x$

59. $8x + 4 > 3x + 4$

60. $7x + 6 \geq 4x + 6$

61. $5x + 7 < 2x + 1$

62. $7x + 2 \geq 4x - 1$

63. $7 - x \leq 3x - 2$

64. $9 - 3x \geq 6 + x$

65. $3(x - 8) < 5x + 6$

66. $9(x - 11) > 13 + 7x$

67. $8(5 - x) \leq 10(8 - x)$

68. $17(3 - x) \geq 3 - 13x$

69. $\frac{1}{2} + \frac{x}{5} > \frac{3}{4}$

70. $\frac{1}{3} + \frac{c}{5} > -\frac{3}{2}$

71. $-\frac{2}{3} \geq \frac{2y}{3} - \frac{3}{4}$

72. $-\frac{2}{9} \geq \frac{5x}{6} - \frac{1}{3}$

Solve each inequality, graph the solution, and write the solution in interval notation.

73. $2 < x - 5 < 5$

74. $3 < x - 2 < 7$

75. $-5 < x + 4 \leq 7$

76. $-9 \leq x + 8 < 1$

77. $0 \leq x + 10 \leq 10$

78. $-8 < x - 8 < 8$

79. $4 < -2x < 10$

80. $-4 \leq -4x < 12$

81. $-3 \leq \frac{x}{2} \leq 5$

82. $-12 < \frac{x}{3} < 0$

83. $3 \leq 2x - 1 < 5$

84. $4 < 3x - 5 \leq 7$

85. $0 < 10 - 5x \leq 15$

86. $1 \leq -7x + 8 \leq 15$

Solve each inequality.

87. $0.6(0.5x - 2.94) < -1.353$

88. $-0.7688 \leq \frac{m}{3.5} - 0.1988$

89. $9(0.05 - 0.3x) + 0.162 \leq 0.081 + 15x$

90. $-1{,}630 \leq \dfrac{b + 312{,}451}{47} < 42{,}616$

APPLICATIONS

91. CALCULATING GRADES A student has test scores of 68%, 75%, and 79% in a government class. What must she score on the last exam to earn a B (80% or better) in the course?

92. OCCUPATIONAL TESTING Before taking on a client, an employment agency requires the applicant to average at least 70% on a battery of four job skills tests. If an applicant scored 70%, 74%, and 84% on the first three exams, what must he score on the fourth test to maintain a 70% or better average?

93. FLEET AVERAGES A car manufacturer produces three models in equal quantities. One model has an economy rating of 17 miles per gallon, and the second model is rated for 19 mpg. If government regulations require the manufacturer to have a fleet average of at least 21 mpg, what economy rating is required for the third model?

94. SERVICE CHARGES When the average daily balance of a customer's checking account falls below $500 in any week, the bank assesses a $5 service charge. The following table shows the daily balances of one customer. What must Friday's balance be to avoid the service charge?

Day	Balance
Monday	$540.00
Tuesday	$435.50
Wednesday	$345.30
Thursday	$310.00

95. HOMEWORK A Spanish teacher requires that students devote no less than 1 hour a day to their homework assignments. Write an inequality that describes the number of minutes m a student should spend each week on Spanish homework.

96. CHILD LABOR A child labor law reads, "The number of hours a full-time student under 16 years of age can work on a weekday shall not exceed 4 hours." Write an inequality that describes the number of hours h such a student can work Monday through Friday.

97. SAFETY CODES The illustration shows the acceptable and preferred angles of pitch or slope for ladders, stairs, and ramps. Use a compound inequality to describe each safe-angle range.

a. Ramps or inclines

b. Stairs

c. Preferred range for stairs

d. Ladders with cleats

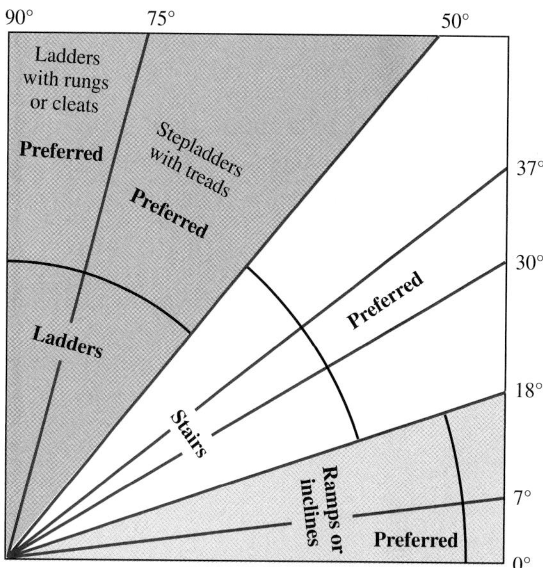

98. WEIGHT CHARTS The following illustration is used to classify the weight of a baby boy from birth to 1 year. Estimate the weight range w for boys in the following classifications, using a compound inequality:

a. 10 months old, "heavy"

b. 5 months old, "light"

c. 8 months old, "average"

d. 3 months old, "moderately light"

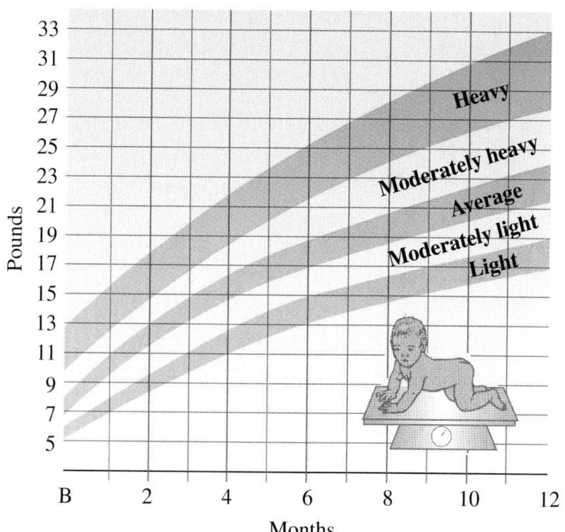

Based on data from *Better Homes and Gardens Baby Book* (Meredith Corp., 1969)

99. LAND ELEVATIONS The land elevations in Nevada range from the 13,143-foot height of Boundary Peak to the Colorado River at 470 feet. Use a compound inequality to express the range of these elevations.

a. In feet

b. In miles (round to the nearest tenth) (*Hint:* 1 mile is 5,280 feet.)

100. COMPARING TEMPERATURES To hold the temperature of a room between 19° C and 22° C, what Fahrenheit temperatures must be maintained? (*Hint:* Fahrenheit temperature F and Celsius temperature C are related by the formula $F = \frac{9C + 160}{5}$.)

101. DRAFTING In the following illustration, the $\pm$ (read "plus or minus") symbol means that the width of a plug a manufacturer produces can range from $1.497 - 0.001$ inches to $1.497 + 0.001$ inches. Write the range of acceptable widths w for the plug and the opening it fits into using compound inequalities.

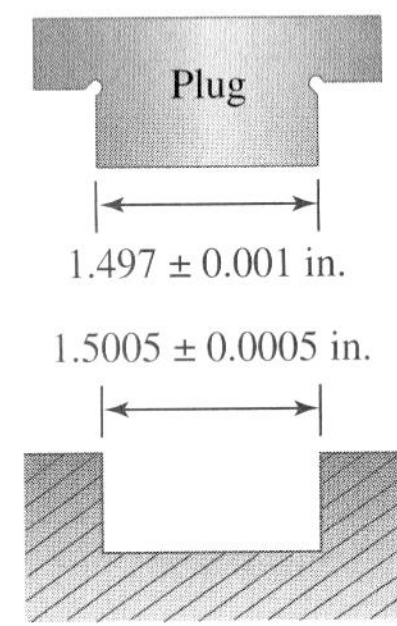

102. COUNTER SPACE In a large discount store, a rectangular counter is being built for the customer service department. If designers have determined that the outside perimeter of the counter (shown in red) needs to be at least 150 feet, use the plan in the illustration to determine the acceptable values for x.

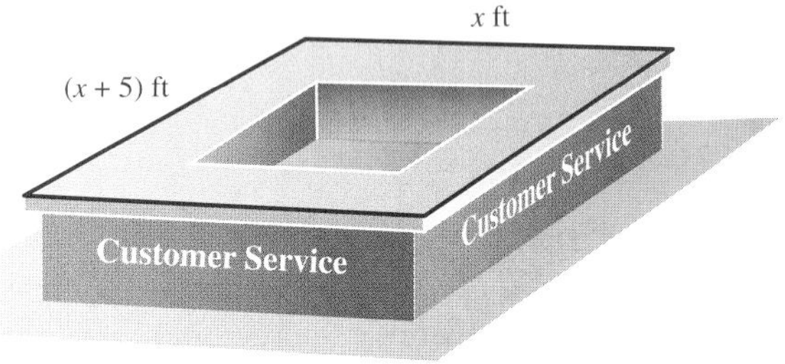

WRITING

103. Explain why multiplying both sides of an inequality by a negative number reverses the direction of the inequality.

104. Explain the use of parentheses and brackets for graphing intervals.

REVIEW *Find each power.*

105. -5^3

106. $(-3)^4$

Complete each input/output table.

107.

x	$x^2 - 3$
-2	
0	
3	

108.

x	$\frac{x}{3} + 2$
-6	
0	
12	

KEY CONCEPT

Simplify and Solve

Two of the most often used instructions in this book are **simplify** and **solve.** In algebra, we *simplify expressions* and we *solve equations and inequalities.*

To simplify an expression, we write it in a less complicated form. To do so, we apply the rules of arithmetic as well as algebraic concepts such as combining like terms, the distributive property, and the properties of 0 and 1.

To solve an equation or an inequality means to find the numbers that make the equation or inequality true, when substituted for its variable. We use the addition, subtraction, multiplication, and division properties of equality or inequality to solve equations and inequalities. Quite often, we must simplify expressions on the left- or right-hand sides of an equation or inequality when solving it.

Use the procedures and the properties that we have studied to simplify the expression in part a and to solve the equation or inequality in part b.

Simplify:	**Solve:**
1. a. $-3x + 2 + 5x - 10$	**b.** $-3x + 2 + 5x - 10 = 4$
2. a. $4(y + 2) - 3(y + 1)$	**b.** $4(y + 2) = 3(y + 1)$
3. a. $\frac{1}{3}a + \frac{1}{3}a$	**b.** $\frac{1}{3}a + \frac{1}{3} = \frac{1}{2}$
4. a. $-(2x + 10)$	**b.** $-2x \geq -10$

5. In the student's work on the right, where was the mistake made? Explain what the student did wrong.

Simplify: $2(x + 3) - x - 12$.

$$\begin{aligned} 2(x + 3) - x - 12 &= 2x + 6 - x - 12 \\ &= x - 6 \\ 0 &= x - 6 \\ 0 + 6 &= x - 6 + 6 \\ 6 &= x \end{aligned}$$

ACCENT ON TEAMWORK

SECTION 8.1

EVALUATING ALGEBRAIC EXPRESSIONS Find five examples of cylinders. Measure and record the diameter d of their bases and their heights h. Express the measurements as decimals. Find the radius r of each base by dividing the diameter by 2. Then find each volume by evaluating the expression $\pi r^2 h$. Round to the nearest tenth of a cubic unit. Present your results in a table of the form shown in the illustration.

Cylinder	d	r	h	Volume
Container of salt	$3\frac{1}{4}$ in. (3.25 in.)	$1\frac{5}{8}$ in. (1.625 in.)	$5\frac{3}{8}$ in. (5.375 in.)	44.6 in.^{3}

SECTION 8.2

THE DISTRIBUTIVE PROPERTY Draw a geometric model on the graph paper below that illustrates why

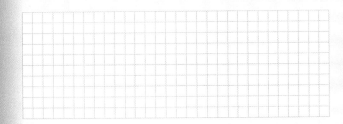

$$5(4 + 2) = 5(4) + 5(2)$$

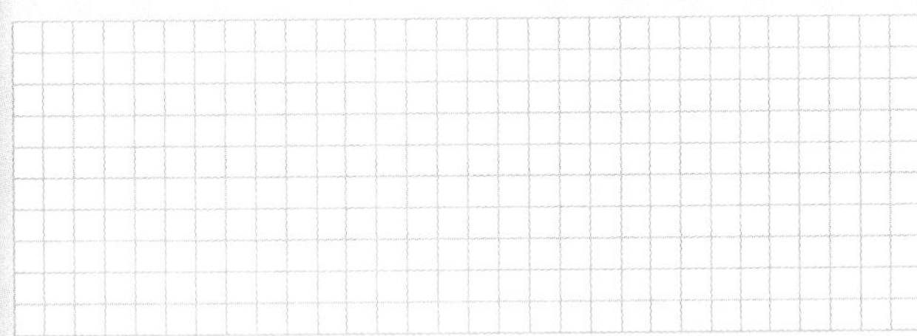

SECTION 8.3

SOLVING EQUATIONS Make a presentation to the class explaining how we undo operations to isolate the variable when solving an equation. As a visual aid, bring in a box, tied shut with string, that contains a toy wrapped in tissue paper. Compare the three-step process a person would use to get to the toy inside the box to the three-step process we could use to solve the equation $\frac{2x}{3} - 4 = 2$.

SOLVING EQUATIONS A scale can be used to illustrate the steps used to solve an equation.

a. What equation is being solved in the illustration? What is the solution?

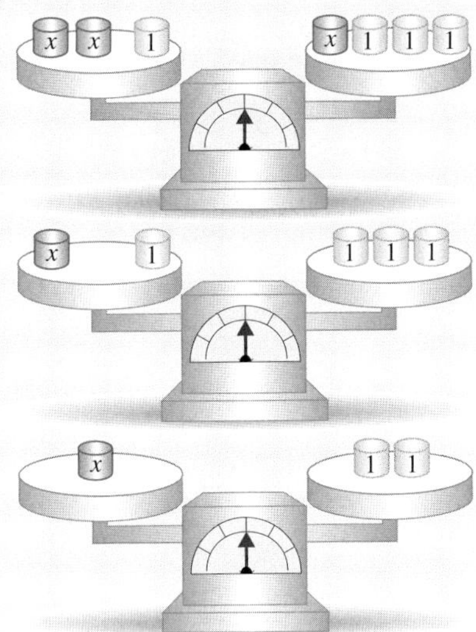

b. Draw a similar series of pictures showing the solution of each of the following equations.

1. $3x + 1 = 2x + 4$ **2.** $2x + 6 = 4x + 2$

SECTION 8.4

GEOMETRY GOURMET Find snack foods that have the shapes of the geometric figures in Table 8-4 and Table 8-5. For example, tortilla chips can be triangular in shape, and malted milk balls are spheres. If you are unable to find a particular shape already available, decide on a way to make a snack in that shape. Make up a tray of snacks to bring to class. Discuss the various shapes of the snacks as you enjoy the food in your groups.

SECTION 8.5

MIXTURES Get several cans of orange juice concentrate and make four pitchers that are 10%, 30%, 50%, and 70% solutions. For example, a 30% solution would consist of three paper cups of concentrate and seven paper cups of water. Pour amounts of each mixture into cups. Have students taste each solution and see whether they can put the mixtures in order from least concentrated to most concentrated.

SECTION 8.6

INEQUALITIES In most states, a person must be at least 16 years old to have a driver's license. We can describe this situation with the inequality $a \geq 16$, where a represents a person's age in years. Think of other situations that can be described using an inequality or a compound inequality.

CHAPTER REVIEW

SECTION 8.1 Algebraic Expressions

CONCEPTS

In order to describe numerical relationships, we need to translate the words of a problem into mathematical symbols.

REVIEW EXERCISES

Write each phrase as an algebraic expression.

1. 25 more than the height h

2. 15 less than the cutoff score s

3. $\frac{1}{2}$ of the time t

4. the product of 6 and x

See the illustration on the right.

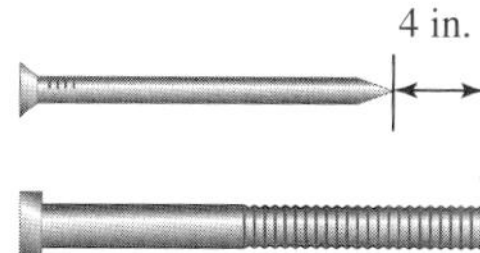

5. If we let n = the length of the nail in inches, write an algebraic expression for the length of the bolt (in inches).

6. If we let b = the length of the bolt in inches, write an algebraic expression for the length of the nail (in inches).

Sometimes we must rely on common sense and insight to find *hidden operations.*

7. How many years are in d decades?

8. If you have x donuts, how many dozen donuts do you have?

9. Five years after a house was constructed, a patio was added. How old, in years, is the patio if the house is x years old?

Number · value = total value

10. Complete the table.

Type of coin	Number	Value (¢)	Total value (¢)
Nickel	6		
Dime	d		

When we replace the variable, or variables, in an algebraic expression with specific numbers and then apply the rules for the order of operations, we are *evaluating* the algebraic expression.

11. Complete the table.

x	$20x - x^3$
0	
1	
−4	

Evaluate each algebraic expression for the given value(s) of the variable(s).

12. $7x^2 - \frac{x}{2}$ for $x = 4$

13. $b^2 - 4ac$ for $b = -10$, $a = 3$, and $c = 5$

14. $2(24 - 2c)^3$ for $c = 9$

15. $\frac{x + y}{-x - z}$ for $x = 19$, $y = 17$, and $z = -18$

16. Use a calculator to find the volume, to the nearest tenth of a cubic inch, of the ice cream waffle cone in the illustration by evaluating the algebraic expression

$$\frac{\pi r^2 h}{3}$$

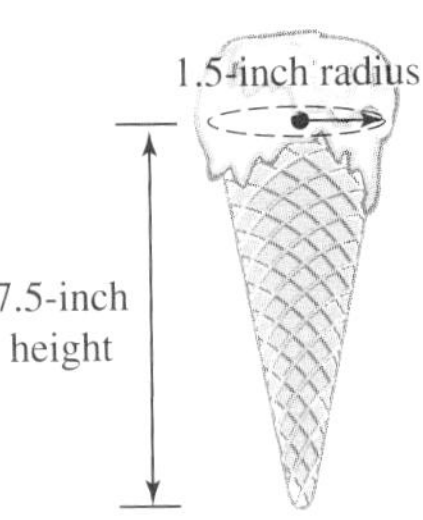

SECTION 8.2 Simplifying Algebraic Expressions

To *simplify* an algebraic expression means to write it in less complicated form.

Simplify each expression.

17. $-4(7w)$

18. $-3r(-5r)$

19. $3(-2x)(-4y)$

20. $0.4(5.2f)$

The *distributive property:*

$a(b + c) = ab + ac$

$a(b - c) = ab - ac$

Write each expression without parentheses.

21. $5(x + 3)$

22. $-2(2x + 3 - y)$

23. $-(a - 4)$

24. $\frac{3}{4}(4c - 8)$

A *term* is a number or a product of a number and one or more variables. Addition signs separate algebraic expressions into terms.

How many terms are in each expression?

25. $3x^2 + 2x - 5$

26. $-12xyz$

In a term, the numerical factor is called the *coefficient.*

Identify the coefficient of each term.

27. $2x - 5$

28. $16x^2 - 5x + 25$

29. $\frac{1}{2}x + y$

30. $9.6t^2 - t$

♡Michelle♡

Like terms are terms with exactly the same variables raised to exactly the same powers.

Simplify each expression by combining like terms.

31. $8p + 5p - 4p$

32. $-5m + 2n - 2m - 2n$

33. $6a + 2b - 8a - 12b$

34. $5(p - 2) - 2(3p + 4)$

35. $x^2 - x(x - 1)$

36. $8a^3 + 4a^3 - 20a^3$

37. Write an algebraic expression in simplified form for the perimeter of the triangle in the illustration.

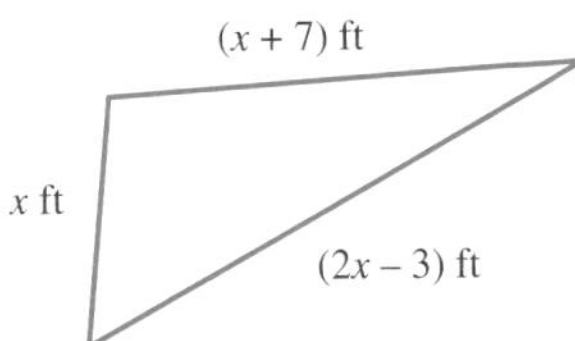

SECTION 8.3 *Solving Equations*

To solve an equation means to find all the values of the variable that, when substituted for the variable, make a true statement.

An equation that is true for all values of its variable is called an *identity.*

An equation that is not true for any values of its variable is called a *contradiction.*

Solve each equation.

38. $5x + 4 = 14$

39. $-1.2y + 0.8 = 2.0$

40. $\frac{n}{5} - 2 = 4$

41. $\frac{b - 5}{4} = -6$

42. $5(2x - 4) - 5x = 0$

43. $-2(x - 5) = 5(-3x + 4) + 3$

44. $\frac{3}{4} = \frac{1}{2} + \frac{d}{5}$

45. $-\frac{2}{3}f = 4$

46. $3(a + 8) = 6(a + 4) - 3a$

47. $2(y + 10) + y = 3(y + 8)$

SECTION 8.4 *Formulas*

A *formula* is an equation that is used to state a known relationship between two or more variables.

Retail price: $r = c + m$
Profit: $p = r - c$
Distance: $d = rt$

Temperature: $C = \frac{5(F - 32)}{9}$

Formulas from geometry:
Square: $P = 4s$, $A = s^2$
Rectangle: $P = 2l + 2w$, $A = lw$

48. Find the markup on a CD player whose wholesale cost is \$219 and whose retail price is \$395.

49. One month, a restaurant had sales of \$13,500 and made a profit of \$1,700. Find the expenses for the month.

50. INDY 500 In 1996, the winner of the Indianapolis 500-mile automobile race averaged 147.956 mph. To the nearest hundredth of an hour, how long did it take him to complete the race?

51. JEWELRY MAKING Gold melts at about 1,065° C. Change this to degrees Fahrenheit.

52. CAMPING Find the perimeter of the air mattress shown below.

53. CAMPING Find the amount of sleeping area on the top surface of the air mattress shown below.

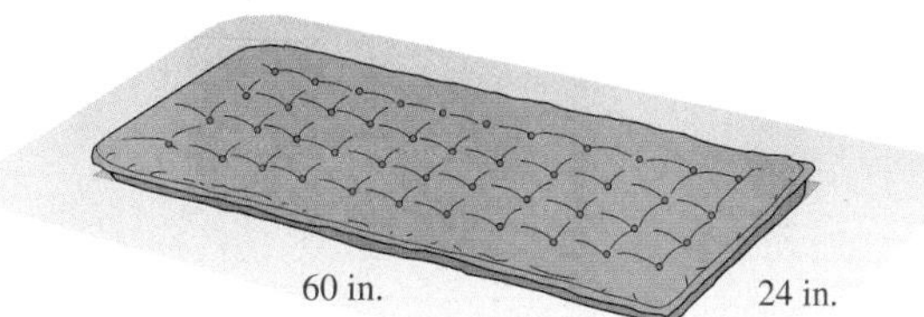

Triangle: $P = a + b + c$
$A = \frac{1}{2}bh$

Trapezoid: $P = a + b + c + d$
$A = \frac{1}{2}h(b + d)$

Circle: $D = 2r$
$C = 2\pi r$
$A = \pi r^2$

Rectangular solid: $V = lwh$

Cylinder: $V = \pi r^2 h$

54. Find the area of a triangle with a base 17 meters long and a height of 9 meters.

55. Find the area of a trapezoid with bases 11 inches and 13 inches long and a height of 12 inches.

56. To the nearest hundredth, find the circumference of a circle with a radius of 8 centimeters.

57. To the nearest hundredth, find the area of the circle in Exercise 70.

58. CAMPING Find the approximate volume of the air mattress in the above illustration if it is 3 inches thick.

59. Find the volume of a 12-foot cylinder whose circular base has a radius of 0.5 feet. Give the result to the nearest tenth.

Pyramid: $V = \frac{1}{3}Bh$

Cone: $V = \frac{1}{3}\pi r^2 h$

Sphere: $V = \frac{4}{3}\pi r^3$

60. Find the volume of a pyramid that has a square base, measuring 6 feet on a side, and a height of 10 feet.

61. HALLOWEEN After being cleaned out, a spherical-shaped pumpkin has an inside diameter of 9 inches. To the nearest hundredth, what is its volume?

Solve each formula for the required variable.

62. $A = 2\pi rh$ for h

63. $P = 2l + 2w$ for l

SECTION 8.5 Problem Solving

To solve problems, use the five-step problem-solving strategy.

1. Analyze the problem.
2. Form an equation.
3. Solve the equation.
4. State the conclusion.
5. Check the result.

64. SOUND SYSTEMS A 45-foot-long speaker wire is to be cut into three pieces. One piece is to be 15 feet long. Of the remaining pieces, one must be 2 feet less than 3 times the length of the other. Find the length of the shorter piece of wire.

65. UTILITY BILLS The electric company charges \$17.50 per month plus 18 cents for every kilowatt hour of energy used. One resident's bill was \$43.96. How many kilowatt hours were used that month?

66. ART HISTORY *American Gothic* was painted in 1930 by American artist Grant Wood. The length of the rectangular painting is 5 inches more than the width. Find the dimensions of the painting if it has a perimeter of $109\frac{1}{2}$ inches.

The sum of the measures of the angles of a triangle is 180°.

67. Find the missing angle measures of the triangle in the illustration.

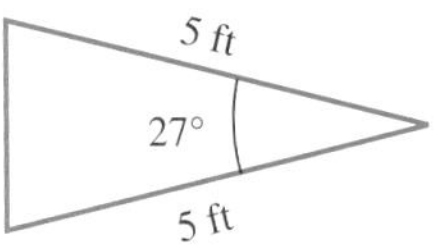

Total value = number · value

Interest = principal · rate · time

$I = Prt$

68. What is the value of x video games each costing \$45?

69. INVESTMENT INCOME A woman has \$27,000. Part is invested for one year in a certificate of deposit paying 7% interest, and the remaining amount in a cash management fund paying 9%. After 1 year, the total interest on the two investments is \$2,110. How much is invested at each rate?

Distance = rate · time

$d = rt$

70. WALKING AND BICYCLING A bicycle path is 5 miles long. A man walks from one end at the rate of 3 mph. At the same time, a friend bicycles from the other end, traveling at 12 mph. In how many minutes will they meet?

The value v of a commodity is its price per pound p times the number of pounds n:

$v = pn$

71. MIXTURES A store manager mixes candy worth 90¢ per pound with gumdrops worth \$1.50 per pound to make 20 pounds of a mixture worth \$1.20 per pound. How many pounds of each kind of candy does he use?

72. SOLUTIONS How much acetic acid is in x gallons of a solution that is 12% acetic acid?

SECTION 8.6 *Inequalities*

An *inequality* is a mathematical expression that contains a $>, <, \geq, \leq,$ or $\neq$ symbol.

A *solution of an inequality* is any number that makes the inequality true.

A *parenthesis* indicates that a number is not on the graph. A *bracket* indicates that a number is included in the graph.

Interval notation can be used to describe a set of real numbers.

Solve each inequality, graph the solution, and use interval notation to write the solution.

73. $3x + 2 < 5$

74. $-5x - 8 > 7$

75. $5x - 3 \geq 2x + 9$

76. $7x + 1 \leq 8x - 5$

77. $5(3 - x) \leq 3(x - 3)$

78. $-\frac{3}{4}x \geq -9$

79. $8 < x + 2 < 13$

80. $0 \leq 2 - 2x < 6$

81. Graph the interval represented by $[-13, \infty)$.

82. SPORTS EQUIPMENT The acceptable weight of Ping-Pong balls used in competition can range from 2.40 to 2.53 grams. Express this range using a compound inequality.

83. SIGNS A large office complex has a strict policy about signs. Any sign to be posted in the building must meet three requirements:

- It must be rectangular in shape.
- Its width must be 18 inches.
- Its perimeter is not to exceed 132 inches.

What possible sign lengths meet these specifications?

CHAPTER 8 TEST

1. A rock band recorded x songs for a CD. Technicians had to delete two songs from the album because of poor sound quality. Express the number of songs on the CD using an algebraic expression.

2. What is the value of q quarters in cents?

3. Complete the table.

x	$2x - \frac{30}{x}$
5	
10	
30	

4. Evaluate $2lw + w^2$ for $l = 4$ and $w = 8$.

5. How many terms are in the expression $4x^2 + 5x - 7$? What is the coefficient of the second term?

6. What property is illustrated below?

$$2(x + 7) = 2x + 2(7)$$

Simplify each expression.

7. $5(-4x)$

8. $-8(-7t)(4t)$

9. $3(x + 2) + 3(4 - x)$

10. $-1.1d^2 - 3.8d^2$

Solve each equation.

11. $12x = -144$

12. $\frac{4}{5}t = -4$

13. $\frac{c}{7} + 6 = -1$

14. $0.3x = 0.5 - 0.2x$

15. $\frac{m}{2} - \frac{1}{3} = \frac{1}{4}$

16. $23 - 5(x + 10) = -12$

17. Solve the equation for the variable indicated.

$$A = P + Prt; \text{ for } r$$

18. On its first night of business, a pizza parlor brought in \$445. The owner estimated his costs that night to be \$295. What was the profit?

19. Find the Celsius temperature reading if the Fahrenheit reading is 14°.

20. PETS The spherical fishbowl shown in the illustration is three-quarters full of water. To the nearest cubic inch, what is the volume of water in the bowl?

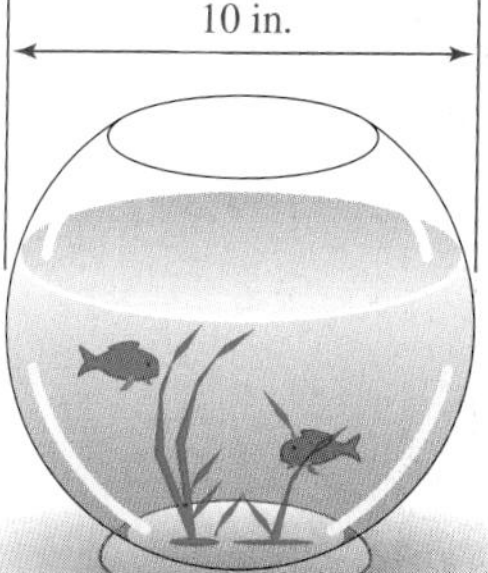

21. TRAVEL TIMES A car leaves Rockford, Illinois, at the rate of 65 mph, bound for Madison, Wisconsin. At the same time, a truck leaves Madison at the rate of 55 mph, bound for Rockford. If the cities are 72 miles apart, how long will it take for the car and the truck to meet?

22. SALT SOLUTIONS How many liters of a 2% brine solution must be added to 30 liters of a 10% brine solution to dilute it to an 8% solution?

23. GEOMETRY If the vertex angle of an isosceles triangle is 44°, find the measure of each base angle.

24. INVESTMENTS Part of $13,750 is invested at 9% annual interest, and the rest is invested at 8%. After one year, the accounts paid $1,185 in interest. How much was invested at the lower rate?

Solve each inequality, graph the solution, and use interval notation to write the solution.

25. $-8x - 20 \leq 4$

26. $-4 \leq 2(x + 1) < 10$

27. After we have solved an equation, how can we check the answer to be sure that it is a solution?

28. What are like terms? Give an example.

CHAPTERS 1–8 CUMULATIVE REVIEW EXERCISES

1. Classify each of the following as an equation or an expression.
   a. $4m - 3 + 2m$ b. $4m = 3 + 2m$
2. Use the formula $t = \frac{w}{5}$ to complete the table.

Weight (lb)	Cooking time (hr)
15	
20	
25	

3. Give the prime factorization of 100.
4. Simplify: $\frac{24}{36}$.
5. Multiply: $\frac{11}{21}\left(-\frac{14}{33}\right)$.
6. COOKING A recipe calls for $\frac{3}{4}$ cup of flour, and the only measuring container you have holds $\frac{1}{8}$ of a cup. How many $\frac{1}{8}$ cups of flour would you need to add to follow the recipe?
7. Add: $\frac{4}{5} + \frac{2}{3}$.
8. Subtract: $42\frac{1}{8} - 29\frac{2}{3}$.
9. Write $\frac{15}{16}$ as a decimal.
10. Multiply: 0.45(100).
11. Evaluate each expression.
    a. $|-65|$ b. $-|-12|$
12. What property of real numbers is illustrated below?
    $x \cdot 5 = 5x$

Classify each number as a natural number, a whole number, an integer, a rational number, an irrational number, and a real number. Each number may have several classifications.

13. 3
14. -1.95
15. $\frac{17}{20}$
16. π
17. Write each product using exponents.
    a. $4 \cdot 4 \cdot 4$ b. $\pi \cdot r \cdot r \cdot h$
18. Perform each operation.
    a. $-6 + (-12) + 8$
    b. $-15 - (-1)$
    c. $2(-32)$
    d. $\frac{0}{35}$
19. Write each phrase as an algebraic expression.
    a. The sum of the width w and 12.
    b. Four less than a number n.
20. SICK DAYS Use the data in the table to find the average (mean) number of sick days used by this group of employees this year.

Name	Sick days	Name	Sick days
Chung	4	Ryba	0
Cruz	8	Nguyen	5
Damron	3	Tomaka	4
Hammond	2	Young	6

21. Complete the table.

x	$x^2 - 3$
−2	
0	
3	

22. Translate to mathematical symbols.

The loudness of a stereo speaker	is	2,000	divided by	the square of the distance of the listener from the speaker

23. LAND OF THE RISING SUN The flag of Japan is a red disc (representing sincerity and passion) on a white background (representing honesty and purity). (See the next page.)
    a. What is the area of the rectangular-shaped flag?
    b. To the nearest tenth of a square foot, what is the area of the red disc?

c. Use the results from parts a and b to find what percent of the area of the Japanese flag is occupied by the red disc.

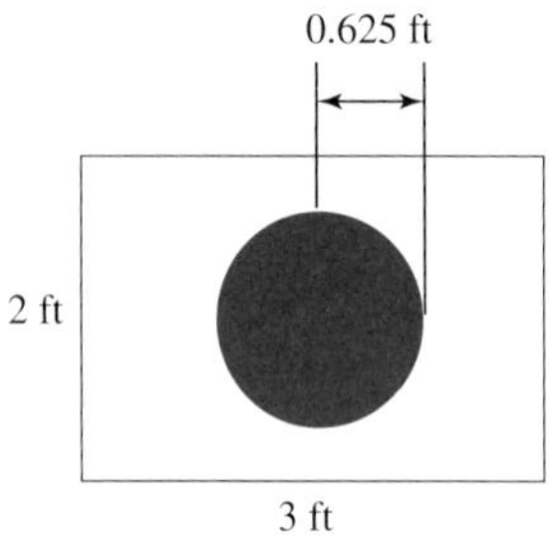

24. 45 is 15% of what number?

Let $x = -5$, $y = 3$, and $z = 0$. Evaluate each expression.

25. $(3x - 2y)z$

26. $\dfrac{x - 3y + |z|}{2 - x}$

27. $x^2 - y^2 + z^2$

28. $\dfrac{x}{y} + \dfrac{y + 2}{3 - z}$

Simplify each expression.

29. $-8(4d)$

30. $5(2x - 3y + 1)$

31. $2x + 3x$

32. $3a + 6a - 17a$

33. $q(q - 5) + 7q^2$

34. $5(t - 4) + 3t$

35. What is the length of the longest side of the triangle in the following illustration?

36. Write an algebraic expression in simplest form for the perimeter of the triangle in the following illustration.

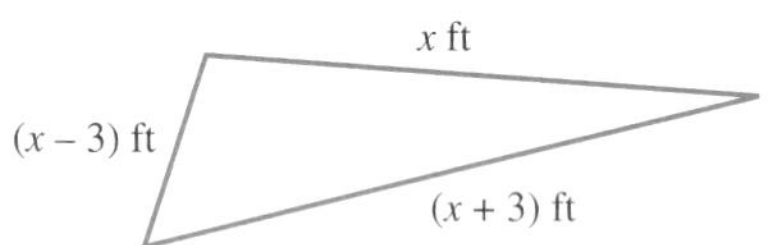

Solve each equation.

37. $3x - 4 = 23$

38. $\dfrac{x}{5} + 3 = 7$

39. $-5p + 0.7 = 3.7$

40. $\dfrac{y - 4}{5} = 3$

41. $-\dfrac{4}{5}x = 16$

42. $-9(n + 2) - 2(n - 3) = 10$

43. $9y - 3 = 6y$

44. $\dfrac{1}{2} + \dfrac{x}{5} = \dfrac{3}{4}$

45. Find the area of a rectangle with sides of 5 meters and 13 meters.

46. Find the volume of a cone that is 10 centimeters tall and has a circular base whose diameter is 12 centimeters. Round to the nearest hundredth.

47. Solve $A = P + Prt$ for t.

48. WORK Physicists say that *work* is done when an object is moved a distance d by a force F. To find the work done, we can use the formula $W = Fd$. Find the work done in lifting the bundle of newspapers shown in the illustration onto the workbench. (*Hint:* The force that must be applied to lift the newspapers is equal to the weight of the newspapers.)

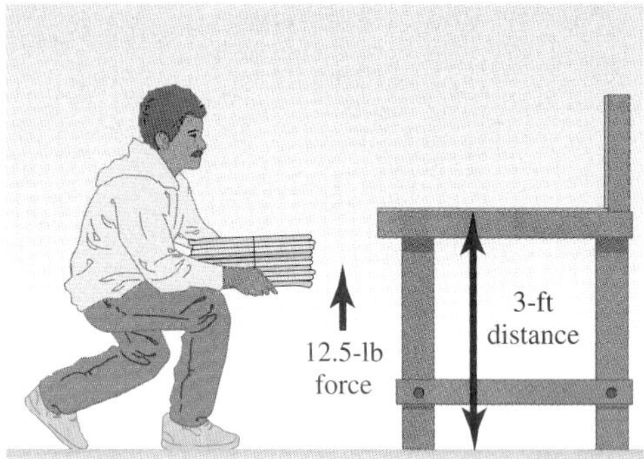

49. WORK See Exercise 48. Find the weight of a 1-gallon can of paint if the amount of work done to lift it onto the workbench is 28.35 foot-pounds.

50. Find the unknown angle measure represented by x.

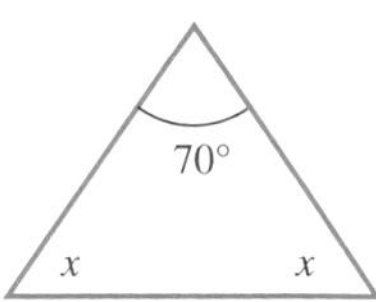

51. INVESTING An investment club invested part of $10,000 at 9% annual interest and the rest at 8%. If the annual income from these investments was $860, how much was invested at 8%?

52. GOLDSMITHING How many ounces of a 40% gold alloy must be mixed with 10 ounces of a 10% gold alloy to obtain an alloy that is 25% gold?

Solve each inequality, graph the solution, and use interval notation to describe the solution.

53. $x - 4 > -6$

54. $-6x \geq -12$

55. $8x + 4 \geq 5x + 1$

56. $-1 \leq 2x + 1 < 5$

CHAPTER 9

Graphs, Linear Equations, and Functions

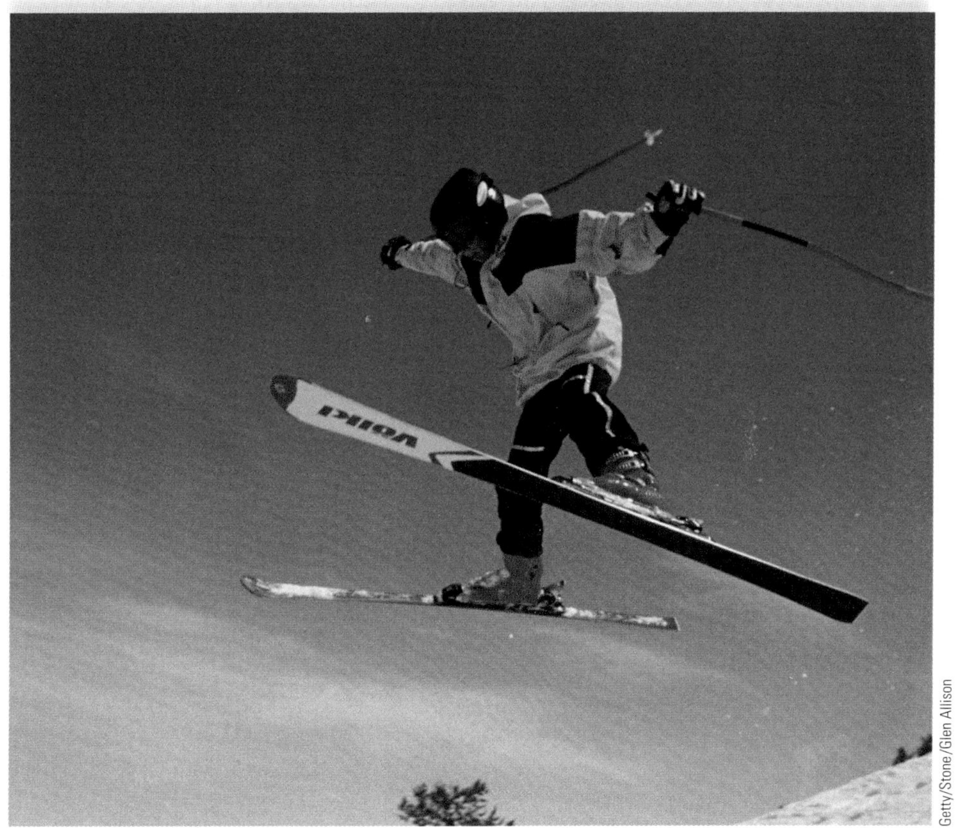

Getty/Stone/Glen Allison

Snow skiing is one of our country's most popular recreational activities. Whether on the gentle incline of a cross-country trip or racing down a near-vertical mountainside, a skier constantly adapts to the steepness of the course. In this chapter, we discuss lines and a means of measuring their steepness, called *slope.* The concept of slope has a wide variety of applications, including roofing, road design, and handicap accessible ramps.

To learn more about the slope of a line, visit *The Learning Equation* on the Internet at http://tle.brookscole.com. (Log-in instructions are in the Preface.) For Chapter 9, the online lessons are:

- *TLE* Lesson 17: Equations Containing Two Variables
- *TLE* Lesson 18: Rate of Change and the Slope of a Line

Check Your Knowledge

1. An ordered pair is a ________ of an equation in two variables if the numbers in the ordered pair satisfy the equation.
2. The point represented by the ordered pair (0, 0) is call the ________.
3. The rate of change of a linear relationship can be found by finding the ______ of the graph of the line.
4. Two different lines that have the same slope are ________.
5. The set of all possible values of the independent variable of a function is called the ________ of the function, and the set of all values of the dependent variable is called the _______ of the function.
6. The graph shows the temperature measured each hour from midnight to 7:00 A.M. in degrees Celsius.
   a. What is the lowest temperature measured?
   b. What was the temperature at 2:00 A.M.?
   c. What was the change in temperature from midnight to 3:00 A.M.?
   d. What was the rate of change in temperature from 5:00 to 6:00 A.M.?
7. Complete the table of solutions for $2x - y = 3$. Then graph the equation.

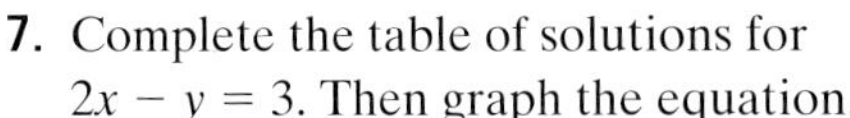

$2x - y = 3$	
x	y
0	
	0
2	

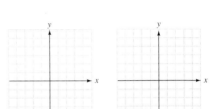

8. Find the slope of the line through the points (2, 3) and (−1, −2).
9. Consider the graph of $3x + y = 1$.
   a. Find the x- and y-intercepts.
   b. Find the slope.
10. Consider $y = x^2 + 1$.
    a. Graph the equation.
    b. Find the y-intercept.
    c. Is the equation linear?
11. Consider $y = 2$.
    a. Graph the equation.
    b. Find the slope.
    c. Find the y-intercept.
12. Graph the line passing through (−2, 1) with slope −1.
13. What is the slope of a horizontal line?
14. What is the slope of a line perpendicular to $3x + y = 1$?
15. A collect telephone call costs $0.50 plus $0.10 per minute. Write a linear equation that gives the cost c of a call lasting m minutes.
16. Determine whether each of the following is the graph of a function.

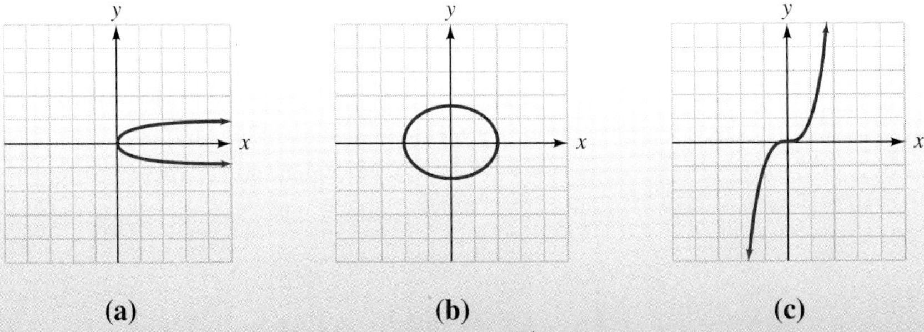

17. Write an equation of the line passing through (0, −3) and (2, 2) in slope–intercept form.
18. If $f(x) = -x^2$, find $f(-2)$.

Study Skills Workshop

MAKING CONNECTIONS

Getty Images

It's often hard to get through new course material completely on your own. It is normal to require help in learning new mathematical concepts, particularly if math is not a strong subject for you. Here are a few suggestions for seeking help.

Instructor Office Hours. Most full-time and some part-time instructors keep regular office hours. The purpose of this time is for instructors to be available to answer students' questions on a drop-in basis. Make sure that you are aware of your instructor's office hours, particularly those hours that are close to times when homework assignments are due. When you visit your instructor during office hours, be as well prepared as possible. It is generally easier for your instructor to help you if you come with a list of identified problems and can point to exactly where in the process you had trouble.

Tutorial Services. Many colleges offer tutorial services for free. Sometimes tutorial assistance is available in a lab setting where you are able to drop-in at your convenience. In some cases, you need to make an appointment to see a tutor in advance. Make sure to seek help as soon as you recognize the need and come to see your tutor with a list of identified problems.

Study Groups. Many students benefit greatly from taking part in a study group composed of classmates. Study groups can be extremely helpful. In order to get the most out of a study group there are a few guidelines that should be followed.

- Keep the group small — a maximum of four students. Make sure that everyone in the group is committed. You may want to schedule days and times to meet.
- Find a place to meet where you can talk and where you have space to spread out; quiet areas in a library will not work well. Possible meeting places are in math labs (if available at your school), designated study areas at your college, coffee shops, or around your kitchen table. If you have a limited amount of time and are comfortable with technology, you may want to use email, chat rooms, or bulletin boards. Sometimes just having a phone number so that you can call someone is helpful when you get stuck on a problem.
- Do not plan to start and finish your whole homework assignment in the study group. You should agree to have at least *attempted* each problem in the assignment before meeting as a group.
- No one person should be responsible for doing all of the work — each person in the group should contribute. The best study groups are those in which the members have complementary abilities.
- In your study group, practice verbalizing problems and explanations. Being able to formulate a question sometimes is helpful in understanding your problem. Being able to explain a solution is the best way to learn a concept.

ASSIGNMENT

1. Find three other students in your class who have a study time compatible with yours.
2. Establish a time to meet together as a group (two days before your next homework assignment is due would be ideal).
3. Before meeting with the group, attempt all of your homework problems. Make a list of all questions that you encountered with homework.
4. Meet with your group and go through everyone's list.
5. After your meeting, evaluate how well the group worked. Share your thoughts with other group members and decide when you will meet next.

Relationships between two quantities can be described by a table, a graph, or an equation.

9.1 Graphing Using the Rectangular Coordinate System

- The rectangular coordinate system • Graphing mathematical relationships
- Reading graphs • Step graphs • The midpoint formula

It is often said, "A picture is worth a thousand words." In this section, we will show how numerical relationships can be described using mathematical pictures called **graphs.** We will also show how graphs are constructed and how we can obtain information from them.

The rectangular coordinate system

When designing the Gateway Arch in St. Louis, shown in Figure 9-1(a), architects created a mathematical model called a **rectangular coordinate graph.** This graph, shown in Figure 9-1(b), is drawn on a grid called a **rectangular coordinate system.** This coordinate system is sometimes called a **Cartesian coordinate system,** after the 17th-century French mathematician René Descartes.

GETTY

(a)

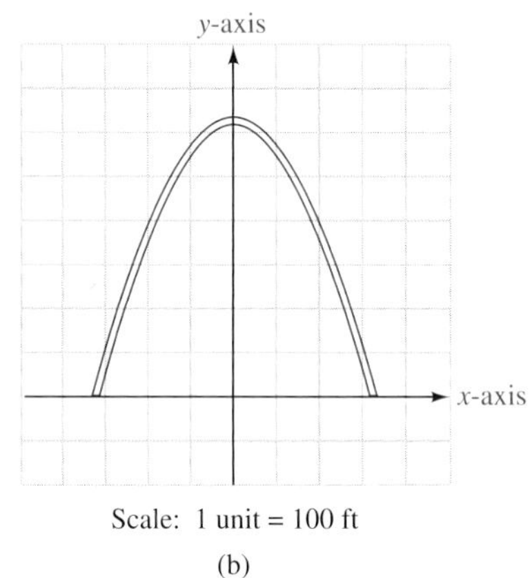

Scale: 1 unit = 100 ft

(b)

FIGURE 9-1

A rectangular coordinate system (see Figure 9-2) is formed by two perpendicular number lines. The horizontal number line is called the ***x*-axis,** and the vertical number line is called the ***y*-axis.** The positive direction on the x-axis is to the right, and the positive direction on the y-axis is upward. The scale on each axis should fit the data. For example, the axes of the graph of the arch shown in Figure 9-1(b) are scaled in units of 100 feet.

! COMMENT If no scale is indicated on the axes, we assume that the axes are scaled in units of 1.

The point where the axes cross is called the **origin.** This is the zero point on each axis. The axes form a **coordinate plane** and divide it into four regions called **quadrants,** which are numbered using Roman numerals as shown in Figure 9-2 on the next page. The axes are not considered to be in any quadrant.

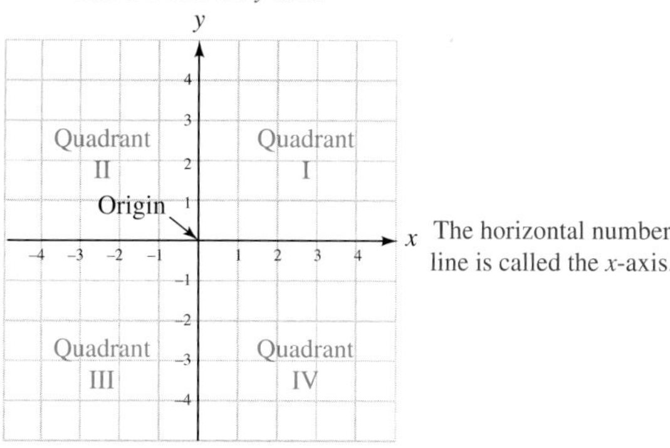

FIGURE 9-2

Each point in a coordinate plane can be identified by a pair of real numbers x and y written in the form (x, y). The first number x in the pair is called the ***x*-coordinate,** and the second number y is called the ***y*-coordinate.** The numbers in the pair are called the **coordinates** of the point. Some examples of such pairs are $(3, -4)$, $\left(-1, \frac{3}{2}\right)$, and $(0, 2.5)$.

$(3, -4)$
↑ ↑
The x-coordinate is listed first. The y-coordinate is listed second.

! COMMENT Don't be confused by this new use of parentheses. The notation $(3, -4)$ represents a point on the coordinate plane, whereas $3(-4)$ indicates a multiplication.

The process of locating a point in the coordinate plane is called **graphing** or **plotting** the point. In Figure 9-3(a), we use two blue arrows to show how to graph the point with coordinates of $(3, -4)$. Since the ***x*-coordinate,** 3, is positive, we start at the origin and move 3 units to the *right* along the x-axis. Since the ***y*-coordinate,** -4, is negative, we then move *down* 4 units and draw a dot. The **graph** of $(3, -4)$ lies in quadrant IV.

In Figure 9-3(a), two red arrows are used to show how to plot the point $(-4, 3)$. We start at the origin, move 4 units to the left along the x-axis, and then move up 3 units and draw a dot. The graph of $(-4, 3)$ lies in quadrant II.

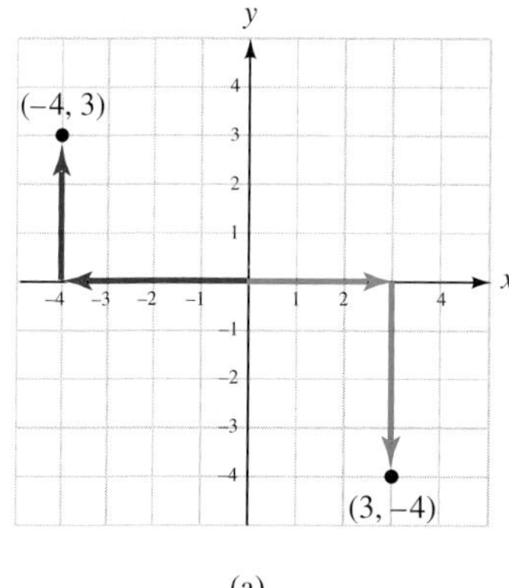

(a)

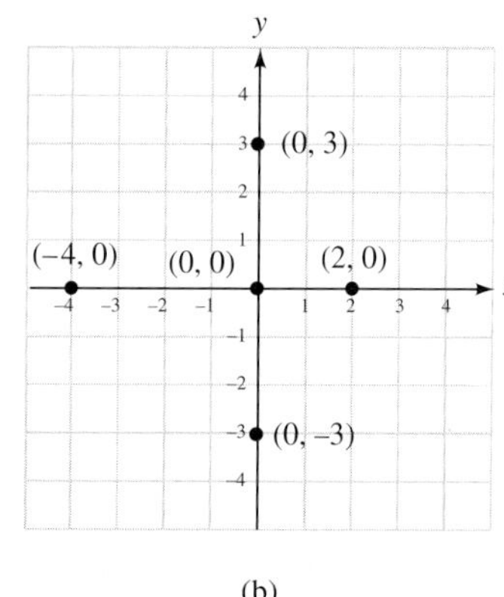

(b)

FIGURE 9-3

! COMMENT Note that the point with coordinates $(3, -4)$ is not the same as the point with coordinates $(-4, 3)$. Since the order of the coordinates of a point is important, we call the pairs **ordered pairs.**

In Figure 9-3(b) on the previous page, we see that the points $(-4, 0)$, $(0, 0)$, and $(2, 0)$ lie on the x-axis. In fact, all points with a y-coordinate of zero will lie on the x-axis. We also see that the points $(0, -3)$, $(0, 0)$, and $(0, 3)$ lie on the y-axis. All points with an x-coordinate of zero lie on the y-axis. We can also see that the coordinates of the origin are $(0, 0)$.

Self Check 1

Plot the points:

a. $(2, -2)$

b. $(-4, 0)$

c. $(1.5, \frac{5}{2})$

d. $(0, 5)$

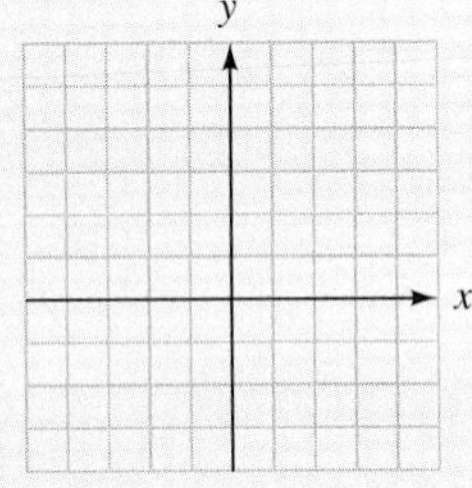

Answers

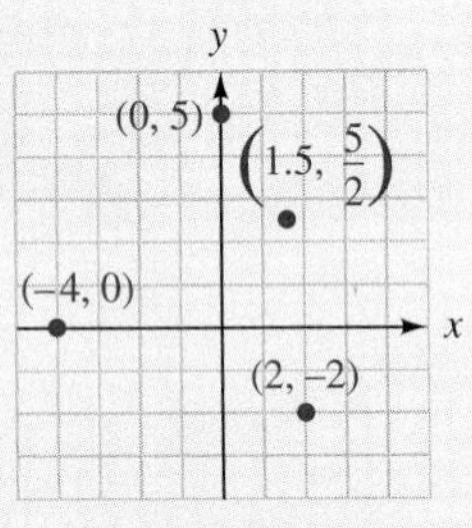

EXAMPLE 1

Plot the points: **a.** $(-2, 3)$, **b.** $\left(-1, -\frac{3}{2}\right)$, **c.** $(0, 2.5)$, and **d.** $(4, 2)$.

Solution See Figure 9-4.

a. To plot $(-2, 3)$, we start at the origin, move 2 units to the *left* on the x-axis, and move 3 units *up*. The point lies in quadrant II.

b. To plot $\left(-1, -\frac{3}{2}\right)$, we start at the origin and move 1 unit to the *left* and $\frac{3}{2}$ $\left(\text{or } 1\frac{1}{2}\right)$ units *down*. The point lies in quadrant III.

c. To graph $(0, 2.5)$, we start at the origin and move 0 units on the x-axis and 2.5 units *up*. The point lies on the y-axis.

d. To graph $(4, 2)$, we start at the origin and move 4 units to the *right* and 2 units *up*. The point lies in quadrant I.

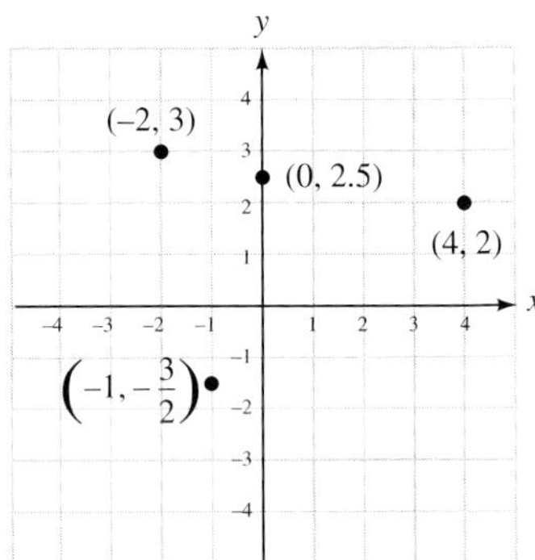

FIGURE 9-4

! COMMENT If no scale is indicated on the axes, we assume that the axes are scaled in units of 1.

EXAMPLE 2 Orbit of Earth.

The circle shown in Figure 9-5 on the next page is an approximate **graph** of the orbit of Earth. The graph is made up of infinitely many points, each with its own x- and y-coordinates. Use the graph to find the coordinates of Earth's position during the months of February, May, August, and December.

Solution To find the coordinates of each position, we start at the origin and move left or right along the x-axis to find the x-coordinate and then up or down to find the y-coordinate.

Month	Position of Earth on graph	Coordinates
February	3 units to the *right*, then 4 units *up*	(3, 4)
May	4 units to the *left*, then 3 units *up*	(−4, 3)
August	3.5 units to the *left*, then 3.5 units *down*	(−3.5, −3.5)
December	5 units to the *right*, no units *up* or *down*	(5, 0)

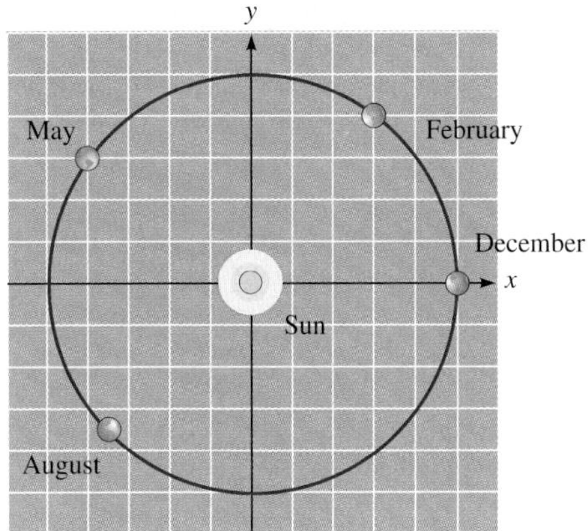

Scale: 1 unit = 18,600,000 mi

FIGURE 9-5

Graphing mathematical relationships

Every day, we deal with quantities that are related:

- The distance that we travel depends on how fast we are going.
- Our weight depends on how much we eat.
- The amount of water in a tub depends on how long the water has been running.

We often use graphs to visualize relationships between two quantities. For example, suppose we know the number of gallons of water that are in a tub at several time intervals after the water has been turned on. We can list that information in a **table.** (See Figure 9-6.)

The information in the table can be used to construct a graph that shows the relationship between the amount of water in the tub and the time the water has been running. Since the amount of water in the tub depends on the time, we will associate *time* with the x-axis and *amount of water* with the y-axis.

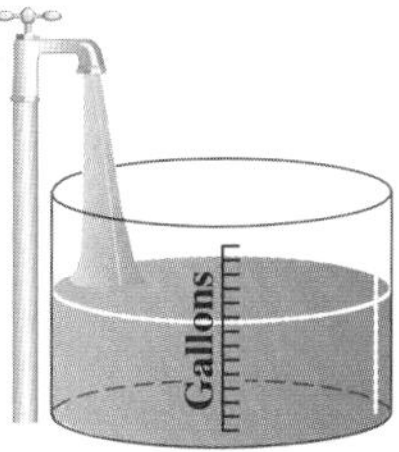

At various times, the amount of water in the tub was measured and recorded in the table.

(a)

Time (min)	Water in tub (gal)		
0	0	→	(0, 0)
1	8	→	(1, 8)
3	24	→	(3, 24)
4	32	→	(4, 32)
↑	↑		↑
x-coordinate	y-coordinate		The data in the table can be expressed as ordered pairs (x, y).

(b)

FIGURE 9-6

To construct the graph in Figure 9-7, we plot the four ordered pairs and draw a straight line through the resulting data points. The y-axis is scaled in larger units (4 gallons) because the data range from 0 to 32 gallons.

From the graph, we can see that the amount of water in the tub steadily increases as the water is allowed to run. We can also use the graph to make observations about the amount of water in the tub at other times. For example, the dashed line on the graph shows that in 5 minutes, the tub will contain 40 gallons of water.

x	y	(x, y)
0	0	(0, 0)
1	8	(1, 8)
3	24	(3, 24)
4	32	(4, 32)

The data can be listed in a table with headings x, y, and (x, y).

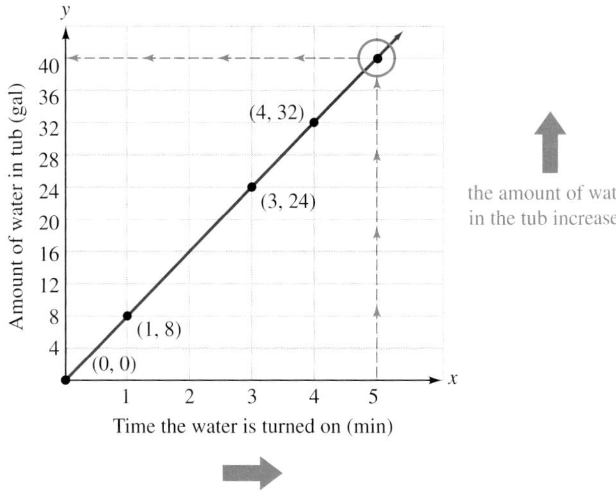

FIGURE 9-7

Reading graphs

We can often read information from a graph, as can be seen in the next example.

Self Check 3

Use the graph in Figure 9-8 to answer the following questions.

a. At what times were there exactly 50 people in the audience?

b. What was the size of the audience that watched the taping?

c. How long did it take for the audience to leave the studio after the taping ended?

EXAMPLE 3 The graph in Figure 9-8 shows the number of people in an audience before, during, and after the taping of a television show. On the x-axis, zero represents the time when taping began. Use the graph to answer the following questions, and record each result in a table.

a. How many people were in the audience when taping began?

b. What was the size of the audience 10 minutes before taping began?

c. At what times were there exactly 100 people in the audience?

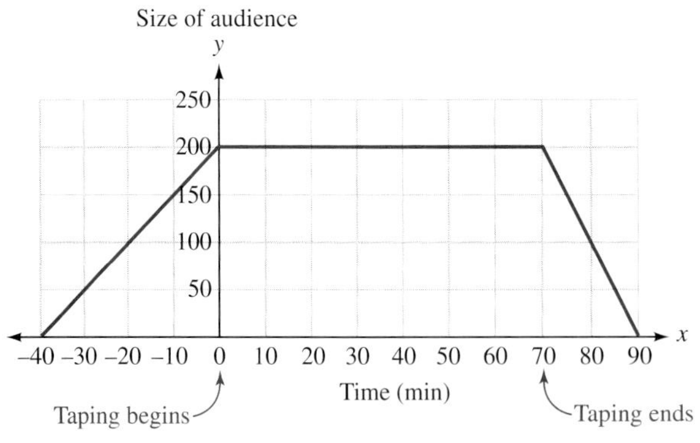

FIGURE 9-8

Solution

a. The time when taping began is represented by 0 on the x-axis. Since the point on the graph directly above 0 has a y-coordinate of 200, the point (0, 200) is on the graph. The y-coordinate of this point indicates that 200 people were in the audience when the taping began. We enter this result in the table at the right.

Time (min) x	Size of audience y
0	200
−10	150
−20	100
80	100

b. Ten minutes before taping began is represented by -10 on the x-axis. Since the point on the graph directly above -10 has a y-coordinate of 150, the point $(-10, 150)$ is on the graph. The y-coordinate of this point indicates that 150 people were in the audience 10 minutes before the taping began. We enter this result in the table.

c. We can draw a horizontal line passing through 100 on the y-axis. This line intersects the graph twice, at $(-20, 100)$ and at (80, 100). So there are two times when 100 people were in the audience. The first time was 20 minutes before taping began (-20), and the second time was 80 minutes after taping began (80). The y-coordinates of these points indicate that there were 100 people in the audience 20 minutes before and 80 minutes after taping began. We enter these results in the table.

Answers **a.** 30 min before and 85 min after taping began, **b.** 200, **c.** 20 min

Step graphs

The graph in Figure 9-9 shows the cost of renting a trailer for different periods of time. For example, the cost of renting the trailer for 4 days is \$60, which is the y-coordinate of the point (4, 60). The cost of renting the trailer for a period lasting over 4 and up to 5 days jumps to \$70. Since the jumps in cost form steps in the graph, we call this graph a **step graph.**

EXAMPLE 4 Use the information in Figure 9-9 to answer the following questions. Write the results in a table.

a. Find the cost of renting the trailer for 2 days.

b. Find the cost of renting the trailer for $5\frac{1}{2}$ days.

c. How long can you rent the trailer if you have \$50?

d. Is the rental cost per day the same?

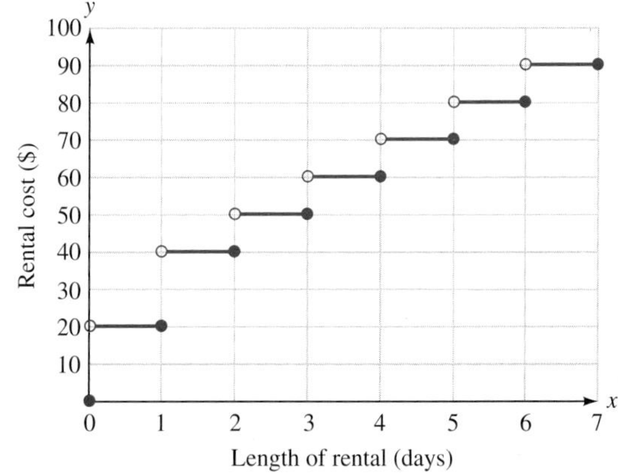

FIGURE 9-9

Solution

a. The solid dot at the end of each step indicates the rental cost for 1, 2, 3, 4, 5, 6, or 7 days. An open circle indicates that that point is not on the graph. We locate 2 days on the x-axis and move up to locate the point on the graph directly above the 2. Since the point has coordinates (2, 40), a 2-day rental would cost \$40. We enter this ordered pair in the table at the left.

b. We locate $5\frac{1}{2}$ days on the x-axis and move straight up to locate the point with coordinates $(5\frac{1}{2}, 80)$, which indicates that a $5\frac{1}{2}$-day rental would cost \$80. We then enter this ordered pair in the table.

c. We draw a horizontal line through the point labeled 50 on the y-axis. Since this line intersects one step in the graph, we can look down to the x-axis to find the x-values that correspond to a y-value of 50. From the graph, we see that the trailer can be rented for more than 2 and up to 3 days for \$50. We write (3, 50) in the table.

Length of rental (days) x	Cost (dollars) y
2	40
$5\frac{1}{2}$	80
3	50

d. No. If we look at the y-coordinates, we see that for the first day, the rental fee is \$20. The second day, the cost jumps another \$20. The third day, and all subsequent days, the cost jumps only \$10.

The midpoint formula

To distinguish between the coordinates of two points on a line, we often use subscript notation. Point $P(x_1, y_1)$ is read as "point P with coordinates of x sub 1 and y sub 1." Point $Q(x_2, y_2)$ is read as "point Q with coordinates of x sub 2 and y sub 2."

If point M in the graph below lies midway between points $P(x_1, y_1)$ and $Q(x_2, y_2)$, point M is called the **midpoint** of segment PQ. To find the coordinates of point M, we find the mean (average) of the x-coordinates and the mean of the y-coordinates of P and Q.

The midpoint formula

The **midpoint** of a line segment with endpoints at $P(x_1, y_1)$ and $Q(x_2, y_2)$ is the point M with coordinates of

$$\left(\frac{x_1 + x_2}{2}, \frac{y_1 + y_2}{2}\right)$$

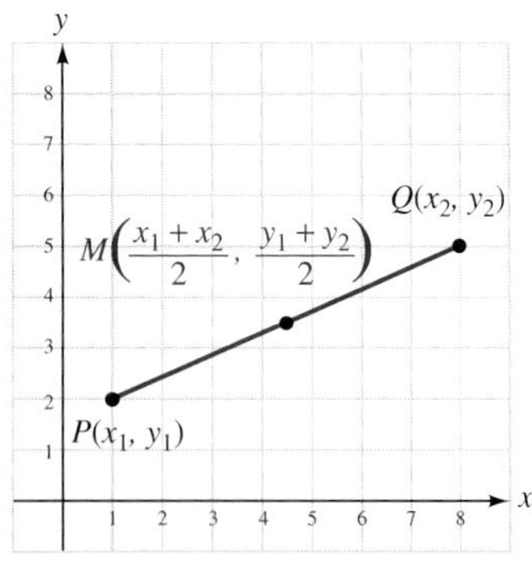

EXAMPLE 5 Find the midpoint of the line segment joining $(-2, 3)$ and $(3, -5)$.

Solution To find the midpoint, we find the mean of the x-coordinates and the mean the y-coordinates to get

$$\frac{x_1 + x_2}{2} = \frac{-2 + 3}{2} \qquad \text{and} \qquad \frac{y_1 + y_2}{2} = \frac{3 + (-5)}{2}$$

$$= \frac{1}{2} \qquad\qquad\qquad = -1$$

The midpoint of segment PQ is the point $M(\frac{1}{2}, -1)$.

Self Check 5
Find the midpoint of the segment joining $(5, -4)$ and $(-3, 5)$.

Answer $(1, \frac{1}{2})$

Section 9.1 STUDY SET

VOCABULARY *Fill in the blanks.*

1. The pair of numbers $(-1, -5)$ is called an ________ pair.
2. In the ordered pair $\left(-\frac{3}{2}, -5\right)$, the -5 is called the ____________.
3. The point with coordinates $(0, 0)$ is called the _______.
4. The x- and y-axes divide the coordinate plane into four regions called __________.
5. The point with coordinates $(4, 2)$ can be graphed on a ____________ coordinate system.
6. The process of locating the position of a point on a coordinate plane is called ______________ the point.

CONCEPTS *Fill in the blanks.*

7. To plot the point with coordinates $(-5, 4.5)$, we start at the _______ and move 5 units to the ______ and then move 4.5 units _____.
8. To plot the point with coordinates $\left(6, -\frac{3}{2}\right)$, we start at the _______ and move 6 units to the ______ and then move $\frac{3}{2}$ units _______.

9. Do $(3, 2)$ and $(2, 3)$ represent the same point?
10. In the ordered pair $(4, 5)$, is the number 4 associated with the horizontal or the vertical axis?
11. In which quadrant do points with a negative x-coordinate and a positive y-coordinate lie?
12. In which quadrant do points with a positive x-coordinate and a negative y-coordinate lie?
13. In the following illustration, fill in the missing coordinate of each highlighted point on the graph of the circle.

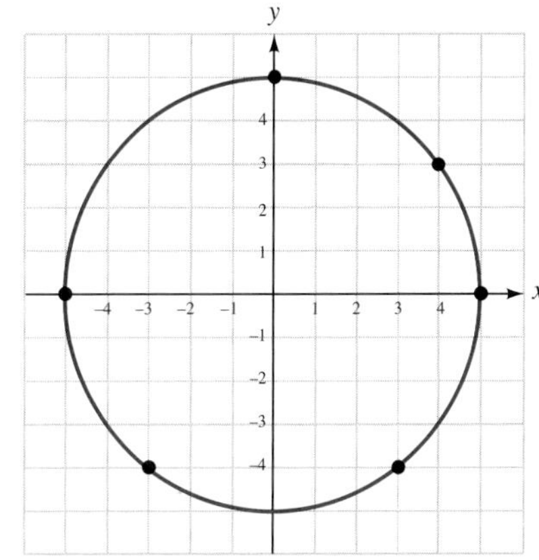

a. $(4, \quad)$
b. $(3, \quad)$
c. $(5, \quad)$
d. $(-3, \quad)$
e. $(-5, \quad)$
f. $(0, \quad)$

14. In the following illustration, fill in the missing coordinate of each point on the graph of the line.

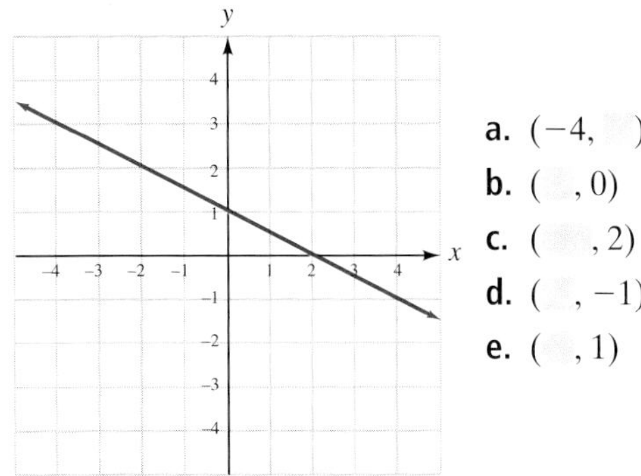

a. $(-4, \quad)$
b. $(\quad, 0)$
c. $(\quad, 2)$
d. $(\quad, -1)$
e. $(\quad, 1)$

The graph in the following Illustration gives the heart rate of a woman before, during, and after an aerobic workout. In Exercises 15–22, use the graph to answer the questions.

15. What information does the point $(-10, 60)$ give us?

16. After beginning her workout, how long did it take the woman to reach her training-zone heart rate?

17. What was the woman's heart rate half an hour after beginning the workout?

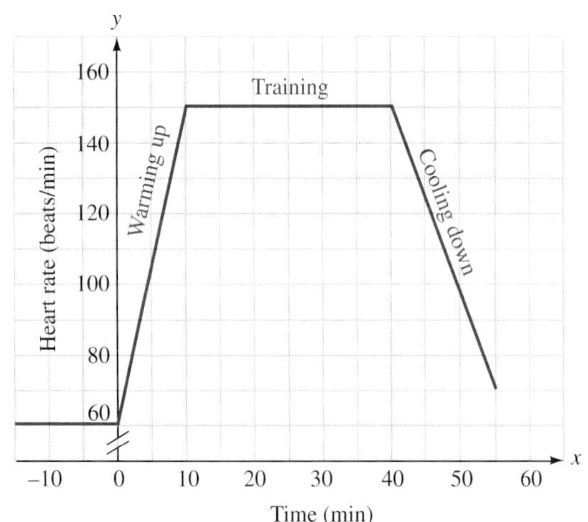

18. For how long did the woman work out at her training zone?

19. At what time was her heart rate 100 beats per minute?

20. How long was her cool-down period?

21. What was the difference in the woman's heart rate before the workout and after the cool-down period?

22. What was her approximate heart rate 8 minutes after beginning?

NOTATION

23. Explain the difference between $(3, 5)$, $3(5)$, and $5(3 + 5)$.

24. In the table, which column contains values associated with the vertical axis of a graph?

x	y
2	0
5	-2
-1	$-\frac{1}{2}$

25. Do these ordered pairs name the same point?

$$\left(2.5, -\tfrac{7}{2}\right), \left(2\tfrac{1}{2}, -3.5\right), \left(2.5, -3\tfrac{1}{2}\right)$$

26. Do these ordered pairs name the same point?

$$(-1.25, 4), \left(-1\tfrac{1}{4}, 4.0\right), \left(-\tfrac{5}{4}, 4\right)$$

PRACTICE *Graph each point on the grid provided.*

27. $(-3, 4)$
$(4, 3.5)$
$\left(-2, -\frac{5}{2}\right)$
$(0, -4)$
$\left(\frac{3}{2}, 0\right)$
$(3, -4)$

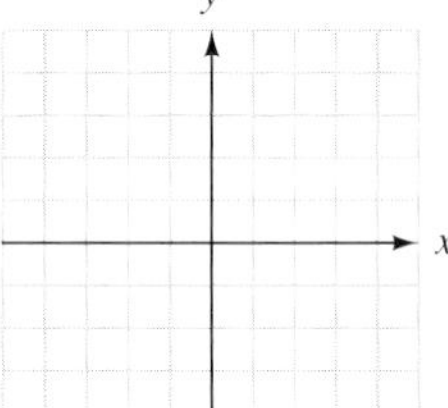

28. $(4, 4)$
$(0.5, -3)$
$(-4, -4)$
$(0, -1)$
$(0, 0)$
$(0, 3)$
$(-2, 0)$

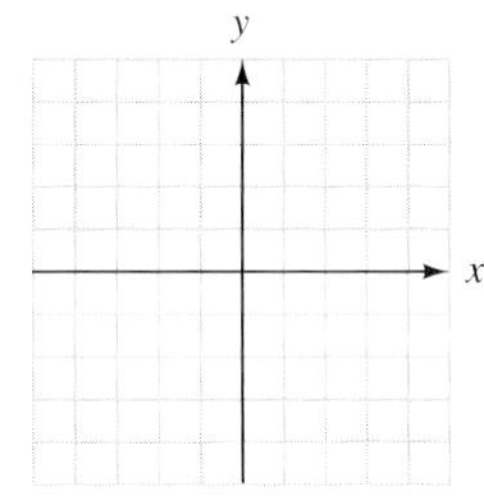

Find the midpoint of a line segment with the given endpoints.

29. $(5, 3)$ and $(7, 9)$

30. $(5, 6)$ and $(7, 10)$

31. $(2, -7)$ and $(-3, 12)$

32. $(-8, 12)$ and $(3, -9)$

33. $(4, 6)$ and $(10, 6)$

34. $(8, -6)$ and $(0, 0)$

APPLICATIONS

35. CONSTRUCTION The graph on the next page shows a side view of a bridge design. Make a table with three columns; label them *rivets*, *welds*, and *anchors*. List the coordinates of the points at which each category is located.

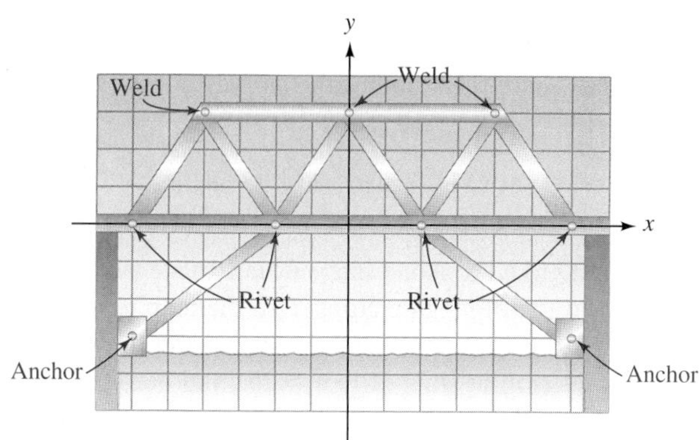

Scale: 1 unit = 8 ft

36. WATER PRESSURE The graph shows how the path of a stream of water changes when the hose is held at two different angles.

a. At which angle does the stream of water shoot up higher? How much higher?

b. At which angle does the stream of water shoot out farther? How much farther?

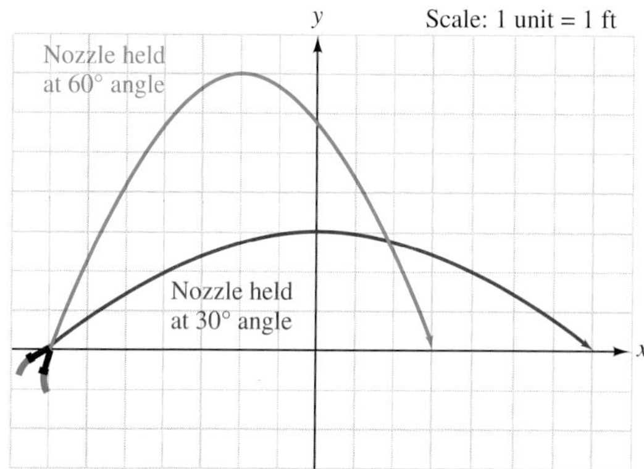

37. GOLF To correct her swing, a golfer is videotaped and then has her image displayed on a computer monitor so that it can be analyzed by a golf pro. Give the coordinates of the points that are highlighted on the arc of her swing.

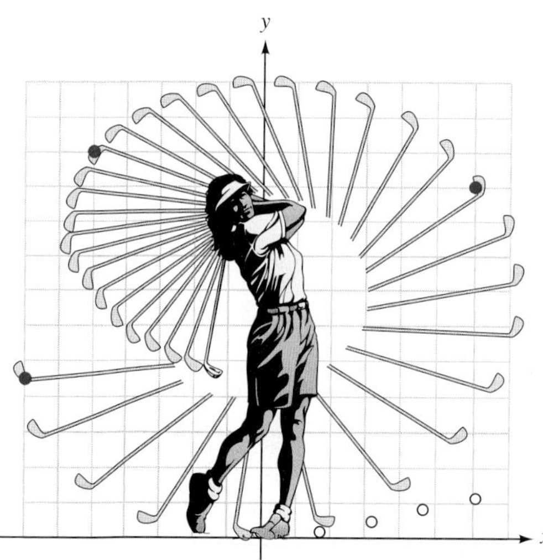

38. MEDICINE Scoliosis is a lateral curvature of the spine that can be more easily detected when a grid is superimposed over an X ray. In the illustration, find the coordinates of the center points of the indicated vertebrae. Note that T3 means the third thoracic vertebra, L4 means the fourth lumbar vertebra, and so on.

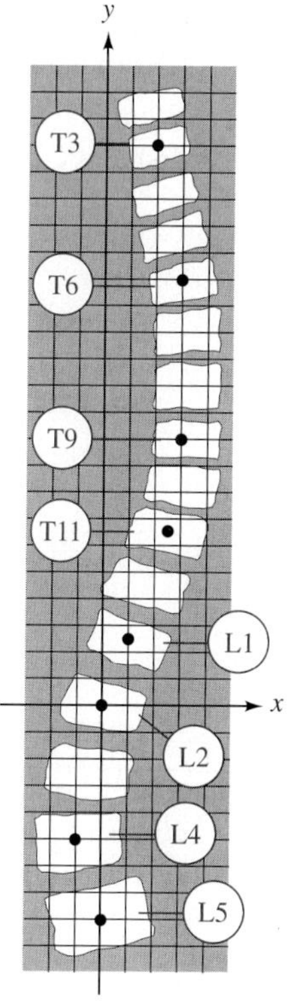

Scale: 1 unit = 0.5 in.

39. VIDEO RENTALS The charges for renting a video are shown in the graph.

a. Find the charge for a 1-day rental.

b. Find the charge for a 2-day rental.

c. What is the charge if a tape is kept for 5 days?

d. What is the charge if a tape is kept for a week?

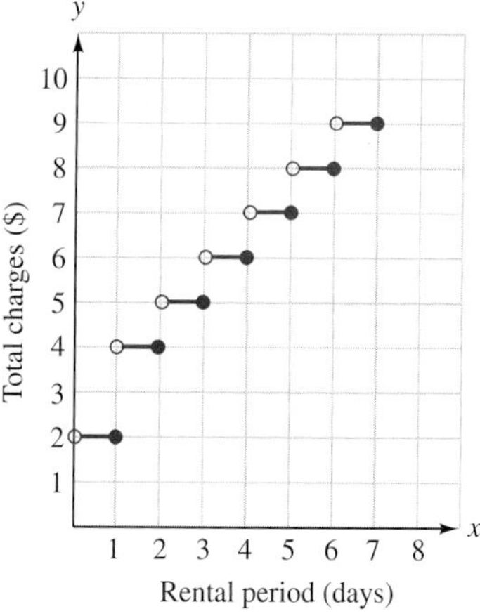

40. POSTAGE RATES The graph on the next page shows the first-class postage rates in 2004 for mailing items weighing up to 5 ounces.

a. Find the postage costs for mailing each of the following letters first class: a 1-ounce letter, a 4-ounce letter, and a $2\frac{1}{2}$-ounce letter.

b. Find the difference in postage for a 3.75-ounce letter and a 4.75-ounce letter.

c. What is the heaviest letter that could be mailed first class for 60¢?

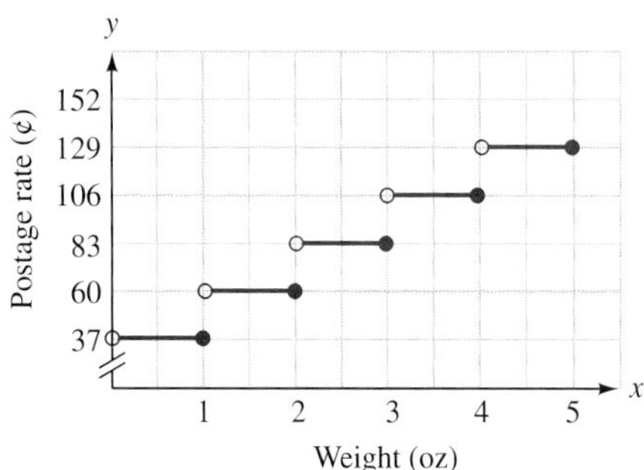

41. GAS MILEAGE The following table gives the number of miles (y) that a truck can be driven on x gallons of gasoline. Plot the ordered pairs and draw a line connecting the points.

x	y
2	10
3	15
5	25

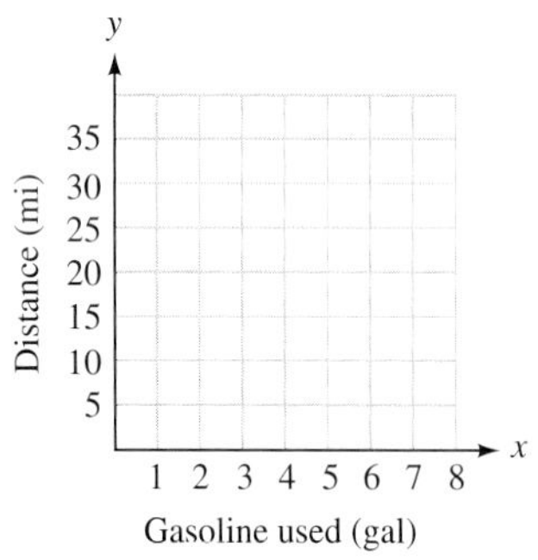

a. Estimate how far the truck can go on 7 gallons of gasoline.

b. How many gallons of gas are needed to travel a distance of 20 miles?

c. How far can the truck go on 6.5 gallons of gasoline?

42. VALUE OF A CAR The following table shows the value y (in thousands of dollars) of a car that is x years old. Plot the ordered pairs and draw a line connecting the points.

x	y
3	7
4	5.5
5	4

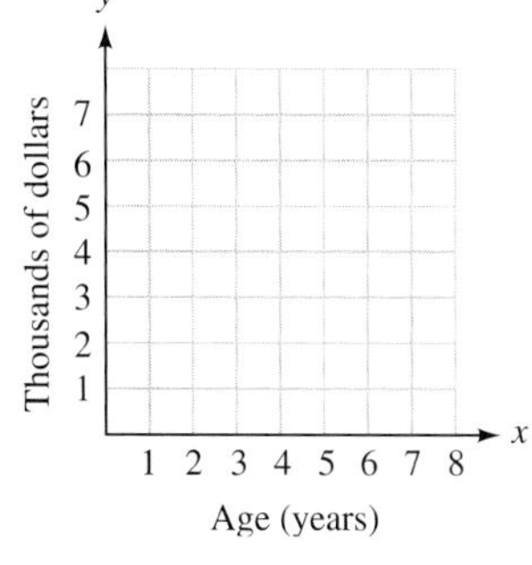

a. What does the point (3, 7) on the graph tell you?

b. Estimate the value of the car when it is 7 years old.

c. After how many years will the car be worth $2,500?

43. ROAD MAPS Road maps usually have a coordinate system to help locate cities. Use the map to locate Rockford, Mount Carroll, Harvard, and the intersection of state Highway 251 and U.S. Highway 30. Express each answer in the form (number, letter).

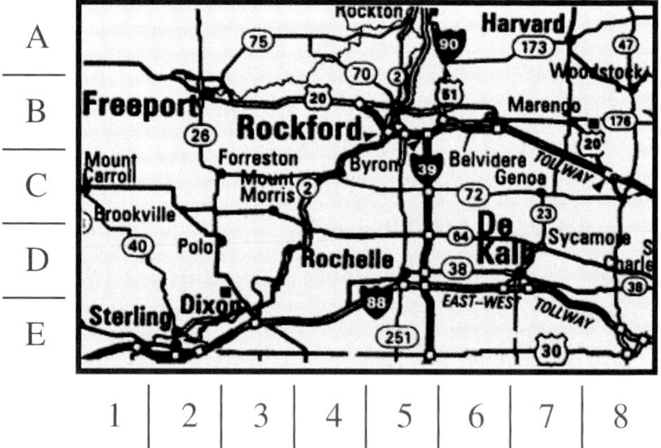

44. BATTLESHIP In the game Battleship, the player uses coordinates to drop depth charges from a battleship to hit a hidden submarine. What coordinates should be used to make three hits on the exposed submarine shown? Express each answer in the form (letter, number).

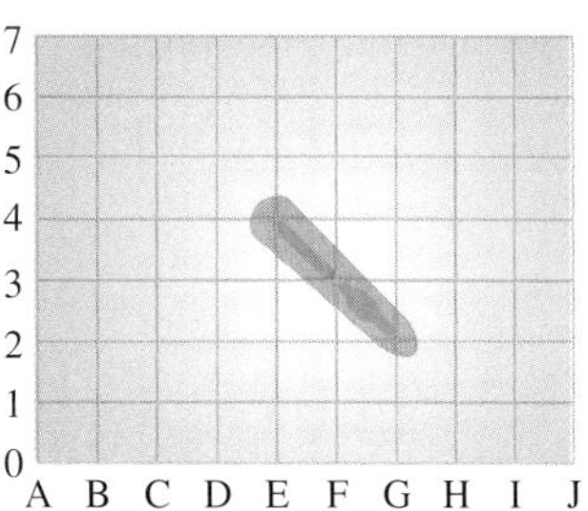

WRITING

45. Explain why the point $(-3, 3)$ is not the same as the point $(3, -3)$.

46. Explain what is meant when we say that the rectangular coordinate graph of the St. Louis Gateway Arch is made up of *infinitely many* points.

47. Explain how to plot the point $(-2, 5)$.

48. Explain why the coordinates of the origin are $(0, 0)$.

REVIEW

49. Evaluate: $-3 - 3(-5)$.
50. Evaluate: $(-5)^2 + (-5)$.
51. What is the opposite of -8?
52. Simplify: $|-1 - 9|$.
53. Solve: $-4x + 7 = -21$.
54. Solve $P = 2l + 2w$ for w.
55. Evaluate $(x + 1)(x + y)^2$ for $x = -2$ and $y = -5$.
56. Simplify: $-6(x - 3) - 2(1 - x)$.

9.2 Equations Containing Two Variables

- Solving equations in two variables
- Constructing tables of solutions
- Graphing equations
- Using different variables

In this section, we will discuss equations that contain two variables. These equations are often used to describe relationships between two quantities. To see a mathematical picture of these relationships, we will construct graphs of their equations.

Solving equations in two variables

We have previously solved equations containing one variable. For example, we can show that the solution of each of the following equations is $x = 3$.

$$2x + 3 = 9, \qquad -5x + 1 = 4 - 6x, \qquad \text{and} \qquad -3(x + 1) = 2x - 18$$

If we graph the solution $x = 3$ on a number line, we get the graph shown in Figure 9-10.

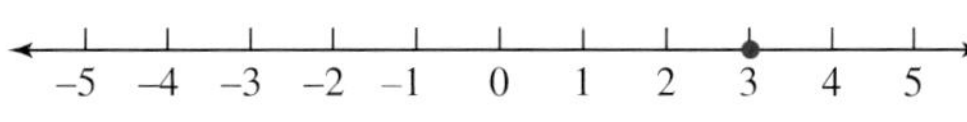

FIGURE 9-10

To describe relationships between two quantities mathematically, we use equations with two variables. Some examples of equations in two variables are

$$y = x - 1, \qquad y = x^2, \qquad y = |x|, \qquad \text{and} \qquad y = x^3$$

Solutions of equations in two variables are ordered pairs. For example, one solution of $y = x - 1$ is the ordered pair (5, 4), because the equation is true when $x = 5$ and $y = 4$.

$y = \mathbf{x} - 1$	This is the original equation.
$\mathbf{4} \stackrel{?}{=} \mathbf{5} - 1$	Substitute 5 for x and 4 for y.
$4 = 4$	Subtract on the right-hand side: $5 - 1 = 4$.

Since $4 = 4$ is a true statement, the ordered pair (5, 4) is a solution, and we say that (5, 4) satisfies the equation.

EXAMPLE 1 Is the ordered pair $(-1, -3)$ a solution of $y = x - 1$?

Solution We substitute -1 for x and -3 for y and see whether the resulting equation is a true statement.

$y = \mathbf{x} - 1$	This is the original equation.
$\mathbf{-3} \stackrel{?}{=} \mathbf{-1} - 1$	Substitute -1 for x and -3 for y.
$-3 = -2$	Perform the subtraction: $-1 - 1 = -2$.

Since $-3 = -2$ is a false statement, $(-1, -3)$ is not a solution.

Self Check 1
Is (9, 8) a solution of $y = x - 1$?

Answer yes

Self Check 2

Is $(-2, 5)$ a solution of $y = x^2$?

Answer no

EXAMPLE 2 Is the ordered pair $(-6, 36)$ a solution of $y = x^2$?

Solution We substitute -6 for x and 36 for y and see whether the resulting equation is a true statement.

$y = x^2$	This is the original equation.
$36 \stackrel{?}{=} (-6)^2$	Substitute -6 for x and 36 for y.
$36 = 36$	Find the power: $(-6)^2 = 36$.

Since the equation $36 = 36$ is true, $(-6, 36)$ is a solution.

Constructing tables of solutions

To find solutions of equations in x and y, we can pick numbers at random, substitute them for x, and find the corresponding values of y. For example, to find some ordered pairs that satisfy the equation $y = x - 1$, we can let $x = -4$ (called the **input value**), substitute -4 for x, and solve for y (called the **output value**).

$y = x - 1$	This is the original equation.
$y = -4 - 1$	Substitute the input -4 for x.
$y = -5$	The output is -5.

$y = x - 1$		
x	y	(x, y)
−4	**−5**	**(−4, −5)**

The ordered pair $(-4, -5)$ is a solution. We list this ordered pair in red in the **table of solutions** (or **table of values**).

To find another ordered pair that satisfies $y = x - 1$, we let $x = -2$.

$y = x - 1$	This is the original equation.
$y = -2 - 1$	Substitute the input -2 for x.
$y = -3$	The output is -3.

$y = x - 1$		
x	y	(x, y)
−4	−5	(−4, −5)
−2	**−3**	**(−2, −3)**

A second solution is $(-2, -3)$, and we list it in the table of solutions.

If we let $x = 0$, we can find a third ordered pair that satisfies $y = x - 1$.

$y = x - 1$	This is the original equation.
$y = 0 - 1$	Substitute the input 0 for x.
$y = -1$	The output is -1.

$y = x - 1$		
x	y	(x, y)
−4	−5	(−4, −5)
−2	−3	(−2, −3)
0	**−1**	**(0, −1)**

A third solution is $(0, -1)$, which we also add to the table of solutions.

If we let $x = 2$, we can find a fourth solution.

$y = x - 1$	This is the original equation.
$y = 2 - 1$	Substitute the input 2 for x.
$y = 1$	The output is 1.

$y = x - 1$		
x	y	(x, y)
−4	−5	(−4, −5)
−2	−3	(−2, −3)
0	−1	(0, −1)
2	**1**	**(2, 1)**

A fourth solution is $(2, 1)$, and we add it to the table of solutions.

If we let $x = 4$, we have

$y = x - 1$ This is the original equation.

$y = \mathbf{4} - 1$ Substitute the input 4 for x.

$y = \mathbf{3}$ The output is 3.

A fifth solution is (4, 3).

Since we can choose any real number for x, and since any choice of x will give a corresponding value of y, it is apparent that the equation $y = x - 1$ has *infinitely many solutions*. We have found five of them: $(-4, -5)$, $(-2, -3)$, $(0, -1)$, $(2, 1)$, and $(4, 3)$.

$y = x - 1$		
x	y	(x, y)
−4	−5	(−4, −5)
−2	−3	(−2, −3)
0	−1	(0, −1)
2	1	(2, 1)
4	**3**	**(4, 3)**

Graphing equations

To graph the equation $y = x - 1$, we plot the ordered pairs listed in the table of solutions on a rectangular coordinate system, as shown in Figure 9-11(a). From the figure, we can see that the five points lie on a line.

In Figure 9-11(b), we draw a line through the points, because the graph of any solution of $y = x - 1$ will lie on this line. The arrowheads show that the line continues forever in both directions. The line is a picture of all the solutions of the equation $y = x - 1$. This line is called the **graph** of the equation.

$y = x - 1$		
x	y	(x, y)
−4	−5	(−4, −5)
−2	−3	(−2, −3)
0	−1	(0, −1)
2	1	(2, 1)
4	3	(4, 3)

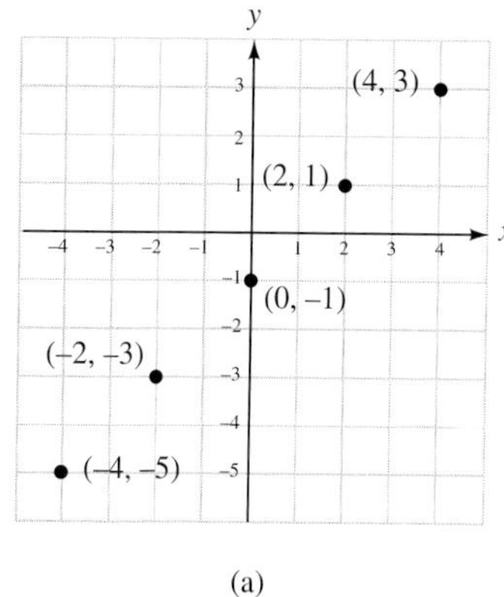

(a)

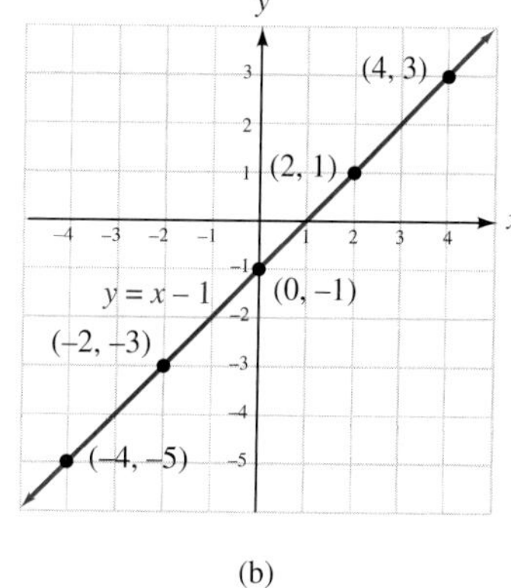

(b)

FIGURE 9-11

To graph an equation in x and y, we follow these steps.

Graphing an equation in x and y

1. Make a table of solutions containing several ordered pairs of numbers (x, y) that satisfy the equation. Do this by picking values for x and finding the corresponding values for y.
2. Plot each ordered pair on a rectangular coordinate system.
3. Carefully draw a line or smooth curve through the points.

Since we will usually choose a number for x and then find the corresponding value of y, the value of y depends on x. For this reason, we call y the **dependent variable** and x the **independent variable.** The value of the independent variable is the input value, and the value of the dependent variable is the output value.

Self Check 3
Graph: $y = -3x + 1$.

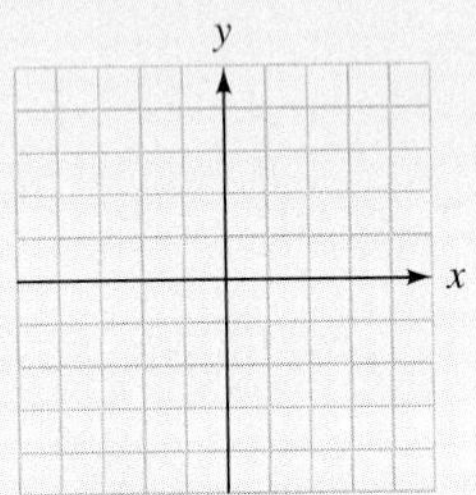

Answer

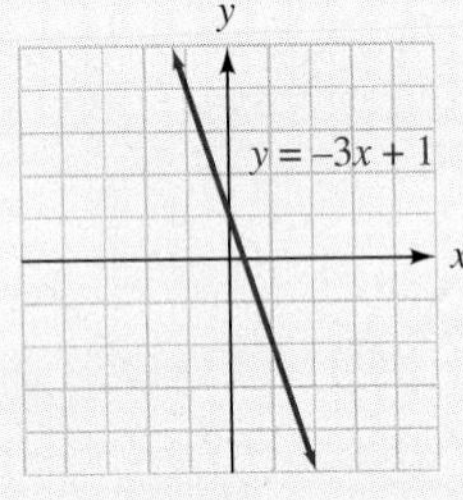

EXAMPLE 3 Graph: $y = -2x - 2$.

Solution To make a table of solutions, we choose numbers for x and find the corresponding values of y. If $x = -3$, we have

$y = -2x - 2$ This is the original equation.

$y = -2(-3) - 2$ Substitute -3 for x.

$y = 6 - 2$ Perform the multiplication: $-2(-3) = 6$.

$y = 4$ Perform the subtraction.

Thus, $x = -3$ and $y = 4$ is a solution. In a similar manner, we find the corresponding y-values for x-values of -2, -1, 0, and 1 and record the results in the table of solutions in Figure 9-12(a). After plotting the ordered pairs, we draw a line through the points to get the graph shown in the Figure 9-12(b).

$y = -2x - 2$		
x	y	(x, y)
-3	4	$(-3, 4)$
-2	2	$(-2, 2)$
-1	0	$(-1, 0)$
0	-2	$(0, -2)$
1	-4	$(1, -4)$

(a)

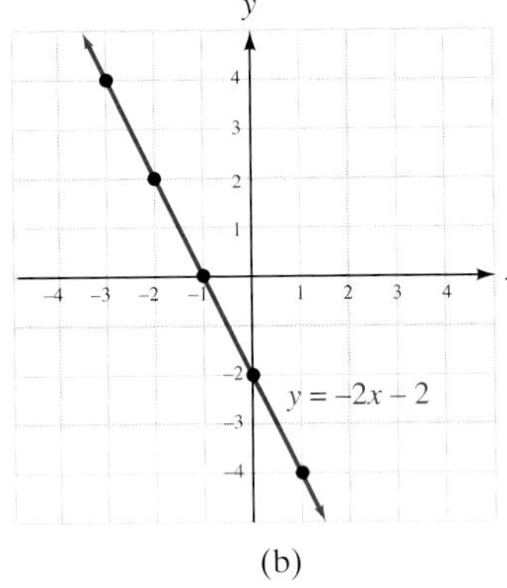

(b)

FIGURE 9-12

Self Check 4
Graph $y = x^2 - 2$ and compare the result to the graph of $y = x^2$. What do you notice?

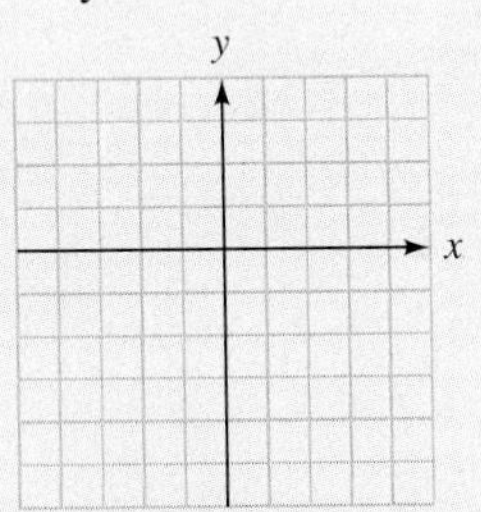

Answer The graph has the same shape, but is 2 units lower.

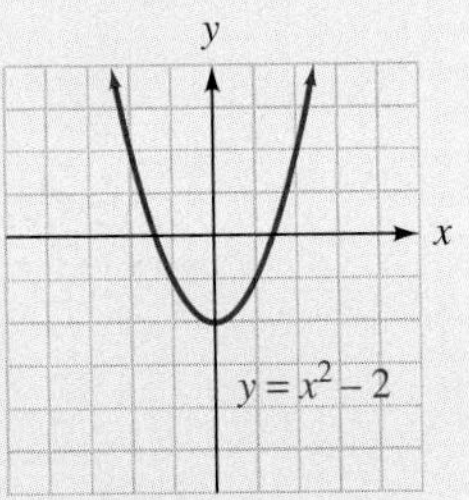

EXAMPLE 4 Graph: $y = x^2$.

Solution To make a table of solutions, we will choose numbers for x and find the corresponding values of y. If $x = -3$, we have

$y = x^2$ This is the original equation.

$y = (-3)^2$ Substitute the input -3 for x.

$y = 9$ The output is 9.

Thus, $x = -3$ and $y = 9$ is a solution. In a similar manner, we find the corresponding y-values for x-values of -2, -1, 0, 1, 2, and 3. If we plot the ordered pairs listed in the table in Figure 9-13 and join the points with a smooth curve, we get the graph shown in the figure, which is called a **parabola.**

$y = x^2$		
x	y	(x, y)
-3	9	$(-3, 9)$
-2	4	$(-2, 4)$
-1	1	$(-1, 1)$
0	0	$(0, 0)$
1	1	$(1, 1)$
2	4	$(2, 4)$
3	9	$(3, 9)$

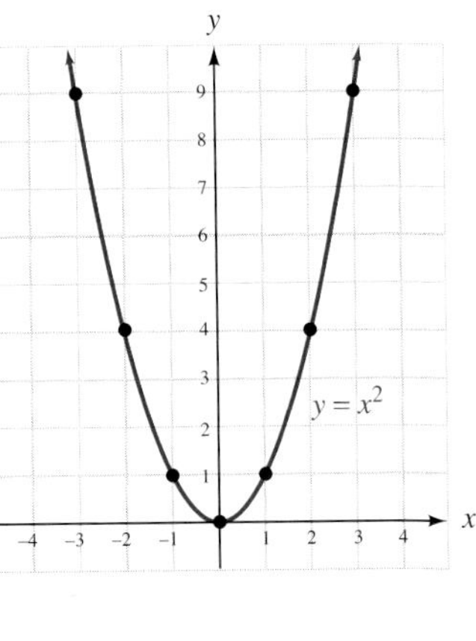

FIGURE 9-13

EXAMPLE 5 Graph: $y = |x|$.

Solution To make a table of solutions, we will choose numbers for x and find the corresponding values of y. If $x = -5$, we have

$y = |x|$ This is the original equation.

$y = |\mathbf{-5}|$ Substitute the input -5 for x.

$y = 5$ The output is 5.

The ordered pair $(-5, 5)$ satisfies the equation. This pair and several others that satisfy the equation are listed in the table of solutions in Figure 9-14. If we plot the ordered pairs in the table, we see that they lie in a "V" shape. We join the points to complete the graph shown in the figure.

$y = \|x\|$		
x	y	(x, y)
-5	5	$(-5, 5)$
-4	4	$(-4, 4)$
-3	3	$(-3, 3)$
-2	2	$(-2, 2)$
-1	1	$(-1, 1)$
0	0	$(0, 0)$
1	1	$(1, 1)$
2	2	$(2, 2)$
3	3	$(3, 3)$

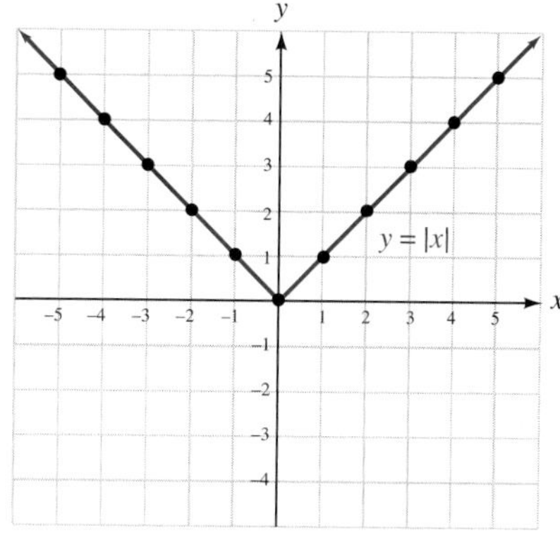

FIGURE 9-14

Self Check 5
Graph $y = |x| + 2$ and compare the result to the graph of $y = |x|$. What do you notice?

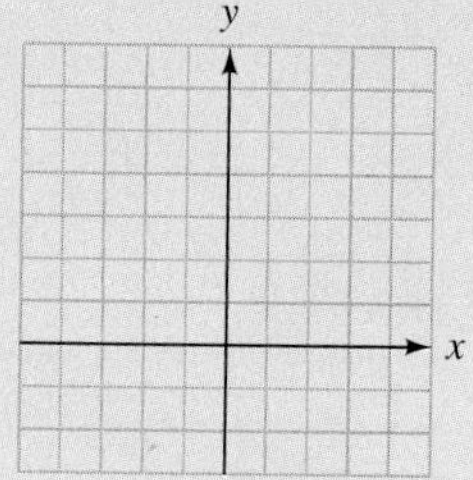

Answer The graph has the same shape, but is 2 units higher.

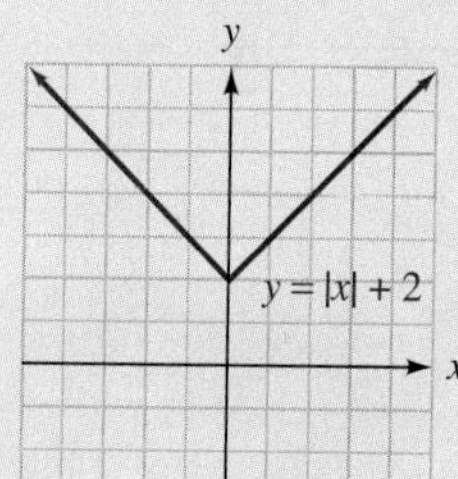

EXAMPLE 6 Graph: $y = x^3$.

Solution If we let $x = -2$, we have

$y = x^3$ This is the original equation.

$y = (\mathbf{-2})^3$ Substitute the input -2 for x.

$y = -8$ The output is -8.

The ordered pair $(-2, -8)$ satisfies the equation. This ordered pair and several others that satisfy the equation are listed in the table of solutions in Figure 9-15 on the next page. Plotting the ordered pairs and joining them with a smooth curve gives us the graph shown in the figure.

Self Check 6
Graph $y = (x - 2)^3$ and compare the result to the graph of $y = x^3$. What do you notice?

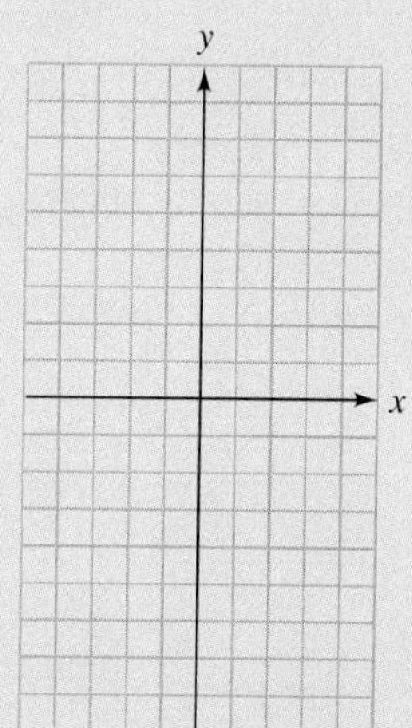

Answer The graph has the same shape but is 2 units to the right.

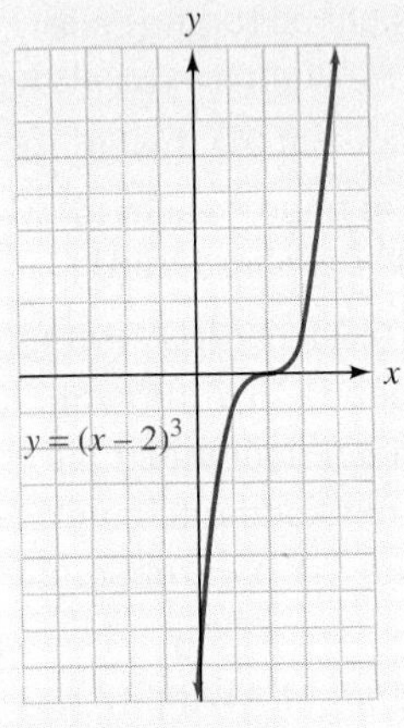

$y = x^3$		
x	y	(x, y)
−2	−8	(−2, −8)
−1	−1	(−1, −1)
0	0	(0, 0)
1	1	(1, 1)
2	8	(2, 8)

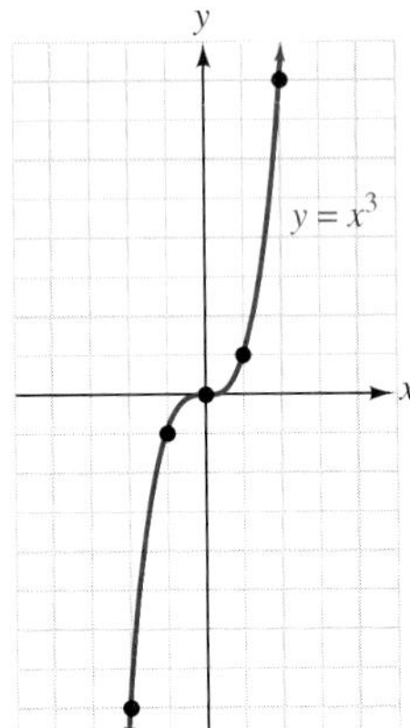

FIGURE 9-15

CALCULATOR SNAPSHOT

Using a graphing calculator to graph an equation

FIGURE 9-16

We have graphed equations by making tables of solutions and plotting points. The task of graphing is much easier when we use a graphing calculator (Figure 9-16). The instructions in this discussion will be general in nature. For specific details about your calculator, please consult your owner's manual.

The viewing window: All graphing calculators have a viewing **window,** used to display graphs. The **standard window** has settings of

$$\text{Xmin} = -10, \qquad \text{Xmax} = 10, \qquad \text{Ymin} = -10, \qquad \text{and} \qquad \text{Ymax} = 10$$

which indicate that the minimum x- and y-coordinates used in the graph will be -10 and that the maximum x- and y-coordinates will be 10.

Graphing an equation: To graph the equation $y = x - 1$ using a graphing calculator, we press the $\boxed{\text{Y} =}$ key and enter the right-hand side of the equation after the symbol Y_1. The display will show the equation

$$\text{Y}_1 = x - 1$$

Then we press the $\boxed{\text{GRAPH}}$ key to produce the graph shown in Figure 9-17.

Next, we will graph the equation $y = |x - 4|$. Since absolute values are always nonnegative, the minimum y-value is zero. To obtain a reasonable viewing window, we set the Ymin value slightly lower, at Ymin $= -3$. We set Ymax to be 10 units greater than Ymin, at Ymax $= 7$. The minimum value of y occurs when $x = 4$. To center the graph in the viewing window, we set the Xmin and Xmax values 5 units to the left and right of 4. Therefore, Xmin $= -1$ and Xmax $= 9$.

After entering the right-hand side of the equation, we obtain the graph shown in Figure 9-18. Consult your owner's manual to learn how to enter an absolute value.

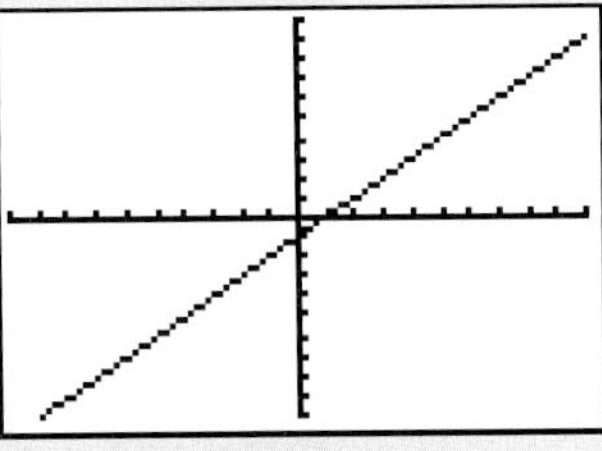

FIGURE 9-17

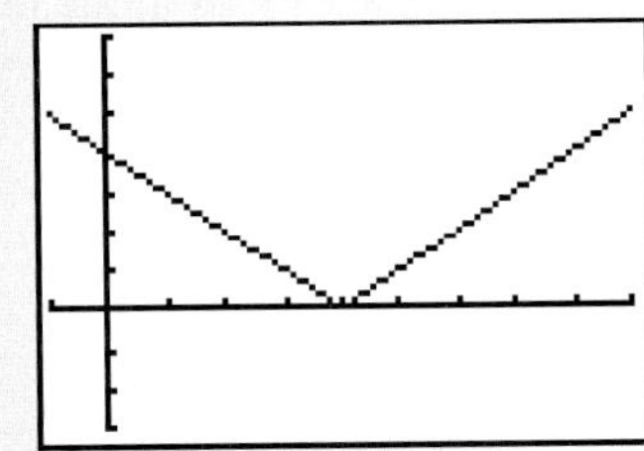

FIGURE 9-18

Changing the viewing window: The choice of viewing windows is extremely important when graphing equations. To show this, let's graph $y = x^2 - 25$ with x-values from -1 to 6 and y-values from -5 to 5.

To graph this equation, we set the x and y window values and enter the right-hand side of the equation. The display will show

$$Y_1 = x^2 - 25$$

Then we press the GRAPH key to produce the graph shown in Figure 9-19(a). Although the graph appears to be a straight line, it is not. Actually, we are seeing only part of a parabola. If we pick a viewing window with x-values of -6 to 6 and y-values of -30 to 2, as in Figure 9-19(b), we can see that the graph is a parabola.

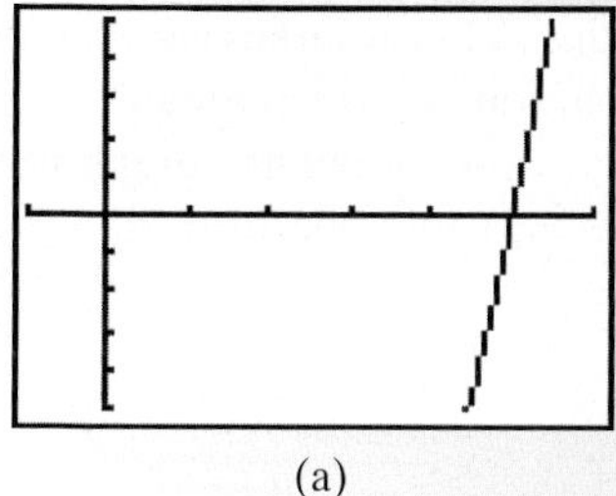

(a) (b)

FIGURE 9-19

Use a graphing calculator to graph each equation. Use a viewing window of x = −5 to 5 and y = −5 to 5.

1. $y = 2.1x - 1.1$

2. $y = 1.12x^2 - 1$

3. $y = |x + 0.7|$

4. $y = 0.1x^3 + 1$

Graph each equation in a viewing window of x = −4 to 4 and y = −4 to 4. Each graph is not what it first appears to be. Pick a better viewing window and find a better representation of the true graph.

5. $y = -x^3 - 8.2$

6. $y = -|x - 4.01|$

7. $y = x^2 + 5.9$

8. $y = -x + 7.95$

Using different variables

We will often encounter equations with variables other than x and y. When we make tables of solutions and graph these equations, we must know which is the independent variable (the input values) and which is the dependent variable (the output values). The independent variable is usually associated with the horizontal axis of the coordinate system, and the dependent variable is usually associated with the vertical axis.

EXAMPLE 7 **Speed limits.** In some states, the maximum speed limit on a U.S. interstate highway is 75 mph. The distance covered by a vehicle traveling at 75 mph depends on the time the vehicle travels at that speed. This relationship is described by the equation $d = 75t$, where d represents the distance (in miles) and t represents the time (in hours). Graph the equation.

FIGURE 9-20

Solution Since d depends on t in the equation $d = 75t$, t is the independent variable (the input) and d is the dependent variable (the output). Therefore, we choose values for t and find the corresponding values of d. Since t represents the time spent traveling at 75 mph, we choose no negative values for t.

If $t = 0$, we have

$d = 75t$ This is the original equation.

$d = 75(\mathbf{0})$ Substitute the input 0 for t.

$d = 0$ Perform the multiplication.

The pair $t = 0$ and $d = 0$, or (0, 0), is a solution. This ordered pair and others that satisfy the equation are listed in the table of solutions shown in Figure 9-21(a). If we plot the ordered pairs and draw a line through them, we obtain the graph shown in Figure 9-21(b). From the graph, we see (as expected) that the distance covered steadily increases as the traveling time increases.

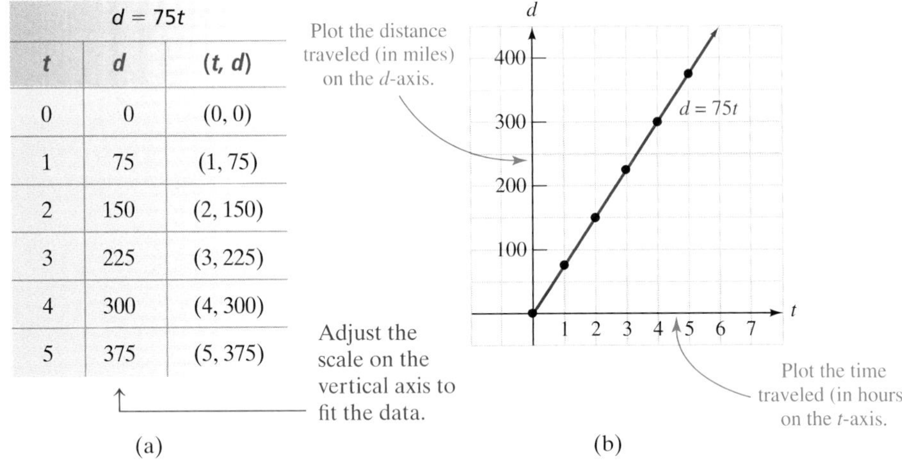

$d = 75t$

t	d	(t, d)
0	0	(0, 0)
1	75	(1, 75)
2	150	(2, 150)
3	225	(3, 225)
4	300	(4, 300)
5	375	(5, 375)

FIGURE 9-21

Section 9.2 STUDY SET

VOCABULARY *Fill in the blanks.*

1. The equation $y = x + 1$ is an equation in ______ variables.
2. An ordered pair is a ________ of an equation if the numbers in the ordered pair satisfy the equation.
3. In equations containing the variables x and y, x is called the ____________ variable and y is called the ____________ variable.
4. When constructing a ________ of solutions, the values of x are the input values and the values of y are the ________ values.

CONCEPTS

5. Consider the equation $y = -2x + 6$.
   a. How many variables does the equation contain?
   b. Does the ordered pair $(4, -2)$ satisfy the equation?
   c. Is $(-3, 12)$ a solution of the equation?
   d. How many solutions does this equation have?

6. To graph an equation, five solutions were found, they were plotted (in black), and a straight line was drawn through them, as shown. From the graph, determine three other solutions of the equation.

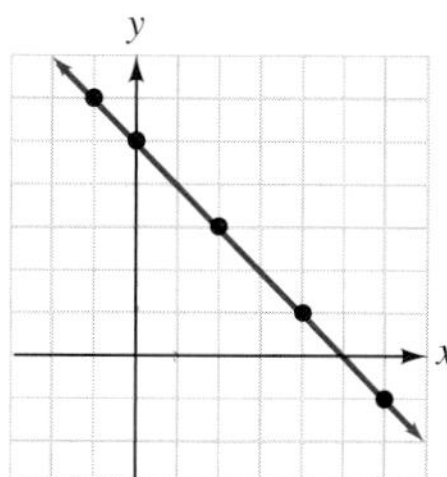

7. The graph of $y = -x + 5$ is shown in Problem 6. Fill in the blanks: Every point on the graph represents an ordered-pair ________ of $y = -x + 5$, and every ordered-pair solution is a ______ on the graph.

8. Consider the graph of an equation shown below.
   a. If the coordinates of point M are substituted into the equation, is the result a true or false statement?
   b. If the coordinates of point N are substituted into the equation, is the result a true or false statement?

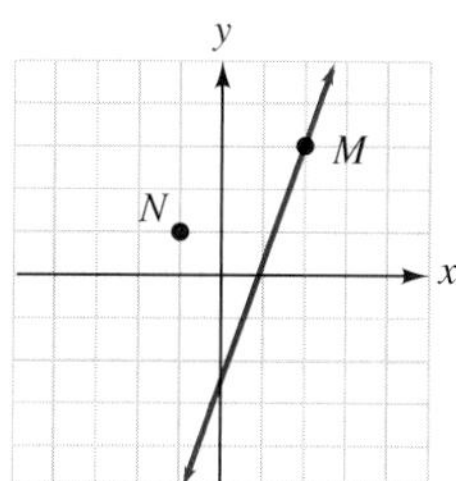

9. Complete the table.

$y = x^3$	
x (inputs)	**y (outputs)**
0	
−1	
−2	
1	
2	

10. What is wrong with the graph of $y = x - 3$ shown below?

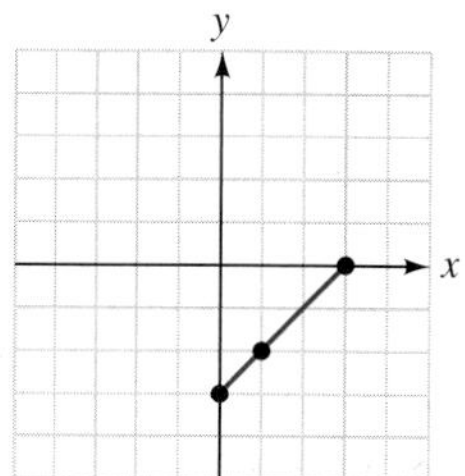

11. To graph $y = -x + 1$, a student constructed a table of solutions and plotted the ordered pairs as shown. Instead of drawing a crooked line through the points, what should he have done?

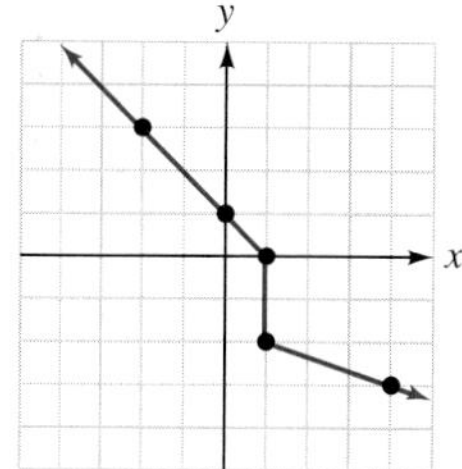

12. To graph $y = x^2 - 4$, a table of solutions is constructed and a graph is drawn, as shown. Explain the error made here.

$y = x^2 - 4$		
x	**y**	**(x, y)**
0	−4	(0, −4)
2	0	(2, 0)

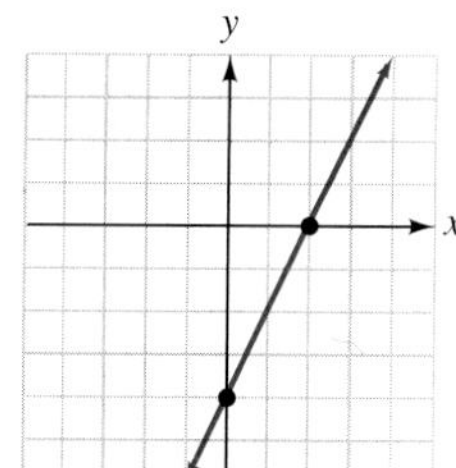

13. Explain the error with the graph of $y = x^2$ shown in the illustration.

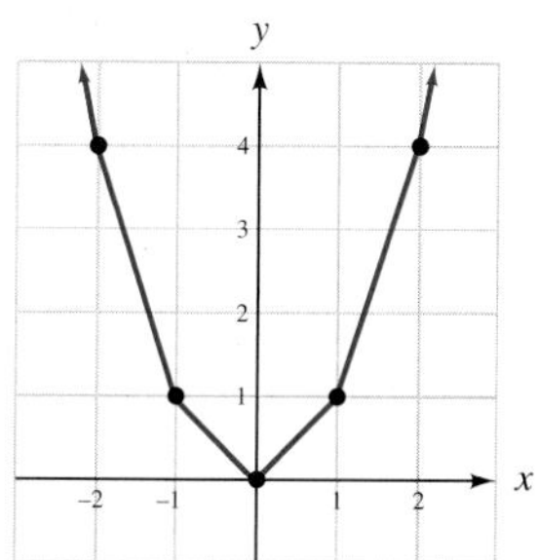

14. Several solutions of an equation are listed in the table of solutions. When graphing them, with what variable should the horizontal and vertical axes of the graph be labeled?

t	s	(t, s)
0	4	(0, 4)
1	5	(1, 5)
2	10	(2, 10)

NOTATION *Complete each solution.*

15. Verify that $(-2, 6)$ satisfies $y = -x + 4$

$y = -x + 4$

$\quad \stackrel{?}{=} -(\quad) + 4$

$6 \stackrel{?}{=} \quad + 4$

$6 = 6$

16. For the equation $y = |x - 2|$, if $x = -3$, find y.

$y = |x - 2|$

$y = |\quad - 2|$

$y = |\quad|$

$y = 5$

PRACTICE *Determine whether the ordered pair satisfies the equation.*

17. $y = 2x - 4$; $(4, 4)$

18. $y = x^2$; $(8, 48)$

19. $y = |x - 2|$; $(4, -3)$

20. $y = x^3 + 1$; $(-2, -7)$

Complete each table.

21.

$y = x - 3$	
x	y
0	
1	
−2	

22.

$y = \|x - 3\|$	
x	$\|x - 3\|$
0	
−1	
3	

23.

$y = x^2 - 3$	
Input	Output
0	
2	
−2	

24.

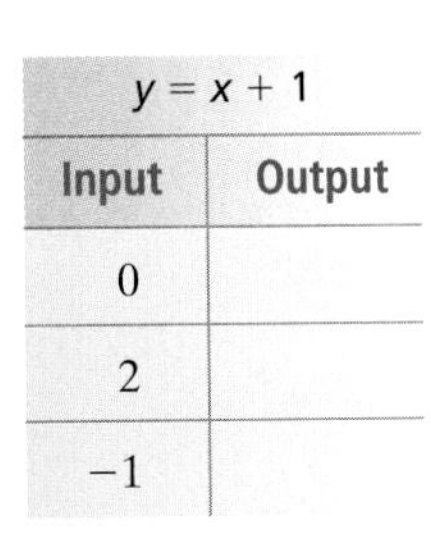

$y = x + 1$	
Input	Output
0	
2	
−1	

Construct a table of solutions and graph each equation.

25. $y = 2x - 3$

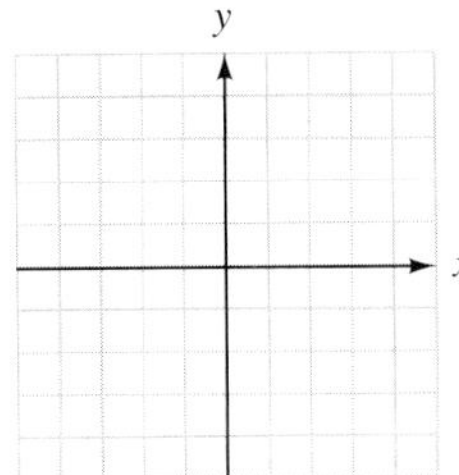

26. $y = 3x + 1$

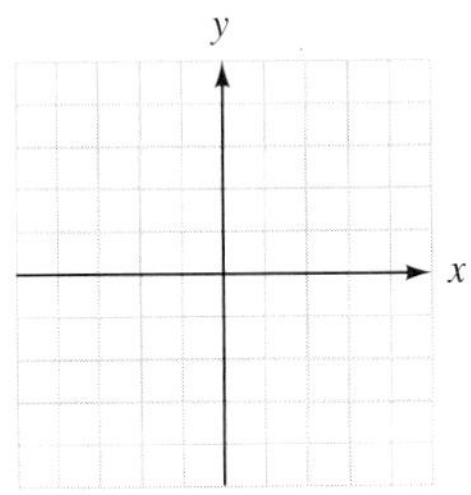

27. $y = -2x + 1$

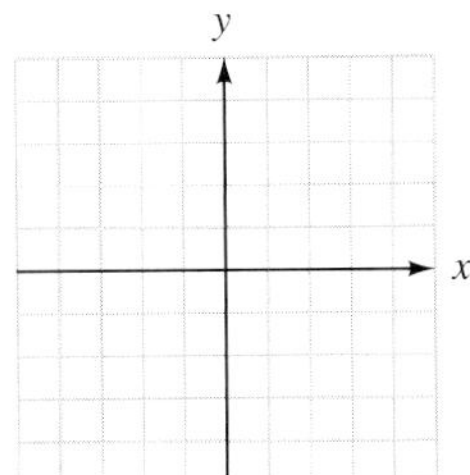

28. $y = -3x + 2$

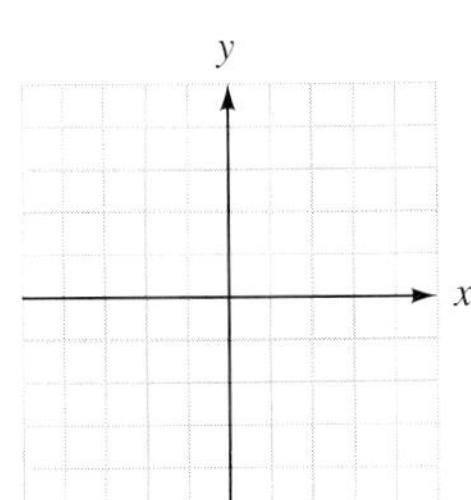

Construct a table of solutions and graph each equation. Compare the result to the graph of $y = x^2$.

29. $y = x^2 + 1$

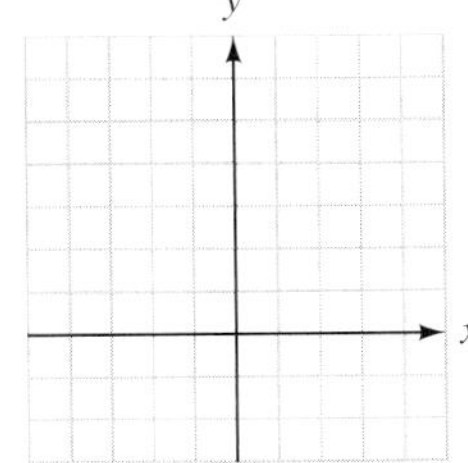

30. $y = -x^2$

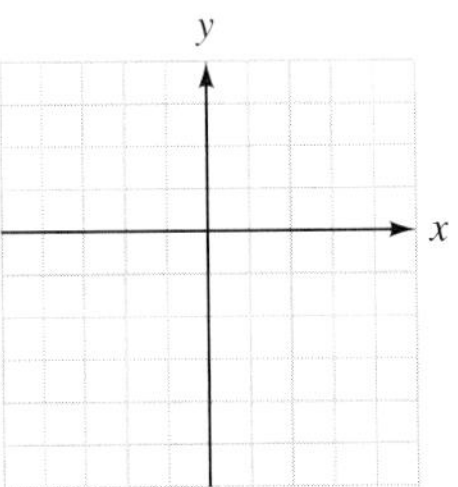

31. $y = (x - 2)^2$

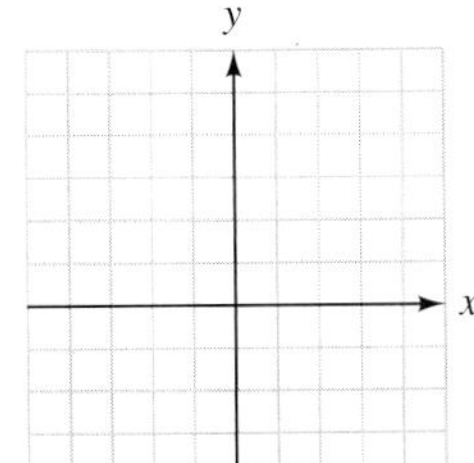

32. $y = (x + 2)^2$

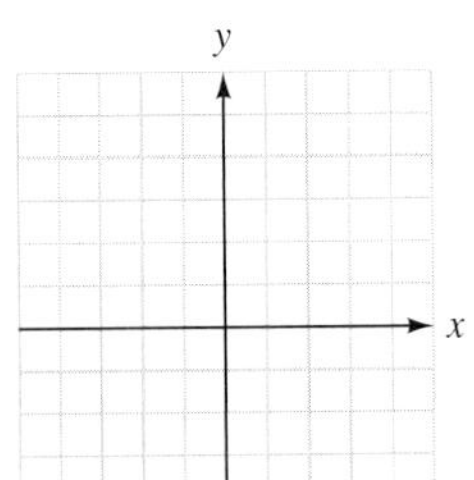

Construct a table of solutions and graph each equation. Compare the result to the graph of $y = |x|$.

33. $y = -|x|$

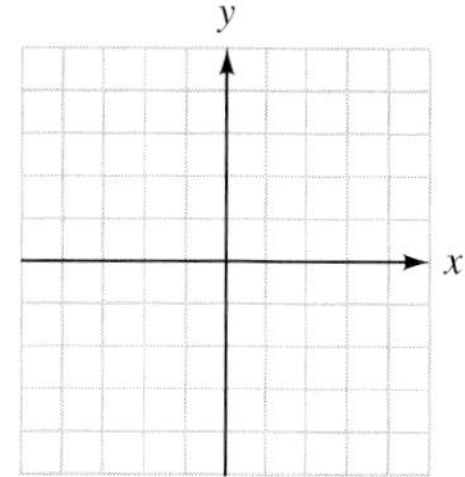

34. $y = |x| - 2$

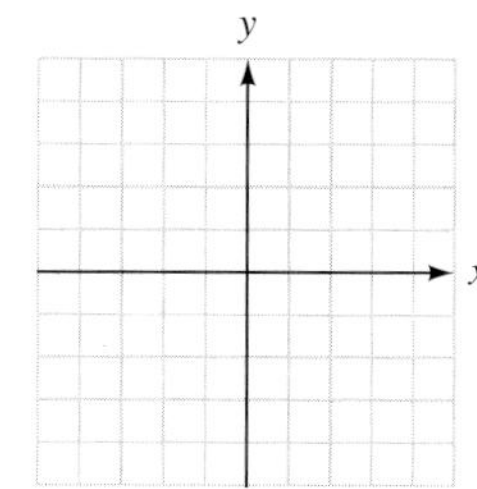

35. $y = |x + 2|$

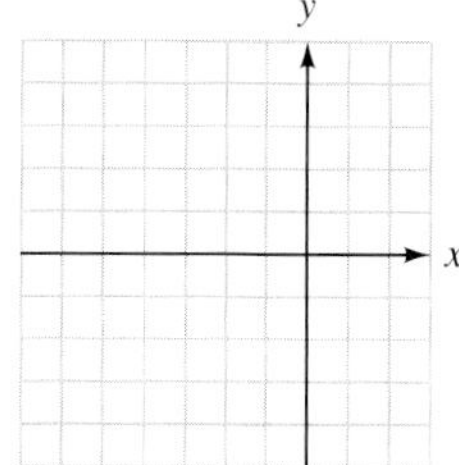

36. $y = |x - 2|$

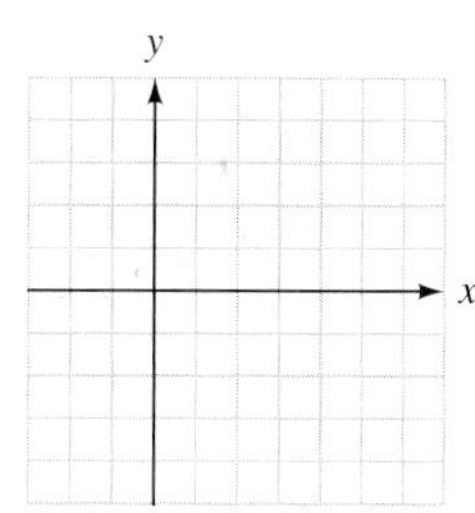

Construct a table of solutions and graph each equation. Compare the result to the graph of $y = x^3$.

37. $y = -x^3$

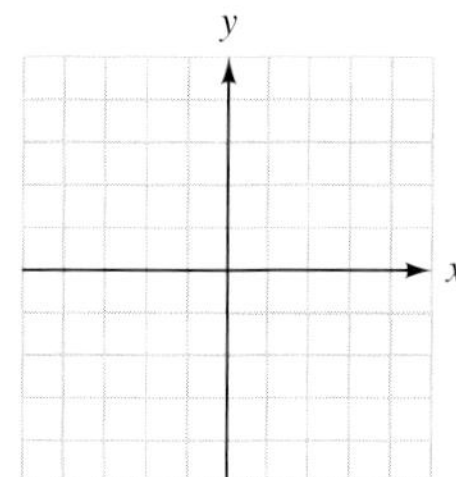

38. $y = x^3 + 2$

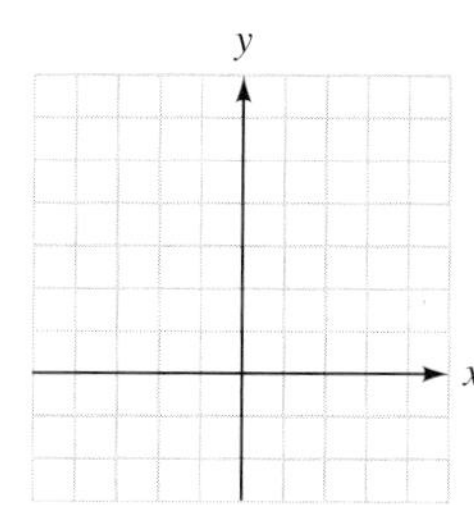

39. $y = x^3 - 2$

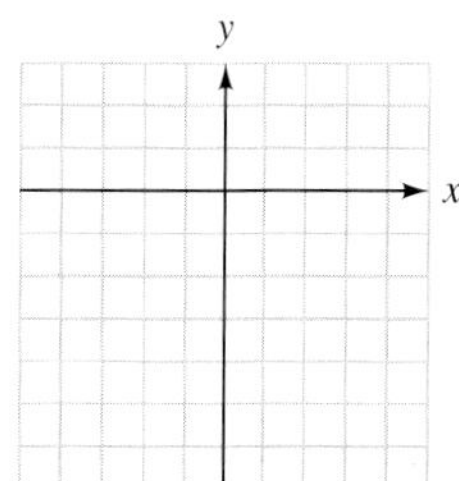

40. $y = (x + 2)^3$

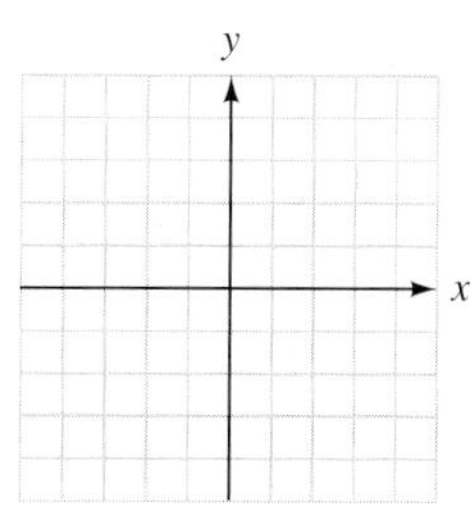

APPLICATIONS

41. EIGHT BALL The illustration shows the path traveled by the 8 ball as it is banked off of a cushion into the right corner pocket. Use the information in the illustration to complete the table.

x	−1	2	5	8
y				

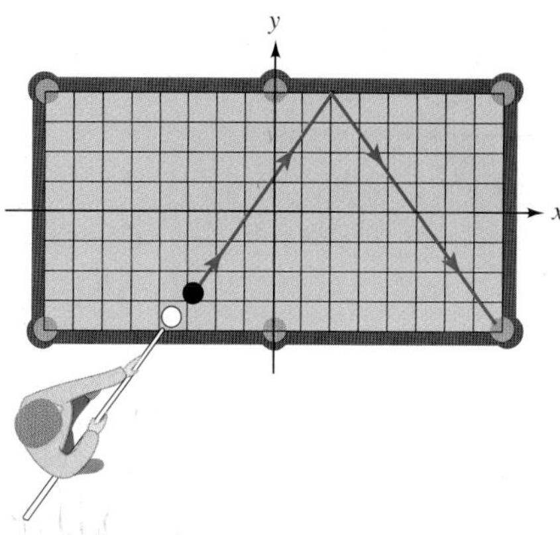

42. TABLE TENNIS The illustration shows the path traveled by a Ping-Pong ball as it bounces off the table. Use the information in the illustration to complete the table.

x	−7	−3	1	3	5
y					

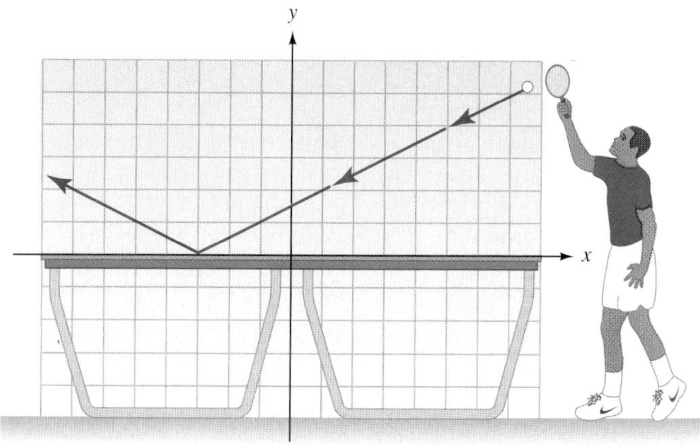

43. SUSPENSION BRIDGES The suspension cables of a bridge hang in the shape of a parabola, as shown. Use the information in the illustration to complete the table.

x	0	2	4	−2	−4
y					

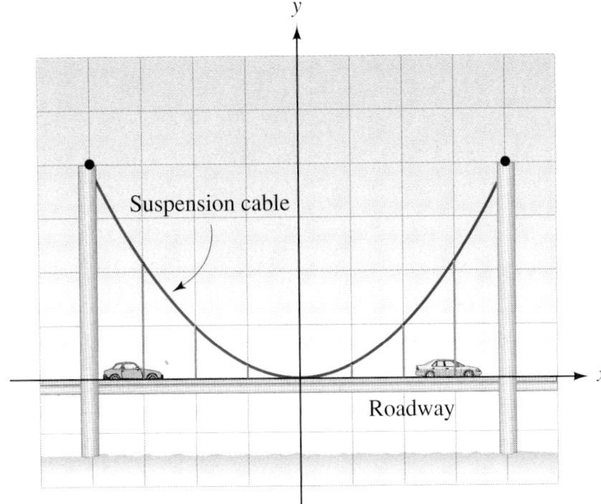

44. FIRE BOATS A stream of water from a high-pressure hose on a fire boat travels in the shape of a parabola. Use the information in the graph to complete the table.

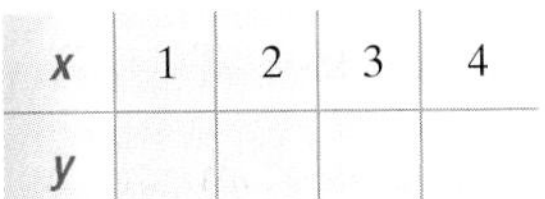

x	1	2	3	4
y				

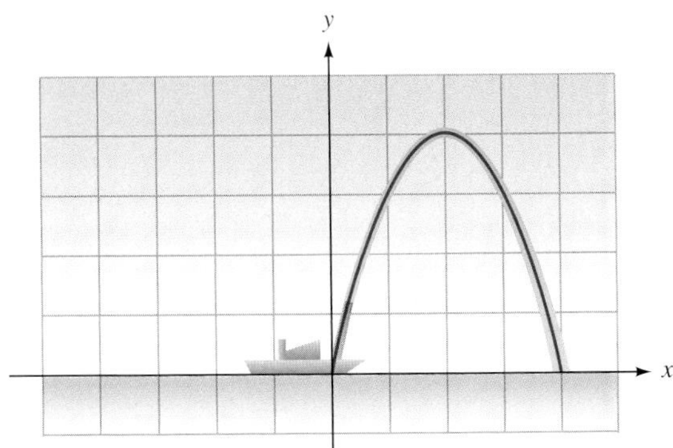

45. MANUFACTURING The graph in the next column shows the relationship between the length l (in inches) of a machine bolt and the cost C (in cents) to manufacture it.

a. What information does the point (2, 8) on the graph give us?

b. How much does it cost to make a 7-inch bolt?

c. What length bolt is the least expensive to make?

d. Describe how the cost changes as the length of the bolt increases.

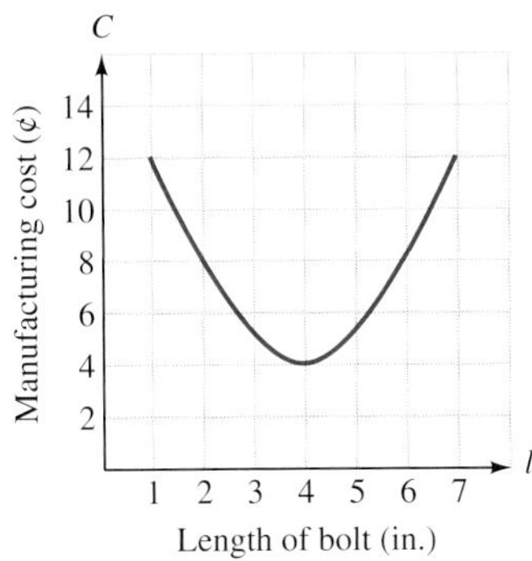

46. SOFTBALL The following graph shows the relationship between the distance d (in feet) traveled by a batted softball and the height h (in feet) it attains.

a. What information does the point (40, 40) on the graph give us?

b. At what distance from home plate does the ball reach its maximum height?

c. Where will the ball land?

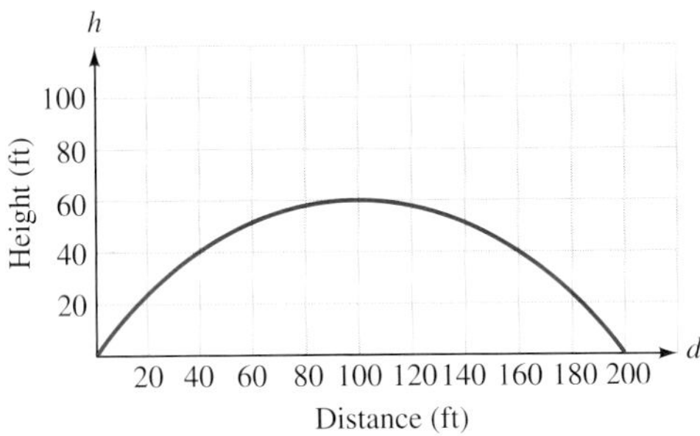

47. MARKET VALUE OF A HOUSE The following graph shows the relationship between the market value v of a house and the time t since it was purchased.

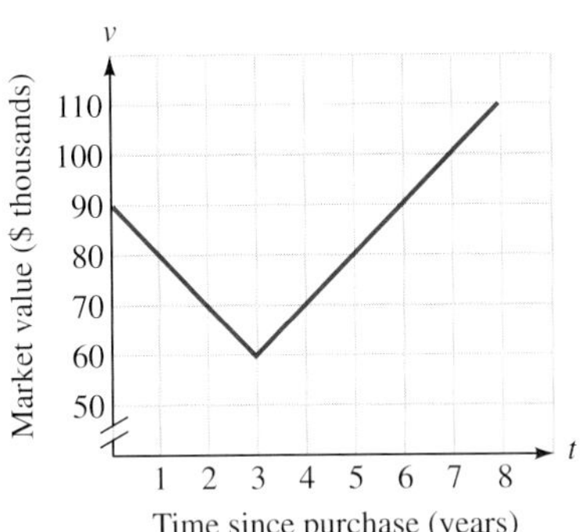

a. What was the purchase price of the house?

b. When did the value of the house reach its lowest point?

c. When did the value of the house begin to surpass the purchase price?

d. Describe how the market value of the house changed over the 8-year period.

48. POLITICAL SURVEYS The following graph shows the relationship between the percent P of those surveyed who rated their senator's job performance as satisfactory or better and the time t she had been in office.

a. When did her job performance rating reach a maximum?

b. When was her job performance rating at or above the 60% mark?

c. Describe how her job performance rating changed over the 12-month period.

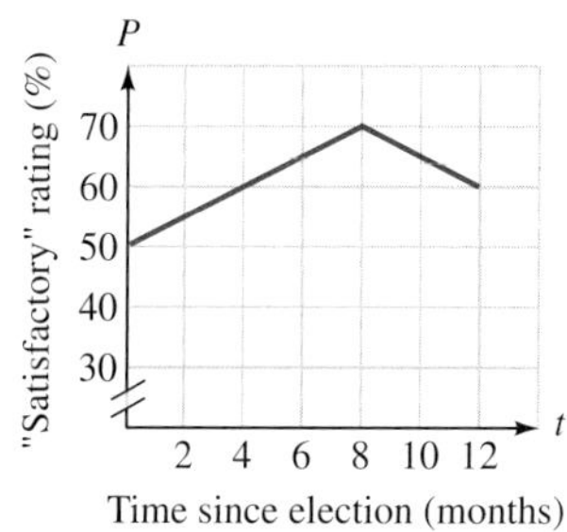

WRITING

49. What is a table of solutions?

50. To graph an equation in two variables, how many solutions of the equation must be found?

51. Give an example of an equation in one variable and an equation in two variables. How do their solutions differ?

52. When we say that $(-2, -6)$ is a solution of $y = x - 4$, what do we mean?

53. On a quiz, students were asked to graph $y = 3x - 1$. One student made the table of solutions on the left. Another student made the one on the right. Which table is incorrect? Or could they both be correct? Explain.

x	y	(x, y)
0	−1	(0, −1)
2	5	(2, 5)
3	8	(3, 8)
4	11	(4, 11)
5	14	(5, 14)

x	y	(x, y)
−2	−7	(−2, −7)
−1	−4	(−1, −4)
1	2	(1, 2)
−3	−10	(−3, −10)
2	5	(2, 5)

54. What does it mean when we say that an equation in two variables has infinitely many solutions?

REVIEW

55. Solve: $\frac{x}{8} = -12$.

56. Combine like terms: $3t - 4T + 5T - 6t$.

57. Is $\frac{x + 5}{6}$ an expression or an equation?

58. What formula is used to find the perimeter of a rectangle?

59. What number is 0.5% of 250?

60. Solve: $-3x + 5 > -7$.

61. Find: $-2.5 - (-2.6)$.

62. Evaluate: $(-5)^3$.

9.3 Graphing Linear Equations

- Linear equations • Solutions of linear equations • Graphing linear equations
- The intercept method • Graphing horizontal and vertical lines
- An application of linear equations

We have previously graphed the equations shown in Figure 9-22. Because the graph of the equation $y = x - 1$ is a line, we call it a *linear equation.* Since the graphs of $y = x^2$, $y = |x|$, and $y = x^3$ are *not* lines, they are *nonlinear equations.*

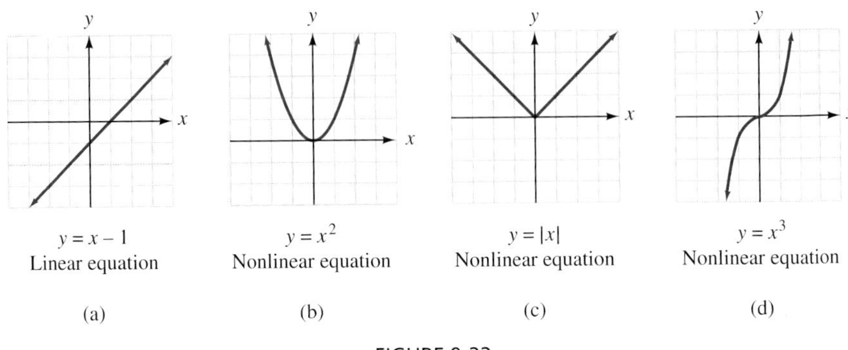

FIGURE 9-22

In this section, we will discuss how to graph linear equations and show how to use their graphs to solve problems.

Linear equations

Any equation, such as $y = x - 1$, whose graph is a straight line is called a **linear equation in x and y.** Some other examples of linear equations are

$$y = \frac{1}{2}x + 2, \quad 3x - 2y = 8, \quad 5y - x + 2 = 0, \quad y = 4, \quad \text{and} \quad x = -3$$

A linear equation in x and y is any equation that can be written in a special form, called **general** (or **standard**) form.

General form of a linear equation

If A, B, and C represent real numbers, the equation

$$Ax + By = C \quad (A \text{ and } B \text{ are not both zero})$$

is called the **general form** (or **standard form**) of the equation of a line.

Whenever possible, we will write the general form $Ax + By = C$ so that A, B, and C are integers and $A \geq 0$. Note that in a linear equation in x and y, the exponents on x and y are 1.

Self Check 1

Which of the following are linear equations and which are nonlinear?

a. $y = |x|$

b. $-x = 6 - y$

c. $y = x$

EXAMPLE 1 Which of the following equations are linear equations?

a. $3x = 1 - 2y$, **b.** $y = x^3 + 1$, and **c.** $y = -\frac{1}{2}x$

Solution

a. Since the equation $3x = 1 - 2y$ can be written in $Ax + By = C$ form, it is a linear equation.

$3x = 1 - 2y$	This is the original equation.
$3x + \mathbf{2y} = 1 - 2y + \mathbf{2y}$	Add $2y$ to both sides.
$3x + 2y = 1$	Simplify the right-hand side: $-2y + 2y = 0$.

Here $A = 3$, $B = 2$, and $C = 1$.

b. Since the exponent on x in $y = x^3 + 1$ is 3, the equation is a nonlinear equation.

c. Since the equation $y = -\frac{1}{2}x$ can be written in $Ax + By = C$ form, it is a linear equation.

$y = -\frac{1}{2}x$	This is the original equation.
$\mathbf{-2}(y) = \mathbf{-2}\left(-\frac{1}{2}x\right)$	Multiply both sides by -2 so that the coefficient of x will be 1.
$-2y = x$	Simplify the right-hand side: $-2(-\frac{1}{2}) = 1$.
$0 = x + 2y$	Add $2y$ to both sides.
$x + 2y = 0$	Write the equation in general form.

Here $A = 1$, $B = 2$, and $C = 0$.

Answers **a.** nonlinear, **b.** linear, **c.** linear

Solutions of linear equations

To find solutions of linear equations, we substitute arbitrary values for one variable and solve for the other.

EXAMPLE 2 Complete the table of solutions for $3x + 2y = 5$.

x	y	(x, y)
7		(7,)
	4	(, 4)

Self Check 2
Complete the table of solutions for $3x + 2y = 5$.

x	y	(x, y)
	−2	(, −2)
5		(5,)

Solution In the first row, we are given an x-value of 7. To find the corresponding y-value, we substitute 7 for x and solve for y.

$3x + 2y = 5$	This is the original equation.
$3(\mathbf{7}) + 2y = 5$	Substitute 7 for x.
$21 + 2y = 5$	Perform the multiplication: $3(7) = 21$.
$2y = -16$	Subtract 21 from both sides: $5 - 21 = -16$.
$y = -8$	Divide both sides by 2.

A solution of $3x + 2y = 5$ is $(7, -8)$.

In the second row, we are given a y-value of 4. To find the corresponding x-value, we substitute 4 for y and solve for x.

$3x + 2y = 5$	This is the original equation.
$3x + 2(\mathbf{4}) = 5$	Substitute 4 for y.
$3x + 8 = 5$	Perform the multiplication: $2(4) = 8$.
$3x = -3$	Subtract 8 from both sides: $5 - 8 = -3$.
$x = -1$	Divide both sides by 3.

Another solution is $(-1, 4)$. The completed table is shown on the right.

x	y	(x, y)
7	**−8**	(7, **−8**)
−1	4	(**−1**, 4)

Answer

x	y	(x, y)
3	−2	(3, −2)
5	−5	(5, −5)

Graphing linear equations

Since two points determine a line, only two points are needed to graph a linear equation. However, we will often plot a third point as a check. If the three points do not lie on a straight line, then at least one of them is in error.

Graphing linear equations

1. Find three pairs (x, y) that satisfy the equation by picking arbitrary numbers for x and finding the corresponding values of y.
2. Plot each resulting pair (x, y) on a rectangular coordinate system. If the three points do not lie on a straight line, check your computations.
3. Draw the straight line passing through the points.

Self Check 3

Graph $y = -3x + 2$ and compare the result to the graph of $y = -3x$. What do you notice?

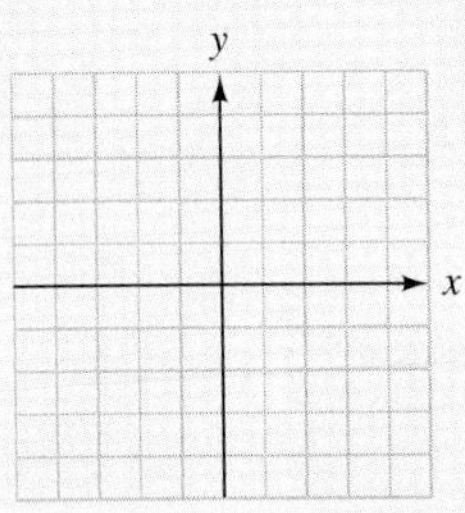

Answer It is a line 2 units above the graph of $y = -3x$.

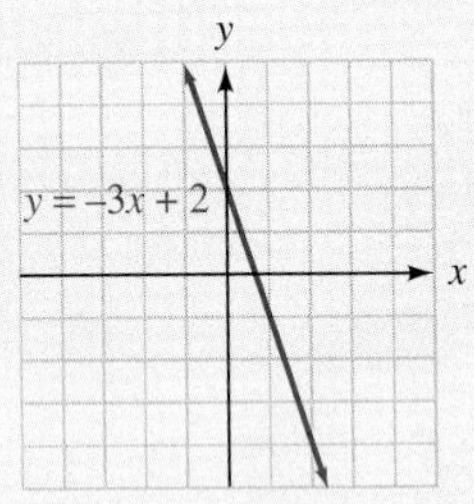

EXAMPLE 3 Graph: $y = -3x$.

Solution To find three ordered pairs that satisfy the equation, we begin by choosing three x-values: -2, 0, and 2.

If $x = -2$	If $x = 0$	If $x = 2$
$y = -3x$	$y = -3x$	$y = -3x$
$y = -3(-2)$	$y = -3(0)$	$y = -3(2)$
$y = 6$	$y = 0$	$y = -6$

We enter the results in a table of solutions, plot the points, and draw a straight line through the points. The graph appears in Figure 9-23. Check this work with a graphing calculator.

$y = -3x$

x	y	(x, y)
-2	6	$(-2, 6)$
0	0	$(0, 0)$
2	-6	$(2, -6)$

This point will serve as a check.

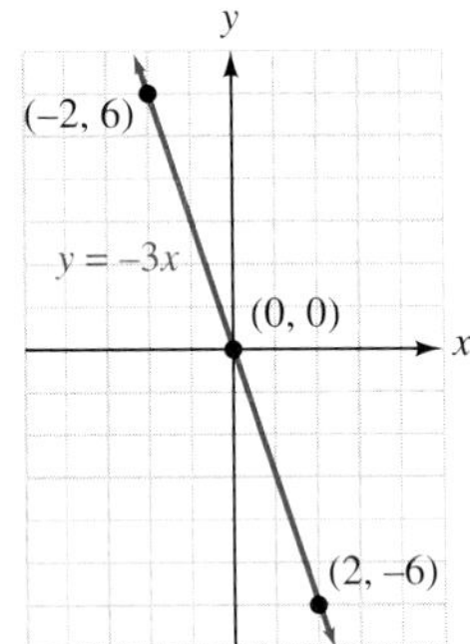

FIGURE 9-23

When graphing linear equations, it is often easier to find solutions of the equation if it is first solved for y.

Self Check 4

Solve $3y = 3 + x$ for y. Then graph the equation

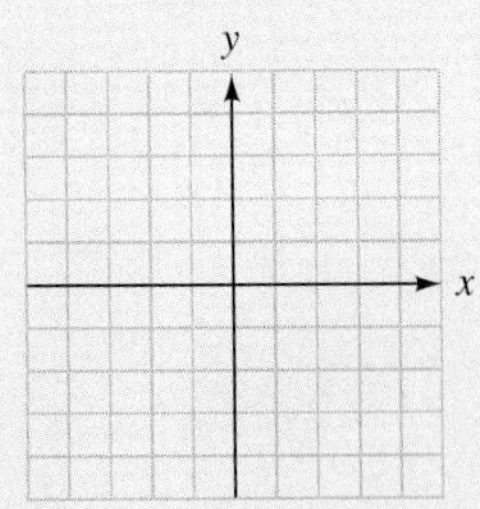

EXAMPLE 4 Graph: $2y = 4 - x$.

Solution To solve for y, we undo the multiplication of 2 by dividing both sides by 2.

$$2y = 4 - x$$

$$\frac{2y}{2} = \frac{4}{2} - \frac{x}{2}$$ On the right-hand side, dividing each term by 2 is equivalent to dividing the entire side by 2: $\frac{4-x}{2} = \frac{4}{2} - \frac{x}{2}$.

$$y = 2 - \frac{x}{2}$$ Simplify: $\frac{4}{2} = 2$.

Since each value of x will be divided by 2, we will choose values of x that are divisible by 2. Three such choices are -4, 0, and 4. If $x = -4$, we have

$$y = 2 - \frac{x}{2}$$

$$y = 2 - \frac{-4}{2} \quad \text{Substitute } -4 \text{ for } x.$$

$$y = 2 - (-2) \quad \text{Divide: } \tfrac{-4}{2} = -2.$$

$$y = 4 \quad \text{Perform the subtraction.}$$

A solution is $(-4, 4)$. This pair and two others satisfying the equation are shown in the table in Figure 9-24. If we plot the points and draw a straight line through them, we will obtain the graph shown in the figure. Check this work with a graphing calculator.

$y = 2 - \frac{x}{2}$

x	y	(x, y)
−4	4	(−4, 4)
0	2	(0, 2)
4	0	(4, 0)

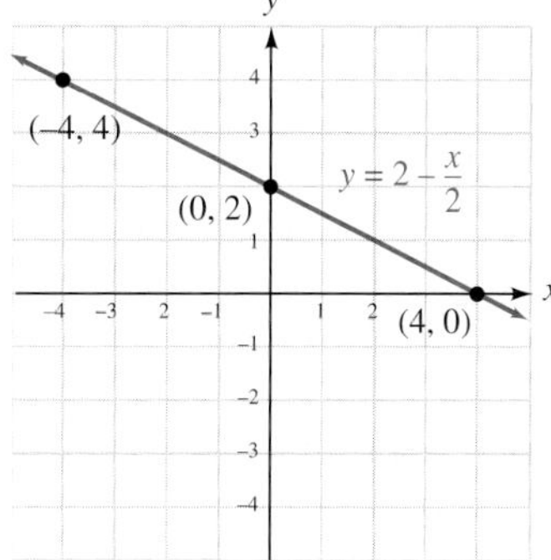

FIGURE 9-24

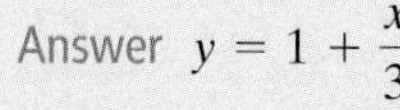

Answer $y = 1 + \frac{x}{3}$

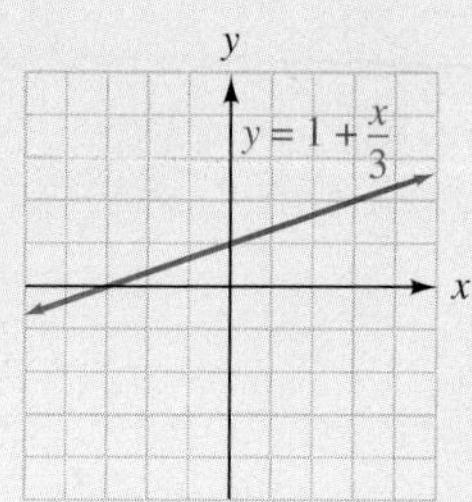

The intercept method

In Figure 9-25, the graph of $3x + 4y = 12$ intersects the y-axis at the point $(0, 3)$; we call this point the ***y*-intercept** of the graph. Since the graph intersects the x-axis at $(4, 0)$, the point $(4, 0)$ is the ***x*-intercept.**

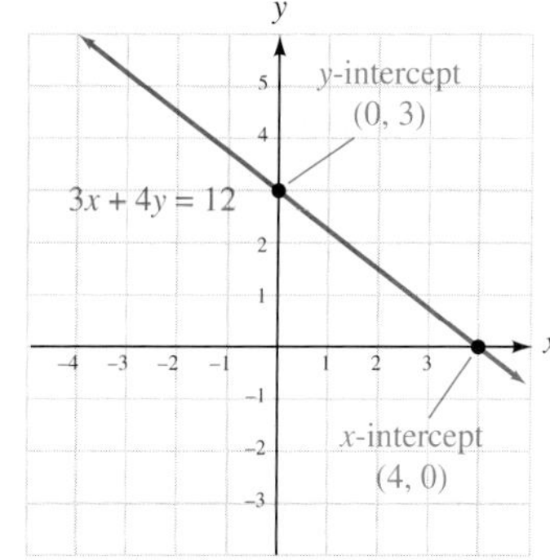

FIGURE 9-25

In general, we have the following definitions.

y- and *x*-intercepts

The ***y*-intercept** of a line is the point $(0, b)$ where the line intersects the y-axis. To find b, substitute 0 for x in the equation of the line and solve for y.

The ***x*-intercept** of a line is the point $(a, 0)$ where the line intersects the x-axis. To find a, substitute 0 for y in the equation of the line and solve for x.

Plotting the x- and y-intercepts of a graph and drawing a straight line through them is called the **intercept method of graphing a line.** This method is useful when graphing equations written in general form.

Self Check 5
Graph $4x + 3y = 6$ using the intercept method.

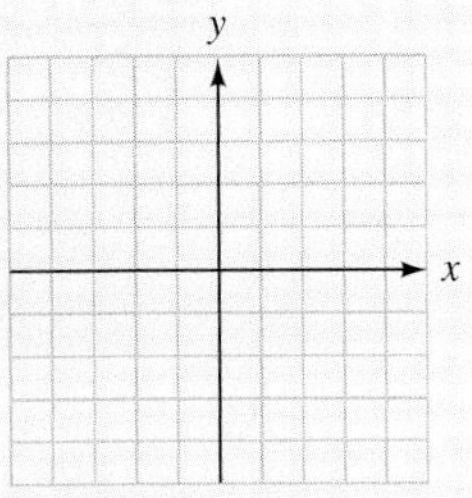

EXAMPLE 5 Graph: $3x - 2y = 8$.

Solution To find the x-intercept, we let $y = 0$ and solve for x.

$$3x - 2y = 8$$

$$3x - 2(\mathbf{0}) = 8 \quad \text{Substitute 0 for } y.$$

$$3x = 8 \quad \text{Simplify the left-hand side: } 2(0) = 0.$$

$$x = \frac{8}{3} \quad \text{Divide both sides by 3.}$$

The x-intercept is $\left(\frac{8}{3}, 0\right)$, which can be written $\left(2\frac{2}{3}, 0\right)$. This ordered pair is entered in the table in Figure 9-26. To find the y-intercept, we let $x = 0$ and solve for y.

$$3x - 2y = 8$$

$$3(\mathbf{0}) - 2y = 8 \quad \text{Substitute 0 for } x.$$

$$-2y = 8 \quad \text{Simplify the left-hand side: } 3(0) = 0.$$

$$y = -4 \quad \text{Divide both sides by } -2.$$

The y-intercept is $(0, -4)$. It is entered in the table below. As a check, we find one more point on the line. If $x = 4$, then $y = 2$. We plot these three points and draw a straight line through them. The graph of $3x - 2y = 8$ is shown in Figure 9-26.

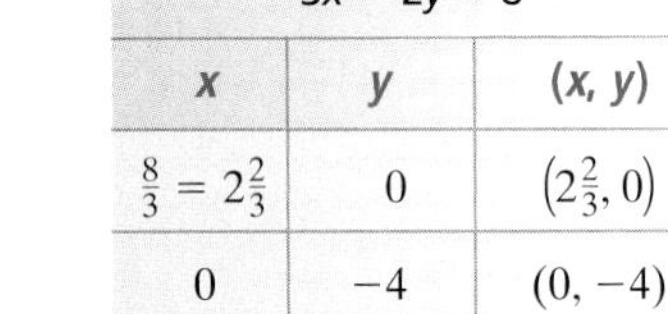

$3x - 2y = 8$		
x	y	(x, y)
$\frac{8}{3} = 2\frac{2}{3}$	0	$\left(2\frac{2}{3}, 0\right)$
0	-4	$(0, -4)$
4	2	$(4, 2)$

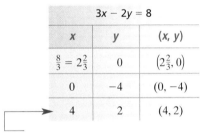

This point serves as a check.

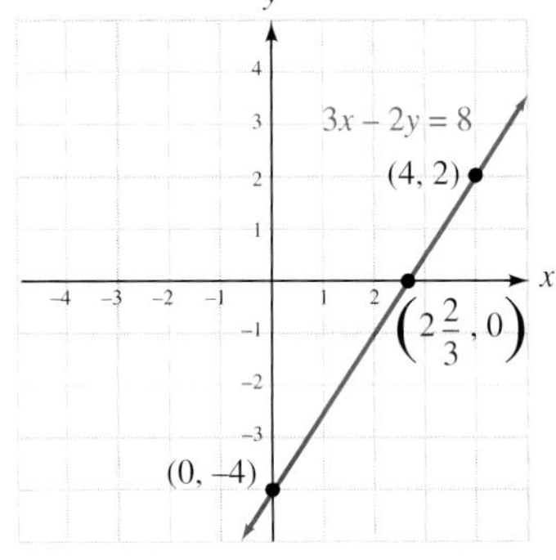

FIGURE 9-26

Answer

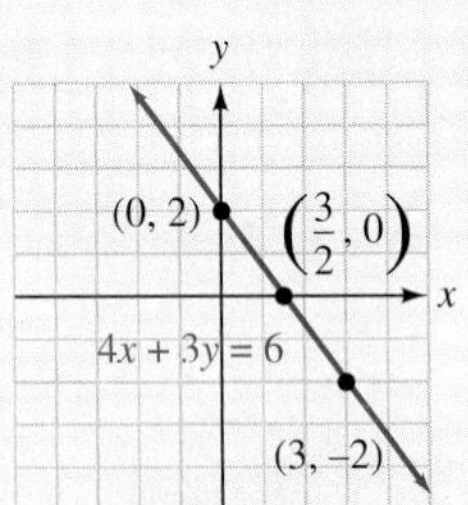

Graphing horizontal and vertical lines

Equations such as $y = 4$ and $x = -3$ are linear equations, because they can be written in the general form $Ax + By = C$.

$$y = 4 \quad \text{is equivalent to} \quad 0x + 1y = 4$$

$$x = -3 \quad \text{is equivalent to} \quad 1x + 0y = -3$$

We now discuss how to graph these types of linear equations.

EXAMPLE 6 Graph: $y = 4$.

Solution We can write the equation in general form as $0x + y = 4$. Since the coefficient of x is 0, the numbers chosen for x have no effect on y. The value of y is always 4. For example, if $x = 2$, we have

$0x + y = 4$ This is the original equation written in general form.

$0(\mathbf{2}) + y = 4$ Substitute 2 for x.

$y = 4$ Simplify the left-hand side: $0(2) = 0$.

The table of solutions shown in Figure 9-27 contains three ordered pairs that satisfy the equation $y = 4$. If we plot the points and draw a straight line through them, the result is a horizontal line. The y-intercept is $(0, 4)$, and there is no x-intercept.

$y = 4$		
x	y	(x, y)
2	4	(2, 4)
−1	4	(−1, 4)
−3	4	(−3, 4)

↑ Note that each y-coordinate is 4.

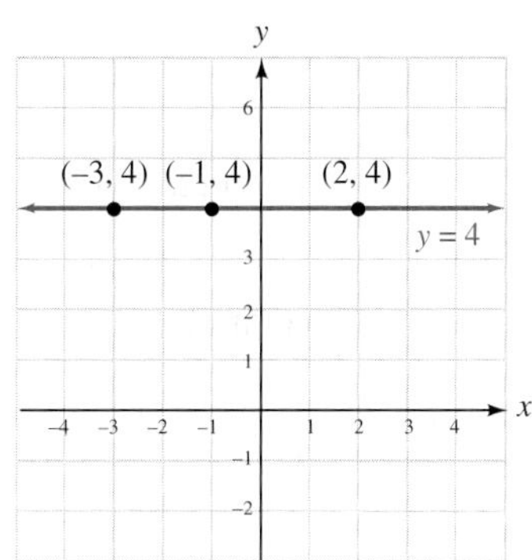

FIGURE 9-27

Self Check 6
Graph: $y = -2$.

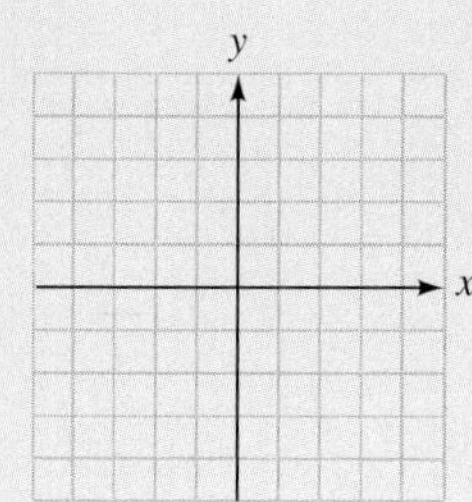

Answer

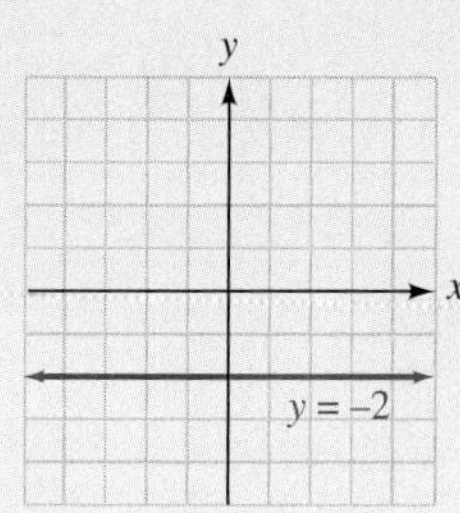

EXAMPLE 7 Graph: $x = -3$.

Solution We can write the equation in general form as $x + 0y = -3$. Since the coefficient of y is 0, the numbers chosen for y have no effect on x. The value of x is always -3. For example, if $y = -2$, we have

$x + 0y = -3$ This is the original equation written in general form.

$x + 0(\mathbf{-2}) = -3$ Substitute -2 for y.

$x = -3$ Simplify the left-hand side: $0(-2) = 0$.

The table of solutions shown in Figure 9-28 contains three ordered pairs that satisfy the equation $x = -3$. If we plot the points and draw a line through them, the result is a vertical line. The x-intercept is $(-3, 0)$, and there is no y-intercept.

$x = -3$		
x	y	(x, y)
−3	−2	(−3, −2)
−3	0	(−3, 0)
−3	3	(−3, 3)

↑ Note that each x-coordinate is −3.

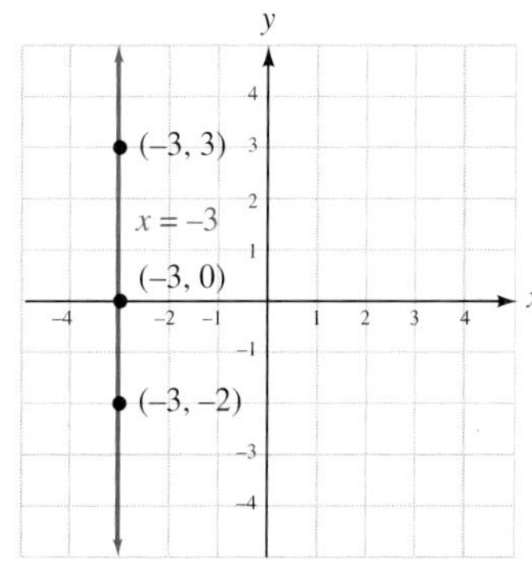

FIGURE 9-28

Self Check 7
Graph: $x = 4$.

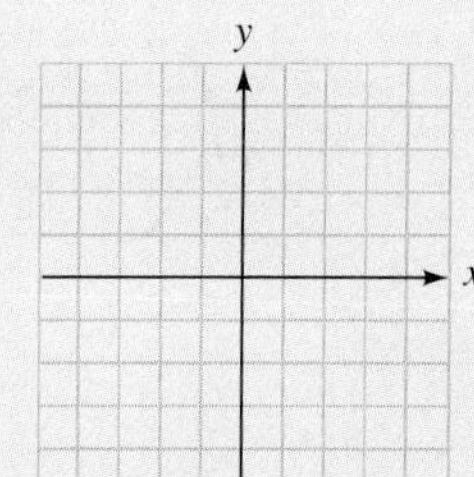

Answer

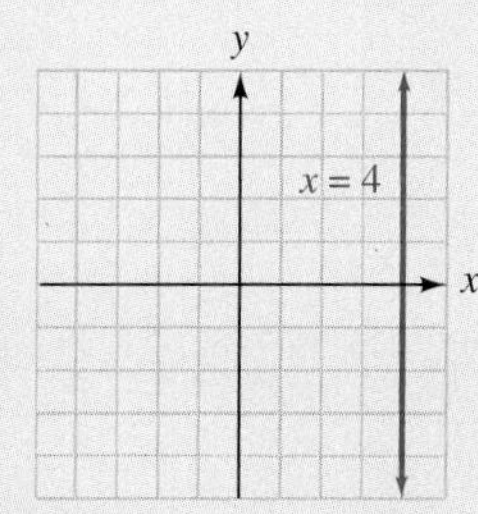

From the results of Examples 6 and 7, we have the following facts.

Equations of horizontal and vertical lines

The equation $y = b$ represents the horizontal line that intersects the y-axis at $(0, b)$. If $b = 0$, the line is the x-axis.

The equation $x = a$ represents the vertical line that intersects the x-axis at $(a, 0)$. If $a = 0$, the line is the y-axis.

An application of linear equations

EXAMPLE 8 **Birthday parties.** A restaurant offers a party package that includes food, drinks, cake, and party favors for a cost of \$25 plus \$3 per child. Write a linear equation that will give the cost for a party of any size, and then graph the equation.

Solution We can let c represent the cost of the party. The cost c is the sum of the basic charge of \$25 and the cost per child times the number of children attending. If the number of children attending is n, at \$3 per child, the total cost for the children is \$$3n$.

The cost	is	the basic \$25 charge	plus	\$3	times	the number of children.
c	$=$	25	$+$	3	$\cdot$	n

For the equation $c = 25 + 3n$, the independent variable (input) is n, the number of children. The dependent variable (output) is c, the cost of the party. We will find three points on the graph of the equation by choosing n-values of 0, 5, and 10 and finding the corresponding c-values. The results are shown in the table.

If $n = 0$

$c = 25 + 3(\mathbf{0})$

$c = 25$

If $n = 5$

$c = 25 + 3(\mathbf{5})$

$c = 25 + 15$

$c = 40$

If $n = 10$

$c = 25 + 3(\mathbf{10})$

$c = 25 + 30$

$c = 55$

$c = 25 + 3n$

n	c
0	25
5	40
10	55

Next, we graph the points and draw a line through them (Figure 9-29). We don't draw an arrowhead on the left, because it doesn't make sense to have a negative number of children attend a party. Note that the c-axis is scaled in units of \$5 to accommodate costs ranging from \$0 to \$65. We can use the graph to determine the cost of a party of any size. For example, to find the cost of a party with 8 children, we locate 8 on the horizontal axis and then move up to find a point on the graph directly above the 8. Since the coordinates of that point are (8, 49), the cost for 8 children would be \$49.

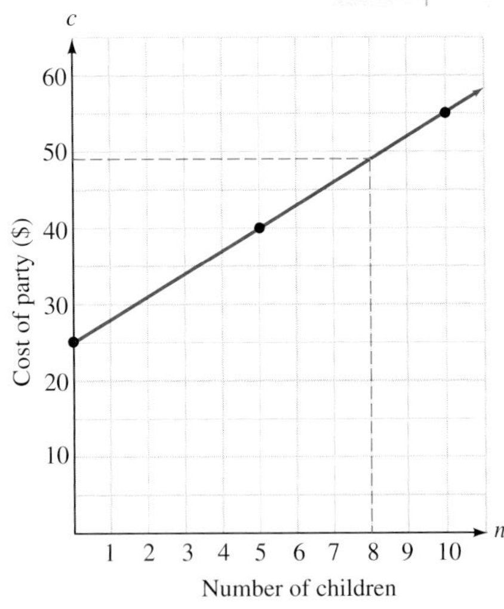

FIGURE 9-29

Section 9.3 STUDY SET

VOCABULARY *Fill in the blanks.*

1. An equation whose graph is a line and whose variables are to the first power is called a ______ equation.
2. The equation $Ax + By = C$ is the ______ form of the equation of a line.
3. The ______ of a line is the point $(0, b)$ where the line intersects the y-axis.
4. The ______ of a line is the point $(a, 0)$ where the line intersects the x-axis.
5. Lines parallel to the y-axis are ______ lines.
6. Lines parallel to the x-axis are ______ lines.

CONCEPTS

7. Classify each equation as linear or nonlinear.
   a. $y = x^3$
   b. $2x + 3y = 6$
   c. $y = |x + 2|$
   d. $x = -2$
   e. $y = -x^2$
8. Classify each of the following as the graph of a linear equation or of a nonlinear equation.

a.

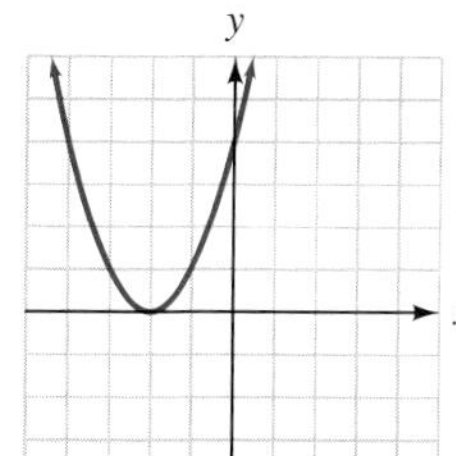

b.

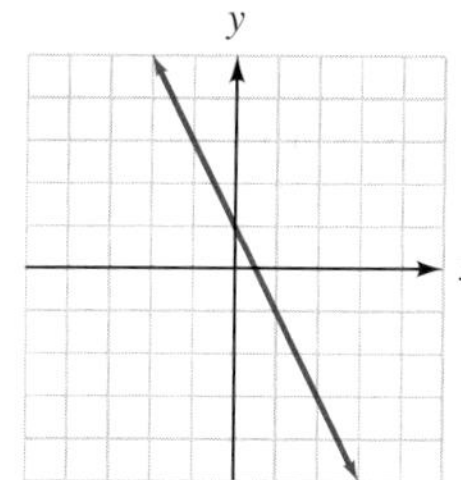

c.

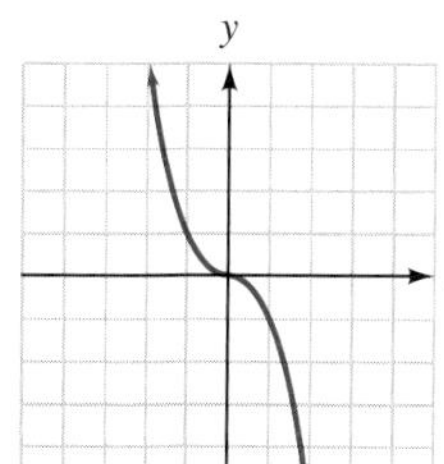

d.

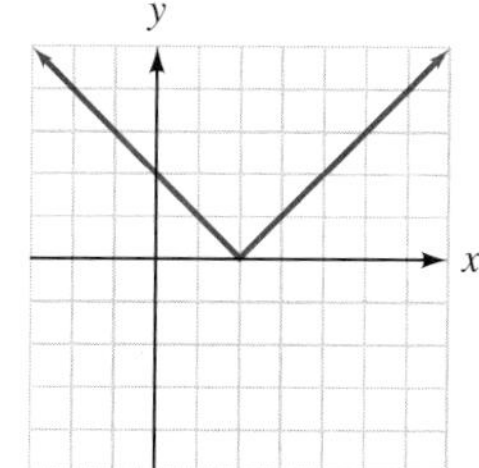

9. Find the power of each variable in the following equations.
   a. $y = 2x - 6$
   b. $y = x^2 - 6$
   c. $y = x^3 + 2$
10. In a linear equation in x and y, what are the exponents on x and y?

Complete each table.

11. $5y = 2x + 10$

x	y
10	
	0
5	

12. $2x + 4y = 24$

x	y
4	
	7
−4	

13. $x - 2y = 4$

x	y
0	
	0
1	

14. $5x - y = 3$

x	y
0	
	0
1	

Consider the graph of a linear equation shown below.

15. Why will the coordinates of point A, when substituted into the equation, yield a true statement?
16. Why will the coordinates of point B, when substituted into the equation, yield a false statement?

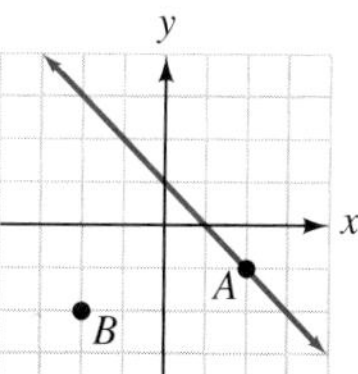

17. A student found three solutions of a linear equation and plotted them as shown. What conclusion can be made?

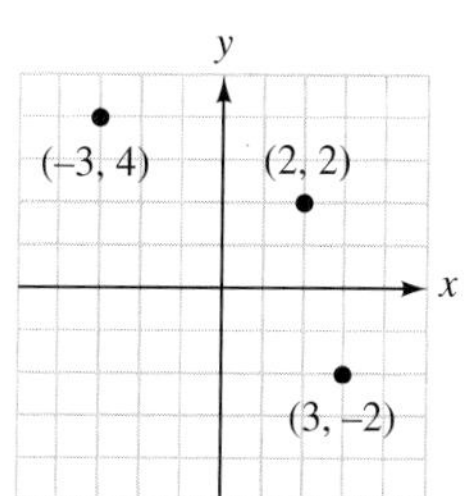

18. How many solutions are there for a linear equation in two variables?

19. Give the x- and y-intercepts of the following graph.

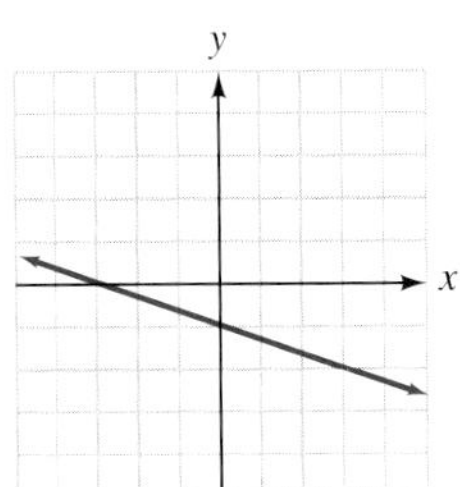

20. On the same coordinate system:

a. Draw the graph of a line with no x-intercept.

b. Draw the graph of a line with no y-intercept.

c. Draw a line with an x-intercept of $(2, 0)$.

d. Draw a line with a y-intercept of $(0, -\frac{5}{2})$.

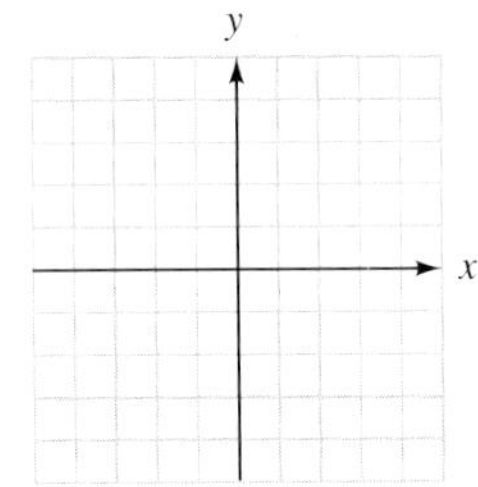

21. Fill in the blanks.

a. To find the y-intercept of the graph of a linear equation, we let ___ = 0 and solve for ___.

b. To find the x-intercept of the graph of a linear equation, we let ___ = 0 and solve for ___.

22. a. What is another name for the line $x = 0$?

b. What is another name for the line $y = 0$?

NOTATION

23. Write each equation in general form.

a. $4x = y + 6$

b. $2y = x$

c. $x - 9 = -3y$

d. $x = 12$

24. Solve each equation for y.

a. $x + y = 8$

b. $2x - y = 8$

c. $3x + \dfrac{y}{2} = 4$

d. $y - 2 = 0$

PRACTICE

Find three solutions of the equation and graph it.

25. $y = -x + 2$

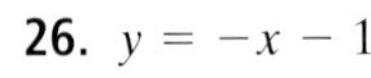

26. $y = -x - 1$

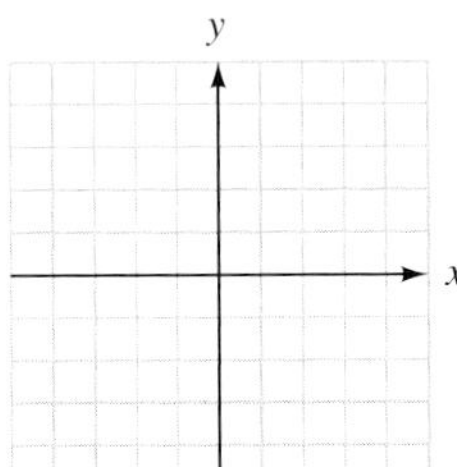

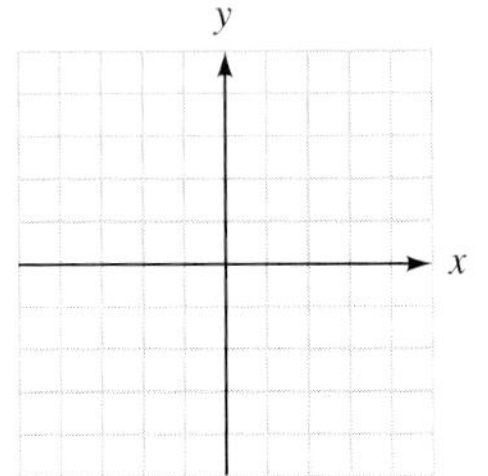

27. $y = 2x + 1$

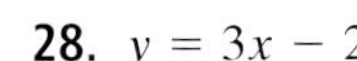

28. $y = 3x - 2$

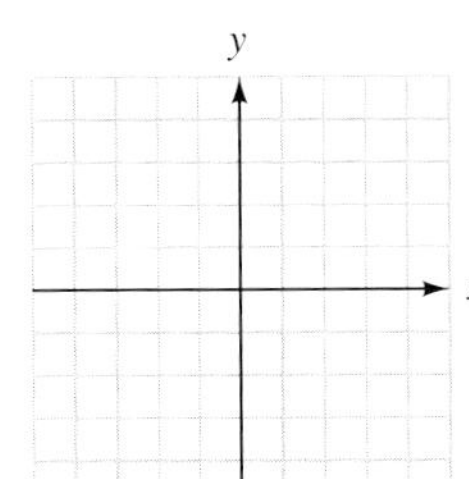

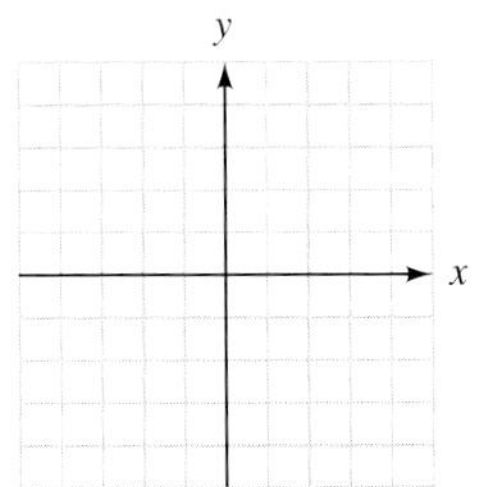

29. $y = x$

30. $y = 3x$

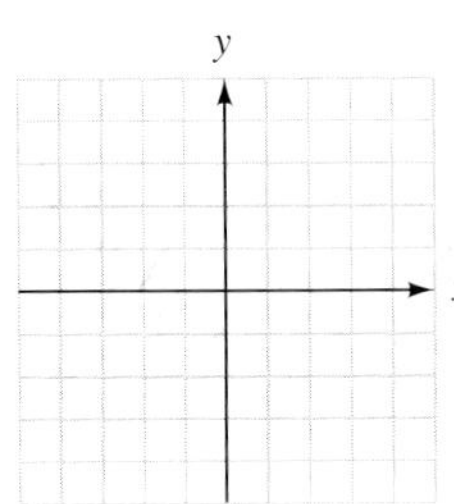

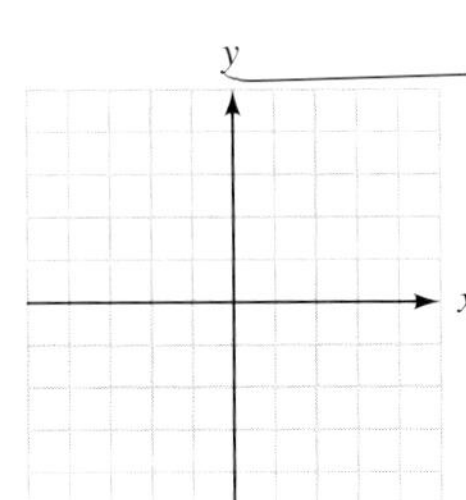

31. $y = -3x$

32. $y = -2x$

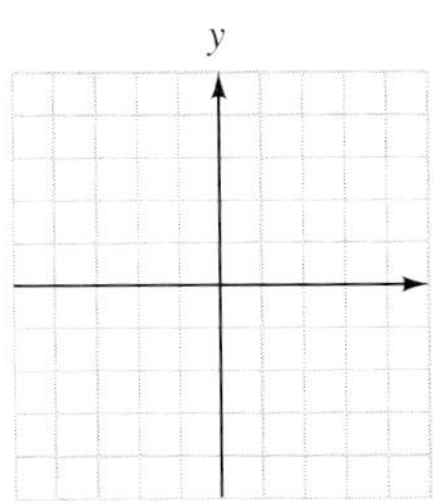

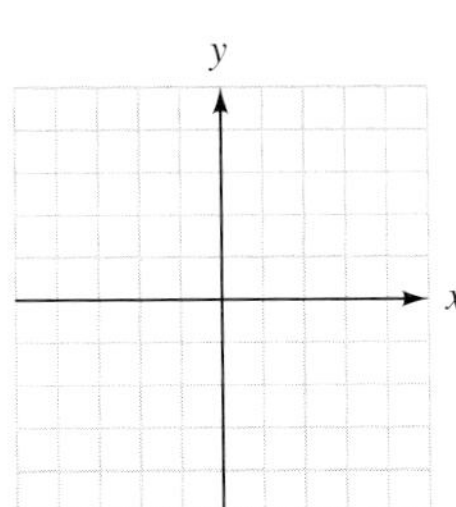

33. $y = \dfrac{x}{3}$

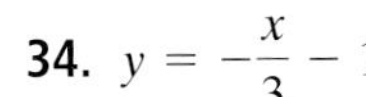

34. $y = -\dfrac{x}{3} - 1$

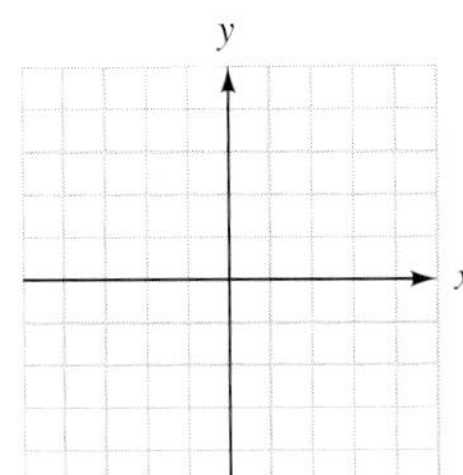

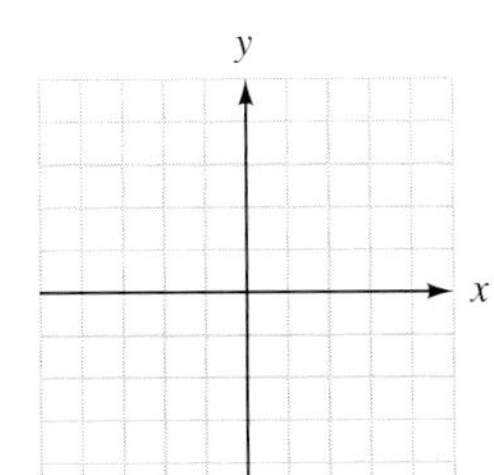

35. $y = -\frac{3}{2}x + 2$

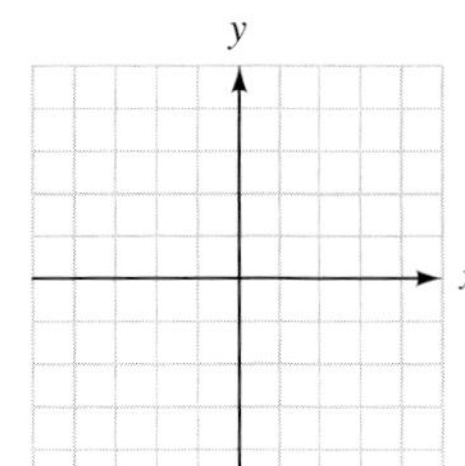

36. $y = \frac{2}{3}x - 2$

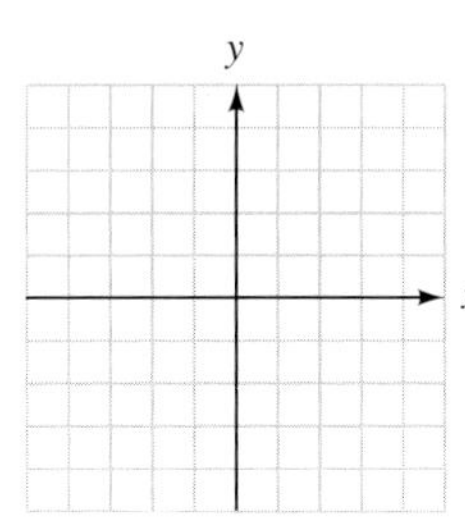

Solve each equation for y, find three solutions of the equation, and then graph it.

37. $2y = 4x - 6$

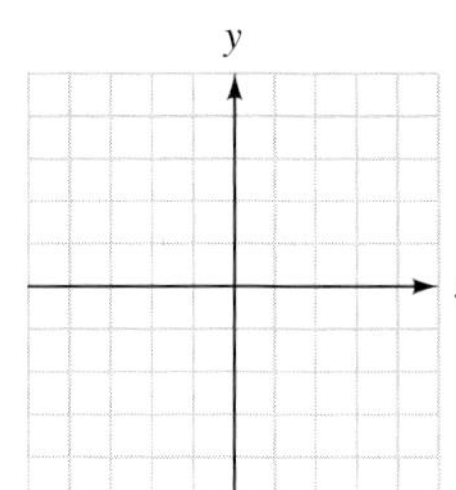

38. $3y = 6x - 3$

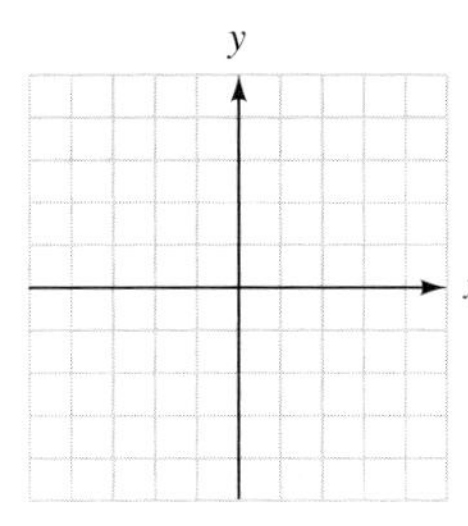

39. $2y = x - 4$

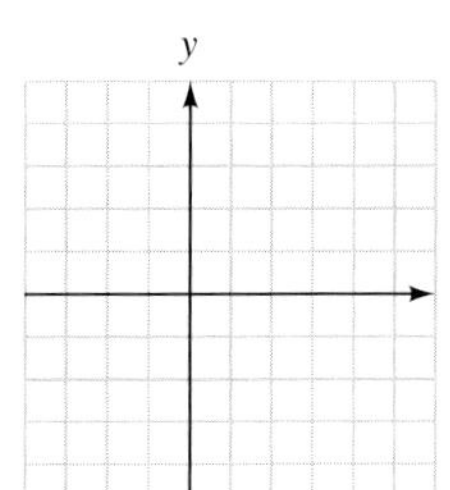

40. $4y = x + 16$

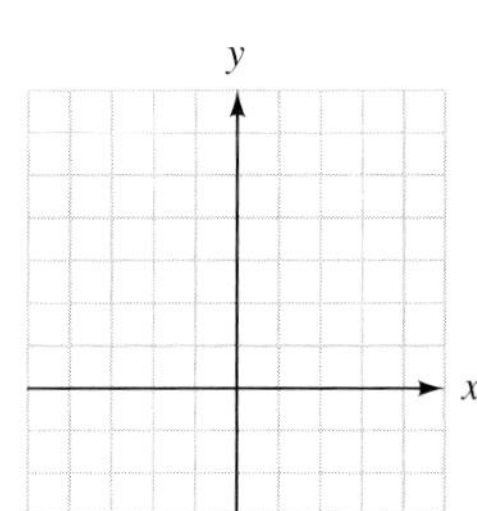

41. $2y + x = -2$

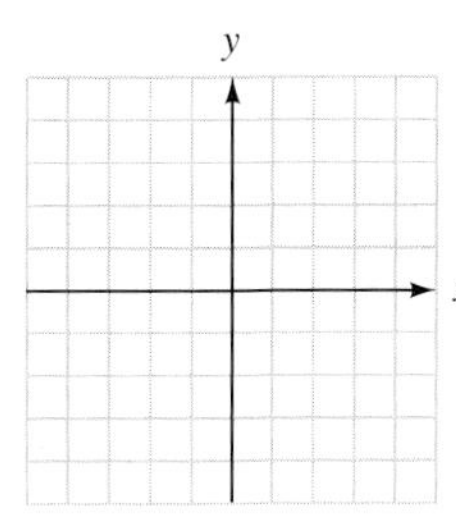

42. $4y + 2x = -8$

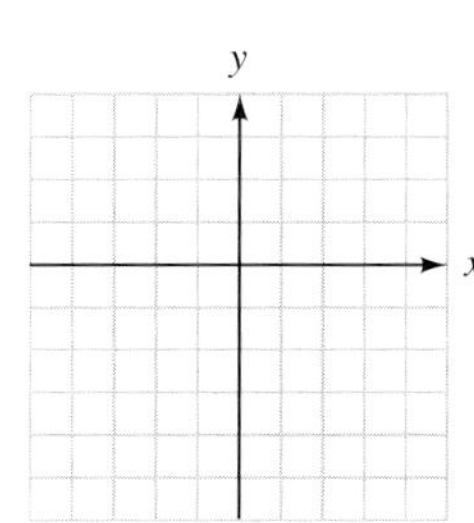

Graph each equation.

43. $y = 4$

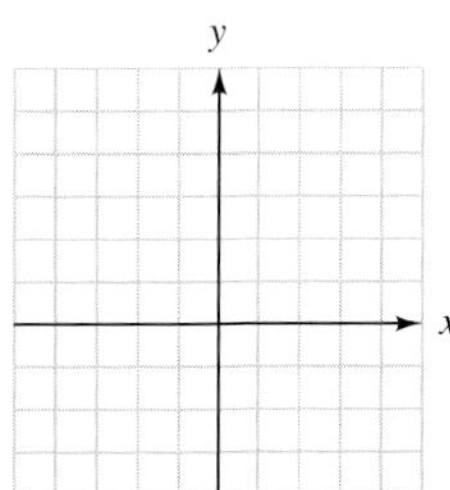

44. $y = -3$

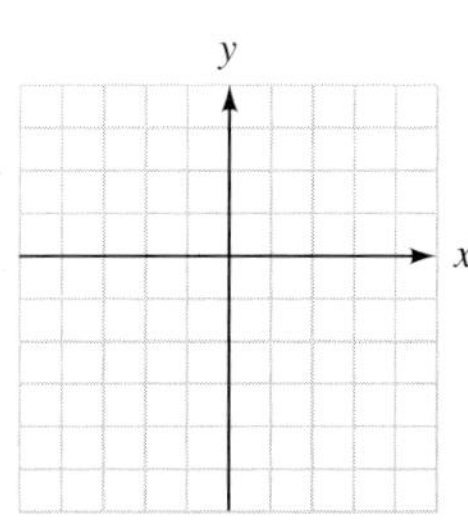

45. $x = -2$

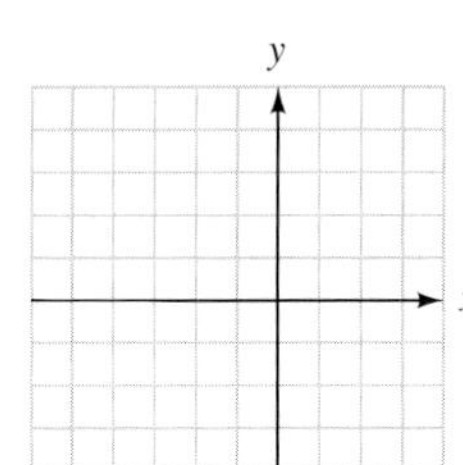

46. $x = 5$

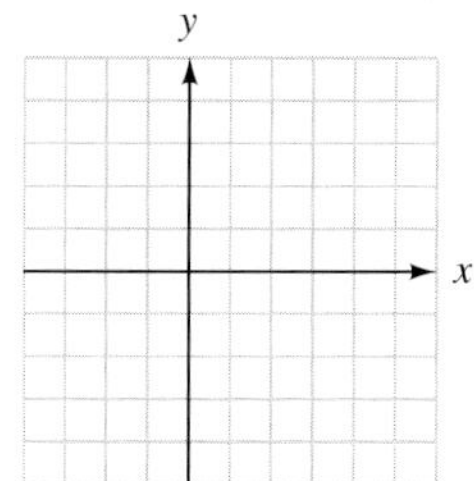

47. $y = -\frac{1}{2}$

48. 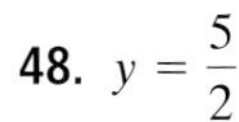 $y = \frac{5}{2}$

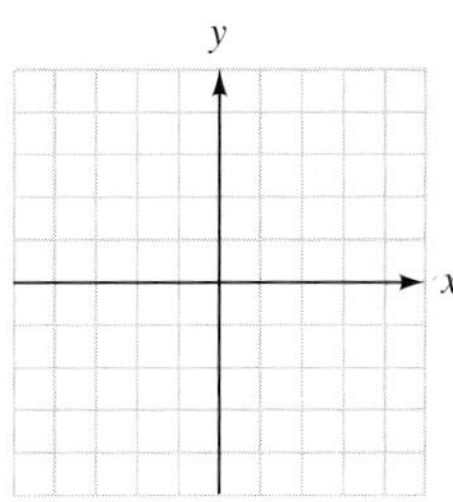

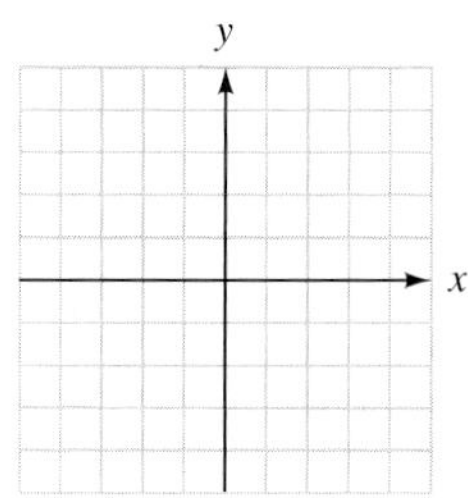

49. $x = \frac{4}{3}$

50. 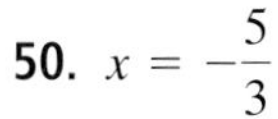 $x = -\frac{5}{3}$

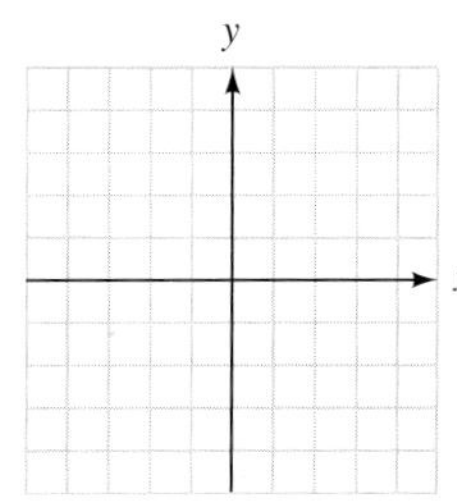

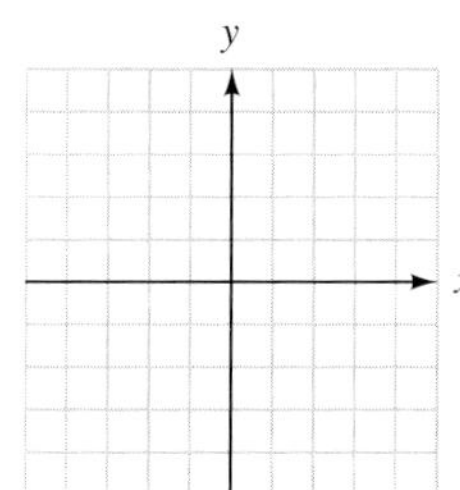

Graph each equation using the intercept method.

51. $2y - 2x = 6$

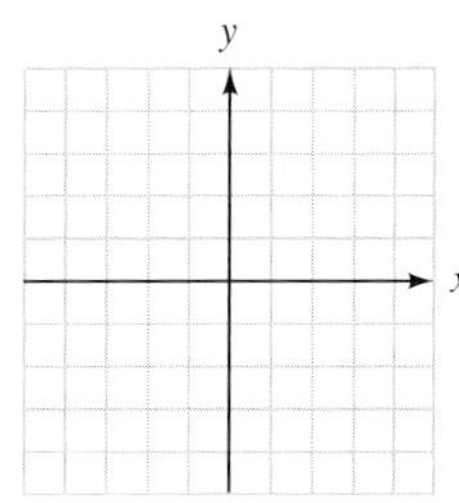

52. $3x - 3y = 9$

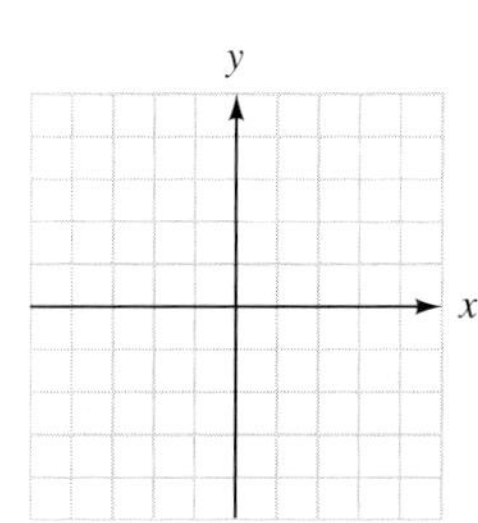

53. $-4y + 9x = -9$

54. $-4y + 5x = -15$

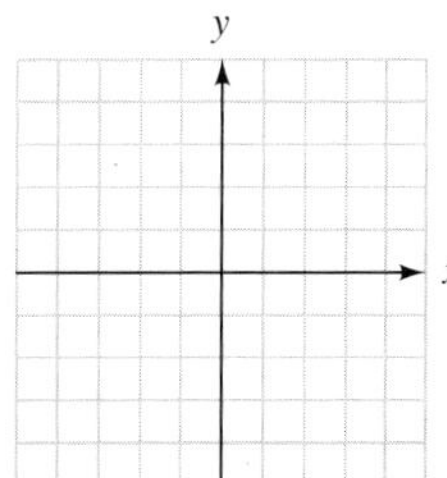

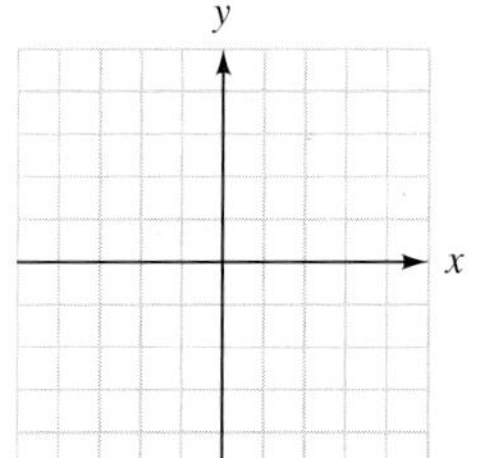

55. $4x + 5y = 20$

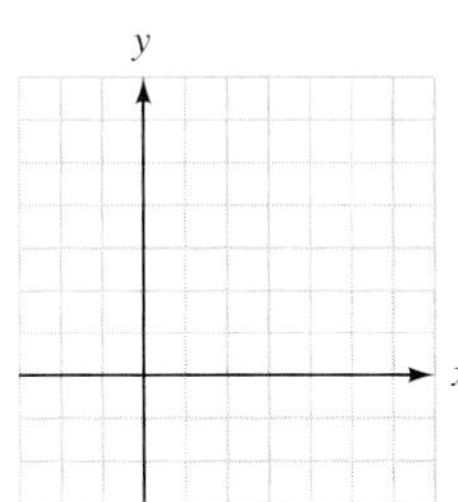

56. $3x + y = -3$

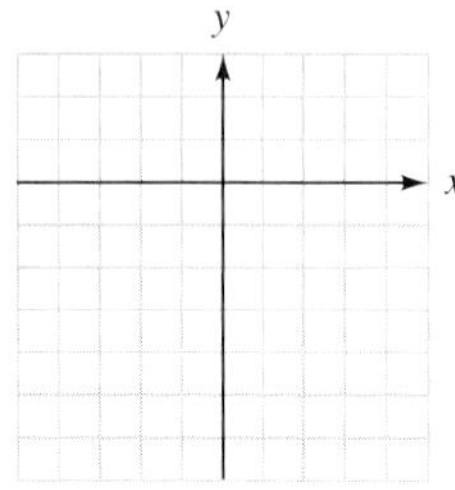

57. $3x + 4y = 12$

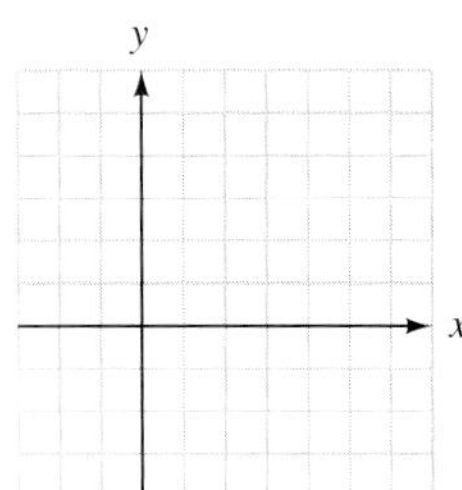

58. $4x - 3y = 12$

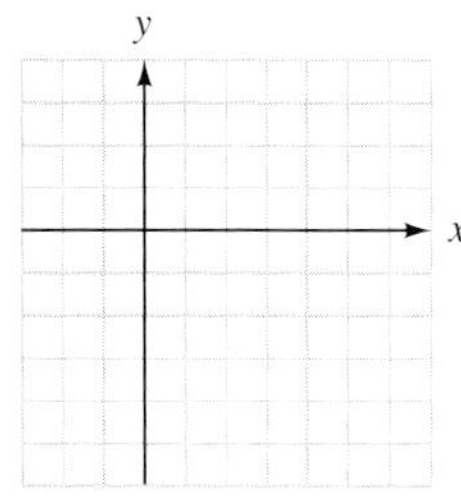

59. $15y + 5x = -15$

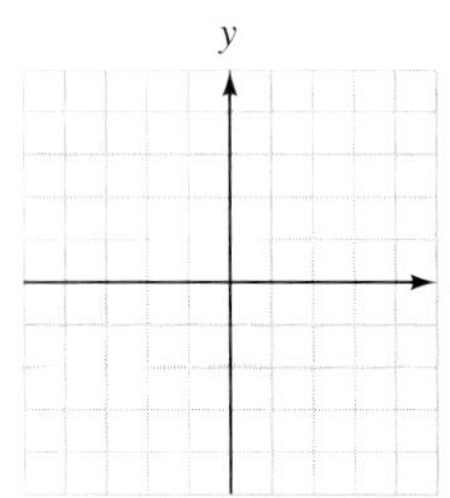

60. $8x + 4y = -24$

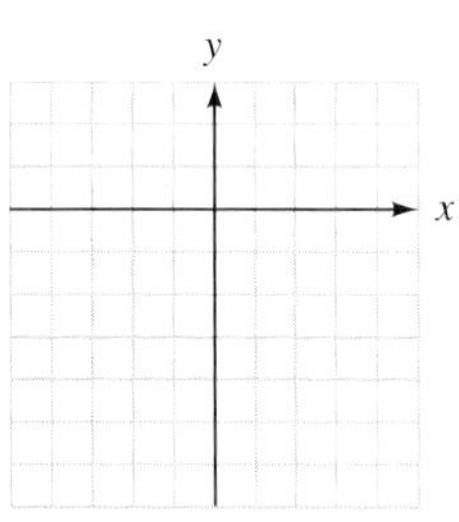

61. $3x + 4y = 8$

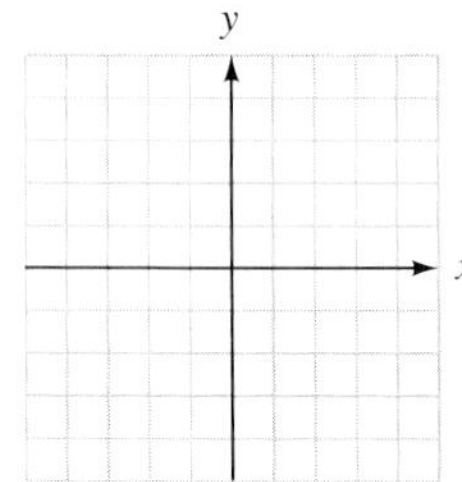

62. $2x + 3y = 9$

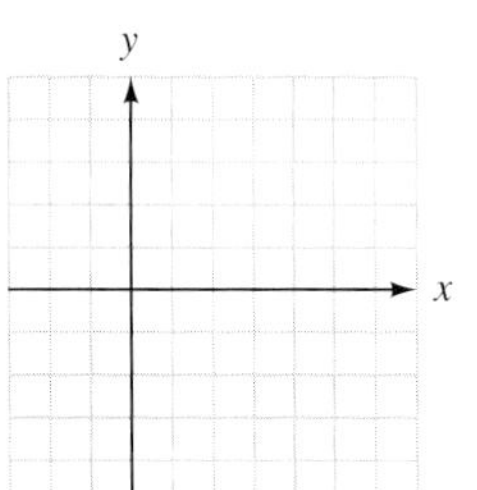

APPLICATIONS

63. EDUCATION COSTS Each semester, a college charges a services fee of \$50 plus \$25 for each unit taken by a student.

   a. Write a linear equation that gives the total enrollment cost c for a student taking u units.

   b. Complete the table of solutions in the illustration in the next column and graph the equation.

   c. Use the graph to find the total cost for a student taking 18 units the first semester and 12 units the second semester.

   d. What does the y-intercept of the line tell you?

u	c
4	
8	
14	

64. GROUP RATES To promote the sale of tickets for a cruise to Alaska, a travel agency reduces the regular ticket price of \$3,000 by \$5 for each individual traveling in the group.

   a. Write a linear equation that would find the ticket price t for the cruise if a group of p people travel together.

   b. Complete the table of solutions in the following illustration and graph the equation.

   c. As the size of the group increases, what happens to the ticket price?

   d. Use the graph to determine the cost of an individual ticket if a group of 25 will be traveling together.

p	t
10	
30	
60	

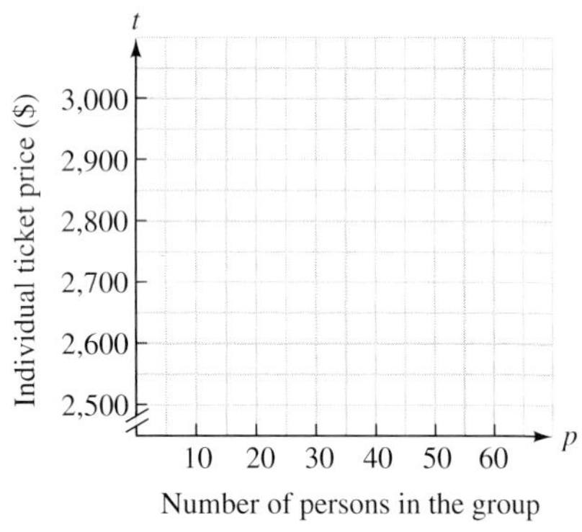

65. PHYSIOLOGY Physiologists have found that a woman's height h in inches can be approximated using the linear equation $h = 3.9r + 28.9$, where r represents the length of her radius bone in inches.

   a. Complete the table of solutions on the next page. Round to the nearest tenth and then graph the equation.

   b. Complete this sentence: From the graph, we see that the longer the radius bone, the . . . .

   c. From the graph on the next page, estimate the height of a woman whose radius bone is 7.5 inches long.

r	h
7	
8.5	
9	

h
65
60
55
Height (in.)
7 8 9 10
r
Length of radius (in.)

66. RESEARCH EXPERIMENTS A psychology major found that the time t (in seconds) that it took a white rat to complete a maze was related to the number of trials n the rat had been given. The resulting equation was $t = 25 - 0.25n$.

a. Complete the table of solutions in the following illustration and graph the equation.

b. Complete this sentence: From the graph, we see that the more trials the rat had, the . . . .

c. From the graph, estimate the time it will take the rat to complete the maze on its 32nd trial.

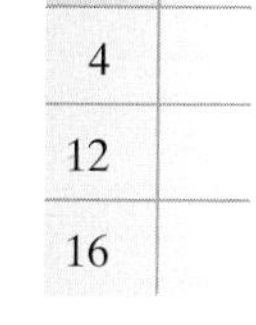

n	t
4	
12	
16	

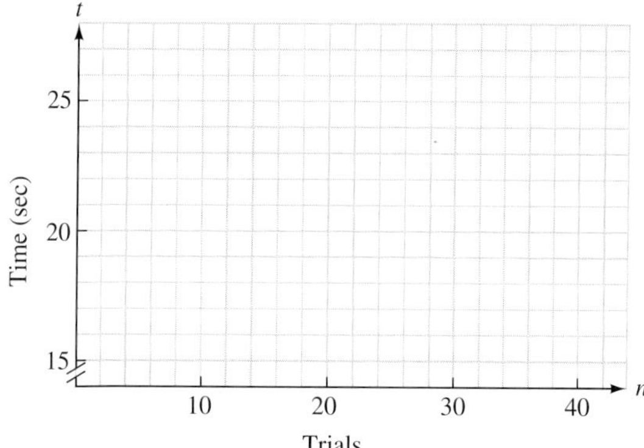

WRITING

67. A linear equation and a graph are two ways of mathematically describing a relationship between two quantities. Which do you think is more informative and why?

68. From geometry, we know that two points determine a line. Explain why it is a good practice when graphing linear equations to find and plot three points instead of just two.

69. How can we tell by looking at an equation whether its graph will be a straight line?

70. Can the x-intercept and the y-intercept of a line be the same point? Explain.

REVIEW

71. Simplify: $-(-5 - 4c)$.

72. Write the set of integers.

73. Solve: $\frac{x + 6}{2} = 1$.

74. Evaluate: $-2^2 + 2^2$.

75. Write a formula that relates profit, revenue, and costs.

76. Find the volume, to the nearest tenth, of a sphere with radius 6 feet.

77. Evaluate: $1 + 2[-3 - 4(2 - 8^2)]$.

78. Evaluate $\frac{x + y}{x - y}$ for $x = -2$ and $y = -4$.

9.4 Rate of Change and the Slope of a Line

- Rates of change • Slope of a line • The slope formula • Positive and negative slope
- Slopes of horizontal and vertical lines • Using slope to graph a line

Since our world is one of constant change, we must be able to describe change so that we can plan for the future. In this section, we will show how to describe the amount of change of one quantity in relation to the amount of change of another quantity by finding a *rate of change.*

Rates of change

The line graph in Figure 9-30(a) shows the number of business permits issued each month by a city over a 12-month period. From the shape of the graph, we can see that the number of permits issued *increased* each month.

For situations such as the one graphed in Figure 9-30(a), it is often useful to calculate a rate of increase (called a **rate of change**). We do so by finding the **ratio** of the change in the number of business permits issued each month to the number of months over which that change took place.

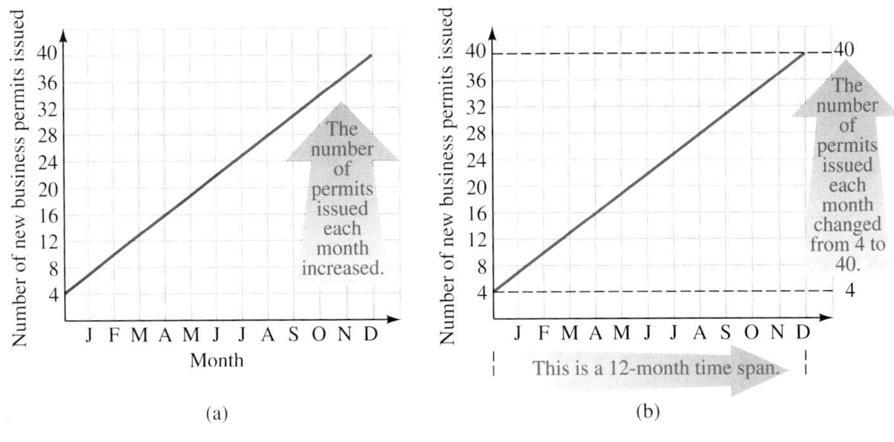

FIGURE 9-30

Ratios and rates

A **ratio** is the quotient of two numbers or the quotient of two quantities with the same units. In symbols, if a and b represent two numbers, the ratio of a to b is $\frac{a}{b}$. Ratios that are used to compare quantities with different units are called **rates.**

In Figure 9-30(b), we see that the number of permits issued prior to the month of January was 4. By the end of the year, the number of permits issued during the month of December was 40. This is a change of $40 - 4$, or 36, over a 12-month period. So we have

$$\text{Rate of change} = \frac{\text{change in number of permits issued each month}}{\text{change in time}}$$

The rate of change is a ratio.

$$= \frac{36 \text{ permits}}{12 \text{ months}}$$

$$= \frac{\overset{1}{\cancel{12}} \cdot 3 \text{ permits}}{\underset{1}{\cancel{12}} \text{ months}}$$

Factor 36 as $12 \cdot 3$ and divide out the common factor of 12.

$$= \frac{3 \text{ permits}}{1 \text{ month}}$$

The number of business permits being issued increased at a rate of 3 per month, denoted as 3 permits/month.

EXAMPLE 1 The graph in Figure 9-31 shows the number of subscribers to a newspaper. Find the rate of change in the number of subscribers over the first 5-year period. Write the rate in simplest form.

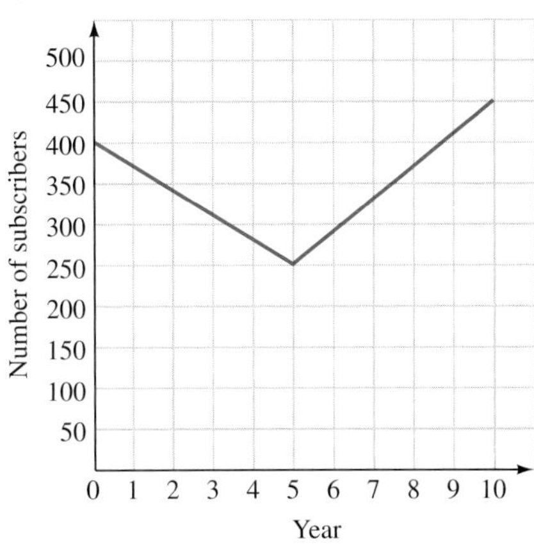

FIGURE 9-31

Self Check 1

Find the rate of change in the number of subscribers over the second 5-year period. Write the rate in simplest form.

Solution We need to write the ratio of the change in the number of subscribers over the change in time.

$$\text{Rate of change} = \frac{\text{change in number of subscribers}}{\text{change in time}}$$ Set up the ratio.

$$= \frac{(250 - 400) \text{ subscribers}}{5 \text{ years}}$$ Subtract the earlier number of subscribers from the later number of subscribers.

$$= \frac{-150 \text{ subscribers}}{5 \text{ years}}$$ $250 - 400 = -150$

$$= \frac{-30 \cdot \overset{1}{\cancel{5}} \text{ subscribers}}{\underset{1}{\cancel{5}} \text{ years}}$$ Factor -150 as $-30 \cdot 5$ and divide out the common factor of 5.

$$= \frac{-30 \text{ subscribers}}{1 \text{ year}}$$

The number of subscribers for the first 5 years *decreased* by 30 per year, as indicated by the negative sign in the result. We can write this as -30 subscribers/year.

Answer 40 subscribers/year

Slope of a line

The **slope** of a nonvertical line is a number that measures the line's steepness. We can calculate the slope by picking two points on the line and writing the ratio of the vertical change (called the **rise**) to the corresponding horizontal change (called the **run**) as we move from one point to the other. As an example, we will find the slope of the line that was used to describe the number of building permits issued and show that it gives the rate of change.

In Figure 9-32 (a modified version of Figure 9-30(a)), the line passes through points $P(0, 4)$ and $Q(12, 40)$. Moving along the line from point P to point Q causes the value of y to change from $y = 4$ to $y = 40$, an increase of $40 - 4 = 36$ units. We say that the *rise* is 36.

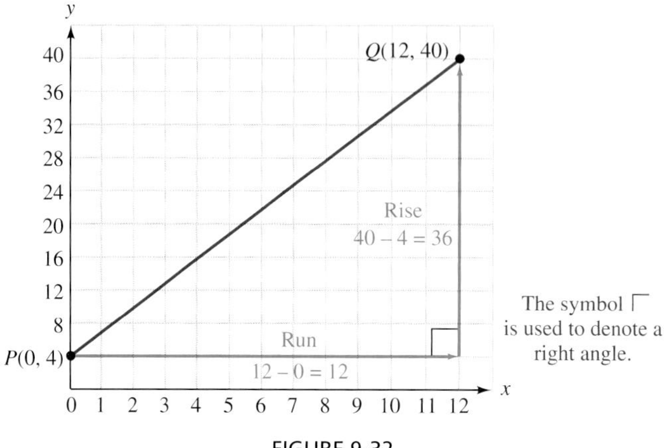

FIGURE 9-32

Moving from point P to point Q, the value of x increases from $x = 0$ to $x = 12$, an increase of $12 - 0 = 12$ units. We say that the *run* is 12. The slope of a line, usually denoted with the letter m, is defined to be the ratio of the change in y to the change in x.

$$m = \frac{\text{change in } y\text{-values}}{\text{change in } x\text{-values}}$$ Slope is a ratio.

$$= \frac{40 - 4}{12 - 0}$$ To find the change in y (the rise), subtract the y-values. To find the change in x (the run), subtract the x-values.

$$= \frac{36}{12}$$ Perform the subtractions.

$$= 3$$ Perform the division.

This is the same value we obtained when we found the rate of change of the number of business permits issued over the 12-month period. Therefore, by finding the slope of the line, we found a rate of change.

Self Check 2
Find the slope of the line shown in Figure 9-33(a) using two points different from those used in Example 2.

EXAMPLE 2 Find the slope of the line shown in Figure 9-33(a).

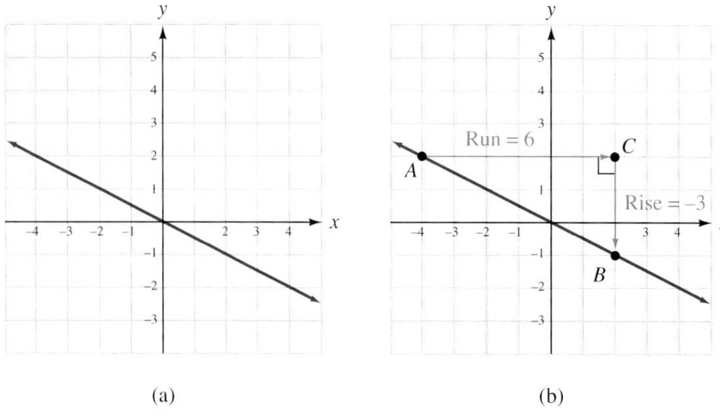

FIGURE 9-33

Solution In Figure 9-33(b), we begin by choosing two points on the line — call them A and B. Then we draw right triangle ABC, having a horizontal leg and a vertical leg. The longest side $\overline{AB}$ of the right triangle is called the **hypotenuse.** As we move from A to B (shown using blue arrows), we move to the right, a run of 6, and then down, a rise of -3. To find the slope of the line, we write a ratio.

$m = \frac{\text{rise}}{\text{run}}$ The slope of a line is the ratio of the rise to the run.

$m = \frac{-3}{6}$ From Figure 9-33(b), the rise is -3 and the run is 6.

$m = -\frac{1}{2}$ Simplify the fraction.

The slope of the line is $-\frac{1}{2}$.

Answer $-\frac{1}{2}$

! COMMENT The answers from Example 2 and the Self Check illustrate an important fact about slope: The same value for the slope of a line will result no matter which two points on the line are used to determine the rise and the run.

The slope formula

The slope of a line can be described in several ways.

$$\text{Slope} = m = \frac{\text{vertical change}}{\text{horizontal change}} = \frac{\text{rise}}{\text{run}} = \frac{\text{change in } y}{\text{change in } x}$$

To distinguish between the coordinates of two points, say points P and Q (see Figure 9-34), we often use **subscript notation.**

- Point P is denoted as $P(x_1, y_1)$. Read as "point P with coordinates of x sub 1 and y sub 1."
- Point Q is denoted as $Q(x_2, y_2)$. Read as "point Q with coordinates of x sub 2 and y sub 2."

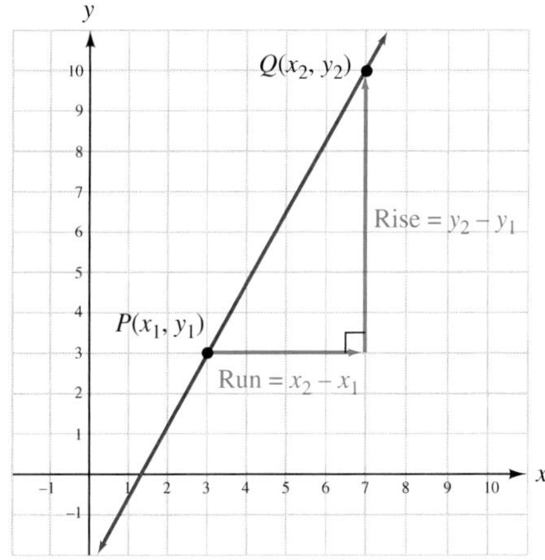

FIGURE 9-34

As a point on the line in Figure 9-34 moves from P to Q, its y-coordinate changes by the amount $y_2 - y_1$ (the rise), while its x-coordinate changes by $x_2 - x_1$ (the run). Since the slope is the ratio $\frac{\text{rise}}{\text{run}}$, we have the following formula for calculating slope.

Slope of a nonvertical line

The **slope** of a nonvertical line passing through points (x_1, y_1) and (x_2, y_2) is

$$m = \frac{y_2 - y_1}{x_2 - x_1}$$

Self Check 3
Find the slope of line l_2 shown in Figure 9-35.

EXAMPLE 3 Find the slope of line l_1 shown in Figure 9-35.

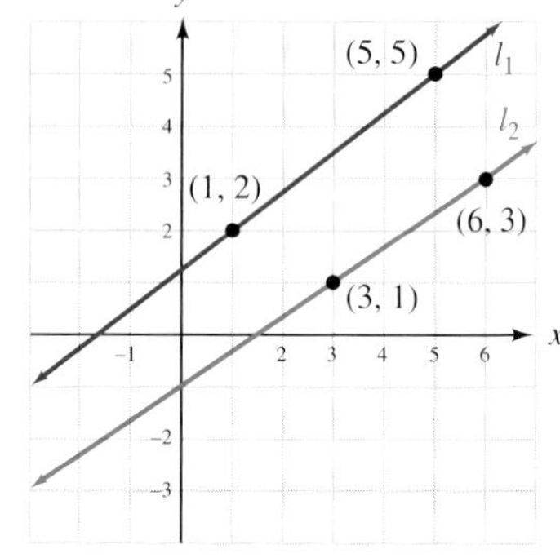

FIGURE 9-35

Solution To find the slope of l_1, we will use two points on the line whose coordinates are given: (1, 2) and (5, 5). If (x_1, y_1) is (1, 2) and (x_2, y_2) is (5, 5), then

$$x_1 = 1 \quad \text{and} \quad x_2 = 5$$
$$y_1 = 2 \qquad\qquad y_2 = 5$$

To find the slope of line l_1, we substitute these values into the formula for slope and simplify.

$$m = \frac{y_2 - y_1}{x_2 - x_1} \quad \text{This is the slope formula.}$$

$$= \frac{5 - 2}{5 - 1} \quad \text{Substitute 5 for } y_2\text{, 2 for } y_1\text{, 5 for } x_2\text{, and 1 for } x_1.$$

$$= \frac{3}{4} \quad \text{Perform the subtractions.}$$

The slope of l_1 is $\frac{3}{4}$. We would have obtained the same result if we had let $(x_1, y_1) = (5, 5)$ and $(x_2, y_2) = (1, 2)$.

$$m = \frac{y_2 - y_1}{x_2 - x_1} = \frac{2 - 5}{1 - 5} = \frac{-3}{-4} = \frac{3}{4}$$

Answer $\frac{2}{3}$

! COMMENT When finding the slope of a line, always subtract the y-values and the x-values in the same order. Otherwise your answer will have the wrong sign:

$$m \neq \frac{y_2 - y_1}{x_1 - x_2} \quad \text{and} \quad m \neq \frac{y_1 - y_2}{x_2 - x_1}$$

THINK IT THROUGH Average Rate of Tuition Increase

"Whatever happens in the future to the economy, whether up or down or more of the same, all current predictions point to a continuing rise over the coming decade in the cost of college education." Daniel Silver in *Show Me the Money,* News-Tribune

The line graphed below approximates the average cost of tuition and fees at U.S. public two-year academic institutions for the years 1990–2003. Find the average rate of increase in cost over this time period by finding the slope of the line.

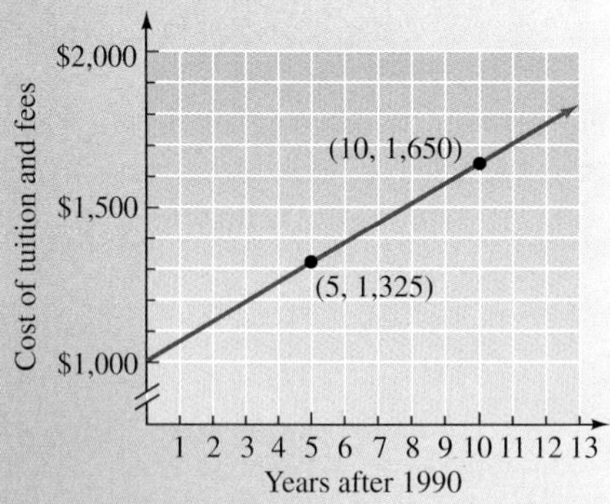

EXAMPLE 4 Find the slope of the line that passes through $(-2, 4)$ and $(5, -6)$ and draw its graph.

Solution Since we know the coordinates of two points on the line, we can find its slope. If (x_1, y_1) is $(-2, 4)$ and (x_2, y_2) is $(5, -6)$, then

$$x_1 = -2 \quad \text{and} \quad x_2 = 5$$
$$y_1 = 4 \qquad\qquad y_2 = -6$$

$m = \dfrac{y_2 - y_1}{x_2 - x_1}$ This is the slope formula.

$m = \dfrac{-6 - 4}{5 - (-2)}$ Substitute -6 for y_2, 4 for y_1, 5 for x_2, and -2 for x_1.

$m = -\dfrac{10}{7}$ Simplify the numerator: $-6 - 4 = -10$. Simplify the denominator: $5 - (-2) = 7$.

The slope of the line is $-\frac{10}{7}$. Figure 9-36 shows the graph of the line. Note that the line falls from left to right — a fact that is indicated by its negative slope.

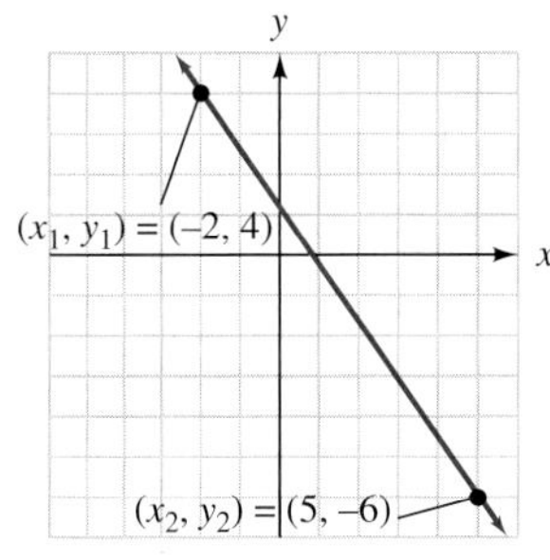

FIGURE 9-36

Self Check 4
Find the slope of the line that passes through $(-1, -2)$ and $(1, -7)$.

Answer $-\frac{5}{2}$

Positive and negative slope

In Example 3, the slope of line l_1 was positive $\left(\frac{3}{4}\right)$. In Example 4, the slope of the line was negative $\left(-\frac{10}{7}\right)$. In general, lines that rise from left to right have a positive slope, and lines that fall from left to right have a negative slope, as shown in Figure 9-37.

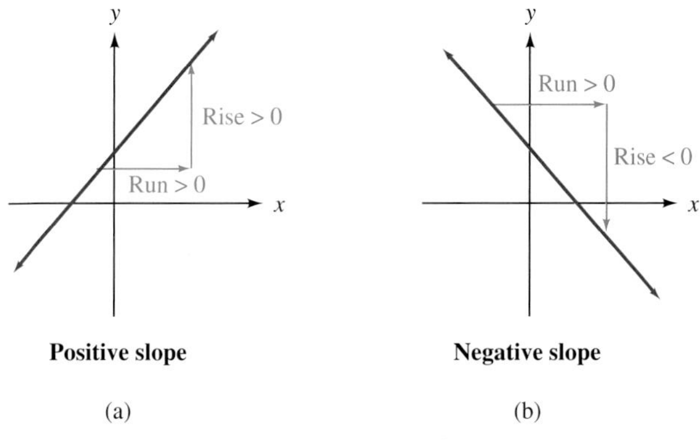

FIGURE 9-37

Slopes of horizontal and vertical lines

In the next two examples, we will calculate the slope of a horizontal line and show that a vertical line has no defined slope.

EXAMPLE 5 Find the slope of the line $y = 3$.

Solution To find the slope of the line $y = 3$, we need to know two points on the line. In Figure 9-38, we graph the horizontal line $y = 3$ and label two points on the line: $(-2, 3)$ and $(3, 3)$.

If (x_1, y_1) is $(-2, 3)$ and (x_2, y_2) is $(3, 3)$, we have

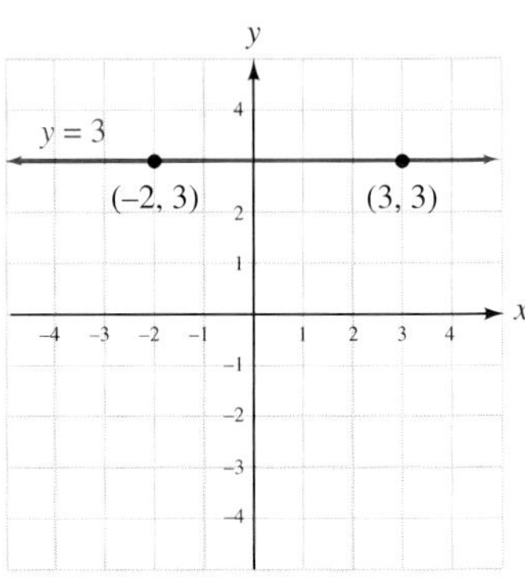

FIGURE 9-38

$$m = \frac{y_2 - y_1}{x_2 - x_1}$$ This is the slope formula.

$$m = \frac{3 - 3}{3 - (-2)}$$ Substitute 3 for y_2, 3 for y_1, 3 for x_2, and -2 for x_1.

$$m = \frac{0}{5}$$ Simplify the numerator and the denominator.

$$m = 0$$

The slope of the line $y = 3$ is 0.

The y-values of any two points on any horizontal line will be the same, and the x-values will be different. Thus, the numerator of

$$\frac{y_2 - y_1}{x_2 - x_1}$$

will always be zero, and the denominator will always be nonzero. Therefore, the slope of a horizontal line is zero.

EXAMPLE 6 If possible, find the slope of the line $x = -2$.

Solution To find the slope of the line $x = -2$, we need to know two points on the line. In Figure 9-39, we graph the vertical line $x = -2$ and label two points on the line: $(-2, -1)$ and $(-2, 3)$.

If (x_1, y_1) is $(-2, -1)$ and (x_2, y_2) is $(-2, 3)$, we have

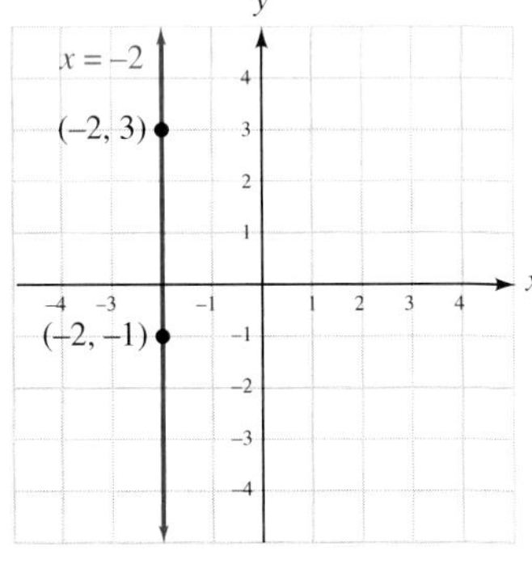

FIGURE 9-39

$$m = \frac{y_2 - y_1}{x_2 - x_1}$$ This is the slope formula.

$$m = \frac{3 - (-1)}{-2 - (-2)}$$ Substitute 3 for y_2, -1 for y_1, -2 for x_2, and -2 for x_1.

$$m = \frac{4}{0}$$ Simplify the numerator and the denominator.

Since division by zero is undefined, $\frac{4}{0}$ has no meaning. The slope of the line $x = -2$ is undefined.

The y-values of any two points on a vertical line will be different, and the x-values will be the same. Thus, the numerator of

$$\frac{y_2 - y_1}{x_2 - x_1}$$

will always be nonzero, and the denominator will always be zero. Therefore, the slope of a vertical line is undefined.

We now summarize the results from Examples 5 and 6.

Slopes of horizontal and vertical lines

Horizontal lines (lines with equations of the form $y = b$) have a slope of 0. (See Figure 9-40a.)

Vertical lines (lines with equations of the form $x = a$) have undefined slope. (See Figure 9-40b.)

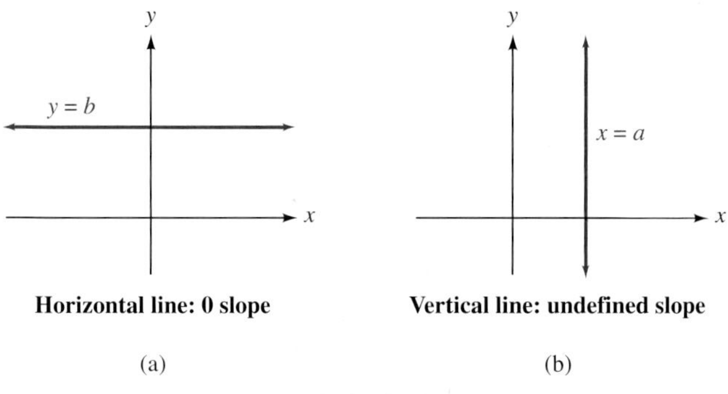

FIGURE 9-40

Using slope to graph a line

We can graph a line whenever we know the coordinates of one point on the line and the slope of the line. For example, to graph the line that passes through $P(2, 4)$ and has a slope of 3, we first plot $P(2, 4)$, as in Figure 9-41. We can express the slope of 3 as a fraction: $3 = \frac{3}{1}$. Therefore, the line *rises* 3 units for every 1 unit it *runs* to the right. We can find a second point on the line by starting at $P(2, 4)$ and moving 1 unit to the right (run) and then 3 units up (rise). This brings us to a point that we will call Q with coordinates $(2 + \mathbf{1}, 4 + \mathbf{3})$ or $(3, 7)$. The required line passes through points P and Q.

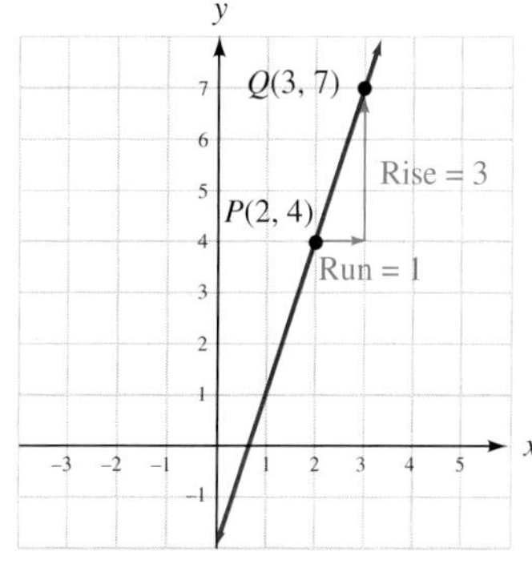

FIGURE 9-41

EXAMPLE 7 Graph the line that passes through the point $(-3, 4)$ with slope $-\frac{2}{5}$.

Solution We plot the point $(-3, 4)$ as shown in Figure 9-42 on the next page. Then, after writing the slope $-\frac{2}{5}$ as $\frac{-2}{5}$, we see that the *rise* is -2 and the *run* is 5. From the point $(-3, 4)$, we can find a second point on the line by moving 5 units to the right (run) and then 2 units down (a rise of -2 means to move down 2 units). This brings us to the point with coordinates of $(-3 + 5, 4 - 2) = (2, 2)$. We then draw a line that passes through the two points.

Self Check 7

Graph the line that passes through the point $(-4, 2)$ with slope -4.

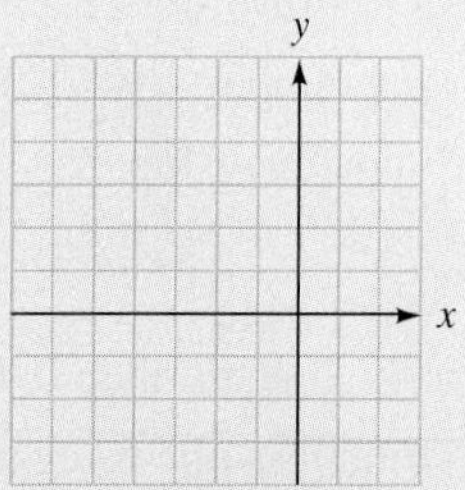

Answer

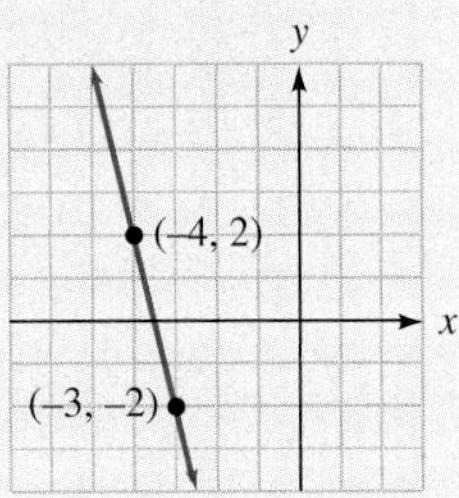

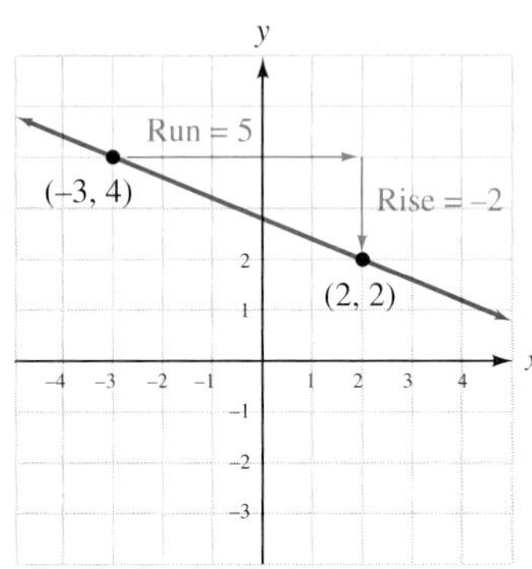

FIGURE 9-42

Section 9.4 STUDY SET

VOCABULARY *Fill in the blanks.*

1. A ______ is the quotient of two numbers.
2. Ratios used to compare quantities with different units are called ______.
3. The ______ of a line is defined to be the ratio of the change in y to the change in x.
4. $m = \frac{\text{\_\_\_\_\_\_ change}}{\text{horizontal change}} = \frac{\text{rise}}{\text{\_\_\_\_\_\_}} = \frac{\text{change in \_\_\_}}{\text{change in \_\_\_}}$
5. The rate of ______ of a linear relationship can be found by finding the slope of the graph of the line.
6. ______ lines have a slope of 0. Vertical lines have ______ slope.

CONCEPTS

7. Which line graphed in the illustration has
   a. a positive slope?
   b. a negative slope?
   c. zero slope?
   d. undefined slope?

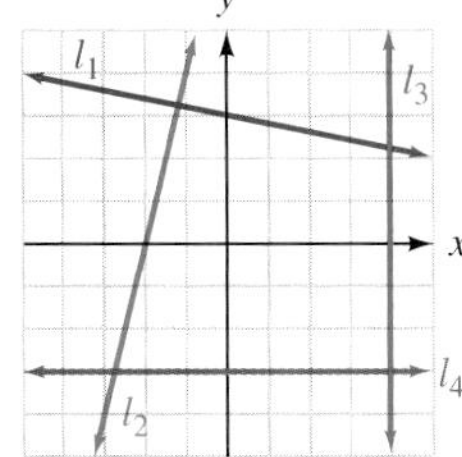

8. For the line graphed in the following illustration:
   a. Find its slope using points A and B.
   b. Find its slope using points B and C.
   c. Find its slope using points A and C.
   d. What observation is suggested by your answers to parts a, b, and c?

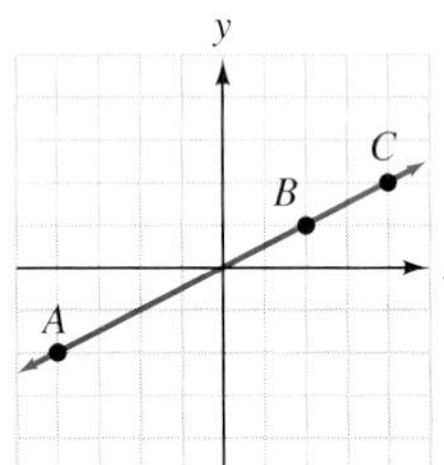

9. Use the information in the table of solutions for a linear equation to determine what the slope of the line would be if it were graphed.

x	y
−4	2
5	−7

10. Fill in the blanks.
    a. A line with positive slope ______ from left to right.
    b. A line with negative slope ______ from left to right.

11. GROWTH RATES Use the graph to find the rate of change of a boy's height during the time shown.

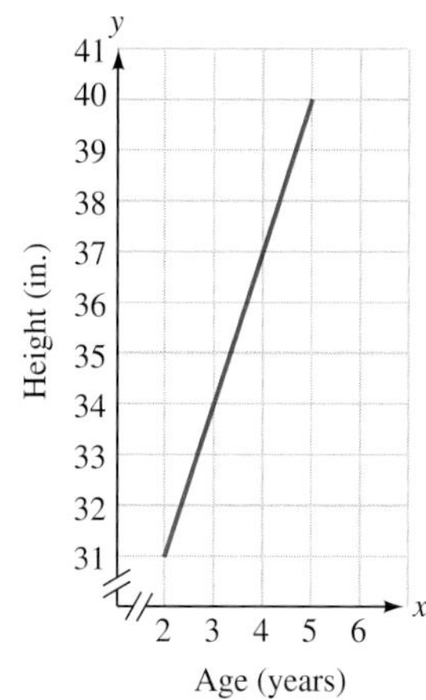

12. IRRIGATION The following graph shows the number of gallons of water remaining in a reservoir as water is discharged from it to irrigate a field. Find the rate of change in the number of gallons of water for the time the field was being irrigated.

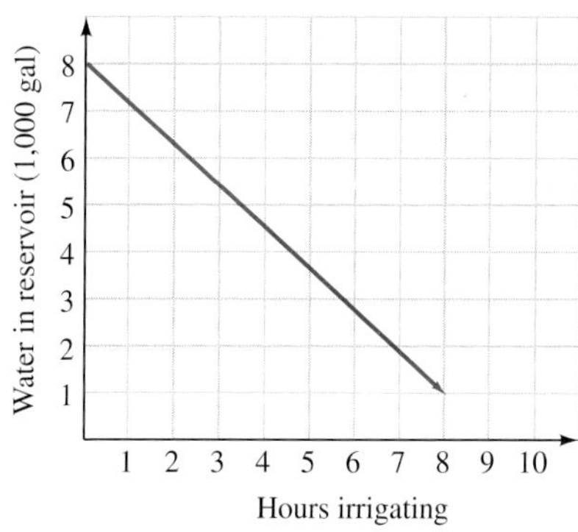

13. DEPRECIATION The following graph shows how the value of some sound equipment decreased over the years. Find the rate of change of its value during this time.

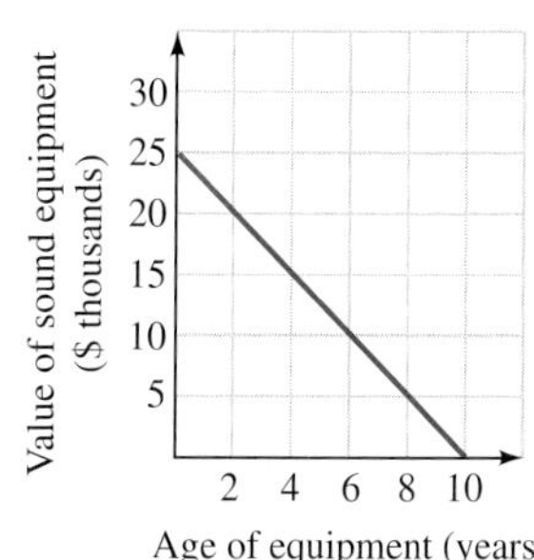

14. WAL-MART The graph in the next column approximates the net sales of Wal-Mart for the years 1991–2002.

a. Find the rate of change in sales for the years 1991–1998.

b. Find the rate of change in sales for the years 1998–2002.

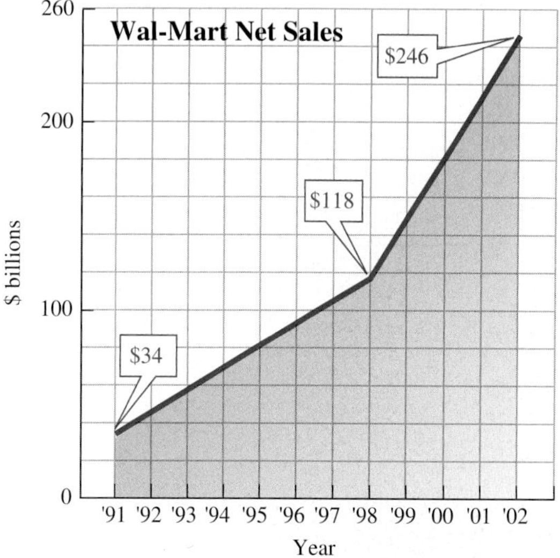

Based on data from Wal-Mart, *USA TODAY* (November 6, 1998), and Hoover's online.

15. THE UNCOLA

a. From the graph, estimate the rate of change in the sales of 7-Up for the years 1996–1997. Interpret this result.

b. From 1998–1999, which noncola had the greatest rate of change in sales?

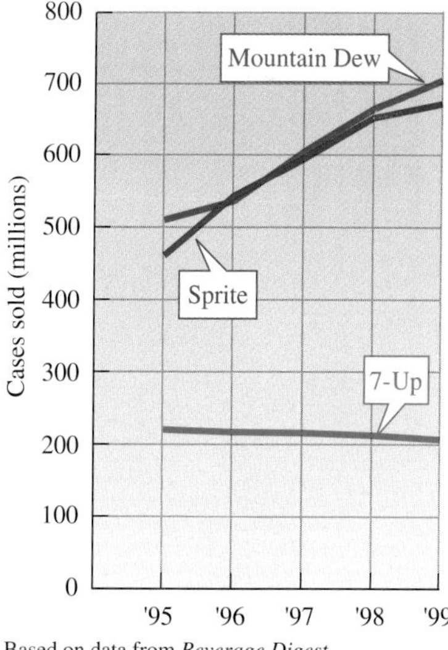

Based on data from *Beverage Digest*

16. COMMERCIAL JETS Examine the graph on the next page and consider trips of more than 7,000 miles by a Boeing 777. Use a rate of change to estimate how the maximum payload decreases as the distance traveled increases. Explain your result in words.

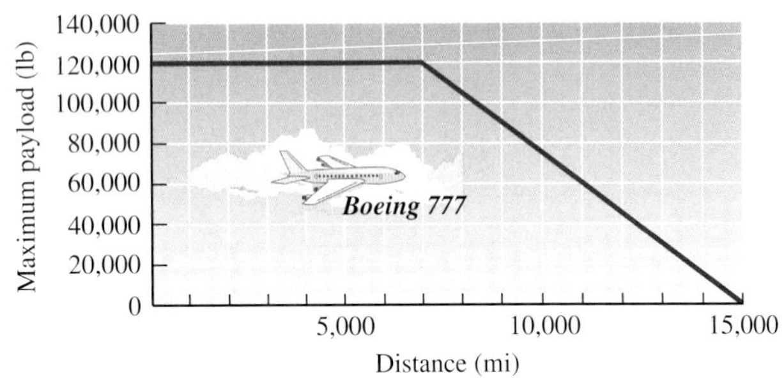

Based on data from Lawrence Livermore National Laboratory and *Los Angeles Times* (October 22, 1998).

NOTATION

17. What is the formula used to find the slope of the line passing through (x_1, y_1) and (x_2, y_2)?

18. Explain the difference between y^2 and y_2.

PRACTICE *Find the slope of the line passing through the given points, when possible.*

19. (2, 4) and (1, 3)
20. (1, 3) and (2, 5)
21. (3, 4) and (2, 7)
22. (3, 6) and (5, 2)
23. (0, 0) and (4, 5)
24. (4, 3) and (7, 8)
25. (−3, 5) and (−5, 6)
26. (6, −2) and (−3, 2)
27. (−2, −2) and (−12, −8)
28. (−1, −2) and (−10, −5)
29. (5, 7) and (−4, 7)
30. (−1, −12) and (6, −12)
31. (8, −4) and (8, −3)
32. (−2, 8) and (−2, 15)
33. (−6, 0) and (0, −4)
34. (0, −9) and (−6, 0)
35. (−2.5, 1.75) and (−0.5, −7.75)
36. (6.4, −7.2) and (−8.8, 4.2)

Find the slope of each line.

37.

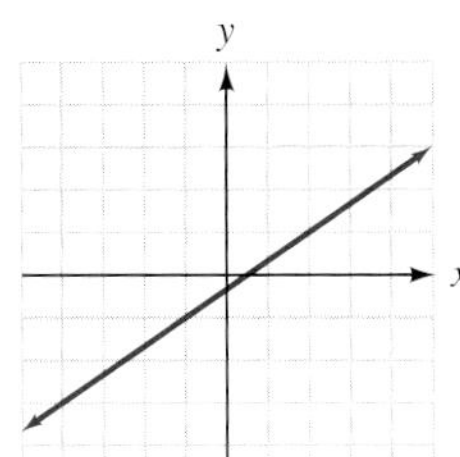

38.

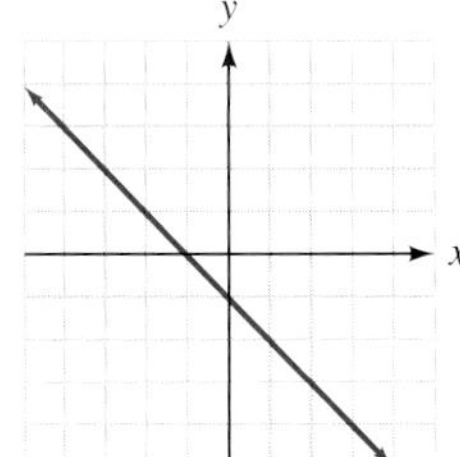

39.

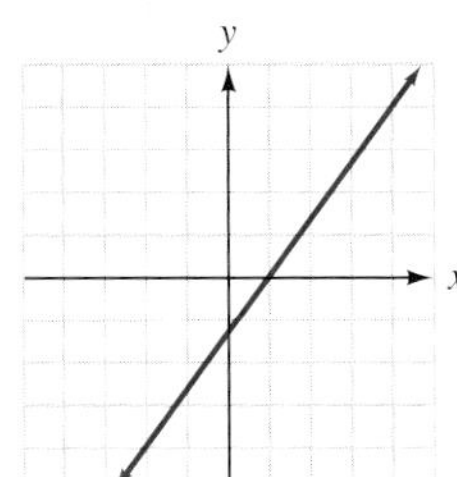

40.

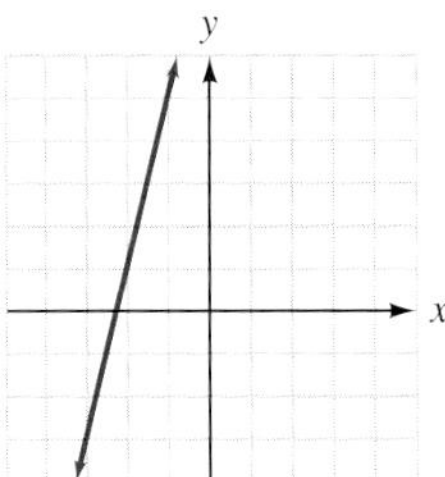

41.

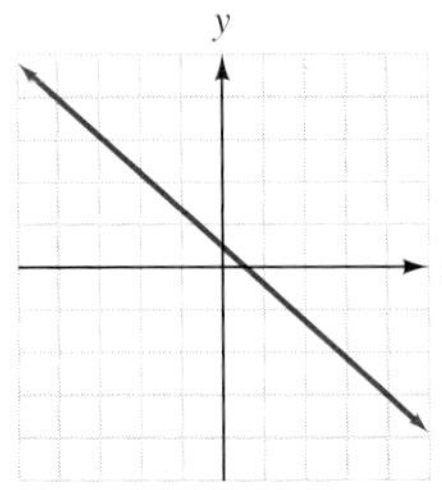

42.

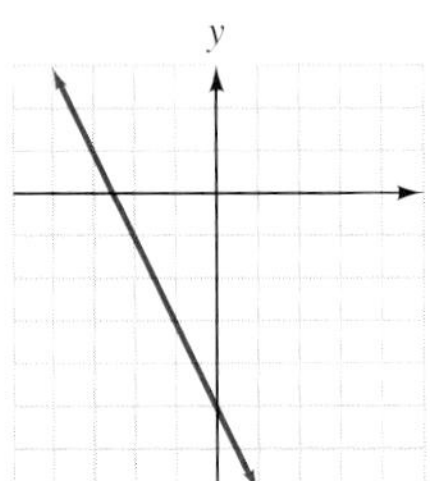

43.

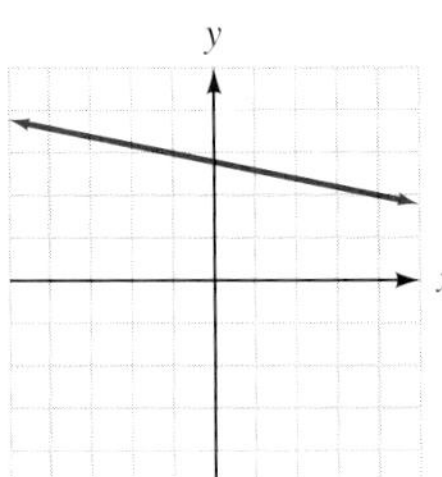

44.

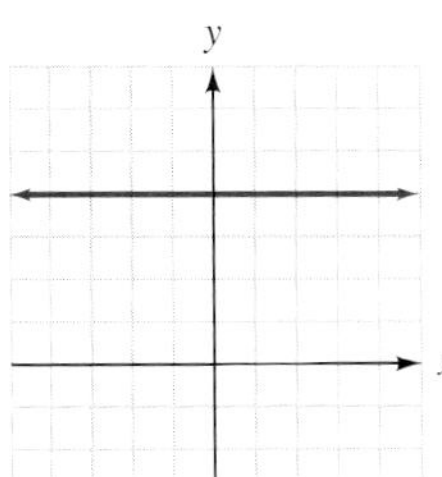

Graph the line that passes through the given point and has the given slope.

45. $(0, 1), m = 2$

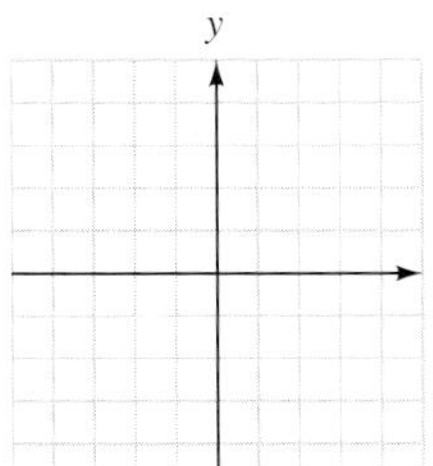

46. $(-4, 1), m = -3$

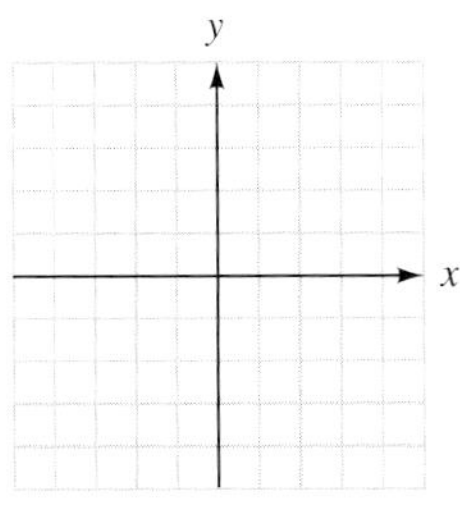

47. $(-3, -3), m = -\frac{3}{2}$

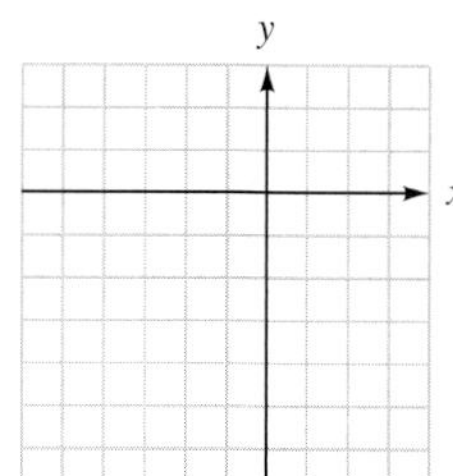

48. $(-2, -1), m = \frac{4}{3}$

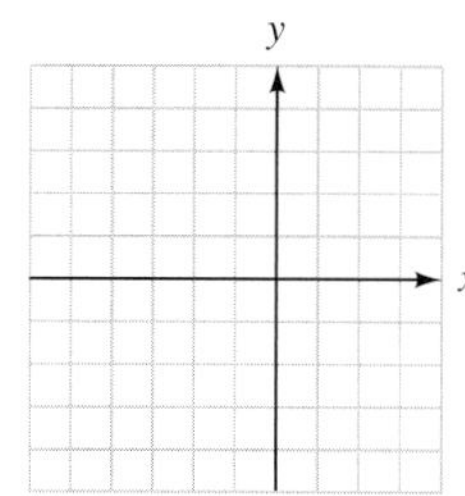

49. $(5, -3), m = \frac{3}{4}$

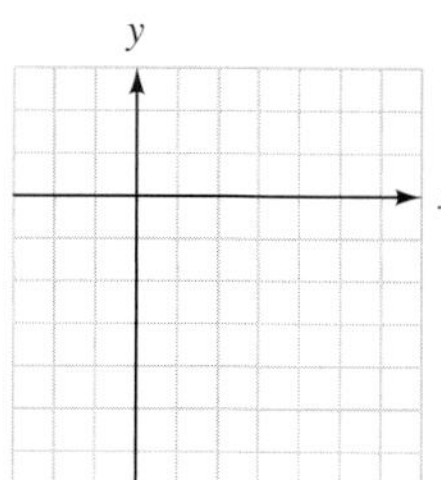

50. $(2, -4), m = \frac{2}{3}$

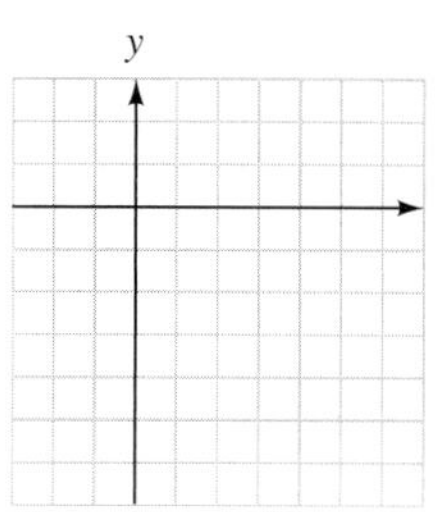

51. $(0, 0), m = -4$

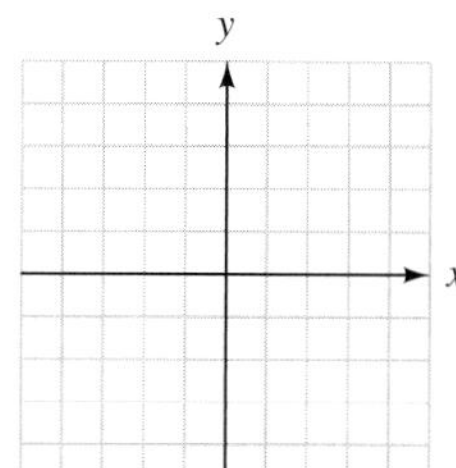

52. $(0, 0), m = 5$

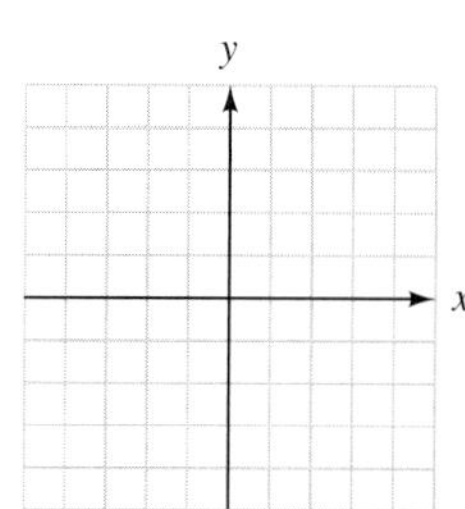

53. $(-5, 1), m = 0$

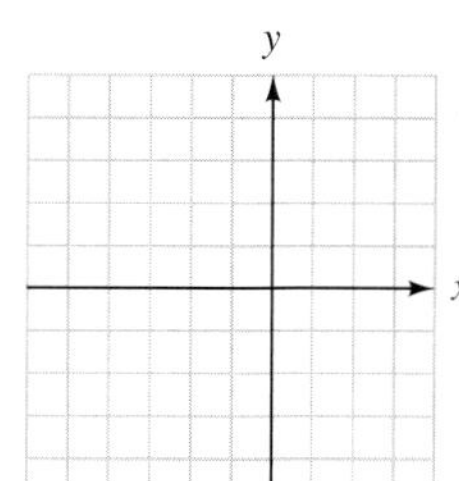

54. $(0, 3)$, undefined slope

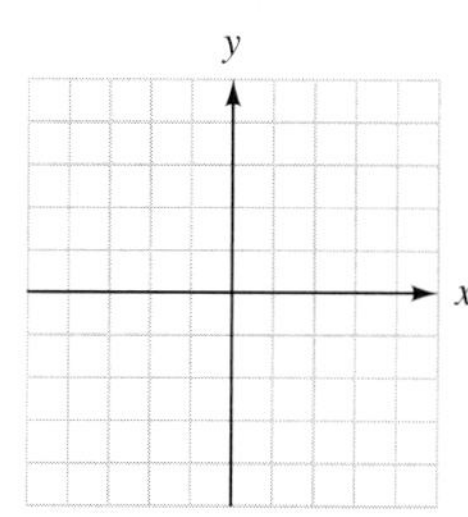

55. $(-1, -4)$, undefined slope

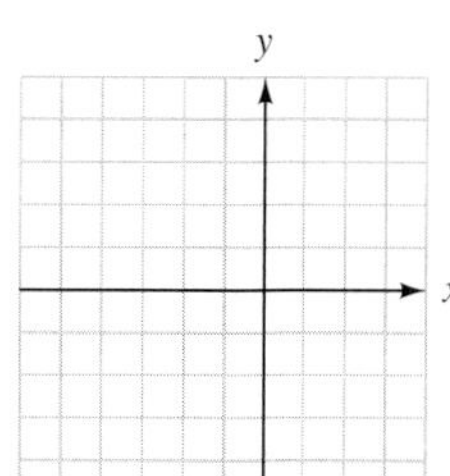

56. $(-3, -2), m = 0$

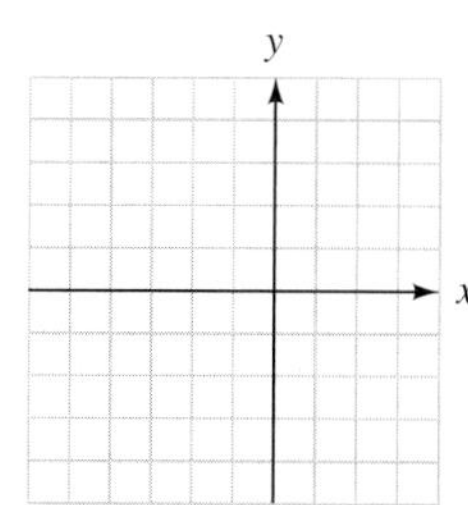

APPLICATIONS

57. POOL DESIGN Find the slope of the bottom of the swimming pool as it drops off from the shallow end to the deep end, as shown.

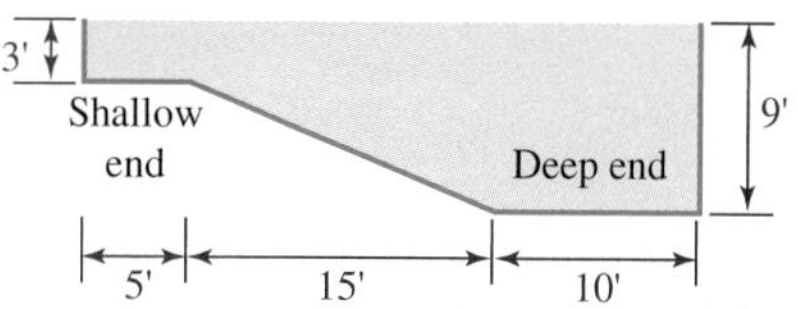

58. DRAINAGE To measure the amount of fall (slope) of a concrete patio slab, a 10-foot-long 2-by-4, a 1-foot ruler, and a level were used. Find the amount of fall in the slab. Explain what it means.

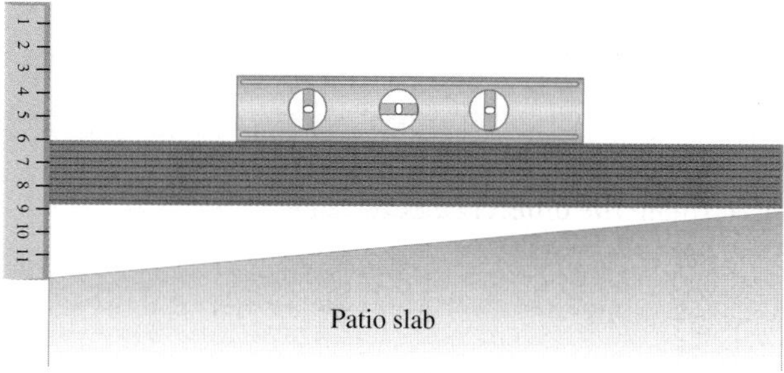

59. GRADE OF A ROAD The vertical fall of the road shown is 264 feet for a horizontal run of 1 mile. Find the slope of the decline and use that fact to complete the roadside warning sign for truckers. (*Hint:* 1 mile = 5,280 feet.)

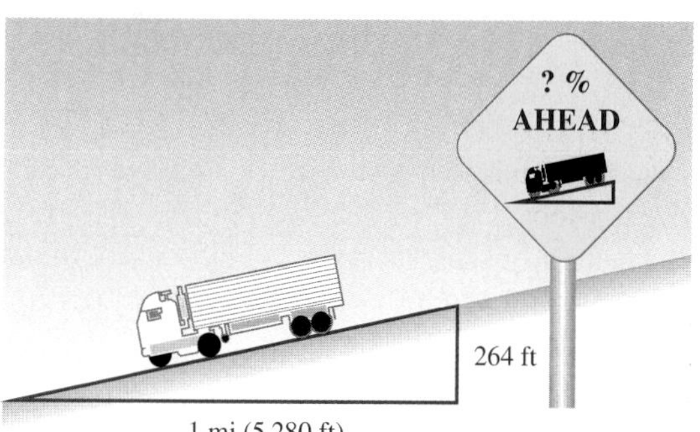

60. TREADMILLS For each height setting listed in the table, find the resulting slope of the jogging surface of the treadmill on the next page. Express each incline as a percent.

Height setting	% incline
2 inches	
4 inches	
6 inches	

61. ACCESSIBILITY The illustration shows two designs to make the upper level of a stadium wheelchair-accessible.

a. Find the slope of the ramp in design 1.

b. Find the slopes of the ramps in design 2.

c. Give one advantage and one drawback of each design.

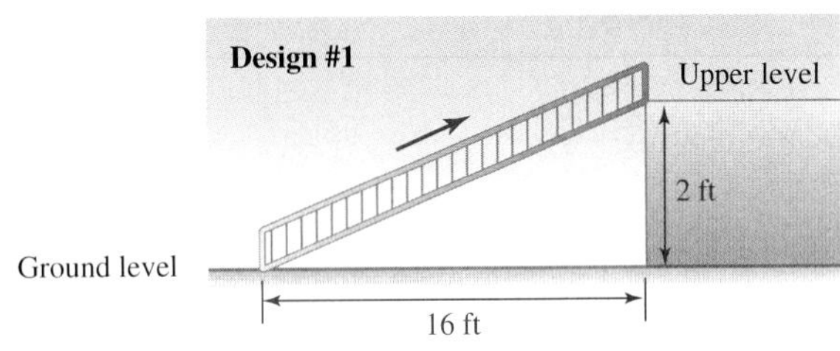

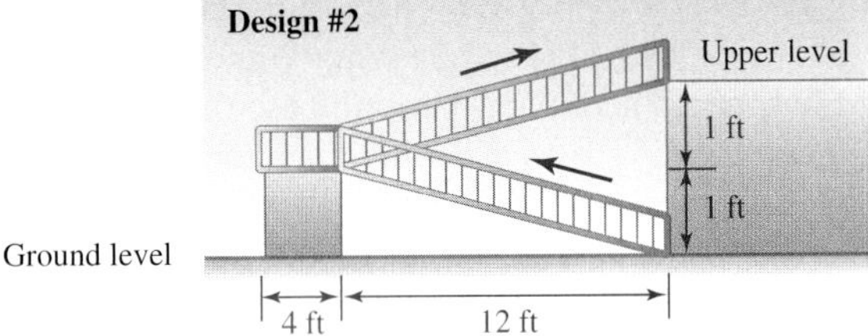

62. ARCHITECTURE Since the slope of the roof of the house shown is to be $\frac{2}{5}$, there will be a 2-foot rise for every 5-foot run. Draw the roof line if it is to pass through the given black points. Find the coordinates of the peak of the roof.

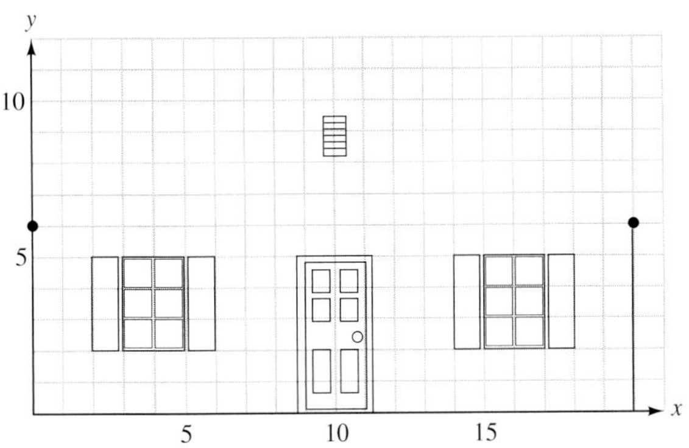

63. ENGINE OUTPUT Use the graph in the next column to find the rate of change in the horsepower (hp) produced by an automobile engine for engine speeds in the range of 2,400–4,800 revolutions per minute (rpm).

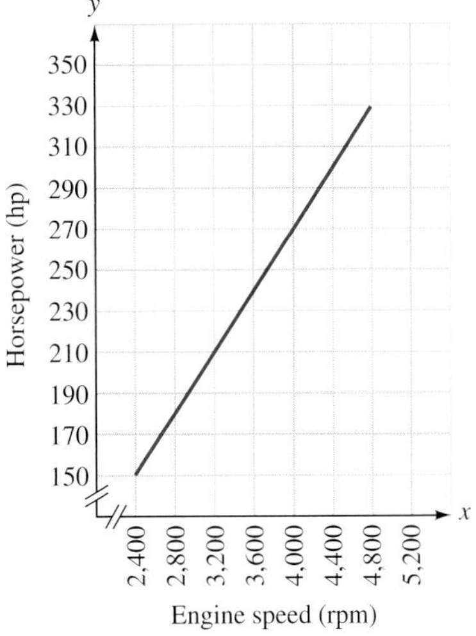

64. UNEMPLOYMENT See the following graph.

a. Between what two years did the unemployment rate increase the most? Find the rate of change for that period of time.

b. Between what two years did the unemployment rate decrease the most? Find the rate of change for that period of time.

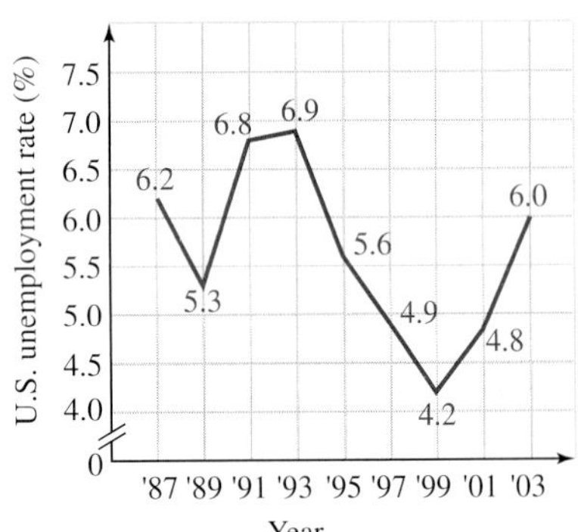

Source: Bureau of Labor Statistics

WRITING

65. Explain why the slope of a vertical line is undefined.

66. How do we distinguish between a line with positive slope and a line with negative slope?

67. Give an example of a rate of change that government officials might be interested in knowing so they can plan for the future needs of our country.

68. Explain the difference between a rate of change that is positive and one that is negative. Given an example of each.

REVIEW

69. In what quadrant does the point $(-3, 6)$ lie?

70. What is the name given the point $(0, 0)$?

71. Is $(-1, -2)$ a solution of $y = x^2 + 1$?

72. What basic shape does the graph of the equation $y = |x - 2|$ have?

73. Is the equation $y = 2x + 2$ linear or nonlinear?

74. Solve: $-3x \le 15$.

9.5 Slope–Intercept Form

- Slope–intercept form of the equation of a line • Parallel lines • Perpendicular lines

Numerical relationships are often described by tables or graphs. For example, various lengths of pipe and their corresponding weights are listed in the table in Figure 9-43. When this information is plotted as ordered pairs, we see that the points lie in a straight line. We say that the relationship between length and weight in this example is *linear.*

Length of pipe (ft) x	Weight of pipe (lb) y
6	120
10	200
14	280
20	400

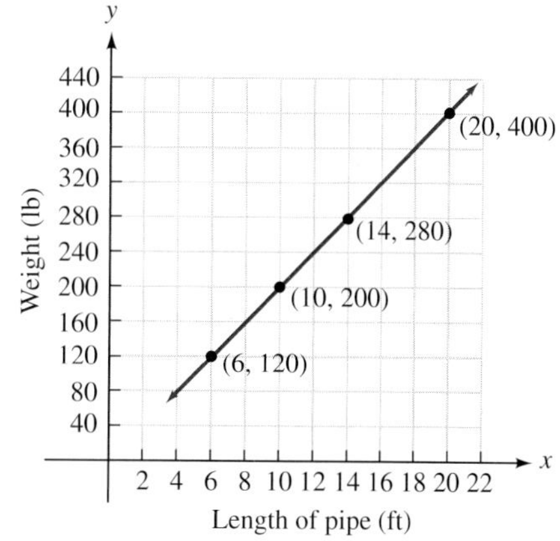

FIGURE 9-43

Figure 9-44 shows a graph of the time a cup of coffee has been sitting on a kitchen counter and its temperature. Since the graph is not a straight line, the relationship between time and temperature in this example is nonlinear.

Time on counter (min) x	Temperature of coffee (°F) y
1	180
5	140
10	110
20	80
30	72
45	70

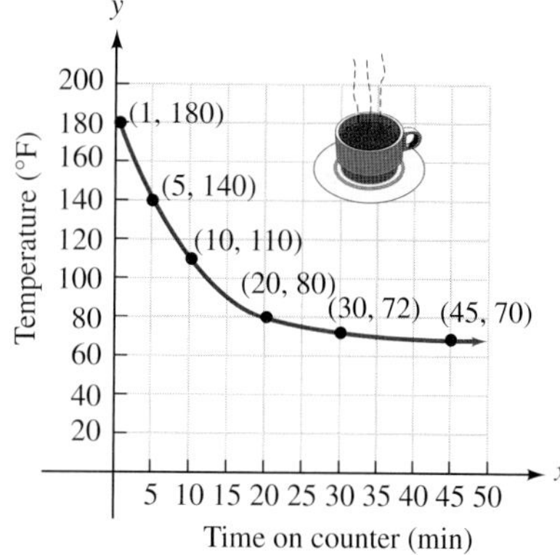

FIGURE 9-44

In this section, we will begin discussing a special type of relationship between two quantities whose graph is a straight line. Our objective is to learn how to write equations in two variables that describe these *linear relationships.*

Slope–intercept form of the equation of a line

The graph of $2x + 3y = 12$ shown in Figure 9-45 enables us to see that the slope of the line is $-\frac{2}{3}$ and that the y-intercept is (0, 4).

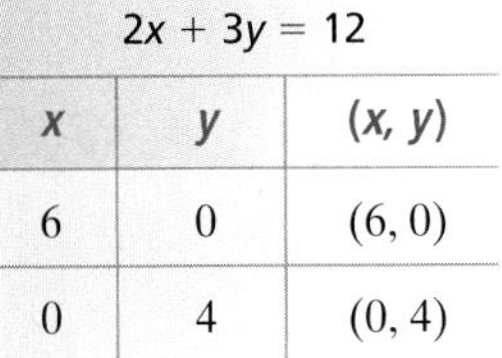

$2x + 3y = 12$		
x	y	(x, y)
6	0	(6, 0)
0	4	(0, 4)

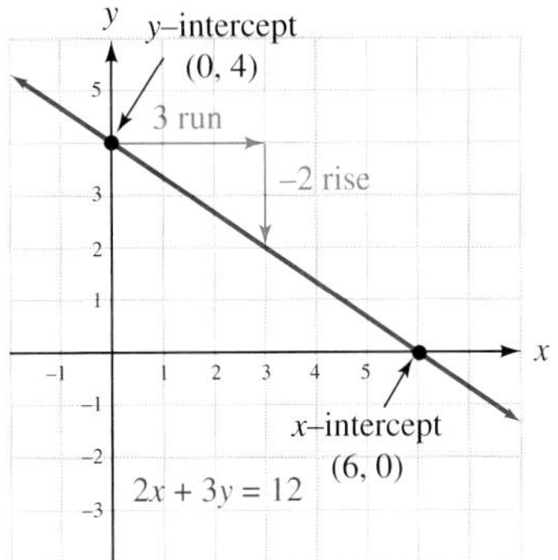

FIGURE 9-45

If we solve the equation for y, we will observe some interesting results.

$2x + 3y = 12$

$3y = -2x + 12$ Subtract $2x$ from both sides.

$\frac{3y}{3} = \frac{-2x}{3} + \frac{12}{3}$ To undo the multiplication by 3, divide both sides by 3. On the right-hand side, dividing each term by 3 is equivalent to dividing the entire side by 3: $\frac{-2x+12}{3} = \frac{-2x}{3} + \frac{12}{3}$.

$y = -\frac{2}{3}x + 4$ Perform the divisions. Rewrite $\frac{-2x}{3}$ as $-\frac{2}{3}x$.

In the equation $y = -\frac{2}{3}x + 4$, the *slope* of the graph $\left(-\frac{2}{3}\right)$ is the coefficient of x, and the constant (4) is the y-coordinate of the *y-intercept* of the graph.

$$y = -\frac{\mathbf{2}}{\mathbf{3}}x + \mathbf{4}$$

↑ The slope of the line. ↑ The y-intercept is (0, 4).

These observations suggest the following form of an equation of a line.

Slope–intercept form of the equation of a line

If a linear equation is written in the form

$$y = mx + b$$

where m and b represent constants, the graph of the equation is a line with slope m and y-intercept $(0, b)$.

Self Check 1
Find the slope and the y-intercept:
a. $y = -5x - 1$
b. $y = \frac{7}{8}x$
c. $y = 5 - \frac{x}{3}$

EXAMPLE 1 Find the slope and the y-intercept of the graph of each equation:
a. $y = 6x - 2$, **b.** $y = -\frac{5}{4}x$, and **c.** $y = \frac{x}{2} + 6$.

Solution

a. If we write the subtraction as the addition of the opposite, the equation will be in $y = mx + b$ form:

$$y = 6x + (-2)$$

Since $m = 6$ and $b = -2$, the slope of the line is 6 and the y-intercept is $(0, -2)$.

b. Writing $y = -\frac{5}{4}x$ in slope–intercept form, we have

$$y = -\frac{5}{4}x + 0$$

Since $m = -\frac{5}{4}$ and $b = 0$, the slope of the line is $-\frac{5}{4}$ and the y-intercept is $(0, 0)$.

c. Since $\frac{x}{2}$ means $\frac{1}{2}x$, we can rewrite $y = \frac{x}{2} + 6$ as

$$y = \frac{1}{2}x + 6$$

We see that $m = \frac{1}{2}$ and $b = 6$, so the slope of the line is $\frac{1}{2}$ and the y-intercept is $(0, 6)$.

Answers **a.** $m = -5$, $(0, -1)$; **b.** $m = \frac{7}{8}$, $(0, 0)$; **c.** $m = -\frac{1}{3}$, $(0, 5)$

! COMMENT If a linear equation is written in the form $y = mx + b$, the slope of the graph is the *coefficient* of x, not the term involving x. For example, it would be incorrect to say that the graph of $y = 5x + 1$ has a slope of $m = 5x$. Its graph has slope $m = 5$.

Prospects for a Teaching Career

THINK IT THROUGH

"While student enrollments are rising rapidly, more than a million veteran teachers are nearing retirement. Experts predict that overall we will need more than 2 million new teachers in the next decade." National Education Association, 2004

Have you ever thought about becoming a teacher? There will be plenty of openings in the future, especially for mathematics and science teachers. The equation

$$y = 865x + 11{,}100$$

approximates the average beginning teacher salary y, where x is the number of years after 1980. Graph the equation. What information about beginning teacher salaries is given by the slope of the line? By the y-intercept? What is the predicted average beginning teacher salary 5 years from now? (Source: American Federation of Teachers)

EXAMPLE 2 Find the slope and the y-intercept of the line determined by $6x - 3y = 9$. Then graph it.

Solution To find the slope and the y-intercept of the line, we write the equation in slope–intercept form by solving for y.

$$6x - 3y = 9$$

$$-3y = -6x + 9 \quad \text{Subtract } 6x \text{ from both sides.}$$

$$\frac{-3y}{-3} = \frac{-6x}{-3} + \frac{9}{-3}$$ To undo the multiplication by -3, divide both sides by -3. On the right-hand side, dividing each term by -3 is equivalent to dividing the entire side by -3: $\frac{-6x + 9}{-3} = \frac{-6x}{-3} + \frac{9}{-3}$.

$$y = 2x - 3$$ Perform the divisions. Here, $m = 2$ and $b = -3$.

From the equation, we see that the slope is 2 and the y-intercept is $(0, -3)$.

Self Check 2

Find the slope and the y-intercept of the line determined by $8x - 2y = -2$. Then graph it.

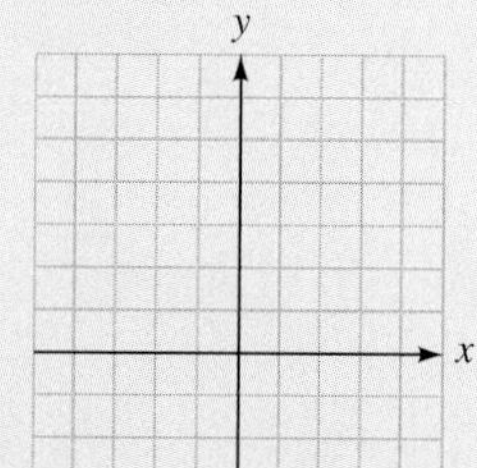

To graph $y = 2x - 3$, we plot the y-intercept $(0, -3)$, as shown in Figure 9-46. Since the slope is $\frac{\text{rise}}{\text{run}} = 2 = \frac{2}{1}$, the line rises 2 units for every unit it moves to the right. If we begin at $(0, -3)$ and move 1 unit to the right (run) and then 2 units up (rise), we locate the point $(1, -1)$, which is a second point on the line. We then draw a line through $(0, -3)$ and $(1, -1)$.

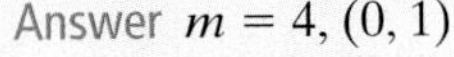
Answer $m = 4$, $(0, 1)$

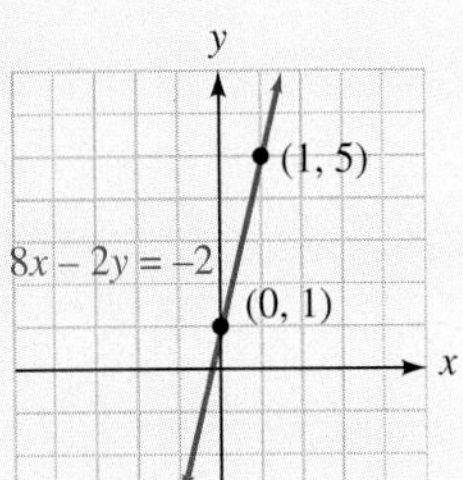

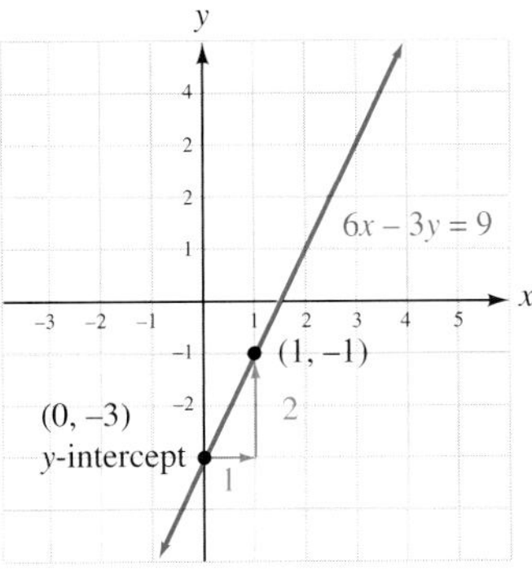

FIGURE 9-46

If we are given the slope and y-intercept of a line, we can write its equation, as in the next example.

EXAMPLE 3 **Limo service.** On weekends, a limousine service charges a fee of \$100, plus 50¢ per mile, for the rental of a stretch limo. Write a linear equation that describes the relationship between the rental cost and the number of miles driven. Graph the result.

Solution To write an equation describing this relationship, we will let x represent the number of miles driven and y represent the cost (in dollars). We can make two observations:

- The cost increases by 50¢ or \$0.50 for each mile driven. This is the *rate of change* of the rental cost to miles driven, and it will be the *slope* of the graph of the equation. Thus, $m = 0.50$.
- The basic fee is \$100. Before driving any miles (that is, when $x = 0$), the cost y is 100. The ordered pair $(0, 100)$ will be the y-intercept of the graph of the equation. So we know that $b = 100$.

We substitute 0.50 for m and 100 for b in the slope–intercept form to get

$$y = \underset{m = 0.50}{\underset{\uparrow}{0.50}}x + \underset{b = 100}{\underset{\uparrow}{100}}$$

Here the cost y depends on x, the number of miles driven.

To graph $y = 0.50x + 100$, we plot its y-intercept, $(0, 100)$, as shown in Figure 9-47. Since the slope is $0.50 = \frac{50}{100} = \frac{5}{10}$, we can start at $(0, 100)$ and locate a second point on the line by moving 10 units to the right (run) and then 5 units up (rise). This point will have coordinates $(0 + 10, 100 + 5)$ or $(10, 105)$. We draw a straight line through these two points to get a graph that illustrates the relationship between the rental cost and the number of miles driven. We draw the graph only in quadrant I, because the number of miles driven is always positive.

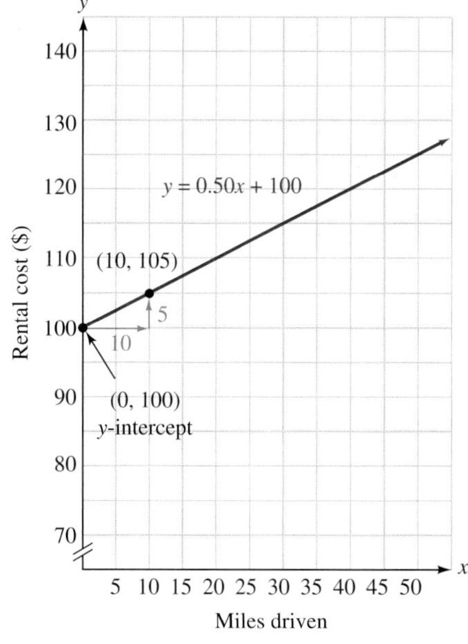

FIGURE 9-47

EXAMPLE 4 Videotapes. A VHS videocassette contains 800 feet of tape. In the long play (LP) mode, it plays 10 feet of tape every 3 minutes. Write a linear equation that relates the number of feet of tape yet to be played and the number of minutes the tape has been playing. Graph the equation.

Solution The number of feet yet to be played depends on the time the tape has been playing. To write an equation describing this relationship, we let x represent the number of minutes the tape has been playing and y represent the number of feet of tape yet to be played. We can make two observations:

- Since the VCR plays 10 feet of tape every 3 minutes, the number of feet remaining is constantly *decreasing*. This rate of change ($-\frac{10}{3}$ feet per minute) will be the slope of the graph of the equation. Thus, $m = -\frac{10}{3}$.
- The cassette tape is 800 feet long. Before any of the tape is played (that is, when $x = 0$), the amount of tape yet to be played is $y = 800$. Written as an ordered pair, we have (0, 800). Thus, $b = 800$.

Writing the equation in slope–intercept form, we have

$$y = -\frac{10}{3}x + 800$$

Its graph is shown in Figure 9-48.

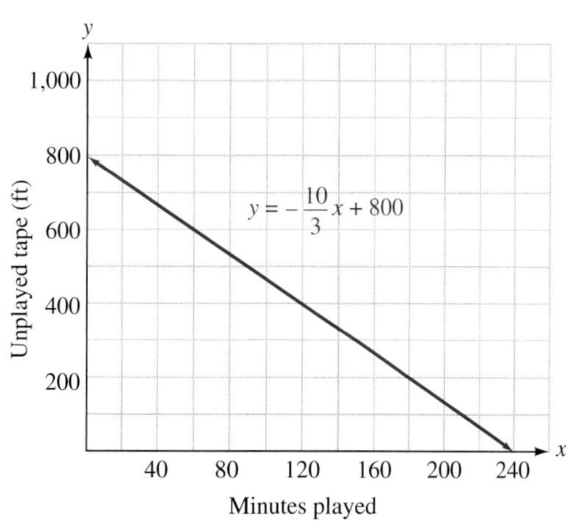

FIGURE 9-48

Self Check 4
In Example 4, let's say that the VCR is in super long play (SLP) mode, which plays 11 feet every 5 minutes. Use Figure 9-48 to graph the equation and then make an observation.

Answer $y = -\frac{11}{5}x + 800$; the graphs have the same y-intercept but different slopes.

Parallel lines

Suppose it costs \$75, plus 50¢ per mile, to rent the limo discussed in Example 3 on a weekday. If we substitute 0.50 for m and 75 for b in the slope–intercept form of a line, we have

$$y = 0.50x + 75$$

The graph of this equation and the graph of the equation

$$y = 0.50x + 100$$

appear in Figure 9-49.

From the figure, we see that the lines, each with slope 0.50, are parallel (do not intersect). This observation suggests the following fact.

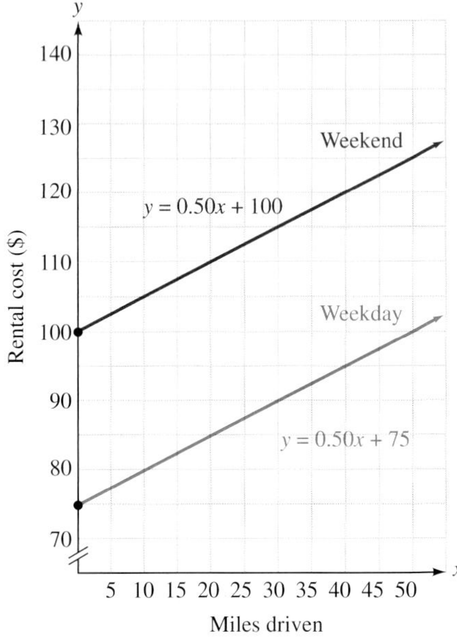

FIGURE 9-49

Slopes of parallel lines

Two different lines with the same slope are parallel.

Self Check 5

Graph $y = \frac{5}{2}x - 2$ and $y = \frac{5}{2}x$ on the same set of axes.

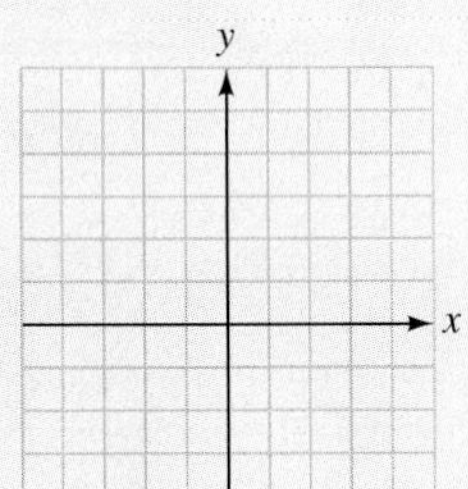

Answer

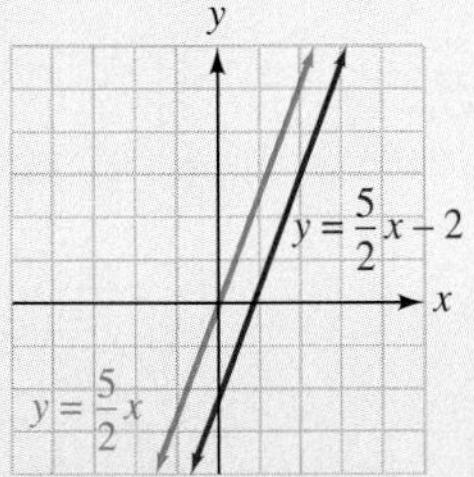

EXAMPLE 5 Graph $y = -\dfrac{2}{3}x$ and $y = -\dfrac{2}{3}x + 3$ on the same set of axes.

Solution The graph of the first equation has a slope of $-\frac{2}{3}$ and a y-intercept of (0, 0). The graph of the second equation has a slope of $-\frac{2}{3}$ and a y-intercept of (0, 3). We graph each equation as in Figure 9-50. Since the lines have the same slope of $-\frac{2}{3}$ and different y-intercepts, they are parallel.

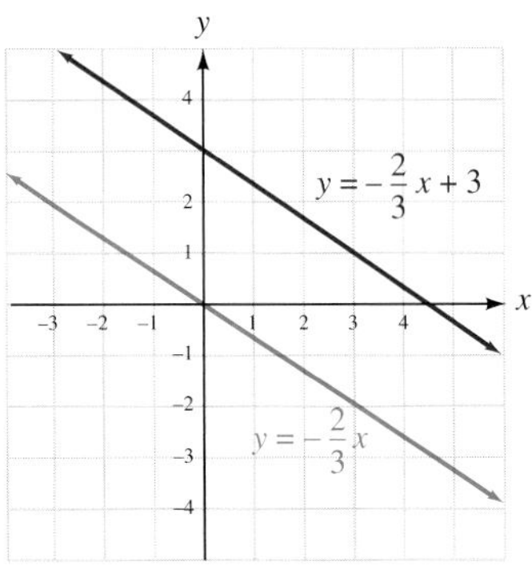

FIGURE 9-50

Perpendicular lines

The two lines shown in Figure 9-51 meet at right angles and are called **perpendicular lines.** In the figure, the symbol ⏋ is used to denote a right angle. Each of the four angles that are formed has a measure of 90°.

The product of the slopes of two (nonvertical) perpendicular lines is -1. For example, the perpendicular lines shown in Figure 9-51 have slopes of $\frac{3}{2}$ and $-\frac{2}{3}$. If we find the product of their slopes, we have

$$\frac{3}{2}\left(-\frac{2}{3}\right) = -\frac{6}{6} = -1$$

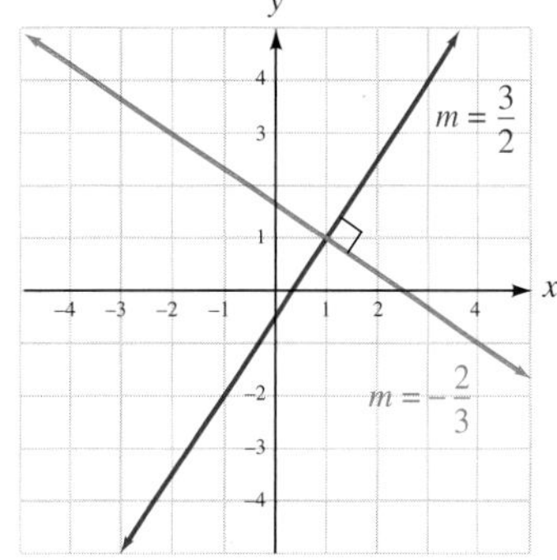

FIGURE 9-51

Two numbers whose product is -1 are called **negative reciprocals.** The numbers $\frac{3}{2}$ and $-\frac{2}{3}$, for example, are negative reciprocals, because their product is -1. The term *negative reciprocal* can be used to relate perpendicular lines and their slopes.

Slopes of perpendicular lines

If two nonvertical lines are perpendicular, their slopes are negative reciprocals.

If the slopes of two lines are negative reciprocals, the lines are perpendicular.

We can also state the fact given above symbolically: If the slopes of two nonvertical lines are m_1 and m_2, then the lines are perpendicular if

$$m_1 \cdot m_2 = -1 \quad \text{or} \quad m_2 = -\frac{1}{m_1}$$

Because a horizontal line is perpendicular to a vertical line, a line with a slope of 0 is perpendicular to a line with no defined slope.

EXAMPLE 6 Determine whether the graphs of $y = -5x + 6$ and $y = \frac{x}{5} - 2$ are parallel, perpendicular, or neither.

Solution The slope of the line $y = -5x + 6$ is -5. The slope of the line $y = \frac{x}{5} - 2$ is $\frac{1}{5}$. (Recall that $\frac{x}{5} = \frac{1}{5}x$.) Since the slopes are not equal, the lines are not parallel. If we find the product of their slopes, we have

$$-5\left(\frac{1}{5}\right) = -\frac{5}{5} = -1$$

Since the product of their slopes is -1, the lines are perpendicular.

Self Check 6

Determine whether the graphs of $y = 4x + 10$ and $y = \frac{1}{4}x$ are parallel, perpendicular, or neither.

Answer neither

Section 9.5 STUDY SET

VOCABULARY *Fill in the blanks.*

1. The equation $y = mx + b$ is called the ______________ form for the equation of a line.
2. The graph of the linear equation $y = mx + b$ has a ____________ of $(0, b)$ and a ________ of m.
3. __________ lines do not intersect.

4. The slope of a line is a ______ of change.

5. The numbers $\frac{5}{6}$ and $-\frac{6}{5}$ are called negative __________. Their product is -1.

6. The product of the slopes of ____________ lines is -1.

CONCEPTS

7. TREE GROWTH Graph the values shown in the following illustration and connect the points with a smooth curve. Does the graph indicate a linear relationship between the age of the tree and its height? Explain your answer.

Age	Height
0	0
5	8
10	15
15	28
20	45
25	62
30	85
35	100
40	112
45	118

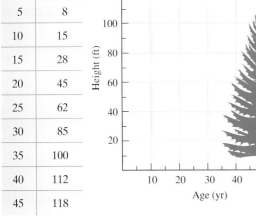

8. See the illustration.

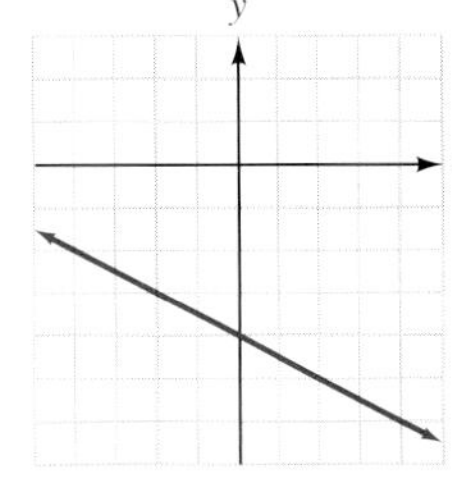

a. What is the slope of the line?

b. What is the y-intercept of the line?

c. Write the equation of the line.

9. NAVIGATION The graph in the next column shows the recommended speed at which a ship should proceed into head waves of various heights.

a. What information does the y-intercept of the graph give?

b. What is the rate of change in the recommended speed of the ship as the wave height increases?

c. Write the equation of the graph.

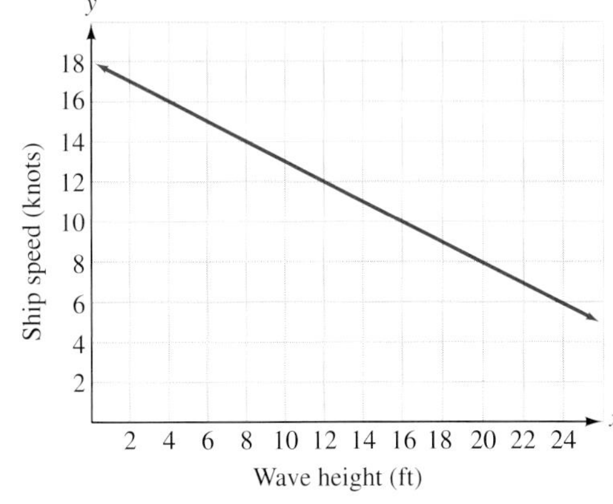

10. In the illustration, the slope of line l_1 is 2.

a. What is the slope of line l_2?

b. What is the slope of line l_3?

c. What is the slope of line l_4?

d. Which lines have the same y-intercept?

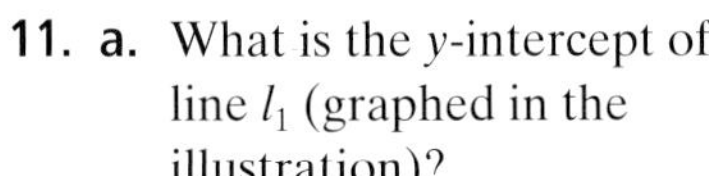

11. a. What is the y-intercept of line l_1 (graphed in the illustration)?

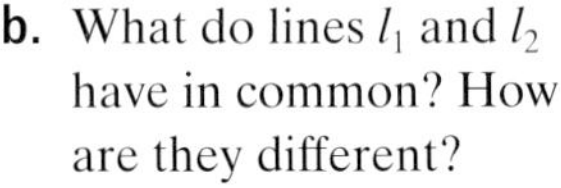

b. What do lines l_1 and l_2 have in common? How are they different?

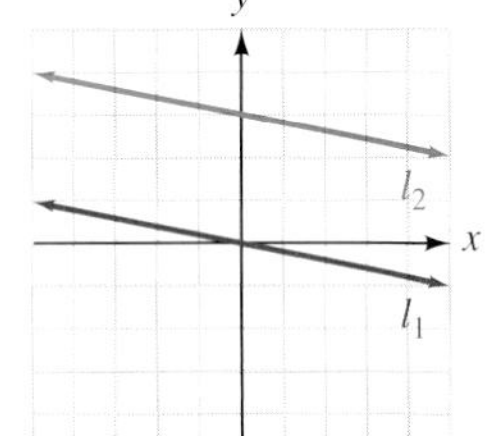

12. Use the graph to determine m and b; then write the equation of the line in slope–intercept form.

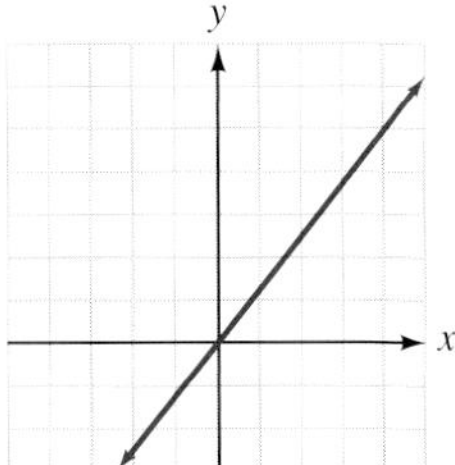

13. What is the slope of the line defined by each equation?

a. $y = \frac{-2x}{3} - 2$ **b.** $y = \frac{x}{4} + 1$

c. $y = 2 - 8x$ **d.** $y = 3x$

e. $y = x$ **f.** $y = -x$

14. Without graphing, determine whether the graphs of each pair of lines are parallel, perpendicular, or neither.

a. $y = 0.5x - 3$; $y = \frac{1}{2}x + 3$

b. $y = 0.75x$; $y = -\frac{4}{3}x + 2$

c. $y = -x$; $y = x$

15. To solve $-2y = 6x - 12$ for y, both sides of the equation were divided by -2. Complete each of the three divisions shown below.

$$\frac{-2y}{-2} = \frac{6x}{-2} - \frac{12}{-2}$$
$$\square = \square + \square$$

16. A graphing calculator was used to graph $y = -2.5x - 1.25$. What important feature of the graph is typed at the bottom of the screen?

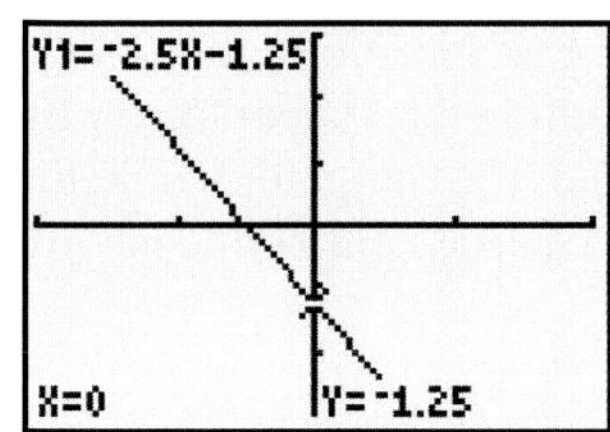

NOTATION *Complete each solution by solving the equation for y. Then find the slope and the y-intercept of its graph.*

17.
$$6x - 2y = 10$$
$$6x - \square - 2y = -6x + 10$$
$$-2y = \square + 10$$
$$\frac{-2y}{\square} = \frac{-6x}{\square} + \frac{10}{\square}$$
$$y = \square - 5$$

The slope is $\square$ and the y-intercept is $\square$.

18.
$$2x + 5y = 15$$
$$2x + 5y - \square = \square + 15$$
$$\square = -2x + 15$$
$$\frac{5y}{\square} = \frac{-2x}{\square} + \frac{15}{\square}$$
$$y = -\frac{2}{5}x + 3$$

The slope is $\square$ and the y-intercept is $\square$.

PRACTICE *Find the slope and the y-intercept of the graph of each equation.*

19. $y = 4x + 2$

20. $y = -4x - 2$

21. $y = \frac{x}{4} - \frac{1}{2}$

22. $4x - 2 = y$

23. $y = \frac{1}{2}x + 6$

24. $y = 6 - x$

25. $6y = x - 6$

26. $6x - 1 = y$

27. $x + y = 8$

28. $x - y = -30$

29. $2x + 3y = 6$

30. $3x - 5y = 15$

31. $3y - 13 = 0$

32. $-5y - 2 = 0$

33. $y = -5x$

34. $y = 14x$

Write an equation of the line with the given slope and y-intercept. Then graph it.

35. $m = 5, (0, -3)$

36. $m = -2, (0, 1)$

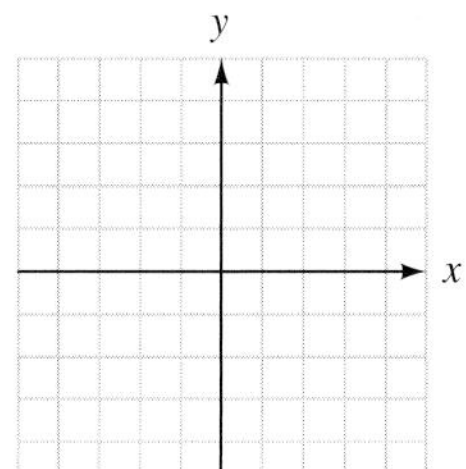

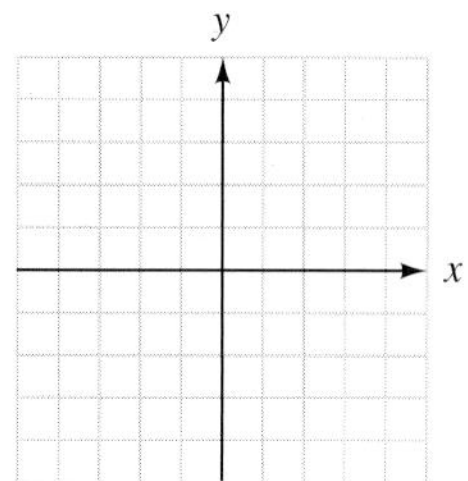

37. $m = \frac{1}{4}, (0, -2)$

38. $m = \frac{1}{3}, (0, -5)$

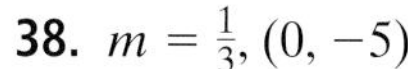

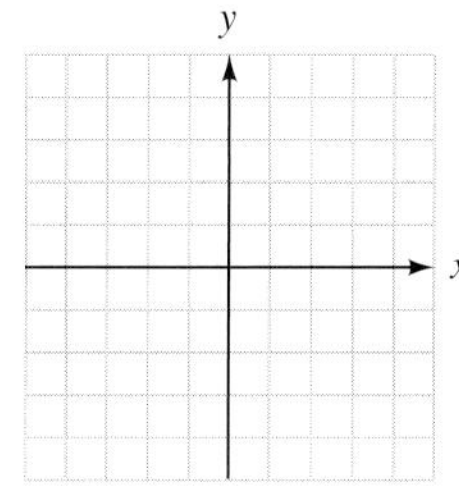

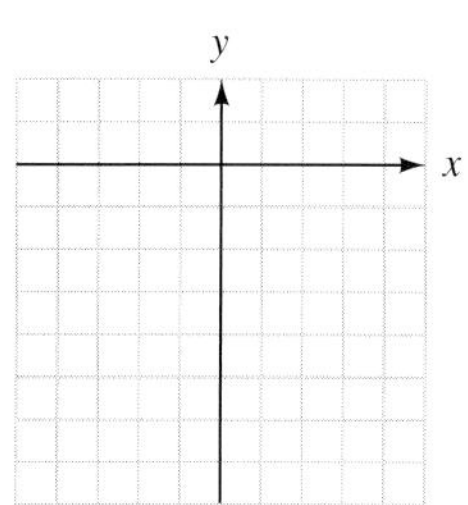

39. $m = -3, (0, 6)$

40. $m = -2, (0, 1)$

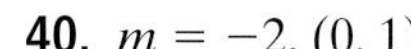

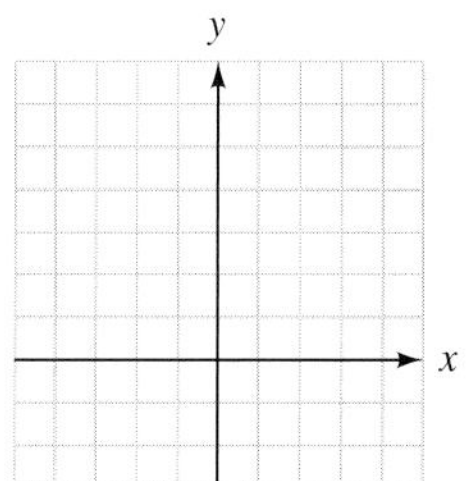

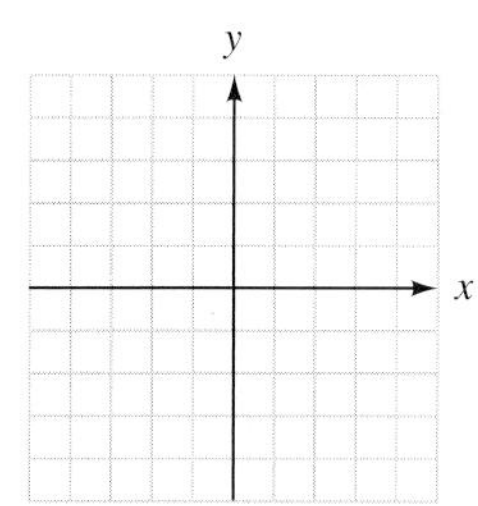

41. $m = -\frac{8}{3}, (0, 5)$

42. $m = -\frac{7}{6}, (0, 2)$

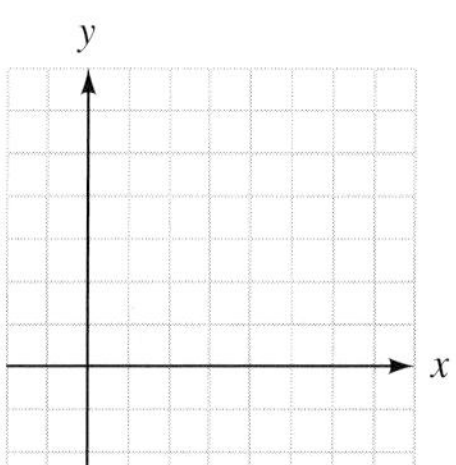

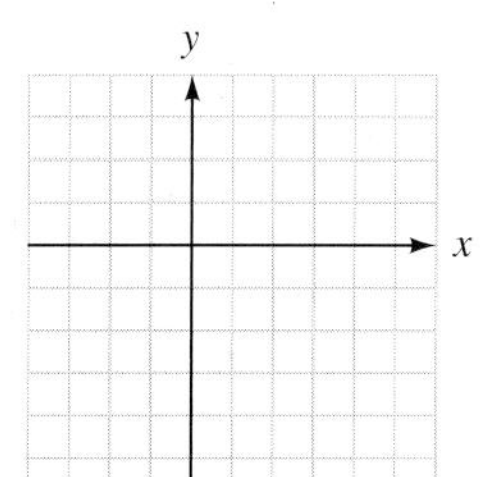

Find the slope and the y-intercept of the graph of each equation. Then graph it.

43. $y = 3x + 3$

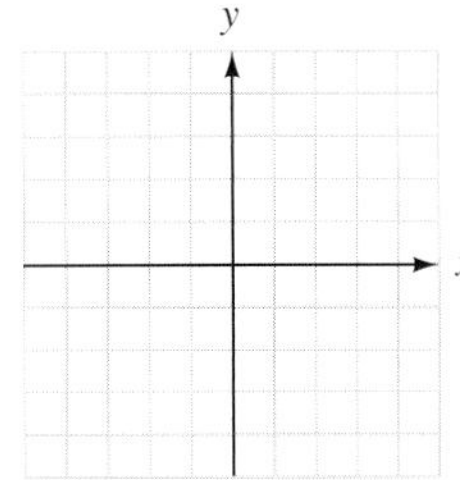

44. $y = -3x + 5$

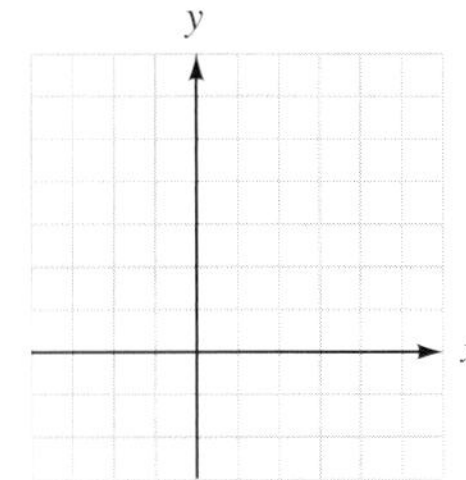

45. $y = -\frac{x}{2} + 2$

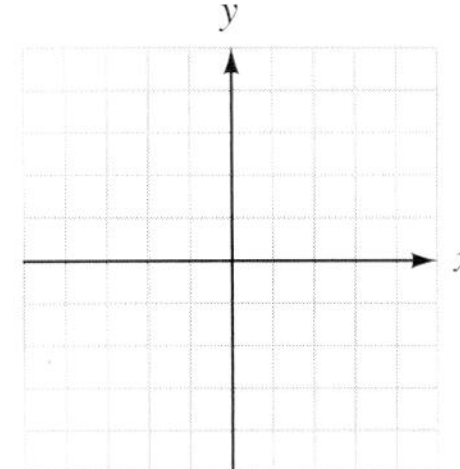

46. $y = \frac{x}{3}$

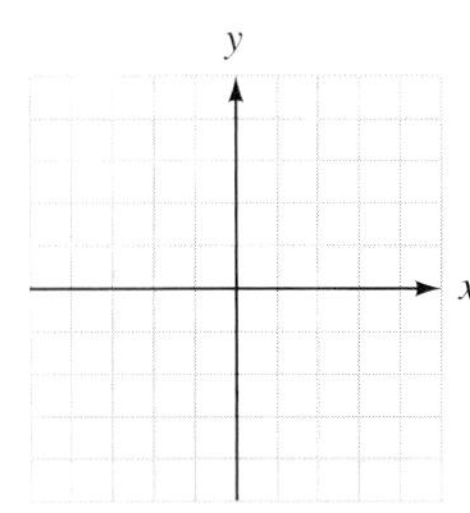

47. $3x + 4y = 16$

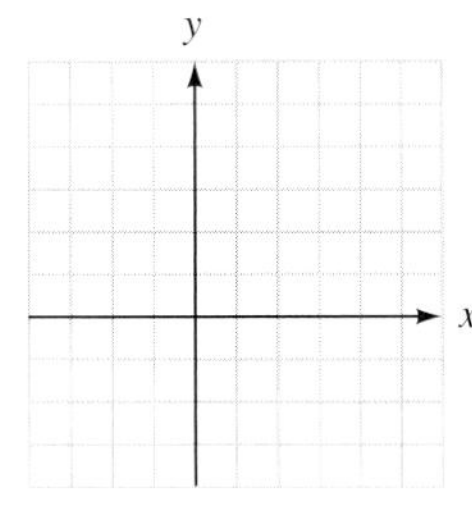

48. $2x + 3y = 9$

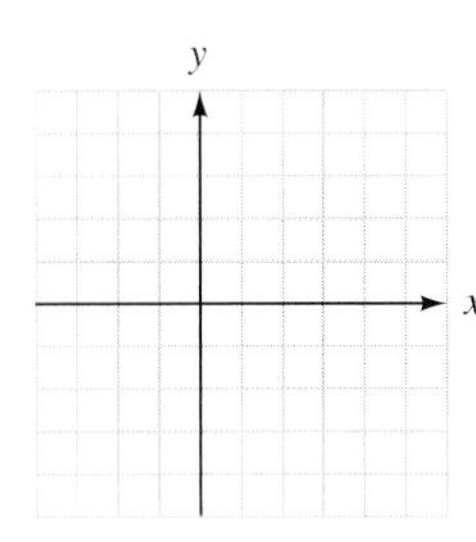

49. $10x - 5y = 5$

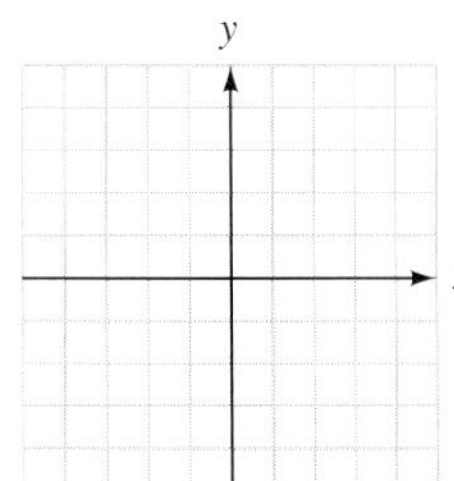

50. $4x - 2y = 6$

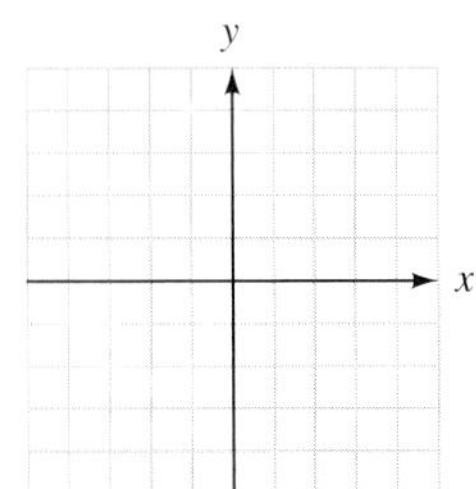

APPLICATIONS

51. PRODUCTION COSTS A television production company charges a basic fee of \$5,000 and then \$2,000 an hour when filming a commercial.

a. Write a linear equation that describes the relationship between the total production costs y and the hours of filming x.

b. Use your answer to part a to find the production costs if a commercial required 8 hours of filming.

52. COLLEGE FEES Each semester, students enrolling at a community college must pay tuition costs of \$20 per unit as well as a \$40 student services fee.

a. Write a linear equation that gives the total fees y to be paid by a student enrolling at the college and taking x units.

b. Use your answer to part a to find the enrollment cost for a student taking 12 units.

53. CHEMISTRY EXPERIMENT The following illustration shows a portion of a student's chemistry lab manual. Use the information to write a linear equation relating the temperature y (in degrees Fahrenheit) of the compound to the time x (in minutes) elapsed during the lab procedure.

Chem. Lab #1 Aug. 13
Step 1: Removed compound from freezer @ −10° F.
Step 2: Used heating unit to raise temperature of compound 5° F every minute.

54. INCOME PROPERTY Use the information in the newspaper advertisement to write a linear equation that gives the amount of income y (in dollars) the apartment owner will receive when the unit is rented for x months.

APARTMENT FOR RENT
1 bedroom/1 bath, with garage
\$500 per month + \$250 nonrefundable security fee.

55. SALAD BAR For lunch, a delicatessen offers a "Salad and Soda" special where customers serve themselves at a well-stocked salad bar. The cost is \$1.00 for the drink and 20¢ an ounce for the salad.

a. Write a linear equation that will find the cost y of a lunch when a salad weighing x ounces is purchased.

b. Graph the equation using the grid on the next page.

c. How would the graph from part b change if the delicatessen began charging $2.00 for the drink?

d. How would the graph from part b change if the cost of the salad changed to 30¢ an ounce?

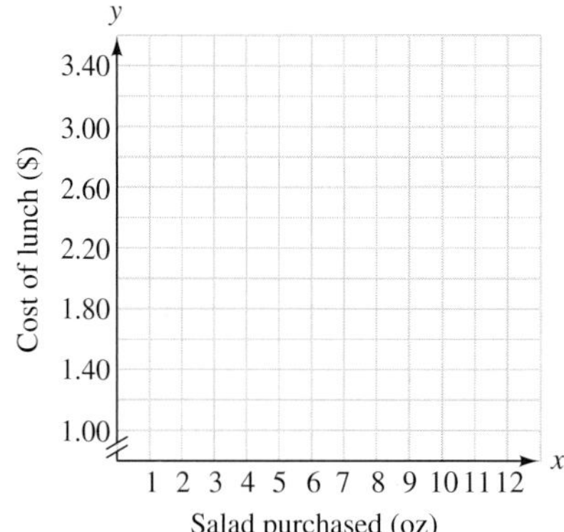

56. SEWING COSTS A tailor charges a basic fee of $20 plus $2.50 per letter to sew an athlete's name on the back of a jacket.

a. Write a linear equation that will find the cost y to have a name containing x letters sewn on the back of a jacket.

b. Graph the equation.

c. Suppose the tailor raises the basic fee to $30. On your graph from part b, draw the new graph showing the increased cost.

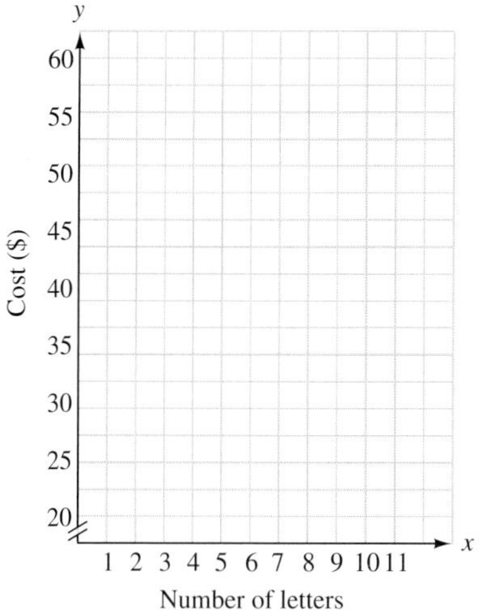

57. EMPLOYMENT SERVICE A policy statement of LIZCO, Inc., is shown in the next column. Suppose a secretary had to pay an employment service $500 to get placed in a new job at LIZCO. Write a linear equation that tells the secretary the actual cost y of the employment service to her x months after being hired.

Policy no. 23452– A new hire will be reimbursed by LIZCO for any employment service fees paid by the employee at the rate of $20 per month.

58. COMPUTER DRAFTING The illustration shows a computer-generated drawing of an automobile engine mount. When the designer clicks the mouse on a line of the drawing, the computer finds the equation of the line. Determine whether the two lines selected in the drawing are perpendicular.

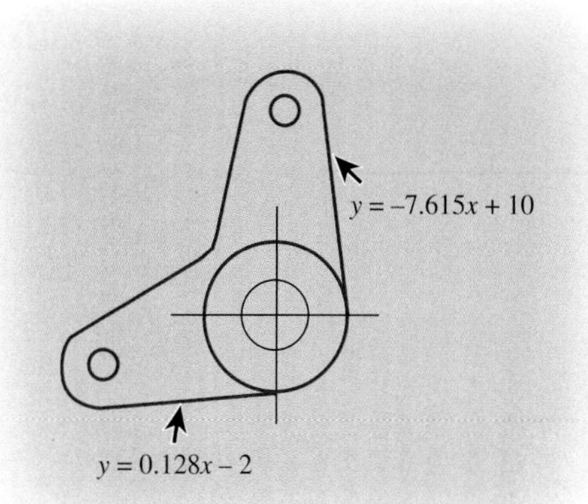

WRITING

59. Explain the advantages of writing the equation of a line in slope–intercept form ($y = mx + b$) as opposed to general form ($Ax + By = C$).

60. Why is $y = mx + b$ called the slope–intercept form of the equation of a line?

61. What is the minimum number of points needed to draw the graph of a line? Explain why.

62. List some examples of parallel and perpendicular lines that you see in your daily life.

REVIEW

63. Find the slope of the line passing through the points $(6, -2)$ and $(-6, 1)$.

64. Is $(3, -7)$ a solution of $y = 3x - 2$?

65. Evaluate: $-4 - (-4)$.

66. Solve: $2(x - 3) = 3x$.

67. To evaluate $[-2(4 - 8) + 4^2]$, which operation should be performed first?

68. Translate to mathematical symbols: four less than twice the price p.

69. What percent of 6 is 1.5?

70. Is -6.75 a solution of $x + 1 > -9$?

9.6 Point–Slope Form; Writing Linear Equations

- Point–slope form of the equation of a line
- Writing the equation of a line through two points
- Horizontal and vertical lines

If we know the slope of a line and its y-intercept, we can use the slope–intercept form to write the equation of the line. If we know the slope and any other point on the line, we can use point–slope form to write the equation of the line.

Point–slope form of the equation of a line

For the line shown in Figure 9-52, suppose we know that it has a slope of 3 and that it passes through the point $P(2, 1)$. If we pick another point on the line and call it $Q(x, y)$, we can find the slope of the line by using the coordinates of points P and Q. Using the slope formula, we have

$$\frac{y_2 - y_1}{x_2 - x_1} = m \quad \text{This is the slope formula.}$$

$$\frac{y - 1}{x - 2} = m \quad \text{Substitute } y \text{ for } y_2\text{, 1 for } y_1\text{, } x \text{ for } x_2\text{, and 2 for } x_1.$$

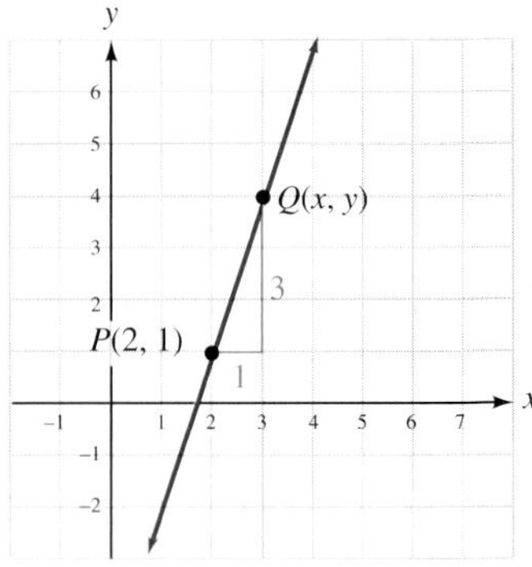

FIGURE 9-52

Since the slope of the line is given to be 3, we can substitute 3 for m in the previous equation.

$$\frac{y - 1}{x - 2} = m$$

$$\frac{y - 1}{x - 2} = 3$$

We then multiply both sides by $x - 2$ to get

$$\frac{y - 1}{x - 2}(x - 2) = 3(x - 2) \quad \text{Clear the equation of the fraction.}$$

$$y - 1 = 3(x - 2) \quad \text{Simplify the left-hand side.}$$

The resulting equation displays the slope of the line and the coordinates of one point on the line:

Slope of the line ↓

$$y - 1 = 3(x - 2)$$

↑ y-coordinate of the point ↑ x-coordinate of the point

In general, suppose we know that the slope of a line is m and that the line passes through the point (x_1, y_1). Then if (x, y) is any other point on the line, we can use the definition of slope to write

$$\frac{y - y_1}{x - x_1} = m$$

If we multiply both sides by $x - x_1$, we have

$$y - y_1 = m(x - x_1)$$

This form of a linear equation is called the **point–slope form.** It can be used to write the equation of a line when the slope and one point on the line are known.

Point–slope form of the equation of a line

If a line with slope m passes through the point (x_1, y_1), the equation of the line is

$$y - y_1 = m(x - x_1)$$

EXAMPLE 1 Write an equation of a line that has a slope of -3 and passes through $(-1, 5)$. Express the result in slope–intercept form.

Solution Since we are given the slope and a point on the line, we will use the point–slope form.

$y - y_1 = m(x - \mathbf{x_1})$ This is the point–slope form.

$y - 5 = -3[x - (\mathbf{-1})]$ Substitute -3 for m, -1 for x_1, and 5 for y_1.

$y - 5 = -3(x + 1)$ Simplify within the brackets.

We can write this result in slope–intercept form, as follows:

$y - 5 = -3x - 3$ Distribute the multiplication by -3.

$y = -3x + 2$ To undo the subtraction of 5, add 5 to both sides: $-3 + 5 = 2$.

In slope–intercept form, the equation is $y = -3x + 2$.

Self Check 1

Write an equation of a line that has a slope of -2 and passes through $(4, -3)$. Write the result in slope–intercept form.

Answer $y = -2x + 5$

EXAMPLE 2 **Temperature drop.** A refrigeration unit lowers the temperature in a railroad car 6° F every 5 minutes. One day, the temperature in a car was 76° F after the cooler had run for 10 minutes. Find a linear equation that describes the relationship between the time the cooler has been running and the temperature in the car.

Graph the equation and use it to find the temperature in the car before the cooler was turned on and the temperature in the car after the cooler had run for 25 minutes.

Solution We will let x represent the time, in minutes, that the cooler was running, and y will represent the air temperature in the car. We can make two observations:

- With the cooler on, the temperature in the railroad car drops 6° every 5 minutes. The rate of change of $-\frac{6}{5}$ degrees per minute is the slope of the graph of the linear equation that we want to find. Thus, $m = -\frac{6}{5}$.
- We know that after the cooler had been running for 10 minutes ($x = 10$), the temperature in the car was 76° ($y = 76$). We can express these facts with the ordered pair (10, 76). This is a point on the graph of the linear equation.

To write the linear equation, we substitute $-\frac{6}{5}$ for m, 10 for x_1, and 76 for y_1, into the point–slope form of the equation of a line.

$y - y_1 = m(x - \mathbf{x_1})$ This is the point–slope form.

$y - 76 = -\frac{\mathbf{6}}{\mathbf{5}}(x - \mathbf{10})$ Substitute: $m = -\frac{6}{5}$, $x_1 = 10$, and $y_1 = 76$.

$y - 76 = -\frac{6}{5}x - \left(-\frac{6}{5}\right)10$ Distribute the multiplication by $-\frac{6}{5}$.

$$y - 76 = -\frac{6}{5}x - (-12)$$ Multiply: $\left(-\frac{6}{5}\right)10 = \left(-\frac{6}{5}\right)\frac{10}{1} = -12$.

$$y - 76 = -\frac{6}{5}x + 12$$ On the right-hand side, change the subtraction to the addition of the opposite.

$$y - 76 + \mathbf{76} = -\frac{6}{5}x + 12 + \mathbf{76}$$ To undo the subtraction of 76, add 76 to both sides.

$$y = -\frac{6}{5}x + 88$$ Perform the additions.

The graph of $y = -\frac{6}{5}x + 88$ is shown in Figure 9-53. From the graph, we see that the temperature in the railroad car before the cooler was turned on was 88° F. This is given by the y-intercept of the graph, (0, 88). If we locate 25 on the x-axis and move straight up to intersect the graph, we will see that the temperature in the car was 58° F. This shows that after the cooler ran for 25 minutes, the temperature was about 58° F.

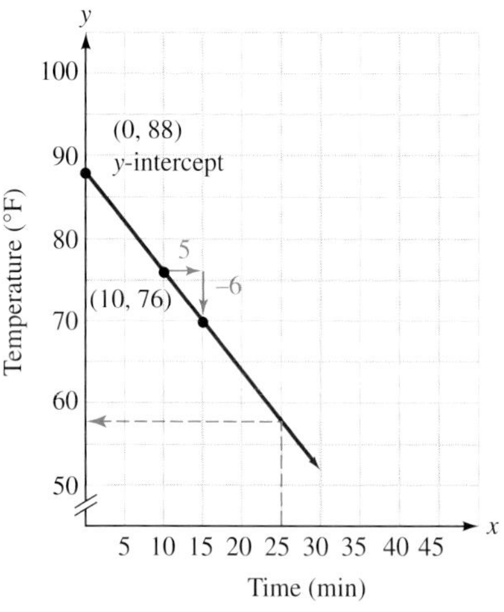

FIGURE 9-53

Writing the equation of a line through two points

In the next example, we will show that it is possible to write the equation of a line when we know the coordinates of two points on the line.

Self Check 3

Write an equation of the line passing through (0, −3) and (2, 1).

EXAMPLE 3 Write an equation of the line passing through (4, 0) and (6, −8).

Solution First we find the slope of the line that passes through (4, 0) and (6, −8).

$$m = \frac{y_2 - y_1}{x_2 - x_1}$$ This is the slope formula.

$$= \frac{\mathbf{-8} - 0}{6 - \mathbf{4}}$$ Substitute −8 for y_2, 0 for y_1, 6 for x_2, and 4 for x_1.

$$= \frac{-8}{2}$$ Simplify.

$$= -4$$

Since the line passes through both (4, 0) and (6, −8), we can choose either point and substitute its coordinates into the point–slope form. If we choose (4, 0), we substitute 4 for x_1, 0 for y_1, and −4 for m and proceed as follows.

$y - y_1 = m(x - x_1)$	This is the point–slope form.
$y - 0 = -4(x - 4)$	Substitute -4 for m, 4 for x_1, and 0 for y_1.
$y = -4x + 16$	Distribute the multiplication by -4.

The equation of the line is $y = -4x + 16$.

Answer $y = 2x - 3$

EXAMPLE 4 Market research. A company that makes a breakfast cereal has found that the number of discount coupons redeemed for its product is linearly related to the coupon's value. In one advertising campaign, 10,000 of the "10¢ off" coupons were redeemed. In another campaign, 45,000 of the "50¢ off" coupons were redeemed. How many coupons can the company expect to be redeemed if it issues a "35¢ off" coupon?

Solution If we let x represent the value of a coupon and y represent the number of coupons that will be redeemed, ordered pairs will have the form

(coupon value, number redeemed)

Two points on the graph of the equation are (10, 10,000) and (50, 45,000). These points are plotted on the graph shown in Figure 9-54. To write the equation of the line passing through the points, we first find the slope of the line.

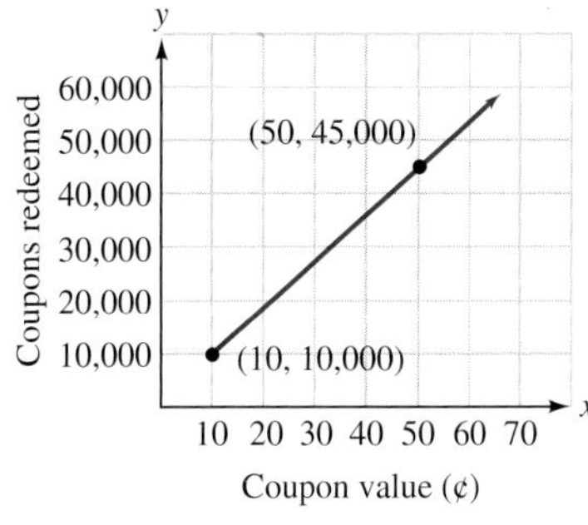

FIGURE 9-54

$$m = \frac{y_2 - y_1}{x_2 - x_1} \quad \text{This is the slope formula.}$$

$$= \frac{45{,}000 - 10{,}000}{50 - 10} \quad \text{Substitute 45,000 for } y_2\text{, 10,000 for } y_1\text{, 50 for } x_2\text{, and 10 for } x_1.$$

$$= \frac{35{,}000}{40}$$

$$= 875$$

We then substitute 875 for m and the coordinates of one known point—say, (10, 10,000)—into the point–slope form of the equation of a line and proceed as follows.

$y - y_1 = m(x - x_1)$	This is the point–slope form.
$y - 10{,}000 = 875(x - 10)$	Substitute for m, x_1, and y_1.
$y - 10{,}000 = 875x - 8{,}750$	Distribute the multiplication by 875.
$y = 875x + 1{,}250$	Add 10,000 to both sides.

To find the expected number of coupons that will be redeemed, we substitute the value of the coupon, 35¢, into the equation $y = 875x + 1{,}250$ and find y.

$y = 875x + 1{,}250$	
$y = 875(35) + 1{,}250$	Substitute 35 for x.
$y = 30{,}625 + 1{,}250$	Perform the multiplication.
$y = 31{,}875$	

The company can expect 31,875 of the 35¢ coupons to be redeemed.

Horizontal and vertical lines

We have graphed horizontal and vertical lines. We will now discuss how to write their equations.

Self Check 5

Write an equation of each line and then graph it:

a. a horizontal line passing through (3, 2)

b. a vertical line passing through $(-1, -3)$

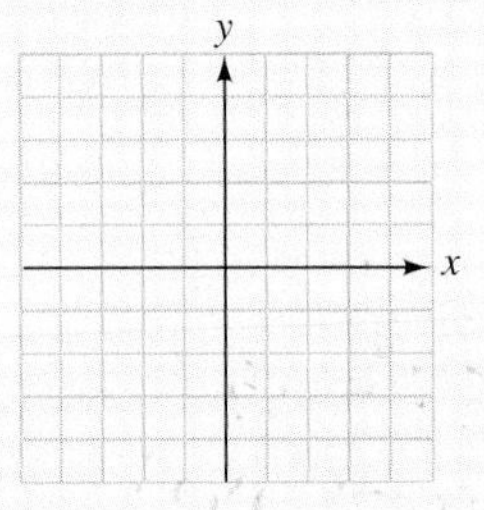

Answers **a.** $y = 2$, **b.** $x = -1$

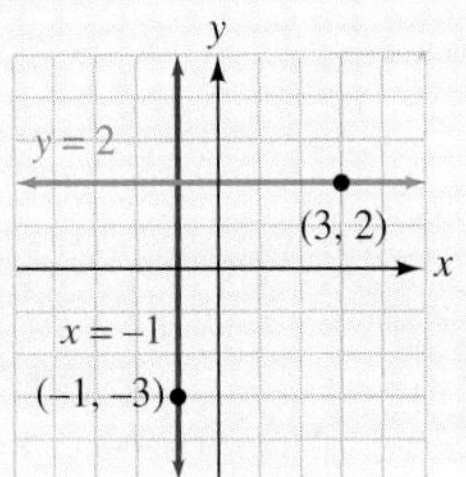

EXAMPLE 5 Write an equation of each line and then graph it: **a.** a horizontal line passing through $(-2, -4)$ and **b.** a vertical line passing through $(1, 3)$.

Solution

a. The equation of a horizontal line can be written in the form $y = b$. Since the y-coordinate of $(-2, -4)$ is -4, the equation of the line is $y = -4$. The graph is shown in Figure 9-55.

b. The equation of a vertical line can be written in the form $x = a$. Since the x-coordinate of $(1, 3)$ is 1, the equation of the line is $x = 1$. The graph is shown in Figure 9-55.

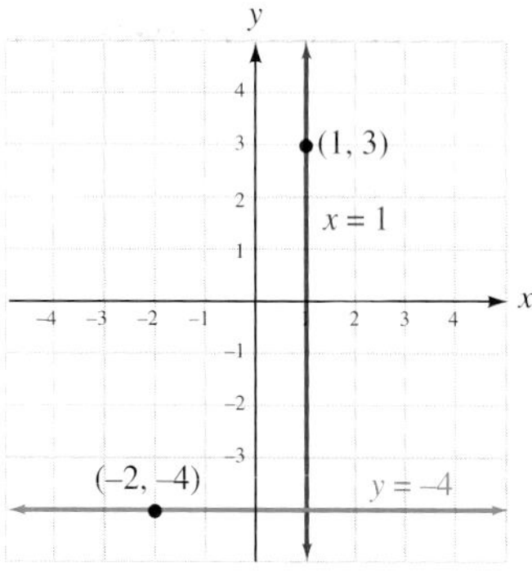

FIGURE 9-55

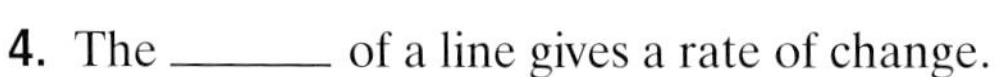

Section 9.6 STUDY SET

VOCABULARY *Fill in the blanks.*

1. $y - y_1 = m(x - x_1)$ is called the ____________ form of the equation of a line.
2. The line in the following illustration _______ through point P.
3. In the following illustration, point P has an ____________ of 2 and a ____________ of -1.
4. The _______ of a line gives a rate of change.

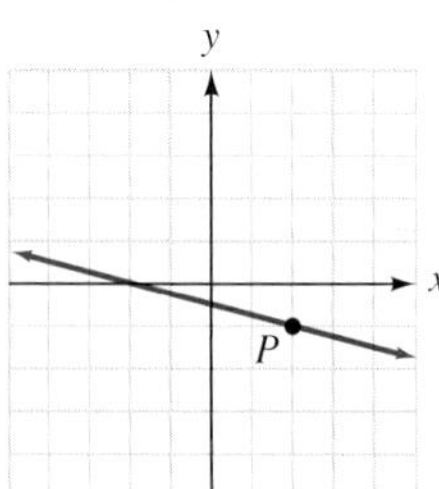

CONCEPTS

5. a. The linear equation $y = 2x - 3$ is written in *slope–intercept* form. What are the slope and the y-intercept of the graph of this line?

   b. The linear equation $y - 4 = 6(x - 5)$ is written in *point–slope* form. What point does the graph of this equation pass through, and what is the line's slope?

6. Is the following statement true or false? The equations

$$y - 1 = 2(x - 2)$$
$$y = 2x - 3$$
$$2x - y = 3$$

all describe the same line.

7. a. Find two points on the line shown below whose coordinates are integers.

   b. What is the slope of the line?

   c. Use your answers to parts a and b to write the equation of the line. Answer in point–slope form.

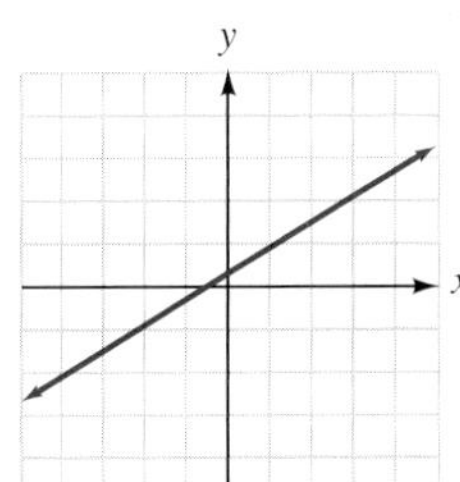

8. In each case, a linear relationship between two quantities is described. If the relationship were graphed, what would be the slope of the line?

   a. The sales of new cars increased by 15 every 2 months.

   b. There were 35 fewer robberies for each dozen police officers added to the force.

   c. Withdrawals were occurring at the rate of \$700 every 45 minutes.

   d. One acre of forest is being destroyed every 30 seconds.

9. In each case, is the given information sufficient to write an equation of the line?

   a. It passes through $(2, -7)$.

   b. Its slope is $-\frac{3}{4}$.

   c. It has the following table of solutions:

x	y
2	3
−3	−6

10. In each case, is the given information sufficient to write the equation of the line?

    a. It is horizontal.

    b. It is vertical and passes through $(-1, 1)$.

    c. It has the following table of solutions:

x	y
4	5

NOTATION

11. Fill in the blank. In $y - y_1 = m(x - x_1)$, we read x_1 as "x ____ one."

12. Write an equation of a horizontal line passing through $(0, b)$.

Write the equation in slope–intercept form.

13.
$$y - 2 = -3(x - 4)$$
$$y - 2 = \square + \square$$
$$y - 2 + \square = -3x + 12 + \square$$
$$y = -3x + 14$$

14.
$$y + 2 = \frac{1}{2}(x + 2)$$
$$y + 2 = \square + \square$$
$$y + 2 - \square = \frac{1}{2}x + 1 - \square$$
$$y = \frac{1}{2}x - 1$$

Complete each solution.

15. Write an equation of the line with slope -2 that passes through the point $(-1, 5)$.

$$y - y_1 = m(x - x_1)$$
$$y - \square = -2[x - (\square)]$$
$$y - 5 = \square - 2$$
$$y = -2x + 3$$

16. Write an equation of the line with slope 4 that passes through the point $(0, 3)$.

$$y - y_1 = m(x - x_1)$$
$$y - \square = 4(x - \square)$$
$$y - 3 = \square$$
$$y = 4x + 3$$

PRACTICE

Use the point–slope form to write an equation of the line with the given slope and point.

17. $m = 3$, passes through $(2, 1)$

18. $m = 2$, passes through $(4, 3)$

19. $m = -\frac{4}{5}$, passes through $(-5, -1)$

20. $m = -\frac{7}{8}$, passes through $(-2, -9)$

Use the point–slope form to first write an equation of the line with the given slope and point. Write the result in slope–intercept form.

21. $m = \frac{1}{5}$, passes through $(10, 1)$

22. $m = \frac{1}{4}$, passes through $(8, 1)$

23. $m = -5$, passes through $(-9, 8)$

24. $m = -4$, passes through $(-2, 10)$

25. $m = -\frac{4}{3}$,

x	y
6	−4

26. $m = -\frac{3}{2}$,

x	y
−2	1

27. $m = -\frac{2}{3}$, passes through $(3, 0)$

28. $m = -\frac{2}{5}$, passes through $(15, 0)$

29. $m = 8$, passes through $(0, 4)$

30. $m = 6$, passes through $(0, -4)$

31. $m = -3$, passes through the origin

32. $m = -1$, passes through the origin

Write an equation of the line that passes through the two given points. Write the result in slope–intercept form.

33. Passes through $(1, 7)$ and $(-2, 1)$

34. Passes through $(-2, 2)$ and $(2, -8)$

35.

x	y
−4	3
2	0

36.

x	y
−1	−4
1	−2

37. Passes through $(5, 5)$ and $(7, 5)$

38. Passes through $(-2, 1)$ and $(-2, 15)$

39. Passes through $(5, 1)$ and $(-5, 0)$

40. Passes through $(-3, 0)$ and $(3, 1)$

41. Passes through $(-8, 2)$ and $(-8, 17)$

42. Passes through $(\frac{2}{3}, 2)$ and $(0, 2)$

Write an equation of the line with the given characteristics.

43. Vertical, passes through $(4, 5)$

44. Vertical, passes through $(-2, -5)$

45. Horizontal, passes through $(4, 5)$

46. Horizontal, passes through $(-2, -5)$

APPLICATIONS

47. POLE VAULT See the following illustration.

a. For each of the four positions of the vault shown, give two points that the pole passes through.

b. Write the equations of the lines that describe the position of the pole for parts 1, 3, and 4 of the jump.

c. Why can't we write a linear equation describing the position of the pole for part 2?

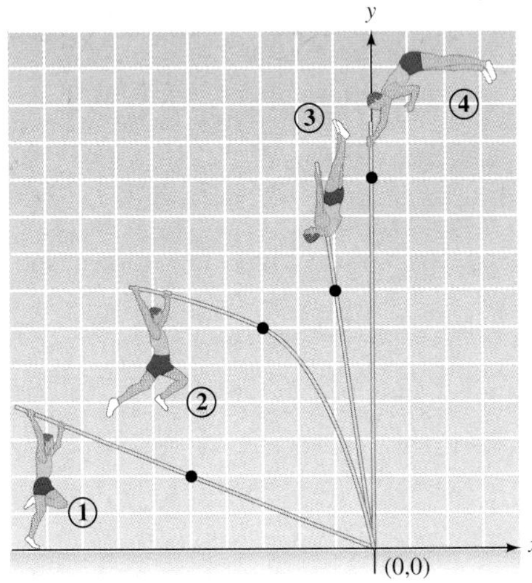

48. FREEWAY DESIGN The graph on the next page shows the route of a proposed freeway.

a. Give the coordinates of the points where the proposed freeway will join Interstate 25 and Highway 40.

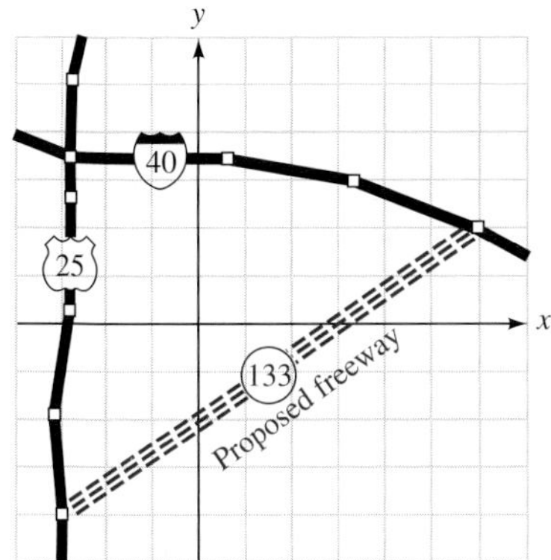

b. Write the equation of the line that mathematically describes the route of the proposed freeway. Answer in slope–intercept form.

49. TOXIC CLEANUP Three months after cleanup began at a dump site, 800 cubic yards of toxic waste had yet to be removed. Two months later, that number had been lowered to 720 cubic yards.

a. Write an equation that mathematically describes the linear relationship between the length of time x (in months) the cleanup crew has been working and the number of cubic yards y of toxic waste remaining.

b. Use your answer to part a to predict the number of cubic yards of waste that will still be on the site 1 year after the cleanup project began.

50. DEPRECIATION To lower its corporate income tax, accountants of a large company depreciated a word processing system over several years using a linear model, as shown in the worksheet.

a. Use the information in the worksheet to write a linear equation relating the years since the system was purchased x and its value y, in dollars.

b. Find the purchase price of the system by substituting $x = 0$ into your answer from part a.

Tax Worksheet

Method of depreciation: *Linear*

Property	Value	Years after purchase
Word processing system	*$60,000*	2
"	*$30,000*	4

51. COUNSELING In the first year of her practice, a family counselor saw 75 clients. In the second year, the number of clients grew to 105. If a linear trend continues, write an equation that gives the number of clients c the counselor will have t years after beginning her practice.

52. HEALTH CARE See the illustration. When the per person health care expenditures in the U.S. for the years 1990–1998 are graphed, the data nearly lie on a straight line. The expenditures can be approximated by the straight line drawn through two of the data points.

a. Use the two highlighted points on the graph to write the equation of the line. Let $x = 0$ represent 1990, $x = 1$ represent 1991, and so on. Answer in slope–intercept form.

b. Use your answer to part a to predict the per person health care expenditure in the year 2020.

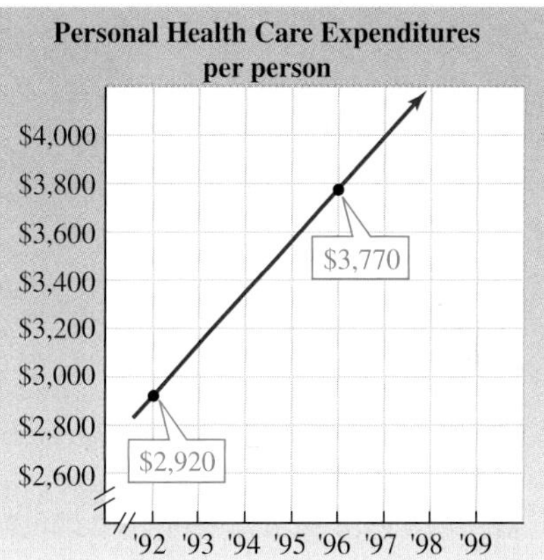

Based on data from the Health Care Financing Administration

53. CONVERTING TEMPERATURES The relationship between Fahrenheit temperature, F, and Celsius temperature, C, is linear.

a. Use the data in the illustration to write two ordered pairs of the form (C, F).

b. Use your answer to part a to write a linear equation relating the Fahrenheit and Celsius scales.

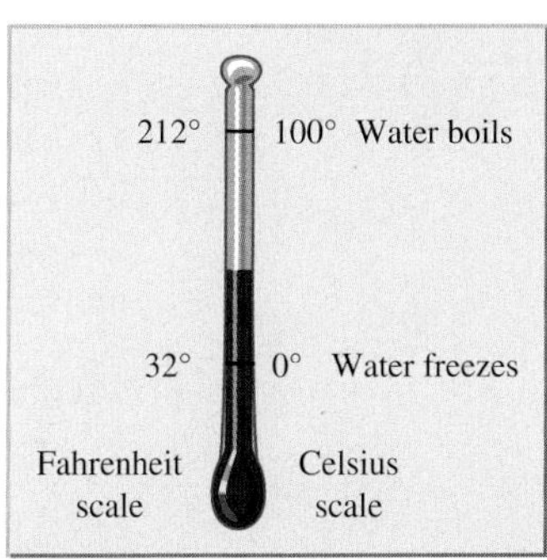

54. TRAMPOLINES The relationship between the circumference of a circle and its radius is linear.

For example, the length l of the protective pad that wraps around a trampoline is related to the radius r of the trampoline. Use the data in the illustration to write a linear equation that approximates the length of pad needed for any trampoline radius.

Radius (ft)	Approximate length of padding (ft)
3	19
7	44

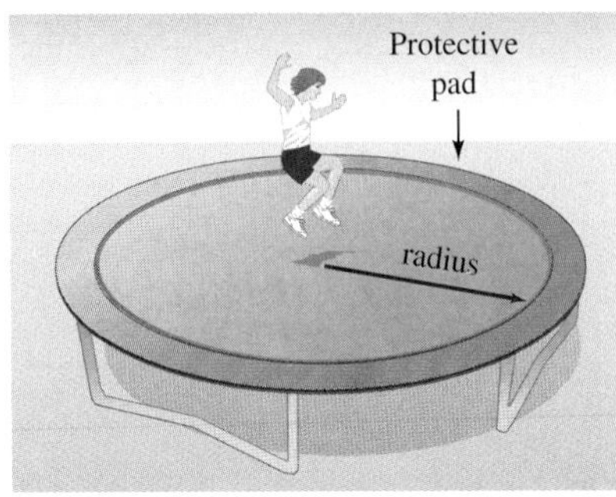

55. AIR-CONDITIONING An air-conditioning unit can lower the air temperature in a classroom 4° every 15 minutes. After the air conditioner had been running for half an hour, the air temperature in the room was 75° F. Write a linear equation relating the time in minutes x the unit had been on and the temperature y of the classroom. (*Hint:* How many minutes are there in half an hour?)

56. AUTOMATION An automated production line uses distilled water at a rate of 300 gallons every 2 hours to make shampoo. After the line had run for 7 hours, planners noted that 2,500 gallons of distilled water remained in the storage tank. Write a linear equation relating the time in hours x since the production line began and the number of gallons y of distilled water in the storage tank.

WRITING

57. Why is $y - y_1 = m(x - x_1)$ called the point–slope form of the equation of a line?
58. If we know two points that a line passes through, we can write its equation. Explain how this is done.
59. If we know the slope of a line and a point it passes through, we can write its equation. Explain how this is done.
60. Think of several points on the graph of the horizontal line $y = 4$. What do the points have in common? How do they differ?

REVIEW

61. Find the slope of the line passing through the points $(2, 4)$ and $(-6, 8)$.
62. Is the graph of $y = x^2$ a straight line?
63. Find the area of a circle with a diameter of 12 feet. Round to the nearest tenth.
64. If a 15-foot board is cut into two pieces and we let x represent the length of one piece (in feet), how long is the other piece?
65. Evaluate: $(-1)^5$.
66. Solve: $\frac{x - 3}{4} = -4$.
67. What is the coefficient of the second term of $-4x^2 + 6x - 13$?
68. Simplify: $(-2p)(-5)(4x)$.

9.7 Functions

- Functions • Domain and range of a function • Function notation
- Graphs of functions • The vertical line test

In everyday life, we see a wide variety of situations where one quantity depends on another:

- The distance traveled by a car depends on its speed.
- The cost of renting a video depends on the number of days it is rented.
- A state's number of representatives in Congress depends on the state's population.

In this section, we will discuss many situations where one quantity depends on another according to a specific rule, called a *function.* For example, the equation

$y = 2x - 3$ sets up a rule where each value of y depends on the choice of some number x. The rule is: *To find y, double the value of x and subtract* 3. In this case, y (the *dependent variable*) depends on x (the *independent variable*).

Functions

We have previously described relationships between two quantities in different ways:

Using words

The number of tires to order	is	two	times	the number of bicycles to be manufactured.

Here words are used to state that the number of bicycle tires to order depends on the number of bicycles to be manufactured.

Using equations

$t = 1{,}500 - d$

This equation describes how the amount of take-home pay t depends on the amount of deductions d.

Using graphs

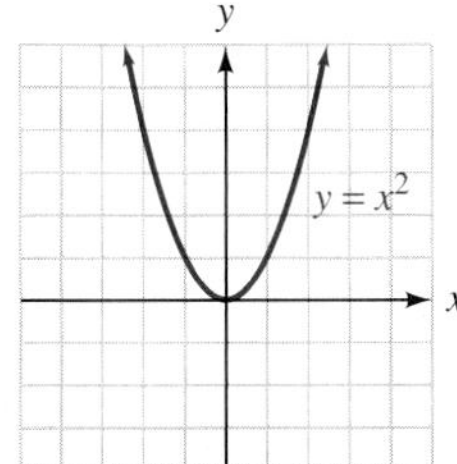

This rectangular coordinate graph shows many ordered pairs (x, y) that satisfy the equation $y = x^2$, where the value of the y-coordinate depends on the value of the x-coordinate.

Using tables

Acres	Schools
400	4
800	8
1,000	10
2,000	20

This table shows that the number of schools needed depends on the size of the housing development.

Two observations can be made about these examples:

- Each one establishes a relationship between two sets of values. For example, the number of bicycle tires that must be ordered *depends* on the number of bicycles to be manufactured.
- In these relationships, each value in one set is assigned a *single* value of a second set. For example, for each number of bicycles to be manufactured, there is exactly one number of tires to order.

Relationships between two quantities that exhibit both of these characteristics are called **functions.**

Functions

A **function** is a rule that assigns to each value of one variable (the **independent variable**) a single value of another variable (the **dependent variable**).

Using the variables x and y, we can restate the previous definition as follows: For y to be a function of x, each value of x must determine exactly one value of y.

Self Check 1

a. Does $y = 2 - x^2$ define a function?

b. Does the table below define a function?

x	y
2	4
1	1
0	0
−1	1
−2	4

Answers **a.** yes, **b.** yes

EXAMPLE 1 **a.** Does $y = 4x + 1$ define a function?

b. Does the table define y as a function of x?

x	y
0	6
5	3
9	1
5	7
10	8

Solution

a. For each value of the independent variable x, we apply the rule: *Multiply x by 4 and add 1.* Since this arithmetic gives a single value of the dependent variable y, the equation defines a function.

b. Since the table assigns two different values of y, 3 and 7, to the x-value of 5, it does not define y as a function of x.

! COMMENT The table in the above Self Check illustrates an important fact about functions. In a function, different values of x can determine the *same* value of y. In the table, x-values of 2 and -2 determine a y-value of 4, and x-values of 1 and -1 determine a y-value of 1. Nevertheless, each value of x determines exactly one value of y, so the table does define a function.

Domain and range of a function

We have seen that functions can be represented by equations in two variables. Some examples of functions are

$$y = 2x - 10, \qquad y = x^2 + 2x - 3, \qquad \text{and} \qquad s = 5 - 16t$$

For a function, the set of all possible values of the independent variable (the inputs) is called the **domain of the function.** The set of all possible values of the dependent variable (the outputs) is called the **range of the function.**

Self Check 2

Find the domain and range of the function $y = -x$.

Answer domain: all real numbers; range: all real numbers

EXAMPLE 2 Find the domain and range of $y = |x|$.

Solution To find the domain of $y = |x|$, we determine which real numbers are allowable inputs for x. Since we can find the absolute value of any real number, the domain is the set of all real numbers. Since the absolute value of any real number x is greater than or equal to zero, the range of $y = |x|$ is the set of all real numbers greater than or equal to zero.

Function notation

When the variable y is a function of x, there is a special notation that we can use to denote the function.

Function notation

The notation $y = f(x)$ denotes that y is a function of x.

The notation $y = f(x)$ is read as "y equals f of x." Note that y and $f(x)$ are two different notations for the same quantity. Thus, the equations $y = 4x + 1$ and $f(x) = 4x + 1$ represent the same relationship.

! COMMENT The symbol $f(x)$ denotes a function. It does not mean "f times x."

The notation $y = f(x)$ provides a way of denoting the value of y that corresponds to some number x. For example, if $f(x) = 4x + 1$, the value of y that is determined when $x = 2$ is denoted by $f(2)$.

$$f(\mathbf{x}) = 4\mathbf{x} + 1 \quad \text{This is the function.}$$
$$f(\mathbf{2}) = 4(\mathbf{2}) + 1 \quad \text{Replace } x \text{ with 2.}$$
$$= 8 + 1$$
$$= 9$$

Thus, $f(2) = 9$.

The letter f used in the notation $y = f(x)$ represents the word *function.* However, other letters can be used to represent functions. For example, $y = g(x)$ and $y = h(x)$ also denote functions involving the variable x.

EXAMPLE 3 For $g(x) = 3 - 2x$ and $h(x) = x^3 - 1$, find **a.** $g(3)$ and **b.** $h(-2)$.

Self Check 3
Find $g(0)$ and $h(4)$ using the functions in Example 3.

Solution

a. To find $g(3)$, we use the function rule $g(x) = 3 - 2x$ and replace x with 3.

$$g(\mathbf{x}) = 3 - 2\mathbf{x}$$
$$g(\mathbf{3}) = 3 - 2(\mathbf{3}) \quad \text{Substitute 3 for } x.$$
$$= 3 - 6 \quad \text{Perform the multiplication.}$$
$$= -3$$

So $g(3) = -3$.

b. To find $h(-2)$, we use the function rule $h(x) = x^3 - 1$ and replace x with -2.

$$h(\mathbf{x}) = \mathbf{x}^3 - 1$$
$$h(\mathbf{-2}) = (\mathbf{-2})^3 - 1 \quad \text{Substitute } -2 \text{ for } x.$$
$$= -8 - 1 \quad \text{Evaluate the power.}$$
$$= -9$$

So $h(-2) = -9$.

Answers **a.** 3, **b.** 63

We can think of a function as a machine that takes some input x and turns it into some output $f(x)$, as shown in Figure 9-56(a). The machine in Figure 3-56(b) turns the input value of -2 into the output value of -9, and we can write $f(-2) = -9$.

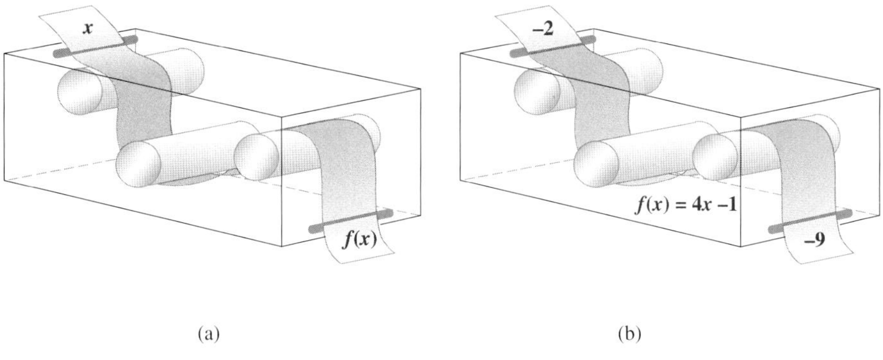

(a) (b)

FIGURE 9-56

CALCULATOR SNAPSHOT

Business profits

Accountants have found that the function $f(x) = -0.000065x^2 + 12x - 278{,}000$ estimates the profit a bowling alley will make when x games are bowled per year. Suppose that management predicts that 90,000 games will be bowled in the upcoming year. The expected profit for that year can be found by evaluating $f(90{,}000)$ on a scientific calculator.

$$f(\mathbf{90{,}000}) = -0.000065(\mathbf{90{,}000})^2 + 12(\mathbf{90{,}000}) - 278{,}000$$

.000065 [+/−] [×] 90000 [x^2] [+] 12 [×] 90000 [−] 278000 [=]

275500

To evaluate $f(90{,}000)$ with a graphing calculator, we enter these numbers and press these keys.

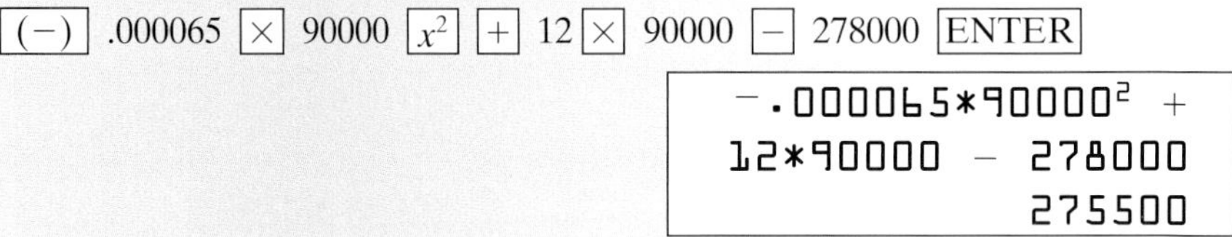

Graphs of functions

We have seen that a function, such as $f(x) = 4x + 1$, assigns to each value of x a single value of y. The ordered pairs (x, y) that a function determines can be shown on a graph. Since $y = f(x)$, the graph of the function $f(x) = 4x + 1$ is the same as the graph of the equation $y = 4x + 1$. We can graph the function by making a **table of values,** plotting the points, and drawing the graph.

To make a table of values for $f(x) = 4x + 1$, we will choose numbers for x and find the corresponding values of $f(x)$. If $x = -1$, we have

$$\begin{aligned} f(\boldsymbol{x}) &= 4\boldsymbol{x} + 1 \\ f(\mathbf{-1}) &= 4(\mathbf{-1}) + 1 && \text{Substitute } -1 \text{ for } x. \\ &= -4 + 1 && \text{Perform the multiplication.} \\ &= -3 \end{aligned}$$

We have found that $f(-1) = -3$. In a similar manner, we find the corresponding values for $f(x)$ for x-values of 0 and 2 and record them in the table of values in Figure 9-57 (next page). If we plot the ordered pairs in the table and draw a straight line through them, we get the graph of the function $f(x) = 4x + 1$ shown in the figure.

$f(x) = 4x + 1$		
x	$f(x)$	$(x, f(x))$
−1	−3	(−1, −3)
0	1	(0, 1)
2	9	(2, 9)

↑ Pick input values from the domain.

↑ Find each corresponding output value: $f(-1), f(0)$, and $f(2)$.

↑ From ordered pairs.

(2, 9)
$f(x) = 4x + 1$
(0, 1)
(−1, −3)

The vertical axis can be labeled y or $f(x)$.

FIGURE 9-57

Any linear equation, except those of the form $x = a$, can be written using function notation by writing it in slope–intercept form ($y = mx + b$) and then replacing y with $f(x)$. We call this type of function a **linear function.**

Figure 9-58 shows the graphs of four basic functions.

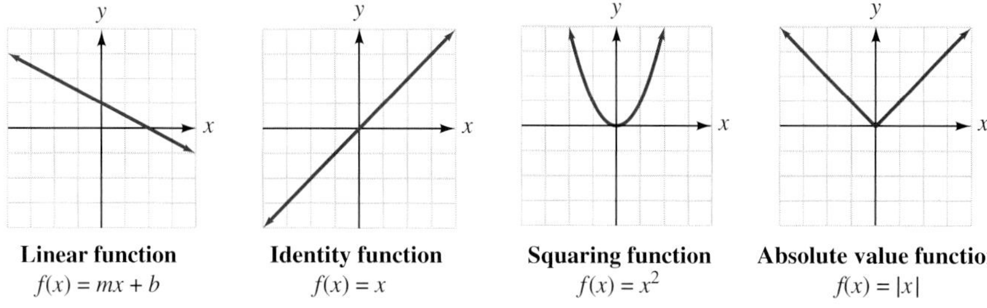

Linear function $f(x) = mx + b$

Identity function $f(x) = x$

Squaring function $f(x) = x^2$

Absolute value function $f(x) = |x|$

FIGURE 9-58

The vertical line test

We can use the **vertical line test** to determine whether a given graph is the graph of a function. If any vertical line intersects a graph more than once, the graph cannot represent a function, because to one value of x, there corresponds more than one value of y. The graph in Figure 9-59(a), shown in red, is not the graph of a function, because the x-value −1 determines three different y-values: 3, −1, and −4.

The graph shown in Figure 9-59(b) does represent a function, because every vertical line that can be drawn intersects the graph exactly once.

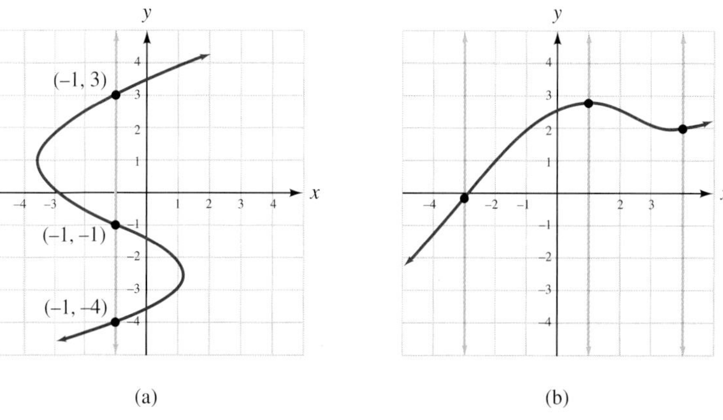

FIGURE 9-59

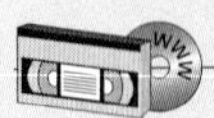

Self Check 4
Which of the following graphs are graphs of functions?

a.

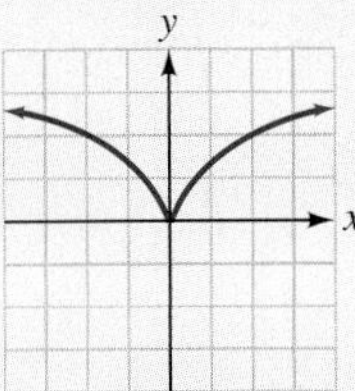

b.

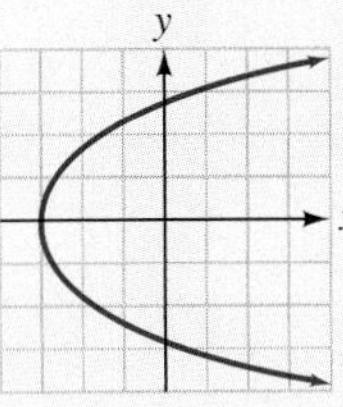

Answers **a.** function, **b.** not a function

EXAMPLE 4 Which of the following graphs in red are graphs of functions?

a.

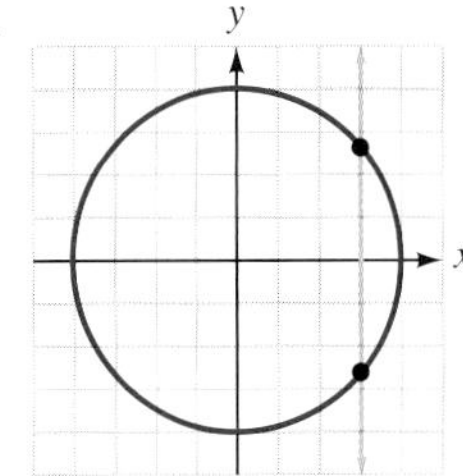

b.

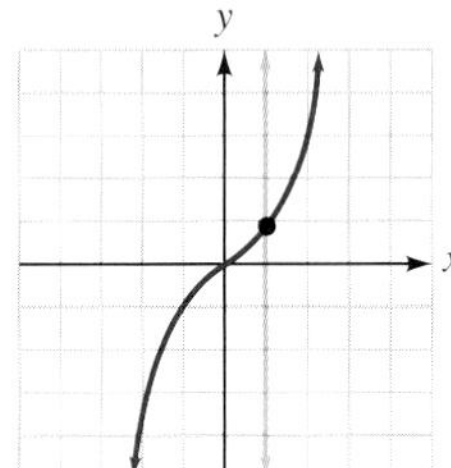

Solution

a. This graph is not the graph of a function, because the vertical line intersects the graph at more than one point.

b. This graph is the graph of a function, because no vertical line will intersect the graph at more than one point.

Section 9.7 STUDY SET

VOCABULARY *Fill in the blanks.*

1. A ________ is a rule that assigns to each value of the independent variable a single value of another variable.
2. The set of all possible input values for a function is called the ________, and the set of all possible output values is called the ________.
3. For $y = 2x + 8$, x is called the ____________ variable, and y is called the ____________ variable.
4. $f(x) = 6 - 5x$ is an example of ________ notation.

CONCEPTS

5. Consider the function $f(x) = x^2$.
   - **a.** If positive real numbers are substituted for x, what type of numbers result?
   - **b.** If negative real numbers are substituted for x, what type of numbers result?
   - **c.** If zero is substituted for x, what number results?
   - **d.** What are the domain and range of the function?
6. Consider the function $g(x) = x^4$.
   - **a.** What type of numbers can be inputs in this function? What is the special name for this set?
   - **b.** What type of numbers will be outputs in this function? What is the special name for this set?
7. Consider the following problems. Fill in the blank so that they ask for the same thing.
   1. In the equation $y = -5x + 1$, find the value of y when $x = -1$.
   2. In the equation $f(x) = -5x + 1$, find ________.
8. A function can be thought of as a machine that converts inputs into outputs. Use the terms *domain, range, input, and output* to label the diagram of a function machine in the illustration. Then find $f(2)$.

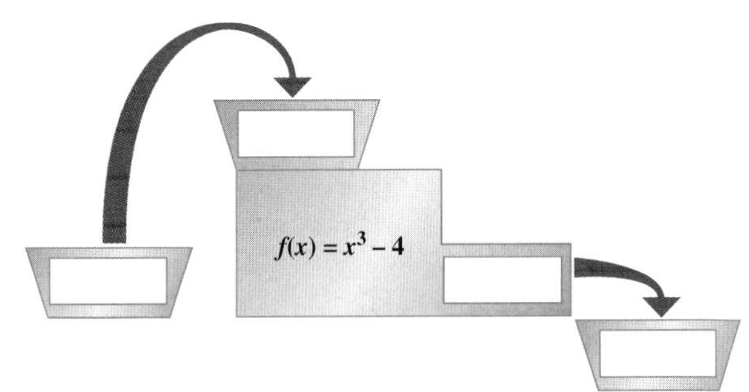

9. See the illustration.

   a. Give the coordinates of the points where the given vertical line intersects the graph.

   b. Is this the graph of a function? Explain your answer.

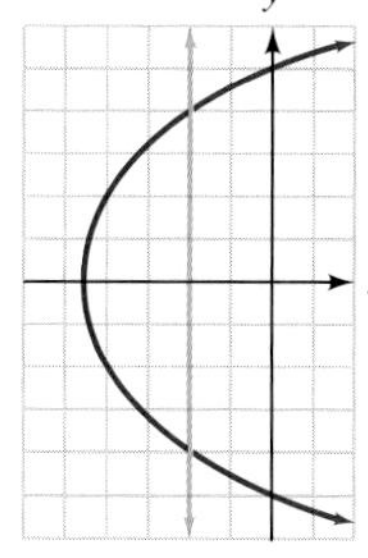

10. A student was asked to determine whether the graph in the illustration is the graph of a function. What is wrong with the following reasoning?

    When I draw a vertical line through the graph, it intersects the graph only once. By the vertical line test, this is the graph of a function.

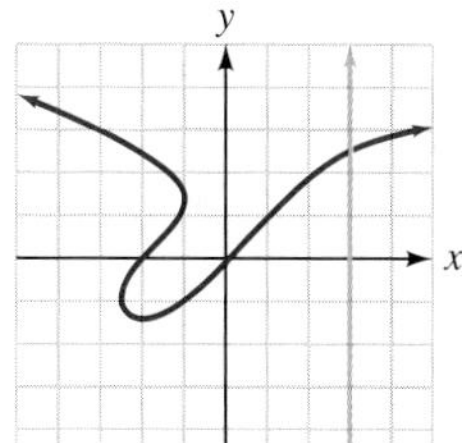

NOTATION

11. Fill in the blanks to make the statements true. The function notation $f(4) = -5$ states that when 4 is substituted for ___ in function f, the result is ___. This fact can be illustrated graphically by plotting the point (___, ___).

12. Fill in the blank: $f(x) = 6 - 5x$ is read as "f ___ x is $6 - 5x$."

13. Fill in the blanks: If $f(x) = 6 - 5x$, then $f(0) = 6$ is read as "f ___ zero ___ 6."

14. Determine whether this statement is true or false: The equations $y = 3x + 5$ and $f(x) = 3x + 5$ are the same.

PRACTICE *Determine whether a function is defined. If it is not, indicate an input for which there is more than one output.*

15. $y = 2x + 10$
16. $y = x - 15$
17. $y = x^2$
18. $y = |x|$
19. $y^2 = x$
20. $|y| = x$
21. $y = x^3$
22. $y = -x$
23. $x = 3$
24. $y = 3$

25.

x	y
1	7
2	15
3	23
4	16
5	8

26.

x	y
−1	1
−3	1
−5	1
−7	1
−9	1

27.

x	y
−4	6
−1	0
0	−3
2	4
−1	2

28.

x	y
30	2
30	4
30	6
30	8
30	10

29.

t	d
3	4
3	−4
4	3
4	−3

30.

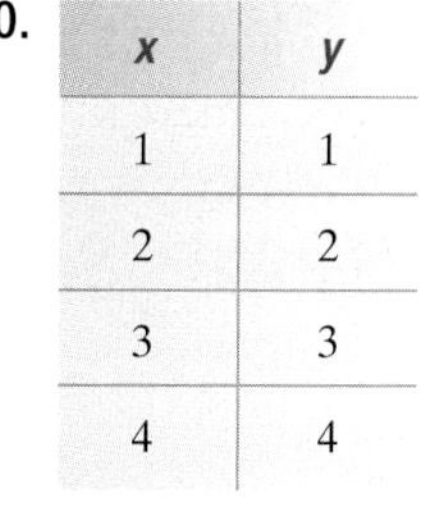

x	y
1	1
2	2
3	3
4	4

31.

32.

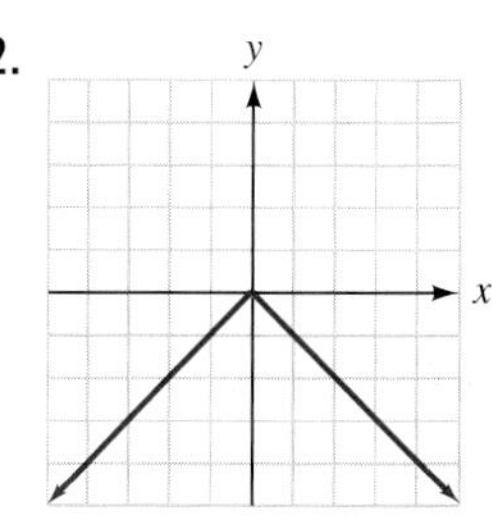

33.

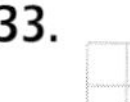

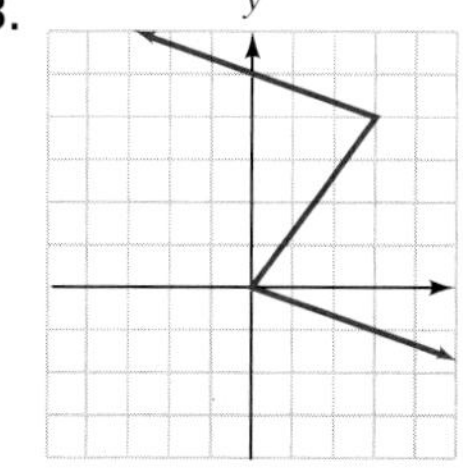

34.

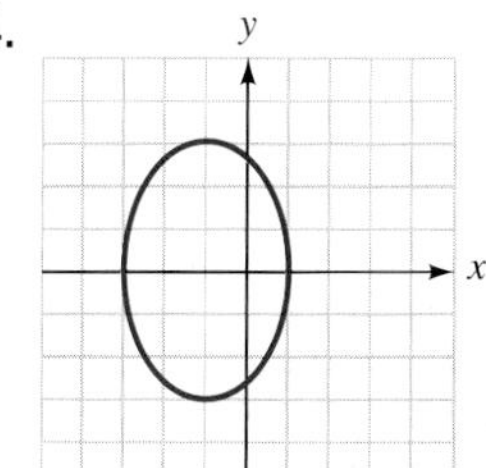

35. 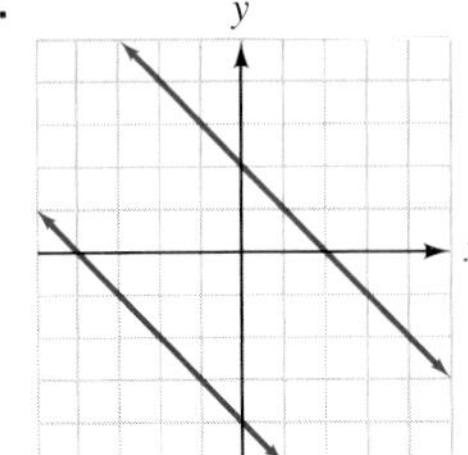

36. 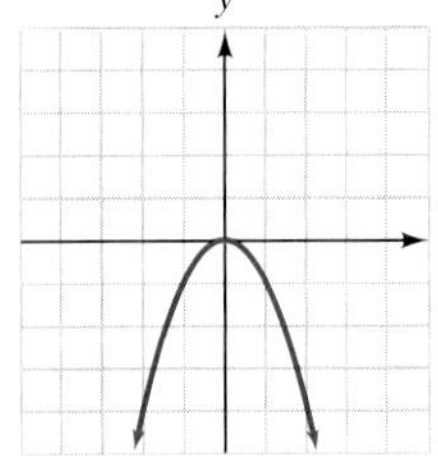

Find the domain and range of the function.

37. $f(x) = x + 1$

38. $f(x) = 3x - 2$

39. $y = x^2$

40. $y = -|x|$

41. $f(x) = x^3$

42. $f(x) = x$

Find each value.

43. $f(x) = 4x - 1$
   a. $f(1)$ b. $f(-2)$
   c. $f\left(\frac{1}{4}\right)$ d. $f(50)$

44. $g(x) = 1 - 5x$
   a. $g(0)$ b. $g(-75)$
   c. $g(0.2)$ d. $g\left(-\frac{4}{5}\right)$

45. $h(t) = 2t^2$
   a. $h(0.4)$ b. $h(-3)$
   c. $h(1{,}000)$ d. $h\left(\frac{1}{8}\right)$

46. $v(t) = 6 - t^2$
   a. $v(30)$ b. $v(6)$
   c. $v(-1)$ d. $v(0.5)$

47. $s(x) = |x - 7|$
   a. $s(0)$ b. $s(-7)$
   c. $s(7)$ d. $s(8)$

48. $f(x) = |2 + x|$
   a. $f(0)$ b. $f(2)$
   c. $f(-2)$ d. $f(-99)$

49. $f(x) = x^3 - x$
   a. $f(1)$ b. $f(10)$
   c. $f(-3)$ d. $f(6)$

50. $g(x) = x^4 + x$
   a. $g(1)$ b. $g(-2)$
   c. $g(0)$ d. $g(10)$

51. If $f(x) = 3.4x^2 - 1.2x + 0.5$, find $f(-0.3)$.

52. If $g(x) = x^4 - x^3 + x^2 - x$, find $g(-12)$.

Complete each table and graph the function.

53. $f(x) = -2 - 3x$

x	f(x)
0	
1	
−1	
−2	

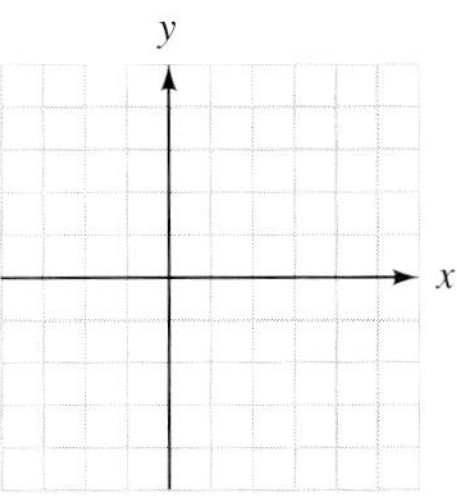

54. $h(x) = |1 - x|$

x	h(x)
0	
1	
2	
3	
−1	
−2	

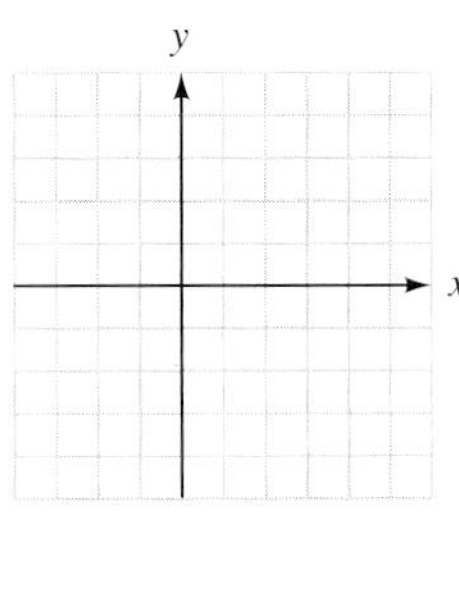

55. $f(x) = \frac{1}{2}x - 2$

x	f(x)
−2	
0	
2	

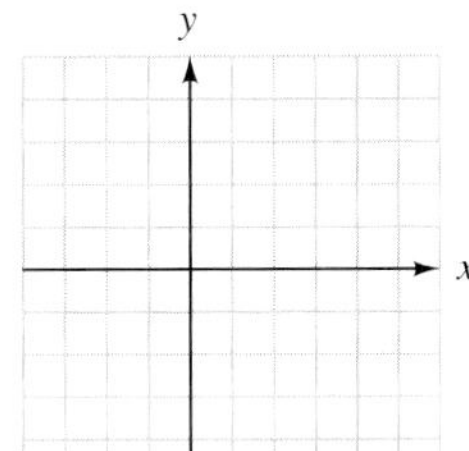

56. $f(x) = -\frac{2}{3}x + 3$

x	f(x)
0	
3	
6	

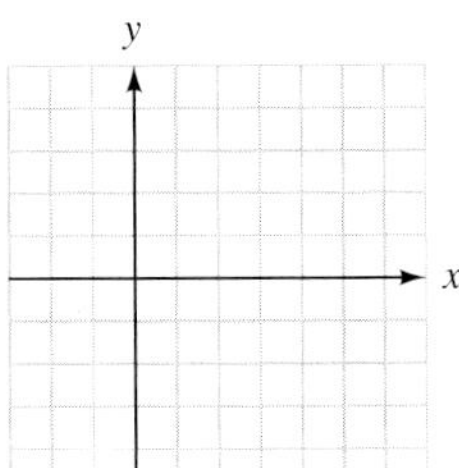

57. $s(x) = 2 - x^2$

x	s(x)
0	
1	
2	
−1	
−2	

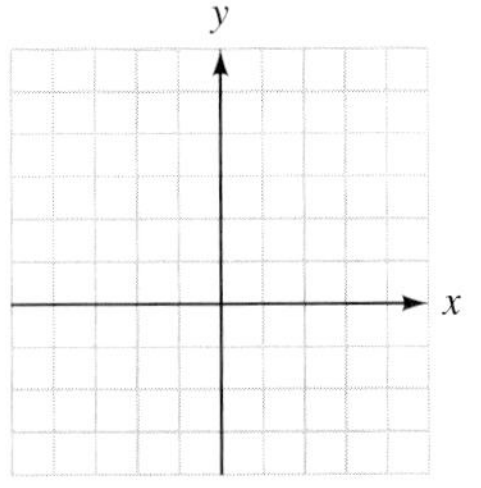

58. $g(x) = 1 + x^3$

x	$g(x)$
0	
1	
2	
−1	
−2	

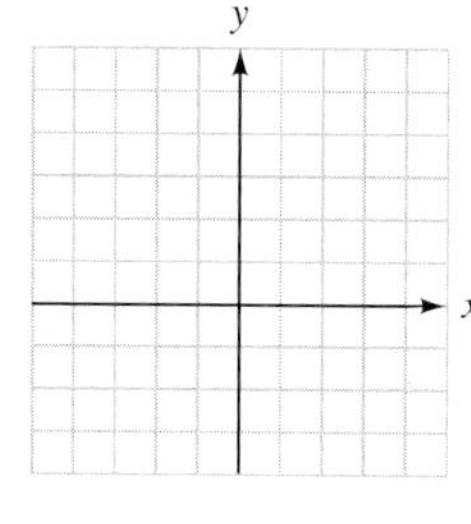

APPLICATIONS

59. REFLECTIONS When a beam of light hits a mirror, it is reflected off the mirror at the same angle that the incoming beam struck the mirror, as shown. What type of function could serve as a mathematical model for the path of the light beam shown here?

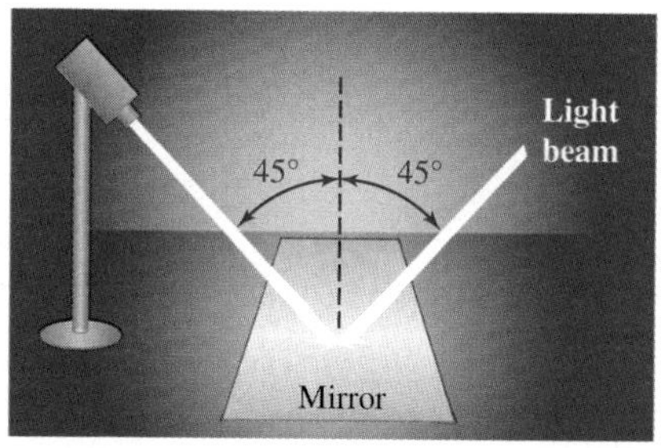

60. MATHEMATICAL MODELS The illustration shows the path of a basketball shot taken by a player. What type of function could be used to mathematically model the path of the basketball?

61. TIDES The illustration in the next column shows the graph of a function f, which gives the height of the tide for a 24-hour period in Seattle, Washington. (Note that military time is used on the x-axis: 3 A.M. = 3, noon = 12, 3 P.M. = 15, 9 P.M. = 21, and so on.)

a. Find the domain of the function.

b. Find $f(3)$.

c. Find $f(6)$.

d. Estimate $f(15)$.

e. What information does $f(12)$ give?

f. Estimate $f(21)$.

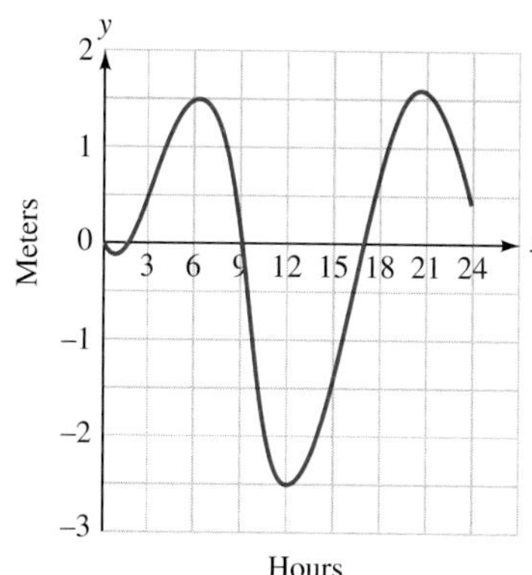

62. SOCCER The illustration shows the graphs of three functions on the same coordinate system: $g(x)$ represents the number of girls, $f(x)$ represents the number of boys, and $t(x)$ represents the total number playing high school soccer in year x.

a. What is the domain of each of these functions?

b. Find $g(87)$, $f(86)$, and $t(93)$.

c. Estimate $g(95)$, $f(95)$, and $t(95)$.

d. For what year x was $g(x) = 75{,}000$?

e. For what year x was $f(x) = 225{,}000$?

f. For what year x was $t(x)$ first greater than 350,000?

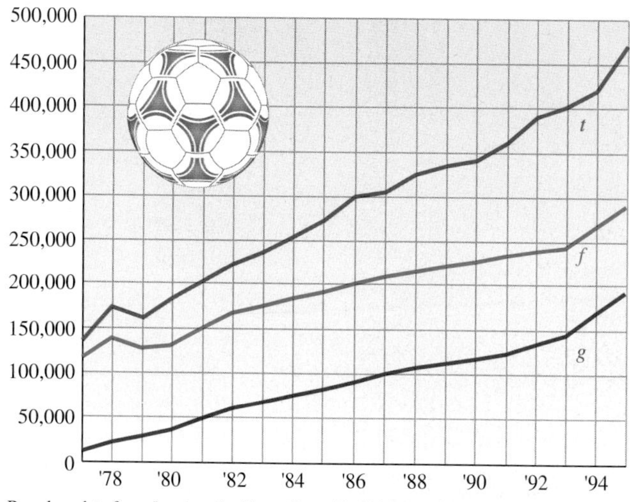

Based on data from *Los Angeles Times* (Dec. 12, 1996), p. A5

63. LAWN SPRINKLERS The function $A(r) = \pi r^2$ can be used to determine the area that will be watered by a rotating sprinkler that sprays out a stream of water r feet. Find $A(5)$, $A(10)$, and $A(20)$. Round to the nearest tenth.

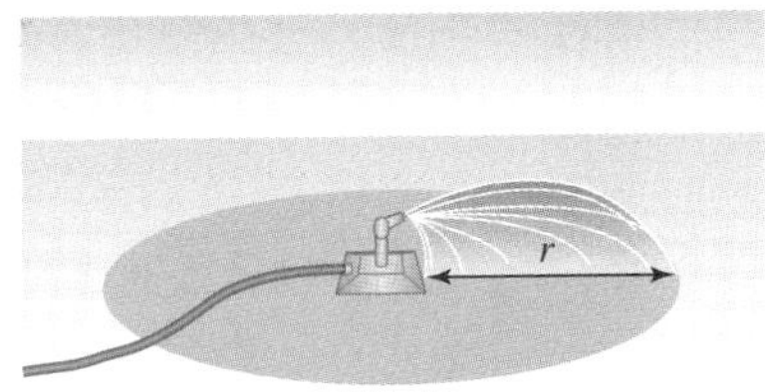

64. PARTS LIST The function

$$f(r) = 2.30 + 3.25(r + 0.40)$$

approximates the length (in feet) of the belt that joins the two pulleys shown. r is the radius (in feet) of the smaller pulley. Find the belt length needed for each pulley in the parts list.

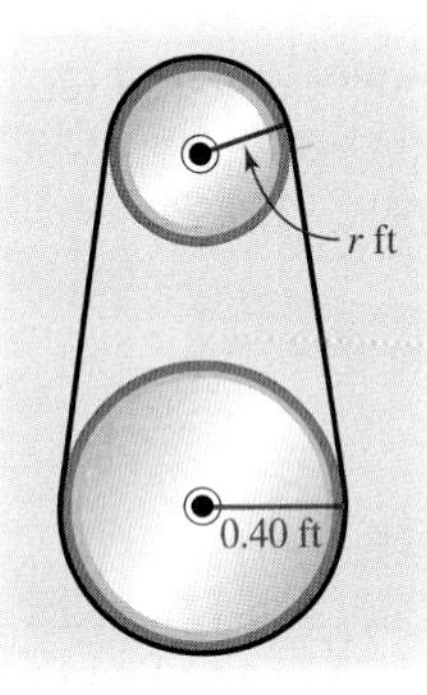

Parts list

Pulley	r	Belt length
P-45M	0.32	
P-08D	0.24	
P-00A	0.18	
P-57X	0.38	

WRITING

65. In the function $y = -5x + 2$, why do you think x is called the *independent* variable and y the *dependent* variable?

66. Explain what a politician meant when she said, "The speed at which the downtown area will be redeveloped is a function of the number of low-interest loans made available to the property owners."

REVIEW

67. Give the equation of the horizontal line passing through $(-3, 6)$.

68. Is -3 a solution of $t^2 - t + 1 = 13$?

69. Write the formula that relates profit, revenue, and costs.

70. What is the word used to represent the perimeter of a circle?

71. Use the distributive property to remove the parentheses in $-3(2x - 4)$.

72. Evaluate $r^2 - r$ for $r = -0.5$.

73. Write an expression for how many eggs there are in d dozen.

74. On a rectangular coordinate graph, what variable is associated with the horizontal axis?

KEY CONCEPT

Describing Linear Relationships

In Chapter 9, we discussed ways to mathematically describe linear relationships between two quantities.

Equations in Two Variables

The general form of the equation of a line is $Ax + By = C$. Two very useful forms of the equation of a line are the slope–intercept form and the point–slope form.

1. Write an equation of a line with a slope of -3 and a y-intercept of $(0, -4)$.

2. Write an equation of the line that passes through $(5, 2)$ and $(-5, 0)$. Answer in slope–intercept form.

Rectangular Coordinate Graphs

The graph of an equation is a picture of all of its solutions (x, y). Important information can be obtained from a graph.

3. Complete the table of solutions for $2x - 4y = 8$. Then graph the equation.

$2x - 4y = 8$	
x	y
0	
	0
-2	

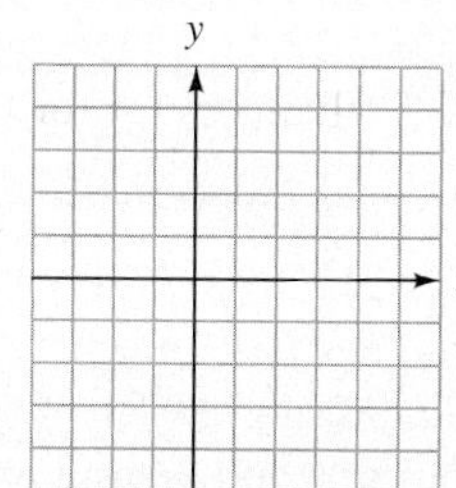

4. See the illustration.

a. What information does the y-intercept of the graph give us?

b. What is the slope of the line and what does it tell us?

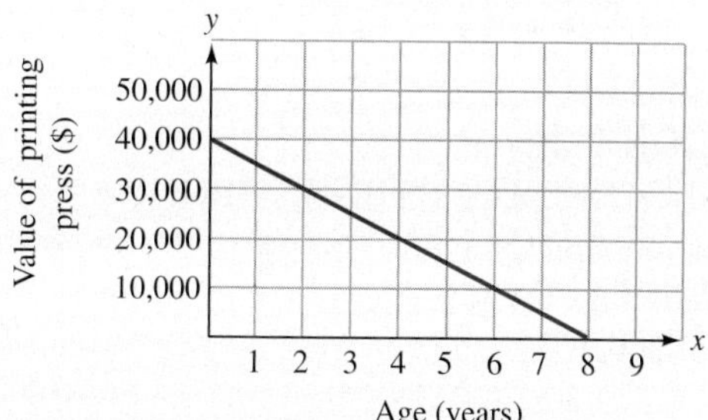

5. Consider the line graphed in the illustration.

a. Find a point on the line.

b. Determine the slope of the line.

c. Write an equation of the line. Express your answer in slope–intercept form.

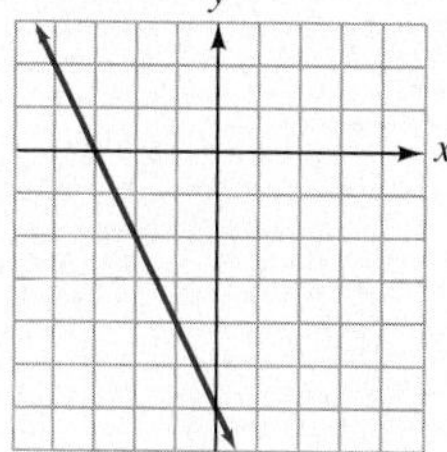

6. Write an equation of the line that passes through $(1, -1)$ and is parallel to the line graphed in the illustration.

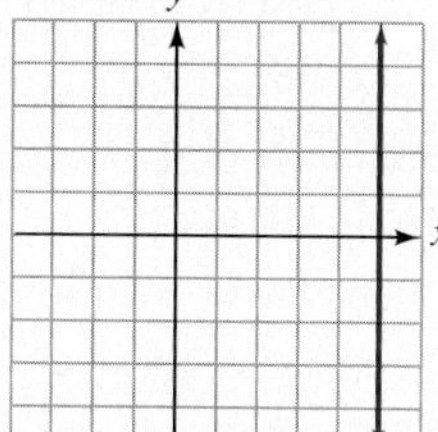

Linear Functions

We can use the notation $f(x) = mx + b$ to describe linear functions.

7. The function $f(x) = 35x + 25$ gives the cost (in dollars) to rent a cement mixer for x days. Find $f(3)$. What does it represent?

8. The function $T(x) = \frac{1}{4}x + 40$ predicts the outdoor temperature T in degrees Fahrenheit using the number of cricket chirps x per minute. Find $T(160)$.

ACCENT ON TEAMWORK

SECTION 9.1

DAILY HIGH TEMPERATURE For a 2-week period, plot the daily high temperature for your city on a rectangular coordinate system. You can normally find this information in a local newspaper. Label the x-axis "observation day" and the y-axis "daily high temperature in degrees Fahrenheit." For example, the ordered pair (3, 72) indicates that on day 3 of the observation period, the high temperature was 72° F. At the end of the 2-week period, see whether any temperature trend is apparent from the graph.

SECTION 9.2

TRANSLATIONS On a piece of graph paper, sketch the graph of $y = |x|$ with a black marker. Using a different color, sketch the graphs of $y = |x| + 2$ and $y = |x| - 2$ on the same coordinate system. On another piece of graph paper, do the same for $y = |x|$ and $y = |x + 2|$ and $y = |x - 2|$. Make some observations about how the graph of $y = |x|$ is "moved" or "translated" by the addition or subtraction of 2. Use what you have learned to discuss the graphs of $y = x^2$, $y = x^2 + 2$, $y = x^2 - 2$, $y = (x + 2)^2$, and $y = (x - 2)^2$.

SECTION 9.3

COMPUTER GRAPHING PROGRAMS If your school has a mathematics computer lab, ask the lab supervisor whether there is a graphing program on the system. If so, familiarize yourself with the operation of the program and then graph each of the equations from Figure 9-22 and from Examples 3–6 in Section 3.3. Print out each graph and compare with those in the textbook.

SECTION 9.4

MEASURING SLOPE Use a tape measure (and a level if necessary) to find the slopes of five objects by finding $\frac{\text{rise}}{\text{run}}$. See the applications in Study Set 3.4 for some ideas about what you can measure. Record your results in a chart like the one shown. List the examples in increasing order of magnitude, starting with the smallest slope.

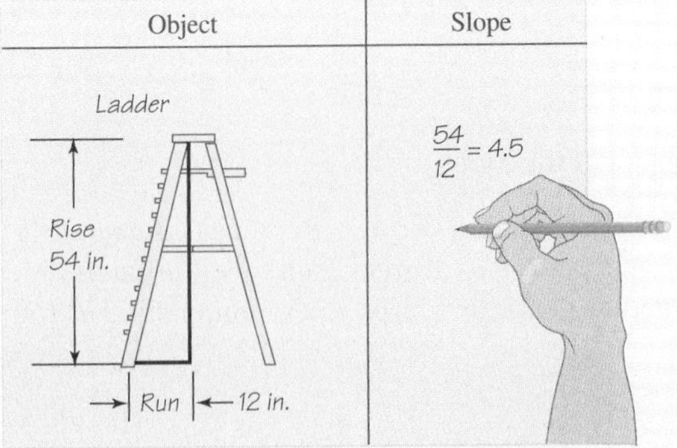

SECTION 9.5

SHOPPING Visit a local grocery store and find the price per pound of bananas. Make a rectangular coordinate graph that could be posted next to the scale in the produce area so that shoppers could determine from the graph the cost of a banana purchase up to 8 pounds in weight. Label the x-axis in quarters of a pound and label the y-axis in cents.

SECTION 9.6

MATCHING GAME Have a student in your group write 10 linear equations on 3 × 5 note cards, one equation per card. Then have the student graph each equation on a separate set of 10 cards. Shuffle each set of cards. Then put all the equation cards on one side of a table and all the cards with graphs on the other side. Work together to match each equation with its proper graph.

SECTION 9.7

FUNCTIONS We have seen that a function can be thought of as a machine that takes some input x and turns it into some output $f(x)$. Write a function that takes an input of 6 and turns it into an output of 19, where

a. only 1 operation is performed to get the output.

b. 2 operations are performed to get the output.

c. 3 operations are performed to get the output.

d. 4 operations are performed to get the output.

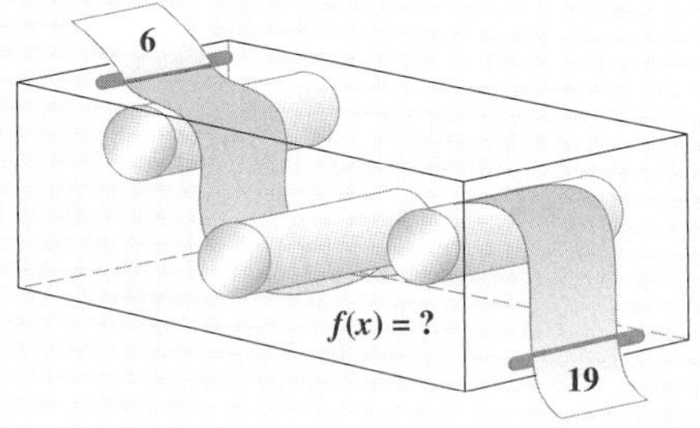

CHAPTER REVIEW

SECTION 9.1 Graphing Using the Rectangular Coordinate System

CONCEPTS

A *rectangular coordinate system* is composed of a horizontal number line called the x-axis and a vertical number line called the y-axis.

The coordinates of the *origin* are (0, 0).

To *graph* ordered pairs means to locate their position on a coordinate system.

The two axes divide the coordinate plane into four distinct regions called *quadrants*.

The midpoint formula:

$$\left(\frac{x_1 + x_2}{2}, \frac{y_1 + y_2}{2}\right)$$

REVIEW EXERCISES

1. Graph the points with coordinates $(-1, 3)$, $(0, 1.5)$, $(-4, -4)$, $\left(2, \frac{7}{2}\right)$, and $(4, 0)$.

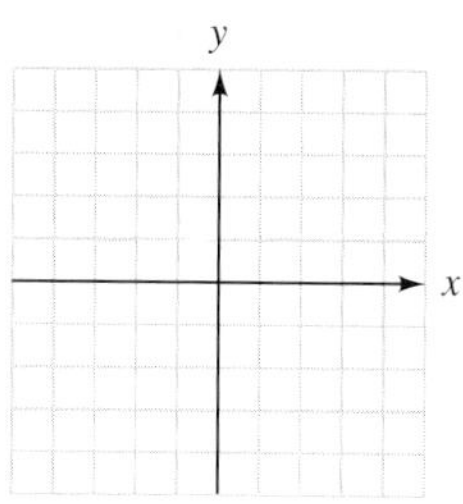

2. Use the graph in the illustration to complete the table.

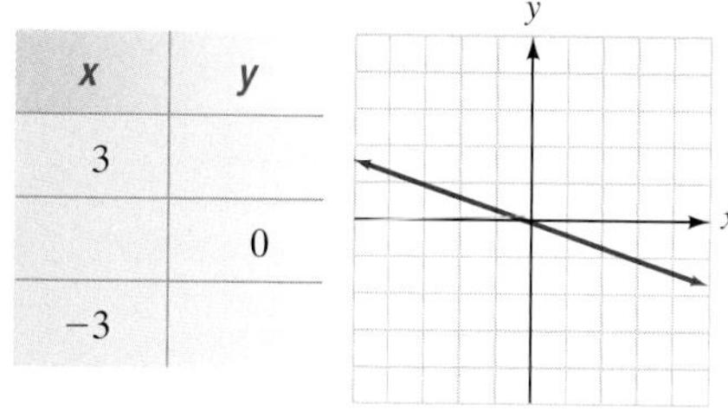

x	y
3	
	0
−3	

3. In what quadrant does the point $(-3, -4)$ lie?

4. SNOWFALL The amount of snow on the ground at a mountain resort was measured once each day over a 7-day period.

a. On the first day, how much snow was on the ground?

b. What was the difference in the amount of snow on the ground when the measurements were taken on the second and third day?

c. How much snow was on the ground on the sixth day?

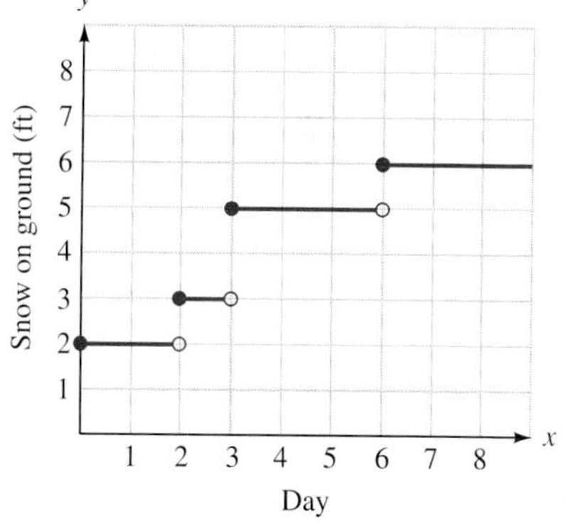

5. COLLEGE ENROLLMENTS The graph gives the number of students enrolled at a college for the period from 4 weeks before to 5 weeks after the semester began.

a. What was the maximum enrollment and when did it occur?

b. How many students had enrolled 2 weeks before the semester began?

c. When was enrollment 2,250?

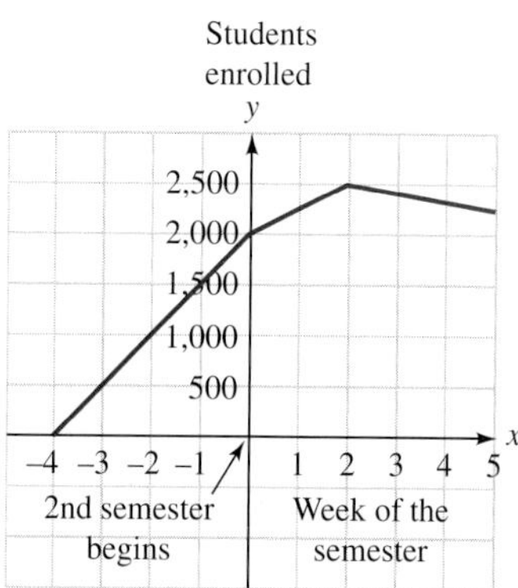

6. Find the midpoint of a line segment with endpoints at $(-3, 7)$ and $(3, -3)$.

SECTION 9.2 *Equations Containing Two Variables*

An ordered pair is a *solution* if, after substituting the values of the ordered pair for the variables in the equation, the result is a true statement.

Solutions of an equation can be shown in a *table of solutions*.

In an equation in x and y, x is called the *independent variable*, or *input*, and y is called the *dependent variable*, or *output*.

7. Check to see whether $(-3, 5)$ is a solution of $y = |2 + x|$.

8. a. Complete the table of solutions and graph the equation $y = -x^3$.

$y = -x^3$		
x	y	(x, y)
−2		
−1		
0		
1		
2		

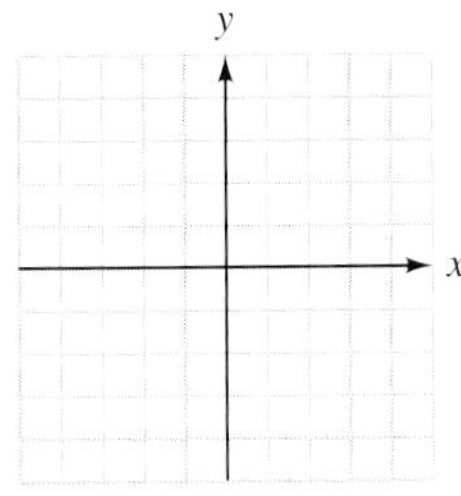

To graph an equation in two variables:

1. Make a table of solutions that contains several solutions written as ordered pairs.
2. Plot each ordered pair.
3. Draw a line or smooth curve through the points.

b. How would the graph of $y = -x^3 + 2$ compare to the graph of the equation given in part a?

9. The graph shows the relationship between the number of oranges O an acre of land will yield if t orange trees are planted on it.

a. If $t = 70$, what is O?

b. What importance does the point (40, 18) on the graph have?

In many application problems, we encounter equations that contain variables other than x and y.

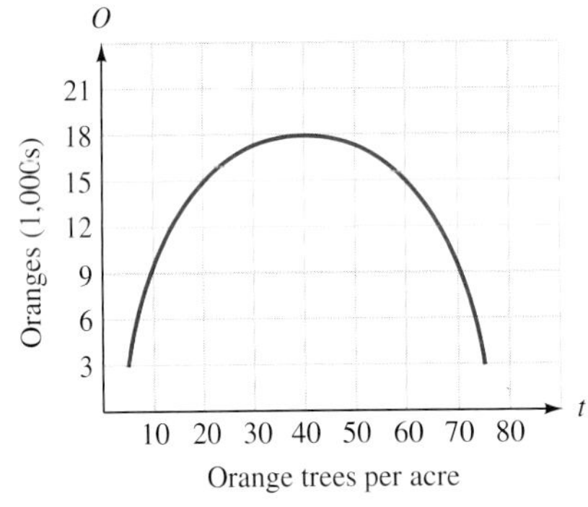

SECTION 9.3 *Graphing Linear Equations*

An equation whose graph is a straight line and whose variables are raised to the first power is called a *linear equation.*

Classify each equation as either linear or nonlinear.

10. $y = |x + 2|$

11. $3x + 4y = 12$

12. $y = 2x - 3$

13. $y = x^2 - x$

The *general* or *standard form* of a linear equation is $Ax + By = C$, where A, B, and C are real numbers and A and B are not both zero.

14. The equation $5x + 2y = 10$ is in general form; what are A, B, and C?

15. Complete the table of solutions for the equation $3x + 2y = -18$.

x	y	(x, y)
−2		(−2,)
	3	(, 3)

To graph a linear equation:

1. Find three (x, y) pairs that satisfy the equation by picking three arbitrary x-values and finding their corresponding y-values.
2. Plot each ordered pair.
3. Draw a straight line through the points.

16. Solve the equation $x + 2y = 6$ for y, find three solutions, and graph it.

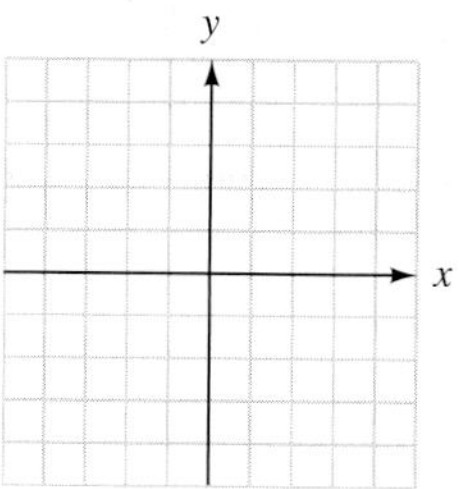

To find the *y-intercept* of a linear equation, substitute 0 for x in the equation of the line and solve for y. To find the *x-intercept* of a linear equation, substitute 0 for y in the equation of the line and solve for x.

17. Graph $-4x + 2y = 8$ by finding its x- and y-intercepts.

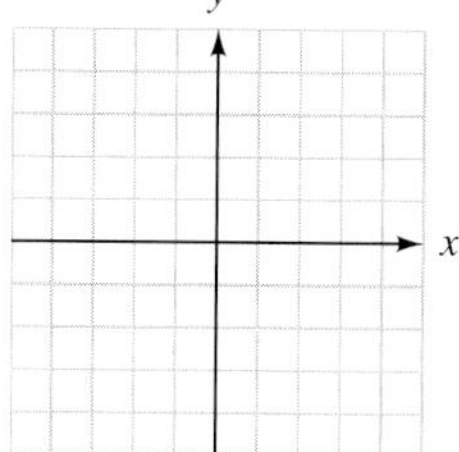

The equation $y = b$ represents the horizontal line that intersects the y-axis at $(0, b)$. The equation $x = a$ represents the vertical line that intersects the x-axis at $(a, 0)$.

Graph each equation.

18. $y = 4$

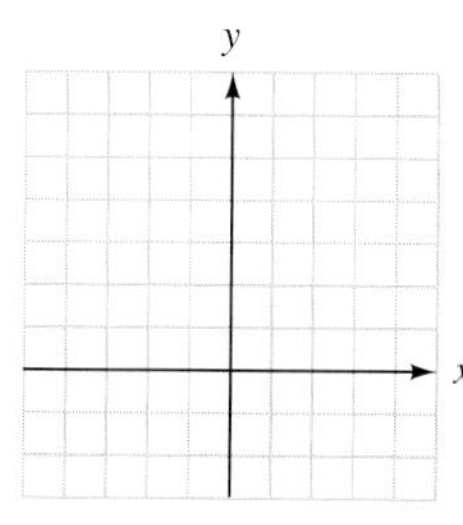

19. $x = -1$

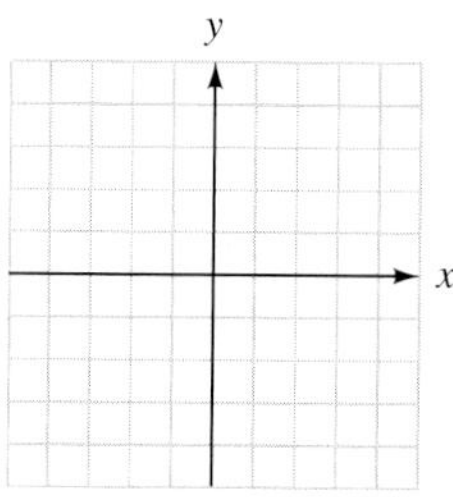

20. Since two points determine a line, only two points are needed to graph a linear equation. Why is it is a good idea to plot a third point?

SECTION 9.4 Rate of Change and the Slope of a Line

The *slope m* of a nonvertical line is a number that measures "steepness" by finding the ratio $\frac{\text{rise}}{\text{run}}$.

$$m = \frac{\text{change in the } y\text{-values}}{\text{change in the } x\text{-values}}$$

In each case, find the slope of the line.

21.

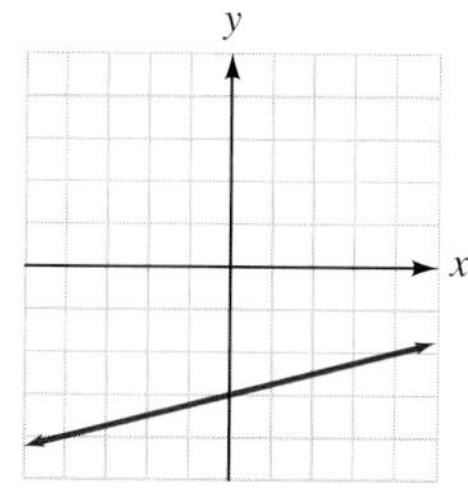

22. The line with the table of solutions shown here.

x	y	(x, y)
2	−3	(2, −3)
4	−17	(4, −17)

If $P(x_1, y_1)$ and $Q(x_2, y_2)$ are two points on a nonvertical line, the slope m of line PQ is

$$m = \frac{y_2 - y_1}{x_2 - x_1}$$

Lines that rise from left to right have a *positive slope,* and lines that fall from left to right have a *negative slope.*

Horizontal lines have a slope of zero. Vertical lines have *undefined slope.*

The slope of a line gives a rate of change.

23. The line passing through the points $(2, -5)$ and $(5, -5)$.

24. The line passing through the points $(1, -4)$ and $(3, -7)$.

25. Graph the line that passes through $(-2, 4)$ and has slope $m = -\frac{4}{5}$.

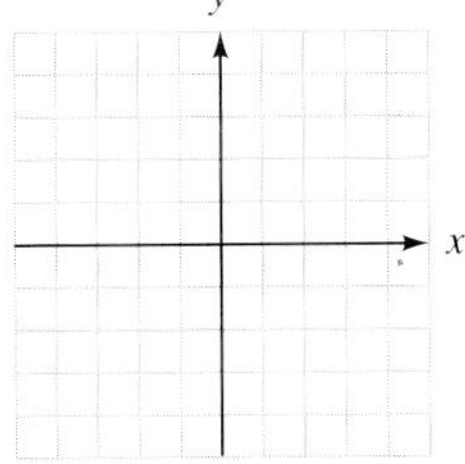

26. TOURISM The graph shows the number of international travelers to the United States from 1986 to 2002, in 2-year increments.

a. Between 2000 and 2002, the largest decline in the number of visitors occurred. What was the rate of change?

b. Between 1986 and 1988, the largest increase in the number of visitors occurred. What was the rate of change?

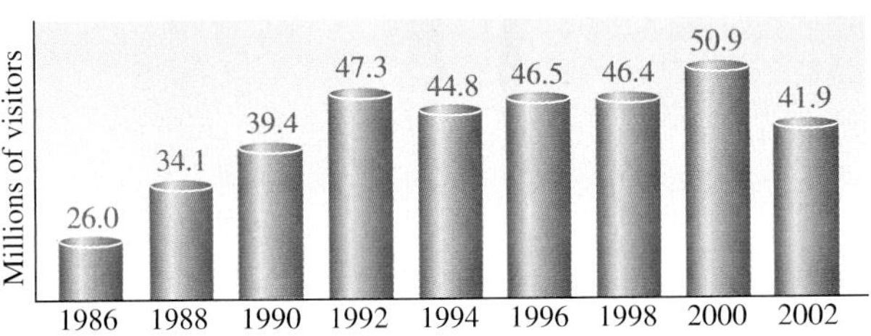

Source: *World Almanac* 2004

SECTION 9.5 *Slope–Intercept Form*

If a linear equation is written in *slope–intercept* form,

$$y = mx + b$$

the graph of the equation is a line with slope m and y-intercept $(0, b)$.

Find the slope and the y-intercept of each line.

27. $y = \frac{3}{4}x - 2$

28. $y = -4x$

29. Find the slope and the y-intercept of the line determined by $9x - 3y = 15$ and graph it.

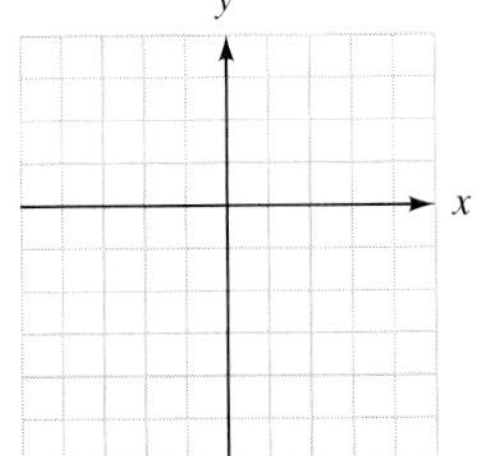

The *rate of change* is the slope of the graph of a linear equation.

30. COPIERS A business buys a used copy machine that, when purchased, has already produced 75,000 copies.

a. If the business plans to run 300 copies a week, write a linear equation that would find the number of copies c the machine has made in its lifetime after the business has used it for w weeks.

b. Use your result to part a to predict the total number of copies that will have been made on the machine 1 year, or 52 weeks, after being purchased by the business.

Two lines with the same slope are *parallel.*

The product of the slopes of *perpendicular* lines is -1.

Without graphing, determine whether the graphs of the given pairs of lines would be parallel, perpendicular, or neither.

31. $\begin{cases} y = -\frac{2}{3}x + 6 \\ y = -\frac{2}{3}x - 6 \end{cases}$

32. $\begin{cases} x + 5y = -10 \\ y = 5x \end{cases}$

SECTION 9.6 Point–Slope Form: Writing Linear Equations

If a line with slope m passes through the point (x_1, y_1), the equation of the line in *point–slope* form is

$$y - y_1 = m(x - x_1)$$

Write an equation of a line with the given slope that passes through the given point. Express the result in slope–intercept form and graph the equation.

33. $m = 3, (1, 5)$

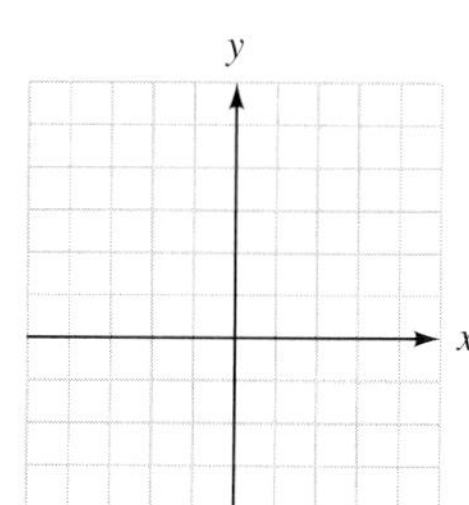

34. $m = -\frac{1}{2}, (-4, -1)$

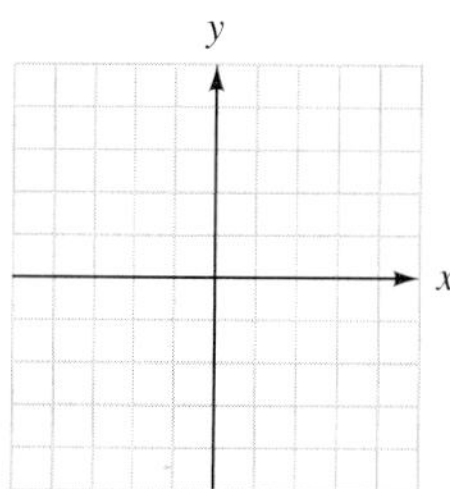

Write an equation of the line with the following characteristics. Express the result in slope–intercept form.

35. passing through (3, 7) and (−6, 1)

36. horizontal, passing through (6, −8)

37. CAR REGISTRATION When it was 2 years old, the annual registration fee for a Dodge Caravan was \$380. When it was 4 years old, the registration fee dropped to \$310. If the relationship is linear, write an equation that gives the registration fee f in dollars for the van when it is x years old.

SECTION 9.7 Functions

A *function* is a rule that assigns to each input value a single output value.

In each case, determine whether a function is defined.

38. $y = 3x - 2$

39. $y^2 = x$

40.

x	2	2	3	4	5	6
y	−2	2	3	−4	5	−6

For a function, the set of all possible values of the independent variable x (the inputs) is called the *domain,* and the set of all possible values of the dependent variable y (the outputs) is called the *range.*

Find the domain and range of each function.

41. $f(x) = x + 10$

42. $y = x^2$

The notation $y = f(x)$ denotes that y is a function of x.

For the function $g(x) = 1 - 6x$, find each value.

43. $g(1)$

44. $g(-6)$

45. $g(0.5)$

46. $g\left(\frac{3}{2}\right)$

Four basic functions are

Linear: $f(x) = mx + b$
Identity: $f(x) = x$
Squaring: $f(x) = x^2$
Absolute value: $f(x) = |x|$

Complete the table and graph the function.

47.

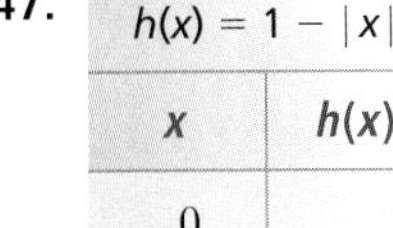

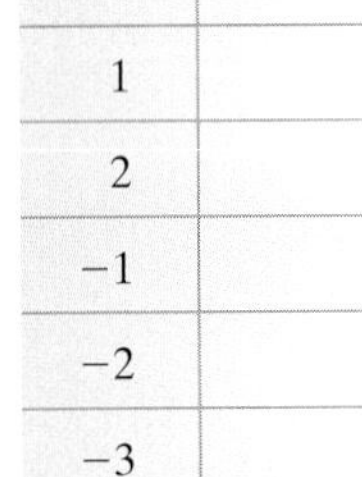

$h(x) = 1 - |x|$

x	$h(x)$
0	
1	
2	
−1	
−2	
−3	

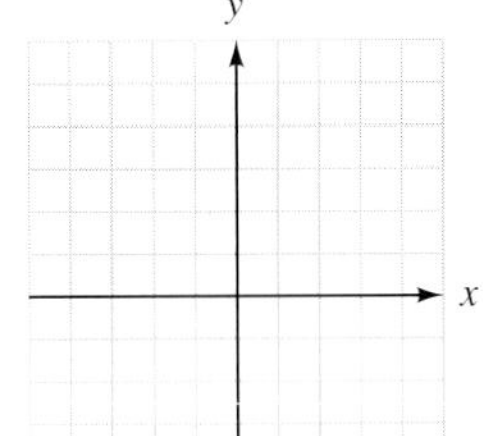

We can use the *vertical line test* to determine whether a graph is the graph of a function.

Determine whether each graph is the graph of a function.

48.

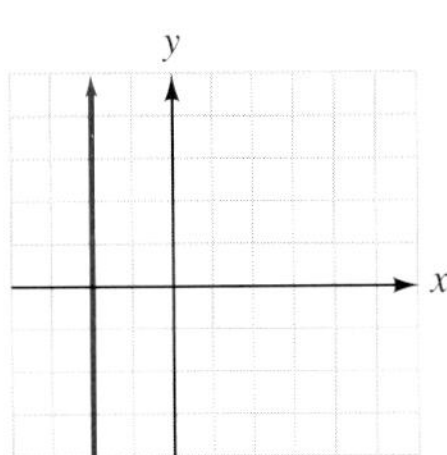

49.

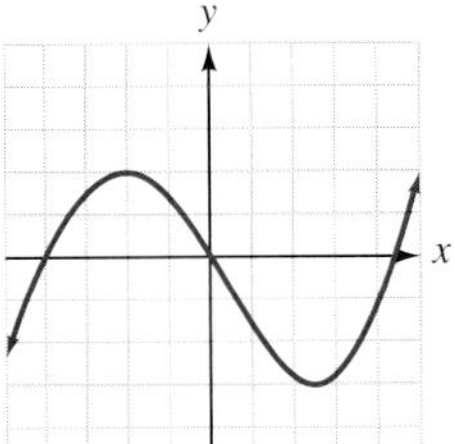

50. The function $f(r) = 15.7r^2$ estimates the volume in cubic inches of a can 5 inches tall with a radius of r inches. Find the volume of the can in the illustration. Round to the nearest tenth.

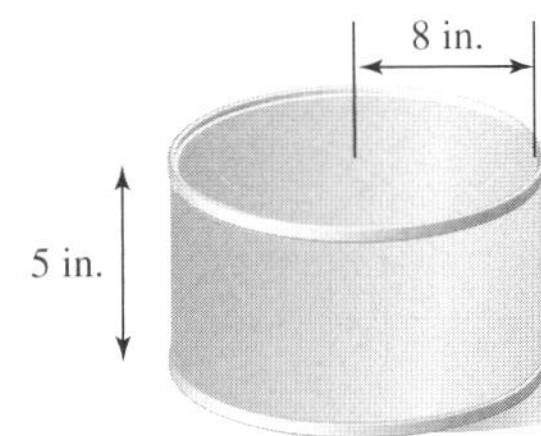

CHAPTER 9 TEST

The graph in the following illustration shows the number of dogs being boarded in a kennel over a 3-day holiday weekend. Use the graph to answer Exercises 1–4.

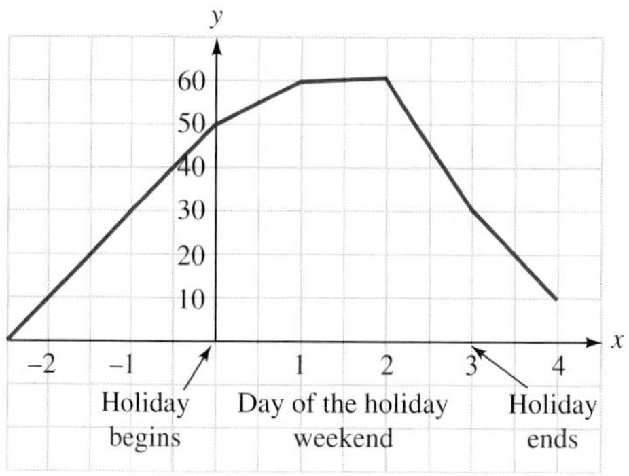

1. How many dogs were in the kennel 2 days before the holiday?

2. What is the maximum number of dogs that were boarded on the holiday weekend?

3. When were there 30 dogs in the kennel?

4. What information does the y-intercept of the graph give?

5. Graph: $y = x^2 - 4$.

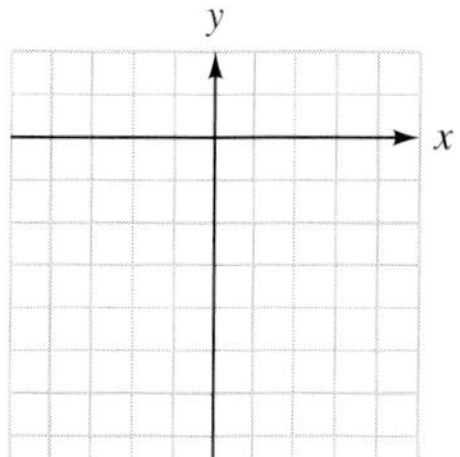

6. Graph: $8x + 4y = -24$.

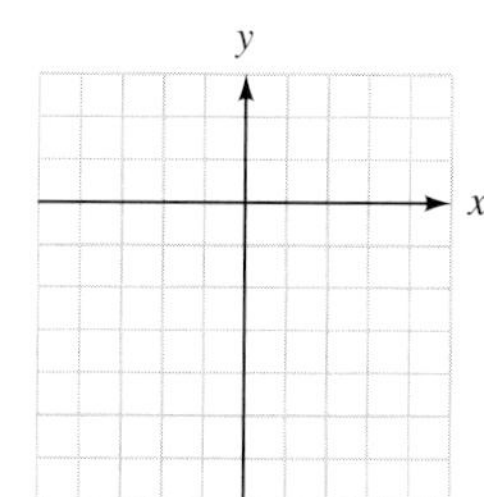

7. Is $(-3, -4)$ the midpoint of $(-6, 2)$ and $(0, -10)$?

8. Is $y = x^3$ a linear equation?

9. What are the x- and y-intercepts of the graph of $2x - 3y = 6$?

10. Find the slope and the y-intercept of $x + 2y = 8$.

11. Graph: $x = -4$.

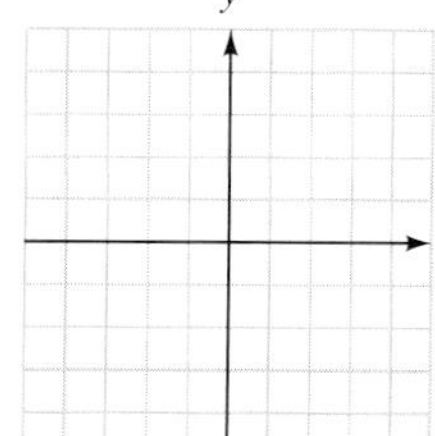

12. Graph the line passing through $(-2, -4)$ having a slope of $\frac{2}{3}$.

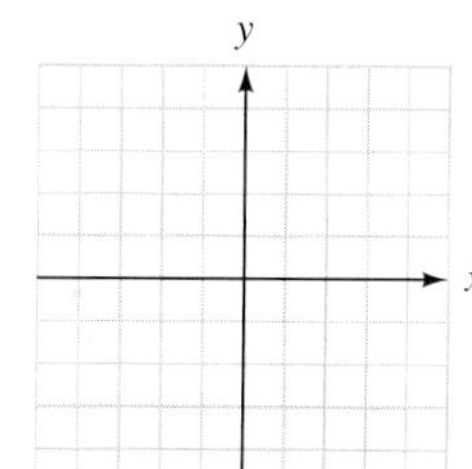

13. What is the slope of the line passing through $(-1, 3)$ and $(3, -1)$?

14. What is the slope of a vertical line?

15. What is the slope of a line that is perpendicular to a line with slope $-\frac{7}{8}$?

16. When graphed, are the lines $y = 2x + 6$ and $6x - 3y = 0$ parallel, perpendicular, or neither?

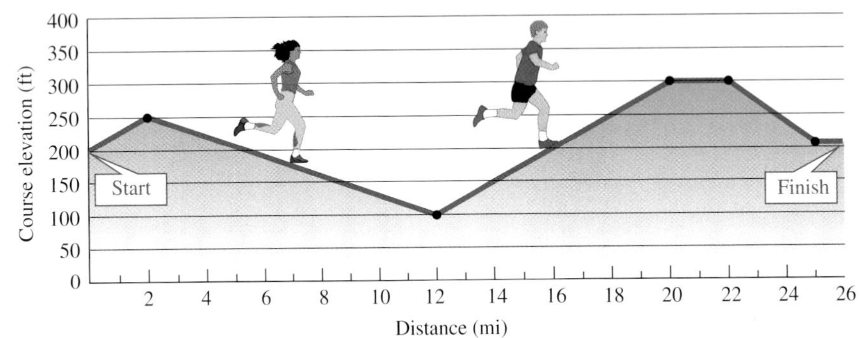

Refer to the graph in the above illustration, which shows the elevation changes in a 26-mile marathon course.

17. Find the rate of change of the decline on which the woman is running.

18. Find the rate of change of the incline on which the man is running.

19. DEPRECIATION After it is purchased, a \$15,000 computer loses \$1,500 in resale value every year. Write a linear equation that gives the resale value v of the computer x years after being purchased.

20. Write an equation of the line passing through $(-2, 5)$ and $(-3, -2)$. Answer in slope–intercept form.

21. Is this the graph of a function?

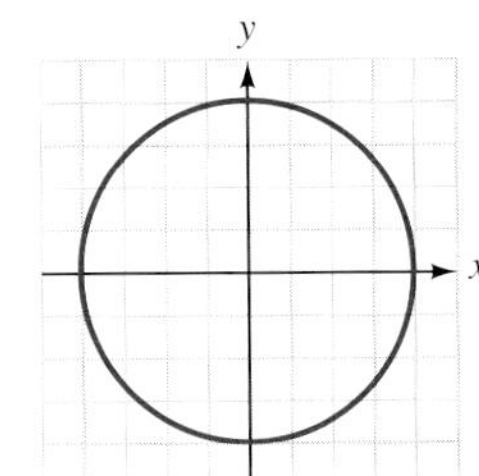

22. Find the domain and range of the function $f(x) = -|x|$.

23. Does the equation $y = 2x - 8$ define a function?

24. If $f(x) = 2x - 7$, find $f(-3)$.

25. If $g(x) = 3.5x^3$, find $g(6)$.

26. Explain what is meant by the statement slope $= \frac{\text{rise}}{\text{run}}$.

CHAPTERS 1–9 CUMULATIVE REVIEW EXERCISES

1. UNITED AIRLINES On May 2, 2003, UAL Corporation, the parent company of United Airlines, announced its 11th straight quarterly loss. (See the graph below.)
   a. During this time span, in which quarter was the loss the greatest?
   b. Estimate the corporation's losses for the year 2002.

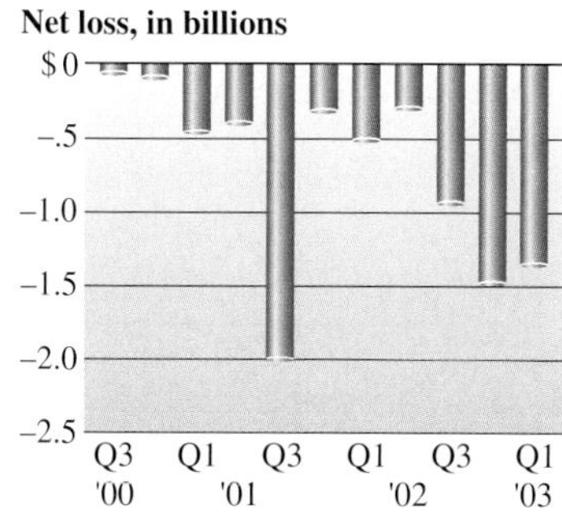

Source: UAL Corporation

2. Give the prime factorization of 108.
3. Write $\frac{1}{250}$ as a decimal.
4. Determine whether each statement is true or false.
   a. Every whole number is an integer.
   b. Every integer is a real number.
   c. 0 is a whole number, an integer, and a rational number.
5. AUTO SALES The following illustration shows the top 5 best-selling vehicles in the United States in the year 2000, as reported by the automakers. Complete the table. Round to the nearest tenth of one percent.

Rank	Vehicle	Units sold 2000	1999	'99 ranks	% change
1	Ford F-Series pickup	876,716	869,001	1	+0.9
2	Chevrolet Silverado pickup	642,119	636,150	2	+0.9
3	Ford Explorer	445,157	428,772	5	
4	Toyota Camry	422,961	448,162	3	
5	Honda Accord	404,515	404,192	6	+0.1

Based on information from Reuters

6. Evaluate each expression.
   a. $12 - 2 \cdot 3$
   b. $\dfrac{(6-5)^4 - (-21)}{-27 + 4^2}$
   c. $19 - 2[(-3.1 + 6.1) \cdot 3]$
   d. $64 - 6[15 - (3)3]$
7. Evaluate $b^2 - 4ac$ for $a = 2$, $b = -8$, and $c = 4$.
8. Suppose x sheets from a 500-sheet ream of paper have been used. How many sheets are left?
9. How many terms does the algebraic expression $3x^2 - 2x + 1$ have? What is the coefficient of the second term?
10. Use the distributive property to remove parentheses.
    a. $2(x + 4)$
    b. $2(x - 4)$
    c. $-2(x + 4)$
    d. $-2(x - 4)$

Simplify each expression.

11. $5a + 10 - a$
12. $-2b^2 + 6b^2$
13. $(a + 2) - (a - 2)$
14. $-y - y - y$

Solve each equation.

15. $3x - 5 = 13$
16. $1.2 - x = -1.7$
17. $\dfrac{2x}{3} - 2 = 4$
18. $\dfrac{y - 2}{7} = -3$
19. $-3(2y - 2) - y = 5$
20. $9y - 3 = 6y$
21. $\dfrac{1}{3} + \dfrac{c}{5} = -\dfrac{3}{2}$
22. $5(x + 2) = 5x - 2$

23. Solve the equation $y = mx + b$ for x

24. Find the perimeter and the area of the gauze pad of the bandage shown.

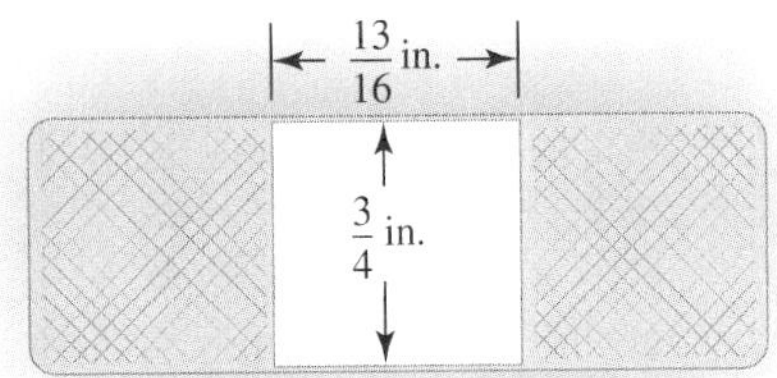

25. If the vertex of an isosceles triangle is 22°, find the measure of each base angle.

26. Complete the table.

Solution	Liters	% acid	Amount of acid
50% solution	x	0.50	
25% solution	$13 - x$	0.25	
30% mixture	13	0.30	

27. ROAD TRIPS A bus, carrying the members of a marching band, and a truck, carrying their instruments, leave a high school at the same time. The bus travels at 60 mph and the truck at 50 mph. In how many hours will they be 75 miles apart?

28. MIXING CANDY Candy corn worth \$1.90 per pound is to be mixed with black gumdrops that cost \$1.20 per pound to make 200 pounds of a mixture worth \$1.48 per pound. How many pounds of each candy should be used?

Solve each inequality, graph the solution set, and write the solution in interval notation.

29. $-\frac{3}{16}x \geq -9$

30. $8x + 4 > 3x + 4$

31. MEDICATION Dosages for a certain medication are shown. What is the dosage for

a. a 5-year-old child?

b. a 9-year-old child?

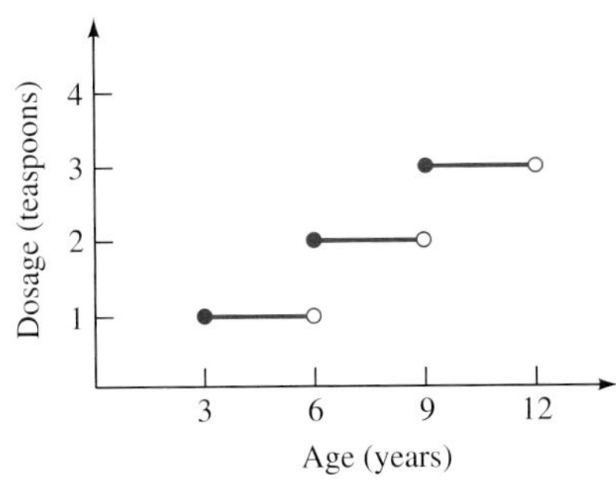

32. Is $(-2, 4)$ a solution of $y = 2x - 8$?

Graph each equation.

33. $y = |x - 2|$

34. $4y + 2x = -8$

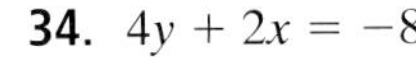

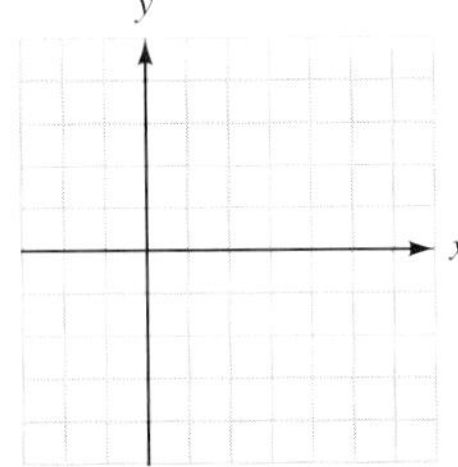

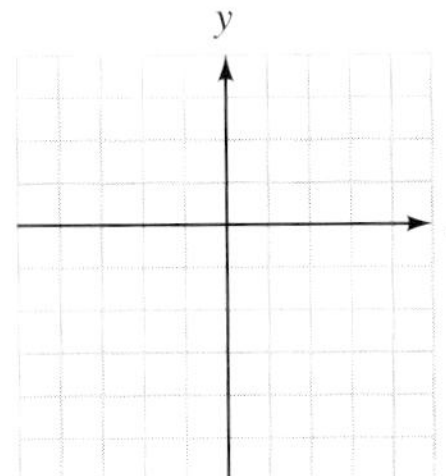

35. What is the slope of the graph of the line $y = 5$?

36. What is the slope of the line passing through $(-2, 4)$ and $(5, -6)$?

37. Find the slope and the y-intercept of the graph of the line described by $4x - 6y = -12$.

38. Write an equation of the line that has slope -2 and y-intercept of $(0, 1)$.

39. Write an equation of the line that has slope $-\frac{7}{8}$ and passes through $(2, -9)$. Express the answer in point–slope form.

40. If $f(x) = x^2 - 3x$, find $f(-2)$.

CHAPTER 10

Exponents and Polynomials

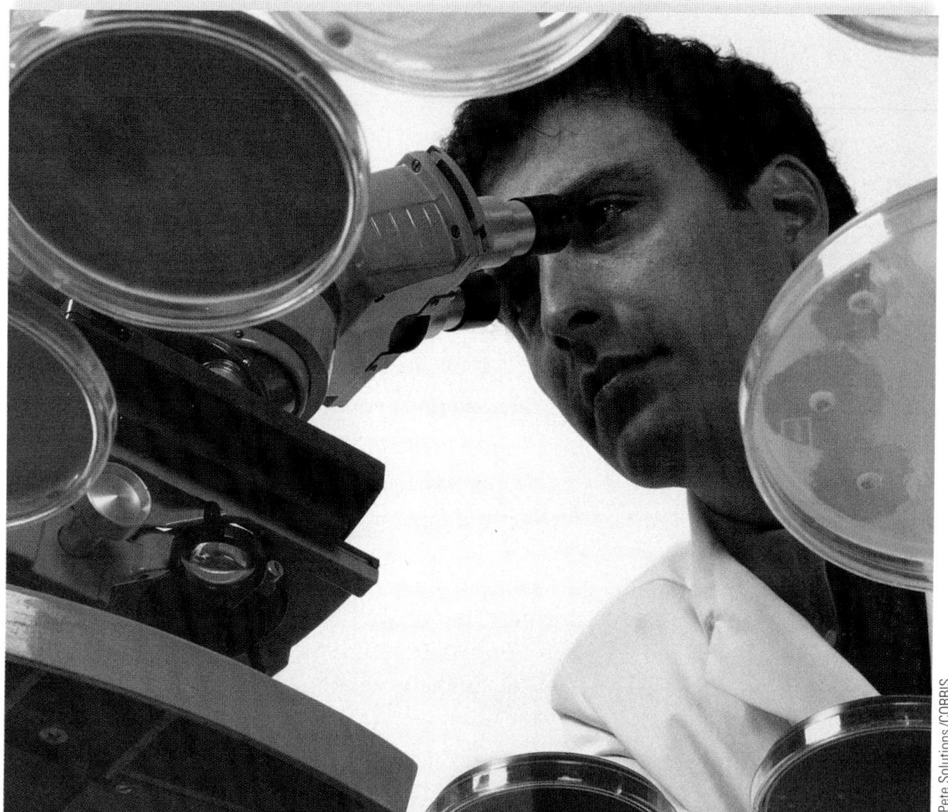

© Pete Solutions/CORBIS

Under certain conditions, bacteria can grow at an incredible rate. To model the growth, scientists use exponents. In this chapter, we will discuss several rules for simplifying expressions involving exponents. We will also see the role exponents play as part of a compact notation that is used to represent extremely large numbers, such as the distance from the Earth to the sun, and extremely small numbers, such as the mass of a proton.

To learn more about exponents visit *The Learning Equation* on the Internet at http://tle.brookscole.com. (The log-in instructions are in the Preface.) For Chapter 10, the online lessons are:

- *TLE* Lesson 19: The Product Rule for Exponents
- *TLE* Lesson 20: The Quotient Rule for Exponents
- *TLE* Lesson 21: Negative Exponents

Check Your Knowledge

1. In the expression x^5, x is the ______, and 5 is the ________.
2. A __________ is a term or a sum of terms in which all variables have whole-numbered exponents.
3. The degree of a polynomial is determined by the ______ with largest degree.
4. Polynomials are normally written in ____________________ powers of one variable — for example, $x^3 + 2x^2 + x + 3$.
5. Write using exponents: $3xxyyyzz$

Simplify:

6. $2x^3y\,(5x^4y^2)$
7. $(2x^2)^3\,(3x^3)^2$

Simplify. Write answers without using parentheses or negative exponents.

8. $(-5)^0$
9. $2x^3y^{-1}x^{-4}y^2$
10. $\dfrac{3x^3y^{-1}}{x^{-4}y^2}$
11. $\left(\dfrac{2x^3y^{-1}}{x^{-4}y^2}\right)^{-3}$

12. A room is $2x + 3$ feet long, $2x + 3$ feet wide, and x feet high.
    a. Write an expression representing the area of the room's floor.
    b. Write an expression representing the volume of the room.
13. Scientific notation:
    a. Write 0.000023 in scientific notation.
    b. Write 6.25×10^6 in standard notation.
14. Consider $2x^2y^5$.
    a. Identify the polynomial as a monomial, a binomial, or a trinomial.
    b. What is the degree of the polynomial?
15. If $f(x) = x^2 + 3x + 4$, find $f(-1)$.
16. Simplify:
    a. $2xy^2 + \dfrac{3x^2y^2}{x} + x(2y)^2$
    b. $(3x^2 - 2x + 1) - (x^2 + x - 1) - 2(x - 1)$

Find the products.

17. $(-3xy^2)\,(-4xyz^2)$
18. $2x(x^2 + 3x - 5)$
19. $(2y - 7)(2y + 7)$
20. $(3y - 5)^2$
21. $(2x + 3)(4x - 7)$
22. $(2x - 3)(x^2 - x + 1)$

23. Solve: $(x - 4)^2 = x^2$.
24. Divide: $2x - 3\overline{)6x^2 + x - 15}$.
25. The area of a rectangular table top is $(4x^2 - 9)$ ft², and the length is $(2x + 3)$ ft. Use division to determine the width.

Getty Images

Study Skills Workshop

HOW TO FORM A STUDY GROUP

In the Study Skills Workshop in Chapter 3, one assignment was to create an interest list about forming a study group. If you have not already done so, you may want to consider this idea again. Peer study groups allow you to ask questions in a nonthreatening environment and also to help fellow students by explaining a solution to a problem to them. Teaching someone else is one of the best ways to learn a topic. If you have already started a group, take some time to evaluate how well it's working and fine-tune the rules under which it operates.

Group Size. Ideally, a study group should be small. Three or four people per group is best, more than four can be too chaotic.

Time and Place. Establish a common meeting time once a week, usually two days before an assignment is due, for all group participants. This can be done by sharing your calendars and comparing study times. Find a meeting place that is convenient and practical for all members. There should be enough room to spread out and a place where you can talk (sometimes the conversation becomes quite animated!). Quiet sections in the library are *not* good meeting sites. Some schools have designated study areas that don't have to be kept quiet, like learning centers or math labs. Coffee shops or restaurants at off-peak times might be good places to meet, as are peoples' homes — a kitchen table can be an excellent place if you have your family's blessing. You may want to schedule an extra day to meet with your group during a test week.

How it Works. Don't expect to start and finish a complete homework assignment within your study group — at least attempt each problem before you meet with the group. During your private study time, make a list of the problems that you had trouble with, noting exactly where in the process you got stuck. Hopefully, someone in your group was able to do that problem correctly. If not, try verbalizing the problem with your group and brainstorm about possible ways to solve it. If there are still unanswered questions at the end of your meeting time, write these down. Plan to attend your instructor's office hour or see your group again the next day with this list.

As you become more familiar with your group, you may decide that it is working well without any formal rules: however, if you notice problems, it is okay to communicate your dissatisfaction to the group and ask whether a few ground-rules might make your group effective and organized.

ASSIGNMENT

1. Are you currently involved with a study group?
   a. If so, evaluate how well it is working. Are there difficulties with the group that need to be addressed? Are there improvements that you would like to see made?
   b. If not, evaluate how well you are doing in the class at this point. If you're not sure, schedule an appointment with your instructor to find out. If you are not doing as well as expected, try forming a study group.
      i. Contact students either before or after your next class meeting to see whether you can find someone who can meet when you're free.
      ii. Find a meeting place and schedule a meeting with your group.
      iii. After the meeting, decide whether it helped. If it didn't, are there things that could be changed to make it work?

In this chapter, we introduce the rules for exponents and use them when performing operations on polynomials.

10.1 Natural-Number Exponents

- The product rule for exponents
- The quotient rule for exponents
- The power rule for exponents
- Power rules for products and quotients

We have used natural-number exponents to indicate repeated multiplication. For example,

$9^2 = 9 \cdot 9 = 81$	Write 9 as a factor 2 times.
$7^3 = 7 \cdot 7 \cdot 7 = 343$	Write 7 as a factor 3 times.
$(-2)^4 = (-2)(-2)(-2)(-2) = 16$	Write -2 as a factor 4 times.
$-2^4 = -(2 \cdot 2 \cdot 2 \cdot 2) = -16$	The $-$ sign in front of 2^4 means the opposite of 2^4.

These examples illustrate a definition for x^n, where n is a natural number.

Natural-Number Exponents

If n is a natural number, then

$$x^n = \overbrace{x \cdot x \cdot x \cdot \cdots \cdot x}^{n \text{ factors of } x}$$

In the **exponential expression** x^n, x is called the **base** and n is called the **exponent.** The entire expression is called a **power of x.**

$$\text{base} \rightarrow x^n \leftarrow \text{exponent}$$

If an exponent is a natural number, it tells how many times its base is to be used as a factor. An exponent of 1 indicates that its base is to be used one time as a factor, an exponent of 2 indicates that its base is to be used two times as a factor, and so on. The base of an exponential expression can be a number, a variable, or a combination of numbers and variables.

$$x^1 = x \qquad (y+1)^2 = (y+1)(y+1) \qquad (-5s)^3 = (-5s)(-5s)(-5s)$$

In this section, we will continue the discussion of exponents as we simplify exponential expressions that are multiplied, divided, and raised to powers. To perform these simplifications, we will use the rules for exponents.

The product rule for exponents

To develop a rule for multiplying exponential expressions with the same base, we consider the product $x^2 \cdot x^3$. Since the expression x^2 means that x is to be used as a factor two times, and the expression x^3 means that x is to be used as a factor three times, we have

$$\begin{aligned} x^2 \cdot x^3 &= \overbrace{x \cdot x}^{2 \text{ factors of } x} \cdot \overbrace{x \cdot x \cdot x}^{3 \text{ factors of } x} \\ &= \overbrace{x \cdot x \cdot x \cdot x \cdot x}^{5 \text{ factors of } x} \\ &= x^5 \end{aligned}$$

In general,

$$x^m \cdot x^n = \overbrace{x \cdot x \cdot x \cdot \cdots \cdot x}^{m \text{ factors of } x} \cdot \overbrace{x \cdot x \cdot x \cdot x \cdot \cdots \cdot x}^{n \text{ factors of } x}$$

$$= \overbrace{x \cdot x \cdot x \cdot x \cdot x \cdot x \cdot \cdots \cdot x \cdot x \cdot x}^{m+n \text{ factors of } x}$$

$$= x^{m+n}$$

This discussion suggests the following rule: *To multiply two exponential expressions with the same base, keep the common base and add the exponents.*

Product rule for exponents

If m and n represent natural numbers, then

$$x^m x^n = x^{m+n}$$

EXAMPLE 1 Simplify each expression: **a.** $9^5(9^6)$, **b.** $x^3 \cdot x^4$, **c.** y^2y^4y, and **d.** $(c^2d^3)(c^4d^5)$.

Solution

a. To simplify $9^5(9^6)$ means to write it in an equivalent form using one base and one exponent.

$9^5(9^6) = 9^{5+6}$ Use the product rule for exponents: Keep the common base, which is 9, and add the exponents.

$= 9^{11}$ Perform the addition.

b. $x^3 \cdot x^4 = x^{3+4}$ Keep the common base x and add the exponents.

$= x^7$ Perform the addition.

c. $y^2y^4y = y^{2+4}y$ Working from left to right, keep the common base y and add the exponents.

$= y^6y$ Perform the addition.

$= y^{6+1}$ Keep the common base and add the exponents.

$= y^7$ Perform the addition.

d. $(c^2d^3)(c^4d^5) = c^2d^3c^4d^5$ Use the associative property of multiplication.

$= c^2c^4d^3d^5$ Change the order of the factors.

$= c^{2+4}d^{3+5}$ Keep the common base c and add the exponents. Keep the common base d and add the exponents.

$= c^6d^8$ Perform the additions.

Self Check 1

Simplify:

a. $7^8(7^7)$

b. $z \cdot z^3$

c. $x^2x^3x^6$

d. $(s^4t^3)(s^4t^4)$

Answers **a.** 7^{15}, **b.** z^4, **c.** x^{11}, **d.** s^8t^7

! COMMENT When simplifying expressions, note the operations that are involved. For example, we cannot simplify $x^3 + x^4$ or $x^3 - x^4$, because x^3 and x^4 are not like terms. However, we can simplify $x^3 \cdot x^4$, because x^3 and x^4 have the same base: $x^3 \cdot x^4 = x^7$.

Furthermore, the expressions $x^2 + y^3$ and $x^2 - y^3$ cannot be simplified, because they do not contain like terms; neither can the expression x^2y^3, because x^2 and y^3 have different bases.

THINK IT THROUGH PIN Code Choices

"According to a Student Monitor LLC survey, ATM debit card ownership among college students has almost doubled from 30 percent to 57 percent in the past four years." *BYU Newsletter*, Oct 2002

In 2002, there were 13.9 billion ATM transactions in the United States. On average, that's more than 38 million a day! Before each transaction, the card owner is required to enter his or her PIN (personal identification number). When an ATM card is issued, many financial institutions have the applicant select a four-digit PIN. There are 10 possible choices for the first digit of the PIN, 10 possible choices for the second digit, and so on. Write the total number of possible choices of a PIN as an exponential expression. Then evaluate the expression.

The quotient rule for exponents

We now consider the fraction

$$\frac{4^5}{4^2}$$

where the exponent in the numerator is greater than the exponent in the denominator. We can simplify this fraction as follows:

$$\frac{4^5}{4^2} = \frac{4 \cdot 4 \cdot 4 \cdot 4 \cdot 4}{4 \cdot 4}$$

$$= \frac{\overset{1}{\cancel{4}} \cdot \overset{1}{\cancel{4}} \cdot 4 \cdot 4 \cdot 4}{\underset{1}{\cancel{4}} \cdot \underset{1}{\cancel{4}}} \quad \text{Divide out the common factors of 4.}$$

$$= 4^3$$

The result of 4^3 has a base of 4 and an exponent $5 - 2$ (or 3). This suggests that *to divide exponential expressions with the same base, we keep the common base and subtract the exponents.*

Quotient rule for exponents

If m and n represent natural numbers, $m > n$, and $x \neq 0$, then

$$\frac{x^m}{x^n} = x^{m-n}$$

Self Check 2
Simplify:

a. $\frac{55^{30}}{55^5}$

b. $\frac{a^5}{a^3}$

c. $\frac{b^{15}c^4}{b^4c}$

EXAMPLE 2 Simplify each expression. Assume that there are no divisions by 0.

a. $\frac{20^{16}}{20^9}$, **b.** $\frac{x^4}{x^3}$, and **c.** $\frac{a^3b^8}{ab^5}$

Solution

a. To simplify $\frac{20^{16}}{20^9}$ means to write it in an equivalent form using one base and one exponent.

$\frac{20^{16}}{20^9} = 20^{16-9}$ Use the quotient rule for exponents: Keep the common base, which is 20, and subtract the exponents.

$= 20^7$ Perform the subtraction.

b. $\frac{x^4}{x^3} = x^{4-3}$ Keep the common base x and subtract the exponents.

$= x^1$ Perform the subtraction.

$= x$

c. $\frac{a^3b^8}{ab^5} = \frac{a^3}{a} \cdot \frac{b^8}{b^5}$

$= a^{3-1}b^{8-5}$ Keep the common base a and subtract the exponents. Keep the common base b and subtract the exponents.

$= a^2b^3$ Perform the subtractions.

Answers **a.** 55^{25}, **b.** a^2, **c.** $b^{11}c^3$

EXAMPLE 3 Simplify: $\frac{a^3a^5a^7}{a^4a}$.

Self Check 3 Simplify: $\frac{b^2b^6b}{b^4b^4}$

Solution We use the product rule for exponents to simplify the numerator and denominator separately and proceed as follows.

$\frac{a^3a^5a^7}{a^4a} = \frac{a^{15}}{a^5}$ In the numerator, keep the common base a and add the exponents. In the denominator, keep the common base a and add the exponents.

$= a^{15-5}$ Use the quotient rule for exponents: Keep the common base a and subtract the exponents.

$= a^{10}$ Perform the subtraction.

Answer b

The power rule for exponents

To find another rule for exponents, we consider the expression $(x^3)^4$, which can be written as $x^3 \cdot x^3 \cdot x^3 \cdot x^3$. Because each of the four factors of x^3 contains three factors of x, there are $4 \cdot 3$ (or 12) factors of x. This product can be written as x^{12}.

$$(x^3)^4 = x^3 \cdot x^3 \cdot x^3 \cdot x^3$$
$$= \overbrace{\underbrace{x \cdot x \cdot x}_{x^3} \cdot \underbrace{x \cdot x \cdot x}_{x^3} \cdot \underbrace{x \cdot x \cdot x}_{x^3} \cdot \underbrace{x \cdot x \cdot x}_{x^3}}^{\text{12 factors of } x}$$
$$= x^{12}$$

In general,

$$(x^m)^n = \overbrace{x^m \cdot x^m \cdot x^m \cdot \cdots \cdot x^m}^{n \text{ factors of } x^m}$$
$$= \overbrace{x \cdot x \cdot x \cdot x \cdot x \cdot x \cdot x \cdot \cdots \cdot x}^{m \cdot n \text{ factors of } x}$$
$$= x^{m \cdot n}$$

This discussion illustrates the following rule: *To raise an exponential expression to a power, keep the base and multiply the exponents.*

Power rule for exponents

If m and n represent natural numbers, then

$$(x^m)^n = x^{m \cdot n} \quad \text{or, more simply,} \quad (x^m)^n = x^{mn}$$

EXAMPLE 4 Simplify each expression: **a.** $(2^3)^7$ and **b.** $(z^8)^8$.

Solution

a. To simplify $(2^3)^7$ means to write it in an equivalent form using one base and one exponent.

$(2^3)^7 = 2^{3 \cdot 7}$ Keep the base of 2 and multiply the exponents.

$= 2^{21}$ Perform the multiplication.

b. $(z^8)^8 = z^{8 \cdot 8}$ Keep the base and multiply the exponents.

$= z^{64}$ Perform the multiplication.

Self Check 4

Simplify each expression:

a. $(5^4)^6$

b. $(y^5)^2$

Answer **a.** 5^{24}, **b.** y^{10}

EXAMPLE 5 Simplify each expression: **a.** $(x^2x^5)^2$ and **b.** $(z^2)^4(z^3)^3$.

Solution

a. We begin by using the product rule for exponents. Then we use the power rule.

$(x^2x^5)^2 = (x^7)^2$ Within the parentheses, keep the base x and add the exponents.

$= x^{14}$ Keep the base x and multiply the exponents.

b. We begin by using the power rule for exponents twice. Then we use the product rule.

$(z^2)^4(z^3)^3 = z^8z^9$ For each power of z raised to a power, keep the base z and multiply the exponents.

$= z^{17}$ Keep the base z and add the exponents.

Self Check 5

Simplify each expression:

a. $(a^4a^3)^3$

b. $(a^3)^3(a^4)^2$

Answers **a.** a^{21}, **b.** a^{17}

Power rules for products and quotients

To develop two more rules for exponents, we consider the expression $(2x)^3$, which is a *power of the product* of 2 and x, and the expression $\left(\frac{2}{x}\right)^3$, which is a *power of the quotient* of 2 and x.

$$(2x)^3 = (2x)(2x)(2x)$$
$$= (2 \cdot 2 \cdot 2)(x \cdot x \cdot x)$$
$$= 2^3x^3$$
$$= 8x^3$$

$$\left(\frac{2}{x}\right)^3 = \left(\frac{2}{x}\right)\left(\frac{2}{x}\right)\left(\frac{2}{x}\right)$$

$= \dfrac{2 \cdot 2 \cdot 2}{x \cdot x \cdot x}$ Multiply the numerators. Multiply the denominators.

$$= \frac{2^3}{x^3}$$
$$= \frac{8}{x^3}$$

These examples illustrate the following rules: *To raise a product to a power, we raise each factor of the product to that power, and to raise a fraction to a power, we raise both the numerator and the denominator to that power.*

Powers of a product and a quotient

If n represents a natural number, then

$$(xy)^n = x^n y^n \qquad \text{and if } y \neq 0, \text{ then} \qquad \left(\frac{x}{y}\right)^n = \frac{x^n}{y^n}$$

EXAMPLE 6 Simplify: **a.** $(3c)^3$, **b.** $(x^2y^3)^5$, and **c.** $(-2a^3b)^2$.

Solution

a. Since $3c$ is the product of 3 and c, the expression $(3c)^3$ is a power of a product.

$(3c)^3 = 3^3c^3$ Use the power rule for products: Raise each factor of the product $3c$ to the 3rd power.

$= 27c^3$ Evaluate 3^3.

b. $(x^2y^3)^5 = (x^2)^5(y^3)^5$ Raise each factor of the product x^2y^3 to the 5th power.

$= x^{10}y^{15}$ For each power of a power, keep the base and multiply the exponents.

c. $(-2a^3b)^2 = (-2)^2(a^3)^2b^2$ Raise each of the three factors of the product $-2a^3b$ to the 2nd power.

$= 4a^6b^2$ Evaluate $(-2)^2$. Keep the base a and multiply the exponents.

Self Check 6

Simplify:

a. $(2t)^4$

b. $(c^3d^4)^6$

c. $(-3ab^5)^3$

Answers **a.** $16t^4$, **b.** $c^{18}d^{24}$, **c.** $-27a^3b^{15}$

EXAMPLE 7 Simplify: $\dfrac{(a^3b^4)^2}{ab^5}$.

Solution

$\dfrac{(a^3b^4)^2}{ab^5} = \dfrac{(a^3)^2(b^4)^2}{ab^5}$ In the numerator, raise each factor within the parentheses to the 2nd power.

$= \dfrac{a^6b^8}{ab^5}$ In the numerator, for each power of a power, keep the base and multiply the exponents.

$= a^{6-1}b^{8-5}$ Keep each of the bases, a and b, and subtract the exponents.

$= a^5b^3$ Perform the subtractions.

Self Check 7

Simplify: $\dfrac{(c^4d^5)^3}{c^2d^3}$.

Answer $c^{10}d^{12}$

EXAMPLE 8 Simplify: **a.** $\left(\frac{4}{k}\right)^3$ and **b.** $\left(\frac{3x^2}{2y^3}\right)^5$.

Solution

a. Since $\frac{4}{k}$ is the quotient of 4 and k, the expresion $\left(\frac{4}{k}\right)^3$ is a power of a quotient.

$\left(\frac{4}{k}\right)^3 = \frac{4^3}{k^3}$ Use the power rule for quotients: Raise the numerator and denominator to the 3rd power.

$= \frac{64}{k^3}$ Evaluate 4^3.

b. $\left(\frac{3x^2}{2y^3}\right)^5 = \frac{(3x^2)^5}{(2y^3)^5}$ Raise the numerator and the denominator to the 5th power.

$= \frac{3^5(x^2)^5}{2^5(y^3)^5}$ In the numerator and denominator, raise each factor within the parentheses to the 5th power.

$= \frac{243x^{10}}{32y^{15}}$ Evaluate 3^5 and 2^5. For each power of a power, keep the base and multiply the exponents.

Self Check 8
Simplify:
a. $\left(\frac{x}{7}\right)^3$
b. $\left(\frac{2x^3}{3y^2}\right)^4$

Answers **a.** $\frac{x^3}{343}$, **b.** $\frac{16x^{12}}{81y^8}$

EXAMPLE 9 Simplify: $\frac{(5b)^9}{(5b)^6}$.

Solution

$\frac{(5b)^9}{(5b)^6} = (5b)^{9-6}$ Keep the common base $5b$, and subtract the exponents.

$= (5b)^3$ Perform the subtraction.

$= 5^3b^3$ Raise each factor within the parentheses to the 3rd power.

$= 125b^3$ Evaluate 5^3.

Self Check 9
Simplify: $\frac{(-2h)^{20}}{(-2h)^{14}}$

Answer $64h^6$

The rules for natural-number exponents are summarized below.

Rules for exponents

If n represents a natural number, then

$$x^n = \overbrace{x \cdot x \cdot x \cdot \cdots \cdot x}^{n \text{ factors of } x}$$

If m and n represent natural numbers and there are no divisions by zero, then

1. $x^m x^n = x^{m+n}$ **2.** $\frac{x^m}{x^n} = x^{m-n}$ **3.** $(x^m)^n = x^{mn}$

4. $(xy)^n = x^n y^n$ **5.** $\left(\frac{x}{y}\right)^n = \frac{x^n}{y^n}$

Section 10.1 STUDY SET

VOCABULARY *Fill in the blanks.*

1. The ______ of the exponential expression $(-5)^3$ is -5. The ______ is 3.
2. The exponential expression x^4 represents a repeated multiplication where x is to be written as a ______ four times.
3. x^n is called a ______ of x.
4. The expression $(2x^2b)^5$ is a power of a ______, and $\left(\frac{2x^2}{b}\right)^5$ is a power of a ______.

CONCEPTS *Fill in the blanks.*

5. $(3x)^4$ means $\square \cdot \square \cdot \square \cdot \square$
6. Using an exponent, $(-5y)(-5y)(-5y)$ can be written as $\square$.
7. $x^m x^n = \square$
8. $(xy)^n = \square$
9. $\left(\frac{a}{b}\right)^n = \square$
10. $(a^b)^c = \square$
11. $\frac{x^m}{x^n} = \square$
12. $x = x^{\square}$
13. $(x^m)^n = \square$
14. $(t^3)^2 = \square \cdot \square$
15. a. Write a power of a product that has two factors.

 b. Write a power of a quotient.
16. a. To simplify $(2y^3z^2)^4$, how many factors within the parentheses must be raised to the fourth power?

 b. To simplify $\left(\frac{y^3}{z^2}\right)^4$ what two expressions must be raised to the fourth power?

Simplify each expression, if possible.

17. a. $x^2 + x^2$ b. $x^2 - x^2$ c. $x^2 \cdot x^2$ d. $\frac{x^2}{1}$
18. a. $x^2 + x$ b. $x^2 - x$ c. $x^2 \cdot x$ d. $\frac{x^2}{x}$
19. a. $x^3 + x^2$ b. $x^3 - x^2$ c. $x^3 \cdot x^2$ d. $\frac{x^3}{x^2}$
20. Simplify each expression, if possible.

 a. $x^3 + y^3$ b. $x^3 - y^3$ c. x^3y^3 d. $\frac{x^3}{y^3}$

Find each area or volume. You may leave π in your answer.

21.

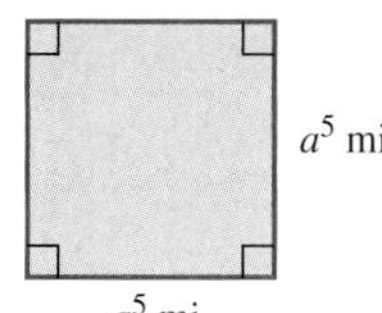

22.

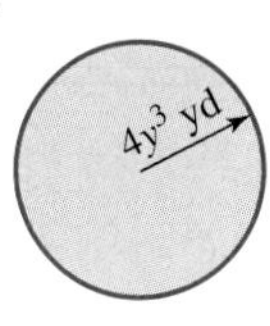

23.

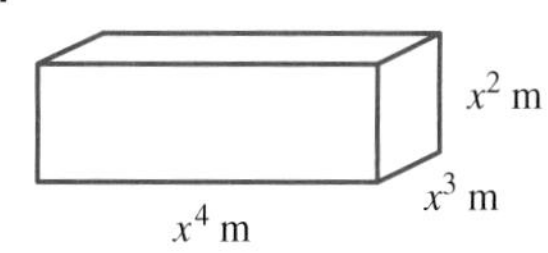

24.

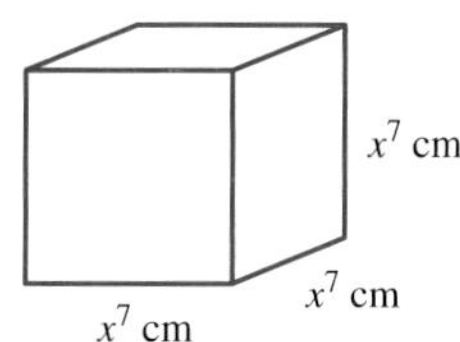

NOTATION *Complete each solution.*

25. $(x^4x^2)^3 = (\square)^3$

 $= x^{18}$

26. $\frac{a^3a^4}{a^2} = \frac{\square}{a^2}$

 $= a^{\square - 2}$

 $= a^5$

Identify the base and the exponent in each expression.

27. 4^3
28. $(-8)^2$
29. x^5
30. $\left(\frac{5}{x}\right)^3$
31. $(-3x)^2$
32. $-x^4$
33. $-\frac{1}{3}y^6$
34. $3.14r^4$

Evaluate each expression.

35. $(-4)^2$
36. $(-5)^2$
37. -4^2
38. -5^2

Write the repeated multiplication that is indicated.

39. x^5
40. $(-7y)^4$
41. $\left(\frac{t^2}{2}\right)^3$
42. c^3d^2

Write each expression using an exponent.

43. $4t(4t)(4t)(4t)$
44. $-5u(-5u)$
45. $-4 \cdot t \cdot t \cdot t$
46. $-5 \cdot u \cdot u$

PRACTICE *Write each expression as an expression involving one base and one exponent.*

47. $12^3 \cdot 12^4$
48. $3^4 \cdot 3^6$
49. $2(2^3)(2^2)$
50. $5(5^5)(5^3)$
51. $a^3 \cdot a^3$
52. $m^7 \cdot m^7$
53. x^4x^3
54. y^5y^2
55. a^3aa^5
56. b^2b^3b
57. $y^3(y^2y^4)$
58. $(y^4y)y^6$
59. $\frac{8^{12}}{8^4}$
60. $\frac{10^4}{10^2}$
61. $\frac{x^{15}}{x^3}$
62. $\frac{y^6}{y^3}$
63. $\frac{c^{10}}{c^9}$
64. $\frac{h^{20}}{h^{10}}$
65. $(3^2)^4$
66. $(4^3)^3$
67. $(y^5)^3$
68. $(b^3)^6$
69. $(m^{50})^{10}$
70. $(n^{25})^4$

Simplify. Assume there are no divisions by 0.

71. $(a^2b^3)(a^3b^3)$
72. $(u^3v^5)(u^4v^5)$
73. $(cd^4)(cd)$
74. $ab^3c^4 \cdot ab^4c^2$
75. $xy^2 \cdot x^2y$
76. $s^8t^2s^2t^7$
77. $\frac{y^3y^4}{yy^2}$
78. $\frac{b^4b^5}{b^2b^3}$
79. $\frac{c^3d^7}{cd}$
80. $\frac{r^8s^9}{rs}$
81. $(x^2x^3)^5$
82. $(y^3y^4)^4$
83. $(3zz^2z^3)^5$
84. $(4t^3t^6t^2)^2$
85. $(x^5)^2(x^7)^3$
86. $(y^3y)^2(y^2)^2$
87. $(uv)^4$
88. $(xy)^3$
89. $(a^3b^2)^3$
90. $(r^3s^2)^2$
91. $(-2r^2s^3)^3$
92. $(-3x^2y^4)^2$
93. $\left(\frac{a}{b}\right)^3$
94. $\left(\frac{r}{s}\right)^4$
95. $\left(\frac{x^2}{y^3}\right)^5$
96. $\left(\frac{u^4}{v^2}\right)^6$
97. $\left(\frac{-2a}{b}\right)^5$
98. $\left(\frac{-2t}{3}\right)^4$
99. $\frac{(6k)^7}{(6k)^4}$
100. $\frac{(-3a)^{12}}{(-3a)^{10}}$
101. $\frac{(a^2b)^{15}}{(a^2b)^9}$
102. $\frac{(s^2t^3)^4}{(s^2t^3)^2}$
103. $\frac{a^2a^3a^4}{(a^4)^2}$
104. $\frac{(aa^2)^3}{a^2a^3}$
105. $\frac{(ab^2)^3}{(ab)^2}$
106. $\frac{(m^3n^4)^3}{(mn^2)^3}$
107. $\frac{(r^4s^3)^4}{(rs^3)^3}$
108. $\frac{(x^2y^5)^5}{(x^3y)^2}$
109. $\left(\frac{y^3y}{2yy^2}\right)^3$
110. $\left(\frac{2y^3y}{yy^2}\right)^3$
111. $\left(\frac{3t^3t^4t^5}{4t^2t^6}\right)^3$
112. $\left(\frac{4t^3t^4t^5}{3t^2t^6}\right)^3$

APPLICATIONS

113. ART HISTORY Leonardo da Vinci's drawing relating a human figure to a square and a circle is shown.
    a. Find an expression that represents the area of the square if the man's height is $5x$ feet.
    b. Find an expression that represents the area of the circle if the distance from his waist to his feet is $3x$ feet. You may leave π in your answer.

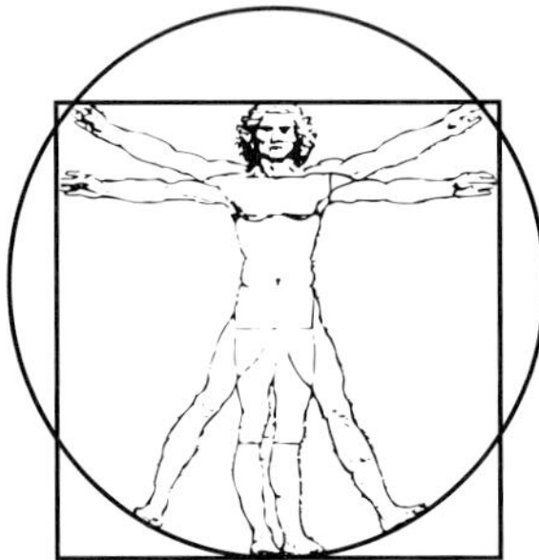

114. PACKAGING Use the illustration to find the volume of the bowling ball and the cardboard box it is packaged in. You may leave π in your answer.

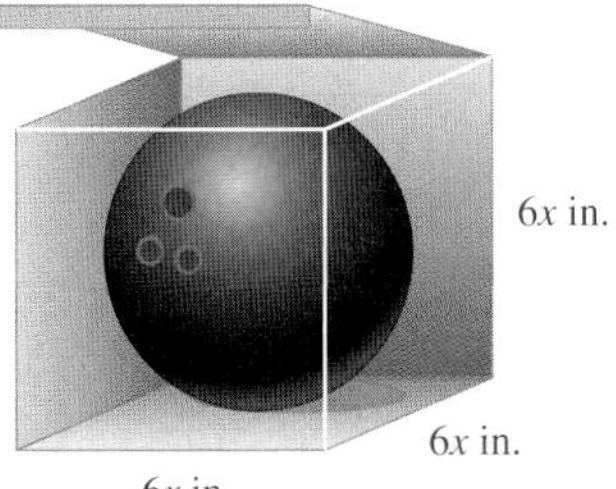

115. BOUNCING BALLS A ball is dropped from a height of 32 feet. Each rebound is one-half of its previous height.

a. Draw a diagram of the path of the ball, showing four bounces.

b. Explain why the expressions $32\left(\frac{1}{2}\right)$, $32\left(\frac{1}{2}\right)^2$, $32\left(\frac{1}{2}\right)^3$, and $32\left(\frac{1}{2}\right)^4$ represent the height of the ball on the first, second, third, and fourth bounces, respectively. Find the heights of the first four bounces.

116. HAVING BABIES The probability that a couple will have n baby boys in a row is given by the formula $\left(\frac{1}{2}\right)^n$. Find the probability that a couple will have four baby boys in a row.

117. COMPUTERS Text is stored by computers using a sequence of eight 0's and 1's. Such a sequence is called a **byte.** An example of a byte is 10101110.

a. Write four other bytes, all ending in 1.

b. Each of the eight digits of a byte can be chosen in *two* ways (either 0 or 1). The total number of different bytes can be represented by an exponential expression with base 2. What is it?

118. INVESTING Guess the answer to the following problem. Then use a calculator to find the correct answer. Were you close?
If the value of 1¢ is to double every day, what will the penny be worth after 31 days?

WRITING

119. Explain the mistake.

$$2^3 \cdot 2^2 = 4^5$$
$$= 1{,}024$$

120. Are the expressions $2x^3$ and $(2x)^3$ equivalent? Explain.

121. Is the operation of raising to a power commutative? That is, is $a^b = b^a$? Explain.

122. When a number is raised to a power, is the result always larger than the original number? Support your answer with some examples.

REVIEW *Match each equation with its graph below.*

123. $y = 2x - 1$

124. $y = 3x - 1$

125. $y = 3$

126. $x = 3$

a.

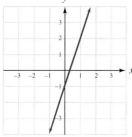

b.

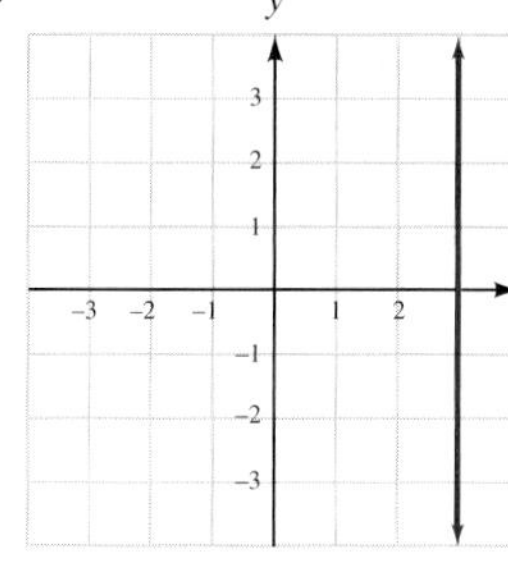

c.

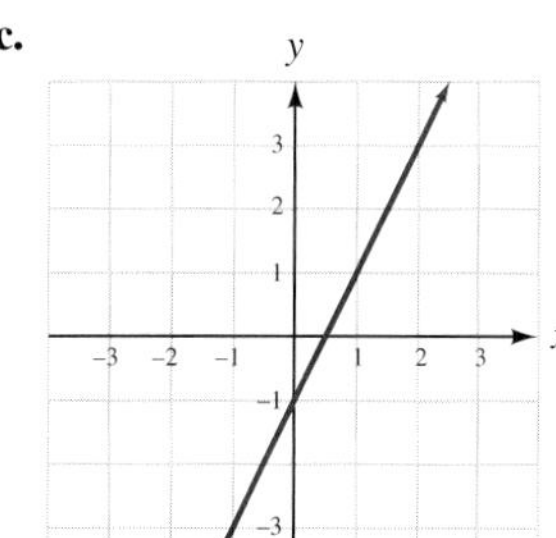

d.

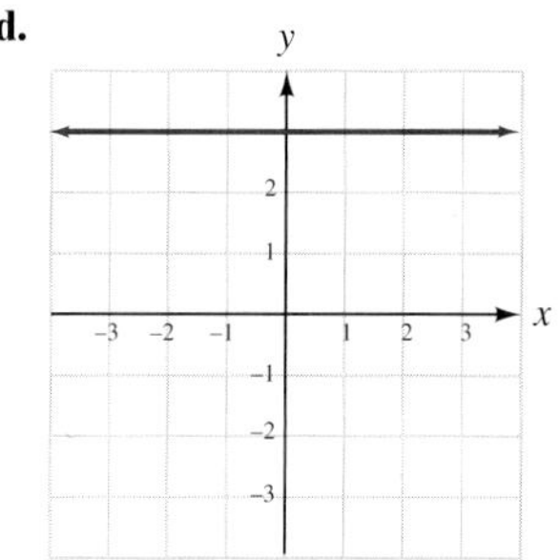

10.2 Zero and Negative Integer Exponents

- Zero exponents • Negative integer exponents • Variable exponents

In the previous section, we discussed natural-number exponents. We now extend the discussion to include exponents that are zero and exponents that are negative integers.

Zero exponents

When we discussed the quotient rule for exponents in the previous section, the exponent in the numerator was always greater than the exponent in the denominator.

We now consider what happens when the exponents are equal. To develop the definition of a zero exponent, we will simplify the expression

$$\frac{5^3}{5^3}$$

in two ways. If we use the quotient rule for exponents, where the exponents in the numerator and denominator are equal, we obtain 5^0. However, by dividing out common factors of 5, we obtain 1.

$$\frac{5^3}{5^3} = 5^{3-3} = \mathbf{5^0} \qquad \frac{5^3}{5^3} = \frac{\overset{1}{\cancel{5}} \cdot \overset{1}{\cancel{5}} \cdot \overset{1}{\cancel{5}}}{\underset{1}{\cancel{5}} \cdot \underset{1}{\cancel{5}} \cdot \underset{1}{\cancel{5}}} = \mathbf{1}$$

These must be equal.

For this reason, we will define 5^0 to be equal to 1. This example suggests the following rule.

Zero exponents

If x represents any nonzero real number, then

$$x^0 = 1$$

Self Check 1

Write each expression without using exponents:

a. $(-0.115)^0$

b. $-5a^0b$

Answers **a.** 1, **b.** $-5b$

EXAMPLE 1 Write each expression without using exponents: **a.** $\left(\frac{1}{13}\right)^0$, **b.** $\frac{x^5}{x^5}$, **c.** $3x^0$, and **d.** $(3x)^0$.

Solution

a. $\left(\frac{1}{13}\right)^0 = 1$

b. $\frac{x^5}{x^5} = x^{5-5}$
$= x^0$
$= 1$

c. $3x^{\mathbf{0}} = 3(\mathbf{1})$ The base is x; the exponent is 0.
$= 3$

d. $(3x)^0 = 1$ The base is $3x$; the exponent is 0.

Parts c and d point out that $3x^0 \neq (3x)^0$.

Negative integer exponents

To develop the definition of a negative exponent, we will simplify the expression

$$\frac{6^2}{6^5}$$

in two ways. If we use the quotient rule for exponents, where the exponent in the numerator is less than the exponent in the denominator, we obtain 6^{-3}. However, by dividing out two factors of 6, we obtain $\frac{1}{6^3}$.

$$\frac{6^2}{6^5} = 6^{2-5} = \mathbf{6^{-3}} \qquad \frac{6^2}{6^5} = \frac{\overset{1}{\cancel{6}} \cdot \overset{1}{\cancel{6}}}{\underset{1}{\cancel{6}} \cdot \underset{1}{\cancel{6}} \cdot 6 \cdot 6 \cdot 6} = \frac{\mathbf{1}}{\mathbf{6^3}}$$

These must be equal.

For this reason, we define 6^{-3} to be equal to $\frac{1}{6^3}$. In general, we have the following rule.

Negative exponents

If x represents any nonzero number and n represents a natural number, then

$$x^{-n} = \frac{1}{x^n}$$

The definition of a negative exponent states that another way to write x^{-n} is to write its reciprocal, changing the sign of the exponent. We can use this definition to write expressions that contain negative exponents as expressions without negative exponents.

EXAMPLE 2 Simplify by using the definition of negative exponents: **a.** 3^{-5} and **b.** $(-2)^{-3}$.

Solution

a. $3^{-5} = \frac{1}{3^5}$ Write the reciprocal of 3^{-5} and change the exponent from -5 to 5.

$= \frac{1}{243}$ Evaluate 3^5.

b. $(-2)^{-3} = \frac{1}{(-2)^3}$ Write the reciprocal of $(-2)^{-3}$ and change the exponent from -3 to 3.

$= -\frac{1}{8}$ Evaluate $(-2)^3$.

Self Check 2

Simplify by using the definition of negative exponents:

a. 4^{-4}

b. $(-5)^{-3}$

Answers **a.** $\frac{1}{256}$, **b.** $-\frac{1}{125}$

! COMMENT A negative exponent does not indicate a negative number. It indicates a reciprocal.

$$4^{-2} = \frac{1}{4^2} \qquad 4^{-2} \neq -16 \qquad 4^{-2} \neq -\frac{1}{4^2}$$

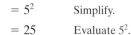

EXAMPLE 3 Simplify by using the definition of negative exponents: **a.** $\frac{1}{5^{-2}}$ and **b.** $\frac{2^{-3}}{3^{-4}}$.

Solution

a. $\frac{1}{5^{-2}} = \frac{1}{\frac{1}{5^2}}$ In the denominator, write the reciprocal of 5^{-2} and change the exponent from -2 to 2.

$= 1 \div \frac{1}{5^2}$ The fraction bar indicates division of 1 by $\frac{1}{5^2}$.

$= 1 \cdot \frac{5^2}{1}$ To divide by $\frac{1}{5^2}$, we multiply by its reciprocal.

$= 5^2$ Simplify.

$= 25$ Evaluate 5^2.

Self Check 3

Simplify by using the definition of negative exponents:

a. $\frac{1}{9^{-1}}$

b. $\frac{8^{-2}}{7^{-1}}$

b. $\dfrac{2^{-3}}{3^{-4}} = \dfrac{\frac{1}{2^3}}{\frac{1}{3^4}}$ In the numerator, write the reciprocal of 2^{-3} and change the exponent from -3 to 3. In the denominator, write the reciprocal of 3^{-4} and change the exponent from -4 to 4.

$= \dfrac{1}{2^3} \cdot \dfrac{3^4}{1}$ To divide by $\frac{1}{3^4}$, we multiply by its reciprocal.

$= \dfrac{3^4}{2^3}$ Multiply the fractions.

$= \dfrac{81}{8}$ Evaluate 3^4 and 2^3.

Answers **a.** 9, **b.** $\dfrac{7}{64}$

The results from Example 3 suggest that we can move factors that have negative exponents between the numerator and denominator of a fraction if we change the sign of their exponents. For example,

$$\frac{3^{-3}}{b^{-1}} = \frac{b^1}{3^3} = \frac{b}{27}$$

Self Check 4
Simplify by using the definition of negative exponents:
a. a^{-5}
b. $\dfrac{r^{-4}}{s^{-5}}$
c. $3y^{-3}$

Answers **a.** $\dfrac{1}{a^5}$, **b.** $\dfrac{s^5}{r^4}$, **c.** $\dfrac{3}{y^3}$

EXAMPLE 4 Simplify by using the definition of negative exponents. Assume that no denominators are zero.

a. x^{-4}, **b.** $\dfrac{x^{-3}}{y^{-7}}$, **c.** $(-2x)^{-2}$, and **d.** $-2x^{-2}$

Solution

a. $x^{-4} = \dfrac{1}{x^4}$

b. $\dfrac{x^{-3}}{y^{-7}} = \dfrac{y^7}{x^3}$

c. $(-2x)^{-2} = \dfrac{1}{(-2x)^2}$

$= \dfrac{1}{4x^2}$

d. $-2x^{-2} = -2\left(\dfrac{1}{x^2}\right)$

$= -\dfrac{2}{x^2}$

The rules for exponents discussed in Section 10.1 (the product, power, and quotient rules) are also true for zero and negative exponents.

Rules for exponents

If m and n represent integers and there are no divisions by zero, then

$$x^m x^n = x^{m+n} \qquad (x^m)^n = x^{mn} \qquad (xy)^n = x^n y^n \qquad \left(\frac{x}{y}\right)^n = \frac{x^n}{y^n}$$

$$x^0 = 1 \quad (x \neq 0) \qquad x^{-n} = \frac{1}{x^n} \qquad \frac{x^m}{x^n} = x^{m-n}$$

EXAMPLE 5 Write $\left(\frac{5}{16}\right)^{-1}$ without using exponents.

Solution

$$\left(\frac{5}{16}\right)^{-1} = \frac{5^{-1}}{16^{-1}}$$ Use the quotient rule for exponents: Raise the numerator and denominator to the -1 power.

$$= \frac{16^1}{5^1}$$ Move the factors that have negative exponents between the numerator and denominator.

$$= \frac{16}{5}$$

Self Check 5

Write $\left(\frac{3}{7}\right)^{-2}$ without using exponents.

Answer $\frac{49}{9}$

EXAMPLE 6 Simplify and write the result without using negative exponents. Assume that no denominators are zero.

a. $(x^{-3})^2$, **b.** $\frac{x^3}{x^7}$, **c.** $(x^3x^2)^{-3}$, **d.** $\frac{y^{-4}y^{-3}}{y^{-20}}$, **e.** $\frac{12a^3b^4}{4a^5b^2}$ and **f.** $\left(-\frac{x^3y^2}{xy^{-3}}\right)^{-2}$.

Solution

a. $(x^{-3})^2 = x^{-6}$
$= \frac{1}{x^6}$

b. $\frac{x^3}{x^7} = x^{3-7}$
$= x^{-4}$
$= \frac{1}{x^4}$

c. $(x^3x^2)^{-3} = (x^5)^{-3}$
$= \frac{1}{(x^5)^3}$
$= \frac{1}{x^{15}}$

d. $\frac{y^{-4}y^{-3}}{y^{-20}} = \frac{y^{-7}}{y^{-20}}$
$= y^{-7-(-20)}$
$= y^{-7+20}$
$= y^{13}$

e. $\frac{12a^3b^4}{4a^5b^2} = 3a^{3-5}b^{4-2}$
$= 3a^{-2}b^2$
$= \frac{3b^2}{a^2}$

f. $\left(-\frac{x^3y^2}{xy^{-3}}\right)^{-2} = (-x^{3-1}y^{2-(-3)})^{-2}$
$= (-x^2y^5)^{-2}$
$= \frac{1}{(-x^2y^5)^2}$
$= \frac{1}{x^4y^{10}}$

Self Check 6

Simplify and write the result without using negative exponents:

a. $(x^4)^{-3}$

b. $\frac{a^4}{a^8}$

c. $\frac{a^{-4}a^{-5}}{a^{-3}}$

d. $\frac{20x^5y^3}{5x^3y^6}$

Answers **a.** $\frac{1}{x^{12}}$, **b.** $\frac{1}{a^4}$, **c.** $\frac{1}{a^6}$, **d.** $\frac{4x^2}{y^3}$

Variable exponents

We can apply the rules for exponents to simplify expressions involving variable exponents.

EXAMPLE 7 Simplify each expression. Assume that there are no divisions by 0.

a. $\frac{6^n}{6^n}$, **b.** $x^{2m}x^{3m}$, and **c.** $\frac{y^{2m}}{y^{4m}}$.

Self Check 7

Simplify each expression:

a. $\frac{7^m}{7^m}$

b. $z^{3n}z^{2n}$

c. $\frac{z^{3n}}{z^{5n}}$

Answers a. 1, b. z^{5n}, c. $\frac{1}{z^{2n}}$

Solution

a. $\frac{6^n}{6^n} = 6^{n-n}$ Keep the common base and subtract the exponents.

$= 6^0$ Combine like terms: $n - n = 0$.

$= 1$

b. $x^{2m}x^{3m} = x^{2m+3m}$ Keep the common base and add the exponents.

$= x^{5m}$ Combine like terms: $2m + 3m = 5m$.

c. $\frac{y^{2m}}{y^{4m}} = y^{2m-4m}$ Keep the base and subtract the exponents.

$= y^{-2m}$ Combine like terms: $2m - 4m = -2m$.

$= \frac{1}{y^{2m}}$ Write the reciprocal of y^{-2m} and change the exponent to $2m$.

CALCULATOR SNAPSHOT

Finding present value

As a gift for their newborn grandson, the grandparents want to deposit enough money in the bank now so that when he turns 18, the young man will have a college fund of \$20,000 waiting for him. How much should they deposit now if the money will earn 6% annually?

To find how much money P must be invested at an annual rate i (expressed as a decimal) to have \$$A$ in n years, we use the formula $P = A(1 + i)^{-n}$. If we substitute 20,000 for A, 0.06 (6%) for i, and 18 for n, we have

$P = A(1 + i)^{-n}$ P is called the **present value.**

$P = \mathbf{20{,}000}(1 + \mathbf{0.06})^{-18}$

To find P with a scientific calculator, we enter these numbers and press these keys.

[(] 1 [+] .06 [)] [y^x] 18 [+/−] [×] 20000 [=]   7006.875823

To evaluate the expression with a graphing calculator, we use the following keystrokes.

20000 [×] [(] 1 [+] .06 [)] [^] [(−)] 18 [ENTER]

```
20000*(1+.06)^-1
8
         7006.875823
```

They must invest approximately \$7,006.88 to have \$20,000 in 18 years.

Section 10.2 STUDY SET

VOCABULARY *Fill in the blanks.*

1. In the exponential expression 8^{-3}, 8 is the ______ and −3 is the ________.

2. In the exponential expression 5^{-1}, the exponent is a ________ integer.

3. Another way to write 2^{-3} is to write its ________ and to change the sign of the exponent:

$$2^{-3} = \frac{1}{2^3}$$

4. In the expression 6^m, the ________ is a variable.

CONCEPTS

5. In parts a and b, fill in the blanks as you simplify the fraction in two different ways. Then complete the sentence in part c.

a. $\frac{6^4}{6^4} = 6^{\square}$

$= 6^{\square}$

b. $\frac{6^4}{6^4} = \frac{\square \cdot \square \cdot \square \cdot \square}{6 \cdot 6 \cdot 6 \cdot 6}$

$= \square$

c. So we define 6^0 to be ☐, and in general, if x is any nonzero real number, then $x^0 =$ ☐.

6. In parts a and b, fill in the blanks as you simplify the fraction in two different ways. Then complete the sentence in part c.

a. $\frac{8^3}{8^5} = 8^{\square}$

$= 8^{\square}$

b. $\frac{8^3}{8^5} = \frac{\square \cdot \square \cdot \square}{8 \cdot 8 \cdot 8 \cdot 8 \cdot 8}$

$= \frac{1}{8^2}$

c. So we define 8^{-2} to be ☐, and in general, if x is any nonzero real number, then $x^{-n} =$ ☐.

Complete each table.

7.

x	3^x
2	
1	
0	
−1	
−2	

8.

x	4^x
2	
1	
0	
−1	
−2	

9.

x	$(-9)^x$
2	
1	
0	
−1	
−2	

10.

x	$(-5)^x$
2	
1	
0	
−1	
−2	

Use the graph to determine the missing y-coordinates in the table and express each y-coordinate as a power of 2.

11.

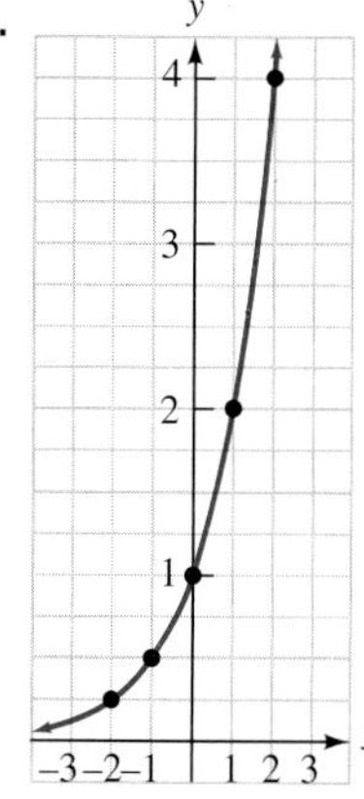

x	y	y as a power of 2
2		$2^{\square}$
1		$2^{\square}$
0		$2^{\square}$
−1		$2^{\square}$
−2		$2^{\square}$

12.

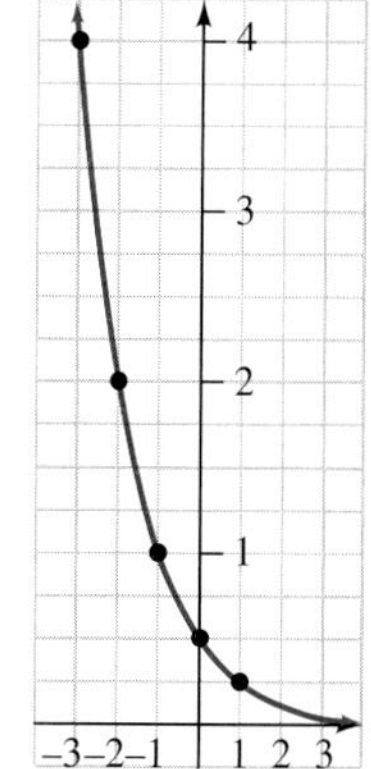

x	y	y as a power of 2
1		$2^{\square}$
0		$2^{\square}$
−1		$2^{\square}$
−2		$2^{\square}$
−3		$2^{\square}$

NOTATION *Complete each solution.*

13. $(y^5y^3)^{-5} = (\square)^{-5}$

$= y^{\square}$

$= \frac{1}{y^{40}}$

14. $\left(\frac{a^2b^3}{a^{-3}b}\right)^{-3} = (a^{2-(-3)}b^{\ \ -1})^{-3}$

$= (a\ \ b\ \ )^{-3}$

$= \frac{1}{(a^5b^2)}$

$= \frac{1}{a^{15}b^6}$

15. In the expression $3x^{-2}$, what is the base and what is the exponent?

16. In the expression $-3x^{-2}$, what is the base and what is the exponent?

17. Determine the base and the exponent and evaluate each expression.

a. -4^2

b. 4^{-2}

c. -4^{-2}

18. Determine the base and the exponent and evaluate each expression.

a. $(-7)^2$

b. $(-7)^{-2}$

c. -7^{-2}

PRACTICE

Simplify each expression. Write each answer without using parentheses or negative exponents.

19. 7^0

20. 9^0

21. $\left(\frac{1}{4}\right)^0$

22. $\left(\frac{3}{8}\right)^0$

23. $2x^0$

24. $(2x)^0$

25. $(-x)^0$

26. $-x^0$

27. $\left(\frac{a^2b^3}{ab^4}\right)^0$

28. $\frac{2}{3}\left(\frac{xyz}{x^2y}\right)^0$

29. $\frac{5}{2x^0}$

30. $\frac{4}{3a^0}$

31. 12^{-2}

32. 11^{-2}

33. $(-4)^{-1}$

34. $(-8)^{-1}$

35. $\frac{1}{5^{-3}}$

36. $\frac{1}{3^{-3}}$

37. $\frac{2^{-4}}{3^{-1}}$

38. $\frac{7^{-2}}{2^{-3}}$

39. -4^{-3}

40. -6^{-3}

41. $-(-4)^{-3}$

42. $-(-4)^{-2}$

43. x^{-2}

44. y^{-3}

45. $-b^{-5}$

46. $-c^{-4}$

47. $(2y)^{-4}$

48. $(-3x)^{-1}$

49. $(ab^2)^{-3}$

50. $(m^2n^3)^{-2}$

51. $2^5 \cdot 2^{-2}$

52. $10^2 \cdot 10^{-4}$

53. $4^{-3} \cdot 4^{-2} \cdot 4^5$

54. $3^{-4} \cdot 3^5 \cdot 3^{-3}$

55. $\left(\frac{7}{8}\right)^{-1}$

56. $\left(\frac{16}{5}\right)^{-1}$

57. $\frac{3^5 \cdot 3^{-2}}{3^3}$

58. $\frac{6^2 \cdot 6^{-3}}{6^{-2}}$

59. $\frac{y^4}{y^5}$

60. $\frac{t^7}{t^{10}}$

61. $\frac{(r^2)^3}{(r^3)^4}$

62. $\frac{(b^3)^4}{(b^5)^4}$

63. $\frac{y^4y^3}{y^4y^{-2}}$

64. $\frac{x^{12}x^{-7}}{x^3x^4}$

65. $\frac{10a^4a^{-2}}{5a^2a^0}$

66. $\frac{9b^0b^3}{3b^{-3}b^4}$

67. $(ab^2)^{-2}$

68. $(c^2d^3)^{-2}$

69. $(x^2y)^{-3}$

70. $(-xy^2)^{-4}$

71. $(x^{-4}x^3)^3$

72. $(y^{-2}y)^3$

73. $(a^{-2}b^3)^{-4}$

74. $(y^{-3}z^5)^{-6}$

75. $(-2x^3y^{-2})^{-5}$

76. $(-3u^{-2}v^3)^{-3}$

77. $\left(\frac{a^3}{a^{-4}}\right)^2$

78. $\left(\frac{a^4}{a^{-3}}\right)^3$

79. $\left(\frac{b^5}{b^{-2}}\right)^{-2}$

80. $\left(\frac{b^{-2}}{b^3}\right)^3$

81. $\left(\frac{4x^2}{3x^{-5}}\right)^4$

82. $\left(\frac{-3r^4r^{-3}}{r^{-3}r^7}\right)^3$

83. $\left(\frac{12y^3z^{-2}}{3y^{-4}z^3}\right)^2$

84. $\left(\frac{6xy^3}{3x^{-1}y}\right)^3$

Simplify each expression. Assume that there are no divisions by 0.

85. $x^{2m}x^m$

86. $y^{3m}y^{2m}$

87. $u^{2m}u^{-3m}$

88. $r^{5m}r^{-6m}$

89. $\frac{y^{3m}}{y^{2m}}$

90. $\frac{z^{4m}}{z^{2m}}$

91. $\frac{x^{3n}}{x^{6n}}$

92. $\frac{x^m}{x^{5m}}$

APPLICATIONS

93. THE DECIMAL NUMERATION SYSTEM Decimal numbers are written by putting digits into place-value columns that are separated by a decimal point. Express the value of each of the columns shown using a power of 10.

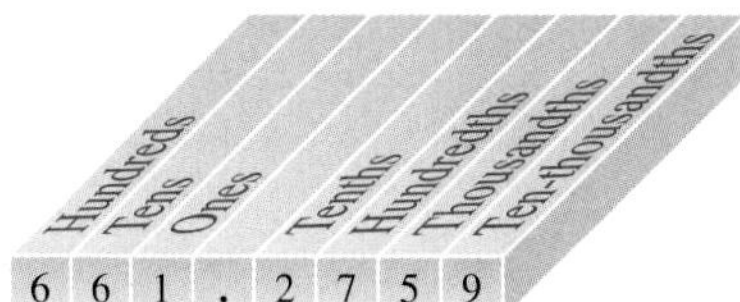

94. UNIT COMPARISONS Consider the relative sizes of the items listed in the table. In the column titled "measurement," write the most appropriate number from the following list. Each number is used only once.

10^0 meter 10^{-1} meter 10^{-2} meter
10^{-3} meter 10^{-4} meter 10^{-5} meter

Item	Measurement (m)
Thickness of a dime	
Height of a bathroom sink	
Length of a pencil eraser	
Thickness of soap bubble film	
Width of a video cassette	
Thickness of a piece of paper	

95. RETIREMENT YEARS How much money should a young married couple invest now at an 8% annual rate if they want to have \$100,000 in the bank when they reach retirement age in 40 years? (See the Accent on Technology in this section for the formula.)

96. BIOLOGY During bacterial reproduction, the time required for a population to double is called the **generation time.** If b bacteria are introduced into a medium, then after the generation time has elapsed, there will be $2b$ bacteria. After n generations, there will be $b \cdot 2^n$ bacteria. Explain what this expression represents when $n = 0$.

WRITING

97. Explain how you would help a friend understand that 2^{-3} is not equal to -8.

98. Describe how you would verify on a calculator that

$$2^{-3} = \frac{1}{2^3}$$

REVIEW

99. IQ TESTS An IQ (intelligence quotient) is a score derived from the formula

$$IQ = \frac{\text{mental age}}{\text{chronological age}} \cdot 100$$

Find the mental age of a 10-year-old girl if she has an IQ of 135.

100. DIVING When you are under water, the pressure in your ears is given by the formula

Pressure = depth · density of water

Find the density of water (in lb/ft^3) if, at a depth of 9 feet, the pressure on your eardrum is 561.6 lb/ft^2.

101. Write the equation of the line having slope $\frac{3}{4}$ and y-intercept -5.

102. Find $f(-6)$ if $f(x) = x^2 - 3x + 1$.

10.3 Scientific Notation

- Scientific notation • Writing numbers in scientific notation
- Changing from scientific notation to standard notation
- Using scientific notation to simplify computations

Scientists often deal with extremely large and extremely small numbers. Two examples are shown in Figure 10-1 on the next page.

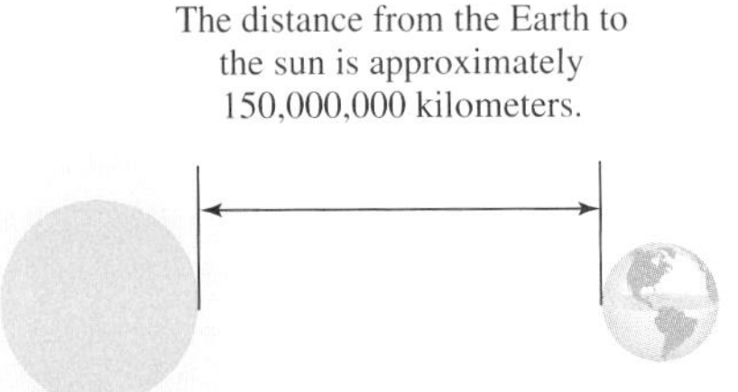

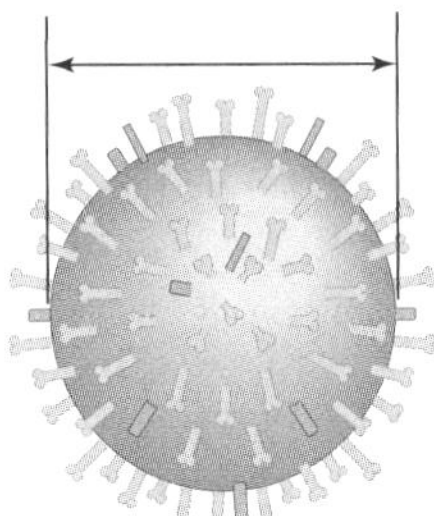

The influenza virus, which causes "flu" symptoms of cough, sore throat, headache, and congestion, has a diameter of 0.00000256 inch.

FIGURE 10-1

The large number of zeros in 150,000,000 and 0.00000256 makes them difficult to read and hard to remember. In this section, we will discuss a notation that will make such numbers easier to use.

Scientific notation

Scientific notation provides a compact way of writing large and small numbers.

> **Scientific notation**
>
> A number is written in **scientific notation** when it is written as the product of a number between 1 (including 1) and 10 and an integer power of 10.

These numbers are written in scientific notation:

3.9×10^{6}, 2.24×10^{-4}, and 9.875×10^{22}

Every number written in scientific notation has the following form:

An integer exponent

$$\underbrace{\square.\square}_{\text{A decimal between 1 and 10}} \times 10^{\square}$$

Writing numbers in scientific notation

EXAMPLE 1 Change 150,000,000 to scientific notation.

Solution We note that 1.5 lies between 1 and 10. To obtain 150,000,000, the decimal point in 1.5 must be moved eight places to the right.

1.50000000

8 places to the right

Because multiplying a number by 10 moves the decimal point one place to the right, we can accomplish this by multiplying 1.5 by 10 eight times. We can show the multiplication of 1.5 by 10 eight times using the notation 10^8. Thus, 150,000,000 written in scientific notation is 1.5×10^8.

Self Check 1
The distance from the Earth to the sun is approximately 93,000,000 miles. Write this number in scientific notation.

Answer 9.3×10^7

EXAMPLE 2 Change 0.00000256 to scientific notation.

Solution We note that 2.56 is between 1 and 10. To obtain 0.00000256, the decimal point in 2.56 must be moved six places to the left.

0 0 0 0 0 2 . 56

6 places to the left

We can accomplish this by dividing 2.56 by 10^6, which is equivalent to multiplying 2.56 by $\frac{1}{10^6}$ (or by 10^{-6}). Thus, 0.00000256 written in scientific notation is 2.56×10^{-6}.

Self Check 2
The *Salmonella* bacterium, which causes food poisoning, is 0.00009055 inch long. Write this number in scientific notation.

Answer 9.055×10^{-5}

EXAMPLE 3 Write **a.** 235,000 and **b.** 0.0000073 in scientific notation.

Solution

a. $235{,}000 = 2.35 \times 10^5$ Because $2.35 \times 10^5 = 235{,}000$ and 2.35 is between 1 and 10

b. $0.0000073 = 7.3 \times 10^{-6}$ Because $7.3 \times 10^{-6} = 0.0000073$ and 7.3 is between 1 and 10

Self Check 3
Write in scientific notation:
a. 17,500
b. 0.657

Answers **a.** 1.75×10^4, **b.** 6.57×10^{-1}

From Examples 1, 2, and 3, we see that in scientific notation, a positive exponent is used when writing a number that is greater than 10. A negative exponent is used when writing a number that is between 0 and 1.

Calculators and scientific notation

CALCULATOR SNAPSHOT

When displaying a very large or a very small number as an answer, most scientific calculators express it in scientific notation. To show this, we will find the values of $(453.46)^5$ and $(0.0005)^{12}$. We enter these numbers and press these keys.

453.46 [y^x] 5 [=] 1.917321395 ^{13}

.0005 [y^x] 12 [=] 2.44140625 $^{-40}$

Since the answers in standard notation require more space than the calculator display has, the calculator gives each result in scientific notation. The first display represents $1.917321395 \times 10^{13}$, and the second represents $2.44140625 \times 10^{-40}$.

If we evaluate the same two expressions using a graphing calculator, we see that the letter E is used when displaying a number in scientific notation.

453.46 [∧] 5 [ENTER]

```
453.46^5
   1.917321395E13
```

.0005 [∧] 12 [ENTER]

```
.0005^12
   2.44140625E-40
```

Self Check 4
Write 85×10^{-3} in scientific notation.

Answer 8.5×10^{-2}

EXAMPLE 4 Write 432.0×10^5 in scientific notation.

Solution The number 432.0×10^5 is not written in scientific notation, because 432.0 is not a number between 1 and 10. To write this number in scientific notation, we proceed as follows:

$$\mathbf{432.0} \times 10^5 = \mathbf{4.32 \times 10^2} \times 10^5 \quad \text{Write 432.0 in scientific notation.}$$
$$= 4.32 \times 10^7 \quad 10^2 \times 10^5 = 10^{2+5} = 10^7.$$

Changing from scientific notation to standard notation

We can change a number written in scientific notation to **standard notation.** For example, to write 9.3×10^7 in standard notation, we multiply 9.3 by 10^7.

$$9.3 \times 10^7 = 9.3 \times 10{,}000{,}000 \quad 10^7 \text{ is equal to 1 followed by 7 zeros.}$$
$$= 93{,}000{,}000$$

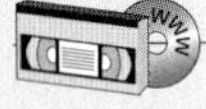

Self Check 5
Write in standard notation:
a. 4.76×10^5
b. 9.8×10^{-3}

Answers a. 476,000, b. 0.0098

EXAMPLE 5 Write **a.** 3.4×10^5 and **b.** 2.1×10^{-4} in standard notation.

Solution

a. $3.4 \times 10^5 = 3.4 \times 100{,}000$
$= 340{,}000$

b. $2.1 \times 10^{-4} = 2.1 \times \frac{1}{10^4}$
$= 2.1 \times \frac{1}{10{,}000}$
$= 2.1 \times 0.0001$
$= 0.00021$

The following numbers are written in both scientific and standard notation. In each case, the exponent gives the number of places that the decimal point moves, and the sign of the exponent indicates the direction that it moves.

$5.32 \times 10^5 = 5\,3\,2\,0\,0\,0.$ The decimal point moves 5 places to the right.

$8.95 \times 10^{-4} = 0.0\,0\,0\,8\,9\,5$ The decimal point moves 4 places to the left.

$9.77 \times 10^0 = 9.77$ There is no movement of the decimal point.

Using scientific notation to simplify computations

Another advantage of scientific notation becomes apparent when we evaluate products or quotients that contain very large or very small numbers.

EXAMPLE 6 **Astronomy.** Except for the sun, the nearest star visible to the naked eye from most parts of the United States is Sirius. Light from Sirius reaches Earth in about 70,000 hours. If light travels at approximately 670,000,000 mph, how far from Earth is Sirius?

Solution We are given the rate at which light travels (670,000,000 mph) and the time it takes the light to travel from Sirius to Earth (70,000 hr). We can find the distance the light travels using the formula $d = rt$.

$d = rt$

$d = \mathbf{670{,}000{,}000(70{,}000)}$	Substitute 670,000,000 for r and 70,000 for t.
$= (6.7 \times 10^8)(7.0 \times 10^4)$	Write each number in scientific notation.
$= (6.7 \cdot 7.0) \times (10^8 \cdot 10^4)$	Group the numbers together and the powers of 10 together.
$= (6.7 \cdot 7.0) \times 10^{8+4}$	Keep the base and add the exponents.
$= 46.9 \times 10^{12}$	Perform the multiplication. Perform the addition.

We note that 46.9 is not between 0 and 1, so 46.9×10^{12} is not written in scientific notation. To answer in scientific notation, we proceed as follows.

$= 4.69 \times 10^1 \times 10^{12}$	Write 46.9 in scientific notation as 4.69×10^1.
$= 4.69 \times 10^{13}$	Keep the base of 10 and add the exponents.

Sirius is approximately 4.69×10^{13} or 46,900,000,000,000 miles from Earth.

Science Majors and Space Travel

THINK IT THROUGH

"The number of U.S. college students earning degrees in science, technology, engineering, and math has fallen over the last 15 years. What a better way to hook our children than with a new space exploration plan?"

Patricia Arnold, Space Foundation, 2004

It has been over 30 years since a U.S. astronaut last walked on the moon. Many educators feel that manned flights to the moon and Mars would ignite a passion for space and science studies among young people. However, the minimum distance Mars is from Earth is 135 times further than the moon is from Earth. Traveling such a long way poses many problems. If the average distance from Earth to the moon is about 2.4×10^5 miles, what is the distance between Earth and Mars? Express the result in scientific notation.

EXAMPLE 7 **Atoms.** Scientific notation is used in chemistry. As an example, we can approximate the weight (in grams) of one atom of the heaviest naturally occurring element, uranium, by evaluating the following expression.

$$\frac{2.4 \times 10^2}{6.0 \times 10^{23}}$$

Solution

$\frac{2.4 \times 10^2}{6.0 \times 10^{23}} = \frac{2.4}{6.0} \times \frac{10^2}{10^{23}}$	Divide the numbers and the powers of 10 separately.
$= \frac{2.4}{6.0} \times 10^{2-23}$	For the powers of 10, keep the base and subtract the exponents.
$= 0.4 \times 10^{-21}$	Perform the division. Then subtract the exponent.
$= 4.0 \times 10^{-1} \times 10^{-21}$	Write 0.4 in scientific notation as 4.0×10^{-1}.
$= 4.0 \times 10^{-22}$	Keep the base and add the exponents.

One atom of uranium weighs 4.0×10^{-22} gram. Written in standard notation, this is 0.00000000000000000000004 g.

Self Check 7

Find the approximate weight (in grams) of one atom of gold by evaluating

$$\frac{1.98 \times 10^2}{6.0 \times 10^{23}}$$

Answer 3.3×10^{-22} g

Entering numbers in scientific notation — CALCULATOR SNAPSHOT

We can evaluate the expression from Example 7 by entering the numbers written in scientific notation, using [EE] the key on a scientific calculator.

2.4 [EE] 2 [÷] 6 [EE] 23 [=]   Display: 4. -22

The result shown in the display means 4.0×10^{-22}.

If we use a graphing calculator, the keystrokes are similar.

2.4 [2nd] [EE] 2 [÷] 6 [2nd] [EE] 23 [ENTER]

Display:
2.4E2/6E23
4E-22

Section 10.3 STUDY SET

VOCABULARY *Fill in the blanks.*

1. A number is written in ______ notation when it is written as the product of a number between 1 (including 1) and 10 and an integer power of 10.
2. The number 125,000 is written in ______ notation.

CONCEPTS *Fill in the blanks using standard notation.*

3. $2.5 \times 10^2 =$ ______
4. $2.5 \times 10^{-2} =$ ______
5. $2.5 \times 10^{-5} =$ ______
6. $2.5 \times 10^5 =$ ______

Fill in the blanks with a power of 10.

7. $387{,}000 = 3.87 \times$ ______
8. $38.7 = 3.87 \times$ ______
9. $0.00387 = 3.87 \times$ ______
10. $0.000387 = 3.87 \times$ ______
11. When we multiply a decimal by 10^5, the decimal point moves ______ places to the ______.
12. When we multiply a decimal by 10^{-7}, the decimal point moves ______ places to the ______.
13. Dividing a decimal by 10^4 is equivalent to multiplying it by ______.
14. Multiplying a decimal by 10^0 does not move the decimal point, because $10^0 =$ ______.
15. When a real number greater than 10 is written in scientific notation, the exponent on 10 is a ______ number.
16. When a real number between 0 and 1 is written in scientific notation, the exponent on 10 is a ______ number.

NOTATION *Complete each solution.*

17. Write 63.7×10^5 in scientific notation.

$$63.7 \times 10^5 = \text{______} \times 10^5$$
$$= 6.37 \times 10^{\text{___}+5}$$
$$= 6.37 \times 10^6$$

18. Simplify: $\dfrac{64{,}000}{0.00004}$.

$$\frac{64{,}000}{0.00004} = \frac{6.4 \times \text{___}}{4 \times \text{___}}$$
$$= \frac{\text{___}}{\text{___}} \times \frac{10^4}{10^{-5}}$$
$$= 1.6 \times 10^{\text{___}-(-5)}$$
$$= 1.6 \times 10^9$$

PRACTICE *Write each number in scientific notation.*

19. 23,000

20. 4,750

21. 1,700,000

22. 290,000

23. 0.062

24. 0.00073

25. 0.0000051

26. 0.04

27. 42.5×10^2

28. 0.3×10^3

29. 0.25×10^{-2}

30. 25.2×10^{-3}

Write each number in standard notation.

31. 2.3×10^2

32. 3.75×10^4

33. 8.12×10^5

34. 1.2×10^3

35. 1.15×10^{-3}

36. 4.9×10^{-2}

37. 9.76×10^{-4}

38. 7.63×10^{-5}

39. 25×10^6

40. 0.07×10^3

41. 0.51×10^{-3}

42. 617×10^{-2}

43. ASTRONOMY The distance from Earth to Alpha Centauri (the nearest star outside our solar system) is about 25,700,000,000,000 miles. Express this number in scientific notation.

44. SPEED OF SOUND The speed of sound in air is 33,100 centimeters per second. Express this number in scientific notation.

45. GEOGRAPHY The largest ocean in the world is the Pacific Ocean, which covers 6.38×10^7 square miles. Express this number in standard notation.

46. ATOMS The number of atoms in 1 gram of iron is approximately 1.08×10^{22}. Express this number in standard notation.

47. LENGTH OF A METER One meter is approximately 0.00622 mile. Use scientific notation to express this number.

48. ANGSTROM One angstrom is 1.0×10^{-7} millimeter. Express this number in standard notation.

Use scientific notation and the rules for exponents to simplify each expression. Give all answers in standard notation.

49. $(3.4 \times 10^2)(2.1 \times 10^3)$

50. $(4.1 \times 10^{-3})(3.4 \times 10^4)$

51. $\dfrac{9.3 \times 10^2}{3.1 \times 10^{-2}}$

52. $\dfrac{7.2 \times 10^6}{1.2 \times 10^8}$

53. $\dfrac{96{,}000}{(12{,}000)(0.00004)}$

54. $\dfrac{(0.48)(14{,}400{,}000)}{96{,}000{,}000}$

Evaluate each expression.

55. $(456.4)^6$

56. $(0.053)^8$

57. $(0.009)^{-6}$

58. 225^{-5}

59. $\left(\dfrac{1}{3}\right)^{-55}$

60. $\left(\dfrac{8}{5}\right)^{50}$

APPLICATIONS

61. WAVELENGTHS Transmitters, vacuum tubes, and lights emit energy that can be modeled as a wave, as shown. Examples of the most common types of electromagnetic waves are given in the table. List the wavelengths in order from shortest to longest.

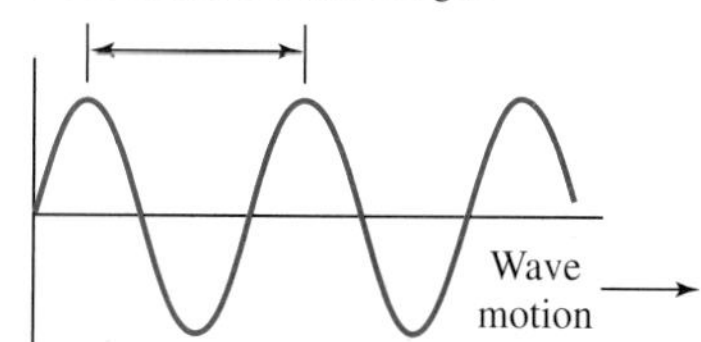
This distance between the two crests
of the wave is called the wavelength.
Wave
motion

Type	Use	Wavelength (m)
visible light	lighting	9.3×10^{-6}
infrared	photography	3.7×10^{-5}
x-ray	medical	2.3×10^{-11}
radio wave	communication	3.0×10^2
gamma ray	treating cancer	8.9×10^{-14}
microwave	cooking	1.1×10^{-2}
ultraviolet	sun lamp	6.1×10^{-8}

62. SPACE EXPLORATION On July 4, 1997, the *Pathfinder,* carrying the rover vehicle called Sojourner, landed on Mars to perform a scientific investigation of the planet. The distance from Mars to Earth is approximately 3.5×10^7 miles. Use scientific notation to express this distance in feet. (*Hint:* 5,280 feet = 1 mile.)

63. PROTONS The mass of one proton is approximately 1.7×10^{-24} gram. Use scientific notation to express the mass of 1 million protons.

64. SPEED OF SOUND The speed of sound in air is approximately 3.3×10^4 centimeters per second. Use scientific notation to express this speed in kilometers per second. (*Hint:* 100 centimeters = 1 meter and 1,000 meters = 1 kilometer.)

65. LIGHT YEARS One light year is about 5.87×10^{12} miles. Use scientific notation to express this distance in feet. (*Hint:* 5,280 feet = 1 mile.)

66. OIL RESERVES As of January 1, 2001, Saudi Arabia was believed to have crude oil reserves of about 2.617×10^{11} barrels. A barrel contains 42 gallons of oil. Use scientific notation to express Saudi Arabia oil reserves in gallons. (Source: *The World Almanac and Book of Facts 2003.*)

67. INTEREST EARNED As of December 2000, the Federal Deposit Insurance Corporation (FDIC) reported that the total insured deposits in U.S. banks and savings and loans was approximately 5.2×10^{12} dollars. If this money was invested at a rate of 4% simple annual interest, how much would it earn in 1 year? Use scientific notation to express the answer. (Source: *The World Almanac and Book of Facts 2003.*)

68. CURRENCY As of June 30, 2002, the U.S. Treasury reported that the number of \$20 bills in circulation was approximately 4.84×10^9. What was the total value of the currency? Use scientific notation to express the answer. (Source: *The World Almanac and Book of Facts 2003.*)

69. THE MILITARY The graph shows the number of U.S. troops for 1983–2001. Estimate each of the following and express your answers in scientific and standard notation.

a. The number of troops in 1993

b. The largest numbers of troops during these years

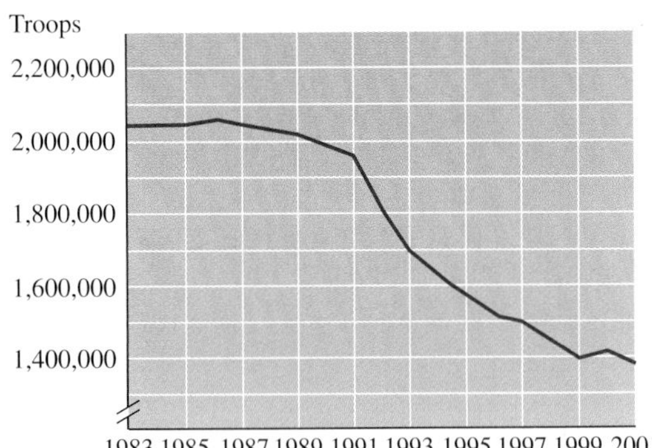

Based on data from the U.S. Department of Defense

70. THE NATIONAL DEBT The graph below shows the growth of the national debt for the fiscal years 1993–2002.

a. Use scientific notation to express the debt as of 1994, 1995, and 2001.

b. In 2002, the population of the United States was about 2.88×10^8. Estimate the share of the debt for each man, woman, and child in the United States. Answer in standard notation.

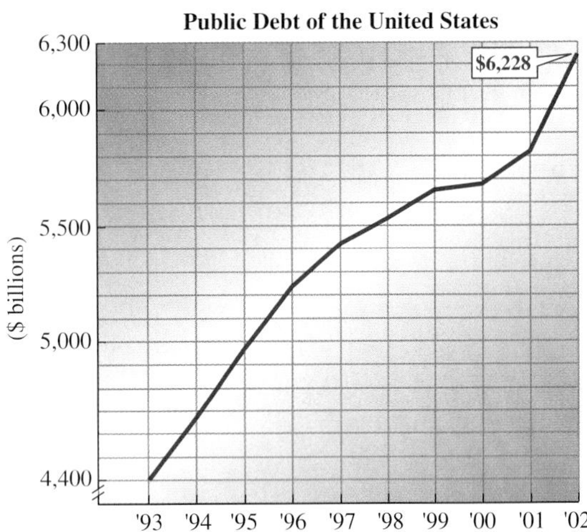

Based on data from the U.S. Department of Defense

WRITING

71. In what situations would scientific notation be more convenient than standard notation?

72. To multiply a number by a power of 10, we move the decimal point. Which way, and how far? Explain.

73. 2.3×10^{-3} contains a negative sign but represents a positive number. Explain.

74. Is this a true statement? $2.0 \times 10^3 = 2 \times 10^3$. Explain.

REVIEW

75. If $y = -1$, find the value of $-5y^{55}$.

76. What is the y-intercept of the graph of $y = -3x - 5$?

Determine which property of real numbers justifies each statement.

77. $5 + z = z + 5$

78. $7(u + 3) = 7u + 7 \cdot 3$

Solve each equation.

79. $3(x - 4) - 6 = 0$

80. $8(3x - 5) - 4(2x + 3) = 12$

10.4 Polynomials

- Polynomials • Monomials, binomials, and trinomials
- Degree of a polynomial • Evaluating polynomial functions

In arithmetic, we learned how to add, subtract, multiply, divide, and find powers of numbers. In algebra, we will learn how to perform these operations on *polynomials.* In this section, we will introduce polynomials, classify them into groups, define their degrees, and show how to evaluate them at specific values of their variables.

Polynomials

Recall that a **term** is a number or a product of a number and one or more variables, which may be raised to powers. Examples of terms are

$$3x, \quad -4y^2, \quad \frac{1}{2}a^2b^3, \quad t, \quad \text{and} \quad 25$$

The **numerical coefficients,** or simply **coefficients,** of the first four of these terms are $3, -4, \frac{1}{2}$, and 1, respectively. Because $25 = 25x^0$, 25 is considered to be the numerical coefficient of the term 25.

Polynomials

A **polynomial** is a term or a sum of terms in which all variables have whole-number exponents. No variable appears in a denominator.

Here are some examples of polynomials:

$$3x + 2, \quad 4y^2 - 2y - 3, \quad -8xy^2, \quad \text{and} \quad a^3 + 3a^2b + 3ab^2 + b^3$$

The polynomial $3x + 2$ has two terms, $3x$ and 2, and we say it is a **polynomial in x.** A single number is called a **constant,** and so its last term, 2, is called the **constant term.**

Since $4y^2 - 2y - 3$ can be written as $4y^2 + (-2y) + (-3)$, it is the sum of three terms, $4y^2$, $-2y$, and -3. It is written in **decreasing** or **descending powers** of y, because the powers on y decrease from left to right.

$-8xy^2$ is a polynomial with just one term. We say that it is a **polynomial in x and y.**

The four-term polynomial $a^3 + 3a^2b + 3ab^2 + b^3$ is written in descending powers of a and **ascending powers** of b.

! COMMENT The expression $2x^3 - 3x^{-2} + 5$ is not a polynomial, because the second term contains a variable with an exponent that is not a whole number. Similarly, $y^2 - \frac{7}{y}$ is not a polynomial, because $\frac{7}{y}$ has a variable in the denominator.

EXAMPLE 1 Determine whether each expression is a polynomial.

a. $x^2 + 2x + 1$, **b.** $3a^{-1} - 2a - 3$, **c.** $\frac{1}{2}x^3 - 2.3x$, and **d.** $\frac{p+3}{p-1}$.

Solution

a. $x^2 + 2x + 1$ is a polynomial.

b. $3a^{-1} - 2a - 3$ is not a polynomial. In the first term, the exponent on the variable is not a whole number.

Self Check 1

Determine whether each expression is a polynomial:

a. $3x^{-4} + 2x^2 - 3$

b. $7.5p^3 - 4p^2 - 3p + 4$

c. $\frac{1}{2}x^3 - 2.3x$ is a polynomial, because it can be written as the sum $\frac{1}{2}x^3 + (-2.3x)$.

d. $\frac{p+3}{p-1}$ is not a polynomial. Variables cannot be in the denominator of a fraction.

Answers **a.** no, **b.** yes

Monomials, binomials, and trinomials

A polynomial with one term is called a **monomial.** A polynomial with two terms is called a **binomial.** A polynomial with three terms is called a **trinomial.** Here are some examples.

Monomials	Binomials	Trinomials
$-6x$	$3u^3 - 4u^2$	$-5t^2 + 4t + 3$
$5x^2y$	$18a^2b + 4ab$	$27x^3 - 6x - 2$
29	$-29z^{17} - 1$	$a^2 + 2ab + b^2$

EXAMPLE 2 Classify each polynomial as a monomial, a binomial, or a trinomial: **a.** $5.2x^4 + 3.1x$, **b.** $7g^4 - 5g^3 - 2$, and **c.** $-5x^2y^3$.

Solution

a. The polynomial $5.2x^4 + 3.1x$ has two terms, $5.2x^4$ and $3.1x$, so it is a binomial.

b. The polynomial $7g^4 - 5g^3 - 2$ has three terms, $7g^4$, $-5g^3$, and -2, so it is a trinomial.

c. The polynomial $-5x^2y^3$ has one term, so it is a monomial.

Self Check 2

Classify each polynomial as a monomial, a binomial, or a trinomial:

a. $5x$

b. $-5x^2 + 2x - 0.5$

c. $16x^2 - 9y^2$

Answers **a.** monomial, **b.** trinomial, **c.** binomial

Degree of a polynomial

The monomial $7x^6$ is called a **monomial of sixth degree** or a **monomial of degree 6,** because the variable x occurs as a factor six times. The monomial $3x^3y^4$ is a monomial of seventh degree, because the variables x and y occur as factors a total of seven times. Here are some more examples:

$2.7a$ is a monomial of degree 1.

$-2x^3$ is a monomial of degree 3.

$47x^2y^3$ is a monomial of degree 5.

8 is a monomial of degree 0, because $8 = 8x^0$.

These examples illustrate the following definition.

Degree of a monomial

If a represents a nonzero constant, the **degree of the monomial** ax^n is n.

The **degree of a monomial** in several variables is the sum of the exponents on those variables.

! COMMENT Note that the degree of ax^n is not defined when $a = 0$. Since $ax^n = 0$ when $a = 0$, the constant 0 has no defined degree.

Because each term of a polynomial is a monomial, we define the degree of a polynomial by considering the degrees of each of its terms.

Degree of a polynomial

The **degree of a polynomial** is determined by the term with the largest degree.

Here are some examples:

$x^2 + 2x$ is a binomial of degree 2, because the degree of its first term is 2 and the degree of its second term is less than 2.

$d^3 - 3d^2 + 1$ is a trinomial of degree 3, because the degree of its first term is 3 and the degree of each of its other terms is less than 3.

$25y^{13} - 15y^8z^{10} - 32y^{10}z^8 + 4$ is a polynomial of degree 18, because its second and third terms are of degree 18. Its other terms have degree less than 18.

EXAMPLE 3 Find the degree of each polynomial:

a. $-4x^3 - 5x^2 + 3x$, **b.** $1.6w - 1.6$, and **c.** $-17a^2b^3 + 12ab^6$.

Solution

a. The trinomial $-4x^3 - 5x^2 + 3x$ has terms of degree 3, 2, and 1. Therefore, its degree is 3.

b. The first term of $1.6w - 1.6$ has degree 1 and the second term has degree 0, so the binomial has degree 1.

c. The degree of the first term of $-17a^2b^3 + 12ab^6$ is 5 and the degree of the second term is 7, so the binomial has degree 7.

Self Check 3

Find the degree of each polynomial:

a. $15p^3 - 15p^2 - 3p + 4$

b. $-14st^4 + 12s^3t$

Answers **a.** 3, **b.** 5

If written in descending powers of the variable, the **lead term** of a polynomial is the term of highest degree. For example, the leading term of $-4x^3 - 5x^2 + 3x$ is $-4x^3$. The coefficient of the leading term (in this case, -4) is called the **lead coefficient.**

Evaluating polynomial functions

Each of the equations below defines a function, because each input x-value determines exactly one output value. Since the right-hand side of each equation is a polynomial, these functions are called **polynomial functions.**

$f(x) = 6x + 4$	$g(x) = 3x^2 + 4x - 5$	$h(x) = -x^3 + x^2 - 2x + 3$
This polynomial has two terms. Its degree is 1.	This polynomial has three terms. Its degree is 2.	This polynomial has four terms. Its degree is 3.

To evaluate a polynomial function for a specific value, we replace the variable in the defining equation with the input value. Then we simplify the resulting expression to find the output. For example, suppose we wish to evaluate the polynomial function

$f(x) = 6x + 4$ for $x = 1$. Then $f(1)$ represents the value of $f(x) = 6x + 4$ when $x = 1$. We find $f(1)$ as follows.

$f(x) = 6x + 4$ This is the given function.
$f(\mathbf{1}) = 6(\mathbf{1}) + 4$ Substitute 1 for x. The number 1 is the input.
$= 6 + 4$ Perform the multiplication.
$= 10$ Perform the addition. 10 is the output.

Thus, $f(1) = 10$.

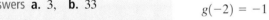

Self Check 4
Consider the function

$$h(x) = -x^3 + x - 2x + 3$$

Find
a. $h(0)$
b. $h(-3)$

Answers **a.** 3, **b.** 33

EXAMPLE 4 Consider the function $g(x) = 3x^2 + 4x - 5$. Find **a.** $g(0)$ and **b.** $g(-2)$.

Solution

a. $g(x) = 3x^2 + 4x - 5$ This is the given function.
$g(\mathbf{0}) = 3(\mathbf{0})^2 + 4(\mathbf{0}) - 5$ To find $g(0)$, substitute 0 for x.
$= 3(0) + 4(0) - 5$ Evaluate the power.
$= 0 + 0 - 5$ Perform the multiplications.
$g(0) = -5$

b. $g(x) = 3x^2 + 4x - 5$ This is the given function.
$g(\mathbf{-2}) = 3(\mathbf{-2})^2 + 4(\mathbf{-2}) - 5$ To find $g(-2)$, substitute -2 for x.
$= 3(4) + 4(-2) - 5$ Evaluate the power.
$= 12 + (-8) - 5$ Perform the multiplications.
$g(-2) = -1$

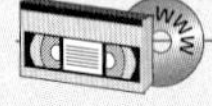

EXAMPLE 5 **Supermarket displays.** The polynomial function

$$f(c) = \frac{1}{3}c^3 + \frac{1}{2}c^2 + \frac{1}{6}c$$

gives the number of cans used in a display shaped like a square pyramid, having a square base formed by c cans per side. Find the number of cans of soup used in the display shown in Figure 10-2.

Solution Since each side of the square base of the display is formed by 4 cans, $c = 4$. We can find the number of cans used in the display by finding $f(4)$.

FIGURE 10-2

$f(c) = \frac{1}{3}c^3 + \frac{1}{2}c^2 + \frac{1}{6}c$ This is the given function.

$f(4) = \frac{1}{3}(4)^3 + \frac{1}{2}(4)^2 + \frac{1}{6}(4)$ Substitute 4 for c.

$= \frac{1}{3}(64) + \frac{1}{2}(16) + \frac{1}{6}(4)$ Find the powers.

$= \frac{64}{3} + 8 + \frac{2}{3}$ Multiply, and then simplify: $\frac{4}{6} = \frac{2}{3}$.

$= \frac{66}{3} + 8$ Add the fractions.

$= 22 + 8$

$= 30$

30 cans of soup were used in the display.

Section 10.4 STUDY SET

VOCABULARY *Fill in the blanks.*

1. A __________ is a term or a sum of terms in which all variables have whole-number exponents.
2. The numerical __________ of the term $-25x^2y^3$ is -25.
3. The ________ of the monomial $3x^7$ is 7.
4. The degree of a polynomial is the same as the ________ of its term with the largest degree.
5. A __________ is a polynomial with one term. A __________ is a polynomial with two terms.
6. A __________ is a polynomial with three terms.
7. $x^3 - 6x^2 + 9x - 2$ is a polynomial written in ____________________ powers of x.
8. For the polynomial $6x^2 + 3x - 1$, the ______ term is $6x^2$, and the lead __________ is 6. The __________ term is -1.
9. The notation $f(x)$ is read as f _____ x.
10. $f(2)$ represents the _______ of a function when $x = 2$.

CONCEPTS *Determine whether each expression is a polynomial.*

11. $x^3 - 5x^2 - 2$
12. $x^{-4} - 5x$
13. $\frac{1}{2x} + 3$
14. $x^3 - 1$
15. $x^2 - y^2$
16. $a^4 + a^3 + a^2 + a$

Classify each polynomial as a monomial, a binomial, a trinomial, or none of these.

17. $3x + 7$
18. $3y - 5$
19. $y^2 + 4y + 3$
20. $3xy$
21. $3z^2$
22. $3x^4 - 2x^3 + 3x - 1$
23. $t - 32$
24. $9x^2y^3z^4$
25. $s^2 - 23s + 31$
26. $2x^3 - 5x^2 + 6x - 3$
27. $3x^5 - x^4 - 3x^3 + 7$
28. x^3
29. $2a^2 - 3ab + b^2$
30. $a^3 - b^3$

Find the degree of each polynomial.

31. $3x^4$
32. $3x^5$
33. $-2x^2 + 3x + 1$
34. $-5x^4 + 3x^2 - 3x$
35. $3x - 5$
36. $y^3 + 4y^2$
37. $-5r^2s^2 - r^3s + 3$
38. $4r^2s^3 - 5r^2s^8$
39. $x^{12} + 3x^2y^3$
40. $17ab^5 - 12a^3b$
41. 38
42. -24

NOTATION *Complete each solution.*

43. If $f(x) = -2x^2 + 3x - 1$, find $f(2)$.

$f(2) = -2(\square)^2 + 3(\square) - 1$

$= -2(\square) + \square - 1$

$= -8 + 6 - \square$

$= \square - 1$

$= -3$

44. If $f(x) = -2x^2 + 3x - 1$, find $f(-2)$.

$$\begin{aligned} f(-2) &= -2(\ \ )^2 + 3(\ \ ) - 1 \\ &= -2(\ \ ) + 3(\ \ ) - 1 \\ &= \ \ + (-6) - 1 \\ &= \ \ - 1 \\ &= -15 \end{aligned}$$

45. Explain why $f(x) = x^3 + 2x^2 - 3$ is called a polynomial function.

46. **a.** Write $x - 9 + 3x^2$ in descending powers of x.

b. Write $-2xy + y^2 + x^2$ in descending powers of x.

PRACTICE *Let $f(x) = 5x - 3$ and find each value.*

47. $f(2)$ **48.** $f(0)$

49. $f(-1)$ **50.** $f(-2)$

51. $f\left(\frac{1}{5}\right)$ **52.** $f\left(\frac{4}{5}\right)$

53. $f(-0.9)$ **54.** $f(-1.2)$

Let $g(x) = -x^2 - 4$ and find each value.

55. $g(0)$ **56.** $g(1)$

57. $g(-1)$ **58.** $g(-2)$

Let $g(x) = -x^2 - 4$ and find each value.

59. $g(1.3)$ **60.** $g(2.4)$

61. $g(-13.6)$ **62.** $g(-25.3)$

Let $h(x) = x^3 - 2x + 3$ and find each value.

63. $h(0)$ **64.** $h(3)$

65. $h(-2)$ **66.** $h(-1)$

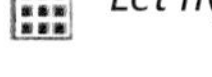

Let $h(x) = x^3 - 2x + 3$ and find each value.

67. $h(0.9)$ **68.** $h(0.4)$

69. $h(-8.1)$ **70.** $h(-7.7)$

Let $f(x) = -x^4 - x^3 + x^2 + x - 1$ and find each value.

71. $f(1)$ **72.** $f(-1)$

73. $f(-2)$ **74.** $f(2)$

APPLICATIONS *Use a calculator to help solve each problem.*

75. PACKAGING To make boxes, a manufacturer cuts equal-sized squares from each corner of a 10 in. × 12 in. piece of cardboard, and then folds up the sides. The polynomial function $f(x) = 4x^3 - 44x^2 + 120x$ gives the volume (in cubic inches) of the resulting box when a square with sides x inches long is cut from each corner. Find the volume of a box if 3-inch squares are cut out.

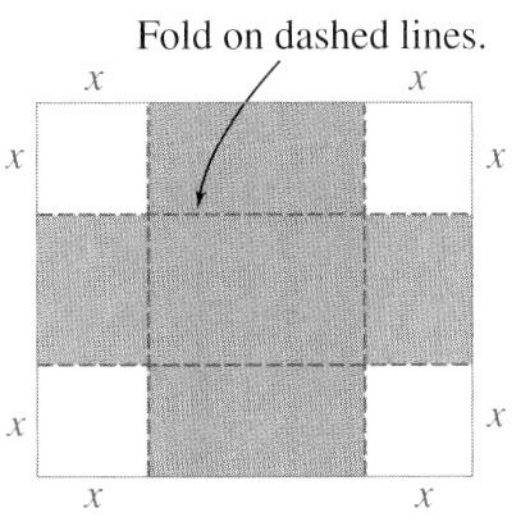

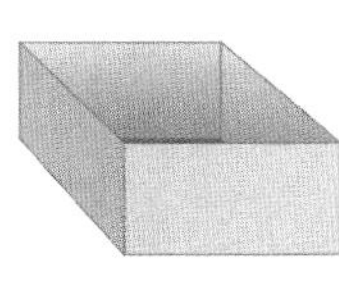

76. MAXIMIZING REVENUE The revenue (in dollars) that a manufacturer of office desks receives is given by the polynomial function

$$f(d) = -0.08d^2 + 100d$$

where d is the number of desks manufactured.

a. Find the total revenue if 625 desks are manufactured.

b. Does increasing the number of desks being manufactured to 650 increase the revenue?

77. WATER BALLOONS Some college students launched water balloons from the balcony of their dormitory on unsuspecting sunbathers. The height in feet of the balloons at a time t seconds after being launched is given by the polynomial function

$$f(t) = -16t^2 + 12t + 20$$

What was the height of the balloons 0.5 second and 1.5 seconds after being launched?

78. STOPPING DISTANCE The number of feet that a car travels before stopping depends on the driver's reaction time and the braking distance, as shown below. For one driver, the stopping distance is given by the polynomial function

$$f(v) = 0.04v^2 + 0.9v$$

where v is the velocity of the car. Find the stopping distance when the driver is traveling at 30 mph.

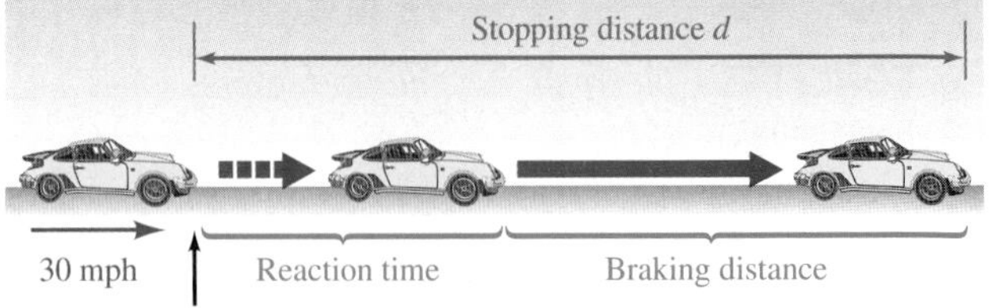

79. SUSPENSION BRIDGES The function

$$f(s) = 400 + 0.0066667s^2 - 0.0000001s^4$$

approximates the length of the cable between the two vertical towers of a suspension bridge, where s is the sag in the cable. Estimate the length of the cable of the following bridge if the sag is 24.6 feet.

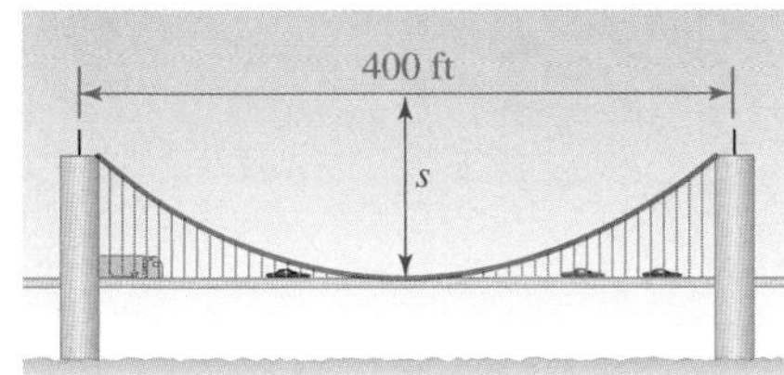

80. PRODUCE DEPARTMENTS Suppose a grocer is going to set up a pyramid-shaped display of cantaloupes like that shown in Figure 10-2 in Example 5. If each side of the square base of the display is made of six cantaloupes, how many will be used in the display?

81. DOLPHINS At a marine park, three trained dolphins jump in unison over an arching stream of water whose path can be described by the polynomial function

$$f(x) = -0.05x^2 + 2x$$

Given the takeoff points for each dolphin, how high must each dolphin jump to clear the stream of water?

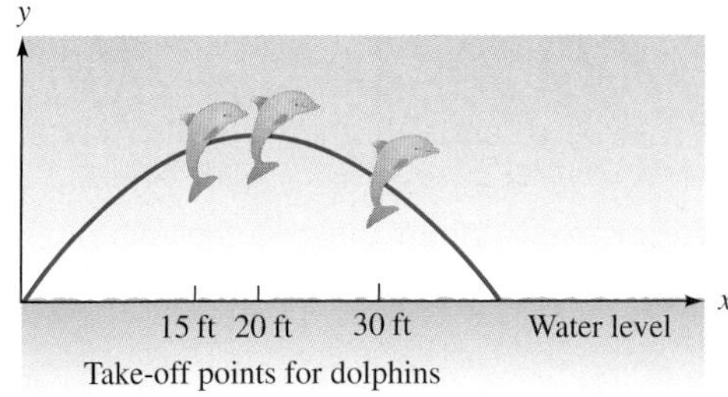

82. TUNNELS The arch at the entrance to a tunnel is described by the polynomial function

$$f(x) = -0.25x^2 + 23$$

What is the height of the arch at the edge of the pavement?

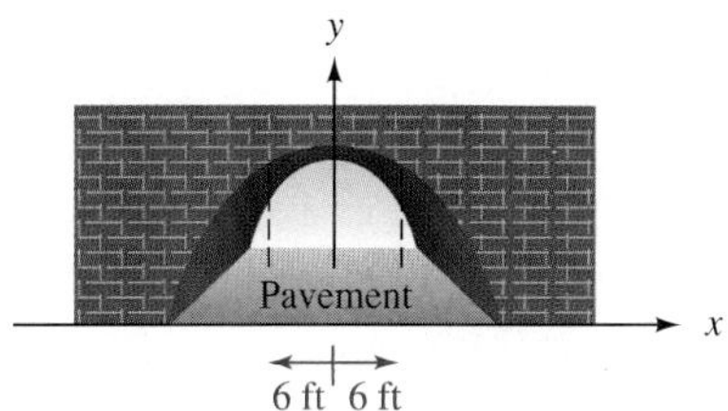

WRITING

83. Describe how to determine the degree of a polynomial.

84. List some words that contain the prefixes *mono, bi,* or *tri*.

REVIEW *Solve each inequality and graph its solution set.*

85. $-4(3y + 2) \le 28$

86. $-5 < 3t + 4 \le 13$

Write each expression without using parentheses or negative exponents.

87. $(x^2x^4)^3$

88. $(a^2)^3(a^3)^2$

89. $\left(\dfrac{y^2y^5}{y^4}\right)^3$

90. $\left(\dfrac{2t^3}{t}\right)^{-4}$

10.5 Adding and Subtracting Polynomials

- Adding monomials • Subtracting monomials • Adding polynomials
- Subtracting polynomials • Adding and subtracting multiples of polynomials
- An application of adding polynomials

In Figure 10-3(a) on the next page, the heights of the Seattle Space Needle and the Eiffel Tower in Paris are given. Using rules from arithmetic, we can find the difference in the heights of the towers by subtracting two numbers.

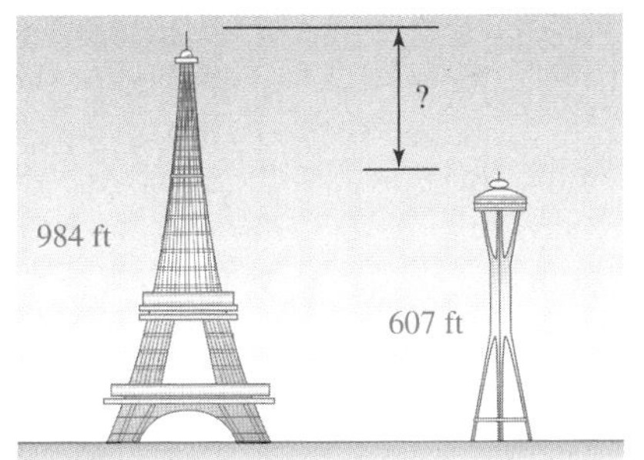

Arithmetic

984 – 607 = 377

The difference in height is 377 feet.

(a)

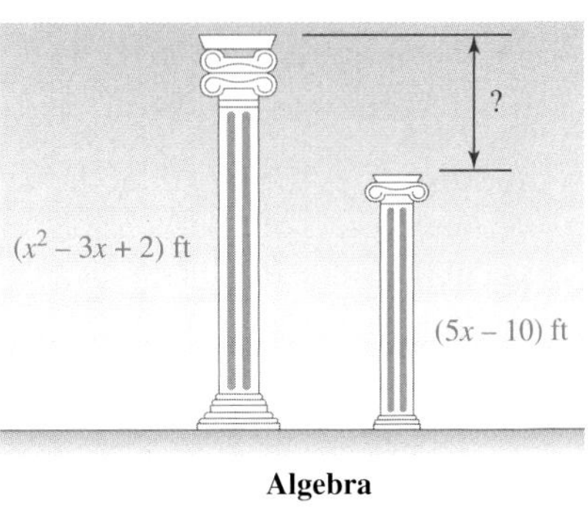

Algebra

$(x^2 - 3x + 2) - (5x - 10) = ?$

(b)

FIGURE 10-3

In Figure 10-3(b), the heights of two types of classical Greek columns are expressed using *polynomials*. To find the difference in their heights, we must subtract the polynomials. In this section, we will discuss the algebraic rules that are used to do this. Since any subtraction can be written in terms of addition, we will consider the procedures used to add polynomials first. We begin with monomials, which are polynomials having just one term.

Adding monomials

Recall that like terms have the same variables with the same exponents:

Like terms	**Unlike terms**
$-7x$ and $15x$	$-7x$ and $15a$
$4y^3$ and $16y^3$	$4y^3$ and $16y^2$
$\frac{1}{2}xy^2$ and $-\frac{1}{3}xy^2$	$\frac{1}{2}xy^2$ and $-\frac{1}{3}x^2y$

Also recall that to combine like terms, we combine their coefficients and keep the same variables with the same exponents. For example,

$$4y + 5y = (4 + 5)y = 9y \quad \text{and} \quad 8x^2 - x^2 = (8 - 1)x^2 = 7x^2$$

Likewise,

$$3a + 4b - 6a + 3b = -3a + 7b \quad \text{and} \quad -4cd^3 + 9cd^3 = 5cd^3$$

These examples suggest that to add like monomials, we simply combine like terms.

EXAMPLE 1 Perform the following additions.

a. $4x^4 + 81x^4 = 85x^4$

b. $-8x^2y^2 + 6x^2y^2 + x^2y^2 = -2x^2y^2 + x^2y^2$ Work from left to right. Combine like terms.

$= -x^2y^2$ Combine like terms.

c. $32c^2 + 10c + 4c^2 = 32c^2 + 4c^2 + 10c$ Write the like terms together.

$= 36c^2 + 10c$ Combine like terms.

Self Check 1

Perform the following additions:

a. $27x^6 + 8x^6$

b. $-12pq^2 + 5pq^2 + 8pq^2$

c. $6a^3 + 15a + a^3$

Answers **a.** $35x^6$, **b.** pq^2, **c.** $7a^3 + 15a$

! COMMENT When performing operations on polynomials, we usually write the terms of the solution in decreasing (or descending) powers of one variable. For instance, in Example 1, part c, the solution was written as $36c^2 + 10c$ instead of $10c + 36c^2$.

Subtracting monomials

To subtract one monomial from another, we add the opposite of the monomial that is to be subtracted.

EXAMPLE 2 Find each difference.

a. $8x^2 - 3x^2 = 8x^2 + (-3x^2)$ Add the opposite of $3x^2$, which is $-3x^2$.

$= 5x^2$ Combine like terms.

b. $6xy - 9xy = 6xy + (-9xy)$

$= -3xy$

c. $-3r - 5 - 4r = -3r + (-5) + (-4r)$ Add the opposite of 5 and $4r$.

$= -3r + (-4r) + (-5)$ Group like terms together.

$= -7r - 5$ Combine like terms. Write the addition of -5 as a subtraction of 5.

Self Check 2
Find each difference:
a. $12m^3 - 7m^3$
b. $-4pq - 27p - 8pq$

Answers **a.** $5m^3$, **b.** $-12pq - 27p$

Adding polynomials

Because of the distributive property, we can remove parentheses enclosing several terms when the sign preceding the parentheses is a + sign. We simply drop the parentheses.

$$+(3x^2 + 3x - 2) = \mathbf{+1}(3x^2 + 3x - 2)$$
$$= \mathbf{1}(3x^2) + \mathbf{1}(3x) + \mathbf{1}(-2) \quad \text{Distribute the multiplication by 1.}$$
$$= 3x^2 + 3x + (-2)$$
$$= 3x^2 + 3x - 2$$

We can add polynomials by removing parentheses, if necessary, and then combining any like terms that are contained within the polynomials.

EXAMPLE 3 Add: $(3x^2 - 3x + 2) + (2x^2 + 7x - 4)$.

Solution

$(3x^2 - 3x + 2) + (2x^2 + 7x - 4)$

$= 3x^2 - 3x + 2 + 2x^2 + 7x - 4$ Drop the parentheses.

$= 3x^2 + 2x^2 - 3x + 7x + 2 - 4$ Write like terms together.

$= 5x^2 + 4x - 2$ Combine like terms.

Self Check 3
Add:
$(2a^2 - a + 4) + (5a^2 + 6a - 5)$.

Answer $7a^2 + 5a - 1$

Problems such as Example 3 are often written with like terms aligned vertically. We can then add the polynomials column by column.

$$\begin{array}{r} 3x^2 - 3x + 2 \\ + \; 2x^2 + 7x - 4 \\ \hline 5x^2 + 4x - 2 \end{array}$$

Self Check 4
Add $4q^2 - 7$ and $2q^2 - 8q + 9$ vertically.

Answer $6q^2 - 8q + 2$

EXAMPLE 4 Add $4x^2 - 3$ and $3x^2 - 8x + 8$.

Solution Since the first polynomial does not have an x-term, we leave a space so that the constant terms can be aligned.

$$\begin{array}{r} 4x^2 - 3 \\ + \; 3x^2 - 8x + 8 \\ \hline 7x^2 - 8x + 5 \end{array}$$

Subtracting polynomials

Because of the distributive property, we can remove parentheses enclosing several terms when the sign preceding the parentheses is a − sign. We simply drop the minus sign and the parentheses, and *change the sign of every term within the parentheses*.

$$\begin{aligned} -(3x^2 + 3x - 2) &= \mathbf{-1}(3x^2 + 3x - 2) \\ &= \mathbf{-1}(3x^2) + (\mathbf{-1})(3x) + (\mathbf{-1})(-2) \\ &= -3x^2 + (-3x) + 2 \\ &= -3x^2 - 3x + 2 \end{aligned}$$

This suggests that the way to subtract polynomials is to remove parentheses, change the sign of each term of the second polynomial, and combine like terms.

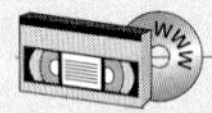

Self Check 5
Find the difference:
$(-2a^2 + 5) - (-5a^2 - 7)$

Answer $3a^2 + 12$

EXAMPLE 5 Find each difference.

a. $(3x - 4) - (5x + 7) = 3x - 4 - 5x - 7$ Change the sign of each term inside $(5x + 7)$.
$= -2x - 11$ Combine like terms.

b. $(3x^2 - 4x - 6) - (2x^2 - 6x) = 3x^2 - 4x - 6 - 2x^2 + 6x$
$= x^2 + 2x - 6$

c. $(-t^3 - 2t^2 - 1) - (-t^3 - 2t^2 + 1) = -t^3 - 2t^2 - 1 + t^3 + 2t^2 - 1$
$= -2$

To subtract polynomials in vertical form, we add the opposite of the **subtrahend** (the bottom polynomial) to the **minuend** (the top polynomial).

Self Check 6
Subtract $2p^2 + 2p - 8$ from $5p^2 - 6p + 7$.

Answer $3p^2 - 8p + 15$

EXAMPLE 6 Subtract $3x^2 - 2x$ from $2x^2 + 4x$.

Solution Since $3x^2 - 2x$ is to be subtracted from $2x^2 + 4x$, we write $3x^2 - 2x$ below $2x^2 + 4x$ in vertical form. Then we change the signs of the terms of $3x^2 - 2x$ and add:

$$\begin{array}{r} 2x^2 + 4x \\ - \; 3x^2 - 2x \\ \hline \end{array} \longrightarrow \begin{array}{r} 2x^2 + 4x \\ + \; -3x^2 + 2x \\ \hline -x^2 + 6x \end{array}$$

EXAMPLE 7 Subtract $12a - 7$ from the sum of $6a + 5$ and $4a - 10$.

Solution We will use brackets to show that $(12a - 7)$ is to be subtracted from the *sum* of $(6a + 5)$ and $(4a - 10)$.

$$[(6a + 5) + (4a - 10)] - (12a - 7)$$

Next, we remove the grouping symbols to obtain

$= 6a + 5 + 4a - 10 - 12a + 7$ Change the sign of each term in $(12a - 7)$.

$= -2a + 2$ Combine like terms.

Self Check 7
Subtract $-2q^2 - 2q$ from the sum of $q^2 - 6q$ and $3q^2 + q$.

Answer $6q^2 - 3q$

Adding and subtracting multiples of polynomials

Because of the distributive property, we can remove parentheses enclosing several terms when a monomial precedes the parentheses. We simply multiply every term within the parentheses by that monomial. For example, to add $3(2x + 5)$ and $2(4x - 3)$, we proceed as follows:

$3(2x + 5) + 2(4x - 3) = 6x + 15 + 8x - 6$ Distribute the multiplication by 3 and by 2.

$= 6x + 8x + 15 - 6$ $15 + 8x = 8x + 15$.

$= 14x + 9$ Combine like terms.

EXAMPLE 8 Use the distributive property to remove parentheses and simplify.

a. $3(x^2 + 4x) + 2(x^2 - 4) = 3x^2 + 12x + 2x^2 - 8$
$= 5x^2 + 12x - 8$

b. $-8(y^2 - 2y + 3) - 4(2y^2 + y - 6) = -8y^2 + 16y - 24 - 8y^2 - 4y + 24$
$= -16y^2 + 12y$

Self Check 8
Remove parentheses and simplify:
$2(a^2 - 3a) + 5(a^2 + 2a)$

Answer $7a^2 + 4a$

An application of adding polynomials

EXAMPLE 9 **Property values.** A house purchased for \$95,000 is expected to appreciate according to the polynomial function $f(x) = 2{,}500x + 95{,}000$, where $f(x)$ is the value of the house after x years. A second house purchased for \$125,000 is expected to appreciate according to the equation $f(x) = 4{,}500x + 125{,}000$. Find one polynomial function that will give the total value of both properties after x years.

Solution The value of the first house after x years is given by the polynomial $2{,}500x + 95{,}000$. The value of the second house after x years is given by the polynomial $4{,}500x + 125{,}000$. The value of both houses will be the sum of these two polynomials.

$$(2{,}500x + 95{,}000) + (4{,}500x + 125{,}000) = 7{,}000x + 220{,}000$$

The total value of the properties is given by the polynomial function $f(x) = 7{,}000x + 220{,}000$.

Section 10.5 STUDY SET

VOCABULARY *Fill in the blanks.*

1. The expression $(x^2 - 3x + 2) + (x^2 - 4x)$ is the sum of two __________.
2. ______ terms have the same variables and the same exponents.
3. "To add or subtract like terms" means to combine their __________ and keep the same variables with the same exponents.
4. Consider

$$\begin{array}{r} 2x^2 + 4x \\ + \quad -3x^2 + 2x \\ \hline -x^2 + 6x \end{array}$$

To add the polynomials, like terms are aligned __________.

CONCEPTS *Fill in the blanks.*

5. To add like monomials, combine like ______.
6. $a - b = a +$ ______
7. To add two polynomials, combine any _____ terms contained in the polynomials.
8. To subtract two polynomials, change the _____ of each term in the second polynomial, and combine like terms.
9. When the sign preceding parentheses is a $-$ sign, we can remove the parentheses by dropping the sign and the parentheses, and ________ the sign of every term within the parentheses.
10. When a monomial precedes parentheses, we can remove the parentheses by __________ every term within the parentheses by that monomial.
11. $-(-2x^2 - 3x + 4) =$ ______
12. $-3(-2x^2 - 3x + 4) =$ ______
13. JETS Find the polynomial representing the length of the passenger jet.

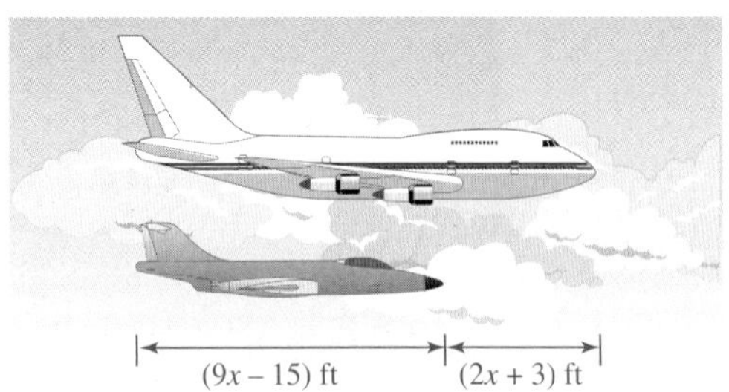

14. WATER SKIING Find the polynomial representing the distance of the water skier from the boat.

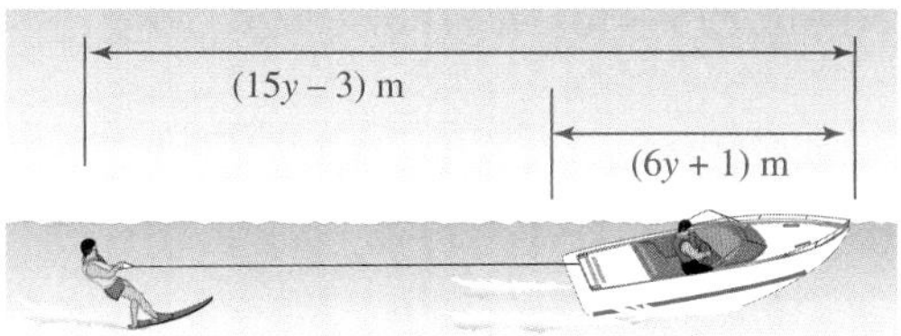

NOTATION *Complete each solution.*

15. $(5x^2 + 3x) - (7x^2 - 2x)$
$= 5x^2 + \underline{\quad} - 7x^2 + \underline{\quad}$
$= 5x^2 - \underline{\quad} + 3x + 2x$
$= -2x^2 + 5x$

16. $4(3x^2 - 2x) - (2x + 4)$
$= 12x^2 - \underline{\quad} - \underline{\quad} - 4$
$= 12x^2 - 10x - 4$

PRACTICE *Simplify each expression, if possible.*

17. $4y + 5y$
18. $-2x + 3x$
19. $8t^2 + 4t^2$
20. $15x^2 + 10x^2$
21. $-32u^3 - 16u^3$
22. $-25x^3 - 7x^3$
23. $1.8x - 1.9x$
24. $1.7y - 2.2y$
25. $\frac{1}{2}st + \frac{3}{2}st$
26. $\frac{2}{5}at + \frac{1}{5}at$
27. $3r - 4r + 7r$
28. $-2b + 7b - 3b$
29. $-4ab + 4ab - ab$
30. $xy - 4xy - 2xy$
31. $(3x)^2 - 4x^2 + 10x^2$
32. $(2x)^4 - (3x^2)^2$

Perform the operations.

33. $(3x + 7) + (4x - 3)$
34. $(2y - 3) + (4y + 7)$
35. $(4a + 3) - (2a - 4)$
36. $(5b - 7) - (3b - 5)$
37. $(2x + 3y) + (5x - 10y)$
38. $(5x - 8y) - (-2x + 5y)$
39. $(-8x - 3y) - (-11x + y)$
40. $(-4a + b) + (5a - b)$
41. $(3x^2 - 3x - 2) + (3x^2 + 4x - 3)$
42. $(3a^2 - 2a + 4) - (a^2 - 3a + 7)$
43. $(2b^2 + 3b - 5) - (2b^2 - 4b - 9)$

44. $(4c^2 + 3c - 2) + (3c^2 + 4c + 2)$

45. $(2x^2 - 3x + 1) - (4x^2 - 3x + 2) + (2x^2 + 3x + 2)$

46. $(-3z^2 - 4z + 7) + (2z^2 + 2z - 1) - (2z^2 - 3z + 7)$

47. $(4.52x^2 + 1.13x - 0.89) + (9.02x^2 - 7.68x + 7.04)$

48. $(0.891a^4 - 0.442a^2 + 0.121a) - (-0.160a^4 + 0.249a^2 + 0.789a)$

Add the polynomials.

49. $\begin{array}{r} 3x^2 + 4x + 5 \\ + \; 2x^2 - 3x + 6 \\ \hline \end{array}$

50. $\begin{array}{r} 2x^3 + 2x^2 - 3x + 5 \\ + \; 3x^3 - 4x^2 - x - 7 \\ \hline \end{array}$

51. $\begin{array}{r} 2x^3 - 3x^2 + 4x - 7 \\ + \; -9x^3 - 4x^2 - 5x + 6 \\ \hline \end{array}$

52. $\begin{array}{r} -3x^3 + 4x^2 - 4x + 9 \\ + \; 2x^3 \qquad\quad + 9x - 3 \\ \hline \end{array}$

53. $\begin{array}{r} -3x^2 + 4x + 25 \\ + \; 5x^2 \qquad - 12 \\ \hline \end{array}$

54. $\begin{array}{r} -6x^3 - 4x^2 + 7 \\ + \; -7x^3 + 9x^2 \quad \\ \hline \end{array}$

Find each difference.

55. $\begin{array}{r} 3x^2 + 4x - 5 \\ - \; -2x^2 - 2x + 3 \\ \hline \end{array}$

56. $\begin{array}{r} 3y^2 - 4y + 7 \\ - \; 6y^2 - 6y - 13 \\ \hline \end{array}$

57. $\begin{array}{r} 4x^3 + 4x^2 - 3x + 10 \\ - \; 5x^3 - 2x^2 - 4x - 4 \\ \hline \end{array}$

58. $\begin{array}{r} 3x^3 + 4x^2 + 7x + 12 \\ - \; -4x^3 + 6x^2 + 9x - 3 \\ \hline \end{array}$

59. $\begin{array}{r} -2x^2y^2 \qquad\quad + 12y^2 \\ - \; 10x^2y^2 + 9xy - 24y^2 \\ \hline \end{array}$

60. $\begin{array}{r} 25x^3 \qquad\quad + 31xz^2 \\ - \; 12x^3 + 27x^2z - 17xz^2 \\ \hline \end{array}$

61. Find the difference when $t^3 - 2t^2 + 2$ is subtracted from the sum of $3t^3 + t^2$ and $-t^3 + 6t - 3$.

62. Find the difference when $-3z^3 - 4z + 7$ is subtracted from the sum of $2z^2 + 3z - 7$ and $-4z^3 - 2z - 3$.

63. Find the sum when $3x^2 + 4x - 7$ is added to the sum of $-2x^2 - 7x + 1$ and $-4x^2 + 8x - 1$.

64. Find the difference when $32x^2 - 17x + 45$ is subtracted from the sum of $23x^2 - 12x - 7$ and $-11x^2 + 12x + 7$.

Simplify each expression.

65. $2(x + 3) + 4(x - 2)$

66. $3(y - 4) - 5(y + 3)$

67. $-2(x^2 + 7x - 1) - 3(x^2 - 2x + 2)$

68. $-5(y^2 - 2y - 6) + 6(2y^2 + 2y - 5)$

69. $2(2y^2 - 2y + 2) - 4(3y^2 - 4y - 1) + 4(y^2 - y - 1)$

70. $-4(z^2 - 5z) - 5(4z^2 - 1) + 6(2z - 3)$

71. $2(ab^2 - b) - 3(a + 2ab) + (b - a + a^2b)$

72. $3(xy + y) - 2(x - 4 + y) + 2(y^3 + y^2)$

Find the polynomial that represents the perimeter of each figure.

73.

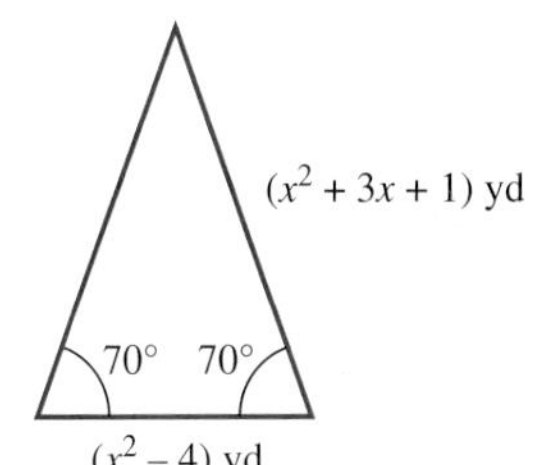

74.

$(2x^2 - 7)$ mi

$(x + 6)$ mi

$(x + 6)$ mi

45° 45°

$(5x^2 + 3x + 1)$ mi

APPLICATIONS

75. GREEK ARCHITECTURE Find the difference in the heights of the columns shown in Figure 10-3(b) at the beginning of this section.

76. CLASSICAL GREEK COLUMNS If the columns shown in Figure 10-3(b) at the beginning of this section were stacked one atop the other, to what height would they reach?

77. AUTO MECHANICS Find the polynomial representing the length of the fan belt shown. The dimensions are in inches. Your answer will involve π.

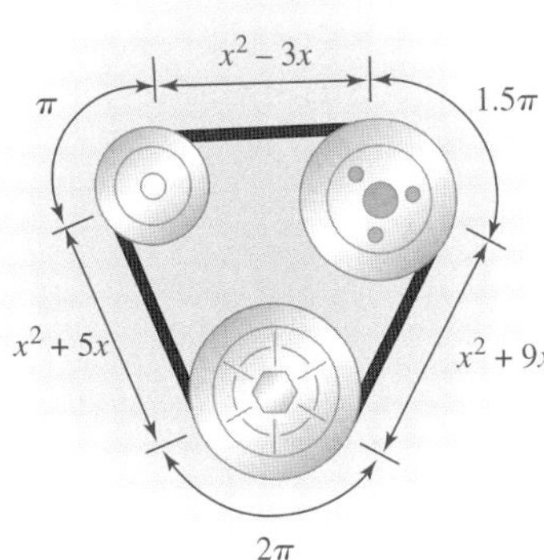

78. READING BLUEPRINTS

a. What is the difference in the length and width of the one-bedroom apartment shown below?

b. Find the perimeter of the apartment.

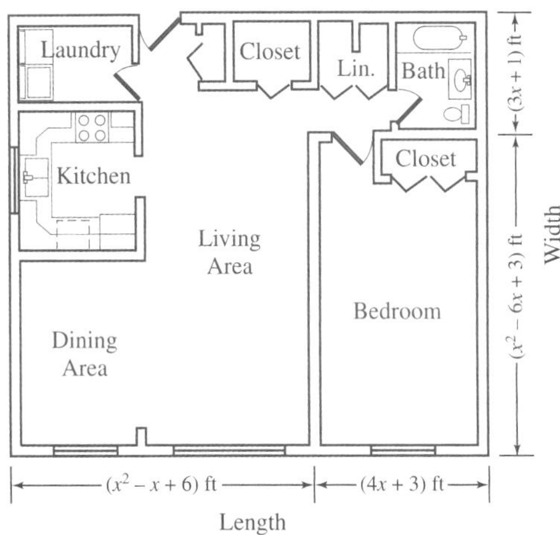

If a house is purchased for \$105,000 *and is expected to appreciate* \$900 *per year, its value y after x years is given by the polynomial function* $f(x) = 900x + 105{,}000$.

79. VALUE OF A HOUSE Find the expected value of the house in 10 years.

80. VALUE OF A HOUSE A second house is purchased for \$120,000 and is expected to appreciate \$1,000 per year.

a. Find a polynomial function that will give the value y of the house in x years.

b. Find the value of this second house after 12 years.

81. VALUE OF TWO HOUSES Find one polynomial function that will give the combined value y of both houses after x years.

82. VALUE OF TWO HOUSES Find the value of the two houses after 20 years by

a. substituting 20 into the polynomial functions $f(x) = 900x + 105{,}000$ and $f(x) = 1{,}000x + 120{,}000$ and adding.

b. substituting into the result of Exercise 81.

A business purchases two computers, one for \$6,600 *and the other for* \$9,200. *The first computer is expected to depreciate* \$1,100 *per year and the second* \$1,700 *per year.*

83. VALUE OF A COMPUTER Write a polynomial function that gives the value of the first computer after x years.

84. VALUE OF A COMPUTER Write a polynomial function that gives the value of the second computer after x years.

85. VALUE OF TWO COMPUTERS Find one polynomial function that gives the combined value of both computers after x years.

86. VALUE OF TWO COMPUTERS In two ways, find the combined value of the two computers after 3 years.

87. NAVAL OPERATIONS Two warning flares are simultaneously fired upward from different parts of a ship. The height of the first flare is $(-16t^2 + 128t + 20)$ feet and the height of the higher-traveling second flare is $(-16t^2 + 150t + 40)$ feet, after t seconds.

a. Find a polynomial that represents the difference in the heights of the flares.

b. In 4 seconds, the first flare reaches its peak, explodes, and lights up the sky. How much higher is the second flare at that time?

88. PIÑATAS Find the polynomial that represents the length of the rope used to hold up the piñata.

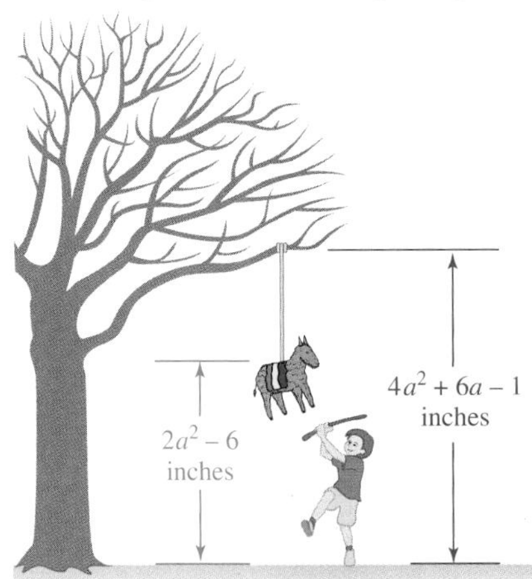

WRITING

89. How do you recognize like terms?

90. How do you add like terms?

91. Explain the concept that is illustrated by the statement

$$-(x^2 + 3x - 1) = -1(x^2 + 3x - 1)$$

92. Explain the mistake made in the solution.

Simplify: $(12x - 4) - (3x - 1)$.

$$\begin{aligned}(12x - 4) - (3x - 1) &= 12x - 4 - 3x - 1\\ &= 9x - 5\end{aligned}$$

REVIEW

93. What is the sum of the measures of the angles of a triangle?

94. What is the sum of the measures of two complementary angles?

95. Solve the inequality $-4(3x - 3) \geq -12$ and graph the solution.

96. CURLING IRONS A curling iron is plugged into a 110-volt electrical outlet and used for $\frac{1}{4}$ hour. If its resistance is 10 ohms, find the electrical power (in kilowatt hours, kwh) used by the curling iron by applying the formula

$$\text{kwh} = \frac{(\text{volts})^2}{1{,}000 \cdot \text{ohms}} \cdot \text{hours}$$

10.6 Multiplying Polynomials

- Multiplying monomials • Multiplying a polynomial by a monomial
- Multiplying a binomial by a binomial • The FOIL method • Special products
- Multiplying a polynomial by a binomial • Multiplying three polynomials
- Multiplying binomials to solve equations

In Figure 10-4(a), the length and width of a dollar bill are given. We can find the area of the bill by multiplying its length and width.

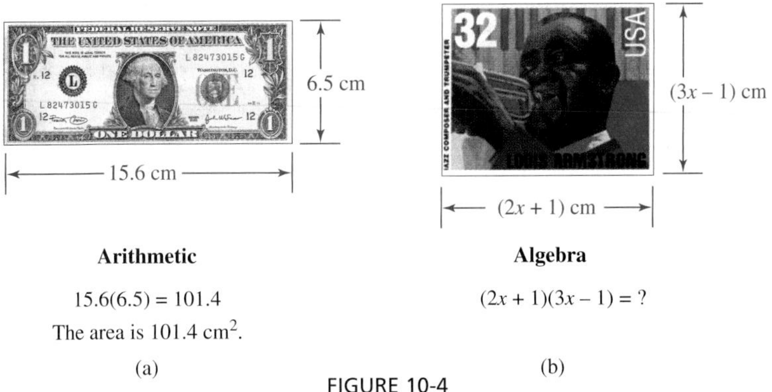

FIGURE 10-4

In Figure 10-4(b), the length and the width of a postage stamp are represented by binomials. To find the area of the stamp, we must multiply the binomials. In this section, we will discuss how to multiply polynomials. We begin with the simplest case, the product of two monomials.

Multiplying monomials

To multiply two monomials, such as $8x^2$ and $-3x^4$, we use the commutative and associative properties of multiplication to group the numerical factors and the variable factors. Then we multiply the numerical factors and multiply the variable factors.

$$\begin{aligned}8x^2(-3x^4) &= 8(-3)x^2x^4\\ &= -24x^6\end{aligned}$$

This example suggests the following rule.

Multiplying monomials

To multiply two monomials, multiply the numerical factors and then multiply the variable factors.

EXAMPLE 1 Multiply: **a.** $3x^4(2x^5)$, **b.** $-2a^2b^3(5ab^2)$, and **c.** $-4y^5z^2(2y^3z^3)(3yz)$.

Solution

a. $3x^4(2x^5) = 3(2)\,x^4x^5$

$= 6x^9$ Multiply the numerical factors, 3 and 2. Multiply the variable factors: $x^4x^5 = x^{4+5} = x^9$.

b. $-2a^2b^3(5ab^2) = -2(5)a^2ab^3b^2$

$= -10a^3b^5$

c. $-4y^5z^2(2y^3z^3)(3yz) = -4(2)(3)y^5y^3yz^2z^3z$

$= -24y^9z^6$

Self Check 1
Multiply:
a. $(5a^2b^3)(6a^3b^4)$
b. $(-15p^3q^2)(5p^3q^2)$

Answers **a.** $30a^5b^7$, **b.** $-75p^6q^4$

Multiplying a polynomial by a monomial

To find the product of a polynomial (with more than one term) and a monomial, we use the distributive property. To multiply $2x + 4$ by $5x$, for example, we proceed as follows:

$\mathbf{5x}(2x + 4) = \mathbf{5x}(2x) + \mathbf{5x}(4)$ Distribute the multiplication by $5x$.

$= 10x^2 + 20x$ Multiply the monomials: $5x(2x) = 10x^2$ and $5x(4) = 20x$.

This example suggests the following rule.

Multiplying polynomials by monomials

To multiply a polynomial with more than one term by a monomial, use the distributive property to remove parentheses and simplify.

EXAMPLE 2 Multiply: **a.** $3a^2(3a^2 - 5a)$ and **b.** $-2xz^2(2x - 3z + 2z^2)$.

Solution

a. $\mathbf{3a^2}(3a^2 - 5a) = \mathbf{3a^2}(3a^2) - \mathbf{3a^2}(5a)$ Distribute the multiplication by $3a^2$.

$= 9a^4 - 15a^3$ Multiply the monomials.

b. $\mathbf{-2xz^2}(2x - 3z + 2z^2)$

$= \mathbf{-2xz^2}(2x) - (\mathbf{-2xz^2})(3z) + (\mathbf{-2xz^2})(2z^2)$ Use the distributive property.

$= -4x^2z^2 - (-6xz^3) + (-4xz^4)$ Multiply the monomials.

$= -4x^2z^2 + 6xz^3 - 4xz^4$

Self Check 2
Multiply:
a. $2p^3(3p^2 - 5p)$
b. $-5a^2b(3a + 2b - 4ab)$

Answers **a.** $6p^5 - 10p^4$,
b. $-15a^3b - 10a^2b^2 + 20a^3b^2$

Multiplying a binomial by a binomial

To multiply two binomials, we must use the distributive property more than once. For example, to multiply $2a - 4$ by $3a + 5$, we proceed as follows.

$(\mathbf{2a - 4})(3a + 5) = (\mathbf{2a - 4})(3a) + (\mathbf{2a - 4})(5)$	Distribute the multiplication by $(2a - 4)$.
$= \mathbf{3a}(2a - 4) + \mathbf{5}(2a - 4)$	Use the commutative property of multiplication.
$= \mathbf{3a}(2a) - \mathbf{3a}(4) + \mathbf{5}(2a) - \mathbf{5}(4)$	Distribute the multiplication by $3a$ and by 5.
$= 6a^2 - 12a + 10a - 20$	Perform the multiplications.
$= 6a^2 - 2a - 20$	Combine like terms.

This example suggests the following rule.

> **Multiplying two binomials**
>
> To multiply two binomials, multiply each term of one binomial by each term of the other binomial and combine like terms.

The FOIL method

We can use a shortcut method, called the **FOIL** method, to multiply binomials. FOIL is an acronym for **F**irst terms, **O**uter terms, **I**nner terms, and **L**ast terms. To use the FOIL method to multiply $2a - 4$ by $3a + 5$, we

1. multiply the **F**irst terms $2a$ and $3a$ to obtain $6a^2$,
2. multiply the **O**uter terms $2a$ and 5 to obtain $10a$,
3. multiply the **I**nner terms -4 and $3a$ to obtain $-12a$, and
4. multiply the **L**ast terms -4 and 5 to obtain -20.

Then we simplify the resulting polynomial, if possible.

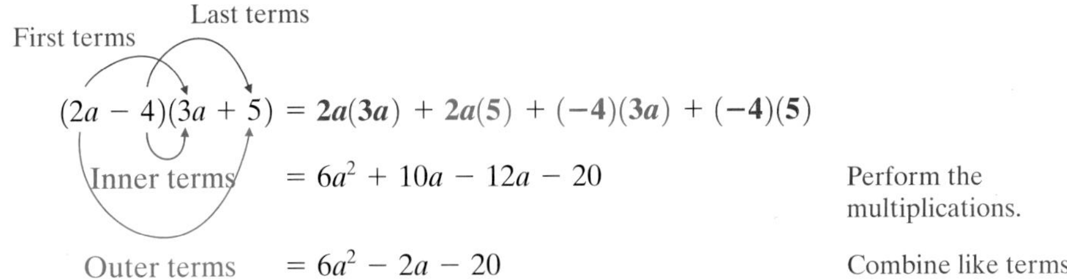

$(2a - 4)(3a + 5) = \mathbf{2a(3a)} + \mathbf{2a(5)} + \mathbf{(-4)(3a)} + \mathbf{(-4)(5)}$

$= 6a^2 + 10a - 12a - 20$ Perform the multiplications.

$= 6a^2 - 2a - 20$ Combine like terms.

EXAMPLE 3 Find each product.

F L

a. $(x + 5)(x + 7) = \mathbf{x(x)} + \mathbf{x(7)} + \mathbf{5(x)} + \mathbf{5(7)}$

I

$= x^2 + 7x + 5x + 35$

O

$= x^2 + 12x + 35$

Self Check 3

Find each product:

a. $(y + 3)(y + 1)$

b. $(2a - 1)(3a + 2)$

c. $(5y - 2z)(2y + 3z)$

b. $(3x + 4)(2x - 3) = \mathbf{3x(2x) + 3x(-3) + 4(2x) + 4(-3)}$

$= 6x^2 - 9x + 8x - 12$

$= 6x^2 - x - 12$

c. $(a - 7b)(a - 4b) = \mathbf{a(a) + a(-4b) + (-7b)(a) + (-7b)(-4b)}$

$= a^2 - 4ab - 7ab + 28b^2$

$= a^2 - 11ab + 28b^2$

d. $(2r - 3s)(2r + t) = \mathbf{2r(2r) + 2r(t) - 3s(2r) - 3s(t)}$

$= 4r^2 + 2rt - 6rs - 3st$ There are no like terms.

Answers **a.** $y^2 + 4y + 3$, **b.** $6a^2 + a - 2$, **c.** $10y^2 + 11yz - 6z^2$

Self Check 4
Simplify:
$(x + 3)(2x - 1) + 2x(x - 1)$.

Answer $4x^2 + 3x - 3$

EXAMPLE 4 Simplify each expression.

a. $3(2x - 3)(x + 1) = 3(2x^2 + 2x - 3x - 3)$ Multiply the binomials.

$= 3(2x^2 - x - 3)$ Combine like terms.

$= 6x^2 - 3x - 9$ Distribute the multiplication by 3.

b. $(x + 1)(x - 2) - 3x(x + 3) = x^2 - 2x + x - 2 - 3x^2 - 9x$

$= -2x^2 - 10x - 2$ Combine like terms.

Special products

Certain products of binomials occur so frequently in algebra that it is worthwhile to learn formulas for computing them. To develop a rule to find the *square of a sum,* we consider $(x + y)^2$.

$(x + y)^2 = (x + y)(x + y)$ In $(x + y)^2$, the base is $(x + y)$ and the exponent is 2.

$= x^2 + xy + xy + y^2$ Multiply the binomials.

$= x^2 + 2xy + y^2$ Combine like terms: $xy + xy = 2xy$.

We note that the terms of this result are related to the terms of the original expression. That is, $(x + y)^2$ is equal to the square of its first term (x^2), plus twice the product of both its terms ($2xy$), plus the square of its last term (y^2).

To develop a rule to find the *square of a difference,* we consider $(x - y)^2$.

$(x - y)^2 = (x - y)(x - y)$

$= x^2 - xy - xy + y^2$ Multiply the binomials.

$= x^2 - 2xy + y^2$ Combine like terms: $-xy - xy = -2xy$.

Again, the terms of the result are related to the terms of the original expression. When we find $(x - y)^2$, the product is composed of the square of its first term (x^2), twice the product of both its terms $(-2xy)$, and the square of its last term (y^2).

The final special product is the product of two binomials that differ only in the signs of the last terms. To develop a rule to find the product of a *sum and a difference,* we consider $(x + y)(x - y)$.

$$(x + y)(x - y) = x^2 - xy + xy - y^2 \quad \text{Multiply the binomials.}$$
$$= x^2 - y^2 \quad \text{Combine like terms: } -xy + xy = 0.$$

The product is the square of the first term (x^2) minus the square of the second term (y^2). The expression $x^2 - y^2$ is called a **difference of two squares.**

Because these special products occur so often, it is wise to memorize their forms.

Special products

$(x + y)^2 = x^2 + 2xy + y^2$ The square of a sum

$(x - y)^2 = x^2 - 2xy + y^2$ The square of a difference

$(x + y)(x - y) = x^2 - y^2$ The product of a sum and difference

EXAMPLE 5 Find **a.** $(t + 9)^2$, **b.** $(8a - 5)^2$, and **c.** $(3y + 4z)(3y - 4z)$.

Solution

a. This is the square of a sum. The terms of the binomial being squared are t and 9.

$$(t + 9)^2 = \underbrace{t^2}_{\text{The square of the first term, } t.} + \underbrace{2(t)(9)}_{\text{Twice the product of both terms.}} + \underbrace{9^2}_{\text{The square of the last term, 9.}}$$

$$= t^2 + 18t + 81$$

b. This is the square of a difference. The terms of the binomial being squared are $8a$ and -5.

$$(8a - 5)^2 = \underbrace{(8a)^2}_{\text{The square of the first term, } 8a.} - \underbrace{2(8a)(5)}_{\text{Twice the product of both terms.}} + \underbrace{5^2}_{\text{The square of the last term, 5.}}$$

$$= 64a^2 - 80a + 25 \quad \text{Use the power rule for products: } (8a)^2 = 8^2a^2 = 64a^2.$$

c. The binomials differ only in the signs of the last terms. This is the product of a sum and a difference.

$$(3y + 4z)(3y - 4z) = (3y)^2 - (4z)^2 \quad \text{This is the square of the first term minus the square of the second term.}$$
$$= 9y^2 - 16z^2 \quad \text{Use the power rule for products twice.}$$

Self Check 5

Find:

a. $(r + 6)^2$

b. $(7g - 2)^2$

c. $(5m - 9n)(5m + 9n)$

Answers **a.** $r^2 + 12r + 36$, **b.** $49g^2 - 28g + 4$, **c.** $25m^2 - 81n^2$

! COMMENT A common error when squaring a binomial is to forget the middle term of the product. For example, $(x + 2)^2 \neq x^2 + 4$ and $(x - 2)^2 \neq x^2 - 4$. Applying the special product formulas, we have $(x + 2)^2 = x^2 + \mathbf{4x} + 4$ and $(x - 2)^2 = x^2 - \mathbf{4x} + 4$.

Multiplying a polynomial by a binomial

We must use the distributive property more than once to multiply a polynomial by a binomial. For example, to multiply $3x^2 + 3x - 5$ by $2x + 3$, we proceed as follows:

$$\begin{aligned}(\mathbf{2x+3})(3x^2+3x-5) &= (\mathbf{2x+3})3x^2 + (\mathbf{2x+3})3x - (\mathbf{2x+3})5 \\ &= 3x^2(2x+3) + 3x(2x+3) - 5(2x+3) \\ &= 6x^3 + 9x^2 + 6x^2 + 9x - 10x - 15 \\ &= 6x^3 + 15x^2 - x - 15\end{aligned}$$

This example suggests the following rule.

Multiplying polynomials

To multiply one polynomial by another, multiply each term of one polynomial by each term of the other polynomial and combine like terms.

It is often convenient to organize the work vertically.

Self Check 6

Multiply:

a. $(3x + 2)(2x^2 - 4x + 5)$

b. $(-2x^2 + 3)(2x^2 - 4x - 1)$

Answers

a. $6x^3 - 8x^2 + 7x + 10$,

b. $-4x^4 + 8x^3 + 8x^2 - 12x - 3$

EXAMPLE 6

a. Multiply:

$$\begin{array}{rrrrl} & 3a^2 & -\ 4a & +\ 7 & \\ & & 2a & +\ 5 & \\ \hline & 15a^2 & -\ 20a & +\ 35 & \text{Multiply } 3a^2 - 4a + 7 \text{ by } 5. \\ 6a^3 & -\ 8a^2 & +\ 14a & & \text{Multiply } 3a^2 - 4a + 7 \text{ by } 2a. \\ \hline 6a^3 & +\ 7a^2 & -\ 6a & +\ 35 & \text{In each column, combine like terms.} \end{array}$$

b. Multiply:

$$\begin{array}{rrrrrl} & & 3y^2 & -\ 5y & +\ 4 & \\ & & -4y^2 & & -\ 3 & \\ \hline & & -9y^2 & +\ 15y & -\ 12 & \text{Multiply } 3y^2 - 5y + 4 \text{ by } -3. \\ -12y^4 & +\ 20y^3 & -\ 16y^2 & & & \text{Multiply } 3y^2 - 5y + 4 \text{ by } -4y^2. \\ \hline -12y^4 & +\ 20y^3 & -\ 25y^2 & +\ 15y & -\ 12 & \end{array}$$

Multiplying three polynomials

When finding the product of three polynomials, we begin by multiplying any two of them, and then we multiply that result by the third polynomial.

Self Check 7

Find the product:

$-2y(y + 3)(3y - 2)$

Answer $-6y^3 - 14y^2 + 12y$

EXAMPLE 7 Find the product: $-3a(4a + 1)(a - 7)$.

Solution First we find the product of $4a + 1$ and $a - 7$. Then we multiply that result by $-3a$.

$$\begin{aligned}-3a(\mathbf{4a+1})(\mathbf{a-7}) &= -3a(\mathbf{4a^2 - 28a + a - 7}) && \text{Multiply the two binomials.} \\ &= -3a(4a^2 - 27a - 7) && \text{Combine like terms.} \\ &= -12a^3 + 81a^2 + 21a && \text{Distribute the multiplication by } -3a.\end{aligned}$$

Multiplying binomials to solve equations

To solve an equation such as $(x + 2)(x + 3) = x(x + 7)$, we can do the multiplication on each side and proceed as follows:

$$(x + 2)(x + 3) = x(x + 7)$$

$$x^2 + 3x + 2x + 6 = x^2 + 7x$$

$$x^2 + 3x + 2x + 6 - \mathbf{x^2} = x^2 + 7x - \mathbf{x^2} \quad \text{Subtract } x^2 \text{ from both sides.}$$

$$5x + 6 = 7x \quad \text{Combine like terms: } x^2 - x^2 = 0 \text{ and } 3x + 2x = 5x.$$

$$6 = 2x \quad \text{Subtract } 5x \text{ from both sides.}$$

$$3 = x \quad \text{Divide both sides by 2.}$$

Check: $(x + 2)(x + 3) = x(x + 7)$

$$(\mathbf{3} + 2)(\mathbf{3} + 3) \stackrel{?}{=} \mathbf{3}(\mathbf{3} + 7) \quad \text{Replace } x \text{ with 3.}$$

$$5(6) \stackrel{?}{=} 3(10) \quad \text{Perform the additions within parentheses.}$$

$$30 = 30$$

EXAMPLE 8 **Dimensions of a painting.** A square painting is surrounded by a border 2 inches wide. If the area of the border is 96 square inches, find the dimensions of the painting.

Analyze the problem

Refer to Figure 10-5, which shows a square painting surrounded by a border 2 inches wide. We know that the area of this border is 96 square inches, and we are to find the dimensions of the painting.

Form an equation

Let $x =$ the length of a side of the square painting. Since the border is 2 inches wide, the length and the width of the outer rectangle are both $(x + 2 + 2)$ inches. Then the outer rectangle is also a square, and its dimensions are $(x + 4)$ by $(x + 4)$ inches. Since the area of a square is the product of its length and width, the area of the larger square is $(x + 4)(x + 4)$, and the area of the painting is $x \cdot x$. If we subtract the area of the painting from the area of the larger square, the difference is 96.

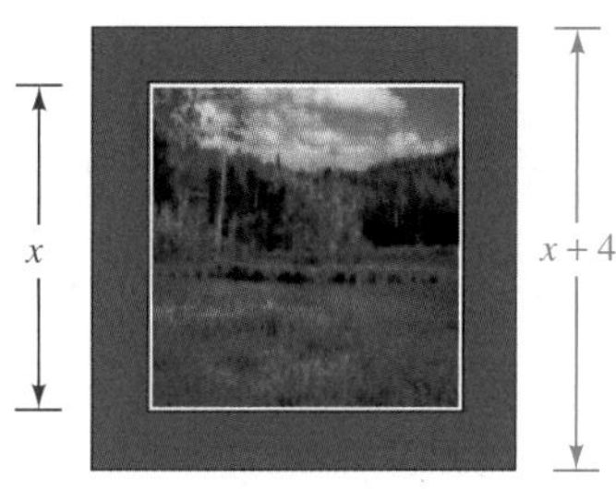

FIGURE 10-5

The area of the large square	minus	the area of the square painting	is	the area of the border.
$(x + 4)(x + 4)$	$-$	$x \cdot x$	$=$	96

Solve the equation

$(x + 4)(x + 4) - x^2 = 96$	$x \cdot x = x^2$.
$x^2 + 8x + 16 - x^2 = 96$	$(x + 4)(x + 4) = (x + 4)^2 = x^2 + 8x + 16$.
$8x + 16 = 96$	Combine like terms: $x^2 - x^2 = 0$.
$8x = 80$	Subtract 16 from both sides.
$x = 10$	Divide both sides by 8.

State the conclusion

The dimensions of the painting are 10 inches by 10 inches.

Check the result

Verify that the 2-inch-wide border of a 10-inch-square painting would have an area of 96 square inches.

Section 10.6 STUDY SET

VOCABULARY *Fill in the blanks.*

1. The expression $(2a - 4)(3a + 5)$ is the product of two ________.
2. The expression $(2a - 4)(3a^2 + 5a - 1)$ is the product of a ________ and a ________.
3. In the acronym FOIL, F stands for ______ terms, O for ______ terms, I for ______ terms, and L for ______ terms.
4. $(x + 5)^2$ is the square of a ______, and $(x - 5)^2$ is the square of a ________.

CONCEPTS *Consider the product* $(2x + 5)(3x - 4)$.

5. The product of the first terms is ▢.
6. The product of the outer terms is ▢.
7. The product of the inner terms is ▢.
8. The product of the last terms is ▢.
9. STAMPS Find the area of the stamp shown in Figure 10-4(b) at the beginning of this section.
10. LUGGAGE Find the volume of the garment bag shown below.

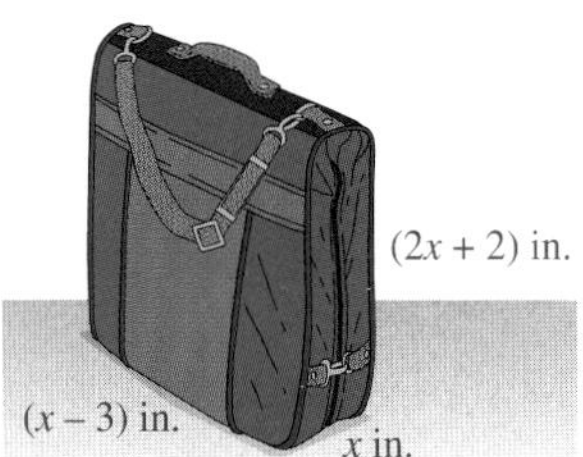

NOTATION *Complete each solution.*

11. $7x(3x^2 - 2x + 5) = \square(3x^2) - \square(2x) + \square(5)$
$= 21x^3 - 14x^2 + 35x$
12. $(2x + 5)(3x - 2) = 2x(3x) - \square(2) + \square(3x) - \square(2)$
$= 6x^2 - \square + \square - 10$
$= 6x^2 + 11x - 10$

PRACTICE *Find each product.*

13. $(3x^2)(4x^3)$
14. $(-2a^3)(3a^2)$
15. $(3b^2)(-2b)(4b^3)$
16. $(3y)(2y^2)(-y^4)$
17. $(2x^2y^3)(3x^3y^2)$
18. $(-5x^3y^6)(x^2y^2)$
19. $(x^2y^5)(x^2z^5)(-3z^3)$
20. $(-r^4st^2)(2r^2st)(rst)$
21. $3(x + 4)$
22. $-3(a - 2)$
23. $-4(t + 7)$
24. $6(s^2 - 3)$
25. $3x(x - 2)$
26. $4y(y + 5)$
27. $-2x^2(3x^2 - x)$
28. $4b^3(2b^2 - 2b)$
29. $3xy(x + y)$
30. $-4x^2z(3x^2 - z)$
31. $2x^2(3x^2 + 4x - 7)$
32. $3y^3(2y^2 - 7y - 8)$
33. $(3x)(-2x^2)(x + 4)$
34. $(-2a^2)(-3a^3)(3a - 2)$
35. $(a + 4)(a + 5)$
36. $(y - 3)(y + 5)$
37. $(3x - 2)(x + 4)$
38. $(t + 4)(2t - 3)$
39. $(2a + 4)(3a - 5)$
40. $(2b - 1)(3b + 4)$

41. $(3x - 5)(2x + 1)$
42. $(2y - 5)(3y + 7)$
43. $(x + 3)(2x - 3)$
44. $(2x + 3)(2x - 5)$
45. $(2t + 3s)(3t - s)$
46. $(3a - 2b)(4a + b)$
47. $(x + y)(x + z)$
48. $(a - b)(x + y)$
49. $(4t - u)(-3t + u)$
50. $(-3t + 2s)(2t - 3s)$

Simplify each expression.

51. $4(2x + 1)(x - 2)$
52. $-5(3a - 2)(2a + 3)$
53. $3a(a + b)(a - b)$
54. $-2r(r + s)(r + s)$
55. $2t(t + 2) + 3t(t - 5)$
56. $3y(y + 2) + (y + 1)(y - 1)$
57. $(x + y)(x - y) + x(x + y)$
58. $(3x + 4)(2x - 2) - (2x + 1)(x + 3)$

Find each special product.

59. $(x + 4)(x + 4)$
60. $(a + 3)(a + 3)$
61. $(t - 3)(t - 3)$
62. $(z - 5)(z - 5)$
63. $(r + 4)(r - 4)$
64. $(b + 2)(b - 2)$
65. $(4x + 5)(4x - 5)$
66. $(5z + 1)(5z - 1)$
67. $(2s + 1)(2s + 1)$
68. $(3t - 2)(3t - 2)$
69. $(x + 5)^2$
70. $(y - 6)^2$
71. $(x - 2y)^2$
72. $(3a + 2b)^2$
73. $(2a - 3b)^2$
74. $(2x + 5y)^2$
75. $(4x + 5y)^2$
76. $(6p - 5q)^2$

Find the area of each figure. You may leave π in your answer.

77.

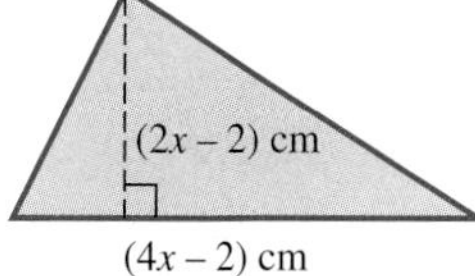

78.

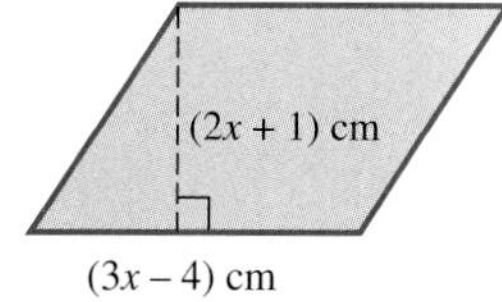

79.

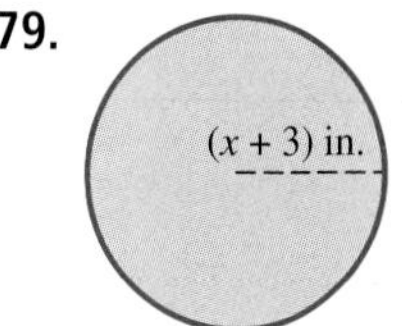

80.

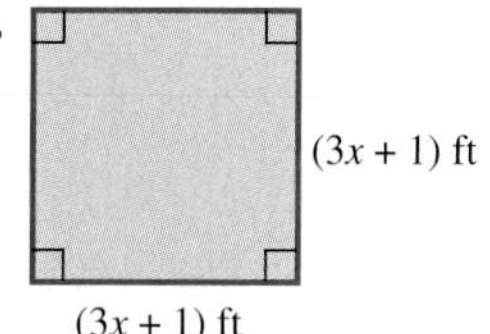

Find each product.

81. $(x + 2)(x^2 - 2x + 3)$
82. $(x - 5)(x^2 + 2x - 3)$
83. $(4t + 3)(t^2 + 2t + 3)$
84. $(3x + 1)(2x^2 - 3x + 1)$
85. $(-3x + y)(x^2 - 8xy + 16y^2)$
86. $(3x - y)(x^2 + 3xy - y^2)$

87. $\begin{array}{r} x^2 - 2x + 1 \\ x + 2 \\ \hline \end{array}$

88. $\begin{array}{r} 5r^2 + r + 6 \\ 2r - 1 \\ \hline \end{array}$

89. $\begin{array}{r} 4x^2 + 3x - 4 \\ 3x + 2 \\ \hline \end{array}$

90. $\begin{array}{r} x^2 - x + 1 \\ x + 1 \\ \hline \end{array}$

Solve each equation.

91. $(s - 4)(s + 1) = s^2 + 5$
92. $(y - 5)(y - 2) = y^2 - 4$
93. $z(z + 2) = (z + 4)(z - 4)$
94. $(z + 3)(z - 3) = z(z - 3)$
95. $(x + 4)(x - 4) = (x - 2)(x + 6)$
96. $(y - 1)(y + 6) = (y - 3)(y - 2) + 8$
97. $(a - 3)^2 = (a + 3)^2$
98. $(b + 2)^2 = (b - 1)^2$

APPLICATIONS

99. TOYS Find the perimeter and the area of the screen of the Etch A Sketch®.

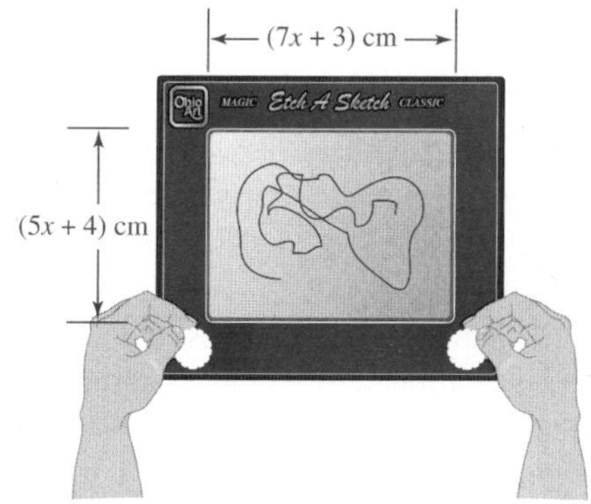

100. SUNGLASSES An ellipse is an oval-shaped closed curve. The area of an ellipse is approximately $3.14ab$, where a is its length and b is its width. Find the polynomial that approximates the total area of the elliptical-shaped lenses of the sunglasses shown.

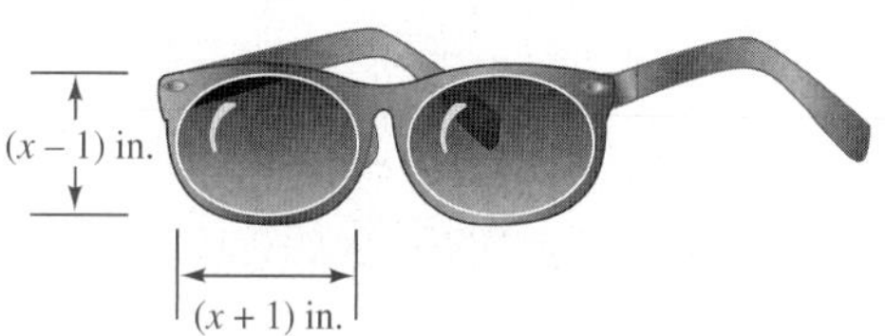

101. GARDENING See the following illustration.

a. What is the area of the region planted with corn? tomatoes? beans? carrots? Use your answers to find the total area of the garden.

b. What is the length of the garden? What is its width? Use your answers to find its area.

c. How do the answers from parts a and b for the area of the garden compare?

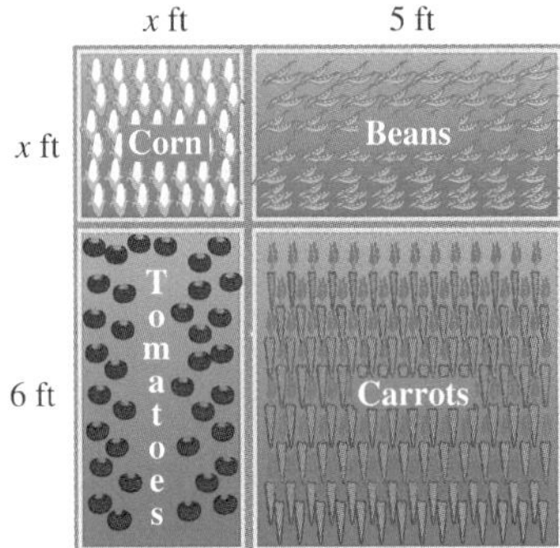

102. PAINTING See the illustration. To purchase the correct amount of enamel to paint these two garage doors, a painter must find their areas. Find a polynomial that gives the number of square feet to be painted. All dimensions are in feet, and the windows are squares with sides of x feet.

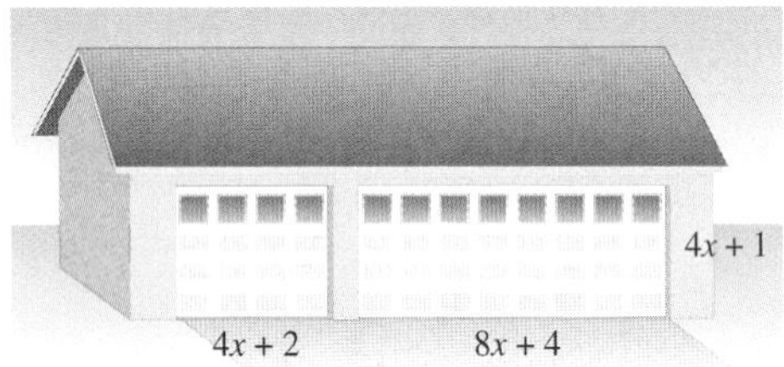

103. INTEGER PROBLEM The difference between the squares of two consecutive positive integers is 11. Find the integers. (*Hint:* Let x and $x + 1$ represent the consecutive integers.)

104. INTEGER PROBLEM If 3 less than a certain integer is multiplied by 4 more than the integer, the product is 6 less than the square of the integer. Find the integer.

105. STONE-GROUND FLOUR The radius of a millstone is 3 meters greater than the radius of another, and their areas differ by 15π square meters. Find the radius of the larger millstone.

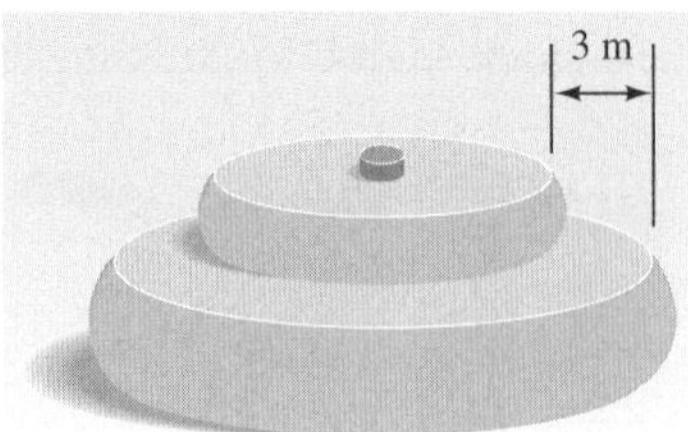

106. BOOKBINDING Two square sheets of cardboard used for making book covers differ in area by 44 square inches. An edge of the larger square is 2 inches greater than an edge of the smaller square. Find the length of an edge of the smaller square.

107. BASEBALL In major league baseball, the distance between bases is 30 feet greater than it is in softball. The bases in major league baseball mark the corners of a square that has an area 4,500 square feet greater than for softball. Find the distance between the bases in baseball.

108. PULLEY DESIGNS The radius of one pulley in the illustration is 1 inch greater than the radius of the second pulley, and their areas differ by 4π square inches. Find the radius of the smaller pulley.

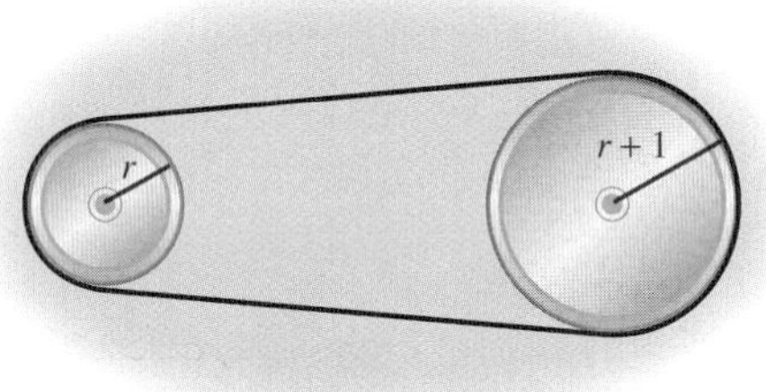

109. SIGNS Find a polynomial that represents the area of the sign.

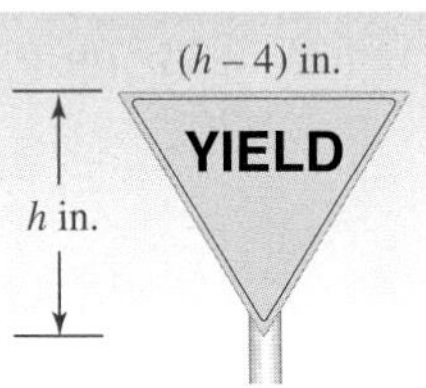

110. BASEBALL Find a polynomial that represents the amount of space within the batting cage.

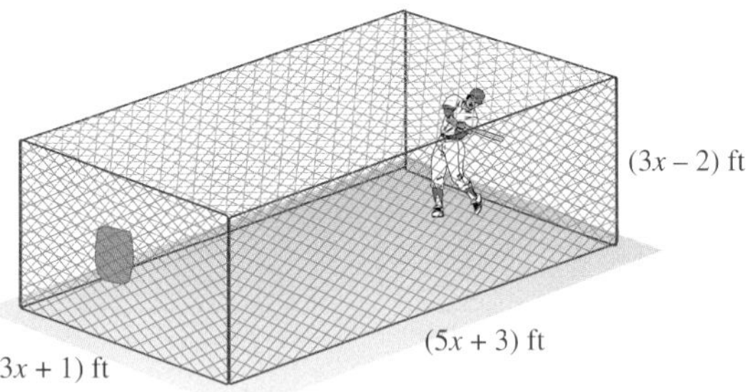

WRITING

111. Describe the steps involved in finding the product of $x + 2$ and $x - 2$.

112. Writing $(x + y)^2$ as $x^2 + y^2$ illustrates a common error. Explain.

REVIEW *Refer to the illustration.*

113. What is the slope of line AB?

114. What is the slope of line BC?

115. What is the slope of line CD?

116. What is the slope of the x-axis?

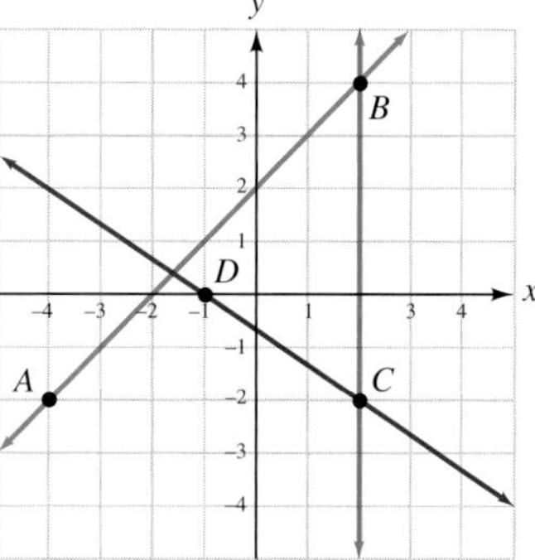

117. What is the y-intercept of line AB?

118. What is the x-intercept of line AB?

10.7 Dividing Polynomials by Monomials

- Dividing a monomial by a monomial
- Dividing a polynomial by a monomial
- An application of dividing a polynomial by a monomial

In this section, we will discuss how to divide polynomials by monomials. We will first divide monomials by monomials and then divide polynomials with more than one term by monomials.

Dividing a monomial by a monomial

Recall that to simplify a fraction, we write both its numerator and denominator as the product of several factors and then divide out all common factors:

$$\frac{4}{6} = \frac{2 \cdot 2}{2 \cdot 3} \quad \text{Factor 4 and 6.}$$

$$= \frac{\overset{1}{\cancel{2}} \cdot 2}{\underset{1}{\cancel{2}} \cdot 3} \quad \text{Divide out the common factor of 2.}$$

$$= \frac{2}{3}$$

$$\frac{20}{25} = \frac{4 \cdot 5}{5 \cdot 5} \quad \text{Factor 20 and 25.}$$

$$= \frac{4 \cdot \overset{1}{\cancel{5}}}{\underset{1}{\cancel{5}} \cdot 5} \quad \text{Divide out the common factor of 5.}$$

$$= \frac{4}{5}$$

We can use the same method to simplify algebraic fractions that contain variables.

$$\frac{3p^2}{6p} = \frac{3 \cdot p \cdot p}{2 \cdot 3 \cdot p} \quad \text{Factor } p^2 \text{ and 6.}$$

$$= \frac{\overset{1}{\cancel{3}} \cdot \overset{1}{\cancel{p}} \cdot p}{2 \cdot \underset{1}{\cancel{3}} \cdot \underset{1}{\cancel{p}}} \quad \text{Divide out the common factors of 3 and } p.$$

$$= \frac{p}{2}$$

To divide monomials, we can use either the preceding method for simplifying arithmetic fractions or the rules for exponents.

Self Check 1

Simplify: $\frac{-5p^2q^3}{10pq^4}$.

Answer $-\frac{p}{2q}$

EXAMPLE 1 Simplify: **a.** $\frac{x^2y}{xy^2}$ and **b.** $\frac{-8a^3b^2}{4ab^3}$.

Solution

By simplifying fractions

a. $\frac{x^2y}{xy^2} = \frac{x \cdot x \cdot y}{x \cdot y \cdot y}$

$= \frac{\overset{1}{\cancel{x}} \cdot x \cdot \overset{1}{\cancel{y}}}{\underset{1}{\cancel{x}} \cdot y \cdot \underset{1}{\cancel{y}}}$

$= \frac{x}{y}$

b. $\frac{-8a^3b^2}{4ab^3} = \frac{-2 \cdot 4 \cdot a \cdot a \cdot a \cdot b \cdot b}{4 \cdot a \cdot b \cdot b \cdot b}$

$= \frac{-2 \cdot \overset{1}{\cancel{4}} \cdot \overset{1}{\cancel{a}} \cdot a \cdot a \cdot \overset{1}{\cancel{b}} \cdot \overset{1}{\cancel{b}}}{\underset{1}{\cancel{4}} \cdot \underset{1}{\cancel{a}} \cdot \underset{1}{\cancel{b}} \cdot \underset{1}{\cancel{b}} \cdot b}$

$= -\frac{2a^2}{b}$

Using the rules for exponents

$\frac{x^2y}{xy^2} = x^{2-1}y^{1-2}$

$= x^1y^{-1}$

$= \frac{x}{y}$

$\frac{-8a^3b^2}{4ab^3} = \frac{-2^3a^3b^2}{2^2ab^3}$

$= -2^{3-2}a^{3-1}b^{2-3}$

$= -2^1a^2b^{-1}$

$= -\frac{2a^2}{b}$

Self Check 2

Simplify: $\frac{-24(h^3p)^5}{20(h^2p^2)^3}$.

Answer $-\frac{6h^9}{5p}$

EXAMPLE 2 Simplify: $\frac{25(s^2t^3)^2}{15(st^3)^3}$. Write the result using positive exponents only.

Solution To divide these monomials, we will use the method for simplifying fractions and several rules for exponents.

$\frac{25(s^2t^3)^2}{15(st^3)^3} = \frac{25s^4t^6}{15s^3t^9}$ Use the power rules for exponents: $(xy)^n = x^ny^n$ and $(x^m)^n = x^{mn}$.

$= \frac{5 \cdot 5 \cdot s^{4-3}t^{6-9}}{5 \cdot 3}$ Factor 25 and 15. Use the quotient rule for exponents: $\frac{x^m}{x^n} = x^{m-n}$.

$= \frac{5 \cdot \overset{1}{\cancel{5}} \cdot s^1t^{-3}}{\underset{1}{\cancel{5}} \cdot 3}$ Divide out the common factors of 5. Perform the subtractions.

$= \frac{5s}{3t^3}$ Use the negative integer exponent rule: $t^{-3} = \frac{1}{t^3}$.

Dividing a polynomial by a monomial

We have discussed the following rules to add and subtract fractions with like denominators.

Adding and subtracting fractions with like denominators

To add (or subtract) fractions with like denominators, we add (or subtract) their numerators and keep the common denominator. In symbols, if a, b, and d represent numbers, and d is not 0,

$$\frac{a}{d} + \frac{b}{d} = \frac{a+b}{d} \quad \text{and} \quad \frac{a}{d} - \frac{b}{d} = \frac{a-b}{d}$$

We can use this rule in reverse to divide polynomials by monomials.

EXAMPLE 3 Divide $9x + 6$ by 3.

Solution

$$\frac{9x + 6}{3} = \frac{9x}{3} + \frac{6}{3}$$ Divide each term of the numerator by the denominator.

$$= 3x + 2$$ Simplify each fraction.

Self Check 3

Simplify: $\frac{4 - 8b}{4}$.

Answer $1 - 2b$

EXAMPLE 4 Divide: $\frac{6x^2y^2 + 4x^2y - 2xy}{2xy}$.

Solution

$$\frac{6x^2y^2 + 4x^2y - 2xy}{2xy}$$

$$= \frac{6x^2y^2}{2xy} + \frac{4x^2y}{2xy} - \frac{2xy}{2xy}$$ Divide each term of the numerator by the denominator.

$$= 3xy + 2x - 1$$ Simplify each fraction.

Self Check 4

Simplify: $\frac{9a^2b - 6ab^2 + 3ab}{3ab}$.

Answer $3a - 2b + 1$

EXAMPLE 5 Divide: $\frac{12a^3b^2 - 4a^2b + a}{6a^2b^2}$.

Solution

$$\frac{12a^3b^2 - 4a^2b + a}{6a^2b^2}$$

$$= \frac{12a^3b^2}{6a^2b^2} - \frac{4a^2b}{6a^2b^2} + \frac{a}{6a^2b^2}$$ Divide each term of the numerator by the denominator.

$$= 2a - \frac{2}{3b} + \frac{1}{6ab^2}$$ Simplify each fraction.

Self Check 5

Simplify: $\frac{14p^3q + pq^2 - p}{7p^2q}$.

Answer $2p + \frac{q}{7p} - \frac{1}{7pq}$

EXAMPLE 6 Simplify: $\frac{(x - y)^2 - (x + y)^2}{xy}$.

Solution

$$\frac{(x - y)^2 - (x + y)^2}{xy}$$

$$= \frac{x^2 - 2xy + y^2 - (x^2 + 2xy + y^2)}{xy}$$ Use the special product rules to square the binomials in the numerator.

$$= \frac{x^2 - 2xy + y^2 - x^2 - 2xy - y^2}{xy}$$ Change the sign of each term within $(x^2 + 2xy + y^2)$.

$$= \frac{-4xy}{xy}$$ Combine like terms.

$$= -4$$ Divide out the common factors of x and y.

Self Check 6

Simplify: $\frac{(x + y)^2 - (x - y)^2}{xy}$.

Answer 4

An application of dividing a polynomial by a monomial

The area of the trapezoid shown in Figure 10-6 is given by the formula $A = \frac{1}{2}h(B + b)$, where B and b are its bases and h is its height. To solve the formula for b, we proceed as follows.

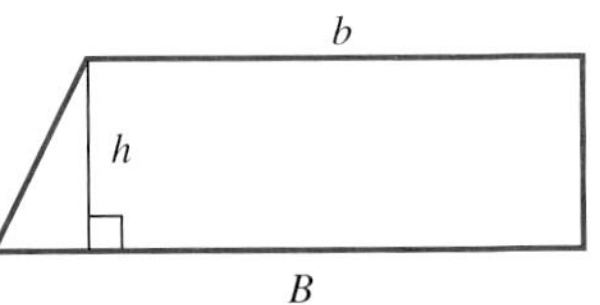

FIGURE 10-6

$$A = \frac{1}{2}h(B + b)$$

$$\mathbf{2} \cdot A = \mathbf{2} \cdot \frac{1}{2}h(B + b)$$ Multiply both sides by 2 to clear the equation of the fraction.

$$2A = h(B + b)$$ Simplify: $2 \cdot \frac{1}{2} = \frac{2}{2} = 1$

$$2A = hB + hb$$ Distribute the multiplication by h.

$$2A - \mathbf{hB} = hB + hb - \mathbf{hB}$$ Subtract hB from both sides.

$$2A - hB = hb$$ Combine like terms: $hB - hB = 0$.

$$\frac{2A - hB}{\mathbf{h}} = \frac{hb}{\mathbf{h}}$$ To undo the multiplication by h, divide both sides by h.

$$\frac{2A - hB}{h} = b$$

Self Check 7

Suppose another student got $b = 2A - \frac{hB}{h}$. Is this result correct?

EXAMPLE 7 Another student worked the previous problem in a different way and got a result of $b = \frac{2A}{h} - B$. Is this result correct?

Solution To determine whether this result is correct, we must show that

$$\frac{2A - hB}{h} = \frac{2A}{h} - B$$

We can do this by dividing $2A - hB$ by h.

$$\frac{2A - hB}{h} = \frac{2A}{h} - \frac{hB}{h}$$ Divide each term of $2A - hB$ by the denominator, which is h.

$$= \frac{2A}{h} - B$$ Simplify the second fraction: $\frac{\overset{1}{\cancel{h}}B}{\underset{1}{\cancel{h}}} = B$.

The results are the same.

Answer no

Section 10.7 STUDY SET

VOCABULARY *Fill in the blanks.*

1. A ________ is an algebraic expression that is the sum of one or more terms containing whole-number exponents.
2. A ________ is a polynomial with one term.
3. A binomial is a polynomial with ____ terms.
4. A trinomial is a polynomial with ____ terms.
5. $\frac{x^m}{x^n} = x^{m-n}$ is a rule for ________.
6. To ________ a fraction, we divide out common factors of the numerator and denominator.

CONCEPTS *Fill in the blanks.*

7. $\dfrac{18x + 9}{9} = \dfrac{18x}{\square} + \dfrac{9x}{\square}$

8. $\dfrac{30x^2 + 12x - 24}{6} = \dfrac{30x^2}{6} \square \dfrac{12x}{6} \square \dfrac{24}{6}$

9. What do the slashes and the small 1's mean?

$$\frac{4}{6} = \frac{\overset{1}{\cancel{2}} \cdot 2}{\underset{1}{\cancel{2}} \cdot 3}$$

10. Complete each rule of exponents.

a. $\dfrac{x^m}{x^n} = \square$

b. $x^{-n} = \square$

11. a. Solve the formula $d = rt$ for t.

b. Use your answer from part a to complete the table.

	r	$\cdot$ t	$=$ d
Motorcycle	$2x$		$6x^3$

12. a. Solve the formula $I = Prt$ for r.

b. Use your answer from part a to complete the table.

	P	$\cdot$ r	$\cdot$ t	$=$ I
Savings account	$8x^3$		$2x$	$24x^6$

13. How many nickels would have a value of $(10x + 35)$ cents?

14. How many twenty-dollar bills would have a value of $\$(60x - 100)$?

NOTATION *Complete each solution.*

15. $\dfrac{a^2b^3}{a^3b^2} = \dfrac{a \cdot a \cdot \square \cdot \square \cdot \square}{\square \cdot \square \cdot \square \cdot b \cdot b}$

$= \dfrac{\overset{1}{\cancel{a}} \cdot \overset{1}{\cancel{a}} \cdot \overset{1}{\cancel{b}} \cdot \overset{1}{\cancel{b}} \cdot \square}{\underset{1}{\cancel{a}} \cdot \underset{1}{\cancel{a}} \cdot \square \cdot \underset{1}{\cancel{b}} \cdot \underset{1}{\cancel{b}}}$

$= \dfrac{b}{a}$

16. $\dfrac{6pq^2 - 9p^2q^2 + pq}{3p^2q}$

$= \dfrac{6pq^2}{\square} - \dfrac{9p^2q^2}{\square} + \dfrac{pq}{\square}$

$= \dfrac{6 \cdot p \cdot q \cdot q}{3 \cdot p \cdot p \cdot q} - \dfrac{\square}{3 \cdot p \cdot p \cdot q} + \dfrac{p \cdot q}{3 \cdot p \cdot p \cdot q}$

$= \dfrac{2q}{p} - 3q + \dfrac{1}{3p}$

PRACTICE *Simplify each fraction.*

17. $\dfrac{5}{15}$

18. $\dfrac{64}{128}$

19. $\dfrac{-125}{75}$

20. $\dfrac{-98}{21}$

21. $\dfrac{120}{160}$

22. $\dfrac{70}{420}$

23. $\dfrac{-3{,}612}{-3{,}612}$

24. $\dfrac{-288}{-112}$

Perform each division.

25. $\dfrac{x^5}{x^2}$

26. $\dfrac{a^{12}}{a^8}$

27. $\dfrac{r^3s^2}{rs^3}$

28. $\dfrac{y^4z^3}{y^2z^2}$

29. $\dfrac{8x^3y^2}{4xy^3}$

30. $\dfrac{-3y^3z}{6yz^2}$

31. $\dfrac{12u^5v}{-4u^2v^3}$

32. $\dfrac{16rst^2}{-8rst^3}$

33. $\dfrac{-16r^3y^2}{-4r^2y^4}$

34. $\dfrac{35xyz^2}{-7x^2yz}$

35. $\dfrac{-65rs^2t}{15r^2s^3t}$

36. $\dfrac{112u^3z^6}{-42u^3z^6}$

37. $\dfrac{x^2x^3}{xy^6}$

38. $\dfrac{x^2y^2}{x^2y^3}$

39. $\dfrac{(a^3b^4)^3}{ab^4}$

40. $\dfrac{(a^2b^3)^3}{a^6b^6}$

41. $\dfrac{15(r^2s^3)^2}{-5(rs^5)^3}$

42. $\dfrac{-5(a^2b)^3}{10(ab^2)^3}$

43. $\dfrac{-32(x^3y)^3}{128(x^2y^2)^3}$

44. $\dfrac{68(a^6b^7)^2}{-96(abc^2)^3}$

45. $\dfrac{-(4x^3y^3)^2}{(x^2y^4)^3}$

46. $\dfrac{(2r^3s^2)^2}{-(4r^2s^2)^2}$

47. $\dfrac{(a^2a^3)^4}{(a^4)^3}$

48. $\dfrac{(b^3b^4)^5}{(bb^2)^2}$

49. $\dfrac{6x + 9}{3}$

50. $\dfrac{8x + 12y}{4}$

51. $\dfrac{5x - 10y}{25xy}$

52. $\dfrac{2x - 32}{16x}$

53. $\dfrac{3x^2 + 6y^3}{3x^2y^2}$

54. $\dfrac{4a^2 - 9b^2}{12ab}$

55. $\dfrac{15a^3b^2 - 10a^2b^3}{5a^2b^2}$

56. $\dfrac{9a^4b^3 - 16a^3b^4}{12a^2b}$

57. $\dfrac{4x - 2y + 8z}{4xy}$

58. $\dfrac{5a^2 + 10b^2 - 15ab}{5ab}$

59. $\dfrac{12x^3y^2 - 8x^2y - 4x}{4xy}$

60. $\dfrac{12a^2b^2 - 8a^2b - 4ab}{4ab}$

61. $\dfrac{-25x^2y + 30xy^2 - 5xy}{-5xy}$

62. $\dfrac{-30a^2b^2 - 15a^2b - 10ab^2}{-10ab}$

Simplify each numerator and perform the division.

63. $\dfrac{5x(4x - 2y)}{2y}$

64. $\dfrac{9y^2(x^2 - 3xy)}{3x^2}$

65. $\dfrac{(-2x)^3 + (3x^2)^2}{6x^2}$

66. $\dfrac{(-3x^2y)^3 + (3xy^2)^3}{27x^3y^4}$

67. $\dfrac{4x^2y^2 - 2(x^2y^2 + xy)}{2xy}$

68. $\dfrac{-5a^3b - 5a(ab^2 - a^2b)}{10a^2b^2}$

69. $\dfrac{(3x - y)(2x - 3y)}{6xy}$

70. $\dfrac{(2m - n)(3m - 2n)}{-3m^2n^2}$

71. $\dfrac{(a + b)^2 - (a - b)^2}{2ab}$

72. $\dfrac{(x - y)^2 + (x + y)^2}{2x^2y^2}$

APPLICATIONS

73. POOL The rack shown in the illustration is used to set up the balls when beginning a game of pool. If the perimeter of the rack, in inches, is given by the polynomial $6x^2 - 3x + 9$, what is the length of one side?

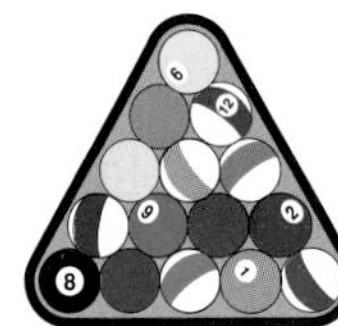

74. CHECKERBOARDS If the perimeter (in inches) of the checkerboard is $12x^2 - 8x + 32$, find an expression that represents the length of one side?

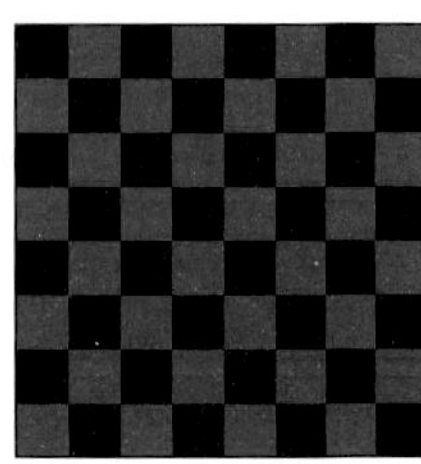

75. AIR CONDITIONING If the volume occupied by the air conditioning unit shown is $(36x^3 - 24x^2)$ cubic feet, find an expression that represents its height.

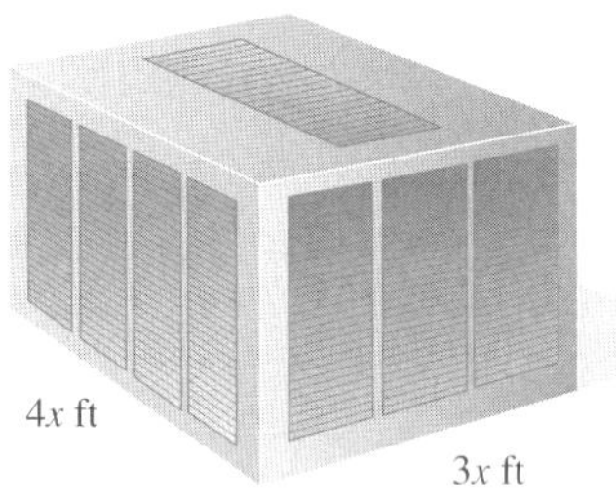

76. MINI-BLINDS The area covered by the mini-blinds is $(3x^3 - 6x)$ square feet. Find an expression that represents the length of the blinds.

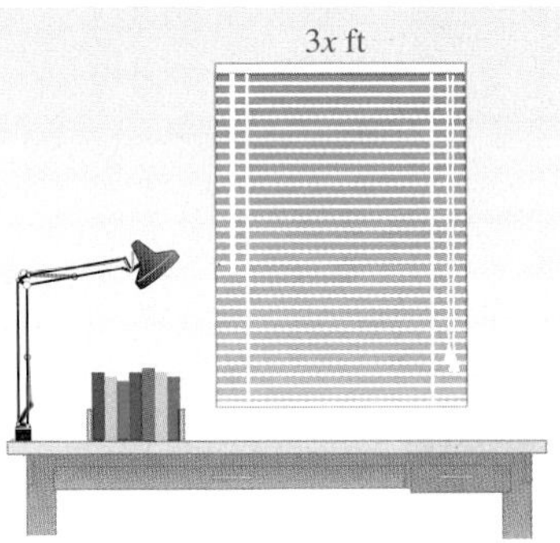

77. CONFIRMING FORMULAS Are these formulas the same?

$$l = \frac{P - 2w}{2} \quad \text{and} \quad l = \frac{P}{2} - w$$

78. CONFIRMING FORMULAS Are these formulas the same?

$$r = \frac{G + 2b}{2b} \quad \text{and} \quad r = \frac{G}{2b} + b$$

79. ELECTRIC BILLS On an electric bill, the following two formulas are used to compute the average cost of x kwh of electricity. Are the formulas equivalent?

$$\frac{0.08x + 5}{x} \quad \text{and} \quad 0.08x + \frac{5}{x}$$

80. PHONE BILLS On a phone bill, the following two formulas are used to compute the average cost per minute of x minutes of phone usage. Are the formulas equivalent?

$$\frac{0.15x + 12}{x} \quad \text{and} \quad 0.15 + \frac{12}{x}$$

WRITING

81. Explain the error.

$$\frac{3x + 5}{5} = \frac{3x + \overset{1}{\cancel{5}}}{\underset{1}{\cancel{5}}}$$
$$= 3x$$

82. Explain how to perform this division.

$$\frac{4x^2y + 8xy^2}{4xy}$$

REVIEW *Identify each polynomial as a monomial, a binomial, a trinomial, or none of these.*

83. $5a^2b + 2ab^2$

84. $-3x^3y$

85. $-2x^3 + 3x^2 - 4x + 12$

86. $17t^2 - 15t + 27$

87. What is the degree of the trinomial $3x^2 - 2x + 4$?

88. What is the numerical coefficient of the second term of the trinomial $-7t^2 + 5t + 17$?

10.8 Dividing Polynomials by Polynomials

- Dividing polynomials by polynomials
- Writing powers in descending order
- Missing terms

In this section, we will discuss how to divide one polynomial by another.

Dividing polynomials by polynomials

To divide one polynomial by another, we use a method similar to long division in arithmetic. We illustrate the method with several examples.

EXAMPLE 1 Divide $x^2 + 5x + 6$ by $x + 2$.

Self Check 1
Divide $x^2 + 7x + 12$ by $x + 3$.

Solution Here the divisor is $x + 2$, and the dividend is $x^2 + 5x + 6$.

Step 1: $x + 2\overline{)x^2 + 5x + 6}$ with quotient x

How many times does x divide x^2? $\frac{x^2}{x} = x$. Write x above the division symbol.

Step 2: $x + 2\overline{)x^2 + 5x + 6}$ with quotient x; under it: $x^2 + 2x$

Multiply each term in the divisor by x. Write the product under $x^2 + 5x$ and draw a line.

Step 3: $x + 2\overline{)x^2 + 5x + 6}$ with quotient x; under it: $x^2 + 2x$, then $3x + 6$

Subtract $x^2 + 2x$ from $x^2 + 5x$. Work vertically, column by column: $x^2 - x^2 = 0$ and $5x - 2x = 3x$.

Bring down the 6.

Step 4: $x + 2\overline{)x^2 + 5x + 6}$ with quotient $x + 3$; under it: $x^2 + 2x$, then $3x + 6$

How many times does x divide $3x$? $\frac{3x}{x} = +3$. Write +3 above the division symbol.

Step 5:
$$x + 2\overline{)x^2 + 5x + 6}$$ with quotient $x + 3$
$$\begin{array}{r} x^2 + 2x \\ \hline 3x + 6 \\ 3x + 6 \\ \hline \end{array}$$

Multiply each term in the divisor by 3. Write the product under $3x + 6$ and draw a line.

Step 6:
$$x + 2\overline{)x^2 + 5x + 6}$$ with quotient $x + 3$
$$\begin{array}{r} x^2 + 2x \\ \hline 3x + 6 \\ 3x + 6 \\ \hline 0 \end{array}$$

Subtract $3x + 6$ from $3x + 6$. Work vertically: $3x - 3x = 0$ and $6 - 6 = 0$.

The quotient is $x + 3$ and the remainder is 0.

Step 7: Check the work by verifying that $(x + 2)(x + 3)$ is $x^2 + 5x + 6$.

$$(x + 2)(x + 3) = x^2 + 3x + 2x + 6$$
$$= x^2 + 5x + 6$$

The answer checks.

Answer $x + 4$

Self Check 2

Divide: $\dfrac{8x^2 + 6x - 3}{2x + 3}$.

EXAMPLE 2 Divide: $\dfrac{6x^2 - 7x - 2}{2x - 1}$.

Solution Here the divisor is $2x - 1$ and the dividend is $6x^2 - 7x - 2$.

Step 1:
$$2x - 1\overline{)6x^2 - 7x - 2}$$ with quotient $3x$

How many times does $2x$ divide $6x^2$? $\dfrac{6x^2}{2x} = 3x$. Write the $3x$ above the division symbol.

Step 2:
$$2x - 1\overline{)6x^2 - 7x - 2}$$ with quotient $3x$
$$\begin{array}{r} 6x^2 - 3x \\ \hline \end{array}$$

Multiply each term in the divisor by $3x$. Write the product under $6x^2 - 7x$ and draw a line.

Step 3:
$$2x - 1\overline{)6x^2 - 7x - 2}$$ with quotient $3x$
$$\begin{array}{r} 6x^2 - 3x \quad \\ \hline -4x - 2 \end{array}$$

Subtract $6x^2 - 3x$ from $6x^2 - 7x$. Work vertically: $6x^2 - 6x^2 = 0$ and $-7x - (-3x) = -7x + 3x = -4x$.

Bring down the -2.

Step 4:
$$2x - 1\overline{)6x^2 - 7x - 2}$$ with quotient $3x - 2$
$$\begin{array}{r} 6x^2 - 3x \quad \\ \hline -4x - 2 \end{array}$$

How many times does $2x$ divide $-4x$? $\dfrac{-4x}{2x} = -2$. Write -2 above the division symbol.

Step 5:
$$2x - 1\overline{)6x^2 - 7x - 2}$$ with quotient $3x - 2$
$$\begin{array}{r} 6x^2 - 3x \quad \\ \hline -4x - 2 \\ -4x + 2 \\ \hline \end{array}$$

Multiply each term in the divisor by -2. Write the product under $-4x - 2$ and draw a line.

Step 6:
$$2x - 1\overline{)6x^2 - 7x - 2}$$ with quotient $3x - 2$
$$\begin{array}{r} 6x^2 - 3x \quad \\ \hline -4x - 2 \\ -4x + 2 \\ \hline -4 \end{array}$$

Subtract $-4x + 2$ from $-4x - 2$. Work vertically: $-4x - (-4x) = -4x + 4x = 0$ and $-2 - 2 = -4$.

Here the quotient is $3x - 2$ and the remainder is -4. It is common to write the answer as either

$$3x - 2 + \frac{-4}{2x - 1} \quad \text{or} \quad 3x - 2 - \frac{4}{2x - 1} \qquad \text{Quotient} + \frac{\text{remainder}}{\text{divisor}}.$$

Step 7: To check, we multiply

$$3x - 2 + \frac{-4}{2x - 1} \quad \text{by} \quad 2x - 1$$

The product should be the dividend.

$$\begin{aligned}(2x - 1)\left(3x - 2 + \frac{-4}{2x - 1}\right) &= (2x - 1)(3x - 2) + (2x - 1)\left(\frac{-4}{2x - 1}\right) \\ &= (2x - 1)(3x - 2) - 4 \\ &= 6x^2 - 4x - 3x + 2 - 4 \\ &= 6x^2 - 7x - 2\end{aligned}$$

Because the result is the dividend, the answer checks.

Answer $4x - 3 + \frac{6}{2x + 3}$

Writing powers in descending order

The division method works best when the terms of the divisor and the dividend are written in descending powers of the variable. This means that the term involving the highest power of x appears first, the term involving the second-highest power of x appears second, and so on. For example, the terms in

$$3x^3 + 2x^2 - 7x + 5$$

have their exponents written in descending order.

If the powers in the dividend or divisor are not in descending order, we use the commutative property of addition to write them that way.

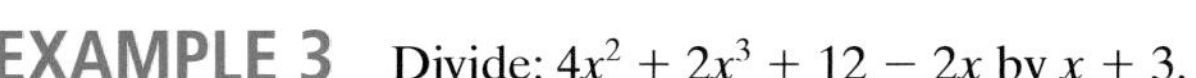

EXAMPLE 3 Divide: $4x^2 + 2x^3 + 12 - 2x$ by $x + 3$.

Solution We write the dividend so that the exponents are in descending order.

$$\begin{array}{r} 2x^2 - 2x + 4 \\ x + 3\overline{)2x^3 + 4x^2 - 2x + 12} \\ \underline{2x^3 + 6x^2} \qquad\qquad \\ -2x^2 - 2x \qquad \\ \underline{-2x^2 - 6x} \qquad \\ 4x + 12 \\ \underline{4x + 12} \\ 0 \end{array}$$

Check: $(x + 3)(2x^2 - 2x + 4) = 2x^3 - 2x^2 + 4x + 6x^2 - 6x + 12$
$= 2x^3 + 4x^2 - 2x + 12$

Self Check 3
Divide $x^2 - 10x + 6x^3 + 4$ by $2x - 1$.

Answer $3x^2 + 2x - 4$

Missing terms

When we write the terms of a dividend in descending powers of x, we must determine whether some powers of the variable are missing. When this happens, we should write such terms with a coefficient of 0 or leave a blank space for them.

Self Check 4

Divide: $\frac{x^2 - 9}{x - 3}$.

Answer $x + 3$

EXAMPLE 4 Divide: $\frac{x^2 - 4}{x + 2}$.

Solution Since $x^2 - 4$ does not have a term involving x, we must either include the term $0x$ or leave a space for it.

$$\begin{array}{r} x - 2 \\ x + 2\overline{)x^2 + \mathbf{0x} - 4} \\ \underline{x^2 + 2x} \\ -2x - 4 \\ \underline{-2x - 4} \\ 0 \end{array}$$

Check: $(x + 2)(x - 2) = x^2 - 2x + 2x - 4$
$= x^2 - 4$

Section 10.8 STUDY SET

VOCABULARY *Fill in the blanks.*

1. In the division $x + 1\overline{)x^2 + 2x + 1}$, the expression $x + 1$ is called the ________ and $x^2 + 2x + 1$ is called the ________.
2. The answer to a division problem is called the ________.
3. If a division does not come out even, the leftover part is called a ________.
4. The powers of x in $2x^4 + 3x^3 + 4x^2 - 7x - 2$ are said to be written in ________ order.

CONCEPTS *Write each polynomial with the powers in descending order.*

5. $4x^3 + 7x - 2x^2 + 6$
6. $5x^2 + 7x^3 - 3x - 9$
7. $9x + 2x^2 - x^3 + 6x^4$
8. $7x^5 + x^3 - x^2 + 2x^4$

Identify the missing terms in each polynomial.

9. $5x^4 + 2x^2 - 1$
10. $-3x^5 - 2x^3 + 4x - 6$

Without performing the division, determine which of the three possible quotients seems reasonable.

11. $\frac{x^4 - 81}{x - 3}$ $x^2 + 3x + 9$
$x^3 + 3x^2 + 9x + 27$
$x^4 + 3x^3 + 9x^2 + 27x + 1$

12. $\frac{8x^3 - 27}{2x - 3}$ $4x^2 + 6x + 9$
$4x^3 - 6x^2 - 9$
$4x^4 - 6x^3 - 9x^2 + 1$

13. a. Solve $d = rt$ for r.
b. Use your answer to part a and the long division method to complete the table.

	r	· t	= d
Subway		$x + 4$	$x^2 + x - 12$

14. a. Solve $I = Prt$ for P.
b. Use your answer to part a and the long division method to complete the table.

	P	· r	· t	= I
Bonds		$x + 4$	1	$x^2 + 7x + 12$

15. Using long division, a student found that

$$\frac{3x^2 + 8x + 4}{3x + 2} = x + 2$$

Use multiplication to see whether the result is correct.

16. Using long division, a student found that

$$\frac{x^2 + 4x - 21}{x - 3} = x - 7$$

Use multiplication to see whether the result is correct.

NOTATION *Complete each division.*

17.

$$\begin{array}{r} \square + 2 \\ x + 2\overline{)x^2 + 4x + 4} \\ \underline{x^2 + \square} \\ \square + 4 \\ \underline{2x + 4} \\ 0 \end{array}$$

18.

$$\begin{array}{r} \square + x - 2 + \square \\ 2x + 1\overline{)2x^3 + 3x^2 - 3x + 5} \\ \underline{\square + x^2} \\ 2x^2 - 3x \\ \underline{2x^2 + \square} \\ \square + 5 \\ \underline{-4x - \square} \\ 7 \end{array}$$

PRACTICE *Perform each division.*

19. Divide $x^2 + 4x - 12$ by $x - 2$.

20. Divide $x^2 - 5x + 6$ by $x - 2$.

21. Divide $y^2 + 13y + 12$ by $y + 1$.

22. Divide $z^2 - 7z + 12$ by $z - 3$.

23. $\dfrac{6a^2 + 5a - 6}{2a + 3}$

24. $\dfrac{8a^2 + 2a - 3}{2a - 1}$

25. $\dfrac{3b^2 + 11b + 6}{3b + 2}$

26. $\dfrac{3b^2 - 5b + 2}{3b - 2}$

Write the terms so that the powers of x are in descending order. Then perform each division.

27. $5x + 3\overline{)11x + 10x^2 + 3}$

28. $2x - 7\overline{)-x - 21 + 2x^2}$

29. $4 + 2x\overline{)-10x - 28 + 2x^2}$

30. $1 + 3x\overline{)9x^2 + 1 + 6x}$

31. $2x - 1\overline{)x - 2 + 6x^2}$

32. $2 + x\overline{)3x + 2x^2 - 2}$

33. $3 + x\overline{)2x^2 - 3 + 5x}$

34. $x - 3\overline{)2x^2 - 3 - 5x}$

Perform each division.

35. $2x + 3\overline{)2x^3 + 7x^2 + 4x - 3}$

36. $2x - 1\overline{)2x^3 - 3x^2 + 5x - 2}$

37. $3x + 2\overline{)6x^3 + 10x^2 + 7x + 2}$

38. $4x + 3\overline{)4x^3 - 5x^2 - 2x + 3}$

39. $2x + 1\overline{)2x^3 + 3x^2 + 3x + 1}$

40. $3x - 2\overline{)6x^3 - x^2 + 4x - 4}$

Perform each division. If there is a remainder, write the answer in quotient + $\dfrac{\text{remainder}}{\text{divisor}}$ form.

41. $\dfrac{2x^2 + 5x + 2}{2x - 3}$

42. $\dfrac{3x^2 - 8x + 8}{3x - 2}$

43. $\dfrac{4x^2 + 6x - 1}{2x + 1}$

44. $\dfrac{6x^2 - 11x + 2}{3x - 1}$

45. $\dfrac{x^3 + 3x^2 + 3x + 1}{x + 1}$

46. $\dfrac{x^3 + 6x^2 + 12x + 8}{x + 2}$

47. $\dfrac{2x^3 + 7x^2 + 4x + 3}{2x + 3}$

48. $\dfrac{6x^3 + x^2 + 2x + 1}{3x - 1}$

49. $\dfrac{2x^3 + 4x^2 - 2x + 3}{x - 2}$

50. $\dfrac{3y^3 - 4y^2 + 2y + 3}{y + 3}$

51. $\dfrac{x^2 - 1}{x - 1}$

52. $\dfrac{x^2 - 9}{x + 3}$

53. $\dfrac{4x^2 - 9}{2x + 3}$

54. $\dfrac{25x^2 - 16}{5x - 4}$

55. $\dfrac{x^3 + 1}{x + 1}$

56. $\dfrac{x^3 - 8}{x - 2}$

57. $\dfrac{a^3 + a}{a + 3}$

58. $\dfrac{y^3 - 50}{y - 5}$

59. $3x - 4\overline{)15x^3 - 23x^2 + 16x}$

60. $2y + 3\overline{)21y^2 + 6y^3 - 20}$

APPLICATIONS

61. FURNACE FILTERS The area of the furnace filter shown is $(x^2 - 2x - 24)$ square inches.

 a. Find an expression for its length.

 b. Find an expression for its perimeter.

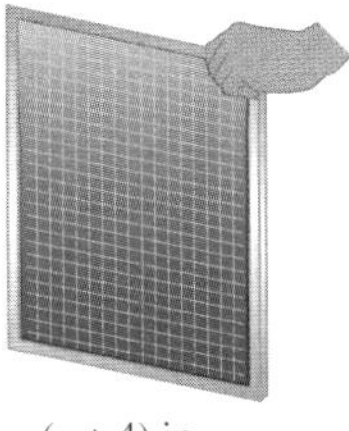

$(x + 4)$ in.

62. SHELF SPACE The formula $V = Bh$ gives the volume of a cylinder where B is the area of the base and h is the height. Find the amount of shelf space that the container of potato chips shown occupies if its volume is $(2x^3 - 4x - 2)$ cubic inches.

63. COMMUNICATIONS See the illustration. Telephone poles were installed every $(2x - 3)$ feet along a stretch of railroad track $(8x^3 - 6x^2 + 5x - 21)$ feet long. How many poles were used?

64. CONSTRUCTION COSTS Find the price per square foot to remodel each of the three rooms listed in the table.

Room	Remodeling cost	Area (ft^2)	Cost (per ft^2)
Bathroom	$\$(2x^2 + x - 6)$	$2x - 3$	
Bedroom	$\$(x^2 + 9x + 20)$	$x + 4$	
Kitchen	$\$(3x^3 - 9x - 6)$	$3x + 3$	

WRITING

65. Explain how the following are related: *dividend, divisor, quotient,* and *remainder.*

66. How would you check the results of a division?

REVIEW *Simplify each expression.*

67. $(x^5x^6)^2$

68. $(a^2)^3(a^3)^4$

69. $3(2x^2 - 4x + 5) + 2(x^2 + 3x - 7)$

70. $-2(y^3 + 2y^2 - y) - 3(3y^3 + y)$

71. What can be said about the slopes of two parallel lines?

72. What is the slope of a line perpendicular to a line with a slope of $\frac{3}{4}$?

KEY CONCEPT

Polynomials

A **polynomial** is a term or a sum of terms in which all variables have whole-number exponents. No variables appear in a denominator.

The Vocabulary of Polynomials

1. Consider $x^3 - 2x^2 + 6x - 8$.

a. Fill in: This is a polynomial in ___. It is written in ___________ powers of x.

b. How many terms does the polynomial have?

c. Give the degree of each term.

d. What is the degree of the polynomial?

e. Give the coefficient of each term.

2. Consider $x^2 + 2xy + y^2$.

a. Fill in: This is a polynomial in ___ and ___. It is written in ___________ powers of x and ___________ powers of y.

b. How many terms does the polynomial have?

c. Give the degree of each term.

d. What is the degree of the polynomial?

e. Give the coefficient of each term.

3. Classify each polynomial as a monomial, binomial, trinomial, or none of these.

a. $x^2 - y^2$

b. $s^2t + st^2 - st + 1$

c. $4y^2 - 10y + 16$

d. $15h$

4. a. Explain why ab is a monomial while $a + b$ is a binomial.

b. Is every term of a polynomial a monomial?

c. The degree of x^2 is 2. What is the degree of 3^2? Explain your answer.

d. For $3x^2 - 4x + 9$, what is the lead term, the lead coefficient, and the constant term?

Operations with Polynomials

Just like numbers in arithmetic, polynomials can be added, subtracted, multiplied, divided, and raised to powers. We have discussed some rules for performing operations with polynomials that have more than one term.

Fill in the blanks.

5. To add two polynomials, remove the parentheses and __________ any like terms.

6. To subtract two polynomials, drop the minus sign and the parentheses, and ________ the sign of every term within the parentheses of the second polynomial. Then combine like terms.

7. To multiply two polynomials, multiply ______ term of one polynomial by ______ term of the other polynomial and combine like terms.

8. To divide two polynomials, use the ______ division method.

Perform the operations.

9. $(2x + 3) + (x - 8)$

10. $(2x + 3) - (x - 8)$

11. $(2x + 3)(x - 8)$

12. $(2x^2 + 3)^2$

13. $(y^2 + y - 6) + (y + 3)$

14. $(y^2 + y - 6) - (y + 3)$

15. $(y^2 + y - 6)(y + 3)$

16. $(y^2 + y - 6) \div (y + 3)$

Polynomial Functions

Polynomial functions can be used to describe such situations as the stopping distance of a car, the appreciation of a house, and the area of a geometric figure.

17. If $f(x) = x^3 - 2x + 5$, find $f(-2)$.

18. STOPPING DISTANCE A vehicle's stopping distance in feet is given by the polynomial function $f(v) = 0.04v^2 + 0.9v$, where v is the velocity. Find the stopping distance when the vehicle is traveling at 40 mph.

ACCENT ON TEAMWORK

SECTION 10.1

RULES FOR EXPONENTS Have a student in your group write each of the five rules for exponents listed on page 662 on separate 3×5 cards. On a second set of cards, write an explanation of each rule using words. On a third set of cards, write a separate example of the use of each rule for exponents. Shuffle the cards and work together to match the symbolic description, the word description, and the example for each of the five rules for exponents.

SECTION 10.2

GRAPHING Complete Table 1, plot the ordered pairs on a rectangular coordinate system, and then draw a smooth curve through the points. Do the same for Table 2, using the same coordinate system. Compare the graphs. How are they similar, and how do they differ?

x	2^x
−2	
−1	
0	
1	
2	
3	

TABLE 1

x	2^{-x}
−2	
−1	
0	
1	
2	
3	

TABLE 2

SECTION 10.3

SCIENTIFIC NOTATION Go to the library and find five examples of extremely large and five examples of extremely small positive numbers. Encyclopedias, government statistics books, and science books are good places to look. Write each number in scientific notation on a separate piece of paper. Include a brief explanation of what the number represents. Present the ten examples in numerical order, beginning with the smallest number first.

SECTION 10.4

POLYNOMIAL FUNCTIONS The height (in feet) of a rock from the floor of the Grand Canyon t seconds after being thrown downward from the rim with an initial velocity of 6 feet per second is given by the polynomial

$$f(t) = -16t^2 - 6t + 5{,}292$$

a. Find $f(0)$ and $f(18)$ and explain their significance.

b. Find $f(3)$, $f(6)$, $f(9)$, $f(12)$, and $f(15)$. Use this information to show the position of the rock for these times on the scale shown in the illustration.

c. Are the distances the rock fell during each 3-second time interval the same?

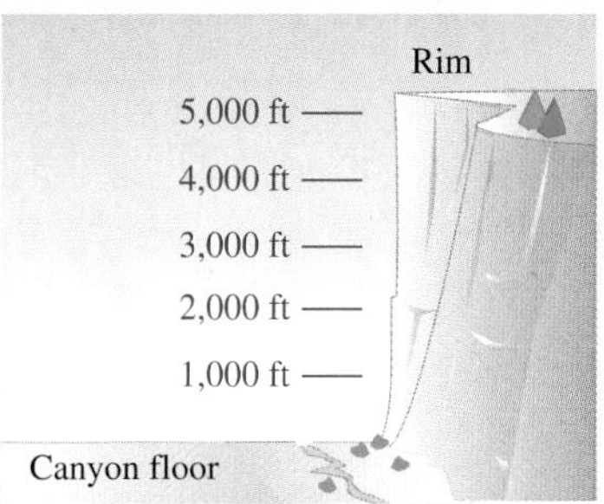

SECTION 10.5

ADDING POLYNOMIALS An old adage is that "You can't add apples and oranges." Give an example of how this concept applies when adding two polynomials.

SECTION 10.6

MULTIPLYING BINOMIALS Recall that the formula for the area of a rectangle is $A = lw$ and the formula for the area of a square is $A = s^2$.

a. Express the area of each of the large figures in the following illustration as a product of two binomials.

b. Find the area of each large figure by finding the sum of the areas of each of its parts.

c. In each case, show the relationship between your answers to part a and part b.

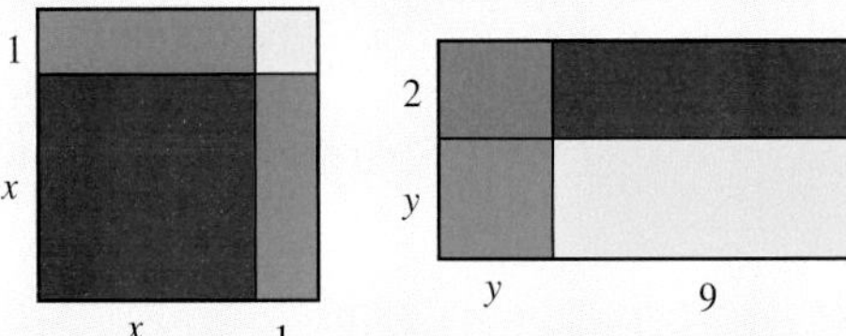

SECTION 10.7

WORKING WITH MONOMIALS For the monomials $15a^3$ and $5a^2$, show, if possible, how they are added, subtracted, multiplied, and divided. If an operation cannot be done, explain why this is so.

SECTION 10.8

WORKING WITH POLYNOMIALS Add, subtract, multiply, and divide the polynomials $6a^2 - 7a + 2$ and $2a - 1$.

CHAPTER REVIEW

SECTION 10.1 Natural-Number Exponents

CONCEPTS

If n represents a natural number, then

$$x^n = \overbrace{x \cdot x \cdot x \cdot \cdots \cdot x}^{n \text{ factors of } x}$$

where x is called the *base* and n is called the *exponent.*

Rules for exponents:
If m and n represent integers, then

$$x^m x^n = x^{m+n}$$
$$(x^m)^n = x^{mn}$$
$$(xy)^n = x^n y^n$$
$$\left(\frac{x}{y}\right)^n = \frac{x^n}{y^n} \quad (y \neq 0)$$
$$\frac{x^m}{x^n} = x^{m-n} \quad (x \neq 0)$$

REVIEW EXERCISES

Write each expression without using exponents.

1. $-3x^4$
2. $\left(\frac{1}{2}pq\right)^3$

Evaluate each expression.

3. 5^3
4. $(-8)^2$
5. -8^2
6. $(5 - 3)^2$

Simplify each expression.

7. x^3x^2
8. $-3y(y^5)$
9. $(y^7)^3$
10. $(3x)^4$
11. $b^3b^4b^5$
12. $-z^2(z^3y^2)$
13. $(-16s)^2s$
14. $(2x^2y)^2$
15. $(x^2x^3)^3$
16. $\left(\frac{x^2y}{xy^2}\right)^2$
17. $\frac{x^7}{x^3}$
18. $\frac{(5y^2z^3)^3}{(yz)^5}$

Find the area or the volume of each figure.

19.

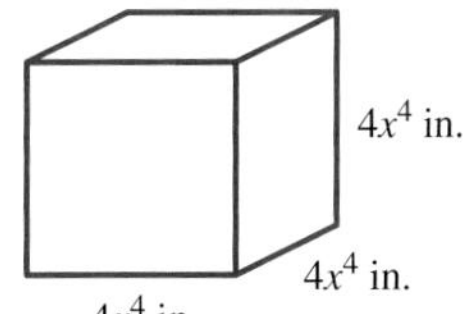

20.

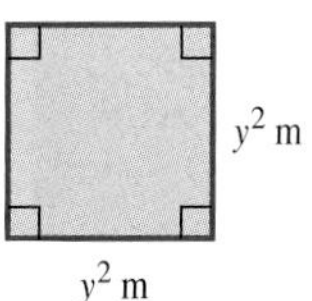

SECTION 10.2 Zero and Negative Integer Exponents

Zero exponents:

$$x^0 = 1 \quad (x \neq 0)$$

Negative integer exponents:

$$x^{-n} = \frac{1}{x^n} \quad (x \neq 0)$$

Write each expression without using negative or zero exponents or parentheses.

21. x^0
22. $(3x^2y^2)^0$
23. $(3x^0)^2$
24. 10^{-3}
25. $\left(\frac{3}{4}\right)^{-1}$
26. -5^{-2}
27. x^{-5}
28. $-6y^4y^{-5}$

29. $\dfrac{x^{-3}}{x^7}$

30. $(x^{-3}x^{-4})^{-2}$

31. $\left(\dfrac{x^2}{x}\right)^{-5}$

32. $\left(\dfrac{3z^4}{z^3}\right)^{-2}$

Write each expression with a single exponent.

33. $y^{3n}y^{4n}$

34. $\dfrac{z^{8c}}{z^{10c}}$

SECTION 10.3 Scientific Notation

A number in written is *scientific notation* if it is written as the product of a number between 1 (including 1) and 10 and an integer power of 10.

Write each number in scientific notation.

35. 728

36. 9,370,000

37. 0.0136

38. 0.00942

39. 0.018×10^{-2}

40. 753×10^3

Write each number in standard notation.

41. 7.26×10^5

42. 3.91×10^{-4}

43. 2.68×10^0

44. 5.76×10^1

Scientific notation provides an easier way to do some computations.

Simplify each fraction by first writing each number in scientific notation. Then perform the arithmetic. Express the result in standard notation.

45. $\dfrac{(0.00012)(0.00004)}{0.00000016}$

46. $\dfrac{(4{,}800)(20{,}000)}{600{,}000}$

47. WORLD POPULATION As of 2000, the world's population was estimated to be 6.08 billion. Write this number in standard notation and in scientific notation.

48. ATOMS The illustration shows a cross section of an atom. How many nuclei, placed end to end, would it take to stretch across the atom?

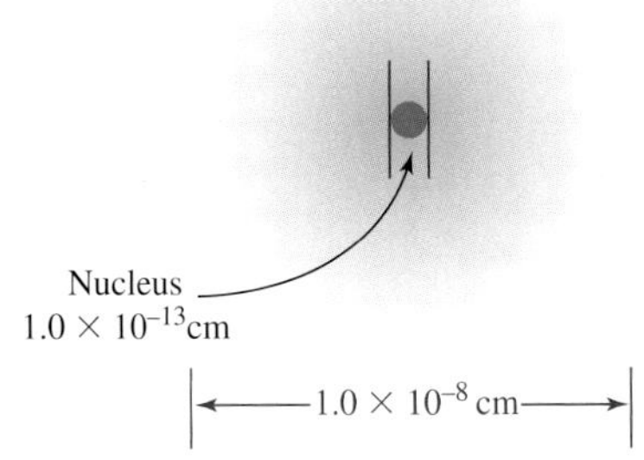

SECTION 10.4 Polynomials

A *polynomial* is a term or a sum of terms in which all variables have whole-number exponents. No variable appears in a denominator.

Determine whether each expression is a polynomial.

49. $x^3 - x^2 - x - 1$

50. $x^{-2} - x^{-1} - 1$

51. $\frac{11}{y} + 4y$

52. $-16x^2y + 5xy^2$

Consider the polynomial $3x^3 - x^2 + x + 10$.

53. How many terms does the polynomial have?

54. What is the leading term?

55. What is the coefficient of the second term?

56. What is the constant term?

The *degree of a monomial* ax^n is n. The *degree of a monomial* in several variables is the sum of the exponents on those variables. The *degree of a polynomial* is the same as the degree of its term with the largest degree.

Find the degree of each polynomial and classify it as a monomial, binomial, trinomial, or none of these.

57. $13x^7$

58. $-16a^2b$

59. $5^3x + x^2$

60. $-3x^5 + x - 1$

61. $9xy^2 + 21x^3y^3$

62. $4s^4 - 3s^2 + 5s + 4$

If $f(x)$ is a polynomial function in x, then $f(3)$ is the value of the function when $x = 3$.

Let $f(x) = 3x^2 + 2x + 1$. *Find each value.*

63. $f(3)$

64. $f(0)$

65. $f(-2)$

66. $f(-0.2)$

67. DIVING The number of inches that the woman deflects the diving board is given by the function

$$f(x) = 0.1875x^2 - 0.0078125x^3$$

where x is the number of feet that she stands from the front anchor point of the board. Find the amount of deflection if she stands on the end of the diving board, 8 feet from the anchor point.

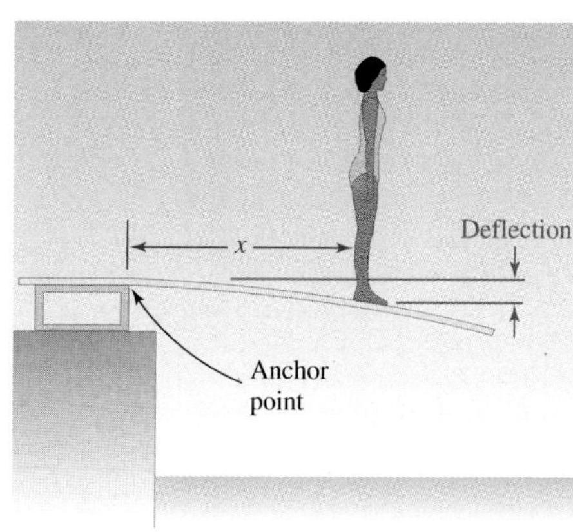

SECTION 10.5 *Adding and Subtracting Polynomials*

When *adding* or *subtracting polynomials,* add or subtract like terms by combining the numerical coefficients and using the same variables and the same exponents.

Simplify each expression.

68. $3x^6 + 5x^5 - x^6$

69. $x^2y^2 - 3x^2y^2$

70. $(3x^2 + 2x) + (5x^2 - 8x)$

71. $3(9x^2 + 3x + 7) - 2(11x^2 - 5x + 9)$

Polynomials can be added or subtracted *vertically.*

Perform the operations.

72. $\begin{array}{r} 3x^2 + 5x + 2 \\ + \quad x^2 - 3x + 6 \\ \hline \end{array}$

73. $\begin{array}{r} 20x^3 \qquad\quad + 12x \\ - \quad 12x^3 + 7x^2 - \; 7x \\ \hline \end{array}$

SECTION 10.6 *Multiplying Polynomials*

To multiply two monomials, first multiply the numerical factors and then multiply the variable factors.

To multiply a polynomial with more than one term by a monomial, multiply each term of the polynomial by the monomial and simplify.

To multiply two binomials, use the *FOIL method:*

F: First

O: Outer

I: Inner

L: Last

Find each product.

74. $(2x^2)(5x)$

75. $(-6x^4z^3)(x^6z^2)$

76. $(2rst)(-3r^2s^3t^4)$

77. $5b^3 \cdot 6b^2 \cdot 4b^6$

78. $5(x + 3)$

79. $x^2(3x^2 - 5)$

80. $x^2y(y^2 - xy)$

81. $-2y^2(y^2 - 5y)$

82. $2x(3x^4)(x + 2)$

83. $-3x(x^2 - x + 2)$

84. $(x + 3)(x + 2)$

85. $(2x + 1)(x - 1)$

86. $(3a - 3)(2a + 2)$

87. $6(a - 1)(a + 1)$

88. $(a - b)(2a + b)$

89. $(-3x - y)(2x + y)$

90. $(x + 3)(x + 3)$

91. $(x + 5)(x - 5)$

92. $(a - 3)^2$

93. $(x + 4)^2$

94. $(-2y + 1)^2$

95. $(y^2 + 1)(y^2 - 1)$

96. $(3x + 1)(x^2 + 2x + 1)$

97. $(2a - 3)(4a^2 + 6a + 9)$

Special products:

$(x + y)^2 = x^2 + 2xy + y^2$

$(x - y)^2 = x^2 - 2xy + y^2$

$(x + y)(x - y) = x^2 - y^2$

Solve each equation.

98. $x^2 + 3 = x(x + 3)$

99. $x^2 + x = (x + 1)(x + 2)$

100. $(x + 2)(x - 5) = (x - 4)(x - 1)$

101. $(x + 5)(3x + 1) = x^2 + (2x - 1)(x - 5)$

To multiply one polynomial by another, multiply each term of one polynomial by each term of the other polynomial, and simplify.

102. APPLIANCES Find the perimeter of the base, the area of the base, and the volume occupied by the dishwasher shown in the illustration.

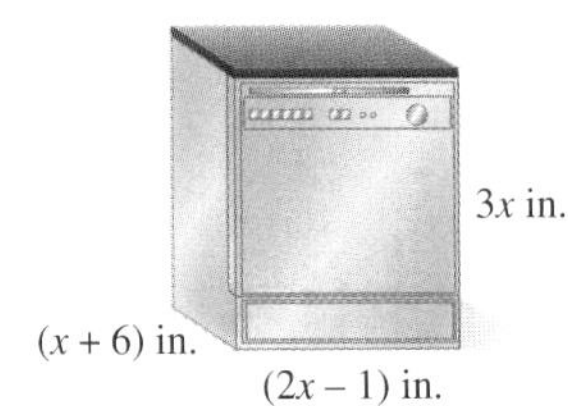

SECTION 10.7 Dividing Polynomials by Monomials

To divide monomials, use the method for simplifying fractions or use the rules for exponents.

Simplify each expression.

103. $\frac{-14x^2y}{21xy^3}$

104. $\frac{(x^2)^2}{xx^4}$

To divide a polynomial by a monomial, divide each term of the numerator by the denominator.

Perform each division. All variables represent positive numbers.

105. $\frac{8x + 6}{2}$

106. $\frac{14xy - 21x}{7xy}$

107. $\frac{15a^2b + 20ab^2 - 25ab}{5ab}$

108. $\frac{(x + y)^2 + (x - y)^2}{-2xy}$

109. SAVINGS BONDS How many \$50 savings bonds would have a total value of \$$(50x + 250)$?

SECTION 10.8 Dividing Polynomials by Polynomials

Long division is used to divide one polynomial by another. When a division has a remainder, write the answer in the form

$$\text{Quotient} + \frac{\text{remainder}}{\text{divisor}}$$

The division method works best when the exponents of the terms of the divisor and the dividend are written in descending order.

When the dividend is missing a term, write it with a coefficient of zero or leave a blank space.

Perform each division.

110. $x + 2\overline{)x^2 + 3x + 5}$

111. $x - 1\overline{)x^2 - 6x + 5}$

112. $\frac{2x^2 + 3 + 7x}{x + 3}$

113. $\frac{3x^2 + 14x - 2}{3x - 1}$

114. $2x - 1\overline{)6x^3 + x^2 + 1}$

115. $3x + 1\overline{)-13x - 4 + 9x^3}$

116. Use multiplication to show that the answer when dividing by $3y^2 + 11y + 6$ by $y + 3$ is $3y + 2$.

117. ZOOLOGY The distance in inches traveled by a certain type of snail in $(2x - 1)$ minutes is given by the polynomial $8x^2 + 2x - 3$. At what rate did the snail travel?

CHAPTER 10 TEST

1. Use exponents to rewrite $2xxxyyyy$.

2. Evaluate: $(3 + 5)^2$.

Write each expression as an expression containing only one exponent.

3. $y^2(yy^3)$

4. $(2x^3)^5(x^2)^3$

Simplify each expression. Write answers without using parentheses or negative exponents.

5. $3x^0$

6. $2y^{-5}y^2$

7. $\dfrac{y^2}{yy^{-2}}$

8. $\left(\dfrac{a^2b^{-1}}{4a^3b^{-2}}\right)^{-3}$

9. What is the volume of a cube that has sides of length $10y^4$ inches?

10. Rewrite 4^{-2} using a positive exponent and then evaluate the result.

11. ELECTRICITY One ampere (amp) corresponds to the flow of 6,250,000,000,000,000,000 electrons per second past any point in a direct current (DC) circuit. Write this number in scientific notation.

12. Write 9.3×10^{-5} in standard notation.

13. Identify $3x^2 + 2$ as a monomial, binomial, or trinomial.

14. Find the degree of the polynomial $3x^2y^3 + 2x^3y - 5x^2y$.

15. If $f(x) = x^2 + x - 2$, find $f(-2)$.

16. Simplify: $(xy)^2 + 5x^2y^2 - (3x)^2y^2$.

17. Simplify: $-6(x - y) + 2(x + y) - 3(x + 2y)$.

18. Subtract: $\begin{array}{r} 2x^2 - 7x + 3 \\ - \quad \underline{3x^2 - 2x - 1} \end{array}$

Find each product.

19. $(-2x^3)(2x^2y)$

20. $3y^2(y^2 - 2y + 3)$

21. $(x - 9)(x + 9)$

22. $(3y - 4)^2$

23. $(2x - 5)(3x + 4)$

24. $(2x - 3)(x^2 - 2x + 4)$

25. Solve: $(a + 2)^2 = (a - 3)^2$.

26. Simplify: $\dfrac{8x^2y^3z^4}{16x^3y^2z^4}$.

27. Simplify: $\dfrac{6a^2 - 12b^2}{24ab}$.

28. Divide: $2x + 3\overline{)2x^2 - x - 6}$.

29. In your own words, explain this rule for exponents:
$x^{-n} = \dfrac{1}{x^n}$

30. A rectangle has an area of $(x^2 - 6x + 5)$ ft^2 and a length of $(x - 1)$ feet. Show how division can be used to find the width of the rectangle. Explain your steps.

CHAPTERS 1–10 CUMULATIVE REVIEW EXERCISES

1. SPORTS CARS The graph shows the Porsche vehicle sales in the United States for the years 1986–2001.
   a. In what year were sales the lowest?
   b. In what year were sales the greatest?
   c. Between what two years was there the greatest increase in sales?

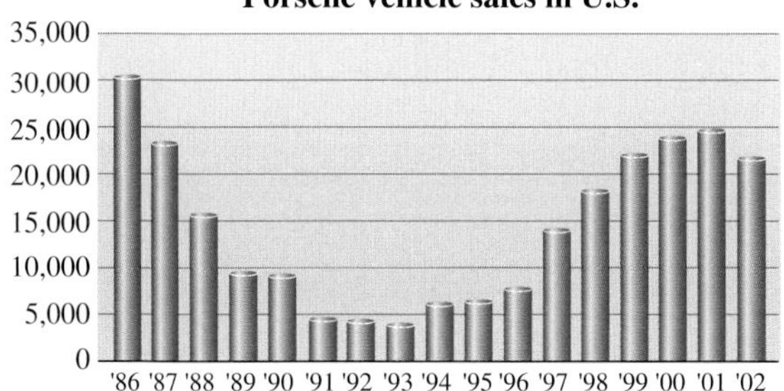

Source: Porsche Cars North America

2. Divide: $\frac{3}{4} \div \frac{6}{5}$.

3. Subtract: $\frac{7}{10} - \frac{1}{14}$.

4. Is π a rational or irrational number?

5. RACING Suppose a driver has completed x laps of a 250-lap race. Write an expression for how many more laps he must make to finish the race.

6. CLINICAL TRIALS In a clinical test of Aricept, a drug to treat Alzheimer's disease, one group of patients took a placebo (a sugar pill) while another group took the actual medication. See the table. Find the number of patients in each group who experienced nausea. Round to the nearest whole number.

Comparison of rates of adverse events in patients		
Adverse event	Group 1: Placebo (number = 315)	Group 2: Aricept (number = 311)
Nausea	6%	5%

Consider the algebraic expression $3x^3 + 5x^2y + 37y$.

7. Find the coefficient of the second term.

8. What is the third term?

Simplify each expression.

9. $3x - 5x + 2y$

10. $3(x - 7) + 2(8 - x)$

11. $2x^2y^3 - xy(xy^2)$

12. $x^2(3 - y) + x(xy + x)$

Solve each equation.

13. $3(x - 5) + 2 = 2x$

14. $\frac{x - 5}{3} - 5 = 7$

Solve each formula for the variable indicated.

15. $A = \frac{1}{2}h(b + B)$; for h

16. $y = mx + b$; for x

Evaluate each expression.

17. $4^2 - 5^2$

18. $(4 - 5)^2$

19. $\frac{-3 - (-7)}{2^2 - 3}$

20. $12 - 2[1 - (-8 + 2)]$

Solve each inequality and graph the solution set. Then describe the solution using interval notation

21. $8(4 + x) > 10(6 + x)$

22. $-9 < 3(x + 2) \leq 3$

Graph each equation.

23. $y = x^2$

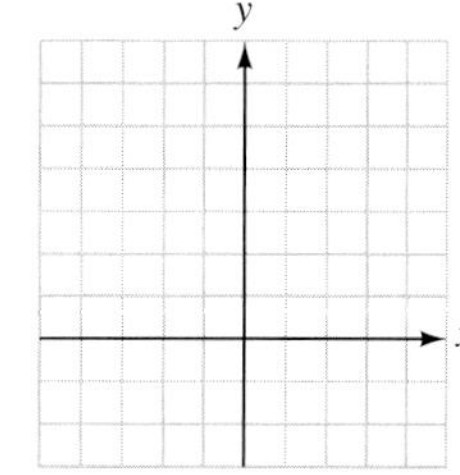

24. $y = |x|$

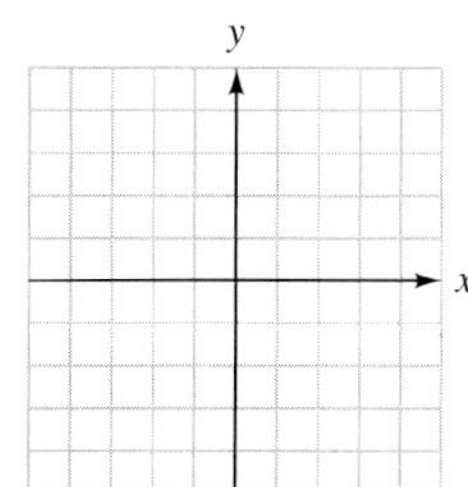

25. $4x - 3y = 12$

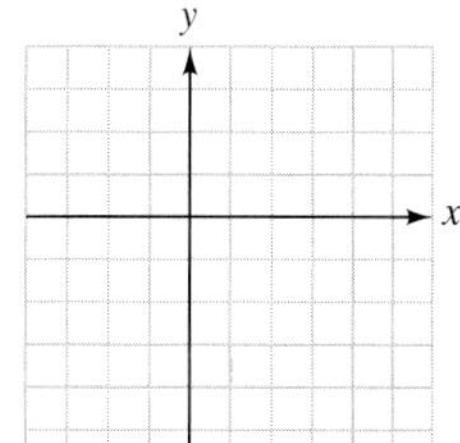

26. $3x = 12$

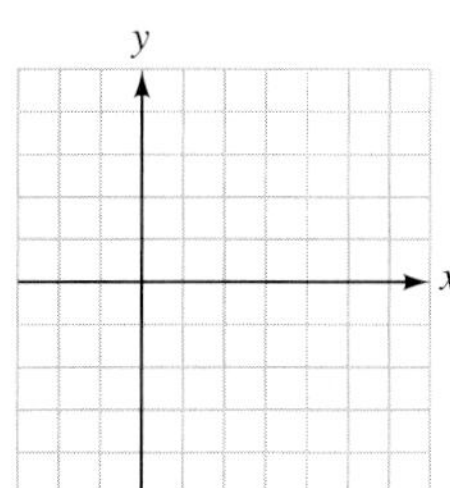

Find the slope of the line with the given properties.

27. Passing through $(-2, 4)$ and $(6, 8)$

28. A line that is horizontal

29. An equation of $y = -4x + 3$

30. An equation of $2x - 3y = 12$

Write the equation of the line with the following properties.

31. Slope $= \frac{2}{3}$, y-intercept $= (0, 5)$

32. Passing through $(-2, 4)$ and $(6, 10)$

33. A horizontal line passing through (2, 4)

34. A vertical line passing through (2, 4)

Are the graphs of the lines parallel or perpendicular?

35. $y = -\frac{3}{4}x + \frac{15}{4}$

$4x - 3y = 25$

36. $y = -\frac{3}{4}x + \frac{15}{4}$

$6x = 15 - 8y$

Determine whether each equation defines a function.

37. $y = x^3 - 4$

38. $x = |y|$

Let $f(x) = 2x^2 - 3$ and find each value.

39. $f(0)$

40. $f(3)$

41. $f(-2)$

42. $f(0.5)$

43. Find the domain and range of the function $f(x) = |x|$.

44. Determine whether the graph is the graph of a function.

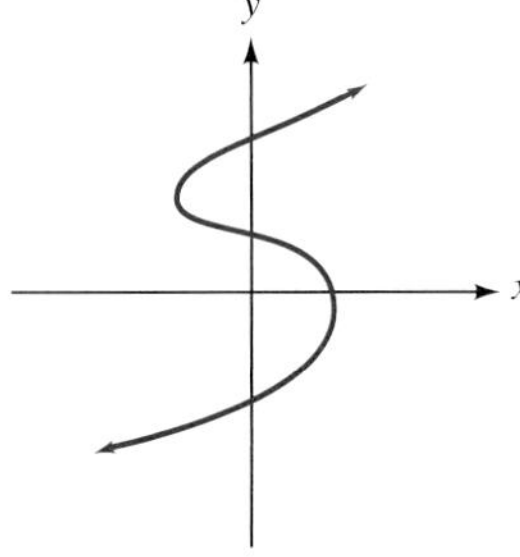

45. Give an example of a positive rate of change and a negative rate of change.

46. Evaluate: -3^2.

Write each expression using one positive exponent.

47. $(y^3y^5)y^6$

48. $(x^3x^4)^2$

49. $\frac{x^3x^4}{x^2x^3}$

50. $\frac{a^4b^0}{a^{-3}}$

51. x^{-5}

52. $(-2y)^{-4}$

53. $(x^{-4})^2$

54. $\left(-\frac{x^3}{x^{-2}}\right)^3$

Write each number in scientific notation.

55. 615,000

56. 0.0000013

Write each number in standard notation.

57. 5.25×10^{-4}

58. 2.77×10^3

Give the degree of each polynomial.

59. $3x^2 + 2x - 5$

60. $-3x^3y^2 + 3x^2y^2 - xy$

61. MUSICAL INSTRUMENTS The gong is a percussion instrument used throughout Southeast Asia. The amount of deflection of the horizontal support (in inches) shown below is given by the polynomial function

$$f(x) = 0.01875x^4 - 0.15x^3 + 1.2x$$

where x is the distance (in feet) that the gong is hung from one end of the support. Find the deflection if the gong is hung in the middle of the support.

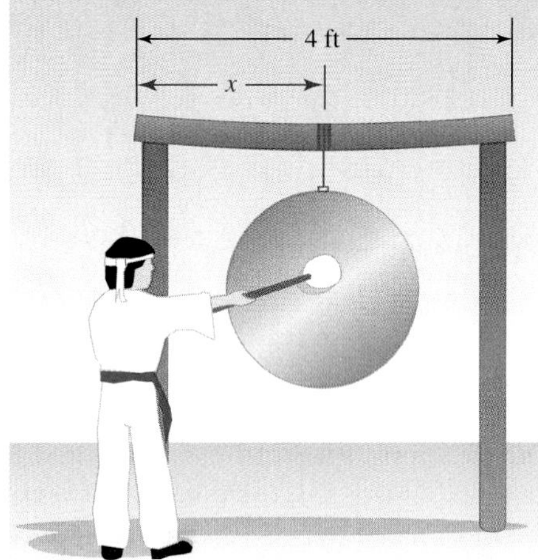

62. Consider the polynomial $2x^2 - 5x + 9$. Determine each of the following.

a. The number of terms

b. The lead term

c. The lead coefficient

d. The degree of the second term

e. The degree of the polynomial

f. The constant term

Perform the operations.

63. $(3x^2 + 2x - 7) - (2x^2 - 2x + 7)$

64. $(2x^2 - 3x + 4) + (2x^2 + 2x - 5)$

65. $-5x^2(7x^3 - 2x^2 - 2)$

66. $(3x^3y^2)(-4x^2y^3)$

67. $(3x - 7)(2x + 8)$

68. $(5x - 4y)(3x + 2y)$

69. $(3x + 1)^2$

70. $(x - 2)(x^2 + 2x + 4)$

71. $\frac{6x^2 - 8x}{2x}$

72. $x - 3\overline{)2x^2 - 5x - 3}$

CHAPTER 11

Factoring and Quadratic Equations

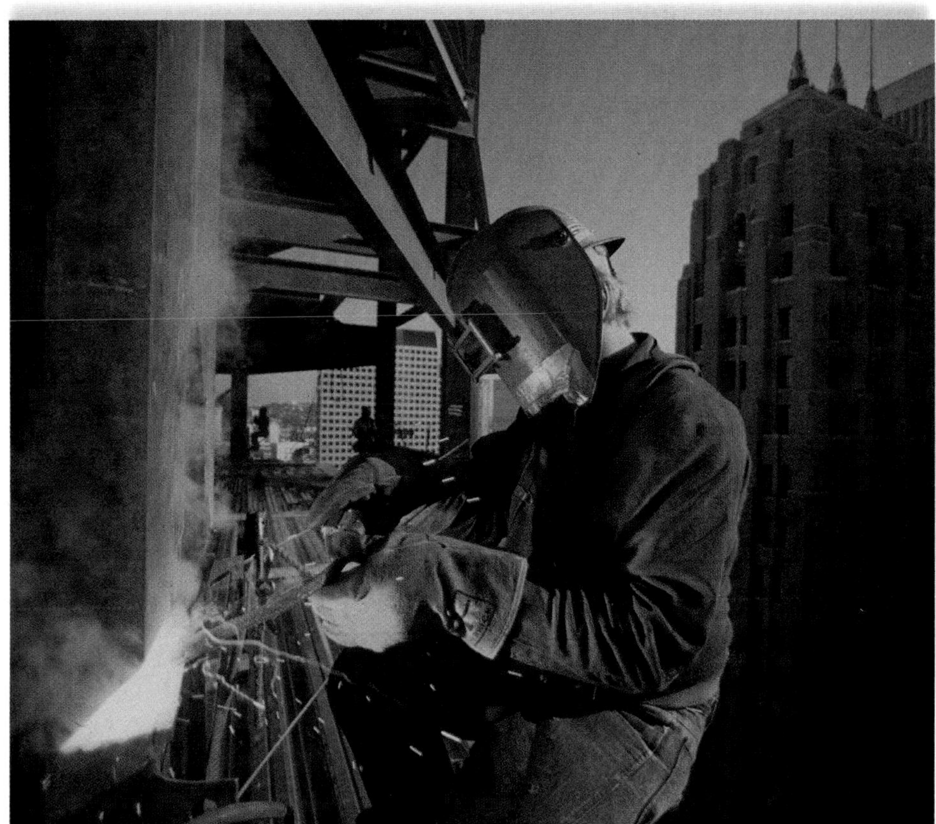

CORBIS

TLE Construction projects come in all shapes and sizes. Whether remodeling a kitchen or erecting a giant skyscraper, algebra and geometry play an important role from start to finish. In this chapter, we explore other applications that involve these branches of mathematics. Using an algebraic process called *factoring,* we will solve applied problems that deal with area, one of the fundamental concepts of geometry.

To learn more about the use of algebra in construction projects, visit *The Learning Equation* on the Internet at http://tle.brookscole.com. (The log-in instructions are in the Preface.) For Chapter 11, the online lessons are:

- *TLE* Lesson 21: Factoring Trinomials and the Difference of Two Squares
- *TLE* Lesson 22: Solving Quadratic Equations by Factoring

Check Your Knowledge

1. A ______ number is a natural number greater than 1 whose only factors are 1 and itself.
2. When we write $2x + 4$ as $2(x + 2)$, we say that we have ______ ______ the GCF, 2.
3. A ______ polynomial cannot be factored using only integers.
4. If the length of the hypotenuse of a right triangle is c and the lengths of the legs of the triangle are a and b, then $c^2 =$ ______.
5. A ______ equation is any equation that can be written in the form $ax^2 + bx + c = 0$, where a, b, and c are real numbers and $a \neq 0$.
6. If the product of two numbers is 0, then at least one of the numbers must be ______.

7. Find the prime factorization of 156.
8. What is the greatest common factor of $8x^3yz^2$ and $20x^5y^2$?

Factor each polynomial completely. If the polynomial cannot be factored, write "prime."

9. $6y^2 + 48y$
10. $3xyz^2 - 6x^2yz^2 + 15x^3y^3z^3$
11. $4x^2 + 49$
12. $9x^2 - 25$
13. $162x^4 - 32y^4$
14. $y^2 + 9y + 18$
15. $x^2 - 8x + 16$
16. $nx + ny - 3x - 3y$
17. $3x^2 + 10x - 8$
18. $27x^3 - 1$

19. The area of a rectangular desktop is $y^2 + 7y + 10$ square units. The length and width can both be written in the form $(y + a)$, where a is a whole number. Find expressions for length and width.

Solve each equation.

20. $(x - 5)(2x - 7) = 0$
21. $16x^2 - 49 = 0$
22. $-16t^2 + 48t = 0$
23. $x^2 - 10x + 16 = 0$
24. $x^2 - 3x = 18$
25. $10x^2 - 3x - 1 = 0$
26. The equation $h = -16t^2 + 80t$ gives the height in feet above the ground, h, of a baseball after t seconds. For what values of t will the height h be 96 feet?
27. A right triangle has legs of lengths x cm and $(x + 1)$ cm. The hypotenuse is $(x + 2)$ cm. Find the lengths of all three sides.

Study Skills Workshop

MIDTERM TUNE-UP

Getty Images

At midterm in your class, take a moment to evaluate your performance — there may still be time to make changes that will significantly impact your final grade.

What Is Your Overall Grade at Midterm? Many instructors give you information on how to calculate your grade based on test, homework, and other scores. Some instructors give formal progress reports to students when tests are returned. Meet with your instructor during office hours if you are not sure how to compute your grade and/or don't know how well you're doing. Based on your midterm grade, ask the following questions:

Should I Continue in the Class? If your midterm grade is a D or F, ask your instructor about your chances for passing the class. In many circumstances, it may be hard to raise your skills to a level that will allow you to be successful in your next math class. It may be best for you to drop the class at this point and spend extra time reviewing the material from your prerequisite class before re-enrolling the next term. (Remember that the most important indicator for success in a math class is the mastery of prerequisite skills!) You may be able to find help in reviewing your prerequisite skills in your college's tutorial center or with software such as that found on the iLrn Web site.

What Can I Do to Improve My Chances of Success? Especially if you find yourself at the C grade level at midterm, you want to do all that you can to improve your skills. Mastery of the skills that you are learning in this class is the most important indicator of success for your next math class! Ask yourself these questions:

1. Am I missing classes?
2. Am I adhering to my study calendar?
3. Am I taking good notes and reworking them to use as a study aid?
4. Am I using suggested techniques and spending the appropriate amount of time studying for tests?
5. Am I participating in a study group? If so, is it working well?
6. Am I seeking a tutor's or my instructor's help when I have unanswered questions?

Even if you have worked hard and are earning an A or B grade at midterm, you may still benefit by examining the above questions and fine-tuning your study routine. Strive to understand concepts as well as to perform routine tasks.

If you are doing well in the course so far, congratulate yourself and get ready to continue to work hard to maintain your grade. If your grade is not where you want it to be, you still have time to improve. Commit to improving your study habits, and schedule a meeting with your instructor to get ideas on how to raise your grade. Most important, don't give up — you have time to increase your success in the second half of the course.

ASSIGNMENT

1. What is your grade at midterm?
2. Respond to the six questions above. Where do you need to make adjustments?
3. Which of the above questions are you doing best with? Which one are you having the most trouble with?
4. What changes can you make that will enable you to do better?
5. What topics have you understood best so far in your class? Which topics are your weakest?

In this chapter, we will factor polynomials. Factoring reverses the process of multiplication. It can be used to simplify expressions and solve equations.

11.1 Factoring Out the Greatest Common Factor and Factoring by Grouping

- Factoring natural numbers • The greatest common factor (GCF)
- Finding the GCF of several monomials • Factoring out the greatest common factor
- Factoring out a negative factor • Factoring by grouping

Recall that we use the distributive property to multiply a monomial and a binomial. For example,

$$\begin{aligned} \mathbf{4y}(3y + 5) &= \mathbf{4y} \cdot 3y + \mathbf{4y} \cdot 5 \\ &= 12y^2 + 20y \end{aligned}$$

In this section, we will reverse the operation of multiplication. Given a polynomial such as $12y^2 + 20y$, we will ask, "What factors were multiplied to obtain $12y^2 + 20y$?" The process of finding the factors of a known product is called **factoring.**

Multiplication	**Factoring**
Given the factors . . . find the product	Given the product . . . find the factors
↓ ↓ ↓	↓ ↓ ↓
$4y(3y + 5) = ?$	$12y^2 + 20y = ?(\ ?\ )$

To begin the discussion of factoring, we consider two methods that can be used to factor natural numbers.

Factoring natural numbers

Because 4 divides 12 exactly, 4 is called a **factor** of 12. The numbers 1, 2, 3, 4, 6, and 12 are the natural-number factors of 12, because each divides 12 exactly.

> A **prime number** is a natural number greater than 1 whose only factors are 1 and itself.

For example, 17 is a prime number, because

1. 17 is a natural number greater than 1, and
2. the only two natural-number factors of 17 are 1 and 17.

The prime numbers less than 50 are

2, 3, 5, 7, 11, 13, 17, 19, 23, 29, 31, 37, 41, 43, and 47

A natural number is said to be in **prime-factored form** if it is written as the product of factors that are prime numbers.

To find the prime-factored form of a natural number, we can use a **factoring tree.** The following examples show two ways to find the prime-factored form of 90 using factoring trees. The process stops when a row of the tree contains only prime-number factors.

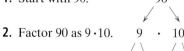

1. Start with 90.	90	**1.** Start with 90.	90
2. Factor 90 as $9 \cdot 10$.	$9 \cdot 10$	**2.** Factor 90 as $6 \cdot 15$.	$6 \cdot 15$
3. Factor 9 and 10.	$3 \cdot 3 \cdot 2 \cdot 5$	**3.** Factor 6 and 15.	$2 \cdot 3 \cdot 3 \cdot 5$

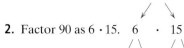

Since the prime factors in either case are $2 \cdot 3 \cdot 3 \cdot 5$, the prime-factored form, or the **prime factorization,** of 90 is $2 \cdot 3^2 \cdot 5$. This example illustrates the **fundamental theorem of arithmetic,** which states that there is only one prime factorization for every natural number greater than 1.

We can also find the prime factorization of a natural number using the **division method.** For example, to find the prime factorization of 42, we begin by choosing the *smallest* prime number that will divide the given number exactly. We continue this process until the result of the division is a prime number.

Step 1: 2 divides 42 exactly. The result is 21, which is not prime. We continue the process.

$$\begin{array}{r|l} 2 & 42 \\ \hline & 21 \end{array}$$

Step 2: We choose the smallest prime number that divides 21. The prime number 2 does not divide 21 exactly, but 3 does. The result is 7, which is prime. We are done.

$$\begin{array}{r|l} 2 & 42 \\ \hline 3 & 21 \\ \hline & 7 \end{array}$$

The prime factorization of 42 is $2 \cdot 3 \cdot 7$.

EXAMPLE 1 Find the prime factorization of 150.

Solution We can use a factoring tree or the division method to find the prime factorization.

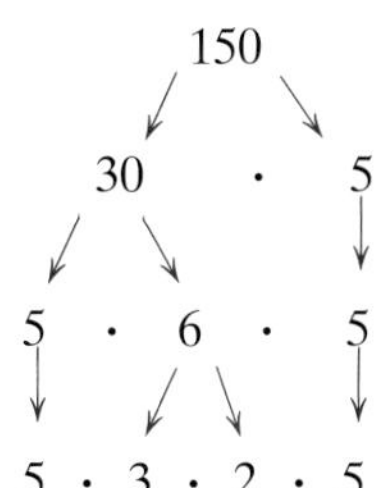

$$\begin{array}{r|l} 2 & 150 \\ \hline 3 & 75 \quad \leftarrow 150 \div 2 \\ \hline 5 & 25 \quad \leftarrow 75 \div 3 \\ \hline & 5 \quad \leftarrow 25 \div 5 \end{array}$$

Using exponents, we can write the prime factorization of 150 as $2 \cdot 3 \cdot 5^2$.

Self Check 1
Find the prime factorization of 225.

Answer $3^2 \cdot 5^2$

The greatest common factor (GCF)

The right-hand sides of the equations

$$90 = \mathbf{2} \cdot \mathbf{3} \cdot 3 \cdot 5$$
$$42 = \mathbf{2} \cdot \mathbf{3} \cdot 7$$

show the prime-factored forms of 90 and 42. The color highlighting indicates that 90 and 42 have one prime factor of 2 and one prime factor of 3 in common. We can conclude that $2 \cdot 3 = 6$ is the largest natural number that divides 90 and 42 exactly, and we say that 6 is their **greatest common factor (GCF).**

$$\frac{90}{\mathbf{6}} = 15 \quad \text{and} \quad \frac{42}{\mathbf{6}} = 7$$

THINK IT THROUGH Inflation and the Price of Goods

"Rising worker productivity in recent years has helped to keep U.S. inflation in check by letting firms produce more goods without adding to wage costs."
Alan Greenspan, Federal Reserve Chairman, 2004

The *inflation factors* in the table below can be used to estimate what items will cost in the future. For example, if a gallon of milk costs \$3.00 today, and if there is an annual inflation rate of 4%, we can find what a gallon of milk will cost in 3 years by multiplying \$3.00 by the inflation factor 1.12. At that rate, a gallon of milk will cost \$3.00 · 1.12, or about \$3.36 in three years.

Go online and search for a Web site that gives the current inflation rate. Round the rate to the nearest percent. Then use the inflation factors in the table below to estimate what the cost of an item that you regularly purchase will be 1, 2, 3, 4, and 5 years in the future. (Assume that the current inflation rate is an annual rate that remains constant.)

Inflation factors

Years	2%	3%	4%	5%	6%	7%
1	1.02	1.03	1.04	1.05	1.06	1.07
2	1.04	1.06	1.08	1.10	1.12	1.14
3	1.06	1.09	1.12	1.16	1.19	1.23
4	1.08	1.13	1.17	1.22	1.26	1.31
5	1.10	1.16	1.22	1.28	1.34	1.40

Self Check 2
Find the GCF for 45, 60, 75.

EXAMPLE 2 Find the greatest common factor of 24, 60, and 96.

Solution We write each prime factorization and highlight the prime factors the three numbers have in common.

$$24 = \mathbf{2} \cdot \mathbf{2} \cdot 2 \cdot 3$$
$$60 = \mathbf{2} \cdot \mathbf{2} \cdot 3 \cdot 5$$
$$96 = \mathbf{2} \cdot \mathbf{2} \cdot 2 \cdot 2 \cdot 2 \cdot 3$$

Since 24, 60, and 96 each have two factors of 2 and one factor of 3, their greatest common factor is $2 \cdot 2 \cdot 3 = 12$.

Answer $3 \cdot 5 = 15$

Finding the GCF of several monomials

The right-hand sides of the equations

$$12y^2 = \mathbf{2} \cdot \mathbf{2} \cdot 3 \cdot y \cdot y$$
$$20y = \mathbf{2} \cdot \mathbf{2} \cdot 5 \cdot y$$

show the prime factorizations of $12y^2$ and $20y$. Since the monomials have two factors of 2 and one factor of y in common, their GCF is

$$2 \cdot 2 \cdot y \quad \text{or} \quad 4y$$

To find the GCF of several monomials, we follow these steps.

Strategy for finding the greatest common factor

1. Find the prime factorization of each monomial.
2. List each common factor the least number of times it appears in any one monomial.
3. Find the product of the factors in the list to obtain the GCF.

EXAMPLE 3 Find the GCF of $10x^3y^2$, $60x^2y$, and $30xy^2$.

Solution

Step 1: Find the prime factorization of each monomial.

$$10x^3y^2 = 2 \cdot 5 \cdot x \cdot x \cdot x \cdot y \cdot y$$
$$60x^2y = 2 \cdot 2 \cdot 3 \cdot 5 \cdot x \cdot x \cdot y$$
$$30xy^2 = 2 \cdot 3 \cdot 5 \cdot x \cdot y \cdot y$$

Step 2: List each common factor the least number of times it appears in any one monomial: 2, 5, x, and y.

Step 3: Find the product of the factors in the list:

$2 \cdot 5 \cdot x \cdot y = 10xy$ The GCF is $10xy$.

Self Check 3
Find the GCF of $20a^2b^3$, $12ab^4$, and $8a^3b^2$.

Answer $4ab^2$

Factoring out the greatest common factor

To factor $12y^2 + 20y$, we find the GCF of $12y^2$ and $20y$ (which is $4y$) and use the distributive property.

$12y^2 + 20y = 4y \cdot 3y + 4y \cdot 5$ Write each term of the polynomial as the product of the GCF, $4y$, and one other factor.

$= 4y(3y + 5)$ $4y$ is a common factor of both terms.

This process is called **factoring out the greatest common factor.**

EXAMPLE 4 Factor: $25 - 5m$.

Solution To find the GCF of 25 and $5m$, we find their prime factorizations.

$$\left.\begin{aligned} 25 &= 5 \cdot 5 \\ 5m &= 5 \cdot m \end{aligned}\right\} \quad \text{GCF} = 5$$

We can use the distributive property to factor out the GCF.

$25 - 5m = 5 \cdot 5 - 5 \cdot m$ Factor each monomial using 5 and one other factor.

$= 5(5 - m)$ Factor out the common factor of 5.

We check by verifying that $5(5 - m) = 25 - 5m$.

Self Check 4
Factor: $18x - 24$.

Answer $6(3x - 4)$

Self Check 5
Factor: $32x^2y^4 + 12x^3y^3$.

Answer $4x^2y^3(8y + 3x)$

EXAMPLE 5 Factor: $35a^3b^2 + 14a^2b^3$.

Solution To find the GCF, we find the prime factorizations of $35a^3b^2$ and $14a^2b^3$.

$$\left.\begin{aligned} 35a^3b^2 &= 5 \cdot 7 \cdot a \cdot a \cdot a \cdot b \cdot b \\ 14a^2b^3 &= 2 \cdot 7 \cdot a \cdot a \cdot b \cdot b \cdot b \end{aligned}\right\} \quad \text{GCF} = 7 \cdot a \cdot a \cdot b \cdot b = 7a^2b^2$$

We factor out the GCF, $7a^2b^2$.

$$\begin{aligned} 35a^3b^2 + 14a^2b^3 &= \mathbf{7a^2b^2} \cdot 5a + \mathbf{7a^2b^2} \cdot 2b \\ &= \mathbf{7a^2b^2}(5a + 2b) \end{aligned}$$

We check by verifying that $7a^2b^2(5a + 2b) = 35a^3b^2 + 14a^2b^3$.

Self Check 6
Factor: $2ab^2c + 4a^2bc - ab$.

Answer $ab(2bc + 4ac - 1)$

EXAMPLE 6 Factor: $4x^3y^2z - 2x^2yz + xz$.

Solution The expression has three terms. We factor out the GCF, which is xz.

$$\begin{aligned} 4x^3y^2z - 2x^2yz + xz &= \mathbf{xz} \cdot 4x^2y^2 - \mathbf{xz} \cdot 2xy + \mathbf{xz} \cdot \mathbf{1} \\ &= \mathbf{xz}(4x^2y^2 - 2xy + \mathbf{1}) \end{aligned}$$

The last term of $4x^3y^2z - 2x^2yz + xz$ has an implied coefficient of 1. When xz is factored out, we must write this coefficient of 1, as shown in blue. We check by verifying that $xz(4x^2y^2 - 2xy + 1) = 4x^3y^2z - 2x^2yz + xz$.

EXAMPLE 7 **Crayons.** The amount of colored wax used to make the crayon shown in Figure 11-1 can be found by computing its volume using the formula

$$V = \pi r^2 h_1 + \frac{1}{3}\pi r^2 h_2$$

Factor the expression on the right-hand side of this equation.

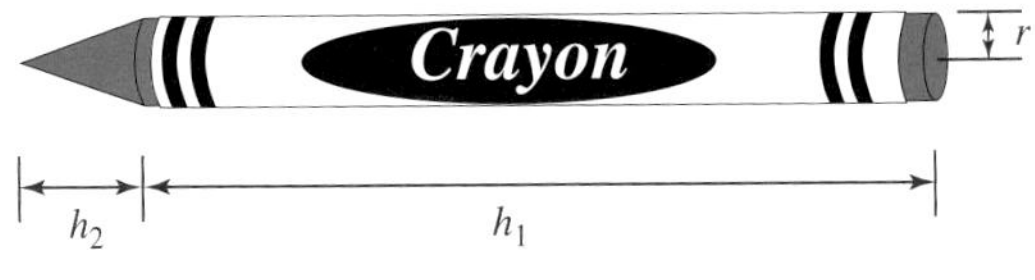

FIGURE 11-1

Solution Each term on the right-hand side of the formula contains a factor of π and r^2.

$$\begin{aligned} V &= \boldsymbol{\pi r^2} h_1 + \frac{1}{3}\boldsymbol{\pi r^2} h_2 \\ &= \boldsymbol{\pi r^2}\left(h_1 + \frac{1}{3}h_2\right) \quad \text{Factor out the GCF, } \pi r^2. \end{aligned}$$

The formula to find the volume of the crayon can be expressed as

$$V = \pi r^2\left(h_1 + \frac{1}{3}h_2\right).$$

EXAMPLE 8 Factor: $x(x + 4) + 3(x + 4)$.

Solution The given expression has two terms:

$$\underbrace{x(x + 4)}_{\text{The first term}} + \underbrace{3(x + 4)}_{\text{The second term}}$$

The GCF of the terms is $x + 4$, which can be factored out.

$$x(\mathbf{x + 4}) + 3(\mathbf{x + 4}) = (\mathbf{x + 4})(x + 3)$$

Self Check 8
Factor: $2y(y - 1) - 7(y - 1)$.

Answer $(y - 1)(2y - 7)$

Factoring out a negative factor

It is often useful to factor out a common factor having a negative coefficient.

EXAMPLE 9 Factor -1 out of $-a^3 + 2a^2 - 4$.

Solution First, we write each term of the polynomial as the product of -1 and another factor. Then we factor out the common factor of -1.

$$\begin{aligned} -a^3 + 2a^2 - 4 &= (\mathbf{-1})a^3 + (\mathbf{-1})(-2a^2) + (\mathbf{-1})4 \\ &= \mathbf{-1}(a^3 - 2a^2 + 4) && \text{Factor out } -1. \\ &= -(a^3 - 2a^2 + 4) && \text{The coefficient of 1 need not be written.} \end{aligned}$$

We check by verifying that $-(a^3 - 2a^2 + 4) = -a^3 + 2a^2 - 4$.

Self Check 9
Factor -1 out of $-b^4 - 3b^2 + 2$.

Answer $-(b^4 + 3b^2 - 2)$

EXAMPLE 10 Factor out the negative (opposite) of the GCF in $-18a^2b + 6ab^2 - 12a^2b^2$.

Solution The GCF is $6ab$. To factor out its negative, we write each term of the polynomial as the product of $-6ab$ and another factor. Then we factor out $-6ab$.

$$\begin{aligned} -18a^2b + 6ab^2 - 12a^2b^2 &= (\mathbf{-6ab})3a - (\mathbf{-6ab})b + (\mathbf{-6ab})2ab \\ &= \mathbf{-6ab}(3a - b + 2ab) \end{aligned}$$

We check by verifying that $-6ab(3a - b + 2ab) = -18a^2b + 6ab^2 - 12a^2b^2$.

Self Check 10
Factor out the negative (opposite) of the GCF in $-27xy^2 - 18x^2y + 36x^2y^2$.

Answer $-9xy(3y + 2x - 4xy)$

Factoring by grouping

Suppose we wish to factor the polynomial

$$ax + ay + cx + cy$$

Although no factor is common to all four terms, there is a common factor of a in $ax + ay$ and a common factor of c in $cx + cy$. We can factor out a and c, and then factor out $x + y$ to obtain

$$\begin{aligned} ax + ay + cx + cy &= a(\mathbf{x + y}) + c(\mathbf{x + y}) \\ &= (\mathbf{x + y})(a + c) && \text{Factor out } x + y. \end{aligned}$$

We can check the result by multiplication.

$$(x + y)(a + c) = ax + cx + ay + cy$$
$$= ax + ay + cx + cy \quad \text{Rearrange the terms.}$$

Thus, $ax + ay + cx + cy$ factors as $(x + y)(a + c)$. This type of factoring is called **factoring by grouping.**

Factoring by Grouping

1. Group the terms of the polynomial so that the first two terms have a common factor and the last two terms have a common factor.
2. Factor out the common factor from each group.
3. Factor out the resulting common binomial factor. If there is no common binomial factor, regroup the terms of the polynomial and repeat steps 2 and 3.

Self Check 11
Factor: $7x - 7y + xy - y^2$.

Answer $(x - y)(7 + y)$

EXAMPLE 11 Factor: $2c - 2d + cd - d^2$.

Solution Since 2 is a common factor of the first two terms and d is a common factor of the last two terms, we have

$$2c - 2d + cd - d^2 = 2(\mathbf{c - d}) + d(\mathbf{c - d}) \quad \text{Factor out 2 from } 2c - 2d \text{ and } d \text{ from } cd - d^2.$$
$$= (\mathbf{c - d})(2 + d) \quad \text{Factor out } c - d.$$

We check by verifying that

$$(c - d)(2 + d) = 2c + cd - 2d - d^2$$
$$= 2c - 2d + cd - d^2 \quad \text{Rearrange the terms.}$$

Self Check 12
Factor: $7bt + 3ct - 7b - 3c$.

Answer $(7b + 3c)(t - 1)$

EXAMPLE 12 Factor: $x^2y - ax - xy + a$.

Solution Since x is a common factor of the first two terms, we can factor it out and proceed as follows.

$$x^2y - ax - xy + a = x(xy - a) - xy + a \quad \text{Factor out } x \text{ from } x^2y - ax.$$

If we factor -1 from $-xy + a$, a common binomial factor $(xy - a)$ appears, which we can factor out.

$$x^2y - ax - xy + a = x(\mathbf{xy - a}) - \mathbf{1}(\mathbf{xy - a})$$
$$= (\mathbf{xy - a})(x - 1) \quad \text{Factor out } xy - a.$$

Check by multiplication.

! **COMMENT** When factoring the expressions in the previous two examples, don't think that $2(c - d) + d(c - d)$ or $x(xy - a) - 1(xy - a)$ are in factored form. For an expression to be in factored form, the result must be a product.

The next example illustrates that when factoring a polynomial, we should always look for a common factor first.

EXAMPLE 13 Factor: $10k + 10m - 2km - 2m^2$.

Solution Since the four terms have a common factor 2, we factor it out first. Then we use factoring by grouping to factor the polynomial within the parentheses. The first two terms have a common factor 5. The last two terms have a common factor $-m$.

$$\begin{aligned} 10k + 10m - 2km - 2m^2 &= 2(5k + 5m - km - m^2) && \text{Factor out the GCF 2.} \\ &= 2[5(\mathbf{k + m}) - m(\mathbf{k + m})] \\ &= 2[(\mathbf{k + m})(5 - m)] && \text{Factor out } k + m. \\ &= 2(k + m)(5 - m) \end{aligned}$$

Use multiplication to check the result.

Self Check 13
Factor: $-4t - 4s - 4tz - 4sz$.

Answer $-4(t + s)(1 + z)$

Section 11.1 STUDY SET

VOCABULARY *Fill in the blanks.*

1. A natural number greater than 1 whose only factors are 1 and itself is called a ______ number.
2. When we write 24 as $2^3 \cdot 3$, we say that 24 has been written in ____________ form.
3. The GCF of several natural numbers is the ______ number that divides each of the numbers exactly.
4. When we write $15x^2 - 25x$ as $5x(3x - 5)$, we say that we have ________ ______ the greatest common factor.
5. The process of finding the individual factors of a known product is called ________.
6. The numbers 1, 2, 3, 4, 6, and 12 are the natural-number ______ of 12.

CONCEPTS *Explain what is wrong with each solution.*

7. Factor: $6a + 9b + 3$.

$$\begin{aligned} 6a + 9b + 3 &= 3(2a + 3b + 0) \\ &= 3(2a + 3b) \end{aligned}$$

8. Prime-factor 100.

$$\begin{array}{r|l} 10 & 100 \\ \hline 5 & 10 \\ \hline & 2 \end{array}$$

$100 = 2 \cdot 5 \cdot 10$

9. Factor out the GCF: $30a^3 - 12a^2$.

$$30a^3 - 12a^2 = 6a(5a^2 - 2a)$$

10. Factor: $ab + b + a + 1$.

$$\begin{aligned} ab + b + a + 1 &= b(a + 1) + (a + 1) \\ &= (a + 1)b \end{aligned}$$

11. What algebraic concept is illustrated in the work shown below?

$$\mathbf{4} \cdot 5x + \mathbf{4} \cdot 3 = \mathbf{4}(5x + 3)$$

12. a. Complete each tree diagram to prime-factor 30.

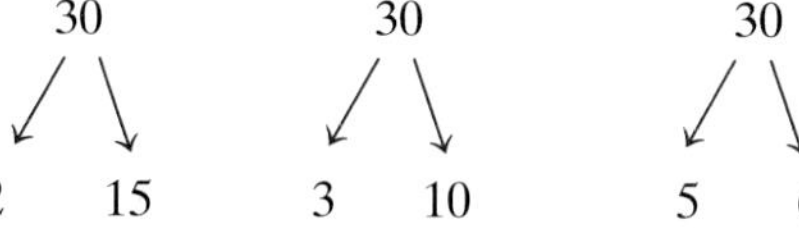

b. Complete the statement: The fundamental theorem of arithmetic states that every natural number greater than 1 has ______ ______ prime factorization.

13. The prime factorizations of three monomials are shown here. Find their GCF.

$3 \cdot 3 \cdot 5 \cdot x \cdot x$

$2 \cdot 3 \cdot 5 \cdot x \cdot y$

$2 \cdot 2 \cdot 3 \cdot x \cdot y \cdot y$

14. Consider the polynomial $2k - 8 + hk - 4h$.
   a. How many terms does the polynomial have?
   b. Is there a common factor of all the terms?
   c. What is the common factor of the first two terms?
   d. What is the common factor of the last two terms?

15. How can we check the answer of the problem shown below?

Factor: $3j^3 + 6j^2 + 2j + 4$.

$$3j^3 + 6j^2 + 2j + 4 = 3j^2(j + 2) + 2(j + 2)$$
$$= (j + 2)(3j^2 + 2)$$

16. List the first 12 prime numbers.

NOTATION *Complete each factorization.*

17. Factor: $b^3 - 6b^2 + 2b - 12$.

$$b^3 - 6b^2 + 2b - 12 = \quad (b - 6) + 2 \quad$$
$$= (b - 6) \quad$$

18. Factor: $12b^3 - 6b^2 + 2b - 2$.

$$12b^3 - 6b^2 + 2b - 2 = \quad (6b^3 - 3b^2 + b - 1)$$

19. In the expression $4x^2y + xy$, what is the coefficient of the last term?

20. Is the following statement true?

$$-(x^2 - 3x + 1) = -1(x^2 - 3x + 1)$$

PRACTICE *Find the prime factorization of each number.*

21. 12 **22.** 24
23. 15 **24.** 20
25. 40 **26.** 62
27. 98 **28.** 112
29. 225 **30.** 144
31. 288 **32.** 968

Complete each factorization.

33. $4a + 12 = \quad (a + 3)$
34. $r^4 + r^2 = r^2(\quad + 1)$
35. $4y^2 + 8y - 2xy = 2y(2y + \quad - \quad)$
36. $3x^2 - 6xy + 9xy^2 = \quad (\quad - 2y + 3y^2)$

Factor out the GCF.

37. $3x + 6$ **38.** $2y - 10$
39. $12x^2 - 6x - 24$
40. $27a^2 - 9a + 45$
41. $t^3 + 2t^2$ **42.** $b^3 - 3b^2$
43. $a^3 - a^2$ **44.** $r^3 + r^2$
45. $24x^2y^3 + 8xy^2$
46. $3x^2y^3 - 9x^4y^3$
47. $12uvw^3 - 18uv^2w^2$
48. $14xyz - 16x^2y^2z$
49. $3x + 3y - 6z$
50. $2x - 4y + 8z$
51. $ab + ac - ad$
52. $rs - rt + ru$
53. $12r^2 - 3rs + 9r^2s^2$
54. $6a^2 - 12a^3b + 36ab$
55. $\pi R^2 - \pi ab$
56. $\frac{1}{3}\pi R^2h - \frac{1}{3}\pi rh$
57. $3(x + 2) - x(x + 2)$
58. $t(5 - s) + 4(5 - s)$
59. $h^2(14 + r) + 14 + r$
60. $k^2(14 + v) - 7(14 + v)$

Factor out −1 from each polynomial.

61. $-a - b$ **62.** $-x - 2y$
63. $-2x + 5y$ **64.** $-3x + 8z$
65. $-3m - 4n + 1$
66. $-3r + 2s - 3$
67. $-3ab - 5ac + 9bc$
68. $-6yz + 12xz - 5xy$

Factor each polynomial by factoring out the negative of the GCF.

69. $-3x^2 - 6x$
70. $-4a^2 + 6a$
71. $-4a^2b^3 + 12a^3b^2$
72. $-25x^4y^3 + 30x^2y^3$
73. $-4a^2b^2c^2 + 14a^2b^2c - 10ab^2c^2$
74. $-10x^4y^3z^2 + 8x^3y^2z - 20x^2y$

Factor by grouping. (Do not combine like terms before factoring.)

75. $2x + 2y + ax + ay$
76. $bx + bz + 5x + 5z$
77. $7r + 7s - kr - ks$
78. $9p - 9q + mp - mq$
79. $xr + xs + yr + ys$
80. $pm - pn + qm - qn$
81. $2ax + 2bx + 3a + 3b$
82. $3xy + 3xz - 5y - 5z$

83. $2ab + 2ac + 3b + 3c$
84. $3ac + a + 3bc + b$
85. $6x^2 - 2x - 15x + 5$
86. $6x^2 + 2x + 9x + 3$
87. $9mp + 3mq - 3np - nq$
88. $ax + bx - a - b$
89. $2xy + y^2 - 2x - y$
90. $2xy - 3y^2 + 2x - 3y$
91. $8z^5 + 12z^2 - 10z^3 - 15$
92. $2a^4 + 2a^3 - 4a - 4$

Factor by grouping. Factor out the GCF first.

93. $ax^3 + bx^3 + 2ax^2y + 2bx^2y$
94. $x^3y^2 - 2x^2y^2 + 3xy^2 - 6y^2$
95. $4a^2b + 12a^2 - 8ab - 24a$
96. $-4abc - 4ac^2 + 2bc + 2c^2$
97. $x^3y - x^2y - xy^2 + y^2$
98. $2x^3z - 4x^2z + 32xz - 64z$

APPLICATIONS

99. PICTURE FRAMING The dimensions of a family portrait and the frame in which it is mounted are shown. Write an algebraic expression that describes
   a. the area of the picture frame.
   b. the area of the portrait.
   c. the area of the mat used in the framing. Express the result in factored form.

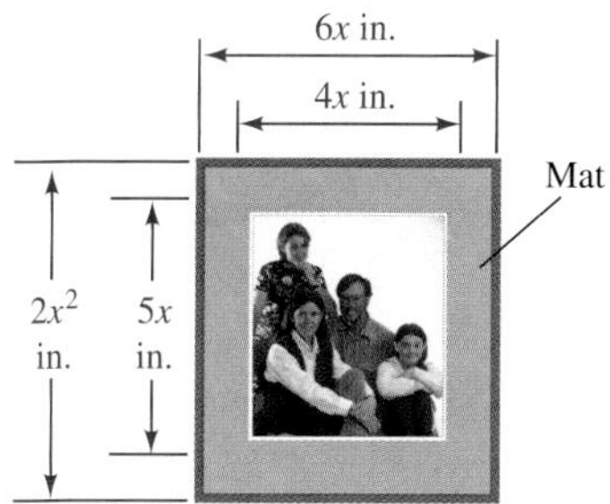

100. REARVIEW MIRRORS The dimensions of the three rearview mirrors on an automobile are given in the illustration in the next column. Write an algebraic expression that gives
   a. the area of the rearview mirror mounted on the windshield.
   b. the total area of the two side mirrors.
   c. the total area of all three mirrors. Express the result in factored form.

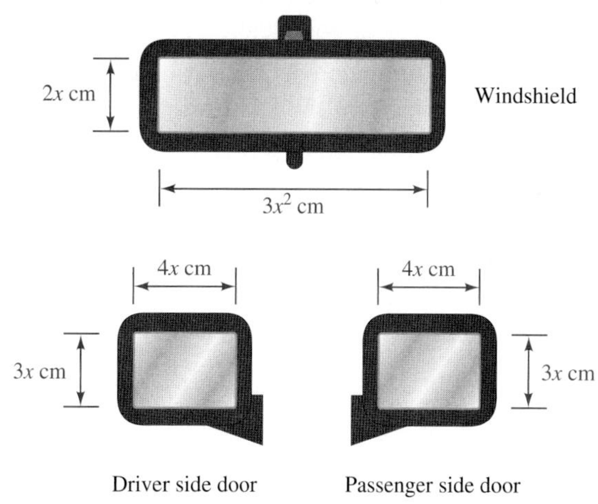

101. COOKING See the following illustration.
   a. What is the length of a side of the square griddle, in terms of r? What is the area of the cooking surface of the griddle, in terms of r?
   b. How many square inches of the cooking surface do the pancakes cover, in terms of r?
   c. Find the amount of cooking surface that is not covered by the pancakes. Express the result in factored form.

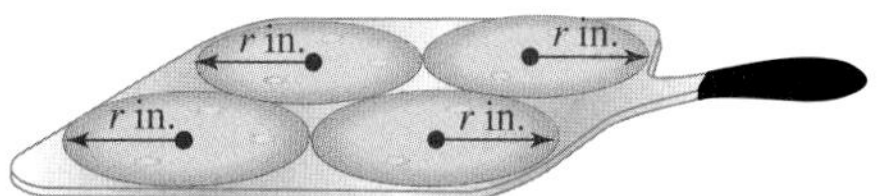

102. AIRCRAFT CARRIERS The rectangular-shaped landing area of $(x^3 + 4x^2 + 5x + 20)$ ft^2 is shown in the illustration. The dimensions of the landing area can be found by factoring. What are the length and width of the landing area?

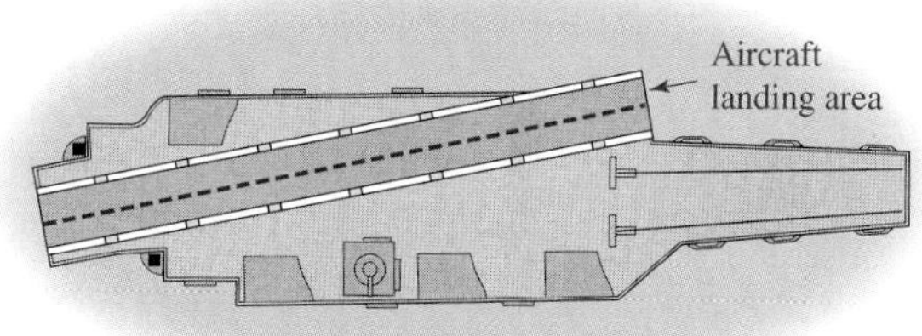

WRITING

103. To add $5x$ and $7x$, we combine like terms: $5x + 7x = 12x$. Explain how this is related to factoring out a common factor.

104. One student commented, "Factoring undoes the distributive property." What do you think she meant? Give an example.

105. If asked to write $ax + ay - bx - by$ in factored form, explain why $a(x + y) - b(x + y)$ is not an acceptable answer.

106. When asked to factor $rx - sy + ry - sx$, a student wrote the expression as $rx + ry - sx - sy$. Then she factored it by grouping. Can the terms be rearranged in this manner? Explain your answer.

REVIEW

107. Simplify: $\left(\frac{y^3y}{2yy^2}\right)^3$.

108. Find the slope of the line passing through the points $(3, 5)$ and $(-2, -7)$.

109. Does the point $(3, 5)$ lie on the graph of the line $4x - y = 7$?

110. Simplify: $-5(3a - 2)(2a + 3)$.

11.2 Factoring Trinomials of the Form $x^2 + bx + c$

- Factoring trinomials that have a lead coefficient of 1
- Multistep factoring
- Prime polynomials

Recall that to multiply two binomials, such as $x + 2$ and $x + 3$, we proceed as follows:

$$(x + 2)(x + 3) = x^2 + 3x + 2x + 6$$
$$= x^2 + 5x + 6$$

In this section, we will reverse the multiplication process. Given a trinomial, such as $x^2 + 5x + 6$, we will ask, "What factors were multiplied to obtain $x^2 + 5x + 6$?" The process of finding the factors of a given trinomial is called *factoring the trinomial.* Since the product of two binomials is often a trinomial, many trinomials factor into the product of two binomials.

Multiplication		**Factoring**	
Given two binomial factors . . .	find the product	Given the product . . .	find the two binomial factors
↓ ↓	↓	↓	↓ ↓
$(x + 2)(x + 3)$	$= ?$	$x^2 + 5x + 6$	$= (\ ?\ )(\ ?\ )$

We will now consider how to factor trinomials of the form $ax^2 + bx + c$, where a (called the **lead coefficient**) is 1.

Factoring trinomials that have a lead coefficient of 1

To develop a method for factoring trinomials, we multiply $(x + a)$ and $(x + b)$.

$$(x + a)(x + b) = x \cdot x + bx + ax + ab \quad \text{Use the FOIL method.}$$
$$= x^2 + ax + bx + ab \quad \text{Write } x \cdot x \text{ as } x^2. \text{ Write } bx + ax \text{ as } ax + bx.$$
$$= x^2 + (a + b)x + ab \quad \text{Factor } x \text{ out of } ax + bx.$$

First term — x^2; Middle term — $(a + b)x$; Last term — ab

The result has three terms. We can see that

- the first term is the product of x and x,
- the last term is the product of a and b, and
- the coefficient of the middle term is the sum of a and b.

We can use these facts to factor trinomials with lead coefficients of 1.

EXAMPLE 1 Factor: $x^2 + 5x + 6$.

Solution Since the first term of the trinomial is x^2, the first term of each binomial factor must be x. To fill in the blanks, we must find two integers whose product is $+6$ and whose sum is $+5$.

$$x^2 + 5x + 6 = (x \quad)(x \quad)$$

The positive factorizations of 6 and the sum of the factors are shown in the table.

Factors of 6	Sum of the factors of 6
1(6)	$1 + 6 = 7$
2(3)	$\mathbf{2 + 3 = 5}$

The last row contains the integers 2 and 3, whose product is 6 and whose sum is 5. To complete the factorization, we enter 2 and 3 as the second terms of the binomial factors.

$$x^2 + 5x + 6 = (x + \mathbf{2})(x + \mathbf{3})$$

To check the result, we verify that $(x + 2)(x + 3)$ is $x^2 + 5x + 6$.

$$\begin{aligned}(x + 2)(x + 3) &= x^2 + 3x + 2x + 6 \\ &= x^2 + 5x + 6\end{aligned}$$

Self Check 1
Factor: $y^2 + 7y + 6$.

Answer $(y + 1)(y + 6)$

! COMMENT When factoring trinomials, the binomial factors can be written in either order. In Example 1, an equivalent factorization is $x^2 + 5x + 6 = (x + 3)(x + 2)$.

EXAMPLE 2 Factor: $y^2 - 7y + 12$.

Solution Since the first term of the trinomial is y^2, the first term of each binomial factor must be y. To fill in the blanks, we must find two integers whose product is 12 and whose sum is -7.

$$y^2 - 7y + 12 = (y \quad)(y \quad)$$

The two-integer factorizations of 12 and the sums of the factors are shown in the following table.

Factors of 12	Sum of the factors of 12
1(12)	$1 + 12 = 13$
2(6)	$2 + 6 = 8$
3(4)	$3 + 4 = 7$
$-1(-12)$	$-1 + (-12) = -13$
$-2(-6)$	$-2 + (-6) = -8$
$\mathbf{-3(-4)}$	$\mathbf{-3 + (-4) = -7}$

Self Check 2
Factor: $p^2 - 5p + 6$.

The last row contains the integers -3 and -4, whose product is 12 and whose sum is -7. To complete the factorization, we enter -3 and -4 as the second terms of the binomial factors.

$$y^2 - 7y + 12 = (y - \mathbf{3})(y - \mathbf{4})$$

To check the result, we verify that $(y - 3)(y - 4)$ is $y^2 - 7y + 12$.

$$(y - 3)(y - 4) = y^2 - 4y - 3y + 12$$
$$= y^2 - 7y + 12$$

Answer $(p - 3)(p - 2)$

Self Check 3
Factor: $p^2 + 3p - 18$.

EXAMPLE 3 Factor: $a^2 + 2a - 15$.

Solution Since the first term of the trinomial is a^2, the first term of each binomial factor must be a. To fill in the blanks, we must find two integers whose product is -15 and whose sum is 2.

$$a^2 + 2a - 15 = (a \quad)(a \quad)$$

The possible factorizations of -15 and the sum of the factors are shown in the following table.

Factors of −15	Sum of the factors of −15
1(−15)	1 + (−15) = −14
3(−5)	3 + (−5) = −2
5(−3)	**5 + (−3)** = 2
15(−1)	15 + (−1) = 14

The third row contains the integers 5 and -3, whose product is -15 and whose sum is 2. To complete the factorization, we enter 5 and -3 as the second binomial factors.

$$a^2 + 2a - 15 = (a + \mathbf{5})(a - \mathbf{3})$$

We can check by multiplying.

$$(a + 5)(a - 3) = a^2 - 3a + 5a - 15$$
$$= a^2 + 2a - 15$$

Answer $(p + 6)(p - 3)$

Self Check 4
Factor: $q^2 - 2q - 24$.

EXAMPLE 4 Factor: $z^2 - 4z - 21$.

Solution Since the first term of the trinomial is z^2, the first term of each binomial factor must be z. To fill in the blanks, we must find two integers whose product is -21 and whose sum is -4.

$$z^2 - 4y - 21 = (z \quad)(z \quad)$$

The factorizations of -21 and the sums of the factors are shown in the following table.

Factors of −21	Sum of the factors of −21
1(−21)	1 + (−21) = −20
3(−7)	**3 + (−7)** = −4
7(−3)	7 + (−3) = 4
21(−1)	21 + (−1) = 20

The second row contains the integers 3 and −7, whose product is −21 and whose sum is −4. To complete the factorization, we enter 3 and −7 as the second terms of the binomial factors.

$$z^2 - 4z - 21 = (z + \mathbf{3})(z - \mathbf{7})$$

We can check by multiplying.

$$\begin{aligned}(z + 3)(z - 7) &= z^2 - 7z + 3z - 21\\ &= z^2 - 4z - 21\end{aligned}$$

Answer $(q + 4)(q - 6)$

The following sign patterns can be helpful when factoring trinomials.

Factoring $x^2 + bx + c$

To factor $x^2 + bx + c$, find two integers whose product is c and whose sum is b.

1. If c is positive, the integers have the same sign.
2. If c is negative, the integers have opposite signs.

When factoring out trinomials of the form $ax^2 + bx + c$, where $a = -1$, we begin by factoring out −1.

EXAMPLE 5 Factor: $-h^2 + 2h + 15$.

Self Check 5

Factor: $-x^2 + 11x - 28$.

Solution We factor out −1 and then factor $h^2 - 2h - 15$.

$$\begin{aligned}-h^2 + 2h + 15 &= \mathbf{-1}(h^2 - 2h - 15) && \text{Factor out } -1.\\ &= -(h^2 - 2h - 15)\\ &= -(h - 5)(h + 3) && \text{Use the integers } -5 \text{ and } 3\text{, because their product is } -15 \text{ and their sum is } -2.\end{aligned}$$

We can check by multiplying.

$$\begin{aligned}-(h - 5)(h + 3) &= -(h^2 + 3h - 5h - 15) && \text{Multiply the binomials first.}\\ &= -(h^2 - 2h - 15)\\ &= -h^2 + 2h + 15\end{aligned}$$

Answer $-(x - 4)(x - 7)$

The trinomials in the next two examples are of a form similar to $x^2 + bx + c$, and we can use the methods of this section to factor them.

EXAMPLE 6 Factor: $x^2 - 4xy - 5y^2$.

Self Check 6

Factor: $s^2 + 6st - 7t^2$.

Solution The trinomial has two variables, x and y. Since the first term is x^2, the first term of each factor must be x.

$$x^2 - 4xy - 5y^2 = (x \quad)(x \quad)$$

To fill in the blanks, we must find two *expressions* whose product is the last term, $-5y^2$, and that will give a middle term of $-4xy$. Two such expressions are $-5y$ and y.

$$x^2 - 4xy - 5y^2 = (x - 5y)(x + y)$$

Check: $$\begin{aligned}(x - 5y)(x + y) &= x^2 + xy - 5xy - 5y^2\\ &= x^2 - 4xy - 5y^2\end{aligned}$$

Answer $(s + 7t)(s - t)$

Multistep factoring

If the terms of a trinomial have a common factor, the GCF should always be factored out before any of the factoring techniques of this section are used. A trinomial is **factored completely** when no factor can be factored further. Always factor completely when you are asked to factor.

Self Check 7
Factor: $4m^5 + 8m^4 - 32m^3$.

Answer $4m^3(m + 4)(m - 2)$

EXAMPLE 7 Factor: $2x^4 + 26x^3 + 80x^2$.

Solution We begin by factoring out the GCF, $2x^2$.

$$2x^4 + 26x^3 + 80x^2 = \mathbf{2x^2}(x^2 + 13x + 40)$$

Next, we factor $x^2 + 13x + 40$. The integers 8 and 5 have a product of 40 and a sum of 13, so the completely factored form of the given trinomial is

$$2x^4 + 26x^3 + 80x^2 = 2x^2(x + 8)(x + 5)$$

Check by multiplying $2x^2$, $x + 8$, and $x + 5$.

Self Check 8
Factor: $-12t + t^3 + 4t^2$.

Answer $t(t - 2)(t + 6)$

EXAMPLE 8 Factor: $-13g^2 + 36g + g^3$ completely.

Solution Before factoring the trinomial, we write its terms in descending powers of g.

$$\begin{aligned} -13g^2 + 36g + g^3 &= g^3 - 13g^2 + 36g && \text{Rearrange the terms.} \\ &= g(g^2 - 13g + 36) && \text{Factor out } g\text{, which is the GCF.} \\ &= g(g - 9)(g - 4) && \text{Factor the trinomial.} \end{aligned}$$

Check by multiplying.

Prime polynomials

If a trinomial cannot be factored using only integers, it is called a **prime polynomial,** or more specifically, a **prime trinomial.**

Self Check 9
Factor: $x^2 - 4x + 6$, if possible.

Answer not possible; prime trinomial

EXAMPLE 9 Factor: $x^2 + 2x + 3$, if possible.

Solution To factor the trinomial, we must find two integers whose product is 3 and whose sum is 2. The possible factorizations of 3 and the sums of the factors are shown in the following table.

Factors of 3	Sum of the factors of 3
1(3)	1 + 3 = 4
−1(−3)	−1 + (−3) = −4

Since two integers whose product is 3 and whose sum is 2 do not exist, $x^2 + 2x + 3$ cannot be factored. It is a prime trinomial.

Section 11.2 STUDY SET

VOCABULARY *Fill in the blanks.*

1. A polynomial, such as $x^2 - x - 6$, that has exactly three terms is called a ________. A polynomial, such as $x - 3$, that has exactly two terms is called a ________.
2. The statement $x^2 - x - 12 = (x - 4)(x + 3)$ shows that $x^2 - x - 12$ ______ into the product of two binomials.
3. Since $10 = (-5)(-2)$, we say -5 and -2 are ______ of 10.
4. A ______ polynomial cannot be factored by using only integers.
5. The _____ coefficient of the trinomial $x^2 - 3x + 2$ is 1, the ________ of the middle term is -3, and the last ______ is 2.
6. A trinomial is factored ________ when no factor can be factored further.

CONCEPTS *Fill in the blanks.*

7. Two factorizations of 4 that involve only positive numbers are ▢ · ▢ and ▢ · ▢. Two factorizations of 4 that involve only negative numbers are ▢(▢) and ▢(▢).
8. Before attempting to factor a trinomial, be sure that the exponents are written in ________ order.
9. Before attempting to factor a trinomial into two binomials, always factor out any ________ factors first.
10. To factor $x^2 + x - 56$, we must find two integers whose ________ is -56 and whose ______ is 1.
11. Two factors of 18 whose sum is -9 are ▢ and ▢.
12. $x^2 + 5x + 3$ cannot be factored because we cannot find two integers whose product is ▢ and whose sum is ▢.
13. Complete the table.

Factors of 8	Sum of the factors of 8
1(8)	
2(4)	
−1(−8)	
−2(−4)	

14. If we use the FOIL method to do the multiplication $(x + 5)(x + 4)$, we obtain $x^2 + 9x + 20$.
    a. What step of the FOIL process produced 20?
    b. What steps of the FOIL process produced $9x$?
15. Find two integers whose
    a. product is 10 and whose sum is 7.
    b. product is 8 and whose sum is -6.
    c. product is -6 and whose sum is 1.
    d. product is -9 and whose sum is -8.
16. Given $x^2 + 8x + 15$:
    a. What is the coefficient of the x^2 term?
    b. What is the last term? What is the coefficient of the middle term?
    c. What two integers have a product of 15 and a sum of 8?
17. To factor a trinomial, a student made a table and circled the correct pair of integers, as shown. Complete the factorization of the trinomial.
    $(x \quad ▢)(x \quad ▢)$

Factors of −6	Sum of the factors of −6
1(−6)	−5
2(−3)	−1
3(−2)	1
6(−1)	5

18. Complete the factorization table.

Factors of −9	Sum of the factors of −9

19. Consider factoring a trinomial of the form $x^2 + bx + c$.
    a. If c is positive, what can be said about the two integers that should be chosen for the factorization?

b. If c is negative, what can be said about the two integers that should be chosen for the factorization?

20. What trinomial has the factorization of $(x + 8)(x - 2)$?

NOTATION *Complete each factorization.*

21. $6 + 5x + x^2 = x^2 + \square + 6$
$= (x + 3)(x + \square)$

22. $-a^2 - a + 20 = \square(a^2 + a - 20)$
$= -(a + 5)(a - \square)$

PRACTICE *Complete each factorization.*

23. $x^2 + 3x + 2 = (x \square 2)(x \square 1)$

24. $y^2 + 4y + 3 = (y \square 3)(y \square 1)$

25. $t^2 - 9t + 14 = (t \square 7)(t \square 2)$

26. $c^2 - 9c + 8 = (c \square 8)(c \square 1)$

27. $a^2 + 6a - 16 = (a \square 8)(a \square 2)$

28. $x^2 - 3x - 40 = (x \square 8)(x \square 5)$

Factor each trinomial. If it can't be factored, write "prime."

29. $z^2 + 12z + 11$

30. $x^2 + 7x + 10$

31. $m^2 - 5m + 6$

32. $n^2 - 7n + 10$

33. $a^2 - 4a - 5$

34. $b^2 + 6b - 7$

35. $x^2 + 5x - 24$

36. $t^2 - 5t - 50$

37. $a^2 - 10a - 39$

38. $r^2 - 9r - 12$

39. $u^2 + 10u + 15$

40. $v^2 + 9v + 15$

41. $s^2 + 11s - 26$

42. $y^2 + 8y + 12$

43. $r^2 - 2r + 4$

44. $m^2 + 3m - 10$

45. $m^2 - m - 12$

46. $u^2 + u - 42$

47. $x^2 + 4xy + 4y^2$

48. $a^2 + 10ab + 9b^2$

49. $m^2 + 3mn - 10n^2$

50. $m^2 - mn - 12n^2$

51. $a^2 - 4ab - 12b^2$

52. $p^2 + pq - 6q^2$

53. $r^2 - 2rs + 4s^2$

54. $m^2 + 3mn - 20n^2$

Factor each trinomial. Factor out −1 first.

55. $-x^2 - 7x - 10$

56. $-x^2 + 9x - 20$

57. $-t^2 - 15t + 34$

58. $-t^2 - t + 30$

59. $-r^2 + 14r - 40$

60. $-r^2 + 14r - 45$

61. $-a^2 - 4ab - 3b^2$

62. $-a^2 - 6ab - 5b^2$

63. $-x^2 + 6xy + 7y^2$

64. $-x^2 - 10xy + 11y^2$

Write each trinomial in descending powers of one variable and factor.

65. $4 - 5x + x^2$

66. $y^2 + 5 + 6y$

67. $10y + 9 + y^2$

68. $x^2 - 13 - 12x$

69. $-r^2 + 2 + r$

70. $u^2 - 3 + 2u$

71. $4rx + r^2 + 3x^2$

72. $a^2 + 5b^2 + 6ab$

73. $-3ab + a^2 + 2b^2$

74. $-13yz + y^2 - 14z^2$

Completely factor each trinomial. Factor out any common monomials first (including −1 if necessary).

75. $2x^2 + 10x + 12$

76. $3y^2 - 21y + 18$

77. $-5a^2 + 25a - 30$

78. $-2b^2 + 20b - 18$

79. $3z^2 - 15z + 12$

80. $5m^2 + 45m - 50$

81. $12xy + 4x^2y - 72y$

82. $48xy + 6xy^2 + 96x$

83. $-4x^2y - 4x^3 + 24xy^2$

84. $3x^2y^3 + 3x^3y^2 - 6xy^4$

APPLICATIONS

85. PETS The cage shown is used for transporting dogs. Its volume is $(x^3 + 12x^2 + 27x)$ in.^{3}. The dimensions of the cage can be found by factoring this expression. If the cage is longer than it is tall and taller than it is wide, write expressions that represent its length, height, and width.

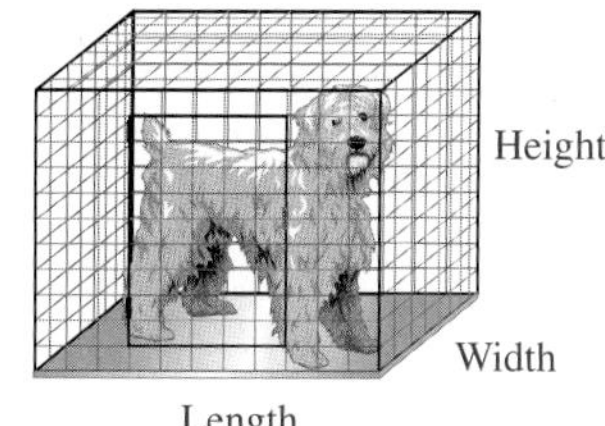

86. CARPOOLING The average rate at which a carpool van travels and the distance it covers are given in the table in terms of t. Factor the expression representing the distance traveled and then complete the table.

Rate (mi/hr)	Time (hr)	Distance traveled (mi)
$t + 11$		$t^2 + 16t + 55$

WRITING

87. Explain what it means when we say that a trinomial is the product of two binomials. Give an example.

88. Are $2x^2 - 12x + 16$ and $x^2 - 6x + 8$ factored in the same way? Explain why or why not.

89. When factoring $x^2 - 2x - 3$, one student got $(x - 3)(x + 1)$, and another got $(x + 1)(x - 3)$. Are both answers acceptable? Explain.

90. Explain how to use multiplication to check the factorization of a trinomial.

91. In the partial solution shown below, a student began to factor the trinomial. Write a note to the student explaining his mistake.

Factor: $x^2 - 2x - 63$.

$(x - \quad)(x - \quad)$

???

92. Explain why the given trinomial is not factored completely.

$$3x^2 - 3x - 60 = 3(x^2 - x - 20)$$

REVIEW *Graph the solution of each inequality on the number line.*

93. $x - 3 > 5$

94. $x + 4 \leq 3$

95. $-3x - 5 \geq 4$.

96. $2x - 3 < 7$

11.3 Factoring Trinomials of the Form $ax^2 + bx + c$

- Observations about multiplying binomials
- The trial-and-check factoring method
- The grouping method

In this section, we will factor trinomials with lead coefficients other than 1, such as $2x^2 + 5x + 3$ and $6a^2 - 17a + 5$. Two methods are used to factor these trinomials. With the first method, educated guesses are made. These guesses are checked by multiplication. The correct factorization is determined by a process of elimination. The second method is an extension of factoring by grouping.

Observations about multiplying binomials

In the work below, we find the products $(2x + 1)(x + 3)$ and $(2x + 3)(x + 1)$. There are several observations that can be made when we compare the results.

$$(2x + 1)(x + 3) = 2x^2 + 6x + x + 3 \qquad (2x + 3)(x + 1) = 2x^2 + 2x + 3x + 3$$
$$= 2x^2 + 7x + 3 \qquad\qquad = 2x^2 + 5x + 3$$

In each case, the result is a trinomial, and

- the first terms are the same ($2x^2$),
- the last terms are the same (3), and
- the middle terms are different ($7x$ and $5x$).

These observations indicate that when the last terms in $(2x + \mathbf{1})(x + 3)$ are interchanged to form $(2x + 3)(x + \mathbf{1})$, only the middle terms of the products are different. This fact is helpful when factoring trinomials using the *trial-and check method.*

The trial-and-check factoring method

To factor a trinomial with a lead coefficient of 1 (say, $x^2 + 4x + 3$), we begin with a factorization of the form

$$x^2 + 4x + 3 = (x \quad)(x \quad)$$

and determine which integers to write in the blanks.

To factor a trinomial with a lead coefficient that is not 1 (say, $2x^2 + 5x + 3$), we begin with a factorization of the form

$$2x^2 + 5x + 3 = (\quad x \quad)(\quad x \quad)$$

We must determine what numbers to write in the blanks. Because there are four blanks, there are more combinations of factors to consider.

Self Check 1
Factor: $3x^2 + 7x + 2$.

EXAMPLE 1 Factor: $2x^2 + 5x + 3$.

Solution Since the first term is $2x^2$, the first terms of the binomial factors must be $2x$ and x. To fill in the blanks, we must find two factors of 3 that will give a middle term of $5x$.

$$(2x \quad)(x \quad)$$

Because each term of the trinomial is positive, we need only consider positive factors of the last term. Since the positive factors of 3 are 1 and 3, there are two possible factorizations.

$$(2x + 1)(x + 3) \qquad \text{or} \qquad (2x + 3)(x + 1)$$

The first possibility is incorrect: When we find the outer and inner products and combine like terms, we obtain an incorrect middle term of $7x$.

Outer: $6x$

$(2x + 1)(x + 3)$ Multiply and add to find the middle term: $6x + x = 7x$.

Inner: x

The second possibility is correct, because it gives a middle term of $5x$.

Outer: $2x$

$(2x + 3)(x + 1)$ Multiply and add to find the middle term: $2x + 3x = 5x$.

Inner: $3x$

Thus,

$$2x^2 + 5x + 3 = (2x + 3)(x + 1)$$

Answer $(3x + 1)(x + 2)$

EXAMPLE 2 Factor: $6a^2 - 17a + 5$.

Self Check 2
Factor: $6x^2 - 7x + 2$.

Solution Since the first term is $6a^2$, the first terms of the binomial factors must be $6a$ and a or $3a$ and $2a$. To fill in the blanks, we must find two factors of 5 that will give a middle term of $-17a$.

$(6a \quad)(a \quad)$ or $(3a \quad)(2a \quad)$

Because the sign of the last term is positive and the sign of the middle term is negative, we need only consider negative factors of the last term. Since the negative factors of 5 are -1 and -5, there are four possible factorizations.

$(6a - 1)(a - 5)$ $-30a - a = -31a$. (outer: $-30a$, inner: $-a$)

$(6a - 5)(a - 1)$ $-6a - 5a = -11a$. (outer: $-6a$, inner: $-5a$)

$(3a - 1)(2a - 5)$ $-15a - 2a = -17a$. (outer: $-15a$, inner: $-2a$)

$(3a - 5)(2a - 1)$ $-3a - 10a = -13a$. (outer: $-3a$, inner: $-10a$)

Only the possibility shown in blue gives the correct middle term of $-17a$. Thus,

$$6a^2 - 17a + 5 = (3a - 1)(2a - 5)$$

Answer $(3x - 2)(2x - 1)$

EXAMPLE 3 Factor: $3y^2 - 7y - 6$.

Self Check 3
Factor: $5a^2 - 23a - 10$.

Solution Since the first term is $3y^2$, the first terms of the binomial factors must be $3y$ and y.

$(3y \quad)(y \quad)$

The second terms of the binomials must be two integers whose product is -6. There are four such pairs: $1(-6)$, $-1(6)$, $2(-3)$, and $-2(3)$. When these pairs are entered, and then reversed, as second terms of the binomials, there are eight possibilities to consider. Four of them can be discarded because they include a binomial whose terms have a common factor. If $3y^2 - 7y - 6$ does not have a common factor, neither can any of its binomial factors.

For the factors −1 and 6:

$(3y - 1)(y + 6)$ $18y - y = 17y$ (outer: $18y$, inner: $-y$)

~~$(3y + 6)(y - 1)$~~ A common factor of 3

For the factors 1 and −6:

$(3y + 1)(y - 6)$ $-18y + y = -17y$ (outer: $-18y$, inner: y)

~~$(3y - 6)(y + 1)$~~ A common factor of 3

For the factors −2 and 3:

$$(3y - 2)(y + 3) \quad 9y - 2y = 7y$$

with products $9y$ and $-2y$.

$$(3y + 3)(y - 2)$$

A common factor of 3

For the factors 2 and −3:

$$(3y + 2)(y - 3) \quad -9y + 2y = -7y$$

with products $-9y$ and $2y$.

$$(3y - 3)(y + 2)$$

A common factor of 3

Only the possibility shown in blue gives the correct middle term of $-7y$. Thus,

$$3y^2 - 7y - 6 = (3y + 2)(y - 3)$$

Check the factorization by multiplication.

Answer $(5a + 2)(a - 5)$

! COMMENT If a trinomial does not have a common factor, the terms of each of its binomial factors will not have a common factor.

Self Check 4
Factor: $4x^2 + 4xy - 3y^2$.

EXAMPLE 4 Factor: $4b^2 + 8bc - 45c^2$.

Solution Since the first term is $4b^2$, the first terms of the factors must be $4b$ and b or $2b$ and $2b$.

$$(4b \quad)(b \quad) \quad \text{or} \quad (2b \quad)(2b \quad)$$

To fill in the blanks, we must find two factors of $-45c^2$ that will give a middle term of $8bc$.

Since $-45c^2$ has many factors, there are many possible combinations for the last terms of the binomial factors. The signs of the factors must be different, because the last term of the trinomial is negative.

If we pick factors of $4b$ and b for the first terms, and $-c$ and $45c$ for the last terms, the multiplication gives an incorrect middle term of $179bc$. So the factorization is incorrect.

$$(4b - c)(b + 45c) \quad 180bc - bc = 179bc.$$

with products $180bc$ and $-bc$.

If we pick factors of $4b$ and b for the first terms and $15c$ and $-3c$ for the last terms, the multiplication gives an incorrect middle term of $3bc$.

$$(4b + 15c)(b - 3c) \quad -12bc + 15bc = 3bc.$$

with products $-12bc$ and $15bc$.

If we pick factors of $2b$ and $2b$ for the first terms and $-5c$ and $9c$ for the last terms, we have

$$\overset{18bc}{\overbrace{(2b - \underbrace{5c)(2b}_{-10bc} + 9c)}} \qquad 18bc - 10bc = 8bc.$$

which gives the correct middle term of $8bc$. Thus,

$$4b^2 + 8bc - 45c^2 = (2b - 5c)(2b + 9c)$$

Check by multiplication.

Answer $(2x + 3y)(2x - y)$

Because some guesswork is often necessary, it is difficult to give specific rules for factoring trinomials with a lead coefficient other than 1. However, the following hints are helpful.

Factoring $ax^2 + bx + c$ $(a \neq 1)$

1. Write the trinomial in descending powers of the variable and factor out any GCF (including -1 if that is necessary to make the lead coefficient positive).
2. Attempt to write the trinomial as *the product of two binomials.* The coefficients of the first terms of each binomial factor must be factors of a, and the last terms must be factors of c.

$$\overset{\text{Factors of } a}{\downarrow \qquad\qquad \downarrow}$$
$$(\;\square\; x + \;\square\;)(\;\square\; x + \;\square\;)$$
$$\underset{\text{Factors of } c}{\uparrow \qquad\qquad \uparrow}$$

3. If the sign of the last term of the trinomial is positive, the signs between the terms of the binomial factors are the same as the sign of the middle term. If the sign of the last term is negative, the signs between the terms of the binomial factors are opposite.
4. Try combinations of coefficients of the first terms and last terms until you find one that gives the middle term of the trinomial. If no combination works, the trinomial is prime.
5. Check the factorization by multiplication.

EXAMPLE 5 Factor: $2x^2 - 8x^3 + 3x$.

Solution We write the trinomial in descending powers of x

$$-8x^3 + 2x^2 + 3x$$

and we factor out the negative of the GCF, which is $-x$.

$$-8x^3 + 2x^2 + 3x = -x(8x^2 - 2x - 3)$$

We must now factor $8x^2 - 2x - 3$. Its factorization has the form

$$(x \;\square\;)(8x \;\square\;) \qquad \text{or} \qquad (2x \;\square\;)(4x \;\square\;)$$

To fill in the blanks, we find two factors of the last term of the trinomial (-3) that will give a middle term of $-2x$. Because the sign of the last term is negative, the signs

Self Check 5
Factor: $12y - 2y^3 - 2y^2$.

within its binomial factors will be different. If we pick factors of $2x$ and $4x$ for the first terms and 1 and -3 for the last terms, we have

$$\overset{-6x}{(2x + 1)(4x - 3)} \quad -6x + 4x = -2x.$$
$$4x$$

which gives the correct middle term of $-2x$, so it is correct.

$$8x^2 - 2x - 3 = (2x + 1)(4x - 3)$$

We can now give the complete factorization.

$$\begin{aligned} -8x^3 + 2x^2 + 3x &= -x(8x^2 - 2x - 3) \\ &= -x(2x + 1)(4x - 3) \end{aligned}$$

Answer $-2y(y + 3)(y - 2)$

Check by multiplication.

The grouping method

The method of factoring by grouping can be used to help factor trinomials of the form $ax^2 + bx + c$. For example, to factor $2x^2 + 5x + 3$, we proceed as follows.

1. We find the product ac: In $2x^2 + 5x + 3$, $a = 2$, $b = 5$, and $c = 3$, so $ac = 2(3) = 6$. This number is called the **key number.**
2. Next, find two numbers whose product is $ac = 6$ and whose sum is $b = 5$. Since the numbers must have a positive product and a positive sum, we consider only positive factors of 6. The correct factors are 2 and 3.

Positive factors of 6	Sum of the factors of 6
$1 \cdot 6 = 6$	$1 + 6 = 7$
$\mathbf{2 \cdot 3 = 6}$	$\mathbf{2 + 3} = 5$

3. Use the factors 2 and 3 as coefficients of two terms to be placed between $2x^2$ and 3:

$$2x^2 + \mathbf{5x} + 3 = 2x^2 + \mathbf{2x} + \mathbf{3x} + 3 \quad \text{Express } 5x \text{ as } 2x + 3x.$$

4. Factor by grouping:

$$\begin{aligned} 2x^2 + 2x + 3x + 3 &= 2x(\mathbf{x + 1}) + 3(\mathbf{x + 1}) && \text{Factor } 2x \text{ out of } 2x^2 + 2x \text{ and 3 out of } 3x + 3. \\ &= (\mathbf{x + 1})(2x + 3) && \text{Factor out } x + 1. \end{aligned}$$

So $2x^2 + 5x + 3 = (x + 1)(2x + 3)$. Verify this factorization by multiplication.

Self Check 6
Factor: $15a^2 + 17a - 4$.

EXAMPLE 6 Factor: $10x^2 + 13x - 3$.

Solution Since $a = 10$ and $c = -3$ in the trinomial, $ac = -30$. We now find two factors of -30 whose sum is 13. Two such factors are 15 and -2. We use these factors as coefficients of two terms to be placed between $10x^2$ and -3.

$$10x^2 + \mathbf{13x} - 3 = 10x^2 + \mathbf{15x} - \mathbf{2x} - 3 \quad \text{Express } 13x \text{ as } 15x - 2x.$$

Finally, we factor by grouping.

$$\begin{aligned} 10x^2 + 15x - 2x - 3 &= 5x(\mathbf{2x + 3}) - 1(\mathbf{2x + 3}) \\ &= (\mathbf{2x + 3})(5x - 1) \end{aligned}$$

Answer $(3a + 4)(5a - 1)$

So $10x^2 + 13x - 3 = (2x + 3)(5x - 1)$. Check the result.

Factoring $ax^2 + bx + c$ by grouping

1. Write the trinomial in descending powers of the variable and factor out any GCF (including -1 if that is necessary to make the lead coefficient positive).
2. Calculate the key number ac.
3. Find two numbers whose product is the key number found in step 2 and whose sum is the coefficient of the middle term of the trinomial.
4. Write the numbers in the blanks of the form shown below, and then factor the polynomial by grouping.

$$ax^2 + \square x + \square x + c$$

5. Check the factorization using multiplication.

EXAMPLE 7 Factor: $12x^5 - 17x^4 + 6x^3$.

Solution First, we factor out the GCF, which is x^3.

$$12x^5 - 17x^4 + 6x^3 = x^3(12x^2 - 17x + 6)$$

To factor $12x^2 - 17x + 6$, we need to find two integers whose product is $12(6) = 72$ and whose sum is -17. Two such numbers are -8 and -9.

$12x^2 \mathbf{-17x} + 6 = 12x^2 \mathbf{- 8x - 9x} + 6$ Express $-17x$ as $-8x - 9x$.

$= 4x(3x - 2) - 3(3x - 2)$ Factor out $4x$ and factor out -3.

$= (3x - 2)(4x - 3)$ Factor out $3x - 2$.

The complete factorization is

$$12x^5 - 17x^4 + 6x^3 = x^3(3x - 2)(4x - 3)$$

Check the result.

Self Check 7
Factor: $21a^4 - 13a^3 + 2a^2$.

Answer $a^2(7a - 2)(3a - 1)$

Section 11.3 STUDY SET

VOCABULARY *Fill in the blanks.*

1. The trinomial $3x^2 - x - 12$ has a ______ coefficient of 3. The ______ term is -12.
2. The numbers 3 and 2 are ________ of the first term of the trinomial $6x^2 + x - 12$.
3. Consider $(x - 2)(5x - 1)$. The product of the ______ terms is $-x$ and the product of the ______ terms is $-10x$.
4. When we write $2x^2 + 7x + 3$ as $(2x + 1)(x + 3)$, we say that we have ________ the trinomial—it has been expressed as the product of two ________.
5. The ________ term of $4x^2 - 7x + 13$ is $-7x$.
6. The polynomial $6x^2 + 2x + 9x + 3$ has four ______
7. The ______ of the middle terms of the polynomial $4a^2 - 12a - a + 3$ is $-13a$.
8. The ______ of the terms of the trinomial $6b^3 - 3b^2 - 12b$ is $3b$.

CONCEPTS *Complete each sentence.*

9. These coefficients must be factors of $\square$.

$$5x^2 + 6x - 8 = (\square x + \square)(\square x + \square)$$

These numbers must be factors of $\square$.

10. The product of these coefficients must be $\square$.

$$3x^2 + 16x + 5 = 3x^2 + \square x + \square x + 5$$

The sum of these coefficients must be $\square$.

A trinomial has been partially factored. Complete each statement that describes the type of integers we should consider for the blanks.

11. $5y^2 - 13y + 6 = (5y \quad)(y \quad)$

Since the last term of the trinomial is ________ and the middle term is ________, the integers must be ________ factors of 6.

12. $5y^2 + 13y + 6 = (5y \quad)(y \quad)$

Since the last term of the trinomial is ________ and the middle term is ________, the integers must be ________ factors of 6.

13. $5y^2 + 7y - 6 = (5y \quad)(y \quad)$

Since the last term of the trinomial is ________, the signs of the integers will be ________.

14. $5y^2 - 7y - 6 = (5y \quad)(y \quad)$

Since the last term of the trinomial is ________, the signs of the integers will be ________.

A trinomial is to be factored by the grouping method. Complete each statement that describes the type of integers we should consider for the blanks.

15. $8c^2 - 11c + 3 = 8c^2 + \quad c + \quad c + 3$

We need to find two integers whose product is ____ and whose sum is ____.

16. $15c^2 + 4c - 4 = 15c^2 + \quad c + \quad c - 4$

We need to find two integers whose product is ____ and whose sum is ____.

NOTATION

17. Write a trinomial of the form $ax^2 + bx + c$

a. where $a = 1$

b. where $a \neq 1$

18. Write the terms of the trinomial $40 - t - 4t^2$ in descending powers of the variable.

PRACTICE *Complete each factorization.*

19. $3a^2 + 13a + 4 = (3a \quad 1)(a \quad 4)$

20. $2b^2 + 7b + 6 = (2b \quad 3)(b \quad 2)$

21. $4z^2 - 13z + 3 = (z \quad 3)(4z \quad 1)$

22. $4t^2 - 4t + 1 = (2t \quad 1)(2t \quad 1)$

23. $2m^2 + 5m - 12 = (2m \quad 3)(m \quad 4)$

24. $10u^2 - 13u - 3 = (2u \quad 3)(5u \quad 1)$

Complete each step of the factorization by grouping.

25. $12t^2 + 17t + 6 = 12t^2 + \quad t + 8t + 6$

$= \quad (4t + 3) + \quad (4t + 3)$

$= (\qquad)(3t + 2)$

26. $35t^2 - 11t - 6 = 35t^2 + \quad t - 21t - 6$

$= 5t(7t + 2) \quad 3(7t \quad 2)$

$= (\qquad)(5t - 3)$

Factor each trinomial, if possible.

27. $2x^2 - 3x + 1$

28. $2y^2 - 7y + 3$

29. $3a^2 + 13a + 4$

30. $2b^2 + 7b + 6$

31. $4z^2 + 13z + 3$

32. $4t^2 - 4t + 1$

33. $6y^2 + 7y + 2$

34. $4x^2 + 8x + 3$

35. $6x^2 - 7x + 2$

36. $15t^2 - 34t + 8$

37. $7x^2 - 9x + 2$

38. $8u^2 - 2u - 15$

39. $2x^2 - 3x - 2$

40. $12y^2 - y - 1$

41. $2m^2 + 5m - 10$

42. $10u^2 - 13u - 6$

43. $10y^2 - 3y - 1$

44. $6m^2 + 19m + 3$

45. $12y^2 - 5y - 2$

46. $10x^2 + 21x - 10$

47. $-5t^2 - 13t - 6$

48. $-16y^2 - 10y - 1$

49. $-16m^2 + 14m - 3$

50. $-16x^2 - 16x - 3$

51. $4a^2 - 4ab + b^2$

52. $2b^2 - 5bc + 2c^2$

53. $6r^2 + rs - 2s^2$

54. $3m^2 + 5mn + 2n^2$

55. $13x^2 + 24xy + 11y^2$

56. $4b^2 + 15bc - 4c^2$

57. $18a^2 + 31ab - 10b^2$

58. $12x^2 + 5xy - 3y^2$

59. $-13x + 3x^2 - 10$

60. $-14 + 3a^2 - a$

61. $15 + 8a^2 - 26a$

62. $16 - 40a + 25a^2$

63. $12y^2 + 12 - 25y$

64. $12t^2 - 1 - 4t$

65. $3x^2 + 6 + x$

66. $25 + 2u^2 + 3u$

67. $2a^2 + 3b^2 + 5ab$
68. $11uv + 3u^2 + 6v^2$
69. $pq + 6p^2 - q^2$
70. $-11mn + 12m^2 + 2n^2$
71. $4x^2 + 10x - 6$
72. $9x^2 + 21x - 18$
73. $-y^3 - 13y^2 - 12y$
74. $-2xy^2 - 8xy + 24x$
75. $6x^3 - 15x^2 - 9x$
76. $9y^3 + 3y^2 - 6y$
77. $30r^5 + 63r^4 - 30r^3$
78. $6s^5 - 26s^4 - 20s^3$
79. $-16m^3n - 20m^2n^2 - 6mn^3$
80. $-84x^4 - 100x^3y - 24x^2y^2$
81. $-28u^3v^3 + 26u^2v^4 - 6uv^5$
82. $-16x^4y^3 + 30x^3y^4 + 4x^2y^5$

APPLICATIONS

83. OFFICE FURNITURE The area of the desktop shown below is given by the expression $(4x^2 + 20x - 11)$ in.^{2}. Factor this expression to find the expressions that represent its length and width. Then determine the difference in the length and width of the desktop.

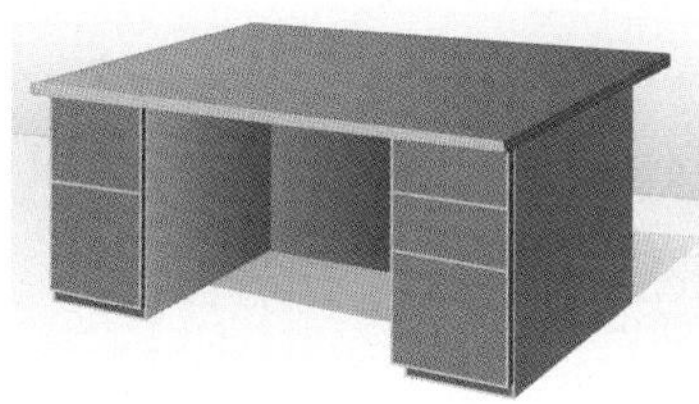

84. STORAGE The volume of the 8-foot-wide portable storage container shown in the next column is given by the expression $(72x^2 + 120x - 400)$ ft^3. If its dimensions can be determined by factoring the expression, find the height and the length of the container.

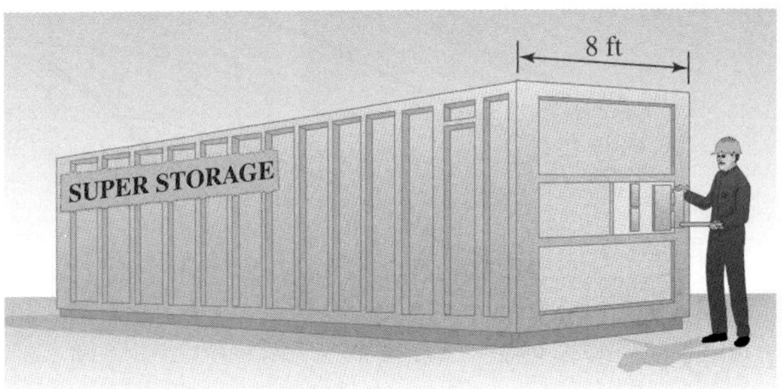

WRITING

85. In the work below, a student began to factor the trinomial. Explain his initial mistake.

Factor: $3x^2 - 5x - 2$.

$(3x - \quad)(x - \quad)$

???

86. Two students factor $2x^2 + 20x + 42$ and get two different answers:

$(2x + 6)(x + 7)$ and $(x + 3)(2x + 14)$

Do both answers check? Why don't they agree? Is either answer completely correct? Explain.

87. Why is the process of factoring $6x^2 - 5x - 6$ more complicated than the process of factoring $x^2 - 5x - 6$?

88. How can the factorization shown below be checked?

$$6x^2 - 5x - 6 = (3x + 2)(2x - 3)$$

REVIEW

89. Simplify: $(x^2x^5)^2$.

90. Simplify: $\frac{(a^3b^4)^2}{ab^5}$.

91. Evaluate: $\frac{1}{2^{-3}}$.

92. Evaluate: 7^0.

11.4 Special Factorizations and a Factoring Strategy

- Factoring perfect square trinomials
- Factoring the difference of two squares
- Multistep factoring
- Factoring the sum and difference of two cubes
- A factoring strategy

We have studied several methods that can be used to factor trinomials. In this section, we will introduce another method that can be used to factor *perfect square trinomials*. We will also develop techniques for factoring three specific types of binomials, called the *difference of two squares*, and the *sum* and *difference of two cubes*.

Factoring perfect square trinomials

In Section 10.6, we saw that the squares of binomials are trinomials.

$(x + y)^2$	$=$	x^2	$+$	$2xy$	$+$	y^2
		↑ This is the square of the first term of the binomial. ↓		↑ This is twice the product of the two terms of the binomial. ↓		↑ This is the square of the last term of the binomial. ↓
$(x - y)^2$	$=$	x^2	$-$	$2xy$	$+$	y^2

Trinomials that are squares of a binomial are called **perfect square trinomials.** Some examples of perfect square trinomials are

$y^2 + 6y + 9$	Because it is the square of $(y + 3)$: $(y + 3)^2 = y^2 + 6y + 9$.
$t^2 - 14t + 49$	Because it is the square of $(t - 7)$: $(t - 7)^2 = t^2 - 14t + 49$.
$4m^2 - 20m + 25$	Because it is the square of $(2m - 5)$: $(2m - 5)^2 = 4m^2 - 20m + 25$.

Self Check 1
Determine which of the following are perfect square trinomials:
a. $y^2 + 4y + 4$
b. $b^2 - 6b - 9$
c. $4z^2 + 4z + 4$

Answers **a.** yes, **b.** no, **c.** no

EXAMPLE 1 Determine whether the following trinomials are perfect square trinomials: **a.** $x^2 + 10x + 25$, **b.** $c^2 - 12c - 36$, and **c.** $25y^2 - 30y + 9$.

Solution

a. To determine whether $x^2 + 10x + 25$ is a perfect square trinomial, we note that

- the first term is the square of x,
- the last term is the square of 5, and
- the middle term is twice the product of x and 5.

Thus, $x^2 + 10x + 25$ is a perfect square trinomial.

b. To determine whether $c^2 - 12c - 36$ is a perfect square trinomial, we note that

- the first term is the square of c, but
- the last term is negative.

Thus, $c^2 - 12c - 36$ is not a perfect square trinomial.

c. To determine whether $25y^2 - 30y + 9$ is a perfect square trinomial, we note that

- the first term is the square of $5y$,
- the last term is the square of -3, and
- the middle term is twice the product of $5y$ and -3.

Thus, $25y^2 - 30y + 9$ is a perfect square trinomial.

Although we can factor perfect square trinomials using techniques discussed earlier in the chapter, we can also factor them by inspecting their terms and applying the special product formulas.

Factoring perfect square trinomials

$$x^2 + 2xy + y^2 = (x + y)^2$$
$$x^2 - 2xy + y^2 = (x - y)^2$$

EXAMPLE 2 Factor: $N^2 + 20N + 100$.

Solution $N^2 + 20N + 100$ is a perfect square trinomial, because:

- The first term N^2 is the square of N: $(N)^2 = N^2$.
- The last term 100 is the square of $\mathbf{10}$: $10^2 = 100$.
- The middle term is twice the product of N and 10: $2(N)(10) = 20N$.

The factored form of the trinomial involves the terms N and 10.

$N^2 + 20N + 100 = (N + \mathbf{10})^2$ The sign in the binomial is the sign of the middle term of the trinomial.

Check by multiplication.

Self Check 2
Factor: $x^2 + 18x + 81$.

Answer $(x + 9)^2$

EXAMPLE 3 Factor: $9x^2 - 30xy + 25y^2$.

Solution $9x^2 - 30xy + 25y^2$ is a perfect square trinomial, because:

- The first term $9x^2$ is the square of $3x$: $(3x)^2 = 9x^2$.
- The last term $25y^2$ is the square of $\mathbf{-5y}$: $(-5y)^2 = 25y^2$.
- The middle term is twice the product of $3x$ and $-5y$: $2(3x)(-5y) = -30xy$.

The factored form of the trinomial involves the terms $3x$ and $-5y$.

$9x^2 - 30xy + 25y^2 = (3x - \mathbf{5y})^2$ The sign in the binomial is the sign of the middle term of the trinomial.

Check by multiplication.

Self Check 3
Factor: $16x^2 - 8xy + y^2$.

Answer $(4x - y)^2$

Factoring the difference of two squares

Whenever we multiply a binomial of the form $x + y$ by a binomial of the form $x - y$, we obtain a binomial of the form $x^2 - y^2$.

$$(x + y)(x - y) = x^2 - xy + xy - y^2 \quad \text{Use the FOIL method.}$$
$$= x^2 - y^2 \quad \text{Combine like terms: } -xy + xy = 0.$$

The binomial $x^2 - y^2$ is called the **difference of two squares,** because x^2 is the square of x and y^2 is the square of y. The difference of the squares of two quantities always factors into the sum of those two quantities multiplied by the difference of those two quantities.

Factoring the difference of two squares

$$x^2 - y^2 = (x + y)(x - y)$$

If we think of the difference of two squares as the square of a **F**irst quantity minus the square of a **L**ast quantity, we have the formula

$$F^2 - L^2 = (F + L)(F - L)$$

and we say: *To factor the square of a First quantity minus the square of a Last quantity, we multiply the First plus the Last by the First minus the Last.*

To factor $x^2 - 9$, we note that it can be written in the form $x^2 - 3^2$ and use the formula for factoring the difference of two squares:

$$\mathbf{F}^2 - \mathbf{L}^2 = (\mathbf{F} + \mathbf{L})(\mathbf{F} - \mathbf{L})$$
$$\downarrow \quad \downarrow \qquad \downarrow \quad \downarrow \quad \downarrow \quad \downarrow$$
$$x^2 - 3^2 = (x + 3)\ (x - 3) \quad \text{Substitute } x \text{ for F and 3 for L.}$$

We can check by verifying that $(x + 3)(x - 3) = x^2 - 9$. Because of the commutative property of multiplication, we can also write this factorization as $(x - 3)(x + 3)$.

To factor the difference of two squares, it is helpful to know the integers that are perfect squares. The number 400, for example, is a perfect square, because $20^2 = 400$. The perfect integer squares through 400 are

1, 4, 9, 16, 25, 36, 49, 64, 81, 100, 121, 144, 169, 196, 225, 256, 289, 324, 361, 400

Expressions containing variables such as $25x^2$ are also perfect squares, because they can be written as the square of a quantity:

$$25x^2 = (5x)^2$$

Self Check 4
Factor: $16a^2 - 1$.

Answer $(4a + 1)(4a - 1)$

EXAMPLE 4 Factor: $25x^2 - 49$.

Solution We can write $25x^2 - 49$ in the form $(5x)^2 - 7^2$ and use the formula for factoring the difference of two squares:

$$\mathbf{F}^2 - \mathbf{L}^2 = (\mathbf{F} + \mathbf{L})(\mathbf{F} - \mathbf{L})$$
$$\downarrow \quad \downarrow \qquad \downarrow \quad \downarrow \quad \downarrow \quad \downarrow$$
$$(5x)^2 - 7^2 = (5x + 7)(5x - 7) \quad \text{Substitute } 5x \text{ for F and 7 for L.}$$

We can check by multiplying.

$$(5x + 7)(5x - 7) = 25x^2 - 35x + 35x - 49$$
$$= 25x^2 - 49$$

Self Check 5
Factor: $9m^2 - 64n^4$.

Answer $(3m + 8n^2)(3m - 8n^2)$

EXAMPLE 5 Factor: $4y^4 - 121z^2$.

Solution We can write $4y^4 - 121z^2$ in the form $(2y^2)^2 - (11z)^2$ and use the formula for factoring the difference of two squares:

$$\mathbf{F}^2 - \mathbf{L}^2 = (\mathbf{F} + \mathbf{L})(\mathbf{F} - \mathbf{L})$$
$$\downarrow \qquad \downarrow \qquad \downarrow \quad \downarrow \quad \downarrow \quad \downarrow$$
$$(2y^2)^2 - (11z)^2 = (2y^2 + 11z)(2y^2 - 11z)$$

Check by multiplying.

Multistep factoring

When factoring a polynomial, we should always factor out the greatest common factor first.

EXAMPLE 6 Factor: $8x^2 - 8$.

Solution We factor out the GCF of 8, and then factor the resulting difference of two squares.

$$8x^2 - 8 = 8(x^2 - 1) \quad \text{The GCF is 8.}$$
$$= 8(x + 1)(x - 1) \quad \text{Think of } x^2 - 1 \text{ as } x^2 - 1^2 \text{ and factor the difference of two squares.}$$

We check by multiplying.

$$8(x + 1)(x - 1) = 8(x^2 - 1) \quad \text{Multiply the binomials first.}$$
$$= 8x^2 - 8 \quad \text{Distribute the multiplication by 8.}$$

Self Check 6
Factor: $2p^2 - 200$.

Answer $2(p + 10)(p - 10)$

Sometimes we must factor a difference of two squares more than once to completely factor a polynomial.

EXAMPLE 7 Factor: $x^4 - 16$.

Solution

$$x^4 - 16 = (x^2 + 4)(x^2 - 4) \quad \text{Factor the difference of two squares.}$$
$$= (x^2 + 4)(x + 2)(x - 2) \quad \text{Factor another difference of two squares: } x^2 - 4.$$

Self Check 7
Factor: $a^4 - 81$.

Answer $(a^2 + 9)(a + 3)(a - 3)$

! COMMENT In Example 7, the binomial $x^2 + 4$ is the **sum of two squares.** If we are limited to integer coefficients, binomials that are the sum of two squares cannot be factored.

Factoring the sum and difference of two cubes

We have seen that the sum of two squares, such as $x^2 + 4$ or $25a^2 + 9b^2$, cannot be factored. However, the sum of two cubes and the difference of two cubes can be factored.

The sum of two cubes		**The difference of two cubes**	
x^3	$+ 8$	a^3	$- 64b^3$
↑	↑	↑	↑
This term is x cubed.	This term is 2 cubed: $2^3 = 8$.	This term is a cubed.	This term is $4b$ cubed: $(4b)^3 = 64b^3$.

To find the formulas for factoring the sum of two cubes and the difference of two cubes, we need to find the following two products:

$$(\mathbf{x} + \mathbf{y})(x^2 - xy + y^2) = (\mathbf{x} + \mathbf{y})x^2 - (\mathbf{x} + \mathbf{y})xy + (\mathbf{x} + \mathbf{y})y^2 \quad \text{Use the distributive property.}$$
$$= x^3 + x^2y - x^2y - xy^2 + xy^2 + y^3$$
$$= x^3 + y^3 \quad \text{Combine like terms.}$$

$$(\mathbf{x} - \mathbf{y})(x^2 + xy + y^2) = (\mathbf{x} - \mathbf{y})x^2 + (\mathbf{x} - \mathbf{y})xy + (\mathbf{x} - \mathbf{y})y^2 \quad \text{Use the distributive property.}$$
$$= x^3 - x^2y + x^2y - xy^2 + xy^2 - y^3$$
$$= x^3 - y^3 \quad \text{Combine like terms.}$$

These results justify the formulas for factoring the **sum and difference of two cubes.**

Factoring the sum and difference of two cubes

$$x^3 + y^3 = (x + y)(x^2 - xy + y^2)$$
$$x^3 - y^3 = (x - y)(x^2 + xy + y^2)$$

If we think of the sum of two cubes as the cube of a **First** quantity plus the cube of a **Last** quantity, we have the formula

$$F^3 + L^3 = (F + L)(F^2 - FL + L^2)$$

In words, we say, *To factor the cube of a **First** quantity plus the cube of a **Last** quantity, we multiply the **First** plus the **Last** by*

- *the **First** squared*
- *minus the **First** times the **Last***
- *plus the **Last** squared.*

The formula for the difference of two cubes is

$$F^3 - L^3 = (F - L)(F^2 + FL + L^2)$$

In words, we say, *To factor the cube of a **First** quantity minus the cube of a **Last** quantity, we multiply the **First** minus the **Last** by*

- *the **First** squared*
- *plus the **First** times the **Last***
- *plus the **Last** squared.*

To factor the sum or difference of two cubes, it's helpful to know the cubes of the numbers from 1 to 10:

1, 8, 27, 64, 125, 216, 343, 512, 729, 1,000

Expressions containing variables such as $64b^3$ are also perfect cubes, because they can be written as the cube of a quantity:

$$64b^3 = (4b)^3$$

Self Check 8
Factor: $h^3 + 27$.

EXAMPLE 8 Factor: $x^3 + 8$.

Solution We think of $x^3 + 8$ as the cube of a **First** quantity, x, plus the cube of a **Last** quantity, 2.

$$x^3 + 8 = x^3 + 2^3$$

Thus, $x^3 + 8$ factors as the product of the sum of x and 2 and the trinomial $x^2 - 2x + 2^2$.

$$\mathbf{F}^3 + \mathbf{L}^3 = (\mathbf{F} + \mathbf{L})(\mathbf{F}^2 - \mathbf{FL} + \mathbf{L}^2)$$
$$\downarrow \quad \downarrow \qquad \downarrow \quad \downarrow \quad \downarrow \qquad \downarrow\downarrow \quad \downarrow$$
$$\mathbf{x}^3 + \mathbf{2}^3 = (\mathbf{x} + \mathbf{2})(\mathbf{x}^2 - \mathbf{x2} + \mathbf{2}^2) \quad \text{Substitute } x \text{ for F and 2 for L.}$$
$$= (x + 2)(x^2 - 2x + 4)$$

We can check by multiplying.

$$(\mathbf{x} + \mathbf{2})(x^2 - 2x + 4) = (\mathbf{x} + \mathbf{2})x^2 - (\mathbf{x} + \mathbf{2})2x + (\mathbf{x} + \mathbf{2})4$$
$$= x^3 + 2x^2 - 2x^2 - 4x + 4x + 8$$
$$= x^3 + 8$$

Answer $(h + 3)(h^2 - 3h + 9)$

EXAMPLE 9 Factor: $a^3 - 64b^3$.

Solution We think of $a^3 - 64b^3$ as the cube of a **F**irst quantity, a, minus the cube of a **L**ast quantity, $4b$.

$$a^3 - 64b^3 = a^3 - (4b)^3$$

Thus, its factors are the difference $a - 4b$ and the trinomial $a^2 + a(4b) + (4b)^2$.

$$\mathbf{F}^3 - \mathbf{L}^3 = (\mathbf{F} - \mathbf{L})(\mathbf{F}^2 + \mathbf{F}\,\mathbf{L} + \mathbf{L}^2)$$

$$\downarrow \quad \downarrow \qquad \downarrow \quad \downarrow \quad \downarrow \quad \downarrow \quad \downarrow \quad \downarrow$$

$$\begin{aligned} a^3 - (4b)^3 &= (a - 4b)[a^2 + a(4b) + (4b)^2] \\ &= (a - 4b)(a^2 + 4ab + 16b^2) \end{aligned}$$

Check by multiplying.

Self Check 9
Factor: $8c^3 - 1$.

Answer $(2c - 1)(4c^2 + 2c + 1)$

Sometimes we must factor out a greatest common factor before factoring a sum or difference of two cubes.

EXAMPLE 10 Factor: $-2t^5 + 250t^2$.

Solution Each term contains the factor $-2t^2$.

$$\begin{aligned} -2t^5 + 250t^2 &= -2t^2(t^3 - 125) && \text{Factor out } -2t^2. \\ &= -2t^2(t - 5)(t^2 + 5t + 25) && \text{Factor } t^3 - 125. \end{aligned}$$

Check by multiplying.

Self Check 10
Factor: $4c^3 + 4d^3$.

Answer $4(c + d)(c^2 - cd + d^2)$

A factoring strategy

When we solve equations and simplify expressions containing polynomials, we usually are not told what type of factoring technique to apply—we have to determine that ourselves. The following strategy is helpful when factoring a random polynomial.

Steps for factoring

1. Factor out all common factors.
2. If a polynomial has two terms, check for the following problem types:
   a. **The difference of two squares:** $x^2 - y^2 = (x + y)(x - y)$
   b. **The sum of two cubes:** $x^3 + y^3 = (x + y)(x^2 - xy + y^2)$
   c. **The difference of two cubes:** $x^3 - y^3 = (x - y)(x^2 + xy + y^2)$
3. If a polynomial has three terms, check for the following problem types:
   a. **A perfect square trinomial:**

   $$x^2 + 2xy + y^2 = (x + y)^2$$
   $$x^2 - 2xy + y^2 = (x - y)^2$$

   b. If the trinomial is not a perfect square, attempt to factor it as a general trinomial using the **trial-and-check method** or **factoring by grouping.**
4. If a polynomial has four or more terms, try **factoring by grouping.**
5. Continue until each individual factor is prime.
6. Check the results by multiplying.

Section 11.4 STUDY SET

VOCABULARY *Fill in the blanks.*

1. The binomial $x^2 - 25$ is called a ________ of two squares.
2. $x^2 + 6x + 9$ is a ________ square trinomial because it is the square of the binomial $(x + 3)$.
3. The binomial $x^3 + 27$ is called a sum of two ________. The binomial $x^3 - 8$ is called a ________ of two cubes.
4. To ________ $4x^2 - 12x + 9$ means to write it as the product of two binomials.

CONCEPTS *Fill in the blanks.*

5. Consider $25x^2 + 30x + 9$.
   a. The first term is the square of ▢.
   b. The last term is the square of ▢.
   c. The middle term is twice the product of ▢ and ▢.
6. Consider $49x^2 - 28xy + 4y^2$.
   a. The first term is the square of ▢.
   b. The last term is the square of ▢.
   c. The middle term is twice the product of ▢ and ▢.
7. To factor the square of a First quantity minus the square of a Last quantity, we multiply the ________ plus the ________ by the ________ minus the ________.
8. If a trinomial is the square of one quantity, plus the square of a second quantity, plus ________ the product of the quantities, it factors into the square of the ________ of the quantities.
9. a. $36x^2 = (\quad)^2$ b. $100x^4 = (\quad)^2$
   c. $27m^3 = (\quad)^3$ d. $a^6 = (\quad)^3$
10. a. $4x^2 - 9 = (\quad)^2 - (\quad)^2$
    b. $8x^3 - 27 = (\quad)^3 - (\quad)^3$
    c. $x^3 + 64y^3 = (\quad)^3 + (\quad)^3$
11. List the first ten perfect integer squares.
12. List the first five perfect integer cubes.
13. Explain why each trinomial is not a perfect square trinomial.
    a. $9h^2 - 6h + 7$
    b. $j^2 - 8j - 16$
    c. $25r^2 + 20r + 16$
14. a. Three incorrect factorizations of $x^2 + 36$ are given below. Show why each is wrong.

    $(x + 6)(x - 6)$
    $(x + 6)(x + 6)$
    $(x - 6)(x - 6)$

    b. Can $x^2 + 36$ be factored using only integer coefficients?

NOTATION *Write each expression as a polynomial in simpler form.*

15. $(6x)^2 - (5y)^2$
16. $(4x)^2 - (9y)^2$
17. $(3a)^2 - 2(3a)(5b) + (5b)^2$
18. $(2s)^2 + 2(2s)(9t) + (9t)^2$

Use an exponent to write each expression in simpler form.

19. $(x + 8)(x + 8)$
20. $(x - 8)(x - 8)$

PRACTICE *Complete each factorization.*

21. $a^2 - 6a + 9 = (a - \quad)^2$
22. $t^2 + 2t + 1 = (t \quad 1)^2$
23. $4x^2 + 4x + 1 = (2x \quad 1)^2$
24. $9y^2 - 12y + 4 = (3y - \quad)^2$

Factor each polynomial.

25. $x^2 + 6x + 9$
26. $x^2 + 10x + 25$
27. $y^2 - 8y + 16$
28. $z^2 - 2z + 1$
29. $t^2 + 20t + 100$
30. $r^2 + 24r + 144$
31. $u^2 - 18u + 81$
32. $v^2 - 14v + 49$
33. $4x^2 + 12x + 9$
34. $4x^2 - 4x + 1$
35. $36x^2 + 12x + 1$
36. $4x^2 - 20x + 25$
37. $a^2 + 2ab + b^2$
38. $a^2 - 2ab + b^2$
39. $16x^2 - 8xy + y^2$
40. $25x^2 + 20xy + 4y^2$

Complete each factorization.

41. $y^2 - 49 = (y + \quad)(y - \quad)$

42. $p^4 - q^2 = (p^2 + q)(\quad - \quad)$

43. $t^2 - w^2 = (\quad + \quad)(t - w)$

44. $49u^2 - 64v^2 = (\quad + 8v)(7u \quad 8v)$

Factor each polynomial, if possible.

45. $x^2 - 16$

46. $x^2 - 25$

47. $4y^2 - 1$

48. $9z^2 - 1$

49. $9x^2 - y^2$

50. $4x^2 - z^2$

51. $16a^2 - 25b^2$

52. $36a^2 - 121b^2$

53. $a^2 + b^2$

54. $121a^2 + 144b^2$

55. $a^4 - 144b^2$

56. $81y^4 - 100z^2$

57. $t^2z^2 - 64$

58. $900 - B^2C^2$

59. $8x^2 - 32y^2$

60. $2a^2 - 200b^2$

61. $7 - 7a^2$

62. $5 - 20x^2$

63. $6x^4 - 6x^2y^2$

64. $4b^2y - 16c^2y$

65. $x^4 - 81$

66. $y^4 - 625$

67. $a^4 - 16$

68. $b^4 - 256$

69. $81r^4 - 256s^4$

70. $16y^8 - 81z^4$

Complete each factorization.

71. $a^3 + 8 = (a + 2)(a^2 - \quad + 4)$

72. $x^3 - 1 = (x - 1)(x^2 + \quad + 1)$

73. $b^3 + 27 = (\qquad)(b^2 - 3b + 9)$

74. $z^3 - 125 = (\qquad)(z^2 + 5z + 25)$

Factor each polynomial.

75. $y^3 + 1$

76. $x^3 - 8$

77. $a^3 - 27$

78. $b^3 + 125$

79. $8 + x^3$

80. $27 - y^3$

81. $s^3 - t^3$

82. $8u^3 + w^3$

83. $a^3 + 8b^3$

84. $27a^3 - b^3$

85. $64x^3 - 27$

86. $27x^3 + 125$

87. $a^6 - b^3$

88. $a^3 + b^6$

89. $x^9 + y^6$

90. $x^3 - y^9$

91. $2x^3 + 54$

92. $2x^3 - 2$

93. $-x^3 + 216$

94. $-x^3 - 125$

95. $64m^3x - 8n^3x$

96. $16r^4 + 128rs^3$

97. $x^4y + 216xy^4$

98. $16a^5 - 54a^2b^3$

99. $81r^4s^2 - 24rs^5$

100. $4m^5n + 500m^2n^4$

APPLICATIONS

101. GENETICS The Hardy–Weinberg equation, one of the fundamental concepts in population genetics, is

$$p^2 + 2pq + q^2 = 1$$

where p represents the frequency of a certain dominant gene and q represents the frequency of a certain recessive gene. Factor the left-hand side of the equation.

102. SPACE TRAVEL The surface area of the spherical part of the spacecraft shown below is given by $(36\pi r^2 - 48\pi r + 16\pi)$ m^2. Factor the expression.

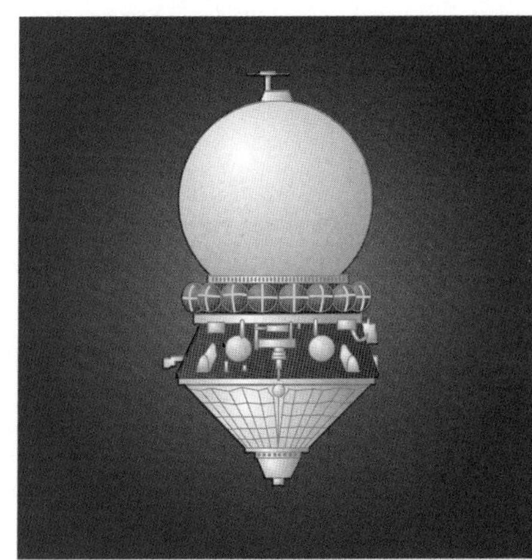

103. PHYSICS The illustration shows a time-sequence picture of a falling apple. Factor the expression, which gives the distance the apple falls during the time interval from t_1 to t_2 seconds.

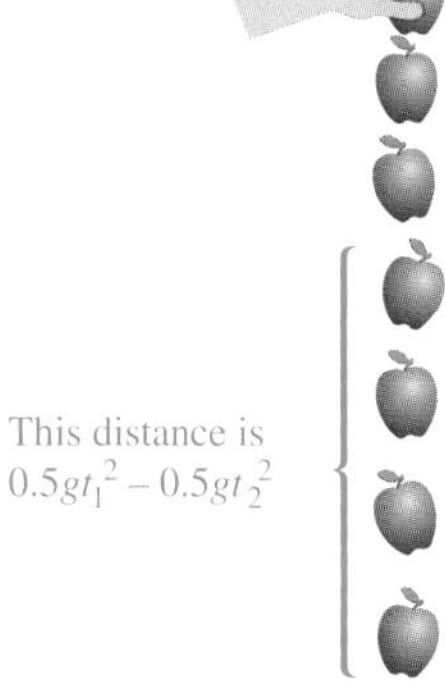

104. DARTS A circular dart board has a series of rings around a solid center, called the bullseye. To find the area of the outer black ring, we can use the formula

$$A = \pi R^2 - \pi r^2$$

Factor the expression on the right-hand side of the equation.

WRITING

105. When asked to factor $x^2 - 25$, one student wrote $(x + 5)(x - 5)$, and another student wrote $(x - 5)(x + 5)$. Are both answers correct? Explain.

106. Write a comment to the student whose work is shown below, explaining the initial error that was made.

Factor: $4x^2 - 16y^2$.

$$(2x + 4y)(2x - 4y)$$

107. Explain why $x^6 - 1$ can be thought of as a difference of two squares or as a difference of two cubes.

108. Why is $a^2 + 2a + 1$ a perfect square trinomial, and why isn't $a^2 + 4a + 1$ a perfect square trinomial?

REVIEW *Perform each division.*

109. $\dfrac{5x^2 + 10y^2 - 15xy}{5xy}$

110. $\dfrac{-30c^2d^2 - 15c^2d - 10cd^2}{-10cd}$

111. $2a - 1\overline{)a - 2 + 6a^2}$

112. $4b + 3\overline{)4b^3 - 5b^2 - 2b + 3}$

ADDITIONAL FACTORING PROBLEMS *Apply the factoring strategy to factor each polynomial completely. If a polynomial is not factorable, write "prime."*

113. $a^2(x - a) - b^2(x - a)$

114. $a^2c + a^2d^2 + bc + bd^2$

115. $70p^4q^3 - 35p^4q^2 + 49p^5q^2$

116. $a^2b^2 - 144$

117. $2ab^2 + 8ab - 24a$

118. $t^4 - 16$

119. $-8p^3q^7 - 4p^2q^3$

120. $8m^2n^3 - 24mn^4$

121. $20m^2 + 100m + 125$

122. $3rs + 6r^2 - 18s^2$

123. $x^2 + 7x + 1$

124. $3a^3 + 24b^3$

125. $-2x^5 + 128x^2$

126. $16 - 40z + 25z^2$

127. $14t^3 - 40t^2 + 6t^4$

128. $-9x^2y^2 + 6xy - 1$

129. $x^2y^2 - 2x^2 - y^2 + 2$

130. $5x^3y^3z^4 + 25x^2y^3z^2 - 35x^3y^2z^5$

131. $8p^6 - 27q^6$

132. $2c^2 - 5cd - 3d^2$

133. $125p^3 - 64y^3$

134. $8a^2x^3y - 2b^2xy$

135. $-16x^4y^2z + 24x^5y^3z^4 - 15x^2y^3z^7$

136. $2ac + 4ad + bc + 2bd$

137. $81p^4 - 16q^4$

138. $6x^2 - x - 16$

139. $4x^2 + 9y^2$

140. $30a^4 + 5a^3 - 200a^2$

141. $54x^3 + 250y^6$

142. $6a^3 + 35a^2 - 6a$

143. $10r^2 - 13r - 4$

144. $21t^3 - 10t^2 + t$

145. $49p^2 + 28pq + 4q^2$

146. $16x^2 - 40x^3 + 25x^4$

11.5 Quadratic Equations

- Quadratic equations • Solving quadratic equations by factoring • Applications

Equations that involve first-degree polynomials, such as $9x - 6 = 0$, are called *linear equations.* Equations that involve second-degree polynomials, such as $9x^2 - 6x = 0$, are called **quadratic equations.** In this section, we will define quadratic equations and learn how to solve many of them by factoring.

Quadratic equations

If a polynomial contains one variable with an exponent to the second (but no higher) power, it is called a **second-degree polynomial.** Equations in which a second-degree polynomial is equal to zero are called **quadratic equations.** Some examples are

$$9x^2 - 6x = 0, \qquad x^2 - 2x - 63 = 0, \qquad \text{and} \qquad 2x^2 + 3x - 2 = 0$$

A **quadratic equation** is an equation that can be written in the form

$$ax^2 + bx + c = 0$$

where a, b, and c represent real numbers, and a is not 0.

To write a quadratic equation such as $21x = 10 - 10x^2$ in $ax^2 + bx + c = 0$ form (called **quadratic form**), we use the addition and subtraction properties of equality to get 0 on the right-hand side.

$$21x = 10 - 10x^2$$

$$\mathbf{10x^2} + 21x = 10 - 10x^2 + \mathbf{10x^2} \qquad \text{Add } 10x^2 \text{ to both sides.}$$

$$10x^2 + 21x = 10 \qquad \text{Combine like terms: } -10x^2 + 10x^2 = 0.$$

$$10x^2 + 21x - 10 = 0 \qquad \text{Subtract 10 from both sides.}$$

When $21x = 10 - 10x^2$ is written in quadratic form, we see that $a = 10$, $b = 21$, and $c = -10$.

The techniques we have used to solve linear equations cannot be used to solve a quadratic equation, because those techniques cannot isolate the variable on one side of the equation. However, we can often solve quadratic equations using factoring and the following property of real numbers.

The zero-factor property

Suppose a and b represent real numbers. Then

If $ab = 0$, then $a = 0$ or $b = 0$.

In words, the zero-factor property states that when the product of two numbers is zero, at least one of them must be zero.

EXAMPLE 1 Solve: $(4y - 1)(y + 6) = 0$.

Solution The left-hand side of the equation is $(4y - 1)(y + 6)$. By the zero-factor property, one of these factors must be 0.

$$4y - 1 = 0 \qquad \text{or} \qquad y + 6 = 0$$

Self Check 1
Solve: $b(5b - 3) = 0$.

We can solve each of the linear equations.

$$4y - 1 = 0 \quad \text{or} \quad y + 6 = 0$$
$$4y = 1 \qquad\qquad y = -6$$
$$y = \frac{1}{4}$$

The equation has two solutions, $\frac{1}{4}$ and -6. To check, we substitute the results for y in the original equation and simplify.

For $y = \frac{1}{4}$

$$(4y - 1)(y + 6) = 0$$
$$\left[4\left(\frac{1}{4}\right) - 1\right]\left(\frac{1}{4} + 6\right) \stackrel{?}{=} 0$$
$$(1 - 1)\left(6\frac{1}{4}\right) \stackrel{?}{=} 0$$
$$0\left(6\frac{1}{4}\right) \stackrel{?}{=} 0$$
$$0 = 0$$

For $y = -6$

$$(4y - 1)(y + 6) = 0$$
$$[4(-6) - 1](-6 + 6) \stackrel{?}{=} 0$$
$$(-24 - 1)(0) \stackrel{?}{=} 0$$
$$-25(0) \stackrel{?}{=} 0$$
$$0 = 0$$

Answer $0, \frac{3}{5}$

Solving quadratic equations by factoring

In Example 1, the right-hand side of the equation was zero, and the left-hand side was in factored form, so we were able to use the zero-factor property immediately. However, to solve many quadratic equations, we must first do the factoring.

Self Check 2
Solve: $5x^2 + 10x = 0$

EXAMPLE 2 Solve: $9x^2 - 6x = 0$.

Solution We begin by factoring the left-hand side of the equation.

$$9x^2 - 6x = 0$$
$$3x(3x - 2) = 0 \quad \text{Factor out the GCF of } 3x.$$

By the zero-factor property, we have

$$3x = 0 \quad \text{or} \quad 3x - 2 = 0$$

We can solve each of the linear equations to get

$$x = 0 \quad \text{or} \quad x = \frac{2}{3}$$

To check, we substitute the results for x in the original equation and simplify.

For $x = 0$

$$9x^2 - 6x = 0$$
$$9(0)^2 - 6(0) \stackrel{?}{=} 0$$
$$0 - 0 \stackrel{?}{=} 0$$
$$0 = 0$$

For $x = \frac{2}{3}$

$$9x^2 - 6x = 0$$
$$9\left(\frac{2}{3}\right)^2 - 6\left(\frac{2}{3}\right) \stackrel{?}{=} 0$$
$$9\left(\frac{4}{9}\right) - 6\left(\frac{2}{3}\right) \stackrel{?}{=} 0$$
$$4 - 4 \stackrel{?}{=} 0$$
$$0 = 0$$

The solutions of $9x^2 - 6x = 0$ are 0 and $\frac{2}{3}$.

Answer $0, -2$

We can use the following steps to solve a quadratic equation by factoring.

Solving quadratic equations by factoring

1. Write the equation in $ax^2 + bx + c = 0$ form.
2. Factor the left-hand side of the equation.
3. Use the zero-factor property to set each factor equal to zero.
4. Solve each resulting linear equation.
5. Check the results in the original equation.

EXAMPLE 3 Solve: $x^2 = 9$.

Solution Before we can use the zero-factor property, we must subtract 9 from both sides to make the right-hand side zero.

$$x^2 = 9$$

$$x^2 - 9 = 0 \quad \text{Subtract 9 from both sides.}$$

$$(x + 3)(x - 3) = 0 \quad \text{Factor the difference of two squares.}$$

$$x + 3 = 0 \quad \text{or} \quad x - 3 = 0 \quad \text{Set each factor equal to zero.}$$

$$x = -3 \quad \Big| \quad x = 3 \quad \text{Solve each linear equation.}$$

Check each possible solution by substituting it into the original equation.

For $x = -3$	For $x = 3$
$x^2 = 9$	$x^2 = 9$
$(-3)^2 \stackrel{?}{=} 9$	$(3)^2 \stackrel{?}{=} 9$
$9 = 9$	$9 = 9$

The solutions of $x^2 = 9$ are -3 and 3.

Self Check 3
Solve: $x^2 - 25 = 0$.

Answer $-5, 5$

EXAMPLE 4 Solve: $x^2 - 2x - 63 = 0$.

Solution In this case, we must factor a trinomial to solve the equation.

$$x^2 - 2x - 63 = 0$$

$$(x + 7)(x - 9) = 0 \quad \text{Factor the trinomial } x^2 - 2x - 63.$$

$$x + 7 = 0 \quad \text{or} \quad x - 9 = 0 \quad \text{Set each factor equal to zero.}$$

$$x = -7 \quad \Big| \quad x = 9 \quad \text{Solve each linear equation.}$$

The solutions are -7 and 9. Check each one.

Self Check 4
Solve: $x^2 + 5x + 6 = 0$.

Answer $-2, -3$

EXAMPLE 5 Solve: $2x^2 + 3x = 2$.

Solution We write the equation in the form $ax^2 + bx + c = 0$ and then solve for x.

$$2x^2 + 3x = 2$$

$$2x^2 + 3x - 2 = 0 \quad \text{Subtract 2 from both sides so that the right-hand side is zero.}$$

$$(2x - 1)(x + 2) = 0 \quad \text{Factor } 2x^2 + 3x - 2.$$

Self Check 5
Solve: $3x^2 - 6 = -7x$.

$$2x - 1 = 0 \quad \text{or} \quad x + 2 = 0$$ Set each factor equal to zero.

$$2x = 1 \qquad x = -2$$ Solve each linear equation.

$$x = \frac{1}{2}$$

Use a check to verify that $\frac{1}{2}$ and -2 are solutions.

Answer $\frac{2}{3}, -3$

Self Check 6
Solve: $x(4x + 12) = -9$.

EXAMPLE 6 Solve: $-4 = x(9x - 12)$.

Solution First, we need to write the equation in the form $ax^2 + bx + c = 0$.

$$-4 = x(9x - 12)$$

$$-4 = 9x^2 - 12x$$ Distribute the multiplication by x.

$$0 = 9x^2 - 12x + 4$$ Add 4 to both sides to make the left-hand side zero.

$$0 = (3x - 2)(3x - 2)$$ Factor the trinomial.

$$3x - 2 = 0 \quad \text{or} \quad 3x - 2 = 0$$ Set each factor equal to zero.

$$3x = 2 \qquad 3x = 2$$ Add 2 to both sides.

$$x = \frac{2}{3} \qquad x = \frac{2}{3}$$ Divide both sides by 3.

The equation has two solutions that are the same. We call $\frac{2}{3}$ a *repeated solution*. Check by substituting it into the original equation.

Answer $-\frac{3}{2}, -\frac{3}{2}$

Self Check 7
Solve: $10x^3 + x^2 - 2x = 0$.

EXAMPLE 7 Solve: $6x^3 + 12x = 17x^2$.

Solution This is not a quadratic equation, because it contains the term x^3. However, we can solve it using factoring and an extension of the zero-factor property.

$$6x^3 + 12x = 17x^2$$

$$6x^3 - 17x^2 + 12x = 0$$ Subtract $17x^2$ from both sides to get 0 on the right-hand side.

$$x(6x^2 - 17x + 12) = 0$$ Factor out the GCF, x.

$$x(2x - 3)(3x - 4) = 0$$ Factor $6x^2 - 17x + 12$.

$$x = 0 \quad \text{or} \quad 2x - 3 = 0 \quad \text{or} \quad 3x - 4 = 0$$ Set each factor equal to zero.

$$2x = 3 \qquad 3x = 4$$ Solve the linear equations.

$$x = \frac{3}{2} \qquad x = \frac{4}{3}$$

This equation has three solutions, $0, \frac{3}{2}$, and $\frac{4}{3}$.

Answer $0, \frac{2}{5}, -\frac{1}{2}$

Applications

The solutions of many problems involve the use of quadratic equations.

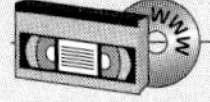

EXAMPLE 8 Softball. A pitcher can throw a fastball underhand at 63 feet per second (about 45 mph). If she throws a ball into the air with that velocity, its height h in feet, t seconds after being released, is given by the formula

$$h = -16t^2 + 63t + 4$$

After the ball is thrown, in how many seconds will it hit the ground?

FIGURE 11-2

Solution When the ball hits the ground, its height will be 0 feet. To find the time that it will take for the ball to hit the ground, we set h equal to 0, and solve the quadratic equation for t.

$$\mathbf{h} = -16t^2 + 63t + 4$$

$$\mathbf{0} = -16t^2 + 63t + 4 \quad \text{Substitute 0 for the height } h.$$

$$0 = -(16t^2 - 63t - 4) \quad \text{Factor out } -1.$$

$$0 = -(16t + 1)(t - 4) \quad \text{Factor } 16t^2 - 63t - 4.$$

$$16t + 1 = 0 \quad \text{or} \quad t - 4 = 0 \quad \text{Set each factor that contains a variable equal to 0.}$$

$$16t = -1 \quad \Big| \quad t = 4 \quad \text{Solve each equation.}$$

$$t = -\frac{1}{16}$$

Since time cannot be negative, we discard the solution $-\frac{1}{16}$. The second solution indicates that the ball hits the ground 4 seconds after being released. Check this answer by substituting 4 for t in $h = -16t^2 + 63t + 4$. You should get $h = 0$.

EXAMPLE 9 Perimeter of a rectangle. Assume that the rectangle in Figure 11-3 has an area of 52 square centimeters and that its length is 1 centimeter more than 3 times its width. Find the perimeter of the rectangle.

$3w + 1$

w | $A = 52 \text{ cm}^2$

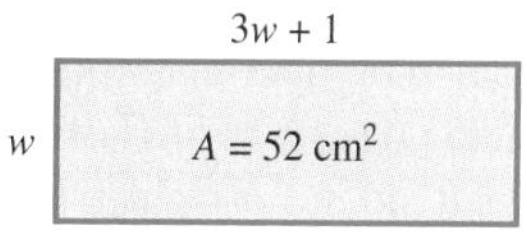

FIGURE 11-3

Analyze the problem

The area of the rectangle is 52 square centimeters. Recall that the formula that gives the area of a rectangle is $A = lw$. To find the perimeter of the rectangle, we need to know its length and width. We are told that its length is related to its width; the length is 1 centimeter more than 3 times the width.

Form an equation

Let w represent the width of the rectangle. Then $3w + 1$ represents its length. Because the area is 52 square centimeters, we substitute 52 for A and $3w + 1$ for l in the formula $A = lw$.

$$\mathbf{A} = lw$$

$$\mathbf{52} = (3w + 1)w$$

Solve the equation

Now we solve the equation for w.

$$52 = (3w + 1)w \quad \text{This is the equation to solve.}$$

$$52 = 3w^2 + w \quad \text{Distribute the multiplication by } w.$$

$$0 = 3w^2 + w - 52 \quad \text{Subtract 52 from both sides to make the left-hand side zero.}$$

$$0 = (3w + 13)(w - 4) \quad \text{Factor the trinomial.}$$

$$3w + 13 = 0 \quad \text{or} \quad w - 4 = 0 \quad \text{Set each factor equal to zero.}$$

$$3w = -13 \quad \Big| \quad w = 4 \quad \text{Solve each linear equation.}$$

$$w = -\frac{13}{3}$$

State the conclusion

Since the width cannot be negative, we discard the solution $-\frac{13}{3}$. Thus, the width of the rectangle is 4, and the length is given by

$$\begin{aligned} 3w + 1 &= 3(\mathbf{4}) + 1 \quad \text{Substitute 4 for } w. \\ &= 12 + 1 \\ &= 13 \end{aligned}$$

The dimensions of the rectangle are 4 centimeters by 13 centimeters. We find the perimeter by substituting 13 for l and 4 for w in the formula for the perimeter of a rectangle.

$$\begin{aligned} P &= 2l + 2w \\ &= 2(\mathbf{13}) + 2(\mathbf{4}) \\ &= 26 + 8 \\ &= 34 \end{aligned}$$

The perimeter of the rectangle is 34 centimeters.

Check the result

A rectangle with dimensions of 13 centimeters by 4 centimeters does have an area of 52 square centimeters, and the length is 1 centimeter more than 3 times the width. A rectangle with these dimensions has a perimeter of 34 centimeters.

The next example involves a right triangle. A **right triangle** is a triangle that contains a 90° angle. The longest side of a right triangle is the **hypotenuse,** which is the side opposite the right angle. The remaining two sides are the **legs** of the triangle. (See Figure 11-4.) The **Pythagorean theorem** provides a formula relating the lengths of the three sides of a right triangle.

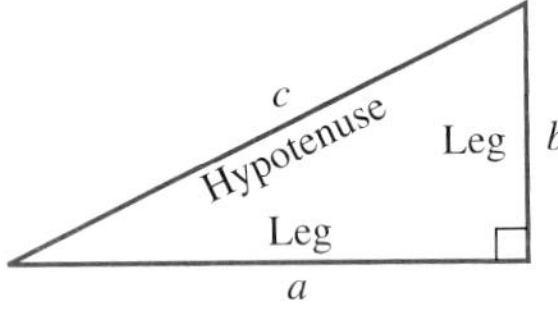

FIGURE 11-4

THINK IT THROUGH — Pythagorean Triples

"Fraternity and sorority membership nationwide is declining, down about 30% in the last decade." *Chronicle of Higher Education, 2003*

The first college social fraternity, Phi Beta Kappa, was founded in 1776 on the campus of The College of William and Mary. However, secret societies have existed since ancient times, and from these roots the essence of today's fraternities and sororities have their foundation. Pythagoras, the Greek mathematician of the 6th century B.C. was the leader of a secret fraternity/sorority called the Pythagoreans. They were a community of men and women that studied mathematics, and in particular, the "magic 3-4-5 triangle." This right triangle is special because the sum of the squares of the lengths of its legs is equal to the square of the length of its hypotenuse: $3^2 + 4^2 = 5^2$ or $9 + 16 = 25$. Today, we call a set of three natural numbers a, b, and c that satisfy $a^2 + b^2 = c^2$ a Pythagorean triple. Show that each list of numbers is a Pythagorean triple.

1. 5, 12, 13 **2.** 7, 24, 25 **3.** 8, 15, 17
4. 9, 40, 41 **5.** 11, 60, 61 **6.** 12, 35, 37

Pythagorean Theorem

If the length of the hypotenuse of a right triangle is c and the lengths of the two legs are a and b, then

$$c^2 = a^2 + b^2$$

EXAMPLE 10 **Right triangles.** The longer leg of a right triangle is 3 units longer than the shorter leg. If the hypotenuse is 6 units longer than the shorter leg, find the lengths of the sides of the triangle.

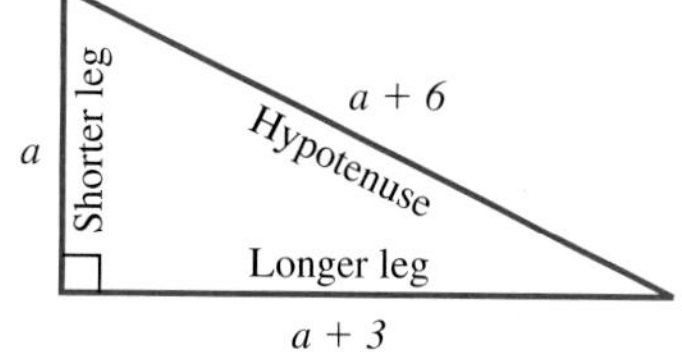

Analyze the problem

We begin by drawing a right triangle and labeling the legs and the hypotenuse.

Form an equation

We let a = the length of the shorter leg. Then the length of the hypotenuse is $a + 6$ and the length of the longer leg is $a + 3$. By the Pythagorean theorem, we have

$$\begin{pmatrix}\text{The length of}\\ \text{the shorter leg}\end{pmatrix}^2 \quad \text{plus} \quad \begin{pmatrix}\text{The length of}\\ \text{the longer leg}\end{pmatrix}^2 \quad \text{equals} \quad \begin{pmatrix}\text{The length of}\\ \text{the hypotenuse}\end{pmatrix}^2$$

$$a^2 \quad + \quad (a + 3)^2 \quad = \quad (a + 6)^2$$

Solve the equation

$$a^2 + (a + 3)^2 = (a + 6)^2$$

$$a^2 + a^2 + 6a + 9 = a^2 + 12a + 36 \qquad \text{Find } (a + 3)^2 \text{ and } (a + 6)^2.$$

$$2a^2 + 6a + 9 = a^2 + 12a + 36 \qquad \text{Combine like terms on the left-hand side.}$$

$$a^2 - 6a - 27 = 0 \qquad \text{Subtract } a^2, 12a, \text{ and 36 from both sides to make the right-hand side 0.}$$

Now solve the quadratic equation for a.

$$a^2 - 6a - 27 = 0$$

$$(a - 9)(a + 3) = 0 \qquad \text{Factor.}$$

$$a - 9 = 0 \quad \text{or} \quad a + 3 = 0 \qquad \text{Set each factor to zero.}$$

$$a = 9 \quad \Big| \quad a = -3 \qquad \text{Solve each equation.}$$

State the conclusion

Since a side cannot have a negative length, we discard the solution -3. Thus, the shorter leg is 9 units long, the hypotenuse is $9 + 6 = 15$ units long, and the longer leg is $9 + 3 = 12$ units long.

Check the result

The longer leg, with length 12, is 3 units longer than the shorter leg, with length 9. The hypotenuse, with length 15, is 6 units longer than the shorter leg. Since these lengths satisfy the Pythagorean theorem, the results check.

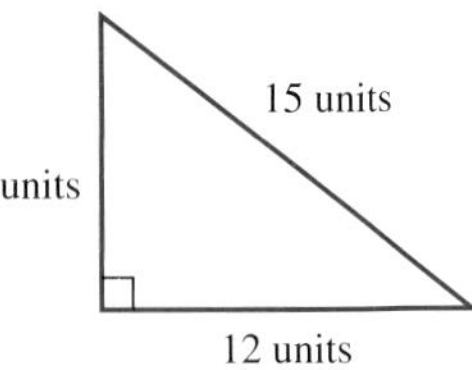

$$9^2 + 12^2 \stackrel{?}{=} 15^2$$
$$81 + 144 \stackrel{?}{=} 225$$
$$225 = 225$$

Section 11.5 STUDY SET

VOCABULARY *Fill in the blanks.*

1. Any equation that can be written in the form $ax^2 + bx + c = 0$ is called a ________ equation.
2. To ________ a binomial or trinomial means to write it as a product.

CONCEPTS *Fill in the blanks.*

3. When the product of two numbers is zero, at least one of them is ______. Symbolically, we can state this: If $ab = 0$, then $a =$ ☐ or $b =$ ☐.
4. We can often use ________ and the zero-factor property to solve quadratic equations.
5. To write a quadratic equation in *quadratic form* means that one side of the equation must be ______ and the other side must be in the form $ax^2 + bx + c$.
6. If the length of the hypotenuse of a right triangle is c and the legs are a and b, then $c^2 =$ ☐.
7. Classify each equation as quadratic or linear.
   a. $3x^2 + 4x + 2 = 0$
   b. $3x + 7 = 0$
   c. $2 = -16 - 4x$
   d. $-6x + 2 = x^2$
8. Check to see whether the given number is a solution of the given quadratic equation.
   a. $x^2 - 4x = 0; x = 4$
   b. $x^2 + 2x - 4 = 0; x = -2$
   c. $4x^2 - x + 3 = 0; x = 1$
9. a. Evaluate $x^2 + 6x - 16$ for $x = 0$.
   b. Factor: $x^2 + 6x - 16$.
   c. Solve: $x^2 + 6x - 16 = 0$.
10. The equation $3x^2 - 4x + 5 = 0$ is written in $ax^2 + bx + c = 0$ form. What are a, b, and c?
11. What is the first step that should be performed to solve each equation?
    a. $x^2 + 7x = -6$
    b. $x(x + 7) = -3$
12. a. How many solutions does the linear equation $2a + 3 = 2$ have?
    b. How many solutions does the quadratic equation $2a^2 + 3a = 2$ have?

NOTATION *Complete each solution to solve the equation.*

13. $7y^2 + 14y = 0$

$\square(y + 2) = 0$

$7y = 0$ or $\square = 0$

$y = 0$ | $y = -2$

14. $12p^2 - p - 6 = 0$

$(\square - 3)(3p + \square) = 0$

$\square = 0$ or $3p + 2 = \square$

$4p = \square$ | $3p = \square$

$p = \frac{3}{4}$ | $p = -\frac{2}{3}$

PRACTICE *Solve each equation.*

15. $(x - 2)(x + 3) = 0$
16. $(x - 3)(x - 2) = 0$
17. $(2s - 5)(s + 6) = 0$

18. $(3h - 4)(h + 1) = 0$
19. $(x - 1)(x + 2)(x - 3) = 0$
20. $(x + 2)(x + 3)(x - 4) = 0$
21. $x(x - 3) = 0$
22. $x(x + 5) = 0$
23. $x(2x - 5) = 0$
24. $x(5x + 7) = 0$
25. $w^2 - 7w = 0$
26. $p^2 + 5p = 0$
27. $3x^2 + 8x = 0$
28. $5x^2 - x = 0$
29. $8s^2 - 16s = 0$
30. $15s^2 - 20s = 0$
31. $x^2 - 25 = 0$
32. $x^2 - 36 = 0$
33. $4x^2 - 1 = 0$
34. $9y^2 - 1 = 0$
35. $9y^2 - 4 = 0$
36. $16z^2 - 25 = 0$
37. $x^2 = 100$
38. $z^2 = 25$
39. $4x^2 = 81$
40. $9y^2 = 64$
41. $x^2 - 13x + 12 = 0$
42. $x^2 + 7x + 6 = 0$
43. $x^2 - 4x - 21 = 0$
44. $x^2 + 2x - 15 = 0$
45. $x^2 - 9x + 8 = 0$
46. $x^2 - 14x + 45 = 0$
47. $a^2 + 8a = -15$
48. $a^2 - a = 56$
49. $2y - 8 = -y^2$
50. $-3y + 18 = y^2$
51. $x^3 + 3x^2 + 2x = 0$
52. $x^3 - 7x^2 + 10x = 0$
53. $k^3 - 27k - 6k^2 = 0$
54. $j^3 - 22j - 9j^2 = 0$
55. $(x - 1)(x^2 + 5x + 6) = 0$
56. $(x - 2)(x^2 - 8x + 7) = 0$
57. $2x^2 - 5x + 2 = 0$
58. $2x^2 + x - 3 = 0$
59. $5x^2 - 6x + 1 = 0$
60. $6x^2 - 5x + 1 = 0$
61. $4r^2 + 4r = -1$
62. $9m^2 + 6m = -1$
63. $-15x^2 + 2 = -7x$
64. $-8x^2 - 10x = -3$
65. $x(2x - 3) = 20$
66. $x(2x - 3) = 14$
67. $(d + 1)(8d + 1) = 18d$
68. $4h(3h + 2) = h + 12$
69. $2x(3x^2 + 10x) = -6x$
70. $2x^3 = 2x(x + 2)$
71. $x^3 + 7x^2 = x^2 - 9x$
72. $x^2(x + 10) = 2x(x - 8)$

APPLICATIONS *An object has been thrown straight up into the air. The formula $h = vt - 16t^2$ gives the height h of the object above the ground after t seconds, when it is thrown upward with an initial velocity v.*

73. TIME OF FLIGHT After how many seconds will the object hit the ground if it is thrown with a velocity of 144 feet per second?

74. TIME OF FLIGHT After how many seconds will the object hit the ground if it is thrown with a velocity of 160 feet per second?

75. OFFICIATING Before a football game, a coin toss is used to determine which team will kick off. The height h (in feet) of a coin above the ground t seconds after being flipped up into the air is given by

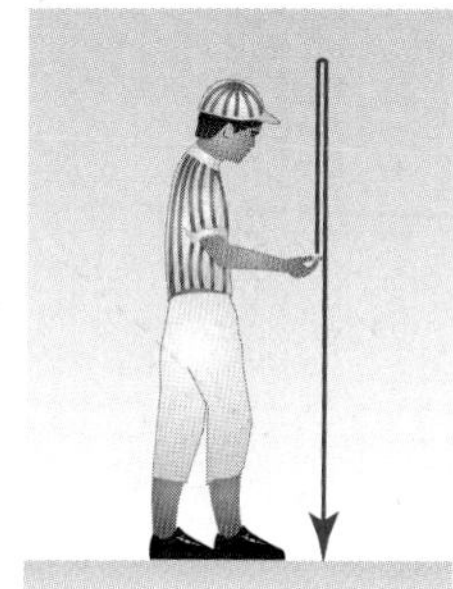

$$h = -16t^2 + 22t + 3$$

How long does a team captain have to call heads or tails if it must be done while the coin is in the air?

76. DOLPHINS The height h in feet reached by a dolphin t seconds after breaking the surface of the water is given by

$$h = -16t^2 + 32t$$

How long will it take the dolphin to jump out of the water and touch the trainer's hand?

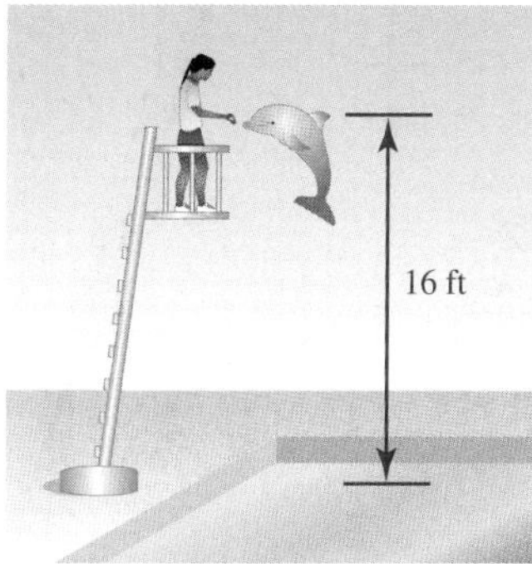

77. EXHIBITION DIVING In Acapulco, Mexico, men diving from a cliff to the water 64 feet below are a tourist attraction. A diver's height h above the water t seconds after diving is given by $h = -16t^2 + 64$. How long does a dive last?

78. FORENSIC MEDICINE The kinetic energy E of a moving object is given by $E = \frac{1}{2}mv^2$, where m is the mass of the object (in kilograms) and v is the object's velocity (in meters per second). Kinetic energy is measured in joules. Examining the damage done to a victim, a police pathologist determines that the energy of a 3-kilogram mass at impact was 54 joules. Find the velocity at impact. (*Hint:* Multiply both sides of the equation by 2.)

79. CHOREOGRAPHY For the finale of a musical, 36 dancers are to assemble in a triangular-shaped series of rows, where each successive row has one more dancer than the previous row. The illustration shows the beginning of such a formation. The relationship between the number of rows r and the number of dancers d is given by

$$d = \frac{1}{2}r(r + 1)$$

Determine the number of rows in the formation. (*Hint:* Multiply both sides of the equation by 2.)

80. CRAFTS The illustration shows how a geometric wall hanging can be created by stretching yarn from peg to peg across a wooden ring. The relationship between the number of pegs p placed evenly around the ring and the number of yarn segments s that criss-cross the ring is given by the formula

$$s = \frac{p(p - 3)}{2}$$

How many pegs are needed if the designer wants 27 segments to criss-cross the ring? (*Hint:* Multiply both sides of the equation by 2.)

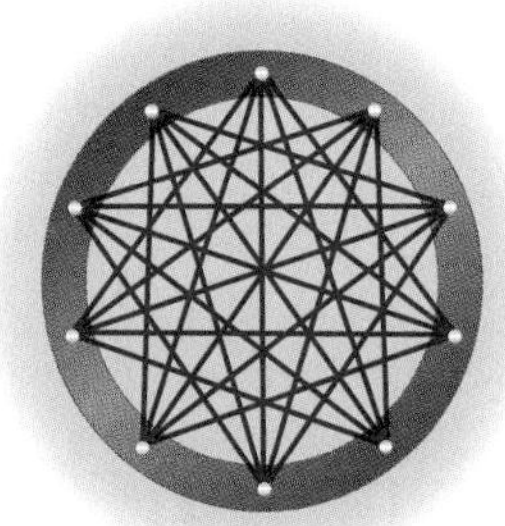

81. CUSTOMER SERVICE At a pharmacy, customers take tickets to reserve their turn for service. If the product of the ticket number now being served and the next ticket number to be served is 156, what number is now being served?

82. HISTORY Delaware was the first state to enter the union and Hawaii was the 50th. If we order the positions of entry for the rest of the states, we find that Tennessee entered the union right after Kentucky, and the product of their order-of-entry numbers is 240. Use the given information to complete this statement.

Kentucky was the ____th state to enter the union.

Tennessee was the ____th state to enter the union.

83. INSULATION The area of the rectangular slab of foam insulation in the illustration is 36 square meters. Find the dimensions of the slab.

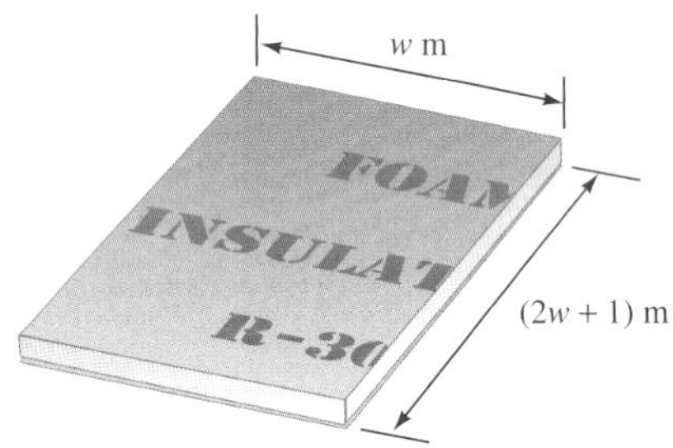

84. SHIPPING PALLETS The length of a rectangular shipping pallet is 2 feet less than 3 times its width. Its area is 21 square feet. Find the dimensions of the pallet.

85. BOATING The inclined ramp of the boat launch shown is 8 meters longer than the rise of the ramp. The run is 7 meters longer than the rise. How long are the three sides of the ramp?

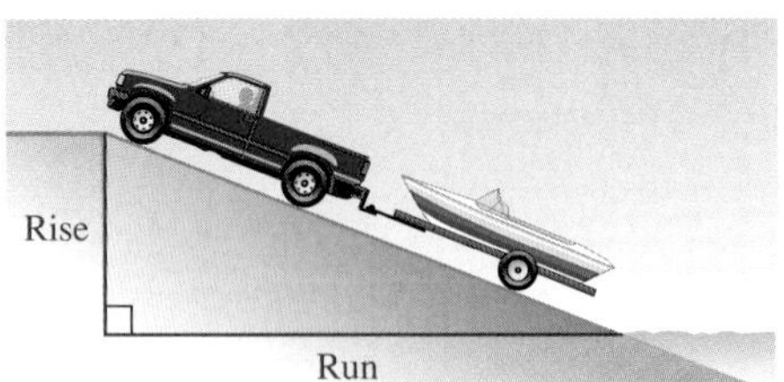

86. CAR REPAIRS To create some space to work under the front end of a car, a mechanic drives it up steel ramps. The ramp in the illustration is 1 foot longer than the back, and the base is 2 feet longer than the back of the ramp. Find the length of each side of the ramp.

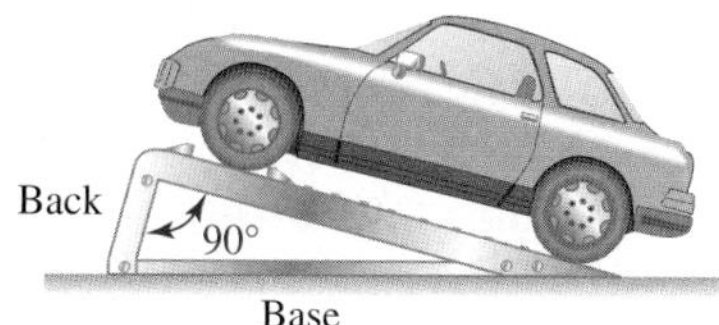

87. GARDENING TOOLS The dimensions (in millimeters) of the teeth of a pruning saw blade are given in the illustration. Find each length.

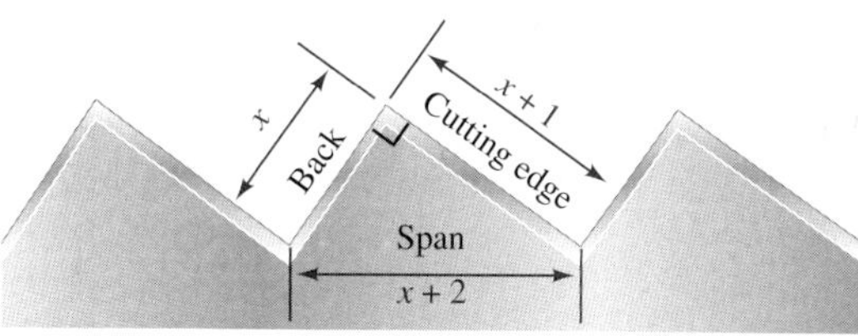

88. HARDWARE An aluminum brace used to support a wooden shelf has a length that is 2 inches less than twice the width of the shelf. The brace is anchored to the wall 8 inches below the shelf. Find the width of the shelf and the length of the brace.

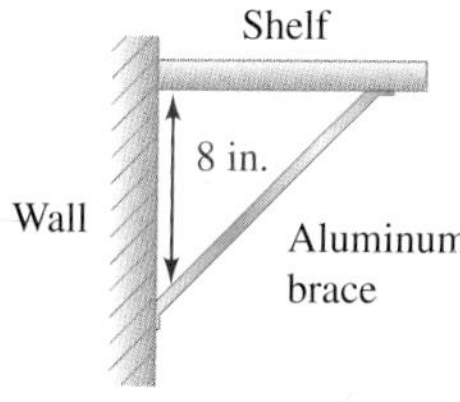

89. DESIGNING A TENT The length of the base of the triangular sheet of canvas above the door of a tent is 2 feet more than twice its height. The area is 30 square feet. Find the height and the length of the base of the triangle.

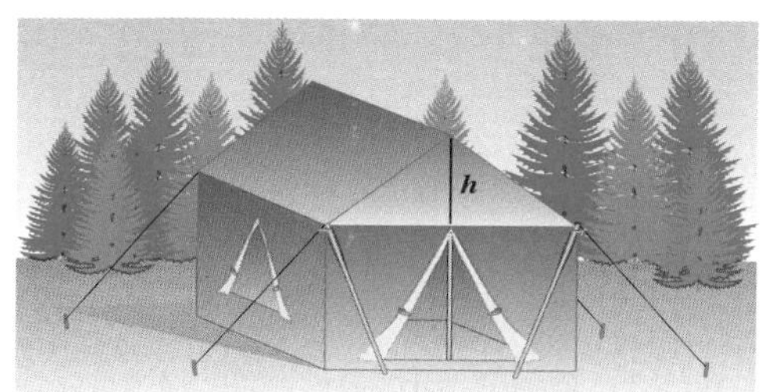

90. DIMENSIONS OF A TRIANGLE The height of a triangle is 2 inches less than 5 times the length of its base. The area is 36 square inches. Find the length of the base and the height of the triangle.

91. TUBING A piece of cardboard in the shape of a parallelogram is twisted to form the tube for a roll of paper towels. The parallelogram has an area of 60 square inches. If its height h is 7 inches more than the length of the base b, what is the circumference of the tube? (*Hint:* The formula for the area of a parallelogram is $A = bh$.)

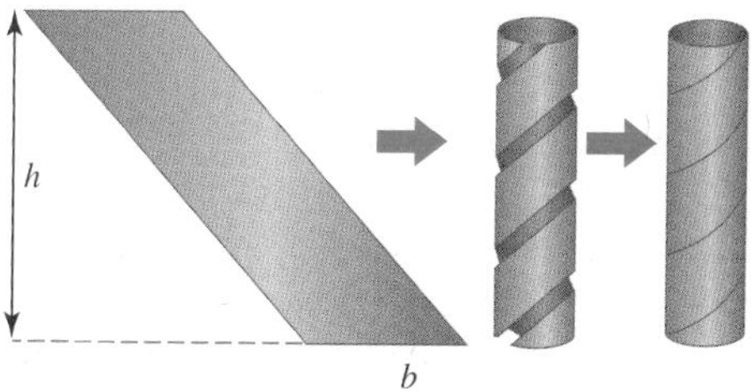

92. SWIMMING POOL BORDERS The owners of the rectangular swimming pool in the illustration want to surround the pool with a crushed-stone border of uniform width. They have enough stone to cover 74 square meters. How wide should they make the border? (*Hint:* The area of the larger rectangle minus the area of the smaller is the area of the border.)

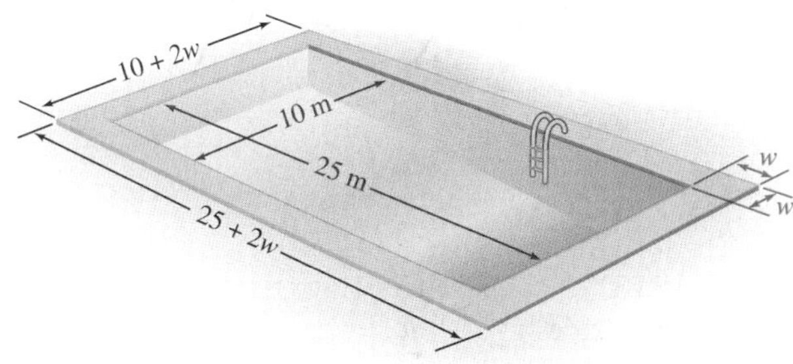

93. HOUSE CONSTRUCTION The formula for the area of a trapezoid is

$$A = \frac{h(B + b)}{2}$$

The area of the trapezoidal truss in the illustration is 24 square meters. Find the height of the trapezoid if one base is 8 meters and the other base is the same as the height. (*Hint:* Multiply both sides of the equation by 2.)

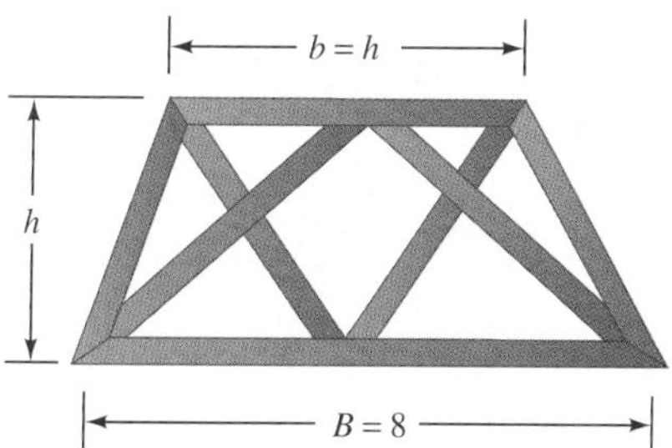

94. VOLUME OF A PYRAMID The volume of a pyramid is given by the formula

$$V = \frac{Bh}{3}$$

where B is the area of its base and h is its height. The volume of the following pyramid is 192 cubic centimeters. Find the dimensions of its rectangular base if one edge of the base is 2 centimeters longer than the other and the height of the pyramid is 12 centimeters.

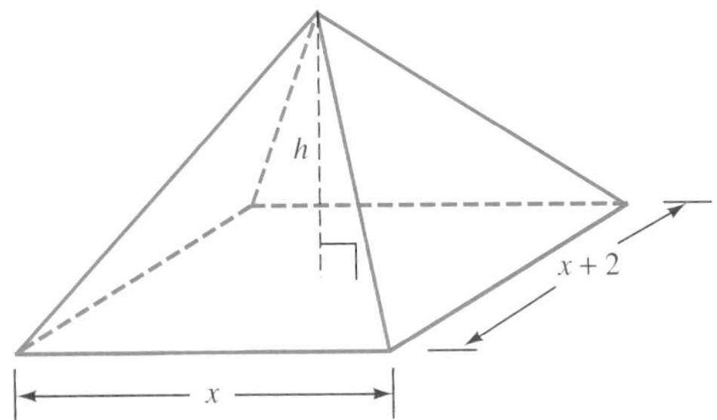

WRITING

95. What is wrong with the logic used to solve $x^2 + x = 6$?

$$x(x + 1) = 6$$
$$x = 6 \quad \text{or} \quad x + 1 = 6$$
$$x = 5$$

The solutions are 6 and 5.

96. Suppose that to find the length of the base of a triangle, you write a quadratic equation and solve it to find $b = 6$ or $b = -8$. Explain why one solution should be discarded.

REVIEW

97. EXERCISE A doctor advises one patient to exercise at least 15 minutes but less than 30 minutes per day. Use a compound inequality to express the range of these times in minutes.

98. SNACKS A bag of peanuts is worth \$0.30 less than a bag of cashews. Equal amounts of peanuts and cashews are used to make 40 bags of a mixture that is worth \$1.05 per bag. How much is a bag of cashews worth?

99. A rectangle is 3 times as long as it is wide, and its perimeter is 120 centimeters. Find its area.

100. INVESTING A woman invests \$15,000, part at 7% annual interest and part at 8% annual interest. If she receives \$1,100 interest per year, how much did she invest at 7%?

KEY CONCEPT

Factoring

Factoring polynomials is the reverse of the process of multiplying polynomials. When we factor a polynomial, we write it as a product of two or more factors.

1. In the following problem, the distributive property is used to multiply a monomial and a binomial

Find $3(x + 9)$.

$$3(x + 9) = 3 \cdot x + 3 \cdot 9$$
$$= 3x + 27$$

Rewrite this so that it becomes a factoring problem. What would you start with? What would the answer be?

2. In the following problem, we multiply two binomials.

Find $(x + 3)(x + 9)$.

$$(x + 3)(x + 9) = x^2 + 9x + 3x + 27$$
$$= x^2 + 12x + 27$$

Rewrite this so that it becomes a factoring problem. What would you start with? What would the answer be?

A factoring strategy

The following flowchart leads you through the correct steps to identify the type(s) of factoring necessary to factor any given polynomial having two or more terms.

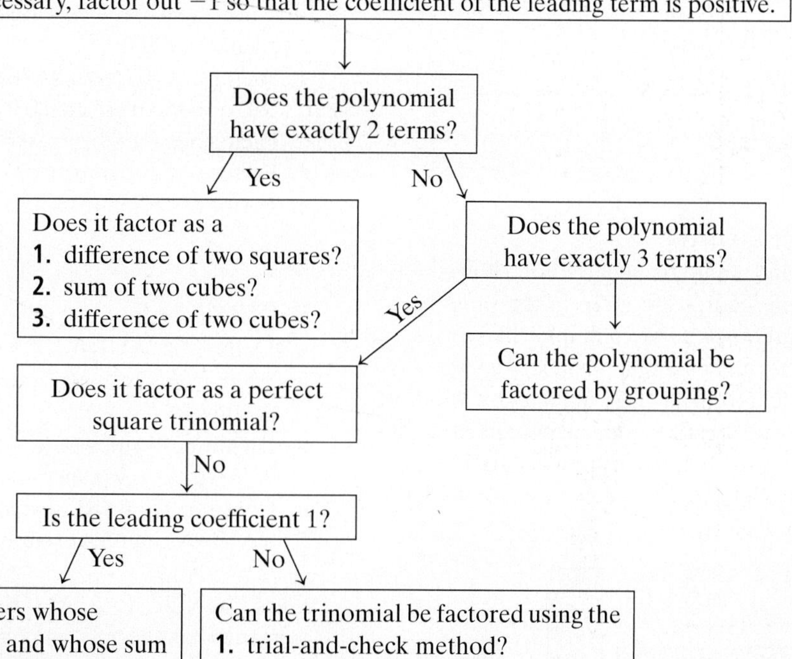

Factor each polynomial completely.

3. $-3a^2 + 21a - 36$

4. $x^2 - 121y^2$

5. $rt + 2r + st + 2s$

6. $v^3 - 8$

7. $6t^2 - 19t + 15$

8. $25y^2 - 20y + 4$

9. $2r^3 - 50r$

10. $46w - 6 + 16w^2$

ACCENT ON TEAMWORK

SECTION 11.1

PRIME NUMBERS We can use a procedure called the **sieve of Eratosthenes** to find all the prime numbers in the set of the first 100 whole numbers. Give each member in your group a copy of the table. Cross out 1, since it is not a prime number by definition. Cross out any numbers divisible by 2, 3, 5, or 7, because they have a factor of 2, 3, 5, or 7 and thus would not be prime. Don't cross out 2, 3, 5, or 7, because they are prime numbers. At the end of this process, you will end up with the first 25 prime numbers.

1	2	3	4	5	6	7	8	9	10
11	12	13	14	15	16	17	18	19	20
21	22	23	24	25	26	27	28	29	30
31	32	33	34	35	36	37	38	39	40
41	42	43	44	45	46	47	48	49	50
51	52	53	54	55	56	57	58	59	60
61	62	63	64	65	66	67	68	69	70
71	72	73	74	75	76	77	78	79	80
81	82	83	84	85	86	87	88	89	90
91	92	93	94	95	96	97	98	99	100

SECTION 11.2

FACTORING TRINOMIALS

a. Four squares and four rectangles are shown. The dimensions of the figures are given in the same units. Find the sum of their areas by combining like terms.

b. Cut out the figures and assemble them into a large rectangle having a length of $(x + 3)$ units and a width of $(x + 1)$ units. Note that these dimensions give the factored form of $x^2 + 4x + 3$, the answer to part a.

c. Make a new model that could be used to find the factored form of $x^2 + 5x + 4$.

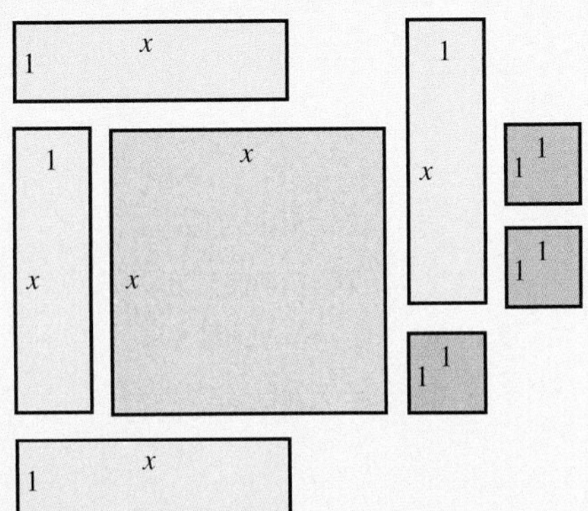

SECTION 11.3

COMPARING METHODS Factor $18x^2 + 3x - 10$ by using the trial-and-check method and by using the grouping method. Which method do you think is better? Explain why.

SECTION 11.4

COMPARING METHODS Show how $x^6 - 1$ can be factored in two ways: first as a difference of two squares and then as a difference of two cubes.

SUMS AND DIFFERENCES OF CUBES Long division can be used to verify the following factoring formulas.

$$x^3 + y^3 = (x + y)(x^2 - xy + y^2)$$
$$x^3 - y^3 = (x - y)(x^2 + xy + y^2)$$

a. Divide: $x + y\overline{)x^3 + y^3}$.
(*Hint:* Write $x^3 + y^3$ as $x^3 + 0x^2y + 0xy^2 + y^3$)

b. Divide: $x - y\overline{)x^3 - y^3}$.

FACTORING Factor each of the following polynomials completely. Begin by factoring out the indicated amount first.

a. Factor out 503 from $44{,}767a - 12{,}072b$.

b. Factor out 0.05 from $0.05t^2 - 0.25t + 0.3$.

c. Factor out -1.8 from $-16.2s^2 - 9s + 7.2$.

d. Factor out 2.5 from $10x^2 - 22.5y^2$.

e. Factor out 6.758 from $182.466b^3 - 54.064c^3$.

SECTION 11.5

SOLVING QUADRATIC EQUATIONS

a. Write $x^2 = 12x$ in quadratic form and then use factoring to solve it.

b. Divide both sides of $x^2 = 12x$ by x to "solve" it. Why don't you get two answers as you did in part a?

c. Consider the division property of equality and explain how it was inappropriately applied in part b.

QUADRATIC EQUATIONS Find a quadratic equation that has the following solutions. To do this, use the factoring method in reverse.

a. 3, 4 **b.** $-5, 1$

c. $\frac{1}{3}, 9$ **d.** $-\frac{2}{5}, -\frac{4}{3}$

CHAPTER REVIEW

SECTION 11.1 Factoring Out the Greatest Common Factor and Factoring by Grouping

CONCEPTS

A *prime number* is a natural number greater than 1 whose only factors are 1 and itself. A natural number is in *prime-factored form* when it is written as the product of prime numbers.

To find the *greatest common factor* (GCF) of several monomials:

1. Prime factor each monomial.
2. List each common factor the least number of times it appears in any one monomial.
3. Find the product of the factors in the list to obtain the GCF.

To *factor by grouping*, arrange the polynomial so that the first two terms have a common factor and the last two terms have a common factor. Factor out the common factor from both groups. Then factor out the resulting common binomial factor.

REVIEW EXERCISES

Find the prime factorization of each number.

1. 35 **2.** 45
3. 96 **4.** 99
5. 2,050 **6.** 4,096

Factor each polynomial completely.

7. $3x + 9y$ **8.** $5ax^2 + 15a$
9. $7s^2 + 14s$ **10.** $\pi ab - \pi ac$
11. $2x^3 + 4x^2 - 8x$ **12.** $x^2yz + xy^2z + xyz$
13. $-5ab^2 + 10a^2b - 15ab$ **14.** $4(x - 2) - x(x - 2)$

Factor out −1 from each polynomial.

15. $-a - 7$ **16.** $-4t^2 + 3t - 1$

Factor by grouping:

17. $2c + 2d + ac + ad$ **18.** $3xy + 9x - 2y - 6$
19. $2a^3 - a + 2a^2 - 1$ **20.** $4m^2n + 12m^2 - 8mn - 24m$

SECTION 11.2 Factoring Trinomials of the Form $x^2 + bx + c$

To *factor a trinomial* of the form $x^2 + bx + c$ means to write it as the product of two binomials.

To factor, $x^2 + bx + c$, find two integers whose product is c, and whose sum is b.

$(x \quad)(x \quad)$

Write the trinomial in descending powers of the variable and factor out −1 before factoring the trinomial.

21. Complete the table.

Factors of 6	Sum of the factors of 6
1(6)	
2( )	
(−6)	
−2(−3)	

If a trinomial cannot be factored using only integer coefficients, it is called a *prime polynomial.*

The *GCF* should always be factored out first. A trinomial is *factored completely* when it is expressed as a product of prime polynomials.

Factor each trinomial, if possible.

22. $x^2 + 2x - 24$

23. $x^2 - 4x - 12$

24. $n^2 - 7n + 10$

25. $t^2 + 10t + 15$

26. $-y^2 + 9y - 20$

27. $10y + 9 + y^2$

28. $c^2 + 3cd - 10d^2$

29. $-3mn + m^2 + 2n^2$

30. Explain how we can check to see whether $(x - 4)(x + 5)$ is the factorization of $x^2 + x - 20$.

Completely factor each trinomial.

31. $5a^2 + 45a - 50$

32. $-4x^2y - 4x^3 + 24xy^2$

SECTION 11.3 Factoring Trinomials of the Form $ax^2 + bx + c$

To factor $ax^2 + bx + c$ using the *trial-and-check* factoring method, we must determine four integers. Use the FOIL method to check your work.

Factors of a

$(\ x + \ )(\ x + \ )$

Factors of c

To factor $ax^2 + bx + c$ using the *grouping* method, we write it as

$ax^2 + \ \ x + \ \ x + c$

Factor each trinomial completely, if possible.

33. $2x^2 - 5x - 3$

34. $10y^2 + 21y - 10$

35. $-3x^2 + 14x + 5$

36. $-9p^2 - 6p + 6p^3$

37. $4b^2 - 17bc + 4c^2$

38. $3y^2 + 7y - 11$

39. ENTERTAINING The rectangular-shaped area occupied by the table setting shown is $(12x^2 - x - 1)$ square inches. Factor the expression to find the binomials that represent the length and width of the table setting.

SECTION 11.4 Special Factorizations and a Factoring Strategy

Special product formulas are used to factor *perfect square trinomials.*

$$x^2 + 2xy + y^2 = (x + y)^2$$
$$x^2 - 2xy + y^2 = (x - y)^2$$

To factor the *difference of two squares*, use the formula

$$\mathbf{F}^2 - \mathbf{L}^2 = (\mathbf{F} + \mathbf{L})(\mathbf{F} - \mathbf{L})$$

Factor each polynomial completely.

40. $x^2 + 10x + 25$

41. $9y^2 - 24y + 16$

42. $-z^2 + 2z - 1$

43. $25a^2 + 20ab + 4b^2$

Factor each polynomial completely, if possible.

44. $x^2 - 9$

45. $49t^2 - 25y^2$

46. $x^2y^2 - 400$

47. $8at^2 - 32a$

48. $c^4 - 256$

49. $h^2 + 36$

To factor the *sum* and *difference of two cubes*, use the formulas

$$F^3 + L^3 = (F + L)(F^2 - FL + L^2)$$

$$F^3 - L^3 = (F - L)(F^2 + FL + L^2)$$

Factor each polynomial completely, if possible.

50. $h^3 + 1$

51. $125p^3 + q^3$

52. $x^3 - 27$

53. $16x^5 - 54x^2y^3$

To factor a random polynomial, use the *factoring strategy* discussed in Section 11.4.

Factor each polynomial completely, if possible.

54. $14y^3 + 6y^4 - 40y^2$

55. $s^2t + s^2u^2 + tv + u^2v$

56. $j^4 - 16$

57. $-3j^3 - 24k^3$

58. $12w^2 - 36w + 27$

59. $121p^2 + 36q^2$

SECTION 11.5 Quadratic Equations

A *quadratic equation* is an equation of the form $ax^2 + bx + c = 0 \quad (a \neq 0)$, where a, b, and c represent real numbers.

Solve each quadratic equation by factoring.

60. $x^2 + 2x = 0$

61. $x(x - 6) = 0$

62. $x^2 - 9 = 0$

63. $a^2 - 7a + 12 = 0$

64. $t^2 + 4t + 4 = 0$

65. $2x - x^2 + 24 = 0$

66. $5a^2 - 6a + 1 = 0$

67. $2p^3 = 2p(p + 2)$

To use the *factoring method* to solve a quadratic equation:

1. Write the equation in $ax^2 + bx + c = 0$ form.
2. Factor the left-hand side.
3. Use the *zero-factor property* (if $ab = 0$, then $a = 0$ or $b = 0$) and set each factor equal to zero.
4. Solve each resulting linear equation.
5. Check the results in the original equation.

68. CONSTRUCTION The face of the triangular preformed concrete panel shown has an area of 45 square meters, and its base is 3 meters longer than twice its height. How long is its base?

The Pythagorean theorem: If the length of the hypotenuse of a right triangle is c and the lengths of the two legs are a and b, then $c^2 = a^2 + b^2$.

69. GARDENING A rectangular flower bed occupies 27 square feet and is 3 feet longer than twice its width. Find its dimensions.

70. TIGHTROPE WALKERS A circus performer intends to walk up a taut cable to a platform atop a pole, as shown. How high above the ground is the platform?

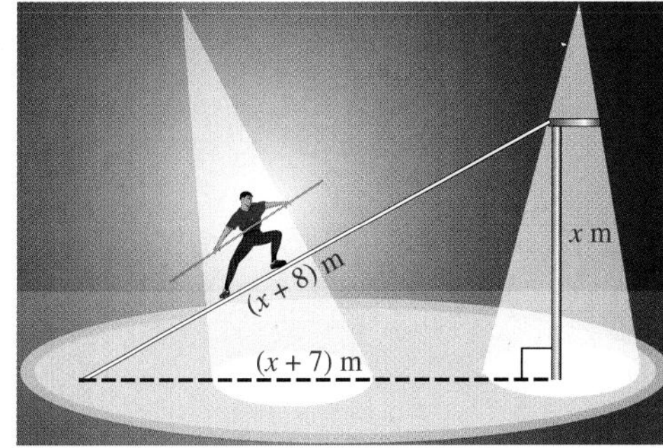

CHAPTER 11 TEST

Find the prime factorization of each number.

1. 196

2. 111

Factor each polynomial completely. If a polynomial cannot be factored, write "prime."

3. $4x + 16$

4. $30a^2b^3 - 20a^3b^2 + 5abc$

5. $q^2 - 81$

6. $x^2 + 9$

7. $16x^4 - 81$

8. $x^2 + 4x + 3$

9. $-x^2 + 9x + 22$

10. $9a - 9b + ax - bx$

11. $2a^2 + 5a - 12$

12. $18x^2 - 60xy + 50y^2$

13. $x^3 + 8$

14. $2a^3 - 54$

15. LANDSCAPING The combined area of the portions of the square lot that the sprinkler doesn't reach is given by $4r^2 - \pi r^2$, where r is the radius of the circular spray. Factor this expression.

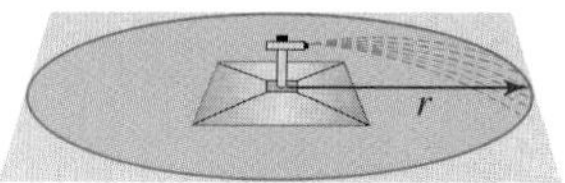

16. CHECKERS The area of the square checkerboard is $25x^2 - 40x + 16$. Find an expression that represents the length of a side.

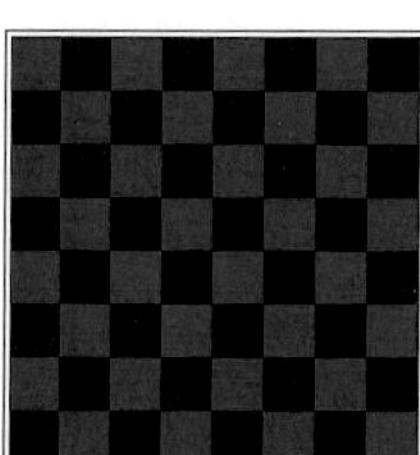

17. What is the greatest common factor of $4a^3b^2$ and $18ab^2$?

18. Factor: $x^2 - 3x - 54$. Show a check of your answer.

Solve each equation.

19. $(x + 3)(x - 2) = 0$

20. $x^2 - 25 = 0$

21. $6x^2 - x = 0$

22. $x^2 + 6x + 9 = 0$

23. $6x^2 + x - 1 = 0$

24. $x^2 + 7x = -6$

25. DRIVING SAFETY Virtually all cars have a blind spot where it is difficult for the driver to see a car behind and to the right. The area of the blind spot shown is 54 square feet. Find the width and length of the blind spot.

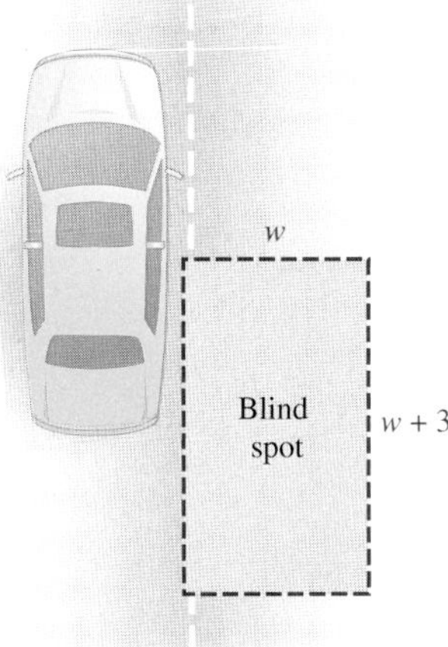

26. What is a quadratic equation? Give an example.

27. Find the length of the hypotenuse of the following right triangle.

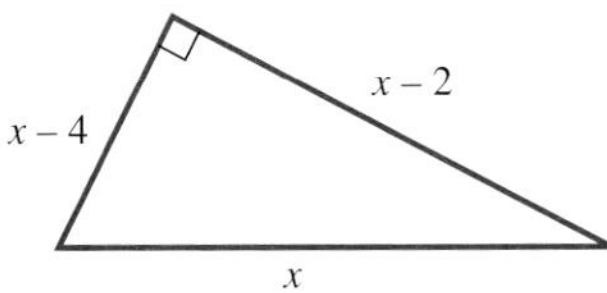

28. If the product of two numbers is 0, what conclusion can be drawn about the numbers?

CHAPTERS 1–11 CUMULATIVE REVIEW EXERCISES

1. HEART RATES Refer to the graph. Determine the difference in the maximum heartbeat rate for a 70-year-old as compared to someone half that age.

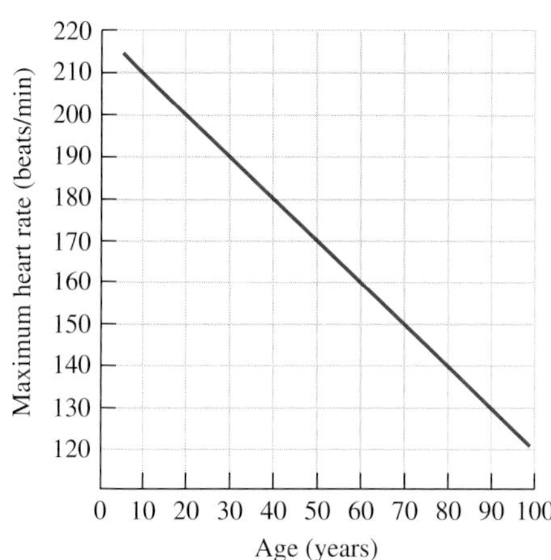

Based on data from *Cardiopulmonary Anatomy and Physiology: Essentials for Respiratory Care*, 2nd ed.

2. Give the prime factorization of 250.

3. Write $\frac{124}{125}$ as a decimal.

4. Determine whether each statement is true or false.
   a. Every integer is a whole number.
   b. Every integer is a rational number.
   c. π is a real number.

5. Find the quotient: $\frac{16}{5} \div \frac{10}{3}$.

6. What is -3 cubed?

Evaluate each expression.

7. $3 + 2[-1 - 4(5)]$

8. $\frac{|-25| - 2(-5)}{9 - 2^4}$

9. Evaluate $\frac{-x - a}{y - b}$ for $x = -2, y = 1, a = 5$, and $b = 2$.

10. Which division is undefined, $\frac{0}{5}$ or $\frac{5}{0}$?

Simplify each expression.

11. $-8y^2 - 5y^2 + 6$

12. $3z + 2(y - z) + y$

Solve each equation.

13. $-(3a + 1) + a = 2$

14. $2 - (4x + 7) = 3 + 2(x + 2)$

15. $\frac{3t - 21}{2} = t - 6$

16. $-\frac{1}{3} - \frac{x}{5} = \frac{3}{2}$

17. Solve $A = P + Prt$ for t.

18. Solve $-\frac{x}{2} + 4 > 5$ and graph the solution.

19. GEOMETRY TOOLS Find the total distance around the outside edge of the protractor. Round to the nearest tenth of an inch.

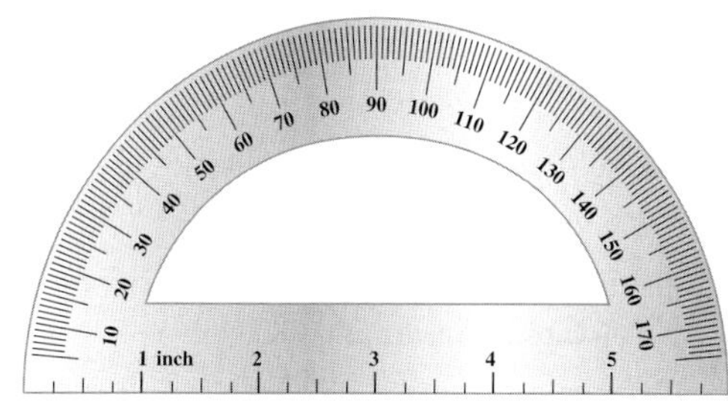

20. What is the formula for simple interest?

21. What is the value of x twenty-dollar bills?

22. PHOTOGRAPHIC CHEMICALS A photographer wishes to mix 6 liters of a 5% acetic acid solution with a 10% solution to get a 7% solution. How many liters of 10% solution must be added?

Graph each equation.

23. $y = (x + 2)^2$

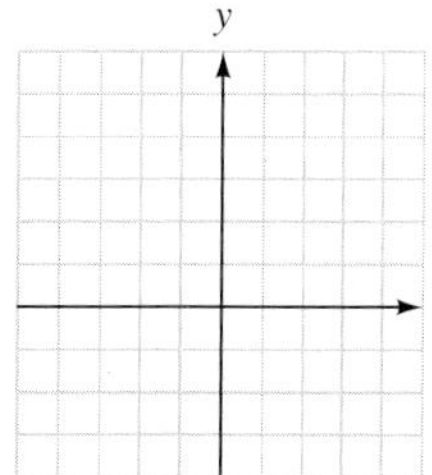

24. $y = |x| - 2$

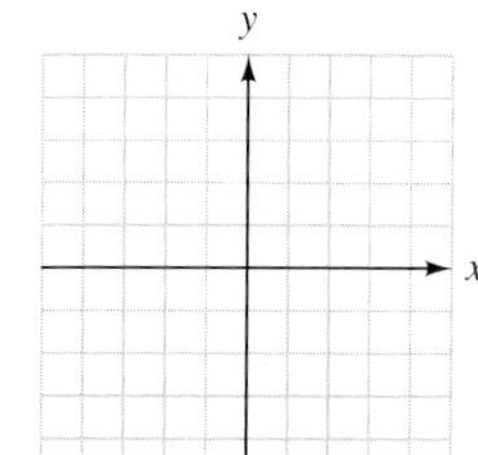

25. Find the slope and the y-intercept of the graph of $3x - 3y = 6$.

26. Write the equation of the line passing through $(-2, 5)$ and $(-3, -2)$. Answer in slope–intercept form.

27. Graph the line passing through $(-4, 1)$ that having slope $m = -3$.

28. Graph: $8x + 4y = -24$.

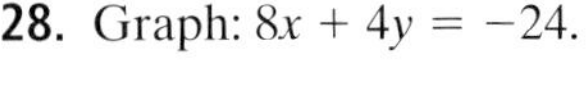

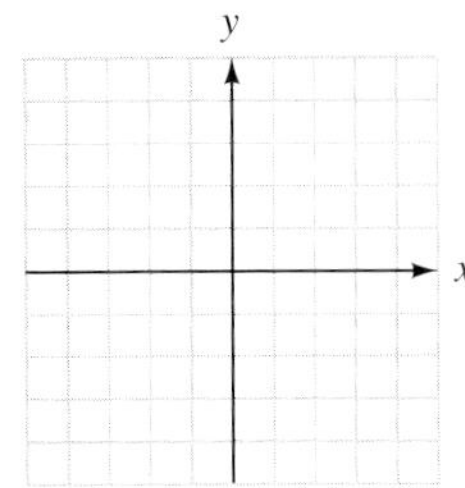

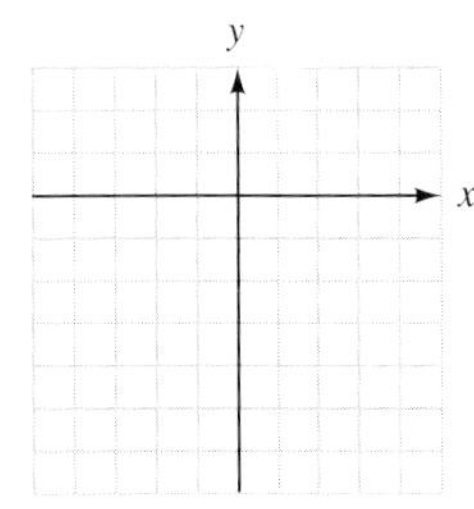

29. If two lines are parallel, what can be said about their slopes?

30. BEVERAGES The illustration shows the annual per-person consumption of coffee and tea in the United States.

a. What was the rate of change in coffee consumption for 1995–2000?

b. Did the per-person consumption of tea change for 1995–2000?

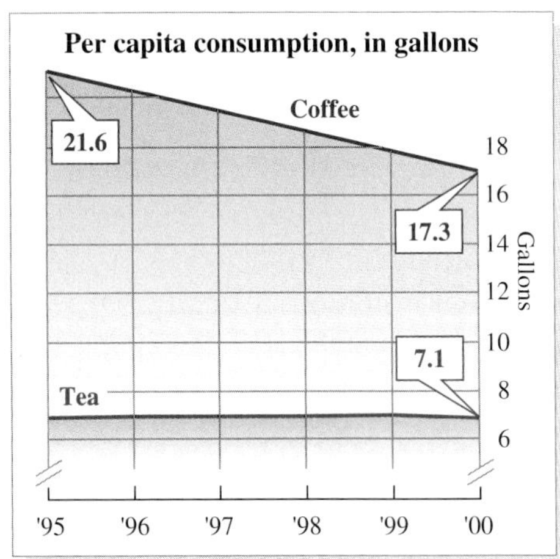

Based on data fom Davenport & Co. and the U.S. Department of Agriculture

31. If $g(x) = 3x - x^3$, find $g(-2)$.

32. Write 1,700,000 in scientific notation.

Simplify each expression. Write each answer without using parentheses or negative exponents.

33. $-y^2(4y^3)$

34. $\dfrac{15(x^2y^5)^5}{21(x^3y)^2}$

35. $\left(\dfrac{b^5}{b^{-2}}\right)^{-2}$

36. $2x^0$

Perform each operation.

37. $(x^2 - 3x + 8) - (3x^2 + x + 3)$

38. $4b^3(2b^2 - 2b)$

39. $(y - 6)^2$

40. $(3x - 2)(x + 4)$

41. $\dfrac{12a^2b^2 - 8a^2b - 4ab}{4ab}$

42. $x - 3\overline{)2x^2 - 5x - 3}$

43. PLAYPENS See the illustration.

a. Find the perimeter of the playpen.

b. Find the area of the floor of the playpen.

c. Find the volume of the playpen.

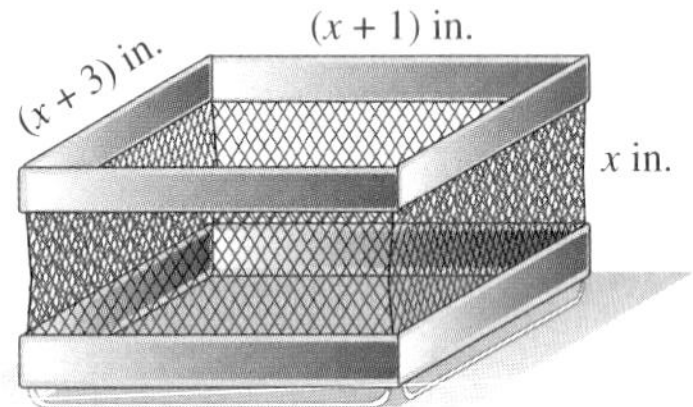

44. What is the degree of the polynomial $7y^3 + 4y^2 + y + 3$?

Factor each polynomial completely.

45. $b^3 - 3b^2$

46. $u^2 - 3 + 2u$

47. $2x^2 - 3x - 2$

48. $9z^2 - 1$

49. $-5a^2 + 25a - 30$

50. $ax + bx + ay + by$

51. $t^3 - 8$

52. $4a^2 - 12a + 9$

Solve each equation.

53. $15s^2 - 20s = 0$

54. $2x^2 - 5x = -2$

CHAPTER 12

Rational Expressions and Equations

© Norbert Schaeffer/CORBIS

TLE In today's fast-paced world, where would we be without technology? Cell phones, computers, CD players, and microwave ovens have become an important part of our daily lives. To design these products, engineers use formulas that involve fractions. Often, these fractions contain variables in their numerators and/or denominators. In this chapter, we will consider such fractions, which are more formally referred to as *rational expressions.*

To learn more about the use of algebra in the field of electronics, visit *The Learning Equation* on the Internet at www.the learning equation.com. (The log-in instructions are in the preface.) For Chapter 12, the online lesson is:

- *TLE* Lesson 23: Adding and Subtracting Rational Expressions

Check Your Knowledge

1. The ______ ________ __________ for a set of fractions is the smallest number that each denominator divides exactly.
2. False solutions that result from multiplying both sides of an equation by a variable are called __________ solutions.
3. A __________ is a statement that two ratios or two rates are equal.
4. The equation $y = \frac{k}{x}$ defines ________ variation.
5. Which values of x make $\frac{2x + 3}{x^2 + 7x}$ undefined?
6. Simplify: $\frac{81xy^2z^3}{45x^3y^2z}$.
7. Simplify: $\frac{x^2 - x - 6}{x^2 - 6x + 9}$.
8. Multiply and simplify: $\frac{x^2 + 7x + 12}{-2x - 6} \cdot \frac{2x}{x^2 - 16}$.
9. Divide and simplify: $\frac{2x^2 + 6x - 20}{x^2 - 25} \div (4x - 8)$.
10. Add: $\frac{3y + 10}{y + 2} + \frac{y - 2}{y + 2}$.
11. Add: $\frac{x}{x + 2} + \frac{1}{x - 1}$.
12. Subtract: $\frac{4}{x^2y} - \frac{5}{xy^2}$.
13. Subtract: $\frac{y}{y - 1} - \frac{1}{1 - y}$.
14. Simplify: $\frac{1 + \frac{1}{x}}{x - \frac{1}{x}}$.
15. Solve: $\frac{y}{2} - 1 = 2 + \frac{y}{5}$.
16. Solve: $\frac{1}{x - 2} - \frac{2}{x + 7} = \frac{4x - 4}{x^2 + 5x - 14}$.
17. Solve: $\frac{x}{x - 1} - \frac{x - 1}{x} = \frac{1}{x^2 - x}$.
18. Solve for W: $L = \frac{P - 2W}{2}$.
19. Solve for a: $\frac{1}{a} + \frac{1}{b} = \frac{1}{c}$.
20. Solve the proportion for t: $\frac{t}{t - 5} = \frac{t + 2}{t - 4}$.
21. A building casts a shadow 40 feet long at the same time that a student who is 5 feet tall casts an 8-foot shadow. How tall is the building?
22. A highway ascending a mountain pass rises 320 feet for each horizontal mile. How many feet does the road level rise in the 12 miles to the top of the pass?
23. With only a small valve open, a large vat fills in 6 hours. With only a larger valve open, the vat can be filled in 4 hours. How long will it take to fill the vat with both valves open?
24. The time in hours, t, it takes to travel from Santa Ana to San Diego varies inversely with the average speed of the trip in miles per hour, s.
    a. Find the constant of variation if $t = 2$ hours when $s = 45$ miles per hour.
    b. How long will the trip take at 60 miles per hour?

Study Skills Workshop

FINDING MATH IN THE WORLD AROUND YOU

Getty Images

This text contains many problems that have applications that directly affect your life, so project-based assignments are particularly relevant. However, some students, even those who have completed research papers in an English class, may feel a little out of their element when doing a research project for a math class. The good news is that researching a project in math follows the same basic steps as writing a paper for any other class. The main idea is that the paper you hand in is your product, but there is a process that you must go through to get your final results.

Process

- *Know your topic* Clearly state your idea and set the bounds you will explore in your project (you might adjust the bounds after doing some research). Decide whether you wish to write a report (a restatement of information that you found from a variety of sources) or an analysis of a topic. An analysis-based research paper is one in which you use others' ideas to make an argument of your own.
- *Do research* A wealth of material is available to explore in both libraries and on the Internet. In fact, once you start looking, you may find lots more information than you can use to write your project!
- *Write your project* This step requires organization and analysis, and may make use of math skills on your part. Examination will help you decide which topics you found in your research belong within the bounds of your project.

Tools for Research

- Explore how to research and write a paper. A good resource for getting started can be found at http://www.ipl.org/div/aplus/.
- Visit your college library or your library's Web site to see what subject matter resources are available there.
- Use Internet resources. Some good ones are "The Electric Library," (http://www.highbeam.com/library/index.asp?) and "Internet Mathematics Library" (http://mathforum.org/library/). Use an Internet search engine to look up specific math topics. To determine whether what you find is a good resource, use "How to Critically Analyze Information Sources" at http://www.library.cornell.edu/olinuris/ref/research/skill26.htm.

Most important, when doing any research project, give yourself ample time to be able to think about the information that you have found and how you want to use it. Depending on the size of the project, give yourself at least a few days after you have found information so that you can analyze and incorporate it into your work.

ASSIGNMENT

1. Choose a topic that you learned this semester that especially interests you. Your project could be a report, an analysis, or a relevant application.
2. Use at least three sources to gather information on your project.
3. Write the details of your project in a paper (3–5 pages).

12.1 Simplifying Rational Expressions

- Evaluating rational expressions
- Simplifying rational expressions
- Division of opposites

Fractions such as $\frac{1}{2}$ and $\frac{3}{4}$ that are the quotient of two integers are *rational numbers*. Fractions such as

$$\frac{3}{2y}, \quad \frac{x}{x+2}, \quad \text{and} \quad \frac{5a^2+6ab+b^2}{25a^2-b^2}$$

where the numerators and denominators are polynomials are called **rational expressions.**

Evaluating rational expressions

To evaluate a rational expression, we replace each variable with a given number value and simplify.

EXAMPLE 1 Find the value of $\frac{2x-1}{x^2+1}$ for $x = -3$.

Solution We replace each x in the expression with -3 and evaluate the numerator and denominator separately.

$$\frac{2x-1}{x^2+1} = \frac{2(-3)-1}{(-3)^2+1} \quad \text{Substitute } -3 \text{ for } x.$$

$$= \frac{-6-1}{9+1} \quad \text{In the numerator, perform the multiplication. In the denominator, evaluate the exponential expression.}$$

$$= -\frac{7}{10}$$

Self Check 1
Evaluate $\frac{3a-7}{a^3+1}$ for $a = 1$.

Answer -2

Since rational expressions indicate division, we must make sure that the denominator of a rational expression is not 0.

EXAMPLE 2 Find the values for x for which each rational expression is undefined: **a.** $\frac{7x}{x-5}$ and **b.** $\frac{x-1}{x^2-x-6}$.

Solution

a. The denominator of $\frac{7x}{x-5}$ will be 0 if we replace x with 5.

$$\frac{7x}{x-5} = \frac{7(5)}{5-5} = \frac{35}{0}$$

Since $\frac{35}{0}$ is undefined, the rational expression is undefined for $x = 5$.

b. The expression $\frac{x-1}{x^2-x-6}$ will be undefined for values of x that make the denominator 0. To find these values, we solve $x^2 - x - 6 = 0$.

$$x^2 - x - 6 = 0 \quad \text{Set the denominator of the rational expression equal to 0.}$$

$$(x-3)(x+2) = 0 \quad \text{Factor the trinomial.}$$

$$x - 3 = 0 \quad \text{or} \quad x + 2 = 0 \quad \text{Set each factor equal to 0.}$$

$$x = 3 \quad | \quad x = -2 \quad \text{Solve each equation.}$$

Self Check 2
Find the values for x for which each expression is undefined.

a. $\frac{x}{x+9}$

b. $\frac{x+7}{x^2-25}$

Since the values 3 and -2 make the denominator 0, the expression is undefined for $x = 3$ or $x = -2$.

Answers **a.** -9, **b.** 5 or -5

Simplifying rational expressions

To simplify a fraction we remove a factor equal to 1. This can be accomplished in two ways. For example, to simplify $\frac{6}{15}$, we proceed as follows:

Method 1

$\frac{6}{15} = \frac{2 \cdot 3}{5 \cdot 3}$ Factor the numerator and the denominator.

$= \frac{2}{5} \cdot \frac{3}{3}$ From Section 1.2, we know that $\frac{a \cdot c}{b \cdot d} = \frac{a}{b} \cdot \frac{c}{d}$.

$= \frac{2}{5} \cdot 1$ A number divided by itself is equal to 1: $\frac{3}{3} = 1$.

$= \frac{2}{5}$ Any number multiplied by 1 remains the same.

Method 2

$\frac{6}{15} = \frac{2 \cdot 3}{5 \cdot 3}$ Factor the numerator and the denominator.

$= \frac{2 \cdot \overset{1}{\cancel{3}}}{5 \cdot \underset{1}{\cancel{3}}}$ Remove the common factor, 3. Replace $\frac{3}{3}$ with the equivalent fraction $\frac{1}{1}$.

$= \frac{2}{5}$ Multiply to find the numerator: $2 \cdot 1 = 2$. Multiply to find the denominator: $5 \cdot 1 = 5$.

When all pairs of factors common to the numerator and denominator of a fraction have been removed, the fraction is **expressed in lowest terms.** Since it usually requires fewer steps, we will use method 2 in the following examples. The generalization of method 2 is called the **fundamental property of fractions.**

Fundamental property of fractions

If a represents a real number and b and c represent nonzero real numbers,

$$\frac{ac}{bc} = \frac{a}{b}$$

Simplifying rational expressions is similar to simplifying fractions. We write the rational expression so that the numerator and denominator have no common factors other than 1.

Simplifying rational expressions

1. Factor the numerator and denominator completely to determine their common factors.
2. Remove factors equal to 1 by replacing each pair of factors common to the numerator and denominator with the equivalent fraction $\frac{1}{1}$.

EXAMPLE 3 Simplify: $\frac{21x^2y}{14xy^2}$.

Solution We look for common factors in the numerator and denominator and remove them.

$\frac{21x^2y}{14xy^2} = \frac{3 \cdot 7 \cdot x \cdot x \cdot y}{2 \cdot 7 \cdot x \cdot y \cdot y}$ Factor the numerator and denominator.

$= \frac{3 \cdot \overset{1}{\cancel{7}} \cdot \overset{1}{\cancel{x}} \cdot x \cdot \overset{1}{\cancel{y}}}{2 \cdot \underset{1}{\cancel{7}} \cdot \underset{1}{\cancel{x}} \cdot y \cdot \underset{1}{\cancel{y}}}$ Replace $\frac{7}{7}$, $\frac{x}{x}$, and $\frac{y}{y}$ with the equivalent fraction $\frac{1}{1}$. This removes the factor $\frac{7 \cdot x \cdot y}{7 \cdot x \cdot y}$, which is equal to 1.

$= \frac{3x}{2y}$ Perform the multiplications in the numerator and in the denominator: $3 \cdot 1 \cdot 1 \cdot x \cdot 1 = 3x$ and $2 \cdot 1 \cdot 1 \cdot y \cdot 1 = 2y$.

Self Check 3
Simplify: $\frac{32a^3b^2}{24ab^4}$

Answer $\frac{4a^2}{3b^2}$

To simplify rational expressions, we often make use of the factoring techniques discussed in the preceding chapter.

Self Check 4
Write

$$\frac{x^2 - 5x}{5x - 25}$$

in lowest terms.

Answer $\frac{x}{5}$

EXAMPLE 4 Write $\frac{x^2 + 3x}{3x + 9}$ in lowest terms.

Solution We note that the terms of the numerator have a common factor of x and the terms of the denominator have a common factor of 3.

$\frac{x^2 + 3x}{3x + 9} = \frac{x(x + 3)}{3(x + 3)}$ Factor the numerator and the denominator.

$= \frac{x\overset{1}{\cancel{(x + 3)}}}{3\underset{1}{\cancel{(x + 3)}}}$ Remove a factor equal to 1 by replacing $\frac{x + 3}{x + 3}$ with $\frac{1}{1}$.

$= \frac{x}{3}$ Multiply in the numerator: $x \cdot 1 = x$. Multiply in the denominator: $3 \cdot 1 = 3$.

Self Check 5
Simplify: $\frac{3x^2 - 8x - 3}{x^2 - 9}$.

Answer $\frac{3x + 1}{x + 3}$

EXAMPLE 5 Simplify: $\frac{x^2 + 13x + 12}{x^2 - 144}$.

Solution The numerator is a trinomial, and the denominator is a difference of two squares.

$\frac{x^2 + 13x + 12}{x^2 - 144} = \frac{(x + 1)(x + 12)}{(x + 12)(x - 12)}$ Factor the numerator and the denominator.

$= \frac{(x + 1)\overset{1}{\cancel{(x + 12)}}}{\underset{1}{\cancel{(x + 12)}}(x - 12)}$ Remove a factor equal to 1 by replacing $\frac{x + 12}{x + 12}$ with $\frac{1}{1}$.

$= \frac{x + 1}{x - 12}$

! COMMENT When simplifying a fraction, remember that only *factors* that are common to the *entire numerator* and the *entire denominator* can be removed. For example, consider the correct simplification

$$\frac{5 + 8}{5} = \frac{13}{5}$$

It would be incorrect to remove the common *term* of 5 in this simplification. Doing so gives an incorrect answer of 9.

$$\frac{5 + 8}{5} = \frac{\overset{1}{\cancel{5}} + 8}{\underset{1}{\cancel{5}}} = \frac{1 + 8}{1} = 9$$

When simplifying rational expressions, it is incorrect to remove terms common to both the numerator and denominator.

$$\frac{\overset{1}{\cancel{x}} + 5}{\underset{1}{\cancel{x}} + 6} \qquad \frac{a^2 - 3\overset{1}{\cancel{a}} + \overset{1}{\cancel{2}}}{\underset{1}{\cancel{a}} + \underset{1}{\cancel{2}}} \qquad \frac{\overset{1}{\cancel{y^2}} - 36}{\underset{1}{\cancel{y^2}} - y - 7}$$

Any number or algebraic expression divided by 1 remains unchanged. For example,

$$\frac{37}{1} = 37, \quad \frac{5x}{1} = 5x, \quad \text{and} \quad \frac{3x + y}{1} = 3x + y$$

In general, we have the following.

Division by 1

For any real number a, $\frac{a}{1} = a$.

EXAMPLE 6 Simplify: $\frac{x^3 + x^2}{x + 1}$.

Solution

$$\frac{x^3 + x^2}{x + 1} = \frac{x^2(x + 1)}{x + 1} \quad \text{Factor the numerator.}$$

$$= \frac{x^2\cancel{(x + 1)}^{1}}{\cancel{x + 1}_{1}} \quad \text{Remove a factor equal to 1 by replacing } \tfrac{x+1}{x+1} \text{ with } \tfrac{1}{1}.$$

$$= \frac{x^2}{1}$$

$$= x^2 \quad \text{Denominators of 1 need not be written.}$$

Self Check 6
Simplify:
$\frac{a^2 + a - 2}{a - 1}$

Answer $a + 2$

EXAMPLE 7 Simplify: $\frac{5(x + 3) - 5}{7(x + 3) - 7}$.

Solution We cannot remove $x + 3$, because it is not a factor of the entire numerator, nor is it a factor of the entire denominator. Instead, we simplify the numerator and denominator, factor the results, and remove any common factors.

$$\frac{5(x + 3) - 5}{7(x + 3) - 7} = \frac{5x + 15 - 5}{7x + 21 - 7} \quad \text{Use the distributive property twice.}$$

$$= \frac{5x + 10}{7x + 14} \quad \text{Combine like terms.}$$

$$= \frac{5(x + 2)}{7(x + 2)} \quad \text{Factor the numerator and the denominator.}$$

$$= \frac{5\cancel{(x + 2)}^{1}}{7\cancel{(x + 2)}_{1}} \quad \text{Remove a factor equal to 1 by replacing } \tfrac{x+2}{x+2} \text{ with } \tfrac{1}{1}.$$

$$= \frac{5}{7}$$

Self Check 7
Simplify:
$\frac{4(x - 2) + 4}{3(x - 2) + 3}$

Answer $\frac{4}{3}$

EXAMPLE 8 Simplify: $\frac{x(x + 3) - 3(x - 1)}{x^2 + 3}$.

Solution We simplify the numerator and look for any common factors to remove in the numerator and denominator.

Self Check 8
Simplify:
$\frac{a(a + 2) - 2(a - 1)}{a^2 + 2}$

$$\frac{x(x+3)-3(x-1)}{x^2+3} = \frac{x^2+3x-3x+3}{x^2+3}$$ Use the distributive property twice in the numerator.

$$= \frac{x^2+3}{x^2+3}$$ Combine like terms in the numerator: $3x - 3x = 0$.

$$= \frac{\overset{1}{\cancel{x^2+3}}}{\underset{1}{\cancel{x^2+3}}}$$ Remove the common factor: $\frac{x^2+3}{x^2+3} = 1$

$$= 1$$

Answer 1

Sometimes a fraction does not simplify. For example, to attempt to simplify

$$\frac{x^2+x-2}{x^2+x}$$

we factor the numerator and the denominator.

$$\frac{x^2+x-2}{x^2+x} = \frac{(x+2)(x-1)}{x(x+1)}$$

Because there are no factors common to the numerator and denominator, this fraction is already in lowest terms.

Division of opposites

If the terms of two polynomials are the same, except for sign, the polynomials are called **opposites (negatives)** of each other. For example, the following pairs of polynomials are opposites of each other:

$x - y$	and	$-x + y$
$2a - 1$	and	$-2a + 1$
$-3x^2 - 2x + 5$	and	$3x^2 + 2x - 5$

Example 9 shows why the quotient of two binomials that are opposites is equal to -1.

Self Check 9
Simplify: $\frac{3p-2}{2-3p}$

EXAMPLE 9 Simplify: $\frac{2a-1}{1-2a}$.

Solution We can rearrange terms in each numerator, factor out -1, and proceed as follows:

$$\frac{2a-1}{1-2a} = \frac{-1+2a}{1-2a}$$ In the numerator, think of $2a - 1$ as $2a + (-1)$. Then change the order of the terms: $2a + (-1) = -1 + 2a$.

$$= \frac{-(1-2a)}{1-2a}$$ In the numerator, factor out -1: $-1 + 2a = -(1 - 2a)$.

$$= \frac{-\overset{1}{\cancel{(1-2a)}}}{\underset{1}{\cancel{1-2a}}}$$ Remove the common factor: $\frac{1-2a}{1-2a} = 1$.

$$= -1$$

Answer -1

In general, we have this important fact.

The quotient of opposites

The quotient of any nonzero expression and its opposite is -1.

! COMMENT Only apply the preceding rule to expressions that are opposites. For example, it would be incorrect to use this rule to simplify $\frac{x+1}{1+x}$. Since $x + 1$ equals $1 + x$ by the commutative property of addition, this is the quotient of a number and itself. The result is 1, not -1.

$$\frac{x+1}{1+x} = \frac{\overset{1}{\cancel{x+1}}}{\underset{1}{\cancel{x+1}}} = 1$$

Section 12.1 STUDY SET

VOCABULARY *Fill in the blanks.*

1. In a fraction, the part above the fraction bar is called the ________, and the part below the fraction bar is called the ________.
2. A fraction that has polynomials in its numerator and denominator, such as $\frac{x+2}{x-3}$, is called a ________ expression.
3. Division by 0 is ________.
4. A fraction is in ________ terms when all common factors of the numerator and denominator have been removed.
5. To ________ a rational expression means we remove factors common to the numerator and denominator.
6. If the terms of two polynomials are the same, except for sign, the polynomials are called ________ of each other.

CONCEPTS

7. What value of x makes each rational expression undefined?

   a. $\frac{x+2}{x}$ b. $\frac{x+2}{x-6}$ c. $\frac{x+2}{x+6}$

8. Fill in the blank: When a ________ factor of the numerator and the denominator of a fraction is removed, the resulting fraction is equivalent to the original fraction.
9. In the following work, what common factor has been removed?

$$\frac{x^2+2x+1}{x^2+4x+3} = \frac{\overset{1}{\cancel{(x+1)}}(x+1)}{(x+3)\underset{1}{\cancel{(x+1)}}} = \frac{x+1}{x+3}$$

10. Simplify each rational expression.

    a. $\frac{x-8}{x-8}$ b. $\frac{x-8}{8-x}$ c. $\frac{x-8}{1}$

11. Explain the error in the following work.

$$\frac{x}{x+2} = \frac{\overset{1}{\cancel{x}}}{\underset{1}{\cancel{x}}+2} = \frac{1}{3}$$

12. What is the first step in the process of simplifying $\frac{3(x+1)+2x+2}{x+1}$?

NOTATION *Complete each solution.*

13. $\frac{x^2+5x-6}{x^2-1} = \frac{(x+\square)(x-1)}{(x+1)(x-\square)}$

    $= \frac{x+6}{\square}$

14. $\frac{5(x+2)-5}{4(x+2)-4} = \frac{5x+\square-5}{4x+\square-4}$

    $= \frac{5x+\square}{4x+\square}$

    $= \frac{5\square}{4(x+1)}$

    $= \frac{5}{\square}$

PRACTICE *Evaluate each expression for $x = 6$.*

15. $\frac{x-2}{x-5}$
16. $\frac{3x-2}{x-2}$
17. $\frac{-2x-3}{x^2-1}$
18. $\frac{x^2-11}{-x-4}$
19. $\frac{x^2-4x-12}{x^2+x-2}$
20. $\frac{x^2-1}{x^3-1}$

Which value(s) of x make each rational expression undefined?

21. $\frac{15}{x-2}$
22. $\frac{5x}{x+5}$
23. $\frac{15x+2}{16}$
24. $\frac{x^2-4x}{25}$
25. $\frac{x+1}{2x-1}$
26. $\frac{-6x}{3x-1}$
27. $\frac{30}{x^2-36}$
28. $\frac{2x-15}{x^2-49}$
29. $\frac{15}{x^2+x-2}$
30. $\frac{x-20}{x^2+2x-8}$

Write each fraction in lowest terms.

31. $\frac{28}{35}$
32. $\frac{14}{20}$
33. $\frac{9}{27}$
34. $\frac{15}{45}$
35. $-\frac{36}{48}$
36. $-\frac{32}{40}$

Simplify each expression. If it is already in lowest terms, so indicate. Assume that no denominators are zero.

37. $\frac{45}{9a}$
38. $\frac{48}{16y}$
39. $\frac{5 + 5}{5z}$
40. $\frac{x + x}{2}$
41. $\frac{(3 + 4)a}{24 - 3}$
42. $\frac{(3 - 18)k}{25}$
43. $\frac{2x}{3x}$
44. $\frac{5y}{7y}$
45. $\frac{6x^2}{4x^2}$
46. $\frac{9xy}{6xy}$
47. $\frac{2x^2}{3y}$
48. $\frac{5y^2}{2y^2}$
49. $\frac{15x^2y}{5xy^2}$
50. $\frac{12xz}{4xz^2}$
51. $\frac{6x + 3}{3y}$
52. $\frac{4x + 12}{2y}$
53. $\frac{x + 3}{3x + 9}$
54. $\frac{2x + 14}{x - 7}$
55. $\frac{x - 7}{7 - x}$
56. $\frac{18 - d}{d - 18}$
57. $\frac{6x - 30}{5 - x}$
58. $\frac{6t - 42}{7 - t}$
59. $\frac{12 - 3x^2}{x^2 - x - 2}$
60. $\frac{-5x + 10}{x^2 - 4x + 4}$
61. $\frac{x^2 + 3x + 2}{x^2 + x - 2}$
62. $\frac{x^2 + x - 6}{x^2 - x - 2}$
63. $\frac{x^2 - 8x + 15}{x^2 - x - 6}$
64. $\frac{x^2 - 6x - 7}{x^2 + 8x + 7}$
65. $\frac{2x^2 - 8x}{x^2 - 6x + 8}$
66. $\frac{3y^2 - 15y}{y^2 - 3y - 10}$
67. $\frac{2 - a}{a^2 - a - 2}$
68. $\frac{4 - b}{b^2 - 5b + 4}$
69. $\frac{x^2 + 3x + 2}{x^3 + x^2}$
70. $\frac{x^2 - 13x + 30}{x^4 - 3x^3}$
71. $\frac{6x^2 - 13x + 6}{3x^2 + x - 2}$
72. $\frac{7x^2 + 20x - 3}{3x^2 + 8x - 3}$
73. $\frac{3x^2 - 14x + 8}{x^2 - 16}$
74. $\frac{4x^2 + 22x + 10}{x^2 - 25}$
75. $\frac{2x^2 - 8}{4x^2 - 7x - 2}$
76. $\frac{3x^2 - 27}{2x^2 - 7x + 3}$
77. $\frac{m^2 - 2mn + n^2}{2m^2 - 2n^2}$
78. $\frac{c^2 - d^2}{c^2 - 2cd + d^2}$
79. $\frac{5x^2 + 2x - 3}{6x^2 - x - 1}$
80. $\frac{3x^2 - 10x - 77}{x^2 - 4x - 21}$
81. $\frac{x^2 - 3(2x - 3)}{9 - x^2}$
82. $\frac{x(x - 8) + 16}{16 - x^2}$
83. $\frac{4(x + 3) + 4}{3(x + 2) + 6}$
84. $\frac{4 + 2(x - 5)}{3x - 5(x - 2)}$
85. $\frac{x^2 - 9}{(2x + 3) - (x + 6)}$
86. $\frac{x^2 + 5x + 4}{2(x + 3) - (x + 2)}$
87. $\frac{y - xy}{xy - x}$
88. $\frac{x^2 + y^2}{x + y}$
89. $\frac{6a - 6b + 6c}{9a - 9b + 9c}$
90. $\frac{3a - 3b - 6}{2a - 2b - 4}$
91. $\frac{15x - 3x^2}{25y - 5xy}$
92. $\frac{xz - 2x}{yz - 2y}$
93. $\frac{a + b - c}{c - a - b}$
94. $\frac{x - y - z}{z + y - x}$

APPLICATIONS

95. ROOFING The *pitch* of a roof is a measure of how steep or how flat the roof is. If pitch $= \frac{\text{rise}}{\text{run}}$, find the pitch of the roof of the cabin shown. Express the result in lowest terms.

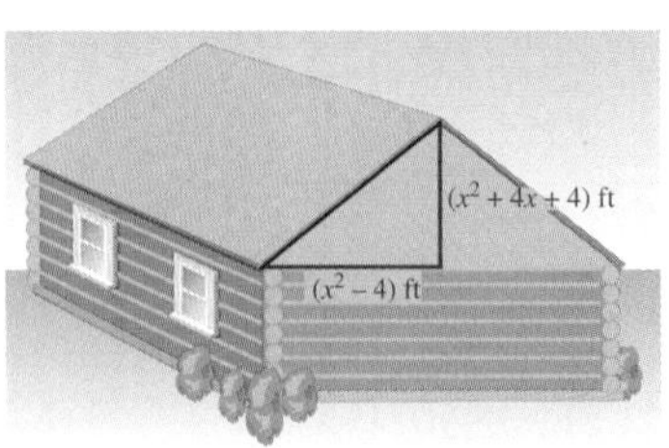

96. GRAPHIC DESIGN A chart of the basic food groups, in the shape of an equilateral triangle, is to be enlarged and distributed to schools for display in their health classes. What is the length of a side of the original design divided by the length of a side of the enlargement? Express the result in lowest terms.

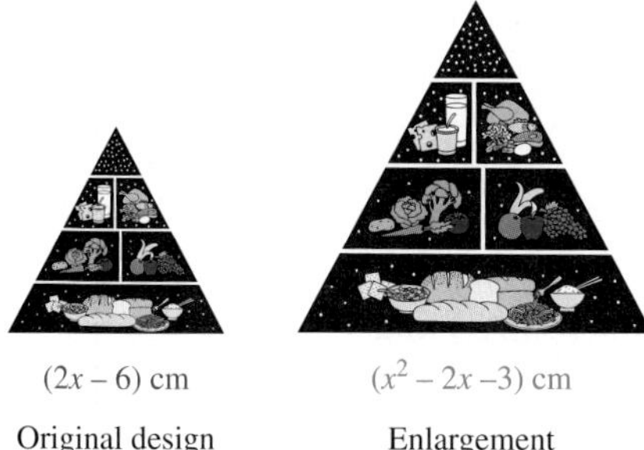

97. WORD PROCESSORS For the word processor shown, the number of words w that can be typed on a piece of paper is given by the formula

$$w = \frac{8{,}000}{x}$$

where x is the font size used. Find the number of words that can be typed on a page for each font size choice shown.

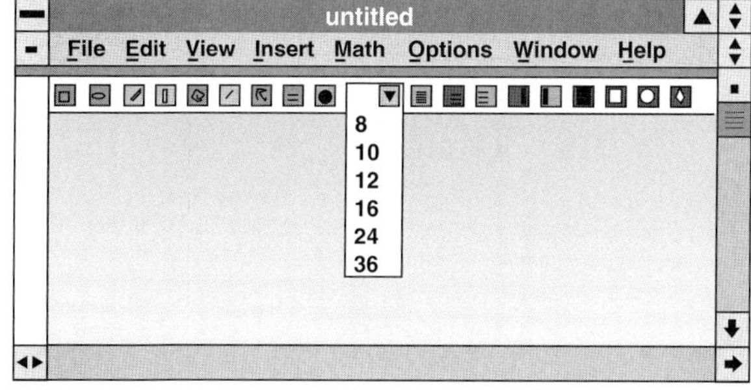

98. ORGAN PIPES The number of vibrations n per second of an organ pipe is given by the formula

$$n = \frac{512}{L}$$

where L is the length of the pipe in feet. How many times per second will a 6-foot pipe vibrate?

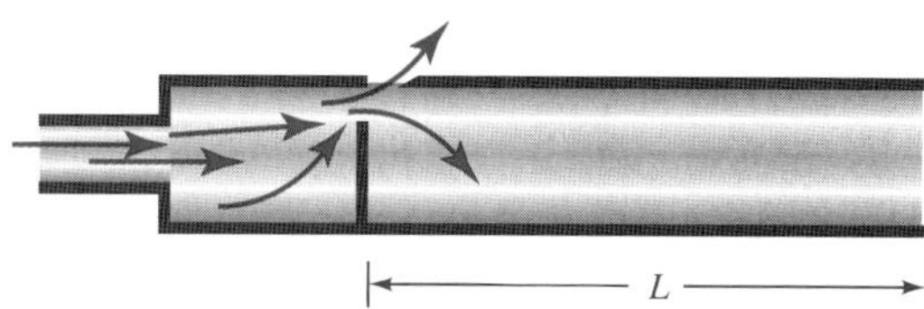

WRITING

99. Explain why $\dfrac{x - 7}{7 - x} = -1$.

100. Explain the difference between a factor and a term. Give several examples.

101. Explain the error.

$$\frac{3\overset{1}{\cancel{(x + 1)}} - x}{\underset{1}{\cancel{x + 1}}} = 3 - x$$

102. Explain why there are no values for x for which $\dfrac{x - 7}{x^2 + 49}$ is undefined.

REVIEW

103. State the associative property of addition using the variables a, b, and c.

104. State the distributive property using the variables x, y, and z.

105. If $ab = 0$, what must be true about a or b?

106. What is the product of a number and 1?

107. What is the opposite of $-\dfrac{5}{3}$?

108. What is the cube of 2 squared?

12.2 Multiplying and Dividing Rational Expressions

- Multiplying rational expressions
- Multiplying a rational expression by a polynomial
- Dividing rational expressions
- Dividing a rational expression by a polynomial
- Combined operations

In this section, we extend the rules for multiplying and dividing numerical fractions to problems involving multiplication and division of rational expressions.

Multiplying rational expressions

To multiply fractions, we multiply their numerators and multiply their denominators. For example,

$\frac{4}{7} \cdot \frac{3}{5} = \frac{4 \cdot 3}{7 \cdot 5}$ Multiply the numerators and multiply the denominators.

$= \frac{12}{35}$ Perform the multiplication in the numerator: $4 \cdot 3 = 12$. Perform the multiplication in the denominator: $7 \cdot 5 = 35$.

In general, we have the following rule.

Rule for multiplying fractions

If a, b, c, and d represent real numbers and b and d are not 0,

$$\frac{a}{b} \cdot \frac{c}{d} = \frac{ac}{bd}$$

We use the same procedure to multiply rational expressions.

Self Check 1
Multiply: $\frac{3x}{4} \cdot \frac{x-3}{5}$

EXAMPLE 1 Multiply: **a.** $\frac{x}{3} \cdot \frac{2}{5}$, **b.** $\frac{7}{9} \cdot \frac{-5}{3x}$, **c.** $\frac{x^2}{2} \cdot \frac{3}{y^2}$, and **d.** $\frac{t+1}{t} \cdot \frac{t-1}{t-2}$.

Solution

a. $\frac{x}{3} \cdot \frac{2}{5} = \frac{x \cdot 2}{3 \cdot 5}$

$= \frac{2x}{15}$

b. $\frac{7}{9} \cdot \frac{-5}{3x} = \frac{7(-5)}{9 \cdot 3x}$

$= \frac{-35}{27x}$

$= -\frac{35}{27x}$

c. $\frac{x^2}{2} \cdot \frac{3}{y^2} = \frac{x^2 \cdot 3}{2 \cdot y^2}$

$= \frac{3x^2}{2y^2}$

d. $\frac{t+1}{t} \cdot \frac{t-1}{t-2} = \frac{(t+1)(t-1)}{t(t-2)}$

Answer $\frac{3x(x-3)}{20}$

Self Check 2
Multiply: $\frac{a^2b^2}{2a} \cdot \frac{9a^3}{3b^3}$

EXAMPLE 2 Multiply: $\frac{35x^2y}{7y^2z} \cdot \frac{z}{5xy}$.

Solution

$\frac{35x^2y}{7y^2z} \cdot \frac{z}{5xy} = \frac{35x^2y \cdot z}{7y^2z \cdot 5xy}$ Multiply the numerators and multiply the denominators.

$= \frac{5 \cdot 7 \cdot x \cdot x \cdot y \cdot z}{7 \cdot y \cdot y \cdot z \cdot 5 \cdot x \cdot y}$ Factor $35x^2$ and factor y^2.

$= \frac{\overset{1}{\cancel{5}} \cdot \overset{1}{\cancel{7}} \cdot \overset{1}{\cancel{x}} \cdot x \cdot \overset{1}{\cancel{y}} \cdot \overset{1}{\cancel{z}}}{\underset{1}{\cancel{7}} \cdot \underset{1}{\cancel{y}} \cdot y \cdot \underset{1}{\cancel{z}} \cdot \underset{1}{\cancel{5}} \cdot \underset{1}{\cancel{x}} \cdot y}$ Simplify by removing the common factors.

$= \frac{x}{y^2}$ Perform the multiplications in the numerator and the denominator.

Answer $\frac{3a^4}{2b}$

EXAMPLE 3 Multiply: $\frac{x^2 - x}{2x + 4} \cdot \frac{x + 2}{x}$.

Solution

$$\frac{x^2 - x}{2x + 4} \cdot \frac{x + 2}{x} = \frac{(x^2 - x)(x + 2)}{(2x + 4)(x)}$$ Multiply the numerators and multiply the denominators.

We now factor the numerator and denominator to see whether this product can be simplified.

$$\frac{x^2 - x}{2x + 4} \cdot \frac{x + 2}{4} = \frac{x(x - 1)(x + 2)}{2(x + 2)x}$$ Factor the numerator: $(x^2 - x) = x(x - 1)$. Factor the denominator: $(2x + 4) = 2(x + 2)$.

$$= \frac{\overset{1}{\cancel{x}}(x - 1)\overset{1}{\cancel{(x + 2)}}}{2\underset{1}{\cancel{(x + 2)}}\underset{1}{\cancel{x}}}$$ Simplify by removing the common factors.

$$= \frac{x - 1}{2}$$

Self Check 3
Multiply: $\frac{x^2 + x}{3x + 6} \cdot \frac{x + 2}{x + 1}$

Answer $\frac{x}{3}$

EXAMPLE 4 Multiply: $\frac{x^2 - 3x}{x^2 - x - 6} \cdot \frac{x^2 + x - 2}{x^2 - x}$.

Solution

$$\frac{x^2 - 3x}{x^2 - x - 6} \cdot \frac{x^2 + x - 2}{x^2 - x}$$

$$= \frac{(x^2 - 3x)(x^2 + x - 2)}{(x^2 - x - 6)(x^2 - x)}$$ Multiply the numerators and multiply the denominators.

$$= \frac{x(x - 3)(x + 2)(x - 1)}{(x + 2)(x - 3)x(x - 1)}$$ Factor the numerator and denominator to see whether the result can be simplified.

$$= \frac{\overset{1}{\cancel{x}}\overset{1}{\cancel{(x - 3)}}\overset{1}{\cancel{(x + 2)}}\overset{1}{\cancel{(x - 1)}}}{\underset{1}{\cancel{(x + 2)}}\underset{1}{\cancel{(x - 3)}}\underset{1}{\cancel{x}}\underset{1}{\cancel{(x - 1)}}}$$ Simplify by removing the common factors.

$$= 1$$

Self Check 4
Multiply: $\frac{a^2 + a}{a^2 - 4} \cdot \frac{a^2 - a - 2}{a^2 + 2a + 1}$

Answer $\frac{a}{a + 2}$

Multiplying a rational expression by a polynomial

Since any number divided by 1 remains unchanged, we can write any polynomial as a fraction by inserting a denominator of 1.

EXAMPLE 5 Multiply: **a.** $\frac{4}{x} \cdot x$, **b.** $63x\left(\frac{1}{7x}\right)$, and **c.** $5a\left(\frac{3a - 1}{a}\right)$.

Solution

a. $\frac{4}{x} \cdot x = \frac{4}{x} \cdot \frac{x}{\mathbf{1}}$ Write x as a fraction: $x = \frac{x}{1}$.

$$= \frac{4 \cdot \overset{1}{\cancel{x}}}{\underset{1}{\cancel{x}} \cdot 1}$$ Multiply the fractions and remove the common factor in the numerator and denominator.

$$= 4$$ Simplify.

Self Check 5
Multiply:

a. $\frac{9}{7y} \cdot 7y$

b. $36b\left(\frac{1}{6b}\right)$

c. $4x\left(\frac{x + 3}{x}\right)$

b. $63x\left(\frac{1}{7x}\right) = \frac{63x}{\mathbf{1}}\left(\frac{1}{7x}\right)$ Write $63x$ as a fraction: $63x = \frac{63x}{1}$.

$= \frac{63x \cdot 1}{1 \cdot 7 \cdot x}$ Multiply the fractions.

$= \frac{9 \cdot \overset{1}{\cancel{7}} \cdot \overset{1}{\cancel{x}} \cdot 1}{1 \cdot \underset{1}{\cancel{7}} \cdot \underset{1}{\cancel{x}}}$ Write $63x$ in factored form as $9 \cdot 7 \cdot x$ and remove the common factors.

$= 9$ Simplify.

c. $5a\left(\frac{3a - 1}{a}\right) = \frac{5a}{\mathbf{1}}\left(\frac{3a - 1}{a}\right)$ Write $5a$ as a fraction: $5a = \frac{5a}{1}$.

$= \frac{5\overset{1}{\cancel{a}}(3a - 1)}{1 \cdot \underset{1}{\cancel{a}}}$ Multiply the fractions and remove the common factor of a.

$= 5(3a - 1)$ Simplify.

$= 15a - 5$ Distribute the multiplication by 5.

Answers **a.** 9, **b.** 6, **c.** $4x + 12$

Self Check 6

Multiply:

$(a - 7) \cdot \frac{a^2 - a}{a^2 - 8a + 7}$

EXAMPLE 6 Multiply: $\frac{x^2 + x}{x^2 + 8x + 7} \cdot (x + 7)$.

Solution

$\frac{x^2 + x}{x^2 + 8x + 7} \cdot (x + 7)$

$= \frac{x^2 + x}{x^2 + 8x + 7} \cdot \frac{x + 7}{\mathbf{1}}$ Write $x + 7$ as a fraction with a denominator of 1.

$= \frac{x(x + 1)(x + 7)}{(x + 1)(x + 7)1}$ Multiply the fractions and factor where possible.

$= \frac{x\overset{1}{\cancel{(x + 1)}}\overset{1}{\cancel{(x + 7)}}}{\underset{1}{\cancel{(x + 1)}}\underset{1}{\cancel{(x + 7)}}1}$ Simplify by removing the common factors.

$= x$

Answer a

Dividing rational expressions

Division by a nonzero number is equivalent to multiplying by its reciprocal. Thus, to divide two fractions, we can invert the divisor (the fraction following the ÷ sign) and multiply. For example,

$\frac{4}{7} \div \frac{\mathbf{3}}{\mathbf{5}} = \frac{4}{7} \cdot \frac{\mathbf{5}}{\mathbf{3}}$ Invert $\frac{3}{5}$ and change the division to a multiplication.

$= \frac{20}{21}$ Multiply the fractions.

In general, we have the following rule.

Rule for dividing fractions

If a represents a real number and b, c, and d represent nonzero real numbers,

$$\frac{a}{b} \div \frac{c}{d} = \frac{a}{b} \cdot \frac{d}{c}$$

We use the same procedures to divide rational expressions.

EXAMPLE 7

Divide: **a.** $\frac{a}{13} \div \frac{17}{26}$ and **b.** $-\frac{9x}{35y} \div \frac{15x^2}{14}$.

Solution

a. $\frac{a}{13} \div \frac{17}{26} = \frac{a}{13} \cdot \frac{26}{17}$ Invert the divisor, which is $\frac{17}{26}$, and change the division to a multiplication.

$= \frac{a \cdot 2 \cdot 13}{13 \cdot 17}$ Multiply the fractions and factor where possible.

$= \frac{a \cdot 2 \cdot \overset{1}{\cancel{13}}}{\underset{1}{\cancel{13}} \cdot 17}$ Simplify by removing the common factors.

$= \frac{2a}{17}$

b. $-\frac{9x}{35y} \div \frac{15x^2}{14} = -\frac{9x}{35y} \cdot \frac{14}{15x^2}$ Multiply by the reciprocal of $\frac{15x^2}{14}$.

$= -\frac{3 \cdot 3 \cdot x \cdot 2 \cdot 7}{5 \cdot 7 \cdot y \cdot 3 \cdot 5 \cdot x \cdot x}$ Multiply the fractions and factor where possible.

$= -\frac{3 \cdot \overset{1}{\cancel{3}} \cdot \overset{1}{\cancel{x}} \cdot 2 \cdot \overset{1}{\cancel{7}}}{5 \cdot \underset{1}{\cancel{7}} \cdot y \cdot \underset{1}{\cancel{3}} \cdot 5 \cdot \underset{1}{\cancel{x}} \cdot x}$ Simplify by removing the common factors.

$= -\frac{6}{25xy}$ Multiply the remaining factors.

Self Check 7

Divide:

$-\frac{8a}{3b} \div \frac{16a^2}{9b^2}$

Answer $-\frac{3b}{2a}$

EXAMPLE 8

Divide: $\frac{x^2 + x}{3x - 15} \div \frac{x^2 + 2x + 1}{6x - 30}$.

Solution

$\frac{x^2 + x}{3x - 15} \div \frac{x^2 + 2x + 1}{6x - 30}$

$= \frac{x^2 + x}{3x - 15} \cdot \frac{6x - 30}{x^2 + 2x + 1}$ Invert the divisor and change the division to multiplication.

$= \frac{x(x + 1) \cdot 2 \cdot 3(x - 5)}{3(x - 5)(x + 1)(x + 1)}$ Multiply the fractions and factor.

$= \frac{x\overset{1}{\cancel{(x + 1)}} \cdot 2 \cdot \overset{1}{\cancel{3}}\overset{1}{\cancel{(x - 5)}}}{\underset{1}{\cancel{3}}\underset{1}{\cancel{(x - 5)}}\underset{1}{\cancel{(x + 1)}}(x + 1)}$ Simplify by removing the common factors.

$= \frac{2x}{x + 1}$

Self Check 8

Divide:

$\frac{z^2 - 1}{z^2 + 4z + 3} \div \frac{z - 1}{z^2 + 2z - 3}$

Answer $z - 1$

Dividing a rational expression by a polynomial

To divide a rational expression by a polynomial, we write the polynomial as a fraction by inserting a denominator of 1, and then we divide the fractions.

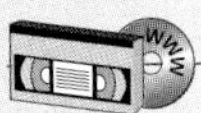

Self Check 9
Divide:
$$\frac{a^2 - b^2}{a^2 + ab} \div (b - a)$$

EXAMPLE 9 Divide: $\frac{2x^2 - 3x - 2}{2x + 1} \div (4 - x^2)$.

Solution

$$\frac{2x^2 - 3x - 2}{2x + 1} \div (4 - x^2)$$

$$= \frac{2x^2 - 3x - 2}{2x + 1} \div \frac{4 - x^2}{1}$$ Write $4 - x^2$ as a fraction with a denominator of 1.

$$= \frac{2x^2 - 3x - 2}{2x + 1} \cdot \frac{1}{4 - x^2}$$ Invert the divisor and change the division to multiplication.

$$= \frac{(2x + 1)(x - 2) \cdot 1}{(2x + 1)(2 + x)(2 - x)}$$ Multiply the fractions and factor where possible.

$$= \frac{\overset{1}{\cancel{(2x + 1)}}\overset{-1}{\cancel{(x - 2)}} \cdot 1}{\underset{1}{\cancel{(2x + 1)}}(2 + x)\underset{1}{\cancel{(2 - x)}}}$$ Remove the common factors. The binomials $x - 2$ and $2 - x$ are opposites: $\frac{x - 2}{2 - x} = -1$.

$$= \frac{-1}{2 + x}$$

$$= -\frac{1}{2 + x}$$

Answer $-\frac{1}{a}$

Combined operations

Unless parentheses indicate otherwise, we perform multiplications and divisions in order from left to right.

Self Check 10
Simplify:
$$\frac{a^2 + ab}{ab - b^2} \cdot \frac{a^2 - b^2}{a^2 + ab} \div \frac{a + b}{b}$$

EXAMPLE 10 Simplify: $\frac{x^2 - x - 6}{x - 2} \div \frac{x^2 - 4x}{x^2 - x - 2} \cdot \frac{x - 4}{x^2 + x}$.

Solution Since there are no parentheses to indicate otherwise, we perform the division first.

$$\frac{x^2 - x - 6}{x - 2} \div \frac{x^2 - 4x}{x^2 - x - 2} \cdot \frac{x - 4}{x^2 + x}$$

$$= \frac{x^2 - x - 6}{x - 2} \cdot \frac{x^2 - x - 2}{x^2 - 4x} \cdot \frac{x - 4}{x^2 + x}$$ Invert the divisor, which is $\frac{x^2 - 4x}{x^2 - x - 2}$, and change the division to a multiplication.

$$= \frac{(x + 2)(x - 3)(x + 1)(x - 2)(x - 4)}{(x - 2)x(x - 4)x(x + 1)}$$ Multiply the fractions and factor.

$$= \frac{(x + 2)(x - 3)\overset{1}{\cancel{(x + 1)}}\overset{1}{\cancel{(x - 2)}}\overset{1}{\cancel{(x - 4)}}}{\underset{1}{\cancel{(x - 2)}}x\underset{1}{\cancel{(x - 4)}}x\underset{1}{\cancel{(x + 1)}}}$$ Simplify by removing the common factors.

$$= \frac{(x + 2)(x - 3)}{x^2}$$

Answer 1

EXAMPLE 11 Simplify: $\frac{x^2+6x+9}{x^2-2x}\left(\frac{x^2-4}{x^2+3x} \div \frac{x+2}{x}\right)$.

Solution We perform the division within the parentheses first.

$$\frac{x^2+6x+9}{x^2-2x}\left(\frac{x^2-4}{x^2+3x} \div \frac{x+2}{x}\right)$$

$$= \frac{x^2+6x+9}{x^2-2x}\left(\frac{x^2-4}{x^2+3x} \cdot \frac{x}{x+2}\right)$$ Invert the divisor and change the division to multiplication.

$$= \frac{(x+3)(x+3)(x-2)(x+2)x}{x(x-2)x(x+3)(x+2)}$$ Multiply the fractions and factor where possible.

$$= \frac{\overset{1}{\cancel{(x+3)}}(x+3)\overset{1}{\cancel{(x-2)}}\overset{1}{\cancel{(x+2)}}\overset{1}{\cancel{x}}}{\underset{1}{\cancel{x}}\underset{1}{\cancel{(x-2)}}x\underset{1}{\cancel{(x+3)}}\underset{1}{\cancel{(x+2)}}}$$ Simplify by removing the common factors.

$$= \frac{x+3}{x}$$

Self Check 11
Simplify: $\frac{x^2-2x}{x^2+6x+9} \div \left(\frac{x^2-4}{x^2+3x} \cdot \frac{x}{x+2}\right)$

Answer $\frac{x}{x+3}$

Section 12.2 STUDY SET

VOCABULARY *Fill in the blanks.*

1. In a fraction, the part above the fraction bar is called the __________.
2. In a fraction, the part below the fraction bar is called the __________.

CONCEPTS *Fill in the blanks.*

3. To multiply fractions, we multiply their __________ and multiply their __________.
4. $\frac{a}{b} \cdot \frac{c}{d} = \square$
5. To write a polynomial in fractional form, we insert a denominator of $\square$.
6. $\frac{a}{b} \div \frac{c}{d} = \frac{a}{b} \cdot \square$
7. To divide fractions, we invert the ________ and ________.
8. The __________ of $\frac{x}{x+2}$ is $\frac{x+2}{x}$.

NOTATION *Complete each solution.*

9. $\frac{x^2+x}{3x-6} \cdot \frac{x-2}{x+1} = \frac{(x^2+x)\square}{\square(x+1)}$
$= \frac{x\square(x-2)}{3\square(x+1)}$
$= \frac{x}{\square}$

10. $\frac{x^2-x}{4x+12} \div \frac{x-1}{x+3} = \frac{x^2-x}{4x+12} \cdot \frac{\square}{\square}$
$= \frac{\square(x+3)}{(4x+12)\square}$
$= \frac{x\square(x+3)}{4\square(x-1)}$
$= \frac{x}{\square}$

PRACTICE *Perform the multiplications. Simplify answers if possible.*

11. $\frac{3}{y} \cdot \frac{y}{2}$
12. $\frac{2}{z} \cdot \frac{z}{3}$
13. $\frac{5y}{7} \cdot \frac{7}{5}$
14. $\frac{4x}{3y} \cdot \frac{3y}{7x}$
15. $\frac{7z}{9z} \cdot \frac{4z}{2z}$
16. $\frac{8}{2x} \cdot \frac{16x}{3x}$
17. $\frac{2x^2y}{3xy} \cdot \frac{3xy^2}{2}$
18. $\frac{2x^2z}{z} \cdot \frac{5x}{z}$
19. $\frac{8x^2y^2}{4x^2} \cdot \frac{2xy}{2y}$
20. $\frac{9x^2y}{3x} \cdot \frac{3xy}{3y}$
21. $-\frac{2xy}{x^2} \cdot \frac{3xy}{2}$
22. $-\frac{3x}{x^2} \cdot \frac{2xz}{3}$
23. $\frac{ab^2}{a^2b} \cdot \frac{b^2c^2}{abc} \cdot \frac{abc^2}{a^3c^2}$
24. $\frac{x^3y}{z} \cdot \frac{xz^3}{x^2y^2} \cdot \frac{yz}{xyz}$

25. $\dfrac{10r^2st^3}{6rs^2} \cdot \dfrac{3r^3t}{2rst} \cdot \dfrac{2s^3t^4}{5s^2t^3}$

26. $\dfrac{3a^3b}{25cd^3} \cdot \dfrac{-5cd^2}{6ab} \cdot \dfrac{10abc^2}{2bc^2d}$

27. $\dfrac{z+7}{7} \cdot \dfrac{z+2}{z}$

28. $\dfrac{a-3}{a} \cdot \dfrac{a+3}{5}$

29. $\dfrac{x-2}{2} \cdot \dfrac{2x}{x-2}$

30. $\dfrac{y+3}{y} \cdot \dfrac{3y}{y+3}$

31. $\dfrac{x+5}{5} \cdot \dfrac{x}{x+5}$

32. $\dfrac{y-9}{y+9} \cdot \dfrac{y}{9}$

33. $\dfrac{5}{m} \cdot m$

34. $p \cdot \dfrac{10}{p}$

35. $4d \cdot \dfrac{3}{2d}$

36. $9x \cdot \dfrac{25}{3x}$

37. $15x\left(\dfrac{x+1}{15x}\right)$

38. $30t\left(\dfrac{t-7}{30t}\right)$

39. $12y\left(\dfrac{y+8}{6y}\right)$

40. $16x\left(\dfrac{3x+8}{4x}\right)$

41. $(x+8)\dfrac{x+5}{x+8}$

42. $(y-2)\dfrac{y+3}{y-2}$

43. $10(h+9)\dfrac{h-3}{h+9}$

44. $r(r-25)\dfrac{r+4}{r-25}$

45. $\dfrac{(x+1)^2}{x+1} \cdot \dfrac{x+2}{x+1}$

46. $\dfrac{(y-3)^2}{y-3} \cdot \dfrac{y-3}{y-3}$

47. $\dfrac{2x+6}{x+3} \cdot \dfrac{3}{4x}$

48. $\dfrac{3y-9}{y-3} \cdot \dfrac{y}{3y^2}$

49. $\dfrac{x^2-x}{x} \cdot \dfrac{3x-6}{3x-3}$

50. $\dfrac{5z-10}{z+2} \cdot \dfrac{3}{3z-6}$

51. $\dfrac{7y-14}{y-2} \cdot \dfrac{x^2}{7x}$

52. $\dfrac{y^2+3y}{9} \cdot \dfrac{3x}{y+3}$

53. $\dfrac{x^2+x-6}{5x} \cdot \dfrac{5x-10}{x+3}$

54. $\dfrac{z^2+4z-5}{5z-5} \cdot \dfrac{5z}{z+5}$

55. $\dfrac{m^2-2m-3}{2m+4} \cdot \dfrac{m^2-4}{m^2+3m+2}$

56. $\dfrac{p^2-p-6}{3p-9} \cdot \dfrac{p^2-9}{p^2+6p+9}$

57. $\dfrac{3x^2+5x+2}{x^2-9} \cdot \dfrac{x-3}{x^2-4} \cdot \dfrac{x^2+5x+6}{6x+4}$

58. $\dfrac{x^2-25}{3x+6} \cdot \dfrac{x^2+x-2}{2x+10} \cdot \dfrac{6x}{3x^2-18x+15}$

Perform each division. Simplify answers when possible.

59. $\dfrac{2}{y} \div \dfrac{4}{3}$

60. $\dfrac{3}{a} \div \dfrac{a}{9}$

61. $\dfrac{3x}{2} \div \dfrac{x}{2}$

62. $\dfrac{y}{6} \div \dfrac{2}{3y}$

63. $\dfrac{3x}{y} \div \dfrac{2x}{4}$

64. $\dfrac{3y}{8} \div \dfrac{2y}{4y}$

65. $\dfrac{4x}{3x} \div \dfrac{2y}{9y}$

66. $\dfrac{14}{7y} \div \dfrac{10}{5z}$

67. $\dfrac{x^2}{3} \div \dfrac{2x}{4}$

68. $\dfrac{z^2}{z} \div \dfrac{z}{3z}$

69. $\dfrac{x^2y}{3xy} \div \dfrac{xy^2}{6y}$

70. $\dfrac{2xz}{z} \div \dfrac{4x^2}{z^2}$

71. $\dfrac{x+2}{3x} \div \dfrac{x+2}{2}$

72. $\dfrac{z-3}{3z} \div \dfrac{z+3}{z}$

73. $\dfrac{(z-2)^2}{3z^2} \div \dfrac{z-2}{6z}$

74. $\dfrac{(x+7)^2}{x+7} \div \dfrac{(x-3)^2}{x+7}$

75. $\dfrac{(z-7)^2}{z+2} \div \dfrac{z(z-7)}{5z^2}$

76. $\dfrac{y(y+2)}{y^2(y-3)} \div \dfrac{y^2(y+2)}{(y-3)^2}$

77. $\frac{x^2 - 4}{3x + 6} \div \frac{x - 2}{x + 2}$

78. $\frac{x^2 - 9}{5x + 15} \div \frac{x - 3}{x + 3}$

79. $\frac{x^2 - 1}{3x - 3} \div \frac{x + 1}{3}$

80. $\frac{x^2 - 16}{x - 4} \div \frac{3x + 12}{x}$

81. $\frac{x^2 - 2x - 35}{3x^2 + 27x} \div \frac{x^2 + 7x + 10}{6x^2 + 12x}$

82. $\frac{x^2 - x - 6}{2x^2 + 9x + 10} \div \frac{x^2 - 25}{2x^2 + 15x + 25}$

83. $\frac{2d^2 + 8d - 42}{d - 3} \div \frac{2d^2 + 14d}{d^2 + 5d}$

84. $\frac{5x^2 + 13x - 6}{x + 3} \div \frac{5x^2 - 17x + 6}{x - 2}$

Perform the operations.

85. $\frac{x}{3} \cdot \frac{9}{4} \div \frac{x^2}{6}$

86. $\frac{y^2}{2} \div \frac{4}{y} \cdot \frac{y^2}{8}$

87. $\frac{x^2}{18} \div \frac{x^3}{6} \div \frac{12}{x^2}$

88. $\frac{y^3}{3y} \cdot \frac{3y^2}{4} \div \frac{15}{20}$

89. $\frac{z^2 - 4}{2z + 6} \div \frac{z + 2}{4} \cdot \frac{z + 3}{z - 2}$

90. $\frac{2}{3x - 3} \div \frac{2x + 2}{x - 1} \cdot \frac{5}{x + 1}$

91. $\frac{x - x^2}{x^2 - 4}\left(\frac{2x + 4}{x + 2} \div \frac{5}{x + 2}\right)$

92. $\frac{2}{3x - 3} \div \left(\frac{2x + 2}{x - 1} \cdot \frac{5}{x + 1}\right)$

93. $\frac{y^2}{x + 1} \cdot \frac{x^2 + 2x + 1}{x^2 - 1} \div \frac{3y}{xy - y}$

94. $\frac{x^2 - y^2}{x^4 - x^3} \div \frac{x - y}{x^2} \div \frac{x^2 + 2xy + y^2}{x + y}$

95. $\frac{x^2 + x - 6}{x^2 - 4} \cdot \frac{x^2 + 2x}{x - 2} \div \frac{x^2 + 3x}{x + 2}$

96. $\frac{x^2 - x - 6}{x^2 + 6x - 7} \cdot \frac{x^2 + x - 2}{x^2 + 2x} \div \frac{x^2 + 7x}{x^2 - 3x}$

APPLICATIONS

97. INTERNATIONAL ALPHABET The symbols representing the letters A, B, C, D, E, and F in an international code used at sea are printed six to a sheet and then cut into separate cards. If each card is a square, find the area of the large printed sheet shown in the illustration.

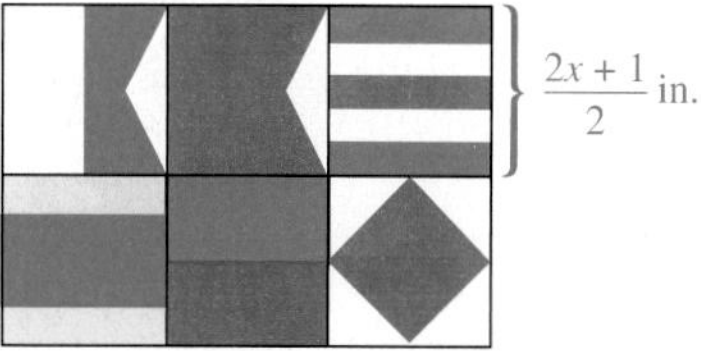

98. PHYSICS The following table contains algebraic expressions for the rate an object travels, and the time traveled at that rate, in terms of a constant k. Complete the table.

Rate (mph)	Time (hr)	Distance (mi)
$\frac{k^2 + k - 6}{k - 3}$	$\frac{k^2 - 9}{k^2 - 4}$	

WRITING

99. Explain how to multiply two fractions and how to simplify the result.

100. Explain why any mathematical expression can be written as a fraction.

101. To divide fractions, you must first know how to multiply fractions. Explain.

102. Explain how to do the division: $\frac{a}{b} \div \frac{c}{d} \div \frac{e}{f}$.

REVIEW *Simplify each expression. Write all answers without using negative exponents.*

103. $2x^3y^2(-3x^2y^4)$

104. $\frac{8x^4y^5}{-2x^3y^2}$

105. $(3y)^{-4}$

106. $x^{3m} \cdot x^{4m}$

Perform the operations and simplify.

107. $-4(y^3 - 4y^2 + 3y - 2) - 4(-2y^3 - y)$

108. $y - 5\overline{)5y^3 - 3y^2 + 4y - 1}$

12.3 Adding and Subtracting Rational Expressions

- Adding and subtracting rational expressions with like denominators
- Combined operations • The LCD
- Adding and subtracting rational expressions with unlike denominators
- Combined operations

In this section, we extend the rules for adding and subtracting numerical fractions to problems involving addition and subtraction of rational expressions.

Adding and subtracting rational expressions with like denominators

To add (or subtract) fractions with a common denominator, we add (or subtract) their numerators and keep the common denominator. For example,

$$\frac{3}{7} + \frac{2}{7} = \frac{3 + 2}{7} = \frac{5}{7} \qquad \frac{3}{7} - \frac{2}{7} = \frac{3 - 2}{7} = \frac{1}{7}$$

In general, we have the following rule.

Adding and subtracting fractions with like denominators

If a, b, and d represent real numbers, and d is not 0,

$$\frac{a}{d} + \frac{b}{d} = \frac{a + b}{d} \quad \text{and} \quad \frac{a}{d} - \frac{b}{d} = \frac{a - b}{d}$$

We use the same procedure to add and subtract rational expressions with like denominators.

Self Check 1

Add:

a. $\frac{x}{7} + \frac{4x}{7}$

b. $\frac{3x}{7y} + \frac{4x}{7y}$

Answers **a.** $\frac{5x}{7}$, **b.** $\frac{x}{y}$

EXAMPLE 1 Perform each addition: **a.** $\frac{x}{8} + \frac{3x}{8}$ and **b.** $\frac{3x + y}{5x} + \frac{x + y}{5x}$.

Solution

a. $\frac{x}{8} + \frac{3x}{8} = \frac{x + 3x}{8}$ Add the numerators and keep the common denominator.

$= \frac{4x}{8}$ Combine like terms: $x + 3x = 4x$.

$= \frac{\overset{1}{\cancel{4}} \cdot x}{\underset{1}{\cancel{4}} \cdot 2}$ Factor the numerator and denominator and remove the common factor, 4.

$= \frac{x}{2}$ Simplify.

b. $\frac{3x + y}{5x} + \frac{x + y}{5x} = \frac{3x + y + x + y}{5x}$ Add the numerators and keep the common denominator.

$= \frac{4x + 2y}{5x}$ Combine like terms.

EXAMPLE 2 Add: $\frac{3x + 21}{5x + 10} + \frac{8x + 1}{5x + 10}$.

Solution Because the fractions have the same denominator, we add their numerators and keep the common denominator.

$$\frac{3x + 21}{5x + 10} + \frac{8x + 1}{5x + 10} = \frac{3x + 21 + 8x + 1}{5x + 10}$$ Add.

$$= \frac{11x + 22}{5x + 10}$$ Combine like terms.

$$= \frac{11\overset{1}{\cancel{(x + 2)}}}{5\underset{1}{\cancel{(x + 2)}}}$$ Simplify the result by factoring the numerator and denominator. Remove the common factor of $x + 2$.

$$= \frac{11}{5}$$

Self Check 2

Add: $\frac{x + 4}{6x - 12} + \frac{x - 8}{6x - 12}$

Answer $\frac{1}{3}$

EXAMPLE 3 Subtract: **a.** $\frac{5x}{3} - \frac{2x}{3}$ and **b.** $\frac{5x + 1}{x - 3} - \frac{4x - 2}{x - 3}$.

Solution In each part, the fractions have the same denominator. To subtract them, we subtract their numerators and keep the common denominator.

a. $$\frac{5x}{3} - \frac{2x}{3} = \frac{5x - 2x}{3}$$

$$= \frac{3x}{3}$$ Combine like terms: $5x - 2x = 3x$.

$$= \frac{x}{1}$$ Remove the common factor of 3.

$$= x$$ Denominators of 1 need not be written.

b. $$\frac{5x + 1}{x - 3} - \frac{4x - 2}{x - 3} = \frac{5x + 1 - (4x - 2)}{x - 3}$$ The second numerator, $4x - 2$, is written within parentheses to make sure that we subtract both of its terms.

$$= \frac{5x + 1 - 4x + 2}{x - 3}$$ Distribute the multiplication by -1: $-(4x - 2) = -4x + 2$.

$$= \frac{x + 3}{x - 3}$$ Combine like terms.

Self Check 3

Subtract: $\frac{2y + 1}{y + 5} - \frac{y - 4}{y + 5}$

Answer 1

Combined operations

To add and/or subtract three or more rational expressions, we follow the rules for the order of operations.

EXAMPLE 4 Simplify: $\frac{3x + 1}{x^2 + x + 1} - \frac{5x + 2}{x^2 + x + 1} + \frac{2x + 1}{x^2 + x + 1}$.

Solution This example combines addition and subtraction. Unless parentheses indicate otherwise, we perform additions and subtractions from left to right.

Self Check 4

Simplify:

$\frac{2a^2 - 3}{a - 5} + \frac{3a^2 + 2}{a - 5} - \frac{5a^2}{a - 5}$

$$\frac{3x+1}{x^2+x+1} - \frac{5x+2}{x^2+x+1} + \frac{2x+1}{x^2+x+1}$$

$$= \frac{3x+1-(5x+2)+2x+1}{x^2+x+1}$$ Combine the numerators and keep the common denominator.

$$= \frac{3x+1-5x-2+2x+1}{x^2+x+1}$$ Distribute the multiplication by -1: $-(5x+2) = -5x-2$.

$$= \frac{0}{x^2+x+1}$$ Combine like terms.

$$= 0$$ If the numerator of a fraction is zero and the denominator is not zero, the fraction's value is zero.

Answer $-\frac{1}{a-5}$

The LCD

Since the denominators of the fractions in the addition $\frac{4}{7} + \frac{3}{5}$ are different, we cannot add the fractions in their present form.

four-sevenths + three-fifths
↑ Different denominators ↑

To add these fractions, we need to find a common denominator. The smallest common denominator (called the **least** or **lowest common denominator**) is usually the easiest one to work with.

> The **least common denominator (LCD)** for a set of fractions is the smallest number that each denominator will divide exactly.

In the addition $\frac{4}{7} + \frac{3}{5}$, the denominators are 7 and 5. The smallest number that 7 and 5 will divide exactly is 35. This is the LCD. We now **build** each fraction into an equivalent fraction with a denominator of 35. To do so, we use the fundamental property of fractions to multiply both the numerator and the denominator of each fraction by some appropriate number.

$$\frac{4}{7} + \frac{3}{5} = \frac{4 \cdot \mathbf{5}}{7 \cdot \mathbf{5}} + \frac{3 \cdot \mathbf{7}}{5 \cdot \mathbf{7}}$$ Since $7 \cdot 5 = 35$, multiply the numerator and denominator of $\frac{4}{7}$ by 5. Since $5 \cdot 7 = 35$, multiply the numerator and denominator of $\frac{3}{5}$ by 7.

$$= \frac{20}{35} + \frac{21}{35}$$ Perform the multiplications.

Now that the fractions have a common denominator, we can add them.

$$\frac{20}{35} + \frac{21}{35} = \frac{20+21}{35} = \frac{41}{35}$$

Self Check 5

Change $\frac{5}{6b}$ into a fraction with a denominator of $30ab$.

EXAMPLE 5 Change each fraction into one with a denominator of $30y$: **a.** $\frac{1}{2y}$, **b.** $\frac{3y}{5}$, and **c.** $\frac{7+x}{10y}$.

Solution To build each fraction, we multiply the numerator and denominator by the factor that makes the denominator $30y$.

a. $\frac{1}{2y} = \frac{1 \cdot \mathbf{15}}{2y \cdot \mathbf{15}} = \frac{15}{30y}$ Multiply numerator and denominator by 15, because $2y \cdot 15 = 30y$.

b. $\frac{3y}{5} = \frac{3y \cdot \mathbf{6y}}{5 \cdot \mathbf{6y}} = \frac{18y^2}{30y}$ Multiply numerator and denominator by $6y$, because $5 \cdot 6y = 30y$.

c. $\frac{7 + x}{10y} = \frac{(7 + x)\mathbf{3}}{(10y)\mathbf{3}} = \frac{21 + 3x}{30y}$ Multiply numerator and denominator by 3, because $10y \cdot 3 = 30y$.

Answer $\frac{25a}{30ab}$

The least common denominator of several rational expressions can be found as follows.

Finding the least common denominator (LCD)

1. Factor each denominator completely.
2. The LCD is a product that uses each different factor obtained in step 1 the greatest number of times it appears in any one factorization.

EXAMPLE 6 Find the LCD of $\frac{5}{24b}$ and $\frac{11}{18b}$.

Solution We list and factor each denominator into the product of prime numbers.

$$24b = 2 \cdot 2 \cdot 2 \cdot 3 \cdot b$$
$$18b = 2 \cdot 3 \cdot 3 \cdot b$$

To find the LCD, we use each of these factors the greatest number of times it appears in any one factorization. We use 2 three times, because it appears three times as a factor of 24. We use 3 twice, because it occurs twice as a factor of 18. We use b once.

$$\begin{aligned} \text{LCD} &= 2 \cdot 2 \cdot 2 \cdot 3 \cdot 3 \cdot b \\ &= 8 \cdot 9 \cdot b \\ &= 72b \end{aligned}$$

Self Check 6

Find the LCD of $\frac{3}{28z}$ and $\frac{5}{21z}$.

Answer $84z$

Adding and subtracting rational expressions with unlike denominators

Recall that the steps used to add (or subtract) fractions that have unlike denominators are as follows.

Adding or subtracting fractions with unlike denominators

1. Find the LCD.
2. Write each fraction as an equivalent fraction whose denominator is the LCD.
3. Add (or subtract) the resulting fractions and simplify the result, if possible.

We use the same procedure to add or subtract rational expressions with unlike denominators.

Self Check 7

Add: $\frac{y}{2} + \frac{6y}{7}$

EXAMPLE 7 Add: $\frac{4x}{7} + \frac{3x}{5}$.

Solution The LCD is 35. We build each fraction so that it has a denominator of 35 and then add the resulting fractions.

$$\frac{4x}{7} + \frac{3x}{5} = \frac{4x \cdot \mathbf{5}}{7 \cdot \mathbf{5}} + \frac{3x \cdot 7}{5 \cdot 7}$$ Multiply the numerator and the denominator of $\frac{4x}{7}$ by 5 and the numerator and denominator of $\frac{3x}{5}$ by 7.

$$= \frac{20x}{35} + \frac{21x}{35}$$ Perform the multiplications.

$$= \frac{41x}{35}$$ Add the numerators and keep the common denominator.

Answer $\frac{19y}{14}$

Self Check 8

Add: $\frac{3}{28z} + \frac{5}{21z}$

EXAMPLE 8 Add: $\frac{5}{24b} + \frac{11}{18b}$.

Solution In Example 6, we saw that the LCD of these fractions is $2 \cdot 2 \cdot 2 \cdot 3 \cdot 3 \cdot b = 72b$. To add them, we first factor each denominator:

$$\frac{5}{24b} + \frac{11}{18b} = \frac{5}{2 \cdot 2 \cdot 2 \cdot 3 \cdot b} + \frac{11}{2 \cdot 3 \cdot 3 \cdot b}$$

In each resulting fraction, we multiply the numerator and the denominator by whatever it takes to build the denominator to the LCD of $2 \cdot 2 \cdot 2 \cdot 3 \cdot 3 \cdot b$.

$$= \frac{5 \cdot \mathbf{3}}{2 \cdot 2 \cdot 2 \cdot 3 \cdot b \cdot \mathbf{3}} + \frac{11 \cdot 2 \cdot 2}{2 \cdot 3 \cdot 3 \cdot b \cdot 2 \cdot 2}$$ Build each fraction to get the common denominator.

$$= \frac{15}{72b} + \frac{44}{72b}$$ Perform the multiplications.

$$= \frac{59}{72b}$$ Add the numerators and keep the common denominator.

Answer $\frac{29}{84z}$

Self Check 9

Add: $\frac{a-1}{9a} + \frac{2-a}{a^2}$

EXAMPLE 9 Add: $\frac{x+4}{x^2} + \frac{x-5}{4x}$.

Solution First we find the LCD.

$$\left.\begin{aligned} x^2 &= x \cdot x \\ 4x &= 2 \cdot 2 \cdot x \end{aligned}\right\} \text{LCD} = x \cdot x \cdot 2 \cdot 2 = 4x^2$$

$$\frac{x+4}{x^2} + \frac{x-5}{4x} = \frac{(x+4)\mathbf{4}}{(x^2)\mathbf{4}} + \frac{(x-5)x}{(4x)x}$$ Build the fractions to get the common denominator, $4x^2$.

$$= \frac{4x+16}{4x^2} + \frac{x^2-5x}{4x^2}$$ Perform the multiplications.

$$= \frac{4x + 16 + x^2 - 5x}{4x^2}$$ Add the numerators and keep the common denominator.

$$= \frac{x^2 - x + 16}{4x^2}$$ Combine like terms.

Answer $\frac{a^2 - 10a + 18}{9a^2}$

Self Check 10

Subtract: $\frac{a}{a-1} - \frac{5}{a}$

EXAMPLE 10 Subtract: $\frac{x}{x+1} - \frac{3}{x}$.

Solution By inspection, the least common denominator is $(x+1)x$.

$$\frac{x}{x+1} - \frac{3}{x} = \frac{x(x)}{(x+1)x} - \frac{3(x+1)}{x(x+1)}$$ Build the fractions to get the common denominator.

$$= \frac{x(x) - 3(x+1)}{x(x+1)}$$ Subtract the numerators and keep the common denominator.

$$= \frac{x^2 - 3x - 3}{x(x+1)}$$ Perform the multiplications in the numerator.

Answer $\frac{a^2 - 5a + 5}{a(a-1)}$

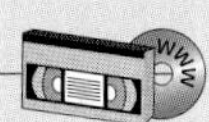

EXAMPLE 11 Subtract: $\frac{a}{a-1} - \frac{2}{a^2-1}$.

Solution We factor $a^2 - 1$ to see that the LCD is $(a+1)(a-1)$.

$$\frac{a}{a-1} - \frac{2}{a^2-1}$$

$$= \frac{a(a+1)}{(a-1)(a+1)} - \frac{2}{(a+1)(a-1)}$$ Build the first fraction to get the LCD.

$$= \frac{a(a+1) - 2}{(a-1)(a+1)}$$ Subtract the numerators and keep the common denominator.

$$= \frac{a^2 + a - 2}{(a-1)(a+1)}$$ Distribute the multiplication by a.

$$= \frac{(a+2)\overset{1}{\cancel{(a-1)}}}{\underset{1}{\cancel{(a-1)}}(a+1)}$$ Simplify the result by factoring $a^2 + a - 2$. Remove the common factor of $a - 1$.

$$= \frac{a+2}{a+1}$$

Self Check 11

Subtract: $\frac{b}{b-2} - \frac{8}{b^2-4}$

Answer $\frac{b+4}{b+2}$

EXAMPLE 12 Subtract: $\frac{2a}{a^2+4a+4} - \frac{1}{2a+4}$.

Solution Find the least common denominator by factoring each denominator.

$$\left.\begin{array}{l} a^2 + 4a + 4 = (a+2)(a+2) \\ 2a + 4 = 2(a+2) \end{array}\right\} \quad \text{LCD} = (a+2)(a+2)2$$

We build each fraction into a new fraction with a denominator of $2(a+2)(a+2)$.

$$\frac{2a}{a^2+4a+4} - \frac{1}{2a+4}$$

$$= \frac{2a}{(a+2)(a+2)} - \frac{1}{2(a+2)}$$ Write the denominators in factored form.

$$= \frac{2a \cdot 2}{(a+2)(a+2)2} - \frac{1(a+2)}{2(a+2)(a+2)}$$ Build each fraction to get a common denominator.

$$= \frac{4a - 1(a+2)}{2(a+2)^2}$$ Subtract the numerators and keep the common denominator. Write $(a+2)(a+2)$ as $(a+2)^2$.

$$= \frac{4a - a - 2}{2(a+2)^2}$$ Distribute the multiplication by -1.

$$= \frac{3a - 2}{2(a+2)^2}$$ Combine like terms.

Self Check 12

Subtract: $\frac{a}{a^2 - 2a + 1} - \frac{1}{6a - 6}$

Answer $\frac{5a+1}{6(a-1)^2}$

Self Check 13

Subtract: $\dfrac{5}{a-b} - \dfrac{2}{b-a}$

Answer $\dfrac{7}{a-b}$

EXAMPLE 13 Subtract: $\dfrac{3}{x-y} - \dfrac{x}{y-x}$.

Solution We note that the second denominator is the opposite (negative) of the first. So we can multiply the numerator and denominator of the second fraction by -1 to get

$$\frac{3}{x-y} - \frac{x}{y-x} = \frac{3}{x-y} - \frac{-\mathbf{1}x}{-\mathbf{1}(y-x)}$$ Multiply numerator and denominator by -1.

$$= \frac{3}{x-y} - \frac{-x}{-y+x}$$ Distribute the multiplication by -1: $-1(y-x) = -y+x$.

$$= \frac{3}{x-y} - \frac{-x}{x-y}$$ $-y+x = x-y$. The fractions now have a common denominator of $x-y$.

$$= \frac{3-(-x)}{x-y}$$ Subtract the numerators and keep the common denominator.

$$= \frac{3+x}{x-y}$$ $-(-x) = x$.

Combined operations

To add and/or subtract three or more rational expressions, we follow the rules for the order of operations.

Self Check 14

Combine: $\dfrac{5}{ab^2} - \dfrac{b}{a} + \dfrac{a}{b}$

Answer $\dfrac{5 - b^3 + a^2b}{ab^2}$

EXAMPLE 14 Perform the operations: $\dfrac{3}{x^2y} + \dfrac{2}{xy} - \dfrac{1}{xy^2}$.

Solution Find the least common denominator.

$$\left.\begin{aligned} x^2y &= x \cdot x \cdot y \\ xy &= x \cdot y \\ xy^2 &= x \cdot y \cdot y \end{aligned}\right\}$$ Factor each denominator.

In any one of these denominators, the factor x occurs at most twice, and the factor y occurs at most twice. Thus,

$$\begin{aligned} \text{LCD} &= x \cdot x \cdot y \cdot y \\ &= x^2y^2 \end{aligned}$$

We build each fraction into one with a denominator of x^2y^2.

$$\frac{3}{x^2y} + \frac{2}{xy} - \frac{1}{xy^2}$$

$$= \frac{3 \cdot y}{x \cdot x \cdot y \cdot y} + \frac{2 \cdot x \cdot y}{x \cdot y \cdot x \cdot y} - \frac{1 \cdot x}{x \cdot y \cdot y \cdot x}$$ Factor each denominator and build each fraction.

$$= \frac{3y + 2xy - x}{x^2y^2}$$ Perform the multiplications and combine the numerators. Write the result over the LCD.

Section 12.3 STUDY SET

VOCABULARY *Fill in the blanks.*

1. The ______ for a set of fractions is the smallest number that each denominator divides exactly.
2. When we multiply the numerator and denominator of a fraction by some number to get a common denominator, we say that we are ______ the fraction.

CONCEPTS *Fill in the blanks.*

3. To add two fractions with like denominators, we add their ______ and keep the ______ ______.
4. To subtract two fractions with ______ denominators, we need to find a common denominator.

NOTATION *Complete each solution.*

5. $\frac{6a-1}{4a+1}+\frac{2a+3}{4a+1}=\frac{6a-1+\square}{4a+1}$

$=\frac{8a+\square}{4a+1}$

$=\frac{2\square}{4a+1}$

$=2$

6. $\frac{x}{2x+1}-\frac{1}{3x}=\frac{x(\square)}{(2x+1)(3x)}-\frac{1(2x+1)}{3x\square}$

$=\frac{x(3x)-1\square}{3x(2x+1)}$

$=\frac{3x^2-\square-\square}{3x(2x+1)}$

$=\frac{(3x+1)(\square)}{3x(2x+1)}$

PRACTICE *Perform each operation. Simplify answers, if possible.*

7. $\frac{x}{9}+\frac{2x}{9}$
8. $\frac{5x}{7}+\frac{9x}{7}$
9. $\frac{2x}{y}+\frac{2x}{y}$
10. $\frac{4y}{3x}+\frac{2y}{3x}$
11. $\frac{4}{7y}+\frac{10}{7y}$
12. $\frac{x^2}{4y}+\frac{x^2}{4y}$
13. $\frac{y+2}{10z}+\frac{y+4}{10z}$
14. $\frac{x+3}{2x^2}+\frac{x+5}{2x^2}$
15. $\frac{3x-5}{x-2}+\frac{6x-13}{x-2}$
16. $\frac{8x-7}{x+3}+\frac{2x+37}{x+3}$
17. $\frac{a}{a^2+5a+6}+\frac{3}{a^2+5a+6}$
18. $\frac{b}{b^2-4}+\frac{2}{b^2-4}$
19. $\frac{35y}{72}-\frac{44y}{72}$
20. $\frac{13t}{99}-\frac{35t}{99}$
21. $\frac{2x}{y}-\frac{x}{y}$
22. $\frac{7y}{5}-\frac{4y}{5}$
23. $\frac{9y}{3x}-\frac{6y}{3x}$
24. $\frac{5r^2}{2r}-\frac{r^2}{2r}$
25. $\frac{6x-5}{3xy}-\frac{3x-5}{3xy}$
26. $\frac{7x+7}{5y}-\frac{2x+7}{5y}$
27. $\frac{3y-2}{2y+6}-\frac{2y-5}{2y+6}$
28. $\frac{5x+8}{3x+15}-\frac{3x-2}{3x+15}$
29. $\frac{2c}{c^2-d^2}-\frac{2d}{c^2-d^2}$
30. $\frac{3t}{t^2-8t+7}-\frac{3}{t^2-8t+7}$
31. $\frac{13x}{15}+\frac{12x}{15}-\frac{5x}{15}$
32. $\frac{13y}{32}+\frac{13y}{32}-\frac{10y}{32}$
33. $-\frac{x}{y}+\frac{2x}{y}-\frac{x}{y}$
34. $\frac{5y}{8x}+\frac{4y}{8x}-\frac{9y}{8x}$
35. $\frac{3x}{y+2}-\frac{3y}{y+2}+\frac{x+y}{y+2}$
36. $\frac{3y}{x-5}+\frac{x}{x-5}-\frac{y-x}{x-5}$
37. $\frac{x+1}{x-2}-\frac{2(x-3)}{x-2}+\frac{3(x+1)}{x-2}$
38. $\frac{3xy}{x-y}-\frac{x(3y-x)}{x-y}-\frac{x(x-y)}{x-y}$

Build each fraction into an equivalent fraction with the indicated denominator.

39. $\frac{25}{4}$; $20x$
40. $\frac{5}{y}$; y^2
41. $\frac{8}{x}$; x^2y
42. $\frac{7}{y}$; xy^2
43. $\frac{3x}{x+1}$; $(x+1)^2$
44. $\frac{5y}{y-2}$; $(y-2)^2$

45. $\frac{2y}{x}; x^2 + x$

46. $\frac{3x}{y}; y^2 - y$

47. $\frac{z}{z - 1}; z^2 - 1$

48. $\frac{y}{y + 2}; y^2 - 4$

49. $\frac{2}{x + 1}; x^2 + 3x + 2$

50. $\frac{3}{x - 1}; x^2 + x - 2$

Several denominators are given. Find the LCD.

51. $2x, 6x$

52. $3y, 9y$

53. $6y, 9xy^2$

54. $6y, 3x^2y$

55. $x^2 - 1, x + 1$

56. $y^2 - 9, y - 3$

57. $x^2 + 6x, x + 6, x$

58. $xy^2 - xy, xy, y - 1$

59. $x^2 - 4x - 5, x^2 - 25$

60. $x^2 - x - 6, x^2 - 9$

Perform the operations. Simplify answers, if possible.

61. $\frac{2y}{9} + \frac{y}{3}$

62. $\frac{8a}{15} - \frac{5a}{12}$

63. $\frac{21x}{14} - \frac{5x}{21}$

64. $\frac{7y}{6} + \frac{10y}{9}$

65. $\frac{4x}{3} + \frac{2x}{y}$

66. $\frac{2y}{5x} - \frac{y}{2}$

67. $\frac{2}{x} - 3x$

68. $14 + \frac{10}{y^2}$

69. $\frac{y + 2}{5y^2} + \frac{y + 4}{15y}$

70. $\frac{x + 3}{x^2} + \frac{x + 5}{2x}$

71. $\frac{x + 5}{xy} - \frac{x - 1}{x^2y}$

72. $\frac{y - 7}{y^2} - \frac{y + 7}{2y}$

73. $\frac{x}{x + 1} + \frac{x - 1}{x}$

74. $\frac{3x}{xy} + \frac{x + 1}{y - 1}$

75. $\frac{x - 1}{x} + \frac{y + 1}{y}$

76. $\frac{a + 2}{b} + \frac{b - 2}{a}$

77. $\frac{x}{x - 2} + \frac{4 + 2x}{x^2 - 4}$

78. $\frac{y}{y + 3} - \frac{2y - 6}{y^2 - 9}$

79. $\frac{x + 1}{x - 1} + \frac{x - 1}{x + 1}$

80. $\frac{2x}{x + 2} + \frac{x + 1}{x - 3}$

81. $\frac{5}{a - 4} + \frac{7}{4 - a}$

82. $\frac{4}{b - 6} - \frac{b}{6 - b}$

83. $\frac{t + 1}{t - 7} - \frac{t + 1}{7 - t}$

84. $\frac{r + 2}{r^2 - 4} + \frac{4}{4 - r^2}$

85. $\frac{2x + 2}{x - 2} - \frac{2x}{2 - x}$

86. $\frac{y + 3}{y - 1} - \frac{y + 4}{1 - y}$

87. $\frac{b}{b + 1} - \frac{b + 1}{2b + 2}$

88. $\frac{4x + 1}{8x - 12} + \frac{x - 3}{2x - 3}$

89. $\frac{2}{a^2 + 4a + 3} + \frac{1}{a + 3}$

90. $\frac{1}{c + 6} - \frac{-4}{c^2 + 8c + 12}$

91. $\frac{x + 1}{2x + 4} - \frac{x^2}{2x^2 - 8}$

92. $\frac{x + 1}{x + 2} - \frac{x^2 + 1}{x^2 - x - 6}$

93. $\frac{2x}{x^2 - 3x + 2} + \frac{2x}{x - 1} - \frac{x}{x - 2}$

94. $\frac{4a}{a - 2} - \frac{3a}{a - 3} + \frac{4a}{a^2 - 5a + 6}$

95. $\frac{2x}{x - 1} + \frac{3x}{x + 1} - \frac{x + 3}{x^2 - 1}$

96. $\frac{a}{a - 1} - \frac{2}{a + 2} + \frac{3(a - 2)}{a^2 + a - 2}$

APPLICATIONS *Refer to the illustration of the funnel.*

97. Find the total height of the funnel.

98. What is the difference between the diameter of the opening at the top of the funnel and the diameter of its spout?

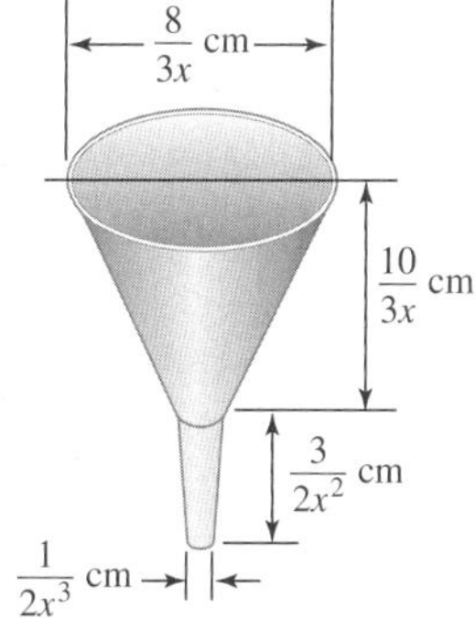

WRITING

99. Explain how to add fractions with the same denominator.

100. Explain how to find a lowest common denominator.

Explain what is wrong with each solution:

101. $\frac{2x+3}{x+5} - \frac{x+2}{x+5} = \frac{2x+3-x+2}{x+5}$

$= \frac{x+5}{x+5}$

$= 1$

102. $\frac{5x-4}{y} + \frac{x}{y} = \frac{5x-4+x}{y+y}$

$= \frac{6x-4}{2y}$

$= \frac{2(3x-2)}{2y}$

$= \frac{3x-2}{y}$

REVIEW *Write each number in prime-factored form.*

103. 49

104. 64

105. 136

106. 315

12.4 Complex Fractions

- Simplifying complex fractions
- Simplifying fractions with terms containing negative exponents

A rational expression whose numerator and/or denominator contain rational expressions is called a **complex rational expression** or, more simply, a **complex fraction.** The expression above the main fraction bar of a complex fraction is the numerator, and the expression below the main fraction bar is the denominator. Two examples are:

$$\frac{\frac{5x}{3}}{\frac{2y}{9}}, \qquad \frac{\frac{x+1}{2}}{x+\frac{1}{x}}$$

← Numerator →
← Main fraction bar →
← Denominator →

In this section, we will simplify complex fractions.

Simplifying complex fractions

To *simplify a complex fraction* means to express it in the form $\frac{P}{Q}$, where P and Q are polynomials that have no common factors. One method for simplifying complex fractions uses the fact that fractions indicate division. For example, to simplify

$$\frac{\frac{5x}{3}}{\frac{2y}{9}}$$

we proceed as follows:

$$\frac{\frac{5x}{3}}{\frac{2y}{9}} = \frac{5x}{3} \div \frac{2y}{9}$$ The main fraction bar indicates division.

$$= \frac{5x}{3} \cdot \frac{9}{2y}$$ Multiply by the reciprocal of $\frac{2y}{9}$.

$$= \frac{5x \cdot 9}{3 \cdot 2y}$$ Multiply the numerators. Multiply the denominators.

$$= \frac{5x \cdot 3 \cdot \overset{1}{\cancel{3}}}{\underset{1}{\cancel{3}} \cdot 2y}$$ Simplify the fraction by removing the common factor 3 in the numerator and denominator.

$$= \frac{15x}{2y}$$

Another method for simplifying complex fractions uses the fundamental property of fractions. For example, to simplify

$$\frac{\frac{3x}{5} + 1}{2 - \frac{x}{5}}$$

we proceed as follows:

$$\frac{\frac{3x}{5} + 1}{2 - \frac{x}{5}} = \frac{5\left(\frac{3x}{5} + 1\right)}{5\left(2 - \frac{x}{5}\right)}$$ Multiply both the numerator and denominator of the complex fraction by 5, the LCD of $\frac{3x}{5}$ and $\frac{x}{5}$.

$$= \frac{5 \cdot \frac{3x}{5} + 5 \cdot 1}{5 \cdot 2 - 5 \cdot \frac{x}{5}}$$ Distribute the multiplication by 5.

$$= \frac{3x + 5}{10 - x}$$ Perform the multiplications.

The methods for simplifying a complex fraction using division and using the LCD are summarized below.

Methods for simplifying complex fractions

Method 1: Write the numerator and denominator of the complex fraction as single fractions. Then divide the fractions and simplify.

Method 2: Multiply the numerator and denominator of the complex fraction by the LCD of the fractions in its numerator and denominator. Then simplify the results, if possible.

EXAMPLE 1 Simplify: $\dfrac{\frac{x}{3}}{\frac{y}{3}}$.

Solution

Method 1

$$\frac{\frac{x}{3}}{\frac{y}{3}} = \frac{x}{3} \div \frac{y}{3}$$

$$= \frac{x}{3} \cdot \frac{3}{y}$$

$$= \frac{3x}{3y}$$

$$= \frac{x}{y}$$

Method 2

$$\frac{\frac{x}{3}}{\frac{y}{3}} = \frac{\mathbf{3}\left(\frac{x}{3}\right)}{\mathbf{3}\left(\frac{y}{3}\right)}$$

The LCD for the fractions in the given complex fraction is 3.

$$= \frac{x}{y}$$

Note that method 1 and method 2 give the same result.

Self Check 1
Simplify:
$\dfrac{\frac{a}{4}}{\frac{5}{b}}$

Answer $\dfrac{ab}{20}$

! COMMENT When simplifying a complex fraction, the same result will be obtained regardless of the method used.

EXAMPLE 2 Simplify: $\dfrac{\frac{x}{x + 1}}{\frac{y}{x}}$.

Solution

Method 1

$$\frac{\frac{x}{x+1}}{\frac{y}{x}} = \frac{x}{x + 1} \div \frac{y}{x}$$

$$= \frac{x}{x + 1} \cdot \frac{x}{y}$$

$$= \frac{x^2}{y(x + 1)}$$

$$= \frac{x^2}{xy + y}$$

Method 2

$$\frac{\frac{x}{x+1}}{\frac{y}{x}} = \frac{\boldsymbol{x(x + 1)}\left(\frac{x}{x + 1}\right)}{\boldsymbol{x(x + 1)}\left(\frac{y}{x}\right)}$$

The LCD for the fractions in the given complex fraction is $x(x + 1)$.

$$= \frac{x^2}{y(x + 1)}$$

$$= \frac{x^2}{xy + y}$$

Self Check 2
Simplify:
$\dfrac{\frac{x}{y}}{\frac{x}{y + 1}}$

Answer $\dfrac{y + 1}{y}$

In general, simplifying using division (method 1) works well when a complex fraction is written, or can be easily written, as a quotient of two single rational expressions.

In general, simplifying using the LCD (method 2) works well when the complex fraction has sums and/or differences in the numerator or denominator.

Self Check 3
Simplify:
$$\frac{\frac{1}{x}+1}{\frac{1}{x}-1}$$

Answer $\frac{1+x}{1-x}$

EXAMPLE 3 Simplify: $\dfrac{1+\frac{1}{x}}{1-\frac{1}{x}}$.

Solution

Method 1

$$\frac{1+\frac{1}{x}}{1-\frac{1}{x}} = \frac{\frac{x}{x}+\frac{1}{x}}{\frac{x}{x}-\frac{1}{x}}$$
$$= \frac{\frac{x+1}{x}}{\frac{x-1}{x}}$$
$$= \frac{x+1}{x} \div \frac{x-1}{x}$$
$$= \frac{x+1}{x} \cdot \frac{x}{x-1}$$
$$= \frac{(x+1)\overset{1}{\cancel{x}}}{\underset{1}{\cancel{x}}(x-1)}$$
$$= \frac{x+1}{x-1}$$

Method 2

$$\frac{1+\frac{1}{x}}{1-\frac{1}{x}} = \frac{x\left(1+\frac{1}{x}\right)}{x\left(1-\frac{1}{x}\right)}$$
$$= \frac{x\cdot 1 + x\cdot\frac{1}{x}}{x\cdot 1 - x\cdot\frac{1}{x}}$$
$$= \frac{x+1}{x-1}$$

Self Check 4
Simplify:
$$\frac{2}{\frac{1}{x+2}-2}$$

Answer $\frac{2(x+2)}{-2x-3}$

EXAMPLE 4 Simplify: $\dfrac{1}{1+\frac{1}{x+1}}$.

Solution We use method 2.

$$\frac{1}{1+\frac{1}{x+1}} = \frac{(\boldsymbol{x+1})\cdot 1}{(\boldsymbol{x+1})\left(1+\frac{1}{x+1}\right)}$$ Multiply the numerator and the denominator of the complex fraction by $x + 1$.

$$= \frac{x+1}{(x+1)1+1}$$ In the denominator, distribute $x + 1$.

$$= \frac{x+1}{x+2}$$ Simplify.

Simplifying fractions with terms containing negative exponents

Many fractions with terms containing negative exponents are complex fractions in disguise.

EXAMPLE 5 Simplify: $\dfrac{x^{-1} + y^{-2}}{x^{-2} - y^{-1}}$.

Solution Write the fraction as a complex fraction and simplify using method 2.

$$\frac{x^{-1} + y^{-2}}{x^{-2} - y^{-1}} = \frac{\frac{1}{x} + \frac{1}{y^2}}{\frac{1}{x^2} - \frac{1}{y}}$$

$$= \frac{x^2y^2\left(\frac{1}{x} + \frac{1}{y^2}\right)}{x^2y^2\left(\frac{1}{x^2} - \frac{1}{y}\right)}$$ Multiply the numerator and denominator by x^2y^2, which is the LCD of the fractions in the numerator and the denominator of the complex fraction.

$$= \frac{xy^2 + x^2}{y^2 - x^2y}$$ Distribute the multiplication by x^2y^2 and simplify.

$$= \frac{x(y^2 + x)}{y(y - x^2)}$$ Attempt to simplify the fraction by factoring the numerator and the denominator. The result cannot be simplified.

Self Check 5
Simplify:
$\dfrac{x^{-2} - y^{-1}}{x^{-1} + y^{-2}}$

Answer $\dfrac{y(y - x^2)}{x(y^2 + x)}$

Section 12.4 STUDY SET

VOCABULARY *Fill in the blanks.*

1. If a fraction has a fraction in its numerator or denominator, it is called a ________ ________.

2. The denominator of the complex fraction $\dfrac{\frac{3}{x} + \frac{x}{y}}{\frac{1}{x} + 2}$ is ____.

CONCEPTS *Fill in the blanks.*

3. To simplify a complex fraction using method 1, we write the numerator and denominator of a complex fraction as ______ fractions and then ______.

4. To simplify a complex fraction using method 2, we multiply the numerator and denominator of the complex fraction by the ______ of the fractions in its numerator and denominator.

NOTATION *Complete each solution.*

5. $\dfrac{\frac{2b - a}{ab}}{\frac{b + 2a}{ab}} = \dfrac{2b - a}{ab} \square \dfrac{b + 2a}{ab}$

$= \dfrac{2b - a}{ab} \cdot \dfrac{\square}{b + 2a}$

$= \dfrac{(2b - a)\square}{ab\square}$

$= \dfrac{\square}{b + 2a}$

6. $\dfrac{\frac{2}{a} - \frac{1}{b}}{\frac{1}{a} + \frac{2}{b}} = \dfrac{\square\left(\frac{2}{a} - \frac{1}{b}\right)}{\square\left(\frac{1}{a} + \frac{2}{b}\right)}$

$= \dfrac{\square - a}{b + 2a}$

PRACTICE *Simplify each complex fraction.*

7. $\dfrac{\frac{2}{3}}{\frac{3}{4}}$

8. $\dfrac{\frac{3}{5}}{\frac{2}{7}}$

9. $\dfrac{\frac{4}{5}}{\frac{32}{15}}$

10. $\dfrac{\frac{7}{8}}{\frac{49}{4}}$

11. $\dfrac{\frac{2}{3} + 1}{\frac{1}{3} + 1}$

12. $\dfrac{\frac{3}{5} - 2}{\frac{2}{5} - 2}$

13. $\dfrac{\frac{1}{2} + \frac{3}{4}}{\frac{3}{2} + \frac{1}{4}}$

14. $\dfrac{\frac{2}{3} - \frac{5}{2}}{\frac{2}{3} - \frac{3}{2}}$

15. $\dfrac{\frac{x}{y}}{\frac{1}{x}}$

16. $\dfrac{\frac{y}{x}}{\frac{x}{xy}}$

17. $\dfrac{\frac{5t^2}{9x^2}}{\frac{3t}{x^2t}}$

18. $\dfrac{\frac{5w^2}{4tz}}{\frac{15wt}{z^2}}$

19. $\dfrac{\frac{1}{x} - 3}{\frac{5}{x} + 2}$

20. $\dfrac{\frac{1}{y} + 3}{\frac{3}{y} - 2}$

21. $\dfrac{\frac{2}{x} + 2}{\frac{4}{x} + 2}$

22. $\dfrac{\frac{3}{x} - 3}{\frac{9}{x} - 3}$

23. $\dfrac{\frac{3y}{x} - y}{y - \frac{y}{x}}$

24. $\dfrac{\frac{y}{x} + 3y}{y + \frac{2y}{x}}$

25. $\dfrac{\frac{1}{x+1}}{1 + \frac{1}{x+1}}$

26. $\dfrac{\frac{1}{x-1}}{1 - \frac{1}{x-1}}$

27. $\dfrac{\frac{x}{x+2}}{\frac{x}{x+2} + x}$

28. $\dfrac{\frac{2}{x-2}}{\frac{2}{x-2} - 1}$

29. $\dfrac{1}{\frac{1}{x} + \frac{1}{y}}$

30. $\dfrac{1}{\frac{b}{a} - \frac{a}{b}}$

31. $\dfrac{\frac{2}{x}}{\frac{2}{y} - \frac{4}{x}}$

32. $\dfrac{\frac{2y}{3}}{\frac{2y}{3} - \frac{8}{y}}$

33. $\dfrac{3 + \frac{3}{x-1}}{3 - \frac{3}{x}}$

34. $\dfrac{2 - \frac{2}{x+1}}{2 + \frac{2}{x}}$

35. $\dfrac{\frac{3}{x} + \frac{4}{x+1}}{\frac{2}{x+1} - \frac{3}{x}}$

36. $\dfrac{\frac{5}{y-3} - \frac{2}{y}}{\frac{1}{y} + \frac{2}{y-3}}$

37. $\dfrac{\frac{2}{x} - \frac{3}{x+1}}{\frac{2}{x+1} - \frac{3}{x}}$

38. $\dfrac{\frac{5}{y} + \frac{4}{y+1}}{\frac{4}{y} - \frac{5}{y+1}}$

39. $\dfrac{\frac{1}{y^2+y} - \frac{1}{xy+x}}{\frac{1}{xy+x} - \frac{1}{y^2+y}}$

40. $\dfrac{\frac{2}{b^2-1} - \frac{3}{ab-a}}{\frac{3}{ab-a} - \frac{2}{b^2-1}}$

41. $\dfrac{x^{-2}}{y^{-1}}$

42. $\dfrac{a^{-4}}{b^{-2}}$

43. $\dfrac{1 + x^{-1}}{x^{-1} - 1}$

44. $\dfrac{y^{-2} + 1}{y^{-2} - 1}$

45. $\dfrac{a^{-2} + a}{a}$

46. $\dfrac{t - t^{-2}}{t^{-1}}$

47. $\dfrac{2x^{-1} + 4x^{-2}}{2x^{-2} + x^{-1}}$

48. $\dfrac{x^{-2} - 3x^{-3}}{3x^{-2} - 9x^{-3}}$

49. $\dfrac{1 - 25y^{-2}}{1 + 10y^{-1} + 25y^{-2}}$

50. $\dfrac{1 - 9x^{-2}}{1 - 6x^{-1} + 9x^{-2}}$

APPLICATIONS

51. GARDENING TOOLS In the illustration, what is the result when the opening of the cutting blades is divided by the opening of the handles? Express the result in simplest form.

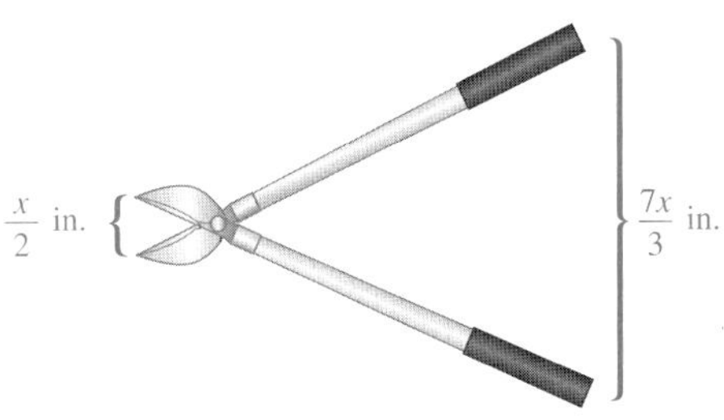

52. EARNED RUN AVERAGE The earned run average (ERA) is a statistic that gives the average number of earned runs a pitcher allows. For a softball pitcher, this is based on a six-inning game. The formula for ERA is

$$\text{ERA} = \dfrac{\frac{\text{earned runs}}{\text{innings pitched}}}{6}$$

Simplify the complex fraction on the right-hand side of the equation.

53. ELECTRONICS In electronic circuits, resistors oppose the flow of an electric current. To find the total resistance of a parallel combination of two resistors, we can use the formula

$$\text{Total resistance} = \frac{1}{\frac{1}{R_1} + \frac{1}{R_2}}$$

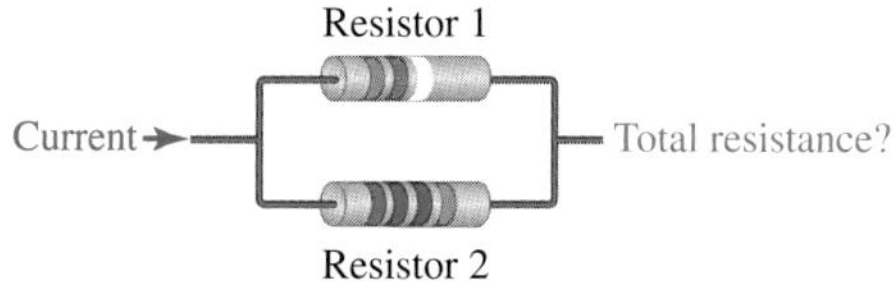

where R_1 is the resistance of the first resistor and R_2 is the resistance of the second. Simplify the complex fraction on the right-hand side of the formula.

54. DATA ANALYSIS Use the data in the table to find the average measurement for the three-trial experiment.

	Trial 1	Trial 2	Trial 3
Measurement	$\frac{k}{2}$	$\frac{k}{3}$	$\frac{k}{2}$

WRITING

55. Explain how to use method 1 to simplify

$$\frac{1 + \frac{1}{x}}{3 - \frac{1}{x}}$$

56. Explain how to use method 2 to simplify the expression in Exercise 55.

REVIEW *Write each expression as an expression involving only one exponent.*

57. $t^3t^4t^2$

58. $(a^0a^2)^3$

59. $-2r(r^3)^2$

60. $(s^3)^2(s^4)^0$

Write each expression without using parentheses or negative exponents.

61. $\left(\frac{3r}{4r^3}\right)^4$

62. $\left(\frac{12y^{-3}}{3y^2}\right)^{-2}$

63. $\left(\frac{6r^{-2}}{2r^3}\right)^{-2}$

64. $\left(\frac{4x^3}{5x^{-3}}\right)^{-2}$

12.5 Rational Equations and Problem Solving

- Solving rational equations
- Extraneous solutions
- Solving formulas
- Applications

In this section, we will solve problems from banking, petroleum engineering, business, electronics, and travel. We will encounter a new type of equation when we write mathematical models of such situations. Since these equations contain one or more rational expressions, they are called **rational equations.**

Solving rational equations

Recall that to solve an equation such as $\frac{x}{6} + \frac{5}{2} = \frac{1}{3}$, we can multiply both sides of the equation by the LCD of the fractions to clear the equation of fractions.

$$\frac{x}{6} + \frac{5}{2} = \frac{1}{3}$$

$$6\left(\frac{x}{6} + \frac{5}{2}\right) = 6\left(\frac{1}{3}\right)$$ Multiply both sides of the equation by the LCD of $\frac{x}{6}$, $\frac{5}{2}$, and $\frac{1}{3}$, which is 6.

$$6 \cdot \frac{x}{6} + 6 \cdot \frac{5}{2} = 6 \cdot \frac{1}{3}$$ Distribute the multiplication by 6.

$$x + 15 = 2$$ Perform the multiplications.

$$x + 15 - \mathbf{15} = 2 - \mathbf{15}$$ To undo the addition of 15, subtract 15 from both sides.

$$x = -13$$ Perform the subtractions.

This method can be used to solve rational equations.

Self Check 1

Solve: $\frac{6}{x} - 1 = \frac{3}{x}$

EXAMPLE 1 Solve: $\frac{4}{x} + 1 = \frac{6}{x}$.

Solution To clear the equation of fractions, we multiply both sides by the LCD of $\frac{4}{x}$ and $\frac{6}{x}$, which is x.

$$\frac{4}{x} + 1 = \frac{6}{x}$$

$$x\left(\frac{4}{x} + 1\right) = x\left(\frac{6}{x}\right)$$

$$x \cdot \frac{4}{x} + x \cdot 1 = x \cdot \frac{6}{x}$$ Distribute the multiplication by x.

$$4 + x = 6$$ Perform each multiplication.

$$x = 2$$ Subtract 4 from both sides.

Check: $\frac{4}{x} + 1 = \frac{6}{x}$

$$\frac{4}{2} + 1 \stackrel{?}{=} \frac{6}{2}$$ Substitute 2 for x.

$$2 + 1 \stackrel{?}{=} 3$$ Simplify.

$$3 = 3$$

Answer 3

Self Check 2

Solve:

$\frac{7}{6} - \frac{2r - 11}{r} = \frac{1}{r}$

EXAMPLE 2 Solve: $\frac{22}{5} - \frac{3a - 1}{a} = \frac{8}{a}$.

Solution We multiply both sides by $5a$, the LCD of the rational expressions in the equation.

$$\frac{22}{5} - \frac{3a - 1}{a} = \frac{8}{a}$$

$$\mathbf{5a}\left(\frac{22}{5} - \frac{3a - 1}{a}\right) = \mathbf{5a}\left(\frac{8}{a}\right)$$

$$\mathbf{5a}\left(\frac{22}{5}\right) - \mathbf{5a}\left(\frac{3a - 1}{a}\right) = \mathbf{5a}\left(\frac{8}{a}\right)$$ Distribute the multiplication by $5a$.

$$22a - 5(3a - 1) = 40$$ Simplify. Note that $3a - 1$ must be written within parentheses.

$$22a - 15a + 5 = 40$$ Distribute the multiplication by -5.

$$7a + 5 = 40$$ Combine like terms: $22a - 15a = 7a$.

$$7a = 35$$ Subtract 5 from both sides.

$$a = 5$$ Divide both sides by 7.

Check: $\frac{22}{5} - \frac{3a - 1}{a} = \frac{8}{a}$

$$\frac{22}{5} - \frac{3(5) - 1}{5} \stackrel{?}{=} \frac{8}{5}$$ Substitute 5 for a.

$$\frac{22}{5} - \frac{14}{5} \stackrel{?}{=} \frac{8}{5}$$

$$\frac{8}{5} = \frac{8}{5}$$

Answer 12

EXAMPLE 3 Solve: $\frac{x+2}{x+3} + \frac{1}{x^2+2x-3} = 1$.

Solution To find the LCD, we must factor the second denominator.

$$\frac{x+2}{x+3} + \frac{1}{x^2+2x-3} = 1$$

$$\frac{x+2}{x+3} + \frac{1}{(x+3)(x-1)} = 1 \quad \text{Factor } x^2+2x-3.$$

To clear the equation of fractions, we multiply both sides by the LCD, which is $(x+3)(x-1)$.

$$(\mathbf{x+3})(\mathbf{x-1})\left[\frac{x+2}{x+3} + \frac{1}{(x+3)(x-1)}\right] = (\mathbf{x+3})(\mathbf{x-1})1$$

Next, we distribute the multiplication by $(x+3)(x-1)$.

$$(\mathbf{x+3})(\mathbf{x-1})\frac{x+2}{x+3} + (\mathbf{x+3})(\mathbf{x-1})\frac{1}{(x+3)(x-1)} = (\mathbf{x+3})(\mathbf{x-1})1$$

$$\begin{aligned}
(x-1)(x+2)+1 &= (x+3)(x-1) && \text{Simplify.}\\
x^2+x-2+1 &= x^2+2x-3 && \text{Multiply the pairs of binomials.}\\
x^2+x-1 &= x^2+2x-3 && \text{Combine like terms.}\\
x-1 &= 2x-3 && \text{Subtract } x^2 \text{ from both sides.}\\
-x-1 &= -3 && \text{Subtract } 2x \text{ from both sides.}\\
-x &= -2 && \text{Add 1 to both sides.}\\
x &= 2 && \text{Multiply (or divide) both sides by } -1.
\end{aligned}$$

Verify that 2 is a solution of the given equation.

Self Check 3
Solve:
$\frac{1}{x+3} + \frac{1}{x-3} = \frac{10}{x^2-9}$

Answer 5

EXAMPLE 4 Solve: $\frac{4}{5} + y = \frac{4y-50}{5y-25}$.

Solution To find the LCD, we must factor $5y-25$.

$$\frac{4}{5} + y = \frac{4y-50}{5y-25}$$

$$\frac{4}{5} + y = \frac{4y-50}{5(y-5)}$$

$$\begin{aligned}
\mathbf{5(y-5)}\left[\frac{4}{5}+y\right] &= \mathbf{5(y-5)}\left[\frac{4y-50}{5(y-5)}\right] && \text{Multiply both sides by the LCD, which is } 5(y-5).\\
4(y-5)+5y(y-5) &= 4y-50 && \text{Distribute } 5(y-5).\\
4y-20+5y^2-25y &= 4y-50 && \text{Distribute 4 and } 5y.\\
5y^2-25y-20 &= -50 && \text{Subtract } 4y \text{ from both sides and rearrange terms.}\\
5y^2-25y+30 &= 0 && \text{Add 50 to both sides.}\\
y^2-5y+6 &= 0 && \text{Divide both sides by 5.}\\
(y-3)(y-2) &= 0 && \text{Factor } y^2-5y+6.
\end{aligned}$$

$$\begin{aligned}
y-3 = 0 \quad &\text{or} \quad y-2=0 && \text{Set each factor equal to zero.}\\
y = 3 \quad & \quad\quad\ \ y=2 && \text{Solve each equation.}
\end{aligned}$$

Verify that 3 and 2 satisfy the original equation.

Self Check 4
Solve:
$\frac{x-6}{3x-9} - \frac{1}{3} = \frac{x}{2}$

Answer 1, 2

Extraneous solutions

If we multiply both sides of an equation by an expression that involves a variable, as we did in the previous examples, we must check the apparent solutions. The next example shows why.

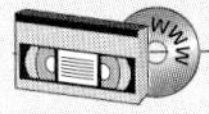

Self Check 5
Solve:
$$\frac{x+5}{x-2} = \frac{7}{x-2}$$

EXAMPLE 5 Solve: $\frac{x+3}{x-1} = \frac{4}{x-1}$.

Solution To clear the equation of fractions, we multiply both sides by the LCD, which is $x - 1$.

$$\frac{x+3}{x-1} = \frac{4}{x-1}$$

$$(\mathbf{x-1})\frac{x+3}{x-1} = (\mathbf{x-1})\frac{4}{x-1} \quad \text{Multiply both sides by } x-1.$$

$$x + 3 = 4 \quad \text{Simplify.}$$

$$x = 1 \quad \text{Subtract 3 from both sides.}$$

Because both sides were multiplied by an expression containing a variable, we must check the apparent solution.

$$\frac{x+3}{x-1} = \frac{4}{x-1}$$

$$\frac{\mathbf{1}+3}{\mathbf{1}-1} \stackrel{?}{=} \frac{4}{\mathbf{1}-1} \quad \text{Substitute 1 for } x.$$

$$\frac{4}{0} \stackrel{?}{=} \frac{4}{0} \quad \text{Simplify.}$$

We have determined that 1 makes both denominators in the original equation 0. Therefore, 1 is not a solution. Since 1 is the only possible solution, and it must be rejected, it follows that $\frac{x+3}{x-1} = \frac{4}{x-1}$ has no solution.

When solving an equation, a possible solution that does not satisfy the original equation is called an **extraneous solution.** In this example, 1 is an extraneous solution.

Answer 2 is extraneous.

Solving formulas

Many formulas are equations that contain rational expressions.

Self Check 6
Solve the formula in Example 6 for r_1.

EXAMPLE 6 The formula $\frac{1}{r} = \frac{1}{r_1} + \frac{1}{r_2}$ is used in electronics to calculate parallel resistances. Solve it for r.

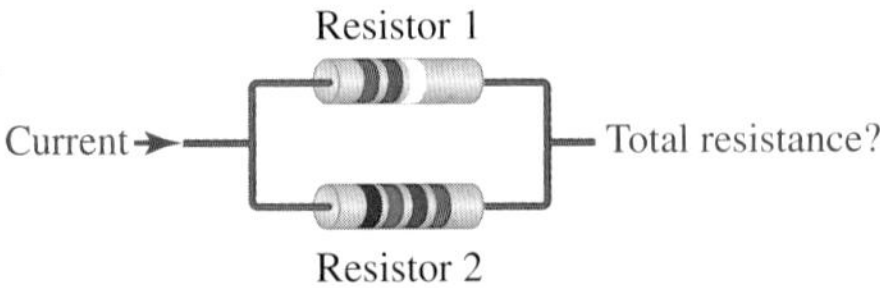

FIGURE 12-1

Solution Clear the equation of fractions by multiplying both sides by the LCD, which is rr_1r_2.

$$\frac{1}{r} = \frac{1}{r_1} + \frac{1}{r_2}$$

$$rr_1r_2\left(\frac{1}{r}\right) = rr_1r_2\left(\frac{1}{r_1} + \frac{1}{r_2}\right) \quad \text{Multiply both sides by } rr_1r_2.$$

$$\frac{rr_1r_2}{r} = \frac{rr_1r_2}{r_1} + \frac{rr_1r_2}{r_2} \quad \text{Distribute the multiplication by } rr_1r_2.$$

$$r_1r_2 = rr_2 + rr_1 \quad \text{Simplify each fraction.}$$

$$r_1r_2 = r(r_2 + r_1) \quad \text{Factor out } r.$$

$$\frac{r_1r_2}{r_2 + r_1} = r \quad \text{To isolate } r\text{, divide both sides by } r_2 + r_1.$$

or

$$r = \frac{r_1r_2}{r_2 + r_1}$$

Answer $r_1 = \dfrac{rr_2}{r_2 - r}$

Applications

EXAMPLE 7 **A number problem.** If the same number is added to both the numerator and the denominator of the fraction $\frac{3}{5}$, the result is $\frac{4}{5}$. Find the number.

Analyze the problem

We are asked to find a number. If we add it to both the numerator and the denominator of $\frac{3}{5}$, we will get $\frac{4}{5}$.

Form an equation

Let n = the unknown number and add n to both the numerator and the denominator of $\frac{3}{5}$. Then set the result equal to $\frac{4}{5}$ to get the equation

$$\frac{3 + n}{5 + n} = \frac{4}{5}$$

Solve the equation

To solve the equation, we proceed as follows:

$$\frac{3 + n}{5 + n} = \frac{4}{5}$$

$$\mathbf{5(5 + n)}\frac{3 + n}{5 + n} = \mathbf{5(5 + n)}\frac{4}{5} \quad \text{Multiply both sides by } 5(5 + n)\text{, which is the LCD of the fractions appearing in the equation.}$$

$$5(3 + n) = (5 + n)4 \quad \text{Simplify.}$$

$$15 + 5n = 20 + 4n \quad \text{Distribute the multiplications by 5 and by 4.}$$

$$15 + n = 20 \quad \text{Subtract } 4n \text{ from both sides.}$$

$$n = 5 \quad \text{Subtract 15 from both sides.}$$

State the conclusion

The number is 5.

Check the result

When we add 5 to both the numerator and denominator of $\frac{3}{5}$, we get

$$\frac{3+5}{5+5} = \frac{8}{10} = \frac{4}{5}$$

The result checks.

We can use rational equations to model shared-work problems. In this case, we assume that the work is being performed at a constant rate by all of those involved.

EXAMPLE 8 Filling an oil tank. An inlet pipe can fill an oil tank in 7 days, and a second inlet pipe can fill the same tank in 9 days. If both pipes are used, how long will it take to fill the tank?

Analyze the problem

The key is to determine what each pipe can do in 1 day. If we add what the first pipe can do in 1 day to what the second pipe can do in 1 day, the sum is what they can do in 1 day, working together.

Since the first pipe can fill the tank in 7 days, it can do $\frac{1}{7}$ of the job in 1 day. Since the second pipe can fill the tank in 9 days, it can do $\frac{1}{9}$ of the job in 1 day. If it takes x days for both pipes to fill the tank, together they can do $\frac{1}{x}$ of the job in 1 day.

Form an equation

Let x = the number of days it will take to fill the tank if both inlet pipes are used. Then form the equation.

What the first inlet pipe can do in 1 day	plus	what the second inlet pipe can do in 1 day	equals	what they can do together in 1 day.
$\frac{1}{7}$	$+$	$\frac{1}{9}$	$=$	$\frac{1}{x}$

Solve the equation

To solve the equation, we proceed as follows:

$$\frac{1}{7} + \frac{1}{9} = \frac{1}{x}$$

$$\mathbf{63x}\left(\frac{1}{7} + \frac{1}{9}\right) = \mathbf{63x}\left(\frac{1}{x}\right)$$ Multiply both sides by $63x$ to clear the equation of fractions.

$$\mathbf{63x}\left(\frac{1}{7}\right) + \mathbf{63x}\left(\frac{1}{9}\right) = \mathbf{63x}\left(\frac{1}{x}\right)$$ Distribute the multiplication by $63x$.

$$9x + 7x = 63$$

$$16x = 63$$ Combine like terms.

$$x = \frac{63}{16}$$ Divide both sides by 16.

State the conclusion

If both inlet pipes are used, it will take $\frac{63}{16}$ or $3\frac{15}{16}$ days to fill the tank.

Check the result

In $\frac{63}{16}$ days, the first pipe fills $\frac{1}{7}(\frac{63}{16})$ of the tank and the second pipe fills $\frac{1}{9}(\frac{63}{16})$ of the tank. The sum of these efforts, $\frac{9}{16} + \frac{7}{16}$, is equal to one full tank.

EXAMPLE 9 Track and field.

A coach can run 10 miles in the same amount of time as his best student-athlete can run 12 miles. If the student can run 1 mile per hour faster than the coach, how fast can the student run?

Analyze the problem

We can use the formula $d = rt$, where d is the distance traveled, r is the rate, and t is the time. If we solve this formula for t, we obtain

$$t = \frac{d}{r}$$

Form an equation

It will take $\frac{10}{r}$ hours for the coach to run 10 miles at some unknown rate of r mph. It will take $\frac{12}{r+1}$ hours for the student to run 12 miles at some unknown rate of $(r + 1)$ mph. We can organize the information of the problem in a table, as shown in Figure 12-2.

	r	$\cdot\ t$	$= d$
Student	$r+1$	$\frac{12}{r+1}$	12
Coach	r	$\frac{10}{r}$	10

FIGURE 12-2

The time it takes the student to run 12 miles	equals	the time it takes the coach to run 10 miles.
$\frac{12}{r+1}$	$=$	$\frac{10}{r}$

Solve the equation

We can solve the equation as follows:

$$\frac{12}{r+1} = \frac{10}{r}$$

$$\mathbf{r(r+1)}\frac{12}{r+1} = \mathbf{r(r+1)}\frac{10}{r} \quad \text{Multiply both sides by } r(r+1).$$

$$12r = 10(r+1) \quad \text{Simplify.}$$

$$12r = 10r + 10 \quad \text{Distribute the multiplication by 10.}$$

$$2r = 10 \quad \text{Subtract } 10r \text{ from both sides.}$$

$$r = 5 \quad \text{Divide both sides by 2.}$$

State the conclusion

The coach can run 5 mph. The student, running 1 mph faster, can run 6 mph.

Check the result

Verify that these results check.

EXAMPLE 10 Banking. At one bank, a sum of money invested for 1 year will earn \$96 interest. If invested in bonds, that money would earn \$108, because the interest rate paid by the bonds is 1% greater than that paid by the bank. Find the bank's rate.

Analyze the problem

This interest problem is based on the formula $I = Prt$, where I is the interest, P is the principal (the amount invested), r is the annual rate of interest and t is the time in years. If we solve this formula for P, we obtain

$$P = \frac{I}{rt}$$

Form an equation

If we let r = the bank's rate of interest, then $r + 0.01$ = the rate paid by the bonds. If a person earns \$96 interest at a bank at some unknown rate r, the principal invested was $\frac{96}{r(1)}$. If a person earns \$108 interest in bonds at some unknown rate $(r + 0.01)$, the principal invested was $\frac{108}{(r + 0.01)(1)}$. We can organize the information of the problem in a table, as shown in Figure 12-3.

	Principal ·	Rate ·	Time	= Interest
Bank	$\frac{96}{r(1)}$	r	1	96
Bonds	$\frac{108}{(r + 0.01)1}$	$r + 0.01$	1	108

FIGURE 12-3

Because the same principal would be invested in either account, we can set up the following equation:

$$\frac{96}{r(1)} = \frac{108}{(r + 0.01)1} \quad \text{or} \quad \frac{96}{r} = \frac{108}{r + 0.01}$$

Solve the equation

We can solve the equation as follows:

$$\frac{96}{r} = \frac{108}{r + 0.01}$$

$$r(r + 0.01) \cdot \frac{96}{r} = r(r + 0.01) \cdot \frac{108}{r + 0.01} \quad \text{Multiply both sides by } r(r + 0.01).$$

$$96(r + 0.01) = 108r$$

$$96r + 0.96 = 108r \quad \text{Distribute.}$$

$$0.96 = 12r \quad \text{Subtract } 96r \text{ from both sides.}$$

$$0.08 = r \quad \text{Divide both sides by 12.}$$

State the conclusion

The bank's interest rate is 0.08, or 8%. The bonds pay 9% interest, a rate 1% greater than that paid by the bank.

Check the result

Verify that these rates check.

Section 12.5 STUDY SET

VOCABULARY *Fill in the blanks.*

1. Equations that contain one or more rational expressions, such as

$$\frac{x+2}{x+3} + \frac{1}{x^2+2x-3} = 1$$

are called ________ __________.

2. To clear an equation of fractions, we multiply both sides by the ______ of the fractions in the equation.

3. If you multiply both sides of an equation by an expression that involves a variable, you must _______ the solution.

4. False solutions that result from multiplying both sides of an equation by a variable are called __________ solutions.

5. In the formula $I = Pr$, I stands for the amount of ________ earned in one year, P stands for the ________, and r stands for the annual interest ______.

6. In the formula $d = rt$, d stands for the ________ traveled, r is the _____, and t is the _____.

CONCEPTS

7. Is 5 a solution of the following equations?

   a. $\dfrac{1}{x-1} = 1 - \dfrac{3}{x-1}$

   b. $\dfrac{x}{x-5} = 3 + \dfrac{5}{x-5}$

8. By what should we multiply both sides of each equation to clear it of fractions?

   a. $\dfrac{1}{x} + 2 = \dfrac{5}{x}$ b. $\dfrac{x}{x-2} - \dfrac{x}{x-1} = 5$

9. The following table shows the length of time it takes each of two hardware store employees to assemble a metal storage shed, working alone.

   a. Complete the table.

	Time to assemble the shed (hr)	Amount of the shed assembled in 1 hr
Marvin	6	
Kyla	5	

   b. If we assume that working together would not change their individual rates, how much of the shed could they assemble in 1 hour if they worked together?

10. When two ice machines are both running, they can fill a supermarket's order in x hours. At this rate, how much of the order do they fill in 1 hour?

11. If the exits at the front of a theater are opened, a full theater can be emptied of all occupants in 6 minutes. How much of the theater is emptied in 1 minute?

12. Solve: $d = rt$.

    a. for r b. for t

13. Solve: $I = Prt$

    a. for r b. for P

14. a. Complete the following table.

	r ·	t	$= d$
Snowmobile	r		4
4 × 4 truck	$r - 5$		3

   b. Complete the following table.

	P ·	r ·	t	$= I$
City Savings		r	1	50
Credit Union		$r - 0.02$	1	75

NOTATION *Complete each solution to solve the equation.*

15.
$$\frac{2}{a} + \frac{1}{2} = \frac{7}{2a}$$
$$\_\_\_\left(\frac{2}{a} + \frac{1}{2}\right) = \_\_\_\left(\frac{7}{2a}\right)$$
$$\_\_\_ \cdot \frac{2}{a} + \_\_\_ \cdot \frac{1}{2} = \_\_\_ \cdot \frac{7}{2a}$$
$$\_\_\_ + a = 7$$
$$4 + a - \_\_\_ = 7 - \_\_\_$$
$$a = 3$$

16.
$$\frac{3}{5} + \frac{7}{a+2} = 2$$
$$\_\_\_\left(\frac{3}{5} + \frac{7}{a+2}\right) = \_\_\_ \cdot 2$$
$$\_\_\_ \cdot \frac{3}{5} + \_\_\_ \cdot \frac{7}{a+2} = \_\_\_ \cdot 2$$
$$3(a+2) + \_\_\_ = 10(a+2)$$
$$3a + \_\_\_ + 35 = 10a + \_\_\_$$
$$3a + \_\_\_ = 10a + 20$$
$$-7a = \_\_\_$$
$$a = 3$$

17. The following work shows both sides of an equation being multiplied by the LCD to clear it of fractions. What was the original equation?
$$5(x-1)\left(\frac{3}{5}\right) + 5(x-1)\left(\frac{7}{x-1}\right) = 5(x-1) \cdot 2$$

18. After solving a rational equation, a student checked her answer and obtained the following:
$$\frac{-1}{0} + \frac{1}{0} = 0$$
What conclusion can be drawn?

PRACTICE *Solve each equation and check the result. If an equation has no solution, so indicate.*

19. $\frac{x}{2} + 4 = \frac{3x}{2}$

20. $\frac{2y}{5} - 8 = \frac{4y}{5}$

21. $\frac{x+1}{3} + \frac{x-1}{5} = \frac{2}{15}$

22. $\frac{3x-1}{6} - \frac{x+3}{2} = \frac{3x+4}{3}$

23. $\frac{3}{x} + 2 = 3$

24. $\frac{2}{x} + 9 = 11$

25. $\frac{5}{a} - \frac{4}{a} = 8 + \frac{1}{a}$

26. $\frac{11}{b} + \frac{13}{b} = 12$

27. $\frac{3}{4h} + \frac{2}{h} = 1$

28. $\frac{5}{3k} + \frac{1}{k} = -2$

29. $\frac{a}{4} - \frac{4}{a} = 0$

30. $0 = \frac{t}{3} - \frac{12}{t}$

31. $\frac{2}{y+1} + 5 = \frac{12}{y+1}$

32. $\frac{3}{p+6} - 2 = \frac{7}{p+6}$

33. $\frac{x}{x-5} - \frac{5}{x-5} = 3$

34. $\frac{3}{y-2} + 1 = \frac{3}{y-2}$

35. $\frac{3r}{2} - \frac{3}{r} = \frac{3r}{2} + 3$

36. $\frac{2p}{3} - \frac{1}{p} = \frac{2p-1}{3}$

37. $\frac{1}{3} + \frac{2}{x-3} = 1$

38. $\frac{3}{5} + \frac{7}{x+2} = 2$

39. $\frac{z-4}{z-3} = \frac{z+2}{z+1}$

40. $\frac{a+2}{a+8} = \frac{a-3}{a-2}$

41. $\frac{v}{v+2} + \frac{1}{v-1} = 1$

42. $\frac{x}{x-2} = 1 + \frac{1}{x-3}$

43. $\frac{a^2}{a+2} - \frac{4}{a+2} = a$

44. $\frac{z^2}{z+1} + 2 = \frac{1}{z+1}$

45. $\frac{7}{q^2-q-2} + \frac{1}{q+1} = \frac{3}{q-2}$

46. $\frac{3}{x-1} - \frac{1}{x+9} = \frac{18}{x^2+8x-9}$

47. $\frac{u}{u-1} + \frac{1}{u} = \frac{u^2+1}{u^2-u}$

48. $\frac{3}{x-2} + \frac{1}{x} = \frac{2(3x+2)}{x^2-2x}$

49. $\frac{n}{n^2-9} + \frac{n+8}{n+3} = \frac{n-8}{n-3}$

50. $\frac{7}{x-5} - \frac{3}{x+5} = \frac{40}{x^2-25}$

51. $\frac{5}{x+4} + \frac{1}{x+4} = x - 1$

52. $\frac{7}{x-3} + \frac{1}{x-3} = x - 5$

53. $\frac{3}{x+1} - \frac{x-2}{2} = \frac{x-2}{x+1}$

54. $\frac{2}{x-1} + \frac{x-2}{3} = \frac{4}{x-1}$

55. $\frac{b+2}{b+3} + 1 = \frac{-7}{b-5}$

56. $\frac{x-4}{x-3} + \frac{x-2}{x-3} = x - 3$

57. $\frac{x}{x-1} - \frac{12}{x^2-x} = \frac{-1}{x-1}$

58. $y + \frac{2}{3} = \frac{2y-12}{3y-9}$

59. $1 - \frac{3}{b} = \frac{-8b}{b^2+3b}$

60. $\frac{5}{4y+12} - \frac{3}{4} = \frac{5}{4y+12} - \frac{y}{4}$

Solve each formula for the indicated variable.

61. $\frac{1}{a} + \frac{1}{b} = 1$ for a

62. $\frac{1}{a} - \frac{1}{b} = 1$ for b

63. $I = \frac{E}{R+r}$ for r

64. $h = \frac{2A}{b+d}$ for A

65. $\frac{a}{b} = \frac{c}{d}$ for d

66. $F = \frac{L^2}{6d} + \frac{d}{2}$ for L^2

Use the given information to find the number or numbers.

67. If the denominator of $\frac{3}{4}$ is increased by a number and the numerator of the fraction is doubled, the result is 1. What is the number?

68. If a number is added to the numerator of $\frac{7}{8}$ and the same number is subtracted from the denominator, the result is 2. What is the number?

69. If a number is added to the numerator of $\frac{3}{4}$ and twice as much is added to the denominator, the result is $\frac{4}{7}$. What is the number?

70. If a number is added to the numerator of $\frac{5}{7}$ and twice as much is subtracted from the denominator, the result is 8. What is the number?

71. The sum of a number and its reciprocal is $\frac{13}{6}$. What is the number and what is its reciprocal?

72. The sum of the reciprocals of two consecutive even integers is $\frac{7}{24}$. What are the integers? (*Hint:* Let x = the first integer and $x + 2$ = the second integer.)

APPLICATIONS

73. OPTICS The focal length f of a lens is given by the formula

$$\frac{1}{f} = \frac{1}{d_1} + \frac{1}{d_2}$$

where d_1 is the distance from the object to the lens and d_2 is the distance from the lens to the image.

Solve the formula for f.

74. OPTICS Solve the formula in Exercise 73 for d_1.

75. MEDICINE Radioactive tracers are used for diagnostic work in nuclear medicine. The **effective half-life** H of a radioactive material in an organism is given by the formula

$$H = \frac{RB}{R+B}$$

where R is the radioactive half-life and B is the biological half-life of the tracer. Solve for R.

76. CHEMISTRY Charles's law describes the relationship between the volume and the temperature of a gas that is kept at a constant pressure. It states that as the temperature of the gas increases, the volume of the gas will increase:

$$\frac{V_1}{V_2} = \frac{T_1}{T_2}$$

Solve for V_2.

77. FILLING POOLS An inlet pipe can fill an empty swimming pool in 5 hours, and another inlet pipe can fill the pool in 4 hours. How long will it take both pipes to fill the pool?

78. FILLING POOLS One inlet pipe can fill an empty pool in 4 hours, and a drain can empty the pool in 8 hours. How long will it take the pipe to fill the pool if the drain is left open?

79. ROOFING HOUSES A homeowner estimates that it will take her 7 days to roof her house. A professional roofer estimates that he could roof the house in 4 days. How long will it take if the homeowner helps the roofer?

80. SEWAGE TREATMENT A sludge pool is filled by two inlet pipes. One pipe can fill the pool in 15 days, and the other can fill it in 21 days. However, if no sewage is added, continuous waste removal will empty the pool in 36 days. How long will it take the two inlet pipes to fill an empty sludge pool?

81. TOURING A woman can bicycle 28 miles in the same time as it takes her to walk 8 miles. If she can ride 10 mph faster than she can walk, how much time should she allow to walk a 30-mile trail? (*Hint:* How fast can she walk?)

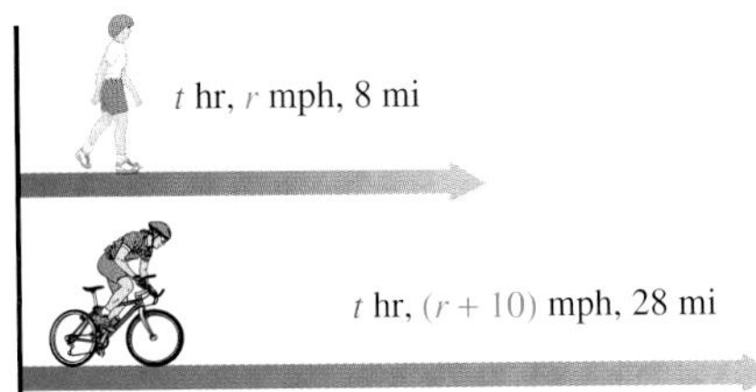

82. COMPARING TRAVEL A plane can fly 300 miles in the same time as it takes a car to go 120 miles. If the car travels 90 mph slower than the plane, find the speed of the plane.

83. BOATING A boat that travels 18 mph in still water can travel 22 miles downstream in the same time as it takes to travel 14 miles upstream. Find the speed of the current in the river.

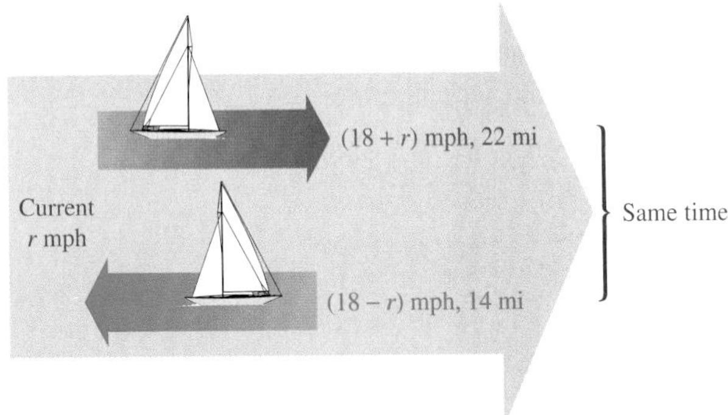

84. WIND SPEED A plane can fly 300 miles downwind in the same time as it can travel 210 miles upwind. Find the velocity of the wind if the plane can fly 255 mph in still air.

85. COMPARING INVESTMENTS Two certificates of deposit (CDs) pay interest at rates that differ by 1%. Money invested for one year in the first CD earns \$175 interest. The same principal invested in the second CD earns \$200. Find the two rates of interest.

86. COMPARING INTEREST RATES Two bond funds pay interest at rates that differ by 2%. Money invested for one year in the first fund earns \$315 interest. The same amount invested in the second fund earns \$385. Find the lower rate of interest.

87. SHARING COSTS Several office workers bought a \$35 gift for their boss. If there had been two more employees to contribute, everyone's cost would have been \$2 less. How many workers contributed to the gift?

88. SALES A dealer bought some radios for a total of \$1,200. She gave away 6 radios as gifts, sold the rest for \$10 more than she paid for each radio, and broke even. How many radios did she buy?

89. SALES A bookstore can purchase several calculators for a total cost of \$120. If each calculator cost \$1 less, the bookstore could purchase 10 additional calculators at the same total cost. How many calculators can be purchased at the regular price?

90. FURNACE REPAIR A repairman purchased several furnace-blower motors for a total cost of \$210. If his cost per motor had been \$5 less, he could have purchased one additional motor. How many motors did he buy at the regular rate?

91. RIVER TOURS A river boat tour begins by going 60 miles upstream against a 5-mph current. There, the boat turns around and returns with the current. What still-water speed should the captain use to complete the tour in 5 hours?

92. TRAVEL TIMES A company president flew 680 miles one way in the corporate jet but returned in a smaller plane that could fly only half as fast. If the total travel time was 6 hours, find the speeds of the planes.

WRITING

93. Explain how you would decide what to do first to solve an equation that involves fractions.

94. Why is it important to check your solutions of an equation that contains fractions with variables in the denominator?

95. In Example 8, one inlet pipe could fill an oil tank in 7 days, and another could fill the same tank in 9 days. We were asked to find how long it would take if both pipes were used. Explain why each of the following approaches is incorrect.

The time it would take to fill the tank

- is the *sum* of the lengths of time it takes each pipe to fill the tank: 7 days + 9 days = 16 days.
- is the *difference* in the lengths of time it takes each pipe to fill the tank: 9 days − 7 days = 2 days.
- is the *average* of the lengths of time it takes each pipe to fill the tank:
$$\frac{7 \text{ days} + 9 \text{ days}}{2} = \frac{16 \text{ days}}{2} = 8 \text{ days}$$

96. Explain the difference between the procedure used to simplify

$$\frac{1}{x} + \frac{1}{3}$$

and the procedure used to solve

$$\frac{1}{x} + \frac{1}{3} = \frac{1}{2}$$

REVIEW *Factor each expression.*

97. $x^2 + 4x$

98. $x^2 - 16y^2$

99. $2x^2 + x - 3$

100. $6a^2 - 5a - 6$

101. $x^4 - 16$

102. $4x^2 + 10x - 6$

12.6 Proportions and Similar Triangles

- Ratios and rates • Proportions • Solving proportions
- Problem solving • Similar triangles

In this section, we will discuss in further detail a problem-solving tool called a *proportion.* A proportion is a type of rational equation that involves two *ratios* or two *rates.*

Ratios and rates

Ratios enable us to compare numerical quantities.

- To prepare fuel for a Lawnboy lawnmower, gasoline must be mixed with oil in the ratio of 50 to 1.
- To make 14-karat jewelry, gold is mixed with other metals in the ratio of 14 to 10.
- In the stock market, winning stocks might outnumber losing stocks in the ratio of 7 to 4.

Ratios

A **ratio** is the quotient of two numbers or the quotient of two quantities that have the same units.

There are three common ways to write a ratio: as a fraction, with the word *to,* or with a colon. For example, the ratio comparing the number of winning stocks to the number of losing stocks mentioned earlier can be written as

$$\frac{7}{4}, \quad 7 \text{ to } 4, \quad \text{or} \quad 7:4$$

Each of these forms can be read as "the ratio of 7 to 4."

When ratios are used to compare quantities with different units, they are called *rates.* For example, if the 495-mile drive from New Orleans to Dallas takes 9 hours, the average rate of speed is the ratio of the miles driven to the length of time the trip takes.

$$\text{Average rate of speed} = \frac{495 \text{ miles}}{9 \text{ hours}} = \frac{55 \text{ miles}}{1 \text{ hour}} \quad \frac{495}{9} = \frac{\cancel{9} \cdot 55}{\cancel{9} \cdot 1} = \frac{55}{1}.$$

Rates

A **rate** is a quotient of two quantities that have different units.

THINK IT THROUGH

Student Loan Calculations

"A consistent majority of students who borrow to pay for their higher education believe they could not have gone to college without student loans. Over 70% agree that student loans were very or extremely important in allowing them access to education after high school." National Student Loan Survey, 2002

Many student loan programs calculate a *debt-to-income ratio* to assist them in determining whether the borrower has sufficient income to repay the loan. A debt-to-income ratio compares an applicant's monthly debt payments (mortgages, credit cards, auto loans, etc.) to their gross monthly income. Most education lenders require borrower debt-to-income ratios of $\frac{2}{5}$ or less, according to the Nellie Mae Debt Management Edvisor. Calculate the debt-to-income ratio for each loan applicant shown below. Then determine whether it makes them eligible for a student loan.

	Applicant #1	Applicant #2	Applicant #3
Monthly debt payments	$250	$1,000	$1,200
Gross monthly income	$1,000	$2,000	$3,000
Debt-to-income ratio			
Is the ratio $\leq \frac{2}{5}$?			

Proportions

Consider the following table, in which we are given the costs of various numbers of gallons of gasoline.

Number of gallons	Cost
2	$3.72
5	$9.30
8	$14.88
12	$22.32
20	$37.20

If we compare the costs to the numbers of gallons purchased, we see that they are equal. In this example, each quotient represents the cost of 1 gallon of gasoline, which is $1.86.

$$\frac{\$3.72}{2} = \$1.86, \quad \frac{\$9.30}{5} = \$1.86, \quad \frac{\$14.88}{8} = \$1.86,$$

$$\frac{\$22.32}{12} = \$1.86, \quad \text{and} \quad \frac{\$37.20}{20} = \$1.86$$

When two ratios or rates (such as $\frac{\$3.72}{2}$ and $\frac{\$9.30}{5}$) are equal, they form a *proportion.*

Proportions

A **proportion** is a mathematical statement that two ratios or two rates are equal.

Some examples of proportions are

$$\frac{1}{2} = \frac{3}{6}, \quad \frac{3 \text{ waiters}}{7 \text{ tables}} = \frac{9 \text{ waiters}}{21 \text{ tables}}, \quad \text{and} \quad \frac{a}{b} = \frac{c}{d}$$

- The proportion $\frac{1}{2} = \frac{3}{6}$ can be read as "1 is to 2 as 3 is to 6."
- The proportion $\frac{3 \text{ waiters}}{7 \text{ tables}} = \frac{9 \text{ waiters}}{21 \text{ tables}}$ can be read as "3 waiters is to 7 tables as 9 waiters is to 21 tables."
- The proportion $\frac{a}{b} = \frac{c}{d}$ can be read as "*a* is to *b* as *c* is to *d*."

In the proportion $\frac{a}{b} = \frac{c}{d}$, *a* and *d* are called the **extremes,** and *b* and *c* are called the **means.** We can show that the product of the extremes (ad) is equal to the product of the means (bc) by multiplying both sides of the proportion by bd and observing that $ad = bc$.

$$\frac{a}{b} = \frac{c}{d}$$

$$bd \cdot \frac{a}{b} = bd \cdot \frac{c}{d}$$ To clear the equation of fractions, multiply both sides by the LCD, which is bd.

$$ad = bc$$ Perform each multiplication and simplify.

Since $ad = bc$, the product of the extremes equals the product of the means.

The fundamental property of proportions

In a proportion, the product of the extremes is equal to the product of the means.

To determine whether an equation is a proportion, we can check to see whether the product of the extremes is equal to the product of the means.

EXAMPLE 1 Determine whether each equation is a proportion: **a.** $\frac{3}{7} = \frac{9}{21}$ and **b.** $\frac{8}{3} = \frac{13}{5}$.

Solution In each case, we check to see whether the product of the extremes is equal to the product of the means.

a. The product of the extremes is $3 \cdot 21 = 63$. The product of the means is $7 \cdot 9 = 63$. Since the products are equal, the equation is a proportion: $\frac{3}{7} = \frac{9}{21}$.

$$3 \cdot 21 = 63 \qquad 7 \cdot 9 = 63$$

$$\frac{3}{7} = \frac{9}{21}$$ The product of the extremes and the product of the means are also known as **cross products.**

Self Check 1

Determine whether the equation is a proportion:

$$\frac{6}{13} = \frac{24}{53}$$

b. The product of the extremes is $8 \cdot 5 = 40$. The product of the means is $3 \cdot 13 = 39$. Since the cross products are not equal, the equation is not a proportion: $\frac{8}{3} \neq \frac{13}{5}$.

$8 \cdot 5 = 40 \qquad 3 \cdot 13 = 39$

$$\frac{8}{3} = \frac{13}{5}$$

Answer no

Solving proportions

Suppose that we know three terms in the proportion

$$\frac{x}{5} = \frac{24}{20}$$

To find the unknown term, we can multiply both sides of the equation by 20 to clear it of fractions, and then solve for x. However, with proportions, it is often easier to simply compute the cross products, set them equal, and solve for the variable.

$$\frac{x}{5} = \frac{24}{20}$$

$20 \cdot x = 5 \cdot 24$ In a proportion, the product of the extremes equals the product of the means.

$20x = 120$ Perform the multiplication: $5 \cdot 24 = 120$.

$\frac{20x}{20} = \frac{120}{20}$ To undo the multiplication by 20, divide both sides by 20.

$x = 6$ Perform the divisions.

Thus, x is 6. To check this result, we substitute 6 for x in $\frac{x}{5} = \frac{24}{20}$ and find the cross products.

$\frac{6}{5} \stackrel{?}{=} \frac{24}{20}$ $\qquad 6 \cdot 20 = 120$, $\quad 5 \cdot 24 = 120$

Since the cross products are equal, this is a proportion. The result, 6, is correct.

Self Check 2
Solve:
$\frac{15}{x} = \frac{25}{40}$

EXAMPLE 2 Solve: $\frac{12}{18} = \frac{3}{x}$.

Solution

$$\frac{12}{18} = \frac{3}{x}$$

$12 \cdot x = 18 \cdot 3$ In a proportion, the product of the extremes equals the product of the means.

$12x = 54$ Multiply: $18 \cdot 3 = 54$.

$\frac{12x}{12} = \frac{54}{12}$ To undo the multiplication by 12, divide both sides by 12.

$x = \frac{9}{2}$ Simplify: $\frac{54}{12} = \frac{9 \cdot \overset{1}{\cancel{6}}}{\underset{1}{\cancel{6}} \cdot 2} = \frac{9}{2}$.

Thus, x is $\frac{9}{2}$. Check the result.

Answer 24

! COMMENT Remember that a cross product is the product of the means or extremes of a *proportion*. For example, it would be incorrect to try to compute cross products to solve the rational equation $\frac{12}{18} = \frac{3}{x} + \frac{1}{2}$. The right-hand side is not a ratio, so the equation is not a proportion.

Solving proportions with a calculator

CALCULATOR SNAPSHOT

To solve the proportion $\frac{3.5}{7.2} = \frac{x}{15.84}$ with a calculator, we can proceed as follows.

$$\frac{3.5}{7.2} = \frac{x}{15.84}$$

$$\frac{3.5(15.84)}{7.2} = x$$ To undo the division by 15.84 and isolate x, multiply both sides of the equation by 15.84.

We can find x by entering these numbers into a scientific calculator.

3.5 [×] 15.84 [÷] 7.2 [=]   7.7

Using a graphing calculator, we enter these numbers and press these keys.

3.5 [×] 15.84 [÷] 7.2 [ENTER]

```
3.5*15.84/7.2
          7.7
```

Thus, x is 7.7.

EXAMPLE 3 Solve: $\frac{2a + 1}{4} = \frac{10}{8}$.

Solution

$$\frac{2a + 1}{4} = \frac{10}{8}$$

$8(2a + 1) = 40$ — In a proportion, the product of the extremes equals the product of the means.

$16a + 8 = 40$ — Distribute the multiplication by 8.

$16a + 8 - \mathbf{8} = 40 - \mathbf{8}$ — To undo the addition of 8, subtract 8 from both sides.

$16a = 32$ — Combine like terms.

$\frac{16a}{\mathbf{16}} = \frac{32}{\mathbf{16}}$ — To undo the multiplication by 16, divide both sides by 16.

$a = 2$ — Perform the divisions.

Thus, a is 2. Check the result.

Self Check 3
Solve:
$$\frac{3x - 1}{2} = \frac{12.5}{5}$$

Answer 2

Problem solving

We can use proportions to solve many real-world problems. If we are given a ratio (or rate) comparing two quantities, the words of the problem can be translated to a proportion, and we can solve it to find the unknown.

Self Check 4
If 9 tickets to a concert cost \$112.50, how much will 15 tickets cost?

EXAMPLE 4 If 6 apples cost \$1.38, how much will 16 apples cost?

Solution
Analyze the problem

We know the cost of 6 apples; we are to find the cost of 16 apples.

Form a proportion

Let c = the cost of 16 apples. If we compare the number of apples to their cost, we know that the two rates are equal.

6 apples is to \$1.38 as 16 apples is to \$c.

$$\begin{array}{r} \text{Number of apples} \rightarrow \\ \text{Cost of the apples} \rightarrow \end{array} \frac{6}{1.38} = \frac{16}{c} \begin{array}{l} \leftarrow \text{Number of apples} \\ \leftarrow \text{Cost of the apples} \end{array}$$

Solve the proportion

$6 \cdot c = 1.38(16)$ In a proportion, the product of the extremes equals the product of the means.

$6c = 22.08$ Perform the multiplication: $1.38(16) = 22.08$.

$\frac{6c}{6} = \frac{22.08}{6}$ To undo the multiplication by 6, divide both sides by 6.

$c = 3.68$ Divide: $\frac{22.08}{6} = 3.68$.

State the conclusion

Sixteen apples will cost \$3.68.

Check the result

If 16 apples are bought, this is about 3 times as many as 6 apples, which cost \$1.38. If we multiply \$1.38 by 3, we get an estimate of the cost of 16 apples: $\$1.38 \cdot 3 = \$4.14$. The result, \$3.68, seems reasonable.

Answer \$187.50

In Example 4, we could have compared the cost of the apples to the number of apples: \$1.38 is to 6 apples as \$c is to 16 apples. This would have led to the proportion

$$\begin{array}{r} \text{Cost of the apples} \rightarrow \\ \text{Number of apples} \rightarrow \end{array} \frac{1.38}{6} = \frac{c}{16} \begin{array}{l} \leftarrow \text{Cost of the apples} \\ \leftarrow \text{Number of apples} \end{array}$$

If we solve this proportion for c, we will obtain the same result: $c = 3.68$.

! COMMENT When solving problems using proportions, we must make sure that the units of both numerators are the same and the units of both denominators are the same. In Example 4, it would be incorrect to write

$$\begin{array}{r} \text{Cost of the apples} \rightarrow \\ \text{Number of apples} \rightarrow \end{array} \frac{1.38}{6} = \frac{16}{c} \begin{array}{l} \leftarrow \text{Number of apples} \\ \leftarrow \text{Cost of the apples} \end{array}$$

EXAMPLE 5 Scale models.

A **scale** is a ratio (or rate) that compares the size of a model, drawing, or map to the size of an actual object. The scale shown in Figure 12-4 indicates that 1 inch on the model carousel is equivalent to 160 inches on the actual carousel. How wide should the model be if the actual carousel is 35 feet wide?

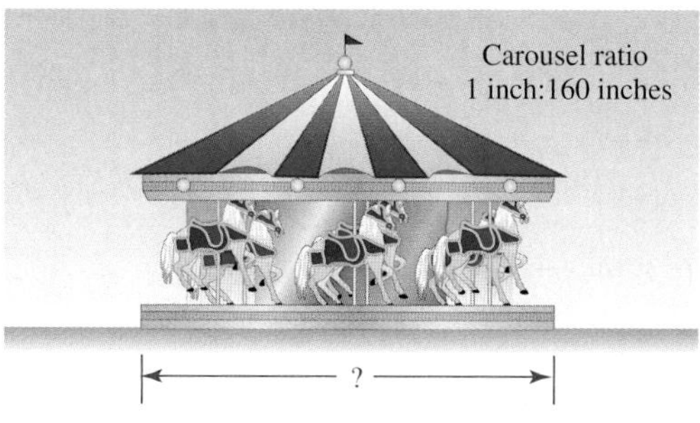

FIGURE 12-4

Solution

Analyze the problem

We are asked to determine the width of the miniature carousel, if a ratio of 1 inch to 160 inches is used. We would like the width of the model to be given in inches, not feet, so we will express the 35-foot width of the actual carousel as $35 \cdot 12 = 420$ inches.

Form a proportion

Let w = the width of the model. The ratios of the dimensions of the model to the corresponding dimensions of the actual carousel are equal.

1 inch is to 160 inches as w inches is to 420 inches.

$$\begin{matrix} \text{model} \rightarrow \\ \text{actual} \rightarrow \end{matrix} \frac{1}{160} = \frac{w}{420} \begin{matrix} \leftarrow \text{model} \\ \leftarrow \text{actual} \end{matrix}$$

Solve the proportion

$420 = 160w$ — In a proportion, the product of the extremes is equal to the product of the means.

$\frac{420}{\mathbf{160}} = \frac{160w}{\mathbf{160}}$ — To undo the multiplication by 160, divide both sides by 160.

$2.625 = w$ — Perform the division: $\frac{420}{160} = 2.625$.

State the conclusion

The width of the miniature carousel should be 2.625 in., or $2\frac{5}{8}$ in.

Check the result

A width of $2\frac{5}{8}$ in. is approximately 3 in. When we write the ratio of the model's approximate width to the width of the actual carousel, we get $\frac{3}{420} = \frac{1}{140}$, which is about $\frac{1}{160}$. The answer seems reasonable.

EXAMPLE 6 **Baking.** A recipe for rhubarb cake calls for $1\frac{1}{4}$ cups of sugar for every $2\frac{1}{2}$ cups of flour. How many cups of flour are needed if the baker intends to use 3 cups of sugar?

Solution

Analyze the problem

The baker needs to maintain the same ratio between the amounts of sugar and flour as is called for in the original recipe.

Self Check 6

How many cups of sugar will be needed to make several cakes that will require a total of 25 cups of flour?

Form a proportion

Let f = the number of cups of flour to be mixed with the 3 cups of sugar. The ratios of the cups of sugar to the cups of flour are equal.

$1\frac{1}{4}$ cups sugar is to $2\frac{1}{2}$ cups flour as 3 cups sugar is to f cups flour.

$$\begin{array}{rcl}\text{Cups sugar} \rightarrow & \dfrac{1\frac{1}{4}}{2\frac{1}{2}} = \dfrac{3}{f} & \leftarrow \text{Cups sugar} \\ \text{Cups flour} \rightarrow & & \leftarrow \text{Cups flour}\end{array}$$

Solve the proportion

$\dfrac{1.25}{2.5} = \dfrac{3}{f}$ Change the fractions to decimals.

$1.25f = 2.5 \cdot 3$ In a proportion, the product of the extremes equals the product of the means.

$1.25f = 7.5$ Perform the multiplication: $2.5 \cdot 3 = 7.5$.

$\dfrac{1.25f}{\mathbf{1.25}} = \dfrac{7.5}{\mathbf{1.25}}$ To undo the multiplication by 1.25, divide both sides by 1.25.

$f = 6$ Divide: $\frac{7.5}{1.25} = 6$.

State the conclusion

The baker should use 6 cups of flour.

Check the result

The recipe calls for about 2 cups of flour for about 1 cup of sugar. If 3 cups of sugar are used, 6 cups of flour seems reasonable.

Answer $12\frac{1}{2}$

Similar triangles

If two angles of one triangle have the same measures as two angles of a second triangle, the triangles have the same shape. Triangles with the same shape are called **similar triangles.** In Figure 12-5, $\triangle ABC \sim \triangle DEF$. (Read the symbol $\sim$ as "is similar to.")

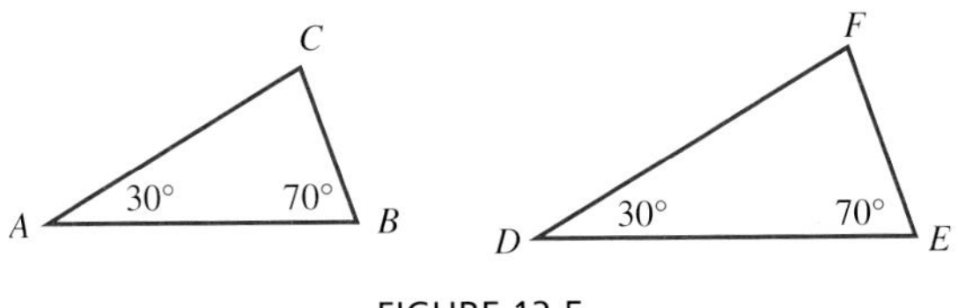

FIGURE 12-5

Property of similar triangles

If two triangles are **similar,** all pairs of corresponding sides are in proportion.

In the similar triangles shown in Figure 12-5, the following proportions are true.

$\dfrac{AB}{DE} = \dfrac{BC}{EF}$, $\dfrac{BC}{EF} = \dfrac{CA}{FD}$, and $\dfrac{CA}{FD} = \dfrac{AB}{DE}$ Read AB as "the length of segment AB."

EXAMPLE 7 A tree casts a shadow 18 feet long at the same time as a woman 5 feet tall casts a shadow 1.5 feet long. Find the height of the tree.

Solution

Analyze the problem

The figure shows the similar triangles determined by the tree and its shadow and the woman and her shadow. Since the triangles are similar, the lengths of their corresponding sides are in proportion. We can use this fact to find the height of the tree.

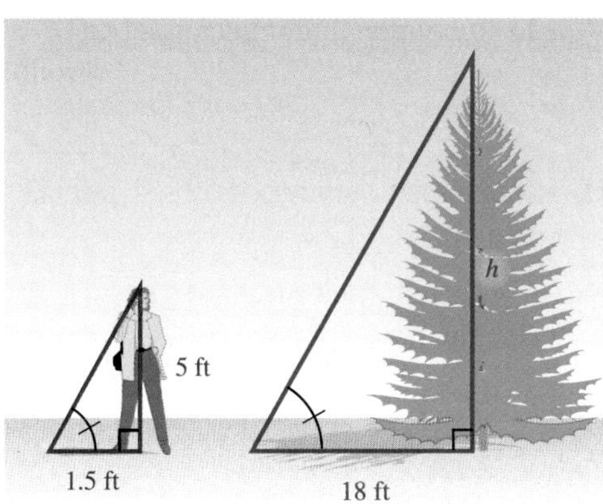

Each triangle has a right angle. Since the sun's rays strike the ground at the same angle, the angles highlighted with a tick mark have the same measure. Therefore, two angles of the smaller triangle have the same measures as two angles of the larger triangle; the triangles are similar.

FIGURE 12-6

Form a proportion

If we let h = the height of the tree, we can find h by solving the following proportion.

$$\frac{h}{5} = \frac{18}{1.5} \qquad \frac{\text{Height of the tree}}{\text{Height of the woman}} = \frac{\text{Length of shadow of the tree}}{\text{Length of shadow of the woman}}$$

Solve the proportion

$1.5h = 5(18)$ In a proportion, the product of the extremes equals the product of the means.

$1.5h = 90$ Perform the multiplication.

$\frac{1.5h}{1.5} = \frac{90}{1.5}$ To undo the multiplication by 1.5, divide both sides by 1.5.

$h = 60$ Divide: $\frac{90}{1.5} = 60$.

State the conclusion

The tree is 60 feet tall.

Check the result

$\frac{18}{1.5} = 12$ and $\frac{60}{5} = 12$. The ratios are the same. The result checks.

Self Check 7

Find the height of the tree in Example 7 if the woman is 5 feet 6 inches tall.

Answer 66 ft

Section 12.6 STUDY SET

VOCABULARY *Fill in the blanks.*

1. A ______ is the quotient of two numbers or the quotient of two quantities with the same units. A ______ is a quotient of two quantities that have different units.

2. A __________ is a mathematical statement that two ratios or two rates are equal.

3. In the proportion $\frac{a}{b} = \frac{c}{d}$, a and d are called the _________ of the proportion. The second and third terms of a proportion are called the ________ of the proportion.

4. The product of the extremes and the product of the means of a proportion are also known as ______ products.

5. If two triangles have the same ______, they are said to be *similar*.

6. If two triangles are ______, their corresponding sides are in proportion.

CONCEPTS *Fill in the blanks.*

7. WEST AFRICA Write the ratio (in fractional form) of the number of red stripes to the number of white stripes on the flag of Liberia.

8. The equation $\frac{a}{b} = \frac{c}{d}$ is a proportion if the cross product ____ is equal to the cross product ____.

9. Is 45 a solution of $\frac{5}{3} = \frac{75}{x}$?

10. Consider $\frac{2}{3} = \frac{x}{15}$.

   a. Solve the proportion by multiplying both sides by the LCD.

   b. Solve the proportion by setting the cross products equal.

11. MINIATURES A high wheeler bicycle is shown below. A model of it is to be made using a scale of 2 inches to 15 inches. The following proportion was set up to determine the height of the front wheel of the model. Explain the error.

$$\frac{2}{15} = \frac{48}{h}$$

12. Two similar triangles are shown in the next column. Fill in the blanks to make the proportions true.

$$\frac{AB}{DE} = \frac{\quad}{EF} \qquad \frac{BC}{\quad} = \frac{CA}{FD} \qquad \frac{CA}{FD} = \frac{AB}{\quad}$$

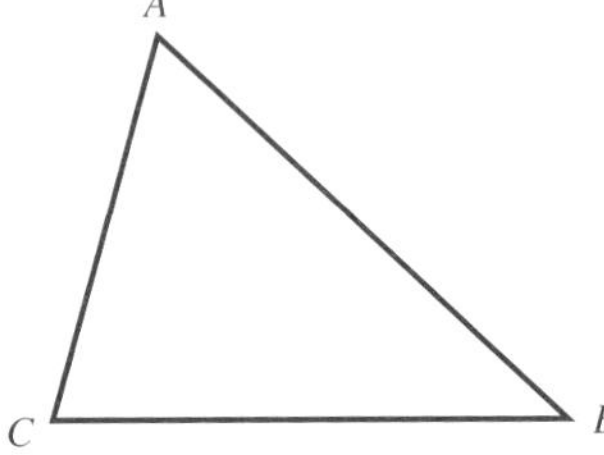

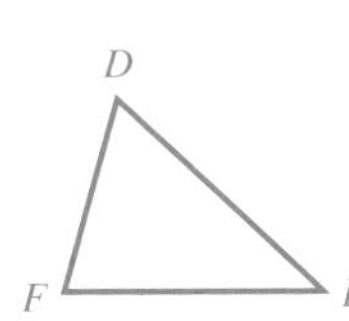

NOTATION *Complete each solution.*

13. Solve for x: $\frac{12}{18} = \frac{x}{24}$.

$$12 \cdot 24 = 18 \cdot \square$$
$$\square = 18x$$
$$\frac{288}{\square} = \frac{18x}{\square}$$
$$16 = x$$

14. Solve for x: $\frac{14}{x} = \frac{49}{17.5}$.

$$14 \cdot \square = 49x$$
$$\square = 49x$$
$$\frac{245}{\square} = \frac{49x}{\square}$$
$$5 = x$$

15. We read "$\triangle ABC$" as "______ ABC."

16. The symbol ~ is read as "______."

PRACTICE *Determine whether each statement is a proportion.*

17. $\frac{9}{7} = \frac{81}{70}$

18. $\frac{5}{2} = \frac{20}{8}$

19. $\frac{7}{3} = \frac{14}{6}$

20. $\frac{13}{19} = \frac{65}{95}$

21. $\frac{9}{19} = \frac{38}{80}$

22. $\frac{40}{29} = \frac{29}{22}$

23. $\frac{10.4}{3.6} = \frac{41.6}{14.4}$

24. $\frac{13.23}{3.45} = \frac{39.96}{11.35}$

Solve each proportion.

25. $\frac{2}{3} = \frac{x}{6}$

26. $\frac{3}{6} = \frac{x}{8}$

27. $\frac{5}{10} = \frac{3}{c}$

28. $\frac{7}{14} = \frac{2}{x}$

29. $\frac{6}{x} = \frac{8}{4}$

30. $\frac{4}{x} = \frac{2}{8}$

31. $\frac{x}{3} = \frac{9}{3}$

32. $\frac{x}{2} = \frac{18}{6}$

33. $\frac{x+1}{5} = \frac{3}{15}$

34. $\frac{x-1}{7} = \frac{2}{21}$

35. $\frac{x+3}{12} = \frac{-7}{6}$

36. $\frac{x+7}{-4} = \frac{1}{4}$

37. $\frac{4-x}{13} = \frac{11}{26}$

38. $\frac{5-x}{17} = \frac{13}{34}$

39. $\frac{2x+1}{18} = \frac{14}{3}$

40. $\frac{2x-1}{18} = \frac{9}{54}$

41. $\frac{y}{4} = \frac{4}{y}$

42. $\frac{2}{3x} = \frac{6x}{36}$

43. $\frac{2}{c} = \frac{c-3}{2}$

44. $\frac{b-5}{3} = \frac{2}{b}$

45. $\frac{2}{x+6} = \frac{-2x}{5}$

46. $\frac{x-1}{x+1} = \frac{2}{3x}$

APPLICATIONS *Set up and solve a proportion. Use a calculator if it is helpful.*

47. GROCERY SHOPPING If 3 pints of yogurt cost $1, how much will 51 pints cost?

48. SHOPPING FOR CLOTHES If shirts are on sale at two for $25, how much will five shirts cost?

49. ADVERTISING In 1997, a 30-second TV ad during the Super Bowl telecast cost $1.2 million. At this rate, what was the cost of a 45-second ad?

50. COOKING A recipe for spaghetti sauce requires four 16-ounce bottles of ketchup to make two gallons of sauce. How many bottles of ketchup are needed to make 10 gallons of sauce?

51. MIXING PERFUME A perfume is to be mixed in the ratio of 3 drops of pure essence to 7 drops of alcohol. How many drops of pure essence should be mixed with 56 drops of alcohol?

52. CPR A first aid handbook states that when performing cardiopulmonary resuscitation on an adult, the ratio of chest compressions to breaths should be 5:2. If 210 compressions were administered to an adult patient, how many breaths should have been given?

53. COOKING A recipe for wild rice soup is shown. Find the amounts of chicken broth, rice, and flour needed to make 15 servings.

Wild Rice Soup

A sumptuous side dish with a nutty flavor

3 cups chicken broth	1 cup light cream
$\frac{2}{3}$ cup uncooked rice	2 tablespoons flour
$\frac{1}{4}$ cup sliced onions	$\frac{1}{8}$ teaspoon pepper
$\frac{1}{2}$ cup shredded carrots	Serves: 6

54. QUALITY CONTROL In a manufacturing process, 95% of the parts made are to be within specifications. How many defective parts would be expected in a run of 940 pieces?

55. QUALITY CONTROL Out of a sample of 500 men's shirts, 17 were rejected because of crooked collars. How many crooked collars would you expect to find in a run of 15,000 shirts?

56. GAS CONSUMPTION If a car can travel 42 miles on 1 gallon of gas, how much gas is needed to travel 315 miles?

57. HIP-HOP According to the *Guinness Book of World Records,* Rebel X.D. of Chicago rapped 674 syllables in 54.9 seconds. At this rate, how many syllables could he rap in 1 minute? Round to the nearest syllable.

58. BANKRUPTCY After filing for bankruptcy, a company was able to pay its creditors only 15 cents on the dollar. If the company owed a lumberyard $9,712, how much could the lumberyard expect to be paid?

59. COMPUTING A PAYCHECK Billie earns $412 for a 40-hour week. If she missed 10 hours of work last week, how much did she get paid?

60. MODEL RAILROADS A model railroad engine is 9 inches long. If the scale is 87 feet to 1 foot, how long is a real engine?

61. MODEL RAILROADS A model railroad caboose is 3.5 inches long. If the scale is 169 feet to 1 foot, how long is a real caboose?

62. NUTRITION The following table shows the nutritional facts about a 10-oz chocolate milkshake sold by a fast-food restaurant. Use the information to complete the table for the 16-oz shake. Round to the nearest unit when an answer is not exact.

	Calories	Fat (gm)	Protein (gm)
10-oz chocolate milkshake	355	8	9
16-oz chocolate milkshake			

63. DRIVER'S LICENSES Of the 50 states, Alabama has one of the largest ratios of licensed drivers to residents. If the ratio is 800:1,000 and Alabama's population is 4,500,000, how many residents of that state have a driver's license?

64. MIXING FUEL The instructions on a can of oil intended to be added to lawnmower gasoline read as follows:

Recommended	Gasoline	Oil
50 to 1	6 gal	16 oz

Are these instructions correct? (*Hint:* There are 128 ounces in 1 gallon.)

65. PHOTO ENLARGEMENT In the illustration, the 3-by-5 photo is to be blown up to the larger size. Find x.

66. BLUEPRINTS The scale for the blueprint shown tells the reader that a $\frac{1}{4}$-inch length ($\frac{1}{4}''$) on the drawing corresponds to an actual size of 1 foot (1′0″). Suppose the length of the kitchen is $2\frac{1}{2}$ inches on the drawing. How long is the actual kitchen?

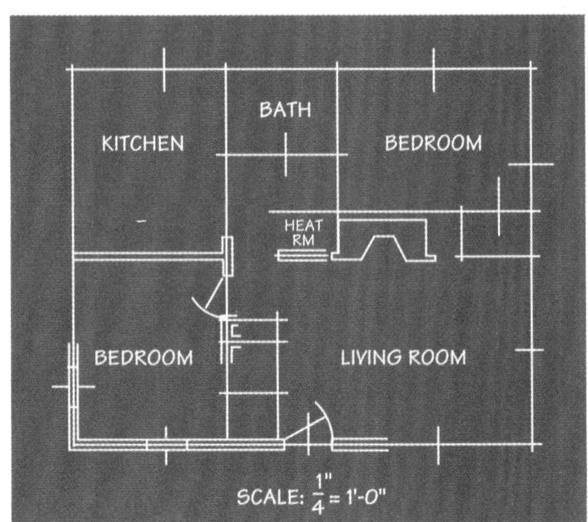

Use similar triangles to solve each problem.

67. HEIGHT OF A TREE A tree casts a shadow of 26 feet at the same time as a 6-foot man casts a shadow of 4 feet (see the illustration in the next column). Find the height of the tree.

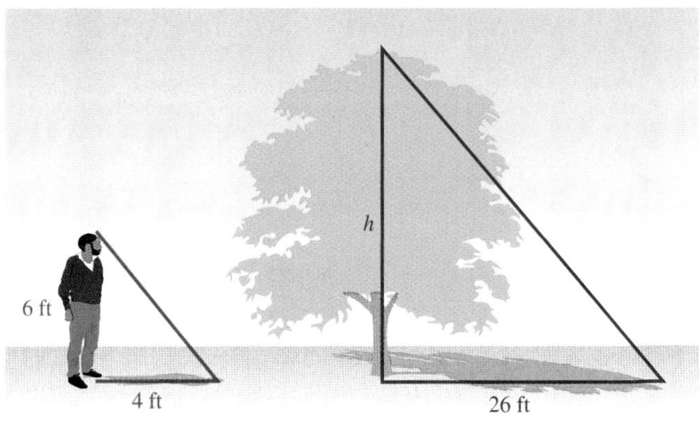

68. HEIGHT OF A BUILDING A man places a mirror on the ground and sees the reflection of the top of a building, as shown. The two triangles in the illustration are similar. Find the height, h, of the building.

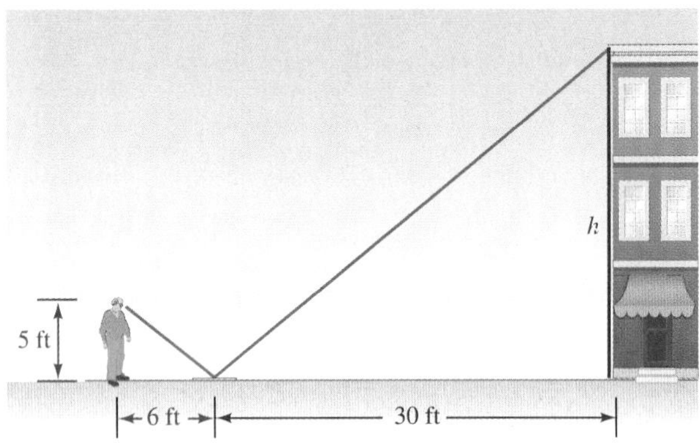

69. WIDTH OF A RIVER Use the dimensions in the illustration to find w, the width of the river. (The two triangles in the illustration are similar.)

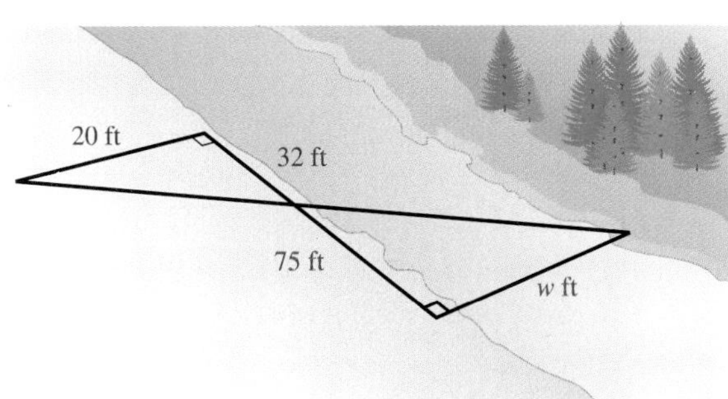

70. FLIGHT PATHS The airplane shown on the next page ascends 100 feet as it flies a horizontal distance of 1,000 feet. How much altitude will it gain as it flies a horizontal distance of 1 mile? (*Hint:* 5,280 feet = 1 mile.)

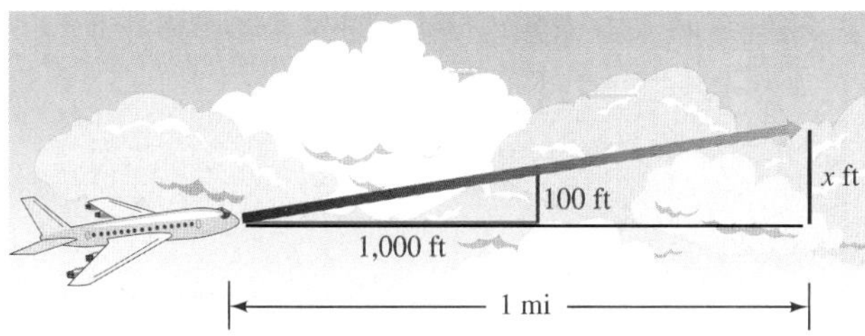

71. FLIGHT PATHS An airplane descends 1,350 feet as it flies a horizontal distance of 1 mile. How much altitude is lost as it flies a horizontal distance of 5 miles?

72. SKI RUNS A ski course falls 100 feet in every 300 feet of horizontal run. If the total horizontal run is $\frac{1}{2}$ mile, find the height of the hill.

WRITING

73. Explain the difference between a ratio and a proportion.

74. Explain how to tell whether $\frac{3.2}{3.7} = \frac{5.44}{6.29}$ is a proportion.

75. Explain why the concept of cross products cannot be used to solve the equation

$$\frac{x}{3} - \frac{3x}{4} = \frac{1}{12}$$

76. Write a problem about a situation you encounter in your daily life that could be solved by using a proportion.

REVIEW

77. Change $\frac{9}{10}$ to a percent.

78. Change $33\frac{1}{3}\%$ to a fraction.

79. Find 30% of 1,600.

80. SHOPPING Maria bought a dress for 25% off the original price of \$98. How much did the dress cost?

81. Find the slope of the line passing through $(-2, -2)$ and $(-12, -8)$.

82. What are the slope and the y-intercept of the graph of $y = 2x - 3$?

12.7 Variation

- Direct variation • Inverse variation

If the value of one quantity depends on the value of another quantity, we can often describe that relationship using the language of variation:

- The sales tax on an item varies with the price.
- The intensity of light varies with the distance from its source.
- The pressure exerted by water on an object varies with the depth of the object beneath the surface.

In this section, we will discuss two types of variation, and we will see how to represent them algebraically using equations.

Direct variation

One type of variation, called **direct variation,** is represented by an equation of the form $y = kx$, where k is a constant (a number). Two variables are said to *vary directly* if one is a constant multiple of the other.

Direct variation

The words *y varies directly with x* mean that

$$y = kx$$

for some constant k, called the **constant of variation.**

Scientists have found that the distance a spring will stretch varies directly with the force applied to it. The more force applied to the spring, the more it will stretch. If d represents the distance stretched and f represents the force applied, this relationship can be expressed by the equation

$$d = kf \quad \text{where } k \text{ is the constant of variation}$$

Suppose that a 150-pound weight stretches a spring 18 inches. (See Figure 12-7.) We can find the constant of variation for the spring by substituting 150 for f and 18 for d in the equation $d = kf$ and solving for k:

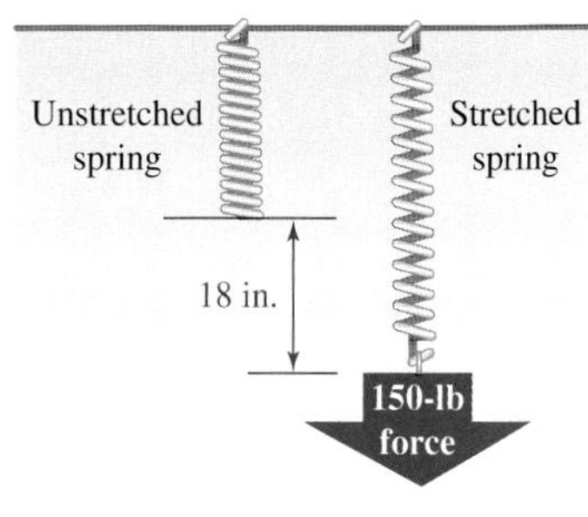

FIGURE 12-7

$$d = kf$$

$$\mathbf{18} = k(\mathbf{150})$$

$$\frac{18}{150} = k \qquad \text{Divide both sides by 150 to isolate } k.$$

$$\frac{3}{25} = k \qquad \text{Simplify the fraction: } \frac{18}{150} = \frac{\overset{1}{\cancel{6}} \cdot 3}{\underset{1}{\cancel{6}} \cdot 25} = \frac{3}{25}.$$

Therefore, the equation describing the relationship between the distance the spring will stretch and the amount of force applied to it is $d = \frac{3}{25}f$. To find the distance that the same spring will stretch when a 50-pound weight is attached, we proceed as follows:

$$d = \frac{3}{25}f \qquad \text{This is the equation describing the direct variation.}$$

$$d = \frac{3}{25}(\mathbf{50}) \qquad \text{Substitute 50 for } f.$$

$$d = 6 \qquad \text{Perform the multiplication.}$$

The spring will stretch 6 inches when a 50-pound weight is attached.

The table in Figure 12-8 shows some other possible values for f and d as determined by the equation $d = \frac{3}{25}f$. When these ordered pairs are graphed and a straight line is drawn through them, it is apparent that as the force f applied to a spring increases, the distance d it stretches increases. Furthermore, the slope of the graph is $\frac{3}{25}$, the constant of variation.

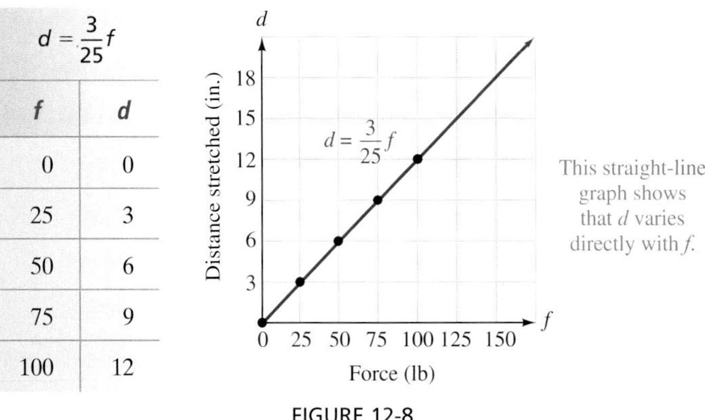

$d = \frac{3}{25}f$

f	d
0	0
25	3
50	6
75	9
100	12

FIGURE 12-8

We can use the following steps to solve variation problems.

Solving variation problems

1. Translate the verbal model into an equation.
2. Substitute the first set of values into the equation from step 1 to determine the value of k.
3. Substitute the value of k into the equation from step 1.
4. Substitute the remaining set of values into the equation from step 3 and solve for the unknown variable.

EXAMPLE 1 The weight of an object on Earth varies directly with its weight on the moon. If a rock weighs 5 pounds on the moon and 30 pounds on Earth, what would be the weight on Earth of a larger rock weighing 26 pounds on the moon?

Solution

Step 1: We let e represent the weight of the object on Earth and m the weight of the object on the moon. Translating the words *weight on Earth varies directly with weight on the moon,* we get the equation

$$e = km$$

Step 2: To find the constant of variation, k, we substitute 30 for e and 5 for m.

$$e = km$$
$$\mathbf{30} = k(\mathbf{5})$$
$$6 = k \quad \text{To undo the multiplication by 5, divide both sides by 5.}$$

Step 3: The equation describing the relationship between the weight of an object on Earth and on the moon is

$$e = 6m$$

Step 4: We can find the weight of the larger rock on Earth by substituting 26 for m in the equation from step 3.

$$e = 6\mathbf{m}$$
$$e = 6(\mathbf{26})$$
$$e = 156$$

The rock would weigh 156 pounds on Earth.

Self Check 1
The cost of a bus ticket varies directly with the number of miles traveled. If a ticket for a 180-mile trip cost $45, what would a ticket for a 1,500-mile trip cost?

Answer $375

Inverse variation

Another type of variation, called **inverse variation,** is represented by an equation of the form $y = \frac{k}{x}$, where k is a constant. Two variables are said to *vary inversely* if one is a constant multiple of the reciprocal of the other.

Inverse variation

The words *y varies inversely with x* mean that

$$y = \frac{k}{x}$$

for some constant k, called the **constant of variation.**

Suppose that the time (in hours) it takes to paint a house varies inversely with the size of the painting crew. As the number of painters increases, the time that it takes to paint the house decreases. If n represents the number of painters and t represents the time it takes to paint the house, this relationship can be expressed by the equation

$$t = \frac{k}{n} \quad \text{where } k \text{ is the constant of variation}$$

If we know that a crew of 8 can paint the house in 12 hours, we can find the constant of variation by substituting 8 for n and 12 for t in the equation $t = \frac{k}{n}$ and solving for k:

$$t = \frac{k}{n}$$

$$12 = \frac{k}{8}$$

$$12 \cdot 8 = k \quad \text{Multiply both sides by 8 to isolate } k.$$

$$96 = k$$

The equation describing the relationship between the size of the painting crew and the time it takes to paint the house is $t = \frac{96}{n}$. We can use this equation to find the time it will take a crew of any size to paint the house. For example, to find the time it would take a 4-person crew, we substitute 4 for n in the equation $t = \frac{96}{n}$.

$$t = \frac{96}{n} \quad \text{This is the equation describing the inverse variation.}$$

$$t = \frac{96}{4} \quad \text{Substitute 4 for } n.$$

$$t = 24$$

It would take a 4-person crew 24 hours to paint the house.

The table in Figure 12-9 shows some possible values for n and t as determined by the equation $t = \frac{96}{n}$. When these ordered pairs are graphed and a smooth curve is drawn through them, it is clear that as the number of painters n increases, the time t decreases.

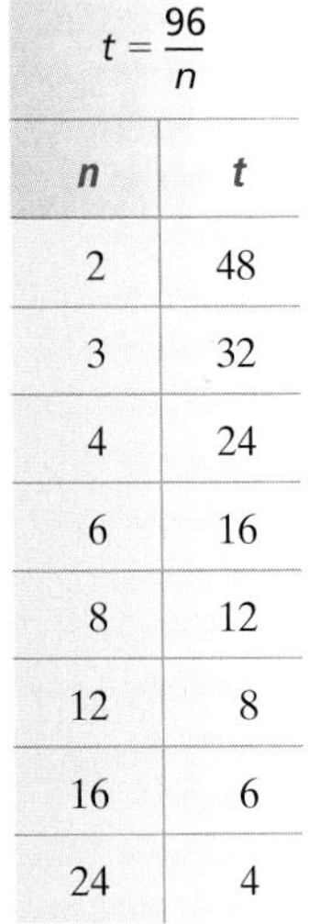

$t = \frac{96}{n}$

n	t
2	48
3	32
4	24
6	16
8	12
12	8
16	6
24	4

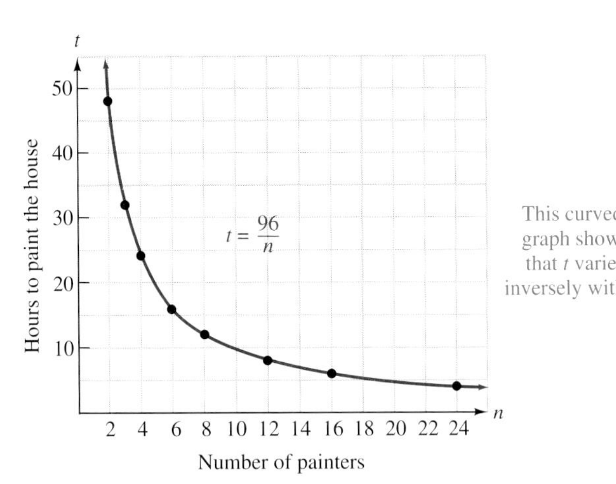

FIGURE 12-9

Self Check 2

How much pressure is needed to compress the gas in Example 2 into a volume of 8 cubic inches?

EXAMPLE 2 **Gas laws.** The volume occupied by a gas varies inversely with the pressure placed on it. That is, the volume decreases as the pressure increases. If a gas occupies a volume of 15 cubic inches when placed under 4 pounds per square inch (psi) of pressure, how much pressure is needed to compress the gas into a volume of 10 cubic inches?

Solution

Step 1: We let V represent the volume occupied by the gas and p represent the pressure. Translating the words *volume occupied by a gas varies inversely with the pressure,* we get the equation

$$V = \frac{k}{p}$$

Step 2: To find the constant of variation, k, we substitute 15 for V and 4 for p.

$$V = \frac{k}{p}$$

$$\mathbf{15} = \frac{k}{\mathbf{4}}$$

$$60 = k \quad \text{Multiply both sides by 4.}$$

Step 3: The equation describing the relationship between the volume occupied by the gas and the pressure placed on it is

$$V = \frac{60}{p}$$

Step 4: We can now find the pressure needed to compress the gas into a volume of 10 cubic inches by substituting 10 for V in the equation and solving for p.

$$V = \frac{60}{p}$$

$$\mathbf{10} = \frac{60}{p}$$

$$10p = 60 \quad \text{To clear the equation of the fraction, multiply both sides by } p.$$

$$p = 6 \quad \text{To undo the multiplication by 10, divide both sides by 10.}$$

It will take 6 psi of pressure to compress the gas into a volume of 10 cubic inches.

Answer 7.5 psi

Study Time vs. Effectiveness

THINK IT THROUGH

"Above all, review regularly and plan to study ahead, so that the night before an exam, all you do is review material. Avoid all-nighters!" *Improve Your Studying Skills,* Counseling Service at The University of North Carolina, Chapel Hill

Each graph below shows a direct or inverse relationship between two components of the educational process as found by researchers in *An Analysis of the Study Time.* (Orlando J. Olivares, Department of Psychology, Bridgewater State College, 2002) For each graph, explain why you agree or disagree with the findings.

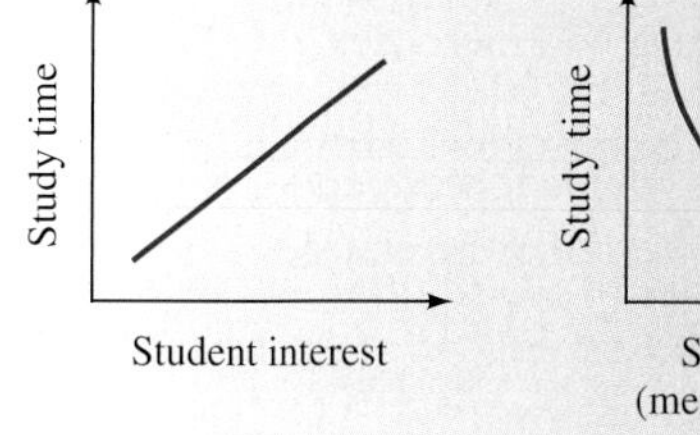

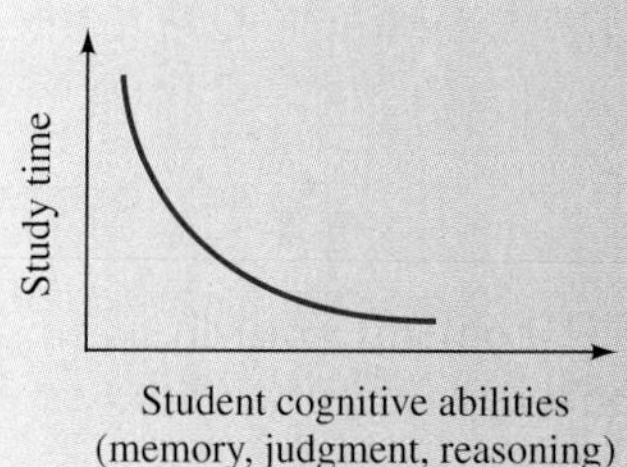

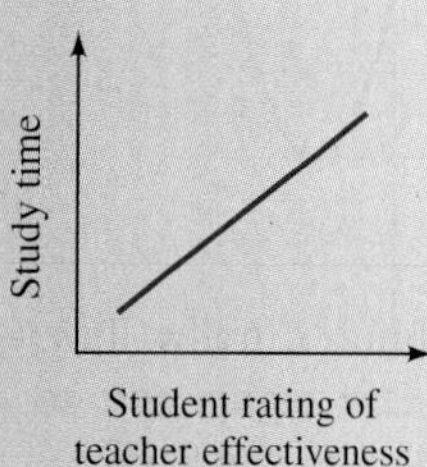

Section 12.7 STUDY SET

VOCABULARY *Fill in the blanks.*

1. The equation $y = kx$ defines ________ variation.
2. The equation $y = \frac{k}{x}$ defines ________ variation.
3. In $y = kx$, the ________ of variation is k.
4. A constant is a ________.

CONCEPTS *Determine whether each graph represents direct variation or inverse variation.*

5.

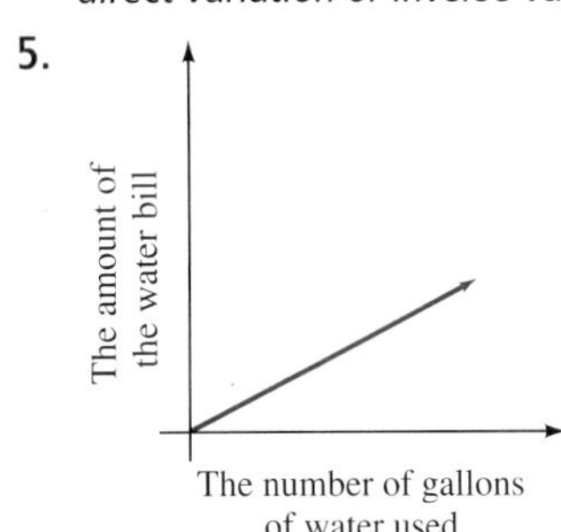

6.

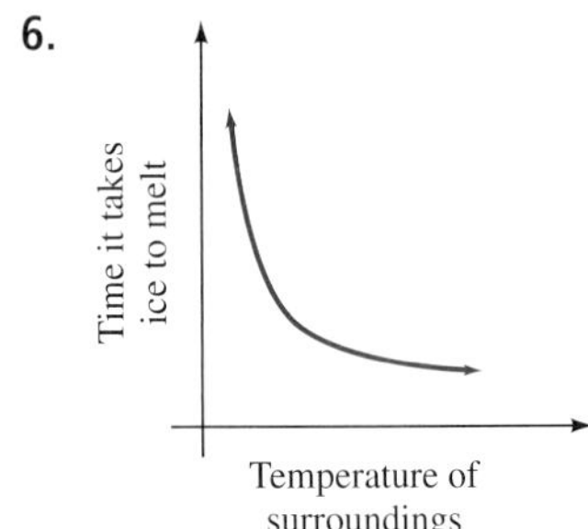

7.

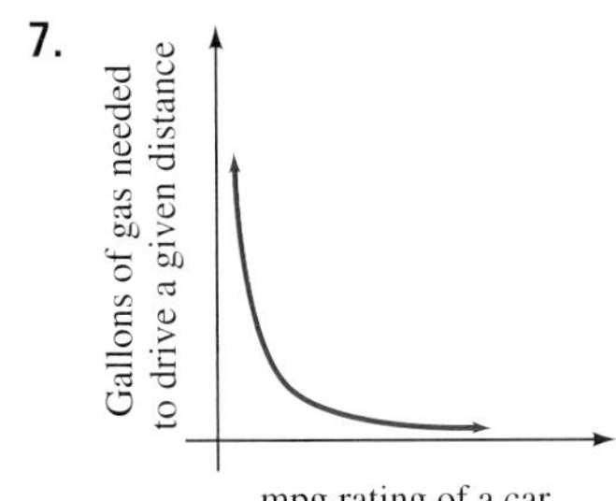

8.

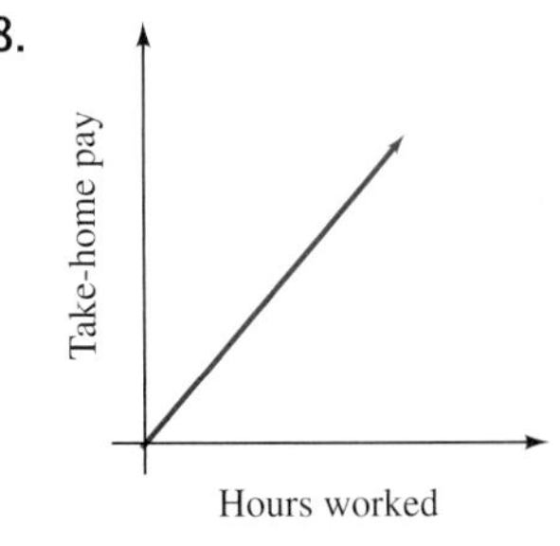

9. Determine whether the equation defines direct variation.
   a. $y = kx$
   b. $y = k + x$
   c. $y = \frac{k}{x}$
   d. $m = kc$

10. a. Translate to mathematical symbols:

A farmer's harvest h	varies directly	with the number of acres planted a.

   b. If the constant of variation for part a is $k = 10{,}000$, what will happen to the size of the harvest as the number of acres planted increases?

11. Express this relationship using an equation: The number of gallons g of paint needed to paint a room varies directly with the number of square feet f to be painted.
12. Express this relationship using an equation: The amount of sales tax t varies directly with the purchase price p of a new car.
13. Assume that t varies directly with s and $t = ks$. If $t = 21$ when $s = 6$, find k.
14. Assume that y varies directly with x and $y = kx$. If $y = 10$ when $x = 2$, find k.
15. Determine whether each equation defines inverse variation.
   a. $y = kx$
   b. $y = \frac{k}{x}$
   c. $y = \frac{x}{k}$
   d. $d = \frac{k}{g}$

16. a. Translate to mathematical symbols:

The time t (in hours) it takes a commuter to drive from her home to her office	varies inversely	with her average speed s (in mph).

   b. If the constant of variation for part a is $k = 30$, what will happen to the time her commute takes her as her average speed increases?

17. Express this relationship using an equation: The number of hot dogs n that a street vendor sells varies inversely with the price p that he charges.
18. a. If y varies directly with x and $k > 0$, what happens to y as x increases?
   b. If y varies inversely with x and $k > 0$, what happens to y as x increases?

19. Assume that y varies inversely with x and $y = \frac{k}{x}$. If $y = 15$ when $x = 10$, find k.

20. Assume that c varies inversely with d and $c = \frac{k}{d}$. If $c = 9$ when $d = 5$, find k.

NOTATION *Complete each solution.*

21. Find f if $d = 21$ and $k = \frac{7}{5}$.

$$d = kf$$
$$21 = \square f$$
$$\square \cdot 21 = \square \cdot \frac{7}{5}f$$
$$15 = f$$

22. Find f if $d = 20$ and $k = 0.75$.

$$d = \frac{k}{f}$$
$$\square = \frac{0.75}{f}$$
$$\square \cdot 20 = \square \cdot \frac{0.75}{f}$$
$$20f = \square$$
$$f = \frac{0.75}{\square}$$
$$f = 0.0375$$

PRACTICE

23. Assume that y varies directly with x. If $y = 10$ when $x = 2$, find y when $x = 7$.

24. Assume that r varies directly with s. If $r = 21$ when $s = 6$, find r when $s = 12$.

25. Assume that l varies directly with m. If $l = 50$ when $m = 200$, find l when $m = 25$.

26. Assume that g varies directly with t. If $g = 3{,}616$ when $t = 8{,}000$, find g when $t = 2{,}405$.

27. Assume that x and y vary directly. If $x = 30$ when $y = 2$, find y when $x = 45$.

28. Assume that n_1 and n_2 vary directly. If $n_1 = 315$ when $n_2 = 3$, find n_2 when $n_1 = 10.5$.

29. Assume that y varies inversely with x. If $y = 8$ when $x = 1$, find y when $x = 8$.

30. Assume that r varies inversely with s. If $r = 40$ when $s = 10$, find r when $s = 15$.

31. Assume that a varies inversely with t. If $a = 600$ when $t = 300$, find a when $t = 15$.

32. Assume that b varies inversely with c. If $b = 0.45$ when $c = 1.6$, find b when $c = 80$.

33. Assume that t_1 and t_2 vary inversely. If $t_1 = 4$ when $t_2 = 5$, find t_2 when $t_1 = 3\frac{1}{3}$.

34. Assume that a and r vary inversely. If $a = 9$ when $r = 7$, find r when $a = \frac{1}{9}$.

APPLICATIONS

35. COMMUTING DISTANCES The distance that a car can travel without refueling varies directly with the number of gallons of gasoline in the tank. If a car can go 360 miles on a full tank of gas (15 gallons), how far can it go on 7 gallons?

36. COMPUTING FORCES The force of gravity acting on an object varies directly with the mass of the object. The force on a mass of 5 kilograms is 49 newtons. What is the force acting on a mass of 12 kilograms?

37. DOSAGES The recommended dose (in milligrams) of Demerol, a preoperative medication given to children, varies directly with the child's weight in pounds. The proper dosage for a child weighing 30 pounds is 18 milligrams. What is the correct dosage for a child weighing 45 pounds?

38. MEDICATIONS To fight ear infections in children, doctors often prescribe Ceclor. The recommended dose in milligrams varies directly with the child's body weight in pounds. The correct dosage for a 20-pound child is 124 milligrams. What is the correct dosage for a 28-pound child?

39. CIDER For the recipe shown, the number of inches of stick cinnamon to use varies directly with the number of servings of spiced cider to be made. How many inches of stick cinnamon are needed to make $2\frac{1}{2}$ dozen servings?

Hot Spiced Cider

8 cups apple cider or apple juice
$\frac{1}{4}$ to $\frac{1}{2}$ cup packed brown sugar
6 inches stick cinnamon
1 teaspoon whole allspice
1 teaspoon whole cloves
8 thin orange wedges or slices (optional)
8 whole cloves (optional) **Makes 8 servings**

40. LUNAR GRAVITY The weight of an object on the moon varies directly with its weight on Earth; 6 pounds on Earth weighs 1 pound on the moon. What would the scale shown in the illustration on the next page register if the astronaut were weighed on the moon?

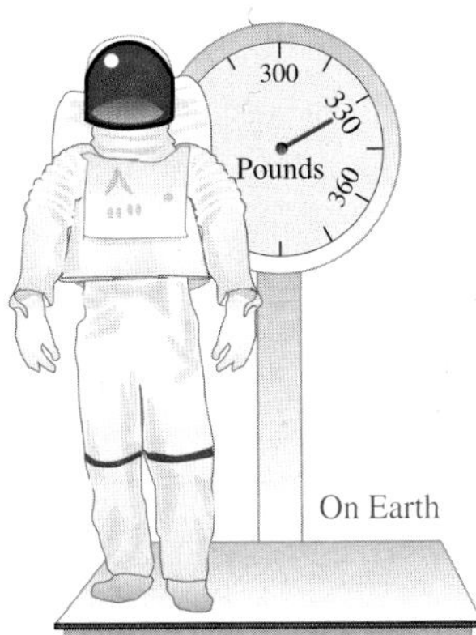

41. COMMUTING TIMES The time it takes a car to travel a certain distance varies inversely with its rate of speed. If a certain trip takes 3 hours at 50 miles per hour, how long will the trip take at 60 miles per hour?

42. GEOMETRY For a fixed area, the length of a rectangle is inversely proportional to its width. A rectangle has a width of 12 feet and a length of 20 feet. If its length is increased to 24 feet, find the width that will maintain the same area.

43. ELECTRICITY The current in an electric circuit varies inversely with the resistance. If the current in the circuit shown is 30 amps when the resistance is 4 ohms, what will the current be for a resistance of 15 ohms?

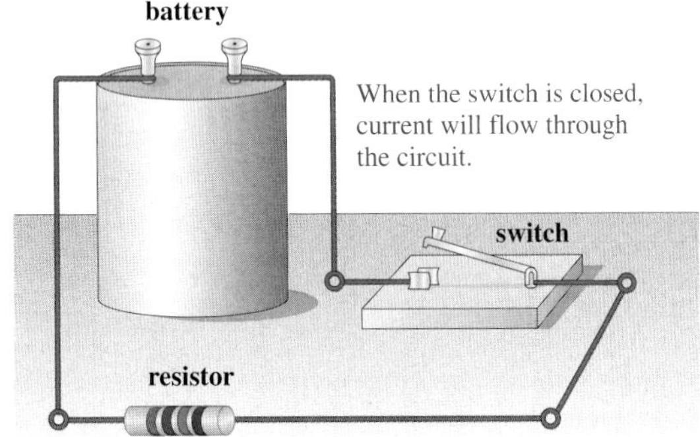

44. FARMING The length of time a given number of bushels of corn will last when feeding cattle varies inversely with the number of animals. If a certain number of bushels will feed 25 cows for 10 days, how long will the feed last for 10 cows?

45. COMPUTING PRESSURES If the temperature of a gas is constant, the volume occupied varies inversely with the pressure. If a gas occupies a volume of 40 cubic meters under a pressure of 8 atmospheres, find the volume when the pressure is changed to 6 atmospheres.

46. DEPRECIATION Assume that the value of a machine varies inversely with its age. If a drill press is worth \$300 when it is 2 years old, find its value when it is 6 years old. How much has the machine depreciated over that 4-year period?

WRITING

47. Give two examples of quantities that vary directly and two that do not.

48. What is the difference between direct variation and inverse variation?

49. What is a constant of variation?

50. Is there a direct variation or an inverse variation between each pair of quantities? Explain why.
    a. The time it takes to type a term paper and the speed at which you type
    b. The time it takes to type a term paper (working at a constant rate) and the length of the term paper

REVIEW *Solve each equation.*

51. $x^2 - 5x - 6 = 0$
52. $x^2 - 25 = 0$
53. $(t + 2)(t^2 + 7t + 12) = 0$
54. $2(y - 4) = -y^2$
55. $y^3 - y^2 = 0$
56. $5a^3 - 125a = 0$
57. $(x^2 - 1)(x^2 - 4) = 0$
58. $6t^3 + 35t^2 = 6t$

KEY CONCEPT

Expressions and Equations

In this chapter, we have discussed procedures for working with **rational expressions** and procedures for solving **rational equations.**

Rational Expressions

The **fundamental property of fractions** is used when simplifying rational expressions and when multiplying and dividing rational expressions: *We can remove factors that are common to the numerator and the denominator of a fraction.*

1. **a.** Simplify: $\frac{2x^2 - 8x}{x^2 - 6x + 8}$.

   b. What common factor was removed?

2. **a.** Multiply: $\frac{x^2 + 2x + 1}{x} \cdot \frac{x^2 - x}{x^2 - 1}$.

   b. What common factors were removed?

The fundamental property of fractions also states that *multiplying the numerator and denominator of a fraction by the same nonzero number does not change the value of the fraction.* We use this concept to "build" fractions when adding or subtracting rational expressions with unlike denominators and when simplifying complex fractions.

3. **a.** Add: $\frac{x}{x + 1} + \frac{x - 1}{x}$.

   b. By what did you multiply the first fraction to rewrite it in terms of the LCD? The second fraction?

4. **a.** Simplify: $\frac{n - 1 - \frac{2}{n}}{\frac{n}{3}}$.

   b. By what did you multiply the numerator and denominator to simplify the complex fraction?

Rational Equations

The multiplication property of equality states that *if equal quantities are multiplied by the same nonzero number, the results are equal quantities.* We use this property when solving rational equations. If we multiply both sides of the equation by the LCD of the rational expressions in the equation, we can clear it of fractions.

5. **a.** Solve: $\frac{11}{b} + \frac{13}{b} = 12$.

   b. By what did you multiply both sides to clear the equation of fractions?

6. **a.** Solve: $\frac{-5}{s^2 + s - 2} + \frac{3}{s + 2} = \frac{1}{s - 1}$.

   b. By what did you multiply both sides to clear the equation of fractions?

7. **a.** Solve: $y + \frac{3}{4} = \frac{3y - 50}{4y - 24}$.

   b. By what did you multiply both sides to clear the equation of fractions?

8. **a.** Solve: $\frac{1}{a} - \frac{1}{b} = 1$ for b.

   b. By what did you multiply both sides to clear the equation of fractions?

ACCENT ON TEAMWORK

SECTION 12.1

CHECKING A SIMPLIFICATION We can use evaluation to check a simplification. To check whether

$$\frac{x^2 - 16}{x + 4} = x - 4$$

have each member of your group evaluate

$$\frac{x^2 - 16}{x + 4} \quad \text{and} \quad x - 4$$

for a given value of x. That is, have one person evaluate both expressions for $x = -5$, have another person evaluate both for $x = -3$, and so on. (Don't use $x = -4$, because the original expression is undefined for this value of x.) The expressions should give identical results for all other values of x. If the evaluations differ for any number, the original expression was not simplified correctly.

Use evaluation to check each of the following simplifications. If an expression was incorrectly simplified, find the correct answer.

a. $\dfrac{x^2 - 4}{x^3 + 8} \stackrel{?}{=} \dfrac{x - 2}{x^2 + 2x + 2}$ **b.** $\dfrac{x^2 + 2x + 1}{x^2 + 4x + 3} \stackrel{?}{=} \dfrac{x + 1}{x + 3}$

c. $\dfrac{x^2 + 2x - 15}{x^2 - 25} \stackrel{?}{=} \dfrac{x - 3}{x - 5}$ **d.** $\dfrac{6x^2 - 13x + 6}{3x^2 + x - 2} \stackrel{?}{=} \dfrac{2x - 3}{x + 2}$

SECTION 12.2

COMBINED OPERATIONS Insert a $\cdot$ symbol or a $\div$ symbol in each blank so that the answer is 1.

$$\frac{x^2 - 2x - 15}{3x^2 - 27} \ \square \ \frac{x^2 - 25}{6x^2 + 45x + 75} \ \square \ \frac{x^2 - x - 6}{2x^2 + 9x + 10}$$

SECTION 12.3

First, add $\dfrac{1}{2x^2} + \dfrac{1}{8x}$

by expressing each fraction in terms of a common denominator $16x^3$. (This is the *product* of their denominators.) Then add the fractions again by expressing each of them in terms of their lowest common denominator. What are one advantage and one disadvantage of each method?

SECTION 12.4

UNIT ANALYSIS Simplify each complex fraction. The units can be divided out just as in the case of common factors.

$$\frac{\dfrac{36 \text{ inches}}{3 \text{ feet}}}{\dfrac{1 \text{ yard}}{3 \text{ feet}}} \qquad \frac{\dfrac{60 \text{ minutes}}{1 \text{ hour}}}{\dfrac{3{,}600 \text{ seconds}}{1 \text{ hour}}} \qquad \frac{\dfrac{4 \text{ quarts}}{1 \text{ gallon}}}{\dfrac{8 \text{ pints}}{1 \text{ gallon}}}$$

SECTION 12.5

SIMPLIFY AND SOLVE The two problems below look similar. Explain why their one-word instructions can't be switched. Write a solution for each problem and then identify the major similarity and the major difference in the solution methods.

Simplify $\dfrac{2}{4x - 4} + \dfrac{3}{x - 1}$ **Solve** $\dfrac{2}{4x - 4} + \dfrac{3}{x - 1} = \dfrac{7}{4}$

SECTION 12.6

PROBLEM SOLVING Problems such as the filling of a water tank are often called *shared-work problems.* For each of the following equations, write a shared-work problem that could be solved using it.

$$\frac{1}{3} + \frac{1}{8} = \frac{1}{x}$$

$$\frac{1}{3} - \frac{1}{8} = \frac{1}{x}$$

$$\frac{1}{3} + \frac{1}{8} - \frac{1}{16} = \frac{1}{x}$$

SECTION 12.7

PI The Greek letter pi (π) represents the ratio of the circumference C of any circle to its diameter d. That is, $\pi = \frac{C}{d}$. Use a tape measure to find the circumference and diameter of various objects that are circular in shape. You can measure anything round: for example, a swimming pool spa, the top of a can, or a ring. Enter your results in a table like the one shown below. Convert each measurement to a decimal and use a calculator to compute the ratio of C to d. Make some observations about your results.

Object	Circumference	Diameter	$\frac{C}{d}$
A quarter	$2\frac{15}{16}$ in. 2.9375 in.	$\frac{15}{16}$ in. 0.9375 in.	3.13333 . . .

COOKING Find a simple recipe for a treat that you can make for your class. Use a proportion to determine the amount of each ingredient needed to make enough for the exact number of people in your class. Write the old and new recipes on separate pieces of poster board. Did the new recipe serve the correct number of people? Share with the class how you made the calculations, as well as any difficulties you encountered.

CHAPTER REVIEW

SECTION 12.1 Simplifying Rational Expressions

CONCEPTS

A *rational expression* is a fraction in which the numerator and denominator are polynomials.

Since division by 0 is undefined, we must make sure that the denominator of a rational expression is not 0.

The fundamental property of fractions:
If b and c are not zero, then

$$\frac{ac}{bc} = \frac{a}{b}$$

When all common factors have been removed, a fraction is in *lowest terms.*

The quotient of any nonzero expression and its opposite is -1.

REVIEW EXERCISES

1. Find the values of x for which the rational expression $\frac{x - 1}{x^2 - 16}$ is undefined.

Write each fraction in lowest terms. If it is already in lowest terms, so indicate.

2. $\frac{10}{25}$
3. $-\frac{12}{18}$

Simplify each rational expression. If it is already in lowest terms, so indicate. Assume that no denominators are zero.

4. $\frac{3x^2}{6x^3}$
5. $\frac{5xy^2}{2x^2y^2}$
6. $\frac{x^2}{x^2 + x}$
7. $\frac{a^2 - 4}{a + 2}$
8. $\frac{3p - 2}{2 - 3p}$
9. $\frac{8 - x}{x^2 - 5x - 24}$
10. $\frac{2x^2 - 16x}{2x^2 - 18x + 16}$
11. $\frac{x^2 + x - 2}{x^2 - x - 2}$
12. Evaluate $\frac{x^2 - 1}{x - 5}$ for $x = -2$.
13. Explain why it would be incorrect to remove the common x's in $\frac{x + 1}{x}$.
14. Simplify: $\frac{4(t + 3) + 8}{3(t + 3) + 6}$.

SECTION 12.2 Multiplying and Dividing Rational Expressions

Rule for multiplying fractions:

$$\frac{a}{b} \cdot \frac{c}{d} = \frac{ac}{bd} \quad \text{where } b, d \neq 0$$

Perform each multiplication and simplify.

15. $\frac{3xy}{2x} \cdot \frac{4x}{2y^2}$
16. $56x\left(\frac{12}{7x}\right)$
17. $\frac{x^2 - 1}{x^2 + 2x} \cdot \frac{x}{x + 1}$
18. $\frac{x^2 + x}{3x - 15} \cdot \frac{6x - 30}{x^2 + 2x + 1}$

Rule for dividing fractions:

$$\frac{a}{b} \div \frac{c}{d} = \frac{a}{b} \cdot \frac{d}{c}$$

where $b, c, d \neq 0$

Perform each division and simplify.

19. $\frac{3x^2}{5x^2y} \div \frac{6x}{15xy^2}$
20. $\frac{x^2 + 5x}{x^2 + 4x - 5} \div \frac{x^2}{x - 1}$

To write the *reciprocal* of a fraction, we invert the fraction.

21. $\dfrac{x^2 - x - 6}{2x - 1} \div \dfrac{x^2 - 2x - 3}{2x^2 + x - 1}$

22. Simplify: $\dfrac{b^2 + 4b + 4}{b^2 + b - 6}\left(\dfrac{b - 2}{b - 1} \div \dfrac{b + 2}{b^2 + 2b - 3}\right)$.

SECTION 12.3 Adding and Subtracting Rational Expressions

Adding and subtracting fractions with like denominators:

$\dfrac{a}{d} + \dfrac{b}{d} = \dfrac{a + b}{d}$ where $d \neq 0$

$\dfrac{a}{d} - \dfrac{b}{d} = \dfrac{a - b}{d}$ where $d \neq 0$

To find the *LCD*, factor each denominator completely. Form a product using each different factor the greatest number of times it appears in any one factorization.

To add or subtract fractions with unlike denominators, first find the LCD of the fractions. Then express each fraction in equivalent form with a common denominator. Finally, add or subtract the fractions.

Perform each operation. Simplify all answers.

23. $\dfrac{x}{x + y} + \dfrac{y}{x + y}$

24. $\dfrac{3x}{x - 7} - \dfrac{x - 2}{x - 7}$

25. $\dfrac{a}{a^2 - 2a - 8} + \dfrac{2}{a^2 - 2a - 8}$

Several denominators are given. Find the lowest common denominator (LCD).

26. $2x^2, 4x$

27. $3y^2, 9x, 6x$

28. $x + 1, x + 2$

29. $y^2 - 25, y - 5$

Perform each operation. Simplify all answers.

30. $\dfrac{x}{x - 1} + \dfrac{1}{x}$

31. $\dfrac{1}{7} - \dfrac{1}{c}$

32. $\dfrac{x + 2}{2x} - \dfrac{2 - x}{x^2}$

33. $\dfrac{2t + 2}{t^2 + 2t + 1} - \dfrac{1}{t + 1}$

34. $\dfrac{x}{x + 2} + \dfrac{3}{x} - \dfrac{4}{x^2 + 2x}$

35. $\dfrac{6}{b - 1} - \dfrac{b}{1 - b}$

36. VIDEO CAMERAS Find the perimeter and the area of the LED screen of the camera.

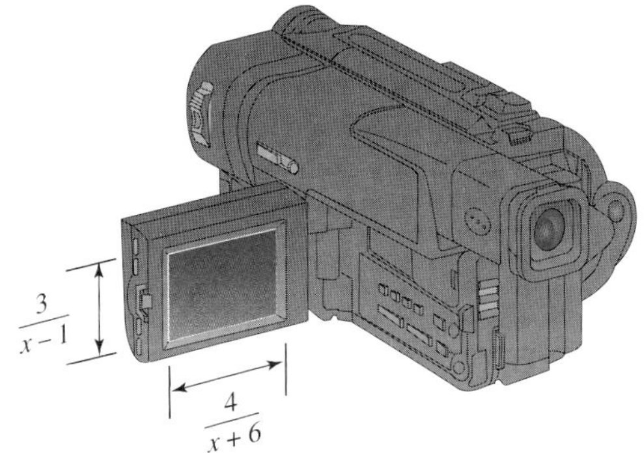

SECTION 12.4 Complex Fractions

Complex fractions contain fractions in their numerators and/or their denominators.

Simplify each complex fraction.

37. $\dfrac{\frac{3}{2}}{\frac{2}{3}}$

38. $\dfrac{\frac{3}{2} + 1}{\frac{2}{3} + 1}$

To simplify a complex fraction, use either of these methods:

1. Write the numerator and denominator of the complex fraction as single fractions, perform the division of the fractions, and simplify.
2. Multiply both the numerator and the denominator of the complex fraction by the LCD of the fractions that appear in the numerator and denominator, then simplify.

39. $\dfrac{\frac{1}{y}+1}{\frac{1}{y}-1}$

40. $\dfrac{1+\frac{3}{x}}{2-\frac{1}{x^2}}$

41. $\dfrac{\frac{2}{x-1}+\frac{x-1}{x+1}}{\frac{1}{x^2-1}}$

42. $\dfrac{x^{-2}+1}{x^{-2}-1}$

SECTION 12.5 *Rational Equations and Problem Solving*

To solve an equation that contains fractions, change it to an equivalent equation without fractions. Do so by multiplying both sides by the LCD of the fractions. Check all solutions.

An apparent solution that does not satisfy the original equation is called an *extraneous* solution.

To solve a problem, follow these steps:

1. Analyze the problem.
2. Form an equation.
3. Solve the equation.
4. State the conclusion.
5. Check the result.

Interest = principal · rate · time

Distance = rate · time

Solve each equation and check all answers.

43. $\dfrac{3}{x}=\dfrac{2}{x-1}$

44. $\dfrac{a}{a-5}=3+\dfrac{5}{a-5}$

45. $\dfrac{2}{3t}+\dfrac{1}{t}=\dfrac{5}{9}$

46. $a=\dfrac{3a-50}{4a-24}-\dfrac{3}{4}$

47. $\dfrac{4}{x+2}-\dfrac{3}{x+3}=\dfrac{6}{x^2+5x+6}$

48. The efficiency E of a Carnot engine is given by the formula

$$E=1-\frac{T_2}{T_1}$$

Solve the formula for T_1.

49. Solve for r_1: $\dfrac{1}{r}=\dfrac{1}{r_1}+\dfrac{1}{r_2}$.

50. NUMBER PROBLEMS If a number is subtracted from the denominator of $\frac{4}{5}$ and twice as much is added to the numerator, the result is 5. Find the number.

51. If a maid can clean a house in 4 hours, how much of the house does she clean in 1 hour?

52. PAINTING HOUSES If a homeowner can paint a house in 14 days and a professional painter can paint it in 10 days, how long will it take if they work together?

53. INVESTMENTS In one year, a student earned $100 interest on money she deposited at a savings and loan. She later learned that the money would have earned $120 if she had deposited it at a credit union, because the credit union paid 1% more interest at the time. Find the rate she received from the savings and loan.

54. EXERCISE A jogger can bicycle 30 miles in the same time that it takes her to jog 10 miles. If she can ride 10 mph faster than she can jog, how fast can she jog?

55. WIND SPEED A plane flies 400 miles downwind in the same amount of time as it takes to travel 320 miles upwind. If the plane can fly at 360 mph in still air, find the velocity of the wind.

SECTION 12.6 *Proportions and Similar Triangles*

A *ratio* is the quotient of two numbers or the quotient of two quantities with the same units.

A *proportion* is a statement that two ratios or two rates are equal.

In the proportion $\frac{a}{b} = \frac{c}{d}$, a and d are the *extremes*, and b and c are the *means*.

In any proportion, the product of the extremes is equal to the product of the means.

56. What is the ratio of the number of teeth of the larger gear to the number of teeth of the smaller gear?

57. Determine whether $\frac{4}{7} = \frac{20}{34}$ is a proportion.

Solve each proportion.

58. $\dfrac{3}{x} = \dfrac{6}{9}$

59. $\dfrac{x}{3} = \dfrac{x}{5}$

60. $\dfrac{x-2}{5} = \dfrac{x}{7}$

61. $\dfrac{2x}{x+4} = \dfrac{3}{x-1}$

62. DENTISTRY The diagram in the illustration was displayed in a dentist's office. According to the diagram, if the dentist has 340 adult patients, how many will develop gum disease?

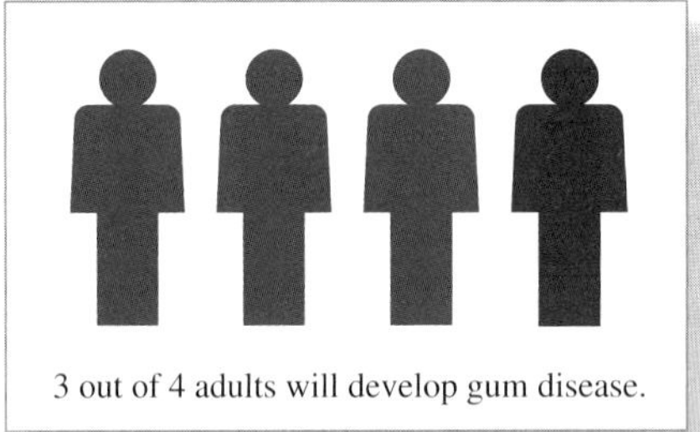

3 out of 4 adults will develop gum disease.

The measures of corresponding sides of *similar triangles* are in proportion.

63. A telephone pole casts a shadow 12 feet long at the same time that a man 6 feet tall casts a shadow of 3.6 feet. How tall is the pole?

SECTION 12.7 *Variation*

Direct variation: As one variable gets larger, the other gets larger as described by the equation $y = kx$, where k is the *constant of variation.*

Inverse variation: As one variable gets larger, the other gets smaller as described by the equation

$$y = \frac{k}{x} \quad \text{where } k \text{ is a constant}$$

64. PROFIT The profit made by a strawberry farm varies directly with the number of baskets of strawberries sold. If a profit of \$500 was made from the sale of 750 baskets, what is the profit when 1,250 baskets are sold?

65. l varies inversely with w. Find the constant of variation if $l = 30$ when $w = 20$.

66. ELECTRICITY For a fixed voltage, the current in an electrical circuit varies inversely with the resistance in the circuit. If a certain circuit has a current of $2\frac{1}{2}$ amps when the resistance is 150 ohms, find the current in the circuit when the resistance is doubled.

67. The graph shows a type of variation. Does it show direct or inverse variation?

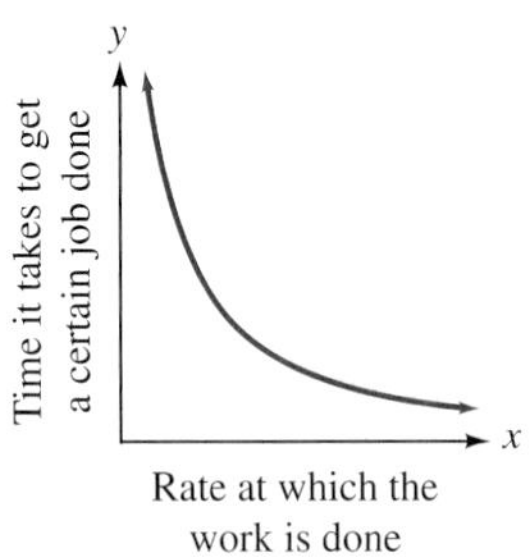

68. Give an example of two quantities that vary directly.

CHAPTER 12 TEST

1. Find the values of x for which $\dfrac{x}{x^2 + x - 6}$ is undefined.

2. Simplify: $\dfrac{48x^2y}{54xy^2}$.

3. Simplify: $\dfrac{2x^2 - x - 3}{4x^2 - 9}$.

4. Simplify: $\dfrac{3(x + 2) - 3}{2x + 3 - (x + 2)}$.

5. Multiply and simplify: $-\dfrac{12x^2y}{15xy} \cdot \dfrac{25y^2}{16x}$.

6. Multiply and simplify: $\dfrac{x^2 + 3x + 2}{3x + 9} \cdot \dfrac{x + 3}{x^2 - 4}$.

7. Divide and simplify: $\dfrac{8x^2}{25x} \div \dfrac{16x^2}{30x}$.

8. Divide and simplify: $\dfrac{x - x^2}{3x^2 + 6x} \div \dfrac{3x - 3}{3x^3 + 6x^2}$.

9. Simplify: $\dfrac{x^2 + x}{x - 1} \cdot \dfrac{x^2 - 1}{x^2 - 2x} \div \dfrac{x^2 + 2x + 1}{x^2 - 4}$.

10. Add: $\dfrac{5x - 4}{x - 1} + \dfrac{5x + 3}{x - 1}$.

11. Subtract: $\dfrac{3y + 7}{2y + 3} - \dfrac{3(y - 2)}{2y + 3}$.

12. Add: $\dfrac{x + 1}{x} + \dfrac{x - 1}{x + 1}$.

13. Subtract: $\dfrac{a + 3}{a - 1} - \dfrac{a + 4}{1 - a}$.

14. Subtract: $\dfrac{2n}{5m} - \dfrac{n}{2}$.

15. Simplify: $\dfrac{1 + \dfrac{y}{x}}{\dfrac{y}{x} - 1}$.

16. $\dfrac{1}{3} + \dfrac{4}{3y} = \dfrac{5}{y}$

17. Solve: $\dfrac{7}{q^2 - q - 2} + \dfrac{1}{q + 1} = \dfrac{3}{q - 2}$.

18. Solve: $\dfrac{2}{3} = \dfrac{2c - 12}{3c - 9} - c$.

19. $\dfrac{9n}{n - 6} = 3 + \dfrac{54}{n - 6}$

20. Solve for B: $H = \dfrac{RB}{R + B}$.

21. Is the equation $\frac{3}{5} = \frac{51}{85}$ a proportion?

22. Solve the proportion: $\frac{y}{y-1} = \frac{y-2}{y}$.

23. HEALTH RISKS A medical newsletter states that a healthy waist-to-hip ratio for men is 19:20 or less. Does the patient shown below fall within the healthy range?

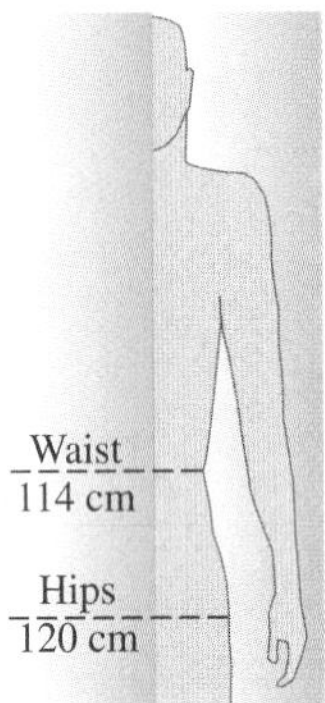

24. FLIGHT PATHS A plane drops 575 feet as it flies a horizontal distance of $\frac{1}{2}$ mile, as shown below. How much altitude will it lose as it flies a horizontal distance of 7 miles?

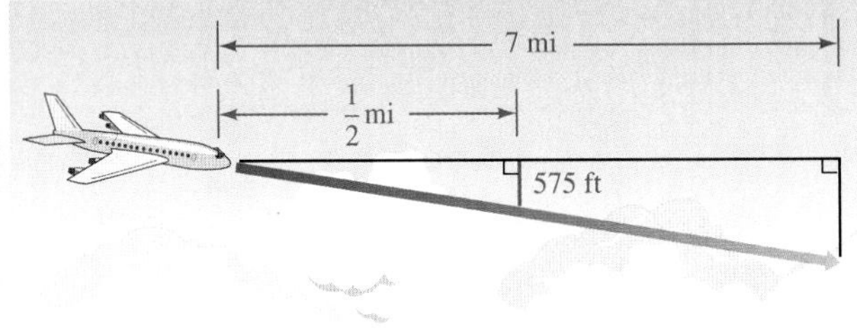

25. POGO STICKS The force required to compress a spring varies directly with the change in the length of the spring. If a force of 130 pounds compresses the spring on the following pogo stick 6.5 inches, how much force is required to compress the spring 5 inches?

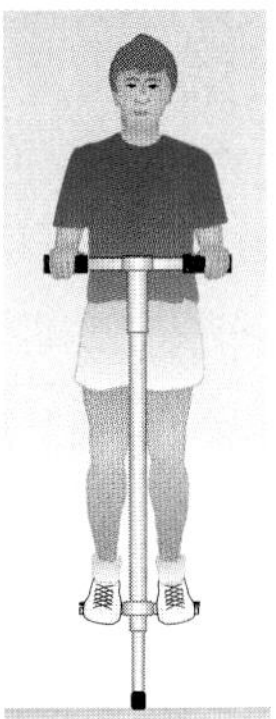

26. If i varies inversely with d, find the constant of variation if $i = 100$ when $d = 2$.

27. CLEANING HIGHWAYS One highway worker can pick up all the trash on a strip of highway in 7 hours, and his helper can pick up the trash in 9 hours. How long will it take them if they work together?

28. BOATING A boat can motor 28 miles downstream in the same amount of time as it can motor 18 miles upstream. Find the speed of the current if the boat can motor at 23 mph in still water.

29. Explain why we can remove the 5's in $\frac{5x}{5}$ and why we can't remove out in $\frac{5+x}{5}$.

30. Explain what it means to clear the following equation of fractions.

$$\frac{u}{u-1} + \frac{1}{u} = \frac{u^2+1}{u^2-u}$$

Why is this a helpful first step in solving the equation?

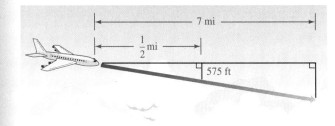

CHAPTERS 1–12 CUMULATIVE REVIEW EXERCISES

1. Evaluate: $9^2 - 3[45 - 3(6 + 4)]$.
2. PAIN RELIEVERS For a 12-month period, Tylenol had sales of \$567,600,000. Use the information in the illustration to determine the total amount of money spent on pain-relieving tablets for that 12-month period.

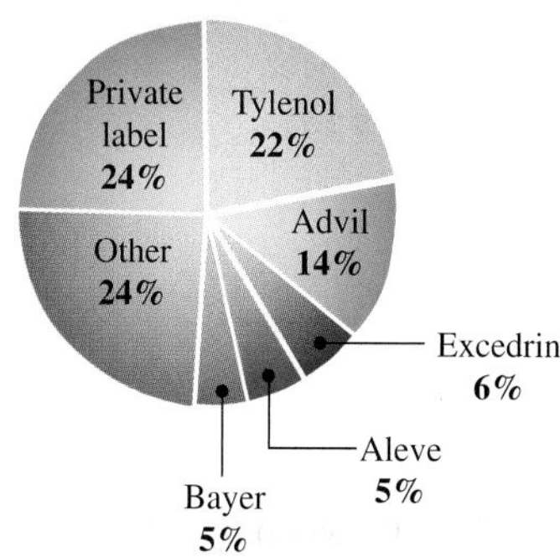

Based on information from *Los Angeles Times* (Sept. 24, 1998)

3. Find the average (mean) test score of a student in a history class with scores of 80, 73, 61, 73, and 98.
4. What is the value in cents of x 35¢ stamps?
5. Solve: $\frac{3}{4} = \frac{1}{2} + \frac{x}{5}$.
6. Change 40° C to degrees Fahrenheit.
7. Find the volume of a pyramid that has a square base, measuring 6 feet on a side, and whose height is 20 feet.
8. Determine whether each statement is true or false.
   a. Every integer is a whole number.
   b. 0 is not a rational number.
   c. π is an irrational number.
   d. The set of integers is the set of whole numbers and their opposites.
9. Solve: $2 - 3(x - 5) = 4(x - 1)$.
10. Simplify: $8(c + 7) - 2(c - 3)$.
11. Solve $A - c = 2B + r$ for B.
12. Solve $7x + 2 \geq 4x - 1$ and graph the solution. Then describe the graph using interval notation.
13. Solve: $\frac{4}{5}d = -4$.
14. BLENDING TEA One grade of tea (worth \$3.20 per pound) is to be mixed with another grade (worth \$2 per pound) to make 20 pounds of a mixture that will be worth \$2.72 per pound. How much of each grade of tea must be used?
15. SPEED OF A PLANE Two planes are 6,000 miles apart, and their speeds differ by 200 mph. If they travel toward each other and meet in 5 hours, find the speed of the slower plane.
16. Graph: $y = 2x - 3$.

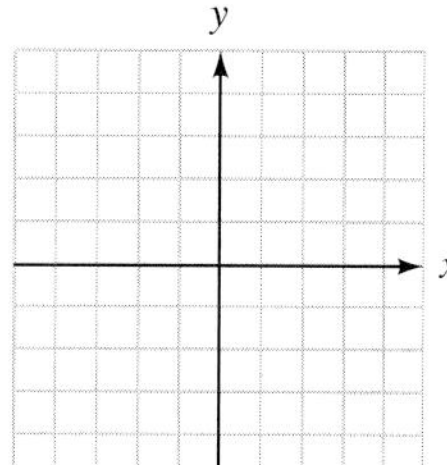

17. Graph: $y = (x + 2)^3$.

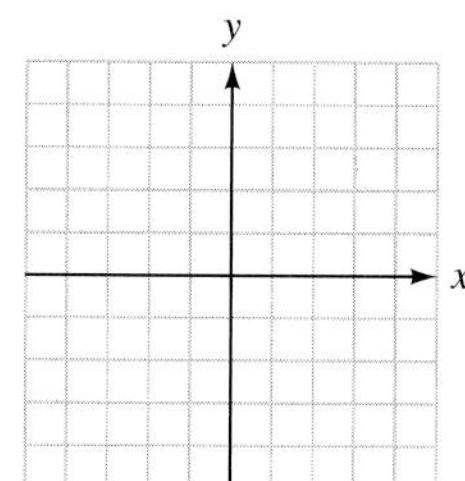

18. Find the slope of the line passing through $(-1, 3)$ and $(3, -1)$.
19. Write the equation of a line that has slope 3 and passes through the point (1, 5).
20. Graph: $3x - 2y = 6$.

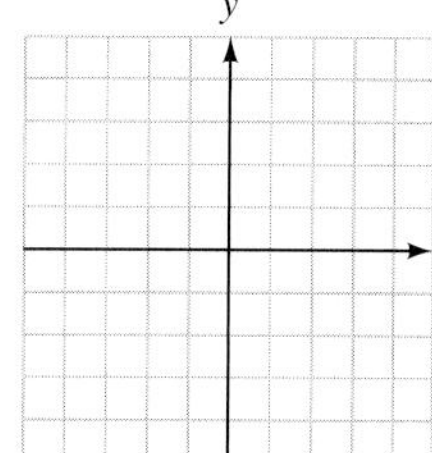

21. Graph: $y = \frac{5}{2}$.

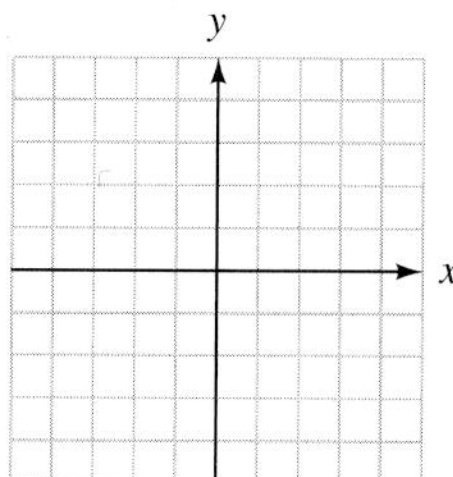

22. What is the slope of a line perpendicular to the line?

    $y = -\frac{7}{8}x - 6$?
23. Is this the graph of a function?

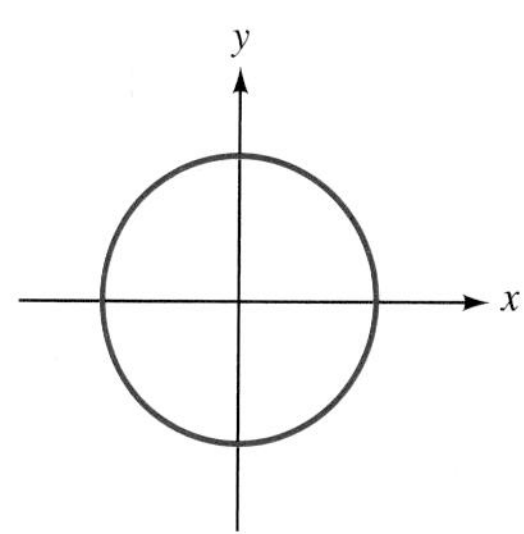

24. CUTTING STEEL The following graph shows the amount of wear (in mm) on a cutting blade for a given length of a cut (in m). Find the rate of change in the length of the cutting blade.

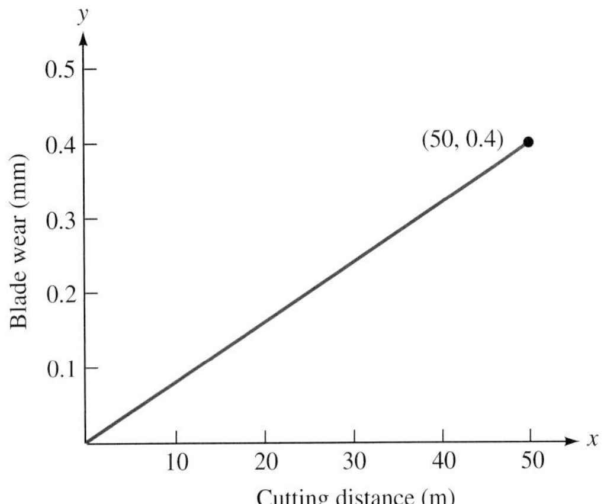

25. Find $f(-4)$ if $f(x) = \dfrac{x^2 - 2x}{2}$.

26. Evaluate: -5^2.

Simplify each expression. Write each answer without using negative exponents.

27. x^4x^3

28. $(x^2x^3)^5$

29. $\left(\dfrac{y^3y}{2yy^2}\right)^3$

30. $\left(\dfrac{-2a}{b}\right)^5$

31. $(a^{-2}b^3)^{-4}$

32. $\dfrac{9b^0b^3}{3b^{-3}b^4}$

33. Write 290,000 in scientific notation.

34. What is the degree of the polynomial $5x^3 - 4x + 16$?

Perform the operations.

35. $(3x^2 - 3x - 2) + (3x^2 + 4x - 3)$

36. $(2x^2y^3)(3x^3y^2)$

37. $(2y - 5)(3y + 7)$

38. $-4x^2z(3x^2 - z)$

39. $\dfrac{6x + 9}{3}$

40. $\dfrac{15(r^2s^3)^2}{-5(rs^5)^3}$

41. LICENSE PLATES The number of different license plates of the form three digits followed by three letters, as shown, is $10 \cdot 10 \cdot 10 \cdot 26 \cdot 26 \cdot 26$. Write this expression using exponents. Then evaluate it.

42. CONCENTRIC CIRCLES The area of the ring between the two concentric circles of radius r and R is given by the formula

$$A = \pi(R + r)(R - r)$$

Perform the multiplication on the right-hand side of the equation.

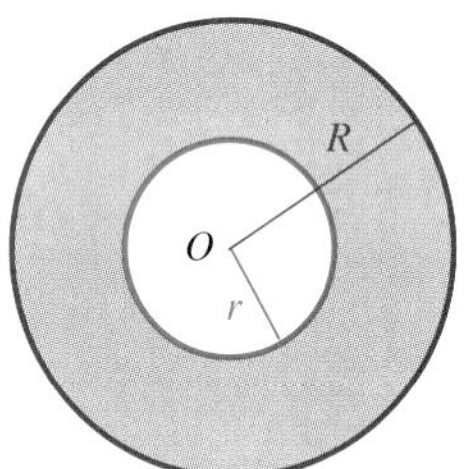

Factor each polynomial completely, if possible.

43. $k^3t - 3k^2t$

44. $2ab + 2ac + 3b + 3c$

45. $2a^2 - 200b^2$

46. $b^3 + 125$

47. $u^2 - 18u + 81$

48. $6x^2 - 63 - 13x$

49. $-r^2 + 2 + r$

50. $u^2 + 10u + 15$

Solve each equation by factoring.

51. $5x^2 + x = 0$

52. $6x^2 - 5x = -1$

53. COOKING The electric griddle shown has a cooking surface of 160 square inches. Find the length and the width of the griddle.

54. For what values of x is the rational expression $\frac{3x^2}{x^2 - 25}$ undefined?

Perform the operations and simplify, if possible.

55. $\frac{x^2 - 16}{x - 4} \div \frac{3x + 12}{x}$

56. $\frac{4}{x - 3} + \frac{5}{3 - x}$

57. $\frac{2 - \frac{2}{x + 1}}{2 + \frac{2}{x}}$

58. $\frac{4a}{a - 2} - \frac{3a}{a - 3} + \frac{4a}{a^2 - 5a + 6}$

Solve each equation.

59. $\frac{7}{5x} - \frac{1}{2} = \frac{5}{6x} + \frac{1}{3}$

60. $\frac{3}{5} + \frac{7}{x + 2} = 2$

61. COMPUTING INTEREST For a fixed rate and principal, the interest earned in a bank account paying simple interest varies directly with the length of time the principal is left on deposit. If an investment earns \$700 in 2 years, how much will it earn in 7 years?

62. HEIGHT OF A TREE A tree casts a shadow of 29 feet at the same time as a vertical yardstick casts a shadow of 2.5 feet. Find the height of the tree.

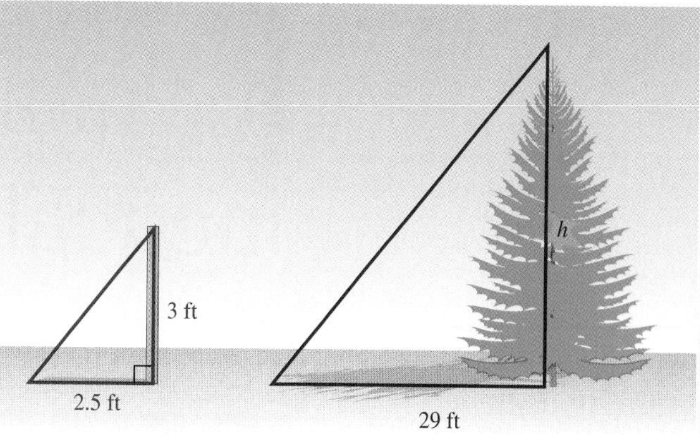

63. DRAINING A TANK If one outlet pipe can drain a tank in 24 hours, and another pipe can drain the tank in 36 hours, how long will it take for both pipes to drain the tank?

64. Explain what it means for two variables to vary inversely.

CHAPTER 13

Solving Systems of Equations and Inequalities

© Getty Images

Managers make many decisions in the day-to-day operation of a business. For example, a theater owner must determine ticket prices, employees' wages, and show times. Algebra can be helpful in making these decisions. When business applications involve two variable quantities, often we can model them using a pair of equations called a *system of equations.*

To learn more about the use of systems of equations, visit *The Learning Equation* on the Internet at http://tle.brookscole.com. (The log-in instructions are in the Preface.) For Chapter 13, the online lesson is

- *TLE* Lesson 24: Solving Systems of Equations by Graphing

Check Your Knowledge

1. An ordered pair that satisfies both equations of a system of linear equations is called a ________ of the system.
2. Systems of equations that have no solutions are called ________ systems.
3. Equations that have the same graph are called ________ equations.
4. To graph a linear inequality, first graph the ________. Then pick a test ________ to determine which half-plane to shade.

Determine whether the ordered pair is a solution of the system of equations.

5. $(1, 1), \begin{cases} x - 3y = -2 \\ x + y = 6 \end{cases}$

6. $(6, -1), \begin{cases} x - y = 7 \\ x + y = 5 \end{cases}$

7. Solve by graphing: $\begin{cases} x - 7y = 6 \\ 2x + 3y = -5 \end{cases}$.

Solve each system by substitution.

8. $\begin{cases} x = 3y + 3 \\ x + y = 3 \end{cases}$

9. $\begin{cases} x - 2y = -2 \\ x + 2y = 6 \end{cases}$

Solve each system by addition.

10. $\begin{cases} 2x - y = 4 \\ 5x + y = 3 \end{cases}$

11. $\begin{cases} 2x + 3y = 1 \\ 3x + 2y = 4 \end{cases}$

Classify each system as consistent or inconsistent and classify the equations as dependent or independent.

12. $\begin{cases} 3x + 4y = 1 \\ 3x + 4y = 5 \end{cases}$

13. $\begin{cases} x + y = 3 \\ x - y = 3 \end{cases}$

14. $\begin{cases} x + 3y = 4 \\ 3x + 9y = 12 \end{cases}$

Solve each system by any method.

15. $\begin{cases} 5x - 2y = -3 \\ 2y = 11 - 3x \end{cases}$

16. $\begin{cases} \frac{1}{2}s - \frac{1}{4}t = 1 \\ \frac{1}{3}s + t = 3 \end{cases}$

17. Julia invested part of \$14,000 at 6% annual interest and the rest at 9%. The annual income from these investments was \$990. Use a system of equations to determine how much she invested at each rate.
18. Peanuts worth \$1.80 per pound are to be mixed with cashews that cost \$2.80 per pound to make 50 pounds of a mixture worth \$2.00 per pound. Use a system of equations to determine how many pounds of each type of nut should be used.

Determine whether the given ordered pair is a solution of $x - 2y < 2$.

19. $(0, 0)$

20. $(4, 1)$

21. Graph the inequality: $2x + y \leq 1$.

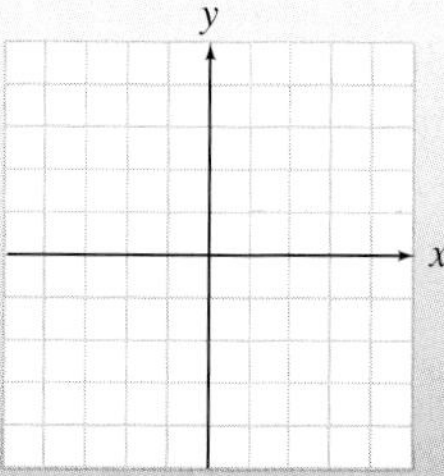

22. Solve the system by graphing: $\begin{cases} 3x + 5y < 8 \\ 5x - 3y \geq 2 \end{cases}$.

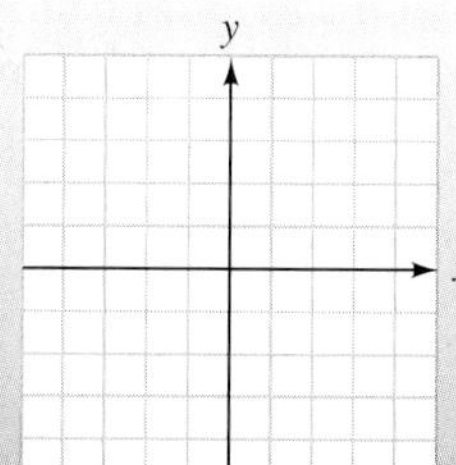

Getty Images

Study Skills Workshop

CAREER EXPLORATION

Your college major can have a direct influence on the math course(s) that you need to take after Intermediate Algebra. If you have plans to major in math, engineering, or the sciences, you will choose a sequence of courses leading to calculus and beyond; if you have a major that is more liberal studies oriented, you will most probably choose a different course or sequence of courses than the math/science majors. Ultimately, your choice of career will affect the math sequence that follows this course. Before the end of this term, it would be wise to have at least a general idea of which direction you would like to go.

How Do You Decide? If you aren't sure of your ultimate career goal, you may want to seek the advice of a counselor, pay a visit to your college's career center, or search the Internet for tools that will help you discover things about yourself and the careers, that might suit your interests. Even if you think you know what career you are aiming for, it's still a good idea to search a little. Personality tests tell you things about your basic makeup but aren't necessarily job or career-oriented; one such test is the Myers-Briggs test which may be available through your college counseling office. Once you know your basic personality type from this test, you may use books such as *Do What You Are*[1] to explore various career choices.

It's probably not a good idea to put all of your hopes on one type of test, though, and there are various free online tests that you can take to help determine your personality type. For example, see: http://www.personalitytype.com/, http://www.keirsey.com/, and http://www.9types.com/. There are also career or vocational tests that you can take online: http://www.review.com/career/careerquizhome.cfm?menuID=0&careers=6, and http://career.missouri.edu/ are examples. If you would rather read than use a computer, a classic book on making career choices is *What Color is Your Parachute? A Practical Manual for Job-Hunters and Career-Changers*.[2] There is also a Web site designed to be used with the book at http://www.jobhuntersbible.com/.

Once You've Decided. After you have a general idea of the career you would like to pursue, talk to a college counselor to find out what type of education is necessary to achieve your goal. Your counselor can tell you which classes you need to take at your particular school and may help you choose a college to which you can transfer, if that is necessary.

ASSIGNMENT

1. Do you have a general career goal in mind? If so, what is it?
2. Take at least two personality tests and two career tests. List the tests that you took and where you found them. Did taking the tests help you in making any decisions about a career choice?
3. Make an appointment and visit a counselor to discuss which classes you should take during your next term and beyond. If you are still unclear about which direction you should take, ask your counselor for a good all-purpose plan. Make a list of classes that your counselor suggests and keep a copy for yourself.

---

[1] Paul D. Tieger and Barbara Barron-Tieger, Little, Brown and Company.

[2] by Dick Bolles, Ten Speed Press Publishers.

To solve many problems, we must use two variables. This requires that we solve a system of equations.

13.1 Solving Systems of Equations by Graphing

- Systems of equations
- The graphing method
- Inconsistent systems
- Dependent equations

Figure 13-1 shows the per capita consumption of chicken and beef in the United States for the years 1985–2002. Plotting both graphs on the same coordinate system makes it easy to compare recent trends. The point of intersection of the graphs indicates that Americans consumed equal amounts of chicken and beef in 1992—about 66 pounds of each, per person.

In this section, we will use a similar graphical approach to solve *systems of equations.*

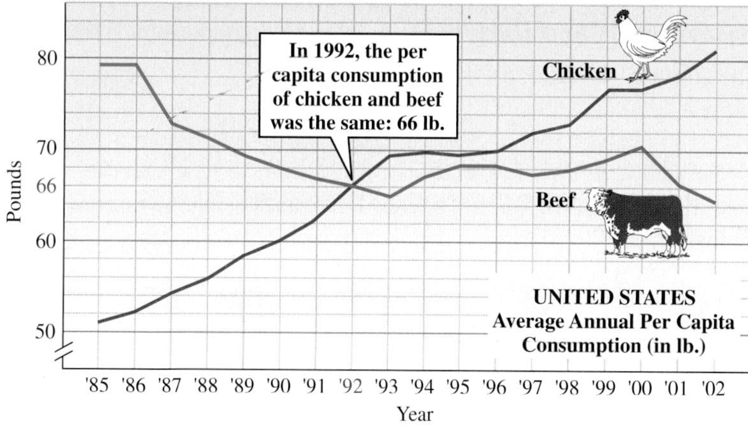

FIGURE 13-1

Systems of equations

We have previously discussed equations that contain two variables, such as $x + y = 3$. Because there are infinitely many pairs of numbers whose sum is 3, there are infinitely many pairs (x, y) that satisfy this equation. Some of these pairs are

$x + y = 3$		
x	y	(x, y)
0	3	(0, 3)
1	2	(1, 2)
2	1	(2, 1)
3	0	(3, 0)

Likewise, there are infinitely many pairs (x, y) that satisfy the equation $3x - y = 1$. Some of these pairs are

$3x - y = 1$		
x	y	(x, y)
0	−1	(0, −1)
1	2	(1, 2)
2	5	(2, 5)
3	8	(3, 8)

Although there are infinitely many pairs that satisfy each of these equations, only the pair (1, 2) satisfies both equations at the same time. The pair of equations

$$\begin{cases} x + y = 3 \\ 3x - y = 1 \end{cases}$$

is called a **system of equations.** Because the ordered pair (1, 2) satisfies both equations simultaneously (at the same time), it is called a **simultaneous solution,** or a **solution of the system of equations.** In this chapter, we will discuss three methods for finding the solution of a system of equations.

The graphing method

To use the graphing method to solve

$$\begin{cases} x + y = 3 \\ 3x - y = 1 \end{cases}$$

we graph both equations on one set of coordinate axes using the intercept method, as shown in Figure 13-2.

$x + y = 3$		
x	y	(x, y)
0	3	(0, 3)
3	0	(3, 0)
2	1	(2, 1)

$3x - y = 1$		
x	y	(x, y)
0	−1	(0, −1)
$\frac{1}{3}$	0	$\left(\frac{1}{3}, 0\right)$
2	5	(2, 5)

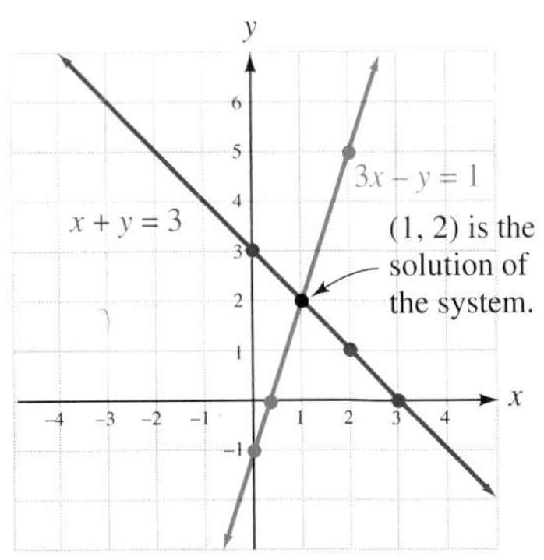

FIGURE 13-2

Although there are infinitely many pairs (x, y) that satisfy $x + y = 3$, and infinitely many pairs (x, y) that satisfy $3x - y = 1$, only the coordinates of the point where their graphs intersect satisfy both equations simultaneously. Thus, the solution of the system is (1, 2).

To check this result, we substitute 1 for x and 2 for y in each equation and verify that the pair (1, 2) satisfies each equation.

Check: **First equation**

$$x + y = 3$$
$$\mathbf{1} + \mathbf{2} \stackrel{?}{=} 3$$
$$3 = 3$$

Second equation

$$3x - y = 1$$
$$3(\mathbf{1}) - \mathbf{2} \stackrel{?}{=} 1$$
$$3 - 2 \stackrel{?}{=} 1$$
$$1 = 1$$

When the graphs of two equations in a system are different lines, the equations are called **independent equations.** When a system of equations has a solution, the system is called a **consistent system.**

To solve a system of equations in two variables by graphing, we follow these steps.

The graphing method

1. Carefully graph each equation.
2. When possible, find the coordinates of the point where the graphs intersect.
3. Check the proposed solution in the equations of the original system.

EXAMPLE 1 Using graphing to solve $\begin{cases} 2x + 3y = 2 \\ 3x = 2y + 16 \end{cases}$.

Solution Using the intercept method, we graph both equations on one set of coordinate axes, as shown in Figure 13-3.

$2x + 3y = 2$		
x	y	(x, y)
0	$\frac{2}{3}$	$(0, \frac{2}{3})$
1	0	$(1, 0)$
−2	2	$(-2, 2)$

$3x = 2y + 16$		
x	y	(x, y)
0	−8	$(0, -8)$
$\frac{16}{3}$	0	$(\frac{16}{3}, 0)$
2	−5	$(2, -5)$

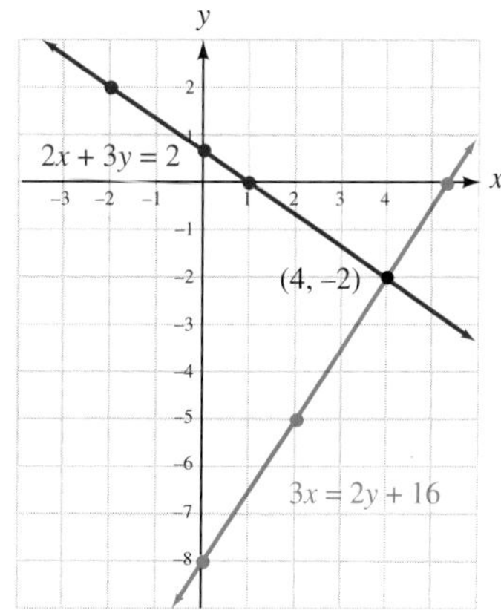
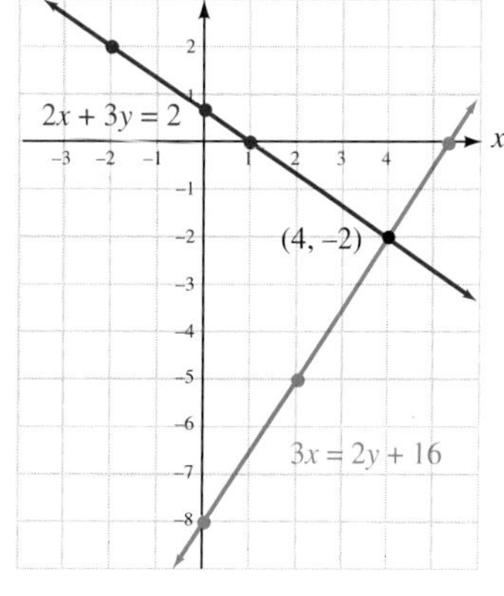

FIGURE 13-3

Although there are infinitely many pairs (x, y) that satisfy $2x + 3y = 2$, and infinitely many pairs (x, y) that satisfy $3x = 2y + 16$, only the coordinates of the point where the graphs intersect satisfy both equations at the same time. From the graph, the solution appears to be $(4, -2)$.

To check, we substitute 4 for x and −2 for y in each equation and verify that the pair $(4, -2)$ satisfies each equation.

Check:

$$\begin{aligned} 2x + 3y &= 2 \\ 2(\mathbf{4}) + 3(\mathbf{-2}) &\stackrel{?}{=} 2 \\ 8 - 6 &\stackrel{?}{=} 2 \\ 2 &= 2 \end{aligned} \qquad \begin{aligned} 3x &= 2y + 16 \\ 3(\mathbf{4}) &\stackrel{?}{=} 2(\mathbf{-2}) + 16 \\ 12 &\stackrel{?}{=} -4 + 16 \\ 12 &= 12 \end{aligned}$$

The equations in this system are independent equations, and the system is a consistent system of equations.

Self Check 1

Solve: $\begin{cases} 2x = y - 5 \\ x + y = -1 \end{cases}$.

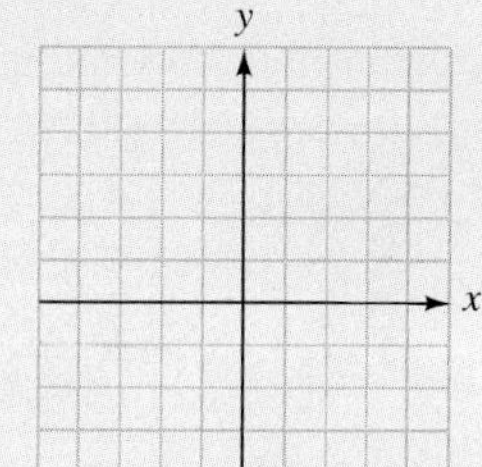

Answer $(-2, 1)$

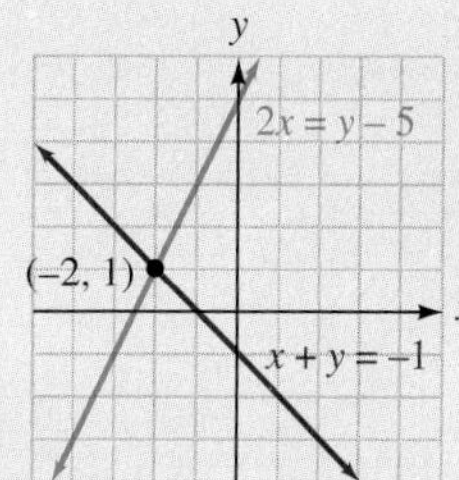

EXAMPLE 2 Solve: $\begin{cases} \dfrac{x}{2} - 1 = \dfrac{y}{2} \\ \dfrac{1}{3}x - \dfrac{1}{2}y = -4 \end{cases}$.

Solution We can multiply both sides of the first equation by 2 to clear it of fractions.

Self Check 2

Solve: $\begin{cases} -\frac{x}{2} = \frac{y}{4} \\ \frac{1}{4}x - \frac{3}{8}y = -2 \end{cases}$.

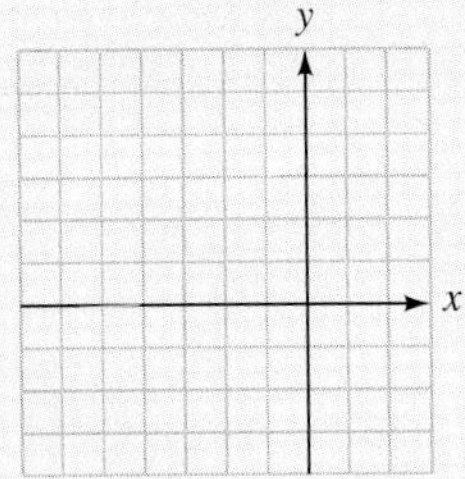

$$-\frac{x}{2} - 1 = \frac{y}{2}$$

$$\mathbf{2}\left(-\frac{x}{2} - 1\right) = \mathbf{2}\left(\frac{y}{2}\right)$$

(1) $\quad -x - 2 = y \qquad$ We will call this Equation 1.

We then multiply both sides of the second equation by 6 to clear it of fractions.

$$\frac{1}{3}x - \frac{1}{2}y = -4$$

$$\mathbf{6}\left(\frac{1}{3}x - \frac{1}{2}y\right) = \mathbf{6}(-4)$$

(2) $\quad 2x - 3y = -24 \qquad$ We will call this Equation 2.

Equations 1 and 2 form the following **equivalent system,** which has the same solutions as the original system:

$$\begin{cases} -x - 2 = y \\ 2x - 3y = -24 \end{cases}$$

In Figure 13-4, we graph $-x - 2 = y$ by plotting the y-intercept $(0, -2)$ and then drawing a slope of -1. We graph $2x - 3y = -24$ using the intercept method. It appears that the point of intersection is $(-6, 4)$.

A check will show that when the coordinates of $(-6, 4)$ are substituted into the two original equations, true statements result. Therefore, the equations are independent and the system is consistent.

$y = -x - 2$

So $m = -1 = \frac{-1}{1}$

and $b = -2$.

2x − 3y = −24		
x	y	(x, y)
0	8	(0, 8)
−12	0	(−12, 0)
−3	6	(−3, 6)

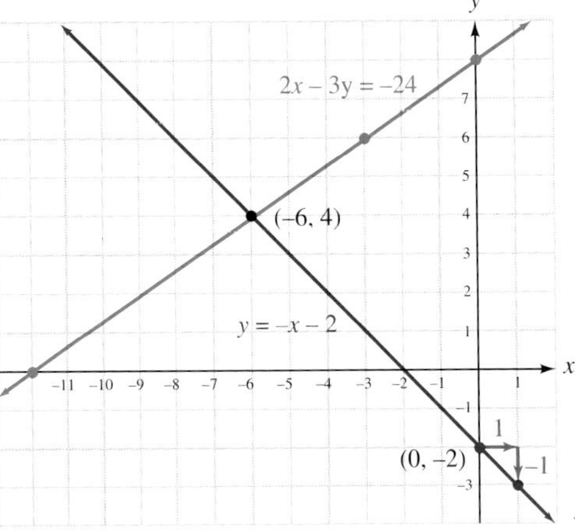

FIGURE 13-4

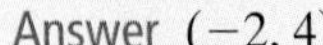

Answer $(-2, 4)$

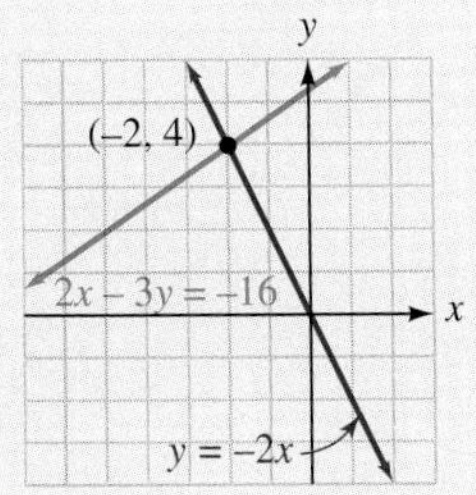

! COMMENT When solving a system of equations, always check your answer by substituting into the *original* equations. Do not check by substituting into the equations of an equivalent system. If an algebraic error was made while finding the equivalent system, an answer that would not satisfy the original system might appear to be correct.

Bridging the Gender Gap in Education

THINK IT THROUGH

"The woman of the 20th century will be the peer of man. In education, in art and science, in literature, in the home, the church, the state, everywhere she will be his acknowledged equal." Susan B. Anthony, 1900

The following graph shows the percent of associate degrees awarded in the U.S. by gender for the years 1970–2002. Determine the point of intersection of the lines and explain its importance.

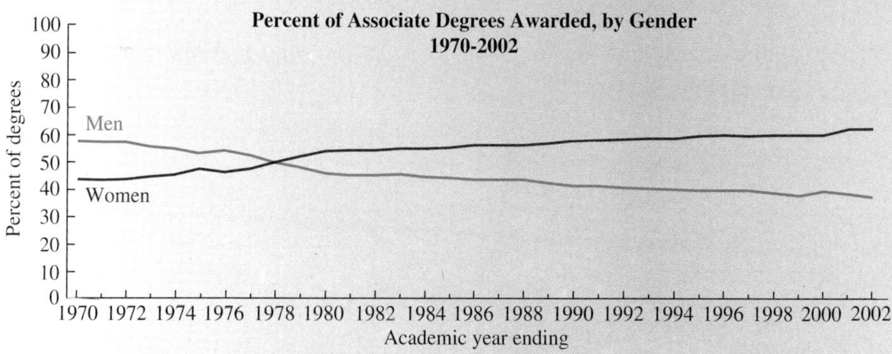

Source: *Chartbook of Degrees Conferred*, National Center for Education Statistics

Solving systems with a graphing calculator

CALCULATOR SNAPSHOT

We can use a graphing calculator to solve the system

$$\begin{cases} 2x + y = 12 \\ 2x - y = -2 \end{cases}$$

However, before we can enter the equations into the calculator, we must solve them for y.

$$\begin{aligned} 2x + y &= 12 \\ y &= -2x + 12 \end{aligned} \qquad \begin{aligned} 2x - y &= -2 \\ -y &= -2x - 2 \\ y &= 2x + 2 \end{aligned}$$

We enter the resulting equations and graph them on the same coordinate axes. If we use the standard window settings, their graphs will look like Figure 13-5(a).

To find the solution of the system, we use the INTERSECT feature that is found on most graphing calculators. With this option, the cursor automatically moves to the point of intersection of the graphs and displays the coordinates of that point. In Figure 13-5(b), we see that the solution is (2.5, 7). Consult your owner's manual for specific keystrokes to use INTERSECT.

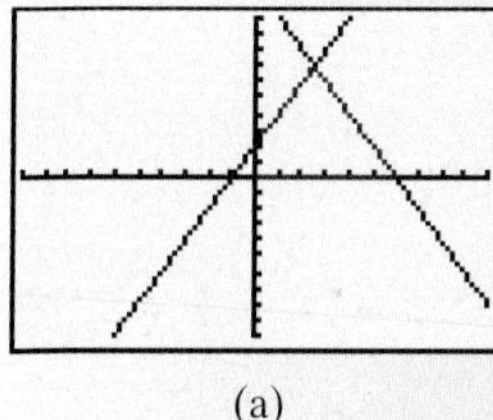

(a)

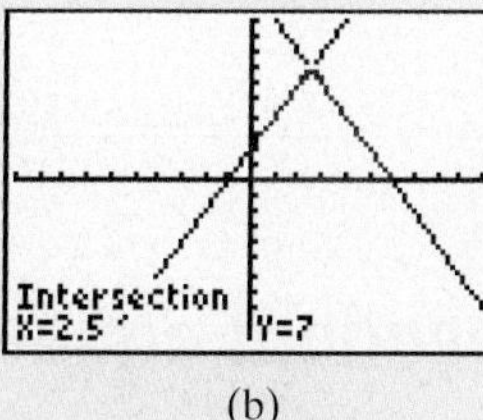

(b)

FIGURE 13-5

Inconsistent systems

Sometimes a system of equations has no solution. Such systems are called **inconsistent systems.**

Self Check 3

Solve: $\begin{cases} y = \frac{3}{2}x \\ 3x - 2y = 6 \end{cases}$.

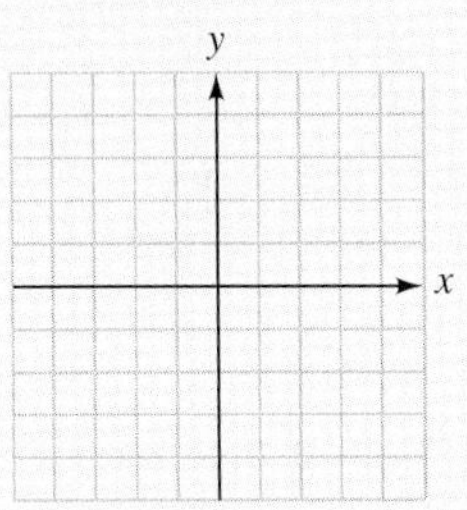

EXAMPLE 3 Solve: $\begin{cases} y = -2x - 6 \\ 4x + 2y = 8 \end{cases}$.

Solution Since $y = -2x - 6$ is written in slope–intercept form, we can graph it by plotting the y-intercept $(0, -6)$ and then drawing a slope of -2. (The run is 1, and the rise is -2.) We graph $4x + 2y = 8$ using the intercept method.

$y = -2x - 6$

So $m = -2 = \frac{-2}{1}$

and $b = -6$.

4x + 2y = 8		
x	y	(x, y)
0	4	(0, 4)
2	0	(2, 0)
1	2	(1, 2)

The system is graphed in Figure 13-6. Since the lines in the figure are parallel, they have the same slope. We can verify this by writing the second equation in slope–intercept form and observing that the coefficients of x in each equation are equal.

$$y = \mathbf{-2}x - 6 \qquad\qquad 4x + 2y = 8$$
$$2y = -4x + 8$$
$$y = \mathbf{-2}x + 4$$

Because parallel lines do not intersect, this system has no solution and is inconsistent. Since the graphs are different lines, the equations of the system are independent.

Answer The lines are parallel; therefore, the system has no solution.

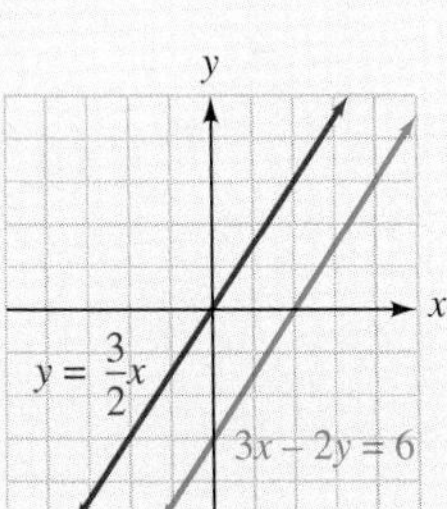

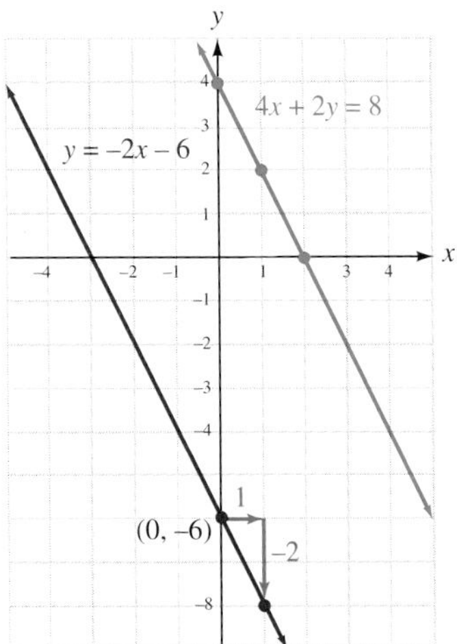

FIGURE 13-6

Dependent equations

Some systems of equations have infinitely many solutions, as we will see in the next example.

EXAMPLE 4 Solve: $\begin{cases} y - 4 = 2x \\ 4x + 8 = 2y \end{cases}$.

Solution We graph both equations on one set of axes, using the intercept method. See Figure 13-7.

$y - 4 = 2x$		
x	y	(x, y)
0	4	(0, 4)
−2	0	(−2, 0)
−1	2	(−1, 2)

$4x + 8 = 2y$		
x	y	(x, y)
0	4	(0, 4)
−2	0	(−2, 0)
−3	−2	(−3, −2)

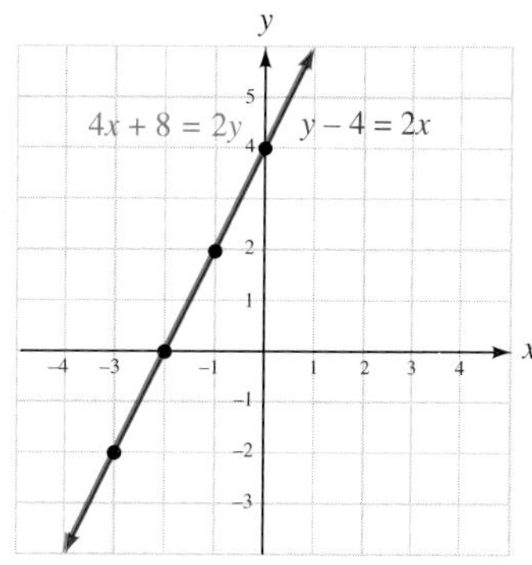

FIGURE 13-7

The lines in Figure 13-7 coincide (they are the same line). Because the lines intersect at infinitely many points, the system has infinitely many solutions. From the graph, we can see that some of the solutions are (0, 4), (−1, 2), and (−3, −2). Equations that have the same graph are called **dependent equations.** Therefore, this system is consistent and its equations are dependent.

Self Check 4

Solve: $\begin{cases} 6x - 2y = 4 \\ y + 2 = 3x \end{cases}$.

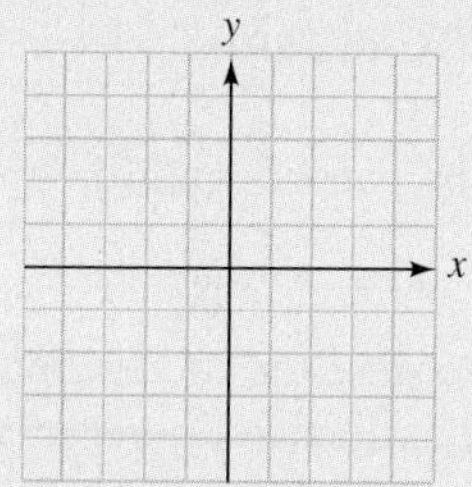

Answer The graphs are the same line. There are infinitely many solutions.

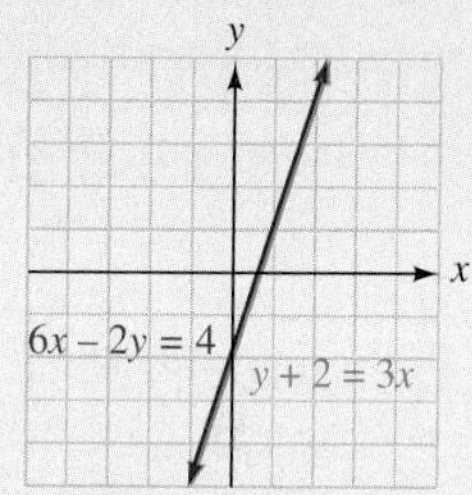

There are three possible outcomes when we solve a system of two linear equations using the graphing method.

Possible graph	If the	Then
	lines are different and intersect,	the equations are independent and the system is consistent. There is one solution.
	lines are different and parallel,	the equations are independent and the system is inconsistent. There are no solutions.
	lines are the same,	the equations are dependent and the system is consistent. There are infinitely many solutions.

Section 13.1 STUDY SET

VOCABULARY *Fill in the blanks.*

1. The pair of equations $\begin{cases} x - y = -1 \\ 2x - y = 1 \end{cases}$ is called a ________ of equations.
2. Because the ordered pair (2, 3) satisfies both equations in Exercise 1, it is called a ________ of the system of equations.
3. When the graphs of two equations in a system are different lines, the equations are called __________ equations.
4. When a system of equations has a solution, the system is called a __________ system.
5. Systems of equations that have no solution are called __________ systems.
6. Equations that have the same graph are called __________ equations.

CONCEPTS *Refer to the following illustration. Determine whether a true or false statement would result if the coordinates of each point are substituted into the equation for the indicated line.*

7. point A, line l_1.
8. point B, line l_2.
9. point A, line l_2.
10. point B, line l_1.
11. point C, line l_1.
12. point C, line l_2.

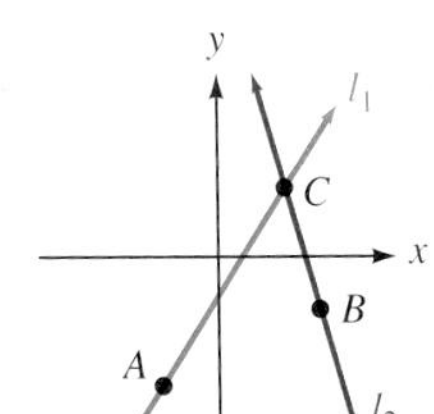

13. The following tables were created to graph the two linear equations in a system. What is the solution of the system?

Equation 1

x	y
0	−5
−5	0
−4	−1
1	−6
2	−7

Equation 2

x	y
0	3
−3	0
−2	1
−4	−1
1	4

14. **a.** To graph $5x - 2y = 10$, we can use the intercept method. Complete the table.

x	y
0	
	0

b. To graph $y = 3x - 2$, we can use the slope and y-intercept. Fill in the blanks.

slope: $\quad = \frac{\quad}{1}$ $\qquad$ y-intercept:

15. How many solutions does the system of equations graphed in the illustration have? Is the system consistent or inconsistent?

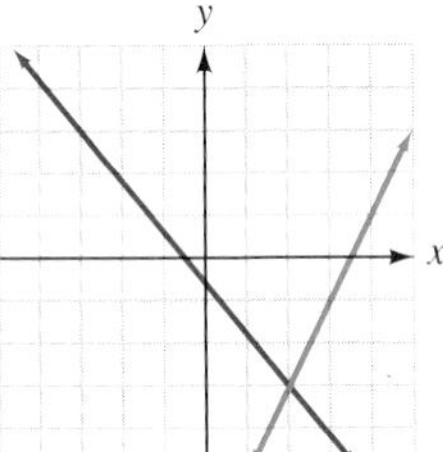

16. How many solutions does the system of equations graphed in the illustration have? Are the equations dependent or independent?

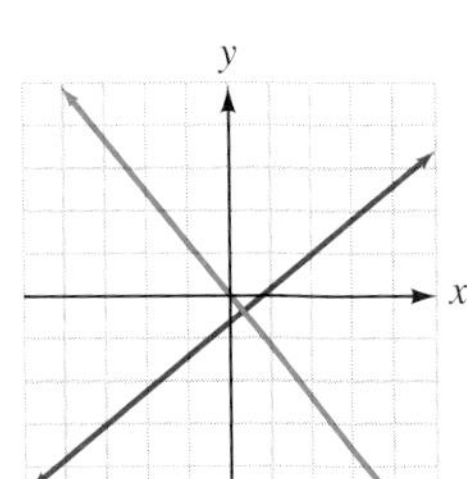

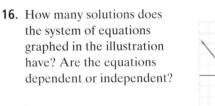

17. The solution of the system of equations graphed on the right is $\left(\frac{2}{5}, -\frac{1}{3}\right)$. Knowing this, can you see any disadvantages to the graphing method?

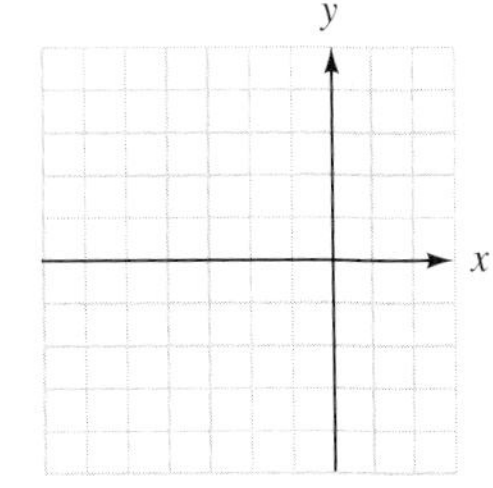

18. Draw the graphs of two linear equations so that the system has
    a. one solution $(-3, -2)$.

b. infinitely many solutions, three of which are $(-2, 0)$, $(1, 2)$, and $(4, 4)$.

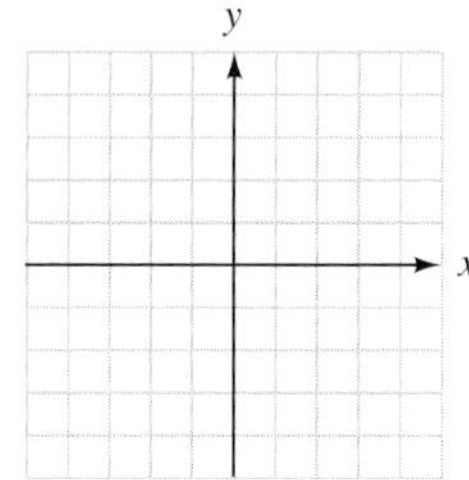

NOTATION *Clear each equation of fractions.*

19.
$$\frac{1}{6}x - \frac{1}{3}y = \frac{11}{2}$$
$$\square\left(\frac{1}{6}x - \frac{1}{3}y\right) = \square\left(\frac{11}{2}\right)$$
$$\square\left(\frac{1}{6}x\right) - 6\left(\square\right) = 6\left(\frac{11}{2}\right)$$
$$x - 2y = 33$$

20.
$$\frac{3x}{5} - \frac{4y}{5} = -1$$
$$\square\left(\frac{3x}{5} - \frac{4y}{5}\right) = \square(-1)$$
$$\square\left(\frac{3x}{5}\right) - 5\left(\square\right) = 5(-1)$$
$$3x - 4y = -5$$

PRACTICE *Determine whether the ordered pair is a solution of the given system.*

21. $(1, 1)$, $\begin{cases} x + y = 2 \\ 2x - y = 1 \end{cases}$

22. $(1, 3)$, $\begin{cases} 2x + y = 5 \\ 3x - y = 0 \end{cases}$

23. $(3, -2)$, $\begin{cases} 2x + y = 4 \\ y = 1 - x \end{cases}$

24. $(-2, 4)$, $\begin{cases} 2x + 2y = 4 \\ 3y = 10 - x \end{cases}$

25. $(-2, -4)$, $\begin{cases} 4x + 5y = -23 \\ -3x + 2y = 0 \end{cases}$

26. $(-5, 2)$, $\begin{cases} -2x + 7y = 17 \\ 3x - 4y = -19 \end{cases}$

27. $\left(\frac{1}{2}, 3\right)$, $\begin{cases} 2x + y = 4 \\ 4x - 11 = 3y \end{cases}$

28. $\left(2, \frac{1}{3}\right)$, $\begin{cases} x - 3y = 1 \\ -2x + 6 = -6y \end{cases}$

29. $\left(-\frac{2}{5}, \frac{1}{4}\right)$, $\begin{cases} x - 4y = -6 \\ 8y = 10x + 12 \end{cases}$

30. $\left(-\frac{1}{3}, \frac{3}{4}\right)$, $\begin{cases} 3x + 4y = 2 \\ 12y = 3(2 - 3x) \end{cases}$

31. $(0.2, 0.3)$, $\begin{cases} 20x + 10y = 7 \\ 20y = 15x + 3 \end{cases}$

32. $(2.5, 3.5)$, $\begin{cases} 4x - 3 = 2y \\ 4y + 1 = 6x \end{cases}$

Solve each system by the graphing method. If the equations of a system are dependent or if a system is inconsistent, so indicate.

33. $\begin{cases} 2x + 3y = 12 \\ 2x - y = 4 \end{cases}$

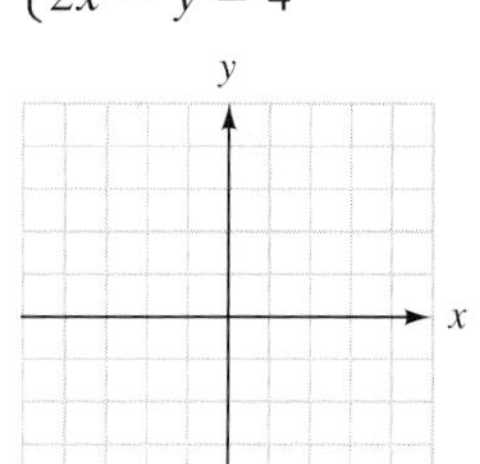

34. $\begin{cases} 5x + y = 5 \\ 5x + 3y = 15 \end{cases}$

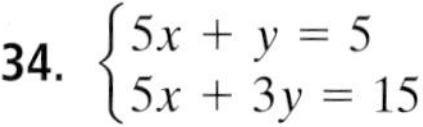

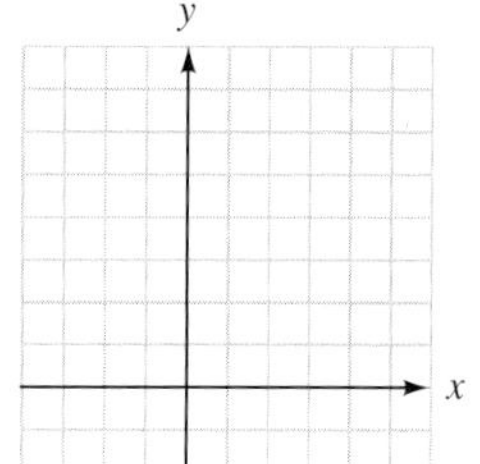

35. $\begin{cases} x + y = 2 \\ y = x - 4 \end{cases}$

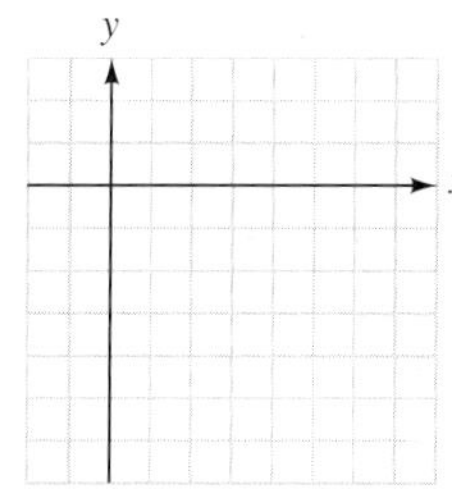

36. $\begin{cases} x + y = 1 \\ y = x + 5 \end{cases}$

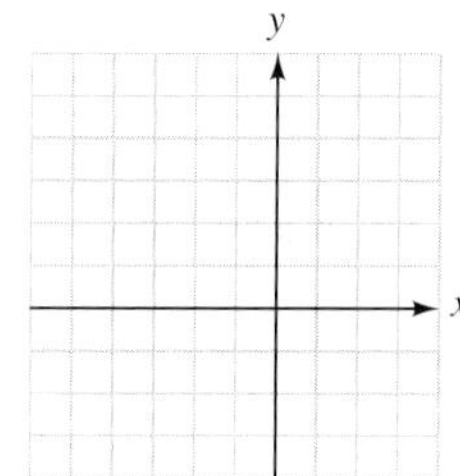

37. $\begin{cases} 3x + 2y = -8 \\ 2x - 3y = -1 \end{cases}$

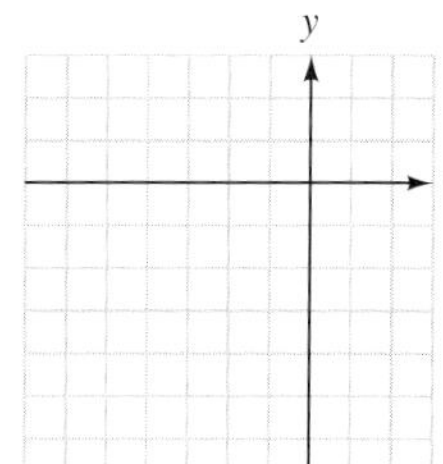

38. $\begin{cases} x + 4y = -2 \\ y = -x - 5 \end{cases}$

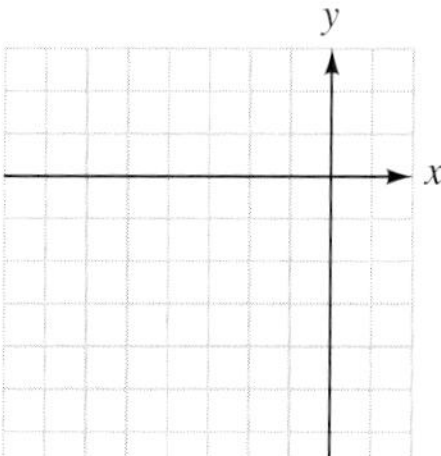

39. $\begin{cases} 4x - 2y = 8 \\ y = 2x - 4 \end{cases}$

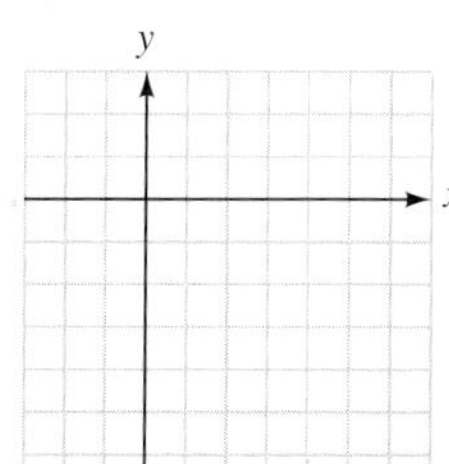

40. $\begin{cases} 3x - 6y = 18 \\ x = 2y + 3 \end{cases}$

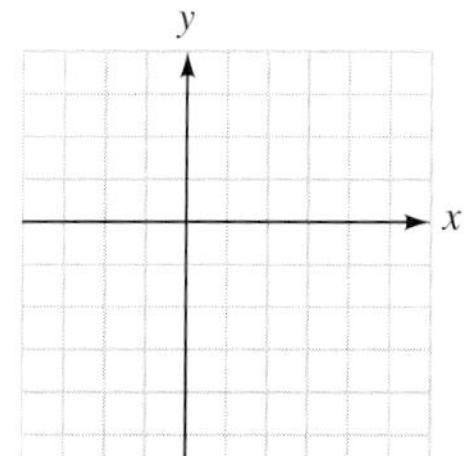

41. $\begin{cases} 2x - 3y = -18 \\ 3x + 2y = -1 \end{cases}$

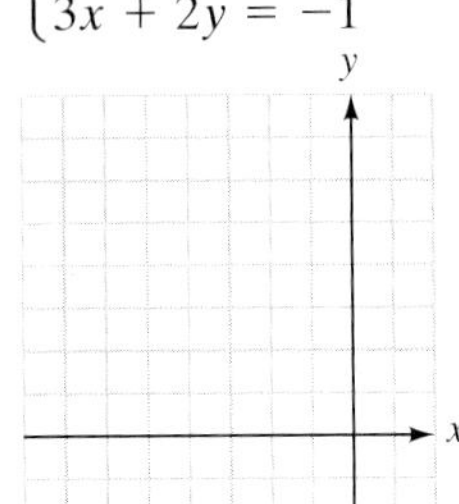

42. $\begin{cases} -x + 3y = -11 \\ 3x - y = 17 \end{cases}$

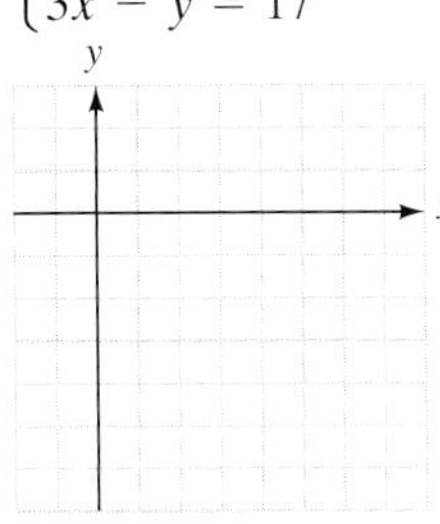

43. $\begin{cases} x = 4 \\ 2y = 12 - 4x \end{cases}$

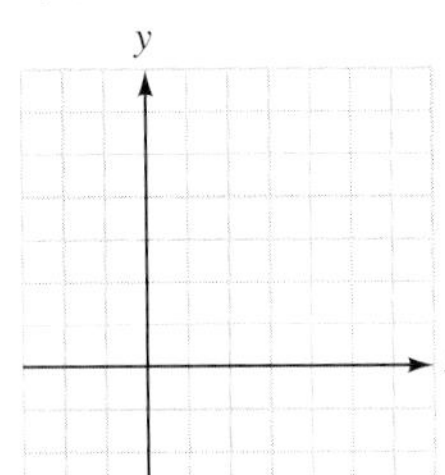

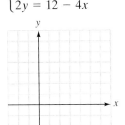

44. $\begin{cases} x = 3 \\ 3y = 6 - 2x \end{cases}$

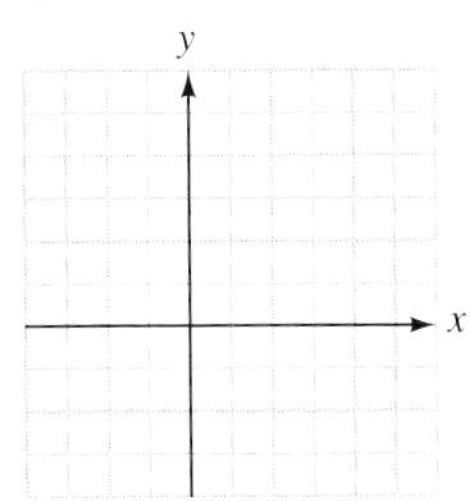

45. $\begin{cases} x + 2y = -4 \\ x - \frac{1}{2}y = 6 \end{cases}$

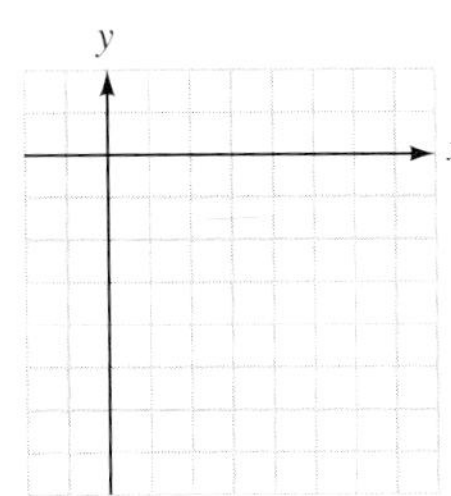

46. $\begin{cases} \frac{2}{3}x - y = -3 \\ 3x + y = 3 \end{cases}$

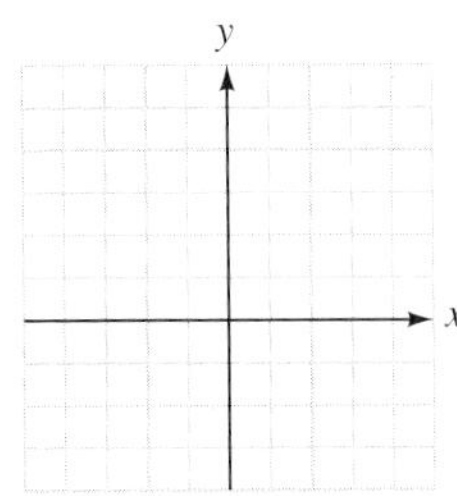

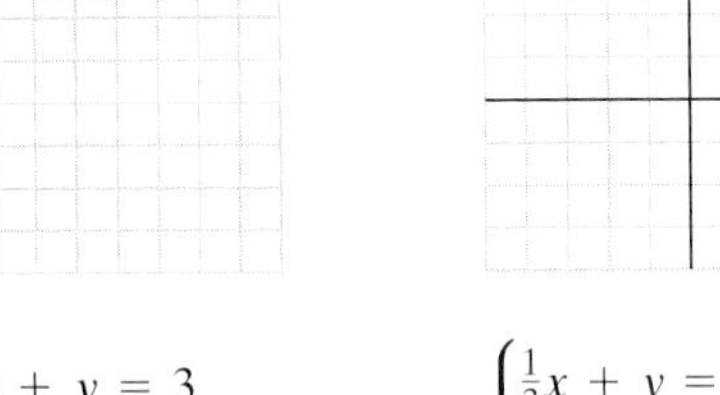

47. $\begin{cases} -\frac{3}{4}x + y = 3 \\ \frac{1}{4}x + y = -1 \end{cases}$

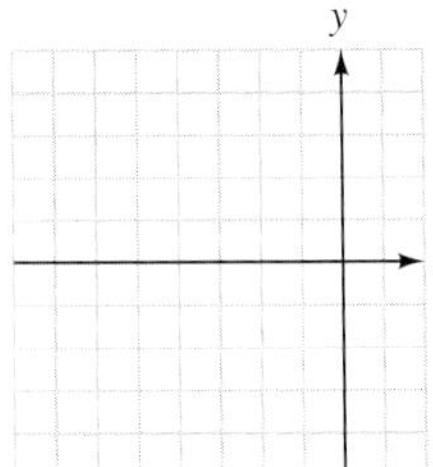

48. $\begin{cases} \frac{1}{3}x + y = 7 \\ \frac{2x}{3} - y = -4 \end{cases}$

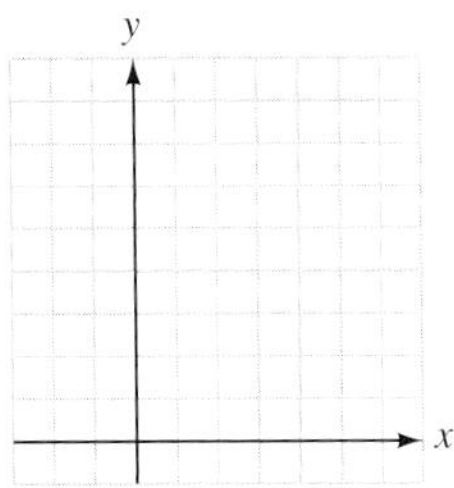

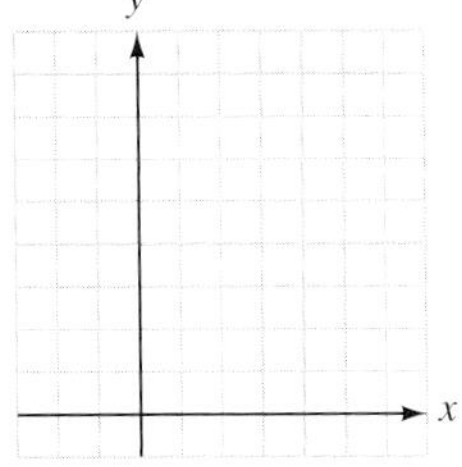

49. $\begin{cases} 2y = 3x + 2 \\ \frac{3}{2}x - y = 3 \end{cases}$

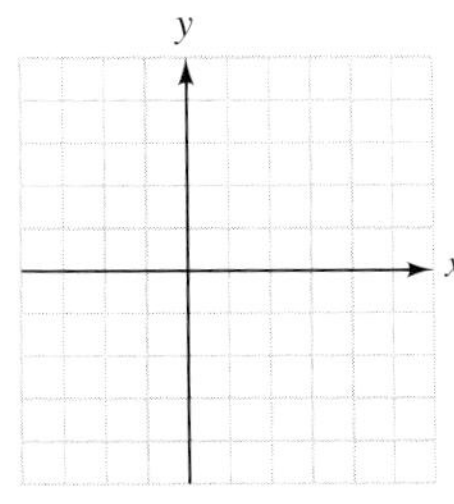

50. $\begin{cases} -\frac{3}{5}x - \frac{1}{5}y = \frac{6}{5} \\ x + \frac{y}{3} = -2 \end{cases}$

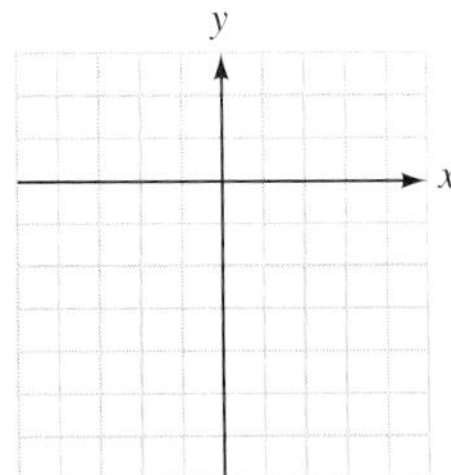

51. $\begin{cases} \frac{1}{3}x - \frac{1}{2}y = \frac{1}{6} \\ \frac{2x}{5} + \frac{y}{2} = \frac{13}{10} \end{cases}$

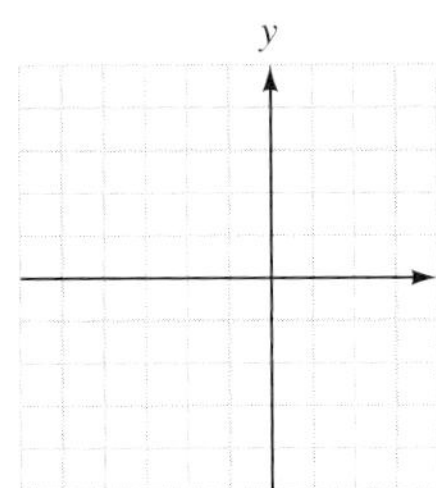

52. $\begin{cases} \frac{3x}{4} + \frac{2y}{3} = -\frac{19}{6} \\ 3y = -x \end{cases}$

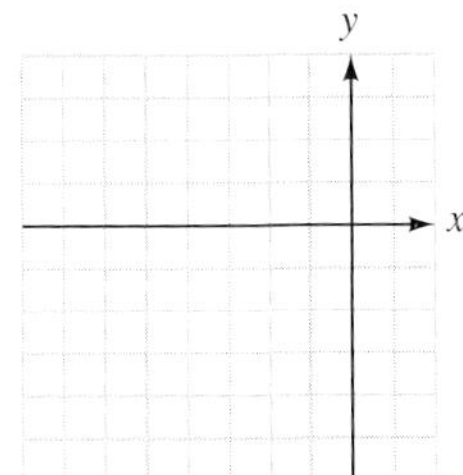

Use a graphing calculator to solve each system, if possible.

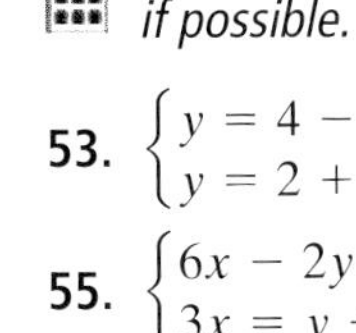

53. $\begin{cases} y = 4 - x \\ y = 2 + x \end{cases}$

54. $\begin{cases} 3x - 6y = 4 \\ 2x + y = 1 \end{cases}$

55. $\begin{cases} 6x - 2y = 5 \\ 3x = y + 10 \end{cases}$

56. $\begin{cases} x - 3y = -2 \\ 5x + y = 10 \end{cases}$

APPLICATIONS

57. TRANSPLANTS See the illustration below.

    a. What was the relationship between the number of donors and those awaiting a transplant in 1989?

    b. In what year were the number of donors and the number waiting for a transplant the same? Estimate the number.

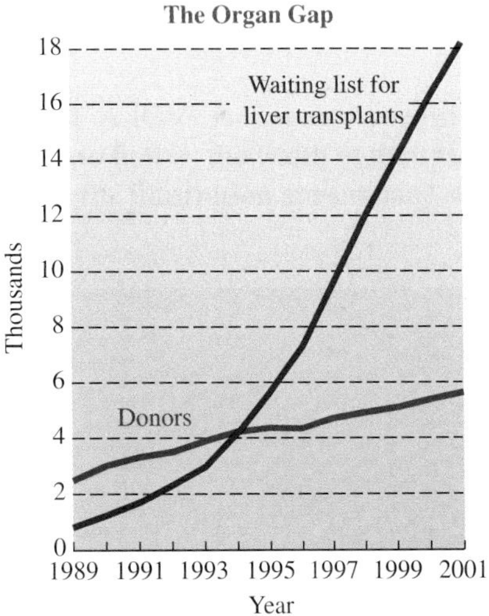

Source: Organ Procurement and Transportation Network

58. DAILY TRACKING POLLS See the illustration on the next page.

    a. Which political candidate was ahead on October 28 and by how much?

    b. On what day did the challenger pull even with the incumbent?

c. If the election was held November 4, who did the poll predict would win, and by how many percentage points?

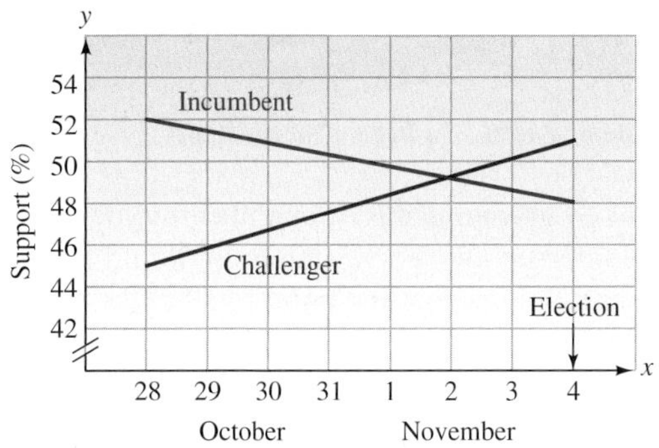

59. LATITUDE AND LONGITUDE See the illustration.

a. Name three American cities that lie on a latitude line of 30° north.

b. Name three American cities that lie on a longitude line of 90° west.

c. What city lies on both lines?

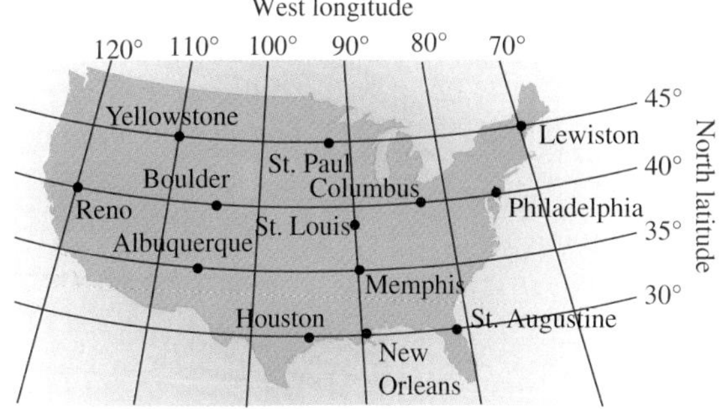

60. ECONOMICS The following graph illustrates the law of supply and demand.

a. Complete this sentence: As the price of an item increases, the *supply* of the item ________.

b. Complete this sentence: As the price of an item increases, the *demand* for the item ________.

c. For what price will the supply equal the demand? How many items will be supplied for this price?

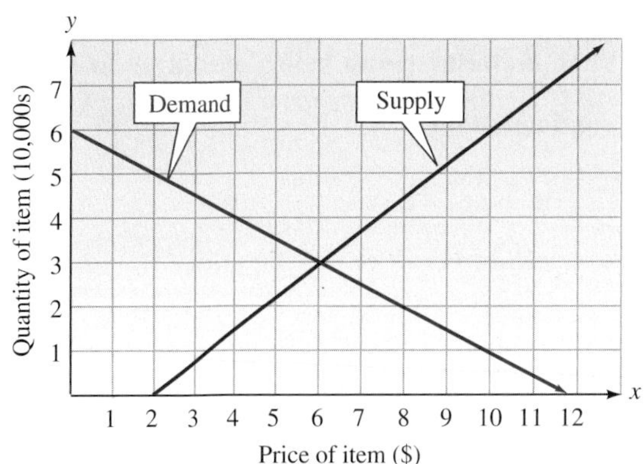

61. AIR TRAFFIC CONTROL The equations describing the paths of two airplanes are $y = -\frac{1}{2}x + 3$ and $3y = 2x + 2$. Graph each equation on the radar screen shown. Is there a possibility of a mid-air collision? If so, where?

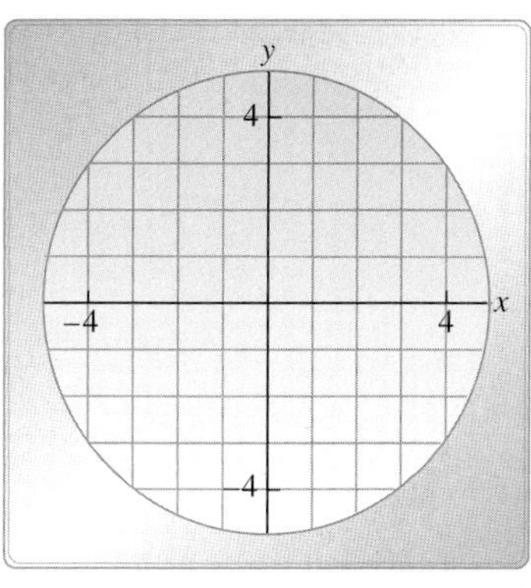

62. TV COVERAGE A television camera is located at $(-2, 0)$ and will follow the launch of a space shuttle, as shown. (Each unit in the illustration is 1 mile.) As the shuttle rises vertically on a path described by $x = 2$, the farthest the camera can tilt back is a line of sight given by $y = \frac{5}{2}x + 5$. For how many miles of the shuttle's flight will it be in view of the camera?

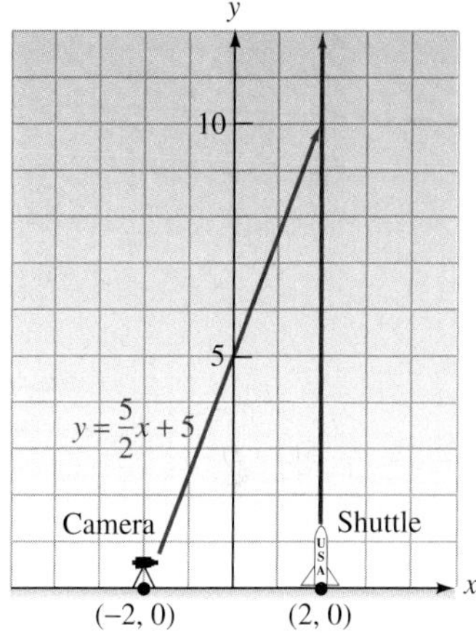

WRITING

63. Look up the word *simultaneous* in a dictionary and give its definition. In mathematics, what is meant by a simultaneous solution of a system of equations?

64. Suppose the solution of a system is $\left(\frac{1}{3}, -\frac{3}{5}\right)$. Do you think you would be able to find the solution using the graphing method? Explain.

REVIEW

65. Are the graphs of the lines $y = 2x - 3$ and $4x - 2y = 8$ parallel, perpendicular, or neither?

66. Are the graphs of the lines $y = 5x$ and $y = -\frac{1}{5}x$ parallel, perpendicular, or neither?

67. If $f(x) = -4x - x^2$, find $f(3)$.

68. In what quadrant does $(-12, 15)$ lie?

69. Write the equation for the y-axis.

70. Does $(1, 2)$ lie on the line $2x + y = 4$?

71. What point does the line with equation $y - 2 = 7(x - 5)$ pass through?

72. Is the word *domain* associated with the inputs or the outputs of a function?

13.2 Solving Systems of Equations by Substitution

- The substitution method • Inconsistent systems • Dependent equations

When solving a system of equations by graphing, it is often difficult to determine the exact coordinates of the point of intersection. For example, it would be virtually impossible to distinguish that two lines intersect at the point $(\frac{1}{16}, -\frac{3}{5})$. In this section, we introduce an algebraic method that finds *exact* solutions. It is called the *substitution method.*

The substitution method

To solve the system

$$\begin{cases} y = 3x - 2 \\ 2x + y = 8 \end{cases}$$

by the **substitution method,** we note that the first equation, $y = 3x - 2$, is *solved for y* (or *y is expressed in terms of x*). Because $y = 3x - 2$, we can substitute $3x - 2$ for y in the equation $2x + y = 8$ to get

$2x + \mathbf{y} = 8$ This is the second equation of the system.

$2x + \mathbf{3x - 2} = 8$ Substitute $3x - 2$ for y.

The resulting equation has only one variable and can be solved for x.

$2x + 3x - 2 = 8$

$5x - 2 = 8$ Combine like terms: $2x + 3x = 5x$.

$5x = 10$ Add 2 to both sides.

$x = 2$ Divide both sides by 5.

We can find y by substituting 2 for x in either equation of the given system. Because $y = 3x - 2$ is already solved for y, it is easier to substitute into this equation.

$y = 3\mathbf{x} - 2$ This is the first equation of the system.

$= 3(\mathbf{2}) - 2$ Substitute 2 for x.

$= 6 - 2$

$= 4$

The solution to the given system is (2, 4).

Check: **First equation**

$y = 3\mathbf{x} - 2$

$\mathbf{4} \stackrel{?}{=} 3(\mathbf{2}) - 2$

$4 \stackrel{?}{=} 6 - 2$

$4 = 4$

Second equation

$2\mathbf{x} + \mathbf{y} = 8$

$2(\mathbf{2}) + \mathbf{4} \stackrel{?}{=} 8$

$4 + 4 \stackrel{?}{=} 8$

$8 = 8$

If we graphed the lines represented by the equations of the given system, they would intersect at the point (2, 4). The equations of this system are independent, and the system is consistent.

To solve a system of equations in x and y by the substitution method, we follow these steps.

The substitution method

1. Solve one of the equations for either x or y. (This step will not be necessary if an equation is already solved for x or y.)
2. Substitute the resulting expression for the variable obtained in step 1 into the remaining equation and solve that equation.
3. Find the value of the other variable by substituting the solution found in step 2 into any equation containing both variables.
4. Check the solution in the equations of the original system.

EXAMPLE 1 Solve: $\begin{cases} 2x + y = -10 \\ x = -3y \end{cases}$.

Solution The second equation, $x = -3y$, tells us that x and $-3y$ have the same value. Therefore, we can substitute $-3y$ for x in the first equation.

$2x + y = -10$	This is the first equation of the system.
$2(\mathbf{-3y}) + y = -10$	Replace x with $-3y$.
$-6y + y = -10$	Perform the multiplication.
$-5y = -10$	Combine like terms.
$y = 2$	Divide both sides by -5.

We can find x by substituting 2 for y in the equation $x = -3y$.

$x = -3y$	This is the second equation of the system.
$= -3(\mathbf{2})$	Substitute 2 for y.
$= -6$	

The solution is $(-6, 2)$.

Check: **First equation** | **Second equation**

$$2x + y = -10 \qquad\qquad x = -3y$$
$$2(-6) + 2 \stackrel{?}{=} -10 \qquad\qquad -6 \stackrel{?}{=} -3(2)$$
$$-12 + 2 \stackrel{?}{=} -10 \qquad\qquad -6 = -6$$
$$-10 = -10$$

Self Check 1

Solve: $\begin{cases} x = -2y \\ 3x - 2y = 8 \end{cases}$.

Answer $(2, -1)$

To find a substitution equation, solve one of the equations of the system for one of its variables. If possible, solve for a variable whose coefficient is 1 or -1 to avoid working with fractions.

EXAMPLE 2 Solve: $\begin{cases} 2x + y = -5 \\ 3x + 5y = -4 \end{cases}$.

Solution We solve one of the equations for one of the variables. Since the term y in the first equation has a coefficient of 1, we solve the first equation for y.

$2x + y = -5$	This is the first equation of the system.
$y = -5 - 2x$	Subtract $2x$ from both sides to isolate y.

Self Check 2

Solve: $\begin{cases} 2x - 3y = 13 \\ 3x + y = 3 \end{cases}$.

We then substitute $-5 - 2x$ for y in the second equation and solve for x.

$3x + 5y = -4$	This is the second equation of the system.
$3x + 5(\mathbf{-5 - 2x}) = -4$	Substitute $-5 - 2x$ for y. Don't forget to write $-5 - 2x$ within parentheses.
$3x - 25 - 10x = -4$	Distribute the multiplication by 5.
$-7x - 25 = -4$	Combine like terms: $3x - 10x = -7x$.
$-7x = 21$	Add 25 to both sides.
$x = -3$	Divide both sides by -7.

We can find y by substituting -3 for x in the equation $y = -5 - 2x$.

$$\begin{aligned} y &= -5 - 2x \\ &= -5 - 2(\mathbf{-3}) \quad \text{Substitute } -3 \text{ for } x. \\ &= -5 + 6 \\ &= 1 \end{aligned}$$

Answer $(2, -3)$

The solution is $(-3, 1)$. Check it in the original equations.

Systems of equations are sometimes written in variables other than x and y. For example, the system

$$\begin{cases} 3a - 3b = 5 \\ 3 - a = -2b \end{cases}$$

is written in a and b. Regardless of the variables used, the procedures used to solve the system remain the same. Unless told otherwise, list the values of the variables of a solution in alphabetical order. Here, the solution should be expressed in the form (a, b).

Self Check 3

Solve: $\begin{cases} 2s - t = 4 \\ 3s - 5t = 2 \end{cases}$.

EXAMPLE 3 Solve: $\begin{cases} 3a - 3b = 5 \\ 3 - a = -2b \end{cases}$.

Solution Since the coefficient of a in the second equation is -1, we will solve that equation for a.

$3 - a = -2b$	This is the second equation of the system.
$-a = -2b - 3$	Subtract 3 from both sides.

To obtain a on the left-hand side, we can multiply (or divide) both sides of the equation by -1.

$\mathbf{-1}(-a) = \mathbf{-1}(-2b - 3)$	Multiply both sides by -1.
$a = 2b + 3$	Perform the multiplications.

We then substitute $2b + 3$ for a in the first equation and proceed as follows:

$3a - 3b = 5$	
$3(\mathbf{2b + 3}) - 3b = 5$	Substitute.
$6b + 9 - 3b = 5$	Distribute the multiplication by 3.
$3b + 9 = 5$	Combine like terms.
$3b = -4$	Subtract 9 from both sides: $5 - 9 = -4$.
$b = -\frac{4}{3}$	Divide both sides by 3.

To find a, we substitute $-\frac{4}{3}$ for b in $a = 2b + 3$ and simplify.

$$\begin{aligned} a &= 2\mathbf{b} + 3 \\ &= 2\left(-\frac{\mathbf{4}}{\mathbf{3}}\right) + 3 && \text{Substitute.} \\ &= -\frac{8}{3} + \frac{9}{3} && \text{Perform the multiplication: } 2\left(-\frac{4}{3}\right) = -\frac{8}{3}. \text{ Write 3 as } \frac{9}{3}. \\ &= \frac{1}{3} && \text{Add the numerators and keep the common denominator.} \end{aligned}$$

The solution is $(\frac{1}{3}, -\frac{4}{3})$. Check it in the original equations.

Answer $\left(\frac{18}{7}, \frac{8}{7}\right)$

EXAMPLE 4 Solve: $\begin{cases} \frac{x}{2} + \frac{y}{4} = -\frac{1}{4} \\ 2x - y = 2 + y - x \end{cases}$.

Self Check 4

Solve: $\begin{cases} \frac{1}{3}x - \frac{1}{6}y = -\frac{1}{3} \\ x + y = -3 - 2x - y \end{cases}$.

Solution It is helpful to rewrite each equation in simpler form before performing a substitution. We begin by clearing the first equation of fractions.

$$\begin{aligned} \frac{x}{2} + \frac{y}{4} &= -\frac{1}{4} \\ \mathbf{4}\left(\frac{x}{2} + \frac{y}{4}\right) &= \mathbf{4}\left(-\frac{1}{4}\right) && \text{Multiply both sides by the LCD, which is 4.} \\ 2x + y &= -1 \end{aligned}$$

We can write the second equation in general form ($Ax + By = C$) by adding x and subtracting y from both sides.

$$\begin{aligned} 2x - y &= 2 + y - x \\ 2x - y \mathbf{+ x - y} &= 2 + y - x \mathbf{+ x - y} \\ 3x - 2y &= 2 && \text{Combine like terms.} \end{aligned}$$

The two results form the following equivalent system, which has the same solution as the original one.

(1) $\begin{cases} 2x + y = -1 \\ 3x - 2y = 2 \end{cases}$
(2)

To solve this system, we solve Equation 1 for y.

$$\begin{aligned} 2x + y &= -1 \\ 2x + y \mathbf{- 2x} &= -1 \mathbf{- 2x} && \text{Subtract } 2x \text{ from both sides.} \\ \textbf{(3)} \quad y &= -1 - 2x && \text{Combine like terms.} \end{aligned}$$

To find x, we substitute $-1 - 2x$ for y in Equation 2 and proceed as follows:

$$\begin{aligned} 3x - 2\mathbf{y} &= 2 \\ 3x - 2(\mathbf{-1 - 2x}) &= 2 && \text{Substitute.} \\ 3x + 2 + 4x &= 2 && \text{Distribute the multiplication by } -2. \\ 7x + 2 &= 2 && \text{Combine like terms.} \\ 7x &= 0 && \text{Subtract 2 from both sides.} \\ x &= 0 && \text{Divide both sides by 7.} \end{aligned}$$

To find y, we substitute 0 for x in Equation 3.

$$y = -1 - 2\mathbf{x}$$
$$y = -1 - 2(\mathbf{0})$$
$$y = -1$$

The solution is $(0, -1)$. Check it in the original equations.

Answer $(-1, 0)$

Inconsistent systems

Self Check 5

Solve: $\begin{cases} 0.1x - 0.4 = 0.1y \\ -2y = 2(2 - x) \end{cases}$.

EXAMPLE 5 Solve: $\begin{cases} 0.01x = 0.12 - 0.04y \\ 2x = 4(3 - 2y) \end{cases}$.

Solution The first equation contains decimal coefficients. We can clear the equation of decimals by multiplying both sides by 100.

$$\begin{cases} x = 12 - 4y \\ 2x = 4(3 - 2y) \end{cases}$$

Since $x = 12 - 4y$, we can substitute $12 - 4y$ for x in the second equation and solve for y.

$2\mathbf{x} = 4(3 - 2y)$	This is the second equation.
$2(\mathbf{12 - 4y}) = 4(3 - 2y)$	Substitute.
$24 - 8y = 12 - 8y$	Distribute.
$24 \neq 12$	Add $8y$ to both sides.

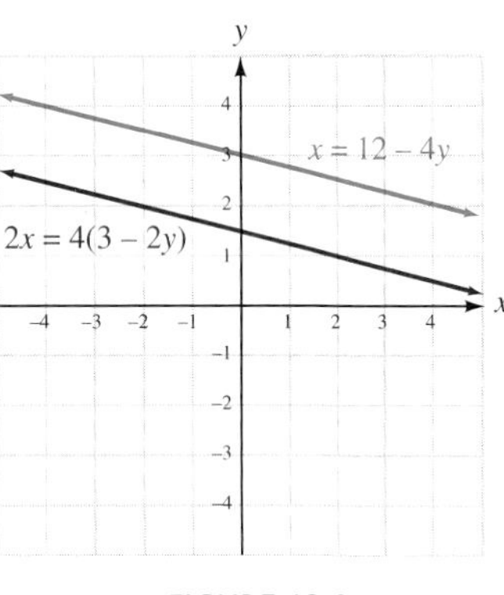

FIGURE 13-8

Here, the terms involving y drop out, and a false result of $24 = 12$ is obtained. This result indicates that the equations are independent and also that the system is inconsistent. As we see in Figure 13-8, when the equations are graphed, the graphs are parallel lines. This system has no solution.

Answer no solution

Dependent equations

Self Check 6

Solve: $\begin{cases} y = 2 - x \\ 3x + 3y = 6 \end{cases}$.

EXAMPLE 6 Solve: $\begin{cases} x = -3y + 6 \\ 2x + 6y = 12 \end{cases}$.

Solution We can substitute $-3y + 6$ for x in the second equation and proceed as follows:

$2\mathbf{x} + 6y = 12$	This is the second equation of the system.
$2(\mathbf{-3y + 6}) + 6y = 12$	Substitute.
$-6y + 12 + 6y = 12$	Distribute the multiplication by 2.
$12 = 12$	Combine like terms.

Although $12 = 12$ is true, we did not find y. This indicates that the equations are dependent. As we see in Figure 13-9, when these equations are graphed, their graphs are identical.

Because any ordered pair that satisfies one equation of the system also satisfies the other, the system has infinitely many solutions. To find some, we substitute 0, 3, and 6 for x in either equation and solve for y. The pairs (0, 2), (3, 1), and (6, 0) are some of the solutions.

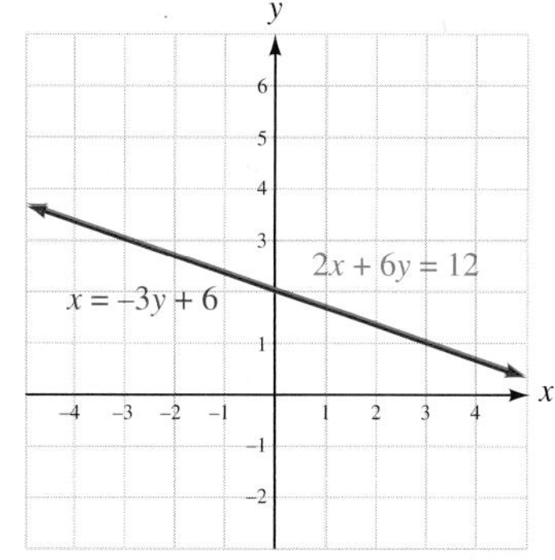

FIGURE 13-9

Answer infinitely many solutions

Section 13.2 STUDY SET

VOCABULARY *Fill in the blanks.*

1. We say that the equation $y = 2x + 4$ is solved for ___ or that y is expressed in _______ of x.
2. "To _______ a solution of a system" means to see whether the coordinates of the ordered pair satisfy both equations.
3. When we write $2(x - 6)$ as $2x - 12$, we are applying the ___________ property.
4. In mathematics, "to __________" means to replace an expression with one that is equivalent to it.
5. A dependent system has __________ many solutions.
6. In the term y, the coefficient is understood to be ___.

CONCEPTS

7. Consider the system $\begin{cases} 2x + 3y = 12 \\ y = 2x + 4 \end{cases}$.
   a. How many variables does each equation of the system contain?
   b. Substitute $2x + 4$ for y in the first equation. How many variables does the resulting equation contain?
8. For each equation, solve for y.
   a. $y + 2 = x$
   b. $2 + x + y = 0$
   c. $2 - y = x$
9. Given the equation $x - 2y = -10$,
   a. solve it for x.
   b. solve it for y.
   c. which variable was easier to solve for, x or y? Explain.
10. Which variable in which equation should be solved for in step 1 of the substitution method?
    a. $\begin{cases} x - 2y = 2 \\ 2x + 3y = 11 \end{cases}$
    b. $\begin{cases} 2x - 3y = 2 \\ 2x - y = 11 \end{cases}$
11. a. Find the error when $x - 4$ is substituted for y.

$x + 2y = 5$	This is the first equation of the system.
$x + 2x - 4 = 5$	Substitute for y: $y = x - 4$.
$3x - 4 = 5$	Combine like terms.
$3x = 9$	Add 4 to both sides.
$x = 3$	Perform the divisions.

    b. Rework the problem to find the correct value of x.
12. A student uses the substitution method to solve the system $\begin{cases} 4a + 5b = 2 \\ b = 3a - 11 \end{cases}$. She finds that $a = 3$. What is the easiest way for her to determine the value of b?

13. Consider the system $\begin{cases} y = 2x \\ 3x + 2y = 6 \end{cases}$.

a. Graph the equations on the same coordinate system. Why is it difficult to determine the solution of the system?

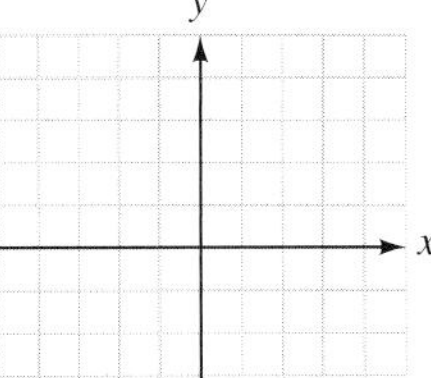

b. Solve the system by the substitution method.

14. Suppose the equation $-2 = 1$ is obtained when a system is solved by the substitution method.

a. Does the system have a solution?

b. Which of the following is a possible graph of the system?

i. 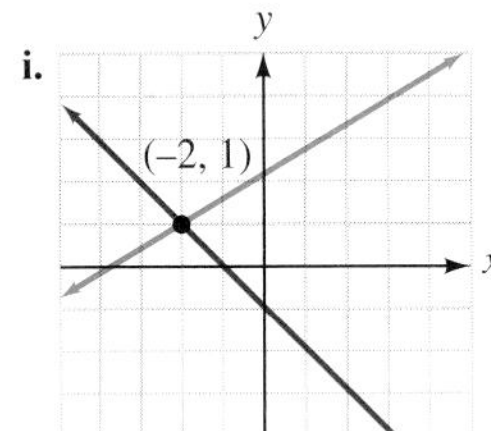

ii. 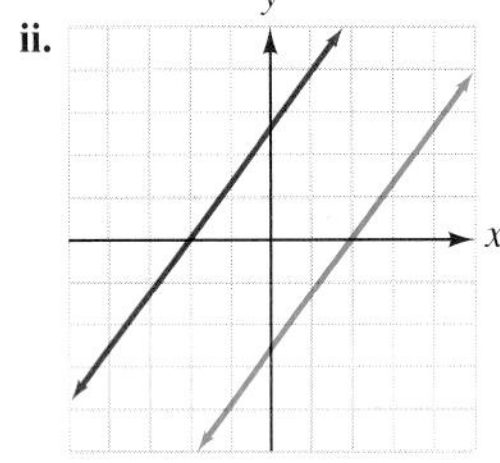

NOTATION *Complete the solution of each system.*

15. Solve: $\begin{cases} y = 3x \\ x - y = 4 \end{cases}$.

$x - y = 4$ This is the second equation.

$x - (\quad) = 4$

$-2x = \quad$

$x = -2$

$y = 3x$ This is the first equation.

$y = 3(\quad)$

$y = -6$

The solution is ______.

16. Solve: $\begin{cases} 2x + y = -5 \\ 2 - 2y = x \end{cases}$.

$2x + y = -5$ This is the first equation.

$2(\quad) + y = -5$

$4 - \quad + y = -5$

$\quad - 3y = -5$

$-3y = \quad$

$y = 3$

$2 - 2y = x$ This is the second equation.

$2 - 2(\quad) = x$

$2 - 6 = x$

$-4 = x$

The solution is ______.

PRACTICE *Use the substitution method to solve each system. If the equations of a system are dependent or if a system is inconsistent, so indicate.*

17. $\begin{cases} y = 2x \\ x + y = 6 \end{cases}$

18. $\begin{cases} y = 3x \\ x + y = 4 \end{cases}$

19. $\begin{cases} y = 2x - 6 \\ 2x + y = 6 \end{cases}$

20. $\begin{cases} y = 2x - 9 \\ x + 3y = 8 \end{cases}$

21. $\begin{cases} y = 2x + 5 \\ x + 2y = -5 \end{cases}$

22. $\begin{cases} y = -2x \\ 3x + 2y = -1 \end{cases}$

23. $\begin{cases} 2a + 4b = -24 \\ a = 20 - 2b \end{cases}$

24. $\begin{cases} 3a + 6b = -15 \\ a = -2b - 5 \end{cases}$

25. $\begin{cases} 2a = 3b - 13 \\ -b = -2a - 7 \end{cases}$

26. $\begin{cases} a = 3b - 1 \\ -b = -2a - 2 \end{cases}$

27. $\begin{cases} r + 3s = 9 \\ 3r + 2s = 13 \end{cases}$

28. $\begin{cases} x - 2y = 2 \\ 2x + 3y = 11 \end{cases}$

29. $\begin{cases} 0.4x + 0.5y = 0.2 \\ 3x - y = 11 \end{cases}$

30. $\begin{cases} 0.5u + 0.3v = 0.5 \\ 4u - v = 4 \end{cases}$

31. $\begin{cases} 6x - 3y = 5 \\ 2y + x = 0 \end{cases}$

32. $\begin{cases} 5s + 10t = 3 \\ 2s + t = 0 \end{cases}$

33. $\begin{cases} 3x + 4y = -7 \\ 2y - x = -1 \end{cases}$

34. $\begin{cases} 4x + 5y = -2 \\ x + 2y = -2 \end{cases}$

35. $\begin{cases} 9x = 3y + 12 \\ 4 = 3x - y \end{cases}$

36. $\begin{cases} 8y = 15 - 4x \\ x + 2y = 4 \end{cases}$

37. $\begin{cases} 0.02x + 0.05y = -0.02 \\ -\frac{x}{2} = y \end{cases}$

38. $\begin{cases} y = -\frac{x}{2} \\ 0.02x - 0.03y = -0.07 \end{cases}$

39. $\begin{cases} b = \frac{2}{3}a \\ 8a - 3b = 3 \end{cases}$

40. $\begin{cases} a = \frac{2}{3}b \\ 9a + 4b = 5 \end{cases}$

41. $\begin{cases} y - x = 3x \\ 2x + 2y = 14 - y \end{cases}$

42. $\begin{cases} y + x = 2x + 2 \\ 6x - 4y = 21 - y \end{cases}$

43. $\begin{cases} 2x - y = x + y \\ -2x + 4y = 6 \end{cases}$

44. $\begin{cases} x = -3y + 6 \\ 2x + 4y = 6 + x + y \end{cases}$

45. $\begin{cases} 3(x - 1) + 3 = 8 + 2y \\ 2(x + 1) = 8 + y \end{cases}$

46. $\begin{cases} 4(x - 2) = 19 - 5y \\ 3(x - 2) - 2y = -y \end{cases}$

47. $\begin{cases} \frac{1}{2}x + \frac{1}{2}y = -1 \\ \frac{1}{3}x - \frac{1}{2}y = -4 \end{cases}$

48. $\begin{cases} \frac{2}{3}y + \frac{1}{5}z = 1 \\ \frac{1}{3}y - \frac{2}{5}z = 3 \end{cases}$

49. $\begin{cases} 5m = \frac{1}{2}n - 1 \\ \frac{1}{4}n = 10m - 1 \end{cases}$

50. $\begin{cases} \frac{2}{3}m = 1 - 2n \\ 2(5n - m) + 11 = 0 \end{cases}$

51. $\begin{cases} \dfrac{6x - 1}{3} - \dfrac{5}{3} = \dfrac{3y + 1}{2} \\ \dfrac{1 + 5y}{4} + \dfrac{x + 3}{4} = \dfrac{17}{2} \end{cases}$

52. $\begin{cases} \dfrac{5x - 2}{4} + \dfrac{1}{2} = \dfrac{3y + 2}{2} \\ \dfrac{7y + 3}{3} = \dfrac{x}{2} + \dfrac{7}{3} \end{cases}$

APPLICATIONS

53. DINING See the following breakfast menu. What substitution from the a la carte menu will the restaurant owner allow customers to make if they don't want hash browns with their country breakfast? Why?

Village Vault Restaurant

Country Breakfast **$5.95**
2 eggs, 3 pancakes, bacon, sausage, hash browns, coffee

A la Carte Menu–Single Servings

Strawberries	$1.25	Melon	$0.95
Croissant	$1.70	Orange juice	$1.65
Hash browns	$0.95	Oatmeal	$1.95
Muffin	$1.30	Ham	$1.80

54. COLLEGE BOARDS Many colleges and universities use the SAT test as one indicator of a high school student's readiness to do college level work. Refer to the graph. For any given year, were the average math score and the average verbal score ever the same?

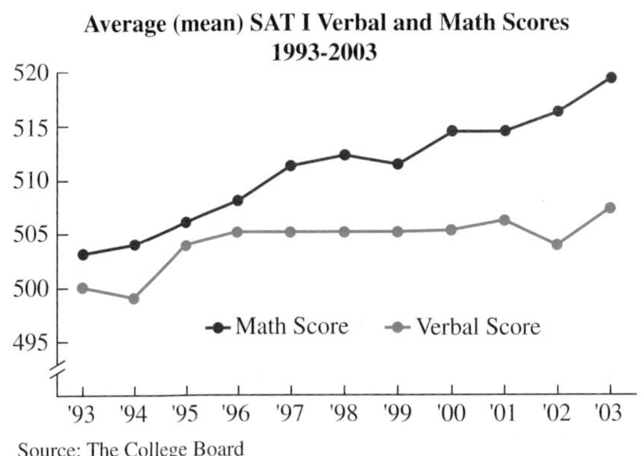

55. DISCOUNT COUPONS In mathematics, the substitution property states:

If $a = b$, then a may replace b or b may replace a in any statement.

Where on the following coupon is there an application of the substitution property? Explain.

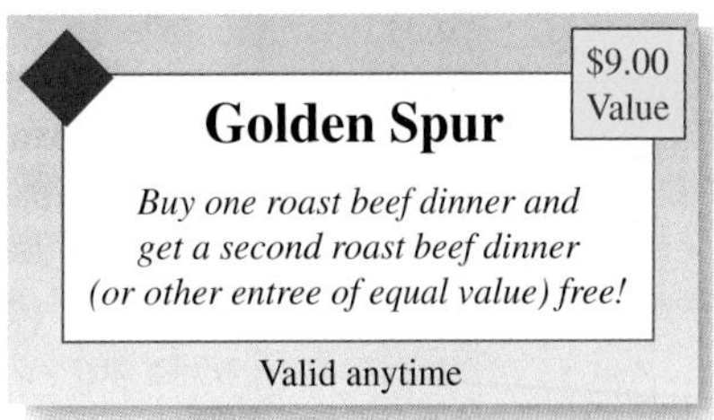

$9.00
Value
Golden Spur
Buy one roast beef dinner and
get a second roast beef dinner
(or other entree of equal value) free!
Valid anytime

56. HIGH SCHOOL SPORTS The equations shown model the number of boys and girls taking part in high school sports. In both models, x is the number of years since 1990, and y is the number of participants. If the trends continue, the graphs will intersect. Use the substitution method to predict the year when the number of boys and girls participating in high school sports will be the same.

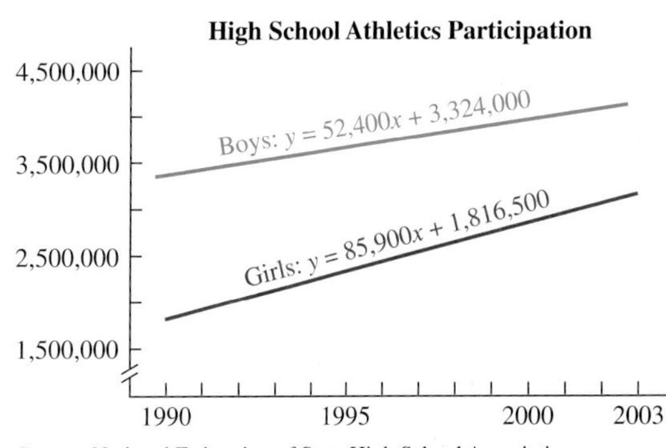

WRITING

57. Explain how to use substitution to solve a system of equations.

58. If the equations of a system are written in general form, why is it to your advantage to solve for a variable whose coefficient is 1 when using the substitution method?

59. When solving a system, what advantages and disadvantages are there with the graphing method? With the substitution method?

60. In this section, the substitution method for solving a system of two equations was discussed. List some other uses of the word *substitution,* or *substitute,* that you encounter in everyday life.

REVIEW

61. What is the slope of the line $y = -\frac{5}{8}x - 12$?
62. If $g(x) = -3x + 9$, find $g(-3)$.
63. Find the y-intercept of $2x - 3y = 18$.
64. Write the equation of the line passing through $(-1, 5)$ with a slope of -3.
65. Can a circle represent the graph of a function?
66. What is the range of the function $f(x) = |x|$?
67. A bowler has found that his score s varies directly with the time t in minutes he practices. Write an equation describing this relationship.
68. On what axis does $(0, -2)$ lie?

13.3 Solving Systems of Equations by Addition

- The addition method
- Inconsistent systems
- Dependent equations

In step 1 of the substitution method for solving a system of equations, we solve one equation for one of the variables. At times, this can be difficult — especially if neither variable has a coefficient of 1 or -1. In cases such as these, we can use another algebraic method called the *addition* or *elimination method* to find the exact solution of the system. This method is based on the addition property of equality: *When equal quantities are added to both sides of an equation, the results are equal.*

The addition method

To solve the system

$$\begin{cases} x + y = 8 \\ x - y = -2 \end{cases}$$

by the **addition method,** we see that the coefficients of y are opposites and then add the left- and right-hand sides of the equations to eliminate the variable y.

$$\begin{aligned} x + y &= 8 \\ \underline{x - y} &\underline{= -2} \end{aligned}$$

Equal quantities, $x - y$ and -2, are added to both sides of the equation $x + y = 8$. By the addition property of equality, the results will be equal.

Now, column by column, we add like terms. The terms y and $-y$ are eliminated.

$$\begin{aligned} \downarrow \quad \downarrow \quad & \downarrow \\ x + y &= 8 \\ \underline{x - y} &\underline{= -2} \\ 2x \quad\;\; &= 6 \end{aligned}$$

Combine like terms: $x + x = 2x$, $y + (-y) = 0$, and $8 + (-2) = 6$.

← Write each result here.

We can then solve the resulting equation for x.

$2x = 6$

$x = 3$ Divide both sides by 2.

To find y, we substitute 3 for x in either equation and solve it for y.

$x + y = 8$ This is the first equation of the system.

$\mathbf{3} + y = 8$ Substitute 3 for x.

$y = 5$ Subtract 3 from both sides.

We check the solution by verifying that (3, 5) satisfies each equation of the system.

To solve a system of equations in x and y by the addition method, we follow these steps.

The addition (elimination) method

1. Write both equations in general form: $Ax + By = C$.
2. If necessary, multiply one or both of the equations by nonzero quantities to make the coefficients of x (or the coefficients of y) opposites.
3. Add the equations to eliminate the terms involving x (or y).
4. Solve the equation resulting from step 3.
5. Find the value of the other variable by substituting the solution found in step 4 into any equation containing both variables.
6. Check the solution in the equations of the original system.

EXAMPLE 1

Solve: $\begin{cases} 5x + y = -4 \\ -5x + 2y = 7 \end{cases}$.

Solution When the equations are added, the terms $5x$ and $-5x$ drop out. We can then solve the resulting equation for y.

$$\begin{array}{r} 5x + y = -4 \\ -5x + 2y = 7 \\ \hline 3y = 3 \end{array}$$

Combine like terms: $5x + (-5x) = 0$, $y + 2y = 3y$, and $-4 + 7 = 3$.

$y = 1$ Divide both sides by 3.

To find x, we substitute 1 for y in either equation. If we use $5x + y = -4$, we have

$5x + \mathbf{y} = -4$ This is the first equation of the system.

$5x + \mathbf{1} = -4$ Substitute 1 for y.

$5x = -5$ Subtract 1 from both sides.

$x = -1$ Divide both sides by 5.

Verify that $(-1, 1)$ satisfies each original equation.

Self Check 1

Solve: $\begin{cases} x + 3y = 7 \\ 2x - 3y = -22 \end{cases}$.

Answer $(-5, 4)$

EXAMPLE 2

Solve: $\begin{cases} 3x + y = 7 \\ x + 2y = 4 \end{cases}$.

Solution If we add the equations as they are, neither variable will be eliminated. We must write the equations so that the coefficients of one of the variables are opposites. To eliminate x, we can multiply both sides of the second equation by $\mathbf{-3}$ to get

$$\begin{cases} 3x + y = 7 \\ \mathbf{-3}(x + 2y) = \mathbf{-3}(4) \end{cases} \longrightarrow \begin{cases} 3x + y = 7 \\ -3x - 6y = -12 \end{cases}$$

Self Check 2

Solve: $\begin{cases} 3x + 4y = 25 \\ 2x + y = 10 \end{cases}$.

The coefficients of the terms $3x$ and $-3x$ are now opposites. When the equations are added, x is eliminated.

$$\begin{aligned} 3x + y &= 7 \\ -3x - 6y &= -12 \\ \hline -5y &= -5 \\ y &= 1 \quad \text{Divide both sides by } -5. \end{aligned}$$

To find x, we substitute 1 for y in the equation $x + 2y = 4$.

$$\begin{aligned} x + 2\mathbf{y} &= 4 && \text{This is the second equation of the original system.} \\ x + 2(\mathbf{1}) &= 4 && \text{Substitute 1 for } y. \\ x + 2 &= 4 && \text{Perform the multiplication.} \\ x &= 2 && \text{Subtract 2 from both sides.} \end{aligned}$$

Answer (3, 4)

Check the solution (2, 1) in the original system of equations.

Self Check 3

Solve: $\begin{cases} 2a + 3b = 7 \\ 5a + 2b = 1 \end{cases}$.

Answer $(-1, 3)$

EXAMPLE 3 Solve: $\begin{cases} 2a - 5b = 10 \\ 3a - 2b = -7 \end{cases}$.

Solution The equations in the system must be written so that one of the variables will be eliminated when the equations are added. To eliminate ***a***, we can multiply the first equation by **3** and the second equation by -2 to get

$$\begin{cases} \mathbf{3}(2a - 5b) = \mathbf{3}(10) \\ -2(3a - 2b) = -2(-7) \end{cases} \longrightarrow \begin{cases} 6a - 15b = 30 \\ -6a + 4b = 14 \end{cases}$$

When these equations are added, the terms $6a$ and $-6a$ are eliminated.

$$\begin{aligned} 6a - 15b &= 30 \\ -6a + 4b &= 14 \\ \hline -11b &= 44 \\ b &= -4 \quad \text{Divide both sides by } -11. \end{aligned}$$

To find a, we substitute -4 for b in the equation $2a - 5b = 10$.

$$\begin{aligned} 2a - 5\mathbf{b} &= 10 && \text{This is the first equation of the original system.} \\ 2a - 5(\mathbf{-4}) &= 10 && \text{Substitute } -4 \text{ for } b. \\ 2a + 20 &= 10 && \text{Simplify.} \\ 2a &= -10 && \text{Subtract 20 from both sides.} \\ a &= -5 && \text{Divide both sides by 2.} \end{aligned}$$

Check the solution $(-5, -4)$ in the original equations.

Self Check 4

Solve: $\begin{cases} \frac{1}{3}x + \frac{1}{6}y = 1 \\ \frac{1}{2}x - \frac{1}{4}y = 0 \end{cases}$.

EXAMPLE 4 Solve: $\begin{cases} \frac{5}{6}x + \frac{2}{3}y = \frac{7}{6} \\ \frac{10}{7}x - \frac{4}{9}y = \frac{17}{21} \end{cases}$.

Solution To clear the equations of fractions, we multiply both sides of the first equation by 6 and both sides of the second equation by 63. This gives the equivalent system

$$\begin{array}{l} \textbf{(1)} \\ \textbf{(2)} \end{array} \begin{cases} 5x + 4y = 7 \\ 90x - 28y = 51 \end{cases}$$

We can solve for x by eliminating the terms involving y. To do so, we multiply Equation 1 by 7 and add the result to Equation 2.

$$\begin{aligned} 35x + 28y &= 49 \\ \underline{90x - 28y} &= \underline{51} \\ 125x &= 100 \end{aligned}$$

$x = \dfrac{100}{125}$ Divide both sides by 125.

$x = \dfrac{4}{5}$ Simplify $\frac{100}{125}$: Remove the common factor of 25.

To solve for y, we substitute $\frac{4}{5}$ for x in Equation 1 and simplify.

$$5x + 4y = 7$$

$$5\left(\frac{4}{5}\right) + 4y = 7$$

$4 + 4y = 7$ Simplify.

$4y = 3$ Subtract 4 from both sides.

$y = \dfrac{3}{4}$ Divide both sides by 4.

Check the solution of $\left(\frac{4}{5}, \frac{3}{4}\right)$ in the original equations.

Answer $\left(\frac{3}{2}, 3\right)$

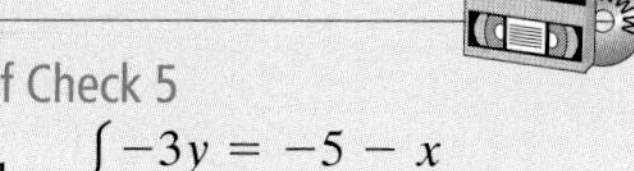

EXAMPLE 5 Solve: $\begin{cases} 2(2x + y) = 13 \\ 8x = 2y - 16 \end{cases}$.

Self Check 5

Solve: $\begin{cases} -3y = -5 - x \\ 3(x - y) = -11 \end{cases}$.

Solution We begin by writing each equation in $Ax + By = C$ form. For the first equation, we need only apply the distributive property. To write the second equation in general form, we subtract $2y$ from both sides.

$$\begin{aligned} \mathbf{2}(2x + y) &= 13 \\ 4x + 2y &= 13 \end{aligned} \qquad \begin{aligned} 8x &= 2y - 16 \\ 8x \mathbf{- 2y} &= 2y - 16 \mathbf{- 2y} \\ 8x - 2y &= -16 \end{aligned}$$

The two resulting equations form the following system.

$$\begin{array}{l} \textbf{(1)} \\ \textbf{(2)} \end{array} \begin{cases} 4x + 2y = 13 \\ 8x - 2y = -16 \end{cases}$$

When the equations are added, the terms involving y are eliminated.

$$\begin{aligned} 4x + 2y &= 13 \\ \underline{8x - 2y} &= \underline{-16} \\ 12x &= -3 \end{aligned}$$

$x = -\dfrac{1}{4}$ Divide both sides by 12 and simplify the fraction: $-\frac{3}{12} = -\frac{1}{4}$.

We can use Equation 1 to find y.

$$4x + 2y = 13$$

$$4\left(-\frac{1}{4}\right) + 2y = 13 \quad \text{Substitute } -\frac{1}{4} \text{ for } x.$$

$$-1 + 2y = 13 \quad \text{Perform the multiplication.}$$

$$2y = 14 \quad \text{Add 1 to both sides.}$$

$$y = 7 \quad \text{Divide both sides by 2.}$$

Verify that $(-\frac{1}{4}, 7)$ satisfies each original equation.

Answer $\left(-3, \frac{2}{3}\right)$

Inconsistent systems

EXAMPLE 6 Solve: $\begin{cases} 3x - 2y = 8 \\ -3x + 2y = -12 \end{cases}$.

Solution We can add the equations to eliminate the term involving x.

$$\begin{array}{r} 3x - 2y = 8 \\ -3x + 2y = -12 \\ \hline 0 = -4 \end{array}$$

Here the terms involving both x and y drop out, and a false result of $0 = -4$ is obtained. This indicates that the equations of the system are independent and that the system is inconsistent. This system has no solution.

Self Check 6

Solve: $\begin{cases} 2t - 7v = 5 \\ -2t + 7v = 3 \end{cases}$.

Answer no solution

Dependent equations

EXAMPLE 7 Solve: $\begin{cases} \frac{2x - 5y}{2} = \frac{19}{2} \\ -0.2x + 0.5y = -1.9 \end{cases}$.

Solution We can multiply both sides of the first equation by **2** to clear it of fractions and both sides of the second equation by **10** to clear it of decimals.

$$\begin{cases} \mathbf{2}\left(\frac{2x - 5y}{2}\right) = \mathbf{2}\left(\frac{19}{2}\right) \\ \mathbf{10}(-0.2x + 0.5y) = \mathbf{10}(-1.9) \end{cases} \longrightarrow \begin{cases} 2x - 5y = 19 \\ -2x + 5y = -19 \end{cases}$$

We add the resulting equations to get

$$\begin{array}{r} 2x - 5y = 19 \\ -2x + 5y = -19 \\ \hline 0 = 0 \end{array}$$

As in Example 6, both x and y drop out. However, this time a true result is obtained. This indicates that the equations are dependent and that the system has infinitely many solutions.

Any ordered pair that satisfies one equation also satisfies the other equation. Some solutions are $(2, -3)$, $(12, 1)$, and $(0, -\frac{19}{5})$.

Self Check 7

Solve: $\begin{cases} \frac{3x + y}{6} = \frac{1}{3} \\ -0.3x - 0.1y = -0.2 \end{cases}$.

Answer infinitely many solutions

Section 13.3 STUDY SET

VOCABULARY *Fill in the blanks.*

1. The ________ of the term $-3x$ is -3.
2. The ________ of 4 is -4.
3. $Ax + By = C$ is the ________ form of the equation of a line.
4. When adding the equations $\begin{array}{r} 5x - 6y = 10 \\ -3x + 6y = 24 \\ \hline \end{array}$ the variable y will be ________.

CONCEPTS

5. If the addition method is to be used to solve this system, what is wrong with the form in which it is written?

$$\begin{cases} 2x - 5y = -3 \\ -2y + 3x = 10 \end{cases}$$

6. Can the system

$$\begin{cases} 2x + 5y = -13 \\ -2x - 3y = -5 \end{cases}$$

be solved more easily using the addition method or the substitution method? Explain.

7. What algebraic step should be performed to clear this equation of fractions?

$$\frac{2}{3}x + 4y = -\frac{4}{5}$$

8. If the addition method is used to solve

$$\begin{cases} 3x + 12y = 4 \\ 6x - 4y = 8 \end{cases}$$

a. By what would we multiply the first equation to eliminate x?

b. By what would we multiply the second equation to eliminate y?

9. Solve: $\begin{cases} 4x + 2y = 2 \\ 3x - 2y = 12 \end{cases}$.

a. by the graphing method.

b. by the addition method.

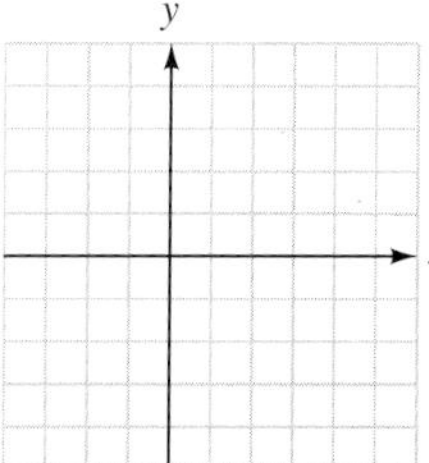

10. a. Suppose $0 = 0$ is obtained when a system is solved by the addition method. Does the system have a solution? Which of the following is a possible graph of the system?

b. Suppose $0 = 2$ is obtained when a system is solved by the addition method. Does the system have a solution? Which of the following is a possible graph of the system?

i 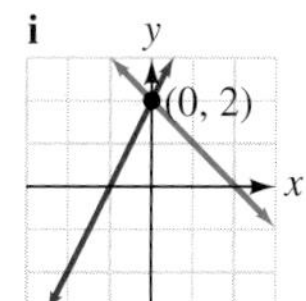

ii 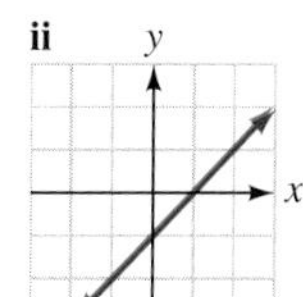

iii 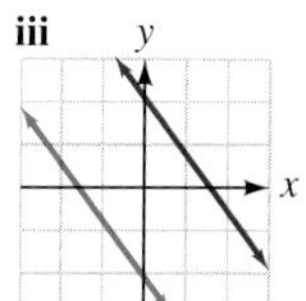

NOTATION *Complete the solution of each system.*

11. Solve: $\begin{cases} x + y = 5 \\ x - y = -3 \end{cases}$.

$$\begin{array}{r} x + y = 5 \\ x - y = -3 \\ \hline \square = 2 \\ x = \square \end{array}$$

$x + y = 5$ This is the first equation.

$(\square) + y = 5$

$y = 4$

The solution is $\square$.

12. Solve: $\begin{cases} x - 2y = 8 \\ -x + 5y = -17 \end{cases}$.

$$\begin{array}{r} x - 2y = 8 \\ -x + 5y = -17 \\ \hline \square = -9 \\ y = \square \end{array}$$

$x - 2y = 8$ This is the first equation.

$x - 2(\square) = 8$

$x + 6 = 8$

$x = 2$

The solution is $\square$.

PRACTICE *Use the addition method to solve each system.*

13. $\begin{cases} x - y = -5 \\ x + y = 1 \end{cases}$

14. $\begin{cases} x + y = 1 \\ x - y = 5 \end{cases}$

15. $\begin{cases} 2r + s = -1 \\ -2r + s = 3 \end{cases}$

16. $\begin{cases} 3m + n = -6 \\ m - n = -2 \end{cases}$

17. $\begin{cases} 2x + y = -2 \\ -2x - 3y = -6 \end{cases}$

18. $\begin{cases} 3x + 4y = 8 \\ 5x - 4y = 24 \end{cases}$

19. $\begin{cases} 4x + 3y = 24 \\ 4x - 3y = -24 \end{cases}$

20. $\begin{cases} 5x - 4y = 8 \\ -5x - 4y = 8 \end{cases}$

Use the addition method to solve each system. If the equations of a system are dependent or if a system is inconsistent, so indicate.

21. $\begin{cases} x + y = 5 \\ x + 2y = 8 \end{cases}$

22. $\begin{cases} x + 2y = 0 \\ x - y = -3 \end{cases}$

23. $\begin{cases} 2x + y = 4 \\ 2x + 3y = 0 \end{cases}$

24. $\begin{cases} 2x + 5y = -13 \\ 2x - 3y = -5 \end{cases}$

25. $\begin{cases} 3x - 5y = -29 \\ 3x + 4y = 34 \end{cases}$

26. $\begin{cases} 3x - 5y = 16 \\ 4x + 5y = 33 \end{cases}$

27. $\begin{cases} 2a - 3b = -6 \\ 2a - 3b = 8 \end{cases}$

28. $\begin{cases} 3a - 4b = 6 \\ 2(2b + 3) = 3a \end{cases}$

29. $\begin{cases} 8x - 4y = 18 \\ 3x - 2y = 8 \end{cases}$

30. $\begin{cases} 4x + 6y = 5 \\ 8x - 9y = 3 \end{cases}$

31. $\begin{cases} 3x + 4y = 12 \\ 4x + 5y = 17 \end{cases}$

32. $\begin{cases} 2x + 11y = -10 \\ 5x + 4y = 22 \end{cases}$

33. $\begin{cases} -3x + 6y = -9 \\ -5x + 4y = -15 \end{cases}$

34. $\begin{cases} -4x + 3y = -13 \\ -6x + 8y = -16 \end{cases}$

35. $\begin{cases} 2x + y = 10 \\ 0.1x + 0.2y = 1.0 \end{cases}$

36. $\begin{cases} 0.3x + 0.2y = 0 \\ 2x - 3y = -13 \end{cases}$

37. $\begin{cases} 2x - y = 16 \\ 0.03x + 0.02y = 0.03 \end{cases}$

38. $\begin{cases} -5y + 2x = 4 \\ -0.02y + 0.03x = 0.04 \end{cases}$

39. $\begin{cases} 6x + 3y = 0 \\ 5y = 2x + 12 \end{cases}$

40. $\begin{cases} 0 = 4x - 3y \\ 5x = 4y - 2 \end{cases}$

41. $\begin{cases} -2(x + 1) = 3y - 6 \\ 3(y + 2) = 10 - 2x \end{cases}$

42. $\begin{cases} 3x + 2y + 1 = 5 \\ 3(x - 1) = -2y - 4 \end{cases}$

43. $\begin{cases} 4(x + 1) = 17 - 3(y - 1) \\ 2(x + 2) + 3(y - 1) = 9 \end{cases}$

44. $\begin{cases} 5(x - 1) = 8 - 3(y + 2) \\ 4(x + 2) - 7 = 3(2 - y) \end{cases}$

45. $\begin{cases} \frac{3}{5}s + \frac{4}{5}t = 1 \\ -\frac{1}{4}s + \frac{3}{8}t = 1 \end{cases}$

46. $\begin{cases} \frac{1}{2}s - \frac{1}{4}t = 1 \\ \frac{1}{3}s + t = 3 \end{cases}$

47. $\begin{cases} \frac{3}{5}x + y = 1 \\ \frac{4}{5}x - y = -1 \end{cases}$

48. $\begin{cases} \frac{1}{2}x + \frac{4}{7}y = -1 \\ 5x - \frac{4}{5}y = -10 \end{cases}$

49. $\begin{cases} \dfrac{x - 3}{2} + \dfrac{y + 5}{3} = \dfrac{11}{6} \\ \dfrac{x + 3}{3} - \dfrac{5}{12} = \dfrac{y + 3}{4} \end{cases}$

50. $\begin{cases} \dfrac{x + 2}{3} = \dfrac{3 - y}{2} \\ \dfrac{x + 3}{2} = \dfrac{2 - y}{3} \end{cases}$

APPLICATIONS

51. EDUCATION The graph shows educational trends during the years 1980–2001 for persons 25 years or older. The equation $9x + 11y = 352$ approximates the percent y who had less than 12 years of schooling. The equation $5x - 11y = -198$ approximates the percent y who had 4 or more years of college. In each case, x is the number of years since 1980. Use the elimination method to determine in what year the percents were equal.

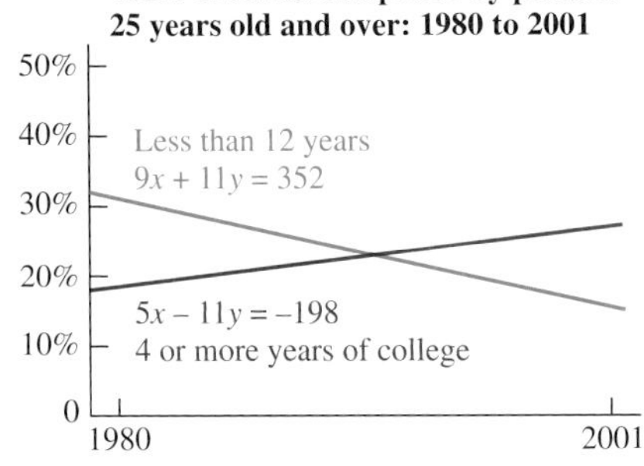

Source: U.S. Department of Commerce, Bureau of the Census

52. EDUCATION Answer Exercise 51 by solving the system of equations using the substitution method. (*Hint:* Solve one equation for $11y$, then substitute for $11y$ in the other equation.)

WRITING

53. Why is it usually to your advantage to write the equations of a system in general form before using the addition method to solve the system?

54. How would you decide whether to use substitution or addition to solve a system of equations?

55. In this section, we discussed the addition method for solving a system of two equations. Some instructors call it the *elimination method.* Why do you think it is known by this name?

56. Explain the error in the following work.

Solve: $\begin{cases} x + y = 1 \\ x - y = 5 \end{cases}$.

$$\begin{aligned} x + y &= 1 \\ x - y &= 5 \\ \hline 2x \quad &= 6 \\ \frac{2x}{2} &= \frac{6}{2} \\ x &= 3 \end{aligned}$$

The solution is 3.

REVIEW

57. Solve: $8(3x - 5) - 12 = 4(2x + 3)$.

58. Solve: $3y + \frac{y + 2}{2} = \frac{2(y + 3)}{3} + 16$.

59. Simplify: $x - x$.

60. Simplify: $3.2m - 4.4 + 2.1m + 16$.

61. Find the area of a triangular-shaped sign with a base of 4 feet and a height of 3.75 feet.

62. Translate to mathematical symbols: *the product of the sum of x and y and the difference of x and y.*

63. What is 10 less than x?

64. Factor: $6x^2 + 7x - 20$.

13.4 Applications of Systems of Equations

- Solving problems using two variables

We have previously formed equations involving one variable to solve problems. In this section, we consider ways to solve problems using two variables.

Solving problems using two variables

We can use the following steps to solve problems involving two unknown quantities.

Problem-solving strategy

1. Read the problem several times and *analyze* the facts. Occasionally, a sketch, table, or diagram will help you visualize the facts of the problem.
2. Pick different variables to represent two unknown quantities. *Form* two equations involving each of the two variables. This will give a system of two equations in two variables.
3. Solve the system of equations using the most convenient method: graphing, substitution, or addition.
4. State the conclusion.
5. Check the result in the words of the problem.

EXAMPLE 1 Photography.

At a school, two picture packages are available, as shown in the illustration. Find the cost of a class picture and the cost of an individual wallet-size picture.

Package 1
1 class picture
10 wallet-size
Only $19
FOOTHILL ELEMENTARY
Package 2
2 class pictures
15 wallet-size
Only $31

Solution

Analyze the problem

- Package 1 contains 1 class picture and 10 wallet-size pictures.
- Package 2 contains 2 class pictures and 15 wallet-size pictures.
- Find the cost of a class picture and the cost of a wallet-size picture.

Form two equations

Let c = the cost of 1 class picture and w = the cost of 1 wallet-size picture. To write an equation that models the first package, we note that (in dollars) the cost of 1 class picture is c and the cost of 10 wallet-size pictures is $10 \cdot w = 10w$.

The cost of 1 class picture	plus	the cost of 10 wallet-size pictures	is	\$19.
c	$+$	$10w$	$=$	19

To write an equation that models the second package, we note that (in dollars) the cost of 2 class pictures is $2 \cdot c = 2c$, and the cost of 15 wallet-size pictures is $15 \cdot w = 15w$.

The cost of 2 class pictures	plus	the cost of 15 wallet-size pictures	is	\$31.
$2c$	$+$	$15w$	$=$	31

The resulting system is $\begin{cases} c + 10w = 19 \\ 2c + 15w = 31 \end{cases}$.

Solve the system

To eliminate c, we proceed as follows:

$$-2c - 20w = -38 \quad \text{Multiply both sides of } c + 10w = 19 \text{ by } -2.$$
$$\underline{2c + 15w = 31}$$
$$-5w = -7 \quad \text{Add the equations to eliminate } c.$$
$$w = 1.4 \quad \text{Divide both sides by } -5. \text{ This is the cost of a wallet-size picture.}$$

To find c, substitute 1.4 for w in the first equation of the original system.

$$c + 10\mathbf{w} = 19$$
$$c + 10(\mathbf{1.4}) = 19 \quad \text{Substitute 1.4 for } w.$$
$$c + 14 = 19 \quad \text{Multiply.}$$
$$c = 5 \quad \text{Subtract 14 from both sides. This is the cost of a class picture.}$$

State the conclusion

A class picture costs \$5 and a wallet-size picture costs \$1.40.

Check the results

Package 1 has 1 class picture and 10 wallets: \$5 + 10(\$1.40) = \$5 + \$14 = \$19. Package 2 has 2 class pictures and 15 wallets: 2(\$5) + 15(\$1.40) = \$10 + \$21 = \$31. The results check.

EXAMPLE 2 **Lawn care.** An installer of underground irrigation systems wants to cut a 20-foot length of plastic tubing into two pieces. The longer piece is to be 2 feet longer than twice the shorter piece. Find the length of each piece.

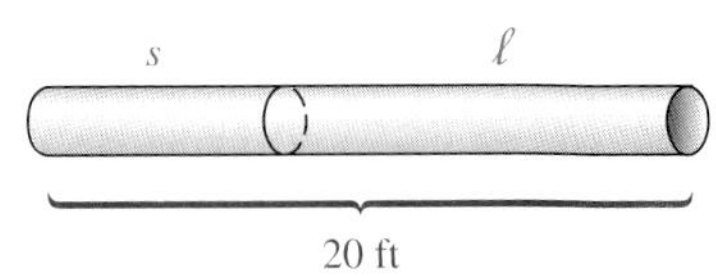

FIGURE 13-10

Analyze the problem

Refer to Figure 13-10, which shows the pipe.

Form two equations

We can let s = the length of the shorter piece and ℓ = the length of the longer piece.

The length of the shorter piece	plus	the length of the longer piece	is	20 feet.
s	$+$	ℓ	$=$	20

Since the longer piece is 2 feet longer than twice the shorter piece, we have

The length of the longer piece	is	2	times	the length of the shorter piece	plus	2 feet.
ℓ	$=$	2	$\cdot$	s	$+$	2

Solve the system

We can use the substitution method to solve the system.

$$\begin{aligned}&(1)\\&(2)\end{aligned}\quad \begin{cases} s + \ell = 20 \\ \boldsymbol{\ell = 2s + 2} \end{cases}$$

$$\begin{aligned} s + \mathbf{2s + 2} &= 20 && \text{Substitute } 2s + 2 \text{ for } \ell \text{ in Equation 1.} \\ 3s + 2 &= 20 && \text{Combine like terms.} \\ 3s &= 18 && \text{Subtract 2 from both sides.} \\ s &= 6 && \text{Divide both sides by 3.} \end{aligned}$$

The shorter piece should be 6 feet long. To find the length of the longer piece, we substitute 6 for s in Equation 2 and find ℓ.

$$\begin{aligned} \ell &= 2s + 2 \\ &= 2(\mathbf{6}) + 2 && \text{Substitute.} \\ &= 12 + 2 && \text{Simplify.} \\ &= 14 \end{aligned}$$

State the conclusion

The longer piece should be 14 feet long, and the shorter piece 6 feet long.

Check the result

The sum of 6 and 14 is 20, and 14 is 2 more than twice 6. The answers check.

College Students and Television Viewing

THINK IT THROUGH

"College students watch a lot less television than other segments of the population. The typical college student watches 14.5 hours of TV a week, compared to 32 hours weekly by the average American." *Media Life, 2002*

According to Nielsen Media Research, the typical high school graduate will have spent 6,000 more hours in front of a TV set than in the classroom. Combined, the television viewing and classroom time totals about 30,000 hours. Use two equations in two variables to find the number of hours spent in the classroom and the number of hours spent watching television by the typical high school graduate.

EXAMPLE 3 **Gardening.** Tom has 150 feet of fencing to enclose a rectangular garden. If the garden's length is to be 5 feet less than 3 times its width, find the area of the garden.

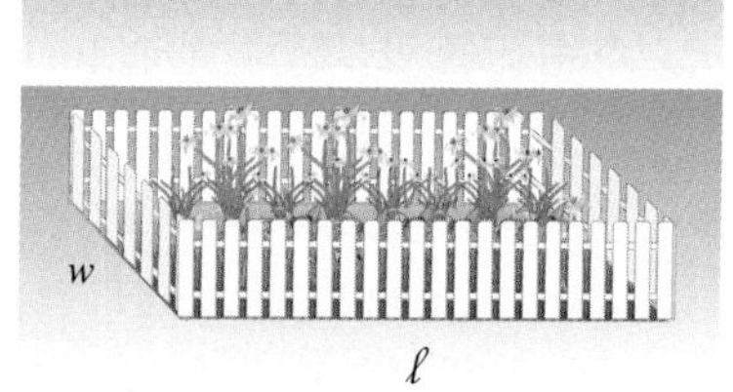

FIGURE 13-11

Analyze the problem

To find the area of a rectangle, we need to know its length and width.

Form two equations

We can let ℓ = the length of the garden and w = its width, as shown in Figure 13-11. Since the perimeter of a rectangle is two lengths plus two widths, we have

2	times	the length of the garden	plus	2	times	the width of the garden	is	150 feet.
2	$\cdot$	ℓ	$+$	2	$\cdot$	w	$=$	150

Since the length is 5 feet less than 3 times the width,

The length of the garden	is	3	times	the width of the garden	minus	5 feet.
ℓ	$=$	3	$\cdot$	w	$-$	5

Solve the system

We can use the substitution method to solve this system.

$$\begin{aligned}&(1)\\&(2)\end{aligned}\quad\begin{cases}2\ell + 2w = 150\\ \ell = \mathbf{3w - 5}\end{cases}$$

$$\begin{aligned}2(\mathbf{3w - 5}) + 2w &= 150 &&\text{Substitute } 3w - 5 \text{ for } \ell \text{ in Equation 1.}\\ 6w - 10 + 2w &= 150 &&\text{Distribute the multiplication by 2.}\\ 8w - 10 &= 150 &&\text{Combine like terms.}\\ 8w &= 160 &&\text{Add 10 to both sides.}\\ w &= 20 &&\text{Divide both sides by 8.}\end{aligned}$$

The width of the garden is 20 feet. To find the length, we substitute 20 for w in Equation 2 and simplify.

$$\begin{aligned}\ell &= 3w - 5\\ &= 3(\mathbf{20}) - 5 &&\text{Substitute.}\\ &= 60 - 5\\ \ell &= 55 &&\text{The length of the garden is 55 feet.}\end{aligned}$$

Now we find the area of the rectangle with dimensions 55 feet by 20 feet.

$$\begin{aligned}A &= \ell w &&\text{This is the formula for the area of a rectangle.}\\ &= \mathbf{55} \cdot \mathbf{20} &&\text{Substitute 55 for } \ell \text{ and 20 for } w.\\ A &= 1{,}100\end{aligned}$$

State the conclusion

The garden covers an area of 1,100 square feet.

Check the result

Because the dimensions of the garden are 55 feet by 20 feet, the perimeter is

$$\begin{aligned}P &= 2\ell + 2w\\ &= 2(\mathbf{55}) + 2(\mathbf{20}) &&\text{Substitute for } \ell \text{ and } w.\\ &= 110 + 40\\ &= 150\end{aligned}$$

It is also true that 55 feet is 5 feet less than 3 times 20 feet. The answers check.

EXAMPLE 4 **Manufacturing.** The setup cost of a machine that mills brass plates is \$750. After setup, it costs \$0.25 to mill each plate. Management is considering the purchase of a larger machine that can produce the same plate at a cost of \$0.20 per plate. If the setup cost of the larger machine is \$1,200, how many plates would the company have to produce to make the purchase worthwhile?

Analyze the problem

We need to find the number of plates (called the **break point**) that will cost equal amounts to produce on either machine.

Form two equations

We can let c = the cost of milling p plates. If we call the machine currently being used machine 1, and the new, larger one machine 2, we can form the two equations.

The cost of making p plates on machine 1	is	the setup cost of machine 1	plus	the cost per plate on machine 1	times	the number of plates p to be made.
c	$=$	750	$+$	0.25	$\cdot$	p

The cost of making p plates on machine 2	is	the setup cost of machine 2	plus	the cost per plate on machine 2	times	the number of plates p to be made.
c	$=$	1,200	$+$	0.20	$\cdot$	p

Solve the system

Since the costs are equal, we can use the substitution method to solve the system

$$\begin{aligned}&(1)\\&(2)\end{aligned}\quad \begin{cases} \mathbf{c = 750 + 0.25p} \\ c = 1{,}200 + 0.20p \end{cases}$$

$\mathbf{750 + 0.25p} = 1{,}200 + 0.20p$	Substitute $750 + 0.25p$ for c in the second equation.
$0.25p = 450 + 0.20p$	Subtract 750 from both sides.
$0.05p = 450$	Subtract $0.20p$ from both sides.
$p = 9{,}000$	Divide both sides by 0.05.

State the conclusion

If 9,000 plates are milled, the cost will be the same on either machine. If more than 9,000 plates are milled, the cost will be cheaper on the larger machine, because it mills the plates less expensively than the smaller machine.

Check the result

We check the solution by substituting 9,000 for p in Equations 1 and 2 and verifying that 3,000 is the value of c in both cases.

If we graph the two equations, we can illustrate the break point. (See Figure 13-12.)

MACHINE 1
$c = 750 + 0.25p$

p	c
0	750
1,000	1,000
5,000	2,000

MACHINE 2
$c = 1{,}200 + 0.20p$

p	c
0	1,200
4,000	2,000
12,000	3,600

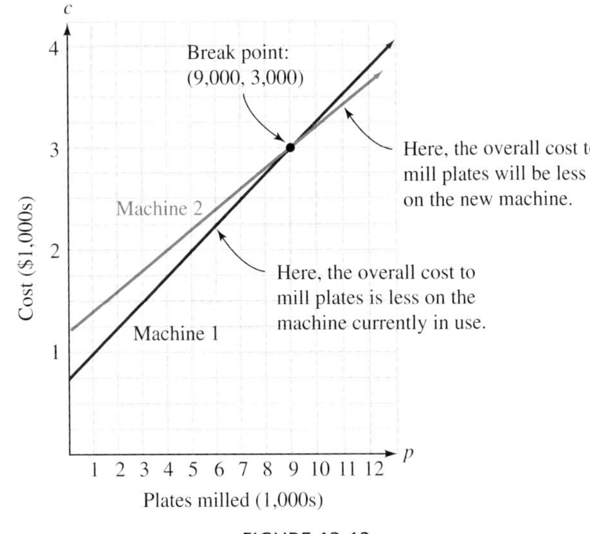

FIGURE 13-12

EXAMPLE 5 **White-collar crime.** Federal investigators discovered that a company secretly moved \$150,000 out of the country to avoid paying corporate income tax on it. Some of the money was invested in a Swiss bank account that paid 8% interest annually. The remainder was deposited in a Cayman Islands account, paying 7% annual interest. The investigation also revealed that the combined interest earned the first year was \$11,500. How much money was invested in each account?

Analyze the problem

We are told that an unknown part of the \$150,000 was invested at an annual rate of 8% and the rest at 7%. Together, the accounts earned \$11,500 in interest.

Form two equations

We can let x = the amount invested in the Swiss bank account and y = the amount invested in the Cayman Islands account. Because the total investment was \$150,000, we have

The amount invested in the Swiss account	+	the amount invested in the Cayman Is. account	is	\$150,000.
x	$+$	y	$=$	150,000

Since the annual income on x dollars invested at 8% is $0.08x$, the income on y dollars invested at 7% is $0.07y$, and the combined income is \$11,500, we have

The income on the 8% investment	+	the income on the 7% investment	is	\$11,500.
$0.08x$	$+$	$0.07y$	$=$	11,500

The resulting system is

$$\begin{cases} x + y = 150{,}000 & \textbf{(1)} \\ 0.08x + 0.07y = 11{,}500 & \textbf{(2)} \end{cases}$$

Solve the system

To solve the system, we use the addition method to eliminate x.

$$\begin{aligned} -8x - 8y &= -1{,}200{,}000 && \text{Multiply both sides on Equation 1 by } -8 \\ 8x + 7y &= 1{,}150{,}000 && \text{Multiply both sides of Equation 2 by 100.} \\ \hline -y &= -50{,}000 \\ y &= 50{,}000 && \text{Multiply (or divide) both sides by } -1. \end{aligned}$$

To find x, we substitute 50,000 for y in Equation 1 and simplify.

$$\begin{aligned} x + y &= 150{,}000 \\ x + \mathbf{50{,}000} &= 150{,}000 && \text{Substitute.} \\ x &= 100{,}000 && \text{Subtract 50,000 from both sides.} \end{aligned}$$

State the conclusion

\$100,000 was invested in the Swiss bank account, and \$50,000 was invested in the Cayman Islands account.

Check the result

$$\begin{aligned} \$100{,}000 + \$50{,}000 &= \$150{,}000 && \text{The two investments total \$150,000.} \\ 0.08(\$100{,}000) &= \$8{,}000 && \text{The Swiss bank account earned \$8,000.} \\ 0.07(\$50{,}000) &= \$3{,}500 && \text{The Cayman Islands account earned \$3,500.} \end{aligned}$$

The combined interest is \$8,000 + \$3,500 = \$11,500. The answers check.

EXAMPLE 6 **Boating.**

A boat traveled 30 kilometers downstream in 3 hours and made the return trip in 5 hours. Find the speed of the boat in still water.

Analyze the problem

Traveling downstream, the speed of the boat will be faster than it would be in still water. Traveling upstream, the speed of the boat will be less than it would be in still water.

Form two equations

We can let s = the speed of the boat in still water and c = the speed of the current. Then the rate of the boat going downstream is $s + c$, and its rate going upstream is $s - c$. We can organize the information as shown in Figure 13-13.

	Rate ·	Time =	Distance
Downstream	$s + c$	3	$3(s + c)$
Upstream	$s - c$	5	$5(s - c)$

FIGURE 13-13

Since each trip is 30 miles long, the Distance column of the table gives two equations in two variables.

$$\begin{cases} 3(s + c) = 30 \\ 5(s - c) = 30 \end{cases}$$

After using the distributive property, we have

$$\begin{cases} 3s + 3c = 30 & (1) \\ 5s - 5c = 30 & (2) \end{cases}$$

Solve the system

To solve this system by addition, we multiply Equation 1 by 5, multiply Equation 2 by 3, add the equations, and solve for s.

$$\begin{aligned} 15s + 15c &= 150 \\ 15s - 15c &= 90 \\ \hline 30s &= 240 \\ s &= 8 \end{aligned}$$

Divide both sides by 30.

State the conclusion

The speed of the boat in still water is 8 kilometers per hour.

Check the result

We leave the check to the reader.

EXAMPLE 7 **Medical technology.** A laboratory technician has one batch of antiseptic that is 40% alcohol and a second batch that is 60% alcohol. She would like to make 8 liters of solution that is 55% alcohol. How many liters of each batch should she use?

Analyze the problem

Some 60% solution must be added to some 40% solution to make a 55% solution.

Form two equations

We can let x = the number of liters to be used from batch 1 and y = the number of liters to be used from batch 2. We then organize the information as shown in Figure 13-14.

	Number of liters of solution ·	% of concentration =	Number of liters of alcohol
Batch 1	x	0.40	$0.40x$
Batch 2	y	0.60	$0.60y$
Mixture	8	0.55	$0.55(8)$

↑ One equation comes from information in this column.

↑ 40%, 60%, and 55% have been expressed as decimals.

↑ Another equation comes from information in this column.

FIGURE 13-14

The information in Figure 13-14 provides two equations.

$$\begin{cases} x + y = 8 & (1) \\ 0.40x + 0.60y = 0.55(8) & (2) \end{cases}$$

(1) The number of liters of batch 1 plus the number of liters of batch 2 equals the total number of liters in the mixture.

(2) The amount of alcohol in batch 1 plus the amount of alcohol in batch 2 equals the amount of alcohol in the mixture.

Solve the system

We can use addition to solve this system.

$$\begin{aligned} -40x - 40y &= -320 && \text{Multiply both sides of Equation 1 by } -40. \\ 40x + 60y &= 440 && \text{Multiply both sides of Equation 2 by } 100. \\ \hline 20y &= 120 \\ y &= 6 && \text{Divide both sides by 20.} \end{aligned}$$

To find x, we substitute 6 for y in Equation 1 and simplify.

$$\begin{aligned} x + y &= 8 \\ x + \mathbf{6} &= 8 && \text{Substitute.} \\ x &= 2 && \text{Subtract 6 from both sides.} \end{aligned}$$

State the conclusion

The technician should use 2 liters of the 40% solution and 6 liters of the 60% solution.

Check the result

The check is left to the reader.

Section 13.4 STUDY SET

VOCABULARY *Fill in the blanks.*

1. A ________ is a letter that stands for a number.
2. An ________ is a statement indicating that two quantities are equal.
3. $\begin{cases} a + b = 20 \\ a = 2b + 4 \end{cases}$ is a ________ of linear equations.
4. A ________ of a system of linear equations satisfies both equations simultaneously.

CONCEPTS

5. For each case in the illustration, write an algebraic expression that represents the speed of the canoe in miles per hour if its speed in still water is x miles per hour.

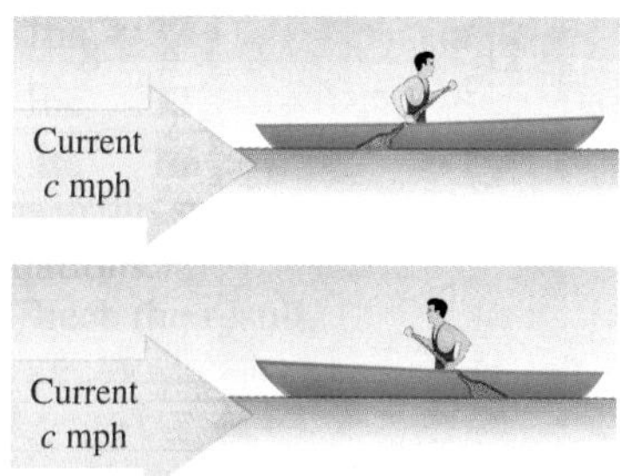

6. See the illustration.
   a. If the contents of the two test tubes are poured into a third test tube, how much solution will the third test tube contain? (mL means milliliters.)
   b. Which is the best estimate of the concentration of the solution in the third test tube: 25%, 35%, or 45% acid solution?

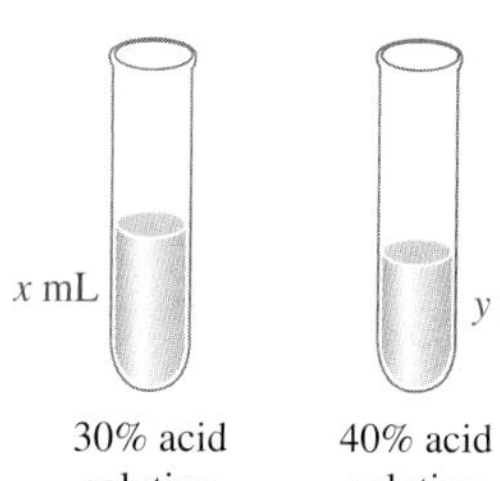

7. Use the information in the table to answer the questions about two investments.

	Principal ·	Rate ·	Time =	Interest
City Bank	x	5%	1 yr	
USA Savings	y	11%	1 yr	

   a. How much money was deposited in the USA Savings account?
   b. What interest rate did the City Bank account earn?
   c. Complete the table.

8. Use the information in the table to answer the questions about a plane flying in windy conditions.

	Rate ·	Time =	Distance
With	$x + y$	3 hr	450 mi
Against	$x - y$	5 hr	450 mi

   a. For how long did the plane fly against the wind?

b. At what rate did the plane travel when flying with the wind?

c. Write two equations that could be used to solve for x and y.

9. a. If a problem contains two unknowns, and if two variables are used to represent them, how many equations must be written to find the unknowns?

b. Name three methods that can be used to solve a system of linear equations.

10. Put the steps of the five-step problem-solving strategy listed below in the correct order.

State the conclusion	Form two equations
Analyze the problem	Check the result
Solve the system	

NOTATION *Write a formula that relates the given quantities.*

11. length, width, area of a rectangle
12. length, width, perimeter of a rectangle
13. rate, time, distance traveled
14. principal, rate, time, interest earned

Translate each verbal model into mathematical symbols. Use variables to represent any unknowns.

15. 2 · (length of pool) + 2 · (width of pool) is 90 yards.

16. \$6 · (number of adults) + \$2 · (number of children) is \$26.

PRACTICE *Use two equations in two variables to find the integers.*

17. One integer is twice another. Their sum is 96.
18. The sum of two integers is 38. Their difference is 12.
19. Three times one integer plus another integer is 29. The first integer plus twice the second is 18.
20. Twice one integer plus another integer is 21. The first integer plus 3 times the second is 33.

APPLICATIONS *Use two equations in two variables to solve each problem.*

21. TREE TRIMMING fully extended, the arm on the tree service truck shown is 51 feet long. If the upper part of the arm is 7 feet shorter than the lower part, how long is each part of the arm?

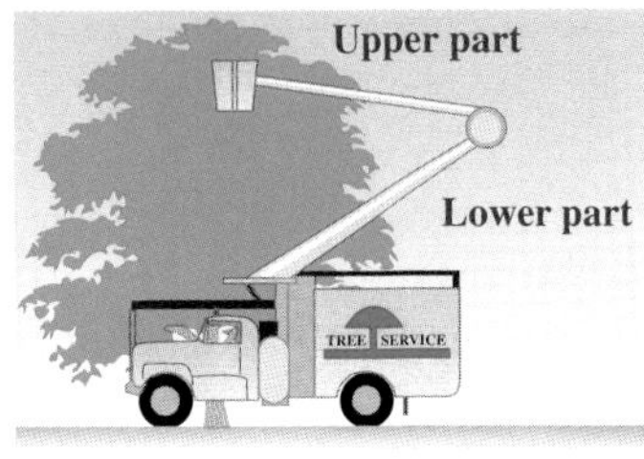

22. TV PROGRAMMING The producer of a 30-minute TV documentary about World War I divided it into two parts. Four times as much program time was devoted to the causes of the war as to the outcome. How long is each part of the documentary?

23. EXECUTIVE BRANCH The salaries of the president and vice president of the United States total \$581,000 a year. If the president makes \$219,000 more than the vice president, find each of their salaries.

24. CAUSES OF DEATH In 1993, the number of Americans dying from cancer was 6 times the number who died from accidents. If the number of deaths from these two causes totaled 630,000, how many Americans died from each cause?

25. BUYING PAINTING SUPPLIES Two partial receipts for paint supplies are shown. How much does each gallon of paint and each brush cost?

COLORFUL PA
Paint and Wal
8 gallons
latex paint @
3 brushes @
Total \$ 270.00

COLORFUL PA
Paint and Wall
6 gallons
latex paint @
2 brushes @
Total \$ 200.00

26. WEDDING PICTURES A photographer sells the two wedding picture packages shown in the illustration.

How much does a 10×14 photo cost? An 8×10 photo?

27. BUYING TICKETS If receipts for the movie advertised below were $1,440 for an audience of 190 people, how many senior citizens attended?

28. SELLING ICE CREAM At a store, ice cream cones cost $0.90 and sundaes cost $1.65. One day the receipts for a total of 148 cones and sundaes were $180.45. How many cones were sold?

29. THE MARINE CORPS The Marine Corps War Memorial in Arlington, Virginia, portrays the raising of the U.S. flag on Iwo Jima during World War II. Find the two angles shown below if the measure of one of the angles is 15° less than twice the other.

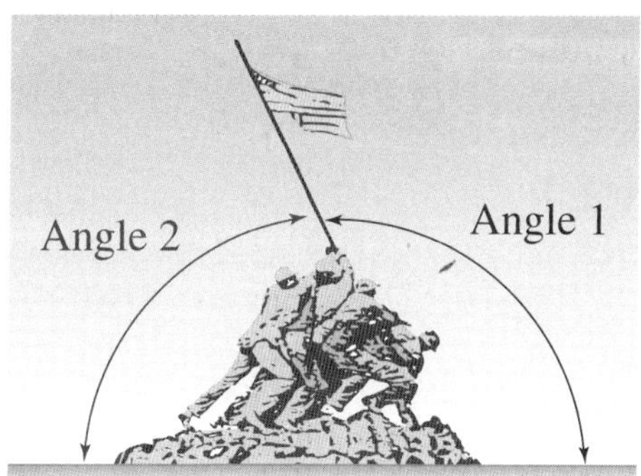

30. PHYSICAL THERAPY To rehabilitate her knee, an athlete does leg extensions. Her goal is to regain a full 90° range of motion in this exercise. Use the information in the illustration to determine her current range of motion in degrees.

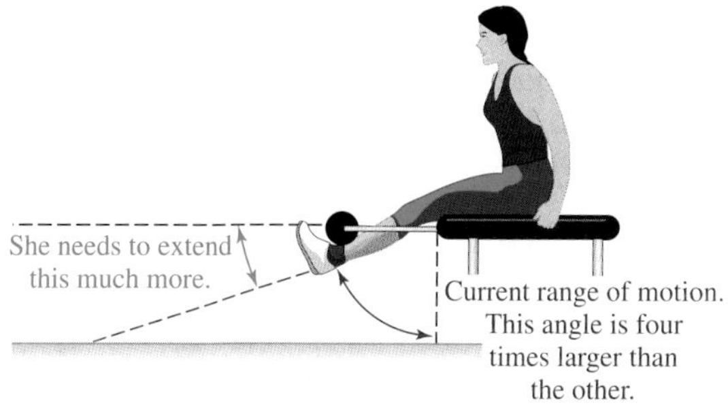

31. THEATER SCREENS At an IMAX theater, the giant rectangular movie screen has a width 26 feet less than its length. If its perimeter is 332 feet, find the area of the screen.

32. GEOMETRY A 50-meter path surrounds the rectangular garden shown below. The width of the garden is two-thirds its length. Find its area.

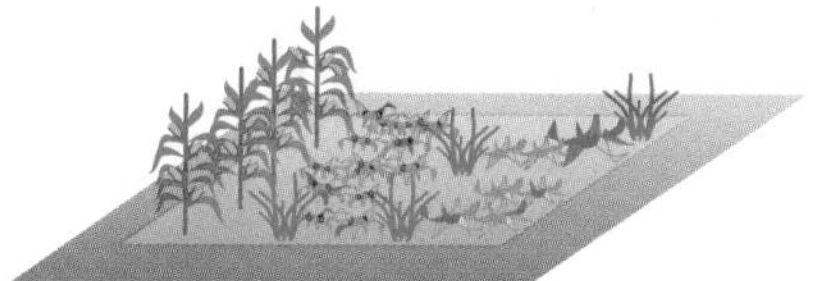

33. MAKING TIRES A company has two molds to form tires. One mold has a setup cost of $1,000, and the other has a setup cost of $3,000. The cost to make each tire with the first mold is $15, and the cost to make each tire with the second mold is $10.

a. Find the break point.

b. Check your result by graphing both equations on the coordinate system in the illustration.

c. If a production run of 500 tires is planned, determine which mold should be used.

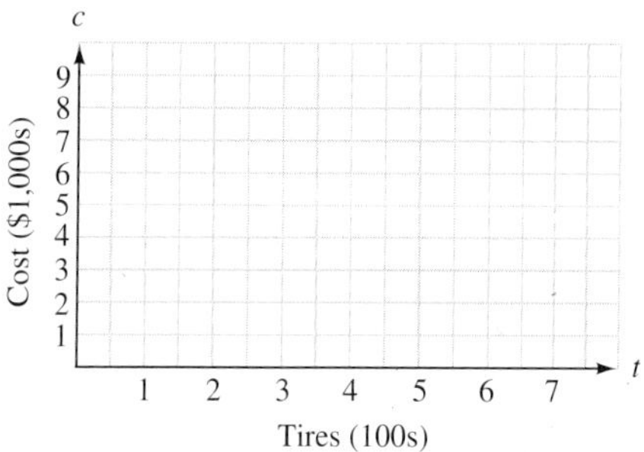

34. CHOOSING A FURNACE A high-efficiency 90+ furnace can be purchased for $2,250 and costs an average of $412 per year to operate in Rockford, Illinois. An 80+ furnace can be purchased for only $1,715, but it costs $466 per year to operate.

a. Find the break point.

b. If you intended to live in a Rockford house for 7 years, which furnace would you choose?

35. STUDENT LOANS A college used a $5,000 gift from an alumnus to make two student loans. The first was at 5% annual interest to a nursing student. The second was at 7% to a business major. If the college collected $310 in interest the first year, how much was loaned to each student?

36. FINANCIAL PLANNING In investing $6,000 of a couple's money, a financial planner put some of it into a savings account paying 6% annual interest. The rest was invested in a riskier mini-mall development plan paying 12% annually. The combined interest earned for the first year was $540. How much money was invested at each rate?

37. THE GULF STREAM The Gulf Stream is a warm ocean current of the North Atlantic Ocean that flows northward, as shown. Heading north with the Gulf Stream, a cruise ship traveled 300 miles in 10 hours. Against the current, it took 15 hours to make the return trip. Find the speed of the current.

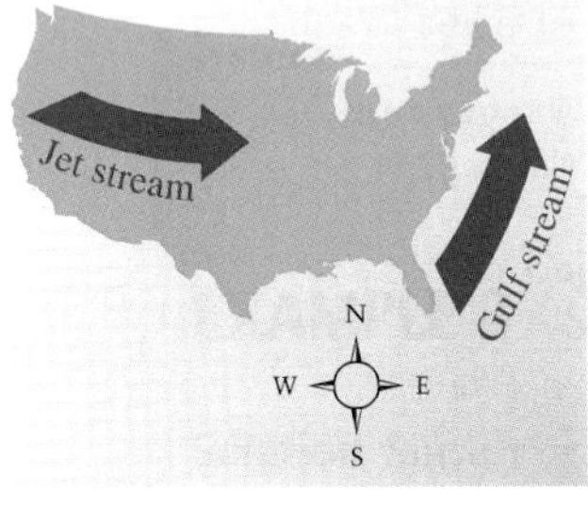

38. THE JET STREAM The jet stream is a strong wind current that flows across the United States, as shown in the illustration above. Flying with the jet stream, a plane flew 3,000 miles in 5 hours. Against the same wind, the trip took 6 hours. Find the airspeed of the plane (the speed in still air).

39. AVIATION An airplane can fly downwind a distance of 600 miles in 2 hours. However, the return trip against the same wind takes 3 hours. Find the speed of the wind.

40. BOATING A boat can travel 24 miles downstream in 2 hours and can make the return trip in 3 hours. Find the speed of the boat in still water.

41. MARINE BIOLOGY A marine biologist wants to set up an aquarium containing 3% salt water. He has two tanks on hand that contain 6% and 2% salt water. How much water from each tank must he use to fill a 16-liter aquarium with a 3% saltwater mixture?

42. COMMEMORATIVE COINS A foundry has been commissioned to make souvenir coins. The coins are to be made from an alloy that is 40% silver. The foundry has on hand two alloys, one with 50% silver content and one with a 25% silver content. How many kilograms of each alloy should be used to make 20 kilograms of the 40% silver alloy?

43. MIXING NUTS A merchant wants to mix peanuts with cashews, as shown in the illustration, to get 48 pounds of mixed nuts that will be sold at $4 per pound. How many pounds of each should the merchant use?

44. COFFEE SALES A coffee supply store waits until the orders for its special coffee blend reach 100 pounds before making up a batch. Coffee selling for $8.75 a pound is blended with coffee selling for $3.75 a pound to make a product that sells for $6.35 a pound. How much of each type of coffee should be used to make the blend that will fill the orders?

45. MARKDOWN A set of golf clubs has been marked down 40% to a sale price of $384. Let r represent the retail price and d the discount. Then use the following equations to find the original retail price.

Retail price − discount = sale price

Discount = discount rate · retail price

46. MARKUP A stereo system retailing at $565.50 has been marked up 45% from wholesale. Let w represent the wholesale cost and m the markup. Then use the following equations to find the wholesale cost.

Wholesale cost + markup = retail price

Markup = markup rate · wholesale cost

WRITING

47. When solving a problem using two variables, why isn't one equation sufficient to find the two unknown quantities?

48. Describe an everyday situation in which you might need to make a mixture.

REVIEW *Graph each inequality.*

49. $x < 4$

50. $x \geq -3$

51. $-1 < x \leq 2$

52. $-2 \leq x \leq 0$

Solve each equation.

53. $x^2 - 4 = 0$

54. $x^2 - 4x = 0$

55. $x^2 - 4x + 4 = 0$

56. $2x^2 + 3x = 2$

13.5 Graphing Linear Inequalities

- Solving linear inequalities
- Graphing linear inequalities
- An application of linear inequalities

The solutions of a linear *equation* in x and y can be expressed as ordered pairs (x, y) and when graphed, the ordered pairs form a line. In this section, we consider linear *inequalities.* Solutions of linear inequalities can also be expressed as ordered pairs and graphed.

Solving linear inequalities

A linear equation in x and y is an equation that can be written in the form $Ax + By = C$. A **linear inequality** in x and y is an inequality that can be written in one of four forms:

$$Ax + By > C, \qquad Ax + By < C, \qquad Ax + By \ge C, \qquad \text{or} \qquad Ax + By \le C$$

where A, B, and C represent real numbers and A and B are not both zero. Some examples of linear inequalities are

$$2x - y > -3, \qquad y < 3, \qquad x + 4y \ge 6, \qquad \text{and} \qquad x \le -2$$

As with linear equations, an ordered pair (x, y) is a solution of an inequality in x and y if a true statement results when the variables in the inequality are replaced by the coordinates of the ordered pair.

EXAMPLE 1 Determine whether each ordered pair is a solution of $x - y \le 5$. Then graph each solution: **a.** $(4, 2)$, **b.** $(0, -6)$, and **c.** $(1, -4)$.

Solution In each case, we substitute the x-coordinate for x and the y-coordinate for y in the inequality $x - y \le 5$. If the ordered pair is a solution, a true statement will be obtained.

a. For $(4, 2)$:

$x - y \le 5$ This is the original inequality.

$4 - 2 \le 5$ Replace x with 4 and y with 2.

$2 \le 5$ This is true.

Because $2 \le 5$ is true, $(4, 2)$ is a solution of the inequality, and we graph it in Figure 13-15.

b. For $(0, -6)$:

$x - y \le 5$ This is the original inequality.

$0 - (-6) \le 5$ Replace x with 0 and y with -6.

$6 \le 5$ This is false.

Because $6 \le 5$ is false, $(0, -6)$ is not a solution.

c. For $(1, -4)$:

$x - y \le 5$ This is the original inequality.

$1 - (-4) \le 5$ Replace x with 1 and y with -4.

$5 \le 5$ This is true.

Because $5 \le 5$ is true, $(1, -4)$ is a solution, and we graph it in Figure 13-15.

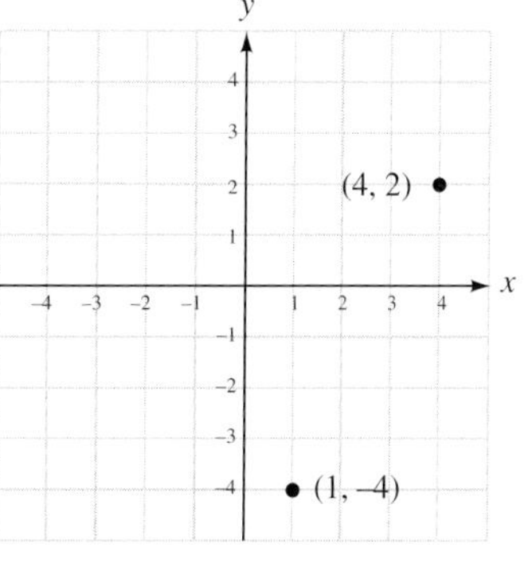

FIGURE 13-15

Self Check 1
Using the inequality in Example 1, determine whether each ordered pair is a solution.
a. $(8, 2)$, **b.** $(4, -1)$, **c.** $(-2, 4)$, and **d.** $(-3, -5)$.

Answers **a.** not a solution, **b.** solution, **c.** solution, **d.** solution

The graph in Figure 13-15 contains some solutions of the inequality $x - y \le 5$. Intuition tells us that there are many more ordered pairs (x, y) such that $x - y$ is less than or equal to 5. How then do we get a complete graph of the solutions of $x - y \le 5$? We address this question in the following discussion.

Graphing linear inequalities

The graph of $x - y = 5$ is a line consisting of the points whose coordinates satisfy the equation. The graph of the inequality $x - y \le 5$ is not a line, but an area bounded by a line, called a **half-plane.** The half-plane consists of the points whose coordinates satisfy the inequality.

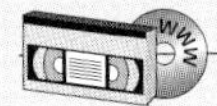

EXAMPLE 2 Graph: $x - y \le 5$.

Solution Since the inequality symbol $\le$ includes an equals sign, the graph of $x - y \le 5$ includes the graph of $x - y = 5$. So we begin by graphing the equation $x - y = 5$, using the intercept method. See Figure 13-16(a).

$x - y = 5$		
x	y	(x, y)
0	−5	(0, −5)
5	0	(5, 0)
6	1	(6, 1)

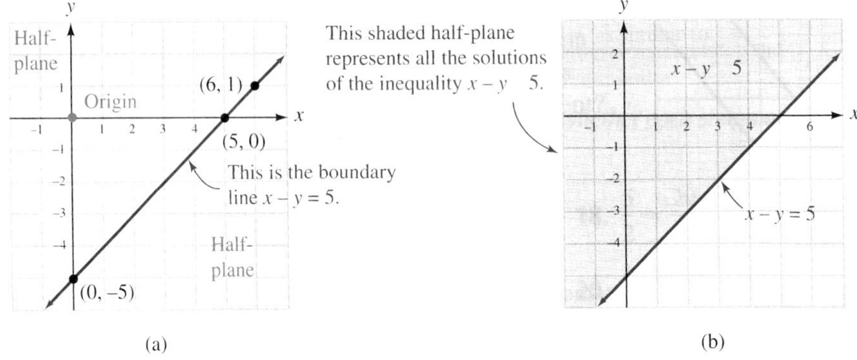

(a) (b)

FIGURE 13-16

Since the inequality $x - y \le 5$ allows $x - y$ to be less than 5, the coordinates of points other than those shown on the line in Figure 13-16(a) satisfy the inequality. For example, the coordinates of the origin (0, 0) satisfy the inequality. We can verify this by letting x and y be zero in the given inequality:

$$x - y \le 5$$
$$\mathbf{0} - \mathbf{0} \le 5 \quad \text{Substitute 0 for } x \text{ and 0 for } y.$$
$$0 \le 5$$

Because $0 \le 5$, the coordinates of the origin satisfy the original inequality. In fact, the coordinates of every point on the *same side* of the line as the origin satisfy the inequality. The graph of $x - y \le 5$ is the half-plane that is shaded in Figure 13-16(b). Since the **boundary line** $x - y = 5$ is included, we draw it with a solid line.

EXAMPLE 3 Graph: $2(x - 3) - (x - y) \geq -3$.

Solution We begin by simplifying the inequality as follows:

$$2(x - 3) - (x - y) \geq -3$$

$$2x - 6 - x + y \geq -3 \quad \text{Use the distributive property.}$$

$$x - 6 + y \geq -3 \quad \text{Combine like terms.}$$

$$x + y \geq 3 \quad \text{Add 6 to both sides.}$$

To graph the inequality $x + y \geq 3$, we graph the boundary line whose equation is $x + y = 3$. Since the graph of $x + y \geq 3$ includes the line $x + y = 3$, we draw the boundary with a solid line. Note that it divides the coordinate plane into two half-planes. See Figure 13-17(a).

To decide which half-plane to shade, we substitute the coordinates of some point that lies on one side of the boundary line into the inequality. If we use the origin (0, 0) for the **test point,** we have

$$x + y \geq 3$$

$$\mathbf{0} + \mathbf{0} \geq 3 \quad \text{Substitute 0 for } x \text{ and 0 for } y.$$

$$0 \geq 3 \quad \text{This is false.}$$

Since $0 \geq 3$ is a false statement, the origin is not in the graph. In fact, the coordinates of *every* point on the origin's side of the boundary line will not satisfy the inequality. However, every point on the other side of the boundary line will satisfy the inequality. We shade that half-plane. The graph of $x + y \geq 3$ is the half-plane that appears in color in Figure 13-17(b).

$x + y = 3$

x	y	(x, y)
0	3	(0, 3)
3	0	(3, 0)
1	2	(1, 2)

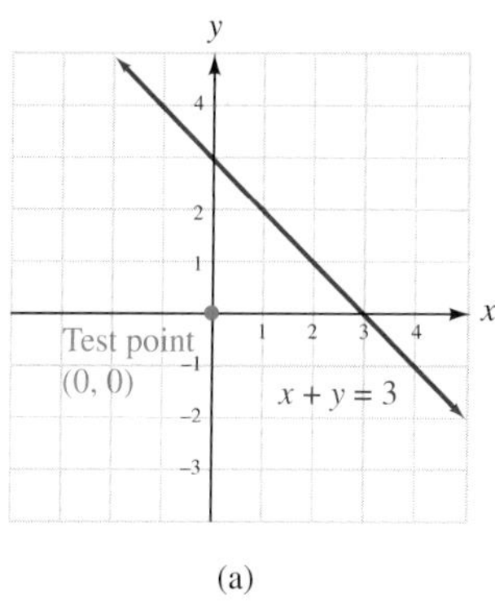

(a)

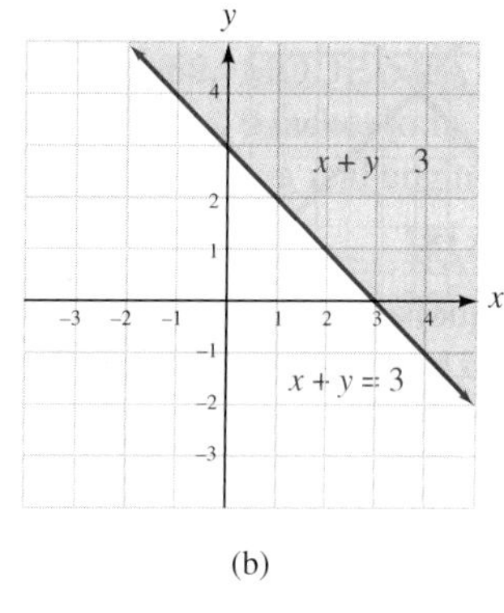

(b)

FIGURE 13-17

Self Check 3
Graph: $3(x - 1) - (2x + y) \leq -5$.

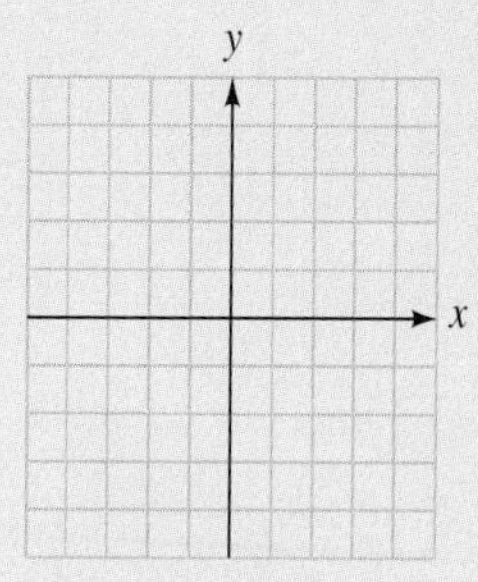

Answer

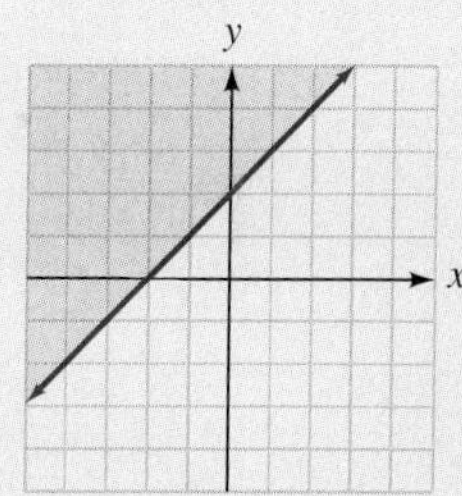

EXAMPLE 4 Graph: $y > 2x$.

Solution To find the boundary line, we graph $y = 2x$. Since the symbol $>$ does not include an equals sign, the points on the graph of $y = 2x$ are not part of the graph of $y > 2x$. We draw the boundary line as a dashed line to show this, as in Figure 13-18(a).

To determine which half-plane to shade, we substitute the coordinates of some point that lies on one side of the boundary line into $y > 2x$. Since the origin is on the boundary, we cannot use it as a test point. The point (2, 0), for example, is below the boundary line. See Figure 13-18(a). To see whether (2, 0) satisfies $y > 2x$, we substitute 2 for x and 0 for y in the inequality.

$$y > 2x$$

$$\mathbf{0} > 2(\mathbf{2}) \quad \text{Substitute 2 for } x \text{ and 0 for } y.$$

$$0 > 4 \quad \text{This is false.}$$

Self Check 4
Graph: $y < 3x$.

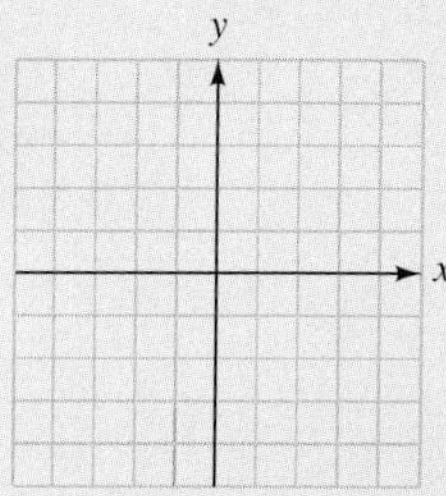

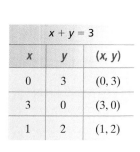

Since $0 > 4$ is a false statement, the point (2, 0) does not satisfy the inequality, and is not on the side of the dashed line we wish to shade. Instead, we shade the other side of the boundary line. The graph of the solution set of $y > 2x$ is shown in Figure 13-18(b).

Answer

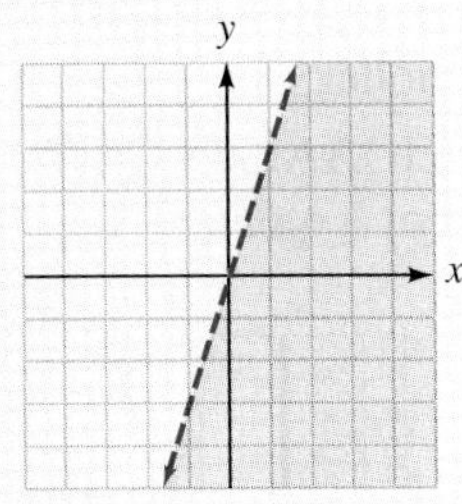

$y = 2x$		
x	y	(x, y)
0	0	(0, 0)
−1	−2	(−1, −2)
1	2	(1, 2)

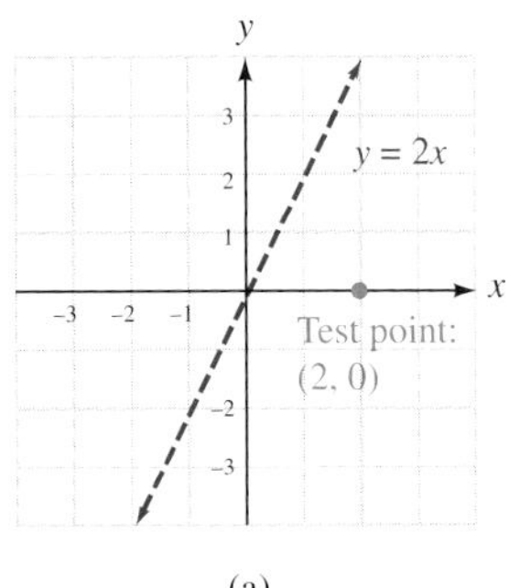

(a)

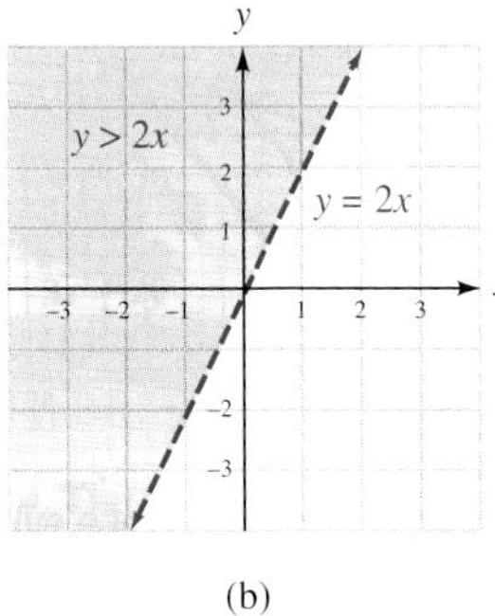

(b)

FIGURE 13-18

Self Check 5
Graph: $2x - y < 4$.

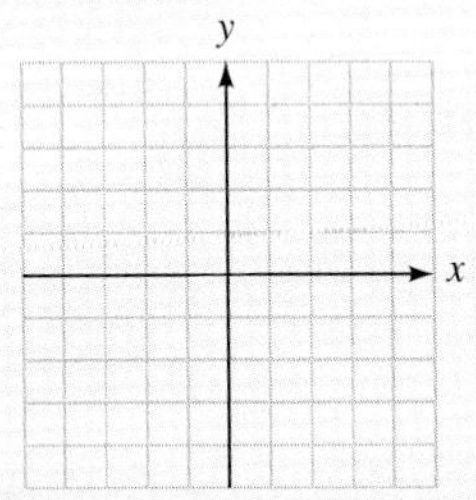

EXAMPLE 5 Graph: $x + 2y < 6$.

Solution We find the boundary by graphing the equation $x + 2y = 6$. We draw the boundary as a dashed line to show that it is not part of the solution. We then choose a test point not on the boundary and see whether its coordinates satisfy $x + 2y < 6$. The origin is a convenient choice.

$x + 2y < 6$
$0 + 2(0) < 6$ Substitute 0 for x and 0 for y.
$0 < 6$ This is true.

Since $0 < 6$ is a true statement, we shade the side of the line that includes the origin. The graph is shown in Figure 13-19.

Answer

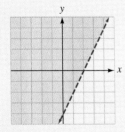

$x + 2y = 6$		
x	y	(x, y)
0	3	(0, 3)
6	0	(6, 0)
4	1	(4, 1)

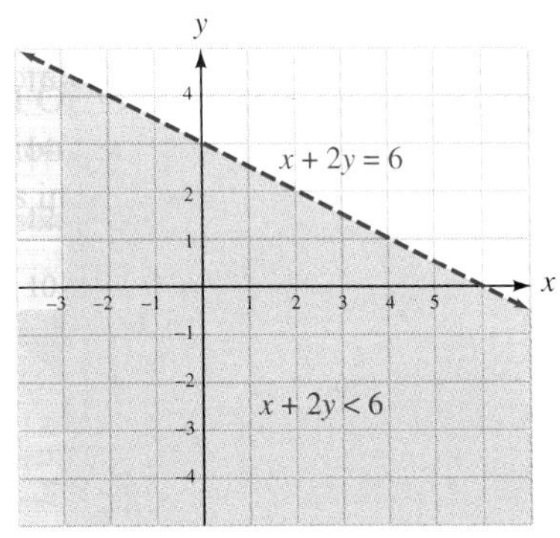

FIGURE 13-19

Self Check 6
Graph: $x \geq 2$.

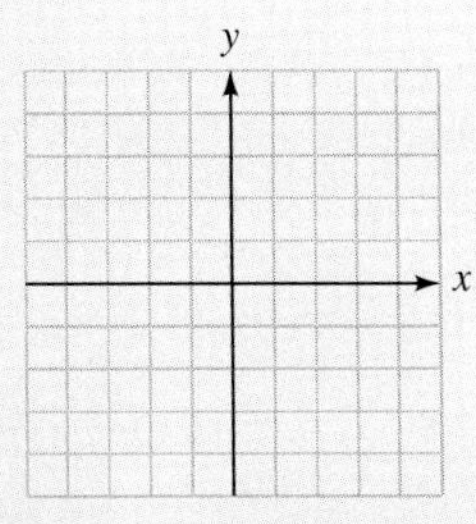

EXAMPLE 6 Graph: $y \geq 0$.

Solution We find the boundary by graphing the equation $y = 0$. We draw the boundary as a solid line to show that it is part of the solution. We then choose a test point not on the boundary and see whether its coordinates satisfy $y \geq 0$. The point (0, 1) is a convenient choice.

$y \geq 0$
$1 \geq 0$ Substitute 1 for y.

Since $1 \geq 0$ is a true statement, we shade the side of the line that includes (0, 1). The graph is shown in Figure 13-20.

$y = 0$		
x	y	(x, y)
1	0	(1, 0)
2	0	(2, 0)
3	0	(3, 0)

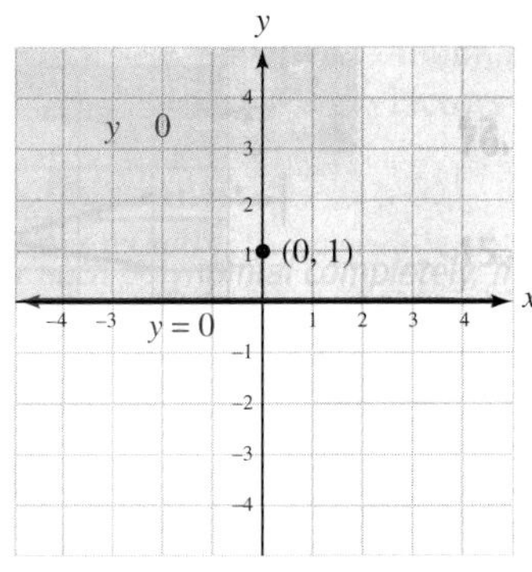

FIGURE 13-20

Answer

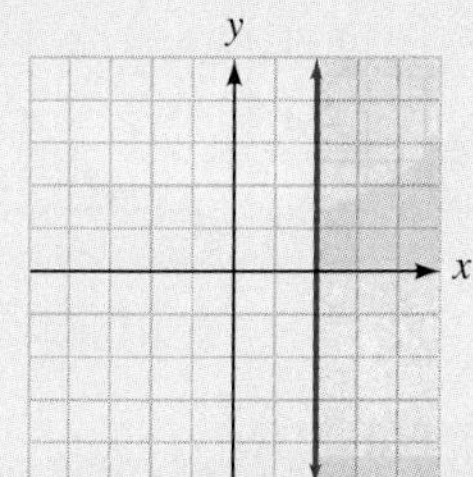

The following is a summary of the procedure for graphing linear inequalities.

Graphing linear inequalities

1. Graph the boundary line of the region. If the inequality allows the possibility of equality (the symbol is either $\leq$ or $\geq$), draw the boundary line as a solid line. If equality is not allowed ($<$ or $>$), draw the boundary line as a dashed line.
2. Pick a test point that is on one side of the boundary line. (Use the origin if possible.) Replace x and y in the inequality with the coordinates of that point. If the inequality is satisfied, shade the side that contains that point. If the inequality is not satisfied, shade the other side of the boundary.

An application of linear inequalities

EXAMPLE 7 Earning money. Carlos has two jobs, one paying \$10 per hour and one paying \$12 per hour. He must earn at least \$240 per week to pay his expenses while attending college. Write an inequality that shows the ways he can schedule his time to achieve his goal.

Solution If we let x represent the number of hours Carlos works on the first job and y the number of hours he works on the second job, we have

The hourly rate on the first job	times	the hours worked on the first job	plus	the hourly rate on the second job	times	the hours worked on the second job	is at least	\$240.
10	$\cdot$	x	$+$	12	$\cdot$	y	$\geq$	240

The graph of the inequality $10x + 12y \geq 240$ is shown in Figure 13-21 on the next page. Any point in the shaded region indicates a possible way Carlos can schedule his time and earn \$240 or more per week. For example, if he works 20 hours on the first job and 10 hours on the second job, he will earn

$$\begin{aligned} \$10(20) + \$12(10) &= \$200 + \$120 \\ &= \$320 \end{aligned}$$

Since Carlos cannot work a negative number of hours, a graph showing negative values of x or y would have no meaning.

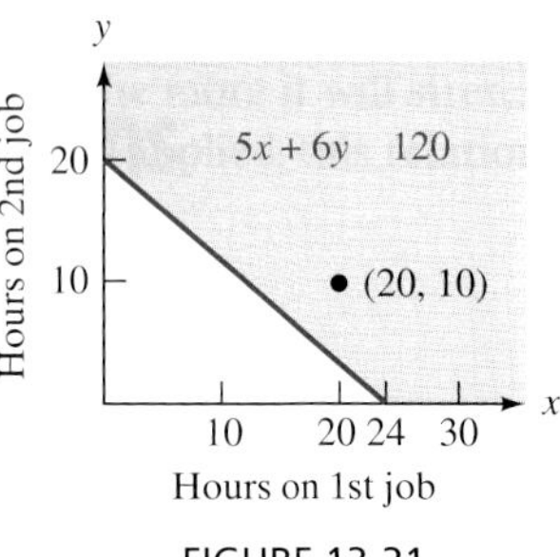

FIGURE 13-21

Section 13.5 STUDY SET

VOCABULARY *Fill in the blanks.*

1. $2x - y \le 4$ is a linear __________ in x and y.
2. The symbol $\le$ means __________ or __________.
3. In the graph in the illustration, the line $2x - y = 4$ is the __________.
4. In the illustration, the line $2x - y = 4$ divides the rectangular coordinate system into two __________.

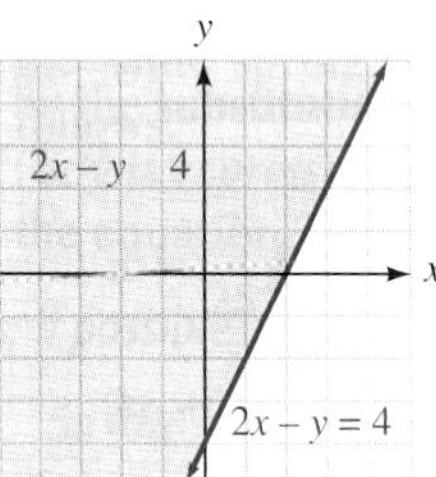

CONCEPTS

5. Determine whether each ordered pair is a solution of $5x - 3y \ge 0$.

   a. $(1, 1)$ b. $(-2, -3)$

   c. $(0, 0)$ d. $\left(\frac{1}{5}, \frac{4}{3}\right)$

6. Determine whether each ordered pair is a solution of $x + 4y < -1$.

   a. $(3, 1)$ b. $(-2, 0)$

   c. $(-0.5, 0.2)$ d. $\left(-2, \frac{1}{4}\right)$

7. Determine whether the graph of each linear inequality includes the boundary line. When graphed, is the boundary line solid or dashed?

   a. $y > -x$ b. $5x - 3y \le -2$

8. If a false statement results when the coordinates of a test point are substituted into a linear inequality, which half-plane should be shaded to represent the solution of the inequality?

9. A linear inequality has been graphed. Determine whether each point satisfies the inequality.

   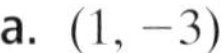

   a. $(1, -3)$
   b. $(-2, -1)$
   c. $(2, 3)$
   d. $(3, -4)$

10. A linear inequality has been graphed. Determine whether each point satisfies the inequality.

   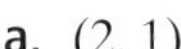

   a. $(2, 1)$
   b. $(-2, -4)$
   c. $(4, -2)$
   d. $(-3, 4)$

11. The boundary for the graph of a linear inequality is shown. Why can't the origin be used as a test point to decide which side to shade?

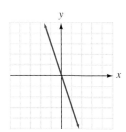

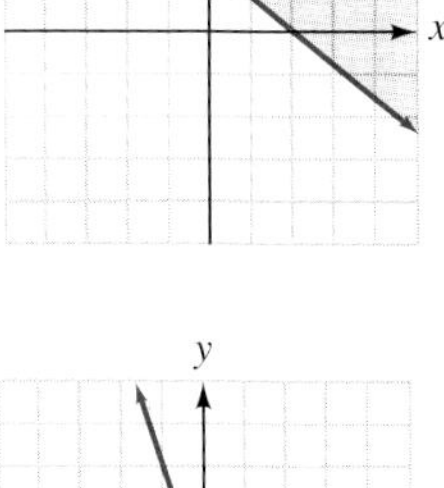

12. To determine how many pallets (x) and barrels (y) a delivery truck can hold, a dispatcher refers to the loading sheet on the right. Can a truck make a delivery of 4 pallets and 10 barrels in one trip?

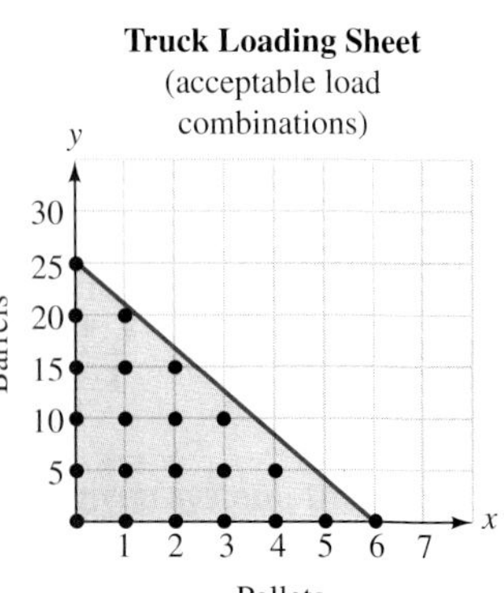

PRACTICE *Complete the graph by shading the correct side of the boundary.*

13. $y \le x + 2$

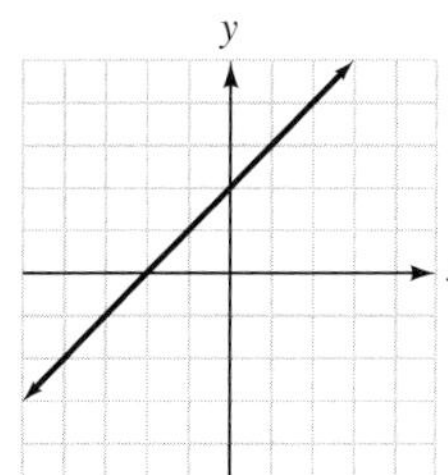

14. $y > x - 3$

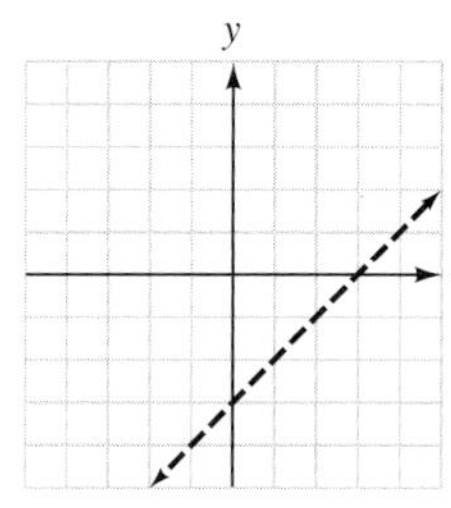

15. $y > 2x - 4$

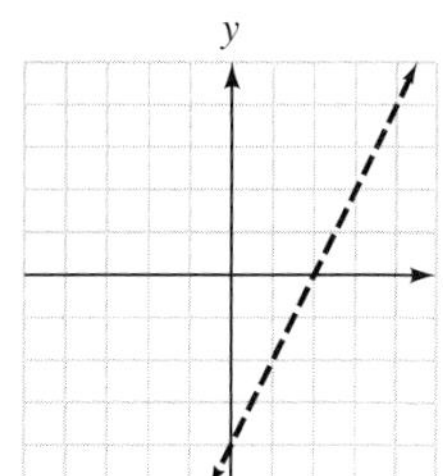

16. $y \le -x + 1$

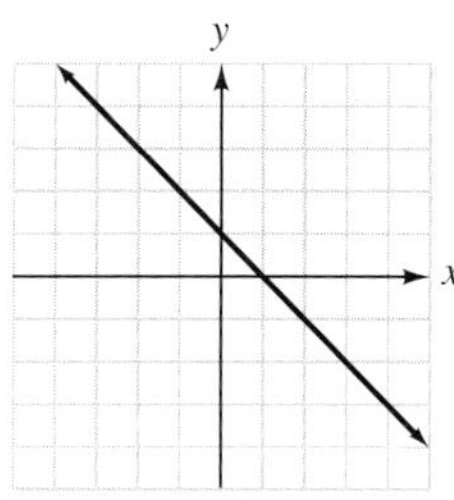

17. $x - 2y \ge 4$

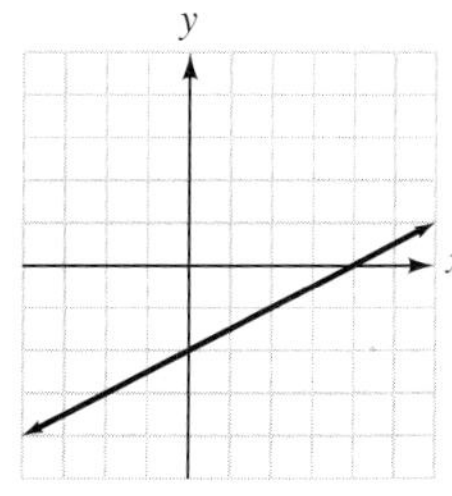

18. $3x + 2y > 12$

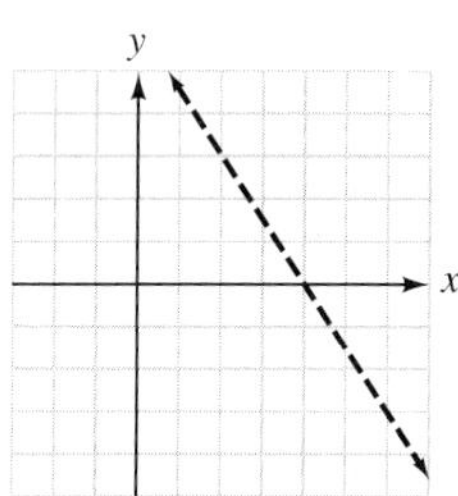

19. $y \le 4x$

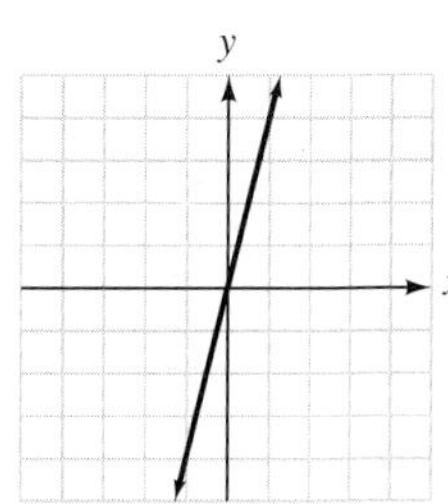

20. $y + 2x < 0$

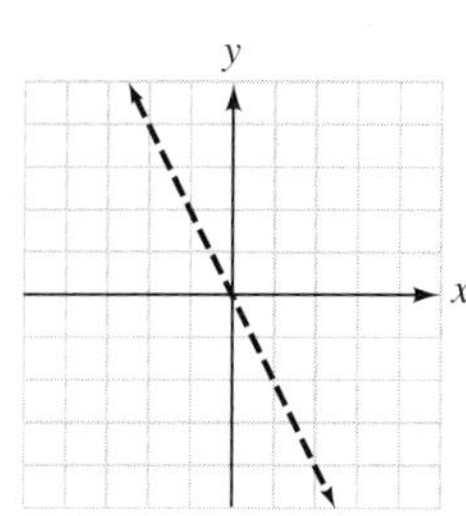

Graph each inequality.

21. $y \ge 3 - x$

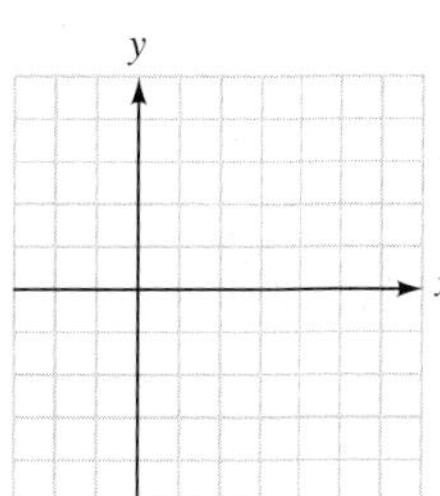

22. $y < 2 - x$

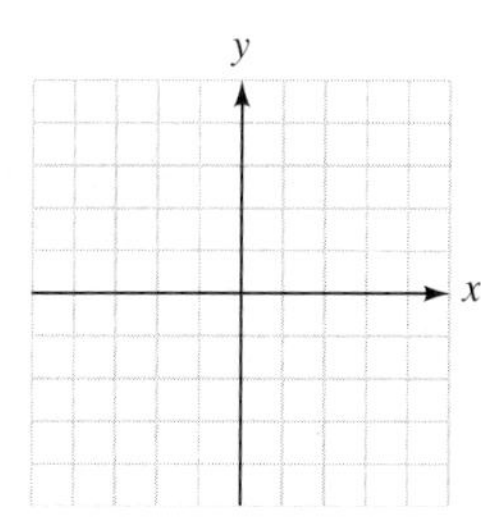

23. $y < 2 - 3x$

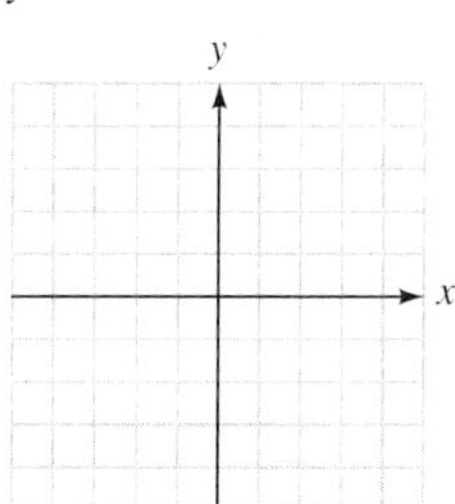

24. $y \ge 5 - 2x$

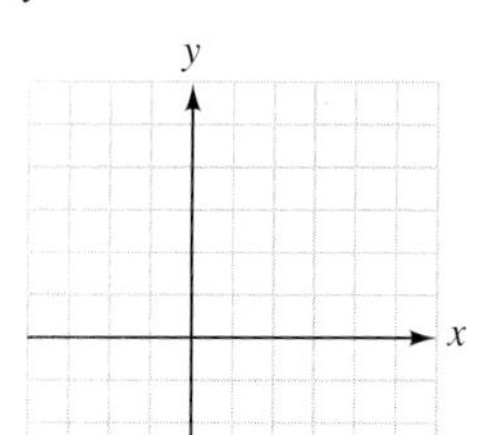

25. $y \ge 2x$

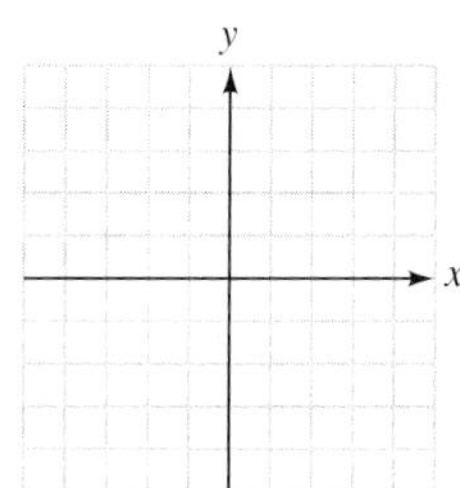

26. $y < 3x$

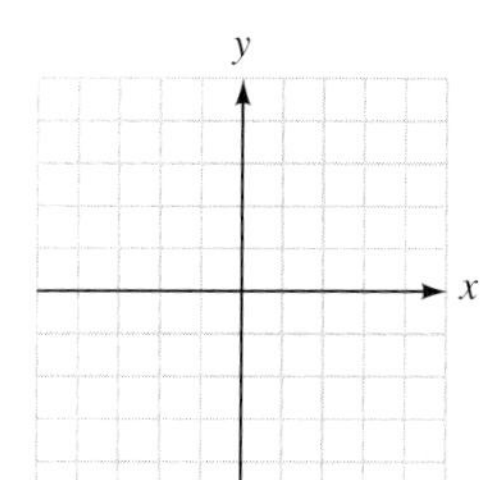

27. $2y - x < 8$

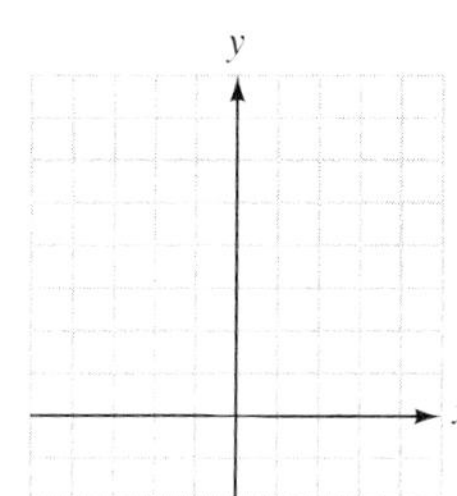

28. $y + 9x \ge 3$

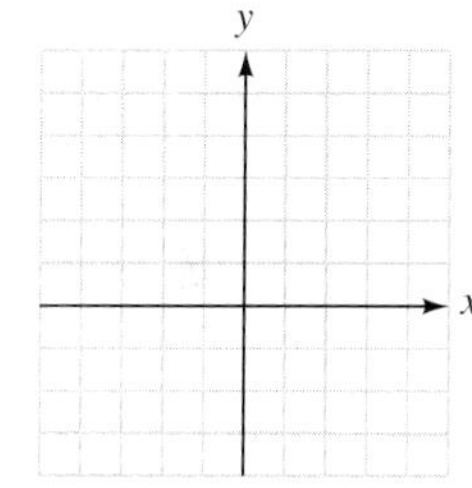

29. $y - x \ge 0$

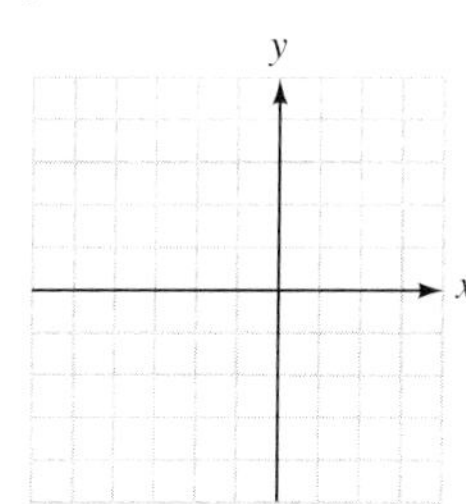

30. $y + x < 0$

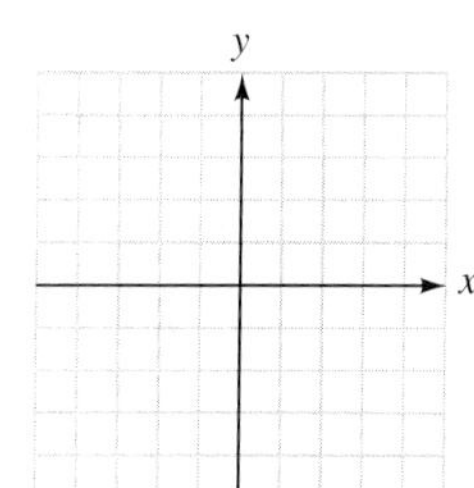

31. $2x + y > 2$

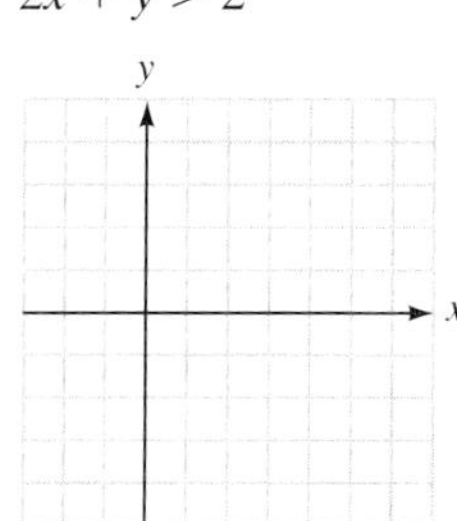

32. $3x - 2y > 6$

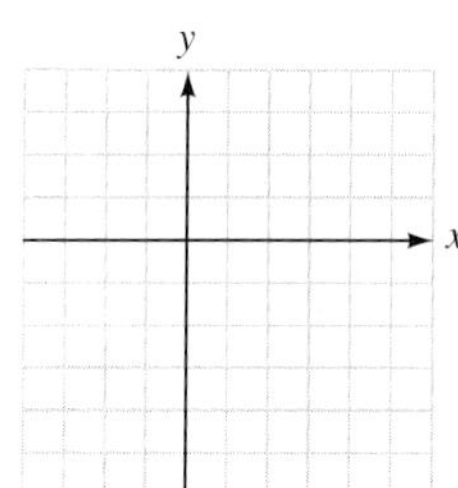

33. $3x - 4y > 12$

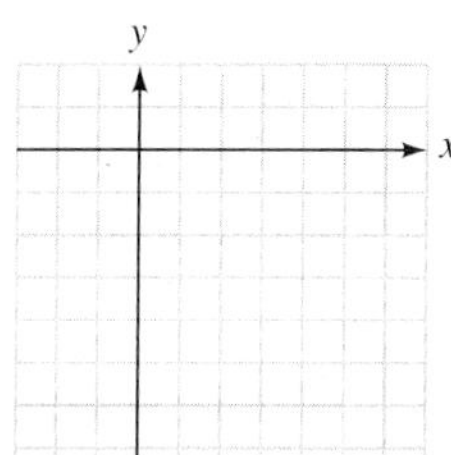

34. $4x + 3y \le 12$

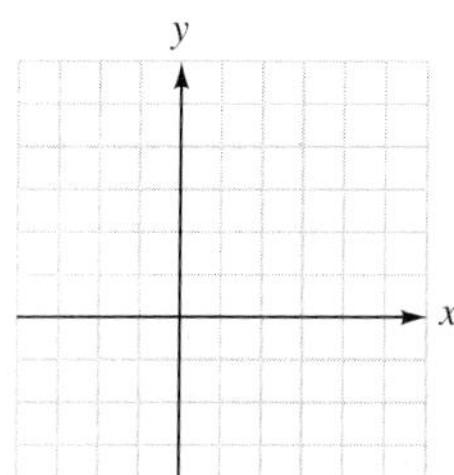

35. $5x + 4y \geq 20$

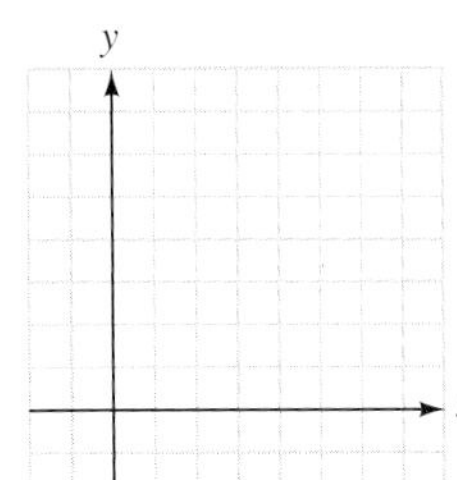

36. $7x - 2y < 21$

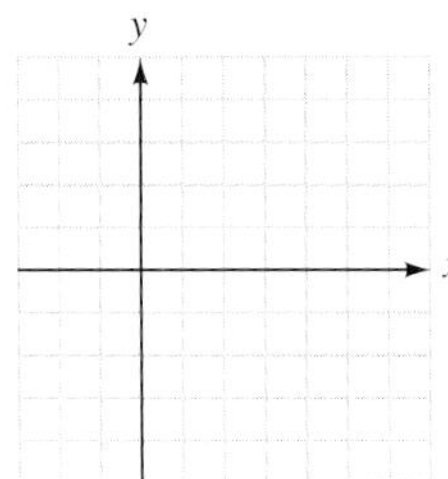

37. $x < 2$

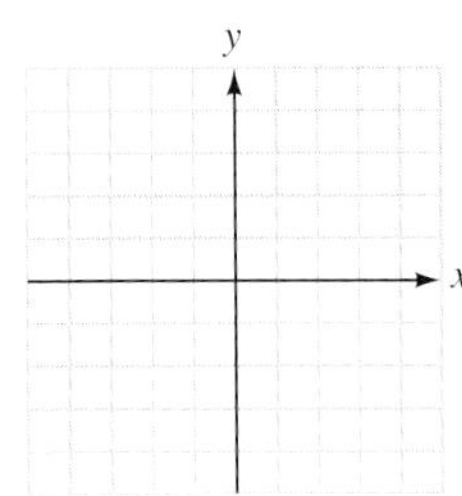

38. $y > -3$

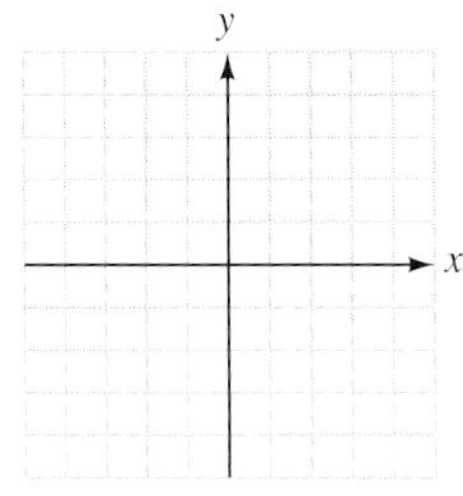

39. $y \leq 1$

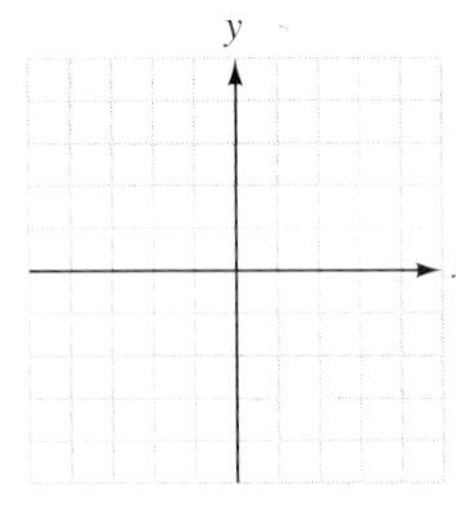

40. $x \geq -4$

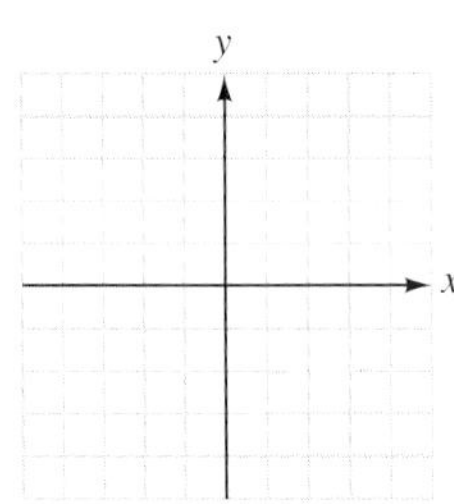

Simplify each inequality and then graph it.

41. $3(x + y) + x < 6$

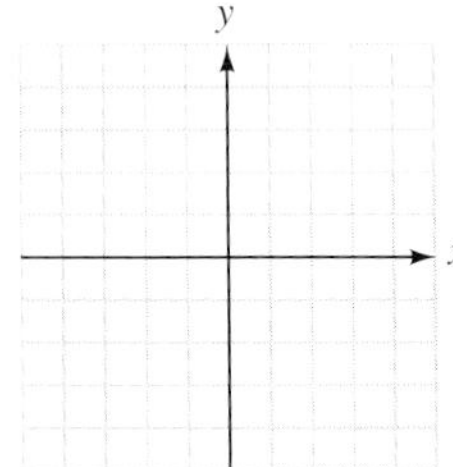

42. $2(x - y) - y \geq 4$

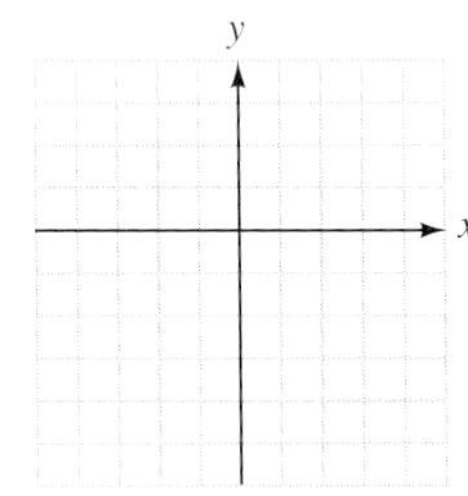

43. $4x - 3(x + 2y) \geq -6y$

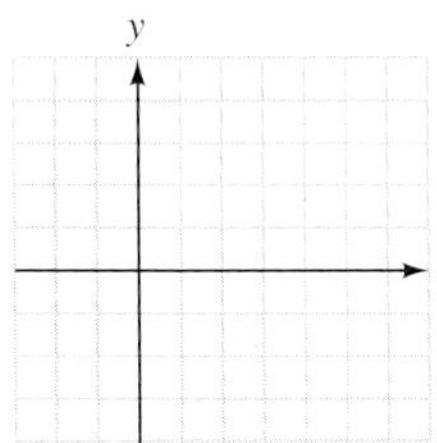

44. $3y + 2(x + y) < 5y$

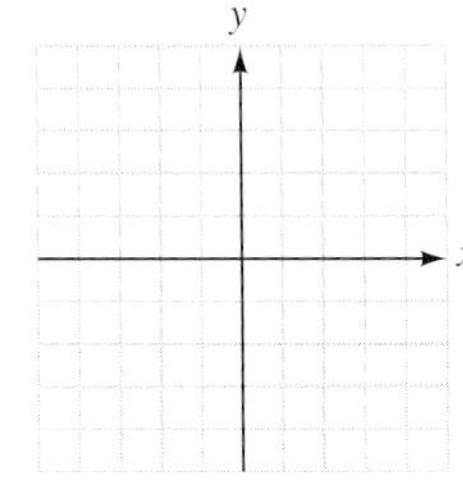

APPLICATIONS

45. NATO In March 1999, NATO aircraft and cruise missiles targeted Serbian military forces that were south of the 44th parallel in Yugoslavia, Montenegro, and Kosovo. Shade the geographic area that NATO was trying to rid of Serbian forces.

46. U.S. HISTORY When he ran for president in 1844, the campaign slogan of James K. Polk was "54-40 or fight!" It meant that Polk was willing to fight Great Britain for the possession of the Oregon Territory north to the 54°40′ parallel, as shown below. In 1846, Polk accepted a compromise to establish the 49th parallel as the permanent boundary of the United States. Shade the area of land that Polk conceded to the British.

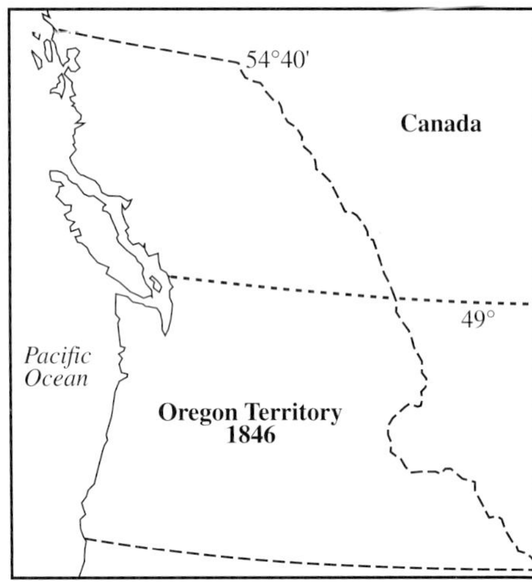

Write an inequality and graph it for nonnegative values of x and y. Then give three ordered pairs that satisfy the inequality.

47. PRODUCTION PLANNING It costs a bakery \$3 to make a cake and \$4 to make a pie. Production costs cannot exceed \$120 per day. Use the illustration on the right to graph an inequality that shows the possible combinations of cakes (x) and pies (y) that can be made.

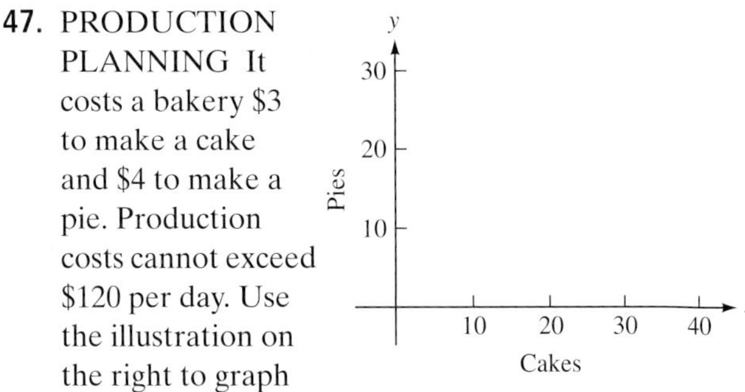

48. HIRING BABYSITTERS Mary has a choice of two babysitters. Sitter 1 charges \$6 per hour, and sitter 2 charges \$7 per hour. If Mary can afford no more than \$42 per week for sitters, use the illustration above to graph an inequality that shows the possible ways that she can hire sitter 1 (x) and sitter 2 (y).

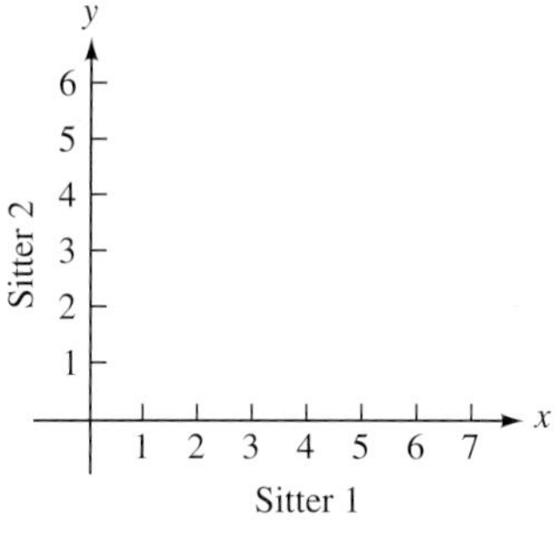

49. INVENTORIES A clothing store advertises that it maintains an inventory of at least \$4,400 worth of men's jackets. If a leather jacket costs \$100 and a nylon jacket costs \$88, use the illustration to graph an inequality that shows the possible ways that leather jackets (x) and nylon jackets (y) can be stocked.

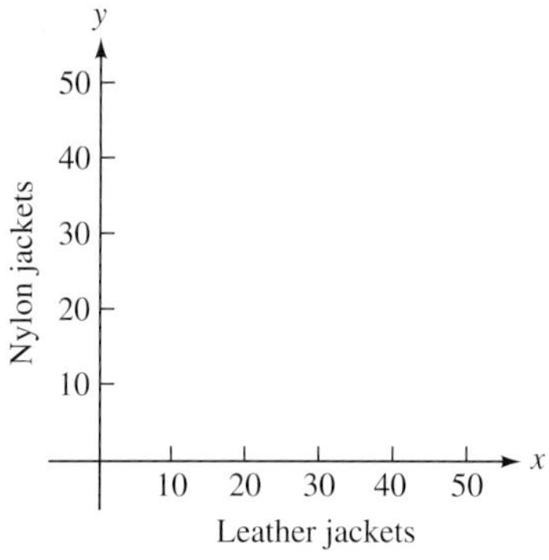

50. MAKING SPORTING GOODS A sporting goods manufacturer allocates at least 2,400 units of production time per day to make baseballs and footballs. If it takes 20 units of time to make a baseball and 30 units of time to make a football, use the illustration to graph an inequality that shows the possible ways to schedule the production time to make baseballs (x) and footballs (y).

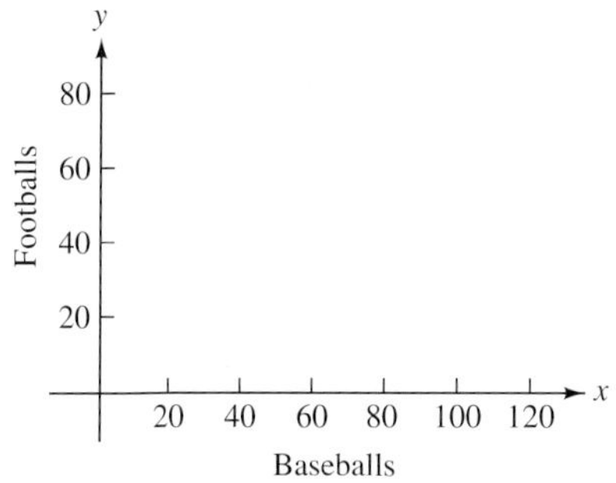

51. INVESTING Robert has up to \$8,000 to invest in two companies. If stock in Robotronics sells for \$40 per share and stock in Macrocorp sells for \$50 per share, use the illustration in the next column to graph an inequality that shows the possible ways that he can buy shares of Robotronics (x) and Macrocorp (y).

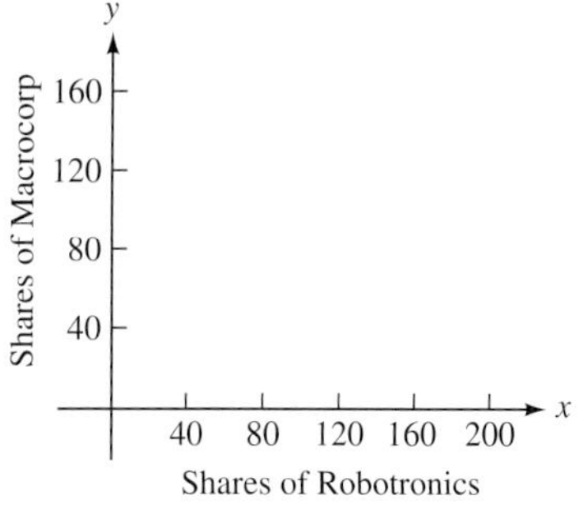

52. BASEBALL TICKETS Tickets to the Rockford Rox baseball games cost \$6 for reserved seats and \$4 for general admission. If nightly receipts must average at least \$10,200 to meet expenses, use the illustration to graph an inequality that shows the possible ways that the Rox can sell reserved seats (x) and general admission tickets (y).

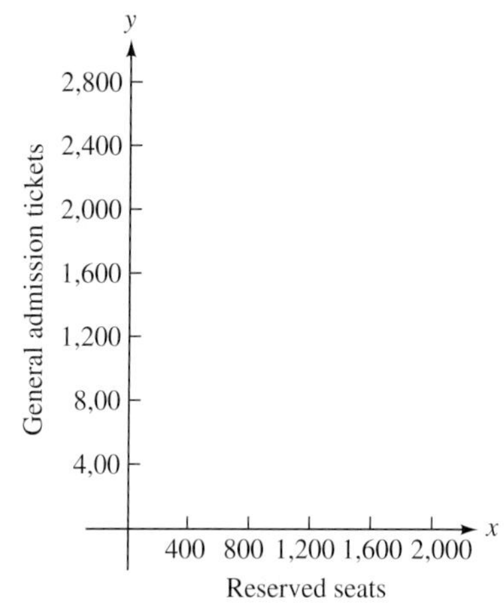

WRITING

53. Explain how to find the boundary for the graph of a linear inequality in two variables.

54. Explain how to decide which side of the boundary line to shade when graphing a linear inequality in two variables.

REVIEW

55. Let $g(x) = 3x^2 - 4x + 3$. Find $g(2)$.

56. Solve: $2(x - 4) \leq -12$.

57. Factor: $x^3 + 27$.

58. Factor: $9p - 9q + mp - mq$.

59. Write a formula relating distance, rate, and time.

60. What is the slope of the line $2x - 3y = 2$?

61. Solve $A = P + Prt$ for t.

62. What is the sum of the measures of the three angles of any triangle?

13.6 Solving Systems of Linear Inequalities

- Systems of linear inequalities
- An application of systems of linear inequalities

We have solved systems of linear *equations* by the graphing method. The solution of such a system is the point of intersection of the straight lines. We now consider how to solve systems of linear *inequalities* graphically. When the solution of a linear inequality is graphed, the result is a half-plane. Therefore, we would expect to find the graphical solution of a system of inequalities by looking for the intersection, or overlap, of shaded half-planes.

Systems of linear inequalities

To solve the **system of linear inequalities**

$$\begin{cases} x + y \geq 1 \\ x - y \geq 1 \end{cases}$$

we first graph each inequality. For instructional purposes, we will graph each inequality on a separate set of axes, although in practice we will draw them on the same axes.

The graph of $x + y \geq 1$ includes the graph of the boundary line $x + y = 1$ and all points above it. Because the boundary is included, we draw it with a solid line, as shown in Figure 13-22(a).

The graph of $x - y \geq 1$ includes the graph of the boundary line $x - y = 1$ and all points below it. Because the boundary is included, it is drawn with a solid line, as shown in Figure 13-22(b).

$x + y = 1$		
x	y	(x, y)
0	1	(0, 1)
1	0	(1, 0)
2	−1	(2, −1)

$x - y = 1$		
x	y	(x, y)
0	−1	(0, −1)
1	0	(1, 0)
2	1	(2, 1)

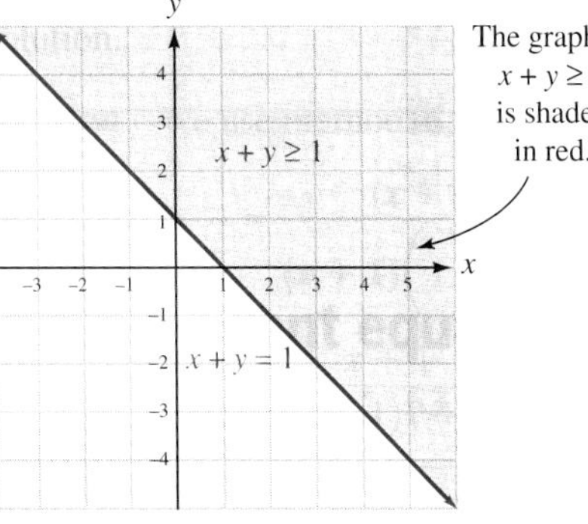

(a)

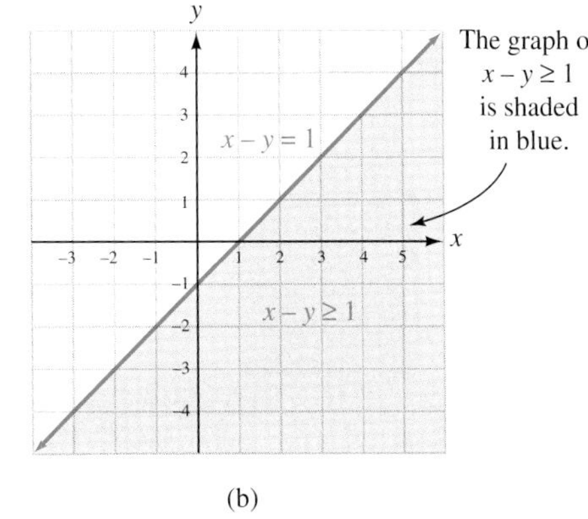

(b)

FIGURE 13-22

In Figure 13-23, we show the result when the inequalities $x + y \geq 1$ and $x - y \geq 1$ are graphed one at a time on the same coordinate axes. The area that is shaded twice represents the set of simultaneous solutions of the given system of inequalities. Any point in the doubly shaded region (shown in purple) has coordinates that satisfy both inequalities of the system.

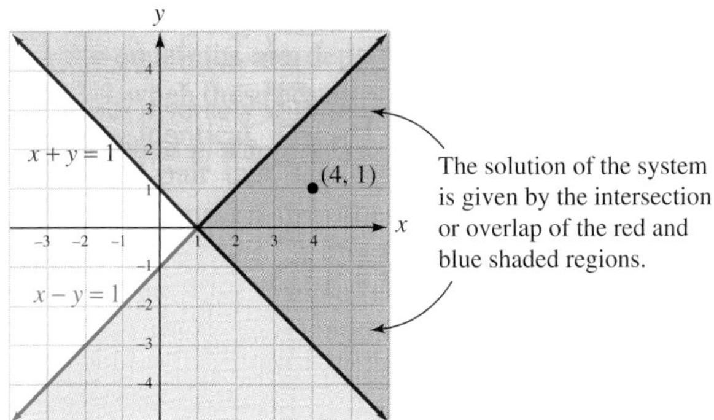

The solution of the system is given by the intersection or overlap of the red and blue shaded regions.

FIGURE 13-23

To see whether this is true, we can pick a point, such as (4, 1), that lies in the doubly shaded region and show that its coordinates satisfy both inequalities.

Check: $x + y \geq 1$ and $x - y \geq 1$

$\mathbf{4} + \mathbf{1} \geq 1$ $\quad$ $\mathbf{4} - \mathbf{1} \geq 1$

$5 \geq 1$ $\quad$ $3 \geq 1$

The resulting true statements verify that (4, 1) is a solution of the system. If we pick a point that is not in the doubly shaded region, its coordinates will fail to satisfy at least one of the inequalities.

In general, to solve systems of linear inequalities, we will follow these steps.

Solving systems of inequalities

1. Graph each inequality in the system on the same coordinate axes.
2. Find the region that is common to every graph.
3. Pick a test point from the region to verify the solution.

EXAMPLE 1 Graph the solution of $\begin{cases} 2x + y < 4 \\ y > 2x + 2 \end{cases}$.

Solution First, we graph each inequality on one set of axes, as shown in Figure 13-24.

2x + y = 4		
x	***y***	***(x, y)***
0	4	(0, 4)
2	0	(2, 0)
1	2	(1, 2)

$y = 2x + 2$

So $m = 2 = \frac{2}{1}$

and $b = 2$.

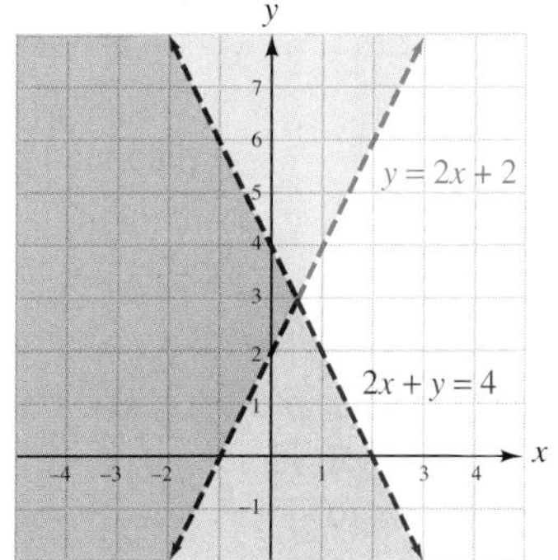

FIGURE 13-24

Self Check 1

Graph the solution of

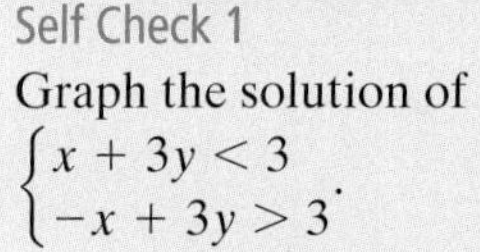

$\begin{cases} x + 3y < 3 \\ -x + 3y > 3 \end{cases}$.

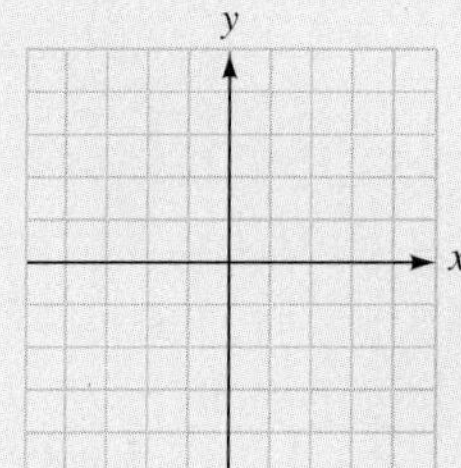

We note that

- The graph of $2x + y < 4$ includes all points below the boundary line $2x + y = 4$. Since the boundary is not included, we draw it as a dashed line.

Answer

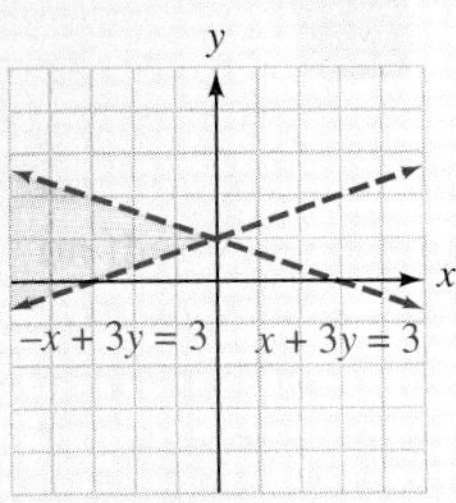

- The graph of $y > 2x + 2$ includes all points above the boundary line $y = 2x + 2$. Since the boundary is not included, we draw it as a dashed line.

The area that is shaded twice (the region in purple) is the solution of the given system of inequalities. Any point in the doubly shaded region has coordinates that will satisfy both inequalities of the system.

Pick a point in the doubly shaded region and show that it satisfies both inequalities.

Self Check 2

Graph the solution of $\begin{cases} y \le 1 \\ x > 2 \end{cases}$.

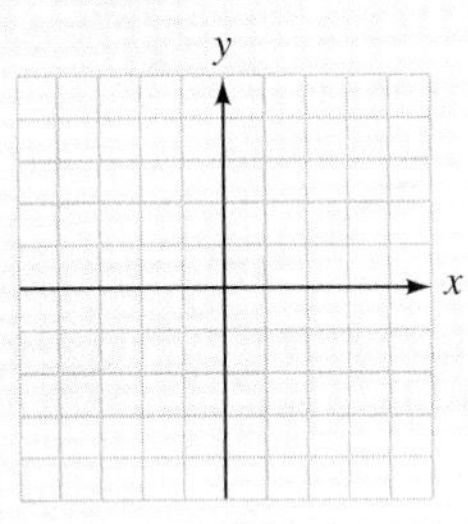

EXAMPLE 2 Graph the solution of $\begin{cases} x \le 2 \\ y > 3 \end{cases}$.

Solution We graph each inequality on one set of axes, as shown in Figure 13-25.

$x = 2$		
x	*y*	(*x*, *y*)
2	0	(2, 0)
2	2	(2, 2)
2	4	(2, 4)

$y = 3$		
x	*y*	(*x*, *y*)
0	3	(0, 3)
1	3	(1, 3)
4	3	(4, 3)

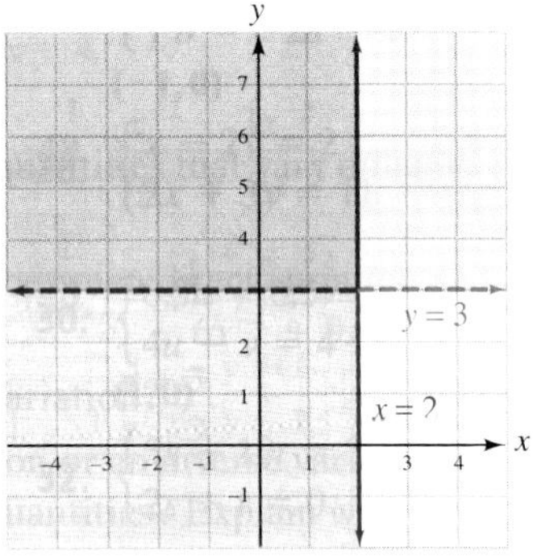

FIGURE 13-25

We note that

- The graph of $x \le 2$ includes all points to the left of the boundary line $x = 2$. Since the boundary is included, we draw it as a solid line.
- The graph $y > 3$ includes all points above the boundary line $y = 3$. Since the boundary is not included, we draw it as a dashed line.

The area that is shaded twice is the solution of the given system of inequalities. Any point in the doubly shaded region (purple) has coordinates that will satisfy both inequalities of the system. Pick a point in the doubly shaded region and show that this is true.

Answer

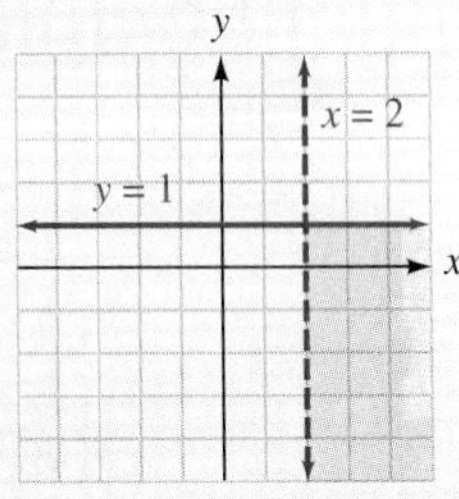

Self Check 3

Graph the solution of $\begin{cases} y \ge -\frac{1}{2}x + 1 \\ y < -\frac{1}{2}x - 1 \end{cases}$.

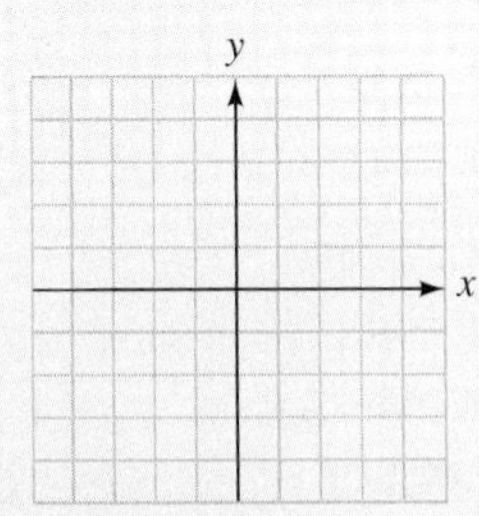

EXAMPLE 3 Graph the solution of $\begin{cases} y < 3x - 1 \\ y \ge 3x + 1 \end{cases}$.

Solution We graph each inequality as shown in Figure 13-26 and make the following observations:

- The graph of $y < 3x - 1$ includes all points below the dashed line $y = 3x - 1$.
- The graph of $y \ge 3x + 1$ includes all points on and above the solid line $y = 3x + 1$.

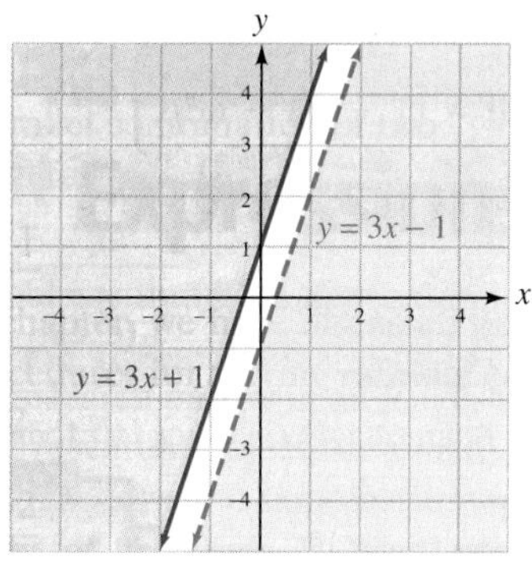

FIGURE 13-26

Because the graphs of these inequalities do not intersect, the solution set is empty. There are no solutions.

Answer no solutions

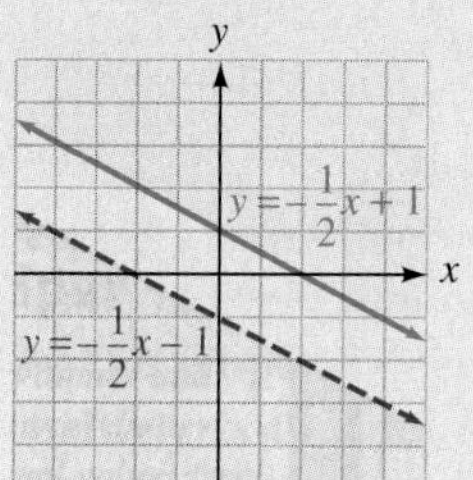

EXAMPLE 4 Graph the solution of $\begin{cases} x \geq 0 \\ y \geq 0 \\ x + 2y \leq 6 \end{cases}$.

Solution We graph each inequality as shown in Figure 13-27 and make the following observations:

- The graph of $x \geq 0$ includes all points on the y-axis and to the right.
- The graph of $y \geq 0$ includes all points on the x-axis and above.
- The graph of $x + 2y \leq 6$ includes all points on the line $x + 2y = 6$ and below.

The solution is the region that is shaded three times. This includes triangle OPQ and the triangular region it encloses.

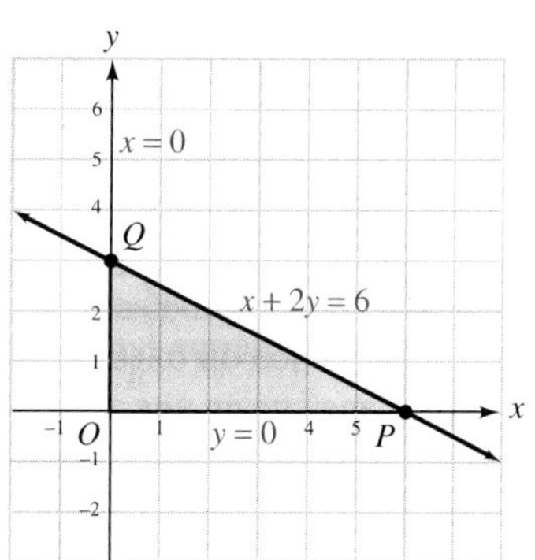

FIGURE 13-27

Self Check 4

Graph the solution of $\begin{cases} x \leq 1 \\ y \leq 2 \\ 2x - y \leq 4 \end{cases}$.

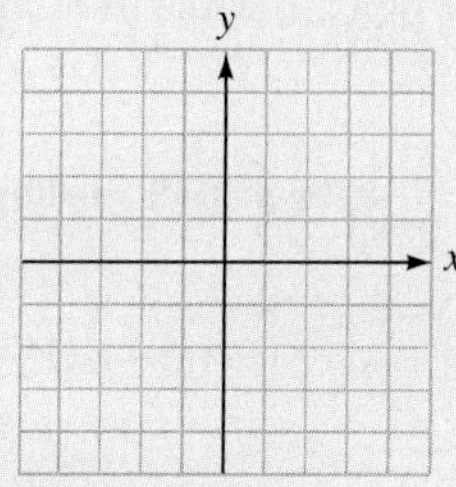

Answer

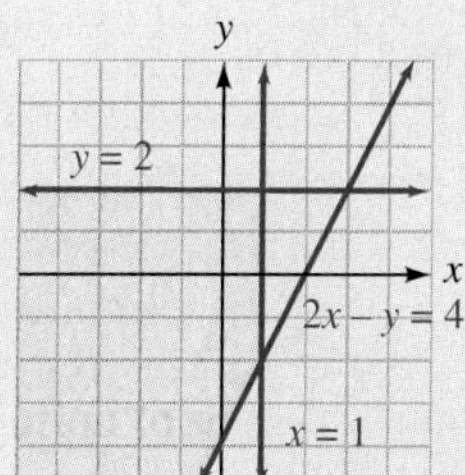

An application of systems of linear inequalities

EXAMPLE 5 **Landscaping.** A homeowner budgets from \$300 to \$600 for trees and bushes to landscape his yard. After shopping around, he finds that good trees cost \$150 and mature bushes cost \$75. What combinations of trees and bushes can he afford to buy?

Analyze the problem

The homeowner wants to spend *at least* \$300 but *not more than* \$600 for trees and bushes.

Form two inequalities

We can let x represent the number of trees purchased and y the number of bushes purchased. We then form the following system of inequalities:

The cost of a tree	times	the number of trees purchased	plus	the cost of a bush	times	the number of bushes purchased	should at least be	\$300.
\$150	$\cdot$	x	$+$	\$75	$\cdot$	y	$\geq$	\$300

The cost of a tree	times	the number of trees purchased	plus	the cost of a bush	times	the number of bushes purchased	should not be more than	\$600.
\$150	$\cdot$	x	$+$	\$75	$\cdot$	y	$\leq$	\$600

Solve the system

We graph the system

$$\begin{cases} 150x + 75y \geq 300 \\ 150x + 75y \leq 600 \end{cases}$$

as shown in Figure 13-28. The coordinates of each point shown in the graph give a possible combination of the number of trees (x) and the number of bushes (y) that can be purchased. These possibilities are

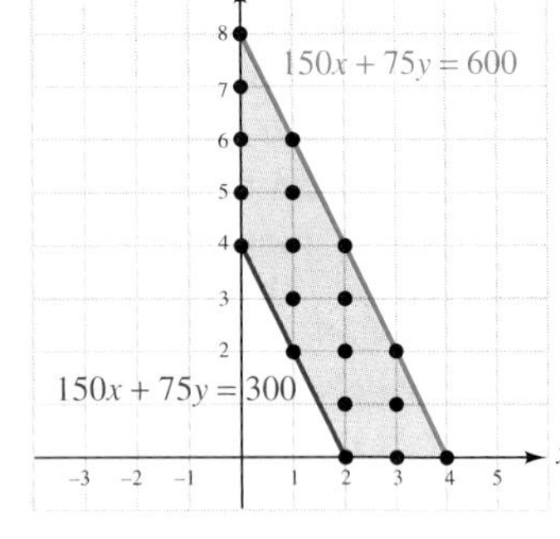

FIGURE 13-28

(0, 4), (0, 5), (0, 6), (0, 7), (0, 8)
(1, 2), (1, 3), (1, 4), (1, 5), (1, 6)
(2, 0), (2, 1), (2, 2), (2, 3), (2, 4)
(3, 0), (3, 1), (3, 2), (4, 0)

Only these points can be used, because the homeowner cannot buy a portion of a tree or a bush.

Section 13.6 STUDY SET

VOCABULARY *Fill in the blanks.*

1. $\begin{cases} x + y > 2 \\ x + y < 4 \end{cases}$ is a system of linear ________.

2. The ________ of a system of linear inequalities is all the ordered pairs that make all inequalities of the system true at the same time.

3. Any point in the doubly ________ region of the graph of the solution of a system of two linear inequalities has coordinates that satisfy both inequalities of the system.

4. To graph a linear inequality such as $x + y > 2$, first graph the boundary. Then pick a test ________ to determine which half-plane to shade.

CONCEPTS

5. In the illustration, the solution of linear inequality 1 was shaded in red, and the solution of linear inequality 2 was shaded in blue. The overlap of the red and the blue regions is shown in purple. Determine whether a true or a false statement results when the coordinates of the given point are substituted into the given inequality.

a. A, inequality 1

b. A, inequality 2

c. B, inequality 1

d. B, inequality 2

e. C, inequality 1

f. C, inequality 2

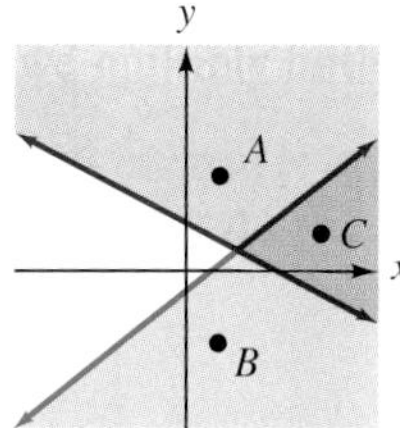

6. Match each equation, inequality, or system with the graph of its solution.

a. $x + y = 2$

b. $x + y \geq 2$

c. $\begin{cases} x + y = 2 \\ x - y = 2 \end{cases}$

d. $\begin{cases} x + y \geq 2 \\ x - y \leq 2 \end{cases}$

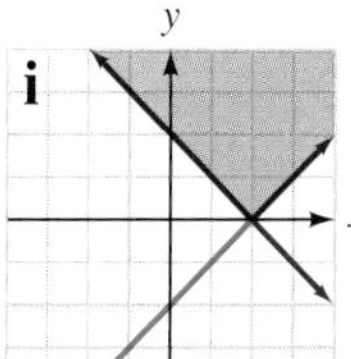

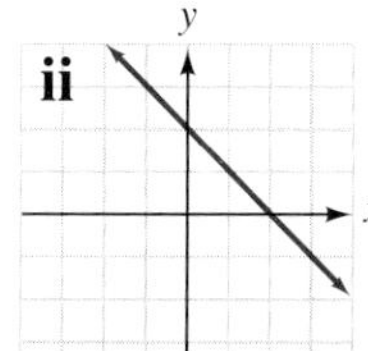

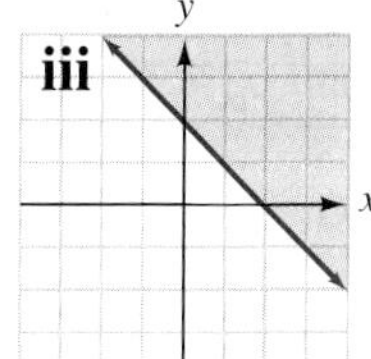

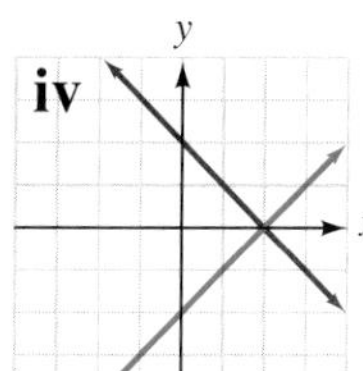

7. The graph of the solution of a system of linear inequalities is shown. Determine whether each point is a part of the solution set.

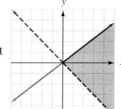

a. $(4, -2)$

b. $(1, 3)$

c. the origin

8. Use a system of inequalities to describe the shaded region in the illustration.

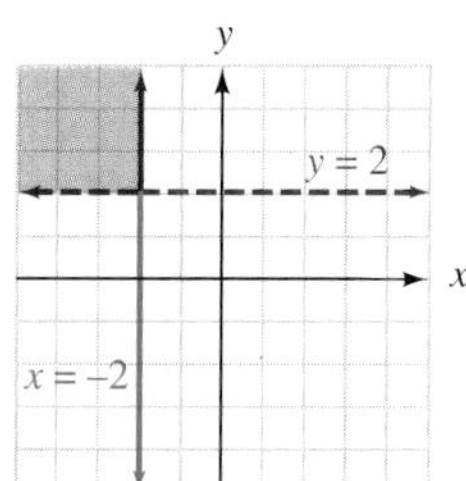

NOTATION

9. Fill in the blank: The graph of the solution of a system of linear inequalities shown can be described as the triangle ______ and the triangular region it encloses.

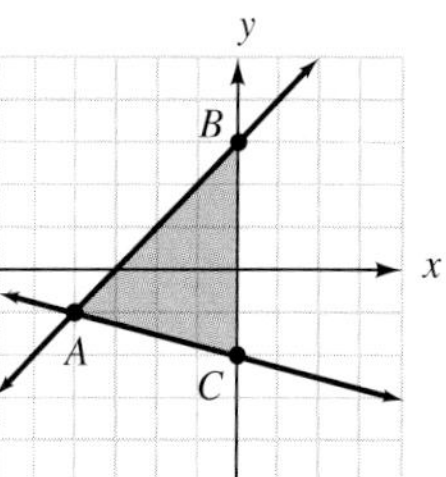

10. Represent each phrase using either $>$, $<$, $\geq$, or $\leq$.

a. is not more than

b. must be at least

c. should not surpass

d. cannot go below

PRACTICE *Graph the solution set of each system of inequalities, when possible.*

11. $\begin{cases} x + 2y \leq 3 \\ 2x - y \geq 1 \end{cases}$

12. $\begin{cases} 2x + y \geq 3 \\ x - 2y \leq -1 \end{cases}$

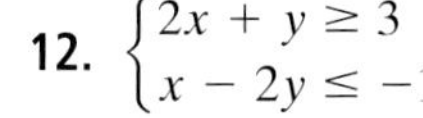

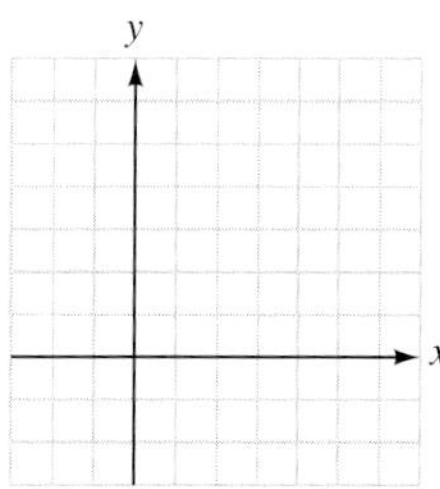

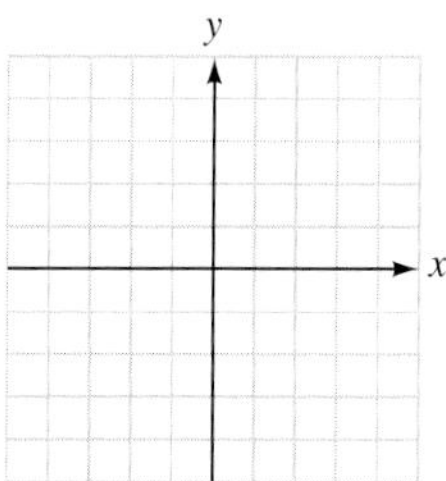

13. $\begin{cases} x + y < -1 \\ x - y > -1 \end{cases}$

14. $\begin{cases} x + y > 2 \\ x - y < -2 \end{cases}$

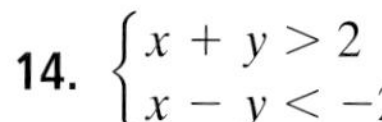

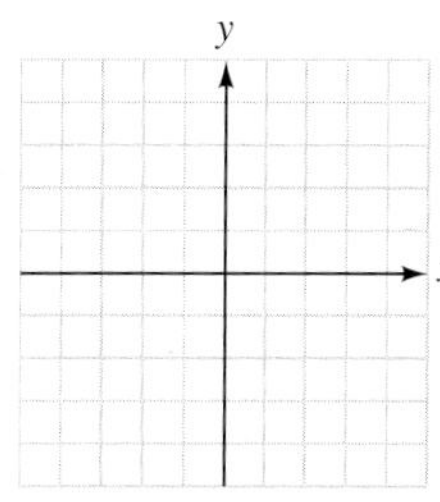

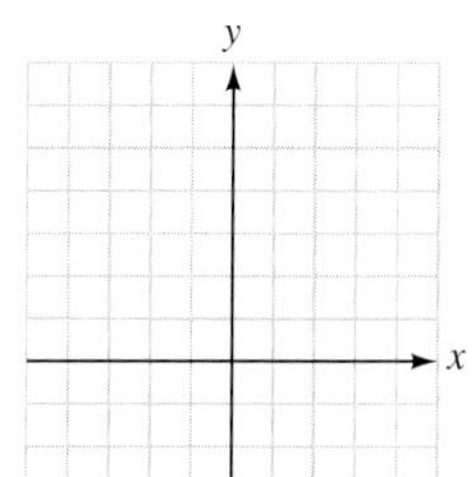

15. $\begin{cases} x > 2 \\ y \le 3 \end{cases}$

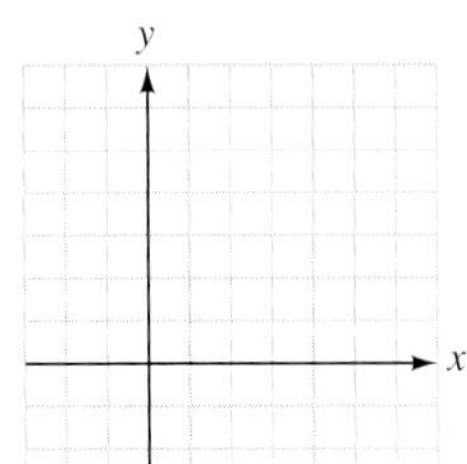

16. $\begin{cases} x \ge -1 \\ y > -2 \end{cases}$

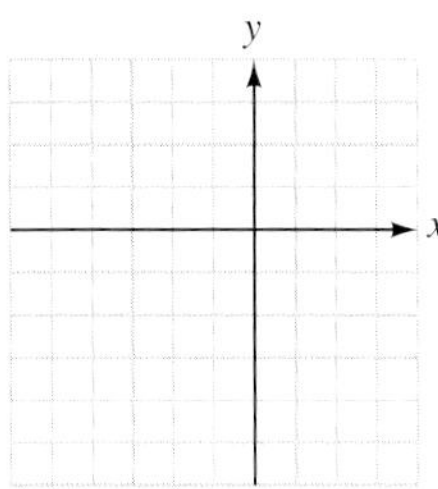

17. $\begin{cases} 2x - 3y \le 0 \\ y \ge x - 1 \end{cases}$

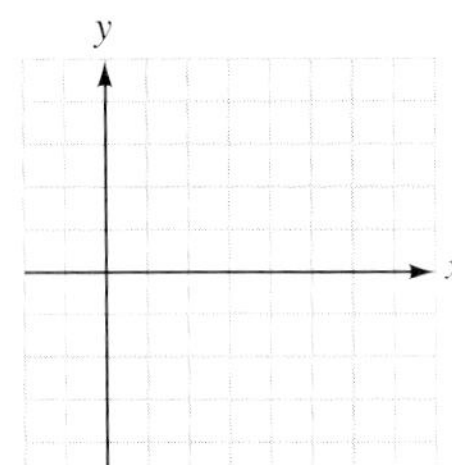

18. $\begin{cases} y > 2x - 4 \\ y \ge -x - 1 \end{cases}$

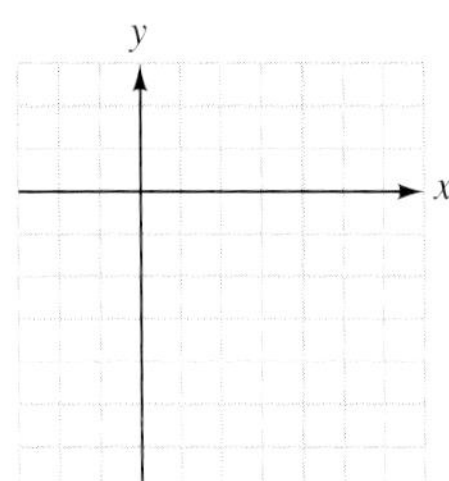

19. $\begin{cases} y < -x + 1 \\ y > -x + 3 \end{cases}$

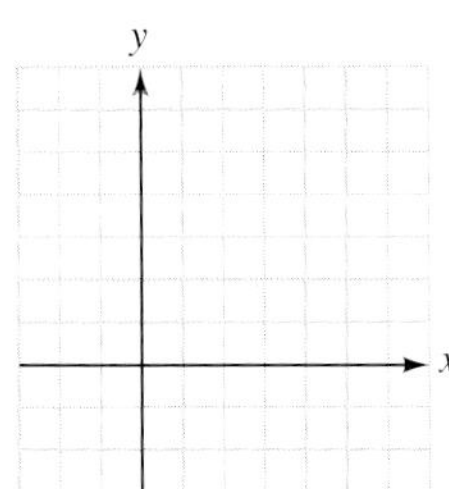

20. $\begin{cases} y > -x + 2 \\ y < -x + 4 \end{cases}$

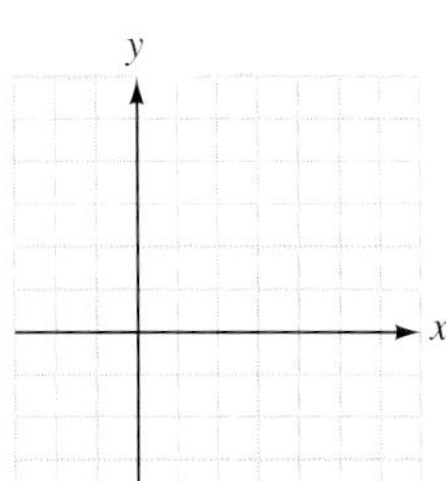

21. $\begin{cases} x > 0 \\ y > 0 \end{cases}$

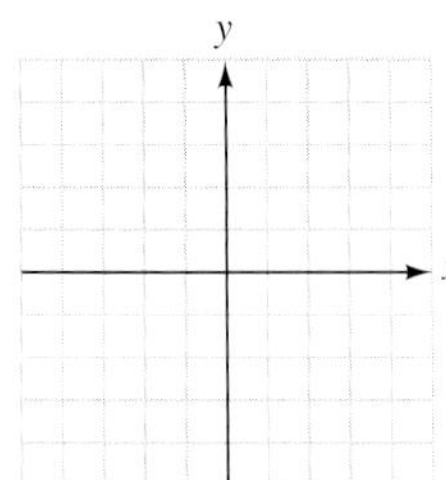

22. $\begin{cases} x \le 0 \\ y < 0 \end{cases}$

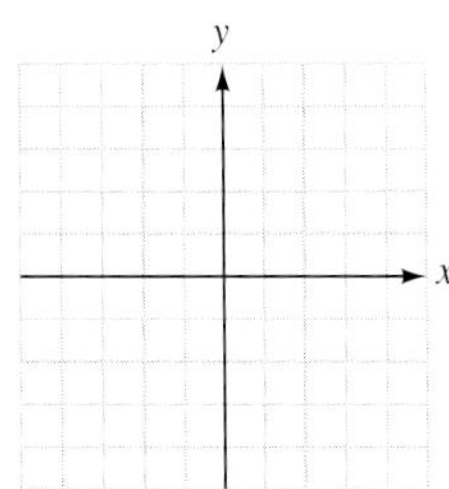

23. $\begin{cases} 3x + 4y > -7 \\ 2x - 3y \ge 1 \end{cases}$

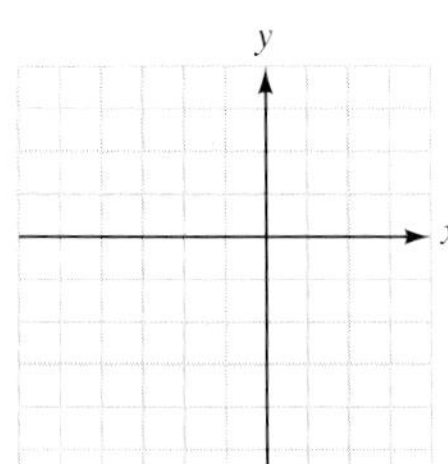

24. $\begin{cases} 3x + y \le 1 \\ 4x - y > -8 \end{cases}$

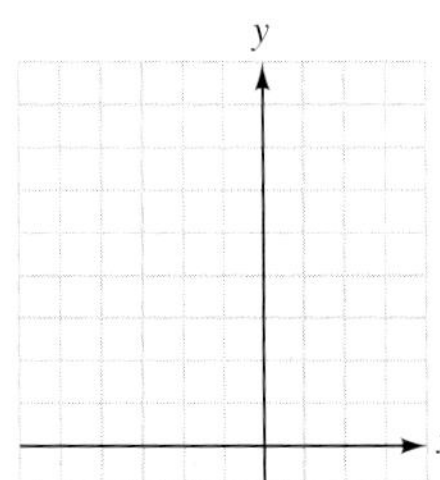

25. $\begin{cases} 2x + y < 7 \\ y > 2(1 - x) \end{cases}$

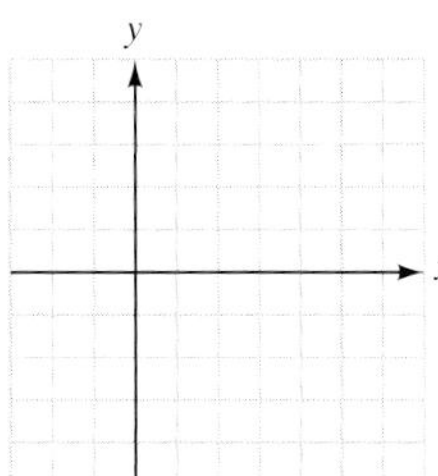

26. $\begin{cases} 2x + y \ge 6 \\ y \le 2(2x - 3) \end{cases}$

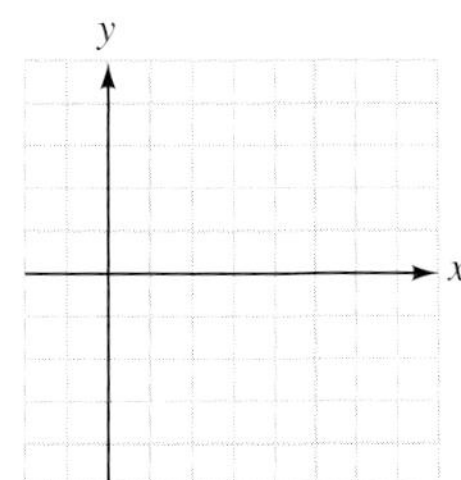

27. $\begin{cases} 2x - 4y > -6 \\ 3x + y \ge 5 \end{cases}$

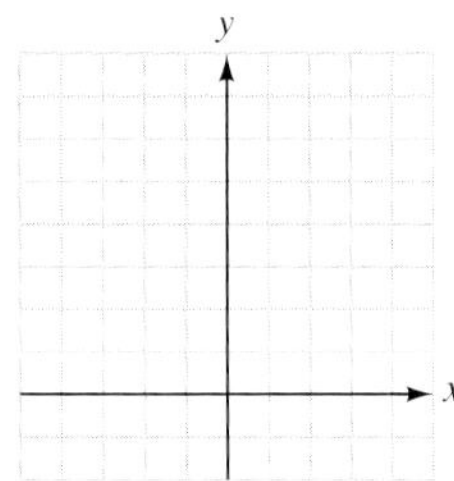

28. $\begin{cases} 2x - 3y < 0 \\ 2x + 3y \ge 12 \end{cases}$

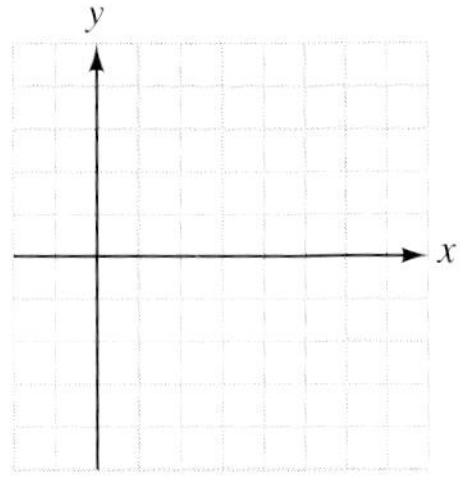

29. $\begin{cases} 3x - y \le -4 \\ 3y > -2(x + 5) \end{cases}$

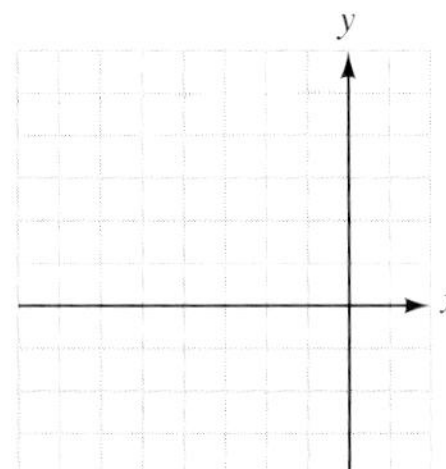

30. $\begin{cases} 3x + y < -2 \\ y > 3(1 - x) \end{cases}$

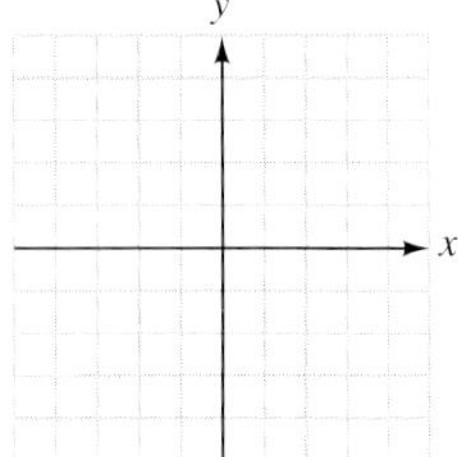

31. $\begin{cases} \frac{x}{2} + \frac{y}{3} \ge 2 \\ \frac{x}{2} - \frac{y}{2} < -1 \end{cases}$

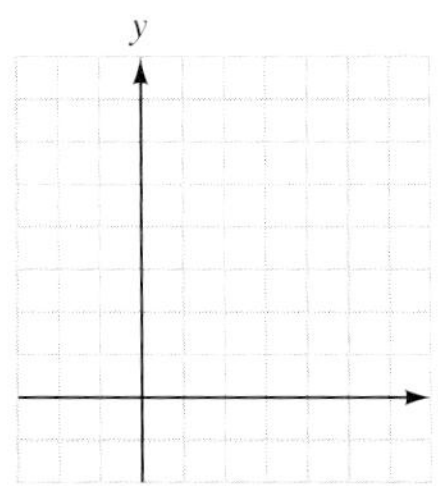

32. $\begin{cases} \frac{x}{3} - \frac{y}{2} < -3 \\ \frac{x}{3} + \frac{y}{2} > -1 \end{cases}$

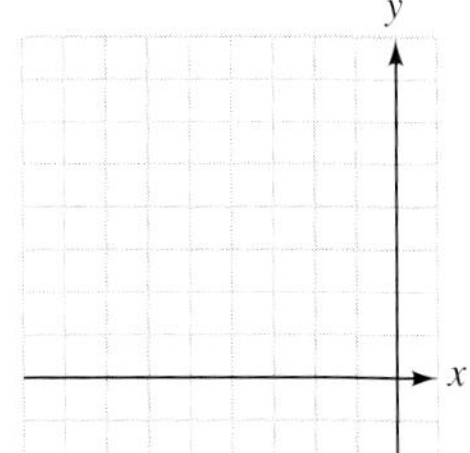

33. $\begin{cases} x \ge 0 \\ y \ge 0 \\ x + y \le 3 \end{cases}$

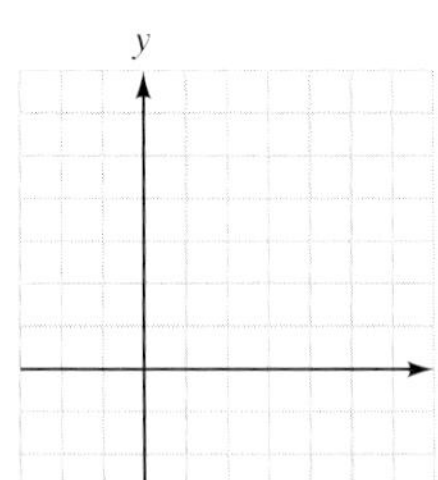

34. $\begin{cases} x - y \le 6 \\ x + 2y \le 6 \\ x \ge 0 \end{cases}$

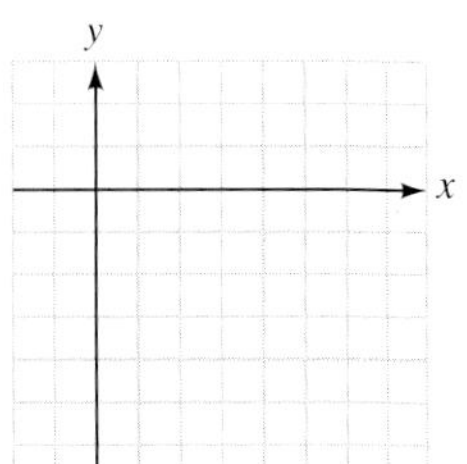

APPLICATIONS

35. BIRDS OF PREY Parts a and b of the illustration show the individual fields of vision for each eye of an owl. In part c, shade the area where the fields of vision overlap—that is, the area that is seen by both eyes.

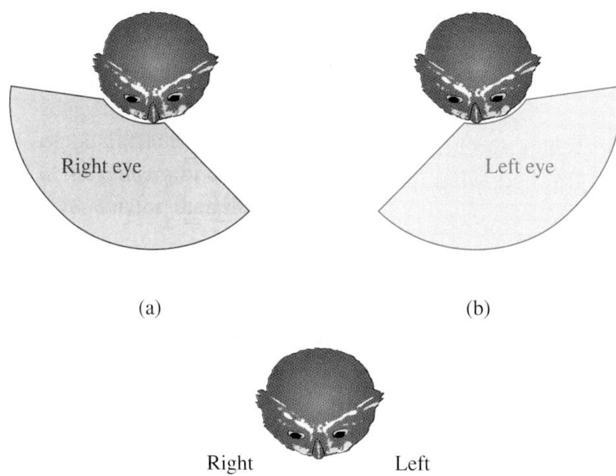

(a) (b)

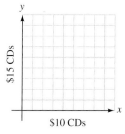

(c)

36. EARTH SCIENCE Shade the area of the Earth's surface that is north of the Tropic of Capricorn and south of the Tropic of Cancer.

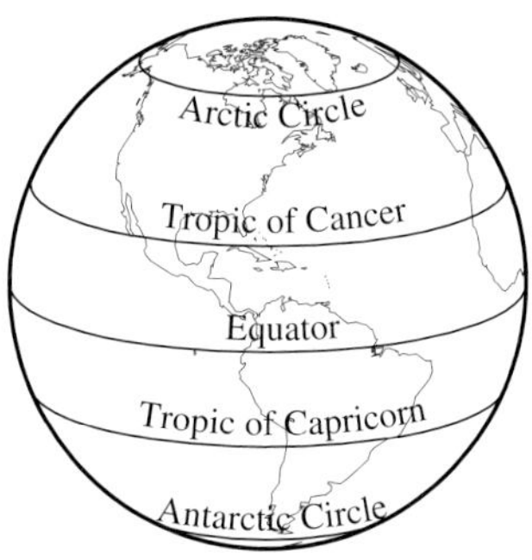

In Exercises 37–40, graph each system of inequalities and give two possible solutions.

37. BUYING COMPACT DISCS Melodic Music has compact discs on sale for either \$10 or \$15. If a customer wants to spend at least \$30 but no more than \$60 on CDs, use the illustration to graph a system of inequalities showing the possible combinations of \$10 CDs (x) and \$15 CDs (y) that the customer can buy.

y
\$15 CDs
x
\$10 CDs

38. BUYING BOATS Dry Boatworks wholesales aluminum boats for \$800 and fiberglass boats for \$600. Northland Marina wants to make a purchase totaling at least \$2,400, but no more than \$4,800. Use the illustration to graph a system of inequalities showing the possible combinations of aluminum boats (x) and fiberglass boats (y) that can be ordered.

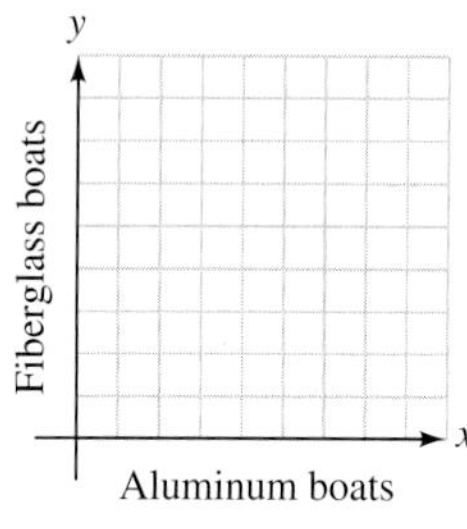

39. BUYING FURNITURE A distributor wholesales desk chairs for \$150 and side chairs for \$100. Best Furniture wants its order to total no more than \$900; Best also wants to order more side chairs than desk chairs. Use the illustration to graph a system of inequalities showing the possible combinations of desk chairs (x) and side chairs (y) that can be ordered.

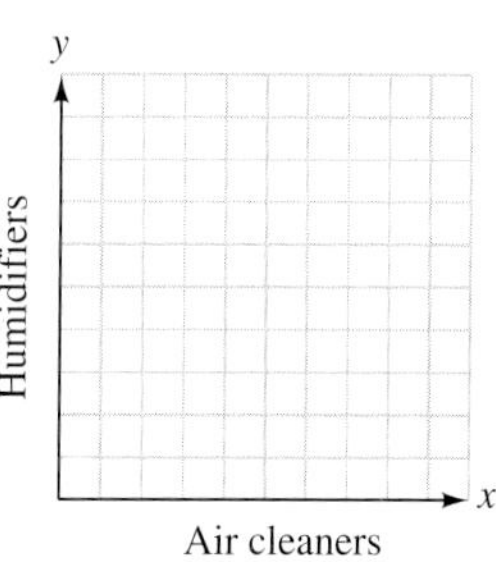

40. ORDERING FURNACE EQUIPMENT J. Bolden Heating Company wants to order no more than \$2,000 worth of electronic air cleaners and humidifiers from a wholesaler that charges \$500 for air cleaners and \$200 for humidifiers. If Bolden wants more humidifiers than air cleaners, use the illustration to graph a system of inequalities showing the possible combinations of air cleaners (x) and humidifiers (y) that can be ordered.

y
Humidifiers
x
Air cleaners

41. PESTICIDES To eradicate a fruit fly infestation, helicopters sprayed an area of a city that can be described by $y \geq -2x + 1$ (within the city limits). Two weeks later, more spraying was ordered over the area described by $y \geq \frac{1}{4}x - 4$ (within the city limits). In the illustration, show the part of the city that was sprayed twice.

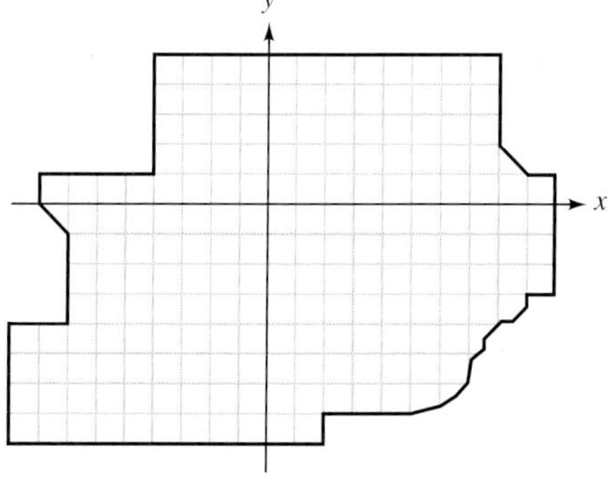

42. REDEVELOPMENT A government agency has declared an area of a city east of First Street, north of Second Avenue, south of Sixth Avenue, and west of Fifth Street as eligible for federal redevelopment funds. See the illustration. Describe this area of the city mathematically using a system of four inequalities, if the corner of Central Avenue and Main Street is considered the origin.

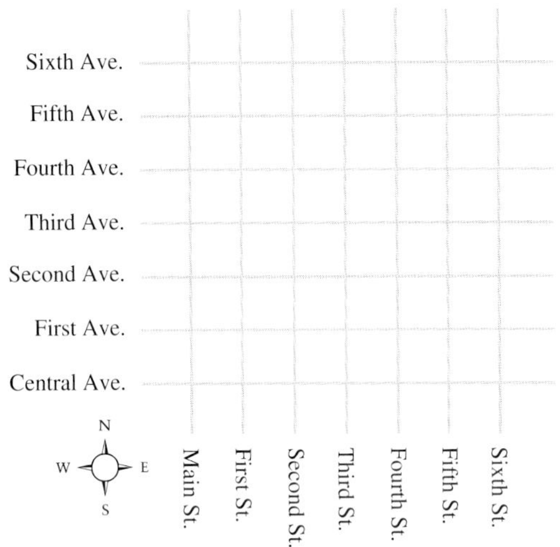

WRITING

43. Explain how to use graphing to solve a system of inequalities.

44. Explain when a system of inequalities will have no solutions.

45. Describe how the graphs of the solutions of these systems are similar and how they differ.

$$\begin{cases} x + y = 4 \\ x - y = 4 \end{cases} \quad \text{and} \quad \begin{cases} x + y \geq 4 \\ x - y \geq 4 \end{cases}$$

46. When a solution of a system of linear inequalities is graphed, what does the shading represent?

REVIEW *Complete each table*

47. $y = 2x^2$

x	y
8	
-2	

48. $t = -|s + 2|$

s	t
-3	
-10	

49. $f(x) = 4 + x^3$

Input	Output
0	
-3	

50. $g(x) = 2x - x^2$

x	$g(x)$
5	
-5	

KEY CONCEPT

Systems of Equations and Inequalities

In Chapter 13, we have solved problems that required the use of two variables to represent two unknown quantities. To find the unknowns, we write a pair of equations (or inequalities) called a **system.**

Solving systems of equations by graphing

A system of linear equations can be solved by graphing both equations and locating the point of intersection of the two lines.

1. FOOD SERVICE The two equations in the table give the fees two different catering companies charge a Hollywood studio for on-location meal service.

Caterer	Setup fee	Cost per meal	Equation
Sunshine	\$1,000	\$4	$y = 4x + 1{,}000$
Lucy's	\$500	\$5	$y = 5x + 500$

Complete the illustration, using the graphing method to find the break point. That is, find the number of meals and the corresponding fee for which the two caterers will charge the studio the same amount.

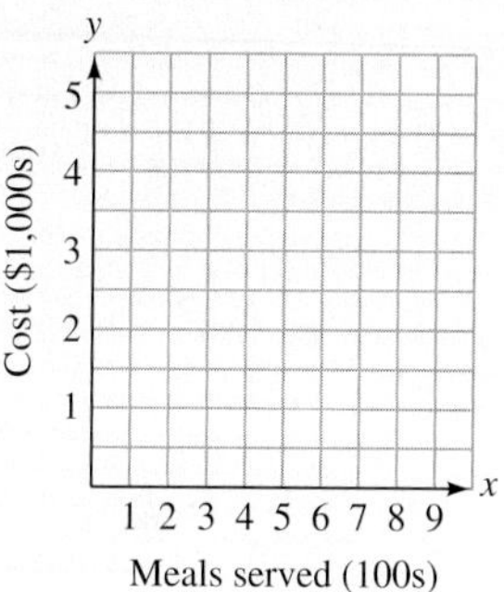

Solving systems of equations by substitution

The substitution method for solving a system of equations works well when a variable in either equation has a coefficient of 1 or −1.

2. Solve by substitution: $\begin{cases} y = 2x - 9 \\ x + 3y = 8 \end{cases}$.

3. Solve by substitution: $\begin{cases} 3x + 4y = -7 \\ 2y - x = -1 \end{cases}$.

Solving systems of equations by addition

With the addition method, equal quantities are added to both sides of an equation to eliminate one of the variables. Then we solve for the other variable.

4. Solve by addition: $\begin{cases} x + y = 1 \\ x - y = 5 \end{cases}$.

5. Solve by addition: $\begin{cases} 2x - 3y = -18 \\ 3x + 2y = -1 \end{cases}$.

Solving systems of inequalities

To solve a system of two linear inequalities, we graph the inequalities on the same coordinate axes. The area that is shaded twice represents the set of solutions.

6. This system of inequalities describes the number of \$20 shirts (x) and \$40 pants (y) a person can buy if he or she plans to spend not less than \$80 but not more than \$120. Using the illustration, graph the system. Then give three solutions.

$$\begin{cases} 20x + 40y \geq 80 \\ 20x + 40y \leq 120 \end{cases}$$

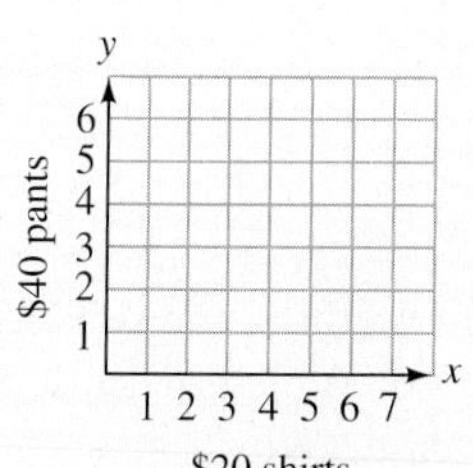

ACCENT ON TEAMWORK

SECTION 13.1

SOLVING SYSTEMS GRAPHICALLY The graphing method was used to solve a system of linear equations. The work is shown below. What are the two equations of the system?

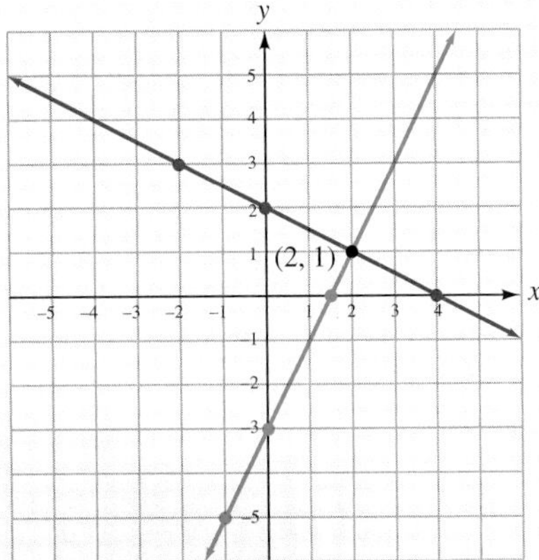

WRITING SYSTEMS OF EQUATIONS In your group, pick a specific ordered pair (x, y) where the x-coordinate and the y-coordinate are integers between -5 and 5. Write a system of two linear equations whose solution is the ordered pair that you picked. Then exchange systems with another group and solve the system that they wrote.

SECTION 13.2

SUBSTITUTION The word *substitution* is used in several ways. Explain how it is used in the context of a sporting event such as a basketball game. Explain how it is sometimes used when ordering food at a restaurant. Explain what is meant by a *substitute* teacher. Finally, explain how the substitution method is used to solve a system such as

$$\begin{cases} y = -2x - 5 \\ 3x + 5y = -4 \end{cases}$$

What is the difference in the mathematical meaning of the word *substitution* as opposed to the everyday usage of the word?

SECTION 13.3

SOLVING SYSTEMS Solve the system

$$\begin{cases} 2x + y = 4 \\ 2x + 3y = 0 \end{cases}$$

using the graphing method, the substitution method, and the addition method. Which method do you think is the best to use in this case? Why?

SECTION 13.4

TWO UNKNOWNS Consider the following problem: A man paid \$89 for two white shirts and four pairs of black socks. Find the cost of a white shirt.

If we let x represent the cost of a white shirt and y represent the cost of a pair of black socks, an equation describing the situation is $2x + 4y = 89$. Explain why there is not enough information to solve the problem.

SECTION 13.5

MATCHING GAMES Have a student in your group write ten linear inequalities on 3×5 note cards, one inequality per card. Then have him or her graph each of the inequalities on separate cards. Mix up the cards and put all the inequality cards on one side of a table and all the cards with graphs on the other side. Work together to match each inequality with its proper graph.

SECTION 13.6

SYSTEMS OF INEQUALITIES The points A, B, C, D, E, F, and G are labeled in the graph of the solution of a system of inequalities. Tell whether the coordinates of each point make the first inequality (whose solution was shown in red) and the second inequality (whose solution was shown in blue) true or false. Use a table of the following form to keep track of your results.

Point	Coordinates	1st inequality	2nd inequality
A			

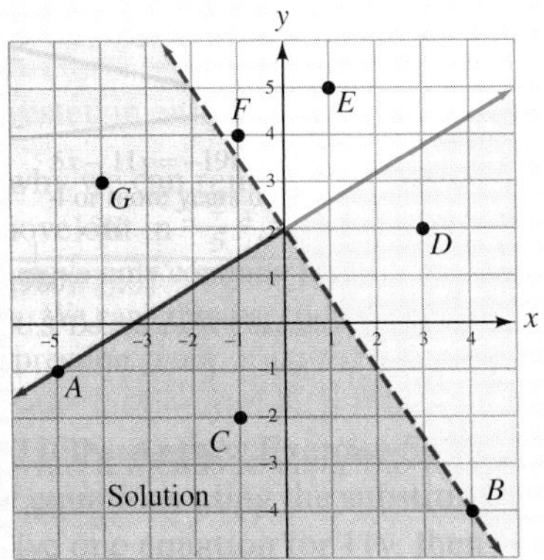

CHAPTER REVIEW

SECTION 13.1 Solving Systems of Equations by Graphing

CONCEPTS

An ordered pair that satisfies both equations of a system is a *solution* of the system.

REVIEW EXERCISES

Determine whether the ordered pair is a solution of the system.

1. $(2, -3)$, $\begin{cases} 3x - 2y = 12 \\ 2x + 3y = -5 \end{cases}$

2. $\left(\frac{7}{2}, -\frac{2}{3}\right)$, $\begin{cases} 4x - 6y = 18 \\ \frac{x}{3} + \frac{y}{2} = \frac{5}{6} \end{cases}$

3. COMPARING STOCKS Refer to the graph below. Estimate the date when Kmart stock and Sears stock sold for the same amount per share. Estimate the price per share at which they were selling on that date.

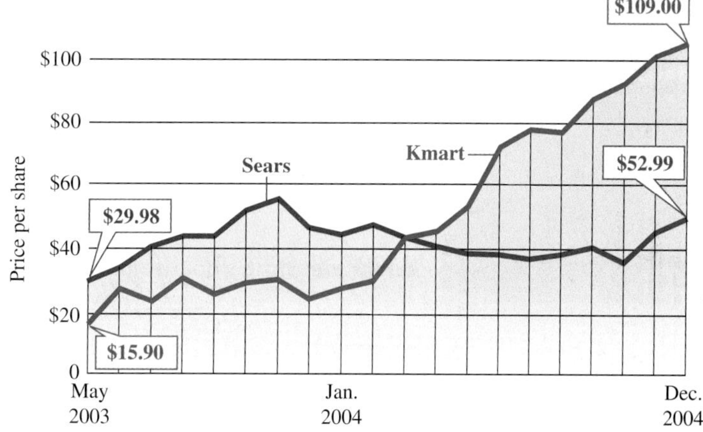

Source: *USA Today*

To solve a system graphically:

1. Carefully graph each equation.
2. If the lines intersect, the coordinates of the point of intersection give the solution of the system.
3. Check the proposed solution in the original equations.

When a system of equations has a solution, it is a *consistent* system. Systems with no solutions are *inconsistent.*

Equations with different graphs are called *independent equations.* If the graphs are the same line, the system has infinitely many solutions. The equations are called *dependent equations.*

Use the graphing method to solve each system. If the equations of a system are dependent, or if a system is inconsistent, so indicate.

4. $\begin{cases} x + y = 7 \\ 2x - y = 5 \end{cases}$

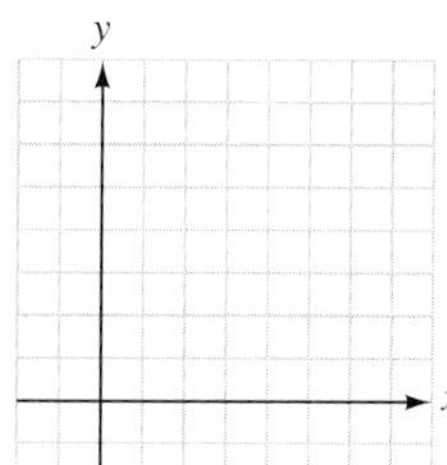

5. $\begin{cases} y = -\frac{x}{3} \\ 2x + y = 5 \end{cases}$

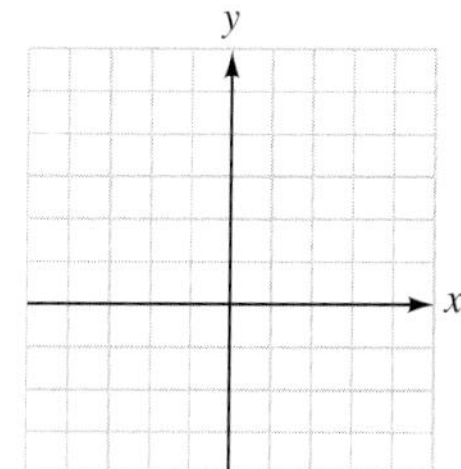

6. $\begin{cases} 3x + 6y = 6 \\ x + 2y = 2 \end{cases}$

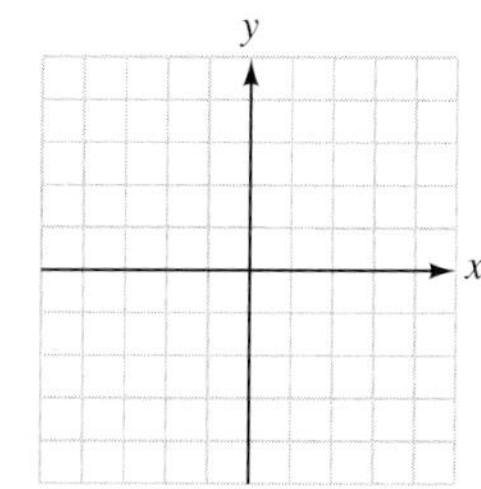

7. $\begin{cases} 6x + 3y = 12 \\ 2x + y = 2 \end{cases}$

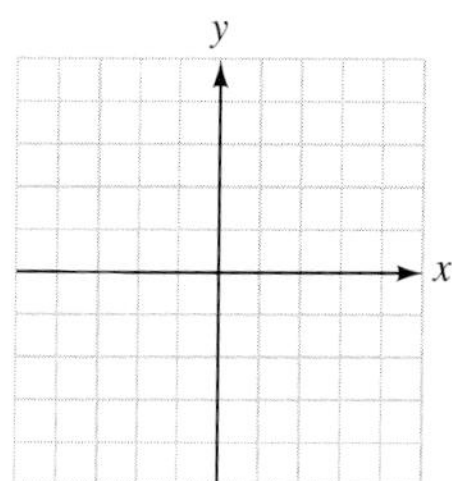

SECTION 13.2 *Solving Systems of Equations by Substitution*

To solve a system of equations in x and y by the *substitution method:*

1. Solve one of the equations for either x or y.
2. Substitute the resulting expression for the variable in step 1 into the other equation, and solve the equation.
3. Find the value of the other variable by substituting the solution found in step 2 in any equation containing x and y.
4. Check the solution in the original equations.

Use the substitution method to solve each system.

8. $\begin{cases} x = y \\ 5x - 4y = 3 \end{cases}$

9. $\begin{cases} y = 15 - 3x \\ 7y + 3x = 15 \end{cases}$

10. $\begin{cases} 0.2x + 0.2y = 0.6 \\ 3x = 2 - y \end{cases}$

11. $\begin{cases} 6(r + 2) = s - 1 \\ r - 5s = -7 \end{cases}$

12. $\begin{cases} 9x + 3y = 5 \\ 3x + y = \dfrac{5}{3} \end{cases}$

13. $\begin{cases} \dfrac{x}{6} + \dfrac{y}{10} = 3 \\ \dfrac{5x}{16} - \dfrac{3y}{16} = \dfrac{15}{8} \end{cases}$

14. In solving a system using the substitution method, suppose you obtain the result of $8 = 9$.

a. How many solutions does the system have?

b. Describe the graph of the system.

c. What term is used to describe the system?

SECTION 13.3 *Solving Systems of Equations by Addition*

To solve a system of equations using the *addition method:*

1. Write each equation in $Ax + By = C$ form.
2. Multiply one or both equations by nonzero quantities to make the coefficients of x (or y) opposites.
3. Add the equations to eliminate the terms involving x (or y).
4. Solve the equation resulting from step 3.
5. Find the value of the other variable by substituting the value of the variable found in step 4 into any equation containing both variables.
6. Check the solution in the original equations.

Solve each system using the addition method.

15. $\begin{cases} 2x + y = 1 \\ 5x - y = 20 \end{cases}$

16. $\begin{cases} x + 8y = 7 \\ x - 4y = 1 \end{cases}$

17. $\begin{cases} 5a + b = 2 \\ 3a + 2b = 11 \end{cases}$

18. $\begin{cases} 11x + 3y = 27 \\ 8x + 4y = 36 \end{cases}$

19. $\begin{cases} 9x + 3y = 15 \\ 3x = 5 - y \end{cases}$

20. $\begin{cases} \dfrac{x}{3} + \dfrac{y + 2}{2} = 1 \\ \dfrac{x + 8}{8} + \dfrac{y - 3}{3} = 0 \end{cases}$

21. $\begin{cases} 0.02x + 0.05y = 0 \\ 0.3x - 0.2y = -1.9 \end{cases}$

22. $\begin{cases} -\dfrac{1}{4}x = 1 - \dfrac{2}{3}y \\ 6(x - 3y) + 2y = 5 \end{cases}$

For each system, tell which method, substitution or addition, would be easier to use to solve the system and why.

23. $\begin{cases} 6x + 2y = 5 \\ 3x - 3y = -4 \end{cases}$

24. $\begin{cases} x = 5 - 7y \\ 3x - 3y = -4 \end{cases}$

SECTION 13.4 *Applications of Systems of Equations*

In this section, we considered ways to solve problems by using *two* variables.

To solve problems involving two unknown quantities:

1. *Analyze* the facts of the problem. Make a table or diagram if necessary.
2. Pick different variables to represent two unknown quantities. *Form* two equations involving the variables.
3. *Solve* the system of equations.
4. *State* the conclusion.
5. *Check* the results.

Use two equations in two variables to solve each problem.

25. CAUSES OF DEATH The number of Americans dying from heart disease is about 4.5 times more than the number dying from a stroke. If the total number of deaths from these causes in 1 year was 880,000, how many deaths were attributed to each?

26. PAINTING EQUIPMENT When fully extended, the ladder shown is 35 feet in length. If the extension is 7 feet shorter than the base, how long is each part of the ladder?

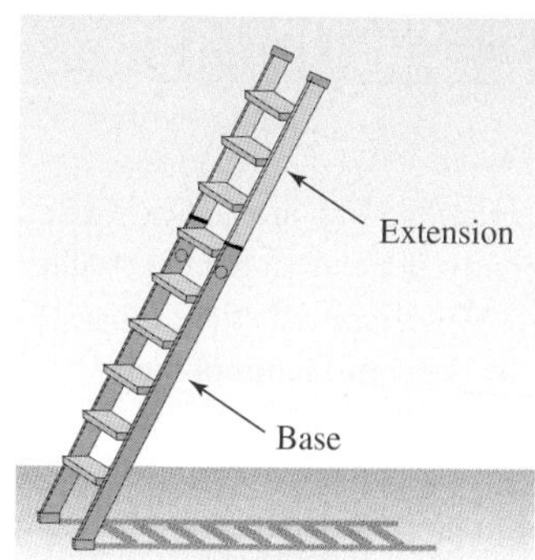

27. CRASH INVESTIGATIONS In an effort to protect evidence, investigators used 420 yards of yellow "Police Line – Do Not Cross" tape to seal off a large rectangular – shaped area around an airplane crash site. How much area will the investigators have to search if the width of the rectangle is three-fourths of the length?

The *break point* of a linear system is the point of intersection of the graph.

28. ENDORSEMENTS A company selling a home juicing machine is contemplating hiring either an athlete or an actor to serve as a spokesperson for a product. The terms of each contract would be as follows:

Celebrity	Base pay	Commission per item sold
Athlete	$30,000	$5
Actor	$20,000	$10

a. For each celebrity, write an equation giving the money ($$y$) the celebrity would earn if x juicers were sold.

b. For what number of juicers would the athlete and the actor earn the same amount?

c. Using the illustration, graph the equations from part a. The company expects to sell over 3,000 juicers. Which celebrity would cost the company the least money to serve as a spokesperson?

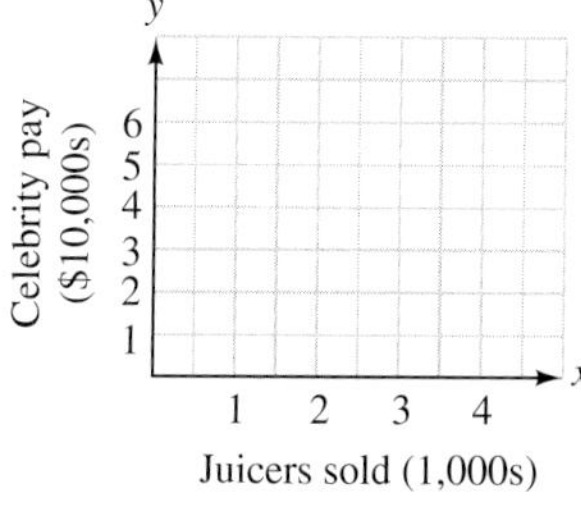

29. CANDY STORES A merchant wants to mix gummy worms worth $3 per pound and gummy bears worth $1.50 per pound to make 30 pounds of a mixture worth $2.10 per pound. How many pounds of each should he use?

30. BOATING It takes a motorboat 4 hours to travel 56 miles down a river, and 3 hours longer to make the return trip. Find the speed of the current.

31. SHOPPING Packages containing 2 bottles of contact lens cleaner and 3 bottles of soaking solution cost $63.40, and packages containing 3 bottles of cleaner and 2 bottles of soaking solution cost $69.60. Find the cost of a bottle of cleaner and a bottle of soaking solution.

32. INVESTING Carlos invested part of \$3,000 in a 10% certificate account and the rest in a 6% passbook account. The total annual interest from both accounts is \$270. How much did he invest at 6%?

33. ANTIFREEZE How much of a 40% antifreeze solution must a mechanic mix with a 70% antifreeze solution if he needs 20 gallons of a 50% antifreeze solution?

SECTION 13.5 Graphing Linear Inequalities

An ordered pair (x, y) is a *solution* of an inequality in x and y if a true statement results when the variables are replaced by the coordinates of the ordered pair.

Determine whether each ordered pair is a solution of $2x - y \leq -4$.

34. $(0, 5)$

35. $(2, 8)$

36. $(-3, -2)$

37. $(\frac{1}{2}, -5)$

To graph a linear inequality:

1. Graph the *boundary line.* Draw a solid line if the inequality contains $\leq$ or $\geq$ and a dashed line if it contains $<$ or $>$.
2. Pick a *test point* on one side of the boundary. Use the origin if possible. Replace x and y with the coordinates of that point. If the inequality is satisfied, shade the side that contains the point. If the inequality is not satisfied, shade the other side.

Graph each inequality.

38. $x - y < 5$

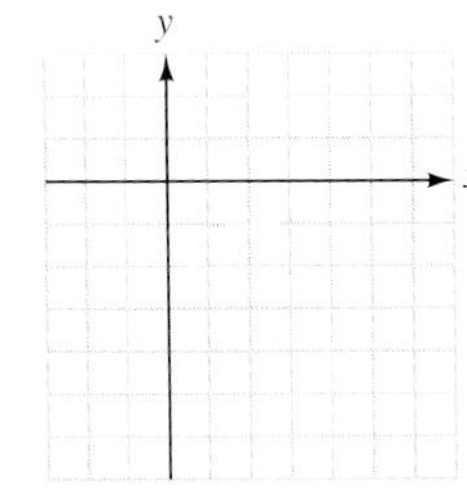

39. $2x - 3y \geq 6$

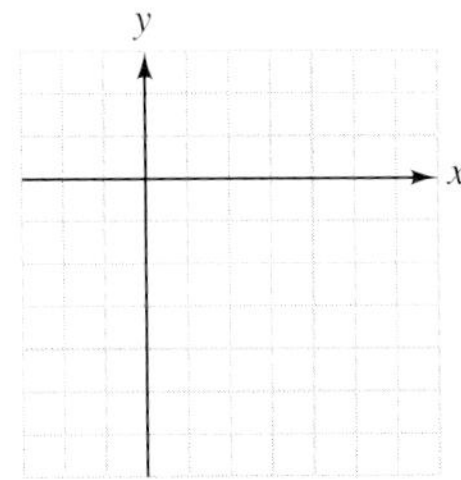

40. $y \leq -2x$

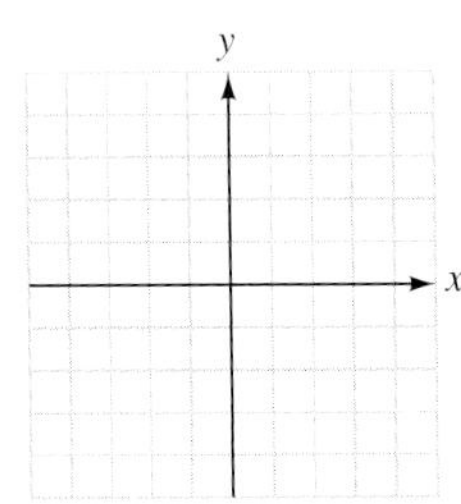

41. $y < -4$

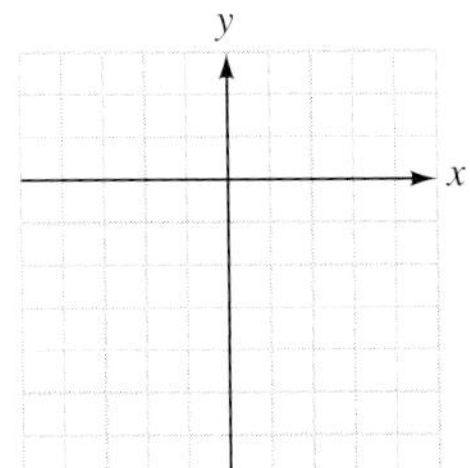

42. In the illustration, the graph of a linear inequality is shown. Would a true or a false statement result if the coordinates of

a. point A were substituted into the inequality?

b. point B were substituted into the inequality?

c. point C were substituted into the inequality?

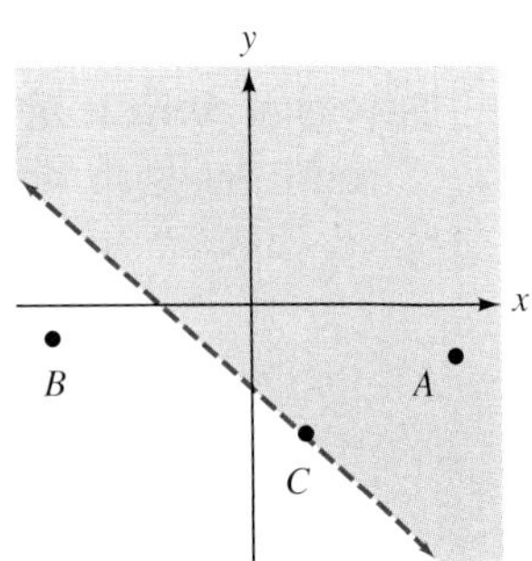

43. WORK SCHEDULES A student told her employer that during the school year, she would be available for up to 30 hours a week, working either 3- or 5-hour shifts. Find an inequality that shows the possible ways to schedule the number of 3-hour (x) and 5-hour shifts (y) she can work, and graph it in the illustration. Give three ordered pairs that satisfy the inequality.

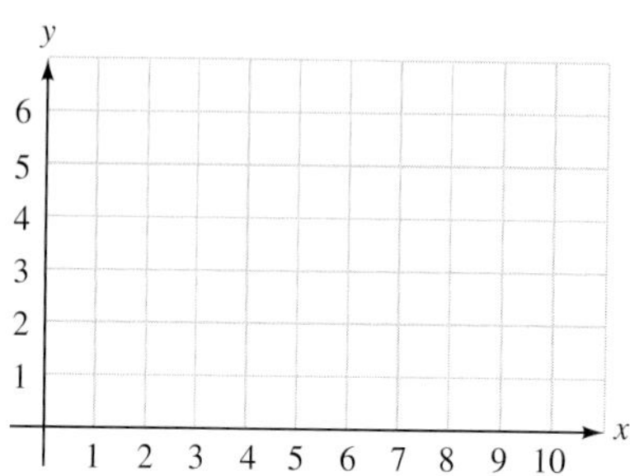

SECTION 13.6 *Solving Systems of Linear Inequalities*

To graph a system of linear inequalities:

1. Graph the individual inequalities of the system on the same coordinate axes.
2. The final solution, if one exists, is that region where all individual graphs intersect.

Solve each system of inequalities.

44. $\begin{cases} 5x + 3y < 15 \\ 3x - y > 3 \end{cases}$

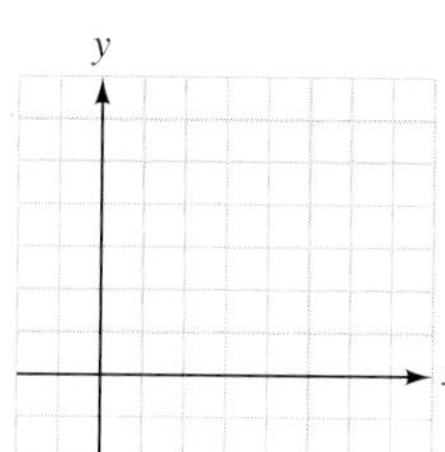

45. $\begin{cases} x \geq 3y \\ y > 3x \end{cases}$

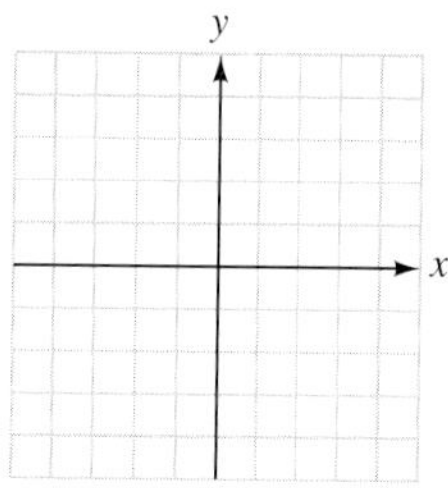

Systems of linear inequalities can be used to solve application problems.

46. GIFT SHOPPING A grandmother wants to spend at least \$40 but no more than \$60 on school clothes for her grandson. If T-shirts sell for \$10 and pants sell for \$20, write a system of inequalities that describes the possible combinations of T-shirts (x) and pants (y) she can buy. Graph the system in the illustration. Give two possible solutions.

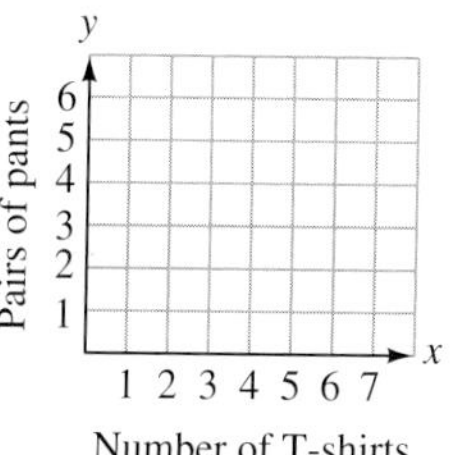

CHAPTER 13 TEST

Determine whether the given ordered pair is a solution of the given system.

1. $(5, 3)$, $\begin{cases} 3x + 2y = 21 \\ x + y = 8 \end{cases}$

2. $(-2, -1)$, $\begin{cases} 4x + y = -9 \\ 2x - 3y = -7 \end{cases}$

3. Solve the system by graphing: $\begin{cases} 3x + y = 7 \\ x - 2y = 0 \end{cases}$

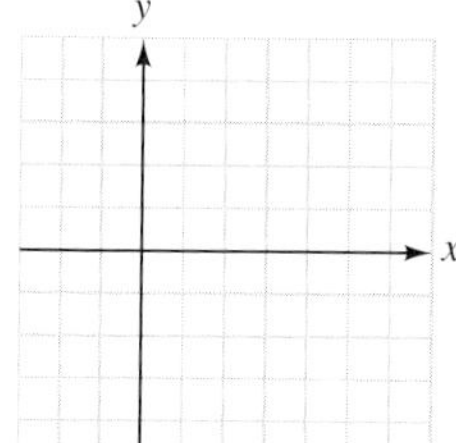

4. To solve a system of two linear equations in x and y, a student used a graphing calculator. From the calculator display shown, determine whether the system has a solution. Explain your answer.

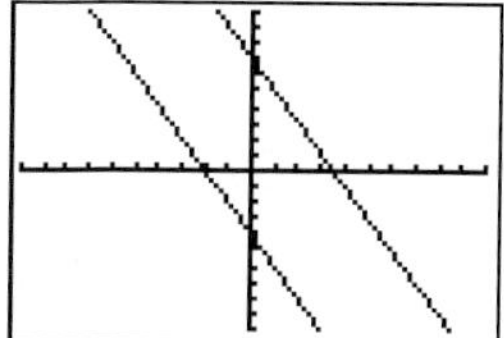

Solve each system by substitution.

5. $\begin{cases} y = x - 1 \\ 2x + y = -7 \end{cases}$

6. $\begin{cases} 3a + 4b = -7 \\ 2b - a = -1 \end{cases}$

Solve each system by addition.

7. $\begin{cases} 3x - y = 2 \\ 2x + y = 8 \end{cases}$

8. $\begin{cases} 4x + 3y = -3 \\ -3x = -4y + 21 \end{cases}$

Classify each system as consistent or inconsistent and classify the equations as dependent or independent.

9. $\begin{cases} x + y = 4 \\ x + y = 6 \end{cases}$

10. $\begin{cases} \dfrac{x}{3} + y = 4 \\ x + 3y = 12 \end{cases}$

11. Which method would be most efficient to solve the following system?

$$\begin{cases} 5x - 3y = 5 \\ 3x + 3y = 3 \end{cases}$$

Explain your answer. (You do not need to solve the system.)

12. FINANCIAL PLANNING A woman invested some money at 8% and some at 9%. The interest for 1 year on the combined investment of \$10,000 was \$840. How much was invested at 9%? Use a system of equations in two variables to solve this problem.

Determine whether the given ordered pair is a solution of $2x - 4y > 8$.

13. $(7, 1)$

14. $(0, -2)$

15. Graph: $x - y > -2$.

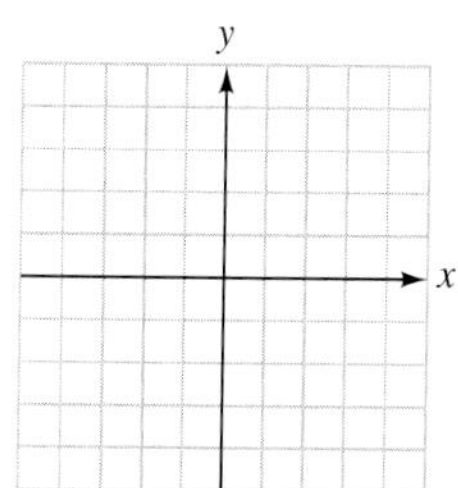

16. Solve by graphing.

$$\begin{cases} 2x + 3y \le 6 \\ x \ge 2 \end{cases}$$

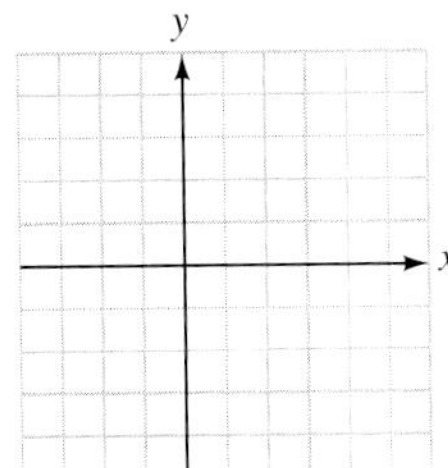

See the following illustration, which shows two different ways in which a salesperson can be paid according to the number of items he or she sells.

17. What is the point of intersection of the graphs? Explain its significance.

18. If a salesperson expects to sell more than 30 items per month, which plan is more profitable?

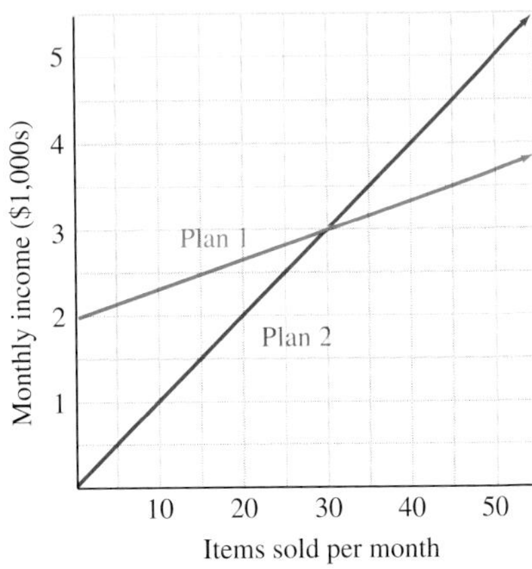

CHAPTERS 1–13 CUMULATIVE REVIEW EXERCISES

1. CANDY SALES The circle graph shows how \$6,300,000,000 in seasonal candy sales for 2002 was spent. Find the candy sales for Halloween.

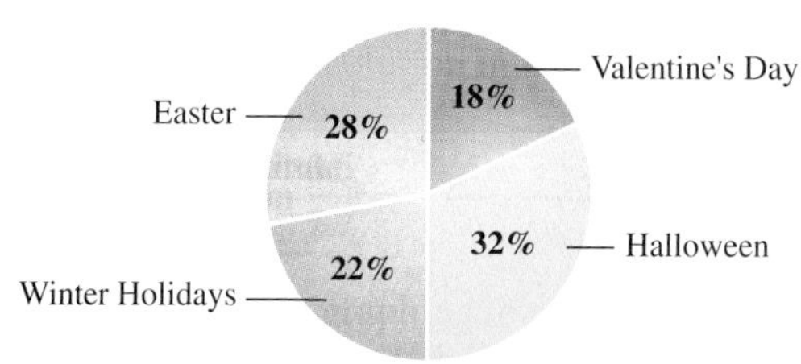

Source: National Confectioners Association

2. List the set of integers.

Evaluate each expression.

3. $3 - 4[-10 - 4(-5)]$

4. $\dfrac{|-45| - 2(-5) + 1^5}{2 \cdot 9 - 2^4}$

5. AIR CONDITIONING Find the volume of air contained in the duct shown below. Round to the nearest tenth of a cubic foot.

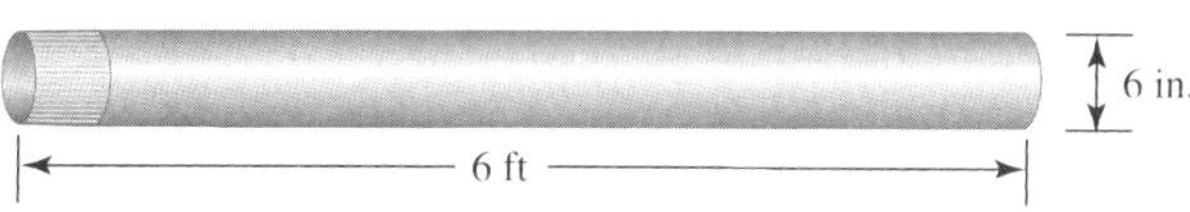

6. Simplify: $3x^2 + 2x^2 - 5x^2$.

Solve each equation.

7. $2 - (4x + 7) = 3 + 2(x + 2)$

8. $\frac{2}{5}y + 3 = 9$

9. Solve $-4x + 6 > 17$, graph the solution set, and describe the graph using interval notation.

10. ANGLE OF ELEVATION Refer to the following illustration. Find x.

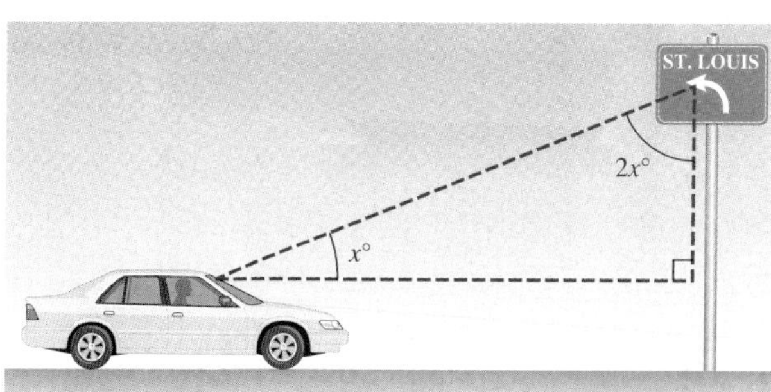

11. STOCK MARKET An investment club invested part of \$45,000 in a high-yield mutual fund that earned 12% annual simple interest. The remainder of the money was invested in Treasury bonds that earned 6.5% simple annual interest. The two investments earned \$4,300 in 1 year. How much was invested in each account?

12. Give the formula for
    a. the perimeter of a rectangle
    b. the area of a rectangle
    c. the area of a circle
    d. the distance traveled

Graph each equation.

13. $y = -x^3$

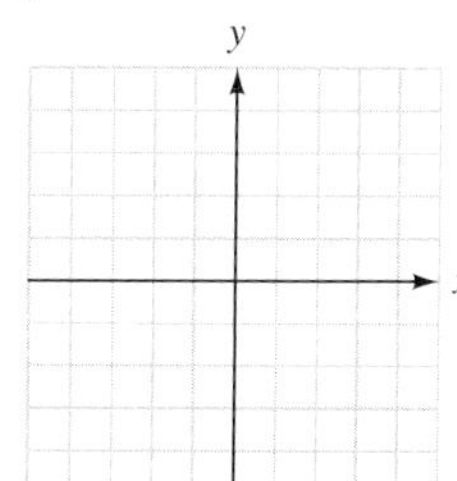

14. $y = -3x + 2$

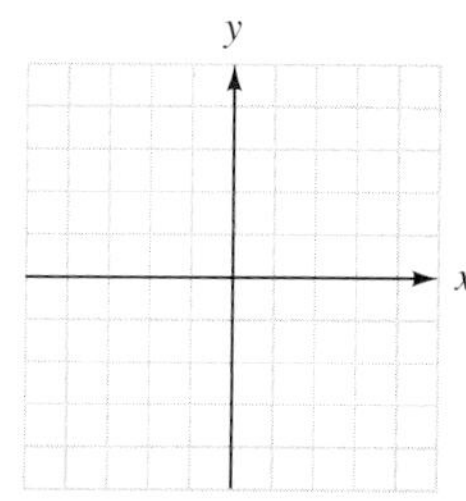

15. $3x + 4y = 8$

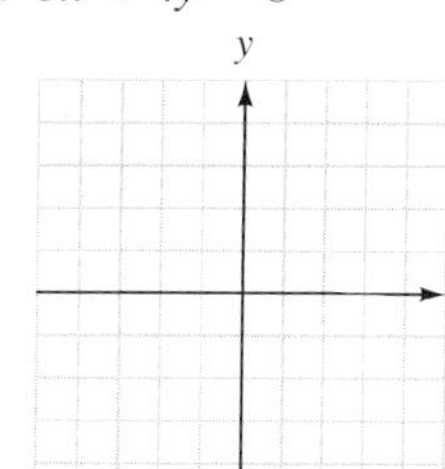

16. $x = -2$

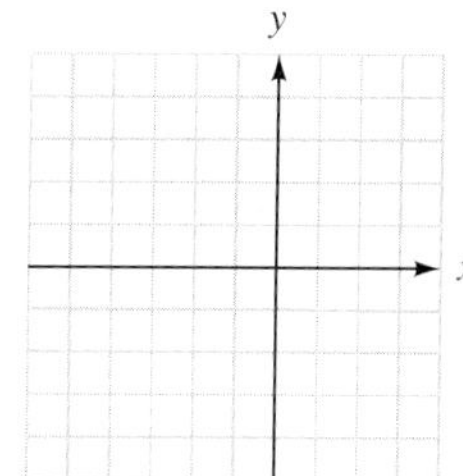

17. Find the slope and y-intercept of the line graphed. Then write the equation of the line.

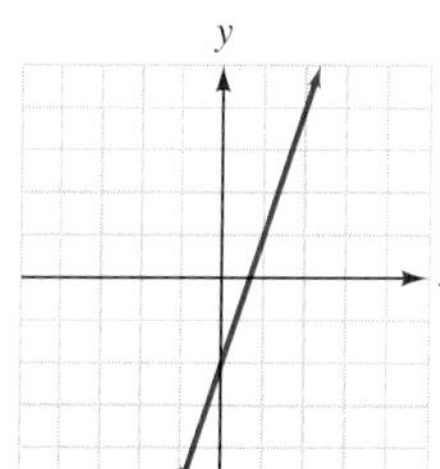

18. Find the slope of the line passing through $(6, -2)$ and $(-3, 2)$.

19. If $f(x) = 4x - 2x^2$, what is $f(-5)$?

20. Write 3,890,000,000 in scientific notation.

Simplify each expression. Write each answer without using parentheses or negative exponents.

21. $(x^5)^2(x^7)^3$

22. $\dfrac{16(aa^2)^3}{2a^2a^3}$

23. $\dfrac{2^{-4}}{3^{-1}}$

24. $(2x)^0$

Perform the operations.

25. $(5x - 8y) - (-2x + 5y)$

26. $2x^2(3x^2 + 4x - 7)$

27. $(c + 16)^2$

28. $(x + 3)(2x - 3)$

29. $\dfrac{2x - 32}{16x}$

30. $3x + 1\overline{)9x^2 + 6x + 1}$

31. Prime-factor 288.

32. Write a polynomial that is a difference of two squares.

Factor each polynomial completely.

33. $12r^2 - 3rs + 9r^2s^2$

34. $u^2 - 18u + 81$

35. $2y^2 - 7y + 3$

36. $x^4 - 81$

37. $t^3 - v^3$

38. $xy - ty + xs - ts$

Solve each equation.

39. $8s^2 - 16s = 0$

40. $x^2 + 2x - 15 = 0$

41. Simplify: $\dfrac{x^2 - 25}{5x + 25}$.

42. Add: $\dfrac{3x}{2y} + \dfrac{5x}{2y}$.

43. Divide: $\dfrac{x^2 - x - 2}{x^2 + x} \div \dfrac{2 - x}{x}$.

44. Subtract: $\dfrac{x + 5}{xy} - \dfrac{x - 1}{x^2y}$.

45. Simplify: $\dfrac{\dfrac{y}{x} + 3y}{y + \dfrac{2y}{x}}$.

46. Solve: $\dfrac{3r}{2} - \dfrac{3}{r} = 3 + \dfrac{3r}{2}$.

47. The triangles shown below are similar. Find a and b.

48. Is $(-5, 2)$ a solution of $\begin{cases} -2x + 7y = 24 \\ 3x - 4y = -19 \end{cases}$?

Solve each system by graphing.

49. $\begin{cases} x + 4y = -2 \\ y = -x - 5 \end{cases}$

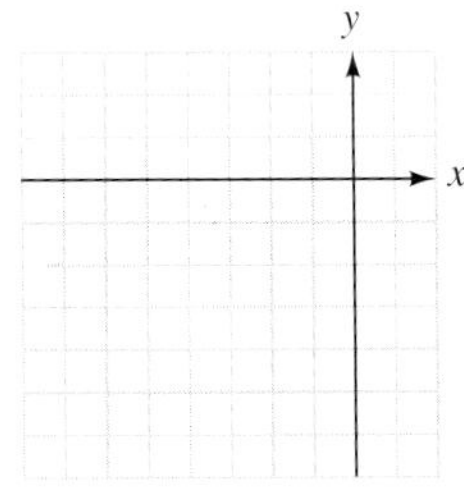

50. $\begin{cases} 2x - 3y < 0 \\ y > x - 1 \end{cases}$

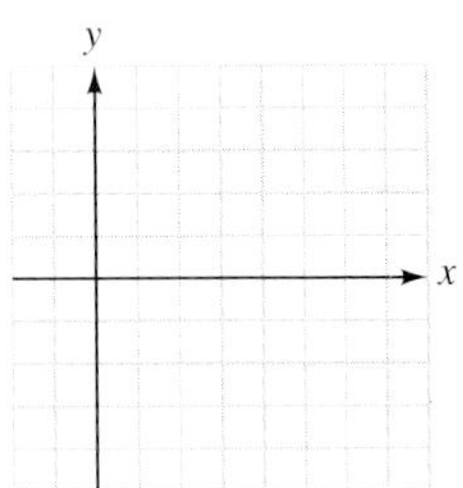

51. Solve $\begin{cases} x - 2y = 2 \\ 2x + 3y = 11 \end{cases}$ by substitution.

52. NUTRITION The table shows per serving nutritional information for egg noodles and rice pilaf. How many servings of each food should be eaten to consume exactly 22 grams of protein and 21 grams of fat?

	Protein (g)	**Fat (g)**
Egg noodles	5	3
Rice pilaf	4	5

CHAPTER 14

Roots and Radicals

© Getty/Alan Pappe

TLE The Parthenon in Athens, Greece, is one of the best examples of the order and harmony for which ancient Greek architecture is known. Built more than 2,500 years ago, its design involves *golden rectangles* of different sizes. Golden rectangles are considered to have the most pleasing shape to the human eye. The exact dimensions of a golden rectangle involve the irrational number $\sqrt{2}$ (the square root of 2).

To learn more about square roots, visit *The Learning Equation* on the Internet at http://tle.brookscole.com. (The log-in instructions are in the Preface.) For Chapter 14, the online lesson is:

- *TLE* Lesson 25: Using Square Roots

Check Your Knowledge

1. The principal square root of a positive real number is a ________ number.
2. The symbol $\sqrt{\ }$ is called a ________ symbol.
3. $4 - \sqrt{3}$ is the __________ of $4 + \sqrt{3}$.
4. The ____________ theorem is a formula that relates the lengths of the sides of a right triangle.

Simplify each radical, if possible. Assume all variables represent positive numbers.

5. $\sqrt{64}$
6. $\sqrt{-16}$
7. $\sqrt{\dfrac{20}{9}}$
8. $-\sqrt[3]{-8}$
9. $\sqrt{18y^4}$
10. $\sqrt{28x^3y^4}$
11. $\sqrt{\dfrac{40x^3}{5y^2}}$
12. $\sqrt[3]{16x^5y^6z^3}$

13. Evaluate $\sqrt{a^2 + b^2}$ for $a = 3$ and $b = 4$.

14. A square measures 2 feet on each side. Express the length of the diagonal of the square in simplified radical form. Then approximate its length to the nearest tenth.

Perform each operation.

15. $3\sqrt{18} - \sqrt{50}$
16. $x\sqrt{27x} + \sqrt{48x^3}$
17. $(2\sqrt{32})(5\sqrt{8})$
18. $\sqrt{2}(\sqrt{3} + \sqrt{8})$
19. $(2 + \sqrt{3})^2$
20. $(\sqrt{x} + 2)(\sqrt{x} - 2)$

21. The playing area of a college football field is 300 feet long and 160 feet wide. Find the length of a diagonal of the playing field.

Rationalize each denominator.

22. $\dfrac{\sqrt{3}}{\sqrt{5x}}$
23. $\dfrac{2}{2 + \sqrt{2}}$

Solve each equation.

24. $\sqrt{x} = 4$
25. $\sqrt{x - 5} - 2 = 1$
26. $\sqrt{2x - 3} - \sqrt{x + 2} = 0$
27. $5 - \sqrt[3]{x + 7} = 2$

28. Find the distance between the points $(2, 5)$ and $(-3, -7)$.
29. Simplify: $27^{2/3}$.
30. Simplify: $(x^{\frac{1}{2}})^{3/2}$.

Study Skills Workshop

FINAL EXAM PREPARATION

Getty Images

Most math instructors give a comprehensive final exam that covers material from the entire course. This is an excellent opportunity for you to get a look at the whole scope of the course and to lay a foundation for your next math course. In many ways, preparing for the final is like preparing for any other test. If you have prepared well throughout the course, the only difference in planning for the final is that it may take a little longer because there is more material to cover. Allow a week, if possible, to prepare for a final exam. The following list of activities will give you a good start.

Review All Tests. Look over all returned tests and make sure you are able to understand how to solve each problem. If you have not already done so, make corrections to all test problems for which you did not receive full credit.

Make a Study Sheet for the Final Exam. Like the sheet that you made to prepare for each test, make up a final exam study sheet (or tape). Keep this handy throughout the week to review whenever possible.

Make a Practice Final Exam. Make a practice final exam like the trial tests you prepared, but covering material for the entire course. Include approximately the same number of problems that the actual final will have, and use problems that have a solution you can check. (You can also use chapter tests on the iLrn Web sites as practice finals.)

Take Your Practice Final. At least 2 days before the final, take the practice final with your book closed, and time yourself. Note which problems are difficult and give special attention to them.

Meet With Your Study Group. After you have reviewed all of your tests and prepared a practice final, meet with your study group to ask questions and compare notes. If possible, make copies of different practice finals so you can see what types of problems others have chosen. Ask your study group for help with difficult problems.

Meet With Your Tutor and/or Instructor. At least a day before the final, make a list of problems that you could not get answered in your study group or that even with study group help are still difficult for you to do. An instructor or tutor may have a different way of doing the problem that is easier for you to understand.

On the Day of Your Final. Make sure you arrive early (check the time in advance—often finals are given at different times than the regular class meeting). Bring all necessary items: Scantrons (if required), scrap-paper, calculators (if allowed), pencils, and erasers. If your instructor agrees, bring a stamped, self-addressed envelope so the instructor can send you the final exam score and course grade.

ASSIGNMENT

1. Make a calendar for the week before your final that includes study time for this final and any other finals that are coming up.
2. Make a study sheet for the final like those that you made for each test.
3. Make a list of your instructor's office hours for the week before the final, and visit your instructor with any unresolved problems.
4. Make a checklist of things that you will need for your final exam. If you are allowed to bring calculators, make sure you have fresh batteries. The night before your final, put all of these items together in one place so that you don't need to stress about finding them right before your final.
5. On the day of the final, relax as much as possible, knowing that you have prepared well.

To solve many applied problems, we must determine what number x must be squared to obtain another number n. We call x the square root of n.

14.1 Square Roots

- Square roots • Approximating square roots • Rational, irrational, and imaginary numbers
- The square root function • The Pythagorean theorem

To find the area A of the square shown in Figure 14-1, we multiply its length by its width.

$$A = l \cdot w$$
$$A = 5 \cdot 5$$
$$= 25$$

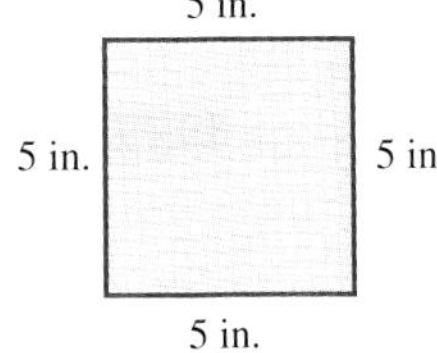

FIGURE 14-1

The area is 25 square inches.

We have seen that the product $5 \cdot 5$ can be denoted by the exponential expression 5^2, where 5 is raised to the second power. Whenever we raise a number to the second power, we are squaring it, or finding its **square.** This example illustrates that the formula for the area of a square with sides of length s is $A = s^2$.

Here are some more squares of numbers:

- The square of 3 is 9, because $3^2 = 9$.
- The square of -3 is 9, because $(-3)^2 = 9$.
- The square of 12 is 144, because $12^2 = 144$.
- The square of -12 is 144, because $(-12)^2 = 144$.
- The square of $\frac{1}{8}$ is $\frac{1}{64}$, because $\left(\frac{1}{8}\right)^2 = \frac{1}{8} \cdot \frac{1}{8} = \frac{1}{64}$.
- The square of $-\frac{1}{8}$ is $\frac{1}{64}$, because $\left(-\frac{1}{8}\right)^2 = \left(-\frac{1}{8}\right)\left(-\frac{1}{8}\right) = \frac{1}{64}$.
- The square of 0 is 0, because $0^2 = 0$.

In this section, we reverse the squaring process and find *square roots* of numbers.

Square roots

Suppose we know that the area of the square shown in Figure 14-2 is 36 square inches. To find the length of each side, we substitute 36 for A in the formula $A = s^2$ and solve for s.

$$A = s^2$$
$$36 = s^2$$

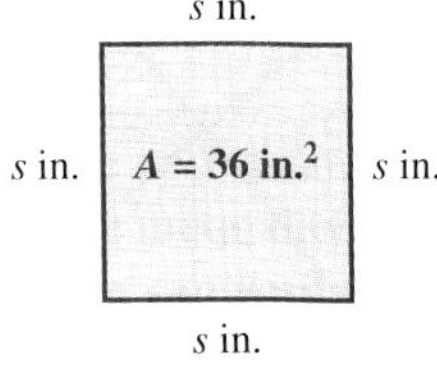

FIGURE 14-2

To solve for s, we must find a positive number whose square is 36. Since 6 is such a number, the sides of the square are 6 inches long. The number 6 is called a *square root* of 36, because 6 is the positive number that we square to get 36.

Here are some more square roots of numbers:

- 3 is a square root of 9, because $3^2 = 9$.
- -3 is a square root of 9, because $(-3)^2 = 9$.
- 12 is a square root of 144, because $12^2 = 144$.
- -12 is a square root of 144, because $(-12)^2 = 144$.

- $\frac{1}{8}$ is a square root of $\frac{1}{64}$, because $\left(\frac{1}{8}\right)^2 = \left(\frac{1}{8}\right)\left(\frac{1}{8}\right) = \frac{1}{64}$.
- $-\frac{1}{8}$ is a square root of $\frac{1}{64}$, because $\left(-\frac{1}{8}\right)^2 = \left(-\frac{1}{8}\right)\left(-\frac{1}{8}\right) = \frac{1}{64}$.
- 0 is a square root of 0, because $0^2 = 0$.

In general, we have the following definition.

Square root

The number b is a **square root** of a if $b^2 = a$.

All positive numbers have two square roots, one positive and one negative. The two square roots of 9 are 3 and -3, and the two square roots of 144 are 12 and -12. The number 0 is the only number that has one square root, which is 0.

The **principal square root** of a positive number is its positive square root. Although 3 and -3 are both square roots of 9, only 3 is the principal square root. The symbol $\sqrt{\ }$, called a **radical symbol,** is used to represent the principal square root of a number, and $-\sqrt{\ }$ is used to represent the negative square root of a number. For example, $\sqrt{9} = 3$ and $-\sqrt{9} = -3$. Likewise, $\sqrt{144} = 12$ and $-\sqrt{144} = -12$.

Principal square root

If $a > 0$, the expression $\sqrt{a}$ represents the **principal** (or positive) **square root** of a.

The principal square root of 0 is 0: $\sqrt{0} = 0$.

The expression under a radical symbol is called the **radicand.** In $\sqrt{9}$, the number 9 is the radicand, and the entire symbol $\sqrt{9}$ is called a **radical.** We read $\sqrt{9}$ as either "the square root of 9" or as "radical 9."

An algebraic expression containing a radical is called a **radical expression.** In this chapter, we will consider radical expressions such as

$$\sqrt{49}, \quad \frac{5}{\sqrt{3}}, \quad -2\sqrt{x+1}, \quad \text{and} \quad \sqrt{28y^2} - 2y\sqrt{63}$$

EXAMPLE 1 Find each square root: **a.** $\sqrt{16}$, **b.** $\sqrt{1}$, **c.** $\sqrt{0.36}$, **d.** $\sqrt{\frac{4}{9}}$, and **e.** $-\sqrt{225}$.

Solution

a. $\sqrt{16} = 4$ Ask: What positive number, when squared, is 16? The answer is 4.

b. $\sqrt{1} = 1$ Ask: What positive number, when squared, is 1? The answer is 1.

c. $\sqrt{0.36} = 0.6$ Ask: What positive number, when squared, is 0.36? The answer is 0.6.

d. $\sqrt{\frac{4}{9}} = \frac{2}{3}$ Ask: What positive number, when squared, is $\frac{4}{9}$? The answer is $\frac{2}{3}$.

e. $-\sqrt{225}$ is the opposite of the square root of 225. Since $\sqrt{225} = 15$, we have

$-\sqrt{225} = -15$ $\quad -\sqrt{225} = -1 \cdot \sqrt{225} = -1 \cdot 15 = -15.$

Self Check 1

Find each square root: **a.** $\sqrt{121}$, **b.** $-\sqrt{49}$, **c.** $\sqrt{0.64}$, **d.** $\sqrt{256}$, **e.** $\sqrt{\frac{1}{25}}$, **f.** $\sqrt{\frac{9}{49}}$

Answers **a.** 11, **b.** -7, **c.** 0.8, **d.** 16, **e.** $\frac{1}{5}$, **f.** $\frac{3}{7}$

Square roots of certain numbers, such as 7, are hard to compute by hand. However, we can approximate $\sqrt{7}$ with a calculator.

Approximating square roots

To find the principal square root of 7, we can enter 7 into a scientific calculator and press the $\sqrt{x}$ key. The approximate value of $\sqrt{7}$ will appear on the display.

$\sqrt{7} \approx 2.6457513$ Read $\approx$ as "is approximately equal to."

Since $\sqrt{7}$ represents the number that, when squared, gives 7, we would expect squares of approximations of $\sqrt{7}$ to be close to 7.

- Rounded to one decimal place, $\sqrt{7} \approx 2.6$ and $(2.6)^2 = 6.76$.
- Rounded to two decimal places, $\sqrt{7} \approx 2.65$ and $(2.65)^2 = 7.0225$.
- Rounded to three decimal places, $\sqrt{7} \approx 2.646$ and $(2.646)^2 = 7.001316$.

CALCULATOR SNAPSHOT Freeway road sign

The sign shown in Figure 14-3 is in the shape of an equilateral triangle, and we can find its height h using the formula

$$h = \frac{\sqrt{3}s}{2}$$

where s is the length of a side of the triangle. In this case, $s = 24$ inches, so we have

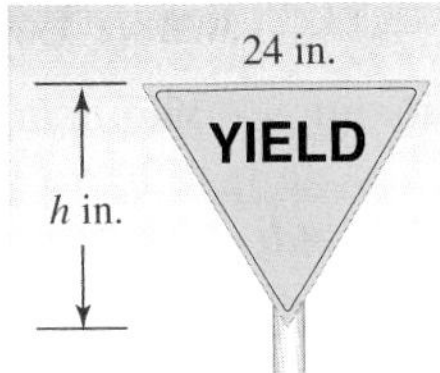

FIGURE 14-3

$$h = \frac{\sqrt{3}(\mathbf{24})}{2} \qquad \sqrt{3}(24) \text{ means } \sqrt{3}\cdot 24.$$

To evaluate this expression with a scientific calculator, we enter these numbers and press these keys.

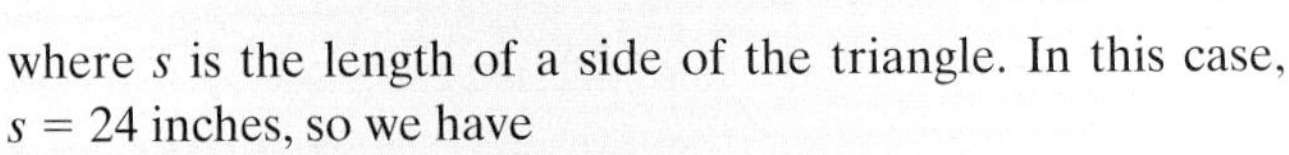

20.784609

To evaluate this expression using a graphing calculator, we press these keys.

2nd √ 3) × 24 ÷ 2 ENTER

√(3)*24/2
20.78460969

The height of the sign is approximately 21 inches.

THINK IT THROUGH Traffic Accidents

"The U.S. Surgeon General reports that life expectancy has improved in the U.S. over the past 75 years for every age group except one: the death rate for 15- to 24-year-olds is higher today than it was 20 years ago. The leading cause of death is drunk/drugged driving." Mothers Against Drunk Driving, MADD

Accident investigators often determine a vehicle's speed prior to braking from the length of the skid marks that it leaves on the street. To do this, they use the formula $s = \sqrt{30Df}$, where s is the speed of the vehicle in mph, D is the skid distance in feet, and f is the drag factor for the road surface. Estimate the speed of each vehicle prior to braking given the following conditions. Round to the nearest mile per hour.

1. Length of skid marks: 71 ft
Road surface: Asphalt, $f = 0.75$
2. Length of skid marks: 133 ft
Road surface: Concrete, $f = 0.90$

Rational, irrational, and imaginary numbers

Whole numbers such as 4, 9, 16, and 49 are called **integer squares,** because each one is the square of an integer. The square root of any integer square is an integer and therefore a rational number:

$$\sqrt{4} = 2, \qquad \sqrt{9} = 3, \qquad \sqrt{16} = 4, \qquad \text{and} \qquad \sqrt{49} = 7$$

The square root of any whole number that is not an integer square is an **irrational number.** For example, $\sqrt{7}$ is an irrational number. Recall that the set of rational numbers and the set of irrational numbers together make up the set of real numbers.

! COMMENT Square roots of negative numbers are not real numbers. For example, $\sqrt{-4}$ is nonreal, because the square of no real number is -4. The number $\sqrt{-4}$ is an example from a set of numbers called **imaginary numbers.** Remember: *The square root of a negative number is not a real number.*

If we attempt to evaluate $\sqrt{-4}$ using a calculator, an error message like the ones shown below will be displayed.

Scientific calculator	**Graphing calculator**
Error	ERR:NONREAL ANS 1:Quit 2:Goto

In this chapter, we will assume that *all radicands under the square root symbols are either positive or zero.* Thus, all square roots will be real numbers.

The square root function

Since there is one principal square root for every nonnegative real number x, the equation $f(x) = \sqrt{x}$ determines a square root function. For example, the value that is determined by $f(x) = \sqrt{x}$ when $x = 4$ is denoted by $f(4)$, and we have $f(4) = \sqrt{4} = 2$.

To graph this function, we make a table of values and plot each ordered pair. Then, from the origin, we draw a smooth curve that passes through the points. In the table, we chose five values for x, 0, 1, 4, 9, and 16, that are integer squares. This made computing $f(x)$ quite simple. The graph appears in Figure 14-4.

$f(x) = \sqrt{x}$

x	$f(x)$	$(x, f(x))$
0	0	(0, 0)
1	1	(1, 1)
4	2	(4, 2)
9	3	(9, 3)
16	4	(16, 4)
↑ Values to be input into $\sqrt{x}$	↑ Output values	↑ Ordered pairs to plot

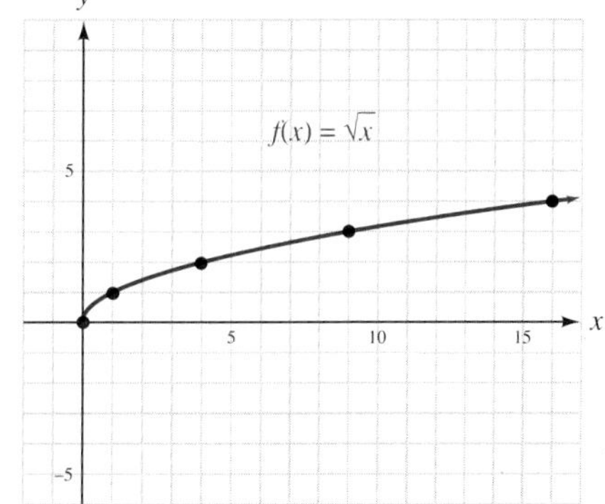

FIGURE 14-4

Self Check 2
Find the period of a pendulum that is 3 feet long.

EXAMPLE 2 **Period of a pendulum.** The *period* of a pendulum is the time required for the pendulum to swing back and forth to complete one cycle. (See Figure 14-5.) The period (in seconds) of a pendulum having length L (in feet) is approximated by the function

$$f(L) = 1.11\sqrt{L}$$ Read $1.11\sqrt{L}$ as "1.11 times $\sqrt{L}$."

Find the period of a pendulum that is 5 feet long.

Solution We substitute 5 for L in the formula and multiply using a calculator.

$$f(L) = 1.11\sqrt{L}$$
$$f(5) = 1.11\sqrt{5}$$ $1.11\sqrt{5}$ means $1.11 \cdot \sqrt{5}$.
$$\approx 2.482035455$$

The period is approximately 2.5 seconds.

Answer about 1.9 sec

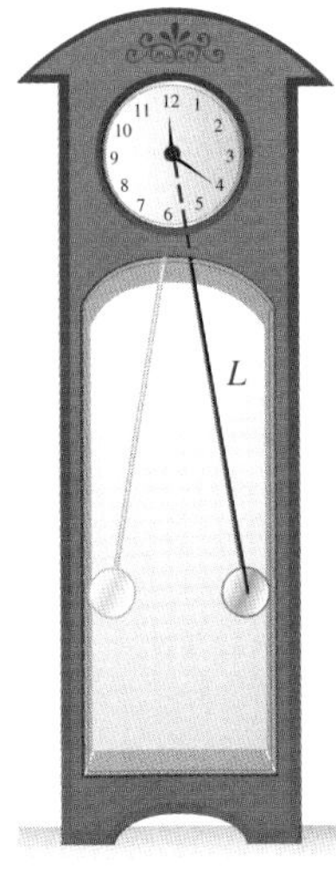

FIGURE 14-5

The Pythagorean theorem

The longest side of a right triangle is the **hypotenuse,** which is the side opposite the right angle. The remaining two sides are the **legs** of the triangle. See Figure 14-6. Recall that the **Pythagorean theorem** provides a formula relating the lengths of the three sides of a right triangle.

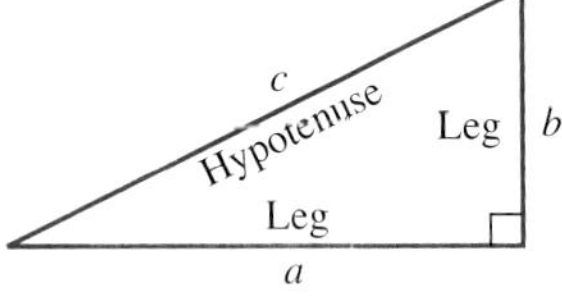

FIGURE 14-6

The Pythagorean theorem

If the length of the hypotenuse of a right triangle is c and the lengths of the two legs are a and b,

$$c^2 = a^2 + b^2$$

Since the lengths of the sides of a triangle are positive numbers, we can use the **square root property of equality** and the Pythagorean theorem to find the length of the third side of any right triangle when the measures of two sides are given.

Square root property of equality

If a and b represent positive numbers, and if $a = b$,

$$\sqrt{a} = \sqrt{b}$$

EXAMPLE 3 **Picture frames.** After gluing together two pieces of picture frame, the maker checks her work by making a diagonal measurement. (See Figure 14-7.) If the sides of the frame form a right angle, what measurement should the maker read on the yardstick?

Solution If the sides of the frame form a right angle, the sides and the diagonal form a right triangle. The lengths of the legs of the right triangle are 15 inches and 20 inches. We can find c, the length of the hypotenuse, using the Pythagorean theorem.

$c^2 = \mathbf{a}^2 + b^2$	This is the Pythagorean theorem.
$c^2 = \mathbf{15}^2 + \mathbf{20}^2$	Substitute 15 for a and 20 for b.
$c^2 = 225 + 400$	$15^2 = 225$ and $20^2 = 400$.
$c^2 = 625$	Perform the addition: $225 + 400 = 625$.

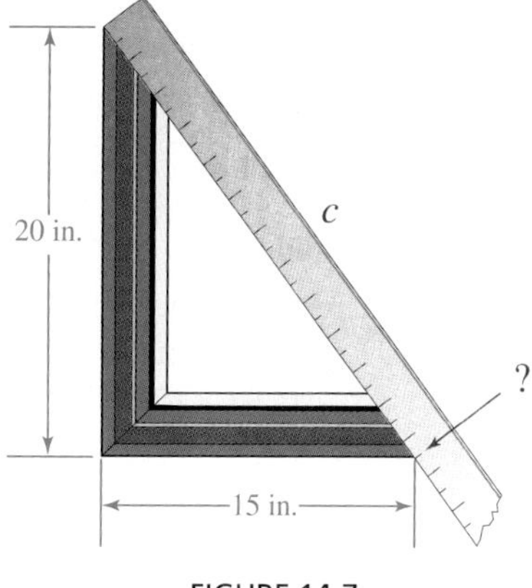

FIGURE 14-7

To find c, we must find a number that, when squared, is 625. There are two such numbers, one positive and one negative. They are called the *square roots* of 625. Since c represents the length of the hypotenuse, c cannot be negative. Thus, we need only determine the positive square root of 625.

$c^2 = 625$	This is the equation to solve.
$\sqrt{c^2} = \sqrt{625}$	To find c, we undo the operation performed on it by taking the positive square root of both sides. Recall that a radical symbol $\sqrt{}$ is used to indicate the positive square root of a number.
$c = 25$	$\sqrt{c^2} = c$ because $(c)^2 = c^2$, and $\sqrt{625} = 25$ because $25^2 = 625$.

The diagonal distance should measure 25 inches. If it does not, the sides of the frame do not form a right angle.

! COMMENT When using the Pythagorean theorem $c^2 = a^2 + b^2$, we can let a represent the length of either leg of the right triangle in question. We then let b represent the length of the other leg. The variable c must always represent the length of the hypotenuse.

EXAMPLE 4 Building a high ropes adventure course. The builder of a high ropes course wants to use a 25-foot cable to stabilize the pole shown in Figure 14-8. To be safe, the ground anchor stake must be farther than 18 feet from the base of the pole. Is the cable long enough to use?

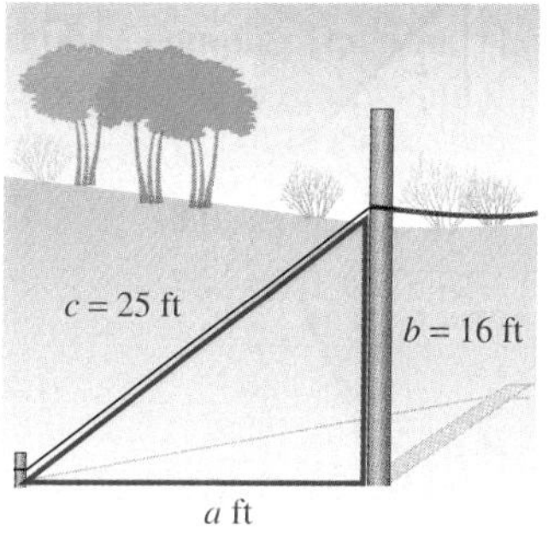

FIGURE 14-8

Solution We use the Pythagorean theorem, with $b = 16$ and $c = 25$, to find a.

$\mathbf{c}^2 = a^2 + b^2$	
$\mathbf{25}^2 = a^2 + \mathbf{16}^2$	Substitute 25 for c and 16 for b.
$625 = a^2 + 256$	$25^2 = 625$ and $16^2 = 256$.
$369 = a^2$	To isolate a^2, subtract 256 from both sides.
$\sqrt{369} = \sqrt{a^2}$	To find a, we undo the operation that is performed on it (squaring) by taking the positive square root of both sides.
$19.209372 \approx a$	Use a calculator to approximate $\sqrt{369}$.

Since the anchor stake will be more than 18 feet from the base, the 25-foot cable is long enough to use.

EXAMPLE 5 Reach of a ladder. A 26-foot ladder rests against the side of a building. If the base of the ladder is 10 feet from the wall, how far up the building will the ladder reach?

Analyze the problem

The wall, the ground, and the ladder form a right triangle, as shown in Figure 14-9. In this triangle, the hypotenuse is 26 feet, and one of the legs is the base-to-wall distance of 10 feet. We can let x = the length of the other leg, which is the distance that the ladder will reach up the wall.

Form an equation

We use the Pythagorean theorem to form the equation.

The hypotenuse squared	is	one leg squared	plus	the other leg squared.
26^2	$=$	10^2	$+$	x^2

Solve the equation

$26^2 = 10^2 + x^2$

$676 = 100 + x^2$ $\quad$ $26^2 = 676$ and $10^2 = 100$.

$676 - 100 = x^2$ $\quad$ To isolate x^2, subtract 100 from both sides.

$576 = x^2$ $\quad$ $676 - 100 = 576$.

$\sqrt{576} = \sqrt{x^2}$ $\quad$ Take the positive square root of both sides.

$24 = x$ $\quad$ $\sqrt{576} = 24$ because $24^2 = 576$.

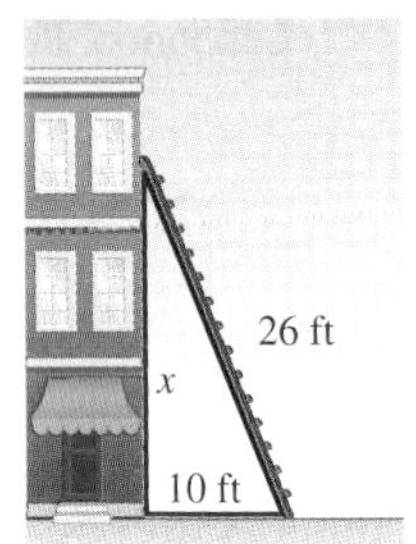

FIGURE 14-9

State the conclusion

The ladder will reach 24 feet up the side of the building.

Check the result

If the ladder reaches 24 feet up the side of the building, we have $10^2 + \mathbf{24}^2 = 100 + 576 = 676$, which is 26^2. The answer, 24, checks.

EXAMPLE 6 Roof design. The gable end of the roof shown in Figure 14-10 is an isosceles right triangle with a span of 48 feet. Find the distance from the eaves to the peak.

Analyze the problem

The two equal sides of the isosceles triangle are the two legs of the right triangle, and the span of 48 feet is the length of the hypotenuse. We can let x = the length of each leg, which is the distance from eaves to peak.

Form an equation

We use the Pythagorean theorem to form the equation.

The hypotenuse squared	is	one leg squared	plus	the other leg squared.
48^2	$=$	x^2	$+$	x^2

Solve the equation

$48^2 = x^2 + x^2$	
$2{,}304 = 2x^2$	$48^2 = 2{,}304$ and $x^2 + x^2 = 2x^2$.
$1{,}152 = x^2$	To isolate x^2, divide both sides by 2.
$\sqrt{1{,}152} = \sqrt{x^2}$	Take the positive square root of both sides.
$33.9411255 \approx x$	Use a calculator to approximate $\sqrt{1{,}152}$.

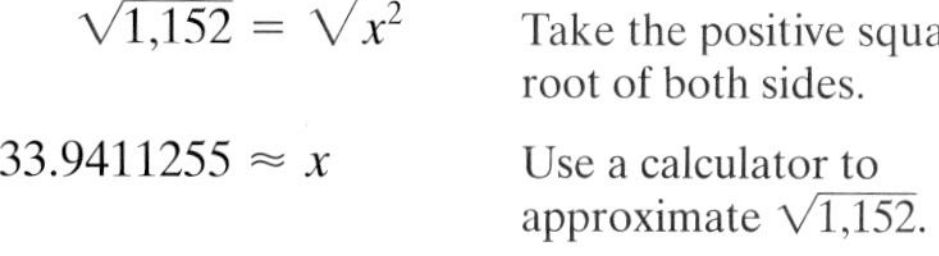

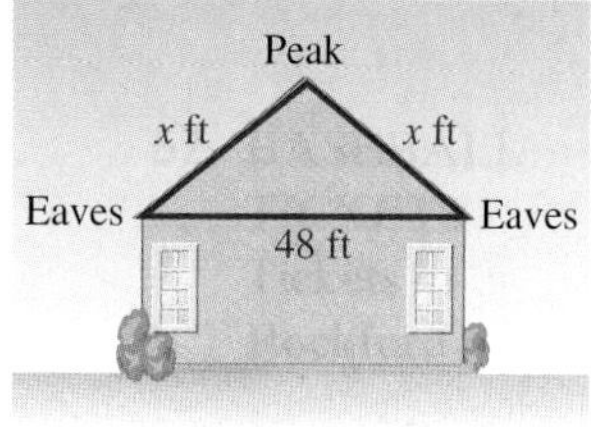

FIGURE 14-10

State the conclusion

The eaves-to-peak distance of the roof is approximately 34 feet.

Check the result

If the eaves-to-peak distance is approximately 34 feet, we have $\mathbf{34^2 + 34^2} = 1{,}156 + 1{,}156 = 2{,}312$, which is approximately 48^2. The answer, 34, seems reasonable.

Section 14.1 STUDY SET

VOCABULARY *Fill in the blanks.*

1. b is a ________ root of a if $b^2 = a$.
2. The symbol $\sqrt{}$ is called a ________ symbol.
3. The principal square root of a positive number is a ________ number.
4. The number under the radical sign is called the ________.
5. If a triangle has a right angle, it is called a ______ triangle.
6. The longest side of a right triangle is called the ____________, and the other two sides are called ______.

CONCEPTS *Fill in the blanks.*

7. The number 25 has ______ square roots. They are ___ and ___.
8. $\sqrt{-11}$ is not a ______ number.
9. If the length of the hypotenuse of a right triangle is c and the legs are a and b, then $c^2 =$ ______.
10.

The hypotenuse squared	is	one leg ______	plus	the other leg ______.

11. If a and b are positive numbers and $a = b$, then $\sqrt{a} =$ ___.
12. 2 is a square ______ of 4, because $2^2 = 4$.
13. To isolate x, what step should be used to undo the operation performed on it? (Assume that x is a positive number.)
    a. $2x = 16$
    b. $x^2 = 16$
14. Graph each number on the number line.

$$\left\{\sqrt{16}, -\sqrt{\tfrac{9}{4}}, \sqrt{1.8}, \sqrt{6}, -\sqrt{23}\right\}$$

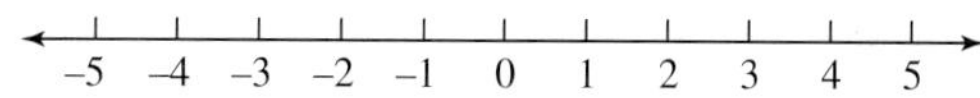

15. Complete the table.
Do not use a calculator.

x	$\sqrt{x}$
0	
$\frac{1}{81}$	
0.16	
36	
400	

16. If $f(x) = \sqrt{x}$, find each value. **Do not use a calculator.**

a. $f(81)$
b. $f(1)$
c. $f(0.25)$
d. $f\left(\frac{1}{121}\right)$
e. $f(900)$

17. a. Use the dashed lines in the following graph to approximate $\sqrt{5}$.
b. Use the graph to approximate $\sqrt{3}$ and $\sqrt{8}$.

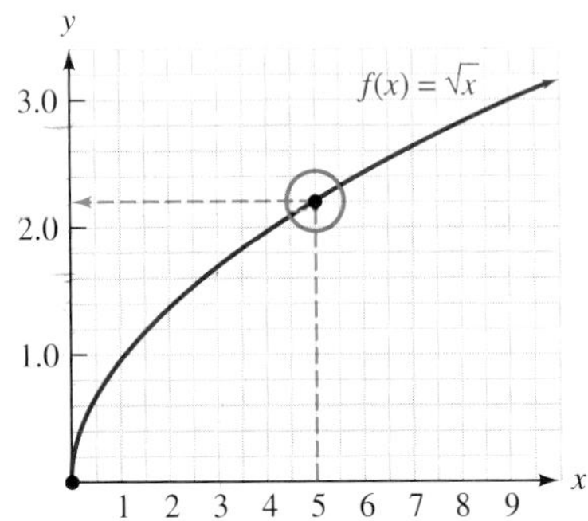

18. A calculator was used to find $\sqrt{-16}$. Explain the message shown on the calculator display.

NOTATION *Complete each solution.*

19. If the legs of a right triangle measure 5 and 12 centimeters, find the length of the hypotenuse.

$c^2 = a^2 + b^2$
$c^2 = \square^2 + \square^2$
$c^2 = 25 + \square$
$c^2 = \square$
$\square = \sqrt{169}$
$c = 13$

20. If the hypotenuse of a right triangle measures 25 centimeters and one leg measures 24 centimeters, find the length of the other leg.

$c^2 = a^2 + b^2$
$\square^2 = \square^2 + b^2$
$625 = \square + b^2$
$\square = b^2$
$\sqrt{49} = \square$
$7 = b$

21. Is the statement $-\sqrt{9} = \sqrt{-9}$ true or false? Explain your answer.

22. Consider the statement $\sqrt{26} \approx 5.1$. Explain why an $\approx$ symbol is used instead of an $=$ symbol.

PRACTICE *Find each square root without using a calculator.*

23. $\sqrt{25}$
24. $\sqrt{49}$
25. $-\sqrt{81}$
26. $-\sqrt{36}$
27. $\sqrt{1.21}$
28. $\sqrt{1.69}$
29. $\sqrt{196}$
30. $\sqrt{169}$
31. $\sqrt{\frac{9}{256}}$
32. $\sqrt{\frac{49}{225}}$
33. $-\sqrt{289}$
34. $-\sqrt{324}$
35. $-\sqrt{2,500}$
36. $-\sqrt{625}$
37. $\sqrt{3,600}$
38. $\sqrt{1,600}$

Use a calculator to evaluate each expression to three decimal places.

39. $\sqrt{2}$
40. $\sqrt{3}$
41. $\sqrt{11}$
42. $\sqrt{53}$
43. $\sqrt{95}$
44. $\sqrt{99}$
45. $\sqrt{428}$
46. $\sqrt{844}$
47. $-\sqrt{9,876}$
48. $-\sqrt{3,619}$
49. $\sqrt{21.35}$
50. $\sqrt{13.78}$
51. $\sqrt{0.3588}$
52. $\sqrt{0.9999}$
53. $-\sqrt{0.8372}$
54. $-\sqrt{0.4279}$
55. $2\sqrt{3}$
56. $3\sqrt{2}$
57. $\frac{2 + \sqrt{3}}{2}$
58. $\frac{2 - \sqrt{3}}{2}$

Determine whether each number in each set is rational, irrational, or imaginary.

59. $\{\sqrt{9}, \sqrt{17}, \sqrt{49}, \sqrt{-49}\}$

60. $\{-\sqrt{5}, \sqrt{0}, \sqrt{-100}, -\sqrt{225}\}$

Complete the table and graph the function.

61. $f(x) = 1 + \sqrt{x}$

x	f(x)
0	
1	
4	
9	
16	

62. $f(x) = -1 + \sqrt{x}$

x	f(x)
0	
1	
4	
9	
16	

63. $f(x) = -\sqrt{x}$

x	f(x)
0	
1	
4	
9	
16	

64. $f(x) = 1 - \sqrt{x}$

x	f(x)
0	
1	
4	
9	
16	

In Exercises 65–72, refer to the following right triangle. Find the length of the unknown side.

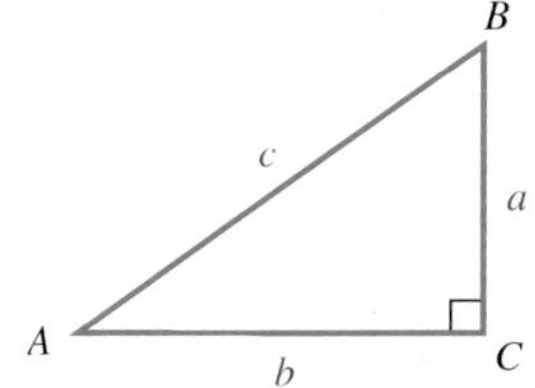

65. Find c if $a = 4$ and $b = 3$.

66. Find c if $a = 5$ and $b = 12$.

67. Find b if $a = 15$ and $c = 17$.

68. Find b if $a = 21$ and $c = 29$.

69. Find a if $b = 16$ and $c = 34$.

70. Find a if $b = 45$ and $c = 53$.

71. Find b if $c = 125$ and $a = 44$.

72. Find c if $a = 176$ and $b = 57$.

APPLICATIONS *Use a calculator to help solve each problem. If an answer is not exact, give it to the nearest tenth.*

73. ADJUSTING A LADDER A 20-foot ladder reaches a window 16 feet above the ground. How far from the wall is the base of the ladder?

74. LINE OF SIGHT A movie viewer in a car parked at a drive-in theater sits 600 feet from the base of the vertical screen. What is the line-of-sight distance for the viewer to the middle of the screen, which is 35 feet above the base?

75. QUALITY CONTROL How can a tool manufacturer use the Pythagorean theorem to verify that the two sides of the carpenter's square shown meet to form a 90° angle?

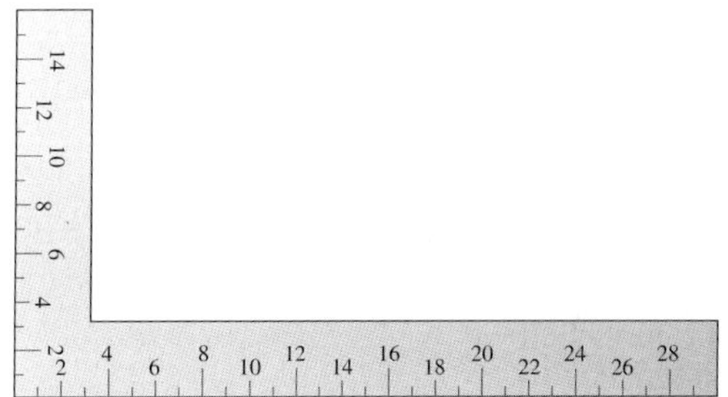

76. GARDENING A rectangular garden has sides of 28 and 45 feet. Find the length of a path that extends from one corner to the opposite corner.

77. BASEBALL A baseball diamond is a square, with each side 90 feet long, as shown. How far is it from home plate to second base?

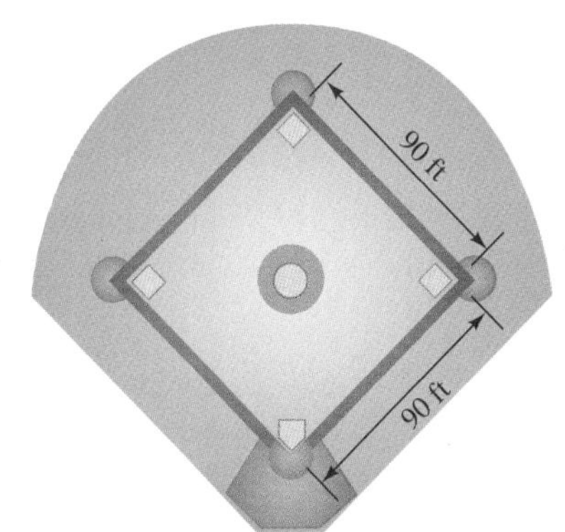

78. TELEVISION The size of a television screen is the diagonal distance from the upper left to the lower right corner. What is the size of the screen shown?

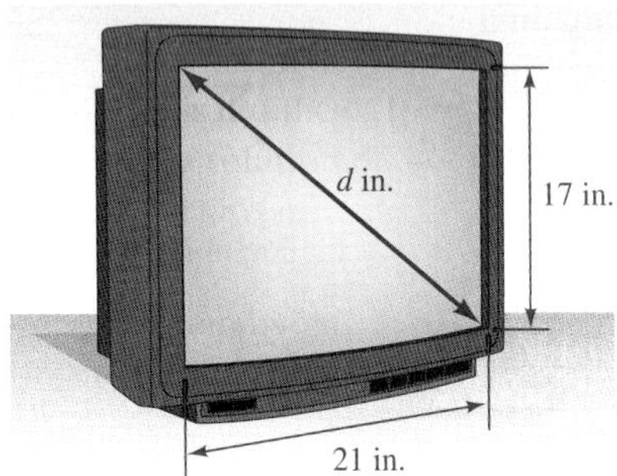

79. FINDING LOCATION A team of archaeologists travels 4.2 miles east and then 4.0 miles north of their base camp to explore some ancient ruins. "As the crow flies," how far from their base camp are they?

80. SHORTCUTS Instead of walking on the sidewalk, students take a diagonal shortcut across the rectangular vacant lot shown. How much distance do they save?

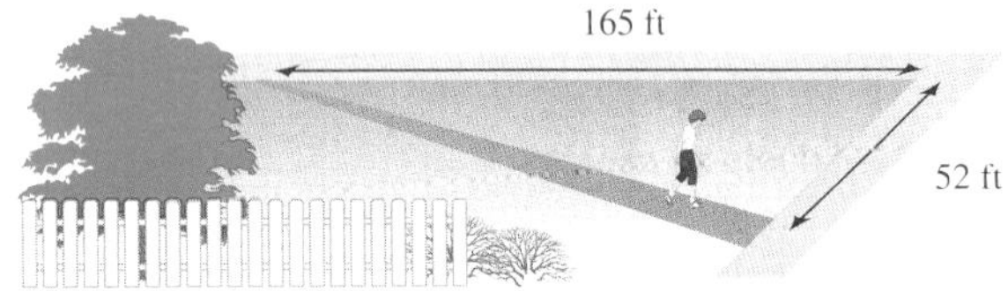

81. FOOTBALL On first down and ten, a quarterback tells his tight end to go out 6 yards, cut 45° to the right, and run 6 yards, as shown. The tight end follows instructions, catches a pass, and is tackled immediately.

a. Find x.

b. Does he gain the necessary 10 yards for a first down?

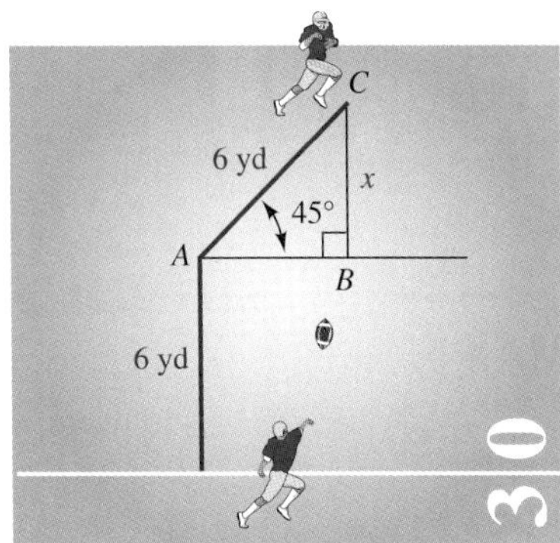

82. GEOMETRY The legs of a right triangle are equal, and the hypotenuse is 2.82843 units long. Find the length of each leg.

83. WRESTLING The sides of a square wrestling ring are 18 feet long. Find the distance from one corner to the opposite corner.

84. PERIMETER OF A SQUARE The diagonal of a square is 3 feet long. Find its perimeter.

85. HEIGHT OF A TRIANGLE Find the area of the isosceles triangle shown.

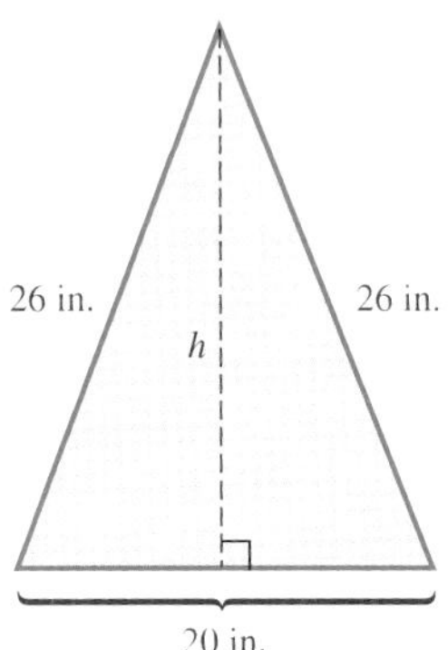

86. INTERIOR DECORATING The following square table is covered by a circular tablecloth. If the sides of the table are 2 feet long, find the area of the tablecloth.

87. DRAFTING Among the tools used in drafting are 30–60–90 and 45–45–90 triangles.

a. Find the length of the hypotenuse of the 45–45–90 triangle if it is $\sqrt{2}$ times as long as a leg.

b. Find the length of the side opposite the 60° angle of the other triangle if it is $\frac{\sqrt{3}}{2}$ times as long as the hypotenuse.

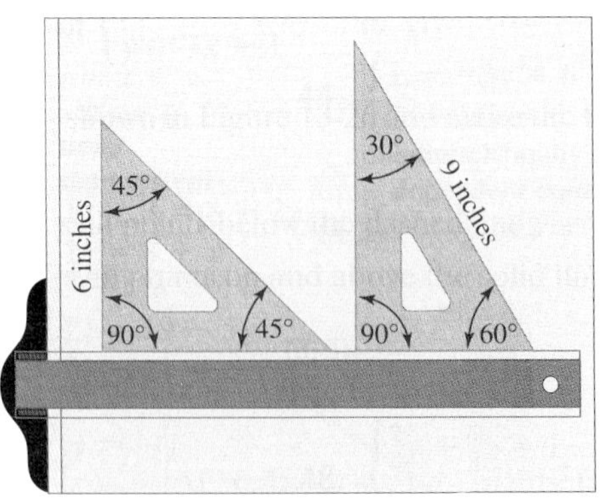

88. ORGAN PIPES The design for a set of brass pipes for a church organ is shown. Find the length of each pipe (to the nearest tenth of a foot), and then find the total length of pipe needed to construct this set.

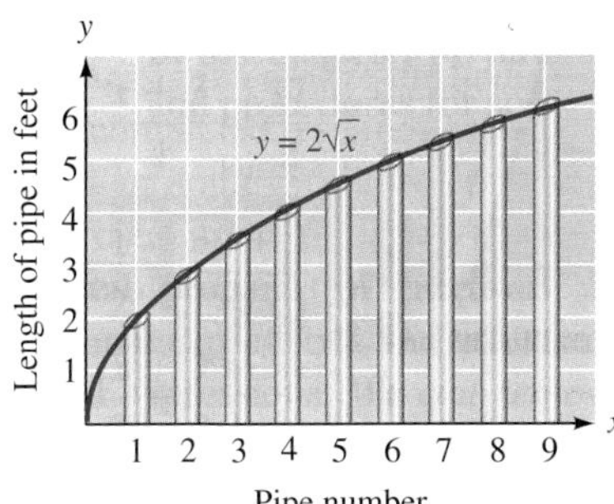

WRITING

89. Explain why the square root of a negative number cannot be a real number.

90. Explain the Pythagorean theorem.

91. Suppose you are told that $\sqrt{10} \approx 3.16$. Explain how another key on your calculator (besides the square root key $\sqrt{}$) could be used to see whether this is a reasonable approximation.

92. Explain the difference between the *square* of a number and the *square root* of a number.

REVIEW

93. Add: $(3s^2 - 3s - 2) + (3s^2 + 4s - 3)$.

94. Subtract: $(3c^2 - 2c + 4) - (c^2 - 3c + 7)$.

95. Multiply: $(3x - 2)(x + 4)$.

96. Divide: $x^2 + 13x + 12$ by $x + 1$.

14.2 Higher-Order Roots; Radicands That Contain Variables

- Cube roots • Approximating cube roots • The cube root function • Higher-order roots
- Radicands that contain variables

To find the volume V of the cube shown in Figure 14-11, we multiply its length, width, and height.

$$V = l \cdot w \cdot h$$
$$V = 5 \cdot 5 \cdot 5$$
$$= 125$$

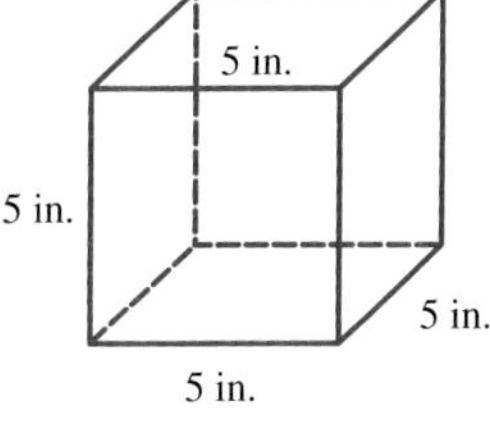

FIGURE 14-11

The volume is 125 cubic inches.

We have seen that $5 \cdot 5 \cdot 5$ can be denoted by the exponential expression 5^3, where 5 is raised to the third power. Whenever we raise a number to the third power, we are cubing it, or finding its **cube.** This example illustrates that the formula for the volume of a cube with each side of length s is $V = s^3$.

Here are some more cubes of numbers:

- The cube of 3 is 27, because $3^3 = 27$.
- The cube of -3 is -27, because $(-3)^3 = -27$.
- The cube of 12 is 1,728, because $12^3 = 1{,}728$.
- The cube of -12 is $-1{,}728$, because $(-12)^3 = -1{,}728$.
- The cube of $\frac{1}{4}$ is $\frac{1}{64}$, because $\left(\frac{1}{4}\right)^3 = \frac{1}{4} \cdot \frac{1}{4} \cdot \frac{1}{4} = \frac{1}{64}$.
- The cube of $-\frac{1}{4}$ is $-\frac{1}{64}$, because $\left(-\frac{1}{4}\right)^3 = \left(-\frac{1}{4}\right)\left(-\frac{1}{4}\right)\left(-\frac{1}{4}\right) = -\frac{1}{64}$.
- The cube of 0 is 0, because $0^3 = 0$.

In this section, we will reverse the cubing process and find **cube roots** of numbers. We will also consider fourth roots, fifth roots, and so on. After graphing the cube root function, we will work with radical expressions having radicands containing variables.

Cube roots

Suppose we know that the volume of the cube shown in Figure 14-12 is 216 cubic inches. To find the length of each side, we substitute 216 for V in the formula $V = s^3$ and solve for s.

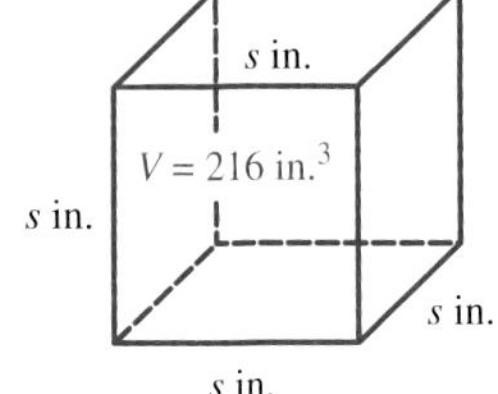

FIGURE 14-12

$$V = s^3$$
$$\mathbf{216} = s^3$$

To solve for s, we must find a number whose cube is 216. Since 6 is such a number, the sides of the cube are 6 inches long. The number 6 is called a *cube root* of 216, because $6^3 = 216$.

Here are more examples of cube roots:

- 3 is a cube root of 27, because $3^3 = 27$.
- -3 is a cube root of -27, because $(-3)^3 = -27$.
- 12 is a cube root of 1,728, because $12^3 = 1{,}728$.
- -12 is a cube root of $-1{,}728$, because $(-12)^3 = -1{,}728$.
- $\frac{1}{4}$ is a cube root of $\frac{1}{64}$, because $\left(\frac{1}{4}\right)^3 = \left(\frac{1}{4}\right)\left(\frac{1}{4}\right)\left(\frac{1}{4}\right) = \frac{1}{64}$.
- $-\frac{1}{4}$ is a cube root of $-\frac{1}{64}$, because $\left(-\frac{1}{4}\right)^3 = \left(-\frac{1}{4}\right)\left(-\frac{1}{4}\right)\left(-\frac{1}{4}\right) = -\frac{1}{64}$.
- 0 is a cube root of 0, because $0^3 = 0$.

In general, we have the following definition.

> The number b is a **cube root** of a if $b^3 = a$.

All real numbers have one real cube root. As the preceding examples show, a positive number has a positive cube root, a negative number has a negative cube root, and the cube root of 0 is 0.

> The **cube root of *a*** is denoted by $\sqrt[3]{a}$. By definition,
>
> $\sqrt[3]{a} = b \quad \text{if} \quad b^3 = a$

EXAMPLE 1 Find each cube root: **a.** $\sqrt[3]{8}$, **b.** $\sqrt[3]{343}$, **c.** $\sqrt[3]{-8}$, **d.** $\sqrt[3]{-125}$, and **e.** $-\sqrt[3]{1000}$.

Solution

a. $\sqrt[3]{8} = 2$, because $2^3 = 8$.

b. $\sqrt[3]{343} = 7$, because $7^3 = 343$.

c. $\sqrt[3]{-8} = -2$, because $(-2)^3 = -8$.

d. $\sqrt[3]{-125} = -5$, because $(-5)^3 = -125$.

e. $-\sqrt[3]{1{,}000}$ is the opposite of the cube root of 1000. Since $\sqrt[3]{1{,}000}$ is 10, we have $-\sqrt[3]{1{,}000} = -10$. $\quad -\sqrt[3]{1{,}000} = -1 \cdot \sqrt[3]{1{,}000} = -1 \cdot 10 = -10.$

Self Check 1
Find each cube root:
a. $\sqrt[3]{64}$, **b.** $\sqrt[3]{-64}$, **c.** $\sqrt[3]{216}$, **d.** $-\sqrt[3]{125}$

Answers **a.** 4, **b.** -4, **c.** 6, **d.** -5

EXAMPLE 2 Find each cube root.

Solution

a. $\sqrt[3]{\frac{1}{8}} = \frac{1}{2}$, because $\left(\frac{1}{2}\right)^3 = \frac{1}{2} \cdot \frac{1}{2} \cdot \frac{1}{2} = \frac{1}{8}$.

b. $\sqrt[3]{-\frac{125}{27}} = -\frac{5}{3}$, because $\left(-\frac{5}{3}\right)^3 = \left(-\frac{5}{3}\right)\left(-\frac{5}{3}\right)\left(-\frac{5}{3}\right) = -\frac{125}{27}$.

Self Check 2

Find each cube root:

a. $\sqrt[3]{\frac{1}{27}}$, **b.** $\sqrt[3]{-\frac{8}{125}}$

Answers **a.** $\frac{1}{3}$, **b.** $-\frac{2}{5}$

Cube roots of numbers such as 7 are hard to compute by hand. However, we can approximate $\sqrt[3]{7}$ with a calculator.

Approximating cube roots

To find $\sqrt[3]{7}$, we can enter 7 into a scientific calculator, press the root key $\sqrt[x]{y}$, enter 3, and press the = key. The approximate value of $\sqrt[3]{7}$ will appear on the calculator's display.

$$\sqrt[3]{7} \approx 1.912931183$$

If your calculator doesn't have a $\sqrt[x]{y}$ key, you can use the y^x key. We will see later that $\sqrt[3]{7} = 7^{1/3}$. To find the value of $7^{1/3}$, we enter 7 into the calculator and press these keys:

7 y^x (1 ÷ 3) =

The display will read 1.912931183.

Since $\sqrt[3]{7}$ represents the number that, when cubed, gives 7, we would expect cubes of approximations of $\sqrt[3]{7}$ to be close to 7.

- Rounded to one decimal place, $\sqrt[3]{7} \approx 1.9$, and $(1.9)^3 = 6.859$.
- Rounded to two decimal places, $\sqrt[3]{7} \approx 1.91$, and $(1.91)^3 = 6.967871$.
- Rounded to three decimal places, $\sqrt[3]{7} \approx 1.913$, and $(1.913)^3 = 7.000755497$.

Numbers such as 8, −27, −64, and 125 are called **integer cubes,** because each one is the cube of an integer. The cube root of any integer cube is an integer and therefore a rational number:

$$\sqrt[3]{8} = 2, \quad \sqrt[3]{-27} = -3, \quad \sqrt[3]{-64} = -4, \quad \text{and} \quad \sqrt[3]{125} = 5$$

Cube roots of integers such as 7 and −10, which are not integer cubes, are irrational numbers. For example, $\sqrt[3]{7}$ and $\sqrt[3]{-10}$ are irrational numbers.

! COMMENT Recall that the square root of a negative number (for example, $\sqrt{-27}$) is not a real number, because no real number squared is equal to a negative number. However, the cube root of a negative number is a real number. For example, $\sqrt[3]{-27} = -3$.

The cube root function

Since every real number has one real-number cube root, there is a cube root function $f(x) = \sqrt[3]{x}$. For example, the value that is determined by $f(x) = \sqrt[3]{x}$ when $x = -8$ is denoted as $f(-8)$, and we have $f(-8) = \sqrt[3]{-8} = -2$.

To graph this function, we substitute numbers for x, compute $f(x)$, plot the resulting ordered pairs, and draw a smooth curve through the points as shown in Figure 14-13. In the table, we chose five values for x, -8, -1, 0, 1, and 8, that are integer cubes. This made computing $f(x)$ quite simple.

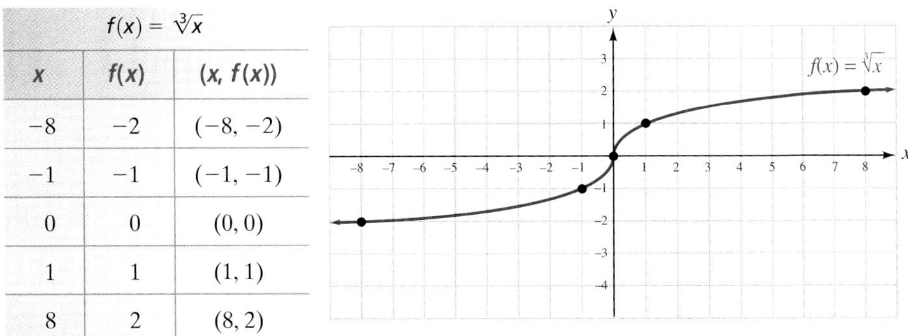

$f(x) = \sqrt[3]{x}$

x	$f(x)$	$(x, f(x))$
-8	-2	$(-8, -2)$
-1	-1	$(-1, -1)$
0	0	$(0, 0)$
1	1	$(1, 1)$
8	2	$(8, 2)$

FIGURE 14-13

CALCULATOR SNAPSHOT

Radius of a water tank

Engineers want to design a spherical tank that will hold 33,500 cubic feet of water, as shown in Figure 14-14. They know that the formula for the radius r of a sphere with volume V is given by the formula

$$r = \sqrt[3]{\frac{3V}{4\pi}} \quad \text{Where } \pi = 3.14159\ldots$$

To use a scientific calculator to find the radius r, we substitute 33,500 for V and enter these numbers and press these keys.

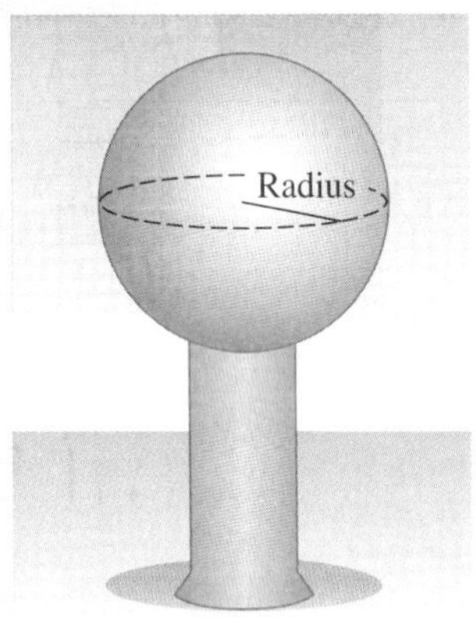

FIGURE 14-14

3 [×] 33500 [÷] [(] 4 [×] [π] [)] [=] [$\sqrt[n]{x}$] 3 [=]   19.99794636

To evaluate this expression using a graphing calculator, we press the [MATH] key. In this mode, arrow down [▼] to highlight the option $\sqrt[3]{\ }$ (and [ENTER]. Then we press the following keys.

3 [×] 33500 [÷] [(] 4 [×] [2nd] [π] [)] [)] [ENTER]

```
∛(3*33500/(4*π)
)        19.99794636
```

Since the result is 19.99794636, the engineers should design a tank with a radius of 20 feet.

Higher-order roots

Just as there are square roots and cube roots, there are also fourth roots, fifth roots, sixth roots, and so on. In general, we have the following definition.

> The ***n*th root of *a*** is denoted by $\sqrt[n]{a}$, and
>
> $\sqrt[n]{a} = b \quad \text{if} \quad b^n = a$
>
> The number n is called the **index** of the radical. In n is an even natural number, a must be positive or zero, and b must be positive.

In the square root symbol $\sqrt{}$, the unwritten index is understood to be 2.

$$\sqrt{a} = \sqrt[2]{a}$$

EXAMPLE 3 Find each root: **a.** $\sqrt[4]{81}$, **b.** $\sqrt[5]{32}$, **c.** $\sqrt[5]{-32}$, and **d.** $\sqrt[4]{-81}$.

Solution

a. $\sqrt[4]{81} = 3$, because $3^4 = 81$.

b. $\sqrt[5]{32} = 2$, because $2^5 = 32$.

c. $\sqrt[5]{-32} = -2$, because $(-2)^5 = -32$.

d. $\sqrt[4]{-81}$ is not a real number, because no real number raised to the fourth power is -81.

Self Check 3
Find each root:
a. $\sqrt[4]{16}$, **b.** $\sqrt[5]{243}$,
c. $\sqrt[5]{-1{,}024}$

Answers **a.** 2, **b.** 3, **c.** -4

EXAMPLE 4 Find each root: **a.** $\sqrt[4]{\frac{1}{81}}$ and **b.** $\sqrt[5]{-\frac{32}{243}}$.

Solution

a. $\sqrt[4]{\frac{1}{81}} = \frac{1}{3}$, because $\left(\frac{1}{3}\right)^4 = \frac{1}{81}$.

b. $\sqrt[5]{-\frac{32}{243}} = -\frac{2}{3}$, because $\left(-\frac{2}{3}\right)^5 = -\frac{32}{243}$.

Self Check 4
Find each root:
a. $\sqrt[4]{\frac{1}{16}}$, **b.** $\sqrt[5]{-\frac{243}{32}}$

Answers **a.** $\frac{1}{2}$, **b.** $-\frac{3}{2}$

Radicands that contain variables

When n is even and $x \geq 0$, we say that the radical $\sqrt[n]{x}$ represents an **even root.** We can find even roots of many quantities that contain variables, provided that these variables represent positive numbers or zero.

EXAMPLE 5 Find each root. Assume that each variable represents a positive number. **a.** $\sqrt{x^2}$, **b.** $\sqrt{x^4}$, **c.** $\sqrt{x^4y^2}$, and **d.** $\sqrt[4]{81x^{12}}$.

Solution

a. $\sqrt{x^2} = x$, because $(x)^2 = x^2$.

b. $\sqrt{x^4} = x^2$, because $(x^2)^2 = x^4$.

c. $\sqrt{x^4y^2} = x^2y$, because $(x^2y)^2 = x^4y^2$.

d. $\sqrt[4]{81x^{12}} = 3x^3$, because $(3x^3)^4 = 81x^{12}$.

Self Check 5
Find each root:
a. $\sqrt{a^4}$ **b.** $\sqrt{m^6n^8}$,
c. $\sqrt[4]{16y^8}$

Answers **a.** a^2, **b.** m^3n^4, **c.** $2y^2$

When n is odd, we say that the radical expression $\sqrt[n]{x}$ represents an **odd root.**

Self Check 6

Find each root:

a. $\sqrt[3]{p^6}$, b. $\sqrt[3]{-27p^9}$,

c. $\sqrt[5]{\frac{1}{32}n^{15}}$

Answers a. p^2, b. $-3p^3$, c. $\frac{1}{2}n^3$

EXAMPLE 6 Find each root: a. $\sqrt[3]{y^3}$, b. $\sqrt[3]{64x^6}$, and c. $\sqrt[5]{x^{10}}$.

Solution

a. $\sqrt[3]{y^3} = y$, because $(y)^3 = y^3$.

b. $\sqrt[3]{64x^6} = 4x^2$, because $(4x^2)^3 = 64x^6$.

c. $\sqrt[5]{x^{10}} = x^2$, because $(x^2)^5 = x^{10}$.

Section 14.2 STUDY SET

VOCABULARY *Fill in the blanks.*

1. If $p^3 = q$, p is called a ______ root of q.
2. If $p^4 = q$, p is called a ________ root of q.
3. We denote the cube root _________ with the notation $f(x) = \sqrt[3]{x}$.
4. If the index of a radical is an even number, the root is called an ______ root.

CONCEPTS *Fill in the blanks.*

5. -3 is a cube ______ of -27, because $(-3)^3 = -27$.
6. $\sqrt[3]{a} = b$ if ▭ $= a$.
7. $\sqrt[3]{-216} = -6$, because (▭)$^3 = -216$.
8. $\sqrt[5]{32x^5} = 2x$, because $(2x)^{\square} = 32x^5$
9. Find each value, if possible.
   a. $\sqrt{-125}$
   b. $\sqrt[3]{-125}$
10. Graph each number on the number line.
   $\{\sqrt[3]{16}, -\sqrt[4]{100}, \sqrt[3]{-1.8}, \sqrt[4]{0.6}\}$

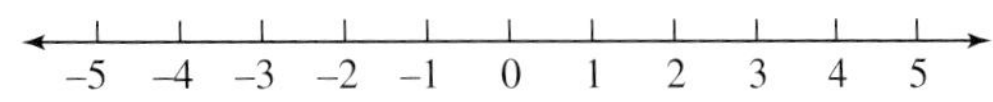

11. If $f(x) = \sqrt[3]{x}$, find each value. **Do not use a calculator.**
   a. $f(1)$
   b. $f\left(-\frac{1}{27}\right)$
   c. $f(125)$
   d. $f(0.008)$
   e. $f(1{,}000)$
12. a. Use the dashed lines in the following graph to approximate $\sqrt[3]{5}$.
   b. Use the graph to approximate $\sqrt[3]{4}$ and $\sqrt[3]{-6}$.

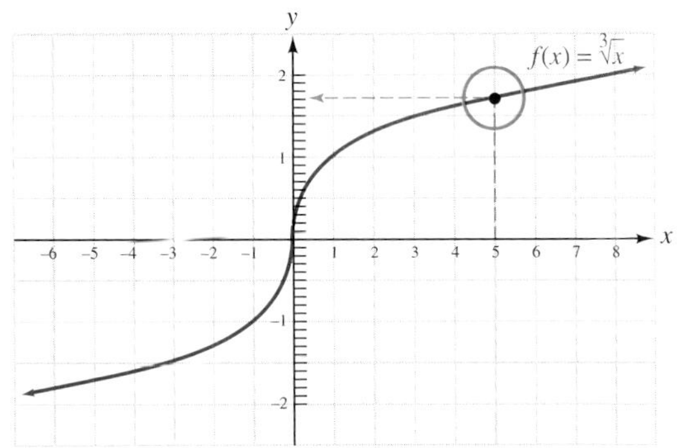

NOTATION *Fill in the blanks.*

13. In the notation $\sqrt[3]{x^6}$, 3 is called the ______ and x^6 is called the _________.
14. $\sqrt{\ }$ is called a ________ symbol.
15. The "understood" index of the radical expression $\sqrt{55}$ is ▭.
16. In reading $f(x) = \sqrt[3]{x}$, we say "f ____ x equals the cube root ____ x."

PRACTICE *Find each value without using a calculator.*

17. $\sqrt[3]{8}$
18. $\sqrt[3]{27}$
19. $\sqrt[3]{0}$
20. $\sqrt[3]{1}$
21. $\sqrt[3]{-8}$
22. $\sqrt[3]{-1}$
23. $\sqrt[3]{-64}$
24. $\sqrt[3]{-27}$
25. $\sqrt[3]{\frac{1}{125}}$
26. $\sqrt[3]{\frac{1}{1{,}000}}$
27. $-\sqrt[3]{-1}$
28. $-\sqrt[3]{-27}$
29. $-\sqrt[3]{64}$
30. $-\sqrt[3]{343}$
31. $\sqrt[3]{729}$
32. $\sqrt[3]{512}$
33. $\sqrt[3]{1{,}000}$
34. $\sqrt[3]{125}$

Use a calculator to find each cube root to the nearest hundredth.

35. $\sqrt[3]{32{,}100}$ **36.** $\sqrt[3]{-25{,}713}$

37. $\sqrt[3]{-0.11324}$ **38.** $\sqrt[3]{0.875}$

Complete the table and graph the function.

39. $f(x) = \sqrt[3]{x} + 1$

x	f(x)
−8	
−1	
0	
1	
8	

40. $f(x) = \sqrt[4]{x}$

x	f(x)
0	
1	
16	

41. $f(x) = -\sqrt[3]{x}$

x	f(x)
−8	
−1	
0	
1	
8	

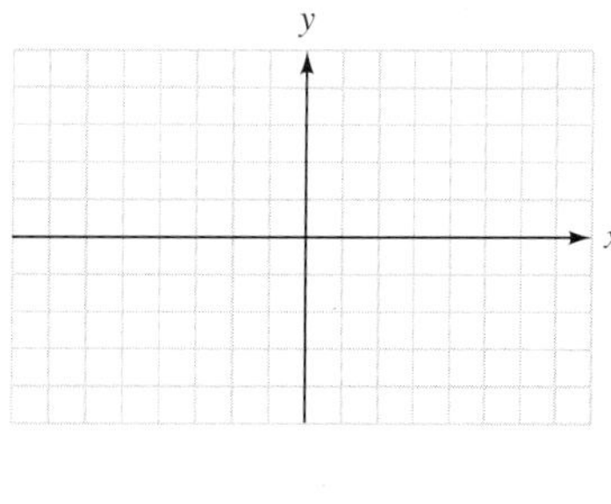

42. $f(x) = \sqrt[4]{x} - 1$

x	f(x)
0	
1	
16	

Find each value without using a calculator.

43. $\sqrt[4]{16}$ **44.** $\sqrt[4]{81}$

45. $-\sqrt[5]{32}$ **46.** $-\sqrt[5]{243}$

47. $\sqrt[6]{1}$ **48.** $\sqrt[6]{0}$

49. $\sqrt[5]{-32}$ **50.** $\sqrt[7]{-1}$

Use a calculator to find each root to the nearest hundredth.

51. $\sqrt[4]{125}$ **52.** $\sqrt[5]{12{,}450}$

53. $\sqrt[5]{-6{,}000}$ **54.** $\sqrt[6]{0.5}$

Find each root. All variables represent positive numbers.

55. $\sqrt{x^2}$ **56.** $\sqrt{y^4}$

57. $\sqrt{x^6}$ **58.** $\sqrt{b^8}$

59. $\sqrt{x^{10}}$ **60.** $\sqrt{y^{12}}$

61. $\sqrt{4z^2}$ **62.** $\sqrt{9t^6}$

63. $-\sqrt{x^4y^2}$ **64.** $-\sqrt{x^2y^4}$

65. $-\sqrt{0.04y^2}$ **66.** $-\sqrt{0.81b^6}$

67. $-\sqrt{25x^4z^{12}}$ **68.** $-\sqrt{100a^6b^4}$

69. $\sqrt{36z^{36}}$ **70.** $\sqrt{64y^{64}}$

71. $-\sqrt{625z^2}$ **72.** $-\sqrt{729x^8}$

73. $\sqrt[3]{y^6}$ **74.** $\sqrt[3]{c^3}$

75. $\sqrt[5]{f^5}$ **76.** $\sqrt[5]{y^{20}}$

77. $\sqrt[3]{27y^3}$ **78.** $\sqrt[3]{64y^6}$

79. $\sqrt[3]{-p^6q^3}$ **80.** $\sqrt[3]{-r^{12}t^6}$

81. $\sqrt[4]{x^4}$ **82.** $\sqrt[4]{x^8}$

APPLICATIONS *Use a calculator to help solve each problem. Give your answers to the nearest hundredth.*

83. PACKAGING A cubical box has a volume of 2 cubic feet. Substitute 2 for V in the formula $V = s^3$ and solve for s to find the length of each side of the box.

84. HOT-AIR BALLOONS If the hot-air balloon shown is in the shape of a sphere, what is its radius? (*Hint:* See the Calculator Snapshot feature in this section.)

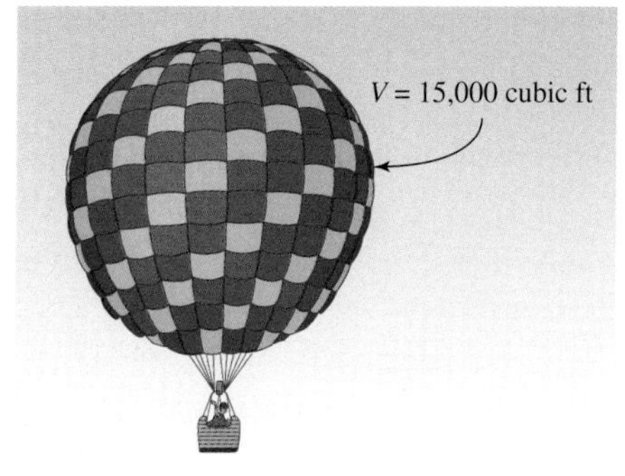

85. WINDMILLS The power generated by a windmill is related to the speed of the wind by the formula

$$S = \sqrt[3]{\frac{P}{0.02}}$$

where S is the speed of the wind (in mph) and P is the power (in watts). Find the speed of the wind when the windmill is producing 400 watts of power.

86. ASTRONOMY In the early 17th century, Johannes Kepler, a German astronomer, discovered that a planet's mean distance R from the sun (in millions of miles) is related to the time T (in years) it takes the planet to orbit the sun by the formula

$$R = 93\sqrt[3]{\frac{T^2}{1.002}}$$

Use the information in the illustration to find R for Mercury, Earth, and Jupiter.

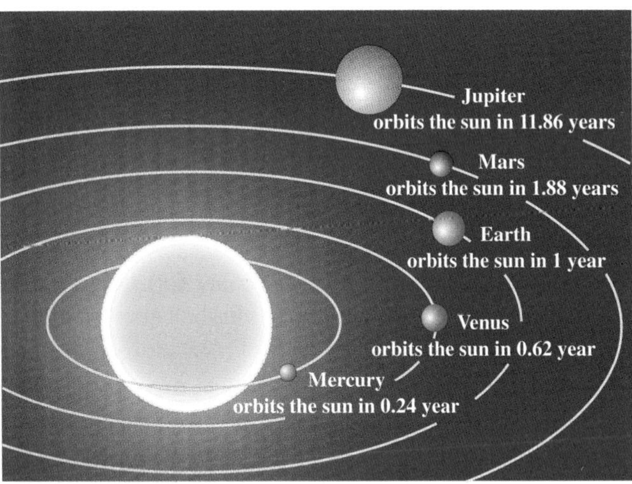
Jupiter
orbits the sun in 11.86 years
Mars
orbits the sun in 1.88 years
Earth
orbits the sun in 1 year
Venus
orbits the sun in 0.62 year
Mercury
orbits the sun in 0.24 year

87. DEPRECIATION The formula

$$r = 1 - \sqrt[n]{\frac{S}{C}}$$

gives the annual depreciation rate r (in percent) of an item that had an original cost of C dollars and has a useful life of n years and a salvage value of S dollars. Use the information in the illustration to find the annual depreciation rate for the new piece of sound equipment.

OFFICE MEMO

To: Purchasing Dept.
From: Bob Kinsell, Engineering Dept. BK
Re: New sound board

We recommend you purchase the new Sony sound board @ \$81K. This equipment does become obsolete quickly but we figure we can use it for 4 yrs. A college would probably buy it from us then. I bet we could get around \$16K for it.

88. SAVINGS ACCOUNTS The interest rate r (in percent) earned by a savings account after n compoundings is given by the formula

$$\sqrt[n]{\frac{V}{P}} - 1 = r$$

where V is the current value and P is the original principal. What interest rate r was paid on an account in which a deposit of \$1,000 grew to \$1,338.23 after 5 compoundings?

WRITING

89. Explain why a negative number can have a real number for its cube root yet cannot have a real number for its fourth root.

90. To find $\sqrt[3]{15}$, we can use the $\sqrt[n]{x}$ key on a calculator to obtain 2.466212074. Explain how a key other than $\sqrt[n]{x}$ can be used to check the validity of this result.

REVIEW *Simplify each expression.*

91. m^5m^2
92. $(-5x^3)(-5x)$
93. $(3^2)^4$
94. r^3rr^5
95. $(x^2x^3)^5$
96. $(3aa^2a^3)^5$
97. $4x^3(6x^5)$
98. $-2x(5x^3)$

14.3 Simplifying Radical Expressions

- The multiplication property of radicals • Simplifying square root radicals
- The division property of radicals • Simplifying cube roots

Square dancing is an American folk dance in which four couples, arranged in a square, perform various moves. Figure 14-15 on the next page shows a group as they move around a square.

If the square shown in the figure has an area of 12 square yards, the length of a side is $\sqrt{12}$ yards. We can use the formula for the area of a square and the concept of square root to show that this is so.

FIGURE 14-15

$\mathbf{A} = s^2$ s is the length of a side of the square.

$\mathbf{12} = s^2$ Substitute 12 for A, the area of the square.

$\sqrt{12} = \sqrt{s^2}$ Take the positive square root of both sides.

$\sqrt{12} = s$ The length of a side of the square is $\sqrt{12}$ yards.

The form in which we express the length of a side of the square depends on the situation. If an approximation is acceptable, we can use a calculator to find that $\sqrt{12} \approx 3.464101615$, and we can then round to a specified degree of accuracy. For example, to the nearest tenth, each side is 3.5 yards long.

If the situation calls for the *exact* length, we must use a radical expression. As you will see in this section, it is common practice to write a radical expression such as $\sqrt{12}$ in *simplified form.* To simplify radicals, we will use the multiplication and division properties of radicals.

The multiplication property of radicals

We introduce the first of two properties of radicals with the following examples:

$$\sqrt{4 \cdot 25} = \sqrt{100} = 10 \qquad \text{and} \qquad \sqrt{4}\sqrt{25} = 2 \cdot 5 = 10$$

Read as "the square root of 4 times the square root of 25."

In each case, the answer is 10. Thus, $\sqrt{4 \cdot 25} = \sqrt{4}\sqrt{25}$. Likewise,

$$\sqrt{9 \cdot 16} = \sqrt{144} = 12 \qquad \text{and} \qquad \sqrt{9}\sqrt{16} = 3 \cdot 4 = 12$$

In each case, the answer is 12. Thus, $\sqrt{9 \cdot 16} = \sqrt{9}\sqrt{16}$. These results illustrate the **multiplication property of radicals.**

The multiplication property of radicals

If a and b represent nonnegative real numbers,

$$\sqrt{ab} = \sqrt{a}\sqrt{b}$$

In words, *the square root of the product of two nonnegative numbers is equal to the product of their square roots.*

Simplifying square root radicals

A square root radical is in **simplified form** when each of the following statements is true.

Simplified form of a square root

1. Except for 1, the radicand has no perfect square factors.
2. No fraction appears in a radicand.
3. No radical appears in the denominator of a fraction.

We can use the multiplication property of radicals to simplify square roots whose radicands have perfect square factors. For example, we can simplify $\sqrt{12}$ as follows:

$\sqrt{12} = \sqrt{4 \cdot 3}$ Factor 12 as $4 \cdot 3$.

$= \sqrt{4}\sqrt{3}$ The square root of $4 \cdot 3$ is equal to the square root of 4 times the square root of 3.

$= 2\sqrt{3}$ Write $\sqrt{4}$ as 2. Read as "2 times the square root of 3" or as "2 radical 3."

The square in Figure 14-15, which we considered in the introduction to this section, has a side length of $\sqrt{12}$ yards. We now see that the *exact* length of a side can be expressed in simplified form as $2\sqrt{3}$ yards.

To simplify more difficult square roots, it is helpful to know the natural-number **perfect squares.** For example, 81 is a perfect square, because it is the square of 9: $9^2 = 81$. The first 20 natural-number perfect squares are

1, 4, 9, 16, 25, 36, 49, 64, 81, 100, 121, 144, 169, 196, 225, 256, 289, 324, 361, 400

Self Check 1
Simplify: $\sqrt{28}$.

Answer $2\sqrt{7}$

EXAMPLE 1 Simplify: $\sqrt{27}$.

Solution Since the greatest perfect square that divides 27 exactly is 9, we will factor 27 as $9 \cdot 3$ and apply the multiplication property of radicals.

$\sqrt{27} = \sqrt{9 \cdot 3}$

$= \sqrt{9}\sqrt{3}$ The square root of a product is equal to the product of the square roots.

$= 3\sqrt{3}$ Find the square root of the perfect square factor: $\sqrt{9} = 3$.

As a check, recall that $\sqrt{27}$ is the number that, when squared, gives 27. If $3\sqrt{3} = \sqrt{27}$, then $(3\sqrt{3})^2$ should be equal to 27.

$(3\sqrt{3})^2 = (3)^2(\sqrt{3})^2$ Use the power of a product rule for exponents: Raise each factor of the product $3\sqrt{3}$ to the second power.

$= 9(3)$ $\sqrt{3}$, when squared, gives 3.

$= 27$

Self Check 2
Simplify: $\sqrt{200}$.

Answer $10\sqrt{2}$

EXAMPLE 2 Simplify: $\sqrt{600}$.

Solution Since the greatest perfect square that divides 600 is 100, we will factor 600 as $100 \cdot 6$ and apply the multiplication property of radicals.

$\sqrt{600} = \sqrt{100 \cdot 6}$

$= \sqrt{100}\sqrt{6}$ The square root of a product is equal to the product of the square roots.

$= 10\sqrt{6}$ Find the square root of the perfect square factor: $\sqrt{100} = 10$.

Check the result.

Variable expressions can also be perfect squares. For example, x^2, x^4, x^6, and x^8 are perfect squares because

$$x^2 = (x)^2, \quad x^4 = (x^2)^2, \quad x^6 = (x^3)^2, \quad \text{and} \quad x^8 = (x^4)^2$$

Perfect squares like these are used to simplify square roots involving variable radicands.

EXAMPLE 3 Simplify: $\sqrt{x^3}$.

Solution To write $\sqrt{x^3}$ in simplified form, we factor x^3 into two factors, one of which is the greatest perfect square that divides x^3. The greatest perfect square that divides x^3 is x^2, so such a factorization is $x^3 = x^2 \cdot x$. We then proceed as follows:

$$\sqrt{x^3} = \sqrt{x^2 \cdot x}$$

$= \sqrt{\mathbf{x^2}}\sqrt{x}$ The square root of a product is equal to the product of the square roots.

$= \mathbf{x}\sqrt{x}$ Find the square root of the perfect square factor: $\sqrt{x^2} = x$.

As a check, recall that $\sqrt{x^3}$, when squared, gives x^3. If $x\sqrt{x} = \sqrt{x^3}$, then $(x\sqrt{x})^2$ should be equal to x^3.

$(x\sqrt{x})^2 = (x)^2(\sqrt{x})^2$ Raise each factor of the product $x\sqrt{x}$ to the 2nd power.

$= x^2(x)$ $\sqrt{x}$, when squared, gives x.

$= x^3$ Keep the base x and add the exponents.

Self Check 3
Simplify: $\sqrt{y^5}$.

Answer $y^2\sqrt{y}$

EXAMPLE 4 Simplify: $-7\sqrt{8m}$.

Solution We first write $\sqrt{8m}$ in simplified form and then multiply the result by -7. By inspection, we see that the radicand, $8m$, has a perfect square factor of 4. We can write $8m$ in factored form as $4 \cdot 2m$.

$$-7\sqrt{8m} = -7\sqrt{4 \cdot 2m}$$

$= -7\sqrt{\mathbf{4}}\sqrt{2m}$ The square root of a product is equal to the product of the square roots.

$= -7(\mathbf{2})\sqrt{2m}$ Find the square root of the perfect square factor: $\sqrt{4} = 2$.

$= -14\sqrt{2m}$ Multiply: $-7(2) = -14$.

Self Check 4
Simplify: $-2\sqrt{50c}$.

Answer $-10\sqrt{2c}$

! COMMENT When writing radical expressions such as $-14\sqrt{2m}$, be sure to extend the radical symbol completely over $2m$, because the expressions $-14\sqrt{2m}$ and $-14\sqrt{2}m$ are not the same. Similar care should be taken when writing expressions such as $\sqrt{3}x$. To avoid any misinterpretation, $\sqrt{3}x$ can be written as $x\sqrt{3}$.

EXAMPLE 5 Simplify: $\sqrt{72x^3}$.

Solution We factor $72x^3$ into two factors, one of which is the greatest perfect square that divides $72x^3$. Since the greatest perfect square that divides $72x^3$ is $36x^2$, such a factorization is $72x^3 = 36x^2 \cdot 2x$. We now use the multiplication property of radicals to get

$$\sqrt{72x^3} = \sqrt{36x^2 \cdot 2x}$$

$= \sqrt{\mathbf{36x^2}}\sqrt{2x}$ The square root of a product is equal to the product of the square roots.

$= \mathbf{6x}\sqrt{2x}$ Find the square root of the perfect square factor: $\sqrt{36x^2} = 6x$.

Self Check 5
Simplify: $\sqrt{48y^3}$.

Answer $4y\sqrt{3y}$

Self Check 6
Simplify: $5q\sqrt{63p^5q^4}$.

Answer $15p^2q^3\sqrt{7p}$

EXAMPLE 6 Simplify: $3a\sqrt{288a^4b^7}$.

Solution We first simplify $\sqrt{288a^4b^7}$. Then we multiply the result by $3a$. To simplify $\sqrt{288a^4b^7}$, we look for the greatest perfect square that divides $288a^4b^7$. Because

- 144 is the greatest perfect square that divides 288,
- a^4 is the greatest perfect square that divides a^4, and
- b^6 is the greatest perfect square that divides b^7,

the factor $144a^4b^6$ is the greatest perfect square that divides $288a^4b^7$.

We can now use the multiplication property of radicals to simplify the radical.

$$3a\sqrt{288a^4b^7} = 3a\sqrt{144a^4b^6 \cdot 2b}$$

$$= 3a\sqrt{\mathbf{144a^4b^6}}\sqrt{2b}$$ The square root of a product is equal to the product of the square roots.

$$= 3a(\mathbf{12a^2b^3})\sqrt{2b}$$ $\sqrt{144a^4b^6} = 12a^2b^3$.

$$= 36a^3b^3\sqrt{2b}$$ Multiply: $3a(12a^2b^3) = 36a^3b^3$.

COMMENT The multiplication property of radicals applies to the square root of the product of two numbers. There is no such property for sums or differences. To illustrate this, we consider these correct simplifications:

$$\sqrt{9 + 16} = \sqrt{25} = 5 \qquad \text{and} \qquad \sqrt{25 - 16} = \sqrt{9} = 3$$

It is incorrect to write

$$\sqrt{9 + 16} = \sqrt{9} + \sqrt{16} = 3 + 4 = 7 \qquad \text{or} \qquad \sqrt{25 - 16} = \sqrt{25} - \sqrt{16} = 5 - 4 = 1$$

Thus, $\sqrt{a + b} \neq \sqrt{a} + \sqrt{b}$ and $\sqrt{a - b} \neq \sqrt{a} - \sqrt{b}$.

The division property of radicals

To introduce the second property of radicals, we consider these examples.

$$\sqrt{\frac{100}{25}} = \sqrt{4} = 2 \qquad \text{and} \qquad \frac{\sqrt{100}}{\sqrt{25}} = \frac{10}{5} = 2$$

Read as "the square root of 100 divided by the square root of 25."

Since the answer is 2 in each case,

$$\sqrt{\frac{100}{25}} = \frac{\sqrt{100}}{\sqrt{25}}$$

Likewise,

$$\sqrt{\frac{36}{4}} = \sqrt{9} = 3 \qquad \text{and} \qquad \frac{\sqrt{36}}{\sqrt{4}} = \frac{6}{2} = 3$$

Since the answer is 3 in each case,

$$\sqrt{\frac{36}{4}} = \frac{\sqrt{36}}{\sqrt{4}}$$

These results illustrate the **division property of radicals.**

The division property of radicals

If a and b represent real numbers, with $a \geq 0$ and $b > 0$,

$$\sqrt{\frac{a}{b}} = \frac{\sqrt{a}}{\sqrt{b}}$$

In words, *the square root of the quotient of two numbers is the quotient of their square roots.*

We can use the division property of radicals to simplify radicals that have fractions in their radicands. For example,

$$\sqrt{\frac{59}{49}} = \frac{\sqrt{59}}{\sqrt{49}}$$

$$= \frac{\sqrt{59}}{7} \qquad \sqrt{49} = 7.$$

EXAMPLE 7 Simplify: $\sqrt{\frac{108}{25}}$.

Solution

$\sqrt{\frac{108}{25}} = \frac{\sqrt{108}}{\sqrt{25}}$ The square root of a quotient is equal to the quotient of the square roots.

$= \frac{\sqrt{36 \cdot 3}}{5}$ Factor 108 using the largest perfect square factor of 108, which is 36. Write $\sqrt{25}$ as 5.

$= \frac{\sqrt{36}\sqrt{3}}{5}$ The square root of a product is equal to the product of the square roots.

$= \frac{6\sqrt{3}}{5}$ This result can also be written as $\frac{6}{5}\sqrt{3}$.

Self Check 7

Simplify: $\sqrt{\frac{20}{81}}$.

Answer $\frac{2\sqrt{5}}{9}$

EXAMPLE 8 Simplify: $\sqrt{\frac{44x^3}{9xy^2}}$.

Solution

$\sqrt{\frac{44x^3}{9xy^2}} = \sqrt{\frac{44x^2}{9y^2}}$ Simplify the fraction by removing the common factor of x: $\frac{44x^3}{9xy^2} = \frac{44x^2\overset{1}{\cancel{x}}}{9\underset{1}{\cancel{x}}y^2} = \frac{44x^2}{9y^2}$.

$= \frac{\sqrt{44x^2}}{\sqrt{9y^2}}$ The square root of a quotient is equal to the quotient of the square roots.

$= \frac{\sqrt{4x^2}\sqrt{11}}{\sqrt{9y^2}}$ Factor $44x^2$ as $4x^2 \cdot 11$. The square root of a product is equal to the product of the square roots.

$= \frac{2x\sqrt{11}}{3y}$ $\sqrt{4x^2} = 2x$ and $\sqrt{9y^2} = 3y$.

Self Check 8

Simplify: $\sqrt{\frac{99b^3}{16a^2b}}$.

Answer $\frac{3b\sqrt{11}}{4a}$

Simplifying cube roots

The multiplication and division properties of radicals are also true for cube roots and higher. To simplify cube roots, we must know the following natural-number **perfect cubes:**

8, 27, 64, 125, 216, 343, 512, 729, 1,000

> **Properties of radicals**
> If a and b represent real numbers,
>
> $$\sqrt[3]{ab} = \sqrt[3]{a}\sqrt[3]{b} \qquad \sqrt[3]{\frac{a}{b}} = \frac{\sqrt[3]{a}}{\sqrt[3]{b}}, \quad \text{provided } b \neq 0.$$

Self Check 9
Simplify: $\sqrt[3]{250}$.

EXAMPLE 9 Simplify: $\sqrt[3]{54}$.

Solution The greatest perfect cube that divides 54 is 27.

$\sqrt[3]{54} = \sqrt[3]{27 \cdot 2}$ Factor 54: $54 = 27 \cdot 2$.

$= \sqrt[3]{\mathbf{27}}\sqrt[3]{2}$ The cube root of a product is equal to the product of the cube roots.

$= \mathbf{3}\sqrt[3]{2}$ Find the cube root of the perfect cube factor: $\sqrt[3]{27} = 3$.

As a check, we note that $\sqrt[3]{54}$ is the number that, when cubed, gives 54. If $3\sqrt[3]{2} = \sqrt[3]{54}$, then $(3\sqrt[3]{2})^3$ will be equal to 54.

$(3\sqrt[3]{2})^3 = (3)^3(\sqrt[3]{2})^3$ Raise each factor of the product $3\sqrt[3]{2}$ to the third power.

$= 27(2)$ $\sqrt[3]{2}$, when cubed, gives 2.

$= 54$

Answer $5\sqrt[3]{2}$

Variable expressions can also be perfect cubes. For example, x^3, x^6, x^9, and x^{12} are perfect cubes because

$$x^3 = (x)^3, \quad x^6 = (x^2)^3, \quad x^9 = (x^3)^3, \quad \text{and} \quad x^{12} = (x^4)^3$$

Perfect cubes like these are used to simplify cube roots involving variable radicands.

Self Check 10
Simplify:
a. $\sqrt[3]{54a^3b^5}$
b. $\sqrt[3]{\frac{27q^5}{64p^3}}$

EXAMPLE 10 Simplify: **a.** $\sqrt[3]{16x^3y^4}$ and **b.** $\sqrt[3]{\frac{64n^4}{27m^3}}$.

Solution

a. We factor $16x^3y^4$ into two factors, one of which is the greatest perfect cube that divides $16x^3y^4$. Since $8x^3y^3$ is the greatest perfect cube that divides $16x^3y^4$, the factorization is $16x^3y^4 = 8x^3y^3 \cdot 2y$.

$\sqrt[3]{16x^3y^4} = \sqrt[3]{8x^3y^3 \cdot 2y}$

$= \sqrt[3]{\mathbf{8x^3y^3}}\sqrt[3]{2y}$ The cube root of a product is equal to the product of the cube roots.

$= \mathbf{2xy}\sqrt[3]{2y}$ Find the cube root of the perfect cube factor: $\sqrt[3]{8x^3y^3} = 2xy$.

b. $\sqrt[3]{\frac{64n^4}{27m^3}} = \frac{\sqrt[3]{64n^4}}{\sqrt[3]{\mathbf{27m^3}}}$ The cube root of a quotient is equal to the quotient of the cube roots.

$= \frac{\sqrt[3]{\mathbf{64n^3}}\sqrt[3]{n}}{\mathbf{3m}}$ In the numerator, use the multiplication property of radicals. In the denominator, $\sqrt[3]{27m^3} = 3m$.

$= \frac{\mathbf{4n}\sqrt[3]{n}}{3m}$ $\sqrt[3]{64n^3} = 4n$.

Answers **a.** $3ab\sqrt[3]{2b^2}$, **b.** $\frac{3q\sqrt[3]{q^2}}{4p}$

Section 14.3 STUDY SET

VOCABULARY *Fill in the blanks.*

1. Squares of integers such as 4, 9, and 16 are called ________ squares.
2. Cubes of integers such as 8, 27, and 64 are called perfect ________.
3. "To ________ $\sqrt{8}$" means to write it as $2\sqrt{2}$.
4. The word *product* is associated with the operation of ____________ and the word *quotient* with ________.

CONCEPTS

5. Fill in the blanks.
   a. The square root of the product of two positive numbers is equal to the ________ of their square roots. In symbols,
   $$\sqrt{ab} = \square$$
   b. The square root of the quotient of two positive numbers is equal to the ________ of their square roots. In symbols,
   $$\sqrt{\frac{a}{b}} = \square$$
6. Which of the perfect squares 1, 4, 9, 16, 25, 36, 49, 64, 81, and 100 is the *largest* factor of the given number?
   a. 20 b. 45
   c. 72 d. 98

Determine what is wrong with each solution.

7. Simplify: $\sqrt{20}$.
$$\begin{aligned}\sqrt{20} &= \sqrt{16+4}\\ &= \sqrt{16}+\sqrt{4}\\ &= 4+2\\ &= 6\end{aligned}$$
8. Simplify: $\sqrt{27}$.
$$\begin{aligned}\sqrt{27} &= \sqrt{36-9}\\ &= \sqrt{36}-\sqrt{9}\\ &= 6-3\\ &= 3\end{aligned}$$
9. A crossword puzzle in a newspaper occupies an area of 28 square inches. See the illustration in the next column.
   a. Express the exact length of a side of the square-shaped puzzle in simplified radical form.
   b. What is the length of a side to the nearest tenth of an inch?

10. See the illustration.
   a. What is the exact length of a side of the cube written in simplified radical form?
   b. What is the length of a side to the nearest tenth of a foot?

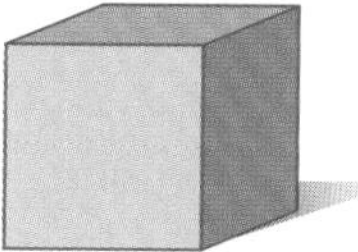

Volume = 40 ft^3

Evaluate the expression $\sqrt{b^2 - 4ac}$ for the given values. Perform the operations within the radical and simplify the radical.

11. $a = 5, b = 10, c = 3$
12. $a = 2, b = 6, c = 1$
13. $a = -1, b = 6, c = 9$
14. $a = 1, b = -2, c = -11$

NOTATION *Complete each solution.*

15.
$$\begin{aligned}\sqrt{80a^3b^2} &= \sqrt{16\cdot \square \cdot a^2\cdot a\cdot b^2}\\ &= \sqrt{16a^2b^2\cdot \square}\\ &= \sqrt{\square}\sqrt{5a}\\ &= 4ab\sqrt{5a}\end{aligned}$$
16.
$$\begin{aligned}\sqrt[3]{\frac{27a^4b^2}{64}} &= \frac{\sqrt[3]{27a^4b^2}}{\square}\\ &= \frac{\sqrt[3]{27a^3\cdot \square}}{\sqrt[3]{64}}\\ &= \frac{\sqrt[3]{\square}\sqrt[3]{ab^2}}{\sqrt[3]{64}}\\ &= \frac{3a\sqrt[3]{ab^2}}{4}\end{aligned}$$
17. What operation is indicated between the two radicals in the expression $\sqrt{4}\sqrt{3}$?
18. Fill in each blank to make a true statement.
   a. $16x^2 = (\quad)^2$ b. $27a^3b^6 = (\quad)^3$
19. Write each expression in a better form.
   a. $\sqrt{5}\cdot 2$ b. $\sqrt{7}a$
   c. $9\sqrt{x^2}\sqrt{6}$ d. $\sqrt{y}\sqrt{25z^4}$

20. a. Explain the difference between $\sqrt{5}x$ and $\sqrt{5x}$.

b. Why do you think it is better to write $\sqrt{5}x$ as $x\sqrt{5}$?

PRACTICE *Simplify each radical. Assume that all variables represent positive numbers.*

21. $\sqrt{20}$ **22.** $\sqrt{18}$
23. $\sqrt{50}$ **24.** $\sqrt{75}$
25. $\sqrt{45}$ **26.** $\sqrt{54}$
27. $\sqrt{98}$ **28.** $\sqrt{147}$
29. $\sqrt{48}$ **30.** $\sqrt{128}$
31. $-\sqrt{200}$ **32.** $-\sqrt{300}$
33. $\sqrt{192}$ **34.** $\sqrt{88}$
35. $\sqrt{250}$ **36.** $\sqrt{1{,}000}$
37. $2\sqrt{24}$ **38.** $3\sqrt{32}$
39. $-2\sqrt{28}$ **40.** $-3\sqrt{72}$
41. $\sqrt{n^3}$ **42.** $\sqrt{x^5}$
43. $\sqrt{4k}$ **44.** $\sqrt{9p}$
45. $\sqrt{12x}$ **46.** $\sqrt{20y}$
47. $6\sqrt{75t}$ **48.** $2\sqrt{24s}$
49. $\sqrt{25x^3}$ **50.** $\sqrt{36y^3}$
51. $\sqrt{a^2b}$ **52.** $\sqrt{rs^4}$
53. $\sqrt{9x^4y}$ **54.** $\sqrt{16xy^2}$
55. $\frac{1}{5}x^2y\sqrt{50x^2y^2}$ **56.** $\frac{1}{5}x^5y\sqrt{75x^3y^2}$
57. $-12x\sqrt{16x^2y^3}$
58. $-4x^5y^3\sqrt{36x^3y^3}$
59. $-\frac{2}{5}\sqrt{80mn^4}$
60. $\frac{5}{6}\sqrt{180ab^6}$

Write each quotient as the quotient of two radicals and simplify.

61. $\sqrt{\frac{25}{9}}$ **62.** $\sqrt{\frac{36}{49}}$
63. $\sqrt{\frac{81}{64}}$ **64.** $\sqrt{\frac{121}{144}}$
65. $\sqrt{\frac{26}{25}}$ **66.** $\sqrt{\frac{17}{169}}$
67. $-\sqrt{\frac{20}{49}}$ **68.** $-\sqrt{\frac{50}{9}}$
69. $\sqrt{\frac{48}{81}}$ **70.** $\sqrt{\frac{27}{64}}$
71. $\sqrt{\frac{32}{25}}$ **72.** $\sqrt{\frac{75}{16}}$

Simplify each expression. All variables represent positive numbers.

73. $\sqrt{\frac{72x^3}{y^2}}$ **74.** $\sqrt{\frac{108b^2}{d^4}}$
75. $\sqrt{\frac{125n^5}{64n}}$ **76.** $\sqrt{\frac{72q^7}{25q^3}}$
77. $\sqrt{\frac{128m^3n^5}{81mn^7}}$ **78.** $\sqrt{\frac{75p^3q^2}{p^5q^4}}$
79. $\sqrt{\frac{12r^7s^7}{r^5s^2}}$ **80.** $\sqrt{\frac{m^2n^9}{100mn^3}}$

Simplify each cube root.

81. $\sqrt[3]{24}$ **82.** $\sqrt[3]{32}$
83. $\sqrt[3]{-128}$ **84.** $\sqrt[3]{-250}$
85. $\sqrt[3]{8x^3}$ **86.** $\sqrt[3]{27x^3}$
87. $\sqrt[3]{-64x^5}$ **88.** $\sqrt[3]{-16x^4}$
89. $\sqrt[3]{54x^3z^6}$ **90.** $\sqrt[3]{-24x^3y^5}$
91. $\sqrt[3]{-81x^2y^3}$ **92.** $\sqrt[3]{81y^2z^3}$
93. $\sqrt[3]{\frac{27m^3}{8n^6}}$ **94.** $\sqrt[3]{\frac{125t^9}{27s^6}}$
95. $\sqrt[3]{\frac{r^4s^5}{1{,}000t^3}}$ **96.** $\sqrt[3]{\frac{54m^4n^3}{r^3s^6}}$

APPLICATIONS *Use a calculator to help solve each problem.*

97. AMUSEMENT PARK RIDES The illustration on the next page shows a pirate ship ride. The time (in seconds) it takes to swing from one extreme to the other is given by

$$t = \pi\sqrt{\frac{L}{32}}$$

a. Find t and express it in simplified radical form. Leave π in your answer.

b. Express your answer to part a as a decimal. Round to the nearest tenth of a second.

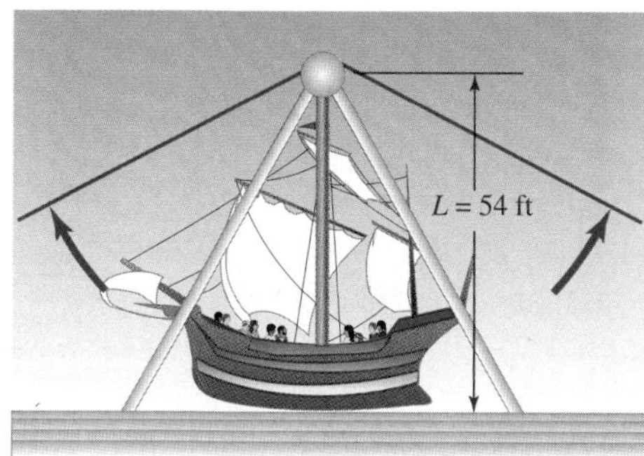

98. HERB GARDENS The perimeter of the herb garden shown below is given by

$$p = 2\pi\sqrt{\frac{a^2 + b^2}{2}}$$

a. Find the length of fencing (in meters) needed to enclose the garden. Express the result in simplified radical form. Leave π in your answer.

b. Express the result from part a as a decimal. Round to the nearest tenth of a meter.

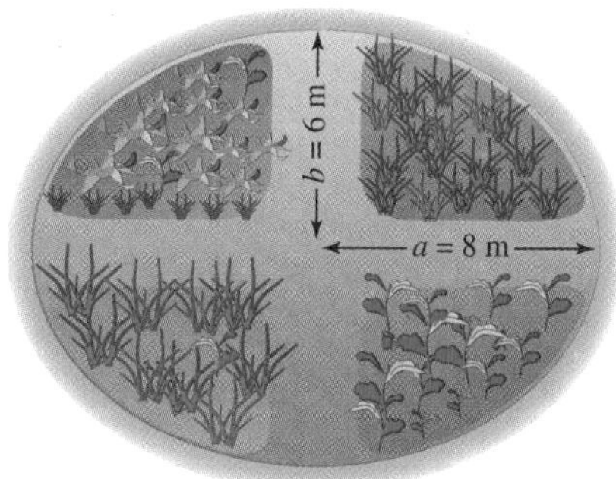

99. ARCHAEOLOGY Framed grids, made up of 20 cm × 20 cm squares, are often used to record the location of artifacts found during an excavation.

a. Use the Pythagorean theorem to determine the *exact* distance between a piece of pottery found at point A and a cooking utensil found at point B.

b. Approximate the distance to the nearest tenth of a centimeter.

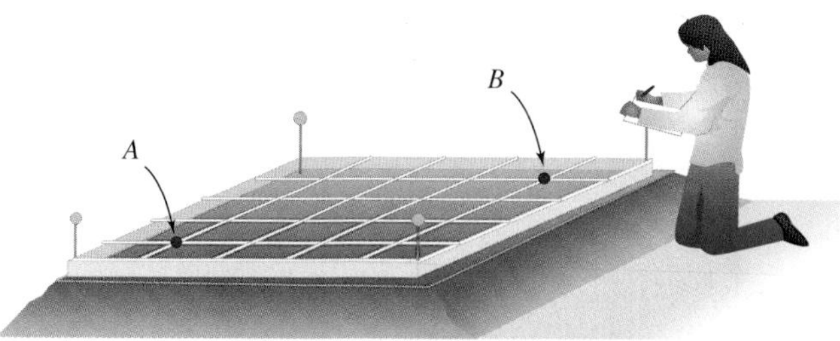

100. ENVIRONMENTAL PROTECTION A new campground is to be constructed 2 miles from a major highway, as shown below. The proposed entrance, although longer than the direct route, bypasses a grove of old-growth redwood trees.

a. Use the Pythagorean theorem to find the length of the proposed entrance road. Express the result as a radical in simplified form.

b. Express the result from part a as a decimal. Round to the nearest hundredth of a mile.

c. How much longer is the proposed entrance as compared to the direct route into the campground?

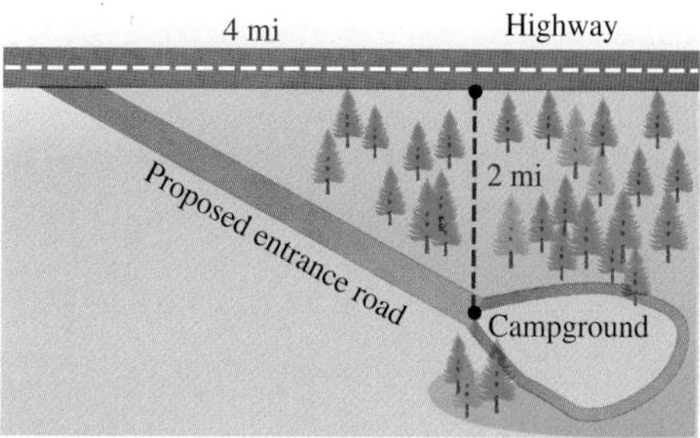

WRITING

101. State the multiplication property of radicals.

102. When comparing $\sqrt{8}$ and $2\sqrt{2}$, why is $2\sqrt{2}$ called simplified radical form?

REVIEW

103. Multiply: $(-2a^3)(3a^2)$.

104. Find the slope of the line passing through $(-6, 0)$ and $(0, -4)$.

105. Write the equation of the line passing through $(0, 3)$ with slope -2.

106. Solve: $-x = -5$.

107. Solve: $-x > -5$.

108. What is the slope of a line perpendicular to a line with a slope of 2?

14.4 Adding and Subtracting Radical Expressions

- Combining like radicals
- Combining expressions containing cube roots

We have discussed how to add and subtract like terms. In this section, we will discuss how to add and subtract expressions that contain like radicals.

Combining like radicals

When adding monomials, we can often combine *like terms.* For example,

$$3x + 5x = (\mathbf{3 + 5})x \quad \text{Use the distributive property.}$$
$$= \mathbf{8}x \quad \text{Perform the addition.}$$

COMMENT The expression $3x + 5y$ cannot be simplified, because $3x$ and $5y$ are not like terms.

It is often possible to combine terms that contain *like radicals.*

Like radicals

Radicals are called **like radicals** when they have the same index and the same radicand.

Like radicals	**Unlike radicals**
$3\sqrt{2}$ and $5\sqrt{2}$	$3\sqrt{\mathbf{2}}$ and $5\sqrt{\mathbf{3}}$
The same index and the same radicand	The same index but different radicands
$5x\sqrt{3y}$ and $-2x\sqrt{3y}$	$5x\sqrt[3]{3y}$ and $-2x\sqrt{3y}$
The same index and the same radicand	The same radicands but a different index

Expressions that contain like radicals can be combined by addition and subtraction. For example, we have

$$3\sqrt{\mathbf{2}} + 5\sqrt{\mathbf{2}} = (3 + 5)\sqrt{\mathbf{2}} \quad \text{Use the distributive property.}$$
$$= 8\sqrt{2} \quad \text{Perform the addition.}$$

Likewise, we can simplify the expression $5x\sqrt{3y} - 2x\sqrt{3y}$.

$$5x\sqrt{\mathbf{3y}} - 2x\sqrt{\mathbf{3y}} = (5x - 2x)\sqrt{\mathbf{3y}} \quad \text{Use the distributive property.}$$
$$= 3x\sqrt{3y} \quad \text{Perform the subtraction: } 5x - 2x = 3x.$$

COMMENT The expression $3\sqrt{\mathbf{2}} + 5\sqrt{\mathbf{3}}$ cannot be simplified, because the radicals are unlike. For the same reason, we cannot simplify $5x\sqrt[3]{3y} - 2x\sqrt{3y}$.

EXAMPLE 1 Simplify: **a.** $\sqrt{6} + 6 + 5\sqrt{6}$ and **b.** $-2\sqrt{m} - 3\sqrt{m}$.

Solution

a. The expression contains three terms: $\sqrt{6}$, 6, and $5\sqrt{6}$. The first and third terms have like radicals, and they can be combined.

Self Check 1
Simplify:
a. $\sqrt{7} + 7 + 7\sqrt{7}$
b. $24\sqrt{m} - 25\sqrt{m}$

$$\sqrt{6} + 6 + 5\sqrt{6} = 6 + (\mathbf{1\sqrt{6} + 5\sqrt{6}})$$ Group the expressions with like radicals. Write $\sqrt{6}$ as $1\sqrt{6}$.

$$= 6 + (\mathbf{1 + 5})\sqrt{\mathbf{6}}$$ Use the distributive property.

$$= 6 + 6\sqrt{6}$$ Perform the addition.

Note that 6 and $6\sqrt{6}$ do not contain like radicals and cannot be combined.

b. The expressions $-2\sqrt{m}$ and $-3\sqrt{m}$ contain like radicals. We can combine them.

$$-2\sqrt{\mathbf{m}} - 3\sqrt{\mathbf{m}} = (-2 - 3)\sqrt{\mathbf{m}}$$ Use the distributive property.

$$= -5\sqrt{m}$$ Perform the subtraction: $-2 - 3 = -5$.

Answers **a.** $7 + 8\sqrt{7}$, **b.** $-\sqrt{m}$

If a sum or difference involves radicals that are unlike, make sure each one is written in simplified form. After doing so, like radicals may result that can be combined.

EXAMPLE 2 Simplify: $3\sqrt{18} + 5\sqrt{8}$.

Solution The radical $\sqrt{18}$ is not in simplified form, because 18 has a perfect square factor of 9. The radical $\sqrt{8}$ is not in simplified form either, because 8 has a perfect square factor of 4. To simplify the radicals and add the expressions, we proceed as follows.

$$3\sqrt{18} + 5\sqrt{8} = 3\sqrt{9 \cdot 2} + 5\sqrt{4 \cdot 2}$$ Factor 18 and 8 using perfect square factors.

$$= 3\sqrt{9}\sqrt{2} + 5\sqrt{4}\sqrt{2}$$ The square root of a product is equal to the product of the square roots.

$$= 3(3)\sqrt{2} + 5(2)\sqrt{2}$$ $\sqrt{9} = 3$ and $\sqrt{4} = 2$.

$$= 9\sqrt{\mathbf{2}} + 10\sqrt{\mathbf{2}}$$ Multiply: $3(3) = 9$ and $5(2) = 10$.

$$= 19\sqrt{\mathbf{2}}$$ To combine like radicals, combine their coefficients: $9 + 10 = 19$.

Self Check 2
Simplify: $2\sqrt{50} + \sqrt{32}$.

Answer $14\sqrt{2}$

EXAMPLE 3 **Orthopedics.** Doctors sometimes use traction to align a broken bone so that a fracture can heal properly. Figure 14-16 shows how traction is applied by fixing a weight, two pulleys, and some stainless steel cable to a broken leg. How many feet of cable are used in the setup shown in the figure?

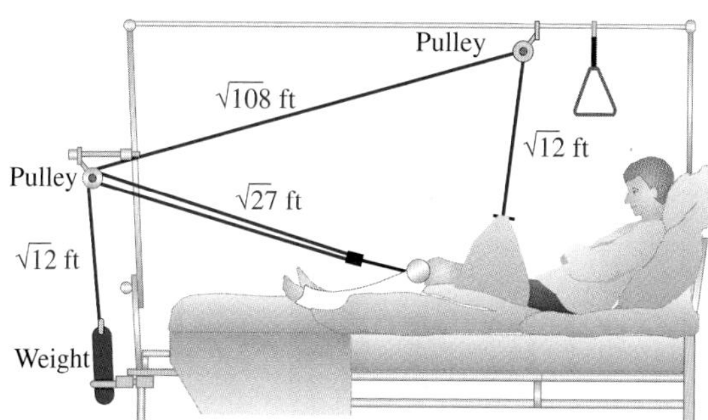

FIGURE 14-16

Solution Two segments of the cable are $\sqrt{12}$ feet long, another is $\sqrt{108}$ feet long, and two others are $\sqrt{27}$ feet long. The total number of feet of cable can be found by adding

$$2\sqrt{12} + \sqrt{108} + 2\sqrt{27}$$

Since $\sqrt{12}$, $\sqrt{108}$, and $\sqrt{27}$ are not like radicals, we cannot perform the addition at this time. First, we need to write each radical in simplified form. Then we can add any expressions that contain like radicals.

$$2\sqrt{12} + \sqrt{108} + 2\sqrt{27}$$

$= 2\sqrt{4 \cdot 3} + \sqrt{36 \cdot 3} + 2\sqrt{9 \cdot 3}$ Factor 12, 108, and 27 using perfect squares.

$= 2\sqrt{4}\sqrt{3} + \sqrt{36}\sqrt{3} + 2\sqrt{9}\sqrt{3}$ The square root of a product is equal to the product of the square roots.

$= 2(2)\sqrt{3} + 6\sqrt{3} + 2(3)\sqrt{3}$ Simplify: $\sqrt{4} = 2$, $\sqrt{36} = 6$ and $\sqrt{9} = 3$.

$= 4\sqrt{\mathbf{3}} + 6\sqrt{\mathbf{3}} + 6\sqrt{\mathbf{3}}$ Perform the multiplications.

$= 16\sqrt{\mathbf{3}}$ To combine like radicals, combine their coefficients: $4 + 6 + 6 = 16$.

The traction setup uses $16\sqrt{3}$ feet of cable.

Self Check 4
Simplify: $\sqrt{12xy^2} + \sqrt{27xy^2}$.

Answer $5y\sqrt{3x}$

EXAMPLE 4 Simplify: $\sqrt{44x^2y} + x\sqrt{99y}$.

Solution We simplify each radical and then add the expressions containing like radicals.

$$\sqrt{44x^2y} + x\sqrt{99y}$$

$= \sqrt{4x^2 \cdot 11y} + x\sqrt{9 \cdot 11y}$ Factor $44x^2y$ and $99y$.

$= \sqrt{4x^2}\sqrt{11y} + x\sqrt{9}\sqrt{11y}$ The square root of a product is equal to the product of the square roots.

$= 2x\sqrt{\mathbf{11y}} + 3x\sqrt{\mathbf{11y}}$ Simplify: $\sqrt{4x^2} = 2x$ and $\sqrt{9} = 3$.

$= 5x\sqrt{\mathbf{11y}}$ To combine like radicals, combine their coefficients: $2x + 3x = 5x$.

Self Check 5
Simplify: $\sqrt{20mn^2} - \sqrt{80m^3}$.

Answer $2n\sqrt{5m} - 4m\sqrt{5m}$

EXAMPLE 5 Simplify: $\sqrt{28x^2y} - 2\sqrt{63y^3}$.

Solution We begin by simplifying each radical.

$$\sqrt{28x^2y} - 2\sqrt{63y^3}$$

$= \sqrt{4x^2 \cdot 7y} - 2\sqrt{9y^2 \cdot 7y}$ Factor $28x^2y$ and $63y^3$.

$= \sqrt{4x^2}\sqrt{7y} - 2\sqrt{9y^2}\sqrt{7y}$ The square root of a product is equal to the product of the square roots.

$= 2x\sqrt{7y} - 2(3y)\sqrt{7y}$ $\sqrt{4x^2} = 2x$ and $\sqrt{9y^2} = 3y$.

$= \mathbf{2x\sqrt{7y} - 6y\sqrt{7y}}$

Since $2x$ and $6y$ are not like terms and therefore cannot be subtracted, the expression does not simplify further.

Self Check 6
Simplify: $\sqrt{75ab} + \sqrt{72ab}$.

Answer $5\sqrt{3ab} + 6\sqrt{2ab}$

EXAMPLE 6 Simplify: $\sqrt{27xy} + \sqrt{20xy}$.

Solution

$\sqrt{27xy} + \sqrt{20xy} = \sqrt{9 \cdot 3xy} + \sqrt{4 \cdot 5xy}$ Factor $27xy$ and $20xy$.

$= \sqrt{9}\sqrt{3xy} + \sqrt{4}\sqrt{5xy}$ The square root of a product is equal to the product of the square roots.

$= 3\sqrt{\mathbf{3xy}} + 2\sqrt{\mathbf{5xy}}$ $\sqrt{9} = 3$ and $\sqrt{4} = 2$.

Since the terms have unlike radicals, the expression does not simplify further.

EXAMPLE 7 Simplify: $\sqrt{8x} + \sqrt{3y} - \sqrt{50x} + \sqrt{27y}$.

Solution We simplify the radicals and then combine like radicals, where possible.

$$\begin{aligned}
&\sqrt{8x} + \sqrt{3y} - \sqrt{50x} + \sqrt{27y} \\
&\quad = \sqrt{4 \cdot 2x} + \sqrt{3y} - \sqrt{25 \cdot 2x} + \sqrt{9 \cdot 3y} && \text{Factor } 8x, 50x, \text{ and } 27y. \\
&\quad = \sqrt{4}\sqrt{2x} + \sqrt{3y} - \sqrt{25}\sqrt{2x} + \sqrt{9}\sqrt{3y} \\
&\quad = 2\sqrt{\mathbf{2x}} + \sqrt{3y} - 5\sqrt{\mathbf{2x}} + 3\sqrt{3y} \\
&\quad = -3\sqrt{\mathbf{2x}} + 4\sqrt{3y} && \text{Combine like radicals: } 2 - 5 = -3 \text{ and } 1 + 3 = 4.
\end{aligned}$$

Self Check 7
Simplify:
$\sqrt{32x} - \sqrt{5y} - \sqrt{200x} + \sqrt{125y}$.

Answer $-6\sqrt{2x} + 4\sqrt{5y}$

Combining expressions containing cube roots

We can extend the concepts used to combine square roots to radicals with higher order.

EXAMPLE 8 Simplify: $\sqrt[3]{81x^4} - x\sqrt[3]{24x}$.

Solution We simplify each radical and then combine like radicals.

$$\begin{aligned}
\sqrt[3]{81x^4} - x\sqrt[3]{24x} &= \sqrt[3]{27x^3 \cdot 3x} - x\sqrt[3]{8 \cdot 3x} && \text{Factor } 81x^4 \text{ and } 24x. \\
&= \sqrt[3]{27x^3}\sqrt[3]{3x} - x\sqrt[3]{8}\sqrt[3]{3x} \\
&= 3x\sqrt[3]{\mathbf{3x}} - 2x\sqrt[3]{\mathbf{3x}} \\
&= x\sqrt[3]{\mathbf{3x}} && \text{Combine like radicals: } 3x - 2x = x.
\end{aligned}$$

Self Check 8
Simplify: $\sqrt[3]{24a^4} + a\sqrt[3]{81a}$.

Answer $5a\sqrt[3]{3a}$

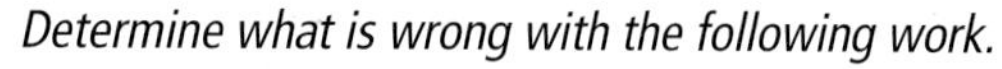

Section 14.4 STUDY SET

VOCABULARY *Fill in the blanks.*

1. Like ________, such as $2\sqrt{3}$ and $5\sqrt{3}$, have the same index and the same radicand.
2. Like ________, such as $2x$ and $5x$, have the same variables with the same exponents.
3. When $\sqrt{8}$ and $\sqrt{18}$ are written in ________ form, the results are the like radicals $2\sqrt{2}$ and $3\sqrt{2}$.
4. The expression $3\sqrt{2} + \sqrt{8} - 2$ contains three ________.

CONCEPTS *Determine whether the expressions contain like radicals.*

5. $5\sqrt{2}$ and $2\sqrt{3}$
6. $7\sqrt{3x}$ and $3\sqrt{3x}$
7. $125\sqrt[3]{13a}$ and $-\sqrt[3]{13a}$
8. $-17\sqrt[4]{5x}$ and $25\sqrt[3]{5x}$

Determine what is wrong with the following work.

9. $7\sqrt{5} - 3\sqrt{2} = 4\sqrt{3}$
10. $12\sqrt{7} + 20\sqrt{11} = 32\sqrt{18}$
11. $7 - 3\sqrt{2} = 4\sqrt{2}$
12. $12 + 20\sqrt{11} = 32\sqrt{11}$

Complete each table.

13.

x	$\sqrt{x} + \sqrt{3}$
3	
12	
27	
48	

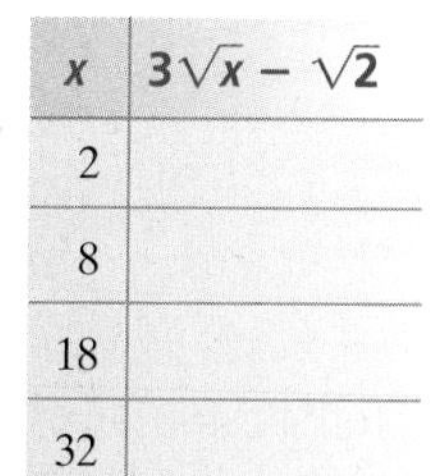

14.

x	$3\sqrt{x} - \sqrt{2}$
2	
8	
18	
32	

NOTATION *Complete each solution.*

15. Subtract: $9\sqrt{5} - 3\sqrt{20}$.

$$\begin{aligned} 9\sqrt{5} - 3\sqrt{20} &= 9\sqrt{5} - 3\sqrt{\quad \cdot 5} \\ &= 9\sqrt{5} - 3\sqrt{\quad}\sqrt{5} \\ &= 9\sqrt{5} - 3\cdot \quad \\ &= 9\sqrt{5} - \quad \\ &= \quad \end{aligned}$$

16. Add: $3\sqrt{80} + 4\sqrt{125}$.

$$\begin{aligned} 3\sqrt{80} + 4\sqrt{125} &= 3\sqrt{\quad \cdot 5} + 4\sqrt{\quad \cdot 5} \\ &= 3\sqrt{16}\quad + 4\sqrt{25}\quad \\ &= 3(\quad)\sqrt{5} + 4(5)\sqrt{5} \\ &= 12\sqrt{5} + \quad\sqrt{5} \\ &= \quad\sqrt{5} \end{aligned}$$

PRACTICE *Simplify each expression. All variables represent positive numbers.*

17. $5\sqrt{7} + 4\sqrt{7}$
18. $3\sqrt{10} + 4\sqrt{10}$
19. $\sqrt{x} - 4\sqrt{x}$
20. $\sqrt{t} - 9\sqrt{t}$
21. $5 + 3\sqrt{3} + 3\sqrt{3}$
22. $\sqrt{5} + 2 + 3\sqrt{5}$
23. $-1 + 2\sqrt{r} - 3\sqrt{r}$
24. $-8 - 5\sqrt{c} + 4\sqrt{c}$
25. $\sqrt{12} + \sqrt{27}$
26. $\sqrt{20} + \sqrt{45}$
27. $\sqrt{18} - \sqrt{8}$
28. $\sqrt{32} - \sqrt{18}$
29. $2\sqrt{45} + 2\sqrt{80}$
30. $3\sqrt{80} + 3\sqrt{125}$
31. $2\sqrt{80} - 3\sqrt{125}$
32. $3\sqrt{245} - 2\sqrt{180}$
33. $\sqrt{20} + \sqrt{180}$
34. $2\sqrt{28} + 7\sqrt{63}$
35. $\sqrt{12} - \sqrt{48}$
36. $\sqrt{48} - \sqrt{75}$
37. $\sqrt{288} - 3\sqrt{200}$
38. $\sqrt{80} - \sqrt{245}$
39. $2\sqrt{28} + 2\sqrt{112}$
40. $4\sqrt{63} + 6\sqrt{112}$
41. $\sqrt{20} + \sqrt{45} + \sqrt{80}$
42. $\sqrt{48} + \sqrt{27} + \sqrt{75}$
43. $\sqrt{200} - \sqrt{75} + \sqrt{48}$
44. $\sqrt{20} + \sqrt{80} - \sqrt{125}$
45. $8\sqrt{6} - 5\sqrt{2} - 3\sqrt{6}$
46. $3\sqrt{2} - 3\sqrt{15} - 4\sqrt{15}$
47. $\sqrt{24} + \sqrt{150} + \sqrt{240}$
48. $\sqrt{28} + \sqrt{63} + \sqrt{18}$
49. $\sqrt{48} - \sqrt{8} + \sqrt{27} - \sqrt{32}$
50. $\sqrt{162} + \sqrt{50} - \sqrt{75} - \sqrt{108}$
51. $\sqrt{2x^2} + \sqrt{8x^2}$
52. $\sqrt{3y^2} - \sqrt{12y^2}$
53. $\sqrt{2d^3} + \sqrt{8d^3}$
54. $\sqrt{3a^3} - \sqrt{12a^3}$
55. $\sqrt{18x^2y} - \sqrt{27x^2y}$
56. $\sqrt{49xy} + \sqrt{xy}$
57. $\sqrt{32x^5} - \sqrt{18x^5}$
58. $\sqrt{27xy^3} - \sqrt{48xy^3}$
59. $3\sqrt{54b^2} + 5\sqrt{24b^2}$
60. $3\sqrt{24x^4y^3} + 2\sqrt{54x^4y^3}$
61. $y\sqrt{490y} - 2\sqrt{360y^3}$
62. $3\sqrt{20x} + 2\sqrt{63y}$
63. $\sqrt{20x^3y} + \sqrt{45x^5y^3} - \sqrt{80x^7y^5}$
64. $x\sqrt{48xy^2} - y\sqrt{27x^3} + \sqrt{75x^3y^2}$
65. $\sqrt[3]{3} + \sqrt[3]{3}$
66. $\sqrt[3]{2} + 5\sqrt[3]{2}$
67. $2\sqrt[3]{x} - 3\sqrt[3]{x}$
68. $4\sqrt[3]{s} - 5\sqrt[3]{s}$
69. $\sqrt[3]{16} + \sqrt[3]{54}$
70. $\sqrt[3]{24} - \sqrt[3]{81}$
71. $\sqrt[3]{81} - \sqrt[3]{24}$
72. $\sqrt[3]{32} + \sqrt[3]{108}$
73. $\sqrt[3]{40} + \sqrt[3]{125}$
74. $\sqrt[3]{3{,}000} - \sqrt[3]{192}$
75. $\sqrt[3]{x^4} - \sqrt[3]{x^7}$
76. $\sqrt[3]{8x^5} + \sqrt[3]{27x^8}$
77. $\sqrt[3]{192x^4y^5} - \sqrt[3]{24x^4y^5}$
78. $\sqrt[3]{24a^5b^4} + \sqrt[3]{81a^5b^4}$
79. $\sqrt[3]{135x^7y^4} - \sqrt[3]{40x^7y^4}$
80. $\sqrt[3]{56a^4b^5} + \sqrt[3]{7a^4b^5}$

APPLICATIONS

81. ANATOMY Determine the length of the patient's arm in the illustration on the next page if he lets it fall to his side.

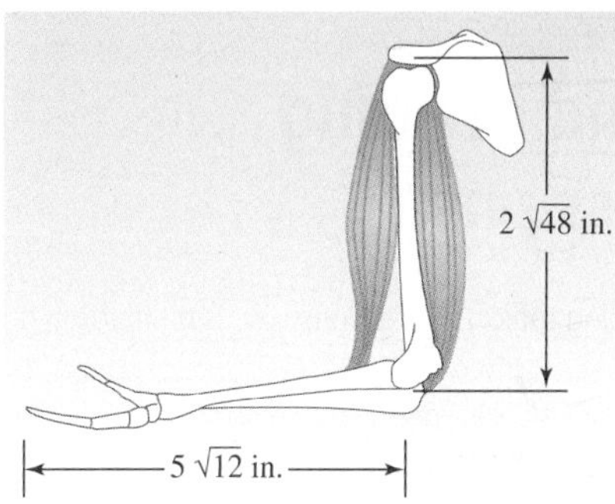

82. PLAYGROUND EQUIPMENT Find the total length of pipe necessary to construct the frame of the swing set shown below.

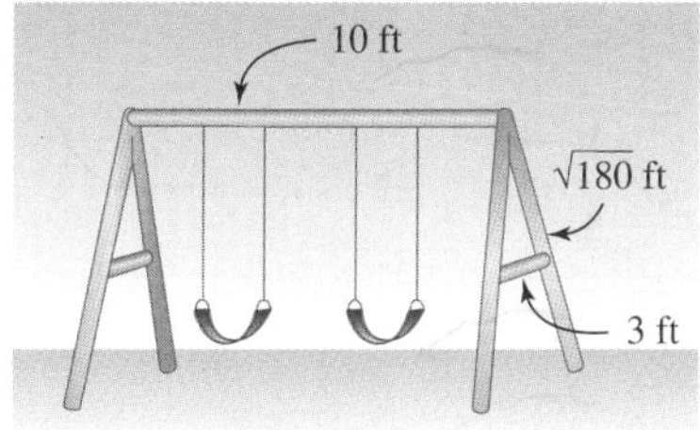

83. BLUEPRINTS What is the length of the motor on the machine shown below?

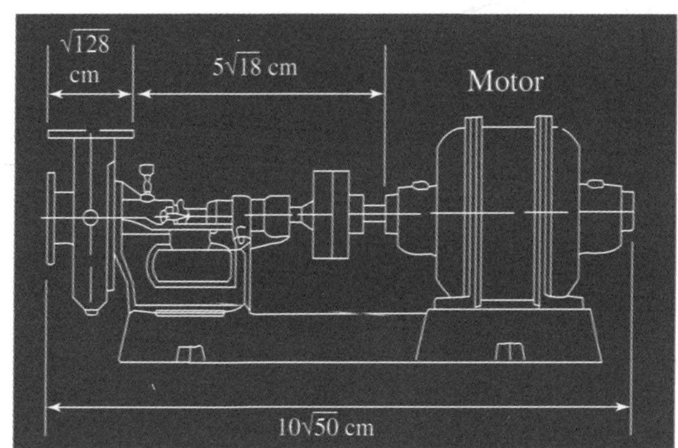

84. TENTS The length of a center support pole for the tents shown below is given by the formula

$$l = 0.5s\sqrt{3}$$

where s is the length of the side of the tent. Find the total length of the four poles needed for the parents' and children's tents.

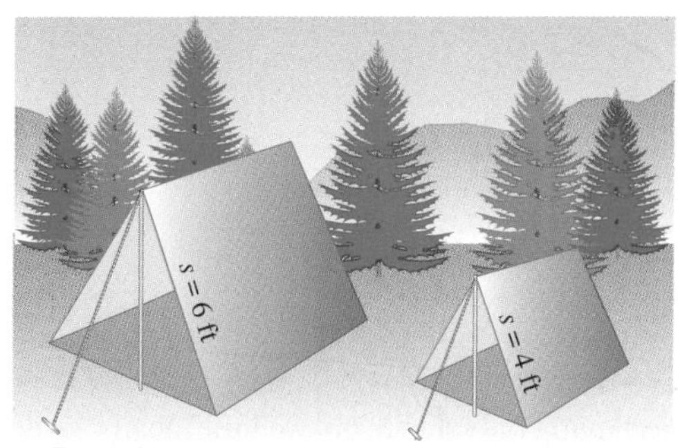

85. FENCING Find the number of feet of fencing needed to enclose the swimming pool complex shown below.

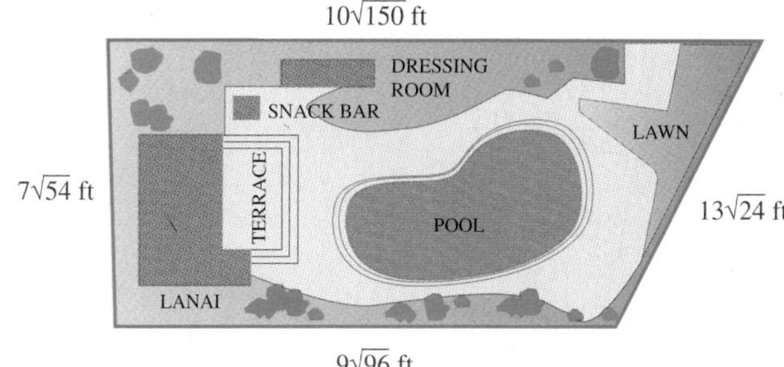

86. HARDWARE Find the difference in the lengths of the arms of the door-closing device shown below.

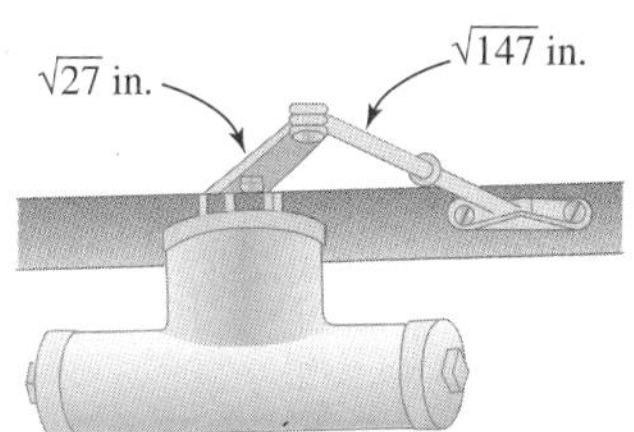

WRITING

87. Explain why $\sqrt{3} + \sqrt{2}$ cannot be combined.

88. Explain why $\sqrt{4x}$ and $\sqrt[3]{4x}$ cannot be combined.

REVIEW *Simplify each expression. Write each answer without using negative exponents.*

89. 3^{-2}

90. $\frac{1}{3^{-2}}$

91. -3^2

92. -3^{-2}

93. x^{-3}

94. $\frac{1}{x^{-3}}$

95. 3^0

96. x^0

14.5 Multiplying and Dividing Radical Expressions

- Multiplying radical expressions • Dividing radical expressions • Rationalizing denominators

In this section, we will discuss the methods used to multiply and divide radical expressions.

Multiplying radical expressions

Recall that the *product of the square roots of two nonnegative numbers is equal to the square root of the product of those numbers.* For example,

$$\sqrt{2}\sqrt{8} = \sqrt{2 \cdot 8} = \sqrt{16} = 4 \qquad \sqrt{3}\sqrt{27} = \sqrt{3 \cdot 27} = \sqrt{81} = 9 \qquad \sqrt{x}\sqrt{x^3} = \sqrt{x \cdot x^3} = \sqrt{x^4} = x^2$$

Likewise, the *product of the cube roots of two numbers is equal to the cube root of the product of those numbers.* For example,

$$\sqrt[3]{2}\sqrt[3]{4} = \sqrt[3]{2 \cdot 4} = \sqrt[3]{8} = 2 \qquad \sqrt[3]{4}\sqrt[3]{16} = \sqrt[3]{4 \cdot 16} = \sqrt[3]{64} = 4 \qquad \sqrt[3]{3x^2}\sqrt[3]{9x} = \sqrt[3]{3x^2 \cdot 9x} = \sqrt[3]{27x^3} = 3x$$

These examples illustrate that radical expressions with the same index can be multiplied.

Self Check 1
Multiply:
a. $\sqrt{5}\sqrt{3}$
b. $\sqrt{8}\sqrt{9}$
c. $\sqrt[3]{6}\sqrt[3]{9}$

EXAMPLE 1 Multiply: **a.** $\sqrt{3}\,\sqrt{2}$, **b.** $\sqrt{6}\,\sqrt{8}$, and **c.** $\sqrt[3]{4}\,\sqrt[3]{10}$.

Solution

a. $\sqrt{3}\sqrt{2} = \sqrt{3 \cdot 2}$ The product of the square roots of two numbers is equal to the square root of the product of those numbers.

$= \sqrt{6}$ Perform the multiplication within the radical.

b. $\sqrt{6}\sqrt{8} = \sqrt{6 \cdot 8}$ The product of two square roots is equal to the square root of the product.

$= \sqrt{48}$ Perform the multiplication within the radical. Note that this radical can be simplified.

$= \sqrt{16}\sqrt{3}$ Factor 48 as $16 \cdot 3$.

$= 4\sqrt{3}$ $\sqrt{16} = 4$.

c. $\sqrt[3]{4}\sqrt[3]{10} = \sqrt[3]{4 \cdot 10}$ The product of two cube roots is equal to the cube root of the product.

$= \sqrt[3]{40}$ Perform the multiplication within the radical.

$= \sqrt[3]{8}\sqrt[3]{5}$ $\sqrt[3]{40} = \sqrt[3]{8 \cdot 5} = \sqrt[3]{8}\sqrt[3]{5}$.

$= 2\sqrt[3]{5}$ $\sqrt[3]{8} = 2$.

Answers **a.** $\sqrt{15}$, **b.** $6\sqrt{2}$, **c.** $3\sqrt[3]{2}$.

To multiply radical expressions having only one term, we multiply the coefficients and multiply the radicals separately and then simplify the result, when possible.

EXAMPLE 2 Multiply: **a.** $3\sqrt{6}$ by $4\sqrt{3}$ and **b.** $-2\sqrt[3]{7x}$ by $6\sqrt[3]{49x^2}$.

Solution The commutative and associative properties enable us to multiply the coefficients and the radicals separately.

a. $3\sqrt{6} \cdot 4\sqrt{3} = 3(4)\sqrt{6}\sqrt{3}$ Write the coefficients together and the radicals together.

$= 12\sqrt{18}$ Multiply the coefficients and multiply the radicals.

$= 12\sqrt{9}\sqrt{2}$ $\sqrt{18} = \sqrt{9 \cdot 2} = \sqrt{9}\sqrt{2}$.

$= 12(3)\sqrt{2}$ $\sqrt{9} = 3$.

$= 36\sqrt{2}$ Perform the multiplication: $12(3) = 36$.

b. $-2\sqrt[3]{7x} \cdot 6\sqrt[3]{49x^2} = -2(6)\sqrt[3]{7x}\sqrt[3]{49x^2}$ Write the coefficients together and the radicals together.

$= -12\sqrt[3]{7x \cdot 49x^2}$ Multiply the coefficients and multiply the radicals.

$= -12\sqrt[3]{343x^3}$ Perform the multiplication within the radical.

$= -12(7x)$ $\sqrt[3]{343x^3} = 7x$.

$= -84x$ Multiply.

Self Check 2
Multiply:
a. $(2\sqrt{2x})(-3\sqrt{3x})$
b. $(5\sqrt[3]{2})(2\sqrt[3]{4})$

Answers **a.** $-6x\sqrt{6}$, **b.** 20

EXAMPLE 3 Find $(2\sqrt{5})^2$.

Solution Recall that a power is used to indicate repeated multiplication.

$(2\sqrt{5})^2 = 2\sqrt{5} \cdot 2\sqrt{5}$ Write $2\sqrt{5}$ as a factor two times.

$= 2(2)\sqrt{5}\sqrt{5}$ Multiply the coefficients and the radicals separately.

$= 4\sqrt{5 \cdot 5}$ The product of two square roots is equal to the square root of the product.

$= 4\sqrt{25}$ Perform the multiplication within the radical.

$= 4 \cdot 5$ $\sqrt{25} = 5$.

$= 20$

Self Check 3
Find $(3\sqrt[3]{-2})^3$

Answer -54

Recall that to multiply a polynomial by a monomial, we use the distributive property. We use the same technique to multiply a radical expression that has two or more terms by a radical expression that has only one term.

EXAMPLE 4 Multiply: **a.** $\sqrt{2x}(\sqrt{6x} + \sqrt{8x})$ and **b.** $\sqrt[3]{3}(\sqrt[3]{9} - 2)$

Solution

a. $\sqrt{2x}(\sqrt{6x} + \sqrt{8x}) = \sqrt{2x}\sqrt{6x} + \sqrt{2x}\sqrt{8x}$ Distribute the multiplication by $\sqrt{2x}$.

$= \sqrt{12x^2} + \sqrt{16x^2}$ The product of two square roots is equal to the square root of the product.

$= \sqrt{4x^2 \cdot 3} + \sqrt{16x^2}$ Factor $12x^2$ as $4x^2 \cdot 3$.

$= \sqrt{4x^2}\sqrt{3} + \sqrt{16x^2}$ The square root of a product is equal to the product of the square roots.

$= 2x\sqrt{3} + 4x$ $\sqrt{4x^2} = 2x$ and $\sqrt{16x^2} = 4x$.

Self Check 4
Multiply:
a. $\sqrt{3}(3\sqrt{6} - \sqrt{3})$
b. $\sqrt[3]{2x}(3 - \sqrt[3]{4x^2})$

b. $\sqrt[3]{3}(\sqrt[3]{9} - 2) = \sqrt[3]{3}\sqrt[3]{9} - 2\sqrt[3]{3}$ Distribute the multiplication by $\sqrt[3]{3}$.

$= \sqrt[3]{27} - 2\sqrt[3]{3}$ The product of two cube roots is equal to the cube root of the product.

$= 3 - 2\sqrt[3]{3}$ $\sqrt[3]{27} = 3$.

Answers **a.** $9\sqrt{2} - 3$, **b.** $3\sqrt[3]{2x} - 2x$

To multiply two binomials, we multiply each term of one binomial by each term of the other binomial and simplify. We multiply two radical expressions, each having two terms, in the same way.

Self Check 5
Multiply:
$(\sqrt{5a} - 2)(\sqrt{5a} + 3)$.

Answer $5a + \sqrt{5a} - 6$

EXAMPLE 5 Multiply: $(\sqrt{3x} + 1)(\sqrt{3x} + 2)$.

Solution

$(\sqrt{3x} + 1)(\sqrt{3x} + 2)$

$= \sqrt{3x}\sqrt{3x} + 2\sqrt{3x} + \sqrt{3x} + 2$ Use the FOIL method.

$= \sqrt{3x}\sqrt{3x} + 3\sqrt{3x} + 2$ Combine like radicals.

$= 3x + 3\sqrt{3x} + 2$ $\sqrt{3x}\sqrt{3x} = (\sqrt{3x})^2 = 3x$.

Self Check 6
Multiply:
$(\sqrt{5} + \sqrt{11})(\sqrt{5} - \sqrt{11})$

Answer -6

EXAMPLE 6 Multiply: $(\sqrt{7} + \sqrt{2})(\sqrt{7} - \sqrt{2})$.

Solution Recall that the product of two binomials that differ only in sign between the terms is the square of the first term minus the square of the second term: $(x + y)(x - y) = x^2 - y^2$. We can use this special product formula to multiply the radical expressions.

$$(\sqrt{7} + \sqrt{2})(\sqrt{7} - \sqrt{2}) = (\sqrt{7})^2 - (\sqrt{2})^2$$
$$= 7 - 2$$
$$= 5$$

! COMMENT Note that the answers to Example 6 and Self Check 6 did not contain any radicals. This will be the case whenever we find the product of radical expressions (containing *square* roots) of this form, which differ only in the sign between the terms.

Self Check 7
Multiply:
$(\sqrt[3]{3x} + 1)(\sqrt[3]{9x^2} - 2)$

Answer
$3x - 2\sqrt[3]{3x} + \sqrt[3]{9x^2} - 2$

EXAMPLE 7 Multiply: $(\sqrt[3]{4x} - 3)(\sqrt[3]{2x^2} + 1)$.

Solution

$(\sqrt[3]{4x} - 3)(\sqrt[3]{2x^2} + 1)$

$= \sqrt[3]{4x}\sqrt[3]{2x^2} + \sqrt[3]{4x} - 3\sqrt[3]{2x^2} - 3$ Use the FOIL method.

$= \sqrt[3]{8x^3} + \sqrt[3]{4x} - 3\sqrt[3]{2x^2} - 3$ The product of two cube roots is equal to the cube root of the product.

$= 2x + \sqrt[3]{4x} - 3\sqrt[3]{2x^2} - 3$ $\sqrt[3]{8x^3} = 2x$.

Dividing radical expressions

To divide radical expressions, we use the division property of radicals. For example, to divide $\sqrt{108}$ by $\sqrt{36}$, we proceed as follows:

$$\frac{\sqrt{108}}{\sqrt{36}} = \sqrt{\frac{108}{36}}$$ The quotient of two square roots is the square root of the quotient.

$$= \sqrt{3}$$ Perform the division within the radical: $108 \div 36 = 3$.

EXAMPLE 8 Divide: $\dfrac{\sqrt{22a^2}}{\sqrt{99a^4}}$.

Solution

$$\frac{\sqrt{22a^2}}{\sqrt{99a^4}} = \sqrt{\frac{22a^2}{99a^4}}$$

$$= \sqrt{\frac{2}{9a^2}}$$ Simplify the radicand: $\dfrac{22a^2}{99a^4} = \dfrac{\overset{1}{\cancel{11}} \cdot 2 \cdot \overset{1}{\cancel{a^2}}}{\underset{1}{\cancel{11}} \cdot 9 \cdot \underset{1}{\cancel{a^2}} \cdot a^2} = \dfrac{2}{9a^2}$.

$$= \frac{\sqrt{2}}{\sqrt{9a^2}}$$ The square root of a quotient is equal to the quotient of the square roots.

$$= \frac{\sqrt{2}}{3a}$$ $\sqrt{9a^2} = 3a$.

Self Check 8
Divide:
$$\frac{\sqrt{30y^9}}{\sqrt{160y^5}}$$

Answer $\dfrac{y^2\sqrt{3}}{4}$

Rationalizing denominators

The length of a diagonal of one of the square tiles shown in Figure 14-17 is 1 foot. Using the Pythagorean theorem, it can be shown that the length of a side of a tile is $\frac{1}{\sqrt{2}}$ feet. Because the expression $\frac{1}{\sqrt{2}}$ contains a radical in its denominator, it is not in simplified form. Since it is often easier to work with a radical expression if the denominator does not contain a radical, we now consider how to change the denominator from a radical that represents an irrational number to a rational number. The process is called **rationalizing the denominator.**

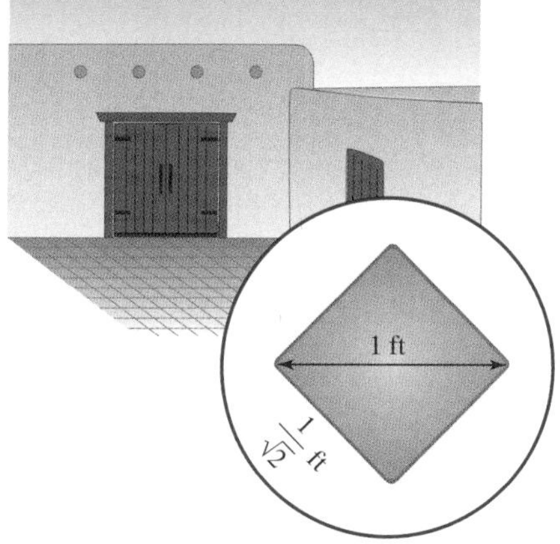

FIGURE 14-17

To rationalize the denominator of $\frac{1}{\sqrt{2}}$, we multiply both the numerator and the denominator by $\sqrt{2}$. Because the expression is multiplied by $\frac{\sqrt{2}}{\sqrt{2}}$, which is 1, the value of $\frac{1}{\sqrt{2}}$ is not changed.

$$\frac{1}{\sqrt{2}} = \frac{1\sqrt{2}}{\sqrt{2}\sqrt{2}}$$ Multiply both numerator and denominator by $\sqrt{2}$.

$$= \frac{\sqrt{2}}{2}$$ In the numerator, $1\sqrt{2} = \sqrt{2}$. In the denominator, $\sqrt{2}\sqrt{2} = (\sqrt{2})^2 = 2$. The denominator is now a rational number.

The length of a side of a patio tile is $\frac{1}{\sqrt{2}} = \frac{\sqrt{2}}{2}$ feet.

This example suggests the following procedure for rationalizing square root denominators.

Rationalizing square root denominators

Multiply the numerator and the denominator by the smallest factor that gives a perfect square radicand in the denominator.

Self Check 9

Rationalize each denominator:

a. $\sqrt{\frac{2}{7}}$

b. $\frac{5}{\sqrt[3]{5}}$

EXAMPLE 9 Rationalize each denominator: **a.** $\sqrt{\frac{5}{3}}$ and **b.** $\frac{2}{\sqrt[3]{3}}$.

Solution

a. The expression $\sqrt{\frac{5}{3}}$ is not in simplified form, because the radicand is a fraction. To write it in simplified form, we use the division property of radicals. Then we use the fundamental property of fractions to rationalize the denominator by multiplying the numerator and the denominator by $\sqrt{3}$.

$$\sqrt{\frac{5}{3}} = \frac{\sqrt{5}}{\sqrt{3}}$$ The square root of a quotient is the quotient of the square roots. Note that the denominator is the irrational number $\sqrt{3}$.

$$= \frac{\sqrt{5}\sqrt{3}}{\sqrt{3}\sqrt{3}}$$ Multiply the numerator and the denominator by $\sqrt{3}$.

$$= \frac{\sqrt{15}}{3}$$ In the numerator, do the multiplication. Simplify in the denominator: $\sqrt{3}\sqrt{3} = (\sqrt{3})^2 = 3$.

b. The denominator contains a cube root. We multiply by the smallest factor that gives an integer cube radicand in the denominator. Since $\sqrt[3]{3}\sqrt[3]{9} = \sqrt[3]{27}$ and 27 is a perfect integer cube, we multiply the numerator and denominator by $\sqrt[3]{9}$ and simplify.

$$\frac{2}{\sqrt[3]{3}} = \frac{2\sqrt[3]{9}}{\sqrt[3]{3}\sqrt[3]{9}}$$

$$= \frac{2\sqrt[3]{9}}{\sqrt[3]{27}}$$ Multiply: $\sqrt[3]{3}\sqrt[3]{9} = \sqrt[3]{27}$.

$$= \frac{2\sqrt[3]{9}}{3}$$ $\sqrt[3]{27} = 3$. The denominator is now a rational number.

Answers **a.** $\frac{\sqrt{14}}{7}$, **b.** $\sqrt[3]{25}$

Self Check 10

Rationalize the denominator and simplify:

$\frac{6\sqrt{z}}{\sqrt{50y}}$

EXAMPLE 10 Rationalize the denominator and simplify: $\frac{5\sqrt{y}}{\sqrt{20x}}$.

Solution To rationalize the denominator, we don't need to multiply the numerator and denominator by $\sqrt{20x}$. To keep the numbers small, we can multiply by $\sqrt{5x}$, because $5x \cdot 20x = 100x^2$, which is a perfect square.

$$\frac{5\sqrt{y}}{\sqrt{20x}} = \frac{5\sqrt{y}\sqrt{5x}}{\sqrt{20x}\sqrt{5x}}$$ Multiply the numerator and denominator by $\sqrt{5x}$.

$$= \frac{5\sqrt{5xy}}{\sqrt{100x^2}}$$ $\sqrt{y}\sqrt{5x} = \sqrt{5xy}$ and $\sqrt{20x}\sqrt{5x} = \sqrt{100x^2}$.

$$= \frac{5\sqrt{5xy}}{10x}$$ $\sqrt{100x^2} = 10x$.

$$= \frac{\overset{1}{\cancel{5}}\sqrt{5xy}}{\underset{1}{\cancel{5}} \cdot 2x}$$ Factor $10x$ and divide out a common factor of 5.

$$= \frac{\sqrt{5xy}}{2x}$$ $\frac{5}{5} = 1$.

Answer $\frac{3\sqrt{2yz}}{5y}$

At times, we will encounter fractions such as $\frac{2}{\sqrt{3} - 1}$, whose denominator has two terms. Note that $\sqrt{3} - 1$ is an irrational number. Because $\sqrt{3} - 1$ has two terms, multiplying it by $\sqrt{3}$ will not make it a rational number. The key to rationalizing this denominator is to multiply the numerator and denominator by $\sqrt{3} + 1$, because the product $(\sqrt{3} + 1)(\sqrt{3} - 1)$ has no radicals. Radical expressions such as $\sqrt{3} + 1$ and $\sqrt{3} - 1$ are called **conjugates** of each other.

EXAMPLE 11

Rationalize the denominator and simplify: $\frac{2}{\sqrt{3} - 1}$.

Solution We rationalize the denominator by multiplying the numerator and denominator by the conjugate of the denominator.

$$\frac{2}{\sqrt{3} - 1} = \frac{2(\sqrt{3} + 1)}{(\sqrt{3} - 1)(\sqrt{3} + 1)}$$ Multiply the numerator and denominator by the conjugate of the denominator, which is $\sqrt{3} + 1$.

$$= \frac{2(\sqrt{3} + 1)}{3 - 1}$$ Use a special product formula: $(\sqrt{3} - 1)(\sqrt{3} + 1) = 3 - 1$.

$$= \frac{2(\sqrt{3} + 1)}{2}$$ Subtract. The denominator is now a rational number.

$$= \sqrt{3} + 1$$ Simplify the fraction by removing the common factor of 2 in the numerator and denominator.

Self Check 11
Rationalize the denominator and simplify: $\frac{3}{\sqrt{2} + 1}$

Answer $3(\sqrt{2} - 1)$

EXAMPLE 12

Rationalize the denominator and simplify: $\frac{\sqrt{x} + 1}{\sqrt{x} - 1}$.

Solution We multiply the numerator and denominator by the conjugate of the denominator, which is $\sqrt{x} + 1$.

Self Check 12
Rationalize the denominator and simplify: $\frac{\sqrt{x} - 1}{\sqrt{x} + 1}$

$$\frac{\sqrt{x}+1}{\sqrt{x}-1} = \frac{(\sqrt{x}+1)(\sqrt{x}+\mathbf{1})}{(\sqrt{x}-1)(\sqrt{x}+\mathbf{1})}$$ Multiply the numerator and denominator by $\sqrt{x}+1$.

$$= \frac{\sqrt{x}\sqrt{x}+\sqrt{x}(1)+1(\sqrt{x})+1}{\sqrt{x}\sqrt{x}+\sqrt{x}(1)-1(\sqrt{x})-1}$$ Perform the multiplications.

$$= \frac{x+2\sqrt{x}+1}{x-1}$$ $\sqrt{x}\sqrt{x} = (\sqrt{x})^2 = x$. Combine like radicals.

Answer $\dfrac{x-2\sqrt{x}+1}{x-1}$

Section 14.5 STUDY SET

VOCABULARY *Fill in the blanks.*

1. The method of changing a radical denominator of a fraction into a rational number is called ________ the denominator.
2. The ________ of the fraction $\frac{4}{\sqrt{3}}$ is 4 and the ________ is $\sqrt{3}$.
3. $3+\sqrt{2}$ is the ________ of $3-\sqrt{2}$.
4. Radical expressions with the same ______ can be multiplied.
5. In the radical expression $3\sqrt{7}$, the number 3 is the ________ of the radical.
6. Nonterminating, nonrepeating decimals such as $\sqrt{2} = 1.414213562\ldots$ and $\sqrt{3} = 1.732050808\ldots$ are ________ numbers.

CONCEPTS *Fill in the blanks.*

7. $\sqrt{a}\cdot\sqrt{b} = \sqrt{}$
8. $\sqrt{\frac{a}{b}} = \frac{}{}$
9. To rationalize the denominator of

   $$\frac{x}{\sqrt{7}}$$

   we multiply the numerator and denominator by ____.
10. To rationalize the denominator of

    $$\frac{x}{\sqrt{x}+1}$$

    we multiply the numerator and denominator by ____.
11. Explain why each expression is not in simplified radical form.

    a. $\sqrt{\frac{3}{4}}$ b. $\frac{1}{\sqrt{10}}$
12. To multiply $2\sqrt{x}$ and $6\sqrt{x}$, we first multiply the ________, then multiply the ________, and simplify the result.
13. Which fractions have a rational denominator and which have an irrational denominator?

    $$\frac{\sqrt{5}}{3},\ \frac{2}{\sqrt{6}},\ -\frac{\sqrt{2}}{8},\ \frac{1+\sqrt{3}}{4},\ \frac{9}{7-\sqrt{10}}$$
14. To multiply $(\sqrt{3}+\sqrt{2})(\sqrt{7}+\sqrt{5})$, we use the FOIL method. What are the

    a. first terms? b. outer terms?

    c. inner terms? d. last terms?

Perform each operation if possible.

15. a. $\sqrt{2}+\sqrt{3}$ b. $\sqrt{2}\cdot\sqrt{3}$

    c. $\sqrt{2}-\sqrt{3}$ d. $\frac{\sqrt{2}}{\sqrt{3}}$

    e. $\sqrt{2}+3\sqrt{2}$ f. $\sqrt{2}\cdot 3\sqrt{2}$

    g. $\sqrt{2}-3\sqrt{2}$ h. $\frac{\sqrt{2}}{3\sqrt{2}}$
16. Find each special product.

    a. $(\sqrt{6}+\sqrt{3})(\sqrt{6}-\sqrt{3})$

    b. $(\sqrt{a}+\sqrt{7})(\sqrt{a}-\sqrt{7})$

NOTATION *Complete each solution.*

17. $(\sqrt{x} + \sqrt{2})(\sqrt{x} - 3\sqrt{2})$
$= \sqrt{x}\ \square - \sqrt{x}(3\sqrt{2}) + \sqrt{2}\ \square - \sqrt{2}(3\sqrt{2})$
$= x - 3\ \square + \sqrt{2x} - 3\sqrt{2}\sqrt{2}$
$= \square - 2\sqrt{2x} - 3(2)$
$= x - 2\sqrt{2x} - 6$

18. $\frac{x}{\sqrt{x} - 2} = \frac{x(\sqrt{x} + 2)}{(\sqrt{x} - 2)\ \square}$
$= \frac{x(\sqrt{x} + 2)}{(\ \square\ )^2 - 2^2}$
$= \frac{x(\sqrt{x} + 2)}{x - 4}$

PRACTICE *Perform each multiplication. All variables represent positive numbers.*

19. $(\sqrt{5})^2$ **20.** $(\sqrt{11})^2$

21. $(3\sqrt{6})^2$ **22.** $(-7\sqrt{2})^2$

23. $\sqrt{2}\sqrt{8}$ **24.** $\sqrt{27}\sqrt{3}$

25. $\sqrt{7}\sqrt{3}$ **26.** $\sqrt{2}\sqrt{11}$

27. $\sqrt{8}\sqrt{7}$ **28.** $\sqrt{6}\sqrt{8}$

29. $3\sqrt{2}\sqrt{x}$ **30.** $4\sqrt{3x}\sqrt{5y}$

31. $\sqrt{x^3}\sqrt{x^5}$ **32.** $\sqrt{a^7}\sqrt{a^3}$

33. $(-5\sqrt{6})(4\sqrt{3})$ **34.** $(6\sqrt{3})(-7\sqrt{2})$

35. $(4\sqrt{x})(-2\sqrt{x})$ **36.** $(3\sqrt{y})(15\sqrt{y})$

37. $\sqrt{8x}\sqrt{2x^3}$ **38.** $\sqrt{27y}\sqrt{3y^3}$

39. $\sqrt{2}(\sqrt{2} + 1)$

40. $\sqrt{5}(\sqrt{5} + 2)$

41. $3\sqrt{3}(\sqrt{27} - 1)$

42. $2\sqrt{2}(\sqrt{8} - 1)$

43. $\sqrt{3}(\sqrt{6} + 1)$

44. $\sqrt{2}(\sqrt{6} - 2)$

45. $\sqrt{x}(\sqrt{3x} - 2)$

46. $\sqrt{y}(\sqrt{y} + 5)$

47. $2\sqrt{x}(\sqrt{9x} + 3)$

48. $3\sqrt{z}(\sqrt{4z} - \sqrt{z})$

49. $(\sqrt{2} + 1)(\sqrt{2} - 1)$

50. $(\sqrt{3} - 1)(\sqrt{3} + 1)$

51. $(2\sqrt{7} - x)(3\sqrt{2} + x)$

52. $(4\sqrt{2} - \sqrt{x})(\sqrt{x} + 2\sqrt{3})$

53. $(\sqrt{6} + 1)^2$

54. $(3 - \sqrt{3})^2$

55. $(\sqrt{2x} + 3)(\sqrt{8x} - 6)$

56. $(\sqrt{5y} - 3)(\sqrt{20y} + 6)$

57. $(-\sqrt[3]{9})^3$

58. $(\sqrt[3]{3})^3$

59. $(2\sqrt[3]{4})(3\sqrt[3]{3})$

60. $(-3\sqrt[3]{3})(\sqrt[3]{5})$

61. $\sqrt[3]{7}(\sqrt[3]{49} - 2)$

62. $\sqrt[3]{5}(\sqrt[3]{25} + 3)$

63. $(\sqrt[3]{2} + 1)(\sqrt[3]{2} + 3)$

64. $(\sqrt[3]{5} - 2)(\sqrt[3]{5} - 1)$

Simplify each expression. Assume that all variables represent positive numbers.

65. $\frac{\sqrt{12x^3}}{\sqrt{27x}}$ **66.** $\frac{\sqrt{32}}{\sqrt{98x^2}}$

67. $\frac{\sqrt{18x}}{\sqrt{25x}}$ **68.** $\frac{\sqrt{27y}}{\sqrt{75y}}$

69. $\frac{\sqrt{196x}}{\sqrt{49x^3}}$ **70.** $\frac{\sqrt{50}}{\sqrt{98z^2}}$

71. $\frac{\sqrt[3]{16x^6}}{\sqrt[3]{54x^3}}$ **72.** $\frac{\sqrt[3]{128a^6}}{\sqrt[3]{16a^3}}$

Rationalize each denominator and simplify. All variables represent positive numbers.

73. $\frac{1}{\sqrt{3}}$ **74.** $\frac{1}{\sqrt{5}}$

75. $\sqrt{\frac{13}{7}}$ **76.** $\sqrt{\frac{3}{11}}$

77. $\frac{9}{\sqrt{27}}$ **78.** $\frac{4}{\sqrt{20}}$

79. $\frac{3}{\sqrt{32}}$

80. $\frac{5}{\sqrt{18}}$

81. $\sqrt{\frac{12}{5}}$

82. $\sqrt{\frac{24}{7}}$

83. $\frac{10}{\sqrt{x}}$

84. $\frac{12}{\sqrt{y}}$

85. $\frac{\sqrt{9y}}{\sqrt{2x}}$

86. $\frac{\sqrt{4t}}{\sqrt{3z}}$

87. $\frac{3}{\sqrt{3} - 1}$

88. $\frac{3}{\sqrt{5} - 2}$

89. $\frac{3}{\sqrt{7} + 2}$

90. $\frac{5}{\sqrt{8} + 3}$

91. $\frac{12}{3 - \sqrt{3}}$

92. $\frac{10}{5 - \sqrt{5}}$

93. $\frac{-\sqrt{3}}{\sqrt{3} + 1}$

94. $\frac{-\sqrt{2}}{\sqrt{2} - 1}$

95. $\frac{5}{\sqrt{3} + \sqrt{2}}$

96. $\frac{3}{\sqrt{3} - \sqrt{2}}$

97. $\frac{\sqrt{x} + 2}{\sqrt{x} - 2}$

98. $\frac{\sqrt{x} - 3}{\sqrt{x} + 3}$

99. $\frac{5}{\sqrt[3]{5}}$

100. $\frac{7}{\sqrt[3]{7}}$

101. $\frac{4}{\sqrt[3]{4}}$

102. $\frac{7}{\sqrt[3]{10}}$

103. $\frac{\sqrt[3]{5}}{\sqrt[3]{2}}$

104. $\frac{\sqrt[3]{2}}{\sqrt[3]{5}}$

APPLICATIONS

105. LAWNMOWERS See the illustration below, which shows the blade of a rotary lawnmower. Use the formula for the area of a circle, $A = \pi r^2$, to find the area of lawn covered by one rotation of the blade. Leave π in your answer.

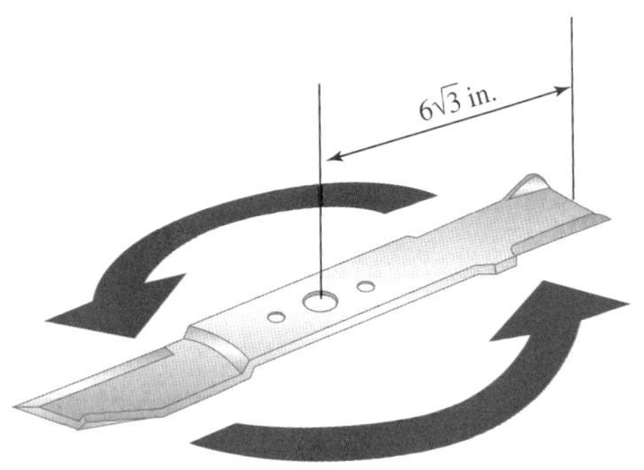

106. AWARDS PLATFORMS Find the total number of cubic feet of concrete needed to construct the Olympic Games awards platforms shown below.

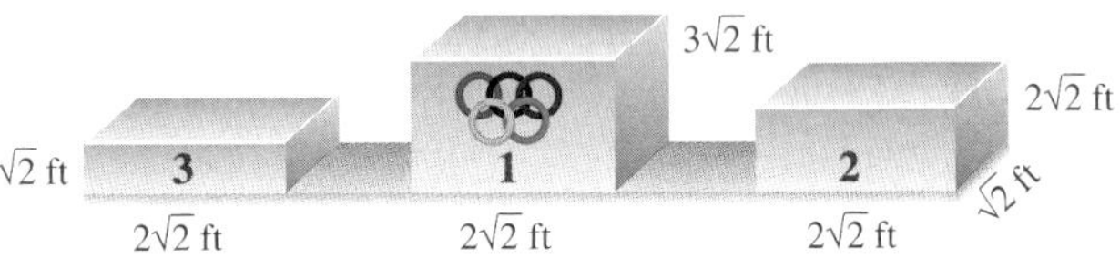

107. AIR HOCKEY Find the area of the playing surface of the air hockey game in the illustration.

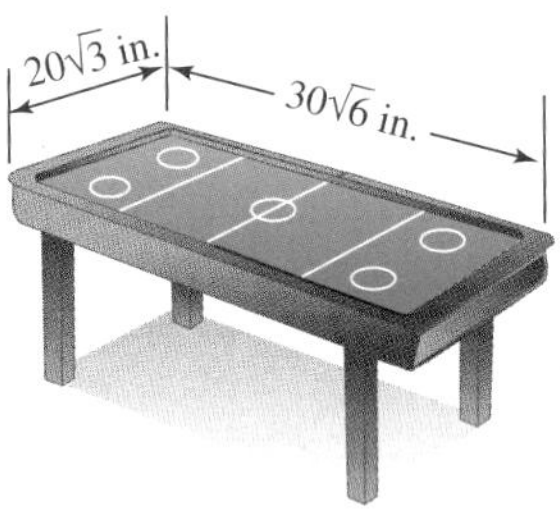

108. PROJECTOR SCREENS To find the length l of a rectangle, we can use the formula

$$l = \frac{A}{w}$$

where A is the area of the rectangle and w is its width. Find the length of the screen shown below if its area is 54 square feet.

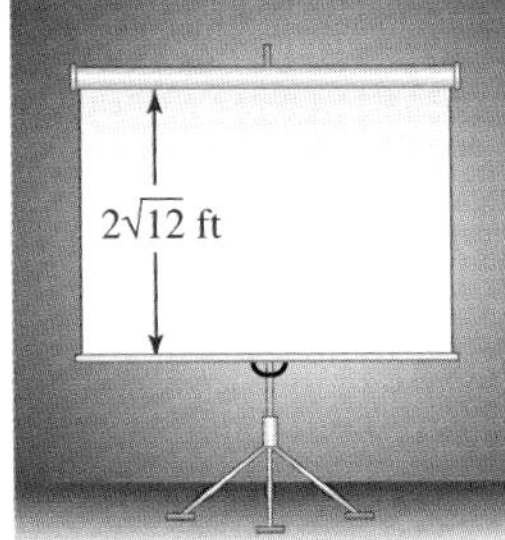

109. COSTUME DESIGNS The pattern for one panel of an 1870s English dress is printed on the 1 in. × 1 in. grid shown on the next page. Find the number of square inches of fabric in the trapezoidal-shaped panel. (*Hint:* Use the Pythagorean theorem to determine the lengths of the sides.)

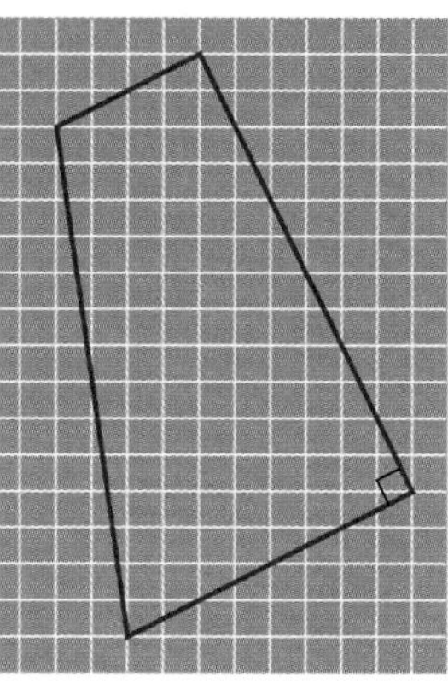

110. SET DESIGNS The director of a stage play requested bright downlighting over the portion of the set shown below. Find the area of the rectangle. (*Hint:* Use the Pythagorean theorem to determine the lengths of the sides.)

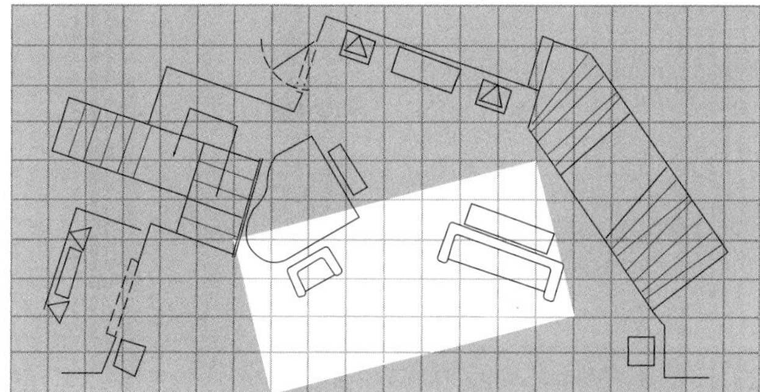

WRITING

111. When rationalizing the denominator of $\frac{5}{\sqrt{6}}$, why must we multiply both the numerator *and* denominator by $\sqrt{6}$?

112. A calculator is used to find decimal approximations for the expressions $\frac{2}{\sqrt{6}}$ and $\frac{\sqrt{6}}{3}$. In each case, the calculator display reads 0.816496581. Explain why the results are the same.

REVIEW

113. Is -2 a solution of $3x - 7 = 5x + 1$?

114. The graph of a line passes through the point (2, 0). Is this the x- or y-intercept?

115. To evaluate the expression $2 - (-3 + 4)^2$, which operation should be performed first?

116. The graph of a straight line rises from left to right. Is the slope of the line positive or negative?

117. Multiply: $(x - 4)(x + 4)$.

118. How far will a car traveling 55 mph go in 3.5 hours?

14.6 Solving Radical Equations; the Distance Formula

- The squaring property of equality • Checking possible solutions
- Solving equations containing one square root • Solving equations containing two square roots
- Solving equations containing cube roots • The distance formula

Many situations can be modeled mathematically by equations that contain radicals. In this section, we will develop techniques to solve such equations.

The squaring property of equality

The equation $\sqrt{x} = 6$ is called a **radical equation,** because it contains a radical expression with a variable radicand. To solve this equation, we isolate x by undoing the operation performed on it. Recall that $\sqrt{x}$ represents the number that, when squared, gives x. Therefore, if we *square* $\sqrt{x}$, we will obtain x.

$$(\sqrt{x})^2 = x$$

Using this observation, we can eliminate the radical on the left-hand side of $\sqrt{x} = 6$ by squaring that side. Intuition tells us that we should also square the right-hand side. This is a valid step, because if two numbers are equal, their squares are equal.

Squaring property of equality

If a and b represent real numbers, with $a = b$, then $a^2 = b^2$.

We can now solve $\sqrt{x} = 6$ by applying the squaring property of equality.

$\sqrt{x} = 6$

$(\sqrt{x})^2 = (6)^2$ Square both sides of the equation to eliminate the radical.

$x = 36$ Simplify each side: $(\sqrt{x})^2 = x$ and $(6)^2 = 36$.

Checking this result, we have

$\sqrt{x} = 6$

$\sqrt{\mathbf{36}} \stackrel{?}{=} 6$ Substitute 36 for x.

$6 = 6$ Simplify the left-hand side: $\sqrt{36} = 6$.

Since we obtain a true statement, 36 is the solution.

Checking possible solutions

If we square both sides of an equation, the resulting equation may not have the same solutions as the original one. For example, consider the equation

$$x = 2$$

The only solution of this equation is 2. However, if we square both sides, we obtain $(x)^2 = 2^2$, which simplifies to

$$x^2 = 4$$

This new equation has solutions 2 and -2, because $2^2 = 4$ and $(-2)^2 = 4$.

The equations $x = 2$ and $x^2 = 4$ are not equivalent equations because they do not have the same solutions. The solution -2 satisfies $x^2 = 4$ but it does not satify $x = 2$. We see that squaring both sides of an equation can produce an equation with solutions that don't satisfy the original one.Therefore, we must check each possible solution in the original equation.

Solving equations containing one square root

To solve an equation containing square root radicals, we follow these steps.

Solving radical equations

1. Whenever possible, isolate a single radical expression on one side of the equation.
2. Square both sides of the equation and solve the resulting equation.
3. Check the possible solutions in the original equation. This step is required.

Self Check 1

Solve: $\sqrt{x - 4} = 9$.

EXAMPLE 1 Solve: $\sqrt{x + 2} = 3$.

Solution To solve the equation $\sqrt{x + 2} = 3$, we note that the radical is already isolated on one side. We proceed to step 2 and square both sides to eliminate the radical.

Since this might produce an equation with more solutions than the original one, we must check each solution.

$$\sqrt{x+2} = 3$$
$$(\sqrt{x+2})^2 = (3)^2 \quad \text{Square both sides.}$$
$$x + 2 = 9 \quad (\sqrt{x+2})^2 = x + 2 \text{ and } 3^2 = 9.$$
$$x = 7 \quad \text{Subtract 2 from both sides.}$$

We check by substituting 7 for x in the original equation.

$$\sqrt{x+2} = 3$$
$$\sqrt{\mathbf{7}+2} \stackrel{?}{=} 3 \quad \text{Substitute 7 for } x.$$
$$\sqrt{9} \stackrel{?}{=} 3 \quad \text{Perform the addition within the radical symbol.}$$
$$3 = 3$$

Since a true statement results, 7 is the solution.

Answer 85

EXAMPLE 2 Solve: $\sqrt{5x+1} + 7 = 3$.

Self Check 2
Solve: $\sqrt{3x-2} + 6 = 1$.

Solution We isolate the radical on one side and proceed as follows:

$$\sqrt{5x+1} + 7 = 3$$
$$\sqrt{5x+1} = -4 \quad \text{Subtract 7 from both sides.}$$
$$(\sqrt{5x+1})^2 = (-4)^2 \quad \text{Square both sides to eliminate the radical.}$$
$$5x + 1 = 16 \quad (\sqrt{5x+1})^2 = 5x + 1 \text{ and } (-4)^2 = 16.$$
$$5x = 15 \quad \text{Subtract 1 from both sides.}$$
$$x = 3 \quad \text{Divide both sides by 5.}$$

We check by substituting 3 for x in the original equation.

$$\sqrt{5\mathbf{x}+1} + 7 = 3$$
$$\sqrt{5(\mathbf{3})+1} + 7 \stackrel{?}{=} 3 \quad \text{Substitute 3 for } x.$$
$$\sqrt{16} + 7 \stackrel{?}{=} 3 \quad \text{Evaluate the expression within the radical symbol.}$$
$$4 + 7 \stackrel{?}{=} 3$$
$$11 \neq 3$$

Since $11 \neq 3$, 3 is not a solution. In fact, the equation has no solution. This was apparent in step 2 of the solution. There is no real number x that could make the nonnegative number $\sqrt{5x+1}$ equal to -4.

Answer no solution

Example 2 shows that squaring both sides of an equation can lead to possible solutions that do not satisfy the original equation. We call such numbers **extraneous solutions.** In Example 2, 3 is an extraneous solution of $\sqrt{5x+1} + 7 = 3$.

EXAMPLE 3 **Height of a bridge.**

The distance d (in feet) that an object will fall in t seconds is given by the formula

$$t = \sqrt{\frac{d}{16}}$$

Self Check 3
If it takes 4 seconds for the stone in Example 3 to hit the water, how high is the bridge?

To find the height of the bridge shown in Figure 14-18, a man drops a stone into the water. If it takes the stone 3 seconds to hit the water, how high is the bridge?

FIGURE 14-18

Solution We substitute 3 for t in the formula and solve for d.

$$t = \sqrt{\frac{d}{16}}$$

$$3 = \sqrt{\frac{d}{16}} \quad \text{Substitute 3 for } t.$$

$$(3)^2 = \left(\sqrt{\frac{d}{16}}\right)^2 \quad \text{Square both sides to eliminate the radical.}$$

$$9 = \frac{d}{16} \quad 3^2 = 9 \text{ and } \left(\sqrt{\frac{d}{16}}\right)^2 = \frac{d}{16}.$$

$$144 = d \quad \text{Multiply both sides by 16.}$$

Answer 256 feet

The bridge is 144 feet above the water. Check this result in the original equation.

Self Check 4
Solve: $b + 4 = \sqrt{b^2 + 6b + 12}$

EXAMPLE 4 Solve: $a + 2 = \sqrt{a^2 + 3a + 3}$.

Solution The radical is isolated on the right-hand side, so we proceed by squaring both sid s to eliminate it.

$$a + 2 = \sqrt{a^2 + 3a + 3}$$

$$(a + 2)^2 = (\sqrt{a^2 + 3a + 3})^2 \quad \text{Square both sides.}$$

$$a^2 + 4a + 4 = a^2 + 3a + 3 \quad \text{Use a special product formula: } (a + 2)^2 = a^2 + 4a + 4. \ (\sqrt{a^2 + 3a + 3})^2 = a^2 + 3a + 3.$$

$$a^2 + 4a + 4 - a^2 = a^2 + 3a + 3 - a^2 \quad \text{To eliminate } a^2 \text{, subtract } a^2 \text{ from both sides.}$$

$$4a + 4 = 3a + 3 \quad \text{Combine like terms: } a^2 - a^2 = 0.$$

$$a + 4 = 3 \quad \text{Subtract } 3a \text{ from both sides.}$$

$$a = -1 \quad \text{Subtract 4 from both sides.}$$

We check by substituting -1 for a in the original equation.

$$a + 2 = \sqrt{a^2 + 3a + 3}$$

$$\mathbf{-1} + 2 \stackrel{?}{=} \sqrt{(\mathbf{-1})^2 + 3(\mathbf{-1}) + 3} \quad \text{Substitute } -1 \text{ for } a.$$

$$1 \stackrel{?}{=} \sqrt{1 - 3 + 3} \quad \text{Within the radical symbol, first find the power, then perform the multiplication.}$$

$$1 \stackrel{?}{=} \sqrt{1} \quad \text{Simplify within the radical symbol.}$$

$$1 = 1$$

The solution is -1.

Answer -2

Sometimes, after clearing an equation of a radical, the result is a quadratic equation.

EXAMPLE 5 Solve: $\sqrt{3 - x} - x = -3$.

Solution To isolate the radical, add x to both sides.

$$\sqrt{3 - x} - x = -3$$

$$\sqrt{3 - x} = x - 3 \quad \text{Add } x \text{ to both sides.}$$

We then square both sides to clear the equation of the radical.

$$(\sqrt{3 - x})^2 = (x - 3)^2 \quad \text{Square both sides.}$$

$$3 - x = (x - 3)(x - 3) \quad \text{The second power indicates two factors of } x - 3.$$

$$3 - x = x^2 - 6x + 9 \quad \text{Multiply the binomials.}$$

To solve the resulting quadratic equation, we write it in standard form so that the left-hand side is 0.

$$3 = x^2 - 5x + 9 \quad \text{Add } x \text{ to both sides.}$$

$$0 = x^2 - 5x + 6 \quad \text{Subtract 3 from both sides.}$$

$$0 = (x - 3)(x - 2) \quad \text{Factor } x^2 - 5x + 6.$$

$$x - 3 = 0 \quad \text{or} \quad x - 2 = 0 \quad \text{Set each factor equal to 0.}$$

$$x = 3 \qquad\qquad x = 2 \quad \text{Solve each equation.}$$

There are two possible solutions to check in the original equation.

***For x* = 3**	***For x* = 2**
$\sqrt{3 - x} - x = -3$	$\sqrt{3 - x} - x = -3$
$\sqrt{3 - \mathbf{3}} - \mathbf{3} \stackrel{?}{=} -3$	$\sqrt{3 - \mathbf{2}} - \mathbf{2} \stackrel{?}{=} -3$
$\sqrt{0} - 3 \stackrel{?}{=} -3$	$\sqrt{1} - 2 \stackrel{?}{=} -3$
$0 - 3 \stackrel{?}{=} -3$	$1 - 2 \stackrel{?}{=} -3$
$-3 = -3$	$-1 = -3$

Since a true statement results when 3 is substituted for x, 3 is a solution. Since a false statement results when 2 is substituted for x, 2 is an extraneous solution.

Self Check 5
Solve:
$\sqrt{x + 4} - x = -2$

Answer 5

Solving equations containing two square roots

In the next example, the equation contains two square roots.

EXAMPLE 6 Solve: $\sqrt{x + 12} = 3\sqrt{x + 4}$.

Solution Note that each radical is isolated on one side of the equation. We begin by squaring both sides to eliminate them.

Self Check 6
Solve:
$\sqrt{x - 4} = 2\sqrt{x - 16}$

$$\sqrt{x + 12} = 3\sqrt{x + 4}$$
$$(\sqrt{x + 12})^2 = (3\sqrt{x + 4})^2 \quad \text{Square both sides.}$$
$$x + 12 = 9(x + 4) \quad (\sqrt{x + 12})^2 = x + 12. \; (3\sqrt{x + 4})^2 = 3^2(\sqrt{x + 4})^2 = 9(x + 4).$$
$$x + 12 = 9x + 36 \quad \text{Distribute the multiplication by 9.}$$
$$-8x = 24 \quad \text{Subtract } 9x \text{ and 12 from both sides.}$$
$$x = -3 \quad \text{Divide both sides by } -8.$$

We check the solution by substituting -3 for x in the original equation.

$$\sqrt{x + 12} = 3\sqrt{x + 4}$$
$$\sqrt{\mathbf{-3} + 12} \stackrel{?}{=} 3\sqrt{\mathbf{-3} + 4} \quad \text{Substitute } -3 \text{ for } x.$$
$$\sqrt{9} \stackrel{?}{=} 3\sqrt{1} \quad \text{Simplify within the radical symbols.}$$
$$3 = 3$$

Answer 20

The solution is -3.

Solving equations containing cube roots

In the next example, we cube both sides of an equation to eliminate a cube root.

Self Check 7
Solve: $\sqrt[3]{3x - 3} = 3$

EXAMPLE 7 Solve: $\sqrt[3]{2x + 10} = 2$.

Solution To undo the operation performed on $2x + 10$, we cube both sides and proceed as follows:

$$\sqrt[3]{2x + 10} = 2$$
$$(\sqrt[3]{2x + 10})^3 = (2)^3 \quad \text{Cube both sides.}$$
$$2x + 10 = 8 \quad (\sqrt[3]{2x + 10})^3 = 2x + 10 \text{ and } (2)^3 = 8.$$
$$2x = -2 \quad \text{Subtract 10 from both sides.}$$
$$x = -1 \quad \text{Divide both sides by 2.}$$

Answer 10

Check the result.

The distance formula

We can use the Pythagorean theorem to derive a formula for finding the distance between two points $P(x_1, y_1)$ and $Q(x_2, y_2)$ on a rectangular coordinate system. The distance d between points P and Q is the length of the hypotenuse of the triangle in Figure 14-19. The two legs have lengths $x_2 - x_1$ and $y_2 - y_1$.

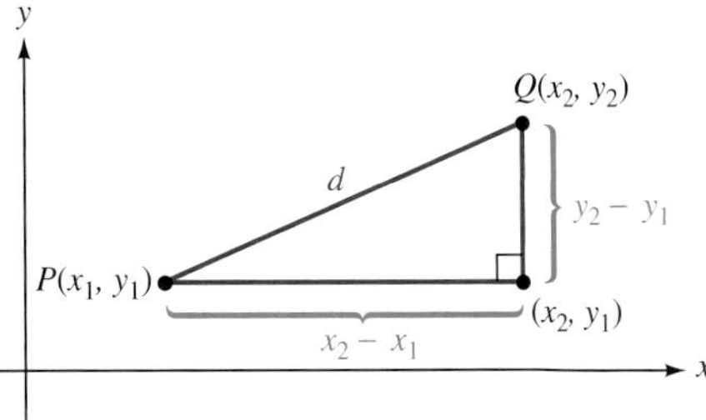

FIGURE 14-19

By the Pythagorean theorem, we have

$$d^2 = (x_2 - x_1)^2 + (y_2 - y_1)^2$$

We can take the positive square root of both sides of this equation to get the **distance formula.**

$$d = \sqrt{(x_2 - x_1)^2 + (y_2 - y_1)^2}$$

The distance formula

The distance d between the points with coordinates (x_1, y_1) and (x_2, y_2) is given by

$$d = \sqrt{(x_2 - x_1)^2 + (y_2 - y_1)^2}$$

EXAMPLE 8 Find the distance between points (1, 2) and (4, 6). (See Figure 14-20.)

Self Check 8

Find the distance between $(-2, 1)$ and $(4, 9)$.

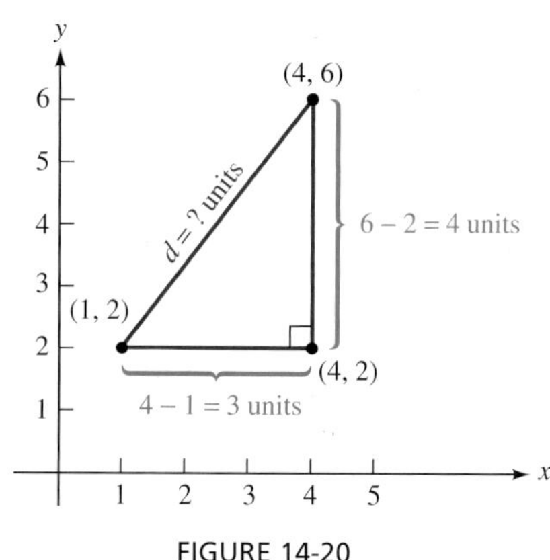

FIGURE 14-20

Solution We use the distance formula and substitute 1 for x_1, 2 for y_1, 4 for x_2, and 6 for y_2. Then we evaluate the expression under the radical symbol.

$d = \sqrt{(x_2 - x_1)^2 + (y_2 - y_1)^2}$

$= \sqrt{(4 - 1)^2 + (6 - 2)^2}$ Substitute.

$= \sqrt{3^2 + 4^2}$ Perform the subtractions within the parentheses first.

$= \sqrt{9 + 16}$ Evaluate the powers.

$= \sqrt{25}$ Perform the addition.

$= 5$ Find the square root.

The distance between the points is 5 units.

Answer 10

EXAMPLE 9 Find the distance between the points $(-4, 5)$ and $(3, -1)$.

Self Check 9

Find the distance between the points $(-1, 2)$ and $(-6, 4)$.

Solution We use the distance formula and substitute -4 for x_1, 5 for y_1, 3 for x_2, and -1 for y_2.

$d = \sqrt{(x_2 - x_1)^2 + (y_2 - y_1)^2}$

$= \sqrt{[3 - (-4)]^2 + (-1 - 5)^2}$ Substitute.

$= \sqrt{7^2 + (-6)^2}$ Perform the subtractions.

$= \sqrt{49 + 36}$ Evaluate the powers.

$= \sqrt{85}$ Perform the addition.

≈ 9.219544457 Use a calculator.

The distance between the points is exactly $\sqrt{85}$ or approximately 9.22 units.

Answer $\sqrt{29} \approx 5.39$

Section 14.6 STUDY SET

VOCABULARY *Fill in the blanks.*

1. A ________ equation contains one or more radical expressions with a variable radicand.
2. To ________ the radical expression in $\sqrt{x} + 1 = 10$ means to get $\sqrt{x}$ by itself on one side of the equation.
3. A false solution that occurs when you square both sides of an equation is called an __________ solution.
4. The squaring property of equality states that if two numbers are equal, their ________ are equal.

CONCEPTS *Fill in the blanks.*

5. The squaring property of equality states that

   If $a = b$, then $a^2 = \square$.

6. The distance formula states that

   $d = \square$

7. To isolate x, what step should be used to undo the operation performed on it? (Assume that x is a positive number.)
   a. $x^2 = 4$
   b. $\sqrt{x} = 4$

8. Simplify each expression.
   a. $(\sqrt{x})^2$
   b. $(\sqrt{x-1})^2$
   c. $(2\sqrt{x})^2$
   d. $(2\sqrt{x-1})^2$
   e. $(\sqrt{2x})^2$
   f. $(\sqrt[3]{x})^3$

In Exercises 9–12, an equation is incorrectly solved. Determine what is wrong with each solution.

9. $$\begin{aligned}\sqrt{x-2} &= 3\\ (\sqrt{x-2})^2 &= 3\\ x-2 &= 3\\ x &= 5\end{aligned}$$

10. $$\begin{aligned}2 &= \sqrt{x-9}\\ (2)^2 &= (\sqrt{x-9})^2\\ 4 &= x-9\\ -5 &= x\\ x &= -5\end{aligned}$$

11. $$\begin{aligned}\sqrt{a+2} - 5 &= 4\\ (\sqrt{a+2} - 5)^2 &= 4^2\\ a+2-25 &= 16\\ a-23 &= 16\\ a &= 39\end{aligned}$$

12. $$\begin{aligned}\sqrt[3]{x+1} &= -2\\ (\sqrt[3]{x+1})^2 &= (-2)^2\\ x+1 &= 4\\ x &= 3\end{aligned}$$

13. a. On the graph below, plot the points $A(-4, 6)$, $B(4, 0)$, $C(1, -4)$, and $D(-7, 2)$.
    b. Draw figure $ABCD$. What type of geometric figure is it?
    c. Find the length of each side of the figure.
    d. Find the perimeter of the figure.

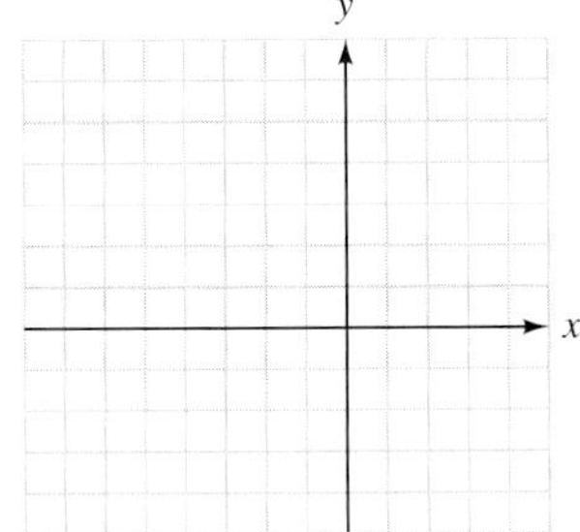

14. a. What type of geometric figure is figure $ABCD$ shown in the following illustration?
    b. Give the coordinates of points A, B, C, and D.
    c. Find the length of each side of the figure.
    d. Find the area of the figure.

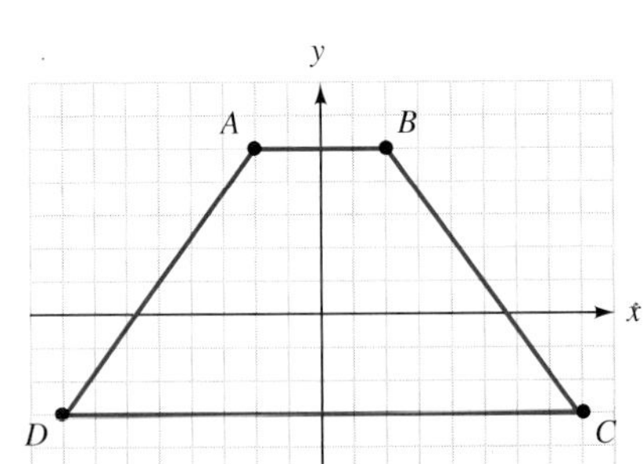

NOTATION *Complete each solution.*

15. Solve: $\sqrt{x-3} = 5$

$(\quad)^2 = \quad^2$

$x - 3 = \quad$

$x = 28$

16. Solve: $\sqrt{2x-18} = \sqrt{x-1}$

$(\quad)^2 = (\sqrt{x-1})^2$

$\quad = x - 1$

$\quad - 18 = -1$

$x = \quad$

PRACTICE *Solve each equation.* ***Check all solutions.***

17. $\sqrt{x} = 3$
18. $\sqrt{x} = 5$
19. $\sqrt{2a} = 4$
20. $\sqrt{3a} = 9$
21. $\sqrt{r} + 4 = 0$
22. $\sqrt{r} + 1 = 0$
23. $-\sqrt{x} = -5$
24. $-\sqrt{x} = -12$
25. $10 - \sqrt{s} = 7$
26. $-4 = 6 - \sqrt{s}$
27. $\sqrt{x+3} = 2$
28. $\sqrt{x-2} = 3$
29. $\sqrt{3-T} = -2$
30. $\sqrt{7+2x} = -4$
31. $\sqrt{6+2x} = 4$
32. $\sqrt{5-T} = 10$
33. $\sqrt{5x-5} - 5 = 0$
34. $\sqrt{6x+19} - 7 = 0$
35. $\sqrt{x+3} + 5 = 12$
36. $\sqrt{x-5} - 3 = 4$
37. $x - 3 = \sqrt{x^2 - 15}$
38. $v - 2 = \sqrt{v^2 - 16}$
39. $\sqrt{3t-9} = \sqrt{t+1}$
40. $\sqrt{a-3} = \sqrt{2a-8}$
41. $\sqrt{10-3x} = \sqrt{2x+20}$
42. $\sqrt{1-2x} = \sqrt{x+10}$
43. $\sqrt{3c-8} - \sqrt{c} = 0$
44. $\sqrt{2x} - \sqrt{x+8} = 0$
45. $x - 1 = \sqrt{x^2 - 4x + 9}$
46. $3d = \sqrt{9d^2 - 2d + 8}$
47. $\sqrt{4m^2 + 6m + 6} = -2m$
48. $\sqrt{9t^2 + 4t + 20} = -3t$
49. $y - 9 = \sqrt{y-3}$
50. $m - 9 = \sqrt{m-7}$
51. $\sqrt{15-3t} = t - 5$
52. $\sqrt{1-8s} = s + 4$
53. $b = \sqrt{2b-2} + 1$
54. $c = \sqrt{5c+1} - 1$
55. $\sqrt{24+10n} - n = 4$
56. $\sqrt{7+6y} - y = 2$
57. $\sqrt{3x+3} = 3\sqrt{x-1}$
58. $2\sqrt{4x+5} = 5\sqrt{x+4}$
59. $2\sqrt{3x+4} = \sqrt{5x+9}$
60. $\sqrt{3x+6} = 2\sqrt{2x-11}$
61. $\sqrt[3]{x} = 7$
62. $\sqrt[3]{x} = -9$
63. $\sqrt[3]{x-1} = 4$
64. $\sqrt[3]{2x+5} = 3$
65. $\sqrt[3]{\frac{1}{2}x - 3} = 2$
66. $\sqrt[3]{x+4} = 1$
67. $\sqrt[3]{7n-1} + 1 = 4$
68. $\sqrt[3]{12m+4} + 2 = 6$

Find the distance between the two points. If an answer contains a radical, give an exact answer and an approximate answer to two decimal places.

69. (3, −4) and (0, 0)
70. (0, 0) and (−6, 8)
71. (2, 4) and (5, 9)
72. (5, 9) and (9, 13)
73. (−2, −8) and (3, 4)
74. (−5, −2) and (7, 3)
75. (6, 8) and (12, 16)
76. (10, 4) and (2, −2)

APPLICATIONS

77. NIAGARA FALLS The distance s (in feet) that an object will fall in t seconds is given by the formula

$$t = \frac{\sqrt{s}}{4}$$

The time it took a stuntman to go over Niagara Falls in a barrel was 3.25 seconds. Substitute 3.25 for t and solve the equation for s to find the height of the waterfall.

78. THE WASHINGTON MONUMENT Gabby Street, a baseball player of the 1920s, was known for once catching a ball dropped from the top of the Washington Monument. If the ball fell for slightly less than 6 seconds before it was caught, find the approximate height of the monument. (*Hint:* See Exercise 77.)

79. PENDULUMS The time t (in seconds) required for a pendulum of length L feet to swing through one back-and-forth cycle, called its **period,** is given by the formula

$$t = 1.11\sqrt{L}$$

The Foucault pendulum in Chicago's Museum of Science and Industry is used to demonstrate the rotation of the Earth. The pendulum completes one cycle in 8.91 seconds. To the nearest tenth of a foot, how long is the pendulum?

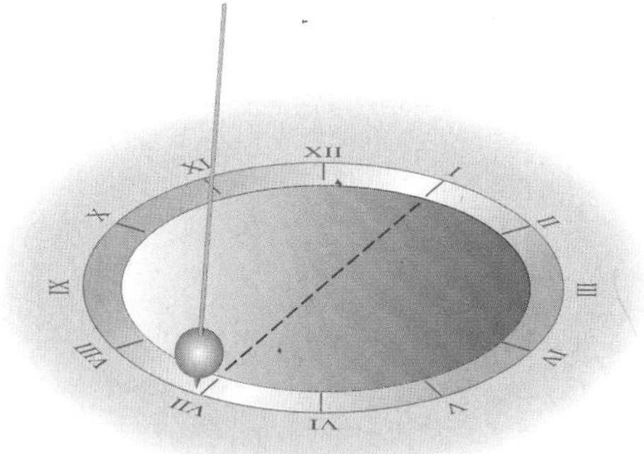

80. POWER USAGE The current I (in amperes), the resistance R (in ohms), and the power P (in watts) are related by the formula

$$I = \sqrt{\frac{P}{R}}$$

Find the power (to the nearest watt) used by a space heater that draws 7 amps when the resistance is 10.2 ohms.

81. ROAD SAFETY The formula $s = k\sqrt{d}$ relates the speed s (in mph) of a car and the distance d of the skid when a driver hits the brakes. On wet pavement, $k = 3.24$. How far will a car skid if it is going 55 mph?

82. ROAD SAFETY How far will the car in Exercise 73 skid if it is traveling on dry pavement? On dry pavement, $k = 5.34$.

83. SATELLITE ORBITS The orbital speed s of an Earth satellite is related to its distance r from the Earth's center by the formula

$$\sqrt{r} = \frac{2.029 \times 10^7}{s}$$

If the satellite's orbital speed is 7×10^3 meters per second, find its altitude a (in meters) above the Earth's surface, as shown.

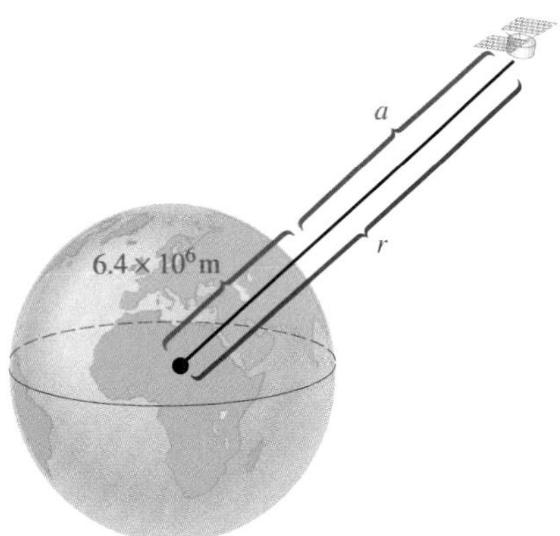

84. HIGHWAY DESIGNS A highway curve banked at 8° will accommodate traffic traveling at speed s (in mph) if the radius of the curve is r (feet), according to the equation $s = 1.45\sqrt{r}$. If highway engineers expect traffic to travel at 65 mph, to the nearest foot, what radius should they specify?

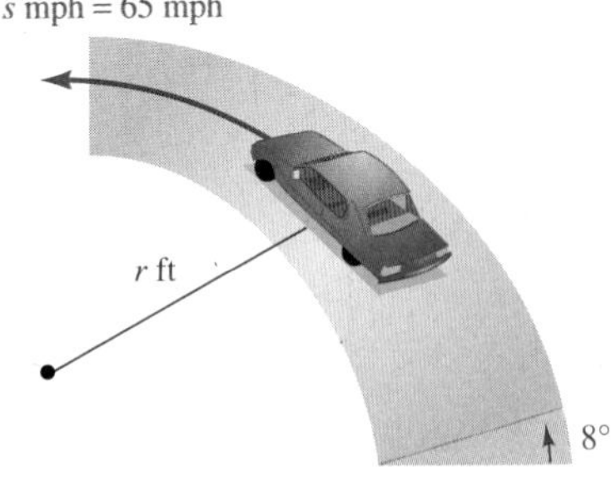

85. GEOMETRY The radius of a cone with volume V and height h is given by the formula

$$r = \sqrt{\frac{3V}{\pi h}}$$

Solve the equation for V.

86. WINDMILLS The power produced by a certain windmill is related to the speed of the wind by the formula

$$s = \sqrt[3]{\frac{P}{0.02}}$$

where P is the power (in watts) and s is the speed of the wind (in mph). How much power will the windmill produce if the wind is blowing at 30 mph?

87. NAVIGATION An oil tanker is to travel from Tunisia to Italy, as shown below. The captain wants to travel a course that is always the same distance from a point on the coast of Sardinia as it is from a point on the coast of Sicily. How far will the tanker be from these points when it reaches

a. position 1?

b. position 2?

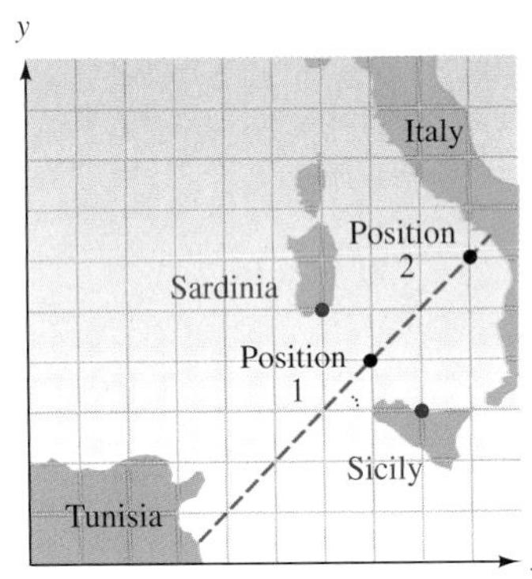

88. DECK DESIGN The plans for a patio deck shown in the illustration call for three redwood support braces directly under the hot tub. Find the length of each support. Round to the nearest tenth of a foot.

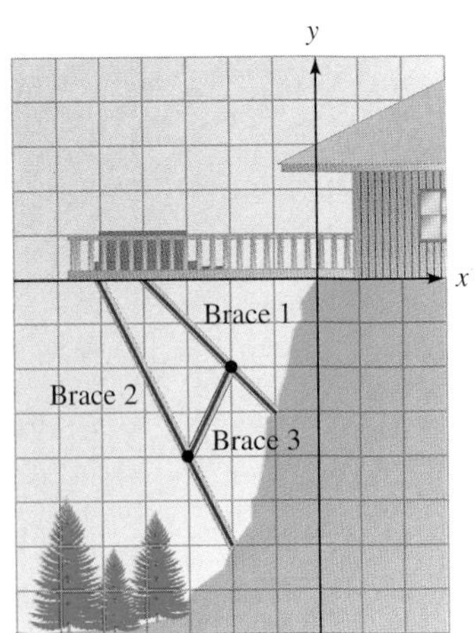

WRITING

89. Explain why a check is necessary when solving radical equations.

90. How would you know, without solving it, that the equation $\sqrt{x+2} = -4$ has no solution?

REVIEW *Perform the operations.*

91. $(3x^2 + 2x) + (5x^2 - 8x)$

92. $(7a^2 + 2a - 5) - (3a^2 - 2a + 1)$

93. $(x + 3)(x + 3)$

94. $x - 1\overline{)x^2 - 6x + 5}$

95. $(3y - 7)^2$

96. $(3y + 7)^2$

14.7 Rational Exponents

- Fractional exponents with numerators of 1
- Fractional exponents with numerators other than 1
- Rules for exponents

We have seen that a positive integer exponent indicates the number of times that a base is to be used as a factor in a product. For example, x^4 means that x is to be used as a factor four times.

$$x^4 = \overbrace{x \cdot x \cdot x \cdot x}^{\text{4 factors of } x}$$

Also, recall the following rules for exponents.

Rules for exponents

If m and n represent natural numbers and there are no divisions by zero, then

$$x^m x^n = x^{m+n} \qquad (x^m)^n = x^{mn} \qquad (xy)^n = x^n y^n \qquad \left(\frac{x}{y}\right)^n = \frac{x^n}{y^n}$$

$$x^0 = 1 \qquad x^{-n} = \frac{1}{x^n} \qquad \frac{x^m}{x^n} = x^{m-n}$$

In this section, we will extend the definition and rules for exponents to cover fractional exponents.

Fractional exponents with numerators of 1

It is possible to raise numbers to fractional powers. To give meaning to **rational** (or **fractional**) **exponents,** we consider $\sqrt{7}$. Because $\sqrt{7}$ is the positive number whose square is 7, we have

$$(\sqrt{7})^2 = 7$$

We now consider the symbol $7^{1/2}$. If fractional exponents are to follow the same rules as integer exponents, the square of $7^{1/2}$ must be 7, because

$$\begin{aligned} (7^{1/2})^2 &= 7^{(1/2)2} && \text{Keep the base 7 and multiply the exponents.} \\ &= 7^1 && \tfrac{1}{2} \cdot 2 = 1. \\ &= 7 \end{aligned}$$

Since $(7^{1/2})^2$ and $(\sqrt{7})^2$ are both equal to 7, we define $7^{1/2}$ to be $\sqrt{7}$. Similarly, we make these definitions.

$$7^{1/3} = \sqrt[3]{7}$$
$$7^{1/7} = \sqrt[7]{7}$$

and so on.

The definition of $x^{1/n}$

If n represents a positive integer greater than 1 and $\sqrt[n]{x}$ represents a real number, then

$$x^{1/n} = \sqrt[n]{x}$$

Self Check 1
Simplify:
a. $81^{1/2}$
b. $125^{1/3}$
c. $(-27)^{1/3}$

Answers **a.** 9, **b.** 5, **c.** -3

EXAMPLE 1 Simplify: **a.** $64^{1/2}$, **b.** $64^{1/3}$, and **c.** $(-64)^{1/3}$.

Solution

a. $64^{1/2} = \sqrt{64} = 8$ The denominator of the fractional exponent is 2. Therefore, we find the square root of the base, 64.

b. $64^{1/3} = \sqrt[3]{64} = 4$ The denominator of the fractional exponent is 3. Therefore, we find the cube root of the base, 64.

c. $(-64)^{1/3} = \sqrt[3]{-64} = -4$ The denominator of the fractional exponent is 3. Therefore, we find the cube root of the base, -64.

Fractional exponents with numerators other than 1

We can extend the definition of $x^{1/n}$ to cover fractional exponents for which the numerator is not 1. For example, because $4^{3/2}$ can be written as $(4^{1/2})^3$, we have

$$4^{3/2} = (4^{1/2})^3 = (\sqrt{4})^3 = 2^3 = 8$$

Because $4^{3/2}$ can also be written as $(4^3)^{1/2}$, we have

$$4^{3/2} = (4^3)^{1/2} = 64^{1/2} = \sqrt{64} = 8$$

In general, $x^{m/n}$ can be written as $(x^{1/n})^m$ or as $(x^m)^{1/n}$. Since $(x^{1/n})^m = (\sqrt[n]{x})^m$, and $(x^m)^{1/n} = \sqrt[n]{x^m}$, we make the following definition.

The definition of $x^{m/n}$

If m and n represent positive integers $(n \neq 1)$ and $\sqrt[n]{x}$ represents a real number, then

$$x^{m/n} = \sqrt[n]{x^m} = (\sqrt[n]{x})^m$$

Self Check 2
Simplify:
a. $16^{3/2}$
b. $(-8)^{4/3}$

EXAMPLE 2 Simplify: **a.** $8^{2/3}$ and **b.** $(-27)^{4/3}$.

Solution These expressions can be simplified in two ways. Using the first method, we take the root of the base and then we find the power. The second method is to find the power first and then take the root.

a. $8^{2/3} = (\sqrt[3]{8})^2$ or $8^{2/3} = \sqrt[3]{8^2}$
$= 2^2$ $= \sqrt[3]{64}$
$= 4$ $= 4$

b. $(-27)^{4/3} = (\sqrt[3]{-27})^4$ or $(-27)^{4/3} = \sqrt[3]{(-27)^4}$
$= (-3)^4$ $= \sqrt[3]{531{,}441}$
$= 81$ $= 81$

Answers **a.** 64, **b.** 16

The work in Example 2 suggests that in order to avoid large numbers, it is usually easier to take the root of the base first and then find the power.

EXAMPLE 3 Simplify: **a.** $125^{4/3}$, **b.** $9^{5/2}$, **c.** $-25^{3/2}$, and **d.** $(-27)^{2/3}$.

Solution

a. $125^{4/3} = (\sqrt[3]{125})^4 = (5)^4 = 625$

b. $9^{5/2} = (\sqrt{9})^5 = (3)^5 = 243$

c. $-25^{3/2} = -(\sqrt{25})^3 = -(5)^3 = -125$

d. $(-27)^{2/3} = (\sqrt[3]{-27})^2 = (-3)^2 = 9$

Self Check 3
Simplify:
a. $100^{3/2}$
b. $(-8)^{2/3}$

Answers **a.** 1,000, **b.** 4

Fractional exponents

CALCULATOR SNAPSHOT

To use a scientific calculator to evaluate an exponential expression containing a fractional exponent, we can use the y^x key. For example, to evaluate $6^{-2/3}$, we enter these numbers and press these keys.

6 [y^x] [(] 2 [+/−] [÷] 3 [)] [=]   0.302853432

So $6^{-2/3} \approx 0.302853432$.

To use a graphing calculator to evaluate $6^{-2/3}$, we press the following keys.

6 [^] [(] [(−)] 2 [÷] 3 [)] [ENTER]   6^(⁻2/3)
.3028534321

Rules for exponents

Because of the way in which $x^{1/n}$ and $x^{m/n}$ are defined, the familiar rules for exponents are valid for rational exponents. The following example illustrates the use of each rule.

EXAMPLE 4 Simplify: **a.** $4^{2/5}4^{1/5}$, **b.** $(5^{2/3})^{1/2}$, **c.** $(3x)^{2/3}$, **d.** $\dfrac{4^{3/5}}{4^{2/5}}$, **e.** $\left(\dfrac{3}{2}\right)^{2/5}$, **f.** $4^{-2/3}$, and **g.** $(5^{1/3})^0$.

Self Check 4
Simplify:
a. $5^{1/3}5^{1/3}$ **b.** $(5^{1/3})^4$
c. $(3x)^{1/5}$ **d.** $\frac{5^{3/7}}{5^{2/7}}$ **e.** $\left(\frac{2}{3}\right)^{2/3}$
f. $5^{-2/7}$ **g.** $(12^{1/2})^0$

Answers **a.** $5^{2/3}$, **b.** $5^{4/3}$, **c.** $3^{1/5}x^{1/5}$, **d.** $5^{1/7}$, **e.** $\frac{2^{2/3}}{3^{2/3}}$, **f.** $\frac{1}{5^{2/7}}$, **g.** 1

Solution

a. $4^{2/5}4^{1/5} = 4^{2/5+1/5} = 4^{3/5}$ $\quad x^m x^n = x^{m+n}$.

b. $(5^{2/3})^{1/2} = 5^{(2/3)(1/2)} = 5^{1/3}$ $\quad (x^m)^n = x^{m \cdot n}$.

c. $(3x)^{2/3} = 3^{2/3}x^{2/3}$ $\quad (xy)^m = x^m y^m$.

d. $\frac{4^{3/5}}{4^{2/5}} = 4^{3/5-2/5} = 4^{1/5}$ $\quad \frac{x^m}{x^n} = x^{m-n}$.

e. $\left(\frac{3}{2}\right)^{2/5} = \frac{3^{2/5}}{2^{2/5}}$ $\quad \left(\frac{x}{y}\right)^n = \frac{x^n}{y^n}$.

f. $4^{-2/3} = \frac{1}{4^{2/3}}$ $\quad x^{-n} = \frac{1}{x^n}$.

g. $(5^{1/3})^0 = 1$ $\quad x^0 = 1$.

We can use the rules for exponents to simplify expressions containing rational exponents.

Self Check 5
Simplify:
a. $25^{-3/2}$
b. $(x^3)^{1/3}$
c. $(x^6y^9)^{-2/3}$

Answers **a.** $\frac{1}{125}$, **b.** x, **c.** $\frac{1}{x^4y^6}$

EXAMPLE 5

Simplify: **a.** $64^{-2/3}$, **b.** $(x^2)^{1/2}$, **c.** $(x^6y^4)^{1/2}$, and **d.** $(27x^{12})^{-1/3}$. All variables represent positive numbers.

Solution

a.
$$64^{-2/3} = \frac{1}{64^{2/3}} = \frac{1}{(64^{1/3})^2} = \frac{1}{4^2} = \frac{1}{16}$$

b.
$$(x^2)^{1/2} = x^{2(1/2)} = x^1 = x$$

c.
$$(x^6y^4)^{1/2} = x^{6(1/2)}y^{4(1/2)} = x^3y^2$$

d.
$$(27x^{12})^{-1/3} = \frac{1}{(27x^{12})^{1/3}} = \frac{1}{27^{1/3}x^{12(1/3)}} = \frac{1}{3x^4}$$

Self Check 6
Simplify:
a. $x^{2/3}x^{1/2}$
b. $\frac{x^{2/3}}{2x^{1/4}}$

EXAMPLE 6

Simplify: **a.** $x^{1/3}x^{1/2}$, **b.** $\frac{3x^{2/3}}{6x^{1/5}}$, and **c.** $\frac{2x^{-1/2}}{x^{3/4}}$.

Solution

a. $x^{1/3}x^{1/2} = x^{2/6}x^{3/6}$ Get a common denominator for the fractional exponents.

$= x^{5/6}$ Keep the base x and add the exponents.

b. $\frac{3x^{2/3}}{6x^{1/5}} = \frac{3x^{10/15}}{6x^{3/15}}$ Get a common denominator for the fractional exponents.

$= \frac{1}{2}x^{10/15-3/15}$ Simplify $\frac{3}{6}$. Keep the base x and subtract the exponents.

$= \frac{1}{2}x^{7/15}$

c. $\dfrac{2x^{-1/2}}{x^{3/4}} = \dfrac{2x^{-2/4}}{x^{3/4}}$ Get a common denominator for the fractional exponents.

$= 2x^{-2/4-3/4}$ Keep the base x and subtract the exponents.

$= 2x^{-5/4}$ Simplify.

$= \dfrac{2}{x^{5/4}}$ $x^{-5/4} = \dfrac{1}{x^{5/4}}$.

Answers **a.** $x^{7/6}$, **b.** $\frac{1}{2}x^{5/12}$

Section 14.7 STUDY SET

VOCABULARY *Fill in the blanks.*

1. A fractional exponent is also called a ________ exponent.

2. In the expression $27^{1/3}$, 27 is called the ______ and the exponent is ___.

CONCEPTS *Complete each rule for exponents.*

3. $x^m x^n =$ ___

4. $(x^m)^n =$ ___

5. $\left(\dfrac{x}{y}\right)^n =$ ___

6. $x^0 =$ ___

7. $x^{-n} =$ ___

8. $\dfrac{x^m}{x^n} =$ ___

9. $x^{1/n} =$ ___

10. $x^{m/n} =$ ___

11. Write $\sqrt{5}$ using a fractional exponent.

12. Write $5^{1/3}$ using a radical.

13. Write $8^{4/3}$ using a radical.

14. Write $(\sqrt{8})^3$ using a fractional exponent.

15. Complete the table.

x	$x^{1/2}$
0	
1	
4	
9	

16. Complete the table.

x	$x^{1/3}$
0	
−1	
−8	
8	

17. Graph each number on the number line.
$\{8^{1/3}, 17^{1/2}, 2^{3/2}, -5^{2/3}\}$

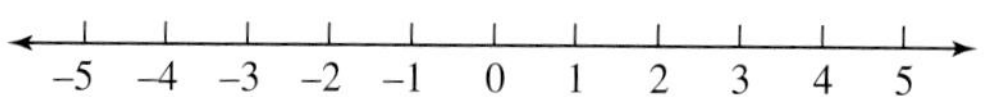

18. Graph each number on the number line.
$\{4^{-1/2}, 64^{-2/3}, (-8)^{-1/3}\}$

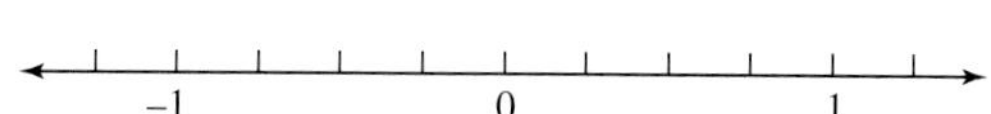

NOTATION *Complete each solution.*

19. Simplify: $(-216)^{4/3}$.

$(-216)^{4/3} = (\sqrt[3]{\quad})^4$

$= (\quad)^4$

$= 1{,}296$

20. Simplify: $\dfrac{3x^{-2/3}}{x^{3/4}}$.

$\dfrac{3x^{-2/3}}{x^{3/4}} = \dfrac{3x^{-8/12}}{\quad}$

$= 3x^{-8/12-\quad}$

$= 3x^{-17/12}$

PRACTICE *Simplify each expression.*

21. $81^{1/2}$

22. $100^{1/2}$

23. $-144^{1/2}$

24. $-400^{1/2}$

25. $\left(\dfrac{1}{4}\right)^{1/2}$

26. $\left(\dfrac{1}{25}\right)^{1/2}$

27. $\left(\dfrac{4}{49}\right)^{1/2}$

28. $\left(\dfrac{9}{64}\right)^{1/2}$

29. $27^{1/3}$
30. $8^{1/3}$
31. $-125^{1/3}$
32. $-1{,}000^{1/3}$
33. $(-8)^{1/3}$
34. $(-125)^{1/3}$
35. $\left(\frac{27}{64}\right)^{1/3}$
36. $\left(\frac{64}{125}\right)^{1/3}$
37. $81^{3/2}$
38. $16^{3/2}$
39. $25^{3/2}$
40. $4^{5/2}$
41. $125^{2/3}$
42. $8^{4/3}$
43. $1{,}000^{2/3}$
44. $27^{2/3}$
45. $(-8)^{2/3}$
46. $(-125)^{2/3}$
47. $\left(\frac{8}{27}\right)^{2/3}$
48. $\left(\frac{49}{64}\right)^{3/2}$

Simplify each expression. Write your answers without using negative exponents.

49. $6^{3/5}6^{2/5}$
50. $3^{4/7}3^{3/7}$
51. $5^{2/3}5^{4/3}$
52. $2^{7/8}2^{9/8}$
53. $(7^{2/5})^{5/2}$
54. $(8^{1/3})^3$
55. $(5^{2/7})^7$
56. $(3^{3/8})^8$
57. $\frac{8^{3/2}}{8^{1/2}}$
58. $\frac{11^{9/7}}{11^{2/7}}$
59. $\frac{5^{11/3}}{5^{2/3}}$
60. $\frac{27^{13/15}}{27^{8/15}}$
61. $4^{-1/2}$
62. $8^{-1/3}$
63. $27^{-2/3}$
64. $36^{-3/2}$
65. $16^{-3/2}$
66. $100^{-5/2}$
67. $(-27)^{-4/3}$
68. $(-8)^{-4/3}$

Simplify each expression. All variables represent positive numbers.

69. $(x^{1/2})^2$
70. $(x^9)^{1/3}$
71. $(x^{12})^{1/6}$
72. $(x^{18})^{1/9}$
73. $x^{5/6}x^{7/6}$
74. $x^{2/3}x^{7/3}$
75. $y^{4/7}y^{10/7}$
76. $y^{5/11}y^{6/11}$
77. $\frac{x^{3/5}}{x^{1/5}}$
78. $\frac{x^{4/3}}{x^{2/3}}$
79. $\frac{x^{1/7}x^{3/7}}{x^{2/7}}$
80. $\frac{x^{5/6}x^{5/6}}{x^{7/6}}$
81. $x^{2/3}x^{3/4}$
82. $a^{3/5}a^{1/2}$
83. $(b^{1/2})^{3/5}$
84. $(x^{2/5})^{4/7}$
85. $\frac{t^{2/3}}{t^{2/5}}$
86. $\frac{p^{3/4}}{p^{1/3}}$
87. $\left(\frac{x^{4/5}}{x^{2/15}}\right)^3$
88. $\left(\frac{y^{2/3}}{y^{1/5}}\right)^{15}$

APPLICATIONS *If an answer is not exact, round to the nearest tenth.*

89. SPEAKERS The formula $A = V^{2/3}$ can be used to find the area A of one face of a cube if its volume V is known. Find the amount of floor space on the dance floor taken up by the speakers shown below if each speaker is a cube with a volume of 2,744 cubic inches.

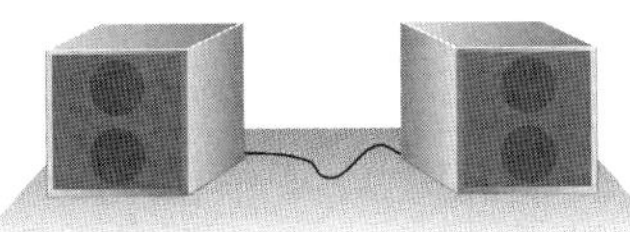

90. MEDICAL TESTS Before a series of X-rays are taken, a patient is injected with a special contrast mixture that highlights obstructions in his blood vessels. The amount of the original dose of contrast material remaining in the patient's bloodstream h hours after it is injected is given by $h^{-3/2}$. How much of the contrast material remains in the patient's bloodstream 4 hours after the injection?

91. HOLIDAY DECORATING Find the length s of each string of colored lights used to decorate an evergreen tree in the manner shown below if $s = (r^2 + h^2)^{1/2}$.

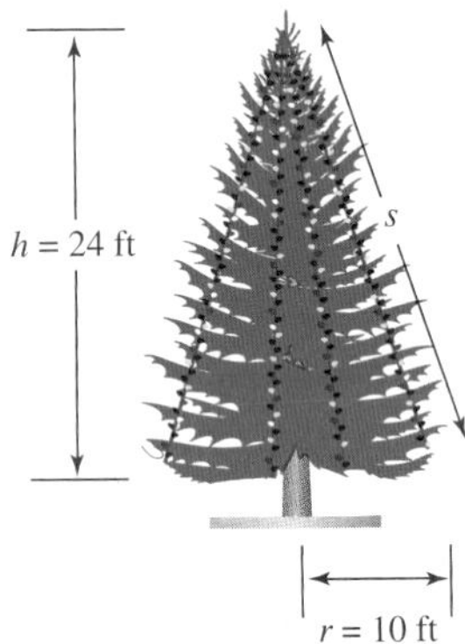

92. VISIBILITY The distance d in miles a person in an airplane can see to the horizon on a clear day is given by the formula $d = 1.22a^{1/2}$, where a is the altitude of the plane in feet. Find d in the illustration.

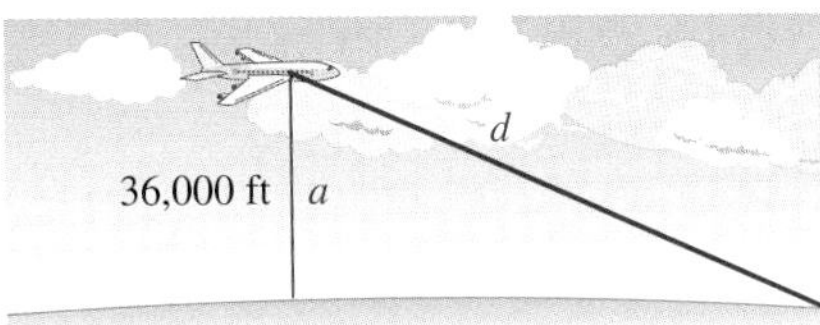

93. TOY DESIGNS Knowing the volume V of a sphere, we can find its radius r using the formula

$$r = \left(\frac{3V}{4\pi}\right)^{1/3}$$

If the volume occupied by a ball is 2π cubic inches, find its radius.

94. EXERCISE EQUIPMENT Find the length l of the incline bench in the illustration, using the formula $l = (a^2 + b^2)^{1/2}$.

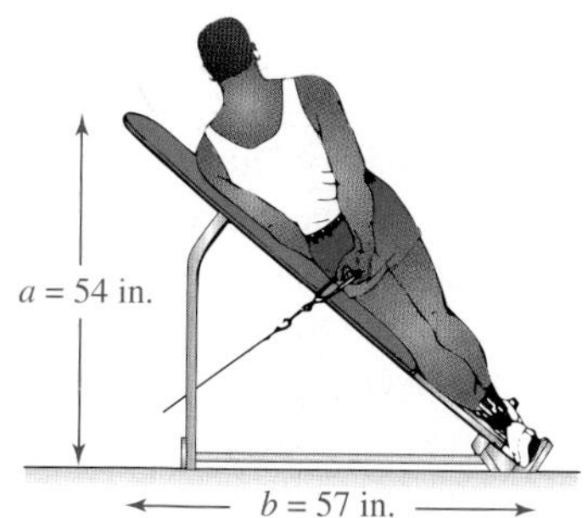

WRITING

95. What is a rational exponent? Give several examples.

96. Explain this statement: *In the expression $16^{3/2}$, the number 3/2 requires that two operations be performed on 16.*

REVIEW *Graph each equation.*

97. $x = 3$

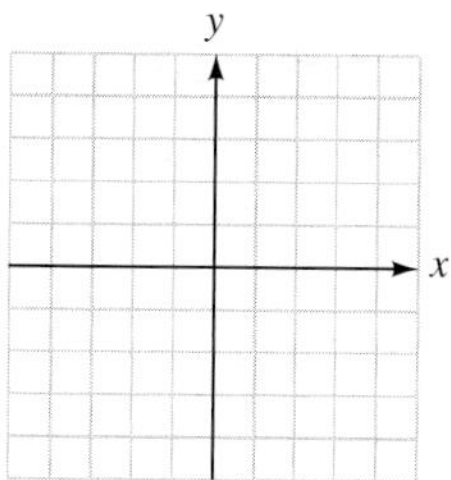

98. $y = -3$

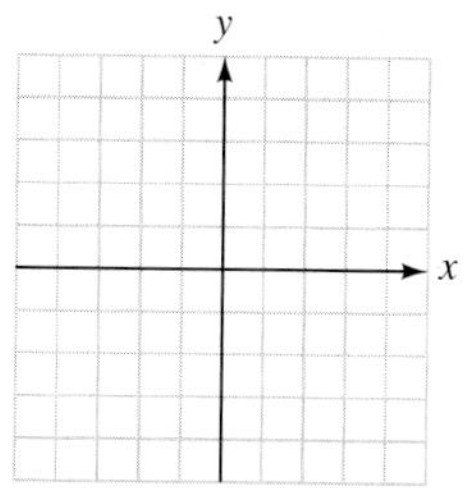

99. $-2x + y = 4$

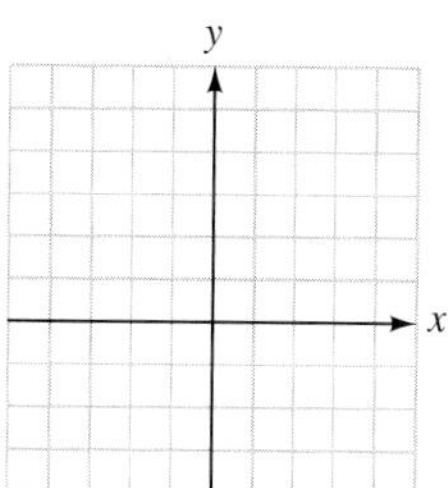

100. $4x - y = 4$

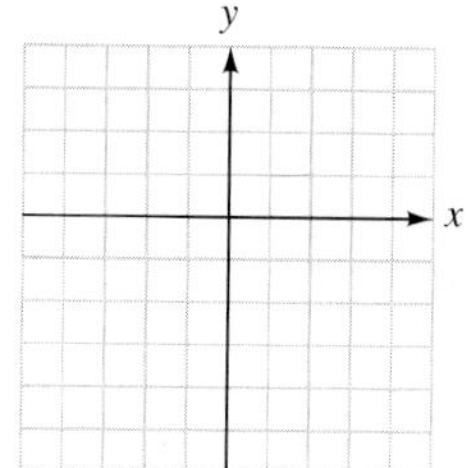

KEY CONCEPT

Inverse Operations

We have performed six operations with real numbers: addition, subtraction, multiplication, division, raising to a power, and finding a root. We have seen that there is a special relationship between *pairs* of operations. That is, subtraction does the opposite of addition, division does the opposite of multiplication, and finding a root does the opposite of raising to a power. Because of this, we call each pair **inverse operations.** Subtraction is the inverse operation of addition, division is the inverse operation of multiplication, and finding a root is the inverse operation of raising to a power.

Solving equations

When solving equations, we use inverse operations to isolate the variable on one side of the equation.

Determine what operation is performed on the variable and what inverse operation should be used to isolate the variable. Then solve the equation.

1. $x + 2 = -4$

2. $x - 5 = 10$

3. $-6x = 24$

4. $\frac{x}{2} = 40$

5. $\sqrt{x} = 7$

6. $x^2 = 169$ (assume $x > 0$)

7. $\sqrt[3]{x} = -2$

8. $x^3 = 64$

When solving equations, we must often undo several operations to isolate the variable. Recall that these operations are undone in the *reverse* order.

Solve each equation and check the result.

9. $-2x - 4 = 6$

10. $\frac{3x}{5} + 3 = 9$

11. $\sqrt{x + 1} = 4$

12. $x^2 = 9$ (assume $x > 0$)

13. $\sqrt{x} - 3 = 5$

14. $\sqrt[3]{x} - 3 = 1$

Applications

We can use the concept of inverse operation to find the length of a side of the cube in the illustration below if we know the area of a face or the volume of the cube.

15. To find the area of a face of the cube, we square the length of a side. How could we find the length of a side, knowing the area of a face?

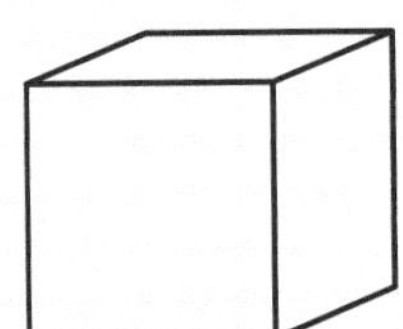

16. To find the volume of the cube, we cube the length of a side. How could we find the length of a side if we knew the volume of the cube?

ACCENT ON TEAMWORK

SECTION 14.1

THE PYTHAGOREAN THEOREM Put 12 knots in a rope, each 1 foot apart, and connect the ends as shown below. Hammer three tent stakes in the ground so that the rope forms a triangle with sides of length 3, 4, and 5 spaces. Make some observations about the triangle. Use the Pythagorean theorem to prove one of your observations.

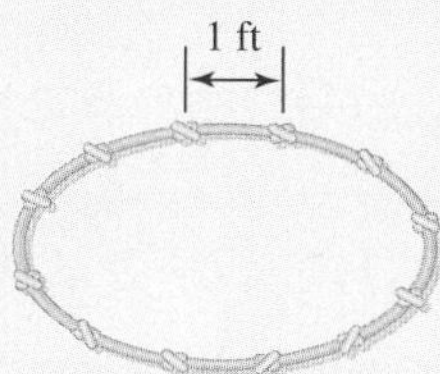

A SPIRAL OF ROOTS For this project, you will need a piece of poster board, a protractor, a yardstick, and a pencil. Begin by drawing an isosceles right triangle near the right margin of the poster board. Label the length of each leg as 1 unit. (See the illustration below.) Use the Pythagorean theorem to determine the length of the hypotenuse. Draw a second right triangle using the hypotenuse of the first triangle as one leg. Draw its second leg with a length of 1 unit. Find the length of the hypotenuse of triangle 2. Continue this process of creating right triangles, using the previous hypotenuse as one leg and drawing a new second leg of length 1 unit each time. Calculate the length of the resulting hypotenuse. What patterns, if any, do you see?

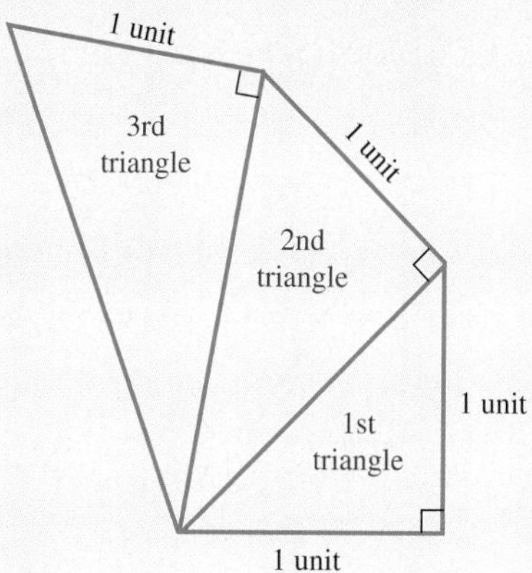

SECTION 14.2

nTH ROOTS Use the $\boxed{\sqrt[x]{y}}$ key on a scientific calculator to approximate $\sqrt{2}$, $\sqrt[3]{2}$, $\sqrt[4]{2}$, $\sqrt[5]{2}$, and $\sqrt[6]{2}$. Do you see any pattern? Explain it in words.

SECTION 14.3

SIMPLIFYING RADICAL EXPRESSIONS Suppose you are the algebra instructor of a student whose work is shown here. Write a note to the student explaining how she could save some steps in simplifying $\sqrt{72}$.

$$\begin{aligned}\sqrt{72} &= \sqrt{4\cdot 18}\\ &= \sqrt{4}\sqrt{18}\\ &= 2\sqrt{18}\\ &= 2\sqrt{9\cdot 2}\\ &= 2\sqrt{9}\sqrt{2}\\ &= 2(3)\sqrt{2}\\ &= 6\sqrt{2}\end{aligned}$$

SECTION 14.4

COMMON ERRORS In each addition and subtraction problem below, tell what mistake was made. Compare each problem to a similar one involving variables to clarify your explanation. For example, compare Problem a to $2x + 3x$ to help you explain the correct procedure that should be used to simplify the expression.

a. $2\sqrt{5} + 3\sqrt{5} = 5\sqrt{10}$ **b.** $30 + 2\sqrt{2} = 32\sqrt{2}$

c. $7\sqrt{3} - 5\sqrt{3} = 2$ **d.** $6\sqrt{7} - 3\sqrt{2} = 3\sqrt{5}$

SECTION 14.5

RATIONALIZING NUMERATORS Some problems in advanced mathematics require that the numerator of a fraction be rationalized. Extend the concepts studied in this section to develop a method to rationalize the numerators of

$$\frac{\sqrt{5}}{3}, \quad \frac{\sqrt{7}}{\sqrt{5}}, \quad \frac{\sqrt{y}}{6y}, \quad \text{and} \quad \frac{\sqrt{3}-\sqrt{2}}{12}$$

SECTION 14.6

SOLVING RADICAL EQUATIONS In this chapter, we solved equations that contained two radicals. The radicals in those equations always had the same index. That is not the case for the following equation:

$$\sqrt{x} = \sqrt[3]{2x}$$

Brainstorm in your group to develop a procedure that can be used to solve this equation. What are its solutions?

SECTION 14.7

GRAPHING Approximate the x- and y-coordinates of the following ordered pairs to the nearest tenth, then graph them on a rectangular coordinate system. (*Hint:* Each quadrant should contain only one point.)

$(\sqrt{2}, 3^{1/2})$ $(-\sqrt{6}, 5^{3/2})$

$(-16^{2/3}, -\sqrt[3]{25})$ $(9^{-1/2}, \sqrt[3]{-10})$

CHAPTER REVIEW

SECTION 14.1 Square Roots

CONCEPTS

The number b is a *square root* of a if $b^2 = a$.

The *principal square root* of a positive number a, denoted by $\sqrt{a}$, is the positive square root of a.

The expression within a *radical symbol* $\sqrt{}$ is called the *radicand.*

If a positive number is not a perfect square, its square root is irrational. Square roots of negative numbers are called *imaginary numbers.*

The Pythagorean theorem: If the length of the hypotenuse of a right triangle is c and the lengths of the two legs are a and b, then $c^2 = a^2 + b^2$.

If a and b are positive numbers, and $a = b$, then $\sqrt{a} = \sqrt{b}$.

REVIEW EXERCISES

1. Fill in the blank: 4 is a ______ root of 16, because $4^2 = 16$.

Find each square root. **Do not use a calculator.**

2. $\sqrt{25}$
3. $\sqrt{49}$
4. $-\sqrt{144}$
5. $-\sqrt{\frac{16}{81}}$
6. $\sqrt{900}$
7. $-\sqrt{0.64}$
8. $\sqrt{1}$
9. $\sqrt{0}$

Use a calculator to approximate each expression to three decimal places.

10. $\sqrt{21}$
11. $-\sqrt{15}$
12. $2\sqrt{7}$
13. $\sqrt{751.9}$

14. Determine whether each number is rational, irrational, or imaginary. Which is not a real number? $\{\sqrt{-2}, \sqrt{68}, \sqrt{81}, \sqrt{3}\}$

Complete the table of values for each function and then graph it.

15. $f(x) = \sqrt{x}$

x	f(x)
0	
1	
4	
9	

16. $f(x) = 2 - \sqrt{x}$

x	f(x)
0	
1	
4	
9	

Refer to the right triangle shown.

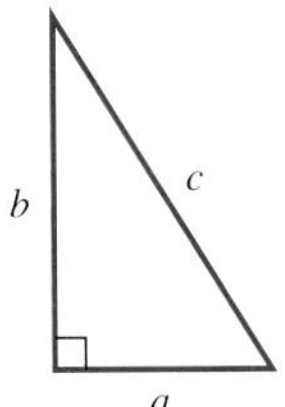

17. Find c where $a = 21$ and $b = 28$.
18. Find b where $a = 1$ and $c = \sqrt{2}$.
19. Find a where $b = 5$ and $c = 7$.

20. THEATER SEATING For the theater seats shown, how much higher is the seat at the top of the incline compared to the one at the bottom?

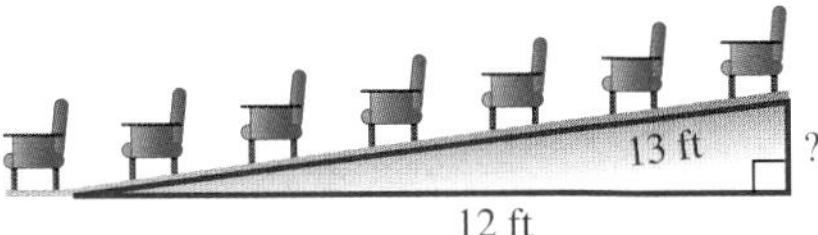

21. ROAD SIGNS To find the maximum velocity a car can safely travel around a curve without skidding, we can use the formula $v = \sqrt{2.5r}$, where v is the velocity in mph and r is the radius of the curve in feet. How should the road sign in the illustration be labeled if it is to be posted in front of a curve with a radius of 360 feet?

SECTION 14.2 Higher-Order Roots; Radicands That Contain Variables

The number b is a *cube root* of a if $b^3 = a$.

The cube root of a is denoted by $\sqrt[3]{a}$. By definition, $\sqrt[3]{a} = b$ if $b^3 = a$.

The number b is an nth root of a if $b^n = a$.

In $\sqrt[n]{a}$, the number n is called the *index* of the radical.

When n is even, we say that the radical $\sqrt[n]{x}$ is an *even root*. When n is odd, $\sqrt[n]{x}$ is an *odd root*.

$$\sqrt{a} = \sqrt[2]{a}$$

22. Fill in the blanks: $\sqrt[3]{125} = 5$, because ______ $= 125$; 5 is called the ______ root of 125.

Find each root. **Do not use a calculator.**

23. $\sqrt[3]{-27}$ **24.** $-\sqrt[3]{125}$ **25.** $\sqrt[4]{81}$ **26.** $\sqrt[5]{32}$

27. $\sqrt[3]{0}$ **28.** $\sqrt[3]{-1}$ **29.** $\sqrt[3]{\dfrac{1}{64}}$ **30.** $\sqrt[3]{1}$

Use a calculator to find each root to three decimal places.

31. $\sqrt[3]{16}$ **32.** $\sqrt[3]{-102.35}$ **33.** $\sqrt[4]{6}$ **34.** $\sqrt[5]{34{,}500}$

Find each root. Each variable represents a positive number.

35. $\sqrt{x^2}$ **36.** $\sqrt{4b^4}$ **37.** $\sqrt{x^4y^4}$ **38.** $-\sqrt{y^{12}}$

39. $\sqrt[3]{x^3}$ **40.** $\sqrt[3]{y^6}$ **41.** $\sqrt[3]{27x^3}$ **42.** $\sqrt[3]{-r^{12}}$

43. DICE Find the length of an edge of one of the dice shown if each one has a volume of 1,728 cubic millimeters.

SECTION 14.3 Simplifying Radical Expressions

The *multiplication property* of radicals: If a and b are positive or zero, then

$$\sqrt{ab} = \sqrt{a}\sqrt{b}$$

Simplified form of a radical:

1. Except for 1, the radicand has no perfect square factors.
2. No fraction appears in the radicand.
3. No radical appears in the denominator.

The *division property* of radicals:

$$\sqrt{\frac{a}{b}} = \frac{\sqrt{a}}{\sqrt{b}} \quad \text{where } b \neq 0$$

Simplify each expression. All variables represent positive numbers.

44. $\sqrt{32}$ **45.** $\sqrt{500}$

46. $\sqrt{80x^2}$ **47.** $-2\sqrt{63}$

48. $-\sqrt{250t^3}$ **49.** $-\sqrt{700z^5}$

50. $\sqrt{200x^2y}$ **51.** $\dfrac{1}{5}\sqrt{75y^4}$

52. $\sqrt[3]{8x^2y^3}$ **53.** $\sqrt[3]{250x^4y^3}$

Simplify each expression. All variables represent positive numbers.

54. $\sqrt{\dfrac{16}{25}}$ **55.** $\sqrt{\dfrac{60}{49}}$

56. $\sqrt[3]{\dfrac{1{,}000}{27}}$ **57.** $\sqrt{\dfrac{242x^4}{169x^2}}$

58. FITNESS EQUIPMENT The length of the sit-up board can be found using the Pythagorean theorem.

a. Find its length. Express the answer in simplified radical form.

b. Express your result to part a as a decimal approximation rounded to the nearest tenth.

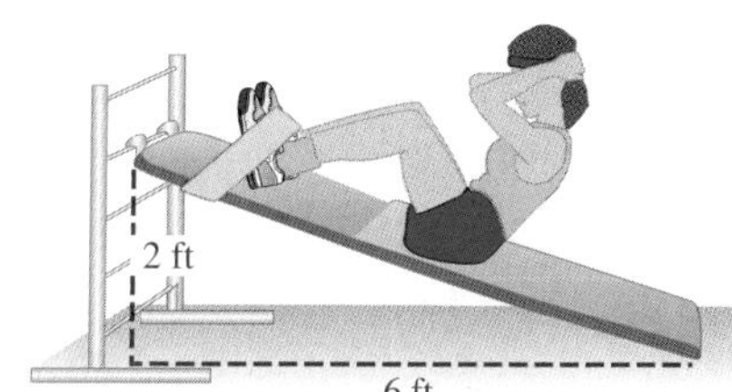

SECTION 14.4 Adding and Subtracting Radical Expressions

Radical expressions can be added or subtracted if they contain like radicals.

Radicals are called *like* radicals when they have the same index and the same radicand.

Perform the operations. All variables represent positive numbers.

59. $\sqrt{2} + \sqrt{8} - \sqrt{18}$

60. $\sqrt{3} + 4 + \sqrt{27} - 7$

61. $5\sqrt{28} - 3\sqrt{63}$

62. $3y\sqrt{5xy^3} - y^2\sqrt{20xy}$

63. $\sqrt[3]{16} + \sqrt[3]{54}$

64. $\sqrt[3]{2{,}000x^3} - \sqrt[3]{128x^3}$

65. Explain why we cannot add $3\sqrt{5}$ and $5\sqrt{3}$.

66. GARDENING Find the difference in the lengths of the two wires used to secure the tree shown.

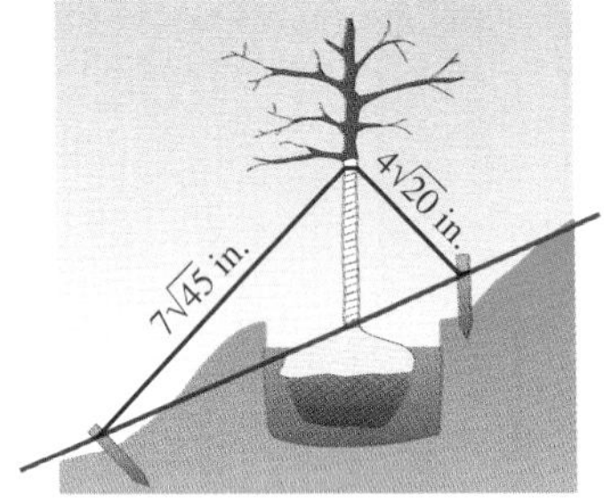

SECTION 14.5 Multiplying and Dividing Radical Expressions

The product of the square roots of two nonnegative numbers is equal to the square root of the product of those numbers.

To multiply radical expressions containing only one term, first multiply the coefficients, then multiply the radicals separately, and simplify the result.

Use the FOIL method to multiply two radical expressions, each having two terms.

If the denominator of a fraction is a square root, *rationalize* the denominator by multiplying the numerator and denominator by some appropriate square root.

If a two-term denominator of a fraction contains square roots, multiply the numerator and denominator by the *conjugate* of the denominator.

Perform the operations.

67. $\sqrt{2}\sqrt{3}$

68. $(-5\sqrt{5})(-2\sqrt{2})$

69. $(3\sqrt{3x})(4\sqrt{6x})$

70. $(\sqrt{15} + 3x)^2$

71. $\sqrt{2}(\sqrt{8} - \sqrt{18})$

72. $(\sqrt{3} + \sqrt{5})(\sqrt{3} - \sqrt{5})$

73. $(\sqrt[3]{4})(2\sqrt[3]{4})$

74. $(\sqrt[3]{3} + 2)(\sqrt[3]{3} - 1)$

75. VACUUM CLEANERS The illustration shows the amount of surface area of a rug suctioned by a vacuum nozzle attachment.

a. Find the perimeter and area of this section of rug. Express the answers in simplified radical form.

b. Express your results to part a as decimal approximations to the nearest tenth.

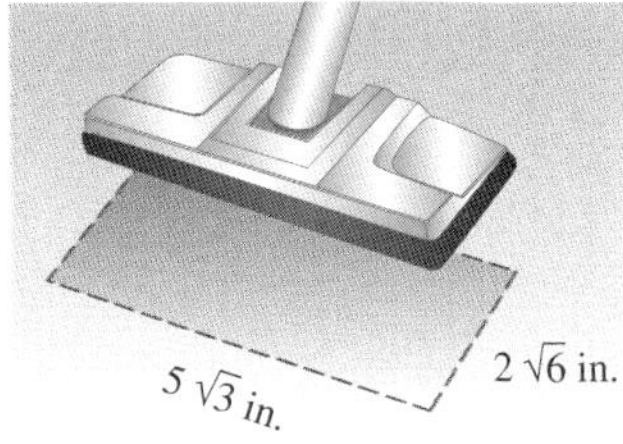

Rationalize each denominator.

76. $\dfrac{1}{\sqrt{7}}$

77. $\sqrt{\dfrac{3}{7}}$

78. $\dfrac{\sqrt{9}}{\sqrt{18}}$

79. $\dfrac{\sqrt{c} - 4}{\sqrt{c} + 4}$

80. $\dfrac{7}{\sqrt{2} + 1}$

81. $\dfrac{8}{\sqrt[3]{16}}$

SECTION 14.6 Solving Radical Equations; the Distance Formula

To solve an equation containing square root radicals:

1. Isolate the radicals.
2. Square both sides and solve the resulting equation.
3. Check the solution. Discard any *extraneous* solutions.

Squaring property of equality:

If $a = b$, then $a^2 = b^2$.

Simplify each expression. All variables represent positive numbers.

82. $(\sqrt{x})^2$ **83.** $(\sqrt[3]{x})^3$ **84.** $(2\sqrt{t})^2$ **85.** $(\sqrt{e-1})^2$

Solve each equation and check all solutions.

86. $\sqrt{x} = 9$

87. $\sqrt{3x+4} + 5 = 3$

88. $\sqrt{24 + 10y} = y + 4$

89. $\sqrt{2(r+4)} = 2\sqrt{r}$

90. $\sqrt{p^2 - 3} = p + 3$

91. $\sqrt[3]{x-1} = 3$

92. FERRIS WHEELS The distance d in feet that an object will fall in t seconds is given by the formula

$$t = \sqrt{\frac{d}{16}}$$

If a person drops a coin from the top of a Ferris wheel and it takes 2 seconds to hit the ground, how tall is the Ferris wheel?

The distance formula:

$$d = \sqrt{(x_2 - x_1)^2 + (y_2 - y_1)^2}$$

Find the distance between the points. If an answer contains a radical, round to the nearest hundredth.

93. $(-7, 12), (-4, 8)$

94. $(-15, -3), (-10, -16)$

SECTION 14.7 Rational Exponents

Real numbers can be raised to fractional powers.

Rational exponents:

$$x^{1/n} = \sqrt[n]{x}$$

$$x^{m/n} = \sqrt[n]{x^m} = (\sqrt[3]{x})^m$$

The rules for exponents can be used to simplify expressions involving rational exponents.

Simplify each expression. Write answers without using negative exponents.

95. $49^{1/2}$ **96.** $(-1{,}000)^{1/3}$ **97.** $36^{3/2}$ **98.** $\left(\frac{8}{27}\right)^{2/3}$

99. $4^{-3/2}$ **100.** $8^{2/3}8^{4/3}$ **101.** $(3^{2/3})^3$ **102.** $(a^4b^8)^{-1/2}$

103. $x^{1/3}x^{2/5}$ **104.** $\frac{t^{3/4}}{t^{2/3}}$ **105.** $\frac{x^{2/5}x^{1/5}}{x^{-2/5}}$ **106.** $\frac{x^{17/7}}{x^{3/7}}$

107. Graph each number on the number line: $\{4^{-1/2}, 12^{1/2}, 9^{1/3}, -2^{2/3}\}$.

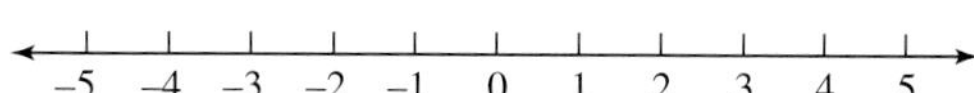

108. DENTISTRY The fractional amount of painkiller remaining in the system of a patient h hours after the original dose was injected into her gums is given by $h^{-3/2}$. How much of the original dose is in the patient's system 16 hours after the injection?

109. Explain why $(-4)^{1/2}$ is not a real number.

CHAPTER 14 TEST

Simplify each radical.

1. $\sqrt{100}$

2. $-\sqrt{\dfrac{400}{9}}$

3. $\sqrt[3]{-27}$

4. $\sqrt{\dfrac{50}{49}}$

5. Evaluate $\sqrt{b^2 - 4ac}$ for $a = 2$, $b = 10$, and $c = 6$. Round to the nearest tenth.

6. A 26-foot ladder reaches a point on a wall 24 feet above the ground. How far from the wall is the ladder's base?

Simplify each expression. Assume that x and y represent positive numbers.

7. $\sqrt{4x^2}$

8. $\sqrt{54x^3}$

9. $\sqrt{\dfrac{18x^2y^3}{2xy}}$

10. $\sqrt[3]{x^6y^3}$

11. A square has an area of 24 square yards.
    a. Express the length of a side of the square in simplified radical form.
    b. Round the length of a side of the square to the nearest tenth.

Perform each operation and simplify.

12. $\sqrt{12} + \sqrt{27}$

13. $\sqrt{8x^3} - x\sqrt{18x}$

14. $(-2\sqrt{8x})(3\sqrt{12x})$

15. $\sqrt{3}(\sqrt{8} + \sqrt{6})$

16. $(\sqrt{2} + \sqrt{3})(\sqrt{2} - \sqrt{3})$

17. $(2\sqrt{x} + 2)(\sqrt{x} - 3)$

18. SEWING A corner of fabric is folded over to form a collar and stitched down as shown below. From the dimensions given in the figure, determine the exact number of inches of stitching that must be made. Then give an approximation to one decimal place. (All measurements are in inches.)

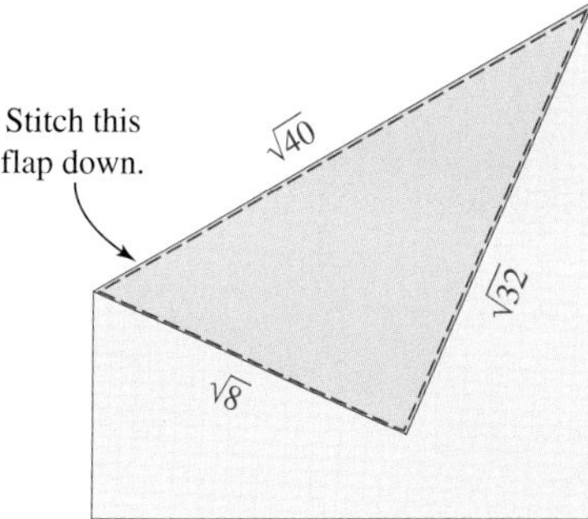

Rationalize each denominator.

19. $\dfrac{2}{\sqrt{2}}$

20. $\dfrac{\sqrt{3x}}{\sqrt{x} + 2}$

Solve each equation.

21. $\sqrt{x} = 15$

22. $\sqrt{2 - x} - 2 = 6$

23. $\sqrt{3x + 9} = 2\sqrt{x + 1}$

24. $x - 1 = \sqrt{x - 1}$

25. $\sqrt{3a + 4} + 2 = 0$

26. $\sqrt[3]{x - 2} = 3$

27. Find the distance between points $(-2, -3)$ and $(-8, 5)$.

28. Complete the table and graph the function. Round to the nearest tenth when necessary.

$f(x) = \sqrt{x}$

x	$f(x)$
0	
1	
2	
3	
4	
5	
6	
7	
8	
9	

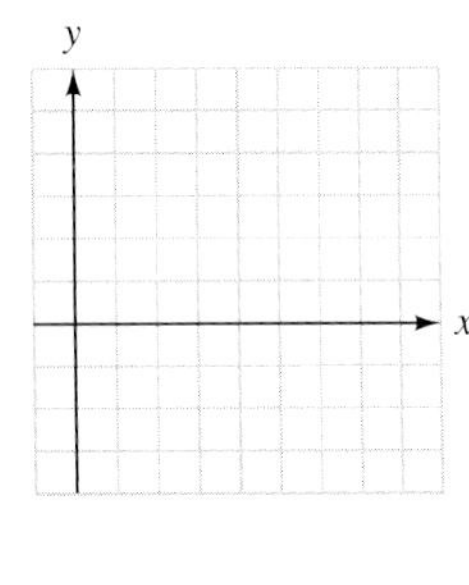

29. Is 0 a solution of the radical equation $\sqrt{3x + 1} = x - 1$? Explain your answer.

30. Explain why we cannot perform the subtraction $4\sqrt{3} - 7\sqrt{2}$.

31. CARPENTRY In the illustration below, a carpenter is using a tape measure to see whether the wall he just put up is perfectly square with the floor. Explain what mathematical concept he is applying. If the wall is positioned correctly, what should the measurement on the tape read?

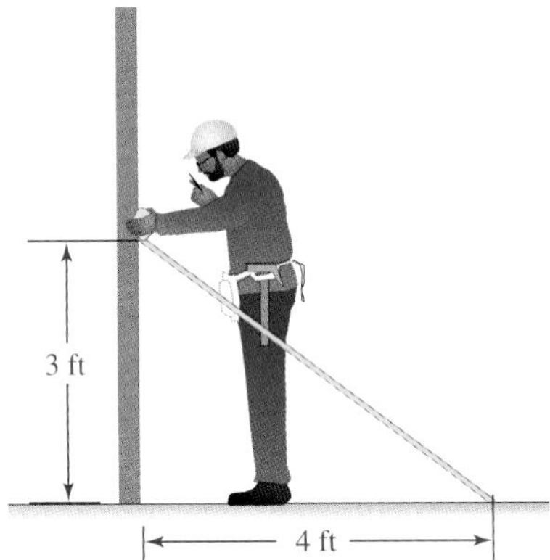

32. Explain why $\sqrt{-9}$ is not a real number.

Simplify each expression.

33. $121^{1/2}$

34. $p^{2/3}p^{4/3}$

CHAPTERS 1–14 CUMULATIVE REVIEW EXERCISES

1. Determine whether each statement is true or false.
   a. All whole numbers are integers.
   b. π is a rational number.
   c. A real number is either rational or irrational.

2. Evaluate:

$$\frac{-3(3 + 2)^2 - (-5)}{17 - 3|-4|}$$

3. BACKPACKS Pediatricians advise that children should not carry more than 20% of their own body weight in backpacks. According to this warning, how much weight can a fifth-grade girl who weighs 85 pounds safely carry in her backpack?

4. SCIENCE The illustration below shows the recent budgets for the National Science Foundation. Determine the percent change for the 1996 budget as compared to the 1995 budget. Round to the nearest tenth of a percent.

In billions		% change
1992	$2.55	8.7%
1993	$2.75	8.0%
1994	$2.99	8.6%
1995	$3.27	9.5%
1996	$3.21	?
1997	$3.30	2.9%
1998	$3.43	3.9%
1999	$3.67	7.1%
2000	$3.91	6.5%

Based on data from the National Science Foundation

5. Simplify: $3p - 6(p + z) + p$.

6. Solve: $2 - (4x + 7) = 3 + 2(x + 2)$.

7. Solve $3 - 3x \geq 6 + x$, graph the solution, and use interval notation to describe the solution.

8. Solve $0 \leq \frac{4 - x}{3} < 2$, graph the solution, and use interval notation to describe the solution.

9. SEARCH AND RESCUE Two search and rescue teams leave base at the same time, looking for a lost boy. The first team, on foot, heads north at 2 mph and the other, on horseback, south at 4 mph. How long will it take them to search a distance of 21 miles between them?

10. BLENDING COFFEES A store sells regular coffee for $4 a pound and gourmet coffee for $7 a pound. Using 40 pounds of the gourmet coffee, the owner makes a blend to put on sale for $5 a pound. How many pounds of regular coffee should he use?

11. SURFACE AREA The total surface area A of a box with dimensions l, w, and h is given by the formula

$$A = 2lw + 2wh + 2lh$$

If $A = 202$ square inches, $l = 9$ inches, and $w = 5$ inches, find h.

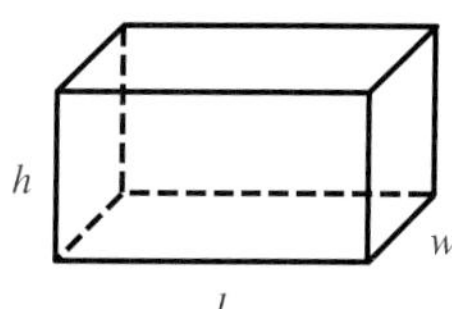

Graph each equation or inequality.

12. $3x - 4y = 12$

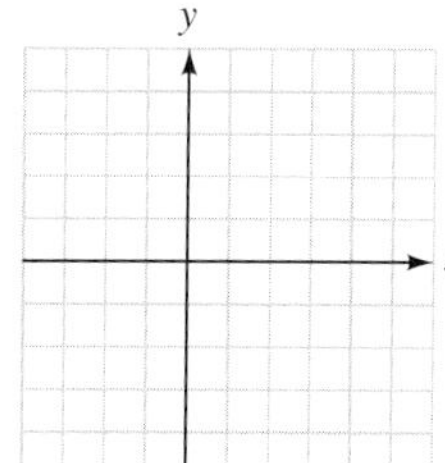

13. 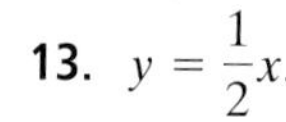 $y = \frac{1}{2}x.$

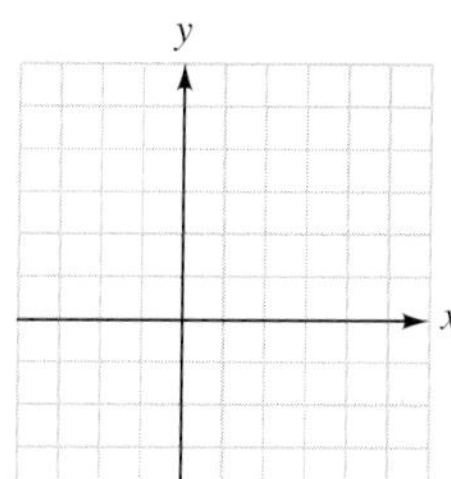

14. $x = 5$

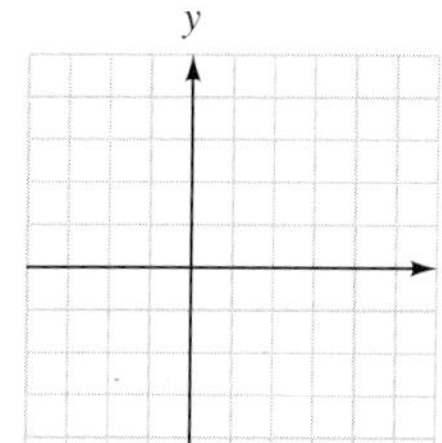

15. 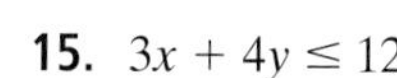 $3x + 4y \leq 12$

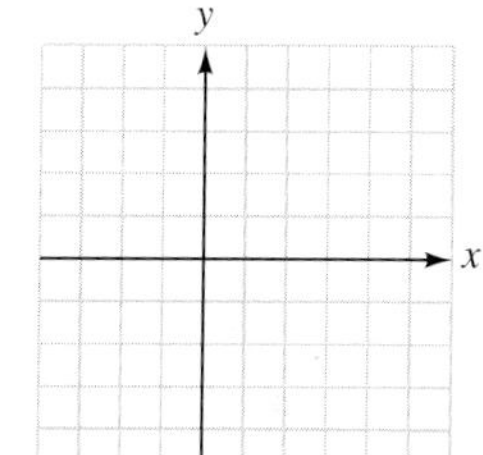

16. Write an equation of the line passing through $(-2, 5)$ and $(4, 8)$. Express the result in slope–intercept form.

17. What is the slope of the line defined by each equation?

a. $y = 3x - 7$ b. $2x + 3y = -10$

18. What is true about the slopes of two

a. parallel lines?

b. perpendicular lines?

19. SHOPPING On the graph, the line approximates the growth in retail sales for U.S. shopping centers during the years 1994–2002. Find the rate of increase in sales by finding the slope of the line.

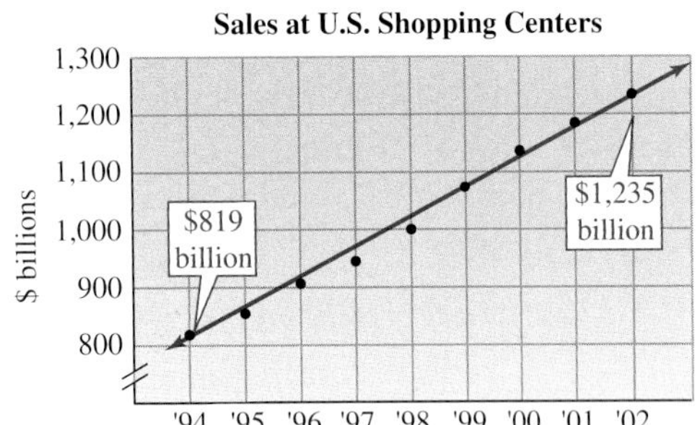

Based on data from International Council of Shopping Centers

20. If $f(x) = x^3 - x + 5$, find $f(-2)$.

21. Complete the table and graph the function. Then give the domain and range of the function.

$f(x) = |1 - x|$

x	$f(x)$
0	
1	
2	
3	
−1	
−2	

22. BOATING The following graph shows the vertical distance from a point on the tip of a propeller to the centerline as the propeller spins. Is this the graph of a function?

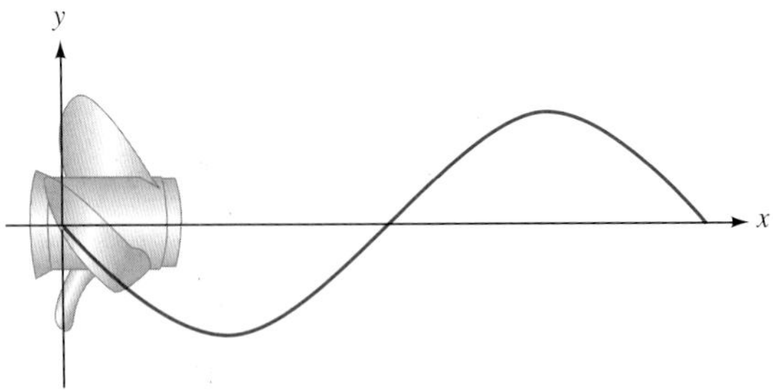

Simplify each expression. Write each answer without using parentheses or negative exponents.

23. $(x^5)^2(x^7)^3$

24. $\left(\frac{a^3b}{c^4}\right)^5$

25. $4^{-3} \cdot 4^{-2} \cdot 4^5$

26. $(a^{-2}b^3)^{-4}$

27. ASTRONOMY The **parsec,** a unit of distance used in astronomy, is 3×10^{16} meters. The distance to Betelgeuse, a star in the constellation Orion, is 1.6×10^2 parsecs. Use scientific notation to express this distance in meters.

28. NCAA MEN'S BASKETBALL The following graph shows the University of Connecticut's lead or deficit during the second half of the 1999 championship game with Duke University.

a. How many x-intercepts does the graph have? Explain their importance.

b. Give the coordinates of the highest point and the lowest point on the graph. What is the importance of each?

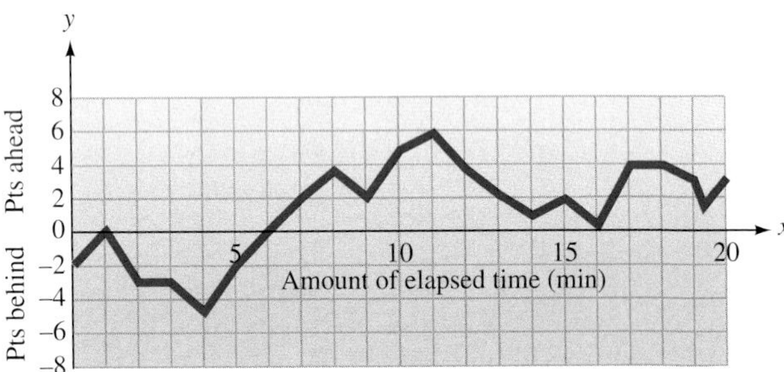

Perform the operations.

29. $(-r^4st^2)(2r^2st)(rst)$

30. $(-3t + 2s)(2t - 3s)$

31. $(3a^2 - 2a + 4) - (a^2 - 3a + 7)$

32. $(y - 6)^2$

33. $\dfrac{4x - 3y + 8z}{4xy}$

34. $2 + x\overline{)3x + 2x^2 - 2}$

Factor each expression completely.

35. $3x^2y - 6xy^2$

36. $2x^2 + 2xy - 3x - 3y$

37. $25p^4 - 16q^2$

38. $3x^3 - 243x$

39. $x^2 - 11x - 12$

40. $a^3 + 8b^3$

41. $6a^2 - 7a - 20$

42. $16m^2 - 20m - 6$

Solve each equation.

43. $x^2 + 3x + 2 = 0$

44. $5x^2 = 10x$

45. $6x^2 - x - 2 = 0$

46. $2y^2 = 12 - 5y$

47. CHILDREN'S STICKERS The following rectangular-shaped sticker has an area of 20 cm^2. The width is 1 cm shorter than the length. Find the length of the sticker.

48. For what value of x is $\dfrac{4x}{x-6}$ undefined?

Simplify each expression.

49. $\dfrac{x^2 + 2x + 1}{x^2 - 1}$

50. $-\dfrac{15a^2}{25a^3}$

Perform the operation(s) and simplify when possible.

51. $\dfrac{p^2 - p - 6}{3p - 9} \div \dfrac{p^2 + 6p + 9}{p^2 - 9}$

52. $\dfrac{x^2y^2}{cd} \cdot \dfrac{d^2}{c^2x}$

53. $\dfrac{x+2}{x+5} - \dfrac{x-3}{x+7}$

54. $\dfrac{3x}{x+2} + \dfrac{5x}{x+2} - \dfrac{7x-2}{x+2}$

55. $\dfrac{3a}{2b} - \dfrac{2b}{3a}$

56. $\dfrac{\frac{1}{x} + \frac{1}{y}}{\frac{1}{x} - \frac{1}{y}}$

Solve each equation.

57. $\dfrac{4}{a} = \dfrac{6}{a} - 1$

58. $\dfrac{a+2}{a+3} - 1 = \dfrac{-1}{a^2 + 2a - 3}$

59. Solve the formula $\dfrac{1}{r} = \dfrac{1}{r_1} + \dfrac{1}{r_2}$ for r.

60. ONLINE SALES A company found that, on average, it made 9 online sales transactions for every 500 hits on its Internet Web site. If the company's Web site had 360,000 hits in one year, how many sales transactions did it have that year?

61. Assume that y varies inversely with x. If $y = 8$ when $x = 2$, find y when $x = 8$.

62. FILLING A POOL An inlet pipe can fill an empty swimming pool in 5 hours, and another inlet pipe can fill the pool in 4 hours. How long will it take both pipes to fill the pool?

Solve each system of equations. If the equations of a system are dependent or if a system is inconsistent, so indicate.

63. $\begin{cases} x = y + 4 \\ 2x + y = 5 \end{cases}$

64. $\begin{cases} \frac{3}{5}s + \frac{4}{5}t = 1 \\ -\frac{1}{4}s + \frac{3}{8}t = 1 \end{cases}$

65. FINANCIAL PLANNING In investing \$6,000 of a couple's money, a financial planner put some of it into a savings account paying 6% annual interest. The rest was invested in a riskier mini-mall development plan paying 12% annually. The combined interest earned for the first year was \$540. How much money was invested at each rate? Use two variables to solve this problem.

66. Graph the solution of $\begin{cases} 3x + 2y \geq 6 \\ x + 3y \leq 6 \end{cases}$.

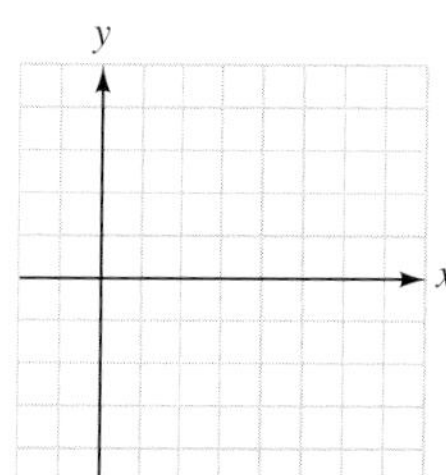

Simplify each expression. All variables represent positive numbers.

67. $\sqrt{\dfrac{49}{225}}$

68. $-\sqrt[3]{-27}$

69. $-12x\sqrt{16x^2y^3}$

70. $\sqrt{48} - \sqrt{8} + \sqrt{27} - \sqrt{32}$

71. $(\sqrt{y} - 4)(\sqrt{y} - 5)$

72. $(-5\sqrt{6})(4\sqrt{3})$

73. $\dfrac{4}{\sqrt{20}}$

74. $\dfrac{\sqrt{x} - 3}{\sqrt{x} + 3}$

75. Solve: $\sqrt{6x + 19} - 5 = 2$.

76. CARGO SPACE How wide a piece of plywood can be stored diagonally in the back of the van shown?

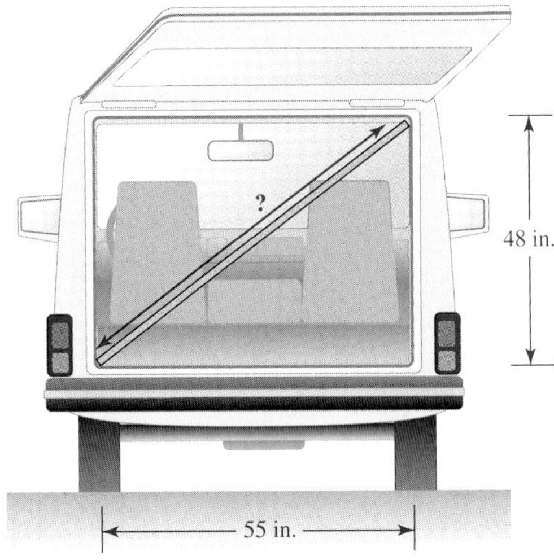

CHAPTER 15

Quadratic Equations

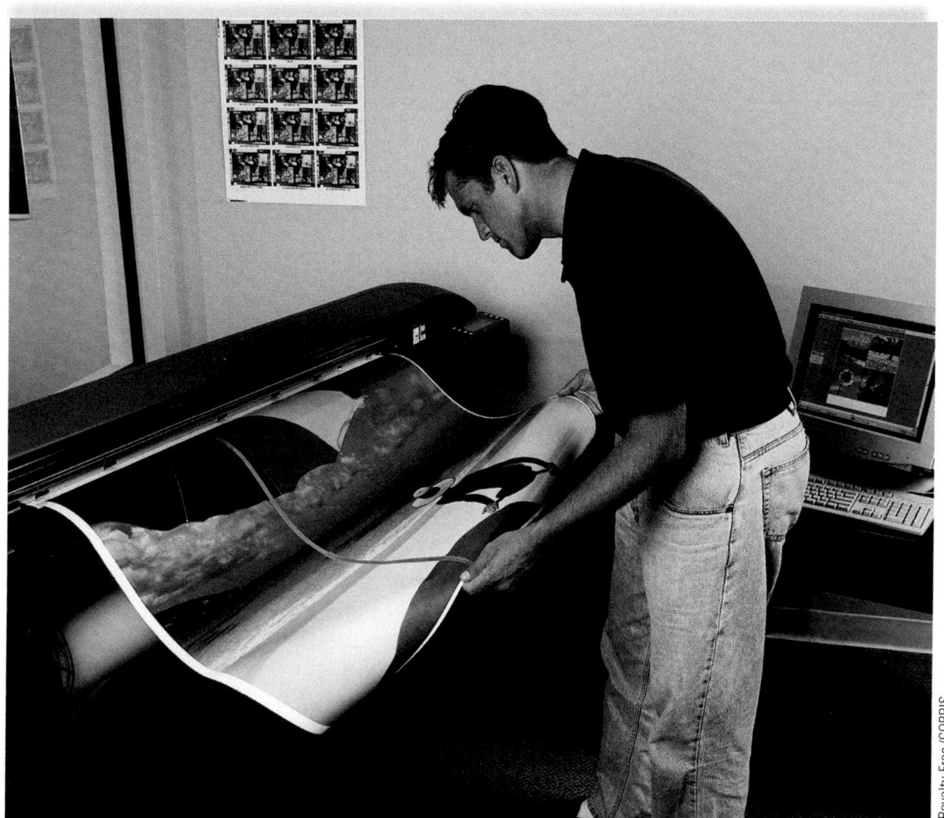

© Royalty-Free/CORBIS

Graphic designers combine their artistic talents with mathematical skills to create attractive advertisements, photography, and packaging. They perform arithmetic computations with fractions and decimals to determine the proper font size and page margins for a design. They apply algebraic concepts such as graphing and equation solving to design page layouts. And they use formulas to create production schedules and calculate costs.

To learn more about the use of algebra in graphic design, visit *The Learning Equation* on the Internet at http://tle.brookscole.com (The log-in instructions are in the Preface.) For Chapter 15, the online lesson is:

- *TLE* Lesson 26: The Quadratic Formula

Check Your Knowledge

1. A ________ equation is an equation that can be written in the form $ax^2 + bx + c = 0$, where $a \neq 0$.
2. To solve a quadratic equation means to find all values of the variable that make the equation ______.
3. The equation $x^2 = c$, where $c > 0$, has ______ real solutions.
4. Completing the _______ and the _________ formula are methods for solving quadratic equations that do not factor.
5. The lowest (or highest) point on a parabola is called the _______.

Solve each equation by the square root method.

6. $y^2 = 49$
7. $x^2 - 18 = 0$
8. $9y^2 + 3 = 19$
9. $(x - 3)^2 = 2$
10. A square sand box has area 20 ft^2. Find the length of a side. Round to the nearest tenth.
11. Find the number required to complete the square on $x^2 + 8x$.
12. Complete the square to solve $x^2 - 6x - 7 = 0$.
13. Complete the square to solve $3x^2 + 2x - 21 = 0$.
14. Complete the square to solve $x^2 - 4x + 2 = 0$. Give the exact solutions and then approximate them. Round to the nearest tenth.

Use the quadratic formula to solve each equation.

15. $x^2 - 2x - 24 = 0$
16. $3x^2 - 5x + 2 = 0$
17. $x^2 + 3x = -4$
18. $x^2 + 4x - 3 = 0$
19. $3x^2 = x + 1$, approximate the solutions to the nearest hundredth
20. A positive number is 6 less than its square. Find the number.
21. The length of a pool table is twice the width and its area is 32 ft^2. Find the width and length.
22. Graph $y = x^2 - 2x - 3$ by finding the vertex, the x- and y-intercepts, and the axis of symmetry.

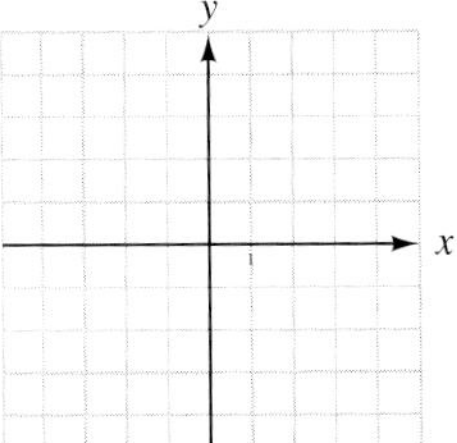

23. A swimming pool is 4 feet longer than it is wide, and its diagonal is 20 feet. Find the width and the length.
24. The graph of $f(x) = x^2 - 2x + 3$ is shown in the illustration below. How many real solutions does $x^2 - 2x + 3 = 0$ have?

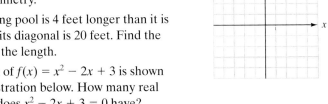

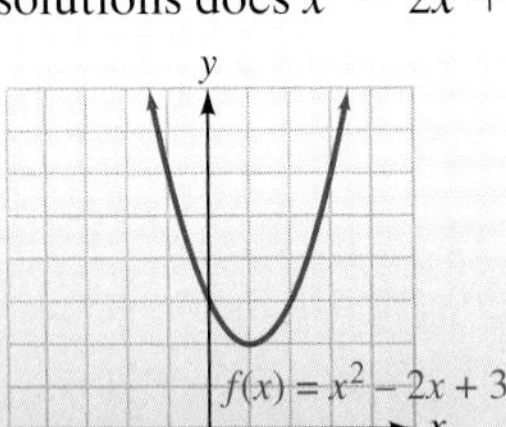

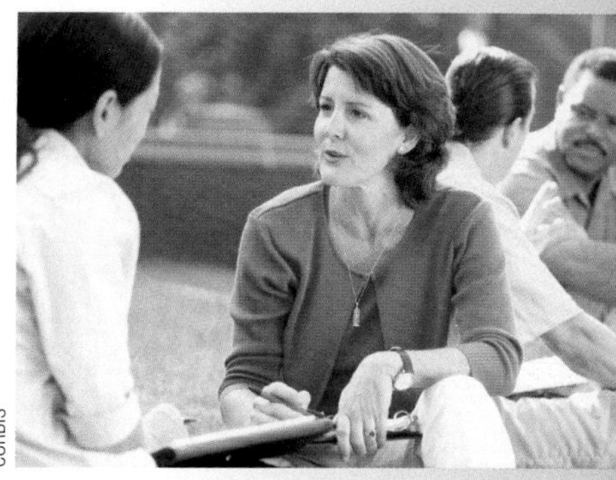
CORBIS

Study Skills Workshop

PREPARING FOR INTERMEDIATE ALGEBRA OR YOUR NEXT MATH COURSE

Because learning math is a skill, we tend to get rusty or forget things if we haven't practiced for a while. Usually there is a time gap between courses, especially if you take a course in the spring and have the whole summer off before taking the next course. Often your Intermediate Algebra course will have a short review component, but you will still be expected to remember some concepts from your Elementary Algebra course. In order to keep these skills honed, plan to spend some time during the semester break to review the concepts learned. The time needed to review will depend on the grade that you received and how well you learned the concepts the first time.

If You Received an A. You may only need to spend a little time once a week reviewing old tests to keep concepts fresh. Enjoy your break—you worked hard to master the concepts!

If You Received a B. You may need to spend a little more time, because there were concepts that you didn't fully master. Take the time during your break to see whether you can understand test problems that you had trouble with during the semester. Also take time to review problems that you understood, so that you don't forget. If your Elementary Algebra textbook came packaged with a tutorial software package, you may want to use this to review material and practice problems with immediate feedback.

If You Received a C. You may want to consider seeing a tutor through your semester break, as there are probably a number of concepts with which you had difficulty. Take advantage of materials that came with your textbook, such as tutorial software packages. In many textbooks there is a set of Summary exercises for each chapter. Try to do as many of these exercises as possible, completing all of the chapters before your new semester. At the beginning of your Intermediate Algebra class, seek tutorial assistance right away to get off to a good start. Analyze your study and homework habits to see whether there is room for improvement, and make sure that you allow enough time in your schedule for your new course.

Your Textbook. Many students are in a hurry to sell their old textbook to their college bookstore as soon as the course is finished. Resist that temptation, as your old book is a good resource for reviewing between semesters and will be a good source of review during your next semester. As you are reviewing for your next class, keep skimming through your old text to look at examples and concepts that you learned. This will help to keep the concepts fresh in your memory.

We have previously solved quadratic equations by factoring. In this chapter, we will discuss three other methods that are used to solve "quadratics."

15.1 Completing the Square

- The square root method • Completing the square
- Solving equations with lead coefficients of 1
- Solving equations with lead coefficients other than 1

Recall that equations that involve second-degree polynomials are called *quadratic equations.*

Quadratic equations

A **quadratic equation** is an equation that can be written in the form

$$ax^2 + bx + c = 0 \quad \text{where } a \neq 0$$

and where a, b, and c represent real numbers. This form is called **quadratic form.**

Some examples of quadratic equations are

$$x^2 + 12x - 13 = 0, \qquad 3q^2 = 3q + 2, \qquad a^2 - 5a = 0, \qquad \text{and} \qquad x^2 = 16$$

We have solved quadratic equations using the factoring method. In this section, we will discuss two other methods for solving them. The first, called the **square root method,** is used when one side of the equation to solve is a quantity squared and the other side is a constant. The second method, called **completing the square,** involves the concept of perfect square trinomials.

The square root method

If $x^2 = 9$, x is a number whose square is 9. Since $3^2 = 9$ and $(-3)^2 = 9$, the equation $x^2 = 9$ has two solutions, $x = \sqrt{9} = 3$ and $x = -\sqrt{9} = -3$. In general, any equation of the form $x^2 = c$, where $c > 0$, has two solutions.

The square root property

If c represents a positive real number, the equation $x^2 = c$ has two solutions:

$$x = \sqrt{c} \qquad \text{or} \qquad x = -\sqrt{c}$$

We can write the previous result with **double-sign notation.** The statement

$$x = \pm\sqrt{c} \qquad \text{(read as "}x\text{ equals positive or negative }\sqrt{c}\text{")}$$

means that $x = \sqrt{c}$ or $x = -\sqrt{c}$.

Self Check 1
Solve: $x^2 = 25$.

EXAMPLE 1 Solve: $x^2 = 16$.

Solution We use the square root method to find that the equation has two solutions.

$x^2 = 16$

$x = \pm\sqrt{16}$ $\quad$ $\pm\sqrt{16}$ means that $x = \sqrt{16}$ or $x = -\sqrt{16}$.

$x = \pm 4$ $\quad$ $\sqrt{16} = 4$.

The solutions of $x^2 = 16$ are 4 and -4.

Check: **For $x = 4$** **For $x = -4$**

$$x^2 = 16 \qquad x^2 = 16$$
$$4^2 \stackrel{?}{=} 16 \qquad (-4)^2 \stackrel{?}{=} 16$$
$$16 = 16 \qquad 16 = 16$$

Answer ± 5

! COMMENT When using the square root method to solve an equation, always write the $\pm$ symbol, or you will lose one of the solutions. For example, consider the equation from Example 1.

$$x^2 = 16$$
$$x = \pm\sqrt{16}$$

↑ If you don't write this symbol, you will lose the solution $-\sqrt{16}$, which is -4.

Note that the equation in Example 1 can also be solved by factoring.

$x^2 = 16$

$x^2 - 16 = 0$ Subtract 16 from both sides.

$(x + 4)(x - 4) = 0$ Factor the difference of two squares.

$x + 4 = 0$ or $x - 4 = 0$

$x = -4$ | $x = 4$

EXAMPLE 2 Solve: $3x^2 - 9 = 0$.

Self Check 2

Solve: $2x^2 - 10 = 0$. Give the exact solutions and approximations to the nearest hundredth.

Solution To solve the equation by the square root method, we first isolate x^2.

$3x^2 - 9 = 0$

$3x^2 = 9$ Add 9 to both sides.

$x^2 = 3$ Divide both sides by 3.

$x = \pm\sqrt{3}$ Use the square root method.

Check: **For $x = \sqrt{3}$** **For $x = -\sqrt{3}$**

$$3x^2 - 9 = 0 \qquad 3x^2 - 9 = 0$$
$$3(\sqrt{3})^2 - 9 \stackrel{?}{=} 0 \qquad 3(-\sqrt{3})^2 - 9 \stackrel{?}{=} 0$$
$$3(3) - 9 \stackrel{?}{=} 0 \qquad 3(3) - 9 \stackrel{?}{=} 0$$
$$9 - 9 \stackrel{?}{=} 0 \qquad 9 - 9 \stackrel{?}{=} 0$$
$$0 = 0 \qquad 0 = 0$$

The solutions of $3x^2 - 9 = 0$ are $\sqrt{3}$ and $-\sqrt{3}$. Note that these are exact solutions. Using a calculator, we can approximate them to the nearest hundredth.

$$\sqrt{3} \approx 1.73 \qquad -\sqrt{3} \approx -1.73$$

Answer $\pm\sqrt{5}$; ± 2.24

EXAMPLE 3 **Hurricanes.** In 1998, Hurricane Mitch dumped heavy rains on Central America, causing extensive flooding in Honduras, Belize, and Guatemala. Figure 15-1 on the next page shows the position of the storm on October 26, at which time the weather service estimated that it covered an area of about 71,000 square miles. What was the diameter of the storm?

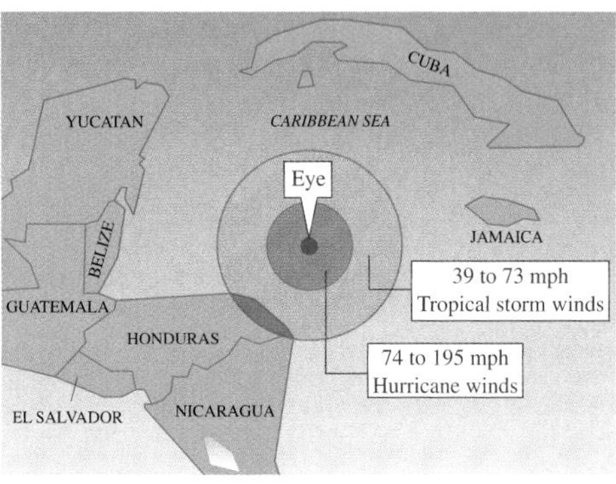

FIGURE 15-1

Solution We can use the formula for the area of a circle to find the *radius* (in miles) of the circular-shaped storm.

$$\mathbf{A} = \pi r^2$$

$$\mathbf{71{,}000} = \pi r^2 \quad \text{Substitute 71,000 for the area } A.$$

$$\frac{71{,}000}{\pi} = r^2 \quad \text{Divide both sides by } \pi \text{ to isolate } r^2.$$

Now we use the square root method to solve for r.

$$r = \sqrt{\frac{71{,}000}{\pi}} \quad \text{or} \quad r = -\sqrt{\frac{71{,}000}{\pi}}$$

Using a calculator to approximate the square root, we have

$r \approx 150.3329702$ The units are miles. The second solution is discarded, because the radius cannot be negative.

If we multiply the radius by 2, we find that the diameter of the storm was about 300 miles.

Self Check 4
Solve: $(x - 3)^2 = 8$.

Answer $3 + 2\sqrt{2} \approx 5.83$, $3 - 2\sqrt{2} \approx 0.17$

EXAMPLE 4 Solve: $(x - 1)^2 = 18$.

Solution

$$(x - 1)^2 = 18$$

$$x - 1 = \pm\sqrt{18} \quad \text{Use the square root method to solve for } x - 1.$$

To solve for x, we undo the subtraction of 1 on the left-hand side by adding 1 to both sides. In this case, when adding (or subtracting) a number on the right-hand side, we customarily write it *in front of* the radical expression.

$$x - 1 = \pm\sqrt{18}$$

$$x - 1 + \mathbf{1} = \mathbf{1} \pm \sqrt{18} \quad \text{Add 1 to both sides, to isolate } x.$$

$$x = 1 \pm \sqrt{18} \quad \text{Simplify the left-hand side.}$$

$$x = 1 \pm 3\sqrt{2} \quad \text{Simplify the radical: } \sqrt{18} = \sqrt{9 \cdot 2} = 3\sqrt{2}.$$

The solutions are $1 \pm 3\sqrt{2}$ (read as "1 plus or minus $3\sqrt{2}$"). We can approximate them to the nearest hundredth.

$1 + 3\sqrt{2} \approx 5.24$ $\qquad$ $1 - 3\sqrt{2} \approx -3.24$

Completing the square

We have previously solved quadratic equations such as $x^2 + 12x - 13 = 0$ using the factoring method.

$$x^2 + 12x - 13 = 0$$

$$(x - 1)(x + 13) = 0 \quad \text{Factor the trinomial } x^2 + 12x - 13.$$

$$x - 1 = 0 \quad \text{or} \quad x + 13 = 0 \quad \text{Set each factor equal to 0.}$$

$$x = 1 \quad \Big| \quad x = -13 \quad \text{Solve each linear equation.}$$

The solutions of $x^2 + 12x - 13 = 0$ are 1 and -13.

Not every quadratic equation can be solved using the factoring method. For example, the trinomial in the equation $x^2 + 4x - 13 = 0$ cannot be factored using techniques we have studied previously. To solve such equations, we can use another method called **completing the square.** It is based on the following **special products:**

$$x^2 + 2bx + b^2 = (x + b)^2 \quad \text{and} \quad x^2 - 2bx + b^2 = (x - b)^2$$

The trinomials $x^2 + 2bx + b^2$ and $x^2 - 2bx + b^2$ are both perfect square trinomials, since each one factors as the square of a binomial. In each trinomial, if we take one-half of the coefficient of the x and square it, we get the third term.

In $x^2 + \mathbf{2b}x + b^2$, if we take $\frac{1}{2}(\mathbf{2b})$, which is b, and square it, we get the third term, b^2.

In $x^2 - \mathbf{2b}x + b^2$, if we take $\frac{1}{2}(\mathbf{-2b}) = -b$ and square it, we get $(-b)^2 = b^2$, which is the third term.

To change a binomial such as $x^2 + 12x$ into a perfect square trinomial, we take *one-half of the coefficient of* x (the 12), *square it,* and *add it* to $x^2 + 12x$.

$$x^2 + 12x + \left[\frac{1}{2}(12)\right]^2 = x^2 + 12x + (6)^2$$

$$= x^2 + 12x + 36$$

This result is a perfect square trinomial, because $x^2 + 12x + 36 = (x + 6)^2$.

EXAMPLE 5 Complete the square and factor the resulting perfect square trinomial: **a.** $x^2 + 4x$, **b.** $x^2 - 6x$, and **c.** $x^2 - 5x$.

Solution

a. Since the coefficient of x is 4, we add the square of one-half of 4.

$$x^2 + 4x + \left[\frac{1}{2}(4)\right]^2 = x^2 + 4x + (2)^2 \quad \frac{1}{2}(4) = 2.$$

$$= x^2 + 4x + 4 \quad \text{Square 2 to get 4.}$$

This perfect square trinomial factors as $(x + 2)^2$.

b. Since the coefficient of x is -6, we add the square of one-half of -6.

$$x^2 - 6x + \left[\frac{1}{2}(-6)\right]^2 = x^2 - 6x + (-3)^2 \quad \frac{1}{2}(-6) = -3.$$

$$= x^2 - 6x + 9 \quad \text{Square } -3 \text{ to get 9.}$$

This perfect square trinomial factors as $(x - 3)^2$.

Self Check 5

Complete the square and factor the resulting perfect square trinomial:

a. $y^2 + 6y$

b. $y^2 - 8y$

c. $y^2 + 3y$

c. Since the coefficient of x is -5, we add the square of one-half of -5.

$$x^2 - 5x + \left[\frac{1}{2}(-5)\right]^2 = x^2 - 5x + \left(-\frac{5}{2}\right)^2 \qquad \frac{1}{2}(-5) = -\frac{5}{2}.$$

$$= x^2 - 5x + \frac{25}{4} \qquad \text{Square } -\frac{5}{2} \text{ to get } \frac{25}{4}.$$

This perfect square trinomial factors as $\left(x - \frac{5}{2}\right)^2$.

Answers

a. $y^2 + 6y + 9 = (y + 3)^2$,
b. $y^2 - 8y + 16 = (y - 4)^2$,
c. $y^2 + 3y + \frac{9}{4} = \left(y + \frac{3}{2}\right)^2$

Solving equations with lead coefficients of 1

If the quadratic equation $ax^2 + bx + c = 0$ has a lead coefficient of 1, it's easy to solve by completing the square.

Self Check 6

Solve: $x^2 + 10x - 4 = 0$. Give the exact solutions and approximations to the nearest hundredth.

EXAMPLE 6 Solve: $x^2 + 4x - 13 = 0$. Give each answer to the nearest hundredth.

Solution Since the coefficient of x^2 is 1, we can complete the square as follows:

$$x^2 + 4x - 13 = 0$$

$$x^2 + 4x = 13 \qquad \text{Add 13 to both sides so that the constant term is on the right-hand side.}$$

We then find one-half of the coefficient of x, square it, and add the result to both sides to make the left-hand side a perfect square trinomial.

$$x^2 + 4x + \left[\frac{1}{2}(4)\right]^2 = 13 + \left[\frac{1}{2}(4)\right]^2 \qquad \text{Since the coefficient of } x \text{ is 4, add the square of one-half of 4 to both sides.}$$

$$x^2 + 4x + 4 = 13 + 4 \qquad \tfrac{1}{2}(4) = 2. \text{ Then square 2 to get 4.}$$

$$(x + 2)^2 = 17 \qquad \text{Factor } x^2 + 4x + 4 \text{ and simplify.}$$

$$x + 2 = \pm\sqrt{17} \qquad \text{Use the square root method to solve for } x + 2.$$

$$x = -2 \pm \sqrt{17} \qquad \text{Subtract 2 from both sides to isolate } x. \text{ Write } -2 \text{ in front of the radical.}$$

We can use a calculator to approximate each solution.

$x = -2 + \sqrt{17}$	or	$x = -2 - \sqrt{17}$
$x \approx -2 + 4.123105626$		$x \approx -2 - 4.123105626$
$x \approx 2.12$		$x \approx -6.12$

Answers $-5 + \sqrt{29}$, $-5 - \sqrt{29}$; 0.39, -10.39

Self Check 7

Solve: $x^2 + 5x = 3$. Approximate the solutions to the nearest hundredth.

EXAMPLE 7 Solve: $x^2 - 7x = 2$.

Solution The constant term is already on the right-hand side. To complete the square on the left-hand side, we find one-half of the coefficient of x and add its square to both sides.

$$x^2 - 7x = 2$$

$$x^2 - 7x + \left[\frac{1}{2}(-7)\right]^2 = 2 + \left[\frac{1}{2}(-7)\right]^2 \qquad \text{Since the coefficient of } x \text{ is } -7, \text{ add the square of one-half of } -7 \text{ to both sides.}$$

$$x^2 - 7x + \frac{49}{4} = 2 + \frac{49}{4} \qquad \frac{1}{2}(-7) = -\frac{7}{2}. \text{ Then square } -\frac{7}{2} \text{ to get } \frac{49}{4}.$$

$$\left(x - \frac{7}{2}\right)^2 = \frac{8}{4} + \frac{49}{4} \qquad \text{Factor the left-hand side. Write 2 as } \frac{8}{4}.$$

$$\left(x - \frac{7}{2}\right)^2 = \frac{57}{4}$$ The fractions have a common denominator. Add them.

$$x - \frac{7}{2} = \pm\sqrt{\frac{57}{4}}$$ Use the square root method to solve for $x - \frac{7}{2}$.

$$x - \frac{7}{2} = \pm\frac{\sqrt{57}}{2}$$ $\sqrt{\frac{57}{4}} = \frac{\sqrt{57}}{\sqrt{4}} = \frac{\sqrt{57}}{2}$.

$$x = \frac{7}{2} \pm \frac{\sqrt{57}}{2}$$ Add $\frac{7}{2}$ to both sides.

$$x = \frac{7 \pm \sqrt{57}}{2}$$ Since the fractions have a common denominator of 2, we can combine them.

The solutions are $\frac{7 \pm \sqrt{57}}{2}$. If we approximate the solutions to the nearest hundredth, we have

$$\frac{7 + \sqrt{57}}{2} \approx 7.27 \qquad \frac{7 - \sqrt{57}}{2} \approx -0.27$$

Answer $\frac{-5 \pm \sqrt{37}}{2}$, 0.54, −5.54

Solving equations with lead coefficients other than 1

If the quadratic equation $ax^2 + bx + c = 0$ has a lead coefficient other than 1, we can make the lead coefficient 1 by dividing both sides of the equation by a.

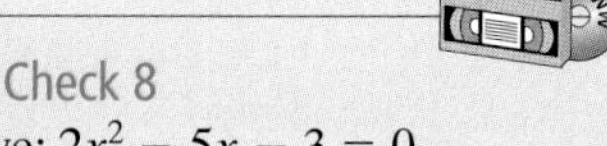

EXAMPLE 8 Solve: $4x^2 + 4x - 3 = 0$.

Self Check 8
Solve: $2x^2 - 5x - 3 = 0$.

Solution We divide both sides by 4 so that the coefficient of x^2 is 1. We then proceed as follows:

$$4x^2 + 4x - 3 = 0$$

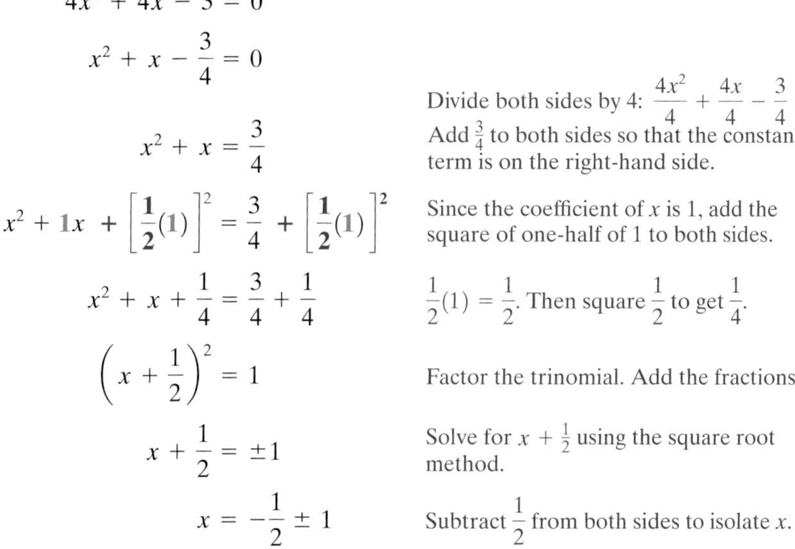

$$x^2 + x - \frac{3}{4} = 0$$ Divide both sides by 4: $\frac{4x^2}{4} + \frac{4x}{4} - \frac{3}{4} = \frac{0}{4}$.

$$x^2 + x = \frac{3}{4}$$ Add $\frac{3}{4}$ to both sides so that the constant term is on the right-hand side.

$$x^2 + \mathbf{1}x + \left[\frac{\mathbf{1}}{\mathbf{2}}(\mathbf{1})\right]^2 = \frac{3}{4} + \left[\frac{\mathbf{1}}{\mathbf{2}}(\mathbf{1})\right]^2$$ Since the coefficient of x is 1, add the square of one-half of 1 to both sides.

$$x^2 + x + \frac{1}{4} = \frac{3}{4} + \frac{1}{4}$$ $\frac{1}{2}(1) = \frac{1}{2}$. Then square $\frac{1}{2}$ to get $\frac{1}{4}$.

$$\left(x + \frac{1}{2}\right)^2 = 1$$ Factor the trinomial. Add the fractions.

$$x + \frac{1}{2} = \pm 1$$ Solve for $x + \frac{1}{2}$ using the square root method.

$$x = -\frac{1}{2} \pm 1$$ Subtract $\frac{1}{2}$ from both sides to isolate x.

$$x = -\frac{1}{2} + 1 \quad \text{or} \quad x = -\frac{1}{2} - 1$$

$$x = \frac{1}{2} \qquad\qquad x = -\frac{3}{2}$$

The solutions are $\frac{1}{2}$ and $-\frac{3}{2}$. Check each one.

Answers 3, $-\frac{1}{2}$

! COMMENT In Example 8, you may have noticed that $4x^2 + 4x - 3$ can be factored. Therefore, we could have solved $4x^2 + 4x - 3 = 0$ by factoring. This example illustrates an important fact: Completing the square can be used to solve any quadratic equation.

The previous examples illustrate that to solve a quadratic equation by completing the square, we follow these steps.

Completing the square to solve a quadratic equation

1. Write the equation in $ax^2 + bx + c = 0$ form. If the coefficient of x^2 is not 1, make it 1 by dividing both sides of the equation by the coefficient of x^2.
2. If necessary, add or subtract a number on both sides of the equation to get the constant term on the right-hand side.
3. Complete the square.
   a. Find half the coefficient of x and square it.
   b. Add that square to both sides of the equation.
4. Factor the perfect square trinomial and combine terms.
5. Solve the resulting quadratic equation using the square root method.
6. Check each solution.

Self Check 9
Solve: $3x^2 - 18x = -12$.

EXAMPLE 9 Solve: $2x^2 - 2 = 4x$.

Solution We write the equation in $ax^2 + bx + c = 0$ form to see whether it can be solved by factoring.

$2x^2 - 4x - 2 = 0$ Subtract $4x$ from both sides to get 0 on the right-hand side.

(1) $x^2 - 2x - 1 = 0$ Divide both sides by 2: $\frac{2x^2}{2} - \frac{4x}{2} - \frac{2}{2} = \frac{0}{2}$.

Since Equation 1 cannot be solved by factoring, we complete the square.

$x^2 - 2x = 1$ Add 1 to both sides.

$x^2 - 2x + \left[\frac{1}{2}(-2)\right]^2 = 1 + \left[\frac{1}{2}(-2)\right]^2$ Since the coefficient of x is -2, add the square of one-half of -2 to both sides.

$x^2 - 2x + 1 = 1 + 1$ $\frac{1}{2}(-2) = -1$. Then square -1 to get 1.

$(x - 1)^2 = 2$ Factor the trinomial and simplify.

$x - 1 = \pm\sqrt{2}$ Use the square root method to solve for $x - 1$.

$x = 1 \pm \sqrt{2}$ Add 1 to both sides.

The solutions are $1 \pm \sqrt{2}$.

Answer $3 \pm \sqrt{5}$

Section 15.1 STUDY SET

VOCABULARY *Fill in the blanks.*

1. If the polynomial in the equation $ax^2 + bx + c = 0$ doesn't factor, we can solve the equation by ________ the square.
2. Since $x^2 + 12x + 36 = (x + 6)^2$, we call the trinomial a perfect ________ trinomial.
3. In the equation $x^2 - 4x + 1 = 0$, the ________ of x is -4.
4. A ________ of an equation is a value of the variable that makes the equation true.

CONCEPTS *Fill in the blanks.*

5. The equation $x^2 = c$, where $c > 0$, has ______ solutions.
6. The solutions of $x^2 = c$, where $c > 0$, are ____ and ____.
7. To complete the square on $x^2 + 8x$, we add the ________ of one-half of 8, which is 16.
8. To complete the square on $x^2 - 10x$, we add the square of ________ of -10, which is 25.
9. If $x^2 = 5$, then $x = \pm$ ____.
10. If $(x - 2)^2 = 7$, then $x - 2 = \pm$ ____.
11. What is the first step if we solve $x^2 - 2x = 35$
    a. by the factoring method?
    b. by completing the square?
12. a. To solve $x^2 - 2x - 1 = 0$, we must complete the square. Why can't we use the factoring method?
    b. Can any quadratic equation be solved by completing the square?
13. Solve $x^2 - 81 = 0$ by using
    a. the square root method.
    b. the factoring method.
14. What is one-half of the given number?
    a. 4 b. -8
    c. 5 d. -7
15. Find one-half of the given number and then square the result.
    a. 6 b. -12
    c. 3 d. -5
16. What is the result when both sides of $2x^2 + 4x - 8 = 0$ are divided by 2?

NOTATION *Complete each solution to solve the equation.*

17.
$$(y - 1)^2 = 9$$
$$y - 1 = \square \quad \text{or} \quad y - 1 = -\sqrt{9}$$
$$\square = 3 \quad \Big| \quad y - 1 = \square$$
$$y = 4 \quad \Big| \quad y = -2$$

18.
$$y^2 + 2y - 3 = 0$$
$$y^2 + 2y = \square$$
$$y^2 + 2y + 1 = 3 + \square$$
$$(y + 1)^2 = \square$$
$$\square = \sqrt{4} \quad \text{or} \quad y + 1 = -\square$$
$$y + 1 = \square \quad \Big| \quad \square = -2$$
$$y = 1 \quad \Big| \quad y = -3$$

19. a. In solving a quadratic equation, a student obtains $x = \pm\sqrt{10}$. How many solutions are represented by this notation? List them.
    b. In solving a quadratic equation, a student obtains $x = 8 \pm \sqrt{3}$. List each solution separately. Then round each one to the nearest hundredth.
20. Solve $x + 1 = \pm\sqrt{2}$ for x.

PRACTICE *Use the square root method to solve each equation.*

21. $x^2 = 1$
22. $r^2 = 4$
23. $x^2 = 9$
24. $x^2 = 32$
25. $t^2 = 20$
26. $x^2 = 0$
27. $3m^2 = 27$
28. $4x^2 = 64$
29. $4x^2 = 16$
30. $5x^2 = 125$
31. $x^2 = \frac{9}{16}$
32. $x^2 = \frac{81}{25}$
33. $(x + 1)^2 = 25$
34. $(x - 1)^2 = 49$
35. $(x + 2)^2 = 81$
36. $(x + 3)^2 = 16$
37. $(x - 2)^2 = 8$
38. $(x + 2)^2 = 50$

Use the square root method to solve each equation. Use a calculator to approximate the solutions. Round to the nearest hundredth.

39. $x^2 = 45.82$
40. $x^2 = 6.05$
41. $(x + 2)^2 = 90.04$
42. $(x - 5)^2 = 33.31$

Factor the trinomial square and use the square root method to solve each equation.

43. $y^2 + 4y + 4 = 4$

44. $y^2 - 6y + 9 = 9$

45. $9x^2 - 12x + 4 = 16$

46. $4x^2 - 20x + 25 = 36$

Complete the square and factor the perfect square trinomial.

47. $x^2 + 2x$ **48.** $x^2 + 12x$

49. $x^2 - 4x$ **50.** $x^2 - 14x$

51. $x^2 + 7x$ **52.** $x^2 + 21x$

53. $a^2 - 3a$ **54.** $b^2 - 13b$

55. $b^2 + \frac{2}{3}b$ **56.** $c^2 - \frac{5}{2}c$

Solve each equation by completing the square.

57. $x^2 + 6x + 8 = 0$ **58.** $x^2 + 8x + 12 = 0$

59. $k^2 - 8k + 12 = 0$ **60.** $p^2 - 4p + 3 = 0$

61. $x^2 - 2x = 15$ **62.** $x^2 - 2x = 8$

63. $g^2 + 5g - 6 = 0$ **64.** $s^2 = 14 - 5s$

65. $2x^2 = 4 - 2x$ **66.** $3q^2 = 3q + 6$

67. $3x^2 + 9x + 6 = 0$ **68.** $3d^2 + 48 = -24d$

69. $2x^2 = 3x + 2$ **70.** $3x^2 = 2 - 5x$

71. $4x^2 = 2 - 7x$ **72.** $2x^2 = 5x + 3$

Solve each equation. Give the exact solutions, and then round them to the nearest hundredth.

73. $x^2 + 4x + 1 = 0$ **74.** $x^2 + 6x + 2 = 0$

75. $x^2 - 2x - 4 = 0$ **76.** $x^2 - 4x = 2$

77. $x^2 = 4x + 3$ **78.** $x^2 = 6x - 3$

79. $4x^2 + 4x + 1 = 20$ **80.** $9x^2 = 8 - 12x$

Write each equation in the form $ax^2 + bx + c = 0$ and solve it by completing the square.

81. $2x(x + 3) = 8$ **82.** $3x(x - 2) = 9$

83. $6(x^2 - 1) = 5x$ **84.** $2(3x^2 - 2) = 5x$

85. $x(x + 3) - \frac{1}{2} = -2$

86. $x[(x - 2) + 3] = 3\left(x - \frac{2}{9}\right)$

APPLICATIONS

87. CAROUSELS In 1999, the city of Lancaster, Pennsylvania, considered installing a classic Dentzel carousel in an abandoned downtown building. After learning that the circular-shaped carousel would occupy 2,376 square feet of floor space and that it was 26 feet high, the proposal was determined to be impractical because of the large remodeling costs. Find the diameter of the carousel to the nearest foot.

88. ESCAPE VELOCITY The speed at which a rocket must be fired for it to leave the Earth's gravitational attraction is called the **escape velocity.** If the escape velocity v_e, in miles per hour, is given by

$$\frac{v_e^2}{2g} = R$$

where $g = 78{,}545$ and $R = 3{,}960$, find v_e. Round to the nearest mi/hr.

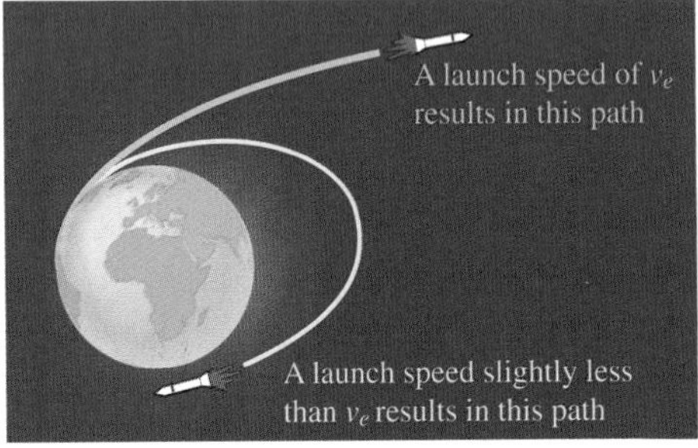
A launch speed of v_e results in this path
A launch speed slightly less than v_e results in this path

89. BICYCLE SAFETY A bicycle training program for children uses a figure-8 course to help them improve their balance and steering. The course is laid out over a paved area covering 800 square feet, as shown below. Find its dimensions.

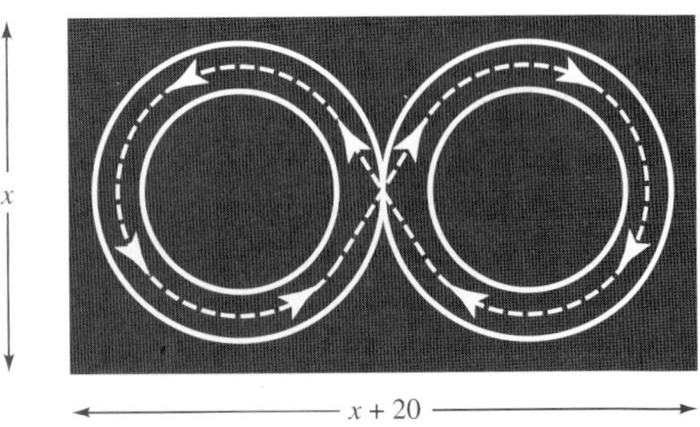
x
x + 20

90. BADMINTON The badminton court shown below occupies 880 square feet of the floor space of a gymnasium. If its length is 4 feet more than twice its width, find its dimensions.

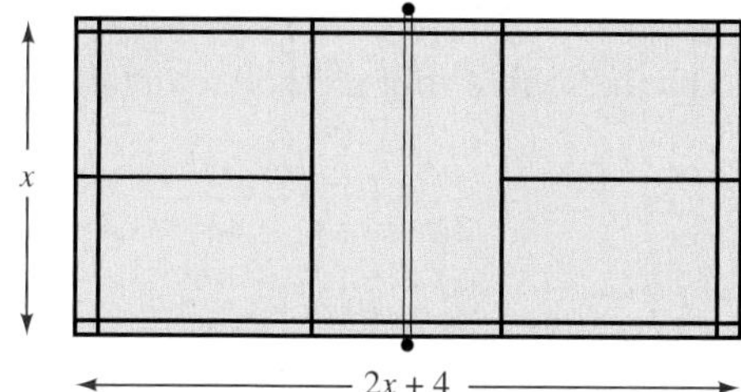

WRITING

91. Explain how to complete the square on $x^2 - 5x$.

92. Explain the error in the following work.

Solve: $x^2 = 28$.

$x = \pm\sqrt{28}$

$x = 2 \pm \sqrt{7}$

93. Rounded to the nearest hundredth, one solution of the equation $x^2 + 4x + 1 = 0$ is -0.27. Use your calculator to check it. How could it be a solution if it doesn't make the left-hand side zero? Explain.

94. Give an example of a perfect square trinomial. Why do you think the word "perfect" is used to describe it?

REVIEW *Perform each operation.*

95. $(y - 1)^2$
96. $(z + 2)^2$
97. $(x + y)^2$
98. $(a - b)^2$
99. $(2z)^2$
100. $(xy)^2$

15.2 The Quadratic Formula

- The quadratic formula
- Quadratic equations with no real solutions
- Applications

We can solve all quadratic equations by completing the square, but the work is often tedious. In this section, we will develop a formula, called the *quadratic formula,* that will enable us to solve quadratic equations with much less effort.

The quadratic formula

We can solve the **general quadratic equation** $ax^2 + bx + c = 0$, where $a \neq 0$, by completing the square.

$$ax^2 + bx + c = 0$$

$$\frac{ax^2}{a} + \frac{bx}{a} + \frac{c}{a} = \frac{0}{a}$$ Divide both sides by a so that the coefficient of x^2 is 1.

$$x^2 + \frac{b}{a}x + \frac{c}{a} = 0$$ Simplify: $\frac{\overset{1}{\cancel{a}}x^2}{\underset{1}{\cancel{a}}} = x^2$. Write $\frac{bx}{a}$ as $\frac{b}{a}x$.

$$x^2 + \frac{b}{a}x = -\frac{c}{a}$$ Subtract $\frac{c}{a}$ from both sides.

Since the coefficient of x is $\frac{b}{a}$, we can complete the square on $x^2 + \frac{b}{a}x$ by adding

$$\left(\frac{1}{2}\cdot\frac{b}{a}\right)^2, \quad \text{which is} \quad \frac{b^2}{4a^2}$$

to both sides:

$$x^2 + \frac{b}{a}x + \mathbf{\frac{b^2}{4a^2}} = \mathbf{\frac{b^2}{4a^2}} - \frac{c}{a}$$

After factoring the perfect square trinomial on the left-hand side, we have

$$\left(x+\frac{b}{2a}\right)\left(x+\frac{b}{2a}\right)=\frac{b^2}{4a^2}-\frac{\mathbf{4a}c}{\mathbf{4a}a}$$ The lowest common denominator on the right-hand side is $4a^2$. Build the second fraction.

$$\left(x+\frac{b}{2a}\right)^2=\frac{b^2-4ac}{4a^2}$$ Subtract the numerators and write the difference over the common denominator.

The resulting equation can be solved by the square root method to obtain

$$x+\frac{b}{2a}=\sqrt{\frac{b^2-4ac}{4a^2}} \quad \text{or} \quad x+\frac{b}{2a}=-\sqrt{\frac{b^2-4ac}{4a^2}}$$

$$x+\frac{b}{2a}=\frac{\sqrt{b^2-4ac}}{\sqrt{4a^2}} \qquad x+\frac{b}{2a}=-\frac{\sqrt{b^2-4ac}}{\sqrt{4a^2}}$$

$$x=-\frac{b}{2a}+\frac{\sqrt{b^2-4ac}}{2a} \qquad x=-\frac{b}{2a}-\frac{\sqrt{b^2-4ac}}{2a}$$

$$x=\frac{-b+\sqrt{b^2-4ac}}{2a} \qquad x=\frac{-b-\sqrt{b^2-4ac}}{2a}$$

These solutions are usually written in one formula called the **quadratic formula.**

Quadratic formula

The solutions of the quadratic equation $ax^2+bx+c=0$ are

$$x=\frac{-b\pm\sqrt{b^2-4ac}}{2a} \quad \text{where } a\neq 0$$

! COMMENT When you write the quadratic formula, draw the fraction bar so that it includes the complete numerator. Do not write

$$x=-b\pm\frac{\sqrt{b^2-4ac}}{2a}$$

Self Check 1
Solve: $x^2+6x+5=0$.

EXAMPLE 1 Solve: $x^2+5x+6=0$.

Solution The equation is written in $ax^2+bx+c=0$ form with $a=1$, $b=5$, and $c=6$.

$$\begin{array}{l}\mathbf{1}x^2+\mathbf{5}x+\mathbf{6}=0\\ \uparrow \quad\ \uparrow \quad \uparrow\\ \mathbf{a}x^2+\mathbf{b}x+\mathbf{c}=0\end{array}$$

To find the solutions, we substitute these values into the quadratic formula and evaluate the right-hand side.

$$x=\frac{-b\pm\sqrt{b^2-4ac}}{2a}$$ This is the quadratic formula.

$$=\frac{-5\pm\sqrt{5^2-4(1)(6)}}{2(1)}$$ Substitute 1 for a, 5 for b, and 6 for c.

$$=\frac{-5\pm\sqrt{25-24}}{2}$$ Evaluate the power and multiply within the radical symbol.

$$=\frac{-5\pm\sqrt{1}}{2}$$ Perform the subtraction within the radical symbol.

$$x=\frac{-5\pm 1}{2}$$ $\sqrt{1}=1$.

This notation represents two solutions. We simplify them separately, first using the + sign and then using the − sign.

$$x = \frac{-5 + 1}{2} \quad \text{or} \quad x = \frac{-5 - 1}{2}$$

$$x = \frac{-4}{2} \qquad x = \frac{-6}{2}$$

$$x = -2 \qquad x = -3$$

The solutions are −2 and −3.

Answer −1, −5

! COMMENT In Example 1, you may have noticed that we could have solved $x^2 + 5x + 6 = 0$ by factoring.

EXAMPLE 2 Solve: $2x^2 = 5x + 3$.

Self Check 2
Solve: $4x^2 - 11x = 3$.

Solution To identify a, b, and c, we must write the equation in quadratic form.

$$2x^2 = 5x + 3$$

$$2x^2 - 5x - 3 = 0 \qquad \text{Subtract } 5x \text{ and 3 from both sides.}$$

In this equation, $a = 2$, $b = -5$, and $c = -3$. To find the solutions, we substitute these values into the quadratic formula and evaluate the right-hand side.

$$x = \frac{-b \pm \sqrt{b^2 - 4ac}}{2a} \qquad \text{This is the quadratic formula.}$$

$$= \frac{-(-5) \pm \sqrt{(-5)^2 - 4(2)(-3)}}{2(2)} \qquad \text{Substitute 2 for } a, -5 \text{ for } b, \text{ and } -3 \text{ for } c.$$

$$= \frac{5 \pm \sqrt{25 - (-24)}}{4} \qquad -(-5) = 5. \text{ Evaluate the power and multiply within the radical symbol.}$$

$$= \frac{5 \pm \sqrt{49}}{4} \qquad \text{Perform the subtraction within the radical symbol: } 25 - (-24) = 25 + 24 = 49.$$

$$= \frac{5 \pm 7}{4} \qquad \sqrt{49} = 7.$$

Thus,

$$x = \frac{5 + 7}{4} \quad \text{or} \quad x = \frac{5 - 7}{4}$$

$$x = \frac{12}{4} \qquad x = \frac{-2}{4}$$

$$x = 3 \qquad x = -\frac{1}{2}$$

The solutions are 3 and $-\frac{1}{2}$. Check each one in the original equation.

Answer $3, -\frac{1}{4}$

EXAMPLE 3 Solve: $3x^2 = 2x + 4$. Round each solution to the nearest hundredth.

Self Check 3
Solve: $2x^2 - 1 = 2x$. Round to the nearest hundredth.

Solution We begin by writing the given equation in $ax^2 + bx + c = 0$ form.

$$3x^2 = 2x + 4$$

$$3x^2 - 2x - 4 = 0 \qquad \text{Subtract } 2x \text{ and 4 from both sides.}$$

In this equation, $a = 3$, $b = -2$, and $c = -4$. To find the solutions, we substitute these values into the quadratic formula and evaluate the right-hand side.

$$x = \frac{-b \pm \sqrt{b^2 - 4ac}}{2a}$$ This is the quadratic formula.

$$= \frac{-(-2) \pm \sqrt{(-2)^2 - 4(3)(-4)}}{2(3)}$$ Substitute 3 for a, -2 for b, and -4 for c.

$$= \frac{2 \pm \sqrt{4 + 48}}{6}$$ $-(-2) = 2$. Simplify within the radical symbol.

$$= \frac{2 \pm \sqrt{52}}{6}$$ Perform the addition within the radical symbol.

$$= \frac{2 \pm 2\sqrt{13}}{6}$$ $\sqrt{52} = \sqrt{4 \cdot 13} = 2\sqrt{13}$.

$$= \frac{\overset{1}{\cancel{2}}(1 \pm \sqrt{13})}{\underset{1}{\cancel{2}} \cdot 3}$$ In the numerator, factor out 2: $2 \pm 2\sqrt{13} = 2(1 \pm \sqrt{13})$. Write 6 as $2 \cdot 3$. Then divide out the common factor of 2.

$$x = \frac{1 \pm \sqrt{13}}{3}$$ Simplify.

The solutions are $\frac{1 \pm \sqrt{13}}{3}$. We can use a calculator to approximate each of them. To the nearest hundredth,

$$\frac{1 + \sqrt{13}}{3} \approx 1.54 \qquad \frac{1 - \sqrt{13}}{3} \approx -0.87$$

Answers $\frac{1 + \sqrt{3}}{2} \approx 1.37$, $\frac{1 - \sqrt{3}}{2} \approx -0.37$

Quadratic equations with no real solutions

The next example shows that some quadratic equations have no real-number solutions.

Self Check 4

Does the equation

$$2x^2 + x + 1 = 0$$

have any real-number solutions?

EXAMPLE 4 Solve: $x^2 + 2x + 5 = 0$.

Solution In this equation $a = 1$, $b = 2$, and $c = 5$. We substitute these values into the quadratic formula.

$$x = \frac{-b \pm \sqrt{b^2 - 4ac}}{2a}$$ This is the quadratic formula.

$$= \frac{-2 \pm \sqrt{2^2 - 4(1)(5)}}{2(1)}$$ Substitute 1 for a, 2 for b, and 5 for c.

$$= \frac{-2 \pm \sqrt{4 - 20}}{2}$$ Evaluate the power and multiply within the radical symbol.

$$x = \frac{-2 \pm \sqrt{-16}}{2}$$ Perform the subtraction within the radical symbol. The result is a negative number, -16.

Since $\sqrt{-16}$ is not a real number, there are no real-number solutions.

Answer no

Applications

We have discussed several methods that are used to solve quadratic equations. To determine the most efficient method for a given equation, we can use the following strategy.

Strategy for solving quadratic equations

1. First, see whether the equation is in a form such that the **square root method** is easily applied.
2. If the square root method can't be used, write the equation in $ax^2 + bx + c = 0$ form.
3. Then see whether the equation can be solved using the **factoring method.**
4. If you can't factor the quadratic, solve the equation by **completing the square** or by the **quadratic formula.**

EXAMPLE 5 **Nutrition.** The poster in Figure 15-2 shows the six basic food groups, as established by the U.S. Department of Agriculture. If the area of the poster is 90 square inches and the base is 3 inches longer than the height, find the length of its base and its height.

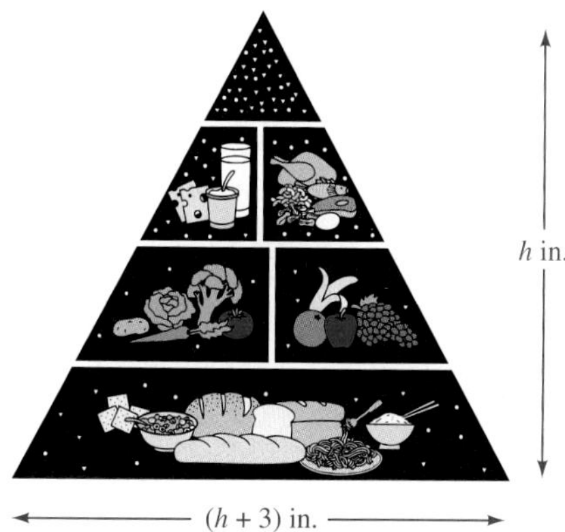

FIGURE 15-2

Analyze the problem

We are given the area of the triangular-shaped poster and asked to find the length of its base and its height.

Form an equation

Since the length of the base is related to the height, we let h = the height of the triangle. Then $h + 3$ = the length of the base. The area of a triangle is given by the formula $A = \frac{1}{2}bh$, which gives the equation

$\frac{1}{2}$	times	the length of the base	times	the height	equals	the area of the triangle.
$\frac{1}{2}$	$\cdot$	$(h + 3)$	$\cdot$	h	$=$	90

Solve the equation

To solve the equation $\frac{1}{2}(h + 3)h = 90$, we first write it in quadratic form.

$$\frac{1}{2}(h + 3)h = 90$$

$$(h + 3)h = 180 \quad \text{Multiply both sides by 2.}$$

$$h^2 + 3h = 180 \quad \text{Distribute the multiplication by } h.$$

$$h^2 + 3h - 180 = 0 \quad \text{Subtract 180 from both sides. The equation is now in quadratic form.}$$

By inspection, we see that -180 has factors of -12 and 15 and that their sum is 3. Therefore, we can use the factoring method to solve the equation.

$(h - 12)(h + 15) = 0$ Factor $h^2 + 3h - 180$.

$h - 12 = 0$ or $h + 15 = 0$ Set each factor equal to 0.

$h = 12$ | $h = -15$ Solve each linear equation.

State the conclusion

When h is 12, the length of the base, $h + 3$, is 15. We discard the solution -15, because the triangle cannot have a negative height. So the length of the base is 15 inches, and the height is 12 inches.

Check the result

With a base of 15 inches and a height of 12 inches, the base of the triangle is 3 inches longer than its height. Its area is $\frac{1}{2}(15)(12) = 90$ square inches. The solution checks.

EXAMPLE 6 **Movie stunts.** As part of a scene in a movie, a stuntman falls from the top of a 95-foot-tall building into a large airbag directly below him on the ground, as shown in Figure 15-3. If an object falls s feet in t seconds, where $s = 16t^2$, and if the bag is inflated to a height of 10 feet, how long will the stuntman fall before making contact with the airbag?

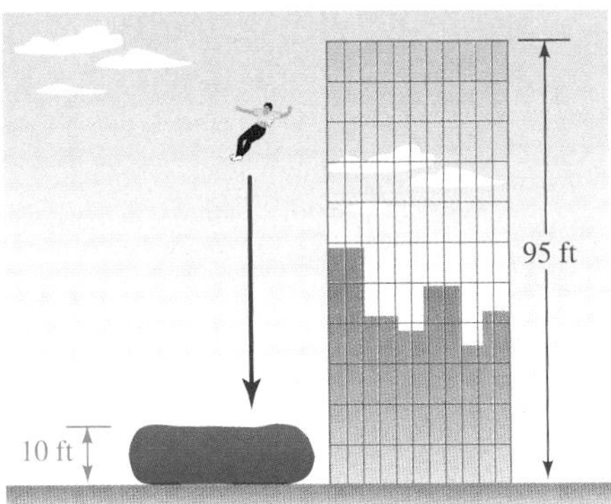

FIGURE 15-3

Solution If we subtract the height of the airbag from the height of the building, we find that the stuntman will fall $95 - 10 = 85$ feet. We substitute 85 for s in the formula and find that the equation is in a form that allows us to use the square root method.

$s = 16t^2$ This is the given formula.

$\mathbf{85} = 16t^2$ Substitute 85 for s.

$\frac{85}{16} = t^2$ Divide both sides by 16.

$\pm\sqrt{\frac{85}{16}} = t$ Use the square root method to solve the equation.

$\pm\frac{\sqrt{85}}{\sqrt{16}} = t$ The square root of a quotient is the quotient of the square roots.

$\pm\frac{\sqrt{85}}{4} = t$ $\sqrt{16} = 4$.

The stuntman will fall for $\frac{\sqrt{85}}{4}$ seconds before making contact with the airbag. To the nearest tenth, this is 2.3 seconds. We discard the other solution, $-\frac{\sqrt{85}}{4}$, because a negative time does not make sense in this context.

EXAMPLE 7 **Televisions.** A television's screen size is measured diagonally. For the 42-inch plasma television shown in Figure 15-4 the screen's height is 16 inches less than its length. What are the height and length of the screen.

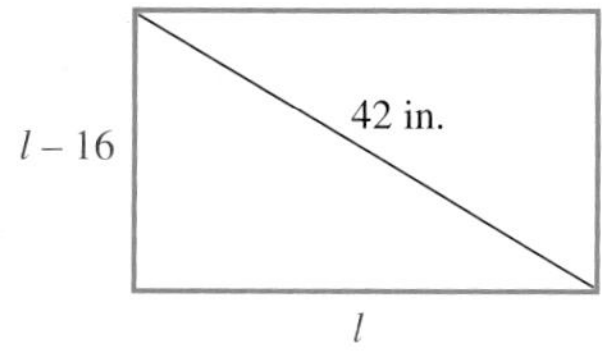

FIGURE 15-4

Analyze the Problem

A sketch of the screen shows that two adjacent sides and the diagonal form a right triangle. The length of the hypotenuse is 42 inches.

Form an equation

If we let l = the length of the screen, then $l - 16$ = the height of the screen. We can use the Pythagorean theorem to form an equation.

$a^2 + b^2 = c^2$	This is the Pythagorean theorem.
$l^2 + (l - 16)^2 = 42^2$	Substitute l for a, $l - 16$ for b, and 42 for c.
$l^2 + l^2 - 32l + 256 = 1{,}764$	Find $(l - 16)^2$ and 42^2.
$2l^2 - 32l - 1{,}508 = 0$	Subtract 1,764 from both sides.
$l^2 - 16l - 754 = 0$	Divide both sides by 2.

Solve the equation

Because of the large constant term, −754, we will not attempt to solve this quadratic equation by factoring. Instead, we will use the quadratic formula, with $a = 1$, $b = -16$, and $c = -754$.

$l = \frac{-b \pm \sqrt{b^2 - 4ac}}{2a}$	In the quadratic formula, replace x with l.
$= \frac{-(-16) \pm \sqrt{(-16)^2 - 4(1)(-754)}}{2(1)}$	Substitute 1 for a, −16 for b, and −754 for c.
$= \frac{16 \pm \sqrt{256 - (-3{,}016)}}{2}$	Evaluate the power and multiply within the radical. Multiply in the denominator.
$= \frac{16 \pm \sqrt{3{,}272}}{2}$	Subtract within the radical.

The solutions are $\frac{16 + \sqrt{3{,}272}}{2}$ and $\frac{16 - \sqrt{3{,}272}}{2}$.

We can use a calculator to approximate the solutions to the nearest tenth.

$$\frac{16 + \sqrt{3{,}272}}{2} \approx 36.6 \qquad \frac{16 - \sqrt{3{,}272}}{2} \approx -20.6$$

Discard the negative solution because the length of the screen cannot be negative.

State the conclusion

The length of the television screen is approximately 36.6 inches. Since the height is $l - 16$, the height is approximately $36.6 - 16$ or 20.6 inches.

Check the result

The sum of the squares of the length of the sides is $(36.6)^2 + (20.6)^2 = 1{,}763.92$. The square of the length of the hypotenuse is $42^2 = 1{,}764$. Since these are approximately equal, the results seem reasonable.

EXAMPLE 8 **Finance.** If $\$P$ is invested at an annual rate r, it will grow to an amount of $\$A$ in n years according to the formula $A = P(1 + r)^n$. What interest rate is needed to make a \$5,000 investment grow to \$5,618 after 2 years?

Solution We can substitute 5,000 for P, 5,618 for A, and 2 for n in the formula and solve for r.

$$A = P(1 + r)^n$$

$$\mathbf{5{,}618} = \mathbf{5{,}000}(1 + r)^2$$

$$5{,}618 = 5{,}000(1 + 2r + r^2) \quad \text{Find } (1 + r)^2.$$

$$5{,}618 = 5{,}000 + 10{,}000r + 5{,}000r^2 \quad \text{Distribute the multiplication by 5,000.}$$

$$0 = 5{,}000r^2 + 10{,}000r - 618 \quad \text{Subtract 5,618 from both sides.}$$

We can use a calculator and solve this equation by the quadratic formula, where $a = 5{,}000$, $b = 10{,}000$, and $c = -618$.

$$r = \frac{-b \pm \sqrt{b^2 - 4ac}}{2a}$$

$$= \frac{-\mathbf{10{,}000} \pm \sqrt{\mathbf{10{,}000}^2 - 4(\mathbf{5{,}000})(\mathbf{-618})}}{2(\mathbf{5{,}000})}$$

$$= \frac{-10{,}000 \pm \sqrt{100{,}000{,}000 + 12{,}360{,}000}}{10{,}000}$$

$$= \frac{-10{,}000 \pm \sqrt{112{,}360{,}000}}{10{,}000}$$

$$= \frac{-10{,}000 \pm 10{,}600}{10{,}000}$$

$$r = \frac{-10{,}000 + 10{,}600}{10{,}000} \quad \text{or} \quad r = \frac{-10{,}000 - 10{,}600}{10{,}000}$$

$$r = \frac{600}{10{,}000} \qquad\qquad r = \frac{-20{,}600}{10{,}000}$$

$$r = 0.06 \qquad\qquad r = -2.06$$

$$r = 6\% \qquad\qquad r = -206\%$$

The required rate is 6%. The rate of −206% has no meaning in this problem.

Section 15.2 STUDY SET

VOCABULARY *Fill in the blanks.*

1. The general ________ equation is $ax^2 + bx + c = 0$.
2. The formula

$$x = \frac{-b \pm \sqrt{b^2 - 4ac}}{2a}$$

is called the ________ formula.
3. To ______ a quadratic equation means to find all the values of the variable that make the equation true.
4. $\sqrt{-16}$ is not a _____ number.

CONCEPTS *Fill in the blanks.*

5. In the quadratic equation $ax^2 + bx + c = 0$, a cannot equal ▢.
6. Before we can determine a, b, and c for $x = 3x^2 - 1$, we must write the equation in ________ form.
7. In the quadratic equation $3x^2 - 5 = 0$, $a =$ ▢, $b =$ ▢, and $c =$ ▢.
8. In the quadratic equation $-4x^2 + 8x = 0$, $a =$ ▢, $b =$ ▢, and $c =$ ▢.
9. The formula for the area of a rectangle is $A =$ ▢, and the formula for the area of a triangle is $A =$ ▢.
10. If a, b, and c are three sides of a right triangle and c is the hypotenuse, then $c^2 =$ ▢.
11. In evaluating the numerator of

$$\frac{-5 \pm \sqrt{5^2 - 4(2)(1)}}{2(2)}$$

what operation should be performed first?
12. Consider the expression

$$\frac{3 \pm 6\sqrt{2}}{3}$$

a. How many terms does the numerator contain?
b. What common factor do the terms have?
c. Simplify the expression.
13. A student used the quadratic formula to solve an equation and obtained

$$x = \frac{-3 \pm \sqrt{15}}{2}$$

a. How many solutions does the equation have?
b. What are they exactly?
c. Approximate them to the nearest hundredth.
14. Write the following steps of the strategy for solving quadratic equations in the proper order.
- Use the quadratic formula.
- Write the equation in $ax^2 + bx + c = 0$ form.
- Use the factoring method.
- Use the square root method.
15. The solutions of a quadratic equation are

$$x = 2 \pm \sqrt{3}$$

Graph them on the number line.

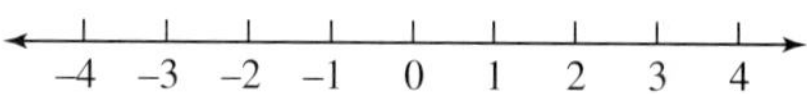

16. The solutions of a quadratic equation are

$$x = \frac{-1 \pm \sqrt{5}}{2}$$

Graph them on the number line.

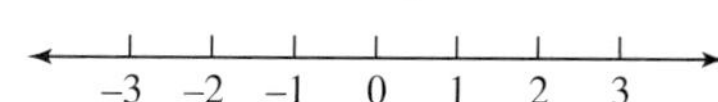

NOTATION *Complete each solution.*

17. Solve: $x^2 - 5x - 6 = 0$.

$$x = \frac{-b \pm \sqrt{b^2 - 4ac}}{2a}$$
$$= \frac{-(\square) \pm \sqrt{(-5)^2 - 4(1)(-6)}}{2(1)}$$
$$= \frac{\square \pm \sqrt{25 + \square}}{2}$$
$$= \frac{5 \pm \sqrt{\square}}{2}$$
$$x = \frac{\square \pm 7}{2}$$
$$x = \frac{5 \square 7}{2} = \square \quad \text{or} \quad x = \frac{5 \square 7}{2} = \square$$

18. Solve: $3x^2 + 2x - 2 = 0$.

$$x = \frac{-b \pm \sqrt{b^2 - 4ac}}{2a}$$
$$= \frac{-2 \pm \sqrt{\square^2 - 4(3)(\square)}}{2(\square)}$$
$$= \frac{-2 \pm \sqrt{4 \,\square\, 24}}{6}$$
$$= \frac{-2 \pm \sqrt{\square}}{6}$$
$$= \frac{-2 \pm \square\sqrt{7}}{6}$$
$$= \frac{\square(-1 \pm \sqrt{7})}{\square \cdot 3}$$
$$x = \frac{-1 \pm \sqrt{7}}{\square}$$

19. What is wrong with the following work?
Solve: $x^2 + 4x - 5 = 0$.

$$x = -4 \pm \frac{\sqrt{16 - 4(1)(-5)}}{2}$$

20. In reading

$$\frac{-b \pm \sqrt{b^2 - 4ac}}{2a}$$

we say, "the ______________ of b, plus or minus the ________ root of b squared minus 4 ________ a times c, all ________ $2a$."

PRACTICE *Change each equation into quadratic form, if necessary, and find the values of a, b, and c.* **Do not solve the equation.**

21. $x^2 + 4x + 3 = 0$
22. $x^2 - x - 4 = 0$
23. $3x^2 - 2x + 7 = 0$
24. $4x^2 + 7x - 3 = 0$
25. $4y^2 = 2y - 1$
26. $2x = 3x^2 + 4$
27. $x(3x - 5) = 2$
28. $y(5y + 10) = 8$
29. $7(x^2 + 3) = -14x$
30. $(2a + 3)(a - 2) = (a + 1)(a - 1)$

Use the quadratic formula to find all real solutions.

31. $x^2 - 5x + 6 = 0$
32. $x^2 + 5x + 4 = 0$
33. $x^2 + 7x + 12 = 0$
34. $x^2 - x - 12 = 0$
35. $2x^2 - x - 1 = 0$
36. $2x^2 + 3x - 2 = 0$
37. $3x^2 + 5x + 2 = 0$
38. $3x^2 - 4x + 1 = 0$
39. $4x^2 + 4x - 3 = 0$
40. $4x^2 + 3x - 1 = 0$
41. $x^2 + 3x + 1 = 0$
42. $x^2 + 3x - 2 = 0$
43. $3x^2 - x = 3$
44. $5x^2 = 3x + 1$
45. $x^2 + 5 = 2x$
46. $2x^2 + 3x = -3$
47. $x^2 = 1 - 2x$
48. $x^2 = 4 + 2x$
49. $3x^2 = 6x + 2$
50. $3x^2 = -8x - 2$

Use the most convenient method to find all real solutions. If a solution contains a radical, give the exact solution and then approximate it to the nearest hundredth.

51. $(2y - 1)^2 = 25$
52. $m^2 + 14m + 49 = 0$
53. $2x^2 + x = 5$
54. $2x^2 - x + 2 = 0$
55. $x^2 - 2x - 1 = 0$
56. $b^2 = 18$
57. $x^2 - 2x - 35 = 0$
58. $x^2 + 5x + 3 = 0$
59. $x^2 + 2x + 7 = 0$
60. $3x^2 - x = 1$
61. $4c^2 + 16c = 0$
62. $t^2 - 1 = 0$
63. $18 = 3y^2$
64. $25x - 50x^2 = 0$

Solve each equation. Round each solution to the nearest tenth.

65. $2.4x^2 - 9.5x + 6.2 = 0$
66. $-1.7x^2 + 0.5x + 0.9 = 0$

APPLICATIONS

67. HEIGHT OF A TRIANGLE The triangle shown has an area of 30 square inches. Find its height.

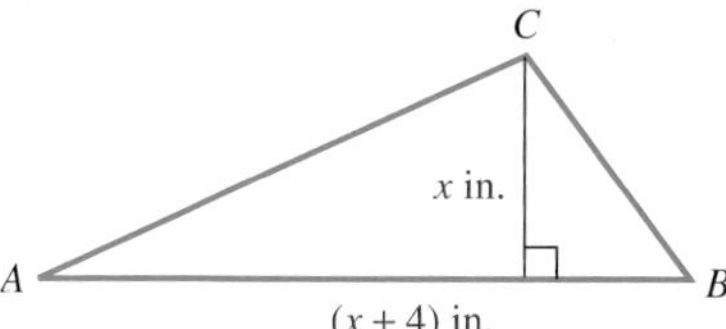

68. BOWLING When the pins for a children's bowling game are set up, they occupy 418 cm^2 of floor space. See the illustration. If the base of the triangular-shaped region is 6 cm longer than twice the height, how wide is the last row of pins?

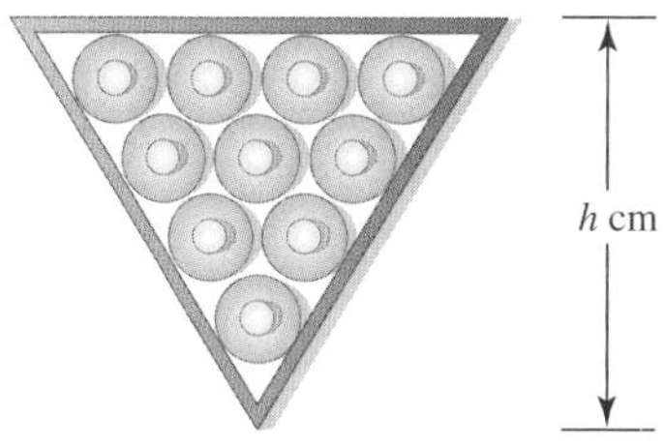

69. FLAGS According to the *Guinness Book of World Records 1998,* the largest flag flown from a flagpole was a Brazilian national flag, a rectangle having an area of 3,102 ft^2. If the flag is 19 feet longer than it is wide, find its width and length.

70. COMICS See the illustration. A comic strip occupies 96 square centimeters of space in a newspaper. The length of the rectangular space is 4 centimeters more than twice its width. Find its dimensions.

71. COMMUNITY GARDENS See the illustration. Residents of a community can work their own 16 ft × 24 ft plot of city-owned land if they agree to the following stipulations:

- The area of the garden cannot exceed 180 square feet.
- A path of uniform width must be maintained around the garden.

Find the dimensions of the largest possible garden.

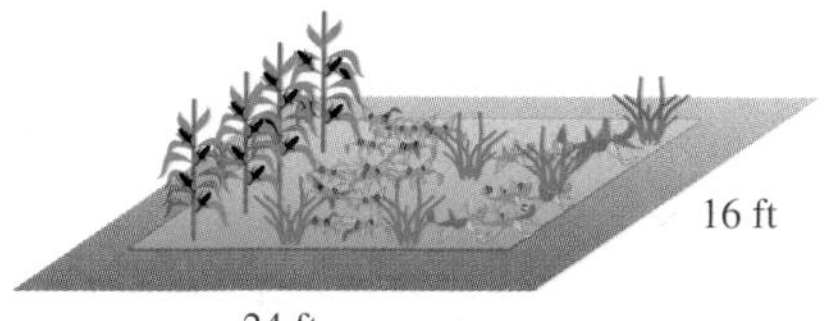

72. DECKING The owner of the pool shown below wants to surround it with a concrete deck of uniform width (shown in gray). If he can afford 368 square feet of decking, how wide can he make the deck?

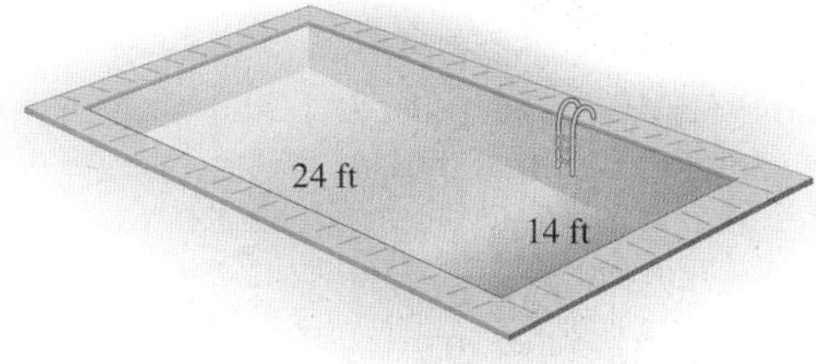

73. DAREDEVILS In 1873, Henry Bellini combined a tightrope walk over the Niagara River with a leap into the churning river below. If the rope was 200 feet above the water, for how many seconds did he fall before hitting the water? Round to the nearest tenth. (*Hint:* Refer to Example 6.)

74. FALLING OBJECTS A tourist drops a penny from the observation deck of a skyscraper 1,377 feet above the ground. How long will it take for the penny to hit the ground? (*Hint:* Refer to Example 6.)

75. ABACUS The Chinese abacus shown consists of a frame, parallel wires, and beads that are moved to perform arithmetic computations. If the frame is 21 centimeters wider than it is high, find its dimensions.

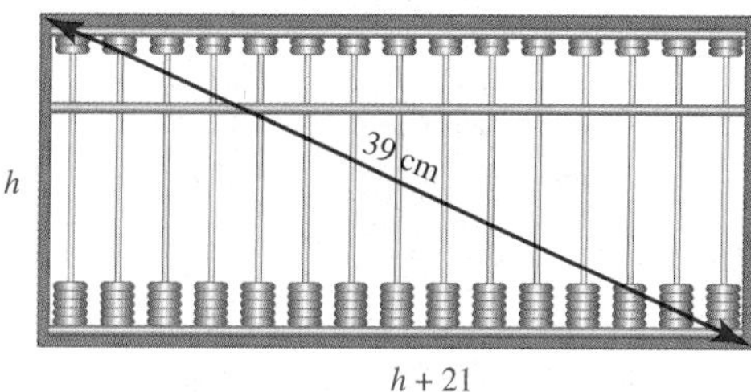

76. SIDEWALKS A 170-meter-long sidewalk from the mathematics building M to the student center C is shown in red in the illustration. However, students prefer to walk directly from M to C. How long are the two segments of the existing sidewalk?

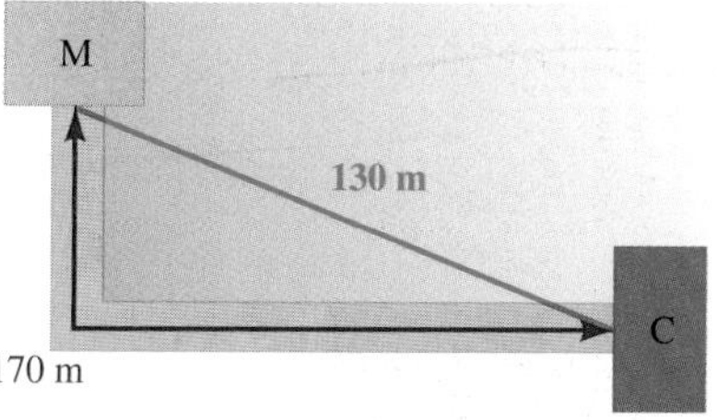

77. NAVIGATION Two boats leave port at the same time, one sailing east and one sailing south. If one boat sails 10 nautical miles more than the other and they are then 50 nautical miles apart, how far does each boat sail?

78. NAVIGATION One plane heads west from an airport, flying at 200 mph. One hour later, a second plane heads north from the same airport, flying at the same speed. When will the planes be 1,000 miles apart?

79. INVESTING We can use the formula $A = P(1 + r)^2$ to find the amount \$$A$ that \$$P$ will become when invested at an annual rate of r% for 2 years. What interest rate is needed to make \$5,000 grow to \$5,724.50 in 2 years?

80. INVESTING What interest rate is needed to make \$7,000 grow to \$8,470 in 2 years? See Exercise 79.

81. MANUFACTURING A firm has found that its revenue for manufacturing and selling x television sets is given by the formula $R = -\frac{1}{6}x^2 + 450x$. How much revenue will be earned by manufacturing 600 television sets?

82. RETAILING When a wholesaler sells n CD players, his revenue R is given by the formula $R = 150n - \frac{1}{2}n^2$. How many players would he have to sell to receive \$11,250? (*Hint:* Multiply both sides of the equation by -2.)

83. METAL FABRICATION A square piece of tin, 12 inches on a side, is to have four equal squares cut from its corners, as shown below. If the edges are then to be folded up to make a box with a floor area of 64 square inches, find the depth of the box.

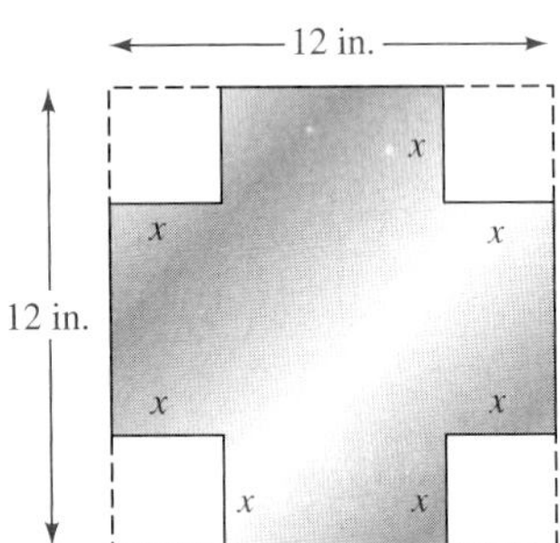

84. MAKING GUTTERS A piece of sheet metal, 18 inches wide, is bent to form the gutter shown. If the cross-sectional area is 36 square inches, find the depth of the gutter.

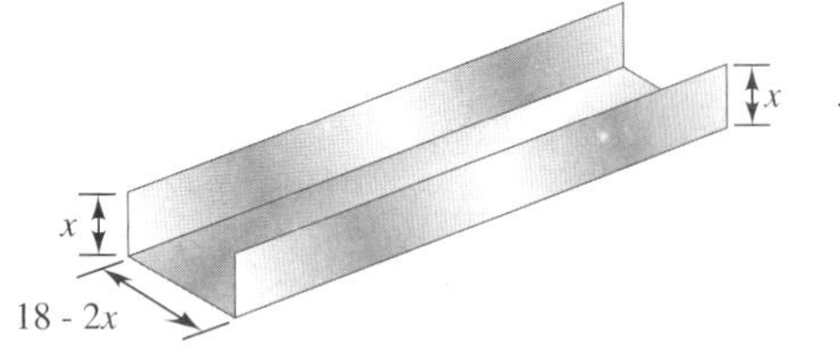

WRITING

85. Do you agree with the following statement? Explain your answer.

> The quadratic formula is the easiest method to use to solve quadratic equations.

86. Explain the meaning of the ± symbol.

87. Use the quadratic formula to solve $x^2 - 2x - 4 = 0$. What is an exact solution, and what is an approximate solution of this equation? Explain the difference.

88. Rewrite in words:

$$x = \frac{-b \pm \sqrt{b^2 - 4ac}}{2a}$$

REVIEW *Solve each equation for the indicated variable.*

89. $A = p + prt$; for r

90. $F = \dfrac{GMm}{d^2}$, for M

Write an equation of the line that has the given properties in slope–intercept form.

91. Slope of $\frac{3}{5}$ and passing through $(0, 12)$

92. Passes through $(6, 8)$ and the origin

Simplify each expression.

93. $\sqrt{80}$

94. $2\sqrt{x^3y^2}$

Rationalize each denominator and simplify.

95. $\dfrac{x}{\sqrt{7x}}$

96. $\dfrac{\sqrt{x} + 2}{\sqrt{x} - 2}$

15.3 Graphing Quadratic Functions

- Quadratic functions • Finding the vertex and the intercepts of a parabola
- A strategy for graphing quadratic functions • Finding a maximum value

In this section, we consider a type of function called a **quadratic function.** When graphing functions in Chapter 9, we constructed a table of values and plotted points. In this section, we will develop a more general strategy for graphing them by analyzing the given function and determining the characteristics of its graph.

Quadratic functions

Quadratic functions are defined by equations of the form $y = ax^2 + bx + c$, where $a \neq 0$, and where the right-hand side is a second-degree polynomial in the variable x. Three examples of quadratic functions are

$$y = x^2 - 3 \qquad y = x^2 - 2x - 3 \qquad y = -2x^2 - 4x + 2$$

We can replace y with the function notation $f(x)$ to express the defining equation in the form $f(x) = ax^2 + bx + c$. For the functions just mentioned, we can write

$$f(x) = x^2 - 3 \qquad f(x) = x^2 - 2x - 3 \qquad f(x) = -2x^2 - 4x + 2$$

In Section 9.2, we constructed the graph of $y = x^2$ (a quadratic function) by plotting points. The result was the **parabola** shown in Figure 15-5.

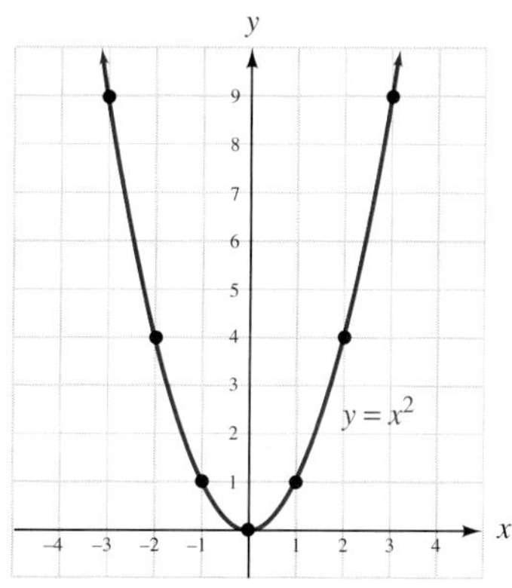

FIGURE 15-5

EXAMPLE 1 Graph: $y = x^2 - 3$. Compare the graph to that of $y = x^2$.

Solution The function is written in $y = ax^2 + bx + c$ form, where $a = 1$, $b = 0$, and $c = -3$. To find ordered pairs (x, y) that satisfy the equation, we pick numbers x and find the corresponding values of y. If we let $x = 3$, we have

$$\begin{aligned} y &= x^2 - 3 \\ &= 3^2 - 3 \quad \text{Substitute 3 for } x. \\ &= 6 \end{aligned}$$

The ordered pair (3, 6) and six others satisfying the equation appear in the table shown in Figure 15-6 on the next page. To graph the equation, we plot each point and draw a smooth curve passing through them. The resulting parabola is the graph of $y = x^2 - 3$. The parabola opens upward, and the lowest point on the graph, called the **vertex of the parabola,** is the point $(0, -3)$.

Self Check 1

Graph: $y = x^2 + 2$. Compare the graph to that of $y = x^2$.

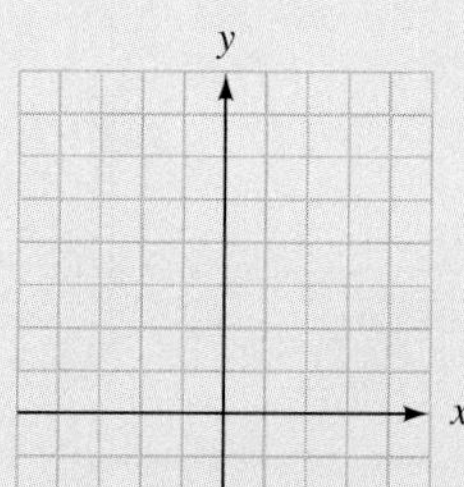

Note that the graph of $y = x^2 - 3$ looks just like the graph of $y = x^2$, except that it is 3 units lower.

$y = x^2 - 3$		
x	y	(x, y)
3	6	(3, 6)
2	1	(2, 1)
1	−2	(1, −2)
0	−3	(0, −3)
−1	−2	(−1, −2)
−2	1	(−2, 1)
−3	6	(−3, 6)

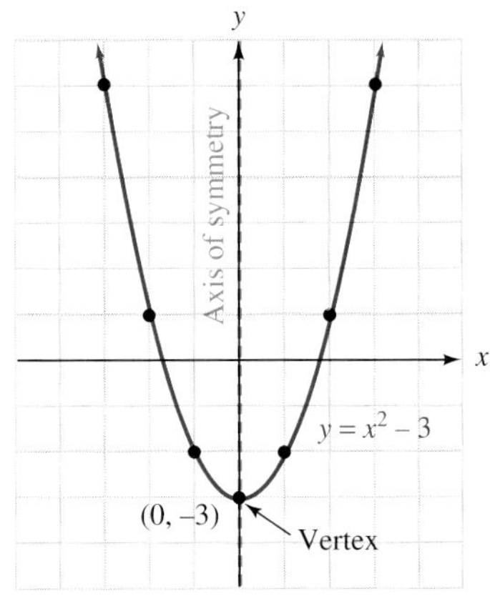

FIGURE 15-6

Answer The graph has the same shape as the graph of $y = x^2$, but it is 2 units higher.

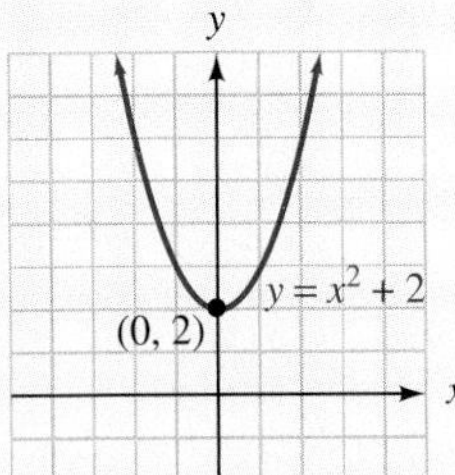

If we draw a vertical line through the vertex of the parabola in Figure 15-6 and fold the graph on this line, the two sides of the graph will match. We call the vertical line the **axis of symmetry.**

Self Check 2
Graph $f(x) = -x^2 - 4x - 4$, find its vertex, and draw its axis of symmetry.

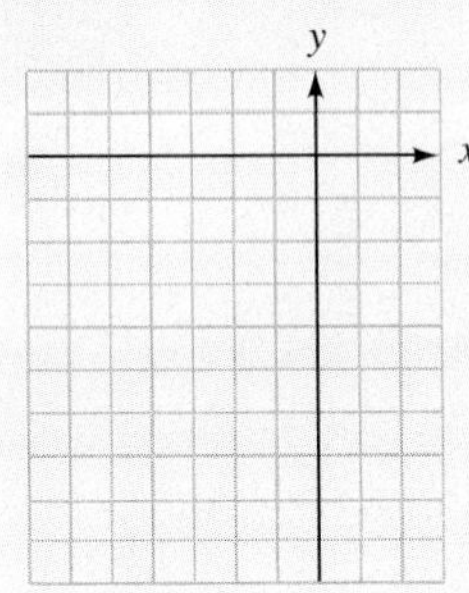

Answer The vertex is at $(-2, 0)$.

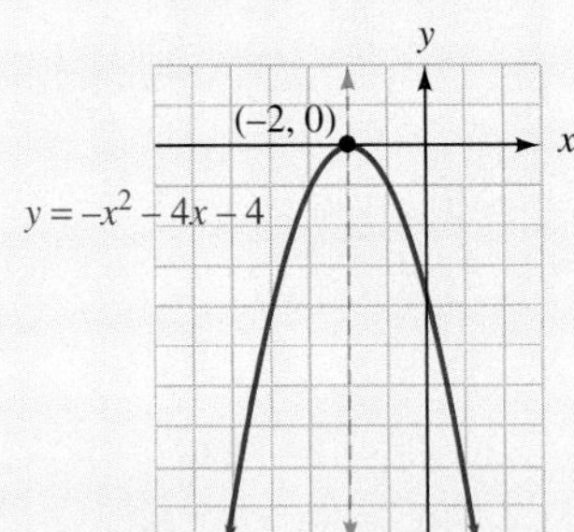

EXAMPLE 2 Graph $f(x) = -2x^2 - 4x + 2$, find its vertex, and draw its axis of symmetry.

Solution The function is written in $f(x) = ax^2 + bx + c$ form, where $a = -2$, $b = -4$, and $c = 2$. We construct the table shown in Figure 15-7, plot the points, and draw the graph.

$f(x) = -2x^2 - 4x + 2$		
x	$f(x)$	$(x, f(x))$
−3	−4	(−3, −4)
−2	2	(−2, 2)
−1	4	(−1, 4)
0	2	(0, 2)
1	−4	(1, −4)

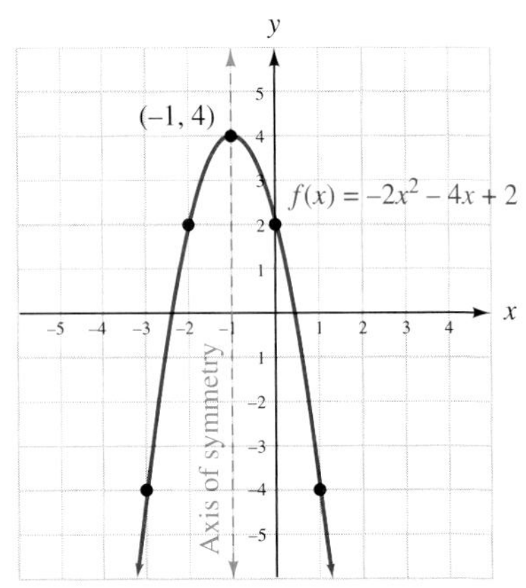

FIGURE 15-7

The parabola opens downward, so its vertex is its highest point, the point $(-1, 4)$.

In Example 1, the coefficient of the x^2 term in $y = x^2 - 3$ is positive $(a = 1)$. In Example 2, the coefficient of the x^2 term in $f(x) = -2x^2 - 4x + 2$ is negative $(a = -2)$. The results of these first two examples illustrate the following fact.

Graphs of quadratic functions

The graph of the function $y = ax^2 + bx + c$ or $f(x) = ax^2 + bx + c$, where $a \neq 0$, is a parabola. It opens upward when $a > 0$ and downward when $a < 0$.

The cup-like shape of a parabola can be seen in a wide variety of real-world settings. Some examples are shown in Figure 15-8.

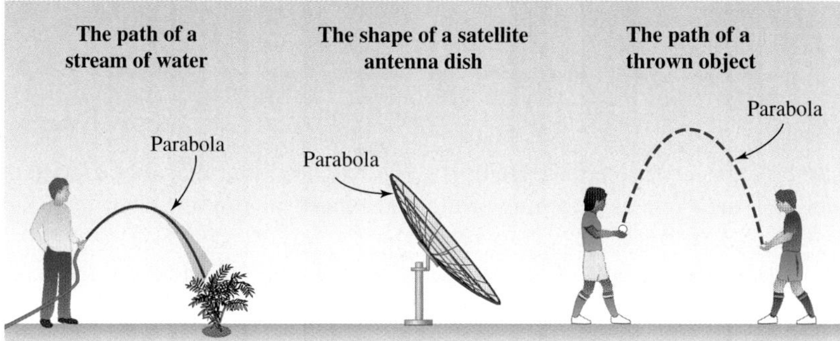

FIGURE 15-8

Finding the vertex and the intercepts of a parabola

It is easier to graph a quadratic function when we know the coordinates of the vertex of its parabolic graph. For a parabola defined by $y = ax^2 + bx + c$ or $f(x) = ax^2 + bx + c$, it can be shown that the x-coordinate of the vertex is given by $-\frac{b}{2a}$. This fact enables us to find the coordinates of its vertex.

Finding the vertex of a parabola

The graph of the quadratic function $y = ax^2 + bx + c$ or $f(x) = ax^2 + bx + c$ is a parabola whose vertex has an x-coordinate of $-\frac{b}{2a}$. To find the y-coordinate of the vertex, substitute $-\frac{b}{2a}$ into the defining equation and find y.

EXAMPLE 3 Find the vertex of the parabola defined by $y = x^2 - 2x - 3$.

Solution For $y = x^2 - 2x - 3$, we have $a = 1$, $b = -2$, and $c = -3$. To find the x-coordinate of the vertex, we substitute the values for a and b into the formula $x = -\frac{b}{2a}$.

$$x = -\frac{b}{2a}$$

$$x = -\frac{-2}{2(1)}$$

$$= 1$$

The x-coordinate of the vertex is 1. To find the y-coordinate, we substitute 1 for x:

$$y = x^2 - 2x - 3$$

$$y = 1^2 - 2(1) - 3$$

$$= 1 - 2 - 3$$

$$= -4$$

The vertex of the parabola is the point $(1, -4)$. See Figure 15-9(a) on the next page.

Self Check 3
Find the vertex of the parabola defined by $y = -x^2 + 6x - 8$.

Answer $(3, 1)$

When graphing quadratic functions, it is often helpful to find the x- and y-intercepts of the parabola.

Self Check 4
Find the x- and y-intercepts of the parabola defined by $y = -x^2 + 6x - 8$.

Answers y-intercept: $(0, -8)$; x-intercepts: $(2, 0)$, $(4, 0)$

EXAMPLE 4 Find the x- and y-intercepts of the parabola defined by $y = x^2 - 2x - 3$.

Solution To find the y-intercept of the parabola, we let $x = 0$ and solve for y.

$$y = x^2 - 2x - 3$$
$$y = 0^2 - 2(0) - 3$$
$$y = -3$$

The parabola passes through the point $(0, -3)$. We note that the y-coordinate of the y-intercept is the same as the value of the constant term c on the right-hand side of $y = x^2 - 2x - 3$.

To find the x-intercepts of the graph, we set y equal to 0 and solve the resulting quadratic equation.

$$y = x^2 - 2x - 3$$
$$0 = x^2 - 2x - 3 \quad \text{Substitute 0 for } y.$$
$$0 = (x - 3)(x + 1) \quad \text{Factor the trinomial.}$$
$$x - 3 = 0 \quad \text{or} \quad x + 1 = 0 \quad \text{Set each factor equal to 0.}$$
$$x = 3 \quad \Big| \quad x = -1$$

Since there are two solutions, the graph has two x-intercepts: $(3, 0)$ and $(-1, 0)$. See Figure 15-9(a).

A strategy for graphing quadratic functions

We can use the characteristics of a parabola to draw its graph. For example, to graph $y = x^2 - 2x - 3$, we note that the coefficient of the x^2 term is positive ($a = 1$). Therefore, the parabola defined by this function opens upward. In Examples 3 and 4, we found that the vertex of the graph of $y = x^2 - 2x - 3$ is at $(1, -4)$ and that the graph has a y-intercept of $(0, -3)$ and x-intercepts of $(3, 0)$ and $(-1, 0)$. See Figure 15-9(a).

We can locate other points on the parabola by noting that the graph has the axis of symmetry shown in Figure 15-9(a). If the point $(0, -3)$, which is 1 unit to the left of the axis of symmetry, is on the graph, the point $(2, -3)$, which is 1 unit to the right of the axis of symmetry, is also on the graph.

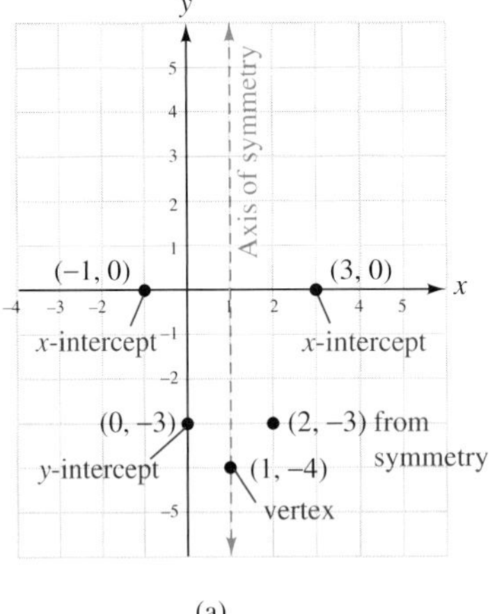

(a)

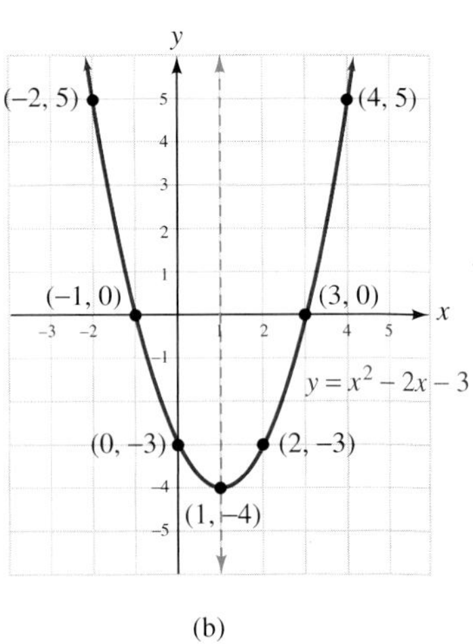

(b)

FIGURE 15-9

We can complete the graph by plotting two more points. If x is -2, then y is 5, and the parabola passes through $(-2, 5)$. Again using symmetry, the parabola must also pass through $(4, 5)$. The completed graph of $y = x^2 - 2x - 3$ is shown in Figure 15-9(b) on the previous page.

$y = x^2 - 2x - 3$		
x	y	(x, y)
-2	5	$(-2, 5)$

Much can be determined about the graph of a quadratic function from the coefficients a, b, and c. This information is summarized below.

Graphing a quadratic function $y = ax^2 + bx + c$

Determine whether the parabola opens upward or downward by examining a.

The x-coordinate of the vertex of the parabola is $x = -\frac{b}{2a}$.

To find the y-coordinate of the vertex, substitute $-\frac{b}{2a}$ for x into the equation and find y.

The axis of symmetry is the vertical line passing through the vertex.

The y-intercept is determined by the value of y when $x = 0$: the y-intercept is $(0, c)$.

The x-intercepts (if any) are determined by the numbers x that make $y = 0$. To find them, solve the quadratic equation $ax^2 + bx + c = 0$.

EXAMPLE 5 Graph: $f(x) = -2x^2 - 8x - 8$.

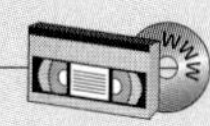

Self Check 5
Graph:
$f(x) = -x^2 + 6x - 8$

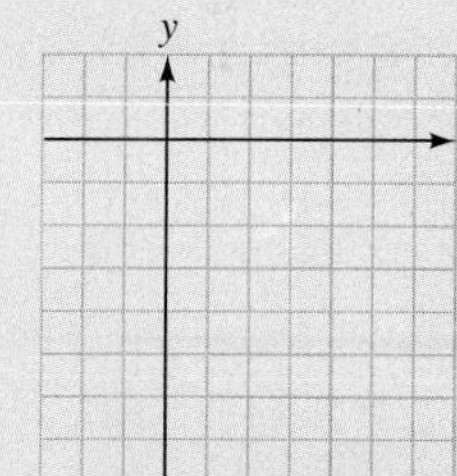

Solution

Step 1 *Determine whether the parabola opens upward or downward.* The equation is in the form $f(x) = ax^2 + bx + c$, with $a = -2$, $b = -8$, and $c = -8$. Since $a < 0$, the parabola opens downward.

Step 2 *Find the vertex and draw the axis of symmetry.* To find the x-coordinate of the vertex, we substitute the values for a and b into the formula $x = -\frac{b}{2a}$.

$$x = -\frac{\mathbf{b}}{2\mathbf{a}}$$

$$x = -\frac{\mathbf{-8}}{2(\mathbf{-2})}$$

$$= -2$$

The x-coordinate of the vertex is -2. To find the y-coordinate, we substitute -2 for x in the equation and find $f(-2)$.

$$f(\mathbf{x}) = -2\mathbf{x}^2 - 8\mathbf{x} - 8$$

$$f(\mathbf{-2}) = -2(\mathbf{-2})^2 - 8(\mathbf{-2}) - 8$$

$$= -8 + 16 - 8$$

$$= 0 \qquad \text{If } f(-2) = 0, \text{ then } y = 0 \text{ for } x = -2.$$

The vertex of the parabola is the point $(-2, 0)$. This point is the blue dot in Figure 15-10.

Step 3 *Find the x- and y-intercepts.* Since $c = -8$, the y-intercept of the parabola is $(0, -8)$. The point $(-4, -8)$, two units to the left of the axis of symmetry, must also be on the graph. We plot both points in black in Figure 15-10 on the next page.

To find the x-intercepts, we set $f(x)$ equal to 0 and solve the resulting quadratic equation.

$$
\begin{aligned}
f(x) &= -2x^2 - 8x - 8 & & \\
\mathbf{0} &= -2x^2 - 8x - 8 & & \text{Set } f(x) = 0. \\
0 &= x^2 + 4x + 4 & & \text{Divide both sides by } -2. \\
&= (x + 2)(x + 2) & & \text{Factor the trinomial.}
\end{aligned}
$$

$$x + 2 = 0 \quad \text{or} \quad x + 2 = 0 \qquad \text{Set each factor equal to 0.}$$

$$x = -2 \quad \Big| \quad x = -2$$

Since the solutions are the same, the graph has only one x-intercept: $(-2, 0)$. This point is the vertex of the parabola and has already been plotted.

Step 4 *Plot another point.* Finally, we find another point on the parabola. If $x = -3$, then $y = -2$. We plot $(-3, -2)$ in Figure 15-10 and use symmetry to determine that $(-1, -2)$ is also on the graph. Both points are in green.

Step 5 Draw a smooth curve through the points, as shown in Figure 15-10.

$f(x) = -2x^2 - 8x - 8$

x	$f(x)$	$(x, f(x))$
−3	−2	(−3, −2)

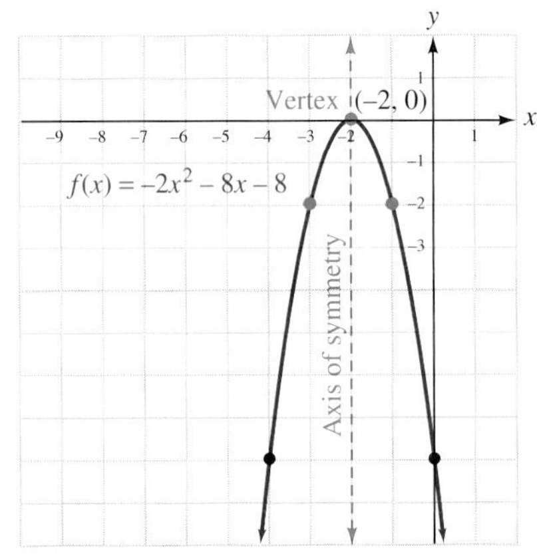

FIGURE 15-10

Answer

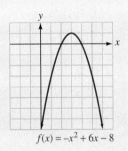

$f(x) = -x^2 + 6x - 8$

COMMENT The number of x-intercepts of the graph of a quadratic function $y = ax^2 + bx + c$ is the same as the number of real solutions of $ax^2 + bx + c = 0$. For example, the graph of $y = x^2 + x - 2$ in Figure 15-11(a) has two x-intercepts, and $x^2 + x - 2 = 0$ has two real-number solutions. In Figure 15-11(b), the graph has one x-intercept, and the corresponding equation has one real-number solution. In Figure 15-11(c), the graph does not have an x-intercept, and the corresponding equation does not have any real-number solutions. Note that the solutions of each equation are given by the x-coordinates of the x-intercepts of each respective graph.

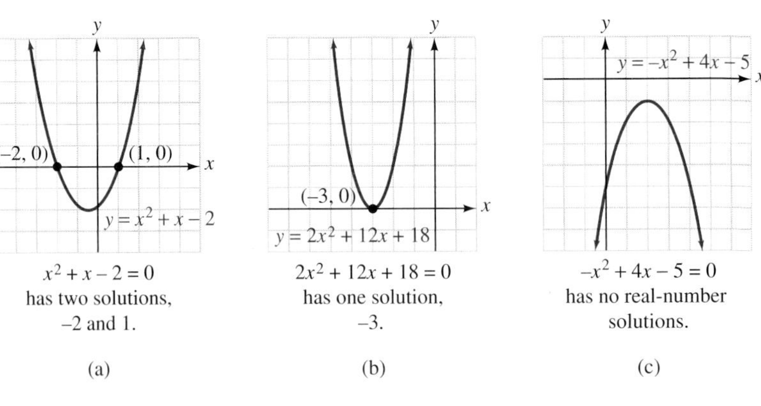

$x^2 + x - 2 = 0$ has two solutions, −2 and 1.

(a)

$2x^2 + 12x + 18 = 0$ has one solution, −3.

(b)

$-x^2 + 4x - 5 = 0$ has no real-number solutions.

(c)

FIGURE 15-11

Finding a maximum value

EXAMPLE 6 **Finding maximum revenue.** A firm manufactures radios. Over the past 10 years, the firm has learned that it can sell x radios at a price of $(200 - \frac{1}{5}x)$ dollars. How many radios should the firm manufacture and sell to maximize its revenue? Find the maximum revenue.

Solution The revenue obtained is the product of the number of radios sold (x) and the price of each radio $(200 - \frac{1}{5}x)$. Thus, the revenue R is given by the function

$$R = x\left(200 - \frac{1}{5}x\right) \quad \text{or} \quad R = -\frac{1}{5}x^2 + 200x$$

Since the graph of this function is a parabola that opens downward, the *maximum* value of R will be the value of R determined by the vertex of the parabola. Because the x-coordinate of the vertex is at $x = -\frac{b}{2a}$, we have

$$x = -\frac{\mathbf{b}}{2\mathbf{a}}$$

$$= -\frac{\mathbf{200}}{2(-\frac{1}{5})} \quad \text{Substitute 200 for } b \text{ and } -\frac{1}{5} \text{ for } a.$$

$$= -\frac{200}{-\frac{2}{5}} \quad \text{Perform the multiplication in the denominator.}$$

$$= (-200)(-\tfrac{5}{2}) \quad \text{Division by } -\tfrac{2}{5} \text{ is the same as multiplication by its reciprocal, which is } -\tfrac{5}{2}.$$

$$= 500$$

If the firm manufactures 500 radios, the maximum revenue will be

$$R = -\frac{1}{5}\mathbf{x}^2 + 200\mathbf{x} \quad \text{This is the revenue formula.}$$

$$= -\frac{1}{5}(\mathbf{500})^2 + 200(\mathbf{500}) \quad \text{Substitute 500 for } x\text{, the number of radios.}$$

$$= 50{,}000$$

The firm should manufacture 500 radios to get a maximum revenue of \$50,000. This fact is verified by examining the graph of $R = -\frac{1}{5}x^2 + 200x$, which appears in Figure 15-12.

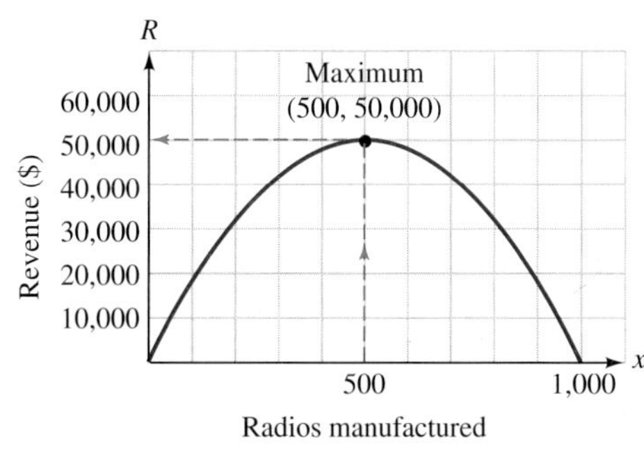

FIGURE 15-12

The "Baby Boom Echo" Generation

THINK IT THROUGH

"The young people who make up the 'baby boom echo' generation are growing older and rapidly starting to fill up our nation's high schools, colleges, and universities." Alliance for Education Excellence, 2002

The following graph shows the surge in births after World War II, nicknamed the "baby boom." It also shows a second surge in births called "baby boom echo"—the children of the baby boom generation.

1. What year during the baby boom had the maximum number of births? Estimate the number.
2. What year during the baby boom echo had the maximum number of births? Estimate the number.
3. For 1947–2007, what year had the minimum number of births? Estimate the number.

Annual Number of births, with Projections: 1947 to 2007

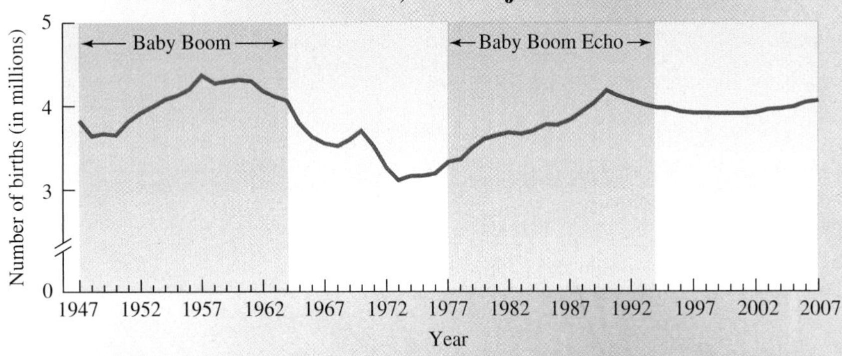

Source: National Center for Education Statistics

Section 15.3 STUDY SET

VOCABULARY *Fill in the blanks.*

1. A function defined by the equation $y = ax^2 + bx + c$, where $a \neq 0$, is called a __________ function.
2. The lowest (or highest) point on a parabola is called the __________ of the parabola.
3. The point where a parabola intersects the y-axis is called the __________.
4. The point (or points) where a parabola intersects the __________ is (are) called the x-intercept(s).
5. For a parabola that opens upward or downward, the vertical line that passes through its vertex and splits the graph into two identical pieces is called the axis of __________.
6. For the graph of $y = ax^2 + bx + c$, the __________ of the x^2 term indicates whether the parabola opens upward or downward.

CONCEPTS *Fill in the blanks.*

7. The graph of $y = ax^2 + bx + c$ opens upward when a ____ 0.
8. The graph of $f(x) = ax^2 + bx + c$ opens downward when a ____ 0.
9. The y-intercept of the graph of $f(x) = ax^2 + bx + c$ is the point ________.
10. The x-coordinate of the vertex of the parabola that results when we graph $y = ax^2 + bx + c$ is $x =$ ____

11. Refer to the graph.

a. What do we call the curve shown there?

b. What are the x-intercepts of the graph?

c. What is the y-intercept of the graph?

d. What is the vertex?

12. The vertex of a parabola is at $(1, -3)$, its y-intercept is $(0, -2)$, and it passes through the point $(3, 1)$, as shown. Draw the axis of symmetry and use it to help determine two other points on the parabola.

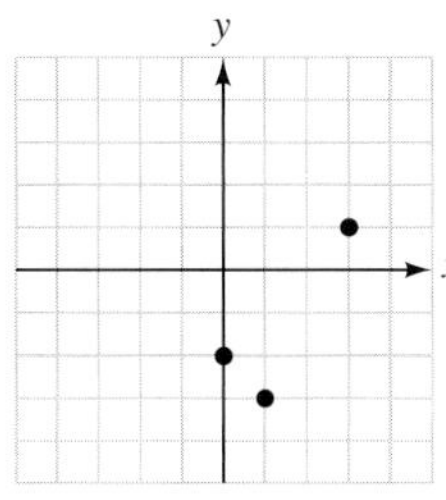

13. Sketch the graphs of parabolas with zero, one, and two x-intercepts.

14. Sketch the graph of a parabola opening upward that doesn't have a y-intercept, if possible.

15. HEALTH CLINICS The number of cases of flu seen by doctors at a county health clinic each week during a 10-week period is described by the quadratic function graphed in the illustration. Write a brief summary report about the flu outbreak. What information about the number of cases of flu does the vertex give?

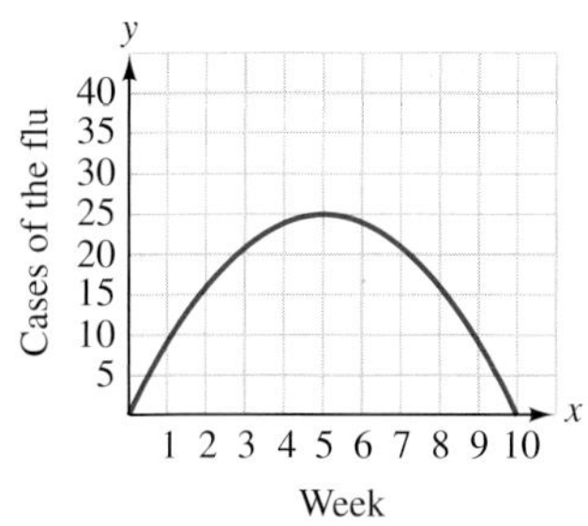

16. COST ANALYSIS A company has found that when it assembles x carburetors in a production run, the manufacturing cost $\$y$ per carburetor is given by the quadratic function graphed in the illustration. What information about the manufacturing cost does the vertex give?

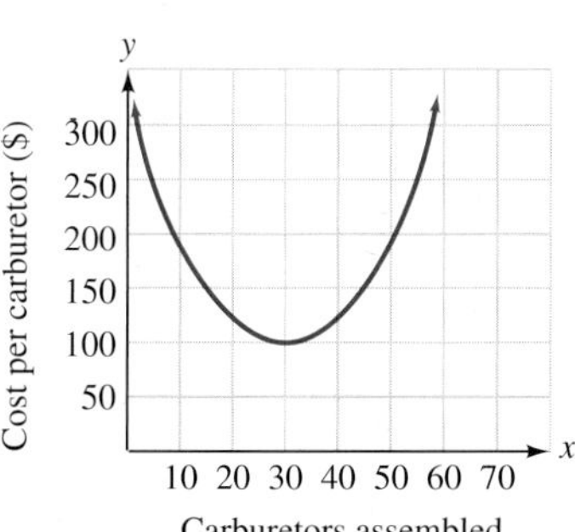

NOTATION

17. Determine whether this statement is true or false: The equations $y = 2x^2 - x - 2$ and $f(x) = 2x^2 - x - 2$ are the same.

18. The function $y = -x^2 + 3x - 5$ is written in $y = ax^2 + bx + c$ form. What are a, b, and c?

19. Consider $y = 3x^2 + 3x - 8$. What is $-\frac{b}{2a}$?

20. Evaluate: $\frac{-12}{2(-3)}$.

PRACTICE *Graph each quadratic function and compare the graph to the graph of $y = x^2$.*

21. $y = x^2 + 1$

22. $y = x^2 - 4$

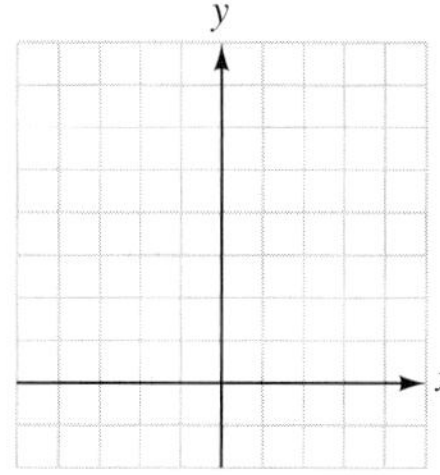

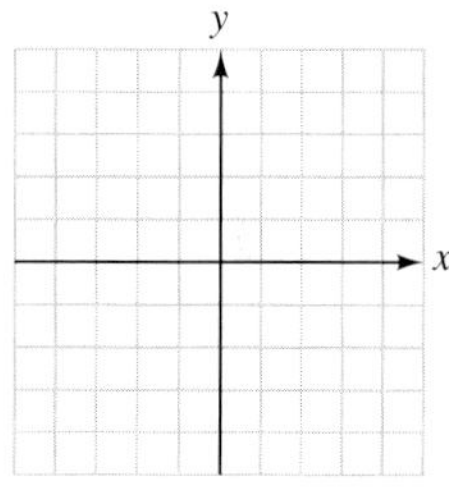

23. $f(x) = -x^2$

24. $f(x) = (x - 1)^2$

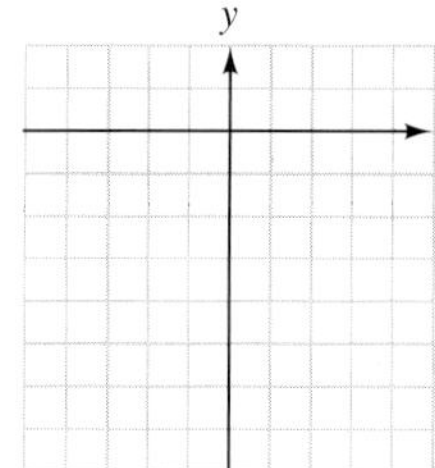

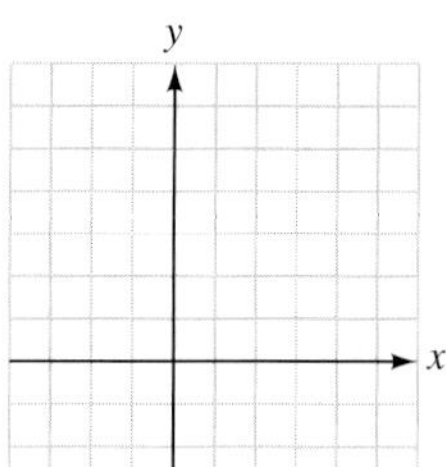

Find the vertex of the graph of each quadratic function.

25. $y = -x^2 + 6x - 8$

26. $y = -x^2 - 2x - 1$

27. $f(x) = 2x^2 - 4x + 1$

28. $f(x) = 2x^2 + 8x - 4$

Find the x- and y-intercepts of the graph of the quadratic function.

29. $f(x) = x^2 - 2x + 1$

30. $f(x) = 2x^2 - 4x$

31. $y = -x^2 - 10x - 21$

32. $y = 3x^2 + 6x - 9$

Graph each quadratic function using the method discussed in Example 5.

33. $y = x^2 - 2x$

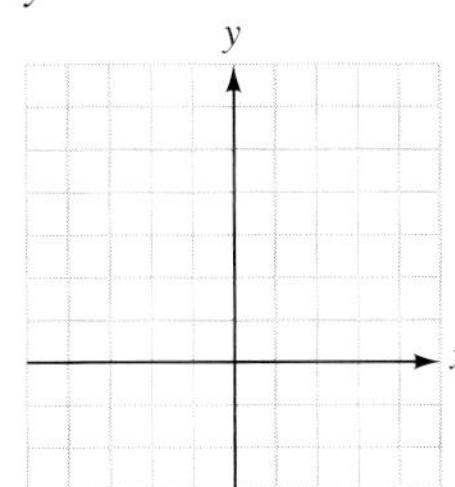

34. $f(x) = -x^2 - 4x$

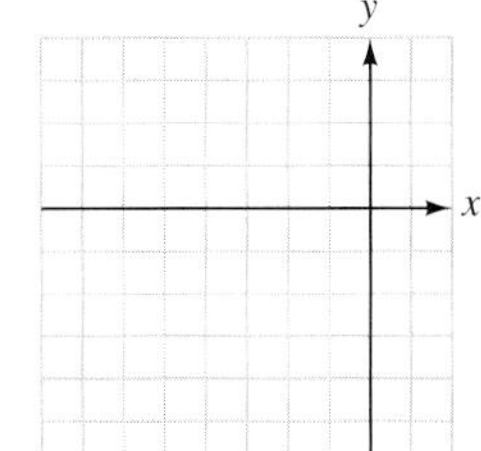

35. $f(x) = -x^2 + 2x$

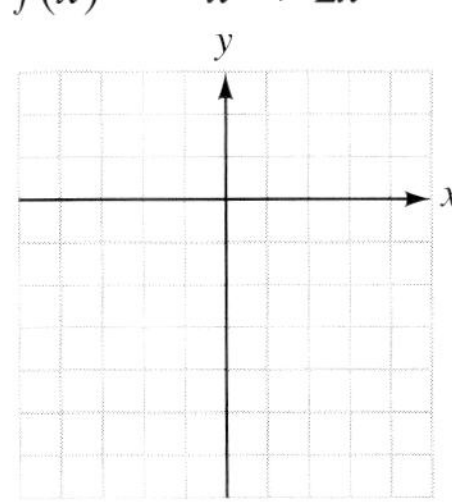

36. $y = x^2 + x$

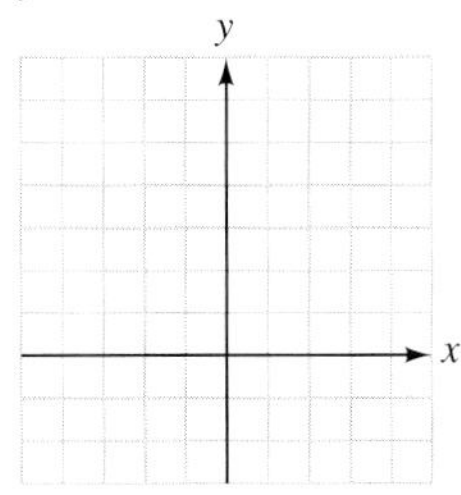

37. $f(x) = x^2 + 4x + 4$

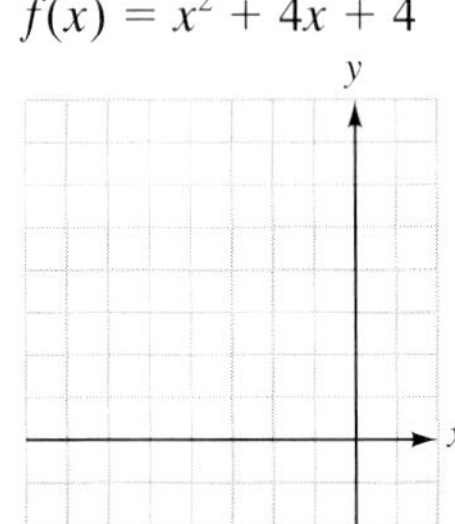

38. $f(x) = x^2 - 6x + 9$

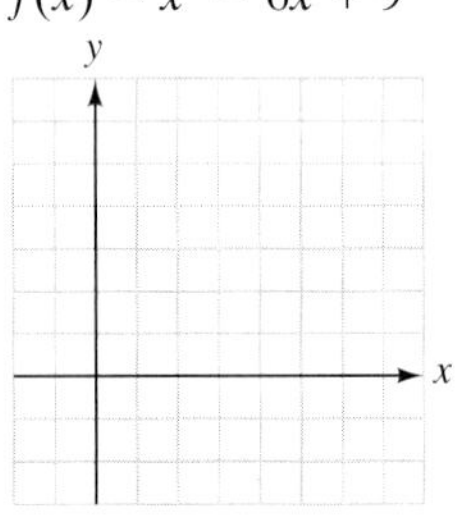

39. $y = -x^2 - 2x - 1$

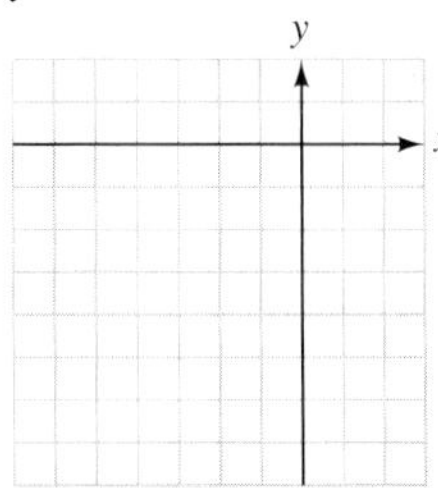

40. $y = -x^2 + 2x - 1$

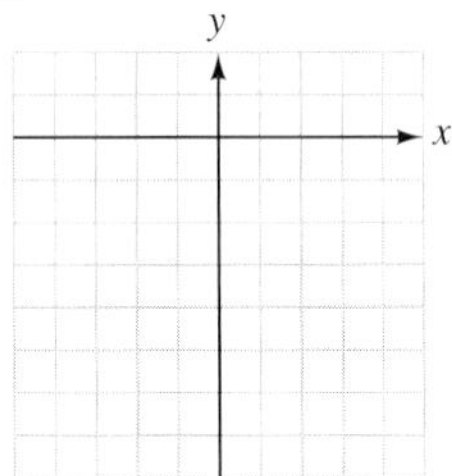

41. $y = x^2 + 2x - 3$

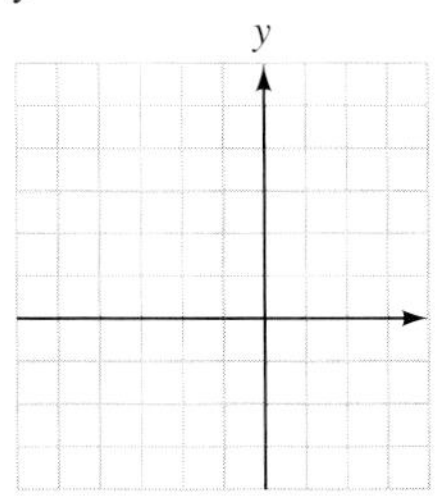

42. $y = x^2 + 6x + 5$

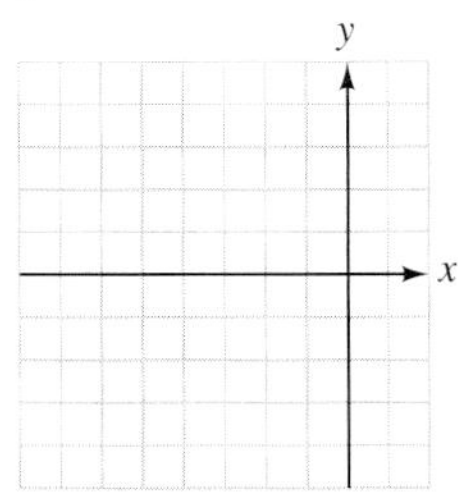

43. $f(x) = 2x^2 + 8x + 6$

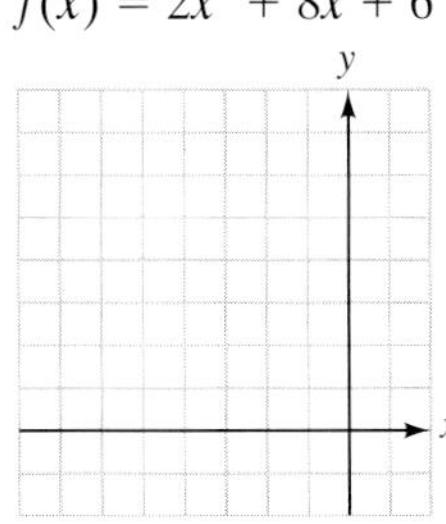

44. $f(x) = 3x^2 - 12x + 9$

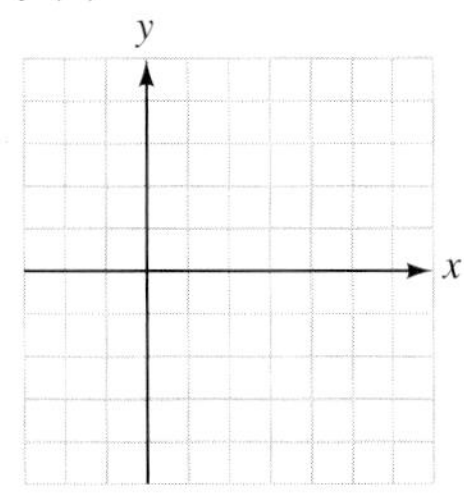

45. $y = x^2 - 2x - 8$

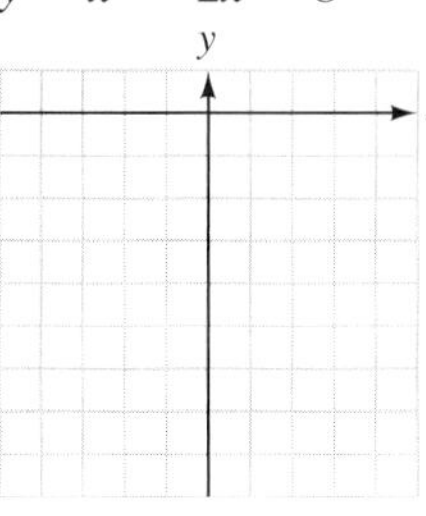

46. $y = -x^2 + 2x + 3$

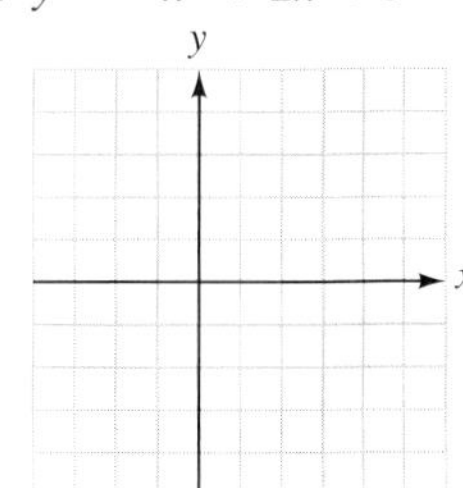

47. $y = x^2 - x - 2$

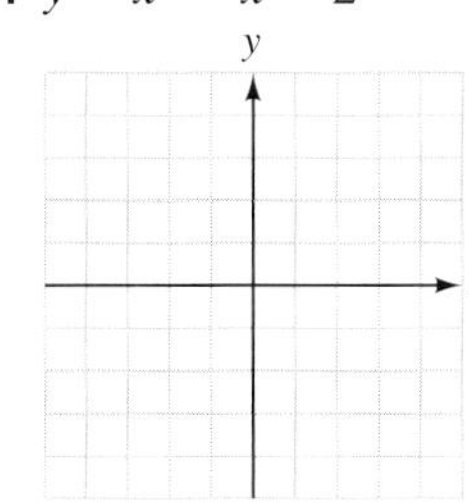

48. $y = -x^2 + 5x - 4$

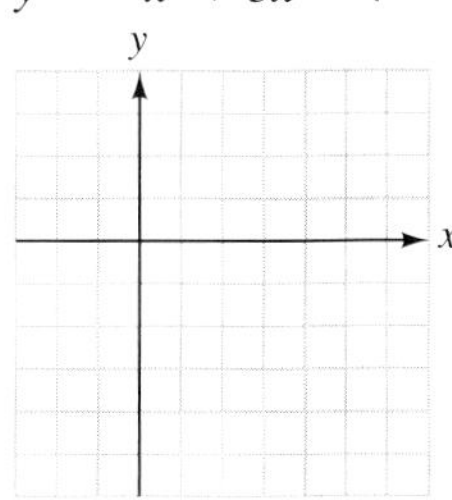

49. $f(x) = 2x^2 + 3x - 2$

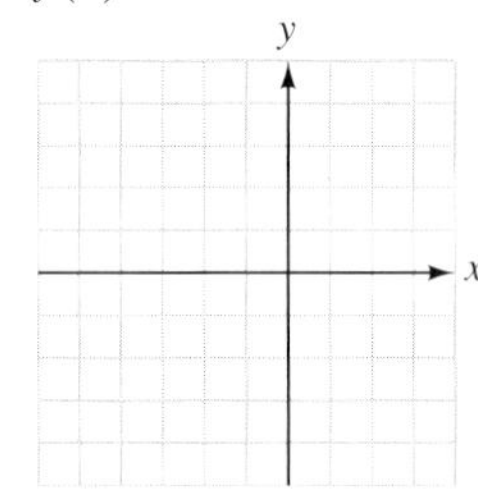

50. $f(x) = 3x^2 - 7x + 2$

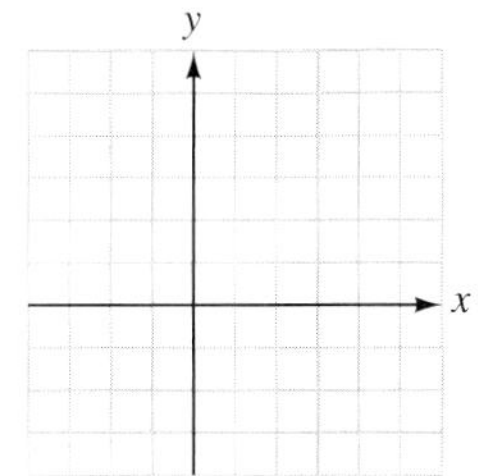

APPLICATIONS

51. TRAMPOLINES The illustration shows how far a trampolinist is from the ground (in relation to time) as she bounds into the air and then falls back down to the trampoline.

a. How many feet above the ground is she $\frac{1}{2}$ second after bounding upward?

b. When is she 9 feet above the ground?

c. What is the maximum number of feet above the ground she gets? When does this occur?

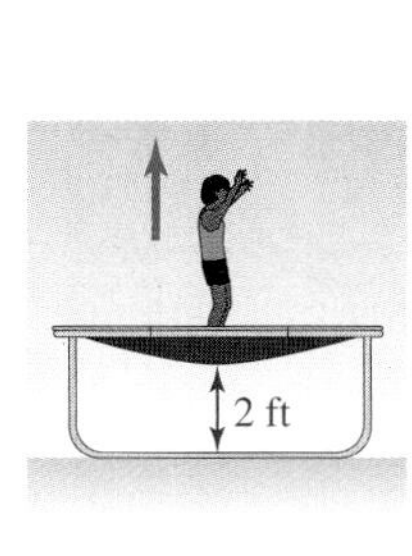

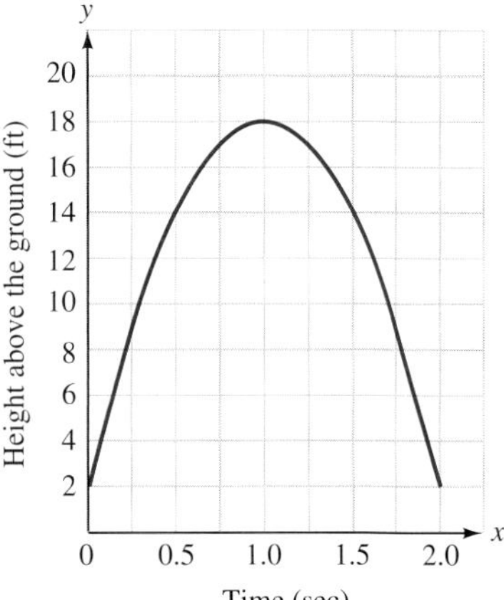

52. MIRRORS A mirror is held against the y-axis of the graph of a quadratic equation. What fact about parabolas does this illustrate?

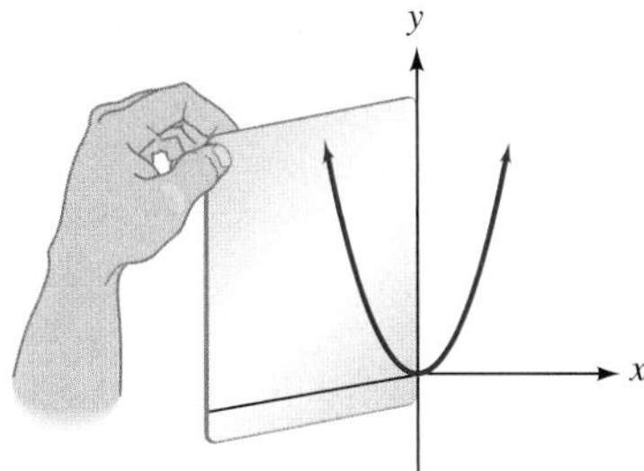

53. BRIDGES The shapes of the suspension cables in certain types of bridges are parabolic. The suspension cable for the bridge shown is described by $y = 0.005x^2$. Finish the mathematical model of the bridge by completing the table of values using a calculator, plotting the points, and drawing a smooth curve through them to represent the cable. Finally, from each plotted point, draw a vertical support cable attached to the roadway.

x	-80	-60	-40	-20	0	20	40	60	80
y									

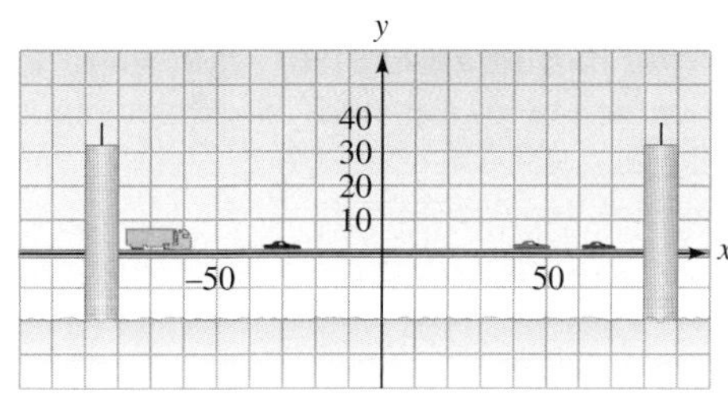

54. SELLING TV SETS A company has found that it can sell x TVs at a price of $\$(450 - \frac{1}{6}x)$.

a. How many TVs must the company sell to maximize its revenue?

b. Find the maximum revenue.

55. SELLING CD PLAYERS A wholesaler sells CD players for \$150 each. However, she gives volume discounts on purchases of 500 to 1,000 units according to the formula $(150 - \frac{1}{10}n)$, where n represents the number of units purchased.

a. How many units would a retailer have to buy for the wholesaler to obtain maximum revenue?

b. Find the maximum revenue.

56. TRACK AND FIELD See the illustration. Sketch the parabolic path traveled by the long-jumper's center of gravity from the take-off board to the landing. Let the x-axis represent the ground.

WRITING

57. Explain why the y-intercept of the graph of $y = ax^2 + bx + c$ is $(0, c)$.

58. Use the example of a stream of water from a drinking fountain to explain the concept of the vertex of a parabola.

59. Explain why parabolas that open left or right are not graphs of functions.

60. **a.** Is it possible for the graph of a parabola that opens upward not to have an x-intercept? Explain.

b. Is it possible for the graph of a parabola that opens downward not to have a y-intercept? Explain.

REVIEW *Simplify each expression.*

61. $\sqrt{12} + \sqrt{27}$

62. $3\sqrt{6y}(-4\sqrt{3y})$

63. $(\sqrt{3} + 1)(\sqrt{3} - 1)$

64. $(\sqrt{x} + 2)^2$

Solve each equation.

65. $\sqrt{6 + 2x} = 4$

66. $\sqrt{1 - 2x} = \sqrt{x + 10}$

KEY CONCEPT

Quadratic Equations

In this chapter, we have studied several ways to solve quadratic equations. We have also graphed quadratic functions and seen that their graphs are parabolas.

What Is a Quadratic Equation?

A quadratic equation can be written in the form $ax^2 + bx + c = 0$, where $a \neq 0$. Determine whether each item is a quadratic equation.

1. $y = 3x + 7$

2. $4(x + 5) = 2x$

3. $2x^2 - 3x + 4 = 0$

4. $y(y - 6) = 0$

5. $a^2 + 7a - 1 > 0$

6. $3y^2 - y + 4$

7. $5 = y - y^2$

8. $|x - 8|$

9. $\sqrt{x + 7} = 4$

10. $x^2 = 16$

11. $\frac{m}{2} - \frac{1}{3} = \frac{1}{4}$

12. $C = \frac{5}{9}(F - 32)$

Solving Quadratic Equations

The techniques we have used to solve linear equations cannot be used to solve a quadratic equation, because those techniques cannot isolate the variable on one side of the equation. Exercises 13–16 show examples of student work to solve quadratic equations. In each case, what did the student do wrong?

13. Solve: $x^2 = 6$.

$$\frac{x^2}{2} = \frac{6}{2}$$
$$x = 3$$

14. Solve: $x^2 - x = 10$.

$$x(x - 1) = 10$$
$$x = 10 \quad \text{or} \quad x - 1 = 10$$
$$x = 11$$

15. Solve: $a^2 = 20$.

$$a = \sqrt{20}$$
$$a = 2\sqrt{5}$$

16. Solve: $x^2 + 5x + 1 = 0$.

$$a = 1 \qquad b = 5 \qquad c = 1$$
$$x = -5 \pm \frac{\sqrt{5^2 - 4(1)(1)}}{2}$$
$$x = -5 \pm \frac{\sqrt{21}}{2}$$

Solve each quadratic equation using the method listed.

17. $4x^2 - x = 0$; factoring method

18. $x^2 + 3x + 1 = 0$; quadratic formula

19. $x^2 = 36$; square root method

20. $a^2 - a - 56 = 0$; factoring method

21. $x^2 + 4x + 1 = 0$; complete the square

22. $(x + 3)^2 = 16$; square root method

Quadratic Functions

The graph of the quadratic function $y = ax^2 + bx + c$ is a parabola. It opens upward when $a > 0$ and downward when $a < 0$.

23. The formula $R = 4x - x^2$ gives the revenue R (in tens of thousands of dollars) that a business obtains from the manufacture and sale of x patio chairs (in hundreds). Graph $R = 4x - x^2$ in the illustration.

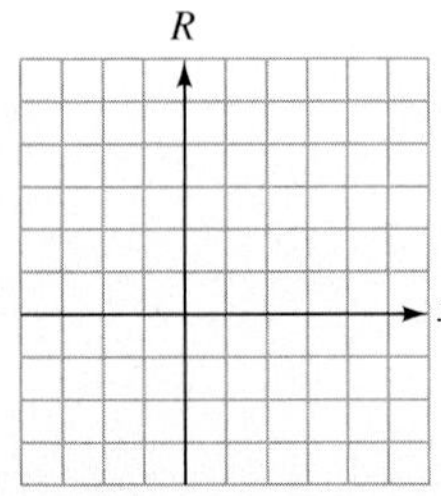

24. Refer to Exercise 23. Find the vertex of the parabola. What is its significance concerning the revenue the business brings in?

ACCENT ON TEAMWORK

SECTION 15.1

COMPLETING THE SQUARE Construct the model in the illustration. Label each piece of the model with its respective area. Show that the total area of the model is $(x^2 + 4x)$ square units. Next, add enough 1×1 squares to make the model a square. How many does it take to do this? Show that the area of the new figure can be expressed as $(x^2 + 4x + 4)$ or $(x + 2)(x + 2)$ square units. Explain how this model demonstrates the process of completing the square on $x^2 + 4x$.

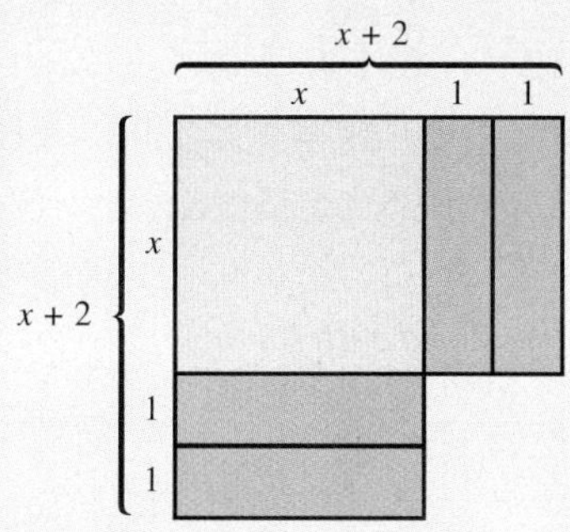

AUTHORING A TEXTBOOK Assign each of the nine examples in Section 15.1 to members of your group. Have them write a new but similar problem for each example, then write a solution complete with an explanation and author notes using the same format as this book. They should also create an accompanying Self Check problem and include the answer. Compile all nine examples into a booklet. Make copies of your booklet for the other members of the class.

SECTION 15.2

PREDICTING SOLUTIONS The expression $b^2 - 4ac$ is called the **discriminant** of the quadratic equation $ax^2 + bx + c = 0$. We can use the discriminant to predict the number of solutions a particular quadratic equation has.

If $b^2 - 4ac > 0$, the equation has two real solutions.
If $b^2 - 4ac = 0$, the equation has one real solution.
If $b^2 - 4ac < 0$, the equation has no real solutions.

For each quadratic equation, evaluate the discriminant to determine how many real-number solutions it has.

a. $4x^2 - 4x + 1 = 0$
b. $6x^2 - 5x - 6 = 0$
c. $5x^2 + x + 2 = 0$
d. $3x^2 + 10x - 2 = 0$
e. $2x^2 = 4x - 1$
f. $9x^2 = 12x - 4$

SOLVING QUADRATIC EQUATIONS Solve the quadratic equation $2x^2 - x - 1 = 0$ using these methods: factoring, completing the square, and the quadratic formula. Write each solution on a separate piece of paper. Under each solution, in two columns, list the advantages and the drawbacks of each method.

SECTION 15.3

PARABOLAS Use a home video camera to make a "documentary" showing examples of parabolic shapes you see in everyday life. Write a script for your video and have a narrator explain the setting, point out the vertex, and tell whether the parabola opens upward or downward in each case.

GRAPHING On one piece of graph paper, graph each of the following quadratic functions.

$$f(x) = x^2 \qquad g(x) = 2x^2 \qquad h(x) = \frac{1}{2}x^2$$

In general, what happens to the graph of $y = ax^2$ as a increases?

MINIMIZING/MAXIMIZING In the business world, it is good for a company to minimize its costs and to maximize its profits. In your group, make a list of quantities that are good to minimize and a list of quantities that are good to maximize.

SOLUTION AND GRAPH Write a quadratic function whose graph has the x-intercepts shown. (*Hint:* Recall that there is a relation between the solution of a quadratic equation and the x-intercepts of the graph of the corresponding quadratic function.)

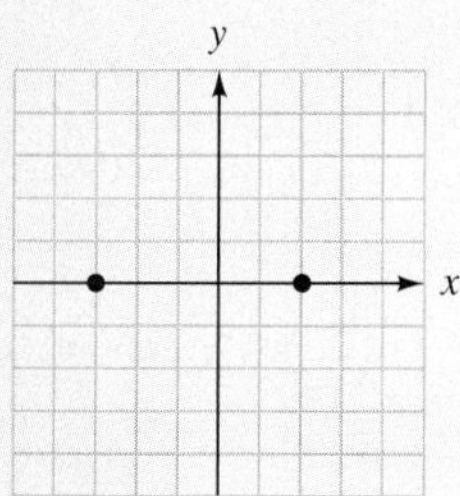

CHAPTER REVIEW

SECTION 15.1 Completing the Square

CONCEPTS

We can use the *square root method* to solve $x^2 = c$, where $c > 0$. The two solutions are $x = \sqrt{c}$ and $x = -\sqrt{c}$, which can be written $x = \pm\sqrt{c}$.

REVIEW EXERCISES

Use the square root method to solve each equation.

1. $x^2 = 25$
2. $x^2 = 400$
3. $2x^2 = 18$
4. $4y^2 = 9$
5. $t^2 = 8$
6. $2x^2 - 1 = 149$

Use the square root method to solve each equation.

7. $(x - 1)^2 = 25$
8. $4(x - 2)^2 = 9$
9. $(x - 8)^2 = 8$
10. $(x + 5)^2 = 75$

Use the square root method to solve each equation. Round each solution to the nearest hundredth.

11. $x^2 = 12$
12. $(x - 1)^2 = 55$

To complete the square on $x^2 + bx$, add the square of one-half of the coefficient of x:

$$x^2 + bx + \left(\frac{1}{2}b\right)^2$$

Complete the square and then factor each perfect square trinomial.

13. $x^2 + 4x$
14. $z^2 - 10z$
15. $t^2 - 5t$
16. $a^2 + \frac{3}{4}a$

The factoring method doesn't always work in solving many quadratic equations. In these cases, we can use a method called *completing the square.*

17. Explain why the quadratic equation $x^2 + 4x + 1 = 0$ can't be solved by the factoring method.

To solve a quadratic equation by completing the square:

1. If necessary, divide both sides of the equation by the coefficient of x^2 to make its coefficient 1.
2. If necessary, get the constant on the right-hand side of the equation.
3. Complete the square and factor the resulting trinomial square.
4. Solve the quadratic equation using the square root method.
5. Check each solution.

Solve each equation by completing the square.

18. $x^2 - 8x + 15 = 0$
19. $x^2 + 5x - 14 = 0$
20. $2x^2 + 5x - 3 = 0$
21. $2x^2 - 2x - 1 = 0$
22. Solve $x^2 + 4x + 1 = 0$ by completing the square. Round each solution to the nearest hundredth.
23. PLAYGROUND EQUIPMENT The tractor tire shown makes a good container for sand. If the circular area that the sandbox covers is 28.3 square feet, what is the radius of the tire? Round to the nearest tenth of a foot.

SECTION 15.2 The Quadratic Formula

For the *general quadratic equation* $ax^2 + bx + c = 0$,

$$x = \frac{-b \pm \sqrt{b^2 - 4ac}}{2a}$$

This is called the *quadratic formula.*

Use the quadratic formula to solve each equation.

24. $x^2 - 2x - 15 = 0$

25. $x^2 - 6x - 7 = 0$

26. $6x^2 - 7x - 3 = 0$

27. $x^2 - 6x + 7 = 0$

28. Use the quadratic formula to solve $3x^2 + 2x - 2 = 0$. Give the solutions in exact form and then rounded to the nearest hundredth.

29. Use the quadratic formula to solve $10x^2 + 2x + 1 = 0$.

30. SECURITY GATES The length of the frame for the iron gate shown is 14 feet longer than the width. A diagonal crossbrace is 26 feet long. Find the width and length of the gate frame.

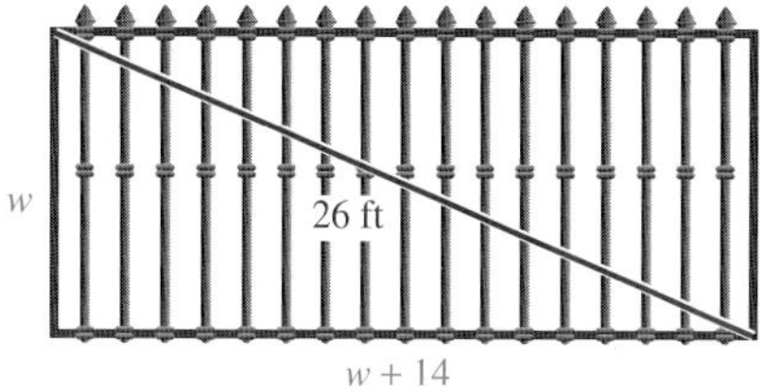

31. THE MILITARY A pilot releases a bomb from an altitude of 3,000 feet. The bomb's height h above the target t seconds after its release is given by the formula

$$h = 3{,}000 + 40t - 16t^2$$

How long will it be until the bomb hits its target?

Strategy for solving quadratic equations:

1. Try the square root method.
2. If it doesn't apply, write the equation in $ax^2 + bx + c = 0$ form.
3. Try the factoring method.
4. If it doesn't work, complete the square or use the quadratic formula.

Use the most convenient method to find all real solutions of each equation.

32. $x^2 + 6x + 2 = 0$

33. $(y + 3)^2 = 16$

34. $x^2 + 5x = 0$

35. $2x^2 + x = 5$

36. $g^2 - 20 = 0$

37. $a^2 = 4a - 4$

38. $a^2 - 2a + 5 = 0$

39. $2c^2 = 800$

SECTION 15.3 Graphing Quadratic Functions

The *vertex* of a parabola is the lowest (or highest) point on the parabola.

A vertical line through the vertex of a parabola that opens upward or downward is its *axis of symmetry.*

See the following graph.

40. What are the x-intercepts of the parabola?

41. What is the y-intercept of the parabola?

42. What is the vertex of the parabola?

43. Draw the axis of symmetry of the parabola on the graph.

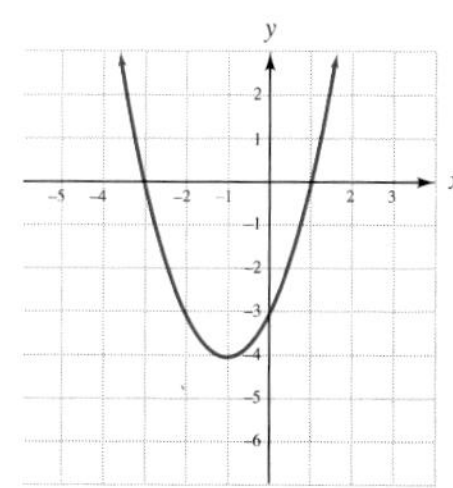

44. What important information can be obtained from the vertex of the parabola in the illustration?

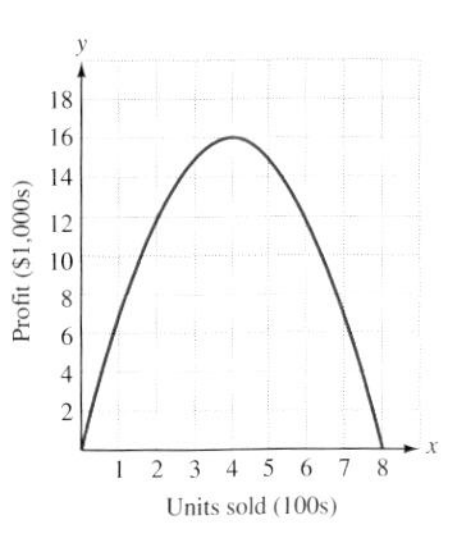

The graph of the quadratic function $y = ax^2 + bx + c$ is a parabola. It opens upward when $a > 0$ and downward when $a < 0$.

Find the vertex of the graph of each quadratic function and tell which direction the parabola opens. **Do not draw the graph.**

45. $y = 2x^2 - 4x + 7$

46. $f(x) = -3x^2 + 18x - 11$

The x-coordinate of the vertex of the parabola $y = ax^2 + bx + c$ is $x = -\frac{b}{2a}$. To find the y-coordinate of the vertex, substitute $-\frac{b}{2a}$ for x in the equation of the parabola and find y.

47. Find the x- and y-intercepts of the graph of $y = x^2 + 6x + 5$.

The x-intercepts of a parabola are determined by solving $ax^2 + bx + c = 0$. The y-intercept is $(0, c)$.

Graph each quadratic function by finding the vertex, x- and y-intercepts, and axis of symmetry of its graph.

48. $y = x^2 + 2x - 3$

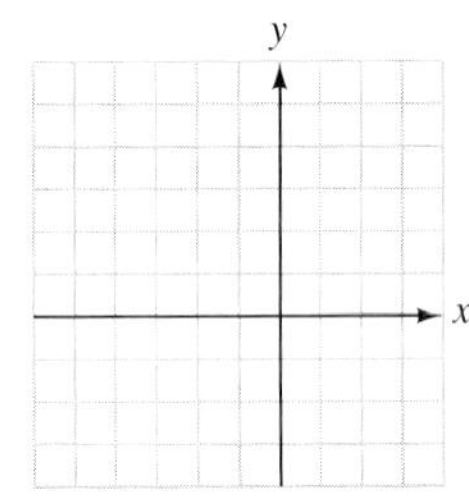

49. $f(x) = -2x^2 + 4x - 2$

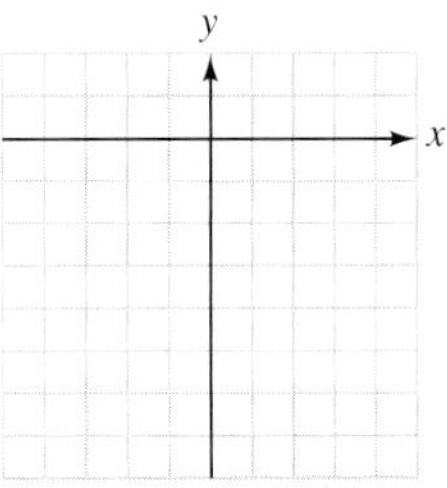

The number of x-intercepts of the graph of a quadratic function $y = ax^2 + bx + c$ is the same as the number of real solutions of $ax^2 + bx + c = 0$.

50. The graphs of three quadratic functions are shown in the illustration. Fill in the blanks.

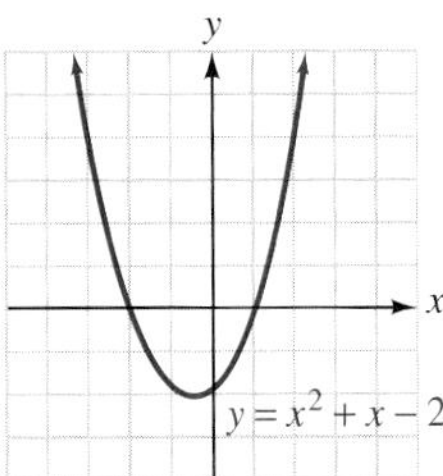

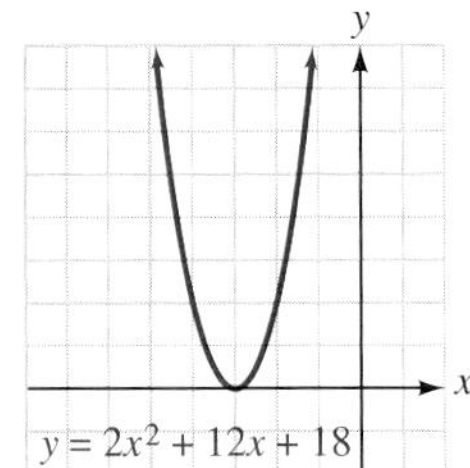

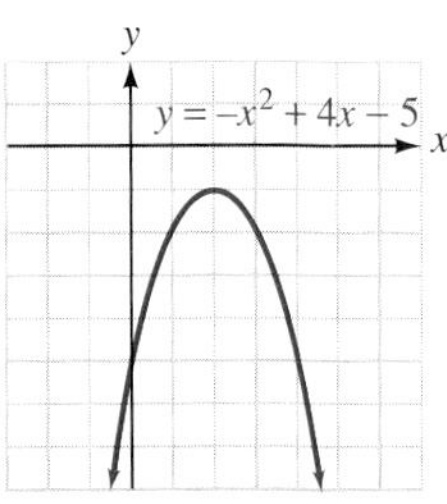

$x^2 + x - 2 = 0$ has ___ real-number solution(s).

$2x^2 + 12x + 18 = 0$ has ___ real-number solution(s).

$-x^2 + 4x - 5 = 0$ has ___ real-number solution(s).

CHAPTER 15 TEST

Solve each equation by the square root method.

1. $x^2 = 16$
2. $u^2 - 24 = 0$
3. $4y^2 = 25$
4. $(x - 2)^2 = 3$
5. ARCHERY The area of the circular archery target shown below is 5,026.5 cm^2. What is the radius of the target? Round to the nearest centimeter.

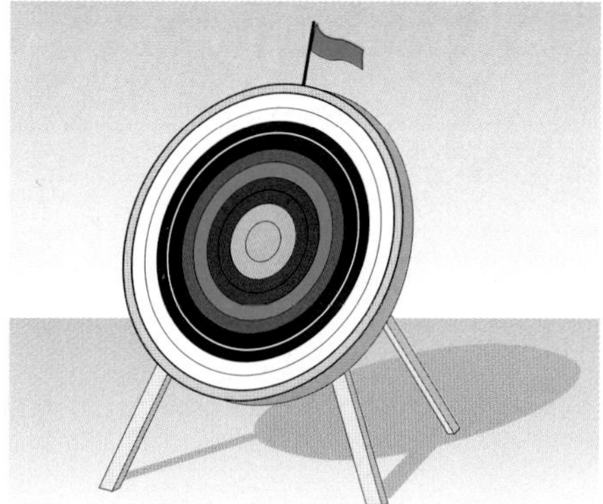

6. Complete the square on $x^2 - 14x$ and then factor the resulting perfect square trinomial.
7. Complete the square to solve $a^2 + 2a - 4 = 0$. Give the exact solutions and then round them to the nearest hundredth.
8. Complete the square to solve $2x^2 = 3x + 2$.

Use the quadratic formula to find all real solutions of each equation.

9. $x^2 + 3x - 10 = 0$
10. $2x^2 - 5x = 12$
11. $x^2 + 5 = 2x$
12. $x^2 = 4x - 2$
13. Solve $3x^2 - x - 1 = 0$ using the quadratic formula. Give the exact solutions, and then approximate them to the nearest hundredth.

14. FLAGS According to the *Guinness Book of World Records 1998,* the largest flag in the world is the American "Superflag," which has an area of 128,775 ft^2. If its length is 5 feet less than twice its width, find its width and length.

15. ADVERTISING When a business runs x advertisements per week on television, the number y of air conditioners it sells is given by the quadratic function graphed in the illustration. What information about sales can be obtained from the vertex?

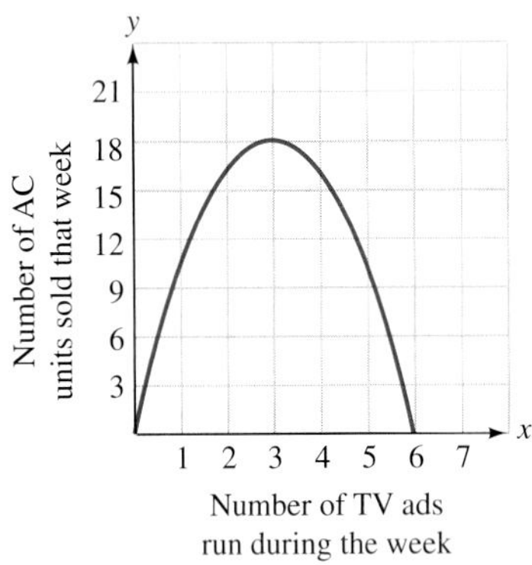

16. Graph the function $y = x^2 + x - 2$ by finding the vertex, x- and y-intercepts, and axis of symmetry.

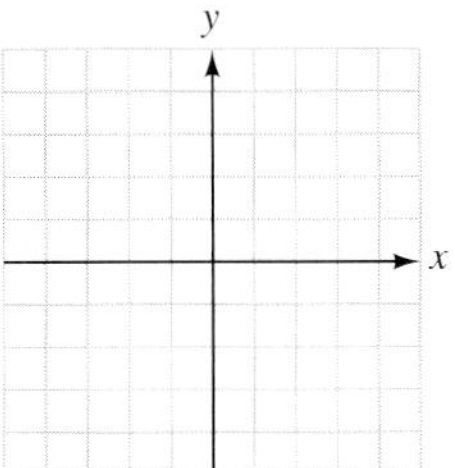

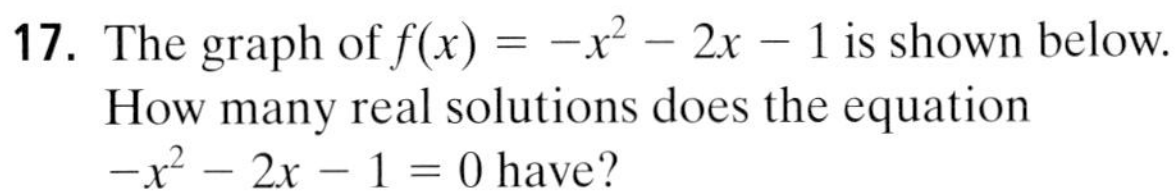

17. The graph of $f(x) = -x^2 - 2x - 1$ is shown below. How many real solutions does the equation $-x^2 - 2x - 1 = 0$ have?

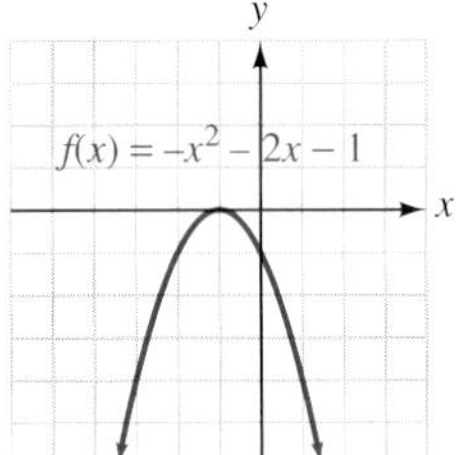

18. Explain the meaning of the $\pm$ symbol.

CHAPTERS 1–15 CUMULATIVE REVIEW EXERCISES

1. Determine whether each statement is true or false.
   a. Every rational number can be written as a ratio of two integers.
   b. The set of real numbers corresponds to all points on the number line.
   c. The whole numbers and their opposites form the set of integers.
2. Evaluate: $-4 + 2[-7 - 3(-9)]$.
3. DRIVING SAFETY In cold climates, salt is spread on roads to melt snow and ice. According to the following graph, when is the accident rate the worst?

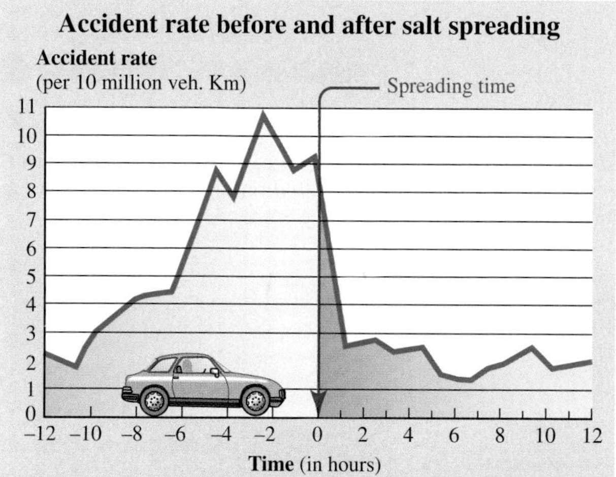

Based on data from the Salt Institute

4. EMPLOYMENT The following headline appeared in a newspaper.

   Xerox to Cut 5,200 Jobs, or 5.3% of Workforce, on Falling Profits

   How many employees did Xerox have at that time?
5. Simplify: $3p - 6(p + z) + p$.
6. Solve: $\frac{5}{6}k = 10$.
7. Solve: $-(3a + 1) + a = 2$.
8. Solve $5x + 7 < 2x + 1$, graph the solution set, and use interval notation to describe the solution.
9. ENTREPRENEURS Last year, a women's professional organization made two small-business loans totaling \$28,000 to young women beginning their own businesses. The money was lent at 7% and 10% simple interest rates. If the annual income the organization received from these loans was \$2,560, what was each loan amount?
10. Evaluate $(x - a)^2 + (y - b)^2$ for $x = -2$, $y = 1$, $a = 5$, and $b = -3$.
11. Evaluate: $|\frac{4}{5} \cdot 10 - 12|$.
12. Find the slope of the line passing through $(-2, -2)$ and $(-12, -8)$.

Graph each equation or inequality.

13. $2y - 2x = 6$

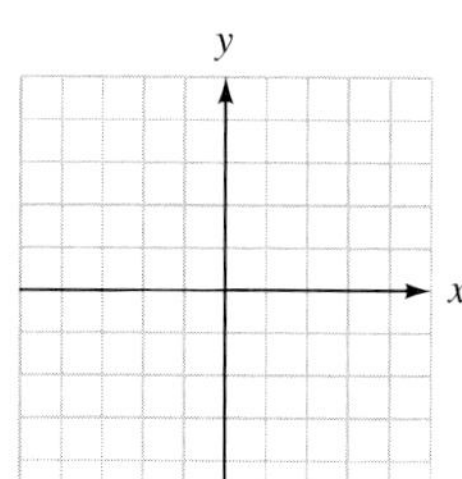

14. $y = -3$

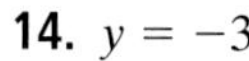

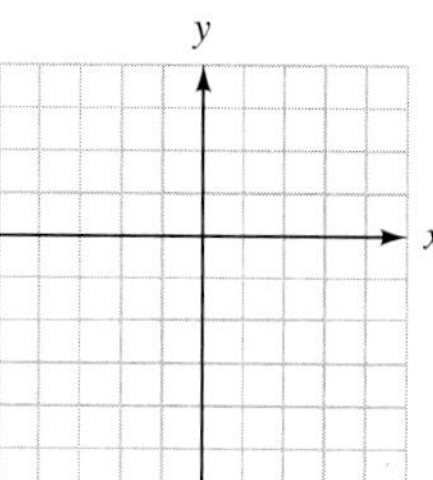

15. $y = -x + 2$

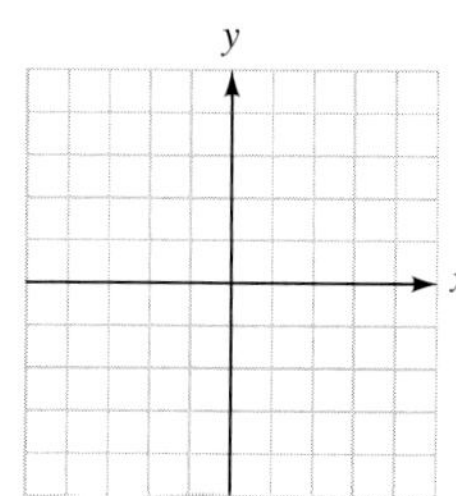

16. $y < 3x$

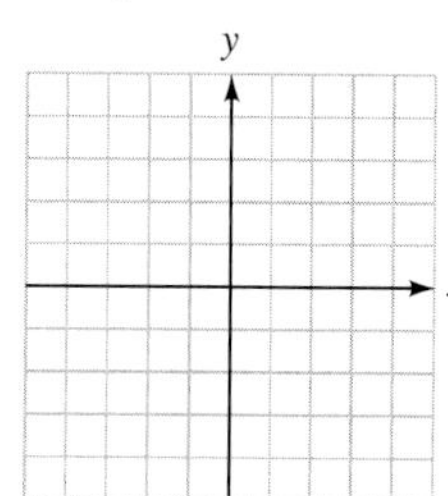

17. Graph the line passing through$(-2, -1)$ and having slope $\frac{4}{3}$.

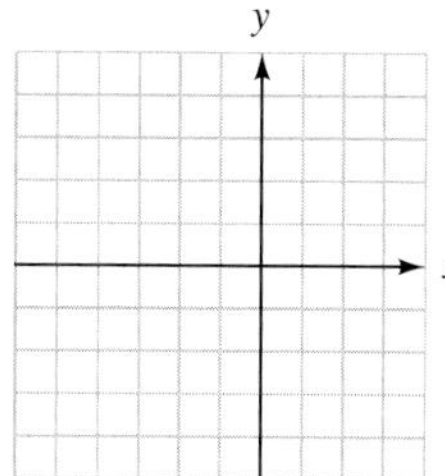

18. Graph: $y = x^3 - 2$.

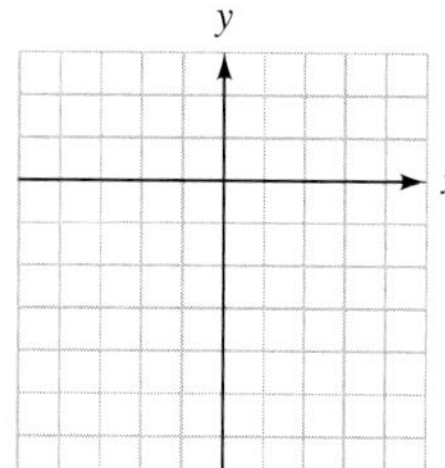

19. Write an equation of the line whose graph has slope $m = -2$ and y-intercept $(0, 1)$.
20. Write an equation of the line whose graph has slope $m = \frac{1}{4}$ and passes through the point $(8, 1)$. Answer in slope–intercept form.
21. What is the slope of the line defined by $4x + 5y = 6$?
22. If $f(x) = 3x^2 + 3x - 8$, find $f(-1)$.

23. Is the word *domain* associated with the input or the output of a function?

24. Is this the graph of a function?

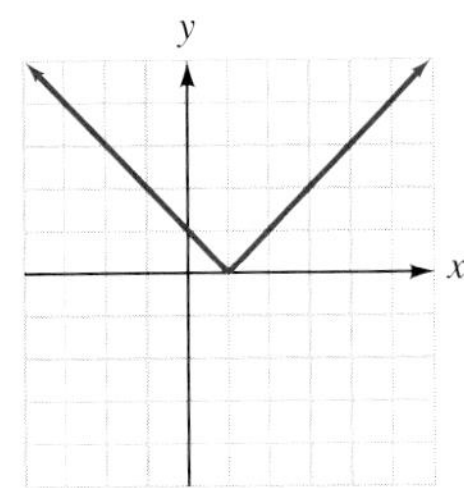

Simplify each expression. Write each answer without using parentheses or negative exponents.

25. $y^3(y^2y^4)$

26. $\left(\dfrac{b^2}{3a}\right)^3$

27. $\dfrac{10a^4a^{-2}}{5a^2a^0}$

28. $\dfrac{(r^2)^3}{(r^3)^4}$

29. POKER The odds against being dealt the hand shown in the illustration are about 2.6×10^6 to 1. Express the odds using standard notation.

30. PAIN RELIEVERS See the illustration. Find the rate of change in the percent of individuals free of headache pain over the given time span after taking Acetaminophen.

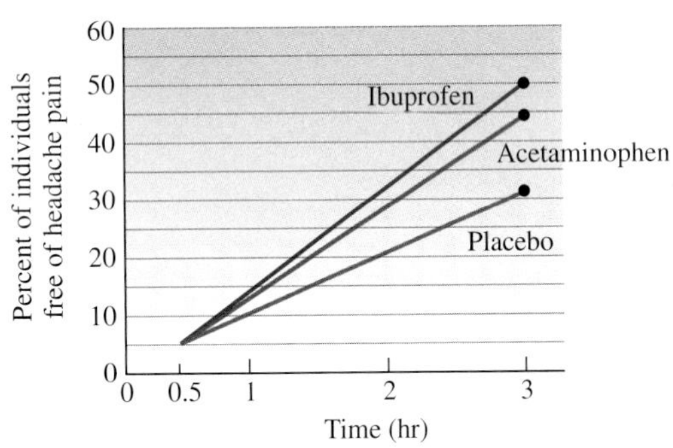

Based on data from *Health and Wellness*, Jones and Bartlett Publishers

Perform the operations.

31. $(-2a^3)(3a^2)$

32. $(2b - 1)(3b + 4)$

33. $(2x + 5y)^2$

34. $x - 3\overline{)2x^2 - 3 - 5x}$

Factor each expression completely.

35. $6a^2 - 12a^3b + 36ab$

36. $2x + 2y + ax + ay$

37. $b^3 + 125$

38. $t^4 - 16$

Solve each equation.

39. $3x^2 + 8x = 0$

40. $15x^2 - 2 = 7x$

41. Write a polynomial that represents the perimeter of the rectangle shown.

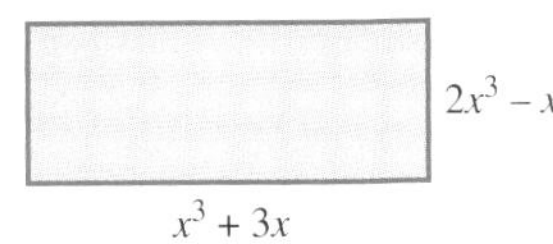

42. TRIANGLES The triangle shown below has an area of 22.5 square inches. Find its height.

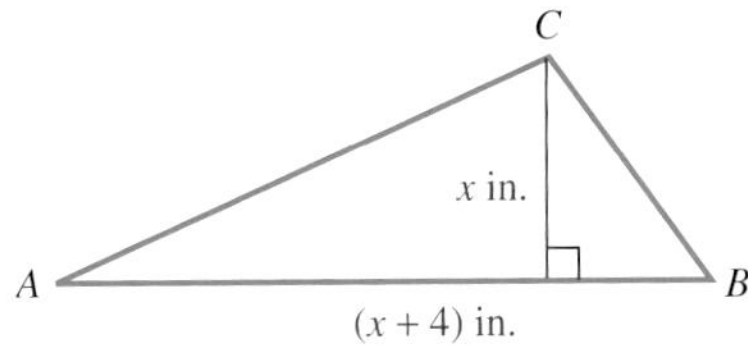

43. For what value of x is $\dfrac{x}{x + 8}$ undefined?

44. Simplify: $\dfrac{3x^2 - 27}{x^2 + 3x - 18}$.

Perform the operations and simplify when possible.

45. $\dfrac{x^2 - x - 6}{2x^2 + 9x + 10} \div \dfrac{x^2 - 25}{2x^2 + 15x + 25}$

46. $\dfrac{x + 3}{x^2} + \dfrac{x + 5}{x^2}$

47. $\dfrac{x}{x - 2} + \dfrac{3x}{x^2 - 4}$

48. $\dfrac{\dfrac{5}{y} + \dfrac{4}{y + 1}}{\dfrac{4}{y} - \dfrac{5}{y + 1}}$

Solve each equation.

49. $\dfrac{3r}{2} - \dfrac{3}{r} = \dfrac{3r}{2} + 3$

50. $\dfrac{7}{q^2 - q - 2} + \dfrac{1}{q + 1} = \dfrac{3}{q - 2}$

51. Solve the formula $\dfrac{1}{a} + \dfrac{1}{b} = 1$ for a.

52. ROOFING A homeowner estimates that it will take him 7 days to roof his house. A professional roofer estimates that he could roof the house in 4 days. How long will it take if the homeowner helps the roofer?

53. LOSING WEIGHT If a person cuts his or her daily calorie intake by 100, it will take 350 days for that person to lose 10 pounds. How long will it take for the person to lose 25 pounds?

54. GEARS The speed of a gear varies inversely with the number of teeth. If a gear with 10 teeth makes 3 revolutions per second, how many revolutions per second will a gear with 25 teeth make?

55. Solve using the graphing method.

$$\begin{cases} x + y = 1 \\ y = x + 5 \end{cases}$$

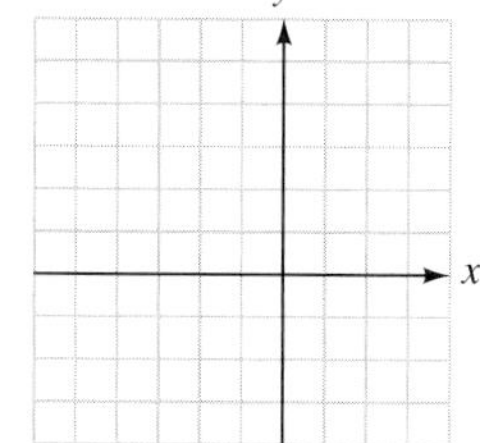

56. Solve using the substitution method.

$$\begin{cases} y = 2x + 5 \\ x + 2y = -5 \end{cases}$$

57. Solve using the addition method.

$$\begin{cases} \frac{3}{5}s + \frac{4}{5}t = 1 \\ -\frac{1}{4}s + \frac{3}{8}t = 1 \end{cases}$$

58. MIXING CANDY How many pounds of each candy shown below must be mixed to obtain 60 pounds of candy that would be worth \$3 per pound? Use two variables to solve this problem.

59. Graph the solution set of

$$\begin{cases} 3x + 4y > -7 \\ 2x - 3y \geq 1 \end{cases}.$$

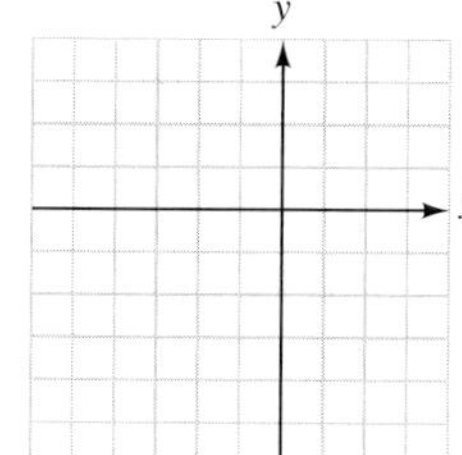

60. DEMOGRAPHICS Refer to the graph in the illustration. To which stage does each of the following descriptions apply?

Stage ___: Rapidly growing population: Births far outnumber deaths.

Stage ___: Stable population: Birth rate drops; births and deaths are more or less equal.

Stage ___: Stable population: Births and deaths are more or less equal.

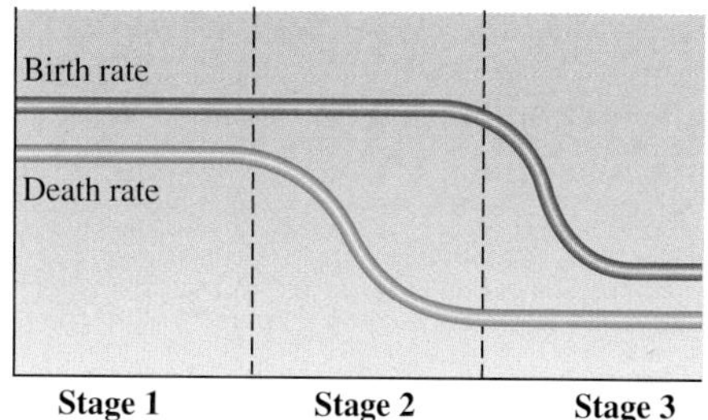

Simplify each expression. All variables represent positive numbers.

61. $\sqrt{50x^2}$

62. $\sqrt[3]{-27y^3}$

63. $3\sqrt{24} + \sqrt{54}$

64. $(\sqrt{2} + 1)(\sqrt{2} - 3)$

65. $\sqrt{\dfrac{72x^3}{y^2}}$

66. $\dfrac{8}{\sqrt{10}}$

67. Solve: $\sqrt{6x + 1} + 2 = 7$.

68. Solve $x^2 + 8x + 12 = 0$ by completing the square.

69. Solve: $t^2 = 75$.

70. STORAGE The diagonal distance across the face of each of the stacking cubes shown below is 15 inches. What is the height of the entire storage arrangement? Round to the nearest tenth of an inch.

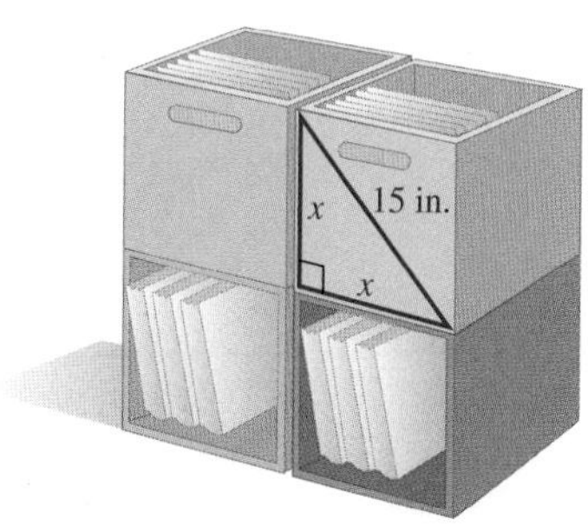

71. Solve $3x^2 - x - 1 = 0$ using the quadratic formula. Give the exact solutions, and then approximate each to the nearest hundredth.

72. QUILTS According to the *Guinness Book of World Records 1998,* the world's largest quilt was made by the Seniors' Association of Saskatchewan, Canada, in 1994. If the length of the rectangular quilt is 11 feet less than twice its width and it has an area of 12,865 ft^2, find its width and length.

73. Find the vertex, the x- and y-intercepts, and the axis of symmetry of the graph of $y = x^2 + 6x + 5$. Then graph the function.

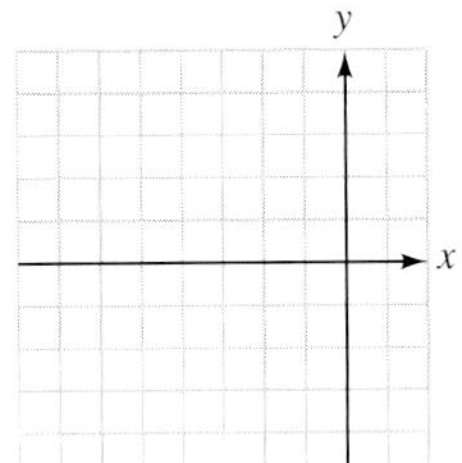

74. POWER OUTPUT The graph below shows the power output (in horsepower, hp) of a certain engine for various engine speeds (in revolutions per minute, rpm).

a. At an engine speed of 3,000 rpm, what is the power output?

b. For what engine speed(s) is the power output 125 hp?

c. For what engine speed does the power output reach a maximum?

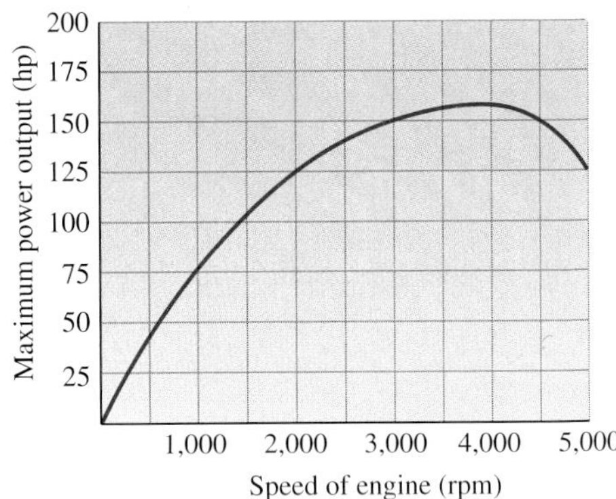

APPENDIX A

Inductive and Deductive Reasoning

- Inductive reasoning • Deductive reasoning

To reason means to think logically. The objective of this appendix is to develop your problem-solving ability by improving your reasoning skills. We will introduce two fundamental types of reasoning that can be applied in a wide variety of settings. They are known as *inductive reasoning* and *deductive reasoning.*

Inductive reasoning

In a laboratory, scientists conduct experiments and observe outcomes. After several repetitions with similar outcomes, the scientist will generalize the results into a statement that appears to be true.

- If I heat water to 212° F, it will boil.
- If I drop a weight, it will fall.
- If I combine an acid with a base, a chemical reaction occurs.

When we draw general conclusions from specific observations, we are using **inductive reasoning.** The next examples show how inductive reasoning can be used in mathematical thinking. Given a list of numbers or symbols, called a *sequence,* we can often find a missing term of the sequence by looking for patterns and applying inductive reasoning.

EXAMPLE 1 **An increasing pattern.** Find the next number in the sequence 5, 8, 11, 14, ...

Solution The terms of the sequence are increasing. To discover the pattern, we find the *difference* between each pair of successive terms.

$8 - 5 = 3$	Subtract the first term from the second term.
$11 - 8 = 3$	Subtract the second term from the third term.
$14 - 11 = 3$	Subtract the third term from the fourth term.

The difference between each pair of numbers is 3. This means that each successive number is 3 greater than the previous one. Thus, the next number in the sequence is $14 + 3$, or 17.

Self Check 1
Find the next number in the sequence $-3, -1, 1, 3, \ldots$.

Answer 5

Self Check 2
Find the next number in the sequence $-0.1, -0.3, -0.5, -0.7 \ldots.$

Answer -0.9

EXAMPLE 2 **A decreasing pattern.** Find the next number in the sequence $-2, -4, -6, -8, \ldots.$

Solution The terms of the sequence are decreasing. Since each successive term is 2 less than the previous one, the next number in the pattern is $-8 - 2$, or -10.

Self Check 3
Find the next entry in the sequence Z, A, Y, B, X, C, ...

Answer W

EXAMPLE 3 **An alternating pattern.** Find the next letter in the sequence A, D, B, E, C, F, D, ...

Solution The letter A is the first letter of the alphabet, D is the fourth letter, B is the second letter, and so on. We can create the following letter–number correspondence.

$$\begin{aligned} A &\to 1 \\ D &\to 4 \\ B &\to 2 \\ E &\to 5 \\ C &\to 3 \\ F &\to 6 \\ D &\to 4 \end{aligned}$$

Add 3.
Subtract 2.
Add 3.
Subtract 2.
Add 3.
Subtract 2.

The numbers in the sequence 1, 4, 2, 5, 3, 6, 4, . . . alternate in size. They change from smaller to larger, to smaller, to larger, and so on.

We see that 3 is added to the first number to get the second number. Then 2 is subtracted from the second number to get the third number. To get successive terms in the sequence, we alternately add 3 to one number and then subtract 2 from that result to get the next number.

If we apply this pattern, the next number in the numerical sequence is $4 + 3$, or 7. The next letter in the original sequence is G, because it is the seventh letter of the alphabet.

Self Check 4
Find the next geometric shape in the sequence below.

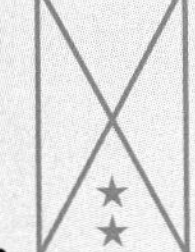
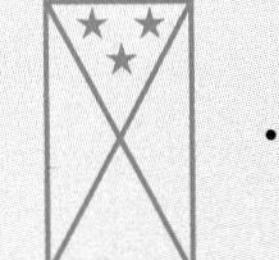
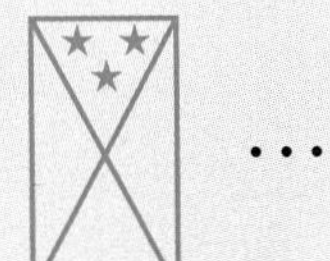

Answer

EXAMPLE 4 **Two patterns.** Find the next geometric shape in the sequence below.

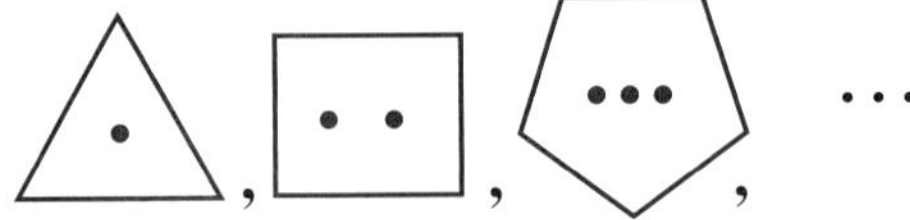

Solution This sequence has two patterns occurring at the same time. The first figure has three sides and one dot, the second figure has four sides and two dots, and the third figure has five sides and three dots. Thus, we would expect the next figure to have six sides and four dots, as shown in Figure A-1.

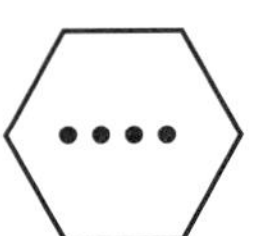

FIGURE A-1

EXAMPLE 5 **A circular pattern.** Find the next geometric shape in the sequence below.

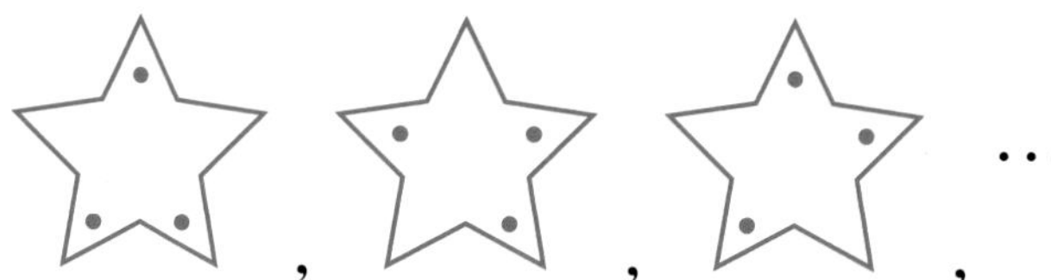

Solution From figure to figure, we see that each dot moves from one point of the star to the next, in a counterclockwise direction. This is a circular pattern. The next shape in the sequence will be the one shown in Figure A-2.

FIGURE A-2

Self Check 5
Find the next geometric shape in the sequence below.

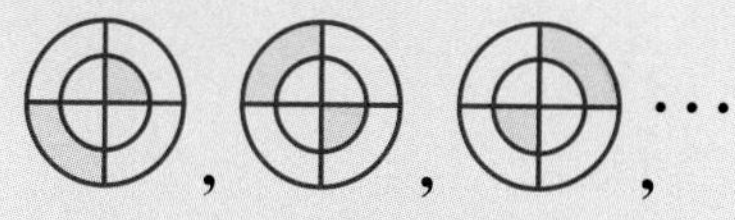

Answer

Deductive reasoning

As opposed to inductive reasoning, **deductive reasoning** moves from the general case to the specific. For example, if we know that the sum of the angles in any triangle is 180°, we know that the sum of the angles of $\triangle ABC$ is 180°. Whenever we apply a general principle to a particular instance, we are using deductive reasoning.

A deductive reasoning system is built on four elements.

1. **Undefined terms:** terms that we accept without giving them formal meaning
2. **Defined terms:** terms that we define in a formal way
3. **Axioms** or **postulates:** statements that we accept without proof
4. **Theorems:** statements that we can prove with formal reasoning

Many problems can be solved by deductive reasoning. For example, suppose that we plan to enroll in an early-morning algebra class and we know that Professors Perry, Miller, and Tveten are scheduled to teach algebra next semester. After some investigating, we find out that Professor Perry teaches only in the afternoon and Professor Tveten teaches only in the evenings. Without knowing anything about Professor Miller, we can conclude that he will be our teacher, since he is the only remaining possibility.

The following examples show how to use deductive reasoning to solve problems.

EXAMPLE 6 **Scheduling classes.** Four professors are scheduled to teach mathematics next semester, with the following course preferences.

1. Professors A and B don't want to teach calculus.
2. Professor C wants to teach statistics.
3. Professor B wants to teach algebra.

Who will teach trigonometry?

Solution The following chart shows each course, with each possible instructor.

Calculus	Algebra	Statistics	Trigonometry
A	A	A	A
B	B	B	B
C	C	C	C
D	D	D	D

Since Professors A and B don't want to teach calculus, we can cross them off the calculus list. Since Professor C wants to teach statistics, we can cross her off every other list. This leaves Professor D as the only person to teach calculus, so we can cross her off every other list. Since Professor B wants to teach algebra, we can cross him off every other list. Thus, the only remaining person left to teach trigonometry is Professor A.

Calculus	Algebra	Statistics	Trigonometry
~~A~~	A	A	A
~~B~~	B	~~B~~	~~B~~
~~C~~	~~C~~	C	~~C~~
D	~~D~~	~~D~~	~~D~~

Self Check 7
Of the 50 cars on a used-car lot, 9 are red, 31 are foreign models, and 6 are red foreign models. If a customer wants to buy an American model that is not red, how many cars does she have to choose from?

EXAMPLE 7 **State flags.** The graph in Figure A-3 gives the number of state flags that feature an eagle, a star, or both. How many state flags have neither an eagle nor a star?

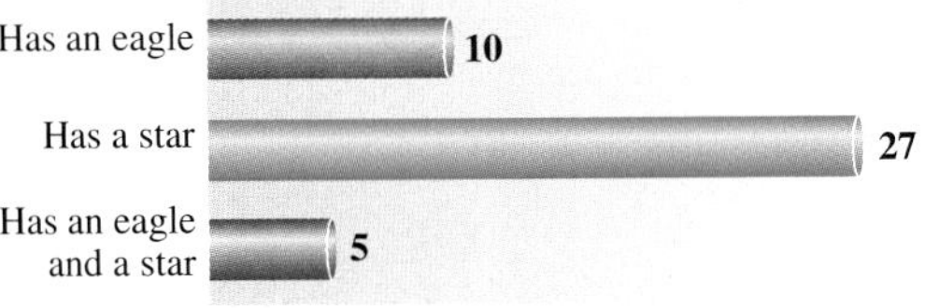

FIGURE A-3

Solution In Figure A-4(a), the intersection (overlap) of the circles is a way to show that there are 5 state flags that have both an eagle and a star. If an eagle appears on a total of 10 flags, then the left circle must contain 5 more flags outside of the intersection. See Figure A-4(b). If a total of 27 flags have a star, the right circle must contain 22 more flags outside the intersection.

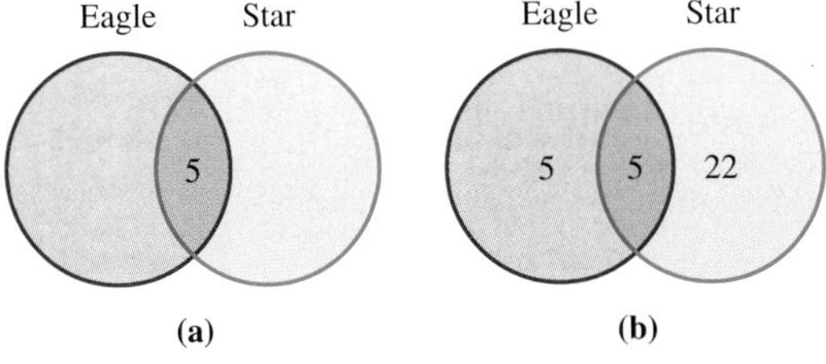

FIGURE A-4

From Figure A-4, we see that 5 + 5 + 22, or 32 flags have an eagle, a star, or both. To find how many flags have neither an eagle nor a star, we subtract this total from the number of state flags, which is 50.

$$50 - 32 = 18$$

There are 18 state flags that have neither an eagle nor a star.

Answer 16

Appendix A STUDY SET

VOCABULARY *Fill in the blanks.*

1. __________ reasoning draws general conclusions from specific observations.
2. __________ reasoning moves from the general case to the specific.

CONCEPTS *Determine whether the pattern shown is increasing, decreasing, alternating, or circular.*

3. 2, 3, 4, 2, 3, 4, 2, 3, 4, . . .
4. 8, 5, 2, −1, . . .
5. −2, −4, 2, 0, 6, . . .
6. 0.1, 0.5, 0.9, 1.3, . . .
7. a, c, b, d, c, e, . . .
8. 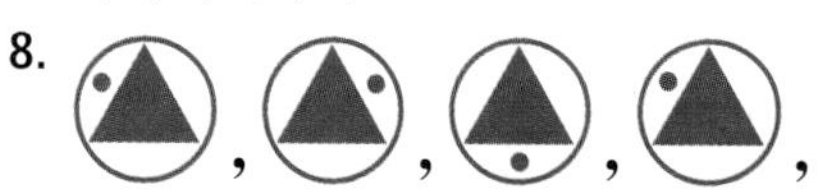

9. ROOM SCHEDULING From the chart, determine what time(s) on a Wednesday morning a practice room in a music building is available. The symbol X indicates that the room has already been reserved.

	M	T	W	Th	F
9 A.M.	X		X	X	
10 A.M.	X	X			X
11 A.M.			X		X

10. QUESTIONNAIRES A group of college students were asked if they were taking a mathematics course and if they were taking an English course. The results are displayed.
    a. How many students were taking a mathematics course and an English course?
    b. How many students were taking an English course but not a mathematics course?
    c. How many students were taking a mathematics course?

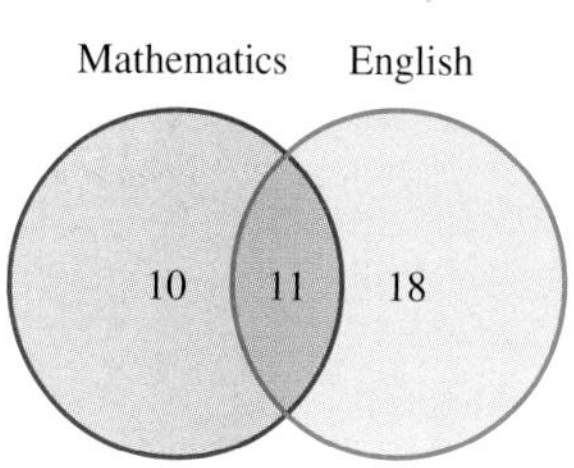

PRACTICE *Find the number that comes next in each sequence.*

11. 1, 5, 9, 13, . . .
12. 15, 12, 9, 6, . . .
13. −3, −5, −8, −12, . . .
14. 5, 9, 14, 20, . . .
15. −7, 9, −6, 8, −5, 7, −4, . . .
16. 2, 5, 3, 6, 4, 7, 5, . . .
17. 9, 5, 7, 3, 5, 1, . . .
18. 1.3, 1.6, 1.4, 1.7, 1.5, 1.8, . . .
19. −2, −3, −5, −6, −8, −9, . . .
20. 8, 11, 9, 12, 10, 13, . . .
21. 6, 8, 9, 7, 9, 10, 8, 10, 11, . . .
22. 10, 8, 7, 11, 9, 8, 12, 10, 9, . . .

Find the figure that comes next in each sequence.

23. 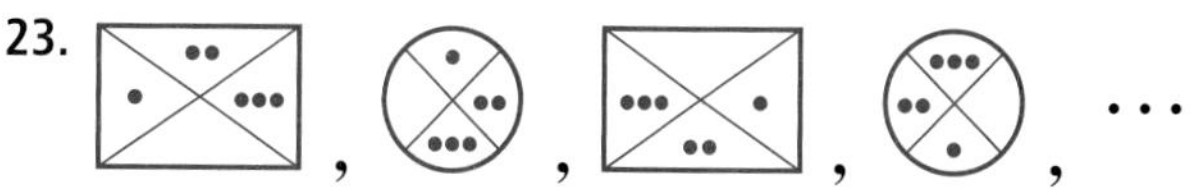

24. 

Find the missing figure in each sequence.

25. 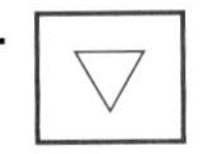, , , ?, 

26. ○, △, ?, 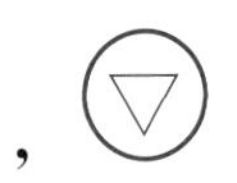, 

Find the next letter or letters in the sequence.

27. A, c, E, g, . . .

28. R, SS, TTT, . . .

29. d, h, g, k, j, n, . . .

30. B, N, C, N, D, . . .

What conclusion(s) can be drawn from each set of information?

31. Four people named John, Luis, Maria, and Paula have occupations as teacher, butcher, baker, and candlestick maker.

1. John and Paula are married.
2. The teacher plans to marry the baker in December.
3. Luis is the baker.

Who is the teacher?

32. In a zoo, a zebra, a tiger, a lion, and a monkey are to be placed in four cages numbered from 1 to 4, from left to right. The following decisions have been made.

1. The lion and the tiger should not be side by side.
2. The monkey should be in one of the end cages.
3. The tiger is to be in cage 4.

In which cage is the zebra?

33. A Ford, a Buick, a Dodge, and a Mercedes are parked side by side.

1. The Ford is between the Mercedes and the Dodge.
2. The Mercedes is not next to the Buick.
3. The Buick is parked on the left end.

Which car is parked on the right end?

34. Four divers at the Olympics finished first, second, third, and fourth.

1. Diver A beat diver B.
2. Diver C placed between divers B and D.
3. Diver B beat diver D.

In which order did they finish?

35. A green, a blue, a red, and a yellow flag are hanging on a flagpole.

1. The blue flag is between the green and yellow flags.
2. The red flag is next to the yellow flag.
3. The green flag is above the red flag.

What is the order of the flags from top to bottom?

36. Andres, Barry, and Carl each have two occupations: bootlegger, musician, painter, chauffeur, barber, and gardener. From the following facts, find the occupations of each man.

1. The painter bought a quart of spirits from the bootlegger.
2. The chauffeur offended the musician by laughing at his mustache.
3. The chauffeur dated the painter's sister.
4. Both the musician and the gardener used to go hunting with Andres.
5. Carl beat both Barry and the painter at monopoly.
6. Barry owes the gardener $100.

APPLICATIONS

37. JURY DUTY The results of a jury service questionnaire are shown. Determine how many of the 20,000 respondents have served on neither a criminal court nor a civil court jury.

Jury Service Questionnaire

997	Served on a criminal court jury
103	Served on a civil court jury
35	Served on both

38. POLLS The following Internet poll shows that 124 people voted for the first choice, 27 people voted for the second choice, and 19 people voted for both the first and the second choice. How many people clicked the third choice, "Neither"?

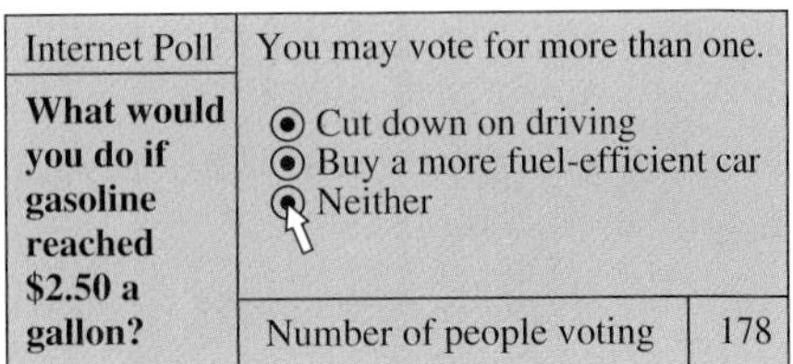

Internet Poll	You may vote for more than one.	
What would you do if gasoline reached $2.50 a gallon?	⦿ Cut down on driving ⦿ Buy a more fuel-efficient car ⦿ Neither	
	Number of people voting	178

39. THE SOLAR SYSTEM The following graph shows some important characteristics of the 9 planets in our solar system. How many planets neither are rocky nor have moons?

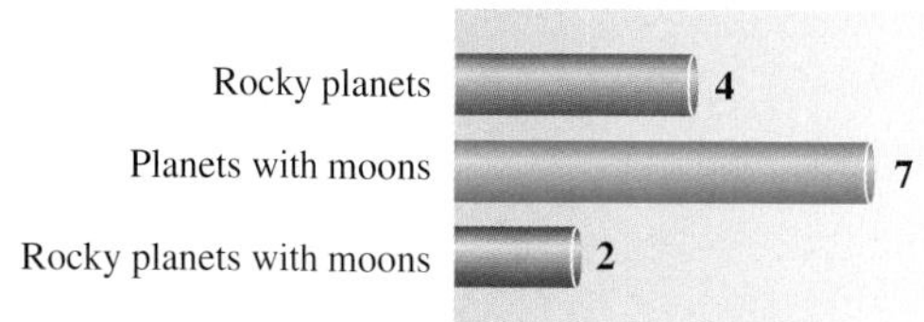

40. Write a problem in such a way that the following diagram can be used to solve it.

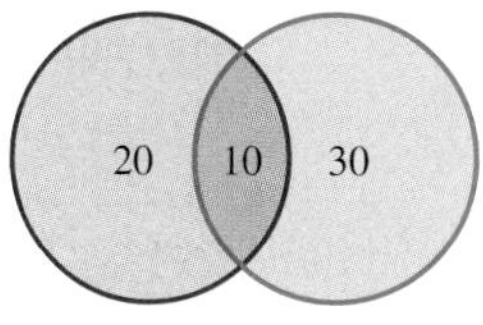

WRITING

41. Describe deductive reasoning.

42. Describe a real-life situation in which you might use deductive reasoning.

43. Describe inductive reasoning.

44. Describe a real-life situation in which you might use inductive reasoning.

APPENDIX B

Measurement

B.1 American Units of Measurement

- American units of length • Converting units of length • American units of weight
- American units of capacity • Units of time

Two common systems of measurement are the American (or English) system and the metric system. We discuss American units in this section and metric units in the next section. Some common American units are *inches, feet, miles, ounces, pounds, tons, cups, pints, quarts,* and *gallons.* These units are used when measuring length, weight, and capacity.

- A newborn baby is 20 inches long.
- The distance from St. Louis to Memphis is 285 miles.
- First-class postage for a letter that weighs less than 1 ounce is 37¢.
- The largest pumpkin ever grown weighed 1,092 pounds.
- Milk is sold in quart and gallon containers.

American units of length

A ruler is one of the most common devices used for measuring distances or lengths. Figure B-1 shows only a portion of a ruler; most rulers are 12 inches (1 foot) long. Since 12 inches = 1 foot, a ruler is divided into 12 equal distances of 1 inch. Each inch is divided into halves of an inch, quarters of an inch, eighths of an inch, and sixteenths of an inch. Several distances are measured using the ruler shown in Figure B-1.

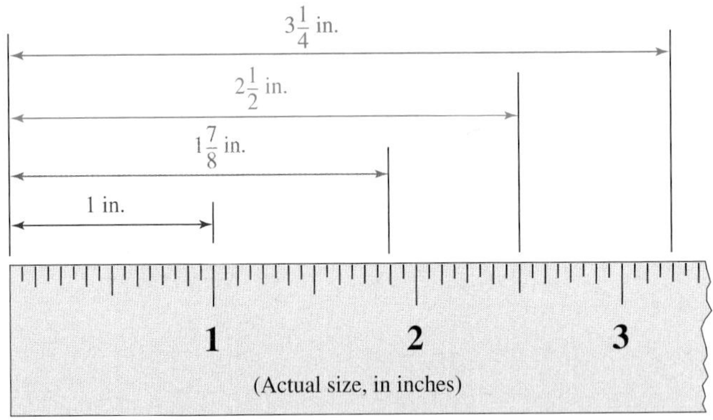

FIGURE B-1

Each point on a ruler, like each point on a number line, has a number associated with it: the distance between the point and 0.

EXAMPLE 1 To the nearest $\frac{1}{4}$ inch, find the length of the nail in Figure B-2.

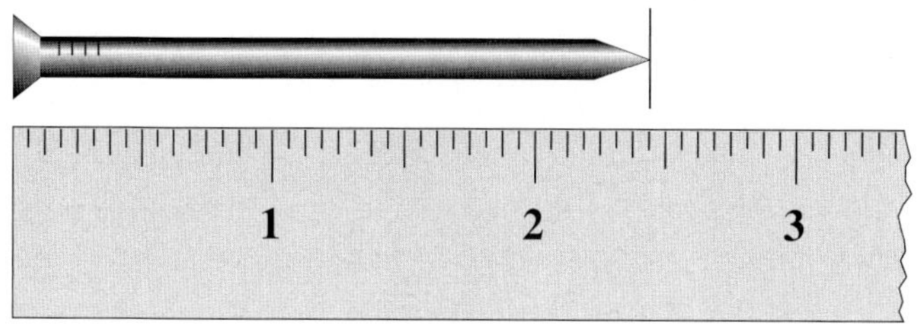

FIGURE B-2

Solution We place the end of the ruler by one end of the nail and note that the other end of the nail is closer to the $2\frac{1}{2}$-inch mark than to the $2\frac{1}{4}$-inch mark on the ruler. To the nearest quarter-inch, the nail is $2\frac{1}{2}$ inches long.

Self Check 1

To the nearest $\frac{1}{4}$ inch, find the width of the circle.

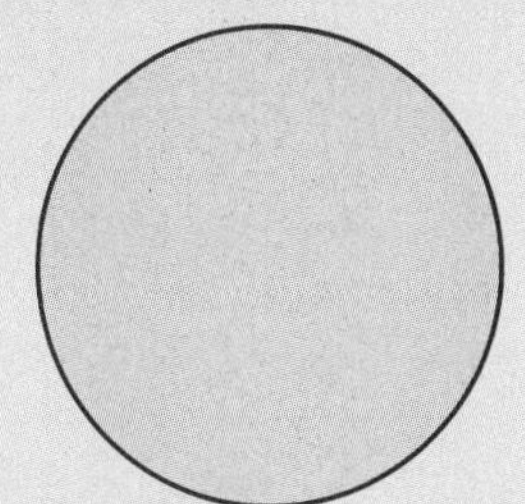

Answer $1\frac{1}{4}$ in.

EXAMPLE 2 To the nearest $\frac{1}{8}$ inch, find the length of the paper clip in Figure B-3.

Solution We place the end of the ruler by one end of the paper clip and note that the other end is closer to the $1\frac{3}{8}$-inch mark than to the $1\frac{1}{2}$-inch mark on the ruler. To the nearest eighth of an inch, the paper clip is $1\frac{3}{8}$ inches long.

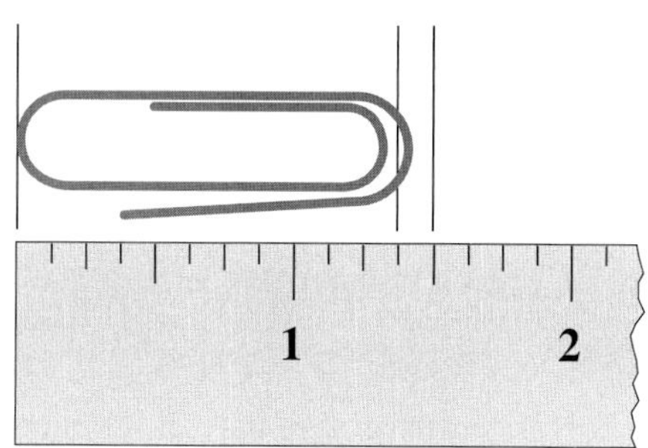

FIGURE B-3

Self Check 2

To the nearest $\frac{1}{8}$ inch, find the length of the jumbo paper clip.

Answer $1\frac{7}{8}$ in.

Converting units of length

American units of length are related in the following ways.

American units of length

1 foot (ft) = 12 inches (in.)	1 yard (yd) = 36 inches
1 yard = 3 feet	1 mile (mi) = 5,280 feet

To convert from one unit to another, we use *unit conversion factors.* To find the unit conversion factor between yards and feet, we begin with this fact:

$$3 \text{ ft} = 1 \text{ yd}$$

If we divide both sides of this equation by 1 yard, we get

$$\frac{3 \text{ ft}}{\mathbf{1\ yd}} = \frac{1 \text{ yd}}{\mathbf{1\ yd}}$$

$$\frac{3 \text{ ft}}{1 \text{ yd}} = 1 \qquad \text{A number divided by itself is 1: } \frac{1 \text{ yd}}{1 \text{ yd}} = 1.$$

The fraction $\frac{3 \text{ ft}}{1 \text{ yd}}$ is called a **unit conversion factor,** because its value is 1. It can be read as "3 feet per yard." Since this fraction is equal to 1, multiplying a length by this fraction does not change its measure; it changes only the *units* of measure.

To convert units of length, we use the following unit conversion factors.

To convert from	Use the unit conversion factor	To convert from	Use the unit conversion factor
feet to inches	$\frac{12\text{ in.}}{1\text{ ft}}$	inches to feet	$\frac{1\text{ ft}}{12\text{ in.}}$
yards to feet	$\frac{3\text{ ft}}{1\text{ yd}}$	feet to yards	$\frac{1\text{ yd}}{3\text{ ft}}$
yards to inches	$\frac{36\text{ in.}}{1\text{ yd}}$	inches to yards	$\frac{1\text{ yd}}{36\text{ in.}}$
miles to feet	$\frac{5{,}280\text{ ft}}{1\text{ mi}}$	feet to miles	$\frac{1\text{ mi}}{5{,}280\text{ ft}}$

Self Check 3
Convert 9 yards to feet.

EXAMPLE 3 Convert 7 yards to feet.

Solution To convert from yards to feet, we must use a unit conversion factor that relates feet to yards. Since there are 3 feet per yard, we multiply 7 yards by the unit conversion factor $\frac{3\text{ ft}}{1\text{ yd}}$.

$$7\text{ yd} = \frac{7\text{ yd}}{1} \cdot \frac{\mathbf{3\text{ ft}}}{\mathbf{1\text{ yd}}} \quad \text{Write 7 yd as a fraction: } 7\text{ yd} = \frac{7\text{ yd}}{1}\text{. Then multiply by 1: } \frac{3\text{ ft}}{1\text{ yd}} = 1.$$

$$= \frac{7\text{ }\overset{1}{\cancel{\text{yd}}}}{1} \cdot \frac{3\text{ ft}}{1\text{ }\underset{1}{\cancel{\text{yd}}}} \quad \text{The units of yards divide out.}$$

$$= 7 \cdot 3\text{ ft}$$

$$= 21\text{ ft} \quad \text{Multiply: } 7 \cdot 3 = 21.$$

Seven yards is equal to 21 feet.

Answer 27 ft

Notice that in Example 3, we eliminated the units of yards and introduced the units of feet by multiplying by the appropriate unit conversion factor. In general, a unit conversion factor is a fraction with the following form:

$$\frac{\text{Unit we want to introduce}}{\text{Unit we want to eliminate}} \quad \begin{matrix}\leftarrow \text{Numerator} \\ \leftarrow \text{Denominator}\end{matrix}$$

Self Check 4
Convert 1.5 feet to inches.

EXAMPLE 4 Convert $1\frac{3}{4}$ feet to inches.

Solution To convert from feet to inches, we must use a unit conversion factor that relates inches to feet. Since there are 12 inches per foot, we multiply $1\frac{3}{4}$ feet by the unit conversion factor $\frac{12\text{ in.}}{1\text{ ft}}$.

$$1\frac{3}{4}\text{ ft} = \frac{7}{4}\text{ ft} \cdot \frac{\mathbf{12\text{ in.}}}{\mathbf{1\text{ ft}}} \quad \text{Write } 1\tfrac{3}{4} \text{ as an improper fraction: } 1\tfrac{3}{4} = \tfrac{7}{4}\text{. Multiply by 1: } \frac{12\text{ in.}}{1\text{ ft}} = 1.$$

$$= \frac{7}{4}\overset{1}{\cancel{\text{ft}}} \cdot \frac{12\text{ in.}}{1\text{ }\underset{1}{\cancel{\text{ft}}}} \quad \text{The units of feet divide out.}$$

$$= \frac{7 \cdot 12}{4 \cdot 1}\text{ in.} \quad \text{Multiply the fractions.}$$

$$= 21\text{ in.} \quad \text{Simplify: } \frac{7 \cdot 12}{4 \cdot 1} = \frac{7 \cdot 3 \cdot \overset{1}{\cancel{4}}}{\underset{1}{\cancel{4}} \cdot 1} = 7 \cdot 3 = 21.$$

$1\frac{3}{4}$ feet is equal to 21 inches.

Answer 18 in.

Sometimes we must use two unit conversion factors to eliminate the given units while introducing the desired units. The following Calculator Snapshot illustrates this concept.

Finding the length of a football field in miles

CALCULATOR SNAPSHOT

A football field (including the end zones) is 120 yards long. To find this distance in miles, we set up the problem so that the units of yards divide out and leave us with units of miles. Since there are 3 feet per yard and 5,280 feet per mile, we multiply 120 yards by $\frac{3 \text{ ft}}{1 \text{ yd}}$ and $\frac{1 \text{ mi}}{5{,}280 \text{ ft}}$.

$$120 \text{ yd} = 120 \text{ yd} \cdot \frac{3 \text{ ft}}{1 \text{ yd}} \cdot \frac{1 \text{ mi}}{5{,}280 \text{ ft}}$$

Use two unit conversion factors: $\frac{3 \text{ ft}}{1 \text{ yd}} = 1$ and $\frac{1 \text{ mi}}{5{,}280 \text{ ft}} = 1$.

$$= \frac{120 \cancel{\text{yd}}}{1} \cdot \frac{3 \cancel{\text{ft}}}{1 \cancel{\text{yd}}} \cdot \frac{1 \text{ mi}}{5{,}280 \cancel{\text{ft}}}$$

Divide out the units of yards and feet.

$$= \frac{120 \cdot 3}{5{,}280} \text{ mi}$$

Multiply the fractions.

We can do this arithmetic using a scientific calculator by entering these numbers and pressing these keys.

120 [×] 3 [÷] 5280 [=]   0.0681818

To the nearest hundredth, a football field is 0.07 mile long.

American units of weight

American units of weight are related in the following ways.

American units of weight

1 pound (lb) = 16 ounces (oz)   1 ton = 2,000 pounds

To convert units of weight, we use the following unit conversion factors.

To convert from	Use the unit conversion factor	To convert from	Use the unit conversion factor
pounds to ounces	$\frac{16 \text{ oz}}{1 \text{ lb}}$	ounces to pounds	$\frac{1 \text{ lb}}{16 \text{ oz}}$
tons to pounds	$\frac{2{,}000 \text{ lb}}{1 \text{ ton}}$	pounds to tons	$\frac{1 \text{ ton}}{2{,}000 \text{ lb}}$

EXAMPLE 5 Convert 40 ounces to pounds.

Solution Since there is 1 pound per 16 ounces, we multiply 40 ounces by the unit conversion factor $\frac{1 \text{ lb}}{16 \text{ oz}}$.

$$40 \text{ oz} = \frac{40 \text{ oz}}{1} \cdot \frac{\mathbf{1\ lb}}{\mathbf{16\ oz}}$$

Write 40 oz as a fraction: $40 \text{ oz} = \frac{40 \text{ oz}}{1}$. Then multiply by 1: $\frac{1 \text{ lb}}{16 \text{ oz}} = 1$.

$$= \frac{40 \cancel{\text{oz}}}{1} \cdot \frac{1 \text{ lb}}{16 \cancel{\text{oz}}}$$

The units of ounces divide out.

$$= \frac{40}{16} \text{ lb}$$

Multiply the fractions.

Self Check 5
Convert 60 ounces to pounds.

There are two ways to complete the solution. First, we can divide out the common factors of the numerator and denominator and write the result as a mixed number.

$$\frac{40}{16}\text{ lb} = \frac{\overset{1}{\cancel{8}}\cdot 5}{\underset{1}{\cancel{8}}\cdot 2}\text{ lb} = \frac{5}{2}\text{ lb} = 2\frac{1}{2}\text{ lb}$$

A second approach is to divide the numerator by the denominator and express the result as a decimal.

$$\frac{40}{16}\text{ lb} = 2.5\text{ lb} \quad \text{Perform the division: } 40 \div 16 = 2.5.$$

Forty ounces is equal to $2\frac{1}{2}$ lb (or 2.5 lb).

Answer $3\frac{3}{4}$ lb = 3.75 lb

Self Check 6
Convert 60 pounds to ounces.

EXAMPLE 6 Convert 25 pounds to ounces.

Solution Since there are 16 ounces per pound, we multiply 25 pounds by the unit conversion factor $\frac{16\text{ oz}}{1\text{ lb}}$.

$$\begin{aligned} 25\text{ lb} &= \frac{25\text{ lb}}{1}\cdot\frac{\mathbf{16\ oz}}{\mathbf{1\ lb}} && \text{Multiply by 1: } \frac{16\text{ oz}}{1\text{ lb}} = 1. \\ &= \frac{25\,\overset{1}{\cancel{\text{lb}}}}{1}\cdot\frac{16\text{ oz}}{1\,\underset{1}{\cancel{\text{lb}}}} && \text{The units of pounds divide out.} \\ &= 25\cdot 16\text{ oz} \\ &= 400\text{ oz} && \text{Multiply: } 25\cdot 16 = 400. \end{aligned}$$

Twenty-five pounds is equal to 400 ounces.

Answer 960 oz

CALCULATOR SNAPSHOT Finding the weight of a car in pounds

A BMW 323Ci convertible weighs 1.78 tons. To find its weight in pounds, we set up the problem so that the units of tons divide out and leave us with units of pounds. Since there are 2,000 pounds per ton, we multiply by $\frac{2{,}000\text{ lb}}{1\text{ ton}}$.

$$\begin{aligned} 1.78\text{ tons} &= \frac{1.78\text{ tons}}{1}\cdot\frac{2{,}000\text{ lb}}{1\text{ ton}} && \text{Multiply by 1: } \frac{2{,}000\text{ lb}}{1\text{ ton}} = 1. \\ &= \frac{1.78\,\overset{1}{\cancel{\text{tons}}}}{1}\cdot\frac{2{,}000\text{ lb}}{1\,\underset{1}{\cancel{\text{ton}}}} && \text{Divide out the units of tons.} \\ &= 1.78\cdot 2{,}000\text{ lb} \end{aligned}$$

We can do this multiplication using a scientific calculator by entering these numbers and pressing these keys.

1.78 [×] 2000 [=] 3560

The convertible weighs 3,560 pounds.

American units of capacity

American units of capacity are related as follows.

American units of capacity

1 cup (c) = 8 fluid ounces (fl oz) 1 pint (pt) = 2 cups (c)

1 quart (qt) = 2 pints (pt) 1 gallon (gal) = 4 quarts (qt)

To convert units of capacity, we use the following unit conversion factors.

To convert from	Use the unit conversion factor	To convert from	Use the unit conversion factor
cups to ounces	$\frac{8\text{ fl oz}}{1\text{ c}}$	ounces to cups	$\frac{1\text{ c}}{8\text{ fl oz}}$
pints to cups	$\frac{2\text{ c}}{1\text{ pt}}$	cups to pints	$\frac{1\text{ pt}}{2\text{ c}}$
quarts to pints	$\frac{2\text{ pt}}{1\text{ qt}}$	pints to quarts	$\frac{1\text{ qt}}{2\text{ pt}}$
gallons to quarts	$\frac{4\text{ qt}}{1\text{ gal}}$	quarts to gallons	$\frac{1\text{ gal}}{4\text{ qt}}$

EXAMPLE 7 **Cooking.** If a recipe calls for 3 pints of milk, how many fluid ounces of milk should be used?

Solution Since there are 2 cups per pint and 8 fluid ounces per cup, we multiply 3 pints by unit conversion factors of $\frac{2\text{ c}}{1\text{ pt}}$ and $\frac{8\text{ fl oz}}{1\text{ c}}$.

$$3\text{ pt} = \frac{3\text{ pt}}{1} \cdot \frac{\mathbf{2\text{ c}}}{\mathbf{1\text{ pt}}} \cdot \frac{\mathbf{8\text{ fl oz}}}{\mathbf{1\text{ c}}}$$ Use two unit conversion factors: $\frac{2\text{ c}}{1\text{ pt}} = 1$ and $\frac{8\text{ fl oz}}{1\text{ c}} = 1$.

$$= \frac{3\,\overset{1}{\cancel{\text{pt}}}}{1} \cdot \frac{2\,\overset{1}{\cancel{\text{c}}}}{1\,\underset{1}{\cancel{\text{pt}}}} \cdot \frac{8\text{ fl oz}}{1\,\underset{1}{\cancel{\text{c}}}}$$ Divide out the units of pints and cups.

$$= 3 \cdot 2 \cdot 8\text{ fl oz}$$

$$= 48\text{ fl oz}$$

Since 3 pints is equal to 48 fluid ounces, 48 fluid ounces of milk should be used.

Self Check 7
How many pints are in 1 gallon?

Answer 8 pt

Units of time

Units of time are related in the following ways.

Units of time

1 minute (min) = 60 seconds (sec) 1 hour (hr) = 60 minutes

1 day = 24 hours

To convert units of time, we use the following unit conversion factors.

To convert from	Use the unit conversion factor	To convert from	Use the unit conversion factor
minutes to seconds	$\frac{60 \text{ sec}}{1 \text{ min}}$	seconds to minutes	$\frac{1 \text{ min}}{60 \text{ sec}}$
hours to minutes	$\frac{60 \text{ min}}{1 \text{ hr}}$	minutes to hours	$\frac{1 \text{ hr}}{60 \text{ min}}$
days to hours	$\frac{24 \text{ hr}}{1 \text{ day}}$	hours to days	$\frac{1 \text{ day}}{24 \text{ hr}}$

Self Check 8
A solar eclipse (eclipse of the sun) can last as long as 450 seconds. Express this time in minutes.

EXAMPLE 8 **Astronomy.** A lunar eclipse occurs when the Earth is between the sun and the moon in such a way that Earth's shadow darkens the moon. (See Figure B-4, which is not to scale.) A total lunar eclipse can last as long as 105 minutes. Express this time in hours.

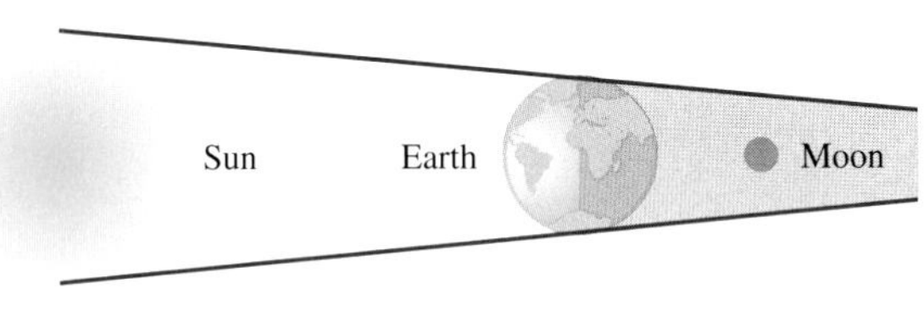

FIGURE B-4

Solution Since there is 1 hour for every 60 minutes, we multiply 105 by the unit conversion factor $\frac{1 \text{ hr}}{60 \text{ min}}$.

$$105 \text{ min} = \frac{105 \text{ min}}{1} \cdot \frac{\mathbf{1\ hr}}{\mathbf{60\ min}}$$ Multiply by 1: $\frac{1 \text{ hr}}{60 \text{ min}} = 1$.

$$= \frac{105 \cancel{\text{min}}}{1} \cdot \frac{1 \text{ hr}}{60 \cancel{\text{min}}}$$ The units of minutes divide out.

$$= \frac{105}{60} \text{ hr}$$ Multiply the fractions.

$$= \frac{\overset{1}{\cancel{3}} \cdot \overset{1}{\cancel{5}} \cdot 7}{2 \cdot 2 \cdot \underset{1}{\cancel{3}} \cdot \underset{1}{\cancel{5}}} \text{hr}$$ Prime factor 105 and 60. Then divide out common factors of the numerator and denominator.

$$= \frac{7}{4} \text{ hr}$$

$$= 1\tfrac{3}{4} \text{ hr}$$ Write $\frac{7}{4}$ as a mixed number.

A total lunar eclipse can last as long as $1\frac{3}{4}$ hours.

Answer $7\frac{1}{2}$ min

Section B.1 STUDY SET

VOCABULARY *Fill in the blanks.*

1. Inches, feet, and miles are examples of American units of ______.
2. A ruler is used for measuring ______.
3. The value of any unit conversion factor is ___.
4. Ounces, pounds, and tons are examples of American units of ______.
5. Some examples of American units of ______ are cups, pints, quarts, and gallons.
6. Some units of ______ are seconds, hours, and days.

CONCEPTS *Fill in the blanks.*

7. 12 in. = ___ ft
8. ___ ft = 1 yd
9. 1 mi = ___ ft
10. 1 yd = ___ in.
11. ___ ounces = 1 pound
12. ___ pounds = 1 ton
13. 1 cup = ___ fluid ounces
14. 1 pint = ___ cups
15. 2 pints = ___ quart
16. 4 quarts = ___ gallon
17. 1 day = ___ hours
18. 2 hours = ___ minutes
19. Determine which measurements the arrows point to on the ruler.

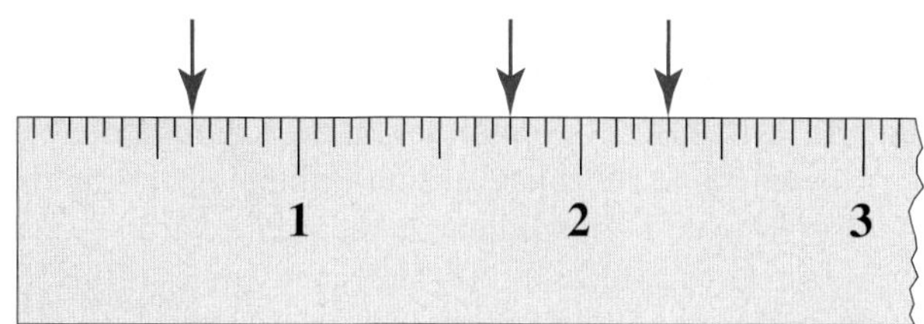

20. Determine which measurements the arrows point to on the ruler, to the nearest $\frac{1}{8}$ inch.

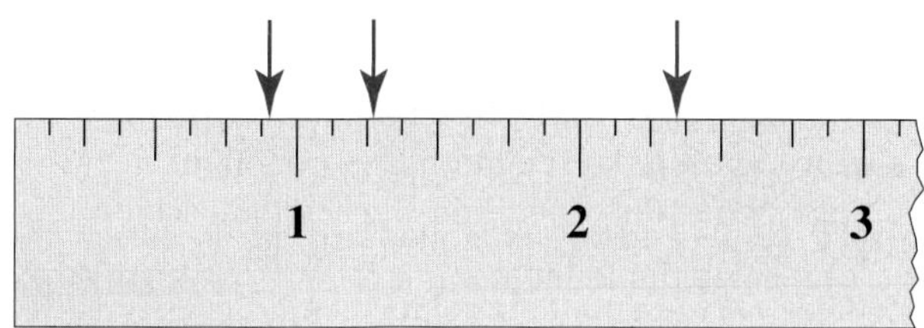

21. Write a unit conversion factor to convert
    a. pounds to tons
    b. quarts to pints
22. Write the two unit conversion factors used to convert
    a. inches to yards
    b. days to minutes
23. Match each item with its proper measurement.
    a. Length of the U.S. coastline
    b. Height of a Barbie doll
    c. Span of the Golden Gate Bridge
    d. Width of a football field

    i. $11\frac{1}{2}$ in.
    ii. 4,200 ft
    iii. 53.5 yd
    iv. 12,383 mi
24. Match each item with its proper measurement.
    a. Weight of the men's shot put used in track and field
    b. Weight of an African elephant
    c. Amount of gold that is worth $500

    i. $1\frac{1}{2}$ oz
    ii. 16 lb
    iii. 7.2 tons
25. Match each item with its proper measurement.
    a. Amount of blood in an adult
    b. Size of the Exxon Valdez oil spill in 1989
    c. Amount of nail polish in a bottle
    d. Amount of flour to make 3 dozen cookies

    i. $\frac{1}{2}$ fluid oz
    ii. 2 cups
    iii. 5 qt
    iv. 10,080,000 gal
26. Match each item with its proper measurement.
    a. Length of first U.S. manned space flight
    b. A leap year
    c. Time difference between New York and Fairbanks, Alaska
    d. Length of Wright Brothers' first flight

    i. 12 sec
    ii. 15 min
    iii. 4 hr
    iv. 366 days

NOTATION *Complete each solution.*

27. Convert 12 yards to inches.

$$\begin{aligned} 12\text{ yd} &= 12\text{ yd}\cdot\frac{\quad\text{ in.}}{1\text{ yd}} \\ &= 12\cdot\quad\text{ in.} \\ &= 432\text{ in.} \end{aligned}$$

28. Convert 1 ton to ounces.

$$\begin{aligned} 1\text{ ton} &= 1\text{ ton}\cdot\frac{\quad\text{ lb}}{1\text{ ton}}\cdot\frac{\quad\text{ oz}}{1\text{ lb}} \\ &= 1\cdot 2{,}000\cdot 16\text{ oz} \\ &= \quad\text{ oz} \end{aligned}$$

29. Convert 12 pints to gallons.

$$\begin{aligned} 12\text{ pt} &= 12\text{ pt}\cdot\frac{1\text{ qt}}{\quad\text{ pt}}\cdot\frac{1\text{ gal}}{\quad\text{ qt}} \\ &= \quad\cdot\frac{1}{2}\cdot\frac{1}{4}\text{ gal} \\ &= 1.5\text{ gal} \end{aligned}$$

30. Convert 37,440 minutes to days.

$$\begin{aligned} 37{,}440\text{ min} &= 37{,}440\text{ min}\cdot\frac{1\text{ hr}}{\quad\text{ min}}\cdot\frac{1\text{ day}}{\quad\text{ hr}} \\ &= \frac{\quad}{60\cdot 24}\text{ days} \\ &= 26\text{ days} \end{aligned}$$

PRACTICE *Use a ruler with a scale in inches to measure each object to the nearest $\frac{1}{8}$ inch.*

31. The width of a dollar bill

32. The length of a dollar bill

33. The length (top to bottom) of this page

34. The length of the word supercalifragilisticexpialidocious

Perform each conversion.

35. 4 feet to inches

36. 7 feet to inches

37. $3\frac{1}{2}$ feet to inches

38. $2\frac{2}{3}$ feet to inches

39. 24 inches to feet

40. 54 inches to feet

41. 8 yards to inches

42. 288 inches to yards

43. 90 inches to yards

44. 12 yards to inches

45. 56 inches to feet

46. 44 inches to feet

47. 5 yards to feet

48. 21 feet to yards

49. 7 feet to yards

50. $4\frac{2}{3}$ yards to feet

51. 15,840 feet to miles

52. 2 miles to feet

53. $\frac{1}{2}$ mile to feet

54. 1,320 feet to miles

55. 80 ounces to pounds

56. 8 pounds to ounces

57. 7,000 pounds to tons

58. 2.5 tons to ounces

59. 12.4 tons to pounds

60. 48,000 ounces to tons

61. 3 quarts to pints

62. 20 quarts to gallons

63. 16 pints to gallons

64. 3 gallons to fluid ounces

65. 32 fluid ounces to pints

66. 2 quarts to fluid ounces

67. 240 minutes to hours

68. 2,400 seconds to hours

69. 7,200 minutes to days

70. 691,200 seconds to days

APPLICATIONS

71. THE GREAT PYRAMID The Great Pyramid in Egypt is about 450 feet high. Express this distance in yards.

72. THE WRIGHT BROTHERS In 1903, Orville Wright made the world's first sustained flight. It lasted 12 seconds, and the plane traveled 120 feet. Express the length of the flight in yards.

73. THE GREAT SPHINX The Great Sphinx of Egypt is 240 feet long. Express this in inches.

74. HOOVER DAM The Hoover Dam in Nevada is 726 feet high. Express this distance in inches.

75. THE SEARS TOWER The Sears Tower in Chicago has 110 stories and is 1,454 feet tall. To the nearest hundredth, express this height in miles.

76. NFL RECORDS Emmit Smith, the former Dallas Cowboys and Arizona Cardinals running back, holds the National Football League record for yards rushing in a career: 18,355. How many miles is this? Round to the nearest tenth of a mile.

77. NFL RECORDS When Dan Marino of the Miami Dolphins retired, it was noted that Marino's career passing total was nearly 35 miles! How many yards is this?

78. LEWIS AND CLARK The trail traveled by the Lewis and Clark expedition is shown on the next page. When the expedition reached the Pacific Ocean, Clark estimated that they had traveled 4,162 miles. (It was later determined that his guess was within 40 miles of the actual distance.) Express Clark's estimate of the distance in feet.

79. WEIGHT OF WATER One gallon of water weighs about 8 pounds. Express this weight in ounces.

80. WEIGHT OF A BABY A newborn baby weighed 136 ounces. Express this weight in pounds.

81. HIPPOS An adult hippopotamus can weigh as much as 9,900 pounds. Express this weight in tons.

82. ELEPHANTS An adult elephant can consume as much as 495 pounds of grass and leaves in one day. How many ounces is this?

83. BUYING PAINT A painter estimates that he will need 17 gallons of paint for a job. To take advantage of a closeout sale on quart cans, he decides to buy the paint in quarts. How many cans will he need to buy?

84. CATERING How many cups of apple cider can be dispensed from a 10-gallon container of cider?

85. SCHOOL LUNCHES Each student attending Eagle River Elementary School receives 1 pint of milk for lunch each day. If 575 students attend the school, how many gallons of milk are used each day?

86. RADIATORS The radiator capacity of a piece of earth-moving equipment is 39 quarts. If the radiator is drained and new coolant put in, how many gallons of new coolant will be used?

87. CAMPING How many ounces of camping stove fuel will fit in the container shown?

88. HIKING A college student walks 11 miles in 155 minutes. To the nearest tenth, how many hours does he walk?

89. SPACE TRAVEL The astronauts of the Apollo 8 mission, which was launched on December 21, 1968, were in space for 147 hours. How many days did the mission take?

90. AMELIA EARHART In 1935, Amelia Earhart became the first woman to fly across the Atlantic Ocean alone, establishing a new record for the crossing: 13 hours and 30 minutes. How many minutes is this?

WRITING

91. Explain how to find the unit conversion factor that will convert feet to inches.
92. Explain how to find the unit conversion factor that will convert pints to gallons.

REVIEW *Round each number as indicated.*

93. 3,673.263; nearest hundred
94. 3,673.263; nearest ten
95. 3,673.263; nearest hundredth
96. 3,673.263; nearest tenth
97. 0.100602; nearest thousandth
98. 0.100602; nearest hundredth
99. 0.09999; nearest tenth
100. 0.09999; nearest one

B.2 Metric Units of Measurement

- Metric units of length • Converting units of length • Metric units of mass
- Metric units of capacity • Cubic centimeters

The metric system is the system of measurement used by most countries in the world. All countries, including the United States, use it for scientific purposes. The metric system, like our decimal numeration system, is based on the number 10. For this reason, converting from one metric unit to another is easier than with the American system.

Metric units of length

The basic metric unit of length is the **meter** (m). One meter is approximately 39 inches, slightly more than 1 yard. Figure B-5 shows the relative sizes of a yardstick and a meterstick.

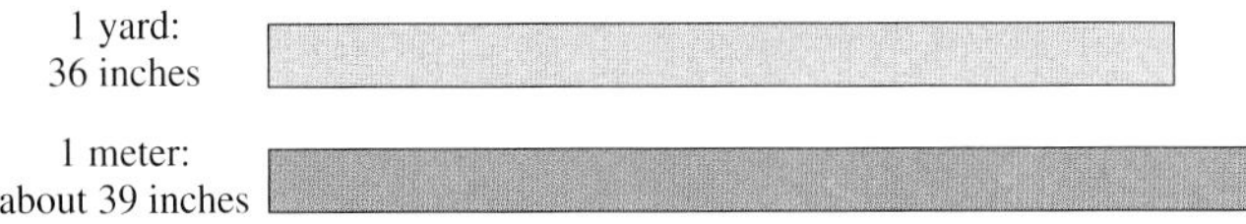

FIGURE B-5

Larger and smaller units are created by adding prefixes to the front of this basic unit, *meter.*

deka means tens

hecto means hundreds

kilo means thousands

deci means tenths

centi means hundredths

milli means thousandths

Metric units of length

1 dekameter (dam) = 10 meters.
1 dam is a little less than 11 yards.

1 hectometer (hm) = 100 meters.
1 hm is about 1 football field long, plus one end zone.

1 kilometer (km) = 1,000 meters.
1 km is about $\frac{3}{5}$ mile.

1 decimeter (dm) = $\frac{1}{10}$ of 1 meter.
1 dm is about the length of your palm.

1 centimeter (cm) = $\frac{1}{100}$ of 1 meter.
1 cm is about as wide as the nail of your little finger.

1 millimeter (mm) = $\frac{1}{1,000}$ of 1 meter.
1 mm is about the thickness of a dime.

Figure B-6 shows a portion of a metric ruler, scaled in centimeters, and a ruler scaled in inches. The rulers are used to measure several lengths.

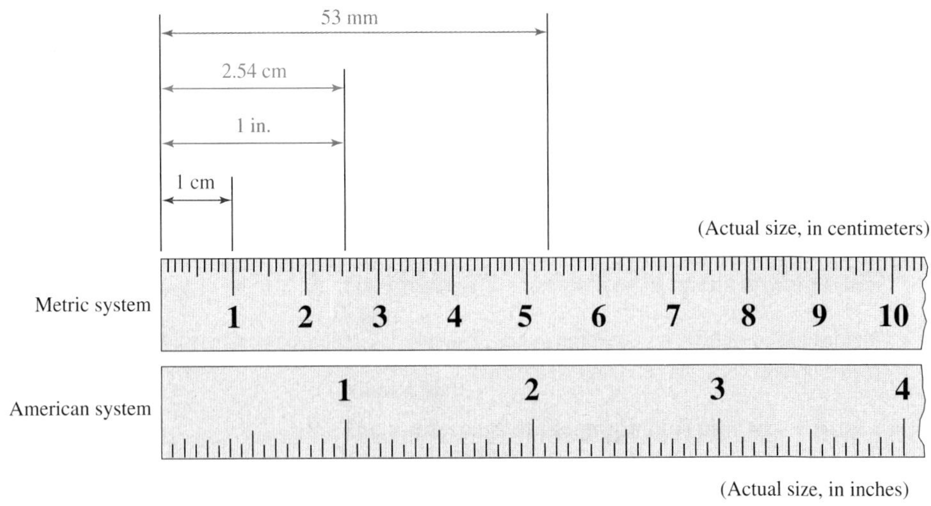

FIGURE B-6

EXAMPLE 1 To the nearest centimeter, find the length of the nail in Figure B-7.

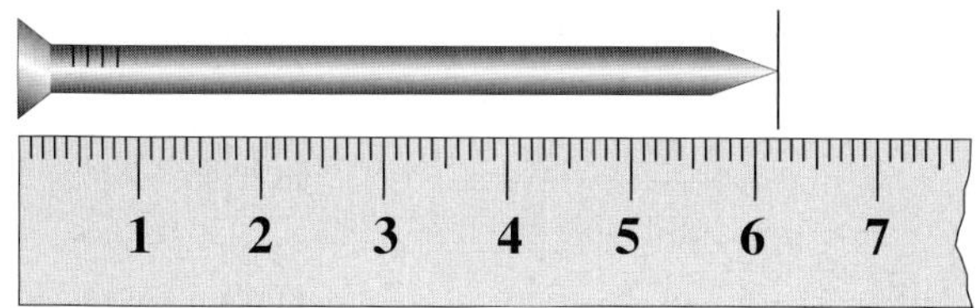

FIGURE B-7

Solution We place the end of the ruler by one end of the nail and note that the other end of the nail is closer to the 6-cm mark than to the 7-cm mark on the ruler. To the nearest centimeter, the nail is 6 cm long.

Self Check 1

To the nearest centimeter, find the width of the circle.

Answer 3 cm

EXAMPLE 2 To the nearest millimeter, find the length of the paper clip in Figure B-8.

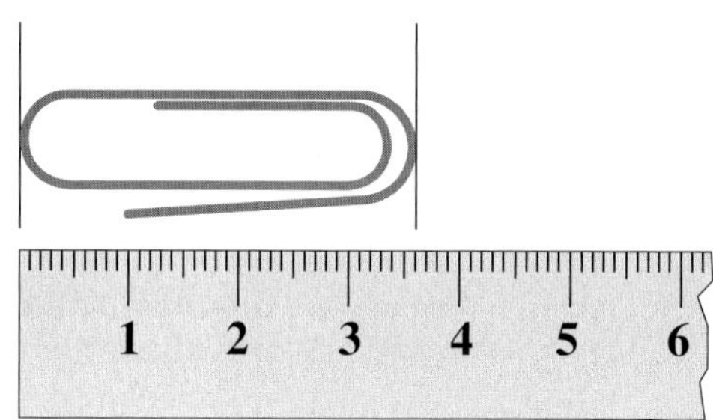

FIGURE B-8

Solution On the ruler, each centimeter has been divided into 10 millimeters. We place the end of the ruler by one end of the paper clip and note that the other end is closer to the 36-mm mark than to the 37-mm mark on the ruler. To the nearest millimeter, the paper clip is 36 mm long.

Self Check 2

To the nearest millimeter, find the length of the jumbo paper clip.

Answer 47 mm

Converting units of length

Metric units of length are related as shown in Table B-1.

Metric units of length

1 kilometer (km) = 1,000 meters	or	1 meter = $\frac{1}{1,000}$ kilometer
1 hectometer (hm) = 100 meters	or	1 meter = $\frac{1}{100}$ hectometer
1 dekameter (dam) = 10 meters	or	1 meter = $\frac{1}{10}$ dekameter
1 decimeter (dm) = $\frac{1}{10}$ meter	or	1 meter = 10 decimeters
1 centimeter (cm) = $\frac{1}{100}$ meter	or	1 meter = 100 centimeters
1 millimeter (mm) = $\frac{1}{1,000}$ meter	or	1 meter = 1,000 millimeters

TABLE B-1

We can use the information in the table to write unit conversion factors that can be used to convert metric units of length. For example, in the table we see that

1 meter = 100 centimeters

From this fact, we can write two unit conversion factors.

$$\frac{1 \text{ m}}{100 \text{ cm}} = 1 \quad \text{and} \quad \frac{100 \text{ cm}}{1 \text{ m}} = 1$$

To obtain the first unit conversion factor, divide both sides of the equation 1 m = 100 cm by 100 cm. To obtain the second unit conversion factor, divide both sides by 1 m.

One advantage of the metric system is that multiplying or dividing by a unit conversion factor involves multiplying or dividing by a power of 10.

Self Check 3
Convert 860 centimeters to meters.

EXAMPLE 3 Convert 350 centimeters to meters.

Solution Since there is 1 meter per 100 centimeters, we multiply 350 centimeters by the unit conversion factor $\frac{1 \text{ m}}{100 \text{ cm}}$.

$$350 \text{ cm} = \frac{350 \text{ cm}}{1} \cdot \frac{\mathbf{1\ m}}{\mathbf{100\ cm}}$$ Multiply by 1: $\frac{1 \text{ m}}{100 \text{ cm}} = 1$.

$$= \frac{350 \not{\text{cm}}}{1} \cdot \frac{1 \text{ m}}{100 \not{\text{cm}}}$$ The units of centimeters divide out.

$$= \frac{350}{100} \text{ m}$$

$$= 3.5 \text{ m}$$ Divide by 100 by moving the decimal point 2 places to the left.

Thus, 350 centimeters = 3.5 meters.

Answer 8.6 m

In Example 3, we converted 350 centimeters to meters using a unit conversion factor. We can also make this conversion by recognizing that all units of length in the metric system are powers of 10 of a meter. Converting from one unit to another is as easy as multiplying by the correct power of 10 or, simply moving a decimal point the correct number of places to the right or left. For example, in the chart below, we see that to convert from centimeters to meters, we move 2 places to the left.

km hm dam **m** dm **cm** mm

To go from centimeters to meters, we must move 2 places to the left.

If we write 350 centimeters as 350.0 centimeters, we can convert to meters by moving the decimal point 2 places to the left.

350.0 centimeters = 3.50 0 meters = 3.5 meters

With the unit conversion factor method or the chart method, we get 350 cm = 3.5 m.

! COMMENT When using a chart to help make a metric conversion, be sure to list the units from largest to smallest when reading from left to right.

EXAMPLE 4 Convert 2.4 meters to millimeters.

Solution Since there are 1,000 millimeters per meter, we multiply 2.4 meters by the unit conversion factor $\frac{1,000 \text{ mm}}{1 \text{ m}}$.

$$2.4 \text{ m} = \frac{2.4 \text{ m}}{1} \cdot \frac{\mathbf{1,000 \text{ mm}}}{\mathbf{1 \text{ m}}} \quad \text{Multiply by 1: } \frac{1,000 \text{ mm}}{1 \text{ m}} = 1.$$

$$= \frac{2.4 \overset{1}{\cancel{\text{m}}}}{1} \cdot \frac{1,000 \text{ mm}}{1 \underset{1}{\cancel{\text{m}}}} \quad \text{The units of meters divide out.}$$

$$= 2.4 \cdot 1,000 \text{ mm}$$

$$= 2,400 \text{ mm} \quad \text{Multiply by 1,000 by moving the decimal point 3 places to the right.}$$

Thus, 2.4 meters = 2,400 millimeters.

We can also make this conversion using a chart.

km hm dam **m** dm cm **mm**

From the chart, we see that we should move the decimal point 3 places to the right to convert from meters to millimeters.

2.4 meters = 2 400. millimeters = 2,400 millimeters

Self Check 4
Convert 5.3 meters to millimeters.

Answer 5,300 mm

EXAMPLE 5 Convert 3.2 kilometers to centimeters.

Solution To convert to centimeters, we set up the problem so that the units of kilometers divide out and leave us with units of centimeters. Since there are 1,000 meters per kilometer and 100 centimeters per meter, we multiply 3.2 kilometers by $\frac{1,000 \text{ m}}{1 \text{ km}}$ and $\frac{100 \text{ cm}}{1 \text{ m}}$.

$$3.2 \text{ km} = \frac{3.2 \overset{1}{\cancel{\text{km}}}}{1} \cdot \frac{\mathbf{1,000} \overset{1}{\cancel{\mathbf{m}}}}{\mathbf{1} \underset{1}{\cancel{\mathbf{km}}}} \cdot \frac{\mathbf{100 \text{ cm}}}{\mathbf{1 \text{ m}}} \quad \text{The units of kilometers and meters divide out.}$$

$$= 3.2 \cdot 1,000 \cdot 100 \text{ cm}$$

$$= 320,000 \text{ cm} \quad \text{Multiply by 1,000 and 100 by moving the decimal point 5 places to the right.}$$

Thus, 3.2 kilometers = 320,000 centimeters.

Using a chart, we see that the decimal point should be moved 5 places to the right to convert kilometers to centimeters.

km hm dam m dm **cm** mm

3.2 kilometers = 3 20000. centimeters = 320,000 centimeters

Self Check 5
Convert 5.15 kilometers to centimeters.

Answer 515,000 cm

Metric units of mass

The **mass** of an object is a measure of the amount of material in the object. When an object is moved about in space, its mass does not change. One basic unit of mass in the metric system is the **gram** (g). A gram is defined to be the mass of water contained in a cube having sides 1 centimeter long. (See Figure B-9 on the next page.)

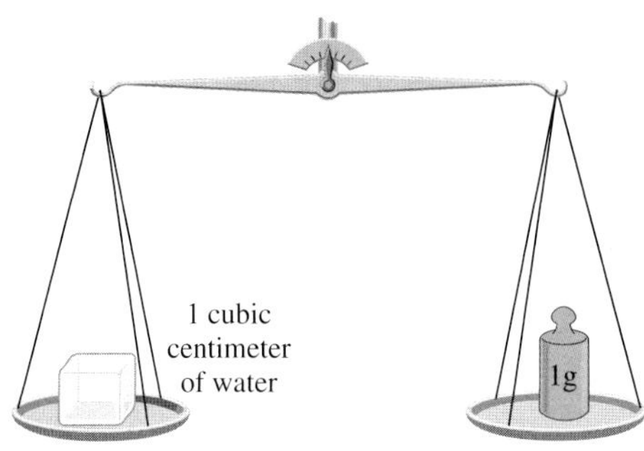

FIGURE B-9

The **weight** of an object is determined by the Earth's gravitational pull on the object. Since gravitational pull on an object decreases as the object gets farther from Earth, the object weighs less as it gets farther from Earth's surface. This is why astronauts experience weightlessness in space. However, since most of us remain near Earth's surface, we will use the words *mass* and *weight* interchangeably. Thus, a mass of 30 grams is said to weigh 30 grams.

Metric units of mass are related as shown in Table B-2.

Metric units of mass

1 kilogram (kg) = 1,000 grams	or	1 gram = $\frac{1}{1,000}$ kilogram
1 hectogram (hg) = 100 grams	or	1 gram = $\frac{1}{100}$ hectogram
1 dekagram (dag) = 10 grams	or	1 gram = $\frac{1}{10}$ dekagram
1 decigram (dg) = $\frac{1}{10}$ gram	or	1 gram = 10 decigrams
1 centigram (cg) = $\frac{1}{100}$ gram	or	1 gram = 100 centigrams
1 milligram (mg) = $\frac{1}{1,000}$ gram	or	1 gram = 1,000 milligrams

TABLE B-2

Here are examples of these units of mass:

- An average bowling ball weighs about 6 kilograms.
- A raisin weighs about 1 gram.
- A certain vitamin tablet contains 450 milligrams of calcium.

We can use the information in Table B-2 to write unit conversion factors that can be used to convert metric units of mass. For example, in the table we see that

1 kilogram = 1,000 grams

From this fact, we can write two unit conversion factors.

$$\frac{1 \text{ kg}}{1{,}000 \text{ g}} = 1 \quad \text{and} \quad \frac{1{,}000 \text{ g}}{1 \text{ kg}} = 1$$

To obtain the first unit conversion factor, divide both sides of the equation 1 kg = 1,000 g by 1,000 g. To obtain the second unit conversion factor, divide both sides by 1 kg.

Self Check 6
Convert 5 kilograms to grams.

EXAMPLE 6 Convert 7.2 kilograms to grams.

Solution To convert to grams, we set up the problem so that the units of kilograms divide out and leave us with the units of grams. Since there are 1,000 grams per 1 kilogram, we multiply 7.2 kilograms by $\frac{1{,}000 \text{ g}}{1 \text{ kg}}$.

$$7.2\text{ kg} = \frac{7.2\text{ kg}}{1} \cdot \frac{\mathbf{1{,}000\text{ g}}}{\mathbf{1\text{ kg}}}$$ Divide out the units of kilograms.

$$= 7.2 \cdot 1{,}000\text{ g}$$

$$= 7{,}200\text{ g}$$ Perform the multiplication by moving the decimal point 3 places to the right.

Thus, 7.2 kilograms = 7,200 grams.

To use a chart to make the conversion, we list the metric units of weight from the largest (kilograms) to the smallest (milligrams).

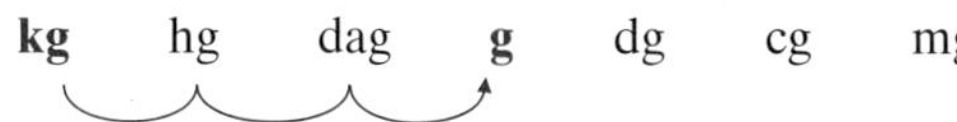

From the chart, we see that we must move the decimal point 3 places to the right to change kilograms to grams.

7.2 kilograms = 7 200. grams = 7,200 grams

Answer 5,000 g

EXAMPLE 7 Medications. A bottle of Verapamil, a drug taken for high blood pressure, contains 30 tablets. If each tablet contains 180 mg of active ingredient, how many centigrams of active ingredient are in the bottle?

Solution Since there are 30 tablets and each one contains 180 mg of active ingredient, there are

$$30 \cdot 180\text{ mg} = 5{,}400\text{ mg}$$

of active ingredient in the bottle. To convert milligrams to centigrams, we multiply 5,400 milligrams by $\frac{1\text{ g}}{1{,}000\text{ mg}}$ and $\frac{100\text{ cg}}{1\text{ g}}$.

$$5{,}400\text{ mg} = \frac{5{,}400\text{ mg}}{1} \cdot \frac{\mathbf{1\text{ g}}}{\mathbf{1{,}000\text{ mg}}} \cdot \frac{\mathbf{100\text{ cg}}}{\mathbf{1\text{ g}}}$$ Divide out the units of milligrams and grams.

$$= \frac{5{,}400 \cdot 100}{1{,}000}\text{ cg}$$ Multiply the fractions.

$$= 540\text{ cg}$$ Simplify

There are 540 centigrams of active ingredient in the bottle.

Using a chart, we see that we must move the decimal point 1 place to the left to convert from milligrams to centigrams.

kg hg dag g dg **cg** **mg**

5,400 milligrams = 540.0 centigrams = 540 centigrams

Self Check 7

One brand name for Verapamil is Isoptin. If a bottle of Isoptin contains 90 tablets, each containing 200 mg of active ingredient, how many centigrams of active ingredient are in the bottle?

Answer 1,800 cg

Metric units of capacity

In the metric system, one basic unit of capacity is the **liter** (L), which is defined to be the capacity of a cube with sides 10 centimeters long. (See Figure B-10.) A liter of liquid is slightly more than 1 quart.

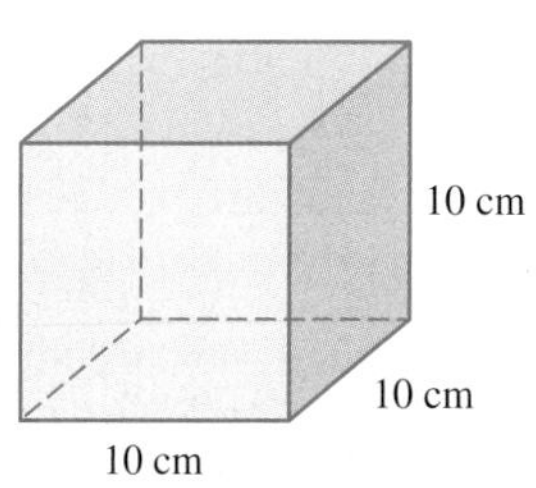

FIGURE B-10

Metric units of capacity are related as shown in Table B-3.

Metric units of capacity		
1 kiloliter (kL) = 1,000 liters	or	1 liter = $\frac{1}{1,000}$ kiloliter
1 hectoliter (hL) = 100 liters	or	1 liter = $\frac{1}{100}$ hectoliter
1 dekaliter (daL) = 10 liters	or	1 liter = $\frac{1}{10}$ dekaliter
1 deciliter (dL) = $\frac{1}{10}$ liter	or	1 liter = 10 deciliters
1 centiliter (cL) = $\frac{1}{100}$ liter	or	1 liter = 100 centiliters
1 milliliter (mL) = $\frac{1}{1,000}$ liter	or	1 liter = 1,000 milliliters

TABLE B-3

Here are examples of these units of capacity:

- Soft drinks are sold in 2-liter plastic bottles.
- The fuel tank of a certain minivan can hold about 75 liters of gasoline.
- Chemists use glass cylinders, scaled in milliliters, to measure liquids.

We can use the information in Table B-3 to write unit conversion factors that can be used to convert metric units of capacity. For example, in the table we see that

1 liter = 100 centiliters

From this fact, we can write two unit conversion factors.

$$\frac{1\text{ L}}{100\text{ cL}} = 1 \quad \text{and} \quad \frac{100\text{ cL}}{1\text{ L}} = 1$$

Self Check 8
How many milliliters are in two 2-liter bottles of cola?

Answer 4,000 mL

EXAMPLE 8 **Soft drinks.** How many centiliters are in three 2-liter bottles of cola?

Solution Three 2-liter bottles of cola contain 6 liters of cola. To convert to centiliters, we set up the problem so that liters divide out and leave us with centiliters. Since there are 100 centiliters per 1 liter, we multiply 6 liters by the unit conversion factor $\frac{100\text{ cL}}{1\text{ L}}$.

$$6\text{ L} = 6\text{ L} \cdot \frac{\mathbf{100\text{ cL}}}{\mathbf{1\text{ L}}} \quad \text{Multiply by 1: } \frac{100\text{ cL}}{1\text{ L}} = 1.$$

$$= \frac{6\,\overset{1}{\cancel{\text{L}}}}{1} \cdot \frac{100\text{ cL}}{1\,\underset{1}{\cancel{\text{L}}}} \quad \text{The units of liters divide out.}$$

$$= 6 \cdot 100\text{ cL}$$

$$= 600\text{ cL}$$

Thus, there are 600 centiliters in three 2-liter bottles of cola.

To make this conversion using a chart, we list the metric units of capacity in order from largest (kiloliter) to smallest (milliliter).

kL hL daL **L** dL **cL** mL

From the chart, we see that we should move the decimal point 2 places to the right to convert from liters to centiliters.

6 liters = 6 00. centiliters = 600 centiliters

Cubic centimeters

Another metric unit of capacity is the **cubic centimeter,** which is represented by the notation cm^3 or, more simply, cc. One milliliter and one cubic centimeter represent the same capacity.

$1 \text{ mL} = 1 \text{ cm}^3 = 1 \text{ cc}$

The units of cubic centimeters are used frequently in medicine. For example, when a nurse administers an injection containing 5 cc of medication, the dosage can also be expressed using milliliters.

$5 \text{ cc} = 5 \text{ mL}$

When a doctor orders that a patient be put on 1,000 cc of dextrose solution, the request can be expressed in different ways.

$1{,}000 \text{ cc} = 1{,}000 \text{ mL} = 1 \text{ liter}$

Section B.2 STUDY SET

VOCABULARY *Fill in the blanks.*

1. *Deka* means ______.
2. *Hecto* means __________.
3. *Kilo* means __________.
4. *Deci* means _______.
5. *Centi* means __________.
6. *Milli* means __________.
7. Meters, grams, and liters are units of measurement in the _______ system.
8. The _______ of an object is determined by the Earth's gravitational pull on the object.

CONCEPTS

9. To the nearest centimeter, determine which measurements the arrows point to on the ruler.

10. To the nearest millimeter, determine which measurements the arrows point to on the ruler.

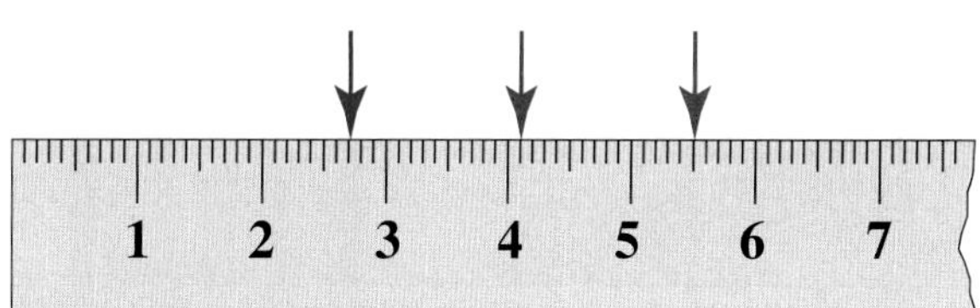

11. Write a unit conversion factor to convert
    a. meters to kilometers
    b. grams to centigrams
    c. liters to milliliters
12. Use the chart to determine how many decimal places and in which direction to move the decimal point when converting the following.
    a. Kilometers to centimeters

       km hm dam m dm cm mm

    b. Milligrams to grams

       kg hg dag g dg cg mg

    c. Hectoliters to centiliters

       kL hL daL L dL cL mL

13. Match each item with its proper measurement.
    a. Thickness of a phone book — i. 6,275 km
    b. Length of the Amazon River — ii. 2 m
    c. Height of a soccer goal — iii. 6 cm

14. Match each item with its proper measurement.

- **a.** Weight of a giraffe
- **b.** Weight of a paper clip
- **c.** Active ingredient in an aspirin tablet

- **i.** 800 kg
- **ii.** 1 g
- **iii.** 325 mg

15. Match each item with its proper measurement.

- **a.** Amount of blood in an adult
- **b.** Cola in an aluminum can
- **c.** Kuwait's daily production of crude oil

- **i.** 290,000 kL
- **ii.** 6 L
- **iii.** 355 mL

16. Of the objects in the illustration, which can be used to measure the following?

- **a.** Millimeters
- **b.** Milligrams
- **c.** Milliliters

Balance

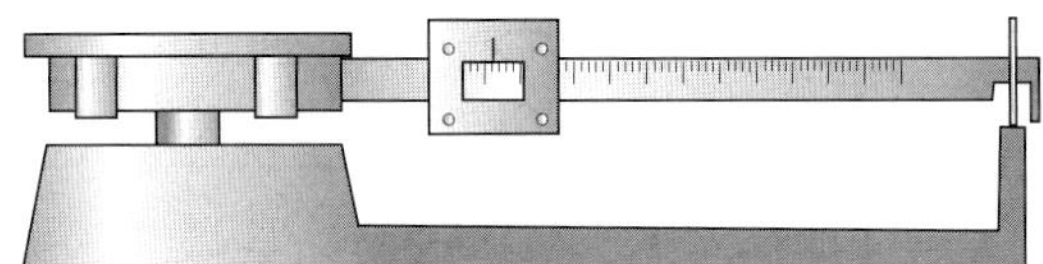

Beaker

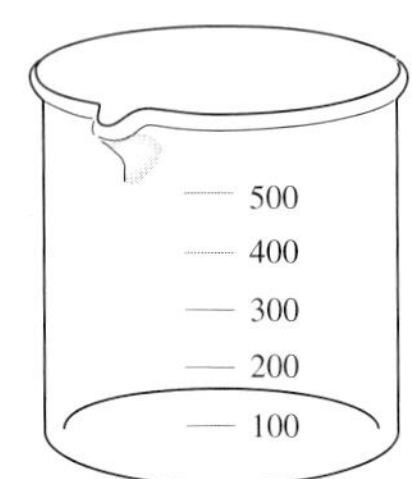

Micrometer

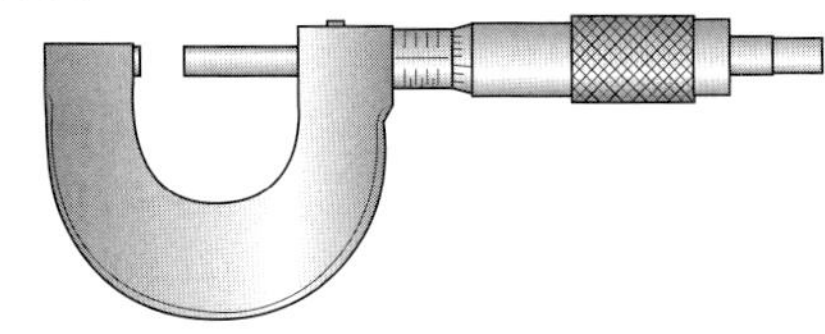

Fill in the blanks.

17. 1 dekameter = ______ meters

18. 1 decimeter = ______ meter

19. 1 centimeter = ______ meter

20. 1 kilometer = ______ meters

21. 1 millimeter = ______ meter

22. 1 hectometer = ______ meters

23. 1 gram = ______ milligrams

24. 100 centigrams = ______ gram

25. 1 kilogram = ______ grams

26. 1 milliliter = ______ cubic centimeter

27. 1 liter = ______ cubic centimeters

28. 1 kiloliter = ______ liters

29. 1 centiliter = ______ liter

30. 1 milliliter = ______ liter

31. 100 liters = ______ hectoliter

32. 10 deciliters = ______ liter

NOTATION *Complete each solution.*

33. Convert 20 centimeters to meters.

$$\begin{aligned} 20\text{ cm} &= 20\text{ cm}\cdot\frac{\underline{\quad}\text{ m}}{100\text{ cm}} \\ &= \frac{20}{\underline{\quad}}\text{ m} \\ &= 0.2\text{ m} \end{aligned}$$

34. Convert 300 centigrams to grams.

$$\begin{aligned} 300\text{ cg} &= 300\text{ cg}\cdot\frac{\underline{\quad}\text{ g}}{100\text{ cg}} \\ &= \frac{\underline{\quad}}{100}\text{ g} \\ &= 3\text{ g} \end{aligned}$$

35. Convert 2 kilometers to decimeters.

$$\begin{aligned} 2\text{ km} &= 2\text{ km}\cdot\frac{\underline{\quad}\text{ m}}{1\text{ km}}\cdot\frac{10\text{ dm}}{\underline{\quad}\text{ m}} \\ &= 2\cdot\underline{\quad}\cdot 10\text{ dm} \\ &= 20{,}000\text{ dm} \end{aligned}$$

36. Convert 3 deciliters to milliliters.

$$\begin{aligned} 3\text{ dL} &= 3\text{ dL}\cdot\frac{1\text{ L}}{\underline{\quad}\text{ dL}}\cdot\frac{\underline{\quad}\text{ mL}}{1\text{ L}} \\ &= \frac{\underline{\quad}\cdot 1{,}000}{10}\text{ mL} \\ &= 300\text{ mL} \end{aligned}$$

PRACTICE *Use a metric ruler to measure each object to the nearest millimeter.*

37. The length of a dollar bill

38. The width of a dollar bill

Use a metric ruler to measure each object to the nearest centimeter.

39. The length (top to bottom) of this page

40. The length of the word antidisestablishmentarianism

Convert each measurement between the given metric units.

41. 3 m = ____ cm

42. 5 m = ____ cm

43. 5.7 m = ____ cm

44. 7.36 km = ____ dam

45. 0.31 dm = ____ cm

46. 73.2 m = ____ dm

47. 76.8 hm = ________ mm

48. 165.7 km = ______ m

49. 4.72 cm = _____ dm

50. 0.593 cm = _______ dam

51. 453.2 cm = _____ m

52. 675.3 cm = _____ m

53. 0.325 dm = _____ m

54. 0.0034 mm = ________ m

55. 3.75 cm = ____ mm

56. 0.074 cm = ____ mm

57. 0.125 m = ____ mm

58. 134 m = ____ hm

59. 675 dam = ______ cm

60. 0.00777 cm = ________ dam

61. 638.3 m = _____ hm

62. 6.77 cm = _____ m

63. 6.3 mm = ____ cm

64. 6.77 mm = _____ cm

65. 695 dm = ____ m

66. 6,789 cm = ____ dm

67. 5,689 m = _____ km

68. 0.0579 km = _____ mm

69. 576.2 mm = _____ dm

70. 65.78 km = _____ dam

71. 6.45 dm = _______ km

72. 6.57 cm = ____ mm

73. 658.23 m = ______ km

74. 0.0068 hm = ______ km

75. 3 g = _____ mg

76. 5 g = ____ cg

77. 2 kg = _____ g

78. 4,000 g = ___ kg

79. 1,000 kg = ________ g

80. 2 kg = ______ cg

81. 500 mg = ____ g

82. 500 mg = ____ cg

83. 3 kL = _____ L

84. 500 mL = ____ L

85. 500 cL = _____ mL

86. 400 L = ___ hL

87. 10 mL = ___ cc

88. 2,000 cc = ___ L

APPLICATIONS

89. SPEED SKATING American Eric Heiden won an unprecedented five gold medals by capturing the men's 500-m, 1,000-m, 1,500-m, 5,000-m, and 10,000-m races at the 1980 Winter Olympic Games in Lake Placid, New York. Convert each race length to kilometers.

90. THE SUEZ CANAL The 163-km-long Suez Canal connects the Mediterranean Sea with the Red Sea. It provides a shortcut for ships operating between European and American ports. Convert the length of the Suez Canal to meters.

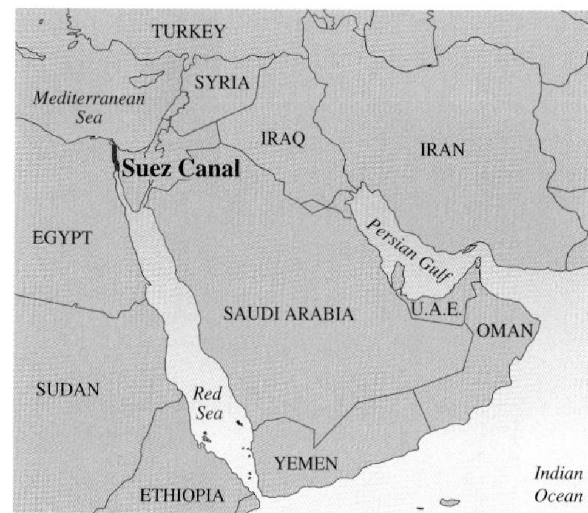

91. THE HANCOCK CENTER The John Hancock Center in Chicago has 100 stories and is 343 meters high. Give this height in hectometers.

92. WEIGHT OF A BABY A baby weighs 4 kilograms. Give this weight in centigrams.

93. HEALTH CARE Blood pressure is measured by a *sphygmomanometer.* The measurement is read at two points and is expressed, for example, as 120/80. This indicates a *systolic* pressure of 120 millimeters of mercury and a *diastolic* pressure of 80 millimeters of mercury. Convert each measurement to centimeters of mercury.

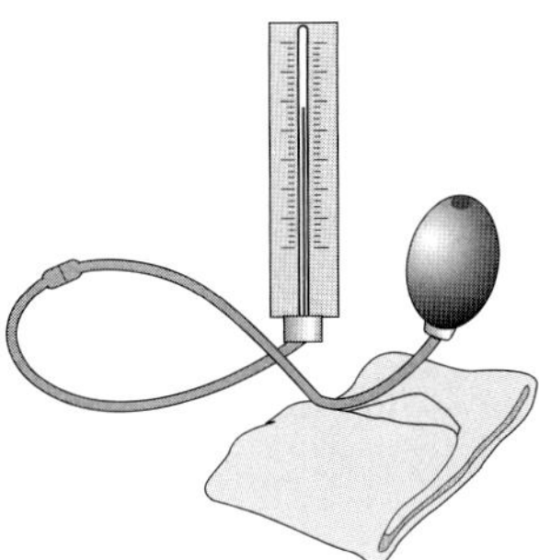

94. JEWELRY A gold chain weighs 1,500 milligrams. Give this weight in grams.

95. CONTAINERS How many deciliters of root beer are in two 2-liter bottles?

96. BOTTLING How many liters of wine are in a 750-mL bottle?

97. BUYING OLIVES The net weight of a bottle of olives is 284 grams. Find the smallest number of bottles that must be purchased to have at least 1 kilogram of olives.

98. BUYING COFFEE A can of Cafe Vienna has a net weight of 133 grams. Find the smallest number of cans that must be packaged to have at least 1 metric ton of coffee. (*Hint:* 1 metric ton = 1,000 kg.)

99. MEDICINE A bottle of hydrochlorothiazine contains 60 tablets. If each tablet contains 50 milligrams of active ingredient, how many grams of active ingredient are in the bottle?

100. INJECTIONS The illustration shows a 3cc syringe. Express its capacity using units of milliliters.

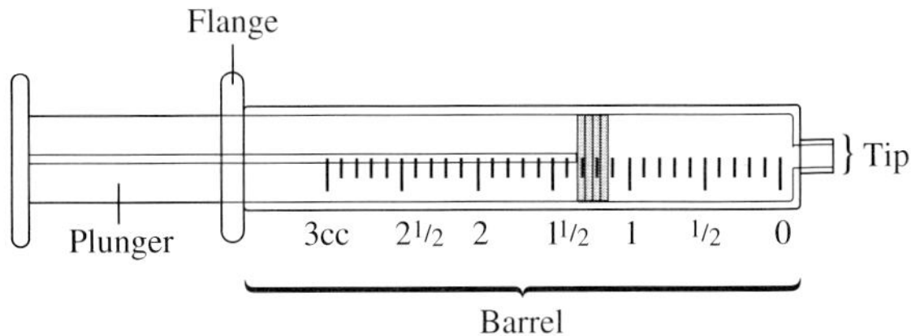

WRITING

101. To change 3.452 kilometers to meters, we can move the decimal point in 3.452 three places to the right to get 3,452 meters. Explain why.

102. To change 7,532 grams to kilograms, we can move the decimal point in 7,532 three places to the left to get 7.532 kilograms. Explain why.

103. A centimeter is one hundredth of a meter. Make a list of other words that begin with the prefix *centi* or *cent* and write a definition for each.

104. List the advantages of the metric system of measurement as compared to the American system. There have been several attempts to bring the metric system into general use in the United States. Why do you think these efforts have been unsuccessful?

REVIEW

105. Find 7% of $342.72

106. $32.16 is 8% of what amount?

107. Divide: $3\frac{1}{7} \div 2\frac{1}{2}$.

108. Simplify: $3\frac{1}{7} + 2\frac{1}{2} \cdot 3\frac{1}{3}$.

B.3 Converting between American and Metric Units

- Converting between American and metric units
- Comparing American and metric units of temperature

It is often necessary to convert between American units and metric units. For example, we must convert units to answer the following questions.

- Which is higher: Pikes Peak (elevation 14,110 feet) or the Matterhorn (elevation 4,478 meters)?
- Does a 2-pound tub of butter weigh more than a 1-kilogram tub?
- Is a quart of soda pop more or less than a liter of soda pop?

In this section, we discuss how to answer such questions.

Converting between American and metric units

We can convert between American and metric units of length using the table below.

Equivalent lengths	
American to metric	**Metric to American**
1 in. ≈ 2.54 cm	1 cm ≈ 0.3937 in.
1 ft ≈ 0.3048 m	1 m ≈ 3.2808 ft
1 yd ≈ 0.9144 m	1 m ≈ 1.0936 yd
1 mi ≈ 1.6093 km	1 km ≈ 0.6214 mi

Read ≈ as "is approximately equal to."

EXAMPLE 1 **Clothing labels.** Figure B-11 shows a label sewn into some pants made in Mexico for sale in the United States. Express the waist size to the nearest inch.

WAIST: 81 cm
INSEAM: 76 cm
RN-80811
SEE REVERSE FOR CARE
MADE IN MEXICO

FIGURE B-11

Solution We need to convert from metric to American units. From the table, we see that there is 0.3937 inch in 1 centimeter. To make the conversion, we substitute 0.3937 inch for 1 centimeter.

81 centimeters = 81 (**centimeters**)

≈ 81(**0.3937 in.**) Substitute 0.3937 inch for 1 centimeter.

≈ 31.8897 in. Perform the multiplication.

To the nearest inch, the waist size is 32 inches.

Self Check 1
Refer to Figure B-11. What is the inseam length, to the nearest inch?

Answer 30 in.

EXAMPLE 2 **Mountain elevations.** Pikes Peak, one of the most famous peaks in the Rocky Mountains, has an elevation of 14,110 feet. The Matterhorn, in the Swiss Alps, rises to an elevation of 4,478 meters. Which mountain is higher?

Solution To make a comparison, the elevations must be expressed in the same units. We will convert the elevation of Pikes Peak, which is given in feet, to meters.

14,110 feet = 14,110 (**feet**)

≈ 14,110(**0.3048 m**) Substitute 0.3048 meter for 1 foot.

≈ 4,300.728 m Perform the multiplication.

Since the elevation of Pikes Peak is about 4,301 meters, we can conclude that the Matterhorn, with an elevation of 4,478 meters, is higher.

Self Check 2
Which is longer: a 500-meter race or a 550-yard race?

Answer the 550-yard race

We can convert between American units of weight and metric units of mass by using the accompanying table.

Equivalent weights and masses	
American to metric	**Metric to American**
1 oz ≈ 28.35 g	1 g ≈ 0.035 oz
1 lb ≈ 0.454 kg	1 kg ≈ 2.2 lb

Self Check 3
Change 20 kilograms to pounds.

Answer 44 lb

EXAMPLE 3 Change 50 pounds to grams.

Solution

$$\begin{aligned} 50 \text{ lb} &= 50(\mathbf{1\ lb}) \\ &= 50(\mathbf{16\ oz}) && \text{Substitute 16 ounces for 1 pound.} \\ &= 50(16)(\mathbf{1\ oz}) \\ &\approx 50(16)(\mathbf{28.35\ g}) && \text{Substitute 28.35 grams for 1 ounce.} \\ &\approx 22{,}680 \text{ g} && \text{Perform the multiplication.} \end{aligned}$$

Thus, 50 pounds is equal to 22,680 grams.

Self Check 4
Who weighs more: a person who weighs 165 pounds or one who weighs 76 kilograms?

Answer the person who weighs 76 kg

EXAMPLE 4 **Packaging.** Does a 2-pound tub of butter weigh more than a 1-kilogram tub?

Solution To decide which contains more butter, we can change 2 pounds to kilograms.

$$\begin{aligned} 2 \text{ lb} &= 2(\mathbf{1\ lb}) \\ &\approx 2(\mathbf{0.454\ kg}) && \text{Substitute 0.454 kilogram for 1 pound.} \\ &\approx 0.908 \text{ kg} && \text{Perform the multiplication.} \end{aligned}$$

Since a 2-pound tub weighs only 0.908 kilogram, the 1-kilogram tub weighs more.

We can convert between American and metric units of capacity by using the accompanying table.

Equivalent capacities	
American to metric	**Metric to American**
1 fl oz ≈ 0.030 L	1 L ≈ 33.8 fl oz
1 pt ≈ 0.473 L	1 L ≈ 2.1 pt
1 qt ≈ 0.946 L	1 L ≈ 1.06 qt
1 gal ≈ 3.785 L	1 L ≈ 0.264 gal

Self Check 5
A student bought a 355-mL can of cola. How many ounces of cola does the can contain?

Answer 12 oz

EXAMPLE 5 **Soft drinks.** A bottle of 7UP contains 750 milliliters. Convert this measure to quarts.

Solution We convert milliliters to liters and then liters to quarts.

$$\begin{aligned} 750 \text{ mL} &= 750 \cancel{\text{mL}} \cdot \frac{1 \text{ L}}{1{,}000 \cancel{\text{mL}}} && \text{Use a unit conversion factor: } \frac{1 \text{ L}}{1{,}000 \text{ mL}} = 1. \\ &= \frac{750}{1{,}000} \text{ L} && \text{The units of mL divide out.} \\ &= \frac{3}{4} \mathbf{L} && \text{Simplify the fraction: } \frac{750}{1{,}000} = \frac{3 \cdot \overset{1}{\cancel{250}}}{4 \cdot \underset{1}{\cancel{250}}} = \frac{3}{4}. \\ &\approx \frac{3}{4}(\mathbf{1.06\ qt}) && \text{Substitute 1.06 quart for 1 liter.} \\ &\approx 0.795 \text{ qt} && \text{Perform the arithmetic.} \end{aligned}$$

The bottle contains 0.795 quart.

From the table of equivalent capacities, we see that 1 liter is equal to 1.06 quarts. Thus, a liter of soda pop is more than a quart of soda pop.

EXAMPLE 6 **Comparison shopping.** A 2-quart bottle of soda pop is priced at \$1.89, and a 1-liter bottle is priced at 97¢. Which is the better buy?

Solution We can convert 2 quarts to liters and find the price per liter of the 2-quart bottle.

$$
\begin{aligned}
2\text{ qt} &= 2(\mathbf{1\ qt}) \\
&\approx 2(\mathbf{0.946\ L}) && \text{Substitute 0.946 liter for 1 quart.} \\
&\approx 1.892\text{ L} && \text{Perform the multiplication.}
\end{aligned}
$$

Thus, the 2-quart bottle contains 1.892 liters. To find the price per liter of the 2-quart bottle, we divide $\frac{\$1.89}{1.892}$.

$$\frac{\$1.89}{1.892} \approx \$0.998942918$$

Since the price per liter of the 2-quart bottle is a little more than 99¢, the 1-liter bottle priced at 97¢ is the better buy.

Self Check 6
Thirty-four fluid ounces of aged vinegar costs \$3.49. A 1-liter bottle of the same vinegar costs \$3.17. Which is the better buy?

Answer the 1-liter bottle

Studying in Other Countries

THINK IT THROUGH

"Over the past decade, the number of U.S. students studying abroad has more than doubled." From *The Open Doors 2003 Report*

In 2001/2002, a record number of 160,920 college students received credit for study abroad. Since students traveling to other countries are almost certain to come into contact with the metric system of measurement, they need to have a basic understanding of metric units.

Suppose a student studying overseas needs to purchase the following school supplies. For each item in red, choose the appropriate metric units.

1. $8\frac{1}{2}$ in. × 11 in. notebook paper:

216 meters × 279 meters　　216 centimeters × 279 centimeters　　216 millimeters × 279 millimeters

2. A backpack that can hold 20 pounds of books:

9 kilograms　　9 grams　　9 milligrams

3. $\frac{3}{4}$ fluid ounce bottle of Liquid Paper correction fluid:

22.5 hectoliters　　2.5 liters　　22.5 milliliters

Comparing American and metric units of temperature

In the American system, we measure temperature using **degrees Fahrenheit** (°F). In the metric system, we measure temperature using **degrees Celsius** (°C). These two scales are shown on the thermometers in Figure B-12. From the figure, we can see that

- 212° F ≈ 100° C   Water boils.
- 32° F ≈ 0° C   Water freezes.
- 5° F ≈ −15° C   A cold winter day
- 95° F ≈ 35° C   A hot summer day

As we have seen, there is a formula that enables us to convert from degrees Fahrenheit to degrees Celsius. There is also a formula to convert from degrees Celsius to degrees Fahrenheit.

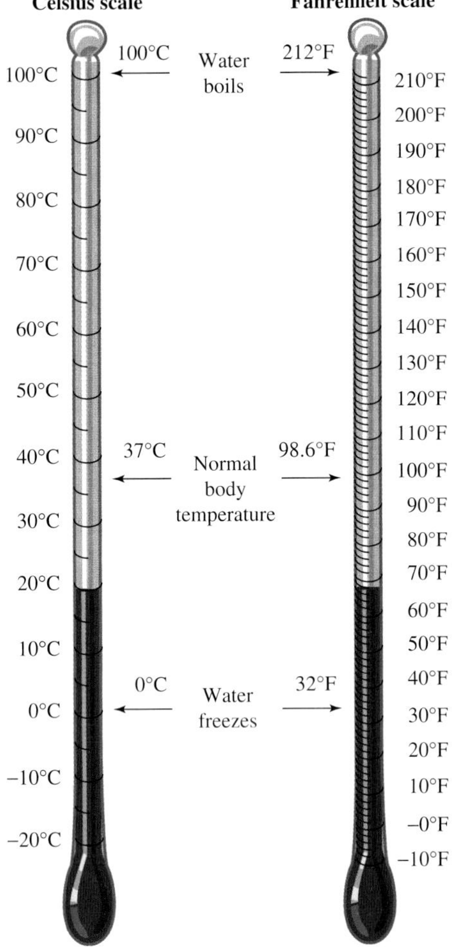

FIGURE B-12

Conversion formulas for temperature

If F is the temperature in degrees Fahrenheit and C is the corresponding temperature in degrees Celsius, then

$$C = \frac{5}{9}(F - 32) \quad \text{and} \quad F = \frac{9}{5}C + 32$$

Self Check 7

Hot coffee is 110° F. To the nearest tenth of a degree, express this temperature in degrees Celsius.

Answer 43.3° C

EXAMPLE 7 **Bathing.** Warm bath water is 90° F. Find the equivalent temperature in degrees Celsius.

Solution We substitute 90 for F in the formula $C = \frac{5}{9}(F - 32)$ and simplify.

$$\begin{aligned} C &= \frac{5}{9}(F - 32) \\ &= \frac{5}{9}(\mathbf{90} - 32) && \text{Substitute 90 for } F. \\ &= \frac{5}{9}(58) && \text{Subtract: } 90 - 32 = 58. \\ &= 32.222\ldots && \text{Perform the arithmetic.} \end{aligned}$$

To the nearest tenth of a degree, the equivalent temperature is 32.2° C.

EXAMPLE 8 A dishwasher manufacturer recommends that dishes be rinsed in hot water with a temperature of 60° C. Express this temperature in degrees Fahrenheit.

Solution We substitute 60 for C in the formula $F = \frac{9}{5}C + 32$ and simplify.

$$F = \frac{9}{5}C + 32$$

$$= \frac{9}{5}(60) + 32 \quad \text{Substitute 60 for } C.$$

$$= \frac{540}{5} + 32 \quad \text{Multiply: } \frac{9}{5}(60) = \frac{540}{5}.$$

$$= 108 + 32 \quad \text{Perform the division.}$$

$$= 140 \quad \text{Perform the addition.}$$

The manufacturer recommends that dishes be rinsed in 140° F water.

Self Check 8
To determine whether a baby has a fever, her mother takes her temperature with a Celsius thermometer. If the reading is 38.8° C, does the baby have a fever? (*Hint:* Normal body temperature is 98.6° F.)

Answer yes

Section B.3 STUDY SET

VOCABULARY *Fill in the blanks.*

1. In the American system, temperatures are measured in degrees ________. In the metric system, temperatures are measured in degrees ________.

2. Inches and centimeters are units used to measure ________. Gallons and liters are units used to measure ________.

CONCEPTS

3. Which is longer?

a. A yard or a meter?

b. A foot or a meter?

c. An inch or a centimeter?

d. A mile or a kilometer?

4. Which is heavier?

a. An ounce or a gram?

b. A pound or a kilogram?

5. Which is the greater unit of capacity?

a. A pint or a liter?

b. A quart or a liter?

c. A gallon or a liter?

6. a. What formula is used for changing degrees Celsius to degrees Fahrenheit?

b. What formula is used for changing degrees Fahrenheit to degrees Celsius?

NOTATION *Complete each solution.*

7. Change 4,500 feet to kilometers.

$$4{,}500 \text{ ft} = 4{,}500(\quad \text{m})$$
$$= \quad \text{m}$$
$$= 1.3716 \text{ km}$$

8. Change 3 kilograms to ounces.

$$3 \text{ kg} = 3(\quad \text{lb})$$
$$= 3(2.2)(\quad \text{oz})$$
$$= 105.6 \text{ oz}$$

9. Change 8 liters to gallons.

$$8 \text{ L} = 8(\quad \text{gal})$$
$$= 2.112 \text{ gal}$$

10. Change 70°C to degrees Fahrenheit.

$$F = \frac{9}{5}C + 32$$
$$= \frac{9}{5}(\quad) + 32$$
$$= \quad + 32$$
$$= 158° \text{ F}$$

PRACTICE *Make each conversion. Since most conversions are approximate, answers will vary depending on the method used.*

11. 3 ft = ____ cm

12. 7.5 yd = ____ m

13. 3.75 m = ____ in.

14. 2.4 km = ____ mi

15. 12 km = ____ ft

16. 3,212 cm = ____ ft

17. 5,000 in. = ____ m
18. 25 mi = ____ km
19. 37 oz = ____ kg
20. 10 lb = ____ kg
21. 25 lb = ____ g
22. 7.5 oz = ____ g
23. 0.5 kg = ____ oz
24. 35 g = ____ lb
25. 17 g = ____ oz
26. 100 kg = ____ lb
27. 3 fl oz = ____ L
28. 2.5 pt = ____ L
29. 7.2 L = ____ fl oz
30. 5 L = ____ qt
31. 0.75 qt = ____ mL
32. 3 pt = ____ mL
33. 500 mL = ____ qt
34. 2,000 mL = ____ gal
35. 50° F = ____ C
36. 67.7° F = ____ C
37. 50° C = ____ F
38. 36.2° C = ____ F
39. −10° C = ____ F
40. −22.5° C = ____ F
41. −5° F = ____ C
42. −10° F = ____ C

APPLICATIONS *Since most conversions are approximate, answers will vary depending on the method used.*

43. THE MIDDLE EAST The distance between Jerusalem and Bethlehem is 8 kilometers. To the nearest mile, give this distance in miles.

44. THE DEAD SEA The Dead Sea is 80 kilometers long. To the nearest mile, give this distance in miles.

45. CHEETAHS A cheetah can run 112 kilometers per hour. Express this speed in mph.

46. LIONS A lion can run 50 mph. Express this speed in kilometers per hour.

47. MOUNT WASHINGTON The highest peak of the White Mountains of New Hampshire is Mount Washington, at 6,288 feet. To the nearest tenth, give this height in kilometers.

48. TRACK AND FIELD Track meets are held on an oval track. One lap around the track is usually 400 meters. However, some older tracks in the United States are 440-yard ovals. Are these two types of tracks the same length? If not, which is longer?

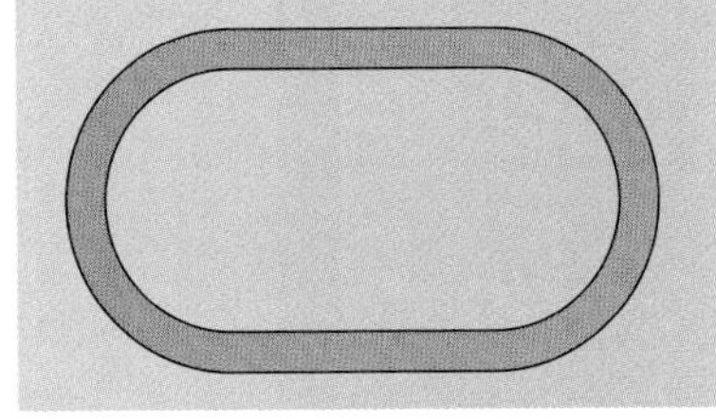

49. HAIR GROWTH When hair is short, its rate of growth averages about $\frac{3}{4}$ inch per month. How many centimeters is this a month?

50. WHALES An adult male killer whale can weigh as much as 12,000 pounds and be as long as 25 feet. Change these measurements to kilograms and meters.

51. WEIGHTLIFTING The table lists the personal best bench press records for two of the world's best powerlifters. Change each metric weight to pounds. Round to the nearest pound.

Name	Hometown	Bench press
Liz Willet	Ferndale, Washington	187 kg
Brian Siders	Charleston, W. Virginia	338 kg

52. WORDS OF WISDOM Refer to the wall hanging. Convert the first metric weight to ounces and the second to pounds. What famous saying results?

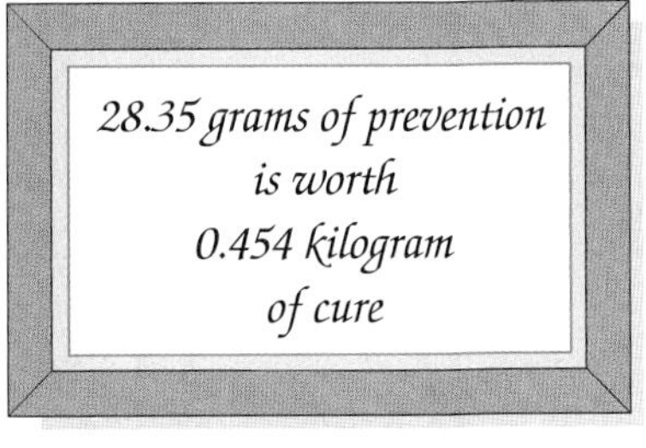
28.35 grams of prevention
is worth
0.454 kilogram
of cure

53. OUNCES AND FLUID OUNCES
   a. There are 310 calories in 8 ounces of broiled chicken. Convert 8 ounces to grams.
   b. There are 112 calories in a glass of fresh Valencia orange juice that holds 8 fluid ounces. Convert 8 fluid ounces to liters.

54. TRACK AND FIELD A shot-put weighs 7.264 kilograms. Give this weight in pounds.

55. POSTAL REGULATIONS You can mail a package weighing up to 70 pounds via priority mail. Can you mail a package that weighs 32 kilograms by priority mail?

56. NUTRITION Refer to the nutrition label for a packet of oatmeal shown. Change each circled weight to ounces.

Nutrition Facts
Serving Size: 1 Packet (46g)
Servings Per Container: 10

Amount Per Serving	
Calories 170	Calories from Fat 20
	% Daily Value
Total fat 2g	**3%**
Saturated fat (0.5g)	**2%**
Polyunsaturated Fat 0.5g	
Monounsaturated Fat 1g	
Cholesterol 0mg	**0%**
Sodium (250mg)	**10%**
Total carbohydrate 35g	**12%**
Dietary fiber 3g	**12%**
Soluble Fiber 1g	
Sugars 16g	
Protein (4g)	

57. COMPARISON SHOPPING Which is the better buy: 3 quarts of root beer for \$4.50 or 2 liters of root beer for \$3.60?

58. COMPARISON SHOPPING Which is the better buy: 3 gallons of antifreeze for \$10.35 or 12 liters of antifreeze for \$10.50?

59. HOT SPRINGS The thermal springs in Hot Springs National Park in central Arkansas emit water as warm as 143° F. Change this temperature to degrees Celsius.

60. COOKING MEAT Meats must be cooked at high enough temperatures to kill harmful bacteria. According to the USDA and the FDA, the internal temperature for cooked roasts and steaks should be at least 145° F, and whole poultry should be 180° F. Convert these temperatures to degrees Celsius. Round up to the next degree.

61. TAKING A SHOWER When you take a shower, which water temperature would you choose: 15° C, 28° C, or 50° C?

62. DRINKING WATER To get a cold drink of water, which temperature would you choose: −2° C, 10° C, or 25° C?

63. SNOWY WEATHER At which temperatures might it snow: −5° C, 0° C, or 10° C?

64. AIR CONDITIONING At which outside temperature would you be likely to run the air conditioner: 15° C, 20° C, or 30° C?

WRITING

65. Explain how to change kilometers to miles.

66. Explain how to change 50° C to degrees Fahrenheit.

67. The United States is the only industrialized country in the world that does not officially use the metric system. Some people claim this is costing American businesses money. Do you think so? Why?

68. What is meant by the phrase *a table of equivalent measures*?

REVIEW *Perform each operation.*

69. $\frac{3}{5} + \frac{4}{3}$

70. $\frac{3}{5} - \frac{4}{3}$

71. $\frac{3}{5} \cdot \frac{4}{3}$

72. $\frac{3}{5} \div \frac{4}{3}$

73. $3.25 + 4.8$

74. $3.25 - 4.8$

75. $3.25 \cdot 4.8$

76. $4.8\overline{)15.6}$

APPENDIX C

Additional Algebraic Topics

C.1 Sets

- Writing sets
- Elements of a set
- Subsets
- Union and intersection of sets
- The complement of a set

When a motorist speaks of purchasing a new *set* of tires or when a child brags about owning a complete *set* of action figures, each is referring to a specific collection of objects. In algebra, we work with sets of numbers. In this section, we will introduce notation that is commonly used with sets. We will also discuss *union* and *intersection,* two important operations that are performed on sets.

Writing sets

A **set** is a collection of objects. The objects that belong to a set are called the **elements** or **members** of the set. One way to specify a set is to list the elements within set **braces** $\{\ \}$. For example, the set containing the elements 1, 2, 3, 4, and 5 is written

$\{1, 2, 3, 4, 5\}$ Elements of a set are separated by commas.

The order in which we list the elements of a set is unimportant. For this reason, the set $\{1, 2, 3, 4, 5\}$ can be written as $\{5, 3, 4, 2, 1\}$ or using any other arrangement of the numbers. Sets that contain the same elements are said to be equal. Thus,

$$\{1, 2, 3, 4, 5\} = \{5, 3, 4, 2, 1\}$$

We often designate sets by using capital letters. For example, we can write

$$A = \{1, 2, 3, 4, 5\}$$

and we say that A is the set consisting of the numbers 1, 2, 3, 4, and 5.

The set $\{1, 2, 3, 4, 5\}$ has five elements. Sets such as this, with a finite (limited) number of elements, are called *finite sets*. A set may also have infinitely many elements. Three examples of *infinite sets* are the set of natural numbers N, the set of whole numbers W, and the set of integers I. When writing these sets, we must list enough elements so that the pattern is apparent. Then we use three dots to indicate that the pattern continues forever.

$$N = \{1, 2, 3, 4, 5, \dots\}$$
$$W = \{0, 1, 2, 3, 4, 5, \dots\}$$
$$I = \{\dots, -3, -2, -1, 0, 1, 2, 3, \dots\}$$

A set may have no elements at all. We call such a set the **empty set** or **null set** and designate it by either of the symbols

$$\{\ \} \quad \text{or} \quad \varnothing$$

The set of all odd numbers that are even is an example of the empty set.

! COMMENT To denote a set with no elements, use { } or $\varnothing$, but not both. It is incorrect to write the empty set as $\{\varnothing\}$ because $\{\varnothing\}$ is not empty. It is a set that contains one element — namely, $\varnothing$.

Elements of a set

To indicate that 3 is an element of the set {1, 2, 3, 4, 5}, we use the symbol $\in$ and we write

$3 \in \{1, 2, 3, 4, 5\}$ Read $\in$ as "is an element (member) of."

Similarly, we can show that 3 is an element of the set of integers by writing $3 \in I$.

To indicate that −7 is not an element of {1, 2, 3, 4, 5}, we draw a slash through the $\in$ symbol.

$7 \notin \{1, 2, 3, 4, 5\}$ Read $\notin$ as "is not an element (member) of."

To show that −7 is not an element of the set of whole numbers, we write $-7 \notin W$.

EXAMPLE 1 Set notation. Place an $\in$ or an $\notin$ in the box to make a true statement.

a. -5 ☐ $\{-9, -6, -3, 0, 3, 6, 9\}$

b. If $C = \left\{\frac{1}{4}, \frac{3}{4}, \frac{5}{4}, \frac{7}{4}, \frac{9}{4}, \frac{11}{4}\right\}$, then $\frac{7}{4}$ ☐ C.

c. 50 ☐ W

Solution

a. Since −5 is not listed in {−9, −6, −3, 0, 3, 6, 9}, we write

$-5 \notin \{-9, -6, -3, 0, 3, 6, 9\}$

b. Since $\frac{7}{4}$ is one of the six elements belonging to set C, we have

$\frac{7}{4} \in C$

c. W denotes the set of whole numbers. Since 50 is a whole number,

$50 \in W$

Self Check 1

Place an $\in$ or an $\notin$ in the box to make a true statement.

a. 9 ☐ {2, 3, 5, 7, 9, 11, 13, 17}

b. If $S = \{-1, 0, 1\}$, then $\frac{1}{2}$ ☐ S.

c. 5.75 ☐ I

Answers **a.** $\in$, **b.** $\notin$, **c.** $\notin$

When discussing sets, it is often helpful to have an all-encompassing set in mind that contains all of the elements of all of the sets under consideration. We call such a set the **universal set** and we use the letter U to represent it. For example, when working with the set of natural numbers, the set of whole numbers, or the set of integers, we normally consider the associated universal set to be the set of real numbers.

Subsets

Since every element of the set of natural numbers is also an element of the whole numbers, we say that the natural numbers are a **subset** of the whole numbers. In symbols, we write this as

$N \subset W$ Read $\subset$ as "is a subset of."

Subset of a set

If every element of set A is also an element of set B, we say that A is a **subset** of B. The notation $A \subset B$ indicates that A is a subset of B.

Since 0 is an element of the whole numbers but it is *not* an element of the natural numbers, the set of whole numbers is not a subset of the natural numbers. To show this, we draw a slash through the $\subset$ symbol.

$W \not\subset N$ Read $\not\subset$ as "is not a subset of."

Self Check 2
Place an $\subset$ or an $\not\subset$ in the box to make a true statement.

a. $\left\{\frac{1}{3}, 3.8\right\} \square \left\{\frac{1}{2}, \frac{1}{3}, \frac{1}{4}, \frac{1}{5}\right\}$

b. $\{-5, 10, -15, 20, -25, \ldots\} \square I$

Answers **a.** $\not\subset$, **b.** $\subset$

EXAMPLE 2 Subsets.

Place an $\subset$ or an $\not\subset$ in the box to make a true statement.

a. $\{2, 4\} \square \{2, 3, 5, 7, 11, 13, 17\}$

b. $I \square W$

c. $\{-4, -8, -16\} \square \{-2, -4, -6, -8, -10, \ldots\}$

d. $N \square N$

Solution

a. $\{2, 4\}$ is *not* a subset of $\{2, 3, 5, 7, 11, 13, 17\}$, because $4 \notin \{2, 3, 5, 7, 11, 13, 17\}$.

$$\{2, 3, 4\} \not\subset \{2, 3, 5, 7, 11, 13, 17\}$$

b. Because many integers (such as the number -1) are not whole numbers,

$$I \not\subset W$$

c. It is apparent from the pattern that -16 is an element of $\{-2, -4, -6, -8, -10, \ldots\}$. Since every element of $\{-4, -8, -16\}$ is an element of $\{-2, -4, -6, -8, -10, \ldots\}$, we have

$$\{-4, -8, -16\} \subset \{-2, -4, -6, -8, -10, \ldots\}$$

d. Obviously, every element of the set of natural numbers is an element of the set of natural numbers. Therefore, $N \subset N$. This illustrates the fact that *every set is a subset of itself.*

! COMMENT The empty set is a subset of every set. This is true because if A is any set, the empty set contains no elements that are not found in A. Therefore, $\varnothing \subset A$.

Illustrations known as **Venn diagrams** can be helpful in showing relationships between sets. For example, the Venn diagram on the right shows that set A is a subset of set B within the context of a universal set U.

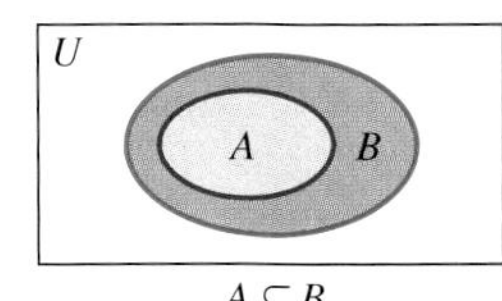

$A \subset B$

Union and intersection of sets

In arithmetic, the operations of addition, subtraction, multiplication, and division take two given numbers and produce a third number. Operations on sets behave in a similar way. Given two sets, these operations produce a third set. To begin the discussion of operations on sets, we will consider the union of two sets.

The Teamsters Union is one of the largest trade organizations in the world. It is the combination of some 700 local unions with members working in such diversified fields as transportation, warehousing, food processing, construction, and corrections. When we find the *union of two sets,* we combine the sets. The operation of union on two sets produces a new third set that is the combination of all of the elements of the two given sets.

Union of two sets

The **union of sets A and B** is the set that contains all of the elements that belong to either set A or set B or both. The notation $A \cup B$, read as "A union B," designates the union of A and B.

EXAMPLE 3 **Union of two sets.** Let $C = \{1, 2, 3\}$ and $D = \{3, 4\}$. Find $C \cup D$.

Solution To form the union of sets C and D, we list the elements of set C and then include any elements from set D that are not already listed.

$\{1, 2, 3\} \cup \{3, 4\} = \{1, 2, 3, 4\}$

Self Check 3
Let $E = \{-2, -1\}$. Find $E \cup W$.

Answer $\{-2, -1, 0, 1, 2, 3, \ldots\}$

One of the most famous street intersections in the country is that of Hollywood and Vine. As you know, an intersection is the place where two streets cross — it's the area that the streets have in common. When we find the *intersection of two sets,* we look for what the two sets have in common. The operation of intersection on two sets produces a new third set that lists all the elements that the two given sets have in common.

Intersection of two sets

The **intersection of sets A and B** is the set that contains only those elements that are in both sets A and B. The notation $A \cap B$, read as "A intersect B," designates the intersection of A and B.

EXAMPLE 4 **Intersection of two sets.** Let $C = \{1, 2, 3\}$, $D = \{3, 4\}$, and $E = \{1, 2, 7, 8, 9\}$. Find **a.** $C \cap D$, **b.** $C \cap E$, and **c.** $D \cap E$.

Solution

a. $C \cap D$ is the set consisting of the elements that are in both $\{1, 2, 3\}$ and $\{3, 4\}$.

$\{1, 2, \mathbf{3}\} \cap \{\mathbf{3}, 4\} = \{3\}$

b. The sets $\{\mathbf{1}, \mathbf{2}, 3\}$ and $\{\mathbf{1}, \mathbf{2}, 7, 8, 9\}$ have the elements 1 and 2 in common. Therefore,

$C \cap E = \{1, 2\}$

c. Since there are no elements common to sets D and E, there are no elements in the intersection. The intersection is the empty set. Two sets such as these, that have no elements in common, are called **disjoint sets.**

$\{3, 4\} \cap \{1, 2, 7, 8, 9\} = \varnothing$

Self Check 4
Let $R = \{-10, -8, 0\}$ and $S = \{-10, 10\}$.
Find **a.** $R \cap S$, **b.** $R \cap N$, and **c.** $I \cap S$.

Answers **a.** $\{-10\}$, **b.** $\varnothing$, **c.** $\{-10, 10\}$

Venn diagrams can be used to illustrate the union and the intersection of sets. The shaded area shown in Figure C-1(a) represents $A \cup B$. The shaded area shown in Figure C-1(b) represents $A \cap B$.

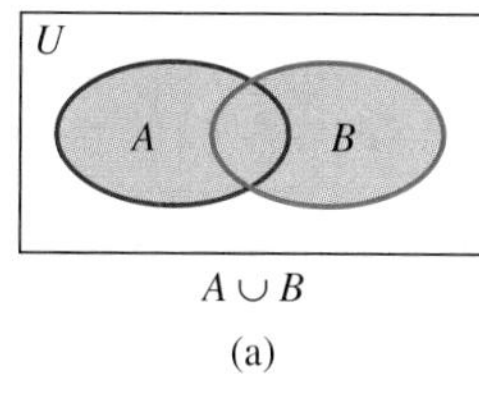

$A \cup B$
(a)

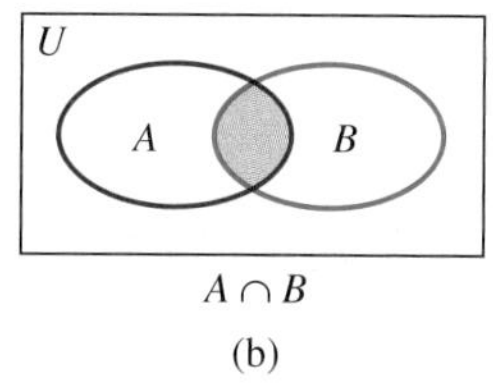

$A \cap B$
(b)

FIGURE C-1

The complement of a set

The manager of a softball team crosses out the names of the players on her team as each one gets in the game. By doing this, at any time during the game, she is aware of who has not yet played. This example illustrates a third operation with sets. It is called finding the **complement** of a set. With this operation, we begin with a set (those who have played in the game) and its associated universal set (all the players on the team). The operation produces a new third set that lists all of the elements of the universal set that are not elements of the given set. In the softball example, that is all those who have not yet played in the game.

Complement of a set

Let U be a universal set of which set A is a subset. The **complement** of set A, denoted A' and read as "A prime," is the set of all elements of U that do not belong to A.

The area shaded in color in the Venn diagram on the right shows the complement of set A with respect to the universal set U.

EXAMPLE 5 **Complement of a set.** If the universal set is $U = \{0, 1, 2, 3, 4, 5, 6, 7, 8, 9, 10\}$ and if $M = \{0, 2, 4, 6, 8, 10\}$, find M'.

Solution The set M' contains the elements of U that are not elements of M. That is,

$$M' = \{1, 3, 5, 7, 9\}$$

Self Check 5
If the universal set is I and if $B = \{-1, -2, -3, -4, -5, \ldots\}$, find B'.

Answer $\{0, 1, 2, 3, 4, 5, \ldots\}$

Section C.1 STUDY SET

VOCABULARY *Fill in the blanks.*

1. A _____ is a collection of objects.
2. The ___________ set does not have any elements.
3. If every element of set S is an element of set K, we say set S is a _______ of set T.
4. The _______ of set S and T is the set that contains all of the elements that belong to either set S or set T or both.
5. The ___________ of set S and T is the set that contains only those elements that are in both sets S and T.
6. The ___________ of set S is the set of all elements of the universal set U that do not belong to S.
7. Two sets that don't ________ are called disjoint sets.
8. A finite set has a limited number of elements; an ________ set does not.

CONCEPTS

9. a. How many elements does the set $\{5, 6, 7, 8, 9\}$ have?
   b. How many elements does the set $\{5, 6, 7, 8, 9, \ldots\}$ have?
10. List the elements of each set within braces.
   a. N
   b. W
   c. I

11. List a subset of {23, 33, 43} that contains
    a. two elements
    b. one element
    c. three elements
    d. no elements
12. Determine whether each statement is true or false.
    a. The set of integers is a subset of the whole numbers.
    b. Every set is a subset of itself.
    c. The empty set is a subset of every set.
    d. {2, 8, 15, 27} = {27, 15, 8, 2}.
13. Fill in the blanks to make a true statement.
    a. The union of two sets is a ____.
    b. The intersection of two sets is a ____.
    c. The complement of a set is a ____.
14. For each problem, copy the Venn diagram below and then shade each set.

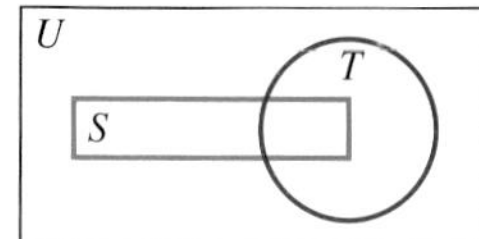

a. $S \cup T$ b. $S \cap T$
c. S' d. T'

15. Determine a set A and a set B such that $A \cap B = \{15\}$ and $A \cup B = \{2, 4, 6, 15\}$.
16. If $H = \{14, 15, 16, 17\}$ and if $H' = \{10, 11, 12, 13\}$ what is U, the universal set which contains H?

NOTATION *Write each expression or statement in words.*

17. $\left\{\frac{1}{2}, \frac{2}{3}, \frac{3}{4}, \frac{4}{5}, \frac{5}{6}, \frac{6}{7}\right\}$
18. $S \subset T$
19. $-1.5 \notin \{-1, -2\}$
20. $\varnothing \subset W$
21. $5 \in S'$
22. $I \not\subset \{\ \}$

PRACTICE *Let $A = \{1, 2, 3\}$ and let $B = \{2, 4, 6\}$. Insert one of the symbols $\in$, $\notin$, $\subset$, or $\not\subset$ in the box to make a true statement.*

23. 1 ☐ A
24. 4 ☐ B
25. {2, 3} ☐ B
26. {2, 3} ☐ A
27. {1} ☐ A
28. {4, 6} ☐ B
29. 3 ☐ A
30. 6 ☐ A
31. A ☐ N
32. B ☐ I
33. $\varnothing$ ☐ A
34. A ☐ A
35. 3 ☐ $\varnothing$
36. { } ☐ B

Let $S = \{-4, -3, -2, -1, 0, 1, 2, 3, 4\}$. Write the subset of S that contains

37. all of the positive numbers in S.
38. all of the negative numbers in S.
39. all of the whole numbers in S.
40. all of the prime numbers in S.
41. all of the numbers less than -3 in S.
42. all of the numbers greater than 4 in S.

Let the universal set be $U = \{-1, 0, 1, 2, 3, 4, 5\}$, $A = \{-1, 3, 4, 5\}$, $B = \{-1, 0, 1\}$, and $C = \{2, 5\}$. Find each set.

43. A'
44. B'
45. C'
46. $\varnothing'$

Let the universal set U be the set of integers, I, and find the complement of each set.

47. {2, 3, 4, 5, . . .}
48. {. . . , −7, −5, −3, −1, 1, 3, 5, 7, . . .}

APPLICATIONS

49. LAW ENFORCEMENT A police report stated that two cars collided in the intersection of First Street and Highland Avenue. In the illustration, shade the area where the accident must have occurred.

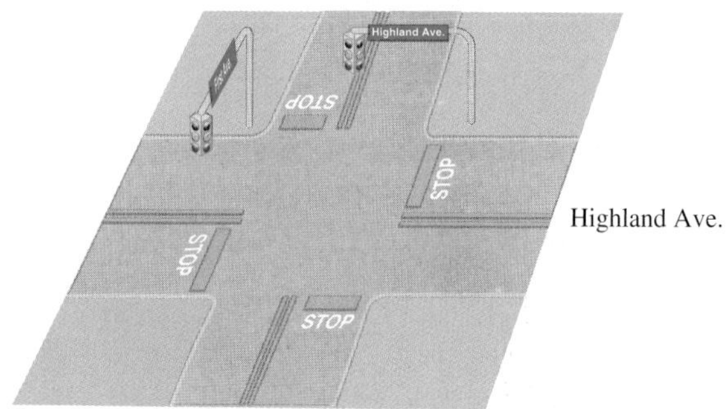

50. AUTO THEFTS The tables on the next page show the three most frequently stolen cars in 1997 and 1998.
    a. Write a set that lists the model/year of the cars that appeared in either list.
    b. Write a set that lists the names of the cars that appeared in both lists.

Cars stolen in 1997		
Rank	Model	Year
1	Camry	89
2	Accord EX	94
3	Camry	90

Cars stolen in 1998		
Rank	Model	Year
1	Camry	89
2	Camry	88
3	Camry	90

51. CELEBRITY EARNINGS The tables below show the highest paid celebrities in 1998 and 1999.

a. Write a set that lists the names of the celebrities that appeared in either list.

b. Write a set that lists the names of the celebrities that appeared in both lists.

1998 Top-earners		1999 Top-earners
Celebrity	Rank	Celebrity
Jerry Seinfeld	1	George Lucas
Larry David	2	Oprah Winfrey
Steven Spielberg	3	Giorgio Armani
Oprah Winfrey	4	David Kelly
James Cameron	5	Tom Hanks

52. GEOGRAPHY Think of the 50 states of the United States as a universal set. What is the complement of the set of states shown in the illustration?

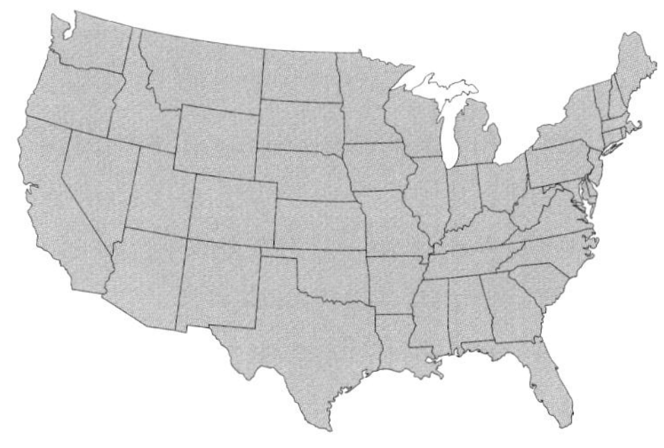

WRITING

53. Explain what is wrong with this statement: $2 \subset \{1, 2, 3\}$.

54. Explain what is wrong with this statement: $\{2\} \in \{1, 2, 3\}$.

55. Explain why the set $\{1, 2, 5, \ldots\}$ is not adequately described.

56. What is a family *reunion?*

57. Explain how to find the union of two sets.

58. A Web site on the Internet was described as:

> An online resource for people who are interested in exploring the *intersection* of new scientific technologies and the American legal system

What do you think this description means?

C.2 Solving Absolute Value Equations and Inequalities

- Absolute value • Equations of the form $|x| = k$ • Equations with two absolute values
- Inequalities of the form $|x| < k$ • Inequalities of the form $|x| > k$

Many quantities in mathematics, science, and engineering are expressed as positive numbers. To guarantee that a quantity is positive, we often use absolute value. In this section, we will work with equations and inequalities involving the absolute value of algebraic expressions. Using the definition of absolute value, we will develop procedures to solve absolute value equations and absolute value inequalities.

Absolute value

Recall that the absolute value of any real number is the distance between the number and zero on the number line. For example, the points shown in Figure C-2 with coordinates of 4 and -4 both lie 4 units from 0. Thus, $|4| = |-4| = 4$.

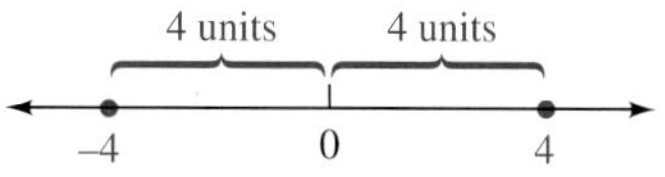

4 units 4 units
-4 0 4

FIGURE C-2

The absolute value of a real number can be defined more formally as follows.

> **Absolute value**
>
> If $x \geq 0$, then $|x| = x$.
>
> If $x < 0$, then $|x| = -x$.

This definition gives a way to associate a nonnegative real number with any real number.

- If $x \geq 0$, then x (which is positive or 0) is its own absolute value.
- If $x < 0$, then $-x$ (which is positive) is the absolute value.

Either way, $|x|$ is positive or 0. That is, $|x| \geq 0$ for all real numbers x.

EXAMPLE 1 Find: **a.** $|9|$, **b.** $|-5.68|$, **c.** $|0|$, and **d.** $2|-8|$.

Solution

a. Since $9 \geq 0$, the number 9 is its own absolute value: $|9| = 9$.

b. Since $-5.68 < 0$, the opposite (negative) of -5.68 is the absolute value:

$$|\mathbf{-5.68}| = -(-5.68) = \mathbf{5.68}$$

c. Since $0 \geq 0$, 0 is its own absolute value: $|0| = 0$.

d. $2|-8| = 2 \cdot |-8|$
$= 2 \cdot 8$
$= 16$

Self Check 1
Find:
a. $|-3|$
b. $|100.99|$
c. $|-2\pi|$
d. $\frac{1}{3}|-6|$

Answers **a.** 3, **b.** 100.99, **c.** 2π, **d.** 2

! COMMENT The placement of a $-$ sign in an expression containing an absolute value symbol is important. For example, $|-19| = 19$, but $-|19| = -19$.

Equations of the form $|x| = k$

The absolute value of a real number represents the distance on a number line from a point to the origin. To solve the **absolute value equation** $|x| = 5$, we must find the coordinates of all points on a number line that are exactly 5 units from zero. See Figure C-3. The only two points that satisfy this condition have coordinates 5 and -5. That is, $x = 5$ or $x = -5$.

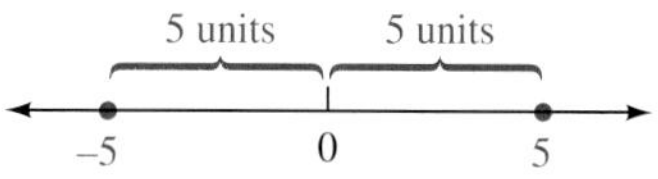

5 units 5 units
-5 0 5

FIGURE C-3

In general, the solution set of the absolute value equation $|x| = k$, where $k \geq 0$, includes the coordinates of the points on the number line that are k units from the origin. (See Figure C-4.)

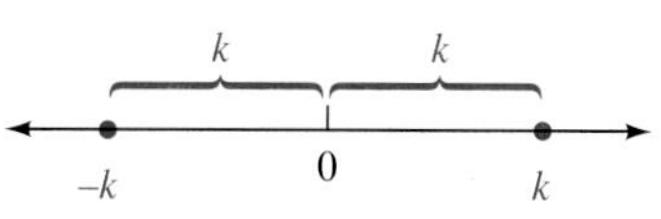

FIGURE C-4

Absolute value equations

If $k \geq 0$, then

$$|x| = k \quad \text{is equivalent to} \quad x = k \text{ or } x = -k$$

Self Check 2
Solve:

a. $|y| = 24$ **b.** $|x| = \frac{1}{2}$

c. $|a| = -1.1$

Answers **a.** 24, −24, **b.** $\frac{1}{2}, -\frac{1}{2}$, **c.** no solution

EXAMPLE 2

Solve: **a.** $|x| = 8$, **b.** $|s| = 0.003$, and **c.** $|c| = -15$.

Solution

a. If $|x| = 8$, then $x = 8$ or $x = -8$.

b. If $|s| = 0.003$, then $s = 0.003$ or $s = -0.003$.

c. The absolute value of a number is either positive or zero but never negative. Therefore, there is no value for c for which $|c| = -15$. This equation has no solution.

The equation $|x - 3| = 7$ indicates that a point on a number line with a coordinate of $x - 3$ is 7 units from the origin. Thus, $x - 3$ can be either 7 or −7.

$$x - 3 = 7 \quad \text{or} \quad x - 3 = -7$$
$$x = 10 \quad \Big| \quad x = -4$$

The solutions of the absolute value equation are 10 and −4. We can graph them on a number line, as shown in Figure C-5. If either of these numbers is substituted for x in $|x - 3| = 7$, the equation is satisfied.

FIGURE C-5

Check:

$$|x - 3| = 7 \qquad |x - 3| = 7$$
$$|\mathbf{10} - 3| \stackrel{?}{=} 7 \qquad |\mathbf{-4} - 3| \stackrel{?}{=} 7$$
$$|7| \stackrel{?}{=} 7 \qquad |-7| \stackrel{?}{=} 7$$
$$7 = 7 \qquad 7 = 7$$

Self Check 3
Solve:

a. $|2x - 3| = 7$

b. $\left|\frac{x}{4} - 1\right| = -3$

Answers **a.** 5, −2, **b.** no solution

EXAMPLE 3

Solve: **a.** $|3x - 2| = 5$ and **b.** $|5 - x| = -10$.

Solution

a. We can solve $|3x - 2| = 5$ by writing and then solving an equivalent compound equation:

$$3x - 2 = 5 \quad \text{or} \quad 3x - 2 = -5$$

Now we solve each equation for x.

$$3x - 2 = 5 \quad \text{or} \quad 3x - 2 = -5$$
$$3x = 7 \quad \Big| \quad 3x = -3$$
$$x = \frac{7}{3} \quad \Big| \quad x = -1$$

Verify that both solutions, $\frac{7}{3}$ and −1, check.

b. For any real number x, $|5 - x|$, will be nonnegative. For this reason, $|5 - x| = -10$ has no solution.

When solving absolute value equations, we want the absolute value isolated on one side. To isolate an absolute value, we can use the equation-solving procedures studied earlier.

EXAMPLE 4 Solve: $\left|\frac{2}{3}x + 3\right| + 4 = 10.$

Self Check 4
Solve: $|0.4x - 2| - 0.6 = 0.4.$

Solution We can isolate $|\frac{2}{3}x + 3|$ on the left-hand side of the equation by subtracting 4 from both sides.

$$\left|\frac{2}{3}x + 3\right| + 4 = 10$$

$$\left|\frac{2}{3}x + 3\right| = 6 \quad \text{Subtract 4 from both sides.}$$

With the absolute value now isolated, we can solve $\left|\frac{2}{3}x + 3\right| = 6$ by writing and then solving an equivalent compound equation.

$$\frac{2}{3}x + 3 = 6 \quad \text{or} \quad \frac{2}{3}x + 3 = -6$$

Now we solve each equation for x.

$$\frac{2}{3}x + 3 = 6 \quad \text{or} \quad \frac{2}{3}x + 3 = -6$$

$$\frac{2}{3}x = 3 \qquad \frac{2}{3}x = -9$$

$$2x = 9 \qquad 2x = -27$$

$$x = \frac{9}{2} \qquad x = -\frac{27}{2}$$

Verify that both solutions, $\frac{9}{2}$ and $-\frac{27}{2}$, check.

Answer 7.5, 2.5

! COMMENT Since the absolute value of a quantity cannot be negative, equations such as $\left|7x + \frac{1}{2}\right| = -4$ have no solution. Their solution sets are empty.

EXAMPLE 5 Solve: $3\left|\frac{1}{2}x - 5\right| - 4 = -4.$

Self Check 5
Solve: $-5\left|\frac{2x}{3} + 4\right| + 1 = 1.$

Solution We first isolate $\left|\frac{1}{2}x - 5\right|$ on the left-hand side.

$$3\left|\frac{1}{2}x - 5\right| - 4 = -4$$

$$3\left|\frac{1}{2}x - 5\right| = 0 \quad \text{Add 4 to both sides.}$$

$$\left|\frac{1}{2}x - 5\right| = 0 \quad \text{Divide both sides by 3.}$$

Since 0 is the only number whose absolute value is 0, the expression $\frac{1}{2}x - 5$ must be 0, and we have

$$\frac{1}{2}x - 5 = 0$$

$$\frac{1}{2}x = 5 \quad \text{Add 5 to both sides.}$$

$$x = 10 \quad \text{Multiply both sides by 2.}$$

Verify that 10 satisfies the original equation.

Answer -6

Equations with two absolute values

The equation $|a| = |b|$ is true when $a = b$ or when $a = -b$. For example,

$$|3| = |3| \qquad \text{or} \qquad |3| = |-3|$$

These are the same number. These numbers are opposites.

In general, the following statement is true.

Equations with two absolute values

If X and Y represent algebraic expressions, the equation $|X| = |Y|$ is equivalent to

$$X = Y \quad \text{or} \quad X = -Y$$

Self Check 6
Solve: $|2x - 3| = |4x + 9|$.

EXAMPLE 6 Solve: $|5x + 3| = |3x + 25|$.

Solution This equation is true when $5x + 3 = 3x + 25$, or when $5x + 3 = -(3x + 25)$. We solve each equation for x.

$$\begin{aligned} 5x + 3 &= 3x + 25 \\ 2x &= 22 \\ x &= 11 \end{aligned} \quad \text{or} \quad \begin{aligned} 5x + 3 &= -(3x + 25) \\ 5x + 3 &= -3x - 25 \\ 8x &= -28 \\ x &= -\frac{28}{8} \\ x &= -\frac{7}{2} \end{aligned}$$

Verify that both solutions, 11 and $-\frac{7}{2}$, check.

Answer $-1, -6$

Inequalities of the form $|x| < k$

To solve the **absolute value inequality** $|x| < 5$, we must find the coordinates of all points on a number line that are less than 5 units from the origin. see Figure C-6. Thus, x is between -5 and 5, and

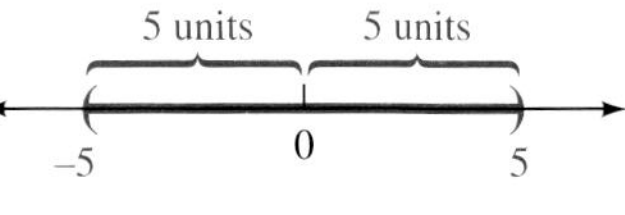

FIGURE C-6

$$|x| < 5 \quad \text{is equivalent to} \quad -5 < x < 5$$

In general, the solution set of the absolute value inequality $|x| < k$ where $k > 0$ includes the coordinates of the points on the number line that are less than k units from the origin. See Figure C-7.

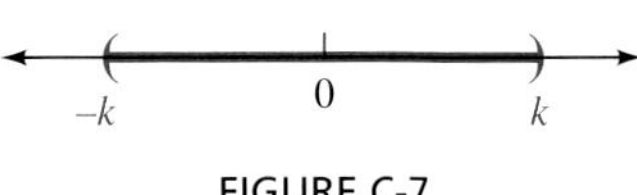

FIGURE C-7

Solving $|x| < k$ and $|x| \leq k$

$\|x\| < k$	is equivalent to	$-k < x < k$	where $k > 0$
$\|x\| \leq k$	is equivalent to	$-k \leq x \leq k$	where $k \geq 0$

EXAMPLE 7 Solve $|2x - 3| < 9$ and graph the solution set.

Solution To solve $|2x - 3| < 9$, we write and then solve an equivalent double inequality.

$$|2x - 3| < 9 \quad \text{is equivalent to} \quad -9 < 2x - 3 < 9$$

Now we solve for x.

$$-9 < 2x - 3 < 9$$
$$-6 < 2x < 12 \quad \text{Add 3 to all three parts.}$$
$$-3 < x < 6 \quad \text{Divide all parts by 2.}$$

Any number between -3 and 6 is in the solution set. This is the interval $(-3, 6)$; its graph is shown in Figure C-8.

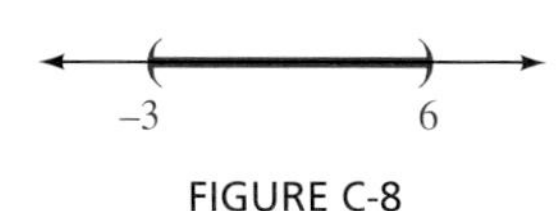

FIGURE C-8

Self Check 7
Solve $|3x + 2| < 4$ and graph the solution set.

Answer $(-2, \frac{2}{3})$

-2 2/3

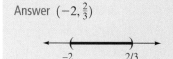

EXAMPLE 8 **Tolerances.**

When manufactured parts are inspected by a quality control engineer, they are classified as acceptable if each dimension falls within a given *tolerance range* of the dimensions listed on the blueprint. For the bracket shown in Figure C-9, the distance between the two drilled holes is given as 2.900 inches. Because the tolerance is ± 0.015 inch, this distance can be as much as 0.015 inch longer or 0.015 inch shorter, and the part will be considered acceptable. The acceptable distance d between holes can be represented by the absolute value inequality $|d - 2.900| \le 0.015$. Solve the inequality and explain the result.

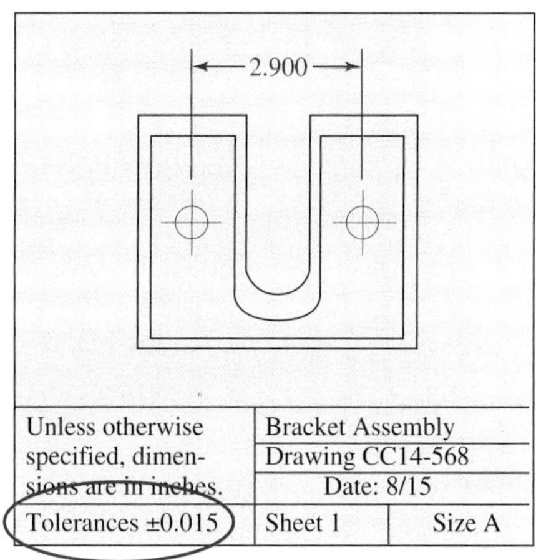

FIGURE C-9

Solution To solve the absolute value inequality, we write and then solve an equivalent double inequality.

$$|d - 2.900| \le 0.015 \quad \text{is equivalent to} \quad -0.015 \le d - 2.900 \le 0.015$$

Now we solve for d.

$$-0.015 \le d - 2.900 \le 0.015$$
$$2.885 \le d \le 2.915 \quad \text{Add 2.900 to all three parts.}$$

The solution set is the interval [2.885, 2.915]. This means that the distance between the two holes should be between 2.885 and 2.915 inches, inclusive. If the distance is less than 2.885 inches or more than 2.915 inches, the part should be rejected.

Inequalities of the form $|x| > k$

To solve the absolute value inequality $|x| > 5$, we must find the coordinates of all points on a number line that are more than 5 units from the origin. See Figure C-10.

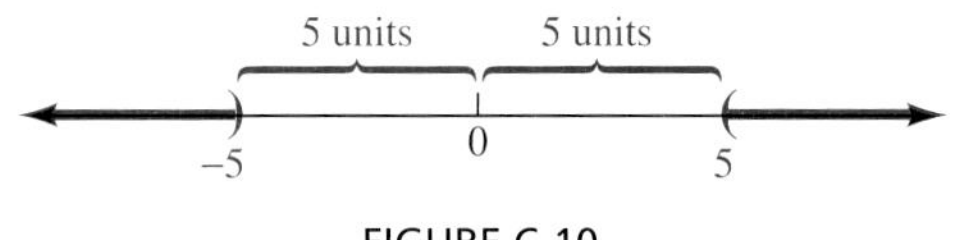

FIGURE C-10

Thus, $x < -5$ or $x > 5$.

In general, the solution set of $|x| > k$ includes the coordinates of the points on the number line that are more than k units from the origin. See Figure C-11. Thus,

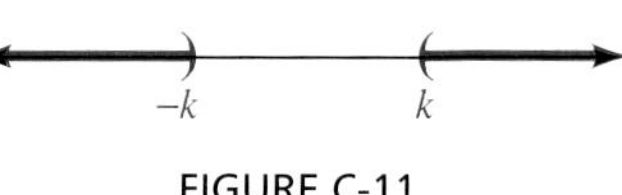

FIGURE C-11

$$|x| > k \quad \text{is equivalent to} \quad x < -k \text{ or } x > k$$

The word *or* indicates an either/or situation. It is only necessary that x satisfy one of the two conditions to be in the solution set.

Solving $|x| > k$ and $|x| \geq k$

If $k \geq 0$, then

$$|x| > k \quad \text{is equivalent to} \quad x < -k \quad \text{or} \quad x > k$$
$$|x| \geq k \quad \text{is equivalent to} \quad x \leq -k \quad \text{or} \quad x \geq k$$

Self Check 9

Solve $\left|\frac{2-x}{4}\right| \geq 1$ and graph the solution set.

Answer $(-\infty, -2] \cup [6, \infty)$

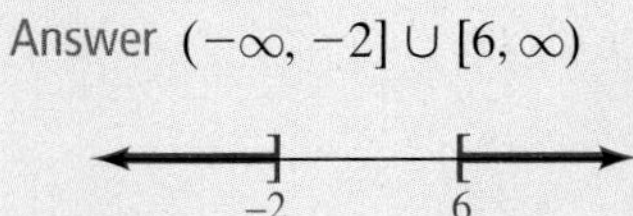

EXAMPLE 9 Solve $\left|\frac{3-x}{5}\right| \geq 6$ and graph the solution set.

Solution To solve $\left|\frac{3-x}{5}\right| \geq 6$, we write and then solve an equivalent compound inequality.

$$\left|\frac{3-x}{5}\right| \geq 6 \quad \text{is equivalent to} \quad \frac{3-x}{5} \leq -6 \quad \text{or} \quad \frac{3-x}{5} \geq 6$$

Now we solve each inequality for x.

$\frac{3-x}{5} \leq -6$	or	$\frac{3-x}{5} \geq 6$	
$3 - x \leq -30$		$3 - x \geq 30$	Multiply both sides by 5.
$-x \leq -33$		$-x \geq 27$	Subtract 3 from both sides.
$x \geq 33$		$x \leq -27$	Divide both sides by −1 and reverse the direction of the inequality symbol.

The solution set is the interval $(-\infty, -27] \cup [33, \infty)$. Its graph appears in Figure C-12.

−27 33

FIGURE C-12

EXAMPLE 10 Solve $\left|\frac{2}{3}x - 2\right| - 3 > 6$ and graph the solution set.

Solution We add 3 to both sides to isolate the absolute value on the left-hand side.

$$\left|\frac{2}{3}x - 2\right| - 3 > 6$$

$$\left|\frac{2}{3}x - 2\right| > 9 \quad \text{Add 3 to both sides to isolate the absolute value.}$$

We then proceed as follows:

$$\frac{2}{3}x - 2 < -9 \quad \text{or} \quad \frac{2}{3}x - 2 > 9$$

$$\frac{2}{3}x < -7 \quad \Big| \quad \frac{2}{3}x > 11 \quad \text{Add 2 to both sides.}$$

$$2x < -21 \quad \Big| \quad 2x > 33 \quad \text{Multiply both sides by 3.}$$

$$x < -\frac{21}{2} \quad \Big| \quad x > \frac{33}{2} \quad \text{Divide both sides by 2.}$$

The solution set is $\left(-\infty, -\frac{21}{2}\right) \cup \left(\frac{33}{2}, \infty\right)$. Its graph appears in Figure C-13.

−21/2 33/2

FIGURE C-13

Self Check 10

Solve $\left|\frac{3}{4}x + 2\right| - 1 > 3$ and graph the solution set.

Answer $(-\infty, -8) \cup \left(\frac{8}{3}, \infty\right)$

−8 8/3

The following summary shows how we can interpret absolute value in three ways. Assume $k > 0$.

Geometric description	Graphic description	Algebraic description
1. $\lvert x\rvert = k$ means that x is k units from 0 on the number line.	−k 0 k	$\lvert x\rvert = k$ is equivalent to $x = k$ or $x = -k$.
2. $\lvert x\rvert < k$ means that x is less than k units from 0 on the number line.	−k 0 k	$\lvert x\rvert < k$ is equivalent to $-k < x < k$.
3. $\lvert x\rvert > k$ means that x is more than k units from 0 on the number line.	−k 0 k	$\lvert x\rvert > k$ is equivalent to $x > k$ or $x < -k$.

Solving absolute value inequalities

CALCULATOR SNAPSHOT

We can also solve absolute value inequalities using a graphing calculator. For example, to solve $|2x - 3| < 9$, we graph the equations $y = |2x - 3|$ and $y = 9$ on the same coordinate system. If we use settings of $[-5, 15]$ for x and $[-5, 15]$ for y, we get the graph shown in Figure C-14.

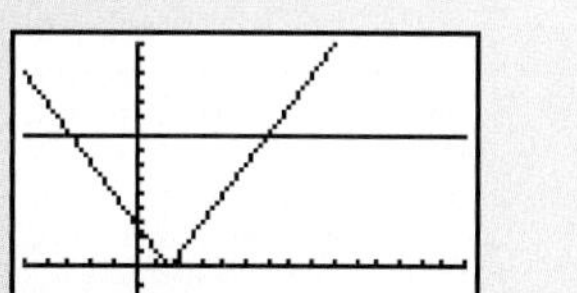

FIGURE C-14

The inequality $|2x - 3| < 9$ will be true for all x-coordinates of points that lie on the graph of $y = |2x - 3|$ and below the graph of $y = 9$. Using the TRACE or INTERSECT feature, we can see that these values of x are in the interval $(-3, 6)$.

Section C.2 STUDY SET

VOCABULARY *Fill in the blanks.*

1. $|2x - 1| = 10$ is an absolute value ________.
2. $|2x - 1| > 10$ is an absolute value ________.
3. To ________ the absolute value in $|3 - x| - 4 = 5$, we add 4 to both sides.
4. $|x| = 2$ is ________ to $x = 2$ or $x = -2$.

CONCEPTS *Fill in the blanks.*

5. $|x| \geq$ ___ for all real numbers x.
6. If $x < 0$, $|x| =$ ___.
7. To solve $|x| > 5$, we must find the coordinates of all points on a number line that are ______ ______ 5 units from 0.
8. To solve $|x| < 5$, we must find the coordinates of all points on a number line that are ______ ______ 5 units from 0.
9. To solve $|x| = 5$, we must find the coordinates of all points on a number line that are ___ units from 0.
10. The equation $|a| = |b|$ is true when $a =$ ___ or when $a =$ ___.
11. Determine whether -3 is a solution of the given equation or inequality.
    a. $|x - 1| = 4$
    b. $|x - 1| > 4$
    c. $|x - 1| \leq 4$
    d. $|5 - x| = |x + 12|$
12. For each absolute value equation or inequality, write an equivalent compound equation or inequality.
    a. $|x| = 8$
    b. $|x| \geq 8$
    c. $|x| \leq 8$
    d. $|5x - 1| = |x + 3|$

NOTATION

13. Match each equation or inequality with its graph.
    a. $|x| = 1$ i.  (graph: open interval from −1 to 1)
    b. $|x| > 1$ ii.  (graph: points at −1 and 1)
    c. $|x| < 1$ iii. (graph: rays left of −1 and right of 1)
14. Match each graph with its corresponding equation or inequality.
    a. (graph: points at −2 and 2) i. $|x| \geq 2$
    b.  (graph: rays left of −2 and right of 2, endpoints included) ii. $|x| \leq 2$
    c.  (graph: closed interval from −2 to 2) iii. $|x| = 2$

Write each compound inequality as an inequality containing absolute value symbols.

15. $-4 < x < 4$
16. $x < -4$ or $x > 4$
17. $x + 3 < -6$ or $x + 3 > 6$
18. $-5 \leq x - 3 \leq 5$

PRACTICE *Find the value of each expression.*

19. $|8|$
20. $|-18|$
21. $-|0.02|$
22. $-|-3.14|$
23. $-\left|-\frac{31}{16}\right|$
24. $-\left|\frac{25}{4}\right|$
25. $|\pi|$
26. $\left|-\frac{\pi}{2}\right|$
27. $5|-5|$
28. $9|-1|$
29. $-\frac{1}{2}|-4|$
30. $-16\left|-\frac{1}{4}\right|$

Solve each equation, if possible.

31. $|x| = 23$
32. $|x| = 90$
33. $|x - 3.1| = 6$
34. $|x + 4.3| = 8.9$
35. $|3x + 2| = 16$
36. $|5x - 3| = 22$
37. $\left|\frac{7}{2}x + 3\right| = -5$
38. $\left|\frac{2x}{3} + 10\right| = 0$
39. $|3 - 4x| = 5$
40. $|8 - 5x| = 18$
41. $2|3x + 24| = 0$
42. $5|x - 21| = -8$
43. $\left|\frac{3x + 48}{3}\right| = 12$
44. $\left|\frac{4x - 64}{4}\right| = 32$
45. $|x + 3| + 7 = 10$
46. $|2 - x| + 3 = 5$
47. $8 = -1 + |0.3x - 3|$
48. $-1 = 1 - |0.1x + 8|$

49. $|2x + 1| = |3x + 3|$

50. $|5x - 7| = |4x + 1|$

51. $|2 - x| = |3x + 2|$

52. $|4x + 3| = |9 - 2x|$

53. $\left|\frac{x}{2} + 2\right| = \left|\frac{x}{2} - 2\right|$

54. $|7x + 12| = |x - 6|$

55. $\left|x + \frac{1}{3}\right| = |x - 3|$

56. $\left|x - \frac{1}{4}\right| = |x + 4|$

Solve each inequality if possible. Write the solution set in interval notation and graph it.

57. $|x| < 4$

58. $|x| < 9$

59. $|x + 9| \le 12$

60. $|x - 8| \le 12$

61. $|3x - 2| < 10$

62. $|4 - 3x| \le 13$

63. $|3x + 2| \le -3$

64. $|5x - 12| < -5$

65. $|x| > 3$

66. $|x| > 7$

67. $|x - 12| > 24$

68. $|x + 5| \ge 7$

69. $|3x + 2| > 14$

70. $|2x - 5| > 25$

71. $|4x + 3| > -5$

72. $|7x + 2| > -8$

73. $|2 - 3x| \ge 8$

74. $|-1 - 2x| > 5$

75. $-|2x - 3| < -7$

76. $-|3x + 1| < -8$

77. $\left|\frac{x - 2}{3}\right| \le 4$

78. $\left|\frac{x - 2}{3}\right| > 4$

79. $|3x + 1| + 2 < 6$

80. $1 + \left|\frac{1}{7}x + 1\right| \le 1$

81. $\left|\frac{1}{3}x + 7\right| + 5 > 6$

82. $-2|3x - 4| < 16$

83. $|0.5x + 1| + 2 \le 0$

84. $15 \ge 7 - |1.4x + 9|$

APPLICATIONS

85. TEMPERATURE RANGES The temperatures on a summer day satisfied the inequality $|t - 78°| \le 8°$, where t is a temperature in degrees Fahrenheit. Solve this inequality and express the range of temperatures as a double inequality.

86. OPERATING TEMPERATURES A car CD player has an operating temperature of $|t - 40°| < 80°$, where t is a temperature in degrees Fahrenheit. Solve the inequality and express this range of temperatures as an interval.

87. AUTO MECHANICS On most cars, the bottoms of the front wheels are closer together than the tops, creating a *camber angle*. This lessens road shock to the steering system. The specifications for a certain car state that the camber angle c of its wheels should be $0.6° \pm 0.5°$.

a. Express the range with an inequality containing absolute value symbols.

b. Solve the inequality and express this range of camber angles as an interval.

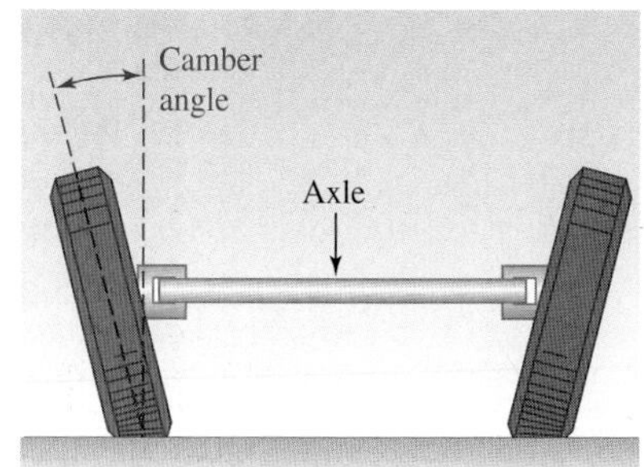

88. STEEL PRODUCTION A sheet of steel is to be 0.250 inch thick with a tolerance of 0.025 inch.

a. Express this specification with an inequality containing absolute value symbols, using x to represent the thickness of a sheet of steel.

b. Solve the inequality and express the range of thickness as an interval.

89. ERROR ANALYSIS In a lab, students measured the percent of copper p in a sample of copper sulfate. The students know that copper sulfate is actually 25.46% copper by mass. They are to compare their results to the actual value and find the amount of *experimental error.*

a. Which measurements shown satisfy the absolute value inequality $|p - 25.46| \leq 1.00$?

b. What can be said about the amount of error for each of the trials listed in part a?

Lab 4 Section A
Title:
"Percent copper (Cu) in copper sulfate ($CuSO_4 \cdot 5H_2O$)"

Results

	% Copper
Trial #1:	22.91%
Trial #2:	26.45%
Trial #3:	26.49%
Trial #4:	24.76%

90. ERROR ANALYSIS See Exercise 89.

a. Which measurements satisfy the absolute value inequality $|p - 25.46| > 1.00$?

b. What can be said about the amount of error for each of the trials listed in part a?

WRITING

91. Explain how to find the absolute value of a given number.

92. Explain why the equation $|x - 4| = -5$ has no solutions.

93. Explain the use of parentheses and brackets when graphing inequalities.

94. Explain the differences between the solution set of $|x| < 8$ and the solution set of $|x| > 8$.

95. Explain how to use the graph to solve $|x - 2| < 3$.

96. Explain how to use the graph to solve $|x - 2| \geq 3$.

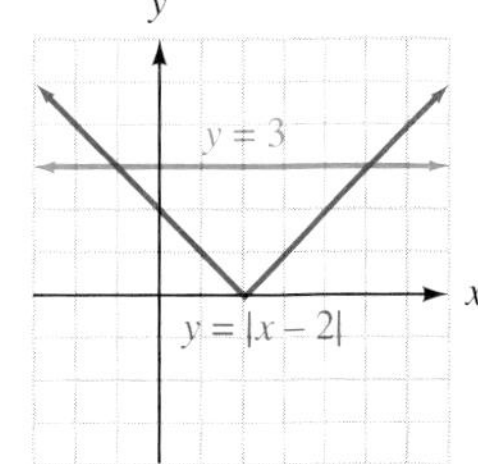

REVIEW

97. RAILROAD CROSSINGS The warning sign in the illustration is to be painted on the street in front of a railroad crossing. If y is 30° more than twice x, find x and y.

98. GEOMETRY Refer to the illustration. What is $2x + 2y$?

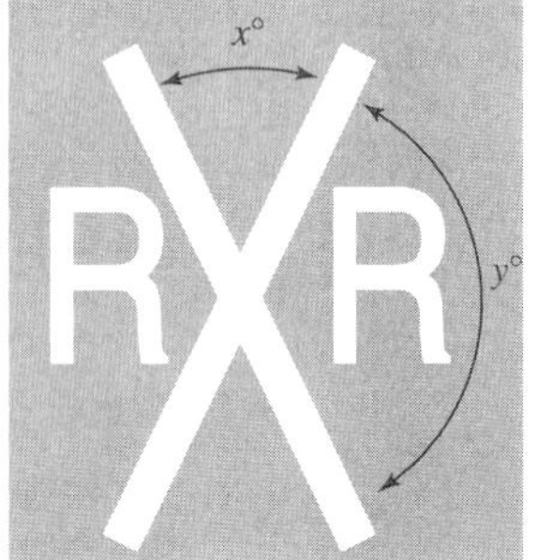

C.3 Quadratic and Other Nonlinear Inequalities

- Solving quadratic inequalities
- Solving rational inequalities
- Graphs of nonlinear inequalities in two variables

If $a \neq 0$, inequalities of the form $ax^2 + bx + c < 0$ and $ax^2 + bx + c > 0$ are called *quadratic inequalities in one variable.* We will begin this section by showing how to solve them by making a sign chart. We will then show how to solve other nonlinear inequalities using the same technique. To conclude, we will show how to find graphical solutions of nonlinear inequalities in two variables.

Solving quadratic inequalities

To solve the inequality $x^2 + x - 6 < 0$, we must find the values of x that make the inequality true. This can be done using a number line. We begin by factoring the trinomial to obtain

$$(x + 3)(x - 2) < 0$$

Since the product of $x + 3$ and $x - 2$ must be less than 0, the values of $x + 3$ and $x - 2$ must be opposite in sign. To find the intervals where this is true, we keep track of their signs by constructing the chart in Figure C-15. The chart shows that

- $x - 2$ is 0 when $x = 2$, is positive when $x > 2$ (indicated with + signs in the figure), and is negative when $x < 2$ (indicated with − signs in the figure).
- $x + 3$ is 0 when $x = -3$, is positive when $x > -3$, and is negative when $x < -3$.

The only place where the values of the binomials are opposite in sign is in the interval $(-3, 2)$. Therefore, the solution set of the inequality can be denoted as

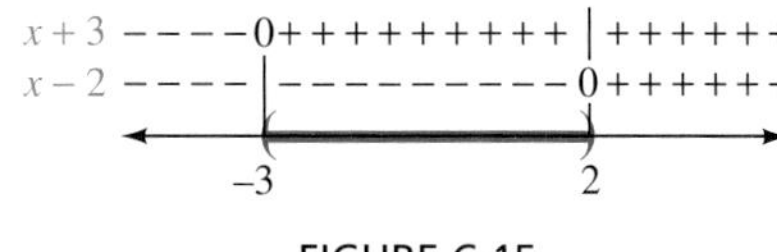

FIGURE C-15

$$-3 < x < 2$$

The graph of the solution set is shown on the number line in Figure C-15.

EXAMPLE 1 Solve: $x^2 + 2x - 3 \geq 0$.

Solution We factor the trinomial to get $(x - 1)(x + 3)$ and construct a sign chart, as in Figure C-16.

- $x - 1$ is 0 when $x = 1$, is positive when $x > 1$, and is negative when $x < 1$.
- $x + 3$ is 0 when $x = -3$, is positive when $x > -3$, and is negative when $x < -3$.

The product of $x - 1$ and $x + 3$ will be greater than 0 when the signs of the binomial factors are the same. This occurs in the intervals $(-\infty, -3)$ and $(1, \infty)$. The numbers -3 and 1 are also included, because they make the product equal to 0. Thus, the solution set is

$$(-\infty, -3] \cup [1, \infty) \qquad \text{or} \qquad x \leq -3 \text{ or } x \geq 1$$

The graph of the solution set is shown on the number line in Figure C-16.

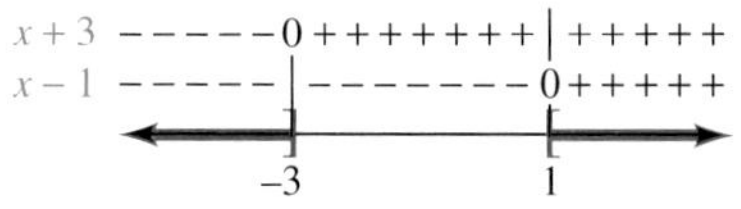

x + 3 -----0+++++++|+++++
x − 1 -----|-------0+++++
−3 1

FIGURE C-16

Self Check 1
Solve $x^2 + 2x - 15 > 0$ and graph the solution set.

Answer $(-\infty, -5) \cup (3, \infty)$

−5 3

Solving rational inequalities

Making a sign chart is useful for solving many inequalities that are neither linear nor quadratic. In the next three examples, we will use a sign chart to solve *rational inequalities*.

Self Check 2

Solve: $\frac{3}{x} > 5$.

Answer $\left(0, \frac{3}{5}\right)$

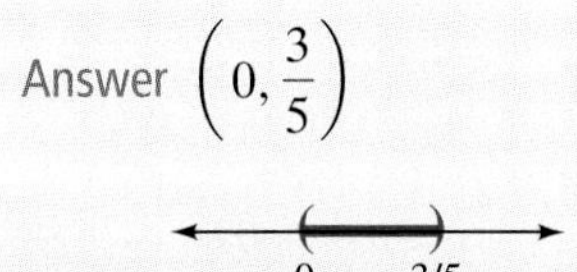

EXAMPLE 2 Solve: $\frac{1}{x} < 6$.

Solution We subtract 6 from both sides to make the right-hand side equal to 0. We then find a common denominator and subtract:

$$\frac{1}{x} < 6$$

$$\frac{1}{x} - 6 < 0 \quad \text{Subtract 6 from both sides.}$$

$$\frac{1}{x} - \frac{6x}{x} < 0 \quad \text{Get a common denominator.}$$

$$\frac{1-6x}{x} < 0 \quad \text{Subtract the numerators and keep the common denominator.}$$

We now make a sign chart, as shown in Figure C-17.

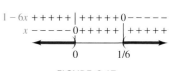

- The denominator x is 0 when $x = 0$, is positive when $x > 0$, and is negative when $x < 0$.
- The numerator $1 - 6x$ is 0 when $x = \frac{1}{6}$, is positive when $x < \frac{1}{6}$, and is negative when $x > \frac{1}{6}$.

1 − 6x + + + + + | + + + + + 0 − − − −
x − − − − − 0 + + + + + | + + + + +
0 1/6

FIGURE C-17

The fraction $\frac{1-6x}{x}$ will be less than 0 when the numerator and denominator are opposite in sign. This occurs in the interval

$$(-\infty, 0) \cup \left(\frac{1}{6}, \infty\right) \quad \text{or} \quad x < 0 \text{ or } x > \frac{1}{6}$$

The graph of this interval is shown in Figure C-17.

! COMMENT Since we don't know whether x is positive, 0, or negative, multiplying both sides of the inequality $\frac{1}{x} < 6$ by x is a three-case situation:

- If $x > 0$, then $1 < 6x$.
- If $x = 0$, then $\frac{1}{x}$ is undefined.
- If $x < 0$, then $1 > 6x$.

If we multiply both sides by x and solve the linear inequality $1 < 6x$, we are considering only one case and will get only part of the answer.

Self Check 3

Solve

$$\frac{x+2}{x^2-2x-3} > 0$$

and graph the solution set.

EXAMPLE 3 Solve: $\frac{x^2-3x+2}{x-3} \geq 0$.

Solution We write the fraction with the numerator in factored form.

$$\frac{(x-2)(x-1)}{x-3} \geq 0$$

To keep track of the signs of $x - 2$, $x - 1$, and $x - 3$, we construct the sign chart shown in Figure C-18. The fraction will be positive in the intervals where all factors are positive, or where exactly two factors are negative. The numbers 1 and 2 are included, because

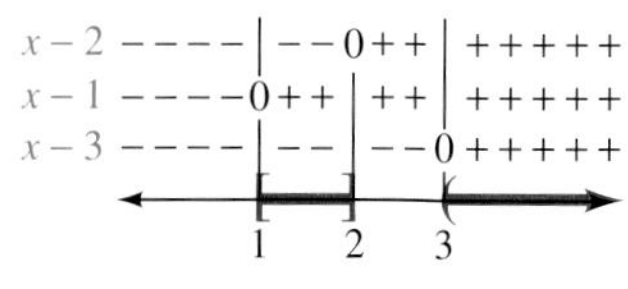

FIGURE C-18

they make the numerator (and thus the fraction) equal to 0. The number 3 is not included, because it gives a 0 in the denominator.

The solution is the interval $[1, 2] \cup (3, \infty)$. The graph appears in Figure C-18.

Answer $(-2, -1) \cup (3, \infty)$

EXAMPLE 4 Solve: $\dfrac{3}{x-1} < \dfrac{2}{x}$.

Solution We subtract $\frac{2}{x}$ from both sides to get 0 on the right-hand side and proceed as follows:

$$\frac{3}{x-1} < \frac{2}{x}$$

$$\frac{3}{x-1} - \frac{2}{x} < 0 \qquad \text{Subtract } \frac{2}{x} \text{ from both sides.}$$

$$\frac{3x}{(x-1)x} - \frac{2(x-1)}{x(x-1)} < 0 \qquad \text{Get a common denominator.}$$

$$\frac{3x - 2x + 2}{x(x-1)} < 0 \qquad \text{Keep the denominator and subtract the numerators.}$$

$$\frac{x+2}{x(x-1)} < 0 \qquad \text{Combine like terms.}$$

We can keep track of the signs of $x + 2$, x, and $x - 1$ with the sign chart shown in Figure C-19. The fraction will be negative in the intervals with either one or three negative factors. The numbers 0 and 1 are not included, because they give a 0 in the denominator, and the number -2 is not included, because it does not satisfy the inequality.

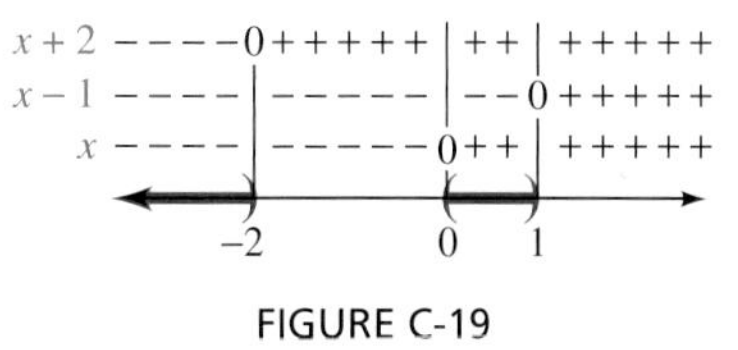

FIGURE C-19

The solution is the interval $(-\infty, -2) \cup (0, 1)$, as shown in Figure C-19.

Self Check 4
Solve

$$\frac{2}{x+1} > \frac{1}{x}$$

and graph the solution set.

Answer $(-1, 0) \cup (1, \infty)$

Solving inequalities graphically

CALCULATOR SNAPSHOT

To approximate the solutions of $x^2 + 2x - 3 \geq 0$ (Example 1) by graphing, we can use the standard window settings of $[-10, 10]$ for x and $[-10, 10]$ for y and graph the quadratic function $y = x^2 + 2x - 3$, as in Figure C-20. The solution of the inequality will be those numbers x for which the graph of $y = x^2 + 2x - 3$ lies above or on the x-axis. We can trace to find that this interval is $(-\infty, -3] \cup [1, \infty)$.

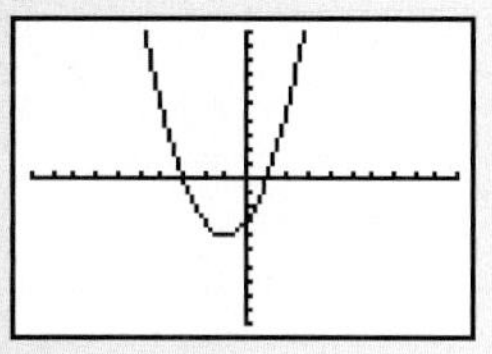

FIGURE C-20

To approximate the solutions of $\dfrac{3}{x-1} < \dfrac{2}{x}$ (Example 4), we first write the inequality in the form

$$\frac{x+2}{x(x-1)} < 0$$

We use window settings of $[-5, 5]$ for x and $[-3, 3]$ for y and graph the function $y = \dfrac{x+2}{x(x-1)}$, as in Figure C-21(a). The solution of the inequality will be those numbers x for which the graph lies below the x-axis.

We can trace to see that the graph is below the x-axis when x is less than -2. Since we cannot see the graph in the interval $0 < x < 1$, we redraw the graph using window settings of $[-1, 2]$ for x and $[-25, 10]$ for y. See Figure C-21(b).

We can now see that the graph is below the x-axis in the interval $(0, 1)$. Thus, the solution of the inequality is the union of two intervals: $(-\infty, -2) \cup (0, 1)$.

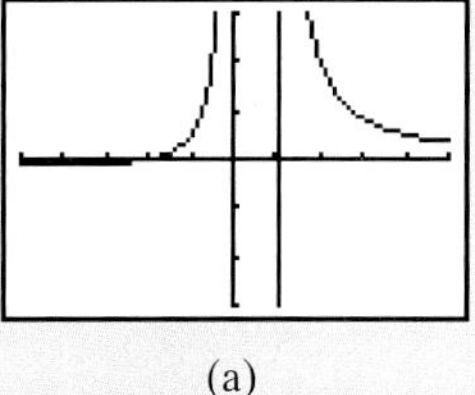
(a)

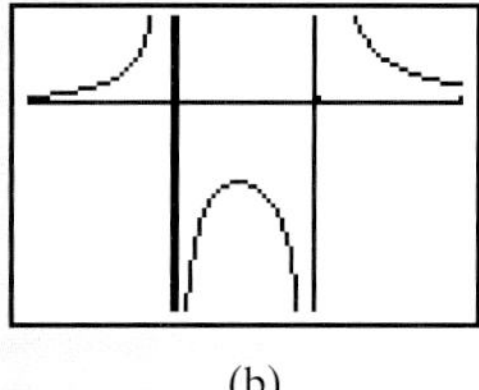
(b)

FIGURE C-21

Graphs of nonlinear inequalities in two variables

We now consider the graphs of nonlinear inequalities in two variables.

Self Check 5
Graph: $y \geq -x^2 + 4$.

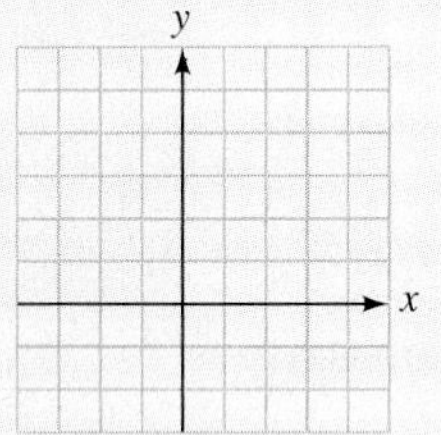

Answer

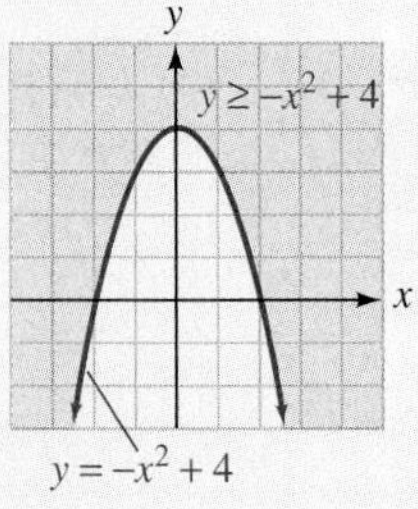

EXAMPLE 5 Graph: $y < -x^2 + 4$.

Solution The graph of $y = -x^2 + 4$ is the parabolic boundary separating the region representing $y < -x^2 + 4$ and the region representing $y > -x^2 + 4$.

We graph the quadratic function $y = -x^2 + 4$ as a dashed parabola, because equality is not permitted. Since the coordinates of the origin satisfy the inequality $y < -x^2 + 4$, the test point $(0, 0)$ is in the graph. The complete graph is shown in Figure C-22.

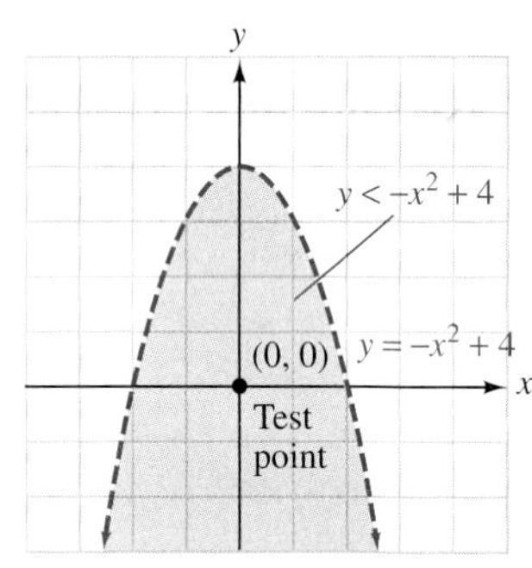

FIGURE C-22

Self Check 6
Graph: $x \geq -|y|$.

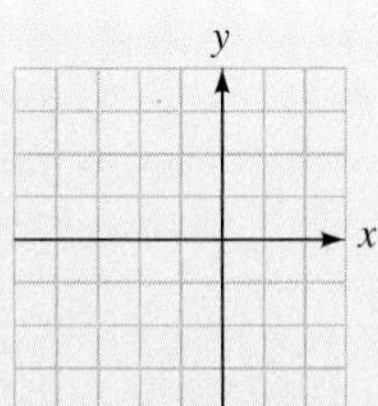

EXAMPLE 6 Graph: $x \leq |y|$.

Solution We first graph $x = |y|$ as in Figure C-23(a). We use a solid line, because equality is permitted. Because the origin is on the graph, we cannot use the origin as a test point. However, any other point, such as $(1, 0)$, will do. We substitute 1 for x and 0 for y into the inequality to get

$$\mathbf{x} \leq |y|$$
$$\mathbf{1} \leq |\mathbf{0}|$$
$$1 \leq 0$$

Since $1 \leq 0$ is a false statement, the test point (1, 0) does not satisfy the inequality and is not part of the graph. Thus, the graph of $x \leq |y|$ is to the left of the boundary.

The complete graph and the test point are shown in Figure C-23(b).

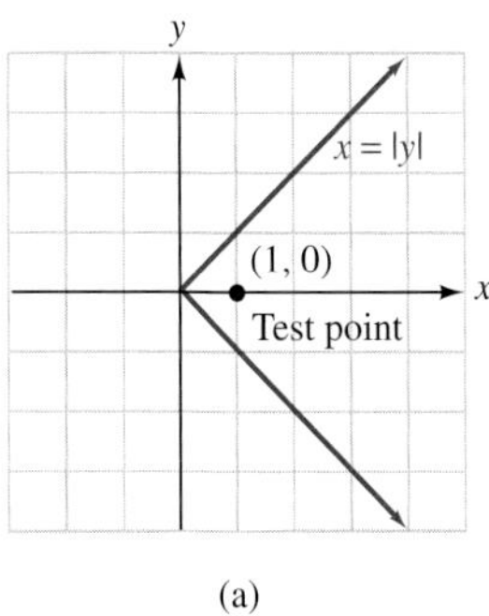

(a)

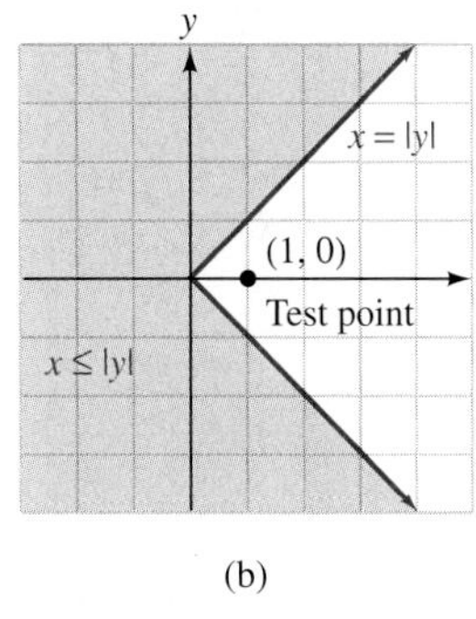

(b)

FIGURE C-23

Answer

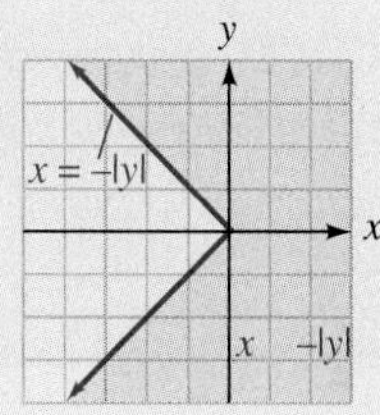

Section C.3 STUDY SET

VOCABULARY *Fill in the blanks.*

1. Any inequality of the form $ax^2 + bx + c > 0$ where $a \neq 0$ is called a ________ inequality.
2. The inequality $y < x^2 - 2x + 3$ is a nonlinear inequality in _____ variables.
3. The ________ (3, 5) represents the real numbers between 3 and 5.
4. To decide which side of the boundary to shade when solving inequalities in two variables, we pick a _____ point.

CONCEPTS *Fill in the blanks.*

5. When $x > 3$, the binomial $x - 3$ is ________ than zero.
6. When $x < 3$, the binomial $x - 3$ is _______ than zero.
7. If $x = 0$, the fraction $\frac{1}{x}$ is ________.
8. To keep track of the signs of factors in a product or quotient, we can use a ______ chart.
9. Estimate the solution of the inequality $x^2 - x - 6 > 0$ using the graph of $y = x^2 - x - 6$, shown on the right.

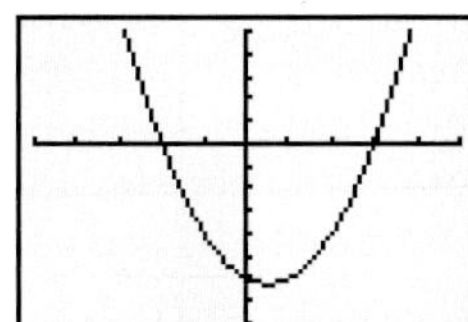

10. Estimate the solution of the inequality $\frac{x-3}{x} \leq 0$ using the graph of $y = \frac{x-3}{x}$, shown below.

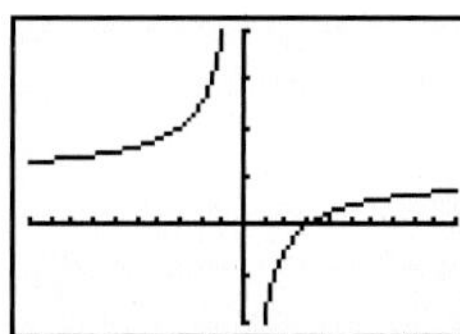

NOTATION *Consider the inequality* $(x + 2)(x - 3) > 0$

11. Since the product of $x + 2$ and $x - 3$ is positive, the values of $x + 2$ and $x - 3$ are both positive or both negative. Fill in each blank with a number.
    a. $x + 2 = 0$ when $x =$
    b. $x + 2 > 0$ when $x >$
    c. $x + 2 < 0$ when $x <$
    d. $x - 3 = 0$ when $x =$
    e. $x - 3 > 0$ when $x >$
    f. $x - 3 < 0$ when $x <$
12. Use the information in Exercise 11 to make a sign chart of the data and graph the solution set.

13. Match each inequality with the corresponding interval notation.

a. $-4 \le x < 5$ **i.** $(-\infty, \infty)$

b. $x \le -4$ or $x > 5$ **ii.** $[-4, \infty)$

c. $x \ge -4$ **iii.** $[-4, 5)$

d. $x < 5$ or $x \ge -4$ **iv.** $(-\infty, -4] \cup (5, \infty)$

14. What is the meaning of the symbol $\cup$?

PRACTICE *Solve each inequality. Give each result in interval notation and graph the solution set.*

15. $x^2 - 5x + 4 < 0$

16. $x^2 - 3x - 4 > 0$

17. $x^2 - 8x + 15 > 0$

18. $x^2 + 2x - 8 < 0$

19. $x^2 + x - 12 \le 0$

20. $x^2 - 8x \le -15$

21. $x^2 + 8x < -16$

22. $x^2 + 6x \ge -9$

23. $x^2 \ge 9$

24. $x^2 \ge 16$

25. $2x^2 - 50 < 0$

26. $3x^2 - 243 < 0$

27. $\frac{1}{x} < 2$

28. $\frac{1}{x} > 3$

29. $-\frac{5}{x} < 3$

30. $\frac{4}{x} \ge 8$

31. $\frac{x^2 - x - 12}{x - 1} < 0$

32. $\frac{x^2 + x - 6}{x - 4} \ge 0$

33. $\frac{6x^2 - 5x + 1}{2x + 1} > 0$

34. $\frac{6x^2 + 11x + 3}{3x - 1} < 0$

35. $\frac{3}{x - 2} < \frac{4}{x}$

36. $\frac{-6}{x + 1} \ge \frac{1}{x}$

37. $\frac{7}{x - 3} \ge \frac{2}{x + 4}$

38. $\frac{-5}{x - 4} < \frac{3}{x + 1}$

39. $\frac{x}{x + 4} \le \frac{1}{x + 1}$

40. $\frac{x}{x + 9} \ge \frac{1}{x + 1}$

41. $(x + 2)^2 > 0$

42. $(x - 3)^2 < 0$

Use a graphing calculator to solve each inequality. Give the answer in interval notation.

43. $x^2 - 2x - 3 < 0$

44. $x^2 + x - 6 > 0$

45. $\frac{x + 3}{x - 2} > 0$

46. $\frac{3}{x} < 2$

Graph each inequality.

47. $y < x^2 + 1$

48. $y > x^2 - 3$

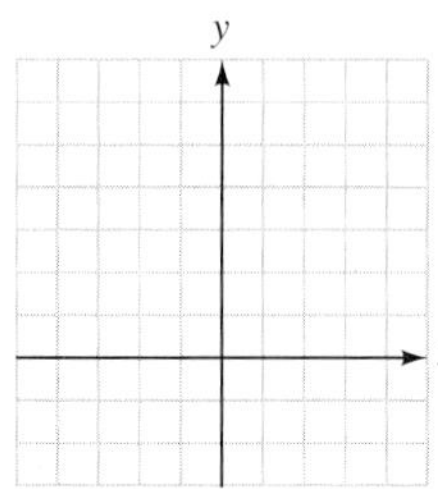

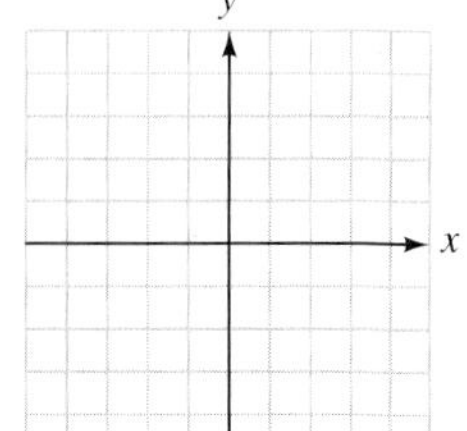

49. $y \leq x^2 + 5x + 6$

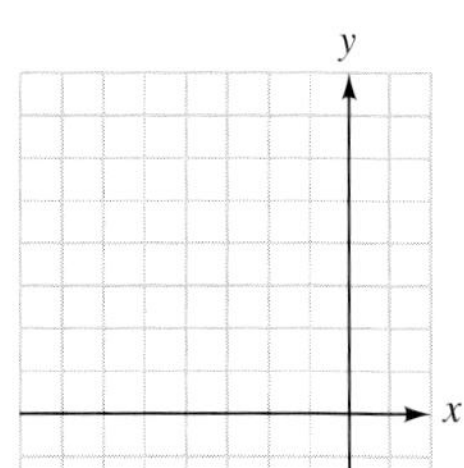

50. $y \geq x^2 + 5x + 4$

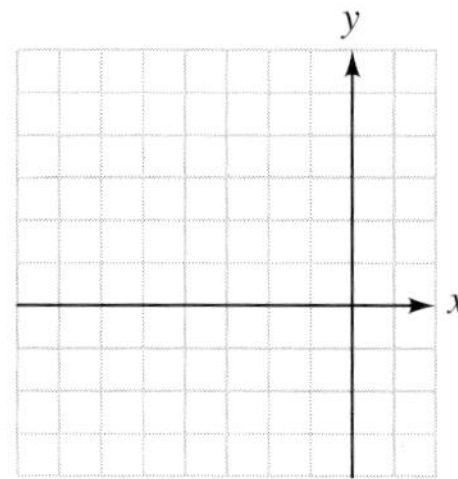

51. $y < |x + 4|$

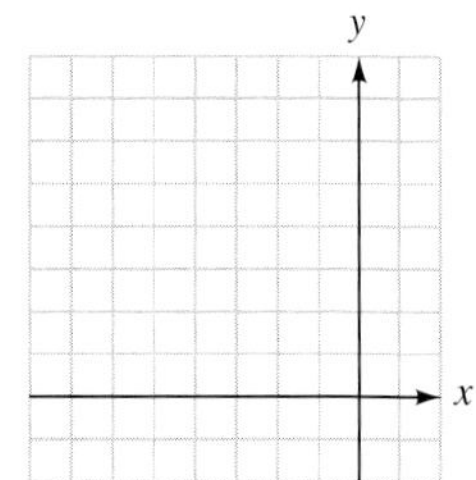

52. $y \geq |x - 3|$

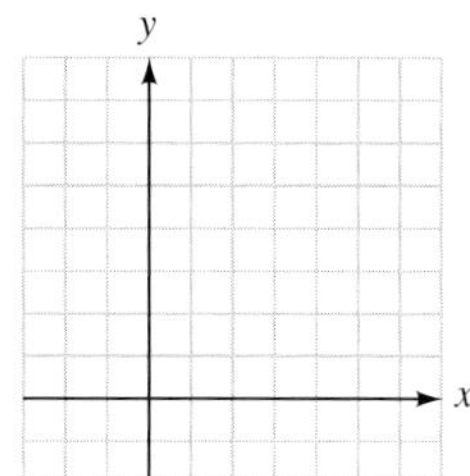

53. $y \leq -|x| + 2$

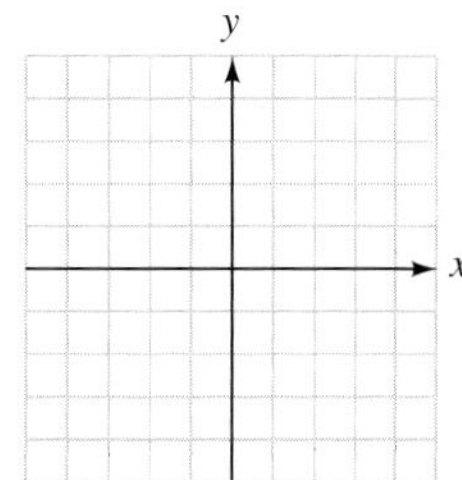

54. $y > |x| - 2$

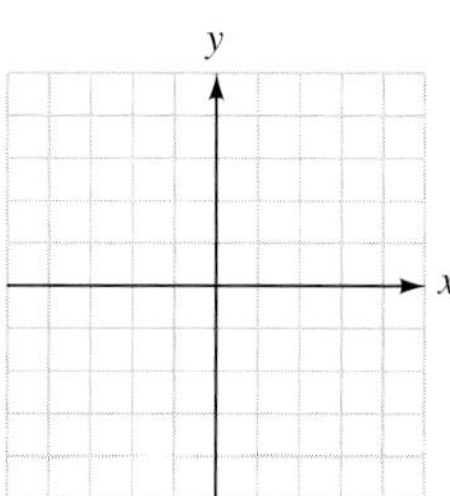

APPLICATIONS

55. SUSPENSION BRIDGES If an x-axis is superimposed over the roadway of the Golden Gate Bridge, with the origin at the center of the bridge as shown in the next column, the length L in feet of a vertical support cable can be approximated by the formula

$$L = \frac{1}{9{,}000}x^2 + 5$$

For the Golden Gate Bridge, $-2{,}100 < x < 2{,}100$. For what intervals along the x-axis are the vertical cables more than 95 feet long?

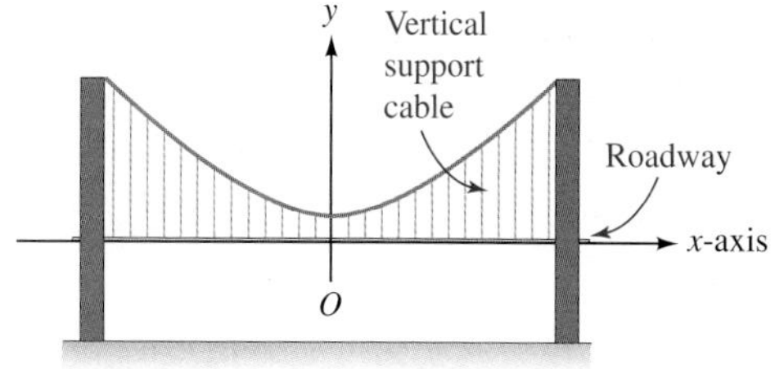

56. MALLS The number of people n in a mall is modeled by the formula

$$n = -100x^2 + 1{,}200x$$

where x is the number of hours since the mall opened. If the mall opened at 9 A.M., when were there 2,000 or more people in it?

WRITING

57. Explain why $(x - 4)(x + 5)$ will be positive only when the signs of $x - 4$ and $x + 5$ are the same.

58. Explain how to find the graph of $y \geq x^2$.

59. The graph of $f(x) = x^2 - 3x + 4$ is shown. Explain why the quadratic inequality $x^2 - 3x + 4 < 0$ has no solution.

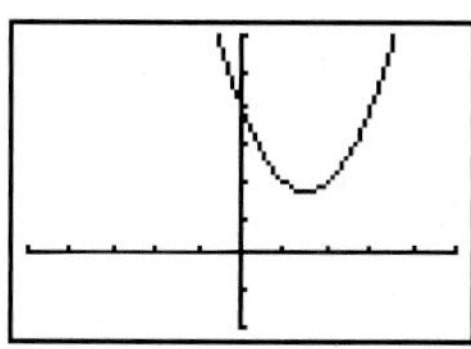

60. What three important facts about the expression $x - 1$ are indicated below?

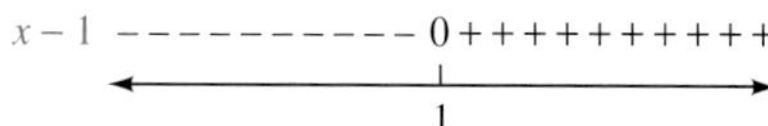

REVIEW *Translate each statement into an equation.*

61. x varies directly with y.

62. y varies inversely with t.

63. t varies jointly with x and y.

64. d varies directly with t and inversely with u^2.

Find the slope of the graph of each linear function.

65. $f(x)= 3x - 4$

66. $f(x) = -x$

C.4 Probability

If we toss a coin, it can land in one of two equally likely ways — either heads or tails. Because one of these two outcomes is heads, we say that the probability of obtaining heads in a single toss is $\frac{1}{2}$. If records show that out of 100 days with weather conditions like today's, 30 have received rain, we say that there is a $\frac{30}{100}$ or 30% probability of rain today. If there are two chances in three that a basketball player will make a free throw, we say that the probability the player will make a free throw is $\frac{2}{3}$.

In this appendix, we will introduce the concept of probability.

Probability

An **experiment** is any process for which the outcome is uncertain. For example, some experiments are

- Tossing a coin
- Rolling dice
- Drawing a card
- Taking a colored marble from a jar

For any experiment, the set of all possible outcomes is called a **sample space.** The sample space for the experiment of tossing a coin twice is the following set of four ordered pairs.

$\{(H, H), (H, T), (T, H), (T, T)\}$ The pair (H, T), for example, represents the outcome "heads on the first coin and tails on the second coin."

An **event** is any set of outcomes of an experiment. For example, if E is the event "getting at least one heads" in the experiment of tossing a coin twice, the event E is the following set of three ordered pairs:

$$E = \{(H, H), (H, T), (T, H)\}$$

Because the outcome of getting at least one heads can occur in 3 ways out of a total of 4 possible ways, we say that the **probability** of event E is $\frac{3}{4}$. In symbols, we write

$$P(E) = P(\text{at least one heads}) = \frac{3}{4}$$ Read $P(E)$ as "the probability of event E."

We can define the probability of an event as follows.

Probability of an event

If E is an event that can occur in n ways out of s possible, equally likely, ways, the **probability of E** is

$$P(E) = \frac{n}{s}$$

Because $0 \le n \le s$, it follows that $0 \le \frac{n}{s} \le 1$. This implies that all probabilities have values from 0 to 1. An event that cannot happen has a probability of 0. An event that is certain to happen has a probability of 1.

A **die** (the singular of *dice*) is a cube with six faces, each containing a number of dots from one to six. See Figure C-24.

FIGURE C-24

EXAMPLE 1 Find the probability of rolling a six on one roll of a fair die.

Solution Since a fair die has 6 faces that are equally likely to appear, and there is 1 way to get a six, we have

$$P(6) = \frac{1}{6}$$

Self Check 1

Find the probability of rolling a 2 or a 5 on one roll of a fair die.

Answer: $\frac{1}{3}$

EXAMPLE 2 Show the sample space of the experiment "rolling two dice one time."

Solution We can list ordered pairs, letting the first number be the result on the first die and the second number the result on the second die. The sample space will be the set containing the following 36 ordered pairs.

$$(1, 1)\ (1, 2)\ (1, 3)\ (1, 4)\ (1, 5)\ (1, 6)$$
$$(2, 1)\ (2, 2)\ (2, 3)\ (2, 4)\ (2, 5)\ (2, 6)$$
$$(3, 1)\ (3, 2)\ (3, 3)\ (3, 4)\ (3, 5)\ (3, 6)$$
$$(4, 1)\ (4, 2)\ (4, 3)\ (4, 4)\ (4, 5)\ (4, 6)$$
$$(5, 1)\ (5, 2)\ (5, 3)\ (5, 4)\ (5, 5)\ (5, 6)$$
$$(6, 1)\ (6, 2)\ (6, 3)\ (6, 4)\ (6, 5)\ (6, 6)$$

Self Check 2
How many pairs in the sample space have a sum of 4?

Answer: 3

EXAMPLE 3 Find the probability of the event "rolling a sum of 7 on one roll of two dice."

Solution The sample space is listed in Example 2. We let E be the set of outcomes that give a sum of 7:

$$E = \{(1, 6), (2, 5), (3, 4), (4, 3), (5, 2), (6, 1)\}$$

Since there are 6 ways to roll a 7 among the 36 equally likely outcomes, we have

$$P(\text{rolling a 7}) = \frac{6}{36} = \frac{1}{6} \quad \text{Simplify the fraction.}$$

Self Check 3
Find the probability of rolling a sum of 4.

Answer: $\frac{1}{12}$

A standard deck of 52 playing cards has two red suits (hearts and diamonds) and two black suits (clubs and spades). Each suit has 13 cards, including a king, a queen, and a jack (called **face cards**), an ace, and cards numbered from 2 to 10. In the next example, we refer to a standard deck of cards.

EXAMPLE 4 Find the probability of drawing an ace from a well-shuffled deck of 52 cards.

Solution Since there are 4 aces in a deck of 52 cards, and each card is equally likely to be drawn, the probability of drawing an ace is

$$P(\text{ace}) = \frac{4}{52} = \frac{1}{13}$$

Self Check 4
Find the probability of drawing a diamond from a well-shuffled deck of 52 cards.

Answer: $\frac{1}{4}$

Section C.4 STUDY SET

VOCABULARY *Fill in the blanks.*

1. An __________ is any process for which the outcome is uncertain.
2. A list of all possible outcomes for an experiment is called a ________ ________.

CONCEPTS *Fill in the blanks.*

3. If an event E can occur in n ways out of s equally likely ways, then $P(E) =$ ____.
4. The probability of any event is always a number from ____ to ____, including ____ and ____.

5. If an event cannot happen, its probability is ▒.
6. If an event is certain to happen, its probability is ▒.

PRACTICE *List the sample space of each experiment.*

7. Rolling a die and tossing a coin
8. Tossing three coins
9. Selecting a letter of the alphabet that is a vowel
10. Guessing a one-digit number

A fair die is rolled once. Find the probability of each event.

11. Rolling a 2
12. Rolling a number greater than 4
13. Rolling a number larger than 1 but less than 6
14. Rolling a number that is an odd number

Two fair dice are rolled twice. Find the probability of each event.

15. Rolling a 6
16. Rolling a 10
17. Rolling a 13
18. Rolling a number from 1 to 13

Balls numbered from 1 to 42 are placed in a jar and stirred. If one is drawn at random, find the probability of each result.

19. The number is less than 20.
20. The number is less than 50.
21. The number is a prime number.
22. The number is less than 10 or greater than 40.

Refer to the spinner in the illustration. If the spinner is spun once, find the probability of each event. Assume that the spinner never stops on a line.

23. The spinner stops on red.
24. The spinner stops on green.
25. The spinner stops on orange.
26. The spinner stops on yellow.

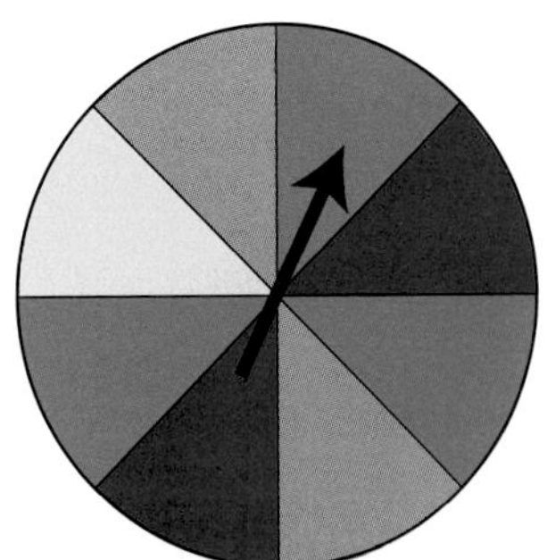

Find the probability of each event.

27. Drawing a black card from a standard card deck
28. Drawing a diamond on one draw from a standard card deck
29. Drawing a face card on one draw from a standard card deck
30. Drawing a jack or a king on one draw from a standard card deck
31. Drawing a red egg from a basket containing 5 red eggs and 7 blue eggs
32. Drawing an orange cube from a bowl containing 5 orange cubes and 1 beige cube

APPLICATIONS *An aircraft has 4 engines. Assume that the probability of an engine failing during a test is $\frac{1}{2}$.*

33. Find the sample space.
34. Find the probability that all engines will survive the test.
35. Find the probability that exactly 1 engine will survive.
36. Find the probability that exactly 2 engines will survive.
37. Find the probability that exactly 3 engines will survive.
38. Find the probability that no engines will survive.
39. Find the sum of the probabilities in Exercises 34 through 38. What do you discover?

A survey of 282 people is taken to determine the opinions of doctors, teachers, and lawyers on a proposed piece of legislation, with the results shown in the table. A person is chosen at random from those surveyed. Refer to the table to find each probability.

	Number that favor	Number that oppose	Number with no opinion	Total
Doctors	70	32	17	119
Teachers	83	24	10	117
Lawyers	23	15	8	46
Total	176	71	35	282

40. The person favors the legislation.
41. A doctor opposes the legislation.
42. A person who opposes the legislation is a lawyer.

C.4 Applied problems involving more than one step

In real life, it is often necessary to use several steps when solving applied problems. Whenever we face a complicated problem, it is a good strategy to break down the problem into smaller problems.

EXAMPLE 1 To landscape his yard, a man bought 8 trees at \$150 each and some bushes at \$36 each. The nursery charged \$35 per hour to plant the trees and bushes and took 7 hours to complete the job. How many bushes were purchased if the total cost was \$2,309?

Solution

Analyze the problem

To solve this problem, we must find two things:

1. The cost of planting of the trees and bushes
2. The number of bushes purchased

Part 1 Form an equation

If we let c = the cost of planting, we can use the following equation to find the cost.

The cost of planting	equals	the cost per hour	times	the number of hours.
c	$=$	35	$\cdot$	7

Solve the equation

$c = 35 \cdot 7$ This is the equation to solve.

$c = 245$ Multiply.

State the conclusion

The cost of planting was \$245.

Part 2 Form an equation

If we let n = the number of bushes purchased, we can use the following equation to find the number of bushes purchased.

The total cost of the job	equals	the cost of planting	plus	the cost of a tree	times	the number of trees	plus	the cost of a bush	times	the number of bushes.
2,309	$=$	245	$+$	150	$\cdot$	8	$+$	36	$\cdot$	n

Solve the equation

$2{,}309 = 245 + 150 \cdot 8 + 36n$ This is the equation to solve.

$2{,}309 = 245 + 1{,}200 + 36n$ Perform the multiplication.

$2{,}309 = 1{,}445 + 36n$ Perform the addition.

$2{,}309 - \mathbf{1{,}445} = 1{,}445 - \mathbf{1{,}445} + 36n$ Subtract 1,445 from both sides.

$864 = 36n$ Simplify.

$\frac{864}{\mathbf{36}} = \frac{36n}{\mathbf{36}}$ Divide both sides by 36.

$24 = n$ Simplify.

State the conclusion

The number of bushes purchased was 24.

Check the result

The trees cost $8 \cdot 150 = \$1{,}200$.
The bushes cost $36 \cdot 24 = \$864$.
The installation cost \$245.
The total cost is $\$1{,}200 + \$864 + \$245 = \$2{,}309$.
The result checks.

EXAMPLE 2 Last year, a woman hosted Thanksgiving dinner and fed 12 people with a 16-pound turkey. This year, she intends to host 21 people and can buy turkey on sale for \$1.29 per pound. If she had the right amount of turkey last year, how much should she expect to spend for turkey this year?

Solution

Analyze the problem

To solve this problem, we must find three things:

1. The amount of turkey needed per person
2. The size of the turkey that will be needed
3. The cost of the turkey

Part 1 Form an equation

Since the woman had the right amount of turkey last year, we can assume that the same portions will be needed this year. If we let a = the amount of turkey needed per person, we can use the following equation to find the amount.

The amount of turkey needed per person	equals	size of turkey last year / number of people fed last year.
a	$=$	$\frac{16}{12}$

Solve the equation

$$a = \frac{16}{12} \quad \text{This is the equation to solve.}$$

$$= \frac{4}{3} \quad \text{Simplify.}$$

State the conclusion

The woman needs $\frac{4}{3}$ pounds of turkey per person.

Part 2 Form an equation

To find the size of the turkey, we can use the following equation:

The size of the turkey	equals	the number of people to be fed	times	the number of pounds required for each person.
s	$=$	21	$\cdot$	$\frac{4}{3}$

Solve the equation

$s = 21 \cdot \frac{4}{3}$

$s = 28$ Perform the multiplication.

State the conclusion

She needs to buy a 28-pound turkey.

Part 3 Form an equation

If we let c = the cost of the turkey, we can use the following equation to find the cost.

The cost of the turkey	equals	the number of pounds needed	times	the price per pound.
c	$=$	28	$\cdot$	1.29

Solve the equation

$c = 28 \cdot 1.29$

$c = 36.12$ Perform the multiplication.

State the conclusion

The cost of the turkey will be \$36.12.

Check the result

For a cost of \$36.12, the woman can buy a 28-pound turkey: $\frac{36.12}{1.29} = 28$.

With a 28-pound turkey, the woman can feed 21 people: $\frac{36.12}{\frac{4}{3}} = 21$.

The answer checks.

Section C.5 STUDY SET

1. To move from an apartment to a house, a couple rented a truck for \$79.95 plus 32¢ per mile. They also purchased 12 medium cardboard boxes at \$1.75 each and 15 large boxes at \$2.15 each. If they drove 107 miles, how much did the move cost?

2. A salesperson is paid \$10.50 per hour plus a commission of 3% on his sales. Find his total pay during a week when he worked 40 hours and had sales of \$12,500.

3. A salesperson is paid a monthly salary of \$2,100 plus a 4% commission on all sales over \$47,000. If her salary for the month was \$2,350, find her total sales.

4. A cell phone company charges \$30 per month for the first 200 minutes of airtime, plus 30¢ per minute for all minutes over 200. Find the bill for a customer who used 430 minutes one month.

5. A homeowner hired a painter to stain his summer home on a time and materials basis. The painter agreed to charge \$20 per hour for labor. If the painter used 12 gallons of stain at \$21.95 per gallon and worked for 24 hours, how much did the job cost?

6. Three students spent a day on the lake water skiing. They rented a boat for the day at a cost of \$140. During the day, they used 11 gallons of gas. After the day was over, they split the costs equally, and each one's share was \$55.32. What was the cost of the gasoline?

7. A rug is 12 feet long. The owner wants to have the perimeter of the rug bound so it won't fray. The cost of binding is \$2.50 per foot. If the total cost of binding was \$110, how wide is the rug?

8. A student purchased 3 shirts at \$23 each, 2 ties at \$12 each, and a pair of slacks at \$25. If the total bill was \$125.08, find the sales tax rate.

9. To play a round of golf, a student paid \$12 for greens fees and \$10 to rent a cart. He also bought 3 golf balls for \$6.95 and a package of tees for \$1.25. He had to pay sales tax on the golf balls and tees. If the total cost of the round was \$30.61, find the sales tax rate.

10. A person bought a car for \$17,500 plus 7% sales tax. In addition, he also had to pay \$9.50 for a title transfer, \$125 for license plates, and a document fee. If the total cost of the car was \$18,867, what was the document fee?

11. A family of four went to the movies. Each person needed a ticket costing \$7.50, had a box of popcorn costing \$3.95, and had a medium drink costing \$2.95. They drove 27 miles roundtrip to the theater in a van that gets 18 miles per gallon. If gasoline costs \$2.20 a gallon, how much did the outing cost?

12. An investor bought two \$50,000 certificates of deposit (CDs), one paying 3% and the other paying 4%. He also bought a long-term CD paying 5%. If his income from the CDs for the first year was \$7,250, how much did he invest in the long-term CD?

APPENDIX D

THEA PRACTICE TEST

1. On Thursday, a concert was held at the Coliseum for which 700 tickets were sold. $\frac{1}{20}$ of those tickets were never picked up from the box office. Of the tickets that remained, $\frac{4}{5}$ were picked up the night of the performance. How many tickets were picked up on Thursday?

   a. 672 tickets b. 532 tickets
   c. 168 tickets d. 35 tickets

2. In a soft-drink manufacturing plant, one machine can fill 12 bottles every 2 seconds. How many bottles can be filled by one machine in 2 hours?

   a. 43,200 bottles b. 1,440 bottles
   c. 86,400 bottles d. 720 bottles

3. Luke got a part-time job as a cashier starting at $6.50 per hour. After 90 days his work was evaluated and he received a 2% raise per hour. In another 6 months he will receive an additional 7% raise. What will be Luke's hourly wage after both raises?

   a. $6.59/hr b. $6.64/hr
   c. $7.09/hr d. $6.96/hr

4. If 5 gallons of gas costs $12.00, how much will it cost to fill up a car that holds 18 gallons?

   a. $40.00 b. $43.00
   c. $43.20 d. $42.30

5. In math class, major tests are counted twice and daily grades are counted once. Elizabeth currently has 92, 88, and 100 as her daily scores. What is the lowest score she can make on the upcoming test and still maintain at least a 90 average?

   a. 90 b. 85
   c. 93 d. 80

6. A new technology stock went on sale on Monday for $4 per share. On Tuesday, the stock price rose $1, on Wednesday the stock price rose another $2, then on both Thursday and Friday the price dropped $1. What was the price of the stock per share at the end of the day on Friday?

   a. $6 b. $4
   c. $5 d. $9

7. Bill and Joan both work full time. Together their salaries total $3,200 per month. According to their budget plan, how much have they allocated for housing and utilities each month?

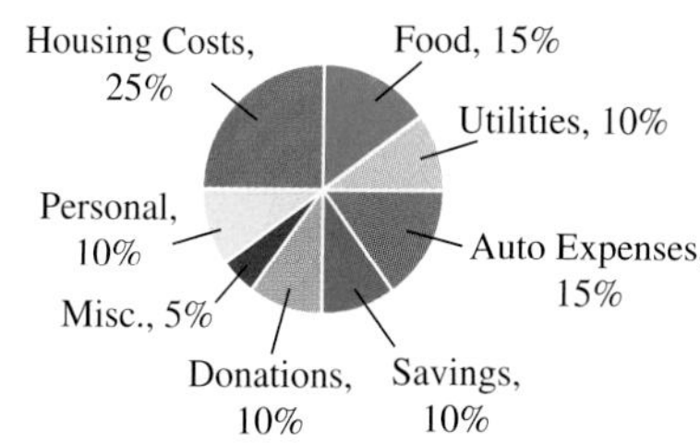

   a. $800 b. $320
   c. $2,080 d. $1,120

8. Light travels at a speed of 300,000 km per second. Express this number in scientific notation.

   a. 3.0×10^6 b. 3.0×10^5
   c. 0.3×10^6 d. 3.0×10^{-5}

9. Simplify: $(3.9 \times 10^{-2})(4.7 \times 10^6)$.

   a. 18.33×10^8 b. 1.833×10^4
   c. 1.833×10^5 d. 1833×10^4

10. Mortgage interest rates have remained at an all-time low for a number of years. However, the lending rates for home loans have recently gradually begun to increase. If time is plotted on the x-axis and the loan rates on the y-axis, which of the following graphs represents mortgage lending rates?

a. 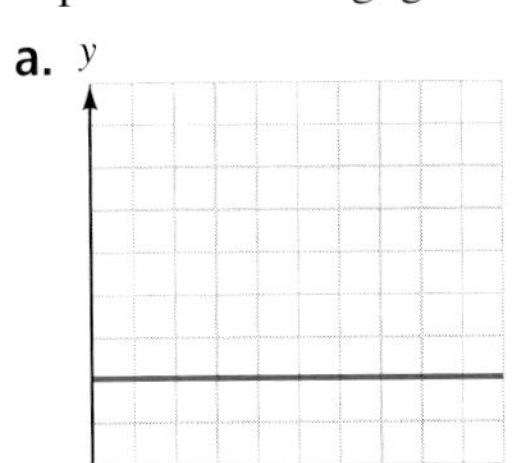

b. 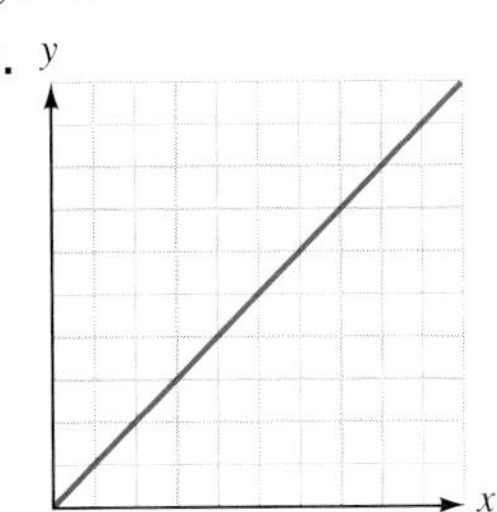

c. 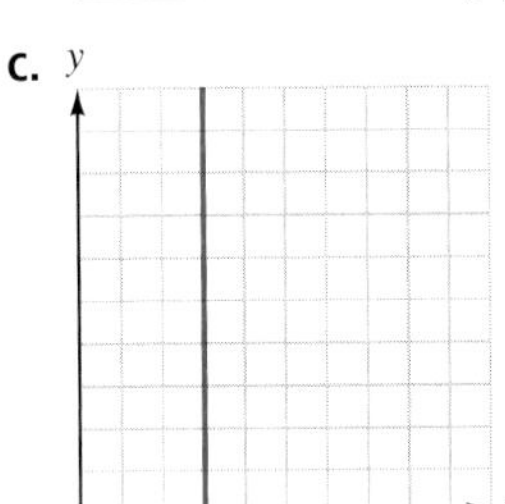

d. 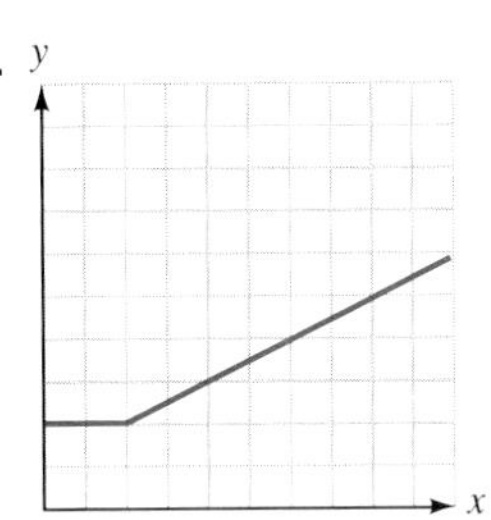

11. Coleman was looking for a reasonably priced used car with low mileage. One weekend, he shopped at three auto dealers and found several cars that interested him. They were priced at \$6,800, \$7,200 and \$7,800. To the nearest dollar, find the mean price of the three automobiles.

a. \$7,200 b. \$7,267

c. \$1,000 d. \$7,500

12. Which of the following equations represents the graphed line?

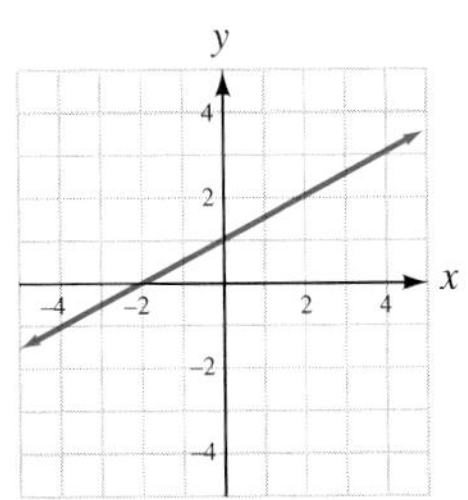

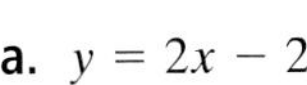

a. $y = 2x - 2$

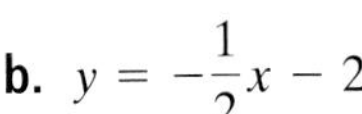

b. $y = -\frac{1}{2}x - 2$

c. $y = \frac{1}{2}x + 1$

d. $y = -2x + 1$

13. Find the slope, if it exists, of the line passing through $(-2, 0)$ and $(1, 4)$.

a. $m = \frac{4}{3}$ b. $m = \frac{3}{4}$

c. $m = 4$ d. $m = -\frac{1}{4}$

14. Which of the following equations passes through $(-3, -4)$ and $(6, 2)$?

a. $y = \frac{3}{2}x + 2$ b. $y = -\frac{2}{3}x + 2$

c. $y = \frac{2}{3}x - 2$ d. $y = -\frac{3}{2}x - 2$

15. What are the coordinates of the y-intercept of the line $3x - 2y = 12$?

a. $(4, 0)$ b. $(0, -6)$

c. $(2, -3)$ d. $(0, 6)$

16. The graph shows the area of a parallelogram remaining constant. The x-axis represents the length of the base and the y-axis represents the height of the triangle. According to the graph, what is the base measurement if the height measures 6 meters?

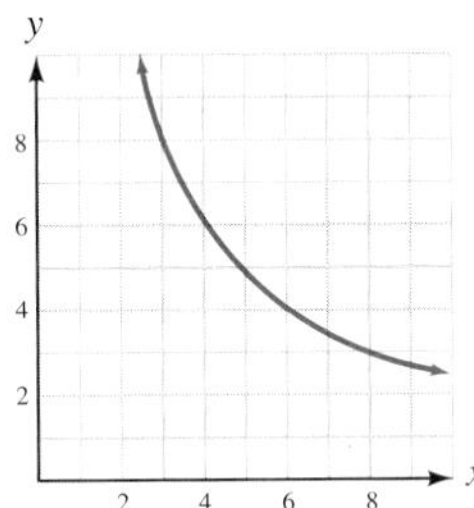

a. 3 meters b. 6 meters

c. 4 meters d. 7 meters

17. Solve $x = \frac{1}{2}(y - 4)$ for y.

a. $y = 2x + 4$ b. $y = 2x - 4$

c. $y = 2x - 2$ d. $y = \dfrac{x - \frac{1}{2}}{4}$

18. If $\frac{1}{5}x - 4 = -3$, what is the value of $2 - 4x$?

a. 22 b. -22

c. -35 d. -18

19. What is the solution of the systems of equations $y = x^2 + 6x + 7$ and $2x + y = -5$?

a. an infinite number of solutions

b. $(6, 7)$ and $(2, -1)$

c. no solution

d. $(-6, 7)$ and $(-2, -1)$

20. Which of the following graphs shows the solution for the system of equations $y = -x^2 + 4$ and $x + y = 3$?

a.

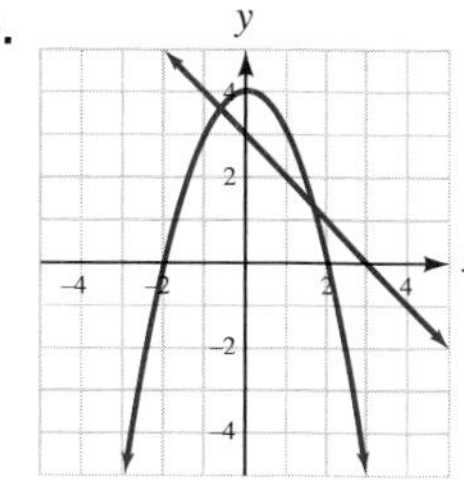

b.

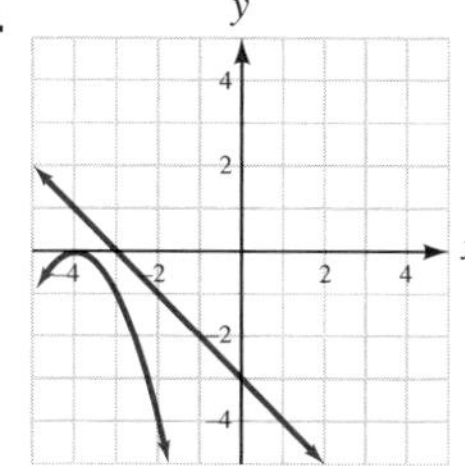

c.

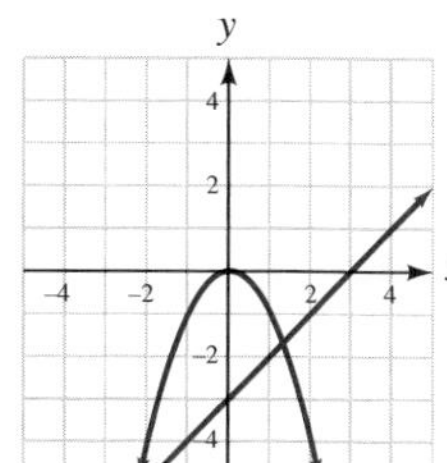

d.

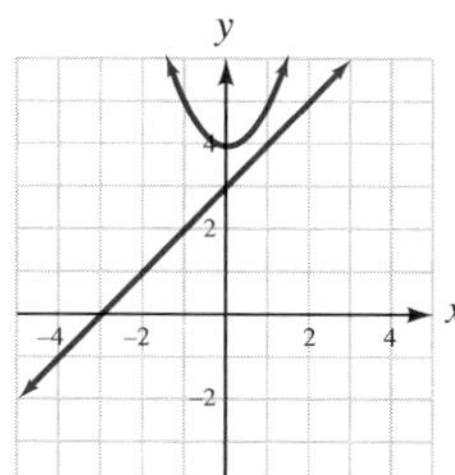

21. Which of the following equations correctly translates this statement? H is eleven less than the square of the quotient of x and the cube of y.

a. $H = 11 - \left(\frac{x}{y}\right)^2$ **b.** $H = \frac{x^2}{y^3} - 1$

c. $H = \left(\frac{x}{y^3}\right)^2 - 11$ **d.** $H - 11 = \frac{x^2}{y^3}$

22. Celeste is a realtor and works on a 3% commission basis of the total sales made during each month. She earned \$6,300 in March by selling two homes. One home sold for twice as much as the other. Determine the selling price of the least expensive home.

a. \$70,000 **b.** \$140,000

c. \$18,900 **d.** \$37,800

23. Which of the following is one factor of $3x^2 - 11x + 6$?

a. $(x - 6)$ **b.** $(3x - 2)$

c. $(3x - 6)$ **d.** $(3x + 2)$

24. Perform the multiplication $(4x - 3)^2$.

a. $16x^2 - 24x + 9$ **b.** $16x^2 - 9$

c. $16x^2 + 9$ **d.** $16x^2 - 24x - 9$

25. Perform the indicated operations assuming all variables are positive. Select the most simplified form of $\sqrt{20x^2} + 3x\sqrt{5} - 4\sqrt{x^2}$.

a. $x\sqrt{5}$ **b.** $5x\sqrt{5} + 4x$

c. $\sqrt{5}$ **d.** $5x\sqrt{5} - 4x$

26. If $f(x) = |3x - 1|$, find $f\left(\frac{1}{3}\right) \cdot f(3)$.

a. 0 **b.** 8

c. 2 **d.** 1

27. Which of the following equations is represented by the graph?

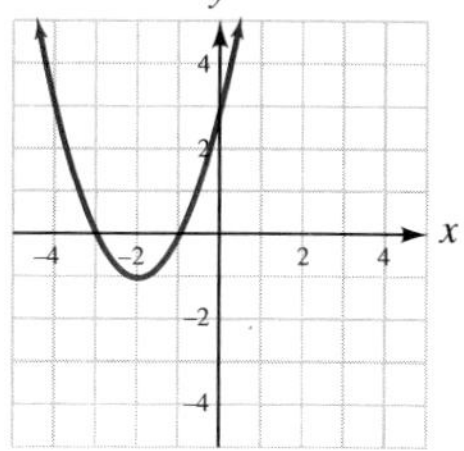

a. $y = x^2 - 2$

b. $y = x^2 + 4x + 3$

c. $y = x^2 + 2$

d. $y = x^2 - 4x + 4$

28. Which of the following inequalities describes the shaded region?

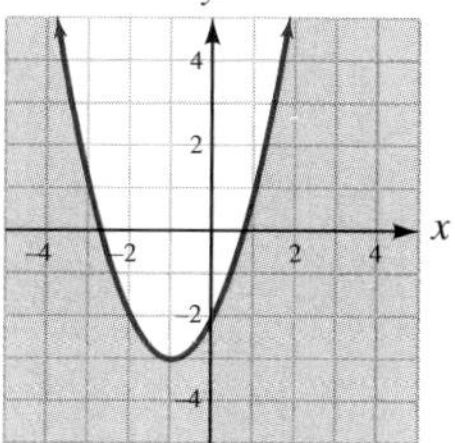

a. $y \le (x + 1)^2 - 3$

b. $y \le (x - 1)^2 + 3$

c. $y \le (x - 1)^2 - 3$

d. $y \le (x + 1)^2 + 3$

29. Which of the following numbers should be added to both sides of the equation $x^2 + 8x = 20$ in order to solve by completing the square?

a. 4 **b.** -16

c. 16 **d.** 8

30. Which of the following expressions appear as a step in solving $5x^2 + x = 3$ using the quadratic formula?

a. $-1 \pm \sqrt{61}$ **b.** $\frac{1 \pm \sqrt{61}}{10}$

c. $\frac{-1 \pm \sqrt{61}}{10}$ **d.** $1 \pm \sqrt{61}$

31. A 12-foot board is cut in three pieces. The third piece is one foot less than twice the first and the second piece is one foot longer than the first piece. What is the length of the longest piece?

a. 3 feet **b.** 4 feet

c. 5 feet **d.** 6 feet

32. Susan has nine coins in her wallet consisting of nickels, dimes, and quarters. The change in her wallet totals \$1.65. The number of quarters she has equals twice the sum of her nickels and dimes. Which set of three equations could be used to determine the number of nickels, dimes, and quarters that Susan has?

a. $n + d + q = 9$
$0.05n + 0.10d + 0.25q = 1.65$
$q = 2(n + d)$

b. $n + d + q = 9$
$n + d + q = 1.65$
$2q = n + d$

c. $n + d + q = 9$
$5n + 10d + 25q = 1.65$
$q = \frac{1}{2}(n + d)$

d. $n + d + q = 9$
$0.05n + 0.10d + 0.25q = 1.65$
$q + 2n = d$

33. Which of the following is the simplified form of $\dfrac{6a^2b - 9ab^2}{3ab}$?

a. $2a + 3b$ **b.** $2a - 3b$

c. $-ab$ **d.** $\dfrac{1}{2a} - \dfrac{1}{3b}$

34. Perform the indicated operation. Select the most simplified form of $\dfrac{x^2 + 4x - 5}{3x^2 + x - 4} \div \dfrac{2x^2 + 11x - 6}{6x^2 + 5x - 4}$.

a. $\dfrac{x + 6}{x + 5}$ **b.** 1

c. $\dfrac{x + 5}{x + 6}$ **d.** $\dfrac{x - 5}{x + 6}$

35. Perform the indicated operation. Select the most simplified form of $\dfrac{a^3b^2c}{(ab^3c^4)^2}$.

a. ab^4c^7 **b.** $\dfrac{a}{b^7c^{15}}$

c. ab^7c^{15} **d.** $\dfrac{a}{b^4c^7}$

36. A silo is filled with grain. It is in the shape of a cylinder with a diameter measurement of 12 ft and a height of 18 ft. What is the volume of the silo to the nearest ft^3?

a. 2,035 ft^3 **b.** 452 ft^3

c. 8,139 ft^3 **d.** 678 ft^3

37. Bill is constructing a ramp that is 26 inches long and is leaning against a raised platform 10 inches above the ground. What is the distance from the point of the ramp's contact with the ground and the platform's base?

a. 28 inches **b.** 24 inches

c. 16 inches **d.** 36 inches

38. If hexagon $ABCDEF$ is similar to hexagon $PQRSTU$, and $BC = 9$, $QR = 15$, and $CD = 12$, what is the measure of RS?

a. 7.2 **b.** 11.25

c. 13 **d.** 20

39. The UR Special Candy factory is manufacturing collector's boxes for Valentine's Day. They are in the shape of a cube, made of tin, with each side measuring 4 inches. If the company is planning to manufacture only 2,000 boxes, how much tin will be needed to complete the entire order (assuming there will be no wasted material)?

a. 192,000 $in.^3$ **b.** 128,000 $in.^3$

c. 32,000 $in.^3$ **d.** 48,000 $in.^3$

40. If $\angle A$ and $\angle C$ are vertical angles and $\angle A$ measures 100°, what is the measure of $\angle C$?

a. $m(\angle C) = 100°$

b. $m(\angle C) = 80°$

c. $m(\angle C) = 50°$

d. not enough information given

41. Complete the statement with the *best* answer choice below.

> If two lines parallel to one another are cut by a transversal, corresponding angles are *always*

a. acute angles.

b. obtuse angles.

c. supplementary angles.

d. congruent angles.

42. Use the clues below to determine which statement that follows must be true.

- All people that wear boots are cowboys.
- Some of the people have black hair.
- All people who have blonde hair love cats.
- People who have black hair love peanuts.
- Jake wears boots.

a. Jake has black hair. **b.** Jake loves peanuts.

c. Jake loves cats. **d.** Jake is a cowboy.

43. Bobby, Danny, Robert and Ken all work in the Plaza Building. One is a banker, one is a realtor, one is a lawyer, and one is a doctor. With the information below, determine who is the banker.

- Bobby and Danny eat lunch with the realtor.
- Robert and Ken live near the banker.
- Danny works on the same floor as the doctor and the banker.

a. Bobby **b.** Danny

c. Robert **d.** Ken

44. Using the following pattern sequence, determine the missing design.

 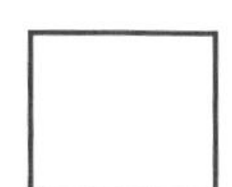 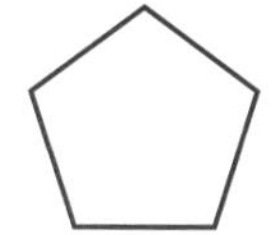 ______

a. 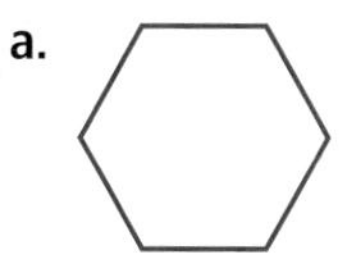 **b.** 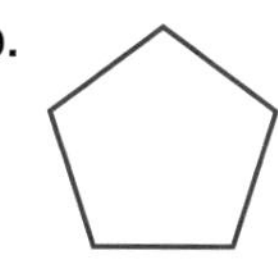

c. 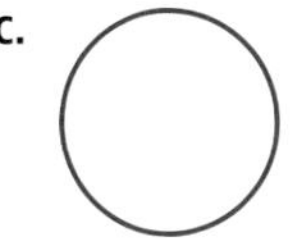 **d.** 

45. Using the following pattern sequence, determine the missing term: A, C, F, J, _____, U.

a. P **b.** Q

c. O **d.** R

46. A rectangular box has dimensions 3 ft $\times$ 4 ft $\times$ 6 ft. To the nearest tenth, what is the length of the longest object that can be put in the box in any position?

a. 13.0 ft **b.** 6.0 ft

c. 7.8 ft **d.** 7.2 ft

47. A 12-oz jar of honey sells for $2.29 and a 26-oz jar for $3.49. Which of the following statements is not true?

a. The better buy per ounce is the 26-oz jar.

b. You can buy two 12-oz jars and spend more for less honey.

c. The smaller jar is about 19¢ per oz.

d. The better buy per ounce is the 12-oz jar.

48. Which of the following dimensions could form a right triangle?

a. 2, 3, 4 **b.** 3, 5, 7

c. 6, 8, 10 **d.** 12, 13, 20

49. Prepared chicken bits are sold by the ounce at the Clucking Chicken House. The empty box weighs 3.1 ounces. Each chicken bit c weighs at least 2.6 ounces. Which inequality best describes the total weight in ounces, t, of a box of chicken bits in terms of c?

a. $t \leq 3.1 + 2.6c$ **b.** $t \geq 5.7c$

c. $t \geq 3.1 + 2.6c$ **d.** $t \geq 3.1c + 2.6$

50. The Gathering Spot pays its wait staff $3.10 per hour plus a portion of the tips received by all the staff. The table shows the number of hours worked by one individual. Which equation best describes the relationship between total pay, p, and hours worked, h?

a. $p = 6.60 + h$

b. $p = (3.10 + 3.50)h$

c. $p = 3.10h + 3.50$

d. $p = 3.10h + 2.50$

Hours worked (h)	Total pay (p)
1	$6.60
2	$9.70
3	$12.80
4	$15.90
5	$19.00

APPENDIX E

THEA TEST MATHEMATICS SKILLS

Basic Mathematics	Text Reference
1. Solving word problems involving integers, fractions, decimals, and units of measurement, including	
• Solving word problems involving integers	Chapter 2
• Solving word problems involving fractions	Chapter 3
• Solving word problems involving decimals (including percents)	Chapters 4 and 5
• Solving word problems involving ratio and proportions	Section 12.6
• Solving word problems involving units of measurement and conversions (including scientific notation)	Section 10.3 and Appendix B
2. Solving problems involving data interpretation and analysis, including	
• Interpreting information from line graphs, bar graphs, pictographs, and pie charts	Throughout, and emphasized in Section 6.1
• Interpreting data from tables	Throughout, and emphasized in Section 6.1
• Recognizing appropriate graphic representations of various data	Throughout, and emphasized in Section 6.2
• Analyzing and interpreting data using measures of central tendency (mean, median, and mode); analyzing and interpreting data using the concept of variability (range)	Section 6.2

Algebra	Text Reference
3. Graphing numbers or number relationships, including	
• Identifying the graph of a given equation	Sections 9.1–9.6
• Identifying the graph of a given inequality	Sections 8.6, 13.5, 13.6, and Appendix C
• Finding the slope and/or intercepts of a given line	Sections 9.3–9.6
• Finding the equation of a line	Sections 9.4–9.6
• Recognizing and interpreting information from the graph of a function (including direct and inverse variation)	Sections 9.7, 12.7

4. Solving one- and two-variable equations, including	
• Finding the value of the unknown in a given one-variable equation	Throughout, and emphasized in Sections 8.1–8.5
• Expressing one variable in terms of a second variable in two-variable equations	Sections 9.2, 9.3, 9.5, 9.6, 13.2
• Solving systems of two equations in two variables (including graphical solutions)	Sections 13.1–13.4
5. Solving word problems involving one and two variables, including	
• Identifying the algebraic equivalent of a stated relationship	Throughout, and emphasized in Chapters 8 and 9
• Solving word problems involving one and two unknowns	Throughout, and emphasized in Chapters 8 and 9
6. Understanding operations with algebraic expressions and functional notations, including	
• Factoring quadratics and polynomials	Sections 11.1–11.4
• Performing operations on and simplifying polynomial expressions	Sections 10.4–10.8
• Performing operations on and simplifying rational expressions	Sections 12.1–12.3
• Performing operations on and simplifying radical expressions	Sections 14.1–14.5
• Applying principles of functions and functional notation	Section 9.7
7. Solving problems involving quadratic equations, including	
• Graphing quadratic functions	Section 15.3
• Graphing quadratic inequalities	Appendix C
• Solving quadratic equations using factoring, completing the square, or the quadratic formula	Sections 15.1 and 15.2
• Solving problems involving quadratic models	Section 11.5, Chapter 15

Geometry

	Text Reference
8. Solving problems involving geometric figures, including	
• Solving problems involving two-dimensional geometric figures (e.g., perimeter and area problems)	Throughout, and emphasized in Sections 7.5, 7.6, 8.4
• Solving problems involving three-dimensional geometric figures (e.g., volume and surface area problems)	Sections 7.7 and 8.4
• Solving problems using the Pythagorean Theorem	Section 7.4
9. Solving problems involving geometric concepts, including	
• Solving problems using principles of similarity and congruence	Sections 7.1–7.4, 12.6
• Solving problems using principles of parallelism and perpendicularity	Sections 7.1–7.4

Problem Solving	
	Text Reference
10. Applying reasoning skills, including	
• Drawing conclusions using inductive reasoning	Appendix A
• Drawing conclusions using deductive reasoning	Appendix A
11. Solving applied problems involving a combination of mathematical skills, including	
• Applying combinations of mathematical skills to solve problems	Throughout
• Applying combinations of mathematical skills to solve a series of related problems	Throughout

APPENDIX F

Answers to Selected Exercises

Chapter 1 Check Your Knowledge (page 2)

1. natural, whole **2.** factors, product, quotient **3.** commutative, associative **4.** prime **5.** 7 thousands + 3 hundreds + 4 tens + 3 ones **6.** 27,500

7.

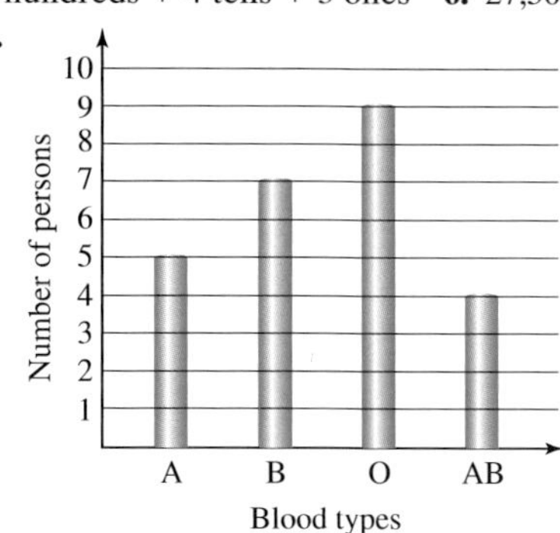

8.

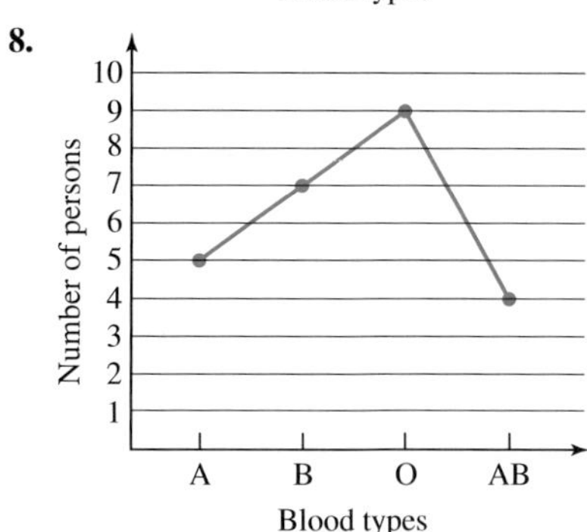

9. > **10.** 5,121 **11.** 58 **12.** 69° **13.** 24,624 **14.** 57 R 34 **15.** 64 ft, 247 ft^2 **16.** $2 \cdot 5^2 \cdot 19$ **17.** 5 **18.** 103 **19.** 3 **20.** 19 **21.** 3 **22.** 2 **23.** 16 **24.** 48 **25.** 63 **26.** 89

Think It Through (page 9)

1. c **2.** b **3.** e **4.** d **5.** a

Study Set Section 1.1 (page 10)

1. set **3.** expanded **5.** number **7.** 3 **9.** 6 **11.** whole numbers

13

15.

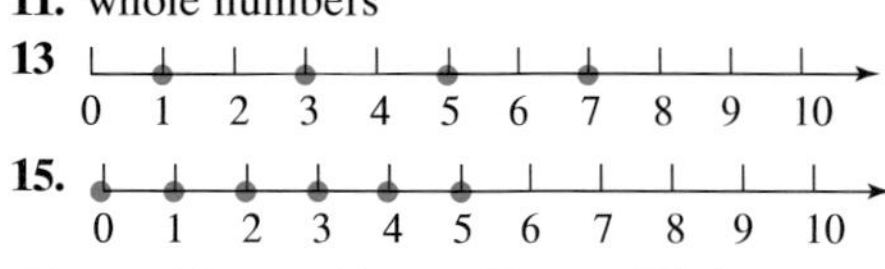

17. > **19.** > **21.** < **23.** > **25.** braces

27. 2 hundreds + 4 tens + 5 ones; two hundred forty-five **29.** 3 thousands + 6 hundreds + 9 ones; three thousand six hundred nine **31.** 3 ten thousands + 2 thousands + 5 hundreds; thirty-two thousand five hundred **33.** 1 hundred thousand + 4 thousands + 4 hundreds + 1 one; one hundred four thousand four hundred one **35.** 425 **37.** 2,736 **39.** 456 **41.** 27,598 **43.** 9,113 **45.** 10,700,506 **47.** 79,590 **49.** 80,000 **51.** 5,926,000 **53.** 5,900,000 **55.** $419,160 **57.** $419,000 **61. a.** the 70s, 7 **b.** the 60s, 9 **c.** the 60s, 12

63.

Gas reserves (trillion cubic ft)
25 50 75 100 125 150 175 200
U.S. Venezuela Canada Argentina Mexico

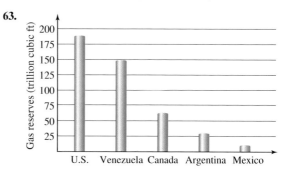

65.

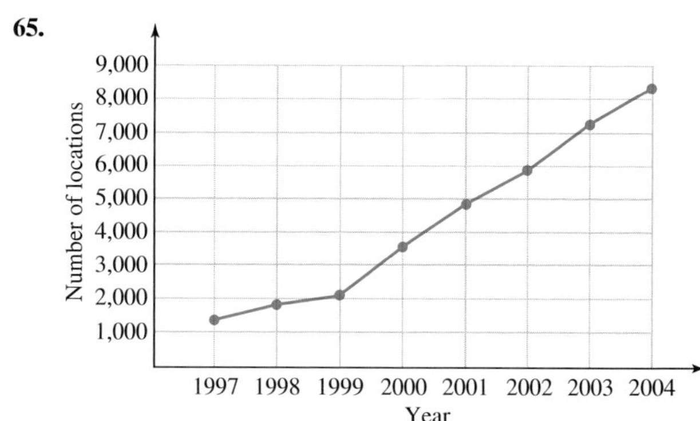

67.

a.

No. 201 March 9, 20 05

Payable to Davis Chevrolet $ 15,601.00

Fifteen thousand six hundred one and $\frac{00}{100}$ DOLLARS

Don Smith

45-365-02

b.

No. 7890	Aug. 12, 20 05
Payable to Dr. Anderson	$ 3,433.00
Three thousand four hundred thirty three and $\frac{00}{100}$	DOLLARS
45-828-02	Juan Decito

69. 1,865,593; 482,880; 1,503; 269; 43,449
71. a. 299,800,000 m/sec **b.** 300,000,000 m/s

Study Set Section 1.2 (page 21)

1. sum, addends **3.** sum **5.** difference, subtrahend, minuend **7.** associative **9.** commutative property of addition **11.** associative property of addition **13. a.** $x + y = y + x$ **b.** $(x + y) + z = x + (y + z)$ **15.** 0 **17.** $4 + 3 = 7$ **19.** parentheses **21.** 47, 52 **23.** 38 **25.** 461 **27.** 111 **29.** 150 **31.** 363 **33.** 979 **35.** 1,985 **37.** 10,000 **39.** 15,907 **41.** 1,861 **43.** 5,312 **45.** 88 ft **47.** 68 in. **49.** 3 **51.** 25 **53.** 103 **55.** 65 **57.** 141 **59.** 0 **61.** 24 **63.** 118 **65.** 958 **67.** 1,689 **69.** 10,457 **71.** 303 **73.** 40 **75.** 110 **77.** $18 **79.** 33 points **81.** $213 **83.** 10,057 mi **85. a.** $147,145 **b.** $161,725 **87.** 91 ft **89.** $6,233,000,000 **91.** 196 in. **95.** 3 thousands + 1 hundred + 2 tens + 5 ones **97.** 6,354,780 **99.** 6,350,000

Study Set Section 1.3 (page 34)

1. multiplication **3.** commutative, associative **5.** divisor, quotient **7.** $4 \cdot 8$ **9.** Multiply its length by its width. **11. a.** 25 **b.** 62 **c.** 0 **d.** 0 **13.** $5 \cdot 12 = 60$ **15. a.** $\times, \cdot, (\ )$ **b.** $\overline{)\ \ }$, ÷, — **17.** square feet **19.** 84 **21.** 324 **23.** 180 **25.** 105 **27.** 7,623 **29.** 1,060 **31.** 2,576 **33.** 20,079 **35.** 2,919,952 **37.** 1,182,116 **39.** 84 in.^{2} **41.** 144 in.^{2} **43.** 8 **45.** 3 **47.** 12 **49.** 13 **51.** 73 **53.** 41 **55.** 205 **57.** 210 **59.** 8 R 25 **61.** 20 R 3 **63.** 30 R 13 **65.** 31 R 28 **67.** $132 **69.** 406 mi **71.** 125,800 **73.** 312 **75.** yes **77.** 72 **79.** 4 **81.** 5 mi **83.** 440 ft **85.** $41 **87.** 9 girls, 24 teams **89.** the square room; the square room **91.** 388 ft^2 **97.** 8 **99.** 872

Study Set Estimation (page 38)

1. no **3.** no **5.** no **7.** approx. 8,900 mi **9.** approx. 30 bags **11.** 1,800,000,000

Study Set Section 1.4 (page 44)

1. factors **3.** factor **5.** composite **7.** prime **9.** base, exponent **11.** $1 \cdot 27$ or $3 \cdot 9$ **13. a.** 44 **b.** 100 **15. a.** 1 and 11 **b.** 1 and 23 **c.** 1 and 37 **d.** They are prime numbers. **17.** yes **19.** 90 **21.** 605 **23.** no **25.** 2 and 5 **27.** 2 **29.** $3 \cdot 5 \cdot 2 \cdot 5$; $5 \cdot 3 \cdot 5 \cdot 2$; they are the same **31.** 13, 8, 7 **33.** 2 **35.** $7 \cdot 7 \cdot 7$ **37.** $3 \cdot 3 \cdot 3 \cdot 3 \cdot 3$ **39.** $5 \cdot 5 \cdot 11$ **41.** 10 **43.** 2^5 **45.** 5^4 **47.** $4^2(5^2)$ **49.** 1, 2, 5, 10 **51.** 1, 2, 4, 5, 8, 10, 20, 40 **53.** 1, 2, 3, 6, 9, 18 **55.** 1, 2, 4, 11, 22, 44 **57.** 1, 7, 11, 77 **59.** 1, 2, 4, 5, 10, 20, 25, 50, 100 **61.** $3 \cdot 13$ **63.** $3^2 \cdot 11$ **65.** $2 \cdot 3^4$ **67.** $2^2 \cdot 5 \cdot 11$ **69.** 2^6 **71.** $3 \cdot 7^2$ **73.** 81 **75.** 32 **77.** 144 **79.** 4,096 **81.** 72 **83.** 3,456 **85.** 12,812,904 **87.** 1,162,213 **89.** 1, 2, 4, 7, 14, 28; $1 + 2 + 4 + 7 + 14 = 28$ **91.** 2^2 square units; 3^2 square units; 4^2 square units **97.** 231,000 **99.** 0 **101.** $A = lw$

Think It Through (page 52)

10, 5, 12, 3, 4

Study Set Section 1.5 (page 52)

1. parentheses, brackets **3.** evaluate **5.** 3; square, multiply, subtract **7.** multiply, subtract **9.** $2 \cdot 3^2 = 2 \cdot 9$; $(2 \cdot 3)^2 = 6^2$ **11.** 4, 20, 8 **13.** 36, 30 **15.** 27 **17.** 2 **19.** 15 **21.** 25 **23.** 5 **25.** 25 **27.** 18 **29.** 813 **31.** 5,239 **33.** 16 **35.** 5 **37.** 49 **39.** 24 **41.** 13 **43.** 10 **45.** 198 **47.** 18 **49.** 216 **51.** 17 **53.** 191 **55.** 3 **57.** 29 **59.** 14 **61.** 64 **63.** 192 **65.** 74 **67.** 137 **69.** 3 **71.** 21 **73.** 11 **75.** 1 **77.** 10,496 **79.** 2,845 **81.** $2(6) + 4(2) + 2(1)$; $22 **83.** $24 + 6(5) + 10(10) + 12(20) + 2(50) + 100$; $594 **85.** brick: $3(3) + 1 + 1 + 3 + 3(5)$; 29; aphid: $3[1 + 2(3) + 4 + 1 + 2]$; 42 **87.** 79° **89.** 5 **91.** 298 **97.** 7,300 **99.** 9,591

Study Set Section 1.6 (page 61)

1. equal, = **3.** check **5.** equivalent **7.** y, c **9.** addition of 6; subtract 6 from both sides **11.** 8, 8, 16, $\stackrel{?}{=}$, 24, 16 **13.** yes **15.** no **17.** yes **19.** yes **21.** yes **23.** yes **25.** no **27.** no **29.** no **31.** yes **33.** 10 **35.** 7 **37.** 3 **39.** 4 **41.** 13 **43.** 75 **45.** 740 **47.** 339 **49.** 3 **51.** 5 **53.** 9 **55.** 10 **57.** 1 **59.** 56 **61.** 84 **63.** 105 **65.** 4 **67.** 12 **69.** 8 **71.** 47 **75.** 94,683,948 **77.** 62 **79.** $218,500 **81.** $180 million **83.** 25 units **85.** $190 **93.** 325,780 **95.** 90 **97.** 3

Study Set Section 1.7 (page 69)

1. division **3.** x **5.** y, z **7.** It is being multiplied by 4. Divide by 4. **9. a.** Subtract 5 from both sides. **b.** Add 5 to both sides. **c.** Divide both sides by 5. **d.** Multiply both sides by 5. **11.** 3, 3, 4, 12, 4 **13.** 1 **15.** 96 **17.** 3 **19.** 6 **21.** 1 **23.** 2 **25.** 14 **27.** 42 **29.** 75 **31.** 39 **33.** 50 **35.** 49 **37.** 10 **39.** 3 **41.** 2 **43.** 1 **45.** 40 **47.** 1,200 **51.** 390 words per minute **53.** 14 **55.** 96 **57.** 32 calls **59.** 55 lb **65.** 48 cm **67.** $2^3 \cdot 3 \cdot 5$ **69.** 72 **71.** 26 mpg

Key Concept (page 73)

1. Let x = the monthly cost to lease the van. **3.** Let x = the width of the field. **5.** Let x = the distance traveled by the motorist. **7.** $a + b = b + a$ **9.** $\frac{b}{1} = b$ **11.** $n - 1 < n$ **13.** $(r + s) + t = r + (s + t)$

Chapter Review (page 75)

1. 0 1 2 3 4 5 6

2. 0 1 2 3 4 5 6

3.

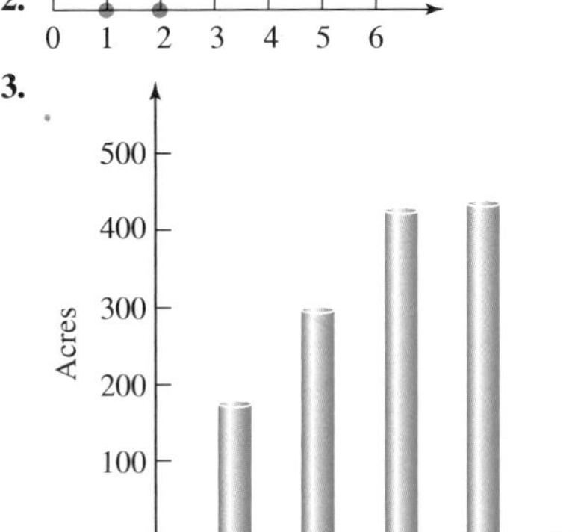

4.
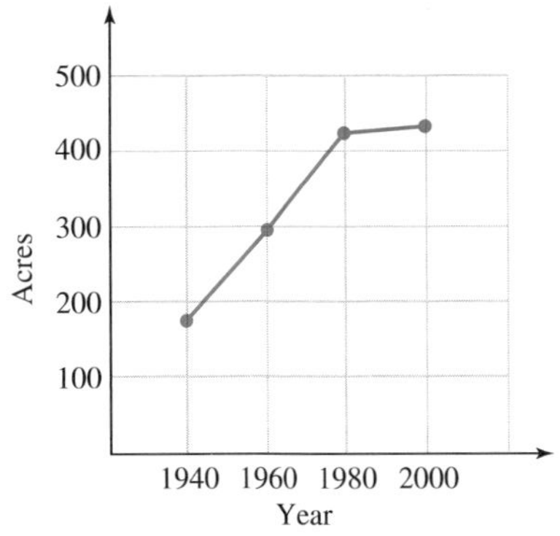

5. 6 **6.** 7 **7.** 5 hundred thousands + 7 ten thousands + 3 hundreds + 2 ones **8.** 3 ten millions + 7 millions + 3 hundred thousands + 9 thousands + 5 tens + 4 ones **9.** 3,207 **10.** 23,253,412 **11.** 16,000,000,000 **12.** > **13.** < **14.** 2,507,300 **15.** 2,510,000 **16.** 2,507,350 **17.** 2,500,000 **18.** 78 **19.** 137 **20.** 55 **21.** 149 **22.** 777 **23.** 2,332 **24.** 518 **25.** 6,000 **26.** 1,010 **27.** 24,986 **28.** commutative property of addition **29.** associative property of addition **30.** 96 in. **31.** 13 **32.** 4 **33.** 11 **34.** 54 **35.** 74 **36.** 2,075 **37.** 4 + 2 = 6 **38.** 5 − 2 = 3 **39.** \$45 **40.** \$785 **41.** \$23,541 **42.** 56 **43.** 56 **44.** 0 **45.** 7 **46.** 560 **47.** 210 **48.** 3,297 **49.** 178,704 **50.** 31,684 **51.** 455,544 **52.** associative property of multiplication **53.** commutative property of multiplication **54.** \$342 **55.** 108 ft, 288 ft^2 **56.** 720 **57.** 2 **58.** 15 **59.** undefined **60.** 0 **61.** 21 **62.** 37 **63.** 19 R 6 **64.** 23 R 27 **65.** 16, 25 **66.** 28 **67.** 1, 2, 3, 6, 9, 18 **68.** 1, 5, 25 **69.** prime **70.** composite **71.** neither **72.** neither **73.** composite **74.** prime **75.** odd **76.** even **77.** even **78.** odd **79.** $2 \cdot 3 \cdot 7$ **80.** $3 \cdot 5^3$ **81.** 6^4 **82.** $5^3 \cdot 13^2$ **83.** 125 **84.** 121 **85.** 200 **86.** 2,700 **87.** 49 **88.** 32 **89.** 75 **90.** 36 **91.** 38 **92.** 24 **93.** 8 **94.** 24 **95.** 53 **96.** 3 **97.** 19 **98.** 7 **99.** 3(6) + 2(5) = 28 **100.** 201 **101.** no **102.** yes **103.** y **104.** t **105.** 9 **106.** 31 **107.** 340 **108.** 133 **109.** 9 **110.** 14 **111.** 120 **112.** 5 **113.** 7 **114.** 985 **115.** \$97,250 **116.** 185 **117.** 4 **118.** 3 **119.** 21 **120.** 14 **121.** 21 **122.** 36 **123.** 315 **124.** 425 **125.** 1 **126.** 144 **127.** 24 in. **128.** \$128

Chapter 1 Test (page 81)

1.

2. 5 thousands + 2 hundreds + 6 tens + 6 ones **3.** 7,507 **4.** 35,000,000

5.
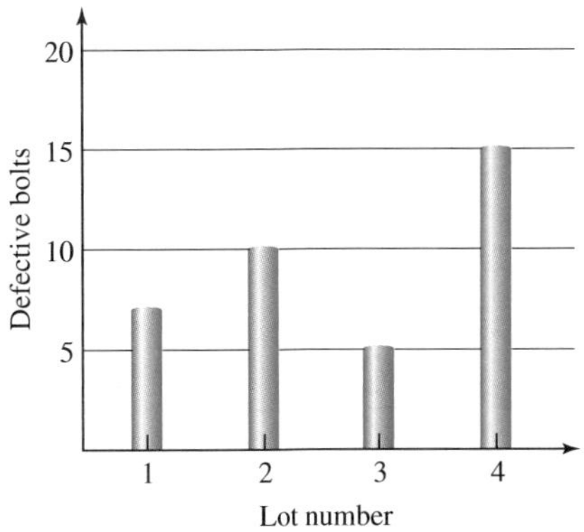

6.
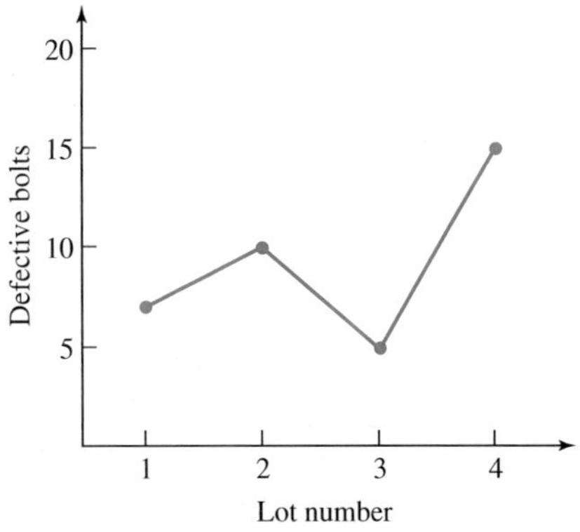

7. > **8.** < **9.** 1,491 **10.** 248 **11.** 58,105 **12.** 942 **13.** \$76 **14.** 1, 2, 4, 5, 10, 20 **15.** 424 **16.** 26,791 **17.** 72 **18.** 114 R 57 **19. a.** 0 **b.** 0 **20. a.** associative property of multiplication **b.** commutative property of addition **21.** 360 ft, 7,875 ft^2 **22.** 47 **23.** 3,456 **24.** $2^2 \cdot 3^2 \cdot 7$ **25.** 29 **26.** 44 **27.** 26 **28.** 39 **29.** yes **30.** 99 **31.** 30 **32.** 11 **33.** 81 **34.** 3,100 **35.** 194 yr **36.** To solve an equation means to find all the values of the variable that, when substituted into the equation, make a true statement.

Chapter 2 Check Your Knowledge (page 84)

1. absolute **2.** identity **3.** opposites (or negatives) **4.** unlike **5.** > **6.** 1 **7. a.** −14 **b.** 0 **c.** −8 **d.** 2 **8. a.** −6 **b.** 6 **c.** −12 **9. a.** −9 **b.** 100 **c.** −24 **10.** 3(−6) = −18 **11. a.** −4 **b.** 30 **c.** undefined **12. a.** 7 **b.** −7 **c.** 12 **13. a.** −9 **b.** 9 **c.** 1 **14. a.** 16 **b.** −13 **c.** 29 **d.** 1 **15.** \$20 **16.** 15 **17.** −\$230

Think It Through (page 89)

\$4,621, \$1,073, \$3,325

Study Set Section 2.1 (page 92)

1. line **3.** graph **5.** inequality **7.** absolute value **9.** integers **11. a.** < **b.** > **c.** <, > **13.** yes **15.** 15 − 8 **17.** 15 > 12 **19. a.** −225 **b.** −10 **c.** −3 **d.** −12,000 **e.** −2 **21.** It is negative. **23.** −4 **25.** −8 and 2 **27.** −7 **29.** 6 −4, −6, −(−6) (answers may vary) **31. a.** −(−8) **b.** |−8| **c.** 8 − 8 **d.** −|−8| **33.** 9 **35.** 8 **37.** 14 **39.** −20 **41.** −6 **43.** 203 **45.** 0 **47.** 11 **49.** 4 **51.** 12

53.
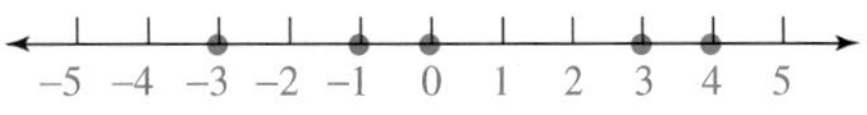

55.
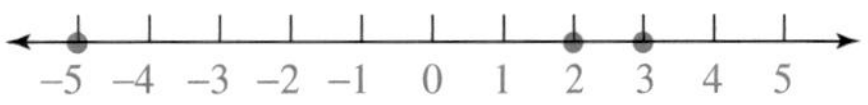

57. < **59.** < **61.** > **63.** > **65.** ≥ **67.** ≥ or ≤ **69.** ≤ **71.** ≤ **73.** 2, 3, 2, 0, −3, −7 **75.** peaks: 2, 4, 0; valleys: −3, −5, −2 **77. a.** −1 (1 below par) **b.** −3 (3 below par) **c.** Most of the scores are below par. **79. a.** −10° to −20°

b. 40° **c.** 10° **81. a.** 200 yr **b.** A.D. **c.** B.C. **d.** the birth of Christ

83.

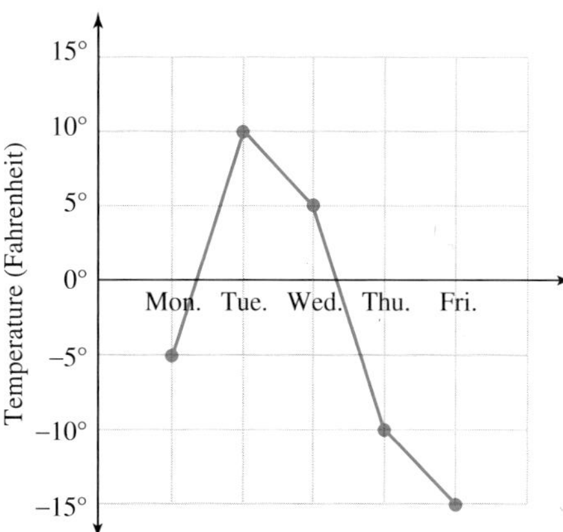

91. 23,500 **93.** 761 **95.** associative property of multiplication

Think It Through (page 100)

decrease expenses, increase income, decrease expenses, increase income, increase income, increase income, decrease expenses, decrease expenses, increase income, decrease expenses

Study Set Section 2.2 (page 103)

1. identity **3.** 3 **5.** −2 **7. a.** yes **b.** yes **9. a.** 7 **b.** 10 **11.** subtract, larger **13.** −18, −19 **15.** 5, 2 **17.** −5 should be within parentheses: −6 + (−5) **19.** 11 **21.** 23 **23.** 0 **25.** −99 **27.** −9 **29.** −10 **31.** 1 **33.** −7 **35.** −20 **37.** 15 **39.** 8 **41.** 2 **43.** −10 **45.** 9 **47.** 8 **49.** −21 **51.** 3 **53.** −10 **55.** −4 **57.** 7 **59.** −21 **61.** −7 **63.** 9 **65.** 0 **67.** 0 **69.** 5 **71.** 0 **73.** −3 **75.** −10 **77.** −1 **79.** −17 **81.** −8,346 **83.** −1,032 **85.** 3G, −3G **87.** no; $70 shortfall each month **89.** 5; 4% risk **91.** −1, 0 **93.** 7 ft over flood stage **95.** about −$2,500 million **101.** 15 ft^2 **103.** 375 mi **105.** 5^3

Study Set Section 2.3 (page 111)

1. difference **3.** adding, opposite **5.** 6 **7.** $(-b)$ **9.** brackets **11.** −8 − (−4) **13.** 7 **15.** no; 8 − 3 = 5, 3 − 8 = −5 **17.** −3, 2, 0 **19.** −2, −10, 6, −4 **21.** 9 **23.** −13 **25.** −10 **27.** −1 **29.** 0 **31.** 8 **33.** 5 **35.** −4 **37.** −4 **39.** −20 **41.** 0 **43.** 0 **45.** −15 **47.** −9 **49.** 3 **51.** 9 **53.** −2 **55.** −10 **57.** −14 **59.** 3 **61.** −8 **63.** −18 **65.** −6 **67.** 10 **69.** −4 **71.** −2,447 **73.** 20,503 **75.** −1,676 **77.** −120 ft **79.** 16 points **81.** −8 **83.** 1,066 ft **85.** −4 yd

87. a.

Bottom Water level Platform

−12 0 25

b. 37 ft

89. No; he will be $244 overdrawn (−244). **95.** 5,990 **97.** 1, 2, 4, 5, 10, 20 **99.** 143 **101.** 3

Study Set Section 2.4 (page 119)

1. factors, product **3.** 3, exponent **5.** unlike **7.** commutative **9.** −9, the opposite of that number **11.** pos · pos, pos · neg, neg · pos, neg · neg **13. a.** negative **b.** positive **15. a.** 3 **b.** 12 **c.** 5 **d.** 9 **e.** 10 **f.** 25 **17. a.** 2, 4; 4, 16; 6, 64 **b.** even **19.** 6, −24 **21.** −5 should be in parentheses: −6(−5) **23.** 54 **25.** −15 **27.** −36 **29.** 56 **31.** −20 **33.** −120 **35.** 0 **37.** 6 **39.** 7 **41.** −23 **43.** −48 **45.** 40 **47.** −30 **49.** −60 **51.** −1 **53.** −18 **55.** 0 **57.** 0 **59.** 60 **61.** 16 **63.** −125 **65.** −8 **67.** 81 **69.** −1 **71.** 1 **73.** 49, −49 **75.** −144, 144 **77.** −59,812 **79.** 43,046,721 **81.** −25,728 **83.** 390,625 **85. a.** plan #1: −30 lb, plan #2: −28 lb **b.** plan #1; the workout time is double that of plan #2 **87. a.** high 2, low −3 **b.** high 4, low −6 **89.** −20° **91.** −20 ft **93.** −$24,330 **99.** 45 **101.** 2,100 **103.** is less than

Study Set Section 2.5 (page 125)

1. quotient, divisor **3.** absolute value **5.** positive **7.** 5(−5) = −25 **9.** 0(?)= −6 **11.** $\frac{-20}{5} = -4$ **13. a.** always true **b.** sometimes true **c.** always true **15.** −7 **17.** 2 **19.** 5 **21.** 3 **23.** −20 **25.** −2 **27.** 0 **29.** undefined **31.** −5 **33.** 1 **35.** −1 **37.** 10 **39.** −4 **41.** −3 **43.** 5 **45.** −4 **47.** −5 **49.** −4 **51.** −542 **53.** −16 **55.** −4° per hour **57.** −1,000 ft **59.** −6 (6 games behind) **61.** −$15 **63.** −$1,740 **69.** 104 **71.** 2 · 3 · 5 · 7 **73.** yes **75.** 142

Study Set Section 2.6 (page 131)

1. order **3.** grouping **5.** 3; power, multiplication, subtraction **7.** multiplication; subtraction **9.** The base of the first exponential expression is 3; the base of the second is −3. **11.** 4, 20, −20, −28 **13.** 9, −36, −42 **15.** −7 **17.** 1 **19.** −21 **21.** −14 **23.** −7 **25.** −5 **27.** 12 **29.** −14 **31.** 30 **33.** 2 **35.** 15 **37.** −42 **39.** −5 **41.** −3 **43.** 4 **45.** 0 **47.** −14 **49.** 19 **51.** 4 **53.** −3 **55.** 25 **57.** −48 **59.** 44 **61.** 91 **63.** 3 **65.** −5 **67.** 17 **69.** 11 **71.** 8 **73.** 112 **75.** −1,707 **77.** −15 **79.** −5 **81.** −35 **83.** −200 **85.** −320 **87.** −9,000 **89.** −1,200 **91.** 19 **93.** 11 yd **95.** about a 60-cent gain **101.** 5,000 **103.** Add the lengths of all its sides. **105.** no

Study Set Section 2.7 (page 139)

1. solve **3.** x **5. a.** 3 **b.** (−3) **7. a.** multiplication by −2 **b.** addition of −6 **c.** mult. by −4, subtraction of 8 **d.** mult. by −5, addition of −6 **9.** simplify **11.** opposite **13. a.** subtraction of 3 **b.** addition of −6 **15.** −13, 7 **17.** 1, 1, −12, −4, −4, 3 **19.** $-10 \cdot x$ **21.** yes **23.** no **25.** −18 **27.** −14 **29.** 5 **31.** −1 **33.** −9 **35.** −14 **37.** −2 **39.** 0 **41.** −8 **43.** 5 **45.** 2 **47.** −1 **49.** 6 **51.** 0 **53.** 6 **55.** −52 **57.** −7 **59.** −4 **61.** −5 **63.** −2 **65.** 10 **67.** −6 **69.** −3 **71.** 3 **73.** −6 **75.** 54 **77.** 30 **79.** −14 **81.** −3 **83.** −2 **85.** −8 **87.** 15 **89.** −75, how many feet the cage was raised, the number of feet the cage was raised, addition, x, −75, −120, −75, 120, 120. The shark cage was raised 45 feet. 45 **91.** 34 **93.** −$435 **95.** 29 points **97.** $5 **99.** 18 ft **101.** zone −8 **105.** 5 · 5 · 5 · 5 · 5 · 5 **107.** 12 **109.** $\frac{16}{8}$

Key Concept (page 143)

1. −5 **3.** −30 **5.** +10 or 10 **7.** −205

9.

−4 −3 −2 −1 0 1 2 3 4

Negatives Positives

11. $x < y$ **13.** Like signs: Add their absolute values and attach their common sign to the sum. Unlike signs: Subtract their absolute values, the smaller from the larger, and attach the sign of the number with the larger absolute value to that result. **15.** Like signs: The quotient is positive. Unlike signs: The quotient is negative.

Chapter Review (page 145)

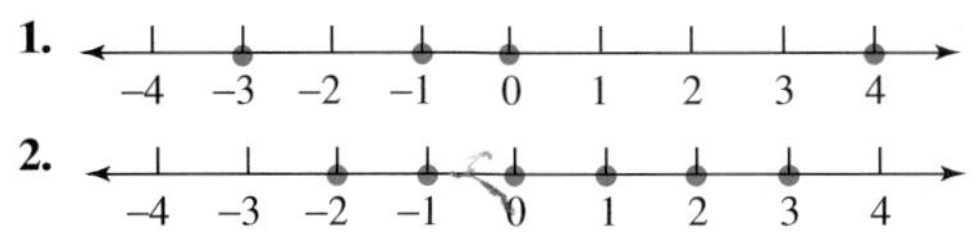

3. < **4.** < **5.** ≥ **6.** ≤ or ≥ **7.** −33 ft **8.** −$1,200 **9.** −10 sec **10.** 4 **11.** 0 **12.** 43 **13.** −12 **14.** negative **15.** the opposite **16.** negative **17.** minus **18.** 12 **19.** −8 **20.** 8 **21.** 0 **22.** 2

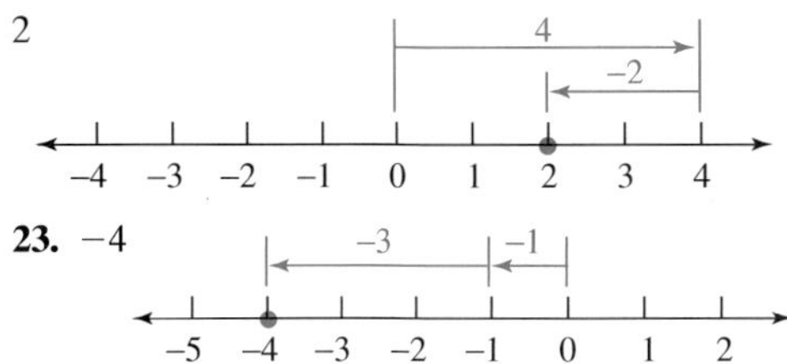

23. −4

24. −10 **25.** −83 **26.** −8 **27.** −1 **28.** 112 **29.** −11 **30.** −3 **31.** −2 **32.** −4 **33.** −20 **34.** 0 **35.** 0 **36.** 11 **37.** −4 **38.** 65 ft **39.** −3 **40.** −21 **41.** 4 **42.** −112 **43.** −6 **44.** 6 **45.** −37 **46.** 30 **47.** adding, opposite **48.** −4 **49.** 15 **50.** 6 **51.** −8 **52.** −77 **53.** −1 **54.** −225 ft **55.** Alaska: 180°; Virginia: 140° **56.** −45 **57.** 18 **58.** −14 **59.** 376 **60.** −100 **61.** 1 **62.** −25 **63.** −150 **64.** −36 **65.** −36 **66.** 0 **67.** 1 **68.** −3, −6, −9 **69.** 25 **70.** −32 **71.** 64 **72.** −64 **73.** negative **74.** first expression: base of 2; second: base of −2; −4, 4 **75.** 5, −3, −15 **76.** −2 **77.** −5 **78.** −8 **79.** 101 **80.** 0 **81.** undefined **82.** 1 **83.** 10 **84.** −2 min **85.** −22 **86.** 4 **87.** −43 **88.** 8 **89.** 41 **90.** 0 **91.** −13 **92.** 32 **93.** 12 **94.** −16 **95.** −4 **96.** 1 **97.** −1 **98.** −4 **99.** −70 **100.** 20 **101.** −7,000 **102.** 1,100 **103.** yes **104.** no **105.** −10 **106.** 12 **107.** −8 **108.** 4 **109.** 15 **110.** −4 **111.** 3 **112.** −2 **113.** −12 **114.** 0 **115.** −$132 **116.** −46° **117.** 121 **118.** $8,200

Chapter 2 Test (page 151)

1. a. > **b.** < **c.** < **2.** {. . . , −3, −2, −1, 0, 1, 2, 3, . . .} **3.** Monroe **4.** −5

−2 −3
−6 −5 −4 −3 −2 −1 0 1 2 3 4 5 6

5. a. −34 **b.** −34 **c.** −8 **6. a.** −13 **b.** −1 **c.** −15 **d.** −150 **7. a.** −70 **b.** −48 **c.** 16 **d.** 0 **8.** (−4)(5) = −20 **9. a.** −8 **b.** undefined **c.** −5 **d.** 0 **10.** $3 million **11.** 154 ft **12. a.** 6 **b.** 7 **c.** −6 **d.** 132 **13. a.** 16 **b.** −16 **c.** −1 **14.** −27 **15.** 1 **16.** −34 **17.** 42 **18.** 4 **19.** −72° **20.** left, −7 **21.** −15 **22.** no **23.** −15 **24.** 16 **25.** −40 **26.** 0 **27.** −5 **28.** 2 **29.** −18 **30.** −$244 **31.** 18 **32.** −4 + (−4) + (−4) + (−4) + (−4) = −20 **33.** The absolute value of a number is the distance from the number to 0 on the number line. Distance is either positive or 0, but never negative. **34.** It is true because 12 = 12.

Cumulative Review Exercises (page 153)

1. 1, 2, 5, 9 **2.** 0, 1, 2, 5, 9 **3.** −2, −1 **4.** −2, −1, 0, 1, 2, 5, 9 **5.** 6 **6.** 3 **7.** 7,326,500 **8.** 7,330,000 **9.** CRF Cable **10.**

Operable nuclear power plants
110 100 90 80 70 60 50 40 30 20 10
1978 1983 1988 1993 1998 2003

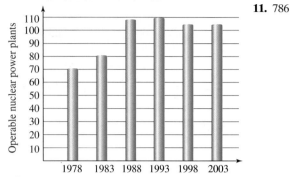

Source: *The World Almanac, 2005*

11. 786 **12.** 3,806 **13.** 4,684 **14.** 13,136 **15.** 104 ft, 595 ft² **16.** 65 **17.** 11,745 **18.** 13 **19.** 307,329 **20.** 467 **21.** 1,728 **22.** 1, 2, 3, 6, 9, 18 **23.** prime, odd **24.** composite, even **25.** even **26.** odd **27.** $2^3 \cdot 3^2 \cdot 7$ **28.** 11^4 **29.** 175 **30.** 38 **31.** 50 **32.** 2 **33.** no **34.** yes **35.** 13 **36.** 53 **37.** 27 **38.** 24 **39.**

−3 −2 −1 0 1 2 3

40.

−4 −3 −2 −1 0 1 2

41. true **42.** 9, −9 **43.** −5 **44.** −14 **45.** −8 **46.** −231 **47.** 24 **48.** −1,715 **49.** 2 **50.** −50 **51.** 26 **52.** −16 **53.** −3 **54.** 4 **55.** 3 **56.** −18 **57.** $126,037 **58.** −279° F

Chapter 3 Check Your Knowledge (page 156)

1. numerator, denominator **2.** lowest **3.** equivalent **4.** reciprocals **5.** mixed **6.** $\frac{5}{12}, \frac{7}{12}$ **7.** $\frac{4}{5}$ **8.** $\frac{3}{8}$ **9.** $-1\frac{1}{2}$ **10.** $\frac{51}{40} = 1\frac{11}{40}$ **11.** $\frac{5}{12}$ **12.** $\frac{30}{36}$ **13.**

$-\frac{9}{8}$ $\frac{1}{9}$ $1\frac{1}{12}$
−2 −1 0 1 2

14. $\frac{5}{8}$ cm² **15.** $313\frac{7}{15}$ **16.** $4\frac{11}{24}$ **17.** $\frac{1}{16}$ **18. a.** $\frac{9}{5}$ **b.** $-\frac{2}{5}$ **19. a.** 72 **b.** $\frac{10}{3}$ **20.** 60 **21.** $\$1\frac{3}{4}$ million **22.** $\$7\frac{1}{2}$ million

Study Set Section 3.1 (page 162)

1. numerator, denominator **3.** proper **5.** equivalent **7.** higher, building **9. a.** 2 **b.** 3 **c.** 5 **d.** 7 **11.** equivalent fractions: $\frac{2}{6} = \frac{1}{3}$ **13. a.** In the first case, 20 and 28 were factored. In the second case, they were prime factored. **b.** yes **15.** The 2's in the numerator and denominator aren't common factors. **17. a.** $\frac{8}{1}$ **b.** $-\frac{25}{1}$ **19.** 3, 2, 2, 3, 2, 3 **21.** $\frac{1}{3}$ **23.** $\frac{1}{3}$ **25.** $\frac{2}{3}$ **27.** $\frac{5}{2}$ **29.** $-\frac{1}{2}$ **31.** $-\frac{6}{7}$ **33.** $\frac{5}{9}$ **35.** $\frac{6}{7}$ **37.** in lowest terms **39.** $\frac{3}{8}$ **41.** $\frac{5}{7}$ **43.** $\frac{4}{5}$ **45.** in lowest terms **47.** in lowest terms **49.** $-\frac{1}{3}$ **51.** $\frac{3}{5}$ **53.** $\frac{5}{4}$ **55.** 2 **57.** $\frac{35}{40}$ **59.** $\frac{28}{35}$ **61.** $\frac{45}{54}$ **63.** $\frac{15}{30}$ **65.** $\frac{4}{14}$ **67.** $\frac{54}{60}$ **69.** $\frac{25}{20}$ **71.** $\frac{6}{45}$ **73.** $\frac{15}{5}$ **75.** $\frac{48}{8}$ **77.** $\frac{36}{9}$ **79.** $-\frac{4}{2}$ **81.** $\frac{3}{5}$ **83.** $-\frac{15}{16}$ in. **85.** $\frac{7}{10}, \frac{1}{8}$

87.

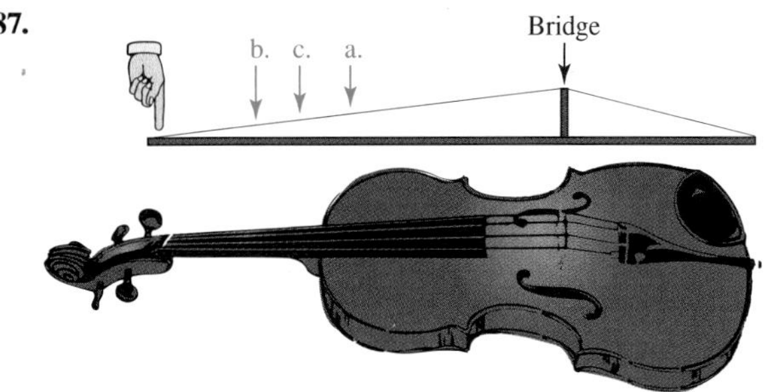

89. one-quarter turn to the left; three-quarters of a turn to the right **91.**

SNACKS
Potato chips
Peanuts
Pretzels
Tortilla chips

93. $\frac{1}{250}$ **99.** -3 **101.** 564,000

Study Set Section 3.2 (page 171)

1. multiply **3.** product **5.** base, height **7.** $\frac{ac}{bd}$
9. **a.** $\frac{1}{4}$ **b.** 12, 1, $\frac{1}{12}$
11. a. negative **b.** positive **13. a.** true **b.** true **c.** false **d.** true **15.** 7, 15, 3, 3 **17.** $\frac{1}{8}$ **19.** $\frac{21}{128}$ **21.** $\frac{4}{7}$ **23.** $\frac{77}{60}$ **25.** $-\frac{1}{5}$
27. $\frac{2}{9}$ **29.** $\frac{2}{3}$ **31.** 1 **33.** $\frac{1}{20}$ **35.** $\frac{1}{30}$ **37.** 15 **39.** -12
41. $\frac{5x}{72}$ **43.** $\frac{b}{40}$ **45.** d **47.** s **49.** $-c$ **51.** $\frac{2x}{9}$ **53.** $\frac{5x}{6}, \frac{5}{6}x$
55. $-\frac{8v}{9}, -\frac{8}{9}v$ **57.** $\frac{4}{9}$ **59.** $\frac{25}{81}$ **61.** $\frac{16}{9}$ **63.** $-\frac{27}{64}$

65.

$\cdot$	$\frac{1}{2}$	$\frac{1}{3}$	$\frac{1}{4}$	$\frac{1}{5}$	$\frac{1}{6}$
$\frac{1}{2}$	$\frac{1}{4}$	$\frac{1}{6}$	$\frac{1}{8}$	$\frac{1}{10}$	$\frac{1}{12}$
$\frac{1}{3}$	$\frac{1}{6}$	$\frac{1}{9}$	$\frac{1}{12}$	$\frac{1}{15}$	$\frac{1}{18}$
$\frac{1}{4}$	$\frac{1}{8}$	$\frac{1}{12}$	$\frac{1}{16}$	$\frac{1}{20}$	$\frac{1}{24}$
$\frac{1}{5}$	$\frac{1}{10}$	$\frac{1}{15}$	$\frac{1}{20}$	$\frac{1}{25}$	$\frac{1}{30}$
$\frac{1}{6}$	$\frac{1}{12}$	$\frac{1}{18}$	$\frac{1}{24}$	$\frac{1}{30}$	$\frac{1}{36}$

67. 15 ft^2 **69.** $\frac{15}{2}$ yd^2 **71.** 290 **73.** 18, 6, and 2 in.
75. $\frac{3}{8}$ cup sugar, $\frac{1}{6}$ cup molasses
77.

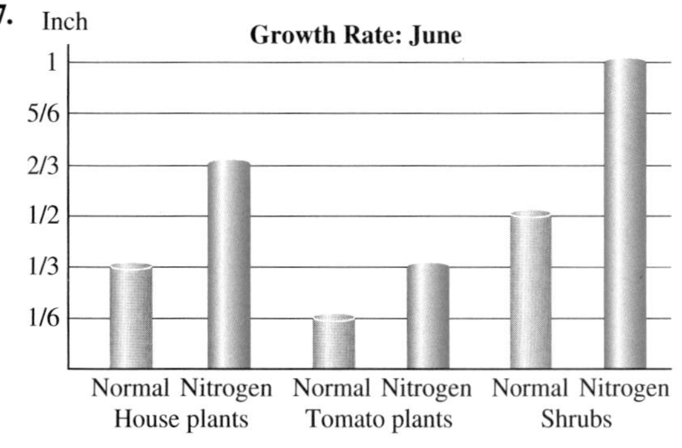

79. 121 in.^{2} **81.** 18 in.^{2} **87.** 987,000 **89.** no

Study Set Section 3.3 (page 179)

1. reciprocals **3.** $\frac{1}{2}, \frac{3}{2}$ **5.** $4 \div \frac{1}{3}$, 12
7. 1 **9. a.** 5 **b.** 5 **c.** $\frac{1}{3}$ **11.** 9, 10, 9, 10, 5, 5, 9, 9, 5
13. $\frac{5}{6}$ **15.** $\frac{27}{16}$ **17.** 1 **19.** $\frac{2}{3}$ **21.** 36 **23.** 50 **25.** $\frac{2}{15}$
27. $\frac{1}{192}$ **29.** $-\frac{27}{8}$ **31.** $-\frac{15}{2}$ **33.** $-\frac{1}{64}$ **35.** 1 **37.** $\frac{8}{15}$ **39.** $\frac{1}{6}$
41. $\frac{13}{8}$ **43.** $-\frac{5}{8}$ **45.** 104 **47.** 56 **49.** route 1 **51. a.** 16 **b.** $\frac{3}{4}$ in. **c.** $\frac{1}{120}$ in. **53.** 7,855 **59.** -4 **61.** 27 **63.** false
65. 637,500

Think It Through (page 185)

$\frac{7}{20}$

Study Set Section 3.4 (page 188)

1. least **3.** higher **5.** denominators, numerators, common
7. The denominators are unlike. **9.** 4 **11. a.** once **b.** twice **c.** three times **13.** 60
15. a. $\frac{1}{3}$

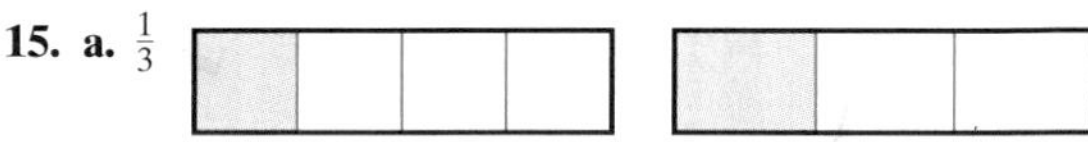

b. $\frac{1}{4}, \frac{1}{3}; \frac{1}{4} = \frac{3}{12}, \frac{1}{3} = \frac{4}{12}$; since $\frac{4}{12} > \frac{3}{12}, \frac{1}{3} > \frac{1}{4}$. **17.** 3, 3, 6, 5, 6, 5
19. 18 **21.** 24 **23.** 40 **25.** 60 **27.** $\frac{4}{7}$ **29.** $\frac{20}{103}$ **31.** $\frac{2}{5}$ **33.** $\frac{8}{7}$
35. $\frac{5}{8}$ **37.** $\frac{9}{20}$ **39.** $\frac{22}{15}$ **41.** $\frac{23}{56}$ **43.** $\frac{1}{12}$ **45.** $\frac{1}{12}$ **47.** $\frac{47}{50}$ **49.** $-\frac{3}{16}$
51. $-\frac{2}{3}$ **53.** $-\frac{23}{24}$ **55.** $-\frac{13}{5}$ **57.** $-\frac{23}{4}$ **59.** $\frac{47}{60}$ **61.** $\frac{3}{4}$ **63.** $\frac{19}{48}$
65. $-\frac{43}{45}$ **67.** $\frac{26}{75}$ **69.** $\frac{17}{54}$ **71.** $\frac{5}{36}$ **73.** $-\frac{17}{60}$ **75. a.** $\frac{7}{32}$ in.
b. $\frac{3}{32}$ in. **77.** $\frac{17}{24}$; no **79.** $\frac{1}{16}$ lb, undercharge **81.** $\frac{4}{5}, \frac{3}{4}, \frac{5}{8}$
83. $\frac{7}{10}$ **85.** $\frac{1}{6}$ hp **91.** $2^2 \cdot 5$ **93.** $A = lw$

The LCM and the GCF (page 193)

1. 15 **3.** 56 **5.** 42 **7.** 18 **9.** 660 **11.** 600 **13.** 72 **15.** 378
17. 3 **19.** 11 **21.** 4 **23.** 25 **25.** 20 **27.** 12 **29.** 6 **31.** 9
33. 360 min (6 hr)

Think It Through (page 196)

$1\frac{1}{3}$

Study Set Section 3.5 (page 198)

1. mixed **3.** graph **5. a.** $-5\frac{1}{2}°$ **b.** $-6\frac{7}{8}$ in.
7. a. $2\frac{2}{3}$ **b.** $1\frac{1}{3}$ **9.** $-\frac{4}{5}, -\frac{2}{5}, \frac{1}{5}$ **11.** $2\frac{1}{2}$
13.

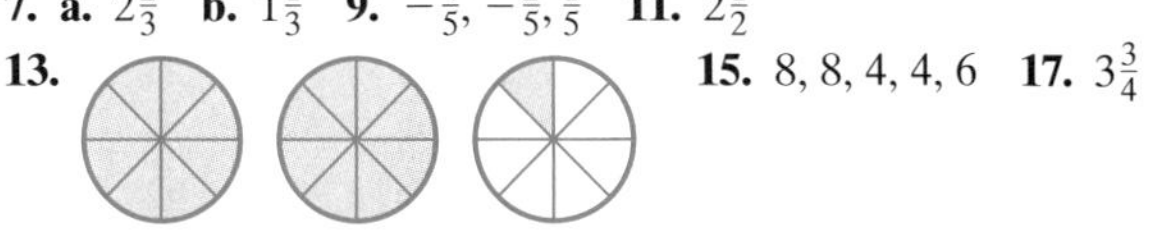

15. 8, 8, 4, 4, 6 **17.** $3\frac{3}{4}$
19. $5\frac{4}{5}$ **21.** $-3\frac{1}{3}$ **23.** $10\frac{7}{12}$ **25.** $\frac{13}{2}$ **27.** $\frac{104}{5}$ **29.** $-\frac{56}{9}$ **31.** $\frac{602}{3}$
33. $-2\frac{8}{9}$, $1\frac{2}{3}$, $\frac{16}{5}$

−5 −4 −3 −2 −1 0 1 2 3 4 5

35. $-\frac{10}{3}$, $-\frac{98}{99}$, $3\frac{1}{7}$

−5 −4 −3 −2 −1 0 1 2 3 4 5

37. $3\frac{4}{7}$
39. $10\frac{1}{2}$ **41.** 14 **43.** $-13\frac{3}{4}$ **45.** $-8\frac{1}{3}$ **47.** $\frac{35}{72}$ **49.** $-1\frac{1}{4}$
51. $\frac{25}{9} = 2\frac{7}{9}$ **53.** $-\frac{64}{27} = -2\frac{10}{27}$ **55.** $1\frac{9}{11}$ **57.** $-\frac{9}{10}$ **59.** 12

61. $\frac{5}{16}$ **63.** $-\frac{2}{3}$ **65.** $2\frac{1}{2}$ **67.** -2 **69.** 64 calories **71.** 357¢ = \$3.57 **73.** 675 **75.** $2\frac{3}{4}$ in., $1\frac{1}{4}$ in. **77.** $42\frac{5}{8}$ in.^{2} **79.** 602 **81.** size 14, slim cut **87.** 72 **89.** 4(8) **91.** -24

Study Set Section 3.6 (page 206)

1. commutative **3.** borrow **5. a.** 76, $\frac{3}{4}$ **b.** $76 + \frac{3}{4}$ **7.** the fundamental property of fractions **9. a.** $10\frac{1}{16}$ **b.** $1{,}290\frac{1}{3}$ **c.** $17\frac{1}{2}$ **d.** $46\frac{1}{5}$ **11.** 70, 39, 70, 39, 7, 5, 7, 5, 35, 35, 31 **13.** $4\frac{2}{5}$ **15.** $5\frac{1}{7}$ **17.** $7\frac{1}{2}$ **19.** $5\frac{11}{30}$ **21.** $1\frac{1}{4}$ **23.** $1\frac{11}{24}$ **25.** $9\frac{3}{10}$ **27.** $3\frac{5}{14}$ **29.** $129\frac{11}{15}$ **31.** $397\frac{5}{12}$ **33.** $273\frac{2}{9}$ **35.** $623\frac{8}{21}$ **37.** $11\frac{1}{30}$ **39.** $101\frac{7}{16}$ **41.** $2\frac{1}{2}$ **43.** $26\frac{7}{24}$ **45.** $10\frac{7}{16}$ **47.** $320\frac{5}{18}$ **49.** $6\frac{1}{3}$ **51.** $\frac{1}{4}$ **53.** $3\frac{12}{35}$ **55.** $3\frac{5}{8}$ **57.** $4\frac{1}{3}$ **59.** $3\frac{7}{8}$ **61.** $53\frac{5}{12}$ **63.** $460\frac{1}{8}$ **65.** $-5\frac{1}{4}$ **67.** $-5\frac{7}{8}$ **69.** $2\frac{3}{4}$ mi **71.** $7\frac{2}{3}$ cups **73.** $48\frac{1}{2}$ ft **75. a.** $16\frac{1}{2}, 16\frac{1}{2}; 5\frac{1}{5}, 5\frac{1}{5}$ **b.** $21\frac{7}{10}$ mi **77. a.** 20¢ **b.** 30¢ **79.** $191\frac{2}{3}$ ft **85.** 7 **87.** 6 **89.** -10

Study Set Section 3.7 (page 214)

1. complex **3.** $\frac{2}{3} \div \frac{1}{5}$ **5.** 15 **7.** negative **9.** 60 **11.** $\frac{3}{4}, \frac{4}{3}$, 4, 4 **13.** $\frac{1}{3}$ **15.** $\frac{31}{45}$ **17.** $\frac{37}{40}$ **19.** $\frac{3}{10}$ **21.** $-1\frac{27}{40}$ **23.** $\frac{3}{4}$ **25.** $-\frac{3}{64}$ **27.** $-1\frac{1}{6}$ **29.** $8\frac{1}{2}$ **31.** $\frac{49}{4}$ **33.** $\frac{121}{16}$ **35.** $8\frac{1}{4}$ in. **37.** $\frac{5}{6}$ **39.** $-1\frac{1}{3}$ **41.** $10\frac{1}{2}$ **43.** $\frac{4}{9}$ **45.** 3 **47.** 5 **49.** -20 **51.** 11 **53.** $\frac{3}{7}$ **55.** $-\frac{3}{8}$ **57.** $8\frac{1}{2}$ **59.** $2\frac{1}{2}, 1\frac{1}{2}, 3\frac{3}{4}; 7\frac{1}{5}, 1\frac{1}{2}, 10\frac{4}{5}; 14\frac{11}{20}$ mi **61.** yes **63.** $10\frac{1}{2}$ mi **65.** 6 sec **71.** 2 **73.** -5 **75.** 8

Study Set Section 3.8 (page 222)

1. reciprocal **3.** least common denominator **5.** Yes; when 40 is substituted for x, the result is a true statement: 25 = 25. **7.** 1 **9. a.** $\frac{4}{5}p$ **b.** $\frac{1}{4}t$ **11.** $\frac{8}{7}, \frac{8}{7}$ **13. a.** true **b.** false **c.** true **d.** true **15.** 28 **17.** -32 **19.** $-\frac{20}{3}$ **21.** $\frac{6}{5}$ **23.** 0 **25.** 30 **27.** $-\frac{14}{5}$ **29.** $\frac{4}{25}$ **31.** $-\frac{1}{2}$ **33.** $\frac{2}{5}$ **35.** $\frac{8}{9}$ **37.** $\frac{1}{3}$ **39.** $\frac{7}{18}$ **41.** $-\frac{5}{8}$ **43.** -1 **45.** $-\frac{5}{4}$ **47.** -36 **49.** $\frac{5}{12}$ **51.** $\frac{8}{9}$ **53.** $-\frac{27}{2}$ **55.** $-\frac{5}{3}$ **57.** $-\frac{24}{7}$ **59.** 12 **61.** 24 **63.** -12 **65.** $\frac{75}{4}$ **67.** $\frac{1}{3}$, 32, customers, the number of customers last year, multiply, $\frac{1}{3}x$, 3, 3, 96. The shop had 96 customers last year. 32 **69.** 20 **71.** 450 **73.** 36 **75.** 8 in. **77.** $\frac{5}{12}$ **83.** -32 **85.** 3 **87.** 13,000,000

Key Concept (page 226)

1. 5, 5, 5, 5, 5 **3.** 7, 7, 7, 7, 7

Chapter Review (page 228)

1. $\frac{7}{24}$ **2.** The figure is not divided into equal parts. **3.** $-\frac{2}{3}, \frac{-2}{3}$ **4.** equivalent fractions: $\frac{6}{8} = \frac{3}{4}$ **5.** The numerator and denominator of the fraction are being divided by 2. **6.** The numerator and denominator of the fraction are being divided by 2. The answer to each division is 1. **7.** $\frac{1}{3}$ **8.** $\frac{5}{12}$ **9.** $-\frac{3}{4}$ **10.** $\frac{11}{18}$ **11.** The numerator and denominator of the original fraction are being multiplied by 2 to obtain an equivalent fraction in higher terms. **12.** $\frac{12}{18}$ **13.** $-\frac{6}{16}$ **14.** $\frac{21}{45}$ **15.** $\frac{36}{9}$ **16.** $\frac{1}{6}$ **17.** $-\frac{14}{45}$ **18.** $\frac{5}{12}$ **19.** $\frac{1}{5}$ **20.** $\frac{21}{5}$ **21.** $\frac{9}{4}$ **22.** 1 **23.** 1 **24.** true **25.** false **26.** $\frac{2}{9}$ **27.** $-\frac{8}{21}s$ **28.** $\frac{1}{21}$ **29.** $-\frac{5}{9}m$ **30.** $\frac{9}{16}$ **31.** $-\frac{125}{8}$ or $-15\frac{5}{8}$ **32.** $\frac{4}{9}$ **33.** $-\frac{8}{125}$ **34.** 30 lb **35.** 60 in.^{2} **36.** 8 **37.** $-\frac{12}{11}$ **38.** $-\frac{1}{6}$ **39.** $\frac{1}{200}$ **40.** $\frac{25}{66}$ **41.** $-\frac{7}{2}$ **42.** $\frac{3}{32}$ **43.** $\frac{5}{2}$ **44.** $\frac{1}{2}$ **45.** $\frac{8}{5}$ **46.** 12 **47.** $\frac{5}{7}$ **48.** $-\frac{6}{5}$ **49.** $\frac{1}{4}$ **50.** $\frac{7}{4} = 1\frac{3}{4}$ **51.** The denominators are not the same. **52.** 90 **53.** $\frac{5}{6}$ **54.** $\frac{1}{40}$ **55.** $-\frac{29}{24}$ **56.** $\frac{20}{7}$ **57.** $-\frac{23}{6}$ **58.** $\frac{47}{60}$ **59.** $\frac{7}{32}$ in. **60.** the second hour **61.** $2\frac{1}{6}$ **62.** $\frac{13}{6}$ **63.** $3\frac{1}{5}$ **64.** $-3\frac{11}{12}$ **65.** 1 **66.** $2\frac{1}{3}$ **67.** $\frac{75}{8}$ **68.** $-\frac{11}{5}$ **69.** $\frac{201}{2}$ **70.** $\frac{199}{100}$

71.

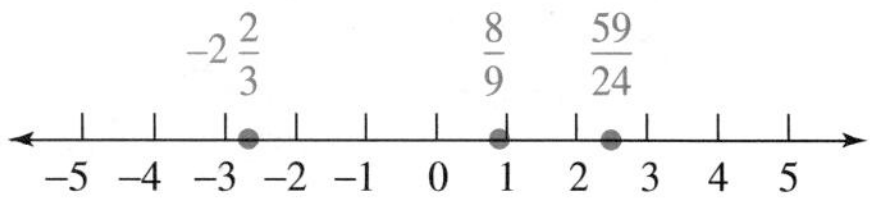

72. $-\frac{3}{10}$ **73.** $\frac{21}{22}$ **74.** 40 **75.** $-2\frac{1}{2}$ **76.** $48\frac{1}{8}$ in. **77.** $3\frac{23}{40}$ **78.** $6\frac{1}{6}$ **79.** $1\frac{1}{12}$ **80.** $1\frac{5}{16}$ **81.** $39\frac{11}{12}$ gal **82.** $182\frac{5}{18}$ **83.** $113\frac{3}{20}$ **84.** $31\frac{11}{24}$ **85.** $316\frac{3}{4}$ **86.** $20\frac{1}{2}$ **87.** $34\frac{3}{8}$ **88.** $\frac{8}{9}$ **89.** $\frac{19}{72}$ **90.** $-\frac{12}{17}$ **91.** $-\frac{2}{5}$ **92.** 24 **93.** 28 **94.** $-\frac{1}{3}$ **95.** $\frac{11}{2}$ **96.** $\frac{57}{8}$ **97.** $-\frac{18}{5}$ **98.** 16 **99.** $\frac{26}{3}$ **100.** 330

Chapter 3 Test (page 233)

1. a. $\frac{4}{5}$ **b.** $\frac{1}{5}$ **2. a.** $\frac{3}{4}$ **b.** $\frac{2}{5}$ **3.** $-\frac{3}{20}$ **4.** 40 **5.** 12 **6.** $-\frac{19}{30}$ **7.** $\frac{21}{24}$ **8.**

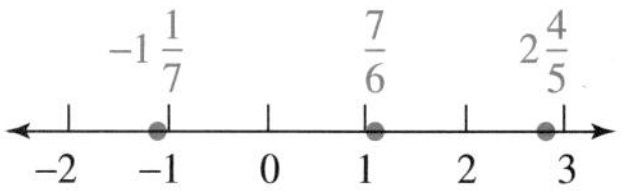

9. \$$1\frac{1}{2}$ million **10.** $261\frac{11}{36}$ **11.** $37\frac{5}{12}$ **12. a.** 0 lb **b.** $2\frac{3}{4}$ in. **c.** $3\frac{3}{4}$ in. **13.** $\frac{11}{7}$ **14.** $11\frac{3}{4}$ in. **15.** perimeter: $53\frac{1}{3}$ in.; area: $106\frac{2}{3}$ in.^{2} **16.** 60 **17.** 12 **18.** $\frac{13}{24}$ **19.** $-\frac{20}{21}$ **20.** $-\frac{5}{3}$ **21.** yes **22.** 42 **23.** $-\frac{36}{5}$ **24.** $\frac{1}{6}$ **25.** $\frac{1}{8}$ **26.** $\frac{9}{2}$ **27.** 108, 36 **28.** numerator, fraction bar, denominator; equal parts of a whole, or a division **29.** When we multiply a number, such as $\frac{3}{4}$, and its reciprocal, $\frac{4}{3}$, the result is 1. **30. a.** simplifying a fraction: dividing the numerator and denominator of a fraction by the same number **b.** equivalent fractions: $\frac{1}{2} = \frac{2}{4}$ **c.** building a fraction: multiplying the numerator and denominator of a fraction by the same number

Cumulative Review Exercises (page 235)

1. 5,434,700 **2.** 5,430,000 **3.** 11,555, 10:30 A.M. **4.** hundred billions **5.** 8,136 **6.** 3,519 **7.** 299,320 **8.** 991 **9.** 450 ft **10.** 11,250 ft^2 **11.** $2^2 \cdot 3 \cdot 7$ **12.** $2 \cdot 3^2 \cdot 5^2$ **13.** $2^3 \cdot 3^2 \cdot 5$ **14.** $2^4 \cdot 3^2 \cdot 5^2$ **15.** 16 **16.** -35 **17.** 2 **18.** 2 **19.** -5 **20.** -5 **21.** -16 **22.** -5 **23.** no **24.** 21 **25.** $\frac{3}{4}$ **26.** $\frac{5}{2}$ **27.** $-\frac{4}{5}$ **28.** $\frac{1}{2}$ **29.** $1\frac{5}{12}$ **30.** $\frac{11}{15}$ **31.** $\frac{23}{6}$ **32.** $-\frac{53}{8}$ **33.** $9\frac{11}{12}$ **34.** $5\frac{11}{15}$ **35.** $\frac{11}{16}$ in. **36.** 30 sec, 60 sec **37.** $\frac{2}{7}$ **38.** $-1\frac{9}{29}$ **39.** $-\frac{17}{15}$ **40.** 4 **41.** -15 **42.** $\frac{8}{3}$ **43.** An expression is a combination of numbers and/or variables with operation symbols. An equation contains an = symbol. **44.** a letter that is used to stand for a number

Chapter 4 Check Your Knowledge (page 238)

1. sum **2.** root **3.** $\frac{21}{250}$ **4.** \$85.80 **5. a.** 354.2782 **b.** 20,004.78 **6. a.** 7.875 ft^2 **b.** 11.5 ft **7.** 3.1 **8. a.** 0.15 **b.** 0.625 **c.** $0.\overline{1}$ **9.** 1.6

10.

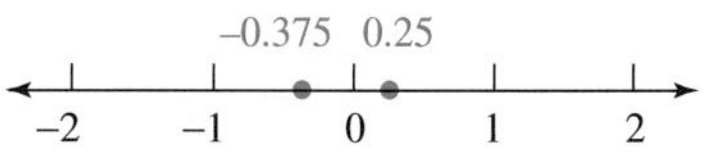

11. $\frac{5}{12}$ **12.** 6.9 **13.**

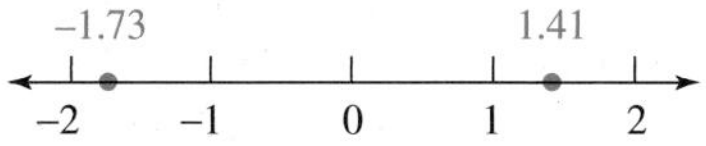

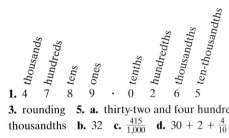

14. a. 19 **b.** $\frac{19}{20}$ **c.** −0.3 **15. a.** > **b.** < **c.** > **16.** 52.6 mph **17.** −9.43 **18.** −9 **19.** 0.8 **20.** 625

Study Set Section 4.1 (page 245)

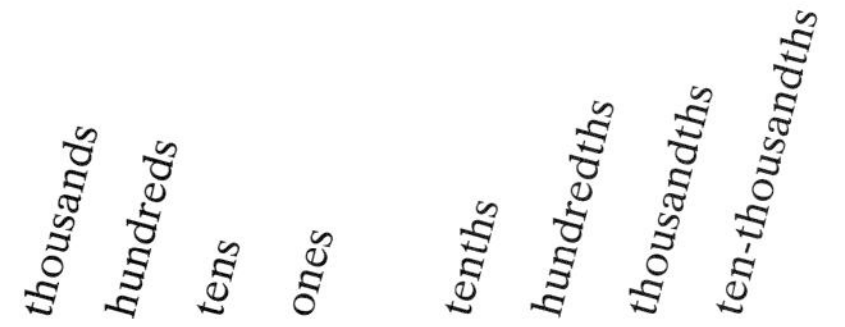

1. 4 7 8 9 · 0 2 6 5

3. rounding **5. a.** thirty-two and four hundred fifteen thousandths **b.** 32 **c.** $\frac{415}{1,000}$ **d.** $30 + 2 + \frac{4}{10} + \frac{1}{100} + \frac{5}{1,000}$

7.

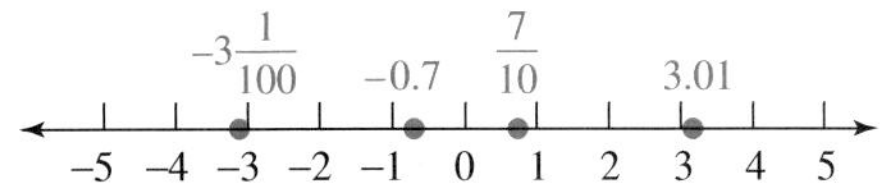

9. a. true **b.** false **c.** true **d.** true **11.** $\frac{47}{100}$, 0.47

13.

15. 9,816.0245

17. fifty and one tenth; $50\frac{1}{10}$ **19.** negative one hundred thirty-seven ten-thousandths; $-\frac{137}{10,000}$ **21.** three hundred four and three ten-thousandths; $304\frac{3}{10,000}$ **23.** negative seventy-two and four hundred ninety-three thousandths; $-72\frac{493}{1,000}$ **25.** −0.39 **27.** 6.187 **29.** 506.1 **31.** 2.7 **33.** −0.14 **35.** 33.00 **37.** 3.142 **39.** 1.414 **41.** 39 **43.** 2,988 **45. a.** $3,090 **b.** $3,090.30 **47.** < **49.** > **51.** 132.64, 132.6401, 132.6499 **53.** $1,025.78

55.

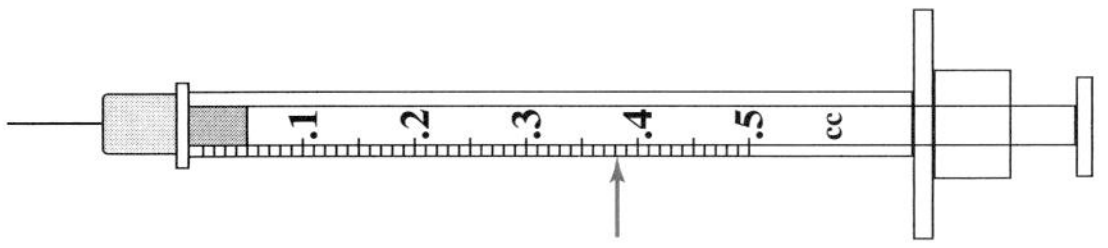

57. a. 0.30 **b.** 1,609.34 **c.** 453.59 **d.** 3.79 **59.** sand, silt, granule, clay **61.** Texas City, Houston, Westport, Galveston, White Plains, Crestline **63.** gold: Patterson; silver: Khorkina; bronze: Zhang **65. a.** Q1, 2004; $0.25 **b.** Q2, 2002; −$0.25 **73.** $164\frac{11}{20}$ **75.** $\frac{1}{8}$ **77.** $\frac{24}{36}$ **79.** −9

Study Set Section 4.2 (page 252)

1. sum **3.** opposite **5.** point **7.** 39.9 **9.** 54.72 **11.** 15.9 **13.** 0.23064 **15.** 288.46 **17.** 58.04 **19.** 9.53 **21.** 70.29 **23.** 4.977 **25.** 0.19 **27.** −10.9 **29.** 38.29 **31.** −14.3 **33.** −0.0355 **35.** −16.6 **37.** 47.91 **39.** 2.598 **41.** 11.01 **43.** 4.1 **45.** 35.85 **47.** −57.47 **49.** 6.2 **51.** 15.2 **53.** 8.03 **55. a.** 53.044 sec **b.** 102.38 **57.** 103.4 in. **59.** 1.8, Texas **61.** 1.74, 2.32, 4.06; 2.90, 0, 2.90 **63.** 43.03 sec **65.** $765.69, $740.69 **67. a.** $101.94 **b.** $55.80 **69.** 8,156.9343 **71.** 1,932.645 **73.** 2,529.0582 **79.** $110\frac{23}{40}$ **80.** $-\frac{5}{6}$

Study Set Section 4.3 (page 261)

1. factors, product **3.** whole, sum **5. a.** $\frac{21}{1,000}$ **b.** $\frac{21}{1,000} = 0.021$. They are the same. **7.** 2.3 **9.** 0.08 **11.** −0.15 **13.** 0.98 **15.** 0.072 **17.** 12.32 **19.** −0.0049 **21.** −0.084 **23.** −8.6265 **25.** 9.6 **27.** −56.7 **29.** 12.24 **31.** −18.183 **33.** 0.024 **35.** −16.5 **37.** 42 **39.** 6,716.4 **41.** −0.56 **43.** 8,050 **45.** 980 **47.** −200 **49.** 0.01, 0.04, 0.09, 0.16, 0.25, 0.36, 0.49, 0.64, 0.81 **51.** 1.44 **53.** 1.69 **55.** −17.48 **57.** 14.24 **59.** 0.84 **61.** −3.872 **63.** 18.72 **65.** 86.49 **67. a.** $12.50, $12,500; $15.75, $1,575 **b.** $14,075 **69.** 0.75 in. **71.** 136.4 lb **73.** $95.20, $123.75, $77.00 **75.** 160.6 m **77.** 0.000000136 in., 0.0000000136 in., 0.00000004 in. **79.** 15.29694 **81.** 631.2722 **83.** $102.65 **89.** 7 **91.** the absolute value of negative three **93.** −1

Think It Through (page 269)

2.86

Study Set Section 4.4 (page 269)

1. dividend, divisor, quotient **3.** whole, right, above **5.** true **7.** 10 **9.** Use multiplication to see whether $2.13 \cdot 0.9 = 1.917$. **11.** yes **13.** moving the decimal points in the divisor and dividend 2 places to the right **15.** 4.5 **17.** −9.75 **19.** 6.2 **21.** 32.1 **23.** 2.46 **25.** −7.86 **27.** 2.66 **29.** 7.17 **31.** 130 **33.** 1,050 **35.** 0.6 **37.** 0.6 **39.** 5.3 **41.** −2.4 **43.** 13.60 **45.** 0.79 **47.** 0.07895 **49.** −0.00064 **51.** 0.0348 **53.** 4.504 **55.** −0.96 **57.** 1,027.19 **59.** 9.1 **61.** 304.07 **63.** 280 **65.** 11 hr later: 6 P.M. **67.** 567 **69.** 1998: $13.00; 2003: $15.35 **71.** 0.37 mi **73.** 7.24 **75.** −3.96 **81.** $\frac{7}{6}$ **83.** {. . . , −3, −2, −1, 0, 1, 2, 3, . . .} **85.** 12

Study Set Estimation (page 273)

1. approx. $240 **3.** approx. 2 cubic feet less **5.** approx. 30 **7.** approx. $330 **9.** approx. $520 **11.** not reasonable **13.** reasonable **15.** reasonable **17.** not reasonable

Study Set Section 4.5 (page 279)

1. repeating **3.** decimal **5. a.** 7 ÷ 8 **b.** numerator **7.** smaller

9.

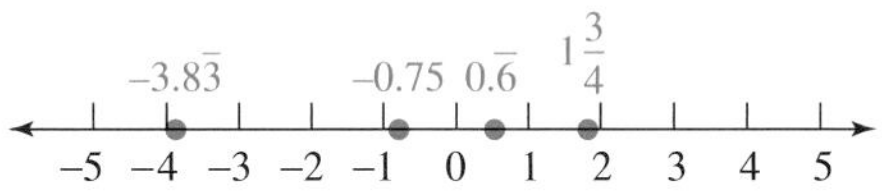

11. a. false **b.** true **c.** true **d.** false **13. a.** no **b.** It is a repeating decimal. **15.** 0.5 **17.** −0.625 **19.** 0.5625 **21.** −0.53125 **23.** 0.55 **25.** 0.775 **27.** −0.015 **29.** 0.002 **31.** $0.\overline{6}$ **33.** $0.\overline{45}$ **35.** $-0.58\overline{3}$ **37.** $0.0\overline{3}$ **39.** 0.23 **41.** 0.38 **43.** 0.152 **45.** 0.370 **47.** 1.33 **49.** −3.09 **51.** 3.75 **53.** −8.67 **55.** 12.6875 **57.** 203.73 **59.** < **61.** < **63.** $\frac{37}{90}$ **65.** $\frac{19}{60}$ **67.** $\frac{3}{22}$ **69.** $-\frac{1}{90}$ **71.** 1 **73.** 0.57 **75.** 5.27 **77.** 0.24 **79.** −2.55 **81.** 0.068 **83.** 7.11 **85.** −1.7 **87.** 4.25 **89.** $0.\overline{2277}$ **91.** 37.2 **93.** 0.0625, 0.375, 0.5625, 0.9375 **95.** $\frac{3}{40}$ in. **97.** 23.4 sec, 23.8 sec, 24.2 sec, 32.6 sec **99.** 93.6 in.^{2} **105.** −1 **107.** $\frac{19}{6}$ **109.** $\frac{2}{3}$

Study Set Section 4.6 (page 285)

1. solve **3.** associative **5.** 2.1(1.7) − 6.3 = −2.73 **7.** 2.3, 2.3, 0.6*s*, 0.6, 0.6 **9.** 1.7 **11.** 7.11 **13.** −11.5 **15.** −0.1 **17.** −4.36 **19.** 1.3 **21.** −8.16 **23.** 22.44 **25.** −21.18 **27.** 0.4 **29.** −2.2 **31.** − 2 **33.** 31 **35.** 1 **37.** 0.3 **39.** 15, 30, 60, signatures, the number of signatures she needs to collect, $0.30, 0.30*x*, 0.30*x*, 0.30*x*, 60, 0.30*x*. She needs to collect 150 signatures to make $60. 150, 150, 45 **41.** $8.6 million **43.** 3.27 **45.** 10.7 **47.** 12.4 mpg **49.** 200 **53.** $\frac{1}{12}$ **55.** $\frac{14}{13}$ **57.** −6 **59.** 12

Study Set Section 4.7 (page 291)

1. root **3.** radical, positive **5.** radicand **7.** 25, 25 **9.** 7^2 **11.** $\frac{3}{4}$ **13.** $\sqrt{6}, \sqrt{11}, \sqrt{23}, \sqrt{27}$ **15. a.** 1 **b.** 0 **17. a.** 2.4 **b.** 5.76 **c.** 0.24

19.

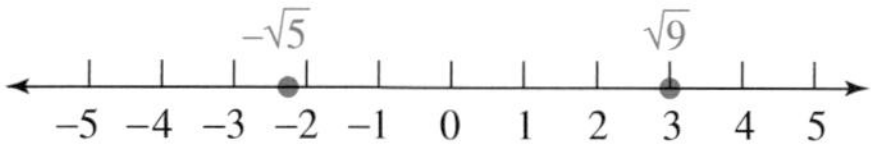

21. a. 4, 5 **b.** 9, 10 **23.** −7, 8 **25.** 4 **27.** −11 **29.** −0.7 **31.** 0.5 **33.** 0.3 **35.** $-\frac{1}{9}$ **37.** $-\frac{4}{3}$ **39.** $\frac{2}{5}$ **41.** 31 **43.** −20 **45.** $-\frac{7}{20}$ **47.** −70 **49.** 2.56 **51.** −3.6 **53.** 1, 1.414, 1.732, 2, 2.236, 2.449, 2.646, 2.828, 3, 3.162 **55.** 37 **57.** 61 **59.** 3.87 **61.** 8.12 **63.** 4.904 **65.** −3.332 **67.** 4,899 **69.** −0.0333 **71. a.** 5 ft **b.** 10 ft **73.** 127.3 ft **75.** 42-inch **83.** subtraction and multiplication **85.** 16 **87.** $\frac{5}{6}$ **89.** 30

Key Concept (page 294)

1. {1, 2, 3, 4, 5, . . .} **3.** {. . . , −3, −2, −1, 0, 1, 2, 3, . . .} **5.** nonterminating, nonrepeating decimals; a number that can't be written as a fraction of two integers **7.** false **9.** false **11.** true **13.** false **15.** true

Chapter Review (page 296)

1. 0.67, $\frac{67}{100}$ **2.**

3. $10 + 6 + \frac{4}{10} + \frac{5}{100} + \frac{2}{1{,}000} + \frac{3}{10{,}000}$ **4.** two and three tenths, $2\frac{3}{10}$ **5.** negative fifteen and fifty-nine hundredths, $-15\frac{59}{100}$ **6.** six hundred one ten-thousandths, $\frac{601}{10{,}000}$ **7.** one one hundred thousandth, $\frac{1}{100{,}000}$

8.

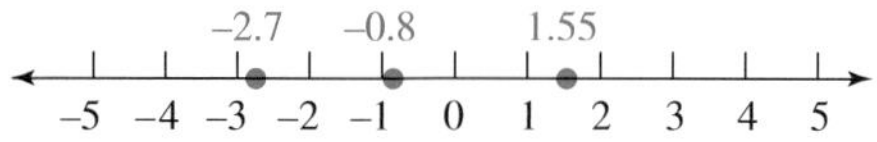

9. Washington, Diaz, Chou, Singh, Gerbac **10.** true **11.** < **12.** > **13.** = **14.** < **15.** 4.58 **16.** 3,706.090 **17.** −0.1 **18.** 88.1 **19.** 66.7 **20.** 45.188 **21.** 15.17 **22.** 27.71 **23.** −7.7 **24.** 3.1 **25.** −4.8 **26.** −29.09 **27.** −25.6 **28.** 4.939 **29.** \$48.21 **30.** 8.15 in. **31.** −0.24 **32.** 2.07 **33.** −17.05 **34.** 197.945 **35.** 0.00006 **36.** 4.2 **37.** 90,145.2 **38.** 2,897 **39.** 0.04 **40.** 0.0225 **41.** 10.89 **42.** 0.001 **43.** −10.61 **44.** 25.82 **45.** 92.38 **46.** 68.62 in.^{2} **47.** 0.07 in. **48.** 1.25 **49.** −10.45 **50.** 1.29 **51.** 4.103 **52.** −2.9 **53.** 0.053 **54.** 63 **55.** 0.81 **56.** 12.9 **57.** −667.3 **58.** 20.22 **59.** \$8.34 **60.** 0.8976 **61.** −0.00112 **62.** 13.95 **63.** 14 **64.** 9.5 **65.** 0.875 **66.** −0.4 **67.** −0.5625 **68.** 0.06 **69.** $0.\overline{54}$ **70.** $-0.\overline{6}$ **71.** 0.58 **72.** 1.03 **73.** > **74.** >

75.

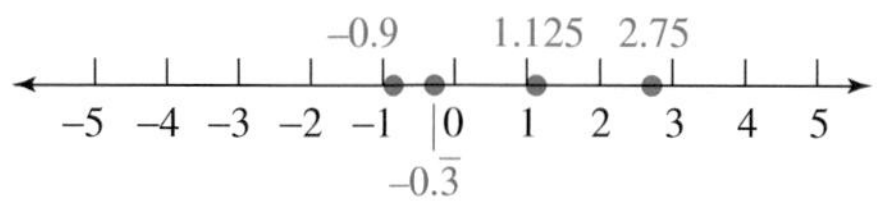

76. $\frac{11}{15}$ **77.** −6.24 **78.** 93 **79.** 39.564 **80.** 33.49 **81.** 34.88 in.^{2} **82.** −18.41 **83.** 4.77 **84.** −5.34 **85.** 17 **86.** yes **87.** 9 **88. a.** radical **b.** 8^2 **89.** 7 **90.** −4 **91.** 10 **92.** 0.3 **93.** $\frac{8}{5}$ **94.** 0.9 **95.** $-\frac{1}{6}$ **96.** 0 **97.** 9 and 10 **98.** It differs by 0.11.

99.

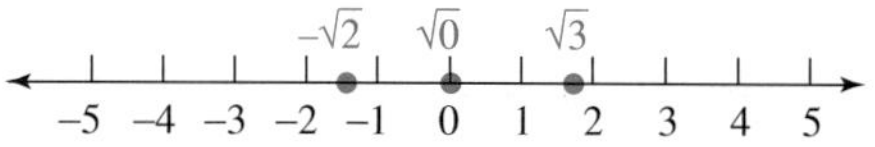

100. −30 **101.** 2.5 **102.** −27 **103.** 1.5 **104.** 4.36

Chapter 4 Test (page 301)

1. $\frac{79}{100}$, 0.79 **2.** Selway, Monroe, Paston, Covington, Cadia **3. a.** sixty-two and fifty-five hundredths; $62\frac{55}{100}$ **b.** eight thousand thirteen one hundred-thousandths; $\frac{8{,}013}{100{,}000}$ **4.** 33.050 **5.** \$208.75 **6. a.** 0.567909 **b.** 0.458 **7.** 1.02 in. **8.** 10.75 **9.** 6.121 **10.** 0.1024 **11.** 14.07 **12.** 1.25 mi^2 **13.** 0.004 in. **14.** 3.588 **15. a.** 0.34 **b.** $0.41\overline{6}$ **16.** −2.29 **17.** $1.\overline{18}$

18.

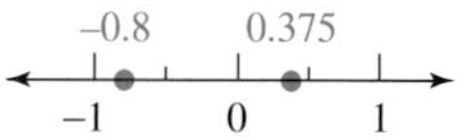

19. $\frac{41}{30}$ **20.** −7 **21.** 6.008 **22.** −0.425 **23.** 0.42 g **24.** 80 **25.** 12^2

26.

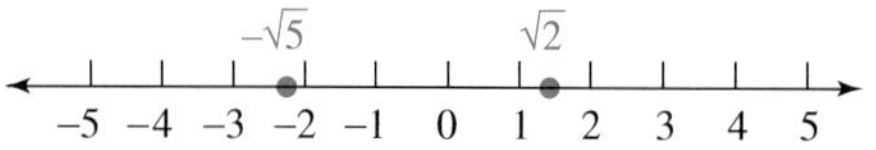

27. 11 **28.** $-\frac{1}{30}$ **29.** > **30.** > **31.** > **32.** < **33.** −0.2 **34.** 1.3

Cumulative Review Exercises (page 303)

1. \$788,000 **2.** $(x + y) + z = x + (y + z)$ **3.** 27 **4.** 1,000 **5.** $11 \cdot 5 \cdot 2^2$ **6.** 1, 2, 4, 5, 10, 20 **7.** {0, 1, 2, 3, 4, 5, . . .} **8.** −13 **9.** adding **10.** 8, −3, 36, −6, 6 **11.** $-15 = -5 \cdot 3$ **12.** −1 **13.** 9 **14.** 30 **15.** 35 **16.** 102 **17.** 3.61 **18.** −\$1,100 **19.** 5 **20.** $\frac{6}{13}$ **21.** equivalent fractions **22.** $\frac{5}{7}$ **23.** $\frac{21}{128}$ **24.** $-\frac{3}{16}$ **25.** $\frac{34}{21}$ **26.** $19\frac{1}{8}$ **27.** $26\frac{7}{24}$ **28.** $-\frac{1}{3}$ **29.** −45 **30.** 8 **31.** 157.5 in.^{2}

32.

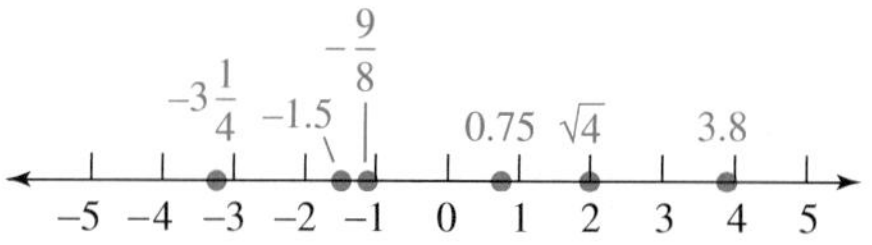

33. 0.001 in. **34.** < **35.** −8.136 **36.** 5.6 **37.** 5,601.2 **38.** 0.0000897 **39.** 47.95 **40.** 33.6 hr **41.** 232.8° C **42.** $0.14\overline{6}$ **43.** −9 **44.** 80 **45.** −6 **46.** $\frac{11}{15}$

Chapter 5 Check Your Knowledge (page 306)

1. one hundred **2.** amount, percent, base **3.** discount **4.** principal **5.** compound, simple **6. a.** 0.75, 75% **b.** 0.625, 62.5% **c.** 1.45, 145% **7. a.** 35%, $\frac{7}{20}$ **b.** 398%, $\frac{199}{50}$ or $3\frac{49}{50}$ **c.** 10.5%, $\frac{21}{200}$ **8. a.** 0.25, $\frac{1}{4}$ **b.** 2 or 2.0, 2 or $\frac{2}{1}$ **c.** 0.005, $\frac{1}{200}$ **9.** 35% **10. a.** 37.5% or $37\frac{1}{2}$% **b.** $83\frac{1}{3}$% **11. a.** $66\frac{2}{3}$% **b.** 66.7% **12.** 325 **13.** 17% **14.** 52 **15.** 250% **16.** \$2.99, \$11.96 **17.** \$29.95, 17% **18.** \$29.94 **19.** \$46.00 **20.** \$1,676.47 **21.** 93% **22.** \$1,045.00 **23.** \$40.71 **24.** 13

Study Set Section 5.1 (page 313)

1. percent **3.** 100, simplify **5.** right **7. a.** 0.84, 84%, $\frac{21}{25}$ **b.** 16% **9.** $\frac{17}{100}$ **11.** $\frac{1}{20}$ **13.** $\frac{3}{5}$ **15.** $\frac{5}{4}$ **17.** $\frac{1}{150}$ **19.** $\frac{21}{400}$ **21.** $\frac{3}{500}$ **23.** $\frac{19}{1{,}000}$ **25.** 0.19 **27.** 0.06 **29.** 0.408 **31.** 2.5 **33.** 0.0079

35. 0.0025 **37.** 93% **39.** 61.2% **41.** 3.14% **43.** 843% **45.** 5,000% **47.** 910% **49.** 17% **51.** 16% **53.** 40% **55.** 105% **57.** 62.5% **59.** 18.75% **61.** $66\frac{2}{3}$% **63.** $8\frac{1}{3}$% **65.** 11.11% **67.** 55.56% **69. a.** $\frac{15}{191}$ **b.** 8% **71. a.** $\frac{9}{22}$ **b.** 41% **73. a.** $\frac{5}{29}$ **b.** 17% **c.** 24% **75.** 5 ft **77.** 0.9944 **79.** as a decimal; 89.6% **81.** torso: 27.5%

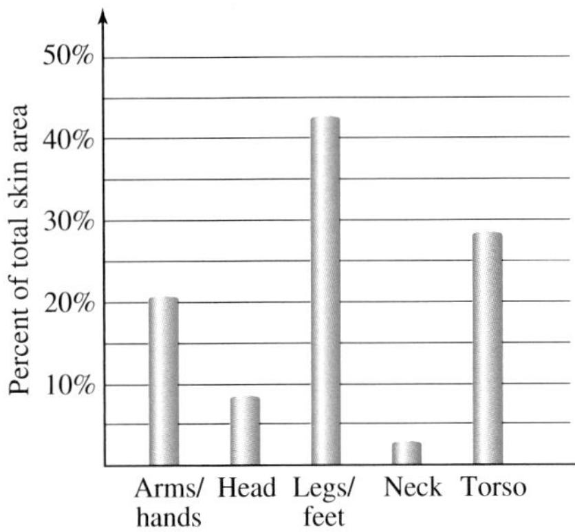

83. 92% **85.** 0.27% **93.** 9 **95.** $\frac{41}{99}$ **97.** 27.075

Solving Percent Problems Using Proportions (page 325)

1. 192 **3.** 225 **5.** 150 **7.** 148% **9.** 700

Think It Through (page 326)

36% are enrolled in college full time; 69% of full-time students read 5 or more assigned textbooks, manuals, or books during the current school year; 38% occasionally

Study Set Section 5.2 (page 326)

1. graph **3.** $x = 0.10 \cdot 50$ **5.** $48 = x \cdot 47$ **7. a.** 0.12 **b.** 0.056 **c.** 1.25 **d.** 0.0025 **9.** more **11. a.** 25 **b.** 100% **c.** 87 **13.** 44% **15. a.** multiply **b.** equals **c.** x (as a variable) **17.** 90 **19.** 80% **21.** 65 **23.** 0.096 **25.** 0.00125% **27.** 44 **29.** 43.5 **31.** 107.1 **33.** 99 **35.** 60 **37.** 31.25% **39.**

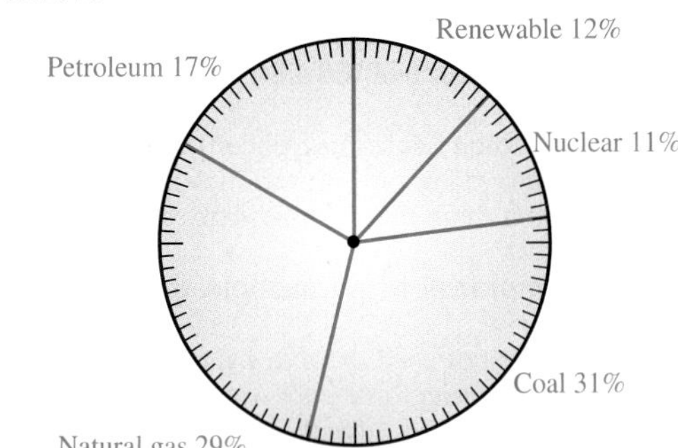

Source: Energy Information Administration

41. 120 **43.** $666 billion **45.** 38,000 = 38K **47.** 24 oz **49.** yes **51.** 30, 12 **53.** 2.7 in. **55.** 5% **57.** yes **63.** 18.17 **65.** 5.001 **67.** 0.008

Think It Through (page 334)

48%, 59%, 49%, 49%, 57%; greatest percent increase is medical assistant

Study Set Section 5.3 (page 336)

1. commission **3.** discount **5.** subtract, original **7.** $42.75 **9.** 8% **11.** $47.34, $2.84, $50.18 **13.** $150 **15.** 8%, 1.2%, 1.45%, 6.2% **17.** 360 hr **19.** 96 calories **21.** 1995–1996; 5% **23.** 10% **25.** 31% **27. a.** 25% **b.** 36% **29.** $5,955 **31.** 1.5% **33.** $12,000 **35.** $39.95, 25% **37.** $187.49 **39.** $349.97, 13% **41.** $3.60, 23%, $11.88 **43.** $76.50 **49.** −50 **51.** 3 **53.** $500 **55.** $-\frac{7}{45}$ **57.** $\frac{10}{7}$ or $1\frac{3}{7}$

Study Set Estimation (page 341)

1. 164 **3.** $60 **5.** $54,000 **7.** 320 lb **9.** 130 **11.** 21 **13.** 18,000 **15.** 3,100

Study Set Section 5.4 (page 347)

1. principal **3.** interest **5.** simple **7. a.** 0.07 **b.** 0.098 **c.** 0.0625 **9.** $1,800 **11. a.** compound interest **b.** $1,000 **c.** 4 **d.** $50 **e.** 1 year **13.** multiplication **15.** $5,300 **17.** $1,472 **19.** $4,262.14 **21.** $10,000, 0.0725, 2 yr, $1,450 **23.** $192, $1,392, $58 **25.** $18.828 million **27.** $755.83 **29.** $1,271.22 **31.** $570.65 **33.** $30,915.66 **39.** $\frac{1}{2}$ **41.** 23.0 **43.** −3 **45.** 50

Key Concept (page 350)

1. 198.4 **3.** 62.5 **5.** 17% **7.** $3,000

Chapter Review (page 352)

1. 39%, 0.39, $\frac{39}{100}$ **2.** 111%, 1.11, $1\frac{11}{100}$ **3.** 61% **4.** $\frac{3}{20}$ **5.** $\frac{6}{5}$ **6.** $\frac{37}{400}$ **7.** $\frac{1}{1,000}$ **8.** 0.27 **9.** 0.08 **10.** 1.55 **11.** 0.018 **12.** 83% **13.** 62.5% **14.** 5.1% **15.** 600% **16.** 50% **17.** 80% **18.** 87.5% **19.** 6.25% **20.** $33\frac{1}{3}$% **21.** $83\frac{1}{3}$% **22.** 55.56% **23.** 266.67% **24.** 63% **25.** 0.1% = $\frac{1}{1,000}$ **26.** amount: 15, base: 45, percent: $33\frac{1}{3}$% **27.** $x = 32\% \cdot 96$ **28.** 200 **29.** 125 **30.** 1.75% **31.** 2,100 **32.** 121 **33.** 30 **34.** 14.4 gal nitro, 0.6 gal methane **35.** 68 **36.** 87% **37.** $5.43 **38.**

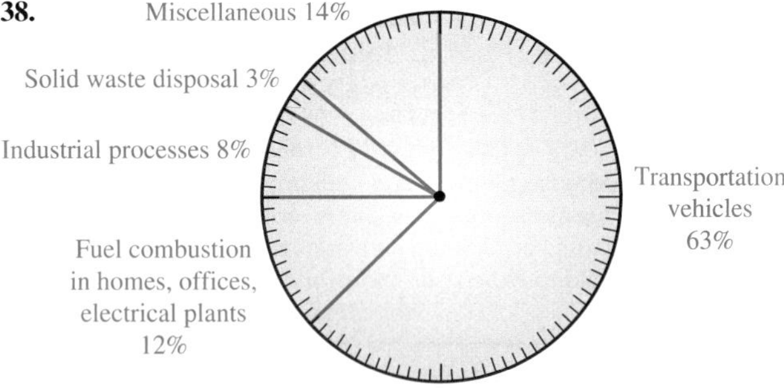

39. 139,531,200 mi^2 **40.** $3.30, $63.29 **41.** 4% **42.** $40.20 **43.** original **44.** 25% **45.** 9.6% **46.** $50, $189.99, 26% **47.** $6,000, 8%, 2 years, $960 **48.** $10,308.22 **49.** $134.69 **50.** $2,142.45 **51.** $6,076.45 **52.** $43,265.78

Chapter 5 Test (page 357)

1. 61%, $\frac{61}{100}$, 0.61 **2.** 199%, $\frac{199}{100}$, 1.99 **3. a.** 0.67 **b.** 0.123 **c.** 0.0975 **4. a.** 25% **b.** 62.5% **c.** 12% **5. a.** 19% **b.** 347% **c.** 0.5% **6. a.** $\frac{11}{20}$ **b.** $\frac{1}{10{,}000}$ **c.** $\frac{5}{4}$ **7.** 23.33% **8.** 60% **9.** $66\frac{2}{3}$% **10.** 25% **11. a.** 1.02 in. **b.** 32.98 in. **12.** 6.5% **13.** $3.81 **14.** 93.9% **15.** 90 **16.** 21 **17.** 144 **18.** 27% **19.** $35.92 **20.** $41,440 **21.** $11.95, $3, 20% **22.** 22% **23.** $150 **24.** $5,079.60 **25.** The phrase "bringing crime down to 37%" is unclear. The question that arises is: 37% of what? **26.** Interest is money that is paid for the use of money.

Cumulative Review Exercises (page 359)

1.

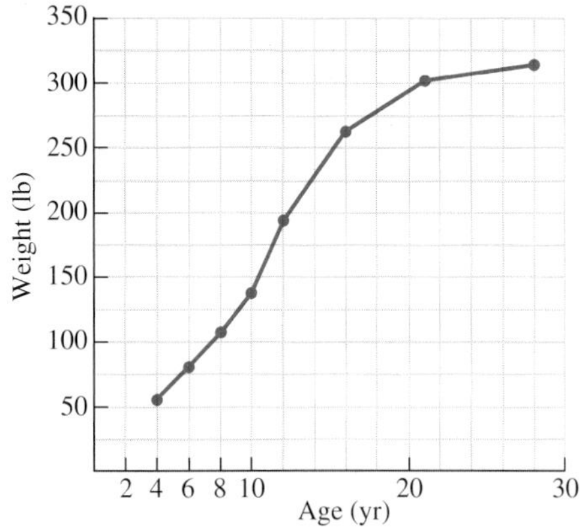

2. If a and b represent numbers, $ab = ba$. **3. a.** 1, 2, 4, 5, 8, 10, 20, 40 **b.** $5 \cdot 2^3$ **4.** $2,106 **5.** 64 ft^2 **6.** −7 **7.** −4 **8.** 55 **9.** −2 **10.** 2 **11.** 15° C **12.** $\frac{4}{11}$ **13.** $\frac{2}{3}$ **14.** $\frac{2}{5}$ **15.** $-\frac{5}{21}$ **16.** $\frac{2}{5}$ **17.** $\frac{34}{21}$ or $1\frac{13}{21}$ **18.** $20\frac{5}{18}$ **19.** −30 **20.** 18 **21.** 70.29 **22.** −8.6265 **23.** 752 **24.** 83.4 **25.** −2.33 **26.** 452.030 **27.** $0.7\overline{3}$ **28.** −11.1 **29.** −29 **30.** 3.5 hr **31.** 29%, $\frac{29}{100}$; 0.473, $\frac{473}{1{,}000}$; 87.5%, 0.875 **32.** 125 **33.** 64 **34.** 8% **35.** $12.00, $87.18 **36.** 0.0018% **37.** $1,450

Chapter 6 Check Your Knowledge (page 362)

1. bar **2.** pictograph **3.** frequency **4.** mean **5.** median **6.** mode **7.** circle graph **8.** 4 **9.** 15 **10.** 20% **11.** O, AB **12.** histogram **13.** 3 **14.** 25 **15.** 40% **16.** 10 **17.** 20% **18.** 7 **19.** 7 **20.** 9 **21.** 7 **22.** mean **23.** mean, median

Study Set Section 6.1 (page 370)

1. a **3.** c **5.** d **7.** bars, equal **9.** $7.35 **11.** $4.01 **13.** $7,895 **15.** $3,110 **17.** nuclear energy **19.** about 30% **21.** coal **23.** 1980 **25.** 1970 **27.** 320 thousand metric tons **29.** reckless driving and failure to yield **31.** reckless driving **33.** seniors **35.** $50 **37.** French and German **39.** English **41.** 51.4% **43.** about 11% **45.** about 49% **47.** 175% **49.** $190 **51.** miners **53.** miners **55.** 1 **57.** 1 **59.** Runner 1 was running; runner 2 was stopped. **61.** 27 **63.** 90 **71.** −7 **73.** $\frac{25}{36}$ **75.** 11, 13, 17, 19, 23, 29 **77.** 4

Think It Through (page 379)

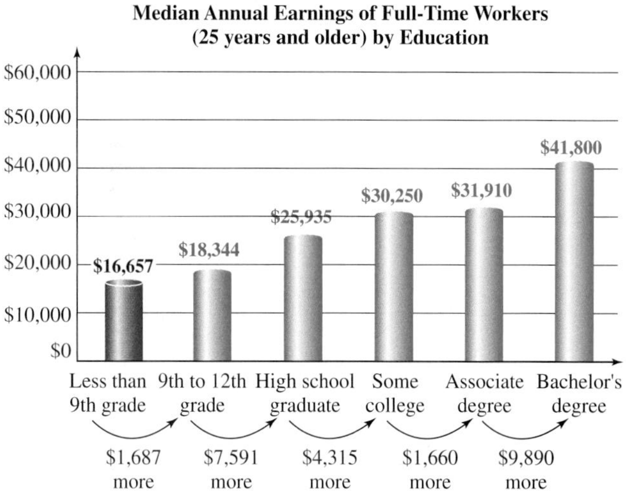

Source: U.S. Census Bureau, June 2004

Study Set Section 6.2 (page 379)

1. mean **3.** median **5.** the number of values **7.** 8 **9.** 35 **11.** 19 **13.** 9 **15.** 6 **17.** 17.5 **19.** 3 **21.** none **23.** 22.7 **25.** about 63¢ **27.** 60¢ **29.** 50¢ **31.** about 61° **33.** 64° **35.** 2,670 mi **37.** 89 mi **39.** Median and mode are 85. **41.** same average (56); sister's scores are more consistent **43.** 22.525 oz, 25 oz **45.** $4.15, $4.19, $4.29, $0.50 **47.** city: mean 43, median 42, mode 42; hwy: mean 48.8, median 49, mode 49 **49. a.** 5.5 **b.** 5.6 **c.** 5.6 **d.** 3.5 **53.** 3^4 **55.** $\frac{1}{6}$ **57.** 6 **59.** $\frac{19}{10} = 1\frac{9}{10}$

Key Concept (page 383)

1. 13.7 **3.** 15 **5.** no **7.** 6

Chapter Review (page 385)

1. −18° **2.** 30 mph **3.** about 3.4 billion **4.** 1997 and 1998 **5.** 1999 and 2000 **6.** 1996 and 1997 **7.** about 830 million **8.** about 865 million **9.** 1987 **10.** about 1,770 million **11.** 180 **12.** 160 **13.** yes **14.** median **15.** 1.2 oz **16.** 1.138 oz **17.** 7.3 microns, 7.2 microns, 6.9 microns, 1.3 microns **18.** $1.45 billion

Chapter 6 Test (page 389)

1. about $1,659 **2.** about $11 **3.** about 4.1% **4.** about 1.2% **5.** about 19% **6.** about 6% **7.** about 270,000 **8.** 1985 **9.** about 7,400 **10.** 65.5% **11.** A **12.** C **13.** E **14.** bicyclist 1 **15.** 7.5 **16.** 7.5 **17.** 5 **18.** mean **19.** 3.6, 3.6, 3.1, 1.6 **20.** Half the families had more debt and half had less debt.

Cumulative Review Exercises (page 391)

1. 358,600,000 gal **2.** 50,000 **3.** 54,604 **4.** 4,209 **5.** 23,115 **6.** 87 **7.** 683 + 459 = 1,142 **8.** 2011 **9.** $4 \cdot 5 = 5 + 5 + 5 + 5 = 20$ **10.** 10,912 in.^{2} **11. a.** 1, 2, 3, 6, 9, 18 **b.** $3^2 \cdot 2$ **12.** 2, 3, 5, 7, 11, 13, 17, 19, 23, 29 **13.** It has factors other than 1 and itself. For example, $27 = 3 \cdot 9$. **14.** 22 **15.** 315 **16.** 6 **17.**

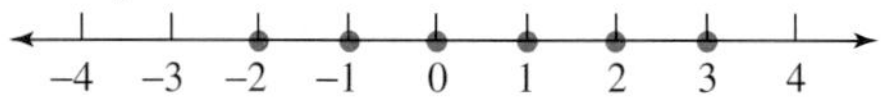

18. 5

19. false **20.**

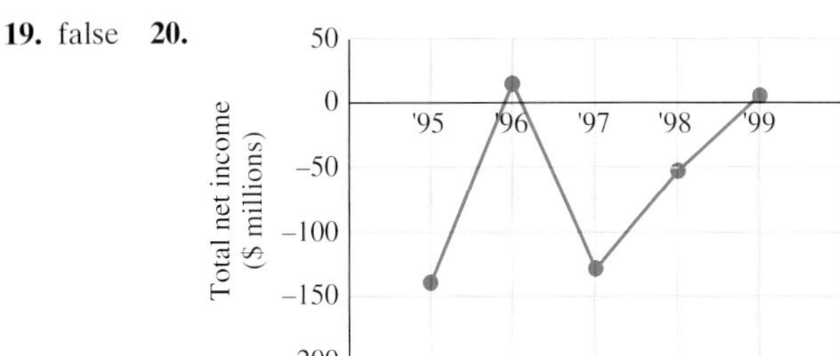

21. −20 **22.** 30 **23.** 125 **24.** 5 **25.** −5 **26.** 429
27. $-3^2 = -(3 \cdot 3) = -9$; $(-3)^2 = (-3)(-3) = 9$ **28.** 1,100° F
29. 800 **30.** 15% **31.** $\frac{5}{0}; \frac{0}{5}$; division by 0 **32.** $\frac{7}{6} = 1\frac{1}{6}$
33. $-\frac{1}{6}$ **34.** $6\frac{3}{4}$ in. **35.** $\frac{8}{35}$ **36.** $\frac{21}{20} = 1\frac{1}{20}$ **37.** 220 **38.** 345
39. 0.744 **40.** 745 **41.** $0.\overline{72}$ **42.** 160 min **43.** 3.1 hr
44. 3.02, 3.005, 2.75

Chapter 7 Check Your Knowledge (page 394)

1. parallel, perpendicular **2.** quadrilateral, triangle **3.** right, hypotenuse **4.** congruent, similar **5.** perimeter, area **6.** radius, circumference, diameter **7.** volume **8. a.** III **b.** I **c.** IV **d.** II **9.** *B* **10.** 14° **11.** 36 **12.** 61° **13. a.** 50° **b.** 130° **c.** 130° **d.** 50° **e.** 75 **14.** 80° **15. a.** 13 ft **b.** 60 ft^2 **16.** 17.55 in.^{2} **17.** 28π ft^2 ≈ 78.5 ft^3 **18.** $10\frac{2}{3}\pi$ ft^3 ≈ 33.5 ft^3 **19.** 200 ft^3 **20.** 160π ft^3 ≈ 502.7 ft^3

Study Set Section 7.1 (page 401)

1. segment **3.** midpoint **5.** protractor **7.** right **9.** 180° **11.** supplementary **13.** true **15.** false **17.** true **19.** true **21.** acute **23.** obtuse **25.** right **27.** straight **29.** true **31.** false **33.** yes **35.** yes **37.** no **39.** true **41.** true **43.** true **45.** true **47.** angle **49.** ray **51.** 3 **53.** 3 **55.** 1 **57.** *B* **59.** 40° **61.** 135° **63.** 10 **65.** 27.5 **67.** 30 **69.** 25 **71.** 60° **73.** 75° **75.** 130° **77.** 230° **79.** 100° **81.** 40° **83.**

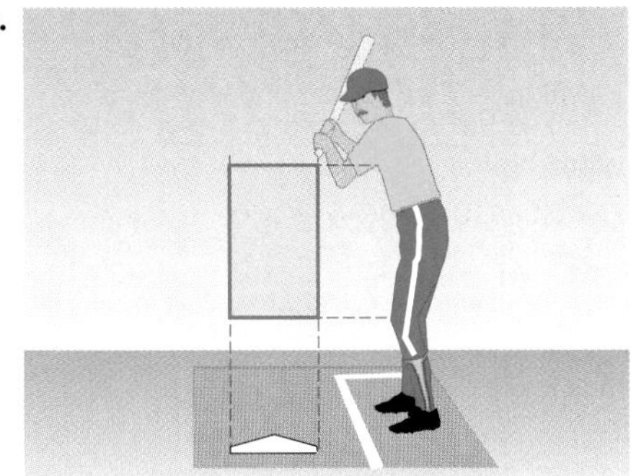

85. 65, 115 **87.** 30° **95.** 16 **97.** $\frac{7}{24}$ **99.** 6

Study Set Section 7.2 (page 409)

1. coplanar **3.** perpendicular **5.** alternate **7.** ∠4 and ∠6, ∠3 and ∠5 **9.** ∠3, ∠4, ∠5, ∠6 **11.** They are parallel. **13.** a right angle **15.** is perpendicular to **17.** m(∠1) = 130°, m(∠2) = 50°, m(∠3) = 50°, m(∠5) = 130°, m(∠6) = 50°, m(∠7) = 50°, m(∠8) = 130° **19.** m(∠*A*) = 50°, m(∠1) = 85°, m(∠2) = 45°, m(∠3) = 135° **21.** 10 **23.** 30 **25.** 40 **27.** 12 **29.** If the stones are level, the plum bob string should pass through the midpoint of the crossbar of the A-frame. **41.** 72 **43.** 45% **45.** yes

Study Set Section 7.3 (page 417)

1. regular **3.** hexagon **5.** octagon **7.** equilateral **9.** hypotenuse **11.** parallelogram **13.** rhombus **15.** isosceles **17.** 4, quadrilateral, 4 **19.** 3, triangle, 3 **21.** 5, pentagon, 5 **23.** 6, hexagon, 6 **25.** scalene triangle **27.** right triangle **29.** equilateral triangle **31.** isosceles triangle **33.** square, rhombus, rectangle **35.** rhombus **37.** rectangle **39.** trapezoid **41.** triangle **43.** 90° **45.** 45° **47.** 90.7° **49.** 70° **51.** 7 cm **53.** 30° **55.** 60° **57.** 720° **59.** 1,440° **61.** 7 sides **63.** 14 sides **69. b.** octagon **c.** triangle **d.** pentagon **71.** pentagon, hexagon **77.** 22 **79.** 40% **81.** 0.10625

Study Set Section 7.4 (page 424)

1. congruent **3.** right **5.** true **7.** false **9.** yes **11.** *a* and *b* represent the lengths of the legs; *c* represents the length of the hypotenuse. **13.** is congruent to **15.** $\overline{DF}$, $\overline{AB}$, $\overline{EF}$, ∠*D*, ∠*B*, ∠*C* **17.** yes, SSS **19.** not necessarily **21.** yes, SSS **23.** yes, SAS **25.** 6 mm **27.** 50° **29.** 5 **31.** 8 **33.** $\sqrt{56}$ **35.** yes **37.** no **39.** 12 ft **41.** 25 in. **43.** 127.3 ft **47.** $1\frac{1}{3}$ **49.** 20 **51.** 9

Think It Through (page 431)

$107\frac{2}{3}$ ft^2

Study Set Section 7.5 (page 434)

1. perimeter **3.** area **5.** square **7.** length 15 in. and width 5 in.; length 16 in. and width 4 in. (answers may vary) **9.** sides of length 5 m **11.** base 5 yd and height 3 yd (answers may vary) **13.** length 5 ft and width 4 ft; length 20 ft and width 3 ft (answers may vary) **15.** $4s$ **17.** square inch **19.** s^2 **21.** triangle **23.** 32 in. **25.** 36 m **27.** 37 cm **29.** 85 cm **31.** $28\frac{1}{3}$ ft **33.** 16 cm^2 **35.** 60 cm^2 **37.** 25 in.^{2} **39.** 169 mm^2 **41.** 80 m^2 **43.** 75 yd^2 **45.** 75 m^2 **47.** 144 **49.** $4,875 **51.** 81 **53.** linoleum **55.** $1,200 **57.** $361.20 **59.** $192 **61.** 111,825 mi^2 **63.** 51 **65.** spot 1: $l = 20$ ft, $w = 10$ ft, 200 ft^2; spot 2: $b_1 = 20$ ft, $b_2 = 16$ ft, $h = 10$ ft, 180 ft^2; spot 3: $b = 28$ ft, $h = 28$ ft, 392 ft^2 **69.** $1\frac{5}{12}$ **71.** $6\frac{1}{12}$ **73.** $1\frac{7}{18}$

Study Set Section 7.6 (page 443)

1. radius **3.** diameter **5.** minor **7.** circumference **9.** $\overline{OA}$, $\overline{OC}$, and $\overline{OB}$ **11.** $\overline{DA}$, $\overline{DC}$ and $\overline{AC}$ **13.** $\widehat{ABC}$ and $\widehat{ADC}$ **15.** Double the radius. **17. a.** 1 in. **b.** 2 in. **c.** 2π in. ≈ 6.28 in. **d.** π in.^{2} ≈ 3.14 in.^{2} **19.** Square 6. **21.** arc *AB* **23.** πD, r **25.** π **27.** 8π **29.** 37.70 in. **31.** 36 m **33.** 25.42 ft **35.** 31.42 m **37.** $A = 28.3$ in.^{2} **39.** 88.3 in.^{2} **41.** 128.5 cm^2 **43.** 27.4 in.^{2} **45.** 66.7 in.^{2} **47.** 3.14 mi^2 **49.** 32.66 ft **51.** 12.73 times **53.** 1.59 ft **55.** 12.57 ft^2; 0.79 ft^2; 6.28% **63.** 90% **65.** five

Study Set Section 7.7 (page 454)

1. volume **3.** cube **5.** surface **7.** cylinder **9.** cone **11.** $V = lwh$ **13.** $V = \frac{4}{3}\pi r^3$ **15.** $V = \frac{1}{3}Bh$ or $V = \frac{1}{3}\pi r^2 h$ **17.** $SA = 2lw + 2lh + 2hw$ **19.** 27 ft^3 **21.** 1,000 dm^3 **23. a.** volume **b.** area **c.** volume **d.** surface area **e.** perimeter **f.** surface area **25. a.** 72 in.^{3} **b.** 18 in.^{2} **c.** 24 in.^{2} **27.** cubic inch **29.** 60 cm^3 **31.** 48 m^3 **33.** 3,053.63 in.^{3} **35.** 1,357.17 m^3 **37.** 314.16 cm^3 **39.** 400 m^3 **41.** 94 cm^2 **43.** 1,256.64 in.^{2} **45.** 576 cm^3 **47.** 335.10 in.^{3} **49.** $\frac{1}{8}$ in.^{3} = 0.125 in.^{3} **51.** 2.125 **53.** 197.92 ft^3 **55.** 33,510.32 ft^3 **57.** 8:1 **63.** -42 **65.** -1

Key Concept (page 457)

1. $d = rt$ **3.** $P = 2l + 2w$ **5.** 210,000 ft^2 **7.** $80.50 **9.** 144 ft **11.** $750, $45, $6,250

Chapter Review (page 459)

1. points C and D, line CD, plane GHI **2.** 5 units **3.** $\angle ABC$, $\angle CBA$, $\angle B$, $\angle 1$ **4.** 48° **5.** $\angle 1$ and $\angle 2$ are acute, $\angle ABD$ and $\angle CBD$ are right angles, $\angle CBE$ is obtuse, and $\angle ABC$ is a straight angle. **6.** obtuse angle **7.** right angle **8.** straight angle **9.** acute angle **10.** 15 **11.** 150 **12. a.** 65° **b.** 115° **13.** 40° **14.** 40° **15.** no **16.** part a **17.** $\angle 4$ and $\angle 6$, $\angle 3$ and $\angle 5$ **18.** $\angle 1$ and $\angle 5$, $\angle 4$ and $\angle 8$, $\angle 2$ and $\angle 6$, $\angle 3$ and $\angle 7$ **19.** $\angle 1$ and $\angle 3$, $\angle 2$ and $\angle 4$, $\angle 5$ and $\angle 7$, $\angle 6$ and $\angle 8$ **20.** $m(\angle 1) = 70°$, $m(\angle 2) = 110°$, $m(\angle 3) = 70°$, $m(\angle 4) = 110°$, $m(\angle 5) = 70°$, $m(\angle 6) = 110°$, $m(\angle 7) = 70°$ **21.** $m(\angle 1) = 60°$, $m(\angle 2) = 120°$, $m(\angle 3) = 130°$, $m(\angle 4) = 50°$ **22.** 40 **23.** 20 **24.** octagon **25.** pentagon **26.** triangle **27.** hexagon **28.** quadrilateral **29.** 3 **30.** 4 **31.** 8 **32.** 6 **33.** isosceles **34.** scalene **35.** equilateral **36.** right triangle **37.** yes **38.** no **39.** 90 **40.** 50 **41.** 50° **42.** It is equilateral. **43.** trapezoid **44.** square **45.** parallelogram **46.** rectangle **47.** rhombus **48.** rectangle **49.** 45° **50.** 11 cm **51.** 15 cm **52.** 40° **53.** 100° **54.** true **55.** false **56.** true **57.** true **58.** 65° **59.** 115° **60.** 360° **61.** 720° **62.** $\angle D$ **63.** $\angle E$ **64.** $\angle F$ **65.** $\overline{DF}$ **66.** $\overline{DE}$ **67.** $\overline{EF}$ **68.** congruent, SSS **69.** congruent, SAS **70.** congruent, ASA **71.** not necessarily congruent **72.** 13 **73.** 15 **74.** 31.3 in. **75.** 72 in. **76.** 9 m **77.** 30 m **78.** 36 m **79.** 9.61 cm^2 **80.** 7,500 ft^2 **81.** 450 ft^2 **82.** 200 in.^{2} **83.** 120 cm^2 **84.** 232 ft^2 **85.** 152 ft^2 **86.** 120 m^2 **87.** 9 ft^2 **88.** 144 in.^{2} **89.** $\overline{CD}$, $\overline{AB}$ **90.** $\overline{AB}$ **91.** $\overline{OA}$, $\overline{OC}$, $\overline{OD}$, $\overline{OB}$ **92.** O **93.** 66.0 cm **94.** 45.1 cm **95.** 254.5 in.^{2} **96.** 130.3 cm^2 **97.** 125 cm^3 **98.** 480 m^3 **99.** 600 in.^{3} **100.** 3,619 in.^{3} **101.** 1,518 ft^3 **102.** 785 in.^{3} **103.** 9,020,833 ft^3 **104.** 35,343 ft^3 **105.** 1,728 in.^{3} **106.** 54 ft^3 **107.** 61.8 ft^2 **108.** 314.2 in.^{2}

Chapter 7 Test (page 469)

1. 4 units **2.** B **3.** true **4.** false **5.** false **6.** true **7.** 50 **8.** 140 **9.** 12 **10.** 45 **11.** 23° **12.** 63° **13.** 70° **14.** 110° **15.** 70° **16.** 40 **17.** 3, 4, 6, 5, 8 **18.** equilateral triangle, scalene triangle, isosceles triangle **19.** 57° **20.** 66° **21.** 30° **22.** 1,440° **23.** $m(\widehat{AB}) = m(\widehat{DC})$, $m(\widehat{AD}) = m(\widehat{BC})$, and $m(\widehat{AC}) = m(\widehat{BD})$ **24.** 130° **25.** 8 in. **26.** 50° **27.** 127.3 ft **28.** 391.6 cm^2 **29.** 83.7 ft^2 **30.** 94.2 ft **31.** 28.3 ft^2 **32.** 159.3 m^3 **33.** 268.1 m^3 **34.** 66.7 ft^3 **35.** The surface area is 6 times the area of one face of the cube.

Cumulative Review Exercises (page 471)

1.

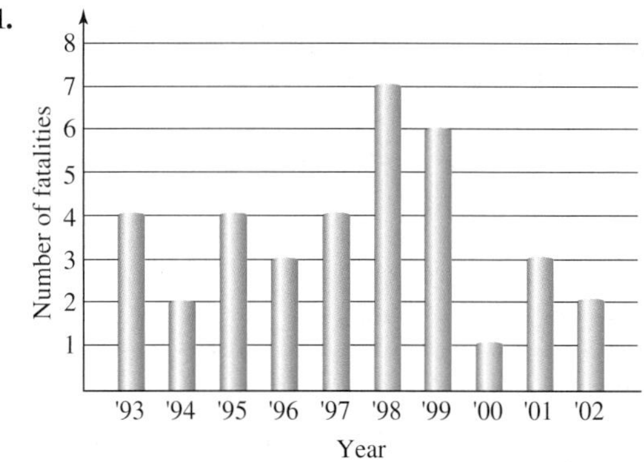

2. $8,995 **3.** 11,022 **4.** 33 **5.** 2,110,000 **6.** $11 \cdot 5 \cdot 2^2$ **7.** 1, 2, 3, 4, 6, 8, 12, 24 **8.** $\{\ldots -3, -2, -1, 0, 1, 2, 3, \ldots\}$ **9.** 13 **10.** -10 **11.** 3 **12.** 5 **13.** -11 **14.** -10 **15.** 5 **16.** 5 **17.** $\frac{5}{4}$ **18.** $142\frac{7}{15}$ **19.** $\frac{3}{20}$ **20.** $13\frac{3}{4}$ cups **21.** $-\frac{11}{20}$ **22.** $\frac{1}{3}$ **23.** $\frac{8}{9}$ **24.** $\frac{15}{2}$ **25.** $\frac{3}{32}$ fluid oz **26.** $\frac{8}{9}$ **27.** $1\frac{9}{29}$ **28. a.** 1998; about 0.6° F **b.** 1986; about -0.4° F

29. $-4\frac{5}{8}$ $-\sqrt{9}$ -0.1 $\frac{2}{3}$ $\frac{3}{2}$ 2.89 $\sqrt{17}$

-5 -4 -3 -2 -1 0 1 2 3 4 5

30. 3.1416 **31.** > **32.** 145.188 **33.** 17.05 **34.** 89,970.8 **35.** 0.053 **36.** -25.6 **37.** 22.3125 **38.** $0.1\overline{3}$ **39.** -9.32 **40.** 97 **41.** 10 **42.** -2 **43.** $\frac{7}{9}$ **44.** 93%, 7% **45.** 67.5 **46.** 120 **47.** 0.57, $\frac{57}{100}$, 0.1%, $\frac{1}{1{,}000}$, $33\frac{1}{3}$%, $0.\overline{3}$

48.

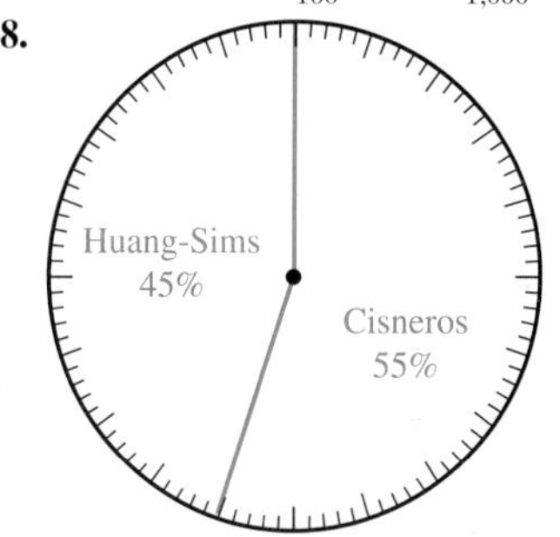

49. $75 **50.** $1,159.38 **51.** 500% **52.** $1,522.50 **53.** $269,390.92 **54.** 90 **55.** more than 0 but less than 90 **56.** 75° **57.** 15° **58.** 50° **59.** 50° **60.** 130° **61.** 50° **62.** 75° **63.** 30° **64.** 105° **65.** 105° **66.** 46, 134 **67.** 540° **68.** 13 m **69.** 42 m, 108 m^2 **70.** 126 ft^2 **71.** 91 in.^{2} **72.** 43.98 cm, 153.94 cm^2 **73.** 98.31 yd^2 **74.** 210 m^3 **75.** 523.60 in.^{3} **76.** 150.80 m^3 **77.** 3.93 ft^3 **78.** 2,124 in.^{2}

Chapter 8 Check Your Knowledge (page 476)

1. 11 **2.** $-\frac{1}{5}$ **3.** $6x + 6y - 12$ **4.** $-2x + 2y + 4$ **5.** $9x$ **6.** $-18a + 26$ **7.** -2 **8.** 72 **9.** 35 **10.** $-\frac{22}{7}$ **11.** 6 **12.** 3 **13.** $y = 2x - 3$ **14.** 42 cm **15.** 37.7 cm **16.** $600, $400 **17.** 5 pints **18.** 25 ft^2 **19.** $721\frac{7}{8}$ in.^{3} **20.** 64 mph **21.** 4 P.M.

22. $x \ge 4$, $[4, \infty)$ **23.** $x \ge 1$, $[1, \infty)$ **24.** $1 < x \le 4$, $(1, 4]$

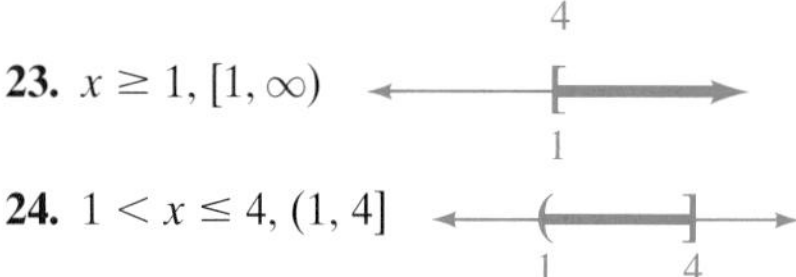

Study Set Section 8.1 (page 485)

1. evaluate **3.** expression, equation **5.** $6 + 20x$; $\frac{6 - x}{20}$ (answers may vary) **7.** We would obtain $34 - 6$; it looks like 34, not 3(4). **9. a.** x = weight of the car; $2x - 500$ = weight of the van **b.** 3,500 lb **11.** 5, 30, 10, 10d, 50, $50(x + 5)$ **13.** 5, 5, 25, 45 **15.** $l + 15$ **17.** $50x$ **19.** $\frac{w}{l}$ **21.** $P + p$ **23.** $k^2 - 2{,}005$ **25.** $J - 500$ **27.** $\frac{1{,}000}{n}$ **29.** $p + 90$ **31.** $35 + h + 300$ **33.** $p - 680$ **35.** $4d - 15$ **37.** $2(200 + t)$ **39.** $|a - 2|$ **41.** 7 less than a number **43.** the product of 7 and a number, increased by 4 **45.** 300; $60h$ **47. a.** $3y$ **b.** $\frac{f}{3}$ **49.** $29x$¢ **51.** $\frac{c}{6}$ **53.** $5b$ **55.** $\$5(x + 2)$ **57.** −1, −2, −28 **59.** 41, 11, 2 **61.** 150, −450 **63.** 0, 0, 5 **65.** 20 **67.** −12 **69.** −5 **71.** 156 **73.** $-\frac{1}{5}$ **75.** 17 **77.** 36 **79.** 230 **81.** 30.5 **83.** 0, 28, 48, 60, 64, 60, 48, 28, 0 **85.** −37° C, −64° C **87.** $1\frac{23}{64}$ in.^{2} **89.** 235 ft^2 **95.** 0 **97.** $\frac{2}{3}$ **99.** 5^4 **101.** 83

Study Set Section 8.2 (page 497)

1. simplify **3.** coefficient, variable **5.** remove **7.** the distributive property **9.** 3, 4; 3, 4 **11. a.** + **b.** − **c.** − **d.** + **e.** − **f.** + **13.** $x + 20 - x = 20$; 20 ft **15. a.** yes **b.** yes **19. a.** no **b.** yes **21. a.** $5x + 1$ **b.** $16t - 6$ **23.** $63m$ **25.** $-35q$ **27.** $5x$ **29.** $6y$ **31.** $20bp$ **33.** $40r^2$ **35.** $5x + 15$ **37.** $-2b + 2$ **39.** $24t - 16$ **41.** $12y - 6$ **43.** $0.4x - 1.6$ **45.** $-2w + 4$ **47.** $r^2 - 10r$ **49.** $-x + 7$ **51.** $34x - 17y + 34$ **53.** $14 - 3p + t$ **55. a.** −1 **b.** −9.9 **c.** $\frac{1}{4}$ **d.** $-\frac{2}{3}$ **57.** −5, 4 **59.** −15 **61.** 50, 2 **63.** 1, −125 **65.** $20x$ **67.** $3x^2$ **69.** 0 **71.** 0 **73.** $3a$ **75.** $-3x$ **77.** t **79.** x **81.** $-16x^2$ **83.** $1.1h$ **85.** $\frac{4}{5}t$ **87.** $0.4r$ **89.** $7z - 15$ **91.** $-3c - 1$ **93.** $7X - 2x$ **95.** $b + 2$ **97.** $-2x^2 + 3x$ **99.** $12x$ **101.** $(4x + 8)$ ft **107.** 0 **109.** 2

Study Set Section 8.3 (page 506)

1. equation **3.** satisfy **5.** reciprocal **7.** subtraction, multiplication **9.** addition, division **11.** $-\frac{5}{4}$ **13.** 30 **15. a.** $2x + 5$ **b.** 2 **c.** 23 **d.** no **19. a.** −1 **b.** 3, 5 **c.** −31 **21.** 6 **23.** $\frac{3}{5}$ **25.** 3.5 **27.** 4 **29.** $-\frac{8}{3}$ **31.** 12 **33.** −12 **35.** 9 **37.** 28 **39.** −19 **41.** 5 **43.** 1 **45.** 3 **47.** $\frac{1}{7}$ **49.** −4 **51.** 1 **53.** −7.2 **55.** 0 **57.** $-\frac{34}{5}$ **59.** $-\frac{55}{6}$ **61.** $-\frac{12}{5}$ **63.** $\frac{2}{15}$ **65.** $\frac{8}{9}$ **67.** $\frac{10}{9}$ **69.** −20 **71.** −41 **73.** 9 **75.** −1 **77.** 3 **79.** all real numbers, identity **81.** no solution, contradiction **83.** no solution, contradiction **85.** all real numbers, identity **87.** −1.238 **89.** 1,645.3 **95.** 8 **97.** 64 **99.** $\frac{1}{64}$ **101.** $16x$

Study Set Section 8.4 (page 515)

1. formula **3.** perimeter **5.** radius **7.** circumference **9. a.** $d = rt$ **b.** $r = c + m$ **c.** $p = r - c$ **d.** $I = Prt$ **e.** $C = 2\pi r$ **11.** 11,176,920 mi, 65,280 mi **13. a.** volume **b.** circumference **c.** area **d.** perimeter **15.** $(2x + 6)$ cm^2 **19. a.** 3.14 **b.** $98 \cdot \pi$ **c.** the radius of the cylinder; the height of the cylinder **21.** 2.5 mph **23.** \$65 million **25.** 3.5% **27.** 4,014° F **29.** \$24.55 **31.** about 132 in. **33.** $R = \frac{E}{I}$ **35.** $w = \frac{V}{lh}$ **37.** $r = \frac{C}{2\pi}$ **39.** $a = 180 - b - c$ **41.** $x = \frac{y - b}{m}$ **43.** $t = \frac{A - P}{Pr}$ **45.** $h = \frac{3V}{\pi r^2}$ **47.** $b = 2x - a$ **49.** $s = C - Dn$ **51.** $c^2 = \frac{E}{m}$ **53.** $a^2 = c^2 - b^2$ **55.** $b = \frac{2A}{h} - d$ or $b = \frac{2A - hd}{h}$ **57.** $y = \frac{1}{3}x + 3$ **59.** $y = -\frac{3}{4}x - 4$ **61.** 212° F, 0° C **63.** 1,174.6, 956.9 **65.** 36 ft, 48 ft^2 **67.** 50.3 in., 201.1 in.^{2} **69.** 56 in., 144 in.^{2} **71.** 2,450 ft^2 **73.** 27.75 in., 47.8125 in.^{2} **75.** 32 ft^2, 128 ft^3 **77.** 348 ft^3 **79.** 254 in.^{2} **81.** $n = \frac{360°}{180° - a}$; 5 sides **87.** 137.76 **89.** 15%

Think It Through (page 524)

\$0.25, \$7.16, \$7.15

Study Set Section 8.5 (page 529)

1. perimeter **3.** vertex **5. a.**

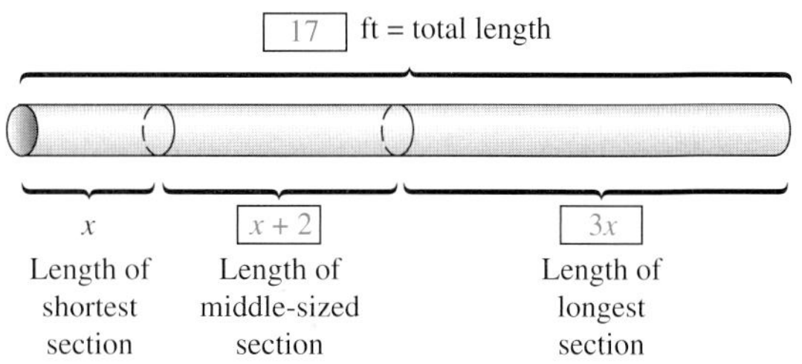

b. 3 ft, 5 ft, 9 ft **9.** \$30,000, 14%, 1 yr **11.** $35t$ mi; t hr, $45t$ mi **13. a.** 16.8 gal **b.** $(x + 42)$ gal **c.** 32% **15.** 6,000 **17.** 22°, 68° **19.** 15 **21.** 4 ft, 8 ft **23.** 7.3 ft, 10.7 ft **25.** Australia: 12 wk; Japan: 16 wk; Sweden: 10 wk **27.** 250 calories in ice cream, 600 calories in pie **29.** in millions of dollars: \$110, \$229, \$189, \$847 **31.** 7 ft, 7 ft, 11 ft **33.** 75 m by 480 m **35.** 20° **37.** 12 **39.** 90 **41.** \$6,000 **43.** \$4,900 **45.** \$7,500 **47.** 2 hr **49.** 65 mph, 45 mph **51.** 4 hr into the flights **53.** 50 **55.** 7.5 oz **57.** 20 **59.** 40 lb lemon drops, 60 lb jelly beans **61.** 80 **67.** $-50x + 125$ **69.** $3x + 3$ **71.** $19p + 11q$

Study Set Section 8.6 (page 542)

1. inequality **3.** solution **5. a.** true **b.** true **c.** false **d.** false **e.** true **f.** false **7.** same **9.** opposite **11. a.** a true statement **b.** a false statement **13. a.** all real numbers greater than 8 **b.**

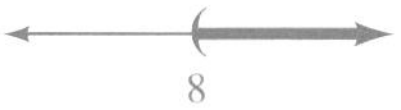

c. $(8, \infty)$ **15.** is less than, is greater than **17.** is not equal to **19.** $x > -2$ **21.** $-2 \le 17$

25.

$(-\infty, 5)$

27.

$(-3, 1]$

29. $x < -1$, $(-\infty, -1)$ **31.** $-7 < x \le 2$, $(-7, 2]$

33. $x > 3$, $(3, \infty)$

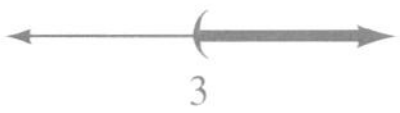

35. $x \ge -10$, $[-10, \infty)$

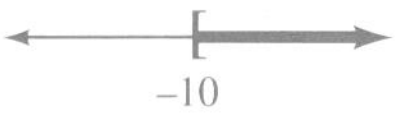

37. $x < -1$, $(-\infty, -1)$

39. $x \le 0.4$, $(-\infty, 0.4]$

41. $x < -2$, $(-\infty, -2)$

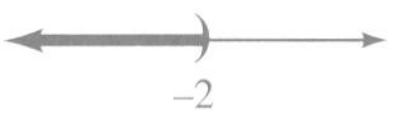

43. $x < -\frac{11}{4}$, $(-\infty, -\frac{11}{4})$

45. $y \le -40$, $(-\infty, -40]$

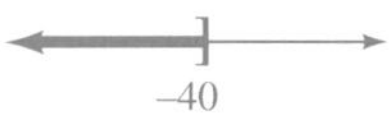

47. $n \le 2$, $(-\infty, 2]$

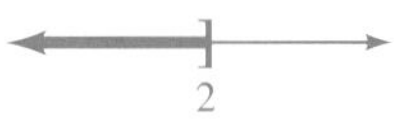

49. $x < 0$, $(-\infty, 0)$

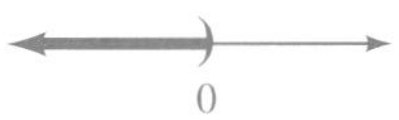

51. $x \geq 3$, $[3, \infty)$ 

53. $x \geq -24$, $[-24, \infty)$ 

55. $x \leq -1$, $(-\infty, -1]$ 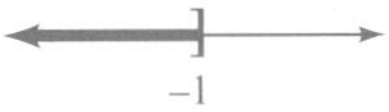

57. $x \geq -13$, $[-13, \infty)$ 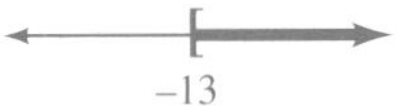

59. $x > 0$, $(0, \infty)$ 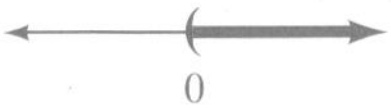

61. $x < -2$, $(-\infty, -2)$ 

63. $x \geq \frac{9}{4}$, $[\frac{9}{4}, \infty)$ 

65. $x > -15$, $(-15, \infty)$ 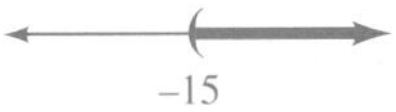

67. $x \leq 20$, $(-\infty, 20]$ 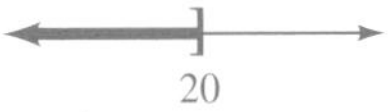

69. $x > \frac{5}{4}$, $(\frac{5}{4}, \infty)$ 

71. $y \leq \frac{1}{8}$, $(-\infty, \frac{1}{8}]$ 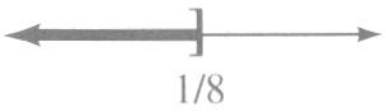

73. $7 < x < 10$, $(7, 10)$ 

75. $-9 < x \leq 3$, $(-9, 3]$ 

77. $-10 \leq x \leq 0$, $[-10, 0]$ 

79. $-5 < x < -2$, $(-5, -2)$ 

81. $-6 \leq x \leq 10$, $[-6, 10]$ 

83. $2 \leq x < 3$, $[2, 3)$ 

85. $-1 \leq x < 2$, $[-1, 2)$ 

87. $x < 1.37$, $(-\infty, 1.37)$ 

89. $x \geq 0.03$, $[0.03, \infty)$ 

91. 98% or better **93.** 27 mpg or better **95.** $m \geq 420$ min **97. a.** $0° < a \leq 18°$ **b.** $18° \leq a \leq 50°$ **c.** $30° \leq a \leq 37°$ **d.** $75° \leq a < 90°$ **99. a.** 470 ft $\leq x \leq$ 13,143 ft **b.** 0.1 mi $\leq x \leq$ 2.5 mi **101.** 1.496 in. $\leq w \leq$ 1.498 in.; 1.5000 in. $\leq w \leq$ 1.5010 in. **105.** -125 **107.** 1, -3, 6

Key Concept (page 547)

1. a. $2x - 8$ **b.** 6 **3. a.** $\frac{2}{3}a$ **b.** $\frac{1}{2}$ **5.** The mistake is on the third line. The student made an equation out of the answer, which is $x - 6$, by writing "0 =" on the left. Then the student solved that equation.

Chapter Review (page 549)

1. $h + 25$ **2.** $s - 15$ **3.** $\frac{1}{2}t$ **4.** $6x$ **5.** $(n + 4)$ in. **6.** $(b - 4)$ in. **7.** $10d$ **8.** $\frac{x}{12}$ **9.** $(x - 5)$ yr **10.** 5, 30, 10, $10d$ **11.** 0, 19, -16 **12.** 110 **13.** 40 **14.** 432 **15.** -36 **16.** 17.7 in.^{3} **17.** $-28w$ **18.** $15r^2$ **19.** $24xy$ **20.** $2.08f$ **21.** $5x + 15$ **22.** $-4x - 6 + 2y$ **23.** $-a + 4$ **24.** $3c - 6$ **25.** 3 **26.** 1 **27.** 2, -5 **28.** 16, -5, 25 **29.** $\frac{1}{2}$, 1 **30.** 9.6, -1 **31.** $9p$ **32.** $-7m$ **33.** $-2a - 10b$ **34.** $-p - 18$ **35.** x **36.** $-8a^3$ **37.** $(4x + 4)$ ft **38.** 2 **39.** -1 **40.** 30 **41.** -19 **42.** 4 **43.** 1 **44.** $\frac{5}{4}$ **45.** -6 **46.** identity, all real numbers **47.** contradiction, no solution **48.** \$176 **49.** \$11,800 **50.** 3.38 hr **51.** 1,949° F **52.** 168 in. **53.** 1,440 in.^{2} **54.** 76.5 m^2 **55.** 144 in.^{2} **56.** 50.27 cm **57.** 201.06 cm^2 **58.** 4,320 in.^{3} **59.** 9.4 ft^3 **60.** 120 ft^3 **61.** 381.70 in.^{3} **62.** $h = \dfrac{A}{2\pi r}$ **63.** $l = \dfrac{P - 2w}{2}$ **64.** 8 ft **65.** 147 **66.** 24.875 in. × 29.875 in. ($24\frac{7}{8}$ in. × $29\frac{7}{8}$ in.) **67.** 76.5°, 76.5° **68.** $\$45x$ **69.** \$16,000 at 7%, \$11,000 at 9% **70.** 20 **71.** 10 lb of each **72.** $0.12x$ gal

73. $x < 1$, $(-\infty, 1)$ 

74. $x < -3$, $(-\infty, -3)$ 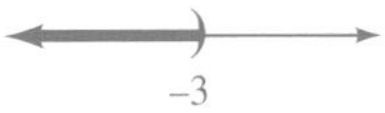

75. $x \geq 4$, $[4, \infty)$ 

76. $x \geq 6$, $[6, \infty)$ 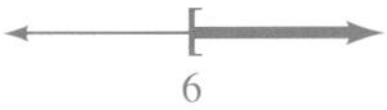

77. $x \geq 3$, $[3, \infty)$ 

78. $x \leq 12$, $(-\infty, 12]$ 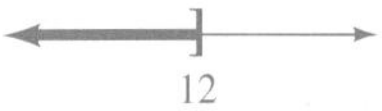

79. $6 < x < 11$, $(6, 11)$ 

80. $-2 < x \leq 1$, $(-2, 1]$ 

81. 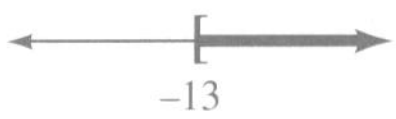

82. $2.40\ g < w < 2.53\ g$ **83.** The sign length must be 48 inches or less.

Chapter 8 Test (page 555)

1. $x - 2$ = number of songs on the CD **2.** $25q$ **3.** 4, 17, 59 **4.** 128 **5.** 3, 5 **6.** the distributive property **7.** $-20x$ **8.** $224t^2$ **9.** 18 **10.** $-4.9d^2$ **11.** -12 **12.** -5 **13.** -49 **14.** 1 **15.** $\frac{7}{6}$ **16.** -3 **17.** $r = \frac{A - P}{Pt}$ **18.** \$150 **19.** $-10°$ C **20.** 393 in.^{3} **21.** $\frac{3}{5}$ hr **22.** 10 **23.** 68° **24.** \$5,250

25. $x \geq -3$, $[-3, \infty)$ 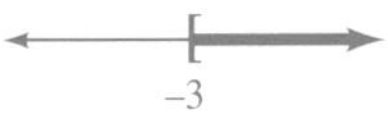

26. $-3 \leq x < 4$, $[-3, 4)$ 

27. Substitute the answer for the variable. If it is a solution, a true statement will result. **28.** Like terms are terms with exactly the same variables, raised to exactly the same powers. $10p^2$ and $6p^2$ are like terms.

Cumulative Review Exercises (page 557)

1. a. expression **b.** equation **2.** 3, 4, 5 **3.** $2 \cdot 2 \cdot 5 \cdot 5 = 2^2 \cdot 5^2$ **4.** $\frac{2}{3}$ **5.** $-\frac{2}{9}$ **6.** 6 **7.** $\frac{22}{15} = 1\frac{7}{15}$ **8.** $12\frac{11}{24}$ **9.** 0.9375 **10.** 45 **11. a.** 65 **b.** -12 **12.** the commutative property of multiplication **13.** natural number, whole number, integer, rational number, real number **14.** rational number, real number **15.** rational number, real number **16.** irrational number, real number **17. a.** 4^3 **b.** $\pi r^2 h$ **18. a.** -10

b. −14 **c.** −64 **d.** 0 **19. a.** $w + 12$ **b.** $n - 4$ **20.** 4
21. 1, −3, 6 **22.** $l = \dfrac{2{,}000}{d^2}$ (answers may vary depending on the variables chosen) **23. a.** 6 ft^2 **b.** 1.2 ft^2 **c.** 20% **24.** 300
25. 0 **26.** −2 **27.** 16 **28.** 0 **29.** $-32d$ **30.** $10x - 15y + 5$
31. $5x$ **32.** $-8a$ **33.** $8q^2 - 5q$ **34.** $8t - 20$ **35.** $(x + 3)$ ft
36. $3x$ ft **37.** 9 **38.** 20 **39.** −0.6 **40.** 19 **41.** −20 **42.** −2
43. 1 **44.** $\frac{5}{4}$ **45.** 65 m^2 **46.** 376.99 cm^3 **47.** $t = \dfrac{A - P}{Pr}$
48. 37.5 ft-lb **49.** 9.45 lb **50.** 55°, 55° **51.** $4,000 **52.** 10 oz
53. $x > -2$, $(-2, \infty)$ 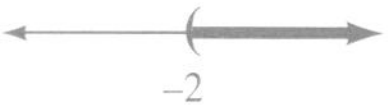

54. $x \le 2$, $(-\infty, 2]$ 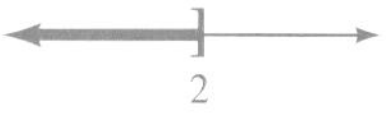

55. $x \ge -1$, $[-1, \infty)$ 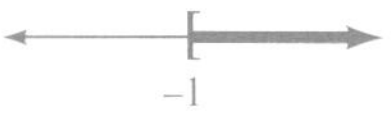

56. $-1 \le x < 2$, $[-1, 2)$ 

Chapter 9 Check Your Knowledge (page 560)

1. solution **2.** origin **3.** slope **4.** parallel **5.** domain, range
6. a. 4° **b.** 6° **c.** −2° **d.** 1° per hour
7.

2x − y = 3	
x	y
0	−3
3/2	0
2	1

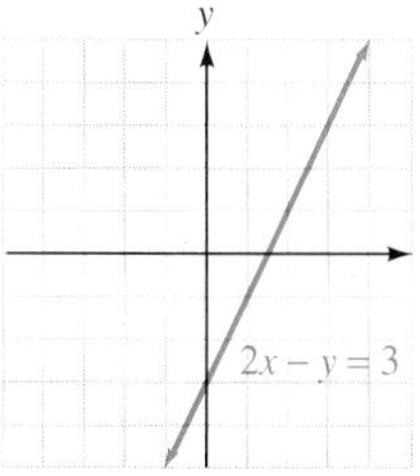

8. $\frac{5}{3}$ **9. a.** $(\frac{1}{3}, 0)$, (0, 1) **b.** −3

10. a. 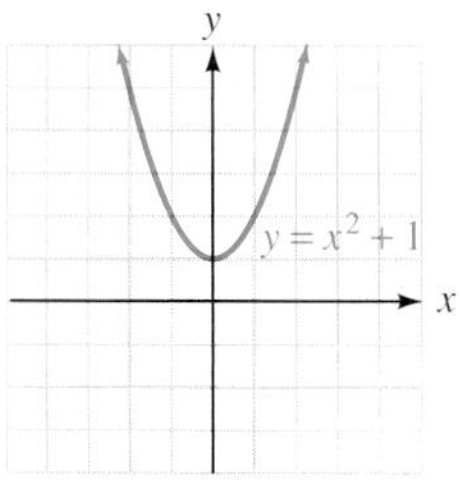

b. (0, 1) **c.** no

11. a. 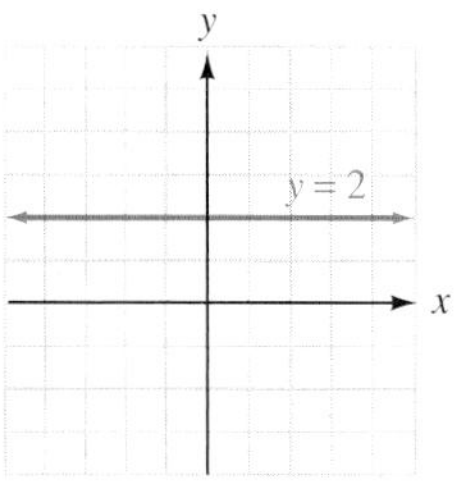

b. 0 **c.** (0, 2)

12. 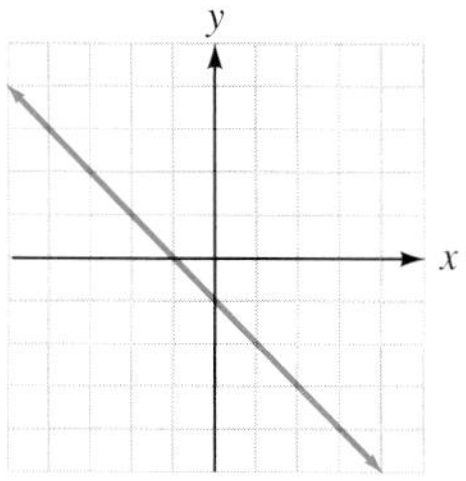

13. 0 **14.** $\frac{1}{3}$ **15.** $c = 0.50 + 0.10m$
16. a. no **b.** no **c.** yes **17.** $y = \frac{5}{2}x - 3$ **18.** $f(-2) = -4$

Study Set Section 9.1 (page 569)

1. ordered **3.** origin **5.** rectangular **7.** origin, left, up
9. no **11.** quadrant II **13. a.** 3 **b.** −4 **c.** 0 **d.** −4 **e.** 0
f. 5 **15.** 10 min before the workout, her heart rate was 60 beats/min. **17.** 150 beats/min **19.** approximately 5 min and 50 min after starting **21.** 10 beats/min faster after cool-down
23. (3, 5) is an ordered pair, 3(5) indicates multiplication, and 5(3 + 5) is an expression containing grouping symbols. **25.** yes

27. 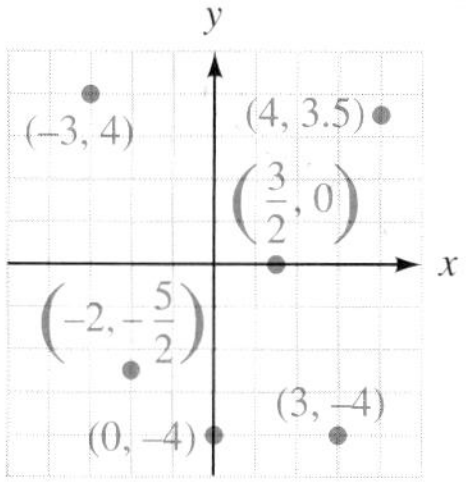

29. (6, 6) **31.** $(-\frac{1}{2}, \frac{5}{2})$ **33.** (7, 6) **35.** rivets: (2, 0), (−6, 0), (−2, 0), (6, 0); welds: (−4, 3), (0, 3), (4, 3); anchors: (−6, −3), (6, −3) **37.** (6, 10), (−7, 4.5), (−5, 11)
39. a. $2 **b.** $4 **c.** $7 **d.** $9
41. a. 35 mi **b.** 4 **c.** 32.5 mi

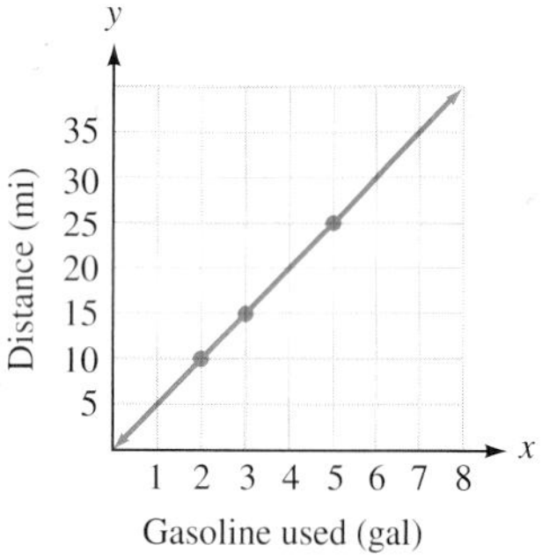

43. Rockford (5, B), Mount Carroll (1, C), Harvard (7, A), intersection (5, E) **49.** 12 **51.** 8 **53.** 7 **55.** −49

Study Set Section 9.2 (page 580)

1. two **3.** independent, dependent **5. a.** 2 **b.** yes **c.** yes **d.** infinitely many **7.** solution, point **9.** 0, −1, −8, 1, 8
11. He should have checked his computations. At least one of his "solutions" is wrong.
13. A smooth curve should be drawn through the points.
17. yes **19.** no **21.** −3, −2, −5 **23.** −3, 1, 1

25. 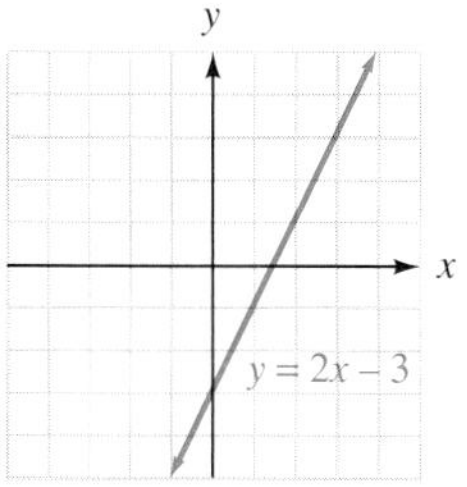

27. 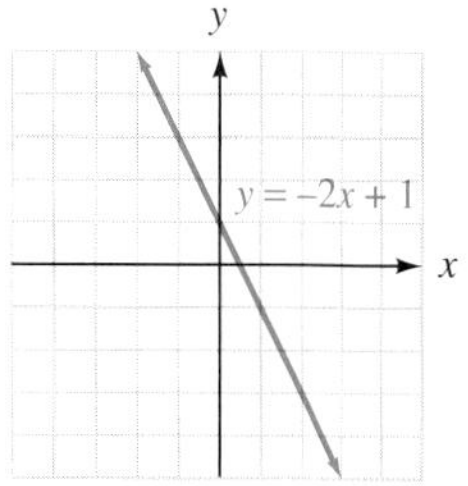

29. 1 unit higher

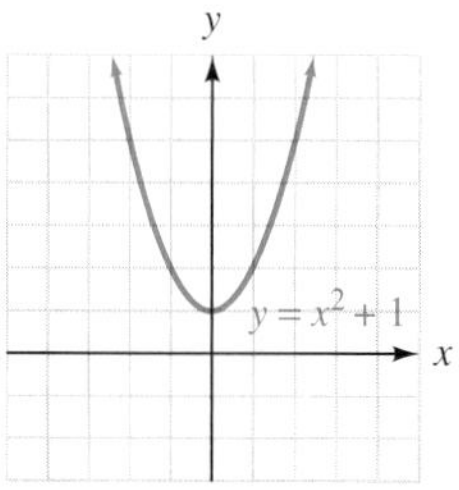

31. 2 units to the right

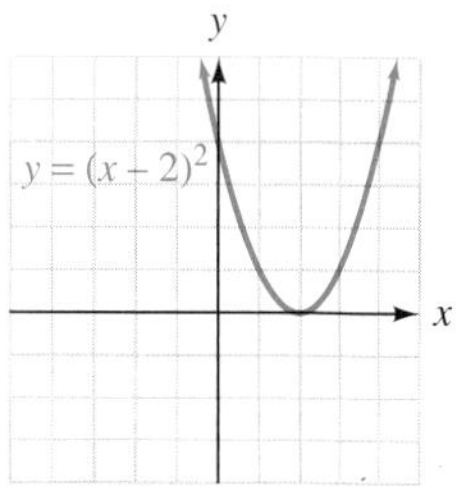

33. It is turned upside down. **35.** 2 units to the left

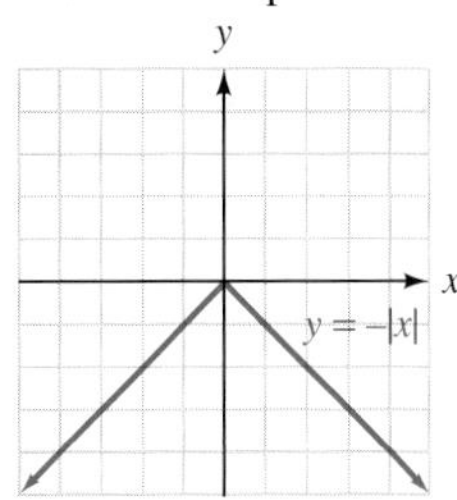

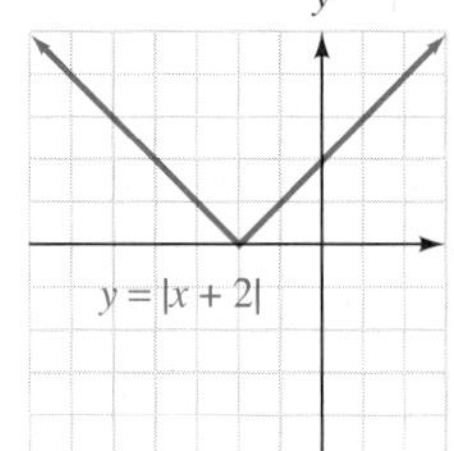

37. It is turned upside down. **39.** 2 units lower

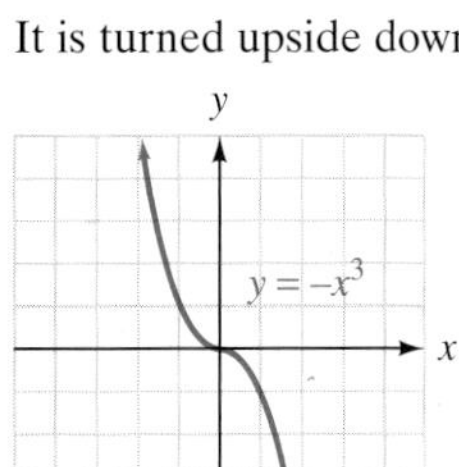

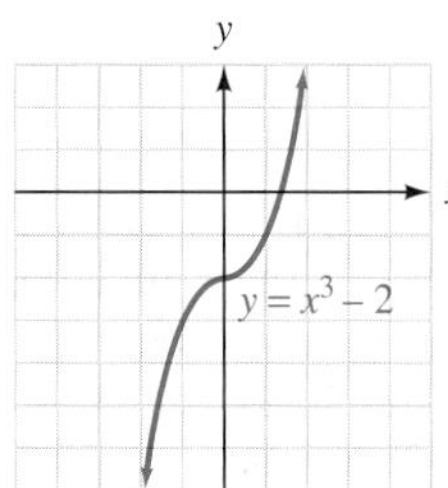

41. 0, 4, 0, −4 **43.** 0, 1, 4, 1, 4 **45. a.** It costs 8¢ to make a 2-in. bolt. **b.** 12¢ **c.** a 4-in. bolt **d.** It decreases as the length increases to 4 in., then increases as the length increases to 7 in. **47. a.** $90,000 **b.** the 3rd yr after being bought **c.** after the 6th yr **d.** It decreased in value for 3 yr, then increased in value for 5 yr. **55.** −96 **57.** an expression **59.** 1.25 **61.** 0.1

Study Set Section 9.3 (page 593)

1. linear **3.** y-intercept **5.** vertical **7. a.** nonlinear **b.** linear **c.** nonlinear **d.** linear **e.** nonlinear **9. a.** y: 1st power; x: 1st power **b.** y: 1st power; x: 2nd power **c.** y: 1st power; x: 3rd power **11.** 6, −5, 4 **13.** −2, 4, $-\frac{3}{2}$ **15.** because A is on the line **17.** The student made a mistake; the points should lie on a straight line. **19.** x-intercept: $(-3, 0)$; y-intercept: $(0, -1)$ **21. a.** x; y **b.** y; x **23. a.** $4x - y = 6$ **b.** $x - 2y = 0$ **c.** $x + 3y = 9$ **d.** $x + 0y = 12$

25.

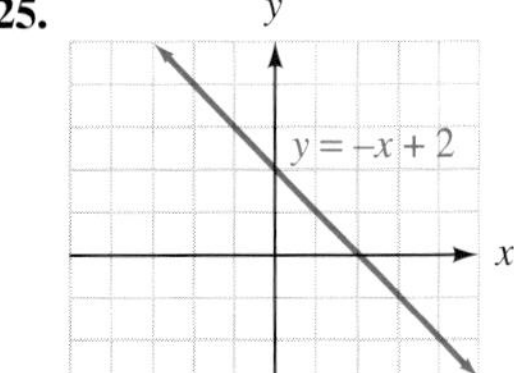

27. **29.**

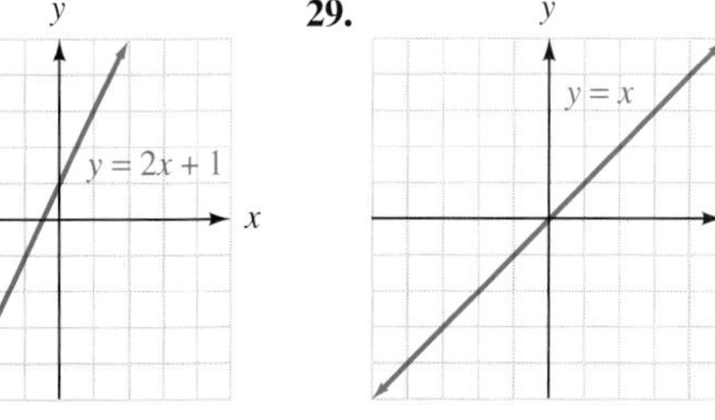
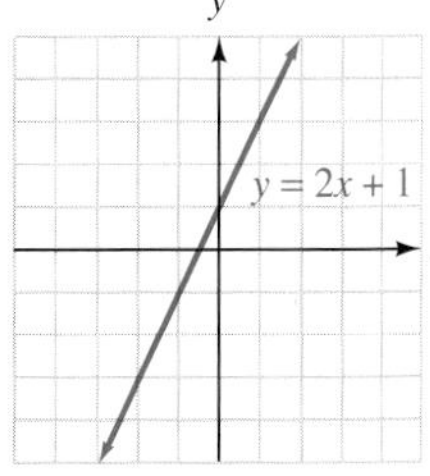

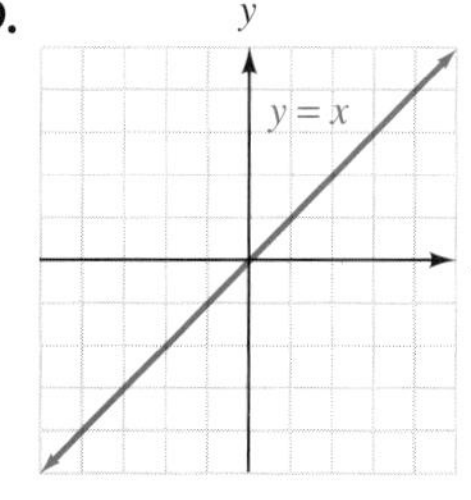

31. **33.**

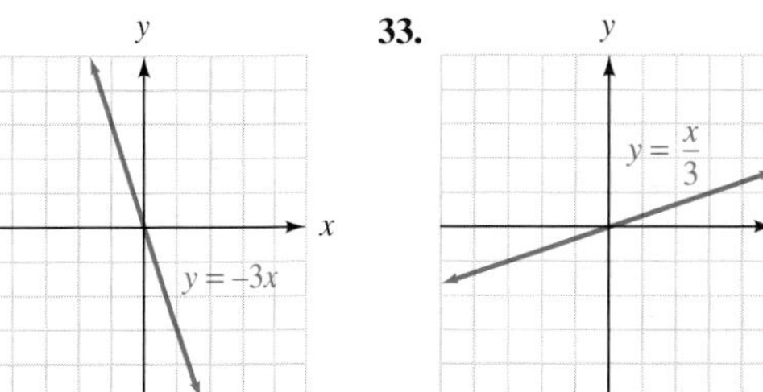
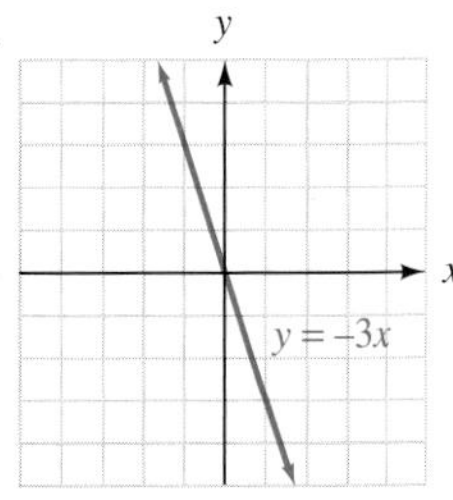

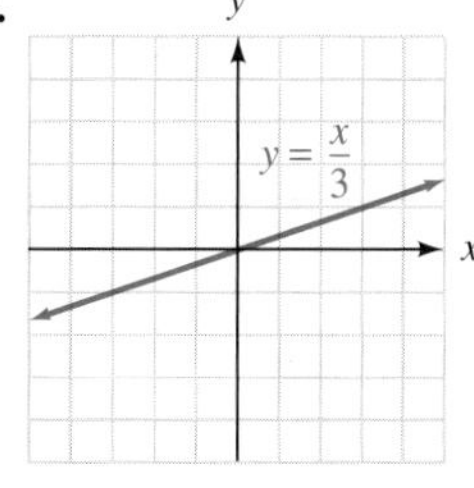

35.

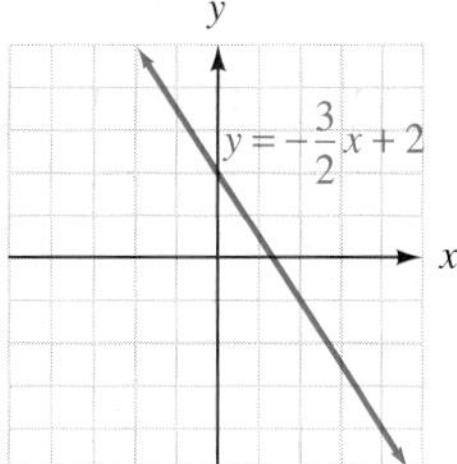

37.

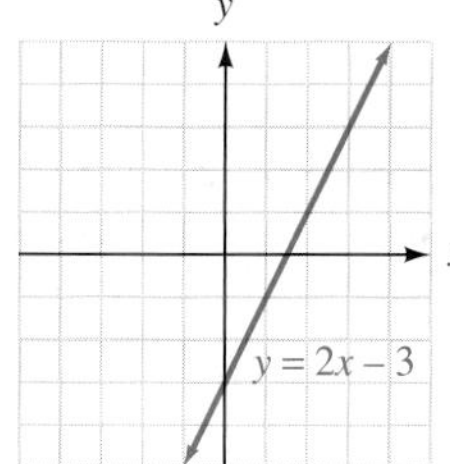

39.

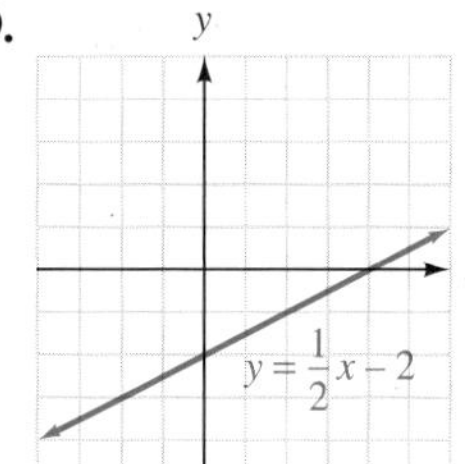

41.

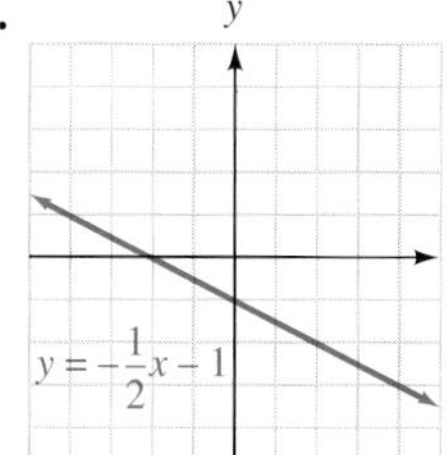

43.

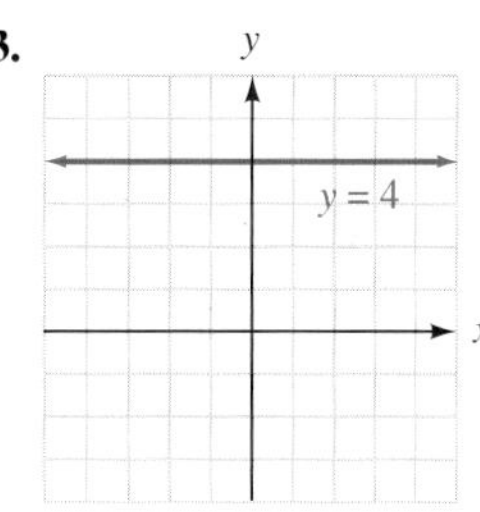

45.

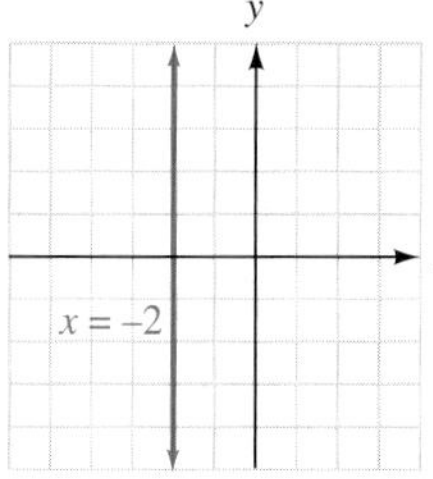

47.

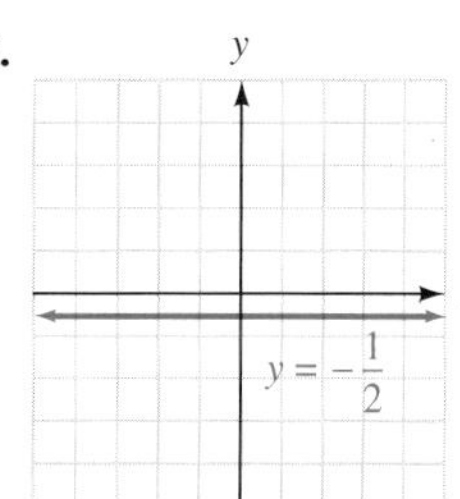

49.

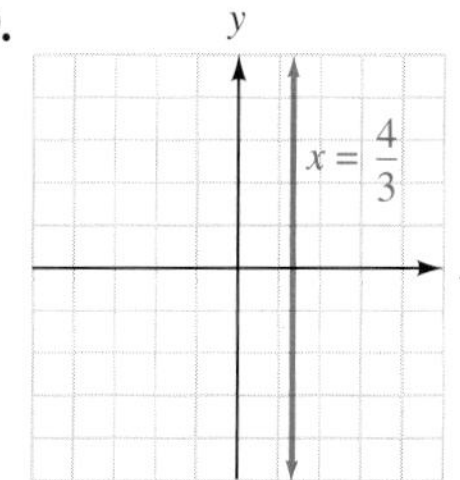

51.

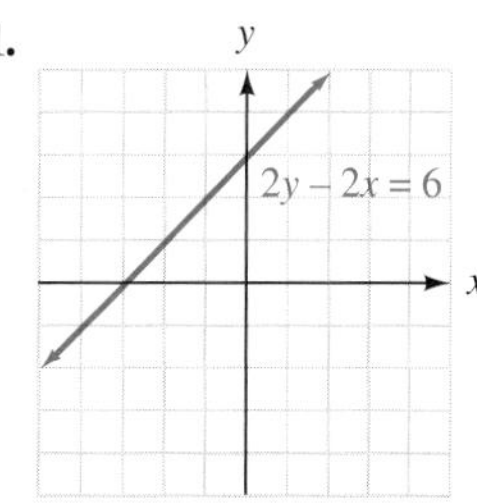

53.

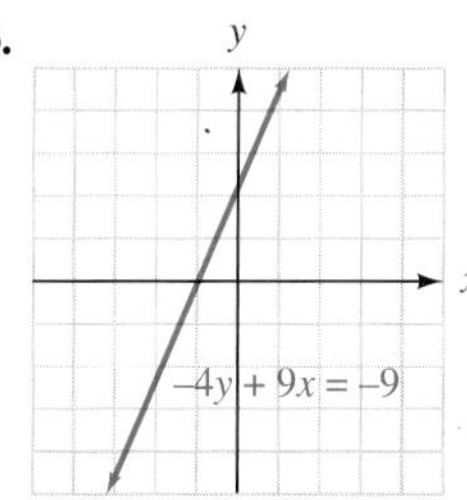

55.

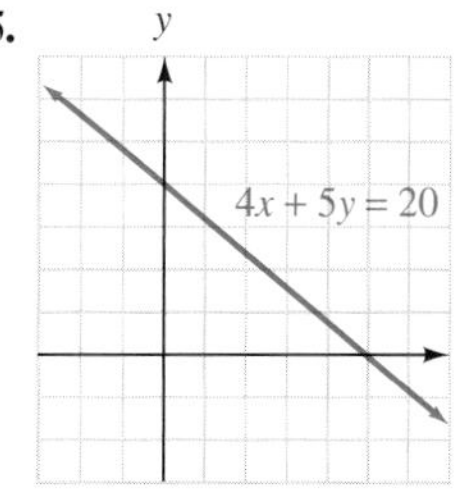

57.

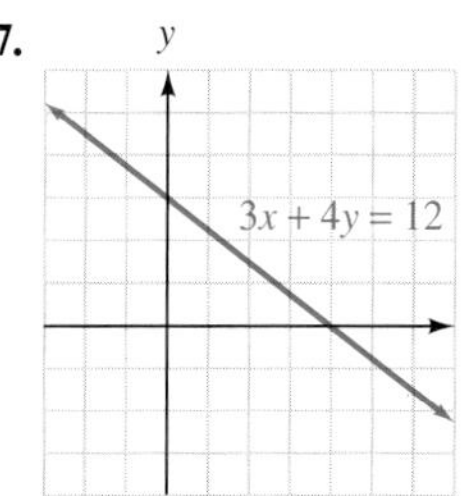

59.

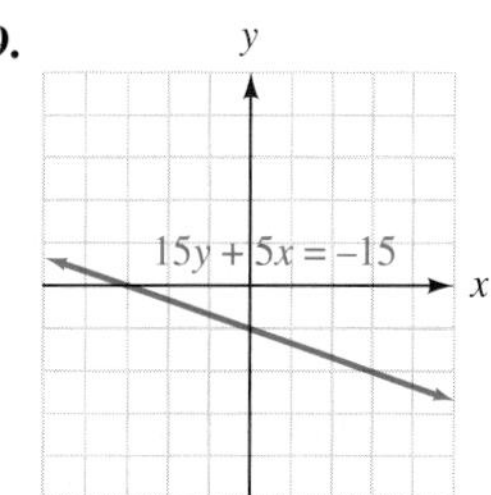

61.

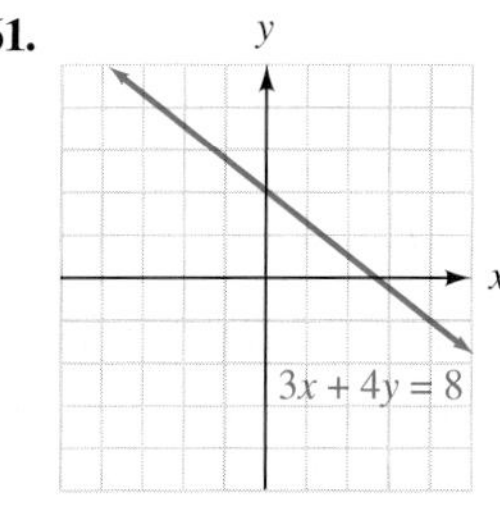

63. **a.** $c = 50 + 25u$ **b.** 150, 250, 400 **c.** \$850 **d.** The service fee is \$50.

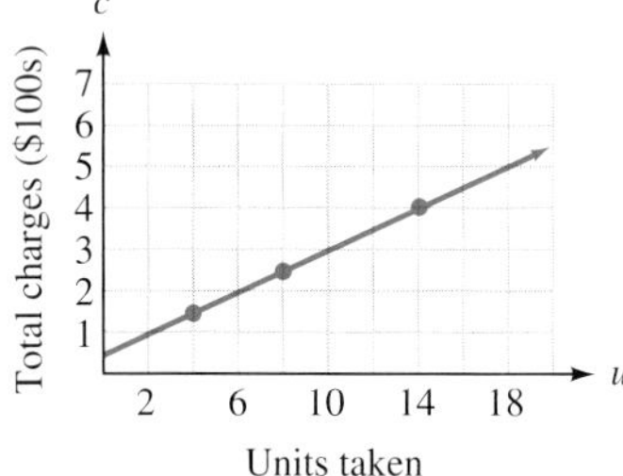

65. **a.** 56.2, 62.1, 64.0

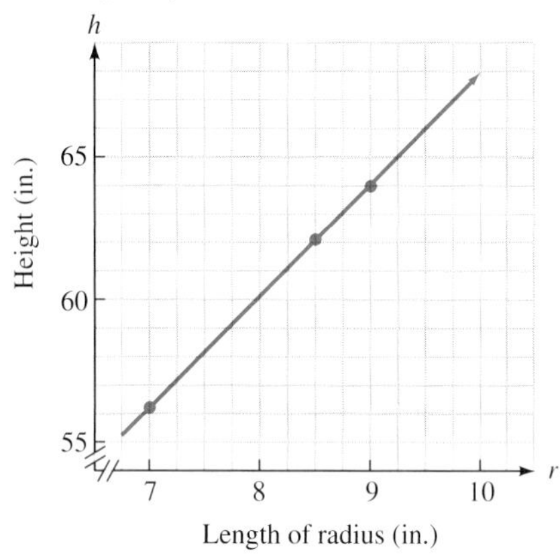

b. taller the woman is **c.** 58 in. **71.** $5 + 4c$ **73.** -4 **75.** profit = revenue − costs **77.** 491

Think It Through (page 602)

\$65 per year

Study Set Section 9.4 (page 606)

1. ratio **3.** slope **5.** change **7.** **a.** l_2 **b.** l_1 **c.** l_4 **d.** l_3 **9.** -1 **11.** 3 in./yr **13.** $-\$2{,}500$/year **15.** **a.** 0; sales of 7-Up were not changing; each year about the same number of cases were sold. **b.** Mountain Dew **17.** $m = \frac{y_2 - y_1}{x_2 - x_1}$ **19.** 1 **21.** -3 **23.** $\frac{5}{4}$ **25.** $-\frac{1}{2}$ **27.** $\frac{3}{5}$ **29.** 0 **31.** undefined **33.** $-\frac{2}{3}$ **35.** -4.75 **37.** $m = \frac{2}{3}$ **39.** $m = \frac{4}{3}$ **41.** $m = -\frac{7}{8}$ **43.** $m = -\frac{1}{5}$

45.

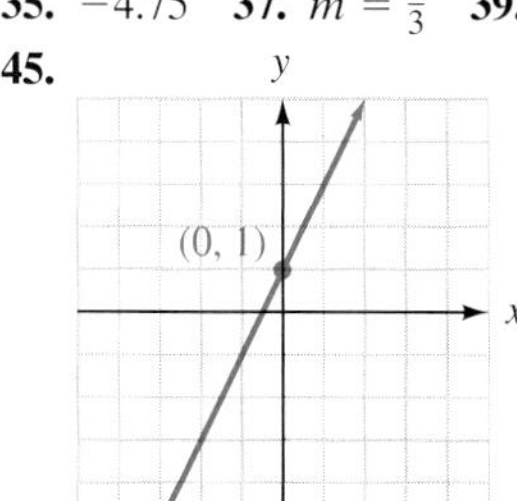

47.

y
x
(−3, −3)

49.

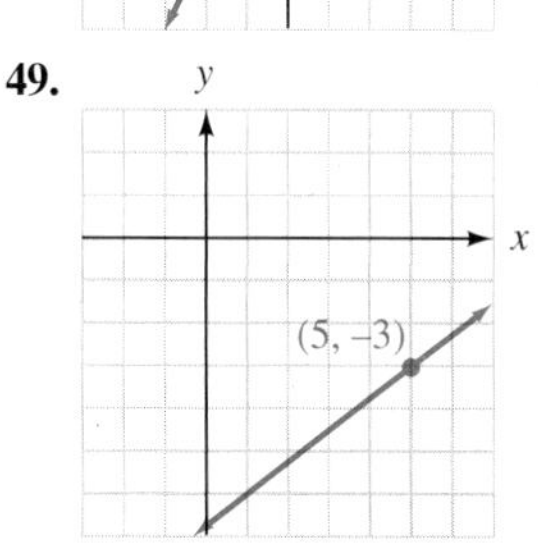

51.

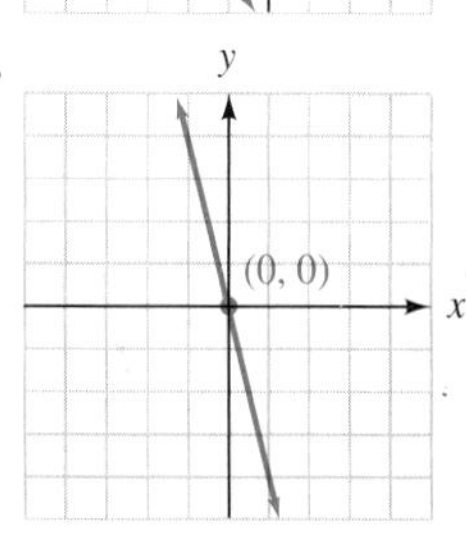

53.

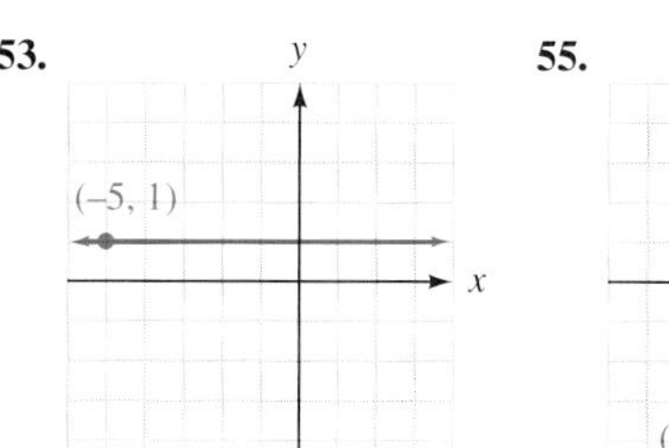

55.

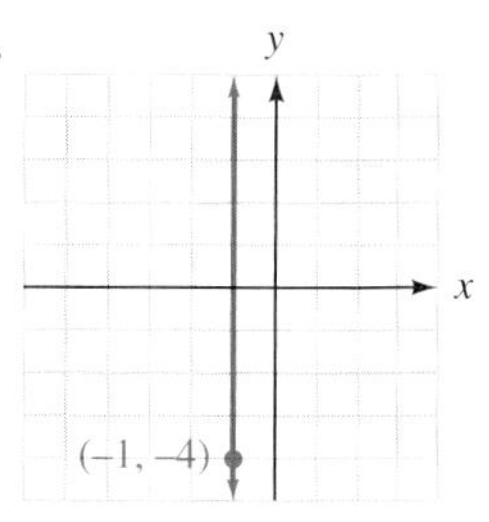

57. $\frac{2}{5}$ **59.** $\frac{1}{20}$; 5% **61.** **a.** $\frac{1}{8}$ **b.** $\frac{1}{12}$ **c.** 1: less expensive, steeper; 2: not as steep, more expensive **63.** 3 hp/40 rpm **69.** quadrant II **71.** no **73.** linear

Think It Through (page 613)

The rate of increase in beginning teacher salary is \$865 per year. In 1980, the average beginning teacher salary was \$11,100.

Study Set Section 9.5 (page 617)

1. slope–intercept **3.** Parallel **5.** reciprocals **7.** No

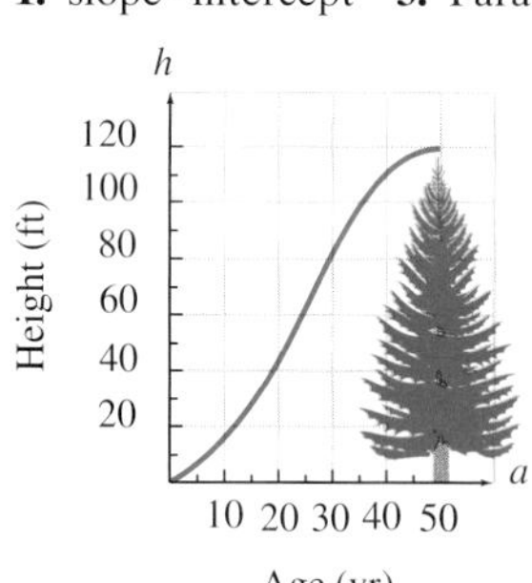

9. **a.** When there are no head waves, the ship could travel at 18 knots. **b.** $-\frac{1}{2}$ knot/ft **c.** $y = -\frac{1}{2}x + 18$ **11.** **a.** (0, 0) **b.** same slope, different y-intercepts **13.** **a.** $-\frac{2}{3}$ **b.** $\frac{1}{4}$ **c.** -8 **d.** 3 **e.** 1 **f.** -1 **15.** $y, -3x, 6$ **19.** 4, (0, 2) **21.** $\frac{1}{4}, (0, -\frac{1}{2})$ **23.** $\frac{1}{2}, (0, 6)$ **25.** $\frac{1}{6}, (0, -1)$ **27.** $-1, (0, 8)$ **29.** $-\frac{2}{3}, (0, 2)$ **31.** $0, (0, \frac{13}{3})$ **33.** $-5, (0, 0)$

35. $y = 5x - 3$

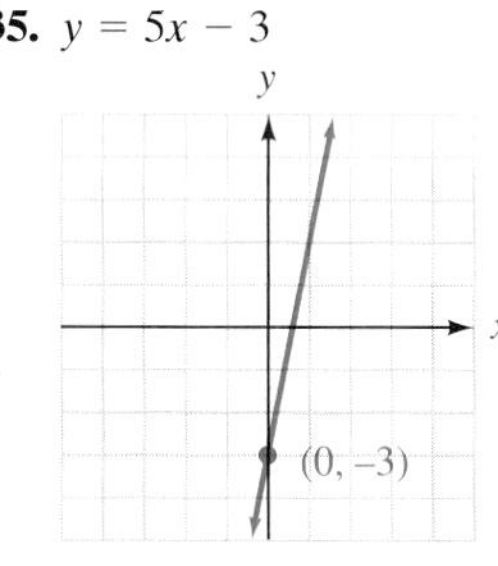

37. $y = \frac{1}{4}x - 2$

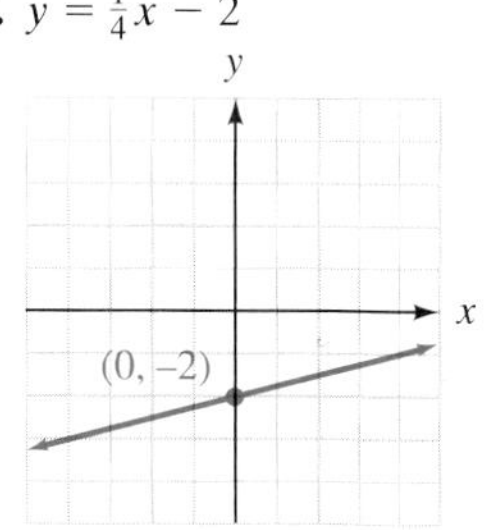

39. $y = -3x + 6$

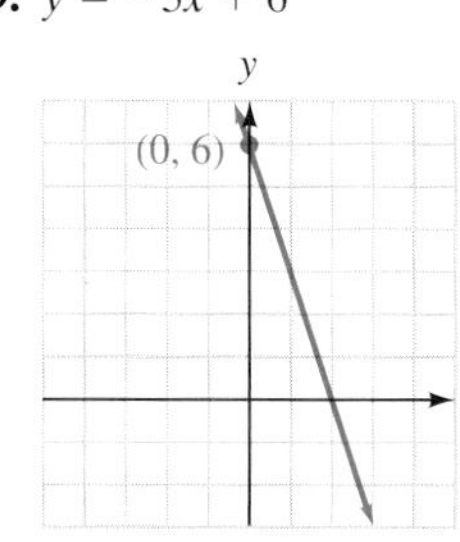

41. $y = -\frac{8}{3}x + 5$

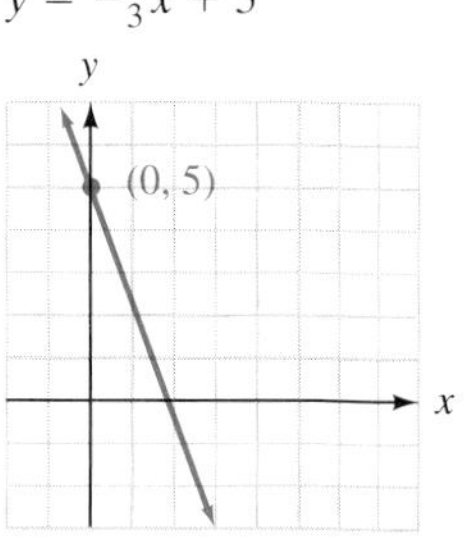

43.

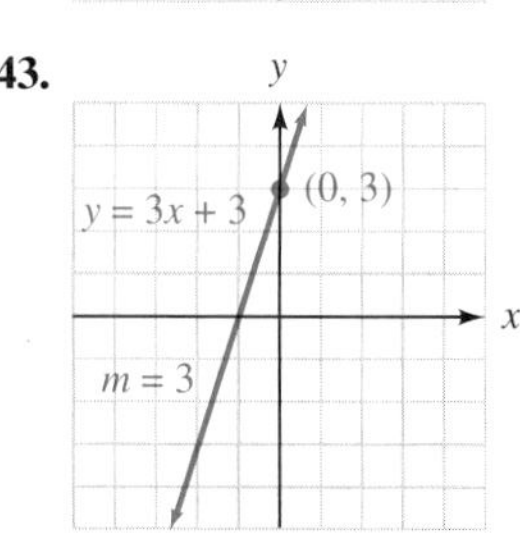

45.

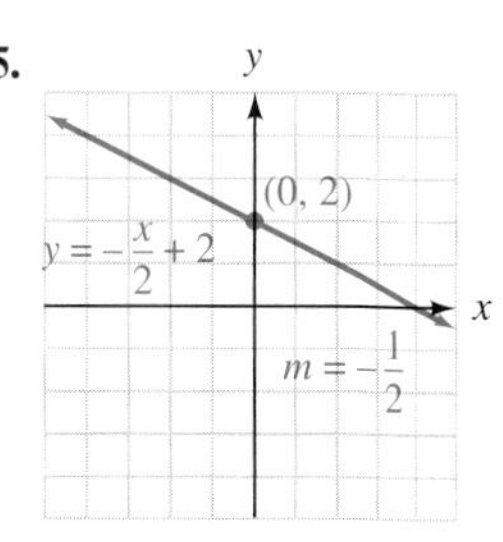

47.

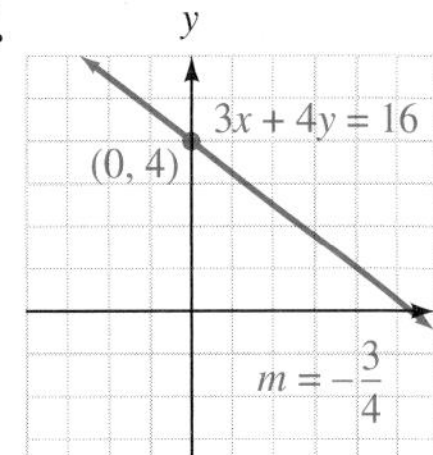

49.

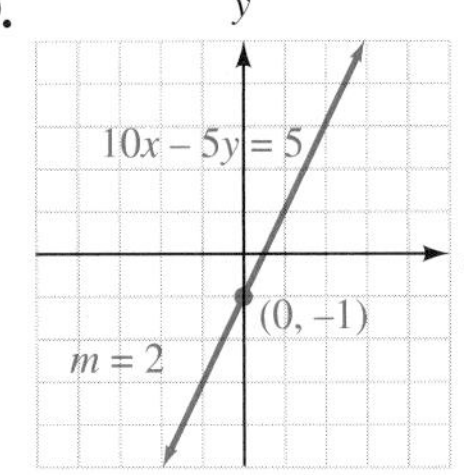

51. a. $y = 2{,}000x + 5{,}000$ **b.** \$21,000 **53.** $y = 5x - 10$
55. a. $y = 0.20x + 1.00$
b.

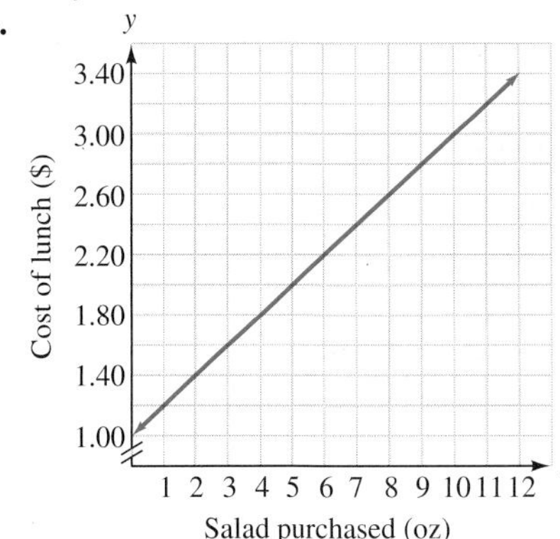

c. same slope; different y-intercept **d.** same y-intercept; steeper slope

57. $y = -20x + 500$ **63.** $-\frac{1}{4}$ **65.** 0 **67.** subtraction
69. 25%

Study Set Section 9.6 (page 626)

1. point–slope **3.** x-coordinate, y-coordinate **5. a.** The slope is 2; the y-intercept is $(0, -3)$ **b.** The graph passes through $(5, 4)$; the slope is 6. **7. a.** $(-4, -2), (3, 2)$ **b.** $\frac{4}{7}$ **c.** $y + 2 = \frac{4}{7}(x + 4)$ or $y - 2 = \frac{4}{7}(x - 3)$ **9. a.** no **b.** no **c.** yes **11.** sub **17.** $y - 1 = 3(x - 2)$ **19.** $y + 1 = -\frac{4}{5}(x + 5)$ **21.** $y = \frac{1}{5}x - 1$ **23.** $y = -5x - 37$ **25.** $y = -\frac{4}{3}x + 4$ **27.** $y = -\frac{2}{3}x + 2$ **29.** $y = 8x + 4$ **31.** $y = -3x$ **33.** $y = 2x + 5$ **35.** $y = -\frac{1}{2}x + 1$ **37.** $y = 5$ **39.** $y = \frac{1}{10}x + \frac{1}{2}$ **41.** $x = -8$ **43.** $x = 4$ **45.** $y = 5$ **47. a.** position 1: $(0, 0), (-5, 2)$; position 2: $(0, 0), (-3, 6)$; position 3: $(0, 0), (-1, 7)$; position 4: $(0, 0), (0, 10)$ **b.** $y = -\frac{2}{5}x, y = -7x, x = 0$ **c.** The pole is not in the shape of a straight line. **49. a.** $y = -40x + 920$ **b.** 440 yd^3 **51.** $c = 30t + 45$ **53. a.** $(0, 32)$; $(100, 212)$ **b.** $F = \frac{9}{5}C + 32$ **55.** $y = -\frac{4}{15}x + 83$ **61.** $-\frac{1}{2}$ **63.** 113.1 ft^2 **65.** −1 **67.** 6

Study Set Section 9.7 (page 636)

1. function **3.** independent, dependent **5. a.** positive numbers **b.** positive numbers **c.** 0 **d.** D: all reals; R: real numbers greater than or equal to 0 **7.** $f(-1)$ **9. a.** $(-2, 4)$, $(-2, -4)$ **b.** No; the x-value −2 is assigned to more than one y-value (4 and −4). **11.** x, −5, $(4, -5)$ **13.** of, is **15.** yes **17.** yes **19.** no; $(4, 2), (4, -2)$ **21.** yes **23.** no; $(3, 1), (3, 2)$ **25.** yes **27.** no; $(-1, 0), (-1, 2)$ **29.** no; $(3, 4), (3, -4)$ or $(4, 3), (4, -3)$ **31.** yes **33.** no; $(3, 4), (3, -1)$ (answers may vary) **35.** no; $(0, 2), (0, -4)$ (answers may vary) **37.** D: all reals; R: all reals **39.** D: all reals; R: real numbers greater than or equal to 0 **41.** D: all reals; R: all reals **43. a.** 3 **b.** −9 **c.** 0 **d.** 199 **45. a.** 0.32 **b.** 18 **c.** 2,000,000 **d.** $\frac{1}{32}$ **47. a.** 7 **b.** 14 **c.** 0 **d.** 1 **49. a.** 0 **b.** 990 **c.** −24 **d.** 210 **51.** 1.166 **53.** −2, −5, 1, 4

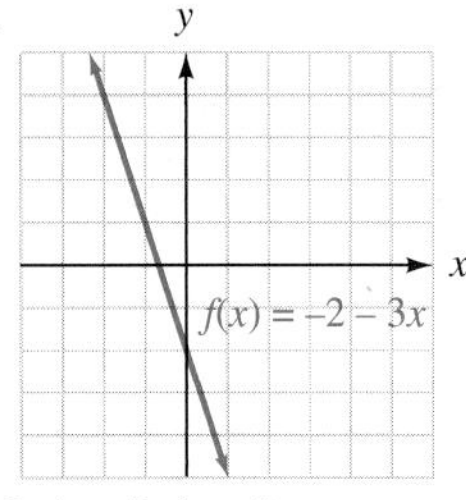

55. −3, −2, −1

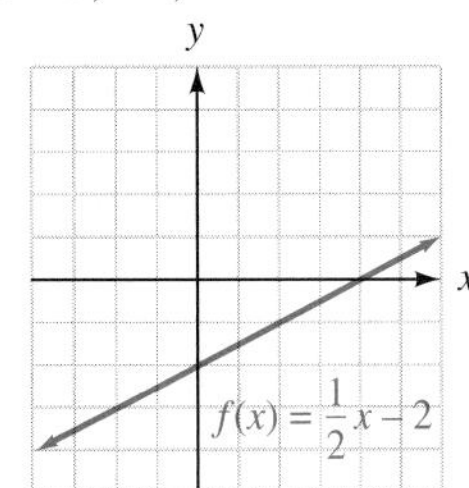

57. 2, 1, −2, 1, −2

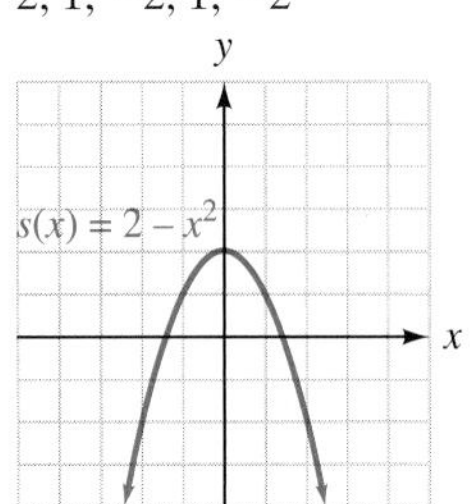

59. $f(x) = |x|$ **61. a.** all real numbers from 0 through 24 **b.** 0.5 **c.** 1.5 **d.** −1.4 **e.** The low-tide mark was −2.5 m. **f.** 1.6 **63.** 78.5 ft^2, 314.2 ft^2, 1,256.6 ft^2 **67.** $y = 6$ **69.** profit = revenue − costs **71.** $-6x + 12$ **73.** $12d$

Key Concept (page 641)

1. $y = -3x - 4$ **3.** −2, 4, −3

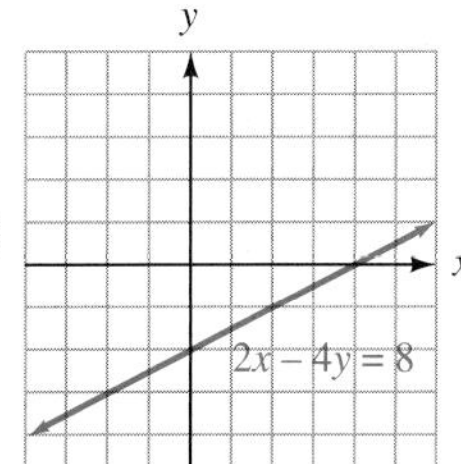

5. a. $(-2, -2)$ (answers may vary) **b.** −2 **c.** $y = -2x - 6$ **7.** 130; the cost to rent the mixer for 3 days

Chapter Review (page 643)

1.

2. −1, 0, 1 **3.** quadrant III **4. a.** 2 ft **b.** 2 ft **c.** 6 ft **5. a.** 2,500; week 2 **b.** 1,000 **c.** 1st week and 5th week **6.** $(0, 2)$ **7.** not a solution

8. a. 8, 1, 0, −1, −8 **b.** It would be 2 units higher.

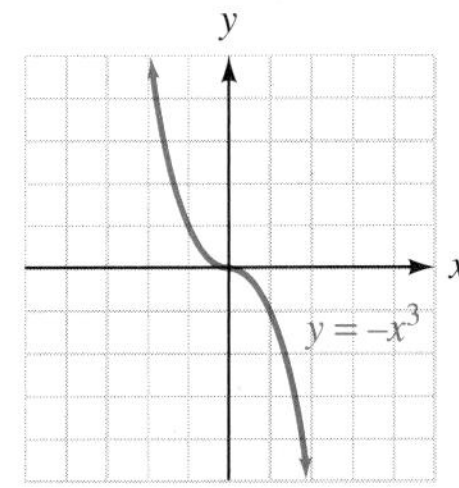

9. a. 9,000 **b.** It tells us that 40 trees on an acre give the highest yield, 18,000 oranges. **10. a.** nonlinear **11.** linear **12.** linear **13.** nonlinear **14.** $A = 5, B = 2, C = 10$ **15.** −6, −6, −8, −8

16.

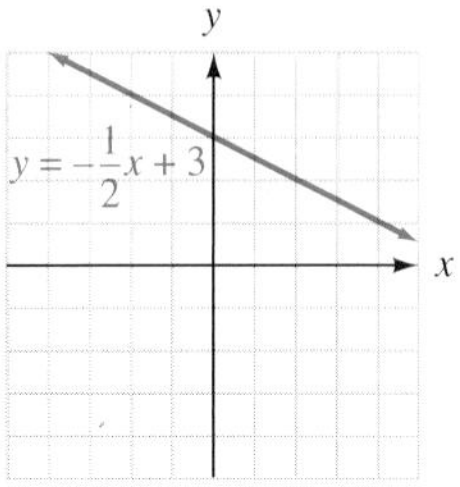

17. x-intercept: $(-2, 0)$; y-intercept: $(0, 4)$

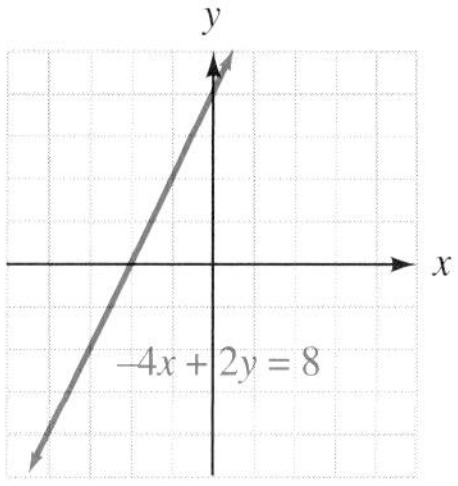

18.

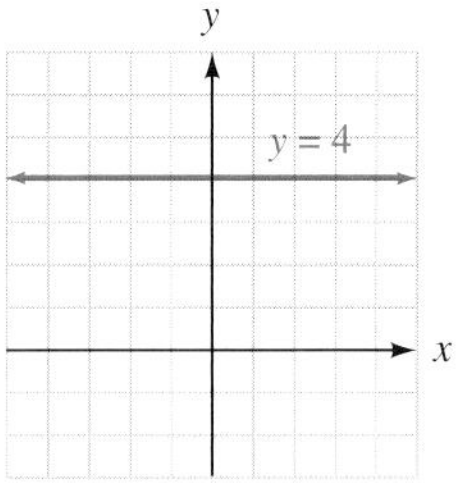

19.

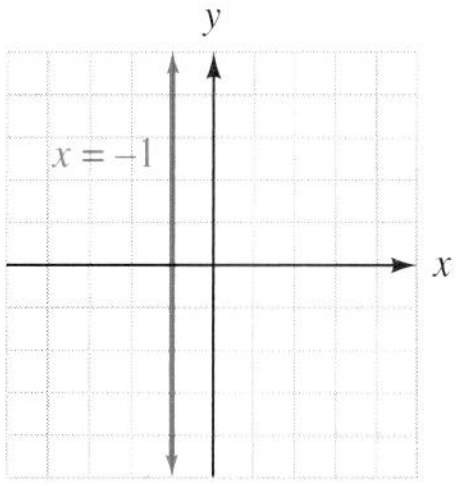

20. If the three points do not lie on a line, then at least one of them is in error. **21.** $\frac{1}{4}$ **22.** -7 **23.** 0 **24.** $-\frac{3}{2}$

25.

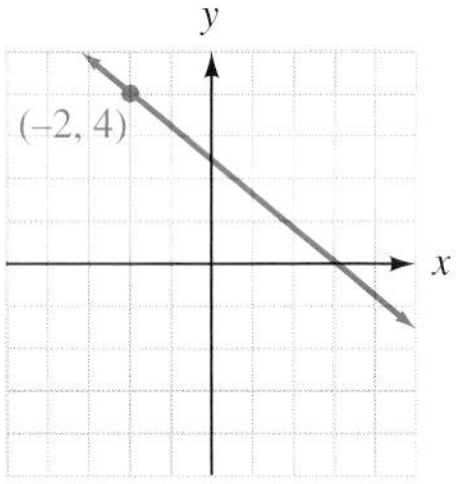

26. a. -4.5 million people/year **b.** 4.05 million people/year **27.** $m = \frac{3}{4}$; y-intercept: $(0, -2)$ **28.** $m = -4$; y-intercept: $(0, 0)$

29. $m = 3$; y-intercept: $(0, -5)$

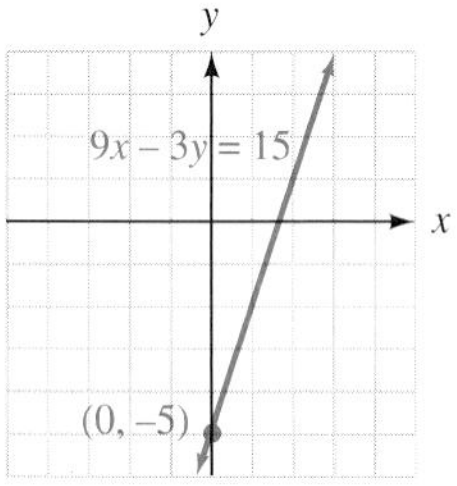

30. a. $c = 300w + 75{,}000$ **b.** 90,600 **31.** parallel **32.** perpendicular

33. $y = 3x + 2$

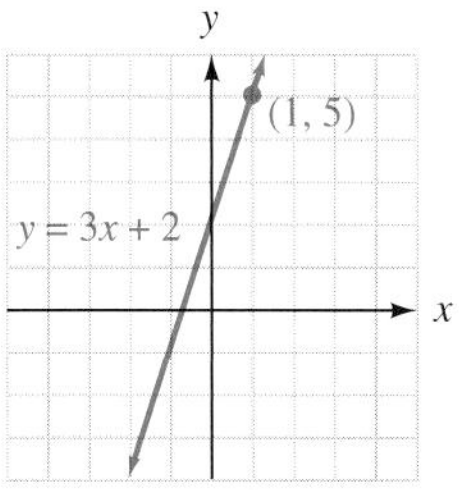

34. $y = -\frac{1}{2}x - 3$

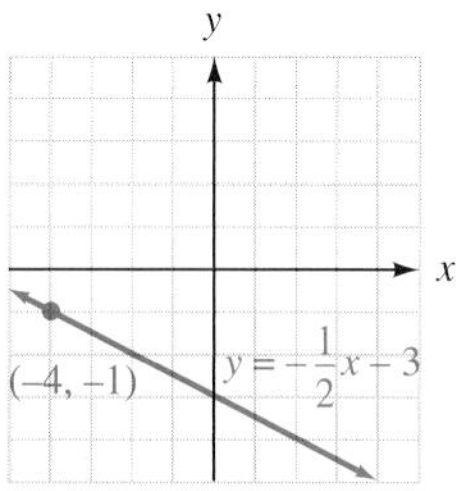

35. $y = \frac{2}{3}x + 5$ **36.** $y = -8$ **37.** $f(x) = -35x + 450$ **38.** yes **39.** no **40.** no **41.** D: all reals; R: all reals

42. D: all reals: R: real numbers greater than or equal to 0 **43.** -5 **44.** 37 **45.** -2 **46.** -8 **47.** 1, 0, -1, 0, -1, -2

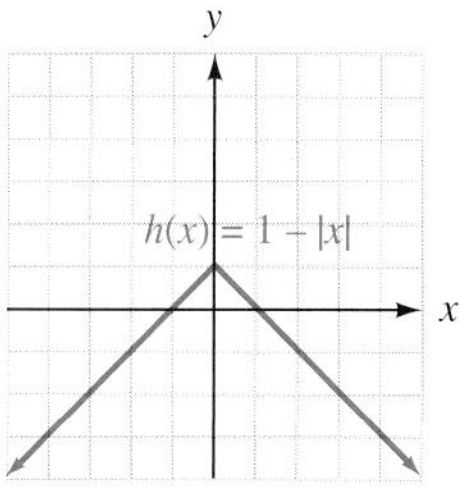

48. no **49.** yes **50.** 1,004.8 in.^{3}

Chapter 9 Test (page 649)

1. 10 **2.** 60 **3.** 1 day before and the 3rd day of the holiday **4.** 50 dogs were in the kennel when the holiday began.

5.

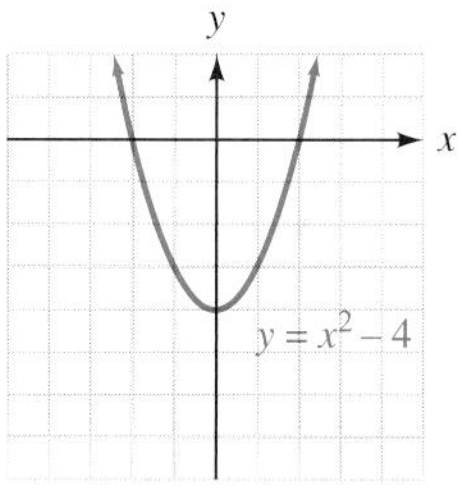

6.

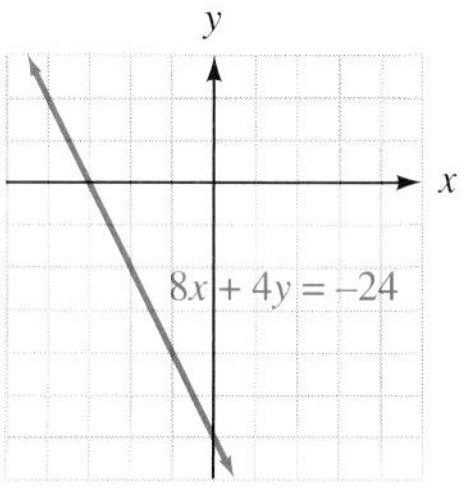

7. yes **8.** no **9.** x-intercept: $(3, 0)$; y-intercept: $(0, -2)$ **10.** $m = -\frac{1}{2}$; $(0, 4)$ **11.**

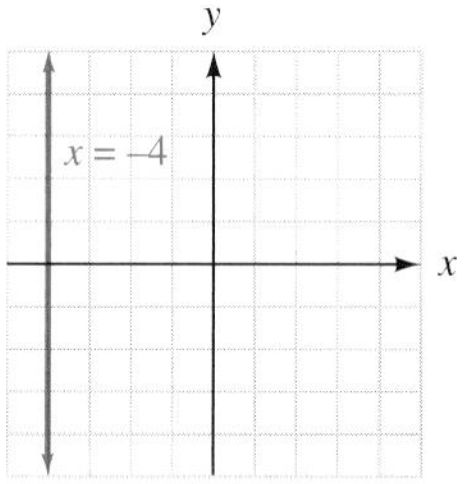

12.

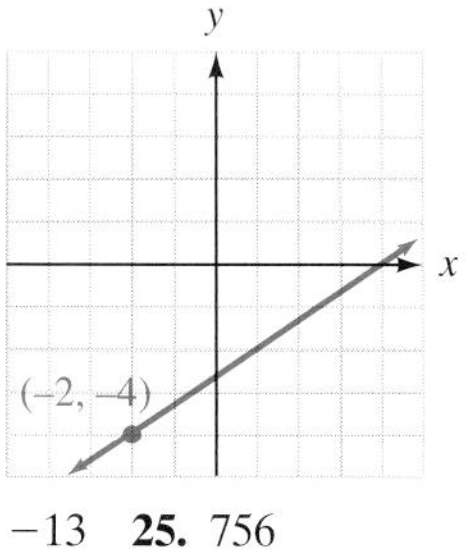

13. -1 **14.** undefined **15.** $\frac{8}{7}$ **16.** parallel **17.** -15 ft per mi **18.** 25 ft per mi **19.** $v = 15{,}000 - 1{,}500x$ **20.** $y = 7x + 19$ **21.** no **22.** D: all reals; R: real numbers less than or equal to 0 **23.** yes **24.** -13 **25.** 756

Cumulative Review Exercises (page 651)

1. a. Q3, 2001 **b.** \$3.1 billion in losses **2.** $2^2 \cdot 3^3$ **3.** 0.004 **4. a.** true **b.** true **c.** true **5.** 3.8, -5.6 **6. a.** 6 **b.** -2 **c.** 1 **d.** 28 **7.** 32 **8.** $500 - x$ **9.** 3, -2 **10. a.** $2x + 8$ **b.** $2x - 8$ **c.** $-2x - 8$ **d.** $-2x + 8$ **11.** $4a + 10$ **12.** $4b^2$ **13.** 4 **14.** $-3y$ **15.** 6 **16.** 2.9 **17.** 9 **18.** -19 **19.** $\frac{1}{7}$ **20.** 1 **21.** $-\frac{55}{6}$ **22.** no solution **23.** $x = \frac{y - b}{m}$ **24.** $3\frac{1}{8}$ in., $\frac{39}{64}$ in.^{2} **25.** 79° **26.** $0.50x$, $0.25(13 - x)$, $0.30(13)$ **27.** 7.5 hr **28.** 80 lb candy corn, 120 lb gumdrops **29.** $x \le 48$,

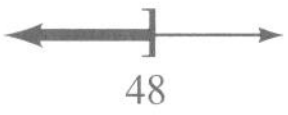

, $(-\infty, 48]$

30. $x > 0$,

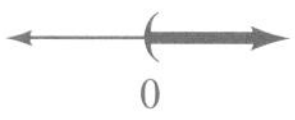

, $(0, \infty)$ **31. a.** 1 tsp **b.** 3 tsp

32. no **33.**

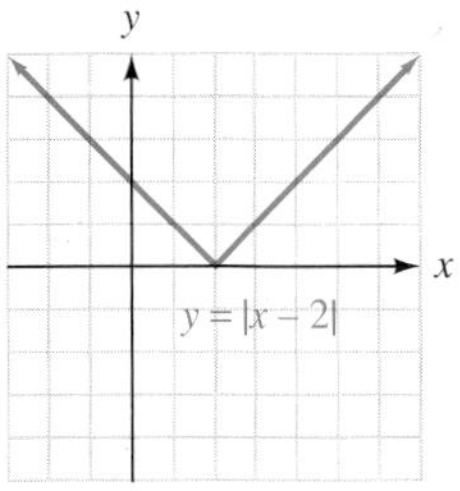

34.

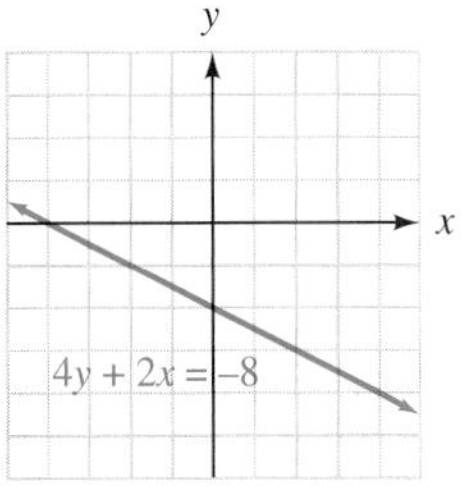

35. 0 **36.** $-\frac{10}{7}$ **37.** $\frac{2}{3}$, (0, 2) **38.** $y = -2x + 1$ **39.** $y + 9 = -\frac{7}{8}(x - 2)$ **40.** 10

Chapter 10 Check Your Knowledge (page 654)

1. base, exponent **2.** polynomial **3.** term **4.** decreasing or descending **5.** $3x^2y^3z^2$ **6.** $10x^7y^3$ **7.** $72x^{12}$ **8.** 1 **9.** $\frac{2y}{x}$ **10.** $\frac{3x^7}{y^3}$ **11.** $\frac{y^9}{8x^{21}}$ **12. a.** $(4x^2 + 12x + 9)$ ft^2 **b.** $(4x^3 + 12x^2 + 9x)$ ft^3 **13. a.** 2.3×10^{-5} **b.** 6,250,000 **14. a.** monomial **b.** 7th degree **15.** $f(-1) = 2$ **16. a.** $9xy^2$ **b.** $2x^2 - 5x + 4$ **17.** $12x^2y^3z^2$ **18.** $2x^3 + 6x^2 - 10x$ **19.** $4y^2 - 49$ **20.** $9y^2 - 30y + 25$ **21.** $8x^2 - 2x - 21$ **22.** $2x^3 - 5x^2 + 5x - 3$ **23.** 2 **24.** $3x + 5$ **25.** $(2x - 3)$ ft

Think It Through (page 658)

$10^4 = 10{,}000$

Study Set Section 10.1 (page 663)

1. base, exponent **3.** power **5.** $3x, 3x, 3x, 3x$ **7.** x^{m+n} **9.** $\frac{a^n}{b^n}$ **11.** x^{m-n} **13.** x^{mn} **15. a.** $(3x^2)^6$ (answers may vary) **b.** $\left(\frac{3a^3}{b}\right)^2$ (answers may vary) **17. a.** $2x^2$ **b.** 0 **c.** x^4 **d.** x^2 **19. a.** doesn't simplify **b.** doesn't simplify **c.** x^5 **d.** x **21.** a^{10} mi^2 **23.** x^9 m^3 **27.** base 4, exponent 3 **29.** base x, exponent 5 **31.** base $-3x$, exponent 2 **33.** base y, exponent 6 **35.** 16 **37.** −16 **39.** $x \cdot x \cdot x \cdot x \cdot x$ **41.** $\left(\frac{t^2}{2}\right)\left(\frac{t^2}{2}\right)\left(\frac{t^2}{2}\right)$ **43.** $(4t)^4$ **45.** $-4t^3$ **47.** 12^7 **49.** 2^6 **51.** a^6 **53.** x^7 **55.** a^9 **57.** y^9 **59.** 8^8 **61.** x^{12} **63.** c **65.** 3^8 **67.** y^{15} **69.** m^{500} **71.** a^5b^6 **73.** c^2d^5 **75.** x^3y^3 **77.** y^4 **79.** c^2d^6 **81.** x^{25} **83.** $243z^{30}$ **85.** x^{31} **87.** u^4v^4 **89.** a^9b^6 **91.** $-8r^6s^9$ **93.** $\frac{a^3}{b^3}$ **95.** $\frac{x^{10}}{y^{15}}$ **97.** $-\frac{32a^5}{b^5}$ **99.** $216k^3$ **101.** $a^{12}b^6$ **103.** a **105.** ab^4 **107.** $r^{13}s^3$ **109.** $\frac{y^3}{8}$ **111.** $\frac{27t^{12}}{64}$ **113. a.** $25x^2$ ft^2 **b.** $9\pi x^2$ ft^2 **115. b.** 16 ft, 8 ft, 4 ft, 2 ft **117. a.** 11000001, 11010001, 11001101, 11000011 (answers may vary) **b.** 2^8 **123.** c **125.** d

Study Set Section 10.2 (page 670)

1. base, exponent **3.** reciprocal **5. a.** 4 − 4, 0 **b.** 6, 6, 6, 6, 1 **c.** 1, 1 **7.** 9, 3, 1, $\frac{1}{3}$, $\frac{1}{9}$ **9.** 81, −9, 1, $-\frac{1}{9}$, $\frac{1}{81}$ **11.** 4, 2^2; 2, 2^1; 1, 2^0; $\frac{1}{2}$, 2^{-1}; $\frac{1}{4}$, 2^{-2} **15.** base x, exponent −2 **17. a.** 4; 2; −16 **b.** 4; −2; $\frac{1}{16}$ **c.** 4; −2; $-\frac{1}{16}$ **19.** 1 **21.** 1 **23.** 2 **25.** 1 **27.** 1 **29.** $\frac{5}{2}$ **31.** $\frac{1}{144}$ **33.** $-\frac{1}{4}$ **35.** 125 **37.** $\frac{3}{16}$ **39.** $-\frac{1}{64}$ **41.** $\frac{1}{64}$ **43.** $\frac{1}{x^2}$ **45.** $-\frac{1}{b^5}$ **47.** $\frac{1}{16y^4}$ **49.** $\frac{1}{a^3b^6}$ **51.** 8 **53.** 1 **55.** $\frac{8}{7}$ **57.** 1 **59.** $\frac{1}{y}$ **61.** $\frac{1}{r^6}$ **63.** y^5 **65.** 2 **67.** $\frac{1}{a^2b^4}$ **69.** $\frac{1}{x^6y^3}$ **71.** $\frac{1}{x^3}$ **73.** $\frac{a^8}{b^{12}}$ **75.** $-\frac{y^{10}}{32x^{15}}$ **77.** a^{14} **79.** $\frac{1}{b^{14}}$ **81.** $\frac{256x^{28}}{81}$ **83.** $\frac{16y^{14}}{z^{10}}$ **85.** x^{3m} **87.** $\frac{1}{u^m}$ **89.** y^m **91.** $\frac{1}{x^{3n}}$ **93.** $10^2, 10^1, 10^0, 10^{-1}, 10^{-2}, 10^{-3}, 10^{-4}$ **95.** approximately \$4,603 **99.** 13.5 yr **101.** $y = \frac{3}{4}x - 5$

Think It Through (page 677)

3.24×10^7 mi

Study Set Section 10.3 (page 678)

1. scientific **3.** 250 **5.** 0.000025 **7.** 10^5 **9.** 10^{-3} **11.** 5, right **13.** 10^{-4} **15.** positive **19.** 2.3×10^4 **21.** 1.7×10^6 **23.** 6.2×10^{-2} **25.** 5.1×10^{-6} **27.** 4.25×10^3 **29.** 2.5×10^{-3} **31.** 230 **33.** 812,000 **35.** 0.00115 **37.** 0.000976 **39.** 25,000,000 **41.** 0.00051 **43.** 2.57×10^{13} mi **45.** 63,800,000 mi^2 **47.** 6.22×10^{-3} mi **49.** 714,000 **51.** 30,000 **53.** 200,000 **55.** $9.038030748 \times 10^{15}$ **57.** $1.881676423 \times 10^{12}$ **59.** $1.74449211 \times 10^{26}$ **61.** g, x, u, v, i, m, r **63.** 1.7×10^{-18} g **65.** 3.099363×10^{16} ft **67.** 2.08×10^{11} dollars **69. a.** 1.7×10^6; 1,700,000 **b.** 1986: 2.05×10^6; 2,050,000 **75.** 5 **77.** commutative property of addition **79.** 6

Study Set Section 10.4 (page 685)

1. polynomial **3.** degree **5.** monomial, binomial **7.** decreasing or descending **9.** of **11.** yes **13.** no **15.** yes **17.** binomial **19.** trinomial **21.** monomial **23.** binomial **25.** trinomial **27.** none of these **29.** trinomial **31.** 4th **33.** 2nd **35.** 1st **37.** 4th **39.** 12th **41.** 0th **45.** because $x^3 + 2x^2 - 3$ is a polynomial **47.** 7 **49.** −8 **51.** −2 **53.** −7.5 **55.** −4 **57.** −5 **59.** −5.69 **61.** −188.96 **63.** 3 **65.** −1 **67.** 1.929 **69.** −512.241 **71.** −1 **73.** −7 **75.** 72 in.^{3} **77.** 22 ft, 2 ft **79.** about 404 ft **81.** 18.75 ft, 20 ft, 15 ft **85.** $y \geq -3$

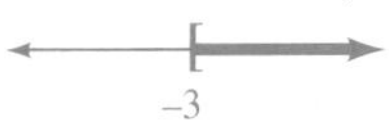

$[-3, \infty)$ **87.** x^{18} **89.** y^9

Study Set Section 10.5 (page 692)

1. polynomials **3.** coefficients **5.** terms **7.** like **9.** changing **11.** $2x^2 + 3x - 4$ **13.** $(11x - 12)$ ft **17.** $9y$ **19.** $12t^2$ **21.** $-48u^3$ **23.** $-0.1x$ **25.** $2st$ **27.** $6r$ **29.** $-ab$ **31.** $15x^2$ **33.** $7x + 4$ **35.** $2a + 7$ **37.** $7x - 7y$ **39.** $3x - 4y$ **41.** $6x^2 + x - 5$ **43.** $7b + 4$ **45.** $3x + 1$ **47.** $13.54x^2 - 6.55x + 6.15$ **49.** $5x^2 + x + 11$ **51.** $-7x^3 - 7x^2 - x - 1$ **53.** $2x^2 + 4x + 13$ **55.** $5x^2 + 6x - 8$ **57.** $-x^3 + 6x^2 + x + 14$ **59.** $-12x^2y^2 - 9xy + 36y^2$ **61.** $t^3 + 3t^2 + 6t - 5$ **63.** $-3x^2 + 5x - 7$ **65.** $6x - 2$ **67.** $-5x^2 - 8x - 4$ **69.** $-4y^2 + 8y + 4$ **71.** $a^2b + 2ab^2 - 6ab - 4a - b$ **73.** $(3x^2 + 6x - 2)$ yd **75.** $(x^2 - 8x + 12)$ ft **77.** $(3x^2 + 11x + 4.5\pi)$ in.

79. \$114,000 **81.** $f(x) = 1{,}900x + 225{,}000$
83. $f(x) = -1{,}100x + 6{,}600$ **85.** $f(x) = -2{,}800x + 15{,}800$
87. a. $22t + 20$ **b.** 108 ft **93.** 180°
95. $x \le 2$ 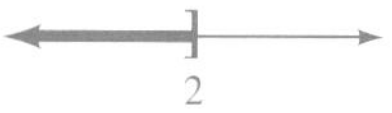 $(-\infty, 2]$

Study Set Section 10.6 (page 702)

1. binomials **3.** first, outer, inner, last **5.** $6x^2$ **7.** $15x$
9. $(6x^2 + x - 1)$ cm^2 **13.** $12x^5$ **15.** $-24b^6$ **17.** $6x^5y^5$
19. $-3x^4y^5z^8$ **21.** $3x + 12$ **23.** $-4t - 28$ **25.** $3x^2 - 6x$
27. $-6x^4 + 2x^3$ **29.** $3x^2y + 3xy^2$ **31.** $6x^4 + 8x^3 - 14x^2$
33. $-6x^4 - 24x^3$ **35.** $a^2 + 9a + 20$ **37.** $3x^2 + 10x - 8$
39. $6a^2 + 2a - 20$ **41.** $6x^2 - 7x - 5$ **43.** $2x^2 + 3x - 9$
45. $6t^2 + 7st - 3s^2$ **47.** $x^2 + xz + xy + yz$
49. $-12t^2 + 7tu - u^2$ **51.** $8x^2 - 12x - 8$ **53.** $3a^3 - 3ab^2$
55. $5t^2 - 11t$ **57.** $2x^2 + xy - y^2$ **59.** $x^2 + 8x + 16$
61. $t^2 - 6t + 9$ **63.** $r^2 - 16$ **65.** $16x^2 - 25$ **67.** $4s^2 + 4s + 1$
69. $x^2 + 10x + 25$ **71.** $x^2 - 4xy + 4y^2$ **73.** $4a^2 - 12ab + 9b^2$
75. $16x^2 + 40xy + 25y^2$ **77.** $(4x^2 - 6x + 2)$ cm^2
79. $(x^2 + 6x + 9)\pi$ in.^{2} **81.** $x^3 - x + 6$
83. $4t^3 + 11t^2 + 18t + 9$ **85.** $-3x^3 + 25x^2y - 56xy^2 + 16y^3$
87. $x^3 - 3x + 2$ **89.** $12x^3 + 17x^2 - 6x - 8$ **91.** -3
93. -8 **95.** -1 **97.** 0 **99.** $(24x + 14)$ cm, $(35x^2 + 43x + 12)$ cm^2 **101. a.** x^2 ft^2, $6x$ ft^2, $5x$ ft^2, 30 ft^2; $(x^2 + 11x + 30)$ ft^2 **b.** $(x + 6)$ ft, $(x + 5)$ ft; $(x^2 + 11x + 30)$ ft^2
c. They are the same. **103.** 5 and 6 **105.** 4 m **107.** 90 ft
109. $(\frac{1}{2}h^2 - 2h)$ in^2. **113.** 1 **115.** $-\frac{2}{3}$ **117.** $(0, 2)$

Study Set Section 10.7 (page 708)

1. polynomial **3.** two **5.** exponents **7.** 9, 9 **9.** In the numerator and denominator, a common factor of 2 was divided out. **11. a.** $t = \frac{d}{r}$ **b.** $3x^2$ **13.** $2x + 7$ **17.** $\frac{1}{3}$ **19.** $-\frac{5}{3}$
21. $\frac{3}{4}$ **23.** 1 **25.** x^3 **27.** $\frac{r^2}{s}$ **29.** $\frac{2x^2}{y}$ **31.** $-\frac{3u^3}{v^2}$ **33.** $\frac{4r}{y^2}$
35. $-\frac{13}{3rs}$ **37.** $\frac{x^4}{y^6}$ **39.** a^8b^8 **41.** $-\frac{3r}{s^9}$ **43.** $-\frac{x^3}{4y^3}$ **45.** $-\frac{16}{y^6}$
47. a^8 **49.** $2x + 3$ **51.** $\frac{1}{5y} - \frac{2}{5x}$ **53.** $\frac{1}{y^2} + \frac{2y}{x^2}$ **55.** $3a - 2b$
57. $\frac{1}{y} - \frac{1}{2x} + \frac{2z}{xy}$ **59.** $3x^2y - 2x - \frac{1}{y}$ **61.** $5x - 6y + 1$
63. $\frac{10x^2}{y} - 5x$ **65.** $-\frac{4x}{3} + \frac{3x^2}{2}$ **67.** $xy - 1$
69. $\frac{x}{y} - \frac{11}{6} + \frac{y}{2x}$ **71.** 2 **73.** $(2x^2 - x + 3)$ in. **75.** $(3x - 2)$ ft
77. yes **79.** no **83.** binomial **85.** none of these **87.** 2

Study Set Section 10.8 (page 714)

1. divisor, dividend **3.** remainder **5.** $4x^3 - 2x^2 + 7x + 6$
7. $6x^4 - x^3 + 2x^2 + 9x$ **9.** $0x^3$ and $0x$ **11.** $x^3 + 3x^2 + 9x + 27$
13. a. $r = \frac{d}{t}$ **b.** $x - 3$ **15.** It is correct. **19.** $x + 6$
21. $y + 12$ **23.** $3a - 2$ **25.** $b + 3$ **27.** $2x + 1$
29. $x - 7$ **31.** $3x + 2$ **33.** $2x - 1$ **35.** $x^2 + 2x - 1$
37. $2x^2 + 2x + 1$ **39.** $x^2 + x + 1$ **41.** $x + 4 + \frac{14}{2x - 3}$
43. $2x + 2 + \frac{-3}{2x + 1}$ or $2x + 2 - \frac{3}{2x + 1}$ **45.** $x^2 + 2x + 1$
47. $x^2 + 2x - 1 + \frac{6}{2x + 3}$ **49.** $2x^2 + 8x + 14 + \frac{31}{x - 2}$
51. $x + 1$ **53.** $2x - 3$ **55.** $x^2 - x + 1$
57. $a^2 - 3a + 10 + \frac{-30}{a + 3}$ or $a^2 - 3a + 10 - \frac{30}{a + 3}$
59. $5x^2 - x + 4 + \frac{16}{3x - 4}$
61. a. $(x - 6)$ in. **b.** $(4x - 4)$ in. **63.** $4x^2 + 3x + 7$ **67.** x^{22}
69. $8x^2 - 6x + 1$ **71.** They are the same.

Key Concept (page 717)

1. a. x, descending **b.** 4 **c.** 3, 2, 1, 0 **d.** 3 **e.** 1, -2, 6, -8
3. a. binomial **b.** none of these **c.** trinomial **d.** monomial
5. combine **7.** each, each **9.** $3x - 5$ **11.** $2x^2 - 13x - 24$
13. $y^2 + 2y - 3$ **15.** $y^3 + 4y^2 - 3y - 18$ **17.** 1

Chapter Review (page 719)

1. $-3 \cdot x \cdot x \cdot x \cdot x$ **2.** $(\frac{1}{2}pq)(\frac{1}{2}pq)(\frac{1}{2}pq)$ **3.** 125 **4.** 64
5. -64 **6.** 4 **7.** x^5 **8.** $-3y^6$ **9.** y^{21} **10.** $81x^4$ **11.** b^{12}
12. $-y^2z^5$ **13.** $256s^3$ **14.** $4x^4y^2$ **15.** x^{15} **16.** $\frac{x^2}{y^2}$ **17.** x^4
18. $125yz^4$ **19.** $64x^{12}$ in.^{3} **20.** y^4 m^2 **21.** 1 **22.** 1 **23.** 9
24. $\frac{1}{1{,}000}$ **25.** $\frac{4}{3}$ **26.** $-\frac{1}{25}$ **27.** $\frac{1}{x^5}$ **28.** $-\frac{6}{y}$ **29.** $\frac{1}{x^{10}}$
30. x^{14} **31.** $\frac{1}{x^5}$ **32.** $\frac{1}{9z^2}$ **33.** y^{7n} **34.** $\frac{1}{z^{2c}}$ **35.** 7.28×10^2
36. 9.37×10^6 **37.** 1.36×10^{-2} **38.** 9.42×10^{-3}
39. 1.8×10^{-4} **40.** 7.53×10^5 **41.** 726,000 **42.** 0.000391
43. 2.68 **44.** 57.6 **45. a.** 0.03 **46.** 160 **47.** 6,080,000,000; 6.08×10^9 **48.** $1.0 \times 10^5 = 100{,}000$ **49.** yes **50.** no **51.** no
52. yes **53.** 4 **54.** $3x^3$ **55.** -1 **56.** 10 **57.** 7th, monomial
58. 3rd, monomial **59.** 2nd, binomial **60.** 5th, trinomial
61. 6th, binomial **62.** 4th, none of these **63.** 34 **64.** 1
65. 9 **66.** 0.72 **67.** 8 in. **68.** $2x^6 + 5x^5$ **69.** $-2x^2y^2$
70. $8x^2 - 6x$ **71.** $5x^2 + 19x + 3$ **72.** $4x^2 + 2x + 8$
73. $8x^3 - 7x^2 + 19x$ **74.** $10x^3$ **75.** $-6x^{10}z^5$ **76.** $-6r^3s^4t^5$
77. $120b^{11}$ **78.** $5x + 15$ **79.** $3x^4 - 5x^2$ **80.** $x^2y^3 - x^3y^2$
81. $-2y^4 + 10y^3$ **82.** $6x^6 + 12x^5$ **83.** $-3x^3 + 3x^2 - 6x$
84. $x^2 + 5x + 6$ **85.** $2x^2 - x - 1$ **86.** $6a^2 - 6$ **87.** $6a^2 - 6$
88. $2a^2 - ab - b^2$ **89.** $-6x^2 - 5xy - y^2$ **90.** $x^2 + 6x + 9$
91. $x^2 - 25$ **92.** $a^2 - 6a + 9$ **93.** $x^2 + 8x + 16$
94. $4y^2 - 4y + 1$ **95.** $y^4 - 1$ **96.** $3x^3 + 7x^2 + 5x + 1$
97. $8a^3 - 27$ **98.** 1 **99.** -1 **100.** 7 **101.** 0
102. $(6x + 10)$ in.; $(2x^2 + 11x - 6)$ in.^{2}; $(6x^3 + 33x^2 - 18x)$ in.^{3}
103. $-\frac{2x}{3y^2}$ **104.** $\frac{1}{x}$ **105.** $4x + 3$ **106.** $2 - \frac{3}{y}$
107. $3a + 4b - 5$ **108.** $-\frac{x}{y} - \frac{y}{x}$ **109.** $x + 5$
110. $x + 1 + \frac{3}{x + 2}$ **111.** $x - 5$ **112.** $2x + 1$
113. $x + 5 + \frac{3}{3x - 1}$ **114.** $3x^2 + 2x + 1 + \frac{2}{2x - 1}$
115. $3x^2 - x - 4$ **117.** $(4x + 3)$ in./min

Chapter 10 Test (page 725)

1. $2x^3y^4$ **2.** 64 **3.** y^6 **4.** $32x^{21}$ **5.** 3 **6.** $\frac{2}{y^3}$ **7.** y^3
8. $\frac{64a^3}{b^3}$ **9.** $1{,}000y^{12}$ in.^{3} **10.** $\frac{1}{4^2}, \frac{1}{16}$ **11.** 6.25×10^{18}
12. 0.000093 **13.** binomial **14.** 5th degree **15.** 0

16. $-3x^2y^2$ **17.** $-7x + 2y$ **18.** $-x^2 - 5x + 4$ **19.** $-4x^5y$
20. $3y^4 - 6y^3 + 9y^2$ **21.** $x^2 - 81$ **22.** $9y^2 - 24y + 16$
23. $6x^2 - 7x - 20$ **24.** $2x^3 - 7x^2 + 14x - 12$ **25.** $\frac{1}{2}$ **26.** $\frac{y}{2x}$
27. $\frac{a}{4b} - \frac{b}{2a}$ **28.** $x - 2$ **30.** $(x - 5)$ ft

Cumulative Review Exercises (page 727)

1. a. 1993 **b.** 1986 **c.** 1996–1997 **2.** $\frac{5}{8}$ **3.** $\frac{22}{35}$ **4.** irrational
5. $250 - x$ **6.** 19, 16 **7.** 5 **8.** $37y$ **9.** $-2x + 2y$ **10.** $x - 5$
11. x^2y^3 **12.** $4x^2$ **13.** 13 **14.** 41 **15.** $h = \frac{2A}{b + B}$
16. $x = \frac{y - b}{m}$ **17.** -9 **18.** 1 **19.** 4 **20.** -2
21. $x < -14$ 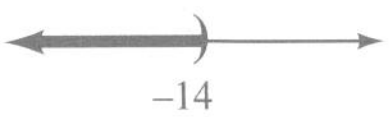 $(-\infty, -14)$

22. $-5 < x \le -1$  $(-5, -1]$

23.
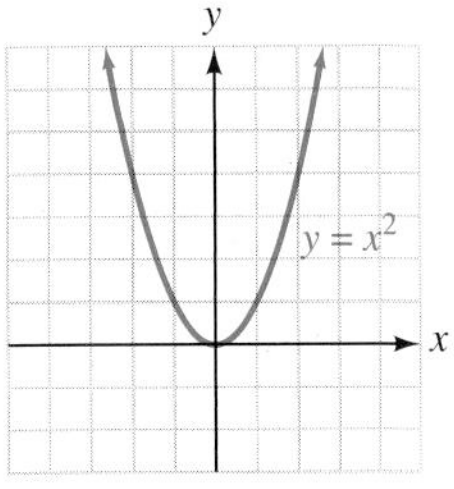

24.
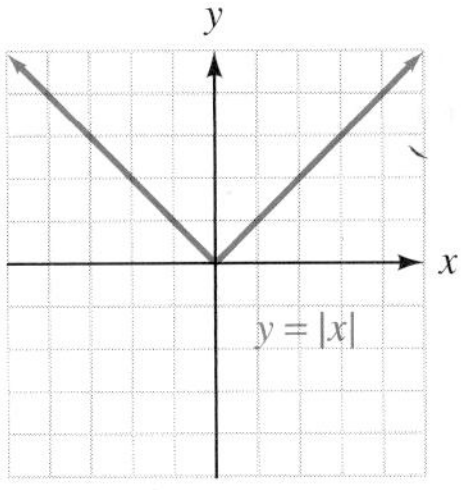

25.
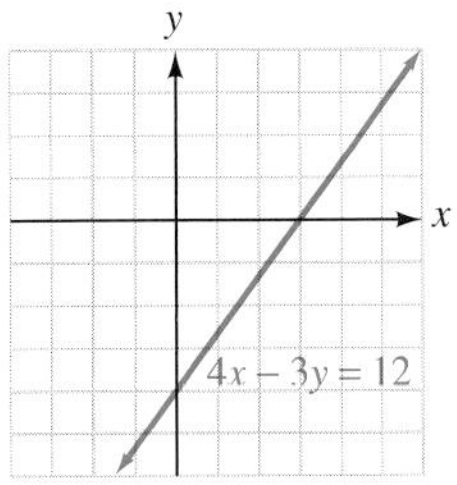

26.
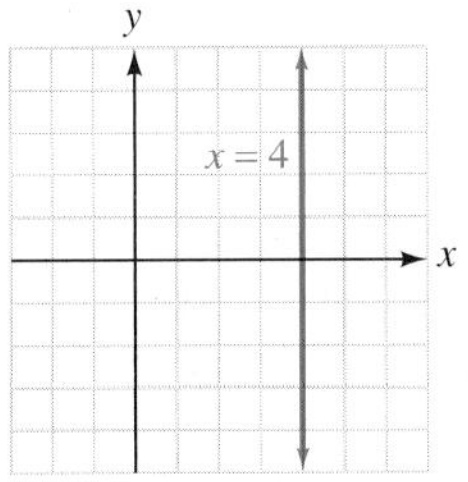

27. $\frac{1}{2}$ **28.** 0 **29.** -4 **30.** $\frac{2}{3}$ **31.** $y = \frac{2}{3}x + 5$
32. $3x - 4y = -22$ **33.** $y = 4$ **34.** $x = 2$ **35.** perpendicular
36. parallel **37.** yes **38.** no **39.** -3 **40.** 15 **41.** 5
42. -2.5 **43.** D: all real numbers, R: real numbers greater than or equal to 0 **44.** no **45.** The temperature is rising at a rate of 3°/hr; the temperature is falling at a rate of -3°/hr.
46. -9 **47.** y^{14} **48.** x^{14} **49.** x^2 **50.** a^7 **51.** $\frac{1}{x^5}$ **52.** $\frac{1}{16y^4}$
53. $\frac{1}{x^8}$ **54.** $-x^{15}$ **55.** 6.15×10^5 **56.** 1.3×10^{-6}
57. 0.000525 **58.** 2,770 **59.** 2 **60.** 5 **61.** 1.5 in.
62. a. 3 **b.** $2x^2$ **c.** 2 **d.** 1 **e.** 2 **f.** 9 **63.** $x^2 + 4x - 14$
64. $4x^2 - x - 1$ **65.** $-35x^5 + 10x^4 + 10x^2$ **66.** $-12x^5y^5$
67. $6x^2 + 10x - 56$ **68.** $15x^2 - 2xy - 8y^2$ **69.** $9x^2 + 6x + 1$
70. $x^3 - 8$ **71.** $3x - 4$ **72.** $2x + 1$

Chapter 11 Check Your Knowledge (page 730)

1. prime **2.** Factored out **3.** prime **4.** $a^2 + b^2$ **5.** quadratic
6. 0 **7.** $2^2 \cdot 3 \cdot 13$ **8.** $4x^3y$ **9.** $6y(y + 8)$
10. $3xyz^2(1 - 2x + 5x^2y^2z)$ **11.** prime **12.** $(3x + 5)(3x - 5)$
13. $2(9x^2 + 4y^2)(3x + 4y)(3x - 4y)$ **14.** $(x + 3)(x + 6)$
15. $(x - 4)^2$ **16.** $(n - 3)(x + y)$ **17.** $(3x - 2)(x + 4)$
18. $(3x - 1)(9x^2 + 3x + 1)$ **19.** $y + 5, y + 2$
20. $\frac{7}{2}, 5$ **21.** $-\frac{7}{4}, \frac{7}{4}$ **22.** 0, 3 **23.** 2, 8 **24.** $-3, 6$ **25.** $-\frac{1}{5}, \frac{1}{2}$
26. 2 seconds, 3 seconds **27.** 3 cm, 4 cm, 5 cm

Study Set Section 11.1 (page 739)

1. prime **3.** largest **5.** factoring **7.** The 0 in the first line should be 1. **9.** The GCF is $6a^2$, not $6a$. **11.** factoring out the GCF **13.** $3x$ **15.** Find $(j + 2)(3j^2 + 2)$. The result, when written in descending powers of j, should be $3j^3 + 6j^2 + 2j + 4$.
19. 1 **21.** $2^2 \cdot 3$ **23.** $3 \cdot 5$ **25.** $2^3 \cdot 5$ **27.** $2 \cdot 7^2$
29. $3^2 \cdot 5^2$ **31.** $2^5 \cdot 3^2$ **33.** 4 **35.** $4, x$ **37.** $3(x + 2)$
39. $6(2x^2 - x - 4)$ **41.** $t^2(t + 2)$ **43.** $a^2(a - 1)$
45. $8xy^2(3xy + 1)$ **47.** $6uvw^2(2w - 3v)$ **49.** $3(x + y - 2z)$
51. $a(b + c - d)$ **53.** $3r(4r - s + 3rs^2)$ **55.** $\pi(R^2 - ab)$
57. $(x + 2)(3 - x)$ **59.** $(14 + r)(h^2 + 1)$ **61.** $-(a + b)$
63. $-(2x - 5y)$ **65.** $-(3m + 4n - 1)$
67. $-(3ab + 5ac - 9bc)$ **69.** $-3x(x + 2)$ **71.** $-4a^2b^2(b - 3a)$
73. $-2ab^2c(2ac - 7a + 5c)$ **75.** $(x + y)(2 + a)$
77. $(r + s)(7 - k)$ **79.** $(r + s)(x + y)$ **81.** $(2x + 3)(a + b)$
83. $(b + c)(2a + 3)$ **85.** $(3x - 1)(2x - 5)$
87. $(3p + q)(3m - n)$ **89.** $(2x + y)(y - 1)$
91. $(2z^3 + 3)(4z^2 - 5)$ **93.** $x^2(a + b)(x + 2y)$
95. $4a(b + 3)(a - 2)$ **97.** $y(x^2 - y)(x - 1)$
99. a. $12x^3$ in.^{2} **b.** $20x^2$ in.^{2} **c.** $4x^2(3x - 5)$ in.^{2}
101. a. $4r$ in.; $16r^2$ in.^{2} **b.** $4\pi r^2$ in.^{2}
c. $16r^2 - 4\pi r^2 = 4r^2(4 - \pi)$ in.^{2} **107.** $\frac{y^3}{8}$ **109.** yes

Study Set Section 11.2 (page 747)

1. trinomial, binomial **3.** factors **5.** lead, coefficient, term
7. 4, 1; 2, 2; $-4, -1$; $-2, -2$ **9.** common **11.** $-6, -3$
13. 9, 6, $-9, -6$ **15. a.** 5, 2 **b.** $-2, -4$ **c.** $3, -2$ **d.** $1, -9$
17. $+3, -2$ **19. a.** They are both positive or both negative. **b.** One will be positive, the other negative.
23. +, + **25.** −, − **27.** +, − **29.** $(z + 11)(z + 1)$
31. $(m - 3)(m - 2)$ **33.** $(a - 5)(a + 1)$ **35.** $(x + 8)(x - 3)$
37. $(a - 13)(a + 3)$ **39.** prime **41.** $(s + 13)(s - 2)$
43. prime **45.** $(m - 4)(m + 3)$ **47.** $(x + 2y)(x + 2y)$
49. $(m + 5n)(m - 2n)$ **51.** $(a - 6b)(a + 2b)$ **53.** prime
55. $-(x + 5)(x + 2)$ **57.** $-(t + 17)(t - 2)$
59. $-(r - 10)(r - 4)$ **61.** $-(a + 3b)(a + b)$
63. $-(x - 7y)(x + y)$ **65.** $(x - 4)(x - 1)$ **67.** $(y + 9)(y + 1)$
69. $-(r - 2)(r + 1)$ **71.** $(r + 3x)(r + x)$ **73.** $(a - 2b)(a - b)$
75. $2(x + 3)(x + 2)$ **77.** $-5(a - 3)(a - 2)$
79. $3(z - 4)(z - 1)$ **81.** $4y(x + 6)(x - 3)$
83. $-4x(x + 3y)(x - 2y)$ **85.** $(x + 9)$ in., $(x + 3)$ in., x in.
93. 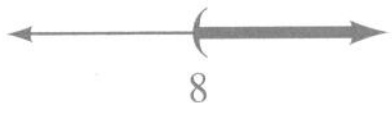

95. 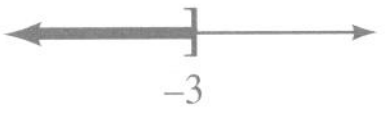

Study Set Section 11.3 (page 755)

1. lead, last **3.** outer, inner **5.** middle **7.** sum **9.** 5, -8
11. positive, negative, negative **13.** negative, different
15. 24, -11 **17. a.** $x^2 + 2x + 3$ (answers may vary)
b. $2x^2 + 2x + 3$ (answers may vary) **19.** +, + **21.** −, −
23. −, + **27.** $(2x - 1)(x - 1)$ **29.** $(3a + 1)(a + 4)$
31. $(z + 3)(4z + 1)$ **33.** $(3y + 2)(2y + 1)$
35. $(3x - 2)(2x - 1)$ **37.** $(7x - 2)(x - 1)$ **39.** $(2x + 1)(x - 2)$
41. prime **43.** $(5y + 1)(2y - 1)$ **45.** $(3y - 2)(4y + 1)$
47. $-(5t + 3)(t + 2)$ **49.** $-(8m - 3)(2m - 1)$
51. $(2a - b)(2a - b)$ **53.** $(3r + 2s)(2r - s)$
55. $(13x + 11y)(x + y)$ **57.** $(18a - 5b)(a + 2b)$
59. $(3x + 2)(x - 5)$ **61.** $(2a - 5)(4a - 3)$
63. $(4y - 3)(3y - 4)$ **65.** prime **67.** $(2a + 3b)(a + b)$
69. $(3p - q)(2p + q)$ **71.** $2(2x - 1)(x + 3)$

73. $-y(y + 12)(y + 1)$ **75.** $3x(2x + 1)(x - 3)$
77. $3r^3(5r - 2)(2r + 5)$ **79.** $-2mn(4m + 3n)(2m + n)$
81. $-2uv^3(7u - 3v)(2u - v)$
83. $(2x + 11)$ in., $(2x - 1)$ in.; 12 in. **89.** x^{14} **91.** 8

Study Set Section 11.4 (page 764)

1. difference **3.** cubes, difference **5. a.** $5x$ **b.** 3 **c.** $5x$, 3
7. First, Last, First, Last **9. a.** $6x$ **b.** $10x^2$ **c.** $3m$ **d.** a^2
11. 1, 4, 9, 16, 25, 36, 49, 64, 81, 100 **13. a.** 7 is not a perfect square. **b.** The sign of the last term must be positive. **c.** The middle term is not twice the product of $5r$ and 4.
15. $36x^2 - 25y^2$ **17.** $9a^2 - 30ab + 25b^2$ **19.** $(x + 8)^2$
21. 3 **23.** + **25.** $(x + 3)^2$ **27.** $(y - 4)^2$ **29.** $(t + 10)^2$
31. $(u - 9)^2$ **33.** $(2x + 3)^2$ **35.** $(6x + 1)^2$ **37.** $(a + b)^2$
39. $(4x - y)^2$ **41.** 7, 7 **43.** t, w **45.** $(x + 4)(x - 4)$
47. $(2y + 1)(2y - 1)$ **49.** $(3x + y)(3x - y)$
51. $(4a + 5b)(4a - 5b)$ **53.** prime **55.** $(a^2 + 12b)(a^2 - 12b)$
57. $(tz + 8)(tz - 8)$ **59.** $8(x + 2y)(x - 2y)$
61. $7(1 + a)(1 - a)$ **63.** $6x^2(x + y)(x - y)$
65. $(x^2 + 9)(x + 3)(x - 3)$ **67.** $(a^2 + 4)(a + 2)(a - 2)$
69. $(9r^2 + 16s^2)(3r + 4s)(3r - 4s)$ **71.** $2a$ **73.** $b + 3$
75. $(y + 1)(y^2 - y + 1)$ **77.** $(a - 3)(a^2 + 3a + 9)$
79. $(2 + x)(4 - 2x + x^2)$ **81.** $(s - t)(s^2 + st + t^2)$
83. $(a + 2b)(a^2 - 2ab + 4b^2)$ **85.** $(4x - 3)(16x^2 + 12x + 9)$
87. $(a^2 - b)(a^4 + a^2b + b^2)$ **89.** $(x^3 + y^2)(x^6 - x^3y^2 + y^4)$
91. $2(x + 3)(x^2 - 3x + 9)$ **93.** $-(x - 6)(x^2 + 6x + 36)$
95. $8x(2m - n)(4m^2 + 2mn + n^2)$
97. $xy(x + 6y)(x^2 - 6xy + 36y^2)$
99. $3rs^2(3r - 2s)(9r^2 + 6rs + 4s^2)$ **101.** $(p + q)^2$
103. $0.5g(t_1 + t_2)(t_1 - t_2)$ **109.** $\frac{x}{y} + \frac{2y}{x} - 3$ **111.** $3a + 2$
113. $(x - a)(a + b)(a - b)$ **115.** $7p^4q^2(10q - 5 + 7p)$
117. $2a(b + 6)(b - 2)$ **119.** $-4p^2q^3(2pq^4 + 1)$
121. $5(2m + 5)^2$ **123.** prime **125.** $-2x^2(x - 4)(x^2 + 4x + 16)$
127. $2t^2(3t - 5)(t + 4)$ **129.** $(y^2 - 2)(x + 1)(x - 1)$
131. $(2p^2 - 3q^2)(4p^4 + 6p^2q^2 + 9q^4)$
133. $(5p - 4y)(25p^2 + 20py + 16y^2)$
135. $-x^2y^2z(16x^2 - 24x^3yz^3 + 15yz^6)$
137. $(9p^2 + 4q^2)(3p + 2q)(3p - 2q)$ **139.** prime
141. $2(3x + 5y^2)(9x^2 - 15xy^2 + 25y^4)$ **143.** prime
145. $(7p + 2q)^2$

Study Set Section 11.5 (page 774)

1. quadratic **3.** zero, 0, 0 **5.** zero **7. a.** quadratic **b.** linear **c.** linear **d.** quadratic **9. a.** -16 **b.** $(x - 2)(x + 8)$ **c.** 2, -8 **11. a.** Add 6 to both sides. **b.** Distribute the multiplication by x. **15.** 2, -3
17. $\frac{5}{2}$, -6 **19.** 1, -2, 3 **21.** 0, 3 **23.** 0, $\frac{5}{2}$ **25.** 0, 7
27. 0, $-\frac{8}{3}$ **29.** 0, 2 **31.** -5, 5 **33.** $-\frac{1}{2}$, $\frac{1}{2}$ **35.** $-\frac{2}{3}$, $\frac{2}{3}$
37. -10, 10 **39.** $-\frac{9}{2}$, $\frac{9}{2}$ **41.** 12, 1 **43.** -3, 7 **45.** 8, 1
47. -3, -5 **49.** -4, 2 **51.** 0, -1, -2 **53.** 0, 9, -3
55. 1, -2, -3 **57.** $\frac{1}{2}$, 2 **59.** $\frac{1}{5}$, 1 **61.** $-\frac{1}{2}$, $-\frac{1}{2}$ **63.** $\frac{2}{3}$, $-\frac{1}{5}$
65. $-\frac{5}{2}$, 4 **67.** $\frac{1}{8}$, 1 **69.** 0, -3, $-\frac{1}{3}$ **71.** 0, -3, -3
73. 9 sec **75.** $\frac{3}{2} = 1.5$ sec **77.** 2 sec **79.** 8 **81.** 12
83. 4 m by 9 m **85.** 5 m, 12 m, 13 m **87.** 3 mm, 4 mm, 5 mm
89. 5 ft, 12 ft **91.** 5 in. **93.** 4 m
97. 15 min $\le t <$ 30 min **99.** 675 cm^2

Key Concept (page 779)

1. Factor $3x + 27$; $3(x + 9)$ **3.** $-3(a - 3)(a - 4)$
5. $(r + s)(t + 2)$ **7.** $(3t - 5)(2t - 3)$ **9.** $2r(r + 5)(r - 5)$

Chapter Review (page 781)

1. $5 \cdot 7$ **2.** $3^2 \cdot 5$ **3.** $2^5 \cdot 3$ **4.** $3^2 \cdot 11$ **5.** $2 \cdot 5^2 \cdot 41$ **6.** 2^{12}
7. $3(x + 3y)$ **8.** $5a(x^2 + 3)$ **9.** $7s(s + 2)$ **10.** $\pi a(b - c)$
11. $2x(x^2 + 2x - 4)$ **12.** $xyz(x + y + 1)$ **13.** $-5ab(b - 2a + 3)$
14. $(x - 2)(4 - x)$ **15.** $-(a + 7)$ **16.** $-(4t^2 - 3t + 1)$
17. $(c + d)(2 + a)$ **18.** $(y + 3)(3x - 2)$ **19.** $(2a^2 - 1)(a + 1)$
20. $4m(n + 3)(m - 2)$ **21.** 3, -1; 7, 5, -7, -5
22. $(x + 6)(x - 4)$ **23.** $(x - 6)(x + 2)$ **24.** $(n - 5)(n - 2)$
25. prime **26.** $-(y - 5)(y - 4)$ **27.** $(y + 9)(y + 1)$
28. $(c + 5d)(c - 2d)$ **29.** $(m - 2n)(m - n)$ **30.** Multiply to see whether $(x - 4)(x + 5) = x^2 + x - 20$. **31.** $5(a + 10)(a - 1)$
32. $-4x(x + 3y)(x - 2y)$ **33.** $(2x + 1)(x - 3)$
34. $(2y + 5)(5y - 2)$ **35.** $-(3x + 1)(x - 5)$
36. $3p(2p + 1)(p - 2)$ **37.** $(4b - c)(b - 4c)$ **38.** prime
39. $(4x + 1)$ in., $(3x - 1)$ in. **40.** $(x + 5)^2$ **41.** $(3y - 4)^2$
42. $-(z - 1)^2$ **43.** $(5a + 2b)^2$ **44.** $(x + 3)(x - 3)$
45. $(7t + 5y)(7t - 5y)$ **46.** $(xy + 20)(xy - 20)$
47. $8a(t + 2)(t - 2)$ **48.** $(c^2 + 16)(c + 4)(c - 4)$ **49.** prime
50. $(h + 1)(h^2 - h + 1)$ **51.** $(5p + q)(25p^2 - 5pq + q^2)$
52. $(x - 3)(x^2 + 3x + 9)$ **53.** $2x^2(2x - 3y)(4x^2 + 6xy + 9y^2)$
54. $2y^2(3y - 5)(y + 4)$ **55.** $(t + u^2)(s^2 + v)$
56. $(j^2 + 4)(j + 2)(j - 2)$ **57.** $-3(j + 2k)(j^2 - 2jk + 4k^2)$
58. $3(2w - 3)^2$ **59.** prime **60.** 0, -2 **61.** 0, 6 **62.** -3, 3
63. 3, 4 **64.** -2, -2 **65.** 6, -4 **66.** $\frac{1}{5}$, 1 **67.** 0, -1, 2
68. 15 m **69.** 3 ft by 9 ft **70.** 5 m

Chapter 11 Test (page 785)

1. $2^2 \cdot 7^2$ **2.** $3 \cdot 37$ **3.** $4(x + 4)$ **4.** $5ab(6ab^2 - 4a^2b + c)$
5. $(q + 9)(q - 9)$ **6.** prime **7.** $(4x^2 + 9)(2x + 3)(2x - 3)$
8. $(x + 3)(x + 1)$ **9.** $-(x - 11)(x + 2)$ **10.** $(a - b)(9 + x)$
11. $(2a - 3)(a + 4)$ **12.** $2(3x - 5y)^2$ **13.** $(x + 2)(x^2 - 2x + 4)$
14. $2(a - 3)(a^2 + 3a + 9)$ **15.** $r^2(4 - \pi)$ **16.** $5x - 4$
17. $2ab^2$ **18.** $(x - 9)(x + 6)$; Multiply the binomials: $(x - 9)(x + 6) = x^2 + 6x - 9x - 54 = x^2 - 3x - 54$. **19.** -3, 2
20. -5, 5 **21.** 0, $\frac{1}{6}$ **22.** -3, -3 **23.** $\frac{1}{3}$, $-\frac{1}{2}$ **24.** -1, -6
25. 6 ft by 9 ft **26.** A quadratic equation is an equation that can be written in the form $ax^2 + bx + c = 0$; $x^2 - 2x + 1 = 0$. (Answers may vary.) **27.** 10 **28.** At least one of them is 0.

Cumulative Review Exercises (page 787)

1. about 35 beats/min difference **2.** $2 \cdot 5^3$ **3.** 0.992
4. a. false **b.** true **c.** true **5.** $\frac{24}{25}$ **6.** -27 **7.** -39
8. -5 **9.** 3 **10.** $\frac{5}{0}$ **11.** $-13y^2 + 6$ **12.** $3y + z$ **13.** $-\frac{3}{2}$
14. -2 **15.** 9 **16.** $-\frac{55}{6}$ **17.** $t = \frac{A - P}{Pr}$
18. $x < -2$, $(-\infty, -2)$

−2

19. 15.4 in. **20.** $I = Prt$ **21.** \$20x **22.** 4 L

23.

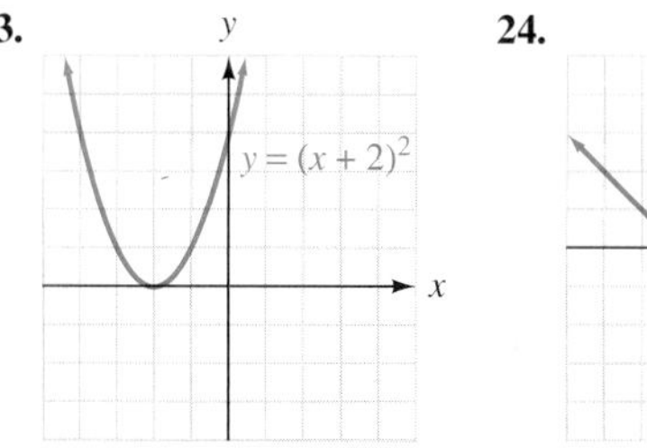

24.

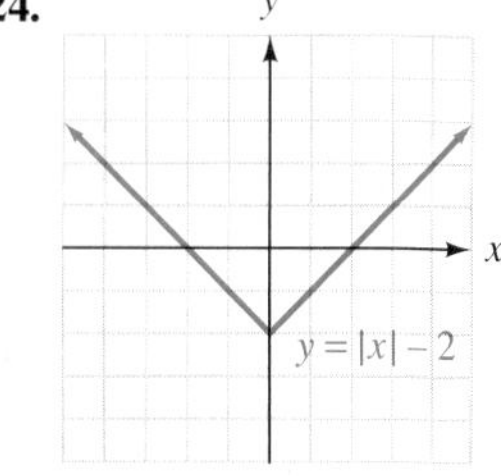

25. 1; (0, -2) **26.** $y = 7x + 19$

27.

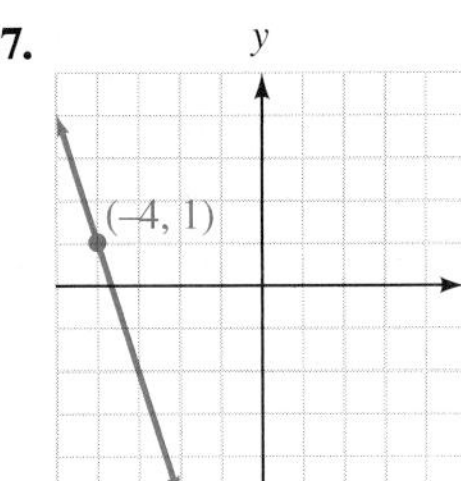

28.

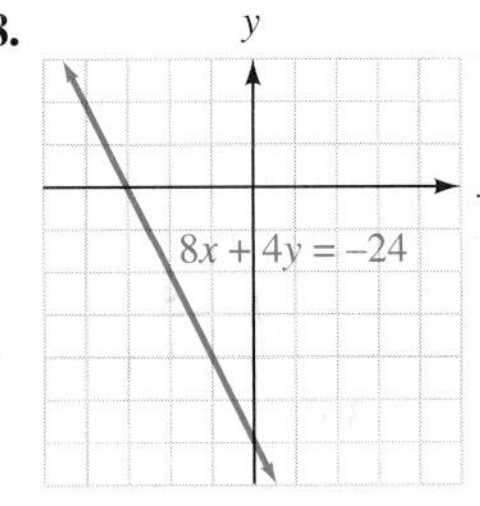

29. They are the same. **30. a.** a decrease of 0.86 gal/year **b.** virtually no change **31.** 2 **32.** 1.7×10^6 **33.** $-4y^5$ **34.** $\frac{5x^4y^{23}}{7}$ **35.** $\frac{1}{b^{14}}$ **36.** 2 **37.** $-2x^2 - 4x + 5$ **38.** $8b^5 - 8b^4$ **39.** $y^2 - 12y + 36$ **40.** $3x^2 + 10x - 8$ **41.** $3ab - 2a - 1$ **42.** $2x + 1$ **43.** $(4x + 8)$ in. **b.** $(x^2 + 4x + 3)$ in.^{2} **c.** $(x^3 + 4x^2 + 3x)$ in.^{3} **44.** 3 **45.** $b^2(b - 3)$ **46.** $(u + 3)(u - 1)$ **47.** $(2x + 1)(x - 2)$ **48.** $(3z + 1)(3z - 1)$ **49.** $-5(a - 3)(a - 2)$ **50.** $(x + y)(a + b)$ **51.** $(t - 2)(t^2 + 2t + 4)$ **52.** $(2a - 3)^2$ **53.** $0, \frac{4}{3}$ **54.** $\frac{1}{2}, 2$

Chapter 12 Check Your Knowledge (page 790)

1. least common denominator **2.** extraneous **3.** proportion **4.** inverse **5.** $-7, 0$ **6.** $\frac{9z^2}{5x^2}$ **7.** $\frac{x + 2}{x - 3}$ **8.** $-\frac{x}{x - 4}$ **9.** $\frac{1}{2(x - 5)}$ **10.** 4 **11.** $\frac{x^2 + 2}{(x + 2)(x - 1)}$ **12.** $\frac{4y - 5x}{x^2y^2}$ **13.** $\frac{y + 1}{y - 1}$ **14.** $\frac{1}{x - 1}$ **15.** 10 **16.** 3 **17.** no solution; 1 is extraneous **18.** $W = \frac{P - 2L}{2}$ **19.** $a = \frac{bc}{b - c}$ **20.** 10 **21.** 25 feet **22.** 3,840 feet **23.** $2\frac{2}{5}$ hr **24. a.** 90 **b.** $1\frac{1}{2}$ hr

Study Set Section 12.1 (page 797)

1. numerator, denominator **3.** undefined **5.** simplify **7. a.** 0 **b.** 6 **c.** −6 **9.** $x + 1$ **11.** x is not a common factor of the numerator and denominator and cannot be removed. **15.** 4 **17.** $-\frac{3}{7}$ **19.** 0 **21.** 2 **23.** none **25.** $\frac{1}{2}$ **27.** −6, 6 **29.** −2, 1 **31.** $\frac{4}{5}$ **33.** $\frac{1}{3}$ **35.** $-\frac{3}{4}$ **37.** $\frac{5}{a}$ **39.** $\frac{2}{z}$ **41.** $\frac{a}{3}$ **43.** $\frac{2}{3}$ **45.** $\frac{3}{2}$ **47.** in lowest terms **49.** $\frac{3x}{y}$ **51.** $\frac{2x + 1}{y}$ **53.** $\frac{1}{3}$ **55.** −1 **57.** −6 **59.** $\frac{-3(x + 2)}{x + 1}$ **61.** $\frac{x + 1}{x - 1}$ **63.** $\frac{x - 5}{x + 2}$ **65.** $\frac{2x}{x - 2}$ **67.** $-\frac{1}{a + 1}$ **69.** $\frac{x + 2}{x^2}$ **71.** $\frac{2x - 3}{x + 1}$ **73.** $\frac{3x - 2}{x + 4}$ **75.** $\frac{2(x + 2)}{4x + 1}$ **77.** $\frac{m - n}{2(m + n)}$ **79.** in lowest terms **81.** $\frac{3 - x}{3 + x}$ or $-\frac{x - 3}{x + 3}$ **83.** $\frac{4}{3}$ **85.** $x + 3$ **87.** in lowest terms **89.** $\frac{2}{3}$ **91.** $\frac{3x}{5y}$ **93.** −1 **95.** $\frac{x + 2}{x - 2}$ **97.** 1,000; 800; about 667; 500; about 333; about 222 **103.** $(a + b) + c = a + (b + c)$ **105.** One of them is zero. **107.** $\frac{5}{3}$

Study Set Section 12.2 (page 805)

1. numerator **3.** numerators, denominators **5.** 1 **7.** divisor, multiply **11.** $\frac{3}{2}$ **13.** y **15.** $\frac{14}{9}$ **17.** x^2y^2 **19.** $2xy^2$ **21.** $-3y^2$ **23.** $\frac{b^3c}{a^4}$ **25.** $\frac{r^3t^4}{s}$ **27.** $\frac{(z + 7)(z + 2)}{7z}$ **29.** x **31.** $\frac{x}{5}$ **33.** 5 **35.** 6 **37.** $x + 1$ **39.** $2y + 16$ **41.** $x + 5$ **43.** $10h - 30$ **45.** $x + 2$ **47.** $\frac{3}{2x}$ **49.** $x - 2$ **51.** x **53.** $\frac{(x - 2)^2}{x}$ **55.** $\frac{(m - 2)(m - 3)}{2(m + 2)}$ **57.** $\frac{x + 1}{2(x - 2)}$ **59.** $\frac{3}{2y}$ **61.** 3 **63.** $\frac{6}{y}$ **65.** 6 **67.** $\frac{2x}{3}$ **69.** $\frac{2}{y}$ **71.** $\frac{2}{3x}$ **73.** $\frac{2(z - 2)}{z}$ **75.** $\frac{5z(z - 7)}{z + 2}$ **77.** $\frac{x + 2}{3}$ **79.** 1 **81.** $\frac{2(x - 7)}{x + 9}$ **83.** $d + 5$ **85.** $\frac{9}{2x}$ **87.** $\frac{x}{36}$ **89.** 2 **91.** $\frac{2x(1 - x)}{5(x - 2)}$ **93.** $\frac{y^2}{3}$ **95.** $\frac{x + 2}{x - 2}$ **97.** $\frac{12x^2 + 12x + 3}{2}$ in.^{2} **103.** $-6x^5y^6$ **105.** $\frac{1}{81y^4}$ **107.** $4y^3 + 16y^2 - 8y + 8$

Study Set Section 12.3 (page 815)

1. LCD **3.** numerators, common denominator **7.** $\frac{x}{3}$ **9.** $\frac{4x}{y}$ **11.** $\frac{2}{y}$ **13.** $\frac{y + 3}{5z}$ **15.** 9 **17.** $\frac{1}{a + 2}$ **19.** $-\frac{y}{8}$ **21.** $\frac{x}{y}$ **23.** $\frac{y}{x}$ **25.** $\frac{1}{y}$ **27.** $\frac{1}{2}$ **29.** $\frac{2}{c + d}$ **31.** $\frac{4x}{3}$ **33.** 0 **35.** $\frac{4x - 2y}{y + 2}$ **37.** $\frac{2x + 10}{x - 2}$ **39.** $\frac{125x}{20x}$ **41.** $\frac{8xy}{x^2y}$ **43.** $\frac{3x(x + 1)}{(x + 1)^2}$ **45.** $\frac{2y(x + 1)}{x^2 + x}$ **47.** $\frac{z(z + 1)}{z^2 - 1}$ **49.** $\frac{2(x + 2)}{x^2 + 3x + 2}$ **51.** $6x$ **53.** $18xy^2$ **55.** $x^2 - 1$ **57.** $x^2 + 6x$ **59.** $(x + 1)(x + 5)(x - 5)$ **61.** $\frac{5y}{9}$ **63.** $\frac{53x}{42}$ **65.** $\frac{4xy + 6x}{3y}$ **67.** $\frac{2 - 3x^2}{x}$ **69.** $\frac{y^2 + 7y + 6}{15y^2}$ **71.** $\frac{x^2 + 4x + 1}{x^2y}$ **73.** $\frac{2x^2 - 1}{x(x + 1)}$ **75.** $\frac{2xy + x - y}{xy}$ **77.** $\frac{x + 2}{x - 2}$ **79.** $\frac{2x^2 + 2}{(x - 1)(x + 1)}$ **81.** $-\frac{2}{a - 4}$ **83.** $\frac{2t + 2}{t - 7}$ **85.** $\frac{4x + 2}{x - 2}$ **87.** $\frac{b - 1}{2(b + 1)}$ **89.** $\frac{1}{a + 1}$ **91.** $-\frac{1}{2(x - 2)}$ **93.** $\frac{x}{x - 2}$ **95.** $\frac{5x + 3}{x + 1}$ **97.** $\frac{20x + 9}{6x^2}$ cm **103.** 7^2 **105.** $2^3 \cdot 17$

Study Set Section 12.4 (page 821)

1. complex fraction **3.** single, divide **7.** $\frac{8}{9}$ **9.** $\frac{3}{8}$ **11.** $\frac{5}{4}$ **13.** $\frac{5}{7}$ **15.** $\frac{x^2}{y}$ **17.** $\frac{5t^2}{27}$ **19.** $\frac{1 - 3x}{5 + 2x}$ **21.** $\frac{1 + x}{2 + x}$ **23.** $\frac{3 - x}{x - 1}$ **25.** $\frac{1}{x + 2}$ **27.** $\frac{1}{x + 3}$ **29.** $\frac{xy}{y + x}$ **31.** $\frac{y}{x - 2y}$ **33.** $\frac{x^2}{(x - 1)^2}$ **35.** $\frac{7x + 3}{-x - 3}$ **37.** $\frac{x - 2}{x + 3}$ **39.** −1 **41.** $\frac{y}{x^2}$ **43.** $\frac{x + 1}{1 - x}$ **45.** $\frac{1 + a^3}{a^3}$ **47.** 2 **49.** $\frac{y - 5}{y + 5}$ **51.** $\frac{3}{14}$ **53.** $\frac{R_1R_2}{R_2 + R_1}$ **57.** t^9 **59.** $-2r^7$ **61.** $\frac{81}{256r^8}$ **63.** $\frac{r^{10}}{9}$

Study Set Section 12.5 (page 831)

1. rational equations **3.** check **5.** interest, principal, rate **7. a.** yes **b.** no **9. a.** $\frac{1}{6}, \frac{1}{5}$ **b.** $\frac{11}{30}$ **11.** $\frac{1}{6}$ **13. a.** $r = \frac{I}{Pt}$ **b.** $P = \frac{I}{rt}$ **17.** $\frac{3}{5} + \frac{7}{x-1} = 2$ **19.** 4 **21.** 0 **23.** 3 **25.** no solution; 0 is extraneous **27.** $\frac{11}{4}$ **29.** −4, 4 **31.** 1 **33.** no solution; 5 is extraneous **35.** −1 **37.** 6 **39.** 1 **41.** 4 **43.** no solution; −2 is extraneous **45.** 1 **47.** 2 **49.** 0 **51.** 2, −5 **53.** −4, 3 **55.** −2, 1 **57.** 3, −4 **59.** 1, −9 **61.** $a = \frac{b}{b-1}$ **63.** $r = \frac{E - IR}{I}$ **65.** $d = \frac{bc}{a}$ **67.** 2 **69.** 5 **71.** $\frac{2}{3}, \frac{3}{2}$ **73.** $f = \frac{d_1 d_2}{d_1 + d_2}$ **75.** $R = \frac{HB}{B - H}$ **77.** $2\frac{2}{9}$ hr **79.** $2\frac{6}{11}$ days **81.** $7\frac{1}{2}$ hr **83.** 4 mph **85.** 7% and 8% **87.** 5 **89.** 30 **91.** 25 mph **97.** $x(x + 4)$ **99.** $(2x + 3)(x - 1)$ **101.** $(x^2 + 4)(x + 2)(x - 2)$

Think It Through (page 836)

$\frac{1}{4}, \frac{1}{2}, \frac{2}{5}$; yes, no, yes

Study Set Section 12.6 (page 843)

1. ratio, rate **3.** extremes, means **5.** shape **7.** $\frac{6}{5}$ **9.** yes **11.** The ratio on the right-hand side should be $\frac{h}{48}$. **15.** triangle **17.** no **19.** yes **21.** no **23.** yes **25.** 4 **27.** 6 **29.** 3 **31.** 9 **33.** 0 **35.** −17 **37.** $-\frac{3}{2}$ **39.** $\frac{83}{2}$ **41.** 4, −4 **43.** 4, −1 **45.** −5, −1 **47.** \$17 **49.** \$1.8 million **51.** 24 **53.** $7\frac{1}{2}, 1\frac{2}{3}, 5$ **55.** 510 **57.** 737 **59.** \$309 **61.** 49 ft, $3\frac{1}{2}$ in. **63.** 3,600,000 **65.** $3\frac{3}{4}$ in. **67.** 39 ft **69.** $46\frac{7}{8}$ ft **71.** 6,750 ft **77.** 90% **79.** 480 **81.** $\frac{3}{5}$

Study Set Section 12.7 (page 852)

1. direct **3.** constant **5.** direct **7.** inverse **9. a.** yes **b.** no **c.** no **d.** yes **11.** $g = kf$ **13.** $\frac{7}{2}$ **15. a.** no **b.** yes **c.** no **d.** yes **17.** $n = \frac{k}{p}$ **19.** 150 **23.** 35 **25.** 6.25 **27.** 3 **29.** 1 **31.** 12,000 **33.** 6 **35.** 168 mi **37.** 27 mg **39.** $22\frac{1}{2}$ **41.** $2\frac{1}{2}$ hr **43.** 8 amps **45.** $53\frac{1}{3}$ m³ **51.** −1, 6 **53.** −2, −3, −4 **55.** 0, 0, 1 **57.** 1, −1, 2, −2

Key Concept (page 855)

1. a. $\frac{2x}{x-2}$ **b.** $x - 4$ **3. a.** $\frac{2x^2 - 1}{x(x+1)}$ **b.** $\frac{x}{x}, \frac{x+1}{x+1}$ **5. a.** 2 **b.** b **7. a.** 2, 4 **b.** $4(y - 6)$

Chapter Review (page 857)

1. 4, −4 **2.** $\frac{2}{5}$ **3.** $-\frac{2}{3}$ **4.** $\frac{1}{2x}$ **5.** $\frac{5}{2x}$ **6.** $\frac{x}{x+1}$ **7.** $a - 2$ **8.** −1 **9.** $-\frac{1}{x+3}$ **10.** $\frac{x}{x-1}$ **11.** in lowest terms **12.** $-\frac{3}{7}$ **13.** x is not a common factor of the numerator and the denominator. **14.** $\frac{4}{3}$ **15.** $\frac{3x}{y}$ **16.** 96 **17.** $\frac{x-1}{x+2}$ **18.** $\frac{2x}{x+1}$ **19.** $\frac{3y}{2}$ **20.** $\frac{1}{x}$ **21.** $x + 2$ **22.** $b + 2$ **23.** 1 **24.** $\frac{2x+2}{x-7}$ **25.** $\frac{1}{a-4}$ **26.** $4x^2$ **27.** $18xy^2$ **28.** $(x + 1)(x + 2)$ **29.** $y^2 - 25$ **30.** $\frac{x^2 + x - 1}{x(x-1)}$ **31.** $\frac{c-7}{7c}$ **32.** $\frac{x^2 + 4x - 4}{2x^2}$ **33.** $\frac{1}{t+1}$ **34.** $\frac{x+1}{x}$ **35.** $\frac{b+6}{b-1}$ **36.** $\frac{14x + 28}{(x+6)(x-1)}$ units, $\frac{12}{(x+6)(x-1)}$ square units **37.** $\frac{9}{4}$ **38.** $\frac{3}{2}$ **39.** $\frac{1+y}{1-y}$ **40.** $\frac{x(x+3)}{2x^2 - 1}$ **41.** $x^2 + 3$ **42.** $\frac{1 + x^2}{1 - x^2}$ **43.** 3 **44.** no solution; 5 is extraneous **45.** 3 **46.** 2, 4 **47.** 0 **48.** $T_1 = \frac{T_2}{1 - E}$ **49.** $r_1 = \frac{rr_2}{r_2 - r}$ **50.** 3 **51.** $\frac{1}{4}$ **52.** $5\frac{5}{6}$ days **53.** 5% **54.** 5 mph **55.** 40 mph **56.** $\frac{3}{2}$ **57.** no **58.** $\frac{9}{2}$ **59.** 0 **60.** 7 **61.** 4, $-\frac{3}{2}$ **62.** 255 **63.** 20 ft **64.** \$833.33 **65.** 600 **66.** 1.25 amps **67.** inverse variation

Chapter 12 Test (page 861)

1. −3, 2 **2.** $\frac{8x}{9y}$ **3.** $\frac{x+1}{2x+3}$ **4.** 3 **5.** $-\frac{5y^2}{4}$ **6.** $\frac{x+1}{3(x-2)}$ **7.** $\frac{3}{5}$ **8.** $-\frac{x^2}{3}$ **9.** $x + 2$ **10.** $\frac{10x - 1}{x - 1}$ **11.** $\frac{13}{2y + 3}$ **12.** $\frac{2x^2 + x + 1}{x(x+1)}$ **13.** $\frac{2a + 7}{a - 1}$ **14.** $\frac{4n - 5mn}{10m}$ **15.** $\frac{x+y}{y-x}$ **16.** 11 **17.** 1 **18.** 1, 2 **19.** no solution; 6 is extraneous **20.** $B = \frac{HR}{R - H}$ **21.** yes **22.** $\frac{2}{3}$ **23.** yes **24.** 8,050 ft **25.** 100 lb **26.** 200 **27.** $3\frac{15}{16}$ hr **28.** 5 mph **29.** We can only remove common factors, as in the first expression. We can't remove common terms, as in the second expression. **30.** We multiply both sides of the equation by the LCD of the rational expressions appearing in the equation. The resulting equation is easier to solve.

Cumulative Review Exercises (page 863)

1. 36 **2.** \$2,580,000,000 **3.** 77 **4.** $35x$¢ **5.** $\frac{5}{4}$ **6.** 104° F **7.** 240 ft³ **8. a.** false **b.** false **c.** true **d.** true **9.** 3 **10.** $6c + 62$ **11.** $B = \frac{A - c - r}{2}$ **12.** $x \geq -1$, $[-1, \infty)$

13. −5 **14.** 12 lb of the \$3.20 tea and 8 lb of the \$2 tea **15.** 500 mph

16.

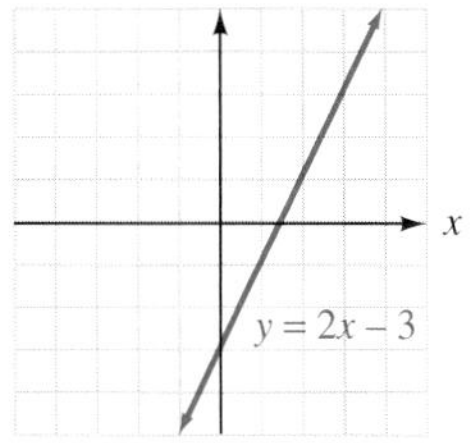

17.

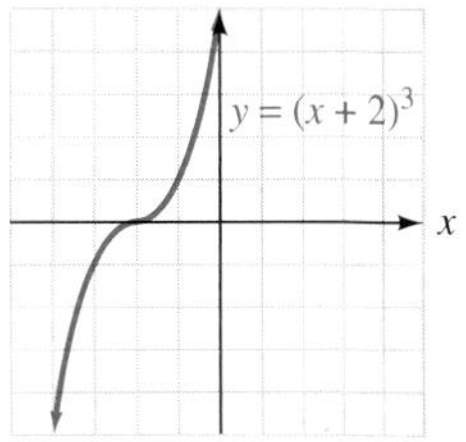

18. −1 **19.** $y = 3x + 2$

20.

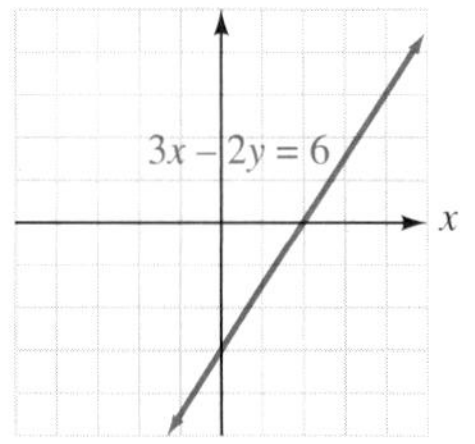

21.

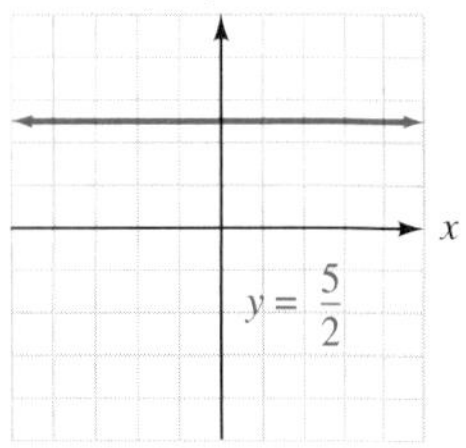

22. $\frac{8}{7}$ **23.** no **24.** 0.008 mm/m **25.** 12 **26.** −25 **27.** x^7

28. x^{25} **29.** $\frac{y^3}{8}$ **30.** $-\frac{32a^5}{b^5}$ **31.** $\frac{a^8}{b^{12}}$ **32.** $3b^2$ **33.** 2.9×10^5
34. 3 **35.** $6x^2 + x - 5$ **36.** $6x^5y^5$ **37.** $6y^2 - y - 35$
38. $-12x^4z + 4x^2z^2$ **39.** $2x + 3$ **40.** $-\frac{3r}{s^9}$
41. $10^3 \cdot 26^3$; 17,576,000 **42.** $A = \pi R^2 - \pi r^2$ **43.** $k^2t(k - 3)$
44. $(b + c)(2a + 3)$ **45.** $2(a + 10b)(a - 10b)$
46. $(b + 5)(b^2 - 5b + 25)$ **47.** $(u - 9)^2$ **48.** $(2x - 9)(3x + 7)$
49. $-(r - 2)(r + 1)$ **50.** prime **51.** $0, -\frac{1}{5}$ **52.** $\frac{1}{3}, \frac{1}{2}$
53. 10 in., 16 in. **54.** 5, −5 **55.** $\frac{x}{3}$ **56.** $-\frac{1}{x - 3}$ **57.** $\frac{x^2}{(x + 1)^2}$
58. $\frac{a}{a - 3}$ **59.** $\frac{17}{25}$ **60.** 3 **61.** \$2,450 **62.** 34.8 ft **63.** $14\frac{2}{5}$
64. One variable is a constant multiple of the reciprocal of the other; $y = \frac{k}{x}$.

Chapter 13 Check Your Knowledge (page 867)

1. solution **2.** inconsistent **3.** dependent **4.** boundary, point
5. no **6.** yes
7. $(-1, -1)$ **8.** $(3, 0)$ **9.** $(2, 2)$ **10.** $(1, -2)$ **11.** $(2, -1)$

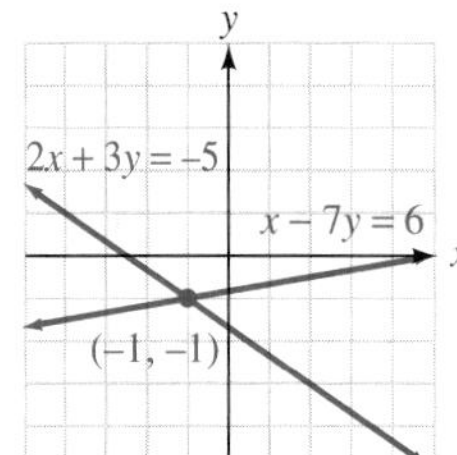

12. inconsistent, independent equations
13. consistent, independent equations
14. consistent, dependent equations
15. $(1, 4)$
16. $(3, 2)$
17. \$9,000 at 6%, \$5,000 at 9%
18. 40 pounds of peanuts, 10 pounds of cashews **19.** yes **20.** no

21.

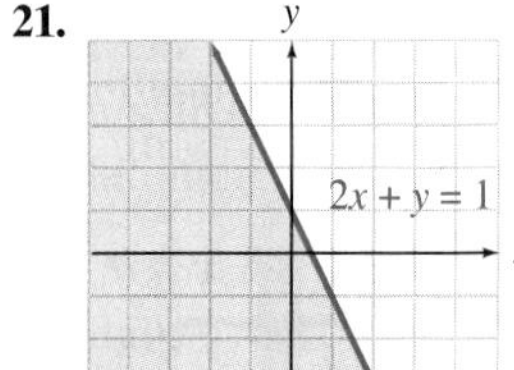

22.

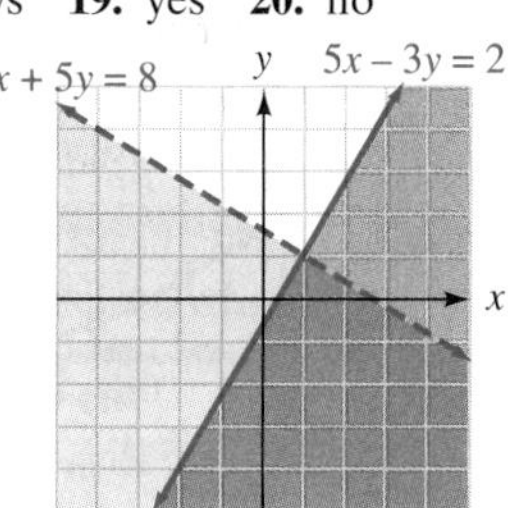

Think It Through (page 873)

(1978, 50); in 1978, 50% of the associate degrees that were awarded went to men and 50% went to women. Since then, the percent awarded to women has increased, while the percent awarded to men has decreased.

Study Set Section 13.1 (page 876)

1. system **3.** independent **5.** inconsistent **7.** true
9. false **11.** true **13.** $(-4, -1)$ **15.** 1 solution; consistent
17. The method is not accurate enough to find a solution such as $(\frac{2}{5}, -\frac{1}{3})$. **21.** yes **23.** yes **25.** no **27.** no **29.** no **31.** yes

33.

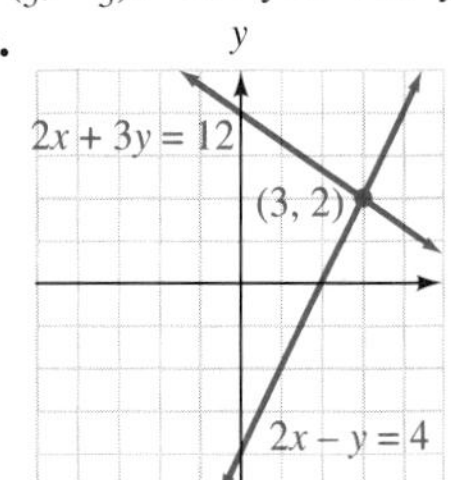

35.

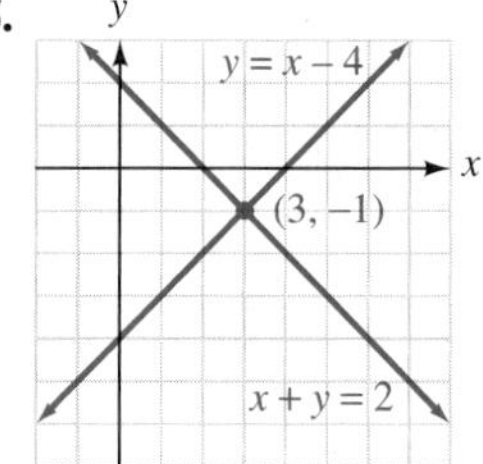

37.

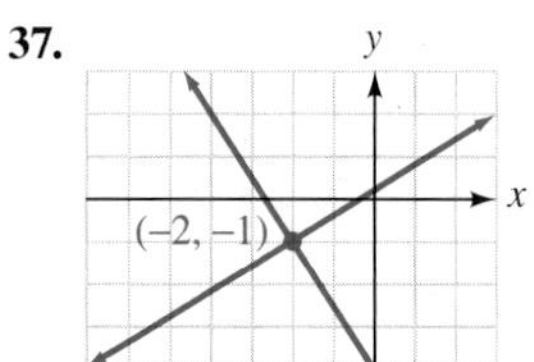

39.

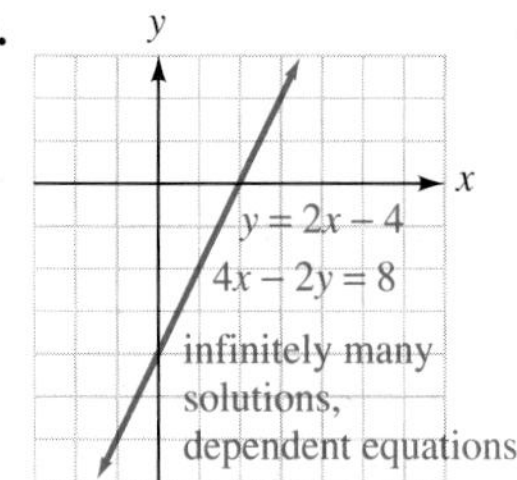

41.

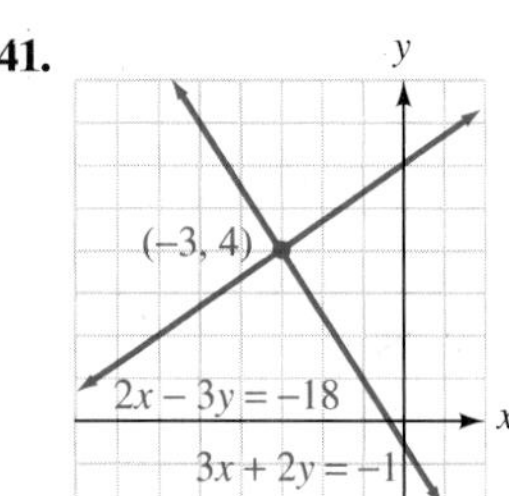

43.

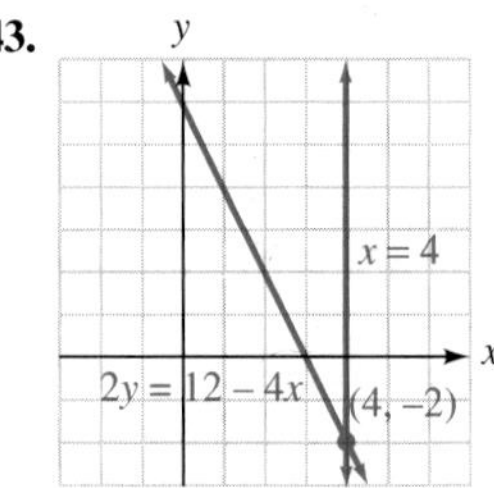

45.

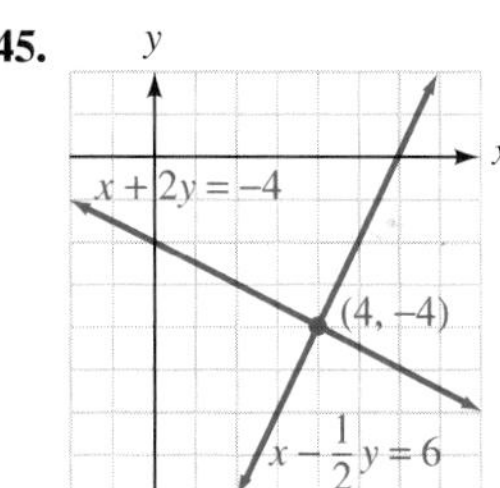

47.

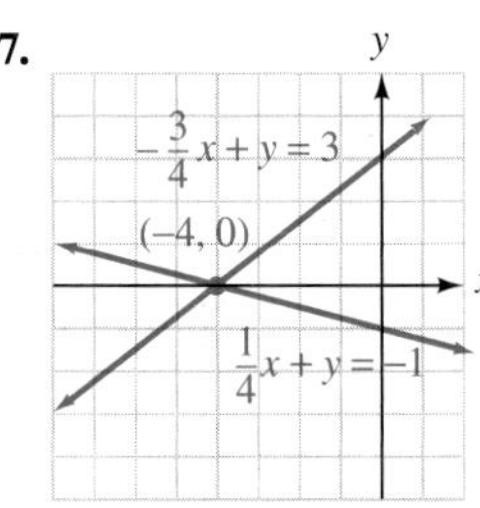

49.

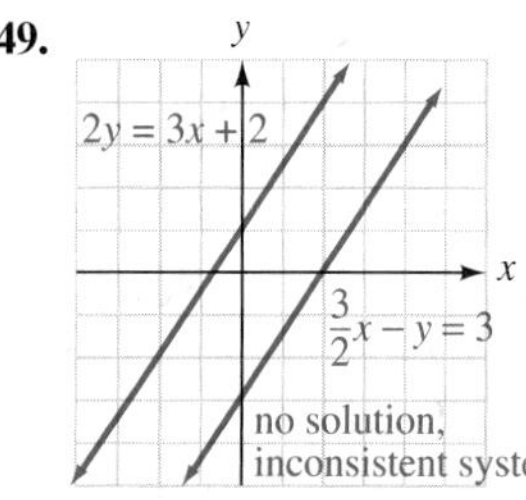

51.

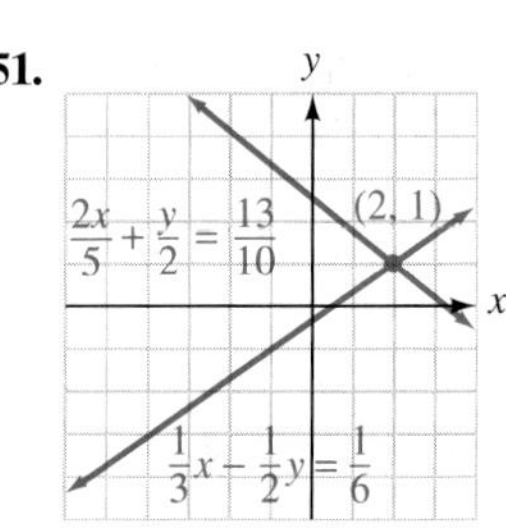

53. $(1, 3)$ **55.** no solution **57. a.** Donors outnumbered those needing a transplant. **b.** 1994; 4,100 **59. a.** Houston, New Orleans, St. Augustine **b.** St. Louis, Memphis, New Orleans **c.** New Orleans **61.** yes, $(2, 2)$ **65.** parallel **67.** −21
69. $x = 0$ **71.** $(5, 2)$

Study Set Section 13.2 (page 885)

1. y, terms **3.** distributive **5.** infinitely **7. a.** 2 **b.** 1
9. a. $x = 2y - 10$ **b.** $y = \frac{x}{2} + 5$ **c.** x; it involved only one step. **11. a.** Parentheses must be written around $x - 4$ in line 2. **b.** $\frac{13}{3}$ **13. a.** The coordinates of the intersection point are not integers.
b. $(\frac{6}{7}, \frac{12}{7})$

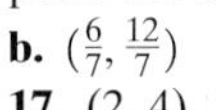

y
x − 2y = 0
x
6x + 3y = 5

17. $(2, 4)$ **19.** $(3, 0)$
21. $(-3, -1)$
23. no solution, inconsistent system
25. $(-2, 3)$

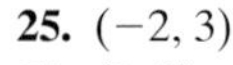

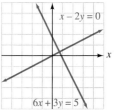

27. $(3, 2)$
29. $(3, -2)$
31. $(\frac{2}{3}, -\frac{1}{3})$ **33.** $(-1, -1)$ **35.** infinitely many solutions, dependent equations **37.** $(4, -2)$ **39.** $(\frac{1}{2}, \frac{1}{3})$ **41.** $(1, 4)$

43. no solution, inconsistent system **45.** (4, 2) **47.** (−6, 4) **49.** $(\frac{1}{5}, 4)$ **51.** (5, 5) **53.** melon, because it's the same price as hash browns **55.** An entry of equal value may be substituted for the second roast beef dinner. **61.** $-\frac{5}{8}$ **63.** (0, −6) **65.** no **67.** $s = kt$

Study Set Section 13.3 (page 893)

1. coefficient **3.** general **5.** The second equation should be written in general form: $3x - 2y = 10$. **7.** Multiply both sides by 15.

9. a.

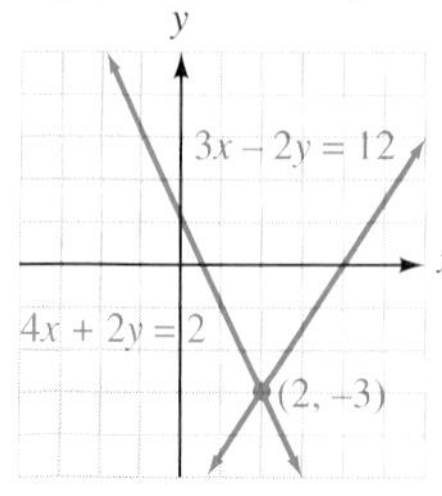

b. (2, −3) **13.** (−2, 3) **15.** (−1, 1) **17.** (−3, 4) **19.** (0, 8) **21.** (2, 3) **23.** (3, −2) **25.** (2, 7) **27.** no solution, inconsistent system **29.** $(1, -\frac{5}{2})$ **31.** (8, −3) **33.** (3, 0) **35.** $(\frac{10}{3}, \frac{10}{3})$ **37.** (5, −6) **39.** (−1, 2) **41.** infinitely many solutions, dependent equations **43.** (4, 0) **45.** (−1, 2) **47.** (0, 1) **49.** (2, 2) **51.** 1991 **57.** 4 **59.** 0 **61.** 7.5 ft^2 **63.** $x - 10$

Think It Through (page 897)

12,000 hr, 18,000 hr

Study Set Section 13.4 (page 903)

1. variable **3.** system **5.** $x + c, x - c$ **7. a.** \$$y$ **b.** 5% **c.** $0.05x, 0.11y$ **9. a.** two **b.** graphing, substitution, addition **11.** $A = lw$ **13.** $d = rt$ **15.** $2l + 2w = 90$ **17.** 32, 64 **19.** 8, 5 **21.** 22 ft, 29 ft **23.** president: \$400,000; vice president: \$181,000 **25.** \$30, \$10 **27.** 40 **29.** 65°, 115° **31.** 6,720 ft^2 **33. a.** 400 tires

b.

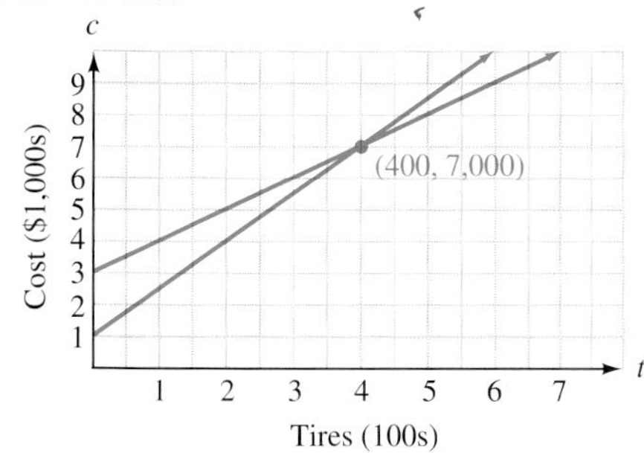

c. the second mold **35.** nursing: \$2,000; business: \$3,000 **37.** 5 mph **39.** 50 mph **41.** 4 L 6% salt water, 12 L 2% salt water **43.** 32 lb peanuts, 16 lb cashews **45.** \$640

49. 4 **51.** −1 2

53. −2, 2 **55.** 2, 2

Study Set Section 13.5 (page 912)

1. inequality **3.** boundary **5. a.** yes **b.** no **c.** yes **d.** no **7. a.** no; dashed **b.** yes; solid **9. a.** no **b.** yes **c.** yes **d.** no **11.** The test point must be on one side of the boundary.

13.

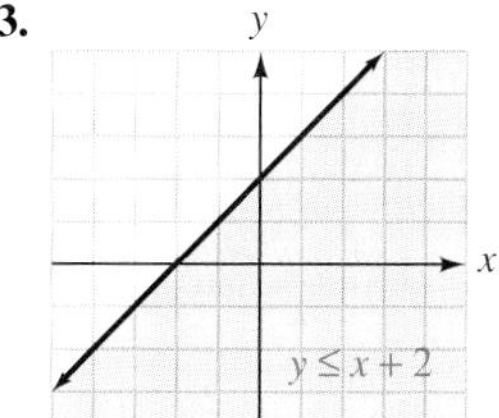

15.

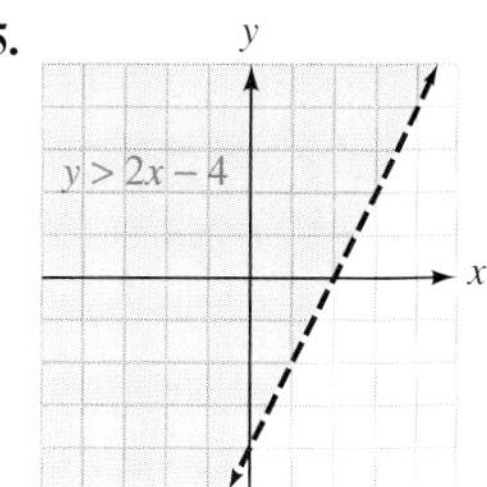

17.

y x x − 2y ≥ 4

19.

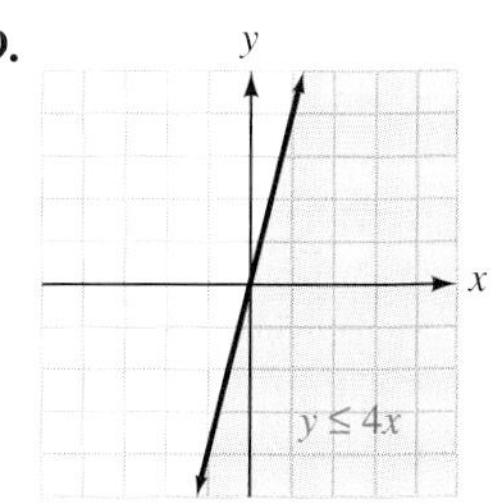

21.

y x y ≥ 3 − x y = 3 − x

23.

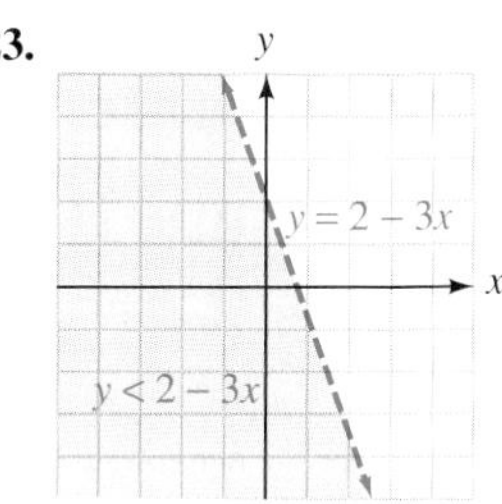

25.

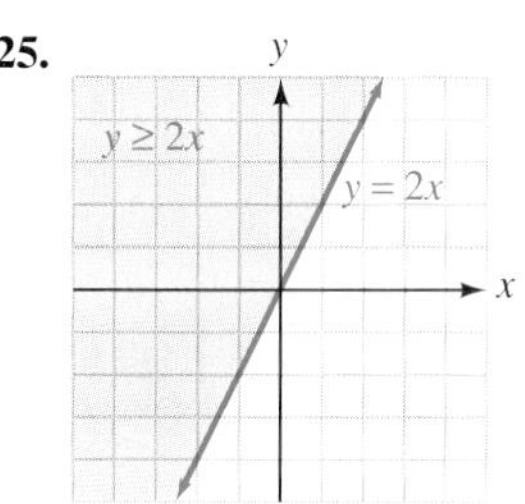

27.

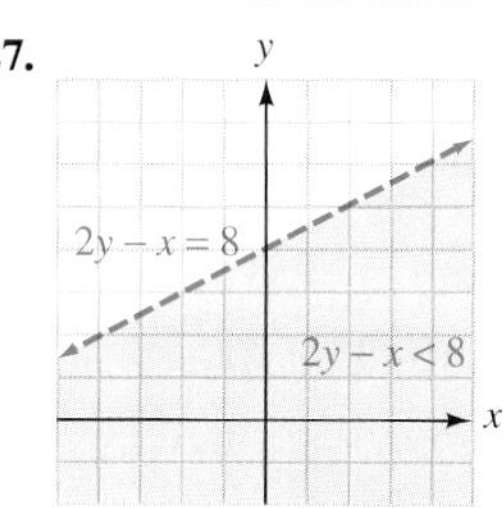

29.

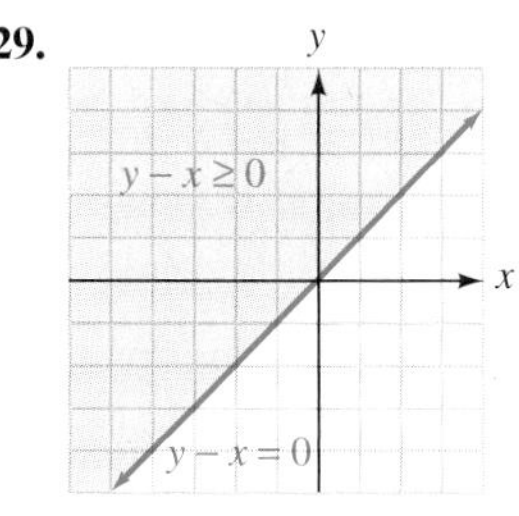

31.

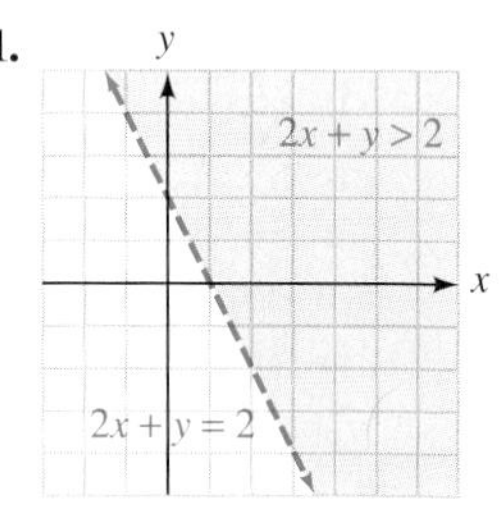

33.

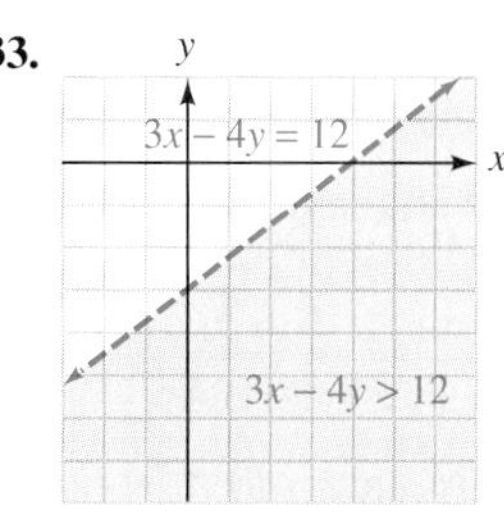

35.

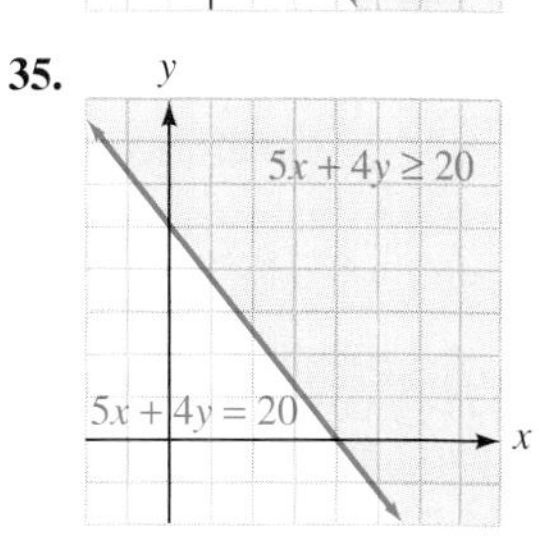

37.

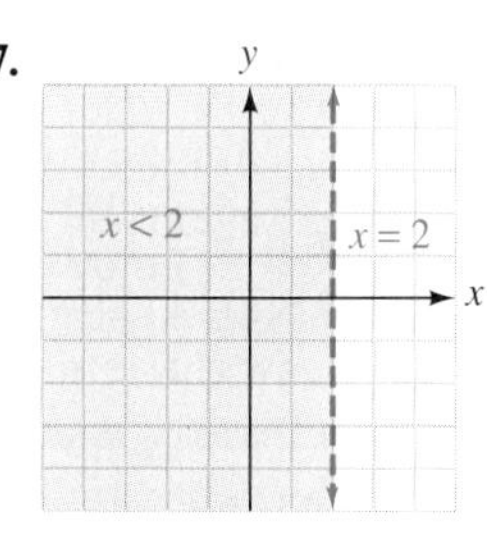

39.

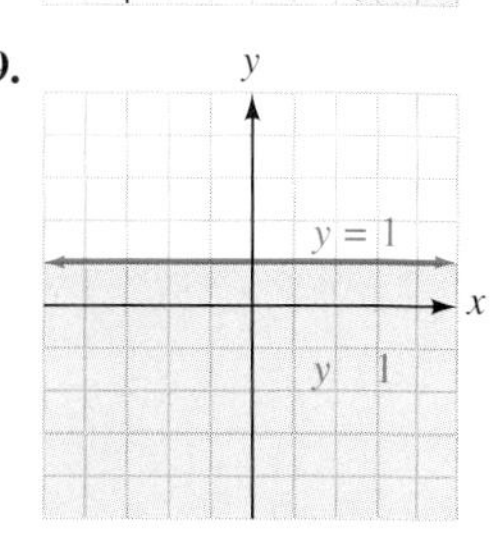

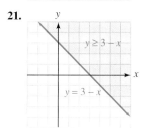

41.

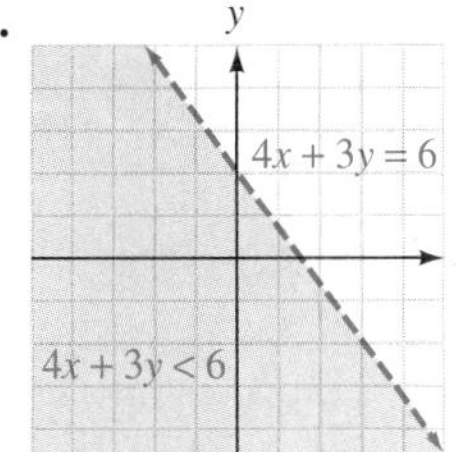

43.

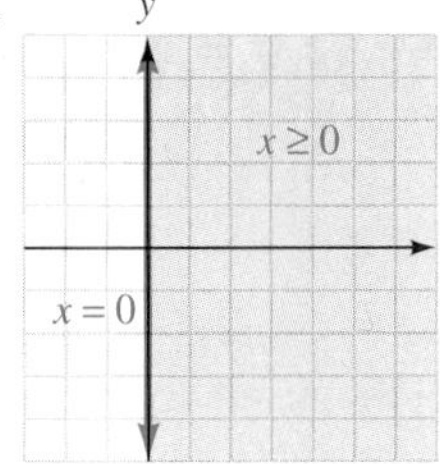

45.

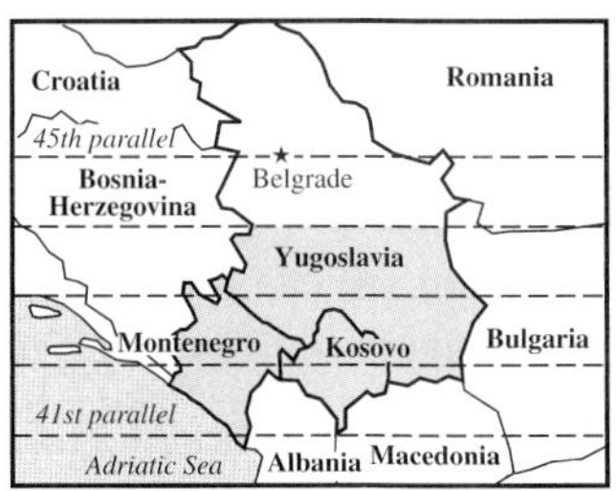

47. (10, 10) (20, 10), (10, 20)

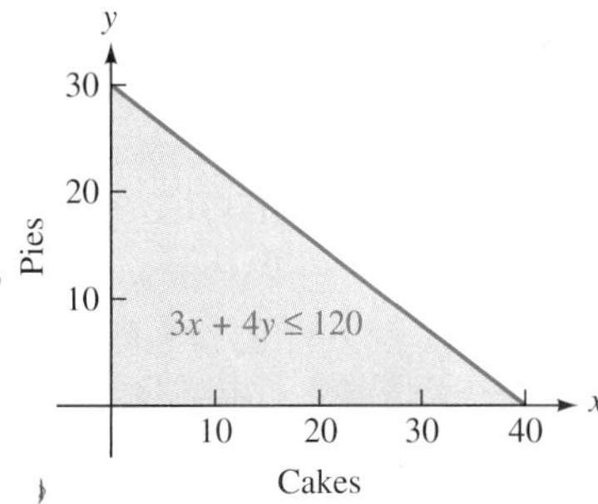

49. (50, 50), (30, 40), (40, 40)

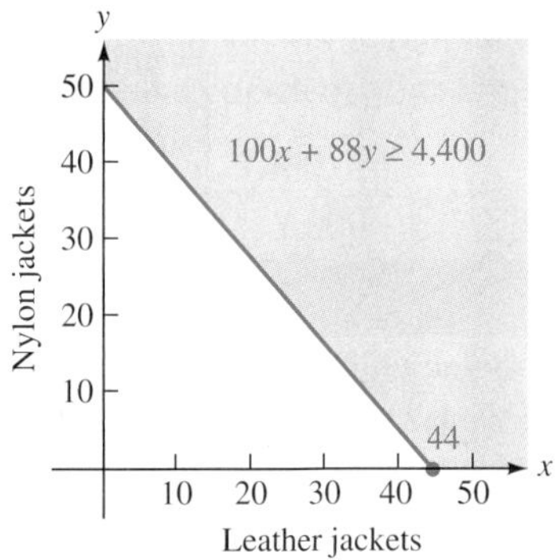

51. (80, 40), (80, 80), (120, 40)

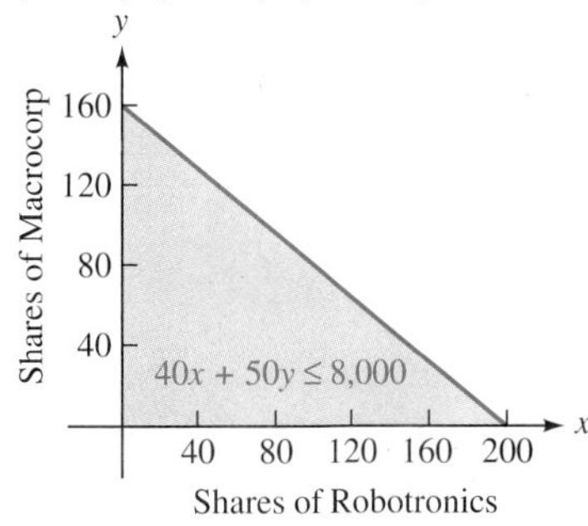

55. 7 **57.** $(x + 3)(x^2 - 3x + 9)$ **59.** $d = rt$ **61.** $t = \frac{A - P}{Pr}$

Study Set Section 13.6 (page 920)

1. inequalities **3.** shaded **5. a.** true **b.** false **c.** false **d.** true **e.** true **f.** true **7. a.** yes **b.** no **c.** no **9.** *ABC*

11.

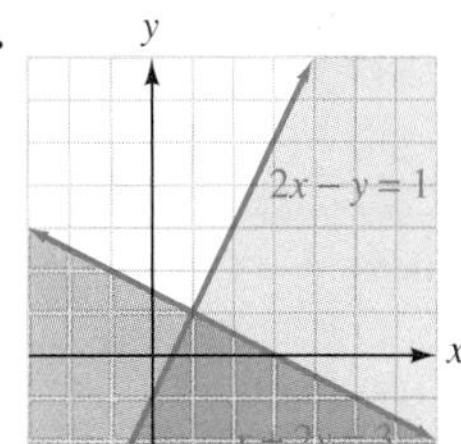

13.

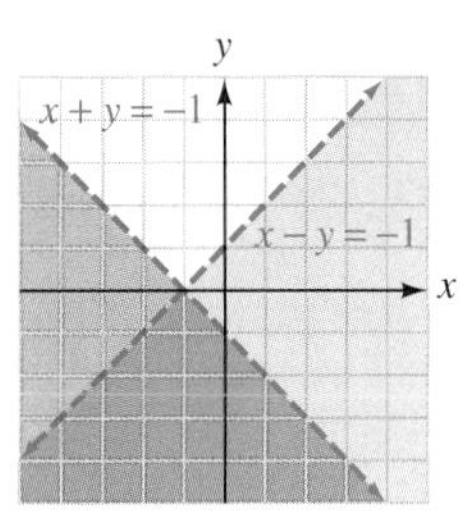

15.

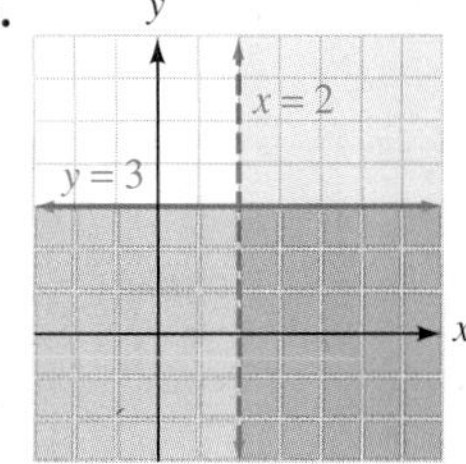

17.

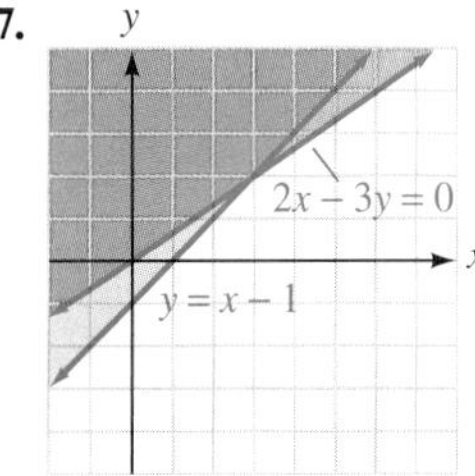

19.

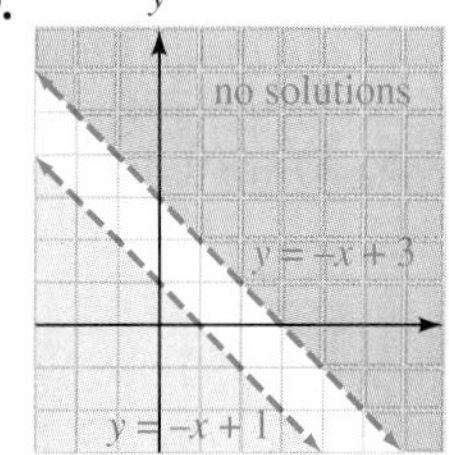

21.

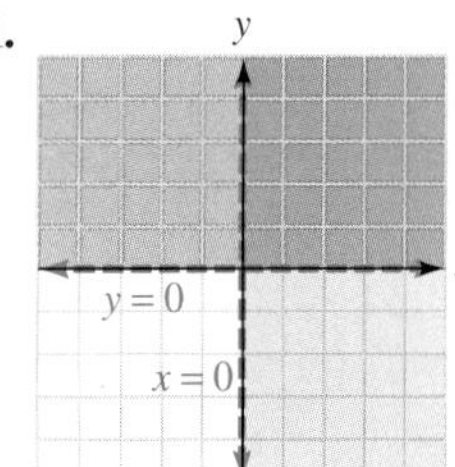

23.

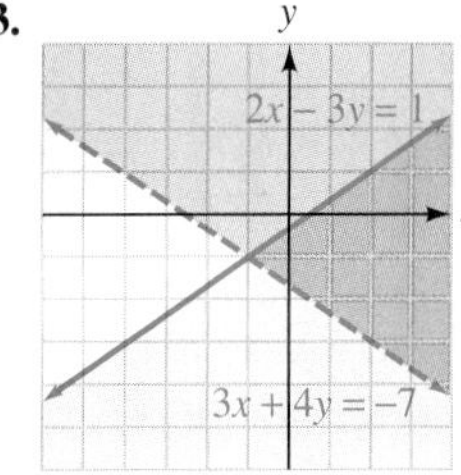

25.

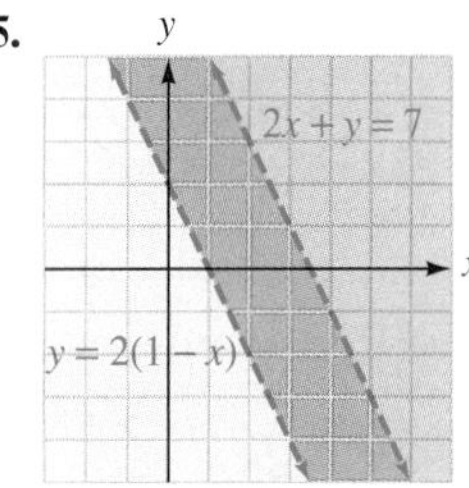

27.

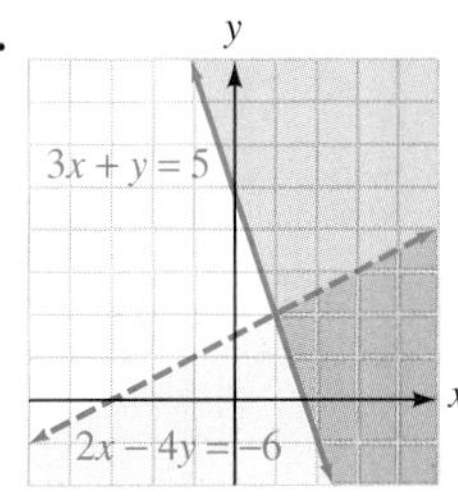

29.

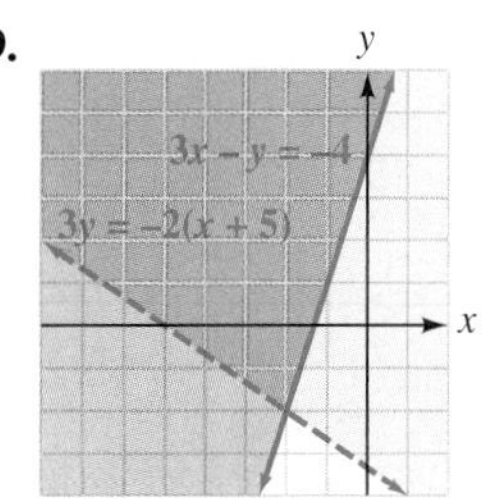

31.

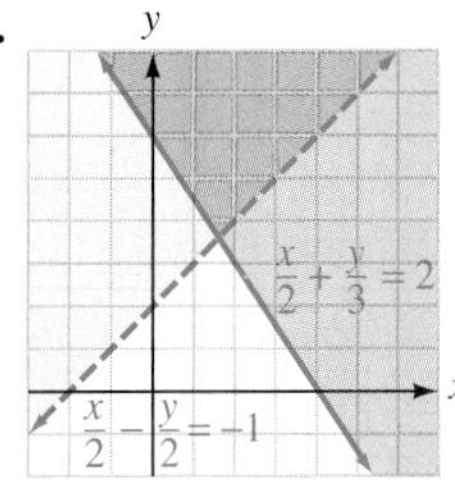

33.

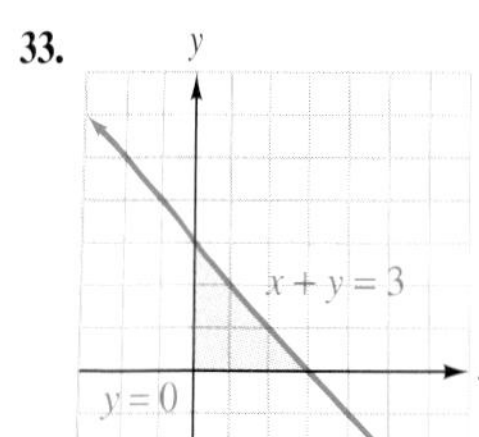

35.

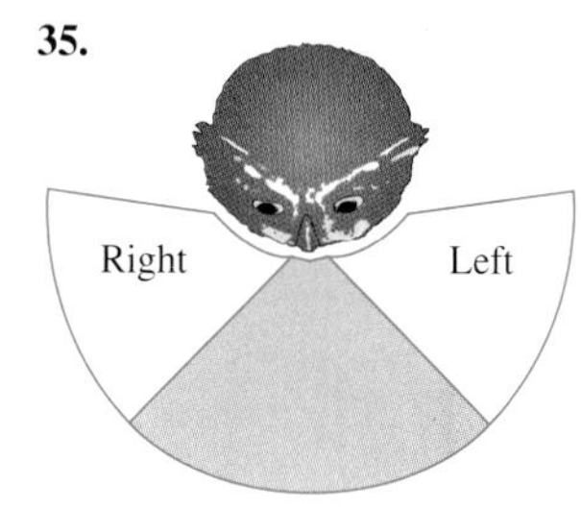

(c)

37. 1 \$10 CD and 2 \$15 CDs; 4 \$10 CDs and 1 \$15 CD

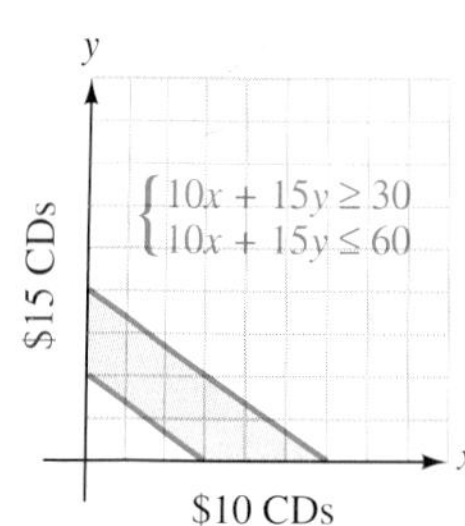

39. 2 desk chairs and 4 side chairs; 1 desk chair and 5 side chairs

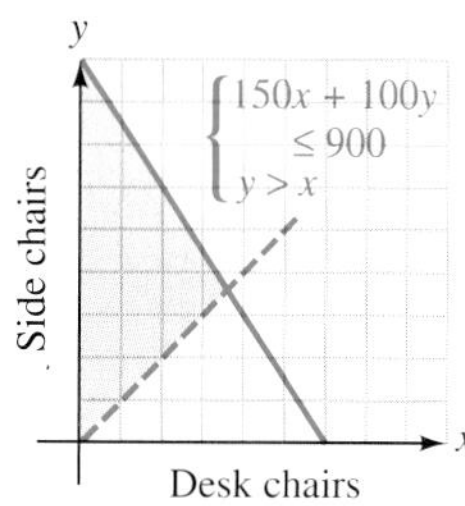

41.

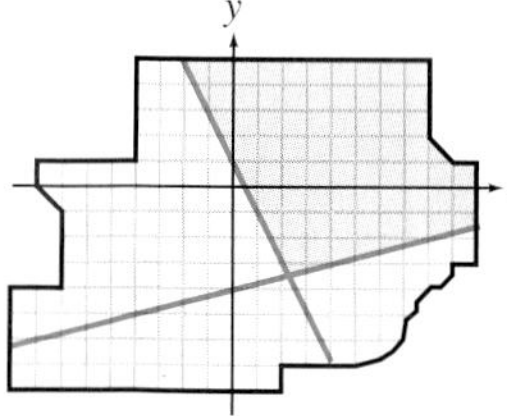

47. 128, 8 **49.** 4, −23

Key Concept (page 925)

1. (500, 3,000)

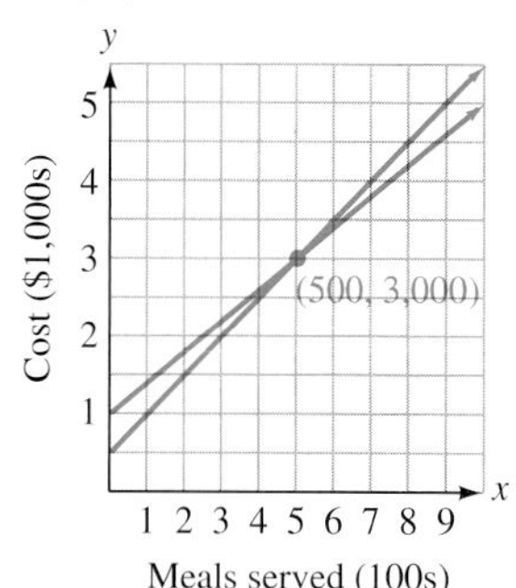

3. $(-1, -1)$ **5.** $(-3, 4)$

Chapter Review (page 927)

1. yes **2.** yes **3.** approx. March 5, about \$42 per share

4.

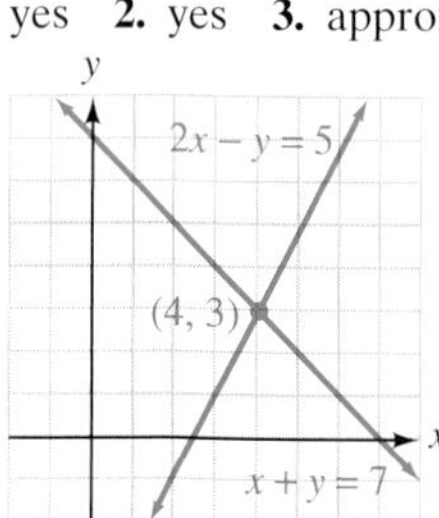

5.

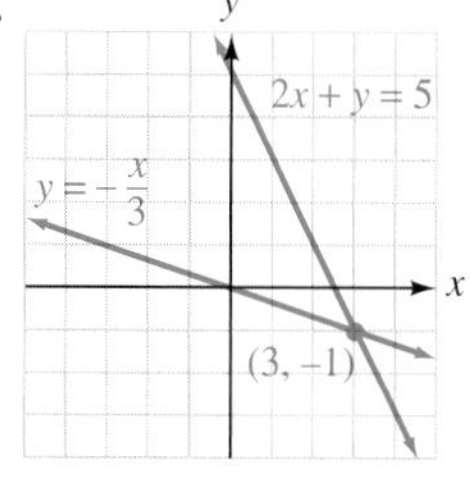

6.

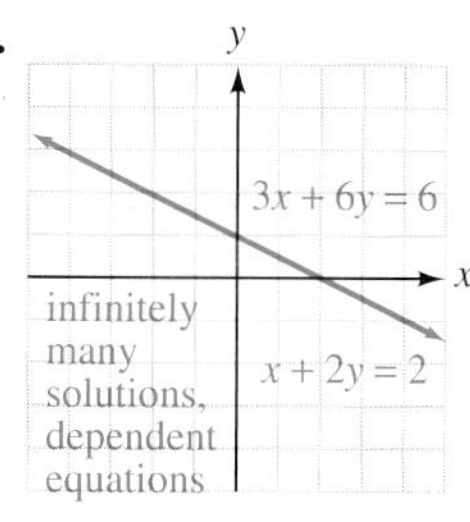

7.

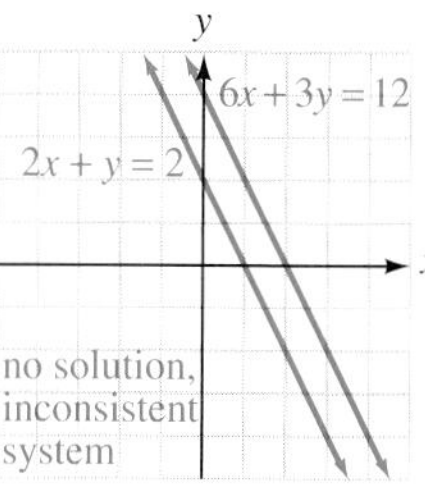

8. (3, 3) **9.** (5, 0) **10.** $\left(-\frac{1}{2}, \frac{7}{2}\right)$ **11.** $(-2, 1)$ **12.** infinitely many solutions, dependent equations **13.** (12, 10) **14. a.** no solutions **b.** two parallel lines **c.** inconsistent system **15. a.** $(3, -5)$ **16.** $\left(3, \frac{1}{2}\right)$ **17.** $(-1, 7)$ **18.** (0, 9) **19.** infinitely many solutions, dependent equations **20.** (0, 0) **21.** $(-5, 2)$ **22.** no solution, inconsistent system **23.** Addition; no variables have a coefficient of 1 or −1. **24.** Substitution; equation 1 is solved for x. **25.** stroke: 160,000; heart disease: 720,000 **26.** base: 21 ft; extension: 14 ft **27.** 10,800 yd^2 **28. a.** $y = 5x + 30{,}000$; $y = 10x + 20{,}000$ **b.** 2,000 **c.** the athlete **29.** 12 lb worms, 18 lb bears **30.** 3 mph **31.** \$16.40, \$10.20 **32.** \$750 **33.** $13\frac{1}{3}$ gal 40%, $6\frac{2}{3}$ gal 70% **34.** yes **35.** yes **36.** yes **37.** no

38.

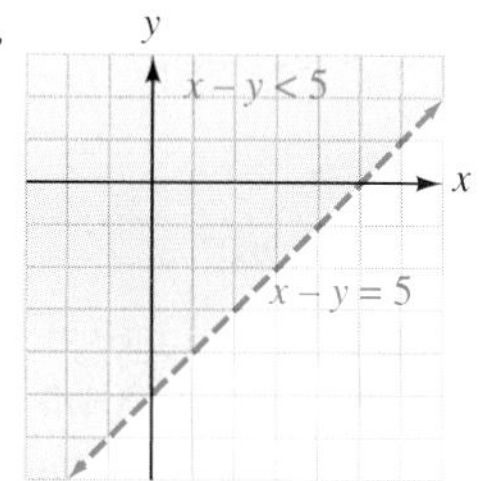

39.

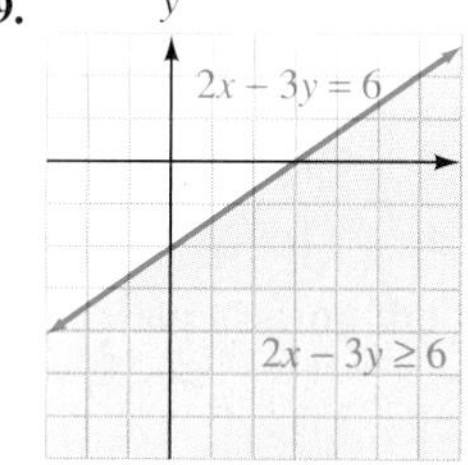

40.

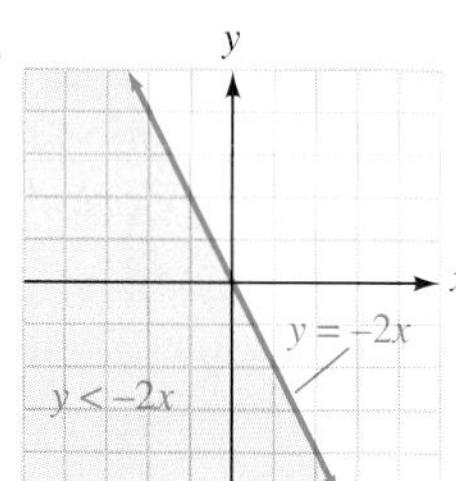

41.

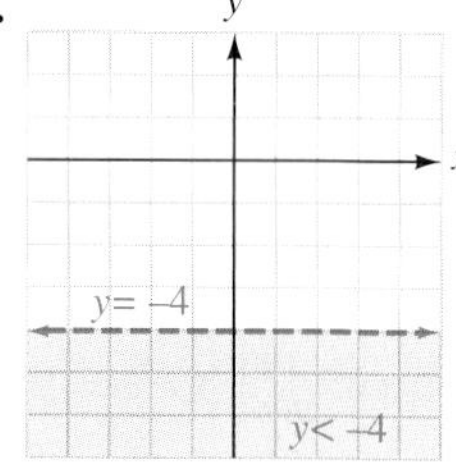

42. a. true **b.** false **c.** false **43.** $3x + 5y \leq 30$; (2, 4), (5, 3), (6, 2) (answers may vary)

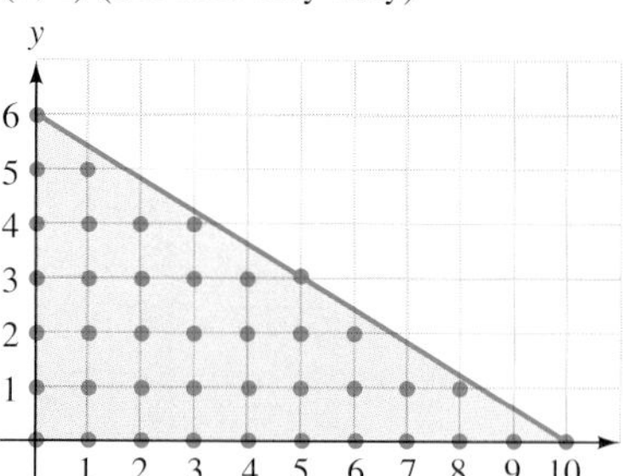

44.

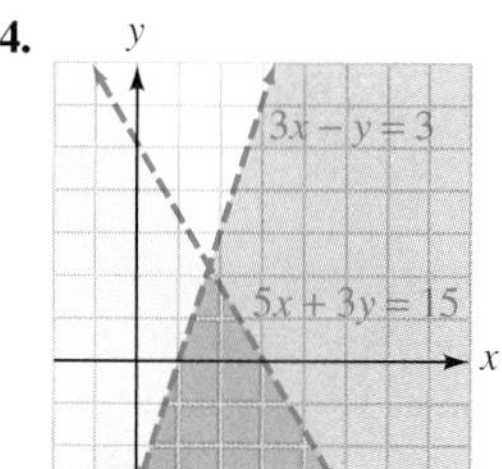

45.

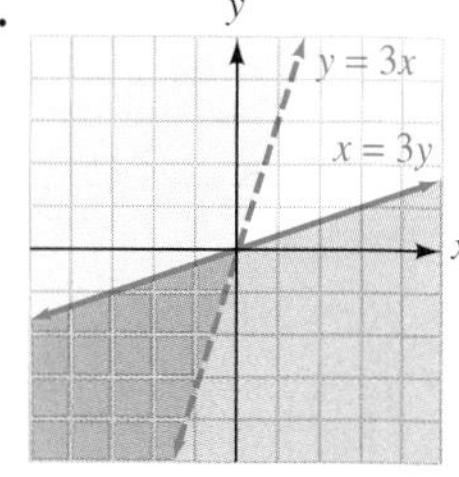

46. $10x + 20y \geq 40$, $10x + 20y \leq 60$; (3, 1): 3 shirts and 1 pair of pants; (1, 2): 1 shirt and 2 pairs of pants (answers may vary)

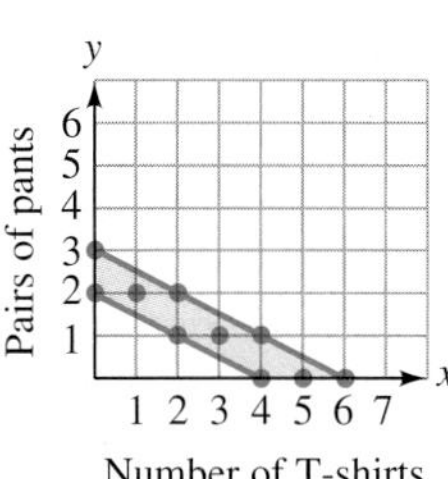

Chapter 13 Test (page 933)

1. yes **2.** no **3.** (2, 1)
4. The lines appear to be parallel. Since the lines do not intersect, the system does not have a solution.
5. $(-2, -3)$ **6.** $(-1, -1)$
7. (2, 4) **8.** $(-3, 3)$
9. inconsistent, independent equations
10. consistent, dependent equations
11. Addition method; the terms involving y can be eliminated easily. **12.** \$4,000 **13.** yes
14. no

15.

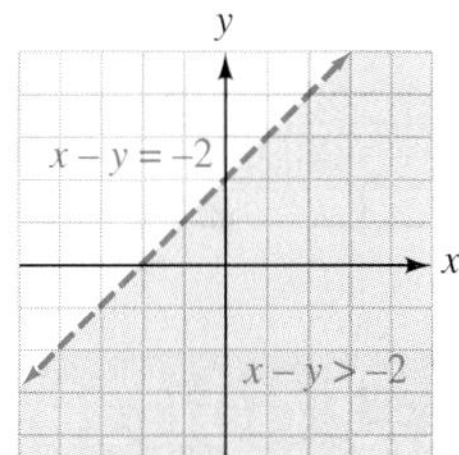

16.

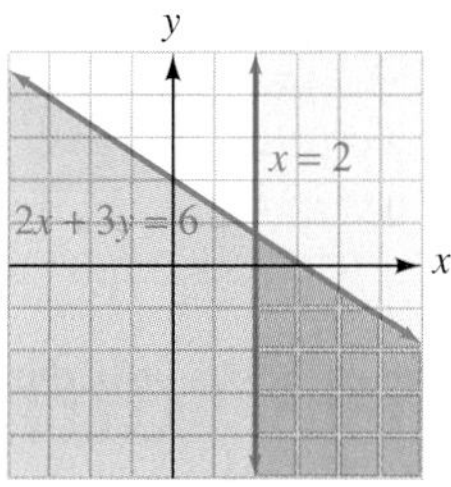

17. (30, 3); if 30 items are sold, the salesperson gets paid the same by both plans, \$3,000. **18.** Plan 2

Cumulative Review Exercises (page 935)

1. 2,016,000,000 **2.** $\{\ldots, -3, -2, -1, 0, 1, 2, 3, \ldots\}$
3. -37 **4.** 28 **5.** 1.2 ft^3 **6.** 0 **7.** -2 **8.** 15 **9.** $x < -\frac{11}{4}$
$(-\infty, -\frac{11}{4})$ **10.** 30
$-11/4$
11. mutual fund: \$25,000; bonds: \$20,000 **12. a.** $P = 2l + 2w$
b. $A = lw$ **c.** $A = \pi r^2$ **d.** $d = rt$

13.

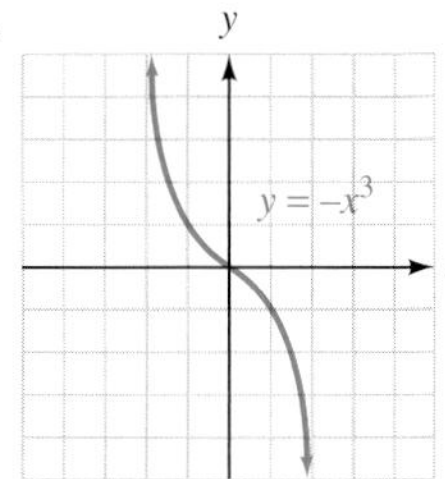

14.

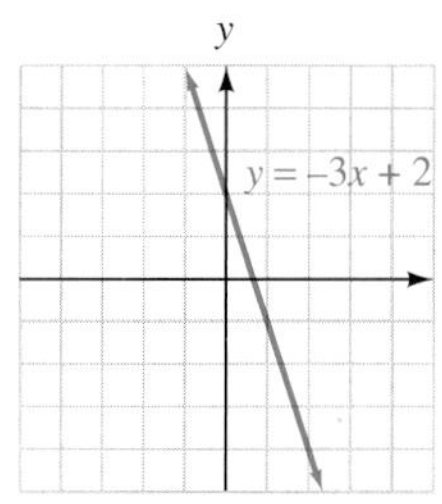

15.

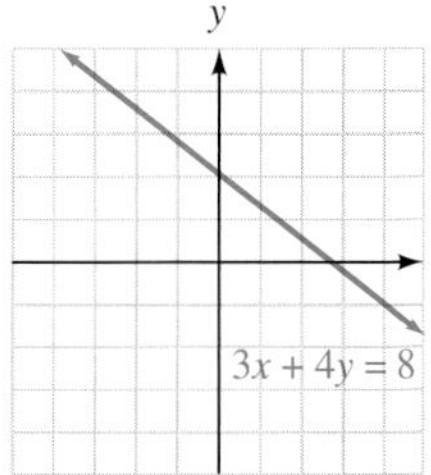

16.

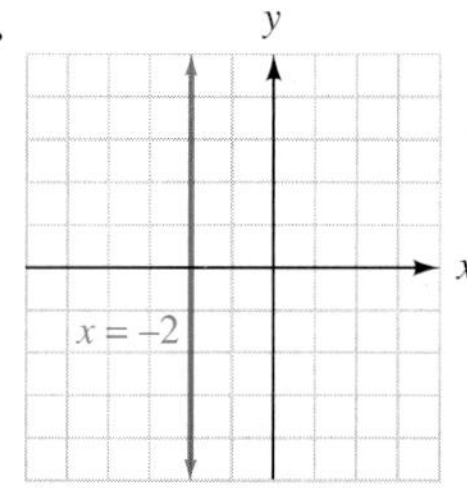

17. $m = 3$, $(0, -2)$; $y = 3x - 2$ **18.** $-\frac{4}{9}$ **19.** -70
20. 3.89×10^9 **21.** x^{31} **22.** $8a^4$ **23.** $\frac{3}{16}$ **24.** 1 **25.** $7x - 13y$
26. $6x^4 + 8x^3 - 14x^2$ **27.** $c^2 + 32c + 256$ **28.** $2x^2 + 3x - 9$

29. $\frac{1}{8} - \frac{2}{x}$ **30.** $3x + 1$ **31.** $2^5 \cdot 3^2$ **32.** $x^2 - 9$ (answers may vary) **33.** $3r(4r - s + 3rs^2)$ **34.** $(u - 9)^2$ **35.** $(2y - 1)(y - 3)$
36. $(x^2 + 9)(x + 3)(x - 3)$ **37.** $(t - v)(t^2 + tv + v^2)$
38. $(x - t)(y + s)$ **39.** 0, 2 **40.** 3, -5 **41.** $\frac{x - 5}{5}$
42. $\frac{4x}{y}$ **43.** -1 **44.** $\frac{x^2 + 4x + 1}{x^2 y}$ **45.** $\frac{3x + 1}{x + 2}$
46. -1 **47.** 16, 8 **48.** no
49. **50.**

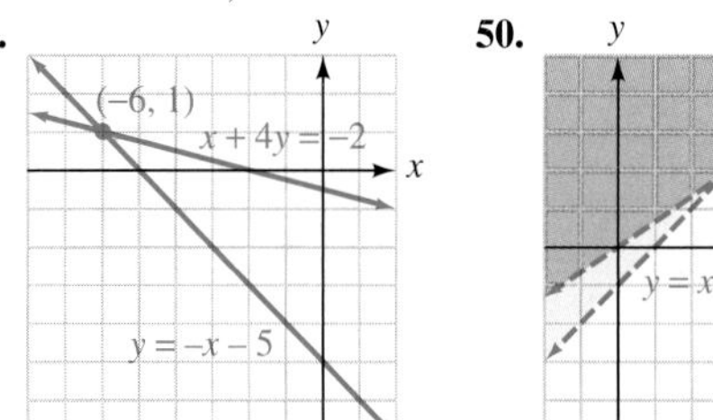

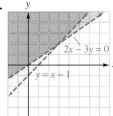

51. (4, 1) **52.** noodles: 2 servings, rice: 3 servings

Chapter 14 Check Your Knowledge (page 938)

1. positive **2.** radical **3.** conjugate **4.** Pythagorean **5.** 8
6. not a real number **7.** $\frac{2\sqrt{5}}{3}$ **8.** 2 **9.** $3y^2\sqrt{2}$ **10.** $2xy^2\sqrt{7x}$
11. $\frac{2x\sqrt{2x}}{y}$ **12.** $2xy^2z\sqrt[3]{2x^2}$ **13.** 5 **14.** $2\sqrt{2}$ ft, 2.8 ft
15. $4\sqrt{2}$ **16.** $7x\sqrt{3x}$ **17.** 160 **18.** $\sqrt{6} + 4$
19. $7 + 4\sqrt{3}$ **20.** $x - 4$ **21.** 340 ft **22.** $\frac{\sqrt{15x}}{5x}$ **23.** $2 - \sqrt{2}$
24. 16 **25.** 14 **26.** 5 **27.** 20 **28.** 13 **29.** 9 **30.** $x^{3/4}$

Think It Through (page 942)

1. 40 mph **2.** 60 mph

Study Set Section 14.1 (page 947)

1. square **3.** positive **5.** right **7.** two, 5, -5 **9.** $a^2 + b^2$
11. $\sqrt{b}$ **13. a.** Divide both sides by 2. **b.** Take the positive square root of both sides. **15.** 0, $\frac{1}{9}$, 0.4, 6, 20
17. a. $\sqrt{5} \approx 2.2$ **b.** $\sqrt{3} \approx 1.7$; $\sqrt{8} \approx 2.8$ **21.** False; $-\sqrt{9} = -3$, $\sqrt{-9}$ is not a real number. **23.** 5 **25.** -9
27. 1.1 **29.** 14 **31.** $\frac{3}{16}$ **33.** -17 **35.** -50 **37.** 60
39. 1.414 **41.** 3.317 **43.** 9.747 **45.** 20.688 **47.** -99.378
49. 4.621 **51.** 0.599 **53.** -0.915 **55.** 3.464 **57.** 1.866
59. rational; irrational; rational; imaginary **61.** 1, 2, 3, 4, 5

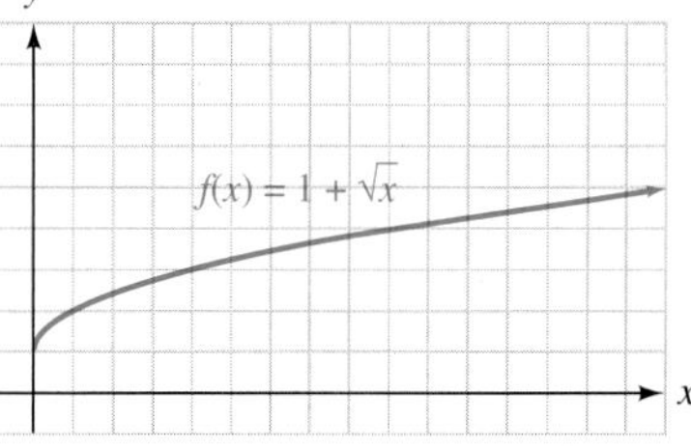

63. 0, -1, -2, -3, -4
65. 5 **67.** 8 **69.** 30
71. 117 **73.** 12 ft
75. The diagonal measurement should be $\sqrt{16^2 + 30^2} = 34$ in.
77. 127.3 ft **79.** 5.8 mi
81. a. $\sqrt{18}$ yd ≈ 4.24 yd
b. yes **83.** 25.5 ft

85. 240 in.^{2} **87. a.** 8.5 in. **b.** 7.8 in. **93.** $6s^2 + s - 5$
95. $3x^2 + 10x - 8$

Study Set Section 14.2 (page 956)

1. cube **3.** function **5.** root **7.** -6 **9. a.** not a real number **b.** -5 **11. a.** 1 **b.** $-\frac{1}{3}$ **c.** 5 **d.** 0.2 **e.** 10 **13.** index, radicand **15.** 2 **17.** 2 **19.** 0 **21.** -2 **23.** -4 **25.** $\frac{1}{5}$ **27.** 1 **29.** -4 **31.** 9 **33.** 10 **35.** 31.78 **37.** -0.48 **39.** $-1, 0, 1, 2, 3$

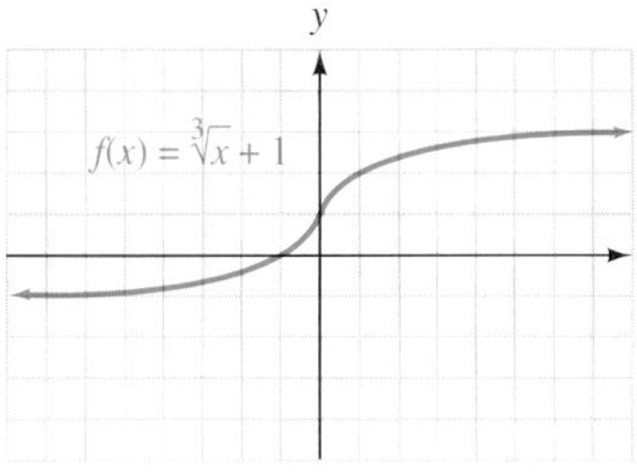

41. $2, 1, 0, -1, -2$
43. 2 **45.** -2
47. 1 **49.** -2
51. 3.34 **53.** -5.70
55. x **57.** x^3
59. x^5 **61.** $2z$
63. $-x^2y$ **65.** $-0.2y$
67. $-5x^2z^6$ **69.** $6z^{18}$
71. $-25z$ **73.** y^2

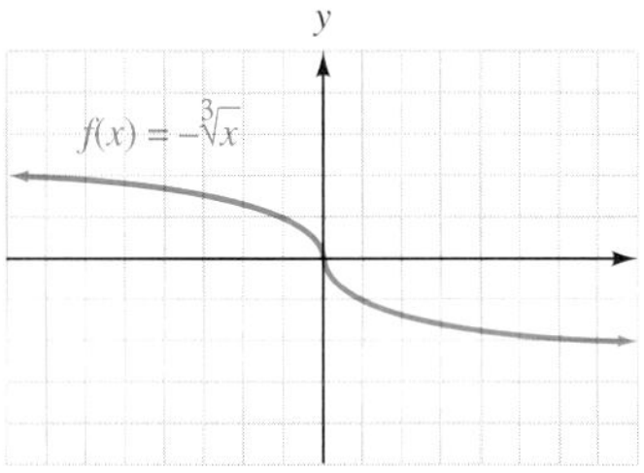

75. f **77.** $3y$ **79.** $-p^2q$ **81.** x **83.** 1.26 ft **85.** 27.14 mph **87.** $33\frac{1}{3}\%$ **91.** m^7 **93.** 3^8 or 9^4 **95.** x^{25} **97.** $24x^8$

Study Set Section 14.3 (page 965)

1. perfect **3.** simplify **5. a.** product, $\sqrt{a}\sqrt{b}$ **b.** quotient, $\frac{\sqrt{a}}{\sqrt{b}}$ **7.** Line 2 is not true. There is no addition property of radicals. **9. a.** $2\sqrt{7}$ in. **b.** 5.3 in. **11.** $2\sqrt{10}$ **13.** $6\sqrt{2}$ **17.** multiplication **19. a.** $2\sqrt{5}$ **b.** $a\sqrt{7}$ **c.** $9x\sqrt{6}$ **d.** $5z^2\sqrt{y}$ **21.** $2\sqrt{5}$ **23.** $5\sqrt{2}$ **25.** $3\sqrt{5}$ **27.** $7\sqrt{2}$ **29.** $4\sqrt{3}$ **31.** $-10\sqrt{2}$ **33.** $8\sqrt{3}$ **35.** $5\sqrt{10}$ **37.** $4\sqrt{6}$ **39.** $-4\sqrt{7}$ **41.** $n\sqrt{n}$ **43.** $2\sqrt{k}$ **45.** $2\sqrt{3x}$ **47.** $30\sqrt{3t}$ **49.** $5x\sqrt{x}$ **51.** $a\sqrt{b}$ **53.** $3x^2\sqrt{y}$ **55.** $x^3y^2\sqrt{2}$ **57.** $-48x^2y\sqrt{y}$ **59.** $-\frac{8n^2\sqrt{5m}}{5}$ **61.** $\frac{5}{3}$ **63.** $\frac{9}{8}$ **65.** $\frac{\sqrt{26}}{5}$ **67.** $-\frac{2\sqrt{5}}{7}$ **69.** $\frac{4\sqrt{3}}{9}$ **71.** $\frac{4\sqrt{2}}{5}$ **73.** $\frac{6x\sqrt{2x}}{y}$ **75.** $\frac{5n^2\sqrt{5}}{8}$ **77.** $\frac{8m\sqrt{2}}{9n}$ **79.** $2rs^2\sqrt{3s}$ **81.** $2\sqrt[3]{3}$ **83.** $-4\sqrt[3]{2}$ **85.** $2x$ **87.** $-4x\sqrt[3]{x^2}$ **89.** $3xz^2\sqrt[3]{2}$ **91.** $-3y\sqrt[3]{3x^2}$ **93.** $\frac{3m}{2n^2}$ **95.** $\frac{rs\sqrt[3]{rs^2}}{10t}$ **97. a.** $\frac{3\pi\sqrt{3}}{4}$ sec **b.** 4.1 sec **99. a.** $60\sqrt{2}$ cm **b.** 84.9 cm **103.** $-6a^5$ **105.** $y = -2x + 3$ **107.** $x < 5$

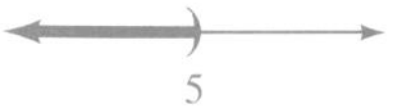

$(-\infty, 5)$

Study Set Section 14.4 (page 971)

1. radicals **3.** simplified **5.** no **7.** yes **9.** The radicals don't have the same radicand, so they can't be combined. **11.** The two terms are not like terms — they cannot be combined. **13.** $2\sqrt{3}, 3\sqrt{3}, 4\sqrt{3}, 5\sqrt{3}$ **17.** $9\sqrt{7}$ **19.** $-3\sqrt{x}$ **21.** $5 + 6\sqrt{3}$ **23.** $-1 - \sqrt{r}$ **25.** $5\sqrt{3}$ **27.** $\sqrt{2}$ **29.** $14\sqrt{5}$ **31.** $-7\sqrt{5}$ **33.** $8\sqrt{5}$ **35.** $-2\sqrt{3}$ **37.** $-18\sqrt{2}$ **39.** $12\sqrt{7}$ **41.** $9\sqrt{5}$ **43.** $10\sqrt{2} - \sqrt{3}$ **45.** $5\sqrt{6} - 5\sqrt{2}$ **47.** $7\sqrt{6} + 4\sqrt{15}$ **49.** $7\sqrt{3} - 6\sqrt{2}$ **51.** $3x\sqrt{2}$ **53.** $3d\sqrt{2d}$ **55.** $3x\sqrt{2y} - 3x\sqrt{3y}$ **57.** $x^2\sqrt{2x}$ **59.** $19b\sqrt{6}$ **61.** $-5y\sqrt{10y}$ **63.** $2x\sqrt{5xy} + 3x^2y\sqrt{5xy} - 4x^3y^2\sqrt{5xy}$ **65.** $2\sqrt[3]{3}$ **67.** $-\sqrt[3]{x}$ **69.** $5\sqrt[3]{2}$ **71.** $\sqrt[3]{3}$ **73.** $24\sqrt[3]{5} + 5$ **75.** $x\sqrt[3]{x} - x^2\sqrt[3]{x}$ **77.** $2xy\sqrt[3]{3xy^2}$ **79.** $x^2y\sqrt[3]{5xy}$ **81.** $18\sqrt{3}$ in. **83.** $27\sqrt{2}$ cm **85.** $133\sqrt{6}$ ft **89.** $\frac{1}{9}$ **91.** -9 **93.** $\frac{1}{x^3}$ **95.** 1

Study Set Section 14.5 (page 980)

1. rationalizing **3.** conjugate **5.** coefficient **7.** $a \cdot b$ **9.** $\sqrt{7}$ **11. a.** The radicand is a fraction. **b.** There is a radical in the denominator. **13.** rational: $\frac{\sqrt{5}}{3}, -\frac{\sqrt{2}}{8}, \frac{1+\sqrt{3}}{4}$, irrational: $\frac{2}{\sqrt{6}}, \frac{9}{7-\sqrt{10}}$ **15. a.** not possible **b.** $\sqrt{6}$ **c.** not possible **d.** $\frac{\sqrt{6}}{3}$ **e.** $4\sqrt{2}$ **f.** 6 **g.** $-2\sqrt{2}$ **h.** $\frac{1}{3}$ **19.** 5 **21.** 54 **23.** 4 **25.** $\sqrt{21}$ **27.** $2\sqrt{14}$ **29.** $3\sqrt{2x}$ **31.** x^4 **33.** $-60\sqrt{2}$ **35.** $-8x$ **37.** $4x^2$ **39.** $2 + \sqrt{2}$ **41.** $27 - 3\sqrt{3}$ **43.** $3\sqrt{2} + \sqrt{3}$ **45.** $x\sqrt{3} - 2\sqrt{x}$ **47.** $6x + 6\sqrt{x}$ **49.** 1 **51.** $6\sqrt{14} + 2x\sqrt{7} - 3x\sqrt{2} - x^2$ **53.** $7 + 2\sqrt{6}$ **55.** $4x - 18$ **57.** -9 **59.** $6\sqrt[3]{12}$ **61.** $7 - 2\sqrt[3]{7}$ **63.** $\sqrt[3]{4} + 4\sqrt[3]{2} + 3$ **65.** $\frac{2x}{3}$ **67.** $\frac{3\sqrt{2}}{5}$ **69.** $\frac{2}{x}$ **71.** $\frac{2x}{3}$ **73.** $\frac{\sqrt{3}}{3}$ **75.** $\frac{\sqrt{91}}{7}$ **77.** $\sqrt{3}$ **79.** $\frac{3\sqrt{2}}{8}$ **81.** $\frac{2\sqrt{15}}{5}$ **83.** $\frac{10\sqrt{x}}{x}$ **85.** $\frac{3\sqrt{2xy}}{2x}$ **87.** $\frac{3\sqrt{3}+3}{2}$ **89.** $\sqrt{7} - 2$ **91.** $6 + 2\sqrt{3}$ **93.** $\frac{\sqrt{3}-3}{2}$ **95.** $5\sqrt{3} - 5\sqrt{2}$ **97.** $\frac{x + 4\sqrt{x} + 4}{x - 4}$ **99.** $\sqrt[3]{25}$ **101.** $2\sqrt[3]{2}$ **103.** $\frac{\sqrt[3]{20}}{2}$ **105.** 108π in.^{2} **107.** $1{,}800\sqrt{2}$ in.^{2} **109.** 90 in.^{2} **113.** no **115.** addition **117.** $x^2 - 16$

Study Set Section 14.6 (page 990)

1. radical **3.** extraneous **5.** b^2 **7. a.** Take the positive square root of both sides. **b.** Square both sides. **9.** On the second line, both sides of the equation were not squared — only the left-hand side. **11.** On the second line, $\sqrt{a+2}$ wasn't isolated before squaring both sides. Also, $(\sqrt{a+2} - 5)^2 \neq a + 2 - 25$.
13. a.

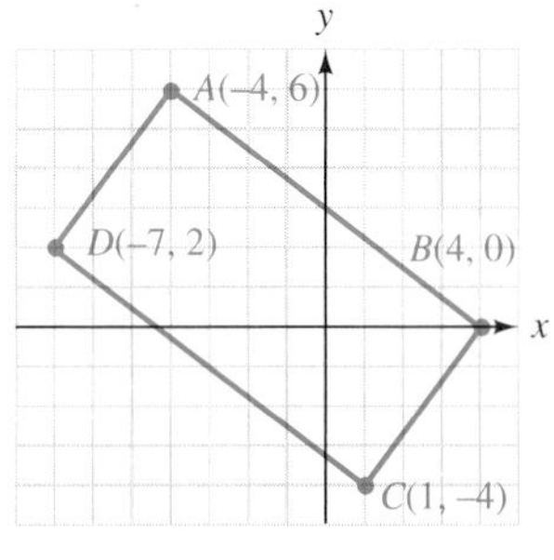

b. rectangle **c.** AB: 10; BC: 5; CD: 10; DA: 5 **d.** 30 units **17.** 9 **19.** 8 **21.** no solution **23.** 25 **25.** 9 **27.** 1 **29.** no solution **31.** 5 **33.** 6 **35.** 46 **37.** 4 **39.** 5 **41.** -2 **43.** 4 **45.** 4 **47.** -1 **49.** 12 **51.** 5 **53.** 1, 3 **55.** $-2, 4$ **57.** 2 **59.** -1 **61.** 343 **63.** 65 **65.** 22 **67.** 4 **69.** 5 **71.** $\sqrt{34}$, 5.83 **73.** 13 **75.** 10 **77.** 169 ft **79.** 64.4 ft **81.** about 288 ft **83.** about 2×10^6 m

85. $V = \frac{\pi r^2 h}{3}$ **87. a.** $\sqrt{2} \approx 1.4$ units **b.** $\sqrt{10} \approx 3.2$ units
91. $8x^2 - 6x$ **93.** $x^2 + 6x + 9$ **95.** $9y^2 - 42y + 49$

Study Set Section 14.7 (page 997)

1. rational **3.** x^{m+n} **5.** $\frac{x^n}{y^n}$ **7.** $\frac{1}{x^n}$ **9.** $\sqrt[n]{x}$ **11.** $5^{1/2}$
13. $(\sqrt[3]{8})^4$ **15.** 0, 1, 2, 3
17.

21. 9 **23.** −12 **25.** $\frac{1}{2}$ **27.** $\frac{2}{7}$ **29.** 3 **31.** −5 **33.** −2
35. $\frac{3}{4}$ **37.** 729 **39.** 125 **41.** 25 **43.** 100 **45.** 4 **47.** $\frac{4}{9}$
49. 6 **51.** 25 **53.** 7 **55.** 25 **57.** 8 **59.** 125 **61.** $\frac{1}{2}$ **63.** $\frac{1}{9}$
65. $\frac{1}{64}$ **67.** $\frac{1}{81}$ **69.** x **71.** x^2 **73.** x^2 **75.** y^2 **77.** $x^{2/5}$
79. $x^{2/7}$ **81.** $x^{17/12}$ **83.** $b^{3/10}$ **85.** $t^{4/15}$ **87.** x^2
89. 392 in.^{2} **91.** 26 ft **93.** 1.1 in.
97.

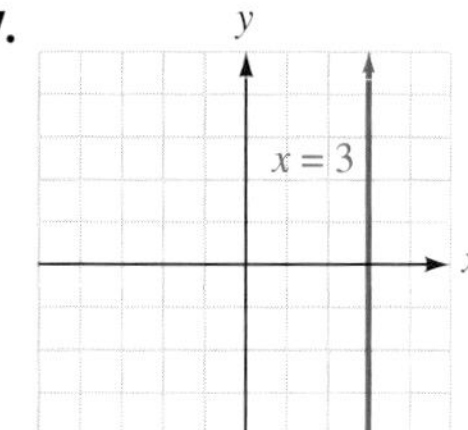

99.

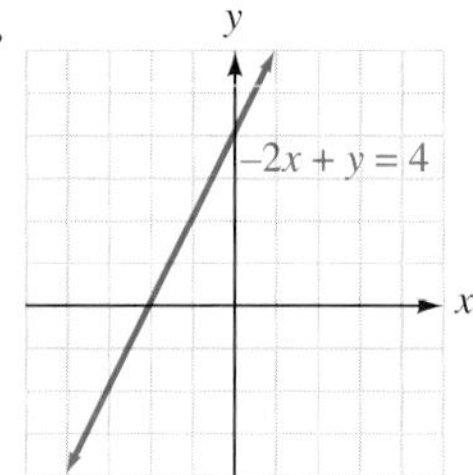

Key Concept (page 1000)

1. addition, subtraction, −6 **3.** multiplication, division, −4
5. square root, square, 49 **7.** cube root, cubed, −8 **9.** −5
11. 15 **13.** 64 **15.** Find the square root of the area.

Chapter Review (page 1002)

1. square **2.** 5 **3.** 7 **4.** −12 **5.** $-\frac{4}{9}$
6. 30 **7.** −0.8 **8.** 1 **9.** 0 **10.** 4.583 **11.** −3.873
12. 5.292 **13.** 27.421 **14.** imag, irr, rat, irr; $\sqrt{-2}$
15. 0, 1, 2, 3

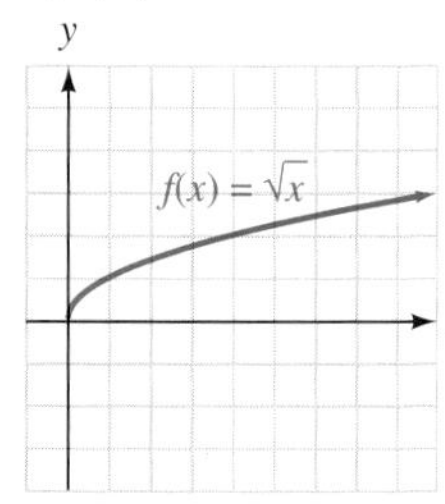

16. 2, 1, 0, −1

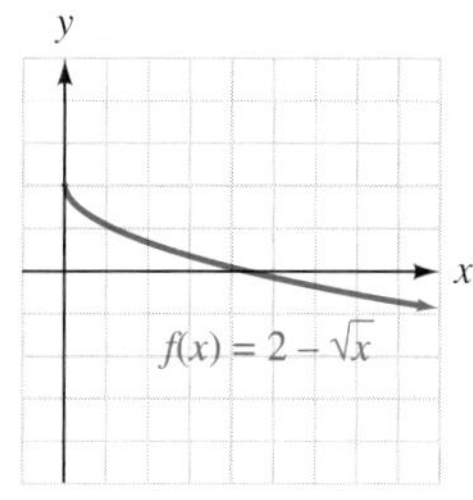

17. 35 **18.** 1 **19.** $2\sqrt{6}$ **20.** 5 ft **21.** 30 mph **22.** 5^3, cube
23. −3 **24.** −5 **25.** 3 **26.** 2 **27.** 0 **28.** −1 **29.** $\frac{1}{4}$
30. 1 **31.** 2.520 **32.** −4.678 **33.** 1.565 **34.** 8.083 **35.** x
36. $2b^2$ **37.** x^2y^2 **38.** $-y^6$ **39.** x **40.** y^2 **41.** $3x$ **42.** $-r^4$
43. 12 mm **44.** $4\sqrt{2}$ **45.** $10\sqrt{5}$ **46.** $4x\sqrt{5}$ **47.** $-6\sqrt{7}$
48. $-5t\sqrt{10t}$ **49.** $-10z^2\sqrt{7z}$ **50.** $10x\sqrt{2y}$ **51.** $y^2\sqrt{3}$
52. $2y\sqrt[3]{x^2}$ **53.** $5xy\sqrt[3]{2x}$ **54.** $\frac{4}{5}$ **55.** $\frac{2\sqrt{15}}{7}$ **56.** $\frac{10}{3}$
57. $\frac{11x\sqrt{2}}{13}$ **58. a.** $2\sqrt{10}$ ft **b.** 6.3 ft **59.** 0
60. $-3 + 4\sqrt{3}$ **61.** $\sqrt{7}$ **62.** $y^2\sqrt{5xy}$ **63.** $5\sqrt[3]{2}$ **64.** $6x\sqrt[3]{2}$
65. They do not contain like radicals — the radicands are different. **66.** $13\sqrt{5}$ in. **67.** $\sqrt{6}$ **68.** $10\sqrt{10}$ **69.** $36x\sqrt{2}$
70. $15 + 6x\sqrt{15} + 9x^2$ **71.** −2 **72.** −2 **73.** $4\sqrt[3]{2}$

74. $\sqrt[3]{9} + \sqrt[3]{3} - 2$ **75. a.** $(4\sqrt{6} + 10\sqrt{3})$in.; $30\sqrt{2}$ in.^{2}
b. 27.1 in.; 42.4 in.^{2} **76.** $\frac{\sqrt{7}}{7}$ **77.** $\frac{\sqrt{21}}{7}$ **78.** $\frac{\sqrt{2}}{2}$
79. $\frac{c - 8\sqrt{c} + 16}{c - 16}$ **80.** $7\sqrt{2} - 7$ **81.** $2\sqrt[3]{4}$ **82.** x **83.** x
84. $4t$ **85.** $e - 1$ **86.** 81 **87.** no solution **88.** −2, 4 **89.** 4
90. −2 **91.** 28 **92.** 64 ft **93.** 5 **94.** 13.93 **95.** 7 **96.** −10
97. 216 **98.** $\frac{4}{9}$ **99.** $\frac{1}{8}$ **100.** 64 **101.** 9 **102.** $\frac{1}{a^2b^4}$
103. $x^{11/15}$ **104.** $t^{1/12}$ **105.** x **106.** x^2
107. $-2^{2/3}$ $4^{-1/2}$ $9^{1/3}$ $12^{1/2}$
−5 −4 −3 −2 −1 0 1 2 3 4 5

108. $\frac{1}{64}$ of the original dose **109.** $(-4)^{1/2} = \sqrt{-4}$; There is no real number that, when squared, gives −4.

Chapter 14 Test (page 1007)

1. 10 **2.** $-\frac{20}{3}$ **3.** −3 **4.** $\frac{5\sqrt{2}}{7}$ **5.** 7.2 **6.** 10 ft **7.** $2x$
8. $3x\sqrt{6x}$ **9.** $3y\sqrt{x}$ **10.** x^2y **11. a.** $2\sqrt{6}$ yd **b.** 4.9 yd
12. $5\sqrt{3}$ **13.** $-x\sqrt{2x}$ **14.** $-24x\sqrt{6}$ **15.** $2\sqrt{6} + 3\sqrt{2}$
16. −1 **17.** $2x - 4\sqrt{x} - 6$ **18.** $(6\sqrt{2} + 2\sqrt{10})$ in., 14.8 in.
19. $\sqrt{2}$ **20.** $\frac{x\sqrt{3} - 2\sqrt{3x}}{x - 4}$ **21.** 225 **22.** −62 **23.** 5
24. 2, 1 **25.** no solution **26.** 29 **27.** 10
28. 0, 1, 1.4, 1.7, 2, 2.2, 2.4, 2.6, 2.8, 3

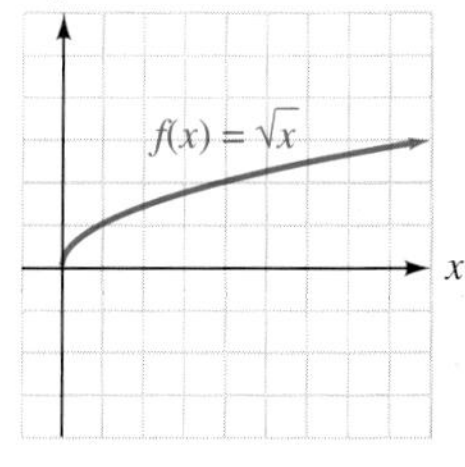

29. No; when 0 is substituted for x, the result is not a true statement: $1 \neq -1$. **30.** The terms do not contain like radicals — the radicands are different. **31.** the Pythagorean theorem; 5 ft **32.** There is no real number that, when squared, gives −9. **33.** 11 **34.** p^2

Cumulative Review Exercises (page 1009)

1. a. true **b.** false **c.** true **2.** −14 **3.** 17 lb
4. −1.8% **5.** $-3p - 6z$ **6.** −2
7. $x \le -\frac{3}{4}, (-\infty, -\frac{3}{4}]$, $-\frac{3}{4}$
8. $-2 < x \le 4, (-2, 4]$, −2 4
9. 3.5 hr **10.** 80 **11.** 4 in.
12.

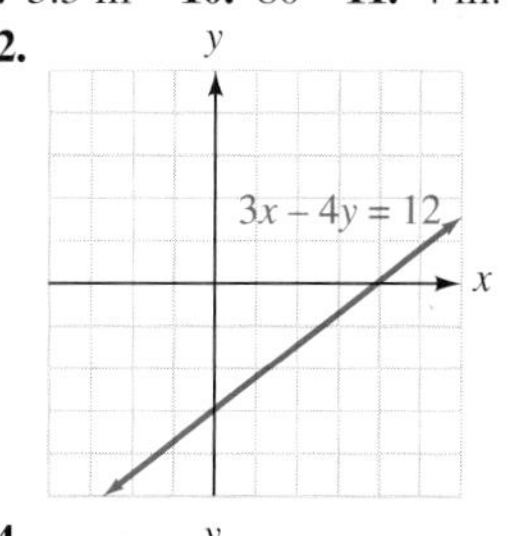

13.

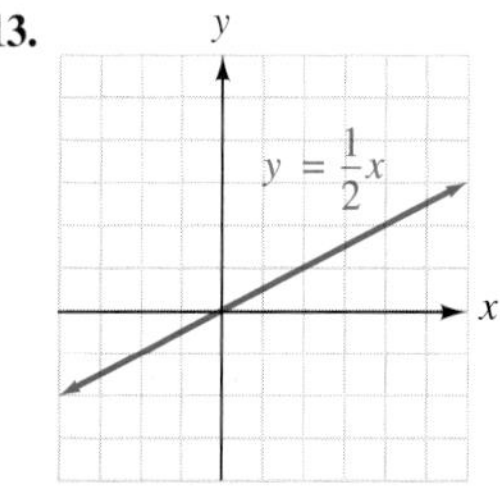

14.

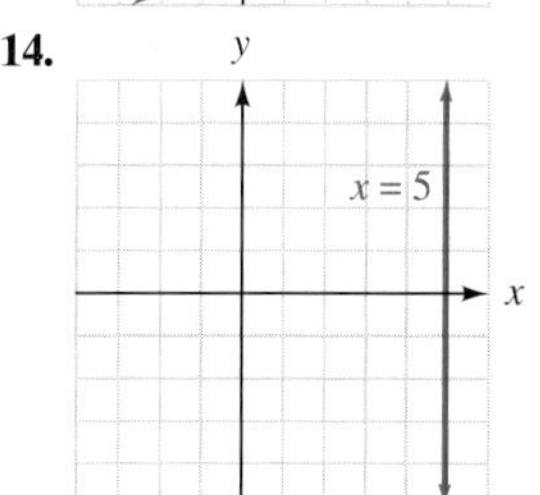

15.

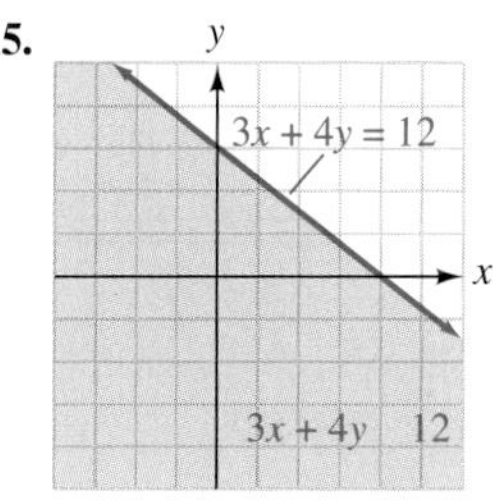

16. $y = \frac{x}{2} + 6$ **17. a.** 3 **b.** $-\frac{2}{3}$ **18. a.** They are the same. **b.** They are negative reciprocals.
19. \$52 billion/yr **20.** -1 **21.** 1, 0, 1, 2, 2, 3

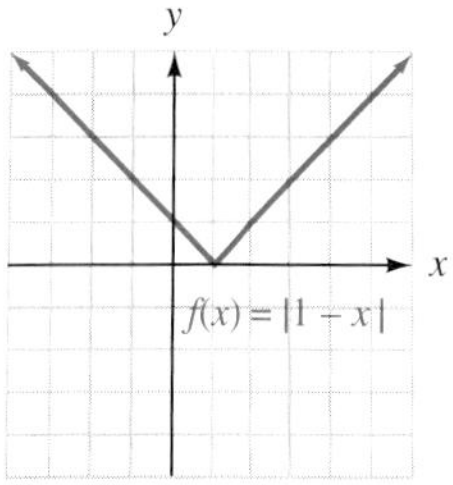

D: all reals; R: all real numbers greater than or equal to 0 **22.** yes **23.** x^{31} **24.** $\frac{a^{15}b^5}{c^{20}}$ **25.** 1 **26.** $\frac{a^8}{b^{12}}$ **27.** 4.8×10^{18} m **28. a.** 3; they indicate that the game was tied 3 times in the second half. **b.** (11, 6); in the second half, UConn had its largest lead (6 points) after 11 minutes had elapsed. (4, −5); in the second half, UConn faced its largest deficit (5 points) after 4 minutes had elapsed.
29. $-2r^7s^3t^4$ **30.** $-6t^2 + 13st - 6s^2$
31. $2a^2 + a - 3$ **32.** $y^2 - 12y + 36$ **33.** $\frac{1}{y} - \frac{3}{4x} + \frac{2z}{xy}$
34. $2x - 1$ **35.** $3xy(x - 2y)$ **36.** $(x + y)(2x - 3)$
37. $(5p^2 + 4q)(5p^2 - 4q)$ **38.** $3x(x + 9)(x - 9)$
39. $(x - 12)(x + 1)$ **40.** $(a + 2b)(a^2 - 2ab + 4b^2)$
41. $(3a + 4)(2a - 5)$ **42.** $2(4m + 1)(2m - 3)$ **43.** $-1, -2$
44. 0, 2 **45.** $\frac{2}{3}, -\frac{1}{2}$ **46.** $\frac{3}{2}, -4$ **47.** 5 cm **48.** 6
49. $\frac{x + 1}{x - 1}$ **50.** $-\frac{3}{5a}$ **51.** $\frac{(p + 2)(p - 3)}{3(p + 3)}$
52. $\frac{xy^2d}{c^3}$ **53.** $\frac{7x + 29}{(x + 5)(x + 7)}$ **54.** 1 **55.** $\frac{9a^2 - 4b^2}{6ab}$
56. $\frac{y + x}{y - x}$ **57.** 2 **58.** 2 **59.** $r = \frac{r_1r_2}{r_2 + r_1}$
60. 6,480 **61.** 2 **62.** $2\frac{2}{9}$ hr **63.** $(3, -1)$
64. $(-1, 2)$ **65.** 6%: \$3,000; 12%: \$3,000
66.

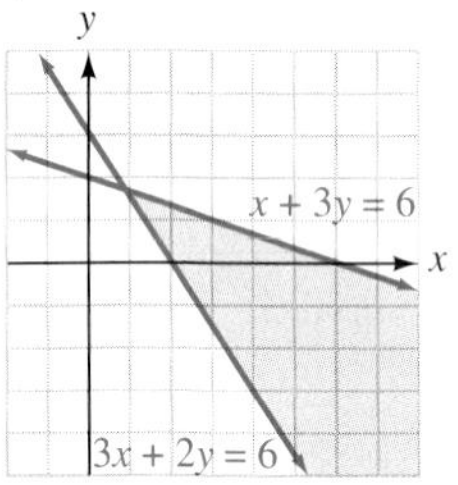

67. $\frac{7}{15}$ **68.** 3 **69.** $-48x^2y\sqrt{y}$ **70.** $7\sqrt{3} - 6\sqrt{2}$
71. $y - 9\sqrt{y} + 20$ **72.** $-60\sqrt{2}$
73. $\frac{2\sqrt{5}}{5}$ **74.** $\frac{x - 6\sqrt{x} + 9}{x - 9}$
75. 5 **76.** 73 in.

Chapter 15 Check Your Knowledge (page 1014)

1. quadratic **2.** true **3.** two **4.** square, quadratic **5.** vertex
6. ± 7 **7.** $\pm 3\sqrt{2}$ **8.** $\pm\frac{4}{3}$ **9.** $3 \pm \sqrt{2}$ **10.** 4.5 ft **11.** 16
12. $-1, 7$ **13.** $-3, \frac{7}{3}$ **14.** $2 \pm \sqrt{2}$; 0.6, 3.4 **15.** $-4, 6$
16. $\frac{2}{3}, 1$ **17.** no real solutions **18.** $-2 \pm \sqrt{7}$
19. $\frac{1 \pm \sqrt{13}}{6}$; −0.43, 0.77 **20.** 3 **21.** 4 ft, 8 ft
22.

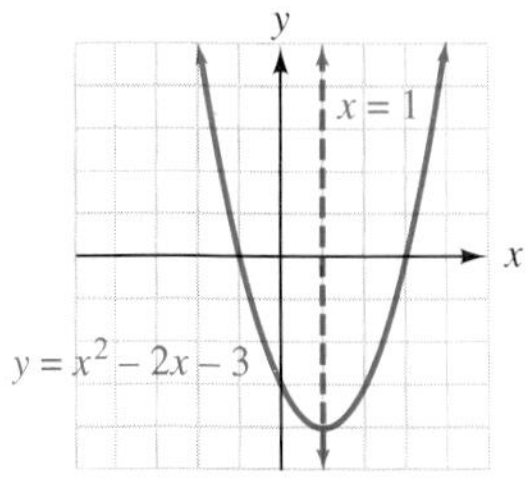

23. 12 ft, 16 ft **24.** none

Study Set Section 15.1 (page 1023)

1. completing **3.** coefficient **5.** two **7.** square **9.** $\sqrt{5}$
11. a. Subtract 35 from both sides. **b.** Add 1 to both sides.
13. a. ± 9 **b.** ± 9 **15. a.** 9 **b.** 36 **c.** $\frac{9}{4}$ **d.** $\frac{25}{4}$
19. a. two; $\sqrt{10}, -\sqrt{10}$ **b.** $8 + \sqrt{3}, 8 - \sqrt{3}$; 9.73, 6.27
21. ± 1 **23.** ± 3 **25.** $\pm 2\sqrt{5}$ **27.** ± 3 **29.** ± 2 **31.** $\pm\frac{3}{4}$
33. $-6, 4$ **35.** $7, -11$ **37.** $2 \pm 2\sqrt{2}$ **39.** ± 6.77
41. 7.49, −11.49 **43.** 0, −4 **45.** $2, -\frac{2}{3}$
47. $x^2 + 2x + 1, (x + 1)^2$ **49.** $x^2 - 4x + 4, (x - 2)^2$
51. $x^2 + 7x + \frac{49}{4}, (x + \frac{7}{2})^2$ **53.** $a^2 - 3a + \frac{9}{4}, (a - \frac{3}{2})^2$
55. $b^2 + \frac{2}{3}b + \frac{1}{9}, (b + \frac{1}{3})^2$ **57.** $-2, -4$ **59.** 2, 6
61. 5, −3 **63.** 1, −6 **65.** 1, −2 **67.** −1, −2 **69.** $2, -\frac{1}{2}$
71. $-2, \frac{1}{4}$ **73.** $-2 \pm \sqrt{3}$; −0.27, −3.73
75. $1 \pm \sqrt{5}$; 3.24, −1.24 **77.** $2 \pm \sqrt{7}$; 4.65, −0.65
79. $\frac{-1 \pm 2\sqrt{5}}{2}$; −2.74, 1.74 **81.** 1, −4 **83.** $\frac{3}{2}, -\frac{2}{3}$
85. $\frac{-3 \pm \sqrt{3}}{2}$ **87.** 55 ft **89.** 20 ft by 40 ft **95.** $y^2 - 2y + 1$
97. $x^2 + 2xy + y^2$ **99.** $4z^2$

Study Set Section 15.2 (page 1033)

1. quadratic **3.** solve **5.** 0 **7.** 3, 0, −5
9. $lw, \frac{1}{2}bh$ **11.** Evaluate 5^2. **13. a.** 2
b. $\frac{-3 + \sqrt{15}}{2}, \frac{-3 - \sqrt{15}}{2}$ **c.** 0.44, −3.44
15.

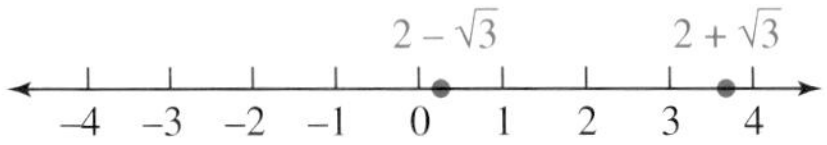

19. The student didn't extend the fraction bar so that it underlines the complete numerator. **21.** $a = 1, b = 4, c = 3$
23. $a = 3, b = -2, c = 7$ **25.** $a = 4, b = -2, c = 1$ **27.** $a = 3, b = -5, c = -2$ **29.** $a = 7, b = 14, c = 21$ **31.** 2, 3
33. −3, −4 **35.** $1, -\frac{1}{2}$ **37.** $-1, -\frac{2}{3}$ **39.** $\frac{1}{2}, -\frac{3}{2}$ **41.** $\frac{-3 \pm \sqrt{5}}{2}$
43. $\frac{1 \pm \sqrt{37}}{6}$ **45.** no real solutions **47.** $-1 \pm \sqrt{2}$
49. $\frac{3 \pm \sqrt{15}}{3}$ **51.** −2, 3 **53.** $\frac{-1 \pm \sqrt{41}}{4}$; −1.85, 1.35
55. $1 \pm \sqrt{2}$; −0.41; 2.41 **57.** −5, 7 **59.** no real solutions
61. −4, 0 **63.** $\pm\sqrt{6}$; ±2.45 **65.** 0.8, 3.1 **67.** 6 in.
69. 47 ft by 66 ft **71.** 10 ft by 18 ft **73.** 3.5 sec **75.** 15 cm by 36 cm **77.** 30 and 40 nautical miles **79.** 7% **81.** \$210,000
83. 2 in. **89.** $r = \frac{A - p}{pt}$ **91.** $y = \frac{3}{5}x + 12$ **93.** $4\sqrt{5}$ **95.** $\frac{\sqrt{7x}}{7}$

Think It Through (page 1044)

1. 1957, 4.3 million **2.** 1990, 4.1 million **3.** 1973, 3.1 million

Study Set Section 15.3 (page 1044)

1. quadratic **3.** y-intercept **5.** symmetry **7.** $a > 0$ **9.** $(0, c)$
11. a. a parabola **b.** (1, 0), (3, 0) **c.** (0, −3) **d.** (2, 1)
15. The most cases of flu (25) were reported the fifth week.
17. true **19.** $-\frac{1}{2}$

21. moved up 1

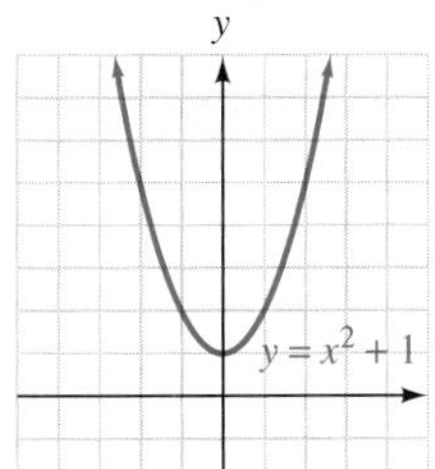

23. opens the opposite direction

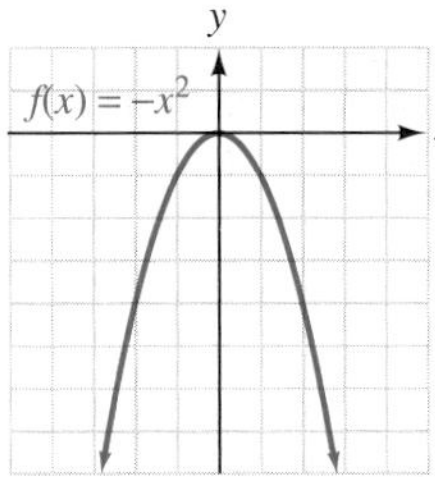

25. (3, 1) **27.** (1, −1) **29.** (1, 0); (0, 1)
31. (−3, 0), (−7, 0); (0, −21)

33.

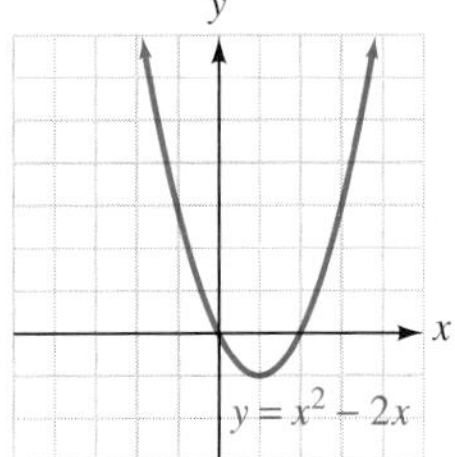

35.

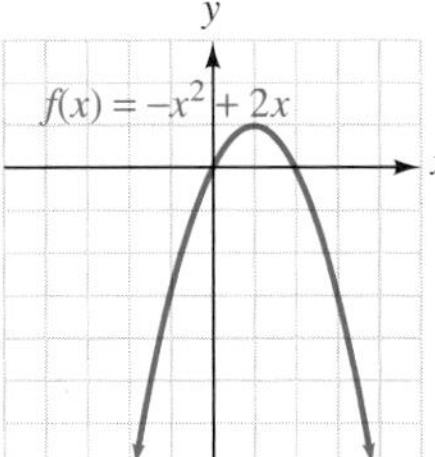

37.

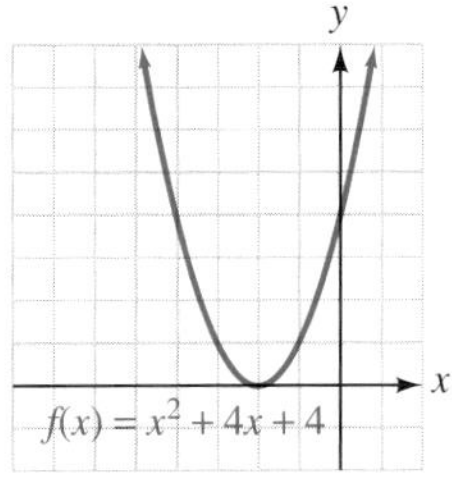

39.

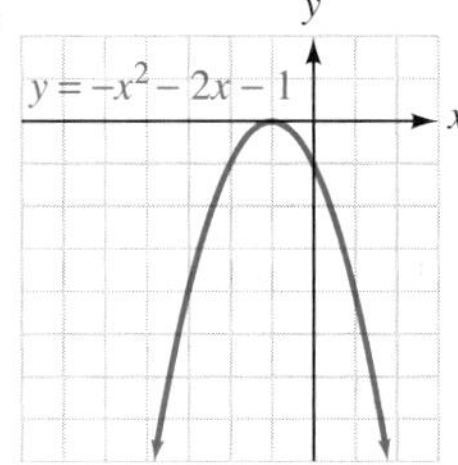

41.

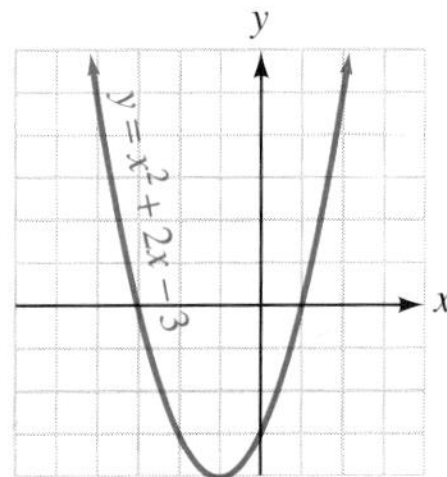

43.

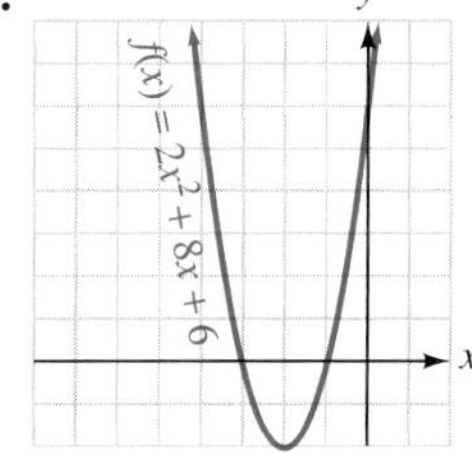

45.

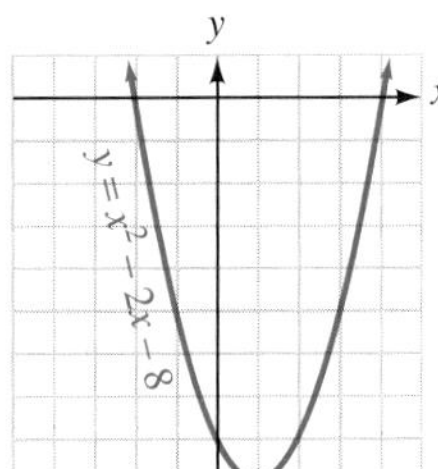

47.

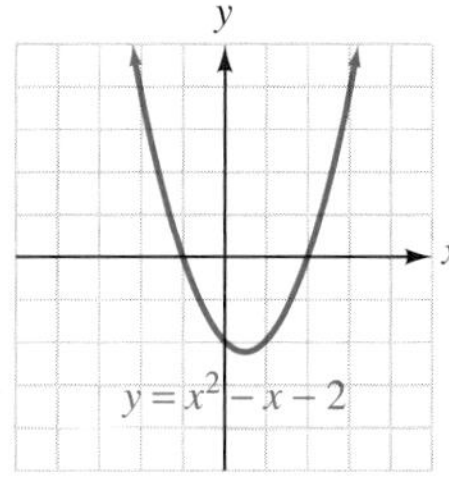

49.

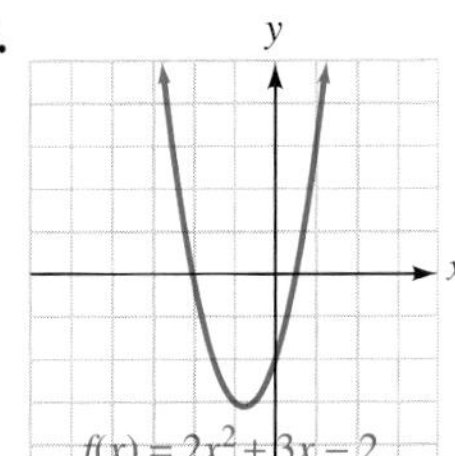

51. a. 14 ft **b.** 0.25 sec and 1.75 sec **c.** 18 ft; 1.0 sec **53.** 32, 18, 8, 2, 0, 2, 8, 18, 32 **55. a.** 750 **b.** \$56,250 **61.** $5\sqrt{3}$ **63.** 2 **65.** 5

Key Concept (page 1048)

1. no **3.** yes **5.** no **7.** yes **9.** no **11.** no **13.** The student divided both sides by 2 and incorrectly thought that $\frac{x^2}{2}$ equals x. The square root method should be used. **15.** The student forgot to write the ± symbol when the square root method was used in Step 2. **17.** $0, \frac{1}{4}$ **19.** ±6 **21.** $-2 \pm \sqrt{3}$

23.

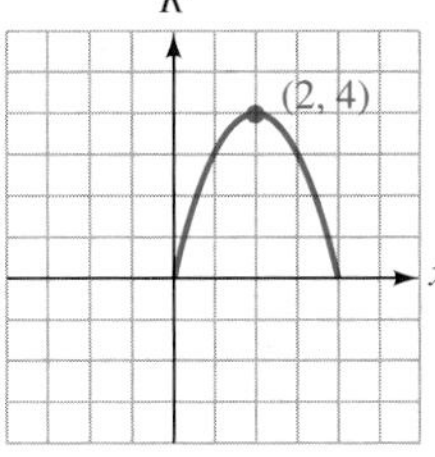

Chapter Review (page 1050)

1. ±5 **2.** ±20 **3.** ±3 **4.** $\pm\frac{3}{2}$ **5.** $\pm 2\sqrt{2}$ **6.** $\pm 5\sqrt{3}$ **7.** −4, 6 **8.** $\frac{7}{2}, \frac{1}{2}$ **9.** $8 \pm 2\sqrt{2}$ **10.** $-5 \pm 5\sqrt{3}$ **11.** ±3.46 **12.** −6.42, 8.42 **13.** $x^2 + 4x + 4, (x + 2)^2$ **14.** $z^2 - 10z + 25, (z - 5)^2$ **15.** $t^2 - 5t + \frac{25}{4}, (t - \frac{5}{2})^2$ **16.** $a^2 + \frac{3}{4}a + \frac{9}{64}, (a + \frac{3}{8})^2$ **17.** $x^2 + 4x + 1$ doesn't factor **18.** 3, 5 **19.** 2, −7 **20.** $\frac{1}{2}, -3$ **21.** $\frac{1 \pm \sqrt{3}}{2}$ **22.** −0.27, −3.73 **23.** 3.0 ft **24.** 5, −3 **25.** 7, −1 **26.** $\frac{3}{2}, -\frac{1}{3}$ **27.** $3 \pm \sqrt{2}$ **28.** $\frac{-1 \pm \sqrt{7}}{3}$, −1.22, 0.55 **29.** no real solutions **30.** 10 ft, 24 ft **31.** 15 sec **32.** $-3 \pm \sqrt{7}$ **33.** 1, −7 **34.** 0, −5 **35.** $\frac{-1 \pm \sqrt{41}}{4}$ **36.** $\pm 2\sqrt{5}$ **37.** 2, 2 **38.** no real solutions **39.** ±20 **40.** (−3, 0), (1, 0) **41.** (0, −3) **42.** (−1, −4) **44.** The maximum profit of \$16,000 is obtained from the sale of 400 units. **45.** (1, 5); upward **46.** (3, 16); downward **47.** (−5, 0), (−1, 0); (0, 5)

48.

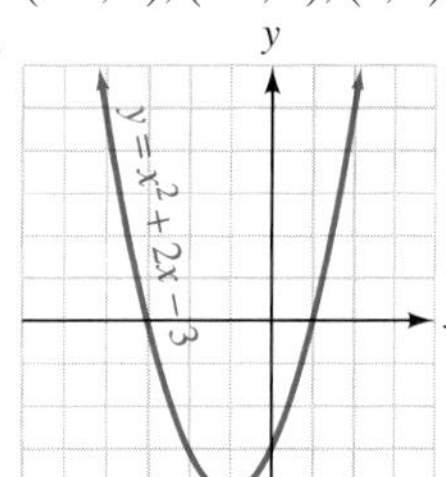

49.

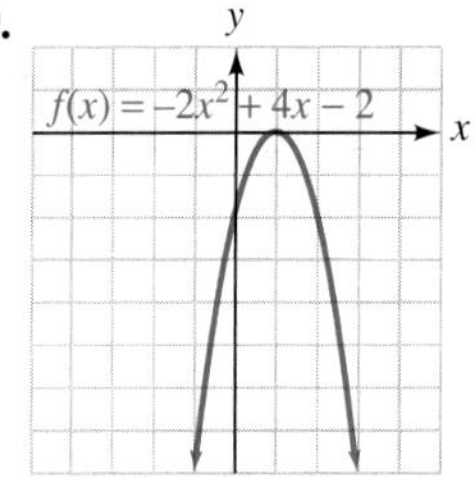

50. 2, 1, 0

Chapter 15 Test (page 1053)

1. ±4, **2.** $\pm 2\sqrt{6}$ **3.** $\pm\frac{5}{2}$ **4.** $2 \pm \sqrt{3}$ **5.** 40 cm **6.** $x^2 - 14x + 49 = (x - 7)^2$ **7.** $-1 \pm \sqrt{5}$; −3.24, 1.24 **8.** $2, -\frac{1}{2}$ **9.** 2, −5 **10.** $-\frac{3}{2}, 4$ **11.** no real solutions **12.** $2 \pm \sqrt{2}$ **13.** $\frac{1 \pm \sqrt{13}}{6}$; −0.43, 0.77 **14.** 255 ft, 505 ft **15.** The most air conditioners sold in a week (18) occurred when 3 ads were run.

16.

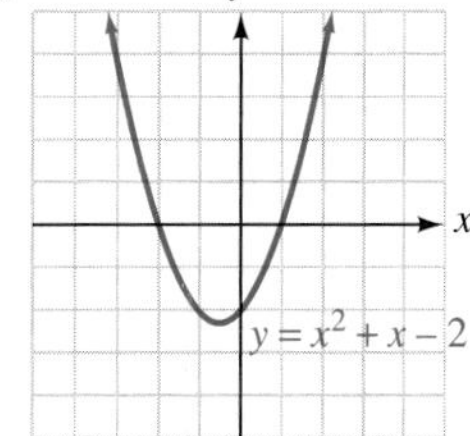

17. 1 **18.** plus or minus

Cumulative Review Exercises (page 1055)

1. a. true **b.** true **c.** true **2.** 36 **3.** 2 hours before salt is spread **4.** about 98,113 **5.** $-2p - 6z$ **6.** 12 **7.** $-\frac{3}{2}$
8. $x < -2$; $(-\infty, -2)$ −2

9. \$8,000 at 7%, \$20,000 at 10% **10.** 65 **11.** 4 **12.** $\frac{3}{5}$

13.

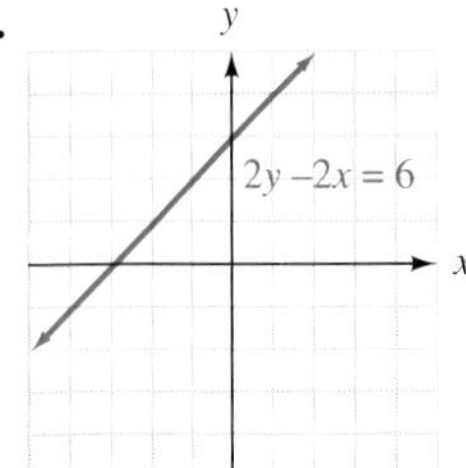

14.

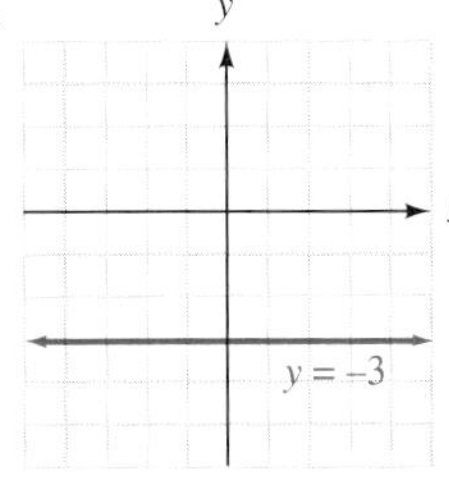

15.

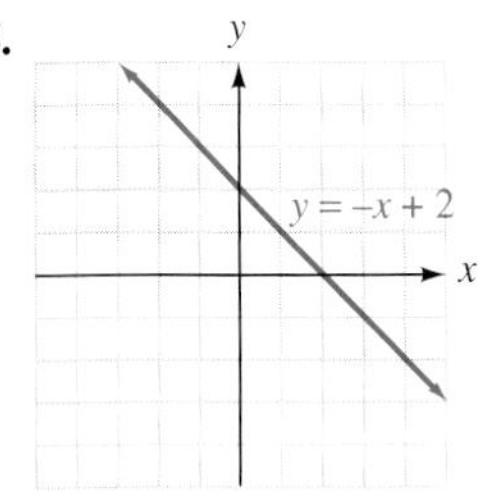

16.

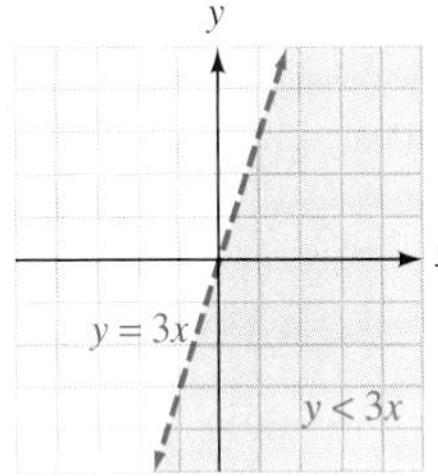

17.

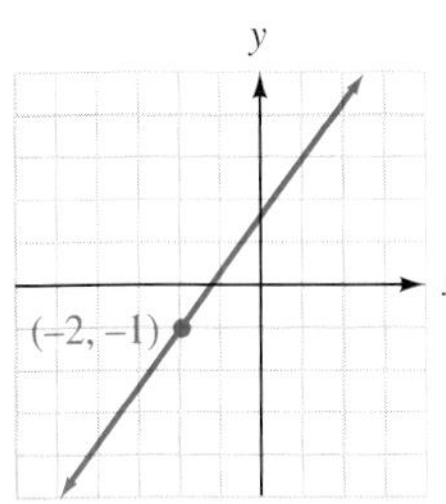

18.

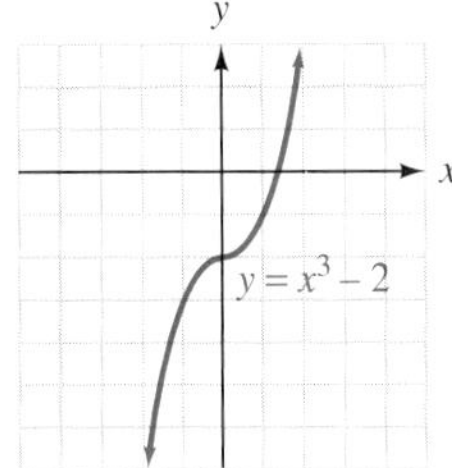

19. $y = -2x + 1$ **20.** $y = \frac{1}{4}x - 1$ **21.** $-\frac{4}{5}$ **22.** -8
23. input **24.** yes **25.** y^9 **26.** $\frac{b^6}{27a^3}$ **27.** 2 **28.** $\frac{1}{r^6}$
29. 2,600,000 to 1 **30.** 16%/hr **31.** $-6a^5$ **32.** $6b^2 + 5b - 4$
33. $4x^2 + 20xy + 25y^2$ **34.** $2x + 1$ **35.** $6a(a - 2a^2b + 6b)$
36. $(x + y)(2 + a)$ **37.** $(b + 5)(b^2 - 5b + 25)$
38. $(t^2 + 4)(t + 2)(t - 2)$ **39.** $0, -\frac{8}{3}$ **40.** $\frac{2}{3}, -\frac{1}{5}$
41. $6x^3 + 4x$ **42.** 5 in. **43.** -8 **44.** $\frac{3(x + 3)}{x + 6}$ **45.** $\frac{x - 3}{x - 5}$
46. $\frac{2x + 8}{x^2}$ **47.** $\frac{x^2 + 5x}{x^2 - 4}$ **48.** $\frac{9y + 5}{4 - y}$ **49.** -1 **50.** 1
51. $a = \frac{b}{b - 1}$ **52.** $2\frac{6}{11}$ days **53.** 875 days **54.** 1.2

55.

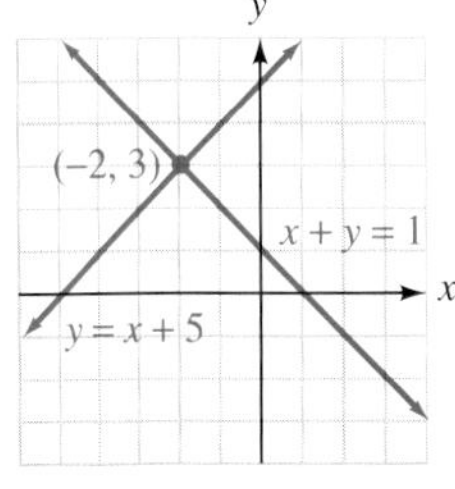

56. $(-3, -1)$ **57.** $(-1, 2)$ **58.** 30 lb of each

59.

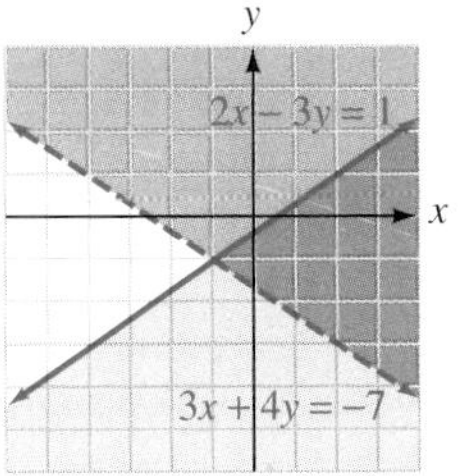

60. 2, 3, 1 **61.** $5x\sqrt{2}$ **62.** $-3y$ **63.** $9\sqrt{6}$ **64.** $-1 - 2\sqrt{2}$
65. $\frac{6x\sqrt{2x}}{y}$ **66.** $\frac{4\sqrt{10}}{5}$ **67.** 4 **68.** $-2, -6$ **69.** $\pm 5\sqrt{3}$
70. 21.2 in. **71.** $\frac{1 \pm \sqrt{13}}{6}$; $-0.43, 0.77$ **72.** 83 ft × 155 ft

73.

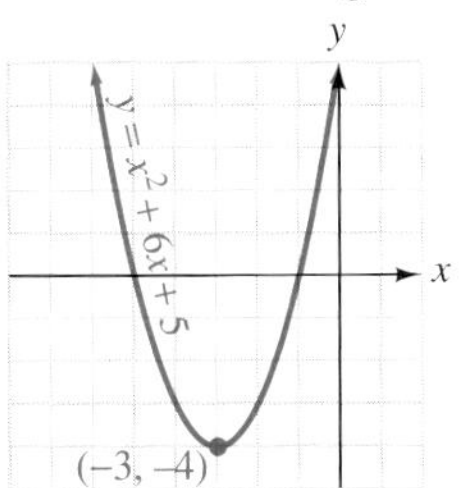

74. a. 150 hp **b.** 2,000 rpm and 5,000 rpm **c.** 4,000 rpm

Study Set Appendix A (page A-5)

1. inductive **3.** circular **5.** alternating **7.** alternating
9. 10 A.M. **11.** 17 **13.** −17 **15.** 6 **17.** 3 **19.** −11 **21.** 9
23. **25.** **27.** I **29.** m **31.** Maria
33. the Mercedes **35.** green, blue, yellow, red **37.** 18,935
39. 0

Study Set Appendix B.1 (page A-15)

1. length **3.** 1 **5.** capacity **7.** 1 **9.** 5,280 **11.** 16 **13.** 8
15. 1 **17.** 24 **19.** $\frac{5}{8}$ in., $1\frac{3}{4}$ in., $2\frac{5}{16}$ in. **21. a.** $\frac{1 \text{ ton}}{2{,}000 \text{ lb}}$ **b.** $\frac{2 \text{ pt}}{1 \text{ qt}}$
23. a. iv **b.** i **c.** ii **d.** iii **25. a.** iii **b.** iv **c.** i **d.** ii
31. $2\frac{5}{8}$ in. **33.** $10\frac{3}{4}$ in. **35.** 48 in. **37.** 42 in. **39.** 2 ft
41. 288 in. **43.** 2.5 yd **45.** $4\frac{2}{3}$ ft **47.** 15 ft **49.** $2\frac{1}{3}$ yd
51. 3 mi **53.** 2,640 ft **55.** 5 lb **57.** 3.5 tons **59.** 24,800 lb
61. 6 pt **63.** 2 gal **65.** 2 pt **67.** 4 hr **69.** 5 days **71.** 150 yd
73. 2,880 in. **75.** 0.28 mi **77.** 61,600 yd **79.** 128 oz **81.** 4.95 tons **83.** 68 **85.** $71\frac{7}{8}$ gal = 71.875 gal **87.** 320 oz **89.** $6\frac{1}{8}$ days = 6.125 days **93.** 3,700 **95.** 3,673.26 **97.** 0.101 **99.** 0.1

Study Set Appendix B.2 (page A-25)

1. tens **3.** thousands **5.** hundredths **7.** metric
9. 1 cm, 3 cm, 6 cm **11. a.** $\frac{1 \text{ km}}{1{,}000 \text{ m}}$ **b.** $\frac{100 \text{ cg}}{1 \text{ g}}$ **c.** $\frac{1{,}000 \text{ milliliters}}{1 \text{ liter}}$
13. a. iii **b.** i **c.** ii **15. a.** ii **b.** iii **c.** i **17.** 10 **19.** $\frac{1}{100}$
21. $\frac{1}{1{,}000}$ **23.** 1,000 **25.** 1,000 **27.** 1,000 **29.** $\frac{1}{100}$ **31.** 1
37. 156 mm **39.** 28 cm **41.** 300 **43.** 570 **45.** 3.1
47. 7,680,000 **49.** 0.472 **51.** 4.532 **53.** 0.0325 **55.** 37.5
57. 125 **59.** 675,000 **61.** 6.383 **63.** 0.63 **65.** 69.5
67. 5.689 **69.** 5.762 **71.** 0.000645 **73.** 0.65823
75. 3,000 **77.** 2,000 **79.** 1,000,000 **81.** 0.5 **83.** 3,000
85. 5,000 **87.** 10 **89.** 0.5 km, 1 km, 1.5 km, 5 km, 10 km

91. 3.43 hm **93.** 12 cm, 8 cm **95.** 40 dL **97.** 4 **99.** 3 g **105.** \$23.99 **107.** $1\frac{9}{35}$

Study Set Appendix B.3 (page A-33)

1. Fahrenheit, Celsius **3. a.** meter **b.** meter **c.** inch **d.** mile **5. a.** liter **b.** liter **c.** gallon **11.** 91.4 **13.** 147.6 **15.** 39,372 **17.** 127 **19.** 1 **21.** 11,350 **23.** 17.5 **25.** 0.6 **27.** 0.1 **29.** 243.4 **31.** 710 **33.** 0.5 **35.** 10° **37.** 122° **39.** 14° **41.** −20.6° **43.** 5 mi **45.** 70 mph **47.** 1.9 km **49.** 1.9 cm **51.** 411 lb; 744 lb **53. a.** 226.8 g **b.** 0.24 L **55.** no **57.** the 3 quarts **59.** 62°C **61.** 28°C **63.** −5°C and 0°C **69.** $\frac{29}{15}$ **71.** $\frac{4}{5}$ **73.** 8.05 **75.** 15.6

Study Set Appendix C.1 (page A-••)

1. set **3.** subset **5.** intersection **7.** intersect **9. a.** five **b.** many **11. a.** {23, 33} or {23, 43} or {33, 43} **b.** {23} or {33} or {43} **c.** {23, 33, 43} **d.** $\varnothing$ **13. a.** set **b.** set **c.** set **15.** $A = \{2, 4, 15\}$, $B = \{6, 15\}$ (answers may vary) **17.** the set containing the elements $\frac{1}{2}, \frac{2}{3}, \frac{3}{4}, \frac{4}{5}, \frac{5}{6}, \frac{6}{7}$ **19.** −1.5 is not an element of the set containing the elements −1 and −2. **21.** 5 is an element of the complement of S. **23.** $\in$ **25.** $\not\subset$ **27.** $\subset$ **29.** $\notin$ **31.** $\subset$ **33.** $\subset$ **35.** $\notin$ **37.** {1, 2, 3, 4} **39.** {0, 1, 2, 3, 4} **41.** {−4} **43.** {0, 1, 2, 4} **45.** {−1, 0, 1, 3, 4} **47.** {. . . , −5, −4, −3, −2, −1, 0, 1}
49.
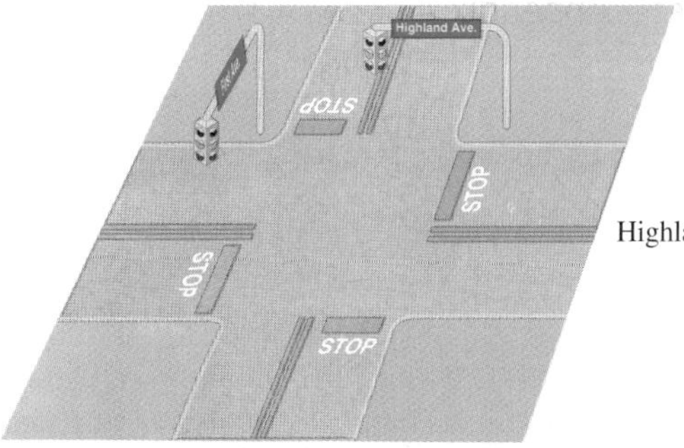

Highland Ave.
First Ave.

51. a. {Seinfeld, David, Spielberg, Winfrey, Cameron, Lucas, Armani, Kelly, Hanks} **b.** {Winfrey}

Study Set Appendix C.2 (page A-50)

1. equation **3.** isolate **5.** 0 **7.** more than **9.** 5 **11. a.** yes **b.** no **c.** yes **d.** no **13. a.** ii **b.** iii **c.** i **15.** $|x| < 4$ **17.** $|x + 3| > 6$ **19.** 8 **21.** −0.02 **23.** $-\frac{31}{16}$ **25.** π **27.** 25 **29.** −2 **31.** 23, −23 **33.** 9.1, −2.9 **35.** $\frac{14}{3}, -6$ **37.** no solution **39.** $2, -\frac{1}{2}$ **41.** −8 **43.** −4, −28 **45.** 0, −6 **47.** 40, −20 **49.** $-2, -\frac{4}{5}$ **51.** 0, −2 **53.** 0 **55.** $\frac{4}{3}$

57. $(-4, 4)$ 

59. $[-21, 3]$ 

61. $(-\frac{8}{3}, 4)$ 

63. no solution **65.** $(-\infty, -3) \cup (3, \infty)$ 

67. $(-\infty, -12) \cup (36, \infty)$ 

69. $(-\infty, -\frac{16}{3}) \cup (4, \infty)$ 

71. $(-\infty, \infty)$ 

73. $(-\infty, -2] \cup [\frac{10}{3}, \infty)$ 

75. $(-\infty, -2) \cup (5, \infty)$ 

77. $[-10, 14]$ 

79. $(-\frac{5}{3}, 1)$ 

81. $(-\infty, -24) \cup (-18, \infty)$ 

83. no solution **85.** $70° \le t \le 86°$ **87. a.** $|c - 0.6°| \le 0.5°$ **b.** [0.1°, 1.1°] **89. a.** 26.45%, 24.76% **b.** It is less than or equal to 1%. **97.** 50°, 130°

Study Set Appendix C.3 (page A-57)

1. quadratic **3.** interval **5.** greater **7.** undefined **9.** $(-\infty, -2) \cup (3, \infty)$ **11. a.** −2 **b.** −2 **c.** −2 **d.** 3 **e.** 3 **f.** 3 **13. a.** iii **b.** iv **c.** ii **d.** i

15. $(1, 4)$ 

17. $(-\infty, 3) \cup (5, \infty)$ 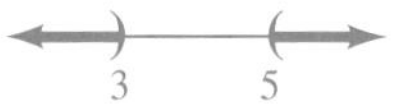

19. $[-4, 3]$ 

21. no solutions **23.** $(-\infty, -3] \cup [3, \infty)$ 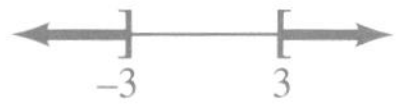

25. $(-5, 5)$ 

27. $(-\infty, 0) \cup (\frac{1}{2}, \infty)$ 

29. $(-\infty, -\frac{5}{3}) \cup (0, \infty)$ 

31. $(-\infty, -3) \cup (1, 4)$ 

33. $(-\frac{1}{2}, \frac{1}{3}) \cup (\frac{1}{2}, \infty)$ 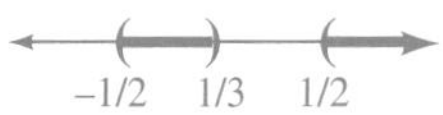

35. $(0, 2) \cup (8, \infty)$ 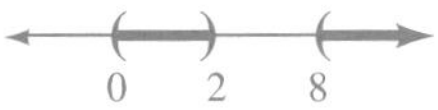

37. $[-\frac{34}{5}, -4) \cup (3, \infty)$ 

39. $(-4, -2] \cup (-1, 2]$ 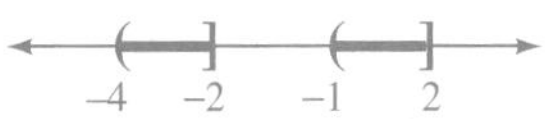

41. $(-\infty, -2) \cup (-2, \infty)$ 

43. $(-1, 3)$ **45.** $(-\infty, -3) \cup (2, \infty)$

47.
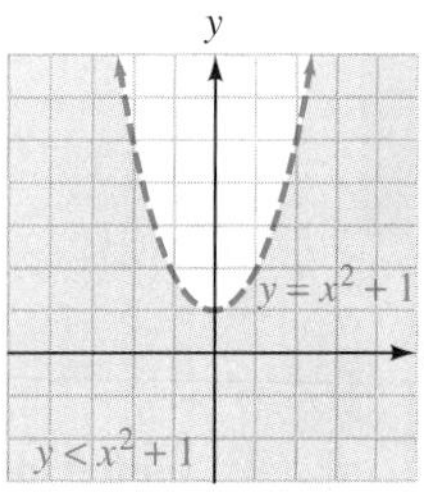

49.
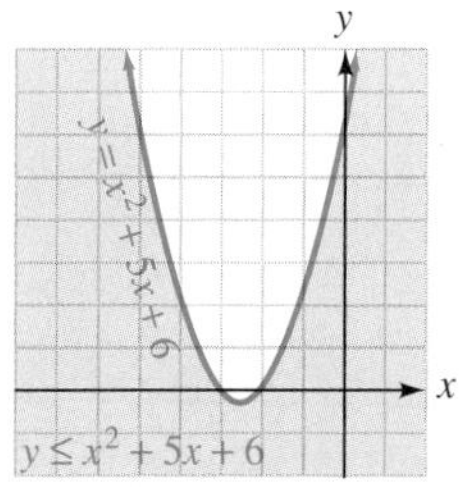

51.

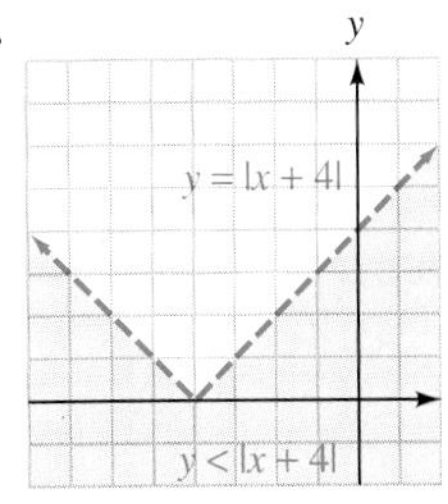

53.

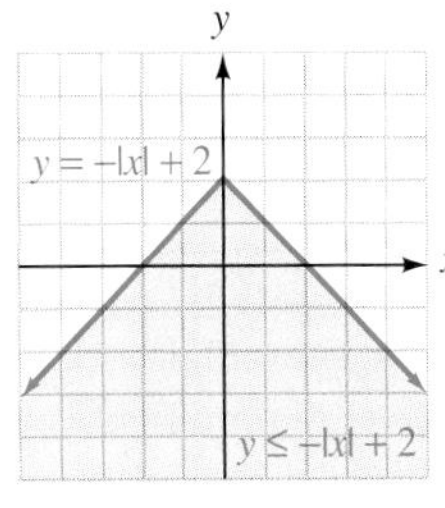

55. $(-2{,}100, -900) \cup (900, 2{,}100)$ **61.** $x = ky$ **63.** $t = kxy$ **65.** 3

Study Set Appendix C.4 (page A-61)

1. experiment **3.** $\frac{n}{s}$ **5.** 0 **7.** {(1, H), (2, H), (3, H), (4, H), (5, H), (6, H), (1, T), (2, T), (3, T), (4, T), (5, T), (6, T)} **9.** {a, e, i, o, u} **11.** $\frac{1}{6}$ **13.** $\frac{2}{3}$ **15.** $\frac{5}{36}$ **17.** 0 **19.** $\frac{19}{42}$ **21.** $\frac{13}{42}$ **23.** $\frac{1}{4}$ **25.** $\frac{1}{4}$ **27.** $\frac{1}{2}$ **29.** $\frac{3}{13}$ **31.** $\frac{5}{12}$ **35.** $\frac{1}{4}$ **37.** $\frac{1}{4}$ **39.** 1 **41.** $\frac{32}{119}$

Study Set Appendix C.5 (page A-65)

1. \$167.44 **3.** \$53,250 **5.** \$743.40 **7.** 10 ft **9.** 5% **11.** \$60.90

THEA Practice Test Appendix D (page A-67)

1. b **2.** a **3.** c **4.** c **5.** b **6.** c **7.** d **8.** b **9.** c **10.** d **11.** b **12.** c **13.** a **14.** c **15.** b **16.** c **17.** a **18.** d **19.** d **20.** a **21.** c **22.** a **23.** b **24.** a **25.** d **26.** a **27.** b **28.** a **29.** c **30.** c **31.** c **32.** a **33.** b **34.** c **35.** d **36.** a **37.** b **38.** d **39.** a **40.** a **41.** d **42.** d **43.** a **44.** a **45.** c **46.** c **47.** d **48.** c **49.** c **50.** c

INDEX